Advances in Neurobiology

Volume 4

Series Editor
Abel Lajtha

For further volumes:
http://www.springer.com/series/8787

In-Young Choi · Rolf Gruetter
Editors

Neural Metabolism *In Vivo*

Springer

Editors
In-Young Choi, Ph.D.
Department of Neurology
Department of Molecular
and Integrative Physiology
Hoglund Brain Imaging Center
University of Kansas Medical Center
Kansas City, KS, USA
ichoi@kumc.edu

Rolf Gruetter, Ph.D.
Laboratory for functional
and metabolic imaging
Ecole Polytechnique Federale de Lausanne
Department of Radiology
University of Lausanne
and Geneva Station 6, CH-1015
Lausanne, Switzerland
rolf.gruetter@epfl.ch

ISSN 2190-5215 e-ISSN 2190-5223
ISBN 978-1-4614-1787-3 e-ISBN 978-1-4614-1788-0
DOI 10.1007/978-1-4614-1788-0
Springer New York Dordrecht Heidelberg London

Library of Congress Control Number: 2011944816

Printed on acid-free paper

Springer is part of Springer Science+Business Media (www.springer.com)

Preface

A greater understanding of the living brain in action has become increasingly feasible through the relentless efforts and devotion of neuroscientists and bioengineers in the past three decades. The development of emerging neuroscience tools and neuroimaging techniques provided us with a new opportunity to integrate real time information regarding the structural organization, function and metabolism of the brain as a whole. The development of this book was truly exciting for us in providing the introduction to the state-of-the-art *in vivo* tools in the neurobiology of the brain, the least understood and most challenging organ. The key advances in neural metabolism in health and disease are indebted to a variety of powerful techniques, namely nuclear magnetic resonance, radiotracer imaging, optical imaging and micro-dialysis.

"Neural Metabolism *In Vivo*" aims to provide a comprehensive overview of neurobiology by presenting the basic principles of up-to-date and cutting-edge technology, as well as their application in assessing the functional, morphological and metabolic aspects of the brain. Investigation of neural activity of the living brain via neurovascular coupling using multimodal imaging techniques extended our understanding of fundamental neurophysiological mechanisms, regulation of cerebral blood flow in connection to neural activity and the interplay between neurons, astrocytes and blood vessels. Constant delivery of glucose and oxygen for energy metabolism is vital for brain function, and the physiological basis of neural activity can be assessed through measurements of cerebral blood flow and consumption of glucose and oxygen.

Dynamic changes in neural metabolism from the basal resting state to the active state accompany the consumption of the major energy substrates of the brain, glucose, and other alternatives. The complexity of the brain is not only reflected in its use of various energy sources as described under the "Cerebral energy metabolism and substrates *in vivo*" section, but also the underlying metabolic processes at the cellular and inter-cellular levels. A close interaction between neuron and astrocytes with regard to oxidative metabolism represents still emerging and controversial research areas since the initial observation of glutamate compartmentation by Dr. Lajtha and others in early 1950s. The working hypothesis of the glutamatergic activity coupled to anaerobic glycolysis is part of the primary neurobiological basis for neural metabolism as presented under the "*In vivo* assessment of metabolic compartmentation between neurons and astrocytes" section, yet to be clearly demonstrated in the future.

This book presents the complex physiological and neurochemical processes of neural metabolism and function in response to various physiological conditions and pharmacological stimulations. Neurochemical detection technologies and quantitative aspects of monitoring cerebral energy substrates and other metabolites in the living brain are described under the "Cerebral metabolism of antioxidants, osmolytes and others *in vivo*" section. Altogether, the advent of new *in vivo* tools has transformed neuroscience and neurobiology research, and demands interdisciplinary approaches as each technology could only approximate a very small fraction of the true complexity of the underlying biological processes. However, translational values of the emerging *in vivo* methods to the application of preclinical to clinical studies cannot be emphasized enough. Thus, it is our hope that advances in our understanding of biochemical, molecular, functional and physiological processes of the brain could eventually help people with neurological problems, which are still dominated by the unknowns.

Kansas City, KS, USA In-Young Choi, Ph.D.
Lausanne, Switzerland Rolf Gruetter, Ph.D.

Acknowledgements

First and foremost, our sincere thanks and appreciation go to all of the authors who put their considerable ruminations and efforts as well as their talents, time and energy to make this work possible. Under one coherent theme of the neural metabolism *in vivo*, 99 contributors provided outstanding overviews of this vast and challenging area, and bore with us with great patience for the past 4 years during the long editing process of this book. We offer our sincere thanks to over 150 reviewers who devoted their time in providing valuable suggestions and constructive comments. Our interaction with prominent scientists from all related disciplines has been a great privilege and an invaluable experience during this endeavor.

We are also eternally thankful to Dr. Abel Lajtha who has been our champion by supporting, encouraging and inspiring us from the start of this project. We express our deep respect to his extraordinary insights, knowledge and contribution to the field of neurochemistry. Our pronounced gratitude goes to Joel Tanzi who helped us through his talent with the editing process. Our gratitude is owed to Drs. Sebastián Cerdán and Phil Lee for their patient and thorough review of many pages of this book. Last but not least, we thank Springer Science for providing us with this unique opportunity and the Springer editors who have worked with us throughout the editing process.

Kansas City, KS, USA — In-Young Choi, Ph.D.
Lausanne, Switzerland — Rolf Gruetter, Ph.D.

Contents

Part I Non-invasive Methods for Neural Metabolism

1 **Introduction to Nuclear Magnetic Resonance (NMR) Methods** 3
Allen W. Song

2 **Hardware Requirements for *In Vivo* Nuclear Magnetic Resonance Studies of Neural Metabolism** 33
Hellmut Merkle, Phil Lee, and In-Young Choi

3 **Diffusion Sensitive Magnetic Resonance Imaging of the Central Nervous System** 65
Jeffrey J. Neil and Joseph J.H. Ackerman

4 **Tools to Enhance the Contrast in MR Images: From Lanthanide Complexes and Iron Oxide Particles to MEMRI** 79
Annemie Van der Linden, Vincent Van Meir, Dario Longo, and Silvio Aime

5 **Principles of ^{1}H NMR Spectroscopy *In Vivo*** 133
Robin A. de Graaf

6 ***In Vivo* Cerebral ^{31}P Magnetic Resonance Spectroscopy** 149
Ernest B. Cady

7 ***In Vivo* ^{13}C Magnetic Resonance Spectroscopy and Metabolic Modeling: Methodology** 181
Pierre-Gilles Henry, Dinesh K. Deelchand, Isabelle Iltis, Malgorzata Marjanska, Gülin Öz, Alexander A. Shestov, and Julien Valette

8 **^{13}C NMR Spectroscopy as a Tool in Neurobiology** 221
Asta Haberg, Ursula Sonnewald, Janniche Hammer, Torun Melø, and Haytham Eloqayli

9 **Overview of Positron Emission Tomography (PET) and Single Photon Emission Computed Tomography (SPECT) Methodologies** ... 255
Paul Cumming and Guido Böning

10 **Fluorodeoxyglucose (FDG) Positron Emission Tomography (PET)** ... 271
Andrei G. Vlassenko and Mark A. Mintun

11 **Neuroreceptor Imaging** ... 305
Habib Zaidi and Miho Shidahara

12 **Optical Approaches to Study Metabolic Consequences of Spreading Depression *In Vivo*** ... 331
C. William Shuttleworth

13 **Microdialysis and Microfiltration: Technology and Cerebral Applications for Energy Substrates** ... 371
Kirsten Huinink, Jakob Korf, and Jan Bert Gramsbergen

Part II *In Vivo* Assessment of Neural Activity Via Neurovascular Coupling

14 **The Coupling of Cerebral Metabolic Rate of Glucose and Cerebral Blood Flow *In Vivo*** ... 417
Steen G. Hasselbalch and Olaf B. Paulson

15 **Neuronal Basis of Non-Invasive Functional Imaging: From Microscopic Neurovascular Dynamics to BOLD fMRI** ... 433
Anna Devor, David A. Boas, Gaute T. Einevoll, Richard B. Buxton, and Anders M. Dale

16 **Imaging of Brain Function Using Deoxyhemoglobin and Magnetic Fields** ... 501
Kâmil Uğurbil

17 **FMRI Using Exogenous Agents and Cerebral Blood Volume Contrast** ... 547
Joseph B. Mandeville

18 **FMRI in Mice: Functional Phenotyping of Transgenic Mouse Lines Based on Hemodynamic Readouts** ... 593
Thomas Mueggler, Christof Baltes, Simone C. Bosshard, Felicitas L. Princz-Kranz, David Ratering, Florence Razoux, Esther Sydekum, and Markus Rudin

19 ***In Vivo* Studies of Brain Oxygen Metabolism and Oxidative Phosphorylation** ... 623
Xiao-Hong Zhu and Wei Chen

Part III Cerebral Energy Metabolism and Substrates *In Vivo*

20 Cerebral Glucose Transport and Homeostasis 655
João M.N. Duarte and Rolf Gruetter

21 Glucose Metabolism During Neural Activation 675
Marianne Fillenz

22 Measurements of Glycogen Metabolism in the Living Brain 699
Rolf Gruetter

23 Functions of Glycogen in the Central Nervous System 707
Richard D. Evans, Angus M. Brown, and Bruce R. Ransom

24 Pyruvate Transport and Metabolism in the Central Nervous System 715
Tiago B. Rodrigues, Alejandra Sierra, Paloma Ballesteros, and Sebastián Cerdán

25 Blood-Brain Barrier Transport of Lactate 755
Gitte Moos Knudsen

26 Cerebral Acetate Metabolism: Towards Its *In Vivo* Assessment 763
Basil Künnecke

27 Brain Fatty Acid Uptake 793
James A. Hamilton, Kellen Brunaldi, Richard P. Bazinet, and Paul A. Watkins

28 The Ketogenic Diet and Brain Metabolism 819
Marc Yudkoff, Yevgeny Daikhin, Oksana Horyn, Ilana Nissim, and Itzhak Nissim

29 Redox Shuttles in the Brain 841
Beatriz Pardo and Laura Contreras

30 Mitochondria in Neurodegeneration 885
Russell H. Swerdlow

Part IV *In Vivo* Assessment of Metabolic Compartmentation Between Neurons and Astrocytes

31 Studies of Brain Metabolism: A Historical Perspective 909
Arne Schousboe

32 Glutamate-Glutamine Cycle and Anaplerosis 921
Gülin Öz, David A. Okar, and Pierre-Gilles Henry

33 Brain Metabolic Compartmentalization, Metabolism Modeling, and Cerebral Activity-Metabolism Relationship 947
Michel Merle and Jean-Michel Franconi

34 Radiotracer Perspective on Cerebral Compartmentation 993
Bruno Weber, Renaud Jolivet, Matthias T. Wyss, and Alfred Buck

35 Nitrogen Metabolism *In Vivo* 1005
Keiko Kanamori and Brian D. Ross

Part V Cerebral Metabolism of Antioxidants, Osmolytes and Others *In Vivo*

36 Glutathione (GSH) Synthesis and Metabolism 1029
Maike M. Schmidt and Ralf Dringen

37 Brain Ascorbate: Protective, Yet Permissive for Redox Signaling 1051
Margaret E. Rice

38 N-Acetyl-Aspartate (NAA) Metabolism 1075
Kishore K. Bhakoo

39 *In Vivo* GABA Metabolism 1095
In-Young Choi and Jun Shen

40 ^{1}H-Magnetic Resonance Spectroscopy of Cerebral Phenylalanine Content and its Transport at the Blood-Brain Barrier 1117
Roland Kreis

41 Measurement of Brain Creatine Metabolism *In Vivo*: Magnetic Resonance Spectroscopy Studies of Transgenic Animals 1135
Arend Heerschap, Christine I.H.C. Nabuurs, Hermien E. Kan, Bé Wieringa, and Dirk Isbrandt

Index 1149

Contributors

Joseph J. H. Ackerman Departments of Chemistry, Radiology and Internal Medicine, Washington University, St. Louis, MO, USA
ackerman@wustl.edu

Silvio Aime Department of Chemistry IFM/Molecular Imaging Center, University of Turin, Turin, Italy
silvio.aime@unito.it

Paloma Ballesteros Laboratory of Organic Synthesis and Molecular Imaging, University at a Distance UNED, Madrid, Spain
pballesteros@ccia.uned.es

Christof Baltes Philips Healthcare, Best, The Netherlands
christof.baltes@philips.com

Richard P. Bazinet Department of Nutritional Sciences, University of Toronto, Toronto, ON, Canada
richard.bazinet@utoronto.ca

Kishore K. Bhakoo Translation Molecular Imaging, Singapore Bioimaging Consortium, Agency for Science, Technology and Research, Biopolis Way, Singapore
kishore_bhakoo@sbic.a-star.edu.sg

David A. Boas Martinos Center for Biomedical Imaging, Massachusetts General Hospital (MGH), Harvard Medical School, Charlestown, MA, USA
dboas@nmr.mgh.harvard.edu

Guido Böning Department of Nuclear Medicine, Ludwig Maximilian University of Munich, Munich, Germany
Guido.Boening@med.uni-muenchen.de

Simone C. Bosshard Institute for Biomedical Engineering, University and Eidgenössische Technische Hochschule (ETH) Zurich, Zurich, Switzerland
bosshard@biomed.ee.ethz.ch

Angus M. Brown University of Nottingham, Queens Medical Centre, Nottingham, UK
ambrown@nottingham.ac.uk

Kellen Brunaldi Department of Physiology and Biophysics, Boston University School of Medicine, Boston, MA, USA
kbrunaldi@gmail.com

Alfred Buck Department of Nuclear Medicine, University Hospital Zurich, Zurich, Switzerland
fred.buck@usz.ch

Richard B. Buxton Department of Radiology, Radiology - Center for Functional MRI, University of California San Diego, La Jolla, CA, USA
rbuxton@ucsd.edu

Ernest B. Cady Department of Medical Physics and Bioengineering, University College London Hospitals Foundation NHS Trust, London, UK
e.cady@medphys.ucl.ac.uk

Sebastián Cerdán Laboratory for Imaging and Spectroscopy by Magnetic Resonance, Institute of Biomedical Research "Alberto Sols" CSIC/UAM, Madrid, Spain
scerdan@iib.uam.es

Wei Chen Center for Magnetic Resonance Research, Department of Radiology, University of Minnesota, School of Medicine, Minneapolis, MN, USA
wei@cmrr.umn.edu

In-Young Choi Department of Neurology, Department of Molecular & Integrative Physiology, Hoglund Brain Imaging Center, University of Kansas Medical Center, Kansas City, KS, USA
ichoi@kumc.edu

Laura Contreras Departamento de Biología Molecular, Centro de Biología Molecular "Severo Ochoa", UAM-CSIC, Universidad Autónoma, Centro de Investigación Biomédica en Red en Enfermedades Raras (CIBERER), ISCIII, Cantoblanco, Madrid, Spain
lcontreras@cbm.uam.es

Paul Cumming Department of Nuclear Medicine, Ludwig Maximilian's University of Munich, Munich, Germany
paul.k.cumming@gmail.com

Yevgeny Daikhin Children's Hospital of Philadelphia and Department of Pediatrics, University of Pennsylvania School of Medicine, Philadelphia, PA, USA

Anders M. Dale Department of Neurosciences, Department of Radiology, University of California San Diego, San Diego, CA, USA
amdale@ucsd.edu

Dinesh K. Deelchand Center for Magnetic Resonance Research, Department of Radiology, University of Minnesota Medical School, Minneapolis, MN, USA
deelc001@umn.edu

Anna Devor Departments of Neurosciences and Radiology, University of California San Diego, La Jolla, CA, USA

Martinos Center for Biomedical Imaging, Massachusetts General Hospital, Harvard Medical School, Charlestown, MA, USA
adevor@ucsd.edu

Ralf Dringen Center for Biomolecular Interactions Bremen, University of Bremen, Bremen, Germany

Center for Environmental Research and Sustainable Technologies, Leobener Strasse, Bremen, Germany

School of Psychology and Psychiatry, Monash University, Clayton, VIC, Australia
ralf.dringen@uni-bremen.de

João M. N. Duarte Laboratory for functional and metabolic imaging, Ecole Polytechnique Federale de Lausanne, Lausanne, Switzerland
joao.duarte@epfl.ch

Gaute T. Einevoll Department of Mathematical Sciences and Technology, Norwegian University of Life Sciences, Aas, Norway
Gaute.Einevoll@umb.no

Haytham M. Eloqayli Department of Neuroscience, Jordan university of Science and Technology (JUST), Irbid, Jordan
hmeloqayli@just.edu.jo

Richard D. Evans School of Biomedical Sciences, Queens Medical Centre, University of Nottingham, Nottingham, UK

Marianne Fillenz Department of Physiology, Anatomy and Genetics, Oxford University, Oxford, UK
marianne.fillenz@dpag.ox.ac.uk

Jean-Michel Franconi Unité de Résonance Magnétique des Systèmes Biologiques, Université Victor Segalen, Bordeaux cedex, France
franconi@rmsb.u-bordeaux2.fr

Robin A. de Graaf Departments of Diagnostic Radiology and Biomedical Engineering, MRRC, Yale University, New Haven, CT, USA
robin.degraaf@yale.edu

Jan Bert Gramsbergen Molecular Neurosciences, South Denmark University, Odense, Denmark
jbgramsbergen@health.sdu.dk

Rolf Gruetter Laboratory for functional and metabolic imaging, Ecole Polytechnique Federale de Lausanne, Department of Radiology, University of Lausanne and Geneva, Station 6, CH-1015, Lausanne, Switzerland
rolf.gruetter@epfl.ch

Asta Haberg Department of Neuroscience, Norwegian University of Science & Technology, Trondheim, Norway
asta.haberg@ntnu.no

James A. Hamilton Department of Physiology and Biophysics, Boston University School of Medicine, Boston, MA, USA
jhamilt@bu.edu

Janniche Hammer Department of Neuroscience, Norwegian University of Science & Technology, Trondheim, Norway
Janniche.Hammer@abbott.com

Steen G. Hasselbalch Neurobiology Research Unit, Copenhagen University Hospital, Rigshospitalet, Copenhagen, Denmark
sgh@nru.dk

Arend Heerschap Department of Radiology, Radboud University Nijmegen Medical Centre, Nijmegen, The Netherlands
A.Heerschap@rad.umcn.nl

Pierre-Gilles Henry Center for Magnetic Resonance Research, Department of Radiology, University of Minnesota Medical School, Minneapolis, MN, USA
henry@cmrr.umn.edu

Oksana Horyn Children's Hospital of Philadelphia and Department of Pediatrics, University of Pennsylvania School of Medicine, Philadelphia, PA, USA

Kirsten Huinink Department of Psychiatry, University Medical Center of Groningen, Groningen, The Netherlands

Isabelle Iltis Center for Magnetic Resonance Research, Department of Radiology, University of Minnesota Medical School, Minneapolis, MN, USA
iltis001@umn.edu

Dirk Isbrandt Zentrum für Molekulare Neurobiologie, University Medical Center Hamburg-Eppendorf, ENP/ZMNH, Hamburg, Germany
isbrandt@enp.org

Renaud Jolivet Institute of Pharmacology and Toxicology, University of Zurich, Zurich, Switzerland
renaud.jolivet@a3.epfl.ch

Hermien Kan Department of Radiology, Radboud University Nijmegen Medical Centre, Nijmegen, The Netherlands
H.E.Kan@lumc.nl

Keiko Kanamori Magnetic Resonance Spectroscopy Laboratory, Huntington Medical Research Institutes, Pasadena, CA, USA
kkanamori@hmri.org

Gitte Moos Knudsen Neurobiology Research Unit, section 9201, Copenhagen University Hospital - Rigshospitalet, Copenhagen, Denmark
gmk@nru.dk; gitte@nru.dk

Jakob Korf Department of Psychiatry, University Medical Center of Groningen, Groningen, The Netherlands
j.korf@psy.umcg.nl

Roland Kreis Department of Clinical Research, University Bern, AMSM, DRNN/DKF, Bern, Switzerland
roland.kreis@insel.ch

Basil Künnecke Magnetic Resonance Imaging and Spectroscopy, CNS Discovery and Translational Area, F. HoffmannLa Roche Ltd, Basel, Switzerland
basil.kuennecke@roche.com

Phil Lee Department of Molecular & Integrative Physiology, Hoglund Brain Imaging Center, University of Kansas Medical Center, Kansas City, KS, USA
plee2@kumc.edu

Annemie Van der Linden Bio-Imaging Lab, Campus Drie Eiken, Building Uc, Department of Biomedical Sciences, University of Antwerp, Antwerp (Wilrijk), Belgium
annemie.vanderlinden@ua.ac.be

Dario Longo Department of Chemistry IFM, Molecular Imaging Center, University of Torino, Turin, Italy
dario.longo@unito.it

Joseph B. Mandeville Athinoula A. Martinos Center for Biomedical Imaging, Massachusetts General Hospital, Charlestown, MA, USA
jbm@nmr.mgh.harvard.edu

Malgorzata Marjanska Center for Magnetic Resonance Research, Department of Radiology, University of Minnesota Medical School, Minneapolis, MN, USA
gosia@umn.edu

Vincent Van Meir Bio-Imaging Lab, Campus Drie Eiken, Building Uc, Department of Biomedical Sciences, University of Antwerp, Antwerp (Wilrijk), Belgium
vincent.vanmeir@ua.ac.be

Torun Melø Department of Neuroscience, Norwegian University of Science & Technology, Trondheim, Norway
torun.melo@ntnu.no

Hellmut Merkle Laboratory of Functional and Molecular Imaging, National Institute of Neurological Disorders and Stroke, National Institutes of Health, Bethesda, MD, USA
merkleh@mail.nih.gov

Michel Merle Unité de Résonance Magnétique des Systèmes Biologiques, Université Victor Segalen, Bordeaux cedex, France
ftc.merle@gmail.com

Mark A. Mintun Department of Radiology, Avid Radiopharmaceuticals, Philadelphia, PA, USA
mintunm@mir.wustl.edu

Thomas Mueggler Pharmaceuticals Division, Preclinical Translational Neuroscience, Basel, Switzerland
thomas.mueggler@roche.com

Christine I. H. C. Nabuurs Department of Radiology, Radboud University Nijmegen Medical Centre, Nijmegen, The Netherlands
christine.nabuurs@maastrichtuniversity.nl

Jeffrey J. Neil Departments of Pediatric Neurology, Pediatrics and Radiology, Washington University School of Medicine, St. Louis, MO, USA
neil@wustl.edu

Ilana Nissim Children's Hospital of Philadelphia and Department of Pediatrics, University of Pennsylvania School of Medicine, Philadelphia, PA, USA

Itzhak Nissim Children's Hospital of Philadelphia and Department of Pediatrics, University of Pennsylvania School of Medicine, Philadelphia, PA, USA
ssitz@mail.med.upenn.edu

David A. Okar Department of Veterans Affairs Medical Center, Department of Biochemistry, Molecular Biology and Biophysics, University of Minnesota, Minneapolis, MN, USA
David.Okar@va.gov

Gülin Öz Center for Magnetic Resonance Research, Department of Radiology, University of Minnesota Medical School, Minneapolis, MN, USA
gulin@cmrr.umn.edu

Beatriz Pardo Departamento de Biología Molecular, Centro de Biología Molecular "Severo Ochoa" UAM-CSIC, Universidad Autónoma, Centro de Investigación Biomédica en Red en Enfermedades Raras (CIBERER), ISCIII, Cantoblanco, Madrid, Spain
bpardo@cbm.uam.es

Olaf B. Paulson Memory Disorders Research Unit and Neurobiology Research Unit, Rigshospitalet, and Danish Research Centre for Magnetic Resonance, Hvidovre Hospital, Faculty of Health Sciences, Copenhagen University, Denmark
olaf.paulson@nru.dk

Felicitas L. Princz-Kranz Institute for Biomedical Engineering, University and Eidgenössische Technische Hochschule (ETH) Zurich, Zurich, Switzerland
princz@biomed.ee.ethz.ch

Bruce R. Ransom Department of Neurology, University of Washington School of medicine, Seattle, WA, USA
bransom@u.washington.edu

David Ratering Institute for Biomedical Engineering, University and Eidgenössische Technische Hochschule (ETH) Zurich, Zurich, Switzerland
ratering@biomed.ee.ethz.ch

Florence Razoux Institute for Biomedical Engineering, University and Eidgenössische Technische Hochschule (ETH) Zurich, Zurich, Switzerland
razoux@hifo.uzh.ch

Margaret E. Rice Department of Neurosurgery, Department of Physiology and Neuroscience, New York University School of Medicine, New York, NY, USA
margaret.rice@nyu.edu

Tiago B. Rodrigues Laboratory for Imaging and Spectroscopy by Magnetic Resonance (LISMAR), Institute of Biomedical Research "Alberto Sols" CSIC/UAM, Madrid, Spain
tiago.rodrigues@cancer.org.uk; tiago.b.rodrigues@gmail.com

Brian D. Ross Magnetic Resonance Spectroscopy Laboratory, Huntington Medical Research Institutes, Pasadena, CA, USA
bdross.hmri@gmail.com

Markus Rudin Institute for Biomedical Engineering, Institute of Pharmacology and Toxicology, University and Eidgenössische Technische Hochschule (ETH) Zurich, University of Zurich, Zurich, Switzerland
rudin@biomed.ee.ethz.ch

Jorgina Satrústegui Department of Molecular Biology, Center for Molecular Biology Severo Ochoa CSIC-UAM, Center for Biomedical Research on Rare Diseases, Universidad Autónoma Madrid Campus of Cantoblanco, Madrid, Spain
jsatrustegui@cbm.uam.es

Maike M. Schmidt Center for Biomolecular Interactions, Center for Environmental Research and Sustainable Technologies, University of Bremen, Bremen, Germany
maike.schmidt@uni-bremen.de

Arne Schousboe Department of Pharmacology and Pharmacotherapy, Faculty of Pharmaceutical Sciences, University of Copenhagen, Copenhagen, Denmark
as@farma.ku.dk

Jun Shen Section on Magnetic Resonance Spectroscopy, National Institute of Mental Health, National Institutes of Health, Molecular Imaging Branch, Mood & Anxiety Disorders Program, Bethesda, MD, USA
shenj@intra.nimh.nih.gov

Alexander A. Shestov Department of Radiology, Center for Magnetic Resonance Research, University of Minnesota Medical School, Minneapolis, MN, USA
shest001@umn.edu

Miho Shidahara Division of Medical Physics, Tohoku University Graduate School of Medicine, Sendai, Miyagi, Japan
shidahara@med.tohoku.ac.jp

C. William Shuttleworth Department of Neurosciences, University of New Mexico School of Medicine, Albuquerque, NM, USA
bshuttleworth@salud.unm.edu

Alejandra Sierra Laboratory for Imaging and Spectroscopy by Magnetic Resonance (LISMAR), Institute of Biomedical Research "Alberto Sols", Madrid, Spain

Allen W. Song Duke University Medical Center, Brain Imaging and Analysis Center, Durham, NC, USA
allen.song@duke.edu

Ursula Sonnewald Department of Neuroscience, Norwegian University of Science & Technology, Trondheim, Norway
ursula.sonnewald@ntnu.no

Russell H. Swerdlow Departments of Neurology, Molecular & Integrative Physiology, University of Kansas School of Medicine, Kansas City, KS, USA
rswerdlow@kumc.edu

Esther Sydekum Institute for Biomedical Engineering, University and ETH Zurich, Zurich, Switzerland
sydekum@biomed.ee.ethz.ch

Kâmil Uğurbil Department of Radiology, Neurosciences, and Medicine Center for Magnetic Resonance Research, University of Minnesota, Medical School, Minneapolis, MN, USA
kamil@cmrr.umn.edu

Julien Valette Center for Magnetic Resonance Research, Department of Radiology, University of Minnesota Medical School, Minneapolis, MN, USA

Andrei G. Vlassenko Department of Radiology, Washington University School of Medicine, St. Louis, MO, USA
andrei@npg.wustl.edu

Paul A. Watkins Kennedy Krieger Institute and Department of Neurology, Johns Hopkins University School of Medicine, Baltimore, MD, USA
watkins@kennedykrieger.org

Bruno Weber Institute of Pharmacology and Toxicology, University of Zurich, Zurich, Switzerland
bweber@pharma.unizh.ch

Bé Wieringa Department of Cell Biology, Radboud University Nijmegen Medical Centre, Nijmegen, The Netherlands
b.Wieringa@ncmls.ru.nl

Matthias T. Wyss Department of Nuclear Medicine, University Hospital Zurich, Zurich, Switzerland
matthias.wyss@usz.ch

Marc Yudkoff Children's Hospital of Philadelphia and Department of Pediatrics, University of Pennsylvania School of Medicine, Philadelphia, PA, USA

Division of Child Development, Rehabilitation and Metabolic Disease, Children's Hospital of Philadelphia, Philadelphia, PA, USA
yudkoff@email.chop.edu

Habib Zaidi Division of Nuclear Medicine and Molecular Imaging, Geneva University Hospital, Geneva, Switzerland

Geneva Neuroscience Center, Geneva University, Geneva, Switzerland

Department of Nuclear Medicine and Molecular Imaging, University Medical Center Groningen, University of Groningen, Groningen, Netherlands
habib.zaidi@hcuge.ch

Xiao-Hong Zhu Center for Magnetic Resonance Research, Department of Radiology, University of Minnesota, School of Medicine, Minneapolis, MN, USA
zhu@cmrr.umn.edu

Part I
Non-invasive Methods for Neural Metabolism

Chapter 1
Introduction to Nuclear Magnetic Resonance (NMR) Methods

Allen W. Song

1.1 Introduction

Magnetic resonance imaging, commonly known as MRI, is perhaps one of the most important inventions in medical imaging and diagnosis. After the first image formation in the early 1970s (Damadian 1971; Lauterbur 1973), MRI has gone through a relatively quiet infancy, largely due to the lack of the necessary hardware. However, even since its first commercialization in the early 1980s, MRI has revolutionized the field of medical imaging, and evolved to be the gold standard for imaging brain and other soft tissues. In over two decades' time, MRI has affected most aspects of clinical diagnosis and treatment.

What makes MRI so powerful and popular? Two compelling reasons may be its ability to produce the clearest and most uniform images of organs, and its complete non-invasiveness. In this chapter, we will attempt to introduce MRI taking a mostly conceptual path, starting from the magnetic field and magnetization (M), signal generation and reception through proton resonance (R), to spatial encoding for generation of images (I). We will also introduce the versatility of MRI contrast generation, which can be used to emphasize different tissue characteristics. To convey both the elegance and the rigor of the MRI theory is a great challenge faced by teachers of MRI. As many prefer to learn by intuition and analogy, this section is written with very limited dependence on detailed equations, in order to provide the many readers a firm conceptual grasp on the key principles of MRI. It is hoped that readers will be able to understand the many basic but key concepts, so as to facilitate their continued reading further into this handbook.

A.W. Song, Ph.D. (✉)
Duke University Medical Center, Brain Imaging and Analysis Center,
Box 3918, Durham, NC 27710, USA
e-mail: allen.song@duke.edu

I.-Y. Choi and R. Gruetter (eds.), *Neural Metabolism In Vivo*,
Advances in Neurobiology 4, DOI 10.1007/978-1-4614-1788-0_1,

1.2 Magnetic Field and Magnetization

1.2.1 Magnetic Field

MRI scanners typically use a strong static magnetic field to align certain nuclei within the human body (most commonly, hydrogen within water molecules) to allow mapping of tissue properties. Early MRI scanners used permanent magnets to generate magnetic fields, which are usually weak and hence lead to low MR signal. Another way of generating a magnetic field was discovered by the Danish physicist Oersted in 1820, when he demonstrated that a current-carrying wire influenced the direction of a compass needle below the wire, redirecting it perpendicularly to the direction of current. This relation was quantified later that year by the French physicists Biot and Savart, who discovered that magnetic field strength was in fact proportional to current strength, so that by adjusting the current in a wire (or sets of wires), one could precisely control field intensity. These findings led to the development of electromagnets, which generate their fields through tight coils of wire. Modern MRI scanners use superconducting electromagnets with wires cooled by cryogens (e.g., liquid helium) to reduce their temperature to near absolute zero.

To generate an extremely large magnetic field, one can inject a huge electric current into the loops of wire. For example, the very large electromagnets used to lift cars in junkyards have magnetic fields on the order of 1 T, similar to that in the center of a MRI scanner. To generate this field, they require enormous electrical power, as much as tens of kilowatts, and thus enormous expense. The use of superconductive materials in MRI scanners allows strong fields to be created without prohibitive cost. Coil windings are typically made of metal alloys such as niobium-titanium, which when immersed in liquid helium can reach temperature at ~4°K (−269°C), less than the critical temperature threshold (10°K) of this particular alloy. In this low-temperature state, the resistance in the wires disappears, thereby enabling a strong and lasting electric current to be generated with no power requirements, which in turn causes a very strong magnetic field.

Combining the precision derived from numerical optimization of the magnetic coil design and the strength afforded by superconductivity, modern MRI scanners can have homogeneous and stable field strengths in the range of 1–9 T for human use and up to 20 T for animal use. In comparison, the earth's magnetic field is approximately 0.00005 T. A typical MRI scanner is shown in Fig. 1.1, the tubular main structure contains the superconductor main magnet, as well as the necessary coils for generating signal and creating images (discussed in detail in the following sections). Since maintaining a field using superconductive wiring requires no electricity, the static fields used in MRI with superconducting magnets are always active, even when no images are being collected. For this reason, these static fields present significant safety challenges, as will be discussed later in this chapter.

Magnetic field creates magnetization in certain objects. A classic demonstration of magnetism can be created by sprinkling some iron filings around a standard bar magnet. The filings clump most densely around the poles of the magnet but also form a series of

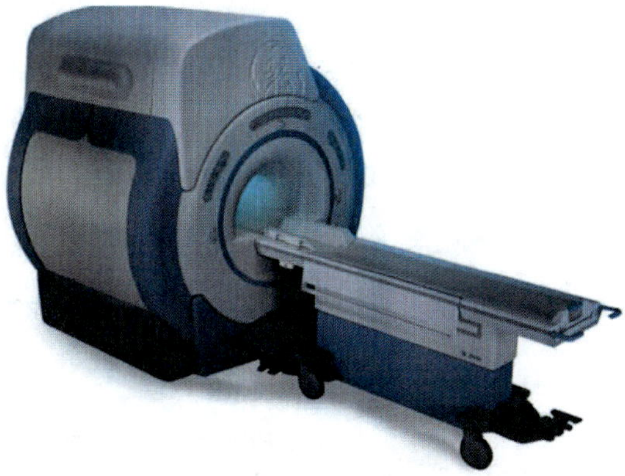

Fig. 1.1 A typical MRI scanner

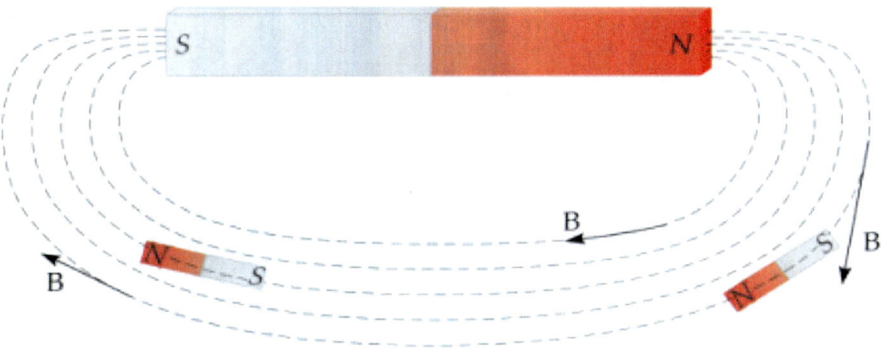

Fig. 1.2 Alignment of bar magnets in an external magnetic field

arcs between the poles. These arcs run along the field lines of the magnet and result from the tendency of individual iron filings to align with the external field. Figure 1.2 presents a schematic illustration of this alignment, which is driven by the principle of energy minimization. Just as massive objects in a gravitational field tend to lower their energy by falling rather than remaining suspended in midair, magnetically susceptible objects in a magnetic field will orient along the field lines rather than across them. For macroscopic objects, like oxygen canisters, the alignment process is known as torsion and presents safety issues. More severely, when placed near magnets where the magnetic field changes are steep, these large magnetic objects will experience tremendous forces attracting them toward the magnets, resulting in the dangerous "projectile" effect. The pattern of field lines is a mathematical description of the contours of the magnetic field, with the density of lines in a location indicating the local strength, or flux, of the magnetic field. In magnetic resonance imaging, the main magnetic field of the scanner is often indicated by the symbol $\mathbf{B}_0$.

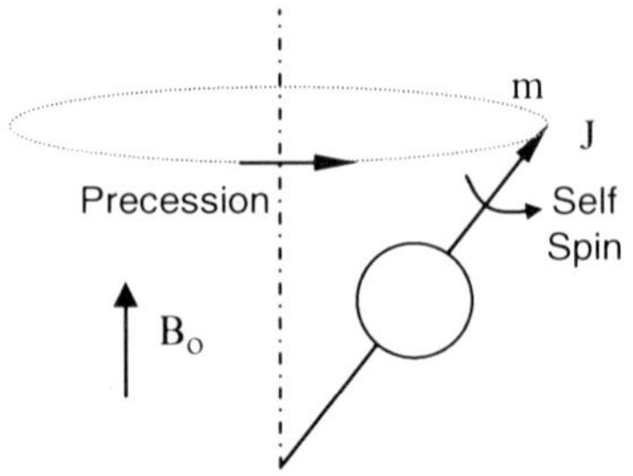

Fig. 1.3 Illustration of precession motion for self-spinning protons

The effect of a magnet on iron filings is intuitive, as they will be magnetized and therefore align with the direction of the magnetic field. What happens if we place biological tissue in a magnetic field?

1.2.2 The Effect of Magnetic Field on Protons in Biological Tissue

To understand the effect of magnetic field on biological tissue, e.g. human body, we will need to first characterize its composition. An average person who weighs 150 lbs. contains approximately 5×10^{27} hydrogen protons (because of the high water content). As such, protons are by far the most dominant nuclei in the human body, and are the main sources of signal for the vast majority of the MRI exams. We will thus focus on hydrogen protons in this introduction chapter. It is worth noting that many other nuclei with odd number of protons and atomic mass (i.e. simultaneous possession of magnetic moment and angular momentum), such as ^{13}C, ^{19}F, ^{23}Na, and ^{31}P, can also generate MR signal.

Under normal conditions, thermal energy causes the proton to spin about itself. The spin motion of a proton has two effects. First, because the proton carries a positive charge, its spin generates an electrical current on its surface, just as a moving electrical charge in a looped wire generates current. This loop current creates a small proton magnetic field and a torque when the proton is placed within an external magnetic field; the strength of the proton magnetic field or the maximum torque (i.e. when placed orthogonally into an external magnetic field) per unit magnetic field strength is called the magnetic moment, or **m**. Second, because the proton has an odd-numbered atomic mass (i.e., 1), its spin results in an angular momentum, or **J**. Both **m** and **J** are vectors pointing in the same direction, along the spin axis.

When placed in a magnetic field, protons, like iron filings, would change their orientation because of their magnetic moment. However, instead of turning to align with the magnetic field, the spinning protons initiate a gyroscopic motion known as precession (Fig. 1.3) because of their simultaneous possession of angular momentum. To understand precession, imagine a spinning top on a desk. The top does not spin perfectly upright; instead its axis of rotation traces a circle perpendicular to the

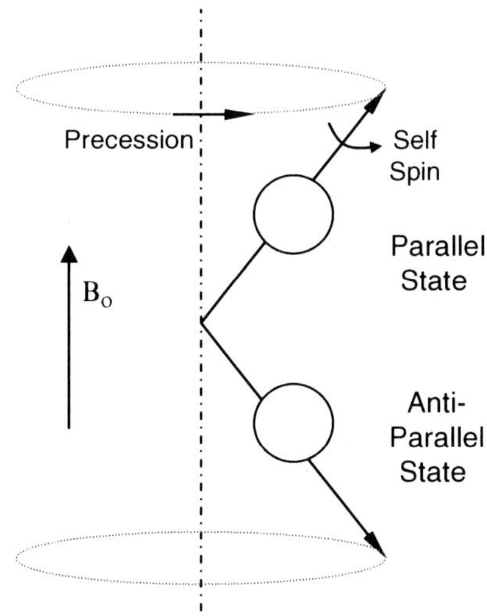

Fig. 1.4 Parallel and antiparallel states of precession

earth's gravitational field. At any moment in time the top is tilted from the vertical, but it does not fall. Why does the top spin at an angle? Spinning objects respond to applied forces by moving their axes in a direction perpendicular to the applied force. A bicycle, for example, is very stable and resists falling over at high speeds, due to the gyroscopic effects of its spinning wheels. When a rider leans to one side, the moving bicycle will not fall but will instead turn in that direction. Similarly, a spinning top turns its axis of rotation at an angle perpendicular to the force exerted by gravity, so that the top precesses in a circle around a vertical axis. Because the precession frequency is determined by the type of nucleus, all protons precess at the same frequency when experiencing the same strength of magnetic field. This characteristic frequency is called the Larmor Frequency.

Interestingly, there are two states for precessing protons: one parallel to the magnetic field and the other antiparallel (Fig. 1.4). Protons in the parallel state have a lower energy level, while protons in the antiparallel state have a higher energy level. The idea of two energy states can be understood by imagining a bar that can rotate around one end. There are two vertical positions for the bar: one balanced above the pivot point and one hanging down from the pivot point. The balanced position is a high-energy state and is not very stable; even a small perturbation may tip the bar over and cause it to fall to the hanging position. The only way to keep the bar in a balanced position is to apply an external force that can counteract gravity. That is, energy must be applied to keep the bar in the high-energy state. The hanging position is much more stable, since it is at the minimum energy level for this system. For protons, the parallel (low-energy) state is slightly more stable, so there will always be more protons in the parallel state than in the antiparallel state, with the relative proportion of the two states dependent upon temperature and the strength of the magnetic field. At room temperature in the earth's magnetic field, roughly equal numbers of protons are in the two energy states, with slightly more in the parallel state.

If the temperature increases, some protons will acquire more energy and jump to the antiparallel state, diminishing but never reversing the already small difference between the two levels. Conversely, if the temperature decreases, protons will possess less energy and even more will stay at the lower energy level. This relationship is described by the Boltzmann distribution $\frac{P_p}{P_a} = e^{\frac{\Delta E}{k_B T}}$, where P_p and P_a are the proportions of protons in parallel and anti-parallel states, respectively, here $P_p + P_a = 1$, T is the absolute temperature, ΔE is the energy difference between the parallel and anti-parallel states, k_B is the so-called Boltzmann constant (1.3806×10^{-23} J/K).

The discrete energy difference between the two states (ΔE) can be measured experimentally. Interestingly, this energy difference is equal to the energy possessed by photons at the precession frequency under the same magnetic field strength. As such, the precessing protons may absorb energy from oscillating electromagnetic fields at the same frequency and change their states. This phenomenon, which was confirmed by the first magnetic resonance experiment carried out by Rabi's laboratory in the 1930s (Rabi et al. 1938), provides a basis for unifying the quantum and classical mechanics concepts in the context of proton excitation discussed later. The precession frequency (ω) is known as the Larmor frequency, and defined by the Larmor equation $\omega = \gamma B_o$ where γ is the gyromagnetic ratio and B_o the magnetic field strength. The gyromagnetic ratio differs for different nuclei, for hydrogen proton, the ratio is 2.67×10^8 radian/T (or 42.58×10^6 cycles/T).

1.2.3 Magnetization of Biological Tissue

It is important to emphasize that MR techniques do not detect just a single hydrogen nucleus, but instead measure the net magnetization of all protons in a volume. Because of the enormous number of protons within even the smallest volume, their transverse components will tend to cancel out, and there will be no net magnetization perpendicular to the main magnetic field (i.e. in the transverse plane). The net magnetization, known as **M**, is thus a vector whose orientation is along the longitudinal direction parallel to the main field, and whose magnitude is proportional to the difference in the number of the proton spins in the parallel and antiparallel states. The more protons at the parallel state, the bigger the **M** (Fig. 1.5), and in turn the larger the MR signal.

Because the proportion of parallel-state protons increases with decreasing temperature, as indicated by the Boltzmann equation, one way to increase the net magnetization is to reduce the temperature. While theoretically possible, this approach is impractical, given that the temperature would need to be lowered by many degrees Celsius to observe a noticeable increase in net magnetization. A more feasible approach is to increase the strength of the external field, based on the Zeeman effect – which states that the net magnetization is proportional to the magnetic field strength. Just as it would be more difficult to lift an object in a stronger gravitational

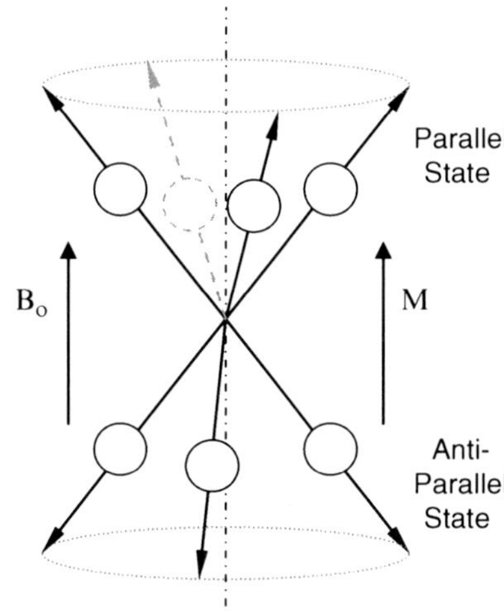

Fig. 1.5 Small difference in parallel and antiparallel protons create a net magnetization M

field than in a weaker gravitational field, it takes more energy to shift from a low-energy state to a high-energy state when the external magnetic field is stronger. Therefore, to increase the net magnetization in a sample (i.e., the proportion of parallel protons), one can place that sample in a very strong magnetic field.

Both the increase in net magnetization and the energy difference between the two states are linearly proportional to the strength of the magnetic field. The combination of these two factors means that the amplitude of the measured MR signal increases quadratically with field strength (e.g., the MR signal is four times as large in a 3 T scanner than in a 1.5 T scanner). Thus, even though the noise in the MR signal increases linearly with field strength, there remains a substantial signal-to-noise advantage to having stronger MRI scanners. For this reason, scanners commonly used for MRI in humans have continually progressed from a field strength of 1.5 T or less in early years, to the wide adoption of 3 T or even higher fields at the present time.

The net magnetization of protons within a volume provides the basis for MR signal generation, but net magnetization itself cannot be measured directly under equilibrium conditions. To understand this principle, recall the analogy of an object whose weight you are trying to estimate. You cannot know the object's weight just by looking at it; instead, you have to lift it. By lifting the object, you perturb its equilibrium state in the gravitational field, and the force observed in reaction to that perturbation allows you to estimate its weight. Measuring net magnetization of protons in a magnetic field is no different; you must perturb the equilibrium state of the protons and then observe how they react to the perturbation. Just as lifting perturbs the position of an object, excitation perturbs the orientation of the net magnetization, which allows its precession to be visible. Because the net magnetization is the sum of all the individual protons, it would naturally precess at the same characteristic frequency, the Larmor frequency.

1.3 Excitation and Reception of MR Signal

Following Rabi's experiments in magnetic resonance, independent groups led by Purcell and Bloch detected MR signal using radiofrequency transmitter and receiver coil (Purcell et al. 1946; Bloch 1946). The phenomenon was then labeled nuclear induction. The early work by these MR physicists earned them respective Nobel Prizes (Rabi in 1944; Purcell and Bloch in 1952), but more importantly, helped establish a core set of physical principles for MR signal generation. These principles are elegant in their simplicity, in that they begin with the properties of atomic nuclei and progress to the signal measured using MRI. Yet, they are also rigorous. In this section, we will introduce the concept of excitation and signal reception using resonating **radiofrequency (RF) coils**, as well as signal relaxation mechanisms.

1.3.1 Radiofrequency Coils

While a strong magnetic field establishes magnetization, it in itself does not produce any MR signal. MR signal is actually produced by the clever use of two types of RF coils, known as transmitter and receiver coils, that resonate at the same frequency as the precessing hydrogen magnetization generated by the main magnet. This process gives the name "resonance" to magnetic resonance imaging. Unlike the static magnetic field, the radiofrequency fields are turned on briefly during periods of the imaging process.

When the RF coils send electromagnetic waves into the body, in a process known as **excitation**, protons at lower energy state (i.e. parallel state) absorb energy and change to higher energy state (i.e. anti-parallel state). When the radiofrequency pulse ends, the protons then release the energy absorbed during excitation. The resulting release of energy can be detected by the RF coils, in a process known as **reception**. One can think of the measurement of MR signal through excitation and reception as analogous to the weighing of an object by lifting and releasing it in a gravitational field. If an object sits motionless on a supporting surface, so that it is in an equilibrium state with respect to gravitational force, we have no information about its weight. To weigh it, we first lift the object to give it potential energy and then release it so that it transfers that energy back into the environment. The amount of energy it releases, whether through impact with a surface or compression of a device like a spring (e.g., in a scale), provides an index of its weight. In the same way, we can perturb the magnetic properties of atomic nuclei (excitation) and then measure the amount of energy returned during their recovery back to an equilibrium state.

The amount of energy that can be transmitted or received by the radiofrequency coils depends upon the distance from the sample being measured. In the case of fMRI, the radiofrequency coils are typically placed immediately around the head, either in a surface coil (a coil that only covers a partial area) or volume coil (a coil that encompasses the entire volume) arrangement. Recent advances in coil design combine the best features of both coil types by using a volume coil for exciting the imaging volume and a set of surface coils for receiving the MR signal. If multiple

Fig. 1.6 An eight-channel array RF coil for head imaging

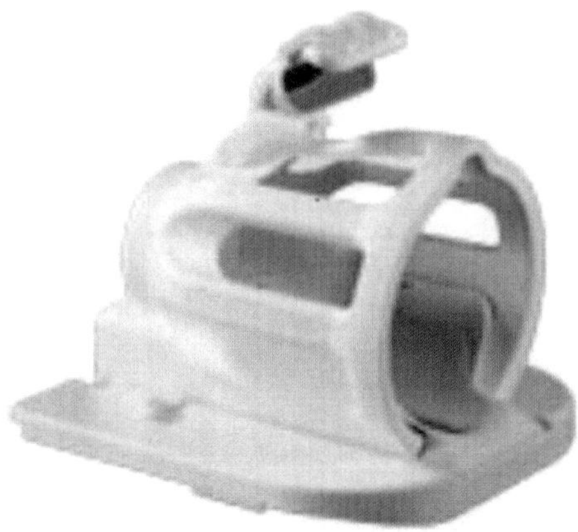

receiver coils are arranged in an overlapping pattern (e.g., phased array coils), the spatial coverage of a single coil can be increased considerably while the high sensitivity of the coils is still maintained. Shown in Fig. 1.6 is an example of a head coil array with eight-channel surface coils. Though sensitivity does change somewhat across the image, the use of multiple receiver coils is an increasingly important technique in fMRI.

1.3.2 Excitation and Signal Reception

We learned that protons can take either a high-energy state or a low-energy state within a magnetic field. The transitions between these two states are sparked by the perturbation of externally applied energy equal to the energy difference between the two states, in the form of the radiofrequency pulse transmission. For example, a proton in the low-energy state can jump to the high-energy state by absorbing energy from the radiofrequency pulse. Such a perturbation changes the probabilistic distributions of the high-energy and low-energy protons, with the rate of change favoring that going from the more populated state to the less populated state. Let's begin with an original condition, where there are more protons at the lower energy level, some protons in the low-energy state absorb the appropriate amount of energy and change to the high-energy state, in a process known as excitation (Fig. 1.7a). At any point when the electromagnetic waves (RF pulses) are turned off, the protons stop changing states. For example, the RF pulse can stop at the time when there are equal numbers at the two states (zero net longitudinal magnetization). Alternatively, in the classical mechanical description, it is the so-called 90 degree RF pulse – as the original net magnetization can be viewed as a vector in the longitudinal direction, and zero net longitudinal magnetization is only possible when the vector turns 90 degrees into the transverse plane. Importantly, the most common radiofrequency pulses for excitation are often designed to fully transition the longitudinal magnetization to the transverse plan, and the most efficient one of such is the 90 degree pulse.

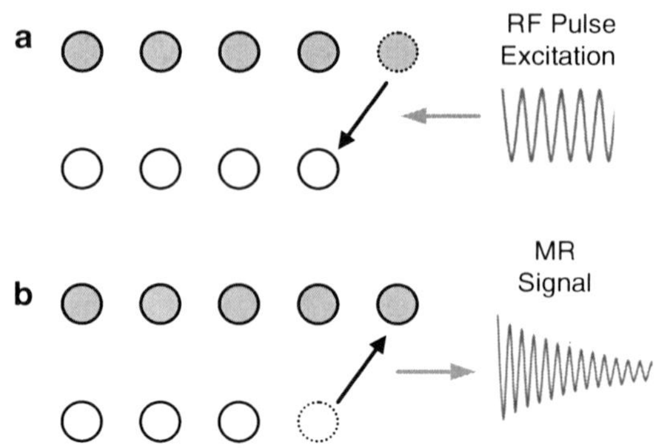

Fig. 1.7 Schematic illustration of proton excitation (**a**) and MR signal generation (**b**)

Since excitation disrupts the thermal equilibrium by creating more high-energy protons than would otherwise be present, immediately after excitation ceases the excess protons at the higher energy level return to the lower level (Fig. 1.7b) to reestablish the equilibrium proportions of energy states. During the time when protons in the high-energy state fall back to the low-energy state, they emit photons back to the system, with energy equal to the energy difference between the two states. Alternatively, in classical mechanics view, the precessing transverse magnetization will broadcast electromagnetic waves oscillating at the Larmor frequency, inducing a sinusoidally-varying voltage in the radiofrequency receiver coil tuned to the same frequency and placed perpendicular to the transverse plane. Because the frequencies of excitation and reception are both at the same Larmor frequency determined by the energy difference between the two states, often the same radiofrequency coil is used for both processes.

The critical concept underlying excitation and reception is that of the change in net magnetization from the longitudinal axis to the transverse plane. When the net magnetization is along the longitudinal axis, the individual protons' precession cannot be measured in detector coils. But, when the net magnetization is tipped to the transverse plane via excitation, its precession now generates an oscillating electric current in receive coils within the scanner.

1.3.3 MR Signal Relaxation Mechanisms

The MR signal detected through receiver coils does not remain stable forever. It changes in two ways after proton excitation and during signal reception: the transverse magnetization quickly loses coherence and the longitudinal magnetization will slowly recover. Together, these changes in the MR signal over time are called relaxation, with the former termed transverse relaxation (or transverse decay), also known as spin-spin relaxation, and the latter longitudinal relaxation (or longitudinal recovery), also known as spin-lattice relaxation. Both mechanisms constrain MR imaging parameters yet allow the generation of different types of images.

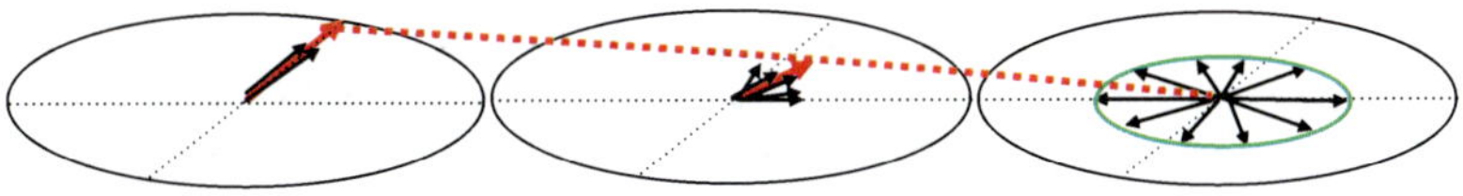

Fig. 1.8 Illustration of transverse (T2) relaxation (decay): the net magnetization vector (in *red*), which is the sum of individual spins (in *black*), decays in time (from *left* to *right*)

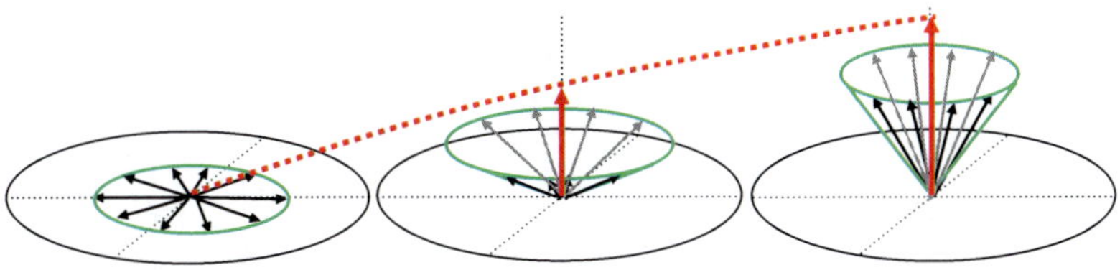

Fig. 1.9 Illustration of longitudinal (T1) relaxation (recovery): the net magnetization vector (in *red*), which is the sum of individual spins (in *black*), recovers in time (from *left* to *right*)

After excitation, the net magnetization is tipped from the longitudinal axis into the transverse plane. Because the net magnetization reflects the vector sum of many individual proton spins, its amplitude depends on the coherence between those spins, specifically that they precess at the same phase and with the same frequency. Yet, over time, they lose coherence. Spatially proximate spins may interact via dipolar coupling, analogously to bumper cars on a track, causing some to precess at higher frequencies, and some lower. This causes a loss of phase coherence among the protons, which leads to an exponential decay in the MR signal that is described by a time constant called T_2.

In addition, any spatial inhomogeneities in the magnetic field will cause different protons to experience different magnetic fields over time, changing their rate of precession. This effect is additive to the T_2 decay, and thus the combined transverse relaxation effect of proton spin-spin interactions and magnetic field inhomogeneity is described by the time constant T_2*. As a result of T_2 or T_2* decay, spins lose coherence relatively quickly (typically within a few tens of milliseconds), resulting in a diminishing net magnetization in the transverse plane (Fig. 1.8).

At the same time while spins are losing coherence due to spin-spin interactions, energy that was absorbed during excitation is emitted as radiofrequency waves that can be detected as MR signal. As the system loses energy, it recovers back to the same state as before the excitation – with the net magnetization pointing along the longitudinal axis (Fig. 1.9). The longitudinal recovery is relatively slow, typically on the order of a few hundreds of milliseconds to a few seconds, and is described by the time constant $\mathbf{T_1}$. The values of T_1 and T_2 for a spin system depend upon the kinds of substance present.

Although the T_1 and T_2 relaxations are plotted as separate processes, which is an approximation based on their largely different time constants (e.g. T_1 is usually one order of magnitude larger than T_2), these two relaxations occur simultaneously. And thus, for practical purposes they can be considered to make separate contributions to the MR signal. Depending upon when an image is acquired during the relaxation process, either T_1 or T_2 (or a combination) will determine the amplitude of the recovered MR signal and thus the intensity of the image. By choosing appropriate imaging parameters, different tissues (e.g. gray vs. white matter) will produce different intensities (and hence contrast), so that investigators can differentiate them for diagnostic or research purposes. In addition, the speed of T_1 recovery influences the rate at which images can be collected, in that the T_1 recovery renews the longitudinal magnetization so that it can be again excited.

1.4 MR Image Formation

The fundamental innovation that made MRI possible was the introduction of superimposed gradient magnetic fields (spatially varying magnetic fields) (Lauterbur 1973; Mansfield 1977). Because of the dependency of the precession frequency on the strength of magnetic field, gradient magnetic fields will lead to spatially dependent distribution of precession frequencies. By differentiating these frequencies according to their spatial locations, we can generate maps (or images) of proton distribution. This innovation has such a large impact in the field of MRI that it has earned Paul Lauterbur and Peter Mansfield a share of the Nobel Prize for Physiology and Medicine in 2003.

1.4.1 Gradient Coils

The combination of a static magnetic field and a radiofrequency coil allows detection of MR signal, but MR signal alone cannot be used to create an image. The fundamental measurement in MRI is merely the amount of current through a coil, which in itself has no spatial information. By introducing magnetic gradients superimposed upon the strong magnetic field, **gradient coils** provide the final component necessary for imaging. The purpose of a gradient coil is to cause the MR signal to become spatially dependent, so that different locations in space contribute differently to the measured signal over time. The gradient fields can be turned on and off, briefly after the excitation process, to provide spatial encoding needed to resolve an image.

To make the recovery of spatial information along any given direction, a gradient coil is used to generate a magnetic field that increases in strength along that particular spatial direction. To resolve all three dimensions, typically three gradient coils are required. Shown in Fig. 1.10 is a schematic illustration of the gradient magnetic fields along x, y and z directions. Note that the gradient direction indicates the direction of change of the magnetic field, not the direction of the field itself (which always points along z).

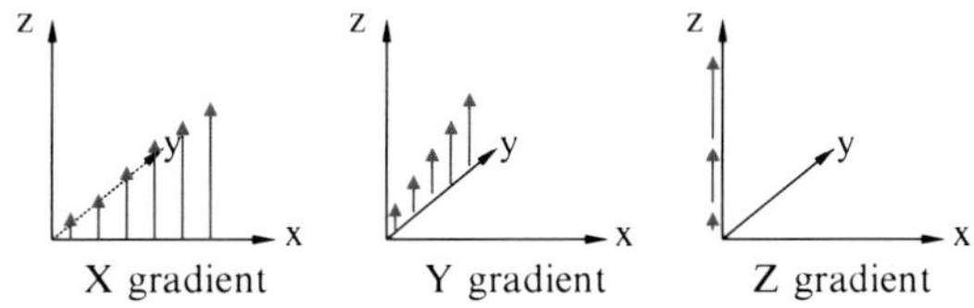

Fig. 1.10 Illustration of gradient fields along x, y and z directions. The vectors (in *blue*) indicate the strength and direction of the magnetic field

The strength of the gradient coil is a function of both the current density and the physical size of the coil. Increasing the current density by increasing the electrical power supplied to the coil produces a stronger gradient field. Reducing the size of the coil, so that a given current travels through a smaller area, also produces a stronger gradient field.

Naturally, since gradient fields along any direction will modulate the precession frequency, one cannot turn on all gradient fields at the same time and hope to resolve spatial information along all directions (as turning on all gradient fields will lead to a single diagonal direction). Instead, a sequence of gradient field actions, along with radiofrequency pulses, is used to create an MR image. We hereby break down MR image formation into the following components, which often occur in temporal order. First, we select a slice (slice selection) to reduce the dimension from 3D to 2D, then we systematically resolve the remaining two dimensions (frequency encoding and phase encoding) within this selected slice to obtain a final image.

1.4.2 Slice Selection

A useful practice in image formation is dimension reduction. Thus, a common first step in producing an MR image is to generate signal only from, for example, a head slice instead of from the whole head - this process is termed slice selection.

The proton excitation principle that we have learned in the last chapter requires a resonance between the precessing proton spins at the Larmor frequency and the RF pulse, that is, the frequency of the RF pulse and spin precession must match each other. In a perfectly uniform magnetic field, all spins within the same field will be excited by the RF excitation pulse at the Larmor frequency, obviously this excitation scheme has no spatial selectivity.

Now let's think of an extreme case where three discrete Larmor frequencies are seen in three contiguous, parallel slabs. In order to select the middle slab, we need to set the frequency of the RF excitation pulse to match that of the middle slab. As such, the spins in the middle slab are "on resonance" with the excitation pulse, thereby changing its state from the longitudinal to the transverse condition and positioning themselves to emit MR signal; the outer slabs are unaffected.

In reality, the frequency distribution in space is not discrete. Indeed, if a spatial gradient field (Gz) is turned on along the slice direction (e.g. z direction), it will result in a continuously changing field as illustrated by arrows (i.e. spin phases) rotating at different rates and angles (Fig. 1.11). Thus a particular slice in space will

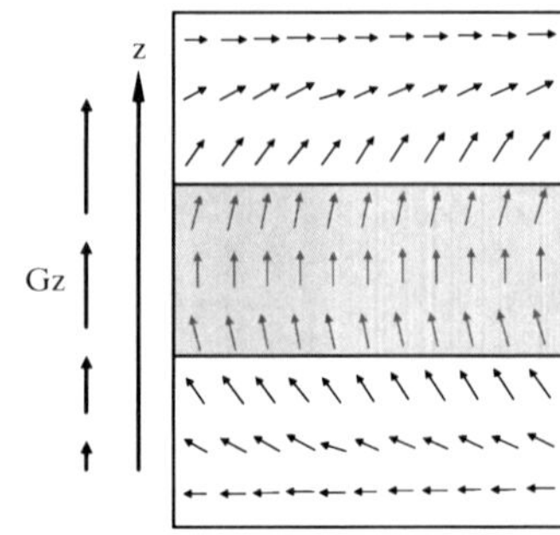

Fig. 1.11 Schematic illustration of slice selection along z direction by matching the frequencies of the RF excitation pulse and proton resonance within the slice

correspond to a band of Larmor frequencies, as shown in dark gray. To match this frequency band, the excitation RF pulse will need to have the same corresponding frequency range as well. Thus, if we know the spatial gradient field along the slice direction, or the slice-selective gradient, as well as the slice thickness, we can determine the resonance frequency range as the product of these two quantities with a scaling constant. Conversely, if we know the bandwidth of the RF excitation pulse, we can calculate the slice thickness. This relationship can be simply described by the following equation: $\omega \pm \frac{\Delta\omega}{2} = \gamma G(x \pm \frac{\Delta x}{2})$, here γ is the gyromagnetic ratio, G is the slice-selection gradient strength, ω and $\Delta\omega$ are the center frequency of the excitation pulse, x and Δx are the position and thickness of the slice.

To this end, we have discussed, conceptually, how slice selection is carried out. The key concept of matching the precessing frequency within the slice (controlled by the spatial gradient) to the frequency of the excitation RF pulse is the basic foundation for slice selection. Immediately after the slice selection, the protons are excited and will undergo T_1 and T_2 relaxation processes. As discussed in the previous chapter, T_2 relaxation will lead to a loss of spin coherence in the transverse plane, and T_1 relaxation will lead to an exponential recovery back to the original state – both resulting in decay of MR signal. As such, the subsequent spatial encoding steps (e.g. frequency encoding) for image formation will need to be quickly carried out after the slice selection.

1.4.3 Frequency Encoding

Once a 2D slice is selected, the next task is to resolve the spatial information within the remaining two dimensions. Because MR signal is acquired from the entire slice, inherently there is no spatial information. To achieve spatial resolution, here we introduce a procedure that involves two intertwined processes, frequency encoding and phase encoding, both taking advantage of the gradient fields. These two processes, however, can often be analyzed independently as they are separated in time. In this section, we will first focus on the concept of frequency encoding.

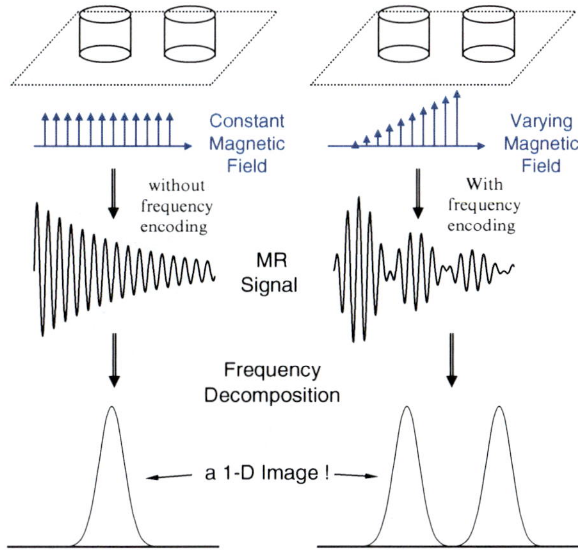

Fig. 1.12 Illustration of the use of gradient field (*right column*) to resolve spatial location

Let's first consider a simple case: imagine we need to make a one dimensional image of two elements, e.g. two thin vials of water. In the absence of gradient field (left column, Fig. 1.12), the resonance frequencies of these two vials are uniform, as a result, the detected analog MR signal from each water-filled vial will oscillate at the same frequency, with a decaying profile due to the aforementioned transverse relaxation mechanism. The frequency spectrum of this single-frequency oscillation will then show a single peak, at the resonance frequency with its width inversely proportional to the transverse relaxation constant. In the presence of a gradient field along the direction of these two vials, however, the two thin vials will have two distinct resonance frequencies, resulting in an oscillating MR signal with a pattern of beat frequency but still with an exponentially decaying envelope due to transverse relaxation. After a frequency decomposition procedure, we can resolve these two resonance frequencies with two peaks (right column, Fig. 1.12). By calibrating the separation of these two peaks and making it correspond to the physical distance between the two vials through the knowledge of the applied gradient field, we just generated a 1D image of these two vials!

While it is straightforward to record the analog MR signal continuously to characterize the two frequency components as demonstrated earlier, it is not always the most efficient way to collect data. To improve the efficiency, often only a limited number of discrete data points are collected, and they are used subsequently to resolve the two discrete frequencies - in fact, only two discrete points are needed. Specifically, let's assume the intensities of the two vials are A and B (which are unknown), the goal of making an MR image is then to find these two quantities. Let's consider collecting data under a spatial gradient G in the *x* direction. Although the protons in these two vials precess at different frequencies, at time 0, they start off at the same phase. This can be better depicted by vectors: considering two vectors

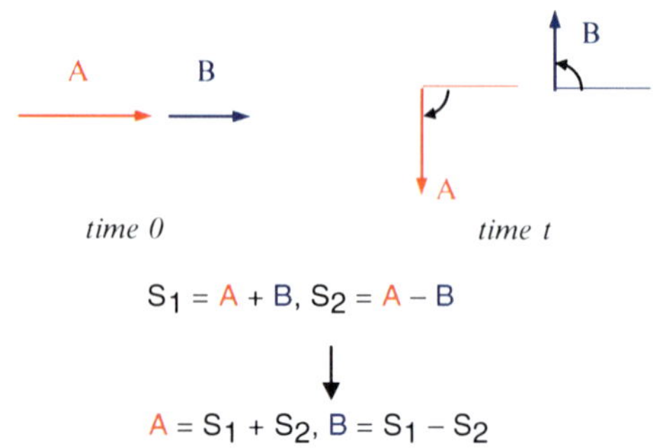

Fig. 1.13 An example of resolving the signal amplitude from two voxels (**a** and **b**) along one dimension

(with intensities A and B), initially pointed along the same direction, which would result in an original total signal $S_0 = A + B$. After some time *t*, different protons at different spatial positions will build up leads and lags in phase because of their different precessing frequencies (frequency × time = phase), and thus, the received signal will have a new intensity C that is governed by the sum of the vectors of A and B. With these two known intensities, and since the precise rotation angles of these two vectors can be determined by the gradient field we applied, we can determine the values of A and B.

To better visualize the frequency encoding process, we can even arrange to acquire data at a specific time point such that the rotation angles of the two vectors are $-\pi/2$ and $\pi/2$, respectively, which will lead to the simplest form of signal representation to solve for the two spatial points (with intensities A and B), as shown in Fig. 1.13.

We can expand our analysis from these two vials to more complex (and realistic) examples. If we have *n* image elements (known as voxels) in space with corresponding intensities (A_1, A_2, … A_n). As usual, these intensities are unknown before the image is formed. We thus need to resolve these intensities to form the final image. From the previous simpler example of resolving two points in space, and by way of extrapolation, we should collect *n* time points under a gradient magnetic field at time t_1, t_2, … t_n, leading to *n* distinct data points with their respective phase angles distributed evenly over a 2π range, leading to *n* uncorrelated signals of S_1, S_2, … S_n. From these *n* known data points and the gradient amplitudes and durations, we will be able to resolve the *n* intensities of each element along this dimension and form a one dimensional image.

1.4.4 Phase Encoding

The final step is then to resolve the second dimension within the selected slice – which will, again, involve the use of a magnetic field gradient. The general term for resolving this second spatial dimension is called phase encoding.

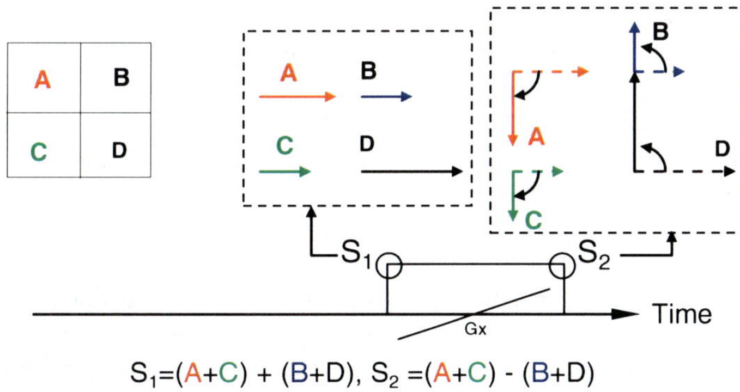

Fig. 1.14 MR data acquisition during frequency encoding, illustrated using a 2×2 matrix

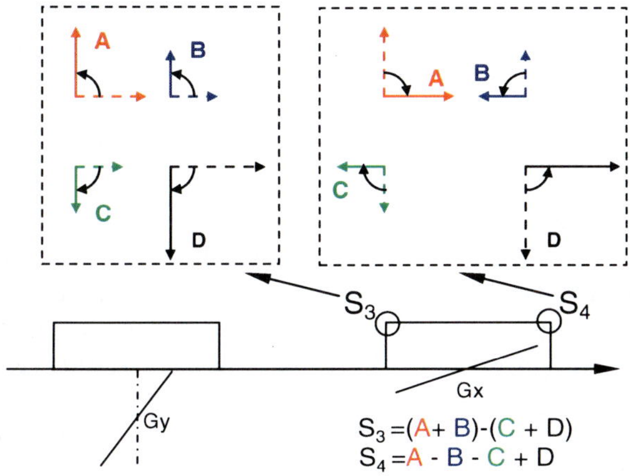

Fig. 1.15 MR data acquisition after phase encoding, illustrated using a 2×2 matrix

For simplicity, we again consider a 2×2 case, with four elements of unknown intensities A, B, C, and D. To resolve the *x* direction, we first perform frequency encoding, as described above. Because there are only two different spatial locations along the *x* direction, we just need to acquire two data points. Similar to the acquisition process in Fig. 1.13, we can collect MR data S_1 and S_2 as illustrated in Fig. 1.14.

We learned previously that we cannot simply turn on the *y* gradient at the same time as the *x* gradient to resolve two dimensional spatial information. Instead, we would first need to complete one gradient, say *y* gradient, before the MR signal is acquired (as opposed to after the MR signal acquisition which serves no purpose). This operation will pre-build phase lags and leads to different voxels according to their respective positions along the *y* direction. The amount of phase accumulations is then carried forward into the frequency encoding process to encode the spatial information in the *y* direction. This process, as illustrated in Fig. 1.15, is the most commonly used method in resolving the second dimension, and known as phase encoding because it builds up a pre-determined amount of phase for a given voxel. The two additional data points, S_3, and S_4, collected after the phase encoding step, along with the previously collected two points, S_1 and S_2, are independent and can be used to resolve four individual intensities A, B, C and D.

While logically it appears simple to directly solve N×N unknowns with N×N known quantities, N being the imaging matrix size, in practice, it can be quite computationally intensive and often impractical for routine use. For example, a 256×256 matrix would require solving 65536 unknowns – which would greatly reduce the image reconstruction speed. Thus, for modern MRI, the two spatial dimensions are resolved not by linear algebra, but rather by a more efficient mathematical process known as the Fourier Transformation (FT) (Ernst et al. 1987). By reorganizing the acquired two-dimensional MRI raw data directly from the scanner, it would become identical to the FT of the actual image. Thus, by carrying out a 2D inverse FT, an MR image can be efficiently resolved. FT is the most prevalent reconstruction technique in modern MRI.

1.4.5 Versatile Contrasts in MR Images

MRI is extraordinary, compared to other methods for medical imaging, in generating versatile and quantitative contrasts in images. Image contrast can be created to emphasize different and unique tissue characteristics. Because of the importance of MRI in specific contrast generation, we introduce in this section several different but important contrast mechanisms. **Static contrasts** are sensitive to the type, number, relaxation and resonance properties of atomic nuclei within a voxel. Typical static contrasts are based on density (e.g., proton density), relaxation time (e.g., T_1, T_2, T_2^*), chemical concentration (e.g. lactate in the brain), and even content of a particular molecular type (e.g., macromolecules). Many of these static contrasts can be used to assess anatomy. In comparison, **motion contrasts** are sensitive to the movement of atomic nuclei. Typical motion contrasts provide information about the dynamic characteristics of the proton pools in the brain, such as blood flow through MR angiography, water diffusion through diffusion-weighted imaging, or capillary irrigation through perfusion-weighted imaging. Many of these motion contrasts can be used to assess function.

In the following sections, we will briefly introduce the various contrasts to appreciate the versatility of MRI in contrast generation. We will begin with several common static contrasts, followed by motion contrasts. More details on the many useful MR contrasts can be found in later chapters, and in several other textbooks on MRI (Stark and Bradley 1992; Haacke et al. 1999).

1.4.5.1 Proton-Density Contrast

One of the simplest forms of MR contrast is proton density imaging. The net magnetization of each voxel is composed of individual protons within that voxel.

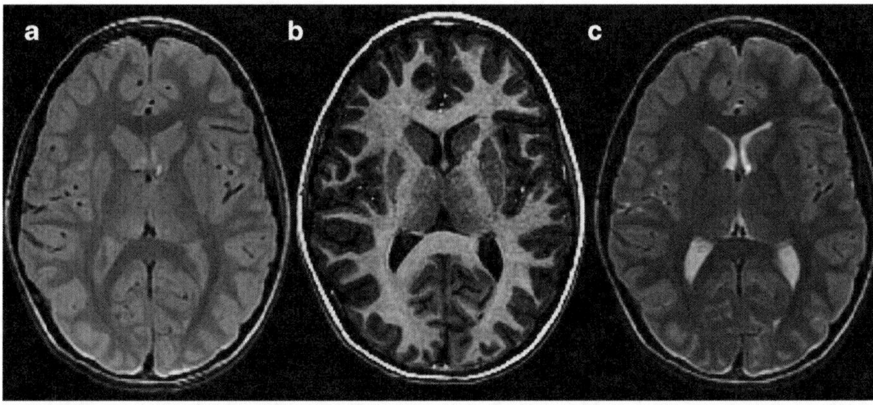

Fig. 1.16 MR Images in a healthy adult with respective weightings on proton density (**a**), T1 (**b**) and T2 (**c**)

Proton-density images, as the name implies, provide contrast based on the sheer number of protons in a voxel, which of course differs across tissue types. To maximize proton-density contrast, researchers use pulse sequences that minimize T_1 and T_2 contrasts. To minimize T_1 contrast, a complete recovery of longitudinal magnetization is needed, which requires a long repetition time (TR); to minimize T_2 contrast, no significant transverse decay is allowed, which requires a very short echo time (TE) to acquire image immediately after excitation.

Proton-density images can be used as high-resolution reference images for determining anatomical structure in the brain. For this reason, they are often an important part of fMRI studies. In addition, the tissue information they provide can be used to improve algorithms for segmenting brain structures on the basis of intensity values. Such segmentation approaches are often important when understanding how damage or atrophy in a region alters its functional properties, such as in the study of disease or aging. Proton-density images are frequently acquired at the same slice locations as T_1- or T_2-weighted images so that complementary anatomical information can be acquired, as shown in Fig. 1.16a.

1.4.5.2 T_1 Contrast

While proton-density images have many uses, other forms of contrast emphasize different features of the brain. One of the commonly used structural contrasts for anatomical images of the brain is T_1 weighting. Images are called T_1 weighted, if the relative signal intensity of voxels within the image depends upon the T_1 value of the tissue. To generate sufficient signal weighting on T_1 recovery, a moderate TR is required. As T_2 decay should be minimized to gain exclusive T_1 contrast, a very short TE is needed. In practice, T_1 weighted images are very useful for brain tissue visualization and segmentation, as it attenuates fluid (e.g. cerebrospinal fluid (CSF)) signal, and differentiates the gray and white matters, as shown in Fig. 1.16b.

To boost T_1 contrast, researchers often use a technique called inversion recovery, which begins the sequence with a 180° inversion pulse rather than the more common 90° pulse. Because the inversion pulse flips the net magnetization to the negative state, it effectively doubles the dynamic range of the signal. This in turn increases the maximal T_1 difference that can be measured between the tissues. Inversion recovery is also useful for selectively eliminating MR signal of a single tissue type. For example, when the magnetization of cerebrospinal fluid (CSF) is recovering from the inverted state to the original state, it will reach a point where there is no longitudinal magnetization (i.e. the zero-crossing point). If images are acquired at this point, there will be no signal from CSF in any voxel. The suppression of CSF allows better assessment of other tissue types, such as gray and white matter.

1.4.5.3 T_2 Contrast

Almost opposite to the T_1 contrast, by using T_2-contrast imaging, images can be created that have maximal signal in fluid-filled regions, which is important for many clinical considerations. Many tumors, arteriovenous malformations (AVMs), and other pathological conditions show up most readily under T_2 contrast. High-resolution T_2 images are also used as anatomical references in fMRI studies (an example is shown in Fig. 1.16c), either in isolation or in conjunction with proton-density or T_1 images in a multi-contrast segmentation algorithm. Thus, common clinical protocols include both T_1- and T_2-weighted images.

In order to allow exclusive T_2 weighting, the longitudinal magnetization needs to be fully recovered to remove any T_1 weighting, which requires a long TR. To remove the influence from unrelated factors such as the magnetic field inhomogeneity, a spin-echo pulse sequence (Hahn 1950), which uses a 180° pulse to refocus any static dephasing process, is usually employed to generate exclusive T_2 contrast. In practice, T_2 weighting is achieved by using a moderate TE, which allows for a moderate T_2 decay.

1.4.5.4 Susceptibility Contrast

MR signal can be influenced by the various susceptibilities of different materials. Susceptibility reflects how the materials are magnetized when placed in the magnetic field. As it is generally known, ferrous metals such as iron are easily magnetized and can retain their magnetism afterwards, making them contraindications for MRI. Many other non-ferrous substances, however, can have negative or positive susceptibility to magnetic fields, and be accordingly labeled as either diamagnetic or paramagnetic. Specifically, they would generate negative and positive magnetic fields when placed in an external magnetic field, but lose their magnetism when taken out the magnetic field. Materials with different susceptibilities disturb the uniformity of the main magnetic field, and can greatly alter MR signal intensities, especially when gradient echo or other T2* sensitive pulse sequences are used.

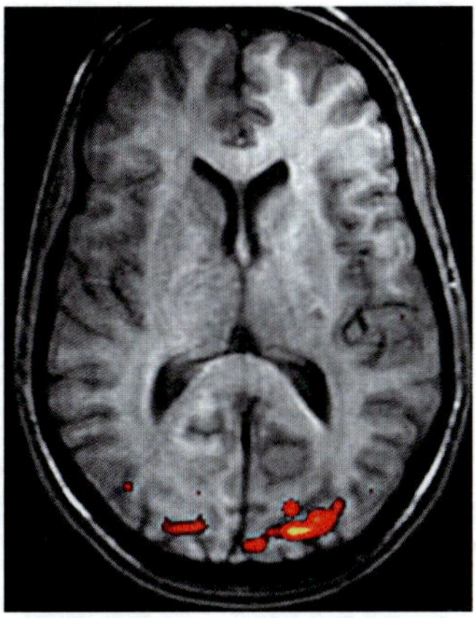

Fig. 1.17 An example of the BOLD signal localized in the primary visual cortex during visual stimulation

One of the primary examples of useful susceptibility contrast is the blood oxygenation level dependent (BOLD) signal (Ogawa et al. 1993) observed in functional MRI (fMRI). Because deoxyhemoglobin in blood is paramagnetic, the dynamic changes of its concentration modulate time-course MR signal intensities upon brain activation, giving rise to the dominant BOLD contrast that are widely used in fMRI research. Shown in Fig. 1.17 is a simple example of the functional BOLD signal in the primary visual cortex during visual stimulation.

While dynamic susceptibility contrast is useful in characterizing hemodynamic responses during brain function, static susceptibility can produce detailed images of brain vasculature. Recent results using susceptibility weighted imaging (SWI) (Haacke et al. 2004) have revealed exquisite details of the venous networks, with dominant contrast generated from the paramagnetic blood within the vasculature. More recent advances in susceptibility contrast at high field have further opened new possibilities to investigate the microanatomical structures of the human brain, with one of the new developments termed susceptibility tensor imaging (Liu 2010). Shown in Fig. 1.18 are three images of tensor element χ_{11}, χ_{23} and χ_{33} from an axial slice of a mouse brain at 7 T. The existence of off diagonal entries and the difference between diagonal entries are clear evidence of magnetic susceptibility anisotropy. The illustration of the brain susceptibility and its anisotropy may provide new ways to characterize the tissue molecular composition and orientation within the human brain.

1.4.5.5 Chemical Contrast

MR signal can also be made sensitive to the types of chemicals, and their concentrations, in the tissue, and thus contrast can be generated from this sensitivity. The principle of chemical contrast is based on the well-known phenomenon called chemical shift. Because protons experience different shielding effects from the surrounding electrons in different molecules, their resonance frequencies vary depending upon which type of molecule they are in. The chemical shift therefore reflects

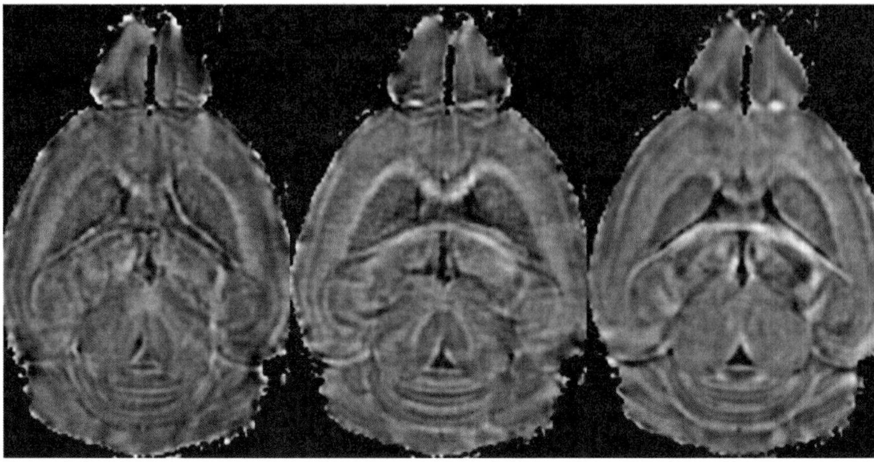

Fig. 1.18 Images of three tensor elements, χ_{11}, χ_{23} and $\chi_{33,}$ in an axial slice of a mouse brain at 7T, demonstrating the anisotropy in magnetic susceptibility of brain tissue

the difference in proton resonance frequencies in different molecules. For example, because protons in fat resonate at a slightly different frequency than those in water, targeted excitation can allow fat selection and fat suppression, which is a very useful practice in clinical MRI.

Chemical shift is also the basis of a different class of MR technique, termed **MR spectroscopy** (MRS), in which the small frequency difference among chemicals are investigated through MR spectra in the frequency domain. As a result, one may be able to quantify important metabolites in the brain such as N-acetyl aspartate (NAA, an index for neuronal viability), creatine (Cr), choline (Cho), Myoinositol (mI), etc. many of these substances play critical roles in regulating brain function, energetics, and metabolism. The ability to quantify these chemicals can be essential in brain function studies.

To resolve the subtle frequency differences of various metabolites, hundreds or thousands of data points are needed within each excitation, making it difficult to accommodate the time needed for spatial encoding. As such, early MRS was acquired in a single-voxel fashion, by using either point-resolved spectroscopy (PRESS) or stimulated echo acquisition mode (STEAM) techniques. Shown in Fig. 1.19 is an example of single-voxel MRS acquired using PRESS.

In recent years, with the improvement in hardware and software to accommodate both frequency and spatial resolutions without an excessively long acquisition time, new MRS techniques that allow for spatial differentiation of the metabolites have been developed. As a result, spatial maps known as chemical shift images (CSI) can be generated that reflect the concentrations of individual chemicals of interest in various regions of interest (Brown and Kincaid 1982).

While ^{1}H MRS has been dominant, other nuclei possessing NMR property (as discussed in earlier sections) can also generate signal. Indeed, recent developments in heteronuclear spectroscopy (e.g. ^{13}C, ^{31}P) have provided specific signal and assessment on cerebral energy metabolism and neurochemistry (Zhu et al. 2009).

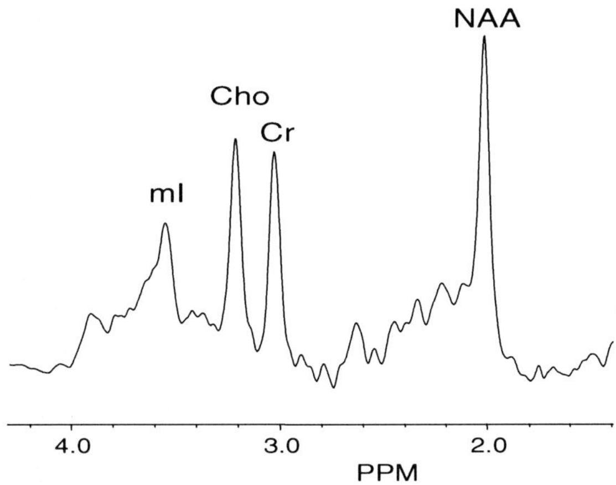

Fig. 1.19 Illustration of a MR spectrum using single-voxel PRESS acquisition, containing various metabolites of interest such as NAA, Cho, Cr, and mI

1.4.5.6 Macromolecular Contrast

In addition to the relaxational mechanisms (e.g. T_1 and T_2) that are based on the evolution of magnetization itself, contrast can also be generated based on the exchange among molecules. For example, the magnetization of the protons in macromolecules (e.g. proteins) can be transferred to those in water molecules, a phenomenon known as magnetization transfer. Many years after its first demonstration (Wolff and Balaban 1989), magnetization transfer is now widely used to generate contrast based on macromolecule concentrations.

The basic principle of magnetization transfer is based on the characteristics of the resonance frequencies of protons in different molecules, due to the various shielding effects of these molecules. In particular, the resonance frequency of protons within water molecules has a fairly narrow range, typically on the order of one hundred Hz; however, protons in macromolecules, albeit at a much lower concentration, can have a range of resonance frequencies that expand across several thousands of Hz. These two distinctly different frequency distributions allow selective excitation of some macromolecule protons, without exciting those in water. If adequate time (e.g. a few tens of milliseconds) is given, the magnetization of the protons in macromolecules will transfer into the protons in free water in the vicinity. If an excitation pulse is given after the transfer to image water, the overall signal in water will experience a reduction because some of the protons in the water already had magnetization from the macromolecules. Because the amount of magnetization transfer is directly proportional to the macromolecule concentration, the reduction seen in water signal can be used to infer macromolecular concentration.

Magnetization transfer contrast can have a significant clinical impact. For example, it can be used to detect changes in myelination in the brain and assess its effect on brain function, as myelin is physically a structure composed of macromolecules (e.g. phospholipids) (shown in Fig. 1.20 is a magnetization transfer-prepared image sensitive to myelin content). As such, magnetization transfer can often be used to

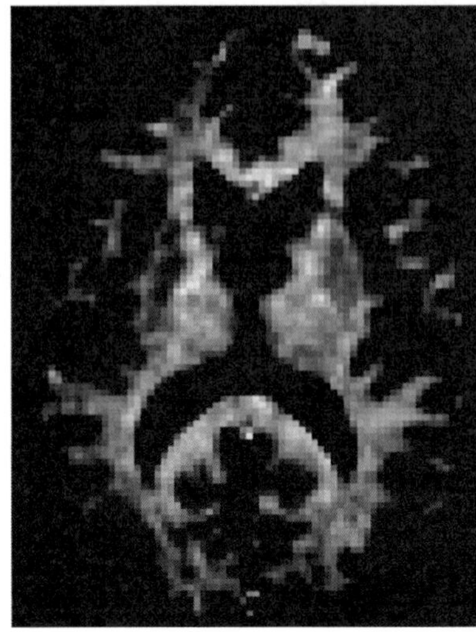

Fig. 1.20 Sample image of myelin content through macromolecular contrast using magnetization transfer

study maturation of myelin in pediatric brains, or investigate myelin disorders such as Multiple Sclerosis (e.g. demyelination) (Grossman 1999).

1.4.5.7 MR Angiography

Magnetic resonance angiography, or MRA, is a common motion-weighted contrast that illustrates the structure of blood vessels using noninvasive MRI. In classic angiography, a contrast agent is injected into the bloodstream through an inserted catheter. X-ray images are then collected with and without the contrast agent present to generate a difference image (i.e., angiogram) that maps the vascular system. Although angiography provides good vascular images, it is a very invasive procedure, requiring both the insertion of foreign substances and exposure to ionizing radiation. Because MRA does not require ionizing radiation, it can noninvasively detect, diagnose, and aid in the treatment of many types of medical problems, including cardiac disorders, stroke, and vascular disease. MRA also complements fMRI studies by identifying major blood vessels that may confound experimental results. If identified, these vessels can be removed from analyses to improve localization of activity to the capillary bed.

MRA can use either exogenous or endogenous contrast. In some clinical settings, exogenous contrast-enhancing agents are used to increase the vessel signal. For a typical contrast-enhanced MRA, a small quantity (or bolus) of a gadolinium-based contrast agent is injected into the patient's bloodstream. The gadolinium itself is not visible on MR images, but it radically shortens the T_1 of nearby blood, allowing the use of specialized pulse sequences with extremely short TR (3–7 ms) and TE (1–3 ms) values. The short TR saturates signal from stationary tissues but not from the gadolinium-enhanced blood, while the short TE minimizes T_2 decay. Depending on

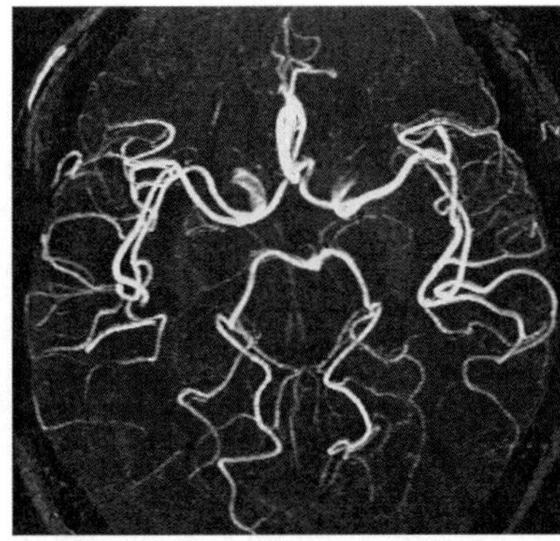

Fig. 1.21 Sample image of MR angiogram using time-of-flight contrast

the delay between bolus injection and image acquisition, the contrast agent may travel through different components of the vascular system, so the images can be calibrated to provide information about arterial or venous networks.

In research settings, MRA is usually obtained using noninvasive endogenous contrast. There are two primary techniques for endogenous contrast MRA (see review by Haacke et al. 1991). The most common is time-of-flight (TOF) MRA, which generates signal based upon blood displacement. The underlying principle of the TOF technique is signal saturation. By repeatedly and frequently applying excitation pulses or gradient pulses to a single imaging plane, the signal within that plane can be suppressed. Thus, static tissues such as gray or white matter, will produce little MR signal and will appear to be very dark on TOF images. Blood vessels, however, are constantly replenished with new blood, hence new protons, from outside the plane. These protons have not experienced the excitation or gradient pulses, and thus they contribute normal MR signal. TOF images are typically acquired in the axial plane and can be reformatted to other planes for ease of viewing. An example of brain angiogram is shown in Fig. 1.21 based on TOF contrast.

A second technique is velocity-encoded phase contrast (VENC-PC) MRA, which uses gradient fields to produce a velocity-dependent phase difference between the vasculature and surrounding tissue. The amount of phase difference accumulated depends on the velocity of the moving blood and the strength and duration of the applied gradient. By acquiring phase differences in each of three orthogonal directions, a map of three-dimensional flow can be created. Typical VENC-PC protocols acquire two sets of images: one with a motion-sensitizing gradient and the other with either no gradient or a gradient in the opposite direction. The difference between these images indicates the magnitude of the phase difference at each voxel, and thus the brightness at each voxel is proportional to flow. Voxels with stationary tissue will not give signal, since there are no phase differences between the images, whereas voxels with rapidly moving blood will produce bright signals due to the large phase differences. The VENC-PC technique, unlike TOF, does not depend upon TR or slice thickness, because it acts upon the blood already present in the imaging slice.

1.4.5.8 Diffusion-Weighted Contrast

The random-walk motion of molecules due to thermodynamic effects is known as diffusion. In gases and liquids, the molecules can move relatively freely, as when a dye spreads through a glass of water or when the smell of freshly baked bread wafts through the house. In solids, however, the motion of molecules is restricted, and thus diffusion is much slower. The abundance of water molecules in the human body makes it possible to perform diffusion-weighted imaging using MRI (Stejskal and Tanner 1965). And, because of the different cellular environment experienced by different water molecules, diffusion-weighted MRI can provide a new dimension of image contrast based on mobility.

Diffusion weighting is the application of controlled gradient magnetic fields to quantify the amplitude and direction of diffusion. The presence of diffusion-weighting gradients further attenuates the MR signal beyond that caused by common T_2 relaxation. Assuming equal, or isotropic, diffusion along all directions, the attenuation effect (A) due to diffusion weighting is given by equation $A = e^{-bD}$, here D is the apparent diffusion coefficient, or ADC, b defines the severity of diffusion weighting (which is governed by the duration and amplitude of the diffusion weighting gradients).

Water molecules in the brain do not always diffuse equally in all directions. In fact, most water is contained within tissue that has considerable structure, such as the long processes of axons or the narrow walls of blood vessels. Unequal, or anisotropic, diffusion refers to the preference in some tissues for water molecules to diffuse in one direction or another. In anisotropic diffusion, the motion of molecules over time does not resemble a sphere, in which molecules move equally in every direction, but instead resembles an ellipsoid whose long axis indicates the fastest axis of diffusion. The diffusion ellipsoid is mathematically described as a three-dimensional tensor (Basser et al. 1994), which is a collection of vector fields governed by three principal axes. One useful image derived from the diffusion tensor is the fractional anisotropy map, which reflects the degree of anisotropy from 0 to 1, with 0 being isotropic, for each voxel. A color coded example of fractional anisotropy map is shown in Fig. 1.22, illustrating the degree of anisotropy (intensity) and directionality (color). The field of diffusion tensor imaging (DTI) has seen marked growth in the past decade as it provides a new way to study structural connectivity of white matter and delineate nerve fiber tracts (Mori et al. 1999).

1.4.5.9 Perfusion-Weighted Contrast

To ensure a constant supply of oxygen, hemoglobin molecules carry oxygen through the bloodstream to tissue and organs (e.g. brain). The irrigation of tissue via blood delivery is known as perfusion, and the family of imaging procedures that measure this process are known as perfusion MRI. Perfusion is expressed as the volume of blood that travels through a tissue mass over time. Unlike the MRA techniques described in the previous section (which are often used for clinical reasons to measure

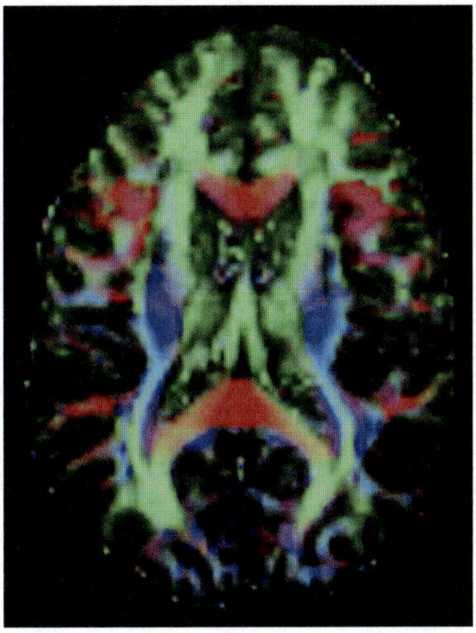

Fig. 1.22 Sample image of fractional anisotropy using diffusion tensor imaging, directionality is color coded (*red*: left – right; *blue*: superior – inferior; *green*: anterior – posterior)

the properties of large blood vessels), perfusion MRI is most frequently used to image blood flow specifically in capillaries and other small vessels.

Similar to MRA, perfusion MRI may use either exogenous or endogenous contrast. Exogenous contrast approaches use intravascular contrast agents that freely perfuse through the vascular system (See review by Prince and Body 1996). The attenuation of the MR signal in each voxel is proportional to the amount of the contrast agent present. Thus, signal changes can be interpreted as a function of perfusion, and images can be created that depict different perfusion properties, such as the relative cerebral blood flow (rCBF), relative cerebral blood volume (rCBV), and mean transit time (mTT). As their names imply, relative blood flow and relative blood volume express changes in how much blood comes into a voxel and how much blood is contained within a voxel, respectively. The mean transit time measures how quickly blood passes through a particular voxel and can indicate brain regions with delayed blood flow. The use of exogenous contrast agents provides very high signal change but has limited use for research because of its invasiveness.

Endogeneous contrast perfusion imaging is noninvasive and has found use in fMRI research. Contrast is generated through the clever use of radiofrequency pulses to magnetically label protons in blood water molecules before they reach the tissue of interest (Detre et al. 1992). This approach is known as arterial spin labeling, or ASL, of which there are two types: continuous and pulsed. Continuous ASL typically uses additional hardware, like a labeling coil, to saturate spins in upstream blood, such as in the carotid arteries of the neck. Following this labeling process, the blood travels to the brain and enters the imaging slice. The images can then be continuously acquired in the brain in the presence of the labeled blood. A drawback of the continuous ASL technique is the requirement for a second transmitter coil to label the inflowing blood. Pulsed ASL uses a single coil both to label blood in a distal labeling plane and to record the MR signal change in the imaging plane.

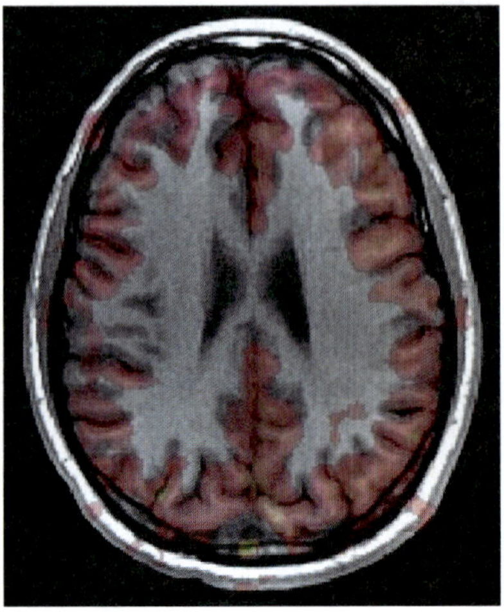

Fig. 1.23 Sample image of perfusion map (in *translucent red*) using pulsed arterial spin labeling

Labeling pulses are broadcast for brief periods and followed by a delay and then image acquisition. The delay period must be calibrated to account for the distance between the labeling plane and imaging plane, so the labeled bolus of blood water will enter the imaging plane during image acquisition and produce quantitative perfusion image (Wong et al. 1998) only related to the labeled blood water, as shown in translucent color in Fig. 1.23.

References

Basser PJ, Mattiello J, Le Bihan D (1994) Estimation of the effective self-diffusion tensor from the NMR spin echo. J Magn Reson B103:247–254

Bloch F (1946) Nuclear induction. Phys Rev 70:460

Brown TR, Kincaid BM (1982) Ugurbil, NMR chemical shift imaging in three dimensions. Proc Natl Acad Sci USA 79(11):3523–3526

Damadian R (1971) Tumor detection by nuclear magnetic resonance. Science 171:1151

Detre J, Leigh JS, Williams DS, Koretsky AP (1992) Perfusion imaging. Magn Reson Med 23(1):37–45

Ernst RR, Bodenhausen G, Wokaun A (1987) Principles of nuclear magnetic resonance in one and two dimensions. Oxford, New York

Grossman RI (1999) Application of magnetization transfer imaging to multiple sclerosis. Neurology 53(5 Suppl 3):S8–S11

Haacke EM, Smith AS, Lin W, Lewin JS, Finelli DA, Duerk JL (1991) Velocity quantification in magnetic resonance imaging. Top Magn Reson Imaging 3(3):34–39

Haacke EM, Brown RW, Thompson MR, Venkatesan R (1999) Magnetic resonance imaging: physical principles and sequence design. Wiley-Liss, New York

Haacke EM, Xu Y, Cheng YC, Reichenbach JR (2004) Susceptibility weighted imaging (SWI). Magn Reson Med 52:612–618

Hahn EL (1950) Spin echoes. Phys Rev 80:580

Lauterbur PC (1973) Image formation by induced local interactions: examples employing NMR. Nature 242:190

Liu C (2010) Susceptibility tensor imaging. Magn Reson Med 63:1471–1477

Mansfield P (1977) Multi-planar image formation using NMR spin echoes. J Phys C 10:L55

Mori S, Crain BJ, Chacko VP, Van Zijl PCM (1999) Three-dimensional tracking of axonal projections in the brain by magnetic resonance imaging. Annal Neurol 45(2):265–269

Ogawa S, Menon RS, Tank DW, Kim SG, Merkle H, Ellermann JM, Ugurbil K (1993) Functional brain mapping by blood oxygenation level-dependent contrast magnetic resonance imaging. A comparison of signal characteristics with a biophysical model. Biophys J 64(3):803–812

Prince MR, Body MR (1996) angiography with gadolinium contrast agents. Magn Reson Imaging Clin N Am 4(1):11–24

Purcell EM, Torrey HC, Pound RV (1946) Resonance absorption by nuclear magnetic moments in a solid. Phys Rev 69:37

Rabi II, Zacharias JR, Millman S, Kusch P (1938) A new method of measuring nuclear magnetic moments. Phys Rev 53:318

Stark DD, Bradley WG (1992) Magnetic resonance imaging, Mosby-Year Book

Stejskal EO, Tanner JE (1965) Spin diffusion measurements: spin echoes in the presence of a time-dependent field gradient. J Chem Phys 42(1):288–292

Wolff SD, Balaban RS (1989) Magnetization transfer contrast (MTC) and tissue water proton relaxation in vivo. Magn Reson Med 10(1):135–144

Wong EC, Buxton RB, Frank LR (1998) Quantitative imaging of perfusion using a single subtraction (QUIPSS and QUIPSS II). Magn Reson Med 39:702–708

Zhu XH, Zhang N, Zhang Y, Lei H, Zhang X, Qiao H, Ugurbil K, Chen W (2009) Advanced in vivo heteronuclear MRS approaches for studying brain bioenergetics driven by mitochondria. Methods Mol Biol 489:317–357

Chapter 2
Hardware Requirements for *In Vivo* Nuclear Magnetic Resonance Studies of Neural Metabolism

Hellmut Merkle, Phil Lee, and In-Young Choi

Abstract Refined technological developments in the field of nuclear magnetic resonance (NMR), within the biomedical environment typically named magnetic resonance (MR), enable to noninvasively obtain biochemical, physiological, morphological, and anatomical information *in vivo* in both clinical and preclinical studies. Currently MR technologies are available for measuring high resolution anatomical images via e.g., T_1- and T_2-weighted magnetic resonance imaging (MRI), microscopic alterations of brain tissue via diffusion tensor imaging (DTI), cerebral blood flow via arterial spin labeling MRI, brain function via the blood oxygen level dependent (BOLD)- MRI, and spatial distribution of neurochemicals via magnetic resonance spectroscopy (MRS), to name a few examples. Furthermore, recent technical advances allow us to combine both NMR and positron emission tomography (PET) technologies, which provide simultaneous acquisition of high resolution anatomical MRI and molecular imaging with radioactive tracers within the magnet, therefore increasing diagnostic values through combining the strength of spatial resolution of MRI and detection sensitivity of PET.

H. Merkle, Ph.D. (✉)
Laboratory of Functional and Molecular Imaging, National Institute of Neurological Disorders and Stroke, National Institutes of Health, Bethesda, MD 20892, USA
e-mail: merkleh@mail.nih.gov

P. Lee, Ph.D.
Department of Molecular & Integrative Physiology, Hoglund Brain Imaging Center, University of Kansas Medical Center, 3901 Rainbow Blvd, Mail Stop 1052, Kansas City, KS 66160, USA
e-mail: plee2@kumc.edu

I.-Y. Choi, Ph.D.
Department of Neurology, Department of Molecular and Integrative Physiology, Hoglund Brain Imaging Center, University of Kansas Medical Center, 3901 Rainbow Blvd, Mail Stop 1052, Kansas City, KS 66160, USA
e-mail: ichoi@kumc.edu

I.-Y. Choi and R. Gruetter (eds.), *Neural Metabolism In Vivo*, Advances in Neurobiology 4, DOI 10.1007/978-1-4614-1788-0_2,

This chapter provides an overview of various configurations and components of MR systems including magnets and gradients. Particular focuses have been employed in explaining the radiofrequency (RF) system, one of the most rapidly develop technologies, from the basic to the state-of-the-art components with various modes of RF system configurations and RF coils.

Keywords Gradients • Magnetic resonance imaging • Magnetic resonance spectroscopy • Magnetic resonance system • Phased array elements • Radio frequency coil • Shims

2.1 Introduction

The principles of magnetic resonance (MR), also called nuclear magnetic resonance (NMR), (Bloch (1946) The nuclear induction experiment. Phys Rev 70:460–473; Abragam (1961) The principles of nuclear magnetism. Clarendon Press, Oxford; Ernst (1966) Nuclear magnetic double resonance with an incoherent radio-frequency field. J Chem Phys 45: 3845; Slichter (1996) Principles of magnetic resonance. Springer, Berlin/New York), rule the basics for magnetic resonance spectroscopy (MRS) and magnetic resonance imaging (MRI) (Lauterbur (1973) Image formation by induced local interaction: examples employing nuclear magnetic resonance. Nature 242: 190–191). MR based technologies have been engineered, further developed and differentiated (Hoult and Richards (1976) The signal-to-noise ratio of the nuclear magnetic resonance experiment. J Magn Reson 24(1): 71–85; Chen and Hoult (1989) Biomedical magnetic resonance technology. Taylor & Francis) for a large variety of *in vivo* applications (Wehrli and Shaw et al (1998) Biomedical magnetic resonance imaging: principles methodology and applications. VCH, New York; Schempp (1998) Magnetic resonance imaging: mathematical foundations and applications. Wiley-Liss, New York; Smith and Lange (1998) Understanding magnetic resonance imaging, CRC, New York; Haacke et al (1999) Magnetic resonance imaging: physical principles and sequence design, Wiley-Liss, New York). Currently, a large number of powerful MR based technologies are available for obtaining biochemical information and high quality *in vivo* images for basic and clinical research within the biomedical community (Ugurbil and Adriany et al (2000) Magnetic resonance studies of brain function and neurochemistry. Annu Rev Biomed Eng 2: 633–660; Shulman and Rothman (eds) (2004) Brain energetics and neuronal activity: applications to fMRI and medicine. Wiley, West Sussex). Furthermore, MR offers a unique opportunity to non-invasively investigate a number of diagnostic facts of the living brain with high temporal and spatial resolution imaging. Anatomy via T_1-weighted and T_2-weighted MRI, morphology via diffusion tensor imaging, spatially-resolved chemical composition for metabolism studies via MRS, cerebral blood flow via arterial spin labeling MRI, and brain function via the blood oxygen level dependent (BOLD) MRI are a few examples of the diverse use of MR technology.

Recent technological developments made it possible to combine both MR and positron emission tomography (PET), which allows simultaneous acquisition of high resolution anatomical MRI and molecular imaging with radioactive tracers within the magnet, therefore increasing diagnostic values. Prototypes of hybrid MR/PET scanners have been built (Jacobs and Cherry (2001) Complementary emerging techniques: high-resolution PET and MRI. Curr Opin Neurobiol 11(5): 621–629; Catana et al (2008) Simultaneous *in vivo* positron emission tomography and magnetic resonance imaging. Proc Natl Acad Sci USA 105(10): 3705–3710) and currently both animal and clinical scanners are commercially available. A variety of parameters obtained through MR and PET allow scientists in basic and/or translational research settings to investigate the living brain in health and disease. In clinical settings, MR technology is widely used to diagnose various disorders including tumors, multiple sclerosis, and stroke.

An MR system consists of an operator console and various hardware subsystems. An operator console includes one or more microprocessors with associated memory and network links, and controls a variety of hardware components. Hardware subsystems include (1) a magnet system with the main magnet and the shim system for homogenizing the main magnetic field, (2) a gradient system for dynamic creation of spatially dependent linear magnetic fields within the three dimensional space, and (3) radiofrequency (RF) system with RF coils and associated RF hardware for transmission and reception of the MR signals.

2.2 Overview of the Basic System for *In Vivo* Magnetic Resonance

MR systems are composed of a magnet system, a gradient system, a radiofrequency (RF) system and a console with microprocessor-based hardware and software for experiment execution and data processing. A variety of electronic hardware components in the magnetic resonance (MR) system are controlled by microprocessors using basic- and application dependent/specific pulse sequence software. From the beginnings of MR technology, sensitivity improvement of MR signal detection has been a major engineering effort (Chen and Hoult 1989; Jin 1998). At present, MR hardware is intensively developed in all components including the magnet, gradient coils, RF components and system controllers with embedded microprocessors. MR systems for human and animal brain research are reviewed elsewhere (Kuperman 2000; Silva and Merkle 2003; Heerschap et al. 2004; Fishbein et al. 2005).

The magnet is a major MR system component and supplies the static magnetic field inside of its cylindrical structure. As MR experiments can only be successful if the magnetic field is extremely homogeneous, MR systems include a shim system for homogenizing the main magnetic field, which consists of amplifiers and a variety of field correction coils mounted on a cylinder located inside the warm magnet bore. The gradient system consists of three sets of coils inside the magnet and high power electronic amplifiers. It allows dynamic creation of magnetic field gradients, well-defined linearly increasing magnetic fields within the three dimensional space as necessary for image generation. The major RF system components are frequency synthesizers, transmit amplifiers, RF coils and associated receive chains. All of

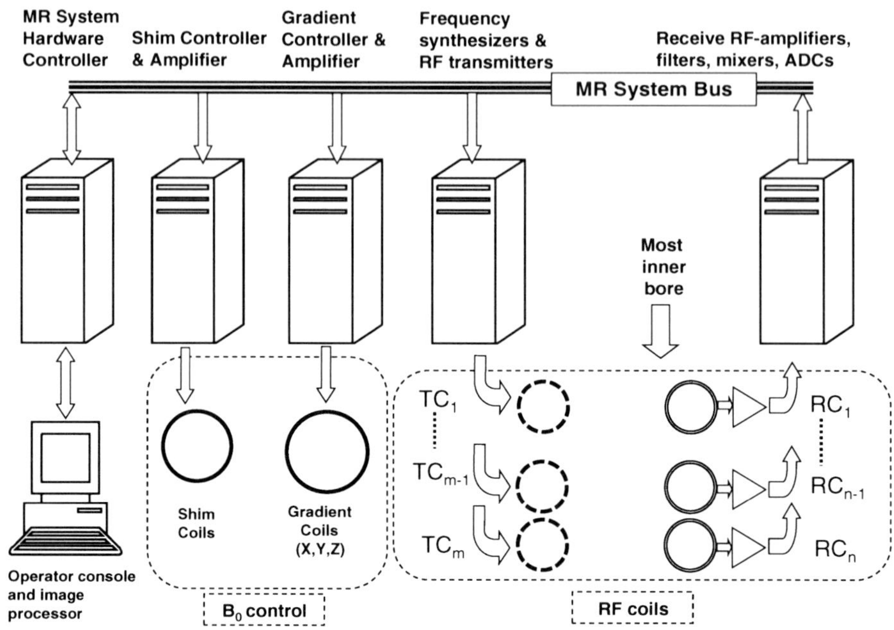

Fig. 2.1 Block diagrams of a MRI/MRS system. All the hardware of the MR system is interconnected via the MR system bus and controlled by a computer. The B_0 field in the volume of interest is homogenized by adjusting electrical currents in the shim coils. The gradient and gradient amplifiers drive current pulses into the gradient coils and produce temporal and spatially well-defined B_0 gradient fields along the x, y and z axes. One or more frequency synthesizers generate radiofrequency pulses of well-defined length, frequency and amplitude modulation. Those are amplified and sent to one or more RF transmit coils (TC_n) that produce B_1 fields within the volume of interest in order to excite nuclear spins. The detection of the NMR signal is achieved with one or more receive coils (RC_n), amplified and sent back to the system where it will be processed to construct images or spectra

these hardware components are interlinked by system hardware and controlled by a dedicated microprocessor (Fig. 2.1).

Recent technological developments made it possible to combine two independently developed imaging modalities, MR and positron emission tomography (PET). Both imaging techniques have been used for research and medical diagnostic purposes, and complement each other. The combined MRI and PET system allows simultaneous acquisition of high resolution anatomical MRI and molecular imaging with PET in the magnet, therefore increasing the diagnostic value. The prototypes of MRI/PET hybrid scanners have been reported since 2001 (Jacobs and Cherry 2001; Catana et al. 2008; Pichler et al. 2008a; Sauter et al. 2010; Wehrl et al. 2011). Clinical scanners are currently commercially available (Siemens, Erlangen, Germany) and animal systems for translational studies will be available in the near future (Bruker Inc., Ettlingen, Germany).

Beside the introduction of the MR scanner, this chapter mainly focuses on describing the conventional and state-of-the art front-end of the MR system namely the RF detection chain, as the RF detection chain largely dominates the overall signal sensitivity of MR systems.

2.3 Magnet System

For *in vivo* metabolism studies, high-field magnets are preferable because MR sensitivity increases with static magnetic field strength, B_0. The magnetic field strength is measured in Tesla (T) or Gauss (G). For comparison, the earth's magnetic field is approximately 0.5 G where 1 T is equal to 10,000 G. Typical low field magnet designs are either C-shaped or open coils supported by large iron structures, whereas high-field magnets are made of superconducting solenoids. To date, the magnetic field strength of clinical scanners ranges from 0.05 up to 3 T. However, within research environments, magnet systems with field strengths of up to 11.7 T are in service and plans are underway to extend beyond this range. *In vivo* translational and animal research environments typically utilize magnetic fields of up to 16 T. In addition, state-of-the-art MRS experiments including the structural determination of molecules or proteins have been performed with a magnetic field strength of up to 22.3 T. The majority of clinical and research MR systems with *in vivo* imaging capability are typically based on magnet designs of horizontally oriented tubes. However, there are some examples of 'wide bore' high field vertical micro imagers and some special 'large bore' vertical magnets for MR studies on non-human primates (Logothetis et al. 1999; Juchem et al. 2004; Pfeuffer et al. 2004a, b; Goense et al. 2008).

Figure 2.2 shows a state-of-the-art 7 T horizontal bore animal-research MR-system with a self-shielded superconducting magnet with a helium cold head, closed-cycle helium refrigerator that reduces liquid helium consumption in the magnet. The magnetic field is generated once during installation by injecting electric currents into the superconducting wire of the solenoid, which is immersed in liquid helium. Since the superconducting solenoid coil alone does not create a sufficiently homogeneous magnetic field, strategically placed steel elements (passive shims) and additional superconducting coils (superconducting shim coils) are added to improve magnetic field homogeneity (Hoult and Lee 1985; Vermilyea 1988; Battocletti et al. 1993; Dorri et al. 1993; Belov et al. 1995). Those superconducting shim coils and passive shims correct the magnetic field homogeneity distortions caused by the site specific magnetic materials including building steel structures.

When a human body, tissue sample or an animal is placed in the homogeneous magnetic field (B_0), the magnetic field (magnetic flux density) within the tissue is distorted depending on the shape and magnetic susceptibility of the tissue. In order to correct the tissue dependent B_0 inhomogeneity within the tissue of interest, external compensation fields, i.e., shim fields, are applied in a spatially dependent manner. The shim fields are produced by electric currents within a network of complex, lower and higher order wire structures that are wrapped around a cylindrical former in the warm bore of the magnet

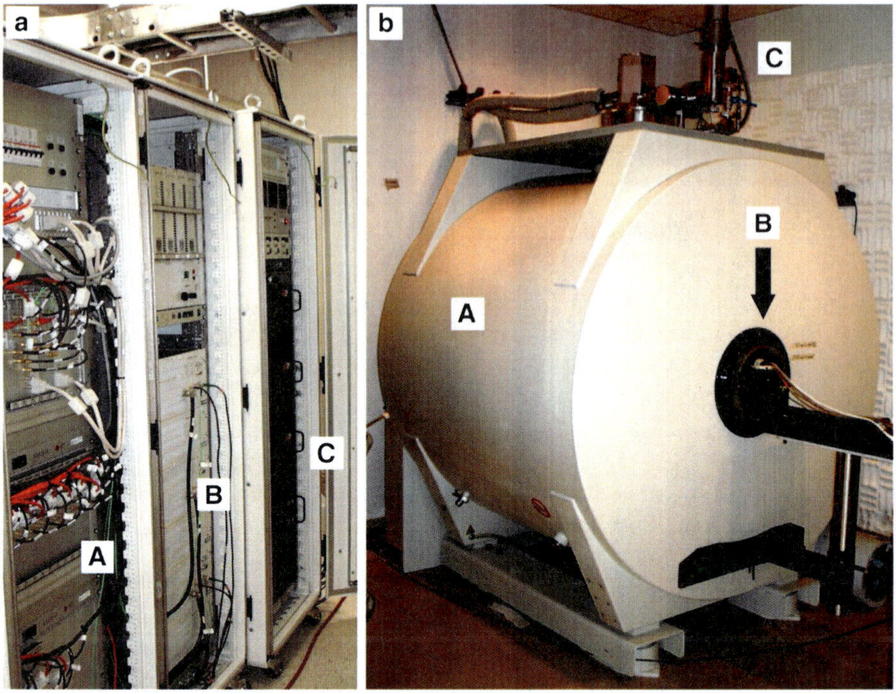

Fig. 2.2 Pictures of a typical animal MR system. System hardware cabinets (**a**) contain (**A**) frequency synthesizers and pulse sequencer, (**B**) shim and RF transmit amplifiers and (**C**) gradient amplifiers and are located adjacent but outside the shielded room of the magnet. Shown in (**b**), a small-footprint, self-shielded, superconducting 7 T horizontal magnet with 26 cm bore diameter (**A**) containing animal holder (**B**), equipped with a helium recovery system (**C**) is housed in an electrically shielded room

(room temperature shim coils). Various approaches for designing and constructing shim coils have been described previously (Golay 1958; Anderson 1961; Hoult and Deslauriers 1994; Forbes and Crozier 2001; Brideson et al. 2002; Forbes and Crozier 2002, 2003, 2004; Forbes et al. 2005). The currents within the shim coils are driven by shim amplifiers and can reach up to 10 Amperes. Typically first order shim corrections for linear B_0 distortions along the major axes (x, y and z) are accomplished by adding constant currents to each of the gradient coils. Local and global shimming is achieved either manually or automatically by executing sophisticated pulse sequence software programs before the actual MR experiments (Gruetter 1993; Shen 2001).

2.4 Gradient System

Spatial information in MRS and MRI is encoded in the MR signals with spatially and temporarily varying magnetic fields that are generated by the magnetic field gradient system as originally observed by Lauterbur (1973). The details of the gradient coil

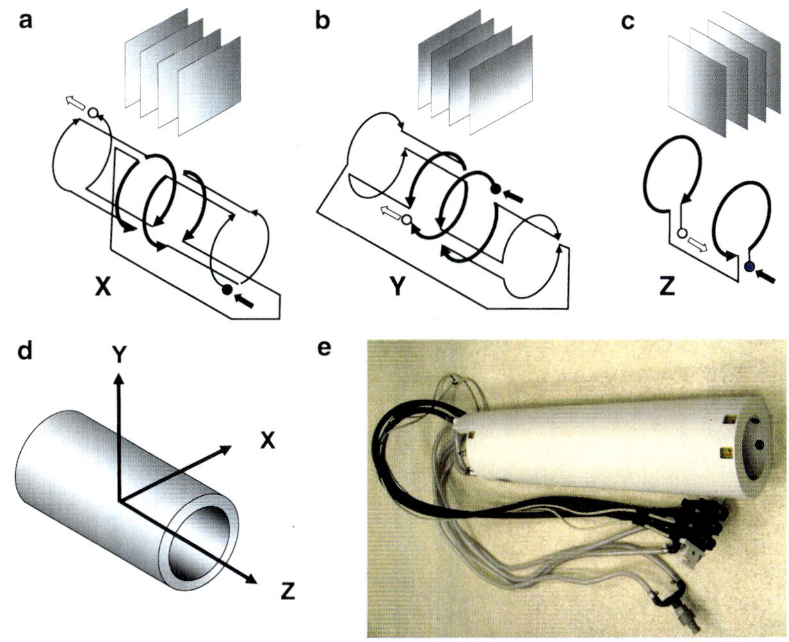

Fig. 2.3 Gradient coils. Top three schematics (**a–c**) are the simplified patterns of wire windings and the generated gradient fields of x-, y-, and z-gradient coils, respectively. All gradient windings are mounted in several layers on a cylindrical former inside the warm bore of the magnet (**d**). The currents from the gradient amplifiers, indicated by block arrows, are sent to the input terminals (*filled circles*). The current in each of the wire structures creates B_0-gradient fields. Thicker lines indicate the structure that produces a useful gradient field at the iso-center as indicated by the shade levels in bundles of squares. A professionally manufactured gradient coil with current leads, temperature sensors and water cooling is shown in (**e**)

construction are reviewed by Turner (1993). The nucleus specific precession frequency of MR (Larmor frequency) increases linearly with the magnetic field B_0, and temporal modulation of B_0 within a well-defined space during a segment of an experiment imparts position dependent information to the MR signals in the localized MRS or MRI. Three sets of gradient coils (Fig. 2.3) are used to produce linear B_0 gradients along the three principal axes in a well-defined area close to the iso-center of the magnet. The gradient coils (Mansfield and Chapman 1986; Turner 1986, 1988; Carlson et al. 1992; Crozier et al. 1994; Eccles et al. 1994; Forbes et al. 2005) are mounted on a cylindrical former in the magnet bore and each set of gradient coils is driven by high current shaped pulses with a typical duration of milliseconds. The pulse waveforms are stored in memory and during the MR experiment recalled and played out sequentially by a software program called the pulse sequence. Gradient pulse waveforms are sent to a set of three high power gradient amplifiers, which amplify the waveforms up to ±1000 Volts and achieve currents of up to several hundred Amperes. The currents in the gradient coils temporarily produce the desired spatially dependent B_0 field within a given volume of the tissue. Each set of gradient coils consists of a primary coil and a shield coil. The currents in the primary coils produce the desired gradient magnetic fields in the tissue whereas shield coils are driven with reverse currents to cancel the magnetic field outside the gradient coils in order to minimize eddy currents in the magnet. Eddy currents are undesirable currents induced within the conductive structures of the magnet by the fringe field of the primary coils, which temporarily distorts the main magnetic field and causes image distortions.

Several wire windings in the gradient coils are mounted on a cylindrical former inside the warm bore of the magnet. Schematics of the structural assembly of the wires that carry currents for x, y and z gradients are shown in Fig. 2.3a–c. Actual winding patterns of gradient coils are optimized to produce linear gradient fields within a specified volume centered at the iso-center (Turner 1993). A real gradient coil picture is shown in Fig. 2.3e. The wires of each gradient coil are terminated with special connectors that can handle high voltages and currents. In addition, wires for temperature sensors and hoses for cooling water are connected to the fully encapsulated gradient module.

Typical gradient strengths of clinical systems are in the range of 20–60 mT/m, whereas those in animal research systems are approximately 200–500 mT/m and are available up to 3000 mT/m. Gradient bore sizes in the human head only or whole body systems range from approximately 45–65 cm in diameter, while those in animal MR systems range from 3–40 cm in diameter.

2.5 Radiofrequency (RF) System

The RF system is composed of RF synthesizers, RF transmit amplifiers, receive amplifiers, preamplifiers, and RF coils. Depending on the specific nucleus and magnetic field, a particular radiofrequency is synthesized by the MR system. RF pulse patterns are generated with a well-defined pulse duration, timing, amplitude, and frequency modulation. Various pulse patterns with associated parameters are originally stored in a specific memory, recalled during an experiment and sent to an RF synthesizer which creates desired RF pulses that are sent to the RF transmit amplifier. The RF synthesizer is also active during reception for processing the signals obtained during an experiment.

The MR experiments require matters containing atoms with an odd number of nucleons which provides non-zero total nuclear spins (quantum mechanical property) in the static magnetic field. When spins are placed in the magnetic field, only a small fraction of the spins are aligned in parallel with the magnetic field and precess at the Larmor frequency which is in the radiofrequency range (Bloch 1946; Abragam 1961; Slichter 1996); for example, at 1 T field strength the net spins of protons (hydrogen atoms) precess at the Larmor frequency of ~42.5 MHz.

RF transmit amplifiers deliver short pulses of RF currents in the order of μs to ms at Larmor frequency to RF coils, also called RF probes or antennas. The RF currents create RF magnetic fields (B_1) in the transverse direction to B_0 within the tissue. The B_1 disturbs the nuclear spins in the tissue from the 'status quo' of Larmor precession, which is in parallel to B_0, to a higher energy level, i.e., the anti-parallel or flipped state. RF coils are also used to detect the signals created by the ensemble of nuclear spins as they return (relax) to the equilibrium state within the B_0 field of the magnet. As all precessing spins relax, they create the B_1 induction fields that are superimposed, also called free induction decay (FID). The FID is an RF signal whose amplitude decays over time in the order of microseconds to seconds depending on the molecular environment of the spins in the tissue. The initial amplitude of FID, its frequency shift off from the Larmor frequency, and decay time are used to characterize tissue and molecules (metabolites) within tissue. The FID is picked up and transformed by the RF probe into a voltage at the Larmor frequency. The RF signals in the RF probe are first amplified with a low-noise low-gain preamplifier and sent to an RF receiver for further processing including amplification, demodulation, digitization, and Fourier transformation.

2.5.1 Modes of Operation and Components of RF System

An MR system can be configured to support a variety of different modes of operation of RF systems. The modes of operation can vary from the simplest single coil transmission and reception mode to the most sophisticated multi-coil (array) transmission and multi-coil (array) reception mode.

Transmit coils (symbolized in this contribution by dashed lines) are used for spin excitation using an RF current in the coil wires that produces a transmit B_1 field. The receive coils (symbolized in this contribution by double lines) probe the magnetic RF field produced by spins. Coils that are used for transmission as well as for reception of the NMR signal are called transmit/receive coils or transceive coils (symbolized in this contribution by solid lines). The details of each mode of operation are described in the following section.

2.5.1.1 Single Coil Transmit/Receive

The most basic RF system configuration consists of a single RF channel operating at one frequency, typically the proton Larmor frequency, and an RF coil for both transmission and reception (Fig. 2.4a). This configuration requires a transmit/receive switch (T/R-S) that directs the RF power and signal flow (Fig. 2.4b). The T/R-S is an electronic device and is controlled or gated by the system during RF transmission and reception via a gating signal to the port (G). High power RF pulses from the transmit amplifier (TA) enter the T/R-S at terminal (T) and exits at the coil port (C), and are applied to the RF coil. During the reception period, the RF signals from the coil are directed to the port (R) with a path connected to a preamplifier which sends the amplified signals to the receiver system for further processing. The strength of this setup lies in its simplicity of operation when combined with an RF coil that produces homogeneous B_1 fields. The single coil setup can be the most desired and sensitive setup in conjunction with RF surface coils (e.g., Fig. 2.14) when inhomogeneous B_1 fields associated with the surface coils are not detrimental. A technology using adiabatic pulses

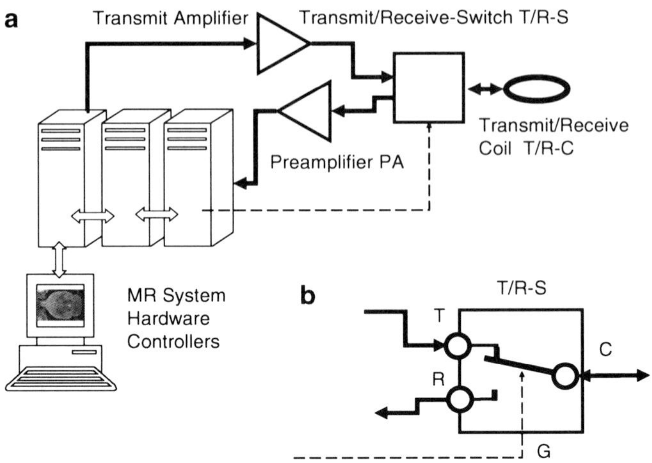

Fig. 2.4 RF system configuration with an RF coil for transmission and reception. The most basic RF system configuration is using a single radiofrequency coil (**a**). The RF pulse is amplified with the transmit amplifier (TA) and sent to the transmit/receive coil (T/R-C) via transmit/receive switch (T/R-S) which is controlled by logic signals from the system (dashed line). During MR signal reception the T/R-S enables the RF path connecting the RF coil and the receiver. The signals from the RF coils are amplified by a low noise preamplifier (PA) and sent to the receiver to enhance system performance. A symbolic schematic of a transmit/receive switch (T/R-S) is shown in (**b**). The T/R-S has three RF ports (*T* input port; *C* coil port; *R* receive port) and a control gate (G). The T/R-S selects the RF pathways to and from the RF coil using PIN diodes and DC control signals

for spin excitation, which provides relatively homogeneous spin excitation, makes this simple surface coil transmit/receive setup more effective and attractive (Bendall 1988; Garwood et al. 1989). The "single coil transmit/receive" configuration is probably the most common MR system configuration installed for animal *in vivo* research.

2.5.1.2 Two Coil TORO

When experiments require well-defined and homogeneous transmit B_1 fields, a solution is to use two independent RF coils, one for **T**ransmission **O**nly and the other for **R**eception **O**nly (TORO). In the TORO system configuration (Fig. 2.5), the transmit coil can be a large volume coil (e.g., Fig. 2.15), which provides homogeneous transmit B_1 fields and can be used in pulse sequences requiring accurate flip angles. The NMR signal can be received using a local coil (surface coil, e.g., Figs. 2.17 and 2.18) placed adjacent to the area of interest. The TORO setup delivers increased reception sensitivity in proximity of the local coil. However, the TORO setup requires detuning of the receive coil while transmitting high power RF signals to the transmit coil, and detuning of the transmit coil while receiving signals. This requirement of detuning of either coil is because the transmit coil and the receive coil would intrinsically interact if they are tuned to the same frequency and their B_1

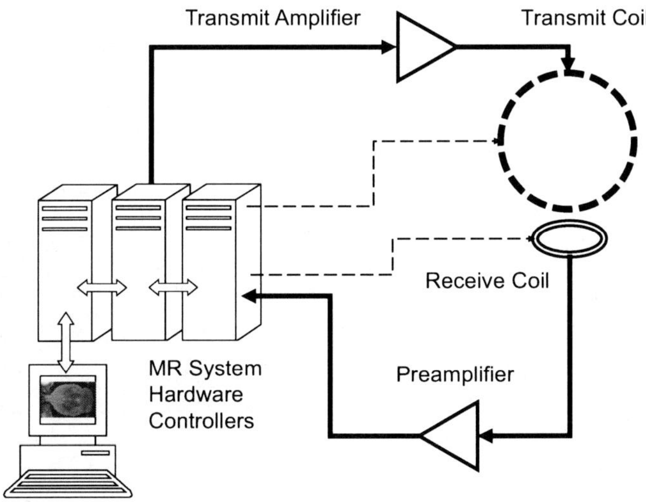

Fig. 2.5 RF system configuration with two RF coils, one for transmission and one for reception (TORO mode). RF systems in the transmit-only and receive-only (TORO) mode provide advantages of homogeneous spin excitation and high sensitivity in signal reception. Typically, RF coils that produce homogeneous B_1 fields in the area of interest (here symbolized by a *dashed circle*) are used as the transmit coil. The transmit coil could be a circumscribing coil (e.g. birdcage coil, saddle or Helmholtz pairs or quadrature combinations) or a partial volume coil. A local coil (e.g., surface coil) is used for reception (*double lines*). Both the transmit coil and the receive coil need to be tuned and de-tuned and vice versa during appropriate segments of time. Timed tuning/detuning is achieved using DC signals from the system (*dashed lines*)

fields are not orthogonally arranged. Detuning requires extra components (e.g., PIN diodes and extra inductors and capacitors) in each coil as well as gating or control signals from the system. TORO configuration is considered as one of standard configurations in human MR systems and has been adopted recently as well within animal research MR systems.

2.5.1.3 Single Coil for Transmission, Array for Reception

The TORO setup as described in the previous paragraph can be further expanded (Fig. 2.6) to obtain greater coverage by replacing the single receive coil with an array of several receive coils, also called phased arrays (Hayes and Roemer 1990; Requardt et al. 1990; Roemer et al. 1990; Wright 1990; Hayes et al. 1991; Wright and Wald 1997; de Zwart et al. 2002; Hardy et al. 2004; Zhu et al. 2004; McDougall and Wright 2005; Niendorf et al. 2006; Wiggins et al. 2006, 2009). However, multiple coils placed adjacent to each other tend to interact not only with the transmit coil but also with each other leading to compromised performance without other means of decoupling. Orthogonal placement of array elements and/or active decoupling methods

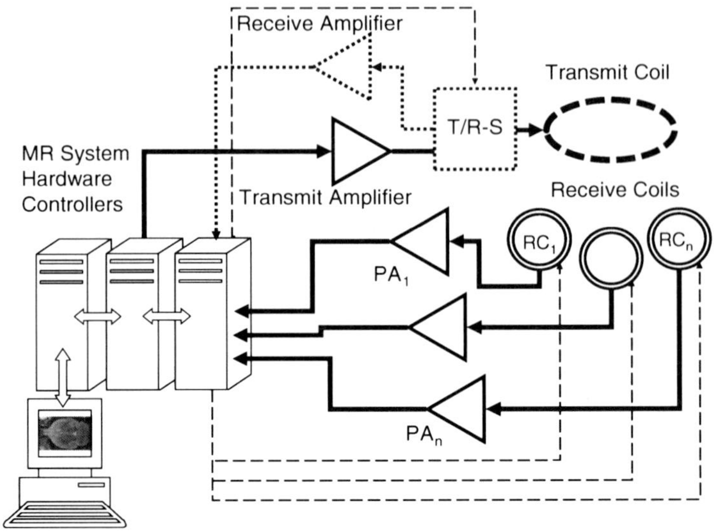

Fig. 2.6 RF system configuration with a transmit coil and N phased receive array elements. An MR system typically has a homogeneous linear or quadrature transmit coil. The modes of operation of can be changed through hardware and software controls and the transmit coil can be used in a transmission only mode or a transmit/receive mode (*dotted* pathway). N receive coil elements with associated inter-array decoupling preamplifiers are utilized for local reception of the RF signals. Receive elements are detuned during transmission and the transmit coil is detuned during reception (*dashed lines*)

using preamplifiers with the high-reflection coefficient is crucial (Reykowski et al. 1995; Duensing et al. 1996; Lee et al. 2002; Wiesinger et al. 2004; Kellman and McVeigh 2005) for proper functionality of the system. After all factors are considered, this setting delivers better coverage of the measurement volume with high reception sensitivity. Currently this system configuration is well-established and nearly exclusively utilized in clinical settings. *In vivo* animal research systems also profit from this approach. In cases that the receive arrays are not necessarily beneficial in specific experiments, the volume-transmit coil can serve as the receive coil as well, as shown in Fig. 2.6 (dotted components). Then the system needs to be configured with a T/R-S, a dedicated receive amplifier and electronic receive path switches (not shown in Fig. 2.6 for simplicity).

2.5.1.4 Transmit/Receive Arrays

Another sophisticated system configuration is using multiple transmit/receive or transceive elements (Fig. 2.7). Each element has its own transmit amplifier, T/R-S and receive amplifier (Adriany et al. 2008, 2010; Driesel et al. 2008; Shajan et al. 2011). All elements are typically mounted on a cylindrical former or other former that conforms to the head shape (e.g., helmet) and are decoupled from each other by geometry. The transceive coil elements can be loops, transverse-electromagnetic

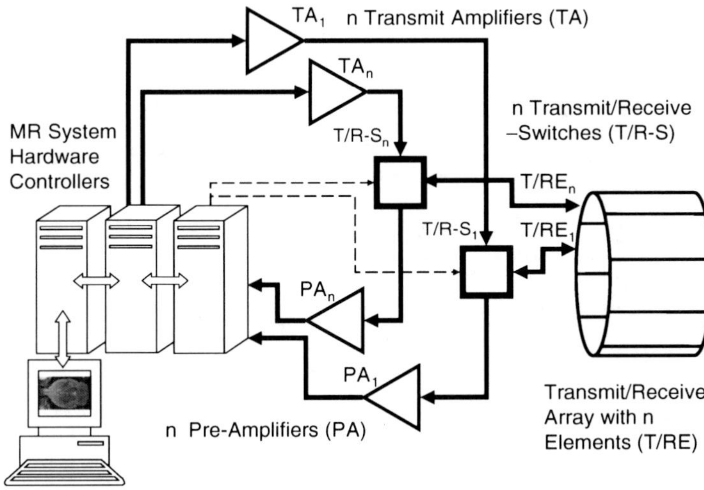

Fig. 2.7 RF system configuration with N transceive elements. When the RF system is configured in multiple (N) transceive mode, the system requires N transmit synthesizers, transmit amplifiers (TA_i), transmit/receive switches ($T/R\text{-}S_i$), transceive elements (T/RE_i), preamplifiers (PA_i), and receive channels

(TEM) resonators, strip-line elements (Kumar and Bottomley 2006; Driesel et al. 2008; Shajan et al. 2011) or a combination of those. The benefits of the multiple transmitters and multiple receivers apply to this setup. The benefits of multiple transmit elements include the ability of B_1 shimming for homogeneous excitation of spins, spatial localization in arbitrary shape, and accelerated imaging through control of the B_1 distribution. The capability of B_1 shimming is especially important at high magnetic fields of 4 T and above where the interaction between B_1 fields and tissue becomes non-linear and B_1 fields are significantly distorted by the presence of tissue (Hetherington et al. 2010; Pan et al. 2010). The B1 shimming is currently an active area of research. The benefits of using multiple receive coils include improved an signal-to-noise ratio (SNR) through close coupling between receive elements and tissue, accelerated imaging through newly developed parallel imaging techniques such as sensitivity encoding for fast MRI (SENSE) (Pruessmann et al. 1999) and generalized auto-calibrating partially parallel acquisitions (GRAPPA) (Griswold et al. 2002). Special consideration is required for decoupling array elements during transmission and reception. Both geometrical decoupling and electronic decoupling with high-reflection-coefficient and low-noise preamplifiers are required for optimal local transmission and reception of the RF signals.

2.5.1.5 TORO Arrays

One of the most sophisticated modes of operation is the combination of transmit arrays and independent receive arrays (Fig. 2.8) (Shajan et al. 2011). Currently this configuration is still a 'work in progress' within research groups that develop animal

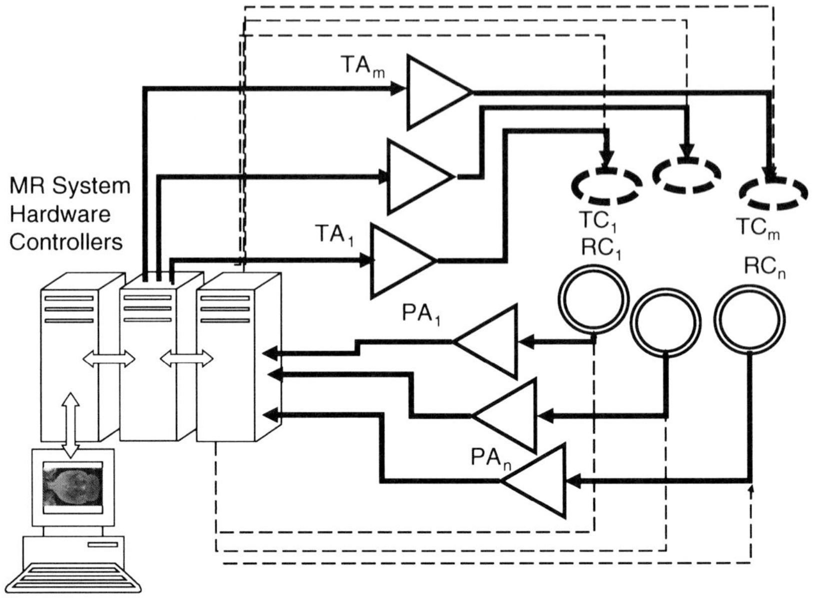

Fig. 2.8 RF system configuration with M transmit coils and N phased array receive elements (TORO arrays). In TORO array mode, the MR system has M transmit channels with independent phase and amplitude control including transmit amplifiers (TA_i) and transmit coils (TC_i) for transmission. For local reception, N receive coil elements (RC_i) with associated inter-array decoupling preamplifiers (PA_j) are utilized. All receive elements are detuned during transmission and the transmit elements are detuned during reception (*dashed lines*)

MR systems and is commercially unavailable. However, this configuration is expected to be essential in the very near future, particularly at high field. This mode of operation requires multiple transmit RF channels with multiple RF synthesizers, waveform generators, and transmit amplifiers, similar to the transmit/receive arrays. This mode provides benefits of both transmit arrays and receive arrays including capability of B_1 shimming and improved receive sensitivity and volume coverage as with transmit/receive arrays. An advantage of this setup over transmit/receive arrays is flexibility of using any number of receive array elements based on the specific signal reception needs, independent of transmit array elements. Each transmit coil element needs to be driven independently from other elements to control local B_1 field distribution in tissue. Although technically challenging, possible solutions for driving each element independently have already been successfully implemented (Hoult et al. 2004a, 2004b; Kurpad et al. 2006; Hoult et al. 2008).

2.5.1.6 Hetero-Nuclear Transmit/Receive

Hetero-nuclear or multi-nuclear MR involves investigation of molecules that contain atoms with odd nuclear spins beside hydrogen, which are often called "X-nuclei".

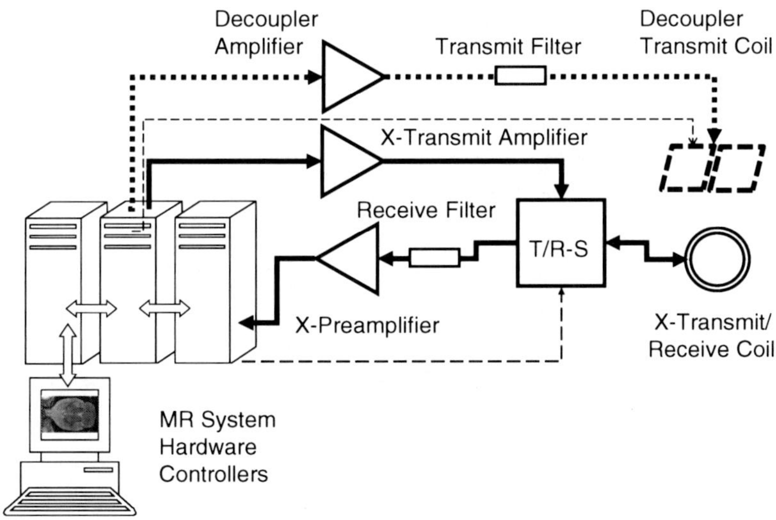

Fig. 2.9 RF system configuration for hetero-nuclear MRI/MRS. The setup for hetero-nuclear MRI/MRS is similar to the homo-nuclear setup (Fig. 2.4a) except the Larmor RF frequency of X-nuclei (e.g. ^{13}C and ^{31}P), which is lower than that of ^{1}H. A band pass filter in the receive pathway is required to suppress unwanted frequency signals that may be picked up with the main RF coil. The decoupler transmit coil tuned to ^{1}H is used for shimming, sensitivity enhancement by the nuclear Overhauser effect (NOE) and spectral simplification by decoupling

Hetero-nuclear experiments require more than one frequency synthesizer and associated RF chains including RF transmit amplifiers, RF receive amplifiers, preamplifiers, and T/R-S. Customized filters are also necessary to avoid interference between RF chains (Fig. 2.9). The RF coil setup for multi (mostly two) nuclear MR experiments is more complex than that for single resonance since the RF coils for both nuclei are simultaneously used for a large fraction of their active time. The probe designs can be volume coils or partial volume coils (Adriany and Gruetter 1997; Gruetter et al. 1998). In the latter case, the coil elements need to be clustered with close proximity to each other in order to provide more homogeneous sensitivity of the volume of interest. Covering the same volume of interest with RF coils of both nuclei provides additional opportunities in enhancing the signal sensitivity through nuclear Overhauser effect (NOE) by transmitting RF power to one RF coil while receiving from the other coil. Nuclei other than hydrogen (proton, ^{1}H) have a different Larmor frequency that is typically lower than that of ^{1}H. Most commonly used and biologically relevant nuclei are carbon (^{13}C) and phosphorous (^{31}P). ^{13}C has about a quarter of the Larmor frequency of ^{1}H and 1.1% of natural abundance, resulting in ~400 times less MR sensitivity compared to ^{1}H without considering the concentration of carbon. 31P has ~40% of the Larmor frequency of ^{1}H, resulting in ~0.4 times of MR sensitivity compared to ^{1}H, without considering the concentration of ^{31}P. As most molecules containing atoms other than hydrogen are much less abundant than hydrogen in biological tissue, less

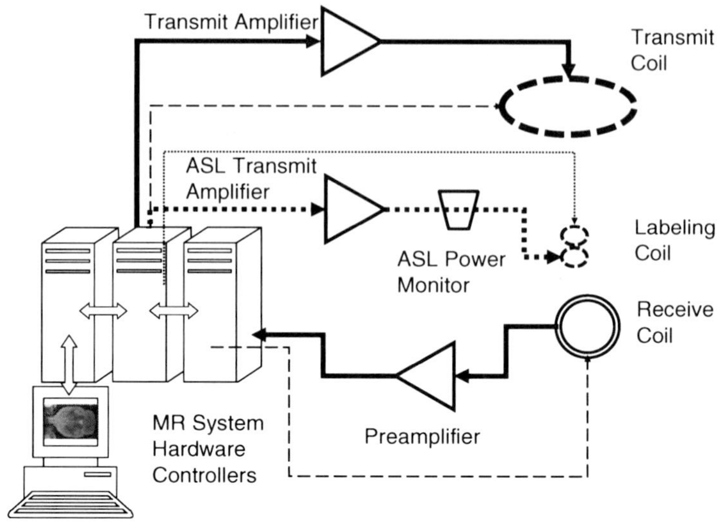

Fig. 2.10 RF system configuration with TORO and arterial spin labeling (ASL). The combined setup of TORO and ASL is similar to that of TORO (Fig. 2.5). To perform arterial spin labeling (ASL), a second independent ^{1}H transmit channel with a frequency synthesizer, an ASL transmit amplifier, and a detunable ASL coil is required. A small ASL coil (*dashed line*) is placed at an area that covers blood vessels such as carotid or vertebral arteries that supply blood to the brain

than 10 mM (other nuclei) vs. ~100 M (^{1}H), the actual MR sensitivity obtainable from other nuclei is far less than that of ^{1}H, e.g., 4 million times less for ^{13}C and 25,000 times less for ^{31}P. A typical MR setup is depicted in Fig. 2.9. The setup for hetero-nuclear MRI/MRS is similar to the homo nuclear setup (Fig. 2.4a) except for the Larmor RF frequency of the X- nuclei of interest, e.g. carbon (^{13}C), phosphorus (^{31}P). A band pass filter in the receive pathway (receive filter) is required in order to suppress signals from unwanted frequencies which may be picked up by the main coil. In addition to the X-nucleus coil, a circumscribing or partial volume ^{1}H coil, i.e., decoupler transmit coil, is placed around the sample. The function of this ^{1}H coil is to acquire MR images, to adjust shim currents for homogenizing B_0, to simplify spectra by collapsing J coupling (decoupling) and to enhance SNR through NOE. Other multi nuclear MR techniques include spectral editing and polarization transfer (Garwood et al. 1989; Merkle et al. 1992). RF wave forms for the ^{1}H "decoupler" are synthesized and phase- and amplitude-modulated in well-defined blocks within the pulse sequence. The waveform is amplified by the decoupler transmit amplifier and sent to the decoupler coil (fat dashed line in Fig. 2.9) through a decoupler transmit filter. The typical setup of X-nucleus experiments can be reversed to indirectly detect ^{1}H MR signals from neurochemicals of interest after transferring polarization from X-nuclei to ^{1}H. One of most common reverse editing experiments is the proton-observed carbon-edited (POCE) experiment. In addition, the setup shown in Fig. 2.9 is not limited to two coil TORO and further improvement is possible by replacing the single receive coil with an array in order to enhance sensitivity

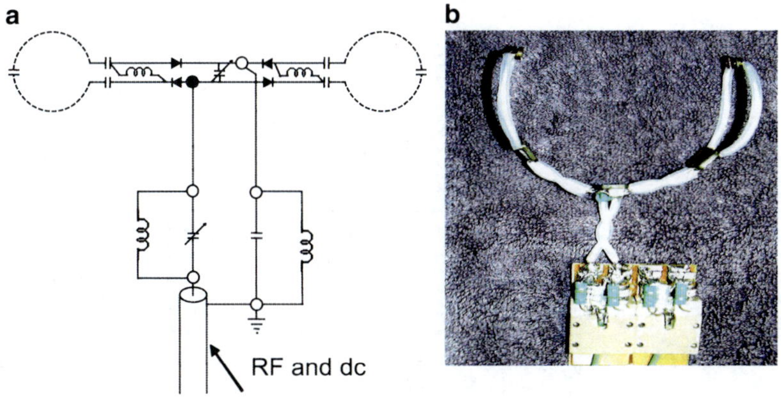

Fig. 2.11 Transmit coil for ASL. A schematic diagram of typical ASL coil is shown in (**a**). The RF and the DC signals combined and fed to the coil through a common coaxial cable. During the period of ASL the in-line PIN diodes are positively biased and allow the RF signals to flow to the coil loops. During other times the PIN diodes are negatively biased and the coil loops become non-resonant and therefore decoupled from other coils in its proximity (e.g., transmit coil and receive coils). An actual ASL coil used to measure cerebral blood flow in macaque monkeys is shown in (**b**)

2.5.1.7 TORO/ASL

Another variation of the setups shown in Figs. 2.5 and 2.6 is the addition of a second ^{1}H channel, a transmit RF chain, and an arterial spin labeling (ASL) coil (Zhang et al. 1992; Zaharchuk et al. 1999; Talagala et al. 2004; Silva 2005). This setup is used to measure cerebral blood flow or perfusion by magnetically labeling spins in feeding arteries to the brain (e.g. carotid and vertebral arteries). This setup requires three independent RF coils: transmit, ASL and receive coils (Fig. 2.10). The ASL coil is typically made in a figure-8 shape or butterfly shape with two coil segments that locally excite spins in the arteries supplying blood to the brain (e.g., Fig. 2.11). The labeling coil is typically placed at an area that covers blood vessels such as carotid or vertebral arteries that supply blood to the brain. Interaction among coils of this multi coil setup has to be strictly avoided using a combination of coil geometry strategic placement and timing of electronic detuning controlled by the system (dashed and dotted lines in Fig. 2.10).

2.5.1.8 TORO/PET

Combining two imaging modalities of MR and PET requires special customization of both MR and PET hardware (Figs. 2.12 and 2.13). (Jacobs and Cherry 2001; Catana et al. 2008; Cherry et al. 2008; Pichler et al. 2008a, b; Antoch and Bockisch 2009; Hicks and Lau 2009; Sauter et al. 2010; Wehrl et al. 2010, 2011). In the MR and PET hybrid scanners, the PET sensor array is located between RF coils and the gradient coils in the main magnet. To avoid distortion of the B_0 field, the PET sensor arrays

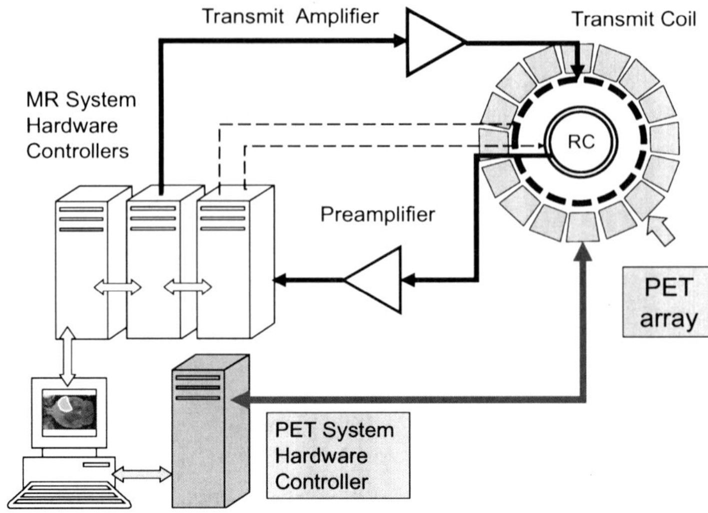

Fig. 2.12 System configuration with combined TORO MR and PET. In the MR and PET combined system, an array of PET sensors is added around the RF coils inside the magnet in addition to MR RF chains in TORO mode. MR hardware and PET hardware are controlled independently, but the images of both MR and PET are combined and displayed on the same monitor. Control lines for TORO operation are shown in *dashed lines*

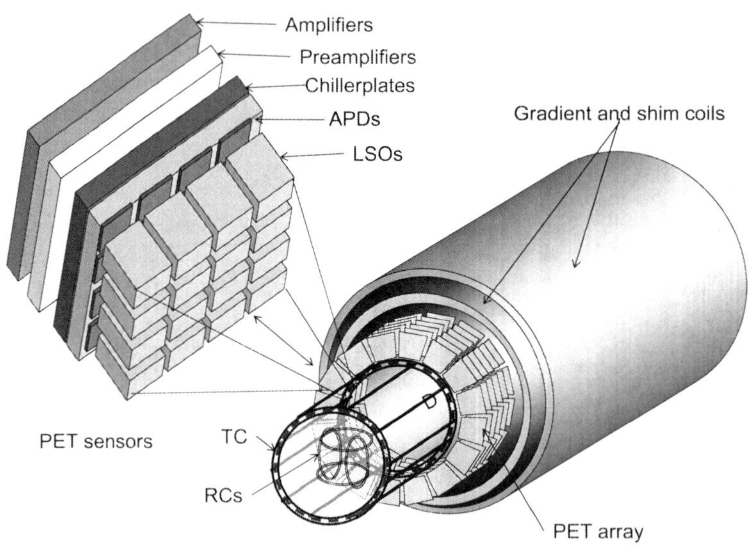

Fig. 2.13 MR and PET integration. To integrate MR and PET, specially designed PET sensors (PS) constructed with non-magnetic components are placed in a layer between the RF coils and gradient/shim coils. A number of individual PET elements (*blocks*) are arranged in z-directional stacks of concentric rings, resulting in a PET detector array. Other magnet inserts include gradient and shim coils (*shaded cylinders*) and an RF transmit coil (TC, *dashed circles*) and RF receive coils (RC, *double lines*). Each PET detector block is constructed in several layers of components that include a pack of lutetium oxyorthosilicate (LSO) scintillation crystals that transform the positron gamma quanta into light quanta. Light is detected by an array of chilled avalanche photodiodes (APD). Output signals of APD are amplified in several stages and sent to the PET controller to be digitized and processed

need to be constructed without any ferromagnetic components and PET interface cables that carry DC currents need to be neutralized by pairing and twisting with cables with opposite currents. The MR RF coils (e.g., transmit coils and receive arrays) and their associated components (e.g. filters and preamplifiers) need to be transparent to the gamma rays from positron decay (511 keV) to avoid scattering and absorption that lead to the unwanted attenuation of gamma rays photons before detection.

2.5.2 *RF Coils*

RF coils are one of the most crucial components of the MR system since they link between the molecular event of magnetic resonance in tissue and the electronic world of the MR system. RF coils have to be placed at close proximity to the sample in order to be sensitive and together with preamplifiers they form the 'front-end' of the MR system. The coils are made with metallic wires that are attached to RF resonance circuits and connected to the system via a transmission line. RF coils can be utilized in a variety of configurations depending on the requirements of the experimental setup described in Sect. 2.5.1 (Wright 1990; Wright and Wald 1997). Recent reviews summarized state-of-the-art small-animal RF coils (Haase et al. 2000; Mispelter et al. 2006; Doty et al. 2007). A single transceive coil is still most commonly utilized in animal research settings, while independent transmit and receive coils (TORO system) are almost exclusively used in clinical settings. The transmit coil used in TORO systems is permanently mounted in the most inner shell of the magnet circumscribing the patient/animal access space. The transmit coil is designed to produce homogeneous transmit B_1 fields. The reception of MR signals is achieved with a receive coil or coil arrays that are strategically placed close to the volume of interest (Fig. 2.14).

2.5.2.1 Partial-Volume and Circumscribing Coils

The simplest front-end configuration (Fig. 2.4) utilizes a single coil for transmission and reception, i.e., a transceive coil. It is important to choose appropriate types and shapes of RF coils based on the requirements of each experiment. MRI experiments typically require a coil that produces a homogeneous B_1 field in order to produce homogeneous image intensity over the whole volume of interest when the image contrast is determined only by characteristics of the given tissue. In some cases, obtaining the desired homogeneity may require some loss of sensitivity in certain areas of the tissue of interest. Certain three-dimensional patterns of coil wires produce quite homogeneous linear B_1 fields with examples of saddle coils (Carlson 1986) or birdcage coils (Hayes 2007 and references therein). Homogeneity of RF coils can be evaluated by field histograms (Yang et al. 1994). If the wire of the coil is bent to a simple loop (Fig. 2.13), placed around a small sample, the B_1 fields produced at the center can be sufficiently homogeneous for a variety experiments, although the sensitivity may be compromised. If, on the other hand, the loop is placed on a surface of a

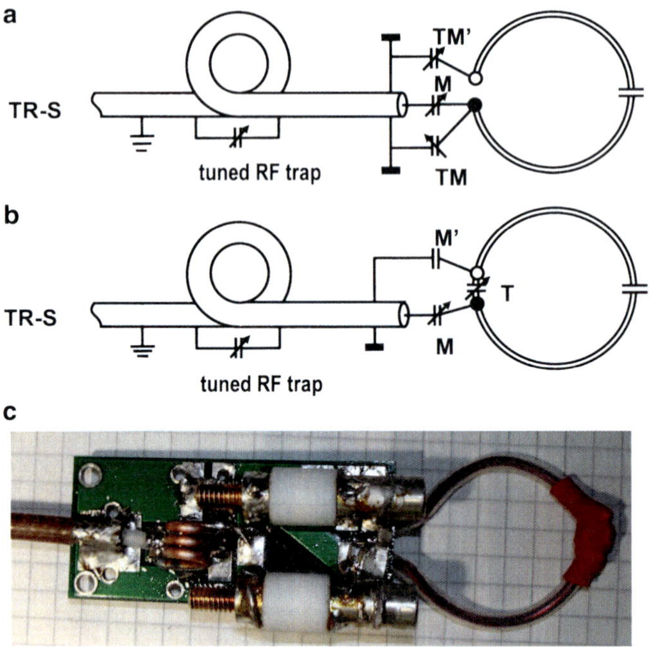

Fig. 2.14 Two simple transceive RF coil circuits. The loop terminals (open and closed circles) of the coil loop are connected to the transmit/receive switch (TR-S) via a tune-match network and a tuned RF trap using 50 Ohm transmission line. The impedance of the coil loops is matched to the characteristic impedance of the transmission line (50 Ohm) with a tune-match network by adjusting match and tune capacitors (M, M', T, T') to minimize RF reflection. An RF coil with a double-balanced tune/match network (two variable tune capacitors and a variable match capacitor) is shown in (**a**). A simplified RF coil with a variable match capacitor and a fixed match capacitor, and an in-loop tune capacitor is shown in (**b**). The system ground and the virtual ground are shown as the three dash symbol and the solid bar symbol, respectively. The virtual ground of the coil is isolated from the system ground by a tuned RF trap. A picture of the actual coil (**b**) is shown in (**c**)

sample as a surface coil, the sensitivity close to the wire is dramatically improved, although the B_1 field in that area is very inhomogeneous. The surface coil or partial volume coil is useful for a variety of MR experiments. Since MR spin precession is a quadrature event (Hoult et al. 1984), a single coil with its associated linearly polarized B_1 fields is unable to interact with spins with full efficiency. Circularly polarized B_1 fields are necessary for full efficiency. A method to create circularly polarized B_1 fields is adding a second linearly polarized B_1 field along the transverse axis with 90° rotation around B_0 using a second coil. A quadrature hybrid (Fig. 2.15a) provides the second RF currents with the 90° rotation, equivalent to phase shifts of 90°, from a single RF input. Using a quadrature hybrid, an MR system with a single channel RF transmitter can benefit from circularly polarized B_1 fields via quadrature RF coils. This setup is more effective during transmission by a factor of 2 (only half the power is necessary to obtain a desired flip angle during excitation) over that of linear surface coils. In addition, this setup provides an improved SNR of ~1.41 (square root of 2)

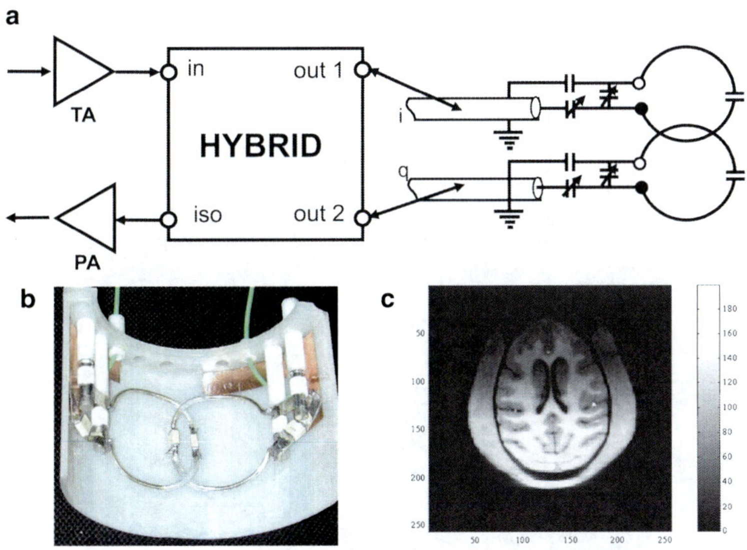

Fig. 2.15 Quadrature hybrid for partial volume quadrature coil with overlapped loops. A quadrature hybrid is utilized to interconnect a single channel RF transmitter and a quadrature RF coil. The quadrature hybrid splits the RF signals from the transmit amplifier equally into two outputs, "**i**" and "**q**" (out1 and out2), with a phase difference of 90 degrees, which are connected to the two circular loop coils (**a**). These two loops produce a partial volume quadrature coil when they are mounted onto a cylindrical former in a way that their axes are orthogonally arranged and the interaction between the loops is nulled by a small overlap (**b**). The resulting B_1 field of the quadrature coil has enhanced homogeneity in its sensitive area close to the structure of the assembly compared to a single surface coil. The MR image of a macaque monkey brain obtained at 4.7 T using the coil demonstrates enhanced homogeneity and good volume coverage (**c**)

times during reception by receiving signals from the two channels, and improved homogeneity of B_1 fields when the coil loops are properly arranged. An example of RF coil setups using a quadrature hybrid and a half-volume quadrature RF coil is shown in Fig. 2.15. Two simple circular surface coil loops are mounted on a plastic former in a partially overlapped arrangement to null the coupling between the two loops (Fig. 2.15b). An MR image acquired using the coil assembly is shown in Fig. 2.15c. Other examples of partial volume transceive coils can be found elsewhere (Choi et al. 2007 and references therein).

When the quadrature principle is applied to already quite homogeneous coils like birdcage coils, the homogeneity of the B_1 field is even more improved. A schematic of a high-pass quadrature birdcage coil and a picture of an actual coil are shown in Fig. 2.16a and b. Figure 2.16c shows examples of MR images of a multi-compartment cylindrical phantom acquired with a birdcage coil at 7 T. Orthogonality between two quadrature channels and the efficiency of creating circularly polarized B_1 fields is demonstrated by the almost complete cancellation of MR signals in the anti-quadrature mode image (Fig. 2.16c, right) compared with the quadrature mode image (Fig. 2.16c, left).

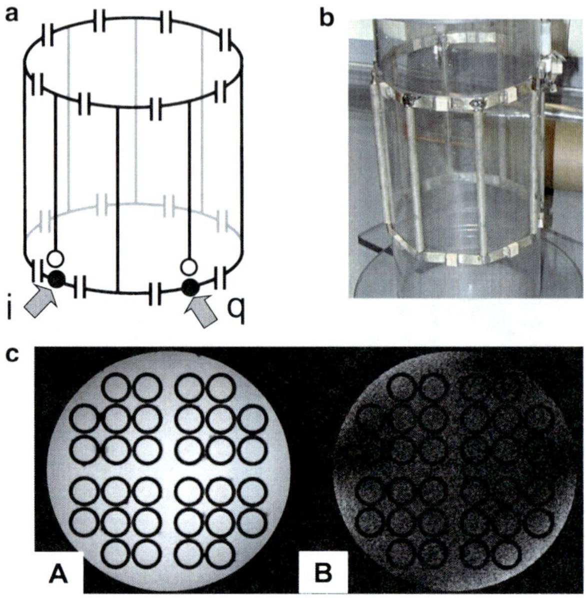

Fig. 2.16 Birdcage coil, a quadrature full volume transceive coil. A high-pass birdcage coil structure with a standard tune/match circuit is shown in (A) and (B). The birdcage produces a very homogeneous internal B_1 field when driven in quadrature mode using a quadrature hybrid. The two shaded arrows in (A) indicate the two quadrature input ports (**i** and **q**) that are connected to the two output ports of the quadrature hybrid. The resulting B_1 field is circularly polarized over a large volume within the coils. MR images of a cylindrical phantom (diameter: 78 mm, height: 100 mm) are acquired at 7 T using the birdcage coil shown in (B). The phantom is filled with water solution of 0.45% NaCl, 1 g/L $CuSO_4$, and polycarbonate tubes (diameter: 10 mm, wall thickness: 1 mm). Fairly homogeneous image contrast across the phantom demonstrates the performance of the quadrature bird cage coil (**C**(a)). The quadrature performance and the efficiency of creating circularly polarized B_1 fields is demonstrated by the almost complete cancellation of MR signals in the anti-quadrature mode image (**C**(b)), which was acquired with the **i** and **q** channels interchanged

2.5.2.2 Receive Only Coil and Receive Only Arrays

Transmission with a homogeneous coil and reception with a local coil (Fig. 2.16) offer the advantage of homogeneous excitation and good sensitivity of surface coils in the volume of interest. In the 1980s, a new technology for receive coils was introduced, driven by the need to compress the time patients have to spend inside a magnet for studies with increased resolution and SNR, particularly in spinal cord MRI (Hayes and Roemer 1990; Requardt et al. 1990; Roemer et al. 1990; Reykowski et al. 1995). An assembly of local receive coils was arranged along the volume of interest such as the spinal cord to have minimal coupling between neighboring elements. This arrangement allows simultaneous or parallel signal acquisition with multiple receivers instead of successive acquisition with one receiver. Coupling between receive coil elements typically results in decreased performance by degrading

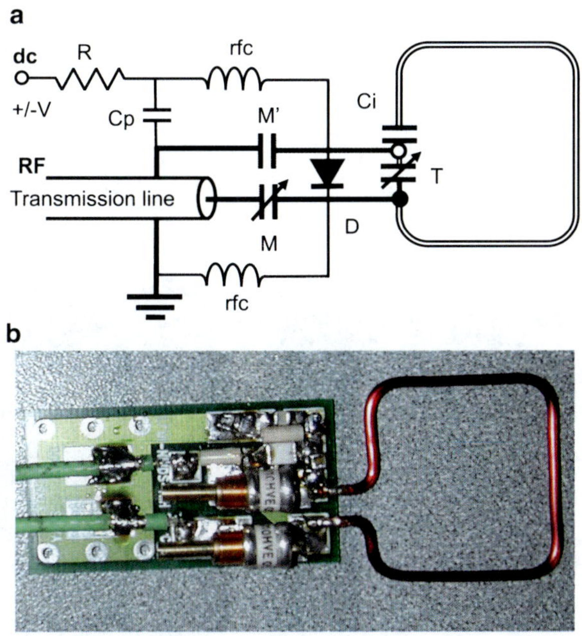

Fig. 2.17 Basic single loop receive coil circuit with detune capability. Basic single loop receive coils with detuning capability, schematic shown in (**a**), can be built based on the designs of simple transceive coils (Fig. 2.14) by adding components required for detuning, which are a PIN diode, resistor, RF chokes, and additional capacitors. The PIN diode is placed in parallel to the terminals of the coil loop. During RF transmission, a DC bias current, limited by a resistor R and applied to the PIN diode, causes an electrical short-circuit of the coil loop and therefore shifts the resonance frequency of the coil outside of the Larmor frequency range. The RF signal leakage through the DC pathway is blocked by radiofrequency chokes (rfc). The large in-loop capacitor (Ci) blocks DC currents from flowing through the loop. Placing Ci further in the loop (e.g. in the middle) allows to tune the circuit at higher frequencies. A picture of an actual coil is shown in (**b**)

the frequency tuning of the coil elements and by coupling of their individual noise floors. To avoid this performance decrease, each receive coil (array element) can be arranged to partially overlap with the next neighbor elements so that total magnetic field flux from one element to another is nulled, thereby achieving decoupling of receive elements. However, this geometrical arrangement can decouple only the nearest neighboring coil elements, but cannot remove coupling between non-neighboring elements. To overcome the limitations of the geometrical decoupling, a new decoupling technology using high reflection coefficient preamplifiers was developed. In this scheme, decoupling among elements is achieved through minimizing electrical currents flowing in each element by connecting each element to a high reflection coefficient preamplifier. This technology enables a more liberal arrangement of receive array elements and wide-spread uses of receive arrays in spinal cord imaging, whole body imaging and brain studies (Fig. 2.17). Basic principles and practical implementations (Roemer et al. 1990; Wright 1990) and more generalized

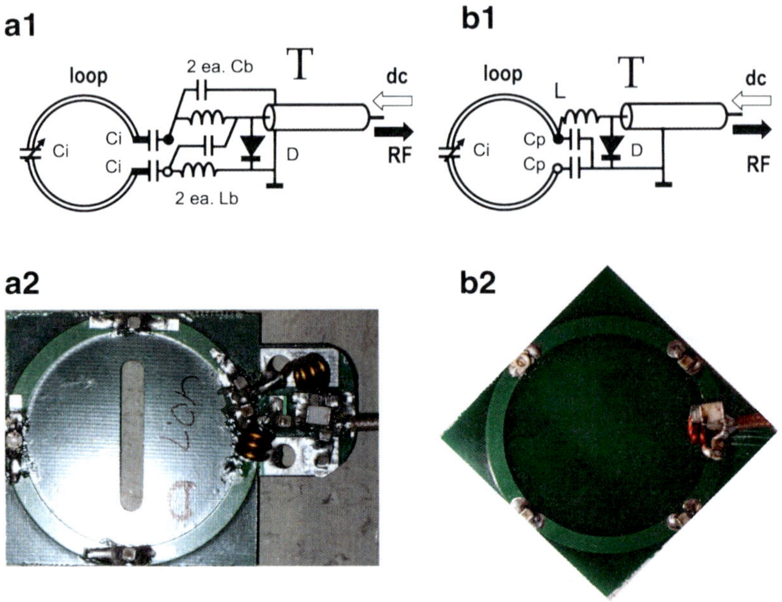

Fig. 2.18 Two designs of phased array element circuits. The loops of phased array elements pick up the RF signals from the sample and guide it via a lattice balun network (**a**) or a split capacitor network and match inductor (**b**) to the transmission line at point T. During transmission, the receive element is de-tuned using a PIN diode that shorts the input to the transmission line and causes high impedance within the coil loop by means of balun elements Lb and Cb in case of (**a**1) or L and Cp in case of (**b**1). The pictures (**a**2 and **b**2) show examples of 50-mm diameter coils with the two designs (**a**1 and **b**1), respectively

theory and relevant applications of receive arrays (Wright and Wald 1997; de Zwart et al. 2002) have been described previously. The number, size and shape of the elements and their arrangement around the areas of interest is dictated by available receiver channels in an MR system and by the requirement of a specific application. Two typical designs of receive elements are shown in Fig. 2.18. The loops in the designs A and B pick up RF signals from the sample and guide it through a lattice balun network (Fig. 2.18a) or a split capacitor network and match inductor (Fig. 2.18b) to the transmission line. During RF transmission the receive element is de-tuned using a PIN diode which shorts the input to the transmission line and causes high impedance within the coil loop by means of balun elements Lb and Cb in (A) or L and Cp in (B). Figure 2.19 shows schematics for the receive chain with preamplifier decoupling (Fig. 2.19a) and examples of receive array arrangement (Fig. 2.19b, c). In the decoupling circuit (Fig. 2.19a), the length of the transmission line between the array element and the preamplifier is crucial for proper decoupling. Figure 2.19b shows a five-element receive array in a linear gapped and overlapped configuration, which is used in a large bore 4.7 T vertical magnet. Actual receive array and MR images of the macaque brain acquired using the array are shown in Fig. 2.20a and b, respectively. Another example of a five-element array in an

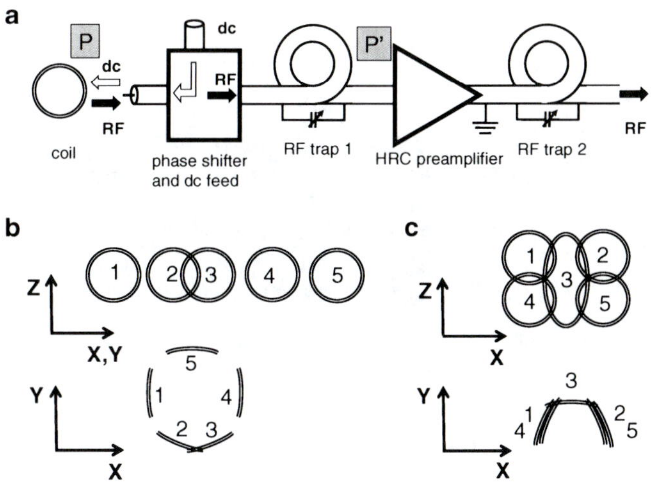

Fig. 2.19 RF receive chain and arrangements for array elements with preamplifier decoupling. In the RF receive chain (**a**), the phase difference of RF signals between the coil element ports (P) and the high reflection coefficient preamplifier (HRC) input port (P') is critical for the performance of the element. The phase difference is adjusted using a phase shifter built from discrete elements inductor and capacitors. At any place between P and P', here shown within the phase shifter, an RF/DC "Bias-T" is inserted to combine RF signals for reception and DC signals for detuning of the coil. The gain factor of the HRC preamplifier is ~28 dB and output impedance is set to 50 Ohm to match those of the transmission line and the system receiver. Array elements can be arranged in a variety of configurations depending on the requirement of including a linear-gapped and overlapped configuration for use in a large bore vertical magnet (**b**) and overlapped oblique configuration for use in a 7 T horizontal magnet for the marmoset brain (**c**)

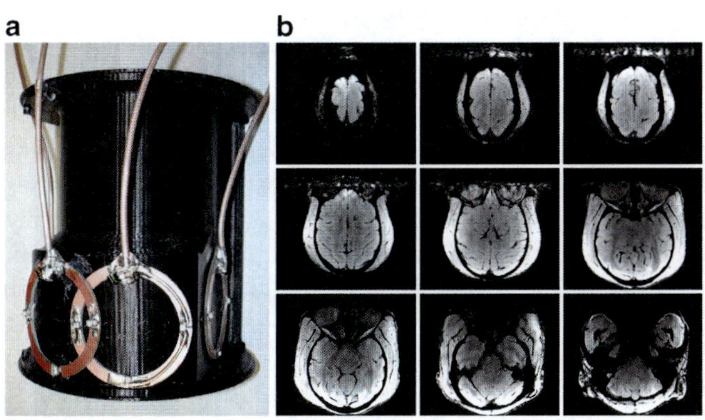

Fig. 2.20 Receive array of linear-gapped and overlapped elements used for macaque monkey brain imaging. An actual constructed receive array of linear-gapped and overlapped elements (Fig. 2.19b) is shown in (**a**). Circular loops in ~50 cm diameter with attached electronics are mounted on a plastic former. The assembly was then placed on a macaque head over the occipital pole of the brain. MR images acquired from a macaque brain using the coil are shown in (**b**). The images show enhanced sensitivity in areas close to the receive elements

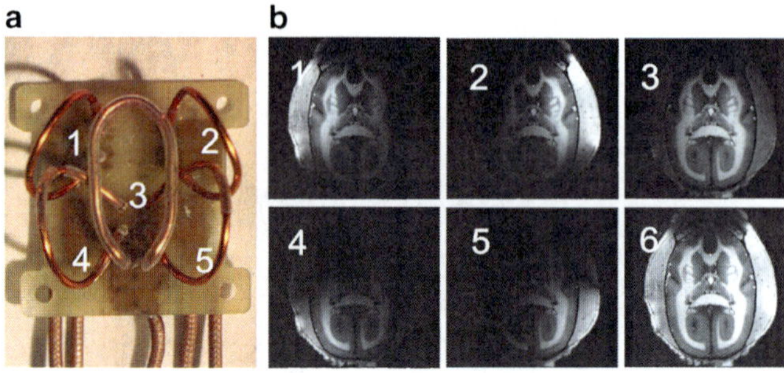

Fig. 2.21 Receive array of overlapped-oblique configuration used for marmoset monkey brain imaging. An actual constructed receive array of overlapped-oblique or basket-cluster configuration (Fig. 2.19c) is shown in (a). The receive array is composed of four circular loops of 25 mm diameter and one elliptical loop of 20 mm × 30 mm. MR images acquired from a marmoset monkey brain using the receive array is shown in (**b**). Reception sensitivity of the array is demonstrated in individual images from each loop (1–5) and the combined image (6)

overlapped oblique configuration is shown in Fig. 2.19c, which is used for marmoset monkeys in a 7 T horizontal magnet. Actual array and MR images acquired using the array are shown in Fig. 2.21a and b, respectively.

2.5.2.3 Hetero-Nuclear Coils

Hetero-nuclear MR coils are built according to the same construction principles as ^{1}H coils, yet the construction is easier due to their lower Larmor frequency. Performance of hetero-nuclear experiments can be increased with techniques that involve direct or indirect nuclear spin interactions between hydrogen and a hetero nucleus in molecules of interest. To this end, the volume of interest in tissue needs to be located within B_1 fields of both Larmor frequencies of X-nucleus and ^{1}H. Consequently, B_1 fields of both frequencies need to be created by independent coils covering the same physical space. Construction of such a hetero- and homo-nuclear coil cluster can be an engineering challenge because the B_1 fields created by the coil wires may couple or interact with each other, and distorts B_1 fields of both frequencies. Such coils can be constructed in either double-tuned configuration that uses a shared coil loop for both frequencies (Fig. 2.21) or a dual-tuned configuration that uses two separate coils (Fig. 2.22). Principles and practical guidelines for constructing hetero-nuclear coils have been described by Mispelter (Mispelter et al. 2006, chapter 6). In order to obtain sensitivity improvement within a hetero-nuclear MR experiment, spins of both nuclei need to be excited or irradiated successively in a short time or at the same time depending on the experimental techniques. Reception of the MR signal can be either at ^{1}H or ^{13}C frequency. For the experimental setup, it is necessary to isolate those two frequencies (e.g., 50 and 200 MHz for ^{1}H and ^{13}C at 4.7 T) using hardware filters that allow only RF signals with the desired frequency

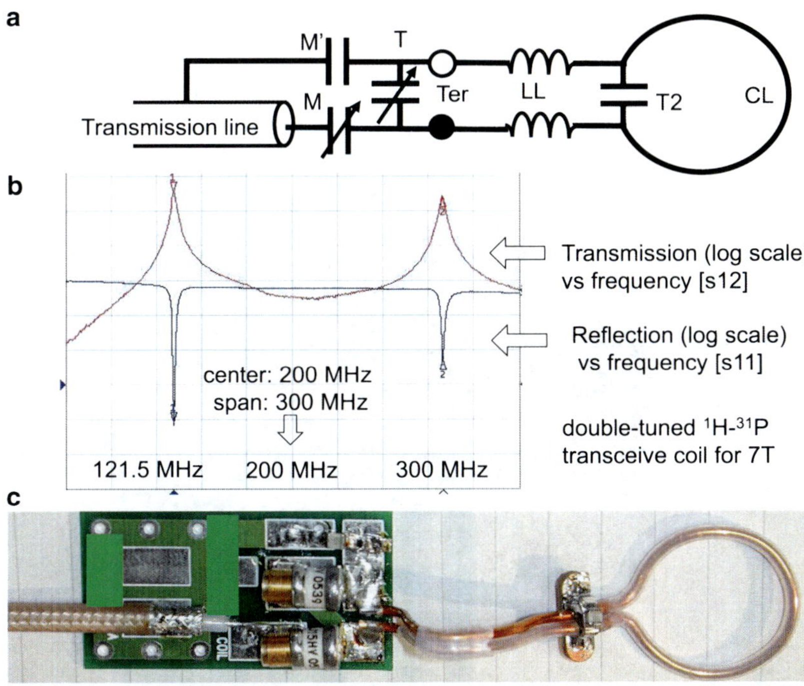

Fig. 2.22 Double tuned transceive RF coil. Double tuned RF coil (**a**) is constructed using a tune/match network with a tune (T), a match (M) and a balance match capacitors (M'), a coil loop (CL), a second tune capacitor (T2) and the parallel lead wire inductance (LL). Measured impedance parameters (s11 and s12) of a constructed coil (e.g., (**c**)) show the impedance of the dual-tuned coil is closely matched to 50 Ohm for both frequencies for X-nucleus and ^{1}H at the same time (121.5 and 300 MHz at 7 T) (**b**). In general, double tuned coils are designed to function at narrow ranges of loading conditions with limited capacity to tune each frequency independently. When the coil cannot be tuned perfectly for both frequencies, the coil is tuned and matched to X-nucleus frequency using tune and match capacitors to preserve the coil sensitivity for X-nucleus

to be directed to the receiver. A careful choice of filters regarding the frequency bandwidth, insertion loss, and power capability is critical to achieve the optimal performance. Depending on the experiment, low-pass, band-pass, high-pass filters, or any combination of them is utilized. The most common hetero-nuclear front-end setup is a ^{13}C MR experiment with a transmit/receive ^{13}C coil for excitation and detection of ^{13}C signals and an ^{1}H transmit coil for NOE and decoupling for improving signal intensity and resolution of ^{13}C MRS (Fig. 2.9). Another common setup uses a ^{1}H transmit/receive coil for excitation and detection and ^{13}C transmit coil for polarization transfer within the molecules of interest. This reverse setup provides an advantage of greater signal sensitivity of up to fourfold by detecting MR signals at ^{1}H frequency. The filter setups in transmit/receive and decoupling pathways need to be exchanged to accommodate the requirements of frequency change (Fig. 2.23).

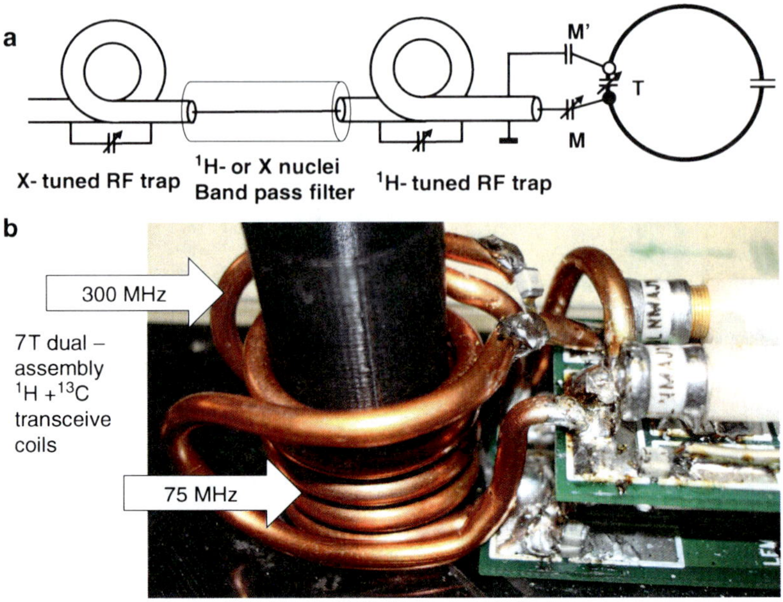

Fig. 2.23 Dual tuned assembly of transceive RF coils for hetero-nuclear studies. Dual tuned RF coil assembly (**a**) uses two independent coils arranged in layers so that they cover the same volume of interest and their B_1 fields are orthogonal. Each transceive coil for the assembly is similar to simple transceive coil except for RF traps for each frequency and an in-line band-pass filter for either X-nucleus or ^{1}H. (**b**) Shows a picture of a dual tuned coil assembly for ^{13}C studies at 7 T. A solenoid coil, tuned to 75 MHz, and an independent saddle coil, tuned at 300 MHz, are combined to cover the same volume of interest

Acknowledgements This work was supported by the intramural program of the Laboratory for Functional and Molecular Imaging (LFMI), National Institute of Neurological Disorders and Stroke (NINDS), National Institutes of Health (NIH), Bethesda, Maryland, USA. We gratefully acknowledge Mark Augath, Max-Planck Institute for Biological Cybernetics, Tuebingen, Germany, for images of Figs. 2.15 and 2.16; Charles Zhu, Neuro Imaging Facility, NINDS, National Eye Institute, National Institute for Mental Health, NIH, for images of Fig. 2.20; Dr. Afonso Silva, Micro Circulation Unit, LFMI, NINDS, NIH, for images of Fig. 2.21.

References

Abragam A (1961) The principles of nuclear magnetism. Clarendon, Oxford

Adriany G, Auerbach EJ et al (2010) A 32-channel lattice transmission line array for parallel transmit and receive MRI at 7 Tesla. Magnetic Resonance Med 63(6):1478–1485

Adriany G, De Moortele PFV et al (2008) A geometrically adjustable 16-channel transmit/receive transmission line array for improved RF efficiency and parallel imaging performance at 7 Tesla. Magn Reson Med 59(3):590–597

Adriany G, Gruetter R (1997) A half-volume coil for efficient proton decoupling in humans at 4 Tesla. J Magn Reson 125(1):178–184

Anderson WA (1961) Electrical current shims for correcting magnetic fields. Rev Sci Instrum 32(3):241–250

Antoch G, Bockisch A (2009) Combined PET/MRI: a new dimension in whole-body oncology imaging? Eur J Nucl Med Mol Imaging 36:113–120
Battocletti JH, Kamal HA et al (1993) Systematic passive shimming of a permanent-magnet for P-31 NMR-spectroscopy of bone-mineral. IEEE Trans Magnetics 29(4):2139–2151
Belov A, Bushuev V et al (1995) Passive shimming of the superconducting magnet for MRI. IEEE Trans Appl Superconductivity 5(2):679–681
Bendall M (1988) Surface coil technology. In: Partain CL, Price RR, Patton JA, Kulkarni MV, James AE (eds) Magnetic resonance imaging. W.B. Saunders, Philadelphia, pp 1201–1268
Bloch F (1946) The nuclear induction experiment. Phys Rev 70:460–473
Brideson MA, Forbes LK et al (2002) Determining complicated winding patterns for shim coils using stream functions and the target-field method. Concepts Magn Reson 14(1):9–18
Carlson JW (1986) Currents and fields of thin conductors in rf saddle coils. Magn Reson Med 3(5):778–790
Carlson JW, Derby KA et al (1992) Design and evaluation of shielded gradient coils. Magn Reson Med 26(2):191–206
Catana C, Procissi D et al (2008) Simultaneous in vivo positron emission tomography and magnetic resonance imaging. Proc Natl Acad Sci USA 105(10):3705–3710
Chen CN, Hoult D (1989) Biomedical magnetic resonance technology. New York and Bristol
Cherry SR, Louie AY et al (2008) The integration of positron emission tomography with magnetic resonance imaging. Proc IEEE 96(3):416–438
Choi IY, Lee SP et al (2007) Simple partial volume transceive coils for in vivo H-1 MR studies at high magnetic fields. Concepts Magn Reson Part B 31B(2):71–85
Crozier S, Forbes LK et al (1994) The design of transverse gradient coils of restricted length by simulated annealing. J Magn Reson Series A 107(1):126–128
de Zwart JA, Ledden PJ et al (2002) Design of a SENSE-optimized high-sensitivity MRI receive coil for brain imaging. Magn Reson Med 47(6):1218–1227
Dorri B, Vermilyea M et al (1993) Passive shimming of MR magnets: algorithm, hardware, and results. Appl Superconductivity, IEEE Trans 3(1):254–257
Doty FD, Entzminger G et al (2007) Radio frequency coil technology for small-animal MRI. NMR Biomed 20(3):304–325
Driesel W, Mildner T et al (2008) A microstrip helmet coil for human brain imaging at high magnetic fields. Concepts Magn Reson B 33B(2):94–108
Duensing GR, Brooker HR et al (1996) Maximizing signal-to-noise ratio in the presence of coil coupling. J Magn Reson B 111(3):230–235
Eccles CD, Crozier S et al (1994) Practical aspects of shielded gradient-coil design for localized in-vivo NMR-spectroscopy and small-scale imaging. Magn Reson Imaging 12(4):621–630
Ernst RR (1966) Nuclear magnetic double resonance with an incoherent radio-frequency field. J Chem Phys 45:3845
Fishbein KW, McGowan JC et al (2005) Hardware for magnetic resonance imaging. In: Filippi M, De Stefano N, Dousset V, McGowan JC (eds) MR imaging in white matter diseases of the brain and spinal cord. Berlin, Heidelberg, p13–28
Forbes LK, Brideson MA et al (2005) A target-field method to design circular biplanar coils for asymmetric shim and gradient fields. IEEE Trans Magn 41(6):2134–2144
Forbes LK, Crozier S (2001) Asymmetric zonal shim coils for magnetic resonance applications. Med Phys 28(8):1644–1651
Forbes LK, Crozier S (2002) A novel target-field method for finite-length magnetic resonance shim coils: II. Tesseral shims. J Phys D 35(9):839–849
Forbes LK, Crozier S (2003) A novel target-field method for magnetic resonance shim coils: III. Shielded zonal and tesseral coils. J Phys D 36(2):68–80
Forbes LK, Crozier S (2004) Novel target-field method for designing shielded biplanar shim and gradient coils. IEEE Trans Magn 40(4):1929–1938
Garwood M, Ugurbil K et al (1989) Magnetic resonance imaging with adiabatic pulses using a single surface coil for RF transmission and signal detection. Magn Reson Med 9(1):25–34
Goense JBM, Ku SP et al (2008) fMRI of the temporal lobe of the awake monkey at 7 T. Neuroimage 39(3):1081–1093

Golay MJE (1958) Field homogenizing coils for nuclear spin resonance instrumentation. Rev Sci Instrum 29(4):313–315
Griswold MA, Jakob PM et al (2002) Generalized autocalibrating partially parallel acquisitions (GRAPPA). Magn Reson Med 47(6):1202–1210
Gruetter R (1993) Automatic, localized in vivo adjustment of all first- and second-order shim coils. Magn Reson Med 29(6):804–811
Gruetter R, Seaquist ER et al (1998) Localized in vivo 13C-NMR of glutamate metabolism in the human brain: initial results at 4 Tesla. Dev Neurosci 20(4–5):380–388
Haacke EM, Brown RW et al (1999) Magnetic resonance imaging: physical principles and sequence design. Wiley-Liss, New York
Haase A, Odoj F et al (2000) NMR probeheads for in vivo applications. Concepts Magn Reson 12(4):361–388
Hardy CJ, Darrow RD et al (2004) Large field-of-view real-time MRI with a 32-channel system. Magn Reson Med 52(4):878–884
Hayes CE (2007) Birdcage resonators: highly homogeneous radiofrequency coils for magnetic resonance. Encyclopedia of Magnetic Resonance, Wiley, New York
Hayes CE, Hattes N et al (1991) Volume imaging with MR phased arrays. Magn Reson Med 18(2):309–319
Hayes CE, Roemer PB (1990) Noise correlations in data simultaneously acquired from multiple surface coil arrays. Magn Reson Med 16(2):181–191
Heerschap A, Sommers MG et al (2004) Nuclear magnetic resonance in laboratory animals. Methods Enzymol 385:41–63
Hetherington HP, Avdievich NI et al (2010) RF shimming for spectroscopic localization in the human brain at 7 T. Magn Reson Med 63(1):9–19
Hicks RJ, Lau EWF (2009) PET/MRI: a different spin from under the rim. Eur J Nucl Med Mol Imaging 36:10–14
Hoult DI, Chen CN et al (1984) Quadrature detection in the laboratory frame. Magn Reson Med 1(3):339–353
Hoult DI, Deslauriers R (1994) Accurate shim-coil design and magnet-field profiling by a power-minimization-matrix method. J Magn Reson Series A 108(1):9–20
Hoult DI, Foreman D et al (2008) Overcoming high-field RF problems with non-magnetic Cartesian feedback transceivers. Magn Reson Mater Phys Biol Med 21(1–2):15–29
Hoult DI, Kolansky G et al (2004a) A ‘hi-fi’ Cartesian feedback spectrometer for precise quantitation and superior performance. J Magn Reson 171(1):57–63
Hoult DI, Kolansky G et al (2004b) The NMR multi-transmit phased array: a Cartesian feedback approach. J Magn Reson 171(1):64–70
Hoult DI, Lee D (1985) Shimming a superconducting NMR imaging magnet with steel. Rev Sci Instrum 56(1):131–135
Hoult DI, Richards RE (1976) The signal-to-noise ratio of the nuclear magnetic resonance experiment. J Magn Reson 24(1):71–85
Jacobs RE, Cherry SR (2001) Complementary emerging techniques: high-resolution PET and MRI. Curr Opin Neurobiol 11(5):621–629
Jin J (1998) Electromagnetic analysis and design in magnetic resonance imaging. CRC, New York
Juchem C, Merkle H et al (2004) Region and volume dependencies in spectral line width assessed by H-1 2D MR chemical shift imaging in the monkey brain at 7 T. Magn Reson Imaging 22(10):1373–1383
Kellman P, McVeigh ER (2005) Image reconstruction in SNR units: a general method for SNR measurement. Magn Reson Med 54(6):1439–1447
Kumar A, Bottomley PA (2006) Optimizing the intrinsic signal-to-noise ratio of MRI strip detectors. Magn Reson Med 56(1):157–166
Kuperman V (2000) Magnetic resonance imaging: physical principles and applications. Academic, San Diego
Kurpad KN, Wright SM et al (2006) RF current element design for independent control of current amplitude and phase in transmit phased arrays. Concepts Magn Reson Part B 29B(2):75–83
Lauterbur PC (1973) Image formation by induced local interaction: examples employing nuclear magnetic resonance. Nature 242:190–191

Lee RF, Giaquinto RO et al (2002) Coupling and decoupling theory and its application to the MRI phased array. Magn Reson Med 48(1):203–213

Logothetis NK, Guggenberger H et al (1999) Functional imaging of the monkey brain. Nat Neurosci 2(6):555–562

Mansfield P, Chapman B (1986) Active magnetic screening of coils for static and time-dependent magnetic-field generation in NMR imaging. J Phys E 19(7):540–545

McDougall MP, Wright SM (2005) 64-channel array coil for single echo acquisition magnetic resonance imaging. Magn Reson Med 54(2):386–392

Merkle H, Wei HR et al (1992) B(1)-Insensitive heteronuclear adiabatic polarization transfer for signal enhancement. J Magn Reson 99(3):480–494

Mispelter J, Lupu M et al (2006) NMR probeheads: for biophysical and biomedical experiments. Imperial College, London

Niendorf T, Hardy CJ et al (2006) Toward single breath-hold whole-heart coverage coronary MRA using highly accelerated parallel imaging with a 32-channel MR system. Magn Reson Med 56(1):167–176

Pan JW, Avdievich N et al (2010) J-refocused coherence transfer spectroscopic imaging at 7 T in human brain. Magn Reson Med 64(5):1237–1246

Pfeuffer J, Juchem C et al (2004a) High-field localized 1H NMR spectroscopy in the anesthetized and in the awake monkey. Magn Reson Imaging 22(10):1361–1372

Pfeuffer J, Merkle H et al (2004b) Anatomical and functional MR imaging in the macaque monkey using a vertical large-bore 7 Tesla setup. Magn Reson Imaging 22(10):1343–1359

Pichler BJ, Judenhofer MS et al (2008) Multimodal imaging approaches: PET/CT and PET/MRI. Handb Exp Pharmacol (185 Pt 1):109–132

Pichler BJ, Judenhofer MS et al (2008b) PET/MRI hybrid imaging: devices and initial results. Eur Radiol 18(6):1077–1086

Pruessmann KP, Weiger M et al (1999) SENSE: sensitivity encoding for fast MRI. Magn Reson Med 42(5):952–962

Requardt H, Offermann J et al (1990) Switched array coils. Magn Reson Med 13(3):385–397

Reykowski A, Wright SM et al (1995) Design of matching networks for low noise preamplifiers. Magn Reson Med 33(6):848–852

Roemer PB, Edelstein WA et al (1990) The NMR phased array. Magn Reson Med 16(2): 192–225

Sauter AW, Wehrl HF et al (2010) Combined PET/MRI: one step further in multimodality imaging. Trends Mol Med 16(11):508–515

Schempp WJ (1998) Magnetic resonance imaging: mathematical foundations and applications. Wiley-Liss, New York

Shajan G, Hoffmann J et al (2011) Design and evaluation of an RF front-end for 9.4 T human MRI. Magn Reson Med 66(2):596–604

Shen J (2001) Effect of degenerate spherical harmonics and a method for automatic shimming of oblique slices. NMR Biomed 14(3):177–183

Shulman RG, Rothman DL (eds) (2004) Brain energetics and neuronal activity: applications to fMRI and medicine. Wiley, West Sussex

Silva AC (2005) Perfusion-based fMRI: insights from animal models. J Magn Reson Imaging 22(6):745–750

Silva AC, Merkle H (2003) Hardware considerations for functional magnetic resonance imaging. Concepts Magn Reson Part A 16A(1):35–49

Slichter CP (1996) Principles of magnetic resonance. Springer, Berlin/New York

Smith RC, Lange RC (1998) Understanding magnetic resonance imaging. CRC, New York

Talagala SL, Ye FQ et al (2004) Whole-brain 3D perfusion MRI at 3.0 T using CASL with a separate labeling coil. Magn Reson Med 52(1):131–140

Turner R (1986) A target field approach to optimal coil design. J Phys D 19(8):L147–L151

Turner R (1988) Minimum inductance coils. J Phys E 21(10):948–952

Turner R (1993) Gradient coil design: a review of methods. Magn Reson Imaging 11(7):903–920

Ugurbil K, Adriany G et al (2000) Magnetic resonance studies of brain function and neurochemistry. Annu Rev Biomed Eng 2:633–660

Vermilyea ME (1988) Method of passively shimming magnetic resonance magnets, Google Patents

Wehrl HF, Judenhofer MS et al (2011) Assessment of MR compatibility of a PET insert developed for simultaneous multiparametric PET/MR imaging on an animal system operating at 7 T. Magn Reson Med 65(1):269–279

Wehrl HF, Sauter AW et al (2010) Combined PET/MR imaging - technology and applications. Technol Cancer Res Treat 9(1):5–20

Wehrli FW, Shaw D et al (1988) Biomedical magnetic resonance imaging: principles methodology and applications. VCH, New York

Wiesinger F, Boesiger P et al (2004) Electrodynamics and ultimate SNR in parallel MR imaging. Magn Reson Med 52(2):376–390

Wiggins GC, Polimeni JR et al (2009) 96-Channel receive-only head coil for 3 Tesla: design optimization and evaluation. Magn Reson Med 62(3):754–762

Wiggins GC, Triantafyllou C et al (2006) 32-channel 3 Tesla receive-only phased-array head coil with soccer-ball element geometry. Magn Reson Med 56(1):216–223

Wright SM (1990) RF coil arrays for magnetic resonance imaging. Engineering in Medicine and Biology Society, 1990. Proceedings of the Twelfth Annual International Conference of the IEEE, Philadelphia

Wright SM, Wald LL (1997) Theory and application of array coils in MR spectroscopy. NMR Biomed 10(8):394–410

Yang QX, Li SH et al (1994) A method for evaluating the magnetic-field homogeneity of a radiofrequency coil by its field histogram. J Magn Reson Series A 108(1):1–8

Zaharchuk G, Ledden PJ et al (1999) Multislice perfusion and perfusion territory imaging in humans with separate label and image coils. Magn Reson Med 41(6):1093–1098

Zhang W, Williams DS et al (1992) Measurement of brain perfusion by volume-localized NMR spectroscopy using inversion of arterial water spins: accounting for transit time and cross-relaxation. Magn Reson Med 25(2):362–371

Zhu Y, Hardy CJ et al (2004) Highly parallel volumetric imaging with a 32-element RF coil array. Magn Reson Med 52(4):869–877

Chapter 3
Diffusion Sensitive Magnetic Resonance Imaging of the Central Nervous System

Jeffrey J. Neil and Joseph J.H. Ackerman

Abstract Diffusion magnetic resonance (MR) imaging offers a wealth of information on the physiologic condition and microstructure of the brain. This chapter starts with a description of the fundamentals of MR and MR diffusion measurements. This is followed by descriptions of: (1) the role of water apparent diffusion coefficient measurements in detecting tissue injury, (2) the use of diffusion anisotropy measurements to assess the development of and injury to cerebral cortex and white matter, and (3) diffusion-based measurements of white matter connections ("tractography"). The chapter concludes with a brief summary of the utility of measurement of diffusion of molecules other than water in the central nervous system.

Keywords Magnetic resonance imaging • Diffusion • Diffusion tensor imaging • Diffusion tractography • Axial diffusivity • Radial diffusivity

3.1 Introduction

In the main, magnetic resonance imaging (MRI) detects the ^{1}H ("proton") nuclear magnetic resonance (NMR) signal from water. While other substances and other nuclides are NMR active, only water is present at sufficient concentration in mammalian tissue

J.J. Neil, M.D., Ph.D. (✉)
Departments of Pediatric Neurology, Pediatrics and Radiology,
Washington University School of Medicine, 660 South Euclid Ave.,
St. Louis, MO 63110, USA
e-mail: neil@wustl.edu

J.J.H. Ackerman, Ph.D.
Departments of Chemistry, Radiology and Internal Medicine, Campus Box 1134,
Washington University, 1 Brookings Drive, St. Louis, MO 63130, USA
e-mail: ackerman@wustl.edu

I.-Y. Choi and R. Gruetter (eds.), *Neural Metabolism In Vivo*,
Advances in Neurobiology 4, DOI 10.1007/978-1-4614-1788-0_3,

to provide the strong signal strength required for routine, robust, time-efficient, imaging applications. Thus, MRI is different from many other imaging modalities, such as positron emission tomography (PET), single photon emission computerized tomography (SPECT), and most applications of optical imaging, in that it does not detect an exogenously administered tracer species but rather the 100% naturally abundant ^{1}H in endogenous tissue water.

Within the medical community, the adjective "nuclear" in "NMR" is generally dropped to form the more politically correct "MR". Importantly however, "nuclear" refers to the fundamental quantum mechanical origin of the phenomenon; specifically, stimulated energy level transitions among different states of nuclear spin (states of different nuclear angular momentum). When placed in a static magnetic field, these nuclear spin states possess different energies and stimulated transitions between these states (between different energy levels) result in detectable absorption or emission of energy.

In marked distinction to the more familiar nuclear disintegration reactions exploited in PET and SPECT, for which very high energy particles are given off as the nucleus of a given isotope decays to form a new isotope, nuclear spin transitions are of extremely low energy, typically in the radio frequency range. Energies in this range are essentially harmless to living systems. Indeed, we are bathed in radio frequency waves emanating from radio, TV, and cell-phone transmissions. Further, static magnetic fields and radio frequency waves readily penetrate most non-metallic objects. Thus, the MR phenomenon is fundamentally noninvasive and nondestructive to living systems, including human.

The low energy nuclear spin states couple weakly to their surrounding environment (formally, to the "lattice"). This results in a particular advantage. Once the spins are perturbed from their thermal equilibrium energy level population distribution, it takes a long time before the thermal equilibrium distribution is reestablished. This return to a state of thermal equilibrium is referred to as "relaxation" and MR relaxation times are very long compared to other familiar spectroscopies. For example, while tissue water ^{1}H MR relaxation times dependend upon tissue type and physiologic state, typical MR relaxation times are on the order of 10–1000 ms (1 ms $=10^{-3}$ s), far longer than optical relaxation times which are measured in nano-, pico-, and even femtoseconds (10^{-9} to 10^{-15} s). A long relaxation time is advantageous because it is precisely during this time when the spins are out of equilibrium that spatial coordinate information is encoded and the image formed. Longer relaxation times confer greater precision in spatial encoding, resulting in the exceptionally clear anatomic definition characteristic of MRI.

The long relaxation times also allow many physical properties of tissue water to be encoded on a voxel by voxel basis in the MR image intensity (contrast generation). For example, images can be created that map water content, blood flow velocity, volumetric perfusion rate, tissue elasticity (stress and strain), tissue displacement, neuronal activation, oxygen extraction fraction, and temperature. Images can also be created in which contrast is determined by the tissue relaxation times. Two relaxation times are most relevant. The exponential relaxation time constant governing the disappearance of the MR signal is referred to as T_2. T_2-weighted contrast typically provides exceptional anatomical definition and is a mainstay of clinical MRI exams.

The exponential relaxation time constant governing reestablishment of the thermal equilibrium spin-state energy-level population distribution is referred to as T_1. T_1-weighted images are often used in concert with administration of exogenous T_1 relaxation agents (contrast agents) to monitor vascular permeability, perfusion, and blood brain barrier integrity. Contrast enhanced T_1-weighted images are a mainstay of oncologic MRI exams.

Another physical property of tissue water, its incoherent displacement motion or diffusion, can be encoded in the MR image and forms the subject of this chapter. The thermal energy present in tissue at 37°C manifests itself in the form of random, incoherent displacements of individual water molecules. These thermally driven stochastic displacements, characteristic of all liquids, are referred to as Brownian motion in honor of Robert Brown, the English botanist who discovered the phenomenon in 1827 upon observing with a microscope the jostling of pollen grains suspended in water. The ongoing, ever continuing, random displacement jumps occur for every member of the vast collection of water molecules that occupy each MR image voxel. The displacement jumps occur with very high frequency, such that molecular collisions occur at a staggering rate. Considering pure water, for example, the rotational correlation time – the average time between two collisions for a given water molecule, collisions which change rotational state – is on the order of 10^{-12} s; put another way, the average water molecule experiences 10^{12} collisions per second. These molecular collisions result in a random walk of all water molecules, i.e., diffusion, even for systems at thermal equilibrium. The net result of this diffusive motion is that, over time, all members of a collection of water molecules move throughout the accessible space. Such an effect can be readily mimicked by carefully placing a small drop of red dye in the center of a beaker of still water. Over time the region of dye grows larger in size (while becoming more dilute) and slowly spreads to the boundaries of the beaker. This is diffusive motion.

Over the typical time scale accessible to the diffusion-sensitive MRI experiment (t_{diff}), 1–100 ms, water diffusion leads to average (root-mean-square, $<X^2>^{1/2}$) displacements of 1–10 μm. This scale matches well with the microarchitectural structural features characteristic of mammalian tissue. These microstructural elements (cell and organelle membranes, for example) serve to hinder and restrict the diffusion of tissue water. Thus, the MRI diffusion experiment serves to probe tissue microstructure and, importantly, changes therein due to physiologic and pathophysiologic challenge. For this reason, diffusion sensitive MRI has become a powerful clinical and preclinical research tool.

When carried out in quantitative fashion, the MRI diffusion experiment results in maps of the apparent diffusion constant (ADC), the coefficient of proportionality between the average displacement distance and time, Eq. 3.1. The term "apparent" is used because the analysis usually employs a mathematics that is strictly true only for free (unhindered, unrestricted) diffusion, a condition clearly violated for water diffusion in tissue.

$$\sqrt{\langle X^2 \rangle} \propto \sqrt{ADC \times t_{diff}} \tag{3.1}$$

The MR encoding of water diffusion employs the same three-axis (x, y, z) magnetic field gradient coils used for anatomic spatial encoding. It follows that it is possible to encode for the diffusion of water along any direction in the laboratory frame of reference by application of the appropriate combination of x-, y-, and z-axis magnetic field gradients. Indeed, for each voxel in an image it is possible to encode the entire diffusion tensor. The diffusion tensor describes the three dimensional diffusion properties of tissue water in a local (voxel-based diffusion-oriented) coordinate frame of reference. In the example of the hypothetical small drop of red dye placed in the middle of a beaker of water, the tendency of the dye to diffuse along any particular direction would be the same as there is no directional preference to diffusive displacement. Likewise, if the MR ADC was determined along different directions for water in the beaker, the result would always be the same. Diffusion in this situation is isotropic (lacks directional preference). But this is not always the case.

Instead of a large beaker of water with a small drop of red dye in its center, consider a long and very narrow glass capillary filled with water. The diffusive motion of water along the axis of the capillary tube is completely unrestricted and unhindered. The ADC measured along this axis would be exactly the same as that measured along any direction in a large beaker of water. However, the same is clearly not true for diffusion measured along a direction radial (orthogonal) to the capillary axis. Here the glass walls of the tube present an impenetrable boundary to diffusion, the displacement motion of water is restricted, and diffusion "appears" slow. Indeed, no matter how long one waits, i.e., no matter how long t_{diff} is, water never appears outside the glass capillary. Thus, ADC_{radial} is much less than ADC_{axial}, and ADC_{radial} appears to decrease as the time allowed for diffusion, t_{diff}, increases. This hypothetical situation is one in which there is extreme diffusion anisotropy, a strong directional preference. Were the water MR diffusion tensor obtained in this case, it would provide (1) the orientation of the capillary tube in the laboratory reference frame (i.e., the angles defining the local coordinate frame – the three eigenvectors) and (2) the two ADCs (ADC_{axial} and ADC_{radial}).

Water diffusion in a narrow glass capillary is not only anisotropic but possesses cylindrical symmetry. Here the measurement of ADC along any direction orthogonal to the capillary axis would yield the same value. While water diffusion in mammalian tissue is often found to be anisotropic, the MR diffusion tensor experiment does not necessarily assume cylindrical symmetry. Indeed, as usually applied, three eigenvectors (v_1, v_2, v_3) and three eigenvalues (λ_1, λ_2, λ_3) are obtained. As noted above, the eigenvectors define the local (voxel-based diffusion coordinate) frame of reference and the three eigenvectors are the ADCs ("diffusivities") along each axis defined by the eigenvectors: λ_1 is the ADC along v_1, λ_2 is the ADC along v_2, and λ_3 is the ADC along v_3. Generally the diffusivities are ordered as $\lambda_1 > \lambda_2 > \lambda_3$. In the special case of cylindrical symmetry, v_1 defines the axial orientation of the cylinder and λ_1 is the diffusivity along the axis of the cylinder, often written as $\lambda_{\parallel}$ and read as "lamda parallel" or "axial diffusivity". Further, for this special case, $\lambda_2 = \lambda_3$ and both are generally written as $\lambda_{\perp}$ and read as "lamda perpendicular" or "radial diffusivity". In practice, water in mammalian tissues rarely presents a case of pure cylindrical symmetry although approximate cylindrical symmetry is not unusual: $\lambda_1 >> \lambda_2 \sim \lambda_3$. Here the average of λ_2 and λ_3 is taken as the radial diffusivity, i.e., $(\lambda_2+\lambda_3)/2=\lambda_\perp$.

Full diffusion tensor imaging presents the problem of displaying the six parameters derived for each voxel (ν_1, ν_2, ν_3, λ_1, λ_2, λ_3) in a single image. Numerous schemes employing color (RBG) and geometrical shapes (diffusion probability ellipsoids) are in use, but will not be reviewed here. However, it is very common to display maps representing the extent of diffusion anisotropy by collapsing the three diffusivities (λ_1, λ_2, λ_3) into a single "summary" parameter, typically either the fractional anisotropy or the relative anisotropy (Armitage and Bastin 2000). Fractional anisotropy (FA) is calculated as follows,

$$FA = \frac{\sqrt{3}}{\sqrt{2}} \frac{\sqrt{(\lambda_1-\lambda)^2+(\lambda_2-\lambda)^2+(\lambda_3-\lambda)^2}}{\sqrt{\lambda_1^2+\lambda_2^2+\lambda_3^2}}, \tag{3.2}$$

where λ is the average diffusivity, $(\lambda_1+\lambda_2+\lambda_3)/3$. Relative anisotropy (RA) is calculated as

$$RA = \frac{1}{\sqrt{6}} \frac{\sqrt{(\lambda_1-\lambda)^2+(\lambda_2-\lambda)^2+(\lambda_3-\lambda)^2}}{\lambda}. \tag{3.3}$$

Both summary parameters seek to represent as a single value, the voxel intensity, the difference between the individual diffusivities and their average. The way in which this works is most readily seen by considering two special cases: isotropic diffusion and diffusion in a long very narrow cylinder (the capillary tube).

In the case of isotropic diffusion $\lambda_1=\lambda_2=\lambda_3$; thus there is no difference between the individual diffusivities and their average, λ. In practice, noise in the MR measurement prevents the situation where $\lambda_1=\lambda_2=\lambda_3$ is found (i.e., there is a noise floor that prevents observation of pure isotropic diffusion by MR). Nevertheless, where diffusion tensor MR finds that water diffusion is well approximated as isotropic, $\lambda_1 \sim \lambda_2 \sim \lambda_3$, the FA or RA would be small and the voxel intensity would be low. In the case of diffusion in a long and very narrow cylinder (the capillary tube), $\lambda_1 > \lambda_2$, $\lambda_3 \sim 0$; thus, there is a substantial difference between the individual diffusivities and their average, λ. As a result, their FA or RA would be large, and the voxel intensity would be high. It follows therefore that FA or RA maps are bright where tissue water diffusion is anisotropic and dark where water diffusion is isotropic. In this manner, the information contained within the full diffusion tensor is compressed into an easily displayed gray scale. However, the true values of the diffusivities and the orientation of the local diffusion frame are lost.

Two final summary parameters worth noting are the sum and the average of the principal diffusivities (λ_1, λ_2, λ_3). The sum $(\lambda_1+\lambda_2+\lambda_3)$ is referred to as the trace of the tensor. The average, λ, is referred to as the mean ADC, i.e., the directionally averaged ADC. If we take the tensor to be represented by an ellipsoidal geometric shape, the orientation of the ellipsoid is given by the three eigenvectors (ν_1, ν_2, ν_3), and the shape of the ellipsoid (rugby ball vs. pancake; long, narrow vs. short, squat; etc.) is given by the length of its axes (λ_1, λ_2, λ_3). Here the trace or mean ADC is seen to reflect the overall size of the ellipsoid (large vs. small rugby ball) independent of its relative dimensions or orientation. Again, employing the trace or mean ADC is a way to compress the information contained within the full diffusion tensor into an easily displayed gray scale. Here bright voxels indicate that tissue water diffusion is, on average, fast, while dark voxels indicate that, on average, diffusion is slow. In this compression scheme, all directional information about water diffusion has been lost.

In summary, MR diffusion measurements exploit and quantify the diffusion of water as a means to infer characteristics of the underlying structural microarchitecture that hinders and restricts the thermally driven stochastic molecular motion of water. The MR "diffusion signal" thus provides information on a spatial scale far smaller than that of normal image voxel dimensions, μm vs. mm. As many tissues present significant structural ordering on the micron scale, the MR diffusion signal has proven extremely powerful in characterizing normal, abnormal, and developing morphology and physiology.

3.2 Information Available Through Diffusion Imaging: ADC

ADC values are typically calculated as the average of the three eigenvalues. The most common clinical application of diffusion imaging is for the acute evaluation of brain injury. As shown in Fig. 3.1, the decrease in ADC values that accompanies stroke serves as an early indicator of brain injury (Moseley et al. 1990), with changes in diffusion images visible within minutes of the onset of injury. In contrast, ischemic injury may not be detectable for days on conventional MR imaging. ADC values decrease following a variety of central nervous system insults in addition to stroke,

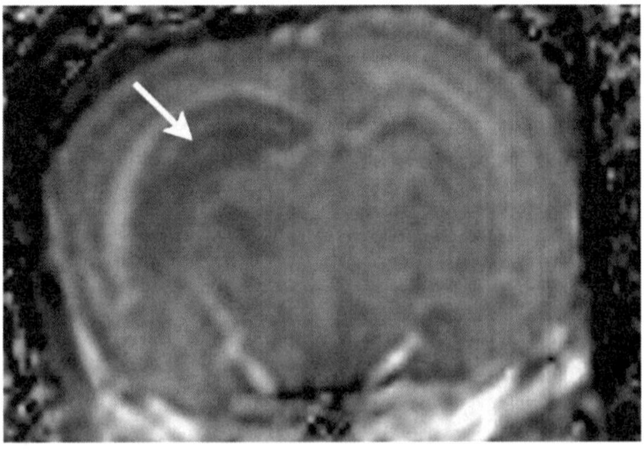

Fig. 3.1 A coronal ADC map from a live, 3.2-g mouse pup taken 24 h after common carotid ligation and 45 min of hypoxia. Darker areas denote lower ADC values. Note the area of injury (*arrow*) which primarily affects the neuronal cell layer of the hippocampus (Image provided courtesy of Drs. Tim West and Greg Lodygensky)

including status epilepticus (Zhong et al. 1993), excitotoxic injury (Dijkhuizen et al. 1999), spreading depression (Busch et al. 1995; Rother et al. 1996; Takano et al. 1996) and trauma (Ito et al. 1996). Thus, a decrease in ADC is a nonspecific finding. In addition, changes in ADC values are sometimes reversible. For example, cerebral vascular occlusion causes local water ADC values to decrease within minutes. If blood flow is restored before permanent injury occurs, ADC values return to normal (Li et al. 1999). Similarly, the decrease in ADC induced by the application of excitotoxins can be reversed by blocking the excitotoxin receptor (Dijkhuizen et al. 1999).

The reduction in ADC values following stroke is not 100% sensitive on the first day after injury. In adults with stroke, for example, the incidence of "diffusion-negative" injury is on the order of 6% (Schlaug et al. 1997; Oppenheim et al. 2000). In newborn infants, the incidence may be as high as 30% (McKinstry et al. 2002b). The explanation for this finding is likely related to the nature of the injury. In animal studies, transient cerebral vascular occlusion (30 min), which is sufficient to cause permanent injury, leads to a delayed decrease in ADC values (Li et al. 2000; Lin et al. 2001; Dijkhuizen et al. 1998). Under these circumstances, ADC values are reduced during occlusion, but return to normal when blood flow is restored. ADC values then decrease again over a time period of approximately 24 h, reaching a nadir at 2 to 4 days after injury. In this case, diffusion imaging does not identify the area of injury for up to 24 h. In contrast, permanent cerebral vascular occlusion leads to a sustained decrease in ADC values within minutes of occlusion. The circumstances of cerebral injury to a newborn infant may be more similar to the animal model of transient occlusion. Infants who are experiencing hypoxia in utero are often delivered quickly by caesarean section, after which blood flow and oxygenation are restored. Vascular occlusion in adults with stroke, on the other hand, lasts for 24–48 h, more closely approximating permanent occlusion. Thus, the mechanism of cerebral injury affects the evolution of diffusion image contrast, even though both injuries result in permanent cerebral infarction.

It should also be kept in mind that ADC values vary with time after injury. As noted above, ADC values reach a nadir at 2 to 4 days after injury and then gradually return to normal over a time period of approximately 1 week – a process known as "pseudonormalization" (Schlaug et al. 1997; Copen et al. 2001; Mukherjee et al. 2000; McKinstry et al. 2002b). After pseudonormalization, ADC values are higher than normal in the area of injury and remain so. Thus, the sensitivity of diffusion measurements for detecting injury varies markedly with time after injury. Overall, the decrease in ADC values is most sensitive for detecting stroke 2 to 4 days after injury, though the majority of injury in adults and children is detectable within minutes of its occurrence.

The biophysical mechanisms behind the decrease in tissue ADC after injury are not well understood. The most-commonly cited (and likely incorrect) hypothesis for the cause of this change is the compartment shift hypothesis. This explanation is based on the assumption that water in the intra- and extracellular spaces has different ADC values – high for extracellular water and low for intracellular water. If water in these two compartments is in fast exchange, the measured overall tissue ADC is a weighted average of the ADC values for water in these two compartments.

When cells swell in response to injury, the proportion of water in the intracellular space increases at the expense of water in the extracellular space. This causes heavier weighting of the overall ADC to that of water in the intracellular compartment, leading to a decrease in the overall tissue water ADC value. While this hypothesis is attractive, the evidence for its validity remains indirect and weak, resting primarily on the close temporal relationship between the injury-induced decrease in water ADC and cell swelling. Estimates of ADC values for water in the intra- and extracellular compartments suggest that their ADC values are not sufficiently different for a shift of water from the extra- to intracellular space to account for the decrease in ADC associated with injury (Duong et al. 1998; Silva et al. 2002). As an alternative hypothesis, some authors have suggested that a decrease in the ADC of intracellular water is responsible for the ADC drop (Neil et al. 1996; Wick et al. 1995). This remains an area of active debate.

The change in water ADC associated with brain injury is not universal, but does occur in other tissues. In skeletal muscle, for example, the decrease in water ADC values following injury is much smaller than that in brain (Heemskerk et al. 2006). In kidney, on the other hand, ADC values decrease following ischemia (Liu and Xie 2003). Similarly, ADC values fall quickly and significantly in prostate following injury caused by heating or cooling (Butts et al. 2003). In addition, diffusion MR methods hold promise for evaluating tumors (Hamstra et al. 2005; Moffat et al. 2006).

ADC values can also increase acutely with some forms of pathology, particularly vasogenic edema. Reversible posterior leukoencephalopathy syndrome (RPLS) is an example. This condition is associated with a variety of clinical conditions, including hypertension or the administration of chemotherapeutic agents (Hinchey et al. 1996). It is characterized by mental status changes, visual disturbance, headache, and seizures. The MRI findings include abnormality of the occipital lobes, with increased ADC values. The change in ADC values has been attributed to vasogenic edema and an increase in the volume of the extracellular space with an associated reduction in the restriction of water displacement. In this case, diffusion tensor imaging is useful for differentiating RPLS from occipital stroke, as stroke is associated with a decrease in ADC values (Ay et al. 1998).

3.3 Information Available Through Diffusion Imaging: Anisotropy

3.3.1 White Matter

In mature brain, anisotropy values are typically higher from regions of white matter (WM) than grey matter (GM), as shown in Fig. 3.2. Further, anisotropy is often taken as representative of the "health" of a white matter tract. One of the first research groups to apply anisotropy in this fashion (Klingberg et al. 2000), showed that diffusion anisotropy in the left temporo-parietal WM was strongly correlated with

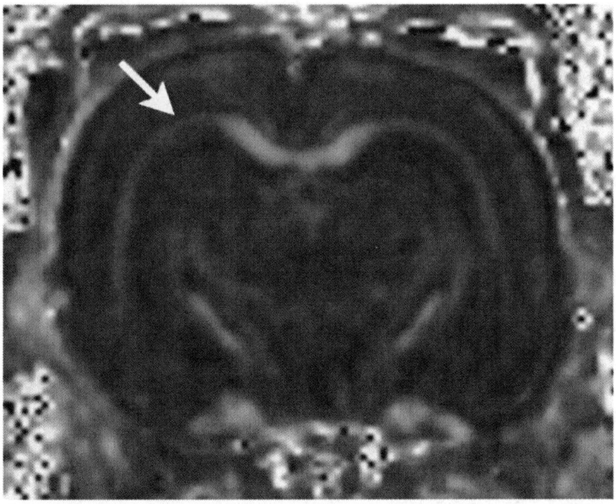

Fig. 3.2 A coronal anisotropy map from a 9-day-old rat pup. Brighter areas denote higher anisotropy. Note that anisotropy values are higher in white matter areas, particularly the corpus callosum and external capsule (*arrow*) (Image provided courtesy of Dr. Greg Lodygensky)

reading scores. This study spawned a series of reports comparing WM anisotropy values to clinical conditions such as schizophrenia, though the validity of such studies remains controversial (Kanaan et al. 2005). The precise relationship between anisotropy and WM health is not yet understood, nor is it clear what parameter (e.g., tract size, anisotropy value) would serve as the best indicator in this application. Despite these limitations, WM diffusion anisotropy measurements are a valuable measure for evaluation of WM diseases such as multiple sclerosis (Goldberg-Zimring et al. 2005).

A more sophisticated approach to evaluating anisotropy data from WM involves analysis of the eigenvalues. The eigenvalue for λ_1, referred to as "axial diffusivity," represents water ADC parallel to axons. The average of the eigenvalues for λ_2 and λ_3, referred to as "radial diffusivity," describes water ADC perpendicular to axons. Data from animal studies show that primary injury to myelin is associated with an increase in radial diffusivity, presumably because there are fewer myelin membranes to hinder water displacements in this direction. Conversely, primary injury to axons, such as Wallerian degeneration, is associated with a decrease in axial diffusivity, presumably because of disruption of the fiber tracts along which water molecules can diffuse (Song et al. 2002, 2005; Sun et al. 2006).

Finally, anisotropy measurement can be applied to evaluation of white matter maturation and injury, particularly in prematurely-born infants, for whom white matter abnormality is common (Woodward et al. 2006). During normal white matter development, the increase in anisotropy takes place in two stages. Surprisingly, the initial increase occurs before the presence of histologically-detectable myelin. This increase has been attributed to changes in axonal structure associated with the pre-myelinating state and has been detected in both animals (Wimberger et al. 1995) and humans (Huppi et al. 1998; Neil et al. 1998). It is followed by a second, larger increase in anisotropy associated with the elaboration of layers of myelin (Mukherjee et al. 2002). Thus, measurement of anisotropy can be used for evaluating aberrant white matter development as well as injury (Neil et al. 2002).

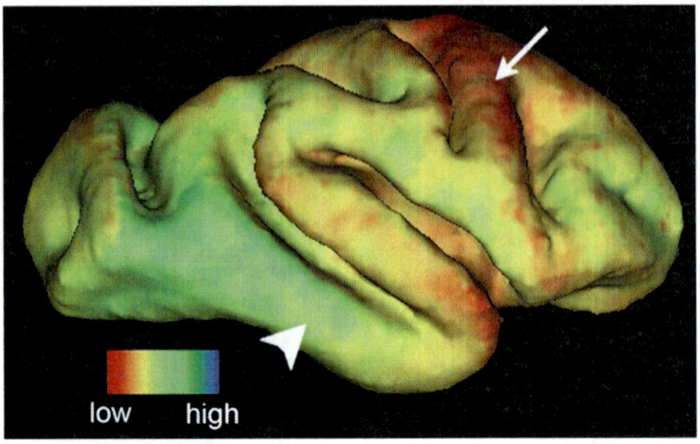

Fig. 3.3 A surface representation of a baboon hemisphere at 125 days gestation (term gestation is 185 days). The color scale shows cortical anisotropy values. Note the heterogeneity of cortical development, with the precentral gyrus appearing more mature (*arrow*, lower anisotropy) than the inferior temporal area (*arrow head*, higher anisotropy)

3.3.2 Grey Matter

In adult brain, anisotropy values from cortical GM are typically low and relatively uninteresting. However, cortical anisotropy values are high during the early stages of cortical development, such as during brain development of prematurely-born human infants, providing information on cortical development. This has been shown in both human (McKinstry et al. 2002a) and animal (Kroenke et al. 2007) studies. Anisotropy values are transiently high during early development because of the radial orientation of apical dendrites of pyramidal cells. This confers larger water displacements orthogonal to the cortical surface than parallel to it. As development progresses, anisotropy values decrease due to the addition of interneurons and elaboration of basal dendrites of pyramidal cells. An example of high cortical anisotropy during cortical development is shown in Fig. 3.3.

3.4 Information Available Through Diffusion Imaging: Tractography

The primary eigenvector, λ_1, is oriented parallel to the orientation of fiber tracts for white matter. As shown in Fig. 3.4, it is possible to follow fiber tracts within the brain by tracking the orientation of this eigenvector from voxel to voxel (Conturo et al. 1999; Xue et al. 1999; Basser et al. 2000). This application of diffusion imaging has not yet seen widespread clinical use, but one potential application is for planning neurosurgical procedures. It may be possible to determine where crucial white matter tracts lie in relation to tumors. This knowledge might be used to alter the neurosurgical approach, helping to minimize the neurologic deficits associated with tumor resection.

One challenge for diffusion tractography is tracking crossing fibers. The tensor formalism presented here is the earliest, and most widely-used, means of representing water displacements in three dimensions. It continues to serve well in a variety

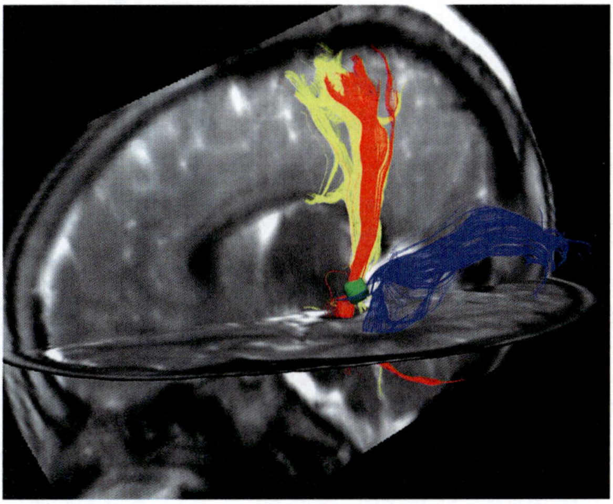

Fig. 3.4 Fiber tracts determined by diffusion tensor imaging and superimposed on a T_2-weighted image. The tracts shown in red and yellow are cortico-spinal tracts; those shown in blue are optic radiations. The lateral geniculate nucleus is shown in green (Image provided courtesy of Dr. Josh Shimony and Mr. Adrian Epstein)

of clinical situations and requires that at least six directionally independent diffusion measurements are made to create this three-vector representation of the diffusion ellipsoid. However, this representation may not suffice for voxels with more complex white matter structure. For example, if two WM tracts cross at 90°, the resultant ellipsoid for this voxel would have a pancake shape, with the plane of the pancake corresponding to the plane of the crossing fibers (remember that the three-vector approach can only describe shapes resembling spheres, rugby balls, and pancakes). Clearly, it would be difficult to follow one of the fiber tracts through such a voxel. Approaches to overcome this limitation are currently under development by several research groups. For example, if diffusion measurements are made in more than six directions [studies with up to 126 directions have been published! (Tuch et al. 2002)], it is possible to apply more sophisticated forms of analysis; in essence, generating shapes more complex than simple ellipsoids. Examples of approaches allowing for more complex shapes include q-space imaging (Tuch 2004; Cohen and Assaf 2002) and spherical harmonic decomposition (Frank 2002). For the example of fiber tracts crossing at 90° given above, a three-dimensional representation of water displacements might look like a cross.

3.5 Diffusion of Molecules Other than ^1H_2O

Though a detailed discussion of these measurements is beyond the scope of this chapter, it is worth noting that it is possible to measure diffusion of molecules other than water. Whereas the concentration of water ^1H in tissues is on the order of 100 M, other metabolites are present in mM concentrations. As a result, these measurements are typically made at much lower spatial resolution than diffusion measurements for water. One advantage of measuring diffusion of molecules other than water is that these molecules can provide both compartmental and cell-type specificity, which is difficult to achieve in measures of water diffusion. For example, the ADC of the metabolite N-acetyl aspartate, which is present in the intracellular space of neurons but not glia and detected by ^{1}H spectroscopy, has been shown to decrease in association with stroke (Wick et al. 1995). Measurements of the diffusion of intracellular ions ^{23}Na (Goodman et al. 2005) and ^{133}Cs (Goodman et al. 2008) show similar changes in diffusion with injury. Finally, it is possible to exogenously administer markers, such as ^{19}F-fluorodeoxyglucose, into a specific compartment and measure its diffusion (Duong et al. 1998).

3.6 Conclusion

Diffusion sensitive MRI is a mainstay of MR imaging in both clinical and research settings. The method provides a wealth of information about the central nervous system. ADC values are a sensitive and early indicator of tissue injury. Diffusion anisotropy reflects underlying tissue microstructure. It has been employed for (1) evaluating the "health" of white matter, (2) following the course of white matter tracts, and (3) assessing cortical grey matter maturity in developing brain.

References

Armitage PA, Bastin ME (2000) Selecting an appropriate anisotropy index for displaying diffusion tensor imaging data with improved contrast and sensitivity. Magn Reson Med 44(1):117–121

Ay H, Buonanno FS, Schaefer PW, Le DA, Wang B, Gonzalez RG, Koroshetz WJ (1998) Posterior leukoencephalopathy without severe hypertension: utility of diffusion-weighted MRI. Neurology 51(5):1369–1376

Basser PJ, Pajevic S, Pierpaoli C, Duda J, Aldroubi A (2000) In vivo fiber tractography using DT-MRI data. Magn Reson Med 44(4):625–632

Busch E, Hoehn-Berlage M, Eis M, Gyngell ML, Hossman KA (1995) Simultaneous recording of EEG, DC potential and diffusion-weighted NMR imaging during potassium induced cortical spreading depression in rats. NMR Biomed 8(2):59–64

Butts K, Daniel BL, Chen L, Bouley DM, Wansapura J, Maier SE, Dumoulin C, Watkins R (2003) Diffusion-weighted MRI after cryosurgery of the canine prostate. Magnetic resonance imaging. J Magn Reson Imaging 17(1):131–135

Cohen Y, Assaf Y (2002) High b-value q-space analyzed diffusion-weighted MRS and MRI in neuronal tissues – a technical review. NMR Biomed 15(7–8):516–542

Conturo TE, Lori NF, Cull TS, Akbudak E, Snyder AZ, Shimony JS, McKinstry RC, Burton H, Raichle ME (1999) Tracking neuronal fiber pathways in the living human brain. Proc Natl Acad Sci USA 96(18):10422–10427

Copen WA, Schwamm LH, Gonzalez RG, Wu O, Harmath CB, Schaefer PW, Koroshetz WJ, Sorensen AG (2001) Ischemic stroke: effects of etiology and patient age on the time course of the core apparent diffusion coefficient. Radiology 221(1):27–34

Dijkhuizen RM, Knollema S, van der Worp HB, Ter Horst GJ, De Wildt DJ, Berkelbach van der Sprenkel JW, Tulleken KA, Nicolay K (1998) Dynamics of cerebral tissue injury and perfusion after temporary hypoxia-ischemia in the rat: evidence for region-specific sensitivity and delayed damage. Stroke 29(3):695–704

Dijkhuizen RM, de Graaf RA, Tulleken KA, Nicolay K (1999) Changes in the diffusion of water and intracellular metabolites after excitotoxic injury and global ischemia in neonatal rat brain. J Cereb Blood Flow Metab 19(3):341–349

Duong TQ, Ackerman JJH, Ying HS, Neil JJ (1998) Evaluation of extra- and intracellular apparent diffusion in normal and globally ischemic rat brain via ^{19}F NMR. Magn Reson Med 40(1):1–13

Frank LR (2002) Characterization of anisotropy in high angular resolution diffusion-weighted MRI. Magn Reson Med 47(6):1083–1099

Goldberg-Zimring D, Mewes AU, Maddah M, Warfield SK (2005) Diffusion tensor magnetic resonance imaging in multiple sclerosis. J Neuroimaging 15(4 Suppl):68S–81S

Goodman JA, Kroenke CD, Bretthorst GL, Ackerman JJ, Neil JJ (2005) Sodium ion apparent diffusion coefficient in living rat brain. Magn Reson Med 53(5):1040–1045

Goodman JA, Ackerman JJ, Neil JJ (2008) Cs+ADC in rat brain decreases markedly at death. Magn Reson Med 59(1):65–72

Hamstra DA, Chenevert TL, Moffat BA, Johnson TD, Meyer CR, Mukherji SK, Quint DJ, Gebarski SS, Fan X, Tsien CI, Lawrence TS, Junck L, Rehemtulla A, Ross BD (2005) Evaluation of the functional diffusion map as an early biomarker of time-to-progression and overall survival in high-grade glioma. Proc Natl Acad Sci USA 102(46):16759–16764

Heemskerk AM, Drost MR, van Bochove GS, van Oosterhout MF, Nicolay K, Strijkers GJ (2006) DTI-based assessment of ischemia-reperfusion in mouse skeletal muscle. Magn Reson Med 56(2):272–281

Hinchey J, Chaves C, Appignani B, Breen J, Pao L, Wang A, Pessin MS, Lamy C, Mas JL, Caplan LR (1996) A reversible posterior leukoencephalopathy syndrome. N Engl J Med 334(8):494–500

Huppi PS, Maier SE, Peled S, Zientara GP, Barnes PD, Jolesz FA, Volpe JJ (1998) Microstructural development of human newborn cerebral white matter assessed in vivo by diffusion tensor magnetic resonance imaging. Pediatr Res 44(4):584–590

Ito J, Marmarou A, Barzo P, Fatouros P, Corwin F (1996) Characterization of edema by diffusion-weighted imaging in experimental traumatic brain injury. J Neurosurg 84(1):97–103

Kanaan RA, Kim JS, Kaufmann WE, Pearlson GD, Barker GJ, McGuire PK (2005) Diffusion tensor imaging in schizophrenia. Biol Psychiatry 58(12):921–929

Klingberg T, Hedehus M, Temple E, Salz T, Gabrieli JD, Moseley ME, Poldrack RA (2000) Microstructure of temporo-parietal white matter as a basis for reading ability: evidence from diffusion tensor magnetic resonance imaging. Neuron 25(2):493–500

Kroenke CD, Van Essen DC, Inder TE, Rees S, Bretthorst GL, Neil JJ (2007) Microstructural changes of the baboon cerebral cortex during gestational development reflected in magnetic resonance imaging diffusion anisotropy. J Neurosci 27(46):12506–12515

Li F, Han SS, Tatlisumak T, Liu KF, Garcia JH, Sotak CH, Fisher M (1999) Reversal of acute apparent diffusion coefficient abnormalities and delayed neuronal death following transient focal cerebral ischemia in rats. Ann Neurol 46(3):333–342

Li F, Silva MD, Liu KF, Helmer KG, Omae T, Fenstermacher JD, Sotak CH, Fisher M (2000) Secondary decline in apparent diffusion coefficient and neurological outcomes after a short period of focal brain ischemia in rats. Ann Neurol 48(2):236–244

Lin SP, Song SK, Miller JP, Ackerman JJ, Neil JJ (2001) Direct, longitudinal comparison of (1)H and (23)Na MRI after transient focal cerebral ischemia. Stroke 32(4):925–932

Liu AS, Xie JX (2003) Functional evaluation of normothermic ischemia and reperfusion injury in dog kidney by combining MR diffusion-weighted imaging and Gd-DTPA enhanced first-pass perfusion. J Magn Reson Imaging 17(6):683–693

McKinstry RC, Mathur A, Miller JP, Ozcan AO, Snyder AZ, Schefft GL, Almli CR, Shiran SI, Conturo TE, Neil JJ (2002a) Radial organization of developing human cerebral cortex revealed by non-invasive water diffusion anisotropy MRI. Cereb Cortex 12(12):1237–1243

McKinstry RC, Miller JH, Snyder AZ, Mathur A, Schefft GL, Almli CR, Shimony JS, Shiran SI, Neil JJ (2002b) A prospective, longitudinal diffusion tensor imaging study of brain injury in newborns. Neurology 59(6):824–833

Moffat BA, Chenevert TL, Meyer CR, McKeever PE, Hall DE, Hoff BA, Johnson TD, Rehemtulla A, Ross BD (2006) The functional diffusion map: an imaging biomarker for the early prediction of cancer treatment outcome. Neoplasia 8(4):259–267

Moseley ME, Cohen Y, Mintorovitch J, Chileuitt L, Shimizu H, Kucharczyk J, Wendland MF, Weinstein PR (1990) Early detection of regional cerebral ischemia in cats: Comparison of diffusion- and T_2-weighted MRI and spectroscopy. Magn Reson Med 14(2):330–346

Mukherjee P, Bahn MM, McKinstry RC, Shimony JS, Cull TS, Akbudak E, Snyder AZ, Conturo TE (2000) Differences between gray matter and white matter water diffusion in stroke: diffusion-tensor MR imaging in 12 patients. Radiology 215(1):211–220

Mukherjee P, Miller JH, Shimony JS, Philip JV, Nehra D, Snyder AZ, Conturo TE, Neil JJ, McKinstry RC (2002) Diffusion-tensor MR imaging of gray and white matter development during normal human brain maturation. AJNR Am J Neuroradiol 23(9):1445–1456

Neil JJ, Duong TQ, Ackerman JJH (1996) Evaluation of intracellular diffusion in normal and globally ischemic rat brain *via* ^{133}Cs NMR. Magn Reso Med 35(3):329–335

Neil JJ, Shiran SI, McKinstry RC, Schefft GL, Snyder AZ, Almli CR, Akbudak E, Aaronovitz JA, Miller JP, Lee BCP, Conturo TE (1998) Normal brain in human newborns: apparent diffusion coefficient and diffusion anisotropy measured using diffusion tensor imaging. Radiology 209(1):57–66

Neil JJ, Miller JP, Mukherjee P, Huppi PS (2002) Diffusion tensor imaging of normal and injured developing human brain. NMR Biomed 15(7–8):543–552

Oppenheim C, Stanescu R, Dormont D, Crozier S, Marro B, Samson Y, Rancurel G, Marsault C (2000) False-negative diffusion-weighted MR findings in acute ischemic stroke. AJNR Am J Neuroradiol 21(8):1434–1440

Rother J, deCrespigny AJ, D'Arceuil H, Moseley ME (1996) MR detection of cortical spreading depression immediately after focal ischemia in the rat. J Cereb Blood Flow Meta 16(2): 214–220

Schlaug G, Siewert B, Benfield A, Edelman RR, Warach S (1997) Time course of the apparent diffusion coefficient (ADC) abnormality in human stroke. Neurology 49(1):113–119

Silva M, Omae T, Helmer K, Li F, Fisher M, Sotak C (2002) Separating changes in the intra- and extracellular water apparent diffusion coefficient following focal cerebral ischemia in the rat brain. Magn Reson Med 48(5):826–837

Song SK, Sun SW, Ramsbottom MJ, Chang C, Russell J, Cross AH (2002) Dysmyelination revealed through MRI as increased radial (but unchanged axial) diffusion of water. Neuroimage 17(3):1429–1436

Song SK, Yoshino J, Le TQ, Lin SJ, Sun SW, Cross AH, Armstrong RC (2005) Demyelination increases radial diffusivity in corpus callosum of mouse brain. Neuroimage 26(1):132–140

Sun SW, Liang HF, Le TQ, Armstrong RC, Cross AH, Song SK (2006) Differential sensitivity of in vivo and ex vivo diffusion tensor imaging to evolving optic nerve injury in mice with retinal ischemia. Neuroimage 32(3):1195–1204

Takano K, Latour LL, Formato JE, Carano RAD, Helmer KG, Hasegawa Y, Sotak CH, Fisher M (1996) The role of spreading depression in focal ischemia evaluated by diffusion mapping. An Neurol 39(3):308–318

Tuch DS (2004) Q-ball imaging. Magn Reson Med 52(6):1358–1372

Tuch DS, Reese TG, Wiegell MR, Makris N, Belliveau JW, Wedeen VJ (2002) High angular resolution diffusion imaging reveals intravoxel white matter fiber heterogeneity. Magn Reson Med 48(4):577–582

Wick M, Nagatomo Y, Prielmeier F, Frahm J (1995) Alteration of intracellular metabolite diffusion in rat brain *in vivo* during ischemia and reperfusion. Stroke 26(10):1930–1933

Wimberger DM, Roberts TP, Barkovich AJ, Prayer LM, Moseley ME, Kucharczyk J (1995) Identification of "premyelination" by diffusion-weighted MRI. J Comput Assis Tomogr 19(1):28–33

Woodward LJ, Anderson PJ, Austin NC, Howard K, Inder TE (2006) Neonatal MRI to predict neurodevelopmental outcomes in preterm infants. N Engl J Med 355(7):685–694

Xue R, van Zijl PCM, Crain BJ, Solaiyappan M, Mori S (1999) In vivo three-dimensional reconstruction of rat brain axonal projections by diffusion tensor imaging. Magn Reson Med 42(6):1123–1127

Zhong J, Petroff OAC, Prichard JW, Gore JC (1993) Changes in water diffusion and relaxation properties of rat cerebrum during status epilepticus. Magn Reson Med 30(2):241–246

Chapter 4
Tools to Enhance the Contrast in MR Images: From Lanthanide Complexes and Iron Oxide Particles to MEMRI

Annemie Van der Linden, Vincent Van Meir, Dario Longo, and Silvio Aime

Abstract In this chapter we have surveyed the basis for the design and modus operandi of MRI contrast agents being paramagnetic metal complexes, including responsive agents, and iron oxide particles. A substantial amount of attention goes to the application of Manganese as a biological Ca^{2+} analogue in Manganese Enhanced MRI (MEMRI). All these agents act by providing contrast enhancements that add molecular, physiological and functional information to the outstanding anatomical resolution of MR images. MEMRI relies on the relaxation enhancement brought about by the administration of Mn^{2+} ions, whereas the development of paramagnetic contrast agents basically involves the use of chelating moieties as the release of the toxic metal ions or iron is of primary concern for the clinical translation of these diagnostic agents. Iron oxides are also considered as prototypes of high sensitive, particulate agents.

Keywords Gd-complexes • Contrast agent • Responsive contrast agent • Iron-oxide particles • Relaxation

4.1 Introduction

The superb spatial resolution and the outstanding capacity for differentiating soft tissues have determined the widespread success of MRI in biomedicine. In 2007, there were about 20,000 MRI scanners operating worldwide, providing about 40 million exams with an annual growth rate (in the last 3 years) of about 5%.

A. Van der Linden, Ph.D. (✉) • V. Van Meir, Ph.D.
Bio-Imaging Lab, Department of Biomedical Sciences, University of Antwerp
Universiteitsplein 1, 2610 Antwerp, Belgium
e-mail: annemie.vanderlinden@ua.ac.be; vincent.vanmeir@ua.ac.be

D. Longo, Ph.D. • S. Aime, Ph.D.
Department of Chemistry IFM/Molecular Imaging Center, University of Torino,
Via Nizza 52, 10126 Turin, Italy
e-mail: dario.longo@unito.it; silvio.aime@unito.it

I.-Y. Choi and R. Gruetter (eds.), *Neural Metabolism In Vivo*,
Advances in Neurobiology 4, DOI 10.1007/978-1-4614-1788-0_4,

The main determinants of the contrast in a MR image are the proton relaxation times T_1 and T_2. When there is poor contrast between healthy and diseased regions, due to a very small variation in relaxation times, the use of contrast agents can be highly beneficial. Already in the early'1980 it was realized that chemicals are able to alter markedly the relaxation times of water protons in the tissues where their distribution may be used as MRI contrast agents. According to whether the dominant effect occurs mainly on T_1 or T_2, MRI contrast agents can be classified as positive or negative agents. Paramagnetic Mn^{2+} and Gd^{3+} chelates are the most common representatives of the T_1-positive agents, whereas iron oxide particles are the most common T_2-negative agents (Toth et al. 2001).

At the pre-clinical level other classes of MRI contrast agents have been proposed such as the CEST agents (CEST=Chemical Exchange Saturation Transfer) (Ward et al. 2000), F-19 containing molecules (Yu et al. 2005a) and the Hyperpolarized molecules (Schroder et al. 2006). CEST agents are chemicals that affect the contrast in a MR image through the transfer of saturated magnetization to the "bulk" water signal. Thus they act as negative agents thanks to the fact that they contain exchangeable protons whose NMR resonance can be suitably irradiated with a proper radiofrequency centered at the absorption of the exchangeable protons. The potential of this class of agent is huge as their mode of action allows the detection of more agents in the same anatomical region simply by shifting the irradiation frequency at the resonance of the given exchangeable proton pool (Terreno et al. 2008). Moreover, in the targeting experiments their use does not require the acquisition of the pre-contrast image prior administration of the agent. In fact the pre-contrast image is acquired just before the application of the CEST procedure simply by disconnecting the second RF field.

F-19 and hyperpolarized molecules differ from the other class of MRI contrast agent because they are not designed to affect the water signal, rather they provide heteronuclear images with no background signal (Caruthers et al. 2006) . The low sensitivity of NMR phenomenon implies that a contrast agent based on F-19 has to contain a very high number of reporting nuclei. This task is basically accomplished by using microemulsion-based particles containing as major compartment fluorinated hydrocarbons (Ruiz-Cabello et al. 2008). In the case of hyperpolarized molecules the sensitivity drawback is overcome by artificially inducing a dramatic change in the Boltzmann population between the spin states involved in the NMR transitions (Golman et al. 2003; Golman and Petersson 2006). Different approaches have been proposed and, if the expectations will be maintained, the hyperpolarization will represent a real breakthrough in the field of "*in vivo*" diagnostic imaging.

As far as concern the contrast agents that have already entered the clinical practice (ca. 1/3 of the MRI scans make use of contrast agents) they belong to the classes of paramagnetic complexes and iron oxide particles. The study of the Central Nervous System (CNS) has been the primary clinical indication for the use of the extra-cellular Gd-based agents. The majority of the pathologies that have been addressed are brain tumours, and three quarters are represented by metastases occurring in patients undergoing treatment for systemic cancers. Other brain diseases, such as multiple sclerosis and cerebral injuries, can also be investigated by contrast-enhanced MRI. There are several other indications for the use of Gd-based contrast agents besides CNS. For instance, in the diagnosis of breast cancer, MRI with contrast agents is becoming an alternative diagnostic procedure to mammography. Particularly interesting (and useful) is the dynamic use of contrast agents (DCE = Dynamic Contrast Enhancement). The breast is imaged repeatedly over the first few minutes following contrast agent administration, and a graph reporting the increase in signal for a selected region of interest is plotted as a function of time (Johansen et al. 2009). The kinetics of the distribution of the contrast agent in the extravascular space are related to the vascular permeability (Choyke et al. 2003). Newly formed vessels display a permeability that is much higher than the normal capillaries and the corresponding areas are therefore characterized by high signal intensity. DCE has been applied to demonstrate the aggressive nature of tumours in several other regions (Cornud et al. 2009; Do et al. 2009).

Although to a smaller extent in respect of Gd-based agents, iron oxide particles have found an important role in imaging. Once administered intravenously, these agents accumulate in the cells of the reticuloendothelial systems. Therefore they report diagnostic responses related to macrophage uptake and in the oncology field they are particularly useful for assessing whether tumour cells have invaded the neighbouring lymph nodes (Reimer and Tombach 1998; Mornet et al. 2004; Boutry et al. 2008).

As the currently used MRI CAs (Contrast Agent) are administered at doses that are much higher than those used for tracers in Nuclear Medicine or for fluorescent dyes in Optical Imaging, much attention has been devoted to their potential toxicity. The concern of Gd-based systems relies on the fact that Gd^{3+} ion is not an essential metal ion and therefore it should never be released from the contrast agent. In fact its release can strongly interfere with Ca^{2+} ions by blocking transmembrane calcium channel or by altering the complex equilibria between cytoplasmatic Ca^{2+} and its storing sink. Therefore Gd^{3+} is administered under the form of very stable complexes that ensure against the "*in vivo*" release of toxic Gd^{3+} ions. Chemistry has identified several octadentate ligands that wrap around the Gd^{3+} ion to form chelates that highly "sequester" it from any interaction with the surrounding biological structures. Its contact with the "environment" occurs basically to the exchange of water molecules that interact with the paramagnetism of the Gd^{3+} ion. Very important is to maintain at least one water molecule directly coordinated to the Gd^{3+} center. In principle, even more stable complexes are attainable by using ligands endowed with a higher coordination ability than DTPA, DOTA, etc. but this would occur at the expenses of the direct coordinated water molecule thus with a net decrease of the relaxation enhancement of the resulting complex. Moreover an intense research is currently pursued in order to find new ligands of denticity six or seven in order to yield highly stable complexes containing two or three directly coordinated water molecules. Of course without having to pay any non acceptable price to the stability of the resulting complexes.

As far as concerns the potential toxicity of MRI-CA based on the manganese and iron, there is a general consensus to think that cells have acquired regulatory pathways that allow them to control the homeostasis of such essential metal ions. This has not yet been definitively proven and certainly it depends on the amounts the cell has to deal with. Moreover, in the case of iron oxide particles, they may move inside the cells when submitted to a magnetic field and thus introduce additional concerns on the possible implications that may lead to interference with the cell machinery.

Besides the search for enhanced sensitivities, the concern on toxicity is one of the reasons for which chemists are still continuing to seek for alternative agents.

4.2 Paramagnetic Metal Complexes

The use of paramagnetic metal complexes is to reduce T_1 in order to obtain an intense signal in short times and better signal-to-noise thanks to the acquisition of a high number of measurements. Since unpaired electrons are remarkably able to reduce T_1 and T_2, the search for positive CAs has been mainly oriented towards paramagnetic metal complexes. The paramagnetic metal ions most extensively investigated are either transition metals or in the lanthanide series.

As far as lanthanides are concerned, the attention has been mainly focused on Gd^{3+} that shows both a high paramagnetism (seven unpaired electrons) and favourable properties in terms of electronic relaxation. Gd is a non-essential metal for living systems as it does not possess any known physiological function and its administration to rodents as free ion is strongly toxic even at low doses (10–20 μmol kg^{-1}). Therefore it's necessary to administrate it under the form of highly stable complexes. Thus an active search has been undertaken in chemical laboratories to identify suitable ligands whose complexes with Gd display very high thermodynamic and kinetic stability. Figure 4.1 reports some representative examples of highly stable Gd^{3+} chelates that have entered the clinical practice as extracellular MRI contrast agents.

Paramagnetic chelates of Mn^{2+} (five unpaired electrons) have also been considered. The main drawback appears to be related to the stability of these complexes. Mn^{2+} is an essential metal, therefore the evolution has selected biological structures able to sequester Mn^{2+} ions with high efficiency. Combined with the fact that Mn^{2+} forms not very stable coordination complexes, it has been difficult to design Mn^{2+} chelates that maintain their integrity when administered to living organisms. Actually $[MnDPDP]^{4-}$ (Fig. 4.2) has entered the clinical practice and it is recommended as a hepatotropic agent (Rummeny and Marchal 1997). It is the only agent that does its job by releasing metal ions to endogenous macromolecules. The huge proton relaxation enhancement brought about by the resulting Mn^{2+} protein adduct is responsible for the MRI visualization of hepatocytes also at the low administered doses of $[MnDPDP]^{4-}$.

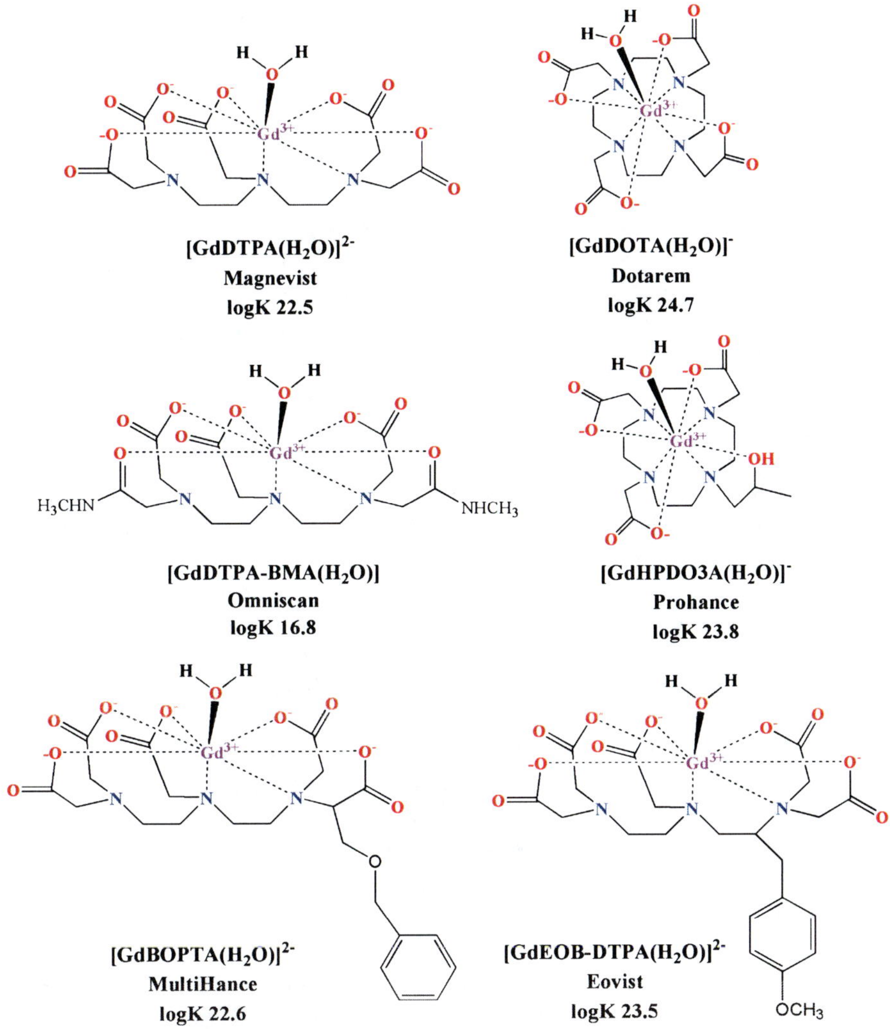

Fig. 4.1 Structures of some Gd^{3+} based MRI contrast agents clinically accepted with brand name and stability constant

4.2.1 Theory of Relaxation Enhancement

The contrast enhancing efficiency of a contrast agent is commonly expressed in terms of relaxivity, i.e. the increase in the solvent proton longitudinal and transverse relaxation rates normalized to a 1 mmol concentration of the chelated paramagnetic metal (Lauffer 1987; Aime et al. 1998).

$$R_i = [CA]r_i + R_{iw} \quad (\mathrm{i} = 1,2) \tag{4.1}$$

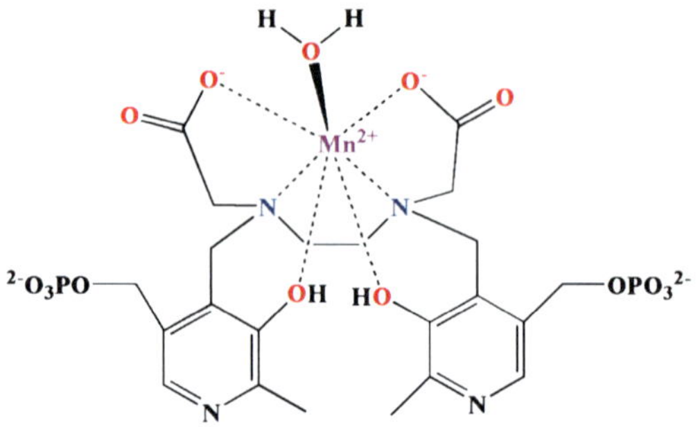

Fig. 4.2 Structure of a Mn^{2+} based MRI contrast agent used in clinical practice

In Eq. 4.1 R_{iw} is the longitudinal (i = 1) or transverse (i = 2) relaxation rate of the bulk water protons, corresponding to that measured for an analogous diamagnetic solution. In practice, R_{iw} coincides with the value determined for pure water under identical conditions of pH, temperature, and observation frequency. Clearly, the above relation strictly holds only for dilute solutions, in the absence of solute-solute interactions and of variation in viscosity.

The relaxation enhancement of water protons in the aqueous solutions of paramagnetic complexes arises from time fluctuation of the dipolar coupling between the electron magnetic moment of the metal ion and the nuclear magnetic moment of the solvent nuclei (Banci et al. 1991). The dipolar interaction involves both the water molecules that belong to the inner coordination sphere of the metal ion (*inner sphere* contribution) and those water molecules that are not directly bound to the paramagnetic center but are either diffusing next to the complex (*outer sphere* contribution) or localized in a well defined position with respect to the metal ion *via* hydrogen-bonding interactions with polar groups of the ligand (*second sphere* contribution). Therefore the total relaxivity is expressed as the sum of three contributions (Aime et al. 1998; Toth et al. 2001).

$$r_i = R_{ip}^{IS} + R_{ip}^{OS} + R_{ip}^{SS} \quad (\mathrm{i} = 1,2) \tag{4.2}$$

The three mechanisms contribute to catalyze the water proton nuclear magnetic relaxation to a different extent that depends on the structural properties of the CA, the temperature and the pH of the aqueous solution, and on the observation frequency. The CAs that are currently used in clinically practice are low molecular weight hydrophilic monoaqua Gd^{3+} complexes (Fig. 4.1) that, at 25°C and 20 MHz, possess a relaxivity of about 4–5 mM^{-1} s^{-1} (Caravan et al. 1999). In this case, the *inner* and *outer sphere* contributions are comparable, whereas the relaxation mechanism due to the water molecules in the second hydratation sphere of the complexes is neglected or simply incorporated into the *outer sphere* contribution. The increase in molecular dimension, the presence of suitable hydrogen-bond acceptor groups on the surface of the complex, the variation in the overall electric charge, and the modification of hydrophilic/hydrophobic domains of the complex are all accompanied by large variations in the relative contributions of the three relaxation pathways and will be discussed in some detail in this chapter.

4.2.1.1 Inner-Sphere Contribution to Relaxivity

This contribution arises from the exchange of the water molecule(s) directly coordinated to the paramagnetic metal center with the solvent molecules. Quantitatively, the inner-sphere contribution to the overall relaxivity is given by the following expression:

$$r_{1p}^{IS} = \frac{[CA]q}{55.6} \frac{1}{T_{1M} + \tau_M} \tag{4.3}$$

where the hydration number q indicates the number of metal-bound water molecules, τ_M is their mean residence lifetime, and T_{1M} is the longitudinal nuclear magnetic relaxation time of the bound water protons. The value of T_{1M} is given by the Solomon-Bloembergen-Morgan (SBM) equations which describe the time fluctuation of the water proton-Gd^{3+} dipolar coupling. Briefly, the modulation of the interaction arises from three main factors: rotation of the complex, described in terms of the correlation time τ_R, the residence lifetime of the bound water (τ_M) and electron paramagnetic relaxation (T_{1e} and T_{2e}). The following equations are commonly used for the analysis of the relaxivity of Gd^{3+} complexes:

$$\frac{1}{T_{1M}} = \frac{2}{15} \frac{\gamma_H^2 g^2 S(S+1)\beta^2}{r_H^6} \left[\frac{3\tau_{C1}}{1+\omega_H^2\tau_{C1}^2} + \frac{7\tau_{C2}}{1+\omega_S^2\tau_{C2}^2} \right] \tag{4.4}$$

$$\frac{1}{\tau_{Ci}} = \frac{1}{\tau_R} + \frac{1}{\tau_M} + \frac{1}{\tau_{ie}} \quad (\mathrm{i} = 1,2) \tag{4.5}$$

$$\left(\frac{1}{T_{1e}}\right)^{ZFS} = \frac{1}{25}\tau_v^2 \{4S(S+1)-3\} \left(\frac{1}{1+\omega_S^2\tau_v^2} + \frac{4}{1+4\omega_S^2\tau_v^2} \right) \tag{4.6}$$

$$\left(\frac{1}{T_{2e}}\right)^{ZFS} = \tau_v^2 \left[\frac{5.26}{1+0.37\omega_S^2\tau_v^2} + \frac{7.18}{1+1.24\omega_S^2\tau_v^2} \right] \tag{4.7}$$

In Eqs. 4.4–4.7 S is the electron spin quantum number, γ_H the proton nuclear magnetogyric ration, g and β the electronic g factor and Bohr magneton, respectively. r_H is the distance between the metal ion and the protons of the coordinated water molecules, ω_H and ω_S the proton and electron Larmor frequencies, respectively, and τ_R is the reorientational correlation time. The longitudinal and transverse electron spin relaxation times, T_{1e} and T_{2e}, are frequency dependent according to Eqs. 4.6 and 4.7, and characterized by the correlation time of the modulation of the zero-field splitting (τ_v) and the mean-square zero-field-splitting energy (Δ^2).

A straightforward approach to increase the inner-sphere relaxivity can be pursued through the increase of the hydration number *q*. This represent a scaling factor in Eq. 4.3 and then a higher number of coordinated water molecules (q > 1) results in an increase of this contribution at any field. The commercial Gd^{3+}-based CA have q = 1 but there are now available some new structures with q = 2 or even 3 still characterized by very high thermodynamic stability (Aime et al. 2000b, 2004 Hajela et al. 2000).

The *inner sphere* relaxivity depends also on other important parameters namely: the mean residence lifetime of the coordinated water(s), the electronic relaxation times and the Larmor frequency. The correlation time for the water proton exchange, τ_M, coincides with the correlation time for the exchange of the entire water molecule near neutral pH and its value covers a large range of values, from few ns, up to several μs. It is a parameter of great relevance because it plays a dual role: it may contribute to the overall correlation time τ_C for the dipolar interaction (Eq. 4.5) when the rotation of the complexes is slowed down following an increase in the molecular dimension of the complexes (larger complexes, micelles, dendrimers or macromolecular adducts) and it controls the efficiency of the transmission of the dipolar interaction to the bulk (Eq. 4.3).

4.2.1.2 Outer-Sphere Contribution to Relaxivity

This contribution arises from the modulation of the dipolar interaction of the paramagnetic center with the water molecule diffusing next to the surface of the complex. The component of the relaxivity due to the r_{ip}^{OS} term is commonly estimated by the expression derived by Freed (Eqs. 4.8 and 4.9) which takes into account the translational diffusive motion of the water molecules, considered as hard spheres, and incorporate the effects of the electron spin relaxation (Freed 1978; Koenig and Brown 1990).

$$r_{ip}^{OS} = C^{OS}\left(\frac{[CA]}{aD}\right)[7J(\omega_S) + 3J(\omega_H)] \tag{4.8}$$

$$J(\omega) = \Re = \left[\frac{1 + 1/4\left(i\omega\tau_d + \frac{\tau_d}{T_{je}}\right)^{1/2}}{1 + \left(i\omega\tau_d + \frac{\tau_d}{T_{je}}\right)^{1/2} + 4/9\left(i\omega\tau_d + \frac{\tau_d}{T_{je}}\right) + 1/9\left(i\omega\tau_d + \frac{\tau_d}{T_{je}}\right)^{3/2}}\right] \tag{4.9}$$

with $j = 1, 2;\quad \tau_d = a^2 / D$

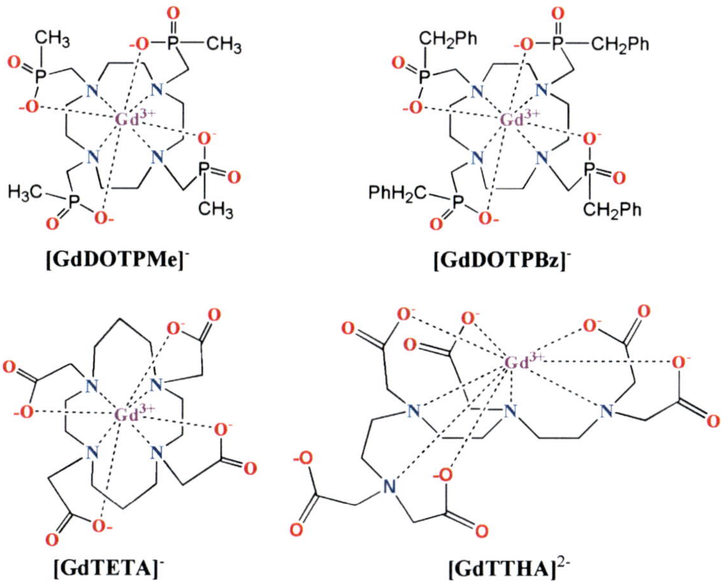

Fig. 4.3 Structures of some Gd^{3+} complexes with q = 0

In the above equations, C^{OS} is a constant (5.8×10^{-10} m^6mol^{-1} s^{-2}) and the non-Lorentzian spectral density function $J(\omega i)$ contain the field dependence on $T_{1,2e}$. At the magnetic field of interest, r_{ip}^{OS} depends primarily on the distance of water closest approach a, related to the molecular dimension and charge distribution of the complex, and on the relative diffusion coefficient of solute and solvent D. It follows that r_{ip}^{OS} assumes a comparable value (above 10 MHz) for all the small gadolinium chelates of clinical interest.

In fact, very similar values of relaxivity (ca. 2.1–2.5 mM^{-1} s^{-1}) have been measured for small Gd^{3+} chelates with q = 0, such as $[GdTTHA]^{3-}$, $[GdTETA]^-$, $[GdDOTPBz]^-$ and $[GdDOTPMe]^-$ (Fig. 4.3) (Aime et al. 1994; Kim et al. 1995; Borel et al. 2001).

The small difference in relaxivity has been accounted for primarily in terms of the different dimensions of the complexes which are reflected in different values of the parameter a. On the other hand, at low magnetic fields the relaxivities differ according to the different values of the electron spin relaxation times of the complexes.

4.2.1.3 Second-Sphere Contribution to Relaxivity

It's clear that Freed's model is not fully adequate for a realistic and physically sound description of the interaction of the outer sphere water molecules with the paramagnetic complexes. The main limitation is the lack of consideration of specific interactions of the solvent molecules with hydrophilic groups of the ligands, i.e. the

Fig. 4.4 Structure of a Gd^{3+}-complex with second-sphere contribution to relaxivity

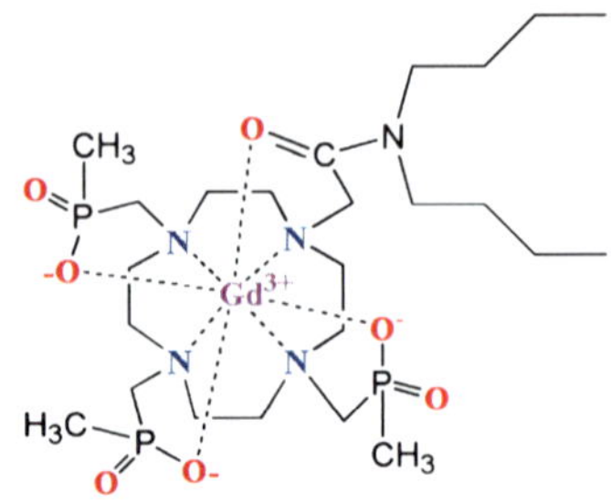

[GdDOTP-MBu$_2$]

formation of a second coordination shell. From a particular point of view, the use of a purely translational diffusion model for the evaluation of the *outer sphere* relaxivity often gives reasonably good results as shown by the analysis of the magnetic field dependence of the relaxivity of q = 0 complexes like $[GdTTHA]^{3-}$, $[GdTETA]^{-}$, and $Gd[DOTPBz]^{-}$ (Fig. 4.3). Presumably, this happens when the residence lifetime of the water molecules in the second hydration shell is close to the value of the self-diffusion of the solvent molecules. In these cases, neglecting r_{ip}^{SS} mostly results in an overestimation of the contribution of r_{ip}^{OS} with perhaps unrealistic values of *a*, often shorter than the molecular radius of the complex. On the other hand, there are cases where the Freed's model completely fails to account for the *outer sphere* relaxivity. A notable example is represented by $[GdDOTP]^{5-}$, a complex structurally similar to $[GdDOTA]^{-}$ but without any coordinated water molecules, as assessed by ^{17}O NMR data. The relaxivity of $[GdDOTP]^{5-}$ is very similar to that of monoaquo complexes $[GdDOTA(H_2O)]^{-}$ and $[GdDTPA(H_2O)]^{2-}$, and therefore, cannot be explained in terms of Freed's equation. The presence of four, highly negatively charged phosphonate groups promotes the formation of strong hydrogen bonds with the solvent molecules with a relatively long residence time. Borel has calculated form MD simulation the presence of 4.3 s sphere water molecules with a τ_M of 56 ps, which is sensibly longer than for the case of $[GdDOTA(H_2O)]^{-}$ and $[GdDTPA(H_2O)]^{2-}$ (Borel et al. 2001). This example clearly shows how the proper choice of the coordinating groups may affect the number and residence lifetime of water molecules in the proximity of the paramagnetic center and enhance this contribution to relaxivity.

Although a clear distinction between r_{ip}^{OS} and r_{ip}^{SS} is impossible, several examples have been reported where a contribution of r_{ip}^{SS} was indeed invoked in order to explain the relaxivity data (Botta 2000). The neutral complex [GdDOTP-Mbu$_2$] has a hydration number q = 0 (Fig. 4.4) but a relaxivity sensibly higher than for $[GdDOTPBz]^{-}$. In this case, the use of the translational diffusion model would give either an *a* value unreasonably too short or parameters describing the diffusional dynamics of the solution physically not meaningful. The relaxivity has been explained assuming the presence of a tightly bound water molecule in the second coordination sphere of the metal ion at a distance of ca. 3.6 Å from the Gd^{3+} ion (Aime et al. 1996b). The interaction was suggested to involve a hydrogen bond between the oxygen and the nitrogen atoms of the carboxamide group, and a water molecule which is to be localized close enough to the paramagnetic center so as to affect the solvent relaxation.

A sizeable contribution of the term r_{ip}^{SS} has also been evidenced in the study of pH-dependence of the relaxivity in the case of the Gd^{3+} complexes with tetraamide DOTA derivatives. These cationic monoaqua complexes are characterized by a slow rate of exchange of the coordinated water molecule, which strongly limits the inner sphere relaxivity around neutral pH. By increasing the pH, a base-catalyzed prototropic exchange occurs that lowers the effective τ_M value and increases the contribution of r_{ip}^{SS}. However, a further increase in the relaxivity is observed at $pH > 10$ and this has been accounted for in terms of the stepwise deprotonation of the amide nitrogen atoms that favors the localization in the proximity of the metal center of a few water molecules, bridging through hydrogen bond interactions, the amide group and the bound water (Aime et al. 1997). Sherry has reported a similar example concerning tetraamide DOTA complexes bearing hydroxypyridyl substituents. At high pH the deprotonation of the amides favors the formation of intramolecular acid–base pair interaction with the phenolic protons accompanied by the formation of a highly organized second hydratation sphere that increases the relaxivity (Woods et al. 2003).

4.3 Iron Oxide Particles

Fe^{3+} having the same number of unpaired electrons as Mn^{2+} has entered the field of MRI contrast agents in the form of iron oxide nanoparticles. These systems correspond to maghemite and magnetite (Fe_2O_3, Fe_3O_4) stabilized by various coating agents and are characterized by a large magnetic moment in the presence of a static external magnetic field. This large magnetic moment is caused by a crystal ordering (spinels) that induces cooperativity between the individual paramagnetic ions of the crystal. The diameter of the iron oxide core is just 3–5 nm thus smaller than a magnetic domain (ca. 30 nm). Consequently they do not behave as ferromagnetic material as their magnetization goes to zero upon removal of the external magnetic field.

Several types of iron oxide particles have been reported containing different coating materials (the most commonly used are dextran and carboxydextran) that endow the particles with quite different sizes which, in turn, results in substantial differences in their MRI properties. The two main classes are represented by (i) superparamagnetic iron oxide particles (SPIO) with a mean diameter of more than 50 nm (Jung and Jacobs 1995) and (ii) ultrasmall superparamagnetic iron oxide particles (USPIO) with a smaller diameter (Reimer et al. 1998). These agents provide excellent (negative) contrast when administered at iron doses as low as 8–15 μmol kg^{-1} body weight. Once administered intravenously, as particles, these agents accumulate in the cells of the reticuloendothelial system. The pharmacodynamic properties of the iron oxide particles are affected by either the size or the overall electric charge. The smaller particles remain in the blood circuit for a time long enough to be considered as blood pool agents for angiographic assays.

The distortion of the magnetic field caused by the superparamagnetic core induces large changes in magnetic susceptibility which, in turn, leads to hypo-intensities in T_2- or T_2*-weighted images. Thus the areas containing particles display fast transverse relaxation rates and low signal intensity ("negative contrast"). Due to the large magnetic susceptibility of an iron oxide particle, the signal void is much larger than the particle size, enhancing detectability at the expenses of resolution. These particles have been used to track "*in vivo*" different cell types (Cunningham et al. 2005), including T lymphocytes (Kircher et al. 2003), macrophages (Oude Engberink et al. 2007), and stem cells (Frangioni and Hajjar 2004).

In general it is now well established that the larger particles are more rapidly and preferentially taken up by macrophages thus their application to assess lymph node functionality is very important (Saksena et al. 2006; Islam and Harisinghani 2009). Moreover, the capacity of nanoparticles to target inflammatory lesions via macrophage labelling is currently under intense scrutiny, in particular, for the characterization of degenerative diseases (Corot et al. 2004). Finally, the conjugation of suitable vectors at the outer surface of the particles endows them with outstanding properties as targeting probes in MR Molecular Imaging applications (Reimer et al. 1990; Boutry et al. 2008).

4.4 Responsive Agents

Responsive agents are systems able to modulate their relaxation enhancement capability as a function of a specific parameter of the microenvironment in which the CA distributes. Although most of the reported responsive agents rely on paramagnetic complexes few examples deal also with iron oxide particles. Basically, to act as a responsive agent a system has to be designed in such a way that at least one determinant of its relaxivity is significantly affected by the changes of the parameters of interest. Paradigmatic examples are reported in the following sub-sections.

4.4.1 pH Responsive Agents

Many pathological events are associated with alterations of pH homeostasis, including tumors and cardiovascular diseases (Aime et al. 1996a; Gerweck and Seetharaman 1996; Lee and Tannock 1998). The task of developing non-invasive methods for measuring pH *in vivo* has stimulated several studies aimed at designing probes sensitive to the hydrogenionic concentration (Sun et al. 2000; Ward and Balaban 2000). The design of a Gd^{3+}-based complex whose relaxivity is pH-dependent requires that at least one of the structural or dynamic parameters determining its relaxivity is made pH dependent.

For instance, it was found that the relaxivity of a series of macrocyclic Gd^{3+} complexes bearing β-arylsulfonamide groups (Lowe et al. 2001) (Fig. 4.5) is markedly

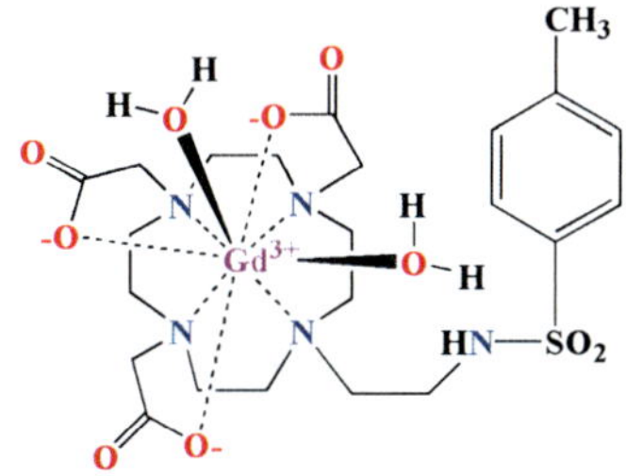

Fig. 4.5 Structure of a pH-responsive Gd^{3+}-complex due to a change in the number of water molecules in the first hydratation sphere

pH-dependent on passing from about 8 s^{-1} mM^{-1} at pH<4 to ca. 2.2 s^{-1} mM^{-1} at pH>8. It has been demonstrated that the observed decrease (about fourfold) of r_1 is the result of a switch in the number of water molecules coordinated to the Gd^{3+} ion from 2 (at low pH values) to 0 (at basic pHs). This corresponds to a change in the coordination ability of the β-arylsulfonamide arm, that binds the metal ion only when it is in the deprotonated form.

Another useful route to pH sensitive probes involves the inclusion of Gd^{3+} complexes, into pH sensitive liposomes (Lokling et al. 2004). The main component of these liposomes is represented by unsaturated phosphatidylethanolamine that is usually "doped" with negatively charged amphiphilic components in order to stabilize the lamellar phase necessary for the formation of the bilayer structure of liposomes. If the acidic group of the amphiphilic component is protonated, then the liposomes destabilizes (transition phase) with the consequent increase of water exchange rate between the inner and outer compartment. On this basis, a relaxometric "off-on" pH switch can be designed by incorporating a Gd^{3+} complex into intact liposomes endowed with a very low water permeability. In this way, the relaxivity of the system will be very low (about 0.5 mM^{-1} s^{-1} at 37°C and 10 MHz), but, the protonation of the acidic components determines a significant relaxation enhancement (almost 4 mM^{-1} s^{-1}). This system has been successively tested in human blood where, unfortunately, it showed to be quite unstable owing to the presence of blood components (Na^+, Ca^{2+}, Mg^{2+} and also HSA) which interfere with the phase transition of the liposomes (Lokling et al. 2001).

4.4.2 Agents Sensitive to Redox Potential

A diagnostic MRI probe sensitive to the *in vivo* redox potential would be very useful for detecting regions with a reduced oxygen partial pressure (P_{O2}), typical symptom of several pathologies including strokes.

Very few Gd^{3+} chelates sensitive to the tissue oxygenation have been so far reported. One possibility is provided by the potential ability of GdDOTP to act as allosteric effector of hemoglobin (Aime et al. 1995). In fact, it has been observed

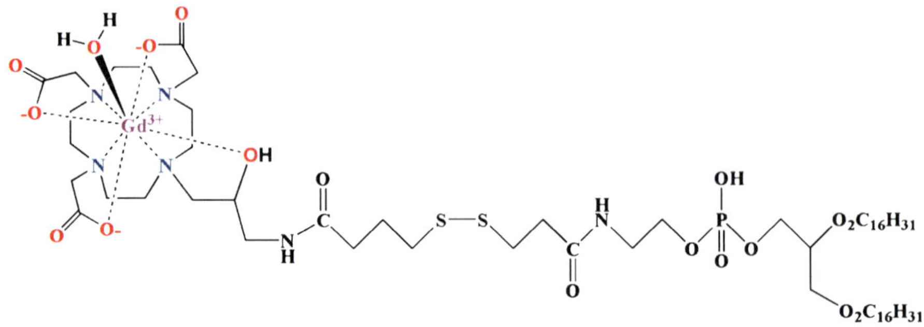

Fig. 4.6 Structure of a redox-potential sensitive Gd^{3+}-complex: the incorporation in the liposome membrane yields a relaxivity of slowly tumbling systems. A decrease of relaxivity is obtained after the cleavage of the S–S bond

that this chelate binds specifically to the T-form of the protein, that is characterized by a lower affinity towards oxygen. The interaction is driven by electrostatic forces and leads to a significant relaxivity enhancement (ca. fivefold) owing to the restricted molecular tumbling of the paramagnetic complex once it is bound to the protein. Although hemoglobin can be considered an excellent indirect target for detecting P_{O2}, the practical applicability of the method suffers for the inability of GdDOTP to enter red blood cells.

Another approach deals with the use of liposomes containing in their membrane an amphiphilic Gd^{3+} complex that has been designed in order to have a radical-sensitive disulfide bridge between the chelate and the lipid moiety (Glogard et al. 2003) (Fig. 4.6). The relaxivity (at 20 MHz and 25°C) of the liposomal paramagnetic agent is 13.6 $mM^{-1}\ s^{-1}$, i.e. twofold higher than that of the free Gd^{3+} complex (r_I of 6.5 $mM^{-1}\ s^{-1}$). Likely, the limited relaxivity enhancement of the "bound" form is due to the rotational flexibility of the complex on the surface of the liposome.

This system has been tested "*in vitro*" by inducing the cleavage of the S-S bond with chemical (by dithiothreitol) or physical (by γ-rays) means. In both cases, the relaxivity decreased from the value of the liposome-bound form to that one of the free Gd^{3+} complex.

Though not based on Gd^{3+} ion, it is worth recalling other paramagnetic probes acting on the relaxation rates of water protons, which have been proposed as responsive agents towards oxygen concentration. Basically, such systems rely on of designing probes containing a metal that exists in two redox states endowed with quite different relaxation enhancement.

A first example is represented by the Mn^{3+}/Mn^{2+} redox switch. The complexes of Mn^{2+} and Mn^{3+} with the water-soluble tetraphenylsulphonate porphyrin (Fig. 4.7) display significantly different r_I values at low magnetic field strength (lower than 1 MHz), but very similar values at the fields used in the clinical practice (>10 MHz) (Aime et al. 2000a). However, the longer electronic relaxation rates of the Mn^{2+} complex makes its relaxivity dependent on the reorientational motion of the chelate. In fact, upon interacting with a poly-β-cyclodextrin, a fourfold enhancement of the

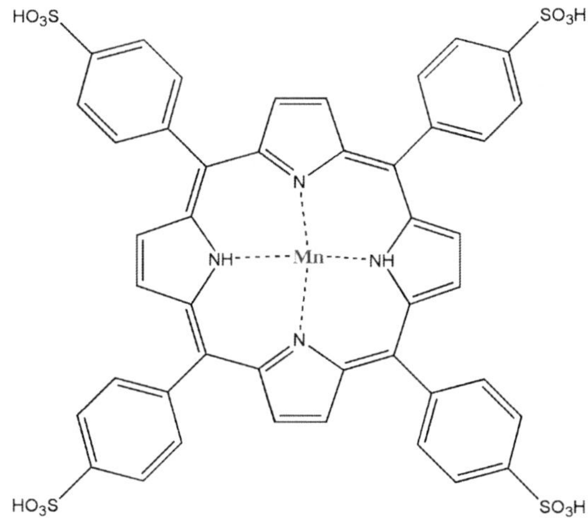

[MnTPPS]

Fig. 4.7 Structure of a Mn-complex: the switch between the two oxidation states Mn^{2+}/Mn^{3+} changes dramatically the relaxometric properties

relaxivity of Mn^{2+}-TPPS at 20 MHz has been detected, whereas a little effect has been observed for the Mn^{3+}-complex. The ability of the Mn^{2+}/Mn^{3+} system to respond to changes in the partial pressure of oxygen has been demonstrated *in vitro*.

Another interesting redox switch deal with the Eu^{3+}/Eu^{2+} couple. Here, Eu^{2+} has the same electronic S_8 ground state of the Gd^{3+} ion, whereas Eu^{3+} ion is a poor relaxing probe owing to its very short electronic relaxation times. Unfortunately, Eu^{2+} is a very unstable cation. Therefore, efforts have been addressed to get more insight into the determinants of thermodynamic and kinetic stability of Eu^{2+} complexes (Seibig et al. 2000). Interestingly, it has been shown that the relaxivity of Eu^{2+}-DOTA is very similar to that of Gd^{3+}-DOTA, but the residence lifetime of the coordinated water in the former is noticeably shorter (τ_M of ca. 0.4 ns vs. 200 ns at 25°C) (Burai et al. 2003).

4.4.3 *Enzyme Responsive Agents*

One possible route to design enzyme responsive agents is to synthesize paramagnetic complexes acting as the substrate for a specific enzyme, whose binding to the active site of the protein can be signalled by the consequent variation in the relaxivity. Along this line of reasoning Lauffer et al. prepared a Gd^{3+} chelate containing a phosphoric ester (Fig. 4.8) sensitive to the attack of the serum alkaline phosphatase (Lauffer et al. 1997). The hydrolysis yields the exposure of an hydrophobic moiety well suitable to bind to HSA. Upon binding, there is an increase of the relaxivity as a consequence of the lengthening of the molecular reorientational time.

This approach was used by the same research group for designing Gd^{3+}-complexes sensitive to TAFI (thrombin-activatable fibrinolysis inhibitor), a carboxypeptidase B involved in the clot degradation (Nivorozhkin et al. 2001).

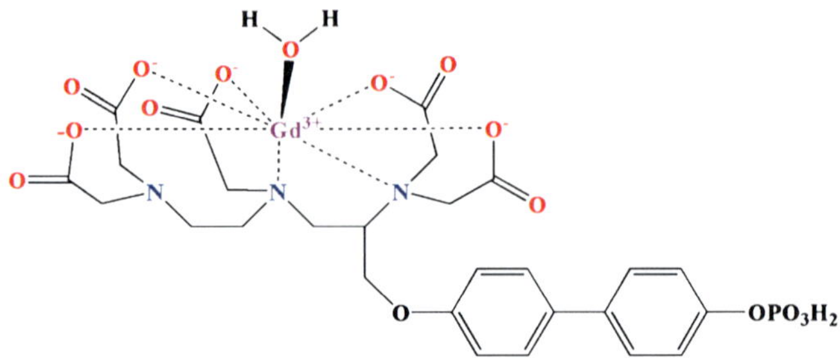

Fig. 4.8 Structure of an enzyme-responsive Gd^{3+}-complex: the binding to HSA is promoted by the action of the serum alkaline phosphatase that cleaves the phosphoric acid yielding a hydrophobic complex

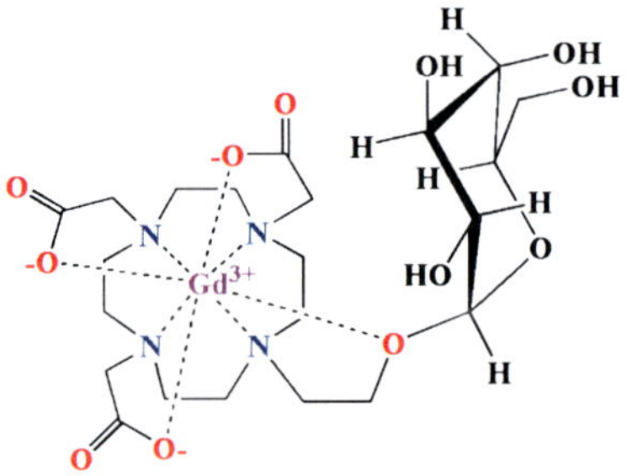

Fig. 4.9 Structure of an enzyme-responsive Gd^{3+}-complex: the relaxivity is increased following the hydratation state change due to the β–galactosidase action

Another example of enzyme responsive agents has been provided by Meade et al. who synthesized a GdDOTA-like system containing a galactopyranosyl substituent capping the metal ion (Fig. 4.9). Hence, the complex shows a low relaxivity (typical $q=0$ complex). In the presence of β-galactosidase, an enzyme widely used as a marker of gene transfection, the sugar moiety is hydrolyzed, and a water molecule can have free access to the paramagnetic center (Moats et al. 1997). This complex has been successfully used for monitoring *in vivo* the gene expression during the development of embryos of *Xenopus laevis* (Louie et al. 2000).

A procedure aimed at being of general applicability has been reported (Bogdanov et al. 2002) which relies on enzyme-mediated polymerization of paramagnetic substrates into oligomers of higher relaxivity. The paramagnetic substrate is represented by a GdDOTA-like complex bearing a catechol functionality on its surface. In the presence of peroxidase, the monomers undergo rapid condensation into paramagnetic oligomers. To give support to the view of a wide applicability of this procedure in MRI, two interesting applications have been reported: (i) visualization of the occurrence of peroxidase activity at nanomolar concentration and (ii) imaging of E-selectin on the surface of endothelial cells through a prior targeting with an anti-E-selectin peroxidase conjugate.

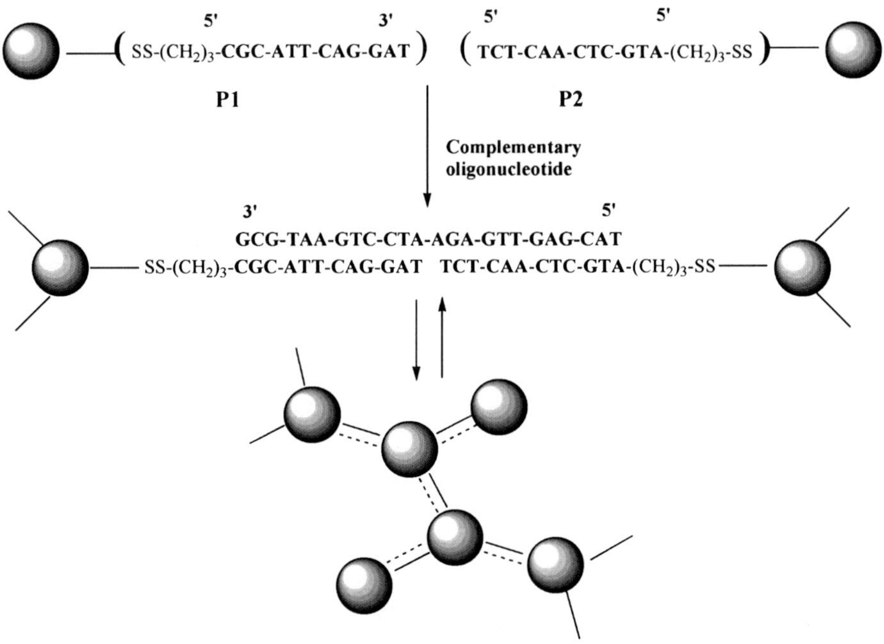

Fig. 4.10 Example of a oligonucleotide sequence reporter agent: the two nanosensors P1 and P2 are formed by oligonuceotides treated with activated nanoparticles. The hybridization of P1 and P2 with a complementary oligonucleotide results in oligomerization and magnetic relaxivity changes

4.4.4 Reporters of Oligonucleotide Sequences

Weissleder and co-workers have shown that is possible to detect specific oligonucleotide sequences by means of MRI probes containing iron oxide particles (Josephson et al. 2001). The particles bear on their outer surface oligonucleotide sequence complementary to that one to be recognized. Hybridization to the target sequence resulted in oligomerization with an increase in T_2-relaxivity (Fig. 4.10).

4.4.5 [Ca^{2+}] -Reporting Agents

Ca^{2+} plays important roles as a carrier of electrical current and as a second messenger in the brain. There is a very large [Ca^{2+}] gradient between the intra- and extracellular compartments, being the extracellular [Ca^{2+}] of the order of 1.5–2 mM whereas the intracellular concentration is only 50–100 nM. During normal brain activity the extracellular [Ca^{2+}] drops ca. 15% but upon maximal stimulation, it drops 30%. Thus much attention has been devoted to the development of Gd^{3+} based agents able to act as MRI-responsive agents towards changes in the extracellular Ca^{2+} concentration. The design of Ca^{2+}-responsive Gd-based complexes was reported by Meade and

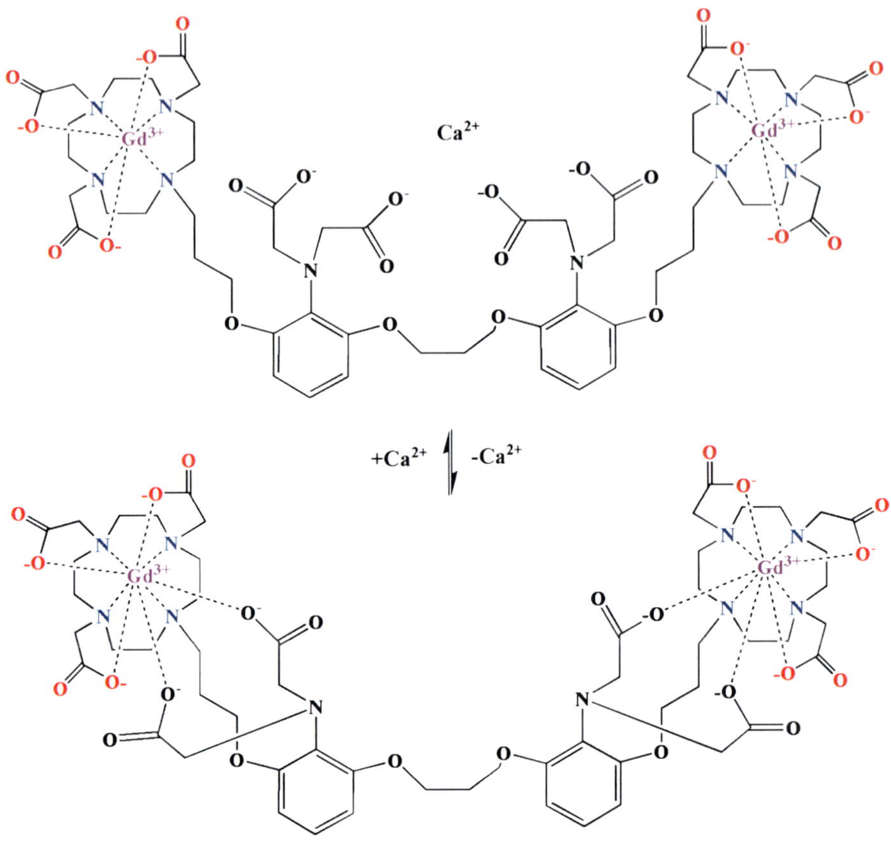

Fig. 4.11 Example of a [Ca^{2+}]-reporting agent: the presence or absence of Ca^{2+} ions induces a conformational change in the Gd^{3+}-complex structure

co-workers in 1999 (Li et al. 1999). They showed that the relaxivity of GdDOPTA (Fig. 4.11) is markedly affected by the presence of Ca^{2+} as the latter ions can compete for the acetate arms that, in the absence of Ca^{2+} coordination, enter the coordination sphere of Gd^{3+} ions. The relaxivity quenching is removed as the acetate groups are coordinated to Ca^{2+} and their positions at the Gd^{3+} center are taken up by water molecules.

Related systems endowed with high selectivity and full reversibility in their interaction with [Ca^{2+}] have been reported (Atanasijevic et al. 2006; Angelovski et al. 2008).

4.5 Manganese-Based MRI Contrast Agents

As anticipated above, Mn^{2+} has attracted much attention since the early days of MRI contrast agents' development because of its high paramagnetism and favourable electronic relaxation time. Mn^{2+} may be hexa- or hepta-coordinated. When coordinated to

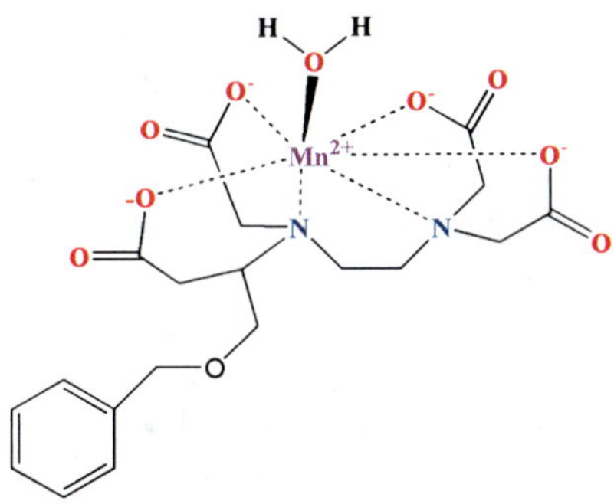

$[Mn(EDTA(BOM))(H_2O)]^{2-}$

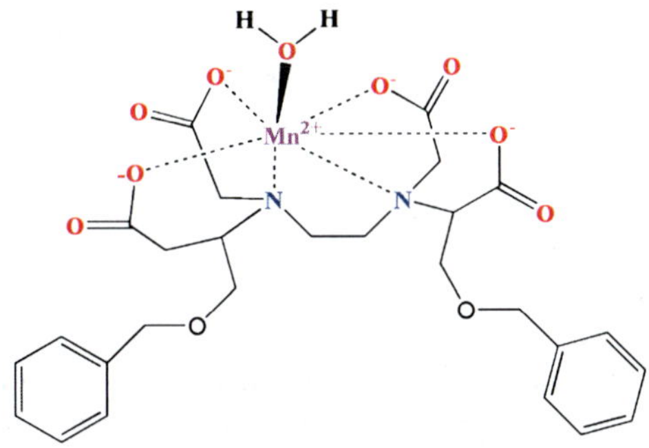

$[Mn(EDTA(BOM)_2)(H_2O)]^{2-}$

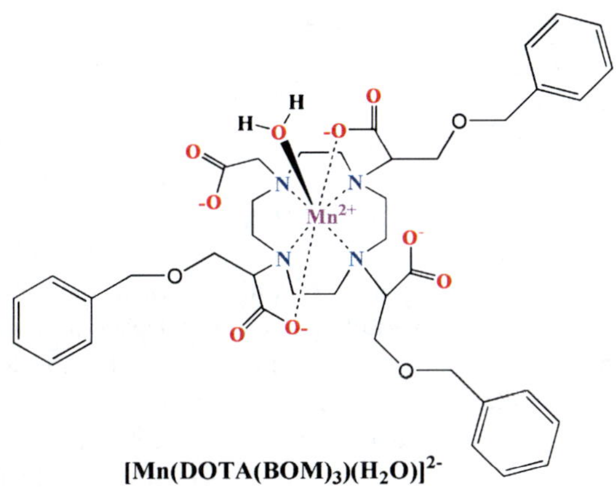

$[Mn(DOTA(BOM)_3)(H_2O)]^{2-}$

Fig. 4.12 Structures of Mn^{2+}-based contrast agents

EDTA (Ethylen Diamino Tetra Acetate) it is hepta-coordinated and then it owns a directly coordinated water. The NMRD profile (Koenig et al. 1984) of $[Mn\text{-}EDTA]^{2-}$ is characterized by the lack of the low-field dispersion, shown by $[Mn(H_2O)_6]^{2+}$, due to the contribution of the contact term to the relaxation enhancement operating at Larmor frequencies lower than 1 MHz. In principle, either a reduction of the Mn^{2+}-water protons coupling constant or a shortening of the electronic relaxation times may account for the observed behavior. ^{17}O measurements have shown that the Mn^{2+}-water oxygen coupling constant doesn't vary appreciably between the $[Mn(H_2O)_6]^{2+}$ and Mn(II) chelates, so it is likely that a significant decrease of T_{1e} in Mn^{2+} chelates is the main reason for the reduction of the contact contribution.

Two derivative containing one or two benzyl-oxy methyl substituents on the acetic arm(s) of EDTA have been reported, Mn[EDTA-BOM] and Mn[EDTA-$(BOM)_2$], respectively (Aime et al. 2002) (Fig. 4.12). Their stability constants are very similar to that reported for $[Mn\text{-}EDTA]^{2-}$ ($\log K = 13.5$) and display slighter higher relaxivities at any magnetic field strength as a result of a longer reorientational correlation time due to the larger molecular size (i.e. longer molecular correlation time τ_R).

The presence of the benzyl-oxymethyl subsituent(s) promotes the binding of these complexes to serum albumin. The non-covalent binding has two important effects: (1) promotes a further relaxation enhancement, mainly as a consequence of the elongation of the molecular reorientational time τ_R and (2) it prolongs the lifetime of the diagnostic agent in the blood stream by limiting the extravasation observed for small-sized CA.

To get an estimate of the contribution arising from the second coordination sphere, a related Mn-DOTA(BOM)$_3$ complex has been synthesized (Aime et al. 1996c) (Fig. 4.12). Now the increased denticity of the ligand does not longer provide room for the entering of a water molecule in the inner coordination sphere of the Mn^{2+} ion. On the other hand, the presence of the three BOM substituents yields a strong binding to Albumin thus allowing us to assign the observed relaxation enhancement as a second sphere contribution. Analogous results were later reported by Caravan et al. on a related Mn-EDTA complex bearing a moiety that allows a strong non-covalent binding to Albumin (Troughton et al. 2004).

In spite of the interesting achievements with these [MnEDTA]-substituted derivatives, the stability of these Mn^{2+} complexes is not high enough to warrant against decomplexation effect in *"in vivo"* applications.

Actually, the release of Mn^{2+} ions is well established for the only approved Mn^{2+} agent, i.e. $[MnDPDP]^{4-}$ (Fig. 4.2) (Padovani et al. 1996). In fact, after intravenous infusion, $[MnDPDP]^{4-}$ dissociates and manganese is released and fixed in the liver and other parenchymatous organs (pancreas, renal cortex). By shortening T_1 relaxation, this contrast agent enhances the hepatic signal on T_1-weighted sequences, with maximum hepatic enhancement occurring at 30–50 min following infusion (5–10 μmol kg^{-1}). $[MnDPDP]^{4-}$ has been used to improve the conspicuity of focal liver lesions (in making the differential diagnosis between hepatocellular lesions and non-hepatocellular ones).

As far as Mn^{3+} ions are concerned the most common vehicle to their stabilization is represented by the coordination by porphyrin ligands. Porphyrins are interesting systems for MRI applications because, in addition to enhance the excretory systems, they show some degree of retention in tumours tissues. $[Mn(TPPS)]^{3-}$ (Fig. 4.7) displays a very good relaxivity ($r_1 = 10.3$ mM^{-1} s^{-1}) at 37°C and 20 MHz (Chen et al. 1984). $[Mn(TPPS)]^{3-}$ has been selected for its good water solubility but it suffers for the drawback of the too high osmolality of its solutions. Therefore, attention was devoted to synthesize (Bradshaw et al. 1998) new water-soluble manganese^{3+} porphyrins having low osmolarity in comparison to $[Mn(TPPS)]^{3-}$ (Fig. 4.13). Polyhydroxylamide substituents have been selected for this purpose because they have been used successfully as non-toxic, water solubilizing substituents for aromatic iodinated X-ray contrast agents.

A detailed assessment of the tumour-seeking capability of Mn^{3+} porphyrin systems has been carried out with the agent reported in Fig. 4.14. During the 24 h period following contrast administration, HOP-8P was consistently cleared from the circulation, liver, kidneys and muscle; however, it was progressively accumulated within the tumor (Takehara et al. 2002).

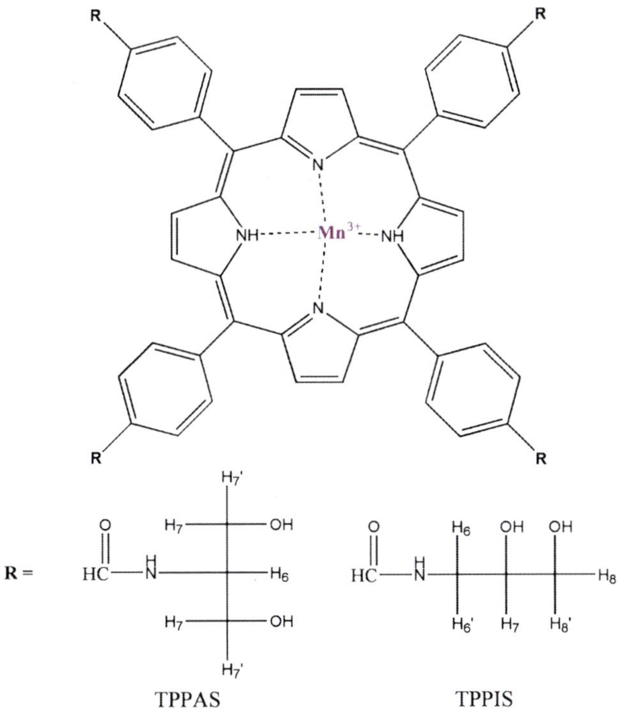

Fig. 4.13 Structures of water-soluble Mn^{3+}-complexes with low osmolarity

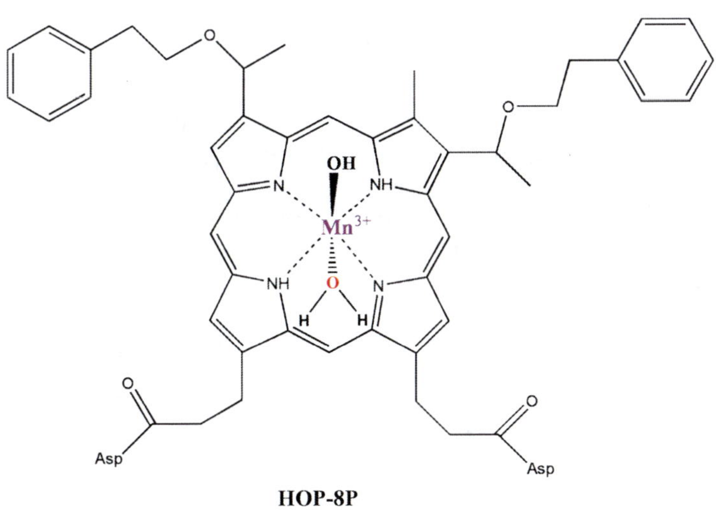

Fig. 4.14 Structures of a tumour-seeking Mn^{3+}-complex

4.5.1 Manganese Ion as Paramagnetic Biological Calcium Ion Analogue

Although limited to animal models the Mn^{2+}- ion can be used as a Ca^{2+} -ion analogue. Despite its' known glia- and neurotoxic effects it has been proven suitable to visualize Ca^{2+} uptake in cells and in excitable cells such as muscle and nervous tissue providing an excellent MRI contrast for studying the activity of excitable cells. By using appropriate concentrations and administration protocols it is increasingly applied in cardiological and neurological animal research. Especially the study of manganese accumulation in the brain is leading to new insights in functional brain anatomy.

The interpretation of the molecular and cellular mechanisms underlying the differential accumulation of Mn^{2+} in the brain after exogenous Mn^{2+} administration are not entirely clear. For example, the use of manganese ions as MRI contrast agent to highlight activity in excitable tissues is thus far largely attributed to its properties to enter cells via voltage gated or ligand gated calcium channels. However, the calcium turnover and (re)distribution in gliacells can also account for much of the activity dependent Mn^{2+} distribution.

The concentration in which Mn^{2+} is administered to the brain together with the administration route determines how much free Mn^{2+} is available in the brain. Most Mn^{2+} will be bound almost immediately to Mn^{2+} -buffering molecules. The free Mn^{2+} ions will be eliminated relatively fast, or distributed over a certain area until all Mn^{2+} is bound. The availability of Mn^{2+} as calcium analogue and thus as a tool for *in vivo* Ca-imaging depends on exchange of Mn^{2+} from its buffered state to the free state and back.

4.5.2 Manganese Administration Routes

For the purpose of inducing contrast in the animal brain, thus far, mainly **systemic injection** techniques have been used (intraperitoneal [i.p.] or subcutaneous [s.c.] or intravenous [i.v.]). This led to contrast enhancements in regions that under natural conditions also contained high Mn^{2+} concentrations (Newland et al. 1989; Watanabe et al. 2002, 2004 Aoki et al. 2004; Kuo et al. 2005; Lee et al. 2005). In Sprague Dawley rats Mn^{2+} distributed under normal conditions among four tested brain regions in the following order: substantia nigra>striatum>hippocampus>frontal cortex in a concentration range of 0.3–0.7 $\mu g\ g^{-1}$ of wet tissue weight (Zheng et al. 1998). In the healthy human brain, highest concentrations (dry weight) were found in the pineal gland (~4 $\mu g\ g^{-1}$), followed by the olfactory bulb (~3 $\mu g\ g^{-1}$) and the caudate nucleus (~2 $\mu g\ g^{-1}$) (Larsen et al. 1979; Bonilla et al. 1982; Duflou et al. 1989) .

Peripheral administration techniques can be useful to obtain a global manganese delivery to the brain non-invasively. Although, a direct access route via the blood exists for most body tissues (e.g. liver, pancreas and heart), this route is largely cut off to the brain by the existence of a blood brain barrier (BBB). After a short systemic exposure, Mn^{2+} is cleared relatively fast from the blood, in the range from minutes (Cotzias et al. 1968) up to 1.83 h (Zheng et al. 2000). The subsequently increased influx into the brain is largely dependent on a fast uptake mechanism that exists at the level of the choroid plexus and the ventricular ependyma, rather than direct uptake through the BBB (Murphy et al. 1991; Rabin et al. 1993).

After **intraventricular injection** Mn^{2+} diffuses into the parenchyma, a process that can last for 24–96 h in rats (Takeda et al. 1994a; Liu et al. 2004). The observed uptake rate and distribution in the brain was similar to that after systemic injection (Dastur et al. 1969; Manghani et al. 1970; Takeda et al. 1994b; Watanabe et al. 2002; Aoki et al. 2004; Lee et al. 2005). Specific stimulation during this period can alter the normal Mn^{2+} distribution pattern of Mn^{2+}-and thus indicate regions of increased activity.(Yu et al. 2005b, 2007).

Alternatively, one can reversibly **break the blood brain barrier** by applying an osmotic shock during the intravenous infusion of Mn^{2+} (Lin and Koretsky 1997). A subsequent stimulation leads to a fast Mn^{2+} accumulation in the activated brain region (Aoki et al. 2002). Mn^{2+} can also be delivered directly into a brain region of interest via its administration into peripheral sensory organs like the olfactory epithelium (Tjalve et al. 1995, 1996; Pautler et al. 1998; Henriksson et al. 1999; Tjalve and Henriksson 1999) and the eye (Pautler et al. 1998; Watanabe et al. 2001), or more invasive **focal brain injections** (Sloot and Gramsbergen 1994; Takeda et al. 1998b; Saleem et al. 2002; Van der Linden et al. 2002; Pautler et al. 2003). This approach has been mostly used to perform anatomical or activation induced tract tracing and is based on the fact that Mn^{2+} is taken up into neurons and subsequently transported via anterograde transport towards the synapse.

A variety of cellular and molecular mechanisms are involved in Mn^{2+} uptake into neurons. On many levels these mechanisms have not been completely resolved. Mn^{2+} is often captured or transported by molecules associated with iron (Fe^{3+}) or calcium (Ca^{2+}) metabolism. This diversity of plausible transport systems obscures the molecular picture of important processes such as manganese accumulation in neurons, their subsequent intracellular distribution and eventual exchange at the synapse. Therefore, the molecular background of these processes is often studied in function of curing and/or preventing manganese toxicity. We will evaluate the multitude of manganese transport mechanisms and discuss their contribution to activity-dependent signal enhancements during MEMRI.

4.5.3 *Molecular and Cellular Correlates of Manganese Uptake and Distribution*

4.5.3.1 Chemical Configurations of Manganese

In biological fluids and tissues, most metals are present largely as complexes with amino acids, peptides, proteins, phospholipids, and other tissue constituents rather than as the free metal cations. Binding of reactive heavy metals to metallothioneins, ferritin, transferrin (Tf), lactoferrin, melanotransferrin, hemosiderin, ceruloplasmin, citrate, ascorbate, glutathione (GSH), cysteine, or other amino acids is a major protective mechanism. Likewise, the biological reactivity of essential metals is regulated by interaction with specific ligands, and in particular with prosthetic groups on proteins (Ballatori 2002). To understand the molecular mechanisms of manganese uptake into cells or neurons one should consider some of the chemical species under which manganese appears.

Thermodynamic modeling of Mn^{2+} in serum suggests it exists in several forms: for example, as an albumin-bound species (84%), as a hydrated ion (6.4%), and in other low molecular weight ligands (1.8%) (Harris and Chen 1994). The calculations based on this model are consistent with the observation of low molecular weight species slightly larger than the Mn^{2+} ion, in plasma (Critchfield and Keen 1992). Approximately 8% of Mn^{2+} in rat blood plasma was found to be in the free fraction, also consistent with the model calculations (Crossgrove and Yokel 2004) According to another model a large portion (approximately 60%) of divalent manganese in the plasma is reported to be non-protein-bound, probably the free ion (May et al. 1977).

In serum, divalent manganese can be oxidized to trivalent manganese by ceruloplasmin or molecular oxygen (Gibbons et al. 1976). Thermodynamic modeling of Mn^{3+} in serum suggests that it is almost 100% bound to Tf (Aisen et al. 1969; Harris and Chen 1994).

4.5.3.2 Manganese Transport across Membranes and Barriers

Several transport systems have been reported for **divalent manganese** uptake: i.e. calcium channels (Drapeau and Nachshen 1984; Narita et al. 1990), Na/Ca exchanger (Frame and Milanick 1991), active calcium uniporter (Gavin et al. 1990), and Na/Mg antiporter (Gavin et al. 1990). Divalent manganese can also be transported into the cell by the divalent metal transporter DMT1 (also known as DCT1 and NRAMP2) (Gunshin et al. 1997; Conrad et al. 2000). **Trivalent manganese**, usually bound to Tf, can be captured by the Tf receptor and subsequently incorporated by endocytosis.

Several studies suggested that Mn^{2+}-ion brain **influx** at the **BBB** is carrier-mediated (Aschner and Aschner 1990; Murphy et al. 1991; Rabin et al. 1993; Aschner and Gannon 1994; Crossgrove et al. 2003) while the identity of the Mn^{2+}-carrier(s) is still unknown. In the brain, capillary endothelia have the highest density of Tf receptors (Jefferies et al. 1984; Pardridge et al. 1987; Connor 1994; Moos 1996), but these seem to play a negligible role in manganese delivery to the brain (Dickinson et al. 1996; Malecki et al. 1998, 1999). It was shown recently that the contribution of DMT1 to Mn^{2+} transport across the BBB might also be limited (Crossgrove and Yokel 2004). Alternative pathways for Mn^{2+} transport across the BBB via store-operated calcium channels, as well as another mechanism at the blood–brain barrier, likely play a role in carrier-mediated Mn^{2+} influx into the brain (Crossgrove and Yokel 2005).

Mn^{2+}-**efflux** across the BBB on the other hand does not appear to occur through a carrier but rather by diffusion (Yokel et al. 2003). However, it has been shown that both the endothelial cells in the BBB as the epithelial cells in de blood – cerebrospinal fluid(CSF) barrier(BCB) express metal transport protein (MEP1, or ferroportin) whose function is to expel the metal from the cells to surrounding environment (Burdo et al. 2001).

Glial cells, particularly **astrocytes** represent a "sink" for brain manganese (Wedler and Denman 1984), and could contribute significantly to signal enhancements after manganese administration. The kinetics of Mn^{2+} uptake have been studied in cultured astrocytes from rat (Aschner et al. 1992) and mixed glia from chick (Wedler et al. 1989). Unlike neurons, astrocytes have the ability to concentrate Mn^{2+} at levels 50-fold higher than the culture media (Wedler et al. 1989; Aschner et al. 1992). The precise transporter(s) for Mn^{2+} into astrocytes is/are unknown. In both reports, saturable Mn^{2+} transport was found to be both competitively and noncompetitively inhibited by Ca^{2+}. Extracellular Mn^{2+} stimulates the rapid component of manganese efflux from rat astrocytes (Aschner et al. 1992). An astrocyte specific manganoprotein critical for ammonia metabolism, glutamine synthetase (GS), is generally thought to account for about 80% of the total brain manganese (Wedler and Denman 1984). Areas of high astrocyte density include the hypothalamus and hippocampus, with low astrocyte density in the cerebral cortex, neostriatum, midbrain, medulla oblongata, and cerebellum (Savchenko et al. 2000). The relatively high influx rates for Mn^{2+} (hippocampus) and manganese-citrate (thalamus/hypothalamus) support a correlation, but the high influx despite low astrocyte density in cortical regions for Mn^{2+}, manganese-citrate, and manganese-Tf and in the cerebellum for manganese-citrate decreases the strength of the putative correlation (Yokel and Crossgrove 2004).

Within the brain parenchyma, the Tf receptor is present in **neurons** at only 10–20% of the level found in capillary endothelia (Connor 1994; Moos 1996). DMT1 is present in high densities in the pyramidal and granule cells of the hippocampus, cerebellar granule cells, the preoptic nucleus and pyramidal cells of the piriform cortex (Gunshin et al. 1997). There is the possibility that this transporter is involved in the uptake of manganese ion and/or low molecular weight ligand-bound manganese by neurons.

In excitable cells such as neurons, excessive Mn^{2+} can be incorporated by L-type voltage gated calcium channels. This was verified by utilizing the drug diltiazem or verapamil which prevents the uptake of Mn^{2+} into cells by blocking calcium channels. This has been verified in the brain as well as the heart (Narita et al. 1990; Simpson et al. 1995; Du et al. 2001; Pautler and Koretsky 2002). Additional support comes from the accumulation of Mn^{2+} in specific brain areas that contain neuronal populations with high spontaneous activity. For instance after a focal injection in the hippocampal region of mice, specific Mn^{2+} enhancement in the dentate–CA3 region as opposed to the CA1–subiculum subfields seems to reflect the local functional activity in relation to intrahippocampal processing (Watanabe et al. 2004). Extracellular recordings (Cohen and Miles 2000) detected spontaneous action potentials from CA3 but not CA1 pyramidal cell populations, while single-channel recordings (Fisher et al. 1990) indicated that low-voltage activated calcium channels are particularly abundant on pyramidal neurons in CA3 but not CA1. Important sensory areas in rodents also provide evidence for local activity induced uptake of Mn^{2+}. The Mn^{2+}-induced MRI signal enhancements in the olfactory and auditory systems of mice after systemic administration are in agreement with 2-deoxyglucose autoradiograms of rat brain, indicating pronounced activity in structures of the olfactory system (Schwartz and Sharp 1978; Astic and Saucier 1981) as well as in the inferior colliculus (Sokoloff et al. 1977; Nelson et al. 1980; Duncan et al. 1999).

4.5.3.3 Intraneuronal Redistribution and Transsynaptic Transport of Manganese

The subcellular distribution of manganese has been recently studied in the basal ganglia of rats (Morello et al. 2008) and in cultured brain cells (Kalia et al. 2008).

In normal conditions Mn^{2+} sub-cellular localization and relative content were the same in neurons and astrocytes of rat striatum and globus pallidus: the Mn^{2+} level was highest in the heterochromatin and in the nucleolus, intermediate in the cytoplasm and lowest in the mitochondria. After chronic treatment, while no utrastructural damage was detected in the neurons and glia cells, the largest rate of Mn^{2+} increase was noted in the mitochondria of astrocytes (+700%), an intermediate rate in the mitochondria of neurons (+200%), and the lowest rate in the nuclei (+100%) of neurons and astrocytes; the Mn^{2+} level in the cytoplasm appeared unchanged (Morello et al. 2008).

These new findings are somewhat surprising with regard to earlier work where most of the Mn^{2+} was associated with superoxide dismutase (Inagaki et al. 1991; Medina et al. 1996), known as MnSOD or SOD2, located in the mitochondria of neurons, or with glutamine synthease, specifically located in the cytoplasm of astrocytes (Norenberg 1979; Wedler and Denman 1984).

After focal injections into the brain it has been demonstrated that fast axonal transport of Mn^{2+} occurs (Sloot and Gramsbergen 1994; Pautler et al. 1998; Takeda et al. 1998b; Van der Linden et al. 2002). Mammalian axons exhibit two major anterograde transport processes with a differential speed. **Slow axonal transport** refers to velocities of 0.01–0.33 mm/h, while the **fast axonal transport** yields velocities in the range of 2–16 mm h^{-1} (Elluru et al. 1995). Since mitochondria are transported via a slow axonal transport mechanism (Grafstein and Forman 1980), it has been concluded that the observed fast axonal transport of Mn^{2+} occurs within vesicles along microtubules via an energy requiring mechanism. This has been demonstrated with the use of the microtubule-disrupting drug, colchicine. Upon disruption of the microtubules, the transport of the Mn^{2+} was halted (Sloot and Gramsbergen 1994; Takeda et al. 1998b; Pautler and Koretsky 2002). Interestingly, although the speed of fast axonal transport is constant, the amount of vesicles that can be transported per unit of time can change according to the neuronal activity (Grafstein and Forman 1980). It was shown in mice lacking the kinesin light chain 1 gene (Bearer et al. 2007), that the fast component of anterograde transport of Mn^{2+}

is affected by conventional kinesin, but also that kinesin is not essential for neuronal transport of Mn^{2+} in the optic tract.

It is likely that Mn^{2+} is sequestered in the endoplasmatic reticulum after neuronal uptake, where it is packaged in vesicles for transport. This has been verified utilizing subcellular fractionation obtained through sucrose gradient centrifugation and photometric sensitive Mn^{2+} assays on subcellular fractions of olfactory bulbs exposed to Mn^{2+} (Pautler and Koretsky 2002). After transport along the microtubules to the synaptic cleft, the Mn^{2+} was then released at the synaptic cleft and taken up by the next neuron in the circuit (Gallez et al. 1998; Takeda et al. 1998b).

Quite interestingly, it has been shown that neurons that have been pre-loaded with Mn^{2+} co-release the ion with glutamate upon stimulation, indicating the possibility that Mn^{2+} is transported within synaptic vesicles (Takeda et al. 1998a). After release in the synaptic cleft Mn^{2+} could then be incorporated in the post-synaptic cells by ligand-gated calcium channels such as NMDA receptors. Fluorescence quenching techniques showed that Mn^{2+} can enter neurons through NMDA receptors (Simpson et al. 1995; Kannurpatti et al. 2000). A pharmacologic disruption of normal synaptic transmission by blocking the NMDA-type glutamate receptors with the antagonist APV (DL-2-amino-5-phosphonovaleric acid) caused a depression of the otherwise pronounced enhancement in the dentate–CA3 region of the dorsal hippocampus by APV on a 3D MRI after 18 h after injection of Mn^{2+}(Watanabe et al. 2004). While Mn^{2+} uptake and transport within the neurons of the visual system occurs independent of electrical activity, Mn^{2+} is not transmitted efficiently across synapses in the absence of electrical activity (Bearer et al. 2007).

At least as important is the fact that manganese is truly transported transsynaptically and not only transneuronally. This is suggested by the observed distribution of Mn^{2+} along the visual system of monkeys after an unilateral injection in the eye (Murayama et al. 2006).

4.5.3.4 Manganese Might Influence Excitable Tissue Activity

Manganese has been found to stimulate adenylate cyclase activity in the brain and other tissues of the body. This is of importance because cyclic-AMP plays a regulatory role in the action of several brain neurotransmitters, by acting as a second messenger within cells in transmitting the messenger hormone (Ashton 1980). Several researchers have demonstrated that manganese influences synaptic neurotransmission at high doses (~mM). Narita et al. (Narita et al. 1990) suggest that Mn^{2+} can enter nerve terminals through the voltage-dependent calcium channel during action potentials in frog motor nerve, enhancing the release of neurotransmitters. Drapeau and Nachshen (1984) also demonstrate that Mn^{2+} permeates presynaptic voltage-dependent calcium channels and induces dopamine release from depolarized nerve terminals. In the absence of extracellular Ca^{2+}, Mn^{2+} induces a long-lasting potentiation of acetylcholine release from cardiac parasympathetic nerve terminals following tetanic nerve stimulation (Kita et al. 1981). These findings indicate that Mn^{2+} can either substitute for Ca^{2+} in the exocytotic process, or induce the release of Ca^{2+} from intracellular stores, possibly the endoplasmic reticulum. Manganese also causes the activation of glutamate-gated cation channels, e.g. N-methyl-aspartate (NMDA) receptor, which contributes to neuronal manganese induced neurodegeneration (Brouillet et al. 1993). Calcium channels in the heart muscle can be blocked by Mn^{2+} and thereby affect cardiac function. However, a high concentration of Mn^{2+} (1 mM) is required to block the calcium channels (Ramos et al. 1996). Manganese also blocks voltage dependent calcium channels and nerve-evoked neurotransmitter release *in vitro*, although the effects of manganese on the voltage-dependent calcium channels and the release of neurotransmitters are controversial (Takeda 2004). After intravenous infusion of 176 mg kg^{-1} $MnCl_2$ brain Mn^{2+} concentrations reach a value of 0.11 mM after 24 h, which declined to 0.03 mM after 35 days (Chuang et al. 2009a).

Note that in most manganese tract tracing studies using focal brain injections the injected concentrations range from 50 to 800 mM. Sixteen nanomoles of $MnCl_2$ appears to be the upper limit for neuronal cell death and 8 nmol for astrogliosis in an approximate tissue volume of 8–11 μL (Canals et al. 2008). When a 2 mM $MnCl_2$ solution was injected via an osmotic pump (constant flow rate of 0.25 μL/h) a total amount of 24 nmol could be injected without inducing neuro- or gliotoxic effects.

4.5.4 Differential Neuronal Tissue Contrast Through MEMRI

It has been demonstrated in mice and rats that an intraperitoneal (i.p.), intravenous (i.v.) or subcutaneous (s.c.) injection of $MnCl_2$ leads to unique MRI contrast revealing the **neuroarchitecture of the brain** (Natt et al. 2002; Watanabe et al. 2002; Aoki et al. 2004; Kuo et al. 2005; Lee et al. 2005). Systemic injected manganese will ultimately reach the brain tissue through uptake into the choroid plexus, which then spreads to the cerebral spinal fluid (CSF) in ventricles and periventricular tissues within 2 h after administration (Aoki et al. 2004). Systemically administered manganese to reveals fine anatomical detail in T_1-weighted MR images of **neonatal mouse brains** (Wadghiri et al. 2004). MEMRI showed to be particularly useful for analyzing early postnatal patterning of the mouse cerebellum correlating qualitatively with granule cell density in the developing cerebellum of different mutant mice. The same group (Deans et al. 2008) expanded this method to **in utero studies of the embryonic CNS** in combination with respiratory gating decreasing motion artifacts and providing an effective means for in utero analysis of CNS phenotypes in a variety of mouse mutants.

4.5.5 Studying Functional Neural Circuitries with MEMRI

Besides tissue contrast enhancement more specific differential applications of MEMRI have been demonstrated in the animal brain and their differences are based on the injection route of Mn^{2+} and the information that is provided. First, due to the fact that manganese ions can enter excitable cells via voltage-gated calcium channels, protocols have been devised that foster accumulation of Mn^{2+} in active areas of the brain. This technique has been referred to as activation induced MEMRI. The second use of MEMRI focuses on the anterograde axonal transport of Mn^{2+}, the so called MEMRI tract tracing which can also be activity dependent.

4.5.5.1 Activation-Induced Manganese-Dependent MRI (AIM)MRI

In **AIMMRI**, $MnCl_2$ is infused intra-arterially after the blood–brain barrier (BBB) is opened with a hyperosmolar agent (Lin and Koretsky 1997). Upon functional stimulation of the brain, Mn^{2+} accumulates in the active region(s) by entering active cells through voltage-gated calcium channels, causing local signal increases in T_1-weighted images. The contrast of AIMMRI however depends strongly on the depth of anaesthesia, and the low levels used in somatosensory stimulation studies can lead to significant non specific accumulation of manganese ion throughout the brain. The method has been used for several other applications such as mapping of the whisker barrels in the rat upon tactile-sense-evoked activity (Weng et al. 2007) and activity of hypothalamic nuclei involved in central osmotic regulation (Morita et al. 2002). The method was also used to explore more complex neural circuitries associated with cognition and fear upon exposing rats to novel/arousing or fear-inducing odors (Chen et al. 2007).

In a **dynamic AIM (DAIM)** paradigm (Aoki et al. 2002), sequential MR scans are acquired during $MnCl_2$ infusion, prior to and following functional stimulation of the brain. Stimulation-specific functional maps are produced using time-course analysis. This method was tested during glutamate administration and electric stimulation of the rat forepaw. It was shown that DAIM maps are better confined to the specific region of brain activated by somatosensory stimulation as compared to AIMMRI. This dynamic version of AIMMRI was used for investigating the effects of pharmacological agents by studying cocaine-induced brain activation in rats (Lu et al. 2007). Dose-dependent, region-specific MEMRI signals were seen mostly in cortical and subcortical mesocorticolimbic structures.

The (D)AIMMRI associated systemic administration of Mn^{2+} in conjunction with the disruption of the blood–brain barrier (BBB) has undoubtedly procedure-related morbidity/mortality and maybe brain activity measured by these methods can diverge from genuine physiological responses. The group of J Bell published a set of studies in which systemic administration of Mn^{2+} without compromising the BBB integrity is applied to detect hypothalamic function associated with feeding and fasting (Kuo et al. 2006), temporal effects of appetite modulating agents on hypothalamic activity in fasted mice (Parkinson et al. 2009) and activity in hypothalamic regions involved in appetite control in mice with different diets (So et al. 2007).

Another approach to obtain information on the activity of neurons without the need to disrupt the BBB pharmacologically is to inject **Mn^{2+} straight into the regions of interest**. The amount of Mn^{2+} ions that entered excitable cells in the injection area is then used as a measure of the activity of that brain region.

The activity of the orbitofrontal cortex (OFC) was investigated, as part of the limbic system involved in the reinforcing effects of drugs of abuse, during opiate administration and withdrawal periods (Sun et al. 2006). $MnCl_2$ was stereotactically injected into the right OFC and 5–6 h later rats were anesthetized, perfused with 10% formalin containing 1% potassium ferrocyanide in order to wash out extracellular Mn^{2+}, and the MRI signal intensity was used as measure for intracellular Mn^{2+} and hence a sign of neuronal activity.

4.5.5.2 Tract Tracing with MEMRI

Tract Tracing

For tract tracing with Mn^{2+} one needs to target the area (inject $MnCl_2$) from which the axonal projections are presumed to start to trace the circuit of interest. In order for the Mn^{2+} to reach the target, the BBB should be circumvented by either using nostril exposure or intravitreal injections of $MnCl_2$ to trace neuronal connections *in vivo* but this approach is limited to the olfactory (Pautler et al. 1998; Pautler and Koretsky 2002) and visual pathways (Watanabe et al. 2001; Thuen et al. 2005). Other brain regions than the olfactory bulb and the visual system can be reached by injecting $MnCl_2$ immediately in the brain. Small focal injections of manganese ions within the central nervous system combined with *in vivo* high resolution MRI allow to delineate neuronal tracts originating from the site of injection and using in some cases the trans-synaptic transport capacity of manganese (Pautler 2004). This was accomplished in the murine striatum and amygdale (Pautler et al. 2003), in the somatosensory cortex (Allegrini and Wiessner 2003; Leergaard et al. 2003) the ventral tegmental area projections to the forebrain (Li et al. 2009) and the spinal cord of the rat (Bilgen et al. 2005, 2006), the basal ganglia of the monkeys in combination with simultaneous injection of the common tract tracing agent, wheat germ agglutinin conjugated to horseradish peroxidase (WGA-HRP) (Saleem et al. 2002) and in the brain of songbirds (Tindemans et al. 2006).

Activity Dependent Tract Tracing

One step further is to perform activity dependent tract tracing as demonstrated in the olfactory system (Pautler and Koretsky 2002). These authors performed a study in which they traced Mn^{2+} from sites of activation in the olfactory epithelium to the olfactory bulb thereby localizing regions within the olfactory bulb that respond to a particular odour. To that end a common odour stimulus (amyl-acetate) and the odour of mice urine, which is a complicated mixture of common odorants and pheromones, were presented to the nostrils of the mice in combination with aerosoled $MnCl_2$. Using the time course of the MRI signal, Koretsky's team generated maps of Mn^{2+} accumulation in the olfactory bulb for both glomerular and mitral cell layers (Chuang et al. 2009b). Results demonstrated that overlapping yet distinct enhancement patterns were produced by exposure to either octanal, acetophenone, or carvone. Notably, areas of Mn^{2+} accumulation in the mitral cell layer were similar to those in the glomerular layer consistent with neural information that passes from specific olfactory sensory neurons to specific mitral cells. Finally, by correlating specific Mn^{2+} signal peaks to genetically labeled glomeruli that are known to be activated by the odorant octanal, these authors showed that MEMRI maps can be resolved at the level of individual glomeruli.

Another relevant example of this approach was accomplished in songbirds to investigate neuroplastic changes in the songbird brain (Van der Linden et al. 2002; Van Meir et al. 2004, 2006) or the impact of hearing songs on specific neuronal populations (Tindemans et al. 2003). In these studies the dynamics of axonal manganese transport were monitored as manganese induced signal intensity (SI) enhancement in the projected areas and translated into a Hill plot (function describing a sigmoid curve) (Fig. 4.15). This so called **Dynamic Manganese Enhanced [DME]MRI** can then be used as a quantitative tool to monitor the activity of the projecting neurons in the injection area. This has resulted in the discovery of gender linked (Van der Linden et al. 2002) or testosterone induced (Van Meir et al. 2004) changes in activity and functional connectivity in two circuits of the song control system of songbirds. These studies used the DMEMRI method to segregate the contribution of different neuronal populations from the activated brain region. In some cases, to follow the dynamics of Mn^{2+} uptake, Mn^{2+} had to be injected through a **permanent non magnetic plastic cannula** implanted in HVC which allowed repeated injections of $MnCl_2$ at exactly the same location. The plastic cannullae was connected to a very long and narrow tubing that allowed stereotaxic injection of very small volumes to be made while the bird is in the magnet so that images could be acquired before, during and after Mn^{2+} injection. In this way DMEMRI can be used to investigate fast changes in activity of specific types of neurons that have been traditionally studied by single cell electrophysiology (Tindemans et al. 2003).

Remodelling of Neuronal Circuitrics

MEMRI was proven to be a very useful method to examine network plasticity in songbirds (for review see (Van der Linden et al. 2004) and de- and regeneration in several pathologies such as stroke (van der Zijden et al. 2007) and Parkinson (Pelled et al. 2007) and disruptions in multisynaptic pathway as the consequences of repeated exposure to psychostimulants (Hsu et al. 2008). A peculiar example of neural circuitry remodelling is axonal plasticity which is recognized to be part of several different pathologic processes and the consequence of hyperactivity during seizures. Sprouting of granule cell axons or mossy fibers is one of the most consistent neuropathologic findings in the hippocampus of animals or humans with temporal lobe epilepsy (Mathern et al. 1995), providing one of the most extensively characterized examples of activity-induced axonal plasticity in the brain. MEMRI was used to characterize this activity-dependent plasticity in the mossy fiber pathway after intraperitoneal kainic acid injection (Nairismagi et al. 2006; Immonen et al. 2008). Enhancement of the MEMRI signal in the dentate gyrus and the CA3 subregion of the hippocampus was evident 3–5 days after injection of $MnCl_2$ into

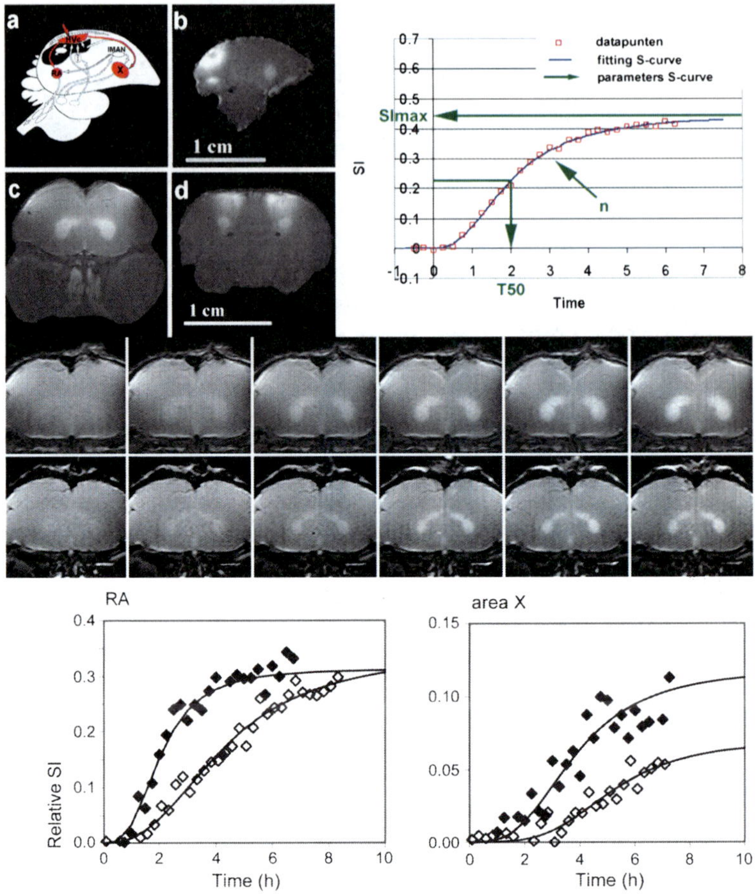

Fig. 4.15 *Top left*: (**a**) Schematic overview of the adult songbird brain showing the song control nuclei (SCN) and their connections in the telencephalon. The red arrow represents the intracerebral injection spot of Manganese. The target is the HVC which gives rise to two pathways which originate from distinct cell populations within HVC. (**b**) Sagittal and (**c**), (**d**) coronal *in vivo* manganese-enhanced MRI of a male starling brain obtained 6 h after $MnCl_2$ injection into HVC showing signal enhancement at the level of area X (**c**) and RA (**d**) (Reprinted from Van der Linden et al. (2002), with permission). *Top right*: Schematic representation of the changes in relative manganese-enhanced signal intensity observed in one target of HVC (RA or area X) as a function of time (up to 6 h) after $MnCl_2$ injection in HVC (0 point). Data are plotted as a function of time and fitted by nonlinear regression to a sigmoid curve. This curve fully defines the kinetics of manganese accumulation in the target using three parameters: the maximal signal intensity (SImax), the time required to reach 50% of this maximum (T50) and the n coefficient that describes the shape of the curve. *Middle panel*: Manganese accumulation in one of the HVC targets, area X, as observed by manganese enhanced MRI of starling brains, displayed as a function of time (*left* 1 h, *right* 6 h) after $MnCl_2$ injection in HVC (Reprinted from Van der Linden et al. (2002), with permission). *Bottom*: Relative Manganese Enhanced Signal Intensity (relative SI) changes in RA (*left*) and area X (*right*) as a function of time after $MnCl_2$ injection in the HVC in a representative canary. *Thick lines* and *full squares* represent data collected when the canary was allowed to listen to canary songs and thin lines and open squares indicate the control situation, without song stimulation, obtained in the same bird. The individual data acquired every 15 min are shown by squares, and the lines through these data points correspond to the sigmoid curves fitted by nonlinear regression as outlined in the text (Reprinted from Tindemans et al. (2003), with permission)

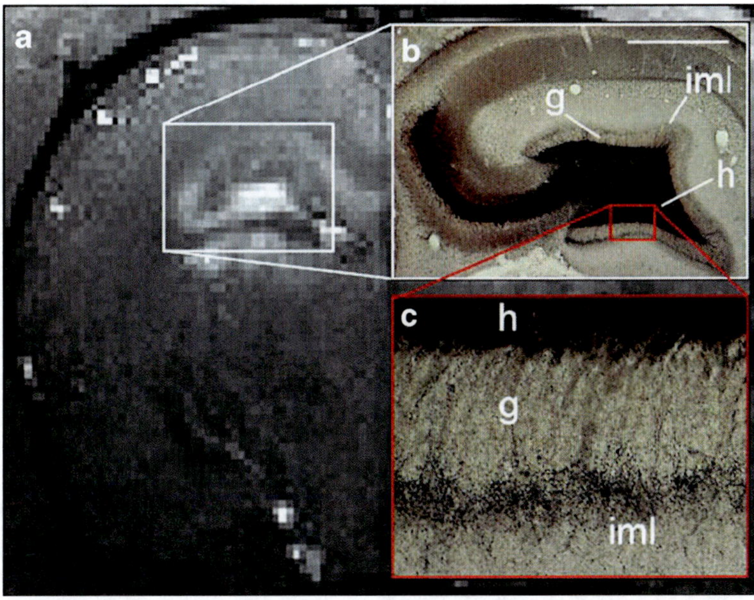

Fig. 4.16 Manganese enhancement in T_1-weighted MRI (**a**) is most pronounced in the same area that appears dark in Timm staining (**b**). Panel **c** is an enlarged field from the inner molecular layer (iml) of panel B showing sprouted mossy fibers in greater detail. The animal was killed 20 days after KA injection. Abbreviations; *g* granule cell layer, *h* hilus. Length of the scale bar in panel B is 1 mm (Reprinted from Nairismagi et al. (2006), with permission)

the entorhinal cortex both in control and kainic acid injected rats. An increase in the number of Mn^{2+} enhanced pixels in the dentate gyrus and CA3 subfield of affected rats correlated ($P<0.05$) with histologically verified mossy fiber sprouting. These data demonstrate that MEMRI can be used to detect specific changes at the cellular level during activity-dependent plasticity *in vivo*. (Fig. 4.16)

Quantifying Neuronal Connectivity

Recently the term **connectivity index (CnI)** was established as a quantitative measure of neuronal connectivity obtained through MEMRI (Canals et al. 2008). The administration strategy was optimized to significantly improve the resolution and at the same time minimize the toxicity. To that end very low concentrations of Mn^{2+} were continuously infused into the target area using osmotic pumps coupled to chronically implanted brain cannulae. Subsequently, connectivity indices (CnI) were obtained based on Mn^{2+} transport in corticofugal somatosensory versus motor pathways. This quantitatively revealed cortical heterogeneities in interhemispheric communication between these two pathways. The Connectivity index (CnI) was calculating through the difference in signal intensity between the baseline images and the images 10 h after Mn^{2+} application in ROIs selected on the basis of the statistical maps (Fig. 4.17).

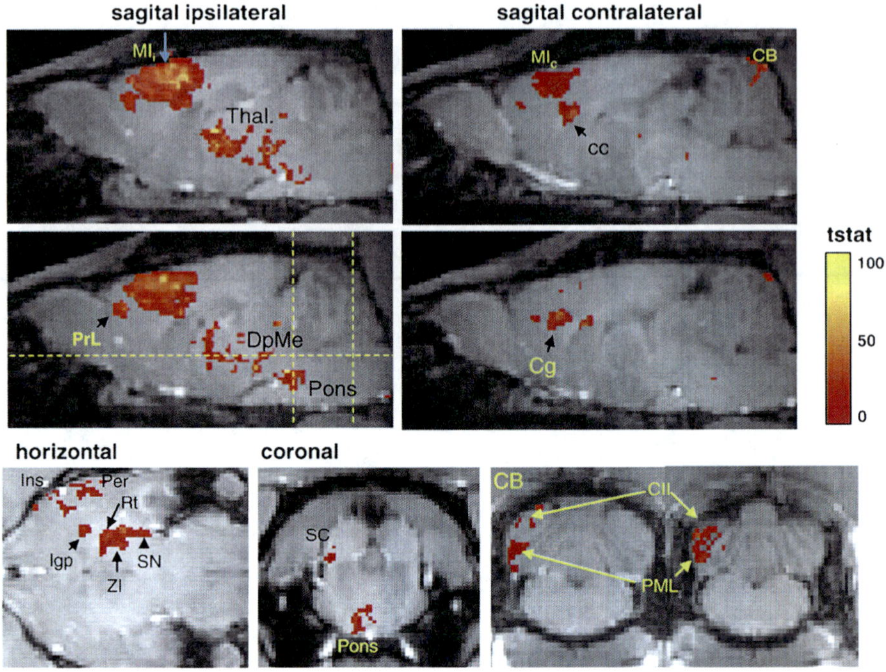

Fig. 4.17 Cortical Mn^{2+} application with microsmotic pumps coupled to chronically implanted cannulae. Sagittal, horizontal and coronal MR images with overlaid thresholded statistical maps (unpaired *t* test comparing the baseline and 48 h scans; $p < 10^{-4}$) illustrating the distribution of Mn^{2+} after 2 days of continuous infusion (10 mM $MnCl_2$ in buffered saline, 0.25 μL/h). The *blue arrow* (*top panel*) indicates the position of the implanted cannula and the *stippled yellow lines* the position of the horizontal and coronal slices in the lower part. Note the extensive connectivity mapping (including polysynaptic) obtained with this method. Abbreviations: *M1c* contralateral Motor Cortex 1, *Cg* cingulate cortex, *PrL* prelimbic cortex, *Ins* insular cortex, *Per* perirhinal cortex, *ZI* zona incerta, *SC* superior colliculus (Reprinted from Canals et al. (2008), with permission)

Manganese Transfer Index, Manganese Transfer Rates and Axonal Transport Rates: Quantitative Readouts from MEMRI

Manganese-enhanced MRI (MEMRI) can also be used to evaluate *in vivo* transneuronal efficiency of manganese ion (substitute for Ca^{2+}) movement, the as such defined '**manganese transfer index or MTI**' as a means to assess overall changes in neuronal function (Serrano et al. 2008) (Fig. 4.18). The MTI was determined in mice upon treatment of pharmacological agents (isoflurane and memantine) and in different APP tg Mice. In all cases Mn^{2+} was administered through the nostrils and a decrease in the MTI value was observed demonstrating the potential utility of the MTI value as an *in vivo* index for the detection of changes in neuronal function in animal models of human disease. Another study (Murayama et al. 2006) used unilateral intravitreal injections in Rhesus monkeys to show that Mn^{2+} not only transpasses 4 synapses but is also sufficiently synapse specific to permit visualization of the lamina of the dorsal lateral geniculate nucleus (dLGN). Interestingly, the **transfer rate of**

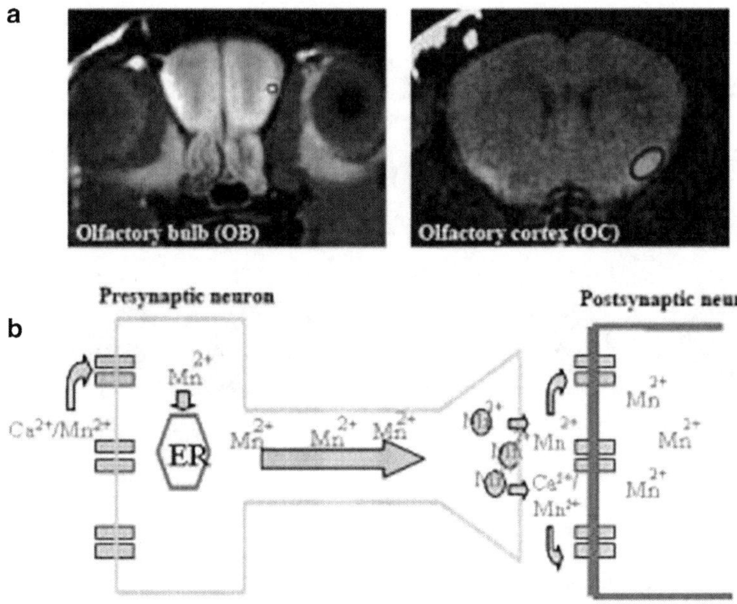

$$MTI = \frac{\text{signal intensity (postsynaptic neuron, eg. OC)}}{\text{signal intensity (presynaptic neuron, eg. OB)}}$$

Fig. 4.18 (**a**) MEMRI at the level of the mouse brain after intranasal lavage with $MnCl_2$. The image demonstrates the region of interest (ROI) selected in the manganese enhanced areas of the olfactory bulb (presynaptic) and olfactory cortex (postsynaptic). (**b**) Cartoon explaining the concept of manganese transfer index (MTI). The MTI value assesses the transneuronal efficiency of Mn_2 ion and is defined as the ratio of the signal intensity in the postsynaptic neuron in relation to the signal intensity in the presynaptic neuron. MTI signal intensity (postsynaptic neuron)/signal intensity (presynaptic neuron) (Reprinted from Serrano et al. (2008), with permission)

Mn^{2+} reflected the well-known axonal size differences between the parvocellular and magnocellular layers of dLGN.

The group of Pautler, one of the pioniers in MEMRI, further developed the Manganese Enhanced MRI (MEMRI) methodology to assess ***in vivo*** **axonal transport rates** (Smith et al. 2007). The rational behind this is the fact that Mn^{2+} is transported anterogradely in vesicles via axonal transport. Modifications in the axonal transport rate as in the case of axonopathy, which is a pronounced attribute of many neurodegenerative diseases, could thus be measured with MEMRI. With their efforts they established an *in vivo* readout for axonal transport rates in animal models for neurological diseases.

4.5.5.3 Study Neural Substrate of Awake Behavior

Besides providing a superior image contrast in the brain, Mn^{2+} injected ip will enhance activated areas enabling to study brain activity during motion accompanied behaviour. Brain activation in ***awake*** **small animals can be monitored by performing**

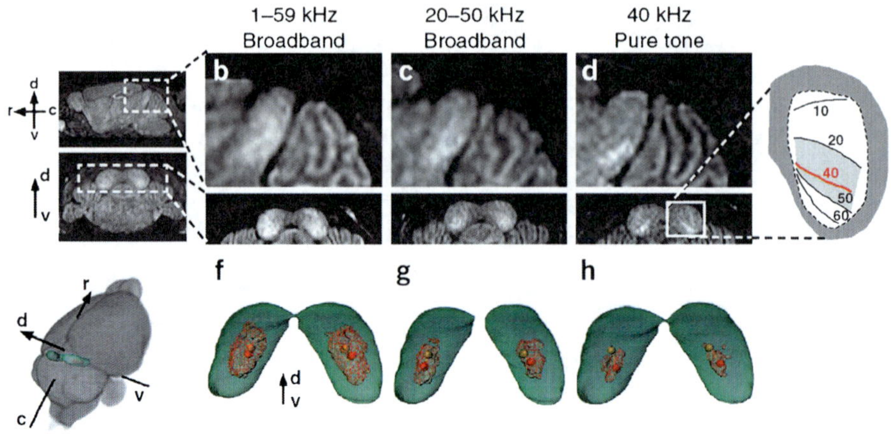

Fig. 4.19 MEMRI was used to map the tonotopic organization of the mouse inferior collilulus. (**a**) Sagittal (*upper*) and coronal (*lower*) images of 21-day-old mice with the inferior collilulus showing obvious differences in mice exposed to defined stimuli. (**b**) After broadband (1–59 kHz) stimulation, enhancement covered most of the rostral–caudal (r-c), ventral–dorsal (v-d) extent of the central nucleus of the inferior collilulus. (**c**) After high-frequency broadband (20–50 kHz) stimulation, enhancement was more restricted to the ventral–caudal region. (**d**) After 40 kHz pure-tone stimulation, enhancement was restricted to an isofrequency band in excellent agreement with electrophysiological maps. (**e**) Averaged, co-registered images (n¼8) were used to extract whole brain (*grey*) and inferior collilulus (*green*) and to generate 3D maps of MEMRI inferior collilulus enhancement (*red*) after stimulation with 1–59 kHz (**f**), 20–50 kHz (**g**) and 40 kHz (**h**) (Reprinted from (Yu et al. (2005a), with permission)

MRI *after* the presumed activity has occurred preceded by a systemic injection of manganese. This minimally invasive approach became feasible due to the low clearance rate from the cell. MEMRI becomes then quite homologue to histological discrimination of IEG expression (cfos) as it highlights areas with prior activity but probably harbours the same drawbacks in terms of specificity. This method has been proven capable of providing a sensitive and effective method for mapping the mouse auditory brainstem (Yu et al. 2005b, 2008). MEMRI was used to map regions of accumulated sound-evoked activity in awake, normally behaving mice resulting in high resolution (100 μm) brain mapping of the tonotopic organization of the mouse inferior colliculus (Fig. 4.19). A similar strategy was applied to study cortical responses to visual stimulation in awake and free-moving rats (Bissig and Berkowitz 2009). Sprague–Dawley rats, with or without systemic injection of $MnCl_2$, were maintained for 8 h in either a visually stimulating environment or darkness. To identify vision-dependent changes in cortical activity, animals were anesthetized and cortices were examined by 3D RARE MEMRI. They could discern layer-specific stimulus-dependant non-invasive MEMRI readout after encoding activity in awake and free moving rats.

4.5.6 *'Non Neuronal Activity' but 'Neuropathology Related' Mn^{2+} Uptake*

Systemic injections of Mn^{2+} can be used to alter the MRI contrast in the brain or to enhance activated brain areas as outlined before. However it was also demonstrated that MEMRI reveals areas where Mn^{2+} accumulation is enhanced as a result of pathology (Chan et al. 2008; Yang et al. 2008). Neonatal hypoxic-ischemic cerebral injury, photothrombotic cortical injury or transient middle cerebral artery occlusion in rats were monitored with and without systemic $MnCl_2$ administration. MEMRI and concomitant histology was performed to unravel enzymatic activities of Mn-superoxide dismutase (Mn-SOD) and glutamine synthetase (GS), which are Mn-binding enzymes against the oxidative stress and glutamate excitotoxicity in neurodegeneration and of which the location could be observed *in vivo* by MEMRI.

In another study (Haapanen et al. 2007), neuropathological CTSD (Cathepsin D) deficient mice were investigated with different MRI methods including MEMRI. The pathology of these mice is characterized by selective neuronal degeneration, gliosis and accumulation of autofluorescent proteinaceous storage material in neurons. MEMRI and histological stainings revealed that the hyperintense signal areas in MEMRI matched perfectly with areas of microglial activation in the brains of these mice at the terminal disease stage. Also in Hypoxic-ischemic injured neonatal rat brain, areas with increased Manganese enhancement corresponded best with areas with high concentrations of activated microglia (Wideroe et al. 2009).

4.6 Clinical Indication for the Currently Used MRI Contrast Agents

MRI agents are still under intense development and novel, highly innovative applications are under scrutiny at the pre-clinical level. It is expected that the efforts to develop markedly more sensitive agents will result into systems suitable for Molecular Imaging investigations, thus making MRI a competitive modality in this arena.

In this concluding section we summarize some of the major clinical applications of the currently used MRI agents.

For soluble systems their biodistribution is determined by properties like size, hydrophobicity, lipophilicity, charge, etc. For particulates their final targets appear essentially cells endowed with macrophagic activity. Also for particulate-based agents size is important and markedly affect the resulting clinical indication, as shown by Endorem (diameter > 50 μm) and Sinerem (diameter < 50 μm). The former agent is typically taken up by macrophages of the RES (most often by Kupffer cells), the latter one is the best system for the visualization of lymph nodes. Thanks to their smaller size, intravenously injected USPIOs are not entrapped during their passage through the lungs, but extravasate in inflamed regions and can be transferred to

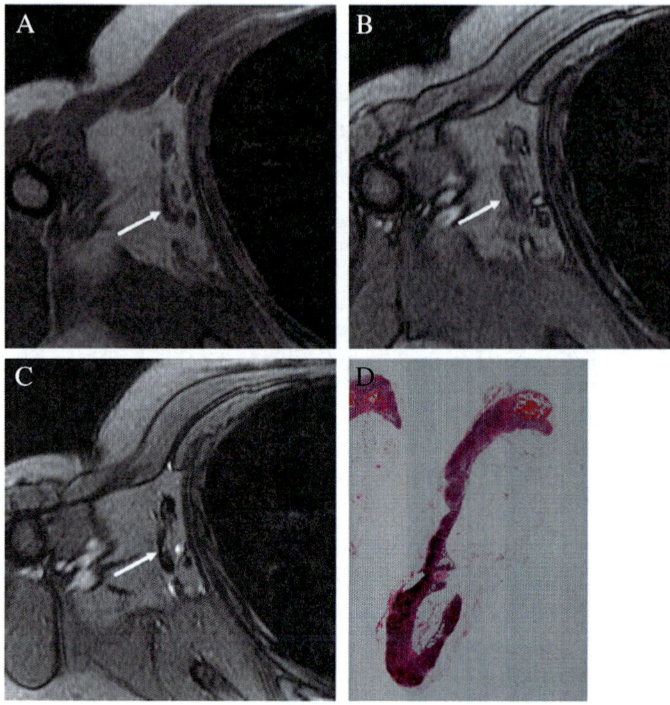

Fig. 4.20 Nonmetastatic lymph nodes on T_1- and T_2*-weighted USPIO images. (**a**) shows the T_1-weighted image. The short axis diameter of the lymph node (*arrow*) was 6.0 mm and was diagnosed as metastatic by conventional criteria; (**b**) the lymph nodes shows high signal intensity before the administration of USPIO (precontrast study); (**c**) the lymph node shows complete darkening after the administration of USPIO (postcontrast study), and was diagnosed as nonmetastatic pathologically in H and E staining (**d**) (Reprinted from Harada et al. (2007), with permission)

lymph nodes through the lymphatic system. Thus the invasion of lymph nodes by metastatic tumour cells is efficiently assessed by the decrease of iron oxide particles uptake. They darken the region homogenously (as a result of the effective T_2/T_2^* shortening) when the tumours cells (that do not possess macrophagic activity) have not yet spread out along the lymphatic system (Figs. 4.20 and 4.21).

As far as the small-sized, hydrophilic Gd^{3+} complexes (that are, by far, the most used MRI CAs) are concerned, their use deals mainly with an enhanced detection of metastases and small tumours and with an improved tumour classification (Caravan et al. 1999; Golman et al. 2003; Caruthers et al. 2006; Terreno et al. 2008). The latter application aims at pursuing the differentiation of vital tumour tissue from central necrosis and from surrounding edema. In particular in brain lesions it exploits the perfusion properties of these agents to assess impairments or absence of the blood–brain barrier.

Very important applications deal with the detection of isointense benign tumours like meningiomas and hamartomas as well as the delineation of tumour and edema in malignant tumours. Absence of enhancement may be of great value as it allows

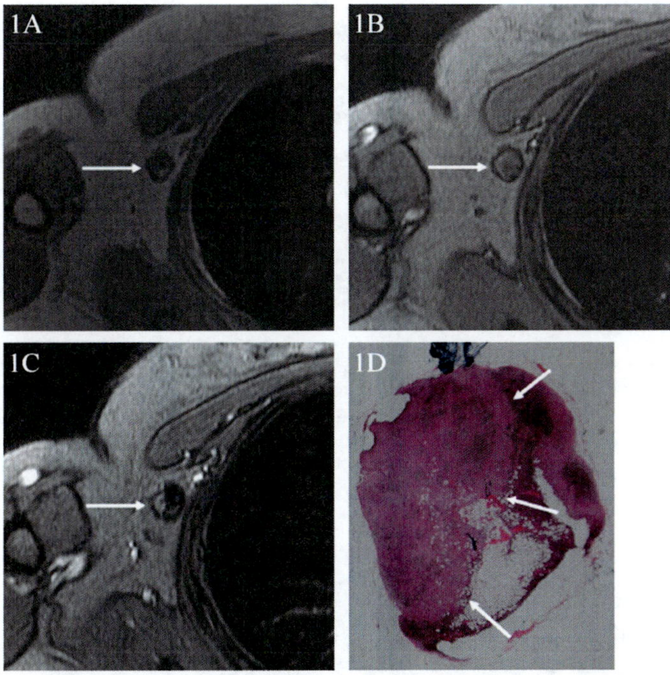

Fig. 4.21 Metastatic lymph nodes on T_1- and T_2*-weighted USPIO images. (**a**) shows the T_1-weighted image. The short axis diameter of the lymph node (*arrow*) was 10.5 mm and was diagnosed correctly as metastatic; (**b**) the lymph node (*arrow*) shows high signal intensity before the administration of USPIO (precontrast study); (**c**) after the administration of USPIO (postcontrast study), the lymph node (*arrow*) showed an irregular darkening pattern; (**d**) histologically confirmed metastatic lymph node. The arrows represents the metastatic areas of the lymph node (Reprinted from Harada et al. (2007), with permission)

to distinguish a low-grade astrocytoma at the cortical surface from a meningioma or small vascular white-matter lesions from metastases (Fig. 4.22).

Although the major diagnostic indication of low molecular weight Gd^{3+} agents deals with the central nervous system (CNS), many applications in other anatomical districts have been reported. Moreover, the simple introduction of an aromatic substituent on the outer surface of GdDTPA yields slightly more lipophilic systems that are excreted not only by kidneys but also via liver. Thus Gd-EOB-DTPA or GdBOPTA (Fig. 4.23) are commercialized as liver agents as they target hepatocytes and are (in part) excreted through the bilary canaliculii.

Interestingly another liver-specific agent, Mn-DPDP, enhance the signal from hepatocytes through a mechanism connected to the release of manganese ions (Fig. 4.24). The observed relaxation enhancement is very good, thus allowing the use of very low dose of Mn-DPDP (10 μmol kg^{-1}) and long lasting (probably because the excretion of Mn^{2+} ions from the hepatocytes follow slow homeostasis related processes).

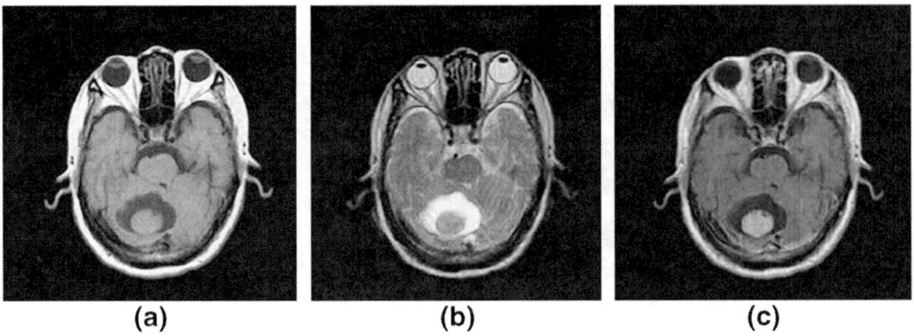

Fig. 4.22 (**a**) Unenhanced, axial, T_1-weighted; (**b**) T_2-weighted; and (**c**) contrast-enhanced, axial, T_1-weighted images of a 50-year-old woman with cystic meningioma of the tentorium of cerebellum. A round mass located inside the cyst showed iso-intensity on both T_1 and T_2-weighted images and moderate enhancement after the injection of Gd-DTPA contrast agents.The arrows represents the metastatic areas of the lymph node (Reprinted from Zhang et al. (2009), with permission)

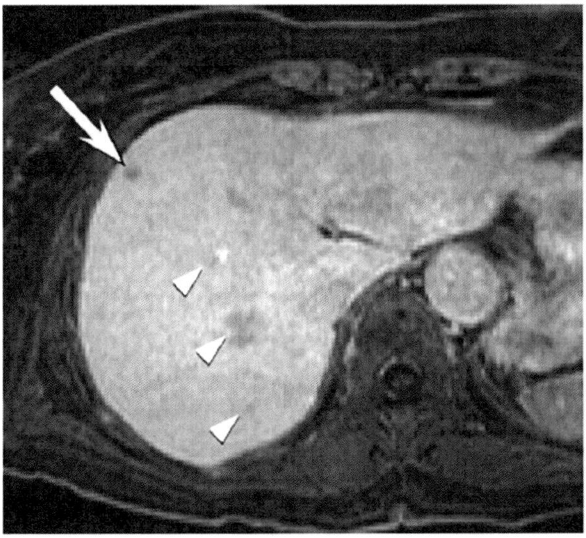

Fig. 4.23 Tiny liver metastases from colon carcinoma. Gadolinium BOPTA enhanced breath-hold fat-suppressed three-dimensional T_1-weighted gradient-recalled-echo MR image obtained 2 h after contrast administration (functional phase), showing intensely enhanced liver parenchyma and relatively hypointense subcentimetric metastasis that were not clearly revealed on gadolinium BOPTA-enhanced portal venous phase or unenhanced T_2-weighted MR images. Note that the other tiny hypointense areas were due to intrahepatic vessels (*arrowheads*) (Reprinted from Kanematsu et al. (2006), with permission)

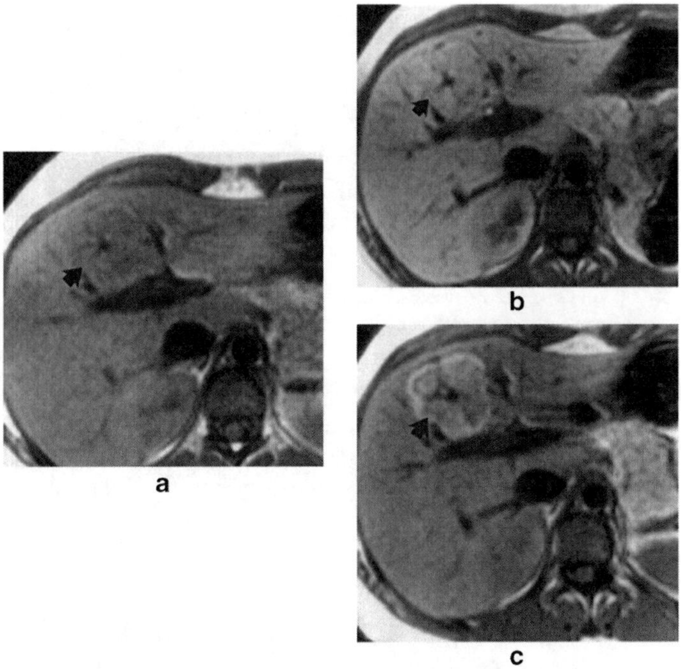

Fig. 4.24 A 35-year-old female patient with a biopsy-proven solitary FNH. The lesion is poorly visualized on (**a**) pre and (**b**) 20 min post MnDPDP T_1-weighted images (*arrow*). (**c**) On delayed T_1-weighted images at 24 h the mass becomes hyperintense to adiacent liver parenchyma improving the lesion conspicuity. A central scar and rim enhancement (*arrow*) are also demonstrated (Reprinted from King et al. (2002), with permission)

MRI-CAs are also useful in the acquisition of angiographic images. For this application, the major requirements are: (a) high longitudinal relaxivity to make T_1 as shorter as possible (and to use reduced doses) and a low transverse relaxivity to avoid spin dephasing effect; (b) prolonged plasma half-life in order to maintain their concentration (and therefore the relaxation enhancement) almost constant during the image acquisition. Different approaches have been tackled in the development of blood pool agents. The one used in the case of Vasovist (Schering) deals with the formation of supramolecular adducts with the circulating albumin (Fig. 4.25). Thus, although the complex maintains a relatively small size, the tight binding to serum albumin endows it with the characteristics of the large macromolecular systems (i.e. high relaxivity and long retention time in the blood circuit).

An alternative approach deals with the development of paramagnetic Gd^{3+} complexes characterized by a large hydrodynamic radius, so they cannot escape the blood compartment in the way usually possible for the corresponding small-sized agents.

Besides reporting on the angiogram of a patient, blood pool agents are under scrutiny as reporters in Dynamic Contrast Enhancements protocols (Fig. 4.26). These procedures deal with the tumours characterization in terms of the permeability of the neo-formed capillaries (neo-angiogenesis) by assessing the "wash in/wash

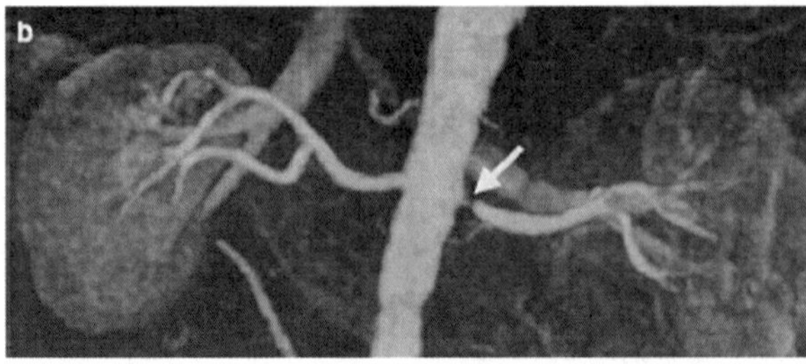

Fig. 4.25 Coronal views of gadofosveset-enhanced MRA, dynamic MIP showing a stenosis in the proximal left renal artery (*arrows*) (Reprinted from McGregor et al. (2008), with permission)

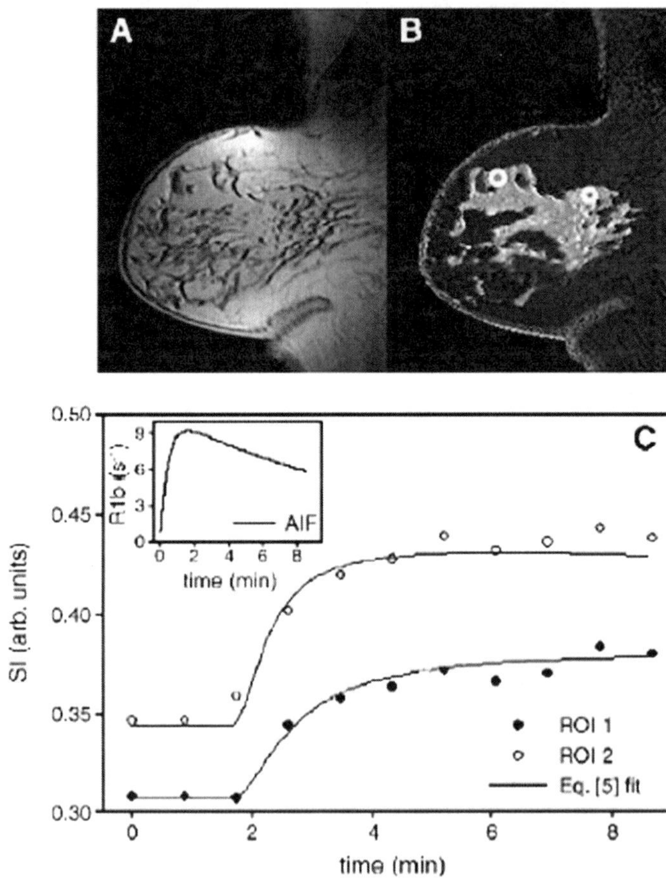

Fig. 4.26 (**a**) A 3.5-min postcontrast T_1-weighted sagittal image. The T_1 map (**b**) readily delineates the pathologic tissue from the healthy-appearing breast tissue. (**c**) Two enhancement curves and the corresponding Eq. 4.5 fit, taken from the two ROI labeled on (**b**). The leftmost *white circle* in (**b**) corresponds to the filled circles in (**c**), which were still enhancing at the end of the 8.5 min study, whereas the *open circles* have reached their maximum enhancement by the ~5 min mark (Reprinted from Yankeelov et al. (2007), with permission)

out" of a xenobiotic (like a MRI CA) from the tumour vasculature. This assessment is of outstanding importance in the monitoring of the therapeutic treatments of breast cancer.

In this chapter we have surveyed two major MRI fields, namely the basis for the design of contrast agents and the use of MEMRI. The two fields developed independently, the latter one being a relatively recent one. Both aim at exploiting the contrast enhancement to add physiological/functional information to the outstanding anatomical resolution of MR images. MEMRI relies on the relaxation enhancement brought about by the administration of Mn^{2+} ions, whereas the development of Gd-based contrast agents basically involves the use of chelating moieties as the release of the toxic metal ions is of primary concern for the clinical transaction of these diagnostic agents. As it is now MEMRI has no chance to be translated to humans though the outstanding achievements obtained in animal studies clearly call for more attention to explore new routes that may conjugate the huge potential of this approach with an acceptable safety profile of the administered Mn^{2+} ions. One may envisage topical administration of the Mn^{2+} payload that substantially limits the systemic toxicity and one may eventually supply vesicles loaded with Mn^{2+} ions that are selectively delivered at the site of interest. The paramagnetic payload may be released (possibly in a triggered mode) by the action of internal (enzyme, pH) or external (heat, ultrasounds) stimuli.

By now one has to note a huge development of the field of MRI contrast agents either in terms of already established clinical applications and even more in the acquisition of innovative proofs of concepts that show the tremendous potential associated to the use of properly designed chemicals to aid the medical diagnosis. Detection of small tumors takes a huge advantage by using small sized Gd- complexes as well as assessing the changes in vascular permeability is outstandingly important for an early evaluation of the therapeutic treatment. Moreover, invasion of lymph nodes by metastatic tumor cells can be accurately visualized by the decreased uptake of iron oxide particles. MR Angiography with contrast agents yields a definitive improvement in the accurate delineation of the blood circuit. Finally, a plethora of experiments has been reported to show how a proper design of the structure of the agent can make it responsive to a specific physico-chemical parameter of the biological milieu. Still much has to be done but unambiguous proofs of concepts have already been acquired that allow to state that, in a not-too-far future, contrast agents will allow to add accurate maps of the relevant physico-chemical parameters to the anatomical MR images.

An entirely different story is the use of the free Mn^{2+} ion in MEMRI that has recently gained field in experimental mouse and transgenic mouse research. It was shown from the many examples reported in this book chapter that it provides an efficient and powerful *in vivo* method for analyzing the brain of normal and genetically engineered mice in terms of assessing plasticity, activity, connectivity and functioning and this in a variety of circuits ranging from sensory up to cognitive networks. One of the biggest benefits of MEMRI which finds no comparison in any other method is however its potential as an *in vivo* non invasive tool to link behaviour, performed in a non restricted environment, with its neuronal substrate. Given the currently acknowledged ease and importance of behavioural phenotyping in neurodegenerative mice models to see whether they mimic the human pathology, we are quite close to inserting MEMRI into protocols for phenotyping the neural substrate of the observed modified behaviour. Its repeated applicability in the same specimen allows unravelling 'how the brain changes the behaviour' but also 'how the behaviour changes the brain'.

In summary, although MEMRI can be included in one of the several ways to alter the contrast in a MR image, the information content brought by this modality goes well beyond what it is currently attainable with contrast agents and it represents an important source of inspiration for the design of more effective responsive agents.

Acknowledgements This was carried out in the frame of the European-funded NoE EMIL (LSHC - 2004–503569) and DiMI (LSHB-CT-2005-512146) of which SA and AVDL are partners.

References

Aime S, Batsanov AS, Botta M, Howard JAK, Parker D, Senanayake K, Williams G (1994) Solution and solid-state characterization of highly rigid, 8-coordinate lanthanide(Iii) complexes of a macrocyclic tetrabenzylphosphinate. Inorg Chem 33:4696–4706

Aime S, Ascenzi P, Comoglio E, Fasano M, Paoletti S (1995) Molecular recognition of R-states and T-states of human adult hemoglobin by a paramagnetic Gd(III) complex by means of the measurement of solvent water proton relaxation rate. J Am Chem Soc 117:9365–9366

Aime S, Botta M, Milone L, Terreno E (1996a) Paramagnetic complexes as novel NMR pH indicators. Chem Commun: 1265–1266

Aime S, Botta M, Parker D, Williams JAG (1996b) Extent of hydration of octadentate lanthanide complexes incorporating phosphinate donors: solution relaxometry and luminescence studies. J Chem Soc-Dalton Trans: 17–23

Aime S, Botta M, Fasano M, Crich SG, Terreno E (1996c) Gd(III) complexes as contrast agents for magnetic resonance imaging: a proton relaxation enhancement study of the interaction with human serum albumin. J Biol Inorg Chem 1:312–319

Aime S, Barge A, Botta M, Parker D, DeSousa AS (1997) Prototropic vs whole water exchange contributions to the solvent relaxation enhancement in the aqueous solution of a cationic Gd3+ macrocyclic complex. J Am Chem Soc 119:4767–4768

Aime S, Botta M, Fasano M, Terreno E (1998) Lanthanide(III) chelates for NMR biomedical applications. Chem Soc Rev 27:19–29

Aime S, Botta M, Gianolio E, Terreno E (2000a) A p(O-2)-responsive MRI contrast agent based on the redox switch of manganese(II/III) - Porphyrin complexes. Angew Chem-Int Edit 39:747–750

Aime S, Botta M, Frullano L, Crich SG, Giovenzana G, Pagliarin R, Palmisano G, Sirtori FR, Sisti M (2000b) [GdPCP2A(H2O)(2)](−): a paramagnetic contrast agent designed for improved applications in magnetic resonance imaging. J Med Chem 43:4017–4024

Aime S, Anelli PL, Botta M, Brocchetta M, Canton S, Fedeli F, Gianolio E, Terreno E (2002) Relaxometric evaluation of novel manganese(II) complexes for application as contrast agents in magnetic resonance imaging. J Biol Inorg Chem 7:58–67

Aime S, Calabi L, Cavallotti C, Gianolio E, Giovenzana GB, Losi P, Maiocchi A, Palmisano G, Sisti M (2004) [Gd-AAZTA](−): a new structural entry for an improved generation of MRI contrast agents. Inorg Chem 43:7588–7590

Aisen P, Aasa R, Redfield AG (1969) Chromium, manganese, and cobalt complexes of transferrin. J Biol Chem 244:4628–4633

Allegrini PR, Wiessner C (2003) Three-dimensional MRI of cerebral projections in rat brain in vivo after intracortical injection of $MnCl_2$. NMR Biomed 16:252–256

Angelovski G, Fouskova P, Mamedov I, Canals S, Toth E, Logothetis NK (2008) Smart magnetic resonance imaging agents that sense extracellular calcium fluctuations. ChemBioChem 9:1729–1734

Aoki I, Tanaka C, Takegami T, Ebisu T, Umeda M, Fukunaga M, Fukuda K, Silva AC, Koretsky AP, Naruse S (2002) Dynamic activity-induced manganese-dependent contrast magnetic resonance imaging (DAIM MRI). Magn Reson Med 48:927–933

Aoki I, Wu YJL, Silva AC, Lynch RM, Koretsky AP (2004) In vivo detection of neuroarchitecture in the rodent brain using manganese-enhanced MRI. Neuroimage 22:1046–1059

Aschner M, Aschner JL (1990) Manganese transport across the blood-brain-barrier - relationship to iron homeostasis. Brain Res Bull 24:857–860

Aschner M, Gannon M (1994) Manganese (Mn) transport across the rat-blood brain barrier - saturable and transferrin-dependent transport mechanisms. Brain Res Bull 33:345–349

Aschner M, Gannon M, Kimelberg HK (1992) Manganese uptake and efflux in cultured rat astrocytes. J Neurochem 58:730–735

Ashton B (1980) Manganese and man. J Orthomol Psychiatry 9:237–249

Astic L, Saucier D (1981) Metabolic mapping of functional-activity in the olfactory projections of the rat - ontogenetic study. Dev Brain Res 2:141–156

Atanasijevic T, Shusteff M, Fam P, Jasanoff A (2006) Calcium-sensitive MRI contrast agents based on superparamagnetic iron oxide nanoparticles and calmodulin. Proc Natl Acad Sci USA 103:14707–14712

Ballatori N (2002) Transport of toxic metals by molecular mimicry. Environ Health Perspect 110:689–694

Banci L, Bertini I, Luchinat C (1991) Nuclear and electron relaxation. VCH, Weinheim

Bearer EL, Falzone TL, Zhang X, Biris O, Rasin A, Jacobs RE (2007) Role of neuronal activity and kinesin on tract tracing by manganese-enhanced MRI (MEMRI). Neuroimage 37(Suppl 1): S37–46

Bilgen M, Dancause N, Al-Hafez B, He YY, Malone TM (2005) Manganese-enhanced MRI of rat spinal cord injury. Magn Reson Imaging 23:829–832

Bilgen M, Peng W, Al-Hafez B, Dancause N, He YY, Cheney PD (2006) Electrical stimulation of cortex improves corticospinal tract tracing in rat spinal cord using manganese-enhanced MRI. J Neurosci Methods 156:17–22

Bissig D, Berkowitz BA (2009) Manganese-enhanced MRI of layer-specific activity in the visual cortex from awake and free-moving rats. Neuroimage 44:627–635

Bogdanov AJ, Matuszewski L, Bremer C, Petrovsky A, Weissleder R (2002) Oligomerization of paramagnetic substrates result in signal amplification and can be used for MR imaging of molecular targets. Mol Imaging 1:16–23

Bonilla E, Salazar E, Villasmil JJ, Villalobos R (1982) The regional distribution of manganese in the normal human-brain. Neurochem Res 7:221–227

Borel A, Helm L, Merbach AE (2001) Molecular dynamics simulations of MRI-relevant Gd-III chelates: direct access to outer-sphere relaxivity. Chem-Eur J 7:600–610

Botta M (2000) Second coordination sphere water molecules and relaxivity of gadolinium(III) complexes: Implications for MRI contrast agents. Eur J Inorg Chem 2000:399–407

Boutry S, Brunin S, Mahieu I, Laurent S, Elst LV, Muller RN (2008) Magnetic labeling of non-phagocytic adherent cells with iron oxide nanoparticles: a comprehensive study. Contrast Media Mol Imaging 3:223–232

Bradshaw JE, Gillogly KA, Wilson LJ, Kumar K, Wan XM, Tweedle MF, Hernandez G, Bryant RG (1998) New non-ionic water-soluble porphyrins: evaluation of manganese(III) polyhydroxylamide porphyrins as MRI contrast agents. Inorg Chim Acta 276:106–116

Brouillet EP, Shinobu L, Mcgarvey U, Hochberg F, Beal MF (1993) Manganese injection into the rat striatum produces excitotoxic lesions by impairing energy-metabolism. Exp Neurol 120:89–94

Burai L, Toth E, Moreau G, Sour A, Scopelliti R, Merbach AE (2003) Novel macrocyclic Eu-II complexes: fast water exchange related to an extreme M-O-water distance. Chem-Eur J 9:1394–1404

Burdo JR, Menzies SL, Simpson IA, Garrick LM, Garrick MD, Dolan KG, Haile DJ, Beard JL, Connor JR (2001) Distribution of divalent metal transporter 1 and metal transport protein 1 in the normal and Belgrade rat. J Neurosci Res 66:1198–1207

Canals S, Beyerlein M, Keller AL, Murayama Y, Logothetis NK (2008) Magnetic resonance imaging of cortical connectivity in vivo. Neuroimage 40:458–472

Caravan P, Ellison JJ, McMurry TJ, Lauffer RB (1999) Gadolinium(III) chelates as MRI contrast agents: structure, dynamics, and applications. Chem Rev 99:2293–2352

Caruthers SD, Neubauer AM, Hockett FD, Lamerichs R, Winter PM, Scott MJ, Gaffney PJ, Wickline SA, Lanza GM (2006) In vitro demonstration using F-19 magnetic resonance to augment molecular imaging with paramagnetic perfluorocarbon nanoparticles at 1.5 Tesla. Invest Radiol 41:305–312

Chan KC, Cai KX, Su HX, Hung VK, Cheung MM, Chiu CT, Guo H, Jian Y, Chung SK, Wu WT, Wu EX (2008) Early detection of neurodegeneration in brain ischemia by manganese-enhanced MRI. Conf Proc IEEE Eng Med Biol Soc 2008:3884–3887

Chen C, Cohen JS, Myers CE, Sohn M (1984) Paramagnetic metalloporphyrins as potential contrast agents in NMR imaging. FEBS Lett 168:70–74

Chen W, Tenney J, Kulkarni P, King JA (2007) Imaging unconditioned fear response with manganese-enhanced MRI (MEMRI). Neuroimage 37:221–229

Choyke PL, Dwyer AJ, Knopp MV (2003) Functional tumor imaging with dynamic contrast-enhanced magnetic resonance imaging. J Magn Reson Imaging 17:509–520

Chuang KH, Koretsky AP, Sotak CH (2009a) Temporal changes in the T1 and T2 relaxation rates (DeltaR1 and DeltaR2) in the rat brain are consistent with the tissue-clearance rates of elemental manganese. Magn Reson Med 61:1528–1532

Chuang KH, Lee JH, Silva AC, Belluscio L, Koretsky AP (2009b) Manganese enhanced MRI reveals functional circuitry in response to odorant stimuli. Neuroimage 44:363–372

Cohen I, Miles R (2000) Contributions of intrinsic and synaptic activities to the generation of neuronal discharges in in vitro hippocampus. J Physiol (London) 524:485–502

Connor JR (1994) Iron regulation in the brain at the cell and molecular-level. Progr Iron Res 356:229–238

Conrad ME, Umbreit JN, Moore EG, Hainsworth LN, Porubcin M, Simovich MJ, Nakada MT, Dolan K, Garrick MD (2000) Separate pathways for cellular uptake of ferric and ferrous iron. Am J Physiol-Gastroint Liver Physiol 279:G767–G774

Cornud F, Beuvon F, Thevenin F, Chauveinc L, Vieillefond A, Descazeaux A, Flam T (2009) Quantitative dynamic MRI and localisation of non-palpable prostate cancer. Prog Urol 19:401–413

Corot C, Petry KG, Trivedi R, Saleh A, Jonkmanns C, Le Bas JF, Blezer E, Rausch M, Brochet B, Foster-Gareau P, Baleriaux D, Gaillard S, Dousset V (2004) Macrophage imaging in central nervous system and in carotid atherosclerotic plaque using ultrasmall superparamagnetic iron oxide in magnetic resonance imaging. Invest Radiol 39:619–625

Cotzias GC, Horiuchi K, Fuenzali S, Mena I (1968) Chronic manganese poisoning - clearance of tissue manganese concentrations with persistence of neurological picture. Neurology 18:376–382

Critchfield JW, Keen CL (1992) Manganese+2 exhibits dynamic binding to multiple ligands in human plasma. Metab Clin Exp 41:1087–1092

Crossgrove JS, Yokel RA (2004) Manganese distribution across the blood-brain barrier III - The divalent metal transporter-1 is not the major mechanism mediating brain manganese uptake. Neurotoxicology 25:451–460

Crossgrove JS, Yokel RA (2005) Manganese distribution across the blood-brain barrier - IV. Evidence for brain influx through store-operated calcium channels. Neurotoxicology 26:297–307

Crossgrove JS, Allen DD, Bukaveckas BL, Rhineheimer SS, Yokel RA (2003) Manganese distribution across the blood-brain barrier I. Evidence for carrier-mediated influx of manganese citrate as well as manganese and manganese transferrin. Neurotoxicology 24:3–13

Cunningham CH, Arai T, Yang PC, McConnell MV, Pauly JM, Conolly SM (2005) Positive contrast magnetic resonance imaging of cells labeled with magnetic nanoparticles. Magn Reson Med 53:999–1005

Dastur DK, Manghani DK, Raghavendran KV, Jeejeebhoy KN (1969) Distribution and fate of Mn54 in rat, with special reference to Cns. Q J Exp Physiol Cogn Med Sci 54:322–331

Deans AE, Wadghiri YZ, Berrios-Otero CA, Turnbull DH (2008) Mn enhancement and respiratory gating for in utero MRI of the embryonic mouse central nervous system. Magn Reson Med 59:1320–1328

Dickinson TK, Devenyi AG, Connor JR (1996) Distribution of injected iron 59 and manganese 54 in hypotransferrinemic mice. J Lab Clin Med 128:270–278

Do RKG, Rusinek H, Taouli B (2009) Dynamic contrast-enhanced mr imaging of the liver: current status and future directions. Magn Reson Imaging Clin N Am 17:339–349

Drapeau P, Nachshen DA (1984) Manganese fluxes and manganese-dependent neurotransmitter release in presynaptic nerve-endings isolated from rat-brain. J Physiol (London) 348:493–510

Du C, MacGowan GA, Farkas DL, Koretsky AP (2001) Calibration of the calcium dissociation constant of Rhod(2) in the perfused mouse heart using manganese quenching. Cell Calcium 29:217–227

Duflou H, Maenhaut W, Dereuck J (1989) Regional distribution of potassium, calcium, and 6 trace-elements in normal human-brain. Neurochem Res 14:1099–1112

Duncan GE, Miyamoto S, Leipzig JN, Lieberman JA (1999) Comparison of brain metabolic activity patterns induced by ketamine, MK-801 and amphetamine in rats: support for NMDA receptor involvement in responses to subanesthetic dose of ketamine. Brain Res 843:171–183

Elluru RG, Bloom GS, Brady ST (1995) Fast axonal-transport of kinesin in the rat visual-system - functionality of kinesin heavy-chain isoforms. Mol Biol Cell 6:21–40

Fisher RE, Gray R, Johnston D (1990) Properties and distribution of single voltage-gated calcium channels in adult hippocampal-neurons. J Neurophysiol 64:91–104

Frame MDS, Milanick MA (1991) Mn and Cd transport by the Na-Ca exchanger of Ferret red-blood-cells. Am J Physiol 261:C467–C475

Frangioni JV, Hajjar RJ (2004) In vivo tracking of stem cells for clinical trials in cardiovascular disease. Circulation 110:3378–3383

Freed JH (1978) Dynamic effects of pair correlation-functions on spin relaxation by translational diffusion in liquids.2. Finite jumps and independent T1 processes. J Chem Phys 68:4034–4037

Gallez B, Baudelet C, Geurts M (1998) Regional distribution of manganese found in the brain after injection of a single dose of manganese-based contrast agents. Magn Reson Imaging 16: 1211–1215

Gavin CE, Gunter KK, Gunter TE (1990) Manganese and calcium efflux kinetics in brain mitochondria - relevance to manganese toxicity. Biochem J 266:329–334

Gerweck LE, Seetharaman K (1996) Cellular pH gradient in tumor versus normal tissue: potential exploitation for the treatment of cancer. Cancer Res 56:1194–1198

Gibbons RA, Dixon SN, Hallis K, Russell AM, Sansom BF, Symonds HW (1976) Manganese metabolism in cows and goats. Biochim Biophys Acta 444:1–10

Glogard C, Stensrud G, Aime S (2003) Novel radical-responsive MRI contrast agent based on paramagnetic liposomes. Magn Reson Chem 41:585–588

Golman K, Petersson JS (2006) Metabolic imaging and other applications of hyperpolarized C-13. Acad Radiol 13:932–942

Golman K, Olsson LE, Axelsson O, Mansson S, Karlsson M, Petersson JS (2003) Silvanus Thompson memorial lecture molecular imaging using hyperpolarized C-13. Br J Radiol 76:S118–S127

Grafstein B, Forman DS (1980) Intracellular-transport in neurons. Physiol Rev 60:1167–1283

Gunshin H, Mackenzie B, Berger UV, Gunshin Y, Romero MF, Boron WF, Nussberger S, Gollan JL, Hediger MA (1997) Cloning and characterization of a mammalian proton-coupled metal-ion transporter. Nature 388:482–488

Haapanen A, Ramadan UA, Autti T, Joensuu R, Tyynela J (2007) In vivo MRI reveals the dynamics of pathological changes in the brains of cathepsin D-deficient mice and correlates changes in manganese-enhanced MRI with microglial activation lift. Magn Reson Imaging 25: 1024–1031

Hajela S, Botta M, Giraudo S, Xu JD, Raymond KN, Aime S (2000) A tris-hydroxymethyl-substituted derivative of Gd-TREN-Me-3,2-HOPO: an MRI relaxation agent with improved efficiency. J Am Chem Soc 122:11228–11229

Harada T, Tanigawa N, Matsuki M, Nohara T, Narabayashi I (2007) Evaluation of lymph node metastases of breast cancer using ultrasmall superparamagnetic iron oxide-enhanced magnetic resonance imaging. Eur J Radiol 63:401–407

Harris WR, Chen Y (1994) Electron-paramagnetic-resonance and difference ultraviolet studies of Mn_2+ binding to serum transferrin. J Inorg Biochem 54:1–19

Henriksson J, Tallkvist J, Tjalve H (1999) Transport of manganese via the olfactory pathway in rats: dosage dependency of the uptake and subcellular distribution of the metal in the olfactory epithelium and the brain. Toxicol Appl Pharmacol 156:119–128

Hsu YH, Chen CCV, Zechariah A, Yen CC, Yang LC, Chang C (2008) Neuronal dysfunction of a long projecting multisynaptic pathway in response to methamphetamine using manganese-enhanced MRI. Psychopharmacology (Berl) 196:543–553

Immonen RJ, Kharatishvili I, Sierra A, Einula C, Pitkanen A, Grohn OHJ (2008) Manganese enhanced MRI detects mossy fiber sprouting rather than neurodegeneration, gliosis or seizure-activity in the epileptic rat hippocampus. Neuroimage 40:1718–1730

Inagaki S, Suzuki K, Taniguchi N, Takagi H (1991) Localization of Mn-superoxide dismutase (Mn-SOD) in cholinergic and somatostatin-containing neurons in the rat neostriatum. Brain Res 549:174–177

Islam T, Harisinghani MG (2009) Overview of nanoparticle use in cancer imaging. Cancer Biomark 5:61–67

Jefferies WA, Brandon MR, Hunt SV, Williams AF, Gatter KC, Mason DY (1984) Transferrin receptor on endothelium of brain capillaries. Nature 312:162–163

Johansen R, Jensen LR, Rydland J, Goa PE, Kvistad KA, Bathen TF, Axelson DE, Lundgren S, Gribbestad IS (2009) Predicting survival and early clinical response to primary chemotherapy for patients with locally advanced breast cancer using DCE-MRI. J Magn Reson Imaging 29:1300–1307

Josephson L, Perez JM, Weissleder R (2001) Magnetic nanosensors for the detection of oligonucleotide sequences. Angew Chem-Int Edit 40:3204–3206

Jung CW, Jacobs P (1995) Physical and chemical-properties of superparamagnetic iron-oxide MR contrast agents - ferumoxides, ferumoxtran. Ferumoxsil Magn Reson Imaging 13:661–674

Kalia K, Jiang W, Zheng W (2008) Manganese accumulates primarily in nuclei of cultured brain cells. Neurotoxicology 29:466–470

Kanematsu M, Kondo H, Goshima S, Kato H, Tsuge U, Hirose Y, Kim MJ, Moriyama N (2006) Imaging liver metastases: review and update. Eur J Radiol 58:217–228

Kannurpatti SS, Joshi PG, Joshi NB (2000) Calcium sequestering ability of mitochondria modulates influx of calcium through glutamate receptor channel. Neurochem Res 25:1527–1536

Kim WD, Kiefer GE, Maton F, Mcmillan K, Muller RN, Sherry AD (1995) Relaxometry, luminescence measurements, electrophoresis, and animal biodistribution of lanthanide(Iii) complexes of some polyaza macrocyclic acetates containing pyridine. Inorg Chem 34:2233–2243

King LJ, Burkill GJC, Scurr ED, Vlavianos P, Murray-Lyons I, Healy JC (2002) MnDPDP enhanced magnetic resonance imaging of focal liver lesions. Clin Radiol 57:1047–1057

Kita H, Narita K, Vanderkloot W (1981) Tetanic stimulation increases the frequency of miniature endplate potentials at the frog neuromuscular-junction in Mn2+–saline, Co2+–saline, and Ni2+–saline solutions. Brain Res 205:111–121

Kircher MF, Allport JR, Graves EE, Love V, Josephson L, Lichtman AH, Weissleder R (2003) In vivo high resolution three-dimensional imaging of antigen-specific cytotoxic T-lymphocyte trafficking to tumors. Cancer Res 63:6838–6846

Koenig SH, Brown RD (1990) Field-cycling relaxometry of protein solutions and tissue: implications for MRI. Prog Nucl Magn Reson Spectrosc 22:487–567

Koenig SH, Baglin C, Brown RD (1984) Magnetic-field dependence of solvent proton relaxation induced by Gd3+ and Mn2+ complexes. Magn Reson Med 1:496–501

Kuo YT, Herlihy AH, So PW, Bhakoo KK, Bell JD (2005) In vivo measurements of T1 relaxation times in mouse brain associated with different modes of systemic administration of manganese chloride. J Magn Reson Imaging 21:334–339

Kuo YT, Herlihy AH, So PW, Bell JD (2006) Manganese-enhanced magnetic resonance imaging (MEMRI) without compromise of the blood-brain barrier detects hypothalamic neuronal activity in vivo. NMR Biomed 19:1028–1034

Larsen NA, Pakkenberg H, Damsgaard E, Heydorn K (1979) Topographical distribution of arsenic, manganese, and selenium in the normal human brain. J Neurol Sci 42:407–416

Lauffer RB (1987) Paramagnetic metal-complexes as water proton relaxation agents for NMR imaging - theory and design. Chem Rev 87:901–927

Lauffer RB, McMurry TJ, Dunham SO, Scott DM, Parmelee DJ, Dumas S (1997) Bioactivated diagnostic imaging contrast agents. In: (EPIX MEDICAL IUURS, Cambridge, MA 02142–1118 (US). ed)

Lee AH, Tannock IF (1998) Heterogeneity of intracellular pH and of mechanisms that regulate intracellular pH in populations of cultured cells. Cancer Res 58:1901–1908

Lee JH, Silva AC, Merkle H, Koretsky AP (2005) Manganese-enhanced magnetic resonance imaging of mouse brain after systemic administration of MnCl2: dose-dependent and temporal evolution of T-1 contrast. Magn Reson Med 53:640–648

Leergaard TB, Bjaalie JG, Devor A, Wald LL, Dale AM (2003) In vivo tracing of major rat brain pathways using manganese-enhanced magnetic resonance imaging and three-dimensional digital atlasing. Neuroimage 20:1591–1600

Li WH, Fraser SE, Meade TJ (1999) A calcium-sensitive magnetic resonance imaging contrast agent. J Am Chem Soc 121:1413–1414

Li YX, Fang F, Wang XX, Lei H (2009) Neuronal projections from ventral tegmental area to forebrain structures in rat studied by manganese-enhanced magnetic resonance imaging. Magn Reson Imaging 27:293–299

Lin YJ, Koretsky AP (1997) Manganese ion enhances T1-weighted MRI during brain activation: an approach to direct imaging of brain function. Magn Reson Med 38:378–388

Liu CH, D'Arceuil HE, de Crespigny AJ (2004) Direct CSF injection of $MnCl_2$ for dynamic manganese-enhanced MRI. Magn Reson Med 51:978–987

Lokling KE, Fossheim SL, Skurtveit R, Bjornerud A, Klaveness J (2001) pH-sensitive paramagnetic liposomes as MRI contrast agents: in vitro feasibility studies. Magn Reson Imaging 19:731–738

Lokling KE, Skurtveit R, Dyrstad K, Klaveness J, Fossheim SL (2004) Tuning the MR properties of blood-stable pH-responsive paramagnetic liposomes. Int J Pharm 274:75–83

Louie AY, Huber MM, Ahrens ET, Rothbacher U, Moats R, Jacobs RE, Fraser SE, Meade TJ (2000) In vivo visualization of gene expression using magnetic resonance imaging. Nat Biotechnol 18:321–325

Lowe MP, Parker D, Reany O, Aime S, Botta M, Castellano G, Gianolio E, Pagliarin R (2001) pH-dependent modulation of relaxivity and luminescence in macrocyclic gadolinium and europium complexes based on reversible intramolecular sulfonamide ligation. J Am Chem Soc 123:7601–7609

Lu HB, Xi ZX, Gitajn L, Rea W, Yang YH, Stein EA (2007) Cocaine-induced brain activation detected by dynamic manganese-enhanced magnetic resonance imaging (MEMRI). Proc Natl Acad Sci USA 104:2489–2494

Malecki EA, Devenyi AG, Beard JL, Connor JR (1998) Transferrin response in normal and iron-deficient mice heterozygotic for hypotransferrinemia; effects on iron and manganese accumulation. Biometals 11:265–276

Malecki EA, Cook BM, Devenyi AG, Beard JL, Connor JR (1999) Transferrin is required for normal distribution of Fe-59 and Mn-54 in mouse brain. J Neurol Sci 170:112–118

Manghani DK, Dastur DK, Jeejeebh KN, Raghaven KV (1970) Effect of stable manganese on fate of radiomanganese in rat with special reference to CNS. Indian J Med Res 58:209–215

Mathern GW, Pretorius JK, Babb TL (1995) Influence of the type of initial precipitating injury and at what age it occurs on course and outcome in patients with temporal-lobe seizures. J Neurosurg 82:220–227

May PM, Linder PW, Williams DR (1977) Computer-Simulation of metal-ion equilibria in biofluids - models for low-molecular-weight complex distribution of calcium(II), magnesium(II), manganese(II), iron(III), copper(II), zinc(II), and lead(II) ions in human-blood plasma. J Chem Soc-Dalton Trans 6:588–595

McGregor R, Vymazal J, Martinez-Lopez M, Neuwirth J, Salgado P, Beregi JP, Peduto A, de la Pena-Almaguer E, Slater GJ, Shamsi K, Parsons EC (2008) A multi-center, comparative, phase 3 study to determine the efficacy of gadofosveset-enhanced magnetic resonance angiography for evaluation of renal artery disease. Eur J Radiol 65:316–325

Medina L, Figueredo-Cardenas G, Reiner A (1996) Differential abundance of superoxide dismutase in interneurons versus projection neurons and in matrix versus striosome neurons in monkey striatum. Brain Res 708:59–70

Moats RA, Fraser SE, Meade T (1997) A "smart" magnetic resonance imaging agent that reports on specific enzymatic activity. Chem Int Ed Engl 36:726–728

Moos T (1996) Immunohistochemical localization of intraneuronal transferrin receptor immunoreactivity in the adult mouse central nervous system. J Comp Neurol 375:675–692

Morello M, Canini A, Mattioli P, Sorge RP, Alimonti A, Bocca B, Forte G, Martorana A, Bernardi G, Sancesario G (2008) Sub-cellular localization of manganese in the basal ganglia of normal and manganese-treated rats An electron spectroscopy imaging and electron energy-loss spectroscopy study. Neurotoxicology 29:60–72

Morita H, Ogino T, Seo Y, Fujiki N, Tanaka K, Takamata A, Nakamura S, Murakami M (2002) Detection of hypothalamic activation by manganese ion contrasted T1-weighted magnetic resonance imaging in rats. Neurosci Lett 326:101–104

Mornet S, Vasseur S, Grasset F, Duguet E (2004) Magnetic nanoparticle design for medical diagnosis and therapy. J Mater Chem 14:2161–2175

Murayama Y, Weber B, Saleem KS, Augath M, Logothetis NK (2006) Tracing neural circuits in vivo with Mn-enhanced MRI. Magn Reson Imaging 24:349–358

Murphy VA, Wadhwani KC, Smith QR, Rapoport SI (1991) Saturable transport of manganese(II) across the rat-blood brain barrier. J Neurochem 57:948–954

Nairismagi J, Pitkanen A, Narkilahti S, Huttunen J, Kauppinen RA, Grohn OHJ (2006) Manganese-enhanced magnetic resonance imaging of mossy fiber plasticity in vivo. Neuroimage 30:130–135

Narita K, Kawasaki F, Kita H (1990) Mn and Mg influxes through Ca channels of motor-nerve terminals are prevented by verapamil in frogs. Brain Res 510:289–295

Natt O, Watanabe T, Boretius S, Radulovic J, Frahm J, Michaelis T (2002) High-resolution 3D MRI of mouse brain reveals small cerebral structures in vivo. J Neurosci Methods 120:203–209

Nelson SR, Howard RB, Cross RS, Samson F (1980) Ketamine-induced changes in regional glucose-utilization in the rat-brain. Anesthesiology 52:330–334

Newland MC, Ceckler TL, Kordower JH, Weiss B (1989) Visualizing manganese in the primate basal ganglia with magnetic-resonance imaging. Exp Neurol 106:251–258

Nivorozhkin AL, Kolodziej AF, Caravan P, Greenfield MT, Lauffer RB, McMurry TJ (2001) Enzyme-activated Gd3+ magnetic resonance imaging contrast agents with a prominent receptor-induced magnetization enhancement. Angew Chem-Int Edit 40:2903–2906

Norenberg MD (1979) Distribution of glutamine synthetase in the rat central nervous system. J Histochem Cytochem 27:756–762

Oude Engberink RD, van der Pol SMA, Döpp ED, de Vries HE, Blezer ELA (2007) Comparison of SPIO and USPIO for in vitro labeling of human monocytes: MR detection and cell function. Radiology 243:467–474

Padovani B, Lecesne R, Raffaelli C, Chevallier P, Drouillard J, Bruneton JN, Lambrechts M, Gordon P (1996) Tolerability and utility of mangafodipir trisodium injection (MnDPDP) at the dose of 5 mumol/kg body weight in detecting focal liver tumors: results of a phase III trial using an infusion technique. Eur J Radiol 23:205–211

Pardridge WM, Eisenberg J, Jing Y (1987) Human blood-brain-barrier transferrin receptor. Metab Clin Exp 36:892–895

Parkinson JRC, Chaudhri OB, Kuo YT, Field BCT, Herlihy AH, Dhillo WS, Ghatei MA, Bloom SR, Bell JD (2009) Differential patterns of neuronal activation in the brainstem and hypothalamus following peripheral injection of GLP-1, oxyntomodulin and lithium chloride in mice detected by manganese-enhanced magnetic resonance imaging (MEMRI). Neuroimage 44:1022–1031

Pautler RG (2004) In vivo, trans-synaptic tract-tracing utilizing manganese-enhanced magnetic resonance imaging (MEMRI). NMR Biomed 17:595–601

Pautler RG, Koretsky AP (2002) Tracing odor-induced activation in the olfactory bulbs of mice using manganese-enhanced magnetic resonance imaging. Neuroimage 16:441–448

Pautler RG, Silva AC, Koretsky AP (1998) In vivo neuronal tract tracing using manganese-enhanced magnetic resonance imaging. Magn Reson Med 40:740–748

Pautler RG, Mongeau R, Jacobs RE (2003) In vivo trans-synaptic tract tracing from the murine striatum and amygdala utilizing manganese enhanced MRI (MEMRI). Magn Reson Med 50:33–39

Pelled G, Bergman H, Ben-Hur T, Goelman G (2007) Manganese-enhanced MRI in a rat model of Parkinson's disease. J Magn Reson Imaging 26:863–870

Rabin O, Hegedus L, Bourre JM, Smith QR (1993) Rapid brain uptake of manganese(Ii) across the blood-brain-barrier. J Neurochem 61:509–517

Ramos KS, Chacon E, Acosta D (1996) Toxic responses of the heart and vascular systems. In: Klaassen CD (ed) Cassarett and Doull's toxicology: the basic science of poisons, 5th edn. McGraw-Hill, New York, pp 487–527

Reimer P, Tombach B (1998) Hepatic MRI with SPIO: detection and characterization of focal liver lesions. Eur Radiol 8:1198–1204

Reimer P, Weissleder R, Lee AS, Wittenberg J, Brady TJ (1990) Receptor imaging - application to MR imaging of liver-cancer. Radiology 177:729–734

Reimer P, Muller M, Marx C, Wiedermann D, Muller R, Rummeny EJ, Ebert W, Shamsi K, Peters PE (1998) T1 effects of a bolus-injectable superparamagnetic iron oxide, SH U 555 A: dependence on field strength and plasma concentration preliminary clinical experience with dynamic T1-weighted MR imaging. Radiology 209:831–836

Ruiz-Cabello J, Walczak P, Kedziorek DA, Chacko VP, Schmieder AH, Wickline SA, Lanza GM, Bulte JWM (2008) In vivo "hot spot" MR imaging of neural stem cells using fluorinated nanoparticles. Magn Reson Med 60:1506–1511

Rummeny EJ, Marchal G (1997) Liver imaging - clinical applications and future perspectives. Acta Radiol 38:626–630

Saksena MA, Saokar A, Harisinghani MG (2006) Lymphotropic nanoparticle enhanced MR imaging (LNMRI) technique for lymph node imaging. Eur J Radiol 58:367–374

Saleem KS, Pauls JM, Augath M, Trinath T, Prause BA, Hashikawa T, Logothetis NK (2002) Magnetic resonance imaging of neuronal connections in the macaque monkey. Neuron 34: 685–700

Savchenko VL, McKanna JA, Nikonenko IR, Skibo GG (2000) Microglia and astrocytes in the adult rat brain: comparative immunocytochemical analysis demonstrates the efficacy of lipocortin 1 immunoreactivity. Neuroscience 96:195–203

Schroder L, Lowery TJ, Hilty C, Wemmer DE, Pines A (2006) Molecular imaging using a targeted magnetic resonance hyperpolarized biosensor. Science 314:446–449

Schwartz WJ, Sharp FR (1978) Autoradiographic maps of regional brain glucose consumption in resting, awake rats using [C-14] 2-deoxyglucose. J Comp Neurol 177:335–359

Seibig S, Toth E, Merbach AE (2000) Unexpected differences in the dynamics and in the nuclear and electronic relaxation properties of the isoelectronic [Eu-II(DTPA)(H2O)](3-) and [Gd-III(DTPA)(H2O)](2-) complexes (DTPA = diethylenetriamine pentaacetate). J Am Chem Soc 122:5822–5830

Serrano F, Deshazer M, Smith KDB, Ananta JS, Wilson LJ, Pautler RG (2008) Assessing transneuronal dysfunction utilizing manganese-enhanced MRI (MEMRI). Magn Reson Med 60:169–175

Simpson PB, Challiss RAJ, Nahorski SR (1995) Divalent-cation entry in cultured rat cerebellar granule cells measured using Mn2+ quench of fura-2 fluorescence. Eur J Neurosci 7:831–840
Sloot WN, Gramsbergen JBP (1994) Axonal-transport of manganese and its relevance to selective neurotoxicity in the rat basal ganglia. Brain Res 657:124–132
Smith KDB, Kallhoff V, Zheng H, Pautler RG (2007) In vivo axonal transport rates decrease in a mouse model of Alzheimer's disease. Neuroimage 35:1401–1408
So PW, Yu WS, Kuo YT, Wasserfall C, Goldstone AP, Bell JD, Frost G (2007) Impact of resistant starch on body fat patterning and central appetite regulation. PLoS One 2:e1309
Sokoloff L, Reivich M, Kennedy C, Desrosiers MH, Patlak CS, Pettigrew KD, Sakurada O, Shinohara M (1977) The [14 C]deoxyglucose method for measurement of local cerebral glucose-utilization: theory, procedure, and normal values in conscious and anesthetized albino rat. J Neurochem 28:897–916
Sun Y, Sugawara M, Mulkern RV, Hynynen K, Mochizuki S, Albert M, Zuo CS (2000) Simultaneous measurements of temperature and pH in vivo using NMR in conjunction with TmDOTP5-. NMR Biomed 13:460–466
Sun N, Li Y, Tian S, Lei Y, Zheng J, Yang J, Sui N, Xu L, Pei G, Wilson FAW, Ma Y, Lei H, Hu X (2006) Dynamic changes in orbitofrontal neuronal activity in rats during opiate administration and withdrawal. Neuroscience 138:77–82
Takeda A (2004) Function and toxicity of trace metals in the central nervous system. Clin Calcium 14:45–49
Takeda A, Sawashita J, Okada S (1994a) Localization is rat-brain of the trace-metals, zinc and manganese, after intracerebroventricular injection. Brain Res 658:252–254
Takeda A, Akiyama T, Sawashita J, Okada S (1994b) Brain uptake of trace-metals, zinc and manganese, in rats. Brain Res 640:341–344
Takeda A, Ishiwatari S, Okada S (1998a) In vivo stimulation-induced release of manganese in rat amygdala. Brain Res 811:147–151
Takeda A, Kodama Y, Ishiwatari S, Okada S (1998b) Manganese transport in the neural circuit of rat CNS. Brain Res Bull 45:149–152
Takehara Y, Sakahara H, Masunaga H, Isogai S, Kodaira N, Sugiyama M, Takeda H, Saga T, Nakajima S, Sakata I (2002) Assessment of a potential tumor-seeking manganese metalloporphyrin contrast agent in a mouse model. Magn Reson Med 47:549–553
Terreno E, Castelli DD, Milone L, Rollet S, Stancanello J, Violante E, Aime S (2008) First ex-vivo MRI co-localization of two LIPOCEST agents. Contrast Media Mol Imaging 3:38–43
Thuen M, Singstad TE, Pedersen TB, Haraldseth O, Berry M, Sandvig A, Brekken C (2005) Manganese-enhanced MRI of the optic visual pathway and optic nerve injury in adult rats. J Magn Reson Imaging 22:492–500
Tindemans I, Verhoye M, Balthazart J, Van der Linden A (2003) In vivo dynamic ME-MRI reveals differential functional responses of RA- and area X-projecting neurons in the HVC of canaries exposed to conspecific song. Eur J Neurosci 18:3352–3360
Tindemans I, Bournans T, Verhoye M, Van der Linden A (2006) IR-SE and IR-MEMRI allow in vivo visualization of oscine neuroarchitecture including the main forebrain regions of the song control system. NMR Biomed 19:18–29
Tjalve H, Henriksson I (1999) Uptake of metals in the brain via olfactory pathways. Neurotoxicology 20:181–195
Tjalve H, Mejare C, Borgneczak K (1995) Uptake and transport of manganese in primary and secondary olfactory neurons in pike. Pharmacol Toxicol 77:23–31
Tjalve H, Henriksson J, Tallkvist J, Larsson BS, Lindquist NG (1996) Uptake of manganese and cadmium from the nasal mucosa into the central nervous system via olfactory pathways in rats. Pharmacol Toxicol 79:347–356
Toth E, Helm L, Merbach AE (2001) The chemistry of contrast agents in medical magnetic resonance imaging. Wiley, Chichester
Troughton JS, Greenfield MT, Greenwood JM, Dumas S, Wiethoff AJ, Wang JF, Spiller M, McMurry TJ, Caravan P (2004) Synthesis and evaluation of a high relaxivity manganese(II)-based MRI contrast agent. Inorg Chem 43:6313–6323

Van der Linden A, Verhoye M, Van Meir V, Tindemans I, Eens M, Absil P, Balthazart J (2002) In vivo manganese-enhanced magnetic resonance imaging reveals connections and functional properties of the songbird vocal control system. Neuroscience 112:467–474

Van der Linden A, Van Meir V, Tindemans I, Verhoye M, Balthazar J (2004) Applications of manganese-enhanced magnetic resonance imaging (MEMRI) to image brain plasticity in song birds. NMR Biomed 17:602–612

van der Zijden JP, Wu O, van der Toorn A, Roeling TP, Bleys RLAW, Dijkhuizen RM (2007) Changes in neuronal connectivity after stroke in rats as studied by serial manganese-enhanced MRI. Neuroimage 34:1650–1657

Van Meir V, Verhoye M, Absil P, Eens M, Balthazart J, Van der Linden A (2004) Differential effects of testosterone on neuronal populations and their connections in a sensorimotor brain nucleus controlling song production in songbirds: a manganese enhanced-magnetic resonance imaging study. Neuroimage 21:914–923

Van Meir V, Pavlova D, Verhoye M, Pinxten R, Balthazart J, Eens M, Van der Linden A (2006) In vivo MR imaging of the seasonal volumetric and functional plasticity of song control nuclei in relation to song output in a female songbird. Neuroimage 31:981–992

Wadghiri YZ, Blind JA, Duan XH, Moreno C, Yu X, Joyner AL, Turnbull DH (2004) Manganese-enhanced magnetic resonance imaging (MEMRI) of mouse brain development. NMR Biomed 17:613–619

Ward KM, Balaban RS (2000) Determination of pH using water protons and chemical exchange dependent saturation transfer (CEST). Magn Reson Med 44:799–802

Ward KM, Aletras AH, Balaban RS (2000) A new class of contrast agents for MRI based on proton chemical exchange dependent saturation transfer (CEST). J Magn Reson 143:79–87

Watanabe T, Michaelis T, Frahm J (2001) Mapping of retinal projections in the living rat using high-resolution 3D gradient-echo MRI with Mn2+–induced contrast. Magn Reson Med 46:424–429

Watanabe T, Natt O, Boretius S, Frahm J, Michaelis T (2002) In vivo 3D MRI staining of mouse brain after subcutaneous application of MnCl2. Magn Reson Med 48:852–859

Watanabe T, Radulovic J, Spiess J, Natt O, Boretius S, Frahm J, Michaelis T (2004) In vivo 3D MRI staining of the mouse hippocampal system using intracerebral injection of MnCl2. Neuroimage 22:860–867

Wedler FC, Denman RB (1984) Glutamine-synthetase - the major Mn(II) enzyme in mammalian brain. Curr Top Cell Regul 24:153–169

Wedler FC, Ley BW, Grippo AA (1989) Manganese(II) dynamics and distribution in glial-cells cultured from chick cerebral-cortex. Neurochem Res 14:1129–1135

Weng JC, Chen JH, Yang PF, Tseng WYI (2007) Functional mapping of rat barrel activation following whisker stimulation using activity-induced manganese-dependent contrast. Neuroimage 36:1179–1188

Wideroe M, Olsen O, Pedersen TB, Goa PE, Kavelaars A, Heijnen C, Skranes J, Brubakk AM, Brekken C (2009) Manganese-enhanced magnetic resonance imaging of hypoxic-ischemic brain injury in the neonatal rat. Neuroimage 45:880–890

Woods M, Zhang SR, Ebron V, Sherry AD (2003) Ph-sensitive modulation of the second hydration sphere in Lanthanide(III) tetraamide - DOTA complexes: A novel approach to smart MR contrast media. Chem-Eur J 9:4634–4640

Yang J, Khong PL, Wang YX, Chu ACY, Ho SL, Cheung PT, Wu EX (2008) Manganese-enhanced MRI detection of neuro degeneration in neonatal hypoxic-ischemic cerebral injury. Magn Reson Med 59:1329–1339

Yankeelov TE, Lepage M, Chakravarthy A, Broome EE, Niermann KJ, Kelley MC, Meszoely I, Mayer IA, Herman CR, McManus K, Price RR, Gore JC (2007) Integration of quantitative DCE-MRI and ADC mapping to monitor treatment response in human breast cancer: initial results. Magn Reson Imaging 25:1–13

Yokel RA, Crossgrove JS (2004) Manganese toxicokinetics at the blood-brain barrier. Res Rep Health Eff Inst 119:7–58, discussion 59–73

Yokel RA, Crossgrove JS, Bukaveckas BL (2003) Manganese distribution across the blood-brain barrier II. Manganese efflux from the brain does not appear to be carrier mediated. Neurotoxicology 24:15–22

Yu JX, Kodibagkar VD, Cui WN, Mason RP (2005a) F-19: A versatile reporter for non-invasive physiology and pharmacology using magnetic resonance. Curr Med Chem 12:819–848

Yu X, Wadghiri YZ, Sanes DH, Turnbull DH (2005b) In vivo auditory brain mapping in mice with Mn-enhanced MRI. Nat Neurosci 8:961–968

Yu X, Sanes DH, Aristizabal O, Wadghiri YZ, Turnbull DH (2007) Large-scale reorganization of the tonotopic map in mouse auditory midbrain revealed by MRI. Proc Natl Acad Sci USA 104:12193–12198

Yu X, Zou J, Babb JS, Johnson G, Sanes DH, Turnbull DH (2008) Statistical mapping of sound-evoked activity in the mouse auditory midbrain using Mn-enhanced MRI. Neuroimage 39: 223–230

Zhang D, Hu LB, Zhen JW, Zou LG, Feng XY, Wang WX, Wen L (2009) MRI findings of intracranial cystic meningiomas. Clin Radiol 64:792–800

Zheng W, Ren S, Graziano JH (1998) Manganese inhibits mitochondrial aconitase: a mechanism of manganese neurotoxicity. Brain Res 799:334–342

Zheng W, Kim H, Zhao QQ (2000) Comparative toxicokinetics of manganese chloride and methylcyclopentadienyl manganese tricarbonyl (MMT) in Sprague-Dawley rats. Toxicol Sci 54: 295–301

Chapter 5
Principles of ^{1}H NMR Spectroscopy *In Vivo*

Robin A. de Graaf

Abstract This chapter gives a brief overview of the principles of ^{1}H NMR spectroscopy when applied to study metabolism *in vivo*. The basic methodology of *in vivo* ^{1}H NMR, such as spatial localization, shimming and water suppression is described. Special attention is given to spectral quantification. The problem of spectral overlap of resonances is addressed through the use of spectral editing methods.

Keywords ^{1}H NMR spectroscopy • Brain • Spatial localization • Water suppression • Spectral quantification • Spectral editing • Shimming

5.1 Introduction

^{1}H NMR spectroscopy is a powerful technique to study metabolism *in vivo*. The hydrogen nucleus provides excellent NMR sensitivity due to the high gyromagnetic ratio (26.752 10^{+7} rad/T/s) and the large natural abundance (99.985%). Since many of the biologically relevant metabolites contain protons, a typical ^{1}H NMR spectrum of brain can hold dozens of (partially overlapping) NMR resonances.

5.1.1 Spectral Information Content

Figure 5.1 shows a typical ^{1}H NMR spectrum of rat brain *in vivo* acquired at 11.75 T, corresponding to 500 MHz for ^{1}H. The excellent spectral resolution achievable at high magnetic fields allows the routine detection of over 15 metabolites (de Graaf

R.A. de Graaf, Ph.D. (✉)
Departments of Diagnostic Radiology and Biomedical Engineering,
Magnetic Resonance Research Center, School of Medicine, Yale University,
TAC, N145, 300 Cedar Street, P. O. Box 208043, New Haven, CT 06520-8043, USA
e-mail: robin.degraaf@yale.edu

I.-Y. Choi and R. Gruetter (eds.), *Neural Metabolism In Vivo*,
Advances in Neurobiology 4, DOI 10.1007/978-1-4614-1788-0_5,

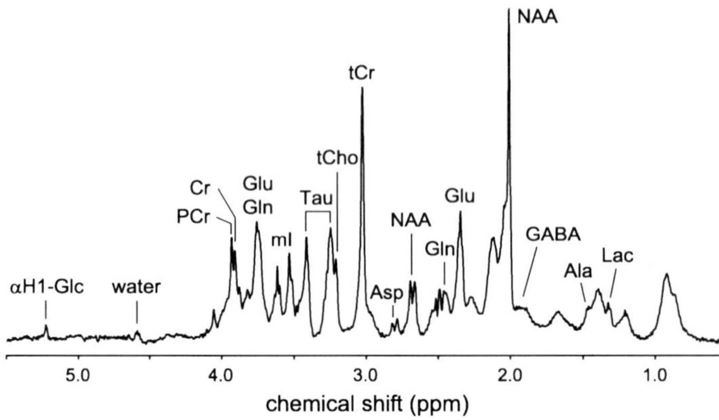

Fig. 5.1 ^{1}H NMR spectrum obtained from rat brain *in vivo* at 11.75 T (TR/TE = 4000/15 ms, 50 μL, 128 averages). The excellent spectral resolution obtainable at high magnetic fields allows the detection of > 15 metabolites

2007; Pfeuffer et al. 1999). The identification of the various metabolites is based on a variety of methods, including prior biochemical knowledge, systemic perturbations, physiological alterations and information obtained from *in vitro* biophysical techniques like gas chromatography–mass spectrometry, high-resolution liquid-state NMR and biochemical assays.

The primary singlet resonances that are visible at almost all echo times ($TE < 3\ T_2$) originate from the methyl protons of *N*-acetyl aspartate (NAA) at 2.01 ppm, total creatine (tCr) at 3.03 ppm which represents the sum of creatine and phosphocreatine and of total choline (tCho) at ~3.2 ppm representing the sum of choline, phosphocholine (PC) and glycerophosphocholine (GPC). The majority of ^{1}H NMR spectroscopic studies reported to date have obtained these three major resonances in longer TE spectra. Changes in NAA levels have been observed in stroke (Bruhn et al. 1989a; Gideon et al. 1992), epilepsy (Hetherington et al. 1995, 2002), multiple sclerosis (Larsson et al. 1991; Arnold et al. 1990) and tumors (Bruhn et al. 1989b; Preul et al. 1996; Majos et al. 2004), whereas tCho level perturbations are commonly observed in cancer (Gillies and Morse 2005). The absolute concentration of tCr is remarkably constant during mild physiological perturbations since creatine and phosphocreatine are interchangeable through the creatine kinase-catalyzed reaction. As a result the tCr resonance is often used as an internal concentration reference (see Sect. 5.3). While these major singlet resonances are readily observable, the metabolic activity of NAA and tCho is relatively low.

Fortunately, a short-TE ^{1}H NMR spectrum holds a wealth of signals originating from metabolites with significant metabolic turnover. These metabolites include the excitatory neurotransmitters glutamate and aspartate, the inhibitory neurotransmitters γ-aminobutyric acid (GABA) and glycine and metabolites involved with intermediary metabolism like glucose, glutamine and lactate. Whereas ^{1}H NMR spectroscopy can be used to obtain the absolute concentrations of these metabolites, indirectly detected ^{13}C NMR or ^{1}H-[^{13}C]-NMR in combination with ^{13}C-enriched substrate infusion can provide the metabolic turnover. Other major metabolites include taurine (Tau), *myo*-inositol (mI) and glutathione (GSH). The biological function of many metabolites is only partially understood. For reviews on cerebral metabolites detectable with ¹H NMR spectroscopy (see de Graaf 2007; Govindaraju et al. 2000).

5.2 Basic *In Vivo* ¹H NMR Data Acquisition

¹H NMR spectra from *in vivo* tissues can offer a wealth of spectral information on a large number of biologically relevant metabolites. However, this information can only be reliably extracted from artifact-free ¹H NMR spectra with high spectral resolution, good sensitivity and which are originating from a well-defined spatial location. The acquisition of high-quality ¹H NMR spectra is thus often synonymous with well-defined spatial localization, an optimized magnetic field homogeneity and a highly suppressed water resonance, whereby the quality of water suppression often strongly depends on the first two factors. Here a brief description will be provided regarding the principles and importance of spatial localization, shimming and water suppression in ¹H NMR spectroscopy.

5.2.1 Spatial Localization

Figure 5.2 shows an unlocalized ¹H NMR spectrum acquired from rat brain *in vivo* at 11.75 T. Besides the large water resonance the spectrum is mainly characterized by lipid resonances which obscure the smaller metabolite signals (Fig. 5.2b). Because the ¹H NMR spectrum originates from a large spatial area (the sensitive volume of the surface coil transceiver), the magnetic field homogeneity can not be optimal for all positions which in turn leads to broad resonance lines (full width at half maximum, FWHM, of water ~ 40 Hz). When the ¹H NMR spectrum is acquired in the presence of spatial localization (Fig. 5.2c, d, volume = $5 \times 4 \times 5$ mm^3, positioned in the rat brain covering cerebral cortex, corpus callosum and hippocampus) the extracranial lipid resonances are eliminated, leaving only the water and metabolite resonances. Since the magnetic field homogeneity over a small, well-defined volume can be optimized more readily, the spectral resolution has greatly increased (FWHM water ~ 14 Hz)

While the development of spatial localization techniques in NMR has a long history (see de Graaf (2007) for review), there are currently only a limited number of single-volume spatial localization techniques in use for ¹H NMR. All of these techniques rely on the principle that a frequency-selective RF pulse executed in the presence of a magnetic field gradient selects a spatially selective slice. When three consecutive RF pulses are applied along three orthogonal magnetic field gradients, the intersection between the three planes defines a single cubic volume. The size and location of the volume are readily modified by changing the magnetic field gradient

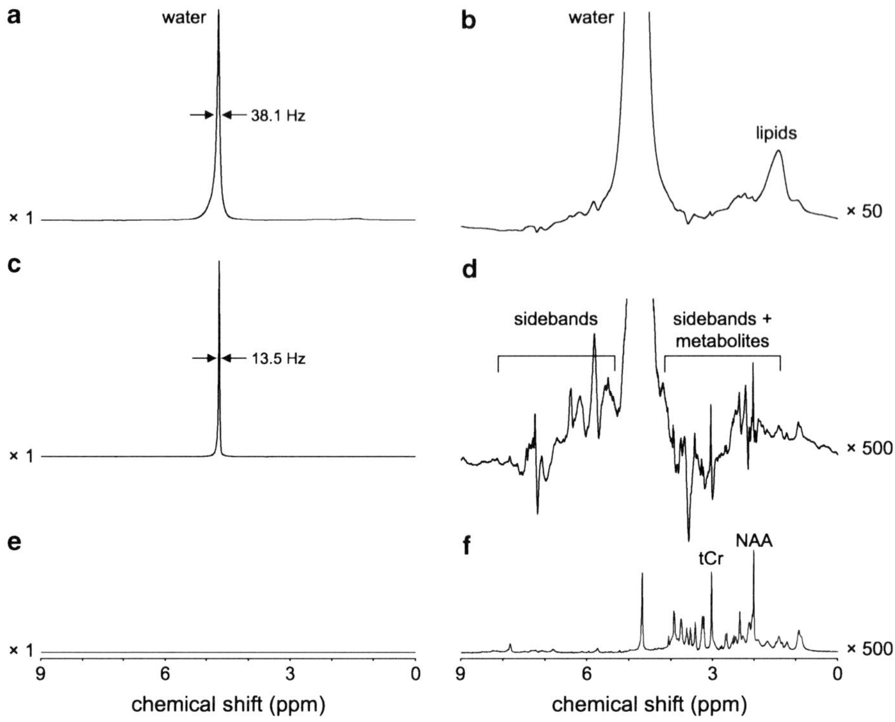

Fig. 5.2 The effects of spatial localization and water suppression on ^{1}H NMR spectra. (**a**, **b**) ^{1}H NMR spectrum acquired from rat head *in vivo* at 11.75 T. Besides the inherent spatial localization provided by the surface coil transceiver, no localization was performed. Besides a broad water resonance, the ^{1}H NMR spectrum is mainly characterized by broad resonances from extracranial lipids. (**c**, **d**) ^{1}H NMR spectrum acquired from rat brain *in vivo* at 11.75 T using LASER localization (100 μL). The extracranial lipid signals have effectively been removed, whereas the limited spatial dimensions provide a greatly improved magnetic field homogeneity. The need for water suppression becomes apparent when the ^{1}H NMR spectrum is zoomed to the metabolite resonances (**d**). Gradient vibrations leads to small modulations of the main magnetic field B_0, which in turn leads to spurious sidebands of the water that overwhelm the much smaller metabolite signals. (**e**, **f**) Water-suppressed (CHESS) ^{1}H NMR spectrum acquired from rat brain *in vivo* at 11.75 T using LASER localization (100 μL). The greatly reduced water resonance no longer generates significant sidebands, thereby allowing the detection of metabolite signals

strength and RF offset, respectively. The different techniques mainly differ in the type of RF pulses used for slice selection. During the STEAM sequence (Frahm et al. (1987), Granot (1986), Kimmich and Hoepfel (1987); Stimulated Echo Acquisition Mode) three 90° pulses are used for volume definition. The advantages of STEAM are that it can maintain a very short echo-time, has only minor RF power deposition and stores the magnetization temporarily along the z axis during the period between the second and third 90° pulse. This fact can be used to achieve additional water suppression during that period and enhance the diffusion-weighting for compounds with short T_2 relaxation constants. The primary disadvantage of STEAM is its inherent signal reduction to 50%. The PRESS method (Bottomley (1984); Bottomley (1987),

Point RESolved Spectroscopy) achieves volume definition by one 90° and two 180° RF pulses. While the minimum echo-time is necessarily longer, PRESS retains full signal intensity and is currently the most popular single-volume localization technique. More recently another localization technique, LASER, (Localization by Adiabatic Spin-Echo Refocusing) has been revitalized (Slotboom et al. 1991; Garwood and DelaBarre 2001). Following a non-selective, adiabatic 90° excitation pulse, 3D localization is achieved by three pairs of adiabatic 180° pulses. The LASER sequence is completely adiabatic, which means that the nutation angle of the RF pulses is independent of the B_1 magnetic field amplitude. This ensures that for a given RF coil, optimal sensitivity is achieved even in the presence of B_1 magnetic field inhomogeneity. Furthermore, since the adiabatic 180° pulses have typically a very large and well-defined bandwidth, the volume definition is excellent with minimal chemical shift displacement. The main disadvantage of LASER is that the requirement for seven RF pulses leads to increased RF power deposition. Furthermore, the minimum echo-time is often significantly longer than for STEAM or PRESS, although the multi-echo CPMG character of the sequence significantly reduces apparent T_2 and scalar coupling related signal losses (de Graaf 2007).

ISIS (image-selected *in vivo* spectroscopy, Ordidge et al. (1986)) achieves spatial localization through an add-subtract scheme in which 180° pulses invert the longitudinal magnetization in a spatially selective slice on alternate scans. Since each spatial direction requires two independent acquisitions to define a spatial slice, ISIS requires eight acquisitions (2×2×2) to achieve 3D volume localization. ISIS is thus not a single-shot localization method, which also represents the main disadvantage of ISIS. Especially when localizing a small volume in a large object, ISIS must subtract a large amount of unwanted signal in order to detect the smaller, desired signal. Any type of motion or system instability can lead to significant subtraction artifacts and a compromised spatial localization. Furthermore, sequence optimization and localized shimming become tedious, time-consuming operations when eight acquisitions are required. As a result, ISIS is almost always combined with outer volume suppression (OVS) methods. OVS is aimed at suppressing the amount of unwanted signal, such that less signal has to be subtracted during the ISIS scheme. The major advantage of ISIS is that it perturbs the longitudinal magnetization, such that any type of NMR sequence can follow the localization part, including a simple non-selective excitation pulse as is desirable for ^{31}P NMR spectroscopy.

In addition to single-volume localization methods there are a number of multi-volume localization techniques, of which MR spectroscopic imaging (MRSI) is the most popular and commonly used (Brown et al. 1982; Maudsley et al. 1983). A typical MRSI scan selects a 1D slice after which the signal distribution within that slice is encoded to allow the generation of a 2D image depicting the signal distribution. MRSI is closely related to MR imaging in that spatial encoding of signal is achieved through a position-dependent phase modulation introduced by so-called phase-encoding gradients. Because MRSI methods are aimed at spatial mapping of different metabolites, the spectroscopic signal is acquired in the absence of a readout gradient. Therefore, in order to spatially encode two dimensions, two independent phase-encoding gradients are required. Following a 2D spatial Fourier transformation and a 1D spectral Fourier transformation, the reconstructed dataset holds NMR spectra from all locations covered by the MRSI field-of-view. The main advantage of MRSI is that (1) it acquires signal from multiple positions, (2) the exact position of the MRSI voxels can be shifted post-acquisition and (3) it is very efficient in terms of information and sensitivity per unit of time. The main downsides of MRSI are that it can be time-consuming to acquire a full 3D dataset and that the voxel definition is not nearly as good as the volume localization achieved by single-volume methods. For more reviews on MRSI acquisition and processing, see de Graaf (2007); Maudsley et al. (2006).

5.2.2 Shimming

The optimization of the magnetic field homogeneity, commonly referred to as shimming, is of crucial importance to almost all aspects of NMR. In NMR the Larmor frequency ω_0 at which the spins precess is directly proportional to the local magnetic field B_0 according to $\omega_0 = \gamma B_0$, where γ is the gyromagnetic ratio of protons. When the magnetic field is spatially homogeneous, the Larmor frequency is the same for all positions. Since the localized signal is simply the sum of the signals from within the 3D localized volume, the total signal thus also precesses at the Larmor frequency, which gives rise to a single Lorentzian line following Fourier transformation. In the case where the magnetic field is spatially inhomogeneous, spins at different spatial locations have different Larmor frequencies. As a result, spins at different locations acquire a different amount of phase following excitation which leads to phase cancellation across a macroscopic volume. Phase cancellation leads to a shorter, apparent T_2 relaxation time constant and hence broader resonance lines and loss of spectral resolution.

One of the simplest ways to improve the magnetic field homogeneity is to restrict the observable signal to a small localized volume (see Fig. 5.2). However, residual magnetic field inhomogeneity is the rule, especially when considering larger volumes and slices. Active shimming based on spherical harmonic functions (Romeo and Hoult 1984) is the standard method to improve the magnetic field homogeneity. Each commercial MR system is equipped with a number of shim coils that produce spatially varying magnetic fields corresponding to one of the low-order spherical harmonic functions. The linear X, Y and Z shim fields, which are typically generated by the magnetic field gradient coils, correspond to the three first-order spherical harmonic functions in which the magnetic field scales linearly with spatial position. Many systems are also equipped with five second-order shim coils, often referred to as Z^2, ZX, ZY, XY and X^2–Y^2. These shim fields have a second-order dependence on spatial position. The crucial 'trick' with active shimming is to find a combination of spherical harmonic fields that best cancels the magnetic field inhomogeneity in the sample. Simple trial-and-error by manual adjustment often works well for first-order shims. However, when higher-order shims are required it is always advisable to use quantitative and often automated methods, such as FASTMAP (Gruetter 1993) or MRI-based magnetic field mapping.

5.2.3 Water Suppression

While the requirements for good spatial localization and magnetic field homogeneity are valid for most nuclei, the presence of a large water resonance is unique to ¹H NMR spectroscopy. Since the water resonance is typically several orders of magnitude larger than the metabolites resonances, its presence leads to baseline distortions and spurious signals that can overwhelm the metabolite signals (Fig. 5.2d). As a result, for most ¹H MRS applications the water signal needs to be suppressed, without disturbing the metabolite resonances. The suppression of a particular resonance in a NMR spectrum requires a difference in a property between the molecule of interest and the interfering compound (e.g. water). This property needs not to be a magnetic one, but it must be reflected in a NMR observable parameter. Thus, besides differences in chemical shift, scalar coupling and T_1 and T_2 relaxation, properties like diffusion and exchange can sometimes also be exploited. Although there is not an universal technique, there are some criteria by which the existing water suppression methods can be evaluated. These criteria include (1) degree of suppression, (2) insensitivity to RF imperfections, (3) ease of phasing the spectra, (4) minimum attainable repetition and echo times, (5) insensitivity to relaxation effects, (6) perturbation of other resonances and (7) degree of distortion of resonances near or at the water frequency.

The most commonly used water suppression techniques employ frequency-selective excitation and/or refocusing (Haase et al. 1985; de Graaf and Nicolay 1998; Mescher et al. 1996; Hore 1983; Tkac et al. 1999). A prominent example is provided by the method of CHEmical Shift Selective (CHESS) water suppression (Haase et al. 1985). CHESS consists of a frequency selective excitation pulse for the water followed by a magnetic field gradient. The water is thus excited onto the transverse plane, where it will be destroyed (dephased) by the magnetic field gradient, while the metabolite resonances reside along the longitudinal axis. Next, the metabolites can be excited onto the transverse plane and be detected (Fig. 5.2e, f). The CHESS pulse sequence element may be repeated a number of times to improve the water suppression. Several related techniques exist, which, for example, do not excite the water (e.g. binomial RF pulses (Hore 1983)), do not refocus the water (e.g. MEGA (Mescher et al. 1996)) or which saturate the water resonance (e.g. pre-saturation). CHESS and all other water suppression techniques have to make compromises between the criteria mentioned above.

5.3 *In Vivo* ¹H NMR Data Processing

5.3.1 Basic Processing Techniques

With modern pulsed NMR techniques the signal is always acquired in the time domain. However, the NMR signal is typically Fourier transformed to the frequency domain to allow for visual discrimination of different resonances. Several operations can be performed prior to and following Fourier transformation to allow for improved sensitivity or spectral resolution, improved baseline or simply to improve the appearance.

These operations may include zero-filling, apodization, Fourier transformation, phasing, baseline correction and residual water removal. Some of these operations, like zero-filling and phasing, are purely cosmetic and do not fundamentally change the acquired data. Others, like baseline correction, do affect the data directly and it should always be considered what their effect on the final, estimated metabolite concentration can be. The final point of pre-processing is to obtain a visually acceptable ^{1}H NMR spectrum. An acceptable ^{1}H NMR spectrum is generally synonymous with (1) well-separated resonances (i.e. a high spectral resolution and magnetic field homogeneity), (2) equal phases for all resonances, (3) no baseline distortions caused by residual water signal or other effects and (4) a good signal-to-noise ratio (SNR). ^{1}H NMR spectra that do not satisfy one or more of these criteria should be scrutinized on their usefulness in attempting spectral quantification.

5.3.2 *Spectral Quantification*

In many cases the purpose of acquiring a ^{1}H NMR spectrum is to obtain metabolite concentrations. Basic NMR theory shows that the intensity of a NMR signal is, in principle, proportional to the Boltzmann equilibrium magnetization and hence to the absolute metabolite concentration. However, there are numerous factors that can affect the intensity of a NMR signal. Many of those factors (e.g. T_1 and T_2 relaxation during TR and TE, respectively) can be quantitatively taken into account. Other factors, like partial NMR visibility, are more difficult to correct.

However, before relaxation-related correction factors can be applied, the resonance areas must be extracted from the ^{1}H NMR spectrum. While there are many different spectral quantification methods available (Naressi et al. 2001; Slotboom et al. 1998), the LCmodel algorithm (Provencher 1993) is currently the most popular for ^{1}H NMR. The LCmodel algorithm approximates the *in vivo* spectrum as a linear combination of model solution or *in vitro* spectra of the individual compounds. The *in vitro* spectra can be obtained through direct measurement, using the same pulse sequence as used for the *in vivo* measurement, or they can be obtained through density matrix calculations (de Graaf 2007). The *in vitro* spectra are then scaled, broadened and shifted such that the sum matches the *in vivo* ^{1}H NMR spectrum.

Figure 5.3 shows a typical LCmodel fit of a short-TE ^{1}H NMR spectrum. Judging from the small difference between the experimental and fitted spectra, the overall mathematical fit is excellent. However, a noticeable complication in fitting short-TE ^{1}H NMR spectra is the presence of an intense baseline originating from cytosolic macromolecules (Behar and Ogino 1993). Since the composition of the baseline is essentially unknown, it can not be obtained *in vitro*. Therefore, the macromolecular baseline is either mathematically modeled or, more accurately, measured *in vivo* through T_1 differences between macromolecules and metabolites (Behar and Ogino 1993; Hwang et al. 1996).

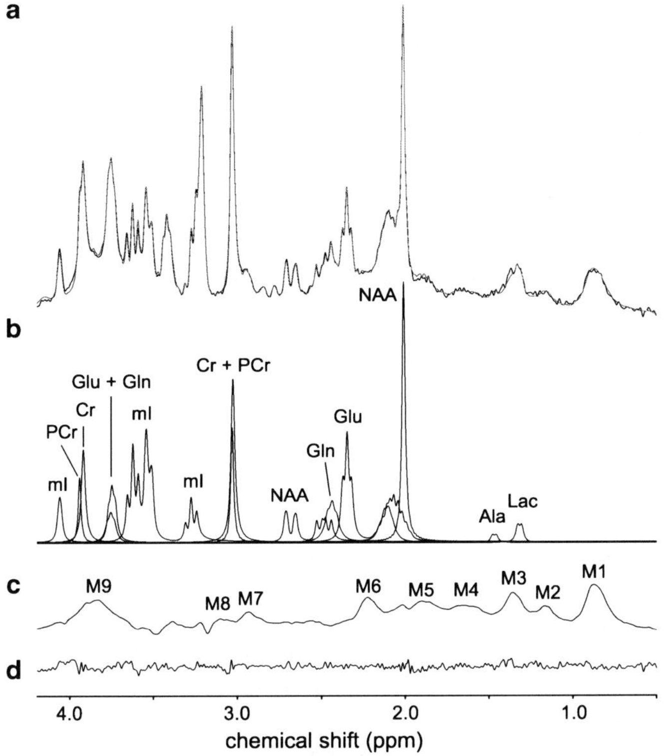

Fig. 5.3 Spectral quantification of ^{1}H NMR spectra using a LCmodel algorithm. (**a**) Experimental ^{1}H NMR spectrum acquired from rat brain *in vivo* at 9.4 T (100 μL, STEAM localization, TR/TE=4000/8 ms, 128 averages) overlaid with a best-fitted NMR spectrum calculated by the LCmodel algorithm. (**b**) ^{1}H NMR spectra of some individual components extracted from the LCmodel fit. (**c**) Experimentally measured macromolecule baseline. Note the close correspondence to the baseline visible in (**a**). (**d**) Residual difference between the experimental and calculated ^{1}H NMR spectra

5.3.3 Metabolite Concentrations

Once the metabolite resonance areas are established, they can be corrected for relaxation-related losses, number of protons and potential pulse sequence-related factors. Since there are so many, often unknown factors affecting the absolute NMR intensity, the corrected resonance areas can not be directly expressed in units of absolute concentration. Instead, the areas are compared to a resonance area of known concentration. This so-called concentration reference can be an endogenous or internal compound, such as water or total creatine, or can be a suitable external compound. The advantage of an internal concentration reference is obvious; all experimental parameters for the compound and the metabolites are identical. However, the main caveat is the uncertainty in the absolute concentration of the compound. While the water and total creatine concentrations are relatively stable, they do change in severe or chronic disorders. The pro's and con's for external concentration referencing are essential opposite to those for internal referencing. While the absolute concentration is exactly known, NMR parameters may be different between metabolite and reference data acquisition. At very high magnetic fields, internal concentration referencing is more desirable as spatially varying RF fields make external concentration referencing unreliable.

5.4 Advanced *In Vivo* ^{1}H NMR Data Acquisition

5.4.1 Spectral Editing – Introduction

The narrow chemical shift dispersion of ^{1}H NMR spectroscopy makes partial or complete overlap of resonances the rule rather than the exception. Prime examples of spectral overlap include lactate and lipids, GABA and total creatine and glutamate and glutamine. In some cases, like for lactate and lipids, the spectral overlap can be eliminated by utilizing the different spatial origins of the compounds. Normal brain tissue does not contain any NMR-observable lipids, such that any lipid resonances must come from extracranial tissues. Adequate spatial localization can then separate cerebral lactate from overlapping lipid signals. However, this strategy does not work when lactate and lipids originate from the same spatial location such as in tumors, infarctions or muscle and liver.

In other cases, like for glutamate and glutamine, the magnetic field dependence of the resonances can be used to achieve separation at high magnetic fields. Glutamate and glutamine are strongly-coupled spin-systems which form complicated and strongly overlapping multiplets at low magnetic fields (i.e. < 3 T). At higher magnetic fields (i.e. > 7 T) the multiplet structures simplify enough to allow complete separation based on the small chemical shift difference of the H4-resonances (see also Fig. 5.1).

In the case of GABA and total creatine or lactate and lipids in tumors, spectral overlap is complete and other methods of separation must be used. Homonuclear spectral editing based on scalar coupling is one of the most powerful methods to separate overlapping resonances, as will be discussed next for GABA.

5.4.2 Homonuclear Spectral Editing

GABA is the major inhibitory neurotransmitter in the mammalian brain and its level varies dynamically during physiological alterations (Epperson et al. 2002; Levy et al. 2002; Babak et al. 2001) and in a number of psychiatric disorders (Brambilla et al. 2003; Petroff et al. 2001; Sanacora et al. 1999, 2004; Goddard et al. 2001). Direct NMR observation of GABA is typically not possible as the three

NMR-observable resonances of GABA overlap with resonances from other, more concentrated, metabolites. GABA-H2 at 2.28 ppm overlaps with glutamate-H4. While the overlap is complete at lower magnetic fields, GABA-H2 and glutamate-H4 become visually separable at high magnetic fields. In short-TE ^{1}H NMR spectra, GABA-H2 is also overlapping with an intense macromolecular resonance (see Fig. 5.3). The GABA-H3 resonance at 1.89 ppm is, in principle, directly observable. However, the GABA-H3 resonance is typically overwhelmed by the 'tail' of the intense NAA resonance at 2.01 ppm. Furthermore, the broad GABA-H3 resonance (quintet multiplet) is difficult to differentiate from the macromolecular baseline. The GABA-H4 resonance at 3.01 ppm overlaps, to a first approximation, only with the singlet resonance of creatine and phosphocreatine (i.e. total creatine, tCr). Since resonances from single spins are readily separated from scalar-coupled spins with homonuclear spectral editing methods, the GABA-H4 resonance is typically observed.

The principle of homonuclear spectral editing is based on the absence and presence of scalar coupling/evolution in uncoupled and scalar-coupled spin-systems, respectively. GABA is a scalar-coupled spin-system in which the protons in the $^{4}CH_2$ group experience the protons in the $^{3}CH_2$ group through their effect on the bonding electrons. As a result of this scalar coupling, the $^{4}CH_2$ resonance is splitted two successive times into a triplet resonance in which the resonances are separated by the scalar coupling constant J. Besides the more complicated spectral pattern of scalar-coupled spin-systems, scalar coupling also leads to scalar evolution during delays following excitation. During a regular spin-echo sequence with echo-time TE = 1/(2 J) the outer two resonances of GABA-H4 become negative. Alternatively, during a spin-echo sequence that is selective for the GABA-H4 resonance, all three GABA-H4 resonances are positive. Since uncoupled spins, like creatine, do not have scalar evolution, they will appear positive during selective and non-selective spin-echo sequences. The difference between the two scans thus subtracts the creatine resonance, thereby revealing the outer two resonances of GABA. This spectral editing technique is referred to as J-difference editing (Rothman et al. 1984; Schupp et al. 1993; Rothman et al. 1993; Mescher et al. 1998) and is only one of many possible homonuclear spectral editing methods (see (de Graaf 2007) for review).

Figure 5.4 shows a practical example of *in vivo* GABA detection by J-difference editing. Figure 5.4a shows a ^{1}H NMR spectrum from rat brain at 11.75 T (100 μL, 512 averages, TE = 1/(2 J) = 68 ms) acquired with a non-selective PRESS sequence. While not directly visible, the outer two GABA-H4 resonances at 3.01 ppm are negative. The ^{1}H NMR spectrum in Fig. 5.4b is acquired in an identical manner with the exception that selective refocusing is applied to GABA-H3 at 1.89 ppm. Because of the finite bandwidth, the NAA resonance at 2.01 ppm is also partially affected. In this scan all three GABA-H4 resonances are positive, such that the difference between (B) and (A) reveals the GABA-H4 multiplet without the overlapping creatine singlet.

From Fig. 5.4c it follows that other resonances are also edited (co-edited). In the case of GABA, the glutamate/glutamine H2 resonance at ~3.75 ppm is partially co-edited, as well as the GABA-H2 resonance at 2.28 ppm. However, this is not a

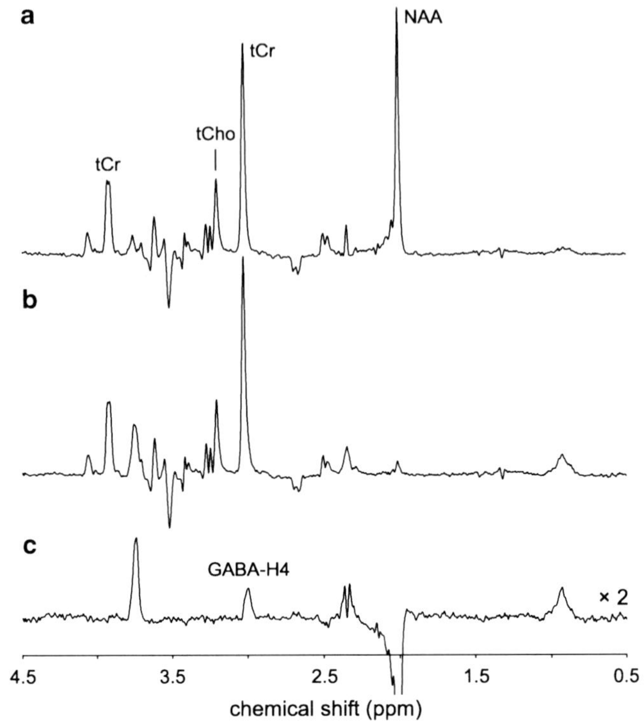

Fig. 5.4 Homonuclear J-difference editing of GABA in rat brain *in vivo* at 11.75 T. (**a**) Control ^{1}H NMR spectrum (100 μL, TE = 1/(2 J) = 68 ms) without selective refocusing of GABA-H3. (**b**) ^{1}H NMR spectrum acquired in the presence of selective GABA-H3 refocusing. (**c**) The difference between the two experiments, i.e. (**b**)–(**a**), reveals the edited GABA-H4 resonance at 3.01 ppm. The total experiment was performed in circa 20 min

problem for GABA-H4 detection simply because these co-edited resonances do not overlap with the peak of interest. Co-editing is a problem in the case of macromolecules because GABA-H4 is directly overlapping with macromolecular resonance M7 which is coupled to resonance M4 at 1.7 ppm (see also Fig. 5.3). Since the editing pulse at 1.89 ppm has a finite bandwidth, it can affect the macromolecular resonance M4 which then leads to co-editing of M7. At low magnetic fields, the contribution of M7 to the edited GABA-H4 resonance is very large and care should be taken during the interpretation of the observed GABA-H4 resonance. At higher magnetic fields (Mescher et al. 1998), or through careful sequence design (Henry et al. 2001), the contribution of M7 can be greatly minimized.

References

Arnold DL, Matthews PM, Francis G, Antel J (1990) Proton magnetic resonance spectroscopy of human brain in vivo in the evaluation of multiple sclerosis: assessment of the load of disease. Magn Reson Med 14:154–159

Babak B, Rothman DL, Petroff OAC, Mason GF, Appel M, Wu C, Duyn JH, Cohen L (2001) Involvement of GABA in rapid experience-dependent plasticity in the human visual cortex. Biol Psych 49:148S

Behar KL, Ogino T (1993) Characterization of macromolecule resonances in the ^{1}H NMR spectrum of rat brain. Magn Reson Med 30:38–44

Bottomley PA (1984) Selective volume method for performing localized NMR spectroscopy. US patent 4 480 228

Bottomley PA (1987) Spatial localization in NMR spectroscopy in vivo. Ann N Y Acad Sci 508:333–348

Brambilla P, Perez J, Barale F, Schettini G, Soares JC (2003) GABAergic dysfunction in mood disorders. Mol Psychiatry 8(8):721–737, 715

Brown TR, Kincaid BM, Ugurbil K (1982) NMR chemical shift imaging in three dimensions. Proc Natl Acad Sci U S A 79(11):3523–3526

Bruhn H, Frahm J, Gyngell ML, Merboldt KD, Hanicke W, Sauter R (1989a) Cerebral metabolism in man after acute stroke: new observations using localized proton NMR spectroscopy. Magn Reson Med 9:126–131

Bruhn H, Frahm J, Gyngell ML, Merboldt KD, Hanicke W, Sauter R, Hamburger C (1989b) Noninvasive differentiation of tumors with use of localized H-1 MR spectroscopy *in vivo*: initial experience in patients with cerebral tumors. Radiology 172:541–548

de Graaf RA (2007) *In vivo* NMR spectroscopy. Principles and techniques. Wiley, Chichester

de Graaf RA, Nicolay K (1998) Adiabatic water suppression using frequency selective excitation. Magn Reson Med 40:690–696

Epperson CN, Haga K, Mason GF, Sellers E, Gueorguieva R, Zhang W, Weiss E, Rothman DL, Krystal JH (2002) Cortical gamma-aminobutyric acid levels across the menstrual cycle in healthy women and those with premenstrual dysphoric disorder: a proton magnetic resonance spectroscopy study. Arch Gen Psychiatry 59(9):851–858

Frahm J, Merboldt KD, Hanicke W (1987) Localized proton spectroscopy using stimulated echoes. J Magn Reson 72:502–508

Garwood M, DelaBarre L (2001) The return of the frequency sweep: designing adiabatic pulses for contemporary NMR. J Magn Reson 153:155–177

Gideon P, Henriksen O, Sperling B, Christiansen P, Olsen TS, Jorgensen HS, Arlien-Soborg P (1992) Early time course of *N*-acetylaspartate, creatine and phosphocreatine, and compounds containing choline in the brain after acute stroke. A proton magnetic resonance spectroscopy study. Stroke 23:1566–1572

Gillies RJ, Morse DL (2005) In vivo magnetic resonance spectroscopy in cancer. Annu Rev Biomed Eng 7:287–326

Goddard AW, Mason GF, Almai A, Rothman DL, Behar KL, Petroff OA, Charney DS, Krystal JH (2001) Reductions in occipital cortex GABA levels in panic disorder detected with ^{1}H-magnetic resonance spectroscopy. Arch Gen Psychiatry 58:556–561

Govindaraju V, Young K, Maudsley AA (2000) Proton NMR chemical shifts and coupling constants for brain metabolites. NMR Biomed 13:129–153

Granot J (1986) Selected volume excitation using stimulated echoes (VEST). Application to spatially localized spectroscopy and imaging. J Magn Reson 70:488–492

Gruetter R (1993) Automatic, localized *in vivo* adjustment of all first- and second-order shim coils. Magn Reson Med 29:804–811

Haase A, Frahm J, Hanicke W, Matthaei D (1985) ^{1}H NMR chemical shift selective (CHESS) imaging. Phys Med Biol 30:341–344

Henry PG, Dautry C, Hantraye P, Bloch G (2001) Brain GABA editing without macromolecule contamination. Magn Reson Med 45(3):517–520

Hetherington HP, Kuzniecky RI, Pan JW, Mason GF, Morawetz R, Harris C, Faught E, Vaughan T, Pohost GM (1995) Proton nuclear magnetic resonance spectroscopic imaging of human temporal lobe epilepsy at 4.1 T. Ann Neurol 38:396–404

Hetherington HP, Pan JW, Spencer DD (2002) ^{1}H and ^{31}P spectroscopy and bioenergetics in the lateralization of seizures in temporal lobe epilepsy. J Magn Reson Imaging 16:477–483

Hore PJ (1983) Solvent suppression in Fourier transform nuclear magnetic resonance. J Magn Reson 55:283–300

Hwang JH, Graham GD, Behar KL, Alger JR, Prichard JW, Rothman DL (1996) Short echo time proton magnetic resonance spectroscopic imaging of macromolecule and metabolite signal intensities in the human brain. Magn Reson Med 35:633–639

Kimmich R, Hoepfel D (1987) Volume-selective multipulse spin-echo spectroscopy. J Magn Reson 72:379–384

Larsson HB, Christiansen P, Jensen M, Frederiksen J, Heltberg A, Olesen J, Henriksen O (1991) Localized *in vivo* proton spectroscopy in the brain of patients with multiple sclerosis. Magn Reson Med 22:23–31

Levy LM, Ziemann U, Chen R, Cohen LG (2002) Rapid modulation of GABA in sensorimotor cortex induced by acute deafferentation. Ann Neurol 52(6):755–761

Majos C, Julia-Sape M, Alonso J, Serrallonga M, Aguilera C, Acebes JJ, Arus C, Gili J (2004) Brain tumor classification by proton MR spectroscopy: comparison of diagnostic accuracy at short and long TE. AJNR Am J Neuroradiol 25:1696–1704

Maudsley AA, Hilal SK, Perman WH, Simon HE (1983) Spatially resolved high resolution spectroscopy by 'four dimensional' NMR. J Magn Reson 51:147–152

Maudsley AA, Darkazanli A, Alger JR, Hall LO, Schuff N, Studholme C, Yu Y, Ebel A, Frew A, Goldgof D, Gu Y, Pagare R, Rousseau F, Sivasankaran K, Soher BJ, Weber P, Young K, Zhu X (2006) Comprehensive processing, display and analysis for *in vivo* MR spectroscopic imaging. NMR Biomed 19:492–503

Mescher M, Tannus A, O'Neil Johnson M, Garwood M (1996) Solvent suppression using selective echo dephasing. J Magn Reson A 123:226–229

Mescher M, Merkle H, Kirsch J, Garwood M, Gruetter R (1998) Simultaneous *in vivo* spectral editing and water suppression. NMR Biomed 11:266–272

Naressi A, Couturier C, Castang I, de Beer R, Graveron-Demilly D (2001) Java-based graphical user interface for MRUI, a software package for quantitation of *in vivo*/medical magnetic resonance spectroscopy signals. Comput Biol Med 31:269–286

Ordidge RJ, Connelly A, Lohman JA (1986) Image-selected in vivo spectroscopy (ISIS). A new technique for spatially selective NMR spectroscopy. J Magn Reson 66:283–294

Petroff OA, Hyder F, Rothman DL, Mattson RH (2001) Homocarnosine and seizure control in juvenile myoclonic epilepsy and complex partial seizures. Neurology 56:709–715

Pfeuffer J, Tkac I, Provencher SW, Gruetter R (1999) Toward an *in vivo* neurochemical profile: quantification of 18 metabolites in short-echo-time ^{1}H NMR spectra of the rat brain. J Magn Reson 141:104–120

Preul MC, Caramanos Z, Collins DL, Villemure JG, Leblanc R, Olivier A, Pokrupa R, Arnold DL (1996) Accurate, noninvasive diagnosis of human brain tumors by using proton magnetic resonance spectroscopy. Nat Med 2:323–325

Provencher SW (1993) Estimation of metabolite concentrations from localized *in vivo* proton NMR spectra. Magn Reson Med 30:672–679

Romeo F, Hoult DI (1984) Magnet field profiling: analysis and correcting coil design. Magn Reson Med 1:44–65

Rothman DL, Behar KL, Hetherington HP, Shulman RG (1984) Homonuclear ^{1}H double-resonance difference spectroscopy of the rat brain *in vivo*. Proc Natl Acad Sci U S A 81:6330–6334

Rothman DL, Petroff OA, Behar KL, Mattson RH (1993) Localized ^{1}H NMR measurements of gamma-aminobutyric acid in human brain *in vivo*. Proc Natl Acad Sci U S A 90:5662–5666

Sanacora G, Mason GF, Rothman DL, Behar KL, Hyder F, Petroff OA, Berman RM, Charney DS, Krystal JH (1999) Reduced cortical gamma-aminobutyric acid levels in depressed patients determined by proton magnetic resonance spectroscopy. Arch Gen Psychiatry 56:1043–1047

Sanacora G, Gueorguieva R, Epperson CN, Wu YT, Appel M, Rothman DL, Krystal JH, Mason GF (2004) Subtype-specific alterations of gamma-aminobutyric acid and glutamate in patients with major depression. Arch Gen Psychiatry 61:705–713

Schupp DG, Merkle H, Ellermann JM, Ke Y, Garwood M (1993) Localized detection of glioma glycolysis using edited ^{1}H MRS. Magn Reson Med 30:18–27

Slotboom J, Mehlkopf AF, Bovee WM (1991) A single-shot localization pulse sequence suited for coils with inhomogneous RF fields using adiabatic slice-selective RF pulses. J Magn Reson 95:396–404

Slotboom J, Boesch C, Kreis R (1998) Versatile frequency domain fitting using time domain models and prior knowledge. Magn Reson Med 39:899–911

Tkac I, Starcuk Z, Choi IY, Gruetter R (1999) *In vivo* ^{1}H NMR spectroscopy of rat brain at 1 ms echo time. Magn Reson Med 41:649–656

Chapter 6
In Vivo Cerebral ^{31}P Magnetic Resonance Spectroscopy

Ernest B. Cady

Abstract This chapter summarises the achievements of *in-vivo* phosphorus magnetic resonance spectroscopy (31P MRS) in the study of brain metabolism. The metabolites detectable by 31P MRS are described as also are the techniques used for spectrum analysis, signal localization and assessment of measures such as intracellular pH and magnesium. Many pathological applications of 31P MRS are also covered.

Keywords Brain • Development • Intracellular pH • Localisation • Magnetic resonance spectroscopy • Metabolism • Pathology • Phosphorus

6.1 Introduction

It is hoped this article will ease navigation towards the keynote literature required to achieve readers' aims. Many techniques described have been applied to human brain: however, appropriate experimental dimension scaling enables application to other mammalian species.

6.1.1 The Phosphorus (^{31}P) Nucleus

The ^{31}P magnetic resonance spectroscopy (MRS) resonant frequency is ~40.5% that of the proton (^{1}H). ^{31}P has intrinsic MRS sensitivity ~6.7% compared to ^{1}H which, coupled with ^{31}P-metabolite resonances mostly resulting from a single nucleus per

E.B. Cady, FInstP, BSc (✉)
Department of Medical Physics and Bioengineering, University College London Hospitals Foundation NHS Trust, Ground Floor, Rosenheim Building, 25 Grafton Way, London WC1E 6DB, UK
e-mail: e.cady@medphys.ucl.ac.uk

I.-Y. Choi and R. Gruetter (eds.), *Neural Metabolism In Vivo*, Advances in Neurobiology 4, DOI 10.1007/978-1-4614-1788-0_6,

molecule, means that ^{31}P signal amplitudes are only ~2% of those for ^{1}H (^{1}H resonances are often from multiple equivalent nuclei). Consequently, to acquire spectra with signal-to-noise ratio (SNR) similar to ^{1}H MRS, many more free induction decays (FID) must be accumulated or with localisation techniques data must be collected from a much larger tissue volume. Some ^{31}P spectrum peaks are multiplets (e.g. doublets): however, tissue heterogeneity and magnetic susceptibility effects engender worse magnetic field homogeneity *in vivo* than that attainable with metabolite solutions *in vitro*; hence, ^{31}P multiplets may not always be resolved *in vivo*.

Over the last decade 1.5 Tesla (T) has been a consensus optimal field strength for clinical magnetic resonance imaging. Recently several major manufacturers have developed 3-T systems with installation specifications similar to those for 1.5 T thereby facilitating a consensus move to higher clinical magnetic field strength. Furthermore, research scanners are now available for human brain studies at 7 T or more (e.g. Lei et al. 2003a). Both metabolite peak separation and signal amplitude increase as magnetic field strength rises. Hence, higher field strength provides an exciting opportunity to compensate for the low intrinsic ^{31}P sensitivity, reduce acquisition times, reduce tissue volumes from which spectra are acquired, and produce spectra with improved metabolite resolution. *In-vivo* experimental scanners at 7 T and above have been available over the last two decades: these have greatly aided interpretation of clinical results and provided important translational research outputs.

6.1.2 The Brain Metabolites Detectable In Vivo

The most important cerebral metabolites detectable by *in-vivo* ^{31}P MRS are nucleotide triphosphate (NTP; mainly adenosine triphosphate (ATP) which is the end-product of energy generation) and phosphocreatine (PCr; an ATP replenisher). The chemical shift of a spectrum peak is its resonant frequency expressed in parts per million (ppm) of the static magnetic field strength (hence, peak positions in spectra acquired at different magnetic field strengths can be compared directly). Intracellular pH (pH_i) can be determined indirectly via the chemical shifts of various ^{31}P metabolites most notably inorganic orthophosphate (Pi) (Moon and Richards 1973; Petroff and Prichard 1985; Pettegrew et al. 1988). Other metabolites easily detected in brain include phosphomonoesters (PME), phosphodiesters (PDE), nicotinamide adenine dinucleotide and its phosphate (NAD and NADP respectively), and sugar phosphates.

Figure 6.1 shows a ^{31}P spectrum acquired *in-vivo* from normal neonatal piglet brain using a simple surface-coil pulse-acquire method (Lorek et al. 1994). The 3 prominent peaks on the right originate mainly from the γ-, α-, and β- ^{31}P-nuclei of magnesium (Mg)-complexed ATP ($ATPMg^{2-}$): adequate ATP levels are essential for maintenance of important cellular processes such as the sodium/potassium pump. However, in mammalian brain other NTPs, mainly guanosine, uridine, and cytidine triphosphates, contribute about a quarter of the signal amplitudes (Chapman et al. 1981). The γ- and α- resonances are doublets whereas β- is a triplet: this is because of coupling between adjacent NTP ^{31}P-nuclei (J ~16 Hz); however, for the reasons given

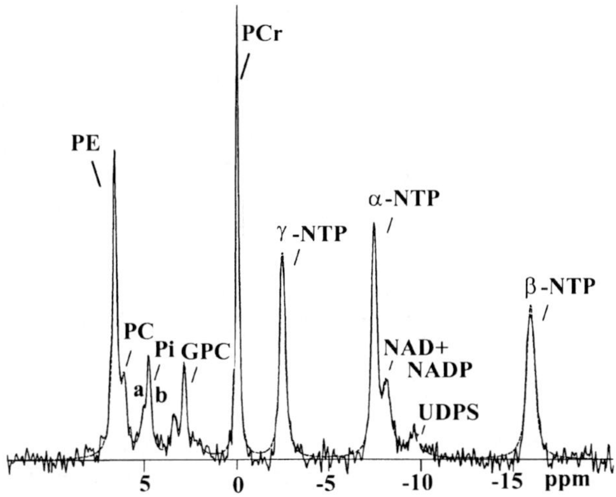

Fig. 6.1 A ^{31}P surface-coil spectrum acquired at 7 T from newborn-piglet brain under normal physiological conditions before transient cerebral hypoxia-ischaemia (192 FIDs summed). Figures 6.1, 6.6, and 6.7 were from the same experiment: spectra are normalised according to the numbers of summed FIDs; hence, signal amplitudes can be directly compared between spectra. Resonance identifications: *PE* phosphoethanolamine, *PC* phosphocholine, Pi_a extracellular inorganic phosphate (Pi), Pi_b intracellular Pi, *GPC* glycerophosphocholine (just to the left is the smaller glycerophosphoethanolamine resonance); *PCr* phosphocreatine, *γ-*, *α-*, and *β-NTP* nucleotide triphosphate (mainly ATP); *NAD* and *NADP* nicotine adenine dinucleotides, and *UDPS* uridine diphosphosugars (Acquired in collaboration with Lorek A, Takei Y, Wyatt JS, Penrice J, Edwards AD, Peebles D, Wylezinska M, Owen-Reece H, Kirkbride V, Cooper CE, Aldridge RF, Roth SC, Brown G, Delpy DT, and Reynolds EOR)

in Sect. 6.1.1 these multiplets may be unresolved *in vivo*. The γ- and α- resonances also include nucleotide diphosphate doublets of mainly adenosine diphosphate (ADP) which is produced by single dephosphorylation of ATP. Immediately to the right of α-NTP is an unresolved NAD doublet which overlaps with NADP (an unresolved doublet of doublets). Further to the right of α-NTP is a small resonance from uridine diphosphosugars (UDPS). On the left is a PME resonance due to phosphoethanolamine (PE) closely to the right of which is a smaller phosphocholine (PC) peak. PE is particularly prominent in spectra from immature brain: PE concentration falls during development (Burri et al. 1988). At ~5 ppm is Pi which clearly shows a lefthand component (a). There is evidence that separate Pi resonances originate from extracellular and cytosolic Pi (Pi_a and Pi_b respectively) with different chemical shifts representing discrete environmental pHs (Robbins et al. 1990; Portman et al. 1991). Cardiac studies suggest that Pi_a may also include a mitochondrial contribution (Garlick et al. 1992) and it has been proposed that further Pi components can be resolved (Gilboe et al. 1998). At ~3 ppm are two small PDE peaks: the larger glycerophosphocholine (GPC); smaller and just to the left glycerophosphoethanolamine (GPE). The spectrum in Fig. 6.1 was acquired at 7 T. At lower field strengths a very broad signal originating from cross-linked PDEs in cell-membrane-bilayer phospholipids (CMBP) underlies

Table 6.1 Approximate chemical shifts of neonatal mammalian brain ^{31}P metabolites resolved *in vivo*

Metabolite resonance	Chemical shift (ppm)	Metabolite resonance	Chemical shift (ppm)
PE	6.80 (.02)	PCr	0.00
PC	6.28 (.05)	γ-NTP	−2.50 (.02)
Pi_a	5.13 (.09)	α-NTP	−7.57 (.01)
Pi_b	4.87 (.04)	NAD/NADP	−8.27 (.02)
GPE	3.53 (.04)	UDPS	−9.87 (.05)
GPC	2.97 (.01)	β-NTP	−16.20 (.03)

Chemical shifts are mean (standard deviation (SD)) relative to PCr at 0 ppm and were obtained from six healthy, anaesthetised piglets under normal physiological conditions as described in Lorek et al. (1994)

the PDE resonances rendering difficult resolution of the mobile PDEs. The CMBPs exhibit chemical-shift anisotropy which greatly reduces the amplitude of this broad resonance at high field strengths (Bates et al. 1989; Murphy et al. 1989). PMEs and PDEs both have roles in cellular membrane synthesis. At 0 ppm there is a singlet peak from PCr: this is often used as a chemical-shift reference since, within the physiological range, its chemical shift is independent of various potential biophysical modulators such as pH_i (Pettegrew et al. 1988). PCr is a high-energy phosphate reserve and assists in maintaining adequate ATP levels (see Sect. 6.2.5). Table 6.1 gives chemical shifts of the ^{31}P brain metabolites in Fig. 6.1.

Numerous additional unresolved resonances of lesser amplitude are identifiable in brain-extract spectra (Glonek et al. 1982; Glonek and Kopp 1985; Pearce and Komoroski 2000). Great care must be taken to account for possible pH modulation of chemical shifts when interpreting *in-vivo* spectra from information provided by extract spectra: e.g. PE was originally misidentified as ribose-5-phosphate (Pettegrew et al. 1986).

6.2 *In-Vivo* ^{31}P Brain Acquisition Methods

The method of choice for a particular application depends critically on the nature of the problem addressed. If regional information is required, MRS localisation methods are essential. However, in other applications less sophistication may be adequate.

6.2.1 *Surface Coils*

A surface coil is simply one or more concentric conducting loops tuned to the appropriate resonant radio frequency (RF) for the nucleus concerned (Ackerman et al. 1980; Haase et al. 2000). With appropriate surface-coil dimensions, positioning, and care to minimise contamination by signal from cranial musculature useful ^{31}P

spectra can be obtained from mammalian brain and sometimes without cranial exposure (e.g. see Fig. 6.1). (^{31}P metabolite concentrations are relatively low in skin (Stubbs et al. 1988) and bone phosphate is immobile and produces an extremely broad resonance easily removed by acquisition (Ackerman et al. 1984) or processing (Vanhamme et al. 1997) techniques). Surface coils have regional sensitivity which decreases with distance: this may result in spectra originating predominantly from brain close to the cranium; however, surface coils often provide better *in-vivo* SNR than other probes.

6.2.2 Localisation Methods

These often use volume coils designed so they completely envelope the head and provide uniform excitation. Localisation enables shaping, sizing, and positioning the volume-of-interest (VOI) spectra are obtained from. However, for the reasons mentioned in the introduction, ^{31}P VOIs or/and acquisition times must be greater than for ^{1}H MRS. However, the development of double-tuned (^{1}H and ^{31}P) and quadrature (Murphy-Boesch et al. 1994) volume coils has improved matters: the latter and ^{1}H-decoupling and the consequent nuclear Overhauser effect (NOE) (Murphy-Boesch et al. 1993) have engendered substantial ^{31}P SNR increases. Sometimes localisation is applied using a volume transmit coil with a surface receive coil: this provides uniform VOI excitation and takes advantage of the superior surface-coil SNR.

6.2.2.1 *Image-Selected In-Vivo Spectroscopy (ISIS)*

ISIS adds or subtracts the signals from a sequence of three slice-selective RF inversion pulses with pulsed orthogonal magnetic field gradients (Ordidge et al. 1986). By adopting an appropriate scheme which disables various inversion pulses and adds or subtracts the resultant signals spectra are obtained originating only from a VOI at the intersection of the orthogonal tissue slices defined by the inversion pulses and the magnetic field gradients. To obtain exact signal cancellation outside the VOI ISIS needs very stable spectrometer sensitivity: hence, it is essential that tissue movement (e.g. breathing) is minimal. ISIS localisation has been improved by the application of saturation RF pulses (OSIRIS) to further suppress signal outside the VOI (Connelly et al. 1988).

6.2.2.2 Point-Resolved Spectroscopy (PRESS)

PRESS is a double spin-echo sequence with one 90° excitation and two 180° "refocussing" slice-selective RF pulses combined with pulsed orthogonal magnetic field gradients (Bottomley 1987). Similar to ISIS, a VOI is defined at the intersection of the three selected slices: however, all three slice-selective pulses are applied in a single acquisition ("single-shot") and signal is localized directly (no add/subtract scheme). Therefore, PRESS localization is less subject to movement-related outer-volume contamination than ISIS. For ^{31}P brain studies the echo time (TE) must be ~12 ms or less otherwise ATP-multiplet phase-modulation (see e.g. Rabenstein and Nakashima 1979) may result in significant signal loss (Jung et al. 1993; Cady 1995; Bluml et al. 1999).

6.2.2.3 Stimulated-Echo Acquisition Mode (STEAM)

This sequence, like PRESS, is single-shot but uses three orthogonal slice-selective 90° pulses which generate a stimulated echo (Frahm et al. 1987). The STEAM signal has half the SNR compared to PRESS. However, STEAM has been used to obtain ^{31}P brain spectra (Merboldt et al. 1990) and as a calibration method for ^{31}P ISIS absolute-quantitation studies of brain *in vivo* (Kreis et al. 2001).

6.2.2.4 Single-Shot, Inversion Recovery (SIRENE)

This is a recently developed, 3-dimensional, non-echo localization technique with potential for application to *in-vivo* ^{31}P brain studies (Choi et al. 2000). Selective pulses first invert magnetisation in slices outside but contiguous with the VOI edges. After an inversion recovery delay, carefully selected to null magnetisation in the selected slices, outer-volume magnetisation is further minimized by 3 pairs of orthogonal slice-selective saturation pulses. VOI magnetization is unaffected by the preparation pulses and is then sampled using a non-selective 90° pulse. Care must be taken to ensure that the selective preparation pulses adequately affect extra-VOI tissue with contamination potential. However, SIRENE is single-shot and, therefore, relatively resilient to tissue movement, and, because it is a non-echo method, there is no multiplet phase-modulation or spin-spin relaxation dependence.

6.2.2.5 Chemical-Shift Imaging (CSI)

PRESS, STEAM, and SIRENE are primarily single VOI methods. CSI acquires information simultaneously from whole brain with spectra localized to multiple VOIs on a "checkerboard" matrix (Fig. 6.2). In the 2-dimensional ^{31}P modality a selective RF pulse defines the required tissue slice: phase-encoding magnetic field gradient pulses are then applied orthogonally within the slice (Bottomley et al. 1988). MRS data are acquired without active gradient pulses and processed by Fourier reconstruction. CSI has been refined by adding ^{1}H decoupling (improves spectroscopic resolution and theoretically increases resonance amplitudes by NOE;

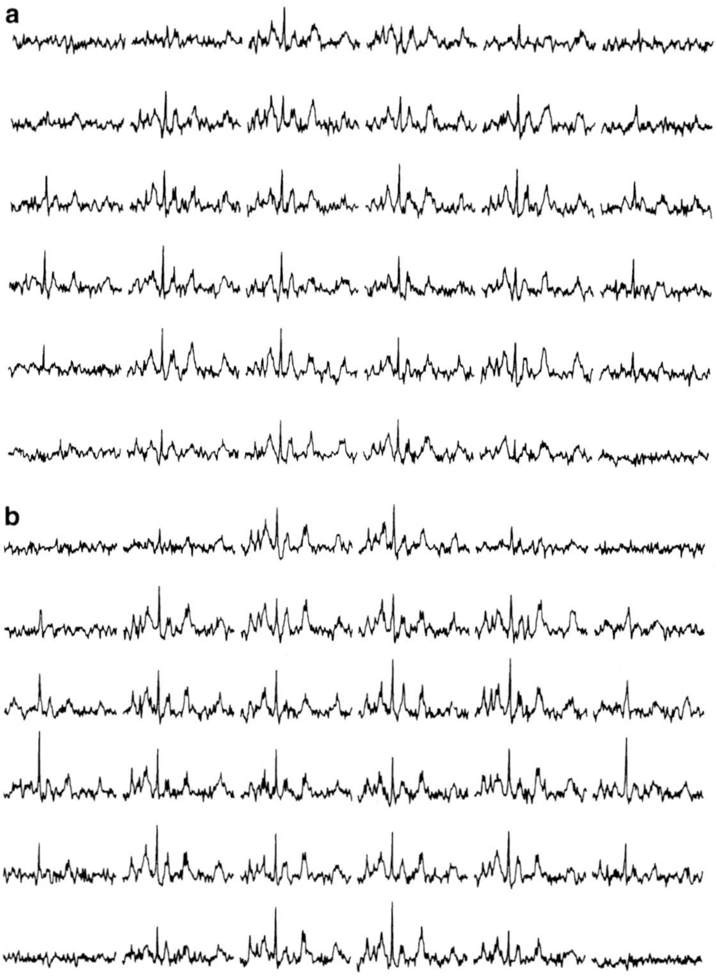

Fig. 6.2 ^{1}H-coupled (**a**) and ^{1}H-decoupled (**b**) CSI spectrum matrices acquired from adult human brain at 1.5 T using 2.6 cm cubic VOIs (17.6 cm^{3}). Due to reduced resonance widths and some NOE the ^{1}H-decoupled spectra (**b**) have better SNR particularly in the PME and PDE region. Automatic analysis of individual CSI spectra enables display of physiological images of e.g. pH_i. Resonance identifications are more clearly displayed in Fig. 6.3 (Reproduced with permission from: Murphy-Boesch et al. (1993). ©John Wiley & Sons Limited)

Fig. 6.3); by phase-encoding in 3 orthogonal directions (3-dimensional CSI; Murphy-Boesch et al. 1993); and by an echo-planar approach to reduce acquisition time (Ulrich et al. 2007). ^{31}P CSI has recently been applied to human brain at 7 T (Lei et al. 2003a). With appropriate data analysis and processing results can be presented as images of individual MRS measures.

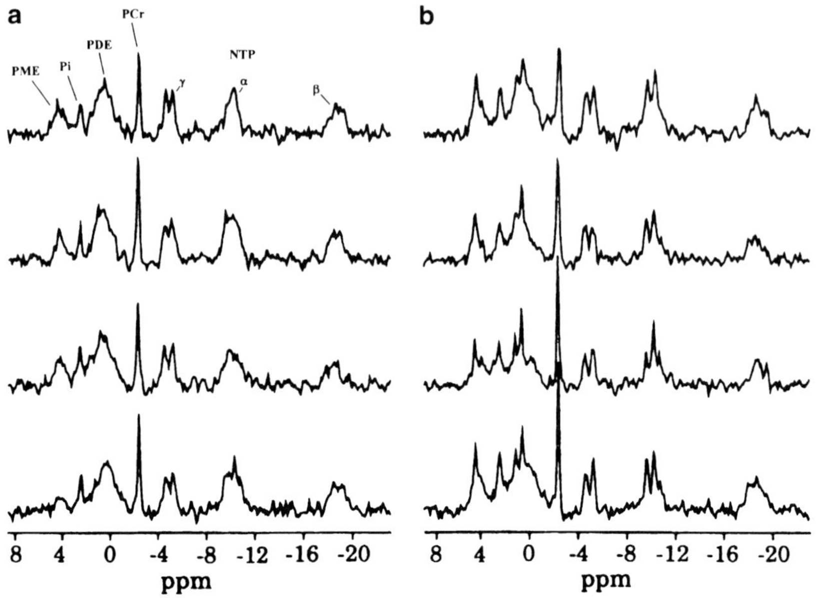

Fig. 6.3 ^{1}H-coupled (**a**) and ^{1}H-decoupled (**b**) spectra acquired at 1.5 T with 3.5 cm cubic VOIs (42.9 cm^3). These spectra illustrate the effects of ^{1}H-decoupling better than those in Fig. 6.2: PME is increased in amplitude with improved resolution; and GPE, GPC, the α-NTP doublet, and NAD/NADP are now well-resolved. Resonance identifications: *PME* phosphomonoesters, *Pi* inorganic phosphate, *PDE* phosphodiesters, *PCr* phosphocreatine, *NTP* nucleotide triphosphate (N.B. In these spectra chemical-shifts are referenced to orthophosphoric acid at 0 ppm – hence, PCr is at –2.52 ppm) (Reproduced with permission from: Murphy-Boesch et al. (1993). ©John Wiley & Sons Limited)

6.2.3 ^{1}H Decoupling

As a result of heteronuclear coupling to neighbouring ^{1}H nuclei many ^{31}P metabolite resonances are unresolved multiplets with small J-couplings ~5–10 Hz (Luyten et al. 1989a). By applying ^{1}H RF power (decoupling) during ^{31}P acquisition these multiplets can be collapsed to singlets with accompanying improvements in spectral resolution and increased SNR due to NOE (Murphy-Boesch et al. 1993; van Sluis et al. 1995). In human studies, so that safety limits for RF-power absorption were not exceeded, special broadband ^{1}H decoupling schemes such as WALTZ-4 and WALTZ-16 have been employed (Shaka et al. 1983). ^{1}H decoupling has been applied to ^{31}P brain studies using both single VOI localization (e.g. ISIS; Luyten et al. 1989a) and CSI (Murphy-Boesch et al. 1993). ^{1}H decoupling significantly reduces the apparent linewidths of PE, PC, GPE, GPC, and α-NTP (Figs. 6.2b and 6.3b) thereby greatly improving spectroscopic resolution near these resonances. In the ISIS studies there was very little NOE SNR increase: this was because WALTZ decoupling was applied only during the acquisition sequence and not during the magnetization recovery delay (TR). However, if low power continuous ^{1}H decoupling is applied during TR full NOE enhancement is obtained while remaining within RF-power

Table 6.2 ^{31}P metabolite T_2s in mature mammalian brain at 1.5, 4.7, and 7 T

Field (T)	PME	Pi	PDE	PCr	γ-NTP	α-NTP	β-NTP
1.5	196	129	3 & 243	241	89 (9)	84 (6)	62 (3)
4.7	84 (17)	62 (3)	nm	169 (7)	40.5 (3.5)	32.7 (0.3)	22.2 (0.3)
7.0	nm	nm	nm	132 (13)	26 (10)	26 (7)	nm

T_2 values (ms) are mean (SD if published); *nm* not measured; 1.5 T (Jung et al. 1993); 4.7 T (Remy et al. 1987); 7 T (Lei et al. 2003a, but also see Mulkern et al. 2003). At 1.5 T two PDE T_2s are given: these correspond to the broad (immobile) and narrow (mobile) PDE constituents respectively

safety limits: significant PE and PCr SNR increases (~30%) have been reported (Figs. 6.2 and 6.3) (Murphy-Boesch et al. 1993). Further increases in PE, GPE, and GPC SNR have been obtained using polarization transfer from ^{1}H nuclei (RINEPT - refocused insensitive nucleus enhancement by polarization transfer) in combination with ^{1}H decoupling and 3-dimensional CSI (Gonen et al. 1997). This method results in CSI spectra with signals other than PE, GPE, and GPC edited out thereby facilitating measurement of these resonances.

6.2.4 ^{31}P Metabolite Relaxation

Cerebral metabolite relaxation times are important not only because they may be of intrinsic pathological interest but also in order to maximise spectrum quality (e.g. SNR) by adjusting TR or TE. The spin–lattice relaxation time (T_1) is a measure of the time required for tissue magnetization to return to equilibrium after the RF pulse sequence. The spin-spin relaxation time (T_2) is a measure of the tissue magnetization decay immediately after an RF pulse sequence due to dephasing incurred by interactions with randomly moving nearby molecules. Both T_1 and T_2 depend on magnetic field strength so care must be taken when comparing results acquired on different spectrometers: T_2 shortens, whilst T_1 lengthens, as field strength increases. Spin-echo methods (e.g. Rabenstein and Nakashima 1979) are used for T_2 measurement. Early studies suggested that ATP T_2s were very short: however, the techniques then used did not account for phase modulation of the ATP multiplets caused by homonuclear coupling. More recent work on brain at 1.5 T (Jung et al. 1993) and 7 T (Lei et al. 2003a, but also see the cautionary comment by Mulkern et al. 2003) has employed pulse sequences inhibiting ^{31}P homonuclear phase-modulation (Table 6.2).

For measurement of ^{31}P brain-metabolite T_1s both saturation recovery and inversion recovery sequences have been used at various magnetic field strengths (e.g. Hida et al. 1992a; Buchli et al. 1994a; Lei et al. 2003a; Table 6.3).

6.2.5 Saturation Transfer

A ^{31}P brain spectrum may appear normal in terms of metabolite concentration ratios. However, for pathological reasons ATP consumption may be reduced and this may be matched by lower ATP generation resulting in an apparently normal ATP signal.

Table 6.3 ^{31}P metabolite T_1s in mature mammalian brain at 1.5, 2, 4.7, and 7 T

Field (T)	PME	Pi	PDE	PCr	γ-NTP	α-NTP	β-NTP
1.5a	2.7 (1.0)	1.5 (0.1)	1.6 (0.1)	3.3 (0.5)	1.4 (0.3)	1.0 (0.1)	1.0 (0.4)
1.5b	1.4 (0.4)	1.5 (0.3)	1.3 (0.1)	3.1 (0.5)	0.7 (0.2)	0.9 (0.2)	0.8 (0.2)
1.5c	1.2 (0.3)	1.2	1.4	2.8	0.9	1.1	1.0
2	1.7 (0.2)	1.4 (0.1)	1.3 (0.1)	2.7 (0.2)	0.6 (0.0)	1.0 (0.2)	0.7 (0.1)
4.7a	4.3 (0.4)	np	np	3.8 (0.4)	np	np	1.6 (0.1)
4.7b	3.8 (0.6)	2.6 (0.6)	np	3.5 (0.4)	1.4 (0.3)	1.3 (0.1)	1.1 (0.2)
7	4.8 (1.0)	3.2 (0.5)	4.1 (1.2) 4.0 (0.3)	3.4 (0.3)	1.3 (0.2)	1.3 (0.1)	1.0 (0.1)

T_1 values (s) are mean (SD if published); np – not published; 1.5 T (a – Luyten et al. 1989b; b – Roth et al. 1989; c – Buchli et al. 1994a); 2 T (Hubesch et al. 1990); 4.7 T (a – Hida et al. 1992a; b – Remy et al. 1987); 7 T (Lei et al. 2003a). At 7 T: PME T_1 is for PE; the PDE T_1s are for GPE and GPC respectively

Such covert changes in metabolic flux can be monitored by saturation transfer (e.g. Le Rumeur 1992). If a particular ^{31}P metabolite resonance is saturated by narrowband RF (i.e. net nuclear magnetization is nulled), then saturated nuclei exchanging chemically retain their nulled magnetization state giving reduced signal from the metabolite they moved to. The creatine kinase (CK) reaction is often monitored via saturation transfer:

$$\mathrm{PCr} + \mathrm{MgADP} + \mathrm{H}^{+} \underset{k_b}{\overset{k_a}{\leftrightarrow}} \mathrm{MgATP} + \mathrm{Cr} \tag{6.1}$$

where k_a and k_b are the forward and reverse reaction rates respectively. (Brain CK reaction stoichiometry is detailed in Roth and Weiner 1991 with an erratum in Roth and Weiner 1995) For example, to monitor reverse CK flux γ-NTP is saturated and the resultant reduction in PCr signal is measured as in Fig. 6.4 (Shoubridge et al. 1982; Corbett and Laptook 1994; Holtzman et al. 1996, 1998; Lei et al. 2003a).

Two spectra must be acquired to account for possible RF saturation effects on PCr: firstly with saturating RF on γ-NTP (Fig. 6.4b); and then with saturating RF the PCr to γ-NTP chemical-shift difference the other side of PCr (Fig. 6.4a). With inhomogeneous RF fields (e.g. using a surface coil), to attain efficient RF saturation, a pulse scheme such as BISTRO should be used (De Graff et al. 1996). Furthermore, the intrinsic PCr T_1 must be known (i.e. that without chemical exchange; Le Rumeur 1992): this can be measured by progressive saturation transfer using a range of saturation durations; hence, measurements can be time consuming. Other metabolic pathways such as Pi to γ-NTP accompanying rephosphorylation of nucleotide diphosphates (NDP) can also be explored (Lei et al. 2003b; Du et al. 2007).

6.3 Analysis of ^{31}P Brain Spectra

The aim of MRS analysis is to obtain biophysical and biochemical information from chemical shifts, peak areas/amplitudes, and relaxation times (see Sect. 6.2.4). Brain-metabolite peak-area/amplitude ratios can be determined e.g. referenced to β-

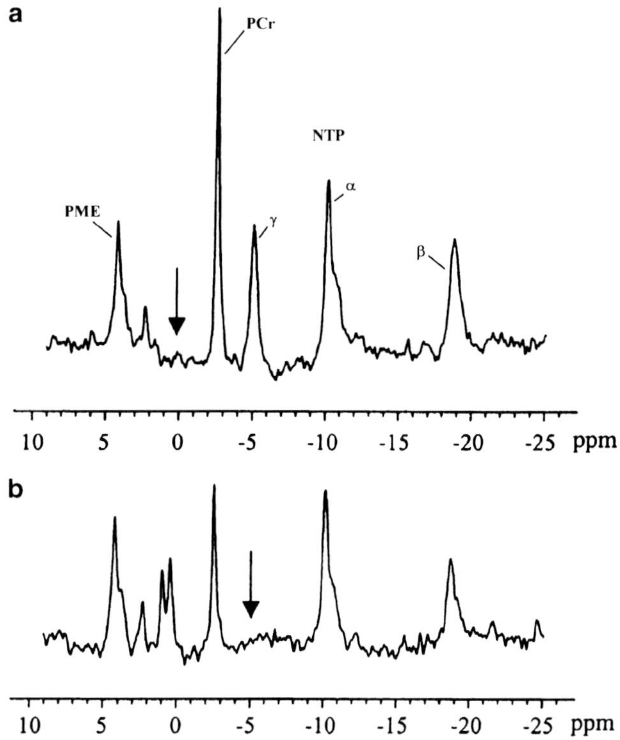

Fig. 6.4 Measurement of the CK forward rate constant in adult human brain at 7 T. Fully-relaxed ^{31}P spectra were acquired from mainly primary visual cortex using a 5-cm diameter surface coil. In spectrum (**b**) γ-NTP (*arrowed*) has been saturated by a 6.8 s BISTRO pulse train with consequent PCr signal decrease due to saturation transfer. In spectrum (**a**) γ-NTP is unsaturated and the BISTRO transmitter frequency shifted symmetrically to the other side of PCr (*arrowed*). Resonance identifications: *PME* phosphomonoesters, *PCr* phosphocreatine, *NTP* nucleotide triphosphate (N.B. In these spectra chemical shifts are referenced to orthophosphoric acid at 0 ppm – hence, PCr is at –2.52 ppm) (Reproduced with permission from Lei et al. (2003a). ©John Wiley & Sons Inc)

NTP: however, tissue metabolites can also be assayed in absolute units (e.g. mmol/kg wet weight of tissue – see Sect. 6.3.1.2). Chemical shifts provide information about pH_i and intracellular metal-ion concentrations in particular that of Mg.

6.3.1 Spectrum Analysis Methods

Ideally an analysis technique must determine 4 measures: peak area in the spectrum (frequency domain) or FID amplitude (time domain), chemical shift, peak width or FID decay rate, and the signal's RF phase. Several methods have been used: the best fit functions statistically modelling the shape of the spectrum peak (frequency domain) or its FID (time domain). State-of-the-art techniques include comprehensive prior knowledge such as the number of resonances sought, starting values for peak amplitudes, chemical shifts, peak widths or FID decay rates, and peak multiplet structures (including J-coupling and component amplitude ratios). A popular method, utilising comprehensive prior knowledge, is the "advanced method for accurate, robust, and efficient spectral fitting" (AMARES) (Vanhamme et al. 1997) although other methods are also available (e.g. Potwarka et al. 1999; Gabr et al. 2006). For accuracy adequate prior knowledge must be available about e.g. NAD and NADP multiplet structure (Carroll et al. 1980) and ATP J-coupling (commonly assumed ~16–17 Hz; Jung et al. 1997; Potwarka et al. 1999). Since ATP J varies with Mg binding an appropriate value must be used (Gupta et al. 1978; Jung et al. 1997). Furthermore, reliable measurement of relatively minor resonances such as Pi in healthy brain is assisted by including in the analysis other neighbouring minor metabolite peaks such as 2,3-diphosphoglycerate (Potwarka et al. 1999).

6.3.1.1 Chemical Shifts, Intracellular pH, and Magnesium

Chemical shifts of several cerebral ^{31}P metabolites depend on pH_i and intracellular Mg concentration (Moon and Richards 1973; Petroff and Prichard 1985; Corbett et al. 1987; Pettegrew et al. 1988; Williams et al. 1993). GPC and α-ATP chemical shifts are effectively pH invariant and that of PCr invariant over the physiological range (Pettegrew et al. 1988). GPC, α-ATP, and PCr (most often) are, therefore, used as internal chemical shift references relative to which pH-dependent chemical shifts of other metabolites are measured. The ^{1}H brain-water peak has also been suggested as a chemical-shift reference for ^{31}P MRS (Cady and Wilkie 1986; Madden et al. 1991).

Pi has often been used to estimate cerebral pH_i. As described earlier, mitochondrial, extra-, and intracellular Pi components have been observed (e.g. Fig. 6.1; Robbins et al. 1990; Portman et al. 1991; Garlick et al. 1992; Gilboe et al. 1998). At high enough magnetic field strengths extra- and intracellular Pi are often resolved and pH_i estimates can be obtained from the latter: however, at lower fields (e.g. 1.5 T) because of poor resolution brain pH_i derived via Pi is a weighted mean of the extra- and intracellular values. Cerebral pH_i is estimated from titrating Pi solutions containing an Mg concentration comparable with cerebral cytosolic (~1.36 mM; Petroff and Prichard 1985). The relationship between cerebral pH_i and Pi chemical shift (δ_{Pi}; PCr at 0 ppm) is given by a Henderson-Hasselbalch equation (e.g. Petroff and Prichard 1985):

$$pH_{pi} = 6.77 + \log_{10}((\delta_{pi} - 3.29 / (5.68 - \delta_{pi})) \tag{6.2}$$

where 3.29 ppm and 5.68 ppm are the acid and alkaline asymptotes respectively. Equation 6.2 is slightly temperature dependent: this must be accounted for if cerebral hyper-/hypothermia is involved.

The PE chemical shift is also pH dependent although its Henderson-Hasselbalch relationship is less ideal than that for Pi: the alkaline asymptote is near normal physiological pH (Corbett et al. 1987; Corbett et al. 1988; Pettegrew et al. 1988):

$$pH_{PE} = 5.625 + \log_{10}\left((\delta_{PE} - 3.190)/(6.946 - \delta_{PE})\right) \tag{6.3}$$

Use of PE may be inappropriate for mature brain because of low signal amplitude: however, PE is very prominent in ^{31}P spectra from developing brain (Burri et al. 1988). PE is almost exclusively intracellular (Westergren et al. 1994): therefore, pH_{PE} may be considered an intracellular pH probe.

The γ- and β-ATP chemical shifts also have strong pH dependences in the physiological range (Pettegrew et al. 1988): however, these also depend significantly on Mg concentration (Cohn and Hughes 1962). To avoid possible Mg modulation of γ- and β-ATP pH estimates a computer model (MAGPAC) of the interrelationships between pH, Mg concentration, and temperature has been devised (Williams et al. 1993; Williams and Smith 1995). Since the MAGPAC pH calculation depends on the chemical-shift differences between γ- and α-ATP and β- and α-ATP, in order to attain useful precision adequate ATP SNR and resolution is essential. Studies of cerebral hypoxia in neonatal rat have demonstrated good correlation between pH_{Pi} and pH_{ATP} (Williams and Smith 1995). In addition to pH, MAGPAC also gives the ratio of the concentrations of Mg_{total} (i.e. ATP-bound plus free) and ATP and the Mg_{free} concentration.

If used in addition to the Pi technique and calibrated under similar conditions, the PE and ATP pH methods may provide further useful information about the state of brain tissue. Unlike Pi, ATP is exclusively intracellular: hence, pH_{ATP} is a purely intracellular probe; furthermore, since necrotic cells contain no ATP, pH_{ATP} exclusively probes surviving cells. As already noted PE is almost exclusively intracellular: pH_{PE} may reflect the intracellular environments of both unlysed necrotic cells and surviving cells. Contrastingly, the observed Pi peak includes extracellular and intracellular components: therefore pH_{Pi} is influenced by the extracellular and intracellular environments; the latter could be in unlysed necrotic cells and surviving cells.

6.3.1.2 Absolute Quantitation of ^{31}P Cerebral Metabolites

Peak-area ratios correspond to metabolite concentration ratios only if $TR \geq 5\ T_1$ (i.e. if spectra have been acquired essentially "fully-relaxed") and if the numbers of magnetically equivalent nuclei contributing to the two peaks are equal. Ideally one would like to assay *in-vivo* metabolites more objectively e.g. in mmol/kg wet weight of tissue. The main problems are: variable MRS probe loading which depends on sample electrical conductivity and volume (increased loading reduces signal); MRS-probe RF-field inhomogeneity; and volume differences between the interrogated tissue and concentration reference. Various approaches have been applied to ^{31}P brain MRS (see Buchli et al. 1994b for a comparative assessment).

Table 6.4 *In-vivo* localized ^{31}P human cerebral metabolite concentrations assayed by various techniques

Tissue	PME	Pi	PDE	PCr	NTP	Method	Reference
Adult central[a]	–	2.6	–	4.3	3.0	ISIS, volume coil, standard/load phantom	Luyten et al. 1989b
Adult inter- hemispheric[a]	4.1	2.0	11.6	4.9	2.3	ISIS, volume coil, standard/load phantom	Roth et al. 1989
Adult parietal[a]	3.4	1.7	11.0	5.1	2.9	ISIS, surface coil, standard/load phantom	Roth et al. 1989
Adult centrum semiovale[a]	3.0 (0.7)	0.7 (0.2)	10.9 (1.8)	2.7 (0.5)	2.9 (0.3)	3D CSI, volume coil, standard/load phantom	Lara et al. 1993
Neonatal central[a]	4.5 (0.7)	0.6 (0.1)	3.2 (0.8)	1.4 (0.2)	1.6 (0.2)	ISIS, volume coil, standard phantom, load matching	Buchli et al. 1994a
Infant central[a]	3.6 (0.9)	0.6 (0.1)	4.2 (0.7)	1.7 (0.3)	1.8 (0.3)	ISIS, volume coil, standard phantom, load matching	Buchli et al. 1994a
Adult centrala	3.5 (0.6)	1.0 (0.2)	11.7 (2.2)	3.4 (0.5)	2.9 (0.4)	ISIS, volume coil, standard phantom, load matching	Buchli et al. 1994a
Neonatal central[b]	5.6 (0.9)	1.4 (0.4)	2.3 (0.6)	2.9 (0.3)	2.8 (0.6)	PRESS, volume coil, internal water	Cady et al. 1996
Adult female supra-ventricular[b]	1.65 (0.14)	0.92 (0.05)	8.02 (0.25)	2.74 (0.03)	2.36 (0.06)	ISIS, volume coil, principle of reciprocity via STEAM	Kreis et al. 2001
Adult male supra-ventricular[b]	1.84 (0.11)	0.98 (0.03)	9.73 (0.63)	2.70 (0.07)	2.46 (0.08)	ISIS, volume coil, principle of reciprocity via STEAM	Kreis et al. 2001

Concentrations (mean SD if published) are

[a]mmol/L or

[b] mmol/kg wet weight tissue

In ^{31}P surface-coil studies, absolute quantitation has been achieved by calibrating the MRS-probe impedance-matching capacitance (increased coil loading requires more capacitance) and matching the spatial dependences of ^{1}H and ^{31}P RF fields (Thulborn and Ackerman 1983; Tofts 1988). An assumed brain-water concentration is then used as a quantitation reference (Cady 1991).

With ISIS 3 approaches have been used: (a) comparing the *in-vivo* spectrum with that from a standard phantom and correcting for differential coil-loading using the signal from a further small phantom always in the MRS probe (Luyten et al. 1989b; Roth et al. 1989); (b) matching *in-vivo* and standard-phantom probe loading by repositioning a saline load (Buchli et al. 1994a); and (c) collecting STEAM spectra to assess the VOI RF field then applying the principal of reciprocity (Kreis et al. 2001; Hoult and Richards 1976). A technique similar to (a) has also been used in 3-dimensional ^{31}P CSI (Lara et al. 1993). Care must be taken using large standard phantoms at high magnetic field strengths: dielectric standing waves may generate a non-uniform RF field in the phantom and RF penetration may be poor (Tofts 1994). With ^{31}P PRESS internal water has been used as a heteronuclear concentration reference (Cady et al. 1996).

Table 6.4 gives ^{31}P cerebral metabolite concentrations obtained by various methods. The large PDE variation probably relates to localization RF-pulse timings: the signal from cross-linked (bound) PDE has a very short T_2 compared to those of mobile GPC and GPE.

6.4 Mammalian Brain Development, Ageing, Regional Metabolism, and Pathology

To interpret pathological *in-vivo* ^{31}P brain spectra normative data are essential including information about regional, developmental, and species variations. Recent ^{31}P MRS technical developments now enable probing of hitherto unexplored aspects of neurology. For example, investigations of human brain metabolism after sleep deprivation reveal NTP and GPC increases which only arise after a subsequent night of sleep (Dorsey et al. 2003).

6.4.1 Development and Ageing

Maturational dependences have been observed in the brains of human and other mammalian species. In both rat and guinea-pig the PCr/NTP concentration ratio increased ~100% between age 4 days and adulthood whereas PME/NTP fell (Tofts and Wray 1985; Hida et al. 1992b). Similar changes have been observed in human brain: notably PCr/Pi, PCr/NTP, and PDE/PME increasing as the brain develops from neonatal to adult (Fig. 6.5; Azzopardi et al. 1989a; Van der Knaap et al. 1990; Buchli et al. 1994a). Mammalian brain development is accompanied by decreasing

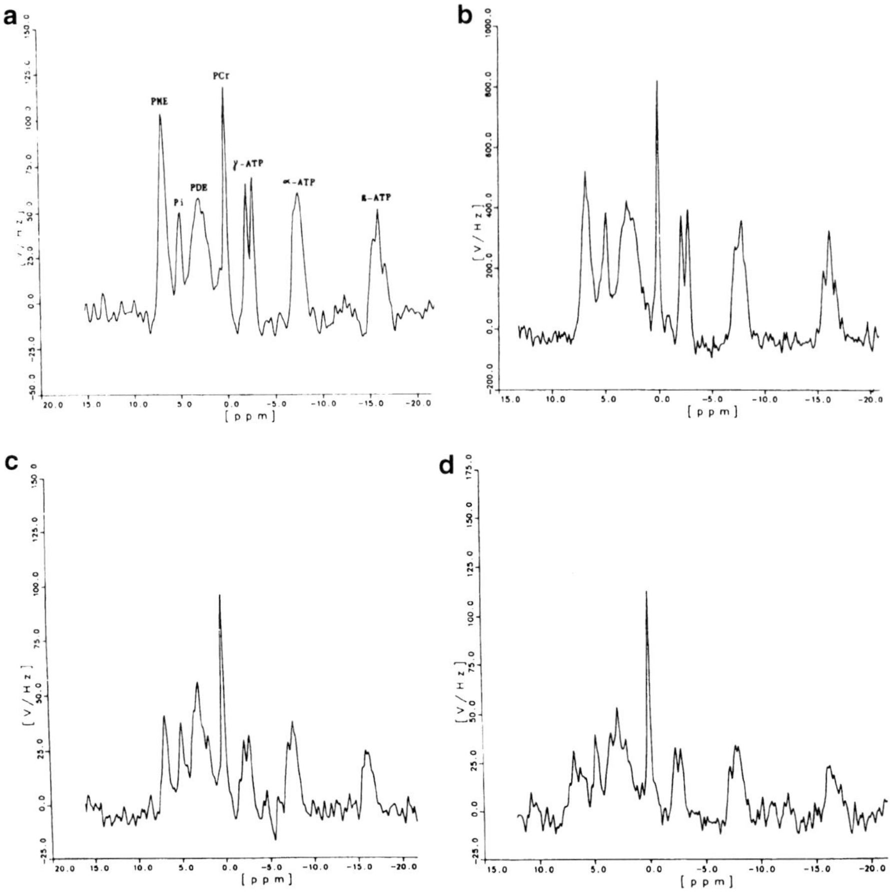

Fig. 6.5 Developmental changes illustrated by ^{31}P-ISIS human-brain spectra at ages 1 month (**a**); 4 months (**b**); 2½ years (**c**); and 15 years (**d**). Smaller brain volumes were sampled in younger subjects and coil efficiency was reduced in older subjects because of greater head volume (increased coil loading): hence, direct comparison of individual metabolite signal amplitudes at different ages is meaningless. However, in relative terms (e.g. compared to ATP) PCr and PDE increase with development whereas PME decreases. Resonance identifications: *PME* phosphomonoesters, *Pi* inorganic phosphate, *PCr* phosphocreatine, *ATP* adenosine triphosphate (including some other nucleotide triphosphates) (Reproduced with permission from: Van der Knaap et al. (1990). ©Radiological Society of North America)

PME concentration and increasing Pi, PDE, PCr, and NTP concentrations (Table 6.4; Buchli et al. 1994a). In rat brain maturational PME decrease is largely due to a fall in PE accompanied by a concomitant increase in phosphatidylethanolamine (Burri et al. 1988). In addition to falling PE human ^{1}H-decoupled ^{31}P-MRS has shown a maturational decline in PC accompanied by increasing GPE and GPC (Bluml et al. 1999). Fetal rat brain has been studied using surface-coil and ISIS techniques (Nakada et al. 1989; Nakada and Kwee 1993).

It has been postulated that high PME in immature brain relates to ongoing membrane development and myelination. The apparent maturational increase in PCr has been interpreted as an increasing energy reserve requirement as the brain develops. Maturational changes in ^{31}P brain spectra indicate the care needed comparing results from altricious and precocious species during perinatal development.

Recent work has investigated ageing: mitochondrial dysfunction increasing with age may underpin many of the characteristics of senescence. A recent 4-T study of healthy human subjects aged 21–84 years revealed a pH_i decrease of about 0.5%/decade, PCr concentration increasing at about 1.1%/decade, and PME concentration decreasing at about 1.7%/decade (Forester et al. 2010). ATP, Pi, PDE, and intracellular magnesium concentrations were age invariant although a trend towards increasing ATP concentration with age (about 1%/decade) only just failed to reach significance. Increased PCr and ATP concentrations and reduced pH_i may reflect altered energy generation in healthy ageing. A further study utilised a polarisation transfer method at 3 T to investigate PME and PDE changes in healthy human subjects aged 16–70 years (Wijnen et al. 2010). Strong inverse relationships were found between age and PE/GPE and PC/GPC.

6.4.2 *Regional Variations*

It is difficult to assess regional variations in the brains of small biological models because of low ^{31}P-MRS SNR and the consequent necessity to often use either surface coils or single VOI localization with large selected volumes. However, recent 3-dimensional CSI of adult human brain has demonstrated higher PCr/NTP and PCr/(total mobile ^{31}P (P_{total})) in gray than white matter whereas PCr/Pi and pH_i were similar (Mason et al. 1998). Further 3-dimensional CSI investigation at 4 T utilising absolute quantitation and multi-compartment regression analysis to assay "pure" cerebral tissue types has revealed 3.5 (0.3), 3.3 (0.4), and 3.8 (0.7) mmol PCr/L brain water (mean (SD)) in gray matter, white matter, and cerebellum respectively (Hetherington et al. 2001): corresponding NTP values were 2.2 (0.3), 3.4 (0.3), and 1.8 (0.6) mmol/L brain water. Gray-matter NTP concentration was less than in white matter ($p<10^{-5}$) and cerebellum NTP concentration was less than in either gray or white matter ($p<.04$ and $<10^{-5}$ respectively). Regional PE, PC, GPE, GPC, and membrane phospholipid concentrations have also been measured by 3-dimensional CSI at 4 T (Jensen et al. 2002).

6.4.3 *Pathological Studies*

The brevity of this chapter renders it impossible to summarise completely all ^{31}P-MRS brain pathology applications in biological analogues and human subjects: however, it is intended to at least illustrate some more important applications.

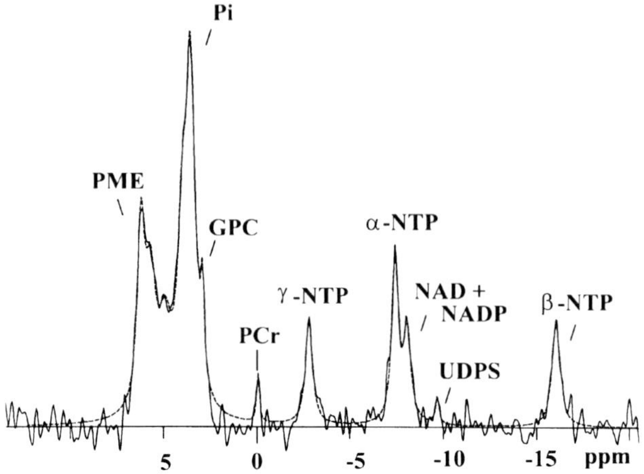

Fig. 6.6 A ^{31}P surface-coil brain spectrum at the end of cerebral hypoxia-ischaemia (64 FIDs summed). Figures 6.1, 6.6, and 6.7 were from the same experiment: spectra are normalised according to the numbers of summed FIDs; hence, signal amplitudes can be directly compared between spectra. The noise level in this hypoxia-ischaemia spectrum is apparently greater than that in Figs. 6.1 and 6.7 because of this normalisation. Compared to Fig. 6.1, PCr and NTP are reduced and Pi greatly increased (and acid-shifted to the right). Within 2 h of resuscitation cerebral energy metabolism apparently recovered to give a spectrum (not shown) similar to that in Fig. 6.1. Resonance identifications as for Fig. 6.1 (Acquired in collaboration with Lorek A, Takei Y, Wyatt JS, Penrice J, Edwards AD, Peebles D, Wylezinska M., Owen-Reece H, Kirkbride V, Cooper CE, Aldridge RF, Roth SC, Brown G, Delpy DT, Reynolds EOR)

6.4.3.1 Hypoxia-Ischaemia

An early cerebral ^{31}P-MRS application was the study of perinatal asphyxia using surface coils (Hope et al. 1984): on resuscitation newborn brain apparently recovers metabolically from the intrapartum energy generation impairment; in biological analogues the latter results in decreased PCr and NTP, increased Pi, and profound intracellular acidosis (see e.g. Williams et al. 1992; Allen et al. 1993; Lorek et al. 1994). However, metabolic recovery was often followed within 12–24 h by ^{31}P-MRS characteristics similar to the acute phase apart from a slight intracellular alkalosis instead of acidosis (Azzopardi et al. 1989b; Martin et al. 1996; Robertson et al. 2002). Reduced PCr/Pi in the first few days of life predicted eventual neurodevelopmental impairment and if NTP was reduced fatality was common. This delayed phase of metabolic abnormality has been termed "secondary energy failure" (SEF) and modelled in the newborn piglet (Lorek et al. 1994). Figure 6.1 (Sect. 6.1.2.) shows a baseline ^{31}P spectrum of newborn piglet brain. Transient hypoxia-ischaemia was then induced by reversible bilateral carotid-artery occlusion and reducing the inspired oxygen fraction initially to 16% and then by further 2% decrements. Serial ^{31}P MRS during the insult revealed developing PCr and NTP depletion accompanied by increased Pi and intracellular acidosis. Figure 6.6 shows a spectrum acquired at

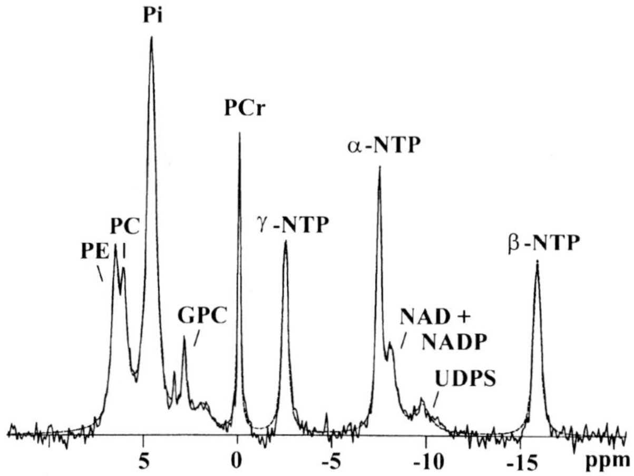

Fig. 6.7 A ^{31}P surface-coil spectrum acquired 33 h after cerebral hypoxia-ischaemia (384 FIDs summed). Figures 6.1, 6.6, and 6.7 were from the same experiment: spectra are normalised according to the numbers of summed FIDs; hence, signal amplitudes can be directly compared between spectra. Although by 2 h after hypoxia-ischaemia cerebral energy metabolism had apparently normalised following resuscitation, at 33 h secondary energy failure was well established: reduced PCr and greatly increased Pi (without acidosis). Resonance identifications as for Fig. 6.1. (Acquired in collaboration with Lorek A, Takei Y, Wyatt JS, Penrice J, Edwards AD, Peebles D, Wylezinska M, Owen-Reece H, Kirkbride V, Cooper CE, Aldridge RF, Roth SC, Brown G, Delpy DT, Reynolds EOR)

the end of hypoxia-ischaemia. On resuscitation metabolic recovery commenced and within ~2 h the ^{31}P spectrum was comparable with baseline. However, by ~24 h SEF was apparent: Fig. 6.7 shows a ^{31}P spectrum acquired ~33 h after resuscitation; compared to the baseline spectrum (Fig. 6.1) and recovery (not shown), amongst less prominent changes PCr has clearly fallen and there is a large Pi increase.

Although ^{31}P MRS suggests brain energy metabolism often recovers apparent normality after resuscitation, there is evidence of injury prior to overt SEF including: raised β-amyloid precursor protein content (Baiden-Amissah et al. 1998); calcium accumulation, mitochondrial swelling with maximal enlargement associated with nuclear chromatin condensation, and apoptosis and necrosis features (Puka-Sundvall et al. 2000); increased lactate dehydrogenase and propidium iodide fluorescence (Graulich et al. 2002); and abnormal white-matter nerve fibres and enhanced white-matter apoptosis (Meng et al. 2005). SEF is probably a consequence of such injury initiated by the primary hypoxia-ischaemia but with metabolic consequences detectable by ^{31}P MRS only many hours later. Recent research suggests supra-baseline PCr recovery in the first hours after perinatal hypoxia-ischaemia is associated with more favourable outcome whereas recovery to a maximum below baseline prognosticates eventual severe injury (Iwata et al. 2008).

A useful feature of the newborn-piglet model is that the serial ^{31}P spectra during hypoxia-ischaemia provide a quantitative insult-severity index (the time-integral of insult and recovery NTP depletion) thereby avoiding the problem of variable biological response (Lorek et al. 1994). Piglet and rat neonatal models have proven useful for testing putative pharmacological and hypothermic therapies (e.g. Thoresen et al. 1995; Adcock et al. 2002; Peeters-Scholte et al. 2003). The piglet model has also shown that brain-water apparent diffusion coefficient (ADC) measured by magnetic resonance imaging (MRI) falls as PCr/Pi and NTP decline during SEF (Thornton et al. 1998): MRI ADC may be a cerebral energetics index with spatial resolution far better than ^{31}P MRS.

In focal subacute and chronic hypoxia-ischaemia (stroke) overall decreases in ^{31}P metabolite concentrations have been observed (Hugg et al. 1992). Such overall concentration decreases are probably associated with, and proportional to, irreversible tissue loss. Furthermore, intracellular Mg concentration in the ischaemic focus appears increased acutely and subacutely and coincides with intracellular acidosis, reduced PCr/(PCr + Pi), and increased Pi/P_{total} (Helpern et al. 1993). During subsequent stroke evolution pH_i normalised by days 2–3, Mg concentration by days 4–10, and PCr/(PCr + Pi) and Pi/P_{total} within 10 days. *In-vivo* ATP is largely bound to Mg (Kuby and Noltmann 1962; Skou 1982). It has been suggested that increased free intracellular Mg during hypoxia-ischaemia is caused by $ATPMg^{2-}$ hydrolysis (Gee et al. 2004). A plasma [Mg] increase could initiate entry into cells via the ATP-independent Na^{+}-Mg^{2+} antiport (Gunzel and Schlue 1996). However, hypoxia-ischaemia by itself does not increase plasma [Mg]. It is likely that the raised free intracellular [Mg] observed in stroke was due to liberation by adenosine-polyphosphate hydrolysis.

6.4.3.2 Epilepsy, Schizophrenia, and Depression

Until recently ^{31}P MRS study of e.g. epilepsy has been hindered by the intrinsic low SNR and consequent requirement for large VOIs. Higher magnetic field strengths, quadrature probes, and ^{1}H decoupling have improved SNR and enabled investigations with smaller VOIs.

In recent epilepsy studies multi-compartment regression analysis which takes into account gray- and white-matter fractions has enabled more detailed investigation of ipsilateral ^{31}P-metabolite changes (Hetherington et al. 2001). Overall ipsilateral hippocampal PCr/ATP was reduced compared to its normal contralateral value (Hetherington et al. 2002; 2004). Furthermore, finer anatomical resolution showed PCr/ATP was less in amygdala than in pes hippocampus (the more anterior part of the hippocampus head) and less in both these structures than in the hippocampal body. There was no correlation between ipsilateral PCr/ATP decrement and either histological neuronal loss or hippocampal volume. It has been suggested that the reduced ipsilateral PCr/ATP was probably related to either: energy production failing to meet demand; or altered relative amounts of different cell types with each type having a discrete PCr/ATP and, hence, changing the ipsilateral ratio. Ipsilateral PCr/ATP did not correlate with neuronal loss (Hetherington et al. 2004) tending to support the former hypothesis and suggesting that successful seizure treatment should reverse the PCr/ATP decrement. Indeed, after surgery or ketogenic-diet treatment ^{31}P-MRS demonstrated reversal of the energetic deficit (Simor et al. 1997;

Pan et al. 1999). The manner by which a ketogenic diet (high fat, low carbohydrate) reverses the energetic abnormality remains unclear (Hetherington et al. 2002): however, it is known that ketones can be an alternative to, or supplement, glucose as an oxidative substrate.

Schizophrenia investigations have mainly focussed on phospholipid changes because of the current membrane-phospholipid hypothesis. First-episode schizophrenics studied at 4 T in thalamus, cerebellum, hippocampus, anterior and posterior cingulate, prefrontal cortex and parieto-occipital cortex showed increased GPC concentration and also raised Pi, PCr, and NTP concentrations (Jensen et al. 2004): CSI studies demonstrated NTP was increased in white matter and decreased in grey matter in the fronto-temporal-striatal region (Jensen et al. 2006). Another CSI study, including multi-episode subjects, reported reduced PME, PDE, Pi, and PCr in caudate nucleus, thalamus, anterior cerebellum, and bilateral prefrontal and medial temporal regions (Smesny et al. 2007). Prefrontal-cortex studies of non-psychotic, but high risk, offspring of parents with schizophrenia or schizoaffective disorder have demonstrated PME and PDE abnormalities (Keshavan et al. 2003). A similar study of children and siblings of schizophrenic parents revealed lower PME/PDE and increased PDE possibly resulting from enhanced phospholipid breakdown (Klemm et al. 2001). It has been suggested that increased PCr and NTP may result from neuronal hyperactivity in early schizophrenia resulting in excitotoxic breakdown of neural-membrane phospholipid and leading to increased PDE (Jensen et al. 2004).

In bipolar disorder (manic depressive illness) reduced PME has been observed in temporal and frontal lobes (Deicken et al. 1995a, b). Furthermore, increased PDE has been observed in both frontal lobes accompanied by an increased frontal-lobe right-to-left PCr ratio (Deicken et al. 1995b). Reduced pH_i has been observed in the frontal lobes and basal ganglia and also in non-localized spectra (Hamakawa et al. 2004). These studies suggest bipolar disorder is associated with alterations in high-energy phosphate and phospholipid metabolism.

6.4.3.3 Degenerative Diseases

^{31}P MRS studies of Alzheimer's disease have indicated that mildly demented patients have increased PME, decreased PCr, and increased oxidative metabolic rate (Pettegrew et al. 1997). High-spectral-resolution ^{31}P MRS of Folch extracts of autopsy tissue has indicated reduced phosphatidylethanolamine (PtdE) and phosphatidylinositol, and increases in sphingomyelin and the plasmalogen derivative of PtdE: diphosphatidylglycerol and phosphatidic acid were reduced in the superior temporal gyrus (Pettegrew et al. 2001). These abnormalities may contribute to altered membrane repair in Alzheimer's disease.

An early study of superior biparietal white and gray matter in Parkinson's disease failed to demonstrate metabolic abnormality (Hoang et al. 1998): however, a later study of non-demented patients found abnormalities in right and left temporoparietal cortex (TPC), occipital cortex, and a central VOI including basal ganglia and brainstem (Hu et al. 2000). In Parkinson's patients TPC Pi/β-NTP was increased bilaterally and in right TPC Pi/P_{tot} was increased: right TPC Pi/β-NTP correlated with reduced performance intelligence quotient (IQ); left TPC Pi/β-NTP correlated with full-scale and verbal IQs. The latter study also included (18)FDG-PET demonstrating bilateral glucose metabolism reductions which, with the ^{31}P-MRS results, suggested impairment of both glycolytic and oxidative energy generation. A recent visual-cortex activation study reported that the combined PCr + β-NTP signal decreased during recovery from activation in Parkinson's disease whereas this increased slightly in controls: it was proposed that the post-activation decrease in Parkinson's disease suggested mitochondrial dysfunction (Rango et al. 2006). A recent proton MRS study has also shown increased lactate (Henchcliffe et al. 2008).

In Huntington's disease ^{1}H-decoupled ^{31}P MRS found no energy metabolism abnormality in superior biparietal white and gray matter (Hoang et al. 1998) in contrast to results from skeletal muscle (Lodi et al. 2000). The latter study revealed reduced muscle ATP/(PCr + Pi) at rest in both symptomatic and presymptomatic patients: compared to normal subjects maximum ATP production rates during recovery from exercise were 44% and 35% less in symptomatic and presymptomatic patients respectively. These abnormalities support a mitochondrial dysfunction role in the pathogenesis of Huntington's disease. Recent ^{31}P-PRESS studies of a murine Huntington's disease model (R6/2) have demonstrated lower Pi/P_{tot} compared to wild-type, age-matched controls (Page et al. 2003).

6.4.3.4 Metabolic Disorders and Substance Abuse

In adult phenylketonuria (PKU) patients under steady state conditions ^{31}P MRS has shown mostly normal metabolism – however, compared to controls Pi and the broad phospholipid signal were decreased (22% and 8% respectively) and ADP and the phosphorylation potential were increased (11% and 22% respectively) (Pietz et al. 2003). The ADP level correlated with that of phenylalanine in both plasma and brain. On phenylalanine loading the phosphorylation potential decreased and the ADP level increased concomitantly with slowed electroencephalogram background activity. Reduced Pi signal could result from less cytosolic Pi or increased binding of Pi to the inner mitochondrial membrane: binding may reduce Pi mobility and render it MRS-invisible: a further possibility is enhanced PCr synthesis. The reduction in broad phospholipid signal was unaccompanied by changes in PME or PDE and did not correlate with white-matter changes detected by imaging.

A genetically-modified murine model of CK deficiency has shown normal high-energy ^{31}P metabolites and pH_i when a single CK isoenzyme was lacking (in 't Zandt et al. 2004): however, in CK double knockouts PCr was almost completely undetectable although there was normal pH_i and Pi. In the same study ^{1}H MRS showed that the double-knockout ^{31}P changes were accompanied by 20–30% increases in creatine and N-acetylaspartate (a putative neuronal marker). Studies of skeletal muscle using the same model suggested that creatine uptake in this tissue was independent of its role in PCr generation (in 't Zandt et al. 2003). It was suggested that, in addition to its phosphorylation role in normal brain and muscle, creatine may have a further, as yet unidentified, metabolic function.

In unchallenged type-I diabetes patients occipital-lobe ^{31}P MRS revealed low PCr/γ-NTP (Bischof et al. 2004): when hypoglycaemia was induced by hyperinsulinaemia, cerebral pH_i increased in both controls and diabetics but became higher in the latter; intracellular Mg fell in both groups. The reduced PCr/γ-NTP in diabetics suggested a chronic alteration of cerebral energy metabolism possibly related to changes in either CK activity or mitochondrial transcription. The greater pH_i increase in diabetics provided further evidence of perturbed cerebral energy metabolism since elevated pH_i facilitates glycolysis.

Increased isolated brain membrane rigidity is related to chronic alcohol exposure. ^{31}P MRS of the centrum semiovale showed that alcohol-dependent heavy drinkers had less broad phospholipid signal (Sect. 6.1.2) than light drinkers (Estilaei et al. 2001a). Further studies of ex-alcoholics (average 31 months abstinence) revealed phospholipid level and T_2 similar to those of light drinkers (Estilaei et al. 2001b). However, the contributions of fast and slow relaxing phospholipid components in the abstainers were still different to those in light drinkers. It was suggested that the reduced broad phospholipid signal in heavy drinkers could be interpreted as decreased myelin or membrane content predating the overt white-matter tissue loss seen in older subjects. Although abstainers showed some reversion to normality, differences in the relative amounts of fast- and slow-relaxing phospholipids suggested on-going, as yet unreversed or permanent, abnormality.

Abnormalities in cerebral energetics and phospholipids have been reported in heroin-dependence. ^{31}P MRS has shown that addicts on methadone-maintenance had ~15% less PCr and ~13% more PDE than age-matched healthy subjects although ATP levels were similar (Silveri et al. 2004): analysis for treatment duration effects showed reduced PCr was apparent only after 8 days of methadone-maintenance and PME was elevated during the period 15–28 days after treatment commenced. It was postulated that the PCr reduction represented cerebral energy metabolism adaptation in response to treatment: the PDE and PME increases were possibly related to membrane proliferation or remodelling.

6.4.3.5 Brain Tumours

There is a vast MRS literature on studies of brain tumour metabolism and monitoring therapy response: it is impossible to give a comprehensive summary here. Many advances have been achieved using ^{1}H MRS (e.g. McKnight 2004). The PME and PDE ^{31}P-MRS resonances are considered potential indices of tumour growth and malignancy because of the respective phospholipid precursor and catabolite roles of their contributory ^{31}P metabolites.

^{31}P ISIS has recently been used to study gliomas, glioblastomas, and meningiomas (Maintz et al. 2002). Meningiomas exhibited the greatest metabolic differences compared to normal brain: they were more alkaline, had less PCr, and increased PDE. Glioblastomas were also alkaline and had decreased PDE/α-NTP. Low malignancy gliomas were slightly alkaline and PDE/α-NTP was more than halved. Tumour spectra displayed clear but inconsistent responses to therapy.

The consistent intracellular alkalosis was contrary to an anticipated acidosis previously postulated to result from glycolytic lactate production. The reason for alkalosis remains unclear: however, it has been suggested that enhanced Na^+/H^+ exchange may be causative (Oberhänsli et al. 1986; Schuldiner and Rozengurth 1982). Maintz et al. were unable to explain their observation of a consistent PDE/α-NTP reduction: altered membrane metabolism might be anticipated; however, there was no concomitant PME change.

Untreated pediatric brain tumours studied by ^{1}H-decoupled 2-dimensional ^{31}P CSI have revealed significant differences amongst tumour types and compared to normal brain tissue (Albers et al. 2005). Six tumour types were studied: primitive neuroectodermal tumor, anaplastic ependymoma, ependymoma, anaplastic astrocytoma, germinoma, and fibrous histocytoma. In primitive neuroectodermal tumor PE/GPE and PC/GPC were both higher than in either normal brain or other tumour types. PC/PE was higher in all tumours relative to controls but was similar between all tumour types. Again, as reported by Maintz et al. tumour pH_i was more alkaline than in normal brain.

6.5 Conclusions

During its early *in-vivo* development localized ^{31}P MRS was handicapped by poor SNR and large VOI dimensions compared to that available with ^{1}H MRS. With improvements such as ^{1}H-decoupling, quadrature probes, and higher magnetic field strengths localised ^{31}P brain spectra can now be acquired from humans and other large mammals with spatial and spectroscopic resolution and SNR adequate for application in fields such as psychiatry, neurodegenerative disease, and oncology. Study of genetically-modified biological analogues also provides significant scenarios for the development of ^{31}P-MRS indices of disease progress and the testing of putative therapies. The combination of ^{1}H and ^{31}P localised MRS now provides a powerful investigatory tool.

References

Ackerman JJH, Grove TH, Wong GG, Gadian DG, Radda GK (1980) Mapping of metabolites in whole animals by ^{31}P NMR using surface coils. Nature 5743:167–170

Ackerman JJH, Evelhoch JL, Berkowitz BA, Kichura GM (1984) Selective suppression of the cranial bone resonance from ^{31}P NMR experiments with rat brain in vivo. J Magn Reson 56:318–322

Adcock KH, Nedelcu J, Loenneker T, Martin E, Walliman T, Wagner BP (2002) Neuroprotection of creatine-supplementation in neonatal rats with transient cerebral hypoxia-ischemia. Dev Neurosci 24:383–388

Albers MJ, Krieger MD, Gonzalez-Gomez I, Gilles FH, McComb JG, Nelson MD, Bluml S (2005) Proton-decoupled ^{31}P MRS in untreated pediatric brain tumors. Magn Reson Med 53:22–29

Allen KL, Busza AL, Proctor E, King MD, Williams SR, Crockard HA, Gadian DG (1993) Controllable graded cerebral ischaemia in the gerbil: studies of cerebral blood flow and energy metabolism by hydrogen clearance and ^{31}P NMR spectroscopy. NMR Biomed 6:181–186

Azzopardi D, Wyatt JS, Hamilton PA, Cady EB, Delpy DT, Hope PL, Reynolds EOR (1989a) Phosphorus metabolites and intracellular pH in the brains of normal and small-for-gestational age infants investigated by magnetic resonance spectroscopy. Pediatr Res 25:440–444

Azzopardi D, Wyatt JS, Cady EB, Delpy DT, Baudin J, Stewart AL, Hope PL, Hamilton PA, Reynolds EOR (1989b) Prognosis of newborn infants with hypoxic-ischemic brain injury assessed by phosphorus magnetic resonance spectroscopy. Pediatr Res 25:445–451

Baiden-Amissah K, Joashi U, Blumberg R, Mehmet H, Edwards AD, Cox PM (1998) Expression of amyloid precursor protein (β-APP) in the neonatal brain following hypoxic-ischaemic injury. Neuropath App Neurobiol 24:346–352

Bates TE, Williams SR, Gadian DG (1989) Phosphodiesters in the liver: the effect of field strength on the ^{31}P signal. Magn Reson Med 12:145–150

Bischof MG, Mlynarik V, Brehm A, Bernroider E, Krssak M, Bauer E, Madl C, Bayerle-Eder M, Waldhausl W, Roden M (2004) Brain energy metabolism during hypoglycaemia in healthy and type I diabetic subjects. Diabetologia 47:648–651

Bluml S, Seymour KJ, Ross BD (1999) Developmental changes in choline- and ethanolamine-containing compounds measured with proton-decoupled ^{31}P MRS in in vivo human brain. Magn Reson Med 42:643–654

Bottomley PA (1987) Spatial localization in NMR spectroscopy in vivo. Ann New York Acad Sci 508:333–348

Bottomley PA, Charles HC, Roemer PB, Flamig D, Engelseth H, Edelstein WA, Mueller OM (1988) Human in vivo phosphate metabolite imaging with ^{31}P NMR. Magn Reson Med 7:319–336

Buchli R, Martin E, Boesiger P, Rumpel H (1994a) Developmental changes of phosphorus metabolite concentrations in the human brain: a ^{31}P magnetic resonance spectroscopy study in vivo. Pediatr Res 35:431–435

Buchli R, Martin E, Boesiger P (1994b) Comparison of calibration strategies for the in vivo determination of absolute metabolite concentrations in the human brain by ^{31}P MRS. NMR Biomed 7:225–230

Burri R, Lazeyras F, Aue WP, Straehl P, Bigler P, Althaus U, Herschkowitz N (1988) Correlation between ^{31}P NMR phosphomonoester and biochemically determined phosphorylethanolamine and phosphatidylethanolamine during development of the rat brain. Dev Neurosci 10:213–221

Cady EB (1991) A reappraisal of the absolute concentrations of phosphorylated metabolites in the human neonatal cerebral cortex obtained by fitting Lorentzian curves to the ^{31}P NMR spectrum. J Magn Reson 91:637–643

Cady EB (1995) Quantitative combined phosphorus and proton PRESS of the brains of newborn human infants. Magn Reson Med 33:557–563

Cady EB, Wilkie DR (1986) Estimation of cerebral intracellular pH by ^{31}P and ^{1}H nuclear magnetic resonance spectroscopy. In: Rolfe P (ed) Neonatal physiological measurements. Butterworths, London, pp 373–381

Cady EB, Wylezinska M, Penrice J, Lorek A, Amess P (1996) Quantitation of phosphorus metabolites in newborn human brain using internal water as reference standard. Magn Reson Imaging 14:293–304

Carroll RC, Edelheit EB, Schmidt PG (1980) Phosphorus nuclear magnetic resonance of bovine platelets. Biochemistry 19:3861–3867

Chapman AG, Westerberg E, Siesjo BK (1981) The metabolism of purine and pyrimidine nucleotides in rat cortex during insulin-induced hypoglycemia and recovery. J Neurochem 36:179–189

Choi I-Y, Tkac I, Gruetter R (2000) Single-shot, three-dimensional "non-echo" localization method for in vivo NMR spectroscopy. Magn Reson Med 44:387–394

Cohn M, Hughes TR (1962) Nuclear magnetic resonance spectra of adenosine di- and triphosphate. II. Effect of complexing with divalent metal ions. J Biol Chem 237:176–181

Connelly A, Counsell C, Lohman JAB, Ordidge RJ (1988) Outer volume suppressed image related in vivo spectroscopy (OSIRIS), a high-sensitivity localization technique. J Magn Reson 78:519–525
Corbett RJT, Laptook AR, Nunnally RL (1987) The use of the chemical shift of the phosphomonoester P-31 magnetic resonance peak for the determination of intracellular pH in the brains of neonates. Neurology 37:1771–1779
Corbett RJ, Laptook AR, Hassan A, Nunnally RL (1988) Quantitation of acidosis in neonatal brain tissue using the 31P NMR resonance peak of phosphoethanolamine. Magn Reson Med 6:99–106
Corbett RJT, Laptook AR (1994) Age-related changes in swine brain creatine kinase-catalyzed ^{31}P exchange measured in vivo using ^{31}P NMR magnetization transfer. J Cereb Blood Flow Metab 14:1070–1077
De Graff RA, Luo Y, Garwood M, Nicolay K (1996) B_1-insensitive, single-shot localization and water suppression. J Magn Reson B 113:35–45
Deicken DF, Weiner MW, Fein G (1995a) Decreased temporal lobe phosphomonoesters in bipolar disorder. J Affect Disord 33:195–199
Deicken RF, Fein G, Weiner MW (1995b) Abnormal frontal lobe phosphorus metabolism in bipolar disorder. Am J Psychiatry 152:915–918
Dorsey CM, Lukas SE, Moore CM, Tartarini WL, Parow AM, Villafuerte RA, Renshaw PF (2003) Phosphorous31 magnetic resonance spectroscopy after total sleep deprivation in healthy adult men. Sleep 26:573–577
Du F, Zhu XH, Qiao H, Zhang X, Chen W (2007) Efficient in vivo 31P magnetization transfer approach for noninvasively determining multiple kinetic parameters and metabolic fluxes of ATP metabolism in the human brain. Magn Reson Med 57:103–114
Estilaei MR, Matson GB, Payne GS, Leach MO, Fein G, Meyerhoff DJ (2001a) Effects of chronic alcohol consumption on the broad phospholipid signal in human brain: an in vivo 31P MRS study. Alcohol Clin Exp Res 25:89–97
Estilaei MR, Matson GB, Payne GS, Leach MO, Fein G, Meyerhoff DJ (2001b) Effects of abstinence from alcohol on the broad phospholipid signal in human brain: an in vivo 31P magnetic resonance spectroscopy study. Alcohol Clin Exp Res 25:1213–1220
Forester BP, Berlow YA, Harper DG, Jensen JE, Lange N, Froimowitz MP, Ravichandran C, Iosifescu DV, Lukas SE, Renshaw PF, Cohen BM (2010) Age-related changes in brain energetics and phospholipid metabolism. NMR Biomed 23:242–250
Frahm J, Merboldt KD, Hanicke W (1987) Localized proton spectroscopy using stimulated echoes. J Magn Reson 72:502–508
Gabr RE, Ouwerkerk R, Bottomley PA (2006) Quantifying in vivo MR spectra with circles. J Magn Reson 179:152–163
Garlick PB, Soboll S, Bullock GR (1992) Evidence that mitochondrial phosphate is visible in ^{31}P spectra of isolated, perfused rat hearts. NMR Biomed 5:29–36
Gee JB, Corbett RJT, Perlman J, Laptook AR (2004) The effects of systemic magnesium sulfate infusion on brain magnesium concentrations and energy state during hypoxia-ischemia in newborn miniswine. Pediatr Res 55:93–100
Gilboe DD, Kintner DB, Anderson ME, Fitzpatrick JH (1998) NMR-based identification of intra- and extracellular compartments of the brain Pi peak. J Neurochem 71:2542–2548
Glonek T, Kopp SJ, Kot E, Pettegrew JW, Harrison WH, Cohen MM (1982) P-31 nuclear magnetic resonance analysis of brain: the perchloric acid extract spectrum. J Neurochem 39:1210–1219
Glonek T, Kopp SJ (1985) Ex vivo P-31 NMR of lens, cornea, heart, and brain. Magn Reson Imaging 3:359–376
Gonen O, Mohebbi A, Stoyanova R, Brown TR (1997) In vivo phosphorus polarization transfer and decoupling from protons in three-dimensional localized nuclear magnetic resonance spectroscopy of human brain. Magn Reson Med 37:301–306
Graulich J, Hoffmann U, Maier RF, Ruscher K, Pomper JK, Ho HK, Gabriel S, Obladen M, Heinemann U (2002) Acute neuronal injury after hypoxia is influenced by the reoxygenation mode in juvenile hippocampal slice cultures. Dev Brain Res 137:35–42

Gunzel D, Schlue WR (1996) Sodium-magnesium antiport in Retzius neurones of the leech Hirudo medicinalis. J Physiol 491:595–608

Gupta RJ, Benovic JL, Rose ZB (1978) The determination of the free-magnesium level in the human red blood cell by ^{31}P NMR. J Biol Chem 253:6172–6176

Haase A, Odoj F, von Kienlin M, Warnking J, Fidler F, Weisser A, Nittka M, Rommel E, Lanz T, Kalusche B, Griswold M (2000) NMR probeheads for in-vivo applications. Concepts Magn Reson 12:361–388

Hamakama H, Murashita J, Yamada N, Inubushi T, Kato N, Kato T (2004) Reduced intracellular pH in the basal ganglia and whole brain measured by 31P-MRS in bipolar disorder. Psychiatry Clin Neurosci 58:82–88

Helpern JA, Vande Linde AM, Welch KM, Levine SR, Schultz LR, Ordidge RJ, Halvorson HR, Hugg JW (1993) Acute elevation and recovery of intracellular [Mg2+] following human focal cerebral ischemia. Neurology 43:1577–1581

Henchcliffe C, Shungu DC, Mao X, Huang C, Nirenberg MJ, Jenkins BG, Beal MF (2008) Multinuclear magnetic resonance spectroscopy for in vivo assessment of mitochondrial dysfunction in Parkinson's disease. Ann NY Acad Sci 1147:206–220

Hetherington HP, Spencer DD, Vaughan JT, Pan JW (2001) Quantitative ^{31}P spectroscopic imaging of human brain at 4 Tesla: Assessment of gray and white matter differences of phosphocreatine and ATP. Magn Reson Med 45:46–52

Hetherington HP, Pan JW, Spencer DD (2002) ^{1}H and ^{31}P spectroscopy and bioenergetics in the lateralization of seizures in temporal lobe epilepsy. J Magn Reson Imaging 16:477–483

Hetherington HP, Kim JH, Pan JW, Spencer DD (2004) ^{1}H and ^{31}P spectroscopic imaging of epilepsy: spectroscopic and histologic correlations. Epilepsia 45(suppl 4):17–23

Hida K, Kwee IL, Nakada T (1992a) T_1 values of phosphomonoester and phosphocreatine of brain show no significant change during development. Magn Reson Med 27:179–182

Hida K, Kwee IL, Nakada T (1992b) In vivo ^{1}H and ^{31}P NMR spectroscopy of the developing rat brain. Magn Reson Med 23:31–36

Hoang TQ, Bluml S, Dubowitz DJ, Moats R, Kopyov O, Jacques D, Ross BD (1998) Quantitative proton-decoupled ^{31}P MRS and ^{1}H MRS in the evaluation of Huntington's and Parkinson's diseases. Neurology 50:1033–1040

Holtzman D, Mulkern R, Tsuji M, Cook C, Meyers R (1996) Phosphocreatine and creatine kinase in piglet cerebral gray and white matter in situ. Devel Neurosci 18:535–541

Holtzman D, Mulkern R, Meyers R, Cook C, Allred E, Khait I, Jensen F, Tsuji M, Laussen P (1998) In vivo phosphocreatine and ATP in piglet cerebral gray and white matter during seizures. Brain Res 783:19–27

Hope PL, Costello AM de L, Cady EB, Delpy DT, Tofts PS, Chu A, Hamilton PA, Reynolds EOR, Wilkie DR (1984). Cerebral energy metabolism studied with phosphorus NMR spectroscopy in normal and birth-asphyxiated infants. Lancet 2: 366-370.

Hoult DI, Richards RE (1976) The signal-to-noise ratio of the nuclear magnetic resonance experiment. J Magn Reson 24:71–85

Hu MT, Taylor-Robinson SD, Chaudhuri KR, Bell JD, Labbe C, Cunningham VJ, Koepp MJ, Hammers A, Morris RG, Turjanski N, Brooks DJ (2000) Cortical dysfunction in non-demented Parkinson's disease patients: a combined (31)P-MRS and (18)FDG-PET study. Brain 123(part 2): 340–352

Hubesch B, Sappey-Marinier D, Roth K, Meyerhoff DJ, Matson GB, Weiner MW (1990) 31P NMR spectroscopy of normal human brain and brain tumors. Radiology 174:401–409

Hugg JW, Matson GB, Twieg DB, Sappey-Marinier D, Weiner MW (1992) Phosphorus-31 MR spectroscopic imaging (MRSI) of normal and pathological human brains. Magn Reson Imaging 10:227–243

in 't Zandt HJ, de Groof AJ, Renema WK, Oerlemans FT, Klomp DW, Wieringa B, Heerschap A (2003) Presence of (phospho)creatine in developing and adult skeletal muscle of mice without mitochondrial and cytosolic muscle creatine kinase isoforms. J Physiol 548:847–858

in 't Zandt HJ, Renema WK, Streijger F, Jost C, Klomp DW, Oerlemans F, Van der Zee CE, Wieringa B, Heerschap A (2004) Cerebral creatine kinase deficiency influences metabolite levels and morphology in the mouse brain: a quantitative in vivo 1H and 31P magnetic resonance study. J Neurochem 90:1321–1330

Iwata O, Iwata S, Bainbridge A, De Vita E, Matsuishi T, Cady EB, Robertson NJ (2008) Supra- and sub-baseline PCr recovery in developing brain after transient hypoxia-ischaemia: relation to baseline energetics, insult severity and outcome. Brain 131:2220–2226

Jensen JE, Drost DJ, Menon RS, Williamson PC (2002) In vivo brain ^{31}P-MRS: measuring the phospholipid resonances at 4 Tesla from small voxels. NMR Biomed 15:338–347

Jensen JE, Miller J, Williamson PC, Neufeld RW, Menon RS, Malla A, Manchanda R, Schaefer B, Densmore M, Drost DJ (2004) Focal changes in brain energy and phospholipid metabolism in first-episode schizophrenia: 31P-MRS chemical shift imaging study at 4 Tesla. Br J Psychiatry 184:409–415

Jensen JE, Miller J, Williamson PC, Neufeld RW, Menon RS, Malla A, Manchanda R, Schaefer B, Densmore M, Drost DJ (2006) Grey and white matter differences in brain energy metabolism in first episode schizophrenia: 31P-MRS chemical shift imaging at 4 Tesla. Psychiatry Res 146:127–135

Jung W-I, Widmaier S, Bunse M, Seeger U, Straubinger K, Schick F, Kuper K, Dietze G, Lutz O (1993) ^{31}P transverse relaxation times of ATP in human brain in vivo. Magn Reson Med 30:741–743

Jung W-I, Staubert A, Widmaier S, Hoess T, Bunse M, van Erckelens F, Dietze G, Lutz O (1997) Phosphorus J-coupling constants of ATP in human brain. Magn Reson Med 37:802–804

Keshavan MS, Stanley JA, Montrose DM, Minshew NJ, Pettegrew JW (2003) Prefrontal membrane phospholipid metabolism of child and adolescent offspring at risk for schizophrenia or schizoaffective disorder; an in vivo 31P MRS study. Mol Psychiatry 8:316–323

Klemm S, Rzanny R, Riehemann S, Volz HP, Schmidt B, Gerhard UC, Schonberg A, Mentzel HJ, Kaiser WA, Blanz B (2001) Cerebral phosphate metabolism in first-degree relatives of patients with schizophrenia. Am J Psychiatry 158:958–960

Kreis R, Slotboom J, Pietz J, Jung B, Boesch C (2001) Quantitation of localized ^{31}P magnetic resonance spectra based on the reciprocity principle. J Magn Reson 149:245–250

Kuby SA, Noltmann EA (1962) ATP-creatine transphosphorylase. In: Boyer PD, Lardy H, Myrback K (eds) The enzymes, vol 6, 2nd edn. Academic, New York, pp 515–603

Lara RS, Matson GB, Hugg JW, Maudsley AA, Weiner MW (1993) Quantitation of in vivo phosphorus metabolites in human brain with magnetic resonance spectroscopic imaging (MRSI). Magn Reson Imaging 11:273–278

Le Rumeur E (1992) Saturation transfer and its application to study biochemical kinetics. In: de Certaines JD, Bovee WMMJ, Podo F (eds) Magnetic resonance spectroscopy in biology and medicine. Pergamon, Oxford, pp 169–173

Lei H, Zhu X-H, Zhang X-L, Ugurbil K, Chen W (2003a) In vivo ^{31}P magnetic resonance spectroscopy of human brain at 7 T: an initial experience. Magn Reson Med 49:199–205

Lei H, Ugurbil K, Chen W (2003b) Measurement of unidirectional Pi to ATP flux in human visual cortex at 7 T by using in vivo ^{31}P magnetic resonance spectroscopy. Proc Nat Acad Sci USA 100:14409–14414

Lodi R, Schapira AH, Manners D, Styles P, Wood NW, Taylor DJ, Warner TT (2000) Abnormal in vivo skeletal muscle energy metabolism in Huntington's disease and dentatorubropallidoluysian atrophy. Ann Neurol 48:72–76

Lorek A, Takei Y, Cady EB, Wyatt JS, Penrice J, Edwards AD, Peebles D, Wylezinska M, Owen-Reece H, Kirkbride V, Cooper C, Aldridge RF, Roth SC, Brown G, Delpy DT, Reynolds EOR (1994) Delayed ("secondary") cerebral energy failure after acute hypoxia-ischemia in the newborn piglet: continuous 48-hour studies by phosphorus magnetic resonance spectroscopy. Pediatr Res 36:99–706

Luyten PR, Bruntink G, Sloff FM, Vermeulen JWAH, van der Heijden JI, den Hollander JA, Heerschap A (1989a) Broadband proton decoupling in human ^{31}P NMR spectroscopy. NMR Biomed 1:177–183

Luyten PR, Groen JP, Vermeulen JWAH, den Hollander JA (1989b) Experimental approaches to image localized human ^{31}P NMR spectroscopy. Magn Reson Med 11:1–21

Madden A, Leach MO, Collins DJ, Payne GS (1991) The water resonance as an alternative pH reference: relevance to in vivo ^{31}P NMR localized spectroscopy studies. Magn Reson Med 19:416–421

Maintz D, Heindel W, Kugel H, Jaeger R, Lackner KJ (2002) Phosphorus-31 MR spectroscopy of normal adult human brain and brain tumours. NMR Biomed 15:18–27

Martin E, Buchli R, Ritter S, Schmid R, Largo RH, Boltshauser E, Fanconi S, Duc G, Rumpel H (1996) Diagnostic and prognostic value of cerebral ^{31}P magnetic resonance spectroscopy in neonates with perinatal asphyxia. Pediatr Res 40:749–758

Mason GF, Chu WJ, Vaughan JT, Ponder SL, Twieg DB, Adams D, Hetherington HP (1998) Evaluation of ^{31}P metabolite differences in human cerebral gray and white matter. Magn Reson Med 39:346–353

McKnight TR (2004) Proton magnetic resonance spectroscopic evaluation of brain tumor metabolism. Semin Oncol 31:605–617

Meng S, Qiao M, Foniok T, Tuor UI (2005) White matter damage precedes that in gray matter despite similar magnetic resonance imaging changes following cerebral hypoxia-ischemia in neonatal rats. Exp Brain Res 166:56–60

Merboldt KD, Chien D, Hanicke W, Gyngell ML, Bruhn H, Frahm J (1990) Localized ^{31}P NMR spectroscopy of the adult human brain in vivo using stimulated-echo (STEAM) sequences. J Magn Reson 89:343–361

Moon RB, Richards JH (1973) Determination of intracellular pH by ^{31}P magnetic resonance. J Biol Chem 248:7276–7278

Mulkern RV, Greenman RL, Rybicki FJ (2003) Comments on ultrahigh field ^{31}P ATP T_2 values. Magn Reson Med 50:654–655

Murphy-Boesch J, Stoyanova R, Srinavasan R, Willard T, Vingeron D, Nelson S, Taylor JS, Brown T (1993) Proton-decoupled ^{31}P chemical shift imaging of the human brain in normal volunteers. NMR Biomed 6:173–180

Murphy-Boesch J, Srinivasan R, Carvajal L, Brown TR (1994) Two configurations of the four-ring birdcage coil for ^{1}H imaging and ^{1} decoupled ^{31}P spectroscopy of the human head. J Magn Reson B 106:103–114

Murphy EJ, Rajagopalan B, Brindle KM, Radda GK (1989) Phospholipid bilayer contribution to ^{31}P NMR spectra in vivo. Magn Reson Med 12:282–289

Nakada T, Kwee IL, Suzuki N, Houkin K (1989) Intrauterine fetal brain NMR spectroscopy: ^{1}H and ^{31}P studies in rats. Magn Reson Med 12:172–180

Nakada T, Kwee IL (1993) ^{31}P localized spectroscopy of fetal brain in utero. Magn Reson Med 29:122–124

Oberhänsli RD, Hilton-Jones D, Bore PJ, Hands LJ, Rampling RP, Radda GK (1986) Biochemical investigation of human tumors in vivo with phosphorus-31 magnetic resonance spectroscopy. Lancet 2(8497):8–11

Ordidge RJ, Connelly A, Lohman JAB (1986) Image selected in-vivo spectroscopy (ISIS). A new technique for spatially-selected NMR spectroscopy. J Magn Reson 66:283–294

Page RA, Cady EB, Priest AN, Bainbridge A, West DA, Thornton JS, Woodman B, Bates GP, Ordidge RJ, Davie C (2003) Localised in-vivo ^{31}P MRS in a mouse model of Huntington's disease. In: Proceedings of the 11th Annual Meeting International Society of Magnetuc Resonance in Medicine, Toronto, Canada, p1306

Pan JW, Bebin EM, Chu WJ, Kiel S, Hetherington HP (1999) Ketosis and epilepsy: ^{31}P spectroscopic imaging at 4.1 T. Epilepsia 40:703–708

Pearce JM, Komoroski RA (2000) Analysis of phospholipid molecular species in brain by ^{31}P NMR spectroscopy. Magn Reson Med 44:215–223

Peeters-Scholte C, Braun K, Koster J, Kops N, Blomgren K, Buonocore G, van Buul-Offers S, Hagberg H, Nicolay K, van Bel F, Groenendaal F (2003) Effects of allopurinol and deferoxamine on reperfusion injury of the brain in newborn piglets after neonatal hypoxia-ischemia. Pediatr Res 54:516–522

Petroff OAC, Prichard JW (1985) Cerebral intracellular pH by ^{31}P nuclear magnetic resonance spectroscopy. Neurology 35:781–788

Pettegrew JW, Kopp SJ, Dadok J, Minshew NJ, Feliksik JM, Glonek T, Cohen MM (1986) Chemical characterization of a prominent phosphomonoester resonance from mammalian brain. ^{31}P and ^{1}H NMR analysis at 4.7 and 14.1 Tesla. J Magn Reson 67:443–450

Pettegrew JW, Withers G, Panchalingam K, Post JFM (1988) Considerations for brain pH assessment by ^{31}P NMR. Magn Reson Imaging 6:135–142

Pettegrew JW, Klunk WE, Panchalingam K, McClure RJ, Stanley JA (1997) Magnetic resonance spectroscopy changes in Alzheimer's disease. Ann NY Acad Sci 826:282–306

Pettegrew JW, Panchalingam K, Hamilton RL, McClure RJ (2001) Brain membrane phospholipid alterations in Alzheimer's disease. Neurochem Res 26:771–782

Pietz J, Rupp A, Ebinger F, Mayatepek E, Boesch C, Kreis R (2003) Cerebral energy metabolism in phenylketonuria: findings by quantitative in vivo 31P MR spectroscopy. Pediatr Res 53:654–662

Portman MA, Lassen NA, Cooper TG, Sills AM, Potchen EJ (1991) Intra- and extracellular pH of the brain in vivo studied by ^{31}P-NMR during hyper- and hypocapnia. J Appl Physiol 71:2168–2172

Potwarka JJ, Drost DJ, Williamson PC (1999) Quantifying ^{1}H decoupled in vivo ^{31}P brain spectra. NMR Biomed 12:8–14

Puka-Sundvall M, Gajkowska B, Cholewinski M, Blomgren K, Lazarewicz JW, Hagberg H (2000) Subcellular distribution of calcium and ultrastructural changes after cerebral hypoxia-ischemia in immature rats. Dev Brain Res 125:31–41

Rabenstein DL, Nakashima TT (1979) Spin-echo Fourier transform nuclear magnetic resonance spectrocopy. Analyt Chem 51:1465A–1474A

Rango M, Bonifati C, Bresolin N (2006) Parkinson's disease and brain mitochondrial dysfunction: a functional phosphorus magnetic resonance spectroscopy study. J Cereb Blood Flow Metab 26:283–290

Remy C, Albrand JP, Benabid AL, Decorps M, Jacrot M, Riondel J, Foray MF (1987) In vivo ^{31}P nuclear magnetic resonance studies of T_1 and T_2 relaxation times in rat brain and in rat brain tumors implanted to nude mice. Magn Reson Med 4:144–152

Robbins RC, Balaban RS, Swain JA (1990) Intermittent hypothermic asanguineous cerebral perfusion (cerebroplegia) protects the brain during prolonged circulatory arrest. J Thorac Cardiovasc Surg 99:878–884

Robertson NJ, Cowan FM, Cox IJ, Edwards AD (2002) Brain alkaline intracellular pH after neonatal encephalopathy. Ann Neurol 52:732–742

Roth K, Hubesch B, Meyerhoff DJ, Naruse S, Gober JR, Lawry TJ, Boska MD, Matson GB, Weiner MW (1989) Noninvasive quantitation of phosphorus metabolites in human tissue by NMR spectroscopy. J Magn Reson 81:299–311

Roth K, Weiner MW (1991) Determination of cytosolic ADP and AMP concentrations and the free energy of ATP hydrolysis in human muscle and brain tissues with ^{31}P NMR spectroscopy. Magn Reson Med 22:505–511, see erratum: 1995. Magn Reson Med 33: 282

Schuldiner S, Rozengurth E (1982) Na^+/H^+ antiport in Swiss 3 T3 cells: mitogenic stimulation leads to cytoplasmic alkalinization. Proc Natl Acad Sci USA 79:7778–7782

Shaka AJ, Keeler J, Freeman R (1983) Evaluation of a new broadband decoupling sequence: Waltz-16. J Magn Reson 53:313–340

Shoubridge EA, Briggs RW, Radda GK (1982) ^{31}P NMR saturation transfer measurements of the steady state rates of creatine kinase and ATP synthetase in the rat brain. FEBS Lett 140: 288–292

Silveri MM, Pollack MH, Diaz CI, Nassar LE, Mendelson JH, Yurgelun-Todd DA, Renshaw PF, Kaufman MJ (2004) Cerebral phosphorus metabolite and transverse relaxation time abnormalities in heroin-dependent subjects at onset of methadone maintenance treatment. Psychiatry Res 131:217–226

Simor T, Chu WJ, Hetherington HP, Kuzniecky RI, Elgavish GA (1997) Tailored temporal lobectomy induced improvements in 4.1 T ^{31}P NMR SI generated phosphorus metabolite indices in temporal lobe epilepsy. In: Proceedings of the 5th Annual Meeting International Society of Magnetic Resonance in Medicine, Vancouver, Canada, p 33

Skou JC (1982) The (Na+ + K+)-ATPase: coupling of the reaction with ATP to the reaction with Na+and K+. Ann N Y Acad Sci 402:169–184

Smesny S, Rosburg T, Nenadic I, Fenk KP, Kunstmann S, Rzanny R, Volz HP, Sauer H (2007) Metabolic mapping using 2D 31P-MR spectroscopy reveals frontal and thalamic metabolic abnormalities in schizophrenia. Neuroimage 35:729–737

Stubbs M, Vanstapel F, Rodrigues LM, Griffiths JR (1988) Phosphate metabolites in rat skin. NMR Biomed 1:50–55

Thoresen M, Penrice J, Lorek A, Cady EB, Wylezinska M, Kirkbride V, Cooper CE, Brown GC, Edwards AD, Wyatt JS, Reynolds EOR (1995) Mild hypothermia after severe transient hypoxia-ischemia ameliorates delayed cerebral energy failure in the newborn piglet. Pediatr Res 37:667–670

Thornton JS, Ordidge RJ, Penrice J, Cady EB, Amess PN, Punwani S, Clemence M, Wyatt JS (1998) Temporal and anatomical variations of brain water apparent diffusion coefficient in perinatal cerebral hypoxic-ischemic injury: relationships to cerebral energy metabolism. Magn Reson Med 39:920–927

Thulborn KR, Ackerman JJH (1983) Absolute molar concentrations by NMR in inhomogeneous B_1. A scheme for analysis of in vivo metabolites. J Magn Reson 55:357–371

Tofts PS (1988) The noninvasive measurement of absolute metabolite concentrations in vivo using surface-coil NMR spectroscopy. J Magn Reson 80:84–95

Tofts PS (1994) Standing waves in uniform water phantoms. J Magn Reson B 104:143–147

Tofts P, Wray S (1985) Changes in brain phosphorus metabolites during post-natal development of the rat. J Physiol 359:417–429

Ulrich M, Wokrina T, Ende G, Lang M, Bachert P (2007) 31P-{1 H} echo-planar spectroscopic imaging of the human brain in vivo. Magn Reson Med 57:784–790

Vanhamme L, van den Boogart A, Huffel SV (1997) Improved method for accurate and efficient quantification of MRS data with use of prior knowledge. J Magn Reson 129:35–43

Van der Knaap M, Van der Grond J, Van Rijen P, Faber J, Valk J, Willemse K (1990) Age-dependent changes in localized proton and phosphorus MR spectroscopy of the brain. Radiology 176:509–515

Van Sluis R, Payne GS, Leach MO (1995) Increased NOE enhancement in ^{1}H decoupled ^{31}P MRS. Magn Reson Med 34:893–897

Westergren I, Nystrom B, Hamberger A, Nordborg C, Johansson BB (1994) Concentrations of amino acids in extracellular fluid after opening of the blood-brain barrier by intracarotid infusion of protamine sulfate. J Neurochem 62:159–165

Wijnen JP, Scheenen TW, Klomp DW, Heerschap A (2010) 31P magnetic resonance spectroscopy imaging with polarisation transfer of phosphomono- and phosphodiesters at 3 T in the human brain: relation with age and spatial difference. NMR Biomed 23:968–976

Williams GD, Palmer C, Roberts RL, Heitjan DF, Smith MB (1992) ^{31}P NMR spectroscopy of perinatal hypoxic-ischemic brain damage: a model to evaluate neuroprotective drugs in immature rats. NMR Biomed 5:145–153

Williams GD, Mosher TJ, Smith MB (1993) Simultaneous determination of intracellular magnesium and pH from the three ^{31}P NMR chemical shifts of ATP. Analyt Biochem 214:458–467

Williams GD, Smith MB (1995) Application of the accurate assessment of intracellular magnesium and pH from the ^{31}P shifts of ATP to cerebral hypoxia-ischemia in neonatal rat. Magn Reson Med 33:853–857

Chapter 7
In Vivo ^{13}C Magnetic Resonance Spectroscopy and Metabolic Modeling: Methodology

Pierre-Gilles Henry, Dinesh K. Deelchand, Isabelle Iltis, Malgorzata Marjanska, Gülin Öz, Alexander A. Shestov, and Julien Valette*

Abstract *In vivo* ^{13}C NMR spectroscopy combined with infusion of ^{13}C-enriched substrates allows non-invasive investigation of brain metabolism in animals and humans. Time courses of ^{13}C label incorporation from the infused substrate(s) into brain metabolites can be analyzed with metabolic models to quantify metabolic rates through specific metabolic pathways such as the neuronal and glial TCA cycles, pyruvate carboxylase, or the glutamate-glutamine cycle. In this chapter, we review the methodology for successful *in vivo* ^{13}C MRS measurements and metabolic modeling studies. The various aspects of this methodology are described with a focus on challenges specific to *in vivo* measurements in the brain. References in this review are current as of March 2008.

Keywords ^{13}C magnetic resonance spectroscopy • Brain metabolism • Metabolic modeling

7.1 Introduction

Carbon-13 nuclear magnetic resonance spectroscopy (^{13}C MRS[1]) allows minimally-invasive regional investigation of metabolism with exquisite chemical specificity. In contrast to radiotracer techniques such as autoradiography and PET, ^{13}C MRS can be used to identify into which molecules ^{13}C atoms are incorporated, and even the exact

**Authors are listed in alphabetical order except for the first author.*

[1] Unless otherwise stated, thete terms '^{13}C MRS' and '^{13}C MR spectra' in this chapter refer to both direct ^{13}C detection (at ^{13}C frequency) and indirect ^{1}H[^{13}C] detection (at ^{1}H frequency).

P.-G. Henry, Ph.D. (✉) • D.K. Deelchand, Ph.D. • I. Iltis, Ph.D. • M. Marjanska, Ph.D. • G. Öz, Ph.D. • A.A. Shestov, Ph.D. • J. Valette, Ph.D.
Center for Magnetic Resonance Research, Department of Radiology,
University of Minnesota Medical School, 2021 6th St SE, Minneapolis, MN 55455, USA
e-mail: henry@cmrr.umn.edu; deelc001@umn.edu; iltis001@umn.edu; gosia@umn.edu; gulin@cmrr.umn.edu; shest001@umn.edu

I.-Y. Choi and R. Gruetter (eds.), *Neural Metabolism In Vivo*,
Advances in Neurobiology 4, DOI 10.1007/978-1-4614-1788-0_7,

carbon position of this ^{13}C label within each molecule. This remarkable specificity provides unique information on biochemical pathways *in vivo*. As a result, in spite of its limitations (relatively low sensitivity, administration of expensive ^{13}C-labeled substrates, technical complexity), ^{13}C MRS has become an indispensable tool to investigate key aspects of brain function such as brain energy metabolism and metabolic cooperation at the cellular level between neurons and astrocytes.

In vivo ^{13}C MRS has been developing steadily over the past 30 years, with constant improvements in instrumentation and methodology, including advent of very high-field large bore magnets for *in vivo* studies, development of new pulse sequences, and analysis of data with more sophisticated metabolic models. The first experiments in bacteria were performed in the late 1970s (Ugurbil et al. 1978; Shulman et al. 1979) soon followed by the first dynamic *in vivo* experiments in liver (Alger et al. 1981). The first *in vivo* studies in the brain were carried out in animals the late 1980s (Rothman et al. 1985; Behar et al. 1986) and in humans in the early 1990s (Beckmann et al. 1991; Rothman et al. 1992; Gruetter et al. 1994) These initial studies focused mostly on glucose metabolism, although initial *in vivo* spectra obtained during ^{13}C-acetate infusion were also reported (Cerdan et al. 1990).

A majority of the initial *in vivo* studies in the brain have focused on the detection of labeled [4-^{13}C]glutamate from [1-^{13}C]glucose (Rothman et al. 1985, 1992; Behar et al. 1986; Fitzpatrick et al. 1990; Beckmann et al. 1991; Inubushi et al. 1993; van Zijl et al. 1993). Glutamate has a high concentration in the brain (~10 mM) and is labeled relatively quickly during infusion of ^{13}C-labeled glucose. Since glutamate is primarily located in the neuronal compartment, measurement of glutamate turnover from glucose primarily reflects TCA cycle activity in neurons. These studies, combined with metabolic modeling, opened the way to *in vivo* measurement of TCA cycle rate in normal brain (Mason et al. 1992a, 1995; Boumezbeur et al. 2005), during functional activation (Hyder et al. 1996, 1997; Chen et al. 2001; Chhina et al. 2001) and in animal models of disease (Henry et al. 2002).

In the mid 1990s, with improvements in sensitivity and NMR methodology, reliable detection of the time course of ^{13}C label incorporation not only into glutamate but also into glutamine became possible, both in rat brain (Sibson et al. 1997) and in human brain (Gruetter et al. 1994, 1996). Since most glutamate is located in neurons, and almost all glutamine is located in astrocytes, it was rapidly recognized that detection of ^{13}C label incorporation simultaneously into glutamate and glutamine opened a new window for the study of compartmentalized brain metabolism. It was shown that the rate of glutamine synthesis can be measured *in vivo* and that it primarily reflects glutamate-glutamine cycling between neurons and astrocytes (Sibson et al. 1997, 1998). Metabolic modeling using two-compartment models were refined in subsequent studies in the past decade (Gruetter et al. 1998, 2001; Shen et al. 1999; Oz et al. 2004). Assessment of the validity of assumptions and of the reliability of such complex metabolic models to determine metabolic rates from experimental *in vivo* ^{13}C MRS data is still a subject of investigation (Shestov et al. 2007b) (see Sects. 7.5.2).

More recently, the range of metabolites than can be detected *in vivo* with sufficient sensitivity for dynamic detection of ^{13}C label incorporation was expanded to include lower concentration metabolites such as aspartate (Pfeuffer et al. 1999b; Gruetter et al. 2001) and GABA (Pfeuffer et al. 1999b; de Graaf et al. 2003b; Henry et al. 2003b; Yang et al. 2005; de Graaf et al. 2006). A slow increase of ^{13}C enrichment in NAA above natural abundance during infusion of ^{13}C glucose was also reported both in animals (Henry et al. 2003a; Choi and Gruetter 2004) and in humans (Moreno et al. 2001a; Henry et al. 2006a), allowing measurement of NAA synthesis rate. Finally, development of specific methods has allowed detection of brain glycogen *in vivo* (Choi et al. 1999; Öz et al. 2007) and has generated renewed interest for glycogen as a substrate for brain energy metabolism (Gruetter 2003).

Numerous ^{13}C studies have been performed in brain extracts, brain slices, or cell cultures, and have provided valuable insights into brain metabolism and compartmentation. In the present chapter, however, we focus on ^{13}C MRS *in vivo*, with particular emphasis on experimental challenges specific to *in vivo* studies and on metabolic modeling. Therefore, most references cited in this chapter describe studies in which ^{13}C MRS measurements were performed *in vivo*. In so doing we do not intend to minimize the important insights provided by the work performed in vitro and ex vivo.

In the next sections, the following experimental aspects of *in vivo* ^{13}C MRS will be examined:

- choice of infused substrate and infusion protocol
- Measurement of *in vivo* ^{13}C MR spectra
- Quantification of *in vivo* ^{13}C MR spectra
- Metabolic modeling

7.2 Choice of Infused Substrate and Infusion Protocol

Without infusion of ^{13}C-labeled substrates, only small natural abundance signals from the most concentrated brain metabolites are observed (mIns, NAA, glutamate, all of which have a concentration of 10 mM, corresponding to 0.11 mM of ^{13}C at natural abundance) (Gruetter et al. 1992; Chen et al. 1998; Henry et al. 2003a).

Infusion of ^{13}C enriched substrates with high isotopic enrichments (typically 50–100%) not only improves the signal-to-noise, but also enables dynamic measurement of ^{13}C incorporation into metabolites located downstream in metabolic pathways, thus providing unique metabolic information (Henry et al. 2006c).[2]

7.2.1 Choice of Infused Substrate

The choice of the metabolic substrate is guided by the following considerations:

- metabolic pathways of interest
- sensitivity
- cost

[2] In principle, the presence of the ^{13}C isotope can alter the rate at which the labeled substrate is metabolized (isotopic effect). In ^{13}C MRS studies, however, it is generally assumed that isotopic effects are negligible, because the presence of ^{13}C atoms does not significantly modify the molecular weight of the substrate.

We review below the substrates most commonly used for *in vivo* studies in the brain:

7.2.1.1 [1-^{13}C]Glucose or [1,6-$^{13}C_2$]Glucose

Glucose is the main source of energy for the brain, and as a result the vast majority of *in vivo* brain studies have used [1-^{13}C]glucose or [1,6-$^{13}C_2$]glucose as a substrate. Both [1-^{13}C] and [1,6-$^{13}C_2$]glucose lead to formation of [3-^{13}C]pyruvate through glycolysis. This [3-^{13}C]pyruvate is then metabolized into the TCA cycle and eventually labels amino acids such as glutamate and glutamine. [3-^{13}C]pyruvate can enter the TCA cycle either through pyruvate dehydrogenase or pyruvate carboxylase (see Fig. 7.1). In the first case (Pyruvate Dehydrogenase), this leads to labeling of the C4 carbon position of glutamate and glutamine in the first turn of the TCA cycle. The C3 and C2 carbon positions are labeled in the second turn. In the second case (pyruvate carboxylase), label from [3-^{13}C]pyruvate is incorporated into the C2 position of glutamate and glutamine in the first turn of TCA cycle.

Double-labeled [1,6-$^{13}C_2$]glucose yields twice higher isotopic enrichment in[3-^{13}C]pyruvate compared to [1-^{13}C]glucose, which is advantageous not only for improved sensitivity, but also for detection of multiply-labeled ^{13}C-^{13}C isotopomers with direct ^{13}C detection (Henry et al. 2003a). However the cost of [1,6-$^{13}C_2$]glucose is ten times that of [1-^{13}C]glucose, and for this reason has been used only in small animal studies.

7.2.1.2 [U-^{13}C]Glucose

A few *in vivo* brain studies have used uniformly labeled [U-^{13}C]glucose (Boumezbeur et al. 2004; Xu and Shen 2006). [U-^{13}C]glucose in principle provides the same improvement in sensitivity as [1,6-$^{13}C_2$]glucose, but it is much cheaper. However, the presence of multiple ^{13}C labeled carbons may complicate spectral analysis. When using indirect ^{1}H[^{13}C] detection without ^{1}H decoupling (Boumezbeur et al. 2004), long-range ^{13}C-^{1}H couplings (typically 2–3 Hz) may slightly affect the lineshape and linewidth of detected resonances. When using direct ^{13}C detection, the use of [U-^{13}C]glucose leads to additional splitting in the spectrum. For example, the C5 carbon of glutamate becomes labeled, resulting in splitting of the C4 resonance. In addition, long-range ^{13}C-^{13}C couplings, although too small to be resolved, cause slight changes in lineshape and linewidth that are dependent on the enrichment of the neighboring carbons. These potential complications notwithstanding, [U-^{13}C] glucose may represent an attractive, cheaper alternative to [1,6-$^{13}C_2$]glucose.

7.2.1.3 [2-^{13}C]Glucose

A few *in vivo* studies have used [2-^{13}C]glucose (Sibson et al. 2001; Mason et al. 2007) in order to specifically investigate pyruvate carboxylase activity. Infusion of [2-^{13}C] glucose leads to the formation of [2-^{13}C]pyruvate. Metabolism of [2-^{13}C]pyruvate

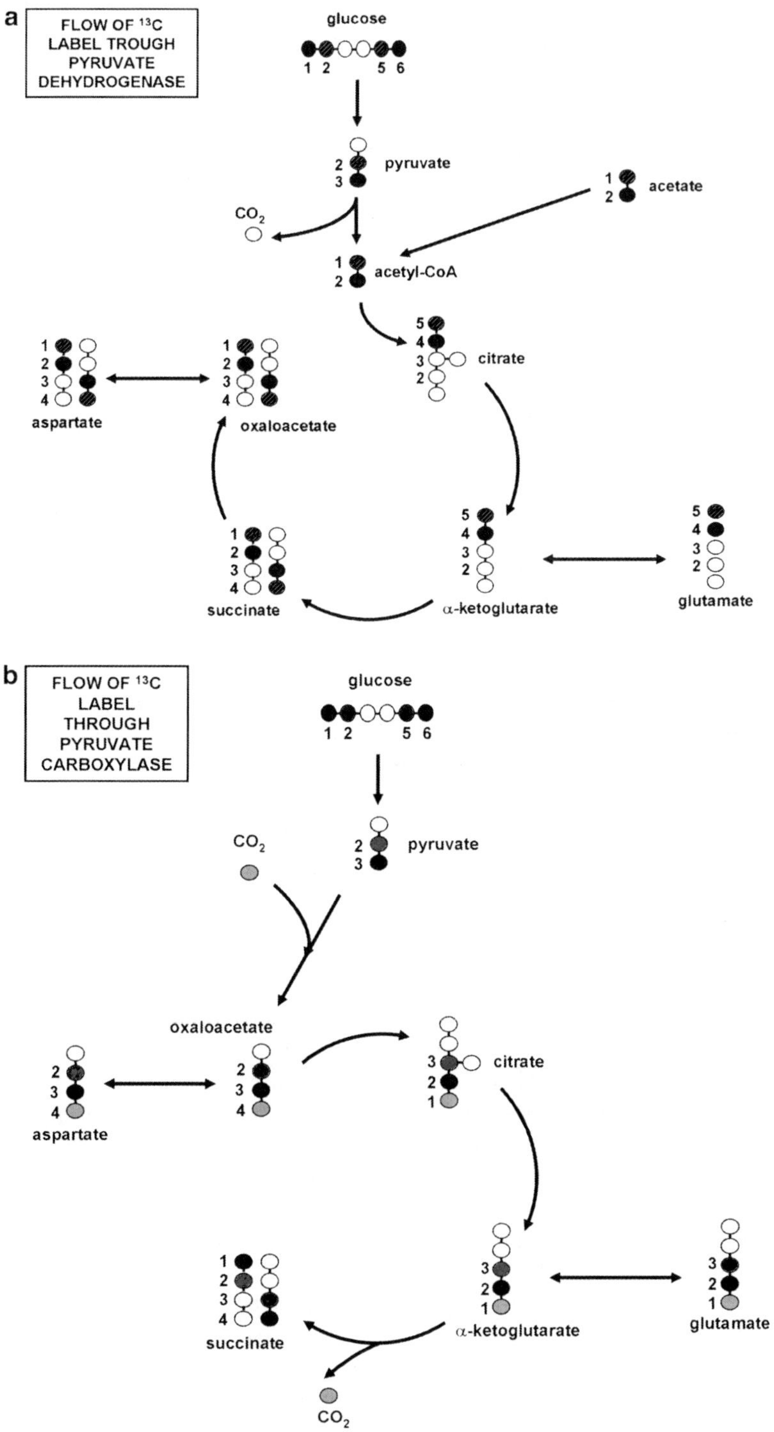

Fig. 7.1 Flow of ^{13}C label from ^{13}C-labeled glucose into brain metabolites. (**a**) ^{13}C label flowing through pyruvate dehydrogenase (only the first turn is shown). Infusion of [2-^{13}C]acetate results in labeling of the same carbons. (**b**) ^{13}C label flowing through pyruvate carboxylase

through pyruvate dehydrogenase leads to the formation of [1-^{13}C]acetylCoA, which is incorporated into [5-^{13}C]glutamate (see Fig. 7.1). In contrast, [2-^{13}C]pyruvate metabolized through astrocytic pyruvate carboxylase leads to the formation of astrocytic [3-^{13}C]glutamine. Therefore the time course of ^{13}C label incorporation from [2-^{13}C]glucose to [3-^{13}C]glutamine directly reflects pyruvate carboxylase activity. However, because the rate of pyruvate carboxylase is relatively small, the isotopic enrichment achieved in [3-^{13}C]glutamine when infusing [2-^{13}C]glucose is low, making ^{13}C MRS signal difficult to detect with good sensitivity.

7.2.1.4 ^{13}C-Labeled Acetate ([1-^{13}C]Acetate, [2-^{13}C]Acetate, [1,2-$^{13}C_2$]Acetate)

Although glucose is its main source of energy under normal physiological conditions, it is well recognized that the brain can also metabolize other substrates such as acetate and lactate. ^{13}C-labeled acetate is an interesting alternative to glucose for studies of compartmentalized brain metabolism. Acetate is metabolized only in glial cells and not in neurons, due to selective transport (Waniewski and Martin 1998). Initial studies with [1,2-$^{13}C_2$]acetate were performed in the 1990s, mostly in extracts (Badar-Goffer et al. 1990; Cerdan et al. 1990; Taylor et al. 1996; Bachelard 1998), but ^{13}C-labeled acetate has recently regained attention as a substrate for *in vivo* studies (Bluml et al. 2002; Lebon et al. 2002; Patel et al. 2005; Mason et al. 2006). Some researchers have used [2-^{13}C]acetate (Lebon et al. 2002; Patel et al. 2005; Mason et al. 2006) resulting in ^{13}C label incorporation into [4-^{13}C]glutamate and [4-^{13}C]glutamine in the first turn of the TCA cycle. Others have used [1-^{13}C] acetate, resulting in labeling of [5-^{13}C]glutamine and [5-^{13}C]glutamate (Bluml et al. 2002). The use of [2-^{13}C]acetate appears preferable because detection of [4-^{13}C] glutamate provides several-fold better sensitivity than detection of [5-^{13}C]glutamate. Indeed, carboxyl carbons do not allow signal enhancement through nuclear Overhauser Enhancement (NOE) or polarization transfer, and have long T_1 relaxation times. One advantage of using [1-^{13}C]acetate is simplicity: [5-^{13}C]glutamate and [5-^{13}C]glutamine signals do not overlap with lipid signals from the scalp and therefore can be detected without spatial localization with a simple pulse-acquire sequence, which can be advantageous for clinical studies (Ross et al. 2003).

Finally, the use of acetate is particularly promising because, as a glial substrate, it may allow increased precision on the measurement of metabolic rates reflecting glial processed such as the glutamine synthetase (glutamate-glutamine cycle), pyruvate carboxylase, or glial TCA cycle. (Shestov et al. 2007b). Monte-Carlo simulations suggest that metabolic modeling of data obtained during simultaneous infusion of [1,2-$^{13}C_2$]acetate (taken up by glia) and [1-^{13}C]glucose or [1,6-$^{13}C_2$]glucose (quantitatively metabolized primarily in neurons because the neuronal TCA cycle rate is much faster than the glial TCA cycle rate) may be advantageous for precise measurement of compartmentalized metabolic fluxes associated with glial processes (Shestov et al. 2007b, 2008; Deelchand et al. 2008a).

7.2.1.5 Other Substrates

A study of ^{13}C label incorporation from [2,4-$^{13}C_2$]β-hydroxybutyrate into brain amino acids glutamate, glutamine and aspartate has been reported *in vivo* (Pan et al. 2002). A few other substrates have been utilized for ex vivo studies (ie the ^{13}C-labeled substrate was infused *in vivo*, but detection of ^{13}C label was done on brain extracts). A non-exhaustive list includes [3-^{13}C]pyruvate (Gonzalez et al. 2005), [3-^{13}C]lactate (Hassel and Brathe 2000; Tyson et al. 2003) and [2,4,6,8-$^{13}C_4$]octanoate (Ebert et al. 2003). Still other substrates were used for in vitro studies on cell cultures of neurons and/or astrocytes.

7.2.2 Infusion Protocol

The infusion protocol is usually designed to rapidly increase the isotopic enrichment of the infused substrate from natural abundance to a high value in order to maximize signal-to-noise. In addition, a large, rapid change in plasma isotopic enrichment of the substrate results in large dynamic changes in isotopic enrichment of metabolites measured by ^{13}C MRS (such as glutamate or glutamine).

7.2.2.1 Route of Administration

Most studies have been performed using intravenous infusion of the ^{13}C-labeled substrate. This allows quick delivery of the ^{13}C-labeled compound, and quick increase of isotopic enrichment.

Although oral administration has been investigated (Moreno et al. 2001b; Mason et al. 2003), and certainly presents advantages in terms of comfort for the subject in human studies, continuous intravenous infusion remains the method of choice. The main reason is that intravenous administration allows better control of the input function, and also typically leads to higher isotopic enrichment than oral administration (and therefore higher sensitivity).

7.2.2.2 Measurement of Input Function

Most studies have attempted to achieve an isotopic steady-state by using an infusion protocol that keeps the concentration and isotopic enrichment of the infused substrate nearly constant in the blood, but this is not absolutely necessary for dynamic ^{13}C studies, as long as the input function is known. Knowledge of this input function (time course of infused substrate concentration and isotopic enrichment in the blood) is critical to interpret the results of ^{13}C experiments.

In most studies, the input function is measured by drawing blood at regular time intervals (with faster sampling at the beginning of the ^{13}C infusion when changes in the concentration and isotopic enrichment of the infused ^{13}C-labeled substrate are rapid) and subsequently measuring substrate concentration and isotopic enrichment using biochemical assays, GC-MS and/or high-resolution NMR.

Note that in principle the input function can also be measured directly in brain tissue from *in vivo* spectra by using the signal of the infused substrate, rendering unnecessary the knowledge of the input function in the blood. Indirect detection, in particular, allows measurement of both concentration and isotopic enrichment *in vivo*. In practice, this is difficult to do for glucose (proximity of water resonance) or lactate (low concentration). In contrast, in studies using [2-^{13}C]acetate, the brain acetate concentration and enrichment can potentially be determined directly from *in vivo* spectra using the CH_3 peak at 1.9 ppm.

7.2.2.3 Recycling of Label

Ideally, the infused ^{13}C substrate is the only source of ^{13}C label into brain metabolism. However, metabolism of the infused ^{13}C substrate (not only in the brain, but also in other parts of the body) may subsequently result in other ^{13}C-labeled molecules being released into the bloodstream, especially during the relatively long infusion times used for *in vivo* studies.

For example, infusion of [2-^{13}C]acetate results in labeling of blood glucose through gluconeogenesis in the liver (Serres et al. 2007). Another example is the progressive increase in isotopic enrichment of blood [3-^{13}C]lactate during infusion of [1-^{13}C] or [1,6-$^{13}C_2$]glucose through isotopic exchange with labeled [3-^{13}C]lactate produced in tissue.

This recycling of ^{13}C label eventually causes ^{13}C label to enter brain metabolism through multiple sources, and therefore may complicate data analysis and interpretation. However, the contribution of label recycling can easily be determined by measuring ^{13}C labeling in blood samples throughout the study, and these additional sources of ^{13}C label can then be taken into account in metabolic modeling.

7.3 Measurement of ^{13}C MR Spectra

The amount of information that can be obtained from MR spectra following infusion of ^{13}C-labeled substrates is limited primarily by sensitivity and to a lesser extent by spectral overlap. These in turn depend on factors such as:

- hardware: strength of main magnetic field (B_0), design of radiofrequency (RF) coil, performance of receiver chain (noise figure), pulse sequence, shimming
- infused substrate and infusion protocol (high enrichments → high signal)
- direct or indirect detection
- size of volume of interest (if the signal is localized to a specific brain region)

Sensitivity is arguably the main limitation of *in vivo* ^{13}C MRS, and optimizing sensitivity is a primary concern for ^{13}C MRS studies. For example, most studies to date have used surface coil to maximize detection sensitivity. In addition, studies have typically been performed in relatively large volumes, although using high magnetic fields has allowed reduction of this volume (Chen et al. 1998; Gruetter et al. 1998; Henry et al. 2003a).

Maximizing sensitivity places strong demands on hardware: high magnetic field, design and construction of RF coil, efficient decoupling without added noise.

The sensitivity threshold for detection of ^{13}C label is very dependent on the specific experimental setup, but even under the best conditions, ^{13}C concentrations below 0.1 mM are usually not detectable, which precludes detection of low concentration metabolites such as glycolytic intermediates or TCA cycle intermediates, even at high isotopic enrichment. Therefore *in vivo* ^{13}C studies are restricted to detection of ^{13}C label incorporation into the most concentrated cerebral metabolites: glutamate, glutamine, aspartate, and more recently GABA, NAA and glycogen.

In this section we will review the different aspects to be considered when measuring ^{13}C MR spectra *in vivo*: hardware requirements, direct vs. indirect detection, localization, decoupling, and power deposition.

7.3.1 Hardware Requirements for ^{13}C NMR Spectroscopy

7.3.1.1 Spectrometer

In principle, ^{13}C MRS requires two radiofrequency channels, one for ^{1}H and one for ^{13}C. This alone represents a significant barrier for routine implementation of ^{13}C MRS on clinical systems, most of which are equipped with only one RF channel. A recent study showed that it is possible to perform ^{1}H-decoupled ^{13}C detection and polarization transfer on a clinical system with only one RF channel (Klomp et al. 2008). In another study, ^{1}H detection without ^{13}C decoupling allowed detection of ^{13}C labeling by subtracting post-infusion spectra from preinfusion spectra, modeling the complex lineshape with density matrix simulations and fitting spectra with linear combination modeling (LCModel) (Boumezbeur et al. 2004).

The advent of high-field *in vivo* magnets has been instrumental in the development of ^{13}C MRS. Sensitivity increases with B_0^{β}, where β lies between 1 (noise dominantly coming from sample) and 1.75 (noise dominantly coming from RF coil). In practice, exact determination of β is difficult, but for ^{13}C and ^{1}H[^{13}C] studies *in vivo*, β is probably closer to one, so that sensitivity increases approximately linearly with magnetic field.

In addition to this increase in sensitivity, high magnetic fields also yield increased spectral dispersion and therefore reduced spectral overlap. This increase in spectral dispersion is beneficial not only for indirect ^{1}H[^{13}C] MRS, but also in direct detection, for example by discriminating glutamate and glutamine C2.

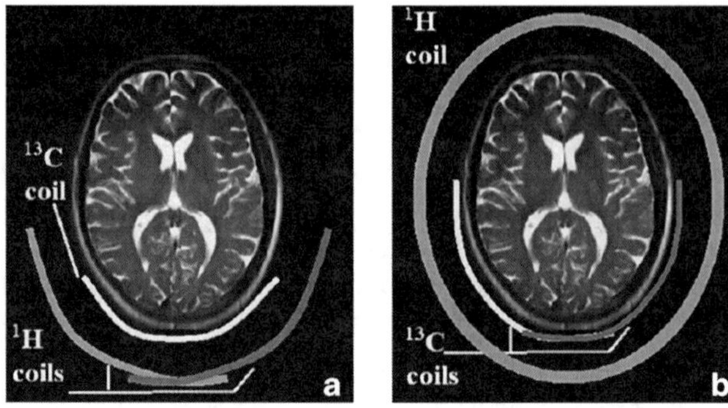

Fig. 7.2 (**a**) The coil concept proposed by Adriany and Gruetter (1997) consists of a circularly polarized 1H surface coil with a linearly polarized ^{13}C coil. (**b**) Transverse overview of the coil concept proposed by Klomp et al. using a birdcage ^{1}H volume coil with a circularly polarized ^{13}C coil positioned close to the human head

7.3.1.2 Coil and Filters

An efficient coil design is critical for ^{13}C studies to ensure sufficient sensitivity and to minimize power deposition. In addition, an optimal coil design is essential to minimize excessive electromagnetic coupling between the ^{1}H and ^{13}C coils, as it increases the risk of injecting excessive voltage into the receiver chain, causing added noise and artifacts. RF filters are usually needed to minimize interferences between ^{1}H and ^{13}C channels (Adriany and Gruetter 1997). Placement of these filters is critical to achieve RF decoupling during acquisition of the MR signal, both for direct ^{13}C detection and for indirect ^{1}H[^{13}C] detection, and the optimal configuration is often site-specific. A typical configuration uses three filters. A first filter is inserted between the detection RF coil and the preamplifier, another filter is added on the decoupler path, and a third filter is inserted after the preamplifier (Adriany and Gruetter 1997).

Due to the limited sensitivity of ^{13}C MRS, most studies have been performed with surface coils. Initial studies have been performed either with concentric single-loop ^{1}H and ^{13}C RF coils or using a single-loop coil for ^{13}C detection and a butterfly for ^{1}H decoupling. In 1997, Adriany *et al.* proposed an improved coil design that allowed ^{13}C MRS studies with surface coils to be performed at high-field (4 T) in human brain within FDA guidelines for power deposition. This design, which has since been adopted by most other groups, consists of a single ^{13}C loop for detection and a quadrature ^{1}H coil for the ^{1}H localization and ^{1}H decoupling. The geometry of the coil assembly minimizes interactions between the ^{1}H and ^{13}C channels and ensures high sensitivity for ^{13}C detection (smaller coil). A similar design can be used in reverse for indirect ^{1}H[^{13}C] detection (Chen et al. 1998).

In addition, a different design was recently proposed which uses a quadrature ^{13}C coil inside a volume ^{1}H coil (Klomp et al. 2006). The use of a quadrature ^{13}C coil potentially yields 40% more signal compared with a single-loop linearly polarized coil. The two different coil designs are shown in Fig. 7.2.

Finally a few studies have been performed with a double-tuned volume coil (Pan et al. 1997; Mason et al. 1999), which has allowed ^{1}H[^{13}C] CSI over an entire slice in human brain (Pan et al. 2000).

7.3.1.3 Shimming

As in all MRS studies, excellent homogeneity of the B_0 magnetic field over the region of interest ("shimming") is required to maximize sensitivity and minimize spectral overlap. Since ^{13}C studies are often performed in a relatively large volume of interest, first-order shimming is usually not sufficient. Strong second order shims (high field scanners may have insufficient 2nd order shim strength capability for optimal shimming) and efficient methods for automated optimization of 2nd order shims (Gruetter and Boesch 1992; Gruetter and Tkac 2000) are necessary to ensure fast, optimal shimming over the entire detected volume.

7.3.1.4 Stability

Stability is another important requirement for ^{13}C studies. Reliable measurement of ^{13}C turnover curves requires excellent stability throughout the relatively long measurement times used for ^{13}C infusion studies. Even when movement is minimized, shimming often degrades over the course of the experiment. Readjustment of B_0 homogeneity during the study can help minimize the corresponding loss in sensitivity.

7.3.1.5 Power Calibration

Like most spectroscopy sequences, RF power needs to be precisely calibrated for optimal performance and to control power deposition. Because no appreciable ^{13}C signal can generally be observed in the brain before infusion of ^{13}C labeled substrate, this RF power calibration is generally performed using an external reference. A small sphere containing ^{13}C-labeled formic acid and positioned at the center of the detection coil is commonly used. (^{13}C formate is often chosen as a reference signal because it is relatively cheap, is available at high-concentration as a liquid, and does not overlap with the resonances of interest).

A simple sequence proposed by Bax allows calibration of both the transmitter and the decoupler power (Bax 1983). With this technique, power calibration can be performed without changing the connection of RF cables, and therefore using the exact same setup as the one used for actual ^{13}C MRS data collection. In addition to accurate calibration of RF power, this technique also makes it possible to verify that both RF channels are functioning properly (eg no defective cable or setup problem) before starting the infusion of ^{13}C label substrate.

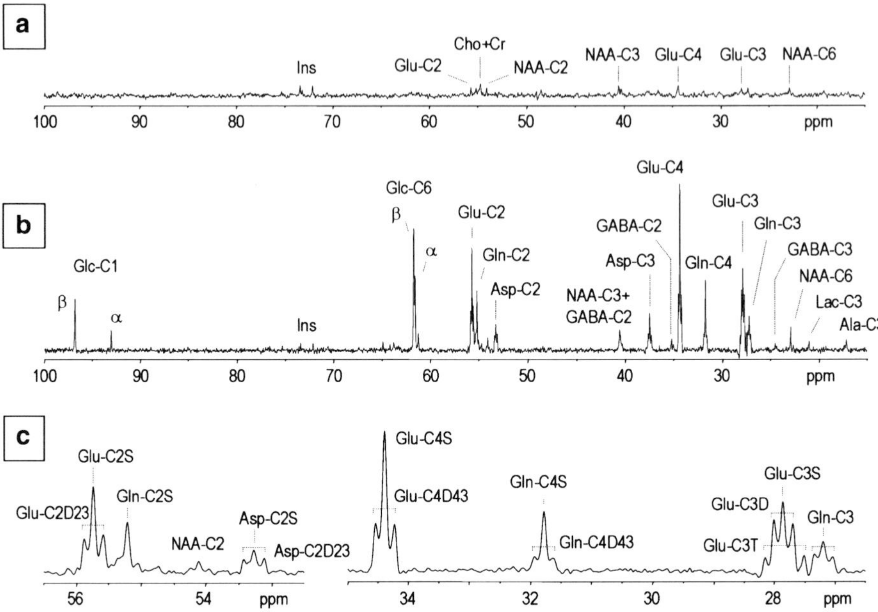

Fig. 7.3 *In vivo* ^{13}C spectra from a 400 μl volume in the rat brain, acquired using the modified DEPT sequence depicted in Fig. 7.4. (**a**) Natural abundance spectrum acquired for 3.5 h (5,120 scans, TR 2.5 s); (**b**) spectrum recorded for 1.8 h (2,560 scans, TR 2.5 s) during an infusion of 70%-enriched [1,6-$^{13}C_2$]glucose. Acquisition was started 1.8 h after the beginning of glucose infusion; (c) expansion of (b) showing the detection of doubly labeled isotopomers of glutamate, glutamine and aspartate. Note the triplet of glutamate (Glu-C3T) at 27.9 ppm, corresponding to glutamate labeled simultaneously at the C2, C3 and C4 positions. Processing consisted of zero-filling, filtering (5 Hz lorentzian line broadening for (**a**) and 2 Hz lorentzian-to-gaussian resolution enhancement for (**b**) and (**c**)), and fast Fourier transform. No baseline correction was applied (Reproduced with permission from Henry et al. 2003b)

7.3.2 *Direct and Indirect Detection*

A key factor that determines which resonances can be detected is the choice of the detected nucleus. Detection at ^{13}C frequency (direct ^{13}C detection) offers the advantage of increased spectral dispersion due to the large chemical-shift range at ^{13}C frequency, while detection at ^{1}H frequency (indirect $^{1}H[^{13}C]$ detection) offers higher sensitivity but also more spectral overlap. Therefore, the choice between direct or indirect detection is a trade-off between sensitivity (signal-to-noise) and spectral overlap (number of resonances than can be measured independently). The choice of a specific technique ultimately depends on the information that is needed for metabolic modeling.

Figures 7.3 and 7.4 show examples of direct and indirect spectra, respectively. These spectra were measured *in vivo* in the rat brain at 9.4 T.

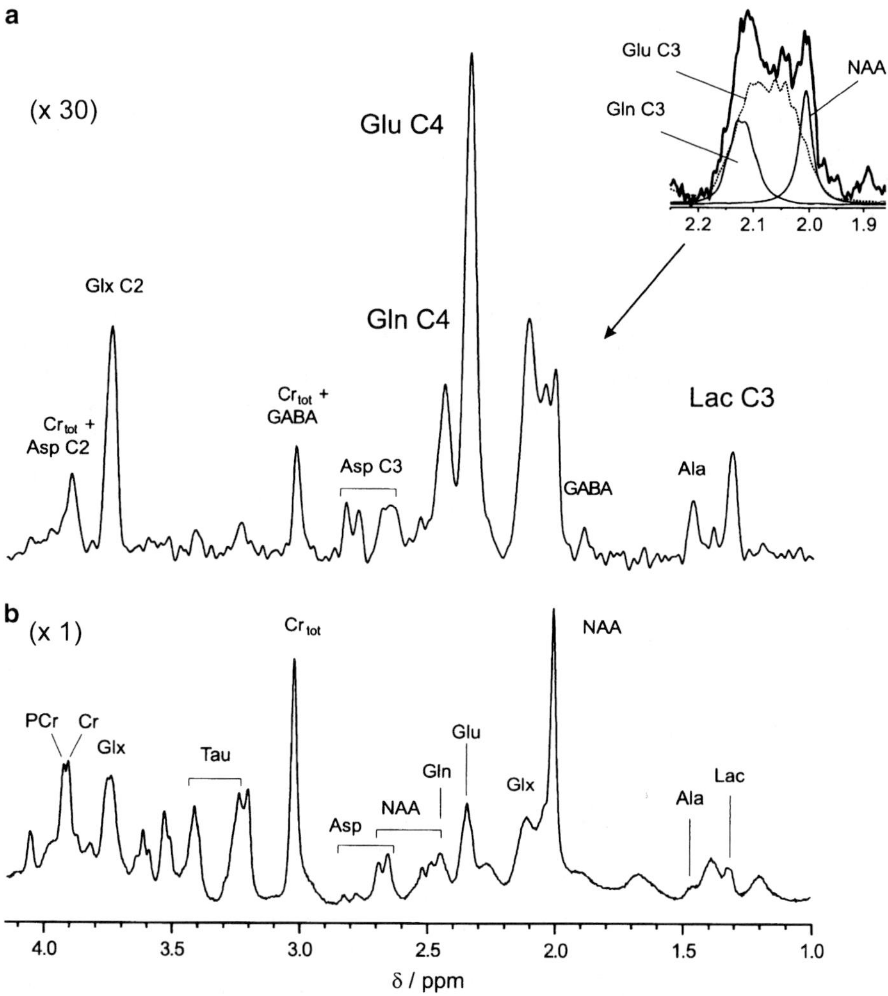

Fig. 7.4 ^{13}C-edited and broadband-decoupled $^{1}H[^{13}C]$ spectrum obtained with the ACED-STEAM sequence from a 137 μl volume in rat brain *in vivo* during infusion of [1-^{13}C] glucose (TE 7.9 ms, TM 20 ms, TR 4 s). (**a**) ^{13}C-edited ^{1}H spectrum obtained by subtracting spectra acquired with the inversion pulse of ACED-STEAM on and off, respectively (Soff – Son, 2,560 transients). Only signals from protons bound to ^{13}C are detected. Peak assignments were based on resonance position, line width, and relative intensity, being consistent with total concentration and fractional enrichment of glutamate. The *inset* illustrates the ability to distinguish ^{13}C label between glutamate C3 and glutamine C3 by comparing the *in vivo* spectrum (*solid line*) to line-broadened solution spectra of Glu, Gln, and NAA (*thin lines*). The spectrum was processed with 5.3 Hz Gaussian linebroadening. (**b**) The reference ^{1}H spectrum gained from the same data set as in a, averaging only the scans when the editing pulse was off (1,280 transients). No apodization was applied (fast Fourier transform only). The normalized vertical scale in (**a**) is 30-fold expanded compared with (**b**). No baseline correction was applied (Reproduced with permission from Pfeuffer et al. 1999b)

7.3.2.1 Sensitivity and Spectral Resolution

The gain in sensitivity provided by indirect detection is often overestimated. The actual gain in sensitivity can be estimated around 2n:1 where n is the proton multiplicity, as discussed in detail in Gruetter et al. (2003). According to this calculation, the gain in sensitivity for CH_2 resonances (eg glutamate $^4CH_2$) is about 4, the gain in sensitivity for CH resonances (eg glucose ^1CH) is about 2, and the gain in sensitivity for CH_3 (eg lactate $^3CH_3$) is about 6 when using indirect detection compared to direct detection with polarization transfer.

Although the gain in sensitivity provided by indirect detection is substantial, it is partly offset by spectral overlap. Consider the example of glutamate C4 and glutamine C4 (two of the most important resonances for metabolic modeling studies). In human brain at 4 T, glutamate C4 and glutamine C4 are well-resolved with direct detection, but cannot be resolved with indirect detection. Even in animal brain at 9.4 T, glutamate C4 and glutamine C4 partially overlap (see Fig. 7.4), requiring line fitting or similar approaches to deconvolute them.

Glutamate C3 (34.37 ppm) and glutamine C3 (27.9 ppm) are easily resolved using direct ^{13}C detection, even at lower field (1.5 T). In contrast, glutamate C3 and glutamine C3 are very difficult to resolve with indirect $^1H[^{13}C]$ detection, even at high-field (Pfeuffer et al. 1999a) (see Fig. 7.4).

Finally, glutamate C2 and glutamine C2 resonate nearly at the same frequency in proton spectra and cannot be separated with indirect detection, whereas they can be detected separately with direct detection.

7.3.2.2 Spectral Editing in Indirect Detection: Semiselective Proton-Observed Carbon-Edited (POCE)

One solution to alleviate spectral overlap when using indirect $^1H[^{13}C]$ detection is to use additional spectral editing to separate overlapping resonances (Henry et al. 2000). Semiselective POCE MRS takes advantage of the increased spectral dispersion at ^{13}C frequency to separate overlapping resonances in the $^1H[^{13}C]$ spectrum (Henry et al. 2000). Using this approach, glutamate C4 and glutamine C3 can be separated at 3 T. The same principle can in principle be used to separate glutamate C4 and glutamine C4.

7.3.2.3 2D Sequences

2D detection in principle combines the sensitivity of indirect detection with the spectral dispersion available at ^{13}C frequency. Several studies have been performed using 2D detection (Inubushi et al. 1993; Watanabe et al. 2000a, b). However, 2D detection has not found widespread use for metabolic studies. This may be due in part to the added complexity and the difficulty to accurately quantify peak volumes in 2D spectra *in vivo*.

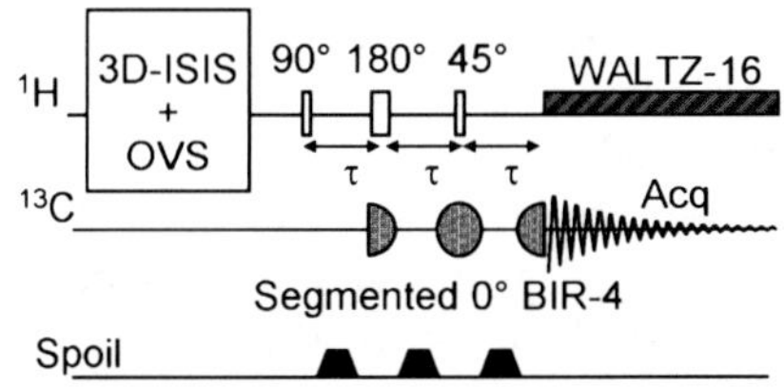

Fig. 7.5 Semi-adiabatic ^{1}H-localized polarization transfer. The ^{13}C part of DEPT is replaced by a segmented adiabatic 0° BIR-4 pulse. Dephasing gradients (2 ms duration, 46 mT/m) are added between the RF pulses in DEPT to eliminate unwanted, offset-dependent coherences. The flip angle of last ^{1}H pulse is set to a nominal 45° to detect signals simultaneously from CH, CH_2 and CH_3 groups. The delay τ is set to $1/(2*J_{CH})$ (Reproduced with permission from Henry et al. 2003a)

7.3.3 *Spatial Localization*

In many cases, spatial localization is highly desirable for ^{13}C and ^{1}H[^{13}C] MRS. Localization makes it possible to study well-defined, functionally specialized regions of the brain and results in superior spectral quality (good B_0 homogeneity, absence of contamination by lipid signals).

7.3.3.1 Localized Direct ^{13}C Detection

When using direct ^{13}C detection without localization (other than that provided by a surface coil), brain ^{13}C spectra are dominated by the strong natural abundance ^{13}C signals from subcutaneous lipids 30 ppm. These natural abundance signals can in principle be removed by subtraction of a natural abundance spectrum acquired prior to infusion of ^{13}C labeled substrate. However, subtraction artifacts usually hamper reliable detection of ^{13}C signals, especially glutamate and glutamine around 30 ppm.

Localized detection was initially achieved using direct localization at ^{13}C frequency using ISIS, which led to the first detection of natural abundance signals in human brain as well as the first detection of glutamine signals from [1-^{13}C]glucose (Gruetter et al. 1992, 1994). However direct localization at ^{13}C frequency suffers from a large chemical-shift displacement artifact due to the large chemical-shift range of ^{13}C spectra, which makes it difficult to achieve broadband detection of the entire ^{13}C spectrum. To overcome this problem, polarization transfer was introduced, in which localization is achieved at ^{1}H frequency, and magnetization is subsequently transferred to carbon using sequences such as INEPT and DEPT. Proton localization followed by polarization transfer has been implemented with ISIS localization (Gruetter et al. 1996, 1998; Shen et al. 1999; Henry et al. 2003a; Nabuurs et al. 2008) (see Fig. 7.5). Single-shot sequences have also been proposed, such as PRESS + DEPT (Yahya and Allen 2005) or single-shot INEPT detection using coherence selection gradients (Li et al. 2005).

Finally, note that the above-mentioned localized sequences are not necessarily suitable for detection of all ^{13}C resonances. For example, ^{13}C resonances in carboxyl groups cannot be detected with polarization transfer due to the absence of attached proton. Polarization transfer or ISIS localization are also not well suited for detection of ^{13}C glycogen due to the short T1 (~100 ms) and T2 (~6 ms). Localization is particularly critical for detection of brain glycogen to eliminate the strong contribution from glycogen in extracerebral muscles. Therefore a specific, non-echo sequence was developed for detection of ^{13}C (Choi et al. 2000; Öz et al. 2005, #1273). This has allowed the detection and quantification of glycogen in the brain in animals (Choi et al. 1999) and humans (Öz et al. 2007).

7.3.3.2 Localized Indirect $^1H[^{13}C]$ Detection

The first localized studies using $^1H[^{13}C]$ detection were performed using ISIS followed by the so-called Proton-Observed Carbon-Edited (POCE) sequence, which is a spin-echo sequence (Rothman et al. 1985) with a ^{13}C inversion pulse applied every other scan.

Later, single shot sequences were proposed, such as ^{13}C-edited PRESS (Chen et al. 1998), ACED-STEAM (Pfeuffer et al. 1999b), ^{13}C-edited J-refocused polarization transfer (Pan et al. 1997) or ^{13}C-edited LASER (Marjanska et al. 2004; Yang et al. 2005). In principle, any ^1H localization sequence can be adapted for $^1H[^{13}C]$ detection by incorporating a ^{13}C pulse and adjusting the timing of the sequence for adequate J-evolution for ^{13}C-editing.

Another way to localize signals is chemical-shift imaging (CSI), a method that does not suffer from chemical-shift displacement artifacts. More precisely, 3D-CSI has no chemical shift displacement artefact whereas slice selection in the presence of a gradient results in a chemical-shift displacement artefact in the direction of the slice selection for in-plane 2D-CSI.

CSI has been employed to obtain maps of ^{13}C glutamate turnover with infusion of ^{13}C label glucose during functional activation (Hyder et al. 1999) and to obtain maps of ^{13}C label incorporation in an entire slice in the human brain *in vivo*, allowing separate measurement of TCA cycle in grey matter and white matter (Pan et al. 2000).

7.3.4 RF Decoupling

Decoupling suppresses the effect of J-coupling during acquisition, resulting in increased signal-to-noise and reduced spectral overlap. However, decoupling is also very challenging *in vivo*, in human in particular. Difficulties related to decoupling are three-fold: first, it requires a large amount of RF power to be sent over a relatively long time during acquisition, and power deposition has to be kept within regulatory guidelines (see Sect. 7.3.6). Second, broadband decoupling schemes have a minimum RF field strength (γB_1) below which they will not achieve adequate decoupling performance. This γB_1 threshold may be difficult to attain in human studies at high field (de Graaf 2005) within FDA guidelines for power deposition. Third, sending large amounts of RF power during acquisition may interfere with the sensitive electronics used for detection of RF signals, resulting in spikes and/or added noise in the detected signals of interest and potentially completely obscuring these signals. These interferences can be minimized by careful design of the RF coil, and the use of appropriate RF filters.

A number of different schemes have been proposed for RF decoupling, such as MLEV, WALTZ, GARP, etc. They differ mostly by the bandwidth over which decoupling is efficiently achieved for a given RF power, as well as the minimum power required (threshold) to achieve decoupling. A number of decoupling schemes were recently evaluated by simulation (de Graaf 2005). It was concluded that, in human applications, only the WALTZ and MLEV decoupling methods provide adequate decoupling performance. For animal studies in which higher RF power levels can be achieved, additional decoupling techniques become available and provide superior performance.

A few studies have been reported without decoupling. Deelchand et al. have shown that taking into account all J-couplings allows quantification of undecoupled ^{13}C spectra with only 40% loss in sensitivity, some of which may be recovered by using a shorter repetition time (Deelchand et al. 2006). This may become a useful approach at very high field (7 T and above) where decoupling is likely to be very difficult in human brain within FDA guidelines. Another example was provided by Boumezbeur et al. who used undecoupled ^{1}H[^{13}C] MRS to detect incorporation of ^{13}C label without a ^{13}C RF channel, taking advantage of the fact that, when no ^{13}C decoupling is used, ^{1}H signals from carbons attached to ^{12}C or to ^{13}C give rise to distinct resonances (Boumezbeur et al. 2004).

7.3.5 Power Deposition (SAR)

Power deposition (specific absorption rate or SAR) is a particular concern for ^{13}C studies because of the use of RF decoupling during the acquisition time. Indeed, in most ^{13}C studies, the bulk of power deposition occurs during decoupling. Although this is usually not an issue for animal studies, power deposition can become a significant limitation in human studies, especially at high magnetic field, due to the necessity to keep power deposition within regulatory guidelines. For example, Food and Drug Administration (FDA) guidelines currently restrict total head power deposition to 3 W/kg averaged over 10 min over the entire head, and local power deposition to 8 W/kg over 1 min in any 1 g of tissue in the head (www.fda.gov/cdrh/ode/guidance/793.html). European Union guidelines similarly restrict local SAR to 10 W/kg averaged over 6 min in any 10 g of tissue in the head (European Union Directive 2004/40/EC).

Careful evaluation of SAR is therefore a necessity before performing experiments in human brain. This evaluation needs to take into account the fact that some RF power is lost (eg in cables). In practice, SAR can be estimated by measuring RF power at the RF coil port, and taking into account the duration and shape of RF pulses. In addition, local SAR can be predicted with prior knowledge of the B_1 distribution (using either assumptions or simulations). With surface coils being used in most studies, the limiting factor is usually local SAR, rather than SAR averaged over the head (van den Bergh et al. 1998).

Carbon-13 MRS stands to benefit from higher magnetic fields. However, high-fields imply additional challenges, which explain in part why no dynamic ^{13}C studies with ^{13}C-labeled substrates have been reported to date in human brain at B_0 magnetic field higher than 4 T. First, power deposition increases with higher magnetic field, making it more difficult to perform RF decoupling within SAR regulatory guidelines. In addition, RF distribution becomes more inhomogeneous at high field, complicating prediction of local SAR. In that context, the use of computer simulations can be a helpful tool to predict possible "hot spots" with a particular RF coil geometry (Collins et al. 2004). In cases where SAR cannot be reduced within safety guidelines, one solution is to shorten the acquisition time (potentially leading to truncation artifacts and decreased spectral resolution), or to lengthen the repetition time, at the expense of S/N per unit time.

7.4 Quantification of ^{13}C MR Spectra

Two steps need to be considered when quantifying time series of ^{13}C MR spectra: relative quantification and absolute quantification.

In a first step, relative quantification is used to convert time series of ^{13}C spectra into ^{13}C turnover curves for ^{13}C resonances of interest (for example, turnover curves for glutamate C4 or glutamine C4). This relative quantification step yields the time course of each ^{13}C resonance in an arbitrary unit. Various methods have been employed for this relative quantification of ^{13}C time courses (measurement of peak height, integration, line fitting, etc.). Spectral fitting with prior knowledge is now the method of choice.

In a second step, absolute quantification is needed to convert time courses expressed in arbitrary units into ^{13}C turnover curves expressed as actual ^{13}C concentrations (or isotopic enrichments). This step is necessary because metabolic modeling requires the fitted data to be expressed in absolute units.

In principle, ^{13}C signals (resonance areas) are directly proportional to ^{13}C concentration for each resonance. For *in vivo* studies, this is often not the case, as the detection efficiency at each ^{13}C resonance may be affected differently by various factors such as:

- off-resonance effects of RF pulses
- imperfections in RF pulse angles or in RF decoupling (especially with surface coils)

– strong coupling effects (both in indirect ^{1}H[^{13}C] detection (Henry et al. 2006b) and in direct ^{13}C detection with polarization transfer (Yahya and Allen 2005))
– differential effects of T_1 relaxation (if spectra are not acquired under fully relaxed conditions. Different carbons have different T_1 relaxation times)
– Effect of proton multiplicity when using polarization transfer

Depending on the method used for absolute quantification, some of these potential sources of bias may be automatically compensated. For other sources, correction factors can be determined on phantom, keeping in mind that relaxation times are often different between in vitro and *in vivo* experimental conditions.

Any uncorrected bias in absolute quantification is likely to cause bias in the determination of metabolic fluxes during metabolic modeling (see metabolic modeling section for more details). For example, when using a one-compartment model to fit glutamate C4 and glutamate C3 time courses, any bias in the relative scaling of the C3 curve relative to the C4 curve may have an impact on the value of fitted metabolic fluxes such as V_{TCA} and V_X (Henry et al. 2002).

7.4.1 Quantification of Direct ^{13}C Spectra

7.4.1.1 Relative Quantification: Spectral Fitting of Direct ^{13}C Spectra

Relative quantification of ^{13}C spectra is usually straightforward because of the lack of spectral overlap between ^{13}C resonances. Traditionally, ^{13}C spectra have been quantified using simple approaches such as measuring peak height, peak integral, or simple line fitting.

Because ^{13}C studies produce large datasets of time series of spectra (eg Fig. 7.6), automated analysis is particularly attractive. Another advantage of automated analysis is to remove potential bias from the investigator. As a result, several automated analysis methods have been proposed in recent years for direct ^{13}C detection (Henry et al. 2003b; Shic and Ross 2003). In particular, it was shown that linear combination modeling with the LCModel software, which was initially developed for quantification of ^{1}H spectra, could be adapted to analyze ^{13}C spectra as well (Henry et al. 2003b). The use of LCModel allows completely automatic analysis of all time series of ^{13}C spectra, including multiplets arising from multiply labeled ^{13}C-^{13}C isotopomers (Figs. 7.7 and 7.8). Using LCModel also allows small phase distortions caused by ^{13}C-^{13}C J-modulation to be taken into account (Henry et al. 2003b).

Strong-coupling effects are not obvious in directly detected ^{13}C spectra, because of the large spectral range at ^{13}C frequency and the smaller value of ^{13}C-^{13}C J-couplings (typically 30–55 Hz). However strong coupling effects can occur when using polarization transfer and may result in different detection efficiency for different resonances across the ^{13}C spectrum (Yahya and Allen 2005).

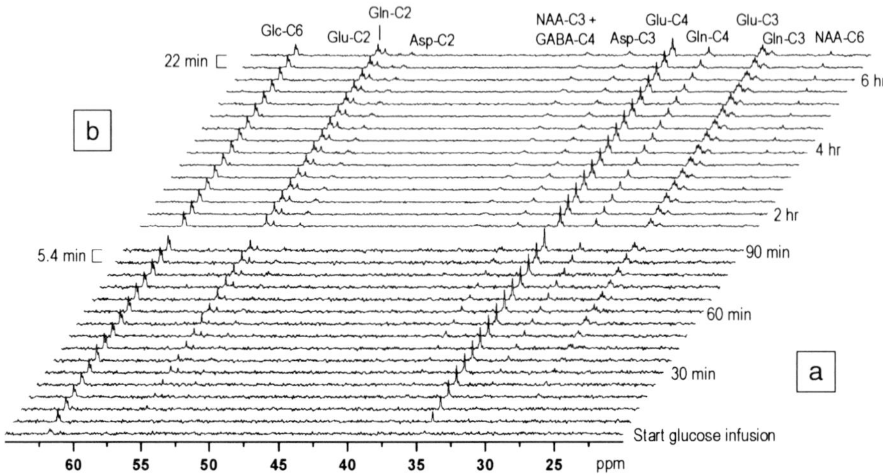

Fig. 7.6 Time course of ^{13}C label incorporation during an infusion of [1,6-$^{13}C_2$]glucose. The stack plot of an expansion of the ^{13}C spectra shows (**a**) the first 90 min with a temporal resolution of 5.4 min (128 scans each, TR 2.5 s), showing label incorporation into glucose C6, glutamate C4, C3, C2, glutamine C4, C3,C2 and aspartate C2 and C3; (**b**) the subsequent 5.5 h with a temporal resolution of 22 min, showing the slower labeling of NAA C2, C3 and C6. Mild lorentzian-to-gaussian resolution enhancement was applied (Reproduced with permission from Henry et al. 2003a)

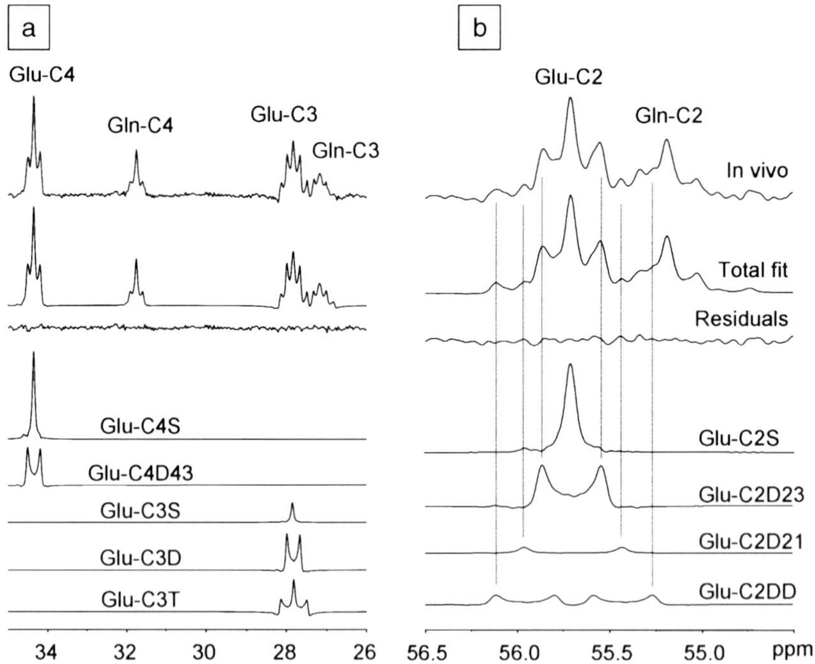

Fig. 7.7 Expansions of Fig. 7.2c showing the fitting of individual isotopomers in glutamate and glutamine. (**a**) Isotopomer analysis of glutamate and glutamine C4 and C3 (**b**) isotopomers analysis of glutamate and glutamine C2. *Vertical dashed lines* indicate the presence of individual isotopomers in the *in vivo* spectrum (Reproduced with permission from Henry et al. 2003b)

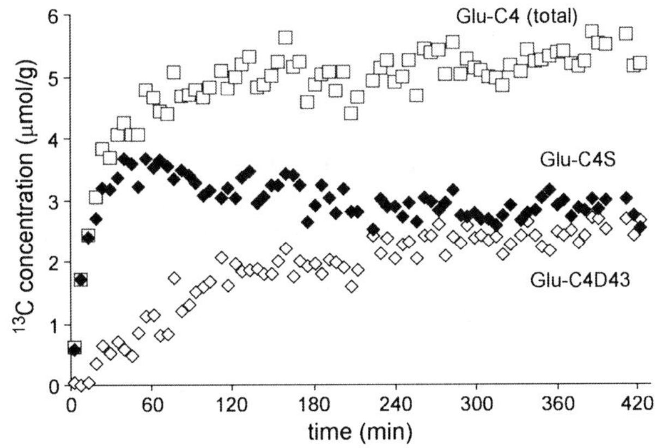

Fig. 7.8 *In vivo* time courses of individual isotopomers (μmol/g of ^{13}C label) during an infusion of [1,6-$^{13}C_2$] glucose. (**a**) Time course of Glu-C4S (*black diamonds*), Glu-C4D43 (*open diamonds*) and total Glu-C4 (=Glu-C4S + Glu-C4D43) (*open squares*) with a time resolution of ~5 min (128 scans, TR 2.5 s) (Reproduced with permission from Henry et al. 2003b)

7.4.1.2 Absolute Quantification of Direct ^{13}C Spectra

As mentioned above, accurate absolute quantification is critical for metabolic modeling. Each turnover curve must reflect the actual isotopic enrichment for the carbon position of interest. Errors in absolute quantification will cause bias in fitted metabolic fluxes or even make it impossible to fit the data properly.

Determination of isotopic enrichment at the end point of the experiment can be done in several ways:

- Internal referencing
  Internal referencing is rarely used in direct ^{13}C MRS due to the lack of a strong internal reference signal. In some studies, however, natural abundance signals from myo-inositol have been used as a reference (based on the observation that there is no significant ^{13}C labeling of myoinositol over the time scale typically used for *in vivo* ^{13}C experiments). In that case, correction factors must be applied for other sources of bias mentioned below.
- External referencing
  With eternal referencing, *in vivo* signals are compared to signals from a phantom of known concentration measured in separate experiment, but with the same setup as the *in vivo* experiment. For example *in vivo* signals from [4-^{13}C]glutamate are compared to signals from a phantom of 100 mM of glutamate (natural abundance). Corrections are made for differences in loading between *in vivo* and in vitro conditions if necessary. Because the same pulse sequence and experimental conditions are used *in vivo* and on phantom, the method inherently compensates for many sources of bias, such as off-resonance effects, effect of multiplicity in polarization transfer, although as stated earlier, one needs to pay a particular attention to the different relaxation times in vitro and *in vivo*, if acquisitions are not performed under fully relaxed conditions.

- Absolute quantification using measurement of isotopic enrichment at end point
  Using the value of isotopic enrichment is probably the most common and straightforward method to convert ^{13}C turnover curves from arbitrary units into units of ^{13}C concentration. Once enrichment is known, ^{13}C concentration can be determined by assuming the total concentration or by measuring it using ^{1}H MRS (^{13}C conc = total conc x FE). Often determination of enrichment is performed at the end point of the experiment, and the entire time course is then scaled accordingly.

  (a) isotopic enrichment measured using POCE
  Even when using direct ^{13}C detection, an additional *in vivo* $^{1}H[^{13}C]$ POCE spectrum can be collected at the end of the experiment (after switching cables) and this spectrum can be quantified (see next section: quantification of indirect $^{1}H[^{13}C]$ spectra) to determine the enrichment of each carbon. However, determination of enrichment may not be possible for every resonance due to spectral overlap. For example, in humans, it is difficult to separate glutamate C4 from glutamine C4 with indirect detection, even at 4 T.

  In animals, $^{1}H[^{13}C]$ POCE NMR can be performed on brain extracts. However, even at 14 T (600 MHz), overlap of certain resonances (eg overlap of glutamate H3 and glutamine H3, or overlap of glutamate H2 and glutamine H2 resonances) makes it difficult to measure enrichment of glutamate and glutamine separately. One way to solve this problem is to perform biochemical separation of amino acids (eg using HPLC) and to measure enrichment in glutamate and in glutamine individually.

  (b) isotopic enrichment measured using isotopomer analysis
  An alternative method to determine isotopic enrichment at the end point is to use isotopomer analysis (Gruetter et al. 2001; Henry et al. 2003a) instead of POCE. When infusing [1-^{13}C] or [1,6-$^{13}C_2$]glucose, the doublet at the glutamate C4 position directly reflects labeling at the C3 position (labeling at the C5 position is negligible) (see Fig. 7.3). Therefore in this particular case ([1-^{13}C]glucose or [1,6-$^{13}C_2$]glucose infusion) the isotopic enrichment at the C3 position can be calculated as: IE(C3) = C4D43/(C4S + C4D43). The enrichment at other carbon positions can then be calculated by comparing signals intensities, and taking into account different detection efficiencies (as measured on phantom). For example, if S(C4) and S(C3) are the measured signal integrals at the C4 and C3 position respectively, and if phantom measurements show that the C3 resonance is detected with 10% less efficiency than the C4 resonance, then the isotopic enrichment of glutamate at the C4 position can be calculated as: IE(C4) = IE(C3) x S(C4)/(S(C3) * 1.1), in which IE(C3) has been determined by isotopomer analysis as explained above. The isotopic enrichment at the C2 position can be determined in the same way.

  Although isotopomer analysis is a powerful method to determine isotopic enrichment, it must be used with caution because it is only valid if there is true randomization of label within the molecule (i.e. for a given molecule, the probability to label a given carbon position is independent of whether other carbon positions are labeled or not). This is the case for glutamate C3 when infusing [1-^{13}C] or [1,6-$^{13}C_2$]glucose. Labeling at the C3 position is independent of whether the C4 carbon position is labeled or not, because acetyl-CoA incorporated into the TCA cycle can be labeled or unlabeled irrespective of labeling at other carbon positions.

However, in other cases, such as infusion of doubly labeled [1,2-$^{13}C_2$] acetate, isotopomer analysis does not allow determination of isotopic enrichment. Indeed, in that case, glutamate C4 and C5 always get labeled together, so that all glutamate labeled at the C4 position will also be labeled at the C5 position. Because all detected ^{13}C signal is a doublet, isotopomer analysis would improperly conclude that enrichment at the C5 position is 100%, irrespective of the actual isotopic enrichment.

7.4.2 *Quantification of Indirect ^{1}H[^{13}C] Spectra*

7.4.2.1 Relative Quantification: Spectral Fitting of Indirect ^{1}H[^{13}C] Spectra

Quantification of indirect detection ^{1}H[^{13}C] spectra is at first sight very similar to quantification of ^{1}H spectra. Sophisticated approaches such as LCModel have become the method of choice for quantification of ^{1}H MRS spectra (Provencher 1993; Pfeuffer et al. 1999a).

However there are a number of particularities that apply only to ^{1}H[^{13}C] spectra compared to quantification of regular ^{1}H spectra. When fitting ^{1}H spectra, each model spectrum in the basis set corresponds to the spectrum of an entire molecule. For example, the basis spectrum for glutamate contains resonances for αCH (3.75 ppm), βCH_2 (2.04 and 2.12 ppm) and γCH_2 (2.34 ppm) groups. In contrast, when fitting edited ^{1}H[^{13}C] spectra, the relative intensities for each group can be different due to different ^{13}C isotopic enrichment at each carbon position. Therefore the peak intensities need to be fitted independently for each CHn group. This can be achieved by using one model spectrum for each CHn group (e.g. for glutamate, three model spectra would be used, for αCH, βCH_2 and γCH_2 respectively) (Pfeuffer et al. 1999b; De Graaf et al. 2003a).

It was recently pointed out that strong coupling effects should be taken into account in the quantification because they may cause serious bias in the quantification of ^{1}H[^{13}C] spectra (Henry et al. 2006b), due to "cross-talk" between CHn groups observed with strong coupling. For example, the signal of glutamate γCH_2 at 2.34 ppm may contain signal contributions from the βCH_2 group as well. These effects can be accounted for by simulating spectra using density matrix simulations (Henry et al. 2006b).

7.4.2.2 Absolute Quantification of Indirect ^{1}H[^{13}C] Spectra

Absolute quantification is relatively straightforward with indirect ^{1}H[^{13}C] detection. Indirect ^{1}H[^{13}C] detection is generally performed by subtracting two spectra. One is a regular ^{1}H spectrum (e.g. obtained with spin-echo), and the second one is a ^{1}H spectrum obtained with an additional inversion pulse at ^{13}C frequency (Rothman et al. 1985). Absolute quantification can be done using water as an internal reference (reference scan without water suppression). Quantification of the unedited spectrum (acquired without ^{13}C inversion) also allows simultaneous determination of total ^{1}H concentrations.

Editing using POCE and similar methods are based on inversion of ^{13}C signals every other scan. Any imperfection in this ^{13}C inversion for a specific resonance (eg due to off-resonance effects) reduces the corresponding signal in the edited $^{1}H[^{13}C]$ spectrum and therefore leads to an underestimation of the isotopic enrichment, unless a correction factor is applied. In practice, several recently proposed sequences achieve close to 100% inversion over a large bandwidth with the use of adiabatic pulses (Pfeuffer et al. 1999b; De Graaf et al. 2003a).

7.5 Metabolic Modeling of ^{13}C Isotopic Turnover Curves

Carbon-13 turnover curves by themselves do not directly allow determination of metabolic rates. In fact, these turnover curves result from ^{13}C label flowing through multiple metabolic pools and multiple metabolic pathways, and therefore depend on the value of multiple metabolic rates as well as on the size of metabolic pools (concentrations). For example, a change in the rate of ^{13}C label incorporation does not necessarily indicate a change in metabolic rate, but may simply reflect a change in the size of metabolic pool(s). (A large metabolic pool will label more slowly than a small pool).

In order to take full advantage of the dynamic information provided by ^{13}C MRS, sophisticated metabolic modeling is used to analyze ^{13}C turnover curves and derive absolute metabolic rates though specific metabolic pathways. The details of this process have been described in several recent reviews (Gruetter 2002; Mason and Rothman 2004; Henry et al. 2006c) and can be divided into the following steps (Henry et al. 2006c):

- choice of metabolic model
- mathematical expression of the model
- fit of model to experimental data
- evaluation of the reliability of the results: fit residuals, sensitivity analysis and Monte-Carlo simulations

7.5.1 Choice of Mathematical Model

The choice of a metabolic model is guided by two main considerations:

1. The model should accurately reflect metabolism of the infused ^{13}C-labeled substrate. For example, when modeling the turnover of [4-^{13}C]glutamate from [1-^{13}C]

glucose, the metabolic model should accurately reflect the steps involved in this metabolic pathway. An inaccurate model is likely to cause a bias in the determination of metabolic rates, and will often (but not always) be evidenced as non-random residuals in the fit.

2. The metabolic model should not be underdetermined. An underdetermined problem has too many degrees of freedom compared to the available experimental data, and will not allow precise determination of metabolic rates. It may even cause the fitting procedure to fail (lack of convergence) or to get trapped in local minima (in which case the obtained fitted values of metabolic rates will be biased and will usually be dependent on initial conditions for the fit). Therefore, in spite of the complexity of brain metabolism, the metabolic model has to be simple enough in order to reliably fit the relatively limited amount of data that can be obtained from ^{13}C NMR (for example, ^{13}C turnover curves cannot be measured for TCA cycle intermediates due to their low concentration). If too many free parameters (unknown fluxes) are present in model, the value of some of these free parameters can be assumed based on literature, in order to reduce the number of unknowns in the model. In that case, it is essential to evaluate carefully the validity of these assumptions and their potential impact on the results, as they may cause bias in the determination of other fluxes in the model.

One nearly systematic assumption in ^{13}C metabolic modeling is that the system under investigation is at metabolic steady-state, i.e. that metabolite concentrations and metabolic rates are constant over the duration of ^{13}C MRS measurements.

In the brain, two main types of metabolic models have been developed:

7.5.1.1 One-Compartment Models

One-compartment models have been proposed to fit glutamate (but not glutamine) turnover curves measured in the brain after infusion of 1-^{13}C glucose (Mason et al. 1992b). An example of current one-compartment model is shown in Fig. 7.9. Glucose is metabolized in the TCA cycle and labels glutamate through exchange with 2-oxoglutarate. From [1-^{13}C]glucose, the C4 of glutamate is labeled on the first turn and C3 is labeled on the second turn of the cycle. Metabolic modeling of glutamate turnover allows determination of the rate of TCA cycle. Since most glutamate is located in neurons, this rate reflects primarily the neuronal TCA cycle rate.

Metabolic modeling of TCA cycle activity was pioneered by Chance et al. (Chance et al. 1983) It was first applied to brain metabolism in the early 1990s in rat brain (Mason et al. 1992b) and human brain (Mason et al. 1995). Later studies have used this one-compartment modeling approach to measure the rate of TCA cycle in a variety of experimental conditions including during functional activation and in animal models of disease (Henry et al. 2002). Studies in the heart have also provided useful insights into one-compartment modeling (Lewandowski et al. 1997; Yu et al. 1997; Jeffrey et al. 1999).

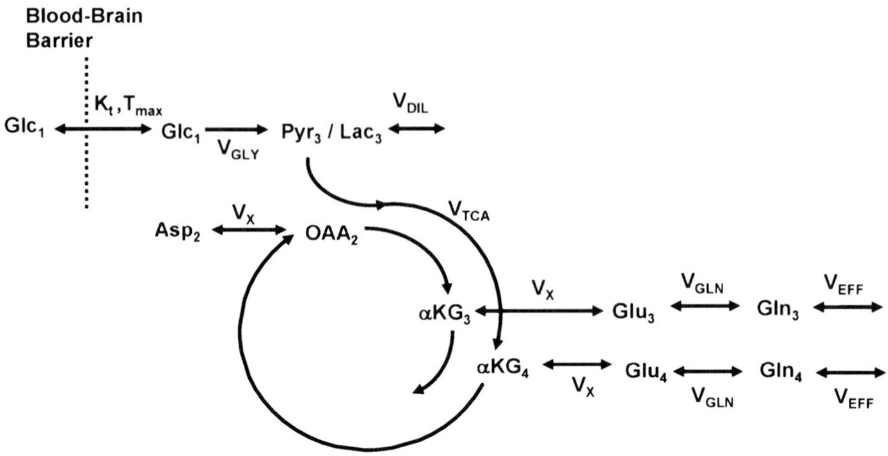

Fig. 7.9 Example of current one-compartment model used to analyze glutamate C4 and glutamate C3 labeling curves measured during infusion of [1-^{13}C]glucose

The most current one-compartment models comprises up to four free parameters (V_{TCA}=TCA cycle rate, V_X=exchange rate between 2-oxoglutarate and glutamate, V_{DIL}=rate of lactate isotopic dilution, and V_{EFF}=combined rate of glutamine isotopic dilution and glutamine efflux). With one-compartment modeling, glutamine ^{13}C time courses are typically not utilized, and the rate of glutamine synthesis (V_{GLN}) is assumed based on literature.

Dilution fluxes are necessary to account for the loss of ^{13}C label along metabolic pathways. For example, glutamate C4 does not reach the same isotopic enrichment as blood glucose during *in vivo* experiments. In experiments using continuous infusion of 70%-enriched [1,6-$^{13}C_2$]glucose, the steady state enrichment of glutamate does not reach 70%, as would be expected if no dilution occurred, but typically reaches around 50%. This loss of label is usually ascribed to isotopic dilution of labeled lactate with unlabeled lactate from the blood, but part of it may also reflect synthesis of unlabeled acetylCoA from unlabeled ketone bodies or acetate. In any case, this dilution is included in the model as a dilution flux V_{DIL} (sometimes also called V_{DILLAC} or V_{OUT}).

Similarly, glutamate C3 does not reach the same isotopic enrichment as glutamate C4, even at isotopic steady-state (Henry et al. 2003c). This can be explained first by isotopic dilution of the C3 carbon position via pyruvate carboxylase, and second by loss of ^{13}C label through exchange of labeled glutamine with unlabeled glutamine from the blood. These two processes may be lumped into the rate of glutamine efflux V_{EFF} in the one-compartment model (Henry et al. 2008).

7.5.1.2 Two-Compartment Models

Brain metabolism is exquisitely compartmentalized, with most glutamate located in neurons and almost all glutamine located in astrocytes. This compartmentation was

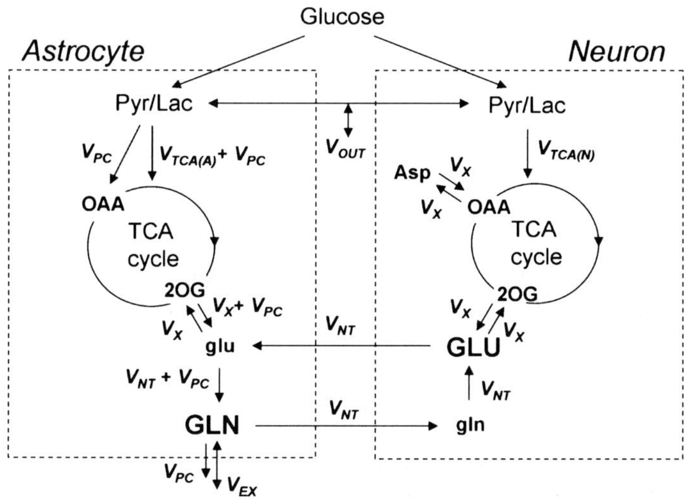

Fig. 7.10 Example of current two-compartment model. The model has seven free parameters: $V_{TCA(N)}$, $V_{TCA(G)}$, V_{NT}, V_{PC}, V_X, V_{OUT}, V_{EX}. See text for details

demonstrated in the 1960s and early 1970s using 14C radioactive tracers (Van den Berg et al. 1969). When not only ^{13}C-glutamate but also ^{13}C-glutamine were first detected by NMR in the brain *in vivo* (Gruetter et al. 1994, 1996), it was quickly recognized as a major progress for the investigation of compartmentalized neuronal-glial metabolism. In seminal studies, Sibson et al. (1997, 1998) showed that analysis of glutamate C4 and glutamine C4 turnover curves could allow measurement of the rate of glutamate-glutamine cycle, and demonstrated that this glutamate-glutamine cycle is a major flux *in vivo*, accounting for most of the flux through glutamine synthetase under normal (normammonemic) conditions. A critical aspect of such measurements is metabolic modeling. As discussed in the next section, the precision and reliability of metabolic modeling with two-compartment models is still a subject of active research.

An example of current two-compartment model is presented in Fig. 7.10. The most up to date modeling studies included up to seven free parameters: neuronal TCA cycle rate (V_{PDH} or $V_{TCA(N)}$), glial TCA cycle rate (V_G or $V_{TCA(A)}$), glutamate glutamine cycle (V_{CYCLE} or V_{NT}), rate of pyruvate carboxylase (V_{PC} or V_{EFFLUX}), exchange rate between 2-oxoglutarate and glutamate (V_X), isotopic dilution of lactate (V_{OUT} or V_{DILLAC}), isotopic dilution of glutamine (V_{EX} or V_{DILGLN}) (Oz et al. 2004). In some studies, a few of these free parameters have been kept constant (with their value assumed based on previous studies or on literature) in order to stabilize the fit. For example, when only the glutamate C4 and glutamine C4 turnover curves are used, the value of V_X has to be assumed (Sibson et al. 1997, 1998; Mason et al. 1999), similar to studies with one-compartment models. When using glucose, fluxes such as V_{PC} or $V_{TCA(A)}$ have also be assumed in several studies (e.g. (Mason et al. 1999; de Graaf et al. 2004; Patel et al. 2005)).

A number of other assumptions are worth pointing out. First, when using [1-^{13}C] glucose or [1,6-^{13}C2]glucose, these models have assumed a single pyruvate/lactate pool for neurons and astrocytes (based on the fact that enrichment in lactate is likely to be the same in neurons and astrocytes). Second, aspartate is assumed to be located primarily in the neuronal compartment. Glutamate is assumed to be predominantly neuronal and glutamine is assumed to be in glia. Typical assumptions for concentrations of glutamate and glutamine are: [GLU(N)] = 9 mM, [GLN(N)] = 0.5 mM, [GLU(A)] = 1 mM, [GLN(A)] = 3.5 mM. It has been shown that, within reasonable limits, variation in the assumed concentrations of glutamate and glutamine does not have significant impact on metabolic rates determined with metabolic modeling (Gruetter et al. 2001).

7.5.2 Estimation of Reliability of Metabolic Models: Residuals, Estimation of Standard Deviation, Monte-Carlo Simulations and Sensitivity Analysis

Evaluation of the reliability of metabolic modeling under the specific experimental conditions of the study is critical to determine how confident one can be in the values of metabolic rates. Unfortunately, this aspect of metabolic modeling studies is often neglected.

The reliability of metabolic modeling can be assessed using a variety of tools:

- **inspection of fit residuals.** Non-random residuals indicate that the metabolic model is not accurate (e.g. a metabolic pathway is missing in the model), or that the experimental data is biased (e.g. if there is a problem with absolute quantification).
- **sensitivity analysis**. Sensitivity analysis consists in varying specific parameters and determining how sensitive metabolic modeling results are to these variations. Therefore it is often used to verify the potential impact of assumptions on the results.
- **Monte-Carlo simulations**. Monte-Carlo simulations are a powerful method to evaluate the reliability of complex models. They consist in repeating the fitting procedures multiple times (often hundreds of times or more), each time with a different noise, to determine the impact of noise on fitted parameters. Monte-Carlo simulations provide a lower bound for the error on fitted metabolic rates. They only take into account the effect of random noise, but not other possible sources of bias, such as inaccurate scaling of turnover curves or incomplete metabolic model. Note that, when performing Monte-Carlo simulations, it is critical to make sure that the algorithm does not get trapped in local minima, which can be verified by performing simulations with different initial conditions for fitted parameters.

7.5.2.1 Reliability of One-Compartment Models

Many studies were performed by modeling the glutamate C4 turnover curve only (Hyder et al. 1996, 1997; Mason et al. 1999). In that case it is not possible to fit all free parameters (underdetermined problem) and an assumption must be made on V_X. The actual value of V_X has been controversial, with initial studies finding that V_X was extremely fast (Mason et al. 1992b, 1995) and several subsequent studies finding that V_X was slower (on the order of V_{TCA}). The current view is that V_X is probably not as fast as initially reported, and that the exact value of V_X is difficult to determine reliably from ^{13}C data. In addition, the Vx flux may change as a function of the metabolic conditions under investigation (Henry et al. 2002). Therefore it is definitely safer to avoid making any assumption on V_X and to analyze both glutamate C4 and C3 curves as opposed to glutamate C4 alone. Analyzing the C4 glutamate curve alone with a wrong assumption on V_X may lead to up to 50% bias on the value of TCA cycle rate (Henry et al. 2002).

A recent study showed that the rate of ^{13}C incorporation into glutamate C4 and glutamate C3 could be described analytically, assuming that intermediate metabolic pools are small and have negligible effect on ^{13}C labeling curves (Uffmann and Gruetter 2007). These developments provide useful insights into metabolic modeling. However, the new analytical formulation does not allow analytic derivation of metabolic rates. These rates must still be obtained using least square fitting, which limits the practical usefulness of this new approach for metabolic modeling.

In conclusion, the determination of (neuronal) TCA cycle rate using a one-compartment model is considered fairly robust as long as both glutamate C4 and C3 curves are fitted simultaneously.

7.5.2.2 Reliability of Two-Compartment Models

Surprisingly, in spite of their complexity, the reliability of two-compartment metabolic models has not been investigated in detail until recently. A Monte-Carlo simulation study (Shestov et al. 2007b) was performed to evaluate the validity of one of the most complete metabolic models published to date (Gruetter et al. 2001). The authors showed that the determination of the glutamate-glutamine cycle rate V_{NT} is relatively unreliable when using [1-^{13}C]glucose or [1,6-^{13}C2]glucose as a substrate. The model by Gruetter et al., like most two-compartment metabolic models used to analyze *in vivo* data to date, does not include a separate parameter for dilution of labeled glutamine by unlabeled glutamine coming from the blood. Such a dilution source appears necessary to account for lower enrichment in glutamine compared to glutamate at isotopic steady-state (Shen et al. 1999; Oz et al. 2004). Our own preliminary results show that including glutamine dilution in the model incidentally leads to more precise estimates of V_{NT} compared to the case when glutamine dilution is omitted. The fact that such a small change in the model has a significant impact on the precision of metabolic rates (as determined by Monte-Carlo simulations) further underscores the importance of performing Monte-Carlo simulations for every study with the specific experimental conditions and the specific metabolic model used in that study.

Two-compartment metabolic models are not only complex, but have also been evolving constantly in the past 10 years. Initial two-compartment modeling was performed with relatively simple models (no astrocytic TCA cycle flux, no independent pyruvate carboxylase flux, assumption of fast exchange rate Vx between 2-oxoglutarate and glutamate, etc.). Two-compartment models were further refined in the past 10 years, with the addition of astrocytic TCA cycle (Gruetter et al. 1998, 2001; Shen et al. 1999), the addition of an independent pyruvate carboxylase flux (Gruetter et al. 1998, 2001; Shen et al. 1999), the recognition that V_X may not be infinitely fast compared to V_{TCA} (Gruetter et al. 2001; Henry et al. 2002) and the incorporation of a glutamine dilution flux (Shen et al. 1999; Oz et al. 2004). Two-compartment modeling continues to be the object of active work and is likely to be further refined in the coming years.

An open question is to determine to which extent separate detection of all carbon resonances of glutamate and glutamine resonances improves the precision of metabolic modeling with two-compartment models. With one-compartment models, it has been shown that glutamate C4 curves are not sufficient to obtain a reliable estimate of V_{TCA} without assumption on the value of V_X. With two-compartment models, however, it is not known whether metabolic modeling with 6 independent time courses (glutamate C4, C3, C2 and glutamine C4, C3, C2), or even 7 time courses when including aspartate, gives more precise estimates of metabolic rates than metabolic modeling with 4 independent time courses (glutamate C4, glutamine C4, glx C3 and glx C2, where glx designates the sum of glutamate and glutamine resonances).This question can be answered with Monte-Carlo simulations similar to those performed in a recent study (Shestov et al. 2007b). As a rule, when the number of fitted time courses is increased, fewer assumptions are required and higher precision and accuracy is obtained on fitted parameters. Fitting too few data with too many free parameters leads to underdetermined problems. For example, fitting only glutamate C4 and glutamine C4 may be unreliable unless sufficient assumptions are used to constrain the fitting procedure. Therefore, which independent time courses are necessary to evaluate metabolic fluxes with good precision must be evaluated carefully for each study. This in turn may affect the choice of the measurement method to be used. For example, glutamate and glutamine C3 can be measured separately only with direct ^{13}C detection, but not with indirect $^{1}H[^{13}C]$ detection.

Finally, a few metabolic modeling studies have been performed with more than two compartments. For example, a few studies were performed to determine compartmentalized metabolic rates in glia and neurons simultaneously multiple tissue types such as grey matter and white matter (Mason et al. 1999; de Graaf et al. 2004). Similarly, a paper recently reported modeling with a 3-compartment model (glutamate-glutamine-GABA) (Patel et al. 2005). The reliability of such a metabolic modeling with more than two compartments has not been investigated in detail using Monte-Carlo simulations.

7.5.3 Development of New "Isotopomer" Models

Most metabolic models used to analyze *in vivo* ^{13}C MRS studies are "positional" models, meaning that they are able to fit time courses of *total* isotopic enrichment at each carbon position (positional enrichment). However, when using direct ^{13}C detection, additional information is available from ^{13}C-^{13}C isotopomers (appearing as multiplets in ^{13}C spectra). Taking advantage of this additional dynamic isotopomer information has been shown to improve the precision of metabolic modeling in the heart (Jeffrey et al. 1999) but no attempt has been made to use dynamic fitting of time courses of ^{13}C-^{13}C isotopomers in the brain. Time-resolved detection and quantification of individual multiplets has recently been reported in the brain *in vivo* (Henry et al. 2003b) (see Fig. 7.8). However, taking this information into account requires development of new metabolic models. Such models are currently under development (Shestov et al. 2007a). Because these models can comprise several hundreds of equations, automatic generation of differential equations is also being investigated (Valette et al. 2008). Preliminary results using this new two-compartment isotopomer model suggest that, indeed, the use of additional information from experimental data improves the precision of metabolic modeling. In particular, analysis of dynamic isotopomer time courses during simultaneous infusion of [1,2-$^{13}C_2$]acetate (taken up by glia) and [1-^{13}C]glucose or [1,6-$^{13}C_2$]glucose (quantitatively metabolized primarily in neurons because the neuronal TCA cycle rate is much faster than the glial TCA cycle rate) appears very promising (Shestov et al. 2007b, 2008; Deelchand et al. 2008a).

7.5.4 Metabolic Modeling in Disease and Animal Models of Disease

Most ^{13}C metabolic modeling studies *in vivo* have been performed in the normal brain. Metabolic modeling of ^{13}C time courses in brain disorders and in animal models of disease is still in its infancy. Only a handful of metabolic studies have been performed *in vivo* in brain disorders.

Clinical ^{13}C MRS studies have been shown to be feasible in a number of disorders such as hepatic encephalopathy or Alzheimer's Disease at 1.5 T (Bluml et al. 2001a, b; Lin et al. 2003; Ross et al. 2003). In most of these initial clinical studies, however, no attempt was made to determine quantitative metabolic rates using metabolic modeling.

A few studies *in vivo* have used metabolic modeling in animal models of disease. For example, the rate of TCA cycle was showed to be reduced in the rat 3NP model of Huntington's disease (Henry et al. 2002). Another study was performed in the MPTP model of Parkinson's Disease (Chassain et al. 2005). A few studies investigated lactate production from glucose in rats with C6 glioma (Ross et al. 1988; Terpstra et al. 1998). In another study, the rate of neurotransmission was shown to be reduced upon administration of antidepressant/antipanic drug phenelzine in normal rats (Yang and Shen 2005; Shen 2006).

In humans, metabolic modeling studies in brain disorders are also just beginning. Recently, the rate of acetate uptake was shown to be higher in diabetic patients than in control subjects during ^{13}C acetate infusion and hypoglycemia. (Mason et al. 2006). The rate of TCA cycle was shown to be unchanged in diabetic patients compared to control (Henry et al. 2008). The ability to simultaneously assess neuronal and glial energy metabolism as well as glutamatergic neurotransmission *in vivo* hold great promise for the investigation of brain disease in animal models and in humans.

A couple of challenges are worth mentioning regarding application of ^{13}C MRS to patients. First, patients often find it more difficult to remain motionless inside the scanner for extended periods of time than do healthy volunteers. Second, most ^{13}C MRS studies to date have been performed in occipital lobe (studies in other brain regions are constrained in part by finding a suitable coil design, and possibly by SAR concerns about the proximity of sensitive tissues such as the eyes), whereas other brain regions are likely to be of interest for investigation in disease.

Finally, it is worth emphasizing again that knowledge of metabolite concentrations is critical for metabolic modeling. Most metabolic modeling studies assume metabolic steady-state, implying that metabolic fluxes and total concentrations of metabolites remain constant for the duration of the experiment. Proton concentrations can be assumed based on literature. However, when comparing a control group and a disease group, this can be a problem as changes in concentrations may not be known. These concentrations can be measured with ^{1}H MRS. In acute animal models, current metabolic modeling approaches may not be applicable if metabolic changes (changes in metabolic fluxes or metabolic concentrations) occur during the time of ^{13}C MRS measurements. In that case, new metabolic models capable of fitting ^{13}C MRS data acquired under non-steady metabolic steady-state would need to be developed. Nearly all studies published to date have assumed metabolic steady-state.

7.5.5 Alternatives to Dynamic Metabolic Modeling

Although this section was concerned primarily with dynamic metabolic modeling, which takes full advantage of *in vivo* ^{13}C MRS data to determine absolute metabolic rates, other analysis methods should be mentioned.

Isotopomer analysis is performed at a single time-point at *isotopic* steady-state, as opposed to the dynamic analysis of ^{13}C isotopic labeling curves described in previous sections (in both cases, however, *metabolic* steady-state is generally assumed over the duration of the ^{13}C infusion). ^{13}C NMR data at isotopic steady-state are used to derive metabolic information. In contrast to dynamic analysis, metabolic fluxes obtained with isotopomer analysis are relative, not absolute. For example, the approach proposed by Lebon et al. (2002) takes advantage of isotopomer analysis at isotopic steady-state to calculate the ratio $V_{NT}/V_{TCA(N)}$. Although isotopomer analysis does not allow determination of absolute metabolic rates, it is simpler than dynamic modeling. The main constraint of this method is the requirement to perform the measurement at isotopic steady-state.

Another, even simpler approach has been used in clinical studies. In such studies, detection of a difference between control and patients using simple analysis methods can be valuable for diagnosis and treatment monitoring. Differences between groups have been measured by simply comparing ^{13}C signals intensities or fitting ^{13}C turnover curves using first-order exponential kinetics. However, these approaches do not directly give information on metabolic rates. For example, differences in the intensity of ^{13}C resonances between control and disease groups may simply reflect differences in pool size. In hepatic encephalopathy, glutamine concentration is elevated, which leads to elevated ^{13}C signals in ^{13}C experiments (Bluml et al. 2001a; Ross et al. 2003). However, this does not necessarily mean that the glutamine synthesis flux is increased. Without metabolic modeling, it is not possible to tell if the change in ^{13}C turnover rate results from a change in concentration or a change in metabolic rate. In contrast, metabolic modeling yields quantitative information on specific metabolic pathways.

7.6 Perspectives and Conclusion

In conclusion, *in vivo* ^{13}C MRS and metabolic modeling have developed steadily in the past 30 years. They have allowed investigation of metabolic pathways *in vivo* that are not accessible non-invasively by any other methods.

Among the unique contributions of ^{13}C MRS to neuroscience, the following deserve to be mentioned:

- Localized non-invasive measurements of brain oxidative glucose consumption (TCA cycle)
- Localized non-invasive measurement of compartmentalized neuronal-glial metabolic rates, including the rate of glutamate-glutamine cycle between neurons and astrocytes, the rate of glial and neuronal TCA cycle, and the rate of pyruvate carboxylasc
- Localized non-invasive measurement of brain glycogen

Progress in NMR instrumentation, and particularly the steady increase in magnetic field strength of *in vivo* NMR spectrometers, has been largely instrumental in the development of ^{13}C MRS. Although NMR instrumentation will continue to improve, providing dynamic ^{13}C time courses with better signal-to-noise and/or temporal resolution, we believe that the main challenges now lie in the following areas of investigation:

- Validation of metabolic models (particularly multicompartmental neuronal-glial models) and evaluation of the reliability of metabolic models. Validation of assumptions.
- Application of these techniques to human patients and to animal models of disease, possibly under non-steady state metabolic conditions (ie fluxes and metabolite concentrations vary during the experiment). This will require investigation of the validity of assumptions in disease state.
- Application of these techniques to evaluating potential treatments (effect of drugs on metabolic rates).

Another exciting development is the advent of hyperpolarized ^{13}C MRS (Ardenkjaer-Larsen et al. 2003; Golman et al. 2006; Golman and Petersson 2006). Hyperpolarization has the potential to increase signal-to-noise by a factor of 10,000 or more. This area of research is just beginning, and the first time courses of hyperpolarized ^{13}C metabolites in the brain were recently reported. (Bhattacharya et al. 2007; Deelchand et al. 2008b; Iltis et al. 2008).

In addition, as a result of the promises of hyperpolarized ^{13}C, more groups are likely to develop an effort in ^{13}C MRS, spurring more progress for the field of ^{13}C MRS as a whole.

Acknowledgements We thank Dee Koski and Chris Nelson for their excellent technical support. This work was supported by NIH P41 RR008079, P30 NS057091, R01 NS038672 (P.G.H.) and the Keck Foundation. The high-resolution NMR facility at the University of Minnesota is supported with funds from the University of Minnesota Medical School, NSF (BIR-961477) and the Minnesota Medical Foundation.

References

Adriany G, Gruetter R (1997) A half-volume coil for efficient proton decoupling in humans at 4 Tesla. J Magn Reson 125:178–184

Alger JR, Sillerud LO, Behar KL, Gillies RJ, Shulman RG, Gordon RE, Shae D, Hanley PE (1981) In vivo carbon-13 nuclear magnetic resonance studies of mammals. Science 214:660–662

Ardenkjaer-Larsen JH, Fridlund B, Gram A, Hansson G, Hansson L, Lerche MH, Servin R, Thaning M, Golman K (2003) Increase in signal-to-noise ratio of >10,000 times in liquid-state NMR. Proc Natl Acad Sci USA 100:10158–10163

Bachelard H (1998) Landmarks in the application of ^{13}C-magnetic resonance spectroscopy to studies of neuronal/glial relationships. Dev Neurosci 20:277–288

Badar-Goffer RS, Bachelard HS, Morris PG (1990) Cerebral metabolism of acetate and glucose studied by ^{13}C-n.m.r. spectroscopy. A technique for investigating metabolic compartmentation in the brain. Biochem J 266:133–139

Bax A (1983) A simple method for the calibration of the decoupler radiofrequency field strength. J Magn Reson 52:76–80

Beckmann N, Turkalj I, Seelig J, Keller U (1991) ^{13}C NMR for the assessment of human brain glucose metabolism in vivo. Biochemistry 30:6362–6366

Behar KL, Petroff OAC, Prichard JW, Alger JR, Shulman RG (1986) Detection of metabolites in rabbit brain by ^{13}C NMR spectroscopy following administration of [1-^{13}C]Glucose. Magn Reson Med 3:911–920

Bhattacharya P, Chekmenev EY, Perman WH, Harris KC, Lin AP, Norton VA, Tan CT, Ross BD, Weitekamp DP (2007) Towards hyperpolarized (13)C-succinate imaging of brain cancer. J Magn Reson 186:150–155

Bluml S, Moreno-Torres A, Ross BD (2001a) [1-^{13}C]glucose MRS in chronic hepatic encephalopathy in man. Magn Reson Med 45:981–993

Bluml S, Moreno A, Hwang JH, Ross BD (2001b) 1-(13)C glucose magnetic resonance spectroscopy of pediatric and adult brain disorders. NMR Biomed 14:19–32

Bluml S, Moreno-Torres A, Shic F, Nguy CH, Ross BD (2002) Tricarboxylic acid cycle of glia in the in vivo human brain. NMR Biomed 15:1–5

Boumezbeur F, Besret L, Valette J, Vaufrey F, Henry PG, Slavov V, Giacomini E, Hantraye P, Bloch G, Lebon V (2004) NMR measurement of brain oxidative metabolism in monkeys using ^{13}C-labeled glucose without a ^{13}C radiofrequency channel. Magn Reson Med 52:33–40

Boumezbeur F, Besret L, Valette J, Gregoire MC, Delzescaux T, Maroy R, Vaufrey F, Gervais P, Hantraye P, Bloch G, Lebon V (2005) Glycolysis versus TCA cycle in the primate brain as measured by combining (18)F-FDG PET and (13)C-NMR. J Cereb Blood Flow Metab 25:1418–1423

Cerdan S, Kunnecke B, Seelig J (1990) Cerebral metabolism of [1,2-$^{13}C_2$]acetate as detected by *in vivo* and *in vitro* ^{13}C NMR. J Biol Chem 265:12916–12926

Chance MC, Seeholzer SH, Kobayashi K, Williamson JR (1983) Mathematical analysis of isotope labeling in the citric acid cycle with applications to ^{13}C NMR studies in perfused rat hearts. J Biol Chem 258:13785–13794

Chassain C, Bielicki G, Donnat JP, Renou JP, Eschalier A, Durif F (2005) Cerebral glutamate metabolism in Parkinson's disease: an in vivo dynamic (13)C NMS study in the rat. Exp Neurol 191:276–284

Chen W, Adriany G, Zhu X-H, Gruetter R, Ugurbil K (1998) Detecting natural abundance carbon signal of NAA metabolite within 12-cm^3 localized volume of human brain ^{1}H-{^{13}C} NMR spectroscopy. Magn Reson Med 40:180–184

Chen W, Zhu XH, Gruetter R, Seaquist ER, Adriany G, Ugurbil K (2001) Study of tricarboxylic acid cycle flux changes in human visual cortex during hemifield visual stimulation using ^{1}H-{^{13}C} MRS and fMRI. Magn Reson Med 45:349–355

Chhina N, Kuestermann E, Halliday J, Simpson LJ, Macdonald IA, Bachelard HS, Morris PG (2001) Measurement of human tricarboxylic acid cycle rates during visual activation by (13)C magnetic resonance spectroscopy. J Neurosci Res 66:737–746

Choi IY, Gruetter R (2004) Dynamic or inert metabolism? Turnover of N-acetyl aspartate and glutathione from D-[1-^{13}C]glucose in the rat brain in vivo. J Neurochem 91:778–787

Choi IY, Tkac I, Ugurbil K, Gruetter R (1999) Noninvasive measurements of [1-^{13}C]glycogen concentrations and metabolism in rat brain in vivo. J Neurochem 73:1300–1308

Choi IY, Tkac I, Gruetter R (2000) Single-shot, three-dimensional "non-echo" localization method for in vivo NMR spectroscopy. Magn Reson Med 44:387–394

Collins CM, Liu W, Wang J, Gruetter R, Vaughan JT, Ugurbil K, Smith MB (2004) Temperature and SAR calculations for a human head within volume and surface coils at 64 and 300 MHz. J Magn Reson Imaging 19:650–656

de Graaf RA (2005) Theoretical and experimental evaluation of broadband decoupling techniques for in vivo nuclear magnetic resonance spectroscopy. Magn Reson Med 53:1297–1306

de Graaf RA, Brown PB, Mason GF, Rothman DL, Behar KL (2003a) Detection of [1,6-$^{13}C_2$]-glucose metabolism in rat brain by in vivo ^{1}H- [^{13}C]-NMR spectroscopy. Magn Reson Med 49:37–46

de Graaf RA, Mason GF, Patel AB, Behar KL, Rothman DL (2003b) In vivo ^{1}H-[^{13}C]-NMR spectroscopy of cerebral metabolism. NMR Biomed 16:339–357

de Graaf RA, Mason GF, Patel AB, Rothman DL, Behar KL (2004) Regional glucose metabolism and glutamatergic neurotransmission in rat brain *in vivo*. Proc Natl Acad Sci USA 101: 12700–12705

de Graaf RA, Patel AB, Rothman DL, Behar KL (2006) Acute regulation of steady-state GABA levels following GABA-transaminase inhibition in rat cerebral cortex. Neurochem Int 48:508–514

Deelchand DK, Ugurbil K, Henry PG (2006) Investigating brain metabolism at high fields using localized ^{13}C NMR spectroscopy without ^{1}H decoupling. Magn Reson Med 55:279–286

Deelchand DK, Nelson C, Shestov AA, Ugurbil K, Henry P-G (2008a) Simultaneous measurement of neuronal and glial metabolism in rat brain in vivo using co-infusion of [1,6-^{13}C2]glucose and [1,2-^{13}C2]acetate. In: Proceedings of the 16th ISMRM, Toronto, ON, Canada

Deelchand DK, Iltis I, Marjanska M, Nelson C, Ugurbil K, Henry P-G (2008b) Localized detection of hyperpolarized [1-^{13}C]pyruvate and its metabolic products in rat brain. In: Proceedings of the 16th ISMRM, Toronto, ON, Canada

Ebert D, Haller RG, Walton ME (2003) Energy contribution of octanoate to intact rat brain metabolism measured by ^{13}C nuclear magnetic resonance spectroscopy. J Neurosci 23:5928–5935

Fitzpatrick SM, Hetherington HP, Behar KL, Shulman RG (1990) The flux from glucose to glutamate in the rat brain in vivo as determined by ^{1}H-observed, ^{13}C-edited NMR spectroscopy. J Cereb Blood Flow Metab 10:170–179

Golman K, Petersson JS (2006) Metabolic imaging and other applications of hyperpolarized ^{13}C. Acad Radiol 13:932–942

Golman K, in't Zand R, Thaning M (2006) Real-time metabolic imaging. Proc Natl Acad Sci USA 103:11270–11275

Gonzalez SV, Nguyen NH, Rise F, Hassel B (2005) Brain metabolism of exogenous pyruvate. J Neurochem 95:284–293

Gruetter R (2002) *In vivo* ^{13}C NMR studies of compartmentalized cerebral carbohydrate metabolism. Neurochem Int 41:143–154

Gruetter R (2003) Glycogen: the forgotten cerebral energy store. J Neurosci Res 74:179–183

Gruetter R, Boesch C (1992) Fast, noniterative shimming of spatially localized signals. *In vivo* analysis of the magnetic field along axes. J Magn Reson 96:323

Gruetter R, Tkac I (2000) Field mapping without reference scan using asymmetric echo-planar techniques. Magn Reson Med 43:319–323

Gruetter R, Rothman DL, Novotny EJ, Shulman RG (1992) Localized ^{13}C NMR spectroscopy of myo-inositol in the human brain *in vivo*. Magn Reson Med 25:204–210

Gruetter R, Novotny EJ, Boulware SD, Mason GF, Rothman DL, Shulman GI, Prichard JW, Shulman RG (1994) Localized ^{13}C NMR spectroscopy in the human brain of amino acid labeling from D-[1-^{13}C]glucose. J Neurochem 63:1377–1385

Gruetter R, Adriany G, Merkle H, Andersen PM (1996) Broadband decoupled, ^{1}H-localized ^{13}C MRS of the human brain at 4 Tesla. Magn Reson Med 36:659–664

Gruetter R, Seaquist ER, Kim S, Ugurbil K (1998) Localized *in vivo* ^{13}C-NMR of glutamate metabolism in the human brain: initial results at 4 Tesla. Dev Neurosci 20:380–388

Gruetter R, Seaquist ER, Ugurbil K (2001) A mathematical model of compartmentalized neurotransmitter metabolism in the human brain. Am J Physiol 281:E100–E112

Gruetter R, Adriany G, Choi I-Y, Henry P-G, Lei H-X, Oz G (2003) Localized *in vivo* ^{13}C NMR spectroscopy of the brain. NMR Biomed 16:313–338

Hassel B, Brathe A (2000) Cerebral metabolism of lactate in vivo: evidence for neuronal pyruvate carboxylation. J Cereb Blood Flow Metab 20:327–336

Henry P-G, Roussel R, Vaufrey F, Dautry C, Bloch G (2000) Semiselective POCE NMR spectroscopy. Magn Reson Med 44:395–400

Henry PG, Lebon V, Vaufrey F, Brouillet E, Hantraye P, Bloch G (2002) Decreased TCA cycle rate in the rat brain after acute 3-NP treatment measured by in vivo ^{1}H-[^{13}C] NMR spectroscopy. J Neurochem 82:857–866

Henry P-G, Tkac I, Gruetter R (2003a) ^{1}H-localized broadband ^{13}C NMR spectroscopy of the rat brain *in vivo* at 9.4 Tesla. Magn Reson Med 50:684–692

Henry P-G, Oz G, Provencher S, Gruetter R (2003b) Toward dynamic isotopomer analysis in the rat brain *in vivo*: automatic quantitation of ^{13}C NMR spectra using LCModel. NMR Biomed 16:400–412

Henry P, Crawford S, Oz G, Ugurbil K, Gruetter R (2003c) Glucose and glial-neuronal metabolism in α-chloralose anesthetized rats measured by in vivo ^{13}C NMR spectroscopy. In: Proceedings of the International Society for Magnetic Resonance in Medicine, Toronto, p 1967

Henry P-G, Oz G, Seaquist ER, Ugurbil K, Gruetter R (2006a) Localized ^{13}C NMR measurement of NAA and GSH turnover in the human brain over multiple days. In: 14th ISMRM, Seattle, WA, p 402

Henry PG, Marjanska M, Walls JD, Valette J, Gruetter R, Ugurbil K (2006b) Proton-observed carbon-edited NMR spectroscopy in strongly coupled second-order spin systems. Magn Reson Med 55:250–257

Henry PG, Adriany G, Deelchand D, Gruetter R, Marjanska M, Oz G, Seaquist ER, Shestov A, Ugurbil K (2006c) In vivo ^{13}C NMR spectroscopy and metabolic modeling in the brain: a practical perspective. Magn Reson Imaging 24:527–539

Henry P-G, Criego AB, Kumar A, Seaquist ER (2008) Brain oxidative glucose metabolism in patients with Type 1 diabetes and hypoglycemia unawareness: an in vivo 13C MRS study. In: American Diabetes Association's 68th Scientific Sessions, San Francisco, CA, pp 19-OR

Hyder F, Chase JR, Behar KL, Mason GF, Siddeef M, Rothman DL, Shulman RG (1996) Increased tricarboxylic acid cycle flux in the rat brain during forepaw stimulation detected with ^{1}H [^{13}C] NMR. Proc Natl Acad Sci USA 93:7612–7617

Hyder F, Rothman DL, Mason GF, Rangarajan A, Behar KL, Shulman RG (1997) Oxidative glucose metabolism in rat brain during single forepaw stimulation: A spatially localized ^{1}H[^{13}C] nuclear magnetic resonance study. J Cereb Blood Flow Metab 17:1040–1047

Hyder F, Renken R, Rothman DL (1999) In vivo carbon-edited detection with proton echo-planar spectroscopic imaging (ICED PEPSI): [3,4-^{13}CH$_2$]glutamate/glutamine tomography in rat brain. Magn Reson Med 42:997–1003

Iltis I, Deelchand DK, Marjanska M, Nelson C, Ugurbil K, Henry P-G (2008) First studies with hyperpolarized [2-13C]pyruvate in the rat brain. In: Proceedings of the 16th ISMRM, Toronto, ON, Canada

Inubushi T, Morikawa S, Kito K, Arai T (1993) ^{1}H-detected in vivo ^{13}C NMR spectroscopy and imaging at 2 T magnetic field: efficient monitoring of ^{13}C-labeled metabolites in the rat brain derived from 1-^{13}C glucose. Biochem Biophys Res Commun 191:866–872

Jeffrey FM, Reshetov A, Storey CJ, Carvalho RA, Sherry AD, Malloy CR (1999) Use of a single (13)C NMR resonance of glutamate for measuring oxygen consumption in tissue. Am J Physiol 277:E1111–1121

Klomp DW, Renema WK, van der Graaf M, de Galan BE, Kentgens AP, Heerschap A (2006) Sensitivity-enhanced ^{13}C MR spectroscopy of the human brain at 3 Tesla. Magn Reson Med 55:271–278

Klomp DW, Kentgens AP, Heerschap A (2008) Polarization transfer for sensitivity-enhanced MRS using a single radio frequency transmit channel. NMR Biomed 21(5):444–452

Lebon V, Petersen KF, Cline GW, Shen J, Mason GF, Dufour S, Behar KL, Shulman GI, Rothman DL (2002) Astroglial contribution to brain energy metabolism in humans revealed by ^{13}C nuclear magnetic resonance spectroscopy: elucidation of the dominant pathway for neurotransmitter glutamate repletion and measurement of astrocytic oxidative metabolism. J Neurosci 22:1523–1531

Lewandowski ED, Yu X, LaNoue KF, White LT, Doumen C, O'Donnell JM (1997) Altered metabolite exchange between subcellular compartments in intact postischemic rabbit hearts. Circ Res 81:165–175

Li S, Chen Z, Zhang Y, Lizak M, Bacher J, Innis RB, Shen J (2005) In vivo single-shot, proton-localized ^{13}C MRS of rhesus monkey brain. NMR Biomed 18:560–569

Lin AP, Shic F, Enriquez C, Ross BD (2003) Reduced glutamate neurotransmission in patients with Alzheimer's disease – an in vivo (13)C magnetic resonance spectroscopy study. MAGMA 16:29–42

Marjanska M, Henry P-G, Gruetter R, Garwood M, Ugurbil K (2004) A new method for proton detected carbon edited spectroscopy using LASER. In: Proceedings of the 12th ISMRM, Kyoto, Japan, p 679

Mason GF, Rothman DL (2004) Basic principles of metabolic modeling of NMR (13)C isotopic turnover to determine rates of brain metabolism in vivo. Metab Eng 6:75–84

Mason GF, Behar KL, Rothman DL, Shulman RG (1992a) NMR determination of intracerebral glucose concentration and transport kinetics in rat brain. J Cereb Blood Flow Metab 12:448–455

Mason GF, Rothman DL, Behar KL, Shulman RG (1992b) NMR determination of the TCA cycle rate and α-ketoglutarate/glutamate exchange rate in rat brain. J Cereb Blood Flow Metab 12:434–447

Mason GF, Gruetter R, Rothman DL, Behar KL, Shulman RG, Novotny EJ (1995) Simultaneous determination of the rates of the TCA cycle, glucose utilization, α-ketoglutarate/glutamate exchange, and glutamine synthesis in human brain by NMR. J Cereb Blood Flow Metab 15:12–25

Mason GF, Pan JW, Chu W-J, Newcomer BR, Zhang Y, Orr R, Hetherington HP (1999) Measurement of the tricarboxylic acid cycle rate in human grey and white matter *in vivo* by ^{1}H-[^{13}C] magnetic resonance spectroscopy at 4.1 T. J. Cereb. Blood Flow Metab 19: 1179–1188

Mason GF, Falk Petersen K, de Graaf RA, Kanamatsu T, Otsuki T, Rothman DL (2003) A comparison of (13)C NMR measurements of the rates of glutamine synthesis and the tricarboxylic acid cycle during oral and intravenous administration of [1-(13)C]glucose. Brain Res Protocols 10:181–190

Mason GF, Petersen KF, Lebon V, Rothman DL, Shulman GI (2006) Increased brain monocarboxylic acid transport and utilization in type 1 diabetes. Diabetes 55:929–934

Mason GF, Petersen KF, de Graaf RA, Shulman GI, Rothman DL (2007) Measurements of the anaplerotic rate in the human cerebral cortex using ^{13}C magnetic resonance spectroscopy and [1-^{13}C] and [2-^{13}C] glucose. J Neurochem 100:73–86

Moreno A, Ross BD, Bluml S (2001a) Direct determination of the N-acetyl-L-aspartate synthesis rate in the human brain by (13)C MRS and [1-(13)C]glucose infusion. J Neurochem 77: 347–350

Moreno A, Bluml S, Hwang JH, Ross BD (2001b) Alternative 1-(13)C glucose infusion protocols for clinical (13)C MRS examinations of the brain. Magn Reson Med 46:39–48

Nabuurs CI, Klomp DW, Veltien A, Kan HE, Heerschap A (2008) Localized sensitivity enhanced in vivo (13)C MRS to detect glucose metabolism in the mouse brain. Magn Reson Med 59:626–630

Öz G, Berkich DA, Henry PG, Xu Y, LaNoue K, Hutson SM, Gruetter R (2004) Neuroglial metabolism in the awake rat brain: CO_2 fixation increases with brain activity. J Neurosci 24: 11273–11279

Öz G, Henry PG, Seaquist ER, Gruetter R (2005). Noninvasive Detection of Brain Glycogen in Humans. Appl Magn Reson 29(1):159–169

Öz G, Seaquist E, Kumar A, Criego A, Benedict L, Rao J, Henry P, Van De Moortele P, Gruetter R (2007) Human brain glycogen content and metabolism: implications on its role in brain energy metabolism. Am J Physiol 292:E946–E951

Pan JW, Mason GF, Vaughan JT, Chu WJ, Zhang Y, Hetherington HP (1997) ^{13}C editing of glutamate in human brain using J-refocused coherence transfer spectroscopy at 4.1 T. Magn Reson Med 37:355–358

Pan JW, Stein DT, Telang F, Lee JH, Shen J, Brown P, Cline G, Mason GF, Shulman GI, Rothman DL, Hetherington HP (2000) Spectroscopic imaging of glutamate C4 turnover in human brain. Magn Reson Med 44:673–679

Pan JW, de Graaf RA, Petersen KF, Shulman GI, Hetherington HP, Rothman DL (2002) [2,4-^{13}C2]-beta-Hydroxybutyrate metabolism in human brain. J Cereb Blood Flow Metab 22:890–898

Patel AB, de Graaf RA, Mason GF, Rothman DL, Shulman RG, Behar KL (2005) The contribution of GABA to glutamate/glutamine cycling and energy metabolism in the rat cortex in vivo. Proc Natl Acad Sci USA 102:5588–5593

Pfeuffer J, Tkac I, Provencher SW, Gruetter R (1999a) Toward an *in vivo* neurochemical profile: quantification of 18 metabolites in short-echo-time (1)H NMR spectra of the rat brain. J Magn Reson 141:104–120

Pfeuffer J, Tkac I, Choi I-Y, Merkle H, Ugurbil K, Garwood M, Gruetter R (1999b) Localized *in vivo* ^{1}H NMR detection of neurotransmitter labeling in rat brain during infusion of [1-^{13}C] D-glucose. Magn Reson Med 41:1077–1083

Provencher S (1993) Estimation of metabolite concentrations from localized *in vivo* proton NMR spectra. Magn Reson Med 30:672–679

Ross BD, Higgins RJ, Boggan JE, Willis JA, Knittel B, Unger SW (1988) Carbohydrate metabolism of the rat C6 glioma. An in vivo ^{13}C and in vitro ^{1}H magnetic resonance spectroscopy study. NMR Biomed 1:20–26

Ross B, Lin A, Harris K, Bhattacharya P, Schweinsburg B (2003) Clinical experience with ^{13}C MRS in vivo. NMR Biomed 16:358–369

Rothman DL, Behar KL, Hetherington HP, den Hollander JA, Bendall MR, Petroff OAC, Shulman RG (1985) ^{1}H-observe/^{13}C-decouple spectroscopic measurements of lactate and glutamate in the rat brain *in vivo*. Proc Natl Acad Sci USA 82:1633–1637

Rothman DL, Novotny EJ, Shulman GI, Howseman AM, Petroff OAC, Mason G, Nixon T, Hanstock CC, Prichard JW, Shulman RG (1992) ^{1}H-[^{13}C] NMR measurements of [4-^{13}C]glutamate turnover in human brain. Proc Natl Acad Sci USA 89:9603–9606

Serres S, Bezancon E, Franconi JM, Merle M (2007) Brain pyruvate recycling and peripheral metabolism: an NMR analysis ex vivo of acetate and glucose metabolism in the rat. J Neurochem 101:1428–1440

Shen J (2006) ^{13}C magnetic resonance spectroscopy studies of alterations in glutamate neurotransmission. Biol Psychiatry 59:883–887

Shen J, Petersen KF, Behar KL, Brown P, Nixon TW, Mason GF, Petroff OAC, Shulman GI, Shulman RG, Rothman DL (1999) Determination of the rate of the glutamate/glutamine cycle in the human brain by *in vivo* ^{13}C NMR. Proc Natl Acad Sci USA 96:8235–8240

Shestov AA, Ugurbil K, Henry P-G (2007a) A new metabolic model for analysis of dynamic ^{13}C isotopomer time courses in the brain. In: Proceedings of the 15th ISMRM, Berlin, Germany, p 196

Shestov AA, Valette J, Ugurbil K, Henry PG (2007b) On the reliability of (13)C metabolic modeling with two-compartment neuronal-glial models. J Neurosci Res 15:3294–3303

Shestov AA, Deelchand DK, Ugurbil K, Henry P-G (2008) Improving the precision of brain ^{13}C metabolic modeling using co-infusion of [1,2-^{13}C2]acetate and [1,6-^{13}C2]glucose. In: Proceedings of the 16th ISMRM, Toronto, ON, Canada

Shic F, Ross B (2003) Automated data processing of [^{1}H-decoupled] ^{13}C MR spectra acquired from human brain in vivo. J Magn Reson 162:259–268

Shulman RG, Brown TR, Ugurbil K, Ogawa S, Cohen SM, den Hollander JA (1979) Cellular applications of 31P and ^{13}C nuclear magnetic resonance. Science 205:160–166

Sibson NR, Dhankhar A, Mason GF, Behar KL, Rothman DL, Shulman RG (1997) *In vivo* ^{13}C NMR measurements of cerebral glutamine synthesis as evidence for glutamate-glutamine cycling. Proc Natl Acad Sci USA 94:2699–2704

Sibson NR, Dhankhar A, Mason GF, Rothman DL, Behar KL, Shulman RG (1998) Stoichiometric coupling of brain metabolism and glutamatergic neuronal activity. Proc Natl Acad Sci USA 95:316–321

Sibson NR, Mason GF, Shen J, Cline GW, Herskovits AZ, Wall JE, Behar KL, Rothman DL, Shulman RG (2001) *In vivo* ^{13}C NMR measurement of neurotransmitter glutamate cycling, anaplerosis and TCA cycle flux in rat brain during [2-^{13}C]glucose infusion. J Neurochem 76:975–989

Taylor A, McLean M, Morris P, Bachelard H (1996) Approaches to studies on neuronal/glial relationships by ^{13}C-MRS analysis. Dev Neurosci 18:434–442

Terpstra M, Gruetter R, High WB, Mescher M, DelaBarre L, Merkle H, Garwood M (1998) Lactate turnover in rat glioma measured by in vivo nuclear magnetic resonance spectroscopy. Cancer Res 58:5083–5088

Tyson RL, Gallagher C, Sutherland GR (2003) ^{13}C-Labeled substrates and the cerebral metabolic compartmentalization of acetate and lactate. Brain Res 992:43–52

Uffmann K, Gruetter R (2007) Mathematical modeling of (13)C label incorporation of the TCA cycle: the concept of composite precursor function. J Neurosci Res 85:3304–3317

Ugurbil K, Brown TR, Den Hollander JA, Glynn P, Shulman RG (1978) High-resolution ^{13}C nuclear magnetic resonance studies of glucose metabolism in *Escherichia coli*. Proc Natl Acad Sci USA 75:3742–3746

Valette J, Shestov AA, Ugurbil K, Henry P-G (2008) ^{13}C isotopomer metabolic modeling: automatic generation of the mathematical model. In Proceedings of the 16th ISMRM, Toronto, ON, Canada

Van den Berg CJ, Krzalic L, Mela P, Waelsch H (1969) Compartmentation of glutamate metabolism in brain. Evidence for the existence of two different tricarboxylic acid cycles in brain. Biochem J 113:281–290 van den Bergh AJ, van den Boogert HJ, Heerschap A (1998) Calibration of the ^{1}H decoupling field strength and experimental evaluation of the specific RF absorption rate in ^{1}H-decoupled human ^{13}C-MRS. Magn Reson Med 39:642–646

van Zijl PCM, Chesnick AS, DesPres D, Moonen CTW, Ruiz-Cabello J, van Gelderen P (1993) In vivo proton spectroscopy and spectroscopic imaging of {1-^{13}C}-glucose and its metabolic products. Magn Reson Med 30:544–551

Waniewski RA, Martin DL (1998) Preferential utilization of acetate by astrocytes is attributable to transport. J Neurosci 18:5225–5233

Watanabe H, Ishihara Y, Okamoto K, Oshio K, Kanamatsu T, Tsukada Y (2000a) 3D localized ^{1}H-13C heteronuclear single-quantum coherence correlation spectroscopy *in vivo*. Magn Reson Med 43:200–210

Watanabe H, Umeda M, Ishihara Y, Okamoto K, Oshio K, Kanamatsu T, Tsukada Y (2000b) Human brain glucose metabolism mapping using multislice 2D ^{1}H-^{13}C correlation HSQC spectroscopy. Magn Reson Med 43:525–533

Xu S, Shen J (2006) In vivo dynamic turnover of cerebral ^{13}C isotopomers from [U-^{13}C]glucose. J Magn Reson 182:221–228

Yahya A, Allen PS (2005) Effect of strong homonuclear proton coupling on localized (13)C detection using PRESS. Magn Reson Med 54:1340–1350

Yang J, Shen J (2005) In vivo evidence for reduced cortical glutamate-glutamine cycling in rats treated with the antidepressant/antipanic drug phenelzine. Neuroscience 135:927–937

Yang J, Li CQ, Shen J (2005) In vivo detection of cortical GABA turnover from intravenously infused [1-^{13}C]D-glucose. Magn Reson Med 53:1258–1267

Yu X, Alpert NM, Lewandowski ED (1997) Modeling enrichment kinetics from dynamic ^{13}C-NMR spectra: theoretical analysis and practical considerations. Am J Physiol 272:C2037–2048

Chapter 8
^{13}C NMR Spectroscopy as a Tool in Neurobiology

Asta Haberg, Ursula Sonnewald, Janniche Hammer, Torun Melø, and Haytham Eloqayli

Abstract Studies of the complex interactions in intermediary metabolism between neurons and glia in cell cultures and animal models are important in order to understand normal physiology, and the origin and development of psychiatric and neurological diseases in humans. This chapter will give examples of how cell culture studies using glutamatergic neurons from cerebellum, GABAergic neurons from cerebral cortex and astrocytes from both areas can be used to probe the importance of the tricarboxylic acid cycle, transamination reactions and glutamine synthesis. The middle cerebral artery occlusion model of ischemic stroke, the pentylene tetrazole model of epileptic seizures, the pilocarpine and kainic acid models of mesial temporal lobe epilepsy in rats and Genetic Absence Epilepsy Rats from Strasbourg (GAERS), will be used to describe what ^{13}C nuclear magnetic resonance spectroscopy analysis of rat brain extracts can tell us about metabolism in the diseased brain.

Keywords Astrocyte • Neuron • Glutamate • GABA • Tricarboxylic acid cycle • Glucose • Acetate • Rat • Human

8.1 Introduction

Nuclear magnetic resonance (NMR) spectroscopy is widely applied to study glial-neuronal interaction (Fig. 8.1). The necessity of this interaction arises from the fact that essential enzymes for glutamate and GABA synthesis are located in

---

A. Haberg, M.D., Ph.D. • U. Sonnewald, Ph.D. (✉) • J. Hammer, Ph.D. • T. Melø, Ph.D.
Department of Neuroscience, Norwegian University of Science & Technology, Trondheim, Norway
e-mail: asta.haberg@ntnu.no; ursula.sonnewald@ntnu.no; Janniche.Hammer@abbott.com; torun.melo@ntnu.no

H. Eloqayli
Department of Neuroscience, Jordan University of Science and Technology (JUST), Irbid, Jordan
e-mail: hmeloqayli@just.edu.jo

I.-Y. Choi and R. Gruetter (eds.), *Neural Metabolism In Vivo*,
Advances in Neurobiology 4, DOI 10.1007/978-1-4614-1788-0_8,

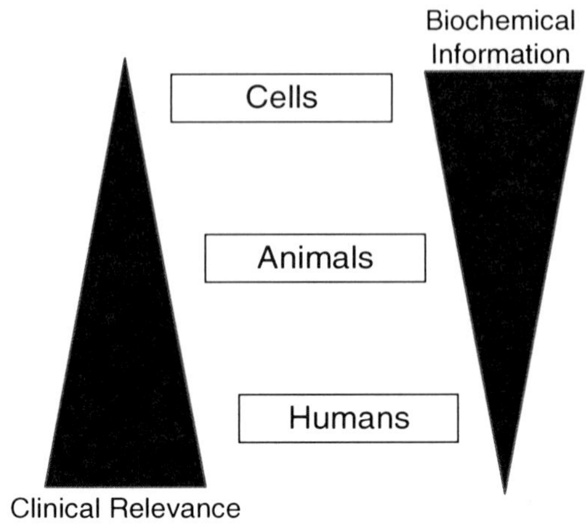

Fig. 8.1 Information from Nuclear Magnetic Resonance Spectroscopy

different cell types. Pyruvate carboxylase, the major anaplerotic enzyme in the brain is present exclusively in astrocytes, the same holds for glutamine synthetase whereas glutamate decarboxylase is mostly in GABAergic neurons (for details see: McKenna et al. 2011).

The nuclei most commonly used in NMR spectroscopy are: ^{1}H, ^{31}P, and ^{13}C. Since ^{1}H and ^{31}P have a high natural abundance, the most common methods of study involve examining differences in the natural abundance of ^{1}H or ^{31}P spectra of samples (subjects) with various metabolic and genetic alterations. As indicated in Fig. 8.1, ^{13}C has a natural abundance of 1.1%. This disadvantage normally makes detection difficult and ^{13}C NMR spectroscopy is thus of limited use for studies on endogenous metabolites unless they occur in large amounts. However, the low natural abundance can be an advantage in that ^{13}C-enriched precursors can be used for pathway mapping with little or no background interference from endogenous metabolites. For detailed description of homonuclear splitting patterns in biological molecules arising from ^{13}C labeled precursors see Cerdan et al. (1990). Quantification of ^{13}C spectra is thoroughly described by Badar-Goffer et al. (1990) and Sonnewald et al. (2011), and a detailed allocation of resonances in biological systems has been done by Barany et al. (1985), Fan (1996) and Fan and Lane (2008).

Bachelard (1998) pioneered the application of ^{13}C NMR spectroscopy to the analysis of metabolic pathways in guinea pig brain slices using [1-^{13}C]glucose both in extracts and during superfusion, whereas Shank et al. (1993) were the first to apply ^{13}C NMR spectroscopy to study glial-neuronal interaction in rat brain extracts. [1-^{13}C]Glucose has also been used for *in vivo* monitoring of cerebral metabolism and how this metabolism can be modeled is described in an excellent paper by Henry et al. (2006). Compartmentation of glutamate metabolism has been studied in brain slices using ^{1}H and ^{13}C NMR spectroscopy (Kauppinen et al. 1994). Using the same system, Rae et al. (2006) showed that the metabolic response to ionotropic glutamate receptor perturbations *in vitro* is a sensitive discriminator of metabolic function. In primary cultures of cortical astrocytes, neurons and co-cultures thereof, glial-neuronal interactions have been studied using ^{13}C labeled substrates since 1991 (Sonnewald et al. 1991). These pioneer studies have unequivocally demonstrated that ^{13}C NMR spectroscopy is a unique and powerful tool for the analysis of brain metabolism and metabolic trafficking between different cerebral compartments and many excellent studies have been performed since.

Studies of the complex interactions in intermediary metabolism between neurons and glia in cell cultures, animal and human brains are important in order to understand normal physiology. Furthermore, progression of psychiatric and neurological diseases is accompanied by derangement in these processes. For example, it has been shown that glutamine synthetase is decreased in hippocampi of patients with mesial temporal lobe epilepsy, and 2-oxoglutarate dehydrogenase in Alzheimer's disease patients (Eid et al. 2004; Gibson et al. 1998). NMR spectroscopy is an excellent technique to analyze disturbances in metabolic processes, and enzyme inhibitors have been used to study the importance of a number of pathways in cell culture experiments and *in vivo*. In rats trifluoroacetic acid which blocks the astrocytic TCA cycle, has been used to show that glutamate synthesis is predominantly neuronal (Hassel et al. 1997). The glutamine synthetase inhibitor methionine sulfoximine (MSO) blocks glutamine synthesis and reduces glutamate labeling by 50% demonstrating the importance of the glutamate – glutamine cycle (Garcia-Espinosa et al. 2004).

8.2 ^{13}C NMR Spectroscopy and Cell Cultures

Cells taken directly from the organism and subsequently grown for at least 24 h *in vitro* are considered to be primary cultures. They are obtained after an initial mechanical and/or enzymatic dissociation of the nervous tissue of neonatal or newborn rat, mouse, or chick. Due to the relative abundance of cells, NMR spectroscopy studies are often performed on cultures consisting of excitatory cerebellar granule cells or inhibitory GABAergic cerebral cortex neurons (Schousboe et al. 1989; Hertz et al. 1989a). The choice of these particular preparations of cultured neurons should be viewed in light of the fact that 90% of the synapses in the brain utilize either glutamate or GABA as the neurotransmitter. As counterparts for the neurons, astrocytes from the same brain region may be maintained in culture (Hertz et al. 1989b). Such cultures of neuronal or glial origin may be kept not only in plastic tissue culture flasks or Petri dishes, but, alternatively, microcarriers or basement membrane gel threads may be used as matrix (Westergaard et al. 1991b; Sonnewald et al. 1992; Alves et al. 1996). This technique allows transfer of the cultures directly to the NMR spectrometer. Additionally, Westergaard et al. (1991a, 1992) have developed a technique to culture neurons on top of preformed layers of astrocytes, so called co-cultures, that have proven useful in studying neuronal-glial interactions. Using cell cultures, it is possible to distinguish intra- and extracellular metabolites, and *in vivo* spectra can be understood on the basis of the characteristics of the separate cell culture types contained in the tissue.

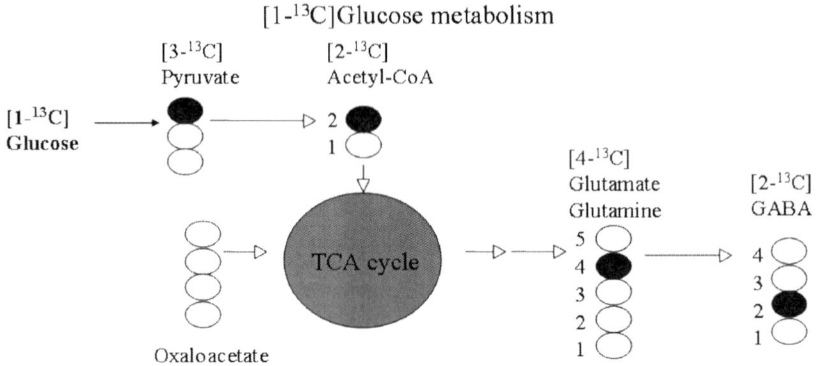

Fig. 8.2 Labeling patterns of amino acids originating from metabolism of [1-^{13}C]glucose. Black circles indicate ^{13}C labeled atoms

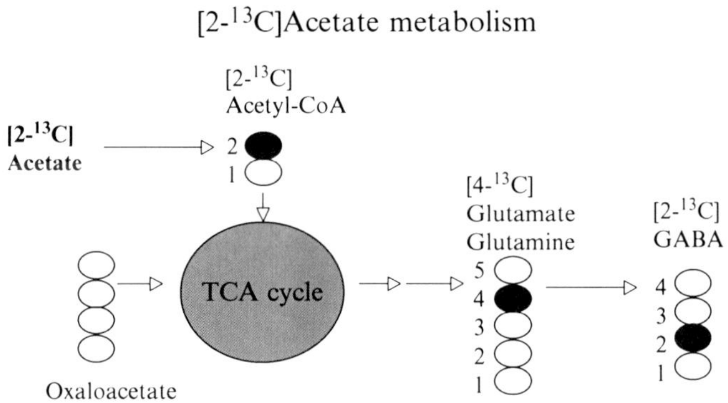

Fig. 8.3 Labeling patterns of amino acids originating from metabolism of [2-^{13}C]acetate. *Black circles* indicate ^{13}C labeled atoms

8.2.1 [1-^{13}C]Glucose and [2-^{13}C]Acetate

Both astrocytes and neurons, the main players in metabolism, can be kept in mono- and co-culture as outlined above. Using cell cultures it is possible to monitor intra- and extracellular metabolites separately. By incubating astrocytes and GABAergic neuronal cultures either alone or as co-cultures with [1-^{13}C]glucose (for labeling pattern see Fig. 8.2 and Sect. 8.2.1) it could be shown that cortical neurons label GABA efficiently and that astrocytes not only export glutamine, but also citrate to the medium (Sonnewald et al. 1991, 1993b). [2-^{13}C]Acetate (for labeling pattern see Fig. 8.3) did not label GABA in mono-cultures, but did so in co-cultures (Sonnewald et al. 1993b). This labeling could be decreased by MSO thereby demonstrating that glutamine is a precursor for GABA (Sonnewald et al. 1993b). Incubation of cortical neurons with [1-^{13}C]glucose has also been used to show that GABA is synthesized in cerebellar neurons which are thought to be largely glutamatergic (Sonnewald et al. 2004).

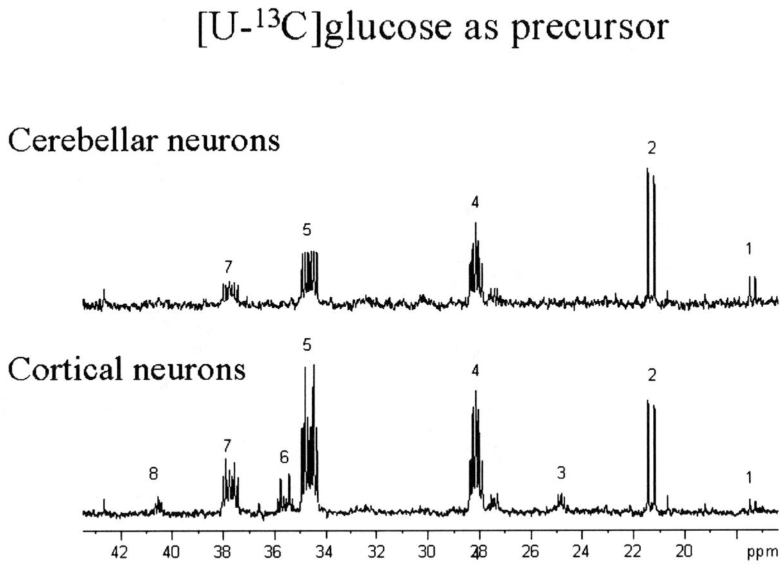

Fig. 8.4 NMR spectra of cell extracts from cultured cerebellar (*top*) and neocortical neurons (*bottom*) incubated in medium containing 12 mM [U-^{13}C]glucose for 4 h after change of medium. Abbreviations: *1*, alanine C-3; *2*, lactate C-3; *3*, GABA C-3; *4*, glutamate C-3; *5*, glutamate C-4; *6*, GABA C-2; *7*, aspartate C-3; *8*, GABA C-4 (from Waagepetersen et al. 2005b)

8.2.2 *[U-^{13}C]Glucose and [U-^{13}C]Lactate*

8.2.2.1 Metabolism in Neurons

The advantage of uniformly labeled precursors is that detection of incorporation of label into metabolites is unambiguous due to ^{13}C-^{13}C spin-spin coupling patterns (Cerdan et al. 1990). The probability of having two ^{13}C atoms in the same molecule from the 1.1% naturally abundant ^{13}C is 0.01%, well below the detection limit for most naturally occurring substances. Uniformly labeled glucose ([U-^{13}C]glucose) and lactate ([U-^{13}C]lactate) have been used to probe neurotransmitter synthesis in both cerebellar and cerebral cortical neurons (Fig. 8.4) (Waagepetersen et al. 2004, 2005a; Bak et al. 2006). It is well known that glutamate exists in a vesicular as well as a cytoplasmic pool and is metabolically closely related to the TCA cycle. Glutamate released during neuronal activity is to a large extent accumulated by astrocytes surrounding the synapse. A compensatory flux from astrocytes to neurons of suitable precursors is obligatory as neurons are incapable of performing a net synthesis of glutamate from glucose (for references see McKenna et al. 2011). Glutamine appears to play a major role in this context. Employing cultured cerebellar granule cells, as a model system for glutamatergic neurons, details of the biosynthetic machinery have been investigated during control (Olstad et al. 2007) and

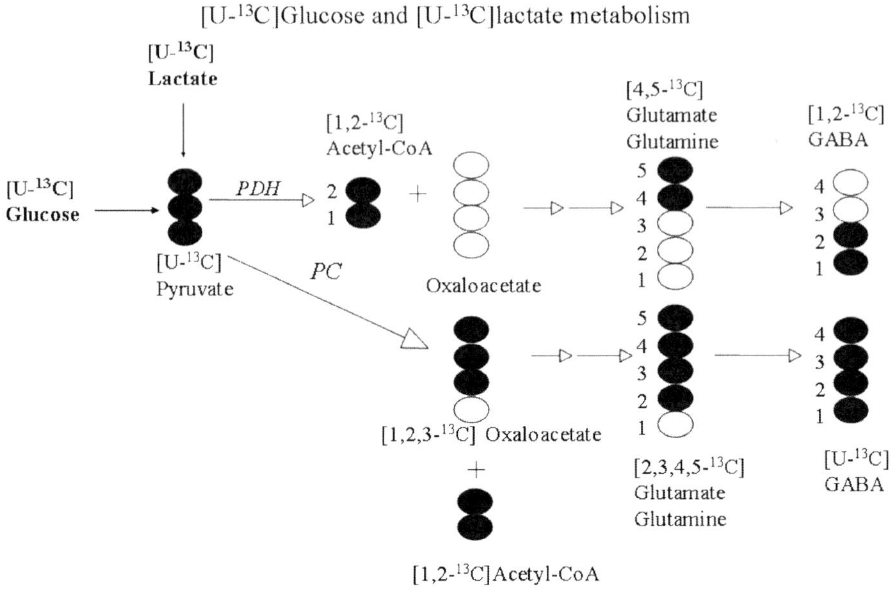

Fig. 8.5 Labeling patterns originating from metabolism of [U-^{13}C]glucose and [U-^{13}C]lactate. Black circles indicate ^{13}C labeled atoms. *Top*: Metabolism of [U-^{13}C]pyruvate involving the pyruvate dehydrogenase (PDH) complex. *Bottom*: [U-^{13}C]pyruvate metabolism via the pyruvate carboxylation pathway

depolarizing conditions inducing vesicular release (Waagepetersen et al. 2005a). [U-^{13}C]Glucose, [U-^{13}C]Glutamine or [U-^{13}C]Glutamate were used as labeled precursors for monitoring metabolic pathways by NMR spectroscopy and/or mass spectrometry (MS), a method complementary to NMR. Surprisingly, the intracellular pool of glutamate was dependent on reuptake of released glutamate in cerebellar granule neurons, disputing the importance of astrocytic glutamate uptake for controlling glutamate levels in the synapse in this brain region.

8.2.2.2 Metabolism in Astrocytes

Pyruvate carboxylation in brain has been linked to the glial specific enzyme, pyruvate carboxylase (PC). However, also malic enzyme is in principle capable of pyruvate carboxylation and this enzyme is present both in neurons and astrocytes. In order to probe in which cell type carboxylation is taking place cerebellar neurons and astrocytes in mono-culture were incubated with [U-^{13}C]glucose plus [U-^{13}C]lactate (Waagepetersen et al. 2001). These labeled precursors can be metabolized into [U-^{13}C]pyruvate and [1,2-^{13}C]acetyl CoA (Fig. 8.5). In the first turn of the TCA cycle, an unlabeled molecule of oxaloacetate condenses with [1,2-^{13}C]acetyl CoA leading to formation of [4,5-^{13}C]glutamate. In astrocytes [4,5-^{13}C]glutamate is converted to [4,5-^{13}C]glutamine, and in GABAergic neurons to [1,2-^{13}C]GABA. [1,2-^{13}C]aspartate

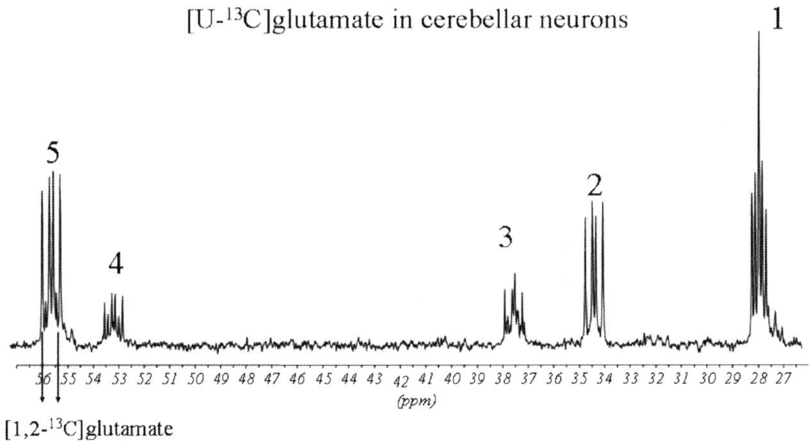

Fig. 8.6 NMR spectrum of cell extract from cultured cerebellar neurons incubated in medium containing 0.25 mM [U-^{13}C]glutamate. Abbreviations: *1*, glutamate C-3; *2*, glutamate C-4; *3*, aspartate C-3; *4*, aspartate C-2; *5*, glutamate C-2

and [3,4-^{13}C]aspartate are formed in equal amounts in the first turn due to the symmetrical succinate molecule. Label can also enter the TCA cycle *via* carboxylation of [U-^{13}C]pyruvate (Fig. 8.5, only labeled acetyl CoA is shown) and thus lead to formation of [1,2,3-^{13}C]oxaloacetate and, when acetyl CoA is labeled, [2,3,4,5-^{13}C] glutamate, [U-^{13}C]GABA and, in astrocytes, [2,3,4,5-^{13}C]glutamine as well. It should be mentioned that the isotopomers generated from pyruvate carboxylation can also be formed in the 3rd turn of the TCA cycle. Therefore, Waagepetersen et al. (2001) used the succinate dehydrogenase inhibitor 3-nitrpropionic acid to block cycling of metabolites. Using this inhibitor, it could be shown that only astrocytes were capable of carboxylation (Waagepetersen et al. 2001). A overview of NMR experiments exploring pyruvate carboxylation is given in Sonnewald and Rae (2010). Also co-cultures of cerebellar neurons and astrocytes have been incubated with uniformly labeled glucose and lactate, showing that metabolism of pyruvate, the precursor for lactate, alanine and acetyl-CoA is highly compartmentalized (Bak et al. 2007).

8.2.3 [U-^{13}C]Glutamate and [3-^{13}C]Glutamate

Glutamate is the most abundant neurotransmitter and derangement in glutamate homeostasis/neurotransmission is linked to a number of neurological disorders like epilepsy, stroke, amylotrophic lateral sclerosis, and Alzheimer's disease. It is therefore important to study glutamate metabolism under normal and pathological conditions. Cell cultures can be incubated with [U-^{13}C]glutamate (Fig. 8.6) and the distribution of label from [U-^{13}C]glutamate can be seen in Fig. 8.7. This approach makes it possible to analyze to which extent [U-^{13}C]glutamate is converted directly

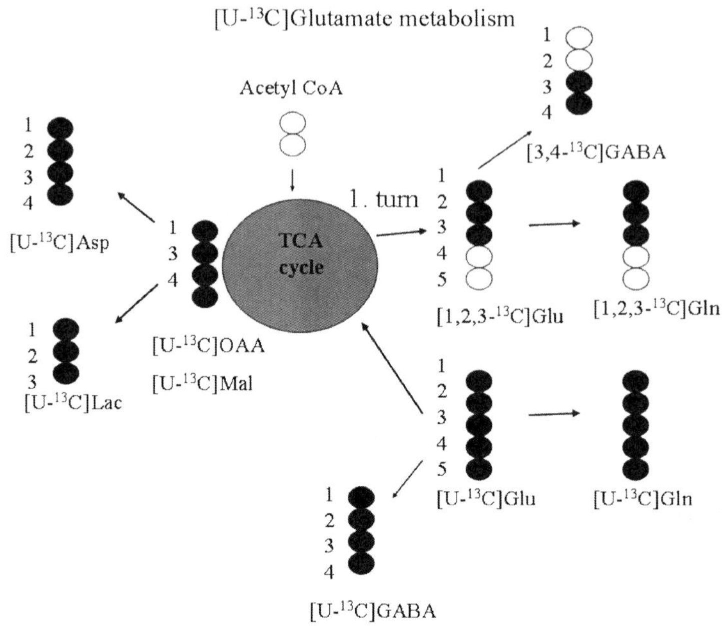

Fig. 8.7 Labeling patterns of amino acids originating from metabolism of [U-^{13}C]glutamate. *Black circles* indicate ^{13}C labeled atoms. Abbreviations: *Asp*, aspartate; *Gln*, glutamine; *Glu*, glutamate; *Lac*, lactate; *Mal*, malate; *OAA*, oxaloacetate

to glutamine or GABA (depending on cell type), or passed through the TCA cycle. Direct conversion will lead to [U-^{13}C]-2-oxoglutarate, [U-^{13}C]GABA (in GABAergic neurons) and [U-^{13}C]glutamine (in astrocytes). [U-^{13}C]-2-oxoglutarat can enter the TCA cycle and lead to the formation of uniformly labeled ([U-^{13}C]) malate and oxaloacetate which in turn can give rise to [U-^{13}C]aspartate and lactate (Fig. 8.7). After one turn of the TCA cycle, the C-4 and C-5 positions in 2-oxoglutarate will be occupied by ^{12}C atoms from unlabeled acetyl-CoA. Glutamate and glutamine derived from this precursor will have ^{12}C atoms in the C-4 and C-5 positions, and GABA will have ^{12}C atoms in the C-1 and C-2 positions (Figs. 8.7 and 8.8).

Cell culture studies have demonstrated that [U-^{13}C]glutamate is metabolized extensively in astrocytes and neurons (Qu et al. 2000b, 2001a, b; Sonnewald et al. 1993a). Astrocytes metabolized [U-^{13}C]glutamate to uniformly labeled aspartate, glutamine, lactate and alanine which were exported to the medium (Sonnewald et al. 1993a). In cerebral cortical neurons [U-^{13}C]glutamate did not label lactate, but in cerebellar granule neurons some labeling of lactate was detected (Sonnewald et al. 1996). However, it cannot be totally excluded that this lactate was formed by the small number of astrocytes present in this type of preparation. Glutamate derived from the first turn of the TCA cycle, i.e. [1,2,3-^{13}C]glutamate (Fig. 8.7) is present in cerebellar and cortical astrocytes and neurons (Sonnewald et al. 1996). [1,2-^{13}C]

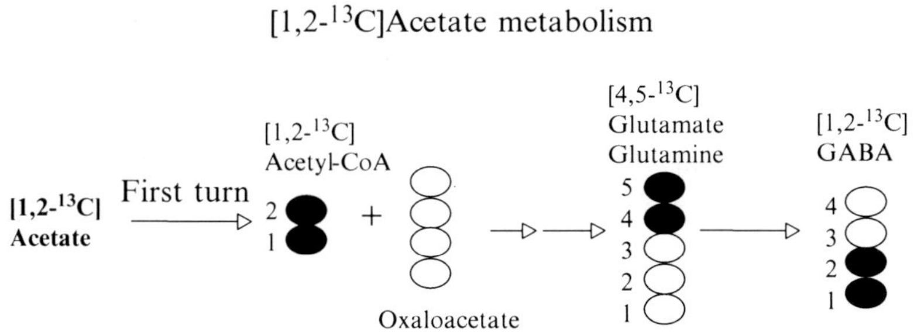

Fig. 8.8 Labeling patterns of amino acids originating from metabolism of [1,2-^{13}C]acetate. *Black circles* indicate ^{13}C labeled atoms

Glutamate which is derived from the second turn of the cycle was only detected in granule neurons (Fig. 8.6) showing that cerebellar neurons have a higher TCA cycle turnover than the other cell types investigated (Sonnewald et al. 1996). McKenna et al. (1996) showed that metabolism of [U-^{13}C]glutamate in astrocytes is concentration dependent. At [U-^{13}C]glutamate levels above 0.2 mM more glutamate entered the TCA cycle after conversion to 2-oxoglutarate than was converted directly to glutamine. As mentioned above, using enzyme inhibitors the significance of certain enzymes for metabolic processes can be probed. In astrocytes it could be shown that the transaminase inhibitor aminooxyacetic acid not only abolished the appearance of aspartate, but also that of the glutamate-isotopomer, [1,2,3-^{13}C]glutamate, derived from the TCA cycle (Westergaard et al. 1996). In this way it could be demonstrated that transamination is necessary for the conversion of 2-oxoglutarate to glutamate, but not for entry of glutamate into the TCA cycle via 2-oxoglutarate since lactate labeling was only slightly decreased. The succinate dehydrogenase inhibitor 3-nitropropionic acid was used to abolish the labeling of aspartate and [1,2,3-^{13}C]glutamate from [U-^{13}C]glutamate by inhibiting TCA cycling in cerebellar granule neurons and cortical astrocytes (Sonnewald et al. 1996; Bakken et al. 1997). As a result lactate was no longer labeled, indicating that malate or oxaloacetate derived from the TCA cycle directly are the precursors, rather than citrate, which could give rise to lactate via the citrate lyase pathway. Furthermore, it was shown that cerebellar neurons were affected by 3-nitropropionic acid at lower concentrations than astrocytes. These findings have been replicated *in vivo* in the rat brain (Hassel and Sonnewald 1995a).

Pyruvate recycling is a pathway for complete oxidation of glutamate. The cellular location and the physiological significance of such recycling have been debated during the last decade (Cerdan et al. 1990; Hassel and Sonnewald 1995b). Using [3-^{13}C] glutamate, pyruvate recycling was probed in neurons, astrocytes and co-cultures of astrocytes and neurons. With the sensitivity at that time, recycling was only detected in astrocytes in mono-culture (Waagepetersen et al. 2001). In a recent study, using [U-^{13}C]glutamate analysis of the labeling pattern of the C-4 position of glutamate and the number of ^{13}C atoms per glutamate molecule showed that pyruvate recycling is clearly present both in astrocytes and neurons (Olstad et al. 2007).

8.2.4 [3-^{13}C]Malate

Malate, specifically labeled with ^{13}C on C-3, was synthesized by chemical means and used to study malate metabolism in cortical astrocytes (Alves et al. 2000). The consumption of malate was only 0.26 μmol/mg of protein, approximately 25-fold lower than the consumption of glucose. [3-^{13}C]Malate in combination with glucose as well as [3-^{13}C]malate alone caused labeling of lactate, glutamine and fumarate which were the three major metabolites released to the medium. Very low and similar levels of isotopic enrichment were detected on C-2 and C-3 of lactate. Glutamine was labeled on C-2 and C-3 to a similar extent as well and labeling on C-4 of glutamine was only detected when glucose was not added. These labeling patterns clearly show that cytosolic malic enzyme is not active in primary astrocytes and demonstrate the occurrence of pyruvate recycling (Alves et al. 2000).

8.3 ^{13}C NMR Spectroscopy and Animal Models

8.3.1 Administration of Labeled Substrates and Factors Relevant to Label Incorporation

In animal studies there are several ways of obtaining ^{13}C enrichment in the compounds of interest. The appropriate approach depends on the hypothesis under investigation.

For *in vivo* ^{13}C NMR spectroscopy studies of metabolic fluxes, a continuous infusion until steady state is required (Sibson et al. 1997, 2001; Cerdan et al. 1990; Henry et al. 2003, for review see Henry et al. 2006). In these experiments plasma glucose concentration and enrichment need to be monitored in parallel with spectra acquisition in order to obtain an input function for the metabolic model used. In some instance this may pose a problem. For instance in small rats and mice obtaining serial blood samples have to be performed in a separate set of animals outside the magnet because the blood volume is very small. Furthermore, lengthy steady state studies may be problematic in many animal models of disease as the condition may progress with time, for instance in acute stroke. Also animals with certain conditions such as chronic epilepsy, may not be fit for lengthy *in vivo* scan protocols. The increased glucose concentration, which accompanies a steady state infusion scheme, may be deleterious in for example stroke and epilepsy models. The effect of hyperglycemia on the underlying condition rather than the primary condition per se is then studied. However, dynamic studies can also be performed in awake animals (Oz et al. 2004). This method requires more animals to be sacrificed, but sensitivity is excellent as ^{13}C NMR spectroscopy analysis of all time points will be performed in tissue extracts. Furthermore, small specific brain regions can be investigated.

There are many instances in which a short intra venous infusion (Serres et al. 2003) or intra venous bolus injection (Haberg et al. 2001) are feasible. Short intervals of label administration lead to short period of increased plasma concentration of the substrate given. This may reduce deleterious and unwanted effects of the labeled substrate. At the same time this "snap shot" approach enables one to differentiate between metabolic processes in different brain regions and due to different pathological processes even though steady state has not been obtained (Haberg et al. 2001, 2007). Indeed, short infusion schemes have been able to elucidate aspects of brain metabolism, not detectable with steady state infusion. Using direct intracarotid bolus injection biphasic labeling in [4-^{13}C]glutamate was found combined with rapid decline in plasma glucose levels. These results suggested that the initial glutamate labeling came from labeled plasma glucose, whereas later glutamate labeling arose from glycogen synthesized in astrocytes from the labeled glucose (Griffin et al. 1999). It is also possible to administer the labeled substrate subcutaneously (Hassel et al. 1995) or intraperitoneally (Shank et al. 1993; Kondziella et al. 2003). These methods may be less reliable than intravenous approaches because variations in local blood flow in the area of injection will influence uptake of the substrate into the circulation. The pro with subcutanouse and intraperitoneal administration is that no anesthesia is required, and that they can be performed by most researchers. Actually results from intraperitoneal injection share some characteristics with that of an intravenous bolus injection, as there is a trend for biphasic labeling of glutamate (Shank et al. 1993) similar to that reported by Griffin et al. (1999).

For studies aimed at investigating synthesis and metabolism of compounds with a very slow conversion rate it is even possible to introduce ^{13}C label through ^{13}C enriched chow (Karelson et al. 2003) or water (Choi and Gruetter 2003). Animal chow can be enriched with ^{13}C labeled glucose, or with ^{13}C enriched biomass (Karelson et al. 2003).

There are also other factors that will influence the results when giving ^{13}C labeled precursors to animals. One important factor is systemic metabolism of labeled substrates, i.e. metabolism of the labeled compound outside the brain. This will give rise to newly labeled compounds that may be substrates for brain metabolism. In the liver label originating from glucose and/or acetate can be incorporated into glucose via gluconeogenesis and thus leads to labeled lactate and glucose. These newly formed compounds can have ^{13}C label in positions other than the original position. Furthermore, the Cori and alanine cycles as well as pyruvate recycling and futile cycling may lead to non-symmetrical scrambling of label in glucose. In plasma, ^{13}C from labeled glucose, acetate and lactate is detected in new positions in glucose

15 min after bolus injection in fasted rats (Haberg et al. 2001, 1998b, a; Qu et al. 2000a; Serres et al. 2003). Label was not detectable in plasma amino acids. Brain amino acids, on the other hand, were labeled. This is due to the fact that conversion of glucose or acetate into amino acid in the brain proceeds at a rate several times higher than in all other organs (Gaitonde et al. 1965; Vrba 1962). Factors like type (continuous versus bolus) and duration of infusion, the metabolic state of the animal (fasted versus non-fasted) and drugs administered will influence the extra-cerebral metabolism of the labeled substrates. Plasma samples should be obtained and analyzed in order to evaluate whether there has been synthesis of significant amounts of newly labeled substrates, which could interfere with the analysis of labeling patterns in the brain. The importance of this phenomenon is illustrated by the results obtained from plasma extract analyses from rats receiving [U-^{13}C]lactate. Fifteen minutes after an intravenous bolus injection of [U-^{13}C]lactate there were approximately equal amounts of [1,2,3-^{13}C]- and [1,2-^{13}C]glucose indicating gluconeogenesis from [U-^{13}C]lactate. It was thus possible that the labeling seen in the cerebral amino acids originated from labeled glucose, not [U-^{13}C]lactate. However, the presence of significantly more label in [U-^{13}C]alanine than in [2,3-^{13}C]alanine demonstrated that [U-^{13}C]lactate did indeed cross the blood-brain barrier, and was metabolized further in the brain (Qu et al. 2000a).

Also anesthesia has important effects on ^{13}C labeling of metabolic intermediates in the brain which need to be taken into account. Surgical procedures demand that the animal is under proper anesthesia. Furthermore, most *in vivo* NMR spectroscopy experiments using animals will require anesthesia. However, anesthetic agents reduce cerebral glucose metabolism (Cremer and Lucas 1971) and thus the incorporation of ^{13}C label from glucose into amino acids (Shank et al. 1993; Hyder et al. 2003). Labeling from ^{13}C acetate is not affected, but astrocytic acetate oxidation is reduced. Moreover, anesthesia affects cortical and subcortical brain regions to different degrees (Hansen et al. 1988; Hoffman et al. 1991; Haberg et al. 2001). It should also be noted that anesthetic compounds influence neurotransmitter release (Haberg et al. 2001), and that some anesthetic agents interact with neurotransmitter receptors (Freo and Ori 2004; Whitehead et al. 2004). In addition, different anesthetic agents depress the cerebral metabolism to varying degrees (Hyder et al. 2003). This implies that the metabolic changes induced by anesthesia will influence the ^{13}C labeling patterns, and must be considered when interpreting ^{13}C NMR spectroscopy results.

8.3.2 ^{13}C Labeled Glucose and Acetate

^{13}C labeled glucose has been used widely in ^{13}C MR spectroscopy studies since it is the main substrate in the adult brain. [1-^{13}C]glucose is the most frequently administrated principally because it is the least expensive. However use of [1,6-^{13}C]glucose will increase the enrichment in the metabolic intermediates with a factor of two, and thus increase the sensitivity. [2-^{13}C]glucose has also been used to explore the importance of astrocytic pyruvate carboxylation and TCA cycle activity (Sibson et al.

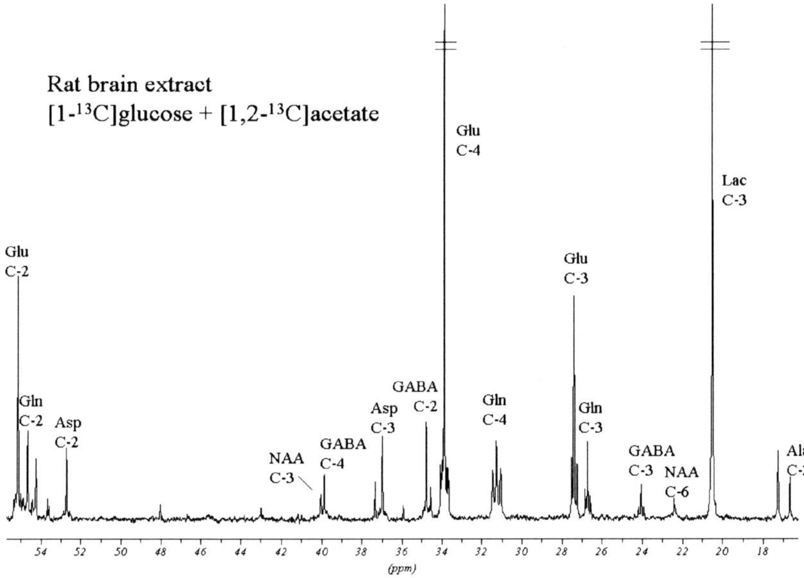

Fig. 8.9 NMR spectrum of rat brain extract. Rats were injected with [1-^{13}C]glucose (3 mM/kg) and [1,2-^{13}C]acetate (3 mM/kg) and decapitated 15 min later. Abbreviations: *ala*, Alanine; *Asp*, aspartate; *Gln*, glutamine; *Glu*, glutamate; *Lac*, lactate; *NAA*, N-acetylaspartate

2001; Mason et al. 2007). [U-^{13}C]glucose is another alternative. It has been shown that glucose enters neurons and astrocytes equally (Nehlig et al. 2004). Still glucose is thought to be metabolized more in the neuronal TCA cycle (Fig. 8.9) (Minchin and Beart 1975b, a; Sonnewald et al. 1991). Using ^{13}C NMR spectroscopy it has been calculated that 65% of acetyl CoA derived from glucose is metabolized in the neuronal TCA cycle *in vivo* (Hassel et al. 1995; Qu et al. 2000a). In the astrocyte glucose may be converted to lactate without entering the TCA cycle and then shuttled to neurons (Serres et al. 2004) (Fig. 8.10).

Acetate labeled in the C-2 or C-1,2 positions are also frequently used. Acetate metabolism is considered to be located to astrocytes. Acetate is selectively taken up by astrocytes by a specialized transport system, which is absent or less active in neurons (Waniewski and Martin 1998).

By simultaneous injection of [1-^{13}C]glucose and [1,2-^{13}C]acetate followed by ^{13}C NMR spectroscopic analysis of brain extracts, information about both neuronal and astrocytic metabolism can be obtained in the same animal (Taylor et al. 1996) (for simplified overview of the metabolic pathways for glucose and acetate see Figs. 8.10 and 8.12). Injection of ^{13}C labeled glucose and acetate leads to efficient labeling of many metabolites as can be seen in the brain extract spectrum in Fig. 8.9. Label from [1-^{13}C]glucose can be quantified by analyzing the singlet peaks in the different compounds. The doublets seen in the spectrum are mostly derived from [1,2-^{13}C] acetate and thus astrocytic metabolism. Simplified schemes of the metabolic

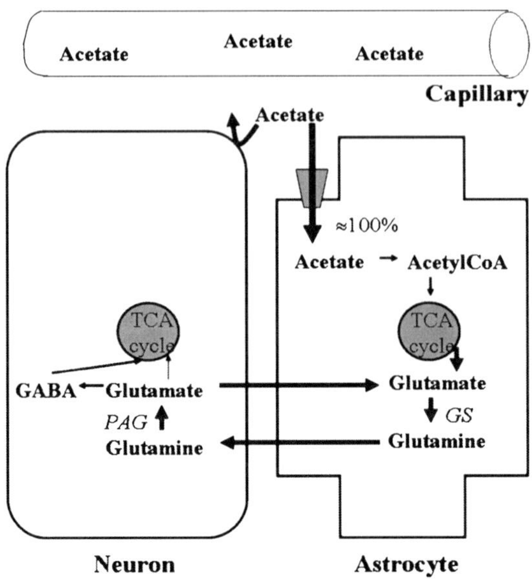

Fig. 8.10 Metabolic pathways for acetate. The neuron depicted is a non-existing neuron, which releases both neurotransmitters glutamate and GABA. Acetate is only taken up by astrocytes due to the presence of a specific acetate transporter on these cells. In the astrocytes acetate is converted to acetyl CoA by acetyl CoA synthetase and can subsequently enter the tricarboxylic acid (TCA) cycle. α-Ketoglutarate can leave the TCA cycle and can be transaminated by aspartate, alanine or branched chain aminotransferases to glutamate. In the astrocytes glutamate is rapidly converted to glutamine by the action of glutamine synthetase (GS). GS is found only in astrocytes and oligodendrocytes. Glutamine is released from astrocytes and then taken up by high affinity transporters present on neurons. Inside the neuron glutamine is converted to glutamate by phosphate activated glutaminase (PAG). Astrocytic glutamine is used specifically to replenish the glutamate neurotransmitter pool in glutamatergic neurons. Glutamate released as neurotransmitter is predominantly taken up by astrocytes. In the astrocyte the neurotransmitter glutamate can either be converted directly to glutamine by GS, or enter the TCA cycle via transamination or oxidative deamination by glutamate dehydrogenase. As schematically illustrated in the neuron, glutamate derived from astrocytic glutamine can be converted to GABA by the action of glutamate decarboxylase (GAD). This enzymatic reaction is considered to take place mostly in GABAergic neurons. GABA released as neurotransmitter is mainly taken up by the GABAergic neuron, but astrocytes possess the capacity to take up and metabolize GABA

pathways of these two substrates are shown in Figs. 8.2 and 8.8. *Via* glycolysis [1-^{13}C]glucose is converted to [3-^{13}C]pyruvate which can be converted to [3-^{13}C]alanine or [3-^{13}C]lactate. Alternatively, pyruvate can enter the TCA cycle *via* [2-^{13}C]acetyl CoA. This will lead to the formation of [4-^{13}C]glutamate and glutamine or [2-^{13}C]GABA. Furthermore, pyruvate can be carboxylated by pyruvate carboxylase (PC) to oxaloacetate which will lead to the synthesis of [2-^{13}C]glutamate and glutamine or [4-^{13}C]GABA. PC activity is illustrated in Fig. 8.5 for [U-^{13}C]glucose (which is analogous to [1-^{13}C]glucose). [1,2-^{13}C]acetate can also be converted to acetyl CoA, however, the product, [1,2-^{13}C]acetyl CoA, has two ^{13}C atoms (Fig. 8.8),

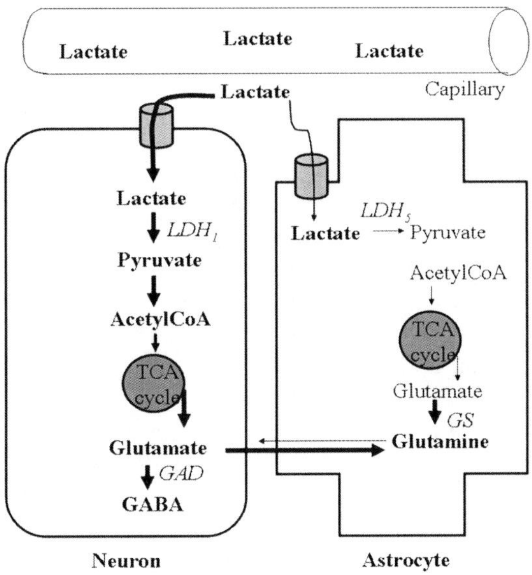

Fig. 8.11 Metabolic pathways for lactate. The neuron depicted is a non-existing neuron, which releases both neurotransmitters glutamate and GABA. Lactate can be taken up by both neurons and astrocytes, but there is an uneven distribution of lactate dehydrogenase (LDH) isoenzymes between the two cell types. Neurons have only the LDH_1 isoform, which favors the production of pyruvate, while astrocytes have the LDH_5 isoform that drives the reaction towards lactate formation and the LDH_1 isoform. Consequently there is limited synthesis of pyruvate from lactate in astrocytes. In neurons, however, lactate is readily converted to pyruvate and subsequently acetyl CoA for entry into the tricarboxylic acid (TCA) cycle. α-Ketoglutarate leaves the neuronal TCA cycle and can be transaminated by aspartate, alanine or branched chain aminotransferases to glutamate. It should be noted that glutamate and GABA de novo synthesis takes place only in astrocytes because the anaplerotic enzyme in brain pyruvate carboxylse (PC) is located in astrocytes (see below). In GABAergic neurons glutamate is converted to GABA by glutamate decarboxylase activity (GAD). Neurotransmitter glutamate or GABA can subsequently be released into the synaptic cleft. Glutamate is predominantly taken up by astrocytes, and GABA by GABAergic neurons. In the astrocyte the neurotransmitter glutamate can either be converted directly to glutamine by glutamine synthetase (GS), or enter the TCA cycle via transamination or oxidative deamination by glutamate dehydrogenase. GABA is mainly taken up by the GABAergic neuron, but astrocytes possess the capacity to take up and metabolize GABA. GABA is metabolized by GABA aminotransferase and succinic-semialdehyde dehydrogenase, which allow four of the five carbon atoms from α-ketoglutarate to re-enter the TCA cycle as succinate. In astrocytes pyruvate from lactate can enter the TCA cycle via either PDH or PC activity. The entry of ^{13}C labeled glucose into the TCA cycle via PC activity gives rise to distinct labeling patterns in glutamate, glutamine and GABA, which can be differentiated from the labeling patterns found when ^{13}C labeled glucose enters the TCA cycle via PDH activity. Hence, in astrocytes either acetyl CoA or oxaloacetate from pyruvate is incorporated into the astrocytic TCA cycle. After additional enzymatic steps α-ketoglutarate is formed and can leave the TCA cycle and be transaminated to glutamate. In the astrocytes glutamate is rapidly converted to glutamine by the action of GS. GS is found only in astrocytes and oligodendrocytes. Glutamine is released from astrocytes and then taken up by high affinity transporters present on neurons. Inside the neuron glutamine is converted to glutamate by phosphate activated glutaminase (PAG). In GABAergic neurons GAD converts glutamate originating from glutamine to GABA

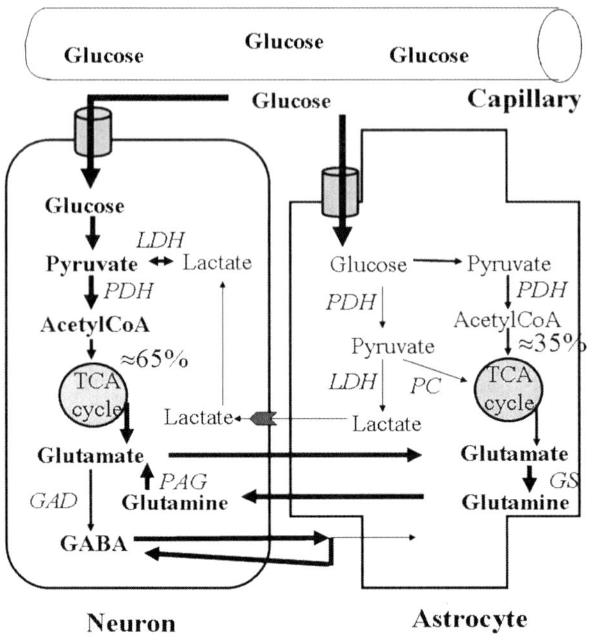

Fig. 8.12 Metabolic pathways for glucose. Glucose is taken up equally by astrocytes and neurons, but at the level of acetyl CoA more glucose enters the neuronal than the astrocytic TCA cycle. In neurons pyruvate from glucose can only be incorporated into the TCA cycle via pyruvate dehydrogenase (PDH) activity. PDH converts pyruvate to Acetyl CoA which condenses with oxaloacetate and produces citrate. α-Ketoglutarate is formed after two additional enzymatic steps in the TCA cycle. α-Ketoglutarate can leave the neuronal TCA cycle for subsequent transamination with aspartate, alanine or branched chain aminotransferases to form glutamate. It should be noted that glutamate and GABA de novo synthesis takes place only in astrocytes sinc the anaplerotic enzyme in brain, pyruvate carboxylase (PC) is located in astrocytes (see below). In GABAergic neurons glutamate is converted to GABA by glutamate decarboxylase activity (GAD). Neurotransmitter glutamate or GABA can subsequently be released into the synaptic cleft. Glutamate is predominantly taken up by astrocytes, and GABA by GABAergic neurons. In the astrocyte the neurotransmitter glutamate can either be converted directly to glutamine by glutamine synthetase (GS), or enter the TCA cycle via transamination or oxidative deamination by glutamate dehydrogenase. GABA is mainly taken up by the GABAergic neuron, but astrocytes possess the capacity to take up and metabolize GABA. GABA is metabolized by GABA aminotransferase and succinic-semialdehyde dehydrogenase, which allows four of the five carbon atoms from α-ketoglutarate to re-enter the TCA cycle as succinate.In the astrocyte pyruvate from glucose can enter the TCA cycle via either PDH or PC activity. The entry of ^{13}C labeled glucose into the TCA cycle via PC activity gives rise to distinct labeling patterns in glutamate, glutamine and GABA, which can be differentiated from the labeling patterns found when ^{13}C labeled glucose enters the TCA cycle via PDH activity. Hence, in astrocytes either acetyl CoA or oxaloacetate from pyruvate is incorporated into the astrocytic TCA cycle. After additional enzymatic steps α-ketoglutarate is formed and can leave the TCA cycle and be transaminated to glutamate. In the astrocytes glutamate is rapidly converted to glutamine by the action of GS. GS is found only in astrocytes and oligodendrocytes. Glutamine is released from astrocytes and then taken up by high affinity transporters present on neurons. Inside the neuron glutamine is converted to glutamate by phosphate activated glutaminase (PAG). Astrocytic glutamine is used specifically to replenish the glutamate neurotransmitter pool in glutamatergic neurons. Glutamine is also taken up by GABAergic neurons, converted to glutamate via PAG and then to GABA by GAD

resulting in doublet formation (Fig. 8.9). Thus [4,5-^{13}C]glutamate and glutamine or [1,2-^{13}C]GABA are formed. Since both acetyl CoA and oxaloacetate can be either labeled or unlabeled, the number of possible isotopomers of the metabolites derived from the TCA cycle is large. Only compounds derived from the first turn are represented in Figs. 8.2 and 8.8. By comparing the doublets with singlets, it can be seen in Fig. 8.9 that glutamine is labeled more from [1,2-^{13}C]acetate (doublet) than from [1-^{13}C]glucose (singlet), the opposite is the case for glutamate and GABA. Lactate, alanine and N-acetylaspartate (NAA) are mainly labeled from glucose.

8.3.3 [U-^{13}C]Lactate

Under normal resting conditions lactate is present in the blood at a concentration of ~1 mM, and it was thought that the brain does not extract appreciable amounts of this potential substrate (Harada et al. 1992). However, it is now widely accepted that lactate is a fuel for neuronal metabolism and that astrocytes play a critical role in the regulation of this meatbolism (Waagepetersen et al. 2009). The astrocyte-neuron lactate shuttle hypothesis (ANLSH) describes the role of astrocytes in shuttling lactate from astrocytes to neurons in response to brain activation (Magistretti et al. 1999; Pellerin et al. 2007). The exact nature of this shuttle has yet to be fully elucidated (Waagepetersen et al. 2009), but an extension of the ANLSH was published by Cerdán et al. (2006) proposing that subcellular compartmentation of pyruvate allows neurons and astrocytes to select between glucose and lactate as alternative substrates, depending on their relative extracellular concentration and the operation of a redox switch. In neuronal cell cultures it has been shown that oxidative metabolism of lactate was more intense than that of glucose, both under resting and depolarizing conditions (Peng et al. 1994). Lactate was also shown to be an important energy fuel in cultured GABAergic neurons, and it extensively labeled amino acids synthesized from TCA cycle intermediates (Waagepetersen et al. 1998).

Several studies have appeared showing that ^{13}C labeled lactate is a useful tool for mapping neuronal metabolism (Fig. 8.11T) (Serres et al. 2003; Qu et al. 2000a; Tyson et al. 2003). In a study by Qu et al. (2000a) fully awake rats were injected intravenously with [U-^{13}C]lactate or [U-^{13}C]glucose followed 15 min later by decapitation. For labeling patterns see Fig. 8.5. Incorporation of label from [U-^{13}C]lactate was seen mainly in glutamate, GABA, glutamine, aspartate, alanine, and lactate. When [U-^{13}C]glucose was the precursor, label incorporation was similar to that observed from [U-^{13}C]lactate, but much larger. However, contribution from pyruvate carboxylase (astrocytic enzyme) was much more pronounced after the glucose injection. This indicated that relatively more pyruvate from lactate than glucose was metabolized in neurons. Surprisingly, the same amount of lactate was synthesized via the TCA cycle in both groups, indicating transfer of neurotransmitters from the neuronal to the astrocytic compartment as Serres et al. (2003) and Hassel et al. (1995) have shown that lactate from the TCA cycle is synthesized primarily in astrocytes. Serres et al. (2003) could come to this conclusion by analyzing ^{13}C enrichment in lactate and Hassel et al. (1995) showed formation of lactate via the astrocytic TCA cycle by documenting lactate formation from [2-^{13}C]acetate.

8.4 ^{13}C NMR Spectroscopic Analyses of Neuronal-Glial Interactions in Focal Cerebral Ischemia

8.4.1 Focal Ischemia

Focal cerebral ischemia or thrombo-embolic occlusion of an intracerebral artery is the cause of approximately 75% of all strokes. In focal cerebral ischemia excessive stimulation of glutamate receptors, glutamate excitotoxicity, is recognized as an important mediator of neuronal death (Rothman and Olney 1986; Choi 1994), whereas increased GABA receptor stimulation in the mature brain is considered neuroprotective (Schwartz-Bloom and Sah 2001). Astrocytes are inextricably involved in the synthesis of both neurotransmitter glutamate and GABA, and any disturbance in the metabolic interaction between astrocytes and neurons may have important implications for the final outcome after focal cerebral ischemia. Indeed, increased astrocytic oxidation of amino acids, glutamate, glutamine and GABA, combined with elevated pyruvate recycling was demonstrated in a model of combined focal and global ischemia in the hemisphere ipsilateral to the focal insult (Pascual et al. 1998). Also glutamate can be released from astrocytes in ischemic conditions, contributing to the excitotoxic damage (for review see Rossi et al. 2007).

In order to study the dynamic changes in the metabolic interactions between astrocytes and neurons during focal cerebral ischemia, we combined *in vivo* injection of labeled [1-^{13}C]glucose and [1,2-^{13}C]acetate in rats subjected to intraluminal occlusion of the middle cerebral artery (MCAO) (Longa et al. 1989). In this model of focal ischemia the cerebral blood flow is most severely reduced in the lateral caudoputamen and lower part of the frontoparietal cortex (Longa et al. 1989; Muller et al. 1995), referred to as the ischemic core. In the upper frontoparietal cortex, cerebral blood flow is moderately reduced (Memezawa et al. 1992; Muller et al. 1995), and this area is considered to represent the penumbra. Reperfusion and/or neuroprotective drug treatment is aimed mainly at salvaging the penumbra. However, it should be noted that focal cerebral ischemia is a dynamic process where the ischemic tissue gradually progresses from reversibly to irreversibly damaged. Moreover, the time-frame for this deterioration is controversial and depends on factors like temperature, reperfusion and individual variations in collateral blood supply. At different time intervals after onset of ischemia rats with permanent or transient MCAO were decapitated and their brains snap frozen. Areas considered representing the ischemic core and penumbra, were dissected and extracted for subsequent *ex vivo* ^{13}C NMR spectroscopy and high performance liquid chromatography (HPLC) analysis. Cerebellum and contralateral hemisphere also undergo changes (Haberg et al. 2009), but these changes are not discussed in this chapter.

8.4.2 Ischemic Core

MCAO disrupted the glutamate-glutamine cycle in the ischemic core (Haberg et al. 1998b, 2001). Incorporation of label from [1-^{13}C]glucose into [4-^{13}C]glutamate, reflecting mainly oxidative metabolism in glutamatergic neurons, declined rapidly after MCAO. The acetate versus glucose utilization ratio which is calculated from the amount of label originating from [1,2-^{13}C]acetate compared to that from [1-^{13}C] glucose (Taylor et al. 1996), i.e. [4,5-^{13}C]glutamate/[4-^{13}C]glutamate or [1,2-^{13}C] GABA/[2-^{13}C]GABA, was significantly increased at all times after MCAO showing that the relative importance of astrocytic glutamine in glutamate production was increased in ischemia (Haberg et al. 1998b, 2001). Thus suggesting that astrocytic glutamine contributes significantly to excitotoxicity in the ischemic core.

In pathological states not only labeling patterns, but also tissue concentrations of the amino acids under investigation may change. The changes in label incorporation and amino acid content may not always correspond, leading to large variations in the percent enrichment of ^{13}C in the amino acids during ongoing pathological processes in the brain. During MCAO there was a steady decline in glutamate content, most likely multifactorial in origin. Glutamate synthesis was significantly reduced, and glutamate consumption elevated since GABA and alanine levels increased with time and astrocytes continued to take up and convert glutamate to glutamine (Haberg et al. 1998b, 2001). However, the reduction in glutamate content far exceeded the aforementioned factors, implying that MCAO caused massive glutamate efflux beyond the capacity of the astrocytic glutamate transporters, and thus glutamate was lost into the systemic circulation and cerebrospinal fluid. This has been observed in patients (Castillo et al. 1996). Glutamate efflux over the blood brain barrier can only occur in instances when the blood brain barrier is damaged, as in ischemia. Under normal conditions only neutral amino acids, like glutamine, can transverse the blood brain barrier. During MCAO GABA content gradually increased, but incorporation of ^{13}C was significantly reduced at all times after MCAO. At 240 min of MCAO no labeling above natural abundance (1.1%) was detectable in GABA. This finding implies that GABA synthesis from glutamate proceeded without functioning mitochondria, and thus death of GABAergic neurons at this time point. GABA synthesis after 240 min of MCAO was therefore direct conversion of glutamate either stored in the GABAergic neurons or originating from other neurons. The amount of glutamate stored in GABAergic neurons is, however, minimal (Ottersen and Storm-Mathisen 1984), which suggests that GABA synthesis took place in the extracellular space. Glutamate decarboxylase, GAD, which is the enzyme converting glutamate to GABA is a small enzyme and exists in to isoforms with molecular masses of approximately 65 and 67 kDa. The GABA accumulation seen during MCAO also implies impaired GABA breakdown. Ischemic conditions have been shown to stimulate GAD activity (Sze 1979; Erecinska et al. 1996), but inhibit GABA transaminase activity (Schousboe et al. 1973; Baxter and Roberts 1961).

Ischemia also immediately affected astrocytic metabolism. The first sign of reduced astrocyte metabolism was the presence of un-metabolized acetate, which was not seen in sham operated rats (Haberg et al. 2001). The enzyme acetyl CoA synthetase, necessary for acetate metabolism, is an ATP dependent enzyme. The glial specific enzymes glutamine synthetase (GS) and pyruvate carboxylase (PC), which are also ATP dependent, had significantly reduced activity at later time points. After 240 min of MCAO PC activity was only detectable in glutamate in significantly reduced amounts, and the acetate/glucose utilization ratio was markedly increased in glutamate, but not in glutamine. These findings suggest that glutamate produced from acetate or glucose in the astrocytic mitochondria was not transported over the mitochondrial membrane into the cytosol, where GS is located. This interpretation is supported by an immunocytochemical observation of mitochondrial glutamate accumulation in astrocytes in global ischemia (Torp et al. 1993).

Reperfusion of the ischemic core 120 min after MCAO did not improve either neuronal or astrocytic TCA cycle activity. Instead a steady decline in mitochondrial metabolism accompanied by lactate accumulation was observed (Haberg et al. 2006). Likewise, the lipid peroxidation inhibitor, U-101033E, considered to penetrate into mitochondrial lipid bilayers and hence prevent free radical damage to mitochondrial membranes and proteins and thereby ensure maintained TCA cycle and respiratory chain activity, did not improve mitochondrial metabolism in the ischemic core. On the contrary, the NMR spectra revealed that the drug caused a further decrease in mitochondrial pyruvate metabolism and more pronounced cytosolic pyruvate metabolism (Haberg et al. 2007).

8.4.3 Penumbra

Neuronal glutamate synthesis from [1-^{13}C]glucose was significantly reduced from onset of ischemia and declined during the experimental period also in the penumbra. The amount of label in [4,5-^{13}C]glutamate, originating from [4,5-^{13}C]glutamine, was halved from 30 min of MCAO and remained at this level for 240 min of MCAO. This finding indicates a non-progressing functional inhibition of glutamine transporters and/or inhibition of the neuronal enzyme phosphate activated glutaminase (PAG) activity. Ammonia accumulation and/or acidosis which are known to occur in ischemia, have been shown to reduced PAG activity (Benjamin 1981; Hogstad et al. 1988; Kvamme et al. 2000). Still, with increasing duration of ischemia glutamine became a more important precursor for glutamate as neuronal glutamate synthesis from glucose continued to decline. Actually the acetate/glucose utilization ratio was doubled between 60 and 240 min of MCAO (Haberg et al. 1998b, 2001). It has been demonstrated that glutamine specifically replenishes the neurotransmitter pool of glutamate (Laake et al. 1995), and it is from this pool glutamate excitotoxicity is considered to originate in the penumbra (Takagi et al. 1993; Obrenovitch 1996). Based on the results from ^{13}C NMR spectroscopy analysis of the ischemic penumbra (Haberg et al. 1998a, b, 2001) the glutamate neurotransmitter pool appear to be at least partly replenished from astrocytic precursors, at the same time as neuronal TCA cycle metabolism was markedly reduced. This constellation may exacerbate glutamate excitotoxicity. Indeed, GS activity inhibition has been shown to reduce infarct size in cortex in rats subjected to MCAO (Swanson et al. 1990).

In the penumbra astrocytic metabolism appeared relatively unaffected by MCAO until 90–120 min of MCAO, when un-metabolized [1,2-^{13}C]acetate appeared in the fully awake rats (Haberg et al. 1998b, 2001). This is considerably later than in the ischemic core, and utilization of [1,2-^{13}C]acetate for glutamine synthesis was at the normal level during the entire 240 min of MCAO, indicating that amino acid synthesis via the TCA cycle was more resistant than oxidative metabolism in astrocytes. The activity of GS appeared unaffected by ischemia, and the glutamine content increased steadily. As mentioned above, GS is an ATP dependent enzyme located in the cytosol. Continued glutamine synthesis implies that ATP was present in sufficient amounts to convert glutamate to glutamine. It is conceivable that ATP was provided partly by glycolysis, perhaps fueled by breakdown of glycogen which takes place in the astrocytic cytosol, and partly by the limited amounts of oxygen and glucose supplied to this area by a reduced, but not abolished blood flow.

Re-establishing blood flow has been shown to salvage tissue in the penumbra region. Using ^{13}C NMR spectroscopy four metabolic events that distinguish successfully reperfused tissue have been identified (Haberg et al. 2007). (1) Improved astrocytic acetate metabolism and oxidation, (2) improved neuronal mitochondrial glucose metabolism (3) improved neuronal utilization of astrocytic PC derived metabolites, and (4) no increase in lactate synthesis. The results from reperfusion and permanent MCAO supplement each other and corroborate the notion that preservation of astrocytic metabolism is essential for neuronal survival and a predictor for recovery. Also supporting the importance of preservation of astrocytic metabolism for neuronal survival are results obtained using the lipid peroxidation inhibitor, U-101033E. In the penumbra this drug decreased both neuronal and astrocytic pyruvate metabolism via the TCA cycle, and at the same time increased the anaerobic glucose metabolism (Haberg et al. 2007). Indeed, a similar compound, tirilazad mesylate (U-74006F; Freedox), has been shown to significantly increase in the number of deaths and the degree of disability in randomized controlled trials in acute stroke in humans (Tirilazad International Steering Committee 2000). The results obtained with ^{13}C NMR spectroscopy stress the importance of other outcome measures than ischemic tissue volume for evaluation of drug efficacy in animal models, which in turn could increase the likelihood of success in clinical trials.

8.5 ^{13}C NMR Spectroscopic Analyses of Neuronal-Glial Interactions in Epilepsy

Epilepsy is one of the most common neurological disorders worldwide with a prevalence of 1%. The main feature of epilepsy is the predisposition to recurrent seizures characterized by increased and synchronous neuronal activity. Seizures are associated with increased release of the excitatory neurotransmitter glutamate (Bradford, 1995), and both glutamate receptor agonists and GABA receptor antagonists have been shown to promote epileptic seizures (Hosford 1995). An imbalance of excitatory and inhibitory functions may be due to disturbed neurotransmitter metabolism (Bradford 1995) and ^{13}C NMR spectroscopy is a useful tool to study neurotransmitter synthesis in addition to glial-neuronal interactions. The following section is a description of how ^{13}C NMR spectroscopy in combination with different animal models can elucidate specific changes underlying epileptogenesis and specific events related to seizures.

8.5.1 The Pentylene Tetrazole Model

Pentylene tetrazole (PTZ) is a chemical convulsant and a $GABA_A$ receptor blocker. Given acutely in high concentration it causes generalized tonic-clonic seizures. To study glial-neuronal interactions in the early postictal stage a single PTZ injection (intra peritoneal) in rats was combined with injections (intra peritoneal) of [1-^{13}C] glucose plus [1,2-^{13}C]acetate 30 min later (Eloqayli et al. 2003). The animals were decapitated 45 min after PTZ injection.

^{13}C NMR spectroscopy analyses of extracts from cortex, subcortex and cerebellum showed that astrocyte metabolism in the PTZ treated rats was unchanged. GABAergic neurons in the cerebellum showed a pronounced decrease of GABA synthesis, which might have led to decreased release of GABA. The GABAergic neurons in cortex and subcortex were unaffected. Glutamate labeling was decreased in all brain regions as a response to PTZ treatment, while the labeling of glutamine was unchanged demonstrating a reduction of the metabolic pool of glutamate and not the transmitter pool. Thus, in this model, which is used extensively in testing potential antiepileptic drugs, mainly neuronal metabolism is affected in the early phase.

Epilepsy is also a common geriatric problem (Stephen and Brodie 2000). Since the majority of epilepsy experiments with animals use young adult rodents, a new experimental model to study effects of seizures in the elderly was designed (Kondziella et al. 2002). This model uses PTZ-kindling in senescence accelerated mice (SAMP8), a genetic model of aging (Fujibayashi et al. 1994; Sato et al. 1994). Kindling is characterized by increased sensitivity to seizures after repeated administration of an initially subconvulsive electrical or chemical stimulus. Continuous application of sub-threshold doses of PTZ induces progressive development of seizures resulting in secondary generalized seizures. To investigate whether seizures induce metabolic changes and whether the effect of the anti-convulsant drug phenobarbital is age dependent, 2-month and 8-month old SAMP8 mice received PTZ, PTZ and phenobarbital or saline (20 intra peritoneal injections) every 48 h over a period of 40 days (Kondziella et al. 2003). The mice received intraperitoneal injections of [1-^{13}C]glucose and [1,2-^{13}C]acetate 15 min before decapitation. Extracts from the whole brain (minus cerebellum) were analyzed.

Several differences between the young and old mice, compared to their respective controls, were observed both in response to PTZ alone and PTZ combined with phenobarbital. As in the above mentioned model, the 8-month old mice treated with

PTZ alone showed reduced glutamate labeling from [1-^{13}C]glucose, indicating decreased mitochondrial activity in the glutamatergic neurons. Impaired astrocytic metabolism was apparent in the 2-month old mice since the synthesis of glutamine from [1,2-^{13}C]acetate was decreased. PTZ activates excitatory mechanisms in brain cells, but PTZ-kindling did not increase the labeling of glutamate. The reduced seizure threshold induced by PTZ kindling could be explained by enhanced density of glutamate-binding sites on excitatory synapses (Schroeder et al. 1999) rather than increased synthesis of glutamate.

In the 2-month-old mice receiving PTZ and phenobarbital the labeling of most metabolites from both [1-^{13}C]glucose and [1,2-^{13}C]acetate was decreased. Since the total amount of amino acids was unchanged this indicated that turnover of metabolites in both neurons and astrocytes was decreased. Unchanged labeling of GABA showed that GABAergic neurons were less sensitive to the effect of phenobarbital. In the 8-month-old mice there was an increase in GABA labeling both from [1-^{13}C]glucose and [1,2-^{13}C]acetate, while the labeling in most of the other metabolites was unchanged. In summary, phenobarbital depressed cerebral metabolism in the young mice while it increased the metabolic activity of the GABAergic neurons in the older mice.

8.5.2 Animal Models of Mesial Temporal Lobe Epilepsy

Mesial temporal lobe epilepsy (MTLE) is one of the most common forms of focal epilepsy. The hallmark of MTLE is hippocampal sclerosis, which is characterized by neuronal loss and gliosis in CA1, CA3, and hilus, and reorganization of synaptic connections (Sommer 1880; de Lanerolle and Lee 2005). Animal models, like the kainic acid and lithium pilocarpine (Li pilo) models, are extensively used to study cellular and molecular mechanisms underlying MTLE.

The lithium-pilocarpine model was used to study neurotransmitter metabolism and neuronal-glial interactions in rats that had developed spontaneous recurrent seizures. Approximately 1 month after treatment, the rats were injected with [1-^{13}C]glucose and [1,2-^{13}C]acetate and extracts from cortex, cerebellum, and hippocampal formation (including hippocampus, amygdala, entorhinal and piriform cortices) were analyzed (Melø et al. 2005).

Decreased amounts of glutamate and glutathione were measured in the hippocampal formation of Li-pilo rats compared with control rats, corresponding well with the neuronal cell loss seen in this region. The amounts of [4-^{13}C]glutamate, [3-^{13}C]-/[2-^{13}C]aspartate and [2-^{13}C]GABA, derived from [1-^{13}C]glucose, were decreased in all brain regions investigated in Li-pilo rats compared with controls. These results suggest that the activity in neuronal mitochondria is lower in the epileptic brain than in controls. This is further supported by the decreased level of NAA seen in the cortex and hippocampal formation of Li-pilo rats in comparison with controls. No changes were detected in glial-neuronal interactions in the hippocampal formation while in cortex and cerebellum the flow of glutamate to astrocytes was decreased, indicating a disturbed glutamate-glutamine cycle. In hippocampus astrocytic metabolism was unchanged despite the pronounced gliosis. The results could indicate that astrocytes in sclerotic tissue have lower mitochondrial activity per cell volume. In cortex and cerebellum, labeled glutamine from [1-^{13}C]glucose was decreased whereas the amount of glutamine was unchanged.

In the kainic acid model, neurotransmitter metabolism and neuronal-glial interactions were investigated before (24 h and 14 days) and after (13 weeks) development of spontaneous seizures. Kainic acid was injected (intra peritoneal) into rats, and after 24 h (Qu et al. 2003) 14 days (Muller et al. 2000; Alvestad et al. 2011) the animals received an injection (intra peritoneal) of [1-^{13}C]glucose plus [1,2-^{13}C]acetate. After 15 min the animals were decapitated and extracts from the whole brain (minus cerebellum) were analyzed.

One day after the kainate injection there was increased labeling of glutamine and glutamate from [1,2-^{13}C]acetate, while labeling from [1-^{13}C]glucose was unchanged. At day 14, however, no changes were observed in the metabolites derived from [1,2-^{13}C]acetate, but labeling of glutamate, GABA, glutamine, aspartate, and succinate from [1-^{13}C]glucose was increased. The tissue concentration of metabolites was unchanged both after 24 h and 14 days. Thus, kainate treatment increased the amino acid turnover in astrocytes after 1 day and increased the turnover in glutamatergic and GABAergic neurons after 14 days. This indicates that early and temporary enhanced astrocytic activity may lead to altered metabolism in neurons with an increased turnover of important amino acids such as GABA and glutamate.

Thirteen weeks after injection of kainic acid, rats with spontaneous seizures were injected intra peritoneal with [1-^{13}C]glucose (Alvestad et al. 2008). Brain extracts from hippocampal formation, entorhinal cortex, and neocortex were analyzed. The amounts of glutmate and NAA and labeling of glutamate were reduced in the hippocampal formation and entorhinal cortex of epileptic rats, indicating neuronal loss. Additionally, mitochondrial dysfunction was detected in surviving glutamatergic neurons in the hippocampal formation. In entorhinal cortex glutamine labeling and concentration were unchanged despite the reduced glutamate content and label, possibly due to decreased oxidative metabolism and conserved flux of glutamate through glutamine synthetase in astrocytes. This mechanism was not operative in the hippocampal formation, where glutamine labeling was decreased. In the neocortex, labelling and concentration of GABA were increased in the epileptic rats, possibly representing a compensatory mechanism. The changes in the hippocampus might be of pathophysiological importance and further studies aiming at resolving metabolic causes and consequences of MTLE are required.

8.6 Genetic Abscencee Epilpsy Rat Model

In the brain of immature Genetic Absence Epilepsy Rats from Strasbourg (GAERS), interactions between glutamatergic neurons and astrocytes appeared normal whereas increased astrocytic metabolism took place in adult GAERS, suggesting that astrocytic alterations could possibly be the cause of seizures (Melø et al. 2007).

8.7 ^{13}C NMR Spectroscopy in Human Pathologies

Human *in vivo* studies using ^{13}C NMR spectroscopy as a tool to investigate basic brain neurochemistry have been perform in healthy volunteers (Beckmann et al. 1991; Gruetter et al. 1992; Rothman et al. 1992; Mason et al. 1995; Oz et al. 2007) and selected patient groups (Rothman et al. 1991; Blüml et al. 2001) since the early 1990s. Most human studies are dynamic studies with ^{13}C labeled glucose, but labeled acetate (Lebon et al. 2002; Blüml et al. 2002) and hydroxybutyrate (Pan et al. 2002) have also been administrated. Studies pertaining to basic neuochemical questions in healthy volunteers are performed with protocols quite similar to those used in steady state studies of animal.

Protocols used in clinical research, on the other hand, often differ from the basic design applied in healthy volunteers and animals. Intravenous infusion times may be shorter, only one or a few spectra may be obtained, labeled substrate may be given orally (Rothman et al. 1991; Moreno et al. 2001a, b; Shic and Ross 2003; Blüml et al. 2001), and resected tissue can be extracted for *in vitro* ^{13}C NMR spectroscopy in order to obtain more detailed neurochemical fingerprinting (Petroff et al. 2002). These adaptations are necessary and/or may be the only possible solution in the clinical setting. Dynamic steady-state ^{13}C requires very cooperative subjects, able to lie still for extended periods before and after infusion of labeled substrate. Furthermore, glucose infusion may in some instances be harmful, for instance in acute stroke and epilepsy. In addition ^{13}C labeled substrates are very expensive. The number of patients investigated with ^{13}C MR spectroscopy thus remains low. The feasibility of clinical ^{13}C MR spectroscopy in metabolic fingerprinting of CNS diseases has been explored in single cases and with few patients. The pathologies cover a wide spectrum; Alzheimer's disease, epilepsy, hypoxia/ischemia, hepatic encephalopathy, leucodystrophy and a variety of metabolic diseases such as Caravan's disease (Rothman et al. 1991; Moreno et al. 2001a, b; Blüml et al. 2001, 2002; Shic and Ross 2003). Two cases of ^{13}C MR spectroscopy in subacute stroke/hypoxia have been reported in the literature (Rothman et al. 1991; Blüml et al. 2001). Rothman et al. (1991) demonstrated protracted elevation of lactate production from glucose in the ischemic tissue while Blüml et al. (2001) showed reduced levels of [4-^{13}C]glutamate synthesis. Both describe reduced levels of NAA. These findings are in line with the above mentioned studies of extracts obtained from ischemic rat brain, but the level of detail is of course not comparable. The neurochemical changes in epilepsy have also been studied *in vivo* (Blüml et al. 2002; Shic and Ross 2003) and in extracts of hippocampi after temporal lobectomy (Petroff et al. 2002). The general trend is disturbed glutamate-glutamine cycling, but the results are conflicting as to the direction of the disturbance, i.e. both up and down regulations were reported.

Compared to other MR methods, for instance functional (fMRI) and diffusion MR imaging (DWI/DTI) that were implemented in humans at the beginning of the 1990s also, *in vivo* ^{13}C MR spectroscopy has yet to become a routine investigation. *In vivo* ^{13}C MR spectroscopy in humans is still limited to relatively few sites with high-field MR scanners and expert interdisplinary research groups. Clinical ^{13}C MR spectroscopy may improve accuracy of diagnosis by offering detailed metabolic profiles of pathological tissues. Furthermore, ^{13}C MR spectroscopy may also be a tool for monitoring treatment effects and also explore new avenues for dietary and/or pharmaceutical intervention. However, the workflow from acquisition to spectrum analysis needs to be simplified in order for this to be a clinically relevant method. It will also be necessary to prove that clinical *in vivo* ^{13}C MR spectroscopy can with high specificity and sensitivity, give the correct diagnosis with additional information relevant for the clinical management of the patient not available with other methods, in order for this costly method to become a part of the standard clinical repertoire.

8.8 Conclusion

From the above it is evident that ^{13}C NMR spectroscopy is a tool not only suitable to study metabolism in the intact brain, but it can also provide valuable information on changes in neuronal, astrocytic and neuronal-astrocytic interactions in the diseased state, and may thus contribute to our understanding of nervous system disease, and provide new avenues to explore in the search for new therapies. The importance of astrocytes for neuronal function and survival is clearly illustrated by numerous reports in the literature and the studies presented here. In KA induced epilepsy glia were affected earlier than neurons and in MCAO the degree of astrocytic metabolic disturbances may be predictive of neuronal survival. Also, ^{13}C NMR spectroscopy offers an opportunity to explore the metabolic effects of drugs, and can thereby increase the likelihood of bringing the best suited drugs to clinical trials. Furthermore, it should be noted that neuronal-astrocytic interactions change with age and appropriate animal models for human disease should be used. It appears likely that ^{13}C NMR spectroscopy will have clinical impact in the future since the dynamic picture of metabolic changes complements the rather static information obtained using ^{1}H NMR spectroscopy and conventional MRI.

References

Alves PM, Flogel U, Brand A, Leibfritz D, Carrondo MJ, Santos H, Sonnewald U (1996) Immobilization of primary astrocytes and neurons for online monitoring of biochemical processes by NMR. Dev Neurosci 18:478–483

Alves PM, Nunes R, Zhang C, Maycock CD, Sonnewald U, Carrondo MJ, Santos H (2000) Metabolism of 3-^{13}C-malate in primary cultures of mouse astrocytes. Dev Neurosci 22:456–462

Alvestad S, Hammer J, Eyjolfsson E, Qu H, Ottersen OP, Sonnewald U (2008) Limbic structures show altered glial-neuronal metabolism in the chronic phase of kainate induced epilepsy. Neurochem Res 33:257–266

Alvestad S, Hammer J, Qu H, Håberg A, Ottersen OP, Sonnewald U (2011) Reduced astrocytic contribution to the turnover of glutamate, glutamine, and GABA characterizes the latent phase in the kainate model of temporal lobe epilepsy. JCBFM 31(8):1675–1686

Bachelard H (1998) Landmarks in the application of ^{13}C-magnetic resonance spectroscopy to studies of neuronal/glial relationships. Dev Neurosci 20:277–288

Badar-Goffer RS, Bachelard HS, Morris PG (1990) Cerebral metabolism of acetate and glucose studied by ^{13}C-n.m.r. spectroscopy. A technique for investigating metabolic compartmentation in the brain. Biochem J 266:133–139

Bak LK, Schousboe A, Sonnewald U, Waagepetersen HS (2006) Glucose is necessary to maintain neurotransmitter homeostasis during synaptic activity in cultured glutamatergic neurons. J Cereb Blood Flow Metab 26:1285–1297

Bak LK, Waagepetersen HS, Melø TM, Schousboe A, Sonnewald U (2007) Complex glutamate labeling from [U-^{13}C]glucose or [U-^{13}C]lactate in co-cultures of cerebellar neurons and astrocytes. Neurochem Res 32:671–680

Bakken IJ, Johnsen SF, White LR, Unsgard G, Aasly J, Sonnewald U (1997) NMR spectroscopy study of the effect of 3-nitropropionic acid on glutamate metabolism in cultured astrocytes. J Neurosci Res 47:642–649

Barany M, Arus C, Chang YC (1985) Natural-abundance ^{13}C NMR of brain. Magn Reson Med 2:289–295

Baxter CF, Roberts E (1961) Elevation of gamma-aminobutyric acid in brain: selective inhibition of gamma-aminobutyric-alpha-ketoglutaric acid transaminase. J Biol Chem 236:3287–3294

Beckmann N, Turkalj I, Seelig J, Keller U (1991) ^{13}C NMR for the assessment of human brain glucose metabolism in vivo. Biochemistry 30:6362–6366

Benjamin AM (1981) Control of glutaminase activity in rat brain cortex in vitro: influence of glutamate, phosphate, ammonium, calcium and hydrogen ions. Brain Res 208:363–377

Blüml S, Moreno A, Hwang JH, Ross BD (2001) 1-^{13}C glucose magnetic resonance spectroscopy of pediatric and adult brain disorders. NMR Biomed 14:19–32

Blüml S, Moreno-Torres A, Shic F, Nguy CH, Ross BD (2002) Tricarboxylic acid cycle of glia in the in vivo human brain. NMR Biomed 15:1–5

Bradford HF (1995) Glutamate, GABA and epilepsy. Prog Neurobiol 47:477–511

Castillo J, Davalos A, Naveiro J, Noya M (1996) Neuroexcitatory amino acids and their relation to infarct size and neurological deficit in ischemic stroke. Stroke 27:1060–1065

Cerdan S, Kunnecke B, Seelig J (1990) Cerebral metabolism of [1,2-$^{13}C_2$]acetate as detected by in vivo and in vitro ^{13}C NMR. J Biol Chem 265:12916–12926

Cerdán S, Rodrigues TB, Sierra A, Benito M, Fonseca LL, Fonseca CP, García-Martín ML (2006) The redox switch/redox coupling hypothesis. Neurochem Int 48:523–530

Choi DW (1994) Glutamate receptors and the induction of excitotoxic neuronal death. Prog Brain Res 100:47–51

Choi IY, Gruetter R (2003) In vivo ^{13}C NMR assessment of brain glycogen concentration and turnover in the awake rat. Neurochem Int 43:317–322

Cremer JE, Lucas HM (1971) Sodium pentobarbitone and metabolic compartments in rat brain. Brain Res 35:619–621

de Lanerolle NC, Lee TS (2005) New facets of the neuropathology and molecular profile of human temporal lobe epilepsy. Epilepsy Behav 10:190–203

Eid T, Thomas MJ, Spencer DD, Runden-Pran E, Lai JC, Malthankar GV, Kim JH, Danbolt NC, Ottersen OP, de Lanerolle NC (2004) Loss of glutamine synthetase in the human epileptogenic hippocampus: possible mechanism for raised extracellular glutamate in mesial temporal lobe epilepsy. Lancet 363:28–37

Eloqayli H, Dahl CB, Gotestam KG, Unsgard G, Hadidi H, Sonnewald U (2003) Pentylenetetrazole decreases metabolic glutamate turnover in rat brain. J Neurochem 85:1200–1207

Erecinska M, Nelson D, Daikhin Y, Yudkoff M (1996) Regulation of GABA level in rat brain synaptosomes: fluxes through enzymes of the GABA shunt and effects of glutamate, calcium, and ketone bodies. J Neurochem 67:2325–2334

Fan TWM (1996) Metabolite profiling by one and two-dimensional NMR analysis of complex mixture. Prog Nucl Magn Reson Spectrosc 528:161–219

Fan TW-M, Lane AN (2008) Structure-based profiling of metabolites and isotopomers by NMR. Prog Nucl Magn Reson Spectrosc 52:69–117

Freo U, Ori C (2004) Effects of anesthesia and recovery from ketamine racemate and enantiomers on regional cerebral glucose metabolism in rats. Anesthesiology 100:1172–1178

Fujibayashi Y, Waki A, Wada K, Ueno M, Magata Y, Yonekura Y, Konishi J, Takeda T, Yokoyama A (1994) Differential aging pattern of cerebral accumulation of radiolabeled glucose and amino acid in the senescence accelerated mouse (SAM), a new model for the study of memory impairment. Biol Pharm Bull 17:102–105

Gaitonde MK, Dahl DR, ElliotT KA (1965) Entry of glucose carbon into amino acids of rat brain and liver in vivo after injection of uniformly 14-C-labelled glucose. Biochem J 94:345–352

Garcia-Espinosa MA, Rodrigues TB, Sierra A, Benito M, Fonseca C, Gray HL, Bartnik BL, Garcia-Martin ML, Ballesteros P, Cerdan S (2004) Cerebral glucose metabolism and the glutamine cycle as detected by in vivo and in vitro ^{13}C NMR spectroscopy. Neurochem Int 45:297–303

Gibson GE, Sheu KF, Blass JP (1998) Abnormalities of mitochondrial enzymes in Alzheimer disease. J Neural Transm 105:855–870

Griffin JL, Rae C, Radda GK, Matthews PM (1999) Delayed labelling of brain glutamate after an intra-arterial [^{13}C]glucose bolus: evidence for aerobic metabolism of guinea pig brain glycogen store. Biochim Biophys Acta 1450:297–307

Gruetter R, Novotny EJ, Boulware SD, Rothman DL, Mason GF, Shulman GI, Shulman RG, Tamborlane WV (1992) Direct measurement of brain glucose concentrations in humans by ^{13}C NMR spectroscopy. Proc Natl Acad Sci USA 89:1109–1112. Erratum in: Proc Natl Acad Sci USA 1992 Dec 15;89(24):12208

Haberg A, Qu H, Bakken IJ, Sande LM, White LR, Haraldseth O, Unsgard G, Aasly J, Sonnewald U (1998a) In vitro and ex vivo ^{13}C-NMR spectroscopy studies of pyruvate recycling in brain. Dev Neurosci 20:389–398

Haberg A, Qu H, Haraldseth O, Unsgard G, Sonnewald U (1998b) In vivo injection of [1-^{13}C] glucose and [1,2-^{13}C]acetate combined with ex vivo ^{13}C nuclear magnetic resonance spectroscopy: a novel approach to the study of middle cerebral artery occlusion in the rat. J Cereb Blood Flow Metab 18:1223–1232

Haberg A, Qu H, Saether O, Unsgard G, Haraldseth O, Sonnewald U (2001) Differences in neurotransmitter synthesis and intermediary metabolism between glutamatergic and GABAergic neurons during 4 hours of middle cerebral artery occlusion in the rat: the role of astrocytes in neuronal survival. J Cereb Blood Flow Metab 21:1451–1463

Haberg A, Qu H, Hjelstuen MH, Sonnewald U (2006) Glutamate and GABA metabolism in transient and permanent middle cerebral artery occlusion in rat: importance of astrocytes for neuronal survival. Neurochem Int 48:531–540

Haberg A, Qu H, Hjelstuen MH, Sonnewald U (2007) Effect of the pyrrolopyrimidine lipid peroxidation inhibitor U-101033E on neuronal and astrocytic metabolism and infarct volume in rats with transient middle cerebral artery occlusion. Neurochem Int 50:932–940

Haberg AK, Qu H, Sonnewald U (2009) Acute changes in intermediary metabolism in cerebellum and contralateral hemisphere following middle cerebral artery occlusion in rat. J Neurochem 109(Suppl 1):174–181

Hansen TD, Warner DS, Todd MM, Vust LJ, Trawick DC (1988) Distribution of cerebral blood flow during halothane versus isoflurane anesthesia in rats. Anesthesiology 69:332–337

Harada M, Okuda C, Sawa T, Murakami T (1992) Cerebral extracellular glucose and lactate concentrations during and after moderate hypoxia in glucose- and saline-infused rats. Anesthesiology 77:728–734

Hassel B, Sonnewald U (1995a) Selective inhibition of the tricarboxylic acid cycle of GABAergic neurons with 3-nitropropionic acid in vivo. J Neurochem 65:1184–1191

Hassel B, Sonnewald U (1995b) Glial formation of pyruvate and lactate from TCA cycle intermediates: implications for the inactivation of transmitter amino acids? J Neurochem 65: 2227–2234

Hassel B, Sonnewald U, Fonnum F (1995) Glial-neuronal interactions as studied by cerebral metabolism of [2-^{13}C]acetate and [1-^{13}C]glucose: an ex vivo ^{13}C NMR spectroscopic study. J Neurochem 64:2773–2782

Hassel B, Bachelard H, Jones P, Fonnum F, Sonnewald U (1997) Trafficking of amino acids between neurons and glia in vivo. Effects of inhibition of glial metabolism by fluoroacetate. J Cereb Blood Flow Metab 17:1230–1238

Henry PG, Tkac I, Gruetter R (2003) ^{1}H-localized broadband ^{13}C NMR spectroscopy of the rat brain in vivo at 9.4 T. Magn Reson Med 50:684–692

Henry PG, Adriany G, Deelchand D, Gruetter R, Marjanska M, Oz G, Seaquist ER, Shestov A, Uğurbil K (2006) In vivo ^{13}C NMR spectroscopy and metabolic modeling in the brain: a practical perspective. Magn Reson Imaging 24:527–539

Hertz E, Yu ACH, Shahar A, Juurlink BHJ, Schousboe A (1989a) Preparation of primary cultures of mouse cortical neurons. In: A dissection and tissue culture manual for the nervous system. Alan R. Liss, New York, pp 183–186

Hertz L, Juurlink BHJ, Hertz E, Fosmark H, Schousboe A (1989b) Preparation of primary cultures of mouse (rat) astrocytes. In: A dissection and tissue culture manual for the nervous system. Alan R. Liss, New York, pp 105–108

Hoffman WE, Edelman G, Kochs E, Werner C, Segil L, Albrecht RF (1991) Cerebral autoregulation in awake versus isoflurane-anesthetized rats. Anesth Analg 73:753–757

Hogstad S, Svenneby G, Torgner IA, Kvamme E, Hertz L, Schousboe A (1988) Glutaminase in neurons and astrocytes cultured from mouse brain: kinetic properties and effects of phosphate, glutamate, and ammonia. Neurochem Res 13:383–388

Hosford DA (1995) Models of primary generalized epilepsy. Curr Opin Neurol 8:121–125

Hyder F, Brown P, Nixon TW, Behar KL (2003) Mapping cerebral glutamate ^{13}C turnover and oxygen consumption by in vivo NMR. Adv Exp Med Biol 530:29–39

Karelson G, Ziegler A, Kunnecke B, Seelig J (2003) Feeding versus infusion: a novel approach to study the NAA metabolism in rat brain. NMR Biomed 16:413–423

Kauppinen RA, Pirttilä TR, Auriola SO, Williams SR (1994) Compartmentation of cerebral glutamate in situ as detected by ^{1}H/^{13}C n.m.r. Biochem J 298:121–127

Kondziella D, Bidar A, Urfjell B, Sletvold O, Sonnewald U (2002) The pentylenetetrazole-kindling model of epilepsy in SAMP8 mice: behavior and metabolism. Neurochem Int 40:413–418

Kondziella D, Hammer J, Sletvold O, Sonnewald U (2003) The pentylenetetrazole-kindling model of epilepsy in SAMP8 mice: glial-neuronal metabolic interactions. Neurochem Int 43:629–637

Kvamme E, Roberg B, Torgner IA (2000) Phosphate-activated glutaminase and mitochondrial glutamine transport in the brain. Neurochem Res 25:1407–1419

Laake JH, Slyngstad TA, Haug FM, Ottersen OP (1995) Glutamine from glial cells is essential for the maintenance of the nerve terminal pool of glutamate: immunogold evidence from hippocampal slice cultures. J Neurochem 65:871–881

Lebon V, Petersen KF, Cline GW, Shen J, Mason GF, Dufour S, Behar KL, Shulman GI, Rothman DL (2002) Astroglial contribution to brain energy metabolism in humans revealed by ^{13}C nuclear magnetic resonance spectroscopy: elucidation of the dominant pathway for neurotransmitter glutamate repletion and measurement of astrocytic oxidative metabolism. J Neurosci 22:1523–1531

Longa EZ, Weinstein PR, Carlson S, Cummins R (1989) Reversible middle cerebral artery occlusion without craniectomy in rats. Stroke 20:84–91

Magistretti PJ, Pellerin L, Rothman DL, Shulman RG (1999) Energy on demand. Science 283:496–497

Mason GF, Gruetter R, Rothman DL, Behar KL, Shulman RG, Novotny EJ (1995) Simultaneous determination of the rates of the TCA cycle, glucose utilization, alpha-ketoglutarate/glutamate exchange, and glutamine synthesis in human brain by NMR. J Cereb Blood Flow Metab 15:12–25

Mason GF, Petersen KF, de Graaf RA, Shulman GI, Rothman DL (2007) Measurements of the anaplerotic rate in the human cerebral cortex using ^{13}C magnetic resonance spectroscopy and [1-^{13}C] and [2-^{13}C] glucose. J Neurochem 100:73–86

McKenna MC, Sonnewald U, Huang X, Stevenson J, Zielke HR (1996) Exogenous glutamate concentration regulates the metabolic fate of glutamate in astrocytes. J Neurochem 66: 386–393

McKenna MC, Dienel GA, Sonnewald U, Waagepetersen HS, Schousboe A (2011) Energy metabolism of the brain. In: Siegel G, Albers W, Brady S, Price D (eds) Basic neurochemistry, vol 8, Elsevier, London, pp 201–226, ISBN 978-0-12-374947-5

Melø TM, Nehlig A, Sonnewald U (2005) Metabolism is normal in astrocytes in chronically epileptic rats: a ^{13}C NMR study of neuronal-glial interactions in a model of temporal lobe epilepsy. J Cereb Blood Flow Metab 25:1254–1264

Melø TM, Bastholm IA, Sonnewald U, Nehlig A (2007) Astrocytes may play a role in the etiology of absence epilepsy: a comparison between immature GAERS not yet expressing seizures and adults. Neurobiol Dis 28:227–235

Memezawa H, Minamisawa H, Smith ML, Siesjo BK (1992) Ischemic penumbra in a model of reversible middle cerebral artery occlusion in the rat. Exp Brain Res 89:67–78

Minchin MC, Beart PM (1975a) Compartmentation of amino acid metabolism in the rat dorsal root ganglion; a metabolic and autoradiographic study. Brain Res 83:437–449

Minchin MC, Beart PM (1975b) Compartmentation of amino acid metabolism in the rat posterior pituitary. J Neurochem 24:881–884

Moreno A, Blüml S, Hwang JH, Ross BD (2001a) Alternative 1-^{13}C glucose infusion protocols for clinical ^{13}C MRS examinations of the brain. Magn Reson Med 46:39–48

Moreno A, Ross BD, Blüml S (2001b) Direct determination of the N-acetyl-L-aspartate synthesis rate in the human brain by ^{13}C MRS and [1-^{13}C]glucose infusion. J Neurochem 771:347–350

Muller TB, Haraldseth O, Jones RA, Sebastiani G, Godtliebsen F, Lindboe CF, Unsgard G (1995) Combined perfusion and diffusion-weighted magnetic resonance imaging in a rat model of reversible middle cerebral artery occlusion. Stroke 26:451–457

Muller B, Qu H, Garseth M, White LR, Aasly J, Sonnewald U (2000) Amino acid neurotransmitter metabolism in neurones and glia following kainate injection in rats. Neurosci Lett 279:169–172

Nehlig A, Wittendorp-Rechenmann E, Lam CD (2004) Selective uptake of [^{14}C]2-deoxyglucose by neurons and astrocytes: high-resolution microautoradiographic imaging by cellular ^{14}C-trajectography combined with immunohistochemistry. J Cereb Blood Flow Metab 24:1004–1014

Obrenovitch TP (1996) Origins of glutamate release in ischaemia. Acta Neurochir Suppl 66:50–55

Olstad O, Qu H, Sonnewald U (2007) Glutamate is preferred over glutamine for intermediary metabolism in cultured cerebellar neurons. J Cereb Blood Flow Metab 50:1004–1013

Ottersen OP, Storm-Mathisen J (1984) Glutamate- and GABA-containing neurons in the mouse and rat brain, as demonstrated with a new immunocytochemical technique. J Comp Neurol 229:374–392

Oz G, Berkich DA, Henry PG, Xu Y, LaNoue K, Hutson SM, Gruetter R (2004) Neuroglial metabolism in the awake rat brain: CO2 fixation increases with brain activity. J Neurosci 24:11273–11279

Oz G, Seaquist ER, Kumar A, Criego AB, Benedict LE, Rao JP, Henry PG, Van De Moortele PF, Gruetter R (2007) Human brain glycogen content and metabolism: implications on its role in brain energy metabolism. Am..J Physiol. Endocrinol Metab 292:E946–51

Pan JW, de Graaf RA, Petersen KF, Shulman GI, Hetherington HP, Rothman DL (2002) [2,4-$^{13}C_2$]-beta-Hydroxybutyrate metabolism in human brain. J Cereb Blood Flow Metab 22: 890–898

Pascual JM, Carceller F, Roda JM, Cerdán S (1998) Glutamate, glutamine, and GABA as substrates for the neuronal and glial compartments after focal cerebral ischemia in rats. Stroke 29:1048–1056, discussion 1056–1057

Pellerin L, Bouzier-Sor AK, Aubert A, Serres S, Merle M, Costala R, Magistretti PJ (2007) Activity-dependent regulation of energy metabolism by astrocytes: an update. Glia 55:1251–1262

Peng L, Zhang X, Hertz L (1994) High extracellular potassium concentrations stimulate oxidative metabolism in a glutamatergic neuronal culture and glycolysis in cultured astrocytes but have no stimulatory effect in a GABAergic neuronal culture. Brain Res 663:168–172

Petroff OA, Errante LD, Rothman DL, Kim JH, Spencer DD (2002) Glutamate-glutamine cycling in the epileptic human hippocampus. Epilepsia 43:703–710

Qu H, Haberg A, Haraldseth O, Unsgard G, Sonnewald U (2000a) ^{13}C MR spectroscopy study of lactate as substrate for rat brain. Dev Neurosci 22:429–436

Qu H, Waagepetersen HS, van Hengel M, Wolt S, Dale O, Unsgard G, Sletvold O, Schousboe A, Sonnewald U (2000b) Effects of thiopental on transport and metabolism of glutamate in cultured cerebellar granule neurons. Neurochem Int 37:207–215

Qu H, Konradsen JR, van HM, Wolt S, Sonnewald U (2001a) Effect of glutamine and GABA on [U-13C]glutamate metabolism in cerebellar astrocytes and granule neurons. J Neurosci Res 66:885–890

Qu H, van der GM, Le MT, Sonnewald U (2001b) The effect of thiopental on glutamate metabolism in mouse cerebellar astrocytes in vitro. Neurosci Lett 304:141–144

Qu H, Eloqayli H, Müller B, Aasly J, Sonnewald U (2003) Glial-neuronal interactions following kainate injection in rats. Neurochem Int 42:101–106

Rae C, Moussa Cel-H, Griffin JL, Parekh SB, Bubb WA, Hunt NH, Balcar VJA (2006) Metabolomic approach to ionotropic glutamate receptor subtype function: a nuclear magnetic resonance in vitro investigation. J Cereb Blood Flow Metab 26:1005–1017

Rossi DJ, Brady JD, Mohr C (2007) Astrocyte metabolism and signaling during brain ischemia. Nat Neurosci 10:1377–1386

Rothman SM, Olney JW (1986) Glutamate and the pathophysiology of hypoxic–ischemic brain damage. Ann Neurol 19:105–111

Rothman DL, Howseman AM, Graham GD, Petroff OA, Lantos G, Fayad PB, Brass LM, Shulman GI, Shulman RG, Prichard JW (1991) Localized proton NMR observation of [3-^{13}C]lactate in stroke after [1-^{13}C]glucose infusion. Magn Reson Med 21:302–307

Rothman DL, Novotny EJ, Shulman GI, Howseman AM, Petroff OA, Mason G, Nixon T, Hanstock CC, Prichard JW, Shulman RG (1992) ^{1}H-[^{13}C] NMR measurements of [4-^{13}C]glutamate turnover in human brain. Proc Natl Acad Sci USA 89:9603–9606

Sato E, Inoue A, Kurokawa T, Ishibashi S (1994) Early changes in glucose metabolism in the cerebrum of senescence accelerated mouse: involvement of glucose transporter. Brain Res 637:133–138

Schousboe A, Wu JY, Roberts E (1973) Purification and characterization of the 4-aminobutyrate–2, ketoglutarate transaminase from mouse brain. Biochemistry 12:2868–2873

Schousboe A, Meier E, Drejer J, Hertz L (1989) Preparation of primary cultures of mouse (rat) cerebellar granule cells. In: A dissection and tissue culture manual for the nervous system. Alan R. Liss, New York, pp 183–186

Schroeder H, Becker A, Schroeder U, Hoellt V (1999) ^{3}H-L-glutamate binding and 3 H-D-aspartate release from hippocampal tissue during the development of pentylenetetrazole kindling in rats. Pharmacol Biochem Behav 62:349–352

Schwartz-Bloom RD, Sah R (2001) gamma-Aminobutyric acid(A) neurotransmission and cerebral ischemia. J Neurochem 77:353–371

Serres S, Bouyer JJ, Bezancon E, Canioni P, Merle M (2003) Involvement of brain lactate in neuronal metabolism. NMR Biomed 16:430–439

Serres S, Bezancon E, Franconi JM, Merle M (2004) Ex Vivo Analysis of Lactate and Glucose Metabolism in the Rat Brain under Different States of Depressed Activity. J Biol Chem 279:47881–47889

Shank RP, Leo GC, Zielke HR (1993) Cerebral metabolic compartmentation as revealed by nuclear magnetic resonance analysis of D-[1-^{13}C]glucose metabolism. J Neurochem 61:315–323

Shic F, Ross B (2003) Automated data processing of [^{1}H-decoupled] ^{13}C MR spectra acquired from human brain in vivo. J Magn Reson 162:259–268

Sibson NR, Dhankhar A, Mason GF, Behar KL, Rothman DL, Shulman RG (1997) In vivo ^{13}C NMR measurements of cerebral glutamine synthesis as evidence for glutamate-glutamine cycling. Proc Natl Acad Sci USA 94:2699–2704

Sibson NR, Mason GF, Shen J, Cline GW, Herskovits AZ, Wall JE, Behar KL, Rothman DL, Shulman RG (2001) In vivo ^{13}C NMR measurement of neurotransmitter glutamate cycling, anaplerosis and TCA cycle flux in rat brain during. J Neurochem 2001(76):975–989

Sommer W (1880) Erkrankung des Ammonshorns als aetiologisches Moment der Epilepsie. Arch Psychiatr Nervenkr 10:631–675

Sonnewald U, Rae C (2010) Pyruvate carboxylation in different model systems studied by ^{13}C MRS. Neurochem Res 35(12):1916–21

Sonnewald U, Westergaard N, Krane J, Unsgard G, Petersen SB, Schousboe A (1991) First direct demonstration of preferential release of citrate from astrocytes using [^{13}C]NMR spectroscopy of cultured neurons and astrocytes. Neurosci Lett 128:235–239

Sonnewald U, Petersen SB, Krane J, Westergaard N, Schousboe A (1992) ^{1}H NMR study of cortex neurons and cerebellar granule cells on microcarriers and their PCA extracts: lactate production under hypoxia. Magn Reson Med 23:166–171

Sonnewald U, Westergaard N, Petersen SB, Unsgard G, Schousboe A (1993a) Metabolism of [U-^{13}C]glutamate in astrocytes studied by ^{13}C NMR spectroscopy: incorporation of more label into lactate than into glutamine demonstrates the importance of the tricarboxylic acid cycle. J Neurochem 61:1179–1182

Sonnewald U, Westergaard N, Schousboe A, Svendsen JS, Unsgard G, Petersen SB (1993b) Direct demonstration by [^{13}C]NMR spectroscopy that glutamine from astrocytes is a precursor for GABA synthesis in neurons. Neurochem Int 22:19–29

Sonnewald U, White LR, Odegard E, Westergaard N, Bakken IJ, Aasly J, Unsgard G, Schousboe A (1996) MRS study of glutamate metabolism in cultured neurons/glia. Neurochem Res 21:987–993

Sonnewald U, Olstad E, Qu H, Babot Z, Cristofòfol R, Suñol C, Schousboe A, Waagepetersen H (2004) First direct demonstration of extensive GABA synthesis in mouse cerebellar neuronal cultures. J Neurochem 91:796–803

Sonnewald U, Schousboe A, Waagepetersen HW (2011) ^{13}C NMR spectroscopy and mass spectrometry analysis of intermediary metabolism in cultured neural cells in Neuromethods. W Walz (Series ed), Aschner M (Volume Eds), Sunol C, Price A. Springer Science-Business Media, New York, pp 403–414, ISBN 978-1-61779-076-8, http://issuu.com/asartehutiimhotep/docs/cell-culturing-techniques

Stephen LJ, Brodie MJ (2000) Epilepsy in elderly people. Lancet 355:1441–1446

Swanson RA, Shiraishi K, Morton MT, Sharp FR (1990) Methionine sulfoximine reduces cortical infarct size in rats after middle cerebral artery occlusion. Stroke 21:322–327

Sze PY (1979) L-Glutamate decarboxylase. Adv Exp Med Biol 123:59–78

Takagi K, Ginsberg MD, Globus MY, Dietrich WD, Martinez E, Kraydieh S, Busto R (1993) Changes in amino acid neurotransmitters and cerebral blood flow in the ischemic penumbral region following middle cerebral artery occlusion in the rat: correlation with histopathology. J Cereb Blood Flow Metab 13:575–585

Taylor A, McLean M, Morris P, Bachelard H (1996) Approaches to studies on neuronal/glial relationships by ^{13}C-MRS analysis. Dev Neurosci 18:434–442

Tirilazad International Steering Committee (2000) Tirilazad mesylate in acute ischemic stroke: A systematic review. Stroke 31:2257–2265

Torp R, Arvin B, Le PE, Chapman AG, Ottersen OP, Meldrum BS (1993) Effect of ischaemia and reperfusion on the extra- and intracellular distribution of glutamate, glutamine, aspartate and GABA in the rat hippocampus, with a note on the effect of the sodium channel blocker BW1003C87. Exp Brain Res 96:365–376

Tyson RL, Gallagher C, Sutherland GR (2003) ^{13}C-Labeled substrates and the cerebral metabolic compartmentalization of acetate and lactate. Brain Res 992:43–52

Vrba R (1962) Glucose metabolism in rat brain in vivo. Nature 195:663–665

Waagepetersen HS, Bakken IJ, Larsson OM, Sonnewald U, Schousboe A (1998) Metabolism of lactate in cultured GABAergic neurons studied by ^{13}C nuclear magnetic resonance spectroscopy. J Cereb Blood Flow Metab 18:109–117

Waagepetersen HS, Qu H, Schousboe A, Sonnewald U (2001) Elucidation of the quantitative significance of pyruvate carboxylation in cultured cerebellar neurons and astrocytes. J Neurosci Res 66:763–770

Waagepetersen H, Melø T, Schousboe A, Sonnewald U (2004) Homeostasis of neuroactive amino acids in cultured cerebellar and neocortical neurons is influenced by environmental cues. J Neurosci Res (in press)

Waagepetersen HS, Qu H, Sonnewald U, Shimamoto K, Schousboe A (2005a) Role of glutamine and neuronal glutamate uptake in glutamate homeostasis and synthesis during vesicular release in cultured glutamatergic neurons. Neurochem Int 47:92–102

Waagepetersen HS, Melø TM, Schousboe A, Sonnewald U (2005b) Homeostasis of neuroactive amino acids in cultured cerebellar and neocortical neurons is influenced by environmental cues. J Neurosci Res 79:97–105

Waagepetersen HS, Qu H, Sonnewald U, Schousboe A (2009) Energy and amino acid neurotransmitter metabolism in astrocytes. In: Parpura V, Haydon PG, (eds), Astrocytes in (patho)physiology of the nervous system, Springer, Boston, pp 177–200

Waniewski RA, Martin DL (1998) Preferential utilization of acetate by astrocytes is attributable to transport. J Neurosci 18:5225–5233

Westergaard N, Fosmark H, Schousboe A (1991a) Metabolism and release of glutamate in cerebellar granule cells cocultured with astrocytes from cerebellum or cerebral cortex. J Neurochem 56:59–66

Westergaard N, Sonnewald U, Petersen SB, Schousboe A (1991b) Characterization of microcarrier cultures of neurons and astrocytes from cerebral cortex and cerebellum. Neurochem Res 16:919–923

Westergaard N, Larsson OM, Jensen B, Schousboe A (1992) Synthesis and release of GABA in cerebral cortical neurons co-cultured with astrocytes from cerebral cortex or cerebellum. Neurochem Int 20:567–575

Westergaard N, Drejer J, Schousboe A, Sonnewald U (1996) Evaluation of the importance of transamination versus deamination in astrocytic metabolism of [U-^{13}C]glutamate. Glia 17:160–168

Whitehead KJ, Rose S, Jenner P (2004) Halothane anesthesia affects NMDA-stimulated cholinergic and GABAergic modulation of striatal dopamine efflux and metabolism in the rat in vivo. Neurochem Res 29:835–842

Chapter 9
Overview of Positron Emission Tomography (PET) and Single Photon Emission Computed Tomography (SPECT) Methodologies

Paul Cumming and Guido Böning

Abstract Molecular imaging with single photon emission computed tomography (SPECT) and positron emission tomography (PET) has revolutionized neuroscience. The external detection of radioactively labeled tracers allows the quantitative analysis of fundamental physiological processes in living brain of experimental animals, and human subjects. For molecular imaging, particular radioisotopes are incorporated into molecules (tracers) intended to serve as markers for cerebral blood flow, blood brain barrier physiology, energy metabolism, neurotransmitter synthesis, and neurotransmitter receptors. These tracers are administered to the subject by inhalation or intravenous injection, and swiftly delivered to brain in the arterial blood. During the processes of partitioning of the tracer across the blood brain barrier and its binding to specific molecular targets in brain, decay events are recorded by the tomograph and reconstructed into an image, or source map. After registration of the map to a brain template, the radioactivity concentration is assigned to specified anatomic regions. Especially in PET, temporal changes in the brain concentration of radioactivity can be analyzed according to principles of compartmental analysis, so as to evaluate the step-wise processes of tracer uptake and entrapment in brain. The goal of this endeavour is to extract from the images physiological information about the relatively unperturbed brain, which can inform us about the processes underlying cognition, healthy aging, and diseases of the central nervous system. However the interpretation of PET and SPECT images is subject to caveats dictated by factors such as the limited spatial resolution of the instruments, and the validity of the kinetic analyses, which can only approximate the true complexity of biological processes.

Keywords Image analysis • Kinetics • PET • SPECT • Tracers

P. Cumming, Ph.D. (✉) • G. Böning, Ph.D.
Department of Nuclear Medicine, Ludwig Maximilian's University of Munich,
Marchioninistrasse 15, 81377 Munich, Germany
e-mail: paul.k.cumming@gmail.com; Guido.Boening@med.uni-muenchen.de

I.-Y. Choi and R. Gruetter (eds.), *Neural Metabolism In Vivo*,
Advances in Neurobiology 4, DOI 10.1007/978-1-4614-1788-0_9,

9.1 General Remarks

Non-invasive molecular and physiological imaging of the living brain emerged as a useful tool for neuroscience research in the 1970s, with the advent of single photon (SPECT) and positron emission tomography (PET). These allied technologies provide images of the radioactivity concentration in a slice or reconstructed 3D volume of the living organism, following administration of a radioactive gas or radiopharmaceutical intended to interact with some physiological process or entity. The radiopharmaceuticals are properly known as tracers when their mass concentration is so low as to have little or no physiological action in the organism. The mapping in brain of an externally detected tracer is a form of autoradiography, where the particular aspect of physiology or pharmacology to be investigated is defined by the biochemical properties of the tracer substance. Thus, tracers can be designed to probe cerebral blood flow (CBF), blood brain barrier (BBB) function, energy metabolism, or receptor pharmacology. The quantitation of these processes by molecular imaging is firmly based on kinetics, which might be considered an extension or special case of the first law of thermodynamics; the brain and its circulation are considered to be a closed system, such that the radioactivity measured in brain is explicable as the outcome of relatively simple physiological processes, acting upon the radioactivity delivered to brain in the arterial blood. Molecular imaging thus differs from classical physiological assays in that the tracer concentration in the reaction medium (here, the brain) is not generally held constant by the experimentalist, but changes continuously according to the partitioning of the tracer between blood and brain, and according to metabolism and elimination of the tracer in peripheral tissues.

Molecular imaging by SPECT and PET is a formidably complex undertaking, entailing the production of radiopharmaceuticals, emission recording and data reconstruction, as well as physiological modelling of the observed signal. Among the most successful neuroscience applications of SPECT and PET include the mapping of CBF, the assay of cerebral energy metabolism using radioactive oxygen or the synthetic glucose analogue [^{18}F]-fluorodeoxyglucose ([^{18}F]-FDG; Chapter 1.1.2), and the imaging of a plethora of neuroreceptors and transporters (Chapter 1.1.3). This chapter outlines the physics underlying tracer imaging, and sketches out the procedures required for image reconstruction, and physiological modelling.

9.2 The Radiation Physics of Molecular Imaging

SPECT and PET are based on the external detection of high energy photons emitted from within the living organism. The tracer is a molecule entering into a particular physiological process, while revealing this interaction by virtue of containing within its structure a synthetic radionuclide, which is the actual source of the radiation to be detected. In general, unstable radionuclides have characteristic modes of spontaneous decay, which can entail emission of a helium nucleus (α), electron ($\beta-$), positron ($\beta+$), or high energy photon (γ), among other particles. These different emissions arise from distinct nuclear decay processes, and have distinct interactions with living organisms, which determine their toxicity, or their usefulness in therapeutics or medical imaging. Thus, α-particles can be enormously energetic, but (due to their charge) travel only short distances through water, or living tissue, such that they are unsuited for autoradiography *in vivo*. The β-particle, commonly known as the electron, arises from conversion of an excess neutron to a proton, as in the decays of Tritium ([^{3}H]) or [^{14}C] (Carbon); it has sufficient energy to interact with photographic emulsions or other detection media, but is unsuited for external detection. The γ-rays (photons) can arise in association with α or β-decays, or in the relaxation of an excited nuclear state; typical γ-emissions with energies of a few hundred keV can travel a considerable distance through solid matter before being absorbed by detectors, as in SPECT imaging. The β+, also known as the positron, is an antimatter electron produced by decay of certain neutron-deficient nuclei. Soon after its formation, the positron encounters an ordinary electron, an interaction which ultimately proves fatal for both particles; the electron-positron pair mutually annihilates and its entire rest mass energy, defined by the famous Einstein equation, is converted into a perfectly matched pair of γ-photons, each with energy very nearly equal to 511 keV. These photons flee the site of annihilation at the speed of light, and at very nearly opposite directions, and are detected along a path which is known as the line of response, in the context of PET scanning. As such, the β+ particle makes itself visible through the coincident activation of two detectors in a circular array by a pair of photons, which is attributed to a single decay event somewhere on the line of response.

The earliest γ-cameras, which were developed in the 1950s, recorded planar images, by passing a detector in a raster-like manner over the subject. Since the radiation field emanating from the subject extends in all directions, planar imagers have heavy lead collimator shielding of the detector, so as to admit only a narrow beam of photons. The detector itself consists of a scintillating crystal such as sodium iodide, sitting upon a photoelectron multiplier, which generates an electrical current proportional the number of photons captured in a given time interval. Despite collimator shielding, uncertainty about the origin of a given photon arises from simple geometry; collimator holes can be made narrow so as to admit photons from a more restricted range of angles, but at the penalty of decreased sensitivity. Furthermore, planar imaging is hardly suited to the needs of neuroscience, since spatial information is lost due to projection of the object onto a plane.

With respect to the history of brain SPECT, the first important radionuclide was [^{133}Xe] (Xenon), an inert gas used for early studies of brain perfusion. It decays to [^{133}Cs] (Cesium), yielding β- and γ-emissions (Table 9.1). Given that Xenon is an inert gas, it cannot be integrated into radiopharmaceuticals, but can (upon inhalation) serve only as a CBF tracer, to which end it was admirably suited due to its high solubility in the lipid-rich environment of the brain. However, use of this relatively long-lived radioactive gas presented problems related to safety and radiation dosimetry, such that [^{133}Xe] was soon supplanted by other radionuclides. Important among these isotopes is [^{131}I] (Iodine) which also yields β- and γ-emissions (Table 9.1). Given its characteristic decay, this radionuclide can serve a dual purpose, with the

Table 9.1 Isotopes used for SPECT and PET imaging. In the case of the PET radionuclides, the mean kinetic energy of the positron, and its corresponding mean free path prior to annihilation are indicated

	Half-life	Main emissions
SPECT		
[^{133}Xe]	5 days	β- at 346 keV γ at 81 keV
[^{131}I]	8 days	β- at 606 keV γ at 364 keV
[^{123}I]	13 h	γ at 27 and 159 keV
[^{99m}Tc]	6 h	γ at 140 keV (monochromatic)
PET		
[^{11}C]: 0.33 MeV, 0.3 mm	20 min	2 × monochromatic γ at 512 keV
[^{15}O]: 0.7 MeV, 1.5 mm	2 min	
[^{18}F]: 0.2 MeV, 0.2 mm	110 min	

β-emission imparting toxicity in radiotherapy for tumours, and the γ-emission serving for detection by SPECT. Thus, just for example, SPECT monitoring of patients under treatment with and [^{131}I]-labelled antibody can reveal regression of lymphomas, and (in conjunction with CT for anatomic delineation), allow precise calculation of absorbed dose of radioactivity in the tumours (Dewaraja et al. 2009). However, the relatively high energy of the γ-emission from [^{131}I] is unfavourable for imaging with collimators, compared to [^{123}I], which has predominant γ-emissions of rather lower energy (Table 9.1). This imparts to [^{131}I] relatively low toxicity, which is especially an issue give the extensive accumulation of Iodine in the thyroid glands.

Radioiodine has been historically important in SPECT imaging, not just for thyroid imaging of [^{123}I]-iodine uptake, but also for neuroscience studies; the chemistry of Iodine lends itself for incorporation into radiopharmaceuticals, many of which have proven invaluable for neuroscience research. One of the first of these was [^{123}I]-IBZM, a benzamide antagonist of brain dopamine D2 receptors on striatal projection neurons (Kung et al. 1990), Its binding in living brain occurs in competition with endogenous dopamine in the interstitial fluid. Based on this property, pharmacological challenge with amphetamine reduces tracer binding in living striatum. This paradigm has revealed elevated releasable dopamine in striatum of patients with schizophrenia (Abi-Dargham et al. 1998), consistent with the hypothesis of hyperactivity of dopamine in that disease. Conversely, pharmacologically-evoked reductions in [^{123}I]-IBZM binding can be used to estimate occupancy at D2 receptors by antipsychotic medications (Meisenzahl et al. 2008), which is a key issue is defining the trade-off between clinical efficacy and extrapyramidal side effects.

The integrity of presynaptic dopamine fibres can be probed in SPECT studies with the cocaine analogue [^{123}I]β-CIT and its fluoro-alkyl derivatives, which bind with varying specificity to the plasma membrane transporters for dopamine, serotonin, and noradrenaline (Okada et al. 1998); [^{123}I]-FP-CIT, now known as

DATSCAN, has found favor due to its relative specificity for dopamine transporters, and its relatively fast kinetics (Booij et al. 1998), such that a steady state is achieved in the course of a few hours following tracer injection. In practice, a static SPECT recording is obtained at a time when steady-state binding prevails in brain. In Europe, DATSCAN is widely employed for routine diagnosis of nigrostriatal degeneration, as in Parkinson's disease, for which it is of comparable sensitivity to [^{18}F]-fluorodopa-PET (Eshuis et al. 2009), to be discussed below. Results of a study with serial DATSCAN investigations has suggested that Parkinson's disease may be over-diagnosed in patients with ambiguous clinical presentation (Marshall et al. 2009). However, DATSCAN is not always adequate for differential diagnosis of Parkinson's disease and other basal ganglia disorders; combined use of DATSCAN with [^{123}I]-metaiodobenzylguanidine ([^{123}I]-MIGB) scintigraphy, which reveals the density of sympathetic innervation of the myocardium, can help distinguish Parkinson's disease from mixed or essential tremor cases (Novellino et al. 2009). A recent DATSCAN study has revealed moderately reduced abundance of dopamine transporters in striatum of never-treated patients with schizophrenia (Mateos et al. 2007), which seems somewhat at odds with the finding of increased effects of amphetamine, noted above. The [^{123}I]-SPECT tracers targeting pre- and post-synaptic elements of dopaminergic pathways present advantages over the analogous PET tracers with respect to ease of analysis, and cost-effectiveness, which are important preconditions for use in prospective studies of disease progression and interventions. However, the generally lower spatial resolution of SPECT, and the imperfect pharmacological selectivity of, for example, DATSCAN, presents certain limitations to its use.

While well-established for pre- and post-synaptic markers of brain dopamine, SPECT with [^{123}I]-tracers is relatively less developed for investigations of serotonergic systems. The serotonin transporter component of DATSCAN binding can be pharmacologically dissected in SPECT studies preceded by treatment with a selective serotonin reuptake inhibitor (van der Wee et al. 2008). Alternately, serotonin-selective transporter ligands such as [^{123}I]-ADAM are now available (Frokjaer et al. 2008); SPECT studies with this tracer have revealed increased serotonin transporters in the brainstem of subjects with borderline personality disorder (Koch et al. 2007), and have shown the occupancy at serotonin transporters in depressed patients under treatment with citalopram (Herold et al. 2006). SPECT tracers for serotonin receptors, as opposed to presynaptic serotonin transporters, are not yet widely available.

The metastable nuclear isomer [^{99m}Tc] (Technetium) presents an alternative to SPECT imaging with [^{123}I]-Iodine. Unlike the preceding radionuclides, it is typically generated at the imaging centre from [^{99}Mo] (Molybdenum), which is adsorbed as molybdate (MoO_4^{2-}) to an ion exchange column, where it decays with a half-life of 66 h. Thus, a single charging of the column with Molybdenum can suffice for several weeks. [^{99m}Tc] undergoes a nuclear isomeric transition with a half-life of 6 h, yielding ordinary [^{99}Tc] and a monochromatic γ-emission of moderate energy. The [^{99m}Tc]-pertechnetate (TcO_4^-) formed in the column is readily eluted on the basis of its lower chemical charge. When then prepared as a chemical complex with hemamethylpropylamine oxime (HMPAO), [^{99m}Tc] is suitable for SPECT studies of CBF, as discussed below. This isotope has also been important in neuroscience and clinical research for SPECT imaging of dopamine transporters with [^{99m}Tc]-TRODAT-1 (Kung et al. 1996), which is more widely used in the United States than is DATSCAN for detecting nigorstriatal degeneration.

The moderately long half-life of SPECT isotopes favors their use at wide-spread imaging centres; a single delivery of isotope or radiopharmaceutical may suffice for 2 weeks or longer. However, PET imaging presents certain advantages over SPECT with respect of sensitivity and spatial resolution, arising from the characteristic double γ-emission derived from positron annihilation. To date, only three radionuclides have found wide use in PET studies of brain: [^{11}C] (Carbon), [^{15}O] (Oxygen), and [^{18}F] (Fluorine). All three radionuclides are produced by bombardment of a target with a beam of high energy particles, usually produced in a cyclotron. [^{11}C] is produced by proton bombardment of the stable Nitrogen isotope ([^{14}N]), and decays with a half-life of 20 min, [^{15}O] is produced by deuteron bombardment of [^{14}N], and decays with a half-life of 2 min, and [^{18}F] is produced by bombardment of a stable Oxygen isotope ([^{18}O]) and decays with a half-life of 110 min. Given these half-lives, only [^{18}F] is suitable for same-day delivery from a central cyclotron facility to the site of use; the requirement for a local cyclotron/radiochemistry facility imposes certain practical limitations on the widespread use of [^{11}C] and [^{15}O], despite their very great utility, and the great diversity of [^{11}C]-tracers for brain PET studies.

9.3 Spatial Resolution

SPECT and PET apparatus consist of arrays of detectors, either arranged in rings, or in blocks rotating about the subject over a range of angles. For SPECT, spatial information about the source of the γ-rays is obtained through the use of lead collimators, which restrict the detection of γ-emissions along a particular line of response defined by the geometry of the collimator. As noted above, this angular information is obtained at the expense of loss of sensitivity, since only select photons can arrive at the detector crystals. For this reason, routine SPECT scans frequently consist of a single, rather lengthy time-frame, recorded during an interval when the distribution and binding of tracer in the target organ is at equilibrium. The greater sensitivity of PET lends itself more readily for dynamic acquisitions, as required for pharmacokinetic analysis. Roughly speaking, the best attainable spatial resolution in molecular imaging is determined by the number and angular size of detector elements; current clinical SPECT apparatus afford spatial resolution of less than 3 mm, or even 1 mm, through the use of very restrictive pinhole collimators, albeit with a considerable penalty in sensitivity arising from the use of very restrictive. The characteristic kinetic energies of positron emission influence the best attainable spatial resolution for PET, since uncertainty about the source of annihilation is determined by the mean distance travelled prior to annihilation, as noted in Table 9.1. The most recent PET equipment affords spatial resolution approaching 2 mm, or even 1 mm in the case of dedicated microPET for animal studies; reconstructed images from cylindrical phantoms of various diameter show the superior resolution obtained by microPET (Fig. 9.1a), which is optimal for the case of [^{18}F] due to the lower kinetic energy of its positron emission.

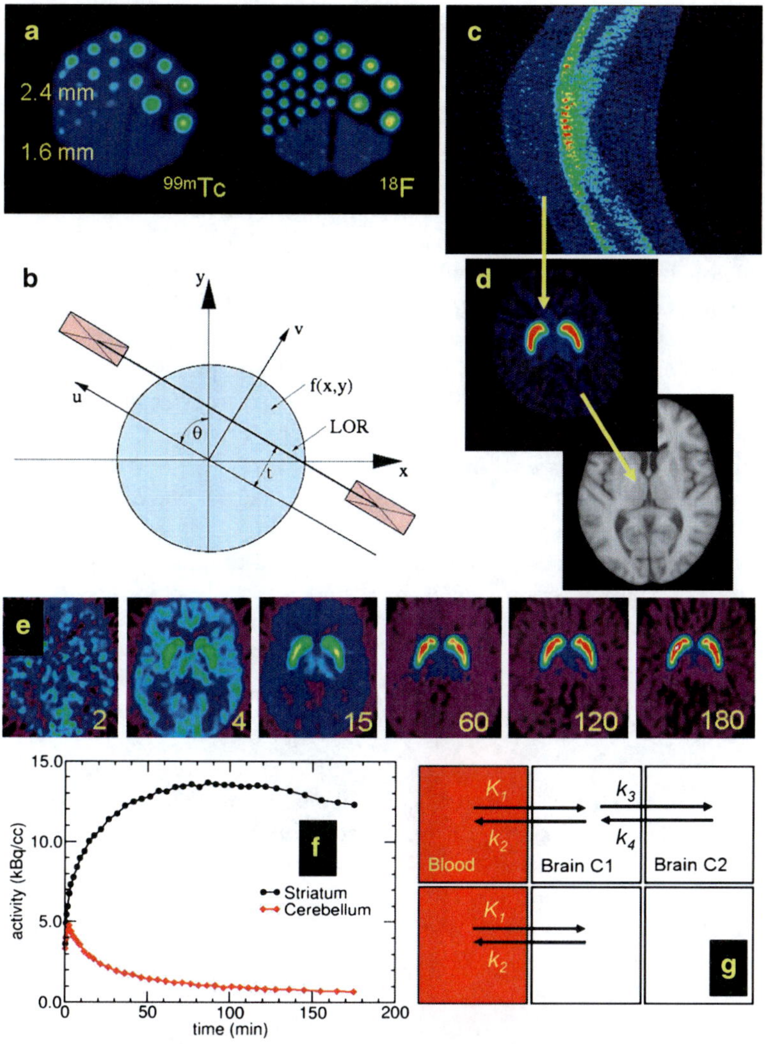

Fig. 9.1 (**a**) Reconstructed cross-sectional images of cylindrical phantom sources containing the SPECT isotope [^{99m}Tc] (*left*) and the PET isotope [^{18}F] (*right*). (**b**) The geometry inherent in the Radon transform and the process of back projection, (**c**) a sinogram for [^{18}F]-fallypride in human brain, and (**d**) the corresponding reconstructed emission image in native space, prior to spatial normalization to an MR-based standard brain template (*grey* scale image). (**e**) Representative frames from a dynamic emission recording with [^{18}F]-fallypride, showing initial tracer deliver to brain followed by the emergence of a heterogeneous binding pattern during the 180 min after tracer injection. (**f**) Time radioactivity curves extracted from the same [^{18}F]-fallypride recording, using ROIs placed over the striatum and cerebellum. (**g**) A schematic diagram depicting the processes determining the temporal changes in radioactivity concentration in a given brain region. The *lower part* of the figure represents the circumstance in a non-binding region, where tracer influx to brain is determined by the magnitude of K_1 (ml g^{-1} min^{-1}) acting about the instantaneous blood concentration, and washout from brain is expressed by the fractional rate constant k_2 (min^{-1}). The *upper part* of the figure represents the case for a brain region in which a radioligand is irreversibly trapped (k_3; min^{-1}), or dissociates from the bound state (k_4; min^{-1}), as in the case of [^{18}F]-fallypride

It is almost axiomatic that neuroscience researchers will chase brain structures of smaller size than afforded by the resolution of their apparatus. A case in point is afforded by the basal ganglia; whereas the caudate and putamen appeared as a single blob in early molecular imaging studies of dopamine markers, researchers now endeavor to investigate dopamine in structures as small as the amygdala or nucleus accumbens. Resolution has important effects on signal recovery, such that it must be conceded that PET and SPECT are really semiquantitative techniques. The loss of signal from a hot brain region to a cold surround, and the intrusion of signal in the target region from adjacent regions constitute the partial volume effects (PVE). Knowing the point-spread function of the detector, which is equivalent to the spatial resolution, it is possible to back-calculate the true radioactivity distribution giving rise to the blurred image. Several methods exist for partial volume correction (Rousset and Zaidi 2006), in one of which a geometric transfer matrix is applied to a segmentation of the emission image into a set of regions assumed to have uniform radioactivity, i.e. cerebrospinal fluid, white matter, grey matter, and specific nuclei such as the caudate and putamen. These procedures are prone to noise magnification, in the manner of a trade-off in favor of accuracy, at the cost of precision. However, partial volume correction may be required to make strong claims about altered tracer uptake in brains which have undergone atrophy, such as occurs in healthy ageing and Alzheimer's dementia.

9.4 Image Reconstruction

For both SPECT and PET, the trick is to reconstruct a source map based on the activation of the detector array. This is accomplished through application of a mathematical transform introduced by the mathematician J. Radon (1917), where r is the analytical description of the 2D measurement of a physical property observed along a line at rotation angle θ and distance t with respect to the center of the field-of-view

$$r(\theta,t) = \int_{LOR} f(x,y)d\bar{l}$$

In both PET and SPECT, f(x,y) in the above equation represent the distribution of the γ-emission from within the patient. The data recorded in a given interval is stored as a matrix, known as the sinogram, which is the discretized representation of the Radon transformed tracer distribution. The sinogram's horizontal coordinate represents parameter t and the vertical coordinate represents the projection angle θ (Fig. 9.1b). The term sinogram refers to the transformation of image space in a sinusoidal trace within the sinogram matrix; the eye can deduce little about the source image from inspection of a typical PET sinogram recorded for the dopamine receptor ligand [^{18}F]-fallypride (Fig. 9.1c). The challenge in tomographic 2D image reconstruction is to approximate the original tracer distribution by estimating f(x,y) from the measured sinogram data. The most famous method for this calculation is filtered back projection algorithm (Budinger et al. 1979), derived from methods developed for radioastronomy, and subsequently extended for microscopic analysis. In filtered back projection, each projection r is Fourier transformed (F{..}), multiplied with a high pass filter H, inverse Fourier (F^{-1}{..}) transformed, and finally back-projected into image space. After performing this calculation for all projection angles θ, these back projections are summed up to form the final image estimate:

$$f(x,y) = \frac{1}{2\pi}\int_0^{\pi} F^{-1}\{F\{r(\theta,t)\}\cdot H\}d\theta.$$

The reconstructed emission image is influenced by some absorption of energy during the passage of γ-rays through the brain, skull, and skin. Therefore, "quantitative" analysis of emission data requires an additional step, known as the attenuation scan. Here, an external X-ray source of known intensity is rotated about the head of the subject, and the attenuation factor is calculated for the emission signal along each line of response in the sinogram.

Although filtered back projection is suitable for routine reconstructions, it is necessarily applicable only for 2D (slice-wise) reconstruction. Considering the lines of response from outside the plane of a given ring of detectors (there are dozens of adjacent rings in modern PET scanners) makes possible more sensitive and higher resolution 3D reconstructions. These are obtained through an iterative reconstruction algorithm, which might be compared to a brute force solution of the mathematics game Sudoku; here, an initial guess of the source map is adjusted through a series of many iterations until it converges to the best possible estimation of the source map. So far, iterative 3D reconstruction has found its main use in research, because of the extensive computational demands arrising from the enormous number of image elements (voxels) and possible lines of response. In our experience, iterative 3D reconstruction of a single state emission frame takes more than 1 h on a dedicated processor. Thus, reconstructing a dynamic [^{18}F]-fallypride emission recording of 20 frames requires approximately 24 h of continuous computation.

However attained, the attenuation-corrected, reconstructed emission image from brain exists in the so-called "native space", which depends on the position and orientation of the head in the aperture of the tomograph (Fig. 9.1d). In general, this image is resampled into a standard coordinate system, so that results obtained on different days and in different individuals can be compared (Fig. 9.1f). This spatial normalization can sometime be obtained by automatic procedures, in which the PET or SPECT data are matched to an MR-based image or some other template in which brain anatomic regions are defined. This approach is based upon the self-similarity of human brains, such that some standard anatomic template can be assumed to serve as a Platonic ideal representation. Naturally, the use of a population-based anatomic atlas presents problems in cases with marked cerebral atrophy or pronounced structural abnormalities. The gold standard for anatomical interpretation of emission data is to register the emission image to the individual's brain MR image, which can be segmented into its component anatomical regions. These regions of interest (ROIs), or 3D volumes of interest (VOIs) serve as cookie-cutters, for extracting the radioactivity signal within the corresponding defined brain region.

9.5 Kinetics of Tracer Uptake

In the case of dynamic emission recordings, a sequence of static images reveals temporal changes in the distribution of radioactivity in brain. Figure 9.1e shows the time course during the minutes and hours following injection of the dopamine D2 receptor ligand [^{18}F]-fallypride. It can be appreciated that the early image frames show nearly homogeneous distribution of the tracer in brain, whereas frames after 15 min show the evolution of contrast between high and low binding regions. By placing anatomical VOIs on the dynamic image sequence, the time-activity curve (TAC) is calculated, showing the temporal changes in radioactivity concentration. In this example, [^{18}F]-fallypride TACs are presented for cerebellum, a non-binding region, and striatum, a region replete with dopamine receptors (Fig. 9.1f). It is the task of physiological modeling to convert these dynamic observations into an estimation of the local magnitude of a physiological process, such as CBF, the activity of an enzyme, or (in this case) the specific binding of radioligand to receptors.

Compartmental models interpret the brain TACs for a given tracer in terms of a finite set of physiological processes, as summarized in Fig. 9.1g. A tracer arriving in brain via arterial blood is presented with the opportunity to enter brain by diffusing across the capillary endothelium of brain, which is characterized by tight junctions between individual capillary cells, constituting the BBB. Whereas water, some gases and lipophilic molecules pass freely into brain, the BBB an absolute barrier to the free diffusion of proteins, and highly charged molecules. Glucose, amino acids, and certain other essential brain nutrients which would otherwise be excluded from brain gain privileged access through the agency of specific transporters resident in the BBB. Whether by simple or facilitated diffusion, the flux of tracer from blood to brain is defined by the product of the instantaneous concentration in arterial plasma, Ca(T), and the unidirectional blood brain clearance (K_l) which has units of CBF (ml g^{-1} min^{-1}). The absolute magnitude of K_l is calculated from the rate of accumulation of radioactivity in a given brain region, interpreted relative to the arterial input, which can be measured in serial blood samples collected following injection or inhalation of the tracer.

A perfect CBF tracer is one which is entirely extracted into brain, i.e. with a K_l equal to the magnitude of CBF, and furthermore is perfectly retained for the time course of the emission recording. As noted above, the radionuclide [^{133}Xe] meets these criteria. Administered by inhalation, it is carried in the arterial blood to the brain, where it is entrapped due to its high lipophilicity. Since its extraction into brain and subsequent retention is so extensive, a SPECT recording made some minutes after inhalation shows the pattern of CBF. However, as noted above, [^{133}Xe] was soon supplanted by injectable CBF tracers such as [^{99m}Tc]-HMPAO, mentioned above.

For this purpose, the [^{99m}Tc]-pertechnetate is reacted with reagents in kit form to produce the stable chemical complex, and purified for intravenous injection. Since [^{99m}Tc]-HMPAO is retained in brain for several hours, a static SPECT image reveals the pattern of CBF which had prevailed at the time of tracer injection some hours earlier. The mechanism for [^{99m}Tc]-HMPAO entrapment in brain is complex; Studies *in vitro* indicate that the preponderant cellular site of trapping is not in neurons, but in glia, in a manner related to the presence of glutathione (Zerarka et al. 2001); long term retention of the tracer in brain thus involves metabolic conversion of the tracer into a hydrophilic metabolite, with more constrained diffusion across plasma membranes. As such, [^{99m}Tc]-HMPAO falls short of perfection as a CBF tracer, since the metabolic dependence of its trapping in brain can sometimes lead to uncoupling from CBF. Indeed, using [^{133}Xe] as a gold standard, regions of hypofixation and hyperfixation of [^{99m}Tc]-HMPAO are evident in post acute infarction of the brain (Ogasawara et al. 2000). However, [^{99m}Tc]-HMPAO deposition in healthy brain tissue is under most circumstances closely coupled to CBF.

Nonetheless, the mechanistic considerations of [^{99m}Tc]-HMPAO deposition and trapping in brain raises the general problem of model identifiability. Almost all kinetic models entail an oversimplication of the real biological circumstances. These simplifications include considering the input measured from a peripheral artery as identical to the brain input, i.e. ignoring delay and dispersion of the bolus, ignoring plasma protein binding of the tracer or the presence of brain-penetrating tracer metabolites, neglecting the presence of additional binding sites in brain, and imperfectly valid assumptions about the reversibility of the binding process in brain. Reduction of the real biological complexity to a manageable number of factors or parameters means that quantitative findings with molecular imaging must always be issued with caveats.

Returning to the general kinetic model, the initial brain uptake of a BBB-permeable tracer is defined in formal kinetic terms only by delivery (K_1), which predominates during the interval of peak plasma concentration in the minutes following intravenous injection. The magnitude of K_1 is less than or equal to the magnitude of CBF, and the ratio of the two is defined as the *extraction fraction* (EF). In general, the EF of a given tracer molecule is determined by its solubility in lipids, although sugars and amino acids can avail themselves of facilitated diffusion via specific carriers, as noted above. Nonetheless, a number of PET tracers proved to have unexpectedly low EF; this is now understood to reflect the action of P-glycoprotein in the BBB, which actively (at the expense of ATP) extrudes its many substrates back across the abluminal membrane. Inhibitors of P-glycoprotein consequently enhance the delivery of certain tracers into brain (e.g. la Fougère et al. 2010).

As noted above, few tracers are irreversibly trapped in brain, but rather diffuse back into circulation at rate determined by a characteristic fractional rate constant (k_2; min^{-1}). The partitioning of [^{15}O]-water, and other tracers devoid of specific binding component, can be schematically represented as in the lower part of Fig. 9.1g with two arrows, describing the fluxes in and out of brain. For quantitative CBF studies with [^{15}O]-water, a series of emission frames is recorded during some 3 min, with concomitant sampling of arterial blood. In general, the two parameters are estimated independently by least squares fitting of the impulse response function relative to the arterial input, with K_l (CBF) being the main parameter of interest. Alternately, the distribution volume at equilibrium, defined as the ratio K_l/k_2, may be the physiological parameter of interest.

When a tracer binds in brain, the model is necessarily more complex, as represented in the upper part of Fig. 9.1g. Here, the trapping of free tracer in brain proceeds at a rate defined by k_3 (min^{-1}). This fractional rate constant may represent the activity of an enzyme, such as hexokinase, which transforms diffusible deoxyglucose into the non-diffusible phosphate (Sokoloff et al. 1977), a model which is extended to [^{18}F]-FDG studies of brain energy metabolism. Alternately, the trapping rate constant can express the activity of DOPA decarboxylase for catalyzing the transformation of diffusible [^{18}F]-fluorodopa into [^{18}F]-fluorodopamine, which is retained within synaptic vesicles of dopamine fibres (Kumakura and Cumming 2009). The rate constant k_3 can also represent the rate constant of association of a radioligand to a receptor; for example, the dopamine D_2 antagonist N-[^{11}C]-methylspiperone ([^{11}C]-NMSP) binds rapidly to its receptors, at a rate defined by the local magnitude of k3, which is proportional to the abundance of receptors (Wong et al. 1986). In a strict sense, none of these three examples are entirely reversible. Thus, [^{18}F]-FDG-phosphate is slowly dephosphorylated, thus returning [^{18}F]-FDG to the precursor compartment (Miyazawa et al. 1993). [^{18}F]-fluorodopamine is eventually decomposed to metabolites, which diffuse out of brain, thus deflecting the TAC downwards (Kumakura and Cumming 2009). Finally, [^{11}C]-NMSP binding eventually approaches an equilibrium due to its slow dissociation from receptors (Rosa-Neto et al. 2004). However, on the time-scale of 90 min PET recordings, these violations of model validity can generally be ignored.

In $K_l/k_2/k_3$ models, individual kinetic parameters are estimated by non-linear regression, or using a robust multi-linear representation of the differential equations defined by the model (Blomqvist, 1984). Alternately, the net influx of tracer to brain can be expressed as the macroparameter K ($ml\ g^{-1}\ min^{-1}$), the composite of the three micro-parameters described above. As a matter of convenience, the magnitude of K can be calculated by linear regression, knowing only the arterial input and the PET recording of uptake in brain (Patlak et al. 1983). In practice, the distribution volume of the tracer, defined as the ratio of brain to blood concentration at a given time, is plotted as a function of a transformation of the arterial input, with the linear regression slope corresponding to the magnitude of K. The product of this flux and the plasma concentration of endogenous dopamine is, roughly speaking, equal to the cerebral metabolic rate for glucose (CMRglc).

For tracers which are very rapidly entrapped in brain, i.e. with k_3 of large magnitude, the magnitude of K approaches the limit imposed by CBF; in other words, some irreversibly trapped tracers may be better suited as tracers of CBF than as tracers for the specific enzymatic process by which it is trapped in brain: this is the case for [^{99m}Tc]-HMPAO, described above. Another instance of this is presented by [^{11}C]-deprenyl, which binds so rapidly to the enzyme monoamine oxidase (MAO), such that its cerebral uptake is close to the upper limit established by tracer deliver, i.e. CBF. While it is consequently difficult to estimate MAO activity (expressed by the magnitude of k_3) with any degree of precision in PET studies using [^{11}C]-deprenyl, is afforded by deuteration of the tracer, which slows the forward trapping reaction to such an extent that k_3 can be separated from tracer delivery (Fowler et al. 1988). In general, the graphical analysis of tracer influx remains linear so long as the assumption of irreversible binding or entrapment is not violated. while this is absolutely true for [^{11}C]-deprenyl, which eventually forms a covalent bond with the MAO enzyme, the slope of influx plot will generally decline at late circulation times, as in the specific cases mentioned above.

Knowing the arterial input and the brain TAC, individual magnitudes of $K_1/k_2/k_3/k_4$ can be estimated for reversibly binding ligands, but with the serious risk of over-specification of the model; there are simply too many parameters at play to be able to obtain stable estimates based on noisy PET data, without imposing some physiological constraints. Since reversible radioligand binding is an equilibrium process; the number of parameters can be reduced by considering the equilibrium in terms of the ratio k_3/k_4, conventionally represented as being proportional to the binding potential (BP). This dimensionless term is proportional to the actual concentration of receptors (B_{max}) in a given brain region, relative to the affinity of the ligand for its binding site (K_D). In practice, stable estimation of BP is dictated by the half-life of the radioisotope relative to the kinetics of the process in question. In the case of [^{11}C]-tracers, the practical limit for a PET recording is usually 100 min, corresponding to five half-lives, whereupon the amount of radioactivity will have declined to only three percent of the original injected dose. Conversely, receptor ligands can be prepared with [^{18}F], which is permissive of much longer PET recordings, due to the half-life of 110 min. The time course of [^{18}F]-fallypride binding in human brain (Fig. 9.1e, f) is such that equilibrium estimates require recordings lasting 3 h (Vernaleken et al. 2008). While this requirement presents no difficulty with respect to sensitivity of detection, it tests the endurance of patients and volunteers, due to the need for prolonged immobility during the scanning session. Some head motion between frames is unavoidable, but can often be corrected by post-processing of the dynamic scan. Alternately, the long recording can be divided into early and late sessions. After a pause following the first part of the recording, the subject returns to the scanner for a second recording, which must subsequently be registered to the initial head position.

The models for tracer uptake and binding represented in Fig. 9.1g are based upon consideration of the arterial input, which is measured by invasive serial sampling. This entails some discomfort for the patients, and can be further complicated by the process of peripheral metabolism. Most PET tracers are substantially metabolized after an hour, usually to hydrophilic species which are unlikely to cross the BBB. Nonetheless, the total arterial blood radioactivity must be corrected for metabolism so as to calculate the "true input", frequently by means of high performance liquid chromatography analysis of selected plasma samples (Cumming et al. 1999). The plasma metabolites of most useful PET and SPECT tracers do not cross the BBB due to their hydrophilicity, but in the particular case of [^{18}F]-fluorodopa, the contribution of a plasma metabolite to the brain signal cannot be ignored (Kumakura and Cumming 2009). On close examination, many tracers have proven to yield some brain-penetrating metabolites, but these can safely be ignored if preclincal studies show that they constitute less than 10% of the brain radioactivity.

While suitable for research protocols, plasma metabolite correction is impractical for large scale clinical studies. Thus, reference tissue methods for the analysis of tracer binding can be very convenient. Here, a brain region devoid of specific binding must be identified, such that the dynamic PET recordings can be transformed into parametric maps of specific binding, without recourse to invasive arterial sampling. In essence, the radioactivity concentration measured in the brain compartment represented in the lower part of Fig. 9.1g serves as a surrogate for calculating the specific binding component defined in the upper part of the figure. The magnitude of BP can be calculated by reference tissue methods employing either linearization (Logan et al. 1990) or non-linear regression (Gunn et al. 1997). The trade-offs between precision and accuracy in the estimation of specific binding differs, such that the method best suited for absolute quantitation may be less sensitive to treatment-induced changes in ligand binding (Cumming et al. 2003). Naturally, the validity of reference tissue methods is predicated upon the assumption that there is no detectable specific binding component in the reference region; this is approximately true in cerebellum for many classes of neuroreceptors.

In summary, the kinetic analysis of the uptake of SPECT and PET tracers always risks the problem of overspecification, wherein the number of kinetic parameters defined in the model exceeds that which can be plausibly evaluated, knowing only the brain concentration of radioactivity, and also (perhaps) the arterial input. Models are always simplifications of the true complexity of biological processes; the challenge inherent in molecular imaging is to strike a balance between the demand for simplicity, and the need for physiological validity. Despite the caveats, modeling is a worthwhile endeavor, as no other present or forseeable *in vivo* imaging method has the process sensitivity or selectivity that is afforded by a good tracer.

References

Abi-Dargham A, Gil R, Krystal J, Baldwin RM, Seibyl JP, Bowers M, van Dyck CH, Charney DS, Innis RB, Laruelle M (1998) Increased striatal dopamine transmission in schizophrenia: confirmation in a second cohort. Am J Psychiatry 155(6):761–767

Blomqvist G (1984) On the construction of functional maps in positron emission tomography. J Cereb Blood Flow Metab 4(4):629–632

Booij J, Habraken JB, Bergmans P, Tissingh G, Winogrodzka A, Wolters EC, Janssen AG, Stoof JC, van Royen EA (1998) Imaging of dopamine transporters with iodine-123-FP-CIT SPECT in healthy controls and patients with Parkinson's disease. J Nucl Med 39(11):1879–1884

Budinger TF, Gullberg GT, Huesman RH (1979) Emission computed tomography. In: Herman GT (ed) Image reconstruction from projections. Springer, Berlin Heidelberg New York, pp 193–199

Cumming P, Yokoi F, Chen A, Deep P, Dagher A, Reutens D, Kapczinski F, Wong DF, Gjedde A (1999) Pharmacokinetics of tracers in human plasma during positron emission tomography. Synapse 34(2):124–134

Cumming P, Rosa-Neto P, Watanabe H, Smith D, Bender D, Clarke PB, Gjedde A (2003) Effects of acute nicotine on hemodynamics and binding of [11C]raclopride to dopamine D2,3 receptors in pig brain. Neuroimage 19(3):1127–1136

Dewaraja YK, Wilderman SJ, Koral KF, Kaminski MS, Avram AM (2009) Use of integrated SPECT/CT imaging for tumor dosimetry in I-131 radioimmunotherapy: a pilot patient study. Cancer Biother Radiopharm 24(4):417–426

Eshuis SA, Jager PL, Maguire RP, Jonkman S, Dierckx RA, Leenders KL (2009) Direct comparison of FP-CIT SPECT and F-DOPA PET in patients with Parkinson's disease and healthy controls. Eur J Nucl Med Mol Imaging 36(3):454–462

Fowler JS, Wolf AP, MacGregor RR, Dewey SL, Logan J, Schlyer DJ, Langstrom B (1988) Mechanistic positron emission tomography studies: demonstration of a deuterium isotope effect in the monoamine oxidase-catalyzed binding of [11C]L-deprenyl in living baboon brain. J Neurochem 51(5):1524–1534

Frokjaer VG, Pinborg LH, Madsen J, de Nijs R, Svarer C, Wagner A, Knudsen GM (2008) Evaluation of the Serotonin Transporter Ligand 123I-ADAM for SPECT Studies on Humans. J Nucl Med 49(2):247–254

Gunn RN, Lammertsma AA, Hume SP, Cunningham VJ (1997) Parametric imaging of ligand-receptor binding in PET using a simplified reference region model. Neuroimage 6(4):279–287

Herold N, Uebelhack K, Franke L, Amthauer H, Luedemann L, Bruhn H, Felix R, Uebelhack R, Plotkin M (2006) Imaging of serotonin transporters and its blockade by citalopram in patients with major depression using a novel SPECT ligand [123I]-ADAM. J Neural Transm 113(5):659–670

Koch W, Schaaff N, Popperl G, Mulert C, Juckel G, Reicherzer M, Ehmer-von Geiso C, Moller HJ, Hegerl U, Tatsch K, Pogarell O (2007) [I-123] ADAM and SPECT in patients with borderline personality disorder and healthy control subjects. J Psychiatry Neurosci 32(4):234–240

Kumakura Y, Cumming P (2009) PET studies of cerebral levodopa metabolism: a review of clinical findings and modeling approaches. Neuroscientist 15(6):635–650

Kung HF, Alavi A, Chang W, Kung MP, Keyes JW Jr, Velchik MG, Billings J, Pan S, Noto R, Rausch A et al (1990) In vivo SPECT imaging of CNS D-2 dopamine receptors: initial studies with iodine-123-IBZM in humans. J Nucl Med 31(5):573–579

Kung HF, Kim HJ, Kung MP, Meegalla SK, Plossl K, Lee HK (1996) Imaging of dopamine transporters in humans with technetium-99m TRODAT-1. Eur J Nucl Med 23(11):1527–1530

la Fougère C, Böning G, Bartmann H, Wängler B, Nowak S, Just T, Wagner E, Winter P, Rominger A, Förster S, Gildehaus FJ, Rosa-Neto P, Minuzzi L, Bartenstein P, Potschka H, Cumming P (2010) Uptake and binding of the serotonin 5-HT1A antagonist [18F]-MPPF in brain of rats: effects of the novel P-glycoprotein inhibitor tariquidar. Neuroimage. 49(2):1406–15

Logan J, Fowler JS, Volkow ND, Wolf AP, Dewey SL, Schlyer DJ, MacGregor RR, Hitzemann R, Bendriem B, Gatley SJ et al (1990) Graphical analysis of reversible radioligand binding from time-activity measurements applied to [N-11C-methyl]-(−)-cocaine PET studies in human subjects. J Cereb Blood Flow Metab 10(5):740–747

Marshall VL, Reininger CB, Marquardt M, Patterson J, Hadley DM, Oertel WH, Benamer HT, Kemp P, Burn D, Tolosa E, Kulisevsky J, Cunha L, Costa D, Booij J, Tatsch K, Chaudhuri KR, Ulm G, Pogarell O, Hoffken H, Gerstner A, Grosset DG (2009) Parkinson's disease is overdiagnosed clinically at baseline in diagnostically uncertain cases: a 3-year European multicenter study with repeat [123I]FP-CIT SPECT. Mov Disord 24(4):500–508

Mateos JJ, Lomena F, Parellada E, Mireia F, Fernandez-Egea E, Pavia J, Prats A, Pons F, Bernardo M (2007) Lower striatal dopamine transporter binding in neuroleptic-naive schizophrenic patients is not related to antipsychotic treatment but it suggests an illness trait. Psychopharmacology (Berl) 191(3):805–811

Meisenzahl EM, Schmitt G, Grunder G, Dresel S, Frodl T, la Fougere C, Scheuerecker J, Schwarz M, Boerner R, Stauss J, Hahn K, Moller HJ (2008) Striatal D2/D3 receptor occupancy, clinical response and side effects with amisulpride: an iodine-123-iodobenzamide SPET study. Pharmacopsychiatry 41(5):169–175

Miyazawa H, Osmont A, Petit-Taboué MC, Tillet I, Travère JM, Young AR, Barré L, MacKenzie ET, Baron JC (1993) Determination of 18F-fluoro-2-deoxy-D-glucose rate constants in the anesthetized baboon brain with dynamic positron tomography. J Neurosci Methods 50(3):263–72

Novellino F, Arabia G, Bagnato A, Cascini GL, Salsone M, Nicoletti G, Messina D, Morelli M, Paglionico S, Giofre L, Restuccia A, Torchia G, Condino F, Quattrone A (2009) Combined use of DAT-SPECT and cardiac MIBG scintigraphy in mixed tremors. Mov Disord 24(15): 2242–2248

Ogasawara K, Ogawa A, Koshu K, Konno H, Suzuki M, Yoshimoto T (2000) Hypofixation and hyperfixation of 99mTc-hexamethyl propyleneamine oxime in subacute cerebral infarction. J Nucl Med 41(5):795–799

Okada T, Fujita M, Shimada S, Sato K, Schloss P, Watanabe Y, Itoh Y, Tohyama M, Nishimura T (1998) Assessment of affinities of beta-CIT, beta-CIT-FE, and beta-CIT-FP for monoamine transporters permanently expressed in cell lines. Nucl Med Biol 25(1):53–58

Patlak CS, Blasberg RG, Fenstermacher JD (1983) Graphical evaluation of blood-to-brain transfer constants from multiple-time uptake data. J Cereb Blood Flow Metab 3(1):1–7

Rosa-Neto P, Gjedde A, Olsen AK, Jensen SB, Munk OL, Watanabe H, Cumming P (2004) MDMA-evoked changes in [11C]raclopride and [11C]NMSP binding in living pig brain. Synapse 53(4):222–33

Rousset OG, Zaidi H (2006) Correction for partial volume effects in emission tomography, *In* Quantitative Analysis in Nuclear Medicine Imaging, Habib Zaidi, Editor, Springer Verlag, Singapore, pp. 236–271

Sokoloff L, Reivich M, Kennedy C, Des Rosiers MH, Patlak CS, Pettigrew KD, Sakurada O, Shinohara M (1977) The [14C]deoxyglucose method for the measurement of local cerebral glucose utilization: theory, procedure, and normal values in the conscious and anesthetized albino rat. J Neurochem 28(5):897–916

van der Wee NJ, van Veen JF, Stevens H, van Vliet IM, van Rijk PP, Westenberg HG (2008) Increased serotonin and dopamine transporter binding in psychotropic medication-naive patients with generalized social anxiety disorder shown by 123I-beta-(4-iodophenyl)-tropane SPECT. J Nucl Med 49(5):757–763

Vernaleken I, Fellows C, Janouschek H, Brocheler A, Veselinovic T, Landvogt C, Boy C, Buchholz HG, Spreckelmeyer K, Bartenstein P, Cumming P, Hiemke C, Rosch F, Schafer W, Wong DF, Grunder G (2008) Striatal and extrastriatal D2/D3-receptor-binding properties of ziprasidone: a positron emission tomography study with [18F]Fallypride and [11C]raclopride (D2/D3-receptor occupancy of ziprasidone). J Clin Psychopharmacol 28(6):608–617

Wong DF, Gjedde A, Wagner HN Jr (1986) Quantification of neuroreceptors in the living human brain. I. Irreversible binding of ligands. J Cereb Blood Flow Metab 6(2):137–146

Zerarka S, Pellerin L, Slosman D, Magistretti PJ (2001) Astrocytes as a predominant cellular site of (99m)Tc-HMPAO retention. J Cereb Blood Flow Metab 21(4):456–468

Chapter 10
Fluorodeoxyglucose (FDG) Positron Emission Tomography (PET)

Andrei G. Vlassenko and Mark A. Mintun

Abstract Fluorodeoxyglucose labeled with [^{18}F] (FDG) is a chemical analog of glucose, where a hydrogen atom of a glucose molecule is replaced with a positron emitter [^{18}F]. Like glucose, FDG is metabolized by hexokinase during glycolysis in the cytosol of the cell; however, unlike glucose-6-phosphate, FDG-6-phosphate is not metabolized further but accumulates intracellularly thus allowing visualization and measurement of local metabolic activity. Relatively long-life of FDG as [^{18}F] compound (110 min) and ability to achieve reliable quantitative estimates of cerebral metabolic rate of glucose and distinguish local changes in FDG uptake with visual inspection and semi-quantitative measures make FDG the most commonly used tracer in PET research and clinical neuroimaging. Besides demonstrating high-quality images of the baseline state of brain activity, FDG also provides accurate measures of the magnitude of changes in brain metabolism in response to various physiological stimulations, pharmacological interventions, and functional tests.

Keywords Accuracy • Acetyl coenzyme A (acetyl CoA) • Activation • Adenosine-5-triphosphate (ATP) • Aerobic glycolysis • Aging • Alignment of images • Alzheimer's disease • Amino acids • Amyloid • Antiepileptic therapy • Astrocytes • Autoradiography • Beta • Brain • Cerebral blood flow (CBF) • Cerebral metabolic rate of glucose (CMRGlc) • Cerebral metabolic rate of oxygen ($CMRO_2$) • Compartment model • Computed tomography (CT) • Co-registration • Cytosol • Dementia • Dementia of Alzheimer's type (DAT) • Dementia with Lewy bodies • Deoxyglucose • Development • Energy metabolism • Epilepsy • Fluorine-18 • Fluorodeoxyglucose (FDG) • Functional brain imaging • Functional magnetic resonance imaging • Glia • Glioma

A.G. Vlassenko, M.D., Ph.D. (✉)
Department of Radiology, Washington University School of Medicine, 4525 Scott Ave, East Building, Campus Box 8225, St. Louis, MO 63110, USA
e-mail: andrei@npg.wustl.edu

M.A. Mintun, M.D.
Department of Radiology, Avid Radiopharmaceuticals, Philadelphia, PA, USA
e-mail: mintunm@mir.wustl.edu

I.-Y. Choi and R. Gruetter (eds.), *Neural Metabolism In Vivo*, Advances in Neurobiology 4, DOI 10.1007/978-1-4614-1788-0_10,

• Glucose • Glucose metabolism • Glucose oxidation • Glucose uptake • Glucose-6-phosphate • Glutamate • Glutamine • Glycolysis • Glycolytic enzymes • Hemiparkinsonism-hemiatrophy syndrome • Hexokinase • High grade tumor • Huntington disease • Hypermetabolism • Hypometabolism • Ketogenic diet • Krebs cycle • Lactate • Lipid biosynthesis • Low-grade tumor • Lumped constant • Magnetic resonance imaging (MRI) • Magnetic resonance spectroscopy (MRS) • Major depressive disorder • Metabolism • Mitochondria • Modeling • Multi-infarct dementia • NADPH • Neural activity • Neural plasticity • Neurodegeneration • Neurodegenerative diseases • Neuroimaging • Neurons • Obesity • Obsessive-compulsive disorder • Oxidative metabolism • Oxidative phosphorylation • Oxygen • Oxygen-to-glucose ratio • Parkinson's disease • Patlak graphical analysis • Pentose phosphate pathway • Photons • Physiological stimulation • Positron emission tomography (PET) • Pyruvate • Quantitative assessment • Radiation necrosis • Radiopharmaceuticals • recurrent tumor • Resolution • Response to therapy • Schizophrenia • Seizures • Semi-quantitative assessment • Sensitivity • Signaling • Single photon emission (computed) tomography (SPECT) • Small animal positron emission tomography • Specificity • Standard uptake value (SUV) • Synaptic activity • Synaptic transmission • Temporal lobe epilepsy • Tracers • Tricarboxylic acid cycle (TCA) • Tumor • Visual cortex • Visual inspection

10.1 FDG as a PET Tracer

18F-Fluorodeoxyglucose (FDG) is used for the measurement of regional cerebral glucose metabolism *in vivo*. This method was derived from the autoradiographic [^{14}C] deoxyglucose method for measuring local cerebral glucose metabolism in animals (Sokoloff et al. 1977) and was first developed for use with single-photon emission computed tomography (Reivich et al. 1979) and later adapted for the use with positron emission tomography (PET) (Phelps et al. 1979).

10.1.1 Main Principles of PET

The main principle of PET is based on coincidence detection of both photons resulting from annihilation due to the interactions of positron, emitted by isotope, with a nearby electron. The registration is provided by paired detectors combined in a circular array. The coincidence data is integrated for each detector pair of the system and tomographic images are reconstructed from the resulting encompassing data set (Ter-Pogossian 1992). A typical PET scanner consists of many radiation detectors and a data acquisition and data processing system (Zaidi and Montandon 2006). The radiation detectors convert the radiation (photon) signal into an electric signal. The data acquisition and data processing system evaluates individual detector-photon interactions, checks for coincidences with other detectors, then stores and processes the signal from the detectors specialized software and reconstructs and displays the PET images.

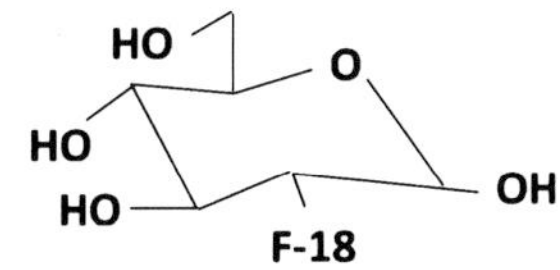

Fig. 10.1 FDG molecule

The positron-emitting isotopes used in PET studies are produced with a particle accelerator (cyclotron). Once the isotope has been made, it is transported to a radiochemistry lab where it is incorporated chemically into the tracer that will be utilized in the study.

10.1.2 Synthesis of FDG

The positron emitters for PET include isotopes of biologically abundant elements such as [^{15}O] (halflife = 2.02 min), [^{11}C] (halflife = 20.2 min), [^{13}N] (halflife = 9.97 min), as well as [^{18}F]. These elements can be substituted for the stable isotopes in biological molecules without changing the physiological characteristics of the molecule. Positron emitters typically have a high specific activity so that minuscule amounts introduced into the body are sufficient to visualize the process. The radiotracers used for PET often are either chemical analogs to natural substrates or a radioactive-labeled natural substrate. An example of chemical analog is FDG, in which hydrogen of a glucose molecule is replaced with a positron emitter, [^{18}F] (Fig. 10.1).

Fluorine-18 has a half-life of 110 min and is commonly produced by the [^{18}O] (p,n) [^{18}F] reaction on a [^{18}O] H_2O target using 11–18 MeV protons. The deoxyglucose molecule is labeled with [^{18}F] by nucleophilic displacement reaction of an acetylated sugar derivative followed by hydrolysis (Saha 2005).

As other PET radiopharmaceuticals, FDG for human use should be tested for chemical purity, radionuclidic purity, radiochemical purity, pH, isotonicity, sterility, and apyrogenicity. According to the United States Pharmacopea specifications, FDG should have pH in the range between 4.5 and 7.5; specific activity not less than 1 Ci (37 GBq)/μmol; radionuclidic purity not less than 99.5%; minimal chemical impurities such as amynopolyether (Kryptofix 2.2.2, should not exceed 50 μg/ml of the sample volume) or 2-chloro-2-deoxy-D-glucose (should not exceed 1 mg per total volume of the batch produced); and radiochemical impurity more than 90% (Saha 2005).

FDG, like all [^{18}F] compounds, has the advantage of a relatively long half-life (110 min) compared to other PET isotopes, which may allow for the synthesis to be performed in one location, the subject injection in another, and the scanning in yet another.

10.2 FDG Model

Interaction of a tracer with tissue *in vivo* is complicated by multiple simultaneous processes; including transport in and out of the tissue via blood flow, diffusion across various membranes, and interactions with transport proteins and metabolic enzymes. One way of describing the cumulative impact of these interactions on the final concentration of the tissue is to define a model that mathematically expresses the most important elements of the interaction while simplifying less relevant aspects to ensure the final expressions have practical solutions.

The model for the measurement of cerebral glucose metabolism with FDG is based on that of [^{14}C] DG (Sokoloff et al. 1977) and includes two compartments: plasma and brain tissue (Fig. 10.2). This model is based on the known biochemistry of glucose and FDG in which both FDG and glucose compete in the plasma for a common carrier to cross the blood-brain barrier, and in the cytosol of the brain cells where they compete for the enzyme hexokinase which phosphorylates them irreversibly to fluorodeoxyglucose-6-phosphate or glucose-6-phosphate. Glucose-6-phosphate undergoes further metabolism eventually: (1) to water and carbon dioxide; or (2) to lactate; or (3) via pentose phosphate pathway (PPP). Fluorodeoxyglucose-6-phosphate cannot be converted to fructose-6-phosphate, it is not a substrate for PPP enzyme glucose-6-phosphate dehydrogenase (Sols and Crane 1954) or any other known enzyme in the brain (Wienhard 2002), and therefore is not metabolized but is trapped and accumulates in brain cells.

The FDG model assumes the following: (1) a steady state of glucose metabolism during measurement, meaning constant rates of transport of glucose between plasma and brain tissue and of phosphorylation of glucose and FDG, as well as constant plasma concentrations of glucose and all other substrates and enzymes of the glycolysis; (2) homogenous tissue compartment with uniform distribution of glucose and FDG, and equal concentrations of glucose and FDG in arterial and capillary plasma; (3) trace amounts of FDG and FDG-6-phosphate.

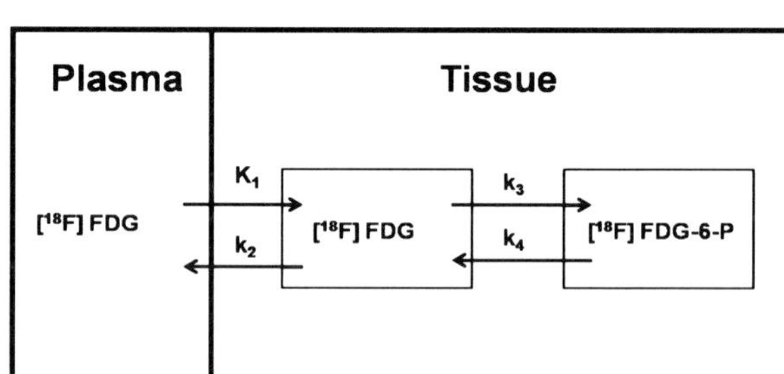

Fig. 10.2 Kinetic model for FDG

10.2.1 Compartmental Model

From the compartmental model describing behavior of glucose and FDG in brain tissue it follows that total [^{18}F] activity $C_i(t)$ in brain tissue is equal to the sum of tissue concentrations of FDG $C_E(t)$ and fluorodeoxyglucose-6-phosphate (FDG-6-PO$_4$) $C_M(t)$, which can be expressed as

$$CE(t) = k1\exp[-(k2+k3)t]\otimes C_p^*(t),$$

$$CM(t) = \frac{k1k3}{k2+k3}[1-\exp[-(k2+k3)t]]\otimes C_p^*(t), \tag{10.1}$$

where $C_p^*(t)$is the capillary plasma FDG concentration as a function of time; k$_1$, k$_2$, and k$_3$ are the first-order rate constants and ⊗ is the convolution operator. The metabolic rate can be expressed as

$$CMRGlc = \frac{Cp}{LC}\frac{k1k3}{k2+k3},$$

where Cp is the concentration of glucose in plasma, and LC is a "lumped constant", the correction factor accounting for the differences in transport and phosphorylation rates between glucose and fluorodeoxyglucose. In the original work by Sokoloff et al. (Sokoloff et al. 1977), it was assumed that FDG-6-PO$_4$ is trapped in the tissue for the whole period of FDG uptake measurement due to very low activity of phosphatase which hydrolyzes it. Later, Phelps et al. (Phelps et al. 1979) allowed possibility of hydrolysis of FDG-6-PO$_4$ to FDG and added a fourth rate constant k$_4$, representing the slow process of dephosphorylation of FDG-6-PO$_4$. Accordingly, Eqs. 10.1 and 10.2 were modified as

$$CE(t) = \frac{k1}{\alpha 2-\alpha 1}[(k4-\alpha 1)\exp[-\alpha 1t]+(\alpha 2-k4)\times\exp[-\alpha 2t]]\otimes C_p^*(t),$$

$$CM(t) = \frac{k1k3}{\alpha 2-\alpha 1}(\exp[-\alpha 1t]-\exp[-\alpha 2t])\otimes C_p^*(t),$$

where

$$\alpha 1 = \left[k2+k3+k4-\sqrt{(k2+k3+k4)^2-4k2k4}\right]/2.$$

$$\alpha 2 = \left[k2+k3+k4+\sqrt{(k2+k3+k4)^2-4k2k4}\right]/2,$$

and

$$Ci(t) = CE(t) + CM(t) = \frac{k1}{\alpha 2 - \alpha 1}[(k3 + k4 - \alpha 1)\exp[-\alpha 1 t] + (\alpha 2 - k3 - k4) \times \exp[-\alpha 2 t]] \otimes C_p^*(t)$$

and metabolic rate is given by Eq. 10.3.

10.2.2 Autoradiographic Method

The most commonly used method of quantitative estimation of the cerebral metabolic rate of glucose (CMRGlc) using FDG PET is called "*in vivo* autoradiographic method", and is based on modifications of Sokoloff's original approach by Phelps et al. (Phelps et al. 1979), when CMRGlc is expressed as

$$CMR_{Glc} \approx \frac{Cp}{LC} \frac{\bar{k}_1 \bar{k}_3}{\bar{k}_2 + \bar{k}_3} \left[\frac{C_i(T) - \bar{C}_E(T)}{\bar{C}_M(T)} \right],$$

where $\bar{k}_1$, $\bar{k}_2$ and $\bar{k}_3$ are mean values of the rate constants in the population of control subjects; $\bar{C}E(T)$ and $\bar{C}M(T)$ are the radioactivity concentrations calculated with $\bar{k}_1$, $\bar{k}_2$ and $\bar{k}_3$ as set parameters; and $Ci(T)$ is the PET-measured tissue radioactivity concentration at time T after FDG injection (Wienhard 2002).

Being accurate for the normal brain studies, nevertheless this approach demonstrated high sensitivity to pathological states when individual rate constants may differ substantially from mean rate constants in the cohort leading to some erroneous values (Huang et al. 1980; Hawkins et al. 1981). Therefore several modifications were proposed leading to the relative independence of CMRGlc estimations from the deviation of the rate constants including normalization method, which use a measured tissue concentration as a correcting factor (Brooks 1982; Rhodes et al. 1983; Hutchins et al. 1984; Wienhard et al. 1985).

10.2.3 Dynamic Methods

An alternative to compartmental models is the multiple time graphical analysis using Patlack plot, which is applicable to both homogeneous and heterogeneous tissues (Patlak et al. 1983; Patlak and Blasberg 1985). Assuming the equilibrium between arterial plasma and tissues and no product loss it is possible to graph the ratio of tracer concentration in tissue $C_i^*(t)/C_p^*(t)$ to concentration in plasma $\int_0^t C_p^*(t')/C_p^*(t)$. Because the FDG graph is characterized by some non-linearity, there is a need in identification of the optimal linear segment of this graph and time to determine CMRGlc, which was found between 45 and 120 min after injection of FDG (Lucignani et al. 1993).

10.2.4 Quantitative Assessment of Glucose Metabolism with FDG PET

Quantitative analysis of FDG PET requires measurement of the time course of radiopharmaceutical in arterial plasma. The total radioactivity should be corrected for the labeled metabolites at several time points. To reduce the risk of side effects of arterial line, arterialized venous blood or a population-based arterial blood curve normalized by two individual arterial samplings were used (Huang et al. 1980; Takikawa et al. 1993). Another promising approach is to estimate an input function with careful segmentation of the internal carotid artery from dynamic FDG PET image and co-registered high quality anatomical MR image (Chen et al. 1998; Wong et al. 2001; Zanotti-Fregonara et al. 2009).

Early quantitative studies of CMRGlc in human brain reported mean value of 5.5 mg glucose/min/100 g ranging from 3.6 to 5.2 mg glucose/min/100 g in white matter and 5.8–10.3 mg glucose/min/100 g in gray matter (Phelps et al. 1979; Reivich et al. 1979; Huang et al. 1980; Kuhl et al. 1980b). For FDG studies, a stable plasma value of glucose (within physiological range) during the time of scanning is necessary, therefore fasting for several hours is an obligatory requirement.

10.2.5 Lumped Constant

The lumped constant (LC) is a unique factor used to correct the differences between FDG and glucose rates of uptake and phosphorylation (Sokoloff et al. 1977). It represents the ratio of metabolic rates of FDG (CMR_{FDG}) and glucose (CMRGlc) and it may be expressed as:

$$LC = \frac{CMR_{FDG}}{CMR_{Glc}} = \left(\frac{\lambda}{\phi}\right)\left(\frac{V_{mFDG}K_{mGlc}}{V_{mGlc}K_{mFDG}}\right),$$

where K_{mFDG} and K_{mGlc} are Michaelis-Menten constants for FDG and glucose, respectively; VmFDG and VmGlc are the maximum velocities in hexokinase reaction for FDG and glucose, respectively; λ is the ratio of the volumes of distribution of FDG and glucose; and ϕ is the proportion of glucose that, once phosphorylated, is further metabolizedand is usually assumed to be 1.0 because of low levels of glucose-6-phosphatase in a healthy brain.

It is very difficult to measure the lumped constant in humans, and only one directly measured study in normal physiological conditions has been done, which provided the mean value of 0.52, while values between 0.42 and 0.65 were derived indirectly from dynamic FDG PET studies (Phelps et al. 1979; Reivich et al. 1985; Brooks et al. 1987; Kuwabara et al. 1990).

LC may be decreased by high plasma and tissue glucose levels (Schuier et al. 1990, 765), and LC may increase substantially, when the glucose supply is limiting, such as in hypoxia or ischemia (Suda et al. 1990); therefore different attempts to stabilize the LC have been suggested (Gjedde et al. 1985; Hawkins et al. 1986; Spence et al. 1998), however, these are not widely used.

10.2.6 Qualitative and Semi-quantitative Assessment

In the diseases resulting in distinct changes in FDG uptake the contrast between the image of the involved area and normal brain tissue may be so obvious that qualitative or semi-quantitative assessment of FDG uptake is often sufficient. Visual inspection is the most typical and simplest approach, which being less precise, nevertheless is widely and successfully used in many clinical studies such as diagnosis, staging and monitoring the treatment effects in brain tumors, or localizing the epileptic foci. Whenever there is a need in some actual value, a standard uptake value may be used (with the correction for weight and height of the patient or surface area, plasma glucose level and injected dose) or tumor-to-non-tumor-tissue (using some reference region or the whole brain) uptake ratio (Zasadny and Wahl 1993; Kim et al. 1994; Vlassenko et al. 2000; Britz-Cunningham et al. 2008). Using the ratio helps avoid many disadvantages of standard uptake value calculations but is also subject to potential influences and alterations (Wienhard 2002). Semi-quantitative approaches are reasonable in situations when the study is focused mostly on changes of FDG (or other PET parameter) and not on the absolute values, such as in response to physiological or pharmacological activations and tasks, which evoke regional but not global alterations. A double injection method has been proposed for repeated quantitative measurements of glucose metabolism on the same day (Chang et al. 1987; Murase et al. 1996), however because of the long halflife of [^{18}F] and the need to account for the residual activity on the second scan, doing FDG studies on different days is a more popular and accurate approach.

10.2.7 Timing of FDG Uptake Measurement

It is typically assumed that FDG PET studies require a 30–40-min equilibration time to approach steady state and any changes in brain metabolism during FDG uptake will alter any precise quantitation of CMRGlc. However, the accumulation of FDG is also known to be heavily weighted to the time immediately after injection (e.g. 0–15 min) when circulating FDG levels are highest (Sokoloff et al. 1977; Crosby and Sokoloff 1983; Gjedde 1987). Moreover, a modification of an experimental

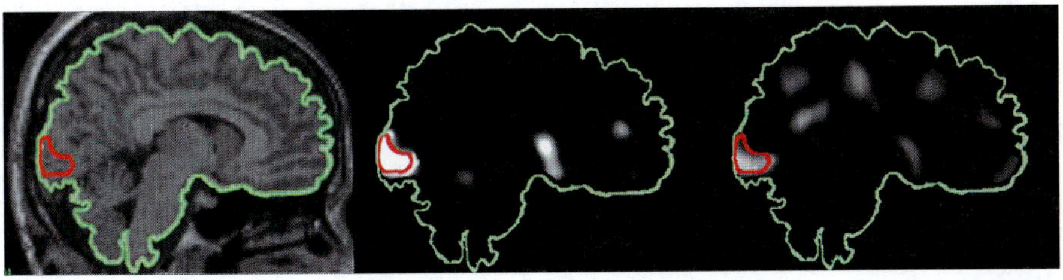

Fig. 10.3 Co-registered sagittal MRI (*left*) and FDG PET (*middle* and *right*) images. PET images are subtractions from short-term (*middle*) and long-term (*right*) activation scans (Reproduced from Vlassenko et al. 2006)

paradigm was designed to evaluate and compare short-term and long-term effects of visual stimulation with FDG PET scans (Vlassenko et al. 2006) when visual stimulation continued for only 5 min of FDG uptake and PET scanning. Such duration of stimulus was chosen in order to achieve a measurable change of FDG uptake (Huang et al. 1981) but keep the stimulus time shorter than what has been associated with changes in the blood flow response to continuous stimulation (Mintun et al. 2002). Interestingly, this short-term visual stimulation evoked a very robust (~28%) increase in FDG uptake in the visual cortex.

10.2.8 Co-registration with MRI and CT

In order to better assess the functional state of some anatomical structures, the registration of PET, MRI and/or CT images has been proposed (Fig. 10.3). This approach allows the reorientation of the PET, MRI or CT slices in such a manner, that images correspond to one another. Fusion or superimposition of images is also possible. Various methods of registration have been tested in models and in clinical studies (Pietrzyk et al. 1994; Woods et al. 1998a, b). Currently, most research and clinical applications that uses anatomical co-registration use MRI obtained separately. Promising results are coming from utilization of in-house or commercially available software packages based on the databases combining normal and clinical images from multiple centers. The American Alzheimer's Disease Neuroimaging (ADNI) and European Network for Standardization of Dementia Diagnosis (NEST-DD) databases, as a combined effort of governmental and private pharmaceutical and non-profit organizations, are aimed at utilization of PET, MR and other methods for measuring the progression of AD at various stages (Walhovd et al. 2010). The results from both databases were comparable and automated FDG PET analysis procedure provided good discrimination power between AD patients and cognitively normal participants, and demonstrated to most accurate for early-onset AD (Haense et al. 2009).

First realization of multi-modality system was a combined PET and CT tomography, which provides practical and effective means to acquire registered anatomical and functional images without moving the patient (Beyer et al. 2000; Townsend 2008). Another novel multimodal approach is combined MR/PET system (Schlemmer et al. 2009). This first human brain MR/PET scanner uses semiconductor-based light detectors, such as avalanche photodiodes instead of conventional photomultipliers (Schlemmer et al. 2008). The image quality obtained with MR/PET is comparable to stand alone systems and allows accurate co-registration of MR anatomy and function (Schlemmer et al. 2008).

10.2.9 Small Animal PET with FDG

Recently, small animal PET scans with FDG became available for visualizing *in vivo* glucose metabolism in primates (Naidoo-Variawa et al. 2007; Machado et al. 2008) and in small rodents (Chatziioannou et al. 1999; Kornblum et al. 2000; Chen et al. 2004; Cheng et al. 2009). Small animal PET imaging with FDG achieves resolution below 2 mm full-width-half-maximum and was used to analyze the regional pattern of glucose metabolism in animal model of diabetes insipidus (Idbaih et al. 2007) and to demonstrate the long-term effects of selective lesions of the brain structures (Machado et al. 2008) and the effects of immobilization stress in rats scanned without anesthesia (Sung et al. 2008). Regional patterns of glucose metabolism were studied in various conditions including mouse model of epilepsy and in pharmacologically-induced tinnitus (Fueger et al. 2006; Mirrione et al. 2006; Paul et al. 2008). The field of experimental neuroimaging has recently made substantial progress and a number of small animal PET systems have been constructed and applied to many studies including those with FDG. While resolution and sensitivity of current scanners are still limited, new small animal PET systems are under development looking for substantial improvement in performance (Levin et al. 2007; Rowland and Cherry 2008).New strategies in imaging system technologies could lead to an order of magnitude increase in molecular sensitivity of PET enabling the measurement of smaller signals associated with various physiological and pathological processes (Levin et al. 2007).

10.3 Physiological Applications of FDG PET

10.3.1 Fate of Glucose in the Brain Cells

In a physiological, normoglycemic state, the brain receives most of its energy from metabolism of glucose (Fig. 10.4). Generation of metabolic energy from glucose starts in the cytosol of the cell with glycolysis, which is a sequence of reactions that

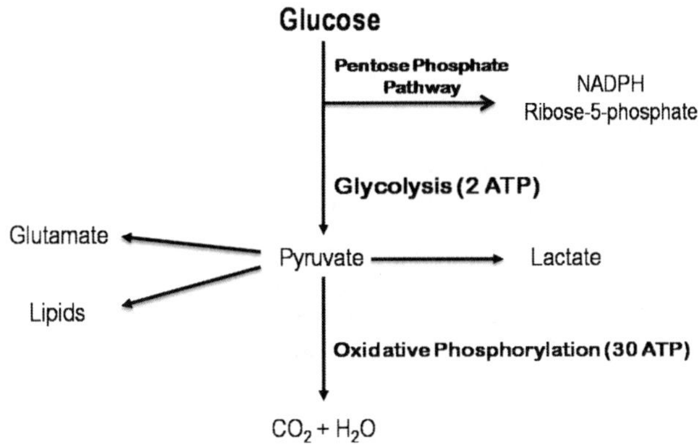

Fig. 10.4 Fate of glucose in the brain cells

converts glucose into pyruvate. Pyruvate then may have a different fate. Most often it enters mitochondria, where it is utilized through the tricarboxylic acid (TCA) cycle and oxidative phosphorylation eventually to CO_2 and H_2O. Mitochondrial processes are much more energetically efficient compared to glycolysis because they provide 30 ATPs per molecule of glucose compared to 2 ATPs generated along the conversion of glucose to pyruvate. However, an important advantage that glycolysis has over oxidative phosphorylation is that it is much faster, which may be of strategic importance especially in accommodating sudden changes in activity (Raichle and Mintun 2006). Also, pyruvate may be utilized in mitochondria not for energy, but for other purposes. As an example, pyruvate may be oxidized by the pyruvate dehydrogenase complex and the acetyl CoA formed may condense with oxaloacetate to produce citrate in the mitochondria. Citrate is then transported out of the mitochondria to the cytosol where it is cleaved to form acetyl CoA for lipid synthesis. In astrocytes, pyruvate may be used for de novo synthesis of glutamate, and carbon flux through the glutamate-glutamine cycle involving pyruvate carboxylation has been demonstrated in the brain (Hertz et al. 2000; Gruetter et al. 2001).

Finally, pyruvate may not be utilized in mitochondria but converted to lactate in the cytosol and leave the cell. Gibbs et al. (Gibbs et al. 1942) concluded from their results that 82% of the glucose is oxidized, while 16% and 2% appear as an output of lactate and pyruvate, respectively. The theoretical value of oxygen-to-glucose ratio (OGI) of 6 has never been shown in actual measurements and an average value for the human brain is approximately 5.3 (Raichle et al. 1970). Scheinberg et al. (1965) did not find any consistent arteriovenous differences of lactate or pyruvate and suggested that the non-oxidized fraction of glucose went into synthesis of non-carbohydrate substrates such as amino acids. Many studies have shown that cerebral activation increases regional cerebral lactate concentration (Richter and Dawson 1948; Kuhr and Korf 1988; Ueki et al. 1988; Prichard et al. 1991). However, it is very hard to verify lactate removal from the brain tissue, because *in vitro* or with magnetic resonance spectroscopy only local content can be evaluated, and it is not possible to differentiate what is removed from what is locally used. Evaluation of arteriovenous difference being the most reliable way and often demonstrating some evidence of lactate removal (Gibbs et al. 1942; Raichle et al. 1970; Madsen et al. 1995; Madsen et al. 1998), nevertheless may be not sensitive enough to detect the small changes in plasma lactate levels evoked by local brain stimulation or during baseline activity of the brain. It is also likely that part of lactate may be utilized in the brain (Pellerin and Magistretti 1994; Magistretti 2000). In experiments with cultured rat neurons and astrocytes, lactate was utilized by both neurons and astrocytes not only as a source of energy but also as a precursor of lipids (Vicario et al. 1993). The rates of lipogenesis from lactate in both neurons and astrocytes were higher than those from glucose, 3-hydroxybutyrate and glutamine (Medina et al. 1996).

Another metabolic fate of glucose in the brain is the PPP, which converts the first product of glycolysis, glucose-6-phosphate eventually to NADPH, a donor for various enzymatic reactions, and to ribose-5-phosphate, a precursor for the synthesis of various nucleotides (Fig. 10.4). Specifically, PPP provides important intermediates for cellular proliferation and turnover of the membrane components including nucleotides for DNA replication and intermediates for fatty acid synthesis (Gaitonde et al. 1987; DeBerardinis et al. 2007). NADPH may be used for cholesterol synthesis, and cholesterol in the brain is produced locally by glial cells (Bjorkhem and Meaney 2004). Moreover, NADPH is essential to regenerate the major antioxidant glutathione, which is highly important for the detoxification of radicals and peroxides in brain cells (Dringen 2000, 2005). In adult brain, under unstressed conditions, the PPP may contribute to 5-10% of the total glucose consumption (Hothersall et al. 1979; Ben-Yoseph et al. 1996). However, if the demand of brain cells for PPP products is increased, the flux of metabolites through the PPP can be substantially increased (Dringen et al. 2007).

Thus, glucose is utilized by the brain for different purposes, and while major metabolism is probably related to high energetic expenses of signaling (Attwell and Laughlin 2001), some of glucose is required for biosynthetic processes and probably for turnover of the membrane components (Marder and Goaillard 2006).

In pathophysiological situations, the relative impact of different pathways of glucose metabolism may be altered. Bolanos et al. (Bolanos et al. 2008) suggested that for neuronal survival during bioenergetic crises and oxidative stress one of the most important control points may be the balance between glucose oxidation through glycolysis and PPP. Earlier, they demonstrated that incubation of rat cortical neurons in presence of excess glucose prevents against mitochondrial damage and neuronal death caused by glutamate (NMDA) receptor activation, however this effect is realized not through increased aerobic glycolysis (because no increase in lactate production was observed) but through intensification of PPP activity (Delgado-Esteban et al. 2000).

Interestingly, in experimental malignant gliomas, hexokinase, the initial enzyme in glycolysis pathway was 40% higher within tumors than the contralateral cerebral tissue, and 2-deoxyglucose had similar increase as observed with autoradiography (Ikezaki et al. 1992). At the same time glucose-6-phosphate dehydrogenase, which is the first enzyme in PPP, was more than 300% higher, whereas enzymes of the energy producing tricarboxylic acid cycle and of the electron transport system were significantly reduced in tumors compared to contralateral cortex suggesting that a substantial part of the increased glucose utilization in glioma cells may be related to elevated activity of the PPP, and not as much to energy production (Ikezaki et al. 1992).

10.3.2 Measurement of Oxidative and Non-oxidative Glucose Metabolism with PET

In an FDG PET study, fluorodeoxyglucose-6-phosphate, which is formed from glucose by hexokinase, essentially is not metabolized further, but is trapped and accumulates in the brain cells. Therefore, FDG uptake or CMRGlc observed with PET represents an overall rate of glucose utilization accomplished by different mechanisms and for different purposes, including glycolysis, oxidative phosphorylation, PPP and other biosynthetic processes, and it is not possible to distinguish the relative impact of each process based only on the measurement of FDG uptake.

The combination of CMRGlc and cerebral metabolic rate of oxygen ($CMRO_2$)PET studies allows assessment of the relationship between oxidative and non-oxidative metabolism of glucose. When the glucose utilization (glycolysis) is non-proportional to oxygen utilization (oxidative phosphorylation) in spite of abundant oxygen in brain cells, such glycolysis is often called aerobic. Aerobic glycolysis may be quantitatively evaluated with PET using oxygen-to-glucose ratio or glycolytic index (GI), which is calculated as the residual after regression of CMRGlc on $CMRO_2$ and corresponds to the amount of glucose utilization that differs from what is predicted by the amount of oxygen utilization. This approach is available for centers which have the ability to measure both these parameters with PET, and preliminary results of ongoing studies demonstrate unique information about the distribution of aerobic glycolysis in the human brain, its response to physiological stimulation and potential link to the pathology (Vaishnavi et al. 2010; Vlassenko et al. 2010).

10.3.3 Changes in Brain Activity as Measured with FDG PET

PET mapping of brain function using sensory, motor, cognitive and emotional tasks is recognized as a highly productive approach, substantially increasing our understanding of the relationship between the structural and functional elements of the brain both in normal conditions and in a variety of pathological states. FDG PET studies have been used for evaluation of brain function during various sensory, motor, behavioral and cognitive tasks (Greenberg et al. 1981; Phelps et al. 1981; Mazziotta et al. 1982; Gur et al. 1987; Fox et al. 1988; Vlassenko et al. 2006).

Besides analysis of physiological aspects of brain functional reserve and plasticity, FDG PET demonstrated high value in addressing many practical clinical issues. Activation studies with specific mental tasks are performed to correlate the level of performance with the level of regional FDG response and to determine how patterns of activation differ between healthy subjects and patients with specific brain disorders.

Impairment of FDG response may be a more specific and sensitive index of functional failure than glucose metabolism measured at the resting state (Pietrini et al. 1997). FDG PET study with visual recognition task was used in assessment of the therapeutic efficacy in patients with Alzheimer's disease (Heiss et al. 1993). FDG PET with motor task demonstrated high sensitivity and specificity for identification of motor-associated brain areas, which was successfully used in pre-surgical planning of tumor resection (Schreckenberger et al. 2001). FDG PET with visual language task provided important information on the plasticity of cortical language networks and effects of deafness on the processing of the visual component of language (Fujiwara et al. 2008). Metabolic changes in response to appetitive food stimuli help to map cortical areas responsible for hunger which contributes to the research in obesity (Wang et al. 2004).

In the human brain, physiological stimulation may demonstrate not only activation but also task-induced decreases in blood flow and MRI signals in several areas, including medial frontal and lateral and medial parietal cortex, that appear to be task-independent, varying little in their location across a wide range of tasks (Raichle et al. 2001; Raichle and Mintun 2006). These areas constitute a "default network" supporting processes more active during passive- than active-task conditions such as monitoring the environment and internal thoughts and emotions and various forms of undirected thought and these areas are chronically active at the resting state of the brain (Raichle et al. 2001). Some of default network regions, including precuneus, posterior cingulate, prefrontal, and lateral parietal cortex are highly glycolytic at the resting state and preliminary data indicate that regional brain deactivation (as well as activation of other brain regions including visual cortex) in response to physiological stimulation is associated with the changes in aerobic glycolysis (Vlassenko et al. 2010).

10.3.4 Changes in Glucose Metabolism During Continuous Physiological Stimulation

The short-term effects of physiological brain stimulation have been extensively studied using positron emission tomography (PET) and functional magnetic resonance imaging (fMRI) (Fig. 10.5). Fox et al. (Fox and Raichle 1986; Fox et al. 1988) used PET to demonstrate that in humans, short-term stimulation of the visual or somatosensory cortex resulted in dramatic increase in blood flow and glucose consumption (approximately 50%) but much less increase in oxygen consumption (~5%). Later, similar findings were found in other studies using PET and fMRI (Kuwabara et al. 1992; Kim and Ugurbil 1997; Davis et al. 1998; Chen et al. 2001; Vlassenko et al. 2006). The blood oxygen level dependent MRI signal, which is dependent on a drop in paramagnetic deoxyhemoglobin in cerebral draining veins, serves as a routine functional MRI signal of the very fact that that blood flow changes are in excess of oxygen consumption changes during acute alteration in brain functional activity.

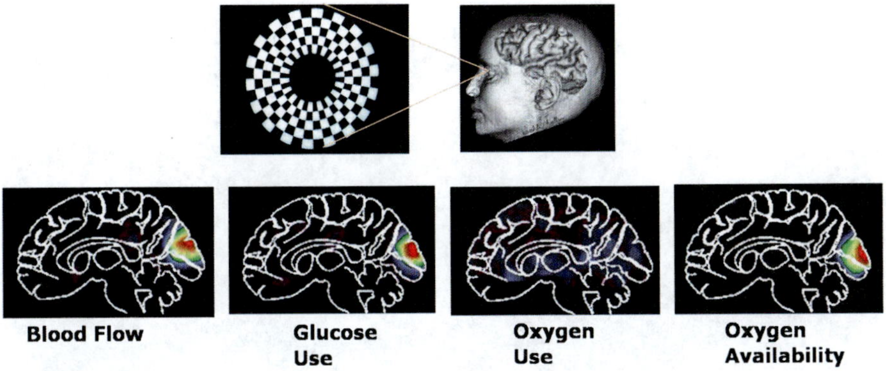

Fig. 10.5 The effects of short-term visual stimulation (Data adapted from Fox et al. 1988; Reproduced from Raichle and Mintun 2006)

However, the behavior and the relationship of glucose and oxygen metabolism and blood flow during prolonged continuous brain stimulation remains the object of study. MR spectroscopy has demonstrated an initial decrease in glucose content with a subsequent turn toward the baseline level after 5 min (Frahm et al. 1996) or 15 min (Chen et al. 1993) of continuous visual stimulation. Comparative assessment of short-term and long-term effects of brain activation using FDG PET is a challenge, because in traditional approaches, the measurement of CMRGlc involves about 40 min of a steady state application of activation (Sokoloff et al. 1977; Fox et al. 1988), but this time constraint is too long to investigate the early changes in CMRGlc. Nevertheless, a modified FDG PET paradigm demonstrated that during the initial period of visual stimulation glucose metabolism increases significantly but after 20 min of continuing stimulation this increase attenuates by ~28% (Vlassenko et al. 2006) in parallel with blood flow response while oxygen utilization increases threefold (Mintun et al. 2002) (Fig. 10.3). This observation may indicate that the energy demands of brain activation in the visual cortex are initially met predominantly by increased aerobic glycolysis but with continuous stimulation it changes back to predominantly oxidative utilization of glucose.

10.4 Clinical Applications of FDG PET

10.4.1 Brain Maturation and Aging

CMRGlc studies with PET and FDG have been used to map the temporal course of "metabolic maturation" of the human brain (Chugani and Phelps 1986; Chugani et al. 1987). These studies demonstrated that the highest metabolic activity at birth is in the phylogenetically older structures of the brain, such as sensorimotor cortices, thalamus, brain stem, and cerebellar vermis. The functions of these old structures

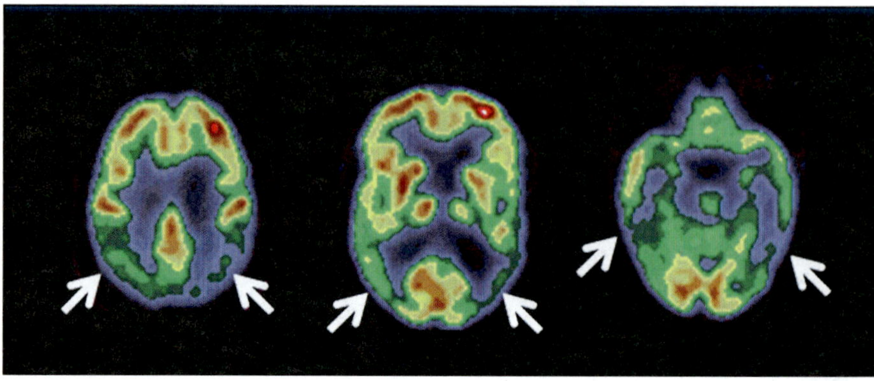

Fig. 10.6 FDG PET in DAT patient

dominate the behavior and primitive intrinsic reflexes of the newborn. As the infant matures in the third month of life, and primitive reflexes become suppressed, increases in CMRGlc occur in parietal, temporal, primary visual cortex, basal ganglia, and cerebellar hemispheres. The frontal cortex was observed to be the last area to metabolically mature, and it proceeded from lateral to medial and inferior to superior. These increases in CMRGlc of the frontal cortex occurred between 6 and 10 months and are associated with rapid cognitive development. By 1 year, the distribution and absolute values of CMRGlc equaled that of the adult.

The changes in FDG uptake with aging are complex and the literature is somewhat contradictory. Difficulty arises with the fact that aging may represent normal aging processes combined with a variety of degenerative and vascular diseases, many of which may not be clinically apparent. It was believed that healthy aging may be associated with diminished glucose metabolism in the temporal, somatosensory, parietal and especially frontal cortex, however precise correction for age-related brain atrophy should be carried out, because it eliminates the variance in CMRGlc substantially and often negates significance in regional differences (Yoshii et al. 1988).

10.4.2 Dementias

In patients with dementia of Alzheimer's type (DAT), FDG PET studies demonstrated specific pattern of hypometabolism in parietal and temporal lobes, especially in posterior cingulate gyrus (Foster et al. 1984; Friedland et al. 1989; Heiss et al. 1991; Ichimiya et al. 1994; Herholz et al. 2002; Li et al. 2008) (Fig. 10.6). FDG PET identified patients with DAT with sensitivity of 94% and sensitivity of 73% (Silverman et al. 2001). However, temporoparietal hypometabolism may be seen in patients with frontotemporal dementia or even Parkinson's disease or stroke (Mazziotta et al. 1992), therefore information from other modalities and clinical testing is often needed for the correct diagnosis. Correction for brain atrophy using

CT or MRI may eliminate metabolic differences between DAT and control subjects, however many studies suggested that even after such correction there is still substantial metabolic deficit (Videen et al. 1988; Chawluk et al. 1990; Labbe et al. 1996; Meltzer et al. 1996; Chetelat et al. 2008).

CMRGlc decreases more rapidly over the time in DAT than in non-demented ageing (Alexander et al. 2002) and magnitude and extent of hypometabolism may correlate with the severity of dementia (Drzezga et al. 2003). In mild amnestic cognitive impairment, glucose metabolism is substantially decreased in the posterior cingulate cortex and to a much lesser extent in temporal and parietal cortex, which is consistent with the hypothesis that mild cognitive impairment is a preclinical DAT state with posterior cingulate cortex being involved in early stages of DAT and parieto-temporal cortex being typical for more advanced disease (Minoshima et al. 1997; Drzezga et al. 2003; Del Sole et al. 2008). Standardized automated analysis of FDG scans in patients with DAT, mild cognitive impairment and dementia with Lewy bodies was suggested as objective and sensitive support to the clinical diagnosis based on the results of international multicenter study (Mosconi et al. 2008).FDG PET was used to evaluate the effects of various pharmacologic interventions and increase in glucose metabolism was reported in some DAT patients who responded to therapy (Nordberg et al. 1992; Mega et al. 2001; Potkin et al. 2001).

FDG PET has been used to differentiate DAT from multi-infarct dementia (MID) and patients with MID demonstrated scattered focal areas of hypometabolism corresponding to lesions visible with MRI (Benson et al. 1983; Duara et al. 1989).

With the development of PET radiotracers having *in vivo* binding to beta-amyloid plaques, such as [^{11}C]PIB (Mathis et al. 2003; Klunk et al. 2004; Mintun et al. 2006), it is now possible to quantify pathological changes in the human brain that were previously restricted to post-mortem study. Recently, increased [^{11}C] PIB binding has been documented not only in DAT patients but also in some age-matched non-demented subjects with binding patterns practically indistinguishable from those of DAT with routine visual examination of images, indicating that these subjects may have a preclinical from of DAT (Fagan et al. 2006; Mintun et al. 2006). Interestingly, there is a high spatial correlation between the pattern of beta-amyloid deposition in non-demented subjects with high [^{11}C] PIB binding and DAT patients and the pattern of aerobic glycolysis as measured using GI in healthy young adults (Vaishnavi et al. 2010). The close spatial relationship between resting aerobic glycolysis in young adults and deposition of beta-amyloid plaques in DAT patients suggests that brain regions shown to have elevation of aerobic glycolysis at rest may spend more time in this state over the lifetime and it may be related to the later development of Alzheimer's pathology.

10.4.3 Brain Tumors

Imaging of tumors of any localization including the brain demonstrates increased glycolysis, which is likely related not only to increased need in ATP but mainly to accelerated usage of glucose for protein synthesis and cellular proliferation.

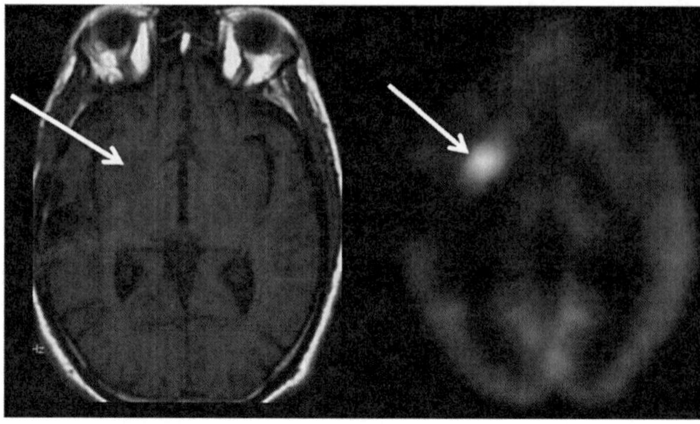

Fig. 10.7 FDG PET and MRI in recurrent brain tumor

Neoplastic degeneration may be associated with increases in glycolysis because of the progressive loss of tricarboxylic cycle (Weber 1977). Also glucose is used to provide carbon backbone to meet the high cell replication rates of tumors through activation of the pentose phosphate shunt (Weber 1977). These two factors increase glucose consumption as neoplastic degeneration progresses. The increase of signals in FDG images delineates even small neoplasm from surrounding tissue.

FDG PET studies were used for grading and differential diagnosis of primary brain tumors, prognostic assessment, treatment planning and evaluation of tumor response to therapy (Di Chiro et al. 1982; Mineura et al. 1986; Alavi et al. 1988; Hanson et al. 1991; Davis et al. 1993; Vlassenko et al. 2000; Basu and Alavi 2008). High-grade tumors show glucose uptake equal or higher than normal cortex, whereas low-grade tumors show glucose uptake equivalent or lower than white matter. It was also demonstrated that FDG uptake is not necessary dependent on the blood-brain barrier breakdown, therefore permitting detection and appropriate classification of tumors that would appear benign based on the absence of contrast enhancement (Davis et al. 1993). Spence et al. (2004) demonstrated better distinction between brain tumor and grey matter when scanning is delayed for 3–8 h after FDG injection, which was due to lower rate constant of FDG-6-phosphate degradation k_4 in tumor compared to contralateral cortex and higher excretion of FDG from normal tissue.

FDG tumor uptake has been proven to be of significant prognostic value. FDG hypermetabolism considered being a negative sign and patients with hypermetabolic tumors consistently demonstrated lower survival in comparison with patients with hypometabolic lesions (Patronas et al. 1985; Mineura et al. 1994). Serial FDG studies can also be helpful in the detection of malignant transformation of previously low-grade tumors (Francavilla et al. 1989). Serial FDG PET studies can determine residual or recurrent tumors after surgery and differentiate the recurrent tumor from radiation necrosis better than MRI with radiation necrosis being hypometabolic and tumor recurrence being hypermetabolic (Patronas et al. 1982; Doyle et al. 1987; Janus et al. 1993) (Figs. 10.7 and 10.8). We must point out, however,

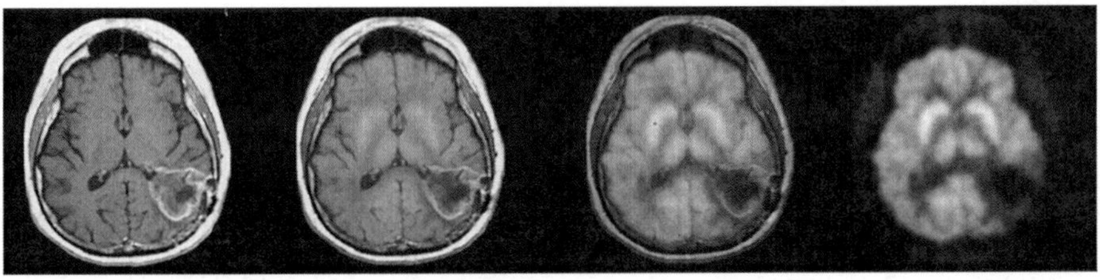

Fig. 10.8 FDG PET and MRI in radiation necrosis

that increased FDG uptake is nonspecific, and it can occur with different tissue abnormalities including inflammatory changes, seizures, and healing processes (Strauss 1996; Smith 1998).

10.4.4 Movement Disorders

In Parkinson's disease (PD), FDG PET data is rather controversial. In early, untreated PD, FDG PET usually demonstrates hypermetabolism in the basal ganglia (Rougemont et al. 1984; Eidelberg et al. 1990), however some studies reported no changes in patients with PD (Kuhl et al. 1982) or even decreases in glucose metabolism in the contralateral basal ganglia in patients with hemiparkinsonism-hemiatrophy syndrome (Przedborski et al. 1993). Dementia associated with PD may demonstrate whole-brain hypometabolism or be undistinguishable from DAT showing parietal hypometabolism (Rougemont et al. 1984).

In Huntington's disease (HD), FDG PET demonstrated hypometabolism in the caudate and putamen, which sometimes precedes the atrophy of these nuclei (Kuhl et al. 1982; Phelps and Mazziotta 1985; Hayden et al. 1986; Kuwert et al. 1989). Cortical hypometabolism has been reported in HD patients and fronto-parietal and temporo-occipital hypometabolism correlated significantly with the duration of the chorea and the severity of dementia (Kuwert et al. 1990). Also, caudate hypometabolism was demonstrated in all stages of HD, whereas the putamen was less involved at the early stage of the disease (Young et al. 1986).

10.4.5 Psychiatric Disorders

FDG PET was used to study functional and energetic abnormalities of schizophrenia and often showed relative frontal hypometabolism with decreased activation in response to various performance tasks (Buchsbaum et al. 1984; Farkas et al. 1984; Wolkin et al. 1988; DeLisi et al. 1989), however some studies failed to demonstrate this pattern (Early et al. 1987; Wiesel et al. 1987; Cleghorn et al. 1989). Glucose metabolism was also decreased in the basal ganglia of unmedicated schizophrenic patients, notably in putamen (Buchsbaum et al. 1992). FDG PET studies have not shown any substantial changes in frontal glucose metabolism in response to neuroleptic treatment (DeLisi et al. 1985; Wolkin et al. 1985; Volkow et al. 1986; Cohen et al. 1988).

In late-onset major depression, FDG PET demonstrates hypometabolism in the prefrontal, cingulate and parietal regions in both hemispheres, and increased metabolism in the occipital pole, cerebellum and basal ganglia in patients compared to controls (Fujimoto et al. 2008). Pretreatment FDG PET studies may have special importance, because hypometabolism of one region, such as anterior cingulate, may be associated with non-responsiveness to the treatment, and hypermetabolism in this very same area may predict positive response to treatment (Mayberg et al. 1997).

In patients with late-age onset of major depression, FDG PET demonstrated global hypometabolism of the same or even greater magnitude as in AD (Kumar et al. 1993). In another study, where bipolar and unipolar depression was compared, bipolar patients were characterized by global decreases in FDG uptake, and unipolar patients did not demonstrate any decrease in global metabolism (Schwartz et al. 1987).

Treatment with selective serotonin reuptake inhibitors resulted in an increase of glucose metabolism in prefrontal and parietal cortex and dorsal anterior cingulate gyrus, interpreted as normalization of previously reduced metabolic activity, and a decrease in insula and hippocampal and parahippocampal regions, interpreted as a decrease from previously elevated metabolic activity compared to healthy controls (Kennedy et al. 2001). In another study, treatment resulted in decreases of glucose metabolism in ventrolateral prefrontal cortex and orbitofrontal cortex in those who clinically responded to treatment compared to non-responders (Brody et al. 1999). Increase in glucose metabolism in prefrontal and some other cortical regions was reported after successful treatment of major depression with other drugs from the same group (Buchsbaum et al. 1997; Mayberg et al. 2000).

In obsessive-compulsive disorder, FDG PET demonstrated increased metabolism in the orbital region of frontal cortex, cingulate gyrus and the caudate nuclei compared to controls (Baxter et al. 1988; Nordahl et al. 1989; Swedo et al. 1989). Treatment with selective serotonin reuptake inhibitors was associated with increase in glucose metabolism in frontal and parietal cortex (Kang et al. 2003).

10.4.6 Epilepsy

Seizure is associated with and characterized by substantial increases in metabolic activity and perfusion in the focus with a tendency to spread and propagate along the progression of the seizure. The comparison between ictal and postictal images, often combining FDG PET and perfusion SPECT studies provides the best localization for the source of a seizure especially in cases when MRI and EEG are inconclusive (Engel et al. 1982b; Theodore et al. 1983; Abou-Khalil et al. 1987; Ho et al. 1995). FDG PET imaging is more sensitive than 99mTc HMPAO or 99mTc ECD SPECT in the interictal state (Lamusuo et al. 1997; Spanaki et al. 1999b). It shows a focal area of hypometabolism, which correlates with the focus of seizure activity (Fig. 10.9).

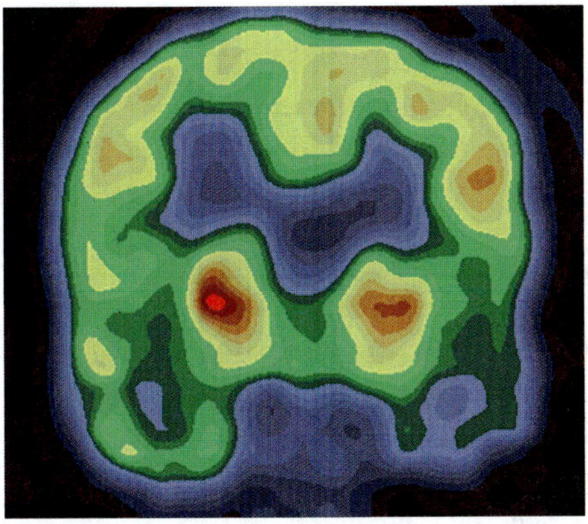

Fig. 10.9 Left temporal hypometabolism on interictal FDG PET image in patient with temporal lobe epilepsy

FDG PET is a useful and reliable tool for accurate localization of seizure foci which may be especially important in planning surgical intervention (Engel et al. 1982b; Theodore et al. 1983; Abou-Khalil et al. 1987). The positive and negative predictive values for MRI and video-EEG monitoring, which ranged from 0.48 to 0.67, were improved to 0.62 to 0.86 in combination with FDG PET in screening for temporal lobe epilepsy (Uijl et al. 2007). Another important application of FDG PET is that it may suggest exclusion criteria such as contraindication to surgery in cases with multiple foci.

Patients appropriate for evaluation by FDG PET have their most common localizations of epileptic foci in temporal and frontal lobes. Sensitivity of FDG PET in detecting the foci of temporal lobe epilepsy is more than 70% in patients with complex partial seizures (Engel et al. 1982a; Theodore et al. 1986b, 1988) and FDG PET was suggested to be a best modality for localizing frontal epileptic foci (Franck et al. 1992).

Glucose uptake and metabolism increase more during epileptic seizures than during any other brain activity (Cornford et al. 2002; Greene et al. 2003). Blood glucose also positively correlates with flurothyl-induced seizures in rats and high glucose may exacerbate human seizure disorders (Schwechter et al. 2003). Brain lactate levels also increase significantly during seizure activity (During et al. 1994; Cornford et al. 2002). Neuronal excitability and epileptic seizures are directly related to rapid glucose utilization and glycolysis however, it is not clear, to what extent enhanced glycolysis is related to the cause or effects of seizure activity. Increased glucose metabolism is often detected in brain epileptic foci upon seizure initiation and then spreads to other brain regions during seizure generalization (Kuhl et al. 1980a; Chugani and Chugani 1999). Although being hypermetabolic during the ictal event, epileptic foci are often hypometabolic between seizures suggesting abnormalities in energy homeostasis (Kuhl et al. 1980a; Chugani and Chugani 1999; Janigro 1999; Vielhaber et al. 2003). Of note, generalized epilepsies associated with increased spike-and-wave discharges on EEG, are not associated with increased glucose utilization measured using FDG PET (Ochs et al. 1987). It is known that some commonly used antiepileptic drugs reduce the CMRGlc, e.g. phenobarbital, primidone, phenytoin, carbamazepine, vugabratin and valproate (Theodore et al. 1986a, b, 1989; Leiderman et al. 1991; Spanaki et al. 1999a).

Interestingly, fasting and a low carbohydrate ketogenic diet have been recognized as an antiepileptic therapy for a broad range of seizure disorders. Successful use of ketogenic diet has been reported in children with refractory seizure disorders (Freeman et al. 1998; Nordli et al. 2001). Younger patients benefit most; 40% of those aged 1–8 years were free of seizures for 12 months, and 80% manifested improved control (Vining et al. 1998). This effect of ketogenic diet may be in part explained by the fact that in contrast to glucose, ketone bodies by-pass cytoplasmic glycolysis and directly enter mitochondria where they are oxidized in the TCA cycle as acetyl-CoA (Nehlig and Pereira 1993; Sato et al. 1995; Veech et al. 2001). This causes significant increases in the TCA cycle metabolites (from citrate to α-ketoglutarate) and improves metabolic efficiency through an increase in the energy of ATP hydrolysis (Sato et al. 1995; Veech et al. 2001). Increased ketone body metabolism could also increase the activity of the sodium pump in both neurons and glia, which could increase the membrane potential thereby decreasing neuronal excitability (Silver and Erecinska 1997; Schwartzkroin 1999; Veech et al. 2001).

10.5 Concluding Remarks

FDG PET is a leading tracer in PET neuroimaging, which has a wide range of research and clinical applications. It shows characteristic pattern for different dementias and is successfully used for grading and differential diagnosis of brain tumors as well as in planning and evaluation of tumor response to the pharmacological, radiation and surgical interventions. FDG PET demonstrates local hypometabolism in major depressive disorder and confirms the effects of medical treatment. In epilepsy, FDG PET is a useful tool for accurate localization of seizure foci and evaluation of the efficacy of the antiepileptic drugs.

Physiologically, FDG uptake and CMRGlc represents an overall rate of glucose utilization related mostly to energetic (glycolysis and oxidative phosphorylation) and also to biosynthetic (PPP, glutamate and lipid synthesis) processes. The combination of CMRGlc and $CMRO_2$ PET studies allows distinguishing oxidative and non-oxidative metabolism of glucose, and non-oxidative part, which is often called aerobic glycolysis, may be visualized and quantitatively evaluated. Local fluctuations of brain activity may be associated with changes in the level of aerobic glycolysis and chronically elevated aerobic glycolysis may be related to later development of pathological disorders.

Improvement of spatial resolution and signal-to-noise ratio with new generations of PET scanners and software; development of new approaches allowing reliable quantitative data without lengthy, complicated, and invasive techniques; and combination of measurement of glucose metabolism with quantitative assessment of oxygen metabolism and other measures of brain functions should further improve our knowledge on the regulation of daily activities of the brain at different physiological and pathological states.

References

Abou-Khalil BW, Siegel GJ, Sackellares JC, Gilman S, Hichwa R, Marshall R (1987) Positron emission tomography studies of cerebral glucose metabolism in chronic partial epilepsy. Ann Neurol 22:480–486

Alavi JB, Alavi A, Chawluk J, Kushner M, Powe J, Hickey W, Reivich M (1988) Positron emission tomography in patients with glioma. A predictor of prognosis. Cancer 62:1074–1078

Alexander GE, Chen K, Pietrini P, Rapoport SI, Reiman EM (2002) Longitudinal PET Evaluation of Cerebral Metabolic Decline in Dementia: A Potential Outcome Measure in Alzheimer's Disease Treatment Studies. Am J Psychiatry 159:738–745

Attwell D, Laughlin SB (2001) An energy budget for signaling in the grey matter of the brain. J Cereb Blood Flow Metab 21:1133–1145

Basu S, Alavi A (2008) Unparalleled contribution of 18F-FDG PET to medicine over 3 decades. J Nucl Med 49:17N-21N, 37N

Baxter LR Jr, Schwartz JM, Mazziotta JC, Phelps ME, Pahl JJ, Guze BH, Fairbanks L (1988) Cerebral glucose metabolic rates in nondepressed patients with obsessive-compulsive disorder. Am J Psychiatry 145:1560–1563

Benson DF, Kuhl DE, Hawkins RA, Phelps ME, Cummings JL, Tsai SY (1983) The fluorodeoxyglucose 18F scan in Alzheimer's disease and multi-infarct dementia. Arch Neurol 40:711–714

Ben-Yoseph O, Boxer PA, Ross BD (1996) Noninvasive assessment of the relative roles of cerebral antioxidant enzymes by quantitation of pentose phosphate pathway activity. Neurochem Res 21:1005–1012

Beyer T, Townsend DW, Brun T, Kinahan PE, Charron M, Roddy R, Jerin J, Young J, Byars L, Nutt R (2000) A combined PET/CT scanner for clinical oncology. J Nucl Med 41:1369–1379

Bjorkhem I, Meaney S (2004) Brain cholesterol: long secret life behind a barrier. Arterioscler Thromb Vasc Biol 24:806–815

Bolanos JP, Delgado-Esteban M, Herrero-Mendez A, Fernandez-Fernandez S, Almeida A (2008) Regulation of glycolysis and pentose-phosphate pathway by nitric oxide: impact on neuronal survival. Biochim Biophys Acta 1777:789–793

Britz-Cunningham SH, Millstine JW, Gerbaudo VH (2008) Improved discrimination of benign and malignant lesions on FDG PET/CT, using comparative activity ratios to brain, basal ganglia, or cerebellum. Clin Nucl Med 33:681–687

Brody AL, Saxena S, Silverman DH, Alborzian S, Fairbanks LA, Phelps ME, Huang SC, Wu HM, Maidment K, Baxter LR Jr (1999) Brain metabolic changes in major depressive disorder from pre- to post-treatment with paroxetine. Psychiatry Res 91:127–139

Brooks RA (1982) Alternative formula for glucose utilization using labeled deoxyglucose. J Nucl Med 23:538–539

Brooks RA, Hatazawa J, Di Chiro G, Larson SM, Fishbein DS (1987) Human cerebral glucose metabolism determined by positron emission tomography: a revisit. J Cereb Blood Flow Metab 7:427–432

Buchsbaum MS, DeLisi LE, Holcomb HH, Cappelletti J, King AC, Johnson J, Hazlett E, Dowling-Zimmerman S, Post RM, Morihisa J et al (1984) Anteroposterior gradients in cerebral glucose use in schizophrenia and affective disorders. Arch Gen Psychiatry 41:1159–1166

Buchsbaum MS, Haier RJ, Potkin SG, Nuechterlein K, Bracha HS, Katz M, Lohr J, Wu J, Lottenberg S, Jerabek PA et al (1992) Frontostriatal disorder of cerebral metabolism in never-medicated schizophrenics. Arch Gen Psychiatry 49:935–942

Buchsbaum MS, Wu J, Siegel BV, Hackett E, Trenary M, Abel L, Reynolds C (1997) Effect of sertraline on regional metabolic rate in patients with affective disorder. Biol Psychiatry 41:15–22

Chang JY, Duara R, Barker W, Apicella A, Finn R (1987) Two behavioral states studied in a single PET/FDG procedure: theory, method, and preliminary results. J Nucl Med 28:852–860

Chatziioannou AF, Cherry SR, Shao Y, Silverman RW, Meadors K, Farquhar TH, Pedarsani M, Phelps ME (1999) Performance evaluation of microPET: a high-resolution lutetium oxyorthosilicate PET scanner for animal imaging. J Nucl Med 40:1164–1175

Chawluk JB, Dann R, Alavi A, Hurtig HI, Gur RE, Resnick S, Zimmerman RA, Reivich M (1990) The effect of focal cerebral atrophy in positron emission tomographic studies of aging and dementia. Int J Rad Appl Instrum B 17:797–804

Chen W, Novotny EJ, Zhu XH, Rothman DL, Shulman RG (1993) Localized 1H NMR measurement of glucose consumption in the human brain during visual stimulation. Proc Natl Acad Sci USA 90:9896–9900

Chen K, Bandy D, Reiman E, Huang SC, Lawson M, Feng D, Yun LS, Palant A (1998) Noninvasive quantification of the cerebral metabolic rate for glucose using positron emission tomography, 18F-fluoro-2-deoxyglucose, the Patlak method, and an image-derived input function. J Cereb Blood Flow Metab 18:716–723

Chen W, Zhu XH, Gruetter R, Seaquist ER, Adriany G, Ugurbil K (2001) Study of tricarboxylic acid cycle flux changes in human visual cortex during hemifield visual stimulation using (1) H-[(13)C] MRS and fMRI. Magn Reson Med 45:349–355

Chen YY, Chien C, Lee TW, Fu YK, Kuo TS, Jaw FS (2004) Dynamic evaluation of [18F]-FDG uptake in the rat brain by microPET imaging. Conf Proc IEEE Eng Med Biol Soc 6:4461–4464

Cheng TE, Yoder KK, Normandin MD, Risacher SL, Converse AK, Hampel JA, Miller MA, Morris ED (2009) A rat head holder for simultaneous scanning of two rats in small animal PET scanners: Design, construction, feasibility testing and kinetic validation. J Neurosci Methods 176:24–33

Chetelat G, Desgranges B, Landeau B, Mezenge F, Poline JB, de la Sayette V, Viader F, Eustache F, Baron JC (2008) Direct voxel-based comparison between grey matter hypometabolism and atrophy in Alzheimer's disease. Brain 131:60–71

Chugani HT, Chugani DC (1999) Basic mechanisms of childhood epilepsies: studies with positron emission tomography. Adv Neurol 79:883–891

Chugani HT, Phelps ME (1986) Maturational changes in cerebral function in infants determined by 18FDG positron emission tomography. Science 231:840–843

Chugani HT, Phelps ME, Mazziotta JC (1987) Positron emission tomography study of human brain functional development. Ann Neurol 22:487–497

Cleghorn JM, Garnett ES, Nahmias C, Firnau G, Brown GM, Kaplan R, Szechtman H, Szechtman B (1989) Increased frontal and reduced parietal glucose metabolism in acute untreated schizophrenia. Psychiatry Res 28:119–133

Cohen RM, Semple WE, Gross M, Nordahl TE (1988) From syndrome to illness: delineating the pathophysiology of schizophrenia with PET. Schizophr Bull 14:169–176

Cornford EM, Shamsa K, Zeitzer JM, Enriquez CM, Wilson CL, Behnke EJ, Fried I, Engel J (2002) Regional analyses of CNS microdialysate glucose and lactate in seizure patients. Epilepsia 43:1360–1371

Crosby G, Sokoloff L (1983) Potential pitfalls in measuring regional cerebral glucose utilization; in reply. Anesthesiology 58:290–292

Davis WK, Boyko OB, Hoffman JM, Hanson MW, Schold SC Jr, Burger PC, Friedman AH, Coleman RE (1993) [18F]2-fluoro-2-deoxyglucose-positron emission tomography correlation of gadolinium-enhanced MR imaging of central nervous system neoplasia. AJNR Am J Neuroradiol 14:515–523

Davis TL, Kwong KK, Weisskoff RM, Rosen BR (1998) Calibrated functional MRI: mapping the dynamics of oxidative metabolism. Proc Natl Acad Sci USA 95:1834–1839

DeBerardinis RJ, Mancuso A, Daikhin E, Nissim I, Yudkoff M, Wehrli S, Thompson CB (2007) Beyond aerobic glycolysis: transformed cells can engage in glutamine metabolism that exceeds the requirement for protein and nucleotide synthesis. Proc Natl Acad Sci USA 104:19345–19350

Del Sole A, Clerici F, Chiti A, Lecchi M, Mariani C, Maggiore L, Mosconi L, Lucignani G (2008) Individual cerebral metabolic deficits in Alzheimer's disease and amnestic mild cognitive impairment: an FDG PET study. Eur J Nucl Med Mol Imaging 35:1357–1366

Delgado-Esteban M, Almeida A, Bolanos JP (2000) D-Glucose prevents glutathione oxidation and mitochondrial damage after glutamate receptor stimulation in rat cortical primary neurons. J Neurochem 75:1618–1624

DeLisi LE, Holcomb HH, Cohen RM, Pickar D, Carpenter W, Morihisa JM, King AC, Kessler R, Buchsbaum MS (1985) Positron emission tomography in schizophrenic patients with and without neuroleptic medication. J Cereb Blood Flow Metab 5:201–206

DeLisi LE, Buchsbaum MS, Holcomb HH, Langston KC, King AC, Kessler R, Pickar D, Carpenter WT Jr, Morihisa JM, Margolin R et al (1989) Increased temporal lobe glucose use in chronic schizophrenic patients. Biol Psychiatry 25:835–851

Di Chiro G, DeLaPaz RL, Brooks RA, Sokoloff L, Kornblith PL, Smith BH, Patronas NJ, Kufta CV, Kessler RM, Johnston GS, Manning RG, Wolf AP (1982) Glucose utilization of cerebral gliomas measured by [18F] fluorodeoxyglucose and positron emission tomography. Neurology 32:1323–1329

Doyle WK, Budinger TF, Valk PE, Levin VA, Gutin PH (1987) Differentiation of cerebral radiation necrosis from tumor recurrence by [18F]FDG and 82Rb positron emission tomography. J Comput Assist Tomogr 11:563–570

Dringen R (2000) Metabolism and functions of glutathione in brain. Prog Neurobiol 62:649–671

Dringen R (2005) Oxidative and antioxidative potential of brain microglial cells. Antioxid Redox Signal 7:1223–1233

Dringen R, Hoepken HH, Minich T, Ruedig C, Gibson GE, Dienel GA (2007) Pentose phosphate pathway and NADPH metabolism. In: Lajtha A (ed) Brain energetics. Integration of molecular and cellular processes. Springer, New York, pp 41–62

Drzezga A, Lautenschlager N, Siebner H, Riemenschneider M, Willoch F, Minoshima S, Schwaiger M, Kurz A (2003) Cerebral metabolic changes accompanying conversion of mild cognitive impairment into Alzheimer's disease: a PET follow-up study. Eur J Nucl Med Mol Imaging 30:1104–1113

Duara R, Barker W, Loewenstein D, Pascal S, Bowen B (1989) Sensitivity and specificity of positron emission tomography and magnetic resonance imaging studies in Alzheimer's disease and multi-infarct dementia. Eur Neurol 29(Suppl 3):9–15

During MJ, Fried I, Leone P, Katz A, Spencer DD (1994) Direct measurement of extracellular lactate in the human hippocampus during spontaneous seizures. J Neurochem 62:2356–2361

Early TS, Reiman EM, Raichle ME, Spitznagel EL (1987) Left globus pallidus abnormality in never-medicated patients with schizophrenia. Proc Natl Acad Sci USA 84:561–563

Eidelberg D, Moeller JR, Dhawan V, Sidtis JJ, Ginos JZ, Strother SC, Cedarbaum J, Greene P, Fahn S, Rottenberg DA (1990) The metabolic anatomy of Parkinson's disease: complementary [18F]fluorodeoxyglucose and [18F]fluorodopa positron emission tomographic studies. Mov Disord 5:203–213

Engel J Jr, Kuhl DE, Phelps ME (1982a) Patterns of human local cerebral glucose metabolism during epileptic seizures. Science 218:64–66

Engel J Jr, Kuhl DE, Phelps ME, Crandall PH (1982b) Comparative localization of epileptic foci in partial epilepsy by PCT and EEG. Ann Neurol 12:529–537

Fagan AM, Mintun MA, Mach RH, Lee SY, Dence CS, Shah AR, Larossa GN, Spinner ML, Klunk WE, Mathis CA, DeKosky ST, Morris JC, Holtzman DM (2006) Inverse relation between in vivo amyloid imaging load and cerebrospinal fluid Abeta42 in humans. Ann Neurol 59:512–519

Farkas T, Wolf AP, Jaeger J, Brodie JD, Christman DR, Fowler JS (1984) Regional brain glucose metabolism in chronic schizophrenia. A positron emission transaxial tomographic study Arch Gen Psychiatry 41:293–300

Foster NL, Chase TN, Mansi L, Brooks R, Fedio P, Patronas NJ, Di Chiro G (1984) Cortical abnormalities in Alzheimer's disease. Ann Neurol 16:649–654

Fox PT, Raichle ME (1986) Focal physiological uncoupling of cerebral blood flow and oxidative metabolism during somatosensory stimulation in human subjects. Proc Natl Acad Sci USA 83:1140–1144

Fox PT, Raichle ME, Mintun MA, Dence C (1988) Nonoxidative glucose consumption during focal physiologic neural activity. Science 241:462–464

Frahm J, Kruger G, Merboldt KD, Kleinschmidt A (1996) Dynamic uncoupling and recoupling of perfusion and oxidative metabolism during focal brain activation in man. Magn Reson Med 35:143–148

Francavilla TL, Miletich RS, Di Chiro G, Patronas NJ, Rizzoli HV, Wright DC (1989) Positron emission tomography in the detection of malignant degeneration of low-grade gliomas. Neurosurgery 24:1–5

Franck G, Maquet P, Sadzot B, Salmon E, Debets R, Dive D, Grisar T, Guillaume D, Van Veelen C, Van Huffelen A et al (1992) Contribution of positron emission tomography to the investigation of epilepsies of frontal lobe origin. Adv Neurol 57:471–485

Freeman JM, Vining EP, Pillas DJ, Pyzik PL, Casey JC, Kelly LM (1998) The efficacy of the ketogenic diet-1998: a prospective evaluation of intervention in 150 children. Pediatrics 102:1358–1363

Friedland RP, Jagust WJ, Huesman RH, Koss E, Knittel B, Mathis CA, Ober BA, Mazoyer BM, Budinger TF (1989) Regional cerebral glucose transport and utilization in Alzheimer's disease. Neurology 39:1427–1434

Fueger BJ, Czernin J, Hildebrandt I, Tran C, Halpern BS, Stout D, Phelps ME, Weber WA (2006) Impact of animal handling on the results of 18F-FDG PET studies in mice. J Nucl Med 47:999–1006

Fujimoto T, Takeuchi K, Matsumoto T, Fujita S, Honda K, Higashi Y, Kato N (2008) Metabolic changes in the brain of patients with late-onset major depression. Psychiatry Res 164:48–57

Fujiwara K, Naito Y, Senda M, Mori T, Manabe T, Shinohara S, Kikuchi M, Hori SY, Tona Y, Yamazaki H (2008) Brain metabolism of children with profound deafness: a visual language activation study by 18F-fluorodeoxyglucose positron emission tomography. Acta Otolaryngol 128:393–397

Gaitonde MK, Jones J, Evans G (1987) Metabolism of glucose into glutamate via the hexose monophosphate shunt and its inhibition by 6-aminonicotinamide in rat brain in vivo. Proc.R.Soc. Lond B Biol. Sci 231:71–90

Gibbs EL, Lennox WG, Nims LF, Gibbs FA (1942) Arterial and venous cerebral blood Arterial-venous differences in man. J Biol Chem 18:325–332

Gjedde A (1987) Does deoxyglucose uptake in the brain reflect energy metabolism? Biochem Pharmacol 36:1853–1861

Gjedde A, Wienhard K, Heiss WD, Kloster G, Diemer NH, Herholz K, Pawlik G (1985) Comparative regional analysis of 2-fluorodeoxyglucose and methylglucose uptake in brain of four stroke patients. With special reference to the regional estimation of the lumped constant J Cereb Blood Flow Metab 5:163–178

Greenberg JH, Reivich M, Alavi A, Hand P, Rosenquist A, Rintelmann W, Stein A, Tusa R, Dann R, Christman D, Fowler J, MacGregor B, Wolf A (1981) Metabolic mapping of functional activity in human subjects with the [18F]fluorodeoxyglucose technique. Science 212:678–680

Greene AE, Todorova MT, Seyfried TN (2003) Perspectives on the metabolic management of epilepsy through dietary reduction of glucose and elevation of ketone bodies. J Neurochem 86:529–537

Gruetter R, Seaquist ER, Ugurbil K (2001) A mathematical model of compartmentalized neurotransmitter metabolism in the human brain. Am J Physiol Endocrinol Metab 281: E100–112

Gur RC, Gur RE, Resnick SM, Skolnick BE, Alavi A, Reivich M (1987) The effect of anxiety on cortical cerebral blood flow and metabolism. J Cereb Blood Flow Metab 7:173–177

Haense C, Herholz K, Jagust WJ, Heiss WD (2009) Performance of FDG PET for detection of Alzheimer's disease in two independent multicentre samples (NEST-DD and ADNI). Dement Geriatr Cogn Disord 28:259–266

Hanson MW, Glantz MJ, Hoffman JM, Friedman AH, Burger PC, Schold SC, Coleman RE (1991) FDG-PET in the selection of brain lesions for biopsy. J Comput Assist Tomogr 15:796–801

Hawkins RA, Phelps ME, Huang SC, Kuhl DE (1981) Effect of ischemia on quantification of local cerebral glucose metabolic rate in man. J Cereb Blood Flow Metab 1:37–51

Hawkins RA, Phelps ME, Huang SC (1986) Effects of temporal sampling, glucose metabolic rates, and disruptions of the blood-brain barrier on the FDG model with and without a vascular compartment: studies in human brain tumors with PET. J Cereb Blood Flow Metab 6: 170–183

Hayden MR, Martin WR, Stoessl AJ, Clark C, Hollenberg S, Adam MJ, Ammann W, Harrop R, Rogers J, Ruth T et al (1986) Positron emission tomography in the early diagnosis of Huntington's disease. Neurology 36:888–894

Heiss WD, Szelies B, Kessler J, Herholz K (1991) Abnormalities of energy metabolism in Alzheimer's disease studied with PET. Ann N Y Acad Sci 640:65–71

Heiss WD, Kessler J, Slansky I, Mielke R, Szelies B, Herholz K (1993) Activation PET as an instrument to determine therapeutic efficacy in Alzheimer's disease. Ann N Y Acad Sci 695:327–331

Herholz K, Salmon E, Perani D, Baron JC, Holthoff V, Frolich L, Schonknecht P, Ito K, Mielke R, Kalbe E, Zundorf G, Delbeuck X, Pelati O, Anchisi D, Fazio F, Kerrouche N, Desgranges B, Eustache F, Beuthien-Baumann B, Menzel C, Schroder J, Kato T, Arahata Y, Henze M, Heiss WD (2002) Discrimination between Alzheimer dementia and controls by automated analysis of multicenter FDG PET. Neuroimage 17:302–316

Hertz L, Yu AC, Kala G, Schousboe A (2000) Neuronal-astrocytic and cytosolic-mitochondrial metabolite trafficking during brain activation, hyperammonemia and energy deprivation. Neurochem Int 37:83–102

Ho SS, Berkovic SF, Berlangieri SU, Newton MR, Egan GF, Tochon-Danguy HJ, McKay WJ (1995) Comparison of ictal SPECT and interictal PET in the presurgical evaluation of temporal lobe epilepsy. Ann Neurol 37:738–745

Hothersall JS, Baquer N, Greenbaum AL, McLean P (1979) Alternative pathways of glucose utilization in brain. Changes in the pattern of glucose utilization in brain during development and the effect of phenazine methosulfate on the integration of metabolic routes. Arch Biochem Biophys 198:478–492

Huang SC, Phelps ME, Hoffman EJ, Sideris K, Selin CJ, Kuhl DE (1980) Noninvasive determination of local cerebral metabolic rate of glucose in man. Am J Physiol 238:E69–82

Huang SC, Phelps ME, Hoffman EJ, Kuhl DE (1981) Error sensitivity of fluorodeoxyglucose method for measurement of cerebral metabolic rate of glucose. J Cereb Blood Flow Metab 1:391–401

Hutchins GD, Holden JE, Koeppe RA, Halama JR, Gatley SJ, Nickles RJ (1984) Alternative approach to single-scan estimation of cerebral glucose metabolic rate using glucose analogs, with particular application to ischemia. J Cereb Blood Flow Metab 4:35–40

Ichimiya A, Herholz K, Mielke R, Kessler J, Slansky I, Heiss WD (1994) Difference of regional cerebral metabolic pattern between presenile and senile dementia of the Alzheimer type: a factor analytic study. J Neurol Sci 123:11–17

Idbaih A, Burlet A, Adle-Biassette H, Boisgard R, Coulon C, Paris S, Marie Y, Donadieu J, Hoang-Xuan K, Ribeiro MJ (2007) Altered cerebral glucose metabolism in an animal model of diabetes insipidus: a micro-PET study. Brain Res 1158:164–168

Ikezaki K, Black KL, Conklin SG, Becker DP (1992) Histochemical evaluation of energy metabolism in rat glioma. Neurol Res 14:289–293

Janigro D (1999) Blood-brain barrier, ion homeostatis and epilepsy: possible implications towards the understanding of ketogenic diet mechanisms. Epilepsy Res 37:223–232

Janus TJ, Kim EE, Tilbury R, Bruner JM, Yung WK (1993) Use of [18F]fluorodeoxyglucose positron emission tomography in patients with primary malignant brain tumors. Ann Neurol 33:540–548

Kang DH, Kwon JS, Kim JJ, Youn T, Park HJ, Kim MS, Lee DS, Lee MC (2003) Brain glucose metabolic changes associated with neuropsychological improvements after 4 months of treatment in patients with obsessive-compulsive disorder. Acta Psychiatr Scand 107:291–297

Kennedy SH, Evans KR, Kruger S, Mayberg HS, Meyer JH, McCann S, Arifuzzman AI, Houle S, Vaccarino FJ (2001) Changes in regional brain glucose metabolism measured with positron emission tomography after paroxetine treatment of major depression. Am J Psychiatry 158:899–905

Kim SG, Ugurbil K (1997) Comparison of blood oxygenation and cerebral blood flow effects in fMRI: estimation of relative oxygen consumption change. Magn Reson Med 38:59–65

Kim CK, Gupta NC, Chandramouli B, Alavi A (1994) Standardized uptake values of FDG: body surface area correction is preferable to body weight correction. J Nucl Med 35:164–167

Klunk WE, Engler H, Nordberg A, Wang Y, Blomqvist G, Holt DP, Bergstrom M, Savitcheva I, Huang GF, Estrada S, Ausen B, Debnath ML, Barletta J, Price JC, Sandell J, Lopresti BJ, Wall A, Koivisto P, Antoni G, Mathis CA, Langstrom B (2004) Imaging brain amyloid in Alzheimer's disease with Pittsburgh Compound-B. Ann Neurol 55:306–319

Kornblum HI, Araujo DM, Annala AJ, Tatsukawa KJ, Phelps ME, Cherry SR (2000) In vivo imaging of neuronal activation and plasticity in the rat brain by high resolution positron emission tomography (microPET). Nat Biotechnol 18:655–660

Kuhl DE, Engel J Jr, Phelps ME, Selin C (1980a) Epileptic patterns of local cerebral metabolism and perfusion in humans determined by emission computed tomography of 18FDG and 13NH3. Ann Neurol 8:348–360

Kuhl DE, Phelps ME, Kowell AP, Metter EJ, Selin C, Winter J (1980b) Effects of stroke on local cerebral metabolism and perfusion: mapping by emission computed tomography of 18FDG and 13NH3. Ann Neurol 8:47–60

Kuhl DE, Phelps ME, Markham CH, Metter EJ, Riege WH, Winter J (1982) Cerebral metabolism and atrophy in Huntington's disease determined by 18FDG and computed tomographic scan. Ann Neurol 12:425–434

Kuhr WG, Korf J (1988) Extracellular lactic acid as an indicator of brain metabolism: continuous on-line measurement in conscious, freely moving rats with intrastriatal dialysis. J Cereb Blood Flow Metab 8:130–137

Kumar A, Newberg A, Alavi A, Berlin J, Smith R, Reivich M (1993) Regional cerebral glucose metabolism in late-life depression and Alzheimer disease: a preliminary positron emission tomography study. Proc Natl Acad Sci USA 90:7019–7023

Kuwabara H, Evans AC, Gjedde A (1990) Michaelis-Menten constraints improved cerebral glucose metabolism and regional lumped constant measurements with [18F]fluorodeoxyglucose. J Cereb Blood Flow Metab 10:180–189

Kuwabara H, Ohta S, Brust P, Meyer E, Gjedde A (1992) Density of perfused capillaries in living human brain during functional activation. Prog Brain Res 91:209–215

Kuwert T, Lange HW, Langen KJ, Herzog H, Aulich A, Feinendegen LE (1989) Cerebral glucose consumption measured by PET in patients with and without psychiatric symptoms of Huntington's disease. Psychiatry Res 29:361–362

Kuwert T, Lange HW, Langen KJ, Herzog H, Aulich A, Feinendegen LE (1990) Cortical and subcortical glucose consumption measured by PET in patients with Huntington's disease. Brain 113(Pt 5):1405–1423

Labbe C, Froment JC, Kennedy A, Ashburner J, Cinotti L (1996) Positron emission tomography metabolic data corrected for cortical atrophy using magnetic resonance imaging. Alzheimer Dis Assoc Disord 10:141–170

Lamusuo S, Ruottinen HM, Knuuti J, Harkonen R, Ruotsalainen U, Bergman J, Haaparanta M, Solin O, Mervaala E, Nousiainen U, Jaaskelainen S, Ylinen A, Kalviainen R, Rinne JK, Vapalahti M, Rinne JO (1997) Comparison of [18F]FDG-PET, [99mTc]-HMPAO-SPECT, and [123I]-iomazenil-SPECT in localising the epileptogenic cortex. J Neurol Neurosurg Psychiatry 63:743–748

Leiderman DB, Balish M, Bromfield EB, Theodore WH (1991) Effect of valproate on human cerebral glucose metabolism. Epilepsia 32:417–422

Levin DC, Rao VM, Frangos AJ, Parker L, Sunshine JH (2007) Recent trends in utilization of vascular ultrasound among radiologists, surgeons, cardiologists, and other physicians. J Am Coll Radiol 4:125–127

Li Y, Rinne JO, Mosconi L, Pirraglia E, Rusinek H, Desanti S, Kemppainen N, Nagren K, Kim BC, Tsui W, de Leon MJ (2008) Regional analysis of FDG and PIB-PET images in normal aging, mild cognitive impairment, and Alzheimer's disease. Eur J Nucl Med Mol Imaging 35(12):2169–2181

Lucignani G, Schmidt KC, Moresco RM, Striano G, Colombo F, Sokoloff L, Fazio F (1993) Measurement of regional cerebral glucose utilization with fluorine-18-FDG and PET in heterogeneous tissues: theoretical considerations and practical procedure. J Nucl Med 34:360–369

Machado CJ, Snyder AZ, Cherry SR, Lavenex P, Amaral DG (2008) Effects of neonatal amygdala or hippocampus lesions on resting brain metabolism in the macaque monkey: a microPET imaging study. Neuroimage 39:832–846

Madsen PL, Hasselbalch SG, Hagemann LP, Olsen KS, Bulow J, Holm S, Wildschiodtz G, Paulson OB, Lassen NA (1995) Persistent resetting of the cerebral oxygen/glucose uptake ratio by brain activation: evidence obtained with the Kety-Schmidt technique. J Cereb Blood Flow Metab 15:485–491

Madsen PL, Linde R, Hasselbalch SG, Paulson OB, Lassen NA (1998) Activation-induced resetting of cerebral oxygen and glucose uptake in the rat. J Cereb Blood Flow Metab 18:742–748

Magistretti PJ (2000) Cellular bases of functional brain imaging: insights from neuron-glia metabolic coupling(1). Brain Res 886:108–112

Marder E, Goaillard JM (2006) Variability, compensation and homeostasis in neuron and network function. Nat Rev Neurosci 7:563–574

Mathis CA, Wang Y, Holt DP, Huang GF, Debnath ML, Klunk WE (2003) Synthesis and evaluation of 11C-labeled 6-substituted 2-arylbenzothiazoles as amyloid imaging agents. J Med Chem 46:2740–2754

Mayberg HS, Brannan SK, Mahurin RK, Jerabek PA, Brickman JS, Tekell JL, Silva JA, McGinnis S, Glass TG, Martin CC, Fox PT (1997) Cingulate function in depression: a potential predictor of treatment response. Neuroreport 8:1057–1061

Mayberg HS, Brannan SK, Tekell JL, Silva JA, Mahurin RK, McGinnis S, Jerabek PA (2000) Regional metabolic effects of fluoxetine in major depression: serial changes and relationship to clinical response. Biol Psychiatry 48:830–843

Mazziotta JC, Phelps ME, Carson RE, Kuhl DE (1982) Tomographic mapping of human cerebral metabolism: auditory stimulation. Neurology 32:921–937

Mazziotta JC, Frackowiak RS, Phelps ME (1992) The use of positron emission tomography in the clinical assessment of dementia. Semin Nucl Med 22:233–246

Medina JM, Tabernero A, Tovar JA, Martin-Barrientos J (1996) Metabolic fuel utilization and pyruvate oxidation during the postnatal period. J Inherit Metab Dis 19:432–442

Mega MS, Cummings JL, O'Connor SM, Dinov ID, Reback E, Felix J, Masterman DL, Phelps ME, Small GW, Toga AW (2001) Cognitive and metabolic responses to metrifonate therapy in Alzheimer disease. Neuropsychiatry Neuropsychol Behav Neurol 14:63–68

Meltzer CC, Zubieta JK, Brandt J, Tune LE, Mayberg HS, Frost JJ (1996) Regional hypometabolism in Alzheimer's disease as measured by positron emission tomography after correction for effects of partial volume averaging. Neurology 47:454–461

Mineura K, Yasuda T, Kowada M, Shishido F, Ogawa T, Uemura K (1986) Positron emission tomographic evaluation of histological malignancy in gliomas using oxygen-15 and fluorine-18-fluorodeoxyglucose. Neurol Res 8:164–168

Mineura K, Sasajima T, Kowada M, Ogawa T, Hatazawa J, Shishido F, Uemura K (1994) Perfusion and metabolism in predicting the survival of patients with cerebral gliomas. Cancer 73: 2386–2394

Minoshima S, Giordani B, Berent S, Frey KA, Foster NL, Kuhl DE (1997) Metabolic reduction in the posterior cingulate cortex in very early Alzheimer's disease. Ann Neurol 42:85–94

Mintun MA, Vlassenko AG, Shulman GL, Snyder AZ (2002) Time-related increase of oxygen utilization in continuously activated human visual cortex. Neuroimage 16:531–537

Mintun MA, Sacco D, Snyder AZ, Couture P, Powers WJ, Hornbeck R, Videen TO, McGee-Minnich L, Perlmutter JS, Mach RH, Morris JC, Raichle ME (2006) Distribution of glycolysis in the resting healthy human brain correlates with distribution of beta-amyloid plaques in Alzheimer's disease. In: 2006 Neuroscience Meeting Planner No. 707.6

Mirrione MM, Schiffer WK, Siddiq M, Dewey SL, Tsirka SE (2006) PET imaging of glucose metabolism in a mouse model of temporal lobe epilepsy. Synapse 59:119–121

Mosconi L, Tsui WH, Herholz K, Pupi A, Drzezga A, Lucignani G, Reiman EM, Holthoff V, Kalbe E, Sorbi S, Diehl-Schmid J, Perneczky R, Clerici F, Caselli R, Beuthien-Baumann B, Kurz A, Minoshima S, de Leon MJ (2008) Multicenter standardized 18F-FDG PET diagnosis of mild cognitive impairment, Alzheimer's disease, and other dementias. J Nucl Med 49: 390–398

Murase K, Kuwabara H, Yasuhara Y, Evans AC, Gjedde A (1996) Mapping of change in cerebral glucose utilization using fluorine-18 fluorodeoxyglucose double injection and the constrained weighted-integration method. IEEE Trans Med Imaging 15:824–835

Naidoo-Variawa S, Hey-Cunningham AJ, Lehnert W, Kench PL, Kassiou M, Banati R, Meikle SR (2007) High-resolution imaging of the large non-human primate brain using microPET: a feasibility study. Phys Med Biol 52:6627–6638

Nehlig A, Pereira dV (1993) Glucose and ketone body utilization by the brain of neonatal rats. Prog Neurobiol 40:163–221

Nordahl TE, Benkelfat C, Semple WE, Gross M, King AC, Cohen RM (1989) Cerebral glucose metabolic rates in obsessive compulsive disorder. Neuropsychopharmacology 2:23–28

Nordberg A, Lilja A, Lundqvist H, Hartvig P, Amberla K, Viitanen M, Warpman U, Johansson M, Hellstrom-Lindahl E, Bjurling P et al (1992) Tacrine restores cholinergic nicotinic receptors and glucose metabolism in Alzheimer patients as visualized by positron emission tomography. Neurobiol Aging 13:747–758

Nordli DR Jr, Kuroda MM, Carroll J, Koenigsberger DY, Hirsch LJ, Bruner HJ, Seidel WT, De VDC (2001) Experience with the ketogenic diet in infants. Pediatrics 108:129–133

Ochs RF, Gloor P, Tyler JL, Wolfson T, Worsley K, Andermann F, Diksic M, Meyer E, Evans A (1987) Effect of generalized spike-and-wave discharge on glucose metabolism measured by positron emission tomography. Ann Neurol 21:458–464

Patlak CS, Blasberg RG, Fenstermacher JD (1983) Graphical evaluation of blood-to-brain transfer constants from multiple-time uptake data. J Cereb Blood Flow Metab 3:1–7

Patlak CS, Blasberg RG (1985) Graphical evaluation of blood-to-brain transfer constants from multiple-time uptake data. Generalizations. J Cereb Blood Flow Metab 5:584–590

Patronas NJ, Di Chiro G, Brooks RA, DeLaPaz RL, Kornblith PL, Smith BH, Rizzoli HV, Kessler RM, Manning RG, Channing M, Wolf AP, O'Connor CM (1982) Work in progress: [18F] fluorodeoxyglucose and positron emission tomography in the evaluation of radiation necrosis of the brain. Radiology 144:885–889

Patronas NJ, Di Chiro G, Kufta C, Bairamian D, Kornblith PL, Simon R, Larson SM (1985) Prediction of survival in glioma patients by means of positron emission tomography. J Neurosurg 62:816–822

Paul AK, Lobarinas E, Simmons R, Wack D, Luisi JC, Spernyak J, Mazurchuk R, Abdel-Nabi H, Salvi R (2008) Metabolic imaging of rat brain during pharmacologically-induced tinnitus. Neuroimage 44(2):312–318

Pellerin L, Magistretti PJ (1994) Glutamate uptake into astrocytes stimulates aerobic glycolysis: a mechanism coupling neuronal activity to glucose utilization. Proc Natl Acad Sci USA 91:10625–10629

Phelps ME, Mazziotta JC (1985) Positron emission tomography: human brain function and biochemistry. Science 228:799–809

Phelps ME, Huang SC, Hoffman EJ, Selin C, Sokoloff L, Kuhl DE (1979) Tomographic measurement of local cerebral glucose metabolic rate in humans with (F-18)2-fluoro-2-deoxy-D-glucose: validation of method. Ann Neurol 6:371–388

Phelps ME, Mazziotta JC, Kuhl DE, Nuwer M, Packwood J, Metter J, Engel J Jr (1981) Tomographic mapping of human cerebral metabolism visual stimulation and deprivation. Neurology 31:517–529

Pietrini P, Dani A, Furey ML, Alexander GE, Freo U, Grady CL, Mentis MJ, Mangot D, Simon EW, Horwitz B, Haxby JV, Schapiro MB (1997) Low glucose metabolism during brain stimulation in older Down's syndrome subjects at risk for Alzheimer's disease prior to dementia. Am J Psychiatry 154:1063–1069

Pietrzyk U, Herholz K, Fink G, Jacobs A, Mielke R, Slansky I, Wurker M, Heiss WD (1994) An interactive technique for three-dimensional image registration: validation for PET, SPECT, MRI and CT brain studies. J Nucl Med 35:2011–2018

Potkin SG, Anand R, Fleming K, Alva G, Keator D, Carreon D, Messina J, Wu JC, Hartman R, Fallon JH (2001) Brain metabolic and clinical effects of rivastigmine in Alzheimer's disease. Int J Neuropsychopharmacol 4:223–230

Prichard J, Rothman D, Novotny E, Petroff O, Kuwabara T, Avison M, Howseman A, Hanstock C, Shulman R (1991) Lactate rise detected by 1H NMR in human visual cortex during physiologic stimulation. Proc Natl Acad Sci USA 88:5829–5831

Przedborski S, Goldman S, Giladi N, Dhawan V, Takikawa S, Hildebrand J, Fahn S, Eidelberg D (1993) Positron emission tomography in hemiparkinsonism-hemiatrophy syndrome. Adv Neurol 60:501–505

Raichle ME, Mintun MA (2006) Brain work and brain imaging. Annu Rev Neurosci 29:449–476

Raichle ME, Posner JB, Plum F (1970) Cerebral blood flow during and after hyperventilation. Arch Neurol 23:394–403

Raichle ME, MacLeod AM, Snyder AZ, Powers WJ, Gusnard DA, Shulman GL (2001) A default mode of brain function. Proc Natl Acad Sci USA 98:676–682

Reivich M, Kuhl D, Wolf A, Greenberg J, Phelps M, Ido T, Casella V, Fowler J, Hoffman E, Alavi A, Som P, Sokoloff L (1979) The [18F]fluorodeoxyglucose method for the measurement of local cerebral glucose utilization in man. Circ Res 44:127–137

Reivich M, Alavi A, Wolf A, Fowler J, Russell J, Arnett C, MacGregor RR, Shiue CY, Atkins H, Anand A (1985) Glucose metabolic rate kinetic model parameter determination in humans: the lumped constants and rate constants for [18F]fluorodeoxyglucose and [11C]deoxyglucose. J Cereb Blood Flow Metab 5:179–192

Rhodes CG, Wise RJ, Gibbs JM, Frackowiak RS, Hatazawa J, Palmer AJ, Thomas DG, Jones T (1983) In vivo disturbance of the oxidative metabolism of glucose in human cerebral gliomas. Ann Neurol 14:614–626

Richter D, Dawson RM (1948) Brain lactate in emotion. Nature 161:205

Rougemont D, Baron JC, Collard P, Bustany P, Comar D, Agid Y (1984) Local cerebral glucose utilisation in treated and untreated patients with Parkinson's disease. J Neurol Neurosurg Psychiatry 47:824–830

Rowland DJ, Cherry SR (2008) Small-animal preclinical nuclear medicine instrumentation and methodology. Semin Nucl Med 38:209–222

Saha GB (2005) Basics of PET imaging: physics, chemistry, and regulations. Springer, New York

Sato K, Kashiwaya Y, Keon CA, Tsuchiya N, King MT, Radda GK, Chance B, Clarke K, Veech RL (1995) Insulin, ketone bodies, and mitochondrial energy transduction. FASEB J 9:651–658

Scheinberg P, Bourne B, Reinmuth OM (1965) Human Cerebral Lactate and Pyruvate Extraction. I Control Subjects Arch Neurol 12:246–250

Schlemmer HP, Pichler BJ, Schmand M, Burbar Z, Michel C, Ladebeck R, Jattke K, Townsend D, Nahmias C, Jacob PK, Heiss WD, Claussen CD (2008) Simultaneous MR/PET imaging of the human brain: feasibility study. Radiology 248:1028–1035

Schlemmer HP, Pichler BJ, Krieg R, Heiss WD (2009) An integrated MR/PET system: prospective applications. Abdom Imaging 34:668–674

Schreckenberger M, Spetzger U, Sabri O, Meyer PT, Zeggel T, Zimny M, Gilsbach J, Buell U (2001) Localisation of motor areas in brain tumour patients: a comparison of preoperative [18F]FDG-PET and intraoperative cortical electrostimulation. Eur J Nucl Med 28:1394–1403

Schuier F, Orzi F, Suda S, Lucignani G, Kennedy C, Sokoloff L (1990) Influence of plasma glucose concentration on lumped constant of the deoxyglucose method: effects of hyperglycemia in the rat. J Cereb Blood Flow Metab 10:765–773

Schwartz JM, Baxter LR Jr, Mazziotta JC, Gerner RH, Phelps ME (1987) The differential diagnosis of depression. Relevance of positron emission tomography studies of cerebral glucose metabolism to the bipolar-unipolar dichotomy. JAMA 258:1368–1374

Schwartzkroin PA (1999) Mechanisms underlying the anti-epileptic efficacy of the ketogenic diet. Epilepsy Res 37:171–180

Schwechter EM, Veliskova J, Velisek L (2003) Correlation between extracellular glucose and seizure susceptibility in adult rats. Ann Neurol 53:91–101

Silver IA, Erecinska M (1997) Energetic demands of the Na+/K+ ATPase in mammalian astrocytes. Glia 21:35–45

Silverman DH, Small GW, Chang CY, Lu CS, Kung De Aburto MA, Chen W, Czernin J, Rapoport SI, Pietrini P, Alexander GE, Schapiro MB, Jagust WJ, Hoffman JM, Welsh-Bohmer KA, Alavi A, Clark CM, Salmon E, de Leon MJ, Mielke R, Cummings JL, Kowell AP, Gambhir SS, Hoh CK, Phelps ME (2001) Positron emission tomography in evaluation of dementia: Regional brain metabolism and long-term outcome. JAMA 286:2120–2127

Smith TA (1998) FDG uptake, tumour characteristics and response to therapy: a review. Nucl Med Commun 19:97–105

Sokoloff L, Reivich M, Kennedy C, Des Rosiers MH, Patlak CS, Pettigrew KD, Sakurada O, Shinohara M (1977) The [14C]deoxyglucose method for the measurement of local cerebral glucose utilization: theory, procedure, and normal values in the conscious and anesthetized albino rat. J Neurochem 28:897–916

Sols A, Crane RK (1954) Substrate specificity of brain hexokinase. J Biol Chem 210:581–595

Spanaki MV, Siegel H, Kopylev L, Fazilat S, Dean A, Liow K, Ben-Menachem E, Gaillard WD, Theodore WH (1999a) The effect of vigabatrin (gamma-vinyl GABA) on cerebral blood flow and metabolism. Neurology 53:1518–1522

Spanaki MV, Spencer SS, Corsi M, MacMullan J, Seibyl J, Zubal IG (1999b) Sensitivity and specificity of quantitative difference SPECT analysis in seizure localization. J Nucl Med 40: 730–736

Spence AM, Muzi M, Graham MM, O'Sullivan F, Krohn KA, Link JM, Lewellen TK, Lewellen B, Freeman SD, Berger MS, Ojemann GA (1998) Glucose metabolism in human malignant gliomas measured quantitatively with PET, 1-[C-11]glucose and FDG: analysis of the FDG lumped constant. J Nucl Med 39:440–448

Spence AM, Muzi M, Mankoff DA, O'Sullivan SF, Link JM, Lewellen TK, Lewellen B, Pham P, Minoshima S, Swanson K, Krohn KA (2004) 18F-FDG PET of gliomas at delayed intervals: improved distinction between tumor and normal gray matter. J Nucl Med 45:1653–1659

Strauss LG (1996) Fluorine-18 deoxyglucose and false-positive results: a major problem in the diagnostics of oncological patients. Eur J Nucl Med 23:1409–1415

Suda S, Shinohara M, Miyaoka M, Lucignani G, Kennedy C, Sokoloff L (1990) The lumped constant of the deoxyglucose method in hypoglycemia: effects of moderate hypoglycemia on local cerebral glucose utilization in the rat. J Cereb Blood Flow Metab 10:499–509

Sung KK, Jang DP, Lee S, Kim M, Lee SY, Kim YB, Park CW, Cho ZH (2008) Neural responses in rat brain during acute immobilization stress: a [F-18]FDG micro PET imaging study. Neuroimage 44(3):1074–1080

Swedo SE, Schapiro MB, Grady CL, Cheslow DL, Leonard HL, Kumar A, Friedland R, Rapoport SI, Rapoport JL (1989) Cerebral glucose metabolism in childhood-onset obsessive-compulsive disorder. Arch Gen Psychiatry 46:518–523

Takikawa S, Dhawan V, Spetsieris P, Robeson W, Chaly T, Dahl R, Margouleff D, Eidelberg D (1993) Noninvasive quantitative fluorodeoxyglucose PET studies with an estimated input function derived from a population-based arterial blood curve. Radiology 188:131–136

Ter-Pogossian MM (1992) The origins of positron emission tomography. Semin Nucl Med 22:140–149

Theodore WH, Newmark ME, Sato S, Brooks R, Patronas N, De La Paz R, DiChiro G, Kessler RM, Margolin R, Manning RG et al (1983) [18F]fluorodeoxyglucose positron emission tomography in refractory complex partial seizures. Ann Neurol 14:429–437

Theodore WH, Bairamian D, Newmark ME, DiChiro G, Porter RJ, Larson S, Fishbein D (1986a) Effect of phenytoin on human cerebral glucose metabolism. J Cereb Blood Flow Metab 6:315–320

Theodore WH, Dorwart R, Holmes M, Porter RJ, DiChiro G (1986b) Neuroimaging in refractory partial seizures: comparison of PET, CT, and MRI. Neurology 36:750–759

Theodore WH, Fishbein D, Dubinsky R (1988) Patterns of cerebral glucose metabolism in patients with partial seizures. Neurology 38:1201–1206

Theodore WH, Bromfield E, Onorati L (1989) The effect of carbamazepine on cerebral glucose metabolism. Ann Neurol 25:516–520

Townsend DW (2008) Positron emission tomography/computed tomography. Semin Nucl Med 38:152–166

Ueki M, Linn F, Hossmann KA (1988) Functional activation of cerebral blood flow and metabolism before and after global ischemia of rat brain. J Cereb Blood Flow Metab 8:486–494

Uijl SG, Leijten FS, Arends JB, Parra J, van Huffelen AC, Moons KG (2007) The added value of [18F]-fluoro-D-deoxyglucose positron emission tomography in screening for temporal lobe epilepsy surgery. Epilepsia 48:2121–2129

Vaishnavi SN, Vlassenko AG, Rundle MM, Snyder AZ, Mintun MA, Raichle ME (2010) Regional aerobic glycolysis in the human brain. Proc Natl Acad Sci USA 107:17757–17762

Veech RL, Chance B, Kashiwaya Y, Lardy HA, Cahill GF, Jr (2001) Ketone bodies, potential therapeutic uses. IUBMB. Life 51:241–247

Vicario C, Tabernero A, Medina JM (1993) Regulation of lactate metabolism by albumin in rat neurons and astrocytes from primary culture. Pediatr Res 34:709–715

Videen TO, Perlmutter JS, Mintun MA, Raichle ME (1988) Regional correction of positron emission tomography data for the effects of cerebral atrophy. J Cereb Blood Flow Metab 8:662–670

Vielhaber S, Von Oertzen JH, Kudin AF, Schoenfeld A, Menzel C, Biersack HJ, Kral T, Elger CE, Kunz WS (2003) Correlation of hippocampal glucose oxidation capacity and interictal FDG-PET in temporal lobe epilepsy. Epilepsia 44:193–199

Vining EP, Freeman JM, Ballaban-Gil K, Camfield CS, Camfield PR, Holmes GL, Shinnar S, Shuman R, Trevathan E, Wheless JW (1998) A multicenter study of the efficacy of the ketogenic diet. Arch Neurol 55:1433–1437

Vlassenko AG, Thiessen B, Beattie BJ, Malkin MG, Blasberg RG (2000) Evaluation of early response to SU101 target-based therapy in patients with recurrent supratentorial malignant gliomas using FDG PET and Gd-DTPA MRI. J Neurooncol 46:249–259

Vlassenko AG, Rundle MM, Mintun MA (2006) Human brain glucose metabolism may evolve during activation: findings from a modified FDG PET paradigm. Neuroimage 33:1036–1041

Vlassenko AG, Vaishnavi NS, Couture L, Sacco D, Shannon BJ, Mach RH, Morris JC, Raichle ME, Mintun MA (2010) Spatial correlation between brain aerobic glycolysis and amyloid-(Aβ) deposition. Proc Natl Acad Sci USA 107:17763–17767

Volkow ND, Brodie JD, Wolf AP, Angrist B, Russell J, Cancro R (1986) Brain metabolism in patients with schizophrenia before and after acute neuroleptic administration. J Neurol Neurosurg Psychiatry 49:1199–1202

Walhovd KB, Fjell AM, Brewer J, McEvoy LK, Fennema-Notestine C, Hagler DJ Jr, Jennings RG, Karow D, Dale AM (2010) Combining MR imaging, positron-emission tomography, and CSF biomarkers in the diagnosis and prognosis of Alzheimer disease. AJNR Am J Neuroradiol 31:347–354

Wang GJ, Volkow ND, Telang F, Jayne M, Ma J, Rao M, Zhu W, Wong CT, Pappas NR, Geliebter A, Fowler JS (2004) Exposure to appetitive food stimuli markedly activates the human brain. Neuroimage 21:1790–1797

Weber G (1977) Enzymology of cancer cells (second of two parts). N Engl J Med 296:541–551

Wienhard K (2002) Measurement of glucose consumption using [(18)F]fluorodeoxyglucose. Methods 27:218–225

Wienhard K, Pawlik G, Herholz K, Wagner R, Heiss WD (1985) Estimation of local cerebral glucose utilization by positron emission tomography of [18F]2-fluoro-2-deoxy-D-glucose: a critical appraisal of optimization procedures. J Cereb Blood Flow Metab 5:115–125

Wiesel FA, Wik G, Sjogren I, Blomqvist G, Greitz T, Stone-Elander S (1987) Regional brain glucose metabolism in drug free schizophrenic patients and clinical correlates. Acta Psychiatr Scand 76:628–641

Wolkin A, Jaeger J, Brodie JD, Wolf AP, Fowler J, Rotrosen J, Gomez-Mont F, Cancro R (1985) Persistence of cerebral metabolic abnormalities in chronic schizophrenia as determined by positron emission tomography. Am J Psychiatry 142:564–571

Wolkin A, Angrist B, Wolf A, Brodie JD, Wolkin B, Jaeger J, Cancro R, Rotrosen J (1988) Low frontal glucose utilization in chronic schizophrenia: a replication study. Am J Psychiatry 145:251–253

Wong KP, Feng D, Meikle SR, Fulham MJ (2001) Simultaneous estimation of physiological parameters and the input function–in vivo PET data. IEEE Trans Inf Technol Biomed 5:67–76

Woods RP, Grafton ST, Holmes CJ, Cherry SR, Mazziotta JC (1998a) Automated image registration: I. General methods and intrasubject, intramodality validation. J Comput Assist Tomogr 22:139–152

Woods RP, Grafton ST, Watson JD, Sicotte NL, Mazziotta JC (1998b) Automated image registration: II. Intersubject validation of linear and nonlinear models. J Comput Assist Tomogr 22:153–165

Yoshii F, Barker WW, Chang JY, Loewenstein D, Apicella A, Smith D, Boothe T, Ginsberg MD, Pascal S, Duara R (1988) Sensitivity of cerebral glucose metabolism to age, gender, brain volume, brain atrophy, and cerebrovascular risk factors. J Cereb Blood Flow Metab 8:654–661

Young AB, Penney JB, Starosta-Rubinstein S, Markel DS, Berent S, Giordani B, Ehrenkaufer R, Jewett D, Hichwa R (1986) PET scan investigations of Huntington's disease: cerebral metabolic correlates of neurological features and functional decline. Ann Neurol 20:296–303

Zaidi H, Montandon ML (2006) The new challenges of brain PET Imaging technology. Curr Med Imaging Rev 2:3–13

Zanotti-Fregonara P, Maroy R, Comtat C, Jan S, Gaura V, Bar-Hen A, Ribeiro MJ, Trebossen R (2009) Comparison of 3 methods of automated internal carotid segmentation in human brain PET studies: application to the estimation of arterial input function. J Nucl Med 50:461–467

Zasadny KR, Wahl RL (1993) Standardized uptake values of normal tissues at PET with 2-[fluorine-18]-fluoro-2-deoxy-D-glucose: variations with body weight and a method for correction. Radiology 189:847–850

Chapter 11
Neuroreceptor Imaging

Habib Zaidi and Miho Shidahara

Abstract Neuroreceptor imaging using PET and SPECT has contributed to clinical neuroscience and diagnosis (e.g. neurodegenerative disease and antipsychotic-drug receptor occupancy). Recent advances in dedicated PET and SPECT instrumentation, disease-specific radioligands, and image analysis techniques contributed to the further development of this field and its widespread application. In this chapter, we introduce the basis concepts of neuroreceptor imaging using typical radioligands, standard modeling and analysis techniques and discuss future research opportunities in the field.

Keywords Kinetic modeling • PET • Quantification • Receptor imaging • SPECT

11.1 Basic Principles of Neuroreceptor Imaging

11.1.1 Neuroimaging Using PET and SPECT

The last quarter century has witnessed the introduction of a variety of powerful medical imaging technologies allowing the visualization of organ structure and function with exquisite detail. Modern functional brain mapping techniques such as positron emission tomography (PET), single-photon emission computed tomography

H. Zaidi, Ph.D. (✉)
Division of Nuclear Medicine and Molecular Imaging, Geneva University Hospital, CH-1211 Geneva, Switzerland

Geneva Neuroscience Center, Geneva University, Geneva, Switzerland

Department of Nuclear Medicine and Molecular Imaging, University Medical Center Groningen, University of Groningen, Groningen, Netherlands
e-mail: habib.zaidi@hcuge.ch

M. Shidahara, Ph.D.
Division of Medical Physics, Tohoku University Graduate School of Medicine, Sendai, Japan
e-mail: shidahara@med.tohoku.ac.jp

I.-Y. Choi and R. Gruetter (eds.), *Neural Metabolism In Vivo*, Advances in Neurobiology 4, DOI 10.1007/978-1-4614-1788-0_11,

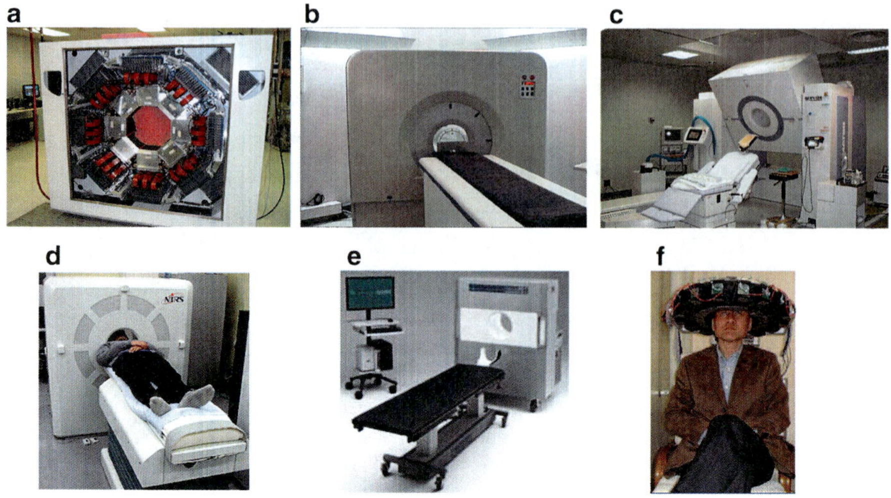

Fig. 11.1 Photographs of dedicated brain PET scanners showing (**a**) the HRRT camera based on LSO scintillation crystals and the phoswich concept, (**b**) the GSO-based PET (G-PET) camera, (**c**) the Hamamatsu SHR-12000 PET scanner based on BGO detector blocks, jPET-D4 brain PET scanner (**d**), the NeuroPET™ (**e**), and the PET-Hat wearable PET system (**f**)

(SPECT), functional magnetic resonance imaging (fMRI), electro-encephalography (EEG), magneto-encephalography (MEG), optical imaging and neuroanatomical tools have been used for assessing the functional organization of the human brain (Gilman 1998; Hammoud et al. 2007). Through these techniques, neuroscience has progressed to a great extent in the understanding of the brain in health and disease.

The specific role of molecular SPECT and PET imaging in the expansion of our understanding of the pathophysiological mechanisms of neurological and psychiatric diseases and in the clinical management of patients is steadily progressing (Jacobs et al. 2003). Both imaging modalities allow *in vivo*, non-invasive three-dimensional (3-D) imaging of regional cerebral blood flow, metabolism and neuroreceptor binding in the brain, amino acid synthesis or amyloid plaque (Heiss and Herholz 2006). Since functional disturbances occur often earlier than structural ones, a faster and more sensitive detection of pathology is possible, compared with anatomical imaging with, for example structural MRI.

During the last two decades, functional brain imaging using SPECT and PET has advanced elegantly and steadily gained importance in the clinical and research arenas. Various clinical indications for molecular neuroimaging using SPECT and PET were adopted and novel applications are being explored. The demand for non-invasive functional, metabolic, and molecular imaging of the brain has stimulated the development of dedicated high-resolution SPECT (Accorsi 2008) and PET (Sossi 2007; Zaidi and Montandon 2006) instrumentation and the synthesis of new radiopharmaceuticals (Fowler et al. 2004; Halldin et al. 2001). Current existing commercial brain PET technology and other dedicated prototype designs constitute state-of-the-art high resolution PET instrumentation dedicated for brain research. Figure 11.1 shows photographs of the some of these designs.

Software- and hardware-based multimodality brain imaging allowing the correlation between anatomical and molecular information has revolutionized clinical diagnosis and now offers unique capabilities for the clinical neuroimaging community and neuroscience researchers at large. To respond to the requirements of emerging clinical and research applications of correlated anatomical and functional brain imaging, several innovative developments in high performance standalone (PET and MRI) and dual-modality PET-MR imaging instrumentation combining both modalities, thus allowing simultaneous or sequential PET and MR brain imaging have been proposed and are currently being assessed in clinical and research settings (Cho et al. 2008; Schlemmer et al. 2008).

This chapter discusses recent advances in neuroreceptor imaging and advanced quantitative imaging procedures in clinical setting. Future opportunities and challenges facing the adoption of neuroreceptor imaging will also be addressed.

11.1.2 Physiological Aspects of Neuroreceptors and Transporters

Receptors are structures, typically proteins, on cellular membranes that, following interaction with specific ligands (first messenger transmitters), activate a signal instigate distinct responses interceded by secondary messengers (G-protein-coupled receptors) as ion channels (ligand-gated ion channels) (Heiss and Herholz 2006). Receptors are usually characterized by their affinity and density and are generally degraded by specific proteases after a functional period. Radioligands are used to visualize and quantify the distribution, density and activity of neuroreceptors.

Apart from indirect cerebral parameters such as metabolism and perfusion, molecular neuroimaging techniques can also offer the unique possibility to image a large set of receptors. Nerve signals pass through synapse between different neurons. This process is a chemical event in which neurotransmitters from the presynaptic nerve terminals are released to the synaptic cleft, and act on the postsynaptic neurotransmitters to induce excitation or inhibition on the target neuron. Through reuptake sites (transporter) on the presynaptic membrane, the surplus in neurotransmitters can be recycled. Changes of important neurotransmitter systems such as the dopamine, GABA (gamma aminobutyric acid), benzodiazepine, choline and serotonin neurotransmitter systems have been demonstrated in many neurological and psychiatric disease stages. For a detailed overview, the reader is referred to Tikofsky et al. (1999).

Receptors play a key role in neurotransmission and neuromodulation and as such are involved in almost all physiological functions of the brain from motor performance to memory, emotion, and pain (Heiss and Herholz 2006). The possibilities offered by molecular imaging to visualize the distribution, density and activity of various receptors open a far-fetched plethora of tools allowing to investigate the physiological activities of functional networks and their disruption by neurologic and psychiatric disorders. In addition, imaging can be efficiently used for assessing selective drug action.

11.2 Radioligands for Neuroreceptor Imaging

11.2.1 Radioligands

The word "ligand" in biochemistry denotes a molecule that binds to a specific protein (e.g. receptor). The distribution, density and affinity of receptors can be visualized and quantified by "radioactive ligands" based on the tracer principle. Over the last three decades, a large number of radioactive ligands have been developed; however, most of them were used only in laboratory animals and primates (e.g. rodents and monkeys). Only a small portion (but not a small number) was introduced and successfully used for human imaging studies (Table 11.1).

Typical isotopes for labeling are ^{11}C (physical half-life $T_{1/2}=20.4$ min) and ^{18}F ($T_{1/2}=109.8$ min) for PET and ^{123}I ($T_{1/2}=13.2$ h) and ^{99m}Tc ($T_{1/2}=6.01$ h) for SPECT. Although the use of short-lived isotopes for PET and SPECT presents the disadvantage in terms of rapid synthesis requirements, their use enable the administration of a reasonable amount of activity for imaging resulting in a relatively low dose to subjects. In the human body, Carbon is contained within every organic molecule and Iodine is contained within several organic molecules which allow isotopic

Table 11.1 Examples of receptor radioligands for human PET and SPECT studies (Adapted from Halldin et al. 2001)

Binding system	Radioligand	Binding system	Radioligand
Dopamine D_1	[^{11}C]NNC 112	Serotonin transporter	[^{11}C]DASB
	[^{11}C]SCH 23390		[^{11}C]McN5652
			[^{123}I]ADAM
			[^{123}I] β-CIT
Dopamine D_2	[^{11}C]Raclopride	Musacrinic	[^{11}C]MNPB
	[^{11}C]FLB 457		[^{123}I]IQNB
	[^{11}C]MNSP		[^{123}I]IQNP
	[^{11}C]MNPA	Nicotinic	Epibatidine
	[^{18}F]fallypride	Opiate	[^{11}C]diprenorphine
	[^{123}I]epidepride		[^{11}C]carfentenyl
	[^{123}I]IBZM		
Dopamine transporter	[^{11}C]PE2I	Peripheral benzodiazepine	[^{11}C]PK 11195
	[^{11}C]β-CFT		[^{11}C]DAA 1106
	[^{11}C]altropane		[^{11}C]AC-5216
	[^{123}I] β-CIT	Benzodiazepine	[^{11}C]flumazenil
	[^{123}I]PE2I		[^{11}C]Ro15 1788
	[^{99m}Tc]altropane		[^{11}C]Ro15 4513
	[^{99m}Tc]TRODAT-1		[^{123}I]iomazenil
Serotonin 5-HT_{1A}	[^{11}C]WAY 100635	Amyloid	[^{11}C]PIB
Serotonin 5-HT_{2A}	[^{11}C]MDL 100907		[^{11}C]BF 227
	[^{11}C]MNSP		[^{11}C]FDDNP
	[^{18}F]altanserin		[^{123}I]IMPY

labeling. Fluorine and Technetium in general do not occur in biologically important compounds. Ligands labeled with ^{11}C offer the great advantage of making repeated examinations in the same subject on the same day feasible owing to its short half-life. On the other hands, the longer half-lives of ^{18}F, ^{99m}Tc, ^{123}I compared to ^{11}C can be attractive for the study of biological processes with relatively slow kinetics. At present, PET represents the most selective and sensitive method for measuring receptor density and interactions *in vivo*. SPECT can also be applied but the sensitivity is by approximately two or three orders of magnitude lower than PET owing to the use of a physical collimation (Rahmim and Zaidi 2008). Longer half-lives may partially compensate for SPECT disadvantages if measurement times over several hours are required. Furthermore, SPECT also enables simultaneous trace of radioligand behavior in the subject through the combination of several isotopes with different emitting energies (Dresel et al. 1999).

Radioligands for neuroreceptor imaging should satisfy the following criteria: stability of labeling radioisotopes, sufficient nanomolar affinity and high selectivity for the specific receptor with low nonspecific binding in brain tissue, rapid permeation through blood-brain barrier (BBB) (Halldin et al. 2001). Furthermore, metabolites in plasma should be avoided to enter the brain through BBB (Pike 2009).

To increase the selectivity of radioligands to specific binding, low non-specific binding is preferable (Pike 2009). The term "non-specific binding" refers to the binding of a radioligand to the lipid bilayer and is generally proportional to the lipophilicity of the ligand (Rosso et al. 2008). The logarithm of the partition coefficient between buffers accounting for ionization at physiological pH and octanol (log P) is often taken as a useful index for the lipophilicity of a compound in the context of biological systems. Conversely, some degree of lipid solubility is needed for passing over the BBB. However, the lipophilic nature of a molecule might also favor binding to plasma proteins, thus reducing the available "free fraction" in blood, which is the fraction of the tracer available for transport into brain tissue. Moreover, lipophilic molecules are also more likely to be extracted and metabolized in lung circulation (Halldin et al. 2001).

11.2.2 Imaging Targets

11.2.2.1 Dopamine Transporters and Receptors

Dopaminergic neurotransmission has a central role in many brain functions. Both pre- and post-synaptic dopaminergic functions can be visualized and quantified through the use of various radioligands (Table 11.1). For the measurement of dopamine D_1 receptor binding, [^{11}C]SCH 23390 (Farde et al. 1987; Halldin et al. 1986) and [^{11}C]NNC 112 (Halldin et al. 1998) are used. On the other hand, dopamine D_2 receptor densities are measured with the low affinity [^{11}C]raclopride in striatum (high D_2 density) and with the high affinity [^{11}C]FLB 457 for extrastriatal low density

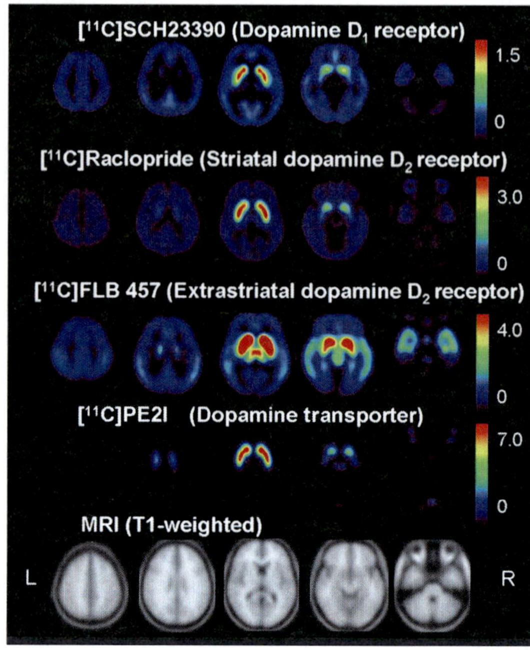

Fig. 11.2 Representative slices of dopaminergic PET imaging using various radioligands: averaged binding potential image of ten young healthy subjects (Data courtesy of Dr. Hiroshi Ito, NIRS)

regions, respectively. Presynaptic dopaminergic function is quantified measuring the dopamine transporter binding with [^{11}C]β-CIT (Farde et al. 1994; Muller et al. 1993), [^{11}C]PE2I (Emond et al. 1997), [^{99m}Tc]TRODAT-1 (Kung et al. 1996), and many others (Brucke et al. 1993) while the endogeneous dopamine synthesis rate can be measured by 6-[^{18}F]fluoro-L-Dopa and L-[b-^{11}C]DOPA (Hartvig et al. 1991). Figure 11.2 demonstrates PET imaging of the dopaminergic system in the human brain that has the highest density in striatum (Ito et al. 2008).

11.2.2.2 Serotonin Transporters and Receptors

Serotonin receptors (5-hydroxytryptamine, 5-HT) and the 5-HT transporter have an important role in affective disorders and for the assessment of the activity of antidepressants. Postsynaptic 5-HT receptor are diversified in 7 major classes containing more than 16 subtypes. Suitable radioligands are only available for 5-HT$_{1A}$ and 5-HT$_{2A}$ subtypes. For the measurement of 5-HT$_{1A}$ receptor binding, the radioligand of choice is [^{11}C]WAY 100635 (Pike et al. 1996). To measure the binding of 5-HT$_{2A}$ receptor, [^{11}C]MNSP (Burns et al. 1984), [^{11}C]MDL 100,907 (Ito et al. 1998b; Lundkvist et al. 1996; Watabe et al. 2000) and [^{18}F]altanserin (Crouzel et al. 1992) are used. Presynaptic 5-HT function, 5-HT transporter binding can be measured with [^{11}C]DASB (Houle et al. 2000; Wilson et al. 2000), [^{123}I] β-CIT (Laruelle et al. 1993), and many other radioligands (Oya et al. 2000; Suehiro et al. 1993). 5-HT transporter belongs to the same family of the transporters for dopamine and

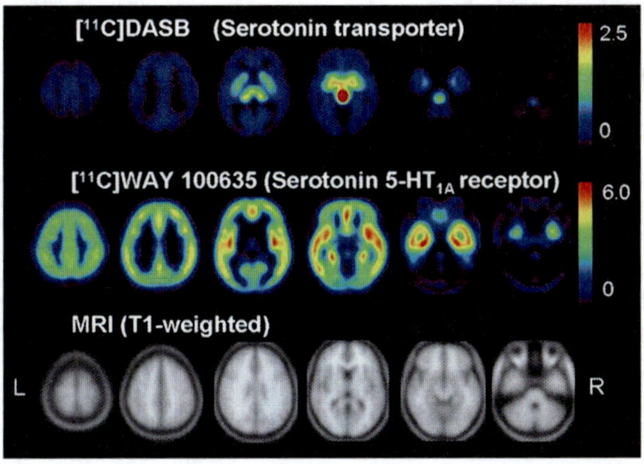

Fig. 11.3 PET images of 5-HT transporter and receptor ligands: averaged binding potential image of 13 young normal subjects (Data courtesy of Dr. Harumasa Takano, NIRS, Japan)

norepinephrine. The distribution and density of a typical 5-HT transporter and receptor is demonstrated in Fig. 11.3.

11.2.2.3 Amyloid Imaging

Excessive amyloid-β (Aβ) deposition in the brain is one of the crucial events in the early pathological state of Alzheimer's disease (AD). PET and SPECT amyloid imaging using a radioligands for Aβ deposit has been paying attention as promising biomarker imaging technique for AD diagnosis (Furumoto et al. 2007). Even though amyloid imaging is not receptor but plaque imaging, here we list several amyloid radioligands, which selectively binds to β-sheet structure of Aβ. [^{11}C]FDDNP is the first radioligand for amyloid imaging (Agdeppa et al. 2003; Shoghi-Jadid et al. 2002), however, [^{11}C]PIB is most widespread. This radioligand is a Thioflavin which is stain analogue to Congo red that sticks to the protein β-sheet (Klunk et al. 2004). Many other radioligands have also been developed, e.g. [^{11}C]BF-227 (Furumoto et al. 2007; Kudo et al. 2007). There are few SPECT radioligands, [123I]IMPY being one of them (Kung et al. 2002). However, the longer half-life of gamma-emitting radionuclides, which allows the delivery of radioligands to virtually all hospitals and facilities, would be strongly beneficial.

11.3 Neuroreceptor Ligand Model

11.3.1 General Considerations

The modeling approach for *in vivo* radioligand binding studies in the brain assumes a compartment model that has at its basis a description of substrate-receptor interactions. The simplest model describing the interaction of a free ligand L with a receptor

R to form the complex RL is the interaction between buffer solution and receptor tissue expressed as follows (Michaelis and Menten 1913):

$$L + R \underset{K_{off}}{\overset{K_{on}}{\rightleftarrows}} RL \tag{11.1}$$

where k_{on} and k_{off} are the association and dissociation constants, respectively. By introducing the equilibrium dissociation constant K_D (= k_{off}/k_{on}) and the total number of receptors B_{max} (= [R]+[RL]), Eq. 11.1 can be written as:

$$[RL] = \frac{B_{max}[L]}{[L] + K_D} \tag{11.2}$$

Separate estimation of B_{max} and K_D requires saturation analysis where measurements are repeated with increasing amount of ligand and the above referenced parameters obtained by fitting the following formula arranged from Eq. 11.2 (Scatchard 1949).

$$\frac{[RL]}{[L]} = \frac{B_{max} - [RL]}{K_D} \tag{11.3}$$

This model can be adapted into the concept of receptor imaging. If the radioligand is administrated in tracer amounts, then [RL] is very small compared with [R], i.e. $B_{max} \approx [R]$ and Eq. 11.2 can be rearranged as follows:

$$BP = \frac{B_{max}}{K_D} = \frac{[RL]}{[L]} \tag{11.4}$$

where BP is the binding potential defined as the ratio of B_{max}/K_D and is equal to the radio of the concentrations of bound and free radioligand. BP is proportional to B_{max} if K_D is constant.

In case of a PET or SPECT study, radioligands are administered intravenously into the human body, go to the heart, and are then delivered to the target organ by arterial blood. The portion of the radioligand that is exchanged with tissue is the one in plasma not bound to plasma proteins (mainly albumin). From plasma, the radiotracer passes through the BBB into the brain and some portion of the radiotracer binds to specific receptors. Dynamic PET or SPECT data acquisitions consisting in a series of time-frames acquired for a time (1–2 h or more in the case of long-half life isotopes and SPECT) provide accurate measurement of the time-course of the tracer. The kinetics of a radioligand depends on the number of receptors in the target organ, its affinity, the non-specific binding, cerebral blood flow, and the concentration of an endogenous competitor. For estimation of detailed physiological parameters (e.g. receptor density and radioligand affinity to a specific receptor, binding

Table 11.2 Typical parameters in receptor modeling *in vivo* (Adapted from Innis et al. 2007)

Abbreviation	Description	Units
B_{max}	Density of receptors *in vitro*	pmol (mg protein)$^{-1}$
B_{avail}	Density of receptors available to bind radioligand *in vivo*	nmol·L^{-1}
BP	*In vitro* binding potential	
BP_{ND}	*In vivo* binding potential	Unitless
f_p	Free fraction in plasma	Unitless
f_{ND}	Free fraction in nondisplaceable compartment	Unitless
K_1	Rate constant for transfer from arterial plasma to tissue	mL·cm^{-3}·min^{-1}
k_2, k_3, k_4	Rate constants	min^{-1}
K_D (k_{off}/k_{on})	Dissociation constant	nmol·L^{-1}
k_{off}	*In vitro* dissociation rate constant	min^{-1}
k_{on}	*In vitro* association rate constant	nmol·L^{-1}·min^{-1}
V_T	Volume of distribution expressed relative to total plasma ligand concentration C_p	mL·cm^{-3}

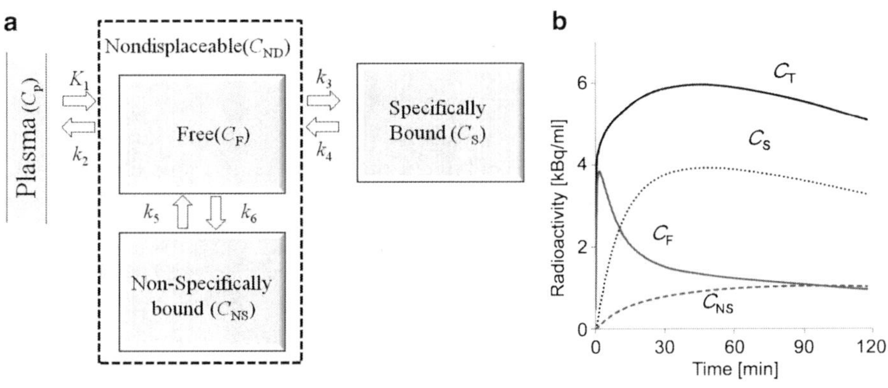

Fig. 11.4 Standard compartmental model for receptor-binding radioligands (**a**) and time courses of radioactivity in each compartment (**b**), where C_T is the total radioactivity concentration in C_S, C_F and C_{NS} compartments, respectively

potential), the application of pharmacokinetic modeling to the tracer time-activity course is required. Table 11.2 summarizes the consensus nomenclature for neuroreceptor imaging that has been adopted since 2007 (Innis et al. 2007).

11.3.2 Standard Reversible Neuroreceptor Model

Depending on each radioligand's kinetics, several compartmental models have been used (Heiss and Herholz 2006; Schmidt and Turkheimer 2002; Watabe et al. 2006; Bentourkia and Zaidi 2007). In this section, we only discuss the standard reversible compartment model shown in Fig. 11.4.

The standard neuroreceptor reversible model in the brain is based on the two-tissue three-compartment model (Fig. 11.4). From arterial blood as a first compartment, the free exchangeable radioligand in plasma passes into the second compartment known as the free compartment. The third compartment is the region of specific binding which we are usually interested to observe. The fourth compartment is a nonspecific binding compartment that exchanges with the free compartment. In practice, for most radioligands, the nonspecific binding compartment is in rapid equilibrium with the free compartment and the two compartments are treated as a single compartment. If the nonspecific binding compartment is considered separately, the compartment model would be much more difficult to solve reliably (Watabe et al. 2006). Exchanges between compartments are governed by simple linear differential equations:

$$\begin{aligned}\frac{dC_{ND}(t)}{dt} &= K_1C_P(t)-(k_2+k_3)C_{ND}(t)+k_4C_S(t)\\ \frac{dC_S(t)}{dt} &= k_3C_{ND}(t)-k_4C_S(t)\\ C_T(t) &= C_{ND}(t)+C_S(t)\end{aligned} \tag{11.5}$$

where K_1 [mL·cm^{-3}·min^{-1}] is the rate constant for transfer from arterial plasma to tissue, k_2, k_3 and k_4 [min^{-1}] are rate constants. C_P [Bq cm^{-3}] is the radioactivity concentration of the unchanged radioligand in plasma (arterial input function), C_{ND} [Bq cm^{-3}] is the radioactivity concentration of non-displaceable radioligand in brain (ND: non-displaceable tissue uptake), including nonspecifically bound and free radioligand, and C_S [Bq cm^{-3}] is the radioactivity concentration of the radioligand specifically bound to receptors. C_T [Bq cm^{-3}] is the radioactivity concentration in a brain region measured by PET or SPECT.

Note that the arterial input function C_P is just the unchanged radioligand in plasma. The free fraction of radioligand in plasma is the fraction of the ligand that is not bound to plasma proteins at equilibrium. The plasma free fraction is referred to as f_P, and the concentration of free exchangeable radioligand in plasma C_{FP} can be calculated as:

$$C_{FP}(t) = f_PC_P(t) \tag{11.6}$$

Actually, for simplicity, unchanged plasma radioactivity concentration (C_P) is used for neuroreceptor modeling. Brain tissues usually have a small fraction of volume occupied by blood (~5%) and Eq. 11.5 may be re-written as:

$$C_T(t) = (1-V_B)[C_{ND}(t)+C_S(t)]+V_BC_a(t) \tag{11.7}$$

where V_B is the fraction of the measured volume occupied by blood. Because the multiplication of blood volumes and radioactivity in whole blood C_a in the human brain are small (Kuhl et al. 1975), the blood volume term is often omitted.

11.3.3 Relation Between the Concept of Receptor Binding and Kinetic Model

The pharmacological parameters described above (e.g. B_{max}, BP) can be linked to the kinetic model parameters. As the unlabeled "cold" ligand is present during the preparation of many radioligands that will compete at binding sites with the labeled tracer, the proportion of labeled tracer to the total amount of injected ligand (specific activity, SA [GBq·μmol^{-1}]) must also be considered in the equations:

$$\begin{aligned} k_3 &= k'_{on} f_{ND} \left(B_{avail} - C_S / SA \right) \\ k_4 &= k'_{off} \end{aligned} \tag{11.8}$$

where k'_{on} [L^{-1}·nmol·min^{-1}] and k'_{off} [L^{-1}·nmol·min^{-1}] are biomolecular association and dissociation rate constants. f_{ND} is the free fraction of radioligand in non-displaceable compartment, B_{avail} [L^{-1}·nmol] is receptor density available to bind radioligand *in vivo*, where B_{avail} can be regarded as (B_{max} – B_{occup}) (B_{occup}: number of receptors occupied by the endogenous ligand). If the amount of total receptor-binding ligand C_S/SA is very small relative to the binding capacity (B_{avail}), it can be disregarded and k_3 is then equal to the first term, i.e. $k'_{on} f_{ND} B_{avail}$.

$$\frac{k_3}{k_4} = f_{ND} \frac{B_{avail}}{K_D} \tag{11.9}$$

Here $K_D = k'_{off}/k'_{on}$. Finally, the binding potential relative to the radioactivity concentration of non-displaceable radioligand in the brain (BP_{ND}), which reflects the ratio of available receptor density (B_{avail}) and affinity (K_D) is defined as follows:

$$BP_{ND} = \frac{k_3}{k_4} \tag{11.10}$$

Full kinetic fits to determine receptor density (B_{avail}) and affinity of radioligand (K_D) are usually obtained from multiple PET/SPECT scans with various SA of radioligands (Delforge et al. 1989; Farde et al. 1989; Gallezot et al. 2008). There has been growing interest in detecting multiple functions simultaneously with single PET/SPECT data acquisition using multiple tracer injections (Ikoma et al. 2009; Koeppe et al. 2001). However, these approaches are still complicated for routine use. Alternatively, it is useful to employ a combination of parameters like BP_{ND} (instead of B_{avail}, K_D, k_3 and k_4) as a macro-parameter to represent the observed data rather than considering the individual parameters. The distribution volume is one of the macro-parameters and the basic concept originated from volume of distribution in pharmacokinetics, which connects the administered dose to the actual initial concentration present in the circulation at the equilibrium. This concept was adopted in the field of *in vivo* imaging. Instead of referring to total administrated dose in the entire body, the distribution volume (V_T) is expressed as the activity of radioligand in the tissue as follows:

$$V_T = \frac{C_T}{C_p} = V_{ND} + V_S = \frac{C_{ND}}{C_p} + \frac{C_S}{C_p} \tag{11.11}$$

where C_{ND} and C_S are the radioactivity concentrations in non-displaceable and specific bound in the tissue, respectively. An important assumption is that the radioactive concentration in V_T must satisfy the equilibrium condition. However, it is difficult to achieve complete secular equilibrium practically where "the rate of change of the concentration in the compartment is zero" due to the limited imaging time (e.g. 1–2 h), physical half-life of isotopes, and other biological reasons. In the case of exchange between compartments in neuroreceptor studies, the term "transient equilibrium", where the maximum turning point of radioactivity concentration, is used (Farde et al. 1989; Ito et al. 1998a). Although at this point the change of concentration is zero, there may still be exchange between compartments (e.g. tissue and blood) and the total loss from the system is not negligible.

An important concept of pharmacokinetics is that the ratio of compartmental rate constant equals selected equilibrium distribution volumes. If C_{ND} and C_S simultaneously reach equilibrium, differentiations in Eq. 11.5, dC_{ND}/dt and dC_S/dt, become zero. After arrangement of Eq. 11.5 with equilibrium conditions, the distribution volumes as expression of rate constant are as follows:

$$\begin{aligned} V_{ND} &= \frac{K_1}{k_2} \\ V_S &= \frac{K_1}{k_2} \cdot \frac{k_3}{k_4} \\ V_T &= V_{ND} + V_S = \frac{K_1}{k_2} \cdot \left(1 + \frac{k_3}{k_4}\right) \end{aligned} \tag{11.12}$$

BP_{ND} are also expressed by the ratio of distribution volumes as follows:

$$BP_{ND} = \frac{k_3}{k_4} = \frac{V_S}{V_{ND}} = \frac{(V_T - V_{ND})}{V_{ND}} \tag{11.13}$$

11.4 Quantification of Receptor Imaging Studies

11.4.1 General Considerations

In standard neuroreceptor compartmental models (Fig. 11.4), the total radioactivity concentration C_T is mathematically expressed as a summation of two compartments as follows:

$$C_T(t) = \frac{K_1}{\alpha_1 - \alpha_2}\left\{(\alpha_1 - k_3 - k_4)e^{-\alpha_1 t} - (\alpha_2 - k_3 - k_4)e^{-\alpha_2 t}\right\} \otimes C_p(t)$$
$$\alpha_{1,2} = \frac{k_2 + k_3 + k_4 \pm \sqrt{(k_2 + k_3 + k_4)^2 - 4k_2 k_4}}{2} \qquad (11.14)$$

For estimation of the K parameters from this nonlinear equation, an iterative nonlinear fitting procedure with an arterial input function (C_p) is necessary. Quantification of the labeled radioligand in the plasma requires to obtain plasma radioactivity concentration, time-course of arterial blood sampling, and plasma separation. Instead of this burden, the large number of parameters introduces instability and variability in the estimates in addition to lengthy calculation time. Therefore, usually the outcome measure in *in vivo* neuroreceptor imaging is BP_{ND} or V_T as macro-parameters rather than k-parameters themselves. Sometimes one of alternative estimation processes, namely "graphical analysis", is used. This graphical analysis is very useful because of rapid calculation and stability in the estimates.

Furthermore, arterial sampling is invasive and technically demanding. If the heart chamber is scanned and imaged simultaneously with the brain, the arterial input function can be directly derived from dynamic images (Iida et al. 1998; Watabe et al. 2005). Alternatively, the input function can be extracted as a component through image processing of dynamic images (Naganawa et al. 2005; Wu et al. 1995). In some applications for receptor studies, the arterial input function can be replaced by a reference to tissue devoid of specific binding sites. A typical derivation of the non-invasive reference technique is the "simplified reference tissue model" (Lammertsma and Hume 1996). Although this approach has several advantages such as non-invasiveness over the method requiring an arterial input function, several assumptions are associated with the simplified technique and as such caution is mandatory for its use.

As a different solution to Eq. 11.5, the experimental paradigm starts with bolus injection of the radiotracer followed by continuous infused administration of the radiotracer to achieve constant concentrations in the tissue and blood (Carson et al. 1993). In this protocol, V_T and BP_{ND} can be easily obtained by calculating ratios between the radioactivity concentration in tissue and blood. The rate of infusion may vary between subjects, and optimal scheduling of the experiments must be sought for the success of the bolus plus infusion paradigm.

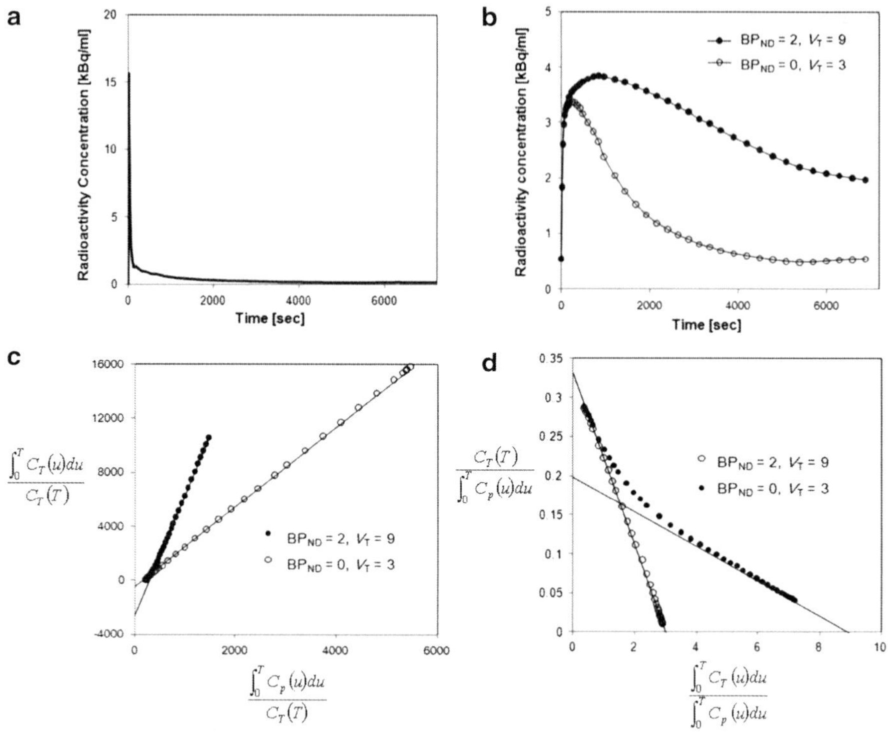

Fig. 11.5 Illustration of time-activity curves and the graphical analysis; (**a**) time-activity curve of plasma input function (C_p) and (**b**) time-activity curves of tissue (C_T) with K_1=0.33, k_2=0.11, k_3=0.08, k_4=0.04 (V_T=9, BP_{ND}=2) and with K_1=0.33, k_2=0.11 (V_T=3, BP_{ND}=0). Two tissue curves are graphically plotted with (**c**) Logan graphical analysis and (**d**) Ito-Yokoi plot

11.4.2 Graphical Analysis

The graphical analysis generally allows quick estimation by graphically fitting a straight line to experimental data or using linear regression analysis. Since it is often difficult to achieve equilibrium conditions, an approximation is used from a graphic representation of kinetic data.

Logan graphical analysis was originally proposed to quantify the total volume of distribution, V_T, for reversibly binding neuroreceptor tracers (Logan et al. 1990). V_T is equivalent to the slope only after reaching equilibrium where the relationship between the two terms of the ratio: the integrated arterial plasma input function (C_p) and the radioactivity concentration in tissue (C_T) becomes linear (Fig. 11.5c). The advantages of using Logan graphical analysis are the rapid determination of V_T and its low variance. V_T is estimated from the following equation:

$$\frac{\int_0^t C_T(u)du}{C_T(t)} = V_T \frac{\int_0^t C_p(u)du}{C_T(t)} + b \qquad t \geq t^*, \tag{11.15}$$

Because V_T and b become constant only after equilibrium time for regression t^*, they are estimated as a slope and an intercept, respectively, using the frames of $t \geq t^*$. The equilibrium time (t^*) for linear regression is different from equilibrium in compartmental models; however, there are still two drawbacks, noise-induced bias in the estimates and objective determination of t^* (Slifstein and Laruelle 2000).

Another approach consists in rearranging Eq. 11.15 into multi-linear form to give Eq. 11.16 (Ichise et al. 2002). This approach is usually called multilinear analysis (MA1).

$$C_T(t) = \beta_1 \int_0^t C_p(u)du + \beta_2 \int_0^t C_T(u)du \qquad t \geq t^* \tag{11.16}$$

$\beta_1 = -V_T/b$ and $\beta_2 = 1/b$ are estimated by linear regression analysis for $t \geq t^*$, and V_T is calculated as $-\beta_1/\beta_2$. Furthermore, Ichise et al. proposed several rigorous approaches, e.g. multilinear analysis (MA2) which does not require t^* (Ichise et al. 2001, 2002, 2003).

Even though both Logan graphical analysis and multilinear analysis provide only estimates for V_T, as a recent progressive work, Ito and Yokoi proposed an alternative graphical analysis for neuroreceptor imaging, which can be used to calculate both V_T and V_{ND} (Ito et al. 2010). In other words, BP_{ND} (= V_T/V_{ND} − 1) can also be estimated as a result. This approach assumes that the time-activity curve is also described by the one-tissue compartment model with two rate constants (Ito et al. 1996; Koeppe et al. 1991). The rate constants, K_i and k_o are the influx and efflux rate constants for the radioligand diffusion through the BBB, respectively, and V_T is defined as K_i/k_o. After integration and arrangement of differential equations based on the two-tissue compartment model given by Eq. 11.5 and the one-tissue compartment model with K_i and k_o, the graphical equilibrium at $t \to 0$ and $t \geq t^*$ will be expressed as follows:

$$Y(t) = \begin{cases} K_1 - k_2 X(t) & t \to 0 \\ K_i - k_o X(t) & t \geq t^* \end{cases}$$

$$X(t) = \frac{\int_0^t C_T(u)du}{\int_0^t C_p(u)du}, \qquad Y(t) = \frac{C_T(u)}{\int_0^t C_p(u)du} \tag{11.17}$$

From Eq. 11.17, both x and y intercepts at each condition ($t \to 0$ and $t \geq t^*$) become as follows:

$$x - \text{intercept} = \begin{cases} V_{ND} = \dfrac{K_1}{k_2} & t \to 0 \\ V_T = \dfrac{k_i}{k_o} & t \geq t^* \end{cases}$$

$$y - \text{intercept} = \begin{cases} K_1 & t \to 0 \\ K_i & t \geq t^* \end{cases} \tag{11.18}$$

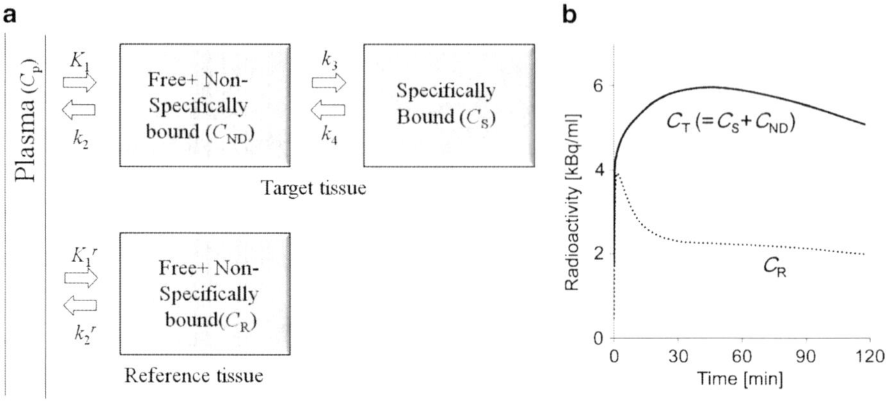

Fig. 11.6 Standard reference tissue model for receptor-binding radioligands (**a**) and time courses of radioactivity in the two compartments C_T and C_R (**b**)

An example plot of this approach is shown in Fig. 11.5d. Advantages of this method can be used to know whether the radioligand binding includes specific binding or not according to the shape of the plot (Ito et al. 2010).

11.4.3 Reference Approach

11.4.3.1 Reference Tissue Region

For most radioligands, reference region with low specific binding can be used as indirect input function to the region containing the specific receptors of interest. A compartment model incorporating the reference region is illustrated in Fig. 11.6 (Blomqvist et al. 1989; Cunningham et al. 1991). Because of low specific binding, radioactivity concentration in the reference region (C_R) is expressed as influx/efflux between only plasma and reference tissue with rate constants K_1^r [mL·cm^{-3}·min^{-1}] and k_2^r [min^{-1}].

$$\frac{dC_R(t)}{dt} = K_1^r \cdot C_p(t) - k_2^r \cdot C_R(t) \tag{11.19}$$

The transient equilibrium that theoretically occurs when the derivative for specific binding ($dC_S(t)/dt$) is 0, that is, the peak point of specific binding, $C_S(t)$, is a simple method for BP_{ND} estimation using reference tissue (Ito et al. 1998a). It follows that $C_S(t)/C_{ND}(t)$ is equal to k_3/k_4 ($=BP_{ND}$) at transient equilibrium (here $C_{ND} \approx C_R$). However, the existence of specific binding in the reference region directly influence the estimates of BP_{ND} though this transient equilibrium method.

11.4.3.2 Simplified Reference Tissue Model (SRTM)

The SRTM yields the binding potential value by eliminating the arterial input function, C_p, arithmetically from model equations by using a TAC from a reference region where specific bindings are negligible, under the assumptions that the distribution volume of the non-displaceable compartment in the target and reference regions are equal, and that a target region can be described with the one-tissue compartment model given by Eq. 11.20.

$$C_T(t) = R_1 \cdot C_R(t) + \left(k_2 - \frac{R_1 k_2}{1 + BP_{ND}}\right) \cdot e^{\left(-\frac{k_2}{1+BP_{ND}}t\right)} \otimes C_R(t)$$

$$R_1 = K_1 / K_1^r \tag{11.20}$$

where C_T and C_R are the radioactivity concentrations in the target and reference regions, respectively. SRTM estimates three parameters, the delivery ratio of the target region to the reference region (R_1), the clearance rate constant of the target region (k_2), and binding potential, referred to as BP_{ND} by non-linear least squares analysis.

11.4.3.3 Noninvasive Graphical Analysis

As a noninvasive Logan plot distribution volume ratio DVR can be estimated as follows (Logan et al. 2001):

$$\frac{\int_0^t C_T(u)du}{C_T(t)} = DVR\left(\frac{\int_0^t C_R(u)du}{C_T(t)} + \frac{C_R(t)/k_2^r}{C_T(t)}\right) + b \qquad t \geq t^*, \tag{11.21}$$

where DVR is defined as the ratio of V_T in the target and reference regions and BP_{ND} is expressed as (DVR − 1). k_2^r is the efflux rate constant k_2 in the reference region. Equation 11.20 can also be extended as a multi-linear formulation. However, determination of k_2^r still remains to be done (Ichise et al. 2003). By using a more strict assumption that C_R is proportional to C_p after equilibrium, Zhou et al. have recently proposed simplified noninvasive graphical plot with reference tissue region (Zhou et al. 2010).

$$\frac{\int_0^t C_T(u)du}{C_R(t)} = DVR\frac{\int_0^t C_R(u)du}{C_R(t)} + b \qquad t \geq t^* \tag{11.22}$$

Even though the assumption of this approach is strict, it is not necessary to define k_2^r in advance and the robustness for noise-induced bias provides stable estimates and this is effective in the case of parametric imaging through pixel-by-pixel estimation process.

11.4.4 Parametric Imaging

In practice, PET measurement C_T(t) has two derivations, at region of interest (ROI) and voxel levels. Parametric imaging indicates voxel-by-voxel estimation of C(t) (Rahmim et al. 2009; Tsoumpas et al. 2008). Due to limited statistical count and temporal resolution of PET and SPECT, noise properties of C_T(t) at the voxel level is poor. Therefore, when parametric imaging is used, the effect of noise on the estimate must be carefully considered. In essence, the effect of noise appears as a noise-induced bias and variance of the estimates and these amounts depend on the kind of analysis used (Ikoma et al. 2008). Furthermore, the number of pixels in the PET image is very large. The computing time for deriving C_T(t) for each pixel is also one of the issues that need to be considered. In other words, time consuming approaches such as non-linear least square fitting, are not desirable for parametric imaging.

Noise-induced bias of V_T in Logan graphical analysis is well-known issue, especially in parametric imaging (Slifstein and Laruelle 2000). The simplified approach derived by Zhou et al. enables stable estimation because the denominator in the graphical analysis has relatively good statistics (i.e. using $C_R(t)$ instead of $C_T(t)$) (Zhou et al. 2010). In order to fasten computing time and improve the stability of the estimates, the SRTM method can be extended using basis function approach (Gunn et al. 1997). As a direct solution, suppressing noise in C_T(t) at each pixel has been investigated (Joshi et al. 2008; Shidahara et al. 2007; Turkheimer et al. 2000). Recent advances in 4D-reconstruction resulting in the generation of parametric images may be a potent approach (Rahmim et al. 2009; Tsoumpas et al. 2008).

11.5 Clinical and Research Applications of Receptor Imaging

11.5.1 Neurodegenerative Disease (e.g. Dopaminergic Function in Parkinson Disease)

Idiopathic Parkinson's disease (IPD) is the most prevalent neurodegenerative disease after dementia, with an estimated prevalence of 150 per 100,000 and incidence of 20 per 100,000. The typical clinical symptoms such as rigidity, hypokinesia, tremor and postural instability, are associated to the degeneration of dopaminergic neurons in the substantia nigra of the brainstem. Although IPD is the most prevalent cause of Parkinsonism, there are other diseases that present with one or more similar symptoms (Van Laere and Zaidi 2006). Recent studies have demonstrated that of the patients that deceased under the diagnosis IPD, up to 25% actually did not have the disease (Lang and Lozano 1998). Various radioligands are currently commercially available to evaluate the pre- and postsynaptic status of the dopaminergic system (Booij et al. 1999). Both ^{123}I labelled iodobenzamide (IBZM) and epidepride, as well as ^{11}C-raclopride can be used as a sensitive differentiator between IPD and other Parkinsonian syndromes, and are predictive for the response on dopaminergic therapy (Schwarz et al. 1997). In Parkinson's disease, the postsynaptic system is normal to upregulated, while in the most Parkinsonian syndromes, of which multi system atrophy (MSA) and progressive supranuclear palsy (PSP) are the most common, there is a decreased postsynaptic binding. The cocaine-analogue ^{123}I-FP-CIT (ioflupane) has a clinical role in the early diagnosis of Parkinson's disease (since abnormalities are present even 5 years before clinical symptomatology), differentiation between IPD and essential tremor, as well as in the differentiation between Alzheimer and Lewy body dementia (Costa et al. 1999). This tracer is very sensitive for presynaptic damage to the striatum and the uptake in the contralateral striatum is correlated to the severity of clinical disease (Seibyl et al. 1995).

Despite controversial clinico-pathological distinctions between Parkinson's disease with dementia (PDD) and DAT, similar patterns of metabolic reduction in the posterior brain were reported previously using PET with ^{18}F-FDG. More specific regional differences were measured between PDD and DAT using accurate and objective brain mapping techniques. PDD showed greater metabolic reduction in the visual cortex and relatively preserved metabolism in the medial temporal cortex (Vander et al. 1997).

Metabolic or perfusion brain imaging is useful in the differential diagnosis of extrapyramidal disorders due to the application of anatomical standardization (Bosman et al. 2003; Spetsieris et al. 1995). Progressive supranuclear palsy (PSP) and corticobasal degeneration (CBD) are neurodegenerative disorders that may be accompanied by dementia and Parkinsonism as clinical symptoms. Measurement of glucose metabolism by PET and a voxel-based analysis is useful to understand the pathophysiology and differentiate IPD and corticobasal degeneration with cognitive impairments (Hosaka et al. 2002).

11.5.2 Receptor Occupancy (e.g. Antipsychotic-Drug Evaluation)

Explicit labeling of striatal receptors with radioligands has been used to determine receptor occupancy during treatment with antipsychotic drugs. The most prevalent application of neuroreceptor imaging in schizophrenia during the last two decades has been the evaluation of receptor occupancy accomplished by antipsychotic drugs. The major spotlight has been on D_2 receptor occupancy although $5HT_{2A}$ and D_1 receptors have also been considered (Laruelle et al. 2003). Several studies have reported the existence of a threshold of occupancy of striatal D_2 receptors beyond which extrapiramidal side effects (EPS) are likely to take place. It has been difficult to establish a relationship between the degree of D_2 receptor occupancy and clinical response. It should be noted that most studies have been carried out at doses reaching more than 50% occupancy and the minimum occupancy needed for observing a therapeutic response still needs to be defined. It has, however, been demonstrated that in few studies carried out with relatively low doses of D_2 receptor antagonists that 50–60% occupancy is needed to monitor a prompt clinical response (Laruelle et al. 2003). It has been determined that antipsychotic action requires a receptor occupancy of 60%, although extrapyramidal side effects occur at ≥80% receptor occupancy (Heiss and Herholz 2006).

Although currently the clinical indications for receptor imaging are limited to predominantly movement disorders, they do show potential for clinical dose titration of medication, early diagnosis of neuropsychiatry disorders, as well as more specific localization of cerebral functional disturbances. Many other applications have been reported in the literature but fall beyond the scope of this review (Heiss and Herholz 2006; Laruelle et al. 2003).

11.6 Summary and Future Perspectives

During the last two decades, molecular neuroreceptor imaging using SPECT and PET has advanced elegantly and steadily gained importance in the clinical and research arenas. High resolution cutting-edge SPECT and PET neuroimaging is poised to advance our understanding of the pathophysiological mechanisms of neurological and psychiatric diseases and in the clinical management of patients. Software- and hardware-based multimodality brain imaging (e.g. PET/MR) has enabled the implementation of sophisticated anatomical-guided quantitative PET procedures that will undoubtedly revolutionize clinical diagnosis and offer unique capabilities for the clinical neuroreceptor imaging community and neuroscience researchers at large.

Frontiers of molecular neuroreceptor imaging research are also being expanded with small-animal brain imaging using dedicated high-resolution SPECT and PET instrumentation in addition to optical imaging and high-field MR imaging (Stout and Zaidi 2008). Most preclinical imaging systems require animals to remain motionless for several minutes to hours in order to obtain useful data, thus some type of anesthesia is necessary. There is considerable interest in imaging without anesthesia, since conscious animals have normal physiological functions compared to those under anesthesia. Several groups have shown that this is indeed feasible under some circumstances (Ferris et al. 2006; Kyme et al. 2008). These observations spurred the development of the rat conscious animal PET scanner (RatCAP), a complete 3D PET scanner designed to image the brain of an awake rat (Vaska et al. 2004), that incorporates the PET system into an integrated, compact arrangement of detector arrays with highly integrated electronics (Woody et al. 2007). Nonetheless, most preclinical neuroreceptor imaging research at present is conducted using anesthesia, and the type and injection route can play a significant role in the sedation and physiological status of the animal. It is expected that the triad molecular neuroscience, genetics and imaging will expand the understanding of the pathophysiology within the psychiatric and neurological clinical diseases in the coming decennia, hopefully leading to improved diagnosis and treatment, and eventually to the implementation of preventive techniques.

Acknowledgements This work was supported by the Swiss National Science Foundation under grants SNSF 31003A-135576 and SNSF 33CM30-124114.

References

Accorsi R (2008) Brain single-photon emission CT physics principles. AJNR Am J Neuroradiol 29:1247–1256

Agdeppa ED, Kepe V, Liu J, Small GW, Huang SC, Petric A, Satyamurthy N, Barrio JR (2003) 2-Dialkylamino-6-acylmalononitrile substituted naphthalenes (DDNP analogs): novel diagnostic and therapeutic tools in Alzheimer's disease. Mol Imaging Biol 5:404–417

Bentourkia M, Zaidi H (2007) Tracer kinetic modeling in PET. PET Clinics 2:267–277

Blomqvist G, Pauli S, Farde L, Eriksson L, Person A, Halldin C (1989) Dynamic models for reversible ligand binding. Kluwer, Dordrecht

Booij J, Tissingh G, Winogrodzka A, van Royen EA (1999) Imaging of the dopaminergic neurotransmission system using single- photon emission tomography and positron emission tomography in patients with Parkinsonism. Eur J Nucl Med 26:171–182

Bosman T, Van Laere K, Santens P (2003) Anatomically standardised (99 m)Tc-ECD brain perfusion SPET allows accurate differentiation between healthy volunteers, multiple system atrophy and idiopathic Parkinson's disease. Eur J Nucl Med Mol Imaging 30:16–24

Brucke T, Kornhuber J, Angelberger P, Asenbaum S, Frassine H, Podreka I (1993) SPECT imaging of dopamine and serotonin transporters with [123I]beta-CIT. Binding kinetics in the human brain. J Neural Transm Gen Sect 94:137–146

Burns HD, Dannals RF, Langstrom B, Ravert HT, Zemyan SE, Duelfer T, Wong DF, Frost JJ, Kuhar MJ, Wagner HN (1984) (3-N-[11C]methyl)spiperone, a ligand binding to dopamine receptors: radiochemical synthesis and biodistribution studies in mice. J Nucl Med 25:1222–1227

Carson RE, Channing MA, Blasberg RG, Dunn BB, Cohen RM, Rice KC, Herscovitch P (1993) Comparison of bolus and infusion methods for receptor quantitation: application to [18F] cyclofoxy and positron emission tomography. J Cereb Blood Flow Metab 13:24–42

Cho ZH, Son YD, Kim HK, Kim KN, Oh SH, Han JY, Hong IK, Kim YB (2008) A fusion PET-MRI system with a high-resolution research tomograph-PET and ultra-high field 7.0 T-MRI for the molecular-genetic imaging of the brain. Proteomics 8:1302–1323

Costa D, Walker Z, Walker R et al (1999) Dementia with Lewy bodies: preliminary data on clinical, pathological and FP-CIT SPECT correlations. Nucl Med Comm 20:467–468

Crouzel C, Guillaume M, Barre L, Lemaire C, Pike VW (1992) Ligands and tracers for PET studies of the 5-HT system–current status. Int J Rad Appl Instrum B 19:857–870

Cunningham VJ, Hume SP, Price GR, Ahier RG, Cremer JE, Jones AK (1991) Compartmental analysis of diprenorphine binding to opiate receptors in the rat in vivo and its comparison with equilibrium data in vitro. J Cereb Blood Flow Metab 11:1–9

Delforge J, Syrota A, Mazoyer BM (1989) Experimental design optimization: theory and application to estimation of receptor model parameters using dynamic positron emission tomography. Phys Med Biol 34:419–435

Dresel S, Kung MP, Huang XF, Plossl K, Hou C, Meegalla SK, Patselas GP, Mu M, Saffer JR, Kung HK (1999) Simultaneous SPECT studies of pre- and postsynaptic dopamine binding sites in baboons. J Nucl Med 40:660–666

Emond P, Garreau L, Chalon S, Boazi M, Caillet M, Bricard J, Frangin Y, Mauclaire L, Besnard JC, Guilloteau D (1997) Synthesis and ligand binding of nortropane derivatives: N-substituted 2beta-carbomethoxy-3beta-(4′-iodophenyl)nortropane and N-(3-iodoprop-(2E)-enyl)-2beta-carbomethoxy-3beta-(3′,4′-disubstituted phenyl)nortropane. New high-affinity and selective compounds for the dopamine transporter. J Med Chem 40:1366–1372

Farde L, Halldin C, Stone-Elander S, Sedvall G (1987) PET analysis of human dopamine receptor subtypes using 11C-SCH 23390 and 11C-raclopride. Psychopharmacology (Berl) 92:278–284

Farde L, Eriksson L, Blomqvist G, Halldin C (1989) Kinetic analysis of central [11C]Raclopride binding to D2-dopamine receptors studied by PET – a comparison to the equilibrium analysis. J Cereb Blood Flow Metab 9:696–708

Farde L, Halldin C, Muller L, Suhara T, Karlsson P, Hall H (1994) PET study of [11C]beta-CIT binding to monoamine transporters in the monkey and human brain. Synapse 16:93–103
Ferris CF, Febo M, Luo F, Schmidt K, Brevard M, Harder JA, Kulkarni P, Messenger T, King JA (2006) Functional magnetic resonance imaging in conscious animals: a new tool in behavioural neuroscience research. J Neuroendocrinol 18:307–318
Fowler JS, Volkow ND, Wang GJ, Ding YS (2004) 2-deoxy-2-[18F]fluoro-D-glucose and alternative radiotracers for positron emission tomography imaging using the human brain as a model. Semin Nucl Med 34:112–121
Furumoto S, Okamura N, Iwata R, Yanai K, Arai H, Kudo Y (2007) Recent advances in the development of amyloid imaging agents. Curr Top Med Chem 7:1773–1789
Gallezot JD, Bottlaender MA, Delforge J, Valette H, Saba W, Dolle F, Coulon CM, Ottaviani MP, Hinnen F, Syrota A, Gregoire MC (2008) Quantification of cerebral nicotinic acetylcholine receptors by PET using 2-[18F]fluoro-A-85380 and the multiinjection approach. J Cereb Blood Flow Metab 28:172–180
Gilman S (1998) Imaging the brain. First of two parts. N Engl J Med 338:812–820
Gunn RN, Lammertsma AA, Hume SP, Cunningham VJ (1997) Parametric imaging of ligand-receptor binding in PET using a simplified reference region model. Neuroimage 6:279–287
Halldin C, Stone-Elander S, Farde L, Ehrin E, Fasth KJ, Langstrom B, Sedvall G (1986) Preparation of 11C-labelled SCH 23390 for the in vivo study of dopamine D-1 receptors using positron emission tomography. Int J Rad Appl Instrum A 37:1039–1043
Halldin C, Foged C, Chou YH, Karlsson P, Swahn CG, Sandell J, Sedvall G, Farde L (1998) Carbon-11-NNC 112: a radioligand for PET examination of striatal and neocortical D1-dopamine receptors. J Nucl Med 39:2061–2068
Halldin C, Gulyas B, Langer O, Farde L (2001) Brain radioligands–state of the art and new trends. Q J Nucl Med 45:139–152
Hammoud DA, Hoffman JM, Pomper MG (2007) Molecular neuroimaging: from conventional to emerging techniques. Radiology 245:21–42
Hartvig P, Agren H, Reibring L, Tedroff J, Bjurling P, Kihlberg T, Langstrom B (1991) Brain kinetics of L-[beta-11C]dopa in humans studied by positron emission tomography. J Neural Transm Gen Sect 86:25–41
Heiss WD, Herholz K (2006) Brain receptor imaging. J Nucl Med 47:302–312
Hosaka K, Ishii K, Sakamoto S, Mori T, Sasaki M, Hirono N, Mori E (2002) Voxel-based comparison of regional cerebral glucose metabolism between PSP and corticobasal degeneration. J Neurol Sci 199:67–71
Houle S, DaSilva JN, Wilson AA (2000) Imaging the 5-HT(1A) receptors with PET: WAY-100635 and analogues. Nucl Med Biol 27:463–466
Ichise M, Meyer JH, Yonekura Y (2001) An introduction to PET and SPECT neuroreceptor quantification models. J Nucl Med 42:755–763
Ichise M, Toyama H, Innis RB, Carson RE (2002) Strategies to improve neuroreceptor parameter estimation by linear regression analysis. J Cereb Blood Flow Metab 22:1271–1281
Ichise M, Liow JS, Lu JQ, Takano A, Model K, Toyama H, Suhara T, Suzuki K, Innis RB, Carson RE (2003) Linearized reference tissue parametric imaging methods: application to [11C]DASB positron emission tomography studies of the serotonin transporter in human brain. J Cereb Blood Flow Metab 23:1096–1112
Iida H, Miura S, Shoji Y, Ogawa T, Kado H, Narita Y, Hatazawa J, Eberl S, Kanno I, Uemura K (1998) Noninvasive quantitation of cerebral blood flow using oxygen-15-water and a dual-PET system. J Nucl Med 39:1789–1798
Ikoma Y, Watabe H, Shidahara M, Naganawa M, Kimura Y (2008) PET kinetic analysis: error consideration of quantitative analysis in dynamic studies. Ann Nucl Med 22:1–11
Ikoma Y, Watabe H, Hayashi T, Miyake Y, Teramoto N, Minato K, Iida H (2009) Quantitative evaluation of changes in binding potential with a simplified reference tissue model and multiple injections of [11C]raclopride. Neuroimage 47:1639–1648
Innis RB, Cunningham VJ, Delforge J, Fujita M, Gjedde A, Gunn RN, Holden J, Houle S, Huang SC, Ichise M, Iida H, Ito H, Kimura Y, Koeppe RA, Knudsen GM, Knuuti J, Lammertsma AA, Laruelle M, Logan J, Maguire RP, Mintun MA, Morris ED, Parsey R, Price JC, Slifstein M,

Sossi V, Suhara T, Votaw JR, Wong DF, Carson RE (2007) Consensus nomenclature for in vivo imaging of reversibly binding radioligands. J Cereb Blood Flow Metab 27:1533–1539

Ito H, Goto R, Koyama M, Kawashima R, Ono S, Sato K, Fukuda H (1996) A simple method for the quantification of benzodiazepine receptors using iodine-123 iomazenil and single-photon emission tomography. Eur J Nucl Med 23:782–791

Ito H, Hietala J, Blomqvist G, Halldin C, Farde L (1998a) Comparison of the transient equilibrium and continuous infusion method for quantitative PET analysis of [11C]raclopride binding. J Cereb Blood Flow Metab 18:941–950

Ito H, Nyberg S, Halldin C, Lundkvist C, Farde L (1998b) PET imaging of central 5-HT2A receptors with carbon-11-MDL 100,907. J Nucl Med 39:208–214

Ito H, Takahashi H, Arakawa R, Takano H, Suhara T (2008) Normal database of dopaminergic neurotransmission system in human brain measured by positron emission tomography. Neuroimage 39:555–565

Ito H, Yokoi T, Ikoma Y, Shidahara M, Seki C, Naganawa M, Takahashi H, Takano H, Kimura Y, Ichise M, Suhara T (2010) A new graphic plot analysis for determination of neuroreceptor binding in positron emission tomography studies. Neuroimage 49:578–586

Jacobs AH, Li H, Winkeler A, Hilker R, Knoess C, Ruger A, Galldiks N, Schaller B, Sobesky J, Kracht L, Monfared P, Klein M, Vollmar S, Bauer B, Wagner R, Graf R, Wienhard K, Herholz K, Heiss WD (2003) PET-based molecular imaging in neuroscience. Eur J Nucl Med Mol Imaging 30:1051–1065

Joshi A, Fessler JA, Koeppe RA (2008) Improving PET receptor binding estimates from Logan plots using principal component analysis. J Cereb Blood Flow Metab 28:852–865

Klunk WE, Engler H, Nordberg A, Wang Y, Blomqvist G, Holt DP, Bergstrom M, Savitcheva I, Huang GF, Estrada S, Ausen B, Debnath ML, Barletta J, Price JC, Sandell J, Lopresti BJ, Wall A, Koivisto P, Antoni G, Mathis CA, Langstrom B (2004) Imaging brain amyloid in Alzheimer's disease with Pittsburgh Compound-B. Ann Neurol 55:306–319

Koeppe RA, Holthoff VA, Frey KA, Kilbourn MR, Kuhl DE (1991) Compartmental analysis of [11C]flumazenil kinetics for the estimation of ligand transport rate and receptor distribution using positron emission tomography. J Cereb Blood Flow Metab 11:735–744

Koeppe RA, Raffel DM, Snyder SE, Ficaro EP, Kilbourn MR, Kuhl DE (2001) Dual-[11C]tracer single-acquisition positron emission tomography studies. J Cereb Blood Flow Metab 21:1480–1492

Kudo Y, Okamura N, Furumoto S, Tashiro M, Furukawa K, Maruyama M, Itoh M, Iwata R, Yanai K, Arai H (2007) 2-(2-[2-Dimethylaminothiazol-5-yl]ethenyl)-6- (2-[fluoro]ethoxy)benzoxazole: a novel PET agent for in vivo detection of dense amyloid plaques in Alzheimer's disease patients. J Nucl Med 48:553–561

Kuhl DE, Reivich M, Alavi A, Nyary I, Staum MM (1975) Local cerebral blood volume determined by three-dimensional reconstruction of radionuclide scan data. Circ Res 36: 610–619

Kung HF, Kim HJ, Kung MP, Meegalla SK, Plossl K, Lee HK (1996) Imaging of dopamine transporters in humans with technetium-99 m TRODAT-1. Eur J Nucl Med 23:1527–1530

Kung MP, Hou C, Zhuang ZP, Zhang B, Skovronsky D, Trojanowski JQ, Lee VM, Kung HF (2002) IMPY: an improved thioflavin-T derivative for in vivo labeling of beta-amyloid plaques. Brain Res 956:202–210

Kyme AZ, Zhou VW, Meikle SR, Fulton RR (2008) Real-time 3D motion tracking for small animal brain PET. Phys Med Biol 53:2651–2666

Lammertsma AA, Hume SP (1996) Simplified reference tissue model for PET receptor studies. Neuroimage 4:153–158

Lang AE, Lozano AM (1998) Parkinson's disease. First of two parts. N Engl J Med 339:1044–1053

Laruelle M, Baldwin RM, Malison RT, Zea-Ponce Y, Zoghbi SS, al-Tikriti MS, Sybirska EH, Zimmermann RC, Wisniewski G, Neumeyer JL et al (1993) SPECT imaging of dopamine and serotonin transporters with [123I]beta-CIT: pharmacological characterization of brain uptake in nonhuman primates. Synapse 13:295–309

Laruelle M, Martinez D, Talbot P, Abi-Dargham A (2003) Molecular imaging in psychiatric disorders. In: Valk PE, Bailey DL, Townsend DW, Maisey MN (eds) Positron emission tomography: basic science and clinical practice. Springer, London, pp 399–426

Logan J, Fowler JS, Volkow ND, Wolf AP, Dewey SL, Schlyer DJ, MacGregor RR, Hitzemann R, Bendriem B, Gatley SJ et al (1990) Graphical analysis of reversible radioligand binding from time-activity measurements applied to [N-11C-methyl]-(−)-cocaine PET studies in human subjects. J Cereb Blood Flow Metab 10:740–747

Logan J, Fowler JS, Volkow ND, Ding YS, Wang GJ, Alexoff DL (2001) A strategy for removing the bias in the graphical analysis method. J Cereb Blood Flow Metab 21:307–320

Lundkvist C, Halldin C, Ginovart N, Nyberg S, Swahn CG, Carr AA, Brunner F, Farde L (1996) [11C]MDL 100907, a radioligland for selective imaging of 5-HT(2A) receptors with positron emission tomography. Life Sci 58:PL187–PL192

Michaelis L, Menten ML (1913) Die kinetic der Invertinwirkung. Biochemistry Zeitschrift 49:333–369

Muller L, Halldin C, Farde L, Karlsson P, Hall H, Swahn CG, Neumeyer J, Gao Y, Milius R (1993) [11C] beta-CIT, a cocaine analogue. Preparation, autoradiography and preliminary PET investigations. Nucl Med Biol 20:249–255

Naganawa M, Kimura Y, Ishii K, Oda K, Ishiwata K, Matani A (2005) Extraction of a plasma time-activity curve from dynamic brain PET images based on independent component analysis. IEEE Trans Biomed Eng 52:201–210

Oya S, Choi SR, Hou C, Mu M, Kung MP, Acton PD, Siciliano M, Kung HF (2000) 2-((2-((dimethylamino)methyl)phenyl)thio)-5-iodophenylamine (ADAM): an improved serotonin transporter ligand. Nucl Med Biol 27:249–254

Pike VW (2009) PET radiotracers: crossing the blood-brain barrier and surviving metabolism. Trends Pharmacol Sci 30:431–440

Pike VW, McCarron JA, Lammertsma AA, Osman S, Hume SP, Sargent PA, Bench CJ, Cliffe IA, Fletcher A, Grasby PM (1996) Exquisite delineation of 5-HT1A receptors in human brain with PET and [carbonyl-11C]WAY-100635. Eur J Pharmacol 301:R5–R7

Rahmim A, Zaidi H (2008) PET versus SPECT: strengths, limitations and challenges. Nucl Med Commun 29:193–207

Rahmim A, Tang J, Zaidi H (2009) Four-dimensional (4D) image reconstruction strategies in dynamic PET: beyond conventional independent frame reconstruction. Med Phys 36: 3654–3670

Rosso L, Gee AD, Gould IR (2008) Ab initio computational study of positron emission tomography ligands interacting with lipid molecule for the prediction of nonspecific binding. J Comput Chem 29:2397–2405

Scatchard G (1949) The attractions of proteins for small molecules and ions. Ann NY Acad Sci 51:660–672

Schlemmer HP, Pichler BJ, Schmand M, Burbar Z, Michel C, Ladebeck R, Jattke K, Townsend D, Nahmias C, Jacob PK, Heiss WD, Claussen CD (2008) Simultaneous MR/PET imaging of the human brain: feasibility study. Radiology 248:1028–1035

Schmidt KC and Turkheimer FE (2002) Kinetic modeling in positron emission tomography. Q J Nucl Med 46:70–85

Schwarz J, Tatsch K, Gasser T, Arnold G, Oertel WH (1997) [123]IBZM binding predicts dopaminergic responsiveness in patients with Parkinsonism and previous dopaminomimetic therapy. Mov Disord 12:898–902

Seibyl JP, Marek K, Quinlan D, Sheff K, Zoghbi S, Zea-Ponce Y, Baldwin RM, Fussell B, Smith EO, Charney DS (1995) Decreased 123Ibeta-CIT striatal uptake correlates with symptom severity in idiopathic Parkinson's disease. Ann Neurol 38:589–598

Shidahara M, Ikoma Y, Kershaw J, Kimura Y, Naganawa M, Watabe H (2007) PET kinetic analysis: wavelet denoising of dynamic PET data with application to parametric imaging. Ann Nucl Med 21:379–386

Shoghi-Jadid K, Small GW, Agdeppa ED, Kepe V, Ercoli LM, Siddarth P, Read S, Satyamurthy N, Petric A, Huang SC, Barrio JR (2002) Localization of neurofibrillary tangles and β-amyloid plaques in the brains of living patients with Alzheimer disease. Am J Geriatr Psychiat 10:24–35

Slifstein M, Laruelle M (2000) Effects of statistical noise on graphic analysis of PET neuroreceptor studies. J Nucl Med 41:2083–2088

Sossi V (2007) Cutting-edge brain imaging with positron emission tomography. PET Clinics 2:91–104

Spetsieris PG, Moeller JR, Dhawan V, Ishikawa T, Eidelberg D (1995) Visualizing the evolution of abnormal metabolic networks in the brain using PET. Comput Med Imaging Graph 19:295–306

Stout DB, Zaidi H (2008) Preclinical multimodality imaging in vivo. PET Clin 3:251–273

Suehiro M, Scheffel U, Ravert HT, Dannals RF, Wagner HN Jr (1993) [11C](+)McN5652 as a radiotracer for imaging serotonin uptake sites with PET. Life Sci 53:883–892

Tikosfky RS, Ichise M, Seibyl JP, Verhoeff NPLG (1999) Functional brain SPECT imaging: 1999 and beyond. Semin Nucl Med 29:193–238

Tsoumpas C, Turkheimer FE, Thielemans K (2008) A survey of approaches for direct parametric image reconstruction in emission tomography. Med Phys 35:3963–3971

Turkheimer FE, Banati RB, Visvikis D, Aston JA, Gunn RN, Cunningham VJ (2000) Modeling dynamic PET-SPECT studies in the wavelet domain. J Cereb Blood Flow Metab 20:879–893

Van Laere K, Zaidi H (2006) Quantitative analysis in functional brain imaging. In: Zaidi H (ed) Quantitative analysis of nuclear medicine images. Springer, New York, pp 435–470

Vander BT, Minoshima S, Giordani B, Foster NL, Frey KA, Berent S, Albin RL, Koeppe RA, Kuhl DE (1997) Cerebral metabolic differences in Parkinson's and Alzheimer's diseases matched for dementia severity. J Nucl Med 38:797–802

Vaska P, Woody CL, Schlyer DJ, Shokouhi S, Stoll SP, Pratte J-F, O'Connor P, Junnarkar SS, Rescia S, Yu B, Purschke M, Kandasamy A, Villanueva A, Kriplani A, Radeka V, Volkow N, Lecomte R, Fontaine R (2004) RatCAP: miniaturized head-mounted PET for conscious rodent brain imaging. IEEE Trans Nucl Sci 51:2718–2722

Watabe H, Channing MA, Der MG, Adams HR, Jagoda E, Herscovitch P, Eckelman WC, Carson RE (2000) Kinetic analysis of the 5-HT2A ligand [11C]MDL 100,907. J Cereb Blood Flow Metab 20:899–909

Watabe H, Jino H, Kawachi N, Teramoto N, Hayashi T, Ohta Y, Iida H (2005) Parametric imaging of myocardial blood flow with 15O-water and PET using the basis function method. J Nucl Med 46:1219–1224

Watabe H, Ikoma Y, Kimura Y, Naganawa M, Shidahara M (2006) PET kinetic analysis–compartmental model. Ann Nucl Med 20:583–588

Wilson AA, Ginovart N, Schmidt M, Meyer JH, Threlkeld PG, Houle S (2000) Novel radiotracers for imaging the serotonin transporter by positron emission tomography: synthesis, radiosynthesis, and in vitro and ex vivo evaluation of (11)C-labeled 2-(phenylthio)araalkylamines. J Med Chem 43:3103–3110

Woody C, Vaska P, Schlyer D, Pratte J-F, Junnarkar S, Park S-J, Stoll S, Purschke M, Southekal S, Kriplani A, Krishnamoorthy S, Maramraju S, Lee D, Schiffer W, Dewey S, Neill J, Kandasamy A, O'Connor P, Radeka V, Fontaine R, Lecomte R (2007) Initial studies using the RatCAP conscious animal PET tomograph. Nucl Instr Meth A 571:14–17

Wu HM, Hoh CK, Choi Y, Schelbert HR, Hawkins RA, Phelps ME, Huang SC (1995) Factor analysis for extraction of blood time-activity curves in dynamic FDG-PET studies. J Nucl Med 36:1714–1722

Zaidi H, Montandon M-L (2006) The new challenges of brain PET imaging technology. Curr Med Imag Rev 2:3–13

Zhou Y, Ye W, Brasic JR, Wong DF (2010) Multi-graphical analysis of dynamic PET. Neuroimage 49(4):2947–2957

Chapter 12
Optical Approaches to Study Metabolic Consequences of Spreading Depression *In Vivo*

C. William Shuttleworth

Abstract Optical imaging methods have been widely used to investigate the function of complex brain tissue. Depending on the approach used, activity can be readily monitored with relatively high temporal resolution throughout large areas of the cortex, or alternatively from individual neurons or sub-cellular compartments. The aim of this chapter is to illustrate the application of optical approaches to provide information about changes in brain metabolism that follow coordinated neuronal firing. The phenomenon of spreading depression (SD) is used to illustrate some of the advantages and challenges of different approaches. SD involves a slowly-propagating coordinated depolarization of neurons and astrocytes and is thought to underlie migraine aura and contribute to enlargement of brain injuries. The stereotyped progression of this event across the cortical surface, coupled with the substantial increase in metabolic demand, has made this response particularly amenable to optical imaging. The discussion will emphasize signals due to changes in mitochondrial redox state, as assessed by changes in levels of NADH autofluorescence. Optical signals associated with changes in blood flow and oxygenation will also be described, since these can significantly impact autofluorescence studies and because the integration of hemodynamic and mitochondrial signals should be valuable for understanding metabolic deficits and neuronal injury that can accompany SD. It is expected that many of the principles discussed here will also apply to less severe metabolic disturbances that accompany reflex activation of discrete neuronal populations.

Keywords Autofluorescence • Intrinsic optical imaging • Ischemia • Metabolism • Mitochondria • NADH • Spreading depression • Stroke

---

C.W. Shuttleworth, Ph.D. (✉)
Department of Neurosciences, University of New Mexico School of Medicine,
Albuquerque, NM 87131, USA
e-mail: bshuttleworth@salud.unm.edu

I.-Y. Choi and R. Gruetter (eds.), *Neural Metabolism In Vivo*,
Advances in Neurobiology 4, DOI 10.1007/978-1-4614-1788-0_12,

12.1 Introduction

Optical imaging approaches have a long history of use for the study of brain function and the range of methods employed includes detection of multiple intrinsic optical signals and responses of many different fluorescent indicators. The power and diversity of optical approaches is well-illustrated by studies of firing behavior within large populations of neurons, which is essential to understanding the function of neuronal circuits. Microelectrodes can be used to record activity of single neurons, but the technical complexity of such recordings makes it difficult to simultaneously measure responses from more than a small number of cells. A number of optical approaches allow indirect assessment of activity in very large populations of neurons. It was recognized in the late 1960s that neuronal depolarization changes the light scattering properties of individual neurons, and this permits measurement of neuronal firing *in vitro* and *in vivo* (Cohen et al. 1968; Tasaki et al. 1968; Wininger et al. 2009). Although such signals are quite small and averaging of repetitive trials is usually required, these types of intrinsic signals can be very useful. The development of fluorescent probes sensitive to membrane voltage changes provided significant increases in signal-noise characteristics (Cohen et al. 1978) and have been used widely into assess circuit activity and changes in excitability with pathophysiological conditions. While voltage-sensitive indicator signals are directly related to neuronal firing, they can be technically challenging because of the rapid time course of membrane voltage changes and spatial resolution is usually compromised by the high acquisition rates that are needed. Partly because of this problem, complementary optical methods that track longer-lasting intracellular consequences of neuronal depolarization (usually Ca^{2+} accumulation) have been very widely used, including recent work with two photon excitation in relatively deep brain structures (Cossart et al. 2005). Thus a range of optical approaches are widely used to assess fundamental changes in CNS activity.

The emphasis of this chapter is on optical imaging of different signals that provide more direct information about changes in brain metabolism that follow coordinated neuronal firing. To provide a framework for discussion of the advantages and challenges of different optical approaches, the chapter will focus on the phenomenon of spreading depression (SD). There are a number of reasons why SD is particularly well-suited for this purpose. As discussed below (Sect. 12.4), SD involves a coordinated depolarization of neurons and astrocytes, that propagates slowly in across the brain and is associated with transient suppression of ongoing synaptic activity. The fact that SD can be used experimentally to deliver a dramatic increase in metabolic demand with high temporal precision has made it very attractive as a test stimulus for development and evaluation of optical methods for at least 40 years. In addition, recent work has provided a new impetus to the study of SD in a range of neurological disorders, suggesting that future work with optical methods could be valuable for development of interventions and evaluation of mechanisms underlying deleterious and possibly protective effects of these events.

Optical methods have long been used to assess changes in tissue mitochondrial function that accompany SD, and these will be emphasized here. Because optical signals associated with changes in blood flow and oxygenation can significantly impact autofluorescence studies of mitochondrial redox potential, optical imaging of hemodynamic signals will also be briefly considered. Recent work has advanced mitochondrial imaging studies to include cellular and subcellular origins of autofluorescence signals, and the integration of these signals with assessment of regional cerebral blood flow is likely to be helpful for understanding the metabolic deficits and neuronal injury that can accompany SD. While SD is an unusually strong stimulus that involves activation of large populations of cells, it is expected that many of the principles discussed here will also apply, albeit with different signal/noise characteristics, to less severe metabolic disturbances that accompany reflex activation of discrete neuronal populations.

12.2 Spreading Depression and Related Events

12.2.1 Basic Mechanisms

Spreading depression (SD) was originally described by Leao as a wave of profound depression of electrical activity that progressed slowly in a wave-like manner across the rabbit cortex (Leao 1944a, 1947; Leao and Morison 1945). Many excellent reviews have discussed the history of Leao's initial discoveries and summarized subsequent studies of SD mechanisms (Marshall 1959; Bures et al. 1984; Martins-Ferreira et al. 2000; Gorji 2001; Somjen 2001, 2005; Teive et al. 2005), however some of the key points will be reiterated here to emphasize why optical methods have been used for a number of decades to examine SD in a range of experimental and clinical settings.

Leao's original discovery came while making recordings from the exposed cortical surface with arrangements of recording electrodes placed on the pia-mater. The original intention of the work was to study mechanisms underlying epileptiform activity, but he found that electrical or mechanical stimulation could result in a "marked, enduring reduction of the spontaneous electrical activity of the cortex". From recordings made from arrangements of multiple electrodes arranged over the cortical surface, it was shown that the depression of electrocorticographic (ECoG) activity propagated slowly in all directions (Fig. 12.1). Leao coined the term "spreading depression" (SD) and, in addition to blocking spontaneous electrical activity, evoked cortical responses were abolished during the depression phase (Leao 1944a). Early studies showed that SD could be generated by both electrical and mechanical stimuli (such as light touch with a glass rod), and also by K^+ or Ca^{2+} applications. In particular, a piece of filter paper soaked in elevated K^+ solutions placed on the brain surface produced a repetitive series of SDs, with a new process commencing even before the surface electrical activity had fully recovered (Leao and Morison 1945). A small plug of filter paper (or similar material) soaked in KCl and placed onto the

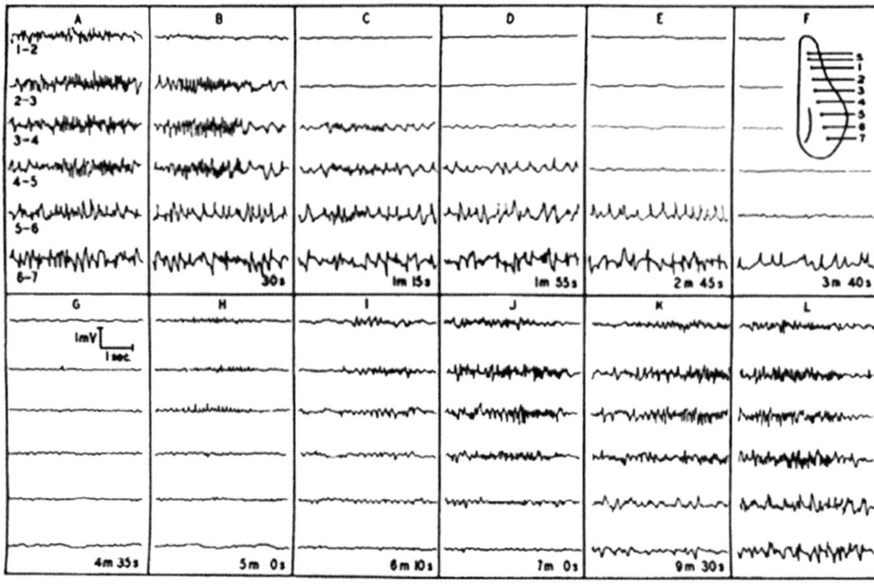

Fig. 12.1 Leao's spreading depression. This figure from Leao's original description of SD (Leao 1944) shows electrocorticographic activity recorded between pairs of electrodes arranged across the rabbit cortex. Electrical stimulation (tetanizing frequency for 3-5 s) applied at the most frontal location (S) induced a slowly-propagating suppression of ECoG activity. By approximately 4 min after the stimulus, the entire cortical region appeared electrically silent, but activity subsequently recovered throughout the region as ionic gradients were restored (From Leao 1944a), with permission)

cortical surface via a small burr hole in the skull is still widely used as a method of generating SD in animal models.

Leao showed that SD can also be described as a slow shift in direct current (DC) potential (Leao 1947), recognized now to be due to sudden opening of ion channels allowing extensive redistribution of ions across the plasma membranes of both neurons and astrocytes. Na^+ and Ca^{2+} accumulation is accompanied by a conspicuous extracellular K^+ elevation, as recorded with ion-sensitive electrodes and Ca^{2+} indicators (Martins-Ferreira et al. 2000; Somjen 2001). Release of glutamate and K^+ together are thought to be responsible for feed-forward nature of the response. Extracellular K^+ concentration increases above about 12 mM are sufficient to initiate SD, and blockers of NMDA-type glutamate receptors are able to prevent propagation of some forms of SD (Somjen 2001).

12.2.2 Metabolic Consequences of SD

The extraordinary ionic shifts that accompany SD provide a trigger for substantial metabolic changes that occur with high temporal precision relative to the progression of SD. In early studies, a significant drop in the partial pressure of oxygen

(PO_2) was reported during the progression of SD (Van Harreveld and Stamm 1952), and decreases in glucose and glycogen content coupled with increased lactate production were also noted (Bures 1956; Krivanek 1958, 1961). Robust increases in glucose utilization (CMR_{glc}) were subsequently recorded throughout the cerebral cortex after SD (Shinohara et al. 1979), and glucose decreases were accompanied by reductions in ATP levels by ~50% following a single SD event in rat cortex (Mies and Paschen 1984).

Because of the well-appreciated extracellular lactate increase following SD, this event has been used for studies of brain lactate release and uptake mechanisms, including efforts to understand the important question of metabolic coupling between neurons and astrocytes (Kocher 1990; Scheller et al. 1992; Cruz et al. 1999). For example, in ^{14}Cglucose studies of SD in rat, it was concluded that SD results in significant efflux of lactate into venous blood, as well as diffusion throughout other brain regions (Cruz et al. 1999).

While the metabolic demands of SD are large, healthy brain tissue has the capacity to recover quite quickly. This recovery was part of Leao's original description of the transient nature of ECoG suppression, as spontaneous electrical activity was recovered within a few minutes *in vivo* (Leao 1944a). Furthermore, when SD is generated by placement of KCl on the brain surface, repetitive events can be generated for hours without apparent injury, despite complete cessation of electrical activity during propagation of each event. Consistent with these observations, large decreases in phosphocreatine levels that occur in the first minutes of a SD event recover to normal levels relatively quickly, together with restoration of extracellular pH (Csiba et al. 1985; Lauritzen et al. 1990; Gault et al. 1994). The lack of neuronal damage produced by repetitive SD events in otherwise healthy brain, has also been demonstrated in detail (Nedergaard and Hansen 1988). As discussed below (Sect. 12.5), significant increases in blood flow accompany SD and the rapid provision of increased oxygen and glucose provides fuel for ATP-dependent pumps involved in restoration of membrane potential. However when SD occurs in conditions where metabolic substrates are limiting, irrecoverable injury can occur (Sect. 12.2.3.2).

12.2.3 Pathophysiological Consequences

12.2.3.1 Migraine

Passage of SD across the visual cortex likely underlies visual aura symptoms that accompany some migraines. Migraine aura comprises a region of visual field loss that moves at about 3 mm/min across the visual field. After a description of aura by Lashley in 1941, Leao discussed the similarity between progression of migraine aura and the propagation rate of SD, and proposed that a progressive depression of activity in visual cortex might underlie aura (Leao and Morison 1945). The scintillating border around the scotoma as aura progresses could be explained by a brief period of increased excitability prior to arrival of a SD wave. There has since been much work to support this idea (Lauritzen 1994; Pietrobon and Striessnig 2003; Eikermann-Haerter and Ayata 2010), with supporting evidence including propagating changes in blood flow and fMRI signals during migraine in human subjects (Olesen et al. 1981; Hadjikhani et al. 2001) and signals corresponding to the slow DC shift being recorded with magnetoencephalography during migraine attacks (Bowyer et al. 2001). While there is accumulating evidence that aura is a consequence of SD, whether or not other symptoms of migraine are due to SD is not clear cut. It is possible that SD is one of a set of downstream consequences of some other initiating stimulus, but it is also possible that SD generates migraine headache. These issues are currently being debated (Ayata 2010; Charles 2010).

Recent work has identified multiple mutations that influence susceptibility to migraine in a disorder termed familial hemiplegic migraine (FHM). Mutations that lead to increased conductance of voltage-dependent calcium or sodium channels have been identified, as well as mutations that lead to decreased Na^{+}/K^{+}/ATPase activity. All of these types of mutation can increase cortical excitability and importantly, in cases where these mutations are expressed in mice, there is a significantly decreased seizure threshold (van den Maagdenberg et al. 2004; Eikermann-Haerter et al. 2009). This evidence suggests that increased susceptibility to SD can explain the symptoms of FHM, and raises the possibility that targeting SD could be useful for treatment and prophylaxis.

12.2.3.2 Brain Injury

While the metabolic demands of SD can be adequately handled by healthy brain, there is considerable evidence to suggest that energy depletion caused by SD can be harmful if tissue metabolism is already partially compromised. A large body of literature has characterized an "anoxic depolarization" that occurs when neurons undergo a large and rapid depolarization following interruption of blood supply, and the injury that rapidly follows if measures are not taken to immediately restore blood supply (Hansen 1985). In clinical situations, it is often difficult to restore perfusion quickly enough to prevent establishment of a core ischemic lesion, but surrounding tissue may still be viable despite a significant perfusion deficit. Such regions have been termed "ischemic penumbra", and the propagation of SD events through penumbral tissue has been implicated in the enlargement of ischemic injuries.

Events that are very similar to SD are generated following focal ischemia, and since they usually originate from edge of the ischemic core, these have been termed peri-infarct depolarizations (PIDs) (see Hossmann 1996). PIDs have been recorded in the hours following stroke injuries and the number of events correlates with the infarct size in animal models (Strong et al. 1983b; Nedergaard and Astrup 1986; Hartings et al. 2003). These events appear to play a role in the enlargement of injury, rather than simply being a secondary consequence of the pathology. Treatments that decrease PID incidence are neuroprotective (Gill et al. 1992; Iijima et al. 1992;

Chen et al. 1993; Mies et al. 1993; Mies et al. 1994; Rawanduzy et al. 1997; Tatlisumak et al. 2000) and experimentally generating SD (which is normally non-injurious) in the context of stroke substantially increases infarct size (Busch et al. 1996; Takano et al. 1996). Depletion of energy substrates by the combination of partial ischemia with the metabolic burden of SD is likely responsible for the increase in injury size (Selman et al. 2004).

An important concept that has emerged is that SD leads to impaired neurovascular coupling. This is based in part on the fact that long-lasting oligemia described following SD is not matched by a long-lasting decrease in regional cortical glucose metabolism following SD (Lauritzen and Diemer 1986) suggesting that the flow decrease is unlikely to be due to decreased tissue metabolic activity. In the vicinity of injured tissue, perfusion decreases (rather than increases) are conspicuous following SD (Back et al. 1994), and this deficiency is likely central to injury produced by SD under a range of conditions.

Results from these and other animal studies provided an additional impetus to test whether SDs are generated in humans following brain injury, and contribute to the enlargement of infarctions. A recent series of reports has appeared from a consortium (Cooperative Study on Brain Injury Depolarisations; COSBID) which has used strip electrodes to make ECoG recordings similar to those originally performed by Leao. Electrodes have been placed after craniotomy, and have led to recordings of SDs for hours and days following large ischemic strokes (Dohmen et al. 2008), subarachnoid hemorrhage (Fabricius et al. 2008; Dreier et al. 2009), and traumatic brain injury (Hartings et al. 2008; Hartings et al. 2009; Sakowitz et al. 2009). In some cases, the passage of SD has been demonstrated to coincide closely with enlargement of the ischemic penumbra, supporting conclusions from animal studies that repetitive SDs are deleterious in the injured brain (Dohmen et al. 2008).

Based on these and other studies, there is strengthened interest in determining the metabolic consequences of SD events *in vivo*. Identifying the relationship between cellular metabolism and vascular coupling is a central part of this goal, and optical imaging approaches can be of significant utility.

12.2.3.3 Terminology

The term "cortical spreading depression" (CSD) has been used to refer specifically to the phenomenon originally described by Leao that propagates across normally-perfused cerebral cortex. The mechanisms underlying wave-like depolarizations triggered by ischemia and by a variety of experimental conditions are very similar in many respects (or possibly identical) to classical spreading depression described by Leao. However since the causes and consequences of the events can be very different, a range of different terminologies have been introduced for post-ischemic depolarizations. The term "spreading depolarization" (also abbreviated as SD) has recently been adopted in the clinical literature to emphasize (1) the similar cellular events involved in propagation and (2) the central aspect of neuronal and glial depolarization that occur in both classical cortical spreading depression, and similar events observed following ischemia (Dreier et al. 2006; Dahlem et al. 2008; Dohmen et al. 2008; Petzold et al. 2008; Hartings et al. 2009). In the following discussion, the term "SD" will be used as a general term for cortical spreading depression and related events occurring in ischemia.

12.3 Autofluorescence Studies

12.3.1 Introduction

In the 1950s and 1960s, Britton Chance and colleagues were largely responsible for demonstrating that changes in mitochondrial function could be assessed in complex tissues, using relatively straightforward optical methods. The term "autofluorescence" simply refers to fluorescence signals that are derived from endogenous fluorophores, rather than from indicators that are experimentally loaded into tissues. While many tissue components can contribute to autofluorescence, nicotinamide adenine dinucleotide and flavoprotein fluorescence emission changes significantly depending on mitochondrial redox state. In the last five decades, many studies have exploited measures of these endogenous fluorophores to assess metabolic consequences of neuronal activation. The reader is referred to a number of excellent reviews that cover the history and range of applications for both NADH (Anderson and Meyer 2002; Chance 2004; Mayevsky and Rogatsky 2007) and flavoprotein signals (Reinert et al. 2007; Shibuki et al. 2007; Tohmi et al. 2009) in studies of nervous system function and dysfunction.

It is not surprising that there is long history utilizing autofluorescence methods to study SD. SD results in a profound metabolic disturbance that can be quite rigorously controlled and, since the event propagates across the brain surface, the event can be accessible with optical methods. NADH autofluorescence has historically been much more widely used than flavoprotein signals to study SD, and the following discussion will therefore concentrate on the NADH signals.

12.3.2 NADH Autofluorescence

Mitochondrial ATP generation occurs as protons flux across the inner mitochondrial membrane, via ATP synthetase. A large electrochemical gradient across the mitochondrial inner membrane is required to power this synthesis, and the gradient is maintained by activity of the electron transport chain. NADH is a principal electron donor that supplies this process. NADH generated within mitochondria by the tricarboxylic acid (TCA) cycle is oxidized to NAD^+ at complex I, and subsequent electron transfer through the electron transport chain culminates in consumption of molecular oxygen at complex IV (cytochrome oxidase) (Fig. 12.2). Thus NADH oxidation is required to maintain the proton gradient during periods of ATP demand,

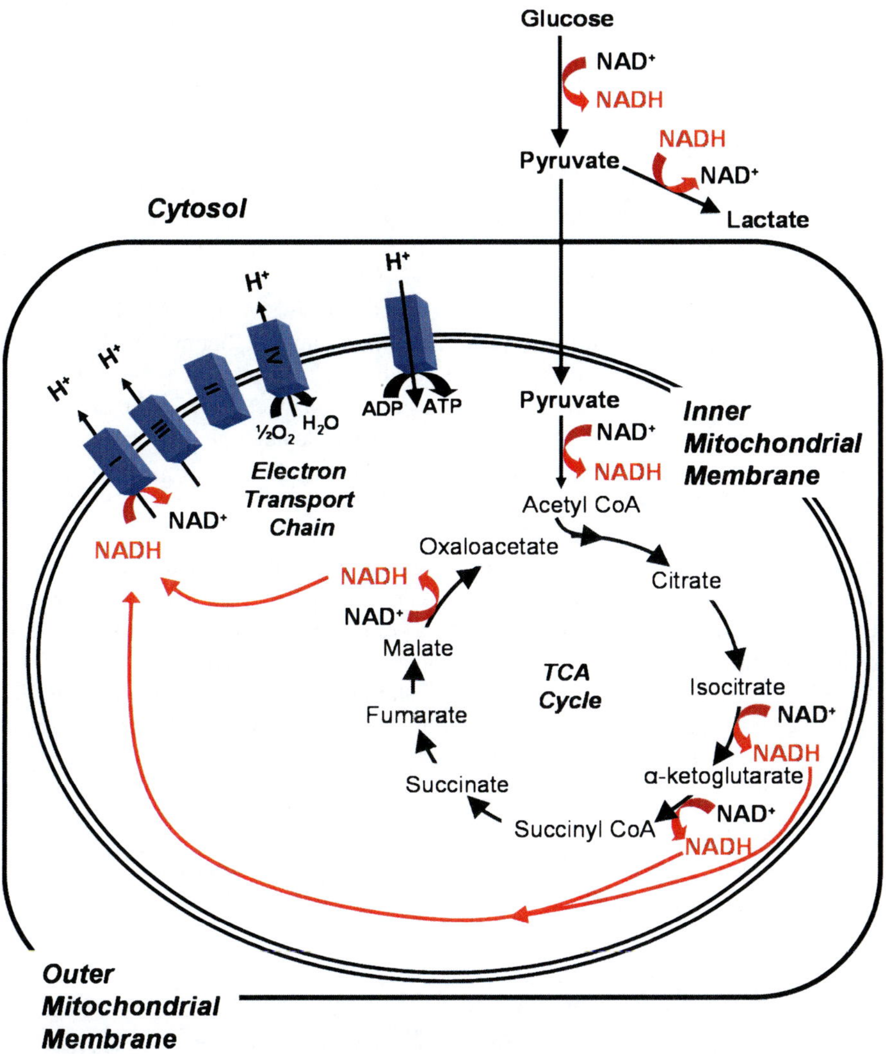

Fig. 12.2 Sources of NADH signals associated with metabolism. When excited at UV wavelengths, NADH is fluorescencent but the oxidized form (NAD^+) is non-fluorescent. NADH fluorescence is also enhanced in the mitochondrial compartment, likely as a consequence of binding to complex I. Decreases in tissue NADH fluorescence can occur as a consequence of increased demand for mitochondrial ATP synthesis and accelerated electron transport. In contrast, when oxygen levels are limiting, mitochondrial NADH fluorescence increases as inhibition of complex IV activity decreases flux through the electron transport chain. In addition to mitochondrial TCA cycle activity, NADH is generated in the cytosol by the glycolytic enzyme glyceraldehyde phosphate dehydrogenase and oxidized to NAD^+ by lactate dehydrogenase. Mitochondrial NADH/NAD^+ transitions appear to dominate NADH autofluorescence transients following synaptic stimulation and SD

and cellular O_2 consumption is normally tightly coupled to NADH/NAD^+ transitions (Nicholls and Ferguson 2002).

NADH is fluorescent, with relatively broad excitation and emission spectra (peak excitation ~340–360 nm and peak emission in the blue range ~430–450 nm; Chance et al. 1962, 1979; Harbig et al. 1976; Aubin 1979). Importantly, early studies confirmed that pure reduced NADH had nearly identical optical characteristics as autofluorescence generated by the same wavelengths in neuronal tissues (including single neurons, axons and intact brain), with the exception of a small wavelength shift due to enzyme binding in some studies (Chance et al. 1962; Terzuolo et al. 1966; Doane 1967; Galeotti et al. 1970; Harbig et al. 1976). While NADH is strongly fluorescent (particularly in mitochondrial compartments, see below), NAD^+ is non-fluorescent and this difference makes it possible to measure changes in the total NADH/NAD^+ ratio in brain tissue relatively easily (Chance et al. 1962).

The fluorescence properties of reduced nicotinamide adenine dinucleotide phosphate (NADPH) are very similar to that of NADH, and therefore it is reasonable to expect that NADPH should contribute to total tissue autofluorescence. To acknowledge the possible contribution of NADPH to compound signals, some groups refer to "NAD(P)H" signals in studies of CNS tissues (Schuchmann et al. 2001; Shuttleworth et al. 2003; Kahraman and Fiskum 2007). However as discussed previously, NADPH levels have long been recognized to be much lower than NADH in brain tissue and since NADPH is not involved in electron transfer in the electron transport chain, it is expected to normally make little contribution to mitochondrial autofluorescence transients (see discussion in Mayevsky and Rogatsky 2007; Shuttleworth 2010).

12.3.3 Technical Considerations for Autofluorescence Imaging of SD

12.3.3.1 Single Photon Penetration and Toxicity

Because no complex loading procedures are required to deliver a fluorescent reporter into target cells, imaging autofluorescence has a fundamental advantage over most other fluorescence methods. However a tradeoff for this convenience is that it can be difficult to detect signal changes from a small number of active cells, if the target cells are surrounded by tissue with high levels of unchanging fluorescence. For this reason, CNS autofluorescence studies are often most effective when populations of synchronously-firing cells are studied. The highly coordinated waves of neuronal depolarization and metabolic increases that occur during SD make this event very well suited for autofluorescence methods, even with imaging systems with quite low spatial resolution.

Optical approaches can be significantly limited by scattering of excitation photons by brain tissue and overlying structures. Scattering is highly dependent on wavelength, and the short wavelengths required for NADH excitation penetrate the intact skull very poorly. For this reason most *in vivo* NADH studies have been done by visualization through open cranial windows with the dura removed. This approach usually limits longitudinal studies, but it is possible to create a stable cranial window using a fixed coverglass that can be used for repeated imaging sessions in the same animal (Holtmaat et al. 2009). Such an approach should be suitable for repeated NADH studies of SD in animal models.

NADH imaging via cranial windows is very effective for many types of SD studies, but even with the skull and dura removed, effective NADH excitation by UV light is limited to superficial layers of the cortex. The effective excitation depth in rat cortex was originally speculated to be restricted to superficial cortical layers (Jobsis et al. 1971) and the degree of attenuation with depth has been modeled recently with Monte Carlo simulations (L'Heureux et al. 2009). Increasing the intensity of UV illumination can increase the effective excitation of deeper layers, but it is important to note that if this is done using conventional optics, superficial tissue will be very strongly excited and complicate the analysis of signals from deeper structures. The large proportion of fluorescence emission derived from shallow structures that do not participate in a response will reduce the signal-noise ratio of responses from deeper structures, and the response may be also impaired by phototoxicity caused by excessive UV absorption in superficial tissues.

12.3.3.2 Two Photon Excitation

Partly because of penetration problems with short wavelength light, multi-photon excitation methods have emerged as useful for NADH imaging studies *in vitro* and *in vivo* (Huang et al. 2002; Zipfel et al. 2003; Rocheleau et al. 2004). Multi-photon excitation is based on the principle that a fluorophore requires an appropriate quantum of energy to become excited to a higher energy state. Excitation energy can be delivered either by a single photon, or by multiple photons of lower energy (i.e. longer wavelengths). Multi-photon excitation is most commonly achieved with pairs of photons of approximately double the wavelength used for single photon excitation (two photon excitation). For NADH excitation, pairs of photons (720–740 nm) are often used, and because these wavelengths are much less scattered than UV illumination, penetration into brain is greatly increased. In addition, because pairs of photons must be absorbed by the fluorophore at very nearly exactly the same time, a very high photon density is achieved at the objective focal point by using a high frequency infrared laser source, and effective two photon excitation falls very steeply as a function of distance from the focal point. As a consequence, only a very small focal plane is effectively excited and this limits photobleaching of fluorophore outside the region of interest. In addition to allowing for substantially improved penetration into the brain, two photon excitation results in much higher spatial resolution, and allows NADH signals to be resolved from individual cells within complex brain tissue (Huang et al. 2002; Zipfel et al. 2003; Kasischke et al. 2004).

Because infrared light is much less attenuated by bone than UV wavelengths, effective two photon excitation of NADH can be achieved without complete removal of a section of skull to form a cranial window. A number of investigators have developed approaches to thin a region of the rodent skull down to a few tens of microns (Yang et al. 2010). This needs to be done without excessive heating of the underlying brain, but can provide a region that is nearly transparent with two photon imaging, suitable for repetitive NADH imaging sessions over a period of days or weeks. It should be noted that while two photon approaches give a considerable advantage in excitation depth, the maximally effective depth of NADH imaging is still limited by scattering and absorption of the NADH emission photons (~450 nm). Thus two photon NADH imaging is still restricted to the order of hundreds of microns below the brain surface, and many interesting deep structures remain out of reach. Surgical approaches to remove overlying tissues, or light guides (fibers or lenses) should be useful to assess NADH dynamics in deeper structures *in vivo* (Jung et al. 2004; Levene et al. 2004; Barretto et al. 2009).

Confocal microscopy using single photon excitation can improve spatial resolution by excluding emission light that is out of the focal plane and has been very effectively used for NADH imaging in a range of settings (e.g. Combs and Balaban 2001; Duchen et al. 2003). However, the excitation of large volumes of tissue with confocal microscopy can lead to significantly more photobleaching of thick preparations than with two photon methods, and furthermore since confocal methods for NADH still rely on UV excitation, penetration through skull and brain tissue remains very limited. For these reasons, single photon confocal approaches have been largely surpassed by multiphoton methods for imaging of NADH in brain slices and *in vivo*.

12.3.3.3 Correction of NADH Signals for Optical Artifacts

Interference from Hemodynamic Responses

It was Leao who first described a dramatic change in cerebral blood flow associated with SD (Leao 1944b). He reported that SD was associated with a wave of vasodilation and increased flow in arteries of all sizes, and concluded that the increase in vessel diameter (in the order of 50–100% increase) was significantly larger than that produced by hypercapnia. Leao also noted that flow in veins was increased, and that they rapidly became red. Leao concluded that the vascular response was secondary to local change in activity neuronal tissue and speculated that the increase in flow probably influenced the activity of the neurons (Leao 1944b). The temporal relationship between pial vessel changes and SD was examined in more detail by Van Harreveld and Stamm (1952) who concluded a wave of vasoconstriction occurs together with the slow potential change and decreased PO_2, and proposed an asphyxial mechanism for the propagation of suppression of surface electrocorticographic activity. The relationship between blood flow and neuronal activity or injury has since been examined in considerable detail using a range of methods, including laser dopper flow and autoradiographic measures. Although there can be significant species differences in responses, experimental SD is often accompanied by a transient increase in cortical blood flow, and a subsequent longer-lasting flow decrease (Lauritzen et al. 1982; Wahl et al. 1987; Piper et al. 1991; Dreier et al. 1998; Busija et al. 2008). In mice, a significant early hypoperfusion has also been noted that coincides with the duration of the DC shift (Ayata et al. 2004; Brennan et al. 2007b). Prominent vascular responses that occur during the progression of SD have been exploited to track (with fMRI and PET approaches) SD events that occur in human subjects during migraine attacks (Woods et al. 1994; Hadjikhani et al. 2001).

As discussed in more detail below (Sect. 12.4.2), the absorption of light by hemoglobin is dependent on the degree of oxygen binding, and occurs over wavelengths that overlap with the NADH emission fluorescence spectrum. Thus hemodynamic changes have the potential to decrease the amount of autofluorescence emission that is detected, even if there are no changes in NADH redox state.

Interference from Cell Swelling

In addition to hemodynamic effects, a substantial tissue swelling response has long been recognized during SD, and this can also be a significant confound to small autofluorescence optical signals.

Van Harreveld first clearly demonstrated a swelling of neurons following SD-like events associated with ischemia in rabbit cortex (Van Harreveld 1957) and subsequently showed that apical dendrites of neurons swelled following SD generated by electrical stimulation in the same preparation (Van Harreveld 1958). The swelling response produces a reliable change in light scattering, and this is of sufficient magnitude that the progression of SD can be seen by the naked eye in some preparations (e.g. Martins-Ferreira and de Castro 1966). The transmission of white (or filtered red light) through thin preparations is also dramatically modified by cellular swelling during SD, and intrinsic optical signals generated by transmitted light have been widely used to track SD in brain slices (Aitken et al. 1999; Jarvis et al. 1999; Anderson and Andrew 2002).

Recent studies have utilized two photon imaging to examine in more detail the optical signals generated by swelling of neurons and astrocytes during SD. In addition to showing a rapid neuronal swelling coincident with SD, a transient loss of dendritic spines was also noted that could contribute to the temporary silencing of electrical activity following SD (Takano et al. 2007). Two photon studies of murine cortical brain slices have demonstrated that astrocytes swell following hyposmotic stress and K^+ exposures (Risher et al. 2009), however when the relative contributions of neurons and astrocytes were specifically studied in relation to SD onset in rat hippocampal slices, it was shown that neurons undergo a rapid and transient swelling, while astrocytes undergo a passive swelling in response to K^+ elevations (Zhou et al. 2010).

In addition to reversible swelling responses due to water movement, stereotypical changes in light transmission can occur as a consequence of neuronal injury. Ischemia or exposure to high concentrations of glutamate receptor agonists can generate a stereotypical pattern of dendrite swelling punctuated by severely constricted regions. This gives the dendrite the appearance of "beads on a string" and is associated with significant rearrangement of structural proteins and irreversible injury (Swann et al. 2000; Hoskison and Shuttleworth 2006; Greenwood et al. 2007). Dendritic beading is prominent when SD occurs in metabolically compromised conditions, and it has been concluded that the extreme structural rearrangement can underlie persistent increases in light scattering (Obeidat and Andrew 1998; Andrew et al. 1999).

Correction from Reflectance Changes

A reflectance subtraction method has been widely used to correct for the impact of swelling and hemodynamic changes on NADH responses. Tissues are illuminated with UV light to excite NADH fluorescence, and the intensity of UV light reflected from the cortical surface is collected together with the NADH fluorescence emission. Subtraction of the reflectance signal generates a corrected NADH fluorescence signal, which was first validated by Jobsis and colleagues for studies of epliptiform activity in perfused cat cortex (Jobsis et al. 1971). A similar correction approach was utilized soon thereafter in cortical slices *in vitro* by Lipton, to report corrected biphasic changes in NADH fluorescence following sustained electrical stimulation (Lipton 1973a, b). The work of Mayevsky and Chance (1974) provided a detailed justification of the subtraction method for SD studies *in vivo*, and showed that simple 1:1 subtraction provides appropriate compensation for vascular effects during SD. The initial subtraction method was later extended by methods to include measurement of fluorescence at the isobestic wavelength of hemoglobin-oxygemoglobin transitions (445 nm) in cat cortex (Harbig et al. 1976). These subtraction methods and their limitations have been discussed in detail relatively recently (Bradley and Thorniley 2006; Mayevsky and Rogatsky 2007). While there are some limitations, correction for reflectance changes have been very important for assessment of NADH responses during SD, since the initial oxidation phase of NADH responses during SD *in vivo* might be completely missed without this correction (see Mayevsky et al. 1974).

The reflectance correction method has been used routinely in NADH studies that utilize fluoremeters to record averaged fluorescence signals from a region of interest on the cortical surface. However, the same approach is not usually applied in CCD-based imaging studies of SD progression across the brain. It has been noted that CCDs lack sensitivity to UV light, making it very difficult to measure meaningful reflectance signals together with the blue light fluorescence emission of NADH (Strong et al. 1996). This limitation requires that hemodynamic and swelling responses are carefully considered in the interpretation of uncorrected NADH signals *in vivo*.

Comparison with Flavoprotein Autofluorescence

Comparison with flavoprotein (FP) autofluorescence signals can be an additional useful test for contamination of NADH autofluorescence signals by changes in tissue swelling or light absorption. Mitochondrial FP fluorescence changes are directly coupled to NADH responses, however the redox sensitivity of the two signals are inverted with respect to each other. Thus mitochondrial oxidation leads to FP fluorescence increases, coincident with NADH fluorescence decreases. Conversely, FP fluorescence decreases and NADH increases occur during transitions to more reduced mitochondrial redox states (see discussion in Shuttleworth 2010). Thus if NADH and FP signals have similar kinetics but are of opposite sign, then it is likely that the responses are both of mitochondrial origin. Conversely, if the fluorescence of both signals change in parallel, it is likely that changes in light absorption and scattering make a major contribution to the signals (Shuttleworth 2010).

The development of a dual-wavelength fluorometer permitted simultaneous NADH/FP measurements to be made (Chance et al. 1971) and scanning measurements of mitochondrial redox ratio have also been achieved (Chance et al. 1979). For CCD-based recording, near-simultaneous imaging can be done by alternating filter sets (Duchen 1992; Mironov and Richter 2001; Shuttleworth et al. 2003; Brennan et al. 2004; Brennan et al. 2007a). However the broad emission spectra of NADH means that it is also possible to using a single filter set (e.g. 505 nm beamsplitter and a single 510/40 nm bandpass filter) to make virtually simultaneous measurements of both fluorphores, if the excitation wavelength is switched quickly with an excitation filter changer of monochromator. This arrangement leads to loss of a great deal of the NADH fluorescence signal, but still allows enough transmission to measure inverted NADH/FP signals (Hepp et al. 2005; Gerich et al. 2006).

In comparison to NADH signals, there have been few reports of FP autofluorescence dynamics during SD *in vivo*. However FP signals have been utilized together with NADH to characterize mechanisms underlying energy depletion in response to spreading depression in the gerbil (Haselgrove et al. 1990), and to track the propagation of a related propagating depolarization in mouse cerebellar cortex (Chen et al. 2005).

12.3.4 NADH Autofluorescence Signals During Normoxic SD

12.3.4.1 Single Site Recordings

In the early 1970s, Jobsis and colleagues utilized a fluorimeter system to make simultaneous fluorescence and reflectance measurements from a region of cat cortex and reported NADH fluorescence dynamics corrected for changes in hemodynamic and/or swelling changes. They reported NADH transients during a range of stimuli, including evoked seizure activity (Jobsis et al. 1971) and graded electrical stimulation (Rosenthal and Jobsis 1971). The same system was soon used by Rosenthal and Somjen for study of mitochondrial activity during SD, elicited by electrical stimulation or elevated K^+ in normoxic cat brain (Rosenthal and Somjen 1973). A significant question at the time was whether brain metabolism was deficient prior to SD, and was responsible for the initiation and propagation of the event. However it was clearly found that there was no NADH fluorescence decrease prior to SD, but a substantial decrease occurred following SD. (Fig. 12.3). This early observation was consistent with the accumulating evidence that SD involved a profound metabolic challenge to brain tissue (see Sect. 12.2.2 above), and the authors discussed possible roles for both astrocytes and neurons in the metabolic signals generated (Rosenthal and Somjen 1973).

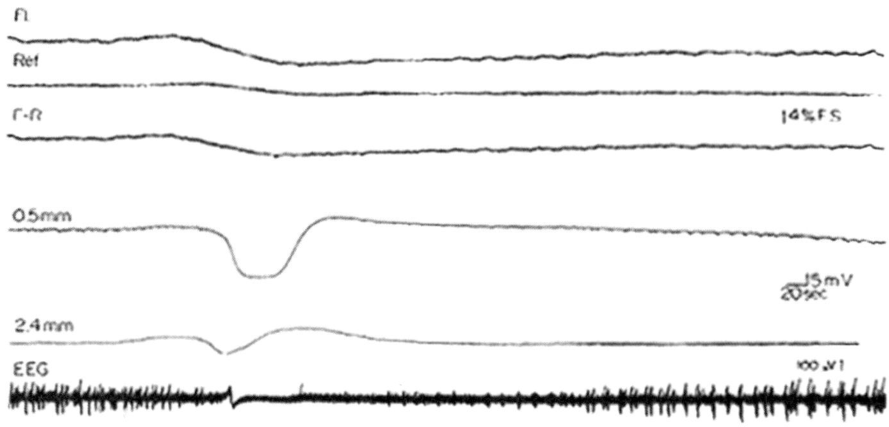

Fig. 12.3 NADH oxidation during SD. This figure is taken from one of the initial descriptions of NADH decreases in cortex during SD (Rosenthal and Somjen 1973). The top trace (FL) is the fluorescence emission from the cortical surface, generated by UV excitation. This record was corrected for changes in UV absorption during SD by subtracting the reflectance signal (Ref) to generate the characteristic signal attributed to NADH oxidation (F-R). The lower three traces are simultaneous electrical recordings of SD, including slow potential changes at two depths under the corticla surface (0.5 and 2.4 mm) and surface EEG activity (From Rosenthal and Somjen 1973, with permission)

At around the same time (Mayevsky and Chance 1974) reported similar large NADH fluorescence decreases during SD, using a newly-developed dual fluorometer coupled to the recording site by a flexible light pipe. The use of small light guides also allowed positioning of NADH and reflectance measurements close to a double-barreled K^+ sensitive microelectrode, and EEG electrodes. From these studies it was confirmed that the durations of NADH decreases were very closely matched with extracellular K^+ increases that occurred during the spread of SD (Mayevsky et al. 1974). NADH oxidation signals were interpreted to be triggered by ATP decreases that occurred after strong activation of plasma membrane Na^+K^+ATPase pumps, and increased O_2 consumption that accompanied increased blood flow (Mayevsky et al. 1974). When the relationship between NADH oxidation and K^+ elevations was examined further, a linear relationship was observed with moderate electrical stimulation in cat cortex, but the NADH oxidation response was much larger for SD than would be expected from that relationship. It was suggested that this might be due to recruitment of an auxiliary source of ADP (i.e. a reserve ATP pool) that is called upon during SD, or possibly a failure of metabolic regulation following SD (Lothman et al. 1975).

While single-site studies of SD *in vivo* have often reported monophasic NADH decreases, this is not always the case. Some studies report that initial the oxidation phase is followed by a large and longer-lasting NAD^+ reduction phase following passage of SD (e.g. Rex et al. 1999). This may be due to differences in measurement methods, but as discussed below (Sect. 12.7.2) differences in oxygen availability between experimental models can also strongly influence the polarity of NADH signals after SD.

12.3.4.2 Full Field Imaging

The application of CCD-based imaging and two photon imaging has been very useful to assess propagation of SD events across the brain surface and cellular mechanisms that contribute to recovery. One of the early studies involved measurement of waves of NADH fluorescence during SD triggered by localized K^+ application onto the surface of the gerbil cortex (Hashimoto et al. 2000). The CCD was placed over a cranial window (2 × 3 mm), and K^+ was applied at a burr hole some distance away to initiate SD events that would spread through the recording field. Images were taken relatively slow frame rate (every 5 s), but it was possible to visualize the propagation of a long-lasting NADH increase across the field of view at a rate consistent with SD. It was curious that an initial NADH oxidation phase was not observed during SD, and it is possible that without adequate correction for reflectance changes (see above), the oxidation phase was obscured by optical interference from hemodynamic changes.

Two photon imaging has recently been employed to examine the spatial distribution of NADH changes following SD evoked by localized injection of KCl under the cortical surface of anesthetized mice (Takano et al. 2007). The high resolution of the images revealed a very interesting heterogeneity of signals that suggested regional differences in oxygen availability (Fig. 12.4). Initial fluorescence decreases were observed throughout the imaging field during SD, but fluorescence increases after SD occurred in an organized pattern that was defined by the vascular architecture. The signals were not corrected for possible heterogeneity in swelling or increased absorption of NADH fluorescence emission, but if these heterogeneity is indeed due to differences in NADH fluorescence, it can be taken to imply that NADH oxidation was limited in tissue where O_2 availability was more limited due to longer diffusion distances from vessels (Takano et al. 2007). This high resolution study also directly demonstrated that if responses were averaged over a whole region imaging, regional heterogeneity was lost. It was also concluded that the NADH transitions following SD do not involve a cellular pattern, and are thus unlikely not to involve neurons and/or astrocytes specifically, but may be a metabolic response that is common to all cell types as a consequence of metabolic demands placed by the SD event (Takano et al. 2007).

12.3.5 Mechanisms Underlying NADH Signals

12.3.5.1 Change in Mitochondrial Redox Potential

Since early studies of NADH and SD, there has been a strong consensus that the signals are due to changes in mitochondrial redox state. Although NADH is involved in cytosolic as well as mitochondrial metabolism, the fluorescence of NADH in mitochondrial compartments appears to dominate fluorescence recordings under most synaptic stimulation conditions (see discussion in Shuttleworth 2010). Thus in

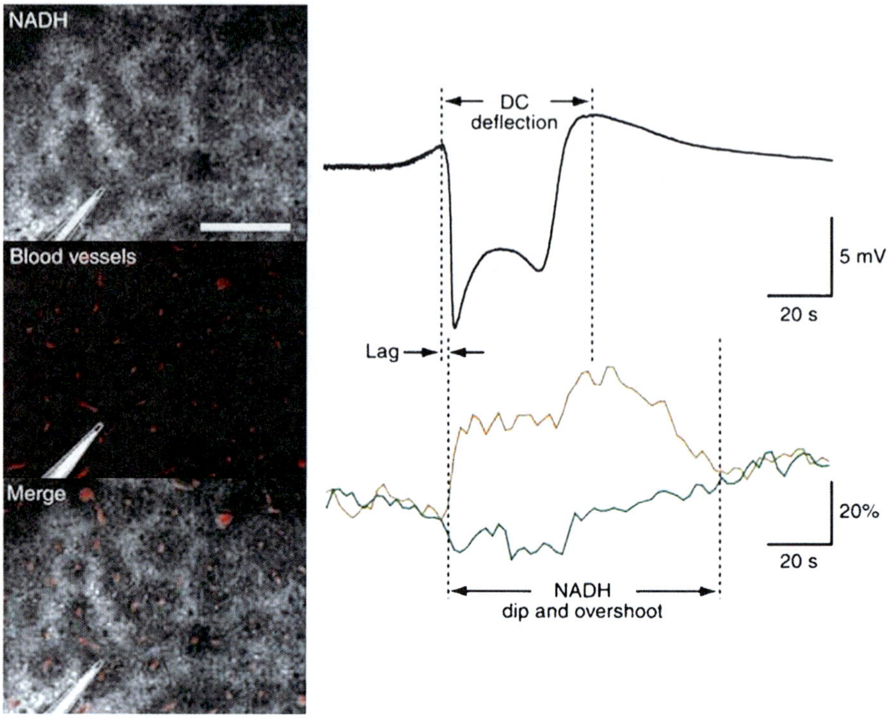

Fig. 12.4 Regional heterogeneity of NADH signals during SD. The recent application of two photon imaging to SD *in vivo* revealed a distinct pattern of NADH dynamics during SD that was determined by proximity of tissue to blood vessels. The figure is modified from Takano et al. (2007), and *left hand panels* show autofluorescence attributed to NADH at the peak of the DC potential shift of SD (*top*), vascular distribution identified following FITC-dextran injection (*middle*). The *merged image* (*bottom*) shows that autofluorescence decreases are concentrated around vessels, while fluorescence increases are prominent at greater distances from the vessels. The longer diffusion distance for O_2 and consequent hypoxia could explain fluorescence increases at sites more distant from vessels during SD. The traces show the DC potential shift (*black*) and the different fluorescence decreases adjacent to the recording electrode, in regions close (*green*) and more distant from a vessel (*brown*) (From Takano et al. 2007)

early studies of surface NADH recordings from cat cortex, it was concluded that mitochondrial dehydrogenase activity, rather than cytosolic NADH transitions, was the prominent source of NADH signals following epileptiform activation (Jobsis et al. 1971).

The classical NADH oxidation observed during normoxic SD was described as a "state 4-3-4 metabolic transition at the mitochondrial level" (Mayevsky et al. 1974). These states refer to different mitochondrial redox potentials described by Chance and Williams, where state 4 is a resting state with mainly reduced NADH, and state 3 is an activation state where ADP levels rise and NADH is oxidized as electron transport is accelerated (Chance and Williams 1955a, b, c). Lothman and co-workers reemphasized that "labile" component of NADH signals is derived almost exclusively from intramitochondrial NADH, and that changes in redox level of NADH measured with fluorescence methods report relative rates of oxidative metabolism (Lothman et al. 1975). Much subsequent work was done to characterize the influence of levels of ischemia, anesthesia, drugs on mitochondrial function using NADH signals during SD signals, and has been reviewed previously (Mayevsky and Rogatsky 2007).

In many studies where tissue NADH simply recovered back to pre-stimulus levels after SD, it could be concluded that mitochondria recovered from state 3 back to state 4 as plasma membrane ionic gradients were restored. However in cases where NADH levels significantly increased after SD, additional explanations are required. As noted above, the availability of oxygen to support mitochondrial metabolism is likely a central issue, as limitations will lead to increased NADH fluorescence and this may be easier to detect with higher resolution measurements (e.g. Takano et al. 2007). However it is also possible that cation accumulation itself leads to increased TCA cycle activity and increased mitochondrial NADH.

SD is associated with very large Ca^{2+} increases (Kunkler and Kraig 2004; Dietz et al. 2008), which are sufficient to cause mitochondrial depolarization (Bahar et al. 2000). Much work has characterized stimulation of mitochondrial dehydrogenases by Ca^{2+}, and led to the suggestion that Ca^{2+} accumulation couples neuronal activity to metabolism increases. In an important study of mechanisms underlying NADH dynamics in isolated neurons, it was concluded that mitochondrial Ca^{2+} accumulation and consequent stimulation of dehydrogenases was responsible for the prominent NADH increases that persist after transient depolarization (Duchen 1992). It is not yet known whether the very large increases in neuronal Ca^{2+} can contribute to NADH increases after SD.

From studies in brain slices, it was also found that NADH increases ("overshoots") could occur following transient neuronal depolarization in the absence of Ca^{2+} (Shuttleworth et al. 2003). The mechanism(s) underlying these responses it not yet known, but on the basis of Na^{+} replacement experiments, it was speculated that metabolic consequences of Na^{+} accumulation could contribute to increased TCA cycle activity. It has also been suggested that large NADH increases after stimulation could occur partly because of inadequate supply of O_2 in the brain slice (Turner et al. 2007). However, this seems unlikely from the responses seen during repetitive challenge with synaptic stimulation at various time points during the "overshoot" phase. If overshoots were due to O_2 depletion, then it should be more difficult to generate a state 4 to state 3 transition, when synaptic stimulation is applied at the peak of NADH overshoots, yet initial NADH oxidation responses are virtually unchanged (Brennan et al. 2007a). The relative contributions of localized O_2 depletion and metabolic consequences of cation loading to NADH increases after the extraordinary metabolic challenges of SD are yet to be determined.

12.3.5.2 Possible Contributions from Astrocyte Metabolism

Astrocyte function is likely and important regulator of initiation and propagation of SD (Somjen 2001), presumably because of the central roles of these cells in buffering extracellular glutamate and K^{+} concentrations (Danbolt 2001; Kofuji and

Newman 2004). Consistent with this assumption, selective inhibition of astrocytic mitochondrial metabolism with the toxin fluorocitrate accelerates the onset of, and consequences of SD (Largo et al. 1997; Lian & Stringer 2004) and prolonged exposures to glial toxins alone can initiate SD accompanied by irreversible neuronal injury (Canals et al., 2008). These findings imply that mitochondrial function in astrocytes is important for the regulation of SD, but the extent to which astrocyte mitochondrial function contributes to NADH autofluorescence signals following SD is currently unclear.

Although mitochondrial inhibitors impair astrocyte function in SD, astrocytes are usually considered to rely heavily on glycolysis for ATP synthesis. Glycolysis depends on cytosolic NAD^+ as a cofactor for activity of glyceraldehyde phosphate dehydrogenase. Conversion of pyruvate to lactate serves to reoxidize NADH to provide the NAD^+ that is required to maintain glycolysis during O_2 depletion. Thus strong activation of glycolysis could potentially contribute to NADH increases if O_2 is available, but significant cytosolic NADH accumulation would not be expected if there were tight coupling between glycolysis and lactate formation.

Glycolysis has been suggested as a candidate for large NADH increases that occur following prolonged periods of electrical stimulation of guinea-pig cortical slices (Lipton 1973b). A more recent two photon imaging study suggested that NADH increases following synaptic stimulation of rat hippocampal slices were restricted to astrocytes (rather than neurons), and that this was in turn due to stimulation of glycolysis by astrocytic glutamate uptake (Kasischke et al. 2004). The latter suggestion was based on spatial and temporal segregation of NADH signals between neurons and astrocytes, and was suggested to be due to stimulation of glycolysis by astrocytic glutamate uptake. This interesting possibility is difficult to test pharmacologically, since the key experiments require block of glycolysis with procedures that maintain substrate for oxidative metabolism. However from subsequent studies in murine hippocampal slices, it was concluded that either glycolytic inhibition or inhibition of astrocytic glutamate uptake did not reduce overshooting NADH increases, over a range of synaptic stimulation durations (Brennan et al. 2006). These results suggested that glycolysis was not necessary to generate robust NADH overshoots, but it was acknowledged that supply of low pyruave concentrations to support oxidative metabolism could artificially shift metabolism from glycolytic to oxidative processes. For this reason, it was useful to note that comparison of NADH and FP responses under the same stimulation conditions (without supplemental pyruvate) supported the conclusion that both components of NADH responses were unlikely to involve glycolysis in murine brain slices (Brennan et al. 2006, 2007a).

These findings are in agreement with the general assumption that mitochondrial metabolism predominates in optical measures of NADH metabolism following activation of brain tissue, and this topic has been reviewed in more detail recently (Shuttleworth 2010). However it may be useful to emphasize that the ability to measure robust mitochondrial NADH transients with fluorimeter and CCD-based imaging systems does not provide evidence that significant accumulation of NADH in the cytosol due to activation of glycolysis does not occur. What these results do suggest is that if such accumulation occurs, it may be too small to detect in comparison to much larger signals generated by NADH/NAD$^+$ transitions that occur in an environment (mitochondria) that leads to much brighter fluorescence (Shuttleworth 2010). Additional high resolution studies, coupled with pharmacological tests for metabolic pathways involved, are needed to directly test whether glycolytic NADH signals can be clearly demonstrated in the cytosol of astrocytes. Discrimination between mitochondrial and cytosolic signals will be of particular interest during SD, when lactate production accelerates.

12.3.5.3 Contributions of Neuronal Compartments to NADH Transients

An important early study of mechanisms underlying NADH transients was undertaken using isolated dorsal root ganglion neurons, and showed that brief depolarization resulted in initial NAD(P)H fluorescence decreases, followed by a longer-lasting overshooting NADH. The fact that these signals were generated by direct depolarization of isolated neurons provided clear evidence that neurons alone could be sufficient to explain both phases of biphasic NAD(P)H transients observed in brain slices or *in vivo* (Duchen 1992). Relatively few studies have dissected the relative contributions of different neuronal compartments to brain autofluorescence transients in complex brain tissue, however it is possible that both presynaptic and postsynaptic neuronal elements could contribute to NADH signals during SD. Mitochondria are densely accumulated in perinuclear regions in nerve cell bodies, but are also found in distributed throughout dendrites and axons (see Kann and Kovacs 2007 for review) and recent work has demonstrated significant trafficking of mitochondria through these processes (Chang and Reynolds 2006; Mironov 2006). Consistent with the idea that dendritic processes can contribute to NADH signals, two photon imaging has shown that localized high K$^+$ produces biphasic NADH transients in single dendritic processes of isolated Purkinje neurons (Hayakawa et al. 2005).

Ionic loading is also expected to place significant metabolic demands on presynaptic compartments following SD presynaptic mitochondrial function may also be expected to contribute to compound NADH transients during these responses. Concentrations of mitochondria have been localized to some presynaptic terminals (Rowland et al. 2000) and pharmacological evidence has implied that mitochondrial Ca^{2+} uptake can modulate presynaptic Ca^{2+} transients (Tang and Zucker 1997; Zucker 1999). Consistent with a role in presynaptic metabolism, NADH transients have recently been documented in a preparation of neurosecretory terminals (Kosterin et al. 2005), and NADH fluorescence decreases in nerve terminals at the neuromuscular junction following extended electrical stimulation (Talbot et al. 2007). Within the CNS, parallel fiber stimulation also leads to some pre-synaptic NADH transients (Diez-Garcia et al. 2005), however it does appear that these are relatively small, and signals are dominated by postsynaptic responses.

12.4 Optical Imaging of Intrinsic Optical and Hemodynamic Signals During SD

12.4.1 Introduction

As discussed above, vascular changes during SD can interfere significantly with autofluorescence measurements, but it has long been appreciated that intrinsic optical signals attributed to vascular dynamics can provide valuable insights into oxygen delivery and recovery following SD (Rosenthal and Somjen 1973). A range of optical methods has been used to assess blood flow and hemoglobin desaturation during SD in diverse experimental settings, and this large body of work will not be reviewed in detail here. The following section will instead introduce some of the main approaches that can provide high resolution imaging of spreading changes in hemoglobin oxygen saturation and blood flow related to SD.

12.4.2 Optical Detection of Vascular Changes

The optical spectra of oxygenated and deoxygenated hemoglobin (Hb) that underlie differences in the color of blood are plotted in Fig. 12.5. Light reflected from

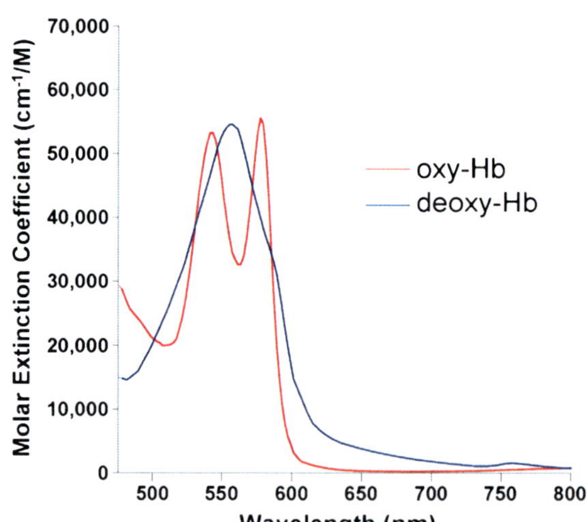

Fig. 12.5 Absorbance spectra of oxyg-hemoglobin and deoxy-hemoglobin. A clear separation in the spectra occurs above 600 nm, so that changes in the amount of reflected red light can be used to assess changes in deoxyhemoglobin levels. Conversely, at green wavelengths (~530–570 nm), the absorption of the two species overlaps to such an extent that changes in reflectance of green light can provide information on changes in local cerebral blood volume, independent of changes in blood oxygen saturation. Since absorption of both species is very low above ~800 nm, changes in infrared signals can be used to assess light scattering changes. Data were obtained from http://omlc.ogi.edu/spectra/hemoglobin/, by S. Prahl, Oregon Medical Laser Center. Replotted with permission

well-oxygenated vessels has a more intense red appearance because deoxygenated Hb continues to absorb light through red wavelengths (>~590 nm), while there is almost no absorption in this range by oxygenated Hb. Based on this difference, imaging at wavelengths above 600 nm has commonly been used to assess blood oxygenation supplying brain tissue. From Fig. 12.5, discrimination between oxy and deoxyHb could also theoretically be achieved at shorter wavelengths but longer wavelengths are favored partly because of good penetration into the brain tissue, including through thinned skull windows. Figure 12.5 also shows that the absorption of oxy and deoxy Hb is identical at some wavelengths (e.g. ~500 nm) and if a range of green wavelengths is assessed (~530–570 nm) absorption of the two species overlaps to such an extent that changes in reflectance of green light can provide information on changes in local cerebral blood volume, independent of changes in blood oxygen saturation (Fujita et al. 2000; Dunn et al. 2005; Vanzetta et al. 2005). Imaging changes in blood oxygenation and volume using red and green light *in vivo* is usually termed intrinsic optical imaging, and has had very wide application for monitoring the regional activation and plasticity of cortical structures (Frostig et al. 1990; Nemoto et al. 1999; Hillman 2007; Prakash et al. 2009).

High resolution optical imaging of brain hemodynamic signals has usually been achieved by using filter-based systems and CCD cameras, but alternative approaches include laser illumination and near-infrared spectroscopic analysis of oxy and deoxy Hb shifts. A more recent development in optical imaging of SD has been the application of laser speckle contrast analysis (LSCA). LSCA was originally developed for studies of flow in retina (reviewed in Briers 2001) but has since become widely used for studies of cerebral blood flow. The method is an extension of single-point laser Doppler flow (LDF) measurements, and is based on changes in the speckled appearance of brain under laser illumination. The speckling is due to a random interference effect, and the image contrast is highest when particles within the image are stationary, or at least moving slowly. When red blood cells move more quickly in an image field, image contrast is reduced. The interference effect requires laser illumination and, like standard laser dopper flow analysis, there remains difficulty in measuring absolute changes in CBF, although a calibration method has been validated in a study of SDs generated by ischemia in cat cortex (Strong et al. 2006). Many studies have utilized open skull preparations, but recent work has begun to utilize LCSA for imaging through thinned skull windows (Parthasarathy et al. 2008), which should be useful for longitudinal studies of SD in rodents.

12.4.3 Signals During Normoxic SD

Figure 12.6 provides an example from a study that examined intrinsic optical responses in rat cortex, using a thinned skull preparation (Ba et al. 2002). SD was initiated by a pin prick, using a 32 gauge needle inserted briefly through a burr hole drilled a few millimeters distant from the imaging site. This is a useful example,

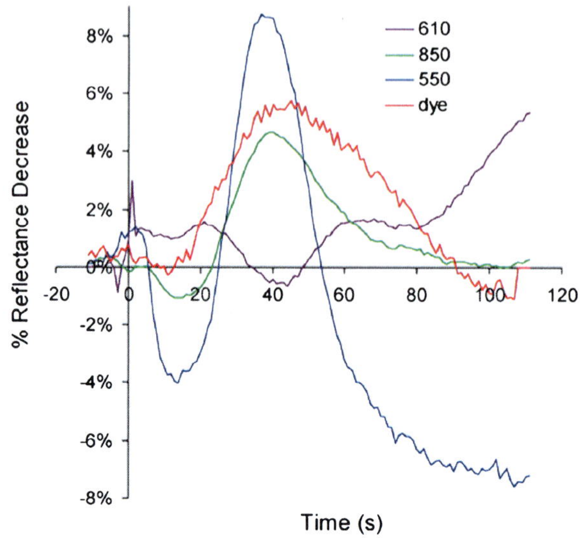

Fig. 12.6 Intrinsic optical signals during SD. This figure is taken from Ba et al. (2002) and illustrates a number of different optical signals recorded from the rat cortical surface, following initiation of SD with a pinprick As noted in the text, intrinsic signals were centered at 550 (*blue trace*), 610 (*purple trace*) and 850 nm (*green trace*) to assess blood volume, deoxgyhemoglobin and light scattering changes, respectively. The red trace is from intravascular injection of an indicator dye Texas Red dextran and plots from different trials have been aligned to form this composite figure. Soon after SD (~40 s on trace) a significant increase in 550 and 850 nm absorption (plotted as reflectance decrease) is seen, and the increase in blood flow at this time point is also visualized as an increase in the intravascular dye signal. The decreased deoxyhemablobin signal (610 nm) at this time point suggests a normal neurovascular coupling response, similar to that observed with reflex stimulation. However at later time points (>60 s on trace), decreased blood flow signals and increased deoxyhemoglobin absorption implies inadequate neurovascular coupling (From Ba et al. 2002, with permission)

because the brain surface was illuminated with white light and reflected light were filtered at three wavelength bands centered at 550 (530–570 nm), 610 (605–615 nm) and 850 (845–855 nm), and signals from an intravascular fluorescent dye (Texas Red dextran) are also shown (Ba et al. 2002). As discussed above, 550 nm reflection was used to assess changes in total hemoglobin concentration, 610 nm for oxy-deoxy Hb shifts, and 850 nm signals for light scattering effects that are not significantly contaminated by Hb absorption. In this report, the authors concluded that a decrease in deoxyhemoglobin occurred soon after SD, as blood volume and flow increase to such a degree that O_2 supply exceeds demand of the tissue, and conclude that these early signals are similar to the normal neurovascular coupling seen following physiological stimulation of somatosensory cortex (Ba et al. 2002). However at later time points post-SD, an increase in 610 nm signals is interpreted as increased O_2 demand being larger than supply (Ba et al. 2002). These and other intrinsic optical imaging studies in a range of preparations are currently being used to investigate regulation of SD propagation and neurovascular coupling (e.g. Chang et al. 2010), and also the propagation of vascular reactivity changes. For example, (Brennan et al. 2007b) recently examined in detail the relationship between the diameter of surface arterioles and intrinsic optical signals recorded from the parenchyma. Interestingly, they found that a dilation response propagated throughout the vascular network significantly faster than the DC shift or parencymal change accompanying SD, and suggested that specific vascular conduction pathways are activated prior to SD (Brennan et al. 2007b).

LSCA of flow changes has emerged as a very useful complementary approach for tracking SD. In an early application, LSCA measurements was used to monitor SD evoked by K^+ in mouse cortex, and provided reliable detection of an initial hypoperfusion phase that coincided with the shift in DC potential (Ayata et al. 2004). LSCA has been successfully combined with simultaneous measurements of intrinsic optical signals following SD in rat cortex (Dunn et al. 2003), and a combination of LSCA and glucose microdialysis was used to characterize the temporal relationship between SD progression and the substantial depletion of extracellular glucose concentration that ensues (Hashemi et al. 2008).

12.5 Optical Imaging of Metabolism During SD in the Context of Brain Ischemia

12.5.1 Introduction

As part of his early descriptions of SD in rabbit cortex, Leao reported the effects of transient common carotid occlusion. The basilar artery was clipped in these animals, so that bilateral carotid artery occlusion would produce a rapid and profound cerebral ischemia without compensation from vertebral artery supply. After about 2–5 min of ischemia, a slow potential shift underlying SD was recorded. Recordings were made from arrangements of five electrodes placed across the cortical surface, and although a propagation rate was not reported, the asynchrony of the responses was consistent with a spreading response over the brain surface (Leao 1947). Leao also described the effects of brief ischemia (1 min) applied to tissue that was undergoing SD generated by electrical stimulation. Ischemia caused a marked prolongation and enlargement of the voltage shift associated with SD (Leao 1947). These observations first raised the idea that ischemia can prolong ionic disruptions of SD, and that ischemia itself can trigger propagating depolarizing events. As noted above in Sect. 12.2, a significant body of work has concentrated on the contributions of SD to ischemic brain injury in animal models, and recent studies have greatly strengthened the suggestion that SD contributes to the evolution of ischemic infarction in human subjects.

Electrophysiological recordings of DC shifts and ECoG activity have been central to efforts to identify the SD events that accompany ischemia (Strong et al. 1983a; Nedergaard and Astrup 1986). From such studies, a relationship between the number of events and infarct volume has been established, and the effectiveness of non-specific inhibitors of SD on reduction of lesion volume has been demonstrated (see Hossmann 1996) However optical imaging has also contributed significantly to understanding of how repetitive SDs contribute to the expansion of injury, partly because of the ability to examine the relationship between regional changes in metabolism and blood flow. As described in Sect. 12.4.1, a substantial increase in blood flow accompanies normoxic SD, and this helps meet increased energy demands. However the ability to increase flow in peri-infarct areas can be greatly reduced. From a variety of experimental approaches, it has been demonstrated that SD events in peri-infarct regions do not elicit appropriate flow increases, and that PO_2 changes are opposite that which is observed in healthy tissue (Back et al. 1994). It is widely presumed that the inappropriate neurovascular coupling that accompanies SD in these regions is a central cause of brain injury (Back 1998). Autofluorescence imaging of NADH has provided a way of assessing regional metabolic status during these events, and when coupled with hemodynamic measures, provides a powerful way of interrogating tissue surrounding infarcts, to predict whether brain tissue will survive or succumb to the repetitive challenges of SD after ischemia.

12.5.2 Influence on NADH Autofluorescence Signals

The relationship between oxygen tension and NADH responses during SD has been extensively investigated by Mayvesky and others, and has provided important support for the idea that NADH autofluorescence can identify vulnerable brain tissue after SD. A large body of work has been summarized elsewhere (Mayevsky and Rogatsky 2007), but one example (Sonn and Mayevsky 2000) will be used to illustrate some key points (Fig. 12.7). NADH autofluorescence was measured from a small region of interest on the rat cortex, together with a laser Doppler flow measurement of blood flow, and electrode recordings of DC potential and extracellular K^+ concentration. These multiple recording probes were embedded in a single rectangular cannula and SD was initiated by KCl delivered from a burr hole a few millimeters away. In normal tissue, the expected NADH decrease was observed during SD (as described in Sect. 12.3.4) which coincided with local flow increases, and this was followed by a longer-lasting fluorescence increase. In contrast, under conditions of hypoxia (inhaled 12% O_2, 88% N_2; decreasing arterial PO_2 from 107 to 57 mmHg), or partial ischemia (permanent bilateral carotid artery occlusion), a quite different NADH response was observed. NADH increases (rather than decreases) were observed, and this switch to a "reduction cycle" coincided with initial local flow decreases (Sonn and Mayevsky 2000). These NADH recordings supported the conclusion that local flow did not compensate for the lack of O_2 during hypoxia and the lack of increased flow was considered to arise either due to lack of capacity for further dilation after ischemia, or release of vasoconstrictor factors (including K^+) during SD (Sonn and Mayevsky 2000).

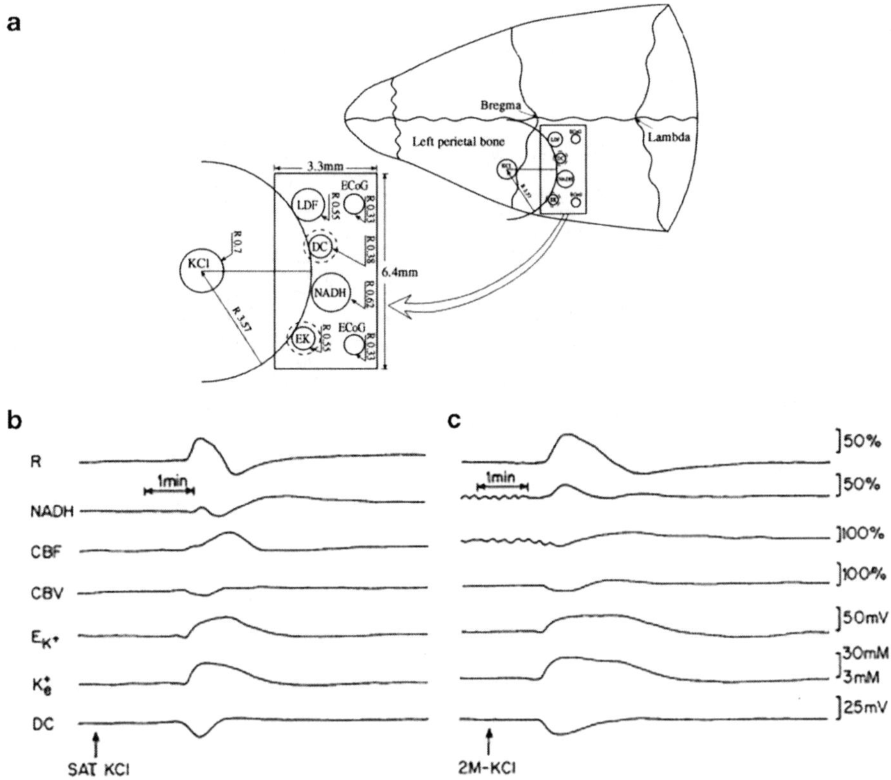

Fig. 12.7 Effects of hypoxia on NADH responses to SD. This composite figure is modified from Sonn and Mayevsky (2000), and shows the recording arrangement for simultaneous NADH fluorescence with laser Doppler flow, extracellular K+, and DC potential recordings from the rat cortex (**a**). Panel **b** shows the typical responses to SD initiated by topical application of KCl. The raw 366 nm reflectance trace (R) is shown together with the corrected NADH fluorescence signal, and emphasizes the importance of this correction to reveal the initial NADH oxidation during SD. Panel **c** shows that the initial NADH oxidation was completely lost and replaced by an initial NADH fluorescence increase when the same stimulus is applied to hypoxic brain (arterial PO_2 reduced from 107 to 57 mmHg). Note that the DC potential shift and extracellular K^+ elevations are also significantly prolonged in the hypoxic condition (From Sonn and Mayevsky 2000, with permission)

12.5.3 Imaging the Spread of SD Events in Relation to Ischemic Infarcts

The application of CCD-based imaging of brain surface NADH autofluorescence signals has proven very useful for the evaluation of site of origin and pattern of spread of SD events in relation to an ischemic infarct. Neuronal injury at an ischemic core is expected to release large amounts of glutamate and K^+ into the extracellular space, and it was proposed that accumulation of these mediators could be

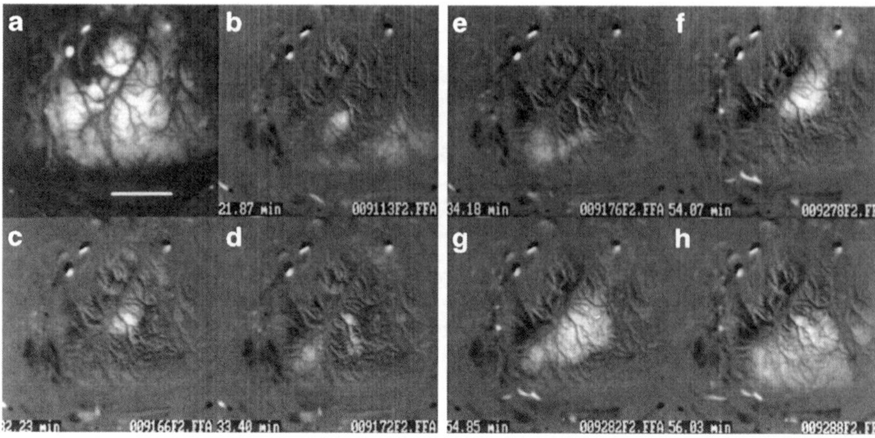

Fig. 12.8 Imaging autofluorescence signals associated with focal ischemic infarction. This figure is taken from a study that demonstrated that waves of autofluorescence increase originated at the site of focal ischemic infarcts and propagated through surrounding tissue. Surface autofluorescence recordings that included NADH (UV excitation, 445–465 nm emission) were made approximately 3 h after generating a middle cerebral artery occlusion in a squirrel monkey. Panel **a** shows a bright-field image of the recording site (scale bar = 5 mm), and **b** shows a small central region of sustained fluorescence increase that appears to correspond to a region of irrecoverably injured tissue. Subsequent frames (**c–e**, 32–34 min from onset of recording) show progression of events around that focal region. The last three frames (**f–h**, 54–56 min after onset of recording) show a larger wave that invades the entire region, including the initial core site (From Strong et al. 2000, with permission)

sufficient to trigger SDs. Alternatively, SD initiation sites could be determined by a pattern of regional blood flow deficiencies, rather than by neuronal injury. Imaging NADH levels from a relatively large field of view that includes the ischemic region has made it possible to address this question.

After an initial report of NADH dynamics from single locations using a fluorimeter (Strong et al. 1988), Strong and co-workers applied a CCD-based approach to image responses following middle cerebral artery occlusion in cats (Strong et al. 1996) (Fig. 12.8). A large cortical area was imaged at a frame rate high enough to capture the spread of a series of complex spreading events in each animal. NADH decreases and increases were both observed, and since these CCD measurements could not be corrected for UV reflectance measures, it was acknowledged that fluorescence decreases were likely a combination of NADH oxidation as well as increases in blood volume as would be expected for SD propagation through healthy tissue (see Sect. 12.4). In addition, these investigators showed that in some regions, clear NADH increases were observed. Again, uncorrected fluorescence increases could be due to decreased blood volume, as well as conversion of NAD^+ to NADH, however in regions where these fluorescence increases were observed, the authors concluded that this represented a conversion to pathologic or "core conditions" (Strong et al. 1996). In addition, the imaging provided evidence that SD events initiated at the edge of ischemic infarcts.

These conclusions were extended in studies of squirrel monkey cortex after middle cerebral artery occlusion. It was concluded that the sign of NADH fluorescence responses could be used to distinguish between periinfarct depolarizations (with risk of promoting damage) and responses in normally perfused cortex (Strong et al. 2000). The demonstration that the propagation of such events could be demonstrated in the gyrencephalic cortex of a primate was also important in strengthening the idea that SDs could occur in higher mammals, in addition to the relatively smooth cortical surface of rodents and cats (Strong et al. 2000). Another important advance has been the demonstration that NADH imaging could be used to clearly evaluate the enlargement of areas of ischemic infarction associated with repetitive SD events (Higuchi et al. 2002). In studies of middle cerebral artery occlusion in rats, sustained NADH fluorescence increases identified the ischemic core, and then propagating NADH increases were observed to originate at the edge of the core. Bands of fluorescence increases (~2–4 mm across) propagated along the margin of the core, and some cases were observed to completely circle the injured region. In about half the cases, areas involved in propagating NADH increases merged with, and could be seen to increase the size of the ischemic core (Higuchi et al. 2002).

The ability to reliably monitor the propagation of periinfarct SDs with NADH is now beginning to be exploited to assess the mechanism of action of drugs and treatments designed to reduce ischemic infarction. For example, CCD-based NADH imaging was recently used suggest that lack of effectiveness of hypothermia against injury produced by permanent vascular occlusion may be a consequence of lack of effect on SD mechanisms (Sasaki et al. 2009).

While the studies above have usually acknowledged the complications of hemodynamic changes for interpretation of NADH signals, it is also possible that pathologic swelling responses modify autofluorescence signals in peri-infarct regions. Recent two photon imaging of neurons has shown dramatic and transient loss of dendritic spines associated with bilateral common carotid artery occlusion in mice (Brown et al. 2008; Murphy et al. 2008) and it is not yet clear whether these changes contribute significantly to NADH signals during ischemia.

Laser speckle contrast imaging (LSCA) has recently been applied to monitor SD events triggered by middle cerebral artery occlusion in cats (Strong et al. 2006). Imaging was performed over a large cortical region, and following the initial drop due to MCAO, flow decreases were observed to propagate in a wave-like manner at a rate (~2–4 mm/min) consistent with progression of peri-infarct depolarizations. DC potential shifts were recorded simultaneously with surface electrodes, and revealed that LSCA measurements failed to detect all electrical events, but it is likely that the detection limit could be improved with higher image acquisition frequency (Strong et al. 2006). It is also noteworthy that some events detected with LSCA did not propagate to the recording electrode, suggesting that a combination of the electrical and optical methods will be useful for comprehensive detection of SDs associated with focal ischemia (Strong et al. 2006). Because this imaging approach can be stable for hours following ischemia surgery, it is likely that LSCA will be of great value for evaluating therapeutic interventions to reduce SD frequency or propagation after ischemia

12.6 Conclusions and Perspectives

Much of the important development of autofluorescence imaging for studies of brain metabolism has relied on spreading depression (SD) as an excellent test response. The stimulus evokes a very strong metabolic challenge and NADH signals that can be detected with good signal-noise characteristics in a variety of experimental settings. In addition, the slow and well-organized propagation of SD makes NADH signals quite easy to capture with conventional CCD cameras, and also by high resolution imaging with two photon excitation methods. The validation of autofluorescence methods in SD studies has provided an important foundation for application of similar optical approaches to other questions of brain pathophysiology. Historically, most such applications have involved NADH imaging, but the complementary flavoprotein autofluorescence signals have recently been applied successfully to *in vivo* studies of plasticity in the rodent cortex, including through thinned skull preparations (Shibuki et al. 2007; Tohmi et al. 2009). It may be useful if more studies can combine NADH and flavoprotein imaging, to assess the relative contributions of mitochondrial and cytosolic metabolism *in vivo*. Such studies could be helpful for addressing questions of metabolic coupling including, if done at sufficiently high resolution, the important questions of compartmentalization of astrocytic and neuronal signals. A useful complement to such studies could be the application of fluorescent analogues of glucose, which have recently been used in brain slice studies to assess differences in glucose uptake between neurons and astrocytes (Barros et al. 2009). Such methods have not yet been applied to synaptic activation *in vivo*.

As a consequence of the large number of SD studies associated with development of NADH methods, much has been learned about the temporal and spatial spread of mitochondrial metabolism after SD. This has provided valuable insights about the spread of the events in relation to ischemic infarcts and, when this has been combined with information about regional blood flow, it has been possible to distinguish tissue that is likely to die following repetitive SDs, and tissue which has sufficient coupling to survive. Much of this work has been done in experimental animal models, but recent evidence that SD events are likely a prominent contributor to some types of human brain injury and it is likely that optical approaches to assess tissue viability could be utilized in clinical settings. As noted above, penetration of light through the skull is a limitation for optical studies, but illumination of relatively large brain areas may be possible in a small number of clinical settings where segments of skull have been removed for decompression after injury. Recent work has demonstrated the effectiveness of small optical probes for recording tissue partial pressure of oxygen, together with electrocorticographic and laser dopper flow probes, all arranged together on strip electrodes and placed under the dura in injured human brain (Dreier et al. 2009). It is possible that other optical signals that have been well-validated for distinguishing at-risk tissue in animal models could also be incorporated into subdural strip arrangements, to identify vulnerable regions in human studies.

The relationship of mitochondrial and hemodynamic signals to other cellular consequences of SD *in vivo* is an important next step. Ca^{2+} overload has long been appreciated as a major contributor to neuronal injury after SD in brain slices (Somjen 2001), but very little is known about the patterns of Ca^{2+} deregulation following repetitive SD *in vivo*. High resolution fluorescence imaging of Ca^{2+} indicators would be helpful to determine whether dendrites serve as initiation sites for neuronal injury after SD, as has been concluded for excitotoxic injury in brain slices (Shuttleworth and Connor 2001). Pathologic accumulations of Zn^{2+} have also recently been considered as a contributor to the initiation and pathologic consequences of SD in brain slices (Dietz et al. 2008, 2009), and it will be of interest to determine whether Zn^{2+} increases are prominent following under similar conditions *in vivo*. As discussed above for NADH autofluorescence (Sect. 12.5.3), such efforts may be most effective with two photon excitation in via open cranial windows or thinned skull preparations. In contrast to autofluorescence methods (which rely no indicator loading procedures), a challenge for fluorescence measurements of any ion indictor is effective loading of organic indicators into the preparations. The use of mice with sensors genetically encoded into subsets of neurons or astrocytes could be very useful in this context.

Recent work has begun to evaluate optical imaging of voltage-sensitive indicators to assess SD propagation in the rat cortex (Obrenovitch et al. 2009; Farkas et al. 2010). These signals have some advantages over metabolic or vascular signals related to SD, since they can be directly related to the initial wave of depolarization associated with SD, rather than downstream consequences. It will be of significant interest to see whether these signals can be used for relatively long-term recordings during focal ischemia, or other injury models, since they could provide valuable high-resolution information about the progression of electrical silencing that is difficult to achieve with conventional electrical recordings.

It is becoming increasingly appreciated that many neurodegenerative diseases are associated with regional or global metabolic dysfunction, and optical imaging methods in animal models can provide valuable insights into mechanisms underlying injury progression. Transfer of these approaches into human subjects is difficult, but may be feasible in some settings, particularly with the development of small optical probes suitable for the types of long-term monitoring that may be needed.

Acknowledgement Supported by NIH grant NS51288

References

Aitken PG, Fayuk D, Somjen GG, Turner DA (1999) Use of intrinsic optical signals to monitor physiological changes in brain tissue slices. Methods 18:91–103

Anderson TR, Andrew RD (2002) Spreading depression: imaging and blockade in the rat neocortical brain slice. J Neurophysiol 88:2713–25

Anderson RE, Meyer FB (2002) In vivo fluorescent imaging of NADH redox state in brain. Methods Enzymol 352:482–94

Andrew RD, Jarvis CR, Obeidat AS (1999) Potential sources of intrinsic optical signals imaged in live brain slices. Methods 18(185–196):179

Aubin JE (1979) Autofluorescence of viable cultured mammalian cells. J Histochem Cytochem 27:36–43

Ayata C (2010) Cortical spreading depression triggers migraine attack: pro. Headache 50:725–730

Ayata C, Shin HK, Salomone S, Ozdemir-Gursoy Y, Boas DA, Dunn AK, Moskowitz MA (2004) Pronounced hypoperfusion during spreading depression in mouse cortex. J Cereb Blood Flow Metab 24:1172–82

Ba AM, Guiou M, Pouratian N, Muthialu A, Rex DE, Cannestra AF, Chen JW, Toga AW (2002) Multiwavelength optical intrinsic signal imaging of cortical spreading depression. J Neurophysiol 88:2726–35

Back T (1998) Pathophysiology of the ischemic penumbra–revision of a concept. Cell Mol Neurobiol 18:621–38

Back T, Kohno K, Hossmann KA (1994) Cortical negative DC deflections following middle cerebral artery occlusion and KCl-induced spreading depression: effect on blood flow, tissue oxygenation, and electroencephalogram. J Cereb Blood Flow Metab 14:12–19

Bahar S, Fayuk D, Somjen GG, Aitken PG, Turner DA (2000) Mitochondrial and intrinsic optical signals imaged during hypoxia and spreading depression in rat hippocampal slices. J Neurophysiol 84:311–24

Barretto RP, Messerschmidt B, Schnitzer MJ (2009) In vivo fluorescence imaging with high-resolution microlenses. Nat Methods 6:511–12

Barros LF, Courjaret R, Jakoby P, Loaiza A, Lohr C, Deitmer JW (2009) Preferential transport and metabolism of glucose in Bergmann glia over Purkinje cells: a multiphoton study of cerebellar slices. Glia 57:962–70

Bowyer SM, Aurora KS, Moran JE, Tepley N, Welch KM (2001) Magnetoencephalographic fields from patients with spontaneous and induced migraine aura. Ann Neurol 50:582–87

Bradley RS, Thorniley MS (2006) A review of attenuation correction techniques for tissue fluorescence. J R Soc Interface 3:1–13

Brennan AM, Connor JA, Shuttleworth CW (2006) NAD(P)H fluorescence transients after synaptic activity in brain slices: predominant role of mitochondrial function. J Cereb Blood Flow Metab 26:1389–1406

Brennan AM, Smith AJ, Connor JA, Shuttleworth CW (2004) Kinetics of intrinsic fluorescence transients in hippocampal slices following synaptic stimulation. Soc Neuroscience Abstracts:197.123

Brennan AM, Connor JA, Shuttleworth CW (2007a) Modulation of the amplitude of NAD(P)H fluorescence transients after synaptic stimulation. J Neurosci Res 85(15):3233–343

Brennan KC, Beltran-Parrazal L, Lopez-Valdes HE, Theriot J, Toga AW, Charles AC (2007b) Distinct vascular conduction with cortical spreading depression. J Neurophysiol 97:4143–51

Briers JD (2001) Laser Doppler, speckle and related techniques for blood perfusion mapping and imaging. Physiol Meas 22:R35–R66

Brown CE, Wong C, Murphy TH (2008) Rapid morphologic plasticity of peri-infarct dendritic spines after focal ischemic stroke. Stroke 39:1286–91

Bures J (1956) Some metabolic aspects of Leao's spreading depression. J Neurochem 1:153–158

Bures J, Buresova O, Krivanek J (1984) The meaning and significance of Leao's spreading depression. An Acad Bras Cienc 56:385–400

Busch E, Gyngell ML, Eis M, Hoehn-Berlage M, Hossmann KA (1996) Potassium-induced cortical spreading depressions during focal cerebral ischemia in rats: contribution to lesion growth assessed by diffusion-weighted NMR and biochemical imaging. J Cereb Blood Flow Metab 16:1090–99

Busija DW, Bari F, Domoki F, Horiguchi T, Shimizu K (2008) Mechanisms involved in the cerebrovascular dilator effects of cortical spreading depression. Prog Neurobiol 86:379–395

Canals S, Larrosa B, Pintor J, Mena MA, Herreras O (2008) Metabolic challenge to glia activates an adenosine-mediated safety mechanism that promotes neuronal survival by delaying the onset of spreading depression waves. J Cereb Blood Flow Metab 28:1835–44

Chance B (2004) Mitochondrial NADH redox state, monitoring discovery and deployment in tissue. Methods Enzymol 385:361–70

Chance B, Williams GR (1955a) Respiratory enzymes in oxidative phosphorylation. I. Kinetics of oxygen utilization. J Biol Chem 217:383–93

Chance B, Williams GR (1955b) Respiratory enzymes in oxidative phosphorylation. III. The steady state. J Biol Chem 217:409–27

Chance B, Williams GR (1955c) Respiratory enzymes in oxidative phosphorylation. II. Difference spectra. J Biol Chem 217:395–07

Chance B, Cohen P, Jobsis F, Schoener B (1962) Intracellular oxidation-reduction states in vivo. Science 137:499–508

Chance B, Mayer D, Legallais V (1971) A dual-wavelength spectrophotometer and fluorometer using interference filters. Anal Biochem 42:494–504

Chance B, Schoener B, Oshino R, Itshak F, Nakase Y (1979) Oxidation-reduction ratio studies of mitochondria in freeze-trapped samples. NADH and flavoprotein fluorescence signals. J Biol Chem 254:4764–71

Chang DT, Reynolds IJ (2006) Mitochondrial trafficking and morphology in healthy and injured neurons. Prog Neurobiol 80:241–68

Chang JC, Shook LL, Biag J, Nguyen EN, Toga AW, Charles AC, Brennan KC (2010) Biphasic direct current shift, haemoglobin desaturation and neurovascular uncoupling in cortical spreading depression. Brain 133:996–1012

Charles A (2010) Does cortical spreading depression initiate a migraine attack? Maybe not. Headache 50:731–33

Chen Q, Chopp M, Bodzin G, Chen H (1993) Temperature modulation of cerebral depolarization during focal cerebral ischemia in rats: correlation with ischemic injury. J Cereb Blood Flow Metab 13:389–94

Chen G, Gao W, Reinert KC, Popa LS, Hendrix CM, Ross ME, Ebner TJ (2005) Involvement of kv1 potassium channels in spreading acidification and depression in the cerebellar cortex. J Neurophysiol 94:1287–98

Cohen LB, Keynes RD, Hille B (1968) Light scattering and birefringence changes during nerve activity. Nature 218:438–41

Cohen LB, Salzberg BM, Grinvald A (1978) Optical methods for monitoring neuron activity. Annu Rev Neurosci 1:171–82

Combs CA, Balaban RS (2001) Direct imaging of dehydrogenase activity within living cells using enzyme-dependent fluorescence recovery after photobleaching (ED-FRAP). Biophys J 80: 2018–28

Cossart R, Ikegaya Y, Yuste R (2005) Calcium imaging of cortical networks dynamics. Cell Calcium 37:451–57

Cruz NF, Adachi K, Dienel GA (1999) Rapid efflux of lactate from cerebral cortex during K+-induced spreading cortical depression. J Cereb Blood Flow Metab 19:380–92

Csiba L, Paschen W, Mies G (1985) Regional changes in tissue pH and glucose content during cortical spreading depression in rat brain. Brain Res 336:167–70

Dahlem MA, Schneider FM, Scholl E (2008) Failure of feedback as a putative common mechanism of spreading depolarizations in migraine and stroke. Chaos 18:026110

Danbolt NC (2001) Glutamate uptake. Prog Neurobiol 65:1–105

Dietz RM, Weiss JH, Shuttleworth CW (2008) Zn2+ influx is critical for some forms of spreading depression in brain slices. J Neurosci 28:8014–24

Dietz RM, Weiss JH, Shuttleworth CW (2009) Contributions of Ca2+ and Zn2+ to spreading depression-like events and neuronal injury. J Neurochem 109(Suppl 1):145–52

Diez-Garcia J, Matsushita S, Mutoh H, Nakai J, Ohkura M, Yokoyama J, Dimitrov D, Knopfel T (2005) Activation of cerebellar parallel fibers monitored in transgenic mice expressing a fluorescent Ca2+ indicator protein. Eur J Neurosci 22:627–35

Doane MG (1967) Fluorometric measurement of pyridine nucleotide reduction in the giant axon of the squid. J Gen Physiol 50:2603–32

Dohmen C, Sakowitz OW, Fabricius M, Bosche B, Reithmeier T, Ernestus RI, Brinker G, Dreier JP, Woitzik J, Strong AJ, Graf R (2008) Spreading depolarizations occur in human ischemic stroke with high incidence. Ann Neurol 63:720–28

Dreier JP, Korner K, Ebert N, Gorner A, Rubin I, Back T, Lindauer U, Wolf T, Villringer A, Einhaupl KM, Lauritzen M, Dirnagl U (1998) Nitric oxide scavenging by hemoglobin or nitric oxide synthase inhibition by N-nitro-L-arginine induces cortical spreading ischemia when K+ is increased in the subarachnoid space. J Cereb Blood Flow Metab 18:978–90

Dreier JP, Woitzik J, Fabricius M, Bhatia R, Major S, Drenckhahn C, Lehmann TN, Sarrafzadeh A, Willumsen L, Hartings JA, Sakowitz OW, Seemann JH, Thieme A, Lauritzen M, Strong AJ (2006) Delayed ischaemic neurological deficits after subarachnoid haemorrhage are associated with clusters of spreading depolarizations. Brain 129:3224–37

Dreier JP, Major S, Manning A, Woitzik J, Drenckhahn C, Steinbrink J, Tolias C, Oliveira-Ferreira AI, Fabricius M, Hartings JA, Vajkoczy P, Lauritzen M, Dirnagl U, Bohner G, Strong AJ (2009) Cortical spreading ischaemia is a novel process involved in ischaemic damage in patients with aneurysmal subarachnoid haemorrhage. Brain 132:1866–81

Duchen MR (1992) Ca(2+)-dependent changes in the mitochondrial energetics in single dissociated mouse sensory neurons. Biochem J 283(Pt 1):41–50

Duchen MR, Surin A, Jacobson J (2003) Imaging mitochondrial function in intact cells. Methods Enzymol 361:353–89

Dunn AK, Devor A, Bolay H, Andermann ML, Moskowitz MA, Dale AM, Boas DA (2003) Simultaneous imaging of total cerebral hemoglobin concentration, oxygenation, and blood flow during functional activation. Opt Lett 28:28–30

Dunn AK, Devor A, Dale AM, Boas DA (2005) Spatial extent of oxygen metabolism and hemodynamic changes during functional activation of the rat somatosensory cortex. Neuroimage 27:279–90

Eikermann-Haerter K, Ayata C (2010) Cortical spreading depression and migraine. Curr Neurol Neurosci Rep 10:167–73

Eikermann-Haerter K, Dilekoz E, Kudo C, Savitz SI, Waeber C, Baum MJ, Ferrari MD, van den Maagdenberg AM, Moskowitz MA, Ayata C (2009) Genetic and hormonal factors modulate spreading depression and transient hemiparesis in mouse models of familial hemiplegic migraine type 1. J Clin Invest 119:99–109

Fabricius M, Fuhr S, Willumsen L, Dreier JP, Bhatia R, Boutelle MG, Hartings JA, Bullock R, Strong AJ, Lauritzen M (2008) Association of seizures with cortical spreading depression and peri-infarct depolarisations in the acutely injured human brain. Clin Neurophysiol 119:1973–84

Farkas E, Bari F, Obrenovitch TP (2010) Multi-modal imaging of anoxic depolarization and hemodynamic changes induced by cardiac arrest in the rat cerebral cortex. Neuroimage 51: 734–42

Frostig RD, Lieke EE, Ts'o DY, Grinvald A (1990) Cortical functional architecture and local coupling between neuronal activity and the microcirculation revealed by in vivo high-resolution optical imaging of intrinsic signals. Proc Natl Acad Sci USA 87:6082–86

Fujita H, Matsuura T, Yamada K, Inagaki N, Kanno I (2000) A sealed cranial window system for simultaneous recording of blood flow, and electrical and optical signals in the rat barrel cortex. J Neurosci Methods 99:71–78

Galeotti T, van Rossum GD, Mayer DH, Chance B (1970) On the fluorescence of NAD(P)H in whole-cell preparations of tumours and normal tissues. Eur J Biochem 17:485–96

Gault LM, Lin CW, LaManna JC, Lust WD (1994) Changes in energy metabolites, cGMP and intracellular pH during cortical spreading depression. Brain Res 641:176–80

Gerich FJ, Hepp S, Probst I, Muller M (2006) Mitochondrial inhibition prior to oxygen-withdrawal facilitates the occurrence of hypoxia-induced spreading depression in rat hippocampal slices. J Neurophysiol 96:492–504

Gill R, Andine P, Hillered L, Persson L, Hagberg H (1992) The effect of MK-801 on cortical spreading depression in the penumbral zone following focal ischaemia in the rat. J Cereb Blood Flow Metab 12:371–79

Gorji A (2001) Spreading depression: a review of the clinical relevance. Brain Res Brain Res Rev 38:33–60

Greenwood SM, Mizielinska SM, Frenguelli BG, Harvey J, Connolly CN (2007) Mitochondrial dysfunction and dendritic beading during neuronal toxicity. J Biol Chem 282:26235–44

Hadjikhani N, Sanchez Del Rio M, Wu O, Schwartz D, Bakker D, Fischl B, Kwong KK, Cutrer FM, Rosen BR, Tootell RB, Sorensen AG, Moskowitz MA (2001) Mechanisms of migraine aura revealed by functional MRI in human visual cortex. Proc Natl Acad Sci USA 98: 4687–92

Hansen AJ (1985) Effect of anoxia on ion distribution in the brain. Physiol Rev 65:101–148

Harbig K, Chance B, Kovach AG, Reivich M (1976) In vivo measurement of pyridine nucleotide fluorescence from cat brain cortex. J Appl Physiol 41:480–88

Hartings JA, Rolli ML, Lu XC, Tortella FC (2003) Delayed secondary phase of peri-infarct depolarizations after focal cerebral ischemia: relation to infarct growth and neuroprotection. J Neurosci 23:11602–10

Hartings JA, Gugliotta M, Gilman C, Strong AJ, Tortella FC, Bullock MR (2008) Repetitive cortical spreading depolarizations in a case of severe brain trauma. Neurol Res 30:876–82

Hartings J, Strong AJ, Fabricius M, Manning A, Bhatia R, Dreier J, Mazzeo AT, Tortella FC, Bullock R (2009) Spreading depolarizations and late secondary insults after traumatic brain injury. J Neurotrauma 26:1857–66

Haselgrove JC, Bashford CL, Barlow CH, Quistorff B, Chance B, Mayevsky A (1990) Time resolved 3-dimensional recording of redox ratio during spreading depression in gerbil brain. Brain Res 506:109–14

Hashemi P, Bhatia R, Nakamura H, Dreier JP, Graf R, Strong AJ, Boutelle MG (2008) Persisting depletion of brain glucose following cortical spreading depression, despite apparent hyperaemia: evidence for risk of an adverse effect of Leao's spreading depression. J Cereb Blood Flow Metab 29:166–175

Hashimoto M, Takeda Y, Sato T, Kawahara H, Nagano O, Hirakawa M (2000) Dynamic changes of NADH fluorescence images and NADH content during spreading depression in the cerebral cortex of gerbils. Brain Res 872:294–300

Hayakawa Y, Nemoto T, Iino M, Kasai H (2005) Rapid Ca2+-dependent increase in oxygen consumption by mitochondria in single mammalian central neurons. Cell Calcium 37:359–70

Hepp S, Gerich FJ, Muller M (2005) Sulfhydryl oxidation reduces hippocampal susceptibility to hypoxia-induced spreading depression by activating BK channels. J Neurophysiol 94: 1091–103

Higuchi T, Takeda Y, Hashimoto M, Nagano O, Hirakawa M (2002) Dynamic changes in cortical NADH fluorescence and direct current potential in rat focal ischemia: relationship between propagation of recurrent depolarization and growth of the ischemic core. J Cereb Blood Flow Metab 22:71–79

Hillman EM (2007) Optical brain imaging in vivo: techniques and applications from animal to man. J Biomed Opt 12:051402

Holtmaat A, Bonhoeffer T, Chow DK, Chuckowree J, De Paola V, Hofer SB, Hubener M, Keck T, Knott G, Lee WC, Mostany R, Mrsic-Flogel TD, Nedivi E, Portera-Cailliau C, Svoboda K, Trachtenberg JT, Wilbrecht L (2009) Long-term, high-resolution imaging in the mouse neocortex through a chronic cranial window. Nat Protoc 4:1128–44

Hoskison MM, Shuttleworth CW (2006) Microtubule disruption, not calpain-dependent loss of MAP2, contributes to enduring NMDA-induced dendritic dysfunction in acute hippocampal slices. Exp Neurol 202:302–12

Hossmann KA (1996) Periinfarct depolarizations. Cerebrovasc Brain Metab Rev 8:195–208

Huang S, Heikal AA, Webb WW (2002) Two-photon fluorescence spectroscopy and microscopy of NAD(P)H and flavoprotein. Biophys J 82:2811–25

Iijima T, Mies G, Hossmann KA (1992) Repeated negative DC deflections in rat cortex following middle cerebral artery occlusion are abolished by MK-801: effect on volume of ischemic injury. J Cereb Blood Flow Metab 12:727–33

Jarvis CR, Lilge L, Vipond GJ, Andrew RD (1999) Interpretation of intrinsic optical signals and calcein fluorescence during acute excitotoxic insult in the hippocampal slice. Neuroimage 10:357–72

Jobsis FF, O'Connor M, Vitale A, Vreman H (1971) Intracellular redox changes in functioning cerebral cortex. I. Metabolic effects of epileptiform activity. J Neurophysiol 34:735–49

Jung JC, Mehta AD, Aksay E, Stepnoski R, Schnitzer MJ (2004) In vivo mammalian brain imaging using one- and two-photon fluorescence microendoscopy. J Neurophysiol 92:3121–33

Kahraman S, Fiskum G (2007) Anoxia-induced changes in pyridine nucleotide redox state in cortical neurons and astrocytes. Neurochem Res 32:799–806

Kann O, Kovacs R (2007) Mitochondria and neuronal activity. Am J Physiol Cell Physiol 292: C641–C657

Kasischke KA, Vishwasrao HD, Fisher PJ, Zipfel WR, Webb WW (2004) Neural activity triggers neuronal oxidative metabolism followed by astrocytic glycolysis. Science 305:99–103

Kocher M (1990) Metabolic and hemodynamic activation of postischemic rat brain by cortical spreading depression. J Cereb Blood Flow Metab 10:564–71

Kofuji P, Newman EA (2004) Potassium buffering in the central nervous system. Neuroscience 129:1045–56

Kosterin P, Kim GH, Muschol M, Obaid AL, Salzberg BM (2005) Changes in FAD and NADH fluorescence in neurosecretory terminals are triggered by calcium entry and by ADP production. J Membr Biol 208:113–24

Krivanek J (1958) Changes of brain glycogen in the spreading EEG-depression of Leao. J Neurochem 2:337–43

Krivanek J (1961) Some metabolic changes accompanying Leao's spreading cortical depression in the rat. J Neurochem 6:183–89

Kunkler PE, Kraig RP (2004) P/Q Ca2+ channel blockade stops spreading depression and related pyramidal neuronal Ca2+ rise in hippocampal organ culture. Hippocampus 14:356–67

Largo C, Ibarz JM, Herreras O (1997) Effects of the gliotoxin fluorocitrate on spreading depression and glial membrane potential in rat brain in situ. J Neurophysiol 78:295–307

Lauritzen M (1994) Pathophysiology of the migraine aura. The spreading depression theory. Brain 117(Pt 1):199–210

Lauritzen M, Diemer NH (1986) Uncoupling of cerebral blood flow and metabolism after single episode of cortical spreading depression in the rat brain. Brain Res 370:405–08

Lauritzen M, Jorgensen MB, Diemer NH, Gjedde A, Hansen AJ (1982) Persistent oligemia of rat cerebral cortex in the wake of spreading depression. Ann Neurol 12:469–74

Lauritzen M, Hansen AJ, Kronborg D, Wieloch T (1990) Cortical spreading depression is associated with arachidonic acid accumulation and preservation of energy charge. J Cereb Blood Flow Metab 10:115–22

Leao AAP (1944a) Spreading depression of activity in the cerebral cortex. J Neurophysiol 7:359–90

Leao AAP (1944b) Pial circulation and spreading depression of activity in the cerebral cortex. J Neurophysiol 7:359–90

Leao AAP (1947) Further observations on the spreading depression of activity in the cerebral cortex. J Neurophysiol 10:409–14

Leao AAP, Morison RS (1945) Propagation of spreading cortical depression. J Neurophysiol 8:33–45

Levene MJ, Dombeck DA, Kasischke KA, Molloy RP, Webb WW (2004) In vivo multiphoton microscopy of deep brain tissue. J Neurophysiol 91:1908–12

L'Heureux B, Gurden H, Pain F (2009) Autofluorescence imaging of NADH and flavoproteins in the rat brain: insights from Monte Carlo simulations. Opt Express 17:9477–90

Lian XY, Stringer JL (2004) Energy failure in astrocytes increases the vulnerability of neurons to spreading depression. Eur J Neurosci 19:2446–54

Lipton P (1973a) Effects of membrane depolarization on light scattering by cerebral cortical slices. J Physiol 231:365–83

Lipton P (1973b) Effects of membrane depolarization on nicotinamide nucleotide fluorescence in brain slices. Biochem J 136:999–1009

Lothman E, Lamanna J, Cordingley G, Rosenthal M, Somjen G (1975) Responses of electrical potential, potassium levels, and oxidative metabolic activity of the cerebral neocortex of cats. Brain Res 88:15–36

Marshall WH (1959) Spreading cortical depression of Leao. Physiol Rev 39:239–79

Martins-Ferreira H, de Castro GO (1966) Light-scattering changes accompanying spreading depression in isolated retina. J Neurophysiol 29:715–26

Martins-Ferreira H, Nedergaard M, Nicholson C (2000) Perspectives on spreading depression. Brain Res Brain Res Rev 32:215–34

Mayevsky A, Chance B (1974) Repetitive patterns of metabolic changes during cortical spreading depression of the awake rat. Brain Res 65:529–33

Mayevsky A, Rogatsky GG (2007) Mitochondrial function in vivo evaluated by NADH fluorescence: from animal models to human studies. Am J Physiol Cell Physiol 292: C615–C640

Mayevsky A, Zeuthen T, Chance B (1974) Measurements of extracellular potassium, ECoG and pyridine nucleotide levels during cortical spreading depression in rats. Brain Res 76:347–49

Mies G, Paschen W (1984) Regional changes of blood flow, glucose, and ATP content determined on brain sections during a single passage of spreading depression in rat brain cortex. Exp Neurol 84:249–58

Mies G, Iijima T, Hossmann KA (1993) Correlation between peri-infarct DC shifts and ischaemic neuronal damage in rat. Neuroreport 4:709–11

Mies G, Kohno K, Hossmann KA (1994) Prevention of periinfarct direct current shifts with glutamate antagonist NBQX following occlusion of the middle cerebral artery in the rat. J Cereb Blood Flow Metab 14:802–07

Mironov SL (2006) Spontaneous and evoked neuronal activities regulate movements of single neuronal mitochondria. Synapse 59:403–11

Mironov SL, Richter DW (2001) Oscillations and hypoxic changes of mitochondrial variables in neurons of the brainstem respiratory centre of mice. J Physiol 533:227–36

Murphy TH, Li P, Betts K, Liu R (2008) Two-photon imaging of stroke onset in vivo reveals that NMDA-receptor independent ischemic depolarization is the major cause of rapid reversible damage to dendrites and spines. J Neurosci 28:1756–72

Nedergaard M, Astrup J (1986) Infarct rim: effect of hyperglycemia on direct current potential and [14C]2-deoxyglucose phosphorylation. J Cereb Blood Flow Metab 6:607–15

Nedergaard M, Hansen AJ (1988) Spreading depression is not associated with neuronal injury in the normal brain. Brain Res 449:395–98

Nemoto M, Nomura Y, Sato C, Tamura M, Houkin K, Koyanagi I, Abe H (1999) Analysis of optical signals evoked by peripheral nerve stimulation in rat somatosensory cortex: dynamic changes in hemoglobin concentration and oxygenation. J Cereb Blood Flow Metab 19:246–59

Nicholls DG, Ferguson SJ (2002) Bioenergetics. Elsevier, London

Obeidat AS, Andrew RD (1998) Spreading depression determines acute cellular damage in the hippocampal slice during oxygen/glucose deprivation. Eur J Neurosci 10:3451–61

Obrenovitch TP, Chen S, Farkas E (2009) Simultaneous, live imaging of cortical spreading depression and associated cerebral blood flow changes, by combining voltage-sensitive dye and laser speckle contrast methods. Neuroimage 45:68–74

Olesen J, Larsen B, Lauritzen M (1981) Focal hyperemia followed by spreading oligemia and impaired activation of rCBF in classic migraine. Ann Neurol 9:344–52

Parthasarathy AB, Tom WJ, Gopal A, Zhang X, Dunn AK (2008) Robust flow measurement with multi-exposure speckle imaging. Opt Express 16:1975–1989

Petzold GC, Haack S, von Bohlen Und Halbach O, Priller J, Lehmann TN, Heinemann U, Dirnagl U, Dreier JP (2008) Nitric oxide modulates spreading depolarization threshold in the human and rodent cortex. Stroke 39:1292–99

Pietrobon D, Striessnig J (2003) Neurobiology of migraine. Nat Rev Neurosci 4:386–98

Piper RD, Lambert GA, Duckworth JW (1991) Cortical blood flow changes during spreading depression in cats. Am J Physiol 261:H96–H102

Prakash N, Uhlemann F, Sheth SA, Bookheimer S, Martin N, Toga AW (2009) Current trends in intraoperative optical imaging for functional brain mapping and delineation of lesions of language cortex. Neuroimage 47(Suppl 2):T116–T126

Rawanduzy A, Hansen A, Hansen TW, Nedergaard M (1997) Effective reduction of infarct volume by gap junction blockade in a rodent model of stroke. J Neurosurg 87:916–20

Reinert KC, Gao W, Chen G, Ebner TJ (2007) Flavoprotein autofluorescence imaging in the cerebellar cortex in vivo. J Neurosci Res 85:3221–32

Rex A, Pfeifer L, Fink F, Fink H (1999) Cortical NADH during pharmacological manipulations of the respiratory chain and spreading depression in vivo. J Neurosci Res 57:359–70

Risher WC, Andrew RD, Kirov SA (2009) Real-time passive volume responses of astrocytes to acute osmotic and ischemic stress in cortical slices and in vivo revealed by two-photon microscopy. Glia 57:207–21

Rocheleau JV, Head WS, Piston DW (2004) Quantitative NAD(P)H/flavoprotein autofluorescence imaging reveals metabolic mechanisms of pancreatic islet pyruvate response. J Biol Chem 279:31780–87

Rosenthal M, Jobsis FF (1971) Intracellular redox changes in functioning cerebral cortex. II. Effects of direct cortical stimulation. J Neurophysiol 34:750–62

Rosenthal M, Somjen G (1973) Spreading depression, sustained potential shifts, and metabolic activity of cerebral cortex of cats. J Neurophysiol 36:739–49

Rowland KC, Irby NK, Spirou GA (2000) Specialized synapse-associated structures within the calyx of Held. J Neurosci 20:9135–44

Sakowitz OW, Kiening KL, Krajewski KL, Sarrafzadeh AS, Fabricius M, Strong AJ, Unterberg AW, Dreier JP (2009) Preliminary evidence that ketamine inhibits spreading depolarizations in acute human brain injury. Stroke 40:e519–e522

Sasaki T, Takeda Y, Taninishi H, Arai M, Shiraishi K, Morita K (2009) Dynamic changes in cortical NADH fluorescence in rat focal ischemia: evaluation of the effects of hypothermia on propagation of peri-infarct depolarization by temporal and spatial analysis. Neurosci Lett 449:61–5

Scheller D, Kolb J, Tegtmeier F (1992) Lactate and pH change in close correlation in the extracellular space of the rat brain during cortical spreading depression. Neurosci Lett 135:83–86

Schuchmann S, Kovacs R, Kann O, Heinemann U, Buchheim K (2001) Monitoring NAD(P)H autofluorescence to assess mitochondrial metabolic functions in rat hippocampal-entorhinal cortex slices. Brain Res Brain Res Protoc 7:267–6

Selman WR, Lust WD, Pundik S, Zhou Y, Ratcheson RA (2004) Compromised metabolic recovery following spontaneous spreading depression in the penumbra. Brain Res 999:167–74

Shibuki K, Hishida R, Kitaura H, Takahashi K, Tohmi M (2007) Coupling of brain function and metabolism: Endogenous flavoprotein fluorescence imaging of neural activities by local changes in energy metabolism. In: Handbook of Neurochemistry & Molecular Neurobiology. 3rd Edition, Vol. 5, Neural Energy Utilization (Gibson G and Dienel G, ed). Springer, New York, pp. 322–42

Shinohara M, Dollinger B, Brown G, Rapoport S, Sokoloff L (1979) Cerebral glucose utilization: local changes during and after recovery from spreading cortical depression. Science 203:188–90

Shuttleworth CW (2010) Use of NAD(P)H and flavoprotein autofluorescence transients to probe neuron and astrocyte responses to synaptic activation. Neurochem Int 56:379–86

Shuttleworth CW, Connor JA (2001) Strain-dependent differences in calcium signaling predict excitotoxicity in murine hippocampal neurons. J Neurosci 21:4225–36

Shuttleworth CW, Brennan AM, Connor JA (2003) NAD(P)H fluorescence imaging of postsynaptic neuronal activation in murine hippocampal slices. J Neurosci 23:3196–08

Somjen GG (2001) Mechanisms of spreading depression and hypoxic spreading depression-like depolarization. Physiol Rev 81:1065–96

Somjen GG (2005) Aristides Leao's discovery of cortical spreading depression. J Neurophysiol 94:2–4

Sonn J, Mayevsky A (2000) Effects of brain oxygenation on metabolic, hemodynamic, ionic and electrical responses to spreading depression in the rat. Brain Res 882:212–16

Strong AJ, Venables GS, Gibson G (1983a) The cortical ischaemic penumbra associated with occlusion of the middle cerebral artery in the cat: 1. Topography of changes in blood flow, potassium ion activity, and EEG. J Cereb Blood Flow Metab 3:86–96

Strong AJ, Tomlinson BE, Venables GS, Gibson G, Hardy JA (1983b) The cortical ischaemic penumbra associated with occlusion of the middle cerebral artery in the cat: 2. Studies of histopathology, water content, and in vitro neurotransmitter uptake. J Cereb Blood Flow Metab 3:97–108

Strong AJ, Gibson G, Miller SA, Venables GS (1988) Changes in vascular and metabolic reactivity as indices of ischaemia in the penumbra. J Cereb Blood Flow Metab 8:79–88

Strong AJ, Harland SP, Meldrum BS, Whittington DJ (1996) The use of in vivo fluorescence image sequences to indicate the occurrence and propagation of transient focal depolarizations in cerebral ischemia. J Cereb Blood Flow Metab 16:367–77

Strong AJ, Smith SE, Whittington DJ, Meldrum BS, Parsons AA, Krupinski J, Hunter AJ, Patel S, Robertson C (2000) Factors influencing the frequency of fluorescence transients as markers of peri-infarct depolarizations in focal cerebral ischemia. Stroke 31:214–22

Strong AJ, Bezzina EL, Anderson PJ, Boutelle MG, Hopwood SE, Dunn AK (2006) Evaluation of laser speckle flowmetry for imaging cortical perfusion in experimental stroke studies: quantitation of perfusion and detection of peri-infarct depolarisations. J Cereb Blood Flow Metab 26:645–53

Swann JW, Al-Noori S, Jiang M, Lee CL (2000) Spine loss and other dendritic abnormalities in epilepsy. Hippocampus 10:617–25

Takano K, Latour LL, Formato JE, Carano RA, Helmer KG, Hasegawa Y, Sotak CH, Fisher M (1996) The role of spreading depression in focal ischemia evaluated by diffusion mapping. Ann Neurol 39:308–18

Takano T, Tian GF, Peng W, Lou N, Lovatt D, Hansen AJ, Kasischke KA, Nedergaard M (2007) Cortical spreading depression causes and coincides with tissue hypoxia. Nat Neurosci 10:754–62

Talbot J, Barrett JN, Barrett EF, David G (2007) Stimulation-induced changes in NADH fluorescence and mitochondrial membrane potential in lizard motor nerve terminals. J Physiol 579:783–98

Tang Y, Zucker RS (1997) Mitochondrial involvement in post-tetanic potentiation of synaptic transmission. Neuron 18:483–91

Tasaki I, Watanabe A, Sandlin R, Carnay L (1968) Changes in fluorescence, turbidity, and birefringence associated with nerve excitation. Proc Natl Acad Sci USA 61:883–88

Tatlisumak T, Takano K, Meiler MR, Fisher M (2000) A glycine site antagonist ZD9379 reduces number of spreading depressions and infarct size in rats with permanent middle cerebral artery occlusion. Acta Neurochir Suppl 76:331–33

Teive HA, Kowacs PA, Maranhao Filho P, Piovesan EJ, Werneck LC (2005) Leao's cortical spreading depression: from experimental "artifact" to physiological principle. Neurology 65:1455–59

Terzuolo CA, Chance B, Handelman E, Rossini L, Schmelzer P (1966) Measurements of reduced pyridine nucleotides in a single neuron. Biochim Biophys Acta 126:361–72

Tohmi M, Takahashi K, Kubota Y, Hishida R, Shibuki K (2009) Transcranial flavoprotein fluorescence imaging of mouse cortical activity and plasticity. J Neurochem 109(Suppl 1):3–9

Turner DA, Foster KA, Galeffi F, Somjen GG (2007) Differences in O2 availability resolve the apparent discrepancies in metabolic intrinsic optical signals in vivo and in vitro. Trends Neurosci 30:390–98

van den Maagdenberg AM, Pietrobon D, Pizzorusso T, Kaja S, Broos LA, Cesetti T, van de Ven RC, Tottene A, van der Kaa J, Plomp JJ, Frants RR, Ferrari MD (2004) A Cacna1a knockin migraine mouse model with increased susceptibility to cortical spreading depression. Neuron 41:701–10

Van Harreveld A (1957) Changes in volume of cortical neuronal elements during asphyxiation. Am J Physiol 191:233–42

Van Harreveld A (1958) Changes in the diameter of apical dendrites during spreading depression. Am J Physiol 192:457–63

Van Harreveld A, Stamm JS (1952) Vascular concomitants of spreading cortical depression. J Neurophysiol 15:487–96

Vanzetta I, Hildesheim R, Grinvald A (2005) Compartment-resolved imaging of activity-dependent dynamics of cortical blood volume and oximetry. J Neurosci 25:2233–44

Wahl M, Lauritzen M, Schilling L (1987) Change of cerebrovascular reactivity after cortical spreading depression in cats and rats. Brain Res 411:72–80

Wininger FA, Schei JL, Rector DM (2009) Complete optical neurophysiology: toward optical stimulation and recording of neural tissue. Appl Opt 48:D218–D224

Woods RP, Iacoboni M, Mazziotta JC (1994) Brief report: bilateral spreading cerebral hypoperfusion during spontaneous migraine headache. N Engl J Med 331:1689–92

Yang G, Pan F, Parkhurst CN, Grutzendler J, Gan WB (2010) Thinned-skull cranial window technique for long-term imaging of the cortex in live mice. Nat Protoc 5:201–08

Zhou N, Gordon GR, Feighan D, Macvicar BA (2010) Transient swelling, acidification, and mitochondrial depolarization occurs in neurons but not astrocytes during spreading depression. Cereb Cortex 20:2614–24

Zipfel WR, Williams RM, Christie R, Nikitin AY, Hyman BT, Webb WW (2003) Live tissue intrinsic emission microscopy using multiphoton-excited native fluorescence and second harmonic generation. Proc Natl Acad Sci USA 100:7075–80

Zucker RS (1999) Calcium- and activity-dependent synaptic plasticity. Curr Opin Neurobiol 9:305–13

Chapter 13
Microdialysis and Microfiltration: Technology and Cerebral Applications for Energy Substrates

Kirsten Huinink, Jakob Korf, and Jan Bert Gramsbergen

Abstract
This chapter is divided into four parts.

1. Sampling methods, detection technologies and quantitative aspects of monitoring cerebral energy substrates in living brain, mainly glucose and lactate, are reviewed, including microdialysis, microfiltration and biosensors. Advances in the field, including metaquant sampling, in situ microsensors, as well as choice of hollow fiber materials for microdialysis and microfiltration are discussed.
2. Experimental and clinical applications of cerebral microdialysis of glucose and lactate are reviewed. Effects of drugs, physiological stimuli and pathological conditions (e.g. hypoxia, ischemia, head or spinal cord trauma) are discussed. For clinical monitoring in neurointensive care low time resolution microdialysis at 0.5–1 min intervals and on-line detection by a biosensor assay may provide immediate information on the metabolic consequences of therapeutic interventions during surgery.
3. The significance of intracerebral trafficking of glucose and lactate in the interstitial space is reviewed and addressed in some detail in an experimental study. In this study ultraslow microdilaysis and reverse dialysis of exogenous glucose or lactate is used to assess the relative glucose and lactate turnover at steady state. It is concluded that only about 12% of the brain's energy substrates traffick through this compartment.
4. Finally, an attempt is made to link our current knowledge of brain energy metabolism to physiological events in the brain. In this context we discuss the concepts

K. Huinink • J. Korf, Ph.D. (✉)
Department of Psychiatry, University Medical Center of Groningen, Groningen, The Netherlands
e-mail: j.korf@psy.umcg.nl

J.B. Gramsbergen, Ph.D.
Molecular Neurosciences, South Denmark University, Odense, Denmark
e-mail: jbgramsbergen@health.sdu.dk

I.-Y. Choi and R. Gruetter (eds.), *Neural Metabolism In Vivo*, Advances in Neurobiology 4, DOI 10.1007/978-1-4614-1788-0_13,

of metabolic and potential brain energy. Potential brain energy, defined as the large ionic gradients over neuronal membranes and storage of glutamate and GABA in synaptic vesicles, is considered a necessary condition for fast neuronal processing, whereas metabolic brain energy (i.e. consumption of glucose and oxygen) serves to restore high levels of potential energy.

Keywords Biosensors • Brain • Clinical microdialysis • Energy substrates • Glucose • *In vivo* • Lactate • Metabolic energy • Microdialysis • Microfiltration • Potential energy • Sampling techniques

13.1 Introduction and Scope

The brain consumes – relative to its weight – a major part of an organism's energy budget. The human brain, for instance, 2–3% of the whole body weight, consumes more than 20% of the energy of the entire organism at physical rest (Attwell and Laughlin 2001). So it is obvious that the unique properties of the brain, e.g. to register, store and integrate sensory information, and to initiate and control behavior, e.g. motor activity and thinking, requires much energy. Already in 1890, the tight coupling between neuronal activities and cerebral blood flow, known as functional hyperemia, was recognized (Roy and Sherrington 1890), and this coupling has been suggested to enable the extra supply of energy substrates. More than 50% of the brain's energy consumption is used for excitatory synaptic activity and action potentials (Attwell and Laughlin 2001). This does not necessarily imply that only glucose directly derived from the circulation is used for excitatory transmission, since – theoretically – also indirect glucose use via lactate uptake from the intercellular space and subsequent consumption by neurons and/or glia may play a role, as will be discussed. Brain energy stores are insufficient to support brain activity for longer than a few minutes (Attwell and Laughlin 2001; Korf et al. 1988; Allen et al. 2005), as was estimated following experimental interruption of cerebral blood flow. If glucose and oxygen are derived from the blood circulation or formed from rapidly turning over cerebral intermediates, always (approximately) six molecules oxygen per molecule glucose are consumed. If there are intermediate substrates leaving the brain (as may occur with lactate), or if the contribution of brain energy reserves (e.g. glycogen, predominantly localized in astroglia) are significant, this ratio becomes less or more than 6, respectively. Under normoxic conditions the ratio is about 5.5, indicating that a small, albeit significant, fraction of the glucose is not aerobically metabolized, either stored as glycogen (Oz et al. 2007; Hertz et al. 2007; Gibbs et al. 2006) or leaving the brain (by the bloodstream) (Leegsma-Vogt et al. 2004; Mintun et al. 2004). The possible importance of glycogen as an *in vivo* energy source must be questioned. For instance 20 min of visual stimulation does not result in detectable glycogen utilization in the visual cortex (Oz et al. 2007). The various cerebral cells, including several types of neurons and glia, are differently positioned towards the blood circulation.

In most current models energy substrates enter the cells following passage of the blood-brain-barrier, consisting of endothelial cells, to glia and some neurons.

Several brain cells have no direct connection to blood vasculature and the energy substrates will reach these cells only indirectly, i.e. their energy supply depends entirely on the transport via cells that have direct access to the blood. Both aerobically or an-aerobically intra-cellular production of energy metabolites leave the cell via trans-membrane transport directly or indirectly, i.e. via other cells, presumably most often glia. Trafficking of substrates between cells is thus considered essential to meet the energy demand of the brain. Accordingly, it has been hypothesized that neuronal energy consumption depends, at least in part, on the release of glucose and lactate from astroglia cells and on subsequent diffusion to neurons via the brain intercellular space (ICS; (Forsyth 1996; Forsyth et al. 1996; Magistretti et al. 1993; Dringen et al. 1993; Pellerin and Magistretti 1994; Pellerin et al. 2007; Hu and Wilson 1997a; McKenna et al. 1998; Schurr and Rigor 1998; Gjedde et al. 2002; Dienel and Hertz 2001).

Monitoring of brain energy substrates requires invasive surgery, as the brain is situated inside the skull. Non-invasive approaches like imaging can only be performed in dedicated locations and is thus unsuitable for frequent or continuous monitoring. Technologies allowing continuous or near-continuous monitoring include biosensor technology, microdialysis (MD), ultrafiltration (UF). In the present text we prefer the term microfiltration (MF) instead of UF as this is more conform the MD terminology. MD has a long history in studying effects of drugs, behavior and cerebral ischemia on the release of neurotransmitters, monoamines, and metabolites such as amino acids and other small endogenous compounds. MF is a sampling technique not used so often but in certain experiments MF is to be preferred over MD.

The importance of estimating glucose and lactate concentration in the ICS has been recognized by several authors (McNay et al. 2000; Lund-Andersen 1979; van der Kuil and Korf 1991; Fellows et al. 1992; Fellows and Boutelle 1993; Fray et al. 1997b; Lowry et al. 1998a; Kuhr et al. 1988; Kuhr and Korf 1988; Shram et al. 1998; Demestre et al. 1997). Partly recovered concentrations of energy substrates as measured with conventional MD (typically using flow rates of 2 μl/min), are sufficient for monitoring relative changes in the extracellular levels of lactate, pyruvate and glucose during experimental interventions. This is indeed the common practice in most studies investigating the effects of drugs (Table 13.1), behavior or physiological conditions (Table 13.2) or changes as a result of hypoxia-ischemia or trauma (Table 13.3). In these studies, MD parameters are allowed to stabilize after insertion of the probe, and then concentrations of the analytes measured in the dialysate are expressed as percent change from baseline (100% = baseline levels). However when the aim is to relate extracellular glucose or lactate levels to absolute turnovers of tissue energy metabolism, the real, absolute concentrations must be known to obtain a meaningful interpretation of the data. The present survey gives examples of studies based on both relative and absolute levels of the energy substrates and metabolites.

There are as yet no reports on either estimates of the turnover rate of glucose and lactate of the ICS or of the proportion of energy substrates that is transferred via this compartment. Quantification of intra-cerebral transport and transport of energy substrates and products to and from the blood circulation may shed light on the

Table 13.1 Effects of drugs on cerebral glucose and/or lactate changes monitored by MD or implanted sensors in experimental animals (rats)

Drug	Drug adm.	Position MD probe	Glucose	Lactate	Assay	References
Pentobarbital	i.p.	Hippocampus Striatum	Increase (+15–76%)	Decrease (−17%)	Off-line On-line glucose sensor	Canal et al. (2005), Osborne et al. (1997), Lonjon et al. (2001)
Ketamine	i.p.	Hippocampus	Increase (+85%)		Off-line	Canal et al. (2005)
Chloral hydrate	i.p.	Striatum	Increase (+150%) No change in response to hypoxia	Hypoxia-induced-increase reduced	On-line, amperometric Implanted glucose sensor	Fellows et al. (1992), Lowry et al. (1998a)
Probenecid (blocks lactate carrier)	Local	Striatum		Increase (+50%)	On-line, fluorometric	Kuhr et al., (1988), Korf (1996), Taylor et al. (1997)
Alpha-cyano-4-hydroxycinnamate (blocks lactate transport)	Local	Cortex		Increase ^{14}C-lactate uptake inhibited	Off-line	Mendelowitsch et al. (2001), Zielke et al. (2007)
High K^+	Local	Striatum N. Accumb	Decrease (−23–70%)	Increase (+70–85%)	On-line, fluorometric Off-line	Taylor et al. (1997), Darbin et al. (2006), Uehara et al. (2007a)
Veratridine	Local	Striatum N. Accumb.	Decrease (−86–94%) Decrease (−75%)	Increase (+100%)	On-line Glucose sensor	Osborne et al. (1997), Fellows et al. (1992), Uehara et al. (2007a)
Tetrodotoxin (TTX)	Local	Striatum N. Accumb	Increase (+34%) Increase (+93%)	Attenuation of ECT-induced increase No change	On-line, amperometric, fluorometric, NMR spectroscopy	Kuhr and Korf (1988a), Fellows et al. (1992), Khandelwal et al. (2004), Uehara et al. (2007a)
Ouabain	Local	Cortex		Inhibits trauma-induced increase	Off-line	Kawamata et al. (1995a)
Glutamate	Local	Cortex	Decrease (−48%)	Increase (+600–700%)	Off-line On-line, amperometric	Alessandri et al. (1996), Mendelowitsch et al. (2001), Ros et al. (2002)
N-methyl-D-aspartate (NMDA)	Local	Striatum		Increase	On-line, fluorometric	Kuhr and Korf (1988b)
Kainic acid (KA)	Local, i.p.	Striatum		Increase	On-line, fluorometric	Kuhr and Korf (1988a), Yager et al. (2002)
D,L-threo-hydroxyaspartate (THA)	Local	Striatum		Increase (+200%)	On-line, amperometric	Fray et al. (1996)
Dihydrokainate (DHK)		Medial Prefrontal Cortex Amygdala		Increased basal release, inhibited foot-shock induced increase	On-line, fluorometric	Uehara et al. (2007b)

AP-5, AP-7 (NMDA receptor antagonists)	Local	Hippocampus Striatum		Inhibits ECS/stress-induced increase	On-line, fluorometric	Kuhr and Korf (1988b), Korf and De (1990)
MK-801 (non-competitive NMDA receptor antagonist)	Local	Hippocampus Striatum		No effect on tail-pinch induced increase	On-line, amperometric	Fellows et al. (1993)
CNQX (AMPA receptor antagonist)	Local	Cerebellum	No change	Inhibits mossy fiber stimulation-induced increase	On-line, amperometric	Caesar et al. (2008)
Kynurenic acid	Local	Cortex		Inhibits trauma- or ischemia- induced increase	Off-line	Katayama et al., (1992), Kawamata et al. (1995a)
Pilocarpine	i.p.	Cortex	Perturbation (±44–48%)	Increase (+379%)	Off-line	Slais et al. (2008)
Isoprenaline	Local	Striatum	Increase	No change	On-line, amperometric	Fray et al. (1996)
Propanolol	Local	Striatum	Decrease (−17%)		On-line, amperometric	Fray et al. (1996)
cAMP	Local	Striatum	Increase		On-line, amperometric	Fray et al. (1996)
Diazepam	i.p.	Amygdala		Attenuation of stress-induced increase	On-line, fluorometric	Uehara et al. (2003, 2005)
Apomorphine	i.p.	N. Accumb.	Increase (+41%)	Increase (+30%)	On-line, fluorometric	Uehara et al. (2007a)
Bromocriptine	i.p.	N. Accumb.	Increase (+110%)	No change	On-line, fluorometric	Uehara et al. (2007a)
Haloperidol	i.p.	Striatum Hippocampus		Increase	On-line, fluorometric	Dijk et al. (1991)
8-OH-DPAT (5-HT1A agonist)	i.p.	Hypothalamus	Decrease		Off-line	Voigt et al. (2004)
Methamphetamine.	i.p.	Striatum Prefrontal cortex		Increase	Off-line	Stephans et al. (1998)
3,4-Methylene-dioxymethamphetamine (MDMA,"Ecstasy")	s.c. i.v.	Striatum	Increase (+39–123%)	Increase (up to +87%)	Off-line On-line, amperometric	Darvesh et al. (2002), Gramsbergen and Cumming (2007), Gudelsky and Yamamoto (2007)
Estradiol	i.v.	Cortex	No change	Attenuation of ischemia or glutamate-induced increase	Off-line	Mendelowitsch et al. (2001), Ritz et al. (2004)
Cyanide, malonate, MPP^+, etc. (mitochondrial toxins)	Local	Cortex Striatum	Decrease (−46%)	Increase (+375–434%)	Off-line NMR	Pazdernik et al. (1994), Jenkins et al. (1996), Clausen et al. (2001)

Table 13.2 Glucose-lactate monitoring following physiological stimuli in rats

Stimilus	Position MD probe	Glucose	Lactate	Time Resolution Assay	References
Tail pinch	Striatum hippocampus Medial prefrontal cortex	Perturbations (first decrease, then increase) ±5–15%	Increase +23–93%	On-line 2.5 min., continuous	Takita et al. (1992), Fellows et al. (1993), Fray et al. (1996), Fillenz and Lowry (1998)
Restraint stress	Striatum Hippocampus Medial prefrontal cortex Amygdala	Perturbations (first decrease, then increase) ±10–125%	Increase +57–135%	On-line 2.5–5 min, continuous	De Bruin et al. (1990), Takita et al. (1992), Fellows et al. (1993), Korf et al. (1993), Elekes et al. (1996), Korf (1996), Uehara et al. (2003)
Acoustic stimulation	Medial prefrontal cortex		Increase +23%	On-line continuous	Takita et al. (1992)
Swimming/exercise	Hippocampus Striatum Hypothalamus	Decrease −26–42%	Increase +12–45%	On-line 2–5 min	Korf (1996), Wang et al. (2002)
Learning and memory	Hippocampus	Decrease −11–48%		Off-line 5 min	McNay et al. (2000, 2001, 2006)
Food shock	Medial prefrontal cortex Amygdala		Increase +50–59%	On-line continuous	Uehara et al. (2005, 2007b)
Electroconvulsive shock	Striatum, hippocampus	Increase two-fold	Increase +47%	On-line 2 min, continuous Off-line 15 min	De Bruin et al. (1990), Krugers et al. (1992), Korf et al. (1993), Korf (1996), Darbin et al. (2006)
Electrophysiological stimulation (climbing fibers)	Cerebellar cortex (molecular layer)	No change	Increased +19–29%	On-line 0.5 min	Caesar et al. (2008)

Table 13.3 Glucose-lactate monitoring in experimental models of acute brain injury

Acute brain injury model	Position MD probe	Glucose	Lactate	Time resolution assay	References
Hypoxia (rat) repetitive	Hippocamp.	Perturb.	Increase +76%	On-line cont.	van der Kuil and Korf (1991), Lowry et al. (1998a), Silver and Erecinska (1994), Jones et al. (2000)
3 min	Striatum	No change Decrease		On-line 1 min. Glucose sensor	
4 min	Cortex	Increase +210%	Increase +1,160%	On-line 3 min	
15 min	Cortex				
Hypoxia (rat)	Hippocamp.	Decrease –48%	Increase +158%	Off-line 30 min.	Harada et al. (1992), Homola et al. (2006), Zoremba et al. (2007)
90 min	Cortex	–75–85%	+153%	Off-line 15 min	
30 min					
Global ischemia 5 min (gerbil) 10 min (rat) 14 min (pig)	Cortex	Decrease –80%	Increase + 200–260%	Off-line 10 min On-line cont. MRS	Katayama et al. (1992), Korf (1996), Valtysson et al. (1998), Corbett et al. (1999)
Focal ischemia (rat, gerbil)	Cortex Striatum	Decrease <–70%	Increase + 100–500%	Off-line 15 min	
(monkey)	Cortex	Core: <–82% Penumbra: –31%	Core: +245% Penumbra: +148%	Off-line 15 min	Fuxe et al. (1992), Alessandri et al. (2000), Lin et al. (2002), Enblad et al. (2001), Frykholm et al. (2005)
Focal ischemia (rat)	Striatum	Decrease <–80%	Increase +320%	On-line 1 min	Gramsbergen et al. (2004), Hopwood et al. (2005)
(cat)	Cortex	–8–38% PID: –5%	+ 51% PID: + 26%	On-line 0.5 min	
Traumatic Brain Injury (rat)	Cortex	No change, decrease (–13–50%) or increase	Increase + 97–600% (+ volume insult)	Off-line 10–30 min	Kawamata et al. (1995b), Lewen and Hillered (1998), Krishnappa et al. (1999), Bentzer et al. (2000), Chen et al. (2000), Marklund et al. (2001), Reinert et al. (2004), Alves et al. (2005), Salci et al. (2006), Thomale et al. (2007)
Traumatic Brain Injury (rat)	Thalamus	No change	Increase (+ hypoxia)	Off-line 15 min	Geeraerts et al. (2006, 2008)
Epilepsy (rat)	Striatum Hippocamp.	See kainic acid, pilocarpine (Table 13.1) and electroconvulsive chok (Table 13.2)			

metabolic organization of brain energy consumption. Current approaches to assess brain energy dynamics include isotope dilution of labelled substrates as detected either with conventional radioactive tracers or with modern imaging technologies, such as magnetic resonance spectroscopy and positron emission tomography (PET). The current review is emphasizing another quantitative approach, which is based on intra-cerebral and intravascular sampling with MD and MF. Both intra-cerebral MD and MF inform only about the fate of substrates in the direct vicinity of the probe and do not inform about the condition of the whole brain. Whole brain metabolism can be studied by measuring arterio-venous differences.

Here we describe the essentials of both the intra-cerebral and intravascular sampling and (biosensor) detection technologies, some approaches for obtaining quantitative *in vivo* results and some new experimental data that allow estimates of the turnover rates of glucose and lactate in the rat brain ICS. We provide an overview of experimental and clinical applications of cerebral glucose/lactate monitoring (mainly conventional MD with off-line or on-line detection).

Finally a general discussion on the implications of recent findings for current theories on intra-cerebral trafficking of energy substrates and products is given.

13.1.1 On Monitoring Methodologies

13.1.1.1 Microdialysis

The basic principle of MD is to mimic the function of capillary blood vessels by perfusing a dialysis tube implanted into the tissue with a physiological fluid. The dialysate reflects the composition of the extracellular fluid containing substances that can diffuse through the pores of the hollow fiber membrane. The driving force of MD is thus the concentration gradients of the various constituents between the extracellular fluid and the perfusion fluid over the membrane of the probe and thus no fluid is withdrawn from the sampling site. The first rather primitive version of the present MD was developed by Delgado et al. (1972) in 1972 and Ungerstedt and Pycock described the first MD for *in vivo* measurements in 1974 (Ungerstedt and Pycock 1974). There are several variables to be considered when applying MD. These include the flow rate and type of the perfusion fluid, length of the hollow fiber membrane, the molecular weight cut-off (MWCO) value of the membrane, and the hollow fiber material. The hollow fiber is the analogue between MD and MF sampling techniques, which will be discussed in the next section.

Conventional MD systems consist of a pump, probe/catheter, and a small vial to collect the dialysate (off-line system) or an analytical device directly coupled to the outlet of the probe (on-line system). The MD probe may be placed inside the brain or in peripheral tissue and fluids (cerebrospinal, bile and blood) or other matrices. Most MD probes are concentric with an inner diameter in the range of 0.15–0.4 mm but there are looped shaped MD probes as well. The perfusate is an aqueous solution that mimics the surrounding biomatrix, so no osmotic differences exist over the probe membrane. The diffusion of the analytes over the probe membrane is dependent on the concentration gradient. While the perfusion fluid passes the membrane, molecules up to a certain molar mass diffuse inwards (recovery) or outwards (delivery) the probe. Thus MD can be used for both collecting substances in the dialysate as well as delivering them into the surrounding tissue (= retro-dialysis).

In far most MD studies, perfusion rates of 0.5–2 μl/min have been used for *in vivo* sampling with relative recoveries far less than 100%. The term relative recovery describes the ratio between the concentration of the substance in the dialysate and (peripheral) tissue. Therefore complicated procedures are required to calculate (or estimate) the real concentration of the analyte in the tissue compartment. There are many different methods for calibration and thus for determination of the relative recovery, method for e.g. of flow rate variation, no-net-flux, dynamic-no-net-flux, retro-dialysis and adding an endogenous reference substance. All these calibration procedures have nicely been reviewed by Plock and Kloft (2005). Quantitative MD, without calibrating the MD device, is also possible by applying an ultraslow (us) perfusion rate, which results in similar concentrations of the analyte in the dialysate and the tissue compartment. Kaptein et al. (1998), Ekberg et al. (2005) and Rosdahl et al. (1998) proved that using usMD the recovery is exactly 100% in contrast to the lower recoveries obtained with conventional MD. Rosdahl described in 1998 one of the first experiments using usMD in human skeletal muscle and adipose tissue. Complete equilibiurm was reached in both tissues using a flow rate 0.16 μl/min. Also the measured concentrations of glucose, glycerol, and urea were in good agreement with expected extracellular tissue-specific levels (Rosdahl et al. 1998). Kaptein et al. compared in anesthetized rats subcutaneous usMD with MF sampling (see also next section) at sampling rates of 100 and 300 nl/min (Kaptein et al. 1998). Glucose and lactate concentrations were exactly the same for both sampling techniques. Ekberg et al. (2005) measured glucose, lactate, pyruvate and glycerol using four different flow rates: 0.3 (= usMD), 1, 2, and 5 μl/min in interstitial fluid of subcutaneous adipose tissue. The conclusion was that the usMD flow rate of 0.3 μl/min resulted in recoveries near 100% with decreasing recoveries at higher perfusion rates.

To make sampling possible at low flow rates, without a substantial loss of perfusion fluid into the tissue (= ultrafiltration fluid loss), some groups add a colloid to the perfusion fluid (Rosdahl et al. 1997; Hamrin et al. 2002). Rosdahl et al. (1997) placed MD probes in the abdominal subcutaneous adipose tissue and femoral muscle testing four different perfusion rates: 0.16, 0.33, 0.66, and 1.6 μl/min in humans. Their perfusion fluid was a modified Krebs-Henseleit bicarbonate buffer (KHB) (Rosdahl et al. 1993) or a KHB supplemented with dextran-70. Without dextran perfusion fluid was lost and adding dextran (40 g/l) to the perfusion fluid, a good balance was observed between the expected and collected volumes (Rosdahl et al. 1997). Hamrin et al. investigated also this phenomenon, adding dextran to the perfusate, but varying the position of the outlet tubing was also investigated. By varying the position of the outlet tubing mechanical/hydrostatic pressure is changed, which is a function of the outlet tubing length and diameter, fluid viscosity, and flow rate as stated by Poiseuille's law. The sample volumes were significantly smaller when the dialysis membranes was influenced by maximal hydrostatic pressure compared with minimal hydrostatic pressure (Hamrin et al. 2002). Both studies used MD probes with a MWCO value of 20 kDa (Rosdahl et al. 1997; Hamrin et al. 2002). Membrane technology is improving, thus hollow fiber membranes with a larger MWCO value can be purchased. MD probes up to 3,000 kDa allows monitoring of a wider range of compounds, including e.g. cytokines, proteins, peptides (Hutchinson et al. 2000, 2005a; Ao and Stenken 2006; Schutte et al. 2004; Rosenbloom et al. 2005; Hillman et al. 2005, 2006a) . However due to the increased porosity of the hollow fiber, ultra-filtration fluid loss can become a problem and over time, the physiology of the tissue being sampled can be altered. Trickler and Miller (2003) proved that adding bovine serum albumin (0–10% m/v BSA) to the perfusion fluid ultrafiltration loss was reduced (MWCO = 20 and 100 kDa using flow rates of 0.5–5 μl/min). Hillman et al. (2005) investigated ultra-filtration fluid loss using an us-flow of 0.3 μl/min with perfusion fluid without and with BSA (3.5%) or with Ringer-dextran 60 (RD60-Braun Medical AB, Stockholm, Sweden). Addition of BSA or dextran reduced fluid loss.

Another problem of using us-flow rates is the very long lag times between sampling and analysis. In addition, evaporation may occur during the long collection times, which concentrates the samples. These disadvantages can be circumvented by using the MetaQuant (MQ) sampling technique (patented by Brainlink B.V.), which allows fast transport of dialysate together with high recoveries of the analytes. In this type of MD probe an extra fluid line is added, which rapidly transports the dialysate sample to the detection device (Cremers and Ebert 2007) (see Fig. 13.1a). A drawback of the additional fluid line is that extra care should be taken to keep the sampling system sterile.

13.1.1.2 Microfiltration

The driving force of the flow rate in MF is a pressure gradient over the probe membrane, pulling sample from the *in vivo* environment. MF samples are undiluted (100% recovery), small (μl or nl size) and sterile. Figure 13.1c is showing a typical MF device schematically. Two different types of MF probes have been developed of which only one is commercially available (Bioanalytical Systems, Lafayette, Ind., USA). The commercial probe consists of one or more hollow fiber loops with various hollow fiber lengths (ranging from 2 to 12 cm). The sampling rate is relatively high, in the range of 0.5–2 μl/h/cm hollow fiber membrane.

A previously described non-commercial probe is much smaller: a concentric probe of 4 cm using a us-sampling rate of 100 nl/min, applied subcutaneously in conscious rats (Moscone et al. 1996). In this study the MF probe was connected to a long capillary, enabling continuous sampling for 30 h and simultaneous storage of the samples in that collection coil. This integrated device is called the ultra-filtration collection device (UCD; Fig. 13.1b, c). With this device time-profiles can be created without manual intervention and the profile can be stored in de collection device until further analysis up to several days without significant diffusion artifacts. The collection coil is made of fused silica tubing, because a significant evaporation

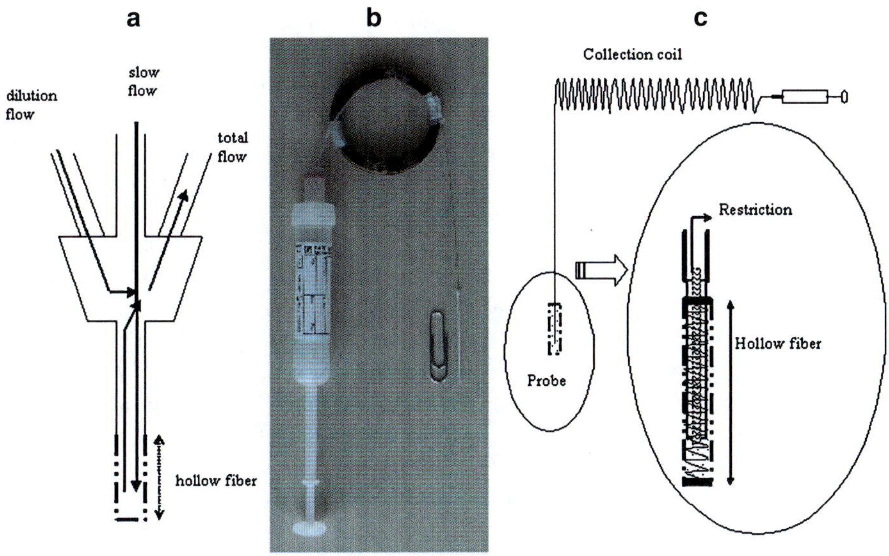

Fig. 13.1 (**a**) Schematic drawing of MQ-MD system. The slow flow into the hollow fiber membrane generates a dialysate, which is rapidly diluted and removed by a second, fast "dilution" flow resulting in a "total flow" out of the probe. (**b**) An ultrafiltration collection device (UCD) consists of a Monovette under-pressure container, a capillary collection tube and a hollow fiber (next to the paperclip). (**c**) A schematic drawing of the UCD with enlargement of the hollow fiber membrane, containing a spiral, which prevents the hollow fiber from collapsing. The spiral extends into the restriction tube, which defines the flow rate in combination with the under-pressure container used

of the collected fluid through several other tested materials was observed. Due to the small size of the probe, the probe can be introduced *in vivo* via adapted needles with minimal tissue damage and without using anesthetics.

The amount of fluid withdrawn from the tissue compartment is usually relatively small, so the chemical balances of the extracellular fluid (ECF) is not or minimally disturbed and in most tissues the body fluid is easily replenished. Because MF concentrations are (nearly) identical to those in the ECF concentrations, the measured levels of the analyte reflect rather accurately the metabolic changes in the intercellular spaces (Moscone et al. 1996). It appeared, however, that the volume of the ECF may differ between animal species, condition of the organism and the investigated organ. For instance, Kissinger et al. mentioned that an anesthetized rat produces almost no MF fluid, whereas a rat, that has free access to water and food and is moving around, produced plenty of MF fluid (Kissinger et al. 2003). Our own studies in anesthetized rats revealed that over a period of 6–7 h and the syringe pump set at a flow rate of 0.3 μl/min, the ultra-filtration flow rate is declining with a subcutaneous placed probe, but remained stable with a probe positioned intraperitoneal. In contrast to most peripheral tissues, MF was not possible in rat brain striatum, not even when applying syringe pump flow rates of 50–100 nl/min (Kaptein et al. 1998). Apparently, the ECF production and/or supply in the rat brain was too low. Whether this limitation applies to the human brain is unlikely, considering the high percentage – as compared to the rat brain – of extracellular volume (see below) of more than 30% of the total brain volume. So in rat (Linhares and Kissinger 1993a; Linhares and Kissinger 1992; Leegsma-Vogt et al. 2001; Linhares and Kissinger 1993b), cat (Janle et al. 1992a), dog (Bidgood and Papich 2005), horse (Davis et al. 2005; Spehar et al. 1998), pig (Tiessen et al. 2001), chicken (Savenije et al. 2003), mouse (Janle et al. 1992b), and sheep (Janle and Sojka 2000; Janle et al. 2001; Sojka et al. 2000) MF has been applied to other tissues, including subcutaneous tissue (Linhares and Kissinger 1992, 1993a, b; Janle et al. 1992a, b, 2001; Davis et al. 2005; Spehar et al. 1998; Bidgood and Papich 2005; Savenije et al. 2003), adipose tissue, blood (Savenije et al. 2003; Tiessen et al. 2001; Leegsma-Vogt et al. 2001, 2003), bone (Janle and Sojka 2000; Janle et al. 2001; Sojka et al. 2000) and muscle (Janle and Sojka 2000; Janle et al. 2001; Spehar et al. 1998; Sojka et al. 2000) to monitor LMW compounds such as lactate (Savenije et al. 2003; Tiessen et al. 1999, 2001; Leegsma-Vogt et al. 2001, 2003), glucose (Janle et al. 1992a, b; Savenije et al. 2003; Tiessen et al. 1999, 2001; Leegsma-Vogt et al. 2001, 2003) and electrolytes like magnesium, calcium, phosphorus, chloride, sodium, potassium (Linhares and Kissinger 1993a; Spehar et al. 1998; Janle and Sojka 2000; Janle et al. 2001) and other analytes/drugs (Linhares and Kissinger 1992, 1993b; Davis et al. 2005; Bidgood and Papich 2005). Only one study described MF in human subcutaneous tissue monitoring glucose and lactate (Tiessen et al. 1999). Most hollow fiber membranes have MWCO values<20 kDa, excluding macromolecules like proteins, peptides and cellular elements. However macro-analytes such as cytokines and peptide hormones may be interesting to sample systemically and/or in local tissues (Huang et al. 2006a, b, c, d; Liu and Huang, 2007; Yang and Huang 2007; Hillman et al. 2006a; Hutchinson et al. 2005a).

13.1.1.3 On-Line Detection: Biosensors

An electrochemical biosensor is defined as a self-contained integrated device, which is capable of providing specific quantitative or semi-quantitative analytical information using a biological recognition element, which is retained in direct spatial contact with an electrochemical transduction element (Thevenot et al. 2001). The biological recognition system translates information from the biochemical domain, usually an analyte concentration, into a chemical or physical output signal with a defined sensitivity. The main purpose of the recognition system is to provide the sensor with a high degree of selectivity for the analyte of interest. The transducer part, also called the detector, sensor, or electrode, of the sensor serves to transfer the signal from the output domain of the recognition system to, mostly, the electrical domain. An electrochemical biosensor is a biosensor with an electrochemical transducer. It is considered to be a chemically modified electrode as a conducting material is coated with a film encapsulating selective proteins. Although a biosensor with different transducer types, e.g. electrochemical, optical, piezoelectric or thermal type, show common features. The amperometric enzymatic glucose biosensors are most extensively studied. Here the enzyme glucose oxidase is immobilized at a charged electrode and glucose concentrations can be monitored by the change in electric current flow caused by the enzyme-catalyzed production of hydrogen peroxide (Abel et al. 1984; Bindra et al. 1991; Cremers and Ebert 2007) or – less often – by the consumption of oxygen (Armour et al. 1990). Biosensors are usually covered with a thin membrane that has several functions, reduction of interference and mechanical protection of the sensing element (Tiessen et al. 2001; Wisniewski et al. 2001; Wisniewski and Reichert 2000), but some are made without membrane (needle-type) and called microsensors (Oldenziel et al. 2006; Hu and Wilson 1997b). Biosensors are most often in the format of a needle coated with the selective enzymes that can be directly inserted in the ICS of any organ or a bodyfluid. The problem is then the reactivity of the tissue: such biosensors become rapidly encapsulated *in vivo*. To avoid such problems the sensor is combined with MD or MF. The direct environment of the sensor remains constant in a relatively clean matrix. The best studied biosensors are sensors for glucose (Tiessen et al. 2001, 2002; Rhemrev-Boom et al. 2002; Dempsey et al. 1997; Fray et al. 1996; Freaney et al. 1997; Lowry et al. 1998b, c; Rabenstein et al. 1996; Perdomo et al. 2000; Wang et al. 2002),(Dong et al. 2003; Jones et al. 2000; Yao et al. 2004; Gramsbergen et al. 2004; Gramsbergen and Cumming 2007; Zhang et al. 2004) and lactate (Tiessen et al. 2001; Gramsbergen et al. 2004; Dempsey et al. 1997; Fray et al. 1996, 1997a; Freaney et al. 1997; Perdomo et al. 2000; Rabenstein et al. 1996; Wang et al. 2002; Dong et al. 2003) (Jones et al. 2000; Yao et al. 2004; Zhang et al. 2004), but sensors for glutamate (Rutherford et al. 2007) (Zhang et al. 2004), and pyruvate (Gajovic et al. 2000) are also described with applications in mammalian brain and blood (Tiessen et al. 2001), via an extracorporeal shunt (Dempsey et al. 1997; Freaney et al. 1997; Rabenstein et al. 1996) or *in vitro* preparations, including single cells, cell cultures or tissue slices (Leegsma-Vogt et al. 2004) and reviewed by Pancrazio et al. (1999). Finally, ion-sensitive electrodes for the influx or efflux of Ca^{2+}, Na^{+}, and K^{+} ions have been widely used in both *in vivo* and *in vitro* preparations (Leegsma-Vogt et al. 2004; Kahlert and Reiser 2004; Kennedy et al. 2002; Buck and Lindner 2001; Smith et al. 1999). Preferably, the biosensor should be applied non-invasively, however such biosensors lack accuracy and researchers are thus focusing on invasive biosensors. Due to the high risk of thrombosis, embolism and septicemia, only a few intravascular sensors have been described (Armour et al. 1990) and tested (Yang et al. 1998; bdel-Hamid et al. 1995). However most glucose sensors developed for continuous monitoring do not measure blood glucose directly, but rely instead on the measurement of the glucose in the interstitial fluid of subcutaneous tissue.

13.1.1.4 Hollow Fiber Materials

The maximum size of particles that are transported over the hollow fiber membrane is defined by the MWCO value of the hollow fiber membrane. Nevertheless the hollow fiber material and build-up, the configuration and the charge of the analyte of interest may have additional effect if the analyte of interest is transported over the hollow fiber membrane. Hollow fibers are made from polymers with different compositions.

The hollow fiber is the interface between the device and the *in vivo* environment and thus can elicit a material-tissue/blood interaction. Introduction of the hollow fiber probe *in vivo* causes little disruption in the tissue, but this might not be the case for implanting the hollow fiber for longer periods of time. The invasiveness may differ depending on the location and the material of the hollow fiber. In general, an inflammatory response can be observed in all tissues. Several articles have addressed this topic in the context of MF, MD sampling and/or implanted biosensors (Wisniewski et al. 2000, 2001; Imsilp et al. 2000; Wisniewski and Reichert 2000; Ben-Yoseph et al. 1994, 1996). Considering the enormous amount of papers published in the field of MD, MF and biosensoring, relatively few studies compared systemically several hollow fiber membranes for biofouling and/or biocompatibility (tissue reactions to the implanted hollow fiber membrane).

Implantation of the MF probe made from polyacrylonitrile in sheep muscle, caused inflammation and degeneration of the surrounding tissue; the longer time elapsed the worse the magnitude of inflammation (Imsilp et al. 2000). No local edema was seen within 2 days implanting MF of the same material intramuscular in horses (Spehar et al. 1998). Subcutaneous placement of polyacrylonitrile MF in rats no inflammatory response was observed 6 weeks after implantation (Janle and Kissinger 1993). Most studies investigating hollow fiber-tissue reactions test subcutaneously in rats, for e.g.: Clark et al. (Clark et al. 2000) tested different hollow fiber membranes and harvested the fibers after 3, 6 and 12 weeks. The membranes made from polysulfone and cellulose diacetate, both with pore size 0.2 μm, were extensively degraded. Even more degraded was the polysulfone fiber with a pore size of 0.65 μm. The mixed ester cellulose with pore size 0.2 μm and regenerated cellulose (MWCO = 3 kDa) retained their integrity even after 12 weeks and degradation did not increase over time (Clark et al. 2000). Although the polysulfone with MWCO = 10 kDa remained intact, it was infiltrated by inflammatory cells and had together with the regenerated cellulose fiber (MWCO = 13 kDa) the thinnest fibrotic capsule formation (Clark et al. 2000). Expanded polytetrafluoroethylene (e-PTFE) was harvested after 4, 10 and 21 days by Zhao et al. (2000). This material was very stable and not degraded (pore size or MWCO value was not mentioned). Brauker et al. (1995) tested the same material in a similar setup with a pore size large enough to be completely penetrated by host cells (0.8–8 μm) comparing this with PTFE membranes with pore size of 0.02 μm. The larger pore size membranes had 80–100-fold more vascular structures.

Only a few studies investigated the brain tissue after implantation of a probe (Benveniste and Diemer 1987; Chen 2006; Shuaib et al. 1990; Khan and Michael 2003). Benveniste and Diemer (1987) investigated cellular reactions to implantation of a dialysis fiber (Diaflo Hollow fiber, MWCO = 50 kDa, Amicon). Within 2 days the tissue adjacent to the hollow fiber (a 50 μm border zone) exhibited edema, minor hemorrhages and accumulations of polymorphononuclear leucocytes followed by astrocyte hypertrophy (e.g. Elekes et al. 1996). Eosinophilic neurons, indicating cell death, were occasionally present within 100–150 μm from the implanted probe, but otherwise normal neuropil surrounded the hollow fiber.

An advantage of subcutaneous and brain sampling is the avoidance of the complement cascade. In MD anticoagulant can be added to the perfusate to reduce or lower the complement cascade, this can't be done with MF.

Several hemodialysis studies have described hollow fiber biofouling interactions (Hillman et al. 2005; von Baeyer et al. 1988; Engstrom et al. 2005). Hollow fibers with smaller pore sizes have greater stability in blood, than membranes with larger pore sizes. Probably proteins can easier clog the hollow fiber membranes with bigger pore size (Hillman et al. 2005; Engstrom et al. 2005). This conclusion may also apply to subcutaneously implanted fibers, not because of clogging, but because cells adhere more easily to larger pores.

For MF studies in pig heart-lactate/glucose changes by myocardial ischemia-, and rat carotid artery and jugular vein-glucose/lactate changes by cerebral ischemia (Leegsma-Vogt et al. 2001)-, heparin was administered intravenously (Tiessen et al. 2001). For intravenous MF sampling in chickens, administration of heparin was not necessary (Savenije et al. 2003). The MF probes in these studies were all made of acrylonitrilsodium methallyl sulfonate with MWCO = 20 kDa.

13.1.2 Applications

13.1.2.1 Levels in Brain ICS and Blood Circulation

A number of studies has estimated absolute concentrations of glucose and/or lactate in rat brain with concentrations ranging from 0.35 to 3.3 mM for glucose (McNay et al. 2000) and 0.35 to 1.1 mM for lactate (Kuhr et al. 1988; Shram et al. 1998; Demestre et al. 1997; Rhemrev-Boom et al. 2001). Some of these studies have directly compared extracellular glucose levels in different brain areas by the zero-net-flux method (based on perfusing with different concentrations of glucose) and showed that striatal levels are lower (0.71 mM) than in hippocampus (about 1.0 mM) (McNay et al. 2000) and that fasting overnight decreased interstitial glucose in hypothalamus from 1.42 to 0.84 mM (Mayer et al. 2006).

For clinical studies, normal levels of glucose, lactate, pyruvate and lactate/pyruvate ratio's have been determined in the human brain: 1.7–2.1 mM for glucose, 2.9–3.1 mM for lactate, 155–166 μM for pyruvate and lactate/pyruvate ratio's of 19–23 and lactate/glucose ratio's of 1.2–1.8 (Schulz et al. 2000; Reinstrup et al. 2000; Langemann et al. 2001; Abi-Saab et al. 2002).The ratio of blood/interstitial glucose has been extensively debated in the literature. Tiessen et al. (2002) studied the relationship of glucose levels in adipose tissue, loose connective tissue of the abdominal subcutis with blood glucose levels in healthy volunteers and found significantly different ratio's. Glucose levels measured in subcutaneous loose connective tissue, but not those in adipose tissue, matched plasma glucose levels (Tiessen et al. 2002). The controversy on the ratio of blood/subcutaneous interstitial glucose may thus be explained by the position of the probe/device in the subcutis. Rhemrev-Boom et al. (2002) refined the technique and developed a portable lightweight biosensor for continuous, subcutaneous glucose monitoring in humans (Rhemrev-Boom et al. 2002). The device was validated for its accuracy, linearity, sensitivity, selectivity and stability during ex vivo and *in vivo* experiments. These biosensors could be used up to 3 days in continuous mode (Rhemrev-Boom et al. 2002).

In the present context the most important conclusion is that brain ICS glucose levels are at least a factor 3 (in human brain) or 5–10 (in rat brain) lower than blood levels, whereas brain ICS lactate is usually somewhat higher than blood lactate levels (in both rat and human brain).

13.1.2.2 Effects of Drugs and Physiological Stimuli

A variety of approaches has been used to study changes of local cerebral glucose/ lactate -reflecting increased metabolism – as a result of increased neuronal activity. Local neuronal activity has been manipulated by various drugs, including pentobarbital or ketamine-xylazine induced anesthesia, local application of depolarizing agents (high K^+, veratridine), agents blocking action potentials (e.g. TTX), glutamate or glutamate receptor agonists (i.e. NMDA, AMPA) or antagonists (e.g. MK801, AP5 or NBQX) or systemic application of convulsive drugs (e.g. kainic acid, pilocarpine) or drugs of abuse (e.g. amphetamines). Effects of these agents and the references are listed in Table 13.1. Other stimuli used include tail pinch, restraint stress, foot shock, electroconvulsive shock, swimming or exercise, learning & memory tasks or local electrophysiological stimulation (Table 13.2). Effects of drugs have been studied in control animals and in animals exposed to stimuli listed in Table 13.2. Finally, MD of glucose and/or lactate has been extensively applied in acute brain injury models to study patterns of metabolic perturbations and effects of neuroprotective treatments (Table 13.3).

Of the conditions listed in Table 13.2 one of the most recent and detailed studies on the coupling between neuronal activity and energy metabolism is the study by Caesar et al. (2008), which specifically addressed the question whether lactate production is triggered by glutamate uptake or glutamate receptor activation using climbing fiber stimulation of cerebellar Purkinje cells. In the latter study glucose and lactate was monitored simultaneously every 30 s using an on-line flow injection-biosensor based detection system. Electrical stimulation of climbing fibers caused rapid lactate increases of about 30% in the molecular layer of the cerebellar cortex, but no changes in dialysate glucose levels within 30 s of stimulation. The stimulus-induced lactate increases could be blocked by the AMPA antagonist CNQX and were independent of glycogen breakdown, indicating that the lactate increases are directly derived from glucose uptake and coupled to neuronal activity through AMPA receptors, which are localized both postsynaptically on neurons (Purkinje cells) and on astrocytes.

Further, effects of drugs of abuse, including the amphetamine derivatives methamphetamine (Stephans et al. 1998) and 3,4-methylenedioxymethamphetamine (MDMA, "ecstasy") (Darvesh et al. 2002; Darvesh and Gudelsky 2003; Gramsbergen and Cumming 2007; Gudelsky and Yamamoto 2007) on brain energy metabolism have been investigated using MD of glucose and lactate. Systemic MDMA administration causes rapid glucose and lactate rises in striatum. The brain glucose changes are related to transient blood glucose elevations, whereas brain lactate remains elevated over a more prolonged period. These metabolic responses seem to be mediated by serotonin and action on 5-HT_2 receptors as discussed in Gramsbergen and Cumming (2007). Others, however, reported decreased blood glucose by high doses of MDMA using female Dark Agouti rats (Soto-Montenegro et al. 2007).

13.1.2.3 Clinical MD: Monitoring Energy Substrates

Over the last two decades cerebral MD of glucose and lactate (and pyruvate and glutamate) has been extensively used for monitoring neurointensive care patients. These studies are still primarily aimed at studying the possible use of MD parameters for the early detection of adverse metabolic changes during or after neurosurgical interventions. Two major categories of patients have been studied: (a) traumatic brain injury (TBI) and (b) subarachnoid hemorrhage (SAH). Several reviews on clinical MD in neurointensive care have appeared (Ungerstedt and Rostami 2004; Hillered et al. 2005, 2006). For clinical decision making during neurological/neurosurgical intensive care, the MD method must fulfill several demands: the MD should be standardized to allow comparison of data between hospitals, the biochemical variables should be analyzed and displayed at the bedside, the data should be quantitative and the biochemical variables should cover important aspects of cerebral metabolism, and each MD probe reflects interstitial biochemical values in small surrounding volume, the locations of the probe in relation of the lesion must be documented and defined (Engstrom et al. 2005).

Hutchinson et al. (2000) investigated different MD hollow fiber lengths, relative recoveries at different perfusion flow rates, the effect of freezing and thawing on glutamate analysis, and compared glucose, lactate, pyruvate and glutamate assays by a bedside enzyme analyzer with results obtained by HPLC analyses. Two MD probes made of polyamide were placed in the frontal cortex of patients with head injury or poor grade subarachnoid hemorrhage. The reference catheter (L = 10 mm, flow rate = 0.3 μl/min) and a probe with different lengths (L = 10 mm or 30 mm, tested flow rates 0.1, 0.3, 0.5, 0.7, 0.9, 1.1, 1.3 and 1.5 μl/min) perfused with Ringer or normal saline. It was concluded that adjacent MD probes of the same length and using the same perfusion fluid, and flow rates revealed equivalent results. Using different compositions of the perfusion fluid or freezing/thawing of the samples caused significant variation in the results. Probe lengths (10 mm) had major effect on the relative recovery which was near 70% at 0.3 μl/min and 30% at 1.0 μl/min. Hillman and coworkers (Hillman et al. 2005, 2006b, 2007; Hutchinson et al. 2005b) performed a series of elegant clinical studies. Patients who became comatose after subarachnoid hemorrhage or traumatic brain injury were inserted with paired MD polyethersulfone membranes (L = 20 mm, flow rate = 0.3 μl/min) with MWCO values of 10 and 100 kDa. The analytes of interest were glucose, glycerol, lactate, and pyruvate (Hutchinson et al. 2005a). In addition, the recovery of the HMW interleukin-6 and total protein values were determined (Hillman et al. 2005). The overall conclusion was that MD probes with a 100 kDa MWCO value can be used in the clinic, allowing sampling of cytokines and other macromolecules (Hillman et al. 2005). Properties of the 20 and 100 kDa hollow fiber were investigated for glucose, glutamate, glycerol, lactate, pyruvate, urea and total proteins (Hillman et al. 2006b). The fluid recovery was similar for both catheters; however significantly more protein was recovered with the 100 kDa catheter. The recoveries for glycerol and pyruvate were similar, while minor differences were observed for glutamate and glucose. A significant difference between the catheter types was observed for lactate (and pyruvate) recovery, which affects the lactate/pyruvate ratio (Hillman et al. 2006b).

The importance of probe location was investigated by Engström et al. (2005). This group compared cerebral energy metabolism, as an indicator of secondary excitotoxic injury and cell membrane degradation close to the focal traumatic lesions (= penumbra zones) and in remote (apparently intact) brain regions of the ipsi-lateral and contra-lateral hemispheres. Except for pyruvate the values of all biochemical variables (glucose, lactate, pyruvate, glutamate, and glycerol) and the lactate/pyruvate (L/P) ratio were significantly different in the penumbra zone when compared with mean values found in "normal" tissue ipsilateral to the parenchymal damage and in contra-lateral normal tissue ($p < 0.001$). In the penumbra zone a slow normalization of the L/P ratio and levels of glutamate and glycerol was observed. In normal tissue these parameters remained within normal limits. Concluding that data obtained from intracerebral MD can be correctly interpreted only if the locations of the catheters as they relate to focal brain lesions are visualized. A "biochemical penumbra zone" surrounds focal traumatic brain lesions. It remains to be proven whether therapeutic interventions can protect the penumbra zone from permanent damage.

In most of the clinical MD studies, samples are analyzed off-line at time intervals of 30 min or more, which makes the technique less effective for clinical monitoring and decision making during surgery or in a neurointensive care setting. A major improvement in this respect has been the introduction of rapid sampling MD (e.g. every 0.5–1 min) coupled to flow injection analysis with flow-through biosensors for glucose and lactate (Bhatia et al. 2006; Parkin et al. 2005), (see also below). In a clinical setting, MD is often combined with other multi-module parameters (e.g. O_2, pH, electrophysiology, PET), thereby providing a broader and more reliable basis for interpretation of neurochemical changes and clinical decision making (Landolt et al. 1994; Zauner et al. 1997a, b; Khaldi et al. 2001; Kett-White et al. 2002; Hlatky et al. 2004; Alves et al. 2005; Clausen et al. 2005; Vespa et al. 2005; Sakowitz et al. 2007). Since hyperglycemia after acute human brain insults (stroke, SAH or head trauma) has been associated with a worse clinical outcome or increased risk of death, a strict insulin-therapy to maintain normal blood glucose levels has been proposed and evaluated in clinical trials. However, maintaining strict euglycemia by glucose-potassium-insulin infusions in stroke patients was not associated with significant clinical benefit (Gray et al. 2007) and intensive insulin therapy to control blood glucose in SAH (Kerner et al. 2007; Schlenk et al. 2008) or head trauma patients (Vespa et al. 2006) caused potentially dangerous decreases in cerebral glucose, despite normoglycemia. In addition, it has been suggested that lower blood glucose levels may trigger peri-infarct depolarizations (PIDs), which may contribute to infarct expansion (Fabricius et al. 2006; Strong et al. 2005). Dynamic metabolic changes related to PIDs have recently been reported in the cat and human brain using rapid sampling MD (0.5–1 min) coupled to flow-injection analysis with enzymatic-amperometric detection (Parkin et al. 2005). Fast metabolic perturbations (within a few seconds) related to spreading-depression-like events cannot be detected by conventional off-line detection techniques, but may be relevant early signs of adverse metabolic events or compromised local cerebral blood flow. A recent review (Hillered et al. 2005) on the potential of cerebral MD monitoring for the early detection of adverse events in neurointensive care was rather positive. It emphasized that no other bedside technology allows the simultaneous monitoring of brain levels of lactate, glucose, glycerol, various amino acids and, last but not least, O_2 and CO_2 (the latter parameters not by MD). Because these levels exert transient changes, a single measurement does not provide clinically useful information, so only continuous monitoring informs adequately on the dynamics of metabolic processes. The current clinical brain monitoring is still in a validation phase rather than a routine diagnostic tool. One of the obstacles for routine use of clinical MD is the too laborious practice of the current monitoring technique, including handling, collecting and analyzing the samples in a 24-h setting, as required in a neurointensive care unit. Introduction of automated high throughput technologies is necessary for a wider application and appreciation of MD and MF (see below) in intensive care medicine.

13.1.2.4 Experimental MF: Monitoring Brain Energy Substrates

Thus far no articles have been published on MF in the human blood or brain. The rat brain experiments were as yet not successful (Kaptein et al. 1998). As emphasized human brain MF may be still an option as the human brain ICF is substantially larger than that of the rat. MF has been used to investigate whole brain metabolism by placing a MF probe in the aorta and jugular vein, thus measuring arterial-venous differences for glucose and lactate, after brain injury and lactate infusion (Leegsma-Vogt et al. 2003). The injury experiments (Leegsma-Vogt et al. 2001) show an increased release of lactate following increased intracranial pressure induced by an inflating balloon. Concomitant uptake of glucose from the blood circulation was decreased. The infusion experiments with lactate showed that the brain may both release and accumulate lactate. The cerebral influx or efflux of lactate depends on the relative concentrations of lactate in the blood and the adjacent brain. The lactate transport is essentially concentration driven, as could be expected considering the molecular properties of the various lactate transporters. Immediately after the infusion there was a net influx of lactate: so the arterial levels were higher than the venous concentration of lactate. The reverse was true when after prompt ending of the infusion of lactate: there was an apparent cerebral release of lactate into the circulation (Leegsma-Vogt et al. 2003). Computer modeling suggests that only a minor fraction of the accumulated lactate is metabolized (Leegsma-Vogt et al. 2004). The results suggest that the majority of lactate moving into the brain is not used as an energy substrate, and that lactate does not replace glucose as an energy source. Instead, the authors propose the concept of a lactate pool in the brain that can be filled and emptied in accordance with the blood lactate concentration, which does not serve as a major substrate for cerebral energy metabolism under normoglycemic conditions at least in anesthetized animals (Leegsma-Vogt et al. 2003, 2004).

13.1.3 Intracerebral Trafficking of Glucose and Lactate

13.1.3.1 Strategy

We give an example of how to estimate ICS steady state levels and the amounts of lactate and glucose diffusing through the ICS of the striatum of freely moving rats. This approach is based on the application of usMD (previous sections), assuming a 100% recovery *in vivo*. Here we propose a simple approach to estimate turnover rates based on the loss of glucose or lactate added to the perfusate at steady state levels. In practice, the decrease in the concentrations of lactate or glucose in the influx and efflux were measured at a constant perfusion rate of 100 nl/min. This approach assumes that higher levels of glucose do not determine the rate of glucose consumption. Such assumption is justified, among others, because acute hyperglycemia does not affect cerebral glucose consumption in conscious rats (Duckrow and Bryan 1987) (Orzi et al. 1988). During slow perfusion rate new steady state levels are achieved rapidly. The difference between the amounts infused (by retrodialysis) and collected per time unit reflects the sum of diffusion and consumption of the substrates in the ICS and brain tissue surrounding the MD-probe. Because passive diffusion may also contribute to the loss of added glucose or lactate, the present approach reveals maximal turnover rates of glucose or lactate in the ICS. Assuming a normal ICS compartment of 18% of total rat brain tissue volume (Lund-Andersen 1979; Klein et al. 1993), we calculated the maximal ICS turnover of glucose and lactate per gram wet weight tissue. On-line usMD of glucose and lactate was performed by application of adapted micro-sensor technology as described previously (Rhemrev-Boom et al. 2001).

13.1.3.2 The Experiment in Brief

Male Wistar rats were anaesthetized, the body temperature was controlled, and the rats were fixated in a stereo-tactic frame. The MD probe was implanted into the left

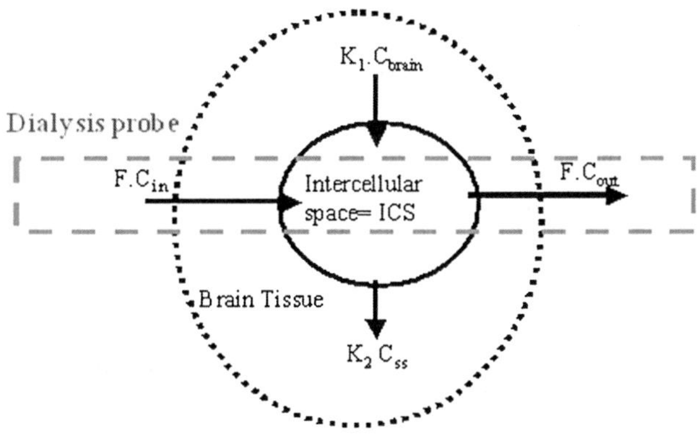

Fig. 13.2 Compartmental model used for the calculations. For definitions of the used parameters see Sect. 13.1.3.3

striatum and secured with screws and dental acrylate. The animals were allowed to recover for 24 h before measuring. Lactate and glucose were monitored continuously using flow-through nanoliter biosensors as described (Rhemrev-Boom et al. 2001). On completion of testing, rats were killed by overdose of sodiumpentobarbital for histological examination of probe placement. Animals showing signs of tissue damage other than probe track, or where the probe was placed outside the striatum were discarded from statistical analysis. Values represent means ± SEM. Statistical significance of the differences was evaluated by using the Mann-Whitney U–test or Student *t* test ($p < 0.05$).

13.1.3.3 Calculations

The calculations are based on the model as defined in Fig. 13.2. The basic idea is that under steady state conditions the outflow of glucose and lactate via the dialysate is the sum of endogenous glucose and lactate released from brain tissue into the ICS and the fraction of exogenous glucose or lactate added to the inflowing perfusate, which did not diffuse into ICS. We assume that the fraction of exogenous glucose or lactate which is released directly into the blood circulation is small, since this is only possible via its cellular transporters (Leegsma-Vogt et al. 2001). Further we assume 100% recovery at usMD (=100 nl/min).

Under steady-state conditions the following equation defines the amounts of lactate or glucose is transporte per time unit into and out of the dialysis probe. At such low perfusion rates we assume that the concentration of glucose and lactate in the (out-flowing) dialysate is the same as that in the ICS, because the diffusion over the probe membrane is not limiting.

$$k_1 * C_{brain} = k_2 * C_{ss} + F * C_{out} = (k_2 + F)C_{ss} \quad (13.1)$$

in which,

k_1 = a transport constant (or composition of various constants; arbitrarily defined) of brain glucose or lactate into the probe;

C_{brain} = apparent (both cellular and intercellular) concentration of the substrates in the brain;

So the term $K_1 * C_{brain}$ equals the amount of lactate or glucose that is released (per L) from brain tissue per time unit reaching the probe

k_2 = transport constant from the ICS into the brain tissue; at usMD we consider the probe as part of the ICS, without limitations of diffusion of glucose or lactate over the probe membrane;

C_{ss} = measured concentration under steady state conditions (without added glucose or lactate to the inflowing perfusate;

C_{out} = C_{ss} concentration glucose or lactate measured in the dialysate when no glucose or lactate was added to the perfusate.

The product $k_2 * C_{ss}$ is the amount of lactate or glucose that diffuses per time unit (minute) from the probe into brain tissue.

$$F = \text{flow rate } \left(\text{in the present experiments, set at 100nl / min}\right);$$

When glucose or lactate are added to the perfusate (C_{in}) and assuming that the rates of inflow and the outflow are identical (there is no loss of fluid to or gain of fluid from brain tissue, which is reasonable because both inflow and outflow were set identical with the perfusion pumps), and (new) steady states are reached, the amounts of the substrates transported or diffused per time unit can be described as:

$$k_1 * C_{brain} + F * C_{in} = \left(k_2 + F\right) * C_{out}, \tag{13.2}$$

C_{in} = concentration glucose or lactate added to the dialysis solution;
Substituting the term $K_1 * C_{brain}$ by combining Eqs. 13.1 and 13.2 results in:

$$\left(k_2 + F\right) * C_{ss} + F * C_{in} = F * C_{out} + k_2 * C_{out}$$

that can be rewritten as:

$$k_2 = F * \left(C_{in} - C_{out} . + C_{ss}\right) / \left(C_{out} - C_{ss}\right) \tag{13.3}$$

The constant k_2 reflects the combined rates of transport out of the extracellular compartment due to diffusion, carrier mediated transport and possibly also metabolism. This rate constant depends on the volume of the probe, which is considered part of the extracellular compartment. The volume of the probe (V_{prob}) is 0.48 μl. The volume of the ICS of the rat striatum is 180 μl/g of brain tissue (18% v/v; e.g. Korf and Postema 1988). We expressed the relative energy metabolism in glucose

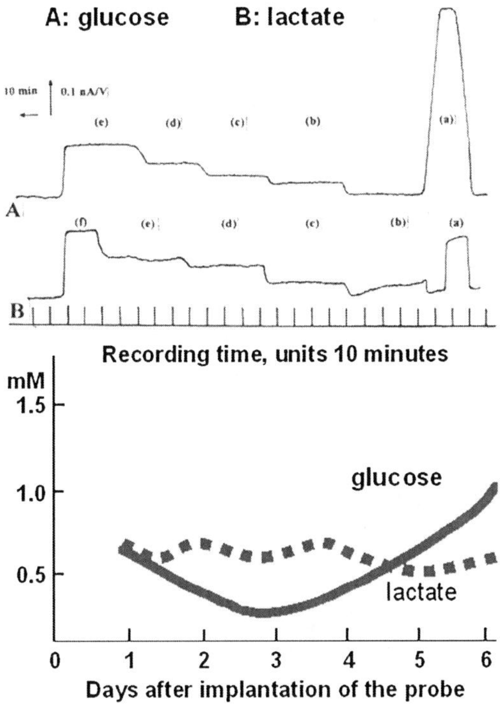

Fig. 13.3 Striatal recordings with ultraslow MD coupled to a microsensor in freely moving rats. **Upper panel**: Representative recordings of glucose (A) and lactate (B) steady state levels following infusion of exogenous glucose or lactate at day 3 or 4 after implantation of the probe. (**a**) standard solutions of 5 mM glucose or 2.5 mM lactate and striatal levels following infusion (via the probe) of glucose 1.0, 2.0, 2.5, 5.0 mM (**b–e**) or lactate 0.5, 1.5, 3.0, 4.0, 5.0 mM (**b–f**). **Lower panel**: Glucose and lactate baseline levels measured up to 6 days following implantation of the probe. Data are mean ± SEM of six rats with SEM between the thickness of the lines

equivalents (in mM/g/min); so the turnover of lactate equals ½ that of glucose. Based upon these assumptions, the flux (or turnover) of glucose or lactate per gram brain tissue (in mM/g/min) can thus be calculated as:

$$\text{Flux (or Turnover ICS)} = (180/0.48)C_{ss}(k_2 + F).$$

The percentage of metabolic trafficking via the ICS (R%) is calculated as the ratio between the calculated Flux (= turnover ICS) and the known Cerebral Metabolic Rate (CMR) for glucose in the rat striatum (0.82 μM/g/min; the mean value of the values given by Barbelivien et al. (1999), Duelli et al. (2000), Horinaka et al. (1997)* 100% as follows:

$$R\ (\%) = [(\text{Turnover ICS in glucose equivalents})/(\text{CMR})] * 100\% \quad (13.4)$$

13.1.3.4 Results

Examples of recordings of typical individual experiments are shown in Fig. 13.3, upper panel. Rat striatal steady state concentrations of ICS (C_{ss}) glucose (n = 10) and lactate (n = 6) were measured up to 6 days following the implantation of a MD probe

Table 13.4 Steady state levels, absolute turnover rates and relative turnover percentages of interstitial glucose and lactate in rat brain. Ultraslow MD was applied in the striatum of freely moving rats 3–4 days after implantation of the dialysis probe. Data are mean ± SEM, n = 6

Analyte	Glucose	Lactate	Total in glucose equivalents
Steady state levels $C_{ss\,(mM)}$	0.23 ± 0.12	0.66 ± 0.36	
Turnover ICS (μMol/g/min)	0.03 ± 0.008	0.13 ± 0.03	0.095 ± 0.02
Relative turnover (%)	3.40 ± 0.99	8.14 ± 2.39[a]	11.54 ± 2.25

[a] In glucose equivalents, different from glucose $p < 0.01$; Mann Whitney-U-test

(Fig. 13.3 lower panel. The lowest levels (±SEM) of glucose were measured at days 2, 3 and 4 and were 0.35 ± 0.07, 0.18 ± 0.04 and 0.31 ± 0.05 mM, respectively. In contrast, the lactate levels (±SEM) did not significantly vary over 4 days; these levels were at day 2, 3 and 4, 0.64 ± 0.03, 0.68 + 0.03 and 0.58 + 0.02 mM, respectively.

The turnover rates of glucose (n = 6) and lactate (n = 6) were estimated by infusing glucose (1.0, 2.0, 2.5 or 5.0 mM) or lactate (0.5, 1.5, 3.0, 4.0 or 5.0 mM) at day 3 and the other substrate at day 4. The lactate or glucose data collected at day 3 (n = 3 for each substrate) and day 4 (n = 3 for each substrate) were pooled and the transport constant k_2 and R (%) were calculated for both glucose and lactate in each individual rat using Eqs. 13.3 and 13.4, respectively. The k_2 values were 0.13/g*min for glucose and 0.2/g*min for lactate. There was no significant correlation between added substrate and calculated turnover rate (R%) of glucose or lactate. The total turnover rate of glucose and lactate expressed in glucose equivalents was estimated to be about 11.5% of total energy consumption, of which 3.4% trafficked as glucose and 8.1% as lactate. These relative turnover rates of glucose and lactate were significantly different (Table 13.4). Thus, more than 88% of the rat brain's energy substrates must be taken up directly from the brain's capillaries or from adjacent glial cells via gap junctions.

13.1.3.5 Discussion of ICF Glucose and Lactate Levels

The concentrations ranging from 0.20 to 0.90 mM for glucose and 0.5 to 0.73 mM for lactate in the striatum of freely moving rats are among the lowest reported (see introduction). The lowest ICS levels of glucose were measured at days 2, 3 and 4 after probe implantation, whereas increasingly higher levels were measured at days 5 and 6. These observations are comparable with those found by Fellows et al. (1992), who reported a 75% reduction of ICS glucose levels over 72 h following implantation of the probe, which were attributed to a disturbed glucose metabolism following implantation of the probe. According to these authors, the blood-brain-barrier is resealed within 2 h, whereas the local cerebral blood flow and glucose utilization in tissue surrounding the probe normalizes within 24 h. Groothuis et al. (1998) reported, however, that the blood-brain-barrier permeability is affected bi-phasically: a prompt increase shortly after insertion, followed by a second increase several days following placement of the probe. Measurements at longer time intervals are not recommended because of gliosis, that starts in the vicinity of the probe already within 2 days after implantation (Elekes et al. 1996; Benveniste and Huttemeier 1990). Although the blood-ICS-barrier may fluctuate over time, we never observed glucose values approaching blood levels, demonstrating that the barrier remained largely intact throughout the experiments. In contrast to glucose, no significant deviations in lactate content were observed over the 4 days after probe implantation. Following implantation the cellular composition of the brain tissue surrounding the probe will change already from day 2 with relatively more glial cells or glial processes as compared to day 0 or 1 as discussed above. Therefore it is unlikely that ICS lactate levels directly reflect local cellular metabolism in this experiment. Rather these steady state levels are determined by the characteristics of the bidirectional transporters and the average intracellular lactate levels. Brain tissue levels of lactate may remain relatively constant, because even under normoxic conditions lactate leaves the brain, particularly when local levels e.g. in glia adjacent to the vascular bed, tend to become higher than blood plasma levels (Leegsma-Vogt et al. 2001).

13.1.3.6 About the Method of Turnover Rate Estimation

A key assumption in our model is that the rate constants of disappearance of glucose or lactate from the dialysate at usMD and from the ICS are similar. Moreover, we assumed that these constants are not affected by the concentrations of the substrates. Indeed, the calculated k_2 values were concentration-independent, thereby suggesting that these values could be applied at steady state perfusion without addition of substrates. The present experimental design does not distinguish between passive diffusion or active metabolism of the added substrates. If diffusion and (already saturated) carrier-mediated transport are the most prominent routes of disappearance of the substrates, the calculated k_2 values should be concentration-independent, in particular at high concentrations of added glucose or lactate (e.g. 5 mM glucose or 5 mM lactate). Direct diffusion from the probe to blood is highly unlikely, considering the large concentration gradient of glucose that has to be overcome at any of the presently applied concentrations of glucose. In case of the lactate perfusion experiments, lactate may diffuse to the blood circulation, but assuming that the blood-brain barrier is intact, only via transport through other cells. The present experiments do not distinguish between transport unrelated to local metabolism or transport of lactate out of the brain. A direct transfer of glucose and lactate to the brain ICS is unlikely, although ICS glucose and lactate levels – after a delay – may follow blood levels as assessed by conventional MD or other techniques (e.g. measuring arterio-venous differences over brain). Fray et al. (1997a) measured reductions in extracellular glucose in rat striatum following exposure to veratridine, but did not give a turnover estimate of ICS glucose. Considering an apparent half-life of extracellular glucose of about 7.5 min, a steady state concentrations of about 0.35 mM and an extracellular space of 20%, we estimate the relative turnover rate in that study about 10% higher than our estimates in the here presented study. Possibly veratridine affects both the size of the extracellular compartment (probably decreased) and cellular processing of glucose (probably increased) as compared to our experiments.

13.1.3.7 Implications of the Extracellular Turnover Rates

The consumption of energy of the rat striatum is estimated to be 0.82 μmol/g/min (Barbelivien et al. 1999; Duelli et al. 2000; Horinaka et al. 1997). The present results suggest that about 12% of the energy metabolites traffic through the ICS, implying that the majority of energy substrates do not traffic through this compartment. Both behavioural and pharmacological investigations show that ICS levels of glucose and lactate can easily be influenced and that these effects might well be related to neuronal and glia activity (see our previous sections). The present study emphasizes that such induced variations does not necessarily reflect cerebral metabolism quantitatively and may only represent a specific brain compartment. Knowledge of contributing cells could be useful for a more precise interpretation of the relationship between metabolic en physiological brain activities. For a further discussion on these subjects see the following sections.

13.1.4 Interpretations and Concepts

13.1.4.1 On the Significance of Lactate Trafficking

In the central nervous system lactate is formed under both aerobic and anaerobic conditions. The question is whether the generated lactate following neural excitation by glutamate is immediately metabolized. According to the so called astrocyte-neuron lactate shuttle hypothesis lactate is formed following glutamate uptake in glia, then released into the intercellular compartment and subsequently used by neurons (Pellerin et al. 2007; Pellerin and Magistretti 1994). In several recent reports (van der Kuil and Korf 1991; Aubert et al. 2005) it has been assumed that lactate serves as a major substrate for oxidative metabolism during nerve cell activation *in vivo*. In this context, the interpretation of the biosensor study of Hu and Wilson (1997a, b) is crucial. In that study the perforant pathway of the rat was simulated for 5-s once or repeatedly and the time course of extracellular lactate, oxygen and glucose in the dentate gyrus of the rat brain was measured with rapidly responding biosensors. A single stimulation showed an initial drop of all the analytes, followed by transient increases. After repeated stimulations extra-cellular lactate was substantially increased, with a concomitant decrease of glucose and little, if any change of oxygen. This study and several others e.g. (De Bruin et al. 1990; Kuhr et al. 1988; van der Kuil and Korf 1991; Krugers et al. 1992; see also Tables 13.1 and 13.2) have shown that lactate is formed under conditions of enhanced neural activity which is -at least in part- associated with lower extra-cellular glucose. The time-course of hippocampus lactate following a single stimulation (lasting 5-s) (Hu and Wilson 1997a, b) and detected with the biosensors is very similar to that seen with continuous flow analysis of micro-dialysates in rats, subjected to a single electroconvulsive stimulus with an intact entorhinal-hippocampal glutamatergic pathway (Krugers et al. 1992). Modelling (Kuhr et al. 1988) shows that such a time-course can be best described by a very rapid (nearly immediate) increase of intracellular lactate that is subsequently released into the extracellular compartment via a carrier mediated process. To appreciate the results of *in vivo* studies, a few cautionary remarks should be made. It should be realized that ICS lactate levels are the net result of appearance and disappearance, which is dependent on the rates of lactate influx and efflux across the blood brain barrier, lactate influx and efflux across the membranes of neurons and glial cells, lactate formation from glucose and glycogen, lactate oxidation in the different brain cells, diffusion to and from the site of the probe or sensor, and also the volume of the ICS in brain and finally, the blood lactate concentration. Similar factors determine ICS glucose and oxygen levels, with the obvious differences in transport capacity and diffusion capability. Changes in extracellular space are emphasized below as an example, but the other factors need also careful consideration. Both electroconvulsive shock (Kuhr et al. 1988) and electrical stimulation (Hu and Wilson 1997a; Hu and Wilson 1997b) of cerebral neuronal pathways are uncommon (and perhaps even artificial) conditions, as in an intact brain it is unlikely that all neurons of a pathway fire synchronously and at the same rate. In our opinion (Korf 2006) the study of Hu and Wilson (1997a) does not support the idea that a substantial proportion of the lactate formed following neural activity, is immediately used as an aerobic substrate. Essential for a quantitative interpretation of the biosensor recordings is that the diffusion of analytes remains constant in the extracellular compartment. Shrinkage of the extra-cellular compartment, as a consequence of neuronal activity and associated swelling, has been reported in models of cerebral ischemia (see e.g. Klein et al. [1993], Zoremba et al. [2007]) and status epilepticus (Slais et al. 2008). Since shrinkage of the extra-cellular space affects lactate diffusion towards and into the dialysis probe, we conclude that the initial dip in lactate as reported in the Hu and Wilson study is probably not due to oxidative metabolism of lactate, but rather to changes in lactate diffusion. Hu and Wilson (1997a, b) reported that the slow transient decrease in the levels of extracellular glucose following stimulation is concomitant with a longer lasting increase in lactate levels and constant oxygen levels. This may indeed point to the possible utilization of lactate instead of glucose as an oxidative substrate 10 min after the onset of stimulation. In any case, however, the possible oxidative metabolism of lactate is a relatively slow pathway.

A consistent interpretation of all the mentioned studies is that most of the glucose used through the glycolytic pathway, proceeds directly through oxidative metabolism, and that some glucose is converted into lactate, in particular during high neuronal activity. *In vivo* estimates of the turnover of lactate and glucose in the freely moving rat suggest that a minor proportion of these substrates traffic through the extra-cellular space (previous section). It is likely that a substantial proportion of cerebrally formed lactate disappears (un-metabolized) via the blood circulation (Leegsma-Vogt et al. 2003). This may also explain the well-known discrepancy between glucose and oxygen uptake and metabolism in brain during activation (Leegsma-Vogt et al. 2004).

Our results (previous section) taken together suggest that under physical rest less than 12% of the glucose and lactate consumed by the brain is transported via the ICS.

It can be argued that our experiments are done at low activity of e.g. the glutaminergic neurotransmission, thus underestimating the contribution of ICS glucose and lactate to energy metabolism in the activated brain. Rat striatal ICS-lactate is increased during physical exercise to a (maximum) increase of only 30% (De Bruin et al. 1990) (see for other ref. Table 13.2). At least part of the increase was mediated by glutamatergic innervations (Korf et al. 1988; Korf and de Boer 1990). These observations suggest that the lactate response is due to both released glutamate and glutamate receptor mediated signaling. A recent paper by Caesar et al. (2008) measuring fast changes of glucose and lactate in the cerebellar cortex upon mossy fiber stimulations is essentially in line with the conclusions by Attwell and Laughlin (2001), since application of the AMPA antagonist NBQX completely blocked the stimulation induced lactate rises. This study (Caesar et al. 2008) emphasizes that glutamate release does not necessarily cause lactate formation in glia, but rather that lactate is primarily formed by activation of the postsynaptic neuron. In neurons postsynaptic to glutamate nerve endings often a high affinity lactate transporter (MCT2) is localized (Bergersen 2007). This high affinity transporter has often been considered to support the hypothesis that neurons can efficiently accumulate lactate, released from glia for aerobic metabolism. Alternatively, we propose that such a high affinity transporter may also serve to export neuronal lactate to the ICS instead, in order to keep intracellular lactate at a low level. A new concept of brain metabolism that acknowledges these findings may include a less strict cellular localization of sequential metabolic processes. Rather, the cellular heterogeneity of the brain should be taken into account to explain simultaneously operating metabolic pathways.

13.1.4.2 Brain Energy and Brain Events in Space and Time

In the introduction we referred to the suggested tight coupling between neuronal activity and cerebral blood flow or increased energy metabolism. In this section we will discuss in some detail the question for which brain processes (increased) energy metabolism is required, and will draw the conclusion that the coupling between energy utilization and brain function may not be as strict as usually assumed. A more complete discussion on this issue has recently been published (Korf and Gramsbergen 2007).

Neurotransmission and Action Potentials

Action potentials and glutamatergic and γ−amino butyric acid (GABA)-ergic neurotransmission are generally considered the fastest processes enabling intercellular communication. A single action potential lasts 2–5 ms and may be followed by a refractory period of about 4 ms. Firing frequencies of cerebral neurons range from less than 1 up to 200 Hz; the latter frequency points to action potentials of 5 ms duration. Fast excitatory and inhibitory neurotransmission by glutamate and GABA are also in the millisecond range. Attwell and Gibb (2005) summarized the kinetic properties and affinities of the AMPA and NMDA receptors and of glutamate transporters, and concluded that glutamate transmission requires only a few milliseconds to proceed. Similarly, the inhibitory action of GABA by opening Cl^- channels via the $GABA_A$ (and $GABA_C$)-receptor requires less than 5 ms. Neurotransmitters are stored in synaptic vesicles. Vesicular glutamate uptake is ATP-dependent (Storm-Mathisen 1977). ATP and Ca^{2+} are required for the fusion of vesicles with the presynaptic membrane (Huang and Bergles 2004) and release of vesicular transmitter is in the range of m*s* (Markram et al. 1997; Felmy et al. 2003). Assuming that released GABA and glutamate are predominantly metabolized and that 1 ATP is required for vesicular uptake and storage, the cerebral (metabolic) turnover of these amino acids indicates how much ATP is required for fast neurotransmission. This has been estimated to be about 2% of the total energy budget of the brain (Attwell and Gibb 2005). If we assume that the amino acid terminals occupy less than 1% of the brain volume, the excitatory nerve terminals use a major proportion of local metabolic energy to keep the vesicles filled up. Similarly, in GABA-ergic terminals a major part of energy metabolism may be required for vesicular transmitter storage. In addition to ionotropic glutamate and GABA receptors, fast neurotransmission is also possible through nicotinic acetylcholine, strychnine-sensitive glycine and $5HT_3$ serotonin receptors (Chebib and Johnston 1999).

The large trans-membrane gradients of ions and fast transmitters allow synaptic release and postsynaptic effects within 2 to 10 ms with a delay in transmitter synthesis and vesicular accumulation.

Nerve Activity and *In Vivo* Metabolic Response

Physiological stimuli or electrical stimulation of nerve fibers *in vivo* or *in vitro* have been used to explore the consequences of enhanced impulse flow on energy metabolism. There is little question, whether during (or following) neuronal activation the consumption of glucose and oxygen is increased.

A decrease in brain (cerebellum and hippocampus) oxygen tensions (presumably reflecting increased consumption) has been observed within a few seconds after excitatory stimulation. Such studies show that the coupling of oxygen consumption and electrophysiological activity is not linear. The energy consumption (glucose and/or oxygen) is only marginally increased at relatively low stimulation frequencies and non-linearly at higher stimulation frequencies. The site of increased energy consumption differs often from that of direct activation (Vafaee and Gjedde 2004). One explanation could be the relatively slow transport of glucose as compared to oxygen. Another possibility is the consumption of lactate or acetate as an alternative energy substrate (Hertz and Dienel 2005; Ames III 2000). Several experiments have shown fast anaerobic glucose metabolism following enhanced neuronal activity (previous section) (Kuhr et al. 1988; Krugers et al. 1992; Hu and Wilson 1997a; Gjedde et al. 2002). Cerebral lactate transporters are essentially bidirectional, thus questioning the idea of a strong coupling between neuronal lactate metabolism and glutamatergic transmission. So the intercellular movement of lactate is mainly concentration-driven and a directed flow of lactate into particular neurons is highly unlikely (Korf 2006). Accordingly, a large proportion of lactate leaves the brain un-metabolized when lactate levels in capillary blood are lower than the lactate levels in adjacent cells.

The Na^+-K^+-membrane-pumps do not always have fast access to mitochondrial energy substrates (e.g. ATP). Thompson et al. (2005) assessed the relationship between the temporal and spatial changes in blood flow and oxygen changes with a locally placed oxygen electrode. After increased focal activity fast negative dips of oxygen tension following visual stimulation in the cat dentate gyrus or cerebral cortex were observed near the stimulating electrode. During widespread activity, the oxygen dip was less pronounced, occurred later and predominantly increases in oxygen tension were seen. The latter changes are rather slow as compared to cellular activity in the visual cortex. As in the case of cerebral blood flow, the late overshoot in metabolic activity occurs always several seconds after the enhanced activity of neurons that triggered these changes. The initial dip in O_2 tension occurring very soon following neuronal activity may be attributed to the utilization of oxygen that is not compensated by the blood supply and diffusion through the tissue. Changes in the cellular and intercellular volume may also explain at least some of the changes in the oxygen signal (Ances 2004; Korf 2006). The dip shows early deoxygenating of hemoglobin within 100 ms, but no changes in CBF or blood volume (Ances 2004). Together, these studies emphasize that after neuronal activation the consumption of freely dissolved oxygen and other energy substrates starts within 100 ms.

Potential Versus Metabolic Energy

The shortest physiological activities involved in inter-neuronal communication in the peripheral and central nervous system are action potentials and excitatory and inhibitory neurotransmission predominantly by glutamate and GABA, lasting 2–10 ms. Activities, such as reflexes, speaking, recognition and complicated information processing of the brain does most often require less than 100 ms. Psychophysiological studies illustrate that fast physiological processes do not primarily depend on extra-cerebral energy substrate recruitment. Moreover, consumption of energy reserves (such as glycogen) is too slow to be available for immediate use. Changes in cerebral glucose uptake and cerebral blood flow, which helps delivering and removing energy substrates to and from the brain, are also relatively slow processes. However, an increased O_2 consumption is already observed within 100 ms following neuronal activity and is therefore most sensitive indicator of increased cerebral energy utilization. The brain is best prepared for fast activities in the *ms* range, when the trans-membrane gradients of the ions (Na^+, K^+ and Cl^-) are kept maximal and when the vesicle stores (of GABA and glutamate) are maximally filled. These concentration gradients are the exclusive prerequisite to perform psycho-physiological functions and may be denoted potential energy of the brain. Brain energy metabolism must be primarily considered a restorative activity: i.e. to restore and maintain the potential energy of the brain at the highest level.

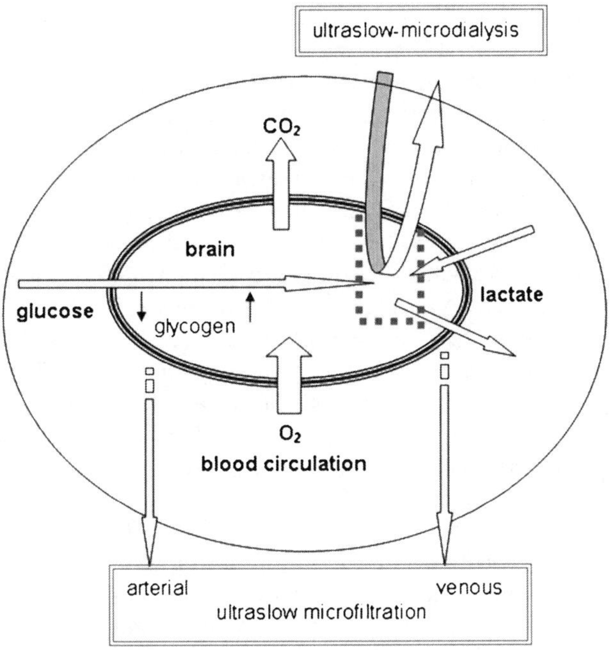

Fig. 13.4 Intercellular trafficking of glucose and lactate as studied with ultraslow intravascular microfiltration (MF) and ultraslow intracerebral microdialysis (MD). Most of the glucose taken up by the brain is directly transported to the oxidative pathways in both neurons and glial cells. A minor part of glucose is stored temporarily as glycogen in astroglia, which can be released as lactate. Lactate produced in the brain, predominantly by astroglia will be oxidized (minor part), predominantly by neurons and oligodendroglia or transported out of the brain (major part). Depending on the concentration gradient of lactate over the blood-brain barrier, lactate enters or leaves the brain

13.1.5 Concluding Remarks

Quantitative MD of glucose and lactate in rat and human brain has revealed much lower glucose levels in the brain ICS than in blood, whereas ICS lactate levels are elevated or similar as compared to blood levels. The various issues discussed in the present paper are shown in Fig. 13.4. In many of the neuronal activation paradigms tested in experimental animals, lactate elevations are faster and more pronounced than glucose perturbations, but the fastest detected changes (e.g. at 30 s intervals) are still slow as compared to the electrophysiological events, which occur on a millisecond time scale. Activation-induced ICS lactate and glucose changes are predominantly mediated through ionotropic glutamate receptors (NMDA and AMPA type). By using usMD coupled to biosensors for glucose and lactate and perfusion with exogenous glucose or lactate, we determined the turnover rates of glucose and lactate in the brain ICS of freely moving rats and found that less than 12% of total glucose metabolism can be attributed to diffusion of glucose (3–4%) and lactate (8%) via the intercellular compartment.

If activated neurons would preferably consume lactate produced by astrocytes, a directed lactate flow from glia to neurons and a high turnover rate would be expected, which is clearly not the case under normal conditions. Steady state levels of lactate, but not of glucose, is essentially independent on the cell types surrounding the probe, suggesting a dominant role of transporters, rather than cellular metabolism. In this review we emphasize that lactate transporters on both neurons and glia are essentially bidirectional and that ICS lactate is derived from both (postsynaptic) neurons and glia. Under normal conditions elevated ICS lactate is predominantly removed by the blood circulation. It remains to be shown whether under conditions of increased neuronal activity (e.g. convulsions or exercise) or limited glucose supply (e.g. hypoglycaemia), ICS lactate turnover rates are increased.

Advances in nanosensor technology may in the future allow simultaneous, real-time recording of intra- and extracellular glucose and lactate at the synaptic level and studies of substrate flow through neuronal or glial transporters. However, such studies can only be performed *in vitro* with no blood supply and abnormal concentrations of energy metabolites in the ICS.

For clinical studies the recently low time resolution MD sampling at 0.5 or 1 min intervals and on-line detection by a FIA-biosensor set-up is most promising, since this may allow early interventions to prevent expansion of ischemic injury. In such studies the position of the probe(s) in relation to the trauma, ischemic core or operated blood vessels (in SAH patients) is of course of paramount importance.

Finally, we discussed brain energy metabolism at a more theoretical level and provided arguments to distinguish potential and metabolic brain energy. Potential brain energy is considered a necessary condition for fast neuronal processing, whereas metabolic brain energy serves to restore high levels of potential energy.

Acknowledgements Support of the Commission of the European Communities, Director General for DG XII Science, Research and Development, Biomed 2 Program PL-972726, and of Dutch Technology Foundation (STW; grants GGN 4680 and GPG 6038) is gratefully acknowledged.

References

Abel P, Muller A, Fischer U (1984) Experience with an implantable glucose sensor as a prerequisite of an artificial beta cell. Biomed Biochim Acta 43(5):577–584

Abi-Saab WM, Maggs DG, Jones T, Jacob R, Srihari V, Thompson J, Kerr D, Leone P, Krystal JH, Spencer DD, During MJ, Sherwin RS (2002) Striking differences in glucose and lactate levels between brain extracellular fluid and plasma in conscious human subjects: effects of hyperglycemia and hypoglycemia. J Cereb Blood Flow Metab 22(3):271–279

Alessandri B, Basciani R, Langemann H, Lyrer P, Pluess D, Landolt H, Gratz O (2000) Chronic effects of an aminosteroid on microdialytically measured parameters after experimental middle cerebral artery occlusion in the rat. J Clin Neurosci 7(1):47–51

Alessandri B, Landolt H, Langemann H, Gregorin J, Hall J, Gratzl O (1996) Application of glutamate in the cortex of rats: a microdialysis study. Acta Neurochir Suppl 67:6–12

Allen NJ, Karadottir R, Attwell D (2005) A preferential role for glycolysis in preventing the anoxic depolarization of rat hippocampal area CA1 pyramidal cells. J Neurosci 25(4):848–859

Alves OL, Bullock R, Clausen T, Reinert M, Reeves TM (2005) Concurrent monitoring of cerebral electrophysiology and metabolism after traumatic brain injury: an experimental and clinical study. J Neurotrauma 22(7):733–749

Ames A III (2000) CNS energy metabolism as related to function. Brain Res Brain Res Rev 34(1–2):42–68

Ances BM (2004) Coupling of changes in cerebral blood flow with neural activity: what must initially dip must come back up. J Cereb Blood Flow Metab 24(1):1–6

Ao X, Stenken JA (2006) Microdialysis sampling of cytokines. Methods 38(4):331–341

Armour JC, Lucisano JY, McKean BD, Gough DA (1990) Application of chronic intravascular blood glucose sensor in dogs. Diabetes 39(12):1519–1526

Attwell D, Gibb A (2005) Neuroenergetics and the kinetic design of excitatory synapses. Nat Rev Neurosci 6(11):841–849

Attwell D, Laughlin SB (2001) An energy budget for signaling in the grey matter of the brain. J Cerebral Blood Flow Metab: Official J Int Soc Cerebral Blood Flow Metab 21(10):1133–1145

Aubert A, Costalat R, Magistretti PJ, Pellerin L (2005) Brain lactate kinetics: modeling evidence for neuronal lactate uptake upon activation. Proc Natl Acad Sci USA 102(45):16448–16453

Barbelivien A, Bertrand N, Besret L, Beley A, MacKenzie ET, Dauphin F (1999) Neurochemical stimulation of the rat substantia innominata increases cerebral blood flow (but not glucose use) through the parallel activation of cholinergic and non-cholinergic pathways. Brain Res 840(1–2):115–124

bdel-Hamid I, Atanasov P, Atanasov P, Wilkins E (1995) Development of a needle-type biosensor for intravascular glucose monitoring. Anal Chim Acta 313(1–2):45–54

Bentzer P, Davidsson H, Grande PO (2000) Microdialysis-based long-term measurements of energy-related metabolites in the rat brain following a fluid percussion trauma. J Neurotrauma 17(5):441–447

Ben-Yoseph O, Boxer PA, Ross BD (1994) Oxidative stress in the central nervous system: monitoring the metabolic response using the pentose phosphate pathway. Dev Neurosci 16(5–6):328–336

Ben-Yoseph O, Boxer PA, Ross BD (1996) Noninvasive assessment of the relative roles of cerebral antioxidant enzymes by quantitation of pentose phosphate pathway activity. Neurochem Res 21(9):1005–1012

Benveniste H, Diemer NH (1987) Cellular reactions to implantation of a microdialysis tube in the rat hippocampus. Acta Neuropathol (Berl) 74(3):234–238

Benveniste H, Huttemeier PC (1990) Microdialysis–theory and application. Prog Neurobiol 35(3):195–215

Bergersen LH (2007) Is lactate food for neurons? Comparison of monocarboxylate transporter subtypes in brain and muscle. Neuroscience 145(1):11–19

Bhatia R, Hashemi P, Razzaq A, Parkin MC, Hopwood SE, Boutelle MG, Strong AJ (2006) Application of rapid-sampling, online microdialysis to the monitoring of brain metabolism during aneurysm surgery. Neurosurgery 584(Suppl 2):ONS-20

Bidgood TL, Papich MG (2005) Plasma and interstitial fluid pharmacokinetics of enrofloxacin, its metabolite ciprofloxacin, and marbofloxacin after oral administration and a constant rate intravenous infusion in dogs. J Vet Pharmacol Ther 28(4):329–341

Bindra DS, Zhang Y, Wilson GS, Sternberg R, Thevenot DR, Moatti D, Reach G (1991) Design and in vitro studies of a needle-type glucose sensor for subcutaneous monitoring. Anal Chem 63(17):1692–1696

Brauker JH, Carr-Brendel VE, Martinson LA, Crudele J, Johnston WD, Johnson RC (1995) Neovascularization of synthetic membranes directed by membrane microarchitecture. J Biomed Mater Res 29(12):1517–1524

Buck RP, Lindner E (2001) Tracing the history of selective ion sensors. Anal Chem 73(3):88A–97A

Caesar K, Hashemi P, Douhou A, Bonvento G, Boutelle MG, Walls AB, Lauritzen M (2008) Glutamate receptor-dependent increments in lactate, glucose and oxygen metabolism evoked in rat cerebellum in vivo. J Physiol 586(5):1337–1349

Canal CE, McNay EC, Gold PE (2005) Increases in extracellular fluid glucose levels in the rat hippocampus following an anesthetic dose of pentobarbital or ketamine-xylazine: an in vivo microdialysis study. Physiol Behav 84(2):245–250

Chebib M, Johnston GA (1999) The 'ABC' of GABA receptors: a brief review. Clin Exp Pharmacol Physiol 26(11):937–940

Chen KC (2006) Effects of tissue trauma on the characteristics of microdialysis zero-net-flux method sampling neurotransmitters. J Theor Biol 238(4):863–881

Chen T, Qian YZ, Di X, Rice A, Zhu JP, Bullock R (2000) Lactate/glucose dynamics after rat fluid percussion brain injury. J Neurotrauma 17(2):135–142

Clark H, Barbari TA, Stump K, Rao G (2000) Histologic evaluation of the inflammatory response around implanted hollow fiber membranes. J Biomed Mater Res 52(1):183–192

Clausen T, Khaldi A, Zauner A, Reinert M, Doppenberg E, Menzel M, Soukup J, Alves OL, Bullock MR (2005) Cerebral acid-base homeostasis after severe traumatic brain injury. J Neurosurg 103(4):597–607

Clausen T, Zauner A, Levasseur JE, Rice AC, Bullock R (2001) Induced mitochondrial failure in the feline brain: implications for understanding acute post-traumatic metabolic events. Brain Res 908(1):35–48

Corbett R, Laptook A, Kim B, Tollefsbol G, Silmon S, Garcia D (1999) Maturational changes in cerebral lactate and acid clearance following ischemia measured in vivo using magnetic resonance spectroscopy and microdialysis. Brain Res Dev Brain Res 113(1–2):37–46

Cremers T, Ebert B (2007) Plasma and CNS concentrations of Gaboxadol in rats following subcutaneous administration. Eur J Pharmacol 562(1–2):47–52

Darbin O, Carre E, Naritoku D, Risso JJ, Lonjon M, Patrylo PR (2006) Glucose metabolites in the striatum of freely behaving rats following infusion of elevated potassium. Brain Res 1116(1):127–131

Darvesh AS, Gudelsky GA (2003) Activation of 5-HT2 receptors induces glycogenolysis in the rat brain. Eur J Pharmacol 464(2–3):135–140

Darvesh AS, Shankaran M, Gudelsky GA (2002) 3,4-Methylenedioxymethamphetamine produces glycogenolysis and increases the extracellular concentration of glucose in the rat brain. J Pharmacol Exp Ther 301(1):138–144

Davis JL, Salmon JH, Papich MG (2005) Pharmacokinetics and tissue fluid distribution of cephalexin in the horse after oral and i.v. administration. J Vet Pharmacol Ther 28(5):425–431

De Bruin LA, Schasfoort EM, Steffens AB, Korf J (1990) Effects of stress and exercise on rat hippocampus and striatum extracellular lactate. Am J Physiol 259(4(Pt 2)):R773–R779

Delgado JM, DeFeudis FV, Roth RH, Ryugo DK, Mitruka BM (1972) Dialytrode for long term intracerebral perfusion in awake monkeys. Arch Int Pharmacodyn Ther 198(1):9–21

Demestre M, Boutelle M, Fillenz M (1997) Stimulated release of lactate in freely moving rats is dependent on the uptake of glutamate. J Physiol 499(Pt 3):825–832

Dempsey E, Diamond D, Smyth MR, Malone MA, Rabenstein K, McShane A, McKenna M, Keaveny TV, Freaney R (1997) In vitro optimisation of a microdialysis system with potential for on-line monitoring of lactate and glucose in biological samples. Analyst 122(2):185–189

Dienel GA, Hertz L (2001) Glucose and lactate metabolism during brain activation. J Neurosci Res 66:824–838

Dijk S, Krugers HJ, Korf J (1991) The effect of theophylline and immobilization stress on haloperidol-induced catalepsy and on metabolism in the striatum and hippocampus, studied with lactography. Neuropharmacology 30(5):469–473

Dong Y, Wang L, Shangguan D, Yu X, Zhao R, Han H, Liu G (2003) Analysis of glucose and lactate in hippocampal dialysates of rats during the operant conditioned reflex using microdialysis. Neurochem Int 43(1):67–72

Dringen R, Wiesinger H, Hamprecht B (1993) Uptake of L-lactate by cultured rat brain neurons. Neurosci Lett 163(1):5–7

Duckrow RB, Bryan RM Jr (1987) Regional cerebral glucose utilization during hyperglycemia. J Neurochem 48(3):989–993

Duelli R, Maurer MH, Staudt R, Heiland S, Duembgen L, Kuschinsky W (2000) Increased cerebral glucose utilization and decreased glucose transporter Glut1 during chronic hyperglycemia in rat brain. Brain Res 858(2):338–347

Ekberg NR, Wisniewski N, Brismar K, Ungerstedt U (2005) Measurement of glucose and metabolites in subcutaneous adipose tissue during hyperglycemia with microdialysis at various perfusion flow rates. Clin Chim Acta 359(1–2):53–64

Elekes O, Venema K, Postema F, Dringen R, Hamprecht B, Korf J (1996) Evidence that stress activates glial lactate formation in vivo assessed with rat hippocampus lactography. Neurosci Lett 208(1):69–72

Enblad P, Frykholm P, Valtysson J, Silander HC, Andersson J, Fasth KJ, Watanabe Y, Langstrom B, Hillered L, Persson L (2001) Middle cerebral artery occlusion and reperfusion in primates monitored by microdialysis and sequential positron emission tomography. Stroke 32(7):1574–1580

Engstrom M, Polito A, Reinstrup P, Romner B, Ryding E, Ungerstedt U, Nordstrom CH (2005) Intracerebral microdialysis in severe brain trauma: the importance of catheter location. J Neurosurg 102(3):460–469

Fabricius M, Fuhr S, Bhatia R, Boutelle M, Hashemi P, Strong AJ, Lauritzen M (2006) Cortical spreading depression and peri-infarct depolarization in acutely injured human cerebral cortex. Brain 129(Pt 3):778–790

Fellows LK, Boutelle MG (1993) Rapid changes in extracellular glucose levels and blood flow in the striatum of the freely moving rat. Brain Res 604(1–2):225–231

Fellows LK, Boutelle MG, Fillenz M (1992) Extracellular brain glucose levels reflect local neuronal activity: a microdialysis study in awake, freely moving rats. J Neurochem 59(6):2141–2147

Fellows LK, Boutelle MG, Fillenz M (1993) Physiological stimulation increases nonoxidative glucose metabolism in the brain of the freely moving rat. J Neurochem 60(4):1258–1263

Felmy F, Neher E, Schneggenburger R (2003) The timing of phasic transmitter release is Ca2+–dependent and lacks a direct influence of presynaptic membrane potential. Proc Natl Acad Sci USA 100(25):15200–15205

Fillenz M, Lowry JP (1998) Studies of the source of glucose in the extracellular compartment of the rat brain. Dev Neurosci 20(4–5):365–368

Forsyth RJ (1996) Astrocytes and the delivery of glucose from plasma to neurons. Neurochem Int 28(3):231–241

Forsyth R, Fray A, Boutelle M, Fillenz M, Middleditch C, Burchell A (1996) A role for astrocytes in glucose delivery to neurons? Dev Neurosci 18(5–6):360–370

Fray AE, Boutelle M, Fillenz M (1997a) Extracellular glucose turnover in the striatum of unanaesthetized rats measured by quantitative microdialysis. J Physiol 504(Pt 3):721–726

Fray AE, Boutelle M, Fillenz M (1997b) Extracellular glucose turnover in the striatum of unanaesthetized rats measured by quantitative microdialysis. J Physiol 504(Pt 3):721–726

Fray AE, Forsyth RJ, Boutelle MG, Fillenz M (1996) The mechanisms controlling physiologically stimulated changes in rat brain glucose and lactate: a microdialysis study. J Physiol 496(Pt 1):49–57

Freaney R, McShane A, Keaveny TV, McKenna M, Rabenstein K, Scheller FW, Pfeiffer D, Urban G, Moser I, Jobst G, Manz A, Verpoorte E, Widmer MW, Diamond D, Dempsey E, de Saez V, Smyth M (1997) Novel instrumentation for real-time monitoring using miniaturized flow systems with integrated biosensors. Ann Clin Biochem 34(Pt 3):291–302

Frykholm P, Hillered L, Langstrom B, Persson L, Valtysson J, Enblad P (2005) Relationship between cerebral blood flow and oxygen metabolism, and extracellular glucose and lactate concentrations during middle cerebral artery occlusion and reperfusion: a microdialysis and positron emission tomography study in nonhuman primates. J Neurosurg 102(6): 1076–1084

Fuxe K, Kurosawa N, Cintra A, Hallstrom A, Goiny M, Rosen L, Agnati LF, Ungerstedt U (1992) Involvement of local ischemia in endothelin-1 induced lesions of the neostriatum of the anaesthetized rat. Exp Brain Res 88(1):131–139

Gajovic N, Beinyamin G, Warsinke A, Scheller FW, Heller A (2000) Operation of a miniature redox hydrogel-based pyruvate sensor in undiluted deoxygenated calf serum. Anal Chem 72(13):2963–2968

Geeraerts T, Friggeri A, Mazoit JX, Benhamou D, Duranteau J, Vigue B (2008) Posttraumatic brain vulnerability to hypoxia-hypotension: the importance of the delay between brain trauma and secondary insult. Intensive Care Med 34(3):551–560

Geeraerts T, Ract C, Tardieu M, Fourcade O, Mazoit JX, Benhamou D, Duranteau J, Vigue B (2006) Changes in cerebral energy metabolites induced by impact-acceleration brain trauma and hypoxic-hypotensive injury in rats. J Neurotrauma 23(7):1059–1071

Gibbs ME, Anderson DG, Hertz L (2006) Inhibition of glycogenolysis in astrocytes interrupts memory consolidation in young chickens. Glia 54(3):214–222

Gjedde A, Marrett S, Vafaee M (2002) Oxidative and nonoxidative metabolism of excited neurons and astrocytes. J Cereb Blood Flow Metab 22(1):1–14

Gramsbergen JB, Cumming P (2007) Serotonin mediates rapid changes of striatal glucose and lactate metabolism after systemic 3,4-methylenedioxymethamphetamine (MDMA, "Ecstasy") administration in awake rats. Neurochem Int 51(1):8–15

Gramsbergen JB, Skjoth-Rasmussen J, Rasmussen C, Lambertsen KL (2004) On-line monitoring of striatum glucose and lactate in the endothelin-1 rat model of transient focal cerebral ischemia using microdialysis and flow-injection analysis with biosensors. J Neurosci Methods 140(1–2):93–101

Gray CS, Hildreth AJ, Sandercock PA, O'Connell JE, Johnston DE, Cartlidge NE, Bamford JM, James OF, Alberti KG (2007) Glucose-potassium-insulin infusions in the management of post-stroke hyperglycaemia: the UK Glucose Insulin in Stroke Trial (GIST-UK). Lancet Neurol 6(5):397–406

Groothuis DR, Ward S, Schlageter KE, Itskovich AC, Schwerin SC, Allen CV, Dills C, Levy RM (1998) Changes in blood-brain barrier permeability associated with insertion of brain cannulas and microdialysis probes. Brain Res 803(1–2):218–230

Gudelsky GA, Yamamoto BK (2007) Actions of 3,4-methylenedioxymethamphetamine (MDMA) on cerebral dopamineric, serotonergic and cholinergic neurons. Pharmacol Biochem Behav 87(4):426–433

Hamrin K, Rosdahl H, Ungerstedt U, Henriksson J (2002) Microdialysis in human skeletal muscle: effects of adding a colloid to the perfusate. J Appl Physiol 92(1):385–393

Harada M, Okuda C, Sawa T, Murakami T (1992) Cerebral extracellular glucose and lactate concentrations during and after moderate hypoxia in glucose- and saline-infused rats. Anesthesiology 77(4):728–734

Hertz L, Dienel GA (2005) Lactate transport and transporters: general principles and functional roles in brain cells. J Neurosci Res 79(1–2):11–18

Hertz L, Peng L, Dienel GA (2007) Energy metabolism in astrocytes: high rate of oxidative metabolism and spatiotemporal dependence on glycolysis/glycogenolysis. J Cereb Blood Flow Metab 27(2):219–249

Hillered L, Persson L, Nilsson P, Ronne-Engstrom E, Enblad P (2006) Continuous monitoring of cerebral metabolism in traumatic brain injury: a focus on cerebral microdialysis. Curr Opin Crit Care 12(2):112–118

Hillered L, Vespa PM, Hovda DA (2005) Translational neurochemical research in acute human brain injury: the current status and potential future for cerebral microdialysis. J Neurotrauma 22(1):3–41

Hillman J, Aneman O, Anderson C, Sjogren F, Saberg C, Mellergard P (2005) A microdialysis technique for routine measurement of macromolecules in the injured human brain. Neurosurgery 56(6):1264–1268

Hillman J, Aneman O, Persson M, Andersson C, Dabrosin C, Mellergard P (2007) Variations in the response of interleukins in neurosurgical intensive care patients monitored using intracerebral microdialysis. J Neurosurg 106(5):820–825

Hillman J, Milos P, Yu ZQ, Sjogren F, Anderson C, Mellergard P (2006a) Intracerebral microdialysis in neurosurgical intensive care patients utilising catheters with different molecular cut-off (20 and 100 kD). Acta Neurochir (Wien) 148(3):319–324

Hillman J, Milos P, Yu ZQ, Sjogren F, Anderson C, Mellergard P (2006b) Intracerebral microdialysis in neurosurgical intensive care patients utilising catheters with different molecular cut-off (20 and 100 kD). Acta Neurochir 148(3):319–324 (Wien)

Hlatky R, Valadka AB, Goodman JC, Robertson CS (2004) Evolution of brain tissue injury after evacuation of acute traumatic subdural hematomas. Neurosurgery 55(6):1318–1323

Homola A, Zoremba N, Slais K, Kuhlen R, Sykova E (2006) Changes in diffusion parameters, energy-related metabolites and glutamate in the rat cortex after transient hypoxia/ischemia. Neurosci Lett 404(1–2):137–142

Horinaka N, Artz N, Cook M, Holmes C, Goldstein DS, Kennedy C, Sokoloff L (1997) Effects of elevated plasma epinephrine on glucose utilization and blood flow in conscious rat brain. Am J Physiol 272(4(Pt 2)):H1666–H1671

Hu Y, Wilson GS (1997a) A temporary local energy pool coupled to neuronal activity: fluctuations of extracellular lactate levels in rat brain monitored with rapid-response enzyme-based sensor. J Neurochem 69(4):1484–1490

Hu Y, Wilson GS (1997b) Rapid changes in local extracellular rat brain glucose observed with an in vivo glucose sensor. J Neurochem 68(4):1745–1752

Huang C, Wang C, Kawai M, Barnes S, Elmets C (2006a) Surfactant sodium lauryl sulfate enhances skin vaccination. Molecular characterization via a novel technique using ultrafiltration capillaries and mass spectrometric proteomics. Mol Cell Proteomics 5(3):523–532

Huang CM, Ananthaswamy HN, Barnes S, Ma Y, Kawai M, Elmets CA (2006b) Mass spectrometric proteomics profiles of in vivo tumor secretomes: capillary ultrafiltration sampling of regressive tumor masses. Proteomics 6(22):6107–6116

Huang CM, Wang CC, Barnes S, Elmets CA (2006c) In vivo detection of secreted proteins from wounded skin using capillary ultrafiltration probes and mass spectrometric proteomics. Proteomics 6(21):5805–5814

Huang CM, Wang CC, Kawai M, Barnes S, Elmets CA (2006d) In vivo protein sampling using capillary ultrafiltration semi-permeable hollow fiber and protein identification via mass spectrometry-based proteomics. J Chromatogr A 1109(2):144–151

Huang YH, Bergles DE (2004) Glutamate transporters bring competition to the synapse. Curr Opin Neurobiol 14(3):346–352

Hutchinson PJ, O'Connell MT, Al-Rawi PG, Maskell LB, Kett-White R, Gupta AK, Richards HK, Hutchinson DB, Kirkpatrick PJ, Pickard JD (2000) Clinical cerebral microdialysis: a methodological study. J Neurosurg 93(1):37–43

Hutchinson PJ, O'Connell MT, Nortje J, Smith P, Al-Rawi PG, Gupta AK, Menon DK, Pickard JD (2005a) Cerebral microdialysis methodology–evaluation of 20 kDa and 100 kDa catheters. Physiol Meas 26(4):423–428

Hutchinson PJ, O'Connell MT, Nortje J, Smith P, Al-Rawi PG, Gupta AK, Menon DK, Pickard JD (2005b) Cerebral microdialysis methodology–evaluation of 20 kDa and 100 kDa catheters. Physiol Meas 26(4):423–428

Imsilp K, Whittem T, Koritz GD, Zachary JF, Schaeffer DJ (2000) Inflammatory response to intramuscular implantation of polyacrylonitrile ultrafiltration probes in sheep. Vet Res 31(6):623–634

Janle E, Cregor M, Sojka JE (2001) Interstitial fluid calcium, magnesium and phosphorus concentrations in bone, muscle and subcutaneous tissue sampled with ultrafiltration probes. Curr Sep 19(3):81–85

Janle EM, Clark T, Ash SR (1992a) Use of an ultrafiltrate sampling probe to control glucose levels in a diabetic cat: case study. Curr Sep 11:3–6

Janle EM, Kissinger PT (1993) Microdialysis and ultrafiltration sampling of small molecules and ions from in vivo dialysis fibers. Am Assoc Clin Chem 147:159–165

Janle EM, Ostroy S, Kissinger PT (1992b) Monitoring the progress of streptozotocin diabetes in the mouse with the ultrafiltrate probe. Curr Sep 11:17–19

Janle EM, Sojka JE (2000) Use of ultrafiltration probes in sheep to collect interstitial fluid for measurement of calcium and magnesium. Contemp Top Lab Anim Sci 39(6):47–50

Jenkins BG, Brouillet E, Chen YC, Storey E, Schulz JB, Kirschner P, Beal MF, Rosen BR (1996) Non-invasive neurochemical analysis of focal excitotoxic lesions in models of neurodegenerative illness using spectroscopic imaging. J Cereb Blood Flow Metab 16(3):450–461

Jones DA, Ros J, Landolt H, Fillenz M, Boutelle MG (2000) Dynamic changes in glucose and lactate in the cortex of the freely moving rat monitored using microdialysis. J Neurochem 75(4):1703–1708

Kahlert S, Reiser G (2004) Glial perspectives of metabolic states during cerebral hypoxia–calcium regulation and metabolic energy. Cell Calcium 36(3–4):295–302

Kaptein WA, Zwaagstra JJ, Venema K, Korf J (1998) Continuous ultraslow microdialysis and ultrafiltration for subcutaneous sampling as demonstrated by glucose and lactate measurements in rats. Anal Chem 70(22):4696–4700

Katayama Y, Kawamata T, Kano T, Tsubokawa T (1992) Excitatory amino acid antagonist administered via microdialysis attenuates lactate accumulation during cerebral ischemia and subsequent hippocampal damage. Brain Res 584(1–2):329–333

Kawamata T, Katayama Y, Hovda DA, Yoshino A, Becker DP (1995a) Lactate accumulation following concussive brain injury: the role of ionic fluxes induced by excitatory amino acids. Brain Res 674(2):196–204

Kawamata T, Katayama Y, Hovda DA, Yoshino A, Becker DP (1995b) Lactate accumulation following concussive brain injury: the role of ionic fluxes induced by excitatory amino acids. Brain Res 674(2):196–204

Kennedy RT, Kauri LM, Dahlgren GM, Jung SK (2002) Metabolic oscillations in beta-cells. Diabetes 51(Suppl 1):S152–S161

Kerner A, Schlenk F, Sakowitz O, Haux D, Sarrafzadeh A (2007) Impact of hyperglycemia on neurological deficits and extracellular glucose levels in aneurysmal subarachnoid hemorrhage patients. Neurol Res 29(7):647–653

Kett-White R, Hutchinson PJ, Al-Rawi PG, Gupta AK, Pickard JD, Kirkpatrick PJ (2002) Adverse cerebral events detected after subarachnoid hemorrhage using brain oxygen and microdialysis probes. Neurosurgery 50(6):1213–1221

Khaldi A, Zauner A, Reinert M, Woodward JJ, Bullock MR (2001) Measurement of nitric oxide and brain tissue oxygen tension in patients after severe subarachnoid hemorrhage. Neurosurgery 49(1):33–38

Khan AS, Michael AC (2003) Invasive consequences of using micro-electrodes and microdialysis probes in the brain. Trends Anal Chem 22(8):503–508

Khandelwal P, Beyer CE, Lin Q, Schechter LE, Bach AC (2004) Studying rat brain neurochemistry using nanoprobe NMR spectroscopy: a metabonomics approach. Anal Chem 76(14): 4123–4127

Kissinger C, Peters S, Zhu Y (2003) New method for automating sample collection from in vivo ultrafiltration probes. Curr Sep 20(3):97–102

Klein HC, Krop-Van GW, Go KG, Korf J (1993) Prediction of specific damage or infarction from the measurement of tissue impedance following repetitive brain ischaemia in the rat. Neuropathol Appl Neurobiol 19(1):57–65

Korf J (1996) Intracerebral trafficking of lactate in vivo during stress, exercise, electroconvulsive shock and ischemia as studied with microdialysis. Dev Neurosci 185–6:405–414

Korf J (2006) Is brain lactate metabolized immediately after neuronal activity through the oxidative pathway? J Cereb Blood Flow Metab 26(12):1584–1586

Korf J, De Boer J (1990) Lactography as an approach to monitor glucose metabolism on-line in brain and muscle. Int J Biochem 22(12):1371–1378

Korf J, De Boer J, Baarsma R, Venema K, Okken A (1993) Monitoring of glucose and lactate using microdialysis: applications in neonates and rat brain. Dev Neurosci 15(3–5):240–246

Korf J, Gramsbergen JB (2007) Timing of potential and metabolic brain energy. J Neurochem 103(5):1697–1708

Korf J, Klein HC, Venema K, Postema F (1988) Increases in striatal and hippocampal impedance and extracellular levels of amino acids by cardiac arrest in freely moving rats. J Neurochem 50(4):1087–1096

Korf J, Postema F (1988) Rapid shrinkage of rat striatal extracellular space after local kainate application and ischemia as recorded by impedance. J Neurosci Res 19(4):504–510

Krishnappa IK, Contant CF, Robertson CS (1999) Regional changes in cerebral extracellular glucose and lactate concentrations following severe cortical impact injury and secondary ischemia in rats. J Neurotrauma 16(3):213–224

Krugers HJ, Jaarsma D, Korf J (1992) Rat hippocampal lactate efflux during electroconvulsive shock or stress is differently dependent on entorhinal cortex and adrenal integrity. J Neurochem 58(3):826–830

Kuhr WG, Korf J (1988) N-methyl-D-aspartate receptor involvement in lactate production following ischemia or convulsion in rats. Eur J Pharmacol 155(1–2):145–149

Kuhr WG, van den Berg CJ, Korf J (1988) In vivo identification and quantitative evaluation of carrier-mediated transport of lactate at the cellular level in the striatum of conscious, freely moving rats. J Cereb Blood Flow Metab 8(6):848–856

Landolt H, Langemann H, Mendelowitsch A, Gratzl O (1994) Neurochemical monitoring and online pH measurements using brain microdialysis in patients in intensive care. Acta Neurochir 60:475–478, Suppl (Wien)

Langemann H, Alessandri B, Mendelowitsch A, Feuerstein T, Landolt H, Gratzl O (2001) Extracellular levels of glucose and lactate measured by quantitative microdialysis in the human brain. Neurol Res 23(5):531–536

Leegsma-Vogt G, van der Werf S, Venema K, Korf J (2004) Modeling cerebral arteriovenous lactate kinetics after intravenous lactate infusion in the rat. J Cereb Blood Flow Metab 24(10):1071–1080

Leegsma-Vogt G, Venema K, Korf J (2003) Evidence for a lactate pool in the rat brain that is not used as an energy supply under normoglycemic conditions. J Cereb Blood Flow Metab 23(8):933–941

Leegsma-Vogt G, Venema K, Postema F, Korf J (2001) Monitoring arterio-venous differences of glucose and lactate in the anesthetized rat with or without brain damage with ultrafiltration and biosensor technology. J Neurosci Res 66(5):795–802

Lewen A, Hillered L (1998) Involvement of reactive oxygen species in membrane phospholipid breakdown and energy perturbation after traumatic brain injury in the rat. J Neurotrauma 15(7):521–530

Lin JY, Chung SY, Lin MC, Cheng FC (2002) Effects of magnesium sulfate on energy metabolites and glutamate in the cortex during focal cerebral ischemia and reperfusion in the gerbil monitored by a dual-probe microdialysis technique. Life Sci 71(7):803–811

Linhares MC, Kissinger PT (1992) Capillary ultrafiltration: in vivo sampling probes for small molecules. Anal Chem 64(22):2831–2835

Linhares MC, Kissinger PT (1993a) Determination of endogenous ions in intercellular fluid using capillary ultrafiltration and microdialysis probes. J Pharm Biomed Anal 11(11–12):1121–1127

Linhares MC, Kissinger PT (1993b) Pharmacokinetic monitoring in subcutaneous tissue using in vivo capillary ultrafiltration probes. Pharm Res 10(4):598–602

Liu YT, Huang CM (2007) In vivo sampling of extracellular {beta}-thymosin by ultrafiltration probes. Ann NY Acad Sci 1112:104–113

Lonjon M, Risso JJ, Palmier B, Negrin J, Darbin O (2001) Effects of hypothermic deep-anaesthesia on energy metabolism at brain and peripheral levels: a multi-probe microdialysis study in free-moving rat. Neurosci Lett 304(1–2):21–24

Lowry JP, Demestre M, Fillenz M (1998a) Relation between cerebral blood flow and extracellular glucose in rat striatum during mild hypoxia and hyperoxia. Dev Neurosci 20(1):52–58

Lowry JP, Miele M, O'Neill RD, Boutelle MG, Fillenz M (1998b) An amperometric glucose-oxidase/poly(o-phenylenediamine) biosensor for monitoring brain extracellular glucose: in vivo characterisation in the striatum of freely-moving rats. J Neurosci Methods 79(1):65–74

Lowry JP, O'Neill RD, Boutelle MG, Fillenz M (1998c) Continuous monitoring of extracellular glucose concentrations in the striatum of freely moving rats with an implanted glucose biosensor. J Neurochem 70(1):391–396

Lund-Andersen H (1979) Transport of glucose from blood to brain. Physiol Rev 59(2):305–352

Magistretti PJ, Sorg O, Yu N, Martin JL, Pellerin L (1993) Neurotransmitters regulate energy metabolism in astrocytes: implications for the metabolic trafficking between neural cells. Dev Neurosci 15:306–312

Marklund N, Clausen F, Lewander T, Hillered L (2001) Monitoring of reactive oxygen species production after traumatic brain injury in rats with microdialysis and the 4-hydroxybenzoic acid trapping method. J Neurotrauma 181(1):1217–1227

Markram H, Lubke J, Frotscher M, Sakmann B (1997) Regulation of synaptic efficacy by coincidence of postsynaptic APs and EPSPs. Science 275(5297):213–215

Mayer CH, Fink H, Rex A, Voigt JP (2006) Changes in extracellular hypothalamic glucose in relation to feeding. Eur J Neurosci 24(6):1695–1701

McKenna MC, Tildon JT, Stevenson JH, Hopkins IB, Huang X, Couto R (1998) Lactate transport by cortical synaptosomes from adult rat brain: characterization of kinetics and inhibitor specificity. Dev Neurosci 20(4–5):300–309

McNay EC, Fries TM, Gold PE (2000) Decreases in rat extracellular hippocampal glucose concentration associated with cognitive demand during a spatial task. Proc Natl Acad Sci USA 97(6):2881–2885

Mendelowitsch A, Ritz MF, Ros J, Langemann H, Gratzl O (2001) 17beta-Estradiol reduces cortical lesion size in the glutamate excitotoxicity model by enhancing extracellular lactate: a new neuroprotective pathway. Brain Res 901(1–2):230–236

Mintun MA, Vlassenko AG, Rundle MM, Raichle ME (2004) Increased lactate/pyruvate ratio augments blood flow in physiologically activated human brain. Proc Natl Acad Sci USA 101(2):659–664

Moscone D, Venema K, Korf J (1996) Ultrafiltrate sampling device for continuous monitoring. Med Biol Eng Comput 34(4):290–294

Oldenziel WH, Dijkstra G, Cremers TI, Westerink BH (2006) In vivo monitoring of extracellular glutamate in the brain with a microsensor. Brain Res 1118(1):34–42

Orzi F, Lucignani G, Dow-Edwards D, Namba H, Nehlig A, Patlak CS, Pettigrew K, Schuier F, Sokoloff L (1988) Local cerebral glucose utilization in controlled graded levels of hyperglycemia in the conscious rat. J Cereb Blood Flow Metab 8(3):346–356

Osborne PG, Niwa O, Kato T, Yamamoto K (1997) On-line, continuous measurement of extracellular striatal glucose using microdialysis sampling and electrochemical detection. J Neurosci Methods 77(2):143–150

Oz G, Seaquist ER, Kumar A, Criego AB, Benedict LE, Rao JP, Henry PG, Van De Moortele PF, Gruetter R (2007) Human brain glycogen content and metabolism: implications on its role in brain energy metabolism. Am J Physiol Endocrinol Metab 292(3):E946–E951

Pancrazio JJ, Whelan JP, Borkholder DA, Ma W, Stenger DA (1999) Development and application of cell-based biosensors. Ann Biomed Eng 27(6):697–711

Parkin M, Hopwood S, Jones DA, Hashemi P, Landolt H, Fabricius M, Lauritzen M, Boutelle MG, Strong AJ (2005) Dynamic changes in brain glucose and lactate in pericontusional areas of the human cerebral cortex, monitored with rapid sampling on-line microdialysis: relationship with depolarisation-like events. J Cereb Blood Flow Metab 25(3):402–413

Pazdernik T, Cross R, Nelson S, Kamijo Y, Samson F (1994) Is there an energy conservation "system" in brain that protects against the consequences of energy depletion? Neurochem Res 191(1):1393–1400

Pellerin L, Bouzier-Sore AK, Aubert A, Serres S, Merle M, Costalat R, Magistretti PJ (2007) Activity-dependent regulation of energy metabolism by astrocytes: an update. Glia 55(12):1251–1262

Pellerin L, Magistretti PJ (1994) Glutamate uptake into astrocytes stimulates aerobic glycolysis: a mechanism coupling neuronal activity to glucose utilization. Proc Natl Acad Sci USA 91(22):10625–10629

Perdomo J, Hinkers H, Sundermeier C, Seifert W, Martinez MO, Knoll M (2000) Miniaturized real-time monitoring system for L-lactate and glucose using microfabricated multi-enzyme sensors. Biosens Bioelectron 15(9–10):515–522

Plock N, Kloft C (2005) Microdialysis–theoretical background and recent implementation in applied life-sciences. Eur J Pharm Sci 25(1):1–24

Rabenstein K, McShane AJ, McKenna MJ, Dempsey E, Keaveny TV, Freaney R (1996) An intravascular microdialysis sampling system suitable for application in continuous biochemical monitoring of glucose and lactate. Technol Health Care 4(1):67–76

Reinert M, Schaller B, Widmer HR, Seiler R, Bullock R (2004) Influence of oxygen therapy on glucose-lactate metabolism after diffuse brain injury. J Neurosurg 101(2):323–329

Reinstrup P, Stahl N, Mellergard P, Uski T, Ungerstedt U, Nordstrom CH (2000) Intracerebral microdialysis in clinical practice: baseline values for chemical markers during wakefulness, anesthesia, and neurosurgery. Neurosurgery 47(3):701–709

Rhemrev-Boom MM, Jonker MA, Venema K, Jobst G, Tiessena R, Korf J (2001) On-line continuous monitoring of glucose or lactate by ultraslow microdialysis combined with a flow-through nanoliter biosensor based on poly(m-phenylenediamine) ultra-thin polymer membrane as enzyme electrode. Analyst 126(7):1073–1079

Rhemrev-Boom MM, Tiessen RG, Jonker AA, Venema K, Vadgama P, Korf J (2002) A lightweight measuring device for the continuous in vivo monitoring of glucose by means of ultraslow microdialysis in combination with a miniaturised flow-through biosensor. Clin Chim Acta 316(1–2):1–10

Ritz MF, Schmidt P, Mendelowitsch A (2004) Acute effects of 17beta-estradiol on the extracellular concentration of excitatory amino acids and energy metabolites during transient cerebral ischemia in male rats. Brain Res 1022(1–2):157–163

Ros J, Jones D, Pecinska N, Alessandri B, Boutelle M, Landolt H, Fillenz M (2002) Glutamate infusion coupled with hypoxia has a neuroprotective effect in the rat. J Neurosci Methods 119(2):129–133

Rosdahl H, Hamrin K, Ungerstedt U, Henriksson J (1998) Metabolite levels in human skeletal muscle and adipose tissue studied with microdialysis at low perfusion flow. Am J Physiol 274(5(Pt 1)):E936–E945

Rosdahl H, Ungerstedt U, Henriksson J (1997) Microdialysis in human skeletal muscle and adipose tissue at low flow rates is possible if dextran-70 is added to prevent loss of perfusion fluid. Acta Physiol Scand 159(3):261–262

Rosdahl H, Ungerstedt U, Jorfeldt L, Henriksson J (1993) Interstitial glucose and lactate balance in human skeletal muscle and adipose tissue studied by microdialysis. J Physiol 471:637–657

Rosenbloom AJ, Sipe DM, Weedn VW (2005) Microdialysis of proteins: performance of the CMA/20 probe. J Neurosci Methods 148(2):147–153

Roy CS, Sherrington CS (1890) On the regulation of the blood-supply of the brain. J Physiol 11(1–2):85–158

Rutherford EC, Pomerleau F, Huettl P, Stromberg I, Gerhardt GA (2007) Chronic second-by-second measures of L-glutamate in the central nervous system of freely moving rats. J Neurochem 102(3):712–722

Sakowitz OW, Stover JF, Sarrafzadeh AS, Unterberg AW, Kiening KL (2007) Effects of mannitol bolus administration on intracranial pressure, cerebral extracellular metabolites, and tissue oxygenation in severely head-injured patients. J Trauma 62(2):292–298

Savenije B, Venema K, Gerritzen MA, Lambooij E, Korf J (2003) Minimally invasive technique based on ultraslow ultrafiltration to collect and store time profiles of analytes. Anal Chem 75(17):4397–4401

Schlenk F, Graetz D, Nagel A, Schmidt M, Sarrafzadeh AS (2008) Insulin-related decrease in cerebral glucose despite normoglycemia in aneurysmal subarachnoid hemorrhage. Crit Care 121:R9

Schulz MK, Wang LP, Tange M, Bjerre P (2000) Cerebral microdialysis monitoring: determination of normal and ischemic cerebral metabolisms in patients with aneurysmal subarachnoid hemorrhage. J Neurosurg 93(5):808–814

Schurr A, Rigor BM (1998) Brain anaerobic lactate production: a suicide note or a survival kit? Dev Neurosci 20(4–5):348–357

Schutte RJ, Oshodi SA, Reichert WM (2004) In vitro characterization of microdialysis sampling of macromolecules. Anal Chem 76(20):6058–6063

Shram NF, Netchiporouk LI, Martelet C, Jaffrezic-Renault N, Bonnet C, Cespuglio R (1998) In vivo voltammetric detection of rat brain lactate with carbon fiber microelectrodes coated with lactate oxidase. Anal Chem 70(13):2618–2622

Shuaib A, Xu K, Crain B, Siren AL, Feuerstein G, Hallenbeck J, Davis JN (1990) Assessment of damage from implantation of microdialysis probes in the rat hippocampus with silver degeneration staining. Neurosci Lett 112(2–3):149–154

Slais K, Vorisek I, Zoremba N, Homola A, Dmytrenko L, Sykova E (2008) Brain metabolism and diffusion in the rat cerebral cortex during pilocarpine-induced status epilepticus. Exp Neurol 209(1):145–154

Smith PJ, Hammar K, Porterfield DM, Sanger RH, Trimarchi JR (1999) Self-referencing, non-invasive, ion selective electrode for single cell detection of trans-plasma membrane calcium flux. Microsc Res Tech 46(6):398–417

Sojka JE, Adams SB, Rohde C, Janie EM (2000) Surgical implantation of ultrafiltration probes in ovine bone and muscle. J Invest Surg 13(5):289–294

Soto-Montenegro ML, Vaquero JJ, Arango C, Ricaurte G, Garcia-Barreno P, Desco M (2007) Effects of MDMA on blood glucose levels and brain glucose metabolism. Eur J Nucl Med Mol Imaging 34(6):916–925

Spehar A, Tiedje L, Sojka JE, Janle EM, Kissinger PT (1998) Recovery of endogenous ions from subcutaneous and intramuscular spaces in horses using ultrafiltrate probes. Curr Sep 17(2):47–51

Stephans SE, Whittingham TS, Douglas AJ, Lust WD, Yamamoto BK (1998) Substrates of energy metabolism attenuate methamphetamine-induced neurotoxicity in striatum. J Neurochem 71(2):613–621

Storm-Mathisen J (1977) Localization of transmitter candidates in the brain: the hippocampal formation as a model. Prog Neurobiol 8(2):119–181

Strong AJ, Boutelle MG, Vespa PM, Bullock MR, Bhatia R, Hashemi P (2005) Treatment of critical care patients with substantial acute ischemic or traumatic brain injury. Crit Care Med 33(9):2147–2149

Takita M, Mikuni M, Takahashi K (1992) Habituation of lactate release responding to stressful stimuli in rat prefrontal cortex in vivo. Am J Physiol 2633(Pt 2):R722–R727

Taylor DL, Urenjak J, Zilkha E, Obrenovitch TP (1997) Effects of probenecid on the elicitation of spreading depression in the rat striatum. Brain Res 764(1–2):117–125

Thevenot DR, Toth K, Durst RA, Wilson GS (2001) Electrochemical biosensors: recommended definitions and classification. Biosens Bioelectron 16(1–2):121–131

Thompson JK, Peterson MR, Freeman RD (2005) Separate spatial scales determine neural activity-dependent changes in tissue oxygen within central visual pathways. J Neurosci 25(39):9046–9058

Tiessen RG, Kaptein WA, Venema K, Korf J (1999) Slow ultrafiltration for continuous in vivo sampling: application for glucose and lactate in man. Anal Chim Acta 379:327–335

Tiessen RG, Rhemrev-Boom MM, Korf J (2002) Glucose gradient differences in subcutaneous tissue of healthy volunteers assessed with ultraslow microdialysis and a nanolitre glucose sensor. Life Sci 70(21):2457–2466

Tiessen RG, Tio RA, Hoekstra A, Venema K, Korf J (2001) An ultrafiltration catheter for monitoring of venous lactate and glucose around myocardial ischemia. Biosens Bioelectron 16(3):159–167

Trickler WJ, Miller DW (2003) Use of osmotic agents in microdialysis studies to improve the recovery of macromolecules. J Pharm Sci 92(7):1419–1427

Uehara T, Kurata K, Sumiyoshi T, Kurachi M (2003) Immobilization stress-induced increment of lactate metabolism in the basolateral amygdaloid nucleus is attenuated by diazepam in the rat. Eur J Pharmacol 459(2–3):211–215

Uehara T, Sumiyoshi T, Itoh H, Kurachi M (2007a) Dopamine D1 and D2 receptors regulate extracellular lactate and glucose concentrations in the nucleus accumbens. Brain Res 1133(1):193–199

Uehara T, Sumiyoshi T, Itoh H, Kurachi M (2007b) Role of glutamate transporters in the modulation of stress-induced lactate metabolism in the rat brain. Psychopharmacology (Berl) 195(2):297–302

Uehara T, Sumiyoshi T, Matsuoka T, Tanaka K, Tsunoda M, Itoh H, Kurachi M (2005) Enhancement of lactate metabolism in the basolateral amygdala by physical and psychological stress: role of benzodiazepine receptors. Brain Res 1065(1–2):86–91

Ungerstedt U, Pycock C (1974) Functional correlates of dopamine neurotransmission. Bull Schweiz Akad Med Wiss 30(1–3):44–55

Ungerstedt U, Rostami E (2004) Microdialysis in neurointensive care. Curr Pharm Des 10(18):2145–2152

Vafaee MS, Gjedde A (2004) Spatially dissociated flow-metabolism coupling in brain activation. Neuroimage 21(2):507–515

Valtysson J, Persson L, Hillered L (1998) Extracellular ischaemia markers in repeated global ischaemia and secondary hypoxaemia monitored by microdialysis in rat brain. Acta Neurochir (Wien) 140(4):387–395

van der Kuil JH, Korf J (1991) On-line monitoring of extracellular brain glucose using microdialysis and a NADPH-linked enzymatic assay. J Neurochem 57(2):648–654

Vespa P, Bergsneider M, Hattori N, Wu HM, Huang SC, Martin NA, Glenn TC, McArthur DL, Hovda DA (2005) Metabolic crisis without brain ischemia is common after traumatic brain injury: a combined microdialysis and positron emission tomography study. J Cereb Blood Flow Metab 25(6):763–774

Vespa P, Boonyaputthikul R, McArthur DL, Miller C, Etchepare M, Bergsneider M, Glenn T, Martin N, Hovda D (2006) Intensive insulin therapy reduces microdialysis glucose values without altering glucose utilization or improving the lactate/pyruvate ratio after traumatic brain injury. Crit Care Med 34(3):850–856

Voigt JP, Nwaiser B, Rex A, Mayer C, Fink H (2004) Effect of 5-HT1A receptor activation on hypothalamic glucose. Pharmacol Res 50(3):359–365

von Baeyer H, Lajous-Petter A, Debrandt W, Hampl H, Kochinke F, Herbst R (1988) Surface reactions on blood contact during haemodialysis and haemofiltration with various membrane types. J Neurosci Methods 36:215–229

Wang L, Dong Y, Yu X, Shangguan DH, Zhao R, Han HW, Liu GQ (2002) Analysis of glucose and lactate in dialysate from hypothalamus of rats after exhausting swimming using microdialysis. Biomed Chromatogr 16(7):427–431

Wisniewski N, Klitzman B, Miller B, Reichert WM (2001) Decreased analyte transport through implanted membranes: differentiation of biofouling from tissue effects. J Biomed Mater Res 57(4):513–521

Wisniewski N, Moussy F, Reichert WM (2000) Characterization of implantable biosensor membrane biofouling. Fresenius J Anal Chem 366(6–7):611–621

Wisniewski N, Reichert M (2000) Methods for reducing biosensor membrane biofouling. Colloids Surf B Biointerfaces 18(3–4):197–219

Yang Q, Atanasov P, Wilkins E (1998) An integrated needle-type biosensor for intravascular glucose and lactate monitoring. Electroanalysis 10(11):752–757

Yang S, Huang CM (2007) Recent advances in protein profiling of tissues and tissue fluids. Expert Rev Proteomics 4(4):515–529

Yao T, Yano T, Nishino H (2004) Simultaneous in vivo monitoring of glucose, L-lactate, and pyruvate concentrations in rat brain by a flow-injection biosensor system with an on-line microdialysis sampling. Anal Chim Acta 510(1):53–59

Zauner A, Doppenberg E, Woodward JJ, Allen C, Jebraili S, Young HF, Bullock R (1997a) Multiparametric continuous monitoring of brain metabolism and substrate delivery in neurosurgical patients. Neurol Res 19(3):265–273

Zauner A, Doppenberg EM, Woodward JJ, Choi SC, Young HF, Bullock R (1997b) Continuous monitoring of cerebral substrate delivery and clearance: initial experience in 24 patients with severe acute brain injuries. Neurosurgery 41(5):1082–1091

Zhang FF, Wan Q, Li CX, Wang XL, Zhu ZQ, Xian YZ, Jin LT, Yamamoto K (2004) Simultaneous assay of glucose, lactate, L-glutamate and hypoxanthine levels in a rat striatum using enzyme electrodes based on neutral red-doped silica nanoparticles. Anal Bioanal Chem 380(4):637–642

Zhao S, Pinholt EM, Madsen JE, Donath K (2000) Histological evaluation of different biodegradable and non-biodegradable membranes implanted subcutaneously in rats. J Craniomaxillofac Surg 28(2):116–122

Zielke HR, Zielke CL, Baab PJ (2007) Oxidation of (14) C-labeled compounds perfused by microdialysis in the brains of free-moving rats. J Neurosci Res 85(14):3145–3149

Zoremba N, Homola A, Rossaint R, Sykova E (2007) Brain metabolism and extracellular space diffusion parameters during and after transient global hypoxia in the rat cortex. Exp Neurol 203(1):34–41

Part II
In vivo Assessment of Neural Activity Via Neurovascular Coupling

Chapter 14
The Coupling of Cerebral Metabolic Rate of Glucose and Cerebral Blood Flow *In Vivo*

Steen G. Hasselbalch and Olaf B. Paulson

Abstract The energy supplied to the brain by metabolic substrate is largely utilized for maintaining synaptic transmission. In this regulation cerebral blood flow and glucose consumption is tightly coupled as well in the resting condition as during activation. Quantification of cerebral blood flow and metabolism was originally performed using the Kety-Schmidt method and this method still represent the gold standard by which subsequent methods have been evaluated. However, in its classical setting, the method overestimates cerebral blood flow. Studies of metabolic changes during activation must take this into account, and subsequent methods for measurement of regional glucose metabolism must be corrected accordingly in order to allow reliable quantitative comparisons of metabolite changes in activation studies. For studies of regional metabolic changes during activation quantification poses further difficulties due to limitation in resolution and partial volume effects.

In contrast to the tight coupling between regional glucose metabolism and cerebral blood flow, there is an uncoupling between flow and oxygen consumption as the latter only increases to a limited extend. The excess glucose uptake is thus not used for aerobic metabolism. Although some of the excess glucose uptake can be explained by lactate production, this phenomenon can still not account for the excess glucose uptake. Thus, more complex metabolic patterns in the brain might be reflected in the excess glucose uptake during activation, and especially temporal relationships must be taken into account.

S.G. Hasselbalch, M.D. (✉)
Neurobiology Research Unit, Copenhagen University Hospital, Rigshospitalet, Copenhagen, Denmark
e-mail: sgh@nru.dk

O.B. Paulson, M.D., D.M.S.C.
Memory Disorders Research Unit and Neurobiology Research Unit, Rigshospitalet, and Danish Research Centre for Magnetic Resonance, Hvidovre Hospital, Faculty of Health Sciences, Copenhagen University, Denmark
e-mail: olaf.paulson@nru.dk

I.-Y. Choi and R. Gruetter (eds.), *Neural Metabolism In Vivo*, Advances in Neurobiology 4, DOI 10.1007/978-1-4614-1788-0_14,

What triggers the flow increase during functional brain activation is not entirely elucidated. The demand for excess glucose uptake may be important and a possible oxygen deficit in tissue distant from the capillaries is probably of minor importance. The mechanism by which cerebral blood flow increases during activation may incorporate changes in glycolytic substrates or local changes in astrocytes or neurons that triggers the production of vasoactive substances.

Keywords Activation of the brain • Astrocyte-neuronal lactate exchange • Blood flow • Brain activation • Cerebral blood flow • Cerebral metabolic rate of glucose • Cerebral metabolic rate of oxygen • Coupling flow and metabolism • Fick's principle • Flow • Functional activation • Glucose • Glutamate • Kety-Schmidt method • Lactate • Lumped constant • Oxygen • Uncoupling flow and metabolism

14.1 Introduction

The energy supplied to the brain by metabolic substrate is largely utilized for maintaining synaptic transmission. In the regulation of cerebral energy supply, cerebral blood flow (CBF) and cerebral metabolic rate for glucose (CMR_{glc}) are tightly coupled in the resting condition as well as during activation. Further, in the resting state, almost all glucose is oxidized. Although the coupling between glucose metabolism and blood flow has been firmly established, several aspects of the cerebral glucose metabolism still remain to be solved. Thus, the spatial and temporal features of glycolysis, glucose oxidation and lactate metabolism are still debated. Irrespective of this, experimental studies points to several putative neurochemical vasodilators that are coupled to an increase in glucose consumption, in an effort to elucidate the observed *in vivo* coupling of CMR_{glc} and CBF. These aspects of the cerebral metabolism are the topic of the present review. It has not been our intention to give a comprehensive review of this vast field, but key aspects of metabolism are reviewed and related to function.

14.2 Quantification of Cerebral Blood Flow

The Fick principle is considered the gold standard for measurement of global cerebral metabolic rate for glucose (CMR_{glc}). The Fick principle states that global CMR_{glc} is equal to cerebral blood flow (CBF) multiplied with arteriovenous differences of glucose over the brain. Quantification of global CBF and CMR_{glc} was originally performed using the Kety-Schmidt method (Kety and Schmidt 1948), and this method still represents the gold standard by which subsequent methods, i.e., positron emission tomography (PET) methods, have been evaluated (Huang et al. 1980; Reivich et al. 1985). Under various physiological conditions, including activation, regional changes occur in both CBF and CMR_{glc}, and if large enough, these changes may be detected by global measurements. Although activation induced relative changes in CBF in specific regions of the brain measured by neuroimaging methods have increased our understanding of the functional organization of the brain, these methods generally fail to quantify the absolute changes in CBF and CMR_{glc}. In the following, arguments are provided to rescale absolute values for global CBF and CMR_{glc} in order to provide a physiologically meaningful background for the complex changes in brain metabolism and CBF that take place during activation.

In its standard application, the Kety-Schmidt technique systematically overestimates global CBF by approximately 20% (Lassen 1959; Madsen et al. 1993), and the "true global CBF" corrected for this overestimation is close to 46 ml × 100 g^{-1} × min^{-1} (Hasselbalch et al. 1998).

More recently, magnetic resonance (MR) perfusion imaging has been applied in the attempt to quantify CBF. Methods such as gadolinium bolus injection and arterial spin labeling have several advantages compared to PET (subjects avoiding radiation and the potential higher spatial resolution), but methodological problems make quantification difficult and may cause both underestimation (Ostergaard et al. 1998) and overestimation (Knutsson et al. 2007) of absolute CBF. Thus, absolute quantification of CBF by MR perfusion imaging still needs further validation.

The purpose of functional neuroimaging during activation is usually to localize specific CBF responses to various activation paradigms. However, if the CBF response is to be quantified, it is important to take into account the spatial resolution of the scanner, which tends to underestimate the CBF response due to the partial volume effect (Hofmann et al. 2009). This is a particular problem for SPECT and PET with spatial resolution in the range of 4–15 mm. Correction for the partial volume effect due to limited spatial resolution may significantly increase the grey matter CBF response to visual stimulation (Law et al. 2000) and if accurate quantification of CBF changes is sought, partial volume correction either by kinetic modeling of PET data (Law et al. 2000) or by co-registration of functional images to structural images (Muller-Gartner et al. 1992) is imperative.

14.3 Quantification of Cerebral Metabolic Rate for Glucose

Using the Kety-Schmidt technique with correction for systematic overestimation of CBF as described above, a very good agreement between several studies has been found: in resting, young subjects global CMR_{glc} is close to 24 μmol × 100 g^{-1} × min^{-1} (Cohen et al. 1967; Madsen et al. 1995; Hasselbalch et al. 1998). Only few studies using other methodologies for calculation of global CMR_{glc} are available for comparison with the Fick principle. Using PET and ^{11}C-labeled glucose as a tracer, Blomqvist and coworkers found an average CMR_{glc} of 26 μmol × 100 g^{-1} × min^{-1}, when loss of tracer by escape of labeled CO_2 and labeled lactate was corrected for (Blomqvist et al. 1990). Other indirect evidence support these studies: Applying a stochiometric relationship between oxygen and glucose metabolism of 5.6, a global CMR_{glc} of 24 μmol × 100 g^{-1} × min^{-1} yields a cerebral metabolic rate for oxygen

(CMR_{O2}) of 134 μmol × 100 g^{-1} × min^{-1}, in line with PET studies using ^{15}O-labeled tracers with CMR_{O2} values ranging from 120 to 130 μmol × 100 g^{-1} × min^{-1} (Mintun et al. 1984; Perlmutter et al. 1987; Leenders et al. 1990).

It should be noted that the global CMR_{glc} value of 24 μmol × 100 g^{-1} × min^{-1} is considerably lower than values of 30–35 μmol × 100 g^{-1} × min^{-1} reported by previous studies using ^{18}F-Fluoro-deoxyglucose (FDG) and PET (Phelps et al. 1979; Reivich et al. 1985; Brooks et al. 1987; Kuwabara et al. 1990). Because the PET-FDG method remains the most widely used method for *in vivo* quantification of regional CMR_{glc}, it is crucial to correctly determine the difference between net uptake of tracer (FDG) and glucose, also known as the lumped constant. The conventional use of a lumped constant in the range of 0.42–0.55 overestimates absolute values for CMR_{glc} obtained with this method. Recent evidence confirms that the lumped constant is close to 0.80, and using this value in PET-FDG experiments, estimates of CMR_{glc} are in accordance with physiologically reasonable levels (Hasselbalch et al. 1998; Graham et al. 2002). Further, it should be noted, that the lumped constant may also vary between different physiological states.

14.4 Coupling of CBF and Cmr_{glc} in the Resting State

More than 100 years ago, Roy and Sherrington proposed that the vascular supply to the brain tissue varied according to the neuronal activity (Roy and Sherrington 1890), and since then the tight coupling of neuronal activity to CBF and CMR_{glc} has formed the basis for measurement of the functional organisation of brain tissue by functional neuroimaging. During the years, it has also become clear that a close coupling between CBF and CMR_{glc} can be demonstrated during activation, whereas the coupling between CBF and CMR_{glc} vs. oxygen metabolism is more complex. The latter problem will be dealt with below, but here we will focus on the basic metabolic pathways that underlie the resting level of CMR_{glc}.

It is generally accepted that metabolic substrates for neuronal activity is supplied by the oxidative metabolism of glucose, because glucose is the major fuel for all brain cells. Measurements of global uptake of glucose and oxygen have been hampered by the problems of correct CBF calculation as described above, generally yielding an overestimation of the absolute levels of both metabolites, but the ratio between glucose and oxygen uptake is quite consistent over many studies when comparing ratios of arteriovenous differences (Kety 1957; Madsen et al. 1995; Hasselbalch et al. 1996). The average ratio of approximately 5.6 obtained by these studies suggests that glucose is almost completely oxidized to CO_2 and H_2O through glycolysis and subsequent TCA cycle. The energy supplied by metabolic substrates is utilized for maintaining synaptic transmission by restoring neurotransmitters and ion gradients across cell membranes (Jueptner and Weiller 1995). Glutamate is the major excitatory neurotransmitter in the brain, and much evidence suggests that glial cells surrounding the synapses clear glutamate from the synaptic cleft, return it in the form of glutamine to the neurons, where glutamate finally is formed again from glutamine (Schousboe et al. 1993; Rothstein, et al. 1994; Magistretti and Pellerin 1996). A study using ^{13}C

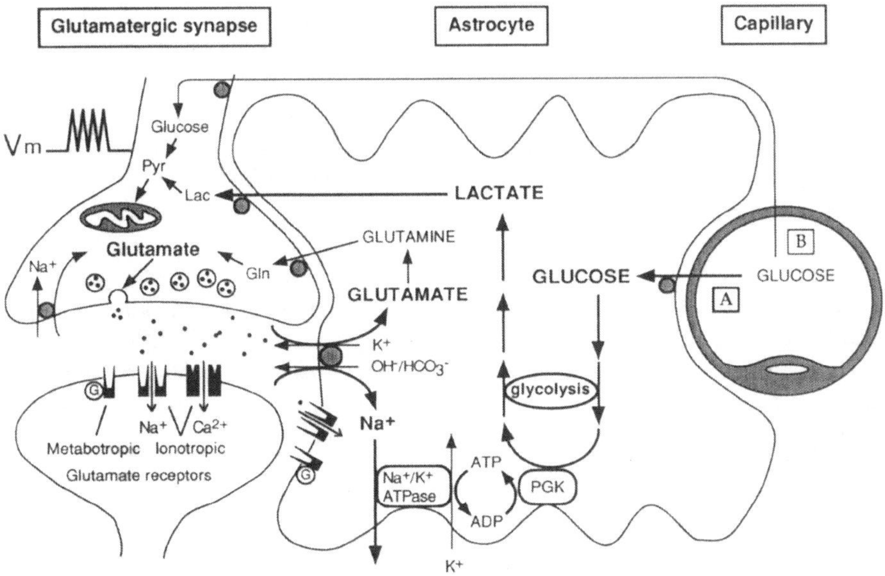

Fig. 14.1 Schematic of the mechanism for glutamate-induced glycolysis in astrocytes during physiological activation. At glutamatergic synapses, glutamate depolarizes neurons by acting at specific receptor subtypes. The action of gutamate is terminated by an efficient glutamate uptake system located primarily in Astrocytes. Glutamate is cotransported with Na^+. resulting in an increase: in $[Na^+]i$, leading to activation of Na^+/K^+-ATPase. The pump fueled by ATP provided by membrane-bound glycolytic enzymes [possibly phosphoglycerate kinase (PGK); see Ref. 22], activates glycolysis – i.e., glucose utilization and lactate production – in astrocytes. Lactate. once released, can be taken up by neurons and serve as an adequate energy substrate. For graphic clarity, only lactate uptake into presynaptic terminals is indicated. However, this process could also occur at the postsynaptic neuron. This model, which summarizes *in vitro* experimental evidence indicating glutamate-induced glycolysis, is taken to reflect cellular and molecular events occurring during activation of a given cortical area [arrow labeled A (activation)]. Direct glucose uptake into neurons under basal conditions is also shown [arrow labeled B (basal conditions)]. Pyr, pyruvate; Lac, lactate; Glu, glutamine; G, guanine nucleotide binding protein (Figure and legend from Pellerin and Magistretti 1994)

MR spectroscopy has shown a remarkable close 1:1 relationship between this glutamate/glutamine cycle and glucose oxidation (Rothman et al. 1999). Thus, these data support the model proposed by Magistretti et al., in which brain glucose metabolism and brain activity are coupled by the glycolytic production of Adenosine triphosphate (ATP) used for pumping glutamate into the glial cells (Fig. 14.1, Magistretti and Pellerin 1999). In this model, astrocytes degrade glucose to lactate to produce small amounts of glycolytic ATP for the glutamate/glutamine cycle. The produced lactate is taken up by the neurons which convert it to pyruvate that enters the TCA cycle to generate large amounts of ATP molecules which are subsequently used for maintenance of neuronal activity, including restoring cellular depolarisation by Na^+/Ka^+-ATPase (for overview, see Magistretti and Pellerin 1999; Pellerin and Magistretti 2004; Pellerin et al. 2007). Thus, coupling of CMR_{glc} and neuronal activity seems to be regulated by the need for energy substrates to maintain synaptic transmission.

More recent studies have, however, shown a more complex metabolic pattern between neurons and astrocytes. Both cell types metabolize glucose via glycolysis, oxidation, and to smaller extent pentose phosphate pathways. In recent years, MR spectroscopy studies have revealed an interesting finding: in resting conditions, the oxidative metabolism in neurons and astrocytes is of similar magnitude, where most of the oxidative metabolism in astrocytes is used for de novo synthesis of carbohydrate substrates, including glutamate (for review, see Hertz et al. 2007). Further, only astrocytes store glucose as glycogen (For review, see Gruetter 2003), and following activation and subsequent glycogen replenishment, the turn-over of glycogen levels in the brain is small (Choi and Gruetter 2003). In neurons, ATP generated from glucose oxidation is used to fuel baseline synaptic transmission activity (for review, see Pellerin and Magistretti 2004). Thus, although it appears that in the resting state, astrocytes and neurons may subserve different metabolic needs, they seem to share a common metabolic pattern with almost complete oxidation of glucose. However, during activation, the metabolic pattern changes significantly.

14.5 Coupling of CBF and CMR_{glc} During Activation

14.5.1 Changes in Metabolic Patterns During Activation

The increased glucose uptake during activation is made possible by increased CBF, and there is a tight coupling between cerebral blood flow and glucose consumption under most physiological conditions, even during functional activation in contrast to an uncoupling between oxygen consumption and blood flow. Thus, in the condition of extreme cerebral activation as seen during the epileptic seizure blood flow increases with at least a factor 2 whereas the arterial venous differences of glucose remains essential unchanged leading to a doubling of the glucose uptake by the brain (Brodersen et al. 1973). In contrast the arterial venous difference of oxygen drops with 30–50% resulting in an only minor increase of oxygen consumption (Brodersen et al. 1973). Also during more normal and physiological activation the same pattern of metabolic changes occurs. This has been demonstrated in PET studies where activation resulted in a parallel increase in blood flow and glucose consumption, but a much less pronounced increase in oxygen consumption (Fox et al. 1988). Similarly, in experimental studies in the rat, more diffuse activation resulted in a decrease of the arteriovenous oxygen difference of about 25% compared to an unchanged arteriovenous difference of glucose (Madsen et al. 1998). Under normal physiological conditions, most of the glucose taken up by the brain is metabolised to water and carbon dioxide. This is reflected by an oxygen/glucose molar consumption rate of close to 6 corresponding to the stoichiometric value of a respiratory coefficient (CO_2 release/oxygen consumption) close to one. The observed decrease of the oxygen/glucose uptake ratio of about 25% that was found in the study by Madsen et al. (1998) indicate that the excess blood flow and excess glucose uptake during activation is not used for increased oxidative glucose metabolism.

14.5.2 What Triggers the Flow and Glucose Consumption Increase During Activation?

It still remains a puzzle why there is an uncoupling of glucose and oxygen metabolism and an increased non-oxidative metabolism during activation. One theory is that despite an increased oxygen tension in the venous outflow from the brain and despite a larger effective capillary diffusion area there is still a lack of oxygen in the tissue areas most distant from the capillaries (Buxton and Frank 1997). According to such a theory the excess lactate production might reflect anaerobic glucose metabolism due to lack of oxygen. Another theory has claimed that part of the excess glucose uptake is cycled through the cerebral glycogen pool resulting in lactate production in order to supply glial cell with energy for a rapid neurotransmitter clearance necessitated by an increased transmitter release during activation (Dringen et al. 1993; Magistretti and Pellerin 1996). This latter theory would request an increased transport of glucose form the blood to the brain. As the extraction of glucose in the cerebrovascular bed is only about 10% a linear flow increase would not increase the glucose supply much according to the Renkin-Crone equation (Knudsen et al. 1990; Kuschinsky and Paulson 1992). However, the increased blood flow will decrease the heterogeneity of the cerebral microcirculation – "slow" capillaries becoming faster - and thereby increase the physiological capillary surface area available for diffusion (Kuschinsky and Paulson 1992; Knudsen 1994). The increased available capillary area for diffusion may be of significance for increased diffusion of both glucose and oxygen to the cerebral tissue. These aspects have been discussed in details in a recent review (Paulson et al. 2010).

14.5.3 The Role of Lactate

According to the astrocyte-neuron lactate shuttle described above, lactate produced from glycolysis in astrocytes is taken up by neurons for energy production (Pellerin and Magistretti 2004). Further, neurons seem to have a preference for metabolizing lactate over glucose (Smith et al. 2003). This model would then predict that during activation glucose uptake is increased into astrocytic end feet that lie in close connection with capillaries, and is metabolized to lactate in astrocytes. Brain levels of lactate may not increase considerably, because lactate is rapidly oxidized in neurons to produce ATP to meet the increased energy demands from neurons. Further, lactate may diffuse rapidly between astrocytes (Medina and Tabernero 2005), for further metabolism in astrocytes more distant from the activated area. Although an abundance of data suggest that this model may in fact describe the fate of glucose uptake, there are some unexplained issues, especially relating to lactate levels in the brain, and the continuous or prolonged uncoupling of oxygen metabolism as discussed below.

Part of the excess glucose taken up seems to be metabolised to lactate since there is a slight excess lactate efflux from the brain during activation (Madsen et al. 1999). Thus, the decrease in oxygen/glucose ratio indicates that the oxidative metabolism account for approximately 60% of the glucose taken up by the brain and of the non-oxidized glucose only approximately 50% seemed to be metabolised to lactate escaping from the brain (Madsen et al. 1999). The metabolic faith of the other part of the excess glucose uptake still remains a puzzle. Lactate diffuses across the blood brain barrier with the same rate constant in both directions and with a permeability that is about 40% lower than that of glucose (Knudsen et al. 1991). Therefore an accumulation of lactate to a major extent in cerebral tissue cannot be expected. Accordingly, during activation in rats the brain lactate levels only nearly doubled during stimulation and glucose levels remained essentially constant (Madsen et al. 1999). This is supported by recent human studies using MR spectroscopy where the lactate concentration in the brain did not reach detectable levels following maximal exercise although the arteriovenous differences for lactate across the brain increased considerably with increasing blood lactate levels (Fig. 14.2, Dalsgaard et al. 2004). Dalsgaard and colleagues found that the metabolic ratio (oxygen/[glucose + ½lactate]) decreased from the expected value of 6 to about 3 and remained low during early recovery, indicating that lactate is not oxidized, at least not during the experimental time (Dalsgaard et al. 2004). In conclusion, previous data support the idea that lactate, both produced peripherally and taken up by the brain as well as produced by astrocytes, is accumulating in the brain during activation, but not to a degree that may explain the discrepancy between oxygen and glucose metabolism. Further, since brain lactate concentration and oxygen consumption do not increase markedly during activation, lactate must be metabolized by other pathways than through the TCA cyclus.

It would be tempting to assume that during activation, part of the excess glucose uptake as well as lactate produced by the brain were metabolised to glycogen. However, experimental studies in the rat reveal that the glycogen content in the brain decreases during activation (Madsen et al. 1999). In order to account for the mismatch between oxygen and glucose metabolism during activation, and also account for the decreasing level of brain glycogen, it must be inferred from the available data that lactate produced by both glycogenolysis and glycolysis is cleared from the activated tissue without further oxidation (Dienel and Cruz 2003). Thus, more complex metabolic patterns in the brain might be reflected in the excess glucose uptake during activation, and there might by a temporal and even spatial relationship between lactate production and subsequent oxidation. Measuring metabolic rates using the Kety-Schmidt technique and arteriovenous differences for glucose, oxygen, and lactate following global mental activation in humans, Madsen and colleagues found that the resetting of the oxygen/glucose ratio (from 6.0 to 5.4) persisted up to at least 40 min after activation (Madsen et al. 1995). However, in similar studies in the rat, during the postactivation period the oxygen/glucose ratio rose to above pre-activation levels, indicating that excess carbohydrates (including lactate) taken up by brain during activation may be oxidized in the postactivation period (Madsen et al. 1998; Madsen et al. 1999), thus "normalizing" the overall stoichiometric relationship between carbohydrates and oxygen, where all carbohydrates needed for energy production are eventually oxidized.

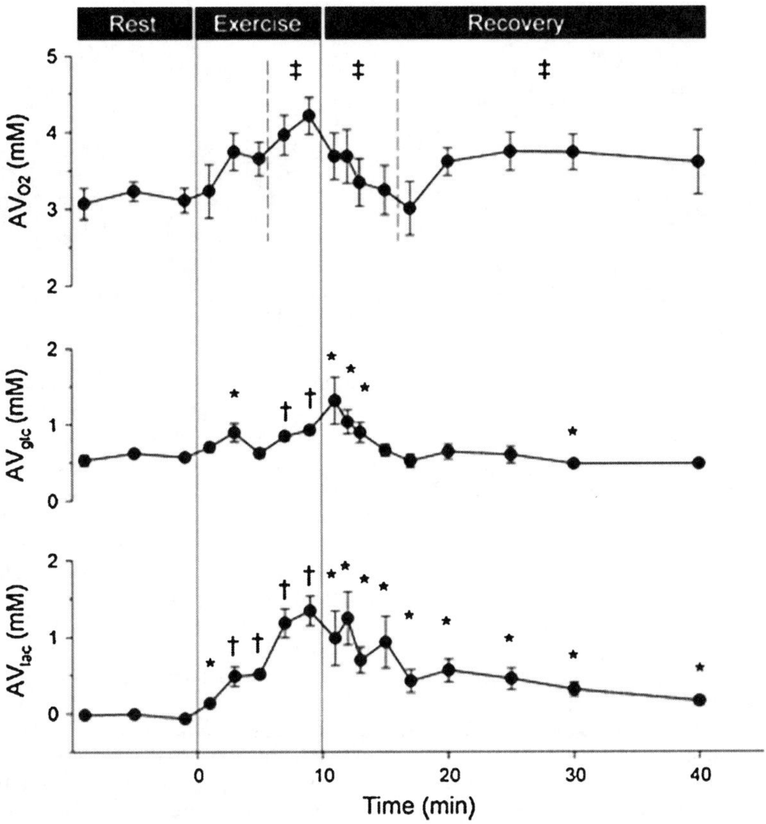

Fig. 14.2 Arteriovenous differences (AV) across the brain during rest, exhaustive arm and leg exercise (Exercise) and during the recovery period. During rest, AV differences for lactate (AV_{lac}) are close to zero, but a significant cerebral uptake of lactate is noted during exercise because of increasing arterial concentrations. During the recovery period, no detectable increase in brain lactate concentrations was found by MRI spectroscopy or in CSF lactate concentrations, suggesting that lactate is taken up by the brain and metabolized when blood lactate increases during exercise (From Dalsgaard et al. 2004)

Recent two-photon fluorescence imaging of nicotinamide adenine dinucleotide (NADH) changes during activation indicate that during the first seconds of the activation there is a very brief glucose oxidation in neurons, followed by a prolonged glycolytic metabolism of glucose in astrocytes (Kasischke et al. 2004). Thus, it seems established that the uncoupling of oxygen and glucose metabolism indeed is caused by non-oxidative metabolism of glucose to lactate, glutamate, and TCA intermediates, which eventually are oxidized during the post-activation period (Fig. 14.3). From studies of astrocytes, it has been suggested that the very thin astrocytic processes that ensheath the capillaries and dendrites of neurons cannot accommodate mitochondria, and that the oxidative capacity of astrocytes is confined to the cell body (Wolff and Chao 2004). Thus, astrocytes may respond immediately to

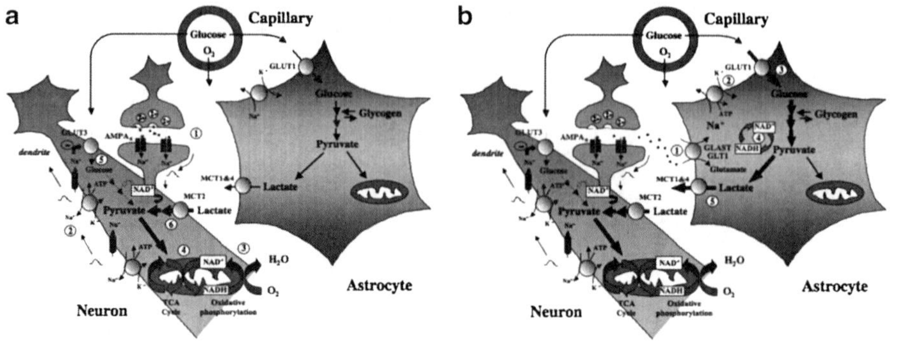

Fig. 14.3 Neuroenergetics upon glutamatergic activation. **A:** Early phase. Activation of glutamatergic afferents leads to synaptic release of glutamate, AMPA receptor (AMPAR) activation and generation of an excitatory postsynaptic potential (EPSP) caused by Na^+ entry within the postsynaptic spine [1]. Depolarization propagates to the dendrite and causes opening of voltage-sensitive Na^+ channels, leading to further Na^+ entry. Reestablishment of ion gradients is accomplished by the Na^+/K^+ ATPase [2] that creates considerable energy expenditure. As a consequence, oxidative phosphorylation is activated [3] and mitochondrial NADH levels first decrease. Then, enhanced TCA cycle activity will ensue [4] to supply NADH for oxidative phosphorylation and support ATP production. As pyruvate utilization in the TCA cycle increases and its cytoplasmic levels decrease, the conditions become favorable for both enhanced glucose and lactate use. Surprisingly, activation of AMPA receptors and coupled Na^+ entry lead to a reduction in glucose uptake and utilization in neurons [5], thus further favoring lactate utilization as preferential oxidative substrate [6]. This would cause a transient drop in extracellular lactate levels as measured *in vivo*. **B:** Late phase. Glutamate released in the synaptic cleft is taken up by astrocytes to be recycled via the specific glutamate transporters GLAST and GLT1 [1]. A large Na^+ influx caused by glutamate uptake takes place and activates the Na^+/K^+ ATPase [2], glucose transport [3] and [4] glucose utilization in astrocytes. The enhancement of aerobic glycolysis in astrocytes first causes a large increase in cytosolic NADH that normalizes with the conversion of pyruvate into lactate and its release via monocarboxylate transporters expressed on astrocytes (mainly MCT1 and 4) [5]. Such a lactate release following glutamatergic activation corresponds to the increase in extracellular lactate levels measured *in vivo*. Lactate produced by astrocytes during this later phase of activation not only replenishes the extracellular pool but also could help sustain neuronal energy needs as activation persists (Figure and Legend modified from Pellerin et al. (2007). For further information and references, please see this review)

changes in synaptic activity but can only do so by glycolytic ATP production (Kasischke et al. 2004). In order to increase the immediate need for energy, astrocytes may also degrade glycogen stores through glycolysis and thus increase ATP production considerably (Cruz and Dienel 2002).

14.5.4 The Neurochemical Coupling of Cerebral Blood Flow and Glucose Metabolism

The close coupling between changes in cerebral blood flow and glucose consumption in the brain during activation suggests that it is the need for an excess glucose uptake that drives the coupling. In the regulation of this response, astrocytes play an important role. Astrocytic endfeets ensheath cerebral arterioles (Straub and Nelson 2007) and link neurons and their synapses with the vasculature. Thereby they are in a strategic position to transmit neuronal signals to the blood vessels. Thus, exposure of astrocytes to glutamate stimulates glucose utilization as well as lactate production (for review, see Pellerin et al. 2007). In genetically modified animals lacking glial glutamate transporters, the increase in CMR_{glc} following activation is reduced (Cholet et al. 2001; Voutsinos-Porche et al. 2003). Further, glutamate rapidly stimulates glucose transport into astrocytes, thus linking glutamatergic synaptic activity to changes in glucose uptake (Loaiza et al. 2003). Zonta et al. (2003) have shown that during synaptic activity, glutamate released from axonal terminals binds to glutamate receptors on neighbouring astrocytic membranes and thus changes the intracellular levels of calcium that provides a proximal stimulus for a variety of vasodilating pathways (Fig. 14.4, for review, see Paulson et al. 2010). Some of these vasodilating substances act at the astrocytic endfeet in contact with arterioles (Zonta et al. 2003). The calcium increase may traverse multiple astrocytes in order to spread the signal to neighbouring areas, and to astrocytic endfeet (Takano et al. 2006). Further, during brain activation, the increase in extracellular potassium may also be transported to astrocytic endfeet, where it (together with calcium) may act as a vasodilator (Paulson and Newman 1987; Filosa et al. 2006). Finally, adenosine produced from the breakdown of ATP in astrocytic processes acts as a strong vasodilator (Dirnagl et al. 1994) and may provide a direct link between energy deficiency and changes in cerebral blood flow.

14.6 Conclusions

A close coupling between cerebral glucose consumption and cerebral blood flow has been established through numerous studies in the past. In contrast to the tight coupling between regional glucose metabolism and cerebral blood flow, there is an uncoupling between flow and oxygen consumption during activation as the latter only increases to a limited extend. The apparent excess glucose uptake is still not completely understood. Although some of the excess glucose uptake can be explained for by lactate production, more complex metabolic patterns in the brain might be reflected in the excess glucose uptake during activation, and especially temporal and spatial relationships must be taken into account. The most likely series of events in the changes in glucose metabolism following activation are a short increase in neuronal glucose oxidation, followed by a more prolonged astrocytic glycolysis, by which ATP can be quickly produced.

The coupling between glucose metabolism and blood flow seems to be driven by the need for local glucose uptake to fuel the synaptic activity. Thus, the possible oxygen deficit in tissue distant from the capillaries is probably of minor importance, and the non-oxidative metabolism during activation may be caused by the astrocytes lacking mitochondria in the tiny processes that encompass synapses. The neurochemical mechanisms by which cerebral blood flow is coupled to glucose metabolism seem to include astrocytes as important mediators that transmit vasoactive substances to endothelial cells.

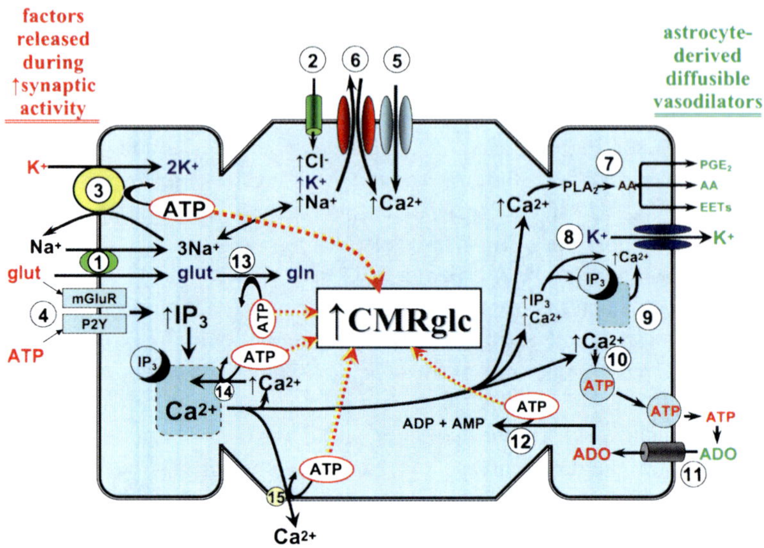

Fig. 14.4 A generalized depiction of the energy cost of neurovascular coupling and its impact on astrocytic glucose utilization (CMRglc). After enhanced synaptic activity, increases in extracellular levels of ATP, glutamate (glut), and K^+ occur (shown in red on the left hand margin of the figure). It has been suggested that these factors, particularly glutamate, arise from pre-synaptic neuronal elements, at least at low to moderate increases in synaptic activity. However, at higher levels of activity, post-synaptic contributions may come into play. Glutamate enters the astrocyte through the glut-3Na^+ cotransporter [1], whereas key K^+ routes of entry include the Na^+-K^+-2Cl^-_ cotransporter [2], and, in exchange for the increased Na^+ from [1] and [2], the Na-K ATPase [3]. Glutamate and ATP can activate their respective metabotropic receptors [4]; leading to IP3-mediated Ca^{2+} release from the ER. Depletion of ER Ca^{2+} promotes Ca^{2+} entry through store-operated Ca^{2+} channels [5]. In addition, the elevated intracellular Na^+ may be exchanged for extracellular Ca^{2+}, through the plasma membrane Na-Ca exchanger operating in reverse mode [6], contributing further to the elevation in intracellular Ca^{2+}. The Ca^{2+} rise provides a proximal stimulus for a variety of vasodilating pathways. Some of the major paracrine vasodilating factors arising from these pathways are shown in green on the right hand margin of the figure. This includes arachidonic acid (AA) derived from Ca^{2+}-activated phospholipase A2 (PLA2) [7]. The AA can be further metabolized, through cyclooxygenase-1 (COX1), to vasodilating prostanoids, primarily PGE2, and/or converted to epoxyeicosatrienoic acids (EETs) by a cytochrome P450 epoxygenase. Arachidonic acid itself can act as a direct vasodilator (Bryan et al. 2006). An increase in K + levels (up to 15–20 mmol/L) in the extracellular milieu provides a potent vasodilating stimulus through acting on smooth muscle inward rectifier K^+ channels. A key conduit for astrocytic release of K^+ is the large-conductance Ca^{2+}-operated K^+ (BKCa) channel [8]. Those channels are well expressed in astrocyte foot processes, along with IP3-sensitive Ca^{2+} storage and release sites [9]. The latter is thought to provide the localized elevations in [Ca^{2+}] necessary for channel activation. Increased cytosolic Ca + may also promote ATP exocytosis [10], although Ca^{2+}-independent and plasma membrane hemichannel routes may also exist. The released ATP is rapidly hydrolyzed to adenosine (ADO) by astrocytic ectonucleotidases. The ADO is returned to the cell through nucleoside transporters [11] to enter the adenosine kinase-mediated adenine nucleotide salvage pathway [12]. There are multiple sites of ATP consumption that, in turn, may trigger increased CMRglc. Some of the key sites that can be directly linked to astrocyte-derived vasodilating factors are shown. Principal among these is the plasma membrane Na-K ATPase [3]. Other ATP-consuming sites depicted in the figure are glutamine (gln) synthase-mediated conversion of glut to gln [13], the ER Ca-ATPase [14], the plasma membrane Ca-ATPase [15], and adenosine kinase-mediated formation of ADP and AMP [12] (From Paulson et al. (2010). For further information and references, please see this review)

References

Blomqvist G, Stone-Elander S, Halldin C, Roland PE, Widen L, Lindqvist M, Swahn CG, Langstrom B, Wiesel FA (1990) Positron emission tomographic measurements of cerebral glucose utilization using [1-11C]D-glucose. J Cereb Blood Flow Metab 10:467–483

Brodersen P, Paulson OB, Bolwig TG, Rogon ZE, Rafaelsen OJ, Lassen NA (1973) Cerebral hyperemia in electrically induced epileptic seizures. Arch Neurol 28:334–338

Brooks RA, Hatazawa J, Di Chiro G, Larson SM, Fishbein DS (1987) Human cerebral gluocse metabolism determined by positron emission tomography: a revisit. J Cereb Blood Flow Metab 7:427–432

Bryan RM Jr, You J, Phillips SC, Andresen JJ, Lloyd EE, Rogers PA, Dryer SE, Marrelli SP (2006) Evidence for two-pore domain potassium channels in rat cerebral arteries. Am J Physiol Heart Circ Physiol Aug 291(2):H770–H780

Buxton RB, Frank LR (1997) A model for the coupling between cerebral blood flow and oxygen metabolism during neural stimulation. J Cereb Blood Flow Metab 17:64–72

Choi IY, Gruetter R (2003) In vivo 13C NMR assessment of brain glycogen concentration and turnover in the awake rat. Neurochem Int 43(4–5):317–322

Cholet N, Pellerin L, Welker E, Lacombe P, Seylaz J, Magistretti P, Bonvento G (2001) Local injection of antisense oligonucleotides targeted to the glial glutamate transporter GLAST decreases the metabolic response to somatosensory activation. J Cereb Blood Flow Metab 21:404–412

Cohen PJ, Alexander SC, Smith TC, Reivich M, Wollman H (1967) Effects of hypoxia and normocarbia on cerebral blood flow and metabolism in conscious man. J Appl Physiol 23:183–189

Cruz NF, Dienel GA (2002) High glycogen levels in brains of rats with minimal environmental stimuli: Implications for metabolic contributions of working astrocytes. J Cereb Blood Flow Metab 22(12):1476–1489

Dalsgaard MK, Quistorff B, Danielsen ER, Selmer C, Vogelsang T, Secher NH (2004) A reduced cerebral metabolic ratio in exercise reflects metabolism and not accumulation of lactate within the human brain. J Physiol 554:571–578

Dienel GA, Cruz NF (2003) Neighborly interactions of metabolically-activated astrocytes in vivo. Neurochem Int 43:339–354

Dirnagl U, Niwa K, Lindauer U, Villringer A (1994) Coupling of cerebral blood flow to neuronal activation: role of adenosine and nitric oxide. Am J Physiol 267:H296–H301

Dringen R, Gebhardt R, Hamprecht B (1993) Glycogen in astrocytes: possible function as lactate supply for neighboring cells. Brain Res 623:208–214

Filosa JA, Bonev AD, Straub SV, Meredith AL, Wilkerson MK, Aldrich RW, Nelson MT (2006) Local potassium signaling couples neuronal activity to vasodilation in the brain. Nat Neurosci 9(11):1397–1403

Fox PT, Raichle ME, Mintun MA, Dence C (1988) Nonoxidative glucose consumption during focal physiologic neural activation. Science 241:462–464

Graham MM, Muzi M, Spence AM, O'Sullivan F, Lewellen TK, Link JM, Krohn KA (2002) The FDG lumped constant in normal human brain. J Nucl Med 43:1157–1166

Gruetter R (2003) Glycogen: the forgotten cerebral energy store. J Neurosci Res 74(2):179–183, Oct 15

Hasselbalch SG, Madsen PL, Hageman LP, Olsen KS, Justesen N, Holm S, Paulson OB (1996) Changes in cerebral blood flow and carbohydrate metabolism during acute hyperketonemia. Am J Physiol 270:E746–E751

Hasselbalch SG, Madsen PL, Knudsen GM, Holm S, Paulson OB (1998) Calculation of the FDG lumped constant by simultaneous measurements of global glucose and FDG metabolism in humans. J Cereb Blood Flow Metab 18:154–160

Hertz L, Peng L, Dienel GA (2007) Energy metabolism in astrocytes: high rate of oxidative metabolism and spatiotemporal dependence on glycolysis/glycogenolysis. J Cereb Blood Flow Metab 27(2):219–249

Hofmann M, Pichler B, Schölkopf B, Beyer T (2009) Towards quantitative PET/MRI: A review of MR-based attenuation correction techniques. Eur J Nucl Med Mol Imaging 36(Suppl 1): S93–S104

Huang SC, Phelps ME, Hoffman EJ, Sideris K, Selin CJ, Kuhl DE (1980) Noninvasive determination of local cerebral metabolic rate of glucose in man. Am J Physiol 238:E69–E82

Jueptner M, Weiller C (1995) Review: does measurement of regional cerebral blood flow reflect synaptic activity? Implications for PET and fMRI. Neuroimage 2:148–156

Kasischke KA, Vishwasrao HD, Fisher PJ, Zipfel WR, Webb WW (2004) Neural activity triggers neuronal oxidative metabolism followed by astrocytic glycolysis. Science 305(5680):99–103, Jul 2

Kety SS (1957) The general metabolism of the brain in vivo. In: Richter D (ed) Metabolism of the nervous system. Pergamon, New York, pp 221–236

Kety SS, Schmidt CF (1948) The nitrous oxide method for quantitative determinations of cerebral blood flow in man: theory, procedure, and normal values. J Clin Invest 27:476–483

Knudsen GM (1994) Application of the double-indicator technique for measurement of blood-brain barrier permeability in humans. Cerebrovasc Brain Metab Rev 6:1–30

Knudsen GM, Pettigrew KD, Paulson OB, Hertz MM, Patlak CS (1990) Kinetic analysis of blood-brain barrier transport of d-glucose in man: quantitative evaluation in the presence of tracer backflux and capillary heterogeneity. Microvasc Res 39(1):28–49

Knudsen GM, Paulson OB, Hertz MM (1991) Kinetic analysis of the human blood-brain barrier transport of lactate and its influence by hypercapnia. J Cereb Blood Flow Metab 11:581–586

Knutsson L, Börjesson S, Larsson EM, Risberg J, Gustafson L, Passant U et al (2007) Absolute quantification of cerebral blood flow in normal volunteers: correlation between xe-133 SPECT and dynamic susceptibility contrast MRI. J Magn Reson Imaging 26(4):913–920

Kuschinsky W, Paulson OB (1992) Capillary circulation in the brain. Cerebrovasc Brain Metab Rev 4:261–286

Kuwabara H, Evans AC, Gjedde A (1990) Michaelis-Menten constraints improved cerebral glucose metabolism and regional lumped constant measurements with [18F]fluorodeoxyglucose. J Cereb Blood Flow Metab 10:180–189

Lassen NA (1959) Cerebral blood flow and oxygen consumption in man. Physiol Rev 39:183–238

Law I, Iida H, Holm S, Nour S, Rostrup E, Svarer C, Paulson OB (2000) Quantitation of regional cerebral blood flow corrected for partial volume effect using O-15 water and PET: II. Normal values and gray matter blood flow response to visual activation. J Cereb Blood Flow Metab 20:1252–1263

Leenders KL, Perani D, Lammertsma AA, Heather JD, Buckingham P, Healy MJ, Gibbs JM, Wise RJ, Hatazawa J, Herold S (1990) Cerebral blood flow, blood volume and oxygen utilization. Normal values and effect of age. Brain 113(Pt 1):27–47

Loaiza A, Porras OH, Barros LF (2003) Glutamate triggers rapid glucose transport stimulation in astrocytes as evidenced by real-time confocal microscopy. J Neurosci 23:7337–7342

Madsen PL, Holm S, Herning M, Lassen NA (1993) Average blood flow and oxygen uptake in the human brain during resting wakefulness: a critical appraisal of the Kety-Schmidt technique. J Cereb Blood Flow Metab 13:646–655

Madsen PL, Hasselbalch SG, Hagemann LP, Olsen KS, Bulow J, Holm S, Wildschiodtz G, Paulson OB, Lassen NA (1995) Persistent resetting of the cerebral oxygen/glucose uptake ratio by brain activation: evidence obtained with the Kety-Schmidt technique. J Cereb Blood Flow Metab 15:485–491

Madsen PL, Linde R, Hasselbalch SG, Paulson OB, Lassen NA (1998) Activation-induced resetting of cerebral oxygen and glucose uptake in the rat. J Cereb Blood Flow Metab 18:742–748

Madsen PL, Cruz NF, Sokoloff L, Dienel GA (1999) Cerebral oxygen/glucose ratio is low during sensory stimulation and rises above normal during recovery: excess glucose consumption during stimulation is not accounted for by lactate efflux from or accumulation in brain tissue. J Cereb Blood Flow Metab 19:393–400

Magistretti PJ, Pellerin L (1996) Cellular mechanisms of brain energy metabolism. Relevance to functional brain imaging and to neurodegenerative disorders. Ann N Y Acad Sci 777: 380–387

Magistretti PJ, Pellerin L (1999) Cellular mechanisms of brain energy metabolism and their relevance to functional brain imaging. Philos Trans R Soc Lond B Biol Sci 354:1155–1163

Medina JM, Tabernero A (2005) Lactate utilization by brain cells and its role in CNS development. J Neurosci Res 79(1–2):2–10

Mintun MA, Raichle ME, Martin WR, Herscovitch P (1984) Brain oxygen utilization measured with O-15 radiotracers and positron emission tomography. J Nucl Med 25(2):177–187

Muller-Gartner HW, Links JM, Prince JL, Bryan RN, McVeigh E, Leal JP, Davatzikos C, Frost JJ (1992) Measurement of radiotracer concentration in brain gray matter using positron emission tomography: MRI-based correction for partial volume effects. J Cereb Blood Flow Metab 12:571–583

Ostergaard L, Smith DF, Vestergaard-Poulsen P, Hansen SB, Gee AD, Gjedde A, Gyldensted C (1998) Absolute cerebral blood flow and blood volume measured by magnetic resonance imaging bolus tracking: comparison with positron emission tomography values. J Cereb Blood Flow Metab 18:425–432

Paulson OB, Newman EA (1987) Does the release of potassium from astrocyte endfeet regulate cerebral blood flow? Science 237:896–898

Paulson OB, Hasselbalch SG, Rostrup E, Knudsen GM, Pelligrino D (2010) Cerebral blood flow response to functional activation. J Cereb Blood Flow Metab 30(1):2–14

Pellerin L, Magistretti PJ (1994) Glutamate uptake into astrocytes stimulates aerobic glycolysis: a mechanism coupling neuronal activity to glucose utilization. Proc Natl Acad Sci USA 91:10625–10629

Pellerin L, Magistretti PJ (2004) Neuroenergetics: calling upon astrocytes to satisfy hungry neurons. Neuroscientist 10:53–62

Pellerin L, Bouzier-Sore AK, Aubert A, Serres S, Merle M, Costalat R, Magistretti PJ (2007) Activity-dependent regulation of energy metabolism by astrocytes: an update. Glia 55(12):1251–1262

Perlmutter JS, Powers WJ, Herscovitch P, Fox PT, Raichle ME (1987) Regional asymmetries of cerebral blood flow, blood volume, and oxygen utilization and extraction in normal subjects. J Cereb Blood Flow Metab 7(1):64–67

Phelps ME, Huang SC, Hoffman EJ, Selin C, Sokoloff L, Kuhl DE (1979) Tomographic measurement of local cerebral glucose metabolic rate in humans with (F-18)2-fluoro-2-deoxy-D-glucose: validation of method. Ann Neurol 6:371–388

Reivich M, Alavi A, Wolf A, Fowler J, Russell J, Arnett C, MacGregor RR, Shiue CY, Atkins H, Anand A et al (1985) Glucose metabolic rate kinetic model parameter determination in humans: the lumped constants and rate constants for [18F]fluorodeoxyglucose and [11C]deoxyglucose. J Cereb Blood Flow Metab 5:179–192

Rothman DL, Sibson NR, Hyder F, Shen J, Behar KL, Shulman RG (1999) In vivo nuclear magnetic resonance spectroscopy studies of the relationship between the glutamate-glutamine neurotransmitter cycle and functional neuroenergetics. Philos Trans R Soc Lond B Biol Sci 354: 1165–1177

Rothstein JD, Martin L, Levey AI, Dykes-Hoberg M, Jin L, Wu D, Nash N, Kuncl RW (1994) Localization of neuronal and glial glutamate transporters. Neuron 13:713–725

Roy CS, Sherrington CS (1890) On the regulation of the blood supply of the brain. J Physiol (Lond) 11:85–108

Schousboe A, Westergaard N, Sonnewald U, Petersen SB, Huang R, Peng L, Hertz L (1993) Glutamate and glutamine metabolism and compartmentation in astrocytes. Dev Neurosci 15: 359–366

Smith D, Pernet A, Hallett WA, Bingham E, Marsden PK, Amiel SA (2003) Lactate: a preferred fuel for human brain metabolism in vivo. J Cereb Blood Flow Metab 23:658–664

Straub SV, Nelson MT (2007) Astrocytic calcium signaling: the information currency coupling neuronal activity to the cerebral microcirculation. Trends Cardiovasc Med 17(6):183–190

Takano T, Tian GF, Peng W, Lou N, Libionka W, Han X, Nedergaard M (2006) Astrocyte-mediated control of cerebral blood flow. Nat Neurosci 9(2):260–267

Voutsinos-Porche B, Bonvento G, Tanaka K, Steiner P, Welker E, Chatton JY, Magistretti PJ, Pellerin L (2003) Glial glutamate transporters mediate a functional metabolic crosstalk between neurons and astrocytes in the mouse developing cortex. Neuron 37:275–286

Wolff J, Chao TI (2004) Cytoarchitectonics of nonneuronal cells in the central nervous system. In: Hertz L (ed) Non-neuronal cells of the nervous system: function and dysfunction. Elsevier, Amsterdam, pp 1–52

Zonta M, Sebelin A, Gobbo S, Fellin T, Pozzan T, Carmignoto G (2003) Glutamate-mediated cytosolic calcium oscillations regulate a pulsatile prostaglandin release from cultured rat astrocytes. J Physiol 553:407–414

Chapter 15
Neuronal Basis of Non-Invasive Functional Imaging: From Microscopic Neurovascular Dynamics to BOLD fMRI

Anna Devor, David A. Boas, Gaute T. Einevoll, Richard B. Buxton, and Anders M. Dale

Abstract With the growing recognition of the complexity of neurovascular coupling, research has focused on the "neurovascular unit", a close association between neurons, astrocytes and blood vessels. A number of experimental tools have been developed for probing the neurovascular unit in animal models, providing the potential for a much deeper understanding of these fundamental physiological mechanisms. In this chapter, we review some of the available experimental and computational methods and present a multi-level conceptual framework for analyzing and interpreting a wide range of experimental measurements. We then discuss

A. Devor, Ph.D. (✉)
Departments of Neurosciences and Radiology, University of California San Diego, La Jolla, CA 92093-0662, USA

Martinos Center for Biomedical Imaging, Massachusetts General Hospital, Harvard Medical School, Charlestown, MA, USA
e-mail: adevor@ucsd.edu

D.A. Boas, Ph.D.
Martinos Center for Biomedical Imaging, Massachusetts General Hospital (MGH), Harvard Medical School, 13th Street Building 149, Charlestown, MA 02129, USA
e-mail: dboas@nmr.mgh.harvard.edu

G.T. Einevoll
Norwegian University of Life Sciences, Department of Mathematical Sciences and Technology, Aas, Norway
e-mail: Gaute.Einevoll@umb.no

R.B. Buxton, Ph.D.
Department of Radiology, University of California San Diego, La Jolla, CA 92093-0677, USA
e-mail: rbuxton@ucsd.edu

A.M. Dale, Ph.D.
Departments of Neurosciences and Radiology, University of California San Diego, San Diego, CA 92103-8756, USA
e-mail: amdale@ucsd.edu

I.-Y. Choi and R. Gruetter (eds.), *Neural Metabolism In Vivo*, Advances in Neurobiology 4, DOI 10.1007/978-1-4614-1788-0_15,

our working hypotheses regarding the regulation of blood flow and neurophysiological correlates of fMRI signals. Finally, we discuss how multimodal imaging, along with valid physiological models, can ultimately be used to obtain quantitative estimates of physiological parameters in health and disease and provide an outlook for the future directions in neurovascular research.

Keywords Imaging • Neurovascular • Neurometabolic • CBF • CMRO2 • LFP • CSD • MUA • fMRI • BOLD • Optical • Microscopy • Hemodynamic • Neurovascular unit • Extracellular potential • Forward modeling • Laminar population analysis

15.1 Introduction

The brain is a highly energy demanding organ, and neurovascular coupling is a critical component linking neuronal activity to blood flow and the delivery of glucose and oxygen for energy metabolism. A better characterization of these connections is important for understanding how these relationships may change with disease. In addition, understanding in a quantitative way how neuronal activity drives changes in blood flow and energy metabolism is critical for laying a solid physiological foundation for interpreting functional neuroimaging studies in humans. In particular, functional magnetic resonance imaging (fMRI) based on the blood oxygenation level dependent (BOLD) response has become a widely used tool for exploring brain function, and yet the physiological basis of this technique is still poorly understood. The primary physiological phenomenon that leads to the BOLD effect is that cerebral blood flow (CBF) increases much more than the cerebral metabolic rate of oxygen ($CMRO_2$), increasing local blood oxygenation and causing a slight increase of the MRI signal due to different magnetic properties of oxy- and deoxyhemoglobin. Although the goal of fMRI studies is usually to assess changes in neuronal activity, the signal measured depends on the relative changes in CBF and $CMRO_2$. For this reason, it is difficult to interpret the magnitude of the BOLD response as a quantitative reflection of the magnitude of the change in neuronal activity without a deeper understanding of how neuronal activity, CBF and $CMRO_2$ are connected. For example, if the BOLD responses to two tasks are different, or the response is different in disease, does that mean the underlying neuronal responses are different or that the vascular or metabolic responses are different?

The simplest picture one could imagine for this physiological coupling is that changes in neuronal activity drive changes in energy metabolism which then drive changes in blood flow. However, a large body of work over the last decade indicates that this simple picture is almost certainly wrong. While molecules produced by increased energy metabolism (CO_2, lactate, H^+, etc.) do have a vasoactive effect, much of the acute CBF response under healthy conditions appears to be driven by molecules related to neuronal signaling. In short, rather than a linear chain of events, blood flow and energy metabolism appear to be driven in parallel by neuronal activity. Because the BOLD response depends on changes in both CBF and $CMRO_2$, a quantitative understanding of exactly which aspects of neuronal activity drive the CBF and $CMRO_2$ changes is critical for interpreting the BOLD response. For example, the primary energy cost of neuronal signaling is thought to be associated with the need to pump ions (particularly sodium) across cell membranes to restore ion gradients that are degraded in neuronal signaling. For this reason $CMRO_2$ may simply respond to the net energy cost of the evoked activity. However, if the CBF response is primarily driven by aspects of synaptic activity in a feed-forward manner, the balance of these changes, and the resulting BOLD response, could differ depending on the details of the changes in neuronal activity.

With the growing recognition of the complexity of neurovascular coupling, research has focused on the "neurovascular unit", a close association between neurons, astrocytes and blood vessels. A number of experimental tools have been developed for probing the neurovascular unit in animal models, providing the potential for a much deeper understanding of these fundamental physiological mechanisms. In this chapter, we review some of the available experimental and computational methods, discuss our working hypotheses regarding the regulation of blood flow and neurophysiological correlates of fMRI signals, and provide an outlook for the future directions in neurovascular research. Our goal is not to provide an exhaustive list of the available methodologies but rather to illustrate how multimodal data can be combined in a theoretical framework in a way where the result is greater than the sum of its parts. Along the way we offer practical guidelines for use of each one of the technologies, describe modeling approaches, and suggest references for further in-depth reading on each subject. Although the main intention is to target an audience of graduate students, the authors hope that this chapter has something to offer to each reader, whether they look for technical advice, a piece of evidence or a conceptual framework. Enjoy the reading!

15.2 Measurement Theory: Measured Parameters and Their Relationship to the Underlying Physiological Variables

Below we provide an overview of our optical and electrophysiological tools and discuss their advantages and shortcomings. The technological aspects of fMRI are described in detail in a number of other chapters in this volume, and also in publications elsewhere (Brown et al. 2007). When considering these methods, it is important to keep in mind that the signals we can measure usually are not direct measurements of the physiological quantities of interest, such as blood flow or energy metabolism. Moreover, different measurement modalities can reflect different aspects of the same physiological process. For example, an increase in neuronal activity can be measured as an increase in spike count, synaptic release, or generation of local field potential (LFP). For this reason, there is a measurement theory associated with each technique that relates the measured signals to more basic physiological quantities, indicated by the arrows in Fig. 15.1. This often involves assumptions about the physical measurements or the underlying physiology, and to use these techniques effectively it is

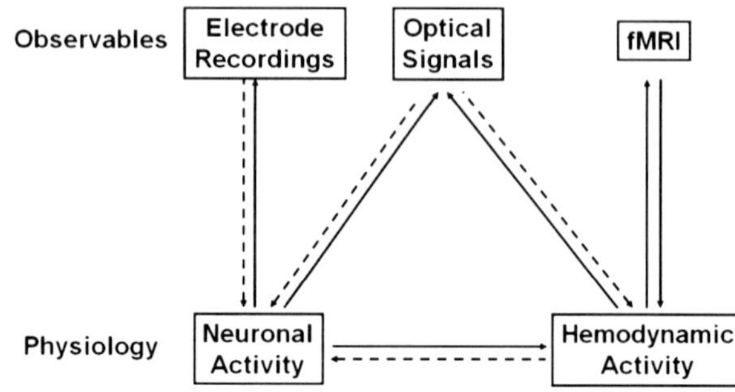

Fig. 15.1 Relationship between measurements to the underlying physiological variables. *Solid lines* indicate mappings that are "well-posed" (i.e., have a unique solution), while *dashed lines* indicate mappings that are "ill-posed" (i.e., have multiple solutions)

important to understand the assumptions and potential limitations (Dale and Halgren 2001). In addition, a recurring theme in this chapter is that the combination of two or more methods in a multi-modal imaging approach often can provide more quantitative or specific information than either technique alone. Examples are the combination of techniques sensitive to blood oxygenation changes (optical or BOLD signals) with a CBF measurement to make possible estimation of $CMRO_2$ changes (see Sect. 15.3.3.2), or measurements of spiking and synaptic response for extraction of activity of specific neuronal populations (see Sect. 15.3.2.3).

15.2.1 Optical Imaging

Optical imaging provides a versatile set of tools for studying cerebral physiology, providing measures of neuronal, metabolic, and vascular processes from the subcellular level up to the intact human brain. Although intrinsic optical changes associated with neuronal and metabolic activity have been recognized for over 50 years, Optical Intrinsic Signal Imaging (OISI) was introduced to measure functional architecture of the cortex *in vivo* only in 1986 (Grinvald et al. 1986). This technique uses a camera to image hemoglobin-induced changes in the absorption of visible light that is reflected from the cortex (either through a cranial window, thinned skull, or intact skull), and provides a spatial resolution ranging from ~10 μm on the cortical surface to hundreds of μm in cortical layers II–III. By measuring the absorption changes at multiple wavelengths of light and given knowledge of the path length of light through the tissue, it is possible to quantify absolute hemoglobin concentration changes (Kohl et al. 2000; Dunn et al. 2003). Blood flow can be imaged with Laser Speckle Contrast Imaging (LSCI) (Dunn et al. 2001; Draijer et al. 2008) by exploiting the dynamic fluctuations that moving red blood cells impose on the random interference pattern of the reflected laser light. By combining measures of blood flow and hemoglobin concentration changes, it is possible to estimate the changes in the $CMRO_2$ during brain activation (Dunn et al. 2005; Royl et al. 2008). These techniques are generally employed in animal models using invasive procedures that usually require thinning of the skull or a cranial window, and readily enable simultaneous electrophysiological measurements (Devor et al. 2003). Near Infrared

Spectroscopy (NIRS) works on a similar principle but enables the measurement of the hemoglobin concentrations through the intact skull of humans by using near infrared light (700–900 nm) (Villringer and Chance 1997; Strangman et al. 2002). The greater penetration depth arises from the fact that the near infrared light is only weakly absorbed by hemoglobin, allowing a detectable portion of the incident light to reach the brain, interrogate the hemoglobin concentrations, scatter back to the surface of the scalp, and be detected. This greater scattering, however, reduces the spatial resolution dramatically ranging from ~5 to 10 mm on the cortex of human infants to ~10–20 mm on the cortex of adult humans. NIRS can be combined with fMRI to estimate evoked changes in $CMRO_2$ (Huppert et al. 2009), and with EEG and MEG (Dale et al. 2000) to explore neurovascular coupling in humans non-invasively (Ou et al. 2009). A major limitation of all of these optical methods is a lack of laminar (i.e. depth) resolution and thus it is not possible to distinguish, e.g., superficial cortical responses from deeper cortical responses.

Significantly improved *in vivo* lateral and depth resolution of the cerebral microvascular network and the response of individual microvessels to brain activation were demonstrated in 1998 with 2-photon laser scanning microscopy (Kleinfeld et al. 1998). While 2-photon microscopy enables ~1 μm lateral and depth resolution of vessel diameter and red blood cell velocity, it is limited by a ~600 μm depth penetration. Optical Coherence Tomography (OCT) shows promise to overcome this issue by using interferometric detection to enhance detection sensitivity enabling a penetration depth of 1 mm or more and increased detection rate (Huang et al. 1991; Drexler et al. 1999). Combined with Doppler detection techniques (Leitgeb et al. 2003; White et al. 2003; Bizheva et al. 2004), it is possible to form volumetric images of red blood cell velocity spanning several cubic millimeters with ~10 μm resolution in ~10 s. This OCT methodology is being used extensively in studies of the retina (Huber et al. 2007) and is finding application in the cortex (Aguirre et al. 2006; Chen et al. 2008; Srinivasan et al. 2009; Srinivasan et al. 2010a; Srinivasan et al. 2010b). Optical imaging is used also for direct measurements of neuronal activity by employing fluorescent sensors of voltage or ionic concentrations (Tsien 1981; Miller 1988; Ross 1989; Grinvald and Hildesheim 2004; Baker et al. 2005). The most popular indicators are voltage- and calcium-sensitive dyes. While both can be applied extrinsically, significant progress has been recently made in development of genetically encoded calcium indicators with sufficient sensitivity for detection of single spikes (Wallace et al. 2008).

Below we choose to focus on a number of microscopic techniques, which have been successfully used by us and others for optical imaging of neuronal, vascular and metabolic activity.

15.2.1.1 Combined Spectral Imaging of Blood Oxygenation and Laser Speckle Imaging of Blood Flow

When OISI is performed using a number of different illumination wavelengths (referred to thereafter as "spectral" imaging), it enables estimation of changes in

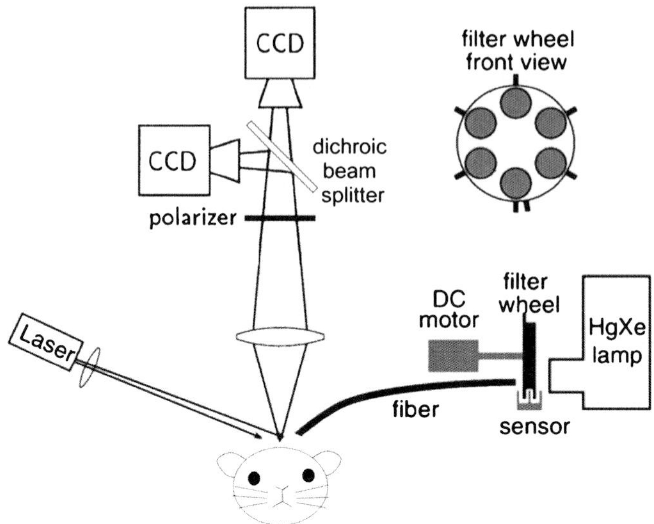

Fig. 15.2 Double-camera setup for simultaneous spectral and laser speckle measurements

oxyhemoglobin (ΔHbO), deoxyhemoglobin (ΔHb) and total hemoglobin (ΔHbT) (Devor et al. 2003; Dunn et al. 2003; Devor et al. 2005; Dunn et al. 2005; Boas et al. 2008). Spectral imaging can be combined with LSCI (referred to thereafter as "laser speckle" imaging) for concurrent 2-dimensional mapping of blood flow (Dunn et al. 2001; Bolay et al. 2002; Dunn et al. 2003; Ayata et al. 2004; Dunn et al. 2005). Simultaneous imaging of blood flow, volume and oxygenation allows for calculation of $CMRO_2$ (see Sect. 15.3.3.2 below).

During simultaneous spectral and laser speckle measurements, the detected light is split via a dichroic mirror, filtered and directed towards two dedicated detectors for imaging of blood oxygenation and speckle contrast respectively (Fig. 15.2). In our system, the filtering is achieved by passing light below 650 nm to the spectral detector, and 780 nm light (FWHM of 10 nm) to the speckle detector.

For spectral imaging n different bandpass filters ($n=6$ in our imager) are placed on a multi-position filter wheel, mounted on a DC motor. In the imager used by Dunn et al. (2003, 2005) and Devor et al. (2003, 2005, 2007, 2008b), the center wavelengths of the filters range between 560 and 610 nm at 10-nm intervals. Illuminating light from a tungsten-halogen light source (Oriel, Spectra-Physics) is directed through the filter wheel, coupled to a 12 mm fiber bundle. Images of ~6×12 mm area are acquired by a cooled 16 bit CCD camera (Cascade 512B, Photometrics). Image acquisition is triggered at ~13 Hz by individual filters in the filter wheel passing through an optic sensor. The image set at each wavelength is averaged across trials and the averaged data are converted to changes in HbO and Hb at each pixel using the modified Beer Lambert relationship:

$$\Delta A(\lambda,t) = \left(\varepsilon_{HbO}(\lambda)\Delta C_{HbO}(t) + \varepsilon_{HbR}(\lambda)\Delta C_{HbR}(t)\right)D(\lambda) \qquad (15.1)$$

where $\Delta A(\lambda,t) = \log(R_o/R(t))$ is the attenuation at each wavelength, R_o and $R(t)$ are the measured reflectance intensities at baseline and time t, respectively, ΔC_{HbO} and ΔC_{Hb} are the changes in concentrations of HbO and Hb, respectively, and ε_{HbO} and ε_{Hb} are the molar extinction coefficients. This equation is solved for ΔC_{HbO} and ΔC_{Hb} using a least-squares approach. The differential pathlength factor, $D(\lambda)$, accounts for the fact that each wavelength travels slightly different pathlengths through the tissue due to the wavelength dependence of scattering and absorption in the tissue, and is estimated using the approach of Kohl et al. (2000) through Monte Carlo simulations of light propagation in tissue. Baseline concentrations of 60 μM and 40 μM are assumed for HbO and Hb, respectively (Mayhew et al. 2000; Jones et al. 2002). Dunn et al. (2005) indicates that the results for relative hemoglobin changes are only weakly sensitive to these assumed baseline values.

A laser diode (785 nm, 80 mW) is used as a light source for speckle imaging. Raw speckle images are acquired by a high-speed (~120 Hz) 8-bit CMOS camera (A602f, Basler) with an exposure time of 5 ms and in-plane resolution of ~20 μm. The speckle contrast is defined as the ratio of the standard deviation to the mean pixel intensities, $\sigma/\langle I \rangle$ within a localized region of the image (Briers 2001). Speckle contrast is calculated in a series of laser speckle images as described in Dunn et al. (2001, 2005) following spatial smoothing using 5×5 pixel sliding window (Dunn et al. 2001; Dunn et al. 2005). Numerous theoretical advances have recently been made in speckle imaging as reviewed in (Draijer et al. 2008).

Spectral imaging in the brain detects not only changes in hemoglobin oxygenation, but also other perfusion- and metabolism-related signals, including changes in cytochrome oxidation and light scattering. However, at visible wavelengths, such as within the 560–610 nm window used in our imager, the hemoglobin in blood is the most significant absorber (Frostig et al. 1990; Grinvald 1992; Narayan et al. 1994; Narayan et al. 1995; Malonek and Grinvald 1996; Nemoto et al. 1999; Vanzetta and Grinvald 1999). At longer wavelengths, at which Hb and HbO have negligible absorbance, light-scattering effects dominate.

The biggest limitation of the spectral and laser speckle imaging, common for all CCD-based optical methods, is the lack of laminar resolution. The signal at every pixel represents a weighted sum of the response though the whole depth of light penetration (with the highest sensitivity to the cortical surface) (Polimeni et al. 2005). As a result, the spatial resolution in the XY plane is somewhat ambiguous: while on the surface it is determined by the size of the CCD chip, it gradually decreases with the cortical depth due to light scattering in the tissue. One can use complementary optical methods such as OCT (Aguirre et al. 2006; Chen et al. 2008; Srinivasan et al. 2009; Srinivasan et al. 2010a; Srinivasan et al. 2010b), LOT (Hillman et al. 2004; Hillman et al. 2007) and 2-photon microscopy (Kleinfeld et al. 1998; Takano et al. 2006; Chaigneau et al. 2007; Devor et al. 2007; Devor et al. 2008b) to gain depth-resolved images.

Fig. 15.3 Line-scan diameter measurement

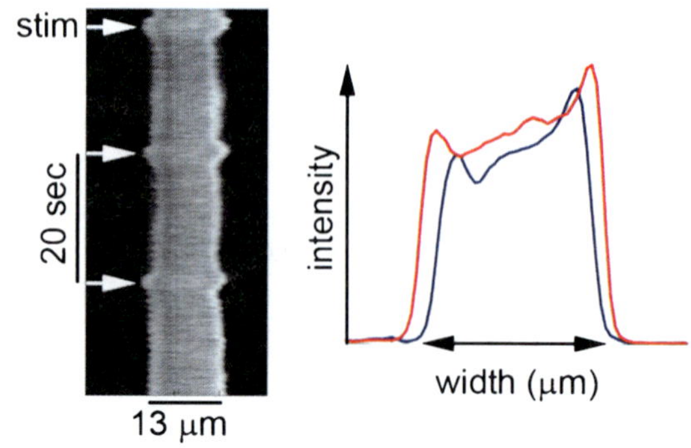

15.2.1.2 Two-Photon Measurements of Single-Vessel Vascular Dynamics

Two-photon microscopy has been employed to image changes in diameter and velocity of blood flow in single arterioles evoked by neuronal, glial and pharmacological stimuli *in vitro* (Simard et al. 2003; Zonta et al. 2003; Cauli et al. 2004; Mulligan and MacVicar 2004; Filosa et al. 2006; Rancillac et al. 2006) and *in vivo* (Faraci and Breese 1993; Kleinfeld et al. 1998; Chaigneau et al. 2003; Kleinfeld and Griesbeck 2005; Takano et al. 2006; Chaigneau et al. 2007; Boas et al. 2008). *In vivo* the cerebral blood flow and vascular diameter changes can be imaged from the cortical surface to as deep as 600 μm below the pia mater of the cortex – down to layer IV in rat cerebral cortex (Kleinfeld et al. 1998). To visualize the vasculature and to track movement of red blood cells (RBCs), dextran-conjugated fluorescent dyes (such as fluorescein or Texas Red) are injected intravenously. In our practice, a 4x air objective (Olympus XLFluor4x/340, NA = 0.28) is used to obtain images of the surface vasculature across the entire cranial window to aid in navigating in the XY-plane. 10x (Zeiss Achroplan, NA = 0.3), Super20x (Olympus, NA = 0.95) and 40x (Olympus, NA = 0.8) water-immersion objectives are used for subsequent high-resolution imaging and line-scan measurements of vessel diameter and RBC velocity.

Measurements of vessel diameter. Absolute vessel diameter is measured by obtaining a planar image stack, or by using continuous line scans to gauge rapid changes over time (Fig. 15.3). The diameter changes are captured by repeated line scans across the vessel that form a space-time image when stacked sequentially (Fig. 15.3, *left*). Diameters are extracted from profile changes (Fig. 15.3, *right*: compare blue (*baseline*) to *red* (*peak dilation*)). By scanning longer distances across multiple vessels, or when using advanced scanning algorithms (Gobel et al. 2007), one can obtain the diameter changes of multiple vessels in a single space-time image. In this case, the scanning direction may not always be perpendicular to the vessel axis for each of the measured vessels when using this approach. However, the fractional diameter change is unaffected and the absolute diameter D can be obtained by $D = D_m \sin\theta$, where D_m is the measured diameter, and θ is the angle between the scan

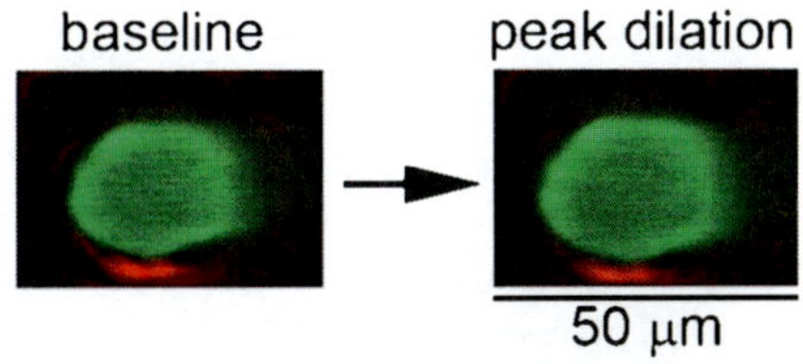

Fig. 15.4 Frame-scan diameter measurement. Intravascular lumen and astrocytic endfeet are in green and red, respectively

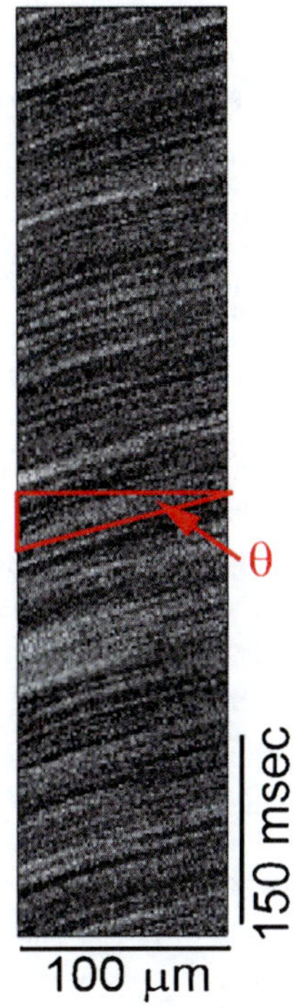

Fig. 15.5 Measurement of RBC velocity

line and the axis of the vessel. In practice, when scanning in a single plane, the number of simultaneously measured vessels is limited to vessels in focus. The diameter can also be estimated from the image time series of a diving vessel obtained in a frame scanning mode by counting the number of pixels above a pre-set intensity threshold (Fig. 15.4). A scan rate of 150 Hz or video-rate frame mode while maintaining diffraction limited spatial resolution is sufficient to capture relatively slow diameter changes.

Measurements of RBC velocity. Intravenously injected dextran-conjugated fluorescent dyes label the plasma leaving RBCs visible as dark shadows on bright fluorescent background (Fig. 15.5). The speed of the RBCs is captured by repeated line scans along the axis of the vessel that form a space-time image when stacked sequentially and leads to the generation of streaks caused by the motion of RBCs. High scan rate (above 1 KHz) is required for capturing of fast arterial flow. The speed of RBCs is given by the reciprocal of the slope of these streaks and the direction of flow is determined from the sign of the slope (Fig. 15.4, velocity = $1/\tan\theta$). An algorithm based on singular value decomposition is used to automate the calculation of speed from the line-scan data (Kleinfeld et al. 1998). Given both measurements of diameter and velocity one can calculate flux (number of RBCs per second) (Devor et al. 2008a) that might have a more direct relationship to the MRI pulsed Arterial Spin Labeling (ASL) measure of blood flow (Brown et al. 2007).

15.2.1.3 Optical Coherence Tomography (OCT)

OCT is an important clinically established biomedical imaging modality that uses broadband near-infrared light to provide depth-resolved, cross-sectional images of tissue to depths > 1 mm and with resolutions in the range of 1–10 μm (Huang et al. 1991; Drexler et al. 1999). With OCT, the light source is split into two beams. One of the beams of light travels through the sample arm and is focused into the tissue by a lens similar to 2-photon microscopy but generally with a smaller numerical aperture. This light is scattered as it travels deeper into the tissue and a portion of this scattered light is collected by the focusing lens. In parallel, the other beam of light travels an equivalent path length in the reference arm but is reflected by a mirror. The light reflected back in the reference arm and the sample arm are mixed and produce an interference pattern. The OCT signal is derived from the interference fringes. This heterodyne method used in OCT allows detection of signals smaller than one part in 10^{-10}, or −100 dB. Such exquisite sensitivity allows deeper imaging than multi-photon microscopy and suggests that OCT may be sensitive to weak changes in scattered light associated with functional activation in the cortex. Depth resolution is achieved by using a broadband light source that has a coherence length on the order of 1–10 μm. When the light has a short coherence length, the interference between the sample and reference arms occurs only when the path length of light in each arm is matched. Thus, by varying the path length of the reference arm, it is possible to probe the amount of light scattered at different depths in the sample with a depth resolution given by the coherence length. This is the classical time-domain OCT that required a moving mirror in the reference arm. Recently, spectral (White et al. 2003; Wojtkowski et al. 2003) and frequency (Leitgeb et al. 2003; Vakoc et al. 2005) domain OCT methodology were introduced that do not require moving parts or a time varying delay line and thus enable up to 100x faster signal acquisition.

This only gives signal along the axis of the optical beam. A small numerical aperture is used to maintain a small lateral profile of the beam over a long axial distance of 500–1,000 μm. A higher numerical aperture would give a lateral resolution of ~1 μm but then the signal could only be acquired over a few micrometers in the axial direction. Cross-sectional images are generated by scanning the optical beam across the tissue in the same fashion that 2-photon microscopy generates images. If the scattering particle in the tissue is moving, such as RBCs, then the scattered light has a Doppler frequency shift that can be detected as a phase shift versus time thus enabling Doppler OCT to image RBC velocity (Leitgeb et al. 2003;

White et al. 2003; Bizheva et al. 2004). Overall, OCT imaging is analogous to ultrasound B-mode imaging, except that it uses light rather than acoustical waves.

Few studies exist to date using OCT imaging to study functional activation in neuronal tissues. Maheswari et al. showed depth resolved stimulus specific profiles of slow processes during functional activation in the cat visual cortex (Maheswari et al. 2003). They attributed these signals to variation in scattering due to localized structural changes such as capillary dilation and cell swelling. This was further supported by recent studies from our group which showed spatial and temporal agreement between the evoked OCT signal changes and the multi-spectral images of hemoglobin changes (Aguirre et al. 2006; Chen et al. 2008). Further, Chen et al. observed a delay in the OCT signal at superficial depths relative to the multi-spectral hemoglobin changes suggestive of retrograde vasodilation (Chen et al. 2008). Evoked scattering changes can be observed also on a fast scale corresponding to propagation of action potentials (Lazebnik et al. 2003; Akkin et al. 2004; Fang-Yen et al. 2004). These and other studies have generated tremendous recent interest in OCT as a potential tool for imaging of both fast and slow functional signals of neuronal and vascular origins.

15.2.1.4 Optical Imaging of pO_2

A recent optical approach exploits O_2-dependent quenching of a phosphorescence molecule for the *in vivo* imaging of O_2 content in living tissue (Dunphy et al. 2002; Smith et al. 2002). The technique is attractive because the calibration between the phosphorescent lifetime and the partial pressure of oxygen (pO_2) is absolute (Dunphy et al. 2002), and it can be applied to measure both intravascular and tissue oxygenation. To obtain local measures of pO_2, the phosphorescence molecule must be introduced into the blood stream or into the tissue. A pulse of light is introduced into the tissue at the appropriate wavelength to excite an electron out of the ground state of the phosphorescent molecule. The excited state electron transitions into a triplet state that has a "forbidden" transition back to the ground state that can take from one to hundreds of microseconds (depending on the molecule), releasing a red-shifted phosphorescent photon in the process. Measuring these phosphorescent photons reveals an exponential decay profile described by a single lifetime. This long lifetime increases the probability of collisions between molecular oxygen and the excited phosphorescent molecule providing a non-radiative decay pathway. The longer the electron stays in the excited state, the greater the probability of it decaying non-radiatively. Thus, oxygen has the effect of decreasing the apparent lifetime of the phosphorescent decay. A greater concentration of oxygen increases the probability of collisions and thus further decreases the phosphorescent lifetime. The lifetime relationship with the partial pressure of oxygen follows the Stern-Volmer relation,

$$\frac{\tau_o}{\tau} = 1 + k\tau_o pO_2 \tag{15.2}$$

where τ and τ_o is the phosphorescent lifetime at a particular partial pressure of oxygen pO_2 and at $pO_2=0$ respectively, and k is the second-order rate constant related to the frequency of collisions of the excited-state phosphor with molecular oxygen.

Intravascular imaging of pO_2 can be performed with a thermoelectrically cooled CCD camera (Sakadzic et al. 2009) or using confocal imaging (Yaseen et al. 2009). The camera frame rate is synchronized with the triggering rate of the pulsed laser used for excitation of phosphorescence. The light is collected by the low magnification infinity corrected objective (e.g., Olympus XL Fluor 4x/340, 0.28 NA) and an image is created on the CCD sensor by a 100 mm focal length tube lens. Ambient light and phosphorescence excitation light are suppressed with a 650 nm high pass filter. The light from the pulsed laser is coupled into the multimode fiber and typically 5–10 mJ/cm^2 is delivered to the brain tissue at an angle with respect to the cranial window surface. For estimation of phosphorescence lifetimes, 4×4 binning of the CCD pixels is used and the CCD exposure time is set to 50 μs. A sequence of 25 frames is acquired with variable delay times with respect to the laser Q-switch opening. The first frame (200 μs before the laser pulse) is used to subtract the background light from the phosphorescence intensity images, and the delays of the remaining 24 frames are divided into three groups: initial 10 acquisitions with 5 μs increments in delay following the laser pulse, followed by 10 acquisitions with 20 μs increments in delay and four acquisitions with 400 μs increments in delay (Sakadzic et al. 2009). In addition to single photon excitation of O_2-sensitive probes, 2-photon excitation is now being employed to achieve better depth penetration and higher spatial localization (Mik et al. 2004; Finikova et al. 2008).

Oxygenation data alone cannot be used to infer $CMRO_2$, because a mere increase in blood flow without a concurrent increase in local O_2 metabolism would raise O_2 concentration in the tissue and blood. However, $CMRO_2$ can be calculated given simultaneous measurements of flow and oxygenation. For example, $CMRO_2$ can be calculated by combining phosphorescence lifetime imaging with laser speckle imaging (Fig. 15.6) (Sakadzic et al. 2009). $CMRO_2$ can also be calculated from simultaneous spectral and laser speckle imaging (Dunn et al. 2003; Dunn et al. 2005), or (on a more macroscopic scale) from interleaved BOLD and ASL fMRI (Brown et al. 2007). In future simultaneous 2-photon imaging of intravascular and tissue pO_2 would allow direct measurements the pO_2 gradient as a function of the depth and distance to the closest capillary, arteriole or venule, and calculation of the microscopic distribution of $CMRO_2$.

15.2.1.5 Optical Imaging of NADH Intrinsic Fluorescence

Optical imaging of intrinsic β-nicotinamide adenine dinucleotide (NADH) tissue fluorescence has drawn some attention in recent years, motivated by the early work of Britton Chance and colleagues (Chance et al. 1962). NADH is the principal electron carrier in glycolysis, the Krebs cycle and the mitochondrial respiratory chain. NADH is generated during glycolysis in the cytosol, shuttled to mitochondria (directly or via electron shuttles) and subsequently oxidized to NAD^+ in the electron

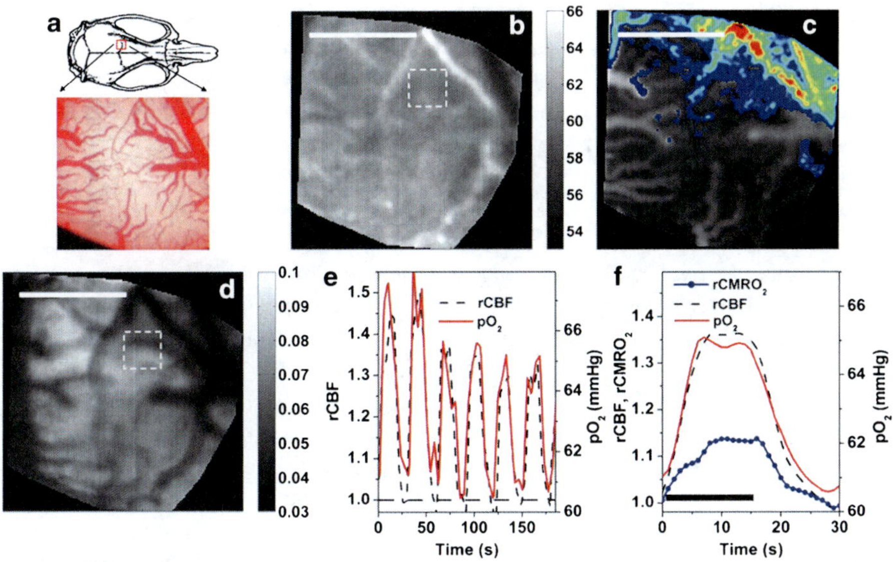

Fig. 15.6 Imaging of pO_2 by measuring the phosphorescence lifetime of an oxygen-sensitive probe Oxyphor R2 during forepaw stimulation. (**a**) Position of the cranial window and photograph of the cortical vasculature. (**b**) Baseline pO_2 map. (**c**) Composite image consisting of phosphorescence intensity (*gray*) and the functional CBF response (*color*). (**d**) Speckle contrast image of baseline flow. (**e**) Time-courses of pO_2 (*red solid curve*) and rCBF (*black dashed curve*) during several stimulation sequences. Both pO_2 and rCBF values were averaged over the area marked by rectangles in (**b**) and (**d**). (**f**) Average pO_2, rCBF, and calculated $rCMRO_2$ averaged over six stimulation trials. Duration of the stimulus is marked by the black horizontal bars in (**e**) and (**f**) Scale bar is 1 mm (Reproduced with permission from (Sakadzic et al. 2009))

transport chain, establishing a potential across the inner mitochondrial membrane, enabling the production of ATP. In mitochondria, the oxidation is followed by regeneration of NADH from NAD^+ during the TCA cycle. In the cytosol, conversion of pyruvate into lactate (with a subsequent secretion of lactate into the extracellular space) decreases the NADH pool but on a slower time scale (Hu and Wilson 1997). Thus, although a close correlation between NADH oxidation and oxygen consumption has been reported (see (Turner et al. 2007) for a recent review), interpretation of the ratio of NADH/NAD^+ can be complex, depending on the balance of $CMRO_2$ and non-oxidative glycolysis.

Since NADH is auto-fluorescent while NAD^+ is not, intrinsic NADH fluorescence serves as an indicator of the cellular redox state. Previous studies *in vivo* established that on a macroscopic level NADH auto-fluorescence of brain tissue decreases in response to stimulation, cortical spreading depression or seizures throughout the duration of the stimulus as far as blood flow is not compromised, and increases in response to hypoxia or ischemia. A recent 2-photon microscopy study in a hippocampal brain slice revealed that astrocytes have a higher resting NADH fluorescence than neurons and respond to stimulation of Schaffer collaterals with an increase in NADH signal (Kasischke et al. 2004). However, NADH behavior *in vitro* might differ from *in vivo* because of the limited O_2 availability in absence blood flow and O_2 carriers (Devor et al. 2009). Future 2-photon studies are required to establish metabolic imaging biomarkers on a single cell level *in vivo*.

15.2.1.6 Voltage-Sensitive Dyes Imaging

Voltage-sensitive dyes (VSD) imaging provides a unique tool for visualizing real-time neuronal activity in space and time (Cohen and Lesher 1986; Grinvald et al. 1988). The dye molecules bind to the external surface of the excitable membranes of neurons and act as molecular transducers that transform changes in membrane potential into optical signals. Following a stimulus, but also during the spontaneous neuronal activity, there is a change in membrane potential in excitable brain tissue that produces a change in the absorption or the emitted fluorescence of the dye. The optical signal represents a spatial integral of membrane potentials over all membranes in a given area. Since dendritic arborizations constitute a large percentage of the total membrane area, voltage-sensitive dye signals reflect potential changes that result mostly from synaptic activity (Ebner and Chen 1995; Ferezou et al. 2006). In common practice, a well is built around a cortical exposure and filled with the dye solution. The dye is left for 1–1.5 h to impregnate the brain. Next, the staining solution in the well is replaced multiple times with fresh buffered saline to remove any unbound dye molecules, and the well is sealed. The cortex is illuminated using an epi-illumination system with appropriate excitation and emission filters (e.g., 630 and 665 nm respectively, using RH 1691) and a dichroic mirror. Fluorescent images are acquired using a cooled CCD camera usually at 200 Hz or above.

Respiration and heartbeat represent dominant sources of noise in VSD imaging experiments (Shoham et al. 1999). Bleaching and photodynamic damage provide an additional constraint for the number of trials that can be acquired to improve the single-to-noise ratio (SNR). To reduce the noise from breathing and heart pulsation, the data acquisition is usually synchronized with ventilation and electrocardiogram, with a subsequent subtraction of blank (no stimulus) trials.

Similar to optical imaging of intrinsic signals, voltage-sensitive optical measurements are done from the cortical surface and do not posses laminar (depth) resolution. The signal at every pixel represents an integral of the response though the whole depth of light penetration. It has been demonstrated that dyes RH-1691, RH-1692, and RH-1838 stain the cortex to a depth of at least 1 mm (Shoham et al. 1999).

15.2.2 Laminar Electrophysiological Recordings

Although VSD and fluorescent ionic indicators, such as calcium-sensitive dyes, become increasingly popular, electrophysiological recordings are still considered "the gold standard" for measurements of neuronal activity. Most of the recordings *in vivo* are performed using extracellular metal or glass microelectrodes.

Electrophysiological recordings with extracellular microelectrodes or microelectrode arrays measure electrical potentials with respect to a distant site – usually a reference electrode that is attached to the skull. In this configuration, the measured potential reflects spikes of multiple neurons superimposed on other lower frequency waves related mostly to synaptic activity. The spiking (multiple unit activity, MUA) and synaptic (local field potential, LFP) activity can be separated by high- and low-pass filtering, respectively. This separation is based on the fact that spikes are fast events lasting ~ 1 ms, whereas synaptic potentials typically range from 10 to 100 ms.

Intracellular recording from single cells can be made *in vivo* using either "sharp" or "patch" configurations. In contrast to extracellular recordings that usually sample many cells in the neighborhood of the electrode tip, intracellular recordings are made from one cell at a time. Until recently, these recordings could not be targeted to a particular cell type, which made studies of infrequent cell types very difficult. Recently, a new approach has been described that uses 2-photon microscopy and cell-type-specific labeling to guide patch electrodes to specific cells - "2-photon guided patch" (Margrie et al. 2003). In principle, this method can be combined with other 2-photon measurements such as measurements of vascular diameters and RBC velocity.

Whereas intracellular recordings provide the ultimate single-cell resolution, they are impractical for estimation of population activity and reconstruction of circuit dynamics. For that reason, MUA and LFP measurements have been extensively employed in studies of neurovascular coupling. However, a growing body of recent data reviewed below (see Sect. 15.4.1) increasingly suggest that cell-type specific release of vascular mediator can play a central role in regulation of blood flow. These data emphasize the need for computational methods for extraction of cell type-specific activity from MUA and LFP that can be validated through intracellular recordings.

The cortical column can be considered as a processing unit of the cerebral cortex (Simons 1978). However, within a given column, processing is organized according to cortical lamina. Different cortical layers contain distinct neuronal types, and cortical connections in different areas have characteristic laminar origins and terminations (Thomson and Bannister 2003). Therefore, depth-resolved (laminar) recordings increase our ability to extract more detailed information on the activity within the cortical circuit. To obtain a laminar depth profile, one can record sequentially, inserting one microelectrode at different depths. However, this method is time-consuming and inevitably inaccurate due to elastic properties of the cortical tissue. An alternative is to use a one-dimentional multielectrode array with multiple contacts spaced at equal intervals (Fig. 15.7) (Ulbert et al. 2001; Devor et al. 2003). Using multielectrode arrays, simultaneous measurements of MUA and LFP are made throughout the entire cortical depth. The depth is estimated based on the contact number when the top contact is positioned at the cortical surface using visual control. In our practice, the recorded potential is amplified and filtered into two signals: a low-frequency LFP part (0.1–500 Hz, sampled at 2 kHz with 16 bits) and a high-frequency MUA part (150–5,000 Hz, sampled at 20 kHz with 12 bits), see (Ulbert et al. 2001) for details. MUA is further digitally bandpass filtered between 750 Hz and 5,000 Hz using a zero phase-shift second order Butterworth filter, and then rectified to provide the MUA. The MUA data is usually smoothed along the time axis using a Gaussian kernel of 1 ms full width at 1/e of peak amplitude. Dense mapping using extracellular

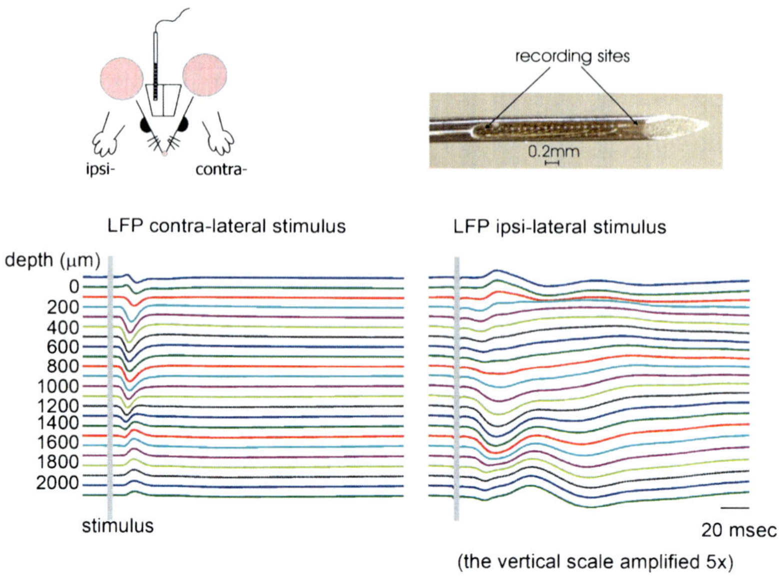

Fig. 15.7 Laminar recordings of LFP in response to contra- and ipsilateral forepaw stimulation

microelectrodes (2–4 MΩ, FHC) is performed prior to insertion of the array to determine the location of the maximal neuronal response.

The size of the laminar electrode (diameter = 300 μm in Ulbert et al. (2001)) is comparable to the size of a cortical column. However, recordings using a laminar electrode and sequential recordings using a single microelectrode at different penetration depths yielded similar results (Rappelsberger et al. 1981). Due to different spatial sensitivity of the recording of spikes and synaptic potentials, LFP signals are recorded from a bigger area. Specifically, MUA represents a weighted sum of the extracellular action potentials of neurons within a sphere of ~ 100 μm radius (Buzsaki 2004; Somogyvari et al. 2005; Pettersen and Einevoll 2008; Pettersen et al. 2008), with the electrode in the center. LFP, on the other hand, appears to reflect a weighted average of dendro-somatic components of synaptic signals of a neuronal population within a few hundred micrometers or more of the electrode tip (Pettersen et al. 2008).

15.3 Modeling Approaches

Our general conceptual framework for modeling the relationship between physiological variables and imaging observables (forward models), as well as the coupling between physiological variables (physiological models) is illustrated in Fig. 15.8. In this framework, spiking activity of each neuronal population is associated with two parallel processes: release of vasoactive messenger molecules and energy metabolism

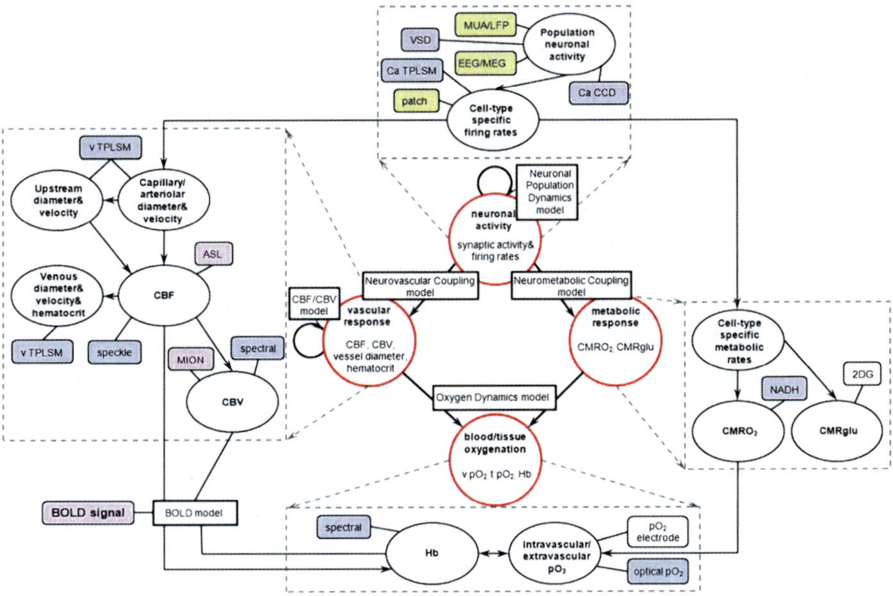

Fig. 15.8 Theoretical framework. Experimental observables are indicated by rounded rectangles (*yellow for electrophysiological; blue for optical; pink for MR-based*). The physiological parameters of interest are indicated by black ovals, and the relationships between physiological variables and observables are indicated by arrows. The computational models, linking physiological parameters and measurements, are indicated by black rectangles. vTPLSM and Ca TPLSM stand for 2-photon laser scanning microscopy measurements of vascular diameters/velocities and calcium indicators, respectively. CBV, cerebral blood volume; MION, MRI contrast agent monocrystalline iron oxide nanocolloid used for measuring of CBV; CMRglu, cerebral metabolic rate of glucose; v pO_2, intravascular pO_2; t pO_2, tissue pO_2; Hb, deoxyhemoglobin

(Neurovascular and Neurometabolic Coupling models). The resulting vascular and metabolic response then lead to spatiotemporal changes in blood flow, volume and oxygenation, as specified by the Oxygen Dynamics model. In the following, we introduce a mathematical modeling framework that is used to capture the relationship between the physiological variables, and show how the physiological parameters can be estimated from the experimental observables.

15.3.1 Multi-modality Integration: A Bayesian Modeling Framework

A key goal of multi-modality integration is to derive estimates of unknown physiological variables with the greatest possible spatial and temporal accuracy, by exploiting the relative strengths of different imaging modalities. Bayesian estimation provides a convenient and general mathematical framework for formulating this integration problem. Specifically, we aim to estimate the spatiotemporal behavior of various electrophysiological variables ('$\boldsymbol{\beta}_E(\mathbf{r},t)$' : [membrane potential, action potentials, trans-synaptic currents]), metabolic variables ('$\boldsymbol{\beta}_M(\mathbf{r},t)$' : [$CMRO_2$]), and vascular/hemodynamic variables ('$\boldsymbol{\beta}_H(\mathbf{r},t)$' : [vessel diameters, flow, volume, oxygen concentration]), from a given set of experimental observables. The relevant observables are **electric/magnetic:** '$\mathbf{Y}_{EM}(\mathbf{r},t)$' = [MUA, LFP, EEG, MEG]; **optical:** '$\mathbf{Y}_{OPT}(\mathbf{r},t)$' = [spectral/speckle, NIRS]; and **magnetic resonance**: '$\mathbf{Y}_{MRI}(\mathbf{r},t)$' = [BOLD, ASL, MION]). These observables reflect the electrical ('$\boldsymbol{\beta}_E(\mathbf{r},t)$'), metabolic ('$\boldsymbol{\beta}_M(\mathbf{r},t)$') and hemodynamic ('$\boldsymbol{\beta}_H(\mathbf{r},t)$') state of the brain at time 't' and location '$\mathbf{r}$' in the brain. Using Bayes' Rule, assuming independent noise processes for the different observables ($\mathbf{Y}_{EM}$, $\mathbf{Y}_{OPT}$, $\mathbf{Y}_{MRI}$), and assuming that $\mathbf{Y}_{EM}$ reflect only electrophysiological parameters, and $\mathbf{Y}_{OPT}$, $\mathbf{Y}_{MRI}$ reflect only hemodynamic and metabolic variables, we get

$$P(\beta_{E,}\beta_{M,}\beta_H \mid \mathbf{Y}_{EM},\mathbf{Y}_{OPT},\mathbf{Y}_{MRI}) \propto P(\mathbf{Y}_{EM} \mid \beta_E)\bullet P(\mathbf{Y}_{OPT} \mid \beta_{M,}\beta_H)\bullet P(\mathbf{Y}_{MRI} \mid \beta_{M,}\beta_H)\bullet P(\beta_{M,}\beta_H \mid \beta_E)\bullet P(\beta_E) \tag{15.3}$$

The left hand side of this equation represents the *a posteriori* probability of the physiological parameters $\boldsymbol{\beta}_E(\mathbf{r},t)$, $\boldsymbol{\beta}_M(\mathbf{r},t)$,and $\boldsymbol{\beta}_H(\mathbf{r},t)$, given all observables '$\mathbf{Y}_i(\mathbf{r},t)$', as well as *a priori* information, from which the *maximum a posteriori probability* (MAP) estimates and confidence intervals for the electrical variables of interest can be obtained (Schmidt et al. 1999; Dale and Halgren 2001; Friston 2005). The term $P(\mathbf{Y}_{EM}|\boldsymbol{\beta}_E)$ represents the forward solution for the electric/magnetic observables, given the electrophysiological parameters. The *a priori* information about the electrophysiological parameters are encoded in $P(\boldsymbol{\beta}_E)$, reflecting the coupling between these parameters (captured by the Neuronal Population Dynamics model in Fig. 15.8). The mathematical relationship between the electric/magnetic observables and electrophysiological parameters is developed in more detail in Sect. 15.3.2, below.

The term $P(\boldsymbol{\beta}_{M,}\boldsymbol{\beta}_H|\boldsymbol{\beta}_E)$ reflects the coupling between electrophysiological activity (e.g., firing and/or synaptic activity of specific cell types), and hemodynamic and metabolic parameters (e.g., $CMRO_2$ and arteriolar dilation; captured by the Neurovascular Coupling, Neurometabolic Coupling, and Oxygen Dynamics models in Fig. 15.8). The terms $P(\mathbf{Y}_{OPT} \mid \boldsymbol{\beta}_M, \boldsymbol{\beta}_H)$ and $P(\mathbf{Y}_{MRI} \mid \boldsymbol{\beta}_{M,}\boldsymbol{\beta}_H)$ represent the optical and MRI forward solutions, respectively. Characterization of these functions, for the measurement modalities and physiological parameters of interest, is detailed in Sect. 15.3.3, below.

15.3.2 Modeling of Neuronal Populations from Laminar Microelectrode Recordings

15.3.2.1 Forward Modeling of Extracellular Potentials

The potentials recorded by extracellular electrodes stem from the ionic electrical currents going through the membranes of neurons and other cells in the vicinity of the electrode contacts. These transmembrane currents force small electrical currents to be driven through the low-resistance extracellular medium which, following Ohms electrical circuit law, lead to spatial variations in the extracellular potential. The low resistance of the extracellular medium implies that the extracellular potential differences will be small, typically less than a millivolt, i.e., much smaller than the typical neuronal membrane potentials of about 70 mV.

The extracellular potentials generated by transmembrane currents can be calculated using *volume conductor theory* (Nunez and Srinivasan 2006). Here the extracellular medium is envisioned as smooth and continuous and transmembrane currents entering and leaving the extracellular space are represented as *volume current sources*. The fundamental relationship between the extracellular potential φ recorded at a position $\boldsymbol{r}$ due to a transmembrane current I_0 at a position $\boldsymbol{r}_0$ is given by (Hämäläinen et al. 1993; Nunez and Srinivasan 2006)

$$\varphi(\boldsymbol{r},t) = \frac{1}{4\pi\theta}\frac{I_0(t)}{|\boldsymbol{r}-\boldsymbol{r}_0|}. \tag{15.4}$$

Here $\boldsymbol{r}$ and $\boldsymbol{r}_0$ are position *vectors*, σ is the *extracellular conductivity*, and the extracellular potential φ is chosen to be zero infinitely far away from the transmembrane current.

Several assumptions lie behind this simple formula, in particular the *quasistatic approximation of Maxwell's equations*. This approximation amounts to omitting time derivatives of the electric and magnetic fields from the original Maxwell's equations. Then the electric field $\boldsymbol{E}$ in the extracellular medium is related to the extracellular potential φ via $\boldsymbol{E} = -\nabla\varphi$. For frequencies inherent in neuronal activity, i.e., less than a few thousand hertz, the quasistatic approximation seems to be well fulfilled (Hämäläinen et al. 1993). Assumptions have also been made about the electrical properties of the extracellular medium: The current density $\boldsymbol{j}$ is assumed to be proportional to the electrical field, i.e., $\boldsymbol{j} = \sigma\boldsymbol{E}$. Further, the extracellular conductivity is assumed to be purely ohmic, i.e., σ has no imaginary part from capacitive effects (Nunez and Srinivasan 2006; Logothetis et al. 2007). Finally, the formula assumes σ to be the same everywhere and also the same in all directions (Logothetis et al. 2007). More discussion on these assumptions, and also ways of generalizing Eq. 15.4 when they do not apply, can be found in (Pettersen et al. 2012).

The formula in Eq. 15.4 forms the basis for so called *forward modeling* of extracellular potentials, i.e., it describes how a transmembrane current due to neuronal activity contributes to this potential. A convenient feature of the forward-modeling scheme is that due to the linearity of Maxwell's equations, the contributions to the extracellular potential from the various neuronal sources add up linearly. Thus the net extracellular potential from activity in an entire neuron can be found simply by adding contributions of the form in Eq. 15.4 from transmembrane currents from all parts of the neuron. In general, compartmental modeling, using simulation tools like NEURON (Carnevale and Hines 2006) and Genesis (Bower and Beeman 1998), must be used to calculate the transmembrane currents acting as sources for the extracellular potential. However, if all transmembrane currents and their spatial positions are known, the extracellular potential can in principle be computed at any point.

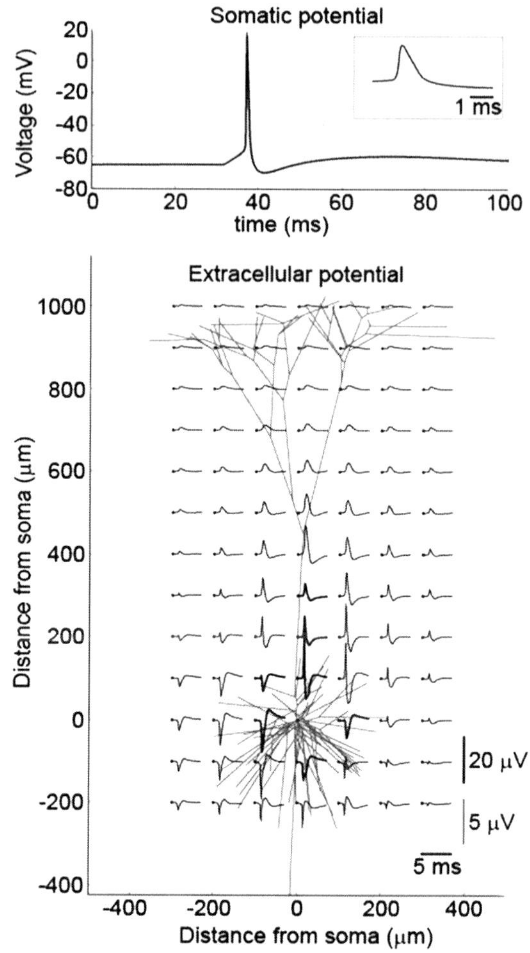

Fig. 15.9 Intracellularly and extracellularly recorded action potentials. The model is a reconstructed pyramidal neuron taken from Mainen and Sejnowski (1996). A synaptic stimuli similar to what is called "stimulus input pattern 1" in Pettersen et al. (2008) is used. Top panel: Membrane potential in soma during an action potential. Inset shows the membrane potential trace in a 5 ms time window around the action potential. Lower panel: Calculated extracellular potentials based on a variant of the forward-modeling formula in Eq. 15.4 assuming an isotropic, homogenous and purely conductive extracellular medium with σ=0.3 S/m. The extracellular potentials are shown for the same 5 ms as in the membrane potential inset in the top panel. All distances are in micrometers

In Fig. 15.9 we illustrate this forward-modeling scheme by showing the calculated extracellular potential for a compartmental neuron model firing an action potential. The top panel shows a characteristic membrane potential trace for an action potential calculated by the simulation tool NEURON using a model pyramidal neuron constructed by Mainen and Sejnowski (Mainen and Sejnowski 1996). This neuron model has several types of active ion channels spread across the neuronal membrane, and the membrane potential trace has a characteristic shape with a fast depolarization followed by an almost equally fast repolarization, and eventually a longer after-hyperpolarization. The corresponding calculated extracellular spike pattern is shown at different positions in the lower panel. These extracellular potentials are found from evaluating a sum over numerous terms (one for each compartment) of the type in Eq. 15.4 where $I_0(t)$ corresponds to transmembrane currents found for each compartment in the NEURON simulation (see (Pettersen et al. 2008) for details). Several features are noteworthy. The extracellular spike has a much lower amplitude than the intracellular action potential. Even close to the soma the amplitude is less than a few tens of microvolts, more than a factor thousand smaller than the intracellular amplitude. Also, the magnitude of the extracellular spike decays rapidly with distance from the neuronal soma. Moreover, not only the size, but also the shape of the extracellular potential vary significantly with position: for example, the shape around the apical (upper) dendrites is typically inverted compared to around the basal (lower) dendrites, and the spikes further away from the soma are generally less sharp. The example in Fig. 15.9 demonstrates that *positive and negative phases of extracellular potentials, such as LFP, cannot be treated as excitation and inhibition* – the sign of the potential changes depending on the location of the recording relative to the activated neuronal population.

Due to the linearity of the forward-modeling scheme the calculation of extracellular potentials can straightforwardly be extended to *populations* of neurons. The computation time grows linearly with the number of neurons, and the calculation of extracellular potentials from joint activity in populations with thousands of morphologically reconstructed neurons can be done on present-day desktop computers (Pettersen et al. 2008).

15.3.2.2 Estimation of Current-Source Density (CSD)

If we introduce the quantity

$$C(\boldsymbol{r},t) \equiv I_0(t)\delta^3(\boldsymbol{r}-\boldsymbol{r}_0), \tag{15.5}$$

where $\delta^3(r)$ is the three-dimensional Dirac δ-function, the forward-modeling formula in Eq. 15.4 can be reformulated as

$$\varphi(\boldsymbol{r},t) = \frac{1}{4\pi\sigma}\iiint_V \frac{C(\boldsymbol{r}',t)}{|\boldsymbol{r}-\boldsymbol{r}'|}d^3\boldsymbol{r}'. \tag{15.6}$$

Here the volume integral goes over all transmembrane currents. The quantity $C(\boldsymbol{r},t)$ corresponds to a *current source density (CSD)*, i.e., the volume density of current entering or leaving the extracellular medium at position $\boldsymbol{r}$ (Nicholson and Freeman 1975; Mitzdorf 1985; Nunez and Srinivasan 2006). A negative $C(\boldsymbol{r},t)$ corresponds to current leaving the extracellular medium and is thus conventionally called a *sink*. Likewise, current entering the extracellular medium is called a *source*.

The CSD distribution from activity in a single or numerous neurons could in principle be described as a sum over point-like CSD contributions as described by Eq. 15.5. However, in practice the CSD is generally considered to be a more coarse-grained measure describing the net transmembrane current entering or leaving a volume a few tens of micrometers across (Nicholson and Freeman 1975). As the

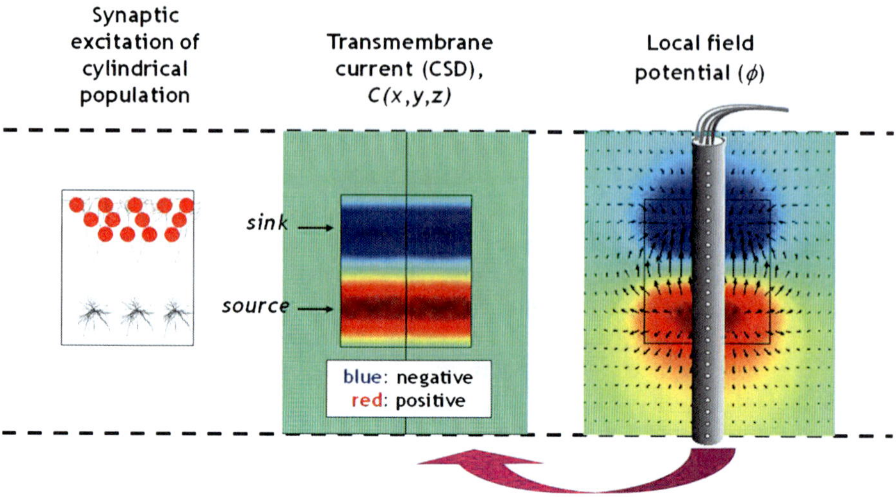

Fig. 15.10 Schematic illustration of principle behind CSD analysis. In the example a cylindrical population of pyramidal neurons receives synaptic excitation in their apical dendrites (*left panel*), resulting in a cylindrically symmetric (*columnar*) CSD distribution with an apical current *sink* accompanied by a basal current *source* due to return currents (*middle panel*). A linear (*laminar*) multielectrode (see Sect. 15.2.2) inserted into the population would measure a corresponding LFP (*right panel*), and the task of CSD analysis is to estimate the CSD distribution $C(\boldsymbol{r},t)$ based on the LFP recordings $\varphi(r_i,t)$, $(i = 1,\ldots,N_c)$ at the N_c electrode contacts

CSD is easier to relate to the underlying neuronal activity than the extracellular potential itself, CSD analysis has become a standard tool for analysis of the low-frequency part (LFP) of such potentials recorded with linear (laminar) multielectrodes (Nicholson and Freeman 1975; Pettersen et al. 2006). The principle behind such analysis is illustrated in Fig. 15.10.

While Eq. 15.6 gives the numerical recipe for calculating the extracellular potential given the CSD, a formula providing the opposite relationship can also be derived. For the case with a position- and direction-independent extracellular conductivity σ one finds (Nicholson and Freeman 1975; Nunez and Srinivasan 2006)

$$\sigma\nabla^2\varphi(\boldsymbol{r},t) = -C(\boldsymbol{r},t). \tag{15.7}$$

This equation, called Poisson's equation, is well known from standard electrostatics where it describes another problem, namely how potentials are generated by electrical charges (with the conductivity σ replaced by the dielectric constant ε (Jackson 1998).

CSD estimation has typically been based on LFP recordings with laminar (linear) multielectrode arrays (see Sect. 15.2.2) with a constant inter-contact distance h inserted perpendicularly to the cortical surface (Rappelsberger et al. 1981; Mitzdorf 1985; Di et al. 1990; Schroeder et al. 2001; Ulbert et al. 2001; Einevoll et al. 2007). Cortical tissue has a prominent laminar structure where changes in the lateral directions are much smaller than in the vertical direction. It has thus been common to assume homogeneous (constant) CSD in the lateral (xy) plane, i.e., perpendicular to the laminar electrode oriented in the z-direction (unlike the example in Fig. 15.10 where the CSD is non-zero only inside a finite column). Variation of the extracellular potential in the x- and y-directions can then be neglected, so that Eq. 15.7 simplifies to its one-dimensional version:

$$\sigma \frac{d^2\phi(z,t)}{dz^2} = -C(z,t). \tag{15.8}$$

The *"standard"* estimator for the CSD at electrode position z_j has thus been (Nicholson and Freeman 1975)

$$C(z_j) = -\sigma \frac{\phi(z_j + h) - 2\phi(z_j) + \phi(z_j - h)}{h^2} \tag{15.9}$$

or variations thereof including additional spatial smoothing filters (Freeman and Nicholson 1975, Ulbert et al. 2001).

The standard estimation formula in Eq. 15.9 has several limitations. For one, the formula only predicts the CSD at the $N_c - 2$ interior contact positions. (This problems gets even more severe with the analogous analysis for two- or three-dimensional cartesian recording grids (Leski et al. 2007) where a higher fraction of the contacts is at the surface.) Secondly, the estimation scheme relies on equidistant electrode contacts and is thus vulnerable to malfunction of individual contacts. Further, the one-dimensional scheme breaks down when the CSD varies significantly in the lateral direction.

In Pettersen et al. (2006) we introduced the *inverse CSD (iCSD)* method to alleviate these problems. The core idea behind iCSD is to exploit the forward-modeling scheme in Eq. 15.6: with an assumed form of the CSD distribution parameterized by N unknown parameters, the forward solution can be calculated and inverted to give estimates of these N parameters based on N recorded potentials. This iCSD approach has several inherent advantages: (1) The method does not rely on a particular regular arrangement of the N electrode contacts recording the LFP signals, but can be straightforwardly be generalized to all sorts of multielectrode geometries (Pettersen et al. 2006; Leski et al. 2007). (2) A priori knowledge, such as estimates of the lateral size of columnar activity or discontinuities and direction dependence of the extracellular conductivity, can be built directly into the iCSD estimator (Pettersen et al. 2006; Einevoll et al. 2007; Pettersen et al. 2008). (3) Unlike the standard CSD method, the iCSD method can also predict CSD at the positions of the boundary electrode contacts.

More detailed information about the iCSD method can be found in Pettersen et al. (2006) and Leski et al. (2007). Moreover, a MATLAB toolbox, *CSDPlotter*, for iCSD estimation based on laminar multielectrodes has been developed and can be downloaded from *http://software.incf.org/*.

15.3.2.3 Extraction of Population-Specific Synaptic Connections and Firing Rates

Even if CSD is a more localized and intuitive measure of neuronal activity than LFP, its interpretation is nevertheless difficult. Transmembrane dendritic currents from multiple populations of neurons contribute to CSD. Ideally one would like to be able to interpret the LFP in terms of activity in individual neuronal populations as this would give more insight into the organization and functioning of the cortical circuits. In a series of papers Barth and collaborators pursued such an scheme by use of principal component analysis (PCA) of the CSD (Barth et al. 1989; Barth et al. 1990; Di et al. 1990; Barth and Di 1991), and in their study of stimulus-evoked data from the rat barrel cortex they identified putative cortical populations in supra- and infragranular layers of the barrel column (Di et al. 1990). PCA is one of several statistical methods where functions of two variables (here electrode contact positions and time) are expanded into sums over spatiotemporally separable functions, i.e., functions that can be written as a product of a spatial function and a temporal function. This can be done in an infinite number of ways, and additional constraints are needed to make the expansion unique. In PCA, for example, the functions (i.e., components) are constrained to be orthogonal both in space and time.

We recently introduced a new analysis method, *Laminar Population Analysis (LPA)*, where we instead of assuming somewhat arbitrary mathematical constraints used *physiological* constraints to specify the population expansion of laminar-electrode data (Einevoll et al. 2007). As a consequence the expansion is not only compatible with the physiology, each component also has a clear physiological interpretation. The basic underlying constraint inherent in LPA is that the observed LFP is evoked by the firing of action potentials in the modelled neuronal populations. An experimental measure of this population firing is obtained from the high-frequency component (>500–750 Hz) of the laminarly recorded extracellular potentials, i.e., the multi-unit activity (MUA) (Schroeder et al. 2001; Ulbert et al. 2001; Pettersen et al. 2008; Blomquist et al. 2009). Thus both the LFP and MUA signals are used in the analysis, unlike in CSD analysis where only the LFP signals are used. The outcomes of LPA are (1) identification of the relevant laminar cortical populations and their vertical spatial position and extent, (2) estimates of the firing-rates of these populations, and (3) estimates of the spatiotemporal LFP signature (and CSD signature) following action-potential firing in the individual populations (Einevoll et al. 2007).

The fundamental equations of LPA analysis are

$$\phi_{\mathrm{MUA}}(z_i,t)=\sum_{n=1}^{N_P} M_n(z_i)r_n(t) \tag{15.10}$$

$$\phi_{\mathrm{LFP}}(z_i,t)=\sum_{n=1}^{N_P} L_n(z_i)(h*r_n)(t) \tag{15.11}$$

In Eq. 15.10 the MUA signal $\varphi_{\text{MUA}}(z_i, t)$, essentially the rectified high-frequency part of the extracellular signal sampled at the various electrode contacts positioned at z_i ($i = 1, \ldots, N_c$), is modelled as as a sum over spatiotemporally separable contributions from several neuronal populations. Here N_p is the number of populations, and $r_n(t)$ represents the firing rate of population n. Modeling studies indeed found this MUA signal to be well correlated with the true population firing rate for stimulus-evoked, trial-averaged data (Pettersen et al. 2008). $M_n(z_i)$ is the MUA spatial profile associated with action-potential firing in population n. This spatial profile will in general depend on the physiological properties of the neurons in the population as well the distribution of their spatial positions in the cortical lamina. The size of the extracellular signature of an action potential decays rapidly with distance from the neuronal soma, and the horizon of visibility is typically less than 100 μm (Somogyvari et al. 2005). The MUA is thus a very localized measure of neuronal firing, and the MUA spatial profile $M_n(z_i)$ appears to be mainly determined by the vertical spread of somas of the neurons belonging to the same population.

Next, in Eq. 15.11 the LFP data is assumed to be driven by the same population firing rates $r_n(t)$ seen in the MUA data: Firing of an action potential of a neuron in population n will lead to postsynaptic transmembrane currents (including both the ligand-gated synaptic currents and consequent return currents) which in turn contribute to the LFP. Consequently, it is assumed that the LFP data can be decomposed into contributions from each of the neuronal populations as implied by Eq. 15.11. Here $(h * r_n)(t)$ is the temporal convolution between $h(t)$ and $r_n(t)$, and $L_n(z_i)$ represents the spatial profile of the contribution to the LFP data following action-potential firing in population n. The temporal coupling kernel $h(t)$ accounts for the temporal delay and spread in the generation of the LFP following firing of neurons in population n. In Einevoll et al. (2007) exponentially decaying coupling kernels were used. A graphical illustration of the principle behind LPA can be found in the left panel of Fig. 15.11.

In Einevoll et al. (2007) the method was applied to stimulus-averaged laminar-electrode data from barrel cortex of anesthetized rat following single whisker deflections. The numerical task was to identify the spatial profiles ($M_n(z_i)$, $L_n(z_i)$), population firing rates $r_n(t)$, and parameters of the temporal coupling kernel $h(t)$ giving the minimum deviations between the model MUA and LFP signals and the experimental data $\varphi_{\text{MUA}}(z_i, t)$ and $\varphi_{\text{MUA}}(z_i, t)$, respectively. The data were found to be well accounted for by a model with four cortical populations: one supragranular, one granular, and two infragranular populations. Further, the spatial LFP population signatures $L_n(z_i)$ were further used to estimate the synaptic connection pattern between the various populations using a new *LFP template-fitting technique* (Einevoll et al. 2007). Here the estimated $L_n(z_i)$ for each laminar population n was decomposed into sums over LFP population templates found by forward modeling with morphologically reconstructed neurons, so that the values of the fitted weights provided specific predictions about the synaptic connections. Results from this analysis are shown in the right panel of Fig. 15.11.

In a recent study we took LPA a step further and used the estimated population firing rates $r_n(t)$ to extract population firing-rate *models* for both thalamocortical and intracortical signal transfer in the rat barrel system (Blomquist et al. 2009). Here the

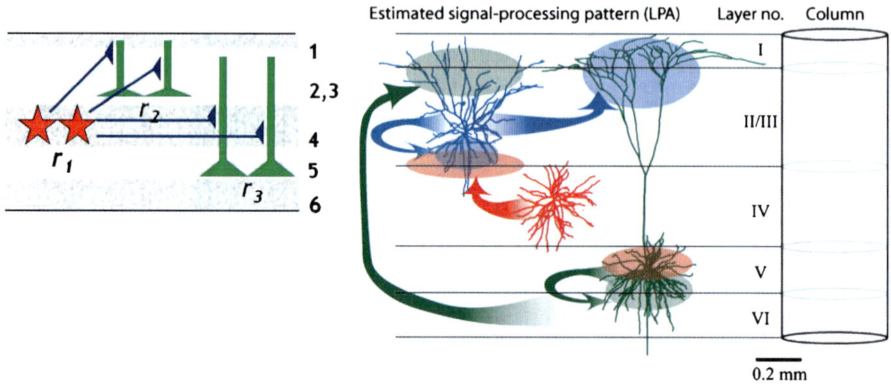

Fig. 15.11 Laminar Population Analysis. *Left panel*: Illustration of principle behind *Laminar Population Analysis (LPA)*. Three populations are considered in the sketch: a granular population of stellate cells with firing rate r_1, a supergranular population of pyramidal neurons with firing rate r_2, and an infragranular population of pyramidal neurons with firing rate r_3. In LPA the population firing rates are assumed to be related to the MUA signal via Eq. 15.10 and also give rise to the LFP signal via their postsynaptic effects, as captured by Eq. 15.11. In the panel this is illustrated for the LFP contribution from the stellate population, i.e., $L_1(z_i)(h * r_1)(t)$. The spatial form of $L_1(z_i)$ will be determined by the total efferent action of the stellate-cell population onto its postsynaptic targets (schematically illustrated with solid lines and triangular synapses). Right panel: Illustration of some results from LPA analysis from Einevoll et al. (2007) on estimation of functional synaptic connection patterns between populations in a barrel column in rat somatosensory cortex. The arrows indicate pre- and postsynaptic populations while the same-colored ellipsoids indicate the region of the dendrites receiving synaptic inputs (assuming that the synaptic connections are predominantly excitatory). For example, the stellate population (*red*) was found to have a strong projection onto the basal dendrites of the supergranular population, and both the supergranular and infragranular populations were found to have a projections onto their own basal dendrites. For more information and results, see Einevoll et al. (2007)

laminar-electrode recordings were supplemented with simultaneous thalamic single-electrode recordings. These experimentally extracted cortical and thalamic firing rates were in turn used to identify population firing-rate models formulated either as integral equations or as more common differential equations. Optimal model structures and model parameters were identified by minimizing the deviation between model firing rates and the experimentally extracted population firing rates. For the thalamocortical transfer the experimental data was found to favor a model with fast feedforward excitation from thalamus to the layer IV laminar population combined with a slower inhibitory process due to feedforward and/or recurrent connections. The intracortical population firing rates were found to exhibit strong temporal correlations and simple feedforward firing-rate models were found to be sufficient to account for the data. Thus while the thalamocortical circuit was found to be optimally stimulated by rapid changes in the thalamic firing rate, the intracortical circuits were found to be low-pass and respond strongest to slowly varying inputs from the cortical layer IV population.

LPA, combined with the abovementioned LFP template-fitting technique and/or the population rate-model extraction method, promise to be a useful tool for extraction of information on neuronal population activity from the new generation of silicon-based multielectrodes (Buzsaki 2004). However, new analysis tools still needs to be developed, and the forward-modeling technique outlined above will be essential not only in the development of the tools, but also in generating model data against which the new tools can be tested.

In the examples above, neuronal populations are defined according to the cortical depth. In the future we hope to extend the approach to cell-type specific populations with the ultimate goal of reconstructing vasoactive transmitter release from each of the relevant neuonal populations across stimulus conditions (see Sect. 15.4.1).

15.3.3 Modeling of Microscopic Vascular Dynamics and Oxygen Consumption

15.3.3.1 Vascular Anatomical Network (VAN) Modeling

Use of reduced or "lumped" models of the vascular and oxygen transport responses is increasingly common in fMRI analyses (reviewed in (Obata et al. 2004)). Although a reasonable fit to measurement data can be obtained with such models (Huppert et al. 2007), the resulting lumped or effective model parameters, e.g., vessel compliance, cannot be directly compared against microscopic properties nor can predictions be made about the spatial pattern of the hemodynamic response. We have developed a dynamic mechanistic *Vascular Anatomical Network (VAN)* model, which incorporates biophysically realistic parameters at the microscopic scale. This approach can be used with either stylistic vascular geometry (Boas et al. 2008), or a realistic geometry derived from 2-photon measurements of an *in vivo* vascular network (Devor et al. 2008b; Fang et al. 2008).

The VAN model represents the vascular tree as a set of branching, cylindrical segments, with a specified topology/geometry, incorporates biophysically realistic parameters at the microscopic scale, and allows a consistent calculation of flow characteristics (resistance, pressure changes, etc.) and O_2 dynamics (hemoglobin saturation in different compartments and average tissue pO_2) (Fang et al. 2008). The activity of specific neurons cause dynamic changes in the diameter of the arterioles/capillaries (that serves as an input to VAN), resulting in CBF changes due to the altered vascular resistance due to these vessel diameter changes. The incorporation of oxygen dynamics enables estimation of intra- and extravascular microscopic oxygenation changes.

The VAN model uses a network of resistors and non-linear capacitors to model the relationship between blood flow, volume, resistance, and pressure (Fig. 15.12). In this framework, the active vessel diameter changes in the arteriolar compartments lead to a change in resistance and, thus a change in CBF proportional to the pressure drop across the compartment. As the arteriolar resistance decreases, pressure shifts downstream into the compliant capillaries and venules resulting in their compliant dilation and secondary blood volume and flow increases. Oxygen dynamics

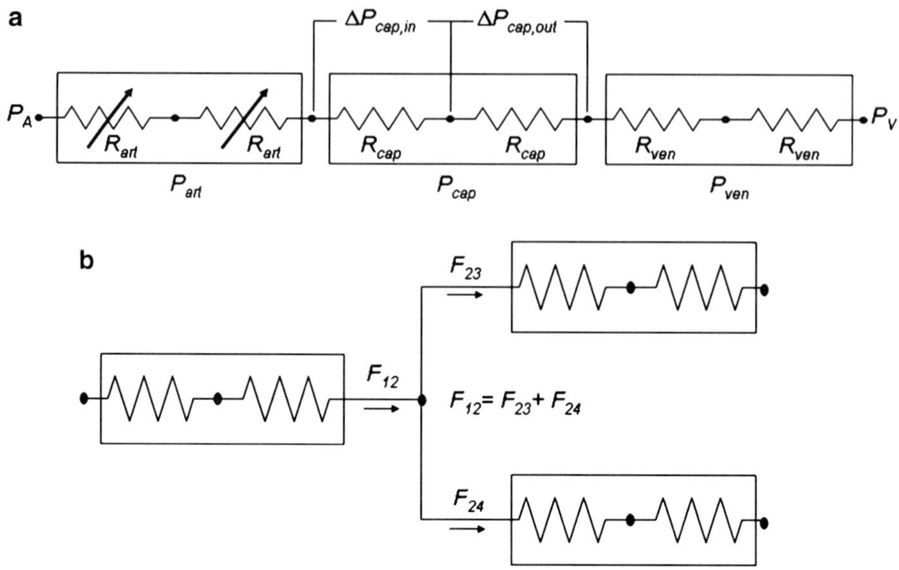

Fig. 15.12 VAN flow-volume dynamics. For more details see (Boas et al. 2008)

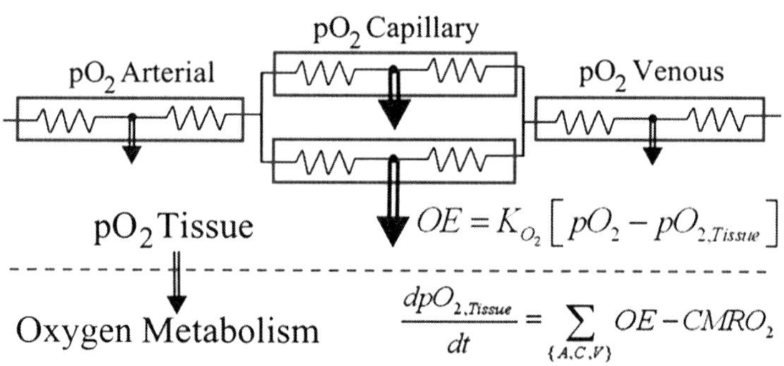

Fig. 15.13 VAN oxygen dynamics. For more details see (Boas et al. 2008)

is incorporated into the network as illustrated in Fig. 15.13. Vessel dilation and constriction, and $CMRO_2$, arising from the firing rates of excitatory and inhibitory neurons, serve as inputs to the model, resulting in predictions of CBF, cerebral blood volume (CBV), and pO_2 in blood and tissue, as functions of time and space.

One question that this VAN model has addressed is the origin of a "surround" hemodynamic negativity that has been observed in response to brain activation (Cox et al. 1993; Woolsey et al. 1996; Takashima et al. 2001; Devor et al. 2005; Devor et al. 2007; Devor et al. 2008b). The question was whether this surround negativity arises from passive redistribution of blood flow to support center increases, or whether neuronal processes are involved that are actively vasoconstricting the vessels. As detailed in (Boas et al. 2008), the VAN model showed that localized arterial dilation induced a chain of passive events in both a center region of interest and in the surrounding vessels such that, concurrent with increases in flow and oxyhemoglobin in the center, there was a passive decrease in flow and oxyhemoglobin in the surround. The VAN model predicted that the observed surround decreases are at least partially due to passive vessel properties that create a local redistribution of blood. However, a comparison with experimental data suggested that the incorporated vessel properties are not sufficient to account for the magnitudes of decreases in the surround. Several factors may contribute to this discrepancy. A likely explanation is active vasoconstriction of arteries and arterioles through activation of the smooth muscles that line them (Cox et al. 1993; Hamel 2004; Lauritzen 2005; Hamel 2006; Devor et al. 2007; Devor et al. 2008b), or active changes in the diameter of capillaries (Peppiatt et al. 2006). Active vasoconstriction could be induced by the release of neurotransmitters or neuropeptides, and may be associated with center-surround electrophysiological patterns of activity consisting of center excitation and surround inhibition (Faraci and Breese 1993; Faraci and Heistad 1998; Takashima et al. 2001; Derdikman et al. 2003; Cauli et al. 2004; Devor et al. 2007; Devor et al. 2008b).

Further evolution of VAN model and validation against experimental data will lead to further insights on the regulation of blood flow by specific vascular branches of the arterioles and capillaries, the role of retrograde vasodilation, the oxygen delivery to the tissue from arterioles versus capillaries versus venules, and flow and dilation induced changes in hematocrit (Hillman et al. 2007) to name a few. While VAN modeling is too complex and ill-constrained to allow routine fitting to experimental data from macroscopic fMRI and optical measures, the interpretation of the "lumped" compartment models is derived from this underlying anatomical detail. Thus, VAN model will illuminate issues with the assumptions of the "lumped" models used to analyze human fMRI and optical imaging data (see Sect. 15.3.3.2 below).

15.3.3.2 Calculation of the Metabolic Rate of O_2

Why Study $CMRO_2$?

In contrast to the diversity of measurement methods that can be applied to study neuronal and vascular activity with high resolution and precision in animal models, human imaging studies are limited by the existing non-invasive tools, mainly fMRI, PET and EEG/MEG. Among them, fMRI has been widely used for cognitive studies and is beginning to make inroads into clinical applications (D'Esposito et al. 2003; Guadagno et al. 2003; Hodics and Cohen 2005; Teasell et al. 2005). However, as was mentioned in the Introduction, the BOLD fMRI signal depends on the balance of the changes in CBF and $CMRO_2$, and therefore is sensitive to the baseline CBF conditions. In other words, the BOLD signal in response to the same increase in neuronal activity is expected to change given different baseline CBF. Therefore, a major focus in the field has been in development methods for extraction of $CMRO_2$ that supposedly provides a better surrogate measure of neuronal activity and is insensitive of the baseline vascular conditions.

The Challenging Task of Measuring $CMRO_2$

In human imaging, the primary physiological variables related to cerebral blood flow and energy metabolism are CBF, cerebral metabolic rate of glucose (CMRGlc) and $CMRO_2$, and a general goal is to be able to quantify these variables in a local brain region. The accepted standard for CBF measurements is a microsphere experiment, in which labeled microspheres are injected arterially (Yang and Krasney 1995). Because the microspheres are too large to pass through the capillaries, they stick in the tissue, and the local concentration of microspheres then directly reflects the local CBF. With an appropriate measurement of the injected arterial bolus, the local CBF can be quantified in absolute units of ml blood/ml tissue-min. For human studies with MRI, ASL methods approach the ideal microsphere experiment because the delivery of magnetically labeled blood is measured only 1–1.5 s after creation of the labeled blood, so there is little time for the labeled blood to pass through the capillary bed and clear from tissue (Buxton 2005).

For CMRGlc methods there is also an accepted standard based on injection of radioactively labeled deoxyglucose (DG), a chemically modified form of glucose (Sokoloff et al. 1977; Phelps et al. 1981) (see also Sect. 15.4.4). Deoxyglucose is similar enough to glucose that the initial stages of uptake from blood and metabolism in the cytosol are similar to glucose. However, when the DG reaches a certain stage in the metabolic pathway the chemical difference between glucose and DG is critical, and the DG is not further metabolized and remains in the tissue. By waiting sufficiently long for unmetabolized DG to diffuse back into blood and clear from the tissue, the remaining radioactive label in tissue directly reflects the rate of metabolism of glucose. As with the CBF measurement with microspheres, an appropriate measurement of the arterial concentration over time makes it possible to measure CMRGlu in absolute units of micromoles/100 ml tissue-min. Technically, this method measures the metabolic rate of DG, and a correction is needed to account for rate constant differences between glucose and DG. Given the need for this correction, at first glance it might seem that a more direct way to measure CMRGlu would be to inject radioactively labeled glucose itself. The essential problem, however, is that measurement of the concentration of the radioactive label does not distinguish between the label in glucose and the label in products of metabolism of glucose (e.g., with carbon-14 the net concentration of label reflects labeled CO_2 as well as glucose). In short, the DG technique is effective because it acts in a similar fashion to microspheres in a CBF measurement: the delivery of the agent reflects the appropriate physiological rate (CBF or CMRGlu), and the agent then sticks in the tissue so that the concentration can be readily measured.

Although O_2 metabolism is a key element of brain function, $CMRO_2$ is much more difficult to measure than CBF or CMRGlu. Currently there is no accepted "gold standard" for such measurements, and no established methods for measuring dynamic $CMRO_2$ changes. In essence, this is because there is no experiment equivalent to the microsphere or DG experiment in which a single measurement reflects $CMRO_2$ alone. For this reason, $CMRO_2$ must be estimated from multiple measurements analyzed within the context of an appropriate mathematical model of the physiology and the measured quantities. This adds a layer of complexity to $CMRO_2$ estimates, and raises associated concerns about the accuracy of the $CMRO_2$ estimates related to the assumptions and limitations of the experimental methods and models used. The following sections review current approaches for combining multi-modal measurements to estimate changes in $CMRO_2$.

The Importance of the O_2 Extraction Fraction

A primary physiological relationship that often underlies methods for estimating $CMRO_2$ is the connection between CBF, $CMRO_2$ and the O_2 extraction fraction (*E*):

$$CMRO_2 = E \cdot CBF \cdot [O_2]_a \tag{15.12}$$

where $[O_2]_a$ is the arterial concentration of O_2. The extraction fraction *E* is defined as the fraction of O_2 delivered to the capillary bed that is extracted from the blood and metabolized in tissue, and so is closely related to the venous O_2 concentration. Note that this relationship follows simply from mass balance and the definition of the terms. The product of CBF and $[O_2]_a$ is the rate of delivery of O_2 to the capillary bed, and multiplying by *E* gives the net rate of metabolism of O_2. Both CBF and $CMRO_2$ are referred to a unit volume (or mass) of tissue. For example, a consistent set of values and associated units for these variables, typical for the human brain, would be: $CMRO_2$ (160 μmol/100 ml tissue-min), CBF (50 ml blood/100 ml tissue-min), *E* (0.4, dimensionless, ranging from 0 to 1), and $[O_2]_a$ (8 mM blood).

Often the fractional change in a variable is the desired quantity, and absolute values may not be essential. For example, in considering a dynamic change from a baseline state due to neuronal activation, it is often useful to consider the CBF normalized to its baseline value (*f*) and the $CMRO_2$ normalized to its baseline value (*m*), related by:

$$m = f \frac{E}{E_0} \tag{15.13}$$

where the subscript "0" denotes values in the baseline state. Note, though, that this relationship is essentially a steady-state relationship, and for rapid changes faster than the capillary transit time, the dynamic definition of *E* would need to be reconsidered. Nevertheless, this basic relationship is usually assumed to be accurate in most applications.

Several approaches for estimating $CMRO_2$ involve combining measurements sensitive to *E* (e.g., OISI and BOLD imaging) with a measurement of the normalized CBF change. The ratio of O_2 extraction fractions is directly related to the venous deoxyhemoglobin concentrations. If Hb_v and HbT_v are the venous deoxyhemoglobin and venous total hemoglobin concentrations, respectively, and Hb_{v0} and HbT_{v0} are the corresponding values in the baseline state, then:

$$\frac{E}{E_0} = \frac{Hb_v / Hb_{v0}}{HbT_v / HBT_{v0}} \tag{15.14}$$

The central goal in estimating $CMRO_2$ changes is then to estimate E/E_0 from the measured data, combine this with a measurement of the normalized CBF change (f), and calculate the relative $CMRO_2$ change from Eq. 15.13.

Estimating $CMRO_2$ with PET

The leading method for measuring $CMRO_2$ in human studies uses positron emission tomography (PET) and molecules labeled with ^{15}O, a positron emitting radionuclide (Mintun et al. 1984). The primary measurement is the uptake of ^{15}O after injection of ^{15}O labeled O_2, providing a measurement that depends on E and CBF. However, such a measurement cannot distinguish between extracted O_2, labeled blood pool O_2 and labeled water produced as a product of O_2 metabolism produced throughout the body and circulated to the brain. The appropriate arterial input curves must distinguish between ^{15}O labeled O_2 and ^{15}O labeled water, requiring frequent arterial blood sampling and separate analysis of whole blood and plasma ^{15}O concentrations. In addition, the study must be repeated with two additional labeled molecules: ^{15}O labeled water to measure CBF, and ^{15}O labeled carbon monoxide to measure blood pool contributions. Taken together, these three measurements provide enough information to estimate $CMRO_2$, with the modeling assumption that all of the extracted O_2 is metabolized. Because of the requirement for three separate studies, this approach measures steady-state $CMRO_2$, and has limited capability for dynamic studies. More recently, variations of this method have been developed to provide a more streamlined approach (Kudomi et al. 2007; Kudomi et al. 2009) and to adapt this method to the challenges of animal studies (Yee et al. 2006). Although technically challenging to apply, the PET method is currently the closest to an accepted standard of $CMRO_2$ measurement.

Estimating $CMRO_2$ Changes with Optical Intrinsic Signal Imaging (OISI)

With the OISI techniques described earlier (Sect. 15.2.1.1), the changes in oxyhemoglobin (ΔHbO), deoxyhemoglobin (ΔHb) and total hemoglobin (ΔHbT) can be measured dynamically, potentially providing the opportunity for measuring dynamic changes in $CMRO_2$. The general approach is to use the measured data to estimate E/E_0, and combine this with a separate measurement of the CBF change to estimate the $CMRO_2$ change from Eq. 15.13 (Mayhew et al. 2000; Jones et al. 2001; Dunn et al. 2005; Royl et al. 2008). There are two challenges related to this approach that require modeling assumptions. The first is that E/E_0 depends on hemoglobin changes normalized to their baseline values, but the measurements are of differences (e.g., ΔHb), and the baseline values Hb_0 and HbT_0 are not known. Typically, assumed values are used for the baseline concentrations (but see (Dunn et al. 2005)). The second challenge is that the relevant hemoglobin concentrations are the venous values (Eq. 15.14), but the measured values sample the full vasculature. Several studies have used the simple assumption that the ratios in Eq. 15.14 for the venous blood are the same as those measured for the total vasculature. Jones and colleagues (Jones et al. 2001) explicitly described the assumptions involved by introducing two new parameters, γ_r and γ_t, defined as:

$$\gamma_r = \frac{\Delta Hb_v / Hb_{v0}}{\Delta Hb / Hb_0} \quad \gamma_t = \frac{\Delta HbT_v / HbT_{v0}}{\Delta HbT / HbT_0} \tag{15.15}$$

The simplest assumption, that $\gamma_r = 1$ and $\gamma_t = 1$ is often used, although more detailed vascular modeling, such as the VAN model described above (Sect. 15.3.3.1), should make possible more realistic estimates of these parameters. In particular, a number of human imaging studies have suggested that the changes in blood volume with activation occur predominantly on the arterial side (Lee et al. 2001; Ito et al. 2005; Kim et al. 2007). Recent 2-photon studies in animals are consistent with implications of these macroscopic fMRI measurements, reporting no change in venous diameters (Hillman et al. 2007; Tian et al. 2008). Therefore, γ_t could be substantially smaller than one.

Estimating $CMRO_2$ Changes with fMRI

The BOLD response also is sensitive to changes in blood oxygenation associated with brain activation. Deoxyhemoglobin is paramagnetic, and distorts the magnetic field around blood vessels, and these field distortions slightly reduce the net MR signal from what it would be if there was no deoxyhemoglobin present. With activation CBF increases more than $CMRO_2$, and the reduction of local deoxyhemoglobin leads to an increase of the MR signal. This BOLD signal change is often modeled as (Davis et al. 1998):

$$\frac{\Delta S}{S_0} = M\left[1 - \frac{Hb_v}{Hb_{v0}}\left(\frac{E}{E_0}\right)^{\beta}\right] \tag{15.16}$$

where S_0 is the baseline signal, M is a scaling parameter that defines the maximum possible BOLD signal change, and β is a parameter usually taken to be equal to 1.5, based on Monte Carlo simulations of the signal decay process due to the deoxyhemoglobin (Boxerman et al. 1995). The venous hemoglobin ratio is taken to be the venous volume ratio, and this is modeled as a power law relationship with the CBF change. Specifically, the assumption used is that the hemoglobin ratio in Eq. 15.16

is equal to f^{α}, with $\alpha = 0.38$ based on early measurements of the relationship between CBF and total blood volume (Grubb et al. 1974). In terms of the normalized changes in CBF and $CMRO_2$, the BOLD signal change is then modeled as:

$$\frac{\Delta S}{S_0} = M\left[1 - f^{\alpha}\left(\frac{m}{f}\right)^{\beta}\right] \tag{15.17}$$

The scaling parameter M depends on baseline state conditions (Hb_{v0} and E_0), as well as the image acquisition method and the magnetic field strength. For this reason, M should be measured rather than assumed. To do this, Davis and colleagues (Davis et al. 1998) introduced the calibrated BOLD method, in which both the BOLD and CBF responses are measured for the neuronal activation experiment and also for a hypercapnia experiment in which the subject breathes a gas with a CO_2 concentration of about 5%. The idea of this calibration experiment is that mild hypercapnia is thought to increase CBF without changing $CMRO_2$. Using Eq. 15.17 with the assumption that $m = 1$, measurement of local BOLD and CBF changes to hypercapnia allow calculation of M, and using this value of M with the measured BOLD and CBF responses to neuronal activation allows calculation of m, the relative $CMRO_2$ change, for the activation experiment.

A key assumption in the calibrated BOLD method is that mild hypercapnia does not alter $CMRO_2$. There is experimental support for the latter assumption (Sicard and Duong 2005), but recently has been questioned (Zappe et al. 2008). High levels of arterial CO_2 depress brain activity, and so the central question is whether the lower levels used in the calibrated BOLD experiment are having a significant effect on $CMRO_2$. One complication in trying to address this central question with animal studies is that the key physiological parameter is arterial CO_2 content, rather than inspired CO_2 content (typically 5% in calibrated BOLD studies). Awake humans often increase ventilation rate during hypercapnia, reducing the effect on arterial CO_2, so for the same inspired CO_2 concentration the resulting change in arterial CO_2 can be larger in an animal experiment in which ventilation is controlled. The issue of $CMRO_2$ changes with hypercapnia needs further study as a potential systematic error in the calibrated BOLD method.

The estimates of $CMRO_2$ change with the calibrated BOLD experiment also depend on the assumed values of the parameters α and β. As in the OISI experiment, there is an assumption about how the venous blood volume change is related to the overall blood volume change. In the context of calibrated BOLD this is described by the parameter α rather than γ_t, but the same underlying assumption is involved. Given the data mentioned above suggesting that most of the blood volume change with activation is on the arterial side, a lower value of α may be more accurate. The parameter β was originally introduced as an approximate description of numerical simulations for deoxyhemoglobin in vessels of different sizes (Boxerman et al. 1995). The same amount of deoxyhemoglobin in a larger vein is more effective at reducing the net MR signal than it is in a smaller capillary, and this leads to the nonlinearity (Ogawa et al. 1993). However, the use of $\beta = 1.5$, rather than 1.0, also fortuitously allows the Davis model to describe additional sources of signal variation – that were not included in the original derivation – due to intravascular signal changes and blood/tissue exchange signal changes (Buxton et al. 2004; Leontiev and Buxton 2007; Leontiev et al. 2007). In addition, the use of the same equation for estimating M and then for estimating m tends to reduce errors related to the assumed values of α and β. For these reasons, the Davis model is likely to be more robust than one would expect based on the original assumptions.

Estimating $CMRO_2$ Changes with Tissue pO_2 Measurements

A somewhat different approach to estimating $CMRO_2$ changes uses measurements of the tissue concentration of O_2 in combination with a measurement of the CBF change (Caesar et al. 2008). Tissue O_2 concentration can be measured dynamically with an appropriate electrode, and the values are usually calibrated as an equivalent partial pressure P_{O2} expressed in units of mm Hg or kilopascals (1 kPa = 7.5 mmHg). A typical value for brain tissue is 25 mmHg, corresponding to a concentration of about 0.03 mM.

The central modeling assumption is that $CMRO_2$ is associated with a passive diffusion of O_2 down a concentration gradient from capillary plasma to the mitochondria:

$$CMRO_2 = k\left[P_{O_2}(cap) - P_{O_2}(tissue)\right] \tag{15.18}$$

where k includes the solubility and diffusion coefficient of O_2 as well as factors related to capillary geometry and capillary density. The usual assumption is that the capillary geometry remains fixed, so that the increased gradient needed to support an increased O_2 flux (increased $CMRO_2$) must come from increasing the difference between capillary and tissue P_{O2} values. This is equivalent to assuming constant k as CBF and $CMRO_2$ change. The capillary P_{O2}, in turn, depends on E and a model for the O_2-hemoglobon saturation curve. Taken together, Eqs. 15.12 and 15.18 provide two equations relating the four variables CBF, $CMRO_2$, E and tissue P_{O2}. Measurement of the change in two of these variables, such as tissue P_{O2} and CBF, then allows calculation of the changes in the other two, $CMRO_2$ and E.

The basic principles outlined above are at the heart of estimating $CMRO_2$ from tissue P_{O2} data, but the details of how this is done can vary considerably. With a distributed model, such as the VAN model described above, the calculations can be done numerically along all of the vessels rather than explicitly calculating average capillary values, and this approach could in principle take into account changes in capillary density as well. Usually, though, the parameter k is assumed to remain constant during dynamic changes in CBF and $CMRO_2$. This is equivalent to assuming that there is no capillary recruitment.

In applying this approach, a central challenge is to properly model the O_2-hemoglobin saturation curve. Most of the O_2 in blood is bound to hemoglobin, but it is the small plasma concentration that drives the diffusion gradient for O_2 flux into tissue. While it is reasonable to assume that O_2 bound to hemoglobin and O_2 dissolved in plasma are in fast equilibrium, this equilibrium curve changes as O_2 is removed from blood and CO_2 is added. The sigmoidal shape of the O_2-hemoglobin saturation curve is often characterized by a parameter P_{50}, the plasma P_{O2} for 50% saturation of hemoglobin. The interesting complexity of O_2 transport is that P_{50} shifts to higher values down the length of the capillary, which has the effect of raising capillary P_{O2} and facilitating unloading of O_2 from hemoglobin. More detailed simulations of O_2 transport are needed to test the accuracy of the assumptions typically used in estimating $CMRO_2$ from tissue pO_2 measurements.

15.3.4 Alternative MRI Methods for Estimating $CMRO_2$

Although the calibrated BOLD approach is currently the most widely used method for estimating $CMRO_2$ changes with MRI data, newer methods under development may provide useful experimental tools for assessing $CMRO_2$. One of the earliest approaches was to use inhalation of O_2 labeled with ^{17}O (Arai et al. 1990; Arai et al. 1991; Fiat and Kang 1992, 1993). Two different experimental approaches have been proposed for using ^{17}O. The earliest exploits a particular magnetic property of ^{17}O: when incorporated into a water molecule, it acts as a relaxation agent, altering the relaxation times of hydrogen nuclei. In this way the presence of ^{17}O labeled water can be detected with standard proton MRI as a change in relaxation rate. However, when the ^{17}O is incorporated in an O_2 molecule it does not have this relaxation property. This creates an interesting scenario in which ^{17}O inhaled as O_2 is not visible with standard proton MRI until the ^{17}O has been incorporated into water by O_2 metabolism in the mitochondria. The rate of appearance of labeled water is then proportional to local $CMRO_2$. To quantify this, however, requires additional measurement of the arterial input function (concentration of the label in arterial blood), and accounting for recirculation of the labeled water. Despite early interest in this approach, it has not been widely used to date due to the complexity involved in addressing these quantitation issues (similar to issues in the PET studies described earlier), combined with the high cost of ^{17}O.

A more recent approach for using ^{17}O exploits the fact that ^{17}O exhibits nuclear magnetic resonance, and so can be directly detected with MR spectroscopy methods (Zhang et al. 2004; Zhu et al. 2005). In this way, the ^{17}O is used as a detectable oxygen label analogous to the use of ^{15}O in PET studies. The same issues of determining the arterial input function remain, but MRI provides ways to do this. Although still hampered by the high cost of ^{17}O, this approach has the potential to provide useful estimates of $CMRO_2$.

Finally, another MRI approach is to measure the local venous O_2 saturation of hemoglobin. The idea is to exploit the fact that the T_2 relaxation time of blood depends on the hemoglobin saturation through the same mechanisms related to the magnetic properties of deoxyhemoglobin that underlie the BOLD effect.

The experimental challenge is to isolate the signal of venous blood and measure the T_2. A recent method called TRUST combines ideas from arterial spin labeling and T_2 measurement to estimate the T_2 in the sagittal sinus, and this is converted to an estimate of hemoglobin saturation with a calibration curve (Lu and Ge 2008; Lu et al. 2008; Xu et al. 2009). The venous saturation then directly gives an estimate of *E*, and combining this with a measurement of CBF using ASL techniques makes possible an estimate of $CMRO_2$. In this initial method looking at the sagittal sinus it is a global estimate of $CMRO_2$, but techniques for extending this approach to local measurements of venous T_2 are under development. A potential challenge for modeling this experiment is that the calibration curve relating blood T_2 with hemoglobin saturation could differ between subjects, particularly due to hematocrit variations, and the potential effect on quantitation accuracy still needs to be investigated.

15.4 What Have We Learned about Neurovascular Coupling from Experimental Measurements of Neuronal, Vascular and Metabolic Activity?

A large number of recent studies have used a wide variety of measurement techniques to address the relationship between neuronal, metabolic and hemodynamic signals on multiple scales. The tools range from non-invasive methods such as fMRI, PET, and EEG/MEG to microscopic imaging of single cells and single vessels. These efforts produced consistent evidence across studies and experimental paradigms and helped settling some of the principles of neurovascular and neurometabolic coupling (Table 15.1). Some other, unexplained or controversial, findings challenge current "working hypotheses" and drive the ongoing investigations. While reviewing this literature, it is constructive to keep in mind the limitations of different measurement methods and other potentially crucial differences in the experimental paradigms between the studies that might have produced seemingly contradicting data. Testing of working hypotheses requires not only collection of high quality empirical measurements but also use of computational analyses to bridge the findings across spatial scales and integrate the components in a multiparameter space of neurovascular physiology.

15.4.1 Cell-Type Specific Regulation of Blood Flow

Of particular note, our recent *in vivo* data, along with *in vitro* data from other labs, increasingly suggest that differential vascular control originates from activation of specific neuronal sub-populations through release of specific vasoactive agents (Table 15.1, findings 1 and 2). *In vitro* (brain slice) studies in cerebral and cerebellar cortices have demonstrated that spiking of specific types of neurons can cause

Table 15.1 Main recent findings in understanding of neurovascular coupling with respect to further questions

Main findings	Questions raised	References
1. Different neuronal cell types cause vasodilation or vasoconstriction of cerebral arterioles *in vitro*.	**Q1.** What are the neuronal cell types that cause vasodilation or vasoconstriction *in vivo*? How is it reflected in the hemodynamic response (that is also sensitive to oxygen consumption)?	(Cauli et al. 2004; Hamel 2004, 2006; Rancillac et al. 2006), see also (Kocharyan et al. 2008)
2. Predominant neuronal inhibition is related to vasoconstriction and negative BOLD fMRI response (at least in sensory cortices).	**Q2.** Are all inhibitory neuronal types vasoactive? Do all of them cause vasoconstriction *in vivo*?	(Shmuel et al. 2006; Alonso et al. 2007; Bressler et al. 2007; Devor et al. 2007; Boas et al. 2008; Devor et al. 2008b)
3. Hemodynamic response is non-linearly coupled to multiunit spiking activity (MUA) and local field potential (LFP); the exact relationship depends on the stimulus.	**Q3.** How can we model the relationship between MUA/LFP to neuronal activity in specific cell types? If we knew this relationship and also the vasoactive effect of each neuronal type (finding 1), could we predict macroscopic vascular response from macroscopic measures of neuronal activity?	(Devor et al. 2003; Jones et al. 2004; Nemoto et al. 2004; Sheth et al. 2004; Devor et al. 2005)
4. Astrocytes release vasodilators *in vivo* and *in vitro*.	**Q4.** Is all vasodilaton mediated by astrocytes?	(Zonta et al. 2003; Mulligan and MacVicar 2004; Takano et al. 2006; Gordon et al. 2008)
5. Astrocytes release vasoconstrictors *in vitro*.	**Q5.** Can astrocytes mediate vasoconstriction *in vivo*? Do they respond to activity of cortical inhibitory interneurons?	(Metea and Newman 2006; Blanco et al. 2008)
6. a) Increase in glucose consumption can happen in presence of vasoconstriction and a decrease in blood flow in areas with predominant neuronal inhibition. b) CBF response is not affected by artificially increasing baseline oxygenation.	**Q6.** Does this finding imply that there is no causal relationship between accumulation of energy metabolites and increases in blood flow? Alternatively, both energy metabolites and neurotransmitters can affect flow but neurotransmitters have a dominant effect.	(Devor et al. 2008b; Lindauer et al. 2009)
7. Arteries/arterioles and capillaries can both dilate and constrict in response to stimulus.	**Q7.** How can we model macroscopic vascular response as a function of microscopic dilation and constriction?	(Devor et al. 2007; Boas et al. 2008; Devor et al. 2008b; Stefanovic et al. 2008)
8. Veins do not change their diameter in response to stimulus (either short or long) but respond by change in blood flow and hematocrit.	**Q8.** Prominent models of fMRI response assume venous dilation (e.g. "Balloon" model (Buxton et al. 1998)). These models need to be revised to reflect new experimental data.	(Vanzetta et al. 2005; Hillman et al. 2007)
9. Vasodlilation propagates along diving arterioles towards the cortical surface *in vivo*.	**Q9.** What is the location of the initial dilation? Is there dependence on the branching order? Does dilation in layer IV precede that in other cortical layers?	(Tian et al. 2008)
10. Exchange of oxygen happens not only from capillaries but also from diving arterioles and their branches.	**Q10.** What is the temporal profile of tissue oxygenation at different distances from the feeding arterioles in response to stimulus? How can we combine these oxygenation profiles with microscopic dilation and constriction (Q7) to model microscopic hemodynamic response?	(Kasischke et al. 2011; Takano et al. 2007; Fang et al. 2008)

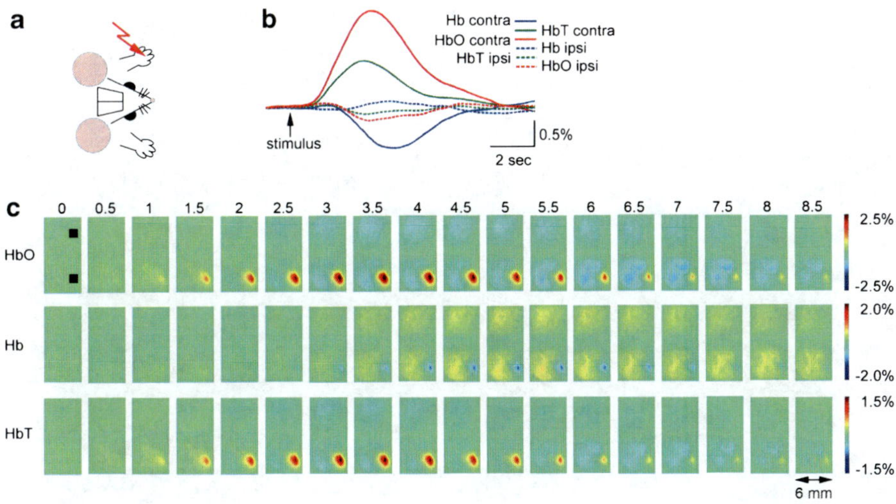

Fig. 15.14 Bilateral spectral imaging of blood oxygenation and volume in response to somatosensory stimulation. (**a**) A schematic representation of bilateral experimental paradigm. SI is exposed on both sides. The signals are imaged bilaterally in the response to stimulation of one (*left*) forepaw. (**b**) Time-courses of contralateral and ipsilateral signal in response to electrical forepaw stimulus (2 s, 3 Hz) extracted from regions of interest (ROIs) shown in (**c**). The contralateral response is shown by solid lines, the ipsilateral response is shown by dotted lines. Colorcode: Hb, blue; HbO, red; HbT, green. (**c**) Spatially smoothed ratio images of Hb, HbO and HbT response. Times in seconds relative to the stimulus onset are denoted above the images. Each image corresponds to the signal during the activation divided by the prestimulus baseline. The black rectangles denote the ROIs centered on the earliest detectable HbT response. For more information see (Devor et al. 2007; Devor et al. 2008b)

dilation or constriction of nearby arterioles (Cauli et al. 2004; Rancillac et al. 2006). *In vivo* studies in the primary somatosensory cortex (SI) in the rat, using spectral and laser speckle optical imaging, VSD and 2-photon microscopy, have demonstrated that a decrease in blood oxygenation and flow (Fig. 15.14) correlated with arteriolar constriction (Fig. 15.15) and increased neuronal inhibition (Fig. 15.16) (Devor et al. 2007; Devor et al. 2008b). This is in agreement with a previous fMRI study that showed a correspondence of negative BOLD to neuronal inhibition (Shmuel et al. 2006). The vasoconstriction was abolished by blocking of neuronal activity (Devor et al. 2008b) and was found on the arterial side while venous diameters stayed unchanged (Hillman et al. 2007; Tian et al. 2008). These findings suggest an active process rather than a passive "blood steal". This conclusion was further supported by a modeling study using VAN (Boas et al. 2008).

In healthy cortical circuits excitation is always coupled to inhibition. In SI, the balance between excitation and inhibition is temporary tilted during a response to a sensory stimulus to generate a net increase in excitation followed by a net increase in inhibition (Simons and Carvell 1989; Moore and Nelson 1998; Pinto et al. 2000; Derdikman et al. 2003). This can be easily observed with VSD that report mostly synaptic polarization changes of apical pyramidal dendrites (Ferezou et al. 2006)

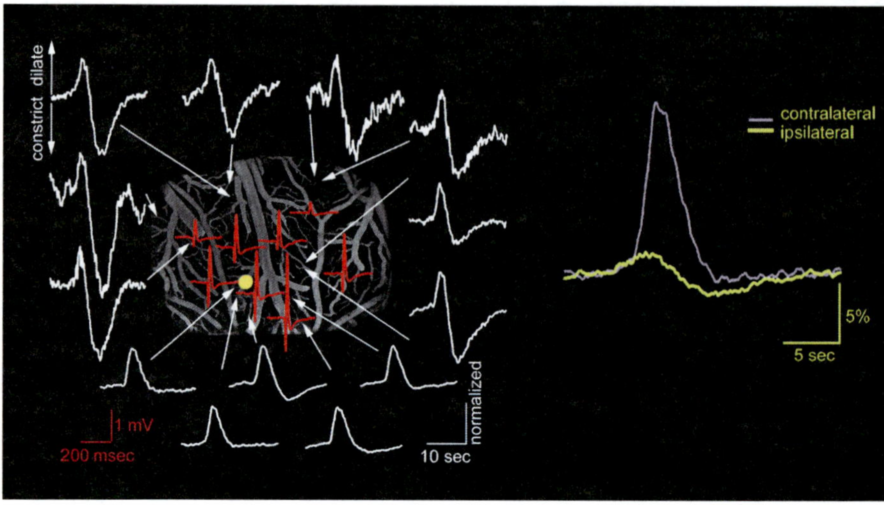

Fig. 15.15 Two-photon imaging of vascular dilation and constriction to a somatosensory stimulus. *On the left:* Diameter changes were measured in single surface arterioles in response to electrical stimulation of contralateral forepaw (1 s, 3 Hz). Increase in the diameter is plotted upward, and decrease is plotted downward. The time-courses are normalized by the peak dilation amplitude. Electrophysiological response to the stimulus was mapped using a ball electrode. Recorded surface potentials (*red traces*) are overlaid on the image of the brain surface vasculature. Both vasodilatation and vasoconstriction are present at all locations. The degree of vasodilation decreases with an increase of distance from the center of neuronal activity (the largest surface potential amplitude). *On the right:* arteriolar diameter changes in response to contralateral stimulus (*blue*) and ipsilateral stimulus (*yellow*) measured at the location marked by a yellow dot on the image of the brain vasculature. For more information see (Devor et al. 2007; Devor et al. 2008b)

(Fig. 15.16). Importantly, neuronal activity is fast, so that both the excitatory and inhibitory phases of the neuronal response occur before the onset of blood flow response. Thus, the vasculature reacts "at once" to the accumulated transmitters released during both phases of neuronal response. This is the main rational behind integration of neuronal measurements over the time window of the response in studies that compared neuronal and hemodynamic measurements over a range of stimulus conditions (Devor et al. 2003; Devor et al. 2005). For example, in SI a sensory stimulus usually elicits an increase in MUA during the VSD depolarization followed by a very small decrease that occurred during the VSD hyperpolarization (Fig. 15.17). The decrease might be below the detection threshold for low-impedance laminar electrodes with large recording contacts, but can be observed using high-impedance electrodes with small tips. However, since the decrease is small, the integral MUA response would indicate an overall increase in neuronal firing (as was discussed above, cell-type specific firing rates cannot be inferred from MUA alone, see Sect. 15.3.2.3).

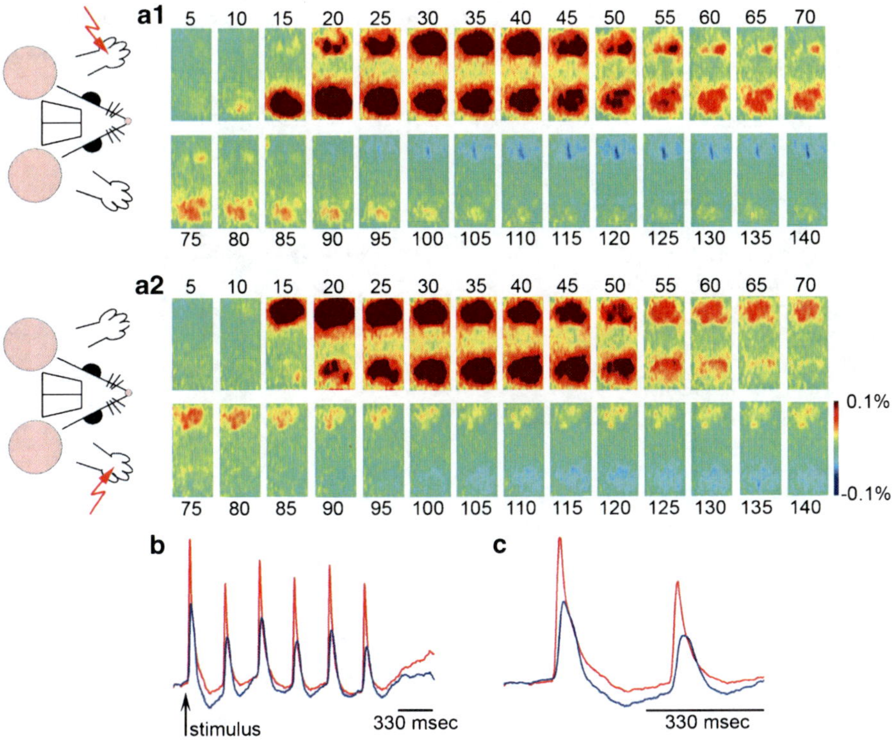

Fig. 15.16 Bilateral VSD imaging of neuronal polarization changes following a somatosensory stimulus. (**a**) Temporal evolution of VSD response to electrical forepaw stimulation (2 s, 3 Hz). The response to the first stimulus in the train is shown. Times relative to the stimulus onset are indicated above the images. A1 and A2 show the response to stimulation of left and right forepaw, respectively. (**b**) Signal time-course extracted from the contralateral (*red*) and ipsilateral (*blue*) hemispheres. The ROIs are shown by red dotted line in A1. (**c**) Zoom in on the response to the first two stimuli. For more information see (Devor et al. 2007; Devor et al. 2008b)

Moreover, synaptic inhibition does not necessarily have to be accompanied by a decrease in firing rate. For example, inhibitory inputs to apical dendrites of layer V pyramidal cells are far enough from the cell body, so that hyperpolarization decays along the membrane before reaching the cell body. Therefore, the hyperpolarization of dendrites in layer I might not affect the generation of action potentials that happens in the initial segment of the axon emerging from the cell body in layer V. Under these conditions, an increase in inhibition would not reduce the firing rate. On contrary, since the inhibitory synaptic inputs result from firing of local inhibitory interneurons, the overall firing might increase.

With respect to the role that neuronal transmitters play in communication of neuronal activity to the nearby arterioles (Cauli et al. 2004; Hamel 2004, 2006; Rancillac et al. 2006), some neurotransmitters are directly vasoactive (Cauli et al. 2004; Hamel 2004, 2006) or trigger a synthesis of vasoactive messengers. For example, the major

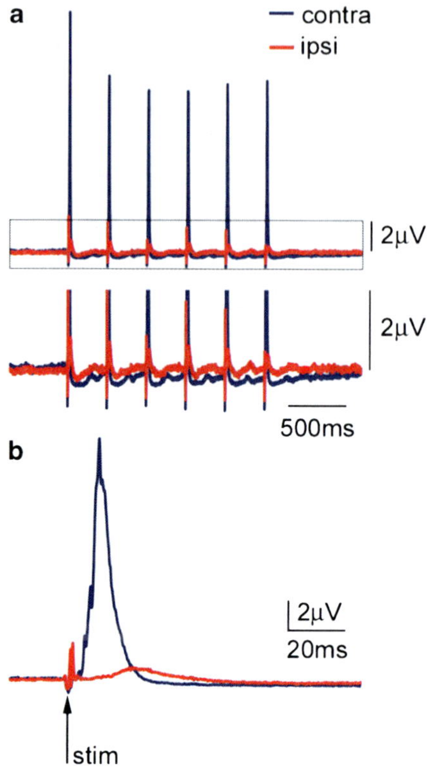

Fig. 15.17 Changes in population firing rate in response to a somatosensory stimulus. (**a**) MUA in response to an electrical forepaw stimulation (2 s, 3 Hz). The response was averaged across the cortical depth to improve signal-to-noise ratio. Responses to contra- and ipsilateral stimuli are superimposed (blue and red, respectively). The part within a rectangle is expanded below. A small decrease in MUA is evident for both stimulus conditions. (**b**) The response to the first stimulus in the train is expanded. Black arrow points to a stimulus artifact indicating stimulus onset

excitatory neurotransmitter, glutamate, causes activation of NMDA receptors that elevate intracellular concentration of calcium and trigger synthesis of a vasodilator nitric oxide (NO) by a neuronal enzyme called nitric oxide synthase (nNOS) (Iadecola and Niwa 2002). Moreover, it has been shown that released glutamate activates astrocytes that can release additional dilators (see Sect. 15.4.3 below) (for recent reviews see (Iadecola and Nedergaard 2007; Koehler et al. 2009)). Thus, neuronal excitation is expected to cause dilation of surround arterioles.

Application of $GABA_A$ agonist muscimol has been reported to cause dilation in brain slices (Fergus and Lee 1997). However, inhibitory interneurons release not only GABA but also other signaling molecules called neuropeptides. In particular, firing in cortical interneurons containing neuropeptides somatostatin and neuropeptide Y was shown to cause arteriolar vasoconstriction *in vitro* (Cauli et al. 2004). The majority of these interneurons, namely Martinotti cells, extend their axons to neighboring cortical columns and thus are in a good position to mediate the phenomenon known as "surround inhibition" (Simons and Carvell 1989; Takashima et al. 2001; Derdikman et al. 2003; Devor et al. 2007). Therefore, activation of these neurons is expected to result in surround vasoconstriction. In this respect, *in vivo* imaging data demonstrated that surround inhibition was correlated with arteriolar vasoconstriction and a decrease in blood oxygenation, or "negative BOLD" (Devor et al. 2007; Boas et al. 2008). Nevertheless, direct *in vivo* evidence showing that selective activation of Martinotti cells, or release of somatostatin, correlates with vasoconstriction is missing.

Assuming that neurotransmitter/neuropeptide release during synaptic transmission might be a critical factor in generation of the hemodynamic response (Cauli et al. 2004; Hamel 2004, 2006; Rancillac et al. 2006), activation of the same cortical circuit by different inputs can lead to different degrees of vasodilation and vasoconstriction, reflecting differential engagement of particular cell types. *In vivo* data are consistent with this hypothesis. Specifically, vasodilation and an increase in blood flow were observed in the forepaw area of SI following stimulation of the contralateral forepaw, while the same area showed vasoconstriction and a decrease in flow following stimulation of the ipsilateral forepaw. In the former, the circuit was excited by thalamic inputs to layer IV, whereas in the latter the ipsilateral SI was activated via transcallosal afferents synapsing below and above layer IV (Wise and Jones 1976; White and DeAmicis 1977; Koralek et al. 1990).

15.4.2 Linearity of Neurovascular Coupling and the Relationship between the Hemodynamic Signals and MUA/LFP

A number of previous studies from our group and others suggest a complex relationship between neuronal firing (MUA) and LFP on the one hand, and the hemodynamic response on the other (Table 15.1, finding 3). These studies have shown that the degree of linearity in neurovascular relationship depends strongly on the stimulus paradigm (Sheth et al. 2004), and that important discrepancies between hemodynamic and electrophysiological measures can be observed under certain conditions (Devor et al. 2008b). Yet, as we have discussed above, it is likely that activation of specific types of neurons as opposed to the dynamics of multiunit spiking or synaptic activity is likely to determine the vascular response.

The following example demonstrates that not only the inverse solution (prediction of MUA and LFP from the hemodynamic response) is an ill-defined problem, but even the forward solution (prediction of the hemodynamic response given the MUA and LFP) can be ambiguous without *a priori* knowledge of stimulus conditions. In this example, paired tactile stimuli were delivered to single whiskers at different time delays (Fig. 15.18a). Electrophysiological recordings of MUA and LFP were performed from layer II/III at the location of the maximal response to stimulation of the second whisker (the "principal barrel" of the second whisker). Spectral imaging of Hb, HbO and HbT were performed simultaneously with the MUA/LFP recordings. In Fig. 15.18b the peak response of Hb, Hb and HbT is plotted against the peak MUA and LFP. In contrast to other studies, the scatter plots in Fig. 15.18b cannot be approximated nether by a linear or any nonlinear continuous function.

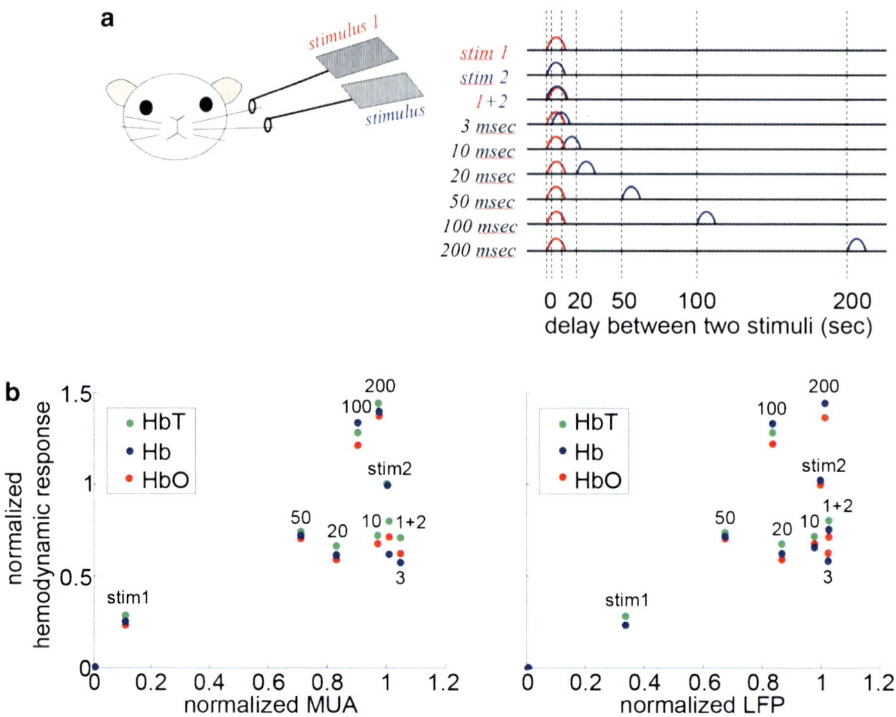

Fig. 15.18 Paired whisker stimulus paradigm with a varying delay. (**a**) Two single neighboring whiskers were stimulated at different delays. Stimulus conditions: first whisker alone, second whisker alone, both whiskers simultaneously, second following first after X msec delay where X = [3 10 20 50 100 200 500 800 1000 1500 2000]. (**b**) Hb (*blue*), HbO (*red*) and HbT (*green*) as a function of MUA (*left*) and LFP (*right*) across stimulus conditions. For each parameter all values are normalized to stimulation of the second whisker alone. Conditions are indicated above data points

What is so different about the paired whisker paradigm that makes it impossible to fit a curve to the relationship between the MUA/LFP and the hemodynamic parameters as has been done in other studies that varied stimulus amplitude, duration, and frequency? As we discussed above, the neuronal response to a sensory stimulus in SI is biphasic, consistent with an increase in excitation followed by an increase in inhibition. In the paired paradigm, the second whisker is deflected before the relaxation of the neuronal response to the first one. Thus, the response to the first whisker sets up neuronal activity, and the response to the second whisker rides on top of this on-going activity, either net excitatory or net inhibitory depending on timing. As different neuronal cell types vary in their frequency response, excitability and relaxation time, stimulation of the second whisker at different delays from the first results in differential engagement of particular cell types and, therefore, in a different hemodynamic response depending on the net neuronal activity. These cell-type specific differences cannot be easily extracted from measurements of MUA and LFP, in particular if only a single-location LFP measurement is available.

In contrast, if we knew the behavior of the neuronal circuit under each one of the stimulus conditions and could reconstruct the activity of each one of the vasoactive cell types, it is conceivable that we could predict the corresponding vasodilation and vasoconstriction.

The issue of linearity of neurovascular coupling is sometimes being confused with the issue of additive properties of hemodynamic responses. Anders Dale and colleagues used one, two and three separate short-lasting visual stimuli to test the former (Dale and Buckner 1997). Individual stimuli were spaced far enough in time to allow relaxation of the neuronal activity between the stimuli. Under these conditions, they were able to confirm that the hemodynamic response to two consecutive stimuli could be modeled as a sum of the responses to each of the stimuli alone. These and other experiments showed that using a rapid presentation, hemodynamic response to each stimulus condition can be recovered as long as neuronal activity recovers between the consecutive stimuli (in the case of the paired whisker paradigm, neuronal activity has to recover between one paired stimulus to the next).

Thus, testing the linearity of neurovascular coupling does not require long interstimulus intervals: the interval just has to be long enough to allow neuronal relaxation. For example, if neuronal activity X evokes hemodynamic response Y, a series of two separate X events with a complete neuronal relaxation between them will evoke a linear sum of Y in the response to the fist X and another Y in the response to the second X. However, if the neuronal activity changes in amplitude to 2X, the hemodynamic response can be either 2Y (linear coupling), <2Y (sublinear), or >2Y (supralinear).

15.4.3 Astrocytes as Neurovascular Mediators

A growing body of literature indicates that astrocytes not only provide a metabolic support for neurons (Magistretti et al. 1999; Rouach et al. 2008) but also release vasoactive molecular mediators ("gliotransmitters") and thus can play a key role in neurovascular communication (Table 15.1, findings 4 and 5) (for recent reviews see (Agulhon et al. 2008; Koehler et al. 2009)). Anatomically, the astrocytes are well-positioned to serve as a bridge between neurons and blood vessels, with processes associated with neuronal synapses and processes that contact the blood vessels (Fig. 15.19). Experiments in brain slices have shown that astrocytes respond to neuronal activity by elevation in intracellular calcium concentration (Simard et al. 2003; Zonta et al. 2003; Gordon et al. 2008). Although astrocytes express receptors for numerous neurotransmitters and neuropeptides, most of the studies have focused on glutamate. Specifically, it has been shown that glutamate activates metabotropic receptors (mGluR) that cause calcium influx from intracellular stores (Porter and McCarthy 1996; Zonta et al. 2003). In addition, activation of astrocytic GABA receptors also has been shown to produce calcium transients (Nilsson et al. 1993). Other receptors, including ACh, NA, VIP, somatostatin and more, are less understood although most of them are also metabotropic and their activation is expected

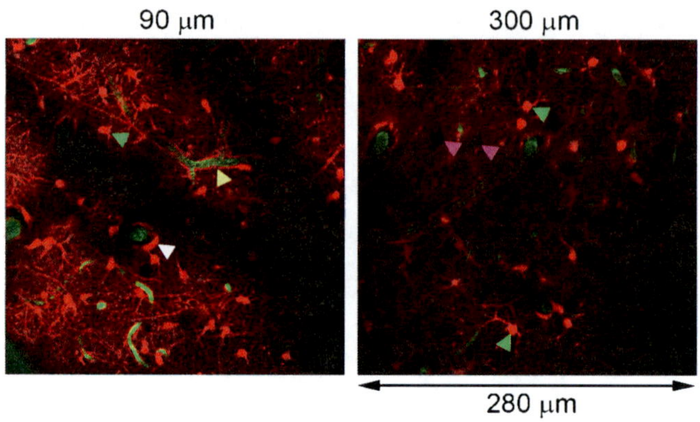

Fig. 15.19 Astrocytic endfeet outline microvasculature. Two single images of astrocytes (labeled with SR101, *red*) and blood vessels (labeled by intravascular fluorescein dextran, *green*) at different depths. Astrocytes (*green arrowheads*) are present throughout the cortical depth; their endfeet wrap around diving arterioles (*white arrowhead*) and capillaries (*yellow arrowhead*). Neurons are not labeled and appear as black holes (*magenta arrowheads*)

to elevate astrocytic calcium concentration. The elevation of intracellular calcium in turn can trigger synthesis and release of gliotransmitters. Both dilators (epoxyeicosatrienoic acids (EETs) and prostaglandins) and constrictors (20-hydroxyeicosatetraenoic acid (20-HETE)) are synthesized in calcium-dependent manner from common precursors (reviewed in (Straub and Nelson 2007)). In fact, slice studies demonstrated that elevation of astrocytic calcium (spontaneous or evoked by neuronal activity) can cause both dilation and constriction of nearby arterioles (Zonta et al. 2003; Mulligan and MacVicar 2004). Since astrocytes are coupled by gap junctions, signals can propagate throughout the astrocytic network and thus might be involved in propagation of vascular response along the vessels. Some of the variability in vascular response observed in slices (dilation or constriction) can be explained by the baseline state of smooth muscle cells in the absence of normal blood pressure and flow (Blanco et al. 2008). An *in vivo* study have has supported the finding that astrocytic calcium increase has vasoactive consequences (usually dilation, but rare case of constriction were also observed) (Takano et al. 2006).

Furthermore, natural sensory stimulation *in vivo* induces transient calcium responses both in neurons and astrocytes (Wang et al. 2006; Schummers et al. 2008). In the majority of these studies, astrocytic calcium transients were significantly delayed with respect to the onset of stimulation, neuronal response, and known dynamics of vascular responses. In fact, a recent study in ferret visual cortex, where both calcium and hemodynamic responses were measured, reported an increase in blood volume preceding an increase in astrocytic calcium by several seconds (Schummers et al. 2008) (see also (Koehler et al. 2009) for further discussion on the discrepancy of astrocytic and blood flow response time scales). Moreover, laser Doppler studies suggest that blocking astrocytic transmission does not affect the blood flow response during the first couple of seconds, but reduces the amplitude of the flow response during prolong stimulation (Liu et al. 2008; Shi et al. 2008). Finally, arteriolar smooth muscle cells express a variety of receptors for neurotransmitters and neuropeptides suggesting that vascular diameters can be regulated independently of astrocytic activity (Hamel 2006).

One study stands apart from others with respect to the kinetics of astrocytic calcium response (Winship et al. 2007). This study reported that a small percent of astrocytes exhibited a fast response to whisker stimulation similar to nearby neurons. The astrocytes in this study were identified based on labeling with SR101. This opens a possibility of heterogeneity within the astrocytic population. The other possibility is that a small percent of SR101 labeled cells were actually neurons. While it has been shown that SR101 exclusively labels astrocytes using topical loading, Winship et al. (2007) injected the dye into the tissue that might have resulted in a small percent of neuronal uptake. The apparent fast response in a small number of astrocytes could also result from a technical error related to the employed subtraction procedure. Nevertheless, it is possible that astrocytes are heterogeneous across and within cortical layers. For example, we know that neurons in layer IV have higher metabolism than in other layers reflected by higher density of cytochrome oxidase (Woolsey et al. 1996). Thus, it is conceivable that astrocytes in layer IV might have faster or more efficient responses to elevated neuronal activity. Noteworthy neurovascular coupling can differ between brain regions, e.g., between SI and the olfactory bulb (Petzold et al. 2008).

15.4.4 Is There a Feedback Vasodilation Signaling Related to Increased Metabolism?

Theoretically, blood flow response can be controlled not only in a feed forward manner via neuronal release, but also by a feedback from accumulation of energy metabolites related to increased neuronal activity. In fact, coupling of blood flow and glucose metabolism is always assumed as a general rule in PET literature (Fox and Raichle 1986; Fox et al. 1988; Raichle and Mintun 2006). However, a recent study, using electrophysiological and optical imaging methods and 2-deoxyglucose (2DG) autoradiography, suggests that blood flow response is not coupled to glucose consumption under all conditions (Devor et al. 2008b). In this study, 2DG uptake was examined bilaterally following stimulation of one (right) forepaw. The results showed a bilateral increase in glucose consumption with a small but significant increase in the ipsilateral SI (Fig. 15.20, red arrows; for statistical quantification see (Devor et al. 2008b), their Fig. 15.5). The authors ruled out potential neuronal deactivation or blood stealing effects by showing that vasoconstriction was observed on arterial side, depended on local neuronal activity, and was observed in the presence of increased MUA (reflecting the spiking response both excitatory and inhibitory cells). Voltage-sensitive dyes imaging data, acquired under the same stimulus conditions, showed that vasoconstriction and negative hemodynamic response can be explained by an increased engagement of synaptic inhibition ((Devor et al. 2008b),

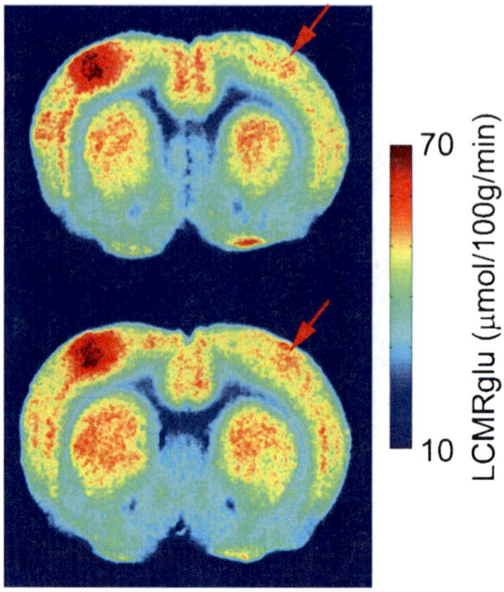

Fig. 15.20 2DG-autoradiography in response to a unilateral stimulation. Consecutive non-adjacent coronal brain sections through the forepaw region of SI are shown. The color scale is expressed in units of local cerebral metabolic rate of glucose, LCMRglu (μmol/100 g/min). *Red arrows* point to a weak ipsilateral label uptake

their Supplementary Fig. 15.2). Moreover, increased inhibition was a common denominator between the ipsilateral results in Devor et al. (2008b) and their previous study considering the "surround" contralateral SI region (Devor et al. 2007). These data are consistent with the idea that some of the inhibitory vasoactive messengers (e.g., somatostatin or neuropeptide Y) can actively cause vasoconstriction and "negative BOLD". Dissociation of vascular and metabolic responses in the ipsilateral SI finding does not rule out the possibility that certain metabolites can have a vasoactive effect (Ido et al. 2004; Mintun et al. 2004; Gordon et al. 2008). Rather, it indicates that blood flow response is not determined by the tissue energy consumption and other factors, such as release of neurotransmitters and neuropeptides, can play a dominant role in the regulation of blood flow (Table 15.1, finding 6).

It has been showed that neuronal stimulation produces adenosine, a potent vasodilator (Rubio et al. 1975). However, most of adenosine release in healthy brain is associated not with ATP utilization but with propagation of vasodilation along arteriolar walls (Iliff et al. 2003; Ohata et al. 2006). Thus, the involvement of adenosine in functional increases in CBF *per se* does not support the "metabolic" hypothesis but rather illustrates the importance of propagation of dilatory response from microvasculature to feeding vessels on the pial surface. Nevertheless, adenosine and energy metabolites, such as a glycolytic product lactate, might play a key role in dilation response under pathological conditions. For instance, it has been demonstrated that extracellular adenosine and lactate facilitate dilation *in vitro* under hypoxia (Gordon et al. 2008).

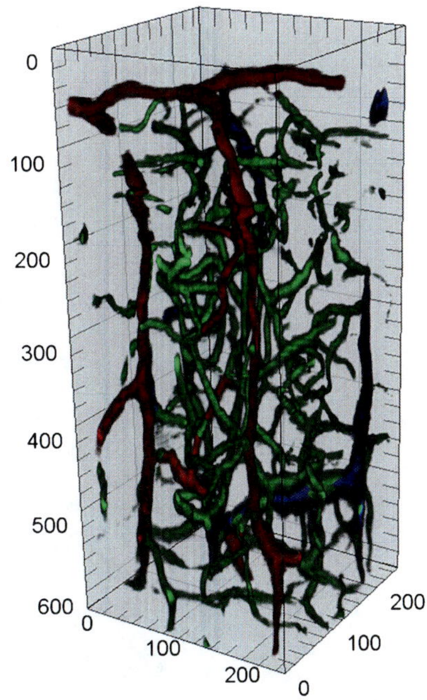

Fig. 15.21 Reconstruction of microvasculature within a 250 × 250 × 600 μm cube of SI cortical tissue. Arterioles are in red, veins in blue and capillaries in green. Top corresponds to the cortical surface

15.4.5 Compartment- and Depth-Specificity of the Vascular Response

In cerebral cortex, pial arteries give rise to smaller arterioles that dive into the cortex at a close to 90°angle. The diving arterioles branch as they dive to feed the local capillary beds. Venules and veins collect the deoxygenated blood from the capillaries and bring it back to the cortical (pial) surface (Fig. 15.21). Thus, measurement modalities with a limited depth penetration and no laminar resolution, such as optical imaging of intrinsic signals, are biased to the largest arteries and veins that lie on the pial surface. However, dilation of feeding arteries and oxygenation changes in draining surface veins are delayed relative to the onset of dilation in tissue microvasculature (Tian et al. 2008). Therefore, not only the initial dip (Malonek and Grinvald 1996) but also the initial changes in HbT can be used for localization of neuronal activity. In fact, examination of the temporal evolution of ratio images (divided by the baseline image) reveals than the first detectable HbO and HbT signals are well localized. As the dilation propagates to the surface, the pial arteries become visible (Fig. 15.22, red arrowheads). The draining veins appear later in images of HbO and Hb (Fig. 15.22, black arrowheads). Speckle contrast images are noisier and, therefore, the surface vasculature is less visible.

In contrast to optical imaging of intrinsic signals, 2-photon microscopy provides the ultimate single-vessel resolution down to ~ 600 μm below the surface. Two-photon

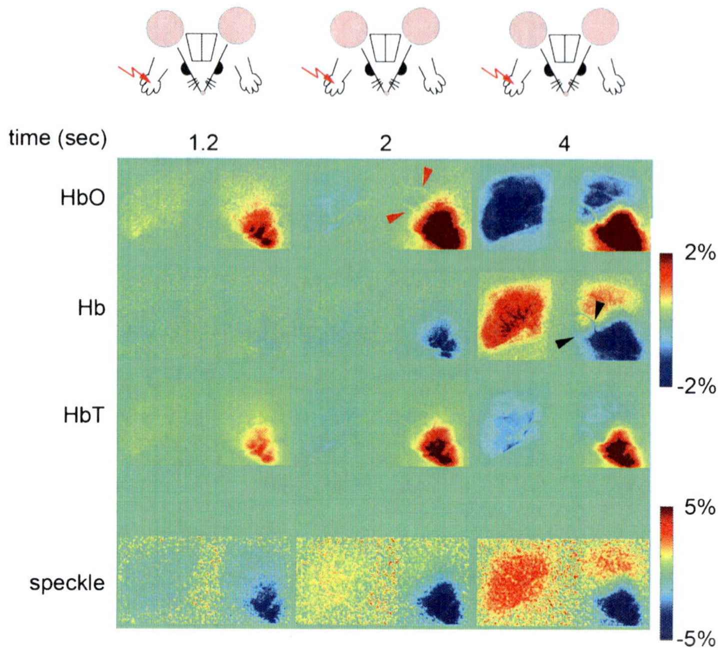

Fig. 15.22 Evolution of intrinsic imaging signals in response to a forepaw stimulus. Bilateral ratio images of HbO, Hb, HbT and speckle contrast are shown. Red and black arrows point to the surface arteries and veins, respectively. Negative change in speckle contrast corresponds to an increase in speed of RBCs (an increase in CBF)

measurements indicate that both arteries/arterioles and capillaries have a biphasic response to stimulus with initial dilation followed by a constriction (Devor et al. 2007; Boas et al. 2008; Devor et al. 2008b; Tian et al. 2008) (Table 15.1, finding 7). Arteriolar walls contain smooth muscle cells and their diameter can be actively controlled by vasoactive mediators (for review see (Iadecola 2004; Lauritzen 2005; Hamel 2006)). Recently, it has been shown in a brain slice preparation that capillary diameters can be actively controlled by perycites - single-cell contractile cells found most often at capillary branching points. This mechanism might be responsible for the capillary dilation observed *in vivo* (Stefanovic et al. 2008; Tian et al. 2008).

Although venules and veins are not surrounded by a wall of smooth muscle cells and cannot dilate actively, it has been proposed that they can passively swell during a functional increase in CBF (Buxton et al. 1998; Mandeville et al. 1999). In fact, a hypothetical increase in venous volume has been used to explain some prominent features of BOLD response, such as "post-stimulus undershoot" (Buxton et al. 1998). Recent experimental studies using 2-photon imaging found no indication of change in venous diameters (Table 15.1, finding 8) (Hillman et al. 2007; Tian et al. 2008). These findings are consistent with observations based on another optical technology, Laminar Optical Tomography (LOT) (Hillman et al. 2004; Hillman et al. 2007), and also with intrinsic (spectral) imaging data (Vanzetta et al. 2005). Interestingly, 2-photon imaging of small surface venules in rat SI provided evidence of an increase in density of RBCs ("hematocrit") in response to a somatosensory stimulus (Hillman et al. 2007).

Because the largest veins run on the pial cortical surface, oxygenation changes detected from the top cortical layer are expected to be delayed relative to deeper layers (Hillman et al. 2007). In addition, our recent *in vivo* 2-photon data demonstrate that dilation propagates to the surface from deeper layers (Tian et al. 2008) (Table 15.1, finding 9). These studies raise a possibility of laminar differences in functional dilation onset. These findings are in agreement with our recent OCT study that observed a delayed response in superficial cortical regions (<200 μm) consistent with retrograde ("upstream") vasodilation (Chen et al. 2008). Importantly, propagation of neuronal activity within a cortical column happens within a couple of milliseconds. The hemodynamic response is much slower: the onset of dilation in response to a brief single-whisker deflection occurs well past the return of neuronal activity to the baseline. For that reason, any laminar hemodynamic/vascular delays are unlikely to be explained by neuronal propagation between cortical layers: on the hemodynamic time scale, neurons respond virtually simultaneously throughout the cortical depth.

15.4.6 Oxygen Availability on the Microscopic Scale

As we discussed in Sect. 15.3.3.2, there are no tools today that allow a direct and dynamic measurement of O_2 consumption. However, O_2 availability has been measured in a number of ways. A number of recent studies have employed polarographic sensors, also called "O_2 electrodes", to measure a partial pressure of O_2 (pO_2) in the tissue (Thompson et al. 2003; Viswanathan and Freeman 2007). These studies demonstrated a drop in tissue pO_2 confined to areas of SI receiving direct thalamic inputs (principal cortical columns). This initial pO_2 drop has been interpreted as an increase in O_2 consumption prior to an increase in CBF, analogous to the "initial dip" often observed by intrinsic optical imaging (Malonek and Grinvald 1996). A followed increase in pO_2 presumably reflected vasodilation and an increase in CBF. While these findings are in general register with intrinsic imaging studies, some timing discrepancies exist. Specifically, the decrease in tissue pO_2 has been reported lasting as long as 10 s (Viswanathan and Freeman 2007), considerably longer than what is usually observed optically (Vanzetta and Grinvald 2001). Some of this can be explained by a slow reaction time of pO_2 sensors. Moreover, the pO_2 measurements can vary considerably depending on micropositioning of the sensor, and sensors need to be calibrated frequently.

Optical measurements of O_2 using phosphorescent lifetime probes are still in infancy, but new and better dyes are being developed and tested (Vanzetta and Grinvald 1999; Dunphy et al. 2002). In contrast to O_2 electrodes, O_2-dependent quenching of phosphorescence can provide a 2-dimensional map of both intravascular

pO_2 if injected i.v. and extravascular pO_2 if microinjected in the tissue. Moreover, phosphorescence lifetime can be combined with 2-dimensional imaging of blood flow using simultaneous laser speckle contrast imaging (Sakadzic et al. 2009). Simultaneous acquisition of spatiotemporal maps of pO_2 and blood flow during functional activation is critical to distinguish between the contribution of CBF and oxygen consumption to functional changes in hemoglobin oxygenation, such as BOLD response. In fact, measurement of pO_2 alone cannot be used to infer neither BOLD response, nor tissue O_2 metabolism. For example, an increase in CBF in areas of baseline neuronal activity (no evoked neuronal response) would lead to an increase in measured tissue pO_2. Thus, one has to measure both O_2 availability and blood flow to predict BOLD or draw conclusions regarding $CMRO_2$.

A couple of recent studies used optical imaging of NADH that provides an indirect tool for looking at processes related to O_2 consumption. Interpretation of NADH images is challenging due to a complex relationship between the cellular processes both producing and oxidizing NADH to a non-fluorescent NAD^+ (for a recent review see (Turner et al. 2007)). Specifically, NADH (and pyruvate) is produced in the cytosol during glycolysis and oxidized during conversion of pyruvate to lactate. NADH is also both produced and oxidized in the mitochondria (during the TCA cycle and oxidative phosphorylation, respectively). If all processes would happen simultaneously and at the same rate, one might expect to see no change in NADH fluorescence. However, thermodynamic studies show that glycolysis occurs faster than the conversion of pyruvate to lactate and empirical optical data show that changes in NADH fluorescence can be observed in neurons and astrocytes following an increase in neuronal activity (Kasischke et al. 2004; Takano et al. 2007). In register with the idea that neurons use mostly oxidative phosphorylation while astrocytes use mostly glycolysis (Magistretti et al. 1999), 2-photon NADH imaging in astrocytes in brain slice preparations revealed an early dip in NADH fluorescence, sensitive to blocking of postsynaptic neuronal activity, followed by an increase in fluorescence in astrocytes (Kasischke et al. 2004). Further *in vivo* 2-photon studies with cellular resolution are needed to elucidate NADH dynamics in neurons and astrocytes under healthy conditions.

In the classical view, the extraction of O_2 and glucose into the tissue occur at the capillary level. However, recent optical data suggest that O_2 travels into the tissue also through the walls of penetrating arterioles and their branches (Hillman et al. 2007; Kasischke et al. 2011; Takano et al. 2007) (Table 15.1, finding 10). Future modeling studies are required to determine to what extent this effect would contribute to the macroscopic BOLD signal (Fig. 15.23).

15.5 Open Questions and Outlook

The methods described above provide a powerful battery of tools for developing a deeper understanding of the fundamental physiological mechanisms linking neuronal activity, blood flow and energy metabolism. A central goal of understanding

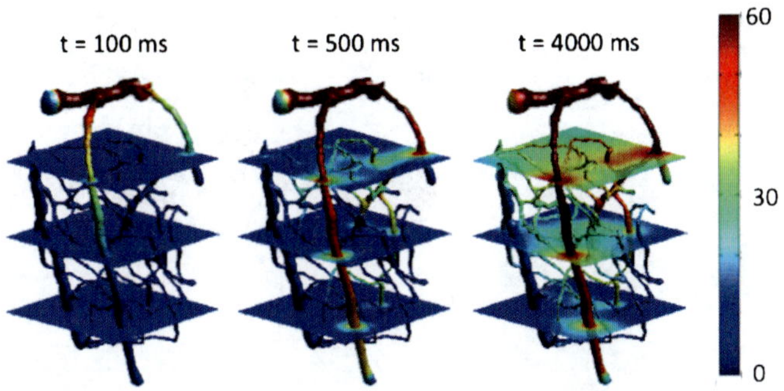

Fig. 15.23 VAN simulation of tissue pO_2. VAN simulation in a 250×250×450 μm subset following an inflow of O_2 from a pial artery at t=0 ms. Baseline pO_2 was 10 mmHg. A steady-state pO_2 is achieved within 4 s

these basic mechanisms is to lay a firm foundation for interpreting noninvasive functional neuroimaging data in humans, with the BOLD effect as a primary example. Given the widespread use of BOLD imaging for brain mapping studies, it is remarkable how small an impact it has had on clinical practice. A key problem is that, because of the physiological complexity of the BOLD response, it is difficult to interpret the magnitude of the response in a quantitative, physiologically meaningful way. For this reason BOLD-fMRI provides a robust index of where neuronal activity is changing, but it is much more difficult to interpret the magnitude, and even polarity, of the change. So it is not surprising that the most promising clinical applications of fMRI to date have been in pre-surgical planning, where the central question is the location of activity changes. However, because the BOLD response is sensitive to key physiological variables (CBF and $CMRO_2$), there is the potential for the development of fMRI into a quantitative probe of physiological function.

Quantitative applications of fMRI require a multi-modal imaging approach. A simple example described earlier is how the combination of a CBF measurement with a BOLD measurement in the calibrated BOLD experiment can yield an estimate of the $CMRO_2$ change, more than either measurement alone provides. The key to achieving such gains with multi-modal imaging, though, is a comprehensive theoretical framework that accurately describes the dependence of each of the measured signals on the underlying physiological variables. To establish that theoretical framework requires systematic studies in animal models to address key questions. For our example of quantitative BOLD imaging, a central concern is the role played by blood volume changes in affecting the signal. Changes on the capillary or venous side could alter the local deoxyhemoglobin content, altering the signal through a BOLD mechanism, but in addition arterial volume changes also could affect the measured signal by exchanging blood and extravascular water carrying different MR signals. To accurately model these effects requires a detailed understanding of volume and hematocrit changes at each stage of the vascular tree (Hillman et al. 2007; Srinivasan et al. 2009). This information can be provided by 2-photon microscopy,

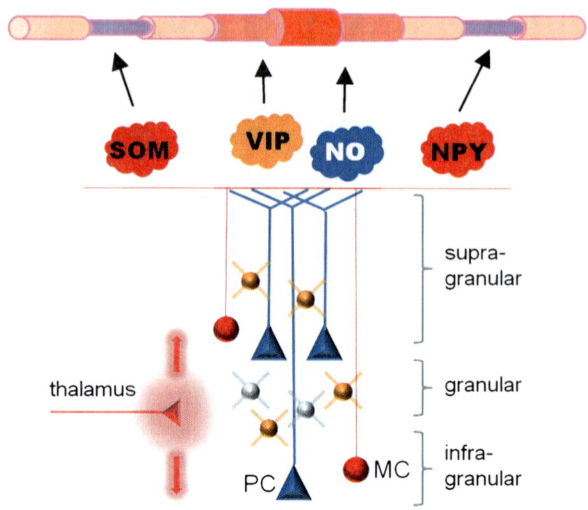

Fig. 15.24 A cartoon of cerebral vasoactive circuit. PC: pyramidal cell; MC: Martinotti cell

and integrated within a multi-scale vascular model, such as the VAN model, to produce macroscopic average behavior appropriate for the spatial scale of imaging. The same approach can be used to address other modeling assumptions noted in Sect. 15.3.

An important area where detailed studies in animal models can shed light on the interpretation of the BOLD response is in understanding the dynamics. Quantitative BOLD applications in human tend to focus on steady-state changes, but the BOLD response exhibits rich dynamics that are still poorly understood. Phenomena such as the initial dip, the post-stimulus undershoot, and the dependence of the time scale of the dynamics on the baseline state are all measurable in human subjects, but still poorly understood in terms of the underlying physiology. Dynamic studies such as those described above will be able to characterize the dynamics of vascular volume at all levels of the vascular tree, as well as the dynamics of CBF and $CMRO_2$ changes. Combined with an appropriate vascular model these measurements will provide a deeper understanding of the origins of these dynamic features of the BOLD response.

According to our current working hypothesis on dynamic regulation of CBF in SI, vasodilation and vasoconstriction result from activation of distinct neuronal subpopulations through release of an array of vasoactive agents (Fig. 15.24). Specifically, the biphasic behavior of arteriolar diameter change observed with 2-photon microscopy is consistent with a "push-pull" mechanism where simultaneously released dilatory and constrictive agents compete as antagonistic forces, with their strengths proportional to the number of released molecules (Devor et al. 2007; Devor et al. 2008b). In agreement with this idea, a simple computational analysis in Devor et al. (2007) shows that the biphasic diameter changes can easily be obtained as a weighted sum of a dilatory and constrictive effects with very similar time-courses and simultaneous onset times.

Figure 15.24 illustrates the idea of the cell-type specific control of cerebral CBF in SI with respect to an incoming sensory input (this figure by no means provides a summary of all the relevant molecular messengers or vasoactive cell types). In SI, thalamic afferents terminate in layer IV (Fig. 15.24, "granular"), and excitation propagates fast up and down the column (red arrows) and to the surround columns. The excitation is always coupled with inhibition because of the wiring of cortical circuits (Lefort et al. 2009). Arteriolar dilation is dominated by excitation in pyramidal cells (PC, Fig. 15.24) and release of glutamate with a subsequent release of NO (via activation of NMDA receptors in NOS-positive neurons), accompanied by release of other dilators such as VIP from VIP-positive interneurons and possibly GABA. Arteriolar constriction, on the other hand, is driven by mainly by activation of somatostatin- and neuropeptide Y-positive inhibitory interneurons, e.g., Martinotti cells (MC, Fig. 15.24).

Cortical columns receiving direct thalamic inputs, respond to a sensory stimulus with a net excitation, while columns in the surround and also those in ipsilateral SI respond with a net inhibition. Although not all inhibitory transmitters are contractile, a net increase in spiking of the inhibitory interneurons might increase the relative strength of the vasoconstrictive force. It has been argued that the hemodynamic response is better correlated with synaptic activity than with spiking (Logothetis 2002). Given that 95% of synapses in primary somatosensory cortex (SI) are local, firing of neurons is expected to be tightly coupled to release of the respective neuro transmitters that drives dilation and constriction (directly or via astrocytes).

Ideally, we would like to directly and simultaneously record neuronal activity for each cell type to reconstruct the release of all the relevant transmitters. Currently this is not feasible using the available experimental tools. However, population-specific activity can be extracted using neuromodeling approaches such as LPA. Moreover, computational methods such as LPA are required for extraction of cell type-specific population activity from macroscopic EEG/MEG data. The general idea is that if we could model the relationship between these macroscopic neuronal measures and activity in specific cell types and given known vasoactive effect of each neuronal population, we could predict macroscopic vascular response from macroscopic measures of neuronal activity.

Although potential vasoactive neuronal cell types have been identified *in vitro* (Cauli et al. 2004), further *in vivo* studies are needed to validate these predictions. In this respect, a recent progress in genetic labeling and remote control of identified neuronal populations will play a central role in unraveling the vasoactive role of cell types with known phenotypes. Inhibitory neurons of the same cell types are coupled by electrotonic junctions and thus discharge synchronously (Beierlein et al. 2000; Mann-Metzer and Yarom 2000; Deans et al. 2001; Beierlein et al. 2003). This synchronous discharge should produce synchronous release of vasoactive mediators leading to vasodilation or vasoconstriction. Thus, one can perform a spike-triggered averaging of vascular responses, with regard to spontaneous activity in a particular neuronal cell type. A complementary approach would be to test the relevant transmitters by local applications *in vivo*. Cell-type specific regulation of CBF would also imply that lessons learned from the cerebral cortex cannot be easily extrapolated to other brain areas with different structures and signaling molecules such as cerebellum or basal ganglia. Indeed, dopaminergic transmission in basal ganglia is likely to play an important vasoactive role. In addition, projections from basal forebrain, raphe nuclei, locus coeruleus, and others might modulate the local vascular tone, CBF baseline and ΔCBF.

Key insights into regulation of CBF can be gained by studying vascular responses across the cortical depth and vascular compartments. In contrast to the previous attempts to differentiate compartment-specific responses while using technologies with low spatial resolution, 2-photon microscopy with the ultimate single-vessel resolution already produces some definite results regarding arteriolar and capillary response to a sensory stimulus under normal conditions (Chaigneau et al. 2003; Chaigneau et al. 2007; Devor et al. 2007; Devor et al. 2008b; Devor et al. 2008a) and redistribution of flow following an intervention (Nishimura et al. 2006; Schaffer et al. 2006; Nishimura et al. 2007). Given the current limitation of 2-photon penetration depth to ~600 μm (layer IV in a mouse), additional technologies are needed to reach the bottom cortical layers. In this respect, recent advances in OCT technology hold promise for imaging of dilation and flow throughout the cortical depth (Aguirre et al. 2006; Chen et al. 2008; Srinivasan et al. 2009; Srinivasan et al. 2010a; Srinivasan et al. 2010b). Specifically, knowledge of where dilation starts (e.g. in the capillary bed in layer IV) and how it propagates throughout the vascular network, will help in screening possible scenarios of neurovascular regulation using experimental or computational approaches.

Another consideration regarding screening possible scenarios of neurovascular regulation is related to time scales and baseline conditions. For example, most astrocytic imaging studies *in vivo* reported slow rises in intracellular calcium with delays considerably longer than known dynamics of hemodynamic/vascular responses (Schummers et al. 2008). It needs to be investigated further if astrocytes are a heterogeneous population with respect to their response kinetics, laminar distribution or metabolic efficiency (Winship et al. 2007). In brain slices, calcium waves have been shown to trigger vasodilation or constriction (Simard et al. 2003). However, because of the extremely slow vascular responses *in vitro*, the slice results are not conclusive in the context of time scales comparison. The other slice-related issue is pathological oxygenation conditions in the absence of hemoglobin circulation. In fact, it has been shown that lowering oxygenation in the bath from the usual 95% to 20% compromises neuronal activity in thin organotypic hippocampal slice cultures (Huchzermeyer et al. 2008). Thus, many of the results obtained in slices, in particular regarding metabolic regulation of vascular diameters, fall into the category of pathophysiology. The metabolic pathways elucidated by these studies are likely to be critical under conditions of stroke and hypoxia (Gordon et al. 2008).

Indeed, in addition to these basic studies providing a firmer foundation for current functional neuroimaging methods, a clear goal is to identify how the coupling of neuronal activity, blood flow, and energy metabolism is altered in disease. Again a multi-modal imaging approach is likely to be essential, because all the normal mechanisms are called into question. With stroke, and a reduction of neurons and proliferation of glial cells, how is the CBF response changed? In neurodegenerative disease, are there early signs of dysfunction in the way CBF or $CMRO_2$ are coupled to neuronal activity? In general, under what circumstances do compromised CBF and $CMRO_2$ responses begin to limit neuronal function? What mechanisms could lead to changes in the dynamics of the BOLD response, due either to disease or to healthy aging?

In future, developments in imaging technology and targeted genetic labeling will allow directly control of firing patterns in identified single cells and cellular populations, while measuring the resultant vascular, metabolic and neuronal response across spatial scales. These future experiments will provide important insights into pathways of neurovascular communication and will identify vasoactive messengers, release of which drives stimulus-evoked hemodynamic signals. The wide range of measurements will need to be integrated within a comprehensive theoretical framework that quantitatively models the steps from activity of specific neuronal cell types to vascular and metabolic changes to BOLD fMRI signals. Ultimately, identification of macro- and microscopic hemodynamic "signatures" of activation in populations of neurons with known phenotype and neurotransmitter content will greatly extend the utility of fMRI as a quantitative probe of physiology for both basic and clinical neuroscience applications.

For translation to human imaging, computational methods are required for neuronal population activity from combined EEG/MEG and fMRI data. The proper interpretation of such multimodal noninvasive imaging measures critically depends on accurate forward models, linking each of the imaging observables to the underlying physiological parameters, along with physiological models, linking the different physiological parameters themselves. Given such models, it should in principle be possible to obtain quantitative spatiotemporal estimates of the relevant physiological parameters, using a Bayesian estimation framework, as outlined above. It should be noted, however, that the accuracy of the resulting estimates will be limited by the accuracy of the models used. Further work is therefore needed to extend and validate the models, under a broad range of normal- and pathophysiological conditions.

Acknowledgements The authors would like you to acknowledge Klas M. Pettersen, Henrik Linden, Ivan C. Teng and Peifang Tian for help with making some of the figures, and the financial support from the following NIH grants: R01NS051188, R01NS057198, R21EB009118 (to Anna Devor), R01EB000790 (to Anders M. Dale), R01NS057476 (to David A. Boas) and R01 NS036722 (to Richard R. Buxton) and Research Council of Norway (eVITA, FRIBIOMOL to Gaute T. Einevoll).

References

Aguirre AD, Chen Y, Fujimoto JG, Ruvinskaya L, Devor A, Boas DA (2006) Depth-resolved imaging of functional activation in the rat cerebral cortex using optical coherence tomography. Opt Lett 31:3459–3461

Agulhon C, Petravicz J, McMullen AB, Sweger EJ, Minton SK, Taves SR, Casper KB, Fiacco TA, McCarthy KD (2008) What is the role of astrocyte calcium in neurophysiology? Neuron 59:932–946

Akkin T, Dave DP, Milner TE, Rylander HG (2004) Detection of neural activity using phase-sensitive optical low-coherence reflectometry. Opt Express 12:2377–2386

Alonso BD, Lowe AS, Dear JP, Lee KC, Williams SC, Finnerty GT (2008) Sensory inputs from whisking movements modify cortical whisker maps visualized with functional magnetic resonance imaging. Cereb Cortex 18(6):1314–1325

Arai T, Nakao S, Mori K, Ishimori K, Morishima I, Miyazawa T, Fritz-Zieroth B (1990) Cerebral oxygen utilization analyzed by the use of oxygen-17 and its nuclear magnetic resonance. Biochem Biophys Res Commun 169:153–158

Arai T, Mori K, Nakao S, Watanabe K, Kito K, Aoki M, Mori H, Morikawa S, Inubushi T (1991) In vivo oxygen-17 nuclear magnetic resonance for the estimation of cerebral blood flow and oxygen consumption. Biochem Biophys Res Commun 179:954–961

Ayata C, Shin HK, Salomone S, Ozdemir-Gursoy Y, Boas DA, Dunn AK, Moskowitz MA (2004) Pronounced hypoperfusion during spreading depression in mouse cortex. J Cereb Blood Flow Metab 24:1172–1182

Baker BJ, Kosmidis EK, Vucinic D, Falk CX, Cohen LB, Djurisic M, Zecevic D (2005) Imaging brain activity with voltage- and calcium-sensitive dyes. Cell Mol Neurobiol 25:245–282

Barth DS, Di S (1991) Laminar excitability cycles in neocortex. J Neurophysiol 65:891–898

Barth DS, Di S, Baumgartner C (1989) Laminar cortical interactions during epileptic spikes studied with principal component analysis and physiological modeling. Brain Res 484: 13–35

Barth DS, Baumgartner C, Di S (1990) Laminar interactions in rat motor cortex during cyclical excitability changes of the penicillin focus. Brain Res 508:105–117

Beierlein M, Gibson JR, Connors BW (2000) A network of electrically coupled interneurons drives synchronized inhibition in neocortex. Nat Neurosci 3:904–910

Beierlein M, Gibson JR, Connors BW (2003) Two dynamically distinct inhibitory networks in layer 4 of the neocortex. J Neurophysiol 90:2987–3000

Bizheva K, Unterhuber A, Hermann B, Povazay B, Sattmann H, Drexler W, Stingl A, Le T, Mei M, Holzwarth R, Reitsamer HA, Morgan JE, Cowey A (2004) Imaging ex vivo and in vitro brain morphology in animal models with ultrahigh resolution optical coherence tomography. J Biomed Opt 9:719–724

Blanco VM, Stern JE, Filosa JA (2008) Tone-dependent vascular responses to astrocyte-derived signals. Am J Physiol Heart Circ Physiol 294:H2855–H2863

Blomquist P, Devor A, Indahl UG, Ulbert I, Einevoll GT, Dale AM (2009) Estimation of thalamocortical and intracortical network models from joint thalamic single-electrode and cortical laminar-electrode recordings in the rat barrel system. PLoS Comput Biol 5:e1000328

Boas DA, Jones SR, Devor A, Huppert TJ, Dale AM (2008) A vascular anatomical network model of the spatio-temporal response to brain activation. Neuroimage 40:1116–1129

Bolay H, Reuter U, Dunn AK, Huang Z, Boas DA, Moskowitz MA (2002) Intrinsic brain activity triggers trigeminal meningeal afferents in a migraine model. Nat Med 8:136–142

Bower JM, Beeman D (1998) The Book of GENESIS: Exploring Realistic Neural Models with the GEneral NEural SImulation System, 2nd edn. Springer, New York

Boxerman JL, Hamberg LM, Rosen BR, Weisskoff RM (1995) MR contrast due to intravascular magnetic susceptibility perturbations. Magn Reson Med 34:555–566

Bressler D, Spotswood N, Whitney D (2007) Negative BOLD fMRI response in the visual cortex carries precise stimulus-specific information. PLoS One 2:e410

Briers JD (2001) Laser Doppler, speckle and related techniques for blood perfusion mapping and imaging. Physiol Meas 22:R35–R66

Brown GG, Perthen JE, Liu TT, Buxton RB (2007) A primer on functional magnetic resonance imaging. Neuropsychol Rev 17:107–125

Buxton RB (2005) Quantifying CBF with arterial spin labeling. J Magn Reson Imaging 22:723–726

Buxton RB, Wong EC, Frank LR (1998) Dynamics of blood flow and oxygenation changes during brain activation: the balloon model. Magn Reson Med 39:855–864

Buxton RB, Uludag K, Dubowitz DJ, Liu TT (2004) Modeling the hemodynamic response to brain activation. Neuroimage 23(Suppl 1):S220–S233

Buzsaki G (2004) Large-scale recording of neuronal ensembles. Nat Neurosci 7:446–451

Caesar K, Offenhauser N, Lauritzen M (2008) Gamma-aminobutyric acid modulates local brain oxygen consumption and blood flow in rat cerebellar cortex. J Cereb Blood Flow Metab 28:906–915

Carnevale NT, Hines ML (2006) The NEURON book. Cambridge University Press, Cambridge

Cauli B, Tong XK, Rancillac A, Serluca N, Lambolez B, Rossier J, Hamel E (2004) Cortical GABA interneurons in neurovascular coupling: relays for subcortical vasoactive pathways. J Neurosci 24:8940–8949

Chaigneau E, Oheim M, Audinat E, Charpak S (2003) Two-photon imaging of capillary blood flow in olfactory bulb glomeruli. Proc Natl Acad Sci USA 100:13081–13086

Chaigneau E, Tiret P, Lecoq J, Ducros M, Knopfel T, Charpak S (2007) The relationship between blood flow and neuronal activity in the rodent olfactory bulb. J Neurosci 27:6452–6460

Chance B, Cohen P, Jobsis F, Schoener B (1962) Intracellular oxidation-reduction states in vivo. Science 137:499–508

Chen Y, Aguirre AD, Ruvinskaya L, Devor A, Boas DA, Fujimoto JG (2008) Optical coherence tomography (OCT) reveals depth-resolved dynamics during functional brain activation. J Neurosci Methods 178(1):162–173

Cohen LB, Lesher S (1986) Optical monitoring of membrane potential: methods of multisite optical measurement. Soc Gen Physiol Ser 40:71–99

Cox SB, Woolsey TA, Rovainen CM (1993) Localized dynamic changes in cortical blood flow with whisker stimulation corresponds to matched vascular and neuronal architecture of rat barrels. J Cereb Blood Flow Metab 13:899–913

D'Esposito M, Deouell LY, Gazzaley A (2003) Alterations in the BOLD fMRI signal with ageing and disease: a challenge for neuroimaging. Nat Rev Neurosci 4:863–872

Dale AM, Buckner RL (1997) Selective averaging of rapidly presented individual trails using fMRI. Hum Brain Mapp 5:329–340

Dale AM, Halgren E (2001) Spatiotemporal mapping of brain activity by integration of multiple imaging modalities. Curr Opin Neurobiol 11:202–208

Dale AM, Liu AK, Fischl BR, Buckner RL, Belliveau JW, Lewine JD, Halgren E (2000) Dynamic statistical parametric mapping: combining fMRI and MEG for high-resolution imaging of cortical activity. Neuron 26:55–67

Davis TL, Kwong KK, Weisskoff RM, Rosen BR (1998) Calibrated functional MRI: mapping the dynamics of oxidative metabolism. Proc Natl Acad Sci USA 95:1834–1839

Deans MR, Gibson JR, Sellitto C, Connors BW, Paul DL (2001) Synchronous activity of inhibitory networks in neocortex requires electrical synapses containing connexin36. Neuron 31:477–485

Derdikman D, Hildesheim R, Ahissar E, Arieli A, Grinvald A (2003) Imaging spatiotemporal dynamics of surround inhibition in the barrels somatosensory cortex. J Neurosci 23:3100–3105

Devor A, Dunn AK, Andermann ML, Ulbert I, Boas DA, Dale AM (2003) Coupling of total hemoglobin concentration, oxygenation, and neural activity in rat somatosensory cortex. Neuron 39:353–359

Devor A, Sakadzic S, Yucel MA, Teng IC, Kasischke KA, Boas DA (2009) In vivo functional NADH imaging with single-cell resolution. In: Society for Neurosciences. Meeting Planner. Chicago, IL: Society for Neuroscience, 2009. Online

Devor A, Shih AY, Tsai PS, Blinder P, Tian P, Teng IC, Kleinfeld D (2008a) Two-photon laser scanning microscopy as a tool to study cortical vasodynamics under normal and ischemic conditions In: Roe AW (ed) Imaging the brain with optical methods. Springer http://www.springer.com/biomed/neuroscience/book/978-1-4419-0451-5

Devor A, Ulbert I, Dunn AK, Narayanan SN, Jones SR, Andermann ML, Boas DA, Dale AM (2005) Coupling of the cortical hemodynamic response to cortical and thalamic neuronal activity. Proc Natl Acad Sci USA 102:3822–3827

Devor A, Tian P, Nishimura N, Teng IC, Hillman EM, Narayanan SN, Ulbert I, Boas DA, Kleinfeld D, Dale AM (2007) Suppressed neuronal activity and concurrent arteriolar vasoconstriction may explain negative blood oxygenation level-dependent signal. J Neurosci 27:4452–4459

Devor A, Hillman EM, Tian P, Waeber C, Teng IC, Ruvinskaya L, Shalinsky MH, Zhu H, Haslinger RH, Narayanan SN, Ulbert I, Dunn AK, Lo EH, Rosen BR, Dale AM, Kleinfeld D, Boas DA (2008b) Stimulus-induced changes in blood flow and 2-deoxyglucose uptake dissociate in ipsilateral somatosensory cortex. J Neurosci 28:14347–14357

Di S, Baumgartner C, Barth DS (1990) Laminar analysis of extracellular field potentials in rat vibrissa/barrel cortex. J Neurophysiol 63:832–840

Draijer M, Hondebrink E, van Leeuwen T, Steenbergen W (2008) Review of laser speckle contrast techniques for visualizing tissue perfusion. Lasers Med Sci 24(4): 639–651

Drexler W, Morgner U, Kartner FX, Pitris C, Boppart SA, Li XD, Ippen EP, Fujimoto JG (1999) In vivo ultrahigh-resolution optical coherence tomography. Opt Lett 24:1221–1223

Dunn AK, Bolay H, Moskowitz MA, Boas DA (2001) Dynamic imaging of cerebral blood flow using laser speckle. J Cereb Blood Flow Metab 21:195–201

Dunn AK, Devor A, Dale AM, Boas DA (2005) Spatial extent of oxygen metabolism and hemodynamic changes during functional activation of the rat somatosensory cortex. Neuroimage 27:279–290

Dunn AK, Devor A, Bolay H, Andermann ML, Moskowitz MA, Dale AM, Boas DA (2003) Simultaneous imaging of total cerebral hemoglobin concentration, oxygenation, and blood flow during functional activation. Opt Lett 28:28–30

Dunphy I, Vinogradov SA, Wilson DF (2002) Oxyphor R2 and G2: phosphors for measuring oxygen by oxygen-dependent quenching of phosphorescence. Anal Biochem 310:191–198

Ebner TJ, Chen G (1995) Use of voltage-sensitive dyes and optical recordings in the central nervous system. Prog Neurobiol 46:463–506

Einevoll GT, Pettersen KH, Devor A, Ulbert I, Halgren E, Dale AM (2007) Laminar population analysis: estimating firing rates and evoked synaptic activity from multielectrode recordings in rat barrel cortex. J Neurophysiol 97:2174–2190

Fang-Yen C, Chu MC, Seung HS, Dasari RR, Feld MS (2004) Noncontact measurement of nerve displacement during action potential with a dual-beam low-coherence interferometer. Opt Lett 29:2028–2030

Fang Q, Sakadzic S, Ruvinskaya L, Devor A, Dale AM, Boas DA (2008) Oxygen advection and diffusion in a three- dimensional vascular anatomical network. Opt Express 16:17530–17541

Faraci FM, Breese KR (1993) Nitric oxide mediates vasodilatation in response to activation of N-methyl-D-aspartate receptors in brain. Circ Res 72:476–480

Faraci FM, Heistad DD (1998) Regulation of the cerebral circulation: role of endothelium and potassium channels. Physiol Rev 78:53–97

Ferezou I, Bolea S, Petersen CC (2006) Visualizing the cortical representation of whisker touch: voltage-sensitive dye imaging in freely moving mice. Neuron 50:617–629

Fergus A, Lee KS (1997) GABAergic regulation of cerebral microvascular tone in the rat. J Cereb Blood Flow Metab 17:992–1003

Fiat D, Kang S (1992) Determination of the rate of cerebral oxygen consumption and regional cerebral blood flow by non-invasive 17O in vivo NMR spectroscopy and magnetic resonance imaging: Part 1. Theory and data analysis methods. Neurol Res 14:303–311

Fiat D, Kang S (1993) Determination of the rate of cerebral oxygen consumption and regional cerebral blood flow by non-invasive 17O in vivo NMR spectroscopy and magnetic resonance imaging. Part 2. Determination of CMRO2 for the rat by 17O NMR, and CMRO2, rCBF and the partition coefficient for the cat by 17O MRI. Neurol Res 15:7–22

Filosa JA, Bonev AD, Straub SV, Meredith AL, Wilkerson MK, Aldrich RW, Nelson MT (2006) Local potassium signaling couples neuronal activity to vasodilation in the brain. Nat Neurosci 9(11):1397–1403

Finikova OS, Lebedev AY, Aprelev A, Troxler T, Gao F, Garnacho C, Muro S, Hochstrasser RM, Vinogradov SA (2008) Oxygen microscopy by two-photon-excited phosphorescence. Chemphyschem 9:1673–1679

Fox PT, Raichle ME (1986) Focal physiological uncoupling of cerebral blood flow and oxidative metabolism during somatosensory stimulation in human subjects. Proc Natl Acad Sci USA 83:1140–1144

Fox PT, Raichle ME, Mintun MA, Dence C (1988) Nonoxidative glucose consumption during focal physiologic neural activity. Science 241:462–464

Freeman JA, Nicholson C (1975) Experimental optimization of current source-density technique for anuran cerebellum. J Neurophysiol 38(2):369–82.

Friston KJ (2005) Models of brain function in neuroimaging. Annu Rev Psychol 56:57–87

Frostig RD, Lieke EE, Ts'o DY, Grinvald A (1990) Cortical functional architecture and local coupling between neuronal activity and the microcirculation revealed by in vivo high-resolution optical imaging of intrinsic signals. Proc Natl Acad Sci USA 87:6082–6086

Gobel W, Kampa BM, Helmchen F (2007) Imaging cellular network dynamics in three dimensions using fast 3D laser scanning. Nat Methods 4:73–79

Gordon GR, Choi HB, Rungta RL, Ellis-Davies GC, Macvicar BA (2008) Brain metabolism dictates the polarity of astrocyte control over arterioles. Nature 456(7223):745–749

Grinvald A (1992) Optical imaging of architecture and function in the living brain sheds new light on cortical mechanisms underlying visual perception. Brain Topogr 5:71–75

Grinvald A, Hildesheim R (2004) VSDI: a new era in functional imaging of cortical dynamics. Nat Rev Neurosci 5:874–885

Grinvald A, Frostig RD, Lieke E, Hildesheim R (1988) Optical imaging of neuronal activity. Physiol Rev 68:1285–1366

Grinvald A, Lieke E, Frostig RD, Gilbert CD, Wiesel TN (1986) Functional architecture of cortex revealed by optical imaging of intrinsic signals. Nature 324:361–364

Grubb RL Jr, Raichle ME, Eichling JO, Ter-Pogossian MM (1974) The effects of changes in PaCO2 on cerebral blood volume, blood flow, and vascular mean transit time. Stroke 5:630–639

Guadagno JV, Calautti C, Baron JC (2003) Progress in imaging stroke: emerging clinical applications. Br Med Bull 65:145–157

Hämäläinen M, Hari R, Ilmoniemi R, Knuutila J, Lounasmaa O (1993) Magnetoencephalography theory, instrumentation, and applications to noninvasive studies of the working human brain. Rev Mod Phys 65:413–497

Hamel E (2004) Cholinergic modulation of the cortical microvascular bed. Prog Brain Res 145:171–178

Hamel E (2006) Perivascular nerves and the regulation of cerebrovascular tone. J Appl Physiol 100:1059–1064

Hillman EM, Boas DA, Dale AM, Dunn AK (2004) Laminar optical tomography: demonstration of millimeter-scale depth-resolved imaging in turbid media. Opt Lett 29:1650–1652

Hillman EM, Devor A, Bouchard MB, Dunn AK, Krauss GW, Skoch J, Bacskai BJ, Dale AM, Boas DA (2007) Depth-resolved optical imaging and microscopy of vascular compartment dynamics during somatosensory stimulation. Neuroimage 35:89–104

Hodics T, Cohen LG (2005) Functional neuroimaging in motor recovery after stroke. Top Stroke Rehabil 12:15–21

Hu Y, Wilson GS (1997) A temporary local energy pool coupled to neuronal activity: fluctuations of extracellular lactate levels in rat brain monitored with rapid-response enzyme-based sensor. J Neurochem 69:1484–1490

Huang D, Swanson EA, Lin CP, Schuman JS, Stinson WG, Chang W, Hee MR, Flotte T, Gregory K, Puliafito CA et al (1991) Optical coherence tomography. Science 254:1178–1181

Huber R, Adler DC, Srinivasan VJ, Fujimoto JG (2007) Fourier domain mode locking at 1050 nm for ultra-high-speed optical coherence tomography of the human retina at 236,000 axial scans per second. Opt Lett 32:2049–2051

Huchzermeyer C, Albus K, Gabriel HJ, Otahal J, Taubenberger N, Heinemann U, Kovacs R, Kann O (2008) Gamma oscillations and spontaneous network activity in the hippocampus are highly sensitive to decreases in pO2 and concomitant changes in mitochondrial redox state. J Neurosci 28:1153–1162

Huppert TJ, Allen MS, Diamond SG, Boas DA (2009) Estimating cerebral oxygen metabolism from fMRI with a dynamic multicompartment Windkessel model. Hum Brain Mapp 30: 1548–1567

Huppert TJ, Allen MS, Benav H, Jones PB, Boas DA (2007) A multicompartment vascular model for inferring baseline and functional changes in cerebral oxygen metabolism and arterial dilation. J Cereb Blood Flow Metab 27:1262–1279

Iadecola C (2004) Neurovascular regulation in the normal brain and in Alzheimer's disease. Nat Rev Neurosci 5:347–360

Iadecola C, Niwa K (2002) Nitric oxide. In: Edvinsson L, Krause DN (eds) Cerebral blood flow and metabolism, 2nd edn. Lippincott Williams & Wilkins, Philadelphia, pp 295–310

Iadecola C, Nedergaard M (2007) Glial regulation of the cerebral microvasculature. Nat Neurosci 10:1369–1376

Ido Y, Chang K, Williamson JR (2004) NADH augments blood flow in physiologically activated retina and visual cortex. Proc Natl Acad Sci USA 101:653–658

Iliff JJ, D'Ambrosio R, Ngai AC, Winn HR (2003) Adenosine receptors mediate glutamate-evoked arteriolar dilation in the rat cerebral cortex. Am J Physiol Heart Circ Physiol 284:H1631–H1637

Ito H, Ibaraki M, Kanno I, Fukuda H, Miura S (2005) Changes in the arterial fraction of human cerebral blood volume during hypercapnia and hypocapnia measured by positron emission tomography. J Cereb Blood Flow Metab 25:852–857

Jackson J (1998) Classical electrodynamics. Wiley, Hoboken, NJ

Jones M, Berwick J, Mayhew J (2002) Changes in blood flow, oxygenation, and volume following extended stimulation of rodent barrel cortex. Neuroimage 15:474–487

Jones M, Berwick J, Johnston D, Mayhew J (2001) Concurrent optical imaging spectroscopy and laser-Doppler flowmetry: the relationship between blood flow, oxygenation, and volume in rodent barrel cortex. Neuroimage 13:1002–1015

Jones M, Hewson-Stoate N, Martindale J, Redgrave P, Mayhew J (2004) Nonlinear coupling of neural activity and CBF in rodent barrel cortex. Neuroimage 22:956–965

Kasischke KA, Vishwasrao HD, Fisher PJ, Zipfel WR, Webb WW (2004) Neural activity triggers neuronal oxidative metabolism followed by astrocytic glycolysis. Science 305:99–103

Kasischke KA, Lambert EM, Panepento B, Sun A, Gelbard HA, Burgess RW, Foster TH, Nedergaard M (2011) Two-photon NADH imaging exposes boundaries of oxygen diffusion in cortical vascular supply regions. J Cereb Blood Flow Metab 31(1):68–81

Kim T, Hendrich KS, Masamoto K, Kim SG (2007) Arterial versus total blood volume changes during neural activity-induced cerebral blood flow change: implication for BOLD fMRI. J Cereb Blood Flow Metab 27:1235–1247

Kleinfeld D, Griesbeck O (2005) From art to engineering? The rise of in vivo mammalian electrophysiology via genetically targeted labeling and nonlinear imaging. PLoS Biol 3:e355

Kleinfeld D, Mitra PP, Helmchen F, Denk W (1998) Fluctuations and stimulus-induced changes in blood flow observed in individual capillaries in layers 2 through 4 of rat neocortex. Proc Natl Acad Sci USA 95:15741–15746

Kocharyan A, Fernandes P, Tong XK, Vaucher E, Hamel E (2008) Specific subtypes of cortical GABA interneurons contribute to the neurovascular coupling response to basal forebrain stimulation. J Cereb Blood Flow Metab 28:221–231

Koehler RC, Roman RJ, Harder DR (2009) Astrocytes and the regulation of cerebral blood flow. Trends Neurosci 32:160–169

Kohl M, Lindauer U, Royl G, Kuhl M, Gold L, Villringer A, Dirnagl U (2000) Physical model for the spectroscopic analysis of cortical intrinsic optical signals. Phys Med Biol 45:3749–3764

Koralek KA, Olavarria J, Killackey HP (1990) Areal and laminar organization of corticocortical projections in the rat somatosensory cortex. J Comp Neurol 299:133–150

Kudomi N, Watabe H, Hayashi T, Iida H (2007) Separation of input function for rapid measurement of quantitative CMRO2 and CBF in a single PET scan with a dual tracer administration method. Phys Med Biol 52:1893–1908

Kudomi N, Hayashi T, Watabe H, Teramoto N, Piao R, Ose T, Koshino K, Ohta Y, Iida H (2009) A physiologic model for recirculation water correction in CMRO2 assessment with 15O2 inhalation PET. J Cereb Blood Flow Metab 29:355–364

Lauritzen M (2005) Opinion: Reading vascular changes in brain imaging: is dendritic calcium the key? Nat Rev Neurosci 6:77–85

Lazebnik M, Marks DL, Potgieter K, Gillette R, Boppart SA (2003) Functional optical coherence tomography for detecting neural activity through scattering changes. Opt Lett 28: 1218–1220

Lee SP, Duong TQ, Yang G, Iadecola C, Kim SG (2001) Relative changes of cerebral arterial and venous blood volumes during increased cerebral blood flow: implications for BOLD fMRI. Magn Reson Med 45:791–800

Lefort S, Tomm C, Floyd Sarria JC, Petersen CC (2009) The excitatory neuronal network of the C2 barrel column in mouse primary somatosensory cortex. Neuron 61:301–316

Leitgeb RA, Schmetterer L, Drexler W, Fercher AF, Zawadzki RJ, Bajraszewski T (2003) Real-time assessment of retinal blood flow with ultrafast acquisition by color Doppler Fourier domain optical coherence tomography. Opt Express 11:3116–3121

Leontiev O, Buxton RB (2007) Reproducibility of BOLD, perfusion, and CMRO2 measurements with calibrated-BOLD fMRI. Neuroimage 35:175–184

Leontiev O, Dubowitz DJ, Buxton RB (2007) CBF/CMRO2 coupling measured with calibrated BOLD fMRI: sources of bias. Neuroimage 36:1110–1122

Leski S, Wojcik DK, Tereszczuk J, Swiejkowski DA, Kublik E, Wrobel A (2007) Inverse current-source density method in 3D: reconstruction fidelity, boundary effects, and influence of distant sources. Neuroinformatics 5:207–222

Lindauer U, Leithner C, Kaasch H, Rohrer B, Foddis M, Fuchtemeier M, Offenhauser N, Steinbrink J, Royl G, Kohl-Bareis M, Dirnagl U (2010) Neurovascular coupling in rat brain operates independent of hemoglobin deoxygenation. J Cereb Blood Flow Metab 30(4):757–768

Liu X, Li C, Falck JR, Roman RJ, Harder DR, Koehler RC (2008) Interaction of nitric oxide, 20-HETE, and EETs during functional hyperemia in whisker barrel cortex. Am J Physiol Heart Circ Physiol 295:H619–H631

Logothetis NK (2002) The neural basis of the blood-oxygen-level-dependent functional magnetic resonance imaging signal. Philos Trans R Soc Lond B Biol Sci 357:1003–1037

Logothetis NK, Kayser C, Oeltermann A (2007) In vivo measurement of cortical impedance spectrum in monkeys: implications for signal propagation. Neuron 55:809–823

Lu H, Ge Y (2008) Quantitative evaluation of oxygenation in venous vessels using T2-Relaxation-Under-Spin-Tagging MRI. Magn Reson Med 60:357–363

Lu H, Zhao C, Ge Y, Lewis-Amezcua K (2008) Baseline blood oxygenation modulates response amplitude: physiologic basis for intersubject variations in functional MRI signals. Magn Reson Med 60:364–372

Magistretti PJ, Pellerin L, Rothman DL, Shulman RG (1999) Energy on demand. Science 283:496–497

Maheswari RU, Takaoka H, Kadono H, Homma R, Tanifuji M (2003) Novel functional imaging technique from brain surface with optical coherence tomography enabling visualization of depth resolved functional structure in vivo. J Neurosci Methods 124:83–92

Mainen ZF, Sejnowski TJ (1996) Influence of dendritic structure on firing pattern in model neocortical neurons. Nature 382:363–366

Malonek D, Grinvald A (1996) Interactions between electrical activity and cortical microcirculation revealed by imaging spectroscopy: implications for functional brain mapping. Science 272:551–554

Mandeville JB, Marota JJ, Ayata C, Zaharchuk G, Moskowitz MA, Rosen BR, Weisskoff RM (1999) Evidence of a cerebrovascular postarteriole windkessel with delayed compliance. J Cereb Blood Flow Metab 19:679–689

Mann-Metzer P, Yarom Y (2000) Electrotonic coupling synchronizes interneuron activity in the cerebellar cortex. Prog Brain Res 124:115–122

Margrie TW, Meyer AH, Caputi A, Monyer H, Hasan MT, Schaefer AT, Denk W, Brecht M (2003) Targeted whole-cell recordings in the mammalian brain in vivo. Neuron 39:911–918

Mayhew J, Johnston D, Berwick J, Jones M, Coffey P, Zheng Y (2000) Spectroscopic analysis of neural activity in brain: increased oxygen consumption following activation of barrel cortex. Neuroimage 12:664–675

Metea MR, Newman EA (2006) Glial cells dilate and constrict blood vessels: a mechanism of neurovascular coupling. J Neurosci 26:2862–2870

Mik EG, van Leeuwen TG, Raat NJ, Ince C (2004) Quantitative determination of localized tissue oxygen concentration in vivo by two-photon excitation phosphorescence lifetime measurements. J Appl Physiol 97:1962–1969
Miller RJ (1988) Calcium signalling in neurons. Trends Neurosci 11:415–419
Mintun MA, Raichle ME, Martin WR, Herscovitch P (1984) Brain oxygen utilization measured with O-15 radiotracers and positron emission tomography. J Nucl Med 25:177–187
Mintun MA, Vlassenko AG, Rundle MM, Raichle ME (2004) Increased lactate/pyruvate ratio augments blood flow in physiologically activated human brain. Proc Natl Acad Sci USA 101:659–664
Mitzdorf U (1985) Current source-density method and application in cat cerebral cortex: investigation of evoked potentials and EEG phenomena. Physiol Rev 65:37–100
Moore CI, Nelson SB (1998) Spatio-temporal subthreshold receptive fields in the vibrissa representation of rat primary somatosensory cortex. J Neurophysiol 80:2882–2892
Mulligan SJ, MacVicar BA (2004) Calcium transients in astrocyte endfeet cause cerebrovascular constrictions. Nature 431:195–199
Narayan SM, Santori EM, Blood AJ, Burton JS, Toga AW (1994) Imaging optical reflectance in rodent barrel and forelimb sensory cortex. Neuroimage 1:181–190
Narayan SM, Esfahani P, Blood AJ, Sikkens L, Toga AW (1995) Functional increases in cerebral blood volume over somatosensory cortex. J Cereb Blood Flow Metab 15:754–765
Nemoto M, Sheth S, Guiou M, Pouratian N, Chen JW, Toga AW (2004) Functional signal- and paradigm-dependent linear relationships between synaptic activity and hemodynamic responses in rat somatosensory cortex. J Neurosci 24:3850–3861
Nemoto M, Nomura Y, Sato C, Tamura M, Houkin K, Koyanagi I, Abe H (1999) Analysis of optical signals evoked by peripheral nerve stimulation in rat somatosensory cortex: dynamic changes in hemoglobin concentration and oxygenation. J Cereb Blood Flow Metab 19: 246–259
Nicholson C, Freeman JA (1975) Theory of current source-density analysis and determination of conductivity tensor for anuran cerebellum. J Neurophysiol 38:356–368
Nilsson M, Eriksson PS, Ronnback L, Hansson E (1993) GABA induces Ca2+ transients in astrocytes. Neuroscience 54:605–614
Nishimura N, Schaffer CB, Friedman B, Lyden PD, Kleinfeld D (2007) Penetrating arterioles are a bottleneck in the perfusion of neocortex. Proc Natl Acad Sci USA 104:365–370
Nishimura N, Schaffer CB, Friedman B, Tsai PS, Lyden PD, Kleinfeld D (2006) Targeted insult to subsurface cortical blood vessels using ultrashort laser pulses: three models of stroke. Nat Methods 3:99–108
Nunez PL, Srinivasan R (2006) Electric fields of the brain: the neurophysics of EEG. Oxford University Press, New York
Obata T, Liu TT, Miller KL, Luh WM, Wong EC, Frank LR, Buxton RB (2004) Discrepancies between BOLD and flow dynamics in primary and supplementary motor areas: application of the balloon model to the interpretation of BOLD transients. Neuroimage 21:144–153
Ogawa S, Menon RS, Tank DW, Kim SG, Merkle H, Ellermann JM, Ugurbil K (1993) Functional brain mapping by blood oxygenation level-dependent contrast magnetic resonance imaging. A comparison of signal characteristics with a biophysical model. Biophys J 64:803–812
Ohata H, Cao S, Koehler RC (2006) Contribution of adenosine A2A and A2B receptors and heme oxygenase to AMPA-induced dilation of pial arterioles in rats. Am J Physiol Regul Integr Comp Physiol 291:R728–R735
Ou W, Nissila I, Radhakrishnan H, Boas DA, Hamalainen MS, Franceschini MA (2009) Study of neurovascular coupling in humans via simultaneous magnetoencephalography and diffuse optical imaging acquisition. Neuroimage 46:624–632
Peppiatt CM, Howarth C, Mobbs P, Attwell D (2006) Bidirectional control of CNS capillary diameter by pericytes. Nature 443:700–704

Pettersen KH, Einevoll GT (2008) Amplitude variability and extracellular low-pass filtering of neuronal spikes. Biophys J. 94(3):784–802.

Pettersen KH, Hagen E, Einevoll GT (2008) Estimation of population firing rates and current source densities from laminar electrode recordings. J Comput Neurosci 24:291–313

Pettersen KH, Linden H, Dale AM, Einevoll GT (2012) Extracellular spikes and CSD. In: Brette R, Destexhe A (eds) Handbook of neural activity measurement Cambridge University Press

Pettersen KH, Devor A, Ulbert I, Dale AM, Einevoll GT (2006) Current-source density estimation based on inversion of electrostatic forward solution: effects of finite extent of neuronal activity and conductivity discontinuities. J Neurosci Methods 154:116–133

Petzold GC, Albeanu DF, Sato TF, Murthy VN (2008) Coupling of neural activity to blood flow in olfactory glomeruli is mediated by astrocytic pathways. Neuron 58:897–910

Phelps ME, Kuhl DE, Mazziota JC (1981) Metabolic mapping of the brain's response to visual stimulation: studies in humans. Science 211:1445–1448

Pinto DJ, Brumberg JC, Simons DJ (2000) Circuit dynamics and coding strategies in rodent somatosensory cortex. J Neurophysiol 83:1158–1166

Polimeni JR, Granquist-Fraser D, Wood RJ, Schwartz EL (2005) Physical limits to spatial resolution of optical recording: clarifying the spatial structure of cortical hypercolumns. Proc Natl Acad Sci USA 102:4158–4163

Porter JT, McCarthy KD (1996) Hippocampal astrocytes in situ respond to glutamate released from synaptic terminals. J Neurosci 16:5073–5081

Raichle ME, Mintun MA (2006) Brain work and brain imaging. Annu Rev Neurosci 29:449–476

Rancillac A, Rossier J, Guille M, Tong XK, Geoffroy H, Amatore C, Arbault S, Hamel E, Cauli B (2006) Glutamatergic control of microvascular tone by distinct GABA neurons in the cerebellum. J Neurosci 26:6997–7006

Rappelsberger P, Pockberger H, Petsche H (1981) Current source density analysis: methods and application to simultaneously recorded field potentials of the rabbit's visual cortex. Pflugers Arch 389:159–170

Ross WN (1989) Changes in intracellular calcium during neuron activity. Annu Rev Physiol 51:491–506

Rouach N, Koulakoff A, Abudara V, Willecke K, Giaume C (2008) Astroglial metabolic networks sustain hippocampal synaptic transmission. Science 322:1551–1555

Royl G, Fuchtemeier M, Leithner C, Megow D, Offenhauser N, Steinbrink J, Kohl-Bareis M, Dirnagl U, Lindauer U (2008) Hypothermia effects on neurovascular coupling and cerebral metabolic rate of oxygen. Neuroimage 40:1523–1532

Rubio R, Berne RM, Bockman EL, Curnish RR (1975) Relationship between adenosine concentration and oxygen supply in rat brain. Am J Physiol 228:1896–1902

Sakadzic S, Yuan S, Dilekoz E, Ruvinskaya S, Vinogradov SA, Ayata C, Boas DA (2009) Simultaneous imaging of cerebral partial pressure of oxygen and blood flow during functional activation and cortical spreading depression. Appl Opt 48:D169–D177

Schaffer CB, Friedman B, Nishimura N, Schroeder LF, Tsai PS, Ebner FF, Lyden PD, Kleinfeld D (2006) Two-photon imaging of cortical surface microvessels reveals a robust redistribution in blood flow after vascular occlusion. PLoS Biol 4:e22

Schmidt DM, George JS, Wood CC (1999) Bayesian inference applied to the electromagnetic inverse problem. Hum Brain Mapp 7:195–212

Schroeder CE, Lindsley RW, Specht C, Marcovici A, Smiley JF, Javitt DC (2001) Somatosensory input to auditory association cortex in the macaque monkey. J Neurophysiol 85:1322–1327

Schummers J, Yu H, Sur M (2008) Tuned responses of astrocytes and their influence on hemodynamic signals in the visual cortex. Science 320:1638–1643

Sheth SA, Nemoto M, Guiou M, Walker M, Pouratian N, Toga AW (2004) Linear and nonlinear relationships between neuronal activity, oxygen metabolism, and hemodynamic responses. Neuron 42:347–355

Shi Y, Liu X, Gebremedhin D, Falck JR, Harder DR, Koehler RC (2008) Interaction of mechanisms involving epoxyeicosatrienoic acids, adenosine receptors, and metabotropic glutamate receptors in neurovascular coupling in rat whisker barrel cortex. J Cereb Blood Flow Metab 28:111–125

Shmuel A, Augath M, Oeltermann A, Logothetis NK (2006) Negative functional MRI response correlates with decreases in neuronal activity in monkey visual area V1. Nat Neurosci 9:569–577

Shoham D, Glaser DE, Arieli A, Kenet T, Wijnbergen C, Toledo Y, Hildesheim R, Grinvald A (1999) Imaging cortical dynamics at high spatial and temporal resolution with novel blue voltage-sensitive dyes. Neuron 24:791–802

Sicard KM, Duong TQ (2005) Effects of hypoxia, hyperoxia, and hypercapnia on baseline and stimulus-evoked BOLD, CBF, and CMRO2 in spontaneously breathing animals. Neuroimage 25:850–858

Simard M, Arcuino G, Takano T, Liu QS, Nedergaard M (2003) Signaling at the gliovascular interface. J Neurosci 23:9254–9262

Simons DJ (1978) Response properties of vibrissa units in rat SI somatosensory neocortex. J Neurophysiol 41:798–820

Simons DJ, Carvell GE (1989) Thalamocortical response transformation in the rat vibrissa/barrel system. J Neurophysiol 61:311–330

Smith AJ, Blumenfeld H, Behar KL, Rothman DL, Shulman RG, Hyder F (2002) Cerebral energetics and spiking frequency: the neurophysiological basis of fMRI. Proc Natl Acad Sci USA 99:10765–10770

Sokoloff L, Reivich M, Kennedy C, Des Rosiers MH, Patlak CS, Pettigrew KD, Sakurada O, Shinohara M (1977) The [14 C]deoxyglucose method for the measurement of local cerebral glucose utilization: theory, procedure, and normal values in the conscious and anesthetized albino rat. J Neurochem 28:897–916

Somogyvari Z, Zalanyi L, Ulbert I, Erdi P (2005) Model-based source localization of extracellular action potentials. J Neurosci Methods 147:126–137

Srinivasan VJ, Sakadzic S, Gorczynska I, Ruvinskaya S, Wu W, Fujimoto JG, Boas DA (2009) Depth-resolved microscopy of cortical hemodynamics with optical coherence tomography. Opt Lett 34:3086–3088

Srinivasan VJ, Sakadzić S, Gorczynska I, Ruvinskaya S, Wu W, Fujimoto JG, Boas DA (2010a) Quantitative cerebral blood flow with optical coherence tomography. Opt Express 18(3): 2477–2494

Srinivasan VJ, Jiang JY, Yaseen MA, Radhakrishnan H, Wu W, Barry S, Cable AE, Boas DA (2010b) Rapid volumetric angiography of cortical microvasculature with optical coherence tomography. Opt Lett 35:43–45

Stefanovic B, Hutchinson E, Yakovleva V, Schram V, Russell JT, Belluscio L, Koretsky AP, Silva AC (2008) Functional reactivity of cerebral capillaries. J Cereb Blood Flow Metab 28:961–972

Strangman G, Culver JP, Thompson JH, Boas DA (2002) A quantitative comparison of simultaneous BOLD fMRI and NIRS recordings during functional brain activation. Neuroimage 17:719–731

Straub SV, Nelson MT (2007) Astrocytic calcium signaling: the information currency coupling neuronal activity to the cerebral microcirculation. Trends Cardiovasc Med 17:183–190

Takano T, Tian GF, Peng W, Lou N, Libionka W, Han X, Nedergaard M (2006) Astrocyte-mediated control of cerebral blood flow. Nat Neurosci 9:260–267

Takano T, Tian GF, Peng W, Lou N, Lovatt D, Hansen AJ, Kasischke KA, Nedergaard M (2007) Cortical spreading depression causes and coincides with tissue hypoxia. Nat Neurosci 10:754–762

Takashima I, Kajiwara R, Iijima T (2001) Voltage-sensitive dye versus intrinsic signal optical imaging: comparison of optically determined functional maps from rat barrel cortex. Neuroreport 12:2889–2894

Teasell R, Bayona NA, Bitensky J (2005) Plasticity and reorganization of the brain post stroke. Top Stroke Rehabil 12:11–26

Thompson JK, Peterson MR, Freeman RD (2003) Single-neuron activity and tissue oxygenation in the cerebral cortex. Science 299:1070–1072

Thomson AM, Bannister AP (2003) Interlaminar connections in the neocortex. Cereb Cortex 13:5–14

Tian P, Teng IC, May LD, Kurz R, Lu K, Scadeng M, Hillman EM, De Crespigny AJ, D'Arceuil HE, Mandeville JB, Marota JJ, Rosen BR, Liu TT, Boas DA, Buxton RB, Dale AM, Devor A (2010) Cortical depth-specific microvascular dilation underlies laminar differences in blood oxygenation level-dependent functional MRI signal. Proc Natl Acad Sci USA 107(34):15246–15251

Tsien RY (1981) A non-disruptive technique for loading calcium buffers and indicators into cells. Nature 290:527–528

Turner DA, Foster KA, Galeffi F, Somjen GG (2007) Differences in O2 availability resolve the apparent discrepancies in metabolic intrinsic optical signals in vivo and in vitro. Trends Neurosci 30:390–398

Ulbert I, Halgren E, Heit G, Karmos G (2001) Multiple microelectrode-recording system for human intracortical applications. J Neurosci Methods 106:69–79

Vakoc B, Yun S, de Boer J, Tearney G, Bouma B (2005) Phase-resolved optical frequency domain imaging. Opt Express 13:5483–5493

Vanzetta I, Grinvald A (1999) Increased cortical oxidative metabolism due to sensory stimulation: implications for functional brain imaging. Science 286:1555–1558

Vanzetta I, Grinvald A (2001) Evidence and lack of evidence for the initial dip in the anesthetized rat: implications for human functional brain imaging. Neuroimage 13:959–967

Vanzetta I, Hildesheim R, Grinvald A (2005) Compartment-resolved imaging of activity-dependent dynamics of cortical blood volume and oximetry. J Neurosci 25:2233–2244

Villringer A, Chance B (1997) Non-invasive optical spectroscopy and imaging of human brain function. Trends Neurosci 20:435–442

Viswanathan A, Freeman RD (2007) Neurometabolic coupling in cerebral cortex reflects synaptic more than spiking activity. Nat Neurosci 10:1308–1312

Wallace DJ, Zum Alten Borgloh SM, Astori S, Yang Y, Bausen M, Kugler S, Palmer AE, Tsien RY, Sprengel R, Kerr JN, Denk W, Hasan MT (2008) Single-spike detection in vitro and in vivo with a genetic Ca(2+) sensor. Nat Methods 5(9):797–804

Wang X, Lou N, Xu Q, Tian GF, Peng WG, Han X, Kang J, Takano T, Nedergaard M (2006) Astrocytic Ca(2+) signaling evoked by sensory stimulation in vivo. Nat Neurosci 9:816–823

White B, Pierce M, Nassif N, Cense B, Park B, Tearney G, Bouma B, Chen T, de Boer J (2003) In vivo dynamic human retinal blood flow imaging using ultra-high-speed spectral domain optical coherence tomography. Opt Express 11:3490–3497

White EL, DeAmicis RA (1977) Afferent and efferent projections of the region in mouse SmL cortex which contains the posteromedial barrel subfield. J Comp Neurol 175:455–482

Winship IR, Plaa N, Murphy TH (2007) Rapid astrocyte calcium signals correlate with neuronal activity and onset of the hemodynamic response in vivo. J Neurosci 27:6268–6272

Wise SP, Jones EG (1976) The organization and postnatal development of the commissural projection of the rat somatic sensory cortex. J Comp Neurol 168:313–343

Wojtkowski M, Bajraszewski T, Targowski P, Kowalczyk A (2003) Real-time in vivo imaging by high-speed spectral optical coherence tomography. Opt Lett 28:1745–1747

Woolsey TA, Rovainen CM, Cox SB, Henegar MH, Liang GE, Liu D, Moskalenko YE, Sui J, Wei L (1996) Neuronal units linked to microvascular modules in cerebral cortex: response elements for imaging the brain. Cereb Cortex 6:647–660

Xu F, Ge Y, Lu H (2009) Noninvasive quantification of whole-brain cerebral metabolic rate of oxygen (CMRO2) by MRI. Magn Reson Med 62:141–148

Yang SP, Krasney JA (1995) Cerebral blood flow and metabolic responses to sustained hypercapnia in awake sheep. J Cereb Blood Flow Metab 15:115–123

Yaseen MA, Srinivasan VJ, Sakadzic S, Wu W, Ruvinskaya S, Vinogradov SA, Boas DA (2009) Optical monitoring of oxygen tension in cortical microvessels with confocal microscopy. Opt Express 17:22341–22350

Yee SH, Lee K, Jerabek PA, Fox PT (2006) Quantitative measurement of oxygen metabolic rate in the rat brain using microPET imaging of briefly inhaled 15O-labelled oxygen gas. Nucl Med Commun 27:573–581

Zappe AC, Uludag K, Oeltermann A, Ugurbil K, Logothetis NK (2008) The influence of moderate hypercapnia on neural activity in the anesthetized nonhuman primate. Cereb Cortex 18: 2666–2673

Zhang N, Zhu XH, Lei H, Ugurbil K, Chen W (2004) Simplified methods for calculating cerebral metabolic rate of oxygen based on 17O magnetic resonance spectroscopic imaging measurement during a short 17O2 inhalation. J Cereb Blood Flow Metab 24:840–848

Zhu XH, Zhang N, Zhang Y, Zhang X, Ugurbil K, Chen W (2005) In vivo 17O NMR approaches for brain study at high field. NMR Biomed 18:83–103

Zonta M, Angulo MC, Gobbo S, Rosengarten B, Hossmann KA, Pozzan T, Carmignoto G (2003) Neuron-to-astrocyte signaling is central to the dynamic control of brain microcirculation. Nat Neurosci 6:43–50

Chapter 16
Imaging of Brain Function Using Deoxyhemoglobin and Magnetic Fields

Kâmil Uğurbil

Abstract In the two decades since its discovery, functional magnetic resonance imaging (fMRI) has seen a revolution in its ability to image brain function, going from early experiments demonstrating relatively course images of activity in the visual cortex, to mapping cortical columns, and to "brain reading" that constructs mental experiences of an individual, all using the fact that we were endowed with a complex paramagnetic molecule sequestered in our blood vessels and that neuronal activity has spatially-specific metabolic and physiologic consequences. These developments owe their success in part to significant improvements in our understanding of underlying mechanisms operative in functional imaging. These mechanisms and how they come into different data collection schemes are reviewed in this chapter. At the same time, we have seen major advances and refinements in instrumentation, which are also touched upon in this chapter, such as the introduction of ultrahigh field instruments with ever increasing capabilities, novel data acquisition strategies, and new image analysis methods, all of which also dramatically improve data quality. Some of these developments have not yet been incorporated into routine use. If history is a guide, however, the fantastically dynamic nature of the MR methodology, and the very large amount of effort committed to this field of research would predict that in a decade or two, fMRI may be performed using totally different approaches and provide substantially better information than the richness we experience with techniques at the cutting edge today.

Keywords Ultrahigh field • 7 Tesla • Orientation columns • Decoding • Encoding • Resting state • fMRI • BOLD • Spin echo • Gradient echo • Brain imaging • Neuroimaging • Prallell imaging • Multiband • Slice acceleration

---

K. Uğurbil, Ph.D. (✉)
Department of Radiology, Neurosciences, and Medicine Center for Magnetic Resonance Research, University of Minnesota Medical School, Minneapolis, MN 55455, USA
e-mail: kamil@cmrr.umn.edu

I.-Y. Choi and R. Gruetter (eds.), *Neural Metabolism In Vivo*, Advances in Neurobiology 4, DOI 10.1007/978-1-4614-1788-0_16,

16.1 Introduction

Most of our understanding of brain function derives from electrophysiology studies that use single and multiple unit recordings. This methodology is precise, but tedious, and the data produced are inherently undersampled because coverage over large areas of the cortex is impractical. Optical imaging techniques provide an alternative approach. However, these methods cannot penetrate more than ~1 mm in depth on the exposed cortical surface when used for high resolution functional mapping and are restricted to a relatively course spatial resolution when performed non-invasively through the intact scalp and skull (Gibson et al. 2005). Radioactively labeled glucose analog 2-[1-^{14}C]-deoxyglucose has been used in early seminal experiments on functional parcellation in the brain (e.g. Kennedy et al. 1975); this method, however, is limited by the availability of differently labeled glucose analogs, and hence, the number of conditions that can be studied in a single animal. Of course, a major drawback with all of these methodologies is that they are invasive, or even terminal, therefore, unsuitable for human brain applications.

Because of the afore-listed limitations, functional magnetic resonance (fMRI) has rapidly emerged as the dominant and virtually indispensible method for studying the human brain, which is endowed with unique capabilities that often cannot be studied in animal models. Since its introduction in 1992 (Bandettini et al. 1992; Kwong et al. 1992; Ogawa et al. 1992), fMRI has fueled explosive developments in our investigation of brain function, achieving depiction of functional activity in three dimensions from whole brain studies with approximately a millimeter resolution to imaging at columnar level (Menon et al. 1997; Cheng et al. 2001; Goodyear and Menon 2001; Fukuda et al. 2006; Yacoub et al. 2007; Yacoub et al. 2008a; Chaimow et al. 2011; Shmuel et al. 2010) in restricted volumes with submillimeter resolution; an example of the latter includes orientation columns that were never previously visualized in the human brain (Yacoub et al. 2008a) (Fig. 16.1). This progress has been possible only with extensive theoretical and experimental studies aimed at understanding the complex set of mechanisms underlying the ultimate coupling between neuronal activity and MRI signals. The drive towards ultrahigh[1] magnetic fields has been critical in this endeavor; such high magnetic fields have provided one of the most important tools to study the mechanisms underlying functional imaging signals and to conduct some of the most advanced applications of fMRI.

Today, functional images in the brain can be obtained using the BOLD mechanism (Bandettini et al. 1992; Kwong et al. 1992; Ogawa et al. 1992), measurement of cerebral blood flow (CBF) changes with arterial spin labeling (ASL) (e.g. Kim 1995; Hoge et al. 1999; Duong et al. 2001; Detre and Wang 2002; Pfeuffer et al. 2002a; Shmuel et al. 2002 and references therein), intravoxel incoherent motion

[1] The terminology is based on that used for classifying radiofrequency (RF) bands. The frequency range 300 MHz to 3 GHz is defined as Ultra High Frequency (UHF) (see http://en.wikipedia.org/wiki/Ultra_high_frequency). The hydrogen nucleus resonance frequency at 7 T is ~300 MHz i.e. in UHF band. Therefore, 7–70 T can be defined as Ultra High Field (UHF).

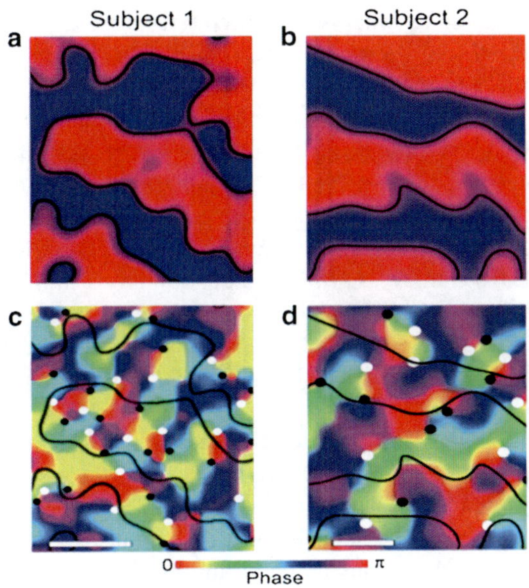

Fig. 16.1 Functional maps of ocular dominance columns (**a**, **b**) and orientation columns (**c**, **d**) in the human brain obtained with SE fMRI at 7 T in two subject. The *black lines* in *c*, and *d* depict the boundaries of the ocular dominance columns seen in *a*, and *b*, respectively, so as to permit the visualization of the relationship between the two columnar organizations. The *black* and *white dots* in *c* and *d* identify the centers of clockwise and counterclockwise rotating pinwheels. The *white bars* (*lower left* hand corner in *c* and *d*) designate 1 mm scale (From Yacoub et al. 2008a)

(Song et al. 2003; Song and Li 2003), diffusion (Yacoub et al. 2008a, b; Zhang et al. 2009) and cerebral blood volume (CBV) alterations (e.g. Belliveau et al. 1991; Mandeville et al. 1998, 2001, 2004; Mandeville and Marota 1999; Harel et al. 2002, 2006a; Kim and Ugurbil 2003; Lu et al. 2003; Zhao et al. 2005, 2006). In this review, the focus is on the BOLD based fMRI, with the other techniques mentioned in the context of their relevance for the BOLD mechanism.

16.2 BOLD-Based Functional Imaging of Brain Activity

The first functional images of the human brain were obtained using an exogenous, gadolinium based contrast agent that was at the time (and still is) available for clinical use (Belliveau et al. 1991). Although this was an exciting development, the approach had many shortcomings: It required two bolus injections of contrast agent separated by many minutes for a single functional map, limited amount of time for data acquisition during a transient bolus passage through the brain, and a limited number of possible contrast agent injections due to saturation. Ultimately human applications of this technique would be limited by safety concerns because gadolinium-containing contrast agents have become a suspected causal factor for nephrogenic systemic fibrosis (NSF) (e.g. Marckmann et al. 2006; Kuo et al. 2007). Because of all these reasons, it would not have been possible with this technique to perform all the incredibly information-rich and complex fMRI studies that are feasible today, and the impact on understanding brain function would have been quite modest. Fortuitously, the exogenous contrast agent based functional imaging approach was rapidly followed by what is now often referred to as blood oxygen level dependent (BOLD) fMRI. This technique relies on intrinsic contrast mechanisms and, as such, does not suffer from the shortcomings of the exogenous contrast based method. It is this development that revolutionized the study of brain function. Today, we know that the fMRI signal is quite complex and has several contributions besides BOLD as originally described (e.g. see Uludag et al. 2009). Nevertheless, it was the BOLD effect itself that motivated some of the investigators involved in the discovery of fMRI to attempt at the first human functional brain mapping. It must also be recognized that such fMRI studies also owe credit to numerous earlier work aimed at mapping brain function using radioactive xenon, and positron emission tomography (PET) technology (see review Raichle 2009). These techniques recognized and relied on the physiological coupling between blood flow and blood volume and neuronal activity. So does the modern fMRI approach.

At the University of Minnesota, the Center for Magnetic Resonance Research, the fMRI effort came about because of the work Seiji Ogawa did in Bell Labs introducing the BOLD effect (Ogawa et al. 1990a, b; Ogawa and Lee 1990). These early experiments, conducted on rats, did not generate functional maps; rather, they demonstrated that metabolic perturbations such as hypoglycemia and graded levels of oxygen in the inhaled gas mixture affected the visibility of venous blood vessels. In their PNAS paper (Ogawa et al. 1990a), however, Ogawa and his coworkers speculated about the use of this method to study oxygen metabolism and even possibly achieve functional imaging in the brain, in a way analogous to the PET approach but, unlike PET, using an endogenous contrast mechanism. The link to oxygen metabolism is what initially attracted my attention to this work as it was presented in conferences. Seiji Ogawa and I started talking about pursuing functional imaging together in the human brain using the 4 T system that was expected at the University of Minnesota in 1989–1990 time frame. When this 4 T became operational in Minnesota, the first experiments conducted on it were functional mapping using visual stimulation. Since we were working for the first time at 4 T, and for the first time with humans, many obstacles had to be overcome including high frequency RF coils, new pulses sequences appropriate for the high field, and regulatory approvals for working at 4 T. Ultimately, experiments like hemifield simulation gave us confidence that maps of brain activity alterations were being generated by gradient recalled echo MR images. The resulting paper was submitted to Nature at the end of 1991 or early 1992. Nature did not send it to scientific review and rejected the paper, deeming it to be not of sufficient general interest. After some delay, the paper was submitted to PNAS in March 1992, within 5 days of submission of a similar paper by Ken Kwong of the MGH group (Kwong et al. 1992).

The BOLD effect originates from the intravoxel magnetic field inhomogeneity induced by paramagnetic deoxyhemoglobin sequestered in red blood cells, which in turn are compartmentalized within the blood vessels. Magnetic susceptibility differences between the deoxyhemoglobin-containing compartments versus the surrounding space devoid of this strongly paramagnetic molecule generate magnetic field gradients across and near the boundaries of these compartments. Therefore, in images sensitized to BOLD contrast, signal intensities are altered if the regional deoxyhemoglobin content is changed. This occurs in the brain because of spatially specific metabolic and hemodynamic responses to enhanced neuronal activity. Regional blood flow (CBF) increases while oxygen consumption rate ($CMRO_2$) in the same area is elevated to a lesser degree, resulting in decreased extraction fraction and lower deoxyhemoglobin content per unit volume of brain tissue. Consequently, signal intensity in a BOLD sensitive image increases in regions of the brain engaged by a "task" or a stimulus relative to a resting, basal state.

16.2.1 Data Acquisition in fMRI

In BOLD contrast, signal is acquired simply after a delay, TE, following excitation. A refocusing pulse may or may not be applied during this delay. When refocusing pulses are not applied, one uses field gradient reversal to form an echo in imaging; consequently, such echoes are called "gradient-recalled" or "gradient" echoes, abbreviated simply as GE. This is the first MR sequence used for functional mapping with the BOLD approach (Bandettini et al. 1992; Kwong et al. 1992; Ogawa et al. 1992). In the presence of a refocusing pulse, the spin-echo (SE) signal (i.e. echo-amplitude) detected after the echo time TE decreases according to $\exp(-TE/T_2)$ where T_2 is the spin-lattice relaxation time associated with the loss of magnetization with time in the transverse plane, the plane perpendicular to the main static magnetic field B_0. Spin echoes can also yield functional information in the brain leading to SE based fMRI. SE fMRI is sensitive only to a subset of the processes that lead to BOLD contrast in the GE experiment, while everything that provides functional signals in SE fMRI also contribute to GE fMRI.

In a gradient echo, signal loss occurs through T_2 relaxation as well as by signal cancellation that arises due to "dephasing" of the magnetization in the presence of magnetic field inhomogeneities. The latter is recoverable with a refocusing pulse and does not contribute to spin-echoes. In a gradient echo image, the appropriate relaxation constant of magnetization in the transverse plane is T_2*. T_2 of pure water is dominated by the magnetic interactions between the two hydrogen atoms, through the dipole-dipole coupling. In the human brain, this is still the case but the water is not in a pure form; rather, it exists in many different environments, such as tightly or weekly bound to macromolecules and metal ions, or free of such binding. The hydrogen atom of the water molecule itself can exchange between water and exchangeable –NH or –OH hydrogens on macromolecules. All of these processes affect the water proton T_2 and impart the contrast that allows the differentiation of one type of tissue from another in MRI. Another mechanism that contributes to T_2 is diffusion of water molecules around magnetic field inhomogeneities; normally in the brain this is not a very large affect though it is present and detectable especially at high magnetic fields (Bartha et al. 2002; Michaeli et al. 2002).

T_2*, on the other hand, has contributions from all of the process that affect T_2 plus the dephasing effect in the presence of magnetic field inhomogeneities.

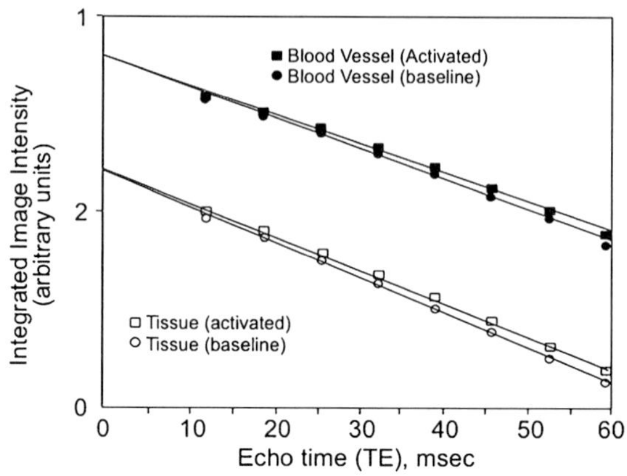

Fig. 16.2 Change in T2* in the visual cortex for the human brain induced by visual activation. Log of signal intensity is plotted as a function of echo delay TE for a gradient recalled echo study. A linear relationship is seen as expected from the exponential signal decay with time constant T2*. The data are shown for the same voxels that were found to be "activated" during the visual stimulations study, both before, and during the visual stimulation period. Some voxels were identified as originating from large venous vessels. Tissue areas had no visible blood vessels at resolution of vessel weighted MR images. A multi-echo sequence was used to collect the data for all echo times simultaneously (From Menon et al. Menon et al. 1993)

In a container of pure water, except near the boundaries of the container where the magnetic field can be non-uniform, T_2 equals T_2*. This is not the case in the brain because of magnetic field inhomogeneities that span scales ranging from microns to many millimeters and even centimeters as encountered around the air filled cavities adjacent to the brain. On the micron to millimeter scale, the inhomogeneities exist around deoxyhemoglobin containing blood vessels ranging from capillaries to large veins; it is this last effect that provides functional mapping signals. Oxygenated blood is diamagnetic and has similar properties to tissue. However, *deoxy*hemoglobin molecule is paramagnetic, and its presence leads to a large susceptibility difference between compartments that contain this molecule and others that are devoid of it. Thus, against a backdrop of numerous contributions to the inherent T_2 and T_2* of brain water, BOLD contrast refers to T_2* or T_2 mechanism arising from magnetic field inhomogeneities generated by magnetic susceptibility difference across the luminal boundaries of blood vessels. During increased neuronal activity, deoxyhemoglobin content is altered in the brain; this change is accompanied with small but measurable decreases in T_2 and T_2* which was experimentally documented in early days of fMRI (Fig. 16.2) (Menon et al. 1993).

Image acquisition can be accomplished using "slices" where a frequency selective RF pulse is employed to restrict the signal origin to a single slice, typically ~1–3 mm in thickness at the present time, though as large as 10 mm slices were not uncommon in early days of fMRI.

In MR imaging terminology, k-space refers to the two or three-dimensional matrix of data points collected during image acquisition. When the image acquired is from a slice only, the k-space data reside in two dimensions, representing the encoding along two orthogonal directions that define a plane perpendicular to the slice selection direction. A 2-dimensional Fourier transform then converts this into an image of the slice. In techniques like FLASH (Haase et al. 1986, and see reviews [Haacke and Tkach 1990; Chien and Edelman 1991; Haacke et al. 1991], one line along a single dimension of k-space is collected after each RF pulse and multiple RF pulses are employed to cover the entire k-space in seconds. In single shot approaches as in Echo Planar imaging (EPI) [Mansfield 1977] or SPIRAL imaging [e.g. Yang et al. 1998]), the entire k-space is covered after a single RF pulse using a train of gradient echoes in ~20–100 ms depending on the available hardware. Segmented EPI schemes (McKinnon 1993; Feinberg and Oshio 1994; Wielopolski et al. 1995a) cover more than one line but less than the entire k-space in a single segment after one RF pulse. They have been particularly useful in high-resolution functional imaging studies (Yacoub et al. 2007, 2008a and references therein) because the higher resolution dictates the acquisition of many more k-space lines and points along each line.

Single shot techniques provide distinct advantages to the others for fMRI. These include the ability to cover the whole brain rapidly, and to partially suppress temporal signal fluctuations in an fMRI time series. The latter touches on a very interesting subject: Namely, in consecutively acquired images from the brain, temporal signal variations contain contributions from and can even be dominated by physiologically induced processes such as blood vessel pulsation, (vasomotion), respiration, and spontaneous neuronal activity (see discussion further on). Of course, the intrinsic signal-to-noise ratio (SNR) of a single image has to be sufficiently good to detect these physiologically induced variations. These temporal instabilities come into single shot versus multiple shot acquisitions in a different way and are much more deleterious for the latter (Hu and Kim 1994; Hu et al. 1995). They can be minimized by post-processing strategies in multiple shot as well as in single shot techniques (e.g. Hu and Kim 1994; Hu et al. 1995; Mitra et al. 1995, 1997; Biswal et al. 1996a; Mitra et al. 1997; Mitra and Pesaran 1999; Glover et al. 2000; Pfeuffer et al. 2002b; Van De Moortele et al. 2002). However, such strategies require intrinsically good SNR and, and in some cases, rapid image acquisition so as to capture these temporal fluctuations accurately.

At higher magnetic fields, it is necessary to acquire single shot images faster because the MR signal disappears faster subsequent to excitation (i.e. the T_2* is shorter) due to increased magnetic field inhomogeneities. A critical development in this respect has been parallel imaging (PI) techniques which utilize the spatial information inherent in multiple surface coils distributed over the sample so as to omit some of the k-space lines as if the image was being acquired with a reduced field of view (FOV). Consequently, the time spent acquiring the signal after the RF pulse is shorter, which for EPI translates into shorter echo trains. When processed without the use of the spatial information from array coil sensitivity profiles, images acquired in this fashion display sections that are folded on top of each other, in other words sections that are "aliased". However, they can be unfolded using parallel imaging techniques. The utility of parallel imaging has been examined for functional imaging

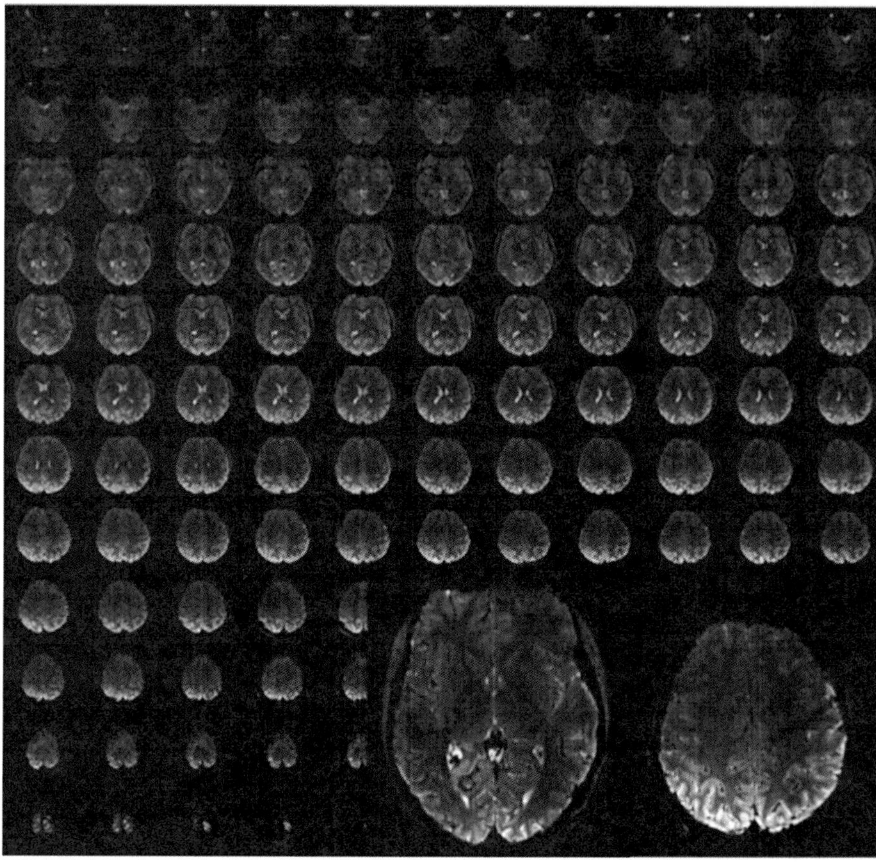

Fig. 16.3 Whole brain EPI images at 7 T. Two slices are shown on the lower right corner as expanded for better visualization of detail: 0.75 mm isotropic resolution echo spacing 0.67 ms, 256 × 256, parallel imaging with reduction factor 4. Partial Fourier = 6/8 TE = 20 ms, 128 slices TR = 6 s, single image (Data generated in CMRR, University of Minnesota by Yacoub et al.)

early on (de Zwart et al. 2006; Moeller et al. 2006) after their introduction and, today, they are typically part of the normal fMRI acquisition, especially at high magnetic fields.

Parallel imaging works better at high magnetic fields (Ohliger et al. 2003; Wiesinger et al. 2004, 2006) and high magnetic fields need the use of parallel imaging relatively more because of the short window available for data acquisition due to the short T_2* (e.g. see (Yacoub et al. 2001, 2003; Uludag et al. 2009) and references therein). Thus, parallel imaging and high fields are synergistic. Partial Fourier (Feinberg et al. 1986) sampling, or sparse data sampling approaches (Liang et al. 2003) are also approaches that reduce the image encoding times. With the use of these techniques, it is feasible today to acquire single shot high-resolution EPI images at high fields such as 7 T. Figure 16.3 illustrates 0.75 mm isotropic GE EPI images obtained at 7 T using a reduction factor, R, of 4 (i.e. fourfold undersampling of the k space, or equivalently, fourfold reduction in FOV, leading, in this case, to fourfold maximum aliasing[2]). Such high resolution images usher in the possibility of investigating the human brain at the level of elementary computational units, such as cortical columns with differentiation even across the cortical layers in the *entire* human brain.

Parallel Imaging, and Partial Fourier sampling reduce the number of phase encoding steps for spatial encoding, and consequently the echo train length after a single RF pulse; however, they do not necessarily reduce *whole brain* image acquisition times significantly in many applications. This is because a physiological contrast preparation period must precede the spatial encoding period for each slice and this contrast preparation period can equal or exceed the time employed for the EPI echo train. 3D echo volume (EVI) (Mansfield et al. 1994) avoids the repetition of the contrast encoding time by following a single contrast preparation period with subsequent 3D volume coverage in a single echo train. However, this approach has limitations in spatial resolution and image quality due to longer echo trains needed to fully encode the volumetric spatial information in the relatively short acquisition period dictated by T_2*, especially at high fields; the consequence is distortions and blurring on two of the 3D image axes, as well as a loss in SNR. Multi-shot (segmented with multiple excitation) 3D EPI approaches that have produced high quality images (Wielopolski et al. 1995b; Poser et al. 2010) overcome this limitation; however, in order to achieve very rapid coverage of the whole brain, high parallel imaging reduction factors must be employed in the two orthogonal phase encoding dimensions in the 3D acquisition, leading to significant SNR losses due to the omitted phase encoding steps; the SNR loss is by factor $g(R_1R_2)^{1/2}$ where R_1 and R_2 represent the reduction factors along the two phase encoding directions, and g is the noise amplification inherent in the unaliasing process. Thus if $R_1=R_2=4$, there is a fourfold loss of SNR and, if such reduction factors are too aggressive relative to the spatial encoding capabilities of the array coil employed, there is an increased penalty paid in high g-factors as well. Echo shifting approachcs, PRESTO (Liu et al. 1993; Golay et al. 2000), increase volume coverage efficiency in fMRI by taking advantage of TE delays to apply additional RF pulses, but are SNR limited and run into restrictions at higher magnetic fields when T_2 and T_2* become inherently short.

In order to bypass these afore-listed limitations, we recently introduced "slice accelerated images" that can achieve significant reductions in whole brain coverage time without substantial or even noticeable degradation in image quality or SNR. The original introduction of this methodology for functional imaging used a multiband

[2] When the FOV is reduced, the image folds and there is aliasing. However, if the FOV is set to be very large, for example more than twice as large as the object, then the use of parallel imaging with a reduction factor of 2, results in no aliasing per se since region of the image devoid of any signal folds onto the regions containing the object. Thus, the degree of maximum aliasing in the object to be imaged is a parameter that conveys the degree of spatial information that must be extracted from the RF coils by PI technique; in appropriate uses of parallel imaging, maximum aliasing is equal to or only marginally smaller than the R factor.

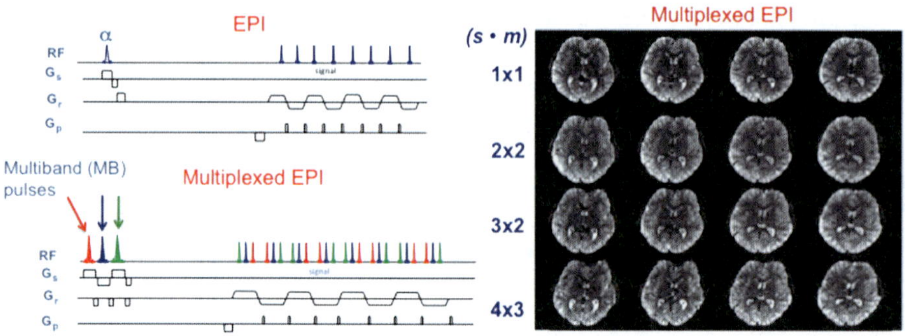

Fig. 16.4 The pulse sequence for regular EPI and Multiplexed EPI (M-EPI) and images obtained with Multiplexed EPI images. EPI pulse sequence generates a single slice image during each readout train, which is repeated by the number of slices to scan the whole brain; replacing the single slice pulse with multiband (MB) pulse in this sequence to excite several slices simultaneously, and then unaliasing them using array coil sensitivity profiles yields the Multiband technique (Moeller et al. 2008, 2010) which requires fewer repeats to scan the whole brain. Multiplexed-EPI (M-EPI) pulse sequence combines the SIR approach with MB technique: SIR consecutively excites s slices ($s=3$ is shown in the pulse sequence diagram with pulses in *red*, *blue* and *green*) and acquires them in a single echo train. Using MB pulses that simultaneously excite m slices for each of the single slice pulses in the SIR approach produces the M-EPI sequence, with a "slice acceleration" of $(s \bullet m)$ leading to $(s \bullet m)$ number of slices collected in a single echo train. Four slices from a 2 mm isotropic resolution, 60 slice, whole brain 3 T data set obtained with the M-EPI technique are shown, together with the $(s \bullet m)$ acceleration factors employed (Adapted from Moeller et al. 2008, 2010; Feinberg et al. 2010)

(MB) approach that excites several slices simultaneously using multibanded pulses and unfolds these aliased slices using parallel imaging principles (Moeller et al. 2008, 2010). Unlike the conventional parallel imaging technique, however, the MB method does *not* suffer from SNR losses associated with omitted phase encoding steps; the g-factor noise amplification due to unaliasing is still present and depends on the adequacy of the spatial information garnered from coil sensitivity profiles relative to the number of slices acquired simultaneously. We subsequently introduced additional slice acceleration (Feinberg et al. 2010) by combining the Multiband approach with the simultaneous image refocused (SIR) scheme (Feinberg et al. 2002) that temporally interleaves signals from several slices within an EPI echo train. In this new approach, referred to as ***Multiplexed***-EPI (M-EPI), if multiband pulses with m bands are used for each of the s RF interleaved pulses in the SIR technique (where *m and* s are positive non-zero integers), the result is simultaneous acquisition of m times s (i.e. $m \bullet s$) slices in a single echo train. Figure 16.4 illustrates the pulse sequence and examples of Multiplexed-EPI images obtained with various "slice acceleration factors".

16.2.2 Mechanism of BOLD-Based Functional Mapping Signals

BOLD contrast reports on the deoxyhemoglobin content in the brain, which is determined by two parameters: (1) the deoxyhemoglobin concentration in blood and

(2) the total amount of deoxyhemoglobin containing blood volume in a given volume of brain tissue. The former is determined by $CMRO_2$/CBF ratio, i.e. the rate at which oxygen is used compared to its delivery rate. Susceptibility gradients generated in and around blood vessels caused by the presence of the paramagnetic deoxyhemoglobin is the source of the BOLD effects. The spatial extent of these gradients depends on vessel diameter. Consequently, blood vessels play a critical role in the coupling between alterations in neuronal activity and MR detectable signals. The nature of the MR detectable signals also depends on parameters other than physiology, namely magnetic field magnitude and the type of pulse sequence used.

16.2.2.1 Extravascular Effects

The BOLD effect has two components: extravascular and intravascular (i.e. blood related). First, let us focus exclusively on the *extravascular space*. When deoxyhemoglobin is present in a blood vessel, magnetic susceptibility of the space within the blood vessel is different than outside the blood vessel. This results in a homogeneous field within the blood vessel[3] but an inhomogeneous field outside. If one considers an infinite cylinder as an approximation for a blood vessel with magnetic susceptibility difference between the inside and the outside of the cylinder, then the magnetic field expressed in angular frequency,[4] at any point in space, will be perturbed from the applied magnetic field ω_o (Springer and Xu 1991). Inside the cylinder, the perturbation, $\Delta\omega^{in}$, will be given by the equation:

$$\Delta\omega^{in} = 2\pi\Delta\chi_O(1-Y)\omega_O[\cos^2(\theta)-1/3] \tag{16.1}$$

At any point outside the cylinder, the magnetic field will vary depending on the distance and orientation relative to the blood vessel and the external magnetic field direction, according to the equation:

$$\Delta\omega^{out} = 2\pi\Delta\chi_O(1-Y)\omega_O\left[r_C/r\right]^2\sin^2(\theta)\cos(2\phi) \tag{16.2}$$

In these equations, $\Delta\chi_O$ is the maximum susceptibility difference expected in the presence of fully deoxygenated blood, Y is the fraction of oxygenated blood present, r_C designates the cylinder radius, r is the distance from the point of interest to the center of the cylinder; θ is the angle the cylinder makes with the static magnetic

[3] In reality, this statement is correct strictly if we treat blood as a homogeneous medium with a magnetic susceptibility that differs from the surrounding tissue. However, in blood, deoxyhemoglobin is compartmentalized within red blood cells. This leads to magnetic field inhomogeneities around the red blood cells, thus to a non-uniform field distribution within the blood. However, the spatial scale of these inhomogeneities is sufficiently small that water diffusion averages them out, effectively resulting in a uniform field.

[4] In the presence of a magnetic field B_0 the angular "resonance" or Larmor frequency ω_0 is given by $\omega_0=\gamma B_0$ where γ is the gyromagnetic ratio which is 2.6751965×10^4 rad s^{-1} G^{-1} or equivalently 42.577 in units of MHz/Tesla for protons in a spherical water sample.

field, and φ is the angle between the line that runs from the point of interest to the cylinder intersecting it perpendicularly and the plane containing the cylinder and the static magnetic field. Note that outside the cylinder, the magnetic field changes rapidly over a distance comparable to two or three times the cylinder radius; at a distance equal to the diameter of the cylinder from the cylinder center, $\Delta\omega^{out}$ is already down to 25% of its value at the cylinder boundary.

In a GE fMRI experiment, images are acquired after a delay TE in order to sensitize the image to magnetic field inhomogeneities. In the SE approach, there is a single 180° "refocusing" pulse in the delay period TE. As previously discussed, intra-voxel magnetic field inhomogeneities will lead to signal loss during the evolution period TE. If the typical diffusion distances of water in tissue during the delay TE are comparable to the distances spanned by the magnetic field gradients within that tissue, than during this delay, the tissue spins will experience a time-averaged magnetic field in the extravascular space where the magnetic field gradients exist. This type of averaging due to motion of the spins is referred to as "dynamic" averaging. Dynamic averaging is what is detected in SE fMRI experiments. GE on the other hand contains additional contributions.

Typical TE values used in GE fMRI experiments depend on the field strength and the specifics of the pulse sequence, but in general range from ~15 ms at ultrahigh fields to ~100 ms at very low fields. Thus, blood vessel size compared to the diffusion distances in this ~15–100 ms time domain becomes a critical parameter in the BOLD effect. In this time scale, diffusion will dynamically average the gradients around small blood vessels (e.g. capillaries) that contain deoxyhemoglobin and, thus, result in a signal decay that will be characterized with a change in apparent T_2 (Ogawa et al. 1993; Kennan et al. 1994; Boxerman et al. 1995a). In a spin-echo experiment with a single refocusing pulse in the middle of the delay period, the phase accumulation that has taken place during the first half of the echo will not be reversed fully because the spins will not be able to trace back their trajectories exactly due to the presence of the diffusion. Of course, applying many refocusing pulses as in a Carr-Purcell pulse train or applying a large B_1 field (relative to the magnitude of the magnetic field inhomogeneity) for spin-locking during this delay will reduce or even eliminate this signal loss due to dynamic averaging. In a gradient echo measurement, dynamic averaging will also occur during the entire delay TE. If the imaging voxel contains only such small blood vessels at a density such that one-half the average distance between them is comparable to or less than diffusion distances (as is the case in the brain where capillaries are separated on the average by ~25–40 μm [Duvernoy et al. 1981; Pawlik et al. 1981]) then the entire signal from the voxel will be affected by dynamic averaging.[5]

In considering the movement of water molecules around blood vessels, we need not be concerned with the exchange that ultimately takes place between intra- and

[5] Capillary density in the brain is not a constant and varies from region to region. In the cat visual cortex, the average distance between capillaries has been reported to be ~25 μm (Pawlik et al. 1981); in contrast, in the human primary motor cortex it was reported to be ~40 μm (Duvernoy et al. 1981).

extravascular water across capillary walls. Typical lifetime of the water in capillaries exceeds 500 ms (Eichling et al. 1974; Paulson et al. 1977a, b), significantly longer than the T_2 and T_2* values in the brain tissue and longer than the period TE typically employed in fMRI studies.

For larger blood vessels, complete dynamic averaging for the entire voxel will not be possible. Instead, there will be "local" or "partial" dynamic averaging over a subsection of the volume spanned by the magnetic field gradients generated by the blood vessel. A water molecule at a given point in space relative to the blood vessel will see a "locally" time-averaged $\Delta\omega^{out}$ and the phase of such spins acquired during the echo time TE will vary across the voxel because this locally time averaged $\Delta\omega^{out}$ is non-uniform in the voxel. Hence signal loss will occur within the voxel due to phase dispersion. This signal loss is referred to as "static averaging" or "static dephasing". In a spin-echo, the static dephasing will be refocused and thus eliminated.

The spin behavior for the extravascular BOLD effect has been examined in numerous modeling studies (e.g. Ogawa et al. 1993; Yablonskiy and Haacke 1994; Boxerman et al. 1995a, b; Kennan et al. 1998; Kiselev and Posse 1999; Fujita 2001; Uludag et al. 2009). An important prediction of modeling studies is that the large and small vessel *extravascular* BOLD effects differ and the microvascular (capillary and post-capillary small venule) contribution varies supralinearly with the external magnetic field. This supralinear dependence however, levels off after about 7–10 T (Uludag et al. 2009). In contrast, the dependence on the external magnetic field is linear for large blood vessels in the static averaging domain.

16.2.2.2 Intravascular Effects

In blood, hemoglobin is also compartmentalized within red blood cells. Thus, when the deoxy form is present, there are field gradients around the red cells. However, because the dimensions are very small compared to diffusion distances, the effect is dynamically averaged and becomes an "apparent" T_2 effect. The dynamic averaging in this case also involves exchange across the red blood cell membrane that is highly permeable to water. The exchange is between two compartments, plasma and the interior of the red blood cell where the magnetic field is significantly different because of the presence of paramagnetic deoxyhemoglobin. Thus, in the presence of deoxyhemoglobin containing red blood cells, apparent T_2 of blood decreases and can be expressed as $1/T_2 = a_0 + kB_0^2(1 - Y)^2$ where a_0 is a field independent term, B_0 is the magnitude of the static magnetic field and k is a constant (see Duong et al. 2003 and references there in). Therefore, the T_2 of *blood* will change when the content of deoxyhemoglobin is altered by elevated neuronal activity and this will lead to a signal change in a T_2 or T_2* weighted image. This effect will be present wherever the content of deoxyhemoglobin has changed, thus both in large and small blood vessels. This effect is a substantial, even the dominant contribution to functional imaging signals at low magnetic fields (van Zijl et al. 1998). It diminishes in importance at high fields because the T_2* and T_2 of blood decreases precipitously with increasing magnetic field magnitude and becomes much less than that of tissue (Duong et al. 2003; Uludag et al. 2009); consequently, in an acquisition where the TE is set approximately equal to tissue T_2* or T_2 for GE and SE fMRI, respectively, blood signals are significantly attenuated and are even diminished beyond detectability.

A complete modeling of fMRI signal changes must consider numerous physiological changes and parameters and basal relaxation rates as well as how the T_2 or T_2* and signal content of extravascular or intravascular space would be effected by changes in brain activity. Such a comprehensive modeling was recently published (Uludag et al. 2009). This paper also presents a good review of what has been done previously.

16.2.2.3 Inflow Effects in BOLD-Based fMRI

During the hemodynamic response that supplies more blood to the area of increased activity in the brain, flow increases must also take place within the vascular tree supplying and draining the activated area. This vascular flow change can lead to signal alterations in large blood vessels in images intended to report on BOLD contrast (e.g. Menon et al. 1993; Duyn et al. 1994; Frahm et al. 1994; Kim et al. 1994; Segebarth et al. 1994). This is not because BOLD contrast itself contains a direct flow effect; rather, image contrast may not be purely of BOLD origin because any repeated, slice-selective image is inherently flow sensitive if the signal within the slice does not attain full relaxation between consecutive signal excitations. In single slice studies, *allowing full relaxation in between RF pulses eliminates this problem completely*; however, this condition often is not satisfied in many studies since it leads to a loss in SNR per unit time. Consequently, such studies essentially obtain images of macrovascular flow rather than the much smaller BOLD changes. This problem was demonstrated with clarity by comparing presumably BOLD-based "functional" images with vessel images in two and three dimensions (Segebarth et al. 1994; Belle et al. 1995).

When full relaxation is allowed for each slice, a macrovascular inflow problem may still be present in multislice studies due to inter slice effects. For example, blood experiencing an RF pulse in one particular slice will subsequently travel to other parts of the brain and, in turn, affect the intensity of signals from a slice sampled at a later time. The sensitivity to this problem will depend on the imaging sequence used, and how the different slices are sampled and the orientation of slices since most macrovascular flow occurs along an inferior-superior direction in the human brain. Experimental evidence, however, indicates that this type of inter-slice flow contribution to BOLD-based multislice imaging is negligible (Howseman et al. 1999).

16.3 Spatial Specificity of BOLD fMRI

Although the neuro- and cognitive- science communities have embraced fMRI with exuberance, caution must be exercised in quantitatively using fMRI data because fMRI maps are based on secondary metabolic and hemodynamic events that follow neuronal activity, and not the electrical activity itself. One of the most important issues confronted is the spatial specificity of the fMRI maps (i.e. how accurate are the maps generated by fMRI compared to actual sites of neuronal activity?).

Neuronal organizations that respond to a highly reduced attribute of an input and perform elementary computations are known to exist in the millimeter to submillimeter spatial scale both tangentially on the cortical surface and across the cortical laminae. The ability to obtain accurate functional maps at this level of organization is imperative if fMRI is to make contributions to brain science beyond its macroscopic circuit level capabilities. This is not just a question of mapping ocular dominance or orientation columns but has important implications for cognitive neurosciences. Columnar clustering is likely to be a fundamental organizational principle in the brain. Clustering, analogous to the well-recognized columns in the early visual areas has been proposed, based on monkey electrophysiology studies, for higher order visual areas such as the middle temporal (MT) for direction of motion (Albright 1984; Albright et al. 1984), and inferio-temporal (IT) cortex for face and object recognition (Fujita et al. 1992; Wang et al. 1996). Imaging such organizations, however, requires submillimeter specificity of mapping signals and sufficiently high image SNR to attain the resolution demands. This is an important technical challenge in fMRI and one our group at the University of Minnesota have been concerned with and worked towards ever since the discovery of fMRI.

In discussing the spatial specificity of the fMRI approach, we must distinguish *single* condition versus *differential* mapping. In addition, one must consider mapping inaccuracies caused by physiology at the level of coupling of blood flow and metabolism to neuronal activity versus those caused by the fMRI mechanisms governing the coupling of MR signal changes to physiological alterations. The latter has been noted to exist virtually ever since the discovery of fMRI (Menon et al. 1993).

Differential mapping refers to functional images obtained by using two analogous but orthogonal activation states designed specifically to eliminate or suppress spatially non-specific signals which are assumed to be common to the orthogonal activation states; the functional image generated corresponds to a subtraction of the two conditions. In contrast, single condition mapping does not rely on a subtraction from a second orthogonal activated state. Mapping ocular dominance columns by stimulating one eye versus the other in an alternating fashion is *differential mapping*. Stimulating one eye and using a dark state as the control would correspond to *single condition mapping*. Even if the imaging signals are *not* specific enough to be confined to the territory of one column, the assumption is that a functional map of the columns can be obtained in the differential image if the stimulus-induced signal changes that are spatially non-specific such as those associated with draining veins for the two different eyes, would be eliminated in the difference. Phase encoded mapping of multiple areas using an activation paradigm that cycles through several different possible stimulations (as employed in the visual system (e.g. Sereno et al. 1995; DeYoe et al. 1996; Engel et al. 1997) is also a kind of differential mapping because it would suppress common, non-modulating components. Although the differential imaging concepts have been used for a long time in optical and MR based functional imaging with the intrinsic assumption that they are effective in cancelling spatially non-specific effects, recent studies demonstrate that this assumption is incorrect and that suppression is not necessarily complete (Yacoub et al. 2007; Shmuel et al. 2010). Therefore, differential mapping alone cannot be counted on to fully eliminate the non-specific signals (Yacoub et al. 2007; Shmuel et al. 2010).

Single condition mapping, on the other hand, is significantly more demanding on the spatial accuracy of the imaging signals than differential imaging. It requires, in fact that that there are no spatially inaccurate components. The ability to perform single condition mapping is important because it is not always feasible to have a priori knowledge of analogous but orthogonal activation conditions.

16.3.1 Specificity of Perfusion Changes Coupled to Neuronal Activity

In fMRI, the degradation of spatial specificity might arise from imprecise spatial-coupling between neuronal activity and the physiological and metabolic events that ultimately yield the functional images. The dominant physiological change induced by alteration in the neuronal activity relevant for fMRI is blood flow (CBF) change. Clearly, when the control of CBF by the vasculature exists at a coarser spatial scale than that governing computational cluster, it will lead to spatial blurring in fMRI. At the time of the introduction of fMRI, this issue was rather controversial. Prevailing concepts suggested that CBF increases far exceed the territorial boundaries of increased neuronal activity. There is, in fact, no a priori requirement that control of CBF match exactly the functional organization spatially as long as CBF change *includes* the functional territory responsible for invoking the altered CBF state. As has been colorfully described before, the brain can water the entire garden for the sake of a single thirsty flower (Malonek and Grinvald 1996); this is adequate as long as the thirsty flowerer gets water. This question was addressed using perfusion based functional imaging or perfusion fMRI.

Images of perfusion or perfusion changes associated with increased neuronal activity can be obtained using arterial spin labeling (ASL) techniques that utilize the water protons in the blood as an endogenous "transient" tag. These methods rely on either continuous (e.g. Detre et al. 1992, 1994; Zhang et al. 1993) or dynamic (i.e. modulated versions of continuous, Barbier et al. 1999, 2001) or pulsed (e.g. Edelman et al. 1994; Kim 1995; Wong et al. 1998) tagging approaches. Efficiency of some of these methods was recently evaluated quantitatively (Pohmann et al. 2010). These ASL techniques can be "tuned" to be selectively sensitive to capillary/tissue level flow and the large vessel effects can be minimized so as to yield accurate perfusion maps, i.e. maps that report on water delivery to the capillary bed and, by exchange across the capillary wall, to surrounding tissue (e.g. Tsekos et al. 1998).

In these MR techniques, a perturbation (the "label") is induced in the population of hydrogen nuclei (spins) of water in blood, outside the tissue of interest; this label is monitored as it shows up in the tissue of interest after a delay (the tagging time) that is long enough to permit arrival into the capillary and tissue but not long enough to reach a new spin equilibrium state. In this case, the amount of "label" detected in the tissue of interest is proportional to blood flow, and increases with elevated neuronal activity. Since arterial side is permeated with fresh blood significantly faster relative to the tissue tagging times, the time to reach the equilibrium state in the large arteries versus the tissue is different. Thus, long tagging times (e.g. ~1.5 to ~2 s) (Tsekos et al. 1998) eliminate arterial component from perfusion images. The long tagging times, however, may lead to the tagged spins appearing in the venous side, thus leading to false "activation" in the veins. At high magnetic fields, the T_2 and T_2* of venous blood is very short (Thulborn et al. 1982; Lee et al. 1999) so that this effect can be selectively eliminated by a brief delay after excitations of spins but before image acquisition. Accordingly, perfusion-based fMRI maps have been shown to yield accurate images that co-localize with Mn^{+2} uptake (Duong et al. 2000a), a marker of calcium dependent synaptic activity (Lin and Koretsky 1997).

In the absence of large vessel contributions, perfusion-based fMRI can be used to examine the critical physiological question related to specificity of blood flow increases. A specific question that can be asked is if perfusion changes are confined accurately to the region of increased neuronal activity in the spatial scale of columnar organizations? Such a study was performed using the iso-orientation columns in the cat visual system (Duong et al. 2001) and demonstrated that while perfusion increases that follow neuronal activation were not "perfectly" localized at the iso-orientation column level, the difference between active and neighboring inactive columns was large and permitted single condition mapping (Fig. 16.5). In other words, the brain does not really water the entire garden for the sake of a single flower; rather, it waters the thirsty flower, and it somewhat sprinkles the garden. To accomplish this, there must exist regulation of blood flow control in the submillimeter (~300–400 μm) scale. This is a fundamentally important result for brain physiology and functional mapping because it demonstrated for the first time that CBF changes are regulated even at the level of capillaries in addition to upstream control at the level of arterioles and arteries that was known to exist. This conclusion was initially considered to be in conflict with results from optical imaging techniques that report on CBV, which is assumed to be correlated with CBF in space and magnitude. However, there is an important difference between optical imaging and MR perfusion imaging data. Namely, the MR methods can be set to eliminate in the image acquisition phase the confounding problem coming from large vessel effects. When the origin of the functional mapping signals is restricted to the spatially accurate capillary/tissue level by the physics of the MR acquisition method, columnar activation is detected. However, the optical methods do not have this selective "tuning" capability and report on *all* CBV changes; thus, they can in fact be dominated by large vessel effects. However, recent optical imaging studies conducted subsequent to the perfusion mapping of iso-orientation columns by perfusion fMRI, have confirmed this fundamental difference and demonstrated that when large vessel effects are taken out, CBV based optical imaging techniques also yield columnar level mapping signals (e.g. Harrison et al. 2002; Sheth et al. 2003; Vanzetta et al. 2004). Subsequently, CBV based fMRI studies, conducted using the extravascular contrast agent MION, have also demonstrated single condition columnar resolution (Zhao et al. 2005; Harel et al. 2006b); these MION based fMRI studies, with their

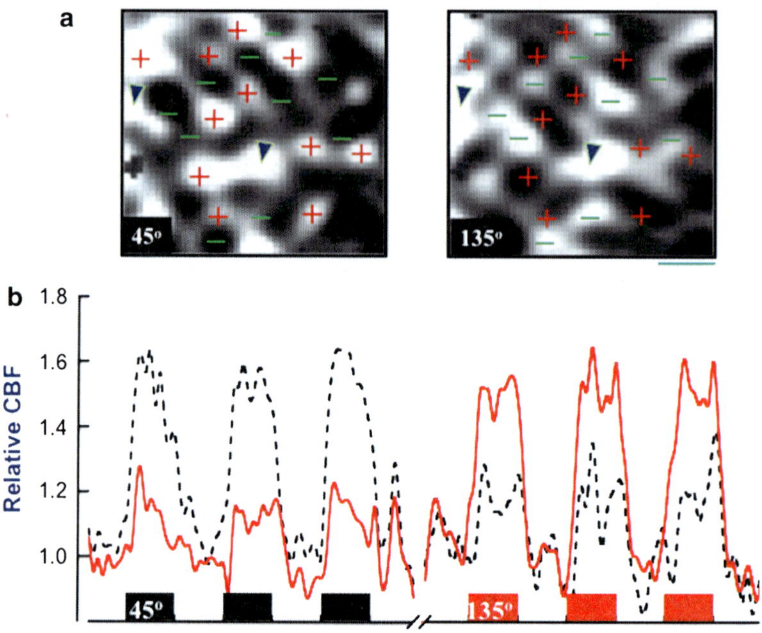

Fig. 16.5 Activation maps of two orthogonal iso-orientation domains in the cat visual cortex obtained separately with perfusion based functional imaging mapping. Panel A, shows perfusion based images obtained with two orthogonal orientations (45° and 135° gratings), demonstrating their complementarity. Each map was acquired as a single condition map where the grating of one orientation was either moving back and forth (activation condition) or static (control condition). Bottom trace (Panel B) shows the blood flow (perfusion) response in all the voxels identified as "activated" either by the 45° or 135° gratings for the two orthogonal stimulations. All voxels identified as "activated" for 45° orientation display a large perfusion increase during stimulation by this orientation gratings and show a smaller but detectable perfusion increase in response to the 135° gratings. The opposite of this is observed for the voxels identified as activated for 135° gratings. A marked perfusion increase (~55%) following 45° or 135° stimulus was observed in the regions tuned to these orientations, while the stimulation with the orthogonal orientation lead to a 3.3 ± 0.6 fold smaller perfusion increase in the same region (Adapted from reference Duong et al. 2001)

superior sensitivity, further revealed that laminar activity can be spatially resolved and that the larger intralaminar signal changes correspond to cortical layer 4 (Lu et al. 2004; Harel et al. 2006a and references therein). These observations further confirm that capillary level control of blood flow and volume in the brain exists and modulates activation-induced alterations in these physiological parameters at the columnar and laminar level.

16.3.2 Spin Echo (SE) Based BOLD Images

Currently, virtually all fMRI studies are carried out using GE, T_2* weighted, BOLD technique. However, with the availability of high magnetic fields, there is increasing justification for considering SE based functional mapping for improved accuracy. BOLD-based functional maps with suppressed large vessel contribution can be obtained with spin echo approach at high but not at low magnetic fields. SE fMRI responds to apparent T_2 (as opposed to T_2*) changes both in the extravascular space around microvasculature (Boxerman et al. 1993; Ogawa et al. 1993), and in blood itself (van Zijl et al. 1998; Ugurbil et al. 1999, 2000). The qualifier "apparent" is used because the fMRI effect is really a modulation of the echo amplitude in a sequence where the signal detected is normally weighted by T_2; but unlike a true T_2 the SE fMRI effect itself is TE dependent because it is caused by diffusion around magnetic field inhomogeneities. Extravascular signals associated with SE fMRI are associated largely with microvasculature and as such should provide improved spatial specificity. The intravascular blood effect, however, can be associated with large and small blood vessels and hence degrades spatial specificity of functional maps. However, the apparent T_2 of venous blood decreases rapidly with magnetic field magnitude (see (Duong et al. 2003; Uludag et al. 2009) and references there in) and is diminished from ~180 ms at 1.5 T (Barth and Moser 1997) to ~6 ms at 9.4 T (Lee et al. 1999), significantly smaller than brain tissue T_2 and the TE values that would be used at such field strengths. Consequently, at very high fields such as 7 T or higher, the functionally *non-specific* blood component becomes negligible in SE studies (Duong et al. 2003; Yacoub et al. 2003; Uludag et al. 2009). Residual large vessel effects are still expected because the extravascular BOLD effect is not exactly zero in SE fMRI for large vessels; rather it is small and significantly less than any effects associated with the vessels the size of capillaries (Uludag et al. 2009).

The *extravascular* SE BOLD effect, which is predominantly of microvascular in origin, is a small effect in the dynamic averaging regime. Because of this, it is reasonable to question whether it is at all detectable at any field strength or whether SE BOLD signals originate predominantly from other mechanisms such as the blood contribution. Certainly, the dominance of the blood effect at lower magnetic fields have been noted, leading to conclusions that the functional mapping signals in SE fMRI may in fact be virtually completely attributable to apparent T_2 changes in blood (van Zijl et al. 1998).

Whether extravascular BOLD effect exists at all was addressed directly and experimentally, rather than through modeling in a study that examined stimulus induced changes on weak but detectable ^{1}H resonances of metabolites, such as N-acetyl aspartate (Fig. 16.6), that are sequestered intracellularly (Zhu and Chen 2001). Any BOLD effect observed on such signals can only be *entirely* extravascular in origin. A BOLD effect on intracellular metabolites was clearly detectable at 4 and 7 T, leading to the clear demonstration that *extravascular spin-echo* BOLD effects exist in the human brain at high magnetic fields.

The blood contribution to SE BOLD fMRI was evaluated using Stejskal-Tanner gradient pair first introduced for diffusion measurements (Stejskal and Tanner 1965). This gradient pair, often referred to as "diffusion weighting" gradients, suppresses blood significantly more than tissue spins in the brain because of several mechanisms. The diffusion of blood water is faster in blood than in tissue, and the gradient pair dephases blood spins due to flow effects. The latter arises because flowing spins acquire a velocity dependent phase in the presence of these gradients.

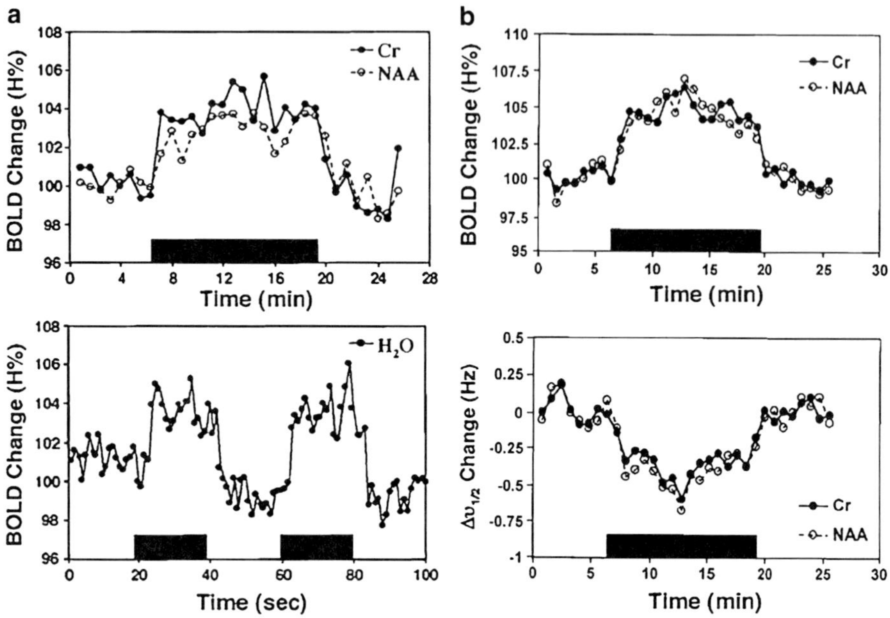

Fig. 16.6 Demonstration of extracellular BOLD effect using intracellular metabolites. (**a**) Time courses of fMRS BOLD changes (%Peak Height) in the NAA and Cr metabolites (*top*) and water (*bottom*) from the same subject. (**b**) Time courses of BOLD and linewidth changes (*top*: %Height; *bottom*: %linewidth) in the NAA and Cr metabolites, respectively, from another subject. The *dark bars* indicate the task periods of visual stimulation (Zhu and Chen 2001)

Flow rates are non-uniform within a blood vessel, leading to phase dispersion and signal cancellation within the vessel. Furthermore, especially for microvasculature, there exists many small vessels with different orientation relative to the direction of the Stejskal-Tanner gradient pair and thus acquire different phases during the gradient pair; since the blood signal detected from the voxel will be a sum of all of these, the net result can be signal cancellation due to phase dispersion again of flowing spins. This is the IVIM mechanism (Le Bihan et al. 1986).

The Stejskal-Tanner pulsed gradients can be used to distinguish between intra- and extra-vascular BOLD effects in functional images by suppressing the blood signals. Such experiments conducted at 1.5 T have concluded that most of the BOLD-based signal increase during elevated neuronal activity is eliminated by bipolar gradients, leading to the conclusion that most of the fMRI signal at 1.5 T arises from *intravascular* or blood related effects (Boxerman et al. 1995b; Song et al. 1996). At 3 T, ~50% of the SE functional signal changes has been attributed to blood using the same type of experiments (Mildner et al. 2001).

The effect of the Stejskal-Tanner gradients on functional imaging signals in the human brain at 4 and 7 T was examined and found to agree well with the modeling predictions (Duong et al. 2003). The dependence on the b- value for diffusion-weighted spin-echo BOLD data, averaged for all subjects, are illustrated in Fig. 16.7

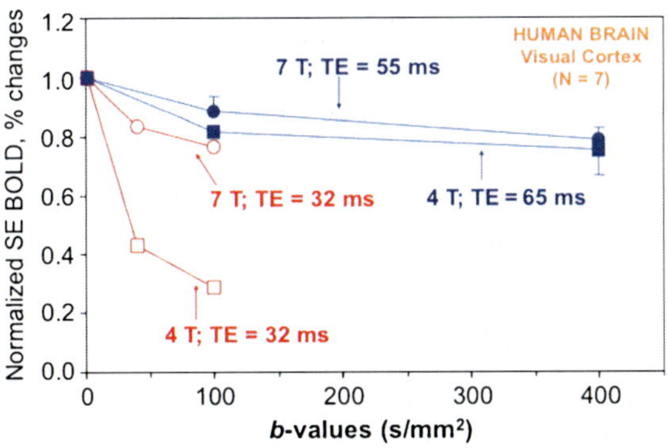

Fig. 16.7 Experimental evaluation of blood contribution to 7 T spin echo Functional Images: Figure shows normalized SE BOLD percent changes as a function of b value for short and long echo times at 4 and 7 T in a region of interest (ROI) defined in the b = 1 map. Percent changes were normalized to the BOLD change at b = 1 s/mm^2 for each subject and averaged for each field. *Closed rectangles* and *circles* indicate SE data with TE of 65 ms at 4 T (n = 7) and with TE of 55 ms at 7 T (n = 4), respectively. *Open rectangles* and *circles* indicate SE data with TE of 32 ms at 4 and 7 T, respectively, for two subjects at each field (six repeated measurements were made for each subject). *Error bars* are standard errors of the means. For long TE, the attenuation was not statistically different between 4 and 7 T (P > 0.05) (From Duong et al. 2003)

for an echo time of 32 ms for both at 4 and 7 T, and for echo times of 65 and 55 ms for 4 and 7 T, respectively. The 32 ms echo time is shorter than tissue T_2 at both fields. Blood contributions are expected to be echo time dependent and diminish with increasing echo time for TE values exceeding the blood T_2. At about 32 ms TE, blood contribution to 7 T is expected to be minimal since T_2 of blood is short (Yacoub et al. 2001). However, for 4 T, where the T_2 of blood is ~20 ms, the blood is still expected to contribute. The 32 ms TE data in Fig. 16.7 emphasize that at this echo time, signal changes associated with activation which are attributable to blood (and thus can be suppressed with diffusion gradients) are a small fraction of the total signal change at 7 T; in contrast, signal changes associated with activation at this echo time arise predominantly from blood at 4 T. At the longer echo times, however, as expected, percent changes at both fields were only slightly reduced by the diffusion-weighting gradients. There is a relatively small but persistent reduction in stimulation-induced signal intensity change in the presence of diffusion gradients that is field independent at long echo times; this is ascribed to the elimination of so called "in-flow" effects that arise because full relaxation was not allowed between consecutive images. At 9.4 T, the effect of the Stejskal-Tanner gradients are similar to 7 T results reported for the human brain above (Lee et al. 1999). In a spin-echo weighted fMRI study conducted in the rat brain (forepaw stimulation, symmetric spin-echo with one 180° pulse), the stimulation-induced signal changes were not altered at all going from very small to very high b values. Therefore, one can conclude that at very high magnetic fields such as 7 T and above, unwanted blood contributions to the BOLD effect are virtually eliminated at typical TE values that correspond to gray matter T_2.

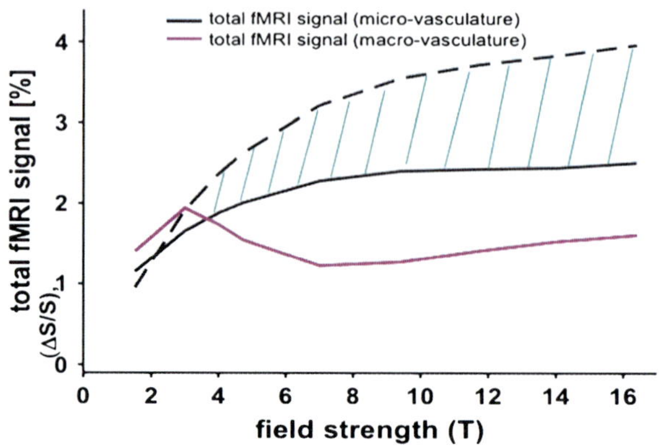

Fig. 16.8 Simulation results for SE fMRI for TE=T_2 as a function of field strength. Total fMRI signal (considering intravascular and extravascular effects) is shown in black for *microvasculature* assuming no stimulus induced CBV change (*dashed line*), and 16% microvascular CBV change. Most probable scenario is between these two extremes. Total fMRI signal for *macrovasculature* is shown in pin*k* for the extreme case of a vessel oriented perpendicularly to the static magnetic field which causes the maximum effect. Microvascular CBV was assumed to be 2.5% and was assumed to be composed of 20% arterioles (diameter =16 μ), 40% capillaries (diameter=5 μ), and 40% venules (diameter =16 μ) (Adapted from Uludag et al. 2009)

Even though both at 4 and 7 T, undesirable blood contributions are suppressed at the TE values that approximate tissue T_2, there is still an advantage for the 7 T over the 4 T for SE fMRI. This advantage is the substantially improved functional CNR. This is expected based on modeling studies (Uludag et al. 2009) (Fig. 16.8) and was also verified experimentally (Yacoub et al. 2003). This CNR enhancement is extremely important in view of the fact that CNR for SE fMRI is low even at 4 T. Note, however, that similarly long echo times are not useful for 1.5 T for suppressing the blood effect because blood T_2 is equal to or exceeds the tissue T_2. At that field magnitude, the blood contribution is substantial at TE values in the vicinity of tissue T_2 (~90 ms), and are increasing with TE until about 150 ms (Duong et al. 2003 and references therein).

The experiments with the "diffusion" weighted gradients confirm the expectation that blood effects are negligible at ultrahigh fields. However, experimentally, we are unable to judge at what level of the vascular tree we were able to eliminate the blood signals from the image using the diffusion gradients. Since the apparent T_2 of venous blood is so short, we expect that at all levels of the venous tree (venules to large veins), blood signals must be suppressed provided we use echo times that are about threefold larger than the apparent venous blood T_2. The intra-capillary blood, however, is unlikely to be fully suppressed. In the capillary, red blood cell density is non-uniform and oxygenation level varies from the arterial value to the venous value from one end of the capillary to the other end. However, this is *immaterial* as to whether capillary blood still contributes or not at high fields to functional maps; in either case, spatial specificity is dictated by the capillary distribution and it is this mapping accuracy one ultimately cares about.

Thus, as in perfusion imaging, SE BOLD techniques are expected to yield columnar level mapping and in fact they have been used to map both ocular dominance and orientation columns in the same subject at 7 T (Yacoub et al. 2008a), revealing the relationship in the organization of these two columnar organizations (Fig. 16.1). That they are capable of distinguishing among different layers has also been shown (Yacoub et al. 2003; Zhao et al. 2004; Harel et al. 2006a). Recent work measuring the point spread function (PSF) in the cat cortex have confirmed this by demonstrating that the full width at half maximum of the PSF is approximately the same for both CBF and SE methods and less than a millimeter (~0.7 mm) (Park et al. 2004 and see discussion further on).

16.3.3 Gradient Echo (T_2*) BOLD Functional Imaging

GE (T_2* weighted) BOLD technique, as previously mentioned, is virtually the only approached employed in fMRI applications. All other fMRI methods are primarily employed by investigators interested in mechanisms and methodology development, or high specificity and high-resolution functional imaging. GE BOLD fMRI, however, suffers from inaccuracies in functional mapping because of large vessel contributions. This is true at all magnetic fields because the *extravascular* BOLD effect associated with large vessels is always present and increase linearly with magnetic field, as previously mentioned. Often one is concerned with the large blood vessels on the pial surface of the brain. That these represent a major confound was realized and documented as early as in 1993 (Menon et al. 1993; Kim et al. 1994). This confound was emphasized again recently (Polimeni et al. 2010); with the availability of high resolution imaging at 7 T, a cortical layer based analysis method was employed in this study (Polimeni et al. 2010) to extract spatial spread of the BOLD response tangential to the cortical surface as a function of cortical depth; it was shown that avoiding surface laminae improved spatial localization by about 40% at a cost of 36% in z-statistic, implying that optimal spatial resolution in functional imaging of the cortex can be achieved using anatomically-informed spatial sampling in mid to deeper layers to avoid large pial vessels. This approach would be adequate for many studies; however, it would not be adequate for layer dependent studies. Not only the superficial layers are sacrificed, GE studies would also have a strong contribution from intracortical veins that run perpendicular to the cortical surface and drain the different layers. These are the blood vessels that can be seen in high resolution T_2* weighted images as dark lines traversing the gray matter perpendicular to the cortical surface. These are the veins that were studied in the original BOLD imaging work conducted in rats (Ogawa and Lee 1990). With the recent introduction of ultrahigh field imaging, including 9.4 T (Vaughan et al. 2006; Budde et al. 2011), that permits the acquisition of very high-resolution human brain images with high T_2* or phase contrast, these intracortical vessels have also been visualized in the human brain (Budde et al. 2011). These intracortical veins are ~50 to ~100 μ in diameter and would be the strongest source of GE fMRI mapping signal within the cortex after the removal of the large pial vessels.

The GE BOLD effect also comes from intravascular and extravascular sources. The intravascular effect originates from the same mechanism as in the SE BOLD, namely the T_2 and, in this case, T_2* changes accompanying the stimulus- or task-induced alterations in deoxyhemoglobin content. However, there is a second mechanism by which blood comes in to the GE BOLD effect (Lai et al. 1993; Haacke et al. 1994). This mechanism is operative when blood occupies a large fraction of the volume of the voxel. When deoxyhemoglobin is present in the blood, the blood water will dynamically average the gradients surrounding the red blood cells and will behave as if it encounters a uniform magnetic field. This will differ from the magnetic field experienced by the rest of the voxel. In the immediate vicinity of the blood vessel, the magnetic field will vary and approach a constant value in tissue distant from the blood vessel. For simplicity, we can neglect the gradients near the blood vessel and consider the voxel to be composed of two large bulk magnetic moments, one associated with blood and the rest with the extravascular volume. These magnetic moments will precess at slightly different frequencies; therefore, the signal from the voxel will decrease with time as the two moments lose phase coherence. In this scenario, the signal can even oscillate as the phase between the two magnetic moments increase and then decrease. Such oscillations in activation-induced fMRI signals have been reported (Diemling et al. 1997). This mechanism cannot be operative when a voxel only contains capillaries since the blood volume is ~2% (Pawlik et al. 1981). This blood-related effect appears to be the main source of fMRI signals at 1.5 T; this explains the reason why there exists very large stimulation induced fractional changes at low magnetic fields (Lai et al. 1993; Haacke et al. 1994). Note that, unlike the extravascular BOLD effect, these blood-related mechanisms do not require that activity related signal change ΔS is dependent on signal amplitude S (or equivalently, $\Delta S/S$ is independent of S).

As in SE fMRI, all blood related effects in GE fMRI diminish dramatically with increasing magnetic field as blood T_2 and, more relevantly for the GE images, blood T_2* decreases and blood contribution to the image decrease in a typical fMRI acquisition unless the TE is set intentionally short. However, unlike SE, the extravascular BOLD effect for large vessels persists and continues to be a source of inaccurate functional mapping signals in GE fMRI. Nevertheless, at higher fields such as 7 T, the microvascular contributions also become large enough so that its contributions is comparable to the large vessel effects (Yacoub et al. 2005). Figure 16.9 displays a histogram of the number of activated voxels, defined by a statistical threshold of $p<0.05$, versus percent signal change for one subject, for three different runs performed on different days at 7 T (Yacoub et al. 2005) for both SE and GE fMRI. The activation in the SE data has a narrow distribution in percent signal change induced by visual stimulation, with only few voxels showing increases larger than 10%. This is consistent with the concept that largely a single blood vessel type contributes to these signal changes. In contrast, the GE BOLD measurements are characterized by a broad distribution of stimulus-induced percent changes. In the GE data, there exists a large concentration of voxels at small percent changes just as in SE series; these would be associated with the microvasculature; however, there are also a significant number of voxels displaying large signal changes ranging up to ~25%.

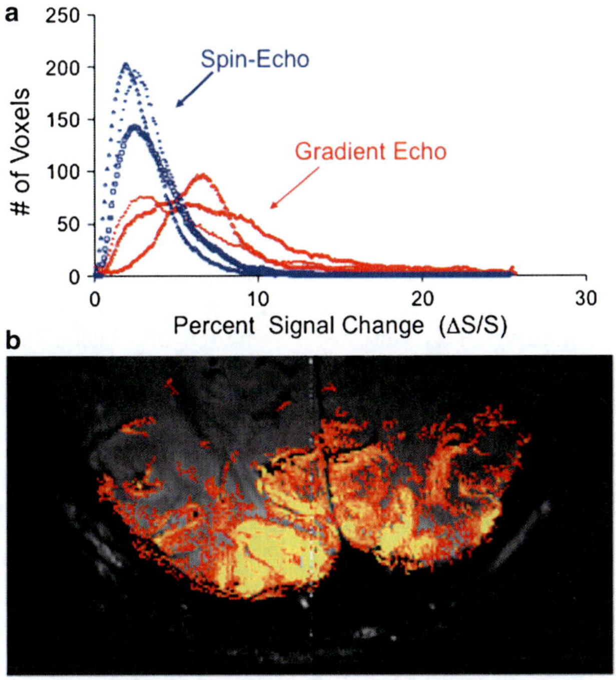

Fig. 16.9 Stimulus-invoked percent signal change at 7 T in the human visual cortex. (**a**) histograms of percent changes of repeated (different days) SE (in *blue*) and GE (in *red*) BOLD studies from the same subject, in the same anatomical location in. (**b**) Percent signal change versus basal image signal intensity (normalized). The data were extracted from functional images obtained with $0.5 \times 0.5 \times 3\ \mathrm{mm}^3$ spatial resolution at 7 T (*b*) (From Yacoub et al. 2005)

These can only come from draining veins as expected from all the modeling studies previously discussed. The presence of large vessel contribution even at 7 T in T_2^* BOLD fMRI images in the human brain was also demonstrated by comparing high resolution perfusion based images in the human brain with T_2^* BOLD fMRI (Pfeuffer et al. 2002a). In 7 T high resolution perfusion images, the activity was found to be confined to the gray matter ribbon; in contrast, the T_2^* BOLD fMRI images showed highest intensity changes not in the cortex but in the cerebral spinal fluid space (CSF) in the sulci (Pfeuffer et al. 2002a) where pial vessels reside. We should add that even with these inaccuracies, still a large number of questions about human cognition can be answered using GE fMRI even at 1.5 T.

The spatial inaccuracies introduced in functional mapping due the large vessel confound present in GE fMRI have been quantified experimentally in several ways. It was rigorously demonstrated that T_2* BOLD fMRI method fails to generate *single-condition* functional images of iso-orientation domains in the cat cortex (Duong et al. 2000b; Kim et al. 2000) at 4.7 T (Duong et al. 2000b). Although the image resolution was sufficiently good to detect columnar organizations associated with orientation preference in these studies, they were *not* observed.

Furthermore, the images were not complementary when orthogonal orientations were used, and the highest "activity" was associated with a large draining vein, the sagittal sinus. Thus, spatial specificity in the submillimeter domain of iso-orientation columns does not exist in T_2* BOLD fMRI. Simultaneous multiple site single unit recordings and fMRI studies on the same animal were performed to evaluate the accuracy of the GE fMRI. The GE fMRI response was found to be proportional to the spiking rate when data from all recording sites was averaged and compared with the BOLD response from the entire area where it was deemed statistically significant. However, this correlation started breaking down when smaller and smaller areas were considered for the averaging of the two distinct measurements, in particular when the area started decreasing below approximately 4×4 mm^2. These studies indicated that the limit of spatial specificity of GE fMRI was in the 2–3 mm range for *single-condition maps* (Toth et al. 2001; Ugurbil et al. 2003). At 1.5 T human brain studies, using a phase-encoding method (Engel et al. 1997), the full width at half maximum (FWHM) of the point spread function (PSF) for GE fMRI was estimated to be 3.5 mm. With a similar approach at 3 T, the PSF at FWHM was reported as 3.9 mm and 3.4 mm for GE and SE fMRI, respectively (Parkes et al. 2005); the latter is consistent with the conclusion that, at this field strength, there is still a lot of intravascular blood BOLD effect (Norris et al. 2002; Duong et al. 2003), which, as discussed before, is a source of inaccurate mapping signals. At 7 T, the PSF at FWHM was measured to be ~2 mm or less in regions devoid of the large veins visible in the image, (Shmuel et al. 2007).

16.4 "Noise" in fMRI and Field Dependence of Functional Contrast-to-Noise Ratio

It is well recognized that the standard deviation of image-to-image fluctuations in an fMRI series (σ_{fMRI}) is usually not dominated by the thermal noise (σ_{Therm}) (Hu and Kim 1994; Hyde et al. 1994; Kim et al. 1994; Hyde et al. 2001; Kruger and Glover 2001; Kruger et al. 2001) that determines the intrinsic SNR of individual MR images. Instead, physiological processes contribute to σ_{fMRI} so that it is equal to ($\sigma^2_{Phys} + \sigma^2_{Thern}$), ignoring the potential contributions from instrument instabilities. These physiological processes include brain motion due to cardiac pulsation, magnetic field perturbations induced by alterations in lung volume, and temporal instabilities in blood flow, blood volume, erythrocyte density, and oxygen consumption caused by vascular events and/or spontaneous activities of neurons. The last process is the source of resting state fMRI that yields information on functional connectivity (discussed next section). All of these contributions may be mediated through BOLD mechanisms but other mechanisms unrelated to BOLD are possible. Usually, respiration dependent contributions, and, occasionally, cardiac pulsation induced instabilities, are suppressed by post-processing (e.g. (Pfeuffer et al. 2002b; Van De Moortele et al. 2002) and references therein).

Recently, it has been postulated that the physiological component of the temporal fluctuations in an fMRI time series, σ_{Phys}, is linearly proportional to signal magnitude S (Hyde et al. 2001; Kruger and Glover 2001; Kruger et al. 2001) and this postulate was supported by experimental data obtained at 1.5 and 3 T for GE based BOLD fMRI. A linear dependence of σ_{Phys} on S is, of course, expected for physiological "noise" that is mediated through the BOLD mechanism; for example, as blood volume, blood flow and/or deoxyhemoglobin concentration in blood fluctuates, so will the basal BOLD effect. In fact, shortly after the introduction of fMRI, it was demonstrated that largest signal fluctuations in a GE fMRI time series occurs in and around large blood vessels, and that this can even be used as a mask to selectively eliminate large vessel regions in functional images (Kim et al. 1994). Recent studies, however, demonstrate that the physiological noise characteristics are somewhat different at 7 T compared to 1.5 or 3 T for GE fMRI (Yacoub et al. 2005). More interestingly, the noise in an SE and GE fMRI time series at 7 T also differ dramatically from each other (Yacoub et al. 2005); this is highly significant since one of the major benefits of the high fields is high specificity functional mapping using SE BOLD effect. The results, however, demonstrate that SE fMRI not only provides gains in accuracy of the functional imaging signals but also substantial gains with increasing magnetic field in contrast-to-noise ratio as well.

A linear dependence of σ_{Phys} on signal intensity S has important implications on the field dependence of CNR in functional imaging. Functional CNR (fCNR) is equal to the activation-induced signal change in the fMRI time series (ΔS) divided by σ_{fMRI}. The complete fMRI signal is quite complex to model since the final functional signals are a dependent on many different parameters (e.g. Uludag et al. 2009), some of which are not reflected in a simple change in T_2 or T_2*. But for those changes such as the *extravascular* or *intravascular* BOLD, functional signal change ΔS can be approximated as $(S_0 e^{-TE \cdot R}(\Delta R \cdot TE))$ for small $1/T_2$ or $1/T_2$* change,[6] and fCNR can thus be expressed as

$$fCNR = \Delta S / \sigma_{fMRI} = (S_0 e^{-TE \cdot R}(\Delta R \cdot TE)) / \sigma_{fMRI} \qquad (16.3)$$

where R is either R_2 or R_2* (i.e. $1/T_2$ or $1/T_2$*) for SE or GE FMRI, respectively. We drop the subscript 2 on the R but it is implicit that we are talking about either R_2 or R_2* depending on whether we are considering a SE or a GE experiment. TE is the echo time, S_o is the signal intensity at TE=0, and Δ represents the stimulus induced changes. Note that $S_0 e^{-TE \cdot R}$ is simply signal intensity S after the echo delay TE.

[6] Signal difference between the activated and the basal state is $S_{activated} - S_{basal} = S_o(e^{-TE \cdot R_{activated}} e^{-TE \cdot R_{basal}}) = S_o e^{-TE \cdot R_{activated}}(1 - e^{-TE \cdot \Delta R})$ and for $TE \cdot \Delta R \ll 1$, one has the approximation $e^{-TE \Delta R} \approx 1 - TE \cdot \Delta R$, where R is either R_2 or R_2*.

At the limit of very high-resolution functional imaging, as employed for example, in mapping columnar organizations, image SNR decreases and we operate in a domain where σ_{Therm} dominates over σ_{Phys}; in this limit

$$fCNR = \Delta S / \sigma_{Therm} \propto SNR_{i0} \cdot e^{-TE \cdot R}(\Delta R \cdot TE) \qquad (16.4)$$

In the above equation, SNR_{i0} is the intrinsic signal-to-noise ratio in the image corresponding to TE = 0. SNR_{i0} increases approximately linearly field magnitude for protons at fields of 1.5 T or above (Vaughan et al. 2001). Based on this information, the fCNR for relatively large veins versus microvasculature can be deduced as a function of TE for the *extravascular* BOLD effect. When thermal noise dominates the temporal noise in an fMRI series (i.e. when Eq. 4 is applicable), the *optimum* CNR is obtained for TE equal to T_2* or T_2 for GE or SE experiments, respectively,[7] in which case, $e^{-TE \cdot R} = e^{-1}$. Thus, fCNR will depend on SNR_{i0}, which we know to increase approximately linearly with field magnitude for 1.5 T or above (Vaughan et al. 2001), and how $TE \cdot \Delta R$ changes with field magnitude. Let us consider SE fMRI first: T_2 changes slowly with magnetic field (Uludag et al. 2009) up to field strengths of ~6 T or even ~7 T. Certainly the T_2 difference between 4 and 7 T is small enough so the same TE can be used for fMRI experiments at the two different fields. Large blood vessels do not contribute significantly to SE fMRI and the functional signals originate predominately from the extravascular BOLD effect associated with microvasculature. For such small vessels, ΔR will increase supralinearly up to field strengths of ~7 T or so and then linearly for higher fields (Uludag et al. 2009). Thus, keeping TE constant, fCNR will increase approximately quadratically or more with the static magnetic field B_0, with the increase with B_0 slowing down as the fields gets much higher than ~7 T.

The situation is different for GE fMRI since T_2* is approximately inversely proportional to B_0 *above* ~4 T, and both large and small vessels contribute to the fMRI signal. Thus, as the field increases, the TE must commensurately decrease. This combined with the fact that ΔR for large vessels increases linearly with field magnitude (Ogawa et al. 1993; Uludag et al. 2009) implies that $\Delta R \cdot TE$ will be field independent for large blood vessels above ~4 T. Because of the supralinear dependence of ΔR *on* B_0 for microvasculature up to ~6 or 7 T, however, $\Delta R \cdot TE$ will increase with B_0 up to ~6 or ~7 T for these blood vessels, and will become approximately field independent above that. These field dependencies will have to be multiplied by the SNR_{i0} that increases approximately linearly with the field magnitude. Thus, under conditions where thermal noise dominates σ_{fMRI}, as in high-resolution imaging, higher magnetic fields provide major advantages for fCNR even when GE fMRI is the method of choice. Not only fCNR increases in a GE experiment, the fCNR for

[7] Optimum TE is calculated simply by solving the equation $\partial CNR / \partial (TE) = 0$ where CNR is given by Eq. 4.

microvasculature increases much more rapidly compared to the undesirable large vessel contributions.

If we are in the limit where σ_{Phys} is much greater than σ_{Therm} (i.e. $\sigma_{fMRI} \sim \sigma_{Phys}$) then predictions of field dependence require that we define σ_{Phys} better: For BOLD-based noise one can write

$$\sigma_{Phys(BOLD)} = c \cdot S \cdot \sigma_{\Delta R} \cdot TE \tag{16.5}$$

where c is a constant and S is the signal intensity, $\sigma_{\Delta R}$ is the standard deviation accounting for the temporal variations in ΔR (Kruger and Glover 2001). For other non-BOLD sources, such as respiration and cardiac pulsation,

$$\sigma_{Phys(non-BOLD)} = c' \cdot \overline{S} \tag{16.6}$$

where $\overline{S}$ is the time averaged signal intensity in the time series. The resultant physiological noise is given by $\sigma_{Phys} = (\sigma^2_{Phys(BOLD)} + \sigma^2_{Phys(non-BOLD)})^{1/2}$ (Kruger and Glover 2001). The CNR is calculated from Eq. 4 with the substitution of σ_{fMRI} with σ_{Phys} and remembering that $S = S_0 e^{-TE \cdot \Delta R}$. For simplicity, the predictions for cases where physiological noise is only BOLD mediated or only non-BOLD mediated can be considered.

If the fMRI noise is exclusively BOLD mediated, in the limit $\sigma_{fMRI} \sim \sigma_{Phys}$, fCNR is just a constant, *independent* of field magnitude, ΔR, and TE for either the SE or GE experiment. No need to buy high field magnets for functional images just to gain fCNR in this case, although one can still argue a utility for high fields based on the accuracy of functional imaging signals. On the other hand, if the noise is purely of non-BOLD variety given by Eq. 6, then fCNR will be proportional to $\Delta R \cdot TE$, predicting an increasing fCNR at longer TE values[8] and for the same TE, increases for fCNR both for small and large blood vessel mediated signals but more so for the former in a GE study.

Thus, when σ_{Phys} dominates, it is possible to be in a scenario where higher magnetic fields provide no gains in fCNR in the context of the noise models considered so far. However, studies conducted at 7 T demonstrated that fMRI "noise" in GE and SE data are fundamentally different at this field strength, and that and unlike GE data, σ_{Phys} at 7 T for SE fMRI was virtually independent of $\overline{S}$ (Yacoub et al. 2005) at any given resolution (i.e. voxel volume) while functionally induced signal changes, ΔS, were proportional to $\overline{S}$. The ratio $\sigma_{Phys}/\sigma_{Term}$ was independent of voxel volume. A more detailed discussion of this behavior can be found in (Yacoub et al. 2005). Note that independence of σ_{Phys} from $\overline{S}$ rules out the explicit formulation of BOLD and non-BOLD mechanisms given by Eqs. 5 and 6 for SE fMRI at 7 T.

[8]One cannot conclude that TE can be increased indefinitely since at some TE, the condition $\sigma_{fMRI} \sim \sigma_{Phys}$ that leads to this prediction will fail, and thermal noise will dominate. But as long as $\sigma_{fMRI} \sim \sigma_{Phys}$ remains valid, longer TE's will yield higher CNR if the source of σ_{Phys} is non-BOLD type given by Eq. 6.

16.5 Resting State fMRI

As discussed in the previous section, shortly after the introduction of fMRI, it was realized that the temporal signal fluctuations in an fMRI time series is not thermal noise per se (except at very high resolution mapping) but arise from physiological processes (e.g. Hu and Kim 1994; Biswal et al. 1995, 1996b; Hu et al. 1995; Mitra et al. 1997). Most of the power in these temporal fluctuations were ascribed to respiration and cardiac cycle effects; however, Biswal et al (1995) provided evidence in 1995 that these fluctuations also contain components that showed temporal correlation among functionally related regions of the brain, suggesting that there exists correlated activity of neuronal origin among regions that have a functional relationship. After several years of dormancy, this early observation led to the recently exploding field of "resting state" fMRI (R-fMRI), garnering interest as a relatively new and powerful method for evaluating regional interactions that occur when a subject is not performing an explicit task (e.g. Lowe et al. 1998; Leopold et al. 2003; Beckmann et al. 2005; Fox et al. 2005, 2006; Salvador et al. 2005; Achard et al. 2006; Damoiseaux et al. 2006; Nir et al. 2006; Dosenbach et al. 2007; Fair et al. 2007a, b; Margulies et al. 2007; Biswal et al. 2010a).

Resting state does not refer to the brain being at rest, which is never the case; rather it simply refers to the subject being at "rest" in the scanner, not performing a directed task or exposed to an external stimulus as in normal fMRI. An fMRI time series collected in such a state contains low-frequency (<0.1 Hz) fluctuations that often display strong correlations among distant gray matter regions (Biswal et al. 1995; Lowe et al. 1998; Damoiseaux et al. 2006; Nir et al. 2006). These low frequency BOLD fluctuations are thought to reflect fluctuations in spontaneous neural activity (Biswal et al. 1995; Leopold et al. 2003; Nir et al. 2006), though the exact mechanisms giving rise to the neural fluctuations remain unclear. R-fMRI correlations were shown to often overlap with known anatomical pathways in the macaque monkey, but they have also been shown sometimes to involve regions that are not directly connected (Vincent et al. 2007). The spatial patterns of R-fMRI correlations were shown to be stable across different states under which the 'resting' state data are collected, such as eyes-open, eyes-closed, and fixation (Fox et al. 2005; Margulies et al. 2007), and across individuals and sessions (Damoiseaux et al. 2006). Regions that co-activate with a seed region in different tasks tend to be positively correlated with the seed region at rest.

While the notion of resting brain oscillations has been known for a long time in neurophysiology (e.g. alpha oscillation emerging in subjects resting with eyes closed were discovered in 1930 by Hans Berger), fMRI has enabled the visualization of networks of correlated fluctuations referred to as resting state networks (RSNs); these networks and, in particular, the default mode network (DMN) (Greicius et al. 2003), are particularly interesting due to their ease of detection and the possibility of interrogating a multitude of functional circuits simultaneously (Biswal et al. 2010a, b). Together with the analysis of the differences in resting state spatial patterns that have proven to be related with age and gender (Esposito et al. 2008; Biswal et al. 2010a, b), the study of the DMN in particular has opened the possibility of using "resting state" paradigms in clinical applications (Greicius et al. 2004; Buckner et al. 2005; Garrity et al. 2007; Sperling et al. 2009; Fox and Greicius 2010).

Recent improvements in the study of R-fMRI involve the use of pulse sequences (Moeller et al. 2008, 2010; Feinberg et al. 2010) targeting *subsecond whole brain* imaging without sacrificing spatial resolution (discussed in section on *Data Acquisition in fMRI*). Virtually all fMRI and diffusion imaging of fiber connections in the human brain rely on EPI which generates a single 2D image in a fraction of a second; however, it requires 2–3 s to acquire multi-slice whole brain coverage for fMRI and even longer for diffusion imaging. Whole brain, 3 T, R-fMRI data obtained with the M-EPI (for **M**ultiplexed-EPI) (Feinberg et al. 2010) sequence in 400 and 800 ms compared to a more conventional 2.5 s TR were used to examine resting state networks. This higher sampling rate was shown to increase the functional sensitivity to RSNs identified through independent component analysis (ICA) by 60% and generally enriched the functional information (Feinberg et al. 2010) (Fig. 16.10).

Until recently, the study of RSNs with R-fMRI has been restricted to 3 T or lower magnetic fields. At such field strengths, R-fMRI studies use voxel volumes of approximately 27 mm^3 or larger (i.e. 3 mm isotropic voxels or larger,) to achieve sufficient minimum SNR and fCNR. Furthermore, typically these studies employ spatial smoothing with large Gaussian kernels (~4–6 mm FWHM) to reduce noise, and often also resort to group averaging, which further amplifies spatial blurring by requiring additional smoothing (~6–8 mm FWHM) of the functional data to compensate for the variability of anatomical structures across subjects. Consequently, at 3 T field, the "partial volume" effects due to both the large effective voxel volume used as well as the non-specific spread of the activation (Yacoub et al. 2003), constrain the spatial accuracy of the resulting fMRI maps. The higher functional specificity, and functional CNR inherent at ultrahigh fields such as 7 T can be exploited to alleviate this limitation. However, this was not done until recently, largely, because of the difficulty of obtaining rapid whole brain coverage at 7 T. Large volume coverage at high resolution confronts technical and methodological challenges at ultrahigh fields. Single shot EPI (Mansfield 1977), the most frequently used method for fMRI, requires shorter acquisition times for each slice (i.e. a shorter echo train length) in order to obtain images that are not excessively contaminated by shifting and/or blurring effects due to decreasing T2 and T2* at 7 T compared to 3 T. The difficulty of this requirement is amplified with increasing resolution as the echo train length is elongated commensurately with resolution. Image segmentation has been used in EPI to reduce the echo train length (McKinnon 1993; Feinberg and Oshio 1994; Wielopolski et al. 1995a); however, this significantly increases the sensitivity to physiological noise and motion (Moeller et al. 2006) and increases substantially the volume acquisition time (TR), the latter leading to prolonged times for the collecting sufficient data in the fMRI time series. With the advent of parallel imaging methods using multi-channel receive coils, as discussed in the section on *Data Acquisition in fMRI*, many of the aforementioned technical limitations at high fields have been significantly alleviated. Despite this, to date only a few whole-brain

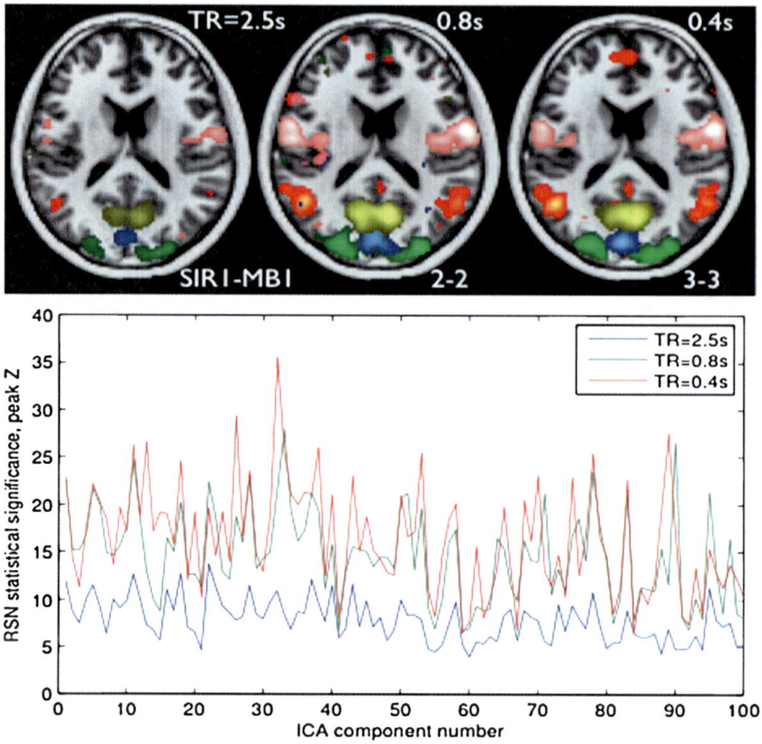

Fig. 16.10 Example RSNs (resting state networks) shown for one subject in different colors from three different 10-min R-fMRI acquisitions with TR of 2.5, 0.8, 0.4 s; whole brain acquisition was performed with the M-EPI technique using 60 slices, 3 mm isotropic resolution at 3 T. The color overlays are z-statistic images, thresholded at z = 4 in all cases. In this central axial slice (Z = +18 mm in MNI152 standard space) the RSNs displayed include visual areas, the default mode network and a sensori-motor network. Lower Plots: plots of peak z-statistics for the 100 components extracted by a 3-session-combined independent component analysis, for one subject using 0.4 and 0.8 s slice accelerated data and unaccelerated acquisitions (Adapted from (Feinberg et al. 2010). Pulse sequences and data acquisition from University of Minnesota, Center for Magnetic Resonance Research, and RSN analysis by Steven Smith and Karla Miller from Oxford University)

applications have been presented at ultrahigh fields and have mostly employed large slices, inter-slice gaps or have sacrificed temporal resolution (Bianciardi et al. 2009a, b, c; Poser and Norris 2009; van der Zwaag et al. 2009a, b; Poser et al. 2010), thus not fully exploiting the full benefit of ultrahigh filed fMRI.

Only recently ultrahigh fields efforts have been focused on RSNs. Figure 16.11 illustrates an example from this work, showing one slice for five RSNs, and multiple slices for one RSN from a whole brain study conducted with 1.5 mm isotropic resolution at 7 T in a single subject. Note the very fine structured depiction of the RSNs largely restricted to the cortical gray matter areas and sulci instead of the large blobs that typically characterize the 3 T RSNs as shown in Fig. 16.10.

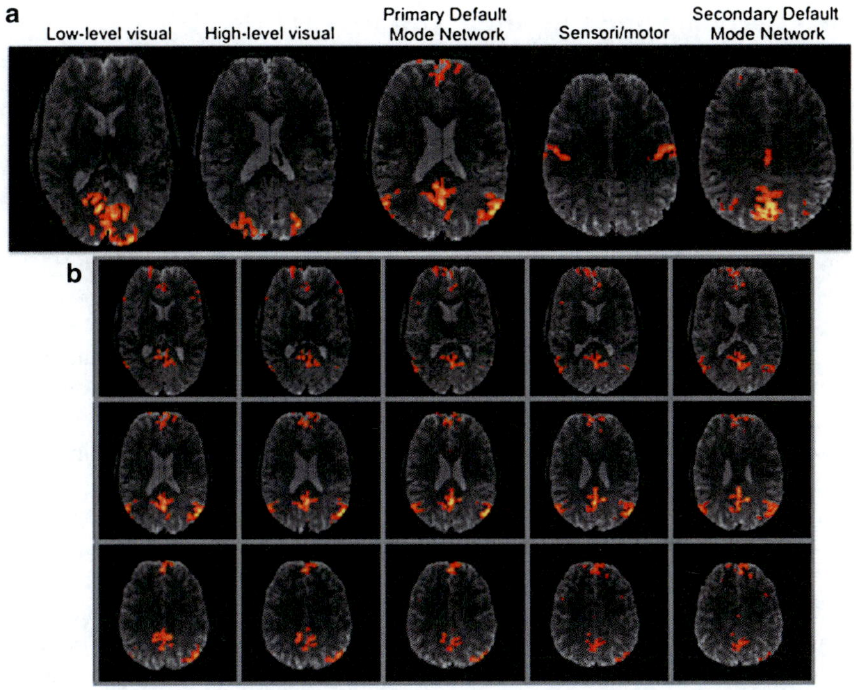

Fig. 16.11 Single subject, whole brain, 1.5 mm isotropic resolution 7 T R-fMRI. RSNs extracted from a 50 component ICA analysis are shown in color superimposed on the native EPI images from the R-fMRI time series, (**a**) Representative single slices from five different RSNs; (**b**) 15 slices from the 90 slice whole brain data for the primary default-mode RSN (Data generated at CMRR, University of Minnesota by Yacoub et al. ICA analysis by Steve Smith, Oxford University)

A more detailed examination of RSNs conducted at 7 T with spatial resolutions ranging from 1 to 3 mm isotropic covering the whole brain without interslice gaps, was recently reported (De Martino et al. 2011). So far, these 7 T studies were not performed with the previously mentioned slice-accelerated sequences (Moeller et al. 2008, 2010; Feinberg et al. 2010) that improve the statistical significance for detection of the RSNs. Rather, conventional slice–selective, single-shot EPI with parallel imaging along the phase encoding dimension was employed. Additional gains will definitely be realized with slice-accelerated sequences in the future. Nevertheless, the potential of ultrahigh fields in R-fMRI is clear. The results showed that the increased fCNR at 7 T permitted the detection of the distribution of spontaneous fluctuations within functionally connected networks in single subjects without the need for spatial smoothing at voxel volumes ranging from 1 to 8 mm^3 (isotropic voxel dimension), significantly smaller than the voxel size of 27 mm^3 commonly used at 3 T.

16.6 Decoding and Encoding with BOLD fMRI

Although a detailed review is outside the scope of this chapter, two recent developments with fMRI are worth mentioning briefly with some references given for the reader to pursue further should they decide to do so. These developments involve the use of multivariate machine learning algorithms to "decode" brain activity patterns and a strategy that can be termed as encoding (Naselaris et al. 2011). Both of these methods offer new possibilities, especially when coupled with the information rich, high resolution, and high specificity functional mapping available at ultrahigh magnetic fields.

Multivariate machine learning algorithms applied to GE fMRI data obtained at 3 T were shown to "decode" visual stimuli (Haxby et al. 2001; Haynes and Rees 2005; Kamitani and Tong 2005). These algorithms were able to reveal the presence of information that is known to be functionally segregated in cortical columns, e.g. ocular dominance, orientation and direction of motion. This was considered a particularly surprising result since the data were obtained with large a voxel size, $3\times3\times3$ mm^3 while these columns have dimensions of ~1 mm; in addition the data were acquired at 3 T with GE fMRI where functional specificity is not expected to be sufficient to image at the columnar level. The "functional images" generated in this process have no resemblance to the actual columnar architecture. Nevertheless, voxels were found that demonstrated a preferential activation corresponding to a unique column, such as an orientation specific column.

The mechanism by which such low resolution imaging decodes information represented at a much finer scale relative to the voxel size has been a source of debate and experimentation. Biased sampling of cortical columns by the large voxels has been hypothesized (Kamitani and Tong 2005, 2006; Haynes and Rees 2006; Kriegeskorte et al. 2006). Biased sampling occurs due to the specific position that a large voxel takes within a fine columnar organization with local variations, even if the overall preferences represented by the columns are distributed equally across the investigated cortical region. Alternatively, draining macroscopic blood vessels can have bias with respect cortical columns and thus could display selective responses (Kamitani and Tong 2005, 2006; Kriegeskorte et al. 2006; Chaimow et al. 2011; Shmuel et al. 2010). The presence of bias in draining vessels with respect to ocular dominance columns was directly demonstrated in high-resolution columnar level mapping at 7 T (Yacoub et al. 2007). Analysis of these data further with "decoding" algorithms revealed that discriminative power was conveyed by both macroscopic blood vessels and coarse-scale organization of the columnar organization, in other words irregularities in the distribution of different columns, in the gray matter areas (Shmuel et al. 2010). A detailed modeling study was performed to examine the latter considering the pattern of cortical columns, the hemodynamic point spread function at 3 T, the voxel size employed in the 3 T decoding studies, and noise (Chaimow et al. 2011). These model studies demonstrated that sub-voxel supra-Nyquist spatial frequencies, including frequencies near the main frequency of the ocular dominance organization (spatial frequency of 0.5 cycles per mm) cannot contribute to the differential contrast in GE fMRI and that the differential functional contrast of local origin is dominated by low-amplitude contributions from low frequencies, associated with irregularities of the cortical pattern (Chaimow et al. 2011). However, decoding due to such irregularities predicted decoding performances much lower than those that have been experimentally observed at 3 T, suggesting the presence of additional mechanisms, such as the presence of biased draining veins implicated in (Shmuel et al. 2010).

The decoding approaches described above rely predominantly on classification of image voxels where a pattern of activity across multiple voxels is used to construct a discrete class of stimulus response. Recent studies have moved beyond classification and demonstrated stimulus *reconstruction* to produce a literal picture of the image that was presented. Reconstruction of geometric stimuli composed of flickering checkerboard patterns was achieved by analyzing the responses of voxels in early visual areas using GE fMRI (Thirion et al. 2006; Miyawaki et al. 2008). The more complex task of reconstruction of natural images was undertaken with significant success using forward encoding models of each voxel (Kay et al. 2008; Naselaris et al. 2009, 2011). These results suggest that it may soon be possible to reconstruct a picture of a person's visual experience from measurements of functional imaging. In this quest, ultrahigh field fMRI is bound to play a critical role since it will provide many more informative voxels with finer, more specific, information content so as to enable the expansion of encoding models and provide substantial increases in discriminative power.

16.7 Conclusions

Our understanding of MR detectable functional signals in the brain has improved significantly over the years since the development of the methodology. At the same time, we have seen major advances and refinements in instrumentation, such as the introduction of ultrahigh field instruments of ever increasing sophistication and refinement, and data acquisition and image analysis methodologies that have significantly impacted data quality. These fantastic improvements today enable mapping of cortical columns, layer specific activation or brain reading. These advances are sure to continue unabated for years to come as new instruments like the 10.5 T system expected at the University of Minnesota and the 11.7 T systems planned for the Intramural Research division of NIH and NeuroSpin Laboratory in France will come on line. With these instruments and the ever improving data analysis methods, MR based functional imaging may be performed in the not so distant future with totally different approaches and provide substantially better information than the already available rich techniques at the cutting edge today.

Acknowledgements The work reviewed in this article conducted at the University of Minnesota, Center for Magnetic Resonance Research, was supported by NIH grants P41 RR007089 and the Keck Foundation.

References

Achard S, Salvador R, Whitcher B, Suckling J, Bullmore E (2006) A resilient, low-frequency, small-world human brain functional network with highly connected association cortical hubs. J Neurosci 26(1):63–72

Albright TD (1984) Direction and orientation selectivity of neurons in visual area MT of the macaque. J Neurophysiol 52(6):1106–1130

Albright TD, Desimone R, Gross CG (1984) Columnar organization of directionally selective cells in visual area MT of the macaque. J Neurophysiol 51(1):16–31

Bandettini PA, Wong EC, Hinks RS, Tikofsky RS, Hyde JS (1992) Time course EPI of human brain function during task activation. Magn Reson Med 25(2):390–397

Barbier EL, Silva AC, Kim HJ, Williams DS, Koretsky AP (1999) Perfusion analysis using dynamic arterial spin labeling (DASL). Magn Reson Med 41(2):299–308

Barbier EL, Silva AC, Kim SG, Koretsky AP (2001) Perfusion imaging using dynamic arterial spin labeling (DASL). Magn Reson Med 45(6):1021–1029

Barth M, Moser E (1997) Proton NMR relaxation times of human blood samples at 1.5 T and implications for functional MRI. Cell Mol Biol (Noisy-le-Grand) 43(5):783–791

Bartha R, Michaeli S, Merkle H, Adriany G, Andersen P, Chen W, Ugurbil K, Garwood M (2002) In vivo 1H2O T2+ measurement in the human occipital lobe at 4 T and 7 T by Carr-Purcell MRI: detection of microscopic susceptibility contrast. Magn Reson Med 47(4):742–750

Beckmann CF, DeLuca M, Devlin JT, Smith SM (2005) Investigations into resting-state connectivity using independent component analysis. Philos Trans R Soc London B Biol Sci 360(1457): 1001–1013

Belle V, Delon-Martin C, Massarell IR, Decety J, Le Bas J, Benabid A, Segebarth C (1995) Intracranial gradient-echo and spin-echo functional MR angiography in humans. Radiology 195(3):739–746

Belliveau JW, Kennedy DN, McKinstry RC, Buchbinder BR, Weisskoff RM, Cohen MS, Vevea JM, Brady TJ, Rosen BR (1991) Functional Mapping of the Human Visual Cortex by Magnetic Resonance Imaging. Science 254:716–719

Bianciardi M, Fukunaga M, van Gelderen P, Horovitz SG, de Zwart JA, Duyn JH (2009a) Modulation of spontaneous fMRI activity in human visual cortex by behavioral state. Neuroimage 45(1):160–168

Bianciardi M, Fukunaga M, van Gelderen P, Horovitz SG, de Zwart JA, Shmueli K, Duyn JH (2009b) Sources of functional magnetic resonance imaging signal fluctuations in the human brain at rest: a 7 T study. Magn Reson Imaging 27(8):1019–1029

Bianciardi M, van Gelderen P, Duyn JH, Fukunaga M, de Zwart JA (2009c) Making the most of fMRI at 7 T by suppressing spontaneous signal fluctuations. Neuroimage 44(2):448–454

Biswal B, Yetkin FZ, Haughton VM, Hyde JS (1995) Functional connectivity in the motor cortex of resting human brain using echo-planar MRI. Magn Reson Med 34(4):537–541

Biswal B, DeYoe EA, Hyde JS (1996a) Reduction of physiological flucuations in fMRI using digital filters. Magn Reson Med 35:117–123

Biswal B, DeYoe AE, Hyde JS (1996b) Reduction of physiological fluctuations in fMRI using digital filters. Magn Reson Med 35(1):107–113

Biswal BB, Mennes M, Zuo XN, Gohel S, Kelly C, Smith SM, Beckmann CF, Adelstein JS, Buckner RL, Colcombe S, Dogonowski AM, Ernst M, Fair D, Hampson M, Hoptman MJ, Hyde JS, Kiviniemi VJ, Kotter R, Li SJ, Lin CP, Lowe MJ, Mackay C, Madden DJ, Madsen KH, Margulies DS, Mayberg HS, McMahon K, Monk CS, Mostofsky SH, Nagel BJ, Pekar JJ, Peltier SJ, Petersen SE, Riedl V, Rombouts SA, Rypma B, Schlaggar BL, Schmidt S, Seidler RD, Siegle GJ, Sorg C, Teng GJ, Veijola J, Villringer A, Walter M, Wang L, Weng XC, Whitfield-Gabrieli S, Williamson P, Windischberger C, Zang YF, Zhang HY, Castellanos FX, Milham MP (2010a) Toward discovery science of human brain function. Proc Natl Acad Sci USA 107(10):4734–4739

Biswal BB, Mennes M, Zuo X-N, Gohel S, Kelly C, Smith SM, Beckmann CF, Adelstein JS, Buckner RL, Colcombe S, Dogonowski A-M, Ernst M, Fair D, Hampson M, Hoptman MJ, Hyde JS, Kiviniemi VJ, Kötter R, Li S-J, Lin C-P, Lowe MJ, Mackay C, Madden DJ, Madsen KH, Margulies DS, Mayberg HS, McMahon K, Monk CS, Mostofsky SH, Nagel BJ, Pekar JJ, Peltier SJ, Petersen SE, Riedl V, Rombouts SARB, Rypma B, Schlaggar BL, Schmidt S, Seidler RD, Siegle GJ, Sorg C, Teng G-J, Veijola J, Villringer A, Walter M, Wang L, Weng X-C, Whitfield-Gabrieli S, Williamson P, Windischberger C, Zang Y-F, Zhang H-Y, Castellanos FX, Milham MP (2010b) Toward discovery science of human brain function. Proc Natl Acad Sci USA 107(10):4734–4739

Boxerman JL, Weisskoff RM, Hoppel BE, Rosen BR (1993) MR contrast due to microscopically heterogeneous magnetic susceptibility: cylindrical geometry. In: 12th annual meeting of the society of magnetic resonance in medicine. International Society for Magnetic Resonance in Medicine, New York

Boxerman JL, Hamberg LM, Rosen BR, Weisskoff RM (1995a) MR contrast due to intravscular magnetic susceptiblity perturbations. Magn Reson Med 34:555–556

Boxerman JL, Bandettini PA, Kwong KK, Baker JR, Davis TL, Rosen BR, Weisskoff RM (1995b) The intravascular contribution to fMRI signal change: Monte Carlo modeling and diffusion-weighted studies in vivo. Magn Reson Med 34(1):4–10

Buckner RL, Snyder AZ, Shannon BJ, LaRossa G, Sachs R, Fotenos AF, Sheline YI, Klunk WE, Mathis CA, Morris JC, Mintun MA (2005) Molecular, structural, and functional characterization of Alzheimer's disease: evidence for a relationship between default activity, amyloid, and memory. J Neurosci 25(34):7709–7717

Budde J, Shajan G, Hoffmann J, Ugurbil K, Pohmann R (2011) Human imaging at 9.4 T using T(2)*-, phase-, and susceptibility-weighted contrast. Magn Reson Med 65(2):544–550

Chaimow D, Yacoub E, Ugurbil K, Shmuel A (2011) Modeling and analysis of mechanisms underlying fMRI-based decoding of information conveyed in cortical columns. Neuroimage 56(2): 627–642

Cheng K, Waggoner RA, Tanaka K (2001) Human ocular dominance columns as revealed by high-field functional magnetic resonance imaging. Neuron 32(2):359–374

Chien D, Edelman RR (1991) Ultrafast imaging using gradient echoes. Magn Reson Q 5(1): 31–56

Damoiseaux JS, Rombouts SA, Barkhof F, Scheltens P, Stam CJ, Smith SM, Beckmann CF (2006) Consistent resting-state networks across healthy subjects. Proc Natl Acad Sci USA 103(37):13848–13853

De Martino F, Esposito F, van de Moortele PF, Harel N, Formisano E, Goebel R, Ugurbil K, Yacoub E (2011) Whole brain high-resolution functional imaging at ultra high magnetic fields: an application to the analysis of resting state networks. Neuroimage 57(3):1031–1044

de Zwart JA, van Gelderen P, Golay X, Ikonomidou VN, Duyn JH (2006) Accelerated parallel imaging for functional imaging of the human brain. NMR Biomed 19(3):342–351

Detre JA, Wang J (2002) Technical aspects and utility of fMRI using BOLD and ASL. Clin Neurophysiol 113(5):621–634

Detre JA, Leigh JS, Williams DS, Koretsky AP (1992) Perfusion imaging. Magn Reson Med 23(1):37–45

Detre JA, Zhang W, Roberts DA, Silva AC, Williams DS, Grandis DJ, Koretsky AP, Leigh JS (1994) Tissue specific perfusion imaging using arterial spin labeling. NMR Biomed 7(1–2):75–82

DeYoe EA, Carman GJ, Bandettini P, Glickman S, Wieser J, Cox R, Miller D, Neitz J (1996) Mapping striate and extrastriate visual areas in human cerebral cortex. Proc Natl Acad Sci USA 93(6):2382–2386

Diemling M, Barth M, Moser E (1997) Quantification of signal changes in gradient recalled echo FMRI. Magn Reson Imaging 15(7):753–762

Dosenbach NU, Fair DA, Miezin FM, Cohen AL, Wenger KK, Dosenbach RA, Fox MD, Snyder AZ, Vincent JL, Raichle ME, Schlaggar BL, Petersen SE (2007) Distinct brain networks for adaptive and stable task control in humans. Proc Natl Acad Sci USA 104(26):11073–11078

Duong TQ, Silva AC, Lee SP, Kim SG (2000a) Functional MRI of calcium-dependent synaptic activity: cross correlation with CBF and BOLD measurements [In Process Citation]. Magn Reson Med 43(3):383–392

Duong QD, Kim D-S, Ugurbil K, S-G K (2000b) Spatio-Temporal Dynamics of BOLD fMRI Signals:Towards mapping Submillimeter Cortical columns Using the early Negative Response. Magn Reson Med 44(2):231–242

Duong TQ, Kim DS, Ugurbil K, Kim SG (2001) Localized cerebral blood flow response at submillimeter columnar resolution. Proc Natl Acad Sci USA 98(19):10904–10909

Duong TQ, Yacoub E, Adriany G, Hu X, Ugurbil K, Kim SG (2003) Microvascular BOLD contribution at 4 and 7 T in the human brain: Gradient-echo and spin-echo fMRI with suppression of blood effects. Magn Reson Med 49(6):1019–1027

Duvernoy HM, Delon S, Vannson JL (1981) Cortical blood vessels of the human brain. Brain Res Bull 7(5):519–579

Duyn JH, Moonen CTW, Van Yperen GH, De Boer RW, Luyten PR (1994) Inflow versus deoxyhemoglobin effects in BOLD functional MRI using gradient echoes at 1.5 T. NMR Biomed 7(1/2):83–88

Edelman RE, Siewer B, Darby DG, Thangaraj V, Nobre AC, Mesulam MM, Warach S (1994) Quantitative mapping of cerebral blood flow and functional localization with echo-planar MR imaging and signal targeting with alternating radio frequency. Radiology 192:513–520

Eichling JO, Raichle ME, Grubb RL, Ter-Pogossian MM (1974) Evidence of the limitations of water as a freely diffusable tracer in brain of the Rheusus monkey. Circ Res 35(3):358–364

Engel SA, Glover GH, Wandell BA (1997) Retinotopic organization in human visual cortex and the spatial precision of functional MRI. Cereb Cortex 7(2):181–192

Esposito F, Aragri A, Pesaresi I, Cirillo S, Tedeschi G, Marciano E, Goebel R, Di Salle F (2008) Independent component model of the default-mode brain function: combining individual-level and population-level analyses in resting-state fMRI. Magn Reson Imaging 26(7):905–913

Fair DA, Dosenbach NU, Church JA, Cohen AL, Brahmbhatt S, Miezin FM, Barch DM, Raichle ME, Petersen SE, Schlaggar BL (2007a) Development of distinct control networks through segregation and integration. Proc Natl Acad Sci USA 104(33):13507–13512

Fair DA, Schlaggar BL, Cohen AL, Miezin FM, Dosenbach NU, Wenger KK, Fox MD, Snyder AZ, Raichle ME, Petersen SE (2007b) A method for using blocked and event-related fMRI data to study "resting state" functional connectivity. Neuroimage 35(1):396–405

Feinberg DA, Oshio K (1994) Phase errors in multi-shot echo planar imaging. Magn Reson Med 32(4):535–539

Feinberg DA, Hale JD, Watts JC, Kaufman L, Mark A (1986) Halving MR imaging time by conjugation: demonstration at 3.5 kG. Radiology 161(2):527–531

Feinberg DA, Reese TG, Wedeen VJ (2002) Simultaneous echo refocusing in EPI. Magn Reson Med 48(1):1–5

Feinberg DA, Moeller S, Smith SM, Auerbach E, Ramanna S, Glasser MF, Miller KL, Ugurbil K, Yacoub E (2010) Multiplexed echo planar imaging for sub-second whole brain FMRI and fast diffusion imaging. PLoS One 5(12):e15710

Fox MD, Greicius M (2010) Clinical applications of resting state functional connectivity. Front Syst Neurosci 4:19

Fox MD, Snyder AZ, Vincent JL, Corbetta M, Van Essen DC, Raichle ME (2005) The human brain is intrinsically organized into dynamic, anticorrelated functional networks. Proc Natl Acad Sci USA 102(27):9673–9678

Fox MD, Snyder AZ, Zacks JM, Raichle ME (2006) Coherent spontaneous activity accounts for trial-to-trial variability in human evoked brain responses. Nat Neurosci 9(1):23–25

Frahm J, Merboldt K-D, Hanicke W, Kleinschmidt A, Boecker H (1994) Brain or vein-oxygenation or flow? On signal physiology in functional MRI of human brain activation. NMR Biomed 7(1/2):45–53

Fujita N (2001) Extravascular contribution of blood oxygenation level-dependent signal changes: a numerical analysis based on a vascular network model. Magn Reson Med 46(4):723–734

Fujita I, Tanaka K, Ito M, Cheng K (1992) Columns for visual features of objects in monkey inferotemporal cortex. Nature 360:343–346

Fukuda M, Moon CH, Wang P, Kim SG (2006) Mapping iso-orientation columns by contrast agent-enhanced functional magnetic resonance imaging: reproducibility, specificity, and evaluation by optical imaging of intrinsic signal. J Neurosci 26(46):11821–11832

Garrity AG, Pearlson GD, McKiernan K, Lloyd D, Kiehl KA, Calhoun VD (2007) Aberrant "default mode" functional connectivity in schizophrenia. Am J Psychiatry 164(3):450–457

Gibson AP, Hebden JC, Arridge SR (2005) Recent advances in diffuse optical imaging. Phys Med Biol 50(4):R1–R43

Glover GH, Li TQ, Ress D (2000) Image-based method for retrospective correction of physiological motion effects in fMRI: RETROICOR. Magn Reson Med 44(1):162–167

Golay X, Pruessmann KP, Weiger M, Crelier GR, Folkers PJ, Kollias SS, Boesiger P (2000) PRESTO-SENSE: an ultrafast whole-brain fMRI technique. Magn Reson Med 43(6):779–786

Goodyear BG, Menon RS (2001) Brief visual stimulation allows mapping of ocular dominance in visual cortex using fMRI. Hum Brain Mapp 14(4):210–217

Greicius MD, Krasnow B, Reiss AL, Menon V (2003) Functional connectivity in the resting brain: a network analysis of the default mode hypothesis. Proc Natl Acad Sci USA 100(1):253–258

Greicius MD, Srivastava G, Reiss AL, Menon V (2004) Default-mode network activity distinguishes Alzheimer's disease from healthy aging: evidence from functional MRI. Proc Natl Acad Sci USA 101(13):4637–4642

Haacke EM, Tkach JA (1990) Fast MR imaging: techniques and clinical applications. AJR Am J Roentgenol 155(5):951–964

Haacke EM, Wieloposki PA, Tkash JA (1991) A comprehensive technical review of short TR, fast, magnetic resonance imaging. Magn Reson Med 3(2):53–170

Haacke EM, Hopkins A, Lai S, Buckley P, Friedman L, Meltzer H, Hedera P, Friedland R, Klein S, Thompson L et al (1994) 2D and 3D high resolution gradient echo functional imaging of the brain: venous contributions to signal in motor cortex studies [published erratum appears in NMR Biomed 1994 Dec;7(8):374]. NMR Biomed 7(1–2):54–62

Haase A, Frahm J, Matthaei D, Hanicke W, Merboldt KD (1986) FLASH imaging. Rapid NMR imaging using low flip angle pulses. J Magn Reson 67:258–266

Harel N, Lee SP, Nagaoka T, Kim DS, Kim SG (2002) Origin of negative blood oxygenation level-dependent fMRI signals. J Cereb Blood Flow Metab 22(8):908–917

Harel N, Lin J, Moeller S, Ugurbil K, Yacoub E (2006a) Combined imaging-histological study of cortical laminar specificity of fMRI signals. Neuroimage 29(3):879–887

Harel N, Ugurbil K, Uludag K, Yacoub E (2006b) Frontiers of brain mapping using MRI. J Magn Reson Imaging 23(6):945–957

Harrison RV, Harel N, Panesar J, Mount RJ (2002) Blood capillary distribution correlates with hemodynamic-based functional imaging in cerebral cortex. Cereb Cortex 12(3):225–233

Haxby JV, Gobbini MI, Furey ML, Ishai A, Schouten JL, Pietrini P (2001) Distributed and overlapping representations of faces and objects in ventral temporal cortex. Science 293(5539): 2425–2430

Haynes JD, Rees G (2005) Predicting the orientation of invisible stimuli from activity in human primary visual cortex. Nat Neurosci 8(5):686–691

Haynes JD, Rees G (2006) Decoding mental states from brain activity in humans. Nat Rev Neurosci 7(7):523–534

Hoge RD, Atkinson J, Gill B, Crelier GR, Marrett S, Pike GB (1999) Linear coupling between cerebral blood flow and oxygen consumption in activated human cortex. Proc Natl Acad Sci USA 96(16):9403–9408

Howseman AM, Grootoonk SG, Porter D, Ramdeen J, Holmes AP, Turner R (1999) The effect of slice order and thickness on fMRI activation data using multislice EPI. Neuroimage 9(4): 363–376

Hu X, Kim S-G (1994) Reduction of physiological noise in functional MRI using navigator echo. Magn Reson Med 31:495–503

Hu X, Le TH, Parrish T, Erhard P (1995) Retrospective estimation and correction of physiological fluctuation in functional MRI. Magn Reson Med 34(2):201–212

Hyde JS, Biswal B, Song AW, Tan SG (1994) Physiological and Instrumental Fluctuations in fMRI data. In: 2nd annual Midwest functional MRI workshop, Madison

Hyde JS, Biswal BB, Jesmanowicz A (2001) High-resolution fMRI using multislice partial k-space GR-EPI with cubic voxels. Magn Reson Med 46(1):114–125

Kamitani Y, Tong F (2005) Decoding the visual and subjective contents of the human brain. Nat Neurosci 8(5):679–685

Kamitani Y, Tong F (2006) Decoding seen and attended motion directions from activity in the human visual cortex. Curr Biol 16(11):1096–1102

Kay KN, Naselaris T, Prenger RJ, Gallant JL (2008) Identifying natural images from human brain activity. Nature 452(7185):352–355

Kennan RP, Zhong J, Gore JC (1994) Intravascular susceptibility contrast mechanisms in tissue. Magn Reson Med 31:9–31

Kennan RP, Scanley BE, Innis RB, Gore JC (1998) Physiological basis for BOLD MR signal changes due to neuronal stimulation: separation of blood volume and magnetic susceptibility effects. Magn Reson Med 40(6):840–846

Kennedy C, Des Rosiers M, Sokoloff L, Reivich M, Jehle J (1975) The ocular dominance columns of the striate cortex as studied by the deoxyglucose method for measurement of local cerebral glucose utilization. Trans Am Neurol Assoc 100:74–77

Kim S-G (1995) Quantification of relative cerebral blood flow change by flow-sensitive alternating inversion recovery (FAIR) technique: application to functional mapping. Magn Reson Med 34:293–301

Kim SG, Ugurbil K (2003) High-resolution functional magnetic resonance imaging of the animal brain. Methods 30(1):28–41

Kim S-G, Hendrich K, Hu X, Merkle H, Ugurbil K (1994) Potential pitfalls of functional MRI using conventional gradient-recalled echo techniques. NMR Biomed 7(1/2):69–74

Kim DS, Duong TQ, Kim SG (2000) High-resolution mapping of iso-orientation columns by fMRI. Nat Neurosci 3(2):164–169

Kiselev VG, Posse S (1999) Analytical model of susceptibility-induced MR signal dephasing: effect of diffusion in a microvascular network. Magn Reson Med 41(3):499–509

Kriegeskorte N, Goebel R, Bandettini P (2006) Information-based functional brain mapping. Proc Natl Acad Sci USA 103(10):3863–3868

Kruger G, Glover GH (2001) Physiological noise in oxygenation-sensitive magnetic resonance imaging. Magn Reson Med 46(4):631–637

Kruger G, Kastrup A, Glover GH (2001) Neuroimaging at 1.5 T and 3.0 T: comparison of oxygenation-sensitive magnetic resonance imaging. Magn Reson Med 45(4):595–604

Kuo PH, Kanal E, Abu-Alfa AK, Cowper SE (2007) Gadolinium-based MR contrast agents and nephrogenic systemic fibrosis. Radiology 242(3):647–649

Kwong KK, Belliveau JW, Chesler DA, Goldberg IE, Weisskoff RM, Poncelet BP, Kennedy DN, Hoppel BE, Cohen MS, Turner R et al (1992) Dynamic magnetic resonance imaging of human brain activity during primary sensory stimulation. Proc Natl Acad Sci USA 89(12): 5675–5679

Lai S, Hopkins AL, Haacke EM, Li D, Wasserman BA, Buckley P, Friedman L, Meltzer H, Hedera P, Friedland R (1993) Identification of vascular structures as a major source of signal contrast in high resolution 2D and 3D functional activation imaging of the motor cortex at 1.5 T: preliminary results. Magn Reson Med 30(3):387–392

Le Bihan D, Breton E, Lallemand D, Grenier P, Cabanis E, Laval-Jeantet M (1986) MR imaging of intravoxel incoherent motions: application to diffusion and perfusion in neurologic disorders. Radiology 161(2):401–407

Lee S-P, Silva AC, Ugurbil K, Kim S-G (1999) Diffusion weighted spin echo fMRI at 9.4 T: microvascular/tissue contribution to BOLD signal changes. Magn Reson Med 42(5):919–928

Leopold DA, Murayama Y, Logothetis NK (2003) Very slow activity fluctuations in monkey visual cortex: implications for functional brain imaging. Cereb Cortex 13(4):422–433

Liang ZP, Madore B, Glover GH, Pelc NJ (2003) Fast algorithms for GS-model-based image reconstruction in data-sharing Fourier imaging. IEEE Trans Med Imaging 22(8):1026–1030

Lin YJ, Koretsky AP (1997) Manganese ion enhances T1-weighted MRI during brain activation: an approach to direct imaging of brain function. Magn Reson Med 38(3):378–388

Liu G, Sobering G, Duyn J, Moonen CT (1993) A functional MRI technique combining principles of echo-shifting with a train of observations (PRESTO). Magn Reson Med 30(6):764–768

Lowe MJ, Mock BJ, Sorenson JA (1998) Functional connectivity in single and multislice echoplanar imaging using resting-state fluctuations. Neuroimage 7(2):119–132

Lu H, Golay X, Pekar JJ, Van Zijl PC (2003) Functional magnetic resonance imaging based on changes in vascular space occupancy. Magn Reson Med 50(2):263–274

Lu H, Patel S, Luo F, Li SJ, Hillard CJ, Ward BD, Hyde JS (2004) Spatial correlations of laminar BOLD and CBV responses to rat whisker stimulation with neuronal activity localized by Fos expression. Magn Reson Med 52(5):1060–1068

Malonek D, Grinvald A (1996) Interactions between electrical activity and cortical microcirculation revealed by imaging spectroscopy: Implication for functional brain mapping. Science 272:551–554

Mandeville JB, Marota JJ (1999) Vascular filters of functional MRI: spatial localization using BOLD and CBV contrast. Magn Reson Med 42(3):591–598

Mandeville JB, Marota JJ, Kosofsky BE, Keltner JR, Weissleder R, Rosen BR, Weisskoff RM (1998) Dynamic functional imaging of relative cerebral blood volume during rat forepaw stimulation. Magn Reson Med 39(4):615–624

Mandeville JB, Jenkins BG, Kosofsky BE, Moskowitz MA, Rosen BR, Marota JJ (2001) Regional sensitivity and coupling of BOLD and CBV changes during stimulation of rat brain. Magn Reson Med 45(3):443–447

Mandeville JB, Jenkins BG, Chen YC, Choi JK, Kim YR, Belen D, Liu C, Kosofsky BE, Marota JJ (2004) Exogenous contrast agent improves sensitivity of gradient-echo functional magnetic resonance imaging at 9.4 T. Magn Reson Med 52(6):1272–1281

Mansfield P (1977) Multi-planar image formation using NMR spin echoes. J Phys C: Solid State Phys 10:L55–L58

Mansfield P, Harvey PR, Stehling MK (1994) Echo-volumar imaging. MAGMA 2:291–294

Marckmann P, Skov L, Rossen K, Dupont A, Damholt MB, Heaf JG, Thomsen HS (2006) Nephrogenic systemic fibrosis: suspected causative role of gadodiamide used for contrast-enhanced magnetic resonance imaging. J Am Soc Nephrol: JASN 17(9):2359–2362

Margulies DS, Kelly AM, Uddin LQ, Biswal BB, Castellanos FX, Milham MP (2007) Mapping the functional connectivity of anterior cingulate cortex. Neuroimage 37(2):579–588

McKinnon GC (1993) Ultrafast interleaved gradient-echo-planar imaging on a standard scanner. Magn Reson Med 30:609

Menon RS, Ogawa S, Tank DW, Ugurbil K (1993) 4 Tesla gradient recalled echo characteristics of photic stimulation- induced signal changes in the human primary visual cortex. Magn Reson Med 30(3):380–386

Menon RS, Ogawa S, Strupp JP, Ugurbil K (1997) Ocular dominance in human V1 demonstrated by functional magnetic resonance imaging. J Neurophysiol 77(5):2780–2787

Michaeli S, Garwood M, Zhu XH, DelaBarre L, Andersen P, Adriany G, Merkle H, Ugurbil K, Chen W (2002) Proton T(2) relaxation study of water, N-acetylaspartate, and creatine in human brain using Hahn and Carr-Purcell spin echoes at 4 T and 7 T. Magn Reson Med 47(4): 629–633

Mildner T, Norris DG, Schwarzbauer C, Wiggins CJ (2001) A qualitative test of the balloon model for BOLD-based MR signal changes at 3 T. Magn Reson Med 46(5):891–899

Mitra PP, Pesaran B (1999) Analysis of dynamic brain imaging data. Biophys J 76(2):691–708

Mitra PP, Thompson DJ, Ogawa S, Hu X, Ugurbil K (1995) Spatio-temporal patterns in fMRI data revealed by principal component analysis and subsequent low pass filtering. In: 3 rd scientific meeting of the Society of Magnetic Resonance, Nice

Mitra PP, Ogawa S, Hu X, Ugurbil K (1997) The nature of spatio-temportal changes in cerebral hemodynamics as manifested in functional magnetic resonance imaging. Magn Reson Med 37(4):511–518

Miyawaki Y, Uchida H, Yamashita O, Sato MA, Morito Y, Tanabe HC, Sadato N, Kamitani Y (2008) Visual image reconstruction from human brain activity using a combination of multi-scale local image decoders. Neuron 60(5):915–929

Moeller S, Van de Moortele PF, Goerke U, Adriany G, Ugurbil K (2006) Application of parallel imaging to fMRI at 7 Tesla utilizing a high 1D reduction factor. Magn Reson Med 56(1): 118–129

Moeller S, Auerbach E, van de Moortele P-F, Adriany G, Ugurbil K (2008) fMRI with 16 fold reduction using multibanded multislice sampling. Proc Int Soc Magn Reson Med 16:2366

Moeller S, Yacoub E, Olman CA, Auerbach E, Strupp J, Harel N, Ugurbil K (2010) Multiband multislice GE-EPI at 7 tesla, with 16-fold acceleration using partial parallel imaging with application to high spatial and temporal whole-brain fMRI. Magn Reson Med 63(5): 1144–1153

Naselaris T, Prenger RJ, Kay KN, Oliver M, Gallant JL (2009) Bayesian reconstruction of natural images from human brain activity. Neuron 63(6):902–915

Naselaris T, Kay KN, Nishimoto S, Gallant JL (2011) Encoding and decoding in fMRI. Neuroimage 56(2):400–410

Nir Y, Hasson U, Levy I, Yeshurun Y, Malach R (2006) Widespread functional connectivity and fMRI fluctuations in human visual cortex in the absence of visual stimulation. Neuroimage 30(4):1313–1324

Norris DG, Zysset S, Mildner T, Wiggins CJ (2002) An investigation of the value of spin-echo-based fMRI using a Stroop color-word matching task and EPI at 3 T. Neuroimage 15(3): 719–726

Ogawa S, Lee TM (1990) Magnetic resonance imaging of blood vessels at high fields: in vivo and in vitro measurments and image simulation. Magn Reson Med 16:9–18

Ogawa S, Lee T-M, Kay AR, Tank DW (1990a) Brain magnetic resonance imaging with contrast dependent on blood oxygenation. Proc Natl Acad Sci USA 87:9868–9872

Ogawa S, Lee T-M, Nayak AS, Glynn P (1990b) Oxygenation-sensitive contrast in magnetic resonance image of rodent brain at high magnetic fields. Magn Reson Med 14:68–78

Ogawa S, Tank DW, Menon R, Ellermann JM, Kim SG, Merkle H, Ugurbil K (1992) Intrinsic signal changes accompanying sensory stimulation: functional brain mapping with magnetic resonance imaging. Proc Natl Acad Sci USA 89(13):5951–5955

Ogawa S, Menon RS, Tank DW, Kim S-G, Merkle H, Ellermann JM, Ugurbil K (1993) Functional brain mapping by blood oxygenation level-dependent contrast magnetic resonance imaging. Biophys J 64:800–812

Ohliger MA, Grant AK, Sodickson DK (2003) Ultimate intrinsic signal-to-noise ratio for parallel MRI: electromagnetic field considerations. Magn Reson Med 50(5):1018–1030

Park JC, Ronen I, Kim D-S, Ugurbil K (2004) Spatial specificity of high resolution GE BOLD and CBF fMRI in the cat visual cortex. Proc Int Soc Magn Reson Med 12:1014

Parkes LM, Schwarzbach JV, Bouts AA, Deckers RH, Pullens P, Kerskens CM, Norris DG (2005) Quantifying the spatial resolution of the gradient echo and spin echo BOLD response at 3 Tesla. Magn Reson Med 54(6):1465–1472

Paulson OB, Hertz MM, Bolwig TG, Lassen NA (1977a) Filtration and diffusion of water across the blood-brain barrier in man. Microvasc Res 13(1):113–124

Paulson OB, Hertz MM, Bolwig TG, Lassen NA (1977b) Water filtration and diffusion across the blood brain barrier in man. Acta Neurol Scand Suppl 64:492–493

Pawlik G, Rackl A, Bing RJ (1981) Quantitative capillary topography and blood flow in the cerebral cortex of cats: an in vivo microscopic study. Brain Res 208(1):35–58

Pfeuffer J, Adriany G, Shmuel A, Yacoub E, Van De Moortele PF, Hu X, Ugurbil K (2002a) Perfusion-based high-resolution functional imaging in the human brain at 7 Tesla. Magn Reson Med 47(5):903–911

Pfeuffer J, Van De Moortele PF, Ugurbil K, Hu X, Glover GH (2002b) Correction of physiologically induced global off-resonance effects in dynamic echo-planar and spiral functional imaging. Magn Reson Med 47(2):344–353

Pohmann R, Budde J, Auerbach EJ, Adriany G, Ugurbil K (2010) Theoretical and experimental evaluation of continuous arterial spin labeling techniques. Magn Reson Med 63(2):438–446

Polimeni JR, Fischl B, Greve DN, Wald LL (2010) Laminar analysis of 7 T BOLD using an imposed spatial activation pattern in human V1. Neuroimage 52(4):1334–1346

Poser BA, Norris DG (2009) Investigating the benefits of multi-echo EPI for fMRI at 7 T. Neuroimage 45(4):1162–1172

Poser BA, Koopmans PJ, Witzel T, Wald LL, Barth M (2010) Three dimensional echo-planar imaging at 7 Tesla. Neuroimage 51(1):261–266

Raichle ME (2009) A brief history of human brain mapping. Trends Neurosci 32(2):118–126

Salvador R, Suckling J, Schwarzbauer C, Bullmore E (2005) Undirected graphs of frequency-dependent functional connectivity in whole brain networks. Philos Trans R Soc London B Biol Sci 360(1457):937–946

Segebarth C, Belle V, Delon C, Massarelli R, Decety J, Le Bas J-F, Decorpts M, Benabid AL (1994) Functional MRI of the human brain: predominance of signals from extracerebral veins. Neuroreport 5:813–816

Sereno MI, Dale AM, Reppas JB, Kwong KK, Belliveau JW, Brady TJ, Rosen BR, Tootell RBH (1995) Borders of Multiple Visual Areas in Humans Revealed by Functional Magnetic Resonance Imaging. Science 268:889–893

Sheth S, Nemoto M, Guiou M, Walker M, Pouratian N, Toga AW (2003) Evaluation of coupling between optical intrinsic signals and neuronal activity in rat somatosensory cortex. Neuroimage 19(3):884–894

Shmuel A, Yacoub E, Pfeuffer J, Van de Moortele PF, Adriany G, Hu X, Ugurbil K (2002) Sustained negative BOLD, blood flow and oxygen consumption response and its coupling to the positive response in the human brain. Neuron 36(6):1195–1210

Shmuel A, Yacoub E, Chaimow D, Logothetis NK, Ugurbil K (2007) Spatio-temporal point-spread function of fMRI signal in human gray matter at 7 Tesla. Neuroimage 35(2):539–552

Shmuel A, Chaimow D, Raddatz G, Ugurbil K, Yacoub E (2010) Mechanisms underlying decoding at 7 T: ocular dominance columns, broad structures, and macroscopic blood vessels in V1 convey information on the stimulated eye. Neuroimage 49(3):1957–1964

Song AW, Li T (2003) Improved spatial localization based on flow-moment-nulled and intra-voxel incoherent motion-weighted fMRI. NMR Biomed 16(3):137–143

Song AW, Wong EC, Tan SG, Hyde JS (1996) Diffusion weighted fMRI at 1.5 T. Magn Reson Med 35(2):155–158

Song AW, Harshbarger T, Li T, Kim KH, Ugurbil K, Mori S, Kim DS (2003) Functional activation using apparent diffusion coefficient-dependent contrast allows better spatial localization to the neuronal activity: evidence using diffusion tensor imaging and fiber tracking. Neuroimage 20(2):955–961

Sperling RA, Laviolette PS, O'Keefe K, O'Brien J, Rentz DM, Pihlajamaki M, Marshall G, Hyman BT, Selkoe DJ, Hedden T, Buckner RL, Becker JA, Johnson KA (2009) Amyloid deposition is associated with impaired default network function in older persons without dementia. Neuron 63(2):178–188

Springer CS, Xu Y (1991) Aspects of bulk magnetic susceptibility in in vivo MRI and MRS. In: Rink PA, Muller RN (eds) New developments in contrast agent research. European Magnetic Resonance Forum, Blonay, pp 13–25

Stejskal EO, Tanner JE (1965) Spin diffusion measurements: spin-echoes in the presence of a time dependent field gradient. J Chem Phys 42:288–292

Thirion B, Duchesnay E, Hubbard E, Dubois J, Poline JB, Lebihan D, Dehaene S (2006) Inverse retinotopy: inferring the visual content of images from brain activation patterns. Neuroimage 33(4):1104–1116

Thulborn KR, Waterton JC, Matthews PM, Radda GK (1982) Oxygenation dependence of the transverse relaxation time of water protons in whole blood at high field. Biochim Biophys Acta 714(2):265–270

Toth LJ, Ronen I, Olman C, Ugurbil K, Kim D-S (2001) Spatial correlation of BOLD activity with neuronal responses. In: Society of Neuroscience. Abstracts

Tsekos NV, Zhang F, Merkle H, Nagayama M, Iadecola C, Kim SG (1998) Quantitative measurements of cerebral blood flow in rats using the FAIR technique: correlation with previous iodo-antipyrine autoradiographic studies. Magn Reson Med 39(4):564–573

Ugurbil K, Hu X, Chen W, Zhu X-H, Kim S-G, Georgopoulos A (1999) Functional Mapping in the human brain using high magnetic fields. Philos Trans R Soc Lond B Biol Sci 354(1387): 1195–1213

Ugurbil K, Adriany G, Andersen P, Chen W, Gruetter R, Hu X, Merkle H, Kim DS, Kim SG, Strupp J, Zhu XH, Ogawa S (2000) Magnetic resonance studies of brain function and neurochemistry. Annu Rev Biomed Eng 2:633–660

Ugurbil K, Toth L, Kim DS (2003) How accurate is magnetic resonance imaging of brain function? Trends Neurosci 26(2):108–114

Uludag K, Muller-Bierl B, Ugurbil K (2009) An integrative model for neuronal activity-induced signal changes for gradient and spin echo functional imaging. Neuroimage 48(1):150–165

Van De Moortele PF, Pfeuffer J, Glover GH, Ugurbil K, Hu X (2002) Respiration-induced B0 fluctuations and their spatial distribution in the human brain at 7 Tesla. Magn Reson Med 47(5):888–895

van der Zwaag W, Francis S, Head K, Peters A, Gowland P, Morris P, Bowtell R (2009a) fMRI at 1.5, 3 and 7 T: characterising BOLD signal changes. Neuroimage 47(4):1425–1434

van der Zwaag W, Marques JP, Hergt M, Gruetter R (2009b) Investigation of high-resolution functional magnetic resonance imaging by means of surface and array radiofrequency coils at 7 T. Magn Reson Imaging 27(8):1011–1018

van Zijl PC, Eleff SM, Ulatowski JA, Oja JM, Ulug AM, Traystman RJ, Kauppinen RA (1998) Quantitative assessment of blood flow, blood volume and blood oxygenation effects in functional magnetic resonance imaging [see comments]. Nat Med 4(2):159–167

Vanzetta I, Slovin H, Omer DB, Grinvald A (2004) Columnar resolution of blood volume and oximetry functional maps in the behaving monkey; implications for FMRI. Neuron 42(5):843–854

Vaughan JT, Garwood M, Collins CM, Liu W, DelaBarre L, Adriany G, Andersen P, Merkle H, Goebel R, Smith MB, Ugurbil K (2001) 7 T vs. 4 T: RF power, homogeneity, and signal-to-noise comparison in head images. Magn Reson Med 46(1):24–30

Vaughan T, DelaBarre L, Snyder C, Tian J, Akgun C, Shrivastava D, Liu W, Olson C, Adriany G, Strupp J, Andersen P, Gopinath A, van de Moortele PF, Garwood M, Ugurbil K (2006) 9.4 T human MRI: preliminary results. Magn Reson Med 56(6):1274–1282

Vincent JL, Larson-Prior LJ, Zempel JM, Snyder AZ (2007) Moving GLM ballistocardiogram artifact reduction for EEG acquired simultaneously with fMRI. Clin Neurophysiol 118(5): 981–998

Wang G, Tanaka K, Tanifuji M (1996) Optical imaging of functional organization in the monkey inferotemporal cortex. Science 272:1665–1668

Wielopolski PA, Manning WJ, Edelman RR (1995a) Single breath-hold volumetric imaging of the heart using magnetization-prepared 3-dimensional segmented echo planar imaging. J Magn Reson Imaging: JMRI 5(4):403–409

Wielopolski PA, Manning WJ, Edelman RR (1995b) Single breath-hold volumetric imaging of the heart using magnetization-prepared 3-dimensional segmented echo planar imaging. J Magn Reson Imaging 5(4):403–409

Wiesinger F, Van de Moortele PF, Adriany G, De Zanche N, Ugurbil K, Pruessmann KP (2004) Parallel imaging performance as a function of field strength–an experimental investigation using electrodynamic scaling. Magn Reson Med 52(5):953–964

Wiesinger F, Van de Moortele PF, Adriany G, De Zanche N, Ugurbil K, Pruessmann KP (2006) Potential and feasibility of parallel MRI at high field. NMR Biomed 19(3):368–378

Wong EC, Buxton RB, Frank LR (1998) Quantitative imaging of perfusion using a single subtraction (QUIPSS and QUIPSS II). Magn Reson Med 39(5):702–708

Yablonskiy DA, Haacke EM (1994) Theory of NMR signal behavior in magnetically inhomogeneous tissues: the static dephasing regime. Magn Reson Med 32(6):749–763

Yacoub E, Shmuel A, Pfeuffer J, Van De Moortele PF, Adriany G, Andersen P, Vaughan JT, Merkle H, Ugurbil K, Hu X (2001) Imaging brain function in humans at 7 Tesla. Magn Reson Med 45(4):588–594

Yacoub E, Duong TQ, Van De Moortele PF, Lindquist M, Adriany G, Kim SG, Ugurbil K, Hu X (2003) Spin-echo fMRI in humans using high spatial resolutions and high magnetic fields. Magn Reson Med 49(4):655–664

Yacoub E, Van De Moortele PF, Shmuel A, Ugurbil K (2005) Signal and noise characteristics of Hahn SE and GE BOLD fMRI at 7 T in humans. Neuroimage 24(3):738–750

Yacoub E, Shmuel A, Logothetis N, Ugurbil K (2007) Robust detection of ocular dominance columns in humans using Hahn Spin Echo BOLD functional MRI at 7 Tesla. Neuroimage 37(4):1161–1177

Yacoub E, Harel N, Ugurbil K (2008a) High-field fMRI unveils orientation columns in humans. Proc Natl Acad Sci USA 105(30):10607–10612

Yacoub E, Uludag K, Ugurbil K, Harel N (2008b) Decreases in ADC observed in tissue areas during activation in the cat visual cortex at 9.4 T using high diffusion sensitization. Magn Reson Imaging 26(7):889–896

Yang Y, Glover GH, van Gelderen P, Patel AC, Mattay VS, Frank JA, Duyn JH (1998) A comparison of fast MR scan techniques for cerebral activation studies at 1.5 tesla [published erratum appears in Magn Reson Med 1998 Mar;39(3):following 505]. Magn Reson Med 39(1):61–67

Zhang W, Williams DS, Koretsky AP (1993) Measurement of rat brain perfusion by NMR using spin labeling of arterial water: in vivo determination of the degree of spin labeling. Magn Reson Med 29(3):416–421

Zhang N, Yacoub E, Zhu XH, Ugurbil K, Chen W (2009) Linearity of blood-oxygenation-level dependent signal at microvasculature. Neuroimage 48(2):313–318

Zhao F, Wang P, Kim SG (2004) Cortical depth-dependent gradient-echo and spin-echo BOLD fMRI at 9.4 T. Magn Reson Med 51(3):518–524

Zhao F, Wang P, Hendrich K, Kim SG (2005) Spatial specificity of cerebral blood volume-weighted fMRI responses at columnar resolution. Neuroimage 27(2):416–424

Zhao F, Wang P, Hendrich K, Ugurbil K, Kim SG (2006) Cortical layer-dependent BOLD and CBV responses measured by spin-echo and gradient-echo fMRI: insights into hemodynamic regulation. Neuroimage 30(4):1149–1160

Zhu XH, Chen W (2001) Observed BOLD effects on cerebral metabolite resonances in human visual cortex during visual stimulation: a functional (1)H MRS study at 4 T. Magn Reson Med 46(5):841–847

Chapter 17
FMRI Using Exogenous Agents and Cerebral Blood Volume Contrast

Joseph B. Mandeville

Abstract The most robust techniques for fMRI employ gradient-echo imaging to measure task-induced changes in MRI signal resulting from endogenous or exogenous paramagnetic contrast agents within the bloodstream. In animal models, blood magnetization can be adjusted to optimal levels using injected agent, and fMRI then reflects dynamic changes in cerebral blood volume (CBV). In the absence of injected agent, blood magnetization, and signal changes are smaller, and blood oxygen level dependent (BOLD) signal reports a more complex physiology that depends upon changes in blood volume, flow, and oxygen utilization. This chapter focuses primarily upon the physics and physiology of dynamic fMRI measurements of CBV using exogenous contrast agent, both as a useful tool for fMRI applications and as a method that contributes to our understanding of BOLD signal. Other recent techniques to measure CBV using endogenous mechanisms also are described briefly, as well as the fMRI method for assessing changes in cerebral blood flow by arterial spin labeling. Together, these techniques are expanding the portfolio of useful fMRI methods beyond BOLD signal and contributing to our understanding of the physiological mechanisms underlying the BOLD phenomenon.

Keywords Arterial spin labeling • ASL • BOLD • CBF • CBV • Cerebral blood flow • Cerebral blood volume • Cerebral metabolic rate of oxygen utilization • CMRO2 • CNR • Contrast to noise ratio • Delayed compliance • Feraheme • fMRI • Hemodynamic response function • Impulse response function • IRON • MION • MRI • Relaxation rate • Relaxation time • Signal to noise ratio • SNR • Undershoot • USPIO • VASO • VERVE

J.B. Mandeville, Ph.D. (✉)
Athinoula A. Martinos Center for Biomedical Imaging, Massachusetts General Hospital, Charlestown, MA 02155, USA
e-mail: jbm@nmr.mgh.harvard.edu

I.-Y. Choi and R. Gruetter (eds.), *Neural Metabolism In Vivo*, Advances in Neurobiology 4, DOI 10.1007/978-1-4614-1788-0_17,

17.1 Introduction

As a tool for assessing brain function, magnetic resonance imaging (MRI) plays multiple roles by providing relatively direct information about metabolic substrates (spectroscopy), or by providing information about neuro-metabolic status that, while indirect, can be obtained with excellent spatial and temporal resolution that is highly desirable for many applications in neuroscience and investigations of pathophysiology. The functional MRI (fMRI) techniques that are most commonly employed today are based upon the assumption that regional brain activity is coupled to the blood supply, as postulated more than a century ago (Roy and Sherrington 1890). Through flow-metabolism coupling, fMRI is thought to represent a summation of competing metabolic processes; as such, changes in brain activity can be localized, but additional information generally is required to interpret findings in terms of the underlying neurotransmission and neurochemistry.

Prior to neuro-imaging, inferences about brain organization and function were derived from lesion and postmortem observations in humans and invasive studies in animal models. Most modern non-invasive imaging methods can be traced back to invasive imaging techniques developed in animal models in the 1970s and earlier. Arterial spin labeling (ASL) by MRI is analogous to the positron emission tomography (PET) method using radiolabeled water (Ter-Pogossian et al. 1969), a method that in turn adapted the radiolabeled diffusible tracers employed by autoradiography (Kety 1960). Bolus measurements of cerebral blood flow (CBF) and volume (CBV) using MRI contrast agents follow the same method employed by x-ray computed tomography using iodinated contrast agent (Axel 1980), which in turn is based upon indicator dilution theory and earlier methods of sampling contrast material (Meier and Zierler 1954). Much of the physics and physiology of the BOLD technique can be related to earlier optical, PET, and MRI studies.

However, despite apparent similarities with other non-invasive imaging modalities, the particular physics associated with MRI modifies many important imaging characteristics relative to other methods. For instance, exogenous contrast agents provide large signal changes associated with CBV by strongly influencing signal in the extravascular space, whereas similarly sized changes in CBV have small effects on PET and CT images that are limited by the small blood volume fraction. Arterial spin labeling can be performed continuously, unlike the PET equivalent using a radioactive label, but quantification of the MRI measurement of CBF is complicated by the rapid decay of the magnetic label.

This chapter focuses on imaging techniques that measure characteristics of the blood supply. A particular emphasis is placed on the use of exogenous contrast agents that enable dynamic measurements of CBV, as this technique is particularly robust and has many connections to blood oxygen level depend (BOLD) signal in terms of both physics and physiology. While BOLD signal is the primary subject of another chapter, it is impossible to discuss CBV techniques without referencing these BOLD connections. Less emphasis is placed on relatively recent methods that employ endogenous contrast mechanisms to assess CBV, as these methods have low sensitivity and are not yet fully developed. However, CBV measurements using endogenous contrast mechanisms in human subjects possess the potential to inform our understanding of coupling relationships between CBV, CBF, and the cerebral metabolic rate of oxygen utilization ($CMRO_2$). Finally, arterial spin labeling (ASL) has proven to be a very useful tool for interpreting BOLD results and will be briefly discussed.

17.1.1 High-CNR Methods

Because fMRI is a sensitivity-limited technique, the contrast to noise ratio (CNR) per unit time is perhaps the most critical characteristic when assessing alternative fMRI strategies. In general, the most sensitive MRI techniques measure changes in the decay rate of transverse magnetization without limiting these rates through the use of refocusing "spin echoes", as discussed in Sect. 17.2. These signals are derived from blood-borne paramagnetic contrast agents, including agents that are injected into the blood stream or proteins that are contained within the red blood cells.

17.1.1.1 IRON fMRI

Initial fMRI investigations of cerebral physiology and functional activation were performed using bolus injections of contrast agent in animal models (Belliveau et al. 1990b; Rosen et al. 1990) and human subjects (Belliveau et al. 1990a; 1991). Gadolinium chelates, the only contrast agents approved in humans at that time, still are used routinely to assess regional CBF and CBV in the clinic. While such agents produce a strong first-pass effect on MRI signal, rapid vascular redistribution (tens of seconds) followed by removal from the blood (1 h) leave them unsuited for continuous imaging of dynamic changes in CBV. Nevertheless, these studies forecast subsequent dynamic measurements of CBV using contrast agents with an extended blood half-life. Many of the basic features of paramagnetic contrast media for fMRI were first described based upon observations from gadolinium studies.

The development of superparamagnetic iron oxide blood pool agents (Weissleder et al. 1990) enabled dynamic measurements of CBV in animal models due to ischemia (Hamberg et al. 1996), respiratory challenges (Kennan et al. 1997), and sensory stimulation (Mandeville et al. 1998; Kennan et al. 1998; van Bruggen et al. 1998). In animal models, this technique provides the practical advantage of very high detection power relative to alternative fMRI techniques like BOLD signal (Mandeville et al. 1998; 2004). While we originally referred to this method as "CBV-fMRI", we now employ the IRON moniker – Increased Relaxation for Optimized NeuroImaging (Chen et al. 2001; Mandeville et al. 2004) – to distinguish this technique from T_1-based measurements of CBV and more recently developed methods that do not rely upon contrast agent.

17.1.1.2 BOLD fMRI

Deoxygenated hemoglobin in the red blood cells is a paramagnetic species (Pauling and Coryell 1936) that acts much like exogenous contrast agents, so the physics of BOLD signal has many similarities to the IRON method, although deoxyhemoglobin produces a much smaller blood magnetization than typical iron doses used for fMRI. The physiology of the two methods is very different. While blood magnetization stays quite constant using the IRON method, blood oxygenation changes during a functional challenge, switching the sign of BOLD signal changes relative to IRON fMRI and producing dependencies of BOLD signal on the cerebral metabolic rate of oxygen utilization ($CMRO_2$) and CBF, in addition to CBV.

Oxygen-induced changes in the transverse relaxation rate of blood were reported early in the history of MRI, leading to a conclusion that "the local availability of oxygen in tissues *in vivo* is reflected in T_2 values" (Thulborn et al. 1981; 1982). However, the implications for imaging were unclear. Optical methods soon reported changes in brain function based upon "intrinsic signals", which were thought to contain a contribution "associated with the conversion of oxyhaemoglobin to haeomoglobin in the capillaries" (Grinvald et al. 1986). Ogawa and colleagues measured changes in "blood oxygen level dependent (BOLD) contrast" due to inspired carbon dioxide in rodents (Ogawa et al. 1990a) and speculated that negative BOLD changes would accompany increases in brain activity "resulting from higher oxygen consumption" (Ogawa et al. 1990b). If not for prior PET observations (Fox and Raichle 1986), the speculation of activity-induced negative BOLD signal would have been correct as postulated but for a different reason: if CBF and $CMRO_2$ exhibited the same magnitude of changes during brain activation, blood oxygenation would remain constant so that negative BOLD signal changes would reflect alterations in CBV, as for the IRON method. Most of the following sections focus explicitly upon IRON fMRI but contain supplemental sections to briefly relate how each imaging characteristic relates to the BOLD method.

17.1.2 Low-CNR Methods

Not every aspect of neurovascular physiology can be measured with high detection power, and the IRON method has not been employed in humans due to a lack of approved contrast agents. Techniques that are not routinely employed for fMRI are nonetheless very important for providing information about physiology.

17.1.2.1 ASL

Blood flow is a responsive index of local brain activity that provides the glucose and oxygen to support metabolism. Arterial spin labeling (ASL) is one of the most important tools for assessing brain function by MRI. This method, developed in the early

1990s concurrent with fMRI (Williams et al. 1992; Detre et al. 1992), was employed together with BOLD signal in the earliest human fMRI measurements (Kwong et al. 1992). However, ASL is seldom employed in traditional brain-mapping studies due its low detection power, and clinical appraisals of basal CBF generally use bolus injections of gadolinium, especially for time-critical applications that cannot afford the averaging time required by ASL.

In many ways, ASL is a seductive technique that possesses most of the characteristics that are desirable for functional brain imaging, except for high detection sensitivity. ASL provides reasonably accurate measurements of absolute CBF in most cases, whereas bolus injections of gadolinium only measure relative regional values of CBF or CBV. For activation studies, ASL signals are localized to the microvessels, where most of the magnetically labeled blood water is extracted through the capillaries into the tissue. Despite these advantages, the limited sensitivity of ASL restricts its usage in brain activation studies to a supporting role for the interpretation of BOLD physiology and the assessment of basal physiology.

17.1.2.2 VASO and VERVE

Two recent techniques have begun to explore the possibility of obtaining quantitative indices of CBV without the use of contrast agent, which would extend dynamic functional assessments of CBV physiology into human subjects for the purpose of informing BOLD signal and potentially other aspects of neurovascular physiology. Each of these methods is challenged with the task of interpreting small signal changes in the face of competing contrast mechanisms, but persistent development efforts are beginning to bear fruit. The VASO (vascular space occupancy) method (Lu et al. 2003) may ultimately provide a reasonably robust index of CBV for targeted studies of physiology, whereas the VERVE (venous refocusing for volume estimation) technique (Stefanovic and Pike 2005) targets a more specific biology by attempting to extract the venous contribution to BOLD signal changes.

17.2 T_1, T_2, and T_2* Contrast

If we exclude several contrast mechanisms like diffusion and chemical shifts (NMR spectroscopy) that are discussed in other chapters, fMRI signals *in vivo* result from changes in the relaxation times (T_1, T_2, or T_2*) of water protons as a result of underlying changes in cerebral physiology. Signal changes can occur from endogenous mechanisms like BOLD signal, or from exogenous contrast agents in the blood stream that magnify the sensitivity to changes in vascular physiology. When imaging normal brain with an intact blood–brain barrier (BBB), contrast agents affect brain water somewhat differently than in solution due to compartmentalization of agent within the small cerebral volume fraction contained inside the vessel walls. In brief, compartmentalization produces the following set of *distinguishing principles* that govern MRI contrast and the technique of choice for particular applications:

1. T_1 contrast agents effect cerebral signal weakly, because the BBB limits exchange of water between the intravascular compartment, containing contrast agent or magnetic labels, and the extravascular compartment, containing the vast majority of water, which is the source of MRI signal.
2. Contrast agents within the blood vessels produce magnetic field gradients that extend into the extravascular tissue. Using T_2* weighting of signal, all water protons in the brain are effected by intravascular agent, so that the influence of CBV on MRI signal is magnified. T_2* includes effects that would be removed by the use of spin echoes, so changes in T_2* always exceed changes in T_2.
3. When using spin echoes to refocus magnetic field inhomogeneities, extravascular water protons are strongly influenced only in the vicinity of microvessels, where magnetic field gradients change rapidly on a spatial scale comparable to the diffusion length. Hence, spin echoes improve microvascular weighting but sacrifice a significant portion of the available signal near larger vessels.

To elucidate upon the first point above, the rate of exchange for water across the BBB is comparable to T_1 values of the brain (Donahue et al. 1996). During the time that longitudinal magnetization.recovers backs to equilibrium, only a small fraction of water protons in the extravascular space are subjected to cross-relaxation by close contact with water that initially relaxed at a different T_1 rate due to the presence of contrast agent. As a consequence, the volume fraction of spins that participates in modifying the MRI signal due to a change in CBV is very limited using T_1 methods. For instance, consider a robust functional stimulus (e.g., strong visual stimulus) that produces a 25% increase in CBV; assuming a 4% basal blood volume fraction in the tissue region responding to the stimulus, only about 1% of the total brain water is displaced by blood water carrying additional contrast agent. Hence, only about 1% of the MRI signal (plus a small additional amount due to transvascular water exchange) senses the additional contrast agent. For this reason, T_1 agents aren't useful for fMRI. Conversely, the low sensitivity of cerebral MRI signal to T_1 agents within the vasculature is a benefit for imaging breakdown of the BBB in pathology, a case where T_1 effects are visualized easily due to extravasation of agent into the extravascular space (Zaharchuk 2007).

In contrast to longitudinal signal, transverse signal requires spin coherence across an image voxel, and this coherence can be disrupted *at a distance* by magnetic field gradients that extend outside the vessels. This effect was postulated (Villringer et al. 1988) based upon physical principles and a very simple observation: the signal attenuation observed in their studies (50%) due to injection of gadolinium into rat brain far exceeded the blood volume fraction, implicating an extravascular source. In fact, the signal to noise ratio (SNR) of gradient-echo sequences can be driven to zero in animal models by injecting large doses of agent, showing that all water protons in the brain are subjected to the magnetic gradients that extend outward from vessels.

From the earliest uses of exogenous agent *in vivo*, it was apparent that gradient echoes and spin echoes produced qualitatively different contrast: gradient echoes

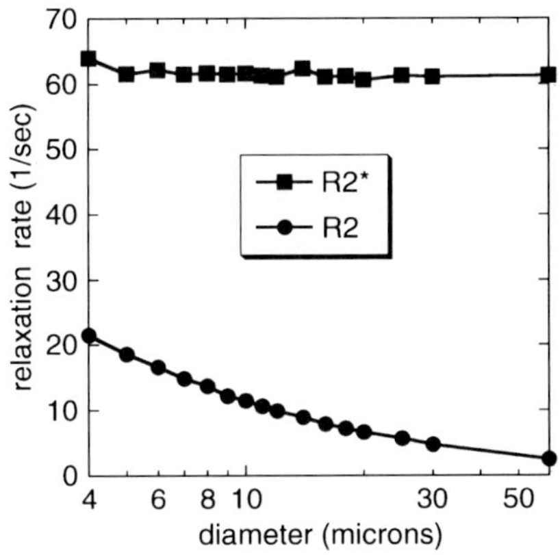

Fig. 17.1 Simulations of gradient-echo (R2*) and spin-echo (R2) relaxation rates versus vessel diameter indicate that gradient echoes have no sensitivity to vessel size, whereas spin echoes are preferentially sensitive to microvessels but have much less overall sensitivity

showed more signal attenuation per unit dose throughout the brain, and differences were particularly pronounced near large vessel regions. Spin-echo measurements were hypothesized to most strongly reflect signal around capillaries (Fisel et al. 1989; Rosen et al. 1991). Given that spin echoes refocus effects of time-independent magnetic field variations, the physics behind the strong capillary weighting for spin-echo measurements is straightforward. In the absence of water diffusion, a spin echo would refocus all phase variations across a voxel, but random field variations sampled by water molecules during diffusion are not refocused. During the few tens of milliseconds over which the NMR signal is sampled during T_2 decay, the average displacement due to water diffusion is no more than a few tens of microns. In order for the magnetic field to produce meaningful changes to the phase of a diffusing water molecule, field variations must occur on a microscopic scale comparable to the diffusion distance. Magnetized vessels produce external magnetic fields with (1) a radial dependence that drops off with a distance comparable to the vessel size, and (2) a strong angular dependence. These dipolar field dependencies implicate microvessels with sizes less than or comparable to the diffusion scale as the dominant sources of signal change when using spin echoes. Conversely, magnetic fields near large vessels appear to be almost static (time-invariant) in the frame of a diffusing water molecule. Near such vessels, phase increments acquired prior to the inversion pulse are reversed and then unwound by the time signal is acquired, so large vessels produce small effects on spin-echo signal.

We can simulate the dependence of relaxation rates upon vessel size for both spin echoes and gradient echoes using standard Monte Carlo methods (Boxerman et al. 1995b). Figure 17.1 shows relaxation rates for a hypothetical voxel containing a physiological CBV distributed in randomly oriented but fixed-diameter vessels, each containing a concentration of contrast agent that would result from systemic injection of about 15 mg of iron per kg of body weight. At such high blood magnetization, gradient-echo measurements have no sensitivity to vessel size, so signal changes can be quantitatively related to total CBV.

Using spin echoes, contrast agents produce the largest influence on MRI signal near the smallest microvessels, and the effect then decreases monotonically with vessel size, as shown in Fig. 17.1 (Mandeville et al. 2007a). This basic dependence, suggested prior to fMRI using exogenous or BOLD contrast (Rosen et al. 1991), underlies claims that spin echoes are preferentially sensitive to microvessels. The strong vessel-size dependence of spin-echo measurements of CBV motivated "vessel size imaging", in which a mean vessel size index can be formed from the ratio of gradient-echo and spin-echo measurements (Dennie et al. 1998; Tropres et al. 2001). The potential advantages, and the clear disadvantages, of spin echoes for fMRI will be discussed later.

17.2.1 BOLD Connection

There are qualitative connections between the simulations of Fig. 17.1 and BOLD signal. At lower levels of blood magnetization appropriate for BOLD signal or for FDA-approved doses of gadolinium, gradient echoes exhibit less sensitivity to capillaries than to larger vessels (Boxerman et al. 1995b). Even at such low levels of blood magnetization, spin echoes are preferentially sensitive to microvessels, and this sensitivity becomes progressively more pronounced as blood magnetization increases (Boxerman et al. 1995b). However, assumptions about vessel size dependencies using either gradient or spin echoes for BOLD imaging should be entertained with caution, as a significant proportion of BOLD signal occurs within the intravascular compartment, particularly at clinical field strengths (Boxerman et al. 1995a; Song et al. 1996). At higher fields, intravascular contributions are reduced for a given echo time, but increased susceptibility artifacts at high field generally force the use of shorter echo times, so that intravascular BOLD contributions remain at high fields. Spin echoes, which are insensitive to susceptibility artifacts, are dominated by intravascular signal at low field (Zhong et al. 1998; Oja et al. 1999), but very high fields like 9.4 T eliminate the intravascular contribution even without using extraordinarily long echo times (Lee et al. 1999). However, spin echoes suffer a multi-fold loss of detection power relative to gradient echoes, a major drawback at any field strength.

17.3 IRON fMRI

17.3.1 Interpretation of T_2 and T_2 Changes*

Although it is common in radiology to speak of MRI contrast in terms of relaxation times, quantitative models are constructed with the inverse quantities, which are rates: $R_2 = 1/T_2$. Several approximations simplify the math. The first is that relaxation rates are additive ($R_2^{TOTAL} = R_2^{ENDOGENOUS} + R_2^{AGENT}$), an assumption that requires a single compartment. This is not strictly true, because the intravascular and extravascular compartments are separated by slow exchange on the scale of the echo time. When agent is first injected into the blood stream, signal from the small intravascular compartment is nulled, leaving only the extravascular compartment as a source of subsequent functional signal changes. However, so long as doses of contrast agent produce sufficiently large changes in MRI signal, the small intravascular compartment can be ignored. A second assumption is that contrast agents increase relaxation rates in proportion to the tissue concentration of contrast agent ($R_2^{AGENT} = k\ C$), where C is the tissue concentration of agent (Rosen et al. 1990). This is a useful concept in terms of connecting bolus infusions, where the time dependence reflects changes in the intravascular concentration of agent at a fixed volume fraction ($R_2^{AGENT}(t) = k\ C_{BLOOD}(t)\ V_0$), to the use of steady-state blood pool contrast agents, where the time dependence reflects dynamic changes in blood volume ($R_2^{AGENT}(t) = k\ C_{BLOOD}\ V(t)$). However, simulations suggest a slightly superlinear dependence of relaxation rate upon blood concentration at very low levels of blood magnetization with a linear dependence upon the blood volume fraction (Boxerman et al. 1995b); the latter dependence is all that matters for IRON fMRI, as long as the blood concentration remains constant.

With these two assumptions in place, MRI signal changes can be related to percentage changes in CBV using the IRON method. Upon injection of contrast agent, a new relaxation process attenuates the signal with a proportionality to the basal blood volume fraction: $R_2^{AGENT}(t=0) = k\ V(0)$. Subsequent functional changes in blood volume are reflected in time-dependent changes in this relaxation rate, $R_2^{AGENT}(t) = k\ V(t)$, such that changes in MRI signal can be directly related to percentage changes in the blood volume fraction, $\Delta V(t)/V$, by normalizing functional responses to the effect produced by injection.

In this formulation, we ignored two additional potential influences on MRI signal: T_1 effects and BOLD signal. As described previously, transvascular water exchange across an intact BBB is slow, so T_1 effects can be ignored even when large doses of contrast agent are used or when the repetition time is short. During a functional response, the BOLD mechanism competes with IRON signal. For a typical sensory stimulus, for instance, BOLD signal changes are positive and IRON signal changes are negative, so changes in relaxation rates oppose each other: $\Delta R_2(t) = \Delta R_2^{IRON}(t) - \Delta R_2^{BOLD}(t)$. Typically, BOLD contamination in IRON signal enters at a level of only 10–20% even at high fields like 9.4 T in animal models (Mandeville et al. 2004), although this may overestimate BOLD effects because the intravascular part of BOLD signal is crushed upon injection of agent, leaving only the extravascular BOLD effect to compete with IRON signal changes. Empirically, drug-induced changes in CBV were found to be quite consistent across a fivefold range of magnetic field when ignoring BOLD effects (Mandeville et al. 2004).

The ability to quantitatively measure a relatively interpretable physiological index (the activation-induced percentage change in CBV using gradient echoes) is a significant advantage of IRON fMRI when juxtaposed relative to the complexities of BOLD signal. IRON results can be compared directly to other modalities that report changes in CBV or CBF (which can be related to CBV), and there is little dependence of this index on magnetic field strength.

The preceding discussion cannot be applied to spin echoes without accounting for the vessel profile shown in Fig. 17.1. When vessels swell, they become less effective at relaxing MRI signal. Hence, relaxation rate ratios systematically underestimate actual percentage changes in microvascular CBV. If we assume that all vessels respond similarly to a stimulus, simulations find that spin-echoes underestimate actual changes in CBV by about 30–40%, and this prediction roughly matches data across most of the brain (Mandeville et al. 2007a). Of course, vessels do not all respond similarly, so model-independent quantification of the percent changes in microvascular CBV is not possible.

17.3.1.1 BOLD Connection

Deoxygenated hemoglobin is a paramagnetic species (Pauling and Coryell 1936; Thulborn et al. 1982; Ogawa and Lee 1990) that acts much like exogenous contrast agents, so BOLD models start from the same perspective described above. However, there are several important differences that complicate quantification of BOLD signal or comparison of BOLD signal across studies.

Firstly, BOLD signal contains an intravascular contribution that generally is dominant at low fields like 1.5 T and progressively diminishes with higher fields. This intravascular contribution can be separated explicitly from extravascular signal or swept up into an effective relaxation rate that has a superlinear dependence upon blood magnetization (Buxton et al. 2004). The relative contribution of intra/extravascular effects is a complicated function of magnetic field strength and echo time.

Secondly, one cannot simply determine the baseline BOLD relaxation rate, which is entangled with other endogenous relaxation processes. Using exogenous agents, the baseline relaxation rate is determined by observing signal attenuation due to injection of agent. There is no analogous method for BOLD signal. An indirect way to determine the baseline BOLD effect is a method that requires inhalation of CO_2 together with combined measurements and BOLD signal, CBF, and optionally CBV (Davis et al. 1998; Mandeville et al. 1999a). Thus, there is no way to routinely compute a physiological index analogous to the IRON method (i.e., the percentage change in deoxygenated hemoglobin). Given these problems with quantification, literature studies typically quote purely statistical quantities, or they report percentage changes in BOLD signal, a quantity that is related to absolute changes in deoxygenated hemoglobin but that has multiple complicated dependencies on specific experimental parameters and the underlying basal physiology.

17.3.2 Detection Power

Detection power generally governs the choice of alternative fMRI strategies. Consider, for example, how rarely ASL is used within the brain mapping community in relation to BOLD signal. Despite the numerous reasons in principle to choose

ASL over BOLD signal (e.g., less uncertainty in terms of interpreting the response, ability to assess the basal physiological state, signal changes more localized to microvessels), the larger CNR of BOLD signal trumps all other arguments, leaving ASL as a niche technique in the fMRI community (albeit an important one).

Similarly, the CNR advantages for IRON fMRI in animal models are so pronounced at typical field strengths that most animal studies would benefit enormously by using exogenous agent in place of BOLD signal. At low magnetic fields like 1.5 or 2 T, IRON boosts CNR relative to BOLD signal roughly sixfold (Mandeville et al. 1998; Vanduffel et al. 2001). Averaging increases CNR only in proportion to the square root of time, meaning that low-field BOLD fMRI requires about 30-fold more averaging to obtain similar results. At higher MRI fields strengths, BOLD detection power improves due to the linear scaling of deoxyhemoglobin magnetization with field strength. At 9.4 T and higher, BOLD signal is much more competitive (within a factor of 2 of IRON CNR), with the comparison depending upon details of the methodology (Mandeville et al. 2004; Zhao et al. 2006).

There are many avenues for improving the absolute CNR in any experiment by increasing the SNR. Because the strongest influence on SNR is the voxel volume, spatial resolution generally is purchased at the expense of detection power. Obviously, this argument falls apart when voxels exceed the size of a focal activation, so that detection power is limited by volumetric dilution of the response. Additionally, physiology and motion contribute noise sources that scale with the magnitude of signal (Kruger and Glover 2001; Kruger et al. 2001; Triantafyllou et al. 2005). When the SNR in individual images exceeds a value of about 100, further improvements in image SNR do not translate to the temporal domain (Triantafyllou et al. 2005). In this regime, CNR does not decrease as much as SNR when voxels become smaller or (presumably) when contrast agent is injected.

Using the IRON method, detection power relative to BOLD signal depends upon a number of considerations that often are counter-intuitive. For instance, contrast agents decrease SNR but increase CNR; detection power for BOLD signal is much more sensitive to basal CBV than IRON signal; the best results using the IRON method are obtained at very short echo times; percentage changes in IRON signal are insensitive to the MRI field strength; the relative ratio of BOLD and IRON CNR depends upon the duration of the applied stimulus. The following sections elaborate upon these points.

17.3.2.1 Echo Time and Dose

To optimize detection power using T_2*-based contrast for fMRI, one always sacrifices SNR for the sake of CNR. Using BOLD signal, for instance, echo times near zero achieve maximal SNR, whereas echo times close to tissue T_2 or T_2* maximize CNR at a much lower SNR. This same principle holds true when using IRON fMRI. At any given echo time, the dose of contrast agent should be adjusted so that SNR falls to about 1/e of the pre-injection value. However, IRON fMRI presents experimenters with the opportunity to adjust two experimental knobs: echo time and dose.

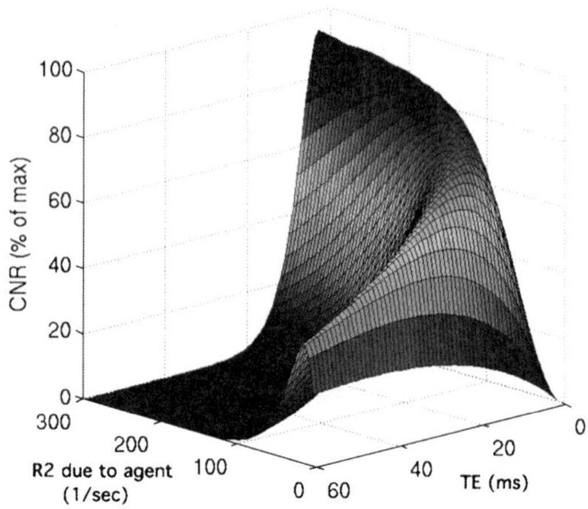

Fig. 17.2 In the two-dimensional space of echo time and contrast agent dose, where dose in the figure is indicated by the effect on the transverse relaxation rate of MRI signal, the functional contrast to noise ratio (CNR) is optimized at short echo times and large doses of contrast agent

In this two-dimensional space, shown in Fig. 17.2, there is a global maximum of CNR corresponding to an unattainable solution: zero TE and an infinite dose of contrast agent.

Fortunately, lengthening the echo time has only a modest effect on the maximum available post-injection CNR, which scales in proportion to the pre-injection SNR as $\exp(-T_E/T_2^*)$. Setting the echo time to $T_2^*/2$, for instance, delivers about 60% of the maximum possible signal change. In animal models where dose limitations are not a concern, short echo times can be used to produce the best possible set of imaging characteristics: large signal changes, minimal susceptibility artifacts, and many slices or excitations per unit time. At high field strengths like 9.4 T, where the T_2^* distribution has a large dispersion relative to lower fields, short echo times can make images insensitive to magnetic susceptibility artifacts, while still delivering excellent detection power by suitably adjusting the dose of contrast agent (Mandeville et al. 2004).

To understand typical agent doses and the corresponding relaxation rates induced by injection in animal models, consider the following examples. For repeated fMRI in non-human primates at 3 T, we typically employ single-shot EPI with echo times in the range 20–25 ms and contrast agent doses of 10–12 mg iron per kg body weight. In these studies, the agent-induced relaxation rate is typically 25–30 1/s (Leite et al. 2002). For pharmacological fMRI using multi-shot imaging at 9.4 T in rodents, we employ very short echo times (5 ms) with very large iron doses (35–40 mg/kg) corresponding to iron-induced relaxation rates of about 150 1/s (Mandeville et al. 2004).

As a "rule of thumb" for determining agent dose, global cerebral signal should be attenuated by about a factor of 2 relative to the pre-injection baseline signal at any echo time. In that case, the brain will exhibit a distribution of signal attenuations similar to that shown by the gray curve in Fig. 17.3 (gray matter corresponds to values greater than about 0.5 on the x axis), so that there is little variation in detection power across the brain.

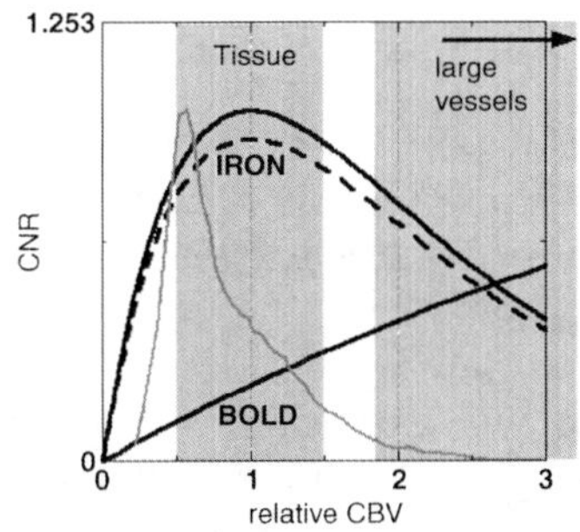

Fig. 17.3 Empirically informed CNR calculations for BOLD and IRON fMRI at 3 T are plotted versus the relative basal blood volume fraction, which is calculated as rCBV = $-\ln(S_{POST}/S_{PRE})$, where S_{POST} and S_{PRE} correspond to signal values prior to and after injection of an exogenous iron oxide blood pool agent. The *dotted line* shows the influence of BOLD contamination in IRON signal. Calculations ignore the different temporal responses and so are appropriate for very long stimuli. The *solid gray line* shows an empirical distribution of basal blood volume as measured in a rat (Mandeville et al. 2007a). Values of rCBV below 0.5 correspond roughly to white matter

17.3.2.2 Blood Volume Fraction

Figure 17.3 shows IRON CNR as a function of basal CBV, with and without inclusion of the concomitant BOLD effect at 3 T. The abscissa in the figure is the relative basal blood volume fraction (rCBV), defined according to the signal attenuation produced by injection of contrast agent: rCBV = $-\ln(S/S_0)$, where S_0 is the pre-injection signal; of course, rCBV is a relative index that is proportional to echo time and dose using this formulation. This index of rCBV is similar to the index used in human clinical studies, where rCBV represents the integral of the regional response profile of the agent-induced relaxation rate following injection of gadolinium. The main difference is that the definition employed in Fig. 17.3 is not normalized by the echo time in order to create a relaxation rate. This definition is convenient for IRON fMRI, because an rCBV value of unity corresponds to the dose and blood volume fraction where detection power is optimal irrespective of the echo time used in the study.

The IRON curves in Fig. 17.3 result from the product of decreasing SNR and increasing percent signal changes versus rCBV: CNR ~ rCBV $\cdot$ e^{-rCBV}. Of course, rCBV is regionally variable in the brain, so CNR cannot be optimized simultaneously for all values of rCBV. However, there is little variation in CNR versus the basal blood volume within a physiological threefold range across brain regions (Mandeville and Marota 1999). The gray line in the figures shows an empirical distribution of gradient-echo rCBV from a recent study by our group in a rat (Mandeville et al. 2007a). Values of rCBV below 0.5 in the figure correspond almost exclusively to white matter. When an appropriate dose of contrast agent is employed, tissue CNR is high and large vessel regions are suppressed. These calculations have been validated empirically across a fivefold range of magnetic field strengths by comparing the ratio of IRON and BOLD CNR versus the basal blood volume fraction (Mandeville et al. 2004).

17.3.2.3 Magnetic Field

Unlike BOLD signal, where blood magnetization is a function of the strength of the magnetic field, the relative signal changes for IRON fMRI have no field dependence aside from the small effect of BOLD contamination. Using a theoretical paramagnetic exogenous contrast agent, dose would be adjusted at each field to achieve the desired signal attenuation, as previously described. The available contrast agents that are most suited to the IRON technique are coated iron oxides like MION (Weissleder et al. 1990) and Feraheme (Coyne 2009). Iron oxides are superparamagnetic, which means that the induced blood magnetization has almost no dependence on field strength above 1.5 T (Shen et al. 1993). Hence, a given dose produces the same effect on MRI signal regardless of field strength.

17.3.2.4 Stimulus Duration

Following stimulation, the time for CBV to reach a maximum response is much longer than for CBF or BOLD signal, as described in more detail in the following section. Thus, the size of the CNR boost relative to BOLD or ASL techniques depends upon the duration of the applied stimulus. For instance, CNR enhancement relative to BOLD signal drops to about a factor of 2 when using short stimuli (e.g., 4 s) that are appropriate for event-related fMRI at 3 T (Leite et al. 2002). A poorly constructed event-related paradigm using periodic stimuli with a short inter-stimulus interval would see a further big loss of CNR due to the low-pass filtering of the slow IRON response. However, randomizing the inter-stimulus interval alleviates any loss of CNR due to overlap of adjacent stimuli (Leite and Mandeville 2006), in the same way as for BOLD signal (Burock et al. 1998). Because the IRON response is slower than the BOLD response, it is even more crucial that event-related stimuli be randomized in time.

BOLD Connection

The physics of BOLD extravascular signal is identical to that for the IRON method, except that blood magnetization generally is much lower. If a very low dose of iron oxide were employed in an experiment, the relationship between CNR and the basal blood volume fraction would be similar to that shown for BOLD signal in Fig. 17.3. When blood magnetization is low enough that regional variations in CBV are not evident in individual images (e.g., the exponential dependence of CNR on SNR is weak), CNR will increase with an almost linear dependence on the basal blood volume fraction. In this contrast-starved regime that is typical for BOLD images, functional responses are much more easily observed in large vessel regions such as draining veins and the superficial layers of cortex (Mandeville and Marota 1999; Zhao et al. 2006). This creates a "brain versus vein" localization problem that is not an issue for IRON fMRI. Note that this argument does not require significant levels

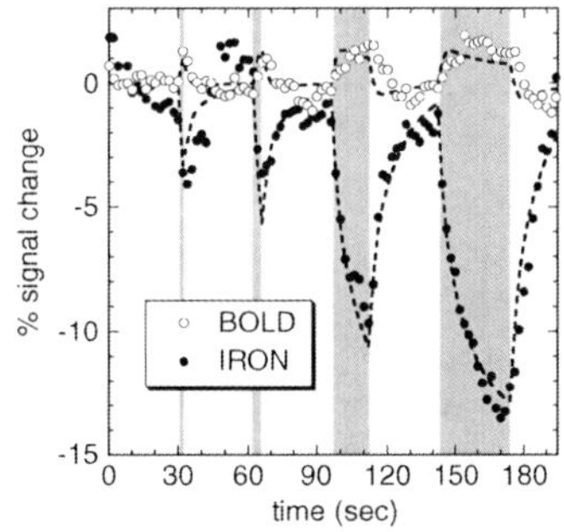

Fig. 17.4 BOLD (*open circles*) and IRON (*closed circles*) data in macaque visual cortex using stimuli of duration 2, 4, 16, and 32 s show the pronounced effect of stimulus duration on IRON signal

of BOLD intravascular signal, which provides a valuable additional source of signal changes but exacerbates the brain-versus-vein conundrum. Loss of spatial localization due to venous drainage of deoxygenated hemoglobin during a stimulus is an effect that is observed not only in large veins, but in regions that bridge activated tissue and large veins (Leite et al. 2002). This effect does not go away at high magnetic fields like 9.4 and 11.7 T (Silva et al. 2007).

17.3.3 Temporal Response

Following our initial observation that CBV responds much more slowly than BOLD signal following a functional stimulus (Mandeville et al. 1998), we compared the temporal responses of CBF and CBV using very high temporal resolution in order to additionally detect the rapid phase of the CBV response that accompanies arterial dilation and the rapid onset of CBF. In fact, CBV initially responds as rapidly as CBF (Mandeville et al. 1999b) and more rapidly than BOLD signal (Marota et al. 1999; Silva et al. 2007), which is delayed by the washout nature of the technique. Prolonged stimulation produces a continued slow rise of CBV with a minimal influence on CBF. Following cessation of stimulation, CBF resolves to baseline more quickly than CBV, which does not return to baseline for 30–60 s. These temporal dynamics for the IRON response now have been observed in many rodent studies (Mandeville et al. 1999b; Lu et al. 2005; Silva et al. 2007; Kida et al. 2007) and in awake non-human primates (Vanduffel et al. 2001; Leite et al. 2002). The IRON response is amenable to a linear model analysis using a hemodynamic response function that contains both a rapid and a slow temporal component (Leite et al. 2002; Leite and Mandeville 2006; Silva et al. 2007).

Figure 17.4 shows typical BOLD and IRON responses in the visual cortex of an awake non-human primate in response to a visual stimulus of variable duration. The response is the summation over all image voxels that were significantly correlated with the stimulus using both methods. During the two longest stimuli (16 and 32 s), the BOLD response reaches a plateau, whereas the IRON response continues to grow.

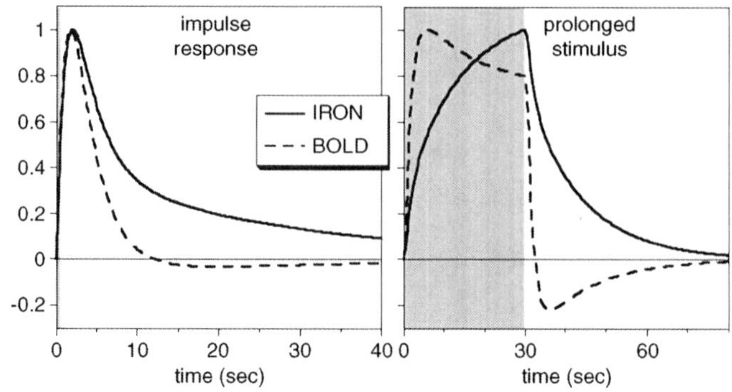

Fig. 17.5 Left: Impulse response functions for IRON and BOLD signal in macaque visual cortex. Right: The response to a long stimulus is predicted by convolving the impulse response with a square function indicating the stimulus timing. For extended stimuli, the slow evolution of IRON signal is more apparent, and BOLD signal shows a pronounced post-stimulus undershoot

As a practical matter of data analysis, the main difference between the BOLD and IRON hemodynamic impulse response functions (IRF) is that sign of the tails are switched (Leite et al. 2002). For a short stimulus, BOLD signal increases with a rapid onset time and then decreases slightly below the baseline level for several tens of seconds after the end of stimulation, as shown in Fig. 17.5 (left). For very short stimuli, the post-stimulus undershoot is a subtle effect that is not detected in all experiments. The response to longer stimuli can be approximately predicted by time-shifted summation of the IRF (Boynton et al. 1996), or alternatively by convolving a parametric IRF with the stimulus timing. For a much longer stimulus (Fig. 17.5, right), the result is a more pronounced post-stimulus undershoot due a slow build-up of the negative tail. The IRON response depicts the same general features, except that (1) the tail of IRF does not switch sign as in the BOLD IRF, and (2) the tail of the IRF is a much more prominent feature in the IRF, so the slow component of the IRON response dominates for extended stimuli. These features of the response also are manifest in more complicated experimental designs using rapidly presented event-related stimuli. For such stimuli, the tail of the IRON response builds into a pronounced offset of the average signal level relative to non-stimulated states, whereas the BOLD tail counteracts baseline shifts that would be otherwise observed (Leite et al. 2002; Leite and Mandeville 2006).

Mathematically, the response of CBV can be simulated by adapting the windkessel concept that is used to describe the compliance of larger arteries in response to periodic pressure waves produced by the heart. In the windkessel formulation (Mandeville et al. 1999b), arteriole dilation in response to a functional stimulus increases intravascular pressure downstream, and this generates dilation of capillaries and venules due to the compliant nature of those vessels. Dilation can be either elastic, meaning that vessels respond rapidly like a balloon, or inelastic due to the presence of an additional physical process with a longer time constant. Due to the biphasic nature of the CBV response, as well as optical experiments that suggest a rapid response of total hemoglobin volume in the capillary compartment, we proposed elastic dilation followed by inelastic expansion of vessels for prolonged stimuli, or "delayed compliance". Optical and fMRI studies using short stimuli have applied or extended this model to accurately describe data for stimuli of short duration without invoking an inelastic response (Friston et al. 2000; Boas et al. 2008), whereas other optical investigators have expanded upon the delayed compliance model by replacing our original ad hoc approach to the inelastic phase with a state variable to govern inelastic dilation (Kong et al. 2004). Within the "balloon" framework for BOLD signal (Buxton et al. 1998b), the CBF-CBV temporal relationship has been modeled using identical equations (Buxton et al. 2004) with a single long time constant, which has the effect of stretching the mean transit time; this approach might be viewed as a purely inelastic model for slow capillary/venous dilation which views the rapid phase in the data as solely due to arterial dilation. Potential mechanisms underlying this response are discussed at the end of the chapter.

17.3.3.1 BOLD Connection

Our data has shown that the temporal mismatch observed between the dynamic responses of CBF and CBV contributes to the BOLD post-stimulus undershoot. Following cessation of electrical forepaw stimulation in rats, the undershoot begins when percent CBF falls below percent CBV, and the undershoot resolves when CBV resolves to baseline (Mandeville et al. 1999a). Using a linear model analysis of fMRI data from the visual cortex of awake non-human primates (Leite et al. 2002), two temporal components were identified that could consistently fit both the CBV and BOLD data in multiple stimulus presentation designs using relative magnitudes that were physiologically reasonable; the CBV response was more heavily weighted by the slow (capillary/venous) response than the fast response, and the influence of CBV (slow response) in BOLD signal was only about one third as large as the influence of blood oxygenation (fast response). Alternate hypotheses about BOLD temporal dynamics are discussed in the last section.

Figure 17.6 compares the integrated amplitude of the BOLD and CBV responses in rat somatosensory cortex during a post-stimulus 20-s window that was delayed by 7 s relative to stimulus cessation (Royl et al. 2001). As a function of stimulus duration, the BOLD undershoot becomes larger, together with the average magnitude of CBV, which has a negative influence on BOLD signal.

Is it possible that the BOLD and IRON responses have such an excellent anti-correlation in the post-stimulus temporal region because the BOLD post-stimulus undershoot is producing an artifact in the IRON response? In other words, could elevated oxygen utilization or a flow undershoot in the post-stimulus region produce elevated levels of deoxygenated hemoglobin, which would add to the negative IRON signal and make it appear that the response was delayed? This potential artifact cannot explain the data. Consider the non-human primate of Leite et al. (2002). Changes in iron-induced relaxation rates were about seven times larger than BOLD changes at the end of a prolonged stimulus, so the BOLD effect is a minor correction to the

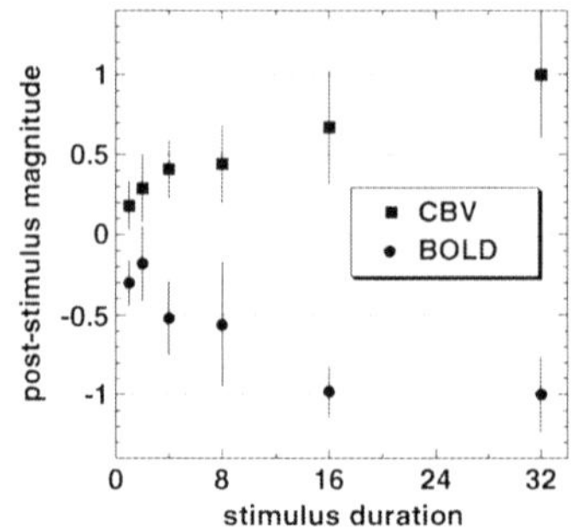

Fig. 17.6 The integrated amplitude of MRI signal during the post-stimulus region for BOLD signal and CBV shows that the magnitude of the BOLD undershoot correlates with the post-stimulus response of CBV, which has a negative effect on BOLD signal (Royl et al. 2001)

IRON response in most circumstances. Moreover, when the BOLD post-stimulus undershoot began about 10 s after stimulus cessation, the IRON-induced relaxation rate was still about three times larger than the maximum positive BOLD response at any time point. In other words, the BOLD effect simply is not strong enough to significantly alter the temporal dynamics of the IRON response when appropriate doses of contrast agent are employed.

Although IRON signal initially responds more rapidly than BOLD signal (Silva et al. 2007), it is fair to say that the IRON method suffers a loss of temporal resolution as compared to BOLD signal, meaning that the slower time-to-peak for the CBV response leads to a greater relative loss of CNR using short or rapidly presented stimuli compared to longer block designs for stimulation.

17.3.4 Spatial Resolution and Localization

Relative to the spatial resolution defined by neuronal ensembles responding to as stimulus, fMRI suffers a loss of spatial resolution that can be understood as convolution with a vascular point spread function (PSF) that introduces blurring (Engel et al. 1997). The 2-deoxyglucose technique revealed that the metabolic response to visual stimuli in macaque visual cortex can be as small as a few hundred microns or less, depending upon the cortical layer (Tootell et al. 1988). However, the ultimate spatial resolution obtainable by fMRI remains somewhat controversial. This issue is confounded by large differences in detection power between alternative methods, by hardware limitations, and by other experimental details.

There have been several oft-cited attempts to define a PSF for fMRI using retinotopy (Engel et al. 1997; Parkes et al. 2005; Shmuel et al. 2007), each reporting a PSF of several millimeters. Not coincidentally, these estimates correspond roughly to the size of the imaging resolution in these studies. Such studies do not probe the "intrinsic" resolution of fMRI, but rather demonstrate the constraints of hardware and techniques employed in typical fMRI studies.

Figure 17.7 illustrates that higher spatial resolution can define a smaller PSF. In this study in awake macaques (Leite et al. 2005), we used isotropic 1.25 mm voxels and parametrically varied the size of small point-like stimuli in order to estimate the

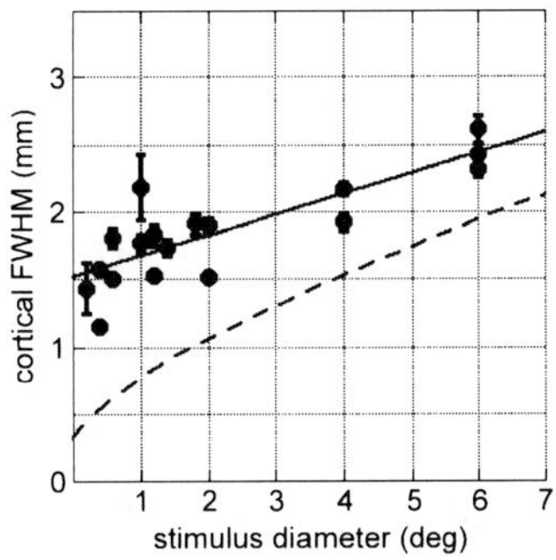

Fig. 17.7 Data show measured full width at half maximum of retinotopic projections of point-like stimuli using IRON fMRI. The solid line fits the data, and the dashed line removes the combined effects of the pixel resolution and T2* blurring to estimate the sum of the neuronal and vascular spreads (Leite et al. 2005)

retinotopic projection of a vanishingly small stimulus, which should correspond to a neuronal response with a minimal spatial spread. As seen in the figure, gradient-echo IRON fMRI can directly measure a spatial spread of no more than 1.5 mm. When one accounts for the finite size of imaging voxels and the blurring function induced by the T_2* decay filter across the spectrum of spatial frequencies during data acquisition, the "intrinsic" point spread of fMRI is projected to be well less than a millimeter (dotted line), and probably just a few hundred microns (Leite 2006).

In addition to spatial resolution, which is a measure of the width of a focal response, spatial localization can be defined as the foci (or centroid) of a response. In turn, localization has aspects that depend upon both the physiology and the physics of a technique. As an example of the differentiation between physiology and physics, consider how ASL can be subtly differentiated from flow. The net flow of blood through the veins is identical to that in arteries and capillaries, and yet very little of the magnetic label for ASL survives into the veins due to water extraction in the capillaries. Hence, small shifts between BOLD and ASL signals will be observed due the venous and capillary weightings, respectively, for the two methods.

The spatial foci of IRON and BOLD fMRI also differ subtly for reasons that largely can be defined from the physics of the methods. BOLD signal has a much stronger dependence on basal CBV, as seen in Fig. 17.3. Regions with high basal CBV exhibit a larger BOLD/IRON signal ratio than regions with low basal CBV, and this is generally true across whole brain (Mandeville et al. 2001). Cortical laminae exhibit a strong gradient of basal CBV that increases toward the superficial layers, and so a focal cortical response due a sensory stimulus produces a BOLD activation pattern that is amplified in superficial cortical layers relative to the IRON response, which is largest in middle cortical layers (Mandeville and Marota 1999; Zhao et al. 2006). In species with cortical folds like non-human primates, the IRON response in sulci can be more easily localized to a specific cortical bank, because the superficial weighting of BOLD signal brings the response centroid near the convergence of the two cortical surfaces (Vanduffel et al. 2001).

The precise spatial relationship between neuronal and fMRI responses is unclear. Some authors have speculated that localization of the CBV response to the middle cortical layers might represent "laminar specificity" for localizing neuronal responses by fMRI (Harel et al. 2006). Alternatively, combined spin-echo and gradient-echo measurements of CBV show that superficial layers of cortex have a higher basal blood volume fraction and a larger mean vessel size than middle cortical layers, suggesting that large but less reactive conduit vessels diminish CBV reactivity near the cortical surface (Mandeville et al. 2007a). Although it is possible and even probably that fMRI eventually will distinguish differential neuronal responses across cortical laminae, there is not yet evidence for this.

17.3.4.1 BOLD Connection

Results are shown for IRON fMRI in Fig. 17.7, but we obtained virtually identical results for BOLD fMRI. fMRI experiments in humans with an isotropic resolution approaching 1 mm will soon be routine as a result of advances in gradient strength technology and parallel imaging, which extend the range of k-space that can be sampled per unit time, and due to the proliferation of higher magnetic field strengths (particularly 7 T), which support higher resolution through a higher intrinsic SNR.

17.3.5 fMRI in Animal Models: Choosing a Technique

There are many disparate reasons for performing fMRI in animal models. Some investigations target specific aspects of physiology or functional coupling (e.g., mechanisms of BOLD signal) and might choose methods less based upon efficacy than on the specific biological information that can be obtained. Another class of experiments – let's call it "brain mapping" – seeks neural correlates related to behavior, state, or stimulus parameters with the presumption that different aspects of the vascular response (CBF, CBV, BOLD signal) provide essentially the same information in terms of neural coupling. Example applications in this class include drug infusions to investigate specific neurotransmitter systems or neuronal plasticity resulting from prolonged drug exposure, sensory stimulation to provide neural correlates of functional recovery following stroke (Dijkhuizen et al. 2003), and investigations of normal sensorimotor/cognitive processing for comparison with human subjects (Tsao et al. 2003) or as the basis for targeted electrophysiological studies (Tsao et al. 2006).

Animal models present an inherent set of difficulties that is beyond the scope of this chapter. The main issues, of course, are anesthesia or motion. Anesthesia blunts the size of signal changes, most generally by inhibiting synaptic transmission but also by shifting vascular baseline physiology. In conscious animals, signal changes may be larger, but noise also is larger due to brain or body motion; the latter produces apparent brain motion by inducing time-dependent changes in the magnetic field. Effective fMRI requires significant attention to the biological/behavioral aspects of an animal model, as well as an understanding of the available fMRI methods. Generally, experimenters desire not only to detect functional changes in obvious regions (e.g., primary sensory cortex) but in higher or connected regions as well, and then to somehow manipulate those responses through stimuli or pretreatment. Obviously, bigger signal changes are better when choosing an fMRI contrast mechanism.

With these considerations, most fMRI applications in animal models come down to a choice between BOLD or IRON signal. At field strengths below 5 T, the IRON method always provides a large CNR gain over BOLD signal. At field strengths approaching 10 T, the choice is more complex. For imaging the effects of drugs, high-resolution whole-brain imaging is desirable with less emphasis on rapid image collection due to the slow evolution of the neural response. In such a case, multi-shot sequences can be employed with short echo times (5 ms) to cover whole brain with minimal susceptibility artifacts at high fields while providing all of the IRON benefits of higher tissue CNR and lower signal from draining venous sinuses. For sensory stimulation studies at 9.4 T, two factors combine to reduce IRON CNR and make the BOLD/IRON comparison virtually a wash in terms of CNR (Zhao et al. 2006): rapid single-shot imaging is beneficial for CNR but requires a somewhat longer minimal echo time (10 ms), and the slower temporal evolution of CBV decreases IRON CNR for short stimuli. The first consideration eventually will be removed by hardware improvements that increase data collection rates through parallel imaging and faster gradients. At that point, event-related fMRI using rapidly presented stimuli might be the main application to benefit from BOLD signal over IRON signal below 10 T.

When using the IRON method, there are many drawbacks and few advantages of spin-echo sequences, so gradient echoes generally should be employed. The microvascular specificity of spin echoes, as shown in Fig. 17.1, might seem to be a potential advantage. However, the lower overall sensitivity of spin echoes shown in the figure implies that much more contrast agent is required in order to maximize spin-echo CNR. Even then, stimulus-induced increases in CBV will produce smaller signal changes than the gradient-echo method due to the shape of the sensitivity curve versus vessel size; as vessels swell during a functional challenge, sensitivity is reduced (Mandeville et al. 2007b). Furthermore, it is not clear that microvascular sensitivity offers much in the context of IRON fMRI. Whereas the BOLD effect occurs in both capillaries and large veins, the largest relative dilation is observed in microvessels, including small arterioles but not large veins (Hillman et al. 2007). Moreover, the IRON method includes a vascular filter that penalizes large blood volume regions in favor of tissue specificity (Fig. 17.3). The largest gradient-echo IRON signal changes, and the largest relative CBV changes, occur in the middle cortical layers (Mandeville and Marota 1999; Zhao et al. 2006), where capillary density is high, rather than in the superficial cortical surface layers, where the blood volume fraction is highest and the average vessel size is larger (Mandevill et al. 2007b). The practical disadvantages of spin-echo IRON fMRI outweigh any theoretical advantages.

17.4 Arterial Spin Labeling

ASL is a variant of the classical method for measuring blood flow based upon freely diffusible labels. If a label that is carried by the blood is completed extracted into the tissue on the first pass through the capillaries, then the regional concentration of label will reflect regional blood flow. Ideally, diffusible tracers should satisfy the condition that the product of capillary surface area (S) and the capillary permeability for the tracer (P) is much greater than CBF (PS>>F). For human brain, the PS product for water is marginally larger than flow, and this is less true in smaller animals, which have higher levels of CBF. Because the first-pass extraction of water is incomplete (~80%), water is an imperfect label that underestimates CBF by escaping the tissue through venous flow. PET measurements found that radiolabeled water underestimated CBF by about 15% relative to the use of a more diffusible tracer, butanol (Herscovitch et al. 1987).

In PET experiments, the wash-in of radioactive label is readily measured, as there is essentially no background. For ASL, the effects of blood flow on signal are very weak, and the background represents all brain water except for the small portion of labeled water that flowed into the region. To isolate the flow effect, images are obtained with and without application of RF tagging, and then the images are subtracted. Typically, tagging proximal blood water produces about a 1% effect on MRI signal at clinical field strengths, but this effect is proportional to blood magnetization and so scales with field strength. Obviously, motion between the control and tagged acquisitions will add noise to the flow signal; since the magnitude of motion-induced artifacts scales with the magnitude of signal, this effect can be reduced somewhat by saturating the image volume for all acquisitions (Ye et al. 2000), although this method is not compatible with combined BOLD/ASL measurements.

In terms of measuring blood flow accurately, the main complication for ASL is the short lifetime of the label, rather than the incomplete extraction of water. The MRI method "tags" water magnetically by inverting longitudinal magnetization proximal to the imaging slice, and then observing the distal effects on MRI signal due to blood water that has flowed into the imaging volume. In contrast to the PET method of labeling water using a radioactive oxygen isotope with a decay constant of about 2 min, the MRI label only survives about 1.5 s, which is the T_1 value for arterial water (Lu et al. 2004a). The short duration of the label is both a blessing and curse. On the one hand, rapid elimination of the label allows for repeated experiments on a time scale of seconds in order to follow dynamic changes in CBF or to improve the quality of basal CBF measurements through averaging. On the other hand, the time scale for delivery of blood from the labeling site to the tissue is comparable with the decay time of the label, so that care must be taken to ensure that the measurements really are reporting CBF, as opposed to regional or temporal variations in arrival times between the labeling site and tissue. Transit times can vary either due to variations in the path lengths of vessels or in the flow velocities.

In order to reduce the sensitivity of ASL measurements to transit times, a time delay is inserted after proximal tagging of water in order to ensure that all labeled water arrives in the imaging volume and is properly counted (Alsop and Detre 1996; Wong et al. 1998b). However, this time delay improves accuracy at the cost of sensitivity, because the additional time leads to loss of ASL signal through T_1 decay. Even with a post-labeling delay, transit time's issues produce systematic errors in measurements of CBF under some conditions. The most obvious examples occurs for stroke or vascular disease in humans, where transit times may be greatly extended as a result of low flow and collateral flow. This cloud may have a silver lining, however, as artifactually low values of flow may improve contrast with normal tissue. A recent study found that ASL detected flow abnormalities in over 40% of patients that had known or suspected vascular disease but normal findings from bolus gadolinium analysis (Zaharchuk et al. 2009).

There are two common variants of ASL that label proximal blood water in fundamentally different ways (Wong et al. 1998a; Williams 2006). Continuous ASL (CASL) tags blood by applying RF radiation to a single slice containing major arteries for extended time periods (1–2 s). This method is conceptually easy to understand, because the amount of labeled water that accumulates in the image volume is proportional to flow and to the time duration of the RF tag, plus a correction factor for decay of the label. Pulsed ASL (PASL) uses a single RF pulse of short duration (milliseconds) to invert water from a proximal slab of tissue. In the absence of any other manipulations of longitudinal magnetization, this method would not measure CBF, because blood is tagged as a volume, rather than as a flow-time product; an upper limit on the ASL signal would occur when all of the tagged blood reached the image volume. To turn PASL into a quantitative method, the spatial tagging scheme is converted to a temporal scheme by saturating the magnetization in the tagging volume after a fixed time delay so that blood remaining in the tagging volume does not contribute to the ASL signal. The selection of PASL tagging widths and timing is crucial for accurate measurements of CBF (Wong et al. 1998b).

Theoretically, CASL provides a large SNR improvement over PASL (Chesler and Kwong 1995; Buxton et al. 1998a), although experimental results in humans have found that actual improvements are less than 50% due to a variety of compensating factors (Wong et al. 1998a; Wang et al. 2002). In animal models on small-bore scanners, the CASL method is almost always employed (e.g., Silva et al. 1995), and this method can be expected to provide larger gains in sensitivity and accuracy relative to PASL as a result of higher CBF and smaller transit times in small animals. Implementation involves separate imaging and labeling RF coils that avoid magnetization transfer effects by physical separation. These small-bore systems always provide a second RF channel for NMR spectroscopy, and this channel can be used to drive the labeling coils using a continuous-wave RF amplifier that is obtained by the user. The CASL implementation is more complicated on human scanners, which don't provide hardware support for a separate RF channel. Instead, implementations use a single RF coil for both labeling and imaging and control off-resonant magnetization transfer using double inversion in the control acquisition (Alsop and Detre 1998). Recent refinements of this method circumvent power limitations and inefficiencies in labeling by using a train of labeling pulses to provide "pseudo-continuous" labeling (Dai et al. 2008).

Given the low detection efficiency for ASL, what is the role of this technique for fMRI? Some authors have argued that the small ASL signal changes may be offset for some applications by less slow-frequency noise in the method (Aguirre et al. 2002) relative to BOLD signal; slow signal drift is particularly problematic for measuring drug-induced responses or other responses that evolve slowly in time. Because ASL is a subtraction method, slow signal drift that is unrelated to physiology will cancel out in the subtraction. As a general method for brain mapping, however, ASL suffers from smaller signal changes, reduced temporal resolution due to the need to acquire two images for subjections, and few slices per unit time due to the labeling period.

Principally, ASL fills two roles in fMRI by (1) providing data on CBF reactivity to inform simultaneously acquired changes in BOLD signal (e.g., Sect. 17.6.1), and (2) measuring basal CBF in cross sectional BOLD studies to aid interpretation of population differences. The latter function is necessary in principle, at least, because BOLD signal depends upon baseline physiology in addition to functional reactivity. Altered flow states, perhaps as a result of drug exposure or aging or other group differences, can modulate BOLD signal by altering the baseline concentration of deoxygenated hemoglobin (see Eq. 17.1).

17.5 Alternative Methods to Measure CBV

Among alternative methods to measure CBV by MRI, this chapter has focused on the IRON technique due the high CNR of that method. T_1-based measurements using exogenous contrast agent are one obvious alternative method. As discussed previously, such an approach would combine the twin disadvantages of requiring exogenous contrast agent and delivering weak signal changes. While alternative CBV techniques provide signal changes much smaller than the IRON or BOLD methods, nevertheless there has been a desire to develop dynamic methods of measuring CBV without exogenous agent in order to clarify mechanisms of BOLD signal in human subjects.

17.5.1 VASO

Vascular space occupancy (VASO) recently became a popular technique due both to the simplicity of implementation and the potential to assess functional CBV changes in humans (Lu et al. 2003; Lu et al. 2004c). This T_1-based method employs a non-selective preparation pulse to globally invert MRI signal, and tissue and blood then relax back to equilibrium at different rates. Image data are acquired as inverted blood goes through the null point of longitudinal signal recovery. The simplest interpretation for this method posits only two compartments (blood and tissue) that exchange blood water rapidly across the BBB during dilation and contraction of vessels (Lu et al. 2003). In this interpretation, the volume fraction missing from the tissue compartment is CBV. Because VASO is such a weak signal, care must be taken to minimize residual BOLD and flow contributions (Donahue et al. 2006).

The original two-compartment model is incomplete, however, because an increase in CBV requires displacement of other volume. Although bidirectional exchange of water by diffusion across the BBB is a rapid process, unidirectional transport of water across the BBB under the influence of hydraulic or osmotic pressure is an extremely slow process that is governed by the Starling equation. During focal brain activation, capillary pressure increases as a consequence of arterial dilation, leading to a further increase in the hydraulic pressure gradient across the BBB. However, the direction of this force is opposite to the direction required by the original VASO interpretation, and the magnitude of water exchange is several orders of magnitude too small to be relevant on the time scale of the vascular response (Leite and Mandeville 2007), according to measurements of the hydraulic conductivity of the BBB (Bradbury 1985). Physiological principles suggest that increases in CBV are compensated by decreases in extravascular fluid, with displacement of cerebrospinal fluid (CSF) presumably providing the largest buffer for CBV variation. In fact, a recent study based upon multi-exponential T_2 fitting for CSF and tissue compartments found that CSF reductions were comparable to CBV elevations (Piechnik et al. 2009).

The extent to which this conceptual objection corrupts the original VASO interpretation is a subject of on-going research. One evaluation of VASO errors found a consistent but insignificant decrease in the CSF fraction in activated regions (Donahue et al. 2006), supporting the authors' original interpretation. A different group reported masking of stimulus-induced activation that might result from dynamic displacement of CSF (Scouten and Constable 2007). However, the most direct evidence that CSF plays a significant role in VASO signal comes from global elevation of CBV by hypercapnia, which produced both positive and negative signal changes using either blood-nulled or CSF-nulled VASO (Scouten and Constable 2008).

Two strategies have been proposed to reduce the influences of CSF displacement on VASO signal. The first employs a dual inversion preparation to null both blood water and CSF (Donahue et al. 2006). The second strategy combines separate blood-nulling and CSF-nulling steps into a combined index (Scouten and Constable 2008). Relative to the original VASO technique, both methods suffer an SNR penalty but hold promise for making VASO an interpretable signal in terms of isolating a CBV-specific mechanism. However, a final difficulty in relating VASO signal to other indices of functional physiology is that changes in VASO signal are related to changes in CBV without providing any information about basal CBV in any region, so it's not possible to report percent changes in CBV without additional information from another method or by assuming global values for basal CBV; the latter method is unlikely to be accurate except possibly for very large regions of interest that might average away regional differences.

The original VASO method reported that CBV resolves more slowly to baseline than CBF or BOLD signal (Lu et al. 2003), consistent with results from IRON fMRI. The authors subsequently reported that the slow response of CBV did not co-localize with BOLD and CBF responses, so the BOLD post-stimulus undershoot should be reinterpreted as due to a post-stimulus elevation of $CMRO_2$ (Lu et al. 2004b). In light of the CSF contamination to the VASO signal that since has been demonstrated, it unclear whether VASO signal changes should co-localize with other methods when CSF signal is not controlled. A very recent VASO study that included CSF-nulling found similar responses for CBF and CBV (Donahue et al.

2009). That study used very long single-shot acquisitions in order to increase the temporal efficiency, but this strategy also may dilute the specificity of the VASO signal by bringing in competing mechanisms.

17.5.2 VERVE

VERVE (venous refocusing for volume estimation) is another recently developed method that differs fundamentally from VASO or the use of exogenous contrast agent, in that VERVE aims to measure changes in the venous component of CBV, rather than total CBV. VERVE is derived from BOLD signal, so "venous" is most directly a reference to blood oxygenation in this case, rather to the anatomical definition of veins, which are differentiated from capillaries by smooth muscle.

The VERVE technique (Stefanovic and Pike 2005) takes advantage of the fact that the spin-echo BOLD signal has as substantially larger intravascular than extravascular contribution at clinical field strengths (Zhong et al. 1998; Oja et al. 1999). To further minimize the extravascular contribution, this method uses a subtraction between acquisitions that employ fast and slow refocusing pulses; this subtraction is the VERVE signal. In the absence of T_2 changes in the blood, activation-induced changes in venous CBV could be estimated in a two-compartment blood-tissue model (CSF is nulled) directly from VERVE signal changes. However, venous T_2 does change during activation, so this must be taken into account. This is accomplished by (1) empirical calibration curves for blood T_2 as a function of oxygenation and refocusing interval, and (2) a model for venous oxygenation changes as a function of changes in CBF (Stefanovic and Pike 2005; Chen and Pike 2009a).

There are a number of potential complications in this method that can affect accuracy and limit the general applicability of the method. The method has low sensitivity because it based upon a subtraction of weak BOLD spin-echo signals. Baseline venous blood oxygenation must be assumed or measured. Changes in venous oxygenation are flow-dependent, and estimation of this effect requires a physiological model; simulations demonstrate that this is perhaps a 20% uncertainty. The degree of error induced by extravascular BOLD changes has not yet been quantitatively estimated.

Despite these complications and drawbacks, the method has reported changes in venous blood volume that are not unreasonable, with about a 10% change for a robust stimulus. However, changes in venous CBV reported by the VERVE method don't scale with CBF (Chen and Pike 2009a). There are several potential reasons that VERVE might underestimate venous CBV changes in a way that becomes progressively larger as CBF increases. Potential reasons include an imperfect compensation for intravascular changes in oxygenation, a breakdown in the assumption that extravascular BOLD effects can be ignored even for large changes in CBF, or simply a dynamic range problem that results from the progressive loss of deoxyhemoglobin, the origin of the VERVE signal, at high levels of CBF. Despite low sensitivity and remaining questions about the quantitative nature of the VERVE signal, this method produced rough agreement with the IRON technique by associating the integrated magnitude of the BOLD post-stimulus undershoot with the integrated magnitude of the CBV response (Chen and Pike 2009b).

17.6 fMRI and Metabolism

17.6.1 Direct Measurement of CMRO$_2$ Reactivity

None of the techniques described in this chapter enable insights into glucose metabolism, but BOLD signal is sensitive to oxygen utilization, so information about CMRO$_2$ should be accessible in principle through the BOLD mechanism. For this purpose, a model is required. BOLD signal can be cast into any number of reasonable physical frameworks that are based upon conservation of mass. At the most basic level, BOLD signal is described by oxygen conservation during transit from oxygenated arteries to partially oxygenated veins due to a diffusive loss of blood oxygen into the tissue due for use in metabolism (Ogawa et al. 1993). Viewing the capillaries as a black box, the difference between oxygen flow into and out of the capillaries is the rate of oxygen delivery to the tissue, $F_{IN} - F_{OUT} = dV/dt$, where flow (F) and volume (V) refer to oxygen rather than blood in this equation. This description, discussed below for measurements of CMRO$_2$ reactivity, ignores transit time phenomena such as the time-dependent washout of deoxygenated hemoglobin. It is a steady state model that is not appropriate for describing BOLD changes on a time scale of a few seconds, but this approximation is not bad for BOLD transients of longer duration like the post-stimulus undershoot. For modeling purposes, a better physical framework would apply mass conservation to the blood as well as to the oxygen in order to allow for transient differences between blood flow and volume; this is the Balloon model (Buxton et al. 1998b). If pronounced changes in hematocrit were discovered to occur during functional activation, separate conservation of the plasma and red cell volumes could further extend this model. All physical models of this type conserve mass but say nothing about the underlying physiology.

To determine CMRO$_2$ reactivity, a measurement model must limit the number of free parameters and be relatively insensitive to any parameters that cannot be directly measured. Consider a simple model of BOLD percentage signal changes in terms of fractional changes in CBV (v), CBF (f), and CMRO$_2$ (m):

$$\left.\frac{\Delta S}{S}\right|_{BOLD} = M\left[1 - v\left(\frac{m}{f}\right)^{\beta}\right] \tag{17.1}$$

The parameter β accounts for a slightly superlinear dependence of the blood relaxation on blood magnetization (Boxerman et al. 1995b). In human studies where it is difficult to dynamically measure CBV, a coupling relationship is assumed with CBF ($v = f^{\alpha}$) so that

$$\left.\frac{\Delta S}{S}\right|_{BOLD} = M\left[1 - m^{\beta} f^{\alpha-\beta}\right] \quad (17.2)$$

The parameter M denotes the maximal possible BOLD signal change. Because BOLD signal reflects the washout of deoxygenated hemoglobin, there is finite headroom for BOLD changes that would be depleted if venous blood became completely oxygenated. M is the product of the echo time and the BOLD-induced relaxation rate (R_2^{BOLD}) during the "resting" or control condition, which in turn depends upon baseline physiology with a proportionality to the local CBV and the ratio of $CMRO_2$ to CBF; the latter is roughly constant across the brain, according to PET measurements (Mazziotta and Phelps 1986). Thus, regional variation in M should roughly reflect regional variations in CBV.

From this description of BOLD signal, it is apparent that the parameter M is playing a role similar to the agent-induced relaxation rate observed upon injection of contrast agent using the IRON method. The latter quantity is readily measurable and is used to normalize signal changes in order to report percentage changes in CBV. For BOLD signal, we can neither estimate changes in $CMRO_2$ nor report percentage changes in deoxyhemoglobin without first finding a way to measure the parameter M. The key insight that enabled BOLD-based measurements of $CMRO_2$ reactivity was that manipulation of blood carbon dioxide levels leads to alterations in blood flow and volume without substantially affecting metabolism, as least on a global basis (Eklöf et al. 1973; Horvath et al. 1994; Yang and Krasney 1995). Therefore, simultaneous BOLD and CBF measurements during hypercapnia can be used to determine M in each image region, and then this value can be used to estimate relative changes in $CMRO_2$ during a metabolically modulated stimulus.

Using this method, experiments in our laboratory were conducted in humans (Davis et al. 1998) and also in rodents (Mandeville et al. 1999a), where IRON fMRI enabled dynamic measurements of CBV as well as CBF. Following the Davis approach, experiments at the Montreal Neurological Institute employed graded hypercapnia and graded visual stimulation to map out an approximately linear coupling in which percentage changes in CBF exceeded percentage changes in $CMRO_2$ (Hoge et al. 1999b). This functional coupling ratio generally has fallen within two to four in healthy human subjects using a variety of stimuli (Davis et al. 1998; Hoge et al. 1999b; Kim et al. 1999; Kastrup et al. 2002; Stefanovic et al. 2004; Uludag et al. 2004; Leontiev et al. 2007).

A number of basic points about the methodology are worth emphasizing. First and foremost, it needs to be emphasized that *details of the model are relatively unimportant if these three assumptions are true*:

Assumption #1): $CMRO_2$ does not change in the region of interest during hypercapnia.

Assumption #2): Hypercapnia and metabolically modulated stimuli produce similar vascular responses in terms of the relationship between CBF and CBV.

The main features of this technique are summarized here:

- While there are several obvious ways to build a more complete theoretical model of BOLD signal, the model presented above enables measurements of $CMRO_2$ reactivity by providing a means to measure M, which depends upon details of the MRI pulse sequence and local physiology. A twofold regional variation in the local blood volume fraction translates into a twofold variation in M, which is such a large range as to render estimations of $CMRO_2$ meaningless when M is not determined empirically on a local basis.
- The method is surprisingly insensitive to uncertainties in α and β due to the "calibration" step provided by the hypercapnia experiment (Davis et al. 1998), particularly when flow changes in the hypercapnic calibration are nearly matched to levels during the metabolically modulated stimulus. When model parameter values are erroneous, the empirical calibration will adjust the model locally to match the data, thereby factoring error into M with little effect on the estimated of $CMRO_2$ reactivity. M reflects the true maximal BOLD signal value (e.g., the resting state BOLD relaxation rate) only to the extent that the model is correct.
- This method fails at very high resolution. The largest BOLD signal changes occur in large draining veins, where ASL signal is small due to the nearly complete extraction of first-pass labeled blood water in the capillaries. For this reason, more accurate estimations of $CMRO_2$ are obtained using large regions of interest or by determining M from pooled subject data as a way of minimizing BOLD/ASL spatial mismatches due to the different vascular weightings of the two techniques (Leontiev et al. 2007).
- Based upon IRON fMRI data obtained in rodents models (Mandeville et al. 1999b) and in non-human primates as reviewed previously, the CBF-CBV coupling relationship (expressed through the parameter α) is not temporally static. CBV continues to increase after CBF and BOLD signal have reached a maximal response. It is important to minimize influences of these transient effects on estimations of $CMRO_2$ reactivity by discounting the first few tens of seconds immediately following onset and cessation of stimulation.

This fMRI technique already has changed some of our previous understandings of flow-metabolism coupling. Of course, PET experiments from the 1980s (Fox and Raichle 1986) discovered that functionally induced changes in $CMRO_2$ were much smaller than changes in CBF and glucose metabolism, an observation that initially was described as "uncoupling" of $CMRO_2$ with CBF and glucose metabolism. Based upon fMRI techniques, it now appears that CBF and $CMRO_2$ couple in very predictable way, but not along the linear trajectory defined by regional CBF-$CMRO_2$ values.

In retrospect, the deviation of $CMRO_2$-CBF coupling from the regional relationship should not have been entirely surprising. As an analogy, regional values of CBF and CBV show a linear coupling; MRI maps of mean transit time, the CBV/CBF ratio, exhibit virtually no contrast in healthy human subjects. However, relative functional changes in CBV are much smaller than corresponding values for CBF (Grubb et al. 1974). Intuitively, these data are reconciled by noting that factors that govern regional relationships, like angiogenesis during development, are not operable on the time scale of functional activation. In other words, CBF increases as a result of vessel dilation during a functional response, rather than by growing new vessels. This same principle appears to hold true for the CBF-$CMRO_2$ relationship. While capillary density and flow can be adjusted regionally on a long-term basis, the oxygen extraction fraction necessarily falls during a functional response (Buxton and Frank 1997) as a consequence of the relatively constant capillary surface area, which increases only nominally by capillary dilation during functional activation.

This fMRI method of determining $CMRO_2$ reactivity will never generalize to routine functional studies as a result of limited ASL sensitivity. However, the method should prove suitably efficient to investigate targeted questions about neurovascular coupling, such as how to view fMRI results using drugs that alter metabolism but also directly affect vascular smooth muscle. Current questions about methodology concern the accuracy of ASL during global challenges that can dramatically alter blood transit times, and the resolution at which the BOLD model of Eq. 17.1 can be applied to both local and global perturbations without a breakdown due to vascular mixing amongst neighboring voxels.

17.6.2 phMRI in Pharmacology

While there is a burgeoning field of "phMRI" in humans, these studies rarely assess the direct effects of drugs, but rather use traditional fMRI stimulation paradigms with and without drug treatment or pretreatment (Wise and Tracey 2006; Anderson et al. 2008). Assessing the direct effects of drug is more difficult from a detection standpoint, because the response evolves slowly and is not amenable to the repeated on-off paradigms that can control signal drift while building statistical power by within-subject averaging. Human BOLD studies designed to characterize the functional response to drug infusion generally have produced disappointing results. This is not surprising; we obtained poor results using BOLD signal for phMRI at low magnetic fields in animal models (Mandeville et al. 2001; Chen et al. 2001), a result that largely motivated our interest in the IRON technique.

Figure 17.8 illustrates the power of this technique using an example that is relevant for this chapter because it demonstrates an interesting result that can be related to previous gold-standard methods for assessing metabolism, and it also demonstrates the challenges for this method in terms of interpretation. The figure maps the response of CBV to an acute infusion of 0.5 mg/kg cocaine in an awake non-human primate (rhesus). The result is interesting because the sign of the response is opposite the sign of the response in rodents using fMRI (Marota et al. 2000), gold-standard flow methods based upon autoradiography (Stein and Fuller 1993), or measurements of flow and glucose metabolism (Sharkey et al. 1991). However, this species difference between rats and rhesus monkeys previously was reported using the 2-deoxyglucose technique to assess cocaine-induced changes in glucose metabolism (Lyons et al. 1996).

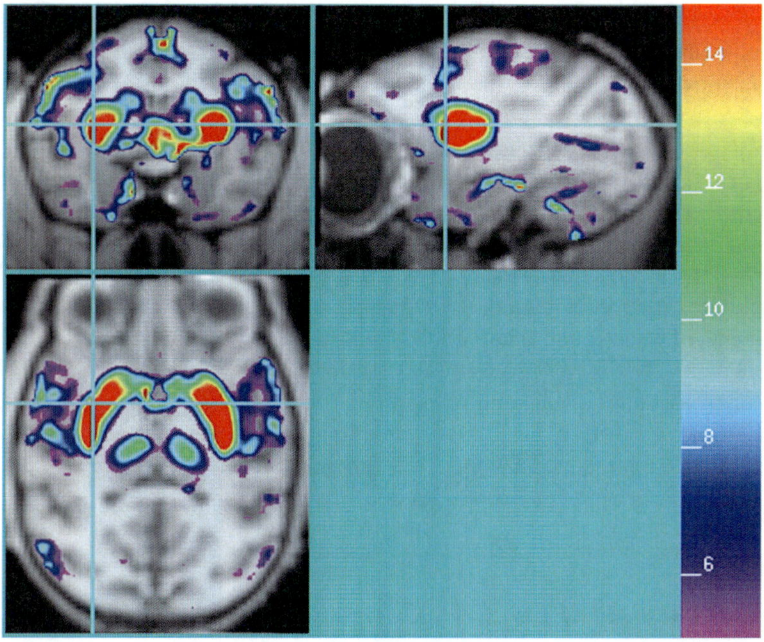

Fig. 17.8 The peak response of CBV in conscious non-human primate following 0.5 mg/kg cocaine infusion is shown using an overlay for negative changes in blood volume. No positive changes in CBV were observed for non-contingent infusion of cocaine, a result opposite to that observed in rodents

This cocaine result also illustrates that fMRI, like metabolism, represents a summation of synaptic consequences that sometimes can be difficult to interpret. Firstly, fMRI measures something very different from PET binding experiments, in that metabolic consequences can be anticipated both at binding sites and in connected circuitry. Even when the drug acts through a single neurotransmitter system like dopamine, post-synaptic receptor binding can be excitatory or inhibitory. Selective D1 and D2 receptor stimulation produce opposite effects on metabolism and fMRI signal (Trugman and James 1993; Marota et al. 2000; Chen et al. 2005), so cocaine might produce opposite effects in the rat and monkey due to different levels of D1 and D2 receptors in these species. Alternatively, cocaine also acts as an indirect serotonergic agonist, producing metabolic modulation through the serotonin system while also potentially reinforce the inhibitory D2 effects. Finally, experiments in conscious animals or human subjects may contain cognitive components in the activation profile.

If fMRI can be viewed as assessing drug-induced metabolism but with the addition of temporal information, what have we gained? One thing we have gained is the ability to employ differential tests in longitudinal studies. The simple assessment of a drug's activation profile represents a differential test relative to the post-drug period, which is not possible with the deoxyglucose approach or with steady-state PET imaging of receptor binding. Repeated infusions can define the proper timing for drug-antagonist-drug studies, providing within-session differential tests of receptor antagonism, an important statistical advantage relative to across-session comparisons. Selective agonists can be used for comparison of regional activation profiles.

phMRI promises to contribute to our understanding of drug effects in the central nervous system, to evaluate longitudinal changes in response to drug as a consequence of repeated exposure (e.g., addiction) or the evolution of pathology (e.g., Parkinson's disease), and even to characterize cognitive components associated with drug intake in properly designed experiments. The information provided by the method is less interpretable than PET binding studies but also more general, in that assessments of brain function are not limited by available agonists or by the density of receptors in a few target regions. Interpreting the fMRI data will require an accumulation of evidence using multiple agonist or antagonists to dissect sources of signal. Studies of drug-induced brain function using the new generation of combined PET/MR scanners should improve our ability to take advantage of the complementary information provided by these different approaches.

17.6.3 When fMRI Does not Correlate with Metabolism

fMRI does not always correlate with metabolism due to a number of factors that generally are well known but warrant brief discussion. Changing blood levels of carbon dioxide associated with changes in respiration represent a major potential confound. Drug infusion (e.g., opiates) or stressful tasks can alter respiration in free-breathing subjects, leading to non-specific dilation or contraction of cerebral vessels. While CBF, CBV, and BOLD signal are all susceptible to this artifact, BOLD signal has the most sensitivity to CO_2. As discussed in Sect. 17.6.1, metabolically modulated stimuli produce changes in $CMRO_2$, whereas hypercapnia does not. As a consequence, hypercapnia-induced BOLD signal changes are about twice as large as those produced by a visual stimulus for matched changes in blood flow (Hoge et al. 1999a). This means that CO_2-induced BOLD signal changes can be expected to swamp metabolically induced BOLD changes, although changing levels of blood CO_2 are a major problem for other fMRI techniques as well.

Changes in blood pressure appear to be a less significant problem as long as pressure is maintained on autoregulatory plateau between pressures of about 60 and 150 mm Hg. Of course, textbooks properly note that the brain autoregulates CBF between within this range by adjusting arterial diameter, so one might expect fMRI based upon CBV to be susceptible to changes in systemic blood pressure. In fact, our experience has shown IRON fMRI to be surprisingly insensitive to changes in blood pressure. For instance, correlation of signal with large changes in blood pressure due to infusion of an opiate agonist revealed an association in the hypothalamus, a region involved in pressure autoregulation, without significant effects elsewhere in the brain (Liu et al. 2007). This result can be understood by recognizing that the brain regulates perfusion pressure distal to arterioles, so that the dominant blood volume fraction in the capillaries and veins sees a constant pressure as long as arteries can compensate for the changes in pressure. Progressive reduction of blood pressure in a rodent model showed that BOLD signal and CBV were very insensitive to pressure changes across the autoregulation range (Zaharchuk et al. 1999). BOLD studies in human subjects support this claim (Liu et al. 2006).

17.7 Future Directions

17.7.1 IRON fMRI in Humans?

BOLD signal will continue to be the dominant fMRI contrast mechanism in humans for the foreseeable future, although ASL increasingly is being used to assess baseline physiology in order to aid interpretation of BOLD results. Is it possible that IRON fMRI will become a viable tool for clinical fMRI application? At this time, the only USPIO (ultra-small superparamagnetic) agent approved for use in humans at a dose efficacious for fMRI is Feraheme, which has been approved for the treatment of iron deficiency anemia in adult patients with chronic kidney disease (Coyne 2009). This agent has a blood half-life of 15 h and is approved at a unit dose of 510 mg.

In order to assess the relative detection power using an agent like Feraheme at a dose that has already passed through clinical trials, we validated theoretical CNR calculations as a function of dose (Mandeville et al. 2004) by using sequential doses of Feraheme in an awake non-human primate. Figure 17.9 summarizes the predicted CNR benefits in humans relative to BOLD signal at 1.5 and 3 T. Because the approved dose of iron is not normalized by patient weight, the shaded dose range in the figure includes subjects within the weight range of 50–100 kg. At 1.5 T, the clinical field strength that dominates the market, CNR should be increased roughly fivefold for the average subject. The benefits at 3 T will be smaller, due to larger BOLD signals at that field strength. At ultra-high field strengths like 7 T, currently a field strength limited to research, current dose limitations in human subjects for these types of agents will prevent a benefit relative to BOLD signal.

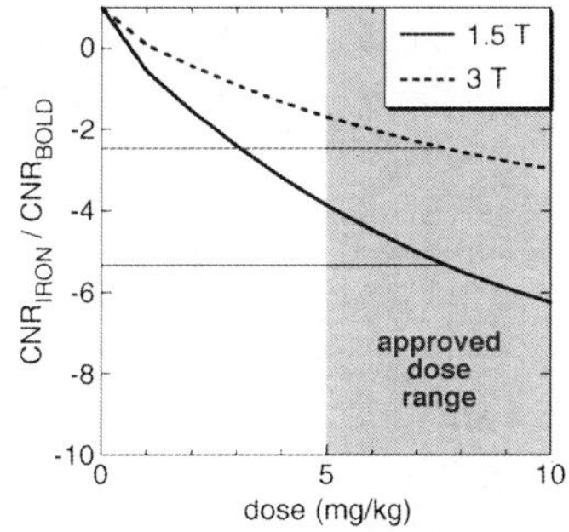

Fig. 17.9 Predicted boosts in CNR at 1.5 and 3 T using the FDA-approved dose of Feraheme, an iron replacement therapy for patients suffering from chronic iron anemia. The *gray* shaded region approximates the weight-normalized dose for the drug, which was approved for adult subjects without regard to weight

BOLD signal has proven to be an excellent research tool in multi-subject studies for mapping brain regions associated with sensory, motor, and cognitive function in health and disease states. For studies of this type, the ethical and practical burdens to recruiting subjects for IRON fMRI would appear to be unmanageable ("Wanted: OCD subjects who also suffer from iron anemia"), unless iron oxide agents of this type receive wider approval that perhaps is conditional on phlebotomy or assessment of blood iron levels. However, it is not clear IRON fMRI would significantly benefit population studies, where across-subject variance ("random effects" in the fMRI parlance) competes with within-subject variance ("fixed effects").

Individualized medicine may be a different story, in that the limitations of BOLD signal become more pronounced when results must be interpreted on a per-patient basis; limited BOLD sensitivity certainly is among the factors that have inhibited the clinical utility of fMRI within individual patients (Jezzard and Buxton 2006). Preoperative brain mapping prior to tumor resection is one application where the risk-benefit tradeoff might work in favor of IRON fMRI. A single dose of Feraheme represents about 10–15% increase in normal total body iron stores of about 60 mg/kg (Barton et al. 2000). However, further studies will be required to clarify neurovascular coupling in the vicinity of a tumor (Sunaert 2006) and to characterize potential complications due to agent extravasation into tissue near a compromised BBB (Enochs et al. 1999).

Pharmacological stimuli represent a more speculative potential application for IRON fMRI in human subjects. IRON fMRI in animal models have shown excellent within-subject results. For instance, dopamine deficits in a monkey model of Parkinson's disease were probed using a dopaminergic stimulus, and the MRI results correlated with both behavior and PET binding potential (Jenkins et al. 2004). The larger installation base and higher resolution of MRI relative to PET could extend drug stimuli into clinical use as a tool to aid diagnosis or follow treatment.

17.7.2 Leveraging the Temporal Response of phMRI

For the vast majority of studies, fMRI provides no differentiation of neural functions based upon temporal response, which is slower than the neuronal response by several orders of magnitude. An exception occurs for pharmacological stimuli, which produce a response that generally evolves so slowly that fMRI directly measures the response of neuronal ensembles. The observation that different regions respond with different kinetics (Marota et al. 2000), and that biphasic responses within an image voxel are observed in some situations (Liu et al. 2007), illustrates that fMRI reflects dynamical responses of the binding targets (receptor or transporter) in addition to drug availability.

Homeostatic feedback mechanisms long have been thought to play some role in governing plastic changes in the brain in response to chronic drug exposure, but emerging evidence suggests that neurotransmitter function may be regulated dynamically even on a timescale of minutes during an acute drug challenge. *In vitro* experiments have identified cellular trafficking of transporters and receptors on this time scale in response to the rate of stimulation due to the presence of agonist (Johnson et al. 2005; Goodkin et al. 2005; van Rijnsoever et al. 2005). In some cases, fMRI temporal dynamics may offer clues about the origins of signal and the functional connections between brain regions, as we proposed based upon fMRI observations using remifentanil (Liu et al. 2007).

From these fMRI and *in vitro* studies, it is apparent that there is more information to be gleaned from the temporal response to the drug that simply the average response magnitude, which is generally reported. It will be interesting to see whether specific temporal components of the signal can be blocked by antagonists or modified by drug exposure history. One can imagine that continued improvements in detection power and analysis strategies will significantly increase the information content of phMRI in the near future by exploiting the temporal response.

17.7.3 Physiology Underlying fMRI

fMRI mechanisms can be divided into three main parts: (1) the physics underlying the signal, (2) vascular and metabolic physiology as they contribute to the signal (e.g., CBF-$CMRO_2$ coupling), and (3) the coupling of flow and other fMRI indices to metabolism and neural processing. The last topic is the subject of many chapters in this volume and won't be discussed here. The physics underlying fMRI signal now is pretty well understood. While we know much of the physiology underpinning fMRI signals, nevertheless a lot of questions remain.

The main hindrance to a more detailed understanding of BOLD signal relates to measurement limitations *in vivo*. BOLD physiology has proven to be a tough nut to crack mainly due our inability to accurately and dynamically measure $CMRO_2$. PET methods cannot address the temporal response of $CMRO_2$ and have their own limitations in terms of accuracy. In MRI or optical studies, $CMRO_2$ is always an invisible variable that can be measured only by inference: essentially, it is the missing oxygen mass not detected by other measurements. This produces an unsatisfying experimental situation that has persisted for almost two decades: $CMRO_2$ can always be invoked to reconcile BOLD data with other measurements (CBF, CBV). Given this situation, it is critical to understand sources of error attached to each of the CBF and CBV methods that are applied in the extraction of $CMRO_2$, and to approach this issue using multiple techniques.

17.7.3.1 What is IRON Signal Measuring?

From a physical viewpoint, we know that changes in IRON signal can be quantitatively related to changes in the cerebral blood plasma volume (CPV). We typically refer to these changes as "CBV" under the assumption that changes in hematocrit are small enough to be ignored, but this may not be true. Presumably, potential changes in hematocrit would be most pronounced in capillaries, where red blood cells stream in single file, and plasma filling during dilation might occur primarily in the slow-velocity regions near the vessel walls. However, data on hematocrit during brain activation currently are unavailable, and the complexities of colloidal flow within the multiple blood volume compartments represent a hindrance to accurate modeling of the response. Recent measurements of CBV in humans using VERVE, a method that measures only the CBV contribution to BOLD signal, produces qualitative agreement with the IRON temporal dynamics that have been reported by us and others during the past decade, supporting a role for CBV in the BOLD post-stimulus undershoot (Chen and Pike 2009b).

However, two-photon microscopy, which enables direct visualization of vessel calipers, has produced results that appear to conflict with interpretation of the IRON temporal response as due to a slow venous response. Using this method in a open-skull rat model, venules showed little change in diameter during brain activation, even while optical imaging found a delayed capillary/venous response in a closed-skull model using the same stimulus (Hillman et al. 2007). In terms of relating these results to fMRI, there are two main concerns. The first is that removal of the skull alleviates intracranial pressure, which may alter vascular dynamics in veins, which operate at low pressure. For this reason, it will be important to determine whether opening the skull model, a requirement for very high-resolution optical imaging, affects details of the vascular dynamics like the slow response of IRON signal and the BOLD post-stimulus undershoot. A second concern is whether vessels in the superficial layers of cortex respond similarly with microvessels in deeper layers. The post-stimulus tail of the IRON response has been shown to be much smaller on the cortical surface than in deep cortical tissue (Yacoub et al. 2006).

One reason to better understand IRON signal is to better understand the physiology underlying BOLD signal, which depends upon the volume of deoxygenated hemoglobin, rather than upon CPV or total CBV. If we overestimate the role played by blood volume in BOLD signal, we underestimate the contribution of $CMRO_2$. In terms of BOLD temporal dynamics, it is clear that CBV (or at least CPV) has temporal dynamics that match elements of BOLD dynamics like the post-stimulus undershoot. In fact, IRON temporal dynamics are perhaps the most reproducible correlate to the BOLD undershoot in the literature, so understanding this signal is a good starting point for understanding BOLD dynamics.

To illustrate how uncertainties in the CBV response affect our understanding of BOLD dynamics, let's constrain the space of possible models by the following two postulates: (1) We can ignore changes in hematocrit, so relative changes in CPV, CBV, and red blood cell volume are equivalent, and (2) the relationship between CBF and $CMRO_2$ is governed by limitations on the diffusion rate of oxygen into the extravascular space. The latter assumption follows the Buxton–Frank model (Buxton and Frank 1997) in which fractional changes in $CMRO_2$ (m) are related to fractional changes in flow (f) with a set point that depends upon the baseline oxygen extraction fraction (E_0):

$$m = (1-g_0)\frac{f}{E_0}\left[1-\frac{(1-E_0)^{v/f}}{1-g_0}\right] \tag{17.3}$$

Fig. 17.10 A hypothetic scenario for the BOLD post-stimulus undershoot in which elevation of $CMRO_2$ in the post-stimulus region shifts the BOLD zero-crossing point relative to the CBF-CBV cross-over point

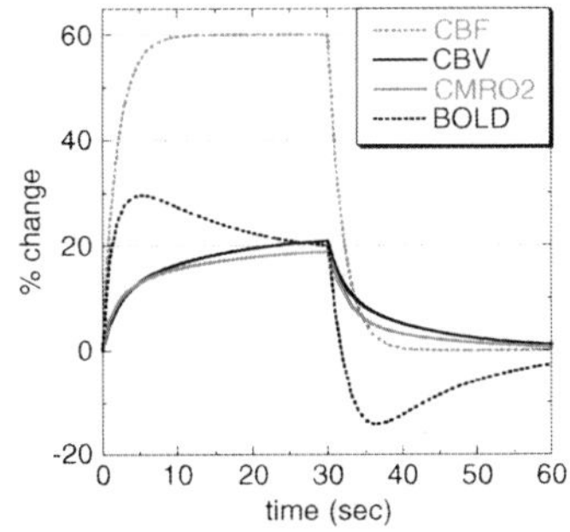

Relative to the original model, this expression includes a tissue oxygen buffer (g_0) that is unchanging for the sake of this discussion (Zheng et al. 2002) and capillary volume expansion expressed as a fractional value (v). A relatively small increase in capillary blood volume (e.g., $v = f^{0.2}$) doubles the CBF:$CMRO_2$ reactivity ratio from 6:1 to 3:1, placing it in much better agreement with published data using the calibrated BOLD method (Sect. 17.6.1). However, even under these relatively restrictive conditions, there are several reasonable hypotheses for the mechanisms underlying the IRON response and the BOLD post-stimulus undershoot that all are consistent with a temporal correlation between BOLD and IRON signals:

Hypothesis #1: *The slow phase of the CBV response is due to stress relaxation in venous smooth muscle, and the BOLD post-stimulus undershoot is attributed to a temporal mismatch between CBF and CBV. Post-stimulus undershoots in CBF may occur for some stimuli as a consequence of neuronal coupling, but this is not a general phenomena, and $CMRO_2$ is temporally linked to CBF. In vitro* experiments in large vessels demonstrate a multi-phase temporal response of diameter in response to a step changes in pressure (Porciuncula et al. 1964), which made this mechanism an attractive solution to the data (Mandeville et al. 1999b). However, stress relaxation seems unlikely to produce such a linear response as shown in the data (Leite et al. 2002; Lu et al. 2005; Silva et al. 2007), and optical microscopy finds minimal changes in venous diameter during the functional response (Hillman et al. 2007).

Hypothesis #2: *The slow phase of the CBV response is due to rate limitations on displacement of CSF within the closed cranium, a mechanism that does not distinguish between capillaries and veins. As a consequence, the BOLD post-stimulus undershoot results from slow decay of both CBV and $CMRO_2$ during the post-stimulus time window after CBF resolves to baseline.* In this scenario, $CMRO_2$ is elevated as a consequence of elevated capillary CBV during the post-stimulus period. Although it seems unlikely that oxygen utilization would be temporally correlated with a mechanical constraint like displacement of CSF, this scenario is presented to highlight the point that the diffusion-limited model does not necessarily imply a temporal coupling of CBF and $CMRO_2$ if capillary CBV follows an altered time course.

This scenario is plotted in Fig. 17.10 for illustrative purposes. Underlying the CBV response is a rapid 50% increase in arterial blood at a 20% blood volume fraction, a slow 20% increase in venous CBV at a 40% blood volume fraction, and a slow 10% increase in capillary CBV at a 40% blood volume fraction. The $CMRO_2$

response was computed from Eq. 17.3, and the BOLD response was computed from Eq. 17.1 and normalized relative to the maximal BOLD response, so that this approximately equals the negative percentage change in deoxygenated hemoglobin.

Hypothesis #3: *The slow phase of the CBV response is due to rate limitations on displacement of CSF. Through a metabolism-flow feedback loop, arteriole diameters adjust flow to temporally regulate $CMRO_2$ to stimulus demands. This requires a post-stimulus undershoot in CBF in response to the post-stimulus elevation of capillary CBV; both effects contribute to the BOLD post-stimulus undershoot, although the mechanical response of CBV is the ultimate cause of the undershoot.* This mechanism temporally couples two phenomena, a slow response of CBV and a post-stimulus undershoot of CBF, which often have been reported in the literature.

The three scenarios above all produce the same multi-phasic CBV response but attribute the BOLD post-stimulus undershoot to either (1) CBV alone, (2) CBV and $CMRO_2$, or (3) CBV and CBF. How might we disentangle these scenarios? In our study in anesthetized rodents (Mandeville et al. 1999a), we observed the post-stimulus undershoot to begin at approximately the "cross-over" point when CBF fell below CBV in the post-stimulus region. Such a simple metric might prove to be more robust than directly calculating $CMRO_2$ through Eq. 17.1, which requires a hypercapnia calibration and which can suffer from error propagated from all the combined measurements. As shown in the simulation of Fig. 17.10, $CMRO_2$ elevation in the post-stimulation region would shift the BOLD undershoot onset relative to the CBF-CBV crossover point.

Not all studies have observed the same temporal dynamics for IRON signal as reported by us (Jin and Kim 2008), and this method is not yet available in human subjects. Evolving techniques like VERVE and VASO may help clarify the physiology underlying the IRON response and its association with BOLD dynamics.

17.7.3.2 Are Changes in CBF and $CMRO_2$ Spatially Variable?

While almost every region of human gray matter has been probed with BOLD signal under some task condition during the past 20 years, a persistent question has been whether or not there are "BOLD silent" regions in the brain or whether some regions intrinsically are harder to map by the BOLD method due to regional variations in the coupling relationship between $CMRO_2$ and CBF. Because BOLD signal in any given brain voxel depends upon the baseline concentration of deoxyhemoglobin, or "M" in the notation of Eq. 17.2, there are expected to be regional differences in the BOLD-CBF coupling ratio and in the BOLD-CBV ratio. From our studies in rats, the measured BOLD-CBV ratio varies by about a factor of 3 across brain, and this variation is completely consistent with the variation in basal CBV, which is easily measured using contrast agent and which should be proportional to "M" (Mandeville et al. 2001). These results suggest that $CMRO_2$-CBF coupling and $CMRO_2$-CBV coupling are regionally invariant.

Hypercapnia calibration in humans has produced a range of results for the coupling ratio of CBF and $CMRO_2$ changes, and much of the variance probably can be ascribed to systematic and statistical measurement errors associated with the methods (Leontiev et al. 2007). A recent study highlights some of the challenges presented by this technique. A combined visual-motor task was used to activate both visual cortex and basal ganglia, and the BOLD to CBF activation ratio was almost threefold smaller in basal ganglia than in visual cortex based upon one criterion for selecting voxels. Unlike our rat study (Mandeville et al. 2001), the hypercapnia calibration technique attributed this variance to the $CMRO_2$-CBF coupling ratio, rather than to "M". While the authors did an extensive review of potential sources of error, the explanation may be rather simple: the method doesn't function well at high resolution. A draining vein from a neighboring voxel may exhibit no oxygenation changes in a voxel of interest for a motor stimulus, but the same vein may dominate BOLD activation in the voxel during a global hypercapnia challenge. Further studies may clarify this issue, but this result demonstrates the challenges of quantifying many aspects of BOLD physiology using available techniques.

17.7.4 Summary

The recent history of fMRI has been one of many incremental improvements that appear small in isolation but ultimately lead to significant improvements in our imaging capabilities. Magnetic field strengths like 7 T that once were labeled as "ultra high" are proliferating. Parallel imaging is only about 10 years old, and already this method is a standard feature on clinical scanners using 32-channel phased array coils or custom-designed arrays with more than 100 elements.

There have been new developments in contrast mechanisms as well. IRON fMRI has become more prevalent in animal studies due to increased availability of contrast agents and continuing demonstrations of efficacy versus other methods, using a wide range of magnetic field strengths and animal models. This method undoubtedly will be extended to human fMRI in the near future, where its long-term role is uncertain. Pseudo-continuous ASL promises to improve the sensitivity and utility of CBF measurements on clinical systems. Persistent efforts to quantify physiologically relevant information using low-SNR endogenous methods like VASO and VERVE are beginning to bear fruit. This broad range of techniques will improve the specificity and accuracy of physiological information provided by fMRI and help clarify remaining uncertainties about the origins of BOLD signal.

Acknowledgements Numerous people contributed to this work by collaboration; I hope their contributions are reflected adequately in the cited publications. There are some collaborators who contributed to work that has been presented in this chapter but not in journal publications, including Georg Royl, Francisca Leite, Ji-Kyung Choi, Bruce Jenkins, Roger Tootell, Wim Vanduffel, and Marge Livingstone.

References

Aguirre GK, Detre JA, Zarahn E, Alsop DC (2002) Experimental design and the relative sensitivity of BOLD and perfusion fMRI. Neuroimage 15:488–500

Alsop DC, Detre JA (1996) Reduced transit-time sensitivity in noninvasive magnetic resonance imaging of human cerebral blood flow. J Cereb Blood Flow Metab 16:1236–1249

Alsop DC, Detre JA (1998) Multisection cerebral blood flow MR imaging with continuous arterial spin labeling. Radiology 208:410–416

Anderson IM, McKie S, Elliott R, Williams SR, Deakin JF (2008) Assessing human 5-HT function in vivo with pharmacoMRI. Neuropharmacology 55:1029–1037

Axel L (1980) Cerebral blood flow determination by rapid-sequence computed tomography: a theoretical analysis. Radiology 137:679–686

Barton JC, McDonnell SM, Adams PC, Brissot P, Powell LW, Edwards CQ, Cook JD, Kowdley KV (2000) Management of hemochromatosis, vol 1. Cambridge University Press, Cambridge, UK

Belliveau JW, Rosen BR, Betteridge D, Kennedy DN, Vevea JM, Johnson KA, Cohen MS, Weisskoff RM, Rzedzian RR, Brady TJ (1990a) Functional NMR imaging of the human brain. In: Proceedings of the ninth annual meeting of the society of magnetic resonance in medicine, New York, SMRM, p 181

Belliveau JW, Rosen BR, Kantor HL, Rzedzian RR, Kennedy DN, McKinstry RC, Vevea JM, Cohen MS, Pykett IL, Brady TJ (1990b) Functional cerebral imaging by susceptibility-contrast NMR. Magn Reson Med 14:538–546

Belliveau JW, Kennedy DN Jr, McKinstry RC, Buchbinder BR, Weisskoff RM, Cohen MS, Vevea JM, Brady TJ, Rosen BR (1991) Functional mapping of the human visual cortex by magnetic resonance imaging. Science 254:716–719

Boas DA, Jones SR, Devor A, Huppert TJ, Dale AM (2008) A vascular anatomical network model of the spatio-temporal response to brain activation. Neuroimage 40:1116–1129

Boxerman JL, Bandettini PA, Kwong KK, Baker JR, Davis TL, Rosen BR, Weisskoff RM (1995a) The intervascular contributions to fMRI signal change: Monte Carlo modeling and diffusion-weighted studies in vivo. Magn Reson Med 34:4–10

Boxerman JL, Hamberg LM, Rosen BR, Weisskoff RM (1995b) MR contrast due to intravascular magnetic susceptibility perturbations. Magn Reson Med 34:555–566

Boynton GM, Engel SA, Glover GH, Heeger DJ (1996) Linear systems analysis of functional magnetic resonance imaging in human V1. J Neurosci 16:4207–4221

Bradbury MWB (1985) The blood-brain barrier: transport across the cerebral endothelium. Circ Res 57:214–222

Burock MA, Buckner RL, Woldorff MG, Rosen BR, Dale AM (1998) Randomized event-related experimental designs allow for extremely rapid presentation rates using functional MRI. Neuroreport 9:3735–3739

Buxton RB, Frank LR (1997) A model for the coupling between cerebral blood flow and oxygen metabolism during neuronal stimulation. J Cereb Blood Flow Metab 17:64–72

Buxton RB, Frank LR, Wong EC, Siewert B, Warach S, Edelman RR (1998a) A general kinetic model for quantitative perfusion imaging with arterial spin labeling. Magn Reson Med 40:383–396

Buxton RB, Wong EC, Frank LR (1998b) Dynamics of blood flow and oxygenation changes during brain activation: the Balloon model. Magn Reson Med 39:855–864

Buxton RB, Uludag K, Dubowitz DJ, Liu TT (2004) Modeling the hemodynamic response to brain activation. Neuroimage 23(Suppl 1):S220–S233

Chen JJ, Pike GB (2009a) BOLD-specific cerebral blood volume and blood flow changes during neuronal activation in humans. NMR Biomed 22:1054–1062

Chen JJ, Pike GB (2009b) Origins of the BOLD post-stimulus undershoot. Neuroimage 46: 559–568

Chen YI, Mandeville JB, Nguyen TV, Talele A, Cavagna F, Jenkins BG (2001) Improved mapping of pharmacologically induced neuronal activation using the IRON technique with Superparamagnetic Iron Blood Pool Agents. J Magn Reson Imaging 14:517–524

Chen YC, Choi JK, Andersen SL, Rosen BR, Jenkins BG (2005) Mapping dopamine D2/D3 receptor function using pharmacological magnetic resonance imaging. Psychopharmacology (Berl) 180(4):705–715

Chesler DA, Kwong KK (1995) An intuitive guide to the T1 based perfusion model. Int J Imaging Syst Tech 6:171–174

Coyne DW (2009) Ferumoxytol for treatment of iron deficiency anemia in patients with chronic kidney disease. Expert Opin Pharmacother 10:2563–2568

Dai W, Garcia D, de Bazelaire C, Alsop DC (2008) Continuous flow-driven inversion for arterial spin labeling using pulsed radio frequency and gradient fields. Magn Reson Med 60:1488–1497

Davis TL, Kwong KK, Weisskoff RM, Rosen BR (1998) Calibrated functional MRI: mapping the dynamics of oxidative metabolism. Proc Natl Acad Sci USA 95:1834–1839

Dennie J, Mandeville JB, Boxerman JL, Packard SD, Rosen BR, Weisskoff RM (1998) NMR imaging of changes in vascular mophology due to tumor angiogenesis. Magn Reson Med 40:793–799

Detre JA, Leigh JS, Williams DS, Koretsky AP (1992) Perfusion Imaging. Magn Reson Med 23:37–45

Dijkhuizen RM, Singhal AB, Mandeville JB, Wu O, Halpern EF, Finklestein SP, Rosen BR, Lo EH (2003) Correlation between brain reorganization, ischemic damage, and neurologic status after transient focal cerebral ischemia in rats: a functional magnetic resonance imaging study. J Neurosci 23:510–517

Donahue KM, Weisskoff RM, Chesler DA, Kwong KK, Bogdanov A Jr, Mandeville JB, Rosen BR (1996) Improving MR quantification of regional blood volume with Intravascular T1 Contrast Agents: accuracy, precision, and water exchange. Magn Reson Med 36:858–867

Donahue MJ, Lu H, Jones CK, Edden RA, Pekar JJ, van Zijl PC (2006) Theoretical and experimental investigation of the VASO contrast mechanism. Magn Reson Med 56:1261–1273

Donahue MJ, Blicher JU, Ostergaard L, Feinberg DA, MacIntosh BJ, Miller KL, Gunther M, Jezzard P (2009) Cerebral blood flow, blood volume, and oxygen metabolism dynamics in human visual and motor cortex as measured by whole-brain multi-modal magnetic resonance imaging. J Cereb Blood Flow Metab 29:1856–1866

Eklöf B, Lassen NA, Nilsson L, Norberg K, Siesjö BK (1973) Blood flow and metabolic rate for oxygen in the cerebral cortex of the rat. Acta Physiol Scand 88:587–589

Engel SA, Glover GH, Wandell BA (1997) Retinotopic organization in human visual cortex and the spatial precision of functional MRI. Cereb Cortex 7:181–192

Enochs WS, Harsh G, Hochberg F, Weissleder R (1999) Improved delineation of human brain tumors on MR images using a long-circulating, superparamagnetic iron oxide agent. J Magn Reson Imaging 9:228–232

Fisel CR, Moore JR, Garrido L, Ackerman JL, Rosen BR, Brady TJ (1989) A general model for susceptibility-based MR contrast. In: Proceedings of the eighth annual meeting of the Society of Magnetic Resonance in Medicine, Amsterdam, p 324

Fox PT, Raichle ME (1986) Focal physiological uncoupling of cerebral blood flow and oxidative metabolism during somatosensory stimulation in human subjects. Proc Natl Acad Sci USA 83:1140–1144

Friston KJ, Mechelli A, Turner R, Price CJ (2000) Nonlinear responses in fMRI: the Balloon model, Volterra kernels, and other hemodynamics. Neuroimage 12:466–477

Goodkin HP, Yeh JL, Kapur J (2005) Status epilepticus increases the intracellular accumulation of GABAA receptors. J Neurosci 25:5511–5520

Grinvald A, Lieke E, Frostig RD, Gilbert CD, Wiesel TN (1986) Functional architecture of cortex revealed by optical imaging of intrinsic signals. Nature 324:361–364

Grubb RL, Raichle ME, Eichling JO, Ter-Pogossian MM (1974) The effects of changes in PaCO2 on cerebral blood volume, blood flow, and vascular mean transit time. Stroke 5:630–639

Hamberg LM, Boccalini P, Stranjalis G, Hunter GJ, Huang Z, Halpern E, Weisskoff RM, Moskowitz MA, Rosen BR (1996) Continuous assessment of relative cerebral blood volume in transient ischemia using steady state susceptibility-contrast MRI. Magn Reson Med 35:168–173

Harel N, Lin J, Moeller S, Ugurbil K, Yacoub E (2006) Combined imaging-histological study of cortical laminar specificity of fMRI signals. Neuroimage 29:879–887

Herscovitch P, Raichle ME, Kilbourn MR, Welch MJ (1987) Positron emission tomographic measurement of cerebral blood flow and permeability-surface area product of water using [15O]water and [11 C]butanol. J Cereb Blood Flow Metab 7:527–542

Hillman EM, Devor A, Bouchard MB, Dunn AK, Krauss GW, Skoch J, Bacskai BJ, Dale AM, Boas DA (2007) Depth-resolved optical imaging and microscopy of vascular compartment dynamics during somatosensory stimulation. Neuroimage 35:89–104

Hoge RD, Atkinson J, Gill B, Crelier GR, Marrett S, Pike GB (1999a) Investigation of BOLD signal dependence on cerebral blood flow and oxygen consumption: the deoxyhemoglobin dilution model. Magn Reson Med 42:849–863

Hoge RD, Atkinson J, Gill B, Crelier GR, Marrett S, Pike GB (1999b) Linear coupling between cerebral blood flow and oxygen consumption in activated human cortex. Proc Natl Acad Sci USA 96:9403–9408

Horvath I, Sandor NT, Ruttner Z, McLaughlin AC (1994) Role of nitric oxide in regulating cerebrocortical oxygen consumption and blood flow during hypercapnia. J Cereb Blood Flow Metab 14:503–509

Jenkins BG, Sanchez-Pernaute R, Brownell AL, Chen YC, Isacson O (2004) Mapping dopamine function in primates using pharmacologic magnetic resonance imaging. J Neurosci 24:9553–9560

Jezzard P, Buxton RB (2006) The clinical potential of functional magnetic resonance imaging. J Magn Reson Imaging 23:787–793

Jin T, Kim SG (2008) Cortical layer-dependent dynamic blood oxygenation, cerebral blood flow and cerebral blood volume responses during visual stimulation. Neuroimage 43:1–9

Johnson LA, Furman CA, Zhang M, Guptaroy B, Gnegy ME (2005) Rapid delivery of the dopamine transporter to the plasmalemmal membrane upon amphetamine stimulation. Neuropharmacology 49:750–758

Kastrup A, Kruger G, Neumann-Haefelin T, Glover GH, Moseley ME (2002) Changes of cerebral blood flow, oxygenation, and oxidative metabolism during graded motor activation. Neuroimage 15:74–82

Kennan RP, Scanley BE, Gore JC (1997) Physiologic basis for BOLD MR signal changes due to hypoxia/hyperoxia: separation of blood volume and magnetic susceptibility effects. Magn Reson Med 37:953–956

Kennan RP, Scanley BE, Innis RB, Gore JC (1998) Physiological basis for BOLD MR signal changes due to neuronal stimulation: separation of blood volume and magnetic susceptibility effects. Magn Reson Med 40:840–846

Kety S (1960) Measurement of local blood flow by the exchange on an inert diffusible substance. Methods Med Res 8:228–236

Kida I, Rothman DL, Hyder F (2007) Dynamics of changes in blood flow, volume, and oxygenation: implications for dynamic functional magnetic resonance imaging calibration. J Cereb Blood Flow Metab 27:690–696

Kim SG, Rostrup E, Larsson HB, Ogawa S, Paulson OB (1999) Determination of relative CMRO2 from CBF and BOLD changes: significant increase of oxygen consumption rate during visual stimulation. Magn Reson Med 41:1152–1161

Kong Y, Zheng Y, Johnston D, Martindale J, Jones M, Billings S, Mayhew J (2004) A model of the dynamic relationship between blood flow and volume changes during brain activation. J Cereb Blood Flow Metab 24:1382–1392

Kruger G, Glover GH (2001) Physiological noise in oxygenation-sensitive magnetic resonance imaging. Magn Reson Med 46:631–637

Kruger G, Kastrup A, Glover GH (2001) Neuroimaging at 1.5 T and 3.0 T: comparison of oxygenation-sensitive magnetic resonance imaging. Magn Reson Med 45:595–604

Kwong K, Belliveau J, Chesler D, Goldberg I, Weisskoff R, Poncelet B, Kennedy D, Hoppel B, Cohen M, Turner R, Cheng H, Brady T, Rosen B (1992) Dynamic magnetic resonance imaging of human brain activity during primary sensory stimulation. Proc Natl Acad Sci 89: 5675–5679

Lee SP, Silva AC, Ugurbil K, Kim SG (1999) Diffusion-weighted spin-echo fMRI at 9.4 T: microvascular/tissue contribution to BOLD signal changes. Magn Reson Med 42:919–928

Leite FP (2006) Detection power, temporal response, and spatial resolution of IRON fMRI in awake, behaving monkeys at 3 Tesla. Massachusetts Institute of Technology, Cambridge, MA

Leite FP, Mandeville JB (2006) Characterization of event-related designs using BOLD and IRON fMRI. Neuroimage 29:901–909

Leite FP, Mandeville JB (2007) Investigating water exchange across the BBB as a mechanism for the slow time constant of blood volume. In: Proceedings of the International Society for Magnetic Resonance in Medicine, Berlin, p 3190

Leite FP, Tsao D, Vanduffel W, Fize D, Sasaki Y, Wald LL, Dale AM, Kwong KK, Orban GA, Rosen BR, Tootell RB, Mandeville JB (2002) Repeated fMRI using Iron Oxide Contrast Agent in Awake, behaving macaques at 3 Tesla. Neuroimage 16:283–294

Leite FP, Vanduffel W, Rosen BR, Mandeville JB (2005) Comparing spatial resolution of IRON and BOLD in awake macaques. In: Proceedings of the International Society for Magnetic Resonance in Medicine, Miami Beach, p 303

Leontiev O, Dubowitz DJ, Buxton RB (2007) CBF/CMRO2 coupling measured with calibrated BOLD fMRI: sources of bias. Neuroimage 36:1110–1122

Liu H, Rainey C, Lauer KK, Piacentine L, Bloom A, Risinger R, Ward BD, Stein E, Li SJ (2006) Peripheral blood pressure changes induced by dobutamine do not alter BOLD signals in the human brain. Neuroimage 30:745–752

Liu CH, Greve DN, Dai G, Marota JJ, Mandeville JB (2007) Remifentanil administration reveals biphasic phMRI temporal responses in rat consistent with dynamic receptor regulation. Neuroimage 34:1042–1053

Lu H, Golay X, Pekar JJ, Van Zijl PC (2003) Functional magnetic resonance imaging based on changes in vascular space occupancy. Magn Reson Med 50:263–274

Lu H, Clingman C, Golay X, van Zijl PC (2004a) Determining the longitudinal relaxation time (T1) of blood at 3.0 Tesla. Magn Reson Med 52:679–682

Lu H, Golay X, Pekar JJ, Van Zijl PC (2004b) Sustained poststimulus elevation in cerebral oxygen utilization after vascular recovery. J Cereb Blood Flow Metab 24:764–770

Lu H, van Zijl PC, Hendrikse J, Golay X (2004c) Multiple acquisitions with global inversion cycling (MAGIC): a multislice technique for vascular-space-occupancy dependent fMRI. Magn Reson Med 51:9–15

Lu H, Soltysik DA, Ward BD, Hyde JS (2005) Temporal evolution of the CBV-fMRI signal to rat whisker stimulation of variable duration and intensity: a linearity analysis. Neuroimage 26: 432–440

Lyons D, Friedman DP, Nader MA, Porrino LJ (1996) Cocaine alters cerebral metabolism within the ventral striatum and limbic cortex of monkeys. J Neurosci 16:1230–1238

Mandeville JB, Marota JJA (1999) Vascular filters of functional MRI: spatial localization using BOLD and CBV contrast. Magn Reson Med 42:591–598

Mandeville JB, Marota JJA, Kosofsky BE, Keltner JR, Weissleder R, Rosen BR, Weisskoff RM (1998) Dynamic functional imaging of relative cerebral blood volume during rat forepaw stimulation. Magn Reson Med 39:615–624

Mandeville JB, Marota JJA, Ayata C, Moskowitz MA, Weisskoff RM, Rosen BR (1999a) An MRI measurement of the temporal evolution of relative CMRO2 during rat forepaw stimulation. Magn Reson Med 42:944–951

Mandeville JB, Marota JJA, Ayata C, Zaharchuk G, Moskowitz MA, Rosen BR, Weisskoff RM (1999b) Evidence of a cerebrovascular post-arteriole windkessel with delayed compliance. J Cereb Blood Flow Metab 19:679–689

Mandeville JB, Jenkins BG, Kosofsky BE, Moskowitz MA, Rosen BR, Marota JJA (2001) Regional Sensitivity and Coupling of BOLD and CBV Changes during Stimulation of Rat Brain. Magn Reson Med 45:443–447

Mandeville JB, Jenkins BG, Chen YC, Choi JK, Kim YR, Belen D, Liu C, Kosofsky BE, Marota JJ (2004) Exogenous contrast agent improves sensitivity of gradient-echo functional magnetic resonance imaging at 9.4 T. Magn Reson Med 52:1272–1281

Mandeville JB, Leite FP, Marota JJ (2007a) Spin-echo MRI underestimates functional changes in microvascular cerebral blood plasma volume using exogenous contrast agent. Magn Reson Med 58:769–776

Mandeville JB, Liu C, Marota JJA (2007) Spin echoes underestimate functional changes in microvascular cerebral blood volume. In: Proceedings of the International Society for Magnetic Resonance in Medicine, Berlin, p 770

Marota JJA, Ayata C, Moskowitz MA, Weisskoff RM, Rosen BR, Mandeville JB (1999) Investigation of the early response to rat forepaw stimulation. Magn Reson Med 41:247–252

Marota JJA, Mandeville JB, Weisskoff RM, Moskowitz MA, Rosen BR, Kosofsky BE (2000) Cocaine activation discriminates dopamergic projections by temporal response: an fMRI study in rat. Neuroimage 11:13–23

Mazziotta JC, Phelps ME (1986) Positron emission tomography studies of the brain. Raven Press, New York

Meier P, Zierler KL (1954) On the theory of the indicator-dilution method for measurement of blood flow and volume. J Appl Physiol 6:731–744

Ogawa S, Lee TM (1990) Magnetic resonance imaging of blood vessels at high fields: in vivo and in vitro measurements and image simulation. Magn Reson Med 16:9–18

Ogawa S, Lee TM, Kay AR, Tank DW (1990a) Brain magnetic resonance imaging with contrast dependent on blood oxygenation. Proc Natl Acad Sci USA 87:9868–9872

Ogawa S, Lee TM, Nayak AS, Glynn P (1990b) Oxygenation-sensitive contrast in magnetic resonance image of rodent brain at high magnetic fields. Magn Reson Med 14:68–78

Ogawa S, Lee RM, Barrere B (1993) The sensitivity of magnetic resonance image signals of a rat brain to changes in the cerebral venous blood oxygenation. Magn Reson Med 29:205–210

Oja JM, Gillen J, Kauppinen RA, Kraut M, van Zijl PC (1999) Venous blood effects in spin-echo fMRI of human brain. Magn Reson Med 42:617–626

Parkes LM, Schwarzbach JV, Bouts AA, Deckers RH, Pullens P, Kerskens CM, Norris DG (2005) Quantifying the spatial resolution of the gradient echo and spin echo BOLD response at 3 Tesla. Magn Reson Med 54:1465–1472

Pauling L, Coryell CD (1936) The magnetic properties and structure of hemoglobin. Oxyhemoglobin and carbon monooxyhemoglobin. Proc Natl Acad Sci USA 22:210–216

Piechnik SK, Evans J, Bary LH, Wise RG, Jezzard P (2009) Functional changes in CSF volume estimated using measurement of water T2 relaxation. Magn Reson Med 61:579–586

Porciuncula CI, Armstrong JGG, Guyton AC, Stone HL (1964) Delayed compliance in external jugular vein of the dog. Am J Physiol 207:728–732

Rosen BR, Belliveau JW, Vevea JM, Brady TJ (1990) Perfusion Imaging with NMR contrast agents. Magn Reson Med 14:249–265

Rosen BR, Belliveau JW, Aronen HJ, Kennedy D, Buchbinder BR, Fischman A, Gruber M, Glas J, Weisskoff RM, Cohen MS, Hochberg FH, Brady TJ (1991) Susceptibility contrast imaging of cerebral blood volume: human experience. Magn Reson Med 22:293–299

Roy CS, Sherrington CS (1890) On the regulation of the blood-supply of the brain. J Physiol (London) 11:85–108

Royl G, Leithner C, Kohl M, Lindauer U, Dirnagl U, Kwong K, Mandeville JB (2001) The BOLD post-stimulus undershoot: fMRI versus imaging spectroscopy. In: Proceedings of the International Society for Magnetic Resonance in Medicine, Glasgow, p 282

Scouten A, Constable RT (2007) Applications and limitations of whole-brain MAGIC VASO functional imaging. Magn Reson Med 58:306–315

Scouten A, Constable RT (2008) VASO-based calculations of CBV change: accounting for the dynamic CSF volume. Magn Reson Med 59:308–315

Sharkey J, McBean DE, Kelly PAT (1991) Acute cocaine administration: effects on local cerebral blood flow and metabolic demand in the rat. Brain Res 548:310–314

Shen T, Weissleder R, Papisov MA, Bogdanov J, Brady TJ (1993) Monocrystalline iron oxide nanocompounds (MION): physiochemical properties. Magn Reson Med 29:599–604

Shmuel A, Yacoub E, Chaimow D, Logothetis NK, Ugurbil K (2007) Spatio-temporal point-spread function of fMRI signal in human gray matter at 7 Tesla. Neuroimage 35:539–552

Silva AC, Zhang W, Williams DS, Koretsky AP (1995) Multi-slice MRI of rat brain perfusion during amphetamine stimulation using arterial spin labeling. Magn Reson Med 33:209–214

Silva AC, Koretsky AP, Duyn JH (2007) Functional MRI impulse response for BOLD and CBV contrast in rat somatosensory cortex. Magn Reson Med 57:1110–1118

Song AW, Wong EC, Tan SG, Hyde JS (1996) Diffusion weighted fMRI at 1.5 T. Magn Reson Med 35:155–158

Stefanovic B, Pike GB (2005) Venous refocusing for volume estimation: VERVE functional magnetic resonance imaging. Magn Reson Med 53:339–347

Stefanovic B, Warnking JM, Pike GB (2004) Hemodynamic and metabolic responses to neuronal inhibition. Neuroimage 22:771–778

Stein EA, Fuller SA (1993) Cocaine's time action profile on regional cerebral blood flow in the rat. Brain Res 626:117–126

Sunaert S (2006) Presurgical planning for tumor resectioning. J Magn Reson Imaging 23:887–905

Ter-Pogossian MM, Eichling JO, Davis DO, Welch MJ, Metzger JM (1969) The determination of regional cerebral blood flow by means of water labeled with radioactive oxygen 15. Radiology 93:31–40

Thulborn KR, Waterton JC, Radda GK (1981) Proton Imaging for in vivo blood flow and oxygen consumption measurements. Journal of Magn Reson 45:188–191

Thulborn KR, Waterton JC, Matthews PM, Radda GK (1982) Oxygenation dependence of the transverse relaxation time of water protons in whole blood at high field. Biochim Biophys Acta 714:265–270

Tootell RBH, Switkes E, Silverman MS, Hamilton SL (1988) Functional anatomy of macaque striate cortex. II. Retinotopic Organization. J Neurosci 8:1531–1568

Triantafyllou C, Hoge RD, Krueger G, Wiggins CJ, Potthast A, Wiggins GC, Wald LL (2005) Comparison of physiological noise at 1.5 T, 3 T and 7 T and optimization of fMRI acquisition parameters. Neuroimage 26:243–250

Tropres I, Gramault S, Vaeth A, Grillon E, Julien C, Payen J-F, Lamalle L, Decorps M (2001) Vessel size imaging. Magn Reson Med 45:397–408

Trugman JM, James CL (1993) D1 dopamine agonist and antagonist effects on regional cerebral glucose utilization in rats with intact dopaminergic innervation. Brain Res 607:270–274

Tsao DY, Vanduffel W, Sasaki Y, Fize D, Knutsen TA, Mandeville JB, Wald LL, Dale AM, Rosen BR, Van Essen DC, Livingstone MS, Orban GA, Tootell RB (2003) Stereopsis activates V3A and caudal intraparietal areas in macaques and humans. Neuron 39:555–568

Tsao DY, Freiwald WA, Tootell RB, Livingstone MS (2006) A cortical region consisting entirely of face-selective cells. Science 311:670–674

Uludag K, Dubowitz DJ, Yoder EJ, Restom K, Liu TT, Buxton RB (2004) Coupling of cerebral blood flow and oxygen consumption during physiological activation and deactivation measured with fMRI. Neuroimage 23:148–155

van Bruggen N, Busch E, Palmer JT, Williams SP, de Crespigny AJ (1998) High-resolution functional magnetic resonance imaging of the rat brain: mapping changes in cerebral blood volume using iron oxide contrast media. J Cereb Blood Flow Metab 18:1178–1183

van Rijnsoever C, Sidler C, Fritschy JM (2005) Internalized GABA-receptor subunits are transferred to an intracellular pool associated with the postsynaptic density. Eur J Neurosci 21:327–338

Vanduffel W, Fize D, Mandeville JB, Nelissen K, Van Hecke P, Rosen BR, Tootell RBH, Orban GA (2001) Visual motion processing investigated using contrast-enhanced fMRI in awake behaving monkeys. Neuron 32:565–577

Villringer A, Rosen BR, Belliveau JW, Ackerman JL, Lauffer RB, Buxton RB, Chao Y, Wedeen VJ, Brady TJ (1988) Dynamic imaging with lanthanide chelates in normal brain: contrast due to magnetic susceptibility effects. Magn Reson Med 6:164–174

Wang J, Alsop DC, Li L, Listerud J, Gonzalez-At JB, Schnall MD, Detre JA (2002) Comparison of quantitative perfusion imaging using arterial spin labeling at 1.5 and 4.0 Tesla. Magn Reson Med 48:242–254

Weissleder R, Elizondo G, Wittenberg K (1990) Ultrasmall superparamagnetic iron oxide. Characterization of a new class of contrast agents for MR imaging. Radiology 175:489–493

Williams DS (2006) Quantitative perfusion imaging using arterial spin labeling. Methods Mol Med 124:151–173

Williams DS, Detre JA, Leigh JS, Koretsky AP (1992) Magnetic resonance imaging of perfusion using spin inversion of arterial water. Proc Natl Acad Sci USA 89:212–216

Wise RG, Tracey I (2006) The role of fMRI in drug discovery. J Magn Reson Imaging 23: 862–876

Wong EC, Buxton RB, Frank LR (1998a) A theoretical and experimental comparison of continuous and pulsed arterial spin labeling techniques for quantitative perfusion imaging. Magn Reson Med 40:348–355

Wong EC, Buxton RB, Frank LR (1998b) Quantitative imaging of perfusion using a single subtraction (QUIPSS and QUIPSS II). Magn Reson Med 39:702–708

Yacoub E, Ugurbil K, Harel N (2006) The spatial dependence of the poststimulus undershoot as revealed by high-resolution BOLD- and CBV-weighted fMRI. J Cereb Blood Flow Metab 26:634–644

Yang SP, Krasney JA (1995) Cerebral blood flow and metabolic responses to sustained hypercapnia in awake sheep. J Cereb Blood Flow Metab 15:115–123

Ye FQ, Frank JA, Weinberger DR, McLaughlin AC (2000) Noise reduction in 3D perfusion imaging by attenuating the static signal in arterial spin tagging (ASSIST). Magn Reson Med 44:92–100

Zaharchuk G (2007) Theoretical basis of hemodynamic MR imaging techniques to measure cerebral blood volume, cerebral blood flow, and permeability. AJNR Am J Neuroradiol 28:1850–1858

Zaharchuk G, Mandeville JB, Bogdonov AA Jr, Weissleder R, Rosen BR, Marota JJA (1999) Cerebrovascular dynamics of autoregulation and hypotension: an MRI study of CBF and changes in total and microvascular cerebral blood volume during hemorrhagic hypotension. Stroke 30:2197–2205

Zaharchuk G, Bammer R, Straka M, Shankaranarayan A, Alsop DC, Fischbein NJ, Atlas SW, Moseley ME (2009) Arterial spin-label imaging in patients with normal bolus perfusion-weighted MR imaging findings: pilot identification of the borderzone sign. Radiology 252:797–807

Zhao F, Wang P, Hendrich K, Ugurbil K, Kim SG (2006) Cortical layer-dependent BOLD and CBV responses measured by spin-echo and gradient-echo fMRI: insights into hemodynamic regulation. Neuroimage 30:1149–1160

Zheng Y, Martindale J, Johnston D, Jones M, Berwick J, Mayhew J (2002) A model of the hemodynamic response and oxygen delivery to brain. Neuroimage 16:617–637

Zhong J, Kennan RP, Fulbright RK, Gore JC (1998) Quantification of intravascular and extravascular contributions to BOLD effects induced by alteration in oxygenation or intravascular contrast agents. Magn Reson Med 40:526–536

Chapter 18
FMRI in Mice: Functional Phenotyping of Transgenic Mouse Lines Based on Hemodynamic Readouts

Thomas Mueggler, Christof Baltes, Simone C. Bosshard, Felicitas L. Princz-Kranz, David Ratering, Florence Razoux, Esther Sydekum, and Markus Rudin

Abstract Considering the increasing number of genetically engineered mouse models of human neurological or psychiatric pathologies there is a rising demand of tools for structural and functional phenotyping. Methods that allow non-invasive assessment of brain function such as functional MRI (fMRI) are highly attractive in this context. This review aims to make the reader aware of the potential of functional MRI (fMRI) in brain activation studies in mice with the focus on experimental aspects of *hemodynamic-based* readouts such as BOLD (blood-oxygen level dependent contrast) or CBV (cerebral blood volume)-based fMRI, and there application to mouse models of CNS disorders. Due to the non-invasiveness nature of MRI functional consequences of disease progression can be investigated in longitudinal studies providing information on disease evolution in the individual that can be combined

T. Mueggler, Ph.D.
Pharmaceuticals Division, Preclinical Translational Neuroscience, CH-4070, Basel, Switzerland
e-mail: thomas.mueggler@roche.com

C. Baltes, Ph.D.
Philips Healthcare, Veenpluis 4-6, Building QR-0135, P.O. Box 10000, NL-5680 DA Best, The Netherlands
e-mail: christof.baltes@philips.com

S.C. Bosshard, MSc. • F.L. Princz-Kranz, Ph.D. • D. Ratering, Ph.D. • F. Razoux • E. Sydekum, Ph.D.
Institute for Biomedical Engineering, University and Eidgenössische Technische Hochschule (ETH) Zurich, CH-8093, Zurich, Switzerland
e-mail: bosshard@biomed.ee.ethz.ch; princz@biomed.ee.ethz.ch; ratering@biomed.ee.ethz.ch; razoux@hifo.uzh.ch; sydekum@biomed.ee.ethz.ch

M. Rudin, Ph.D. (✉)
Institute for Biomedical Engineering, University and Eidgenössische Technische Hochschule (ETH) Zurich, Zurich, Switzerland

Institute of Pharmacology and Toxicology, University of Zurich, Zurich, Switzerland
e-mail: rudin@biomed.ee.ethz.ch

I.-Y. Choi and R. Gruetter (eds.), *Neural Metabolism In Vivo*, Advances in Neurobiology 4, DOI 10.1007/978-1-4614-1788-0_18,

with more traditional pharmacological readouts (i.e. behavioural assessments). FMRI studies investigating transgenic models of brain disorders characterizing the disease phenotype and assessing the efficacy of a therapeutic intervention are therefore warranted and it is anticipated that the field will rapidly grow.

Keywords Alzheimer's disease • Anesthesia • BOLD • Brain • Cerebral blood volume • Functional magnetic resonance imaging • Mouse • Physiological monitoring • Transgenic

Abbreviations

Aβ	amyloid β-peptide
AD	Alzheimer's disease
ADC	apparent diffusion coefficient
AMPA-R	glutamatergic α-amino-3 hydroxy-5-methylisoxazole-4 propionic acid receptor
APP	amyloid precursor protein
ASL	arterial spin labeling
BOLD	blood oxygenation-level dependent contrast
BP	blood pressure (arterial blood pressure)
CAA	cerebral amyloid angiopathy
CBF	cerebral blood flow
CBV	cerebral blood volume
$CMRO_2$	cerebral metabolic rate for oxygen
CNR	contrast-to-noise ratio
CNS	central nervous system
D_2R	dopamine-D_2-receptor
DT	dopamine transporter
DTI	diffusion tensor imaging
EPI	echo-planar-imaging
FMRI	functional magnetic resonance imaging
GE	gradient-echo
$GABA_AR$	-amino butyric acid type A receptor
MCA-O	middle cerebral artery occlusion
NMDA-R	glutamatergic N-methyl-D-aspartate receptor
nAChR	nicotinergic acetyl-cholin receptor
PCASL	pseudo-continuous arterial spin labeling
PET	positron emission tomography
RF	radiofrequency
rCBV	relative cerebral blood volume
SE	spin-echo
SMC	smooth muscle cell
SNR	signal-to-noise ratio
T_2 (R_2)	transverse relaxation time

T_2* (R_2*) effective transverse relaxation time
T_1 (R_1) longitudinal relaxation time
5-HT-R 5-hydroxy-tryptamine (serotonergic) receptor

18.1 Introduction

Functional MRI (fMRI) probes secondary changes associated with neuronal activation such as regional changes in cerebral blood volume (CBV; Belliveau et al. 1990), cerebral blood flow (CBF; Detre et al. 1992; Kwong et al. 1992), and blood oxygenation (BOLD: blood oxygenation level-dependent contrast mechanism; Ogawa et al. 1990). Such fMRI methods have emerged as a powerful tool for investigating the functional organization in the mammalian brain (Ogawa et al. 1998) and are widely applied for mapping human brain function both under normal and pathological conditions (Jezzard and Buxton 2006). In parallel, experimental biomedical applications of these *hemodynamic-based* fMRI techniques (here referred to as fMRI) have been developed at a rapid pace providing relevant information on neuronal function in animal models of human disease covering a wide range of species from non-human primates (Dubowitz et al. 1998) to mice (Mueggler et al. 2001).

Alternatively, a *non-hemodynamic* fMRI imaging approach referred to as manganese-enhanced MRI (MEMRI) has been brought forward to map brain function (Lin and Koretsky 1997). Besides using Mn^{2+} (a T_1 relaxation agent) to image brain architecture or to track neuronal pathways it can be used to demarcate active regions of the brain (activity-induced manganese-enhanced MRI, AIM MRI) – based on its ability to enter cells through voltage-gated calcium channels. The interested reader is referred to recent reviews covering a broad range of its applications (i.e. from structural to functional applications, (van der Linden et al. 2004; Silva et al. 2004; Silva and Bock 2008).

Aim of the current review is to report on methodological aspects of *hemodynamic-based* fMRI approaches in small animals and its application to mouse models of CNS disorders. A principal problem in such fMRI studies, in particular those involving rodents, is the choice of a suitable stimulation paradigm. There are limited possibilities of challenging an anaesthetized animal. Most often used paradigms include sensory stimulation of the paw (Hyder et al. 1994; Gyngell et al. 1996; Kerskens et al. 1996; Mandeville et al. 1998), direct cortical stimulation (Austin et al. 2003), pharmacological stimulation i.e. administration of a neuro-active drug (Silva et al. 1995; Burdett et al. 1995; Marota et al. 2000, for review see Steward et al. 2005). These protocols have been shown to evoke significant focal or global hemodynamic responses in the central nervous system (CNS). They have further been applied to study the brain function under normal and pathological conditions covering a broad range of animal species and models of human disorders (i.e. focal ischemia, neurodegeneration) (for review see Weber et al. 2006a; van der Linden et al. 2007), to investigate functional connectivity of different brain circuits (Schwarz et al. 2007) and to elucidate the underlying mechanisms leading to the changes in the fMRI signal (Logothetis and Pfeuffer 2004).

18.2 Experimental Biomedical Applications of *Hemodynamic-Based* fMRI Techniques

18.2.1 A Sensitive Indicator of Brain Pathology

The value of rodent fMRI studies has been demonstrated for instance in studies of focal cerebral ischemia in rats as models of human embolic stroke. Functional studies in the rat middle cerebral artery occlusion (MCA-O) model revealed a lack of functional response in cortical areas that have been preserved from becoming infarcted by cytoprotective therapy (Reese et al. 2000b). Follow-up study demonstrated functional recovery in some but not all animals (Sauter et al. 2002; Rudin et al. 2003). Similarly, in studies using the transient MCA-O model both structural images and CBV maps recorded during the resting state did not reveal any abnormality due to the focal cerebral ischemia. However, the assessment of CBV changes following pharmacological stimulation displayed a compromised hemodynamic responsiveness. There was a region-specific negative correlation between the fMRI signal and the duration of the ischemia (Reese et al. 2000b). These experiments clearly demonstrated that *structural integrity of brain parenchyma is a necessary yet not sufficient precondition for functional integrity*. For the ischemia models this implies that reduction of infarct volume does not necessarily translate into improved functional outcome, a paradigm that had been widely assumed valid when evaluating cytoprotective therapy. This example shows that fMRI experiments yield relevant information for the characterization of the brain status that must be considered when analyzing therapy efficacy. In fact, there is evidence that functional read-outs are more sensitive indicators of tissue pathology than morphological parameters.

18.2.2 A non-invasive Tool to Phenotype Genetically Engineered Mice

Mice have been widely used in biomedical research because of their genetic similarity to humans. The rapid advances in transgenic and gene-targeting technologies enabling the generation of genetically altered strains for concept validation and/or as humanized disease models. These properties combined with their small body weight requiring only low amounts of compounds to be tested per animal, high reproductive rate and cost effectiveness render mice an attractive species for the evaluation of novel therapy concepts. During the past decades transgenic manipulation of mice allowed the linking of specific genes to molecular, physiological and morphological phenotypes both in the normal and diseased state. Transgenic strains that mimic pathological features of brain disorders have become valuable tools to study the aetiology of a variety of CNS diseases. Such tools are increasingly used in preclinical studies aiming at disease-modifying therapies (Scorer 2001; Tarditi et al. 2009).

The pathophysiological status of genetically modified animals is commonly assessed by post-mortem analysis of brain tissue samples. *Non-invasive*, quantitative phenotyping would offer significant advantages: disease progression can be monitored in an individual in a longitudinal fashion and combined with more 'clinical' readouts such as the assessment of behaviour. High-throughput mouse phenotyping procedures using imaging modalities are currently developed by multiple labs (Paulus et al. 2001; Johnson et al. 2002; Dazai et al. 2004). These approaches aim at the characterization of *morphological phenotypes,* also using fixed specimen. Yet, as already discussed functional readouts frequently provide more subtle information on pathological tissue transformation and should not be discarded, even if they cannot be standardized or even automated to the extent of structural studies in fixed specimen. In this article we report on basic aspects of fMRI studies in mice and their application for phenotyping of transgenic models of neurodegenerative diseases.

18.3 Experimental Aspects of *Hemodynamic-Based* fMRI Techniques – from Rats to Mice

18.3.1 Physiological Control

18.3.1.1 Anaesthesia Protocols for fMRI

Non-invasive functional MRI studies in animals are generally performed under anesthesia, with alpha-chloralose, isoflurane or halothane being the most frequently used narcotics for sensory or pharmacological fMRI in rodents. Propofol has been among the first anaesthetics used for sensory fMRI studies (Scanley et al. 1997). Their effects on BOLD and/or CBV response during activation under different stimulation paradigms have been investigated in several human and animal studies (Table 18.1). Further anaesthetics used in rodent fMRI are pentobarbital (Huang et al. 1996; Kuo et al. 2005) and urethane (Forman et al. 1998; Xu et al. 2000; Xu et al. 2003; Kuriwaki et al. 2004). Recently, medetomidine (domitor®), an α_2-adrenoreceptor agonist with sedative, anxiolytic and analgetic properties has been introduced as an independent anesthesia for functional studies using sensory paradigms in the rat and mouse (Ramos-Cabrer et al. 2005; Van Camp et al. 2006; Weber et al. 2006b; Zhao et al. 2008; Adamczak et al. 2010). Compared to α-chloralose or urethane anesthesia the medetomidine-protocol is fully reversible thus allowing longitudinal studies similar to studies with the volatile anesthetics isoflurane or halothane. The choice of the optimal anesthesia is further influenced by the fact that various anesthetics affect neuronal activity (Jenkins et al. 2006; Rojas et al. 2006; Masamoto et al. 2007) as well as the coupling of the activity to blood flow and metabolism differently (Crosby et al. 1983; Ueki et al. 1992; Lindauer et al. 1993, for review see Jenkins et al. 2006).

Table 18.1 BOLD, CBV-MION and ASL fMRI in rodents for commonly used anesthetic agents: effect of anesthesia and initial reports

Anaesthetic agent	Challenge	Studies investigating dose-dependent effect of anaesthetic on fMRI response[a]	Initial reports of different fMRI readouts for various anesthetics in rats and mice: BOLD	CBV-MION[b]	ASL
α-chloralose[c]	Sensory	el. forepaw stimulation (R), Peeters et al. (2001), Hyder et al. (2002)	el. forepaw stimulation (R), Hyder et al. (1994)	el. forepaw stimulation (R), van Bruggen et al. (1998)	el. forepaw stimulation (R), Kerskens et al. (1996)
		direct cortical stimulation (R) , Austin et al. 2005	el. hindpaw stimulation (M), Ahrens and Dubowitz (2001)		
	CNS drug	Not reported	D_2R antagonist (R), Preece et al. (2001)	nAChR agonist (R), Skoubis et al. (2006)	Monoamine releaser (R), Bruns et al. (2009)
			NMDA-R antagonist, 5-$HT_{2b,2c}$ agonist (R), Houston et al. (2001)		
Isoflurane	Sensory	Tactile stimulation (H), Antonigni et al. (1997)	el. forepaw stimulation (R), Liu et al. (2004)	el. forepaw stimulation (R), Reese et al. (2000b)	el. forepaw stimulation (R), Liu et al. (2004)
		el. forepaw stimulation (R), Masamoto et al. (2007)		el. hindpaw stimulation (M), Mueggler et al. (2003)	
	CNS drug	Dopaminergic agonist (P), Zhang et al. (2000)	AMPA-R potentiator (R), Jones et al. (2005)	$GABA_A$ antagonist (R), Reese et al. (2000a)	NMDA-R antagonist (R), Risterucci et al. (2005)
		$GABA_A$ antagonist (R), Abo et al. (2004)		$GABA_A$ antagonist (M), Mueggler et al. (2001)	Dopaminerigc agonist (R), Schmidt et al. (2006)

Halothane	Sensory	Direct cortical stimulation (R), Austin et al. (2005)	el. forepaw stimulation (R), Mandeville et al. (1998)	el. forepaw stimulation (R) Mandeville et al. 1998	Not reported
	CNS drug	Not reported	NMDA-R antagonist (R), Burdett et al. (1995)	Dopamine re-uptake inhibitor (R), Marota et al. (2000)	Monoamine releaser (R), Silva et al. (1995)
			Monoamine releaser, DT antagonist (R), Chen et al. (1997)	Monoamine releaser (R), Nguyen et al. (2000)	
Propofol	Sensory	el. hindpaw stimulation (R), Lathi et al. (1999)	Median nerve stimulation (R), Scanley et al. (1997)	el. forepaw stimulation (R), Kennan et al. (1998)	Not reported
	CNS drug	Not reported	Not reported	Not reported	Monoamine releaser (R), Bruns et al. (2009)

[a] Clinical *and* preclinical studies using different fMRI readouts. Some studies include the awake state

[b] Some authors refer to as IRON

[c] Induction of anesthesia most frequently by halothane

Abbreviations. AMPA-R: glutamatergic α-amino-3 hydroxy-5-methylisoxazole-4 propionic acid receptor; D_2R: dopamine-D_2 receptor; nAChR: nicotinergic acetyl-cholin receptor; $GABA_A$: γ-amino butyric acid type A receptor; NMDA-R: glutamatergic N-methyl-D-aspartate receptor; (H): clinical studies; (M): mouse studies; (P): non-human primate studies; (R): rat studies

For pharmacological fMRI studies one has to consider potential interaction of the anaesthetic agent with the targeted receptor system. Direct action of the anaesthetic on the neurotransmitter system of interest has to be avoided or minimized in order not to mask or even abolish any specific drug induced neuronal activity. FMRI readouts such as BOLD or CBV do in most cases not only report on a single neurotransmitter or receptor system but rather comprise a complex interaction of primary and secondary, more downstream effects. Thus, the fMRI response might even be affected by an anaesthetic that does not directly interact with the receptor system under investigation. Comparing the action of a particular pharmacological stimulation among different anaesthetic regimes may help to better understand the specific action of the investigated compound. In a similar way we also do have to draw the attention on potential systemic side effects of the applied narcotics as they may have profound effect on the hemodynamic system (i.e. arterial blood pressure (BP), respiratory pattern, and vasodilatation) which then may compromise the expected fMRI readout. To overcome some of the potential side-effects of anaesthetic drug the feasibility of performing fMRI in the rat under awake and restrained conditions has been demonstrated (Lahti et al. 1999; Peeters et al. 2001; Sicard et al. 2003). Such experimental conditions which would allow investigating sub-cortical and higher order cognitive functions thereby approaching clinical studies comprise however the intrinsic problems arising from motion artefacts and restraint stress (Peeters et al. 2001) of an awake animal. Procedures to customise rodents to the magnet environment (i.e. scanner noise) are challenging and require time-consuming training periods (King et al. 2005). Current studies aim at characterizing the differences and similarities in physiological and functional measures between awake and anesthetized conditions which could help to better understand the fMRI signals and improve study designs (Duong 2007).

18.3.1.2 Monitoring and Control of Physiological Parameters

A second critical aspect in rodent fMRI studies is tight physiological control in order to achieve stable baseline conditions and robust functional responses. The control of respiration and monitoring of arterial blood pH, blood gas partial pressure of oxygen (pO2) and carbon dioxide (pCO2) in rat fMRI experiments are important as these factors influence the BOLD signal per se. Previous studies have shown that hypercapnia and apnoea produce fMRI responses (Sicard et al. 2003; Kannurpatti and Biswal 2004), and that artificial ventilation can unmask the BOLD response to a drug compared to spontaneous breathing, especially when using a drug such as heroin which causes marked respiratory depression (Xu et al. 2000). Mechanical ventilation provides stable values that can be maintained in the normal physiological range by adaptation of both the ventilation rate and volume based on periodic determination of blood-gas levels. Paralysing the mechanically ventilated animal facilitates the control of a regular breathing pattern with the additional benefit of reducing motion artefacts also in stereotactically fixed animals.

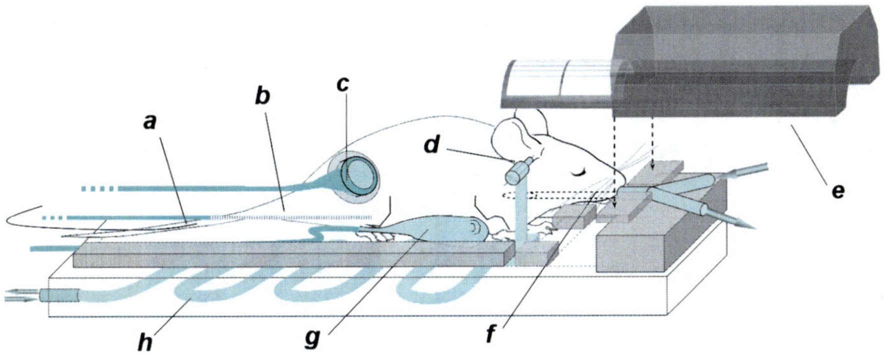

Fig. 18.1 *Schematic drawing of an experimental setup for mouse fMRI*. The anesthetized (1–2% Isoflurane) and endotracheally intubated (polyethylene tubing ID=0.58) mouse is positioned prone on a cradle made from Plexiglas with an in-built warm water circulation for maintaining body temperature. The tracheal tubus is shortened to 24 mm and tightly fixed to the ventilator's Y-connector. Home-built ear bars secure proper head fixation. Tail vein cannulation for subsequent drug infusion is performed using a 30-gauge needle (Microlance3, 0,3x13) which is connected to an infusion pump via polyethylene tubing. Body temperature is monitored using a fluor-optic rectal temperature probe (Quasys, Cham, Switzerland). For blood gas monitoring a combined transcutaneous pCO2/pO2 electrode (TCM4®, Radiometer Copenhagen, Denmark) is attached to the self-adhesive fixation ring on the shaved and cleaned skin above the upper hind limb. (***a***) tail vein cannula; (***b***) rectal temperature probe; (***c***) TCM4® electrode; (***d***) a-traumatic ear bars; (***e***) RF surface coil; (***f***) endotracheal tube; (***g***) pressure sensor connected to a spirometer pressure transducer (ADInstruments, Spechbach, Germany); (***h***) animal support with in-built warm water circulation

Physiological monitoring *and* control in mice (Fig. 18.1) during an fMRI session is demanding: due to their small total blood volume (1.5–2 mL), blood samples of sufficient quantity for blood gas analysis (>70 μL) cannot be collected repeatedly. Repetitive blood sampling can result in blood pressure changes and subsequently influence the BOLD signal by artificially induced changes in CBF and CBV (Kalisch et al. 2001). Therefore, non-invasive blood gas devices either based on transcutaneous (Mueggler et al. 2001) or end-tidal measurements (Thal and Plesnila 2007) are required for continuous blood gas monitoring. Another important parameter to be controlled is body temperature which rapidly drops in the anesthetized mouse. Body temperature does have substantial effects on the metabolic oxygen consumption rate (critical for BOLD-fMRI) and the hemodynamic system in general. Via feedback control mechanism regional brain activity occurs when body temperature changes as recently shown in a rat fMRI study (Vanhoutte et al. 2006). The administration volume for subsequent systemic infusions (anaesthetic, neuromuscular blocking agent, contrast agents, drug infusion) has to be kept at a minimum as larger volumes may interact with the cardiac work load and thus affecting again the hemodynamic system. Moreover centrally-mediated changes in cardiovascular parameters, including BP can occur upon administration of a psychoactive drug potentially introducing confounds in the fMRI response. However, by mimicking the rapid BP changes typically observed with acute drug challenges Gozzi et al. (2007) found that CBF autoregulation was maintained with no changes in local relative CBV (rCBV) over a BP range of 60–120 mmHg suggesting that microvascular rCBV changes are negligible within the autoregulatory range. Another aspect relevant to the animal's physiological state is the position of the animal in the magnet. Due to their small dimensions, mice have been frequently studied in narrow bore vertical high-field magnets commonly used for chemical structure elucidation. Yet, the unnatural vertical posture leads to potential adverse effects on physiology (increased cardiac workload, orthostatic-syndrome) that might affect the outcome during a long acquisition time (Wiesmann et al. 2001; Foley et al. 2005) and horizontal positioning is in any case preferable.

18.3.2 BOLD Versus CBV- and Perfusion-Based fMRI Methods

Several fMRI methods that probe the hemodynamic response to functional brain activation have been described (for review see Buxton 2002), their methodological aspects extensively reviewed and the reader is referred to these references (Buxton et al. 2004; Nair 2005).

Application of BOLD-based imaging exploits the endogenous contrast due to paramagnetic deoxyhemoglobin and hence is completely non-invasive (Ogawa et al. 1990). Yet, fMRI signals due to changes in blood oxygenation are small, of the order of less than 5% of the baseline MRI signal. This is a critical issue considering the high demands on spatial resolution and, correspondingly, the relatively low contrast-to-noise ratio (CNR) commonly associated with rodent imaging as compared to clinical applications. The use of higher magnetic field strengths will increase CNR and hence favour the use of BOLD based fMRI readouts. On the other hand, we have to keep in mind that static susceptibility artefacts due to tissue/air interfaces will also become more severe at higher B_0 values. Cryogenic radiofrequency (RF) probes (Ratering et al. 2008; Baltes et al. 2009) represent an attractive alternative for mouse fMRI experiments to enhance not only image signal-to-noise (SNR), but also temporal SNR and BOLD sensitivity, respectively, at identical B_0 by reducing thermal noise of the receive chain (Baltes et al. 2010).

To achieve local changes in the level of blood oxygenation of a few percent, the magnitude of changes in local CBF and CBV must be significantly larger (Buxton and Frank 1997). Hence, quantitative assessment of CBF and/or CBV yields superior sensitivity as compared to BOLD based methods. This concept has been utilized by several groups by using intravascular contrast agents such as superparamagnetic iron oxide nanoparticles (SPIO) that are slowly cleared from circulation to continuously monitor local CBV (Mandeville et al. 1998; Reese et al. 2000a; Chen et al. 2001). Increases in rCBV will lead to local enhancements in the total amount of particles per voxel, which then leads to an increase in the transverse relaxation rates (R_2, R_2*). Assuming that the particle concentration in the blood has reached a steady state, changes in R_2 (R_2*) directly reflect changes in rCBV. Unfortunately, during CNS activation the signal changes induced by the intravascular contrast agent in the voxel due to a CBV increase and by the changes in blood oxygenation levels are of opposite sign and tend to compensate each other (Scheffler et al. 1999). This may become more obvious when using higher magnetic fields, as the BOLD effect scales with the field (Gati et al. 1997), while the relaxivity effects due to the intravascular contrast agent do not increase at field strength above 1.5 T. This may limit the use of CBV-based fMRI methods at high magnetic field strengths (typically for magnetic fields higher than 7 T). Nevertheless, the utility of iron oxide contrast agents for gradient echo functional MRI at 9.4 T has been evaluated recently using pharmacological and sensory stimuli in the rat (Mandeville et al. 2004; Keilholz et al. 2006; Lu et al. 2007). The combination of short echo times and high iron doses resulted in improved sensitivity as compared to the BOLD method. In fact, rCBV showed excellent agreement with results obtained at much lower field strengths (Mandeville et al. 2004).

As a third surrogate marker of neuronal activity – alternative to BOLD- or CBV-based readouts – CBF can be assessed using MRI. Because CBF is a uniquely defined physiological parameter fMRI techniques based on flow contrast have the advantage (i.e. compared to BOLD) of being specific to tissue signal change providing a direct and quantitative measure of brain activation. Prior to the use of endogenous water as a perfusion tracer, CBF could be measured following the passage of an exogenous injected tracer (i.e. gadolinium chelates) (Villringer et al. 1988). Obviously those bolus tracking techniques are not suited for repetitive measurements or even continuous CBF monitoring. This disadvantage has prompted the development of *tracer-free* perfusion measurement techniques based on arterial spin labelling (ASL) (Williams et al. 1993). The assessment of regional CBF has first been demonstrated in the rodent brain (Detre et al. 1992; Silva et al. 1995; Kerskens et al. 1996). Magnitude, temporal and spatial characteristics of CBF changes upon brain activation have been investigated upon sensory stimulation in the rat and compared to conventional BOLD and CBV based signal changes (for review see Silva 2005).

As mentioned above, the relationship between BOLD and CBF changes is highly non-linear, especially at higher CBF values leading to a higher sensitivity of CBF-based methods. Regarding spatial characteristics it has been shown that CBF-based fMRI is less affected by confounds from venous drainage common to BOLD (Silva and Kim 1999; Duong et al. 2000; Mandeville et al. 2001). Moreover ASL-MRI is advantageous to study brain activity because of its stable noise characteristics and its low inter-subject variability (Detre and Wang 2002; Turner and Jones 2003). This makes perfusion imaging an attractive method to study brain activity, in particular at baseline or during relatively slow processes where the BOLD approach would suffer from higher physiological noise (i.e. studying long-term effects on regional blood perfusion both at rest and during drug challenge in rodent pharmacological fMRI) (Silva et al. 1995; Risterucci et al. 2005; Weidensteiner et al. 2009). The combination of advanced pseudo-continuous ASL (PCASL) with fast GE-EPI readouts enabled temporal resolutions at the order of conventional BOLD acquisitions (Silva and Kim 1999; 2000).

18.3.3 Stimulation Paradigms in Rodents

One could conclude that the use of anesthesia in rodent fMRI studies significantly limits the number of possibilities to probe the functional neuroarchitecture. For the same reason, the use of fMRI in cognitive neurosciences is often wrongly perceived to be limited to studies on conscious primates. However, the need to better understand the underlying molecular and physiological mechanisms of functional brain activation and its interpretations in functional MR images has become obvious and fMRI in small animal models will contribute to this knowledge (Nair 2005; van der Linden et al. 2007). By providing appropriate biophysical models, fMRI in small animals allow a better interpretation of the image contrast, the creation of new forms of contrast (e.g. by using intravascular agents providing CBV and CBF measurements) and an improved design of fMRI techniques per se (e.g. arterial spin labeling methods to measure CBF). Stimulation paradigms that have been applied to anaesthetized animals are pharmacological, physiological or sensory challenges.

18.3.3.1 Pharmacological and Physiological Stimulation in Mice

Administration of neuroactive drug will lead to local or global activation of nerve cells and correspondingly to hemodynamic changes (Villringer and Dirnagl 1995). The concept has been applied by several groups to study compound effects in rats (Silva et al. 1995; Chen et al. 1997; Reese et al. 2000a; Xu et al. 2000) or to use a well characterized stimulus in order to identify dysfunctional brain areas (Reese et al. 2000b). Some of the compounds used (e.g. bicuculline) produced large rCBV changes of the order of 30–40% of baseline values, hence pharmacological stimulation was considered attractive to test the feasibility of fMRI studies in mice. In addition, the CBV-fMRI method has been selected as it provides higher CNR at 4.7 T.

Magnetite nanoparticles administered as an intravascular contrast agent allowed continuous monitoring of rCBV values with a temporal resolution of 30 s. Intravenous infusion of bicuculline, a competitive γ-aminobutyric acid ($GABA_A$) receptor antagonist (Krnjevic 1970) prompted region-specific increases in local rCBV values, which were found to be dose-dependent (Fig. 18.2, Mueggler et al. 2001). The results obtained in mice are comparable to the activation pattern observed in fMRI studies in the rat (Reese et al. 2000a). Blood pCO_2 values, which can cause unspecific changes of CBV, were monitored on-line during the experiment with a transcutaneous monitoring device and did not vary significantly during or following bicuculline infusion (Mueggler et al. 2001). Potential confounding factors for stimuli-induced distinct hemodynamic changes (e.g. increase in systemic BP, K^+ and H^+ activities in the brain) have been reported during epileptic seizure induced by high-dose bicuculline (Meldrum and Nilsson 1976; Leniger-Follert 1984; Blennow et al. 1985; Kloiber et al. 1993) and emphasize the need to carefully adjust experimental conditions such as management of anesthesia (Abo et al. 2004), dose and rate of drug infusion and the importance of physiological monitoring in fMRI studies.

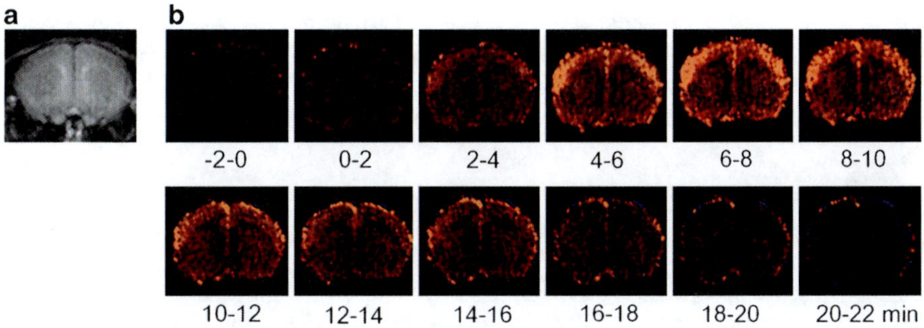

Fig. 18.2 *Pharmacological stimulation protocol*. Stimulation in BALB/c mice using $GABA_A$ antagonist bicuculline prompts significant region-specific responses. (**a**) Transverse T_2-weighted section through the mouse brain at the level of caudate putamen as structural reference image. (**b**) fMRI difference images displaying changes in rCBV. Bicuculline was infused in the interval 0–10 min following the protocol described in Mueggler et al. (2001). Large rCBV changes of 30–40% with regard to baseline values are observed in cerebral cortex

The feasibility of mapping region-specific and dose-dependent changes in serotonergic transmission in the adult mouse brain using a 200-MHz cryogenic RF detector for CBV-fMRI has been demonstrated (Mueggler et al. 2010). Inhibition of the 8-OH-DPAT-induced CBV response by co-injection with the selective 5-HT_{1A} receptor antagonist WAY-100635 or in 5-HT_{1A} knock-out mice, suggests that 5-HT_{1A} receptors are the primary targets of the agonist. This method opens novel perspectives for investigating 5-HT_{1A} receptor functions in mouse models of human pathologies resulting from a dysfunction of the 5-HT_{1A} receptor or the serotonergic system, including depression and anxiety. A methodological similar approach has recently been applied to investigate the neuronal substrate involved in conditioned fear response. By combining correlation analysis of CBV-fMRI data with cell- and tissue-specific pharmacogenetic inhibition Gozzi and colleagues were able to identify downstream circuits of the amygdala that control fear expression in mice (Gozzi et al. 2010).

Despite the lower CNR of BOLD- compared to the CBV-fMRI method at lower field strengths BOLD experiment have been carried out in mice during pharmacological stimulation. Kuriwaki and colleagues used D_2R agonist and antagonist as pharmacological fMRI challenge at 4.7 T to investigate a genetic mouse model of schizophrenia (see below) (Kuriwaki et al. 2004).

In addition to functional receptor imaging (pharmacological stimulation paradigms), fMRI can be used to study the effects of physiological, neuron-independent stimuli. For example, systemic infusion of the carbonic anhydrase blocker acetazolamide which directly triggers a local hemodynamic response via the vascular compartment (endothelium-independent) can be used to probe e.g. the cerebro-vascular reactivity. Substantial CBV increases upon such a physiological stimulus have been measured in mice (Mueggler et al. 2002). Application of this to characterize mouse models of neurodegeneration will be described later. In order to elucidate the role astrocytic NAAG signaling in the control of local cerebral microcirculation the effect of a specific blocker of N-acetylaspartylglutamate (NAAG), a dipeptide derivative of N-acetylaspartate (NAA) and glutamate (Glu) and selective mGluR3 agonist on the BOLD response after systemic application has been studied in mice at 7 T (Baslow et al. 2005).

18.3.3.2 Sensory Stimulation in Mice

Studies by autoradiography have shown that electrical stimulation of the forepaw in rats anesthetized with α-chloralose increases local CBF and glucose metabolism in the sensory-motor cortex (Ueki et al. 1988, 1992). Since then this paradigm has been applied in a number of fMRI studies in both α-chloralose (for review see Buxton 2002) or isoflurane anaesthetized rats (Reese et al. 2000b; Sauter et al. 2002; Liu et al. 2004; Liebscher et al. 2005; Sydekum et al. 2009; Ghosh et al. 2010). Stimulation of the forepaw led to distinct activation in the appropriate contralateral somatosensory cortical area representing the forepaw.

Similarly, robust CBV responses have been obtained in the respective cortical region of isoflurane-anaesthetized and mechanically ventilated mice during unilateral electrical stimulation of the hind paw (Mueggler et al. 2003). The CBV-fMRI signal response was found to be correlated with the current amplitude of the stimulus in the range 0.5–2 mA. Such a correlation of cerebral hemodynamic response with the intensity of the stimulus has been previously described in rats during median nerve stimulation (Backes et al. 2000). Similarly, the dependence of the CBV change on the current amplitude is in line with results derived from a study, in which CBF was assessed using Doppler flowmetry and ASL-MRI in α-chloralose anaesthetized rats. Forepaw stimulation revealed changes in local CBF that increased with the amplitude of the electrical current (0.1–2.5 mA). At higher current amplitudes, the focal CBF responses were accompanied by increases of the mean arterial BP, which were interpreted as indication of painful stimuli (Silva and Kim 1999). In fact, current levels of 3–10 mA have been used as pain models (Strangman and Walker 1999). Such conditions involve the recruitment of nociceptive Aδ-and C-fibers, which are part of the spinal-thalamic tract (Chang and Shyu 2001). The feasibility of investigating somatosensory and noxious signal processing in the isoflurane-anesthetized mouse using fMRI has recently been demonstrated (Bosshard et al. 2010). Electrical fore- and hindpaw stimulation was performed with different current amplitudes ranging from 0.5 to 2 mA. BOLD signal changes were observed in primary and secondary somatosensory cortices, the thalamus and the insular cortex, important regions composing the neuronal substrate of sensory and nociceptive processing.

Isoflurane has long been known to act as a potent vasodilator (Hansen et al. 1988; Matta et al. 1999); yet, the extent to which isoflurane affects cerebral hemodynamic parameters has been discussed controversially. Basal CBF modulates the stimulus-evoked fMRI response and high basal CBF due to vasodilatory effects of isoflurane might result in a "ceiling effect" reducing the amplitude of the stimulus-induced BOLD and/or CBF changes. However, there is substantial experimental evidence that despite the potential drawbacks of isoflurane anesthesia i.e. depressing excitatory synaptic transmission (Antognini et al. 1997; Antkowiak and Helfrich-Forster 1998; Larsen and Langmoen 1998; Detsch et al. 1999) and vasodilatory effects, robust fMRI responses can be observed during electrical paw stimulation in both rats (Sauter et al. 2002; Reese et al. 2000b; Sydekum et al. 2009; Ghosh et al. 2010) and mice (Mueggler et al. 2003; Bosshard et al. 2010) under artificial ventilation. A recent study emphasizes the requirement to optimize stimulation parameters i.e. stimulation frequency for a given anesthetic in order to maximize the fMRI signal change (Masamoto et al. 2007). Application of higher frequencies than commonly used for fMRI under α-chloralose (<5 Hz) clearly improved the BOLD signal change under isoflurane (maximum BOLD change at around 12 Hz) without the need of higher current amplitudes.

As already discussed above the use of high magnetic fields increases the CNR of BOLD-fMRI due to increased sensitivity to magnetic susceptibility changes. In fact, using conventional BOLD-fMRI protocols at 9.4 and 11.7 T, respectively, signal changes have been observed in the murine primary visual and somatosensory cortex during visual- and electrical hindpaw stimulation, respectively of freely breathing anaesthetized animals (Huang et al. 1996; Ahrens and Dubowitz 2001; Nair and Duong 2004). Drug induced neuronal activity has been visualized in the olfactory bulb of mice using BOLD-fMRI (Xu et al. 2003). Odor maps obtained in urethane anaesthetized mice allowed the visualization and quantification of activity patterns in response to olfactory stimulation by aldehyde homologues and esters. This study illustrates the potential of *non-invasive* fMRI to test different odorants in the same animal, as previously shown for the rat (Yang et al. 1998).

In conclusion BOLD, CBV- and flow-based *noninvasive* functional MRI using different stimulation paradigms is feasible in such small structures as the murine brain and open the way to a broad range of applications in genetically engineered mouse models of CNS disorders.

18.4 Application of *Hemodynamic-Based* fMRI to Murine Models of CNS Disorders

Due to technical demands mentioned above (i.e. sensitivity issues and requirements to control mouse physiology are more stringent than in rat fMRI) only a few applications of fMRI in murine models of CNS disorders have been reported so far from two application areas: neurodegenerative and psychiatric disorders.

18.4.1 Mouse Models of Neurodegeneration – a Comparison of Structural and Functional Readouts

In neurodegenerative diseases such as Alzheimer's disease (AD), early symptoms comprise functional (cognitive) deficits, which appear before gross-morphological alterations of associated brain regions (Jack et al. 1999). This led to the hypothesis

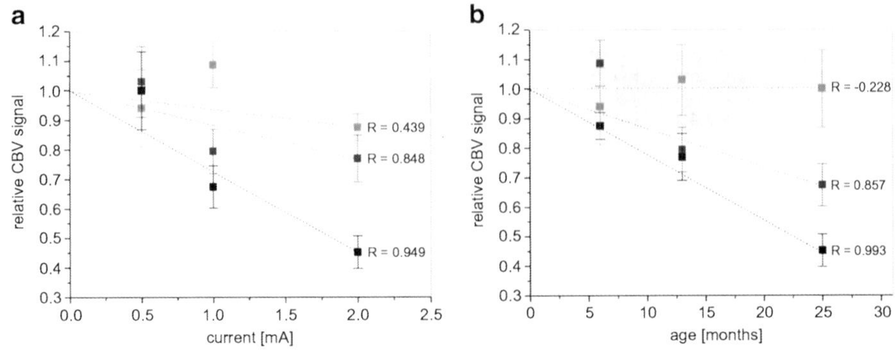

Fig. 18.3 *Functional phenotyping of APP23 transgenic mice*. Somatosensory stimulation of hind paws of APP23 transgenic mice relative to age matched wild-type mice. (**a**) Dependence of fMRI signal on amplitude of stimulation current. The symbols indicate 6 (*light gray symbol*), 13 months (*gray*), and 25 months (*black*). (**b**) Age dependence of fMRI signal. The symbols indicate 0.5 (*light gray symbol*), 1 (*gray*), and 2 mA (*black*). The data indicated that a compromised somatosensory response becomes detectable in APP23 mice of an age between 10 and 13 months for stimulation amplitude of 2 mA

that functional neuroimaging might detect subtle alterations of neuronal function before gross anatomical abnormalities become detectable in structural images. Therefore functional readouts would be of high diagnostic relevance. The various transgenic mouse models available today, which manifest characteristic neuropathological and behavioral features of AD application, offer an attractive basis for the evaluation of this concept.

A well characterized AD model is the APP23 transgenic mouse that overexpresses human amyloid precursor protein (APP). The animals exhibit elevated brain levels of amyloid beta peptide (Aβ) and develop characteristic neuropathological features of AD such as Aβ deposits (referred to as neuritic plaques; Sturchler-Pierrat et al. 1997; Sturchler-Pierrat and Staufenbiel 2000). These pathomorphological alterations occur in age-dependent manner and are associated with behavioral and learning deficits (Kelly et al. 2003; van Dam et al. 2003). FMRI was applied to APP23 mice of various ages to evaluate whether the response to a standardized challenge is different as compared to age matched wild-type animals. The stimuli applied were those previously described, i.e. pharmacological stimulation using the $GABA_A$ receptor antagonist bicuculline, physiological stimulation by inducing hypercapnia using the carbonic anhydrase inhibitor acetazolamide, and peripheral sensory activation using electrical stimulation of the hind paws. All three stimulation paradigms evoked CBV responses in 15 and 24 months old APP23 mice that were significantly reduced when compared to those of age-matched control littermates (Fig. 18.3, Mueggler et al. 2002, 2003). In young animals of 6 to 8 months age, there was no difference between the transgenic and wild-type group. The acetazolamide experiments indicated that the reduced response to the different stimuli might be due to a compromised vascular reserve capacity caused by the severe cerebral amyloid angiopathy (CAA) found in this model (Winkler et al. 2001) – and in AD patients (Burgermeister et al. 2000). Perivascular Aβ deposits impair the ability of the cerebral arteriolar and/or capillary compartment to effectively regulate CBF. As fMRI probes the hemodynamic changes associated with brain activity and not the neuronal activation per se, it is currently unclear to what extent the reduced response to the $GABA_A$ receptor inhibition and the sensory input is caused by impaired neural excitability or by the lack of cerebrovascular reactivity.

Assuming that the impaired vascular reactivity reflects CAA related vasculopathies, the acetazolamide challenge might detect early phases of CAA, which involve focal discontinuities of the smooth muscle cell (SMC) layer, which later progresses to a dramatic loss of SMCs in the tunica media of amyloid-loaded vessels. CAA has not been observed in young mice of an age of 8 months (Burgermeister et al. 2000); in the framework of this hypothesis the normal response observed in 6-month-old APP23 mice as compared to controls might be explained by the absence of CAA at this age. The impact of vascular deposits or even direct effects of Aβ(1–42) on properties of cerebral vascular SMCs have been examined in different models of AD (Iadecola et al. 1999; Christie et al. 2001; Niwa et al. 2002a, b) highlighting the growing awareness of the role of soluble Aβ species in early pathophysiological alterations in AD pathology. In particular, it has been suggested that elevated Aβ leads to the production of reactive oxygen species, which might compromise endothelium-dependent, nitric-oxide (NO) mediated vasodilatory response. In APP23 mice parenchymal Aβ levels have been reported to be increased already in 6-months-old mice (Sommer et al. 2000). At this age, however, the general vascular response as derived from fMRI studies was found normal, indicating that the CBV-fMRI response observed in the mice correlates rather with CAA than with parenchymal Aβ levels. In line with this hypothesis are recent findings in a novel APP-based mouse model (arcAβ; Knobloch et al. 2007) of a reduced CBV response to acetazolamide both with regard to vascular reactivity and extent of vasodilatation in brain regions with high CAA with normal response pattern relative to wildtype in brain areas displaying only little CAA (Princz-Kranz et al. 2010). These findings indicate that vascular β-amyloid deposits reduce the capacity of cerebral blood vessels to dilate upon demand, supporting the hypothesis that vascular β-amyloid contributes to hypoperfusion and neurological deficits observed in AD and CAA. We have, however, to keep in mind that significant hypoperfusion assessed by ASL-MRI has also been observed in different brain regions in a mouse model of AD which does not display profound CAA (PS2APP; Weidensteiner et al. 2009).

Clearly, non-invasive visualization of the parenchymal deposits (and cerebrovascular deposits) under *in vivo* conditions using MRI or other imaging modalities would be of high diagnostic relevance and several attempts have been reported to visualize plaques by exploiting their intrinsic contrast to the surrounding brain parenchyma both in tissue specimen from humans (Benveniste et al. 1999) and transgenic mice (Fig. 18.4, Rudin et al. 2003; Zhang et al. 2004; Vanhoutte et al. 2005). Although measurement time recently approached a range acceptable to *in vivo* imaging (Jack et al. 2005) the sensitivity of this MR readout remains challenging and has to be further improved to meet the requirements to spatial resolution moving either to high-field systems or by the application of cryogenic RF probes currently being validated under *in-vivo* conditions (Ratering et al. 2008;

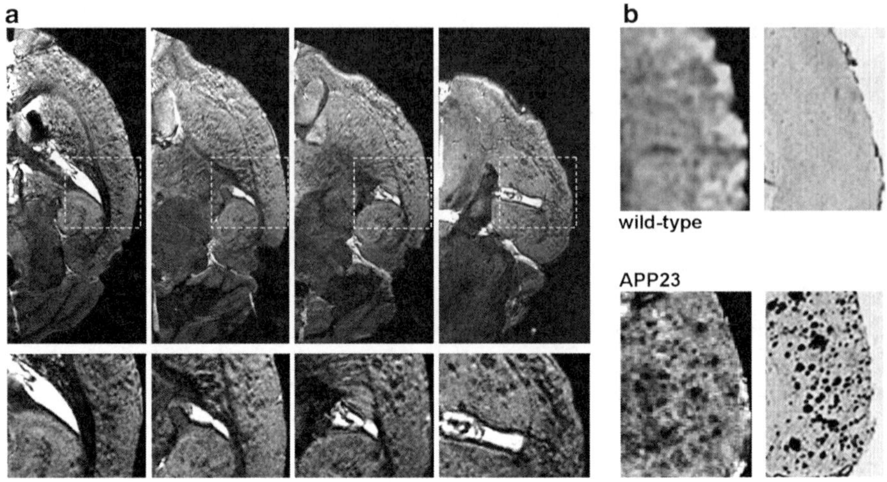

Fig. 18.4 Structural phenotyping: Visualization of β-amyloid plaques in the brain of 25 months old APP23 mouse that over-expresses human APP. (**a**) Four horizontal sections (out of 128) of a three-dimensional MRI data set of a fixed mouse brain. The *bottom* part shows enlarged inserts. Note the *dark structures* predominantly in cerebral cortex and hippocampus that are attributed to the occurrence of β-amyloid plaques. (**b**) Comparison of MRI and immunohistochemical readouts (NT12 antibody, Mueggler et al. 2004)

Baltes et al. 2009). An alternative concept is to amplify plaque related signals by administering Aβ–specific contrast agents. This approach uses labeled Aβ(1–40) peptide, known for its high binding affinity to Aβ plaques (Wadghiri et al. 2003). Intra-arterial injection of magnetically labeled (gadolinium chelates or SPIO) Aβ(1–40) enabled the *in vivo* detection of Aβ plaques. A disadvantage of the approach is that it requires breaking of the blood brain barrier (e.g. with mannitol) to allow the bulky reporter molecules to penetrate the brain parenchyma. In this regard, approaches that use plaque-specific ligands for positron-emission tomography (PET), (Shoghi-Jadid et al. 2002) or fluorescent dyes (Bacskai et al. 2003; Hintersteiner et al. 2005) seem more promising as those compounds readily cross the blood–brain barrier (BBB). BBB permeability can be increased as demonstrated by labeling amyloid plaques *in vivo* in Tg2476 and PSAPP mice with a probe, putrescine-gadolinium-amyloid-β peptide (PUT-Gd-Aβ), which due to the presence of the putrescine moiety is transported across the blood–brain-barrier following intravenous injection, causing specific plaque enhancement in T_1-weigthed images (Poduslo et al. 2002).

Direct visualization and quantification of the plaque load is considered highly attractive for diagnosis, disease staging, and therapy evaluation. Nevertheless, in view of the difficulties of direct visualization of the primary morphological hallmarks of AD a number of MRI methods that probe secondary structural changes have been applied to phenotype AD mouse models. Several readouts being based on volumetric measurements of brain atrophy (Mueggler 2002; Redwine et al. 2003; von Kienlin et al. 2005), changes in MR relaxation rate (R_2) (Helpern et al. 2004; Falangola et al. 2007), cerebrovascular flow disturbances (Beckmann et al. 2003), proton diffusion properties of the brain parenchyma as assessed by diffusion-weighted MRI (Mueggler et al. 2004) and diffusion-tensor imaging (DTI) (Song et al. 2004; Sun et al. 2005) have been reviewed recently (Mueggler 2006). A common feature of these various morphological phenotyping approaches is their lack of sensitivity. Reliable deviations from normal values (i.e. brain atrophy, apparent diffusion coefficient ADC, T_2, vascular defects) are only observed in advanced disease states, i.e. in old animals. This limits the value of these readouts for early diagnosis and as prognostic tools for monitoring disease evolution or therapeutic interventions. Whether fMRI readouts probing the brain functional architecture are superior in detecting early pathological alterations (Mueggler et al. 2002, 2003; Princz-Kranz et al. 2010) remains to be shown and requires longitudinal studies monitoring disease progression and evaluating the efficacy of disease modifying therapeutic interventions (e.g. in preclinical Aβ immunotherapy).

18.4.2 FMRI in D_2R Knock-Out Mouse Model of Schizophrenia

Functional approaches may be even more relevant when studying models of psychiatric diseases, for which there is frequently no clear neuro-anatomical correlate. In order to gain a better understanding of the etiology and pathophysiology of this disorder animal models of schizophrenia have been developed. These models are classically divided into two broad categories. The *neuropharmacological* models are related to specific neurotransmitter systems (i.e. dopamine or glutamate). They rely on the use of psychotomimetic substances such as DR agonists or NMDA-antagonists to produce schizophrenic-like symptoms in animals. Hence they postulate a dysfunction of these systems is underlying the disease. The *pathophysiological* models are particularly related to symptomatology and treatment. They rely on deficits induced by abnormal neurodevelopment, dysfunction of cortical glutamatergic systems or genetic susceptibility.

In both types of models, similar schizophrenia-like symptoms are generated such as hyperlocomotion, sensorimotor deficits or deficits in latent inhibition, impaired performance in cognition and memory tasks, and altered social and reward processes (Lipska and Weinberger 2000; Le Pen et al. 2002). Dopamine alterations in the brain of schizophrenic patients were among the first pathological observations in schizophrenia research, and, for this reason, the classical type of antipsychotics had preferential affinity for dopamine receptors (i.e. D_2R and D_3R). Targeting of dopamine receptors has been associated with amelioration of positive psychotic symptoms and existing evidence indicates that dopamine antagonism is required for antipsychotic activity. PET studies have shown that a minimum occupancy of 60% of striatal D_2R is required to obtain clinical response. Several studies have associated D_2 genetic variants with response to classical antipsychotics, but also with response to atypical antipsychotics such as risperidone, which shows a strong D_2R binding (for review see Arranz and de Leon 2007).

Recently, questions concerning the role of the dopamine-D_2-receptor (D_2R) in schizophrenia have been addressed in a genetically modified mouse model using fMRI in combination with a pharmacological challenge (Kuriwaki et al. 2004). Regional brain activity was analyzed on the basis of perfusion-weighted MR images obtained in D_2R-knockout and wild-type mice after systemic infusion of both a D_2R agonist and antagonist. The results suggested that antipsychotic D_2R antagonists specifically affect brain regions that are associated with the pathophysiology of schizophrenia in humans such as the hippocampal formation, thalamus, anterior cingulated cortex, caudate putamen, nucleus accumbens and central nucleus of the amygdale (Carlsson and Carlsson 1990; Shenton et al. 2001).

Similar to the application in models of neurodegeneration, the combination of fMRI and genetically engineered mice offers great potential to evaluate efficacy as well as cerebral side effects of novel therapeutics in models of psychiatric disorders.

18.5 Conclusion

FMRI studies in mice are feasible with adequate temporal and spatial resolution. While at low magnetic field strengths CBV-based methods using intravascular contrast agents yield superior sensitivity as compared to BOLD-based methods the situation is less clear for high B_0 values. Susceptibility effects scale linearly with the field strength while relaxation effects of paramagnetic and superparamagnetic contrast agent become saturated. As the two effects are of opposite sign they tend to compensate each other (Scheffler et al. 1999).

Obviously, the requirements on spatial resolution in mouse fMRI have a negative impact on the sensitivity. Use of higher magnetic field strengths to enhance sensitivity is inherently associated with increased susceptibility artefacts occurring at tissue interfaces, which due to the small dimensions of the mouse brain may potentially mask brain structures. These highly non-linear artefacts cannot be accounted for by shimming. The use of cryogenic probe detector devices may emerge as a cost efficient solution for increasing the sensitivity in BOLD-fMRI studies as compared to investing into higher magnetic field strengths (Baltes et al. 2010). High CNR of fMRI signals on the other hand depends critically on stable baseline conditions. A dominant noise source in fMRI is physiological noise, which may lead to increased variability in data and potentially confounding activation. Hence, close monitoring of physiological parameters such as blood gases, body temperature, and heart rate is mandatory allowing for corrective measures if required. Artificial ventilation in combination with muscle relaxing agents has yielded superior results as compared to spontaneously breathing animals (Nair and Duong 2004).

An important aspect in fMRI studies is the stimulation paradigm applied. The fact that such studies are commonly carried out in anaesthetized animals limits the number of possibilities. Potential protocols are sensory (visual, auditory, olfactory or peripheral sensory input) and passive motor stimulation, pharmacological challenges using neuro-active compounds, and physiological stimulation such as the induction of hypercapnia leading either to focal or global CNS responses. Application of these paradigms to anaesthetized mice has resulted in robust fMRI signals. For instance, administration of the $GABA_A$ receptor antagonist bicuculline led to CBV changes of up to 30–40% (Mueggler et al. 2001).

FMRI does not measure the neural activity per se, but rather the physiological consequence thereof. Thus, the lack of an fMRI signal may indicate the lack of neuronal activity or an impairment of the metabolic and hemodynamic coupling. Similarly, the observation of an fMRI signal following pharmacological stimulation might reflect a direct effect of the compound on the cerebral metabolic activity or on CBV not involving neuronal activity at all. This has to be kept in mind, in particular when analyzing data obtained from models of brain pathologies in which neurovascular coupling might be impaired.

Quantification is an important aspect when considering applications in pharmacological research. This is clearly possible in well controlled animal models that use standardized stimulation paradigm. For focal stimulation paradigms, quantitative analysis can be based on analysis of the fMRI signal amplitude in a defined ROI or by measuring the spatial extent of the activated region (number of voxels). Reproducibility of fMRI signal amplitudes for an electrical paw stimulation experiment was found to be reasonable with a coefficient of variation of 15% for the rat (forepaw, Sauter et al. 2002) and 20% for the mouse (hind paw, Mueggler et al. 2003). For global pattern, densitometric analysis for selected regions of interest has proven to be useful (Mueggler et al. 2002). Further analysis may direct towards resolving networks of correlated responses in the pharmacologically activated brain as recently demonstrated upon dopaminergic and serotonergic challenges in the rat (Schwarz et al. 2007). Application of network analysis to receptor knock-out mouse models would be obviously highly attractive as specific interruptions of a given neuronal pathway by the down regulation of a receptor or receptor-subunit could be investigated (Gozzi et al. 2010).

In view of the increasing number of genetically engineered mouse models of human neurological or psychiatric pathologies there is an increasing demand of tools for structural and functional phenotyping. Methods that allow non-invasive assessment of brain function such as fMRI are highly attractive in this context. We have discussed functional characterization of mouse models of neurodegeneration in some detail: a significantly compromised response to a standardized challenge (sensory, pharmacological, and physiological) has been observed in transgenic as compared to wild-type animals. The proper design of the test paradigms yields valuable information on mechanisms underlying the fMRI changes. In the APP23 example, the impaired response to a hypercapnic challenge indicates that CAA has to be considered in the analysis of the altered response to sensory and pharmacological stimuli.

Due to its non-invasiveness nature functional consequences of disease progression can be investigated in longitudinal studies providing information on disease evolution in the individual that can be combined with more traditional pharmacological readouts (i.e. behavioural assessments). Further studies investigating the potential of fMRI readouts in transgenic models of CNS disorders to characterize the disease phenotype and to assess the efficacy of a therapeutic intervention are therefore warranted and it is anticipated that the field will rapidly grow.

References

Abo M, Suzuki M, Senoo A et al (2004) Influence of isoflurane concentration and hypoxia on functional magnetic resonance imaging for the detection of bicuculline-induced neuronal activation. Neurosignals 13(3):144–149

Adamczak JM, Farr TD, Seehafer JU, Kalthoff D, Hoehn M (2010) High field BOLD response to forepaw stimulation in the mouse. Neuroimage 51(2):704–712

Ahrens ET, Dubowitz DJ (2001) Peripheral somatosensory fMRI in mouse at 11.7 T. NMR Biomed 14(5):318

Antkowiak B, Helfrich-Forster C (1998) Effects of small concentrations of volatile anesthetics on action potential firing of neocortical neurons in vitro. Anesthesiology 88(6):1592

Antognini JF, Buonocore MH, Disbrow EA et al (1997) Isoflurane anesthesia blunts cerebral responses to noxious and innocuous stimuli: a fMRI study. Life Sci 61(24):PL 349–PL 354

Arranz MJ, de Leon J (2007) Pharmacogenetics and pharmacogenomics of schizophrenia: a review of last decade of research. Mol Psychiatry 12(8):707–747

Austin VC, Blamire AM, Grieve SM et al (2003) Differences in the BOLD fMRI response to direct and indirect cortical stimulation in the rat. Magn Reson Med 49(5):838–847

Austin VC, Blamire AM, Allers KA et al (2005) Confounding effects of anesthesia on functional activation in rodent brain: a study of halothane and alpha-chloralose anesthesia. Neuroimage 24(1):92–100

Backes WH, Mess WH, van Kranen-Mastenbroek V et al (2000) Somatosensory cortex responses to median nerve stimulation: fMRI effects of current amplitude and selective attention. Clin Neurophysiol 111(10):1738

Bacskai BJ, Hickey GA, Skoch J et al (2003) Four-dimensional multiphoton imaging of brain entry, amyloid binding, and clearance of an amyloid-beta ligand in transgenic mice. Proc Natl Acad Sci USA 100(21):12462–12467

Baltes C, Radzwill N, Bosshard S, Marek D, Rudin M (2009) Micro MRI of the mouse brain using a novel 400 MHz cryogenic quadrature RF probe. NMR Biomed 22(8):834–842

Baltes C, Bosshard S, Mueggler T, Ratering D, Rudin M (2010) Increased blood oxygen level-dependent (BOLD) sensitivity in the mouse somatosensory cortex during electrical forepaw stimulation using a cryogenic radiofrequency probe. NMR Biomed. doi:doi: 10.1002/nbm.1613

Baslow MH, Dyakin VV, Nowak KL, Hungund BL, Guilfoyle DN (2005) 2-PMPA, a NAAG peptidase inhibitor, attenuates magnetic resonance BOLD signals in brain of anesthetized mice: evidence of a link between neuron NAAG release and hyperemia. J Mol Neurosci 26(1):1–15

Beckmann N, Schuler A, Mueggler T et al (2003) Age-dependent cerebrovascular abnormalities and blood flow disturbances in APP23 mice modeling Alzheimer's disease. J Neurosci 23(24):8453–8459

Belliveau JW, Rosen BR, Kantor HL et al (1990) Functional cerebral imaging by susceptibility-contrast NMR. Magn Reson Med 14(3):538

Benveniste H, Einstein G, Kim KR et al (1999) Detection of neuritic plaques in Alzheimer's disease by magnetic resonance microscopy. Proc Natl Acad Sci USA 96(24):14079

Blennow G, Nilsson B, Siesjo BK (1985) Influence of reduced oxygen availability on cerebral metabolic changes during bicuculline-induced seizures in rats. J Cereb Blood Flow Metab 5(3):439–445

Bosshard SC, Baltes C, Wyss MT, Mueggler T, Weber B, Rudin M (2010) Assessment of brain responses to innocuous and noxious electrical forepaw stimulation in mice using BOLD fMRI. Pain 151(3):655–663

Bruns A, Künnecke B, Risterucci C, Moreau JL, von Kienlin M (2009) Validation of cerebral blood perfusion imaging as a modality for quantitative pharmacological MRI in rats. Magn Reson Med 61(6):1451–1458

Burdett NG, Menon DK, Carpenter TA et al (1995) Visualisation of changes in regional cerebral blood flow (rCBF) produced by ketamine using long TE gradient-echo sequences: preliminary results. Magn Reson Imaging 13(4):549–553

Burgermeister P, Calhoun ME, Winkler DT, Jucker M (2000) Mechanisms of cerebrovascular amyloid deposition. Lessons from mouse models. Ann NY Acad Sci 903:307–316

Buxton RB (2002) Introduction to functional magnetic resonance imaging: principles and techniques. Cambridge University Press, Cambridge

Buxton RB, Frank LR (1997) A model for the coupling between cerebral blood flow and oxygen metabolism during neural stimulation. J Cereb Blood Flow Metab 17(1):64

Buxton RB, Uludag K, Dubowitz DJ et al (2004) Modeling the hemodynamic response to brain activation. Neuroimage 23(Suppl 1):S220–S233

Carlsson M, Carlsson A (1990) Interactions between glutamatergic and monoaminergic systems within the basal ganglia–implications for schizophrenia and Parkinson's disease. Trends Neurosci 13(7):272–276

Chang C, Shyu BC (2001) A fMRI study of brain activations during non-noxious and noxious electrical stimulation of the sciatic nerve of rats. Brain Res 897(1–2):71

Chen YC, Galpern WR, Brownell AL et al (1997) Detection of dopaminergic neurotransmitter activity using pharmacologic MRI: correlation with PET, microdialysis, and behavioral data. Magn Reson Med 38(3):389–398

Chen YC, Mandeville JB, Nguyen TV et al (2001) Improved mapping of pharmacologically induced neuronal activation using the IRON technique with superparamagnetic blood pool agents. J Magn Reson Imaging 14(5):517

Christie R, Yamada M, Moskowitz M et al (2001) Structural and functional disruption of vascular smooth muscle cells in a transgenic mouse model of amyloid angiopathy. Am J Pathol 158(3):1065

Crosby G, Crane AM, Jehle J et al (1983) The local metabolic effects of somatosensory stimulation in the central nervous system of rats given pentobarbital or nitrous oxide. Anesthesiology 58(1):38–43

Dazai J, Bock NA, Nieman BJ et al (2004) Multiple mouse biological loading and monitoring system for MRI. Magn Reson Med 52(4):709–715

Detre JA, Wang J (2002) Technical aspects and utility of fMRI using BOLD and ASL. Clin Neurophysiol 113(5):621–634

Detre JA, Leigh JS, Williams DS et al (1992) Perfusion imaging. Magn Reson Med 23(1):37

Detsch O, Vahle-Hinz C, Kochs E et al (1999) Isoflurane induces dose-dependent changes of thalamic somatosensory information transfer. Brain Res 829(1–2):77

Dubowitz DJ, Chen DY, Atkinson DJ, Grieve KL, Gillikin B, Bradley WG Jr, Andersen RA (1998) Functional magnetic resonance imaging in macaque cortex. Neuroreport 9(10):2213–2218

Duong TQ (2007) Cerebral blood flow and BOLD fMRI responses to hypoxia in awake and anesthetized rats. Brain Res 1135(1):186–194

Duong TQ, Silva AC, Lee SP et al (2000) Functional MRI of calcium-dependent synaptic activity: cross correlation with CBF and BOLD measurements. Magn Reson Med 43(3):383–392

Falangola MF, Dyakin VV, Lee SP et al (2007) Quantitative MRI reveals aging-associated T2 changes in mouse models of Alzheimer's disease. NMR Biomed 20(3):343–351

Foley LM, Hitchens TK, Kochanek PM, Melick JA, Jackson EK, Ho C (2005) Murine orthostatic response during prolonged vertical studies: effect on cerebral blood flow measured by arterial spin-labeled MRI. Magn Reson Med 54(4):798–806

Forman SD, Silva AC, Dedousis N et al (1998) Simultaneous glutamate and perfusion fMRI responses to regional brain stimulation. J Cereb Blood Flow Metab 18(10):1064–1070

Gati JS, Menon RS, Ugurbil K et al (1997) Experimental determination of the BOLD field strength dependence in vessels and tissue. Magn Reson Med 38(2):296

Ghosh A, Haiss F, Sydekum E et al (2010) Rewiring of hindlimb corticospinal neurons after spinal cord injury. Nat Neurosci 13(1):97–104

Gozzi A, Ceolin L, Schwarz A et al (2007) A multimodality investigation of cerebral hemodynamics and autoregulation in pharmacological MRI. Magn Reson Imaging 25(6):826–833

Gozzi A, Jain A, Giovanelli A et al (2010) A neural switch for active and passive fear. Neuron 67(4):656–666

Gyngell ML, Bock C, Schmitz B et al (1996) Variation of functional MRI signal in response to frequency of somatosensory stimulation in alpha-chloralose anesthetized rats. Magn Reson Med 36(1):13

Hansen TD, Warner DS, Todd MM et al (1988) Distribution of cerebral blood flow during halothane versus isoflurane anesthesia in rats. Anesthesiology 69(3):332

Helpern JA, Lee SP, Falangola MF et al (2004) MRI assessment of neuropathology in a transgenic mouse model of Alzheimer's disease. Magn Reson Med 51(4):794–798

Hintersteiner M, Enz A, Frey P et al (2005) In vivo detection of amyloid-beta deposits by near-infrared imaging using an oxazine-derivative probe. Nat Biotechnol 23(5):577–583

Houston GC, Papadakis NG, Carpenter TA et al (2001) Mapping of brain activation in response to pharmacological agents using fMRI in the rat. Magn Reson Imaging 19(7):905–919

Huang W, Plyka I, Li H et al (1996) Magnetic resonance imaging (MRI) detection of the murine brain response to light: temporal differentiation and negative functional MRI changes. Proc Natl Acad Sci USA 93(12):6037–6042

Hyder F, Behar KL, Martin MA et al (1994) Dynamic magnetic resonance imaging of the rat brain during forepaw stimulation. J Cereb Blood Flow Metab 14(4):649–655

Hyder F, Rothman DL, Shulman RG (2002) Total neuroenergetics support localized brain activity: implications for the interpretation of fMRI. Proc Natl Acad Sci USA 99(16):10771–10776

Iadecola C, Zhang F, Niwa K et al (1999) SOD1 rescues cerebral endothelial dysfunction in mice overexpressing amyloid precursor protein. Nat Neurosci 2(2):157

Jack CR Jr, Petersen RC, Xu YC et al (1999) Prediction of AD with MRI-based hippocampal volume in mild cognitive impairment. Neurology 52(7):1397

Jack CR Jr, Wengenack TM, Reyes DA et al (2005) In vivo magnetic resonance microimaging of individual amyloid plaques in Alzheimer's transgenic mice. J Neurosci 25(43):10041–10048

Jenkins BG, Choi JK, Mandeville JB et al (2006) Pharmacological Magnetic Resonance Imaging (phMRI). In: Beckmann N (ed) In-vivo MR techniques in drug discovery and development. Taylor & Francis, New York, pp 171–220

Jezzard P, Buxton RB (2006) The clinical potential of functional magnetic resonance imaging. J Magn Reson Imaging 23(6):787–793

Johnson GA, Cofer GP, Fubara B et al (2002) Magnetic resonance histology for morphologic phenotyping. J Magn Reson Imaging 16(4):423–429

Jones N, O'Neill MJ, Tricklebank M et al (2005) Examining the neural targets of the AMPA receptor potentiator LY404187 in the rat brain using pharmacological magnetic resonance imaging. Psychopharmacology (Berl) 180(4):743–751

Kalisch R, Elbel GK, Gossl C, Czisch M, Auer DP (2001) Blood pressure changes induced by arterial blood withdrawal influence bold signal in anesthesized rats at 7 Tesla: implications for pharmacologic MRI. Neuroimage 14:891–898

Kannurpatti SS, Biswal BB (2004) Effect of anesthesia on CBF, MAP and fMRI-BOLD signal in response to apnea. Brain Res 1011:141–147

Keilholz SD, Silva AC, Raman M et al (2006) BOLD and CBV-weighted functional magnetic resonance imaging of the rat somatosensory system. Magn Reson Med 55(2):316–324

Kelly PH, Bondolfi L, Hunziker D et al (2003) Progressive age-related impairment of cognitive behavior in APP23 transgenic mice. Neurobiol Aging 24(2):365–378

Kennan RP, Scanley BE, Innis RB et al (1998) Physiological basis for BOLD MR signal changes due to neuronal stimulation: separation of blood volume and magnetic susceptibility effects. Magn Reson Med 40(6):840

Kerskens CM, Hoehn-Berlage M, Schmitz B et al (1996) Ultrafast perfusion-weighted MRI of functional brain activation in rats during forepaw stimulation: comparison with T2 -weighted MRI. NMR Biomed 9(1):20

King JA, Garelick TS, Brevard ME et al (2005) Procedure for minimizing stress for fMRI studies in conscious rats. J Neurosci Methods 148(2):154–160

Kloiber O, Bockhorst K, Hoehn-Berlage M et al (1993) Effect of hypoxia on bicuculline seizures of rat: NMR spectroscopy and bioluminescence imaging. NMR Biomed 6(5):333–338

Knobloch M, Konietzko U, Krebs DC, Nitsch RM (2007) Intracellular Aβ and cognitive deficits precede β-amyloid deposition in transgenic arcAβ mice. Neurobiol Aging 28(9):1297–1306
Krnjevic K (1970) Glutamate and gamma-aminobutyric acid in brain. Nature 228(5267):119–124
Kuo CC, Chen JH, Tsai CY et al (2005) BOLD signals correlate with ensemble unit activities in rat's somatosensory cortex. Chin J Physiol 48(4):200–209
Kuriwaki J, Nishijo H, Kondoh T et al (2004) Comparison of brain activity between dopamine D2 receptor-knockout and wild mice in response to dopamine agonist and antagonist assessed by fMRI. Neurosignals 13(5):227–240
Kwong KK, Belliveau JW, Chesler DA et al (1992) Dynamic magnetic resonance imaging of human brain activity during primary sensory stimulation. Proc Natl Acad Sci USA 89(12):5675
Lahti KM, Ferris CF, Li F et al (1999) Comparison of evoked cortical activity in conscious and propofol-anesthetized rats using functional MRI. Magn Reson Med 41(2):412–416
Larsen M, Langmoen IA (1998) The effect of volatile anaesthetics on synaptic release and uptake of glutamate. Toxicol Lett 100–101:59–64
Le Pen G, Gaudet L, Mortas P et al (2002) Deficits in reward sensitivity in a neurodevelopmental rat model of schizophrenia. Psychopharmacol (Berl) 161(4):434–441
Leniger-Follert E (1984) Mechanisms of regulation of cerebral microflow during bicuculline-induced seizures in anaesthetized cats. J Cereb Blood Flow Metab 4(2):150–165
Liebscher T, Schnell L, Schnell D, Scholl J, Schneider R, Gullo M, Fouad K, Mir A, Rausch M, Kindler D, Hamers FP, Schwab ME (2005) Nogo-A antibody improves regeneration and locomotion of spinal cord-injured rats. Ann Neurol 58(5):706–719
Lin YJ, Koretsky AP (1997) Manganese ion enhances T1-weighted MRI during brain activation: an approach to directly imaging of brain function. Magn Reson Med 83(3):378–388
Lindauer U, Villringer A, Dirnagl U (1993) Characterization of CBF response to somatosensory stimulation: model and influence of anesthetics. Am J Physiol 264(4 Pt 2):H1223–H1228
Lipska BK, Weinberger DR (2000) To model a psychiatric disorder in animals: schizophrenia as a reality test. Neuropsychopharmacology 23(3):223–239
Liu ZM, Schmidt KF, Sicard KM et al (2004) Imaging oxygen consumption in forepaw somatosensory stimulation in rats under isoflurane anesthesia. Magn Reson Med 52(2):277–285
Logothetis NK, Pfeuffer J (2004) On the nature of the BOLD fMRI contrast mechanism. Magn Reson Imaging 22(10):1517–1531
Lu H, Scholl CA, Zuo Y, Stein EA, Yang Y (2007) Quantifying the blood oxygenation level dependent effect in cerebral blood volume-weighted functional MRI at 9.4 T. Magn Reson Med 58(3):616–621
Mandeville JB, Marota JJ, Kosofsky BE et al (1998) Dynamic functional imaging of relative cerebral blood volume during rat forepaw stimulation. Magn Reson Med 39(4):615
Mandeville JB, Jenkins BG, Kosofsky BE et al (2001) Regional sensitivity and coupling of BOLD and CBV changes during stimulation of rat brain. Magn Reson Med 45(3):443
Mandeville JB, Jenkins BG, Chen YC et al (2004) Exogenous contrast agent improves sensitivity of gradient-echo functional magnetic resonance imaging at 9.4 T. Magn Reson Med 52(6):1272–1281
Marota JJ, Mandeville JB, Weisskoff RM et al (2000) Cocaine activation discriminates dopaminergic projections by temporal response: an fMRI study in Rat. Neuroimage 11(1):13
Masamoto K, Kim T, Fukuda M et al (2007) Relationship between neural, vascular, and BOLD signals in isoflurane-anesthetized rat somatosensory cortex. Cereb Cortex 17(4):942–950
Matta BF, Heath KJ, Tipping K et al (1999) Direct cerebral vasodilatory effects of sevoflurane and isoflurane. Anesthesiology 91(3):677–680
Meldrum BS, Nilsson B (1976) Cerebral blood flow and metabolic rate early and late in prolonged epileptic seizures induced in rats by bicuculline. Brain 99(3):523–542
Mueggler T (2002) Thesis, Functional MRI of the Murine Brain. Application to a Transgenic Model of Alzheimer's Disease. University of Basel, Switzerland
Mueggler T (2006) Magnetic resonance imaging and spectorscopy in transgenic mice modeling Alzheimer's disease. In: Beckmann N (ed) In-vivo MR techniques in drug discovery and development. Taylor & Francis, New York, pp 95–110

Mueggler T, Baumann D, Rausch M et al (2001) Bicuculline-induced brain activation in mice detected by magnetic resonance imaging. Magn Reson Med 46:292
Mueggler T, Sturchler-Pierrat C, Baumann D et al (2002) Compromised hemodynamic response in amyloid precursor protein transgenic mice. J Neurosci 22(16):7218
Mueggler T, Baumann D, Rausch M et al (2003) Age-dependent impairment of somatosensory response in the amyloid precursor protein 23 transgenic mouse model of Alzheimer's disease. J Neurosci 23(23):8231–8236
Mueggler T, Meyer-Luehmann M, Rausch M et al (2004) Restricted diffusion in the brain of transgenic mice with cerebral amyloidosis. Eur J Neurosci 20(3):811–817
Mueggler T, Razoux F, Russig H et al (2010) Mapping of CBV changes in 5-HT(1A) terminal fields by functional MRI in the mouse brain. Eur Neuropsychopharmacology 21(4):344–353
Nair DG (2005) About being BOLD. Brain Res Brain Res Rev 50(2):229–243
Nair G, Duong TQ (2004) Echo-planar BOLD fMRI of mice on a narrow-bore 9.4 T magnet. Magn Reson Med 52(2):430–434
Nguyen TV, Brownell AL, Iris CYC et al (2000) Detection of the effects of dopamine receptor supersensitivity using pharmacological MRI and correlations with PET. Synapse 36(1):57
Niwa K, Kazama K, Younkin L et al (2002a) Cerebrovascular autoregulation is profoundly impaired in mice overexpressing amyloid precursor protein. Am J Physiol Heart Circ Physiol 283(1):H315
Niwa K, Kazama K, Younkin SG et al (2002b) Alterations in cerebral blood flow and glucose utilization in mice overexpressing the amyloid precursor protein. Neurobiol Dis 9(1):61–68
Ogawa S, Lee TM, Kay AR et al (1990) Brain magnetic resonance imaging with contrast dependent on blood oxygenation. Proc Natl Acad Sci USA 87(24):9868
Ogawa S, Menon RS, Kim SG et al (1998) On the characteristics of functional magnetic resonance imaging of the brain. Annu Rev Biophys Biomol Struct 27:447–474
Paulus MJ, Gleason SS, Easterly ME et al (2001) A review of high-resolution X-ray computed tomography and other imaging modalities for small animal research. Lab Anim (NY) 30(3):36–45
Peeters RR, Tindemans I, De SE et al (2001) Comparing BOLD fMRI signal changes in the awake and anesthetized rat during electrical forepaw stimulation. Magn Reson Imaging 19(6):821
Poduslo JF, Wengenack TM, Curran GL et al (2002) Molecular targeting of Alzheimer's amyloid plaques for contrast-enhanced magnetic resonance imaging. Neurobiol Dis 11(2):315–329
Preece M, Mukherjee B, Huang CL et al (2001) Detection of pharmacologically mediated changes in cerebral activity by functional magnetic resonance imaging: the effects of sulpiride in the brain of the anaesthetised rat. Brain Res 916(1–2):107–114
Princz-Kranz FL, Mueggler T, Knobloch M, Nitsch RM, Rudin M (2010) Vascular response to acetazolamide decreases as a function of age in the arcA beta mouse model of cerebral amyloidosis. Neurobiol Dis 40(1):284–292
Ramos-Cabrer P, Weber R, Wiedermann D et al (2005) Continuous noninvasive monitoring of transcutaneous blood gases for a stable and persistent BOLD contrast in fMRI studies in the rat. NMR Biomed 18(7):440–446
Ratering D, Baltes C, Nordmeyer-Massner J, Marek D, Rudin M (2008) Performance of a 200-MHz cryogenic RF probe designed for MRI and MRS of the murine brain. Magn Reson Med 59(6):1440–1447
Redwine JM, Kosofsky B, Jacobs RE et al (2003) Dentate gyrus volume is reduced before onset of plaque formation in PDAPP mice: a magnetic resonance microscopy and stereologic analysis. Proc Natl Acad Sci USA 100(3):1381–1386
Reese T, Bjelke B, Porszasz R et al (2000a) Regional brain activation by bicuculline visualized by functional magnetic resonance imaging. Time-resolved assessment of bicuculline-induced changes in local cerebral blood volume using an intravascular contrast agent. NMR Biomed 13(1):43
Reese T, Porszasz R, Baumann D et al (2000b) Cytoprotection does not preserve brain functionality in rats during the acute post-stroke phase despite evidence of non-infarction provided by MRI. NMR Biomed 13(6):361

Risterucci C, Jeanneau K, Schoppenthau S et al (2005) Functional magnetic resonance imaging reveals similar brain activity changes in two different animal models of schizophrenia. Psychopharmacol (Berl) 180(4):724–734

Rojas MJ, Navas JA, Rector DM (2006) Evoked response potential markers for anesthetic and behavioral states. Am J Physiol Regul Integr Comp Physiol 291(1):R189–R196

Rudin M, Mueggler T, Allegrini PR et al (2003) Characterization of CNS disorders and evaluation of therapy using structural and functional MRI. Anal Bioanal Chem 377(6):973–981

Sauter A, Reese T, Porszasz R et al (2002) Recovery of function in cytoprotected cerebral cortex in rat stroke model assessed by functional MRI. Magn Reson Med 47(4):759

Scanley BE, Kennan RP, Cannan S et al (1997) Functional magnetic resonance imaging of median nerve stimulation in rats at 2.0 T. Magn Reson Med 37(6):969

Scheffler K, Seifritz E, Haselhorst R et al (1999) Titration of the BOLD effect: separation and quantitation of blood volume and oxygenation changes in the human cerebral cortex during neuronal activation and ferumoxide infusion. Magn Reson Med 42(5):829

Schmidt KF, Febo M, Shen Q et al (2006) Hemodynamic and metabolic changes induced by cocaine in anesthetized rat observed with multimodal functional MRI. Psychopharmacol (Berl) 185(4):479–486

Schwarz AJ, Gozzi A, Reese T et al (2007) In vivo mapping of functional connectivity in neurotransmitter systems using pharmacological MRI. Neuroimage 34(4):1627–1636

Scorer CA (2001) Preclinical and clinical challenges in the development of disease-modifying therapies for Alzheimer's disease. Drug Discov Today 6(23):1207

Shenton ME, Dickey CC, Frumin M, McCarley RW (2001) A review of MRI findings in schizophrenia. Schizophr Res 49(1–2):1–52

Shoghi-Jadid K, Small GW, Agdeppa ED et al (2002) Localization of neurofibrillary tangles and beta-amyloid plaques in the brains of living patients with Alzheimer disease. Am J Geriatr Psychiatry 10(1):24–35

Sicard K, Shen Q, Brevard ME et al (2003) Regional cerebral blood flow and BOLD responses in conscious and anesthetized rats under basal and hypercapnic conditions: implications for functional MRI studies. J Cereb Blood Flow Metab 23(4):472–481

Silva AC (2005) Perfusion-based fMRI: insights from animal models. J Magn Reson Imaging 22(6):745–750

Silva AC, Bock NA (2008) Manganese-enhanced MRI: an exceptional tool in translational neuroimaging. Schizophr Bull 34(4):595–604

Silva AC, Kim SG (1999) Pseudo-continuous arterial spin labeling technique for measuring CBF dynamics with high temporal resolution. Magn Reson Med 42(3):425–429

Silva AC, Zhang W, Williams DS et al (1995) Multi-slice MRI of rat brain perfusion during amphetamine stimulation using arterial spin labeling. Magn Reson Med 33(2):209–214

Silva AC, Lee SP, Iadecola C et al (2000) Early temporal characteristics of cerebral blood flow and deoxyhemoglobin changes during somatosensory stimulation. J Cereb Blood Flow Metab 20(1):201–206

Silva AC, Lee JH, Aoki I, Koretsky AP (2004) Manganese-enhanced magnetic resonance imaging (MEMRI): methodological and practical considerations. NMR Biomed 17(8):532–543

Skoubis PD, Hradil V, Chin CL et al (2006) Mapping brain activity following administration of a nicotinic acetylcholine receptor agonist, ABT-594, using functional magnetic resonance imaging in awake rats. Neuroscience 137(2):583–591

Sommer B, Sturchler-Pierrat C, Abramowski D et al (2000) Transgenic approaches to model Alzheimer's disease. Rev Neurosci 11(1):47

Song SK, Kim JH, Lin SJ et al (2004) Diffusion tensor imaging detects age-dependent white matter changes in a transgenic mouse model with amyloid deposition. Neurobiol Dis 15(3):640–647

Steward CA, Marsden CA, Prior MJ et al (2005) Methodological considerations in rat brain BOLD contrast pharmacological MRI. Psychopharmacol (Berl) 180(4):687–704

Strangman NM, Walker JM (1999) Cannabinoid WIN 55,212-2 inhibits the activity-dependent facilitation of spinal nociceptive responses. J Neurophysiol 82(1):472

Sturchler-Pierrat C, Staufenbiel M (2000) Pathogenic mechanisms of Alzheimer's disease analyzed in the APP23 transgenic mouse model. Ann NY Acad Sci 920:134–139

Sturchler-Pierrat C, Abramowski D, Duke M et al (1997) Two amyloid precursor protein transgenic mouse models with Alzheimer disease-like pathology. Proc Natl Acad Sci USA 94(24): 13287–13292

Sun SW, Song SK, Harms MP et al (2005) Detection of age-dependent brain injury in a mouse model of brain amyloidosis associated with Alzheimer's disease using magnetic resonance diffusion tensor imaging. Exp Neurol 191(1):77–85

Sydekum E, Baltes C, Ghosh A, Mueggler T, Schwab ME, Rudin M (2009) Functional reorganization in rat somatosensory cortex assessed by fMRI: elastic image registration based on structural landmarks in fMRI images and application to spinal cord injured rats. Neuroimage 44(4): 1345–1354

Tarditi A, Caricasole A, Terstappen G (2009) Therapeutic targets for Alzheimer's disease. Expert Opin Ther Targets 13(5):551–567

Thal SC, Plesnila N (2007) Non-invasive intraoperative monitoring of blood pressure and arterial pCO2 during surgical anesthesia in mice. J Neurosci Methods 159(2):261–267

Turner R, Jones T (2003) Techniques for imaging neuroscience. Br Med Bull 65:3–20

Ueki M, Linn F, Hossmann KA (1988) Functional activation of cerebral blood flow and metabolism before and after global ischemia of rat brain. J Cereb Blood Flow Metab 8(4):486

Ueki M, Mies G, Hossmann KA (1992) Effect of alpha-chloralose, halothane, pentobarbital and nitrous oxide anesthesia on metabolic coupling in somatosensory cortex of rat. Acta Anaesthesiol Scand 36(4):318

Van Bruggen N, Busch E, Palmer JT et al (1998) High-resolution functional magnetic resonance imaging of the rat brain: mapping changes in cerebral blood volume using iron oxide contrast media. J Cereb Blood Flow Metab 18(11):1178

Van Camp N, Verhoye M, De Zeeuw CI et al (2006) Light stimulus frequency dependence of activity in the rat visual system as studied with high-resolution BOLD fMRI. J Neurophysiol 95(5):3164–3170

Van Dam D, D'Hooge R, Staufenbiel M et al (2003) Age-dependent cognitive decline in the APP23 model precedes amyloid deposition. Eur J Neurosci 17(2):388–396

Van der Linden A, Van Meir V, Tindemans I, Verhoye M, Balthazart J (2004) Applications of manganese-enhanced magnetic resonance imaging (MEMRI) to image brain plasticity in song birds. NMR Biomed 17(8):602–612

Van der Linden A, Van Camp N, Ramos-Cabrer P et al (2007) Current status of functional MRI on small animals: application to physiology, pathophysiology, and cognition. NMR Biomed 20(5):522–545

Vanhoutte G, Dewachter I, Borghgraef P, Van Leuven F, Van der Linden A (2005) Noninvasive in vivo MRI detection of neuritic plaques associated with iron in APP[V717I] transgenic mice, a model for Alzheimer's disease. Magn Reson Med 53(3):607–613

Vanhoutte G, Verhoye M, Van der Linden A (2006) Changing body temperature affects the T2* signal in the rat brain and reveals hypothalamic activity. Magn Reson Med 55(5):1006–1012

Villringer A, Dirnagl U (1995) Coupling of brain activity and cerebral blood flow: basis of functional neuroimaging. Cerebrovasc Brain Metab Rev 7(3):240

Villringer A, Rosen BR, Belliveau JW et al (1988) Dynamic imaging with lanthanide chelates in normal brain: contrast due to magnetic susceptibility effects. Magn Reson Med 6(2):164

von Kienlin M, Kunnecke B, Metzger F et al (2005) Altered metabolic profile in the frontal cortex of PS2APP transgenic mice, monitored throughout their life span. Neurobiol Dis 18(1):32–39

Wadghiri YZ, Sigurdsson EM, Sadowski M et al (2003) Detection of Alzheimer's amyloid in transgenic mice using magnetic resonance microimaging. Magn Reson Med 50(2):293–302

Weber R, Ramos-Cabrer P, Hoehn M (2006a) Present status of magnetic resonance imaging and spectroscopy in animal stroke models. J Cereb Blood Flow Metab 26(5):591–604

Weber R, Ramos-Cabrer P, Wiedermann D et al (2006b) A fully noninvasive and robust experimental protocol for longitudinal fMRI studies in the rat. Neuroimage 29(4):1303–1310

Weidensteiner C, Metzger F, Bruns A, Bohrmann B, Kuennecke B, von Kienlin M (2009) Cortical hypoperfusion in the B6.PS2APP mouse model for Alzheimer's disease: comprehensive phenotyping of vascular and tissular parameters by MRI. Magn Reson Med 62(1):35–45

Wiesmann F, Neubauer S, Haase A et al (2001) Can we use vertical bore magnetic resonance scanners for murine cardiovascular phenotype characterization? Influence of upright body position on left ventricular hemodynamics in mice. J Cardiovasc Magn Reson 3(4):311–315

Williams DS, Detre JA, Zhang W et al (1993) Fast serial MRI of perfusion in the rat brain using spin inversion of arterial water. Bull Mag Reson 15:60

Winkler DT, Bondolfi L, Herzig MC et al (2001) Spontaneous hemorrhagic stroke in a mouse model of cerebral amyloid angiopathy. J Neurosci 21(5):1619

Xu H, Li SJ, Bodurka J et al (2000) Heroin-induced neuronal activation in rat brain assessed by functional MRI. Neuroreport 11(5):1085

Xu F, Liu N, Kida I et al (2003) Odor maps of aldehydes and esters revealed by functional MRI in the glomerular layer of the mouse olfactory bulb. Proc Natl Acad Sci USA 100(19):11029–11034

Yang X, Renken R, Hyder F et al (1998) Dynamic mapping at the laminar level of odor-elicited responses in rat olfactory bulb by functional MRI. Proc Natl Acad Sci USA 95(13):7715–7720

Zhang Z, Andersen AH, Avison MJ et al (2000) Functional MRI of apomorphine activation of the basal ganglia in awake rhesus monkeys. Brain Res 852(2):290–296

Zhang J, Yarowsky P, Gordon MN et al (2004) Detection of amyloid plaques in mouse models of Alzheimer's disease by magnetic resonance imaging. Magn Reson Med 51(3):452–457

Zhao F, Zhao T, Zhou L et al (2008) BOLD study of stimulation-induced neural activity and resting-state connectivity in medetomidine-sedated rat. Neuroimage 39(1):248–260

Chapter 19
In Vivo Studies of Brain Oxygen Metabolism and Oxidative Phosphorylation

Xiao-Hong Zhu and Wei Chen

Abstract The greatest merit of *in vivo* magnetic resonance spectroscopy (MRS) methodology used in brain research is its ability to noninvasively determine cerebral metabolites, metabolic rates and their dynamic changes caused by physiological or pathological perturbations. The utility of *in vivo* MRS is further enhanced with high/ultrahigh magnetic field strength instruments because of significant gains in detection sensitivity and improvements in spectral resolution. Recent progress of *in vivo* ^{31}P and ^{17}O MRS methodologies at high field has further demonstrated great potential and promise for *in vivo* measurements of crucial metabolic rates with regard to two essential chemical processes: cerebral oxygen utilization and oxidative phosphorylation of ADP. These metabolic rates are tightly linked to each other and are ultimately regulated by ATP consumption. This chapter provides a brief review of the new developments in *in vivo* ^{31}P and ^{17}O MRS techniques and their applications to the study of cerebral oxidative metabolism in supporting brain activity and function.

Keywords Brain • Brain function • Brain metabolism • Cerebral bioenergetics • CMR_{ATP} • $CMRO_2$ • *In vivo* ^{17}O MRS • *In vivo* ^{31}P MRS • Magnetic field

X.-H. Zhu, Ph.D. • W. Chen, Ph.D. (✉)
Center for Magnetic Resonance Research, Department of Radiology, University of Minnesota, School of Medicine, Minneapolis, MN, USA
e-mail: zhu@cmrr.umn.edu; wei@cmrr.umn.edu

I.-Y. Choi and R. Gruetter (eds.), *Neural Metabolism In Vivo*, Advances in Neurobiology 4, DOI 10.1007/978-1-4614-1788-0_19,

19.1 Cerebral Oxidative Metabolism, Brain Bioenergetics and Function

19.1.1 Oxygen Metabolism and Oxidative Phosphorylation

The brain is a typical aerobic organ, which consumes oxygen and glucose extensively to provide a mechanism for chemical energy storage in the form of the adenosine triphosphate (ATP) molecule. This chemical energy is essential to maintain the unceasing electrophysiological activities in a resting brain; and to supply additional energy for supporting increased neuronal activity during brain activation and task performance (Siesjo 1978; Raichle 1987; Boyer 1999; Clarke and Sokoloff 1999; Stock et al. 1999; Attwell and Laughlin 2001; Shulman et al. 2004; Chen and Zhu 2005; Zhu et al. 2005; Hyder et al. 2006; Raichle and Mintun 2006). The cerebral energy balance between demand and supply is regulated by a number of crucial biochemical reactions, which constitute a metabolic network that supports normal brain function.

Figure 19.1 illustrates a simplified network showing the major metabolic processes occurring in various sub-cellular compartments and the associated hemodynamic processes as well. Oxygen and glucose are the major substrates for brain metabolism and they are continuously supplied by the cerebral blood flow (CBF) through the brain capillary bed. Glucose is first converted to two pyruvate molecules in the cytosol, then the pyruvate enters the mitochondria and is metabolized oxidatively to form acetyl Co-A. The latter is then oxidized via the tricarboxylic acid cycle to generate reducing equivalents (NADH and $FADH_2$), which enter the electron transport chain and their metabolism generates an electrochemical potential gradient across the mitochondrial inner membrane. Because the terminal component of the electron transport chain, cytochrome oxidase, uses oxygen as its electron recipient, the function of the electron transport pathway is dependent on the availability of diffusion delivered oxygen molecules. Reaction 1 (or **R1**) in Fig. 19.1 reflects the net reaction to consuming oxygen molecules in the mitochondria.

The aforementioned electrochemical potential gradient supplies the energy to drive the conversion of adenosine diphosphate (ADP) to ATP via the F_1F_0-ATP_{ase} (i.e., Reaction 2 or **R2** in Fig. 19.1). Thus, oxygen metabolism is tightly coupled to the oxidative production of ATP from inorganic phosphate (Pi) and ADP under physiological condition. If oxygen availability is non-limiting, oxidative phosphorylation comprises ~90% of ATP production (Rolfe and Brown 1997). In contrast, ATP utilization by neuronal energy consuming processes mainly occurs in the cytosol and results in the regeneration of ADP and Pi (i.e., Reaction 5 or **R5** in Fig. 19.1). This ATP utilization is particularly essential for maintaining and restoring the Na^+/K^+ ion gradients across the cellular membranes via the enzyme of Na^+/K^+-ATP_{ase} as well as for supporting the neuronal signaling processes (e.g., neuronal firing, neurotransmitter cycling) during resting and activated brain states (Rolfe and Brown 1997; Magistretti et al. 1999; Shulman et al. 1999; Attwell and Laughlin 2001).

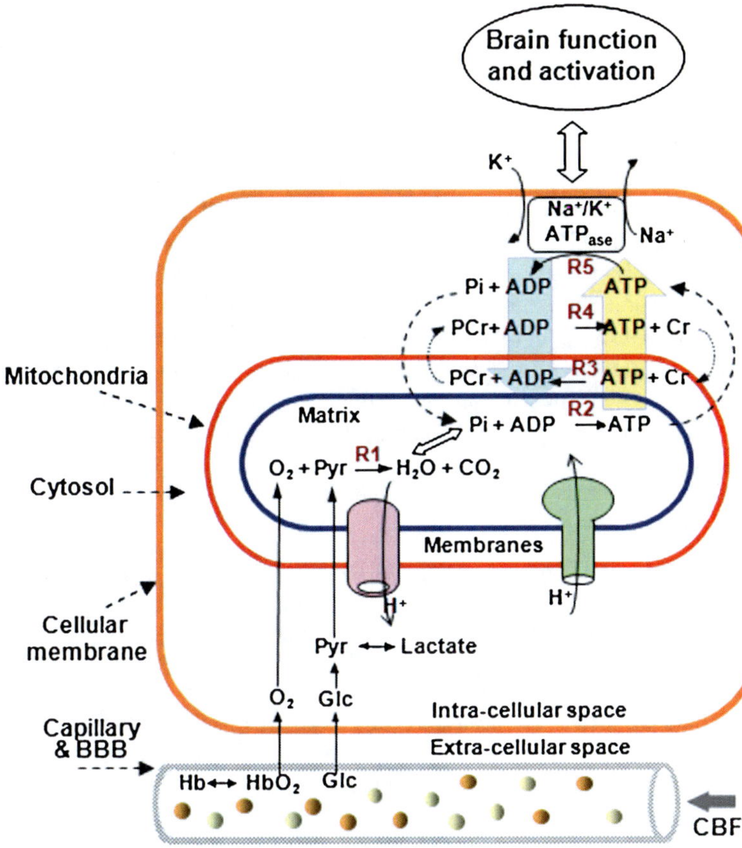

Fig. 19.1 Schematic diagram of major brain network involving metabolisms and hemodynamics occurring in the capillary, subcellular compartments including the mitochondria and cytosol spaces. Oxygen dissociates from hemoglobin (Hb), crosses the blood brain barrier (BBB) and enters mitochondria, then it is metabolized with pyruvate (Pyr) which is converted from glucose (Glc). This oxygen utilization via Reaction 1 (**R1**) is, in general, coupled with two ATP reactions for generating ATP (**R2**) and consuming ATP (**R5**). Other two paired CK reactions (**R3** vs. **V4**) have roles in transporting and buffering ATP molecules. These four ATP relevant reactions constitutes a chemical exchange network PCr ↔ ATP ↔ Pi. The listed reactions are essential to support brain activity and function. For simplicity, multiple cellular compartments (e.g., neuron vs. astrocyte) are not considered in this figure

The high ATP energy demand in the aerobic brain causes extremely rapid chemical cycling ATP, ADP and Pi, which requires rapid transportation of these phosphate molecules between the cytosolic and mitochondrial sub-compartments. This energy transportation is partially accomplished by another phosphate compound of phosphocreatine (PCr) and the reversible creatine kinase (CK) reaction, thus, maintaining a stable ATP level throughout the brain cells (Wallimann and Hemmer 1994; Saks et al. 1996; Kemp 2000). There are at least two coupled CK reactions: one occurring in the mitochondrial intermembrane space (i.e., Reaction 3 or **R3** in Fig. 19.1) and another occurring in the cytosol space (i.e., Reaction 4 or **R4** in Fig. 19.1). They play an important role in carrying the ATP molecules generated in the mitochondria into the cytosol for energy utilization, and then to distribute ATP throughout the cytosol and return ADP and Pi to mitochondria so that ATP generation is sustained and its cellular concentration remains constant. Therefore, PCr plays a vital role for energy transportation among sub-cellular compartments. Because of its role in ATP/ADP transport, the rates of CK reactions are expected to be much faster than the rates of ATP synthesis and utilization.

Four ATP-related reactions (two ATP_{ase} reactions and two CK reactions, i.e., **R2–R5** in Fig. 19.1) constitute a complex ATP metabolic network. They are tightly coupled and integrated with the oxygen metabolism and the hemodynamic process as depicted in Fig. 19.1 for controlling the dynamics of ATP production and utilization, which play a central role in the cerebral bioenergetics and function in normal and diseased brains.

19.1.2 Cerebral Oxidative Metabolism, Brain Function and Diseases

Oxidative metabolism is essential to normal brain function. For instance, the human brain can consume ~20% of the total body oxygen consumption even though the brain is only 2% of the body weight. Moreover, the cerebral ATP concentration is relatively low (~3 mM), resulting in a total ATP weight of 2 g in the entire human brain assuming an average 1.35 kg brain weight for adults. In contrast, the oxidative ATP synthesis rate through the ATP_{ase} reaction is very high in the human brain (8–9 μmol/g/min) (Lei et al. 2003a; Du et al. 2007). These numbers can be used to estimate that the total ATP weight utilized by the resting human brain in a single day could be five times of the human brain weight, an amazing fact! Such a high ATP utilization demand only can be satisfied via rapid ATP synthesis from Pi and ADP and efficient energy transportation among sub-cellular compartments.

Normal brain function relies on intact cerebral oxidative metabolism (Siesjo 1978; Raichle 1987; Boyer 1999; Clarke and Sokoloff 1999; Stock et al. 1999; Attwell and Laughlin 2001; Shulman et al. 2004; Chen and Zhu 2005; Zhu et al. 2005; Hyder et al. 2006; Raichle and Mintun 2006). Deficits or abnormalities in cerebral oxidative metabolism have been linked to numerous brain disorders and neurodegenerative diseases such as: Schizophrenia, Alzheimer's disease, Huntington's disease, Parkinson's disease, mitochondrial dysfunction and aging problems (e.g., Frackowiak et al. 1988; Beal 1992; Wong-Riley et al. 1997; Maurer et al. 2000; 2001). One line of evidence is the observation that the activity of cytochrome oxidase, the key mitochondrial enzyme that catalyzes the reduction of oxygen to form water, is significantly impaired in Schizophrenic (Maurer et al. 2001) and Alzheimer's patients (Wong-Riley et al. 1997; Maurer et al. 2000). Another line of evidences comes from the studies of diseases caused by mitochondrial DNA mutations suggesting that a variety of degenerative processes in Parkinson's and Alzheimer's disease may be associated with defects in mitochondrial oxidative phosphorylation (Wallace 1992; 1999).

Development of neuroimaging methods able to noninvasively detect and quantify the cerebral metabolic rates of oxygen utilization ($CMRO_2$) and oxidative ATP formation (CMR_{ATP}), have become important for studying and understanding the relation between oxidative metabolism and cerebral bioenergetics and the implication of abnormalities of these processes on brain function and diseases. *In vivo* ^{31}P and ^{17}O magnetic resonance spectroscopy (MRS) or MRS imaging (MRSI) methods may provide valuable neuroimaging tools for this goal.

One approach to directly and noninvasively measuring the rate of oxygen metabolism occurring in the mitochondria (i.e., **R1** in Fig. 19.1) is the use of *in vivo* ^{17}O MRSI for imaging $CMRO_2$ (e.g., Mateescu et al. 1988; Pekar et al. 1991; Fiat et al. 1992; Zhu et al. 2002), which has been described in great details in a review article (Zhu et al. 2005) and the references cited therein. The basic idea underlying this approach is to apply *in vivo* ^{17}O MRSI for imaging the dynamic change of the ^{17}O-labeled metabolic water ($H_2{}^{17}O$) produced through the oxidative metabolism of inhaled $^{17}O_2$ gas, and using this information to determine $CMRO_2$.

In vivo ^{31}P MRS provides an alternative tool for directly detecting a number of high-energy phosphate (HEP) compounds, metabolites involving in phosphate lipid metabolism, and tissue pH. This method has been widely applied to study both normal and diseased brains, and has led to a large number of publications (e.g., Ackerman et al. 1980; Burt and Ribolow 1984; 1994; Chance et al. 1987; Weiner 1987; Brown et al. 1989; Ramadan et al. 1989; Young et al. 1989; Hossmann 1991; Nioka et al. 1991a; b; Welch et al. 1992; Gadian et al. 1993; Altura and Altura 1994; Duncan 1997; Pettegrew et al. 1997; Ross et al. 1997; Riehemann et al. 2000; Hetherington et al. 2002; Jensen et al. 2002; Lei et al. 2003b; Chaumeil et al. 2009). The general discussion regarding *in vivo* ^{31}P MRS is covered in another chapter in this book.

Although the steady-state concentrations of cellular HEP metabolites are closely linked to the ATP metabolism, they are relatively stable under physiological conditions. This fact has been attributed to the rigorous regulation of the ATP metabolic reactions and their kinetics (see Fig. 19.1) for maintaining a stable cellular level of ATP and a high Gibbs free energy (ΔG) in the brain. Thus, the measurements of the kinetic network of ATP metabolism (i.e., reaction rates or fluxes) should logically be more sensitive for quantifying the cerebral bioenergetics and its change under varied brain activity states as compared to those of steady-state brain metabolite concentrations.

The main focus of this chapter is to discuss the newly developed *in vivo* ^{31}P and ^{17}O MRS approaches at high field for quantifying and studying the cerebral oxidative metabolic rates noninvasively; and several applications in brain research are also described herein.

19.2 High Field ^{17}O and ^{31}P MRS Studies of *In Vivo* Brain Metabolism

19.2.1 Challenges of In Vivo ^{31}P and ^{17}O MRS and Promise at High/Ultrahigh Fields

In practice, *in vivo* ^{31}P and ^{17}O MRS applications face many challenges due to their unfavorable NMR sensitivities. For instance, the phosphate metabolites detected by *in vivo* ^{31}P MRS and the metabolic $H_2{}^{17}O$ detected by *in vivo* ^{17}O MRS are in very low concentrations (in millimole range), which are several thousand times lower than the tissue water content that is commonly detected by ^{1}H MRI, resulting in a much poor signal-to-noise ratio (SNR). Thus, a large number of signal averaging is usually required to achieve reasonable spectra quality. The sensitivity challenge has significantly limited the reliability, applicability, spatial and temporal resolutions of *in vivo* ^{31}P and ^{17}O MRS for general applications including clinical research. One possible solution is the use of high/ultrahigh field MRS. In addition to the sensitivity gain at high fields, the spectral resolution of *in vivo* MRS is also significantly improved and many overlapped resonance peaks from different metabolites seen at low fields become resolvable at high/ultrahigh fields. We have systematically investigated the high-field advantages that can benefit *in vivo* ^{31}P and ^{17}O MRS for brain research.

19.2.1.1 Improvement of *In Vivo* ^{17}O MRS at High/Ultrahigh Fields

^{17}O is a stable oxygen isotope with a spin quantum number of 5/2, which is detectable by ^{17}O NMR, and it possesses an electric quadrupolar moment. The natural abundance of ^{17}O is only 0.037%, which is almost 30 times lower than that of ^{13}C and 2,700 times lower than that of ^{1}H. Moreover, the magnetogyric ratio of ^{17}O is 7.4 times lower than that of ^{1}H. These facts contribute to the low inherent ^{17}O NMR sensitivity that is the major obstacle for the development and application of *in vivo* ^{17}O MRS in biomedical research.

The relaxation mechanism of ^{17}O spin is distinct from other NMR-detectable spins such as ^{1}H and ^{31}P. The ^{17}O quadrupolar moment can interact with local electric field gradients and the temporal fluctuations of this interaction induced by molecular motion dominate the ^{17}O relaxation processes and determine both the longitudinal relaxation time (T_1) and the transverse relaxation time (T_2) of $H_2{}^{17}O$, thus, the T_1, T_2 and the apparent T_2 (T_2*) values of ^{17}O water spins are expected to be insensitive to the magnetic field strength (B_0) (Abragam 1961; Zhu et al. 2005). We have experimentally investigated and compared the ^{17}O longitudinal and transverse relaxation times of the natural abundance water in the rat brain at both 4.7T and 9.4T and confirmed that the ^{17}O relaxation times are field independent (Zhu et al. 2001).

Recent experimental evidence has indicated that the field-independence of ^{17}O relaxivity can further extend to a much higher magnetic field, such as 17.6T (Blackband et al. 2003).

The field independence of ^{17}O relaxivity reveals that ^{17}O sensitivity gain at higher fields is not compromised by the relaxation times, and this could lead to a substantial sensitivity gain when the field strength increases. Our experimental results showed approximately a fourfold SNR gain in the *in vivo* ^{17}O MRS signal measured in the rat brain at 9.4T compared to 4.7T (Zhu et al. 2001). The significant sensitivity gain provides adequate SNR for acquiring the three-dimension (3D) ^{17}O MRSI of the natural abundance tissue $H_2^{17}O$ (~20 mM) from a small brain of rat with a temporal resolution of ~10 s at 9.4T (see one example shown in Fig. 19.2a). This advantage is crucial for developing an ^{17}O-MR Based neuroimaging modality able to imaging $CMRO_2$ *in vivo*.

19.2.1.2 Improvement of *In Vivo* ^{31}P MRS at High/Ultrahigh Fields

It has been demonstrated that both SNR and the quality of *in vivo* ^{31}P MRS acquired in the human brain are improved at higher fields (Hardy et al. 1988; Boska et al. 1990; Hetherington et al. 2001; Lei et al. 2003b). Recently, we have quantitatively measured and compared the SNR of PCr resonance peak in the *in vivo* ^{31}P MRS acquired from the human brain at 4T and 7T (Qiao et al. 2006). It was found that the SNR is substantially improved at 7T. Another interesting observation of this study is that the linewidth broadening of the phosphate resonance peak with increased field strength is not linearly correlated with the field strength and it is smaller than the magnitude of field strength increase (Qiao et al. 2006). Moreover, increasing field strength linearly increases the chemical shift dispersion of *in vivo* ^{31}P MRS. The combination of these two facts (i.e., large increase in chemical shift dispersion and relatively small increase in linewidth broadening) improves the ^{31}P spectral resolution at 7T. Figure 19.2b demonstrates a typical *in vivo* ^{31}P spectrum acquired from the human occipital lobe at 7T showing excellent detection sensitivity and spectral resolution (Lei et al. 2003b). Besides ATP, PCr and Pi molecules, other detectable phosphate compounds in the human brain ^{31}P MRS include: uridine diphospho (UDP) sugar (an important precursor for glycogen metabolism), diphospho nicotinamide adenine dinucleotides (NAD) which is involved in oxidative metabolism and oxygen radical formation, and four resolved phosphate compounds of glycerophosphoethanolamine (GPE), glycerophosphocholine (GPC), phosphoethanolamine (PE) and phosphocholine (PC), which are actively involved in the membrane phospholipid metabolism through phospholipid biosynthetic enzymes. The improvement in the spectral resolution at 7T makes it possible to resolve many adjacent phosphate metabolite resonance peaks (e.g., α-ATP vs. NAD) and it is especially helpful for resolving the Pi peak from other adjacent resonance peaks. The well-resolved Pi peak is crucial for measuring the oxidative ATP synthetic rate and this topic will be discussed further in this chapter.

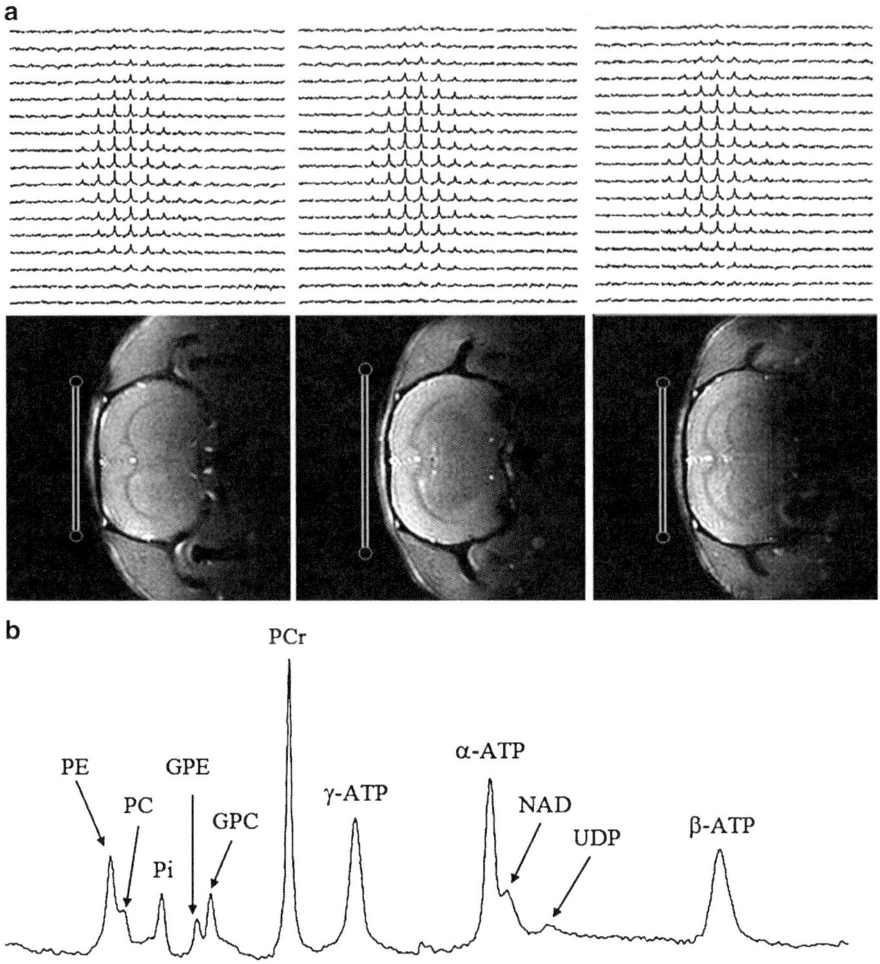

Fig. 19.2 (**a**) 3D ^{17}O MRSI of natural abundance $H_2{}^{17}O$ (*top row*) and corresponding ^{1}H anatomical images (*bottom row*) of rat brain acquired at 9.4T with a total sampling time of ~10 s. The ^{17}O RF surface coil position and cross sections are indicated in the anatomic images. (**b**) A typical ^{31}P MR spectrum acquired from the human occipital lobe at 7T with a total sampling time of 6.4 min. The spectrum is characterized by excellent spectral resolution and NMR sensitivity, and a large number of well-resolved resonance peaks from phosphoethanolamine (PE); phosphocholine (PC); inorganic phosphate (Pi); glycerophosphoethanolamine (GPE); glycerophosphocholine (GPC); phosphocreatine (PCr); adenosine triphosphate (ATP); diphospho nicotinamide adenine dinucleotides (NAD) and uridine diphospho (UDP) sugar (Adapted from Lei et al. 2003b)

19.2.2 In Vivo ^{17}O MRSI Approach for Imaging $CMRO_2$

Historically, *in vivo* ^{17}O NMR methodology has been used to study brain function and cerebral bioenergetics by measuring either CBF or $CMRO_2$ (Hopkins and Barr

1987; Mateescu et al. 1988; 1989; Arai et al. 1990; 1991; Kwong et al. 1991; Pekar et al. 1991; 1995; Fiat and Kang 1992; 1993; Fiat et al. 1992; 1993; Ronen and Navon 1994; Reddy et al. 1995; 1996; Stolpen et al. 1997; Ronen et al. 1998; de Crespigny et al. 2000; Zhu et al. 2001; 2002; Mateescu 2003; Chen et al. 2004; Zhang et al. 2004; Zhu et al. 2005; 2006). Nevertheless, the most important application is to measure and image $CMRO_2$ through monitoring the dynamic changes of metabolically generated $H_2{}^{17}O$ in the brain tissue from inhaled ^{17}O-labeled oxygen gas (Zhu et al. 2005 and references cited therein). There are at least two NMR approaches for monitoring brain $H_2{}^{17}O$ *in vivo*: a direct approach based on ^{17}O NMR detection, and an indirect approach by using ^1H MRI to measure the changes in T_2- or $T_{1\rho}$-weighted proton MRI signals caused by the ^{17}O-^1H scalar coupling and proton chemical exchange (Zhu et al. 2005) and references cited therein). Recent studies suggest that the *in vivo* ^{17}O NMR approach at high/ultrahigh fields is advantageous for quantifying and imaging $CMRO_2$; this approach will be discussed in detail.

19.2.2.1 Theory and Practice of $CMRO_2$ Measurements Using *In Vivo* ^{17}O MRSI

In general, there is a similarity between *in vivo* ^{17}O MRS approach and the positron emission tomography (PET) approach (Ter-Pogossian et al. 1970; Lenzi et al. 1981; Mintun et al. 1984; Fox et al. 1988) in measuring $CMRO_2$. Both approaches introduce oxygen gas with different isotope labels: $^{17}O_2$ for ^{17}O MRS and $^{15}O_2$ for PET. After gas exchange in the lung, the inhaled O_2 molecules quickly bind to hemoglobin (Hb) in the blood, forming O_2-hemoglobin complexes (HbO_2) (see Fig. 19.1). Through the feeding arteries and arterioles, the HbO_2 complex enters the brain capillaries. The O_2 molecules dissociate from hemoglobin in the capillaries, then diffuse across the brain blood barrier (BBB), and subsequently diffuse into the brain tissue (extra- and then intra-cellular space), and finally enter mitochondria where they are metabolized by cytochrome oxidase to produce H_2O according to the following chemical reaction (Siesjo 1978):

$$4\,\text{protons} + O_2 \rightarrow 2H_2O$$

According to this reaction (equivalent to **R1** in Fig. 19.1), one oxygen molecule produces two isotope labeled water molecules ($H_2{}^{17}O$ for ^{17}O MRS and $H_2{}^{15}O$ for PET), which can move out of the mitochondria (traversing the opposite pathway as O_2 entry) and finally be washed out from the brain through venules and veins. The dynamic change of the isotope labeled water in the brain is tightly linked to the cerebral oxygen utilization rate and it provides the vital signal source for determining $CMRO_2$ for both PET and *in vivo* ^{17}O MRS.

Though historically, the principles underlying the *in vivo* ^{17}O MRS approach for measuring $CMRO_2$ were adopted from the well-established PET approach, there are fundamental distinctions between these two neuroimaging modalities. For instance, PET is unable to distinguish the ^{15}O signal contribution of $^{15}O_2$ substrates from the

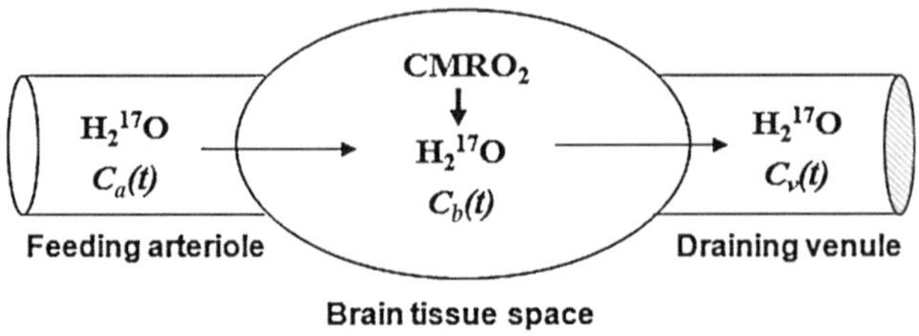

Fig. 19.3 Schematic illustration of a "complete model" describing three parallel processes of the ^{17}O-labeled metabolic water ($H_2{}^{17}O$) occurring in the brain when the ^{17}O-labeled oxygen gas is introduced via an inhalation. In this model, only the metabolic $H_2{}^{17}O$ is considered because the ^{17}O-labeled O_2 is invisible to *in vivo* ^{17}O NMR. $C_a(t)$, $C_b(t)$ and $C_v(t)$ stand for the $H_2{}^{17}O$ concentration in arteriole, brain tissue and venule, respectively, as a function of the $^{17}O_2$ inhalation time

metabolically generated $H_2{}^{15}O$ products. This complication limits the robustness of PET technique for imaging $CMRO_2$ (Mintun et al. 1984; Zhu et al. 2005). In contrast, *in vivo* ^{17}O MRS approach specifically detects the metabolically generated $H_2{}^{17}O$ without confounding signals from the $^{17}O_2$ molecules because the invisibility of $^{17}O_2$ in an *in vivo* ^{17}O spectrum (Zhu et al. 2005). This unique NMR specificity significantly simplifies the *in vivo* ^{17}O MRS methodology for measuring and quantifying $CMRO_2$ (Zhang et al. 2004; Zhu et al. 2005). In addition, ^{17}O is a nonradioactive and stable isotope, and is safe in general.

However, the dynamic change of the metabolically generated $H_2{}^{17}O$ concentration in the brain during an $^{17}O_2$ inhalation reflects an interplay between three parallel processes (see Fig. 19.3): (i) cerebral oxygen utilization for generating $H_2{}^{17}O$ in the brain tissue, (ii) cerebral blood perfusion that determines $H_2{}^{17}O$ washout from the brain, and (iii) blood recirculation brings metabolically generated $H_2{}^{17}O$ in the entire body back to the brain as a result of blood recirculation. The contributions of all three processes have to be considered in quantifying $CMRO_2$ according to the mass balance of the isotope labeled $H_2{}^{17}O$ in the brain tissue during an $^{17}O_2$ gas inhalation (Pekar et al. 1991; Zhu et al. 2002; 2005; Zhang et al. 2004):

$$\frac{dCb(t)}{dt} = 2\alpha f_1 CMRO2 + mCBF\left(f_2 Ca(t) - \frac{nCb(t)}{\lambda}\right), \qquad (19.1)$$

where $C_a(t)$, $C_b(t)$ and $C_v(t)$ are the metabolic $H_2{}^{17}O$ concentrations in excess of the natural abundance $H_2{}^{17}O$ concentration in the arterial blood, brain tissue and venous blood; respectively, as a function of $^{17}O_2$ inhalation time (t); . is the ^{17}O enrichment fraction of the oxygen atoms in the inhaled $^{17}O_2$ gas; λ is the brain/blood partition coefficient (Herscovitch and Raichle 1985). The factor of 2 counts for the fact that one O_2 converts to two H_2O molecules through oxidative metabolism; f_1 and f_2 are two unit conversion factors (Zhu et al. 2002; Zhang et al. 2004). The correction factor *m* is used in Eq. 19.1 to account for the water permeability restriction across BBB (Herscovitch et al. 1987), and *n* is another correction factor that accounts for the permeability restriction occurring when the metabolically generated $H_2{}^{17}O$

molecules inside the mitochondria across the mitochondrial membranes (Zhu et al. 2002; Zhang et al. 2004). The mathematical model as described by Eq. 19.1 requires a number of invasive and noninvasive measurements for determining CBF, *n*, $C_b(t)$ and $C_a(t)$ (Zhu et al. 2002; Zhang et al. 2004). These parameters can be used to solve the differential equation of Eq. 19.1 and to calculate the $CMRO_2$ value at each time point measured during $^{17}O_2$ gas inhalation. The quantification approach based on Eq. 19.1 is a complete model that requires all parameters for precisely determining $CMRO_2$ (Zhu et al. 2002; 2005; Zhang et al. 2004).

19.2.2.2 Imaging $CMRO_2$ in a Rat Brain Using *In Vivo* ^{17}O MRSI at High Field

We have applied 3D *in vivo* ^{17}O MRSI to image $CMRO_2$ using the rat model at 9.4T. The values of $C_b(t)$, CBF and *n* measured from each ^{17}O MRSI voxel and $C_a(t)$ measured during each ^{17}O inhalation in the same animal were used to calculate the absolute $CMRO_2$ value as a function of inhalation time according to Eq. 19.1. Figure 19.4a shows one example of $CMRO_2$ time course from a representative ^{17}O MRSI voxel with a temporal resolution of 11 s (Zhu et al. 2002). It is evident that the $CMRO_2$ values are independent of $^{17}O_2$ inhalation time if the first two $CMRO_2$ values (which are characterized by relatively large signal fluctuations) were excluded. These $CMRO_2$ value were averaged for improving measurement accuracy. The same calculation can be applied to all ^{17}O MRSI voxels for generating 3D $CMRO_2$ images in the rat brain (Zhu et al. 2002; Chen et al. 2004). Figure 19.4b demonstrates three adjacent $CMRO_2$ images in the coronal orientation from a representative rat brain. The averaged $CMRO_2$ value in the rat brains anesthetized with α-chloralose was 2.19 ± 0.14 μmol/g/min (n = 7) (Zhu et al. 2002), this result is consistent with the reported values in the literature, which were measured by different methods under similar physiological conditions (Hyder et al. 2000; Nakao et al. 2001; Yee et al. 2006).

19.2.2.3 Feasibility for Establishing a Non-invasive ^{17}O-MR Based $CMRO_2$ Imaging Method

The major technical limitation of the complete model for imaging $CMRO_2$ using *in vivo* ^{17}O MRSI is the requirement of invasive measurements (e.g., for determining CBF, $C_a(t)$ and *n*). This could significantly limit the potential of this *in vivo* approach for broad biomedical applications; especially in humans. Thus, it is crucial to examine the feasibility of developing a completely noninvasive ^{17}O-MR Based approach for imaging $CMRO_2$. Attempts have been made to simplify the experimental procedures and the models for determining $CMRO_2$ based on a number of approximations (Pekar et al. 1991; 1995; Fiat and Kang 1992; 1993; Fiat et al. 1992; Zhang et al. 2004). We will discuss one of these models, in which invasive measurements are eliminated by using a simplified model based on the Taylor's and Polynomial

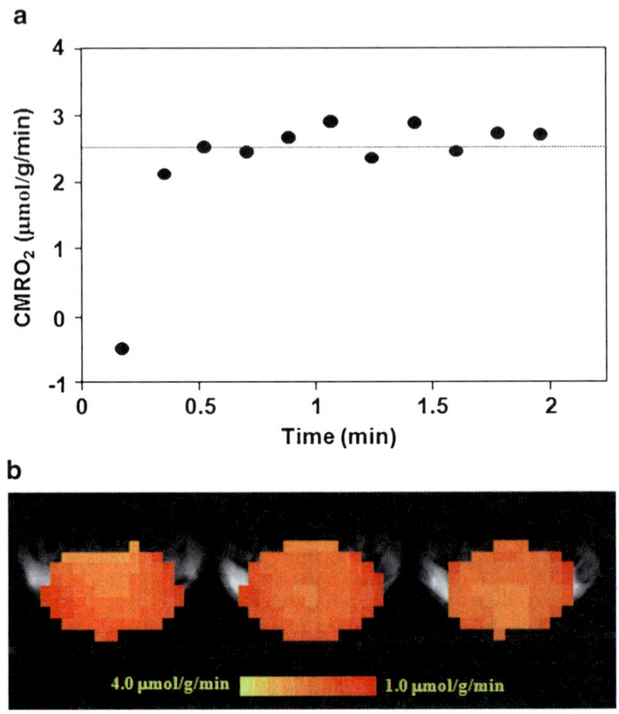

Fig. 19.4 (**a**) Plot of the calculated $CMRO_2$ values using the complete model as a function of $^{17}O_2$ inhalation time. (**b**) Three-dimensional coronal $CMRO_2$ images of rat brain measured by *in vivo* ^{17}O MRS approach during a 2-min $^{17}O_2$ inhalation (Adapted from Zhu et al. 2002)

Theorems (Zhang et al. 2004). Briefly, this simplified model only requires one noninvasive measurement of brain metabolic $H_2{}^{17}O$ concentration as a function of $^{17}O_2$ inhalation time, this measured $C_b(t)$ time course can be fitted to a polynomial function, and the first-order (or linear) coefficient of a_1 is directly proportional to $CMRO_2$ according to the following relationship (Zhang et al. 2004)

$$CMRO_2 = \frac{a_1}{2\alpha f_1}. \tag{19.2}$$

We have demonstrated that the $CMRO_2$ value quantified via the complete model is not statistically different from that of simplified model and single noninvasive measurement of $C_b(t)$ in the rat brain (Zhang et al. 2004).

Recently, we have further examined the simplified model for determining $CMRO_2$ under varied physiological conditions (Zhu et al. 2007). In this study, we compared the rat $CMRO_2$ results obtained with the complete model versus the simplified model using linear fitting of $C_b(t)$ during normothermia (37°C) and during mild

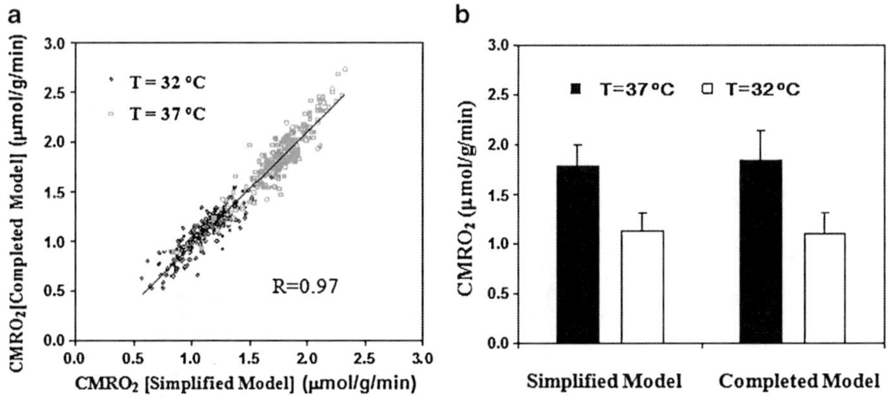

Fig. 19.5 (**a**) Voxel based $CMRO_2$ calculation and comparison using the completed and simplified models from a representative rat (total voxel number used was 224 for 32°C and 254 for 37°C). (**b**) Averaged $CMRO_2$ values in the same rat brain at normothermia (37°C) and hypothermia (32°C) condition, calculated with simplified and completed model, respectively (Adapted from Zhu et al. 2007)

hypothermia (32°C). The latter condition is known to significantly decrease both CBF and $CMOR_2$. Figure 19.5 demonstrates an excellent consistency between the $CMRO_2$ results obtained with both models (Fig. 19.5b) at both brain temperature conditions (Zhu et al. 2007). These results support the validity of the simplified ^{17}O approach for imaging $CMRO_2$ applicable in a wide physiological range using small animal models. Nevertheless, one should be cautious that the approximation of this simple model may require further validation or modification when apply to large animals or humans.

19.2.2.4 Imaging $CMRO_2$ Change Using *In Vivo* ^{17}O MRSI Approach at High Field

One advantage of the ^{17}O-MR Based $CMRO_2$ imaging approach is the ability to perform repeated $CMRO_2$ measurements during a single experimental session. This is due to the fact that the metabolic $H_2{}^{17}O$ concentration in the brain can quickly reach a new steady-state level after the cessation of $^{17}O_2$ inhalation (see Fig. 19.6a), which allows repeated $CMRO_2$ measurements in the same subject with an excellent reproducibility (Zhu et al. 2007). This capability is particularly attractive for studies aiming at $CMRO_2$ changes induced by physiological or pathological perturbations in which multiple measurements are required under different conditions (e.g., control vs. stimulation for studying brain function). Therefore, the combination of the simplified model and ultrahigh-field *in vivo* ^{17}O MRS may potentially provide an alternative neuroimaging modality for studying the central role of oxidative metabolism in brain function and neurological diseases (Chen et al. 2004; Zhu et al. 2005).

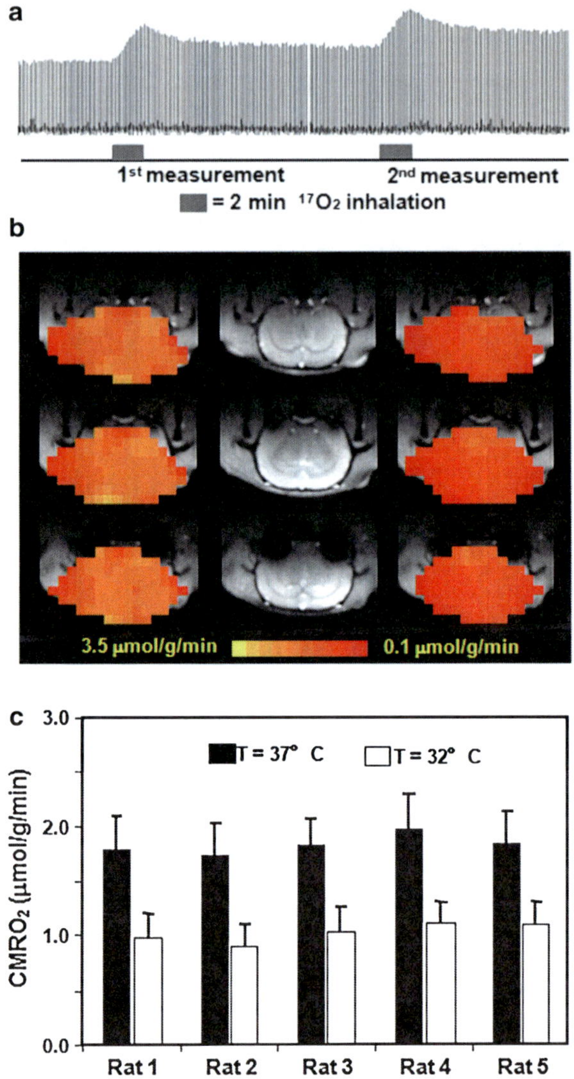

Fig. 19.6 (**a**) Stacked plots of $H_2^{17}O$ spectra from a representative voxel of 3D ^{17}O MRSI data acquired before, during and after two consecutive 2-min $^{17}O_2$ inhalation in a representative rat brain. (**b**) 3D $CMRO_2$ maps of a representative rat brain obtained at normothermia (*left column*) and hypothermia (*right column*), and their corresponding anatomic images (*middle column*). (**c**) Summary of $CMRO_2$ results measured at normothermia and hypothermia conditions (n = 5) (Adapted from Zhu et al. 2007)

It is well documented that the basal $CMRO_2$ is sensitive to the brain temperature (Siesjo 1978; Erecinska et al. 2003). However, most studies reported in the literature are based on global $CMRO_2$ measurements using the Kety-Schmidt method (Kety and Schmidt 1945; 1948), and are limited by the lack of spatial information with regard to regional $CMRO_2$. In a noninvasive $CMRO_2$ imaging study using 3D *in vivo* ^{17}O MRSI at 9.4T for quantifying absolute $CMRO_2$ values in the rat brain at normal brain temperature (37°C) (i.e., normothermia) and mild hypothermia (32°C) conditions, we examined this issue (Zhu et al. 2007). Figure 19.6b displays three representative slices of 3D $CMRO_2$ images from a rat brain under normothermic and hypothermic conditions. These images clearly show that a significant reduction of $CMRO_2$ across the entire brain is induced by lowing brain temperature several degrees. The metabolic activity suppression occurring during hypothermia was consistently observed in all five rats studied (Fig. 19.6c), resulting in an average of 45% $CMRO_2$ reduction compared to normothermic condition (Zhu et al. 2007). These results indicate that the *in vivo* ^{17}O MRS approaches are capable of determining dynamic $CMRO_2$ changes and their spatial distributions. Hence, these methods hold promise for biomedical applications, and especially so to provide a better understanding of the crucial roles of oxidative metabolism in brain bioenergetics, function and dysfunction.

19.2.3 In Vivo ^{31}P MRS Approach for Measuring Cerebral ATP Metabolic Rates

The mechanisms involved in the regulation of ATP production and ATP utilization play a fundamental role in cerebral bioenergetics, brain function and neurodegenerative diseases. Two important chemical processes contributing to brain ATP metabolism are ATP_{ase} and CK reactions. They are coupled together and constitute a three-^{31}P-spin chemical exchange system involving ATP, PCr and Pi (i.e., PCr ↔ ATP ↔ Pi summarizing four reactions of **R2–R5** in Fig. 19.1). One vital function of this exchange network is to maintain a stable cellular ATP concentration that ensures a continuous energy supply for sustaining electrophysiological activity in the brain. Logically, measurements of ATP metabolic fluxes should be more sensitive to brain activity and its energy state and their changes than that of steady-state ATP and other HEP concentrations. Therefore, flux measurements should provide a useful index reflecting cerebral bioenergetics under various brain states. It is therefore essential to find a noninvasive and reliable technique capable of assessing the cellular exchange network of PCr ↔ ATP ↔ Pi in the in situ brain. The sole approach to achieving this goal is the use of *in vivo* ^{31}P MRS combined with magnetization transfer (MT) methods (Frosen and Hoffman 1963; Alger and Shulman 1984; Degani et al. 1985; Ugurbil 1985a; 1985b; Lei et al. 2003a; Weiss et al. 2005; Du et al. 2007). However, to completely determine the kinetics and fluxes involved in the PCr ↔ ATP ↔ Pi exchange network requires extensive measurements and information including: three steady-state phosphate metabolite concentrations: [ATP], [PCr] and [Pi]; three intrinsic T_1 values: T_1^{ATP},T_1^{PCr} and T_1^{Pi}; four pseudo-first order chemical reaction rate constants: forward and reverse rate constants for the CK reaction (k_1 and k_{-1}) and the ATP_{ase} reaction (k_{-2} and k_2), respectively, as defined in Fig. 19.7 (Du et al. 2007). The products of the rate constants and their related phosphate concentrations can provide four ATP metabolic fluxes along both forward and reverse reaction directions in the PCr ↔ ATP ↔ Pi exchange network (see Fig. 19.7).

Several *in vivo* ^{31}P MT methods have been developed such as conventional two-spin magnetization saturation transfer (CST), inversion recovery transfer and two-dimensional chemical exchange spectroscopy (Shoubridge et al. 1982; Balaban et al.

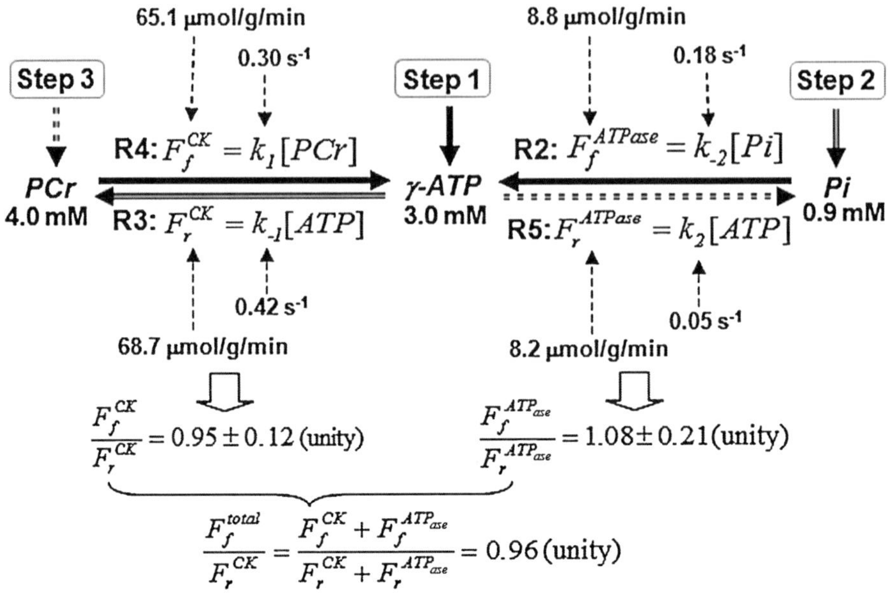

Fig. 19.7 Chart showing the three step measurements of *in vivo* ^{31}P MSS MT approach for determining the entire ATP kinetic network and associated metabolic rate constants and fluxes, and the measurement results from the human visual cortex. Step 1 measures two forward reactions (**R2** and **R4**) along the *solid arrows*. Step 2 measures the indirectly coupled reverse reaction (**R3**) along the *double-line arrows*. Step 3 measures the another reverse reaction (**R5**) along the *dotted-line arrows*. All results point to that the flux ratios satisfy the chemical equilibrium condition if the chemical exchange system of PCr↔γ-ATP↔Pi is treated as a three-spin exchange system

1983; Alger and Shulman 1984; Ugurbil 1985a, b). They have been applied to physiological studies of ATP metabolism in a variety of organs, including heart and brain (Shoubridge et al. 1982; Ugurbil 1985ab; Degani et al. 1987; Bottomley and Hardy 1992; Sauter and Rudin 1993; Chen et al. 1997; Braunova et al. 2000; Lei et al. 2003a; b; Joubert et al. 2004; Mlynarik et al. 2005; Weiss et al. 2005; Du et al. 2007; Chaumeil et al. 2009). Among those methods, the CST method is most commonly used in biomedical research, perhaps due to its methodological simplicity and high efficiency. This method is particularly useful in measuring the forward rate constants and fluxes in the chemical exchange system of PCr↔ATP↔Pi (i.e., Pi→ATP for the ATP_{ase} reaction, and PCr→ATP for the CK reaction), when a frequency-selective RF saturation pulse train is applied to fully saturate the γ-ATP resonance peak. However, the measurements of the reverse CK flux (i.e., ATP→PCr) using CST (i.e., by saturating PCr) resulted in an inequality between the forward and reverse CK fluxes (Matthews et al. 1982; Bittl et al. 1987; Mora et al. 1992), which is paradoxical to the fact that the CK fluxes into and out of the PCr pool must be equal because the CK reaction is under chemical equilibrium condition. One possible explanation is that the ATP is involved in other chemical exchange reactions besides the CK reaction, for example, the ATP hydrolysis reaction (Ugurbil et al. 1986). Consequently, neglecting ATP hydrolysis may lead to an error in estimating the reverse CK flux.

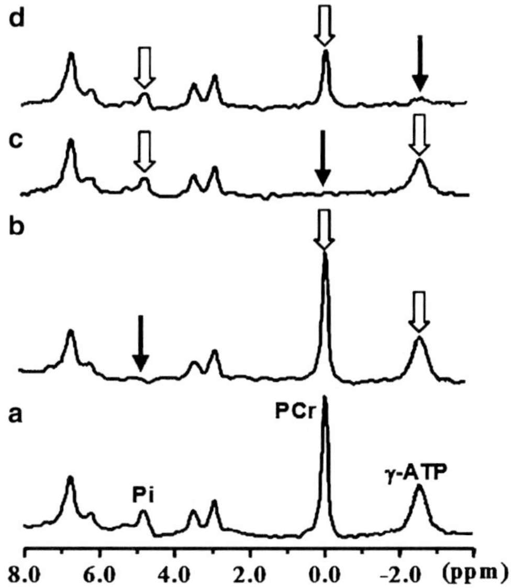

Fig. 19.8 *In vivo* ^{31}P MSS MT measurements and four spectra acquired from a representative human occipital lobe in the absence (**a**) and presence of complete RF saturation on a selected resonance peak of (**b**) Pi (Step 2), (**c**) PCr (Step 3) and (**d**) γ-ATP (Step 1), respectively. The *narrow arrows* indicate the saturation sites and the wide arrows indicate the signal reductions due to the magnetization transfer effect (Adapted from Du et al. 2007)

Therefore, it is necessary to consider PCr↔ATP↔Pi as a three-spin chemical exchange system in order to accurately determine all kinetic parameters; in particular, the reverse rate constants and fluxes (Ugurbil 1985a; b; Ugurbil et al. 1986; Spencer et al. 1988; Du et al. 2007).

In this chapter, we briefly discuss a newly introduced ^{31}P MT approach which is able to noninvasively study the PCr↔ATP↔Pi exchange network in the brain explicitly through the *in vivo* measurements of the following fluxes associated with four coupled ATP reactions in different sub-cellular compartments (Du et al. 2007):

1. The forward flux ($F_f^{ATP_{ase}}$) of ATP_{ase} reaction (**R2** in Figs. 19.1 and 19.7) occurring in mitochondria;
2. The forward flux (F_f^{CK}) of CK reaction (**R4** in Figs. 19.1 and 19.7) occurring in cytosol;
3. The reverse flux ($F_r^{ATP_{ase}}$) of ATP_{ase} reaction (**R5** in Figs. 19.1 and 19.7) occurring in cytosol;
4. The reverse flux (F_r^{CK}) of CK reaction (**R3** in Figs. 19.1 and 19.7) occurring in mitochondria.

This *in vivo* ^{31}P MT approach is based on the Multiple Single-site Saturation (MSS) MT method and the details regarding this method has been reported (Du et al. 2007). The MSS-MT approach requires a total of four *in vivo* ^{31}P spectra: one control spectrum (Fig. 19.8a) in the absence of RF saturation and three single-site RF saturated spectra with the saturation frequency selectively sitting on Pi (Fig. 19.8b), PCr (Fig. 19.8c) and γ-ATP (Fig. 19.8d), respectively. The quantification of the ATP metabolic fluxes relies on solving three coupled Bloch equations based on the three-spin chemical exchange model through three measurement steps (Frosen and Hoffman 1963; Du et al. 2007).

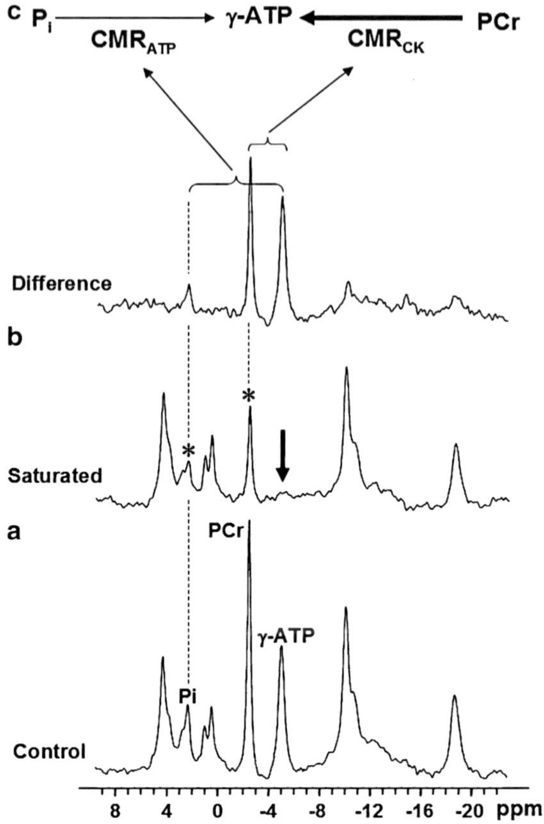

Fig. 19.9 *In vivo* ^{31}P spectra acquired from a health human occipital lobe in the absence (**a**) and presence (**b**) of complete γ-ATP saturation, and the difference spectrum (**c**) between the two. The intensity reduction of Pi can be used to determine the forward rate constant and flux for the ATPase reaction, and the intensity reduction of PCr can be used to determine the forward rate constant and flux for the CK reaction (Adapted from Lei et al. 2003a)

Step 1: *Steady-state saturation of γ-ATP for determining forward CK and* ATP_{ase} *fluxes*

Step 1 is the same as the CST approach which applies a frequency-selective RF pulse train for completely saturating the γ-ATP resonance peak with a sufficiently long saturation time resulting in steady-state, reduced magnetizations of PCr and Pi (see Fig. 19.8a vs. Fig. 19.8d). Under this circumstance, the three-spin chemical exchange system of PCr ↔ ATP ↔ Pi can be treated as two independent two-spin chemical exchange systems (i.e., PCr ↔ ATP and ATP ↔ Pi); consequently, the forward rate constants and fluxes for both the CK reaction (i.e., **R4** of PCr → ATP in Figs. 19.1 and 19.7) and ATP_{ase} reaction (i.e., **R2** of Pi → ATP in Figs. 19.1 and 19.7) can be explicitly determined. Specifically, the relative reduction in the PCr intensity (also see Fig. 19.9) and the predetermined T_1^{PCr} can be used to calculate the forward rate constant for the CK reaction (k_1); and the forward flux, F_f^{CK} given by $k_1 \times [PCr]$ (Lei et al. 2003a; b; Du et al. 2007). Similarly, the reduction in the Pi intensity and T_1^{Pi} can be used to determine both the forward rate constant (k_{-2}) and flux ($F_f^{ATP_{ase}}$) for the ATP_{ase} reaction. These forward metabolic fluxes reflect the ATP production rates catalyzed by CK and ATP_{ase}, respectively.

Step 2: *Steady-state saturation of Pi for determining CK reverse rate constant and flux*
The second step of the *in vivo* ^{31}P MSS-MT measurements is to apply a frequency-selective RF pulse train for completely saturating the Pi resonance peak with a sufficiently long saturation time resulting in steady-state, moderately reduced magnetizations of PCr and γ-ATP (see Fig. 19.8a vs. Fig. 19.8b). For this case, the relative reductions in both PCr and γ-ATP signal intensities, the predetermined T_1^{PCr} and the k_1 value measured by Step 1 can be used to calculate the reverse CK rate constant (k_{-1}) and flux, F_r^{CK}, given by $k_{-1} \times [ATP]$ (Du et al. 2007).

Step 3: *Steady-state saturation of PCr for determining ATP_{ase} reverse rate constant and flux*
The last step of the *in vivo* ^{31}P MSS-MT measurements is to apply a frequency-selective RF pulse train for completely saturating the PCr resonance peak with a sufficiently long saturation time resulting in steady-state, reduced magnetizations of Pi and γ-ATP (see Fig. 19.8a vs. Fig. 19.8c). For this case, the relative reductions in both Pi and γ-ATP signal intensities, the predetermined T_1^{Pi} and the k_{-2} value measured by Step 1 can be used to calculate the reverse ATP_{ase} rate constant (k_2) and flux, $F_r^{ATP_{ase}}$, given by $k_2 \times [ATP]$ (Du et al. 2007).

Therefore, by combining these three measurement steps, one is able to measure all kinetic parameters which determine the entire chemically exchange network of PCr ↔ ATP ↔ Pi including all four ATP metabolic fluxes for both forward and reverse reaction directions. Figure 19.7 summarizes the *in vivo* ^{31}P MSS-MT measurement results from the human brain study at 7T indicating several important observations (Du et al. 2007). There is no a statistically significant difference between the forward and reverse exchange fluxes for both the CK reaction (i.e., F_f^{CK} vs. F_r^{CK}; $p=0.38$) and the ATP_{ase} reaction (i.e., $F_f^{ATP_{ase}}$ vs. $F_r^{ATP_{ase}}$; $p=0.46$); thus, the ratios between the forward and reverse fluxes for the CK reaction ($F_f^{CK} / F_r^{CK} = 0.95 \pm 0.12$) and the ATP_{ase} reaction ($F_f^{ATP_{ase}} / F_r^{ATP_{ase}} = 1.08 \pm 0.21$) are close to unity (Du et al. 2007). In general, the CK fluxes reflect the non-oxidative phosphorylation and the ATP_{ase} fluxes, in contrast, reflect oxidative phosphorylation by ATP synthase and ATP de-phosphorylation by the numerous cytosolic ATP_{ase} enzymes that utilize ATP to support energy requiring processes. Both of these phosphorylation pathways can contribute to the total ATP production flux (i.e., $F_f^{total} = F_f^{CK} + F_f^{ATP_{ase}} = 73.9\,\mu mol/g/min$) and the total ATP utilization flux (i.e., $F_r^{total} = F_r^{CK} + F_r^{ATP_{ase}} = 76.9\,\mu mol/g/min$). The ratio between F_f^{total} and F_r^{total} is again close to unity (=0.96) indicating that the total ATP production flux equals to the total ATP utilization flux in the brain. These results conclude that the fluxes measured by the *in vivo* ^{31}P MSS-MT approach satisfy the chemical equilibrium conditions for the CK and ATP synthetic and utilizations reactions in the human brain, and indicate that the MSS-MT approach is able to explicitly determine the four rate constants in the PCr ↔ ATP ↔ Pi kinetic network and their associated ATP metabolic fluxes in the brain noninvasively. Nevertheless, such equal forward and reverse flux relations were not obtainable when the three-spin PCr ↔ ATP ↔ Pi exchange system was treated as two independent two-spin change systems (i.e., PCr ↔ ATP and ATP ↔ Pi) even when the same experimental MSS-MT data were analyzed (Du et al. 2007). These findings clearly indicate that the chemical exchange system of PCr ↔ ATP ↔ Pi has to be treated as a three-spin exchange system in order to accurately determine the reverse rate constants and fluxes for both the CK and ATP_{ase} reactions.

The PCr ↔ ATP ↔ Pi exchange system in the normal brain likely operates in chemical equilibrium (or near equilibrium) state under most circumstances. As long as the equilibrium condition is satisfied, only two forward fluxes: F_f^{ATPase}, which is further defined as the cerebral metabolic rate of ATP production (CMR_{ATP}) through the F_1F_0-ATP_{ase}, and F_f^{CK}, need to be measured (Du et al. 2007). In this case, only two *in vivo* ^{31}P spectra (one control and another with full saturation of γ-ATP as illustrated in Fig. 19.9) are practically required to determine the entire PCr ↔ ATP ↔ Pi exchange network (Du et al. 2007). Thus, this simplification can make the *in vivo* ^{31}P MT measurement more robust and efficient and it permits rapid acquisition of ATP metabolic flux measurements and the changes in these fluxes induced by either physiological or pathological perturbations.

19.3 Essential Roles of *In Vivo* MRS in Brain Bioenergetics and Function

Despite low NMR detection sensitivity (relative to ^{1}H MRS), researches in the past decades have demonstrated the unique merits of *in vivo* heteronuclear MRS for a broad range of biomedical applications. Two of them (i.e., ^{31}P and ^{17}O) are discussed in this chapter in great details owing to their unique specificity for measuring the metabolic rates related to the brain oxidative metabolism. Each of them provides a powerful tool for noninvasively studying metabolic pathways in the brain through the measurements of metabolite concentrations and biochemical reaction fluxes. The combination of different *in vivo* multinuclear MRS approaches with MRI should significantly enhance the MR capability to perform comprehensive studies concerning brain physiology, neurochemistry, bioenergetics and their relations to brain function and dysfunction (Zhu et al. 2009a).

As aforementioned, both *in vivo* ^{31}P and ^{17}O MRS methods can be used to noninvasively study brain oxidative metabolism and bioenergetics. Each of these methods has unique features for probing the metabolic functions in mitochondria. Specifically, the *in vivo* ^{17}O MRSI approach is useful for mapping the absolute $CMRO_2$ within a relatively short $^{17}O_2$ inhalation time (a few minutes). The $CMRO_2$ images should reflect the total metabolic rate of oxygen utilization (i.e., **R1** in Fig. 19.1) and its spatial distribution in the brain. On the other hand, the *in vivo* ^{31}P MT approach is powerful for measuring all the cerebral ATP metabolic fluxes associated with four ATP reactions (i.e., **R2**–**R5** in Fig. 19.1) occurring in different sub-cellular compartments. All of these measured physiological parameters can be linked to brain ATP generation and transport. Nevertheless, CMR_{ATP} is probably most closely linked to the cerebral oxidative phosphorylation rate, which is crucial to the study of brain bioenergetics during basal state, active or dysfunctional states. However, it is still debated about whether the forward ATP_{ase} flux measured by *in vivo* ^{31}P MT truly reflects the net rate of oxidative phosphorylation in the mitochondria. The ATP synthesis reaction catalyzed by the coupled activities of glycolytic enzymes glyceraldehyde-3-phosphate dehydrogenase (GAPDH) and phosphoglycerate kinase (PGK) have been shown to make a major contribution to the forward Pi → ATP reaction flux measured by ^{31}P MT in E. coli (Mitsumori et al. 1988), yeast (Campbell et al. 1987), liver (Thoma and Ugurbil 1987), and the myocardium (Kingsley-Hickman et al. 1987). For instance, in the perfused rat heart, GAPDH/PGK mediated exchange approximates ~50% of the Pi → ATP flux measured by the ^{31}P MT at low myocardial work states; the relative contribution of this enzyme couple to total flux decreases as the rate of ATP expenditure increases (Kingsley-Hickman et al. 1987). Under these circumstances, *only when* the GAPDH/PGK effect was eliminated either directly (using the glycolytic inhibitor, iodoacetate) or indirectly by eliminating exogenous and endogenous sources of glucose, did the value of the myocardial Pi → ATP flux determined by the *in vivo* ^{31}P MRS-MT measurement strongly correlate with the rate of cardiac oxygen consumption and yield a biologically realistic P:O ratio (~2.4) (Kingsley-Hickman et al. 1987). These observations reveal the complexity of biological systems and potential limitations of *in vivo* ^{31}P MRS-MT approaches for directly measuring the ATP metabolic flux related to oxidative phosphorylation in organs such as heart and liver.

Interestingly, in contrast to the observations in the myocardium, yeast, E. coli and liver, the Pi → ATP flux measured by *in vivo* ^{31}P MRS-MT in the human brain is close to the net oxidative ATP synthesis rate which can be estimated from the measured $CMRO_2$ data and a realistic estimate of the P:O ratio (Lei et al. 2003b; Du et al. 2007). The $CMRO_2$ value in the human occipital lobe has been measured to be 1.71 μmol/g/min by PET (Fox et al. 1988). The ATP synthesis rate attributed by oxidative phosphorylation can be calculated from multiplying this $CMRO_2$ value by the P:O ratio of ~2.5 (Hinkle 2005) and factor of 2, resulting in an estimated value of 8.6 μmol/g/min. This result is almost identical to the CMR_{ATP} value of 8.8 μmol/g/min, which was directly measured by *in vivo* ^{31}P MRS-MT (Du et al. 2007). A similar relation was also evident in the animal studies that showed a tight correlation between the measured Pi → ATP flux and the estimated oxidative phosphorylation rate in the rat brain (Shoubridge et al. 1982; Du et al. 2008; Chaumeil et al. 2009). These comparison results, thus, suggest that in the brain, the Pi → ATP flux measured by the *in vivo* ^{31}P MT approach approximates the net oxidative ATP synthesis rate and this flux should provide a useful index of the rate of production of the vast majority of ATP synthesized to support brain function (Attwell and Laughlin 2001; Lei et al. 2003a; Hyder et al. 2006; Du et al. 2007). Hence, *in vivo* ^{31}P MT is an attractive method for studying the central role of oxidative phosphorylation playing in brain activity. However, the underlying mechanism of the discrepancy in the ^{31}P MT measurements between the brain and other organs (e.g., heart and liver) remains to be determined.

One relevant question in the brain function and neuroimaging research is how much extra brain energy is needed during the brain stimulation and/or task performance. This topic is still under intense debate, especially, regarding to how much $CMRO_2$ is increased by brain stimulation (Fox and Raichle 1986; Roland et al. 1987; Fox Fox et al. 1988; Prichard et al. 1992; Marrett et al. 1993; Ribeiro et al. 1993; Hyder et al. 1996; Davis et al. 1998; Hoge et al. 1999; Kim et al. 1999; Vafaee et al. 1999; Chen et al. 2001; Shulman et al. 2001; Kasischke et al. 2004; Hyder et al. 2006; Raichle and Mintun 2006). Perhaps, one even more important question is how much the rate of ATP utilization is increased when the brain activity level increases.

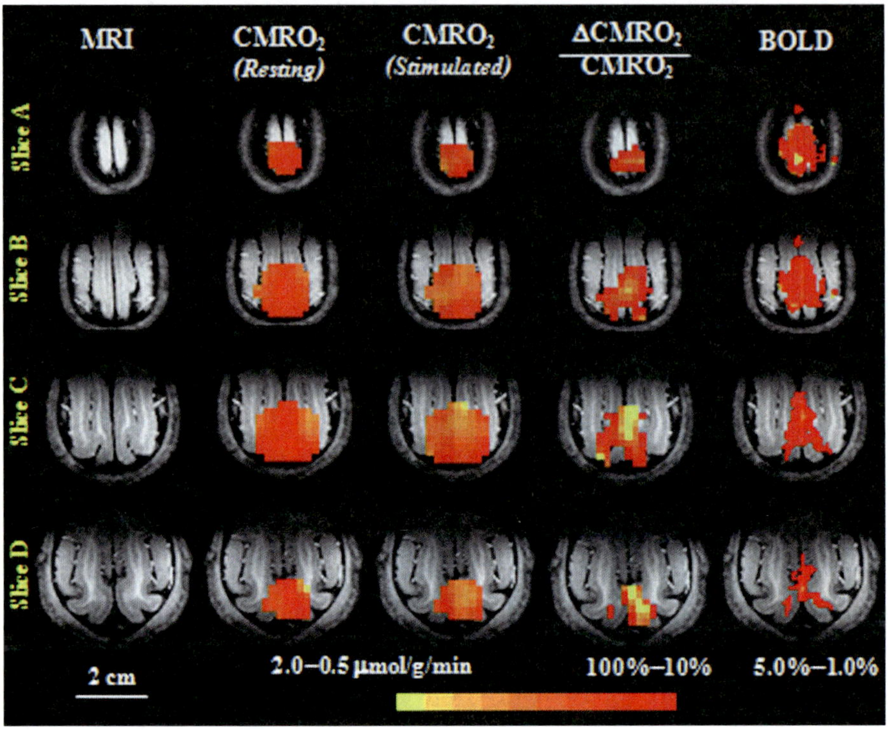

Fig. 19.10 Three-dimensional $CMRO_2$ maps of cat visual cortex obtained at resting (second column from the left) and visual stimulated (third column) conditions; 3D functional $CMRO_2$ activation maps showing percent changes of $CMRO_2$ elevated by visual stimulation (fourth column); BOLD-based functional MRI (fMRI) maps (fifth column); and anatomic brain images (first column) in four adjacent image slices from a representative cat brain (Adapted from Zhu et al. 2009b)

The answers to these questions are central for understanding the mechanisms underlying modern neuroimaging techniques including two most popular methods: fMRI and PET (Ogawa et al. 1990; Bandettini et al. 1992; Blamire et al. 1992; Frahm et al. 1992; Kwong et al. 1992; Ogawa et al. 1992; Barinaga 1997; Hyder et al. 2006; Raichle and Mintun 2006). One major hurdle for addressing these questions has been the development of sophisticated and robust neuroimaging methods for directly, noninvasively assessing the absolute $CMRO_2$ and CMR_{ATP} and the changes in these measurements induced by brain activation. The *in vivo* ^{31}P and ^{17}O MRS approaches as described in this chapter may fill this methodological gap and make a significant contribution in understanding the relation between brain bioenergetics and function. This notion is convincingly supported by several lines of evidence provided by recent research.

One example was the use of the *in vivo* ^{17}O MRSI approach to successfully map $CMRO_2$ in the cat brain under resting and visual stimulation conditions as demonstrated in Fig. 19.10 (Zhu et al. 2009b). A significant $CMRO_2$ increase was observed

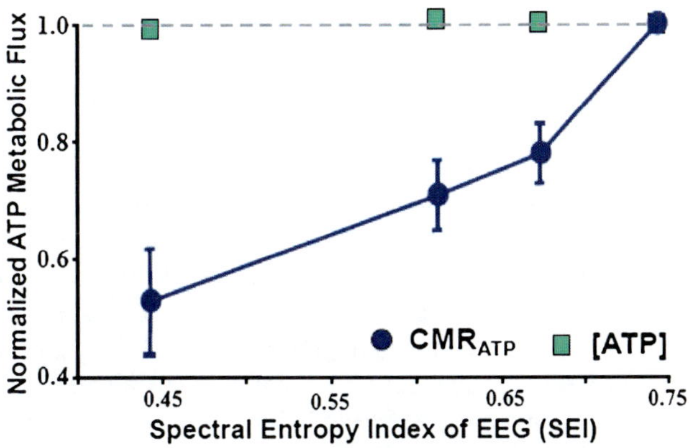

Fig. 19.11 Correlations of the rat EEG brain activity level versus CMR_{ATP} and cerebral ATP concentration, respectively, measured under varied brain states using different anesthetics and/or doses The EEG signal was quantified by the spectral entropy index (Adapted from Du et al. 2008)

in the activated cat visual cortex and the change was spatially coincided with the functional MRI mapping results based on the blood oxygenation dependent level (BOLD) contrast (Ogawa et al. 1990; Bandettini et al. 1992; Kwong et al. 1992; Ogawa et al. 1992). This functional metabolic imaging observation indicates the central role of oxidative ATP production in supporting the evoked brain activity during activation.

Another interesting finding is related to the *in vivo* ^{31}P MRS-MT as applied to the rat brain to study the quantitative correlation between the CMR_{ATP} and the variation in brain activity levels induced by anesthesia (i.e., from mild anesthesia to the isoelectric state characterized by the absence of EEG activity) (Du et al. 2008). In this study, a strong correlation was found between CMR_{ATP} measured by the *in vivo* ^{31}P MRS-MT approach and the brain activity quantified by EEG signal (Du et al. 2008). In contrast, the brain ATP concentration remained constant across the entire range (Fig. 19.11). This result again reveals a tight correlation between the oxidative ATP synthesis rate and brain bioenergetic demand under varied brain activity states. Therefore, the measurement of Pi → ATP flux should provide a sensitive index for quantifying the brain bioenergetics.

19.4 Conclusion

Although, in reality, *in vivo* heteronuclear MRS has not been widely applied in biomedical research and clinic as compared to ^{1}H MRI/MRS due to its incomparable detection sensitivity, the increasing in the high/ultrahigh MRI/MRS scanners availability, the improvement in MR technology and the associated increases in sensitivity have substantially stimulated *in vivo* heteronuclear MRS research and methodological development (Ugurbil et al. 2006). Significant progress has been made in advancing *in vivo* heteronuclear MRS in brain research. Two interesting applications of the *in vivo* ^{17}O MRSI approach and/or ^{31}P MRS-MT approach in the near future are: (i) the study of the dynamic changes in oxygen utilization and oxidative ATP metabolism during activation in normal brains and (ii) exploration of the possibility of detecting abnormalities in oxidative metabolic fluxes during the progression of brain diseases prior to the occurrence of HEP concentration changes. The outcomes of these applications will be essential for a better understanding of brain function and dysfunction. The combination of $CMRO_2$ and CMR_{ATP} imaging methods should provide a powerful tool for studying the decoupled metabolism processes between oxygen utilization and ADP phosphorylation that may be associated with many neurodegenerative diseases. Finally, *in vivo* ^{17}O and ^{31}P MRS can be readily combined with many other MRI modalities for imaging brain anatomy, perfusion, diffusion and BOLD etc. within the same scanning session. This combined MRS/MRI strategy can make it particularly powerful in neuroscience research.

Acknowledgments The authors thank Drs. Fei Du, Nanyin Zhang, Hao Lei, Yi Zhang, Hellmut Merkle, Peter Andersen, Gregor Adriany, Arthur From and Kamil Ugurbil, Mr. John Strupp for their technical assistance, support and scientific discussion. The part of the reviewed work was supported by NIH grants of NS041262, NS041262S1, EB002632, NS039043, EB000329, NS057560, NS070839, P30NS057091 and P41 RR08079, the W.M. Keck Foundation and the MIND Institute.

References

Abragam A (1961) The principles of nuclear magnetism. Oxford University Press, London

Ackerman JJH, Grove TH, Wong GG, Gadian DG, Radda GK (1980) Mapping of metabolites in whole animals by ^{31}P NMR using surface coils. Nature 283:167–170

Alger JR, Shulman RG (1984) NMR studies of enzymatic rates *in vitro* and *in vivo* by magnetization transfer. Quart Rev Biophys 17:83–124

Altura BM, Altura BT (1994) Role of magnesium and calcium in alcohol-induced hypertension and strokes as probed by in vivo television microscopy, digital image microscopy, optical spectroscopy, 31P-NMR, spectroscopy and a unique magnesium ion-selective electrode. Alcohol Clin Exp Res 18:1057–1068

Arai T, Nakao S, Mori K, Ishimori K, Morishima I, Miyazawa T, Fritz-Zieroth B (1990) Cerebral oxygen utilization analyzed by the use of oxygen-17 and its nuclear magnetic resonance. Biochem Biophys Res Commun 169:153–158

Arai T, Mori K, Nakao S, Watanabe K, Kito K, Aoki M, Mori H, Morikawa S, Inubushi T (1991) In vivo oxygen-17 nuclear magnetic resonance for the estimation of cerebral blood flow and oxygen consumption. Biochem Biophys Res Commun 179:954–961

Attwell D, Laughlin SB (2001) An energy budget for signaling in the grey matter of the brain. J Cereb Blood Flow Metab 21:1133–1145

Balaban RS, Kantor HL, Ferretti JA (1983) In vivo flux between phosphocreatine and adenosine triphosphate determined by two-dimensional phosphorous NMR. J Biol Chem 258:12787–12789

Bandettini PA, Wong EC, Hinks RS, Tikofsky RS, Hyde JS (1992) Time course EPI of human brain function during task activation. Magn Reson Med 25:390–397

Barinaga M (1997) What makes brain neurons run. Science 276:196–198

Beal MF (1992) Does impairment of energy metabolism result in excitotoxic neuronal death in neurodegenerative illnesses? Ann Neurol 31:119–130

Bittl JA, DeLayre J, Ingwall JS (1987) Rate equation for creatine kinase predicts the in vivo reaction velocity: ^{31}P NMR surface coil studies in brain, heart, and skeletal muscle of the living rat. Biochemistry 26:6083–6090

Blackband SJ, Thelwall PE, Chen W (2003) Field dependence of ^{17}O T_1, T_2 and SNR – in vitro and in vivo studies at 4.7, 11 and 17.6 Tesla. In: Proceedings of the international society for magnetic resonance in medicine, Toronto, p 504

Blamire AM, Ogawa S, Ugurbil K, Rothman D, McCarthy G, Ellermann JM, Hyder F, Rattner Z, Shulman RG (1992) Dynamic mapping of the human visual cortex by high-speed magnetic resonance imaging. Proc Natl Acad Sci USA 89:11069–11073

Boska MD, Hubesch B, Meyerhoff DJ, Twieg DB, Karczmar GS, Matson GB, Weiner MW (1990) Comparison of ^{31}P MRS and ^{1}H MRI at 1.5 and 2.0 T. Magn Reson Med 13:228–238

Bottomley PA, Hardy CJ (1992) Mapping creatine kinase reaction rates in human brain and heart with 4 Tesla saturation transfer ^{31}P NMR. J Magn Reson 99:443–448

Boyer PD (1999) What makes ATP synthase spin? Nature 402(247):249

Braunova Z, Kasparova S, Mlynarik V, Mierisova S, Liptaj T, Tkac I, Gvozdjakova A (2000) Metabolic changes in rat brain after prolonged ethanol consumption measured by ^{1}H and ^{31}P MRS experiments. Cell Mol Neurobiol 20:703–715

Brown GG, Levine SR, Gorell JM, Pettegrew JW, Gdowski JW, Bueri JA, Helpern JA, Welch KM (1989) In vivo ^{31}P NMR profiles of Alzheimer's disease and multiple subcortical infarct dementia. Neurology 39:1423–1427

Burt CT, Ribolow HJ (1984) A hypothesis: noncyclic phosphodiesters may play a role in membrane control. Biochem Med 31:21–30

Burt CT, Ribolow H (1994) Glycerol phosphorylcholine (GPC) and serine ethanolamine phosphodiester (SEP): evolutionary mirrored metabolites and their potential metabolic roles. Comp Biochem Physiol Biochem Mol Biol 108:11–20

Campbell SL, Jones KA, Shulman RG (1987) ^{31}P NMR saturation-transfer measurements in Saccharomyces cerevisiae: characterization of phosphate exchange reactions by iodoacetate and antimycin A inhibition. Biochemistry 26:7483–7492

Chance B, Leigh JS Jr, Nioka S, Sinwell T, Younkin D, Smith DS (1987) An approach to the problem of metabolic heterogeneity in brain: ischemia and reflow after ischemia. Ann N Y Acad Sci 508:309–320

Chaumeil MM, Valette J, Guillermier M, Brouillet E, Boumezbeur F, Herard AS, Bloch G, Hantraye P, Lebon V (2009) Multimodal neuroimaging provides a highly consistent picture of energy metabolism, validating 31P MRS for measuring brain ATP synthesis. Proc Natl Acad Sci USA 106:3988–3993

Chen W, Zhu XH (2005) Dynamic study of cerebral bioenergetics and brain function using in vivo multinuclear MRS approaches. Concept Magn Reson A 27A:84–121

Chen W, Zhu X-H, Adriany G, Ugurbil K (1997) Increase of creatine kinase activity in the visual cortex of human brain during visual stimulation: a ^{31}P NMR magnetization transfer study. Magn Reson Med 38:551–557

Chen W, Zhu XH, Gruetter R, Seaquist ER, Ugurbil K (2001) Study of oxygen utilization changes of human visual cortex during hemifield stimulation using ^{1}H-{^{13}C} MRS and fMRI. Magn Reson Med 45:349–355

Chen W, Zhu XH, Ugurbil K (2004) Imaging Cerebral Metabolic Rate of Oxygen Consumption ($CMRO_2$) using ^{17}O NMR Approach at Ultra-high Field. In: Shulman RG, Rothman DL (eds) Brain energetics and neuronal activity. Wiley, New York, pp 125–146

Clarke DD, Sokoloff L (1999) Circulation and energy metabolism of the brain. In: Siegel GJ et al (eds) Basic neruochemistry: molecular, cellular and medical aspects. Lippincott-Raven, Philadephia, pp 633–669

Davis TL, Kwong KK, Weisskoff RM, Rosen BR (1998) Calibrated functional MRI: mapping the dynamic of oxidative metabolism. Proc Natl Acad Sci USA 95:1834–1839

de Crespigny AJ, D'Arceuil HE, Engelhorn T, Moseley ME (2000) MRI of focal cerebral ischemia using ^{17}O-labeled water. Magn Reson Med 43:876–883

Degani H, Laughlin M, Campbell S, Shulman RG (1985) Kinetics of creatine kinase in heart: a ^{31}P NMR saturation- and inversion-transfer study. Biochemistry 24:5510–5516

Degani H, Alger JR, Shulman RG, Petroff OA, Prichard JW (1987) ^{31}P magnetization transfer studies of creatine kinase kinetics in living rabbit brain. Magn Reson Med 5:1–12

Du F, Zhu XH, Qiao H, Zhang X, Chen W (2007) Efficient in vivo ^{31}P magnetization transfer approach for noninvasively determining multiple kinetic parameters and metabolic fluxes of ATP metabolism in the human brain. Magn Reson Med 57:103–114

Du F, Zhu XH, Zhang Y, Friedman M, Zhang N, Ugurbil K, Chen W (2008) Tightly coupled brain activity and cerebral ATP metabolic rate. Proc Natl Acad Sci USA 105:6409–6414

Duncan JS (1997) Imaging and epilepsy. Brain 120(Pt 2):339–377

Erecinska M, Thoresen M, Silver IA (2003) Effects of hypothermia on energy metabolism in Mammalian central nervous system. J Cereb Blood Flow Metab 23:513–530

Fiat D, Kang S (1992) Determination of the rate of cerebral oxygen consumption and regional cerebral blood flow by non-invasive ^{17}O in vivo NMR spectroscopy and magnetic resonance imaging: Part 1. Theory and data analysis methods. Neurol Res 14:303–311

Fiat D, Kang S (1993) Determination of the rate of cerebral oxygen consumption and regional cerebral blood flow by non-invasive ^{17}O in vivo NMR spectroscopy and magnetic resonance imaging. Part 2. Determination of $CMRO_2$ for the rat by ^{17}O NMR, and $CMRO_2$, rCBF and the partition coefficient for the cat by ^{17}O MRI. Neurol Res 15:7–22

Fiat D, Ligeti L, Lyon RC, Ruttner Z, Pekar J, Moonen CT, McLaughlin AC (1992) In vivo ^{17}O NMR study of rat brain during $^{17}O_2$ inhalation. Magn Reson Med 24:370–374

Fiat D, Dolinsek J, Hankiewicz J, Dujovny M, Ausman J (1993) Determination of regional cerebral oxygen consumption in the human: ^{17}O natural abundance cerebral magnetic resonance imaging and spectroscopy in a whole body system. Neurol Res 15:237–248

Fox PT, Raichle ME (1986) Focal physiological uncoupling of cerebral blood flow and oxidative metabolism during somatosensory stimulation in human subjects. Proc Natl Acad Sci USA 83:1140–1144

Fox PT, Raichle ME, Mintun MA, Dence C (1988) Nonoxidative glucose consumption during focal physiologic neural activity. Science 241:462–464

Frackowiak RS, Herold S, Petty RK, Morgan-Hughes JA (1988) The cerebral metabolism of glucose and oxygen measured with positron tomography in patients with mitochondrial diseases. Brain 111(Pt 5):1009–1024

Frahm J, Bruhn H, Merboldt KD, Hanicke W (1992) Dynamic MR imaging of human brain oxygenation during rest and photic stimulation. J Magn Reson Image 2:501–505

Frosen S, Hoffman RA (1963) Study of moderately rapid chemical exchange by means of nuclear magnetic double resonance. J Chem Phys 39:2892–2901

Gadian DG, Williams SR, Bates TE, Kauppinen RA (1993) NMR spectroscopy: current status and future possibilities. Acta Neurochir Suppl (Wien) 57:1–8

Hardy CJ, Bottomley PA, Roemer PB, Redington RW (1988) Rapid ^{31}P spectroscopy on a 4-T whole-body system. Magn Reson Med 8:104–109

Herscovitch A, Raichle ME (1985) What is the correct value for the brain-blood partition coefficient for water? J Cereb Blood Flow Metab 5:65–69

Herscovitch P, Raichle ME, Kilbourn MR, Welch MJ (1987) Positron emission tomographic measurement of cerebral blood flow and permeability-surface area product of water using [15O] water and [11C]butanol. J Cereb Blood Flow Metab 7:527–542

Hetherington HP, Spencer DD, Vaughan JT, Pan JW (2001) Quantitative ^{31}P spectroscopic imaging of human brain at 4 Tesla: assessment of gray and white matter differences of phosphocreatine and ATP. Magn Reson Med 45:46–52

Hetherington HP, Pan JW, Spencer DD (2002) ^{1}H and ^{31}P spectroscopy and bioenergetics in the lateralization of seizures in temporal lobe epilepsy. J Magn Reson Imaging 16:477–483

Hinkle PC (2005) P/O ratios of mitochondrial oxidative phosphorylation. Biochim Biophys Acta 1706:1–11

Hoge RD, Atkinson J, Gill B, Crelier GR, Marrett S, Pike GB (1999) Investigation of BOLD signal dependence on cerebral blood flow and oxygen consumption: the deoxyhemoglobin dilution model [In Process Citation]. Magn Reson Med 42:849–863

Hopkins AL, Barr RG (1987) Oxygen-17 compounds as potential NMR T_2 contrast agents: enrichment effects of $H_2{}^{17}O$ on protein solutions and living tissues. Magn Reson Med 4: 399–403

Hossmann KA (1991) Studies of experimental cerebral ischemia with NMR spectroscopy. Arzneimittelforschung 41:292–298

Hyder F, Chase JR, Behar KL, Mason GF, Siddeek M, Rothman DL, Shulman RG (1996) Increase tricarboxylic acid cycle flux in rat brain during forepaw stimulation detected with ^1H-{^{13}C} NMR. Proc Natl Acad Sci USA 93:7612–7617

Hyder F, Kennan RP, Kida I, Mason GF, Behar KL, Rothman D (2000) Dependence of oxygen delivery on blood flow in rat brain: a 7 tesla nuclear magnetic resonance study. J Cereb Blood Flow Metab 20:485–498

Hyder F, Patel AB, Gjedde A, Rothman DL, Behar KL, Shulman RG (2006) Neuronal-glial glucose oxidation and glutamatergic-GABAergic function. J Cereb Blood Flow Metab 26:865–877

Jensen JE, Al-Semaan YM, Williamson PC, Neufeld RW, Menon RS, Schaeffer B, Densmore M, Drost DJ (2002) Region-specific changes in phospholipid metabolism in chronic, medicated schizophrenia: ^{31}P-MRS study at 4.0 Tesla. Br J Psychiatry 180:39–44

Joubert F, Mateo P, Gillet B, Beloeil JC, Mazet JL, Hoerter JA (2004) CK flux or direct ATP transfer: versatility of energy transfer pathways evidenced by NMR in the perfused heart. Mol Cell Biochem 256–257:43–58

Kasischke KA, Vishwasrao HD, Fisher PJ, Zipfel WR, Webb WW (2004) Neural activity triggers neuronal oxidative metabolism followed by astrocytic glycolysis. Science 305:99–103

Kemp GJ (2000) Non-invasive methods for studying brain energy metabolism: what they show and what it means. Dev Neurosci 22:418–428

Kety SS, Schmidt CF (1945) The determination of cerebral blood flow in man by the use of nitrous oxide in low concentrations. Am J Physiol 143:53–66

Kety SS, Schmidt CF (1948) Nitrous oxide method for the quantitative determination of cerebral blood flow in man: theory, procedure and normal values. J Clin Invest 27:476–483

Kim SG, Rostrup E, Larsson HB, Ogawa S, Paulson OB (1999) Determination of relative CMRO2 from CBF and BOLD changes: significant increase of oxygen consumption rate during visual stimulation. Magn Reson Med 41:1152–1161

Kingsley-Hickman PB, Sako EY, Mohanakrishnan P, Robitaille PM, From AH, Foker JE, Ugurbil K (1987) ^{31}P NMR studies of ATP synthesis and hydrolysis kinetics in the intact myocardium. Biochemistry 26:7501–7510

Kwong KK, Hopkins AL, Belliveau JW, Chesler DA, Porkka LM, McKinstry RC, Finelli DA, Hunter GJ, Moore JB, Barr RG, Rosen BR (1991) Proton NMR imaging of cerebral blood flow using $H_2{}^{17}O$. Magn Reson Med 22:154–158

Kwong KK, Belliveau JW, Chesler DA, Goldberg IE, Weisskoff RM, Poncelet BP, Kennedy DN, Hoppel BE, Cohen MS, Turner R, Cheng HM, Brady TJ, Rosen BR (1992) Dynamic magnetic resonance imaging of human brain activity during primary sensory stimulation. Proc Natl Acad Sci USA 89:5675–5679

Lei H, Ugurbil K, Chen W (2003a) Measurement of unidirectional Pi to ATP flux in human visual cortex at 7 Tesla using in vivo ^{31}P magnetic resonance spectroscopy. Proc Natl Acad Sci USA 100:14409–14414

Lei H, Zhu XH, Zhang XL, Ugurbil K, Chen W (2003b) In vivo ^{31}P magnetic resonance spectroscopy of human brain at 7 T: an initial experience. Magn Reson Med 49:199–205

Lenzi GL, Jones T, Frackowiak RS (1981) Positron emission tomography: state of the art in neurology. Prog Nucl Med 7:118–137

Magistretti PJ, Pellerin L, Rothman DL, Shulman RG (1999) Energy on demand. Science 283:496–497

Marrett S, Fujita H, Meyer E, Ribeiro L, Evans A, Kuwabara H, Gjedde A (1993) Stimulus specific increase of oxidative metabolism in human visual cortex. In: Uemura K (ed) Quantification of brain function in tracer kinetics and image analysis in brain PET. Elsevier Science, New York, pp 217–228

Mateescu GD (2003) Functional oxygen-17 magnetic resonance imaging and localized spectroscopy. Adv Exp Med Biol 510:213–218

Mateescu GD, Yvars GM, Dular T (1988) Water, Ions and O-17 Magnetic Resonance Imaging. In: Lauger P, Packer L, Vasilescu V (eds) Water and ions in biological systems. Birkhauser Verlag, Basel/Boston/Berlin, pp 239–250

Mateescu GD, Yvars G, Pazara DI, Alldridge NA, LaManna JC, Lust DW, Mattingly M, Kuhn W (1989) ^{17}O-^{1}H Magnetic Resonance Imaging in Plants, Animals, and Materials. In: Baillie TA, Jones JR (eds) Synthesis and application of isotopically labeled compounds. Elsevier, Amsterdam, pp 499–508

Matthews PM, Bland JL, Gadian DG, Radda GK (1982) A ^{31}P-NMR saturation transfer study of the regulation of creatine kinase in the rat heart. Biochim Biophys Acta 721:312–320

Maurer I, Zierz S, Moller HJ (2000) A selective defect of cytochrome c oxidase is present in brain of Alzheimer disease patients. Neurobiol Aging 21:455–462

Maurer I, Zierz S, Moller H (2001) Evidence for a mitochondrial oxidative phosphorylation defect in brains from patients with schizophrenia. Schizophr Res 48:125–136

Mintun MA, Raichle ME, Martin WR, Herscovitch P (1984) Brain oxygen utilization measured with O-15 radiotracers and positron emission tomography. J Nucl Med 25:177–187

Mitsumori F, Rees D, Brindle KM, Radda GK, Campbell ID (1988) ^{31}P-NMR saturation transfer studies of aerobic Escherichia coli cells. Biochim Biophys Acta 969:185–193

Mlynarik V, Zbyn S, Brehm A, Bischof M, Roden M (2005) An optimized protocol for measuring rate constant of creatine kinase reaction in human brain by ^{31}P NMR saturation transfer. In: Proceedings of the 13th international society for magnetic resonance in medicine, p 2767

Mora BN, Narasimhan PT, Ross BD (1992) ^{31}P magnetization transfer studies in the monkey brain. Magn Reson Med 26:100–115

Nakao Y, Itoh Y, Kuang TY, Cook M, Jehle J, Sokoloff L (2001) Effects of anesthesia on functional activation of cerebral blood flow and metabolism. Proc Natl Acad Sci USA 98:7593–7598

Nioka S, Smith DS, Mayevsky A, Dobson GP, Veech RL, Subramanian H, Chance B (1991a) Age dependence of steady state mitochondrial oxidative metabolism in the in vivo hypoxic dog brain. Neurol Res 13:25–32

Nioka S, Zaman A, Yoshioka H, Masumura M, Miyake H, Lockard S, Chance B (1991b) ^{31}P magnetic resonance spectroscopy study of cerebral metabolism in developing dog brain and its relationship to neuronal function. Dev Neurosci 13:61–68

Ogawa S, Lee T-M, Kay AR, Tank DW (1990) Brain magnetic resonance imaging with contrast dependent on blood oxygenation. Proc Natl Acad Sci USA 87:9868–9872

Ogawa S, Tank DW, Menon R, Ellermann JM, Kim S-G, Merkle H, Ugurbil K (1992) Intrinsic signal changes accompanying sensory stimulation: functional brain mapping with magnetic resonance imaging. Proc Natl Acad Sci USA 89:5951–5955

Pekar J, Ligeti L, Ruttner Z, Lyon RC, Sinnwell TM, van Gelderen P, Fiat D, Moonen CT, McLaughlin AC (1991) In vivo measurement of cerebral oxygen consumption and blood flow using ^{17}O magnetic resonance imaging. Magn Reson Med 21:313–319

Pekar J, Sinnwell T, Ligeti L, Chesnick AS, Frank JA, McLaughlin AC (1995) Simultaneous measurement of cerebral oxygen consumption and blood flow using ^{17}O and ^{19}F magnetic resonance imaging. J Cereb Blood Flow Metab 15:312–320

Pettegrew JW, Klunk WE, Panchalingam K, McClure RJ, Stanley JA (1997) Magnetic resonance spectroscopic changes in Alzheimer's disease. Ann N Y Acad Sci 826:282–306

Prichard J, Rothman D, Novotny E, Petroff O, Kuwabara T, Avison M, Howseman A, Hanstock C, Shulman RG (1992) Lactate rise detected by ^{1}H NMR in human visual cortex during physiologic stimulation. Proc Natl Acad Sci USA 88:5829–5831

Qiao H, Zhang X, Zhu XH, Du F, Chen W (2006) In vivo ^{31}P MRS of human brain at high/ultrahigh fields: a quantitative comparison of NMR detection sensitivity and spectral resolution between 4 T and 7 T. Magn Reson Imaging 24:1281–1286

Raichle ME (1987) Circulatory and metabolic correlates of brain function in normal humans. In: Mountcastle VB, Plum F, Geiger SR (eds) Handbook of physiology-the nervous system. American Physiological Society, Bethesda, pp 643–674

Raichle ME, Mintun MA (2006) Brain work and brain imaging. Annu Rev Neurosci 29:449–476

Ramadan NM, Halvorson H, Vande-Linde A, Levine SR, Helpern JA, Welch KM (1989) Low brain magnesium in migraine. Headache 29:590–593

Reddy R, Stolpen AH, Leigh JS (1995) Detection of ^{17}O by proton T_1 rho dispersion imaging. J Magn Reson B 108:276–279
Reddy R, Stolpen AH, Charagundla SR, Insko EK, Leigh JS (1996) ^{17}O-decoupled 1H detection using a double-tuned coil. Magn Reson Imaging 14:1073–1078
Ribeiro L, Kuwabara H, Meyer E, Fujita H, Marrett S, Evans A, Gjedde A (1993) Cerebral blood flow and metabolism during nonspecific bilateral visual stimulation in normal subjects. In: Uemura K (ed) Quantification of brain function in tracer kinetics and image analysis in brain PET. Elsevier Science, New York, pp 229–236
Riehemann S, Volz HP, Smesny S, Hubner G, Wenda B, Rossger G, Sauer H (2000) Phosphorus 31 magnetic resonance spectroscopy in schizophrenia research. Pathophysiology of cerebral metabolism of high-energy phosphate and membrane phospholipids. Nervenarzt 71:354–363
Roland PE, Ericksson L, Stone-Elander S, Widen L (1987) Does mental activity change the oxidative metabolism of the brain? J Neurosci 7:2373–2389
Rolfe DF, Brown GC (1997) Cellular energy utilization and molecular origin of standard metabolic rate in mammals. Physiol Rev 77:731–758
Ronen I, Navon G (1994) A new method for proton detection of $H_2{}^{17}O$ with potential applications for functional MRI. Magn Reson Med 32:789–793
Ronen I, Merkle H, Ugurbil K, Navon G (1998) Imaging of $H_2{}^{17}O$ distribution in the brain of a live rat by using proton-detected ^{17}O MRI. Proc Natl Acad Sci USA 95:12934–12939
Ross BM, Moszczynska A, Blusztajn JK, Sherwin A, Lozano A, Kish SJ (1997) Phospholipid biosynthetic enzymes in human brain. Lipids 32:351–358
Saks VA, Ventura-Clapier R, Aliev MK (1996) Metabolic control and metabolic capacity: two aspects of creatine kinase functioning in the cells. Biochim Biophys Acta 1274:81–88
Sauter A, Rudin M (1993) Determination of creatine kinase parameters in rat brain by NMR magnetization transfer. J Biol Chem 268:13166–13171
Shoubridge EA, Briggs RW, Radda GK (1982) ^{31}P NMR saturation transfer measurements of the steady state rates of creatine kinase and ATP synthetase in the rat brain. FEBS Lett 140:289–292
Shulman RG, Rothman DL, Hyder F (1999) Stimulated changes in localized cerebral energy consumption under anesthesia. Proc Natl Acad Sci USA 96:3245–3250
Shulman RG, Hyder F, Rothman DL (2001) Cerebral energetics and the glycogen shunt: neurochemical basis of functional imaging. Proc Natl Acad Sci USA 98:6417–6422
Shulman RG, Rothman DL, Behar KL, Hyder F (2004) Energetic basis of brain activity: implications for neuroimaging. Trends Neurosci 27:489–495
Siesjo BK (1978) Brain energy metabolism. Wiley, New York, pp 101–110
Spencer RG, Balschi JA, Leigh JS Jr, Ingwall JS (1988) ATP synthesis and degradation rates in the perfused rat heart. 31P-nuclear magnetic resonance double saturation transfer measurements. Biophys J 54:921–929
Stock D, Leslie AG, Walker JE (1999) Molecular architecture of the rotary motor in ATP synthase. Science 286:1700–1705
Stolpen AH, Reddy R, Leigh JS (1997) ^{17}O-decoupled proton MR spectroscopy and imaging in a tissue model. J Magn Reson 125:1–7
Ter-Pogossian MM, Eichling JO, Davis DO, Welch MJ (1970) The measure in vivo of regional cerebral oxygen utilization by means of oxyhemoglobin labeled with radioactive oxygen-15. J Clin Invest 49:381–391
Thoma WJ, Ugurbil K (1987) Saturation-transfer studies of ATP-Pi exchange in isolated perfused rat liver. Biochim Biophys Acta 893:225–231
Ugurbil K (1985a) Magnetization Transfer Measurements of Creatine Kinase and ATPase Rates in Intact Hearts. Circulation 72:IV94–IV96
Ugurbil K (1985b) Magnetization Transfer Measurements of Individual Rate Constants in the Presence of Multiple Reactions. J Magn Reson 64:207–219
Ugurbil K, Petein M, Maiden R, Michurski S, From A (1986) Measurement of an individual rate constant in the presence of multiple exchanges: Application to myocardial creatine kinase rates. Biochemistry 25:100–108
Ugurbil K, Adriany G, Akgün C, Andersen P, Chen W, Garwood M, Gruetter R, Henry P-G, Marjanska M, Moeller S, Van de Moortele P-F, Prüssmann K, Tkac I, Vaughan JT, Wiesinger F,

Yacoub E, Zhu X-H (2006) High magnetic fields for imaging cerebral morphology, function and biochemistry. In: Robitaille PMLaB LJ (ed) Biological magnetic resonance: ultra high field magnetic resonance imaging. Springer, New York, pp 285–342

Vafaee MS, Meyer E, Marrett S, Paus T, Evans AC, Gjedde A (1999) Frequency-dependent changes in cerebral metabolic rate of oxygen during activation of human visual cortex. J Cereb Blood Flow Metab 19:272–277

Wallace DC (1992) Mitochondrial genetics: a paradigm for aging and degenerative diseases? Science 256:628–632

Wallace DC (1999) Mitochondrial diseases in man and mouse. Science 283:1482–1488

Wallimann T, Hemmer W (1994) Creatine kinase in non-muscle tissues and cells. Mol Cell Biochem 133–134:193–220

Weiner MW (1987) NMR spectroscopy for clinical medicine. Animal models and clinical examples. Ann N Y Acad Sci 508:287–299

Weiss RG, Gerstenblith G, Bottomley PA (2005) ATP flux through creatine kinase in the normal, stressed, and failing human heart. Proc Natl Acad Sci USA 102:808–813

Welch KM, Levine SR, Martin G, Ordidge R, Vande Linde AM, Helpern JA (1992) Magnetic resonance spectroscopy in cerebral ischemia. Neurol Clin 10:1–29

Wong-Riley M, Antuono P, Ho KC, Egan R, Hevner R, Liebl W, Huang Z, Rachel R, Jones J (1997) Cytochrome oxidase in Alzheimer's disease: biochemical, histochemical, and immunohistochemical analyses of the visual and other systems. Vision Res 37:3593–3608

Yee SH, Lee K, Jerabek PA, Fox PT (2006) Quantitative measurement of oxygen metabolic rate in the rat brain using microPET imaging of briefly inhaled ^{15}O-labelled oxygen gas. Nucl Med Commun 27:573–581

Young RS, Chen B, Petroff OA, Gore JC, Cowan BE, Novotny EJ Jr, Wong M, Zuckerman K (1989) The effect of diazepam on neonatal seizure: in vivo ^{31}P and ^{1}H NMR study. Pediatr Res 25:27–31

Zhang N, Zhu XH, Lei H, Ugurbil K, Chen W (2004) Simplified methods for calculating cerebral metabolic rate of oxygen based on ^{17}O magnetic resonance spectroscopic imaging measurement during a short $^{17}O_2$ inhalation. J Cereb Blood Flow Metab 24:840–848

Zhu XH, Merkle H, Kwag JH, Ugurbil K, Chen W (2001) ^{17}O relaxation time and NMR sensitivity of cerebral water and their field dependence. Magn Reson Med 45:543–549

Zhu XH, Zhang Y, Tian RX, Lei H, Zhang N, Zhang X, Merkle H, Ugurbil K, Chen W (2002) Development of ^{17}O NMR approach for fast imaging of cerebral metabolic rate of oxygen in rat brain at high field. Proc Natl Acad Sci USA 99:13194–13199

Zhu XH, Zhang N, Zhang Y, Zhang X, Ugurbil K, Chen W (2005) In vivo ^{17}O NMR approaches for brain study at high field. NMR Biomed 18:83–103

Zhu XH, Zhang Y, Zhang N, Ugurbil K, Chen W (2007) Noninvasive and three-dimensional imaging of $CMRO_2$ in rats at 9.4 T: reproducibility test and normothermia/hypothermia comparison study. J Cereb Blood Flow Metab 27:1225–1234

Zhu XH, Du F, Zhang N, Zhang Y, Lei H, Zhang X, Qiao H, Ugurbil K, Chen W (2009a) Advanced In Vivo Heteronuclear MRS Approaches for Studying Brain Bioenergetics Driven by Mitochondria. Methods Mol Biol 489:317–357

Zhu XH, Zhang N, Zhang Y, Ugurbil K, Chen W (2009b) New insights into central roles of cerebral oxygen metabolism in the resting and stimulus-evoked brain. J Cereb Blood Flow Metab 29:10–18

Part III
Cerebral Energy Metabolism and Substrates *In Vivo*

Chapter 20
Cerebral Glucose Transport and Homeostasis

João M.N. Duarte and Rolf Gruetter

Abstract Cerebral glucose homeostasis is maintained by the equilibrium between glucose transport across the blood-brain-barrier (BBB), which occurs through facilitative carriers, and glucose consumption mainly initiated by hexokinase phosphorylation. Glucose concentrations can be quantified non invasively by nuclear magnetic resonance (NMR) spectroscopy and such measurements, through appropriate mathematical modelling, allow to determine the kinetics of glucose transport and metabolism. This chapter summarizes the insights gained into brain glucose transport from the measurement of the brain glucose concentration, particularly reviewing state of the art NMR methods for non invasive determination of glucose homeostasis and discussing the employment of mathematical models of glucose transport.

Keywords Blood-brain-barrier • Glucose • GLUT1 • NMR

20.1 Introduction

Under normal physiological conditions, adult brain function relies on glucose supply from blood stream (Siesjo 1978), which occurs through facilitative glucose carrier proteins (principally GLUT1) located at the BBB (Simpson et al. 2007). The cerebral

J.M.N. Duarte, Ph.D. (✉)
Laboratory for functional and metabolic imaging, Ecole Polytechnique
Federale de Lausanne, Lausanne, Switzerland
e-mail: joao.duarte@epfl.ch

R. Gruetter, Ph.D.
Laboratory for functional and metabolic imaging, Ecole Polytechnique
Federale de Lausanne, Department of Radiology, University of Lausanne and Geneva,
Station 6, CH-1015, Lausanne, Switzerland
e-mail: rolf.gruetter@epfl.ch

I.-Y. Choi and R. Gruetter (eds.), *Neural Metabolism In Vivo*,
Advances in Neurobiology 4, DOI 10.1007/978-1-4614-1788-0_20,

concentration of glucose can be measured *in vivo* using nuclear magnetic resonance (NMR) techniques and, thereby, glucose homeostasis in the brain has been vastly determined in humans and rodents using non-invasive NMR spectroscopy (e.g. Choi et al. 2002; Duarte et al. 2009a; Gruetter et al. 1998b; Lei et al. 2010). However, appropriate mathematical models are required to distinguish the different components of glucose transport and metabolism involved in cerebral glucose homeostasis (Duarte et al. 2009b; Gruetter et al. 1998a). Models of glucose transport across the BBB were first developed based on standard irreversible Michaelis-Menten kinetics (e.g. Lund-Andersen 1979; Gruetter et al. 1996b). Nonetheless, since brain glucose can reach concentrations that are higher than the K_M of glucose carriers, reversible Michaelis-Menten kinetics has been introduced to describe the transport process (Cunningham et al. 1986; Gruetter et al. 1998b). Although this reversible Michaelis-Menten kinetics can reliably describe brain glucose transport from the steady-state brain glucose concentration (e.g. Gruetter et al. 1998a; Choi et al. 2002), it has not yet been taken as a general model because the affinity constant of glucose transport is not precisely estimated (e.g. Choi et al. 2002; Lei and Gruetter 2006; Seaquist et al. 2001) and does not take into account the iso-inhibition or trans-acceleration of the carrier protein (Barros et al. 2007; Duarte et al. 2009b). In fact, the apparent affinity constant (K_t) estimated by the reversible Michaelis-Menten model is generally much lower than the K_M for glucose transport determined *in vitro* (e.g. Gould et al. 1991; Lee et al. 1997). Therefore, a conformational four-state exchange kinetic model of simple solute carriers (e.g. Cuppoletti and Segel 1975) has been proposed to describe glucose transport across the BBB and to account for these features of glucose transport (Barros et al. 2007; Duarte et al. 2009b).

20.2 Brain Glucose Can Be Directly Measured *In Vivo*

Quantitative measurements of cerebral glucose transport have been reported in studies using radioactive tracers, namely in the human brain (e.g. Brooks et al. 1986; Feinendegen et al. 1986; Gutniak et al. 1990; Blomqvist et al. 1991). The intracellular glucose concentration depends on metabolic as well as on transport rates and thus contains information whether glucose supply is adequate for demand. It is well established that brain glucose concentration is an important parameter in the analysis of experiments using radiolabeled glucose or glucose analogs. Free glucose is measurable in animal brain (e.g. Mason et al. 1992; Holden et al. 1991; Gjedde and Diemer 1983; Silver and Erecinska 1994) and, under physiological conditions, it is also measurable by NMR in the human brain (Gruetter et al. 1992a, 1996a, 1998a; Frahm et al. 1996).

Glucose signals can be observed using either ^{13}C or ^{1}H NMR spectroscopy (Fig. 20.1). Brain glucose concentrations have been quantified *in vivo* using different NMR methods. An early study quantified whole brain glucose in rats using ^{13}C NMR spectroscopy by using the post-mortem cerebral lactate signal as the internal reference (Mason et al. 1992). Non-invasive quantification of the signals of α- and β-glucose C1 measured by ^{13}C NMR spectroscopy was reported in human brain

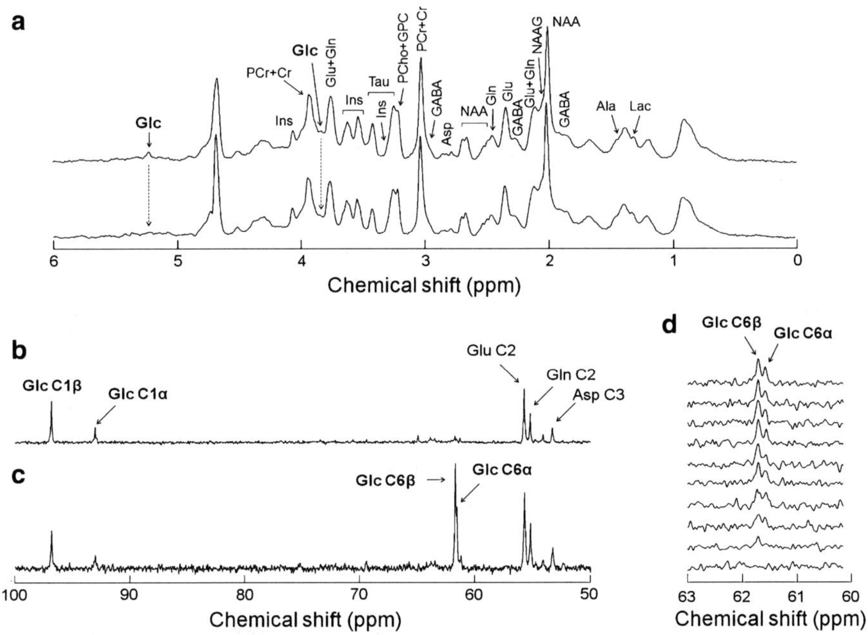

Fig. 20.1 Typical *in vivo* ^{1}H and ^{13}C NMR spectra from the rat brain. **Panel A** shows ^{1}H NMR spectra acquired during 3 min from a volume of 120 μL in the rat brain at euglycaemia (*bottom*) and at hyperglycaemia after glucose infusion (*top*). An extensive neurochemical profile including glucose (Glc) can be measured. **Panel B** shows ^{13}C NMR spectra acquired for ~3 h upon infusion of [1-^{13}C]glucose. **Panels C** and **D** show ^{13}C NMR spectra acquired upon infusion of [1,6-^{13}C] glucose for ~1 h (**C**) or for ~5 min sequentially after onset of infusion. ^{13}C NMR spectra were all acquired from a volume of 320 μL in the rat brain. See Duarte et al. (2011) for details on these spectroscopy methods

using the external reference method (Gruetter et al. 1992a). External referencing is based on repeating the identical measurement in a phantom containing an aqueous solution of a known concentration of natural abundance glucose. This technique is robust, since the calibration measurement is designed to mimic the *in vivo* experiment in as many aspects as experimentally can be controlled.

The direct observation by ^{13}C has the advantage that the signals from α- and β-glucose can be unequivocally observed. The disadvantage lies in the lower sensitivity, thereby requiring longer accumulation times, larger regions of interest and infusion of ^{13}C labeled glucose to increase sensitivity. To minimize potential partial volume effects from extra-cerebral tissue and from large vessels, localized ^{13}C NMR measurements have been performed (Gruetter et al. 1992a). These studies also showed that brain glucose concentration is approximately 1 μmol/g at euglycemia in the human brain. Later, it was demonstrated that, at hyperglycemia, ^{13}C NMR can also be used to quantify brain glucose without infusion of ^{13}C labeled glucose (Gruetter et al. 1998a, b).

Observation of glucose in the more sensitive ^{1}H spectrum is difficult due to overlap with resonances from other compounds (Gruetter et al. 1992b; Gruetter et al.

1996b). To overcome such limitations, changes in brain glucose signal at 3.4 ppm using difference editing have been measured (Gruetter et al. 1992b), which has been used to dynamically investigate glucose transport kinetics (Gruetter et al. 1996b; van Zijl et al. 1997). Alternatively, ^{1}H NMR detection of glucose has also been achieved by exploiting homonuclear J couplings in glucose (Keltner et al. 1998), and by observing the ^{1}H resonances of H1 bound to the C1 of [1-^{13}C]glucose (van Zijl et al. 1993; Inubushi et al. 1993). Although H1 resonance of α-glucose is sufficiently separated to be observed directly without further editing means (Gruetter et al. 1996a, 1998a), high magnetic fields and appropriate algorithms for spectral analysis (e.g. LCModel) provide a reliable determination of cerebral glucose concentrations (Duarte et al. 2009a; Lei et al. 2010), able to detect small modifications of cerebral glucose concentration upon distinct functional states at euglycaemia (Mangia et al. 2007).

Determination of the brain glucose concentration by NMR is highly consistent among different techniques used and with biochemical measurements (compared in Duarte et al. 2009b). Using creatine as an internal concentration standard (Gruetter et al. 1998a) minimizes the overall study time and is robust, when taking into account variations in the gray/white matter composition of the volume studied (Hetherington et al. 1996; Kreis 1997). The glucose peak of the H1 of α-D-glucose at 5.22 ppm (Fig. 20.1A) has been quantified using this approach and the quantification has been compared to that using the external reference method discussed above with very good agreement (e.g. Gruetter et al. 1998a). In recent studies, the water signal from the volume where glucose was quantified was used as reference for quantification (e.g. Duarte et al. 2009a; Lei et al. 2010).

The cerebral blood volume depends on subject's physiology (Shockley and LaManna 1988) and typically is 50 μL/g or less, suggesting that blood glucose is a minor contribution to the overall NMR signal at euglycaemia. Therefore, the glucose signal detected by NMR reflects glucose in the brain aqueous compartment. However, at high plasma glucose concentrations the contribution of blood glucose to the NMR signal can be significant (Duarte et al. 2009a).

In summary, NMR quantifications of brain glucose performed to date have been very consistent, when measured over similar plasma glucose concentrations. The unique ability of *in vivo* NMR spectroscopy to quantify brain glucose naturally provides a unique measurement capability that is complementary to the highly sensitive brain mapping techniques of glucose phosphorylation, namely using radiolabeled glucose analogs.

20.3 The Physical Distribution Space of Glucose Is the Brain's Aqueous Phase

Glucose transport across membranes is mediated by a family of membrane-spanning carrier proteins (e.g. Mueckler 1994). There are several glucose carrier isoforms present in the brain (reviewed in Simpson et al. 2007). However, the predominant glucose transporter (GLUT) proteins involved in cerebral glucose utilization are GLUT1 and GLUT3 (reviewed in Simpson et al. 2007). GLUT1 is present in all brain cells including the endothelial cells of the capillaries and has very low neuronal expression *in vivo*, while GLUT3 almost restricted to neurons (Leino et al. 1997; Vannucci et al. 1997). Therefore, GLUT1, the ubiquitous high-affinity transporter present in erythrocyte membranes, is considered the principal carrier of glucose at the BBB, with some contributions from other transporters (Pardridge et al. 1990; Maher et al. 1993).

While the distribution of GLUT1 in the luminal and abluminal membranes of the endothelium is on the same order of magnitude, glucose transporter proteins are more concentrated in the membrane of brain cells than in membranes of the endothelium (Barros et al. 2007; Simpson et al. 2007). Accordingly, the diffusion across the membranes of astrocytes and neurons is not limiting for the global rate of glucose transport and allows fast equilibration inside brain's aqueous phase. Thus interstitial space and brain cells are traditionally considered as a single kinetic compartment, consistent with similar intra- and extracellular glucose concentrations (Poitry-Yamate et al. 2009; Pfeuffer et al. 2000). Together with the fact that luminal and abluminal membranes of the endothelium have similar amounts of glucose carrier proteins, and that glucose transport at the BBB approaches symmetry (Barros et al. 2007; Duarte et al. 2009b), the BBB can be modeled as a single membrane system in terms of glucose transport (Duarte et al. 2009b; Gruetter et al. 1998a).

The determination of kinetic constants from the relation between brain and plasma glucose concentration depends on the assumption that the rate-limiting step for glucose transport into the brain cells is at the BBB. Alternatively, the rate-limiting step could be at the cellular membrane. In this scenario the glucose is confined to extracellular space at near-equilibrium with plasma glucose. Intracellular glucose would be zero and transport at the cell membranes would be rate-limiting for metabolism. It is generally assumed that brain glucose is almost evenly distributed between the intra- and extracellular compartment. Brain glucose concentration changes measured after a rapid increase in plasma glucose concentration are delayed in a manner that implies a large distribution volume of brain glucose (Gruetter et al. 1996b).

Further investigation into the distribution of brain glucose in the brain was performed by measuring the diffusion behaviour of glucose by NMR. In model systems it has been shown that extra- and intracellular signals can be separated based on a different diffusion attenuation of the MR signal (van Zijl et al. 1991; Pfeuffer et al. 1998a, b). *In vivo*, most metabolites are predominantly intracellular, with the notable exception of glucose whose diffusion-weighted signal was more attenuated than that of intracellular metabolites, consistent with ~20% extracellular volume and glucose being approximately evenly distributed between intra- and extracellular space (Pfeuffer et al. 1998c).

Therefore, the glucose concentration in brain largely reflects the relative transport capacity at the BBB.

20.4 Several Models Exist to Describe Brain Glucose Transport Kinetics at Steady-State

The process of glucose transport in the brain is saturable (Crone 1965) but not rate limiting for the glycolytic pathway (Betz et al. 1976). Glucose selectively binds to facilitative glucose carriers that mediate solute movement in both directions of the BBB. This facilitated diffusion is dependent on the concentration gradient, which is a result of intracellular glucose phosphorylation. Non-carrier mediated diffusion of glucose molecules through the plasma membrane is negligible (e.g. Barros et al. 2007; Duarte et al. 2009b; Gruetter et al. 1998a).

Glucose transport across the BBB can be described as a three compartment system (e.g. Cunningham et al. 1986; Duarte et al. 2009b; Gruetter et al. 1998b; Mason et al. 1992): the BBB separates the blood circulation compartment from the brain, which is kinetically separated from the metabolic pool where glucose is phosphorylated and consumed. In such a model, the physical distribution space of glucose equals the brain water phase (Gjedde and Diemer 1983; Lund-Andersen 1979; Pappenheimer and Setchell 1973), which implies that glucose is evenly distributed in the intra- and extracellular spaces and that transport across cellular membranes is fast compared to glucose consumption rate (see also above).

As the BBB can be simplified to behave as a single membrane (because there is a relatively small distribution space for glucose inside endothelial cells and the rate of transport is fast when compared to glucose consumption), transport across the BBB is described as the sum unidirectional transport rates and consumption rate CMR_{glc}.

$$\frac{dG_{brain}}{dt} = T_{in} - T_{out} - CMR_{glc} \tag{20.1}$$

20.4.1 *Irreversible Michaelis-Menten Model of Glucose Transport*

The description of glucose transport was initially based on the standard Michaelis-Menten model, reviewed by Lund-Andersen (1979), which assumes classical Michaelis-Menten kinetics to describe the unidirectional fluxes across the lumenal and ablumenal membrane, i.e. the net uptake rate is given by

$$v = T_{in} - T_{out} = \frac{T_{max} G_{plasma}}{K_t + G_{plasma}} - \frac{T_{max} G_{brain}}{K_t + G_{brain}} \tag{20.2}$$

Generally, the kinetic constants for influx and efflux across the BBB are assumed to be the same, i.e. glucose transport at the BBB is symmetric. Furthermore, cerebral glucose consumption is assumed to be constant at euglycemia and above, implied from arterio-venous difference measurements (Pappenheimer and Setchell 1973; Boyle et al. 1994).

In accordance, the 1 μmol/g of brain glucose at euglycemia (e.g. Gruetter et al. 1992a) is well above the K_M of brain hexokinase ranging from 30 to 67 μM (Grossbard and Schimke 1966; Schimke and Grossbard 1968; Sprengers et al. 1983; Gjedde 1982; Buschiazzo et al. 1970).

From Eq. 20.1, the steady-state brain glucose concentration in the standard Michaelis-Menten model is given by (e.g. Gruetter et al. 1996b, 1998a):

$$G_{brain} = V_d \frac{\left(\frac{T_{max}}{CMR_{glc}} - 1\right) G_{plasma} - K_t}{\frac{T_{max}}{CMR_{glc}} + 1 + \frac{G_{plasma}}{K_t}} \quad (20.3)$$

Brain glucose G_{brain} is given in μmol/g and the rates T_{max} and CMR_{glc} are in μmol/g min, whereas plasma glucose G_{plasma} and the apparent affinity constant K_t are in mM. V_d represents the physical distribution space of glucose. With a blood volume of 35 μL/g (Shockley and LaManna 1988) and a dry weight of 20%, and assuming that glucose occupies the entire aqueous phase (see above), V_d is ~0.77 mL/g (Lund-Andersen 1979).

It is obvious from Eq. 20.3 that at saturating plasma glucose concentrations, i.e. when G_{plasma} is much larger than K_t, G_{brain} approaches an upper limit of $V_d K_t (T_{max}/CMR_{glc} - 1)$.

Most localized glucose transport kinetic measurements in human brain using the standard model (summarised in Table 20.1), have been highly consistent with an apparent Michaelis-Menten constant K_t of 4–5 mM for transport and an average maximal transport rate T_{max} of approximately 1 μmol/g min, in accordance to data from studies using radiotracers (Brooks et al. 1986; Feinendegen et al. 1986; Blomqvist et al. 1991). However, the standard Michaelis-Menten model also predicted that brain glucose should be below 5 μmol/g for plasma glucose concentrations below 30 mM. However, the brain glucose concentrations detected by NMR spectroscopy are typically ~9 μmol/g at plasma glucose concentrations of 30 mM in humans (Gruetter et al. 1998a) as well as in rodents (Duarte et al. 2009a). At such high plasma glucose concentrations, glucose efflux from the brain is substantial and inhibits its own carrier-mediated uptake.

20.4.2 Reversible Michaelis-Menten Model of Glucose Transport

Brain glucose concentrations substantially above the K_M of GLUT1 indicate that product formation is not unidirectional, i.e. the reverse reaction may proceed at significant rates at steady-state. Glucose binding to the transporter at the ablumenal

Table 20.1 Ratio of apparent maximum transport rate (T_{max}) to cerebral metabolic rate of glucose (CMR_{glc}) and plasma glucose concentration at half maximal transport rate (K_t) and determined non-invasively by NMR spectroscopy and using generally steady-state kinetics (exceptions are identified). Values are accompanied by reported standard error

	Anaesthesia	T_{max}/CMR_{glc}	K_t (mM)	Transport kinetics	References
Human brain		3.6 ± 0.1	4.9 ± 0.9	Standard MM	Gruetter et al. 1992a
		2.5 ± 0.1	4.8 ± 2.4	Standard MM	Gruetter et al. 1996a*
		3.2 ± 0.1	6.2 ± 0.9	Standard MM	de Graaf et al. 2001
		2.1 ± 0.1^{WM}	1.7 ± 0.9^{WM}	Standard MM	de Graaf et al. 2001
		2.3 ± 0.1	0.6 ± 2.0	Rev. MM	Gruetter et al. 1998a
		2.2 ± 0.2	−1.0 ± 2.1	Rev. MM	Seaquist et al. 2001
		2.2 ± 0.3^{WM}	2.0 ± 2.5^{WM}	Rev. MM	Seaquist et al. 2001
		3.9 ± 0.2^{WM}	7.3 ± 1.1^{WM}	Rev. MM	de Graaf et al. 2001
		1.8 ± 0.1	1.1 ± 0.7	Rev. MM	de Graaf et al. 2001
		2.7 ± 1.2	2.1 ± 5.2	Four-state conformational	Duarte et al. 2009b
Rat brain	α-chloralose	11.8 ± 1.6	4.7 ± 0.1	Standard MM	Choi et al. 2001
		2.7 ± 0.2	3.3 ± 0.1	Rev. MM	Choi et al. 2001
		2.7 ± 0.2	n.d.	Rev. MM	Choi et al. 2002
		2.6 ± 0.1	n.d.	Rev. MM	Lei and Gruetter 2006
		1.9 ± 0.1	2.5 ± 1.2	Rev. MM	Lei et al. 2010
		1.8 ± 0.1	1.2 ± 1.3	Rev. MM	Duarte et al. 2009a
		2.0 ± 0.5	1.9 ± 2.6	Four-state conformational	Duarte et al. 2009a [R]
		2.9 ± 0.3	3.5 ± 1.0	Four-state conformational	Duarte et al. 2009b
		1.9 ± 0.1	1.6 ± 1.1	Four-state conformational	Lei et al. 2010[R]
	Enflurane	5.8 ± 0.8	13.9 ± 2.7	Standard MM	Mason et al. 1992
	Pentobarbital or thiopental	1.9 ± 0.8	n.d.	Rev. MM	Cunningham et al. 1986**
		5.0 ± 0.2	n.d.	Rev. MM	Choi et al. 2002
		2.8 ± 0.2	2.8 ± 0.2	Rev. MM	Lei et al. 2010
		2.8 ± 0.2	4.6 ± 1.4	Four-state conformational	Lei et al. 2010[R]

n.d., not determined; ^{WM}determined in white mater tissue; *non-steady-state measurements; **radioisotope study; ^{R}data re-evaluated with conformational model as described in Duarte et al. 2009b membrane may partially inhibit the unidirectional influx. Such behaviour can be expected for the glucose transporter GLUT1, since it has been shown that cytochalasin B, which binds to the efflux binding site, increases the K_i for maltose binding at the sugar influx site of the carrier (Carruthers and Helgerson 1991). When the product formation is not unidirectional, reversible Michaelis-Menten kinetics are applicable. Such a situation amounts to replacing the apparent Michaelis constant in the standard model K_t by the term $K_t + G_{brain}$ for product formation (transport in, T_{in}), and by $K_t + G_{plasma}$ for the reverse reaction T_{out}, conversely (Cunningham et al. 1986; Gruetter et al. 1998a). The net velocity is thus given by

$$v = T_{in} - T_{out} = \frac{T_{max}\left(G_{plasma} - G_{brain}\right)}{K_t + G_{plasma} + G_{brain}} \tag{20.4}$$

This can be interpreted as reduced affinity for the forward reaction (influx) when substantial substrate (brain glucose, G_{brain}) is present, a situation that would result in apparent asymmetric kinetic properties, when assessed using the standard model. At steady state, the reversible Michaelis-Menten model results in the following relationship between brain and plasma glucose:

$$G_{brain} = V_d \frac{\left(\frac{T_{max}}{CMR_{glc}} - 1\right) G_{plasma} - K_t}{\frac{T_{max}}{CMR_{glc}} + 1} \tag{20.5}$$

which implies that brain glucose is a linear function of plasma glucose (Gruetter et al. 1998a). The linearity is also directly illustrated in Fig. 20.2. ^{13}C NMR spectroscopy studies for glucose determination also demonstrated that transport and consumption of brain glucose can be predicted by the reversible model (Duarte et al. 2011; Gruetter et al. 2001).

A linear brain glucose to plasma glucose relationship is consistent with earlier reports suggesting the presence of a high-affinity low-capacity transport system (Gjedde 1981) or a non-saturable transport mechanism (Pardridge 1983), since both mechanisms give rise to a linear relationship as well. Therefore we cannot, based on our measurements, exclude the presence of such a mechanism. However, it is interesting to note that a similar analysis of kinetic tracer experiments in rat brain suggested that the difference of the two models might manifest itself in the standard model as such a non-saturable component (Cunningham 1986).

Some investigators have measured glucose transport from the plasma to brain glucose relationship and reported a K_t of 9–14 mM, using the standard model (e.g. Mason et al. 1992; Holden et al. 1991; Gjedde and Christensen 1984; Gjedde 1992). It is interesting to note that these studies measured a K_t that is markedly higher than the K_M measured in erythrocyte model systems (where GLUT1 is the main operating carrier) against a zero intracellular glucose concentration (Carruthers 1990).

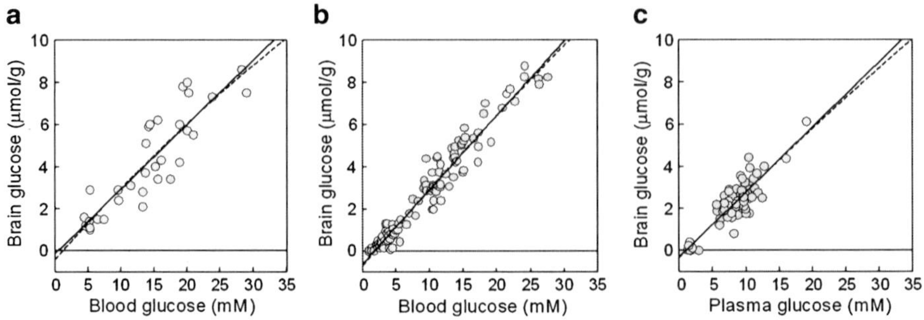

Fig. 20.2 Brain glucose concentration as function of plasma glucose concentration in three different studies: (**a**) human brain, in Gruetter et al. (1998a); (**b**) rat brain (under α-chloralose anaesthesia), in Choi et al. (2002); (**c**) rat brain, in Morgenthaler et al. (2006). The data sets were analysed with the reversible (*solid lines*) and the conformational four-state exchange (*dashed lines*) models. In these experimental conditions and for the range of glucose concentrations studied, the reversible Michaelis-Menten model is clearly able to describe the brain-blood glucose relationship, with the advantage of requiring the estimation of only two kinetic parameters. Figure was adapted from Duarte et al. (2009b). Fit results are included in Table 20.1

Such high apparent K_t can be taken as an indication of an almost linear brain-blood glucose relationship. However, as shown in Table 20.1, apparent K_t for glucose transport has been reported in a wide range of values when using the reversible Michaelis-Menten model, both in human or rodent brains. This suggests that, at least in some experimental conditions, there is a component of glucose transport that is not being accounted for.

20.4.3 Four-State Conformational Model of Glucose Transport

As described above, models of glucose transport across the BBB were developed based on standard Michaelis-Menten kinetics (e.g. Lund-Andersen 1979). Then, since brain glucose can reach concentrations similar to or higher than the K_M of the glucose carrier, reversible Michaelis-Menten kinetics has been introduced to describe the transport process (Cunningham et al. 1986; Gruetter et al. 1998a). The reversible Michaelis-Menten kinetics of glucose transport implies a near linear relation between brain and plasma glucose, as it has been experimentally determined in several studies in both humans (e.g. Gruetter et al. 1998a) and rats (e.g. Choi et al. 2002). The reversible Michaelis-Menten relationship (Eq. 20.5) reliably describes the relationship of brain to plasma glucose but does not account for trans-acceleration or asymmetry (Barros et al. 2007). Conversely, a conformational four-state exchange kinetic model (Fig. 20.3) of simple solute carriers (e.g. Cuppoletti and Segel 1975) can account trans-acceleration, product inhibition and asymmetry (Carruthers 1990), and was used in a recently developed multi-compartmental theoretical model of brain

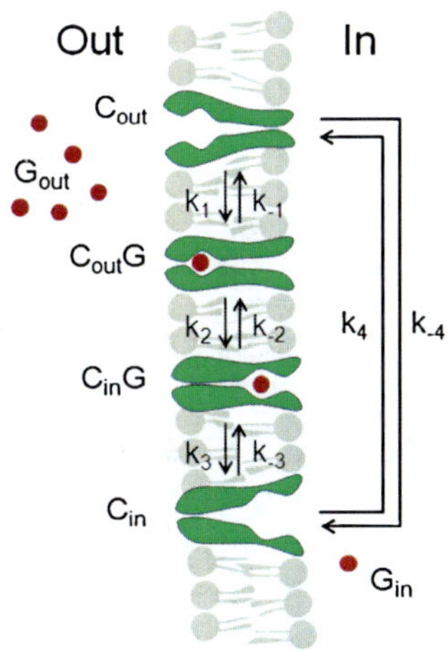

Fig. 20.3 Scheme of the alternating four-state conformational kinetics of the glucose carrier: when loaded, it can isomerise between conformations that favour glucose release to either the outer ($C_{out}G$) or the inner ($C_{in}G$) side of the membrane; in the free form, it exists in two inter-converting isomers C_{out} or C_{in} that are ready to bind glucose either of the membrane sides. The rate constants k_1 and k_{-3} define glucose binding while k_{-1} and k_3 define its dissociation from the carrier. The rate constants k_2 and k_{-2} or k_4 and k_{-4} reflect the isomerisation of the loaded or unloaded carrier, respectively. A fast isomerisation of the unloaded carrier, i.e. for k_4 and k_{-4} much larger than the other rate constants, is equivalent to a simple reversible Michaelis-Menten kinetics (Figure was adapted from Duarte et al. 2009b)

glucose transport (Barros et al. 2007), and further reduced to a simple equation relating brain to plasma glucose (Duarte et al. 2009b).

Unlike the reversible Michaelis-Menten model, this four-site exchange mechanism assumes that the free carrier after glucose release to the brain's interstice is not conformationally the same that binds glucose outside. Thus, an isomerisation equilibrium exists between the two states. The four-state conformational model that is depicted in Fig. 20.3 predicts reverse glucose transport (and thus impediment of glucose uptake by the occupied carrier, like the reversible model) and that the presence of substantial amounts of glucose in the interstice prevents the carrier from acquiring a conformation that binds to glucose from plasma.

In this case (Duarte et al. 2009b), the net velocity of transport is given by

$$v = T_{in} - T_{out} = \frac{T_{max}\left(G_{plasma} - G_{brain}\right)}{K_t + G_{plasma} + G_{brain} + \frac{G_{plasma}G_{brain}}{K_{ii}}} \tag{20.6}$$

where K_{ii} denotes the iso-inhibition constant (in mM) that reflects the ability of glucose to inhibit the translocation of carrier isoforms between the two faces of the membrane. At steady-state, a single analytical expression based on the aforementioned conformational four-state exchange model of carriers can be used to describe steady-state glucose transport across the BBB:

$$G_{brain} = V_d \frac{\left(\frac{T_{max}}{CMR_{glc}} - 1\right) G_{plasma} - K_t}{\frac{T_{max}}{CMR_{glc}} + 1 + \frac{G_{plasma}}{K_{ii}}} \quad (20.7)$$

It is apparent that when K_{ii} equals K_t, an expression formally equivalent to Eq. 20.3, i.e. standard Michaelis-Menten model, is obtained. We further note that if K_{ii} largely exceeds G_{plasma}, an expression approximating Eq. 20.5 is obtained. This means that the reversible Michaelis-Menten model is equivalent to the four-state conformational model with fast exchange between the two unloaded carrier conformations (Duarte et al. 2009b).

It was recently shown that such transport mechanism can be used describe *in vivo* brain glucose homeostasis at steady-state (Barros et al. 2007; Duarte et al. 2009b). The apparent Michaelis constant K_t was found to be on the order of a few mM (Table 20.1), as determined with the reversible model (Duarte et al. 2009b; Gruetter et al. 1998a). T_{max} was also found in the range of that determined by the other models (Duarte et al. 2009b). In this model, the so-called trans-acceleration (Carruthers 1990) is defined by the apparent iso-inhibition constant (K_{ii}). However, since K_{ii} was found to be much larger (above 50 mM) than the concentrations of brain glucose typically observed, the transport mechanism based on the reversible Michaelis-Menten kinetics can fully describe glucose transport under normal physiological conditions (Duarte et al. 2009b). In fact, the fit of experimental data with the conformational model was not statistically different with the four-state conformational model and the reversible Michaelis-Menten model previously proposed (Duarte et al. 2009b). However, in metabolic conditions where high glucose concentrations are observed (for example under eventual uncontrolled diabetes, see Duarte et al. 2009a), the inhibition constant K_{ii} may be important in describing glucose transport at the BBB.

20.5 Brain Glucose Homeostasis Depends on Transport and Consumption Rates

Brain glucose concentrations at steady-state are well defined by the ratio of T_{max} to CMR_{glc}. However, cerebral glucose consumption is regional- and activity-dependent (e.g. Linde et al. 1999; Sokoloff et al. 1977; Ueki et al. 1992), as is summarised in Table 20.2. This may compromise the evaluation of transport rates when CMR_{glc} is unknown. For example, barbiturates are known to depress energy metabolism. These anaesthetics inhibit the oxidation of NADH in the respiratory chain (Aldridge and

Table 20.2 Cerebral glucose metabolic rate (CMR_{glc}) in the rat brain, under different anaesthesia conditions. When available, data shown as mean ± SEM of n experiments. Cortex and hippocampus are depicted to illustrate regional variations of CMR_{glc}

Anaesthesia	CMR_{glc} (μmol/g/min)	References
Conscious rat	1.07 ± 0.03 to 1.57 ± 0.05 (n = 6)[a]	Sokoloff et al. 1977
	0.79 ± 0.01 (n = 6)[b]	Sokoloff et al. 1977
	1.07 ± 0.03 to 1.59 ± 0.06 (n = 15)[a]	Archer et al. 1990
	0.77 ± 0.03 to 0.91 ± 0.04 (n = 15)[b]	Archer et al. 1990
	0.66 ± 0.06 (n = 8)[c]	Linde et al. 1999
	1.03 ± 0.02 to 1.08 ± 0.03 (n = 6)[a]	Nakao et al. 2001
α-chloralose	0.44 ± 0.07 to 0.52 ± 0.09 (n = 4)[a]	Ueki et al. 1992
	0.37 ± 0.03 (n = 5)[a]	Nakao et al. 2001
	0.50 ± 0.02 (n = 5)[c]	Duarte et al. 2011
Halothane	0.67 ± 0.10 (n = 5) to 0.74 ± 0.08 (n = 4)[c]	Gjedde 1980
	0.84 ± 0.03 (n = 8)[a]	Hansen et al. 1989
	0.90 ± 0.04 (n = 8)[b]	Hansen et al. 1989
	0.65 ± 0.08 to 0.66 ± 0.09 (n = 3)[a]	Ueki et al. 1992
	0.96 ± 0.13 (n = 8)[c]	Linde et al. 1999
Isoflurane	0.54 ± 0.02 (n = 8)[a]	Hansen et al. 1989
	0.66 ± 0.02 (n = 8)[b]	Hansen et al. 1989
	0.37 ± 0.07 to 0.48 ± 0.06 (n = 3)[a]	Archer et al. 1990
	0.42 ± 0.07 to 0.63 ± 0.12 (n = 3)[b]	Archer et al. 1990
Nitrous oxide	0.64 ± 0.02 to 0.95 ± 0.04 (n = 5)[a]	Sakabe et al. 1985
	0.67 ± 0.03 (n = 5)[b]	Sakabe et al. 1985
	0.68 ± 0.05 to 0.69 ±0.04 (n = 10)[a]	Pelligrino et al. 1990
	0.46 ± 0.02 (n = 10)[b]	Pelligrino et al. 1990
	0.94 ± 0.13 to 1.07 ±0.16 (n = 3)[a]	Ueki et al. 1992
Pentobarbital	0.38 ± 0.02 to 0.53 ± 0.02 (n = 6)[a]	Sakabe et al. 1985
	0.52 ± 0.02 (n = 6)[b]	Sakabe et al. 1985
	* 0.25 ± 0.01 to 0.33 ± 0.02 (n = 5)[a]	Sakabe et al. 1985
	* 0.38 ± 0.02 (n = 5)[b]	Sakabe et al. 1985
	0.53 to 0.57[a]	Cunningham et al. 1986
	0.42[b]	Cunningham et al. 1986
	0.40 ± 0.03 to 0.55 ± 0.05 (n = 5)[a]	Archer et al. 1990
	0.36 ± 0.03 to 0.37 ± 0.05 (n = 5)[b]	Archer et al. 1990
	0.46 ± 0.06 to 0.48 ± 0.04 (n = 3)[a]	Ueki et al. 1992
	0.33 ± 0.04 (n = 10)[c]	Linde et al. 1999
	* 0.33 ± 0.03 to 0.38 ± 0.04 (n = 5)[c]	Choi et al. 2002
Thiopental	0.64 ± 0.03 to 0.81 ± 0.03 (n = 8)[a]	Sokoloff et al. 1977
	0.56 ± 0.01 (n = 8)[b]	Sokoloff et al. 1977

[a] Values obtained for cerebral cortex; [b] values for hippocampus; [c] values for the whole brain; * experiment under proved pentobarbital-induced cerebral isoelectricity

Parker 1960; Chance et al. 1963), inhibit glucose phosphorylation by hexokinase (Bielicki and Krieglstein 1976) and thus reduce CMR_{glc} (Choi et al. 2002; Linde et al. 1999; Strang and Bachelard 1973; Sokoloff et al. 1977). In addition, barbiturates also inhibit the capability of GLUT1 to transport glucose (Honkanen et al. 1995; Haspel et al. 1999; Stephenson et al. 2000). While a ^{13}C NMR spectroscopy study under deep pentobarbital anaesthesia (at a dose able to induce isoelectricity) reported CMR_{glc} of 0.33–0.38 μmol/g/min for the whole rat brain (Choi et al. 2002), CMR_{glc} was ranging from 0.42 to 0.57 μmol/g/min at a lower pentobarbital dose (Cunningham et al. 1986).

Glucose transport kinetics as measured *in vivo* by NMR spectroscopy has been mostly determined under steady-state conditions, compatible with sufficient measurement time required for glucose quantification. Although in most conditions steady-state transport models describe the glucose transport mechanism *in vivo*, they measure the ratio of the maximum transport rate to glucose consumption rate. A constant CMR_{glc} is then assumed to allow determination of T_{max} (e.g. Gruetter et al. 1998a; Duarte et al. 2009a, b; Lei et al. 2010). Gruetter et al. (1996b) measured glucose transport kinetics using ^{1}H NMR difference spectroscopy after a rapid increased in plasma glucose concentration, i.e. by displacing glucose homeostasis from steady-state equilibrium. This method had the advantage of quantifying ^{1}H NMR spectra with only glucose resonances and allowed the determination of both T_{max} and CMR_{glc} using, at that time, a standard irreversible Michaelis-Menten kinetics for the glucose carriers at the BBB (Gruetter et al. 1996b). Alternatively, upon infusion of ^{13}C-enriched glucose, dynamic ^{13}C NMR spectroscopy allows measuring the time course of signals from cerebral glucose and thus one can also determine both T_{max} and CMR_{glc} (e.g. Duarte et al. 2011).

In summary, while steady-state determination of brain glucose provides a measure of T_{max}/CMR_{glc}, the two fluxes involved in glucose homeostasis can be determined by dynamic NMR measures of glucose (Fig. 20.1) either using both ^{1}H (Gruetter et al. 1996b) and ^{13}C NMR spectroscopy (Duarte et al. 2011; Gruetter et al. 2001). These studies found a T_{max} of 0.80–1.16 μmol/g/min and CMR_{glc} of 0.32–0.35 in humans (Gruetter et al. 1996b, 2001). In the rat brain, T_{max} and CMR_{glc} were 0.91 ± 0.03 and 0.50 ± 0.02 μmol/g/min, respectively (Duarte et al. 2011).

CMR_{glc} is generally taken as invariable above euglycaemia but this assumption is not applicable at low plasma glucose levels. Under hypoglycaemia, brain glucose concentration is largely reduced to the magnitude of the K_M of hexokinase and, therefore, glucose transport across the BBB becomes rate limiting for metabolism, as occurs in GLUT1 deficiency (discussed in Barros et al. 2007). To overcome the reduced glucose supply, a situation of hypoglycaemia mobilizes brain glycogen reserves and increases cerebral blood flow (Choi et al. 2001). In line with this, chronic hypoglycaemia leads to an adaptative increase in the density of the glucose carrier GLUT1 and the rate of glucose uptake (McCall et al. 1986; Pelligrino et al. 1990; Kumagai et al. 1995; Simpson et al. 1999). Accordingly, T_{max}/CMR_{glc} in rats submitted to chronic hypoglycaemia was reported to be 62% higher than in controls (Lei and Gruetter 2006). Conversely, glucose transport kinetics at the BBB (Duarte et al. 2009a), [^{14}C]glucose uptake (Simpson et al. 1999) and brain GLUT1 or GLUT3 densities (Duelli et al. 2000; Simpson et al. 1999) are not altered by chronic hyperglycaemia. Likewise, previous studies in humans reported that poorly controlled diabetes did not affect brain glucose concentration (Seaquist et al. 2005) or glucose transport and metabolism (Fanelli et al. 1998).

Functional cerebral activation induces metabolic fluctuations that involve support from the gliovascular unit to the stimulated neurons. Acute energy demand from activated neurons leads to enhanced local glucose metabolism and, although glucose transport is also stimulated, it remains limiting for metabolism (Barros et al. 2007). Therefore, like under acute hypoglycaemia, brain activation is accompanied by reduction of brain glucose concentrations (Frahm et al. 1996; Chen et al. 1993; Merboldt et al. 1992; Mangia et al. 2007). Also, in this case, CMR_{glc} cannot be assumed constant.

References

Aldridge WN, Parker VH (1960) Barbiturates and oxidative phosphorylation. Biochem J 76:47–56

Archer DP, Elphinstone MG, Pappius HM (1990) The effect of pentobarbital and isoflurane on glucose metabolism in thermally injured rat brain. J Cereb Blood Flow Metab 10:624–630

Barros LF, Bittner CX, Loaiza A, Porras OH (2007) A quantitative overview of glucose dynamics in the gliovascular unit. Glia 55:1222–1237

Betz AL, Gilboe DD, Drewes LR (1976) The characteristics of glucose transport across the blood brain barrier and its relation to cerebral glucose metabolism. Adv Exp Med Biol 69:133–149

Bielicki L, Krieglstein J (1976) Inhibition of glucose phosphorylation in rat brain by thiopental. Naunyn Schmiedebergs. Arch Pharmacol 293:25–29

Blomqvist G, Gjedde A, Gutniak M, Grill V, Widén L, Stone-Elander S, Hellstrand E (1991) Facilitated transport of glucose from blood to brain in man and the effect of moderate hypoglycemia on cerebral glucose utilization. Eur J Nucl Med 18:834–837

Boyle PJ, Nagy RJ, O'Connor AM, Kempers SF, Yeo RA, Qualls C (1994) Adaptation in brain glucose uptake following recurrent hypoglycemia. Proc Natl Acad Sci USA 91:9352–9356

Brooks DJ, Gibbs JS, Sharp P, Herold S, Turton DR, Luthra SK, Kohner EM, Bloom SR, Jones T (1986) Regional cerebral glucose transport in insulin-dependent diabetic patients studied using [11C]3-O-Methyl-D-Glucose and positron emission tomography. J Cereb Blood Flow Metab 6:240–244

Buschiazzo PM, Terrell EB, Regen DM (1970) Sugar transport across the blood-brain barrier. Am J Physiol 219:1505–1513

Carruthers A (1990) Facilitated diffusion of glucose. Physiol Rev 70:1135–1176

Carruthers A, Helgerson AL (1991) Inhibitions of sugar transport produced by ligands binding at opposite sides of the membrane. Evidence for simultaneous occupation of the carrier by maltose and cytochalasin B. Biochemistry 30:3907–3915

Chance B, Williams GR, Hollunger G (1963) Inhibition of electron and energy transfer in mitochondria. I. Effects of Amytal, thiopental, rotenone, progesterone, and methylene glycol. J Biol Chem 238:418–431

Chen W, Novotny EJ, Zhu XH, Rothman DL, Shulman RG (1993) Localized 1H NMR measurement of glucose consumption in the human brain during visual stimulation. Proc Natl Acad Sci USA 90:9896–9900

Choi IY, Lee SP, Kim SG, Gruetter R (2001) In vivo measurements of brain glucose transport using the reversible Michaelis-Menten model and simultaneous measurements of cerebral blood flow changes during hypoglycemia. J Cereb Blood Flow Metab 21:653–663

Choi IY, Lei H, Gruetter R (2002) Effect of deep pentobarbital anesthesia on neurotransmitter metabolism in vivo: on the correlation of total glucose consumption with glutamatergic action. J Cereb Blood Flow Metab 22:1343–1351

Crone C (1965) Facilitated transfer of glucose from blood into brain tissue. J Physiol 181:103–113

Cunningham VJ (1986) The influence of transport and metabolism on brain glucose content. Ann N Y Acad Sci 481:161–173

Cunningham VJ, Hargreaves RJ, Pelling D, Moorhouse SR (1986) Regional blood-brain glucose transfer in the rat: a novel double-membrane kinetic analysis. J Cereb Blood Flow Metab 6:305–314

Cuppoletti J, Segel IH (1975) Kinetic analysis of active membrane transport systems: equations for net velocity and isotope exchange. J Theor Biol 53:125–144

de Graaf RA, Pan JW, Telang F, Lee JH, Brown P, Novotny EJ, Hetherington HP, Rothman DL (2001) Differentiation of glucose transport in human brain gray and white matter. J Cereb Blood Flow Metab 21:483–492

Duarte JMN, Carvalho RA, Cunha RA, Gruetter R (2009a) Caffeine consumption attenuates neurochemical modifications in the hippocampus of streptozotocin-induced diabetic rats. J Neurochem 111:368–379

Duarte JMN, Morgenthaler FD, Lei H, Poitry-Yamate C, Gruetter R (2009b) Steady-state brain glucose transport kinetics re-evaluated with a four-state conformational model. Front Neuroenergetics 1:6

Duarte JMN, Lanz B, Gruetter R (2011) Compartmentalised cerebral metabolism of [1,6-13C] glucose determined by in vivo 13C NMR spectroscopy at 14.1 T. Front Neuroenergetics 3:3

Duelli R, Maurer MH, Staudt R, Heiland S, Duembgen L, Kuschinsky W (2000) Increased cerebral glucose utilization and decreased glucose transporter Glut1 during chronic hyperglycemia in rat brain. Brain Res 858:338–347

Fanelli CG, Dence CS, Markham J, Videen TO, Paramore DS, Cryer PE, Powers WJ (1998) Blood-to-brain glucose transport and cerebral glucose metabolism are not reduced in poorly controlled type 1 diabetes. Diabetes 47:1444–1450

Feinendegen LE, Herzog H, Wieler H, Patton DD, Schmid A (1986) Glucose transport and utilization in the human brain: Model using carbon-11 methylglucose and positron emission tomography. J Nucl Med 27:1867–1877

Frahm J, Kruger G, Merboldt KD, Kleinschmidt A (1996) Dynamic uncoupling and recoupling of perfusion and oxidative metabolism during focal brain activation in man. Magn Reson Med 35:143–148

Gjedde A (1980) Rapid steady-state analysis of blood-brain glucose transfer in rat. Acta Physiol Scand 108:331–339

Gjedde A (1981) High- and low-affinity transport of D-glucose from blood to brain. J Neurochem 36:1463–1471

Gjedde A (1982) Calculation of cerebral glucose phosphorylation from brain uptake of glucose analogs in vivo: a re-examination. Brain Res 257:237–274

Gjedde A (1992) Blood-Brain Glucose Transfer. In: Bradbury M (ed) Physiology and pharmacology of the blood-brain barrier. Springer, New York, pp 65–117

Gjedde A, Christensen O (1984) Estimates of Michaelis-Menten constants for the two membranes of the brain endothelium. J Cereb Blood Flow Metab 4:241–249

Gjedde A, Diemer NH (1983) Autoradiographic determination of regional brain glucose content. J Cereb Blood Flow Metab 3:303–310

Gould GW, Thomas HM, Jess TJ, Bell GI (1991) Expression of human glucose transporters in Xenopus oocytes: kinetic characterization and substrate specificities of the erythrocyte, liver, and brain isoforms. Biochemistry 30:5139–5145

Grossbard L, Schimke RT (1966) Multiple hexokinases of rat tissues. Purification and comparison of soluble forms. J Biol Chem 241:3546–3560

Gruetter R, Novotny EJ, Boulware SD, Rothman DL, Mason GF, Shulman GI, Shulman RG, Tamborlane WV (1992a) Direct measurement of brain glucose concentrations in humans by 13C NMR spectorscopy. Proc Natl Acad Sci USA 89:1109–1112

Gruetter R, Rothman DL, Novotny EJ, Shulman GI, Prichard JW, Shulman RG (1992b) Detection and assignment of the glucose signal in 1H NMR spectra of the human brain. Magn Reson Med 26:183–188

Gruetter R, Novotny EJ, Boulware SD, Rothman DL, Mason GF, Shulman GI, Shulman RG, Tamborlane WV (1993) Non-invasive measurements of the cerebral steady-state glucose concentration and transport in humans by 13C magnetic resonance. In: Drewes L, Betz A (eds) Frontiers in cerebral vascular biology: transport and its regulation. Plenum, New York, pp 35–40

Gruetter R, Garwood M, Ugurbil K, Seaquist ER (1996a) Observation of resolved glucose signals in 1H NMR spectra of the human brain at 4 Tesla. Magn Reson Med 36:1–6

Gruetter R, Novotny EJ, Boulware SD, Rothman DL, Shulman RG (1996b) 1H NMR studies of glucose transport in the human brain. J Cereb Blood Flow Metab 16:427–438

Gruetter R, Ugurbil K, Seaquist ER (1998a) Steady-state cerebral glucose concentrations and transport in the human brain. J Neurochem 70:397–408

Gruetter R, Seaquist E, Kim S-W, Ugurbil K (1998b) Localized in vivo 13C NMR of glutamate metabolism. Initial results at 4 Tesla. Dev Neurosci 20:380–388

Gruetter R, Seaquist ER, Ugurbil K (2001) A mathematical model of compartmentalized neurotransmitter metabolism in the human brain. Am J Physiol Endocrinol Metab 281: E100–E112

Gutniak M, Blomqvist G, Widen L, Stone-Elander S, Hamberger B, Grill V (1990) D-[U-11C] glucose uptake and metabolism in the brain of insulin-dependent diabetic subjects. Am J Physiol 258:E805–E812

Haspel HC, Stephenson KN, Davies-Hill T, El-Barbary A, Lobo JF, Croxen RL, Mougrabi W, Koehler-Stec EM, Fenstermacher JD, Simpson IA (1999) Effects of barbiturates on facilitative glucose transporters are pharmacologically specific and isoform selective. J Membr Biol 169:45–53

Hansen TD, Warner DS, Todd MM, Vust LJ (1989) The role of cerebral metabolism in determining the local cerebral blood flow effects of volatile anesthetics: evidence for persistent flow-metabolism coupling. J Cereb Blood Flow Metab 9:323–328

Hetherington HP, Pan JW, Mason GF, Adams D, Vaughn MJ, Twieg DB, Pohost GM (1996) Quantitative H-1 spectroscopic imaging of human brain at 4.1 T using image segmentation. Magn Reson Med 36:21–29

Holden JE, Mori K, Dienel GA, Cruz NF, Nelson T, Sokoloff L (1991) Modeling the dependence of hexose distribution volumes in brain on plasma glucose concentration: implications for estimation of the local 2-deoxyglucose lumped constant. J Cereb Blood Flow Metab 11: 171–182

Honkanen RA, McBath H, Kushmerick C, Callender GE, Scarlata SF, Fenstermacher JD, Haspel HC (1995) Barbiturates inhibit hexose transport in cultured mammalian cells and human erythrocytes and interact directly with purified GLUT-1. Biochemistry 34:535–544

Inubushi T, Morikawa S, Kito K, Arai T (1993) 1H-detected in vivo 13C NMR spectroscopy and imaging at 2T magnetic field: efficient monitoring of 13C-labeled metabolites in the rat brain derived from 1-13C-glucose. Biochem Biophys Res Commun 191:866–872

Keltner JR, Wald LL, Ledden PJ, Chen YC, Matthews RT, Küestermann EH, Baker JR, Rosen BR, Jenkins BG (1998) A localized double-quantum filter for the in vivo detection of brain glucose. Magn Reson Med 39:651–656

Kreis R (1997) Quantitative localized 1H MR spectroscopy for clinical use. Prog NMR Spectroscopy 31:155–195

Kumagai AK, Kang YS, Boada RJ, Pardridge WM (1995) Upregulation of blood–brain barrier GLUT1 glucose transporter protein and mRNA in experimental chronic hypoglycemia. Diabetes 44:1399–1404

Lee WJ, Peterson DR, Sukowski EJ, Hawkins RA (1997) Glucose transport by isolated plasma membranes of the bovine blood-brain barrier. Am J Physiol 272:C1552–C1557

Lei H, Gruetter R (2006) Effect of chronic hypoglycaemia on glucose concentration and glycogen content in rat brain: a localized 13C NMR study. J Neurochem 99:260–268

Lei H, Duarte JM, Mlynarik V, Python A, Gruetter R (2010) Deep thiopental anesthesia alters steady-state glucose homeostasis but not the neurochemical profile of rat cortex. J Neurosci Res 88:413–419

Leino RL, Gerhart DZ, van Bueren AM, McCall AL, Drewes LR (1997) Ultrastructural localization of GLUT 1 and GLUT 3 glucose transporters in rat brain. J Neurosci Res 49:617–626
Linde R, Schmalbruch IK, Paulson OB, Madsen PL (1999) The Kety-Schmidt technique for repeated measurements of global cerebral blood flow and metabolism in the conscious rat. Acta Physiol Scand 165:395–401
Lund-Andersen H (1979) Transport of glucose from blood to brain. Physiol Rev 59:305–352
Maher F, Vannucci SJ, Simpson IA (1993) Glucose transporter isoforms in brain: absence of GLUT3 from the blood-brain barrier. J Cereb Blood Flow Metab 13:342–345
Mangia S, Tkác I, Gruetter R, Van de Moortele PF, Maraviglia B, Uğurbil K (2007) Sustained neuronal activation raises oxidative metabolism to a new steady-state level: evidence from 1H NMR spectroscopy in the human visual cortex. J Cereb Blood Flow Metab 27:1055–1063
Mason GF, Behar KL, Rothman DL, Shulman RG (1992) NMR determination of intracerebral glucose concentration and transport kinetics in rat brain. J Cereb Blood Flow Metab 12:448–455
McCall AL, Fixman LB, Fleming N, Tornheim K, Chick W, Ruderman NB (1986) Chronic hypoglycemia increases brain glucose transport. Am J Physiol 251:E442–E447
Merboldt KD, Bruhn H, Hänicke W, Michaelis T, Frahm J (1992) Decrease of glucose in the human visual cortex during photic stimulation. Magn Reson Med 25:187–194
Morgenthaler FD, Koski DM, Kraftsik R, Henry PG, Gruetter R (2006) Biochemical quantification of total brain glycogen concentration in rats under different glycemic states. Neurochem Int 48:616–622
Mueckler M (1994) Facilitative glucose transporters. Eur J Biochem 219:713–725
Nakao Y, Itoh Y, Kuang TY, Cook M, Jehle J, Sokoloff L (2001) Effects of anesthesia on functional activation of cerebral blood flow and metabolism. Proc Natl Acad Sci USA 98:7593–7598
Pappenheimer JR, Setchell BP (1973) Cerebral glucose transport and oxygen consumption in sheep and rabbits. J Physiol 283:529–551
Pardridge WM (1983) Brain metabolism: a perspective from the blood-brain barrier. Physiol Rev 63:1481–1535
Pardridge WM, Boado RJ, Farrell CR (1990) Brain-type glucose transporter (GLUT-1) is selectively localized to the blood-brain barrier. Studies with quantitative western blotting and in situ hybridization. J Biol Chem 265:18035–18040
Pelligrino DA, Segil LJ, Albrecht RF (1990) Brain glucose utilization and transport and cortical function in chronic vs. acute hypoglycemia. Am J Physiol 259:E729–E735
Pfeuffer J, Flogel U, Dreher W, Leibfritz D (1998a) Restricted diffusion and exchange of intracellular water – theoretical modelling and diffusion time dependence of h-1 nmr measurements on perfused glial cells. NMR Biomed 11:19–31
Pfeuffer J, Flogel U, Leibfritz D (1998b) Monitoring of cell volume and water exchange time in perfused cells by diffusion-weighted h-1 nmr spectroscopy. NMR Biomed 11:11–18
Pfeuffer J, Tkac I, Gruetter R (1998c) Diffusion weighted in vivo 1H NMR spectroscopy of brain glucose and metabolites. In: 15th annual Meeting ESMRMB, Geneva, p 36
Pfeuffer J, Tkác I, Gruetter R (2000) Extracellular-intracellular distribution of glucose and lactate in the rat brain assessed noninvasively by diffusion-weighted 1H nuclear magnetic resonance spectroscopy in vivo. J Cereb Blood Flow Metab 20:736–746
Poitry-Yamate C, Lei H, Gruetter R (2009) The rate-limiting step for glucose transport into the hypothalamus is across the blood-hypothalamus interface. J Neurochem 109(s1):38–45
Sakabe T, Tsutsui T, Maekawa T, Ishikawa T, Takeshita H (1985) Local cerebral glucose utilization during nitrous oxide and pentobarbital anesthesia in rats. Anesthesiology 63:262–266
Schimke RT, Grossbard L (1968) Studies on isozymes of hexokinase in animal tissues. Ann N Y Acad Sci 151:332–350
Seaquist ER, Damberg GS, Tkac I, Gruetter R (2001) The effect of insulin on in vivo cerebral glucose concentrations and rates of glucose transport/metabolism in humans. Diabetes 50:2203–2209

Seaquist ER, Tkac I, Damberg G, Thomas W, Gruetter R (2005) Brain glucose concentrations in poorly controlled diabetes mellitus as measured by high-field magnetic resonance spectroscopy. Metabolism 54:1008–1013

Shockley RP, LaManna JC (1988) Determination of rat cerebral cortical blood volume changes by capillary mean transit time analysis during hypoxia, hypercapnia and hyperventilation. Brain Res 454:170–178

Siesjo BK (1978) Utilisation of substrates by brain tissues. In: Brain energy metabolism. Wiley, New York, pp 101–130

Silver IA, Erecinska M (1994) Extracellular glucose concentration in mammalian brain: Continues monitoring of changes during increased neuronal activity and upon limitation in oxygen supply during normo-, hypo- and hyperglycemic animals. J Neurosci 14:5068–5076

Simpson IA, Appel NM, Hokari M, Oki J, Holman GD, Maher F, Koehler-Stec EM, Vannucci SJ, Smith QR (1999) Blood-brain barrier glucose transporter: effects of hypo- and hyperglycemia revisited. J Neurochem 72:238–247

Simpson IA, Carruthers A, Vannucci SJ (2007) Supply and demand in cerebral energy metabolism: the role of nutrient transporters. J Cereb Blood Flow Metab 27:1766–1791

Sokoloff L, Reivich M, Kennedy C, Des Rosiers MH, Patlak CS, Pettigrew KD, Sakurada O, Shinohara M (1977) The [14C]deoxyglucose method for the measurement of local cerebral glucose utilization: theory, procedure, and normal values in the conscious and anesthetized albino rat. J Neurochem 28:897–916

Sprengers ED, Koenderman AH, Staal GE (1983) Mitochondrial and cytosolic hexokinase from rat brain: one and the same enzyme? Biochim Biophys Acta 755:112–118

Stephenson KN, Croxen RL, El-Barbary A, Fenstermacher JD, Haspel HC (2000) Inhibition of glucose transport and direct interactions with type 1 facilitative glucose transporter (GLUT-1) by etomidate, ketamine, and propofol: a comparison with barbiturates. Biochem Pharmacol 60:651–659

Strang RH, Bachelard HS (1973) Rates of cerebral glucose utilization in rats anaesthetized with phenobarbitone. J Neurochem 20:987–996

Ueki M, Mies G, Hossmann KA (1992) Effect of alpha-chloralose, halothane, pentobarbital and nitrous oxide anesthesia on metabolic coupling in somatosensory cortex of rat. Acta Anaesthesiol Scand 36:318–322

van Zijl PC, Moonen CT, Faustino P, Pekar J, Kaplan O, Cohen JS (1991) Complete separation of intracellular and extracellular information in NMR spectra of perfused cells by diffusion-weighted spectroscopy. Proc Natl Acad Sci USA 88:3228–3232

van Zijl PCM, Chesnick AS, DesPres D, Moonen CTW, Ruiz-Cabello J, Van Gelderen P (1993) In vivo proton spectroscopy and spectroscopic imaging of [1-13C]-glucose and its metabolic products. Magn Reson Med 30:544–551

van Zijl PC, Davis D, Eleff SM, Moonen CT, Parker RJ, Strong JM (1997) Determination of cerebral glucose transport and metabolic kinetics by dynamic MR spectroscopy. Am J Physiol 273:E1216–E1227

Vannucci SJ, Maher F, Simpson IA (1997) Glucose transporter proteins in brain: delivery of glucose to neurons and glia. Glia 21:2–21

Chapter 21
Glucose Metabolism During Neural Activation

Marianne Fillenz

Abstract The energy consumption of the brain is derived from the oxidation of glucose. The basal oxygen/glucose ratio (O/G) in the brain is 5.5. While 90% of glucose is oxidised, a small fraction undergoes glycolysis and gives rise to lactate. The difference between the delivery of oxygen and its utilisation is the oxygen extraction fraction (OEF). Although brain regions differ in their energy consumption under basal conditions there is little or no variation in the OEF. On activation there is an increase in energy production and a decrease in the O/G and an increase in lactate. There are various theories about the nature of the changes during the increased energy production; one of these is the astrocyte-neuron-lactate-shuttle theory according to which neurones use lactate rather than glucose for their increased energy production. An understanding of the causal connections between the changes requires the elucidation of their temporal relationships. In the present review the *in vivo* changes during activation of oxygen, glucose, glycogen and lactate are examined. The evidence suggests that the additional lactate is not used locally as a substrate for neuronal metabolism. It is proposed that the substantial rise in extracellular lactate is not derived from the uptake of synaptically released glutamate but the reuptake of glutamate released from astrocytes. The release of glutamate is the result of the stimulation of metabotropic glutamate receptors on the astrocytic membrane. Various roles have been proposed for the astrocytic glutamate. The lactate resulting from the reuptake of the astrocytic glutamate has no further function and is removed by diffusion or release into the circulation.

Keywords Astrocytes • Glucose • Glutamate • Lactate • Oxygen

M. Fillenz, DSc. (✉)
Department of Physiology, Anatomy and Genetics, Oxford University, Oxford, UK
e-mail: marianne.fillenz@dpag.ox.ac.uk

I.-Y. Choi and R. Gruetter (eds.), *Neural Metabolism In Vivo*,
Advances in Neurobiology 4, DOI 10.1007/978-1-4614-1788-0_21,

21.1 Introduction

Early estimates of the brain's energy consumption based on the Kety and Schmidt method and on the Fick principle demonstrated that the brain accounts for 20% of energy consumption, measured as its oxygen utilisation, although it represents 2% of body weight. At least 75% of the oxygen is consumed by neurones with glial cells consuming a much smaller percentage (Siesjö 1978). Glucose is the principal source of energy for the brain. Most of the glucose is oxidised to CO_2 and H_2O *in vivo*. The oxygen/glucose ratio (O/G) with complete oxidation is 6; expressed as a percentage, the normal value is around 90%, a minor fraction of up to 15% being converted to lactic acid (Cohen et al. 1964). The function of the brain is the transmission of information, which is mediated by the propagation of action potentials and the setting up of synaptic potentials by neurotransmitters. Both these processes involve the movement of ions, along electrochemical gradients. The maintenance of these electrochemical gradients is the major source of the brain's energy requirements (Ames 2000; Erecinska and Silver 1989, 1994; Silver and Erecinska 1997). The most important enzyme in the restoration of ionic gradients is Na/K ATPase. Inhibition of Na/K ATPase by ouabain produces a substantial reduction in oxygen consumption. Using anatomical and physiological data, the energy expenditure of different components of excitatory signalling in the grey matter of rodent brain has been calculated. According to these calculations the energy consumptions are 47% for action potentials, 34% for synaptic potentials, maintenance of the resting potential 13% and glutamate recycling only 3% (Attwell and Laughlin 2001). The figures for the primate cortex are 52% for excitatory postsynaptic potentials, 33% for action potentials, 5% for transmitter release and 4% for transmitter recycling (Lennie 2003). Glutamate is the main excitatory transmitter. The rate of glutamate neurotransmitter cycling can be calculated using nuclear resonance magnetic (NMR) spectroscopy (Sibson et al. 1997). The flux of ^{13}C label from 1-^{13}C glucose into glutamate provides an *in vivo* measurement of the tricarboxylic acid rate from which oxidative glucose consumption can be derived (Mason et al. 1995). Under mild anaesthesia more than 80% of total glucose oxidation accounts for glutamatergic synaptic activity (Sibson et al. 1998). The main inhibitory transmitter is γ-amino-butyric acid (GABA). A number of attempts have been made to estimate the contribution of GABA transmission to energy metabolism. With the use of proton resonance spectroscopy (MRS) the estimated contribution to neuronal glucose oxidation of GABA was calculated to be 18% as against 82% for glutamate in rats under halothane anaesthesia (Patel et al. 2005). As with glutamate, the action of GABA is terminated by GABA transporters, which belong to a Na+/Cl^--dependent family of neurotransmitter transporters (Gadea and Lopez-Colome 2001). GABA uptake stimulates energy consuming Na/K/ATPase. GABA transporter-evoked Na influx is very brief, produces very little activation of the Na/K pump, and does not lead to an increase in glucose uptake or a reduction in ATP (Chatton et al. 2003). Thus inhibition is not metabolically demanding (Waldfogel et al. 2000).

The changes in local blood flow that accompany changes in brain activity were reported well over a 100 years ago (Roy and Sherrington 1890). These result in increased glucose delivery to the brain. It is only much more recently that it has been shown that the accompanying increase in oxygen utilisation is smaller than the increase in blood flow and glucose delivery (Raichle 1998). This disparity in glucose delivery and oxygen consumption is expressed as the O/G ratio and is the basis of some of the techniques used to monitor neuronal activity. There are a number of such techniques, both direct and indirect (Dale and Halgren 2001).

All these techniques measure changes compared to a baseline. One problem is the definition of such a baseline. An attempt has been made to define the baseline based on brain metabolism. The O/G ratio measured under basal conditions expresses the fact that most, but not all, the glucose delivered to the brain by the blood is completely oxidised. With increased activity in the brain the increase in glucose delivery of glucose is not matched by a similar increase in oxygen utilisation. The difference between oxygen delivery and oxygen utilisation has been defined as the oxygen extraction fraction (OEF). Positron emission tomography (PET) is the only technique that can measure blood flow and oxygen utilisation in absolute figures. The change in OEF measured by PET can therefore be used as a measure of the level of neuronal activity. Although there are wide differences in the energy consumption between grey matter and white matter in the brain and between different brain areas, measurement of the OEF with PET in human subjects, lying quietly with eyes closed, revealed that the OEF shows remarkable uniformity across brain regions. This uniform OEF is an index of the intrinsic activity of the brain and has been called its default activity (Gusnard and Raichle 2001; Raichle and Snyder 2007; Raichle 2001). This intrinsic activity is also present under anaesthesia (Vincent et al. 2007). The default activity includes rhythmic activity monitored with either functional magnetic resonance imaging (fMRI) or electro-encephalography (EEG). The significance of this rhythmic activity, previously regarded as noise, has only recently become appreciated (Fox and Raichle 2007). With EEG, neuronal networks in the mammalian forebrain demonstrate several oscillatory bands covering frequencies from 0.5 to 500n Hz. Different frequencies are associated with different neuronal functions (Buzsaki and Draguhn 2004). These rhythmic activities are the outcome of intrinsic cellular conductances and local circuit properties.

Superimposed on this complex pattern of basal activity are the changes that occur in response to neuronal activation. These often consist of changes in the pattern of activity brought about by changes in the membership of cell assemblies as well as changes in the net level of activity (Buzsaki 2006; Buzsáki et al. 2007). The increase in energy consumption, although only around 5%, has a different pattern from basal metabolism.

In what follows the evidence for the changes in oxygen, glucose, lactate and glycogen during neuronal activation is examined.

21.2 Astrocyte-Neurone-Lactate-Shuttle (ANLS)

There have been a number of theories about the nature of the metabolism of activated neurons. One such model is the astrocyte-neurone-lactate-shuttle (ANLS) theory (Magistretti et al. 1999; Pellerin et al. 1998). This theory proposes that in an activated brain area the main metabolic substrate for neuronal metabolism is lactate not glucose. According to this model glutamate released from presynaptic terminals is taken up by astrocytes, which is an energy-requiring process. It therefore leads to an increased uptake of glucose into astrocytes, where it undergoes glycolysis, results in the formation of lactate, which is exported and taken up by neurones. Here it is converted to pyruvate and enters the tricarboxylic cycle to form ATP. This theory has been challenged by a number of authors on a number of grounds (Chih et al. 2001; Chih and Roberts 2003; Dienel and Cruz 2003, 2004; Dienel and Hertz 2001; Fillenz 2005; Korf 2006; Leegsma-Vogt et al. 2003).

The ANLS theory is based almost exclusively on *in vitro* experiments, that completely lack the temporal dimension. In order to elucidate the causal connection between the changes during neuronal activation their temporal sequence has to be established, which can only be provided by *in vivo* experiments.

21.3 Time Course of Oxygen Changes

Activation of a brain area is triggered by the arrival of the afferent input and the initiation of postsynaptic potentials in the neurones of the area. These electrophysiological events occur on a millisecond time scale (Korf and Gramsbergen 2007). However, because they involve the influx of Na^+ they disturb the ionic equilibrium and lead to the immediate activation of the enzymes that restore these equilibria. The most important of these is Na/K ATPase, which restores the Na^+ equilibrium. The increase in oxygen and glucose consumption results in changes in the extracellular compartment, which is the immediate source of these energy substrates. A number of techniques have been used for measuring brain tissue oxygen (Ndubuizu and LaManna 2007). Simultaneous measurement of tissue oxygen and single cell neuronal activity in the cat visual cortex with a dual-purpose microelectrode showed a biphasic response in oxygen tension: an immediate decrease, which coincided with the neuronal discharge, followed by an increase. The highest neuronal firing frequency produced the largest decrease and the smallest delayed increase in tissue oxygen. The rise in tissue oxygen occurred over a larger area including areas where no spike discharge was recorded; this rise in tissue oxygen is attributed to an increase in local cerebral blood flow. These results suggest two competing mechanisms - increased consumption and increased delivery – which partially overlap in time, and as a result, each response reduces the size of the other (Thompson et al. 2003). In similar experiments measurements of tissue oxygen were combined with recording of local field potentials and cerebral blood flow using Laser-Doppler flowmetry during electrical stimulation of climbing fibres of the Purkinje cells of the rat cerebellum. Increased synaptic activity was accompanied by an increase in oxygen consumption and a decrease in tissue oxygen, followed by an increase in both cerebral blood flow and tissue oxygen. Inhibitors of neuronal nitric oxide synthase enhance the initial reduction of tissue oxygen and attenuate the increase in blood flow and oxygen increase. This confirms the partial overlap between the increase in oxygen consumption and the increase in oxygen delivery that accompanies the increased cerebral blood flow (Offenhauser et al. 2005). A number of studies using similar techniques have confirmed that increased oxygen consumption precedes the increase in cerebral blood flow (Ances et al. 2001a, b; Masamoto et al. 2003).

This pattern has also been confirmed by the use of oxygen-dependent phosphorescence (Vanzetta and Grinvald 1999), and the use of optical imaging (Jones et al. 2001; Mayhew et al. 2000) as well as the use of fMRI, which is based on the ratio of reduced to oxyhaemoglobin, and shows an initial negative dip which signals an increase in reduced haemoglobin (Duong et al. 2000; Yacoub et al. 1999, 2001). Others however, have failed to find this 'initial dip' (Lindauer et al. 2001; Marota et al. 1999; Silva et al. 1999). A number of explanations have been offered for this discrepancy (Buxton 2001; Vanzetta and Grinvald 2001).

There is some disagreement concerning the role of the increase in cerebral blood flow in the delivery of oxygen. According to one view there are negligible levels of oxygen in brain tissue and the increase in blood flow is necessary for the delivery of oxygen (Buxton and Frank 1997; Hyder et al. 1998; Vafaee and Gjedde 2000). This model is contradicted by the demonstration that there is an increase in oxygen consumption that precedes the increase in blood flow (see above). When changes in cerebral blood flow are suppressed by using a vasodilator, neuronal activation leads to a decrease in tissue oxygen tension that is linearly related to the summation of evoked field potentials. This suggests that an increase in cerebral blood flow (CBF) is not required for an increase in oxygen consumption (Masamoto et al. 2009; Vazquez et al. 2008).

21.4 Changes in Glucose

There has been some debate concerning the basal concentration of glucose in the extracellular fluid (ECF). Calculations based on total glucose content of the brain have suggested figures between 2 and 2.5 mM (Silver and Erecinska 1994). Microdialysate concentrations are very much lower but the glucose concentration in microdialysate does not represent the true extracellular concentration, since it depends on the *in vivo* recovery. With the zero-net flux method (Lönnroth et al. 1987), the true extracellular concentration, using microdialysis, can be determined. This has yielded values for extracellular glucose concentration in freely moving rats of between 0.350 mM (Fray et al. 1997) and 0.500 mM (Fellows et al. 1992) for the striatum, 1.0 mM for the hippocampus (McNay and Gold 1999) and 0.6 mM for the cortex (Netchiporouk et al. 2001). Using an implanted glucose sensor ECF concentration was 2.5 mM in the hippocampus (Hu and Wilson 1997b) and 0.350 mM in the striatum (Lowry et al. 1998b). ECF glucose concentration varies with species of rat and brain region (McNay and Gold 1999). Values for extracellular glucose in human subjects range from 1.7 to 2.1 mM (Abi-Saab et al. 2002; Langemann et al. 2001; Reinstrup et al. 2000). The values for extracellular glucose represent the balance between delivery and utilisation. Drugs that increase neuronal activity, such as veratridine, lead to a decrease, and drugs that decrease neuronal activity, such as tetrodotoxin (TTX) and anaesthetics, to an increase extracellular glucose (Canal et al. 2005; Fellows et al. 1992; Netchiporouk et al. 2001; Osborne et al. 1997). This may account for the higher values for human subjects, which are all determined under anaesthesia mM (Abi-Saab et al. 2002; Langemann et al. 2001; Reinstrup et al. 2000).

The development of the 2-deoxyglucose (2DG) technique by Sokoloff (Sokoloff 1981), led to the demonstration that physiologically increased neuronal activity was accompanied by an increased uptake of 2DG (Sokoloff 1980). The distribution of labelled 2DG was mapped ex vivo using autoradiography. These experiments demonstrated that 2DG uptake was absent in dorsal root ganglia, which have no synapses, and was maximal in brain areas rich in synaptic complexes. The spatial resolution was insufficient to determine whether the uptake occurred into astrocytes or neurones. *In vitro* experiments with retina and cultured astrocytes all suggested that glucose was taken up by astrocytes rather than neurones (Tsacopoulos and Magistretti 1996). However, a recent microautoradiographic study demonstrated that under basal conditions around half the ^{14}C-2DG was taken up into neurones and half into astrocytes; uptake during activation has not yet been measured (Nehlig and Coles 2007; Nehlig et al. 2004; Wittendorp-Rechenmann et al. 2002).

Using PET with fluoro-18-deoxy-glucose, it is possible to measure the uptake of glucose *in vivo*. There is general agreement that activation leads to a reduction in O/G ratio, although the degree of reduction varied between studies depending on the nature of the stimuli for neuronal activation (Dienel 2004b). Similar results were obtained in experiments on rats in which arterio-venous differences were measured (Madsen et al. 1998, 1999; Schmalbruch et al. 2002). The rate of glucose utilisation using radiolabelled glucose is always lower than the rate measured using 2DG. The reason is that 2DG does not proceed beyond phosphorylation by hexokinase and the resulting non-metabolisable phosphorylated products are trapped in the tissue. Labelled glucose, by contrast, gives rise to many labelled compounds, which are not retained in activated tissue. Measurement of cerebral metabolic rates of glucose (CMRglc) in parallel studies using 2DG and [6-^{14}C] glucose demonstrated an underestimate of 50% CMRglc with labelled glucose (Cruz et al. 2007).

These techniques have spatial but no temporal resolution. With the use of microdialysis it became possible to study the time course of changes in glucose in response to physiological activation. With this technique, a 5 min period of immobilisation stress led to a 10% reduction in glucose followed by a rise to 125% of basal by 15 min after the end of the stress (van der Kuil and Korf 1991). Application of 5 min tail pinch caused a brief decrease in striatal extracellular glucose, during the period of stimulation, followed by a delayed increase with a return to basal levels, over a period of 30 min (Fray et al. 1996). During a spatial memory task in the rat, hippocampal glucose decreased by between 11% and 30%, depending on the complexity of the task (McNay et al. 2000, 2001).

Similar results were obtained with an implanted enzyme-based glucose sensor; this technique has a high time resolution, since glucose measurement is continuous. Application of the 5 min tail pinch showed an immediate decrease of 5% glucose during stimulation and a delayed, long lasting increase of 11% after the end of stimulation (Lowry and Fillenz 1997). A 5 s electrical stimulation at 40 Hz of the perforant path produced a 20% reduction of glucose lasting 10 s in the dentate gyrus of the rat hippocampus; the decrease was followed by a 20% increase in glucose lasting around 100 s (Hu and Wilson 1997b).

These changes in glucose raise the question of the source of extracellular glucose. Although in the rat the level of glucose in the blood is 10 times higher than in the extracellar compartment of the brain, there seems to be no rapid transfer of glucose between the two compartments. In the behavioural experiments an intraperitoneal injection of 250 mg/kg glucose produced no change in the basal level of extracellular glucose but completely blocked the decrease produced by the behavioural task (McNay et al. 2000). This suggests that the systemically administered glucose entered a store outside the extracellular compartment, from where it could be recruited for neuronal use. Intravenous injection of glucose also produces only a very slow increase in extracellular glucose (Bequet et al. 2001).

The 5-min tail pinch produced a biphasic change in glucose: an early decrease accompanied by an increase blood flow (signalled by the increase in tissue oxygen) and a delayed, prolonged increase in glucose. The early increase in blood flow fails to prevent the decrease in ECF glucose, whereas the delayed increase in ECF glucose occurs when regional cerebral blood flow (rCBF) has returned to basal levels (Lowry and Fillenz 1997). Infusion of veratridine through a dialysis probe produces a steep fall in extracellular glucose (Lowry et al. 1998a). Using the zero-net flux method of Lönnroth, one can determine the true change in extracellular concentration as well as the *in vivo* recovery, which measures the balance between supply and utilisation of glucose. By using a modification of the zero-net-flux method (Olson and Justice 1993; Smith et al. 1992) the time course of changes of *in vivo* recovery and thus changes in the utilisation or turnover of glucose can be measured. Since astrocytes derive glucose from the blood stream the increased turnover of extracellular glucose measured by this technique is an index of neuronal utilisation of glucose. The infusion of veratridine causes a steep increase in turnover rate followed by a decrease, as the availability of glucose has been exhausted when the extracellular level falls to zero (Fray et al. 1997). Again blood glucose appears to be unable to compensate for the rapid depletion of extracellular glucose. This suggests that there is no direct uptake from the blood vascular system into the extracellular compartment.

Astrocytic glycogen is a possible site for the storage of the systemically administered glucose, as well as the source of the late rise in glucose, following neuronal activation.

Studies with cultured astrocytes and brain slices have demonstrated that a number of neurotransmitters, positively coupled to adenylate cyclase, cause glycogen breakdown (Hertz et al. 1998; Sorg and Magistretti 1991). In these *in vitro* experiments, glycogenolysis resulted in the release of lactate (Dringen et al. 1993).

Striatal astrocytes have a high density of $ß_2$ adrenoceptors (Sporn and Molinoff 1976), whose stimulation leads to an increase in intracellular cyclic adenosine monophosphate (cAMP) (Segal et al. 1981). A 5 min infusion of 50 μM isoprenaline, a $ß_2$ receptor agonist, produces an increase in extracellular glucose; this increase is blocked if the β_2 antagonist, propranolol, included in the perfusion medium. An increase in glucose is also produced by the addition of 50 μM 8br-cAMP, a membrane permeable analog of cAMP (Fray et al. 1996). In the presence of propranolol, the delayed increase in glucose in response to tail pinch, seen with microdialysis, is abolished (Fray et al. 1996). Infusion of propranolol through a microdialysis probe implanted next to the glucose sensor completely abolished the delayed rise and produces a greater and more prolonged decrease in glucose during the period of stimulation (Fillenz and Lowry 1998). This supports the hypothesis that the late rise in glucose is derived from the breakdown of glycogen.

The nucleus of the locus coeruleus, which provides the noradrenergic innervation of the forebrain, has a tonic discharge whose frequency is increased by sensory stimulation (Aston-Jones and Cohen 2005). The late increase in glucose is therefore likely to be due to an increase in the release of noradrenaline, stimulation of astrocytic β_2 adrenoceptors and the rise of extracellular glucose similar to that seen the infusion of isoprenaline.

The export of glucose resulting from glycogenolysis requires glucose-6-phosphatase. There is considerable controversy concerning the role of this enzyme in brain. Immunocytochemical techniques demonstrated the presence of glucose-6-phosphatase in brain astrocytes (Bell et al. 1993). More recently it has been shown that the brain glucose-6-phosphatase is an isophorm, which differs from the enzyme found in liver and kidney (Guionie et al. 2003; Middleditch et al. 1997). What is open to question is the degree of *in vivo* activity of the enzyme. A quantitative estimation of *in vivo* dephosphorylation of glucose-6-phosphate in rat brain based on the difference in the rates ^{3}H and ^{14}C utilisation produced a rate of dephosphorylation, which was 35% of the rate of glucose utilisation (Huang and Veech 1986). This finding was challenged by a later study where the results were attributed to contamination of the precursor pool and incomplete recovery of end products (Dienel et al. 1988). Experiments with cultured astrocytes also produced evidence for dephosphorylation in one study (Forsyth et al. 1996), but failed to do so in another (Gotoh et al. 2001). No explanation was found for this difference. The hypothesis of glycogen as the source of extracellular glucose needs further evidence for the activity of glucose-6-phosphatase *in vivo*.

21.5 Changes in Lactate

Using PET during brain activation in humans first revealed the uncoupling of cerebral blood flow and oxidative metabolism (Fox and Raichle 1986; Fox et al. 1988). This implied that some of the glucose was undergoing glycolysis, although there was no direct evidence at this stage for the formation of lactate. In animal experiments using microdialysis, various forms of mild stress including immobilisation, handling and cold exposure were found to lead to a rise in dialysate lactate, which reached a peak ranging from 20% to 30% above baseline, 6–7 min after the onset of a stress of 5 min duration (Schasfoort et al. 1988). In another study a 5-min tail pinch caused an immediate rise in lactate, which continued beyond the end of the period of stimulation and reached a peak of 179% of baseline in the striatum and 190% of baseline in the hippocampus. This was followed by a very slow return to baseline levels over a period of 40 min (Fellows et al. 1993; Fray et al. 1996). The use of MRS in human subjects confirmed the increase in lactate during neuronal activation (Prichard et al. 1991; Sappey-Marinier et al. 1992).

Neuronal activation is accompanied by changes in dihydronicotinamide adenine dinucleotide (NADH), an index of oxidative metabolism and glycolysis. Electrical stimulation of hippocampal slices leads to a biphasic change in NADH (Brennan et al. 2006, 2007; Foster et al. 2005; Kasischke et al. 2004; Shuttleworth et al. 2003). The initial decrease in NADH coincides with the decrease in tissue oxygen and is localised to neurons. It is taken to represent oxidative phosphorylation in neuronal mitochondria (Kasischke et al. 2004). The later increase in NADH has been localised to astrocytes, but although attributed to glycolysis in astrocytes by some (Kasischke et al. 2004), its significance is still controversial (Brennan et al. 2006, 2007; Kann and Kovács 2007). The NADH overshoot, which follows the initial NADH decrease, starts 10 s after the onset of stimulation and reaches a peak at 30 s; it is still above basal 500 s after the onset of the stimulus. The glycolytic nature of the overshoot is corroborated by analytical measurement of the concurrent increase in lactate (Kasischke et al. 2004).

These experiments demonstrate a rise in lactate that extends well beyond the period of increased neuronal activity but provide no evidence for the subsequent fate of lactate. For neuronal activity to be sustained, the electrochemical gradient for Na^+, decreased by the influx of Na that accompanies both synaptic potentials and action potentials, has to be restored which means the immediate activation of the energy-dependent Na/K-ATPase-mediated Na^+ extrusion. The energy supply required by such immediate activation is demonstrated by the increase in oxygen utilisation and the decrease in NADH. It has also been confirmed by PET measurements of oxidative metabolism in human visual cortex (Gjedde and Marrett 2001). If lactate contributed to energy production this would be signalled by a decrease, as is the case with glucose, not an increase in the extracellular level of lactate. There are only two reports of activity-related decreases in lactate. In one study using time-resolved proton magnetic resonance spectroscopy, changes in lactate were recorded in the human visual cortex. A 1 s stimulation was followed by a reduction in lactate 5 s after the end of stimulation; there was no increase in lactate (Mangia et al. 2003). In the second study changes in glucose and lactate resulting from electrical stimulation of the perforant path were measured with sensors implanted in the dentate gyrus of the rat brain. A 5 s period of electrical stimulation led to an immediate 7% reduction in lactate lasting 10–12 s after the end of stimulation followed by an increase to 200% which remained above basal levels for 14 min (Hu and Wilson 1997a).

Two important questions remain: one is the origin of the rise in lactate and the other is the fate of the released lactate. There is at present no agreed explanation for the increase in lactate. Early experiments with cultured mouse cerebral cortical astrocytes demonstrated that the uptake of glutamate stimulated the activity of Na/K/ATPase, increased the utilisation of glucose and increased the release of lactate

(Pellerin and Magistretti 1994, 1996, 1997). These findings were the basis of the ANLS theory (Magistretti and Pellerin 1999; Magistretti et al. 1999). The hypothesis that release of lactate was a consequence of the uptake of glutamate was confirmed in *in vivo* experiments. Infusion of glutamate through the dialysis probe led to an increase in lactate (Demestre et al. 1997), and local infusion of inhibitors of glutamate transporters reduce (Fray et al. 1996), or abolish (Demestre et al. 1997), the stimulation-induced rise in lactate.

There are alternatives to the view that the rise in lactate is due to the uptake of glutamate released from presynatptic terminals. Astrocytes have complex three-dimensional structures (Bushong et al. 2002; Ventura and Harris 1999) and contain nearly all the glycogen in the brain (Pfeiffer et al. 1992). Astrocytes have a number of functions, which require energy; in addition to the uptake of glutamate, astrocytes buffer extracellular K^+, synthesise glutamate and degrade both glutamate and GABA. The narrow astrocytic surface extensions called peripheral astrocytic processes or PAPs are too narrow to contain mitochondria and their sources of energy are glycolysis and glycogenolysis. Glycogenolysis provides three molecules of ATP against the two molecules provided by glycolysis. The end product of both glycolysis and glycogenolysis is lactate, which has to be cleared to avoid feedback inhibition. The end product of energy metabolism of PAPs is therefore another possible source of extracellular lactate (Hertz et al. 2007). A recent paper suggests another possible source of lactate. Electrical stimulation of climbing fibres of cerebellar Purkinje cells produced a 30% rise in lactate. AMPA receptor blockade, which abolished evoked postsynaptic activity, blood flow increases and increases in oxygen consumption, abolished this rise in lactate. This suggests, although it does not prove, that lactate may have been derived from neuronal metabolism (Caesar et al. 2008). The derivation of lactate from neuronal glycolysis has also been suggested by experiments with hippocampal slices (Schurr and Payne 2007) and stimulation of the human visual cortex (Gjedde and Marrett 2001).

The next question is the fate of the released lactate. The ANLS hypothesis postulates that lactate is taken up by neurones as their energy substrate. The greater increase in glucose consumption than oxygen utilisation during neuronal activation indicates that if lactate is produced it is not oxidised by neighbouring cells. Similarly the large underestimation of metabolic activation by labelled glucose compared with deoxyglucose (Ackermann and Lear 1989; Collins et al. 1987) indicates that the diffusible products of glucose metabolism, of which lactate is the most important, are rapidly cleared from the activated tissue (Cruz et al. 2007). The labelled lactate could diffuse in the extracellular compartment, be lost to the blood stream or diffuse through the syncitial astrocytic network coupled by gap junctions (Ball et al. 2007). The reduction of the spatial spread of labelled metabolites by gap junction blockers provides evidence for the latter pathway (Cruz et al. 2007).

There is extensive evidence, which argues against the theory that during physiological activation neurones use lactate rather than glucose for their energy requirements. There is however also evidence that neurones are able to use lactate and do so under a number of circumstances. Thus the use of lactate during reoxygenation after hypoxia fuels recovery (Schurr 2006). Similarly in animal models of brain injury lactate has a neuroprotectice role (Ros et al. 2001).

21.6 Glycogen

The role of glycogen in maintaining brain function is at present not clear. Glycogen represents the largest store of glucose equivalents in the brain and is present in concentrations that exceed by several times total brain glucose (Gruetter 2003). Recent studies suggest that previous values of 3 μmol/g had greatly underestimated brain glycogen content. Measurements of brain glycogen are critically dependent on the combined effect of tissue extraction and the level of stimulation of the animal. The values for rat brain glycogen concentration obtained with novel extraction methods are 8–12 μmol/g (Cruz and Dienel 2002). However, a technique that avoids the difficulties of extraction methods is the recently developed *in vivo* ^{13}C NMR spectroscopy; this has yielded values for total brain glycogen in the awake normoglycaemic rat of 3.3 ± 0.8 μmol/g (Choi and Gruetter 2003). Under basal conditions turnover rates of total glycogen are very slow; measurements using *in vivo* NMR give turnover times of 7–10 h (Choi and Gruetter 2003; Gruetter 2003). However, the use of an auto-radiographic method to label brain glycogen with ^{14}C glucose demonstrated increased utilisation of glycogen in response to sensory stimulation. Tactile stimulation of the rat face and vibrissae for a period of 10 min led to a loss of ^{14}C-labelled glycogen in the contralateral sensory cortex (Swanson et al. 1992). This demonstrated a link between neuronal activity and local glycogen utilisation. Since glycogen is restricted to astrocytes, the authors were the first to suggest a close coupling between glial energy metabolism and neuronal activity. More recently, glycogen was measured in ethanol-insoluble pellets after labelling with ^{14}C glucose. This technique gives the high resting values for brain glycogen. After 60 s of generalised sensory stimulation there was a 16% decrease in labelled glycogen but no change in total quantity of glycogen (Cruz and Dienel 2002). This agrees with the view that there is an outer layer of labile glycogen surrounding a core of stable glycogen (Watanabe and Passonneau 1973). In a recent paper glycogen utilisation during sensory stimulation of unanaesthetised rats was re-examined. Glycogen utilization was assessed by change in concentration, release of ^{14}C from pre-labelled glycogen and increase in utilization of blood glucose (CMR (glc)) in the presence of a glycogen phosphorylase inhibitor. Glycogen level fell, ^{14}C release increased and in the presence of the inhibitor of glycogenolysis, there was a regionally selective compensatory increases in CMR(glc) over and above the activation-induced rise in vehicle-treated rats. The contribution of glycogen to metabolism varied with brain structures. The metabolic fate of glycogen is unknown but adding glycogen to the fuel consumed during activation would further add to the reduction in oxygen/glucose ratio (Dienel et al. 2007).

One hypothesis for the role of glycogen in neuronal activity is the hypothesis of the glycogen shunt (Shulman et al. 2001a, b). According to this hypothesis, during high frequency neuronal firing, the energy produced by glycolysis of glucose derived from blood is insufficient for clearing synaptically released glutamate; it is therefore supplemented by energy derived from glycogen breakdown. This is less efficient method of ATP production since glycogen synthesis uses on molecule of ATP; as a result, excess lactate is produced, which is effluxed into the blood circulation. This mechanism is taken to account for the decrease in the oxygen-to-glucose index (O/G) during neuronal activation.

An alternative role for glycogen is as a source of energy in the laminae of the fine astrocytic processes that surround synaptic complexes. These processes are 50–100 nm wide and contain glycogen granules but no mitochondria (Derouiche and Frotscher 2001). It is suggested that the local astrocytic energy demands associated with rapid excitatory transmission such as K^+ uptake, Na^+-dependent glutamate transport glutamate and glutamine synthesis are met by glycogenolysis and the production of lactate (Dienel and Cruz 2003) (see above).

A further function for astrocytic glycogen is consolidation of memory based on experiments with chicks. These studies demonstrate that learning is dependent on glycogenolysis stimulated by noradrenaline, glycogen acting as a source of glutamate (Gibbs et al. 2006, 2008).

Finally astrocytic glycogen is a possible site for the storage of the systemically administered glucose, which prevents the decrease in extracellular glucose during behavioural experiments (Gibbs et al. 2006; McNay et al. 2001), as well as the source of the late rise in glucose, following neuronal activation (Lowry and Fillenz 1997) (see above).

21.7 Astrocytic Release of Glutamate

There is widespread agreement that neuronal activation leads to a decrease in the O/G ratio, which is accompanied by an increase in extracellular lactate. That there is no increase in oxygen consumption corresponding to the increase in lactate means that the lactate is not used locally but is dispersed throughout the brain or discharged into the blood stream. It also implies that more glucose is taken up than is required for local energy requirements. Since much, if not all the lactate, results from the uptake of glutamate, it raises the question of the origin of this glutamate. According to the ANLS hypothesis the glutamate taken up is the glutamate released from presynaptic terminals. The astrocytes surrounding the synaptic complex represent a very efficient uptake mechanism whose role is to terminate the action of the released glutamate, prevent desensitisation of the receptors and enable the next action potential to exert its effect (Marcaggi and Attwell 2004). There has been extensive electrophysiological investigation to determine how much overflow there is between adjacent synapses. Such studies have demonstrated that glutamate spillover is variable and very limited (Barbour 2001; Diamond 2001, 2002; Kullmann et al. 1999; Kullmann and Asztely 1998, 2000; Piet et al. 2004; Rusakov 2001). In spite of this, basal concentration of glutamate in the extracellular compartment varies between 2 and 3 μM (Baker et al. 2002; Day et al. 2006; Herrera-Marschitz et al. 1996; Miele et al. 1996a). This concentration is increased by physiological stimulation (Miele et al. 1996b; Moghaddam 1993). Addition of TTX to the perfusion medium, which produces dramatic reductions in the concentration of amine transmitters, has no effect on basal glutamate levels (Biggs et al. 1995; Timmerman and Westerink 1997) or stimulated release of glutamate (Miele et al. 1996b). This shows that the extracellular glutamate does not represent impulse-dependent release of glutamate.

Recent evidence has revealed release of stimulated glutamate from a source other than presynaptic nerve terminals. Astrocytes have on their surface membrane a number of G-protein coupled receptors, including metabotropic glutamate receptors, whose stimulation leads to rise of $[Ca^{2+}]_i$, released from intracellular stores (Cornell-Bell et al. 1990). Astrocytes are coupled by gap junctions and form networks through which Ca^{2+} waves propagate. Such Ca^{2+} waves can be set up in cultured astrocytes by mild mechanical stimulation and a number of neurotransmitters (Verkhratsky and Kettenmann 1996). In acutely isolated hippocampal slices loaded with a Ca^{2+}-sensitive dye the response of astrocytes to electrical stimulation was monitored with confocal microscopy (Porter and McCarthy 1996). Stimulation of Schaffer collaterals leads to an increase of Ca^{2+} in astrocytes of the stratum radiatum, identified by the astrocytic marker glial fibrillary acidic protein (GFAP). The astrocytic response is blocked by TTX, and by the selective glutamate metabotropic receptor antagonist, alpha-methyl-4-carboxyphenylglycie (MCPG). These results show that the astrocytic response is triggered by glutamate released from the terminals of the Schaffer collaterals that acts on glutamate metabotropic receptors on the astrocytic membrane. A number of other neurotransmitters, which include Acetylcholine, Histamine and Adrenaline, also lead to a rise in astrocytic Ca^2 (Araque et al. 2002; Muyderman et al. 2001; Shelton and McCarthy 2000). There have been numerous studies of the Ca^{2+} waves and their method of propagation (Scemes and Giaume 2006). Since all these experiments were carried out *in vitro* it was important to discover whether this phenomenon also occurred *in vivo*. Using topical application of the Ca^2-sensitive dye Fluo-4 AM and two-photon laser scanning microscopy on the cerebral cortex of anaesthetised rats, cytosolic calcium fluctuation in astrocyte populations of superficial cortical layers were monitored. Spontaneous Ca^{2+} events in individual astrocytes were similar to those observed *in vitro*. Increased neuronal discharge was associated with increased astrocytic Ca^{2+} activity in individual cells and a robust coordination of Ca^{2+} signals in neighbouring astrocytes (Hirase et al. 2004). Using the same technique, in a more recent study, whisker stimulation in adult mice evoked increases in astrocytic cytosolic Ca^{2+} in the barrel cortex. Changes in Ca^{2+} were a direct function of the sum of local field potentials, occurred seconds after the onset of neuronal activity and were inhibited by metabotropic glutamate receptor antagonists. Iontophoresis of AMPA and NMDA receptor antagonists, which produced a marked reduction in the excitatory postsynaptic currents, had no effect on the increases in Ca^{2+}. This implied that the rises in Ca^{2+} were triggered by glutamate release from presynaptic terminals (Wang et al. 2006).

Experiments with astrocytes in culture have shown that the rise in astrocytic Ca^{2+} leads to the release of glutamate (Pasti et al. 1997, 2001). Although early evidence was indirect and depended on the effects on the response of neurones co-cultured with the astrocytes, there are now direct measurements of glutamate release from cultured astrocytes (Bezzi et al. 1998; Jeftinija et al. 1996; Jeremic et al. 2001). This Ca^{2+}-dependent release shares many of the characteristics of vesicular exocytosis in nerve terminals (Araque et al. 2000; Bergersen and Gundersen 2009; Hepp et al. 1999; Montana et al. 2006). A number of possible alternative mechanisms for glutamate release have been suggested (Malarkey and Parpura 2008; Parpura et al. 2004).

The spatial spread of Ca^{2+} waves and of glutamate release is measured using the Ca^{2+} indicator fluo-3 and an enzyme-linked assay for glutamate. In confluent astrocytic cultures a mechanical stimulus, applied to a single astrocyte, sets up a Ca^{2+} wave, which travels at 10–30 μm/s. The same stimulus produces a wave of glutamate-dependent fluorescence, which propagates at around 26 μm/s and causes a localised accumulation of glutamate of 1–100 μM. A comparison of the spread of glutamate ejected from a pipette with that released from astrocytes demonstrates that the astrocytic release of glutamate is regenerative (Innocenti et al. 2000).

There has been intensive investigation of the various mechanisms for astrocytic glutamate release and extensive speculation of the role of the glutamate released from astrocytes (Araque et al. 1998a, b, 1999a; Haydon 2001; Santello and Volterra 2008; Volterra and Steinhauser 2004). Less consideration has been given to the fate of the released glutamate and its metabolic consequences. There is however one elegant demonstration in cultured astrocytes that electrical or mechanical stimulation of one astrocyte set up Ca^{2+} and Na^{+} waves. The Ca^{2+} waves have a higher propagation velocity than the Na^{+} waves. Na^{+} waves are strongly inhibited by glutamate uptake inhibitors or drugs that reduce the extracellular concentration of glutamate. The Na^{+} wave is closely paralleled by a wave of increased glucose uptake (Bernardinelli et al. 2004). It would appear therefore that stimulation of astrocytes, which *in vivo* results from the action of synaptically released glutamate on metabotropic glutamate receptors, sets up Ca^{2+} waves, which in turn lead to the release of glutamate. This glutamate is taken up by astrocytes. Such glutamate uptake stimulates Na/K ATPase, leads to the uptake of glucose, glycolysis and the release of lactate. This corresponds to the delayed release of lactate that accompanies neuronal activation. It occurs too late to contribute to neuronal metabolism, has no further function and is dispersed in the various ways discussed above. Further evidence that excess lactate derived from astrocytic uptake of glutamate does not contribute to neuronal metabolism comes from the finding that local injection of antisense oligonucleotides, targeted to the glial high affinity glutamate transporter (GLAST), reduces glucose uptake by 50% but leaves neuronal activity unaffected (Cholet et al. 2001).

It would appear therefore that the uptake of glucose in excess of oxygen consumption serves to support the uptake of glutamate released from astrocytes. The glutamate released from astrocytes has a variety of functions, which includes modulation of ion channels in postsynaptic neurones and control of transmitter release by action on presynaptic receptors (Araque et al. 1998a, b, 1999a, b; Araque and Perea 2004; Carmignoto and Fellin 2006; Dienel 2004a; Fellin et al. 2006; Haydon 2001; Newman and Volterra 2004; Ni et al. 2007; Volterra and Steinhauser 2004; Zhang and Haydon 2005), but has to be removed in order to protect neurones from excitotoxicity. The resulting lactate on the other hand has no further function and is removed from the activated brain area.

21.8 Conclusion

Evidence from *in vivo* experiments, which demonstrates the temporal sequence of events, suggests that under physiological conditions neuronal activation leads to an increase in the uptake and utilisation of glucose. The additional neuronal energy supply is derived from the oxidative phosphorylation of glucose. However the uptake of glucose in excess of oxygen utilisation represents the release of lactate derived from the uptake of glutamate released from astrocytes. This astrocytic glutamate modulates synaptic transmission but the lactate resulting from the reuptake of this glutamate has no further function.

References

Abi-Saab WM, Maggs DG, Jones T, Jacob R, Srihari V, Thompson J, Kerr D, Leone P, Krystal JH, Spencer DD, During MJ, Sherwin RS (2002) Striking differences in glucose and lactate levels between brain extracellular fluid and plasma in conscious human subjects: effects of hyperglycemia and hypoglycemia. J Cereb Blood Flow Metab 22:271–279

Ackermann RF, Lear JL (1989) Glycolysis-induced discordance between glucose metabolic rates measured with radiolabeled fluorodeoxyglucose and glucose. J Cereb Blood Flow Metab 9:774–785

Ames A 3rd (2000) CNS energy metabolism as related to function. Brain Res Brain Res Rev 34:42–68

Ances BM, Buerk DG, Greenberg JH, Detre JA (2001a) Temporal dynamics of the partial pressure of brain tissue oxygen during functional forepaw stimulation in rats. Neurosci Lett 306:106–110

Ances BM, Wilson DF, Greenberg JH, Detre JA (2001b) Dynamic changes in cerebral blood flow, O_2 tension, and calculated cerebral metabolic rate of O_2 during functional activation using oxygen phosphorescence quenching. J Cereb Blood Flow Metab 21:511–516

Araque A, Perea G (2004) Glial modulation of synaptic transmission in culture. Glia 47:241–248

Araque A, Parpura V, Sanzgiri RP, Haydon PG (1998a) Glutamate-dependent astrocyte modulation of synaptic transmission between cultured hippocampal neurons. Eur J Neurosci 10:2129–2142

Araque A, Sanzgiri RP, Parpura V, Haydon PG (1998b) Calcium elevation in astrocytes causes an NMDA receptor-dependent increase in the frequency of miniature synaptic currents in cultured hippocampal neurons. J Neurosci 18:6822–6829

Araque A, Parpura V, Sanzgiri RP, Haydon PG (1999a) Tripartite synapses: glia, the unacknowledged partner. Trends Neurosci 22:208–215

Araque A, Sanzgiri RP, Parpura V, Haydon PG (1999b) Astrocyte-induced modulation of synaptic transmission. Can J Physiol Pharmacol 77:699–706

Araque A, Li N, Doyle RT, Haydon PG (2000) SNARE protein-dependent glutamate release from astrocytes. J Neurosci 20:666–673

Araque A, Martin ED, Perea G, Arellano JI, Buno W (2002) Synaptically released acetylcholine evokes Ca^{2+} elevations in astrocytes in hippocampal slices. J Neurosci 22:2443–2450

Aston-Jones G, Cohen JD (2005) An integrative theory of locus coeruleus-norepinephrine function: adaptive gain and optimal performance. Annu Rev Neurosci 28:403–450

Attwell D, Laughlin SB (2001) An energy budget for signaling in the grey matter of the brain. J Cereb Blood Flow Metab 21:1133–1145

Baker DA, Xi ZX, Shen H, Swanson CJ, Kalivas PW (2002) The origin and neuronal function of in vivo nonsynaptic glutamate. J Neurosci 22:9134–9141

Ball KK, Gandhi G, Thrash J, Cruz NF, Dienel GA (2007) Astrocytic connexin distributions and rapid, extensive dye transfer via gap junctions in the inferior colliculus: Implications for [^{14}C] glucose metabolite trafficking. J Neurosci Res 85:3267–3283

Barbour B (2001) An evaluation of synapse independence. J Neurosci 21:7969–7984

Bell JE, Hume R, Busuttil A, Burchell A (1993) Immunocytochemical detection of the microsomal glucose-6-phosphatase in human brain astrocytes. Neuropathol Appl Neurobiol 19:429–435

Bequet F, Gomez Merino D, Berthelot M, Guezennec CY (2001) Exercise-induced changes in brain glucose and serotonin revealed by microdialysis in rat hippocampus: effect of glucose supplementation. Acta Physiol Scand 173:223–230

Bergersen LH, Gundersen V (2009) Morphological evidence for vesicular glutamate release from astrocytes. Neuroscience 158:260–265

Bernardinelli Y, Magistretti PJ, Chatton JY (2004) Astrocytes generate Na^+-mediated metabolic waves. Proc Natl Acad Sci USA 101:14937–14942

Bezzi P, Carmignoto G, Pasti L, Vesce S, Rossi D, Rizzini BL, Pozzan T, Volterra A (1998) Prostaglandins stimulate calcium-dependent glutamate release in astrocytes. Nature 391:281–285

Biggs CS, Fowler LJ, Whitton PS, Starr MS (1995) Impulse-dependent and tetrodotoxin-sensitive release of GABA in the rat's substantia nigra measured by microdialysis. Brain Res 684:172–178

Brennan AM, Connor JA, Shuttleworth CW (2006) NAD(P)H fluorescence transients after synaptic activity in brain slices: predominant role of mitochondrial function. J Cereb Blood Flow Metab 26:1389–1406

Brennan AM, Connor JA, Shuttleworth CW (2007) Modulation of the amplitude of NAD(P)H fluorescence transients after synaptic stimulation. J Neurosci Res 85:3233–3243

Bushong EA, Martone ME, Jones YZ, Ellisman MH (2002) Protoplasmic astrocytes in CA1 stratum radiatum occupy separate anatomical domains. J Neurosci 22:183–192

Buxton RB (2001) The elusive initial dip. NeuroImage 13:953–958

Buxton RB, Frank LR (1997) A model for the coupling between cerebral blood flow and oxygen metabolism during neural stimulation. J Cereb Blood Flow Metab 17:64–72

Buzsaki G (2006) Rhythms of the Brain. Oxford University Press, Oxford

Buzsaki G, Draguhn A (2004) Neuronal oscillations in cortical networks. Science 304:1926–1929

Buzsáki G, Kaila K, Raichle M (2007) Inhibition and brain work. Neuron 56:771–783

Caesar K, Hashemi P, Douhou A, Bonvento G, Boutelle MG, Walls AB, Lauritzen M (2008) Glutamate receptor-dependent increments in lactate, glucose and oxygen metabolism evoked in rat cerebellum in vivo. J Physiol (Lond) 586:1337–1349

Canal CE, McNay EC, Gold PE (2005) Increases in extracellular fluid glucose levels in the rat hippocampus following an anesthetic dose of pentobarbital or ketamine-xylazine: an in vivo microdialysis study. Physiol Behav 84:245–250

Carmignoto G, Fellin T (2006) Glutamate release from astrocytes as a non-synaptic mechanism for neuronal synchronization in the hippocampus. J Physiol 99:2–3

Chatton JY, Pellerin L, Magistretti PJ (2003) GABA uptake into astrocytes is not associated with significant metabolic cost: implications for brain imaging of inhibitory transmission. Proc Natl Acad Sci USA 100:12456–12461

Chih CP, Roberts EL Jr (2003) Energy substrates for neurons during neural activity: a critical review of the astrocyte-neuron lactate shuttle hypothesis. J Cereb Blood Flow Metab 23:1263–1281

Chih CP, Lipton P, Roberts EL Jr (2001) Do active cerebral neurons really use lactate rather than glucose? Trends Neurosci 24:573–578

Choi IY, Gruetter R (2003) In vivo ^{13}C NMR assessment of brain glycogen concentration and turnover in the awake rat. Neurochem Int 43:317–322

Cholet N, Pellerin L, Welker E, Lacombe P, Seylaz J, Magistretti PJ, Bonvento G (2001) Local injection of antisense oligonucleotides targeted to the glial glutamate transporter GLAST decreases the metabolic response to somatosensory activation. J Cereb Blood Flow Metab 21:404–412

Cohen PJH, Wollman SC, Alexander PE, Behar MG (1964) Cerebral carbohydrate metabolism in man during halothane anaesthesia. Anaesthesiology 25:185–191

Collins RC, McCandless DW, Wagman IL (1987) Cerebral glucose utilization: comparison of [^{14}C] deoxyglucose and [6-^{14}C]glucose quantitative autoradiography. J Neurochem 49:1564–1570

Cornell-Bell AH, Finkbeiner SM, Cooper MS, Smith JS (1990) Glutamate induces calcium waves in cultured astrocytes: long-range glial signaling. Science 247:470–473

Cruz NF, Dienel GA (2002) High glycogen levels in brains of rats with minimal environmental stimuli: implications for metabolic contributions of working astrocytes. J Cereb Blood Flow Metab 22:1476–1489

Cruz NF, Ball KK, Dienel GA (2007) Functional imaging of focal brain activation in conscious rats: impact of [(14)C]glucose metabolite spreading and release. J Neurosci Res 85: 3254–3266

Dale AM, Halgren E (2001) Spatiotemporal mapping of brain activity by integration of multiple imaging modalities. Curr Opin Neurobiol 11:202–208

Day BK, Pomerleau F, Burmeister JJ, Huettl P, Gerhardt GA (2006) Microelectrode array studies of basal and potassium-evoked release of L-glutamate in the anesthetized rat brain. J Neurochem 96:1626–1635

Demestre M, Boutelle M, Fillenz M (1997) Stimulated release of lactate in freely moving rats is dependent on the uptake of glutamate. J Physiol 499:825–832

Derouiche A, Frotscher M (2001) Peripheral astrocyte processes: monitoring by selective immunostaining for the actin-binding ERM proteins. Glia 36:330–341

Diamond JS (2001) Neuronal glutamate transporters limit activation of NMDA receptors by neurotransmitter spillover on CA1 pyramidal cells. J Neurosci 21:8328–8338

Diamond JS (2002) A broad view of glutamate spillover. Nat Neurosci 5:291–292

Dienel G (2004a) Glial biology: functional interactions among glia and neurons. Neurochem Int 45:189–190

Dienel GA (2004b) Lactate muscles its way into consciousness: fueling brain activation. Am J Physiol 287:R519–R521

Dienel GA, Cruz NF (2003) Neighborly interactions of metabolically-activated astrocytes in vivo. Neurochem Int 43:339–354

Dienel GA, Cruz NF (2004) Nutrition during brain activation: does cell-to-cell lactate shuttling contribute significantly to sweet and sour food for thought? Neurochem Int 45:321–351

Dienel GA, Hertz L (2001) Glucose and lactate metabolism during brain activation. J Neurosci Res 66:824–838

Dienel GA, Nelson T, Cruz NF, Jay T, Crane AM, Sokoloff L (1988) Over-estimation of glucose-6-phosphatase activity in brain in vivo. Apparent difference in rates of [2-3H]glucose and [U-^{14}C]glucose utilization is due to contamination of precursor pool with ^{14}C-labeled products and incomplete recovery of ^{14}C-labeled metabolites. J Biol Chem 263:19697–19708

Dienel GA, Schmidt K, Cruz NF (2007) Astrocyte activation in vivo during graded photic stimulation. J Neurochem 103:1506–1522

Dringen R, Gebhardt R, Hamprecht B (1993) Glycogen in astrocytes: possible function as lactate supply for neighboring cells. Brain Res 623:208–214

Duong TQ, Kim DS, Ugurbil K, Kim SG (2000) Spatiotemporal dynamics of the BOLD fMRI signals: toward mapping submillimeter cortical columns using the early negative response. Magnetic Resonance in Medicine 44:231–242

Erecinska M, Silver IA (1989) ATP and brain function. J Cereb Blood Flow Metab 9:2–19

Erecinska M, Silver IA (1994) Ions and energy in mammalian brain. Prog Neurobiol 43:37–71

Fellin T, Pascual O, Haydon PG (2006) Astrocytes coordinate synaptic networks: balanced excitation and inhibition. Physiology 21:208–215

Fellows LK, Boutelle MG, Fillenz M (1992) Extracellular brain glucose levels reflect local neuronal activity: a microdialysis study in awake, freely moving rats. J Neurochem 59:2141–2147

Fellows LK, Boutelle MG, Fillenz M (1993) Physiological stimulation increases nonoxidative glucose metabolism in the brain of the freely moving rat. J Neurochem 60:1258–1263

Fillenz M (2005) The role of lactate in brain metabolism. Neurochem Int 47:413–417

Fillenz M, Lowry JP (1998) Studies of the source of glucose in the extracellular compartment of the rat brain. Dev Neurosci 20:365–368

Forsyth RJ, Bartlett K, Eyre J (1996) Dephosphorylation of 2-deoxyglucose 6-phosphate and 2-deoxyglucose export from cultured astrocytes. Neurochem Int 28:243–250

Foster KA, Beaver CJ, Turner DA (2005) Interaction between tissue oxygen tension and NADH imaging during synaptic stimulation and hypoxia in rat hippocampal slices. Neuroscience 132:645–657

Fox PT, Raichle ME (1986) Focal physiological uncoupling of cerebral blood flow and oxidative metabolism during somatosensory stimulation in human subjects. Proc Natl Acad Sci USA 83:1140–1144

Fox MD, Raichle ME (2007) Spontaneous fluctuations in brain activity observed with functional magnetic resonance imaging. Nat Rev Neurosci 8:700–711

Fox PT, Raichle ME, Mintun MA, Dence C (1988) Nonoxidative glucose consumption during focal physiologic neural activity. Science 241:462–464

Fray AE, Forsyth RJ, Boutelle MG, Fillenz M (1996) The mechanisms controlling physiologically stimulated changes in rat brain glucose and lactate: a microdialysis study. J Physiol 496:49–57

Fray AE, Boutelle M, Fillenz M (1997) Extracellular glucose turnover in the striatum of unanaesthetized rats measured by quantitative microdialysis. J Physiol 504:721–726

Gadea A, Lopez-Colome AM (2001) Glial transporters for glutamate, glycine, and GABA III. Glycine transporters. J Neurosci Res 64:218–222

Gibbs ME, Anderson DG, Hertz L (2006) Inhibition of glycogenolysis in astrocytes interrupts memory consolidation in young chickens. Glia 54:214–222

Gibbs ME, Hutchinson D, Hertz L (2008) Astrocytic involvement in learning and memory consolidation. Neurosci Biobehav Rev 32:927–944

Gjedde A, Marrett S (2001) Glycolysis in neurons, not astrocytes, delays oxidative metabolism of human visual cortex during sustained checkerboard stimulation in vivo. J Cereb Blood Flow Metab 21:1384–1392

Gotoh J, Kuang TY, Nakao Y, Cohen DM, Melzer P, Itoh Y, Pak H, Pettigrew K, Sokoloff L (2001) Regional differences in mechanisms of cerebral circulatory response to neuronal activation. Am J Physiol 280:H821–H829

Gruetter R (2003) Glycogen: the forgotten cerebral energy store. J Neurosci Res 74:179–183

Guionie O, Clottes E, Stafford K, Burchell A (2003) Identification and characterisation of a new human glucose-6-phosphatase isoform. FEBS Lett 551:159–164

Gusnard DA, Raichle ME (2001) Searching for a baseline: functional imaging and the resting human brain. Nat Rev Neurosci 2:685–694

Haydon PG (2001) GLIA: listening and talking to the synapse. Nat Rev Neurosci 2:185–193

Hepp R, Perraut M, Chasserot Golaz S, Galli T, Aunis D, Langley K, Grant NJ (1999) Cultured glial cells express the SNAP-25 analogue SNAP-23. Glia 27:181–187

Herrera-Marschitz M, You ZB, Goiny M, Meana JJ, Silveira R, Godukhin OV, Chen Y, Espinoza S, Pettersson E, Loidl CF, Lubec G, Andersson K, Nylander I, Terenius L, Ungerstedt U (1996) On the origin of extracellular glutamate levels monitored in the basal ganglia of the rat by in vivo microdialysis. J Neurochem 66:1726–1735

Hertz L, Peng L, Lai JCK (1998) Functional studies in cultured astrocytes. Methods - A companion to Method Enzymol 16:293–310

Hertz L, Peng L, Dienel GA (2007) Energy metabolism in astrocytes: high rate of oxidative metabolism and spatiotemporal dependence on glycolysis/glycogenolysis. J Cereb Blood Flow Metab 27:219–249

Hirase H, Qian L, Barthó P, Buzsáki G (2004) Calcium dynamics of cortical astrocytic networks in vivo. Plos Biol 2:0494–0499

Hu Y, Wilson GS (1997a) A temporary local energy pool coupled to neuronal activity: fluctuations of extracellular lactate levels in rat brain monitored with rapid-response enzyme-based sensor. J Neurochem 69:1484–1490

Hu Y, Wilson GS (1997b) Rapid changes in local extracellular rat brain glucose observed with an in vivo glucose sensor. J Neurochem 68:1745–1752

Huang MT, Veech RL (1986) Glucose-6-phosphatase activity in brain. Science 234:1128–1129

Hyder F, Shulman RG, Rothman DL (1998) A model for the regulation of cerebral oxygen delivery. J Appl Physiol 85:554–564

Innocenti B, Parpura V, Haydon PG (2000) Imaging extracellular waves of glutamate during calcium signaling in cultured astrocytes. J Neurosci 20:1800–1808

Jeftinija SD, Jeftinija KV, Stefanovic G, Liu F (1996) Neuroligand-evoked calcium-dependent release of excitatory amino acids from cultured astrocytes. J Neurochem 66:676–684

Jeremic A, Jeftinija K, Stevanovic J, Glavaski A, Jeftinija S (2001) ATP stimulates calcium-dependent glutamate release from cultured astrocytes. J Neurochem 77:664–675

Jones M, Berwick J, Johnston D, Mayhew J (2001) Concurrent optical imaging spectroscopy and laser-Doppler flowmetry: the relationship between blood flow, oxygenation, and volume in rodent barrel cortex. NeuroImage 13:1002–1015

Kann O, Kovács R (2007) Mitochondria and neuronal activity. Am J Physiol, Cell Physiol 292:C641–C657

Kasischke KA, Vishwasrao HD, Fisher PJ, Zipfel WR, Webb WW (2004) Neural activity triggers neuronal oxidative metabolism followed by astrocytic glycolysis. Science 305:99–103

Korf J (2006) Is brain lactate metabolized immediately after neuronal activity through the oxidative pathway? J Cereb Blood Flow Metab 26:1584–1586

Korf J, Gramsbergen JB (2007) Timing of potential and metabolic brain energy. J Neurochem 103:1697–1708

Kullmann DM, Asztely F (1998) Extrasynaptic glutamate spillover in the hippocampus: evidence and implications. Trends Neurosci 21:8–14

Kullmann DM, Asztely F (2000) Spillover and synaptic cross talk mediated by glutamate and GABA in the mammalian brain. Extrasynaptic glutamate spillover in the hippocampus: evidence and implications. Prog Brain Res 125:339–351

Kullmann D, Min M, Asztely F, Rusakov D (1999) Extracellular glutamate diffusion determines the occupancy of glutamate receptors at CA1 synapses in the hippocampus. Philos Trans R Soc Lond, B, Biol Sci 354:395–402

Langemann H, Alessandri B, Mendelowitsch A, Feuerstein T, Landolt H, Gratzl O (2001) Extracellular levels of glucose and lactate measured by quantitative microdialysis in the human brain. Neurol Res 23:531–536

Leegsma-Vogt G, Venema K, Korf J (2003) Evidence for a lactate pool in the rat brain that is not used as an energy supply under normoglycemic conditions. J Cereb Blood Flow Metab 23:933–941

Lennie P (2003) The cost of cortical computation. Curr Biol 13:493–497

Lindauer U, Royl G, Leithner C, Kühl M, Gold L, Gethmann J, Kohl-Bareis M, Villringer A, Dirnagl U (2001) No evidence for early decrease in blood oxygenation in rat whisker cortex in response to functional activation. NeuroImage 13:988–1001

Lönnroth P, Jansson PA, Smith U (1987) A microdialysis method allowing characterization of interecellular water space in humans. Am J Physiol 253:652–654

Lowry JP, Fillenz M (1997) Evidence for uncoupling of oxygen and glucose utilization during neuronal activation in rat striatum. J Physiol 498:497–501

Lowry JP, Demestre M, Fillenz M (1998a) Relation between cerebral blood flow and extracellular glucose in rat striatum during mild hypoxia and hyperoxia. Dev Neurosci 20:52–58

Lowry JP, O'Neill RD, Boutelle MG, Fillenz M (1998b) Continuous monitoring of extracellular glucose concentrations in the striatum of freely moving rats with an implanted glucose biosensor. J Neurochem 70:391–396

Madsen PL, Linde R, Hasselbalch SG, Paulson OB, Lassen NA (1998) Activation-induced resetting of cerebral oxygen and glucose uptake in the rat. J Cereb Blood Flow Metab 18:742–748

Madsen PL, Cruz NF, Sokoloff L, Dienel GA (1999) Cerebral oxygen/glucose ratio is low during sensory stimulation and rises above normal during recovery: excess glucose consumption during stimulation is not accounted for by lactate efflux from or accumulation in brain tissue. J Cereb Blood Flow Metab 19:393–400

Magistretti PJ, Pellerin L (1999) Cellular mechanisms of brain energy metabolism and their relevance to functional brain imaging. Philos Trans R Soc Lond, B, Biol Sci 354:1155–1163

Magistretti PJ, Pellerin L, Rothman DL, Shulman RG (1999) Energy on demand. Science 283: 496–497

Malarkey EB, Parpura V (2008) Mechanisms of glutamate release from astrocytes. Neurochem Int 52:142–154

Mangia S, Garreffa G, Bianciardi M, Giove F, Di Salle F, Maraviglia B (2003) The aerobic brain: lactate decrease at the onset of neural activity. Neuroscience 118:7–10

Marcaggi P, Attwell D (2004) Role of glial amino acid transporters in synaptic transmission and brain energetics. Glia 47:217–225

Marota JJ, Ayata C, Moskowitz MA, Weisskoff RM, Rosen BR, Mandeville JB (1999) Investigation of the early response to rat forepaw stimulation. Magn Reson Med 41:247–252

Masamoto K, Omura T, Takizawa N, Kobayashi H, Katura T, Maki A, Kawaguchi H, Tanishita K (2003) Biphasic changes in tissue partial pressure of oxygen closely related to localized neural activity in guinea pig auditory cortex. J Cereb Blood Flow Metab 23:1075–1084

Masamoto K, Vazquez A, Wang P, Kim SG (2009) Brain tissue oxygen consumption and supply induced by neural activation: determined under suppressed hemodynamic response conditions in the anesthetized rat cerebral cortex. Adv Exp Med Biol 645:287–292

Mason GF, Gruetter R, Rothman DL, Behar KL, Shulman RG, Novotny EJ (1995) Simultaneous determination of the rates of the TCA cycle, glucose utilization, alpha-ketoglutarate/glutamate exchange, and glutamine synthesis in human brain by NMR. J Cereb Blood Flow Metab 15:12–25

Mayhew J, Johnston D, Berwick J, Jones M, Coffey P, Zheng Y (2000) Spectroscopic analysis of neural activity in brain: increased oxygen consumption following activation of barrel cortex. NeuroImage 12:664–675

McNay EC, Gold PE (1999) Extracellular glucose concentrations in the rat hippocampus measured by zero-net-flux: effects of microdialysis flow rate, strain, and age. J Neurochem 72:785–790

McNay EC, Fries TM, Gold PE (2000) Decreases in rat extracellular hippocampal glucose concentration associated with cognitive demand during a spatial task. Proc Natl Acad Sci USA 97:2881–2885

McNay EC, McCarty RC, Gold PE (2001) Fluctuations in brain glucose concentration during behavioral testing: dissociations between brain areas and between brain and blood. Neurobiol Learn Mem 75:325–337

Middleditch C, Marcolongo P, Benedetti A, Burchell A (1997) Glucose-6-phosphatase activity: a determinant of brain microsomal intactness. Biochem Soc Trans 25:183S

Miele M, Berners M, Boutelle MG, Kusakabe H, Fillenz M (1996a) The determination of the extracellular concentration of brain glutamate using quantitative microdialysis. Brain Res 707:131–133

Miele M, Boutelle MG, Fillenz M (1996b) The source of physiologically stimulated glutamate efflux from the striatum of conscious rats. J Physiol 497:745–751

Moghaddam B (1993) Stress preferentially increases extraneuronal levels of excitatory amino acids in the prefrontal cortex: comparison to hippocampus and basal ganglia. J Neurochem 60:1650–1657

Montana V, Malarkey EB, Verderio C, Matteoli M, Parpura V (2006) Vesicular transmitter release from astrocytes. Glia 54:700–715

Muyderman H, Angehagen M, Sandberg M, Bjorklund U, Olsson T, Hansson E, Nilsson M (2001) Alpha 1-adrenergic modulation of metabotropic glutamate receptor-induced calcium oscillations and glutamate release in astrocytes. J Biol Chem 276:46504–46514

Ndubuizu O, LaManna JC (2007) Brain tissue oxygen concentration measurements. Antioxid Redox Signal 9:1207–1219

Nehlig A, Coles J (2007) Cellular pathways of energy metabolism in the brain: is glucose used by neurons or astrocytes? Glia 55:1238–1250

Nehlig A, Wittendorp-Rechenmann E, Lam CD (2004) Selective uptake of [^{14}C]2-deoxyglucose by neurons and astrocytes: high-resolution microautoradiographic imaging by cellular ^{14}C-trajectography combined with immunohistochemistry. J Cereb Blood Flow Metab 24:1004–1014

Netchiporouk L, Shram N, Salvert D, Cespuglio R (2001) Brain extracellular glucose assessed by voltammetry throughout the rat sleep-wake cycle. Eur J Neurosci 13:1429–1434

Newman EA, Volterra A (2004) Glial control of synaptic function. Glia 47:207–208

Ni Y, Malarkey E, Parpura V (2007) Vesicular release of glutamate mediates bidirectional signaling between astrocytes and neurons. J Neurochem 103:1273–1284

Offenhauser N, Thomsen K, Caesar K, Lauritzen M (2005) Activity-induced tissue oxygenation changes in rat cerebellar cortex: interplay of postsynaptic activation and blood flow. J Physiol 565:279–294

Olson RJ, Justice JB Jr (1993) Quantitative microdialysis under transient conditions. Anal Chem 65:1017–1022

Osborne PG, Niwa O, Kato T, Yamamoto K (1997) On-line, continuous measurement of extracellular striatal glucose using microdialysis sampling and electrochemical detection. J Neurosci Methods 77:143–150

Parpura V, Scemes E, Spray DC (2004) Mechanisms of glutamate release from astrocytes: gap junction "hemichannels", purinergic receptors and exocytotic release. Neurochem Int 45:259–264

Pasti L, Volterra A, Pozzan T, Carmignoto G (1997) Intracellular calcium oscillations in astrocytes: a highly plastic, bidirectional form of communication between neurons and astrocytes in situ. J Neurosci 17:7817–7830

Pasti L, Zonta M, Pozzan T, Vicini S, Carmignoto G (2001) Cytosolic calcium oscillations in astrocytes may regulate exocytotic release of glutamate. J Neurosci 21:477–484

Patel AB, de Graaf RA, Mason GF, Rothman DL, Shulman RG, Behar KL (2005) The contribution of GABA to glutamate/glutamine cycling and energy metabolism in the rat cortex in vivo. Proc Natl Acad Sci USA 102:5588–5593

Pellerin L, Magistretti PJ (1994) Glutamate uptake into astrocytes stimulates aerobic glycolysis: a mechanism coupling neuronal activity to glucose utilization. Proc Natl Acad Sci USA 91: 10625–10629

Pellerin L, Magistretti PJ (1996) Excitatory amino acids stimulate aerobic glycolysis in astrocytes via an activation of the Na+/K+ATPase. Dev Neurosci 18:336–342

Pellerin L, Magistretti PJ (1997) Glutamate uptake stimulates Na+, K+-ATPase activity in astrocytes via activation of a distinct subunit highly sensitive to ouabain. J Neurochem 69: 2132–2137

Pellerin L, Pellegri G, Bittar PG, Charnay Y, Bouras C, Martin JL, Stella N, Magistretti PJ (1998) Evidence supporting the existence of an activity-dependent astrocyte-neuron lactate shuttle. Dev Neurosci 20:291–299

Pfeiffer B, Meyermann R, Hamprecht B (1992) Immunohistochemical co-localization of glycogen phosphorylase with the astroglial markers glial fibrillary acidic protein and S-100 protein in rat brain sections. Histochemistry 97:405–412

Piet R, Vargová L, Syková E, Poulain DA, Oliet SH (2004) Physiological contribution of the astrocytic environment of neurons to intersynaptic crosstalk. Proc Natl Acad Sci USA 101:2151–2155

Porter JT, McCarthy KD (1996) Hippocampal astrocytes in situ respond to glutamate released from synaptic terminals. J Neurosci 16:5073–5081

Prichard J, Rothman D, Novotny E, Petroff O, Kuwabara T, Avison M, Howseman A, Hanstock C, Shulman R (1991) Lactate rise detected by 1H NMR in human visual cortex during physiologic stimulation. Proc Natl Acad Sci USA 88:5829–5831

Raichle ME (1998) Behind the scenes of functional brain imaging: a historical and physiological perspective. Proc Natl Acad Sci USA 95:765–772

Raichle ME (2001) Cognitive neuroscience. Bold insights. Nature 412:128–130

Raichle M, Snyder AZ (2007) A default mode of brain function: a brief history of an evolving idea. NeuroImage 37:1083–1090

Reinstrup P, Ståhl N, Mellergård P, Uski T, Ungerstedt U, Nordström CH (2000) Intracerebral microdialysis in clinical practice: baseline values for chemical markers during wakefulness, anesthesia, and neurosurgery. Neurosurgery 47:701–710

Ros J, Pecinska N, Alessandri B, Landolt H, Fillenz M (2001) Lactate reduces glutamate-induced neurotoxicity in rat cortex. J Neurosci Res 66:790–794

Roy CS, Sherrington CS (1890) On the regulation of the blood supply of the brain. J Physiol 11:85–108
Rusakov DA (2001) The role of perisynaptic glial sheaths in glutamate spillover and extracellular Ca(2+) depletion. Biophys J 81:1947–1959
Santello M, Volterra A (2008) Synaptic modulation by astrocytes via Ca(2+)-dependent glutamate release. Neuroscience 22(1):253
Sappey-Marinier D, Calabrese G, Fein G, Hugg JW, Biggins C, Weiner MW (1992) Effect of photic stimulation on human visual cortex lactate and phosphates using 1H and 31P magnetic resonance spectroscopy. J Cereb Blood Flow Metab 12:584–592
Scemes E, Giaume C (2006) Astrocyte calcium waves: what they are and what they do. Glia 54:716–725
Schasfoort EM, De Bruin LA, Korf J (1988) Mild stress stimulates rat hippocampal glucose utilization transiently via NMDA receptors, as assessed by lactography. Brain Res 475:58–63
Schmalbruch IK, Linde R, Paulson OB, Madsen PL (2002) Activation-induced resetting of cerebral metabolism and flow is abolished by beta-adrenergic blockade with propranolol. Stroke 33:251–255
Schurr A (2006) Lactate: the ultimate cerebral oxidative energy substrate? J Cereb Blood Flow Metab 26:142–152
Schurr A, Payne RS (2007) Lactate, not pyruvate, is neuronal aerobic glycolysis end product: an in vitro electrophysiological study. Neuroscience 147:613–619
Segal M, Greenberger V, Hofstein R (1981) Cyclic AMP-generating systems in rat hippocampal slices. Brain Res 213:351–364
Shelton MK, McCarthy KD (2000) Hippocampal astrocytes exhibit Ca^{2+}-elevating muscarinic cholinergic and histaminergic receptors in situ. J Neurochem 74:555–563
Shulman RG, Hyder F, Rothman DL (2001a) Cerebral energetics and the glycogen shunt: neurochemical basis of functional imaging. Proc Natl Acad Sci USA 98:6417–6422
Shulman RG, Hyder F, Rothman DL (2001b) Lactate efflux and the neuroenergetic basis of brain function. NMR Biomed 14:389–396
Shuttleworth CW, Brennan AM, Connor JA (2003) NAD(P)H fluorescence imaging of postsynaptic neuronal activation in murine hippocampal slices. J Neurosci 23:3196–3208
Sibson NR, Dhankhar A, Mason GF, Behar KL, Rothman DL, Shulman RG (1997) In vivo 13C NMR measurements of cerebral glutamine synthesis as evidence for glutamate-glutamine cycling. Proc Natl Acad Sci USA 94:2699–2704
Sibson NR, Dhankhar A, Mason GF, Rothman DL, Behar KL, Shulman RG (1998) Stoichiometric coupling of brain glucose metabolism and glutamatergic neuronal activity. Proc Natl Acad Sci USA 95:316–321
Siesjö BK (1978) Brain Energy Metabolism. John Wiley & Sons, Chichester
Silva AC, Lee SP, Yang G, Iadecola C, Kim SG (1999) Simultaneous blood oxygenation level-dependent and cerebral blood flow functional magnetic resonance imaging during forepaw stimulation in the rat. J Cereb Blood Flow Metab 19:871–879
Silver IA, Erecinska M (1994) Extracellular glucose concentration in mammalian brain: continuous monitoring of changes during increased neuronal activity and upon limitation in oxygen supply in normo-, hypo-, and hyperglycemic animals. J Neurosci 14:5068–5076
Silver IA, Erecinska M (1997) Energetic demands of the Na+/K+ATPase in mammalian astrocytes. Glia 21:35–45
Smith AD, Olson RJ, Justice JB Jr (1992) Quantitative microdialysis of dopamine in the striatum: effect of circadian variation. J Neurosci Methods 44:33–41
Sokoloff L (1980) Local cerebral energy metabolism: its relationship to local functional activity and blood flow. Bull Schweiz Akad Med Wiss 36:71–91
Sokoloff L (1981) The deoxyglucose method for the measurement of local glucose utilization and the mapping of local functional activity in the central nervous system. Int Rev Neurobiol 22:287–333
Sorg O, Magistretti PJ (1991) Characterization of the glycogenolysis elicited by vasoactive intestinal peptide, noradrenaline and adenosine in primary cultures of mouse cerebral cortical astrocytes. Brain Res 563:227–233

Sporn J, Molinoff P (1976) beta-Adrenergic receptors in rat brain. J Cyclic Nucleotide Res 2:149–161

Swanson RA, Morton MM, Sagar SM, Sharp FR (1992) Sensory stimulation induces local cerebral glycogenolysis: demonstration by autoradiography. Neuroscience 51:451–461

Thompson JK, Peterson MR, Freeman RD (2003) Single-neuron activity and tissue oxygenation in the cerebral cortex. Science 299:1070–1072

Timmerman W, Westerink BH (1997) Brain microdialysis of GABA and glutamate: what does it signify? Synapse 27:242–261

Tsacopoulos M, Magistretti PJ (1996) Metabolic coupling between glia and neurons. J Neurosci 16:877–885

Vafaee MS, Gjedde A (2000) Model of blood-brain transfer of oxygen explains nonlinear flow-metabolism coupling during stimulation of visual cortex. J Cereb Blood Flow Metab 20:747–754

van der Kuil JH, Korf J (1991) On-line monitoring of extracellular brain glucose using microdialysis and a NADPH-linked enzymatic assay. J Neurochem 57:648–654

Vanzetta I, Grinvald A (1999) Increased cortical oxidative metabolism due to sensory stimulation: implications for functional brain imaging. Science 286:1555–1558

Vanzetta I, Grinvald A (2001) Evidence and lack of evidence for the initial dip in the anesthetized rat: implications for human functional brain imaging. NeuroImage 13:959–967

Vazquez AL, Masamoto K, Kim SG (2008) Dynamics of oxygen delivery and consumption during evoked neural stimulation using a compartment model and CBF and tissue $P(O_2)$ measurements. NeuroImage 42:49–59

Ventura R, Harris KM (1999) Three-dimensional relationships between hippocampal synapses and astrocytes. J Neurosci 19:6897–6906

Verkhratsky A, Kettenmann H (1996) Calcium signalling in glial cells. Trends Neurosci 19:346–352

Vincent JL, Patel GH, Fox MD, Snyder AZ, Baker JT, Van Essen DC, Zempel JM, Snyder LH, Corbetta M, Raichle ME (2007) Intrinsic functional architecture in the anaesthetized monkey brain. Nature 447:83–86

Volterra A, Steinhauser C (2004) Glial modulation of synaptic transmission in the hippocampus. Glia 47:249–257

Waldfogel D, van Gelderen P, Muellbacher W, Ziemann U, Immisch I, Hallett M (2000) The relative metabolic demand of inhibition and excitation. Nature 406:995–998

Wang X, Lou N, Xu Q, Tian GF, Peng W, Han X, Kang J, Takano T, Nedergaard M (2006) Astrocytic Ca^{2+} signaling evoked by sensory stimulation in vivo. Nat Neurosci 9:816–823

Watanabe H, Passonneau JV (1973) Factors affecting the turnover of cerebral glycogen and limit dextrin in vivo. J Neurochem 20:1543–1554

Wittendorp-Rechenmann E, Lam CD, Steibel J, Lasbennes F, Nehlig A (2002) High resolution tracer targeting combining microautoradiographic imaging by cellular C-14-trajectography with immunohistochemistry: A novel protocol to demonstrate metabolism of C-14 2-deoxyglucose by neurons and astrocytes. J Trace Microprobe Tech 20:505–515

Yacoub E, Le TH, Ugurbil K, Hu X (1999) Further evaluation of the initial negative response in functional magnetic resonance imaging. Magn Reson Med 41:436–441

Yacoub E, Shmuel A, Pfeuffer J, Van De Moortele PF, Adriany G, Ugurbil K, Hu X (2001) Investigation of the initial dip in fMRI at 7 Tesla. NMR Biomed 14:408–412

Zhang Q, Haydon PG (2005) Roles for gliotransmission in the nervous system. J Neural Transm 112:121–125

Chapter 22
Measurements of Glycogen Metabolism in the Living Brain

Rolf Gruetter

Abstract To maintain brain function, the combustion of glucose for the generation of energy (ATP) is essential, yet the brain maintains fuel storage in the form of glial glycogen. Although brain glycogen levels are lower than those in the liver, they exceed the normal free glucose concentration in brain several-fold. Under normoxic conditions, when a partial glucose supply deficit is present, brain glycogen can provide fuel for extended periods of time. Brain glycogen metabolism is active, but comparatively slow, consistent with a role of an emergency fuel with neuroprotective effects. The classical analysis with biochemical means requires tissue extraction; brain glycogen metabolism can be measured directly in the living brain only by ^{13}C NMR spectroscopy. This chapter reviews the insights gained into brain glycogen metabolism in the living brain to-date and reviews potential roles for brain glycogen in the context of neuro-glial metabolism, hypoglycemia and neuroprotection.

Keywords Energy metabolism • Glutamate neurotransmission • Glycogen • Hypoglycemia • Ischemia

22.1 Brain Glycogen Concentrations Can Be Measured Non-invasively

Glycogen represents the largest store of glucose equivalents in the brain and is present in concentrations on the order of 5 μmol glucosyl units/g wet weight in humans and rodents. These concentrations exceed the normal total brain glucose concentration

---

R. Gruetter, Ph.D. (✉)
Laboratory for functional and metabolic imaging, Ecole Polytechnique Federale de Lausanne, Department of Radiology, University of Lausanne and Geneva, Station 6, CH-1015, Lausanne, Switzerland
e-mail: rolf.gruetter@epfl.ch

I.-Y. Choi and R. Gruetter (eds.), *Neural Metabolism In Vivo*, Advances in Neurobiology 4, DOI 10.1007/978-1-4614-1788-0_22,

(~1 μmol/g in humans and ~2 μmol/g in rodents due to the slightly higher plasma glucose concentration) by several-fold.

Although brain glycogen content can be assessed using biochemical extraction using carefully controlled rapid extraction procedures (see below), such measurements do not allow for *in vivo* study of its metabolism, nor are they applicable to human studies. In contrast to glucose metabolism, which can be traced using the trapping of radio-active phosphorylated deoxyglucose in cells, no such alternative exists for glycogen. To-date the only method available for assessing glycogen metabolism non-destructively is localized ^{13}C NMR spectroscopy (Oz et al. 2007; Morgenthaler et al. 2008). Localized ^{13}C NMR spectroscopy is non-invasive and thus brain glycogen metabolism can be followed in a single animal or human for several hours.

Here, we will provide a short account of the salient elements required for successful measurement of brain glycogen using NMR, without providing details on the background of ^{13}C NMR methodology, as this is covered elsewhere in the volume, see chapter 20 in this volume.

The measurement of brain glycogen faces three major obstacles: Its low concentration, short relaxation times and the relatively high concentration of glycogen in muscle surrounding the brain. With the use of tailored NMR methods that eliminate the intense signal from extra-cerebral muscle tissue using outer volume suppression, localization can be achieved without signal loss from the measured volume of interest in the brain (Choi et al. 1999; Oz et al. 2007). The choice of a dedicate localization method is important given the short relaxation times of glycogen (on the order of a few ms for T_2 and on the order of 100s of ms for T_1). While T_2 ~6 ms is approximately independent of the magnetic field strength, T_1 increases with the static magnetic field strength, ranging from 300 ms at 4.7 T to 540 ms at 14.1 T (Overloop et al. 1996).

In addition, due the low natural abundance of the ^{13}C isotope (1.1%) and the concentration of a few mM of brain glycogen its ^{13}C NMR signal must be increased by infusion of ^{13}C labeled glucose to increase the concentration of measurable ^{13}C label in glycogen. The incorporation of ^{13}C label from glucose into brain glycogen reflects naturally metabolism (turnover due to simultaneously active synthesis and degradation) as well as net concentration changes. The effect of turnover can be minimized by administration of ^{13}C labeled glucose for approximately 24 h prior to the measurement, which achieves isotopic steady-state, whereas the subsequent administration of differently labeled glucose (such as 1,6-$^{13}C_2$ glucose) allows the simultaneous measurement of glycogen turnover.

Due to lack of an internal reference signal in the ^{13}C spectrum, quantification procedures mainly rely on the phantom displacement or external reference method. This approach to quantification relies on repeating the measurement previously performed *in vivo* on a container filled with a glycogen reference solution (phantom) of known concentration. From the signal corrected for instrumental variations, the *in vivo* signal intensity can be converted to concentration units yielding μmol/g wet weight glucosyl concentrations (Morgenthaler et al. 2008).

Although the direct measurement of glycogen metabolism in the living brain by NMR is mandatory in humans, more available alternatives exist for the study of e.g. the rodent brain using careful extraction methods (vide infra). However, such biochemical extraction methods that measure a longitudinal change in brain glycogen content and metabolism require a substantially higher number of cross-sectional measurements. Therefore, the cost of NMR measurements is at least in part offset by increased cost in the labor required for tissue extraction, homogenization and biochemical analysis. In addition, an order of magnitude increase is necessary in the number of animals that need to be sacrificed.

22.2 Brain Glycogen Is an Energy Reserve

Based on a total glucose metabolic rate ranging between 0.3 μmol/g/min in humans to 1.4 μmol/g/min in rodents, the ~5 μmol/g of glycogen can sustain such a rate of glycolysis fully for at most 10 min. In fact, upon biochemical extraction, glycogen is completely depleted under agonal conditions within minutes due to the rapidly increased rate of anaerobic glycolysis (Choi et al. 1999; Swanson 1992). This rapid postmortem depletion of brain glycogen has limited the precise measurement of its content using biochemical means: only in the past decade rapid focused microwave fixation or equivalent careful experimentation lead to consistent brain glycogen concentration typically on the order of 5 μmol/g wet weight or even higher (Cruz and Dienel 2002; Kong et al. 2002). It would thus be easy to conclude that glycogen in brain serves a minimal role as a fuel reservoir.

However, hypoglycemia is distinct from total ischemia in that only glucose supply is limited, but not that of oxygen. Thus glycogen can be efficiently metabolized to form more than 30 ATP per glucosyl unit. A brain glycogen concentration on the order of several μmol/g can survive longer periods of sustained hypoglycemia, as shall be illustrated with the following: When plasma glucose concentrations approach K_t, the glucose transport capacity decreases and eventually becomes rate-limiting for metabolism when its concentration approaches the K_m of hexokinase (~50 μM). However, even when the brain glucose concentration is zero, glucose is still transported across the BBB *in vivo*: For example, assuming a Michealis-Menten constant on the order of K_t ~3 mM, and an apparent relative maximal glucose transport rate of $T_{max}/CMR_{glc} = 2.7$ (see chapter 20), glucose transport provides ~90% of the resting cerebral glucose metabolic rate, CMR_{glc}, at 1.5 mM plasma glucose concentrations. Assuming a basal glucose consumption of 0.4 μmol/g/min, a 10% glucose supply deficit corresponds to 0.04 μmol/g/min, which would thus completely deplete brain glycogen only after about 120 min of such hypoglycemia. Note that this does not critically depend on the specific values of glucose transport kinetics nor the precise glycogen concentration assumed.

We further note an important difference between the *in vivo* condition and in situ experiments: Both slice preparations as well as cell cultures lack blood supply.

Therefore, when using a superfusate devoid of glucose is equivalent not to hypoglycemia, but aglycemia, i.e. a situation that corresponds to zero glucose concentration in the blood *in vivo*.

Experimentally, it is in the surviving rodent indeed very difficult to completely eliminate all brain glycogen during 2 h of hypoglycemia alone. The presence of substantial brain glycogen levels even after periods of profound hypoglycemia demonstrates that brain glycogen is a significant store of glucose equivalents, protecting the brain in periods of limited glucose supply under normoxic conditions.

In this context it is of interest to note that only extended periods of hours of profound hypoglycemia lead to permanent brain damage in humans, which is consistent with the persistence of brain glycogen during moderate hypoglycemia.

Evidence suggests that glycogen degradation typically starts at low brain glucose concentrations, the point at which brain glucose content becomes rate-limiting for metabolism (Choi et al. 2003) and thus points to insufficient glucose phosphorylation being important in activating cerebral defenses against a deficiency in fuel supply, such as glycogen degradation. Together with the apparent stability of glycogen in the non-stimulated brain at eu-or hyperglycemia (see below), this suggests that intracellular brain glucose may also play a regulatory role in cerebral glycogen metabolism as in other organs.

Consistent with a role for brain glucose in affecting glycogen levels in brain is the observation during chronic hypoglycemia that brain glycogen concentrations remain unchanged: Since blood-brain barrier glucose transport is upregulated (Simpson et al. 1999) leading to elevated brain glucose concentrations, brain cells are exposed to a glucose concentration in chronic hypoglycemia that corresponds to the normoglycemic brain glucose concentration (Lei and Gruetter 2006).

Note that some studies have suggested that brain glycogen levels may be reduced by focal activation (see below). This may for instance indicate that glial glucose content can transiently become rate limiting for metabolism, and is consistent with the hypothesis that glial glycolyis is augmented during glutamatergic activation (Magistretti and Pellerin 1996).

22.3 Glycogen Metabolism in Neuroprotection and Activation

Studies have reported increased glycogen metabolism upon activation of the noradrenergic system (Morgenthaler et al. 2009) and during global physiologic activation in rodents (Dienel et al. 2002), although in humans no effect of focal visual stimulation has been discernible (Oz et al. 2003). It has been proposed that increased cycling of glucose units in and out of glycogen would account for the discrepancy in glucose and oxygen metabolism observed during visual stimulation (Shulman et al. 2001). However, under normal awake conditions the turnover rate of glycogen (simultaneous activity of glycogen phosphorylase and synthase) is on the order of a few% of the resting glucose metabolic rate in humans (Oz et al. 2007) and thus cannot quantitatively account for the reported mismatch in human glucose and oxygen metabolic rates.

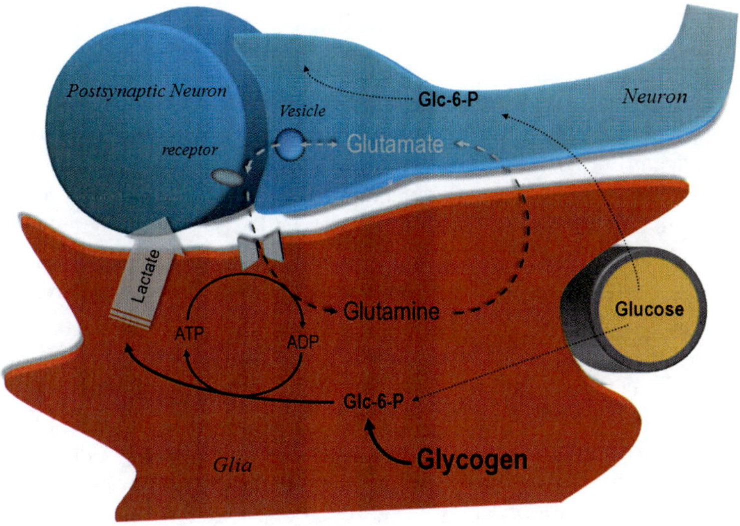

Fig. 22.1 Scheme depicting a potential role for glial glycogen to maintain glutamate uptake into glia when fuel supply is limited. When brain glucose concentration becomes rate-limiting for metabolism, such as during hypoglycemia or during short ischemic bouts, astrocytic glycogenolyis can provide the necessary glucosyl units to generate ATP in the glial compartment (*thick solid lines*), necessary for glutamate uptake into the astrocyte and conversion to glutamine (*dashed lines* indicate the glutamate-glutamine cycle). Glycogen can thus help reduce the glutamate concentration in the synaptic cleft under conditions of limited substrate supply. Glycogen breakdown can further provide fuel to neurons in the form of lactate, which can diffuse to neurons, thereby further reducing the energy deficit in the neuronal compartment when oxygen is present

Astrocytes are a key player in the mechanism of the major excitatory neurotransmitter glutamate (Fig. 22.1): Detoxification of the synaptic cleft is achieved by rapid uptake into the adjacent astrocyte and conversion to glutamine, which diffuses back to the presynaptic neuron where it is back-converted to glutamate by phosphate-activated glutaminase, thereby completing the glutamate-glutamine cycle. As sole energy reservoir of brain cells, the glial localization of brain glycogen implies a neuroprotective role: Since the uptake of glutamate and subsequent conversion to glutamine requires ATP (Magistretti and Pellerin 1996), astrocytes can maintain uptake of glutamate and conversion to glutamine during short ischemic periods, leading to accumulation of glutamine during transient ischemia (Berthet et al. 2011). The role of brain glycogen as an important neuroprotective energy reserve, likely used to minimize glutamatergic exctitotoxicty (Fig. 22.1), is supported by increased neuroprotection of elevated brain glycogen in astrocyte-neuronal co-cultures and in the postnatal rat (Brucklacher et al. 2002; Swanson and Choi 1993), consistent with neuroprotective effects of glycogen on axons in aglycemia (Brown et al. 2003) and that the final depolarization occurs when nerve glycogen levels are depleted (Allen et al. 2005).

22.4 Brain Glycogen and Disturbances of Glucose Metabolism

In contrast to other tissues, bulk brain glucose metabolism is considered largely insulin-insensitive. However, as in other tissues, brain glycogen metabolism and possibly concentrations are affected by plasma insulin and/or glucose concentrations (Choi et al. 2003; Morgenthaler et al. 2006). Insulin can exert an effect on brain glycogen metabolism by transport across the BBB, by insulin receptors in the brain and possibly at the endothelial cell. Whatever the precise mechanism, its insulin-sensitivity makes brain glycogen a likely player in disturbances of systemic glucose metabolism, such as those encountered in diabetes.

Following a single hypoglycemic episode, brain glycogen increased above the basal level and beyond. This rebound or super-compensation appeared unabated several hours after restoring glycemia in many studies, reaching levels up to three times the normal brain glycogen concentration (Canada et al. 2011; Choi et al. 2003; Lei et al. 2007). Such supercompensation of brain glycogen has also been observed following sleep deprivation (Kong et al. 2002) or hypoxic-ischemia (Brucklacher et al. 2002). Because of the importance of brain glycogen as a fuel reserve in the brain, such excess brain glycogen is likely to have a neuroprotective effect through the subsequent release of fuel from glycogen in the form of lactate, which can be metabolized by neurons (Fig. 22.1).

22.5 Human Brain Glycogen Metabolism

Brain glycogen concentrations increase with anesthesia or depressed electrical activity (Swanson 1992; Choi et al. 2002) and brain glycogen metabolism has been reported to be changed during focal activation (Dienel et al. 2002). Consistent with these observations, brain glycogen metabolism in the rat brain was very slow, whether under light (Choi et al. 1999) or deep anesthesia (Choi et al. 2002). However, in the absence of strong focal stimulation, bulk brain glycogen turnover was very slow even in the awake rat brain, on the order of 0.5 μmol/g/h (Morgenthaler et al. 2009).

Human brain glycogen metabolism is comparatively slow with a flux through glycogen synthase of 0.1–0.2 μmol/g/h in the awake, resting condition (Oz et al. 2007). Cycling of glucose in and out of glycogen was ~0.5% of CMR_{glc} in the conscious, unstimulated human brain at euglycemia and above, where the brain glucose concentration is well above the K_m of hexokinase. As a consequence, a brain glycogen pool of a the ~4 μmol/g wet weight measured in human brain (Oz et al. 2007; Dalsgaard et al. 2007) is expected to have a turnover time on the order of days in human brain.

Nonetheless brain glycogen becomes important in any metabolic state where glucose phosphorylation transiently cannot meet demand. Such conditions could occur during prolonged focal activation, sleep deprivation, seizures and mild transient hypoxia.

22.6 Conclusions

Glycogen is a viable and important store of glucose equivalents in the brain, whose metabolism, which is accessible by ^{13}C NMR spectroscopy, is affected by glucose homeostasis, hormones, neurotransmitters and second messengers. The sensitivity of brain glycogen metabolism to insulin, glucose and the profound effect of hypoglycemia on brain glycogen metabolism point to the potential involvement of brain glycogen in hypoglycemia. Further, its ability to provide glial ATP to maintain uptake of glutamate implies a neuroprotective role for glycogen in many conditions of transiently limited fuel supply.

References

Allen NJ, Karadottir R, Attwell D (2005) A preferential role for glycolysis in preventing the anoxic depolarization of rat hippocampal area CA1 pyramidal cells. J Neurosci 25(4):848–859

Berthet C, Lei H, Gruetter R, Hirt L (2011) Early predictive biomarkers for lesion after transient cerebral ischemia. Stroke 42(3):799–805

Brown AM, Tekkok SB, Ransom BR (2003) Glycogen regulation and functional role in mouse white matter. J Physiol 549(Pt 2):501–512

Brucklacher RM, Vannucci RC, Vannucci SJ (2002) Hypoxic preconditioning increases brain glycogen and delays energy depletion from hypoxia-ischemia in the immature rat. Dev Neurosci 24(5):411–417

Canada SE, Weaver SA, Sharpe SN, Pederson BA (2011) Brain glycogen supercompensation in the mouse after recovery from insulin-induced hypoglycemia. J Neurosci Res 89(4):585–591

Choi IY, Tkac I, Ugurbil K, Gruetter R (1999) Noninvasive measurements of [1-(13)C]glycogen concentrations and metabolism in rat brain in vivo. J Neurochem 73(3):1300–1308

Choi IY, Lei H, Gruetter R (2002) Effect of deep pentobarbital anesthesia on neurotransmitter metabolism in vivo: on the correlation of total glucose consumption with glutamatergic action. J Cereb Blood Flow Metab 22(11):1343–1351

Choi IY, Seaquist ER, Gruetter R (2003) Effect of hypoglycemia on brain glycogen metabolism in vivo. J Neurosci Res 72(1):25–32

Cruz NF, Dienel GA (2002) High glycogen levels in brains of rats with minimal environmental stimuli: implications for metabolic contributions of working astrocytes. J Cereb Blood Flow Metab 22(12):1476–1489

Dalsgaard MK, Madsen FF, Secher NH, Laursen H, Quistorff B (2007) High glycogen levels in the hippocampus of patients with epilepsy. J Cereb Blood Flow Metab 27(6):1137–1141

Dienel GA, Wang RY, Cruz NF (2002) Generalized sensory stimulation of conscious rats increases labeling of oxidative pathways of glucose metabolism when the brain glucose-oxygen uptake ratio rises. J Cereb Blood Flow Metab 22(12):1490–1502

Kong J, Shepel PN, Holden CP, Mackiewicz M, Pack AI, Geiger JD (2002) Brain glycogen decreases with increased periods of wakefulness: implications for homeostatic drive to sleep. J Neurosci 22(13):5581–5587

Lei H, Gruetter R (2006) Effect of chronic hypoglycaemia on glucose concentration and glycogen content in rat brain: a localized 13C NMR study. J Neurochem 99(1):260–268

Lei H, Morgenthaler F, Yue T, Gruetter R (2007) Direct validation of in vivo localized 13C MRS measurements of brain glycogen. Magn Reson Med 57(2):243–248

Magistretti PJ, Pellerin L (1996) Cellular mechanisms of brain energy metabolism. Relevance to functional brain imaging and to neurodegenerative disorders. Ann N Y Acad Sci 777:380–387

Morgenthaler FD, Koski DM, Kraftsik R, Henry PG, Gruetter R (2006) Biochemical quantification of total brain glycogen concentration in rats under different glycemic states. Neurochem Int 48(6–7):616–622

Morgenthaler FD, van Heeswijk RB, Xin L, Laus S, Frenkel H, Lei H, Gruetter R (2008) Non-invasive quantification of brain glycogen absolute concentration. J Neurochem 107(5):1414–1423

Morgenthaler FD, Lanz BR, Petit JM, Frenkel H, Magistretti PJ, Gruetter R (2009) Alteration of brain glycogen turnover in the conscious rat after 5h of prolonged wakefulness. Neurochem Int 55(1–3):45–51

Overloop K, Vanstapel F, Van Hecke P (1996) 13C-NMR relaxation in glycogen. Magn Reson Med 36(1):45–51

Oz G, Henry PG, Seaquist ER, Gruetter R (2003) Direct, noninvasive measurement of brain glycogen metabolism in humans. Neurochem Int 43(4–5):323–329

Oz G, Seaquist ER, Kumar A, Criego AB, Benedict LE, Rao JP, Henry PG, Van De Moortele PF, Gruetter R (2007) Human brain glycogen content and metabolism: implications on its role in brain energy metabolism. Am J Physiol Endocrinol Metab 292(3):E946–E951

Shulman RG, Hyder F, Rothman DL (2001) Cerebral energetics and the glycogen shunt: neurochemical basis of functional imaging. Proc Natl Acad Sci USA 98(11):6417–6422

Simpson IA, Appel NM, Hokari M, Oki J, Holman GD, Maher F, Koehler-Stec EM, Vannucci SJ, Smith QR (1999) Blood-brain barrier glucose transporter: effects of hypo- and hyperglycemia revisited. J Neurochem 72(1):238–247

Swanson RA (1992) Physiologic coupling of glial glycogen metabolism to neuronal activity in brain. Can J Physiol Pharmacol 70(Suppl):S138–S144

Swanson RA, Choi DW (1993) Glial glycogen stores affect neuronal survival during glucose deprivation in vitro. J Cereb Blood Flow Metab 13(1):162–169

Chapter 23
Functions of Glycogen in the Central Nervous System

Richard D. Evans, Angus M. Brown, and Bruce R. Ransom

Abstract Glycogen is present in the mammalian brain and has the intriguing feature of being predominantly located in astrocytes. Its importance was discounted for decades because of the presumption that it was found in concentrations too low to be meaningful. Recent studies have begun a full re-evaluation of the physiological role of brain glycogen and many of the findings have been unexpected. Moreover, they have helped expand our whole concept of brain energy metabolism. Among the new concepts is evidence that astrocyte glycogen can be broken down to an energy substrate, lactic acid most likely, that traverses the extracellular space and supports energy metabolism in adjacent neural elements under conditions of need. There is now strong evidence in favor of rapid coupling between active neurons and nearby astrocytes in terms of energy metabolism. Active neurons 'communicate' their energy needs, which are promptly met by astrocytes via glycogen breakdown. We are still learning the rules about this altruism among brain cells.

Keywords Glycogen • White matter • Astrocyte • Lactate • Glucose • Aglycemia

---

R.D. Evans • A.M. Brown, Ph.D. (✉)
School of Biomedical Sciences, Queens Medical Centre, University of Nottingham, Nottingham, NG7 2UH, UK
e-mail: ambrown@nottingham.ac.uk

B.R. Ransom, M.D., Ph.D.
Department of Neurology, University of Washington School of medicine, 1959 NE Pacific Street, Box # 356465, Seattle, WA 98195, USA
e-mail: bransom@u.washington.edu

I.-Y. Choi and R. Gruetter (eds.), *Neural Metabolism In Vivo*, Advances in Neurobiology 4, DOI 10.1007/978-1-4614-1788-0_23,

23.1 Peripheral Glycogen

The role of glycogen as an energy reserve in liver and skeletal muscle is central to the concept of energy storage and utilization. During periods when calorific intake exceeds demand excess carbohydrate is stored as glycogen in the liver, and liberated during periods of famine, when the glycogen is metabolised and released into the systemic circulation as glucose to maintain euglycemia, primarily to support brain function. In skeletal muscle glycogen metabolism supports muscle function with the by-product of glycolytic metabolism, lactate, released into the systemic circulation. Although the presence of glycogen in the brain has been illustrated both by electron microscopy and by biochemical assay, compared to liver and skeletal muscle brain glycogen occurs at fairly low levels, thus a role as a conventional energy reserve was ignored. In the last decade however, there has been a resurgence of interest in brain glycogen, and assigning potential roles to brain glycogen based on recent data is the subject of this chapter.

23.2 Glycogen Metabolism

Glycogen is an efficient molecule in which to store excess glucose. Glucose molecules are hydrolysed, forming glucosyl molecules, during incorporation into the glycogen molecule thus reducing osmotic pressure. Glucosyl molecules are added to the glycogen molecule sequentially forming a 'string of pearls' interrupted by side chains every ten or so glucosyl molecules. Glucosyl molecules are added to glycogen via the action of glycogen synthase and metabolised by the actions of glycogen phosphorylase (Stryer 1995). Glycogen synthase is located in astrocytes as expected given the predominantly astrocytic location of astrocytes, but is also found in neurones (Pellegri et al. 1996), whereas glycogen phosphorylase is found exclusively in astrocytes (Pfeiffer-Guglielmi et al. 2003).

23.3 Tissue Culture Studies of Astrocytic Glycogen

The re-emergence of interest in brain glycogen can be dated to the early 1990s when two papers, which described novel aspects of brain glycogen function, were published. The first of these papers used autoradiographic techniques to describe a decrease in glycogen levels in the barrel cortex in response to whisker stimulation (Swanson et al. 1992). The importance of this study was the demonstration of a link between sensory stimulation, a physiological function, and glycogen metabolism. The second study described a cell culture procedure where the duration of survival of neurones in culture was determined by the presence of astrocytes in the culture. However the key aspect was not merely the presence of the astrocytes but their glycogen content (Swanson and Choi 1993). The combined message of these two papers, that astrocytic glycogen nurtures neurons and that neuronal activity resulted in glycogen metabolism re-energized a dormant field.

23.4 Rodent Optic Nerve Model

The data we will describe in this chapter derive from our studies on the rodent optic nerve, a central white matter tract. This preparation is ideal in the study of glycogen as it is devoid of neuronal cell bodies and lacks the complications of glutamatergic synapses, thus simplifying data interpretation. The stimulus evoked compound action potential (CAP) can be recorded using suction electrodes to provide a quantifiable index of axon conduction *in vitro* (Stys et al. 1991). In addition biochemical assay can be used to quantify glycogen content, thus these techniques, when used in combination, can be used to assess the relationship between glycogen content and axon conduction. An additional advantage of the preparation is that it allows imposition of high frequency stimulus, our experimental correlate for increased tissue energy demand that, due to the morphological nature of the tissue, imposes the energy burden throughout the entire tissue. Thus the result of high frequency stimulus is to increase the energy demand along the entire length of the nerve, rather than in localised areas, as would occur in stimulating the Schaffer collaterals in the hippocampus, for example.

The initial experiments were carried out in the rat optic nerve and showed that in the absence of exogenously applied glucose axon conduction was sustained for up to 30 min (Ransom and Fern 1997), compared to the conduction failure that occurs after a few minutes in the same preparation when exposed to anoxia (Stys et al. 1992). This delayed axon conduction failure suggested an endogenous energy reserve was present within the nerve that sustained axon function in the absence of glucose. The most obvious candidate was glycogen, which was subsequently identified by transmission electron microscopy and localised to astrocytes (Wender et al. 2000).

23.5 CNS Glycogen

The presence of glycogen in the optic nerve raised the issue of its function. In a series of experiments in the rat optic nerve glycogen was shown to be labile, its concentration influenced by not only the ambient glucose levels, but also sensitive to the neurotransmitter nor-epinephrine. In addition lactate was identified as the glycogen derived transportable conduit, which acts as a metabolic shuttle between the astrocyte and the axon (Wender et al. 2000). Although glycogen does support axon conduction during periods of glucopenia, whose clinical correlate is insulin-induced hypoglycemia (Frier and Fisher 2007), this is an iatrogenic condition and is unlikely to have any relevance to glycogen's physiological function. This is compounded by

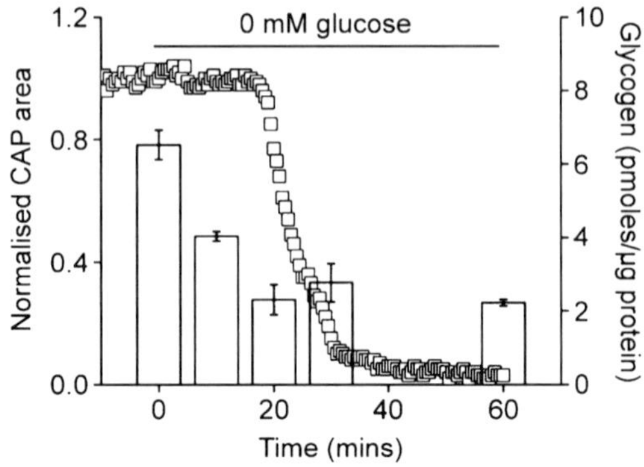

Fig. 23.1 Introduction of glucopenia results in a decrease in the glycogen content (*right* y-axis) in mouse optic nerve. The delayed latency to CAP failure (*left* y-axis) corresponds to the time the nadir of glycogen content is reached. *Open squares* denote CAP area and *open columns* denote glycogen concentration

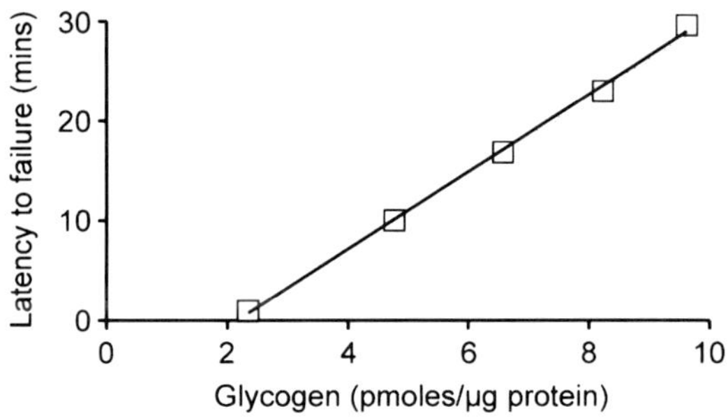

Fig. 23.2 The glycogen content at the onset of glucopenia predicts the latency to CAP failure in mouse optic nerve. Glycogen content stated as pmol of glucose molecules in glycogen

a few simple calculations; the concentration of glycogen in the rodent brain is about 6 μmol g^{-1} wet tissue, and the metabolic rate of glucose is about 0.75 μmol g^{-1} min^{-1} (Dienel 2009), thus in the absence of glucose glycogen would only support function for about 8 min, a very limited period. We were prompted to conduct experiments to investigate the role of glycogen under physiological conditions using the mouse optic nerve. This preparation offers the advantage of decreased diffusion distances between the bath compartment and extracellular fluid, a key factor in studies of this kind, where larger structures may present such an impenetrable diffusion barrier that no glucose is present in the core of the tissue (Baltan Tekkök et al. 2003). Initial studies in the mouse replicated the results from rat, namely that glycogen in labile, its content dependent upon the concentration of glucose in the bathing artificial cerebrospinal fluid (aCSF) (Fig. 23.2), that glycogen is depleted during glucopenia (Fig. 23.1), and that lactate acts as conduit between astrocytes and axons (Brown et al. 2003). It is pertinent at this point to describe studies that show, or at least imply, a role for glycogen under physiological conditions. In the rat glycogen accumulates during sleep, but decreases with sleep deprivation, implying that the awake brain constantly turns over glycogen (Kong et al. 2002). In rodents introduction of general anesthetics results in increased levels of brain glycogen, with the degree of the increase related to the depth and length of the anaesthesia (Folbergrova et al. 1970; Hutchins and Rogers 1970; Nelson et al. 1968; Rosengarten 1970). However, tissue culture studies revealed that anesthetics had no effect on the glycogen content in astrocyte cultures, implying that the increases in glycogen that were seen under *in vivo* conditions were a result of anaesthesia-induced attenuation of neuronal activity resulting in a decreased demand for astrocytic glycogen (Swanson et al. 1989). The consensus opinion drawn from these experiments is that physiological neuronal activity promotes metabolism of glycogen, but during periods when neurones are quiescent, such as sleep or anaesthesia, the reduced neuronal activity elicits a decreased demand for metabolism of astrocytic glycogen and leads to glycogen accumulation. Recent data have shown that maternal under-nutrition results in elevated cortical glycogen in neonatal pigs, implying a role for glycogen in the pathology associated with maternal under-nutrition. (Lizarraga-Mollinedo et al. 2010).

23.6 Hypoglycemia and Glycogen

It is intuitive to imagine that if glycogen supports neural function during periods of glucopenia, where no exogenously applied glucose is present, then under hypoglycemic conditions, where glucose is present but not in sufficient quantities to meet the energy demands of the tissue, glycogen could be metabolised to meet the shortfall in glucose supply. The upper threshold for hypoglycaemia is considered to be 4 mM glucose in the arterial circulation (Frier and Fisher 2007), which would correspond to a CSF blood glucose concentration of about 2.6 mM (Gruetter et al. 1998). We therefore incubated optic nerves for 2 h in concentrations of glucose between 0.5 and 2 mM while continuously monitoring axon conduction. Only at 2 mM glucose was the CAP fully supported for 2 h; at the lower glucose concentrations the CAP fell after short periods of time (Brown et al. 2003). To investigate whether glycogen was supporting function in the presence of 2 mM glucose, we pre-treated nerves for 20 min in 0 mM glucose to deplete glycogen content. On reintroduction of aCSF containing 2 mM glucose there was a rapid and progressive decrease in the CAP area, indicating that 2 mM glucose, in the absence of glycogen, is unable to fully support axon conduction (Brown et al. 2003). Thus, during periods of hypoglycaemia, when ambient glucose is unable to fully meet the energy requirements of the tissue, glycogen provides supplemental energy substrate. The form that this supplemental energy substrate takes was subsequently investigated. There is convincing data that this substrate is lactate (Dringen et al. 1993, 1995; Pellerin 2003), thus we focused on lactate as the most likely candidate. We used cinnemate, a blocker of monocarboxylate transport at the monocarboxylate transporter (Dringen et al. 1993, 1995) to interrupt cell-to-cell transfer of lactate, as we hypothesize that lactate must be transported out of the astrocyte and into the axon via the extracellular fluid by monocarboxylate transporters (MCTs). Nerves that were perfused with 10 mM glucose showed no alteration in the CAP area on introduction of cinnemate, whereas nerves that were perfused with 2 mM glucose showed a reversible decrease in the CAP area on introduction of cinnemate (Brown et al. 2003). Complementary studies were carried out using D-lactate, the enantiomer of L-lactate, which although metabolically inert is transported at the MCT, and can thus act as a competitive antagonist for L-lactate transport (Jackson and Halestrap 1996). While introduction of D-lactate had no effect on the CAP area in nerves perfused with 10 mM glucose, in nerves perfused with 2 mM glucose the CAP rapidly fell to zero (Baltan Tekkök et al. 2005). These experiments indicate that in the presence of hypoglycaemic concentrations of glucose, i.e. concentrations of glucose that are not capable in themselves of meeting the energy demands of the tissue, glycogen is metabolised to lactate in the astrocyte, which is subsequently shuttled to the axon via the extracellular fluid, where it acts as a supplemental energy source to fully support axon conduction (Baltan Tekkök et al. 2005).

23.7 Increased Tissue Energy Demand and Glycogen

That glycogen is metabolised in the presence of glucose is compatible with the concept of glycogen metabolism under physiological conditions in the presence of normoglycaemic concentrations of glucose. A likely occasion for glycogen metabolism under these conditions is during increased tissue energy demand. To simulate this condition *in vitro* we imposed a period of 100 Hz stimulus on the tissue, i.e. a CAP was evoked every 10 ms; at stimulus frequencies higher than this the CAP area fell probably because the stimuli occurred during the refractory period of the previous CAP (Fig. 23.3). The optic nerve is an ideal tissue for this type of experiments as it is capable is generating CAPs at these high frequencies for extended periods of time, which grey matter cannot. Imposing a 15 min period of 100 Hz stimulus on the optic nerve in the presence of 10 mM glucose led to a transient increase in CAP area followed by a sustained fall in the CAP area below baseline, indicative of a failure of all axons in the nerve to conduct. However if the aCSF was switched to one containing 30 mM glucose the CAP fully recovered to baseline levels (Brown et al. 2003). This indicates that there is no threshold of energy demand above which the nerve will not conduct, rather that under conditions of increased energy demand CAP conduction can be maintained as long as there is an increase in the energy supply. This was highlighted in nerves pre-incubated in 2 mM glucose for 2 h to deplete

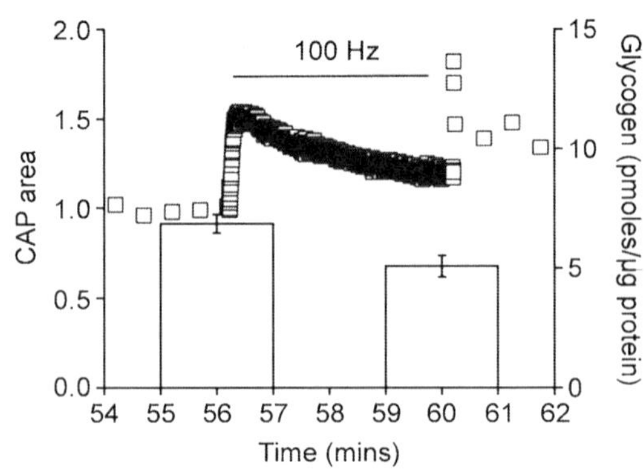

Fig. 23.3 The CAP profile is maintained above baseline during imposition of 100 Hz stimulus for 4 min, but the glycogen content significantly decreases ($p<0.05$). *Open squares* denote CAP area and open columns denote glycogen concentration

glycogen levels. During a 4-min period of 100 Hz stimulus the CAP fell rapidly to zero indicating that 2 mM glucose in the absence of exogenous glycogen could not support this level of energy demand (Brown et al. 2003). The ability to support the CAP during periods when 100 Hz stimulus was imposed was tested under a variety of conditions. Blocking uptake of lactate via MCTs into the axon with D-lactate, resulted in a fall in the CAP area (Baltan Tekkök et al. 2005). Pre-depletion of glycogen with a period of glucopenia, resulted in rapid CAP failure in the presence of 10 mM glucose, indicating that in normoglycaemic conditions supplementary glycogen-derived energy substrate is required to support CAP conduction (Brown et al. 2003). Nerves incubated in 10 mM glucose and subjected to 100 Hz stimulus showed a significant fall in glycogen content complementing the electrophysiological data indicating glycogen metabolism under normoglycaemic conditions when tissue energy demand is increased (Brown et al. 2003).

23.8 Conclusion

In conclusion glycogen is present in the mammalian brain, although at relatively low concentrations. Glycogen acts to delay conduction failure during periods of glucose deprivation, suggesting that it could be a promising clinical target to stave off the cognitive failure that occurs during insulin-induced hypoglycaemia that is a common occurrence in type-1 diabetic patients. In addition glycogen supports axon conduction during periods of increased energy demand strongly suggesting that glycogen may play a role in normal physiological activity and act as an energy buffer when tissue energy demand is not matched by tissue glucose delivery.

References

Baltan Tekkök S, Brown AM, Ransom BR (2003) Axon function persists during anoxia in mammalian white matter. J Cereb Blood Flow Metab 23:1340–1348

Baltan Tekkök S, Brown AM, Westenbroek RE, Pellerin L, Ransom BR (2005) Transfer of glycogen-derived lactate from astrocytes to axons via specific monocarboxylate transporters supports mouse optic nerve activity. J Neurosci Res 81:644–652

Brown AM, Baltan Tekkök S, Ransom BR (2003) Glycogen regulation and functional role in mouse white matter. J Physiol 549(2):501–512

Dienel GA (2009) Energy metabolism in the brain. In: Byrne JH, Roberts JL (eds) From Molecules to Networks. Elsevier, London, pp 49–110

Dringen R, Gebhardt R, Hamprecht B (1993) Glycogen in astrocytes: possible function as lactate supply for neighboring cells. Brain Res 623:208–214

Dringen R, Peters H, Wiesinger H, Hamprecht B (1995) Lactate transport in cultured glial cells. Dev Neurosci 17:63–69

Folbergrova J, Lowry OH, Passonneau JV (1970) Changes in metabolites of the energy reserves in individual layers of mouse cerebral cortex and subjacent white matter during ischaemia and anaesthesia. J Neurochem 17:1155–1162

Frier BM, Fisher BM (2007) Hypoglycaemia in Clinical Diabetes. John Wiley and Sons, Ltd, New York

Gruetter R, Ugurbil K, Seaquist ER (1998) Steady-state cerebral glucose concentrations and transport in the human brain. J Neurochem 70:397–408

Hutchins DA, Rogers KJ (1970) Physiological and drug-induced changes in the glycogen content of mouse brain. Br J Pharmacol 39:9–25

Jackson VN, Halestrap AP (1996) The kinetics, substrate, and inhibitor specificity of the monocarboxylate (lactate) transporter of rat liver cells determined using the fluorescent intracellular pH indicator, 2',7'-bis(carboxyethyl)-5(6)-carboxyfluorescein. J Biol Chem 271:861–868

Kong J, Shepel PN, Holden CP, Mackiewicz M, Pack AI, Geiger JD (2002) Brain glycogen decreases with increased periods of wakefulness: implications for homeostatic drive to sleep. J Neurosci 22:5581–5587

Lizarraga-Mollinedo E, Fernandez-Millan E, de Miguel-Santos L, Martinez-Honduvilla CJ, Alvarez C, Escriva F (2010) Early undernutrition increases glycogen content and reduces the activated forms of GSK3, AMPK, p38 MAPK, and JNK in the cerebral cortex of suckling rats. J Neurochem 112:123–133

Nelson SR, Schulz DW, Passonneau JV, Lowry OH (1968) Control of glycogen levels in brain. J Neurochem 15:1271–1279

Pellegri G, Rossier C, Magistretti PJ, Martin JL (1996) Cloning, localization and induction of mouse brain glycogen synthase. Brain Res Mol Brain Res 38:191–199

Pellerin L (2003) Lactate as a pivotal element in neuron-glia metabolic cooperation. Neurochem Int 43:331–338

Pfeiffer-Guglielmi B, Fleckenstein B, Jung G, Hamprecht B (2003) Immunocytochemical localization of glycogen phosphorylase isozymes in rat nervous tissues by using isozyme-specific antibodies. J Neurochem 85:73–81

Ransom BR, Fern R (1997) Does astrocytic glycogen benefit axon function and survival in CNS white matter during glucose deprivation? Glia 21:134–141

Rosengarten H (1970) Glycogen level and oxygen consumption of rabbit brain tissue slices under ethyl ether and halothane anesthesia. Agressologie 11:159–162

Stryer L (1995) Biochemistry. W.H. Freeman & Co, New York

Stys PK, Ransom BR, Waxman SG (1991) Compound action potential of nerve recorded by suction electrode: a theoretical and experimental analysis. Brain Res 546:18–32

Stys PK, Waxman SG, Ransom BR (1992) Ionic mechanisms of anoxic injury in mammalian CNS white matter: role of Na^+ channels and Na^+-Ca^{2+} exchanger. J Neurosci 12:430–439

Swanson RA, Choi DW (1993) Glial glycogen stores affect neuronal survival during glucose deprivation *in vitro*. J Cereb Blood Flow Metab 13:162–169

Swanson RA, Sagar SM, Sharp FR (1989) Regional brain glycogen stores and metabolism during complete global ischaemia. Neurol Res 11:24–28

Swanson RA, Morton MM, Sagar SM, Sharp FR (1992) Sensory stimulation induces local cerebral glycogenolysis: demonstration by autoradiography. Neuroscience 51:451–461

Wender R, Brown AM, Fern R, Swanson RA, Farrell K, Ransom BR (2000) Astrocytic glycogen influences axon function and survival during glucose deprivation in central white matter. J Neurosci 20:6804–6810

Chapter 24
Pyruvate Transport and Metabolism in the Central Nervous System

Tiago B. Rodrigues, Alejandra Sierra, Paloma Ballesteros, and Sebastián Cerdán

Abstract We review the transport and enzyme systems involved in cerebral pyruvate metabolism, combining the information derived from recent genome sequencing technologies and *in vivo* or *in vitro* ^{13}C and (^{13}C, ^{2}H) NMR approaches. Emphasis is placed on the role of subcellular compartmentation in the metabolic coupling between neurons and glial cells during glutamatergic neurotransmission. Proton-linked monocarboxylate transport through the plasma membrane utilizes the monocarboxylate transporters MCT as coded by the family of SLC16 genes, of which MCT4 and MCT2 are found in astrocytes and neurons, respectively. Cytosolic metabolism of monocarboxylates is kinetically compartmented in neurons and astrocytes with two different pyruvate pools originated from extracellular glucose or monocarboxylates, respectively. Mitochondrial transport of pyruvate is mediated by the pyruvate carrier PyC, a member of the mitochondrial carrier proteins, coded by the SLC25 gene family. Intramitochondrial oxidation of pyruvate in the cerebral tricarboxylic acid cycle is mainly controlled by pyruvate dehydrogenase, an ubiquitous multienzyme complex. Pyruvate carboxylase, an exclusively astrocytic enzyme, plays a fundamental anaplerotic role replenishing the glutamate and GABA pools involved in neurotransmission. Recent results obtained with cerebral mitochondria, primary cultures of neu-

T.B. Rodrigues, Ph.D.
Laboratory for Imaging and Spectroscopy by Magnetic Resonance (LISMAR),
Institute of Biomedical Research "Alberto Sols" CSIC/UAM, Madrid, Spain
e-mail: tiago.rodrigues@cancer.org.uk; tiago.b.rodrigues@gmail.com

A. Sierra • S. Cerdán, Ph.D. (✉)
Laboratory for Imaging and Spectroscopy by Magnetic Resonance (LISMAR),
Institute of Biomedical Research "Alberto Sols" CSIC/UAM, Calle Arturo Duperier 4, E-28029, Madrid, Spain
e-mail: scerdan@iib.uam.es

P. Ballesteros, Ph.D.
Laboratory of Organic Synthesis and Molecular Imaging,
University at a Distance UNED, Paseo Senda del Rey 9, E-28040, Madrid, Spain
e-mail: pballesteros@ccia.uned.es

I.-Y. Choi and R. Gruetter (eds.), *Neural Metabolism In Vivo*,
Advances in Neurobiology 4, DOI 10.1007/978-1-4614-1788-0_24,

rons and astrocytes or in the *in vivo* brain with (^{13}C,^{2}H) NMR have revealed the presence of two kinetically different glutamate pools in neurons and astrocytes. The subcellular compartmentation of glutamate and pyruvate were not considered in previous interpretations of metabolic coupling between neurons and glial cells *in vivo*. To account for these findings we proposed a novel redox coupling mechanism incorporating intracellular glutamate and monocarboxylate compartmentation. Transcellular redox coupling is based on: (a) the intracellular coupling of glycolysis and oxidation in neurons and astrocytes through NAD(P)/NAD(P)H redox switches, (b) the transcellular coupling of NAD(P)/NAD(P)H redox states in neurons and glial cells through the intercellular exchange of monocarboxylate reducing equivalents and (c) the glutamate-glutamine cycle, exchanging only the cytosolic (or vesicular) pools of glutamate and glutamine in both neural cells.

Keywords (^{13}C,^{2}H) NMR • Gene cloning • Glycolysis • Monocarboxylate transport • Redox coupling • Tricarboxylic acid cycle

24.1 Introduction

Pyruvate plays a pivotal role in the metabolism of the Central Nervous System (CNS), linking the anaerobic and aerobic portions of glucose catabolism and providing the carbon skeletons that couple the vital energy exchange between neurons and astrocytes (Denton and Halestrap 1979; Sokoloff 1989, 1992). Pyruvate is known to be an essential nutrient for neural cells in culture and has important neuroprotective actions, contributing also the building blocks for many anabolic activities in the brain (Desagher et al. 1997; Izumi et al. 1997; Matsumoto et al. 1994). Among these functions, cerebral pyruvate oxidation in the mitochondria entails vital bioenergetic importance, having a pronounced influence on the general physiology of the individual (Dienel 2002). An adult human brain of 1,400 grams consumes approximately 55 mg of glucose min^{-1} and produces 110 mg of pyruvate min^{-1} approximately. Oxidation of these pyruvate molecules to CO_2 requires approximately 49 mL of O_2 min^{-1}. This indicates that approximately 20% of the total oxygen used by the adult body is consumed by the brain, an organ that accounts for only 2% of the body weight (Nicholls 2007; Okada and Lipton 2007).

The cerebral metabolism of pyruvate, and that of its precursor glucose, has been reviewed several times. Classical approaches involved the use of radioactive isotopes or optical methods which require the isolation and purification of the enzymes, transport systems or metabolites involved to investigate their *in vitro* kinetics (Bachelard 1989; Clark and Lai 1989; Siesjo 1982; Sokoloff 1989). This reductionist approach provided fundamental information on the operation of the CNS, despite the limitations imposed by the minute amounts of proteins present in neural tissues and by the fact that this methodology is restricted to cerebral extracts or *post mortem* biopsies.

An increasing amount of evidences indicate that neurons and glial cells depict important differences in pyruvate transport and metabolism (Bittar et al. 1996; Laughton et al. 2000; Pellerin et al. 1998b). Astrocytes have been proposed to have a limited oxidative capacity, dedicating their intracellular pyruvate mainly to lactate formation. In contrast, neurons are normally thought to oxidize primarily their intracellular pyruvate to CO_2 and water. These two different pyruvate fates in both cell types have been proposed to be metabolically coupled by a lactate transfer from astrocytes to neurons supporting glutamatergic neurotransmission (Pellerin and Magistretti 1994). In general, it is becoming increasingly apparent that both types of neural cells are metabolically dependent of each other, making it difficult to understand neuronal metabolism *in vivo* without considering the glial contributions and viceversa.

The impressive development during the past three decades of powerful non-invasive tools for studying the human brain has had a tremendous impact on our ability to investigate brain function. Positron Emission Tomography (PET) with (^{18}F)-2-deoxyglucose (FDG) was the first technique used to investigate *non invasively* cerebral glucose uptake in animals and man with moderate resolution (Wienhard 2002). Similarly, functional Magnetic Resonance Imaging (fMRI) allowed indirectly the investigation of the hemodynamic and blood oxygenation changes associated to sensory or motor stimulation (Heeger and Ress 2002). Despite its importance, FDG uptake or fMRI provide no information on the pathways and metabolic interactions underlying the cerebral activation process, hence further advances in this field would require additional methodologies.

From this perspective, genome cloning and sequencing, as well as the development of novel ^{13}C Nuclear Magnetic Resonance (NMR) strategies have overcome many of the limitations of the traditional approaches. In particular, sequencing of the human and mouse genomes has provided a comprehensive view of the different isoforms of enzymes and transporters present in the brain at the gene level, without the need to isolate and purify the corresponding proteins (Consortium IHGS 2001; Consortium MGS 2002). However, the genomic approaches do not provide information on potential post-translational modifications and thus may not allow the comprehensive investigation of the *in vivo* performance of the genes sequenced or cloned. In this respect ^{13}C NMR, and more recently (^{13}C,^{2}H) NMR, technologies have become more helpful, providing the quantitative assessment of transport steps, metabolic fluxes and subcellular compartmentation of pyruvate oxidation in a plethora of neural systems ranging from primary cell cultures to the intact rodent or human brain (Cerdan 2003; Gruetter et al. 2003; Morris and Bachelard 2003; Shulman et al. 2004).

In this review, we address cerebral pyruvate metabolism from an integrative perspective, combining the classical neurochemical approach, involving isolation and purification of the different transport and enzymes, with more recent knowledge on their *in vivo* operation as derived from ^{13}C and (^{13}C, ^{2}H) NMR approaches. We place emphasis on the subcellular compartmentation of pyruvate and glutamate in neural cells and how this can modify the traditional interpretations of neuroglial coupling during glutamatergic neurotransmission. Several recent reviews complement and extend the information presented here (Cerdan et al. 2006; De Meirleir 2002; Garcia-Espinosa et al. 2004; Gruetter 2002; Magistretti and Pellerin 1999; Rodrigues and Cerdan 2005a; Rothman et al. 2003; Schurr 2002).

24.2 Overview of Pyruvate Metabolism

Figure 24.1 provides a general overview of pyruvate oxidation in the brain and its interactions with other cerebral metabolic pathways. Pyruvate oxidation involves; (a) the transport of glucose or monocarboxylates through the blood brain barrier (BBB) and the plasma membrane of neural cells, (b) the production of cytosolic pyruvate and reducing equivalents through the glycolytic pathway, (c) the transport of cytosolic pyruvate and reducing equivalents to the mitochondrial matrix by the carriers of the inner mitochondrial membrane and finally, (d) the terminal oxidation of pyruvate in the mitochondrial tricarboxylic acid (TCA) cycle.

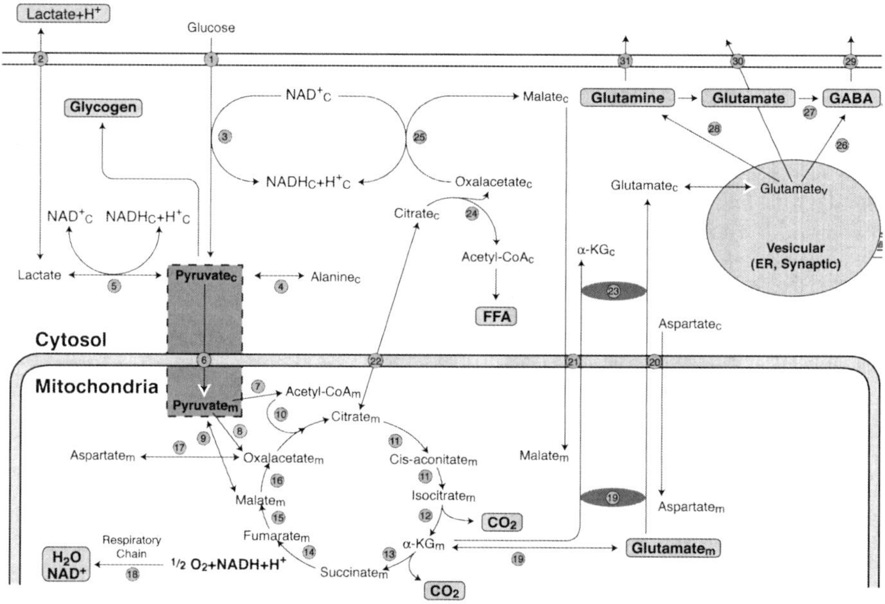

Fig. 24.1 Overview of pyruvate transport and metabolism. Intercellular compartmentation is not considered for simplicity. 1 GLUT transporters of the plasma membrane, 2 MCT transporters of the plasma membrane, 3 glycolysis, 4 alanine aminotransferase, 5 lactate dehydrogenase, 6 pyruvate carrier (PyC) of the inner mitochondrial membrane, 7 pyruvate dehydrogenase complex, 8 pyruvate carboxylase, 9 malic enzyme, 10 citrate synthase, 11 aconitase, 12 isocitrate dehydrogenase, 13 α-ketoglutarate dehydrogenase, 14 succinate dehydrogenase, 15 fumarase, 16 malate dehydrogenase, 17 aspartate aminotransferase (mitochondrial), 18 respiratory chain, 19 aspartate aminotransferase + glutamate dehydrogenase (mitochondrial), 20 aspartate/glutamate exchanger, 21 α-ketoglutarate/malate exchanger (OGC), 22 tricarboxylate carrier (CIC), 23 aspartate aminotransferase (cytosolic), 24 citrate lyase, 25 malate dehydrogenase (cytosolic), 26 glutamate decarboxylase (vesicular), 27 glutamate decarboxylase (cytosolic), 28 glutamine synthetase, 29 GABA transporters, 30 glutamate transporters, 31 glutamine transporters. Subscripts c, m and v refer to cytosolic, mitochondrial or vesicular (synaptosomal) compartments. *FFA* Free Fatty Acids

Glucose and monocarboxylates are transported through the plasma membrane by glucose and monocarboxylate transporters, respectively (Fig. 24.1, processes 1 and 2) (Dwyer et al. 2002; Halestrap and Meredith 2004; Palmieri 2004). Glycolytic pyruvate production from glucose in the cytosol, generates two molecules of NADH and two of ATP (Dienel 2002) (process 3). Cytosolic pyruvate may then be transaminated to alanine by alanine aminotransferase (ALT, process 4; E.C. 2.6.1.2), reduced to lactate by lactate dehydrogenase (LDH, process 5; E.C. 1.1.1.27), or transported to the mitochondrial matrix (process 6) and oxidized (processes 7–16) (Denton and Halestrap 1979). Under normoxic conditions, the mitochondrial respiratory chain is able to reoxidize completely the reducing equivalents (NADH and $FADH_2$) generated in aerobic glycolysis and the TCA cycle (process 18). Since the mitochondrial membrane is impermeable to NADH, the transfer of cytosolic reducing equivalents through the inner mitochondrial membrane requires the operation of specific shuttle mechanisms (processes 20–22) (Palmieri 1994; Palmieri et al. 1996). Oxygen limitations in hypoxic or ischemic conditions may prevent the respiratory chain to reoxidize completely the reducing equivalents generated by glycolysis or during TCA cycle activity. This decreases the mitochondrial and cytosolic NAD(P)/NAD(P)H ratio and causes cytosolic pyruvate to be reduced to lactate and extruded (processes 5 and 2) (Bachelard et al. 1974). Under these conditions, the oxidative production of ATP is reduced and flux through the glycolytic pathway is upregulated, resulting in increased glycerol-3-P levels and increased lactate and alanine productions (Ben-Yoseph et al. 1993; Siesjo 1982).

Mitochondrial pyruvate metabolism plays additional roles. Pyruvate is the main precursor for fatty acid synthesis in the CNS (Moore 2001). In this process, intramitochondrial citrate is transported to the cytosol, in exchange for malate (process 22), to produce cytosolic oxalacetate (OAA) and acetyl-CoA by citrate lyase (process 24, E.C. 4.1.3.6) in order to support fatty acid synthesis (Palmieri 1994, 2004). Cytosolic pyruvate can also be produced from mitochondrial TCA cycle intermediates, as malate, through the combined actions of phosphoenolpyruvate carboxykinase (E.C. 4.1.1.32) and pyruvate kinase (E.C. 2.7.1.40) or the malic enzyme reaction (ME; E.C. 1.1.1.40) in the process known as pyruvate recycling (process 9), a pathway thought to support fatty acid synthesis and free-radical scavenging activities of neural cells (Cerdan et al. 1990; Cruz et al. 1998; Pascual et al. 1998). Finally, mitochondrial pyruvate is a crucial precursor for cerebral anaplerosis, helping to replace amino acids and TCA cycle intermediates liberated to the extracellular space during neurotransmission. Anaplerosis involves a net increase in OAA molecules produced through pyruvate carboxylase (PC; E.C. 6.4.1.1) in astrocytes and, possibly, through the reversal of ME activity in the neurons (process 8 or 9) (Gamberino et al. 1997; Hassel 2000). Closely associated to this anaplerotic role, pyruvate, or its direct products: lactate or alanine, may produce glucose and glycogen in astroglial cells, in a similar way to that observed in liver (Dringen et al. 1993; Schmoll et al. 1995). The following sections address in more detail these aspects.

24.3 Transport of Pyruvate and Monocarboxylates through the Blood Brain Barrier and the Plasma Membrane of Neural Cells.

Pyruvate and other monocarboxylates are classically known to pass the BBB through facilitated diffusion (Cremer et al. 1979; Crone and Sorensen 1970). Transport of monocarboxylates through the plasma membrane was attributed initially to non ionic diffusion, until it was demonstrated that it could be inhibited by α-cyano-4-hydroxycinamate and organomercurials (Halestrap and Price 1999). This led to the identification of the first member of the family of monocarboxylate-proton cotransporter proteins, MCT1 (Enerson and Drewes 2003). MCTs were shown to operate under near equilibrium conditions, catalyzing the reversible cotransport of the monocarboxylate and one proton. In the case of pyruvate the following equation applies,

$$\text{Pyruvate}_{\text{out}} + \text{H}^{+}_{\text{out}} \leftrightarrow \text{Pyruvate}_{\text{in}} + \text{H}^{+}_{\text{in}} \tag{24.1}$$

where the subscripts $_{\text{-in}}$ or $_{\text{-out}}$ refer to intra- or extracellular concentrations, respectively. A similar reaction can be written for other monocarboxylates including lactate, acetate and acetoacetate. Under near equilibrium conditions the relative concentration of pyruvate molecules in the intra- and extracellular spaces is believed to be dictated by the transmembrane pH gradient,

$$\text{Pyruvate}_{\text{in}} / \text{Pyruvate}_{\text{out}} = \text{H}^{+}_{\text{out}} / \text{H}^{+}_{\text{in}} \tag{24.2}$$

If the intracellular pH value is 0.1 units less than that of the extracellular space, intracellular monocarboxylate levels will be 80% that of extracellular concentrations. However, under ischemic or hypoxic conditions when the intracellular pH may be 1.0 unit above extracellular values, intracellular monocarboxylate levels will be an order of magnitude higher than the extracellular concentrations. These important changes in monocarboxylate distribution through the plasma membrane reflect the reversibility of MCTs. Its kinetic parameters are known to be constrained by the Haldane relationship

$$(V_{\text{max}} / K_{\text{m}})_{\text{influx}} = (V_{\text{max}} / K_{\text{m}})_{\text{efflux}} \tag{24.3}$$

where V_{max} of influx or efflux can be stimulated by either decreasing the pH in the same side of the monocarboxylate or by increasing it in the opposite, respectively (Halestrap and Price 1999).

The monocarboxylate transporter proteins are coded by the SCL16 gene family (Halestrap and Meredith 2004). Up to fourteen, relatively homologous, monocarboxylate transporters are coded by this gene family of which only four have been isolated, purified and overexpressed in *Xenopus oocytes*, to our knowledge. Although the three

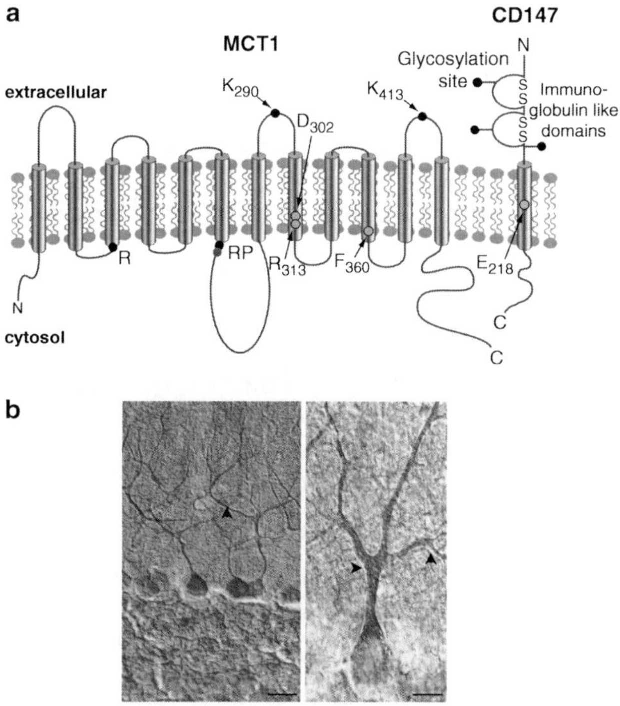

Fig. 24.2 Proposed topology and cellular localization of the monocarboxylate transporter family (MCT). (**a**) Transmembrane domains of the MCT1. CD147, the ancillary protein that associates with MCT1 and MCT4, is also shown. (**b**) Photomicrographies at high magnification of the selective expression of MCT2 in the Purkinje cell bodies and dendrites at P21. Scale bar in the *left panel* is 60 μm and in the *right panel* is 20 μm. *Arrowheads* indicate immunonegative glial lamellae ensheathing the Purkinje cell stem dendrites (Reproduced from Halestrap and Meredith (2004) and Rafiki et al. (2003). With permission of the publisher)

dimensional structure of these transmembrane proteins has not been determined, it is possible to infer some of their structural properties from their primary sequence. The common topology involves up to twelve transmembrane domains, with both C- and N-termini facing the intracellular side (Fig. 24.2a). In addition, at least MCT1 and MCT4 (and probably MCT3) require the ancillary protein CD147 for their correct expression in the plasma membrane (Kirk et al. 2000). MCT1, MCT2 and MCT4 are heterogeneously expressed in brain (Fayol et al. 2004; Rafiki et al. 2003). MCT1 is almost ubiquitously expressed, being found in endothelial cells, neurons, astrocytes as well as in glioma cells. Conversely, MCT2 is expressed only in neurons and MCT4 is expressed in the endothelial or glial cells (Mac and Nalecz 2003) (Fig. 24.2b). The kinetic properties of MCT1, MCT2 and MCT4 are summarized in Table 24.1.

The pattern of expression and the kinetic properties of these transporters support the classical model of vectorial lactate trafficking between neurons and astrocytes (Pellerin et al. 1998b). The low affinity of astrocytic MCT4 favors lactate extrusion

Table 24.1 Kinetic properties of monocarboxylate transporters from the central nervous system, as expressed in Xenopus oocytes (Taken from Halestrap and Price (1999) and (Rafiki et al. (2003))

	Km (mM)		
Substrate/inhibitor	MCT1	MCT2	MCT4
Pyruvate	1.0	0.1	28.0
Lactate	3.5	0.5	150.0
β-hydroxybutyrate	12.5	1.2	
Acetoacetate	5.5	0.8	

from astrocytes, a situation similar to that found in highly glycolytic tissues such as the white fibers of muscle. In contrast, the higher affinity of neuronal MCT2 favors lactate uptake, as found primarily in oxidative tissues as heart and the red fibers of muscle (Brooks 2002). Thus, heterogeneous distribution of MCTs with different kinetic properties supports astrocytic lactate production and neuronal lactate consumption (Pellerin et al. 1998a).

MCT expression of the different isoforms depicts important changes during ontogenic development (Leino et al. 1999; Vannucci and Simpson 2003). Neuronal MCT2 is expressed in most brain areas including hippocampus and cerebellum, being most abundant during the first 3 weeks after birth and decreasing progressively later. This pattern appears to be complementary to that of the glucose transporters GLUT1, endothelial, and GLUT3, neuronal (Vannucci et al. 1997). Both GLUT1 and GLUT3 increase progressively after birth, while MCT2 decreases. In contrast, MCT4 expression is low at birth and increases progressively through development. The temporal pattern of expression of MCTs and GLUTs parallel the decrease in lactate utilization and the increase in glucose metabolism observed during cerebral maturation (Clark et al. 1993, 1994; Dwyer et al. 2002; Medina et al. 1996; Vannucci et al. 1997; Vannucci and Simpson 2003).

24.3.1 (^{13}C, ^{2}H) Methodologies to Investigate Monocarboxylate Transport

Classical methods to investigate pyruvate and lactate transport in neural cells or subcellular organelles included mainly radioactive or fluorescent methods. These techniques provided important insight into the kinetics of net uptake or release of the different monocarboxylates (Halestrap and Price 1999). More recently, it has become possible for the first time to determine simultaneously monocarboxylate flux in both directions of monocarboxylate transport using (^{13}C, ^{2}H) NMR as illustrated in Fig. 24.3. The method is based on the fact that, in incubation media containing $^{2}H_2O$, lactate molecules entering the cell, could exchange very rapidly their H2 proton by a deuteron (Fig. 24.3A). This process occurs only in the cytosol because of the LDH equilibrium, and is not observed under similar conditions in the incubation media without cells (Rodrigues and Cerdan 2005b; Rodrigues et al. 2005).

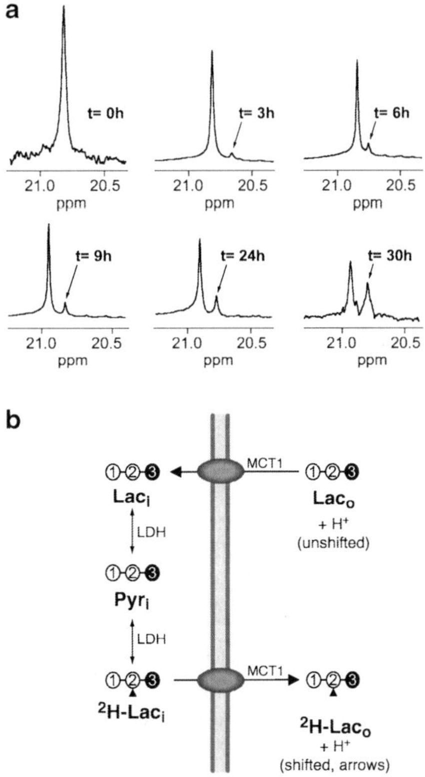

Fig. 24.3 Lactate recycling through the plasma membrane of C6 cells as detected by (^{13}C, ^{2}H) NMR. (**a**) C6 cells were incubated with 5 mM (3-^{13}C) lactate in Krebs-Henseleit buffer containing 50% ^2H_2O (vol/vol). Aliquots from the incubation medium were taken at different times and analyzed by ^{13}C NMR. Two singlet resonances derived from the methyl groups of (3-^{13}C) and (3-^{13}C, 2-^{2}H) lactate are detected at 20.9 and 20.8 ppm, respectively. The latter is derived from recycled lactate. B: Mechanism of H2 deuteration of (3-^{13}C) lactate. Only (3-^{13}C) lactate molecules passing through the cytosolic compartment are deuterated in H2 by the lactate dehydrogenase equilibrium. *Black circles* ^{13}C, *black triangles* ^{2}H (Adapted from (Rodrigues et al. 2005). Reproduced with permission of the publisher)

Figure 24.3A shows an illustrative example of the time course of changes in the lactate C3 resonance from (3-^{13}C) and (3-^{13}C, 2-^{2}H) lactates in the incubation medium. The C6 cells were incubated with 5 mM (3-^{13}C) lactate in Krebs-Henseleit buffer containing 50% ^2H_2O (vol/vol). The singlet resonance from (3-^{13}C) lactate decreases linearly, revealing the net consumption of this precursor. Simultaneously, the shifted singlet signal originated from the methyl group of (3-^{13}C, 2-^{2}H) lactate increases exponentially, revealing a faster (3-^{13}C, 2-^{2}H) lactate extrusion. Both resonances can be easily resolved because the H2 deuteron induces an upfield isotopic shift in the C3 resonance of the same lactate molecule. This allows distinguishing easily by (^{13}C, ^{2}H) NMR the (3-^{13}C, 2-^{2}H) lactate molecules leaving the cell, after cytosolic deuteration through the LDH equilibrium.

24.4 Cytosolic Metabolism of Pyruvate

Pyruvate is produced in the cytosol of neural cells either from glucose, through the glycolytic pathway, or from extracellular lactate, alanine or pyruvate, after transport and equilibration in the LDH or ALT reactions (cf. Fig. 24.1). Cytosolic pyruvate may then be removed to the mitochondrial matrix through the monocarboxylate

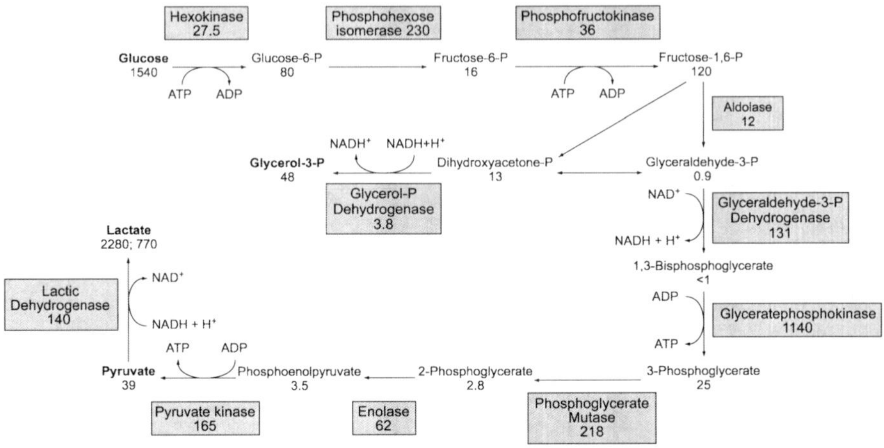

Fig. 24.4 The cerebral glycolytic pathway. Numbers below the enzymes and metabolites refer to the maximal velocity (μmol g^{-1} wet weight. min^{-1}) and concentration (nmol. g^{-1} wet weight) as measured in brain homogenates and acid extracts, respectively (Taken from Clarke et al. (1989b). Reproduced with permission of the publisher

carriers of the inner mitochondrial membrane, or extruded to the extracellular space in the form of lactate or alanine through the corresponding MCTs or amino acid transporters. Cytosolic pyruvate may also be produced from malate in the ME reaction.

The glycolytic pathway (Fig. 24.4) is a sequence of ten enzyme catalyzed reactions degrading glucose to two molecules of pyruvate:

$$\text{Glucose} + 2\text{NAD}^+ + 2\text{ADP} + 2\text{P}_i \rightarrow 2\ \text{Pyruvate} + 2\text{NADH} + 2\text{H}^+ + 2\text{ATP} + 2\text{H}_2\text{O} \quad (24.4)$$

Hexokinase (HK, E.C. 2.7.1.1), phosphofructokinase (PFK, E.C. 2.7.1.11) and pyruvate kinase (PK, E.C. 2.7.1.40) are able to modulate glycolytic flow in most tissues. However, the classical work of Lowry and Saktor showed that in brain, aerobic glycolysis is regulated primarily at the PFK and HK steps (Passonneau and Lowry 1964; Sacktor et al. 1966). A schematic description of this pathway is given in Fig. 24.4, with maximal activities of the enzymes and concentrations of metabolites involved (Clarke et al. 1989a). Glycolytic flux *in vivo* represents approximately 5% of the maximal possible flux through the pathway, revealing the important metabolic capacity of adaptation of this pathway and the need for a strict regulation. Literature data on the kinetic properties and structure of the main glycolytic enzymes are broad. In this review, we will focus exclusively on PK, LDH and ALT, the main enzymes involving pyruvate as product or substrate.

24.4.1 Pyruvate Kinase

PK catalyzes the irreversible phosphoryl transfer from phosphoenolpyruvate to ADP, yielding ATP and pyruvate.

$$\text{Phosphoenolpyruvate} + \text{ADP} \rightarrow \text{Pyruvate} + \text{ATP} \qquad (24.5)$$

PK exists in the cytosol of mammalian cells as tetramers of subunits 58–62 kDa. There are four isoenzymes PK-L, PK-R, PK-M_1 and PK-M_2 types, respectively (Yamada and Noguchi 1999a, b). These isoenzymes are coded by two genes, the L-gene coding for PK-R and PK-L and the M-gene codifying the PK-M_1 and M_2 isoforms by alternative splicing (Takenaka et al. 1989). The isoform found in brain is PK-M_1, the same found in skeletal muscle and heart. The brain isoform is not activated by fructose 2,6-bisphosphate or phosphorylation by protein kinase A, like the isoforms of liver PK-L, or such as erythrocytes PK-R and the widely distributed PK-M_2 (Yamada and Noguchi 1999a, b). PK-M_1 is characterized by Km's for PEP and ATP of 0.08 and 3–3.5 mM, respectively. The ontogenic and regional expression of PK-M_1 parallels those of other key glycolytic enzymes, as HK and PFK. PK-C expression increases progressively after birth and is more abundant in the cerebral cortex than in basal ganglia, pons and medulla (Clark and Lai 1989).

24.4.2 Lactate Dehydrogenase

LDH represents the end point of anaerobic glycolysis, catalyzing the NADH-dependent reduction of pyruvate to lactate and NAD^+, in a reversible manner (Everse and Kaplan 1973).

$$\text{Pyruvate} + \text{NADH} + H^+ \leftrightarrow \text{Lactate} + \text{NAD}^+ \qquad (24.6)$$

Cerebral LDH appears in five different isoforms, derived from two parental homotetramers and three hybrids. These are: LDH1 (also known as LDH-H or LDH-A, isolated from heart) or LDH5 (also known as LDH-M or LDH-B, isolated from muscle) (Bonavita et al. 1964). The remaining isoforms arise from different combinations of H and M subunits (approximately 36 kDa), H4 (LDH1), H3M (LDH2), H2M2 (LDH3), HM3 (LDH4), M4 (LDH5). LDH1 is detected only in neurons and its kinetic properties favor mainly pyruvate formation (Bittar et al. 1996; Laughton et al. 2000) while LDH5 is present mainly in astrocytes and its kinetic properties favor lactate formation (Dawson et al. 1964; Richard 1963). Thus, the distribution of LDH isoenzymes is also adequate to support kinetically the astrocyte to neuron lactate shuttle (Pellerin et al. 1998a). Cerebral LDH is believed to be cytosolic in most tissues, with the exception of the testis, where it has a mitochondrial distribution. Nonetheless, recent evidence indicates that at least part of hepatic and cardiac

LDH may be mitochondrial (Brooks et al. 1999). LDH1 and LDH5 isoenzymes are also heterogeneously distributed within the different cerebral structures, as revealed by in situ m-RNA hybridization (Laughton et al. 2000). LDH also experiences important changes during ontogeny, increasing progressively after birth, in parallel with other glycolytic enzymes and transporters (Medina et al. 1996).

24.4.3 Alanine Aminotransferase

ALT catalyzes the glutamate or aspartate dependent transamination of pyruvate to alanine. Its activity in the brain is relatively low, existing as cytosolic and mitochondrial forms (Ruscak et al. 1982). The Km's (mM) for alanine, glutamate, α-ketoglutarate and pyruvate are very different for the mitochondrial (5.1, 6.6, 0.7 and 0.4) and cytosolic (30.3, 4.3, 0.7 and 0.6) isoforms. ALT has higher specific activity in cortical astrocytes with 17 nmol mg $protein^{-1}$ min^{-1}, than in glutamatergic and GABAergic neurons, 4.7 and 4.5 nmol mg $protein^{-1}$ min^{-1}, respectively (Westergaard et al. 1993). These results led to suggest an intercellular cycle for alanine in glutamatergic neurons, where alanine can be used as a metabolic fuel (Schousboe et al. 2003; Waagepetersen et al. 2000). In this case, it has been proposed that alanine could serve as a carrier of ammonia nitrogen from the neuronal to the astrocytic compartment, compensating for the lactate flux operating in the opposite direction (Schousboe et al. 2003; Waagepetersen et al. 2000). This intercellular cycle was also suggested in GABAergic neurons (Zwingmann et al. 2000).

24.4.4 Malic Enzyme

ME catalyzes the reversible $NAD(P)^+$ dependent oxidative decarboxylation of malate to pyruvate (Chang and Tong 2003). Three isoforms of ME have been described: a cytosolic $NADP^+$-dependent (c-NADP-ME), a mitochondrial $NADP^+$-dependent (m-NADP-ME) and a mitochondrial NAD^+-dependent (m-NAD-ME). The genes for these enzymes have been cloned (Chou et al. 1994; Loeber et al. 1991) and crystal structures are available (Yang et al. 2000). In the brain, ME exists in cytosolic and mitochondrial isoforms. The cytosolic isoenzyme appears to be expressed in astrocytes (Vogel et al. 1998a) while the mitochondrial is expressed in neurons (Vogel et al. 1998b). ME expression increases during ontogeny, reaching a peak during synaptogenic maturation in the second week and declining slightly thereafter (Cruz et al. 1998). Indeed, much larger activities of ME have been found in synaptosomes than in primary cultures of neurons or astrocytes (McKenna et al. 2000). The functions of cytosolic and mitochondrial MEs may be different. While the cytosolic isoenzyme may work as a lipogenic enzyme providing NADPH for lipid synthesis, the mitochondrial isoform could provide NADPH for detoxification processes. These considerations are of interest when interpreting the intercellular compartmentation of the pyruvate recycling system (Cerdan et al. 1990). While this system was originally described in the adult brain as a neuronal process, it was not possible initially to demonstrate its activity in primary cultures of neurons, but it appeared with high activity in cultures of astrocytes (Hassel and Sonnewald 1995; Waagepetersen et al. 2002). However, it was shown that both neurons and astrocytes express the two isoforms of ME and that its expression increases importantly during synaptogenesis and myelination (Cruz et al. 1998). Later the presence of ME in neuronal cultures was also demonstrated (Olstad et al. 2007). Therefore, it is now accepted that both neurons and astrocytes in the adult brain express the mitochondrial and cytosolic isoforms of ME and are consequently able to perform pyruvate recycling. These isoenzymes have different ontogenic properties and functions as redox modulators in neurons and astrocytes, as control of free radical damage and support of lipogenesis, respectively.

24.4.5 Cytosolic Compartmentation of Pyruvate as Detected by ^{13}C NMR

Figure 24.5 illustrates the existence of two kinetically different intracellular pools of pyruvate in primary cultures of neurons and astrocytes (Cruz et al. 2001; Zwingmann et al. 2001). Neurons incubated with mixtures containing 1 mM (1,2-$^{13}C_2$) glucose and increasing (3-^{13}C) pyruvate concentrations produced exclusively (3-^{13}C) lactate, rather than the predicted mixture of (3-^{13}C) and (2,3-$^{13}C_2$) lactate (Fig. 24.5a). This indicated that the cytosolic pools of (3-^{13}C) and (1,2-$^{13}C_2$) pyruvates derived from (3-^{13}C) pyruvate and (1,2-$^{13}C_2$) glucose did not equilibrate rapidly in the neuronal cytosol. One pool, denoted as P_p, is derived from extracellular pyruvate (or other monocarboxylates as lactate) and has a fast turnover, while the other pool denoted as P_g, is originated by glycolysis from extracellular glucose and turns over slower (Fig. 24.5B). As a result of fast MCT transport, changes in extracellular lactate levels are rapidly transferred to cytosolic lactate concentrations. Increased cytosolic lactate may completely switch off glycolysis by competition for NAD^+ with glyceraldehyde-3-phosphate dehydrogenase (GAPDH; E.C. 1.2.1.12). An important implication of this is that glycolytic flux may be modulated or even aborted by changes in the extracellular redox state, such as those occurring during neuronal activation.

Astrocytes have been shown to be able to produce glucose and glycogen from three carbon precursors in a similar way to that described in the liver (Bell et al. 1993; Dringen et al. 1993; Hevor 1994; Schmoll et al. 1995). This indicates that astrocytes are equipped with the battery of enzymes, bypassing the irreversible steps of glycolysis. Indeed, astrocytes express phosphoenolpyruvate carboxykinase (E.C. 4.1.1.32) (Schmoll et al. 1995) and PC (Shank et al. 1985) as bypasses to PK, fructose-1, 6-bisphosphatase (E.C. 3.1.3.11) as a bypass to PFK (Loffler et al. 2001) and glucose 6-phosphatase (E.C. 3.1.3.9) (Bell et al. 1993) as the bypass to HK. Notably, it has been recently possible to decrease astrocytic glycogen synthesis by administering antisense c-DNAs against fructose-1,6-bisphosphatase and glycogen synthase (E.C. 2.4.1.21) (Bernard-Helary et al. 2002a). The function of cerebral glycogen is unclear,

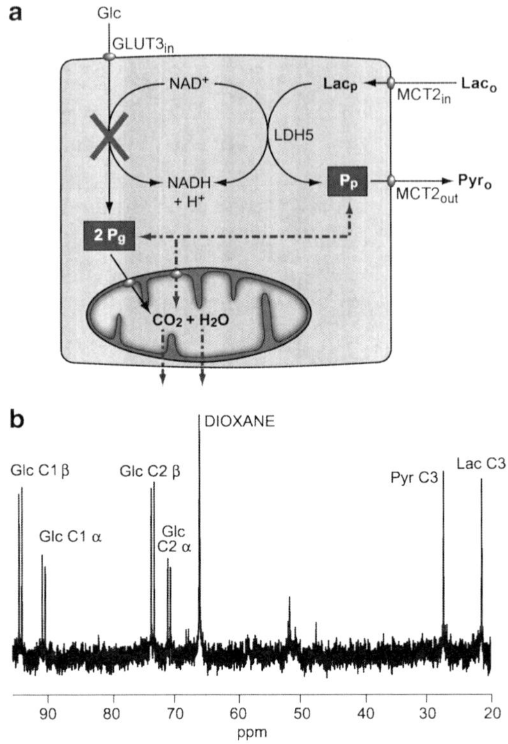

Fig. 24.5 Intracellular compartmentation of pyruvate in primary cultures of neurons. (**a**) Two kinetically different pyruvate pools exist in neurons. The fast turnover P_p pool is derived from extracellular lactate or pyruvate while the slow turnover P_g pool is derived from glucose. In the presence of high lactate concentrations, glycolytic flow may be reduced or stopped by competition with lactate dehydrogenase for cytosolic NAD^+. *Glc* glucose, *Pyr* pyruvate, *Lac* lactate. (**b**) ^{13}C NMR spectrum of neurons medium after incubation with 1 mM (1,2-$^{13}C_2$) glucose and 1 mM (3-^{13}C) pyruvate. Note that only (3-^{13}C) lactate is produced, as derived from (3-^{13}C) pyruvate. No (2,3-$^{13}C_2$) lactate is produced from (1,2-$^{13}C_2$) glucose (Reproduced from Cruz et al. (2001). With permission of the publisher)

but it could play an important role as glucose storage and delivery system in hypoglycemic and hypoxic situations or during seizure activity (Bernard-Helary et al. 2002b; Magistretti and Allaman 2007). The *in vivo* compartmentation of the cerebral pyruvate pool was demonstrated recently (Rodrigues et al. 2009). It was found that the (2-^{13}C) lactate and (3-^{13}C) lactate isotopomers obtained after co-infusion of (2-^{13}C) pyruvate and (1-^{13}C) glucose, in partially deuterated rats, depicted different deuteration patterns in both the C3 and C2 carbons, a finding inconsistent with the homogeneous deuteration of a rapidly mixing, single, intracellular pyruvate pool.

Finally, the use of novel ^{13}C hyperopolarization strategies is expected to provide important advances in cerebral pyruvate metabolism. Golman and colleagues demonstrated a 10.000 times sensitivity gain in ^{13}C NMR when using ^{13}C hyperopolarized substrates, as obtained with the Dynamic Nuclear Polarization method (Ardenkjaer-Larsen et al. 2003). It became possible to hyperpolarize (1-^{13}C) pyruvate and observe by *in vivo* ^{13}C NMR its transformation in (1-^{13}C) lactate, (1-^{13}C) alanine and $^{13}CO_3H^-$ in tumor cells and implanted tumors (Day et al. 2007). The enormous sensitivity gain of the method allowed the obtention of metabolic images of the spatial distribution of pyruvate, lactate or alanine using fast spectroscopic imaging methods (Golman et al. 2006). More quantitative studies are expected to provide additional evidence on the dynamics of monocarboxylate transport and metabolism in the seconds range.

24.5 Mitochondrial Transport

24.5.1 General Characteristics of Mitochondrial Carrier Proteins

Pyruvate and other cytosolic metabolites are transported through the inner mitochondrial membrane by the mitochondrial carrier family (MCF) of proteins (LaNoue and Schoolwerth 1979; Palmieri 1994, 2004; Palmieri et al. 1996; Satrustegui et al. 2007). Their kinetic properties are summarized in Table 24.2.

MCF are integral proteins of the inner mitochondrial membrane coded in man by SLC25 genes family (Palmieri 2004). Approximately 35 mRNA transcripts of this family have been observed, although only ten MCs have been purified, sequenced and reconstituted. The main members of the MCF include: the ATP translocators (AAC), the phosphate exchanger (PiC), the α-ketoglutarate/malate exchanger (OGC), the dicarboxylate carrier (DIC), the citrate/malate exchanger (CIC), the aspartate/glutamate exchanger (AGC), the carnitine and ornithine carriers (CAC and ORC), the uncoupling proteins (UCP) and the pyruvate carrier (PyC). The primary structure of transmembrane MCF proteins is relatively constant (Fig. 24.6). It depicts a tripartite sequence with three tandemly repeated domains of approximately 100 amino acids in length. Each domain contains two hydrophobic portions, believed to span the membrane as α-helices, separated by hydrophilic regions with the N and C terminals facing the intracellular side. Most isolated MC proteins have been

Table 24.2 Kinetic parameters of purified and reconstituted mitochondrial carrier proteins (Taken from Palmieri et al. (1996))

Carrier	Substrate	Km (mM) external	Km (mM) internal	Vmax (μmol/min g)	Activation energy (kJ)
AAC	ATP	0.01	0.025	Up to 7.1	160/62
	ADP	0.01			
PiC	Pi/Pi	1.8 ± 0.1	9.4 ± 0.1	92	64
	Pi/OH⁻		11.2 ± 0.9	53	
OGC	α-ketoglutarate	0.31 ± 0.08	0.17 ± 0.06	9.5 ± 1.5	54
	Malate	1.36 ± 0.14	0.71 ± 0.18	10.0 ± 1.1	
DIC	Malate	0.49 ± 0.05	0.92 ± 0.16	6.0 ± 1.6	96
	Malonate	0.54 ± 0.10	n.d.	5.9 ± 0.9	
	Phosphate	1.41 ± 0.35	0.93 ± 0.18	6.0 ± 1.4	
CIC	Citrate	0.032 ± 0.004	0.027 ± 0.004	10.5 ± 1.9	70
	Malate	0.25 ± 0.03	0.060 ± 0.006	11.5 ± 1.8	
AGC	Aspartate	0.12 ± 0.02	2.8 ± 0.6	n.d. *	77
	Glutamate	0.25	3		
CAC	Carnitine (exchange)	1.1 ± 0.1	10 ± 0.8	1.7 ± 1.3	133
	Carnitine (uniport)	0.53 ± 0.12	2.2 ± 0.4	0.2 ± 0.09	115
ORC	Ornithine	0.16 ± 0.02	4	3.2 ± 0.7	89
PyC	Pyruvate	0.24 ± 0.02	n.d	114 ± 21	n.d.

All data are from studies using hydrophobic chromatography for reconstitution, except those for the AAC, *Vmax value has not been determined with purified protein. *n.d.* not determined

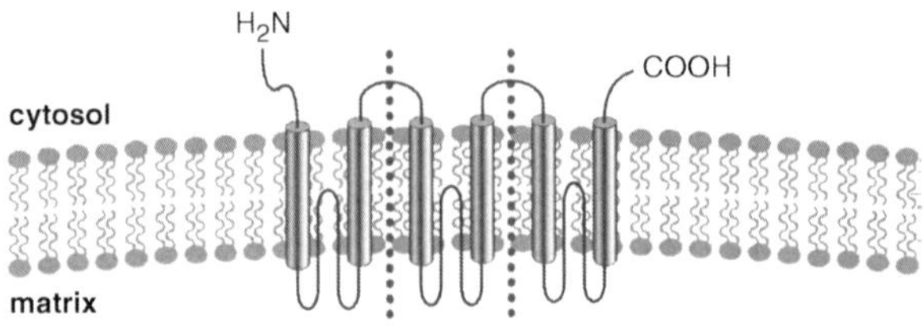

Fig. 24.6 Topological model of mitochondrial carrier monomers. Six helices traverse the inner mitochondrial membrane with the C- and N-termini facing the cytosol. The whole sequence is divided into three similar domains each with two transmembrane helices. Within each domain the two helices are connected by a long hydrophilic matrix loop (A, B and C respectively) which is assumed to protrude into the membrane (Adapted from Palmieri (2004). With permission of the publisher)

shown to be homodimers, probably consisting of a total of twelve transmembrane segments, as in the majority of carrier proteins.

Figure 24.7 summarizes the operation of different shuttle systems that exchange cytosolic and mitochondrial metabolites related to pyruvate metabolism and transfer of reducing equivalents, through the inner mitochondrial membrane.

24.5.2 The Pyruvate Carrier

Cytosolic pyruvate is transported through the inner mitochondrial membrane by the pyruvate carrier (PyC). This was one of the first to be isolated and purified (Bolli et al. 1989) but its precise precursor gene remains unidentified. The mitochondrial pyruvate transporter is a member of the MCF, which is completely unrelated to the monocarboxylate transporters MCT of the plasma membrane, presenting entirely distinct properties. Its Km for cytosolic pyruvate is 0.24 mM and its Vmax 114 $\mu mol\ mg^{-1}\ min^{-1}$, the largest transport capacity found among mitochondrial transporters. This makes PyC activity not to be limiting for the subsequent oxidative metabolism of pyruvate. The Km values for PyC are similar to those of LDH or ALT and similar to MCT1, suggesting that pyruvate transport to the mitochondria competes effectively with its cytosolic metabolism, the preferred process being determined by additional redox or energetic factors.

24.5.3 Other Related Carriers

The aspartate/glutamate mitochondrial carriers (AGC) have been recently identified (del Arco and Satrustegui 1998, 2004; Palmieri et al. 2001; Satrustegui et al. 2007). Two isoforms AGC1 (coded by SLC25A12) and AGC2 (coded by SLC25A13) exist, known as aralar1 and citrin, which belong to a subfamily of Ca^{2+} binding MCs. AGC1 is found mainly in excitable tissues whereas AGC2 is the only isoform expressed in liver. Aspartate is transported as an anion while glutamate is

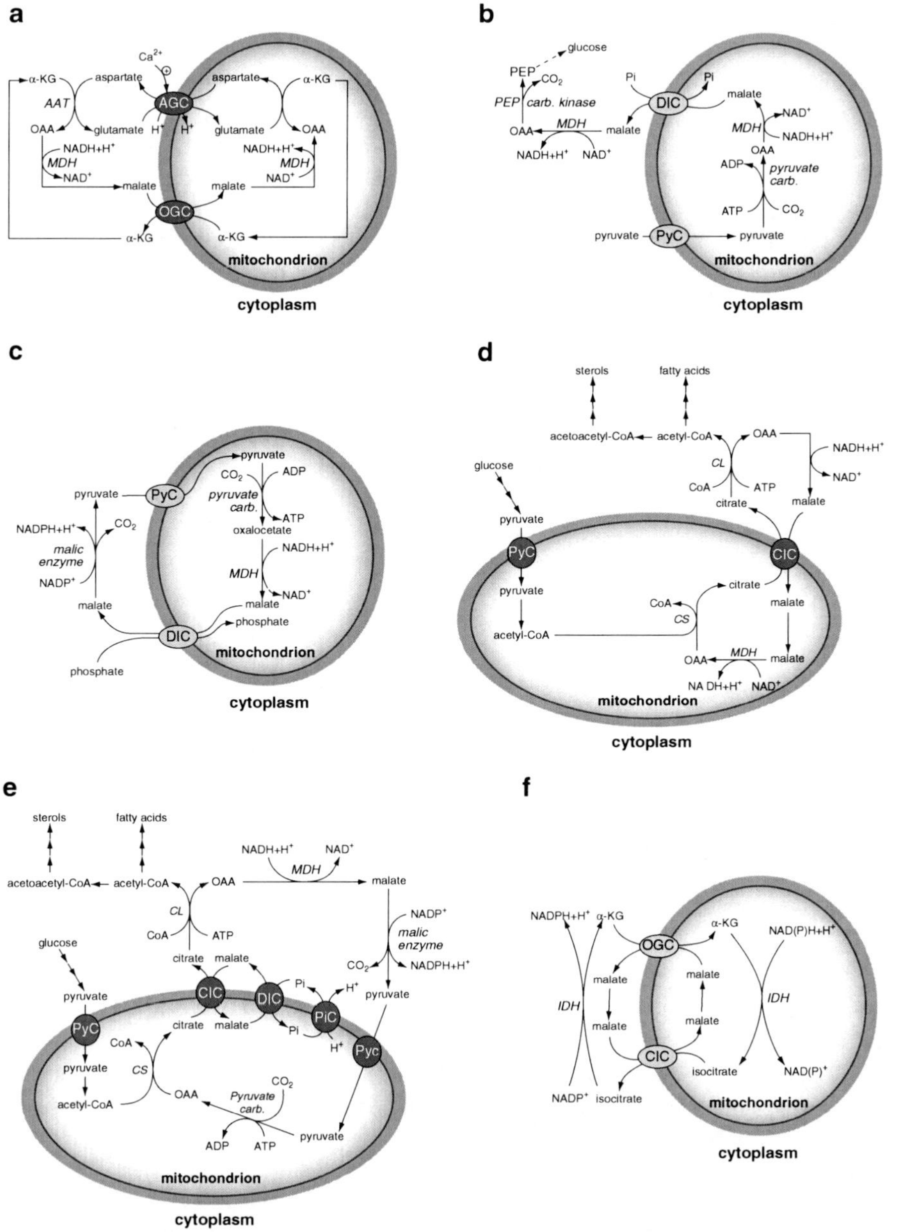

Fig. 24.7 The mitochondrial transporters present in neural cells and their roles. The schemes do not show all the mitochondrial transporters nor all the metabolic pathways in which individual carriers are involved. (**a**) The malate-aspartate shuttle for the transfer of cytosolic NADH to the mitochondrial matrix. (**b**) Extrusion of NADH to the cytosol for gluconeogenesis from pyruvate in astrocytes. (**c**) The pyruvate-malate shuttle; extrusion of reducing equivalents (NADPH) to the cytosol. (**d**) The citrate-malate shuttle; extrusion of citrate to the cytosol for acetyl-CoA generation and sterols and fatty acid synthesis. (**e**) The citrate-pyruvate shuttle; simultaneous extrusion of mitochondrial citrate, NAD^+ and NADPH for cytosolic sterol and fatty acid synthesis. (**f**) The α-ketoglutarate-isocitrate shuttle. Extrusion of mitochondrial isocitrate for cytosolic NADPH synthesis (Adapted from Palmieri (2004). Reproduced with permission of the publisher)

cotransported with one H^+, making AGC an electrogenic carrier. AGC plays a fundamental role in the malate-aspartate shuttle which transfers NADH from cytosol to mitochondria of neural cells. AGC is required also for an adequate supply of aspartate from the mitochondria to the cytosol as well as for the regulation of cytosolic NADH/NAD^+ ratio. Interestingly enough, aralar1 and citrin distributions have been found to vary with development and to be heterogeneous between neurons and astrocytes (Ramos et al. 2003). In the adult brain, aralar1 expression is much higher in neurons than in astrocytes, suggesting that mitochondrial reoxidation of NADH in the astroglia may probably rely on the contribution of a different mitochondrial carrier system for glutamate.

The glutamate carriers GC1 (coded by the SLC25A22 gene) and GC2 (coded by the SLC25A18 gene) catalyze the transport of L-glutamate to the intramitochondrial space either with one H^+ or in exchange with an OH^-. Both forms differ clearly in their kinetics. GC1 has a very high Km for external glutamate (4–5 mM), whereas GC2 has a much lower Km (0.2 mM). However, the Vmax of GC1 is significantly higher than that of GC2. Since glutamate is cotransported with one H^+, its distribution across the inner mitochondrial membrane is dependent on ΔpH, being favored in energized mitochondria.

The α-ketoglutarate carrier OGC (SLC25A11) also plays a fundamental role in the malate-aspartate shuttle. It catalyzes the electroneutral exchange of α-ketoglutarate for other dicarboxylates, mainly malate, being susceptible to inhibition by impermeable substrate analogs as phthalonic acid as well as by mercurials and maleimides. The dicarboxylate carrier DIC (SLC25A10) catalyzes the exchange of some dicarboxylates, such as malate and succinate, inorganic phosphate and inorganic sulphur-containing compounds. It is inhibited by butylmalonate. There are two binding sites on each side of the membrane, specific for dicarboxylates and phosphate, respectively. Km's for malate and phosphate are approximately 0.4 and 1.5 mM, respectively. DIC is used mainly in gluconeogenesis from pyruvate (and amino acids) and ureagenesis.

The tricarboxylate (or citrate) carrier CIC (SLC25A1) catalyzes the electroneutral exchange of citrate or isocitrate for, either another tricarboxylate, a dicarboxylate (malate) or phosphoenolpyruvate (PEP). Citrate malate exchange is electroneutral because it involves only the singly protonated form of citrate ($citrate^{2-}$). CIC is necessary for fatty acid and sterol biosynthesis. It exports citrate from the mitochondria to the cytosol, where it is cleaved by citrate lyase to OAA and acetyl-CoA which is used for fatty acid synthesis. Cytosolic OAA is reduced to malate, which can be decarboxylated to pyruvate by ME with production of cytosolic pyruvate and NADPH + H^+. This carrier is inhibited by 1,2,3-benzenetricarboxylate.

The ATP/ADP carrier (AAC) is the most abundant protein in mitochondria and was the first MC purified, reconstituted and sequenced already in 1982. There are three isoforms of AAC carriers found in man and other organisms. AAC1 (coded by SLC25A4) is expressed in heart and skeletal muscle, AAC2 (coded by SLC25A5) is expressed ubiquitously at a level depending on the respiratory activity and AAC3 (coded by SLC25A6) is expressed weakly in most tissues (Aquila et al. 1982). It is inhibited specifically by carboxyatractyloside and bongkrekic acid.

The main role of PiC is to transport inorganic phosphate into the mitochondrial matrix to support oxidative phosphorylation. PiC achieves this goal either by proton cotransport or by OH^-exchange with cytosolic Pi. The phosphate carrier PiC exists as a single gene in man (SLC25A3), generating two isoforms PiC-A and PiC-B by alternative splicing. These forms differ considerably in their kinetic properties. PiC-A has a Km for Pi on the external surface that is three times that of PiC-B (2.20 vs. 0.78 mM), respectively. The Km values of the internal surface are much higher, around 8.5 and 6.5 mM for PiC-A and PiC-B, respectively. Tissue specificity of PiC resembles that of AAC with one heart-isoform, expressed only in muscle, and one liver-isoform, expressed ubiquitously.

24.5.4 Intracellular Glutamate Compartmentation as Detected by (^{13}C,^{2}H) NMR

Classical evidences support fast α-ketoglutarate/glutamate exchange through the cytosolic and mitochondrial isoforms of cerebral aspartate aminotransferase (c.f. Fig. 24.1, processes 19 and 23; E.C. 2.6.1.1) (Erecinska and Silver 1990). Less is known, however, on the *in situ* activity of the neuronal or glial dicarboxylate carriers (c.f. Fig. 24.1, process 21 and Fig. 24.7A) or the glutamate/aspartate exchanger (c.f. Fig. 24.1, process 20 and Fig. 24.7A), which are responsible for α-ketoglutarate/glutamate exchange between mitochondria and cytosol of neural cells. However, recent experiments with purified mitochondrial preparations from brain incubated with (U-^{14}C) glutamate, have revealed that the exchange of α-ketoglutarate/glutamate may be slower than the TCA cycle flux *in vitro* (Berkich et al. 2005). It was concluded that a significant portion of α-ketoglutarate remained within the mitochondrial matrix without equilibration with the extramitochondrial pool, suggesting the presence of two slowly exchanging pools of glutamate in these preparations. In agreement with these findings, numerous experiments had previously revealed intracellular glutamate compartmentation in primary cultures of neurons or astrocytes (Sonnewald et al. 2004; Waagepetersen et al. 2003).

Hydrogen turnover experiments, as monitored by (^{13}C,^{2}H) NMR, have more recently disclosed the first evidence of intracellular glutamate compartmentation in the neuronal and glial compartments of the adult rat brain *ex vivo* (Cerdan et al. 2009; Sierra et al. 2004). Hydrogen turnover occurs in a much faster timescale than carbon turnover allowing the resolution of faster metabolic processes. (^{13}C, ^{2}H) NMR is ideally suited to investigate hydrogen turnover because it determines the extent of deuterium replacements in individual hydrogens of ^{13}C labeled isotopomers (Garcia-Martin et al. 2001). This is possible because deuterium substitutions induce readily observable isotopic shifts and heteronuclear couplings in the ^{13}C resonances from vicinal or geminal ^{13}C labeled carbons. In this context, the stereospecific exchange of the H3 hydrogens of glutamate and glutamine by deuterons from heavy water, as detected in the C2 resonances of the corresponding (2-^{13}C) isotopomers, has been found to provide important information of the trafficking of

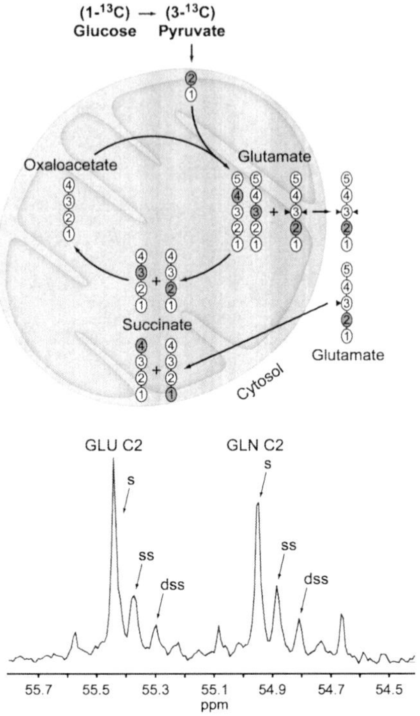

Fig. 24.8 The turnover of the H3 hydrogens of cerebral glutamate and glutamine reveals subcellular compartmentation. A fast deuterium exchange occurs in the $H3_{proS}$ *or* $H3_{proR}$ hydrogens of pre-existing molecules of (2-^{13}C) glutamate during the metabolism of (1-^{13}C) glucose, catalyzed by cytosolic isocitrate dehydrogenase (NADP$^+$) or aconitase, respectively. A slower double exchange of the $H3_{proR}$ *and* $H3_{proS}$ occurs only in newly formed (2-^{13}C) glutamate molecules in the TCA cycle (*upper panel*). Additive isotopic shifts make it possible to distinguish between perprotonated (2-^{13}C) glutamate (GLU C2) or glutamine (GLN C2) (singlets, s), (2-^{13}C, 3-^{2}H) glutamate or glutamine (shifted singlets, ss) and (2-^{13}C, 3,3′-$^2H_2$) glutamate or glutamine (doubly shifted singlet, dss) by high resolution (^{13}C, ^{2}H) NMR (*lower panel*). Deuterons are indicated by *black triangles*. ^{13}C atoms are shown as *grey circles* (Adapted from Cerdan et al. (2009). With permission of the publisher)

glutamate/α-ketoglutarate between cytosol and mitochondria (Cerdan et al. 2003, 2009; Garcia-Martin et al. 2002; Rodrigues and Cerdan 2005a) (Fig. 24.8).

In the perfused liver, it has been possible to show, that the exchange of the H3 hydrogens can be resolved in two different kinetic components (Garcia-Martin et al. 2002). A fast exchange of either the $H3_{proR}$ *or* $H3_{proS}$ hydrogens occurs first on pre-existing (2-^{13}C) glutamate or glutamine molecules. This fast process substitutes *only one* of the H3 hydrogens, in a way that substitution of $H3_{proR}$ precludes the deuterium substitution of the $H3_{proS}$ and *vice versa*. This fast process may be catalyzed by either cytosolic aconitase (E.C. 4.2.3.1) or isocitrate dehydrogenase (NADP$^+$; E.C. 1.1.1.42). A slower double deuteration of the $H3_{proR}$ *and* $H3_{proS}$ occurs later, but it is *only* possible on newly synthesized (2-^{13}C) glutamate molecules during TCA cycle activity. Thus, (^{13}C,^{2}H) NMR allows resolving the fast turnover of cytosolic (2-^{13}C, 3-^{2}H) glutamate and the slower turnover of (2-^{13}C, 3,3′-$^{2}H_2$) glutamate produced in the mitochondrial TCA cycle. Figure 24.8B shows that a similar situation may occur in the brain where it is possible to distinguish the (2-^{13}C, s), (2-^{13}C, 3-^{2}H, ss) and (2-^{13}C, 3,3′-$^{2}H_2$, dss) glutamate and glutamine isotopomers during (2-^{13}C) acetate infusion. The fact that the fast and slow components of H3 exchange can be resolved by (^{13}C, ^{2}H) NMR and that the cytosolic (2-^{13}C, 3-^{2}H) glutamate and mitochondrial (2-^{13}C, 3,3′-$^{2}H_2$) glutamate can be observed simultaneously clearly suggests that the exchange of glutamate across the inner mitochondrial membrane is slow in the cytosolic H3 hydrogen exchange timescale. Taken together, the experiments with isolated cerebral mitochondria, primary cultures of neurons and glial cells and (^{13}C, ^{2}H) NMR of brain extracts indicate that α-ketoglutarate/glutamate exchange through the mitochondrial membrane of neural cells is slower than previously thought, creating two kinetically different intracellular glutamate pools in both cell types.

24.6 Intramitochondrial Pyruvate Oxidation and the Tricarboxylic Acid Cycle

Once transported to the mitochondrial matrix, pyruvate may enter the TCA cycle through the pyruvate dehydrogenase complex (PDHC; E.C. 1.2.4.1) or through PC.

24.6.1 Pyruvate Dehydrogenase

The PDHC plays a key role in the oxidative metabolism of pyruvate, catalyzing the first irreversible step, the decarboxylation of pyruvate to produce CO_2, acetyl-CoA and NADH. Mammalian PDHC is composed of multiple copies of six different enzymes (Fig. 24.9A) (Cullingford et al. 1994): 20–30 copies of pyruvate dehydrogenase (E1, E.C. 1.2.4.1.), 60 copies of dihydrolipoamide acetyltransferase (E2, E.C. 2.3.1.12), 6 copies of dihydrolipoamide dehydrogenase (E3, E.C. 1.8.1.4), 1–3 copies of a regulatory E1-specific kinase, some copies of a loosely bound phospho-E1 phosphatase and 6 copies of a poorly characterized component protein X. E1 is heterotrimeric, having two active sites that catalyze the thiamine dependent decarboxylation of pyruvate and the acetylation of the lipoyl moieties of E2. The three dimensional structure of PDHC remains to be elucidated. It is most probably, however, very similar to its highly homologous branched-chain α-ketoacid dehydrogenase (E.C. 1.2.4.4) which is known. This structure shows a high degree of contact and interaction within the different subunits (Fig. 24.9A) (Aevarsson et al. 2000).

PDHC activity is strictly regulated both by product inhibition and by phosphorylation/dephosphorylation of the E1 subunit by E1-specific kinases/phosphatases (Fig. 24.9B) (Patel and Korotchkina 2001). E1 can be phosphorylated in three sites by at least, four isoforms of pyruvate dehydrogenase kinase (PDK; E.C. 2.7.1.99) 1–4. Phosphorylation of E1 by PDK inhibits the enzyme. This process is stimulated

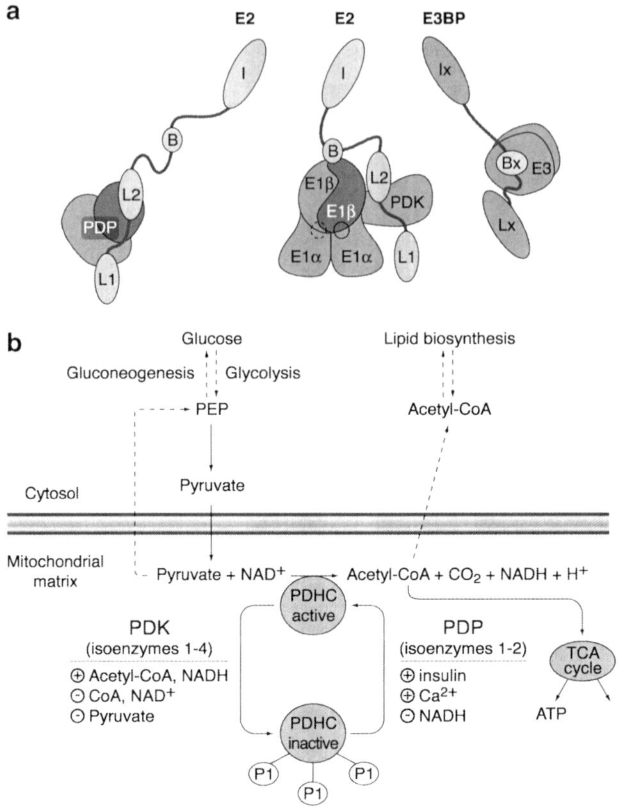

Fig. 24.9 Tentative structure (a) and regulation (b) of the Pyruvate Dehydrogenase multienzyme complex. (**a**) Schematic representation of the domains of the PDHC components and their sites of interactions. Only two molecules of E2 and one molecule of each of E1, E3, E3BP, PDK (pyruvate dehydrogenase kinase) and PDP (pyruvate dehydrogenase phosphatase) are shown. *I* inner domain, *B* binding domain, *L2* and *L1* lipoyl domains. E3BP consists of: *Ix* inner domain, *Bx* binding domain, *Lx* lipoyl domain. The *black* and *dotted circles* denote the position of the active sites. (**b**) Regulation of PDC activity by interconversion between active (dephosphorylated) and inactive (phosphorylated) forms, as catalyzed by PDPs (phosphatases) and PDKs (kinases). The three phosphorylation sites of E1-PDC are denoted as P1, P2 and P3 (Adapted from Patel and Korotchkina (2001). Reproduced with permission of the publisher)

by increased acetyl-CoA or NADH concentrations and inhibited by increased CoA or pyruvate concentrations. Dephosphorylation of E1 by pyruvate dehydrogenase phosphatase (PDP; E.C. 3.1.3.43) activates the enzyme, being stimulated by insulin and Ca^{2+} and inhibited by NADH.

PDHC is more abundant in the cerebral cortex and grey matter than in basal ganglia, probably revealing the larger oxidative capacity of the cortex (Butterworth and Giguere 1984). PDK isoforms also present a heterogeneous distribution in the brain (Nakai et al. 2000). PDK-1, PDK-2 and PDK-4 are found in all cerebral structures, with PDK-2 being the most abundant isoenzyme. PDK-1 is higher in cerebellum and cerebral cortex than in medulla oblongata and hippocampus.

Like many other oxidative and glycolytic enzymes, PDHC is subjected to important ontogenic variations. It increases sharply after birth, reaching a maximum in adulthood (Butterworth and Giguere 1984). This increase is thought to be mediated by translational or post-translational mechanisms (Cullingford et al. 1994). Cerebral PDHC increase, however, appears to lag behind that of glycolytic enzymes during postnatal development, suggesting that it may constitute the rate limiting step in the implementation of fully aerobic cerebral respiration.

24.6.2 Pyruvate Carboxylase

PC catalyzes the ATP- and bicarbonate-dependent carboxylation of pyruvate to produce OAA. PC is apparently a homotetramer α_4 of approximately 120–130 kDa (Attwood 1995; Attwood and Keech 1984; Jitrapakdee and Wallace 1999). A single PC gene exists in rat and human. In the rat, the PC gene consists of 19 coding exons and four introns, spanning 40 kb. Alternative splicing produced up to five different mature transcripts which contain the same coding region but different non coding sequences (Jitrapakdee et al. 1996a, b). In humans, the PC gene has been localized in the long arm of chromosome 11. Although 3D structures are not available yet, sequencing of cDNA genes coding for PC indicates that this enzyme contains three functional domains: a biotin carboxylation domain (N-terminal region), a transcarboxylation domain (central region) and a biotin carboxyl carrier domain (C-terminal region). PC activity in brain is significantly smaller than in liver, being upregulated by acetyl-CoA and Mg^{2+}.

Cerebral PC plays an important anaplerotic function, providing the TCA cycle with OAA molecules synthesized *de novo* (Gamberino et al. 1997). These additional molecules of OAA allow replenishing the cycle from loses of carbon skeletons caused by exocytosis and neurotransmission. However, PC is found only in astrocytes (Shank et al. 1985) while the loss of metabolites by neurotransmission occurs mainly in the neurons. This implies that to accomplish the replenishment of the neuronal pools, an astrocyte to neuron shuttle must exist to transfer to the neurons the newly formed carbon skeletons synthesized in the astrocytes. This role is accomplished mainly by glutamine, a metabolite produced by glutamine synthetase (GS; E.C. 6.3.1.2) *only* in the astrocytes (Martinez-Hernandez et al. 1977).

Notably, PC activity has been reported recently in neurons (Hassel 2000, 2001; Hassel and Brathe 2000). Because of the lack of the PC protein in these cells, it has been proposed to be caused by ME activity, operating in reverse.

24.6.3 Pyruvate Oxidation in the Cerebral Tricarboxylic Acid Cycles as Detected by ^{13}C NMR

In the last decade, a variety of ^{13}C NMR approaches have provided new information on the operation of the neuronal and glial TCA cycles and their interaction during cerebral activation (Cruz and Cerdan 1999; Garcia-Espinosa et al. 2004; Gruetter 2002;

Rodrigues and Cerdan 2005a, 2007; Rothman et al. 2003; Shulman et al. 2004; Sonnewald et al. 2004; Waagepetersen et al. 2003). Progress in this area has been mainly derived from the combined use of ^{13}C labeled substrates that monitor mainly neuronal or glial metabolism and *in vivo* or high-resolution ^{13}C NMR methods to monitor the kinetics of ^{13}C enrichments and isotopomer populations of individual carbons of cerebral glutamate, glutamine and GABA (Bachelard et al. 1993; Cruz et al. 1998; Gruetter et al. 2003; Lapidot and Gopher 1994; Rodrigues et al. 2007; Rothman et al. 2003).

Figure 24.10 illustrates the fundamentals of the ^{13}C NMR approach by presenting the ^{13}C labeling patterns of glutamate carbons derived from (1-^{13}C) glucose and the corresponding *in vivo* ^{13}C NMR spectra (Fig. 24.10a and c). (1-^{13}C) glucose originates by glycolysis an equimolar mixture of (3-^{13}C) and unlabeled pyruvate molecules. Further metabolism of (3-^{13}C) pyruvate through PDHC originates (2-^{13}C) acetyl-CoA and in the first turn of the cycle, (4-^{13}C) α-ketoglutarate. Fast α-ketoglutarate/glutamate exchange has been traditionally assumed, a circumstance that results in a single intracellular α-ketoglutarate/glutamate pool and allows considering glutamate labeling as a direct reflection of α-ketoglutarate labeling. On this basis, it has been possible to calculate the total cerebral TCA cycle flux in rodent or human brain from the kinetics of (4-^{13}C) glutamate production, as detected in consecutive ^{13}C NMR acquisitions during (1-^{13}C) glucose infusions (Fig. 24.10c). Under these conditions, the neuronal glutamate pool is believed to account for approximately 90% of total cerebral glutamate and the neuronal TCA cycle flux can be approximated from the kinetics of ^{13}C enrichment of total cerebral glutamate (Fitzpatrick et al. 1990; Mason et al. 1992b, 1995; Sibson et al. 1998a, 2001).

With (1-^{13}C) glucose as a substrate the turnover of glutamate C4 reflects exclusively PDHC activity. The turnover of glutamate C2 includes labeling derived directly from PC as well as from the second and subsequent turns of the cycle as fed from PDHC. It is possible, however, to correct for the latter by comparing the multiplicities of the C2 and C3 carbons. It becomes possible then to calculate the relative activities of PDHC/PC from the relative ^{13}C labelings of glutamate C4/C2 after appropriate corrections. PC flux has been shown to account for approximately 20% of the total cerebral TCA cycle flux (Cruz and Cerdan 1999; Rodrigues and Cerdan 2005a; Zwingmann and Leibfritz 2003).

The astroglial TCA cycle has been investigated using (2-^{13}C) glucose, a substrate that labels the cerebral TCA cycle through PC, an exclusively glial enzyme (Sibson et al. 2001) (Fig. 24.10b). In this case, (2-^{13}C) pyruvate is produced first by glycolysis, followed by (3-^{13}C) OAA and (2-^{13}C) α-ketoglutarate/glutamate in the first turn of the cycle, originating eventually equimolar mixtures of (1-^{13}C) and (4-^{13}C) succinate, fumarate and malate which are lost in further turns by decarboxylation. Consecutive ^{13}C NMR spectra of the *in vivo* brain, provide in this case the kinetics of enrichment in glutamate and glutamine C3, which can be approximated to the astroglial cycle flux. Under these conditions, labeling in glutamine C3 is faster than in glutamate C3, reflecting the faster turnover of the glial glutamate pool (Fig. 24.10d, e) (Sibson et al. 2001).

More frequently, the astroglial cycle has been investigated using ^{13}C labeled acetate as substrate (Badar-Goffer et al. 1990, 1992; Bluml et al. 2002; Cerdan et al.

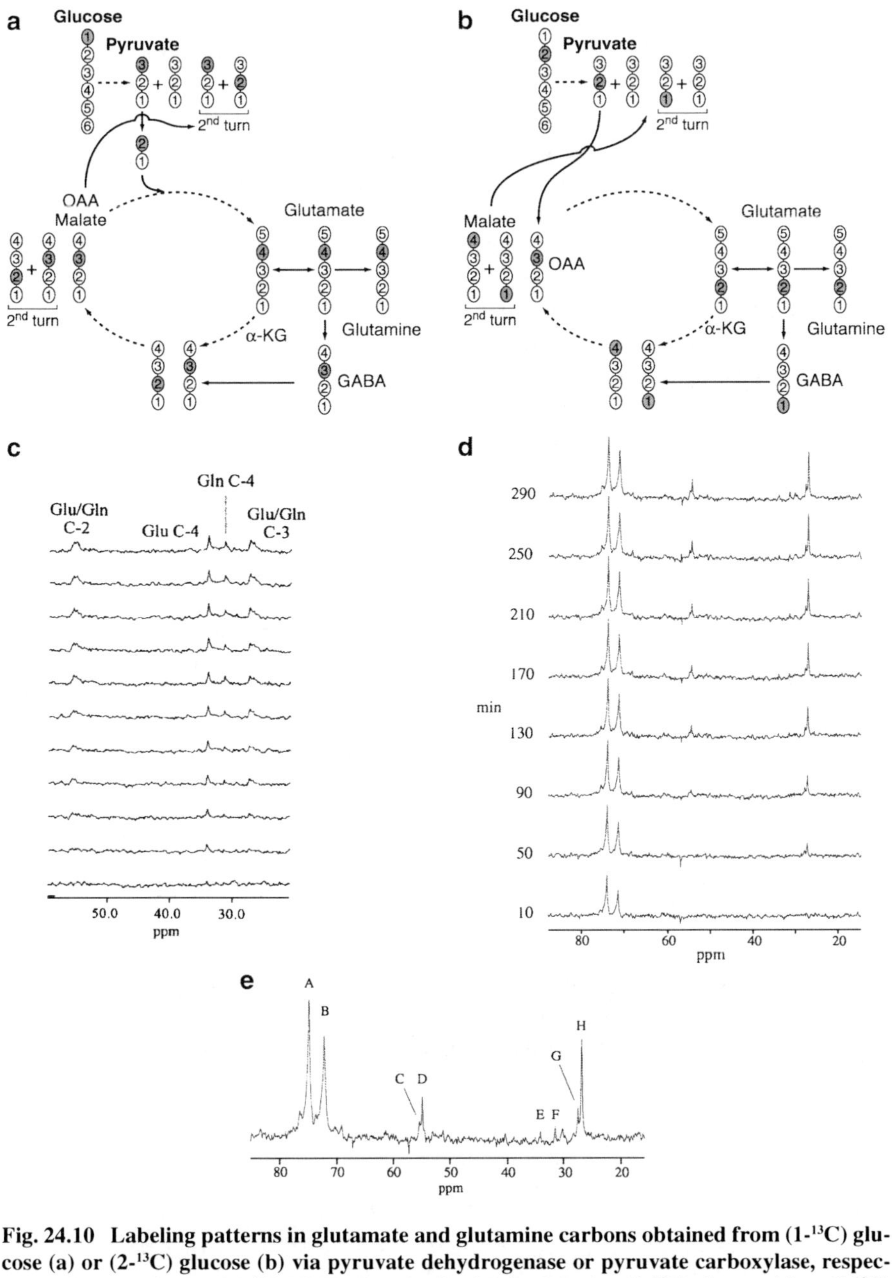

Fig. 24.10 Labeling patterns in glutamate and glutamine carbons obtained from (1-^{13}C) glucose (a) or (2-^{13}C) glucose (b) via pyruvate dehydrogenase or pyruvate carboxylase, respectively. Consecutive *in vivo* ^{13}C NMR of rat brain obtained during (1-^{13}C) glucose (c) or (2-^{13}C) glucose (d) infusions. (e) Summed ^{13}C NMR spectra from the last 60 min of infusion with (2-^{13}C) glucose. Resonances include: A glucose C2β, B glucose C2α, C glutamate C2, D glutamine C2, E glutamate C4, F glutamine C4, G glutamate C3, and H glutamine C3. *Glu* glutamate, *Gln* glutamine, *α-KG* α-ketoglutarate, *OAA* oxalacetate (Reproduced from Sibson et al. (1998a) and Sibson et al. (2001). Reproduced with permission of the publisher)

1990; Cruz and Cerdan 1999; Chapa et al. 1995; Lebon et al. 2002; Ross et al. 2003). Acetate is only transported to glial cells (Waniewski and Martin 1998), being activated to acetyl-CoA by acetate thiokinase (E.C. 6.2.1.1) and entering the TCA cycle directly at the citrate synthase (E.C. 2.3.3.1) step, bypassing the PC and PDHC steps. The ^{13}C labeling patterns of glutamate or glutamine from (2-^{13}C) or (1-^{13}C) acetate are identical to those from (1-^{13}C) or (2-^{13}C) glucose. However, complete acetate oxidation differs from that of glucose, because it does not involve transfer of reducing equivalents from cytosol to mitochondria and bypasses the PDHC and PC steps. Acetate has been shown to be the substrate of choice to investigate ME activity and the pyruvate recycling system. In particular, the use of (1,2-$^{13}C_2$) acetate provided the first evidence for the existence of this pathway and introduced the use of ^{13}C multilabeled substrates and isotopomer analysis in cerebral biochemistry. Along these lines, (1,2-$^{13}C_2$) acetate and (1,2-$^{13}C_2$) glucose have been shown to provide a wealth of information *in vitro* on the compartmentation of the TCA cycle in neurons and glial cells, in physiological and pathological situations (Cruz and Cerdan 1999).

Table 24.3 summarizes the values for the TCA cycle flux obtained using different *in vivo* and *in vitro* ^{13}C NMR approaches, as compared to earlier values obtained with radioactive isotope methods and more recent ^{13}C isotopomer approaches. In general, TCA cycle fluxes calculated for the normoxic adult rat or human brain were in the range 0.5–1.4 μmol min^{-1} g^{-1}. More accurate measurements provided recently values in the vicinity of 0.5 μmol min^{-1} g^{-1} for the normoxic brain under basal conditions. This value may decrease to virtually undetectable levels with increasing states of anesthesia (Sibson et al. 1998b) or increase by 25–50% with different cerebral activation protocols (Chhina et al. 2001; Hyder et al. 1996, 1997). TCA cycle fluxes in the glial compartment of rodent or human brain are in the range 0.14–0.4 μmol min^{-1} g^{-1}, representing approximately 20% of the total cerebral TCA cycle activity, matching well with the relative contribution of PC to the total cycle flux. Table 24.3 also shows sizes and turnover rates of the glutamate compartments associated to the neuronal ("large") and glial ("small") glutamate pools. No direct measurements exist to our knowledge of these pool sizes. The values shown are based on the assumption that the small glutamate pool accounts for approximately 10% of total cerebral glutamate during (1-^{13}C) glucose metabolism. Recently it has been possible to determine quantitatively the relative labeling of the large and small glutamate pools *ex vivo* using (1-^{13}C) glucose and (2-^{13}C, 2-$^{2}H_3$) acetate as substrates (Chapa et al. 2000).

24.6.4 Subcellular Compartmentation of Pyruvate and Glutamate and the Traditional Metabolic Coupling Hypothesis. The Redox Switch/Redox Coupling Hypothesis

The detection of intracellular pyruvate and glutamate compartmentations presents important implications for the traditional interpretations of metabolic coupling between neurons and glia during glutamatergic neurotransmission (Magistretti and

Table 24.3 Tricarboxylic acid cycle and glutamate/glutamine exchange between the neuronal and glial compartments of the adult mammalian brain as calculated with different models and methodologies (Adapted from Cruz and Cerdan (1999). Reproduced with permission of the publisher)

Process/model	Garfinkel[a]	Van den Berg and Garfinkel[b]	Kunnecke et al.[c] Preece and Cerdan[d]	Mason et al.[e], Sibson et al.[f,g], Shen et al.[h], Lebon et al.[i]	Gruetter et al.[j]	Bluml et al.[k]
Total cerebral TCA cycle Flux (μmol min^{-1} g^{-1})	1.05	1.5	1.4	1.6 or 0.7[e], 0.6[f], 0.2–1.0[g], 0.8[h]	0.6	0.84
Neuronal TCA cycle Flux (μmol min^{-1} g^{-1})	0.40	1.2	1.0	1.6[e], 0.6[f], 0.2–1.0[g], 0.8[h]	0.6	0.70
Glial TCA cycle Flux (μmol min^{-1} g^{-1})	0.65	0.3	0.4	0.14[i]		0.14
Size (μmol g^{-1})/turnover (per min) of large glutamate pool	8.8/21.7	7.0/5.7	5.8/5.8	n.a.		n.a.
Size (μmol g^{-1})turnover(per min) of small glutamate pool	1.7/2.6	1.25/4.16	0.5/1.25	n.d./7.7[l]		n.a.
Transfer of neuronal glutamate to glial compartment (μmol min^{-1} g^{-1})	0.08[m]	0.14	0.1	0.21[f], 0.0–0.4[g], 0.32[h], 0.3[i]	0.3 (0.2)[n]	n.a..
Transfer of glial glutamine to neuronal compartment (μmol min^{-1} g^{-1})	0.45	n.a.	0.1	0.21[f], 0.0–0.4[g], 0.32[h], 0.3[i]	0.3 (0.2)[k,n]	n.a.

[a]Calculated from specific radioactivity measurements in brain extracts obtained after intracranial injections of various radioactive precursors including (U-^{14}C) glutamate, (U-^{14}C) aspartate, $^{14}CO_3H^-$ and $^{15}NH_4$ acetate (Garfinkel 1966)

[b]Calculated from specific radioactivity measurements in glutamate, glutamine and aspartate from mouse brain extracts prepared after intraperitoneal injections of ^{14}C labeled glucose and acetate (Van den Berg and Garfinkel 1971)

[c]Relative flux values were calculated as described in (Kunnecke et al. 1993) and from the relative ^{13}C isotopomer populations in glutamate, glutamine and GABA measured by high-resolution ^{13}C NMR in rat brain extracts after infusion of (1,2-$^{13}C_2$) acetate (Chapa et al. 1995)

[d] bsolute flux values were determined from the relative values described in note c by measuring the absolute rate of GABA accumulation induced by vigabatrin, a selective inhibitor of GABA transaminase (Preece and Cerdan 1996)

[e]Determined *in vivo* from the kinetics of ^{13}C enrichment in glutamate and glutamine C4 carbons from rat or human brain during infusion of (1-^{13}C) glucose, respectively (Mason et al. 1992a, 1995)

[f] Determined *in vivo* from the kinetics of ^{13}C labeling in glutamate and glutamine C4 (Sibson et al. 1997)

[g]Determined *in vivo* from the kinetics of ^{13}C enrichment in glutamate and glutamine C4, under different conditions of morphine, α-chloralose and pentobarbital anesthesia (Sibson et al. 1998a)

[h]Determined in the human brain *in vivo* during (1-^{13}C) glucose infusion. Glutamate/glutamine exchange concluded to be stoichiometric 1:1 with CMRglc (Shen et al. 1999)

[i]Determined in the human brain during metabolism of (2-^{13}C) acetate (Lebon et al. 2002)

[j]Determined in the human brain during (1-^{13}C) glucose infusion with a different model than that used in f-i (Gruetter et al. 2001)

[k]Determined in the human brain during (1-^{13}C) acetate metabolism (Bluml et al. 2002)

[l]Determined *in vivo* as the inverse of the rate constant of ^{13}C labeling in cerebral glutamate C4 during (1-^{13}C) glucose infusions (Mason et al. 1992b)

[m] Originally proposed as an α-ketoglutarate exchange between the large and small compartments (Garfinkel 1966)

[n]Only a fraction of glutamine synthesis considered to be derived from neurotransmitter glutamate. Glutamate/glutamine exchange concluded non stoichiometric with CMRglc, approaching 0.4 rather than the value of 1 concluded in f–i (Gruetter et al. 2001)

n.a.: not applicable, n.d.: not detectable

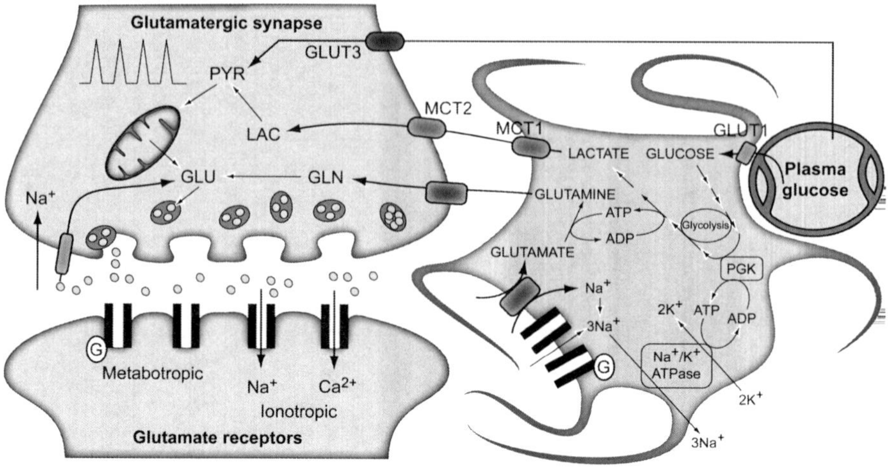

Fig. 24.11 The traditional metabolic coupling hypothesis between neurons and glial cells during glutamatergic neurotransmission. Glutamate released to the synaptic cleft during glutamatergic neurotransmission is co-transported with Na^+ to the astrocytes. Astroglial Na^+ is exchanged by extracellular K^+ through the Na^+/K^+ ATPase, consuming one ATP molecule. Astrocytic glutamate produces glutamine through glutamine synthetase, consuming one additional ATP molecule. Lactate produced exclusively in astroglial glycolysis to support these energy demands, is extruded to the extracellular medium, taken up by the surrounding neurons and oxidized as their main metabolic fuel. Note the apparent stoichiometric coupling between glutamate-glutamine cycling and glucose uptake as well as the exclusive *glycolytic* or *oxidative* metabolisms in astrocytes and neurons, respectively. MCT1 and MCT2: monocarboxylate transporters 1 and 2. *GLUT 1* and *GLUT 3* glucose transporters 1 and 3, *PGK* phosphoglycerate kinase (Adapted with permission from Tsacopoulos and Magistretti (1996))

Pellerin 1997, 1999; Tsacopoulos and Magistretti 1996) (Fig. 24.11). The classical metabolic coupling hypothesis indicates that neurotransmitter glutamate released to synaptic cleft after an action potential is recaptured mainly by the high affinity glutamate transporters of surrounding astrocytes in the neuropil. Glutamate is cotransported to the astrocyte cytosol together with three Na^+ atoms and transformed into glutamine by GS. The three Na^+ atoms are extruded to the extracellular space through the Na^+/K^+ ATPase (E.C. 3.6.3.9), at the expense of one ATP molecule, and GS consumes one additional ATP molecule. These two ATP molecules were originally proposed to be derived exclusively from the degradation of one glucose molecule in the astrocyte through the glycolytic pathway. Glutamine is subsequently extruded to the extracellular space and recaptured by the neurons to regenerate neurotransmitter glutamate. In this process, one glucose molecule from plasma is consumed and two molecules of astrocytic lactate are generated and exported to the neurons to become their main metabolic fuel. Thus, glutamine production appears to be stoichiometrically coupled to glucose uptake with a 1:1 stoichiometry, anaerobic metabolism of glucose being confined to the astrocytes and pyruvate oxidation remaining as an exclusively neuronal process. These findings matched well with early ^{13}C NMR results which determined the cerebral TCA cycle and the glutamine cycle fluxes from a minimal mathematical model, assuming that (4-^{13}C) glutamate or (4-^{13}C) glutamine turnover curves reflected the neuronal TCA cycle or the glutamine cycle fluxes, respectively (Sibson et al. 1998a). Glutamine was then proposed to be the major precursor of cerebral glutamate and the glutamine cycle was found to be stoichiometrically coupled to glucose uptake, accounting for 60–80% of the energy derived from glucose metabolism (Rothman et al. 2003).

However, compelling evidences have accumulated since then indicating that a significant portion of the energy used to synthesize glutamine is derived from the astroglial TCA cycle (Garcia-Espinosa et al. 2003), that up to 40% of cerebral glutamate is derived from alternative sources to glutamine (Garcia-Espinosa et al. 2003), that glutamine cycling may not present a 1:1 stoichiometry with glucose uptake (Gruetter et al. 2001; McKenna 2007) and that different kinetic pools of pyruvate, lactate, glutamate, glutamine and GABA exist in neurons and astrocytes (Cruz et al. 2001; Rodrigues et al. 2005; Sierra et al. 2004; Zwingmann et al. 2001). These findings suggest that the coupling mechanisms between neuronal and glial metabolisms are more complex than initially envisioned.

Figure 24.12 depicts our present conception of metabolic coupling between neurons and glial cells during glutamatergic neurotransmission (Cerdan et al. 2006; Ramirez et al. 2007). Our proposal is based on the operation of intracellular redox switches (as the one described in section 5.5) and the transcellular coupling of oxidative and non oxidative metabolisms in neurons and glial cells through the exchange of monocarboxylate reducing equivalents. Briefly, following presynaptic glutamate release, astrocytic transporters incorporate glutamate and three Na^+ to the astrocytic cytosol, the Na^+ ions being removed later through the plasma membrane Na^+/K^+ ATPase. The energetic cost of this process results in decreased astrocytic ATP/ADP concentrations, a circumstance that stimulates *both* astroglial glycolysis and astroglial TCA cycle. Together, astrocytic glycolysis and the TCA cycle contribute the energy required by GS, with a predominant contribution of the oxidative metabolism. However, the energy demands during glutamatergic neurotransmission eventually exceed the reduced capacity of the astrocytic TCA cycle, resulting in a net activation of the glycolic flux and a net production of astrocytic lactate, which is rapidly extruded to the extracellular space. Extracellular lactate is taken up by neurons, their cytosolic redox being reduced to a point where neuronal glycolysis is inhibited at the GAPDH step. Under these conditions, extracellular lactate is mainly consumed by oxidation in the neurons (and probably also in the astrocytes), until its extracellular concentration decreases to the preactivation levels, setting up the stage for a new glutamatergic event.

The present redox switch/redox coupling hypothesis accounts for essentially all experimental findings *in vivo* and *in vitro*. In particular, it allows for: (a) the simultaneous operation of both glial and neuronal glycolysis and TCA cycles during neuronal activation, (b) the fact that neurons and astrocytes may use glucose or lactate as complementary, or even alternative substrates, depending on the extracellular redox state and availability, (c) the stoichiometric or non stoichiometric coupling of the glutamate cycle and glucose uptake, (d) the intracellular glutamate compartmentation in neurons and astrocytes and (e) the intracellular compartmentation of cytosolic monocarboxylates in neural cells (Cruz et al. 2005; Dienel and Hertz 2005). The transcellular redox switch/redox coupling hypothesis mimics the intracellular

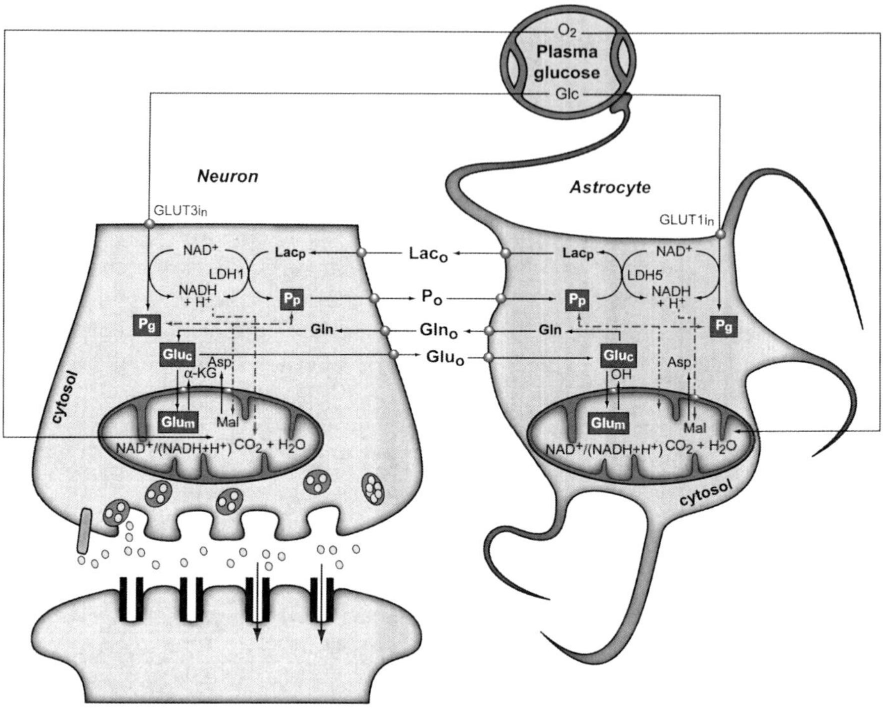

Fig. 24.12 The subcellular compartmentation of pyruvate and glutamate and the redox switch/redox coupling hypothesis. Two pools of pyruvate exist in neurons and astrocytes derived from extracellular monocarboxylates (Pp) or glucose (Pg). A lactate/pyruvate redox shuttle is able to transfer continuously lactate from astrocytes to neurons, taking advantage of the kinetics of plasma membrane transporters and lactate dehydrogenase isoenzymes. High cytosolic lactate concentration inhibits neuronal glycolysis at the glyceraldehyde-3-phosphate dehydrogenase step by competition with cytosolic NAD^+, favoring the oxidation of extracellular lactate. Neuronal pyruvate is transferred back to the astrocyte to close the transcellular exchange of reducing equivalents. Two α-ketoglutarate/glutamate pools exist in neurons and astrocytes, associated probably to cytosolic and mitochondrial compartments. Exchange of α-ketoglutarate/glutamate between mitochondria and cytosol appears to be slow in the H3 glutamate hydrogen exchange timescale and dependent of the cytosolic and mitochondrial $NAD(P)^+/NAD(P)H$ ratios, as determined by the malate-aspartate shuttle. Both glycolysis and oxidative astrocytic metabolism contribute the energy for glutamine production in the astrocytes, indicating that the neuroglial coupling involves both transcellular and intracellular redox coupling mechanisms that allow the simultaneous operation of glycolysis and oxidation in astroglial cells. *Asp* aspartate, *Glc* glucose, *Gln* glutamine, *Glu* glutamate, *GLUT1* and *GLUT3* glutamate transporters 1 and 3, *α-KG* α-ketoglutarate, *Lac* lactate, *LDH1* and *LDH5* lactate dehydrogenase 1 and 5, *Mal* malate (Reproduced with permission from Ramirez et al. (2007))

coupling mechanisms existing between cytosolic glycolysis and the TCA cycle by the transfer of reducing equivalents through the inner mitochondrial membrane. However, during transcellular redox coupling, reducing equivalents are transferred between neural cells in the form of the monocarboxylates lactate and pyruvate (Arco and Satrustegui 2005; McKenna et al. 2006).

24.7 Concluding Remarks

We examined individually the main transport and enzyme systems involved in cerebral pyruvate transport and metabolism combining the information derived from functional ^{13}C NMR approaches and that available from gene sequencing and cloning techniques. The genes from most enzyme and transport systems have been cloned and sequenced and in many cases, the 3D structure of the corresponding proteins has been resolved. However, the relationship between gene expression and 3D structure with physiological performance and metabolic regulation remains difficult to integrate at present. Our study contributes in this direction. In particular, significant progress has been made recently in the determination of the *in vivo* activities of transport and enzyme systems using ^{13}C and (^{13}C, ^{2}H) NMR methods. These data have disclosed novel aspects of cerebral biochemistry, including notably subcellular compartmentation of monocarboxylates, glutamate and glutamine. Although earlier evidences addressed the role of subcellular compartmentation *in vitro*, its role in cerebral metabolism *in vivo* remain more difficult to address. In this review we outline the "Redox switch/Redox coupling hypothesis", a recent proposal that accounts for the subcellular compartmentation of monocarboxylates and glutamate, making neuroglial coupling to rely on the subcellular compartmentation of these metabolites and the operation of intracellular redox switches and intercellular redox shuttles. This process mimics, at the transcellular level, the intracellular coupling between glycolysis and respiration through the metabolite shuttles exchanging redox equivalents between mitochondria and cytosol of neural cells.

Acknowledgements This work was supported in part by grants SAF 2001–2245, SAF 2004–03197, SAF 2008–01397 and S-BIO/0179/2006 to S.C and P.B. T.B.R. was a fellow from FCT-Portugal (SFRH/BPD/26881/2006). A.S. was a predoctoral fellow from UNED. Authors are indebted to Mr. Javier Pérez for the final drafting of the illustrations.

References

Aevarsson A, Chuang JL, Wynn RM, Turley S, Chuang DT, Hol WGJ (2000) Cristal structure of human branched-chain a-ketoacid dehydrogenase and the molecular basis of multienzyme complex deficiency in maple syrup disease. Structure 8:277–291

Aquila H, Misra D, Eulitz M, Klingenberg M (1982) Complete amino acid sequence of the ADP/ATP carrier from beef heart mitochondria. Hoppe Seylers Z Physiol Chem 363:345–349

Arco AD, Satrustegui J (2005) New mitochondrial carriers: an overview. Cell Mol Life Sci 62:2204–2227

Ardenkjaer-Larsen JH, Fridlund B, Gram A, Hansson G, Hansson L, Lerche MH, Servin R, Thaning M, Golman K (2003) Increase in signal-to-noise ratio of > 10,000 times in liquid-state NMR. Proc Natl Acad Sci USA 100:10158–10163

Attwood PV (1995) The structure and the mechanism of action of pyruvate carboxylase. Int J Biochem Cell Biol 27:231–249

Attwood PV, Keech DB (1984) Pyruvate carboxylase. Curr Top Cell Regul 23:1–55

Bachelard HS (1989) Measurement of carbohydrates and their derivatives in neuronal tissues. In: Boulton AA, Baker GB, Butterworth RF (eds) Carbohydrates and energy metabolism. Humana Press, Clifton, pp 133–154

Bachelard HS, Lewis LD, Ponten U, Siesjo BK (1974) Mechanisms activating glycolysis in the brain in arterial hypoxia. J Neurochem 22:395–401

Bachelard H, Badar-Goffer R, Ben-Yoseph O, Morris P, Thatcher N (1993) Studies on metabolic regulation using NMR spectroscopy. Dev Neurosci 15:207–215

Badar-Goffer RS, Bachelard HS, Morris PG (1990) Cerebral metabolism of acetate and glucose studied by ^{13}C-n.m.r. spectroscopy. A technique for investigating metabolic compartmentation in the brain. Biochem J 266:133–139

Badar-Goffer RS, Ben-Yoseph O, Bachelard HS, Morris PG (1992) Neuronal-glial metabolism under depolarizing conditions. A ^{13}C-n.m.r. study. Biochem J 282(Pt 1):225–230

Bell JE, Hume R, Busuttil A, Burchell A (1993) Immunocytochemical detection of the microsomal glucose-6-phosphatase in human brain astrocytes. Neuropathol Appl Neurobiol 19:429–435

Ben-Yoseph O, Badar-Goffer RS, Morris PG, Bachelard HS (1993) Glycerol 3-phosphate and lactate as indicators of the cerebral cytoplasmic redox state in severe and mild hypoxia respectively: a ^{13}C- and ^{31}P-n.m.r. study. Biochem J 291(Pt 3):915–919

Berkich DA, Xu Y, Lanoue KF, Gruetter R, Hutson SM (2005) Evaluation of brain mitochondrial glutamate and alpha-ketoglutarate transport under physiologic conditions. J Neurosci Res 79:106–113

Bernard-Helary K, Ardourel M, Magistretti P, Hevor T, Cloix JF (2002a) Stable transfection of cDNAs targeting specific steps of glycogen metabolism supports the existence of active gluconeogenesis in mouse cultured astrocytes. Glia 37:379–382

Bernard-Helary K, Ardourel MY, Hevor T, Cloix JF (2002b) In vivo and in vitro glycogenic effects of methionine sulfoximine are different in two inbred strains of mice. Brain Res 929:147–155

Bittar PG, Charnay Y, Pellerin L, Bouras C, Magistretti PJ (1996) Selective distribution of lactate dehydrogenase isoenzymes in neurons and astrocytes of human brain. J Cereb Blood Flow Metab 16:1079–1089

Bluml S, Moreno-Torres A, Shic F, Nguy CH, Ross BD (2002) Tricarboxylic acid cycle of glia in the in vivo human brain. NMR Biomed 15:1–5

Bolli R, Nalecz KA, Azzi A (1989) Monocarboxylate and alpha-ketoglutarate carriers from bovine heart mitochondria. Purification by affinity chromatography on immobilized 2-cyano-4-hydroxycinnamate. J Biol Chem 264:18024–18030

Bonavita V, Ponte F, Amore G (1964) Lactate dehydrogenase isoenzymes in the nervous tissue. Iv. An ontogenetic study on the rat brain. J Neurochem 11:39–47

Brooks GA (2002) Lactate shuttles in nature. Biochem Soc Trans 30:258–264

Brooks GA, Dubouchaud H, Brown M, Sicurello JP, Butz CE (1999) Role of mitochondrial lactate dehydrogenase and lactate oxidation in the intracellular lactate shuttle. Proc Natl Acad Sci USA 96:1129–1134

Butterworth RF, Giguere JF (1984) Pyruvate dehydrogenase activity in regions of the rat brain during postnatal development. J Neurochem 43:280–282

Cerdan S (2003) 13C NMR and cerebral biochemistry. NMR Biomed 16:301–302

Cerdan S, Kunnecke B, Seelig J (1990) Cerebral metabolism of [1,2-13C2]acetate as detected by in vivo and in vitro ^{13}C NMR. J Biol Chem 265:12916–12926

Cerdan S, Rodrigues TB, Ballesteros P, Lopez P, Perez-Mayoral E (2003) The subcellular metabolism of water and its implications for magnetic resonance image contrast. In: Belton PS, Gil AM, Webb GA, Rutledge D (eds) Magnetic resonance in food science. Royal Society of Chemistry, Oxford, pp 121–135

Cerdan S, Rodrigues TB, Sierra A, Benito M, Fonseca LL, Fonseca CP, Garcia-Martin ML (2006) The redox switch/redox coupling hypothesis. Neurochem Int 48:523–530

Cerdan S, Sierra A, Fonseca LL, Ballesteros P, Rodrigues TB (2009) The turnover of the H3 deuterons from (2-^{13}C) glutamate and (2-^{13}C) glutamine reveals subcellular trafficking in the brain of partially deuterated rats. J Neurochem 109(Suppl 1):63–72

Chang GG, Tong L (2003) Structure and function of malic enzymes, a new class of oxidative decarboxylases. Biochemistry 42:12721–12733

Chapa F, Kunnecke B, Calvo R, Escobar del Rey F, Morreale de Escobar G, Cerdan S (1995) Adult-onset hypothyroidism and the cerebral metabolism of (1,2-$^{13}C_2$) acetate as detected by ^{13}C nuclear magnetic resonance. Endocrinology 136:296–305

Chapa F, Cruz F, Garcia-Martin ML, Garcia-Espinosa MA, Cerdan S (2000) Metabolism of (1-^{13}C) glucose and (2-^{13}C, 2-$^{2}H_3$) acetate in the neuronal and glial compartments of the adult rat brain as detected by [^{13}C, ^{2}H] NMR spectroscopy. Neurochem Int 37:217–228

Chhina N, Kuestermann E, Halliday J, Simpson LJ, Macdonald IA, Bachelard HS, Morris PG (2001) Measurement of human tricarboxylic acid cycle rates during visual activation by ^{13}C magnetic resonance spectroscopy. J Neurosci Res 66:737–746

Chou WY, Huang SM, Liu YH, Chang GG (1994) Cloning and expression of pigeon liver cytosolic NADP$^+$-dependent malic enzyme cDNA and some of its abortive mutants. Arch Biochem Biophys 310:158–166

Clark JB, Lai JCK (1989) Glycolytic, tricarboxylic acid cycle and related enzymes in brain. In: Boulton AA, Baker GB, Butterworth RF (eds) Carbohydrates and energy metabolism. Humana Press, Clifton, pp 233–281

Clark JB, Bates TE, Cullingford T, Land JM (1993) Development of enzymes of energy metabolism in the neonatal mammalian brain. Dev Neurosci 15:174–180

Clark JB, Bates TE, Almeida A, Cullingford T, Warwick J (1994) Energy metabolism in the developing mammalian brain. Biochem Soc Trans 22:980–983

Clarke DD, Lajhta AL, Maker HS (1989a) Intermediary metabolism. In: Siegel G, Agranoff B, Albers RW, Molinoff P (eds) Basic neurochemistry. Raven, New York, pp 541–564

Clarke DD, Lajtha AL, Maker HS (1989b) Intermediary metabolism. In: Siegel G, Agranof B, Albers RW, Molinoff P (eds) Basic neurochemistry. Raven, New York

Consortium IHGS (2001) Initial sequencing and analysis of the human genome. Nature 409:860–921

Consortium MGS (2002) Initial sequencing and comparative analysis of the mouse genome. Nature 420:520–562

Cremer JE, Cunningham VJ, Pardridge WM, Braun LD, Oldendorf WH (1979) Kinetics of blood-brain barrier transport of pyruvate, lactate and glucose in suckling, weanling and adult rats. J Neurochem 33:439–445

Crone C, Sorensen SC (1970) The permeability of the blood-brain barrier to lactate and pyruvate. Acta Physiol Scand 80:47A

Cruz F, Cerdan S (1999) Quantitative ^{13}C NMR studies of metabolic compartmentation in the adult mammalian brain. NMR Biomed 12:451–462

Cruz F, Scott SR, Barroso I, Santisteban P, Cerdan S (1998) Ontogeny and cellular localization of the pyruvate recycling system in rat brain. J Neurochem 70:2613–2619

Cruz F, Villalba M, Garcia-Espinosa MA, Ballesteros P, Bogonez E, Satrustegui J, Cerdan S (2001) Intracellular compartmentation of pyruvate in primary cultures of cortical neurons as detected by ^{13}C NMR spectroscopy with multiple ^{13}C labels. J Neurosci Res 66:771–781

Cruz NF, Lasater A, Zielke HR, Dienel GA (2005) Activation of astrocytes in brain of conscious rats during acoustic stimulation: acetate utilization in working brain. J Neurochem 92:934–947

Cullingford TE, Clark JB, Phillips IR (1994) The pyruvate dehydrogenase complex: cloning of the rat somatic E1 alpha subunit and its coordinate expression with the mRNAs for the E1 beta, E2, and E3 catalytic subunits in developing rat brain. J Neurochem 62:1682–1690

Dawson DM, Goodfriend TL, Kaplan NO (1964) Lactic dehydrogenases: functions of the two types rates of synthesis of the two major forms can be correlated with metabolic differentiation. Science 143:929–933

Day SE, Kettunen MI, Gallagher FA, Hu DE, Lerche M, Wolber J, Golman K, Ardenkjaer-Larsen JH, Brindle KM (2007) Detecting tumor response to treatment using hyperpolarized ^{13}C magnetic resonance imaging and spectroscopy. Nat Med 13:1382–1387

De Meirleir L (2002) Defects of pyruvate metabolism and the Krebs cycle. J Child Neurol 17(Suppl 3):3S26-33, discussion 23S33-24

del Arco A, Satrustegui J (1998) Molecular cloning of Aralar, a new member of the mitochondrial carrier superfamily that binds calcium and is present in human muscle and brain. J Biol Chem 273:23327–23334

del Arco A, Satrustegui J (2004) Identification of a novel human subfamily of mitochondrial carriers with calcium-binding domains. J Biol Chem 279:24701–24713

Denton RM, Halestrap AP (1979) Regulation of pyruvate metabolism in mammalian tissues. Essays Biochem 15:37–77

Desagher S, Glowinski J, Premont J (1997) Pyruvate protects neurons against hydrogen peroxide-induced toxicity. J Neurosci 17:9060–9067

Dienel GA (2002) Energy generation in the central nervous system. In: Edvinsson L, Krause DN (eds) Cerebral blood flow and metabolism. Lippincott Williams & Wilkins, Philadelphia, pp 141–171

Dienel GA, Hertz L (2005) Astrocytic contributions to bioenergetics of cerebral ischemia. Glia 50:362–388

Dringen R, Schmoll D, Cesar M, Hamprecht B (1993) Incorporation of radioactivity from [^{14}C] lactate into the glycogen of cultured mouse astroglial cells. Evidence for gluconeogenesis in brain cells. Biol Chem Hoppe Seyler 374:343–347

Dwyer DS, Vannucci SJ, Simpson IA (2002) Expression, regulation, and functional role of glucose transporters (GLUTs) in brain. Int Rev Neurobiol 51:159–188

Enerson BE, Drewes LR (2003) Molecular features, regulation, and function of monocarboxylate transporters: implications for drug delivery. J Pharm Sci 92:1531–1544

Erecinska M, Silver IA (1990) Metabolism and role of glutamate in mammalian brain. Prog Neurobiol 35:245–296

Everse J, Kaplan NO (1973) Lactate dehydrogenases: structure and function. Adv Enzymol Relat Areas Mol Biol 37:61–133

Fayol L, Baud O, Monier A, Pellerin L, Magistretti P, Evrard P, Verney C (2004) Immunocytochemical expression of monocarboxylate transporters in the human visual cortex at midgestation. Brain Res Dev Brain Res 148:69–76

Fitzpatrick SM, Hetherington HP, Behar KL, Shulman RG (1990) The flux from glucose to glutamate in the rat brain in vivo as determined by ^{1}H-observed, ^{13}C-edited NMR spectroscopy. J Cereb Blood Flow Metab 10:170–179

Gamberino WC, Berkich DA, Lynch CJ, Xu B, LaNoue KF (1997) Role of pyruvate carboxylase in facilitation of synthesis of glutamate and glutamine in cultured astrocytes. J Neurochem 69:2312–2325

Garcia-Espinosa MA, Garcia-Martin ML, Cerdan S (2003) Role of glial metabolism in diabetic encephalopathy as detected by high resolution ^{13}C NMR. NMR Biomed 16:440–449

Garcia-Espinosa MA, Rodrigues TB, Sierra A, Benito M, Fonseca C, Gray HL, Bartnik BL, Garcia-Martin ML, Ballesteros P, Cerdan S (2004) Cerebral glucose metabolism and the glutamine cycle as detected by in vivo and in vitro ^{13}C NMR spectroscopy. Neurochem Int 45:297–303

Garcia-Martin ML, Ballesteros P, Cerdan S (2001) The metabolism of water in cells and tissues as detected by NMR methods. Prog Nucl Mag Res Spectroscopy 39:41–77

Garcia-Martin ML, Garcia-Espinosa MA, Ballesteros P, Bruix M, Cerdan S (2002) Hydrogen turnover and subcellular compartmentation of hepatic [2-^{13}C]glutamate and [3-13C]aspartate as detected by ^{13}C NMR. J Biol Chem 277:7799–7807

Garfinkel D (1966) A simulation study of the metabolism and compartmentation in brain of glutamate, aspartate, the Krebs cycle, and related metabolites. J Biol Chem 241:3918–3929

Golman K, in't Zandt R, Thaning M (2006) Real-time metabolic imaging. Proc Natl Acad Sci USA 103:11270–11275

Gruetter R (2002) In vivo 13C NMR studies of compartmentalized cerebral carbohydrate metabolism. Neurochem Int 41:143–154

Gruetter R, Seaquist ER, Ugurbil K (2001) A mathematical model of compartmentalized neurotransmitter metabolism in the human brain. Am J Physiol Endocrinol Metab 281: E100–E112

Gruetter R, Adriany G, Choi IY, Henry PG, Lei H, Oz G (2003) Localized in vivo 13C NMR spectroscopy of the brain. NMR Biomed 16:313–338

Halestrap AP, Meredith D (2004) The SLC16 gene family-from monocarboxylate transporters (MCTs) to aromatic amino acid transporters and beyond. Pflugers Arch 447:619–628

Halestrap AP, Price NT (1999) The proton-linked monocarboxylate transporter (MCT) family: structure, function and regulation. Biochem J 343(Pt 2):281–299

Hassel B (2000) Carboxylation and anaplerosis in neurons and glia. Mol Neurobiol 22:21–40

Hassel B (2001) Pyruvate carboxylation in neurons. J Neurosci Res 66:755–762

Hassel B, Brathe A (2000) Cerebral metabolism of lactate in vivo: evidence for neuronal pyruvate carboxylation. J Cereb Blood Flow Metab 20:327–336

Hassel B, Sonnewald U (1995) Glial formation of pyruvate and lactate from TCA cycle intermediates: implications for the inactivation of transmitter amino acids? J Neurochem 65:2227–2234

Heeger DJ, Ress D (2002) What does fMRI tell us about neuronal activity. Nat Rev Neurosci 3:142–151

Hevor TK (1994) Some aspects of carbohydrate metabolism in the brain. Biochimie 76:111–120

Hyder F, Chase JR, Behar KL, Mason GF, Siddeek M, Rothman DL, Shulman RG (1996) Increased tricarboxylic acid cycle flux in rat brain during forepaw stimulation detected with 1H[13C] NMR. Proc Natl Acad Sci USA 93:7612–7617

Hyder F, Rothman DL, Mason GF, Rangarajan A, Behar KL, Shulman RG (1997) Oxidative glucose metabolism in rat brain during single forepaw stimulation: a spatially localized ^{1}H[^{13}C] nuclear magnetic resonance study. J Cereb Blood Flow Metab 17:1040–1047

Izumi Y, Katsuki H, Zorumski CF (1997) Monocarboxylates (pyruvate and lactate) as alternative energy substrates for the induction of long-term potentiation in rat hippocampal slices. Neurosci Lett 232:17–20

Jitrapakdee S, Wallace JC (1999) Structure, function and regulation of pyruvate carboxylase. Biochem J 340(Pt 1):1–16

Jitrapakdee S, Booker GW, Cassady AI, Wallace JC (1996a) Cloning, sequencing and expression of rat liver pyruvate carboxylase. Biochem J 316(Pt 2):631–637

Jitrapakdee S, Walker ME, Wallace JC (1996b) Identification of novel alternatively spliced pyruvate carboxylase mRNAs with divergent 5′-untranslated regions which are expressed in a tissue-specific manner. Biochem Biophys Res Commun 223:695–700

Kirk P, Wilson MC, Heddle C, Brown MH, Barclay AN, Halestrap AP (2000) CD147 is tightly associated with lactate transporters MCT1 and MCT4 and facilitates their cell surface expression. EMBO J 19:3896–3904

Kunnecke B, Cerdan S, Seelig J (1993) Cerebral metabolism of [1,2-$^{13}C_2$]glucose and [U-$^{13}C_4$]3-hydroxybutyrate in rat brain as detected by ^{13}C NMR spectroscopy. NMR Biomed 6:264–277

LaNoue KF, Schoolwerth AC (1979) Metabolite transport in mitochondria. Annu Rev Biochem 48:871–922

Lapidot A, Gopher A (1994) Cerebral metabolic compartmentation. Estimation of glucose flux via pyruvate carboxylase/pyruvate dehydrogenase by ^{13}C NMR isotopomer analysis of D-[U-^{13}C] glucose metabolites. J Biol Chem 269:27198–27208

Laughton JD, Charnay Y, Belloir B, Pellerin L, Magistretti PJ, Bouras C (2000) Differential messenger RNA distribution of lactate dehydrogenase LDH-1 and LDH-5 isoforms in the rat brain. Neuroscience 96:619–625

Lebon V, Petersen KF, Cline GW, Shen J, Mason GF, Dufour S, Behar KL, Shulman GI, Rothman DL (2002) Astroglial contribution to brain energy metabolism in humans revealed by ^{13}C nuclear magnetic resonance spectroscopy: elucidation of the dominant pathway for neurotransmitter glutamate repletion and measurement of astrocytic oxidative metabolism. J Neurosci 22:1523–1531

Leino RL, Gerhart DZ, Drewes LR (1999) Monocarboxylate transporter (MCT1) abundance in brains of suckling and adult rats: a quantitative electron microscopic immunogold study. Brain Res Dev Brain Res 113:47–54

Loeber G, Infante AA, Maurer-Fogy I, Krystek E, Dworkin MB (1991) Human NAD$^+$-dependent mitochondrial malic enzyme. cDNA cloning, primary structure, and expression in Escherichia coli. J Biol Chem 266:3016–3021

Loffler T, Al-Robaiy S, Bigl M, Eschrich K, Schliebs R (2001) Expression of fructose-1, 6-bisphosphatase mRNA isoforms in normal and basal forebrain cholinergic lesioned rat brain. Int J Dev Neurosci 19:279–285

Mac M, Nalecz KA (2003) Expression of monocarboxylic acid transporters (MCT) in brain cells. Implication for branched chain alpha-ketoacids transport in neurons. Neurochem Int 43: 305–309

Magistretti PJ, Allaman I (2007) Glycogen: a Trojan horse for neurons. Nat Neurosci 10: 1341–1342

Magistretti PJ, Pellerin L (1997) Metabolic coupling during activation. A cellular view. Adv Exp Med Biol 413:161–166

Magistretti PJ, Pellerin L (1999) Cellular mechanisms of brain energy metabolism and their relevance to functional brain imaging. Philos Trans R Soc Lond B Biol Sci 354:1155–1163

Martinez-Hernandez A, Bell KP, Noremberg MD (1977) Glutamine synthase: glila localization in brain. Science 195:1356–1358

Mason GF, Behar KL, Rothman DL, Shulman RG (1992a) NMR determination of intracerebral glucose concentration and transport kinetics in rat brain. J Cereb Blood Flow Metab 12:448–455

Mason GF, Rothman DL, Behar KL, Shulman RG (1992b) NMR determination of the TCA cycle rate and alpha-ketoglutarate/glutamate exchange rate in rat brain. J Cereb Blood Flow Metab 12:434–447

Mason GF, Gruetter R, Rothman DL, Behar KL, Shulman RG, Novotny EJ (1995) Simultaneous determination of the rates of the TCA cycle, glucose utilization, alpha-ketoglutarate/glutamate exchange, and glutamine synthesis in human brain by NMR. J Cereb Blood Flow Metab 15:12–25

Matsumoto K, Yamada K, Kohmura E, Kinoshita ATH (1994) Role of pyruvate in ischemia like conditions on cultured neurons. Neurol Res 16:460–464

McKenna MC (2007) The glutamate-glutamine cycle is not stoichiometric: fates of glutamate in brain. J Neurosci Res 85:3347–3358

McKenna MC, Stevenson JH, Huang X, Tildon JT, Zielke CL, Hopkins IB (2000) Mitochondrial malic enzyme activity is much higher in mitochondria from cortical synaptic terminals compared with mitochondria from primary cultures of cortical neurons or cerebellar granule cells. Neurochem Int 36:451–459

McKenna MC, Waagepetersen HS, Schousboe A, Sonnewald U (2006) Neuronal and astrocytic shuttle mechanisms for cytosolic-mitochondrial transfer of reducing equivalents: current evidence and pharmacological tools. Biochem Pharmacol 71:399–407

Medina JM, Tabernero A, Tovar JA, Martin-Barrientos J (1996) Metabolic fuel utilization and pyruvate oxidation during the postnatal period. J Inherit Metab Dis 19:432–442

Moore SA (2001) Polyunsaturated fatty acid synthesis and release by brain-derived cells in vitro. J Mol Neurosci 16:195–200, discussion 215–121

Morris P, Bachelard H (2003) Reflections on the application of ^{13}C-MRS to research on brain metabolism. NMR Biomed 16:303–312

Nakai N, Obayashi M, Nagasaki M, Sato Y, Fujitsuka N, Yoshimura A, Miyazaki Y, Sugiyama S, Shimomura Y (2000) The abundance of mRNAs for pyruvate dehydrogenase kinase isoenzymes in brain regions of young and aged rats. Life Sci 68:497–503

Nicholls D (2007) Bioenergetics. In: Lajhta A, Gibson G, Dienel G (eds) Brain energetics. integration of molecular and cellular processes, 3rd edn. Springer, New York, pp 3–16

Okada Y, Lipton P (2007) Glucose, oxidative energy metabolism, and neural function in brain slices—glycolysis plays a key role in neural activity. In: Lajhta A, Gibson G, Dienel G (eds) Brain energetics. integration of molecular and cellular processes, 3rd edn. Springer, New York, pp 17–39

Olstad E, Olsen GM, Qu H, Sonnewald U (2007) Pyruvate recycling in cultured neurons from cerebellum. J Neurosci Res 85:3318–3325

Palmieri F (1994) Mitochondrial carrier proteins. FEBS Lett 346:48–54

Palmieri F (2004) The mitochondrial transporter family (SLC25): physiological and pathological implications. Pflugers Arch 447:689–709

Palmieri F, Bisaccia F, Capobianco L, Dolce V, Fiermonte G, Iacobazzi V, Indiveri C, Palmieri L (1996) Mitochondrial metabolite transporters. Biochim Biophys Acta 1275:127–132

Palmieri L, Pardo B, Lasorsa FM, del Arco A, Kobayashi K, Iijima M, Runswick MJ, Walker JE, Saheki T, Satrustegui J et al (2001) Citrin and aralar1 are Ca^{2+}-stimulated aspartate/glutamate transporters in mitochondria. EMBO J 20:5060–5069

Pascual JM, Carceller F, Roda JM, Cerdan S, Pascual JM, Carceller F, Roda JM, Cerdan S (1998) Glutamate, glutamine, and GABA as substrates for the neuronal and glial compartments after focal cerebral ischemia in rats. Stroke 29:1048–1056, discussion 1056–1047

Passonneau JV, Lowry OH (1964) The role of phosphofructokinase in metabolic regulation. Adv Enzyme Regul 2:265–274

Patel MS, Korotchkina LG (2001) Regulation of mammalian pyruvate dehydrogenase complex by phosphorylation: complexity of multiple phosphorylation sites and kinases. Exp Mol Med 33:191–197

Pellerin L, Magistretti PJ (1994) Glutamate uptake into astrocytes stimulates aerobic glycolysis: a mechanism coupling neuronal activity to glucose utilization. Proc Natl Acad Sci USA 91:10625–10629

Pellerin L, Pellegri G, Bittar PG, Charnay Y, Bouras C, Martin JL, Stella N, Magistretti PJ (1998a) Evidence supporting the existence of an activity-dependent astrocyte-neuron lactate shuttle. Dev Neurosci 20:291–299

Pellerin L, Pellegri G, Martin JL, Magistretti PJ (1998b) Expression of monocarboxylate transporter mRNAs in mouse brain: support for a distinct role of lactate as an energy substrate for the neonatal vs. adult brain. Proc Natl Acad Sci USA 95:3990–3995

Preece NE, Cerdan S (1996) Metabolic precursors and compartmentation of cerebral GABA in vigabatrin-treated rats. J Neurochem 67:1718–1725

Rafiki A, Boulland JL, Halestrap AP, Ottersen OP, Bergersen L (2003) Highly differential expression of the monocarboxylate transporters MCT2 and MCT4 in the developing rat brain. Neuroscience 122:677–688

Ramirez BG, Rodrigues TB, Violante IR, Cruz F, Fonseca LL, Ballesteros P, Castro MM, Garcia-Martin ML, Cerdan S (2007) Kinetic properties of the redox switch/redox coupling mechanism as determined in primary cultures of cortical neurons and astrocytes from rat brain. J Neurosci Res 85:3244–3253

Ramos M, del Arco A, Pardo B, Martinez-Serrano A, Martinez-Morales JR, Kobayashi K, Yasuda T, Bogonez E, Bovolenta P, Saheki T et al (2003) Developmental changes in the Ca^{2+} – regulated mitochondrial aspartate-glutamate carrier aralar1 in brain and prominent expression in the spinal cord. Brain Res Dev Brain Res 143:33–46

Richard J (1963) Lactic dehydrogenase isoenzymes in the nervous system. Actions and relations to hydrosoluble proteins and mineral elements. Ann Soc R Sci Med Nat Brux 16:185–223, CONCL

Rodrigues TB, Cerdan S (2005a) ^{13}C MRS: an outstanding tool for metabolic studies. Concepts Magn Reson, Part A 27A:1–16

Rodrigues TB, Cerdan S (2005b) A fast and sensitive ^{1}H NMR method to measure the turnover of the H2 hydrogen of lactate. Magn Reson Med 54:1014–1019

Rodrigues TB, Cerdan S (2007) The cerebral tricarboxylic acid cycles. In: Dienel G, Gibson G, Lajtha A (eds) Brain energetics from genes to cells, integration of molecular and cellular processes. Springer, New York, pp 61–93

Rodrigues TB, Gray HL, Benito M, Garrido S, Sierra A, Geraldes CF, Ballesteros P, Cerdan S (2005) Futile cycling of lactate through the plasma membrane of C6 glioma cells as detected by (^{13}C, ^{2}H) NMR. J Neurosci Res 79:119–127

Rodrigues TB, Granado N, Ortiz O, Cerdan S, Moratalla R (2007) Metabolic interactions between glutamatergic and dopaminergic neurotransmitter systems are mediated through D_1 dopamine receptors. J Neurosci Res 85:3284–3293

Rodrigues TB, Lopez-Larrubia P, Cerdan S (2009) Redox dependence and compartmentation of [^{13}C]pyruvate in the brain of deuterated rats bearing implanted C6 gliomas. J Neurochem 109(Suppl 1):237–245

Ross B, Lin A, Harris K, Bhattacharya P, Schweinsburg B (2003) Clinical experience with ^{13}C MRS in vivo. NMR Biomed 16:358–369

Rothman DL, Behar KL, Hyder F, Shulman RG (2003) In vivo NMR studies of the glutamate neurotransmitter flux and neuroenergetics: implications for brain function. Annu Rev Physiol 65:401–427

Ruscak M, Orlicky J, Zubor V, Hager H (1982) Alanine aminotransferase in bovine brain: purification and properties. J Neurochem 39:210–216
Sacktor B, Wilson JE, Tiekert CG (1966) Regulation of glycolysis in brain, in situ, during convulsions. J Biol Chem 241:5071–5075
Satrustegui J, Pardo B, Del Arco A (2007) Mitochondrial transporters as novel targets for intracellular calcium signaling. Physiol Rev 87:29–67
Schmoll D, Fuhrmann E, Gebhardt R, Hamprecht B (1995) Significant amounts of glycogen are synthesized from 3-carbon compounds in astroglial primary cultures from mice with participation of the mitochondrial phosphoenolpyruvate carboxykinase isoenzyme. Eur J Biochem 227:308–315
Schousboe A, Sonnewald U, Waagepetersen HS (2003) Differential roles of alanine in GABAergic and glutamatergic neurons. Neurochem Int 43:311–315
Schurr A (2002) Lactate, glucose and energy metabolism in the ischemic brain (Review). Int J Mol Med 10:131–136
Shank RP, Bennett GS, Freytag SO, Campbell GL (1985) Pyruvate carboxylase: an astrocyte-specific enzyme implicated in the replenishment of amino acid neurotransmitter pools. Brain Res 329:364–367
Shen J, Petersen KF, Behar KL, Brown P, Nixon TW, Mason GF, Petroff OA, Shulman GI, Shulman RG, Rothman DL (1999) Determination of the rate of the glutamate/glutamine cycle in the human brain by in vivo ^{13}C NMR. Proc Natl Acad Sci USA 96:8235–8240
Shulman RG, Rothman DL, Behar KL, Hyder F (2004) Energetic basis of brain activity: implications for neuroimaging. Trends Neurosci 27:489–495
Sibson NR, Dhankhar A, Mason GF, Behar KL, Rothman DL, Shulman RG (1997) In vivo ^{13}C NMR measurements of cerebral glutamine synthesis as evidence for glutamate-glutamine cycling. Proc Natl Acad Sci USA 94:2699–2704
Sibson NR, Dhankhar A, Mason GF, Rothman DL, Behar KL, Shulman RG (1998a) Stoichiometric coupling of brain glucose metabolism and glutamatergic neuronal activity. Proc Natl Acad Sci USA 95:316–321
Sibson NR, Shen J, Mason GF, Rothman DL, Behar KL, Shulman RG (1998b) Functional energy metabolism: in vivo ^{13}C-NMR spectroscopy evidence for coupling of cerebral glucose consumption and glutamatergic neuronalactivity. Dev Neurosci 20:321–330
Sibson NR, Mason GF, Shen J, Cline GW, Herskovits AZ, Wall JE, Behar KL, Rothman DL, Shulman RG (2001) In vivo ^{13}C NMR measurement of neurotransmitter glutamate cycling, anaplerosis and TCA cycle flux in rat brain during. J Neurochem 76:975–989
Sierra A, Lopes da Fonseca L, Ballesteros P, Cerdan S (2004) Quantitative modelling of H3 hydrogen turnover in (2-^{13}C) glutamate and (2-^{13}C) glutamine during (2-^{13}C) acetate metabolism in the adult rat brain. Proceedings of the XXI Annual Meeting of the European Society for Magnetic Resonance in Medicine and Biology. Copenhagen, p 36
Siesjo BK (1982) Lactic acidosis in the brain: occurrence, triggering mechanisms and pathophysiological importance. Ciba Found Symp 87:77–100
Sokoloff L (1989) Circulation and energy metabolism of the brain. In: Siegel G, Agranoff B, Albers RW, Molinoff P (eds) Basic neurochemistry. Raven, New York, pp 565–590
Sokoloff L (1992) The brain as a chemical machine. Prog Brain Res 94:19–33
Sonnewald U, Schousboe A, Qu H, Waagepetersen HS (2004) Intracellular metabolic compartmentation assessed by ^{13}C magnetic resonance spectroscopy. Neurochem Int 45:305–310
Takenaka M, Noguchi T, Inoue H, Yamada K, Matsuda T, Tanaka T (1989) Rat pyruvate kinase M gene. Its complete structure and characterization of the 5′-flanking region. J Biol Chem 264:2363–2367
Tsacopoulos M, Magistretti PJ (1996) Metabolic coupling between glia and neurons. J Neurosci 16:877–885
Van den Berg CJ, Garfinkel D (1971) A stimulation study of brain compartments. Metabolism of glutamate and related substances in mouse brain. Biochem J 123:211–218
Vannucci SJ, Simpson IA (2003) Developmental switch in brain nutrient transporter expression in the rat. Am J Physiol Endocrinol Metab 285:E1127–E1134

Vannucci SJ, Maher F, Simpson IA (1997) Glucose transporter proteins in brain: delivery of glucose to neurons and glia. Glia 21:2–21

Vogel R, Hamprecht B, Wiesinger H (1998a) Malic enzyme isoforms in astrocytes: comparative study on activities in rat brain tissue and astroglia-rich primary cultures. Neurosci Lett 247: 123–126

Vogel R, Jennemann G, Seitz J, Wiesinger H, Hamprecht B (1998b) Mitochondrial malic enzyme: purification from bovine brain, generation of an antiserum, and immunocytochemical localization in neurons of rat brain. J Neurochem 71:844–852

Waagepetersen HS, Sonnewald U, Larsson OM, Schousboe A (2000) A possible role of alanine for ammonia transfer between astrocytes and glutamatergic neurons. J Neurochem 75:471–479

Waagepetersen HS, Qu H, Hertz L, Sonnewald U, Schousboe A (2002) Demonstration of pyruvate recycling in primary cultures of neocortical astrocytes but not in neurons. Neurochem Res 27:1431–1437

Waagepetersen HS, Sonnewald U, Schousboe A (2003) Compartmentation of glutamine, glutamate, and GABA metabolism in neurons and astrocytes: functional implications. Neuroscientist 9:398–403

Waniewski RA, Martin DL (1998) Preferential utilization of acetate by astrocytes is attributable to transport. J Neurosci 18:5225–5233

Westergaard N, Varming T, Peng L, Sonnewald U, Hertz L, Schousboe A (1993) Uptake, release, and metabolism of alanine in neurons and astrocytes in primary cultures. J Neurosci Res 35:540–545

Wienhard K (2002) Measurement of glucose consumption using [^{18}F]fluorodeoxyglucose. Methods 27:218–225

Yamada K, Noguchi T (1999a) Nutrient and hormonal regulation of pyruvate kinase gene expression. Biochem J 337(Pt 1):1–11

Yamada K, Noguchi T (1999b) Regulation of pyruvate kinase M gene expression. Biochem Biophys Res Commun 256:257–262

Yang Z, Floyd DL, Loeber G, Tong L (2000) Structure of a closed form of human malic enzyme and implications for catalytic mechanism. Nat Struct Biol 7:251–257

Zwingmann C, Leibfritz D (2003) Regulation of glial metabolism studied by ^{13}C-NMR. NMR Biomed 16:370–399

Zwingmann C, Richter-Landsberg C, Brand A, Leibfritz D (2000) NMR spectroscopic study on the metabolic fate of [3-^{13}C]alanine in astrocytes, neurons, and cocultures: implications for glia-neuron interactions in neurotransmitter metabolism. Glia 32:286–303

Zwingmann C, Richter-Landsberg C, Leibfritz D (2001) ^{13}C isotopomer analysis of glucose and alanine metabolism reveals cytosolic pyruvate compartmentation as part of energy metabolism in astrocytes. Glia 34:200–212

Chapter 25
Blood-Brain Barrier Transport of Lactate

Gitte Moos Knudsen

Abstract Monocarboxylates cross the blood-brain barrier by means of non-saturable and saturable carrier-mediated diffusion; the latter is mediated through the monocarboxylate transporters. The contribution of non-saturable diffusion across cell membranes is substantial and constitutes as much as 50% of the total amount of lactate transport at normal lactate levels. At high lactate levels the transport is dominated by the non-saturable transport and the equilibration between blood and brain lactate levels is almost instantaneous.

The clinical impact of monocarboxylates (MCT) function to the overall maintenance of cerebral metabolic homeostasis is important and this pertains particularly to disorders such as brain ischemia, meningtitis or head injury.

Keywords Blood-brain barrier • Brain activation • Cerebral metabolic rate of lactate • Lactate • Meningitis • Monocarboxylate transporters

25.1 Monocarboxylates are Transported Across the BBB by the MCT Transporter

Monocarboxylates such as lactate, pyruvate, and ketone bodies cross mammalian cell membranes by facilitated diffusion through a family of integral membrane transport proteins, the monocarboxylate transporters (MCTs), which catalyse the proton-linked transport of the monocarboxylates. The family now comprises fourteen unique members of the MCT family (Halestrap and Meredith 2004, p. 447) and orthologs to each have been identified in a variety of species (Enerson and Drewes 2003, p. 92).

G.M. Knudsen, M.D., DMSc. (✉)
Neurobiology Research Unit, Section 9201, Copenhagen University Hospital - Rigshospitalet, 9 Blegdamsvej, DK-2100, Copenhagen, Denmark
e-mail: gmk@nru.dk; gitte@nru.dk

I.-Y. Choi and R. Gruetter (eds.), *Neural Metabolism In Vivo*,
Advances in Neurobiology 4, DOI 10.1007/978-1-4614-1788-0_25,

Four MCTs (MCT1-MCT4), all are expressed in the brain, have been functionally characterized. Each isoform possesses unique biochemical properties such as kinetic constants and sensitivity to known MCT inhibitors.

MCT1 is ubiquitously expressed, and catalyses either net transport of one monocarboxylate with another one, or with one proton. In the brain, the transporter is particularly localised to astrocytes, blood vessels, ependymocytes, and glia limitans. *In vitro*, a subset of neurons exhibits a weak but significant MCT1 expression (Pellerin et al. 2005, p. 79). In the rodent brain, MCT1 undergoes a profound developmental regulation. The MCT1's are also present in the brain endothelial cells and are thus an integrated part of the blood-brain barrier (BBB).

In contrast to the rodent brain where it is particularly expressed on neuronal dendrites, MCT2 is poorly expressed in human brain. Km for MCT2 mediated lactate transport is relatively low compared to normal plasma levels, about 0.7 mM.

MCT3 is mainly expressed in the retinal pigment epithelium and choroid plexus epithelia, in contrast to the apically located MCT1. It has a Km for lactate transport in the order of 6 mM. MCT4 is also expressed on glial cells, perhaps exclusively, at least in the rodent brain.

Several fold changes in the expression of MCTs may be evoked by altered physiological conditions, yet the molecular mechanisms underlying the regulation of MCTs are poorly understood. Post-translational regulation of MCT1 and MCT4 occurs, in part, by interaction with CD147, an accessory protein that is necessary for trafficking, localization, and functional expression of these transporters (Enerson and Drewes 2003, p. 92).

In addition to the facilitated BBB-transport a substantial transport via non-saturable, non-stereospecific diffusion has been demonstrated in rats. Whereas the MCT mediated component in rats has been shown to have a low efficiency and moderate capacity with transport affinity coefficients between 6 and 14 mM and transport maxima of 23–40 μmol/100 g/min, then the diffusional component ranged from 0.020 to 0.036 ml/g/min (LaManna et al. 1993, p. 614). This means that at normal plasma lactate concentrations (1–1.5 mM) the two influx components contribute approximately equally to the total transport from the blood to the brain, the permeability-surface area, PS1 of about 0.06–0.08 ml g-1 min−1. Presumably, for transport out of the brain and under normal physiological circumstances, the facilitated diffusion constitutes a similar fraction (Knudsen et al. 1990, p. 10) since the brain extracellular concentration of lactate is in the same order of magnitude as plasma concentrations, as determined with cerebral microdialysis (Fig. 25.1).

By means of the double-indicator method (Knudsen et al. 1994, p. 266) Knudsen et al. investigated BBB transport in humans and confirmed the values obtained in rats; the average unidirectional extraction of L-lactate is 15%, PS_1 is 0.08 ml g-1 min-1, and the transport from the brain to the blood (PS_2) is of the same order of magnitude (Knudsen et al. 1991, p. 11). This means that in comparison to, e.g., glucose L-lactate is relatively easily taken up by the human brain, and that the mechanism by which it crosses the blood-brain barrier is equilibrative.

In rats, intravenous sodium lactate infusion with blood lactate levels of up to 2.3 mmol/l is associated with a threefold increase in whole brain lactate, as determined by magnetic resonance spectroscopy. Brain concentrations mirrored

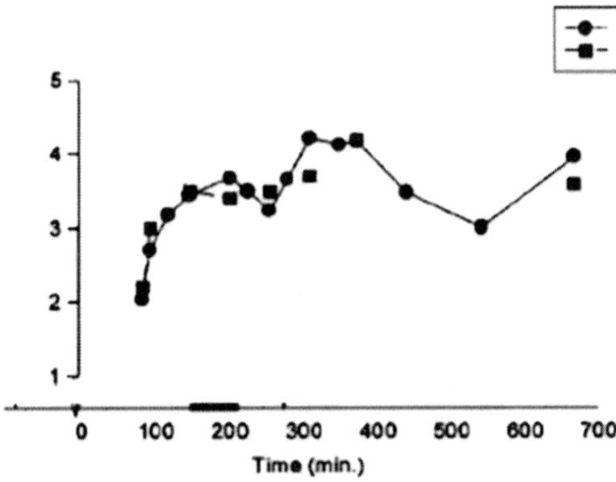

Fig. 25.1 Development of arterial and cerebral lactate levels, measured simultaneously, in a patient with acute liver failure (Modified from Tofteng and Larsen (2002))

plasma lactate levels but in this study, the rise and decline was less steep in the brain (Dager et al. 1992, p. 51). There are also more recent data available from humans, demonstrating how rapid plasma lactate levels equilibrate with brain interstitial lactate levels, as measured with microdialysis. An example of this is given in Fig. 25.1.

When a model of nutrient transport in the mammalian brain incorporated known concentrations and kinetic parameters, one of the main outcomes was the appreciation of the role of the basal lamina as a post-endothelial cell "reservoir" and conduit for diffusion of glucose and metabolites directly to the interstitium, not through the glial end-feet that are in close contact with the endothelial cells' abluminal side (Simpson et al. 2007, p. 27).

It has for a long time been known that in humans, under in hypercapnia, as arterial pH decrease from normal levels to about 7.26, the BBB PS_1 for L-lactate increases considerably, by 110% (Knudsen et al. 1991, p. 11). This enhancement was thought to take place through the concomitant decrease in pH, presumably through the saturable MCT because the fraction of ionized lactate that converts into unionized lactate (and thereby becomes sufficiently lipophilic to cross the BBB) is negligible. This observation was later corroborated by experimental studies showing that the MCT1 mediated transport of lactate can be stimulated by decreasing pH from 8 to 6 (Halestrap and Meredith 2004, p. 447).

25.2 Lactate Net Flux and CMR_{lac}

Under normal physiological circumstances, the human brain exhibits a small global net efflux of lactate, in the order of 0.02–0.05 μmol/g/min (Moller et al. 2004, p. 35, 2002a, p. 22), as measured from arteriovenous differences times cerebral blood flow. That is, under normal conditions, the brain has a small net production of lactate.

In awake rats, infusion with [U-(13)C]lactate has been found to result in label incorporation similar to that observed from [U-(13)C]glucose, although at a much reduced level (Qu et al. 2000, p. 22). In the same study, it was also found that gluconeogenesis takes place from [U-(13)C]lactate. Contributions from the glial enzyme pyruvate carboxylase were detectable in glutamine, glutamate and GABA, and were comparatively more pronounced in the glucose group. This indicated that relatively more pyruvate from lactate than glucose was metabolized in neurons (Qu et al. 2000, p. 22).

From NMR spectroscopy data it has been shown that *in vivo* glucose and lactate distributes relatively evenly between the intra- and extracellular brain compartment (Pfeuffer et al. 2000, p. 20). In the adult rat brain, the calculated rate of cellular lactate transport was found to be two orders of magnitude faster than transport of lactate across the BBB, and passive diffusion of lactate was in this study not found to contribute significantly across either cell or blood-brain barrier(Kuhr et al. 1988, p. 8).

25.3 The Brain Consumes Lactate

During brain activation, lactate is produced aerobically and after ischemia, it may even serve as an obligatory aerobic energy substrate. It has been also hypothesized, based on *in vitro* studies, that lactate, produced by glia in large amounts during activation and/or ischemia/hypoxia, is transported via specific glial and neuronal monocarboxylate transporters into neurons for aerobic utilization. On the basis of *in vitro* experiments it has been claimed that lactate is an efficient energy substrate for neurons that contributes to maintain synaptic transmission, particularly during neuronal activation. In contrast to what is commonly stated (e.g., (Pellerin et al. 1998, p. 20)), as evidenced above, lactate does indeed cross the BBB relatively easy (Knudsen et al. 1994, p. 266), and blood-borne lactate may thus constitute a significant source of energy under activation. There is biochemical evidence to support that lactate is interchangeable with glucose to support oxidative metabolism in cortical neurons. This set of data is consistent with the existence of an activity-dependent astrocyte-neuron lactate shuttle for the supply of energy substrates to neurons (Shulman et al. 2001, p. 14).

In the resting brain, energy is primarily supplied by the oxidation of glucose, but there are exceptions to this rule. For example, in healthy people starved for three days, ketone bodies account for approximately one-fourth of the cerebral energy requirements (Hasselbalch et al. 1994, p. 14), and blood beta-hydroxybutyrate (beta-OHB) concentration and beta-OHB influx into the brain is increased by > 10 times(Hasselbalch et al. 1995, p. 268), whereas lactate consumption is unaltered. This shift in metabolism is directly coupled to circulating blood concentrations of ketone bodies and the immediate oxidation of ketone bodies induces a decrease in cerebral glucose uptake in spite of an adequate glucose supply (Hasselbalch et al. 1996, p. 270).

There is not really any evidence that increased blood lactate levels *per se* leads to increased brain lactate consumption. An increase in blood lactate, e.g. as seen under anesthesia, does not alter that there still is a small net brain lactate efflux. However,

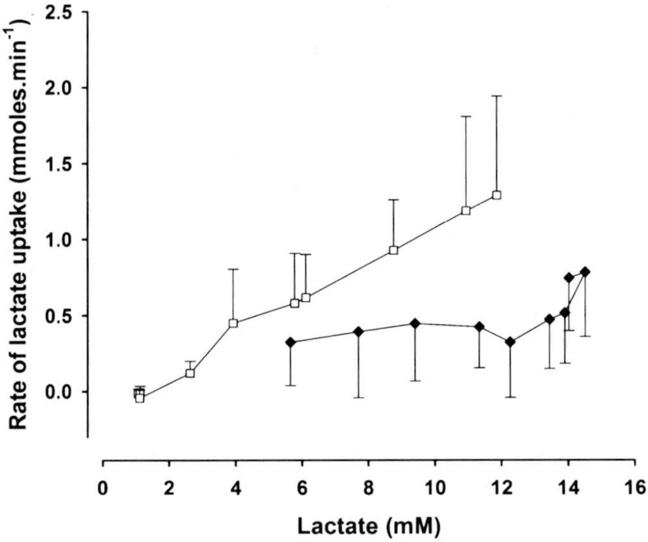

Fig. 25.2 Lactate uptake in the human brain, as a function of plasma lactate levels. □, during rest (0–15 min), and exercise (15–30 min); ♦, during recovery (30–60 min) in young healthy volunteers (Reproduced from Quistorff et al. (2008)) Values are given as means ± SD

under intense physical exercise during which anaerobic metabolism occurs in the muscles and arterial lactate is elevated, brain takes up lactate in amounts that may supersede glucose uptake, as illustrated in Fig. 25.2 (Quistorff et al. 2008).

The oxygen-to-glucose index (OGI), which is the ratio of metabolic rates of oxygen to glucose, CMR(O2)/CMR(glc), diverges from the theoretical value of 6 as brain activity increases. *In vivo* measurements of humans have demonstrated that the brain lactate concentration increases as the brain becomes activated but also that the observed changes in brain lactate decreases in amplitude over time, as the stimulus is presented over and over again, and gradually adapted to a value not higher than 0.1 μmol/g (Mangia et al. 2007, p. 85). The idea that during hypoglycemia, astrocyte glycogen breaks down to lactate that is then transferred to adjacent neurons or axons where it is used aerobically as fuel (Brown and Ransom 2007, p. 55) has been around for a while. This view is not, however, at odds with the notion that blood constitutes a considerable source of lactate that, due to the rapid equilibrative uptake across the BBB, can supply neurons with energy.

25.4 Clinical Relevance

Rapid transport of lactic acid transport across the plasma membrane and the BBB is of fundamental importance, particularly in hypoxic conditions where a build-up of lactate takes place. Apart from the most obvious one, namely stroke, two more examples of pathological conditions where cerebral lactate consumption may take place will be given.

Head-injured patients achieving a more favorable outcome differed from those with an unfavorable outcome by having a higher cerebral metabolic rate of oxygen, a higher rate of abnormal brain lactate uptake relative to arterial lactate levels, and lesser degrees of BBB damage (Glenn et al. 2003, p. 23). This suggests that the brain's ability to use lactate as a fuel may be a key outcome predictor.

The cerebral metabolic rate of lactate (CMRlac) is significantly more negative in patients with meningitis than in healthy subjects, indicating an increased cerebral lactate production in meningitis (Moller et al. 2002a, p. 46). Hyperventilation induces an increase in lactate production and efflux, from 0.02 to 0.12 (Moller et al. 2002b, p. 46) and at the same time, microdialysis studies in humans have revealed that lactate concentrations within the brain remain constant (Tofteng and Larsen 2002, p. 8) suggesting that increased cerebral levels of lactate does not increase CMRlac.

25.5 Conclusion

Because of the physiological importance of monocarboxylates (MCT) to the overall maintenance of cerebral metabolic homeostasis, the function of MCTs is significant to several pathologies that occur with disease, such as in brain ischemia or head injury. It should be emphasized, however, that the contribution of non-saturable diffusion across cell membranes is substantial and that this transport mechanism constitutes as much as 50% of the total amount of lactate transport. At high lactate levels the transport is dominated by the non-saturable transport and the equilibration between blood and brain lactate levels is almost instantaneous.

References

Brown AM, Ransom BR (2007) Astrocyte glycogen and brain energy metabolism. Glia 55(12):1263–1271

Dager SR, Marro KI, Richards TL, Metzger GD (1992) Localized magnetic resonance spectroscopy measurement of brain lactate during intravenous lactate infusion in healthy volunteers. Life Sci 51(12):973–985

Enerson BE, Drewes LR (2003) Molecular features, regulation, and function of monocarboxylate transporters: implications for drug delivery. J Pharm Sci 92(8):1531–1544

Glenn TC, Kelly DF, Boscardin WJ, McArthur DL, Vespa P, Oertel M, Hovda DA et al (2003) Energy dysfunction as a predictor of outcome after moderate or severe head injury: indices of oxygen, glucose, and lactate metabolism. J Cereb Blood Flow Metab 23(10):1239–1250

Halestrap AP, Meredith D (2004) The SLC16 gene family-from monocarboxylate transporters (MCTs) to aromatic amino acid transporters and beyond. Pflugers Arch 447(5):619–628

Hasselbalch SG, Knudsen GM, Jakobsen J, Hageman LP, Holm S, Paulson OB (1994) Brain metabolism during short-term starvation in humans. J Cereb Blood Flow Metab 14(1):125–131

Hasselbalch SG, Knudsen GM, Jakobsen J, Hageman LP, Holm S, Paulson OB (1995) Blood-brain barrier permeability of glucose and ketone bodies during short-term starvation in humans. Am J Physiol 268(6 Pt 1):E1161–E1166

Hasselbalch SG, Madsen PL, Hageman LP, Olsen KS, Justesen N, Holm S, Paulson OB (1996) Changes in cerebral blood flow and carbohydrate metabolism during acute hyperketonemia. Am J Physiol 270(5 Pt 1):E746–E751

Knudsen GM, Paulson OB, Hertz MM (1991) Kinetic analysis of the human blood-brain barrier transport of lactate and its influence by hypercapnia. J Cereb Blood Flow Metab 11(4):581–586

Knudsen GM, Pettigrew KD, Patlak CS, Hertz MM, Paulson OB (1990) Asymmetrical transport of amino acids across the blood-brain barrier in humans. J Cereb Blood Flow Metab 10(5):698–706

Knudsen GM, Pettigrew KD, Patlak CS, Paulson OB (1994) Blood-brain barrier permeability measurements by double-indicator method using intravenous injection. Am J Physiol 266(3 Pt 2): H987–H999

Kuhr WG, van den Berg CJ, Korf J (1988) In vivo identification and quantitative evaluation of carrier-mediated transport of lactate at the cellular level in the striatum of conscious, freely moving rats. J Cereb Blood Flow Metab 8(6):848–856

LaManna JC, Harrington JF, Vendel LM, bi-Saleh K, Lust WD, Harik SI (1993) Regional blood-brain lactate influx. Brain Res 614(1–2):164–170

Mangia S, Tkac I, Logothetis NK, Gruetter R, Van de Moortele PF, Ugurbil K (2007) Dynamics of lactate concentration and blood oxygen level-dependent effect in the human visual cortex during repeated identical stimuli. J Neurosci Res 85(15):3340–3346

Moller K, Qvist T, Tofteng F, Sahl C, Sonderkaer S, Dethloff T, Knudsen GM, Larsen FS (2004) Cerebral blood flow and metabolism during infusion of norepinephrine and propofol in patients with bacterial meningitis. Stroke 35(6):1333–1339

Moller K, Strauss GI, Qvist J, Fonsmark L, Knudsen GM, Larsen FS, Krabbe KS, Skinhoj P, Pedersen BK (2002a) Cerebral blood flow and oxidative metabolism during human endotoxemia. J Cereb Blood Flow Metab 22(10):1262–1270

Moller K, Strauss GI, Thomsen G, Larsen FS, Holm S, Sperling BK, Skinhoj P, Knudsen GM (2002b) Cerebral blood flow, oxidative metabolism and cerebrovascular carbon dioxide reactivity in patients with acute bacterial meningitis. Acta Anaesthesiol Scand 46(5):567–578

Pellerin L, Bergersen LH, Halestrap AP, Pierre K (2005) Cellular and subcellular distribution of monocarboxylate transporters in cultured brain cells and in the adult brain. J Neurosci Res 79(1–2):55–64

Pellerin L, Pellegri G, Bittar PG, Charnay Y, Bouras C, Martin JL, Stella N, Magistretti PJ (1998) Evidence supporting the existence of an activity-dependent astrocyte-neuron lactate shuttle. Dev Neurosci 20(4–5):291–299

Pfeuffer J, Tkac I, Gruetter R (2000) Extracellular-intracellular distribution of glucose and lactate in the rat brain assessed noninvasively by diffusion-weighted 1 H nuclear magnetic resonance spectroscopy in vivo. J Cereb Blood Flow Metab 20(4):736–746

Qu H, Haberg A, Haraldseth O, Unsgard G, Sonnewald U (2000) (13)C MR spectroscopy study of lactate as substrate for rat brain. Dev Neurosci 22(5–6):429–436

Quistorff B, Secher NH, Van Lieshout JJ (2008) Lactate fuels the human brain during exercise. FASEB J 22(10):3443–3449

Shulman RG, Hyder F, Rothman DL (2001) Lactate efflux and the neuroenergetic basis of brain function. NMR Biomed 14(7–8):389–396

Simpson IA, Carruthers A, Vannucci SJ (2007) Supply and demand in cerebral energy metabolism: the role of nutrient transporters. J Cereb Blood Flow Metab 27(11):1766–1791

Tofteng F, Larsen FS (2002) Monitoring extracellular concentrations of lactate, glutamate, and glycerol by in vivo microdialysis in the brain during liver transplantation in acute liver failure. Liver Transpl 8(3):302–305

Chapter 26
Cerebral Acetate Metabolism: Towards Its *In Vivo* Assessment

Basil Künnecke

Abstract Acetate as precursor and tracer for cerebral metabolism has received high interest since the late 1950s when it was first shown with radiolabels that, among others, acetate incorporates very differently into cerebral metabolites than glucose. Subsequent work has taken advantage of acetate's exclusive cerebral uptake into glial cells in order to probe metabolic compartmentation and the interplay of the intricately and inextricably intermingled glial and neuronal tissues in intact brain. The present review takes a three-pronged approach to outline the current understanding of cerebral acetate uptake, metabolism and the learning derived thereof with regard to the integral metabolism in mammalian brain. (1) Acetate-based tracer modalities including radiography, positron emission tomography (PET) and magnetic resonance spectroscopy (MRS) together with corresponding labelling concepts ranging from specific (radio-)activity to advanced multi-labelling strategies used for assessing cerebral metabolism are presented. (2) Translational aspects and efforts in moving the assessment of cerebral acetate metabolism from cell cultures and ex vivo tissue toward its non-invasive detection in situ in the brain of living animals and man are then discussed. (3) The original notion of cerebral substrate selection is complemented with current data on metabolic compartmentation and substrate trafficking to build a comprehensive, though in part still controversial view on cerebral metabolism. A collective of studies, which have utilised acetate to explore cerebral metabolism in health and disease, are put into perspective with this latter notion. Metabolism in brain tumours and cerebral ischemia as well as alterations in cerebral metabolism brought about by primarily extra-cerebral disorders such as diabetes and thyroid hormone deficits are discussed. Finally, acetate's involvement in psychiatric and

B. Künnecke, Ph.D. (✉)
Magnetic Resonance Imaging and Spectroscopy, CNS Discovery and Translational Area, F. Hoffmann-La Roche Ltd, Grenzacherstrasse 124, CH-4070, Basel, Switzerland
e-mail: basil.kuennecke@roche.com

I.-Y. Choi and R. Gruetter (eds.), *Neural Metabolism In Vivo*,
Advances in Neurobiology 4, DOI 10.1007/978-1-4614-1788-0_26,

neurological disorders receives particular focus as this area is largely dominated by newer work addressing neurotransmitter balances and receptor involvement from a glial stance and is offering potential links to functional and behavioural data.

Keywords 11C • 13C • 14C • 2H • 3H • Isotope • Isotopomer • Labelling • Magnetic resonance spectroscopy • MRS • Positron emission tomography • PET • Mass-spectrometry • MS • Astrocyte • Glia • Neuron • Acetate • Glutamate • Glutamine • Tricarboxylicacid cycle • Cell culture • Mouse • Rat • Monkey • Man • Diabetes • Epilepsy • Schizophrenia • Stroke

26.1 Introduction

In 1959, Lajhta et al. (1959) provided first evidence for the existence of two distinct glutamate compartments in mammalian brain. In subsequent radio-labelling experiments, these glutamate pools were shown to be kinetically different and were associated with metabolism in the tricarboxylic acid cycles located in astroglia and neurons, respectively. In this context, cerebral metabolism of numerous putative substrates was studied by supplying brain tissue with radio-labelled precursors including [^{14}C]-glucose, [^{14}C]-glutamate, [^{14}C]-glutamine, and [^{14}C]-acetate. Two distinct classes of cerebral substrates were identified based on the resulting relative specific (radio-)activity (RSA) in cerebral glutamine versus glutamate. Glucose consistently gave rise to RSA values <1 whereas acetate was identified as a representative of the substrates yielding an RSA > 1. This initial qualitative classification was further refined over time and acetate in particular was shown to be almost exclusively metabolised in astrocytes. More recently, the preferential utilisation of acetate by astrocytes was attributed to cellular differences in the transport of acetate (Waniewski and Martin 1998). Acetate uptake by astrocytes has been suggested to be mediated by a carrier similar, if not identical to the monocarboxylate transporter (MCT) or more specifically its subtype 1 (Pierre et al. 2000; Pierre and Pellerin 2005; Hosoi et al. 2009).

Since the early days of intense exploration of cerebral metabolism in the 1960s, acetate has rapidly been recognised as a valuable tool to be used particularly in combination or opposed to glucose for assessing the intricate interrelationship of the different cell populations which make up mammalian brain. This review sets out to provide an overview on tracing strategies and detection modalities applied to elucidate cerebral acetate metabolism. The second part will shed light on the current notion of intra-cerebral compartmentation and its assessment in cell cultures, animal models and man. In the third section, numerous examples will be given, which illustrate metabolite cycling and trafficking in the mammalian brain with explicit reference to acetate in its roles as tracer and cerebral substrate. The forth part will address alterations of cerebral acetate metabolism under pathophysiological conditions. Finally, an outlook to potential future developments and applications of acetate for elucidating and assessing cerebral metabolism will be given.

26.2 Labelling Strategies with Acetate

26.2.1 Acetate as Precursor and Tracer of Cerebral Metabolism

Acetate has extensively been used as a metabolic tracer and precursor complementary to glucose in order to elucidate different aspects of cerebral metabolism including the synthesis and recycling of the major excitatory and inhibitory neurotransmitters glutamate and gamma-aminobutyric acid (GABA), respectively. Whereas glucose is considered the primary fuel for cerebral metabolism, other substrates including ketone bodies and acetate may be used by the brain. Under physiological conditions, however, acetate has been reported to be present in blood plasma at a concentration of as low as 700 μM in rats or 50 μM in humans (Tyce et al. 1981; Blüml et al. 2002). Innumerable tracer experiments have shown that even under these conditions acetate is readily taken up and metabolised by the brain (Muir et al. 1986; Lear and Ackermann 1990). The monocarboxylate transporter (MCT) has been suggested to be implicated in the transport of acetate into the mammalian brain (Waniewski and Martin 1998; Gonda and Quastel 1966). Cremer et al. (1976) showed that during ontogenesis the MCT in the brains of baby rats has a capacity several times higher than in the adult brain and it was demonstrated that the developing brain indeed uses substrates other than glucose, including acetate, more readily than the brains of adults (Van den Berg 1970). At the other end, ageing was reported to have little effect on acetate uptake and metabolism in rat brain beyond a somewhat delayed metabolism of glutamate in the so called "small" (i.e. glial) compartment (De Koning-Verest 1980).

Data accrued to date indicate that acetate uptake into brain tissue is mediated by the MCT subtype 1 (MCT1) or based on more recent data, though with less evidence, also by subtype 4 (Pierre et al. 2000; Hosoi et al. 2009). Both MCTs are thought to be exclusively located in the astroglia (Pierre and Pellerin 2005). As a consequence, acetate enters with high preference the glial compartment and may thus serve as a specific tracer for glial metabolism, as described below. Apart from the clear-cut compartmentation of the transporter for acetate, further aspects of cerebral substrate selection need to be considered. Common monocarboxylic acids such as lactate or the acetate analogue fluoroacetate compete with the transport of acetate and thus modulate cerebral acetate metabolism (Waniewski and Martin 1998; De Koning-Verest 1980). Yet, more recent radio-labelling studies suggested that acetate metabolism scales with astrocyte metabolism and that MCT1 is not the rate-limiting step in acetate metabolism under different conditions including pharmacological modulation of brain activity, metabolic inhibitors and ischaemia (Hosoi et al. 2004, 2007, 2009; Hirose et al. 2009).

Classic radio-labelling- and PET approaches generally introduce labelled molecules in trace amounts. In contrast, the less sensitive NMR and in particular its *in vivo* variant, MRS, rather have used labelled acetate as proper substrate. While this procedure is typically required to boost the detectability of endogenous metabolites, it also increases plasma acetate concentration many fold from low submillimolar to millimolar levels (Tyce et al. 1981; Blüml et al. 2002). Yet, there is ample evidence that despite the increased availability of acetate the labelling mismatch between glutamate and glutamine previously reported by radiotracer methodology is maintained (Hassel et al. 1997; Cerdan et al. 1990; Haberg et al. 2000; Hammer et al. 2001). Finally, it needs to be emphasised that extra-cerebral metabolism of acetate, in particular hepatic conversion of labelled acetate to glucose, provide newly labelled cerebral substrates that may confound data analysis.

26.2.2 Radio-Labelling

Radio-labels, in particular radioactive carbon-14 and hydrogen-3 have found wide-spread application in elucidating cerebral acetate metabolism and have tremendously advanced and shaped our current understanding of cerebral biochemistry. Specific metabolite tagging is an imperative prerequisite in order for acetate to be effective as metabolic marker beyond simple measures of metabolite accumulation or depletion. Thus, all procedures adhere to the same general concept of labelling: One or several atoms of a molecular species are replaced with tags which later can be used to readily identify these molecular entities or their derivatives among all others that were not (or to a lesser extent) involved in the corresponding part of metabolism. With radio-labels, radioactivity, i.e. nuclear decay events which are usually captured with scintillation counters, serve as tags. This procedure is highly sensitive but (bio-) chemically rather unspecific and thus bulk radioactivity in the tissue under investigation only provides a gross measure of substrate accumulation and incorporation. Hence, assessment of metabolism requires metabolites to be extracted, chemically separated and their relative specific radioactivity (RSA) to be determined separately. More detailed elucidation of metabolic precursor-product relationships rely on precise knowledge of the intra-molecular location and distribution of the label(s). Such an analysis, however, necessitates tedious carbon-by-carbon degradation of the different metabolites. Metabolic analyses based on radio-labelling and scintillation counting most often have been restricted to the analysis of one labelled entity at a time. Van den Berg and Ronda (1972) demonstrated ways to partly remedy this obvious limitation by taking advantage of the different radiation bands of carbon-14 and hydrogen-3 isotopes. They used mixtures of [^{14}C]-acetate and [^{3}H]-acetate and double- labelled [^{3}H,^{14}C]-acetate in order to simultaneously assess the isotope effect (see below) and the metabolic fates of different parts of the acetate molecule.

26.2.3 Stable Isotopes and Nuclear Magnetic Resonance

With the advent of nuclear magnetic resonance (NMR) spectroscopy in the biochemical arena, labelling strategies established for radiotracers were initially transferred basically unrevised to this new modality with the only change being the replacement of carbon-14 and hydrogen-3 with the stable (non-radioactive), though magnetically active isotopes carbon-13 and hydrogen-2, respectively (Cerdan et al. 1990; Künnecke 1995; Chapa et al. 2000). However, NMR spectroscopy in conjunction with stable isotope labelling provides several distinct advantages over corresponding radio-tracer approaches: (a) NMR spectroscopy yields inherent chemical separation of different metabolites and even of single carbons (hydrogens) within metabolites. Analysis of specific labelling and positional labelling thus doesn't require biochemical separation of individual metabolites and their tedious carbon-by-carbon degradation, respectively. (b) NMR as such is completely non-destructive and can thus not only be applied *in vitro* but has been extensively used *in vivo* on cell cultures, experimental animals, and man, where it is dubbed magnetic resonance spectroscopy (MRS). (c) With NMR and MRS, full advantage can be taken of advanced labelling strategies. Whereas in nuclear biochemistry multi-labelled compounds have been used mainly to enhance the specific activity of molecules (Van den Berg 1970; Berl et al. 1968), magnetic resonance in addition readily allows with such multi-labelled precursors different entities within the same molecule to be assess simultaneously (Künnecke and Cerdan 1989; Künnecke 1995; Cerdan et al. 1990; Künnecke et al. 1993). Moreover, unique information on intra-molecular label distribution and thus on the metabolic history of observed metabolites is obtained from so-called spin-spin coupling patterns, a property only available by magnetic resonance (Chapa et al. 2000; Künnecke and Cerdan 1989; Künnecke 1995; Cerdan et al. 1990; Künnecke et al 1993). In brief, covalently bound, magnetically active nuclei interact with each other which in turn leads to typical splitting of the nuclei's resonances into several lines. For example, a -^{13}C–^{13}C- moiety features two pairs of doublets in the NMR spectrum as long as the covalent bond is maintained. Whenever the bond is broken, these doublets revert to singlets. By analysing these spin coupling patterns and their underlying isotope-isomer (isotopomer) distribution, the origin and subsequent fate of molecular moieties can be followed. Whereas carbon-13 with a natural abundance of as much as 1.1% allows NMR spectra to be conveniently obtained without the need for labelling, this very same background signal may become a confound in labelling studies. Spin coupling due to multiple labelling and its spectral manifestation turn out very helpful for distinguishing exogenously supplied label from the singlet resonances of the natural abundance background signals (n.b. ^{13}C-^{13}C homonuclear spin couplings may also occur under natural abundance conditions, however their incidence is negligible with 0.01%).

26.2.4 Autoradiography and Positron Emission Tomography

Quantitative autoradiography not only allows radio tracer accumulation to be quantified but provides maps of tracer distribution and local accumulation in tissue sections (Lear and Ackermann 1990; Cruz et al. 2005; Hosoi et al. 2004). Positron emission tomography (PET) picks up the notion of autoradiography and makes it applicable under *in vivo* conditions. PET is a highly sensitive nuclear imaging method which reports the local abundance of molecular entities harbouring a positron emitter such as ^{11}C or ^{18}F. Over the past decades, PET of the brain has found wide-spread application mainly in receptor occupancy and glucose turnover studies. These latter metabolic analyses have in common the "feed and trap" concept. Since PET cannot distinguish labelled metabolites from their parent, metabolic precursors intended as PET tracers are usually required to be chemically modified such that their ultimate metabolism is inhibited and tracer accumulation ensues (Inoue et al. 2006; Momosaki et al. 2007). PET studies with acetate have significantly deviated from this paradigm in that unmodified acetate was used (Spence et al. 2003; Liu et al. 2006; Tsuchida et al. 2008; Yamamoto et al. 2008). Under these conditions, PET in essence assesses the incorporation of label, which is supplied at trace amounts, into the diverse metabolic pools accessible by the labelled precursor. Metabolic turnover is inferred from label accumulation or incorporation but tracer metabolism with its intermediates cannot be followed in detail. The main advantages of PET, however, are its very high sensitivity and its translational applicability *in vivo* in animal models and man. Yet, the spatial resolution power of PET is limited to the millimetre scale which restricts its effectiveness in small animals and alternative but related methods such as local β-scintillation have been used (Wyss et al. 2009).

26.2.5 Mass Spectrometry and Isotope Effects

Mass spectrometry (MS) represents a valuable and sensitive alternative to NMR for analysing *in vitro* (e.g. in extracts) the incorporation of stable isotope tracers. MS differentiates labelled from "unlabelled" metabolites based on their increased molecular masses due to the presence of heavy atoms, e.g. carbon-13, introduced as label. Preceding chromatography is commonly used for increasing the power of MS to unambiguously distinguish a vast variety of constituents in complex (bio-)chemical mixtures. Molecular fragmentation, on the other hand, allows the intra-molecular distribution of label to be assessed similarly to NMR. MS has only rarely been used in the context of cerebral metabolism of acetate (Yudkoff et al. 2005; Sonnewald et al. 1993). Its underlying principle, i.e. the use of isotopes with altered masses, pinpoints an intrinsic though unresolved issue present in all labelling studies with isotope tracers, be it with radioactive or stable isotopes. Increased (decreased) molecular mass can alter the reactivity of the labelled metabolite in (bio-)chemical reactions and thus lead to a preferential metabolism of unlabelled precursors. Though certain enzymes have been reported to be particularly sensitive to this kinetic isotope effect for the replacement of hydrogen with deuterium or tritium, the small percentage increase/decrease in mass for carbon-13, carbon-14 and carbon-11 is likely to be negligible in most cases (Van den Berg and Ronda 1972).

26.3 Compartmentation and Metabolism: From Cell Cultures to Man

26.3.1 Concept of Cerebral Compartmentation

Metabolic compartmentation and the interrelationship of glutamine with its precursor glutamate have extensively been analysed and debated. Figure 26.1 provides a simplified schematic representation of the current understanding of cerebral

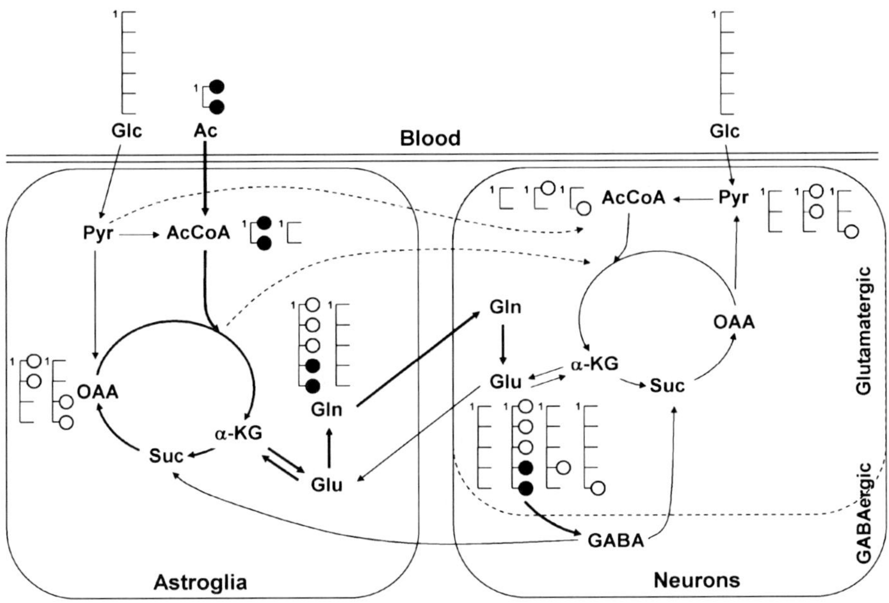

Fig. 26.1 Schematic representation of the current understanding of cerebral metabolism of acetate and glucose, and its compartmentation in astrocytes, glutamatergic- and GABAergic neurons. At least two kinetically different tricarboxylic acid cycle exist in mammalian brain, located presumably in astrocytes and neurons. The glial compartment is characterised by the access of exogenous acetate and glucose, as well as by the exclusive activity of glutamine synthase and pyruvate carboxylase. The neuronal compartments lack the metabolism of acetate and the activities of glutamine synthase and pyruvate carboxylase. Instead, neurons exhibit glutaminase activity, are the site of GABA production and harbour a pyruvate recycling system. The compartments are metabolically connected via the blood supply and with each other at different metabolic levels, in particular by the fluxes of glutamine, glutamate and GABA. Bold arrows highlight the most prominent pathways of cerebral acetate metabolism. Dashed lines designate less established metabolic fluxes. *Filled circles* indicate individual carbons positions which are directly labelled by the incorporation of carbon-13 nuclei supplied by [1,2-$^{13}C_2$] acetate. *Open circles* signify label propagation upon metabolic dilution and recycling after approximately 1.5 turns of the TCA cycles

metabolism that shall be used as a framework for further discussion. It exemplifies the cerebral metabolism of [1,2-$^{13}C_2$]-acetate. Filled circles indicate the ^{13}C-labelling after 1.5 turns of the TCA cycles. Briefly, two metabolically distinct compartments were proposed that were later attributed to astroglia and neurons, respectively (Lajhta et al. 1959; O'Neal et al. 1966; Van den Berg et al. 1966, 1969; Berl and Clarke 1983). Here, we further subdivide the neuronal compartment into glutamatergic and GABAergic neurons (adapted from Chowdhury et al. 2007). These compartments are metabolically connected with the blood supply and with each other at different metabolic levels. All compartments extract glucose from the blood. In addition, the glial compartment contains the MCT1 and MCT4, which, among others, enable acetate uptake. Neurons and glia both comprise the enzymatic machinery for glycolysis and tricarboxylic acid (TCA) cycle activity from which glutamate is derived. The glial compartment is further characterised by the presence of glutamine synthase (EC 6.3.1.2) and pyruvate carboxylase (EC 6.4.1.1), and thus is the exclusive site for cerebral glutamine synthesis and anaplerosis (Norenberg and Martinez-Hernandez 1979; Martinez-Hernandez et al. 1977; Shank et al. 1985; Yu et al. 1983). The neurons, on the other hand, show glutaminase activity (EC 3.4.1.2) and are the site of GABA production due to the action of glutamate decarboxylase (EC 4.1.1.15) (Sze 1979). The principal entry point for acetate into cerebral metabolism in mammals after its transport into glia via MCT1 (and potentially also via MCT4) is its conversion to acetyl-CoA by acetyl-CoA ligase (EC 6.2.1.1). The Acetyl-CoA then enters the TCA cycle by condensing with oxaloacetate to form citrate. From there carbons of acetate reach glial glutamate by transamination of α-ketoglutarate formed in the TCA cycle. Glutamate, in turn, serves as substrate for the synthesis of glial glutamine by glutamine synthase. α-Ketoglutarate that is not transaminated is further processed in the TCA cycle to ultimately form oxaloacetate and close the cycle. During TCA cycle turnover, two carbons are lost primarily in the form of CO_2 in order to keep the mass balance with the freshly introduced carbons of acetylCoA.

26.3.2 Cell Cultures and Brain Slices

Numerous labelling studies with acetate have essentially corroborated the notion of cerebral compartmentation and metabolic interaction established earlier (Lajhta et al. 1959; O'Neal et al. 1966; Van den Berg et al. 1966, 1969; Berl and Clarke 1983; Van den Berg and Garfinkel 1971). Glutamine was found to be labelled more rapidly and more strongly than its direct precursor glutamate. The availability of relatively pure cultures of primary astrocytes and neurons has enabled directed investigation of metabolic activities in the segregated putative metabolic compartments (Sonnewald et al. 1993; Brand et al. 1997; Hassel et al. 1994; Haberg et al. 1998a; Sonnewald et al. 1998). Using inhibitors, Hosoi et al. (2009) investigated the mechanism of acetate uptake in astrocytes. P-(Chloromercuri) benzenesulfonic acid, a putative selective MCT1 inhibitor, completely blocked acetate uptake whereas α-cyano-3-hydroxycinnamate (a selective MCT2 inhibitor) did not lead to any inhibition, thus underpinning the importance of MCT1 in cerebral acetate metabolism. In numerous studies the metabolic fate of labelled acetate was followed further. Whereas labelled glutamate and aspartate were detected in neurons, glutamine was the main product of glial acetate metabolism. Co-cultures or sandwich co-cultures (neurons growing on top of astrocytes) of primary neurons and astrocytes showed a completely different picture that could not be explained by simple additive combination of the two cellular constituents, but rather resembled results obtained from brain slices or *in vivo* studies. Badar-Goffer et al. (1990, 1992) and Rae et al. (2003) investigated cerebral compartmentation in brain slices, which they used as a tractable model of the *in vivo* situation.

26.3.3 Rodents

In vivo metabolism of acetate has been extensively studied in rodent animal models. Generally, brains were extracted and mostly semi-quantitative investigations of cerebral metabolism were performed by *in vitro* modalities. The labelling states of cerebral metabolites derived from metabolism of acetate as tracer and substrate were compared with separate experiments where labelled glucose alone or in combination with labelled acetate was administered. These former studies were limited in as far as metabolism of unlabelled co-substrates remained largely undetected whereas in the latter studies the concomitant use of two labelled substrates usually gave rise to confounding labelling patterns. Elucidation of cerebral metabolism by comparing the effect of different fuels has thus remained semi-quantitative at most.

A novel approach pioneered in brain by Cerdan et al. (1990) has been based on isotopomer analysis upon labelling with multiple-labelled substrates (Cerdan et al. 1990; Künnecke et al. 1993; Chapa et al. 1995, 2000; Nishima et al. 2004; Cruz et al. 1998; Pascual et al. 1998; Garcia-Espinosa et al. 2003; Rodrigues et al. 2007). Figure 26.2 depicts the aliphatic portion of a ^{13}C-NMR spectrum of a perchloric acid extract obtained from rat brain after infusion of double-labelled [1,2-$^{13}C_2$]-acetate. Numerous resonances of pertinent cerebral metabolites including glutamate, glutamine and GABA are discerned. Most of the resonances show typical multiplet structures due to homonuclear ^{13}C–^{13}C spin couplings introduced by the contiguously double labelled acetate. For better understanding of the value of multiple-labelling, Fig. 26.3 shows close-ups of the glutamate C4 and glutamine C4 resonances. It is immediately apparent that the labelling pattern of glutamine is distinctly different from that of its metabolic precursor glutamate, thus imperatively supporting the notion of metabolic compartmentation in brain. In more detail, the two labels in [1,2-$^{13}C_2$]-acetate were simultaneously transferred to the C5–C4 moiety of glial glutamate and thereafter to glial glutamine, giving rise to typical doublet resonances in the ^{13}C-NMR spectrum. Transfer of labelled material from the glia to neurons eventually lead to the large pool of neuronal glutamate being labelled. However, since direct labelling from acetate is absent in neurons, portions of glutamate derived from neuronal metabolism remain unlabelled or, more precisely, labelled at the level of natural abundance. This is borne out by the additional central singlet observed for the glutamate C4 resonance. Quantitative analysis revealed that astroglia and neurons contributed 1.9 mmol/g and 5.7 mmol/g to the total pool of cerebral glutamate in anaesthetised rats, thus fully supporting glutamate compartmentation into a small and a large pool associated with astroglia and neurons, respectively (Künnecke et al. 1993).

The concept of multiple labelling has been further extended and successfully put into practice. Chapa et al. (2000) used carbon-13-deuterium co-labelled [2-^{13}C, 2-$^2H_3$]-acetate in order to obtain more detailed information on metabolite cycling. Sonnewald and co-workers used [1,2-$^{13}C_2$]-acetate in combination with [1-^{13}C]-glucose, which essentially follows the labelling concept of Cerdan et al. (1990) but provides additional labelling to the neuronal compartment beyond the natural abundance of carbon-13 (Haberg et al. 1998a, b, 2000, 2001, 2006, 2007; Kondziella et al. 2003; Eloqayli et al. 2001, 2004; Melo et al. 2005, 2006a, b, 2007,

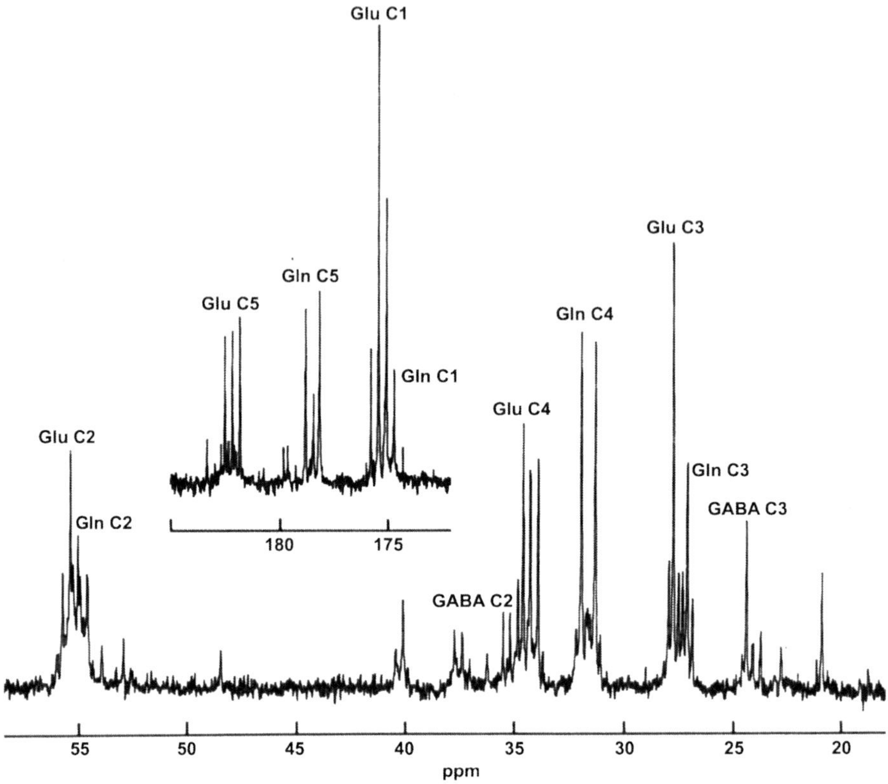

Fig. 26.2 Typical ^{13}C-NMR spectrum of a perchloric acid extract obtained from rat brain after metabolism of [1,2-$^{13}C_2$] acetate. Only the aliphatic and carboxylic regions are shown. Numerous resonances are seen especially those of cerebral glutamate, glutamine and GABA. These resonances all show multiplet character due to the incorporation of contiguous carbon-13 labels upon metabolism of [1,2-$^{13}C_2$] acetate

Müller et al. 2000, Brenner et al. 2007; Brenner et al. 2005; Eyjolfsson et al. 2006; Kondziella et al. 2006; Hammer et al. 2001). These labelling data served as basis for the analysis of cerebral metabolism and compartmentation under different states and pathological conditions as discussed below.

26.3.4 Monkey and Man

Reports on truly non-invasive assessment of cerebral acetate metabolism in situ are very few. Behar and co-workers (Patel et al. 2005; de Graaf et al. 2006) investigated glutamate/glutamine and GABA/glutamine cycling in rat brain by *in vivo* MRS. Carbon-13 labelled acetate was used to discriminate the glial contribution to cerebral

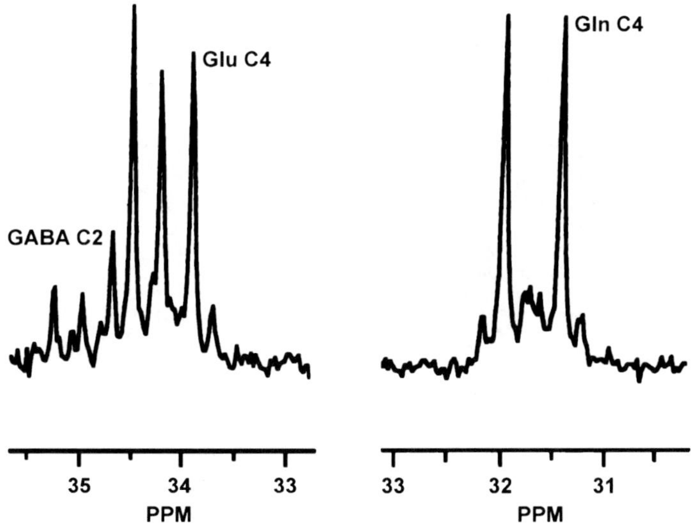

Fig. 26.3 Comparison of the ^{13}C-spin coupling patterns observed by ^{13}C-NMR in GABA C2 and glutamate C4 (*left*), and glutamine C4 (*right*) in rat brain upon metabolism of [1,2-$^{13}C_2$] acetate. The glutamate and glutamine resonances are scaled to the same height in order to emphasise the different labelling patterns. All three resonances are characterised by prominent doublets brought about by spin coupling of the two contiguous carbon-13 labels introduced by [1,2-$^{13}C_2$] acetate. Glutamate and GABA also feature distinct singlet resonances which are indicative of alternate pathways for glutamate (GABA) synthesis that do not involve direct conversion of glutamine and glutamate

metabolism as assessed with labelled glucose. Likewise, Yang et al. (2007) determined the extent of cortical GABA-glutamine cycling in vigabatrin-treated rats from the *in vivo* magnetic resonance signal of [2-^{13}C]-GABA derived from [2-^{13}C]-acetate. They also demonstrated that *in vivo* acetate metabolism can be followed in monkey brain in situ (Li et al. 2005; Yang et al. 2007). Label incorporation into cerebral glutamine, glutamate, aspartate and NAA were observed. Lebon et al. (2002) and Blüml et al. (2002) extended the *in vivo* assessments of glial metabolism to the human brain. [2-^{13}C]-acetate and [1-^{13}C]-acetate, respectively, were used to quantitatively elucidate the extent of astroglial contribution to brain energy metabolism. Blüml et al. (2002) determined the mean rate of acetate oxidation from the ^{13}C-enrichment of bicarbonate detected by ^{13}C-MRS on four subjects as ~0.13 μmol/g/min. This latter flux was estimated to be approximately 20% of the total cerebral TCA cycle rate. Under similar conditions, Lebon et al. (2002) reported a somewhat lower contribution of 14%. Metabolic modelling further provided a quantitative estimate of glutamate compartmentation. Analysis of their *in vivo* data suggested an astroglial glutamate pool size of 0.7 ± 0.5 mM in human brain. This low contribution of astroglial glutamate to total cerebral glutamate of approximately 7.2 mM (Kreis et al. 2002 and references therein) compares favourably with similar investigations carried out in rat brain which reported a ~1.9 mM contribution of glial glutamate (Künnecke et al. 1993). Mason et al. (2006) investigated the monocarboxylic acid

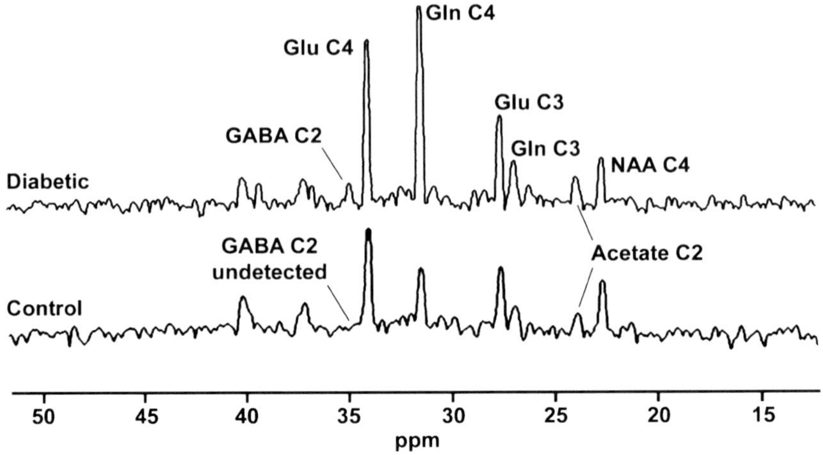

Fig. 26.4 Representative *in vivo* ^{13}C-MR spectra of human brain obtained in subjects with and without type 1 diabetes. The spectra were acquired over the final 45 min of the study during hypoglycaemia and [2-^{13}C] acetate infusion. The subject with type 1 diabetes (*upper trace*) had significantly higher labelling in glutamate C4 and glutamine C4, and the acetate C2 signal was also higher than in the healthy subject (*bottom trace*). Other metabolites clearly visible in the spectra include gamma-aminobutyric acid (GABA C2), glutammine (Gln C3 and Gln C4), glutamate (Glu C3 and Glu C4), and N-acetyl aspartate (originally labelled NAA C4 but reflects NAA C6 instead) (Copyright ©2006 American Diabetes Association from Mason et al. (2006). Reprinted with permission from The American Diabetes Association)

transport and utilisation in type 1 diabetes as compared to healthy controls. Figure 26.4 shows representative *in vivo* ^{13}C-MR spectra from subjects with and without type 1 diabetes during the final 45 min of [2-^{13}C] acetate infusion. The subject with type 1 diabetes (upper trace) had significantly higher labelling in glutamate C4 and glutamine C4, and the acetate C2 signal was also higher than the in healthy subject (bottom trace). These results were interpreted to be consistent with an approximately twofold upregulation of MCT in the brain of well-controlled diabetic patients. The authors hypothesised that this upregulation may play an important role in maintaining cerebral energy metabolism during hypoglycemia similar to what has been reported for brain metabolism in early postnatal phase (van den Berg 1970; Cremer et al. 1976).

26.4 Cerebral Metabolite Cycles and Interactions

26.4.1 Neuronal and Glial Tricarboxylic Acid Cycles

In the normal adult mammalian brain, aerobic metabolism in the TCA cycles of astroglia and neurons provides the primary source of energy and precursors for major neurotransmitters including glutamate and GABA. Cerebral TCA cycle activity has largely been explored with labelled glucose as tracer and fuel. Notwithstanding, the glutamine C3/C4 labelling ratio, which was taken as a measure for second-to-first turn TCA labelling, was found to be significantly smaller with labelled acetate than with labelled glucose or lactate as substrates. Tyson et al. (2003) hence concluded that the TCA cycle in astroglia had a lower turnover than that in neurons. Steady state labelling of key metabolites introduced by [2-^{13}C]-acetate as precursor was employed as a constraint in fitting a three-compartment model including astroglia, glutamatergic- and GABAergic neurons to the labelling kinetics of [^{13}C]-glucose (Chowdhury et al. 2007). Using this approach, Chowdhury et al. (2007) demonstrated in rat a strong increase in cortical neuronal TCA cycle activity from 0.24 μmol/g/min to 0.71 μmol/g/min during post-natal development between day 10 and day 30. Twofold higher metabolic activity was found in awake as compared to anaesthetised rats. Experiments with labelled acetate pointed to a substantial enhancement of the oxidative metabolism notably in the astrocytic compartment of awake animals (Serres et al. 2008). Blüml et al. (2002) assessed the extent of glial TCA cycle activity in humans based on the principle of tracer dilution in bicarbonate, i.e. the end-product of oxidative metabolism in the TCA cycle. At steady-state, the rate of $^{13}CO_2$ production is equal to the rate of [^{13}C]-acetate metabolism in the (predominantly glial) TCA cycle. Unlabelled substrates, primarily glucose, are concomitantly metabolised in the glial as well as the neuronal compartment and thus contribute unlabelled material to the production of CO_2. On the premise that unlabelled CO_2 is primarily formed in neurons, the steady-state fractional enrichment of cerebral bicarbonate hence directly reflects the contribution of astroglial metabolism to overall TCA cycle activity. The rate of glial acetate oxidation determined in this way was 20% or 0.13 μmol/g/min when assuming a total cerebral TCA cycle rate of 0.7 μmol/g/min (Blüml et al. 2002). Alternatively, astroglial TCA cycle activity in human brain was derived from metabolic simulation of labelling kinetics observed in humans upon [^{13}C]-acetate metabolism. Based on metabolic interrelations similar to those shown in Fig. 26.2 and a neuronal TCA cycle rate of 0.8 μmol/g/min, Lebon et al. (2002) calculated a glial TCA cycle rate of 0.14 μmol/g/min, which accounted for ~14% of total brain oxygen consumption. These latter studies provide quantitative insight with rather high consistency and essentially corroborate pioneering work carried out in the 1970s (Van den Berg and Garfinkel 1971) and newer data obtained indirectly with labelled glucose (Shen et al. 1999). Wyss et al. (2009) then investigated semi-quantitatively the relationship between stimulation-induced increase in the cerebral workload and astrocytic oxidative metabolism in human brain. In this latter study, PET was used to elucidate the time course of CO_2 production from [1-^{11}C]-acetate. Stimulation of the visual cortex significantly increased retention of label which was assumed to be directly related to an increased (astrocytic) acetate metabolism and to an increased contribution of astroglia to total cerebral oxygen consumption. Corroborative evidence was obtained in rats: Wyss et al. (2009) reported increased acetate consumption as seen by local β-scintillation for a variety of stimuli whereas Cruz et al. (2005) visualised by autoradiography a modest though distinct increase in (presumably oxidative) acetate metabolism during acoustic stimulation. In a rat model of unilateral spreading cortical depression induced by topical application of KCl to the dura, Dienel et al. (2001b) found by autoradiography highest label incorporation from [^{14}C]-acetate in regions adjacent to the KCl application site. This finding is consistent with a K^+-induced activation of glial metabolism.

26.4.2 Neurotransmitter Cycles: Glutamine, Glutamate and GABA

Strong evidence has accrued that glial glutamine is the major precursor for the biosynthesis of the excitatory pool of neurotransmitter glutamate in neurons. In parallel, glial glutamine is essential for the maintenance of neuronal GABA pool. When [1,2-$^{13}C_2$]-acetate was provided *in vivo* to rats as glial substrate, the double label was readily transferred to neuronal [4,5-$^{13}C_2$]-glutamate and also to [1,2-$^{13}C_2$]-GABA (Cerdan et al. 1990). Triple labelling with [1,2-$^{13}C_2$]-acetate+[1-^{13}C]-glucose corroborated these findings (Hassel et al. 1997). Inhibition of cerebral acetate metabolism by fluoroacetate essentially lead to an abolished transfer of the twin label of acetate to glutamate and GABA. Conversely, massive glial uptake of transmitter glutamate was implied by extensive mono-labelling of astroglial glutamine from [1-^{13}C]-glucose (Hassel et al. 1997). Brain slices incubated with [2-^{13}C]-acetate in presence of an excess of unlabelled glutamine showed a marked reduction of label incorporation into glutamate and GABA as a consequence of label dilution at the level of glial-neuronal glutamine transport (Rae et al. 2003). In sandwich co-cultures of primary astrocytes and neurons incorporation of label from [2-^{13}C]-acetate into GABA was distinctly reduced by the glutamine synthetase inhibitor methionine sulfoximine (Sonnewald et al. 1993). Chowdhury et al. (2007) investigated glutamatergic and GABAergic neurotransmitter cycling in rat cerebral cortex during postnatal development. Time course analysis of label incorporation from glucose backed with steady state enrichment data from [2-^{13}C]-acetate indicated that glutamatergic and GABAergic neurons undergo ontogenic changes involving large increases in neurotransmitter cycling and oxidative energy metabolism. Yang et al. (2007) quantified *in vivo* the cortical GABA-glutamine cycling rate in rats. On the premise of a constant TCA cycle rate, the GABA-glutamine cycling flux rate was determined to be ~0.03 μmol/g/min in animals treated with vigabatrin, an inhibitor of the catabolic enzyme GABA transaminase (EC 2.6.1.19). Under similar conditions de Graaf et al. (2006) found approximately twofold higher rate of GABA-synthesis in the cortex of vigabatrin treated and drug naive rats. Notwithstanding, these latter cycling rates constitute only a small fraction to the overall neurotransmitter cycling (~0.32 μmol/g/min) as reported previously (Lebon et al. 2002). Corroborative evidence was obtained in a study by Patel et al. (2005) in which the percentage contribution of GABA-glutamine cycling to the total neurotransmitter (GABA+glutamate) cycling was assessed in rat cortex under conditions of halothane anaesthesia (23%) and pentobarbital-induced isoelectricity (below detection). Whereas astroglia undisputedly take a dominant role in the clearance of extracellular glutamate present after pre-synaptic release, glial and neuronal transport are implicated in the clearance of GABA. Although quantitative information is still missing, it is thought that neuronal GABA reuptake prevails because of the higher density of GABA transporters in neurons as compared to astroglia (Schousboe 2003). Hence, total (neuronal-glial and synaptic-neuronal) GABA cycling may be underestimated by the current assessments.

26.4.3 Acetylcholine

Apart from the persistent focus on the major neurotransmitters glutamate and GABA, early radiotracer studies elucidated the synthesis of acetylcholine or more precisely the origin of its acetyl moiety in mammalian brain. In cortex slices of rat brain incubated with double-labelled [2-^{14}C,2-^{3}H]-acetate in presence of unlabelled glucose, Sollenberg and Sörbo (1970) found that the ^{3}H/^{13}C labelling ratio was the same in de novo synthesised acetylcholine and in the substrate acetate. These results together with those obtained with doubly labelled [6-^{14}C,6-^{2}H]-glucose were thought to indicate that citrate, and with today's knowledge of cerebral compartmentation more precisely neuronal citrate, confers the precursor for the acetyl moiety in acetylcholine. Data obtained in labelling studies with a variety of different precursors indicated that acetate is poorer a substrate for acetylcholine than pyruvate (Dolezal and Tucek 1981; Nakamura et al. 1970; Tucek and Cheng 1974). Since blood-borne pyruvate is preferentially used in the neuronal compartment, this finding may be taken as confirmation of a mostly neuronal origin of the acetyl moiety in acetylcholine. Tucek and Cheng (1974), however, suggested that acetate and pyruvate do not supply the same acetyl-CoA pool for ultimate synthesis of acetylcholine in cholinergic neurons. A more recent labelling study indicated that extracellular acetate, presumably from acetylcholine release and subsequent degradation during acetylcholinergic neurotransmission is recycled and re-incorporated as acetyl moiety in acetylcholine (Carroll 1997). Despite these investigations, the precise pathway and molecular format by which acetyl moieties generated in the glial compartment are transferred to cholinergic neurons for the synthesis of acetylcholine has remained elusive.

26.4.4 Anaplerosis

Anaplerosis serves to replenish TCA cycle intermediates and compensate for their loss of carbon skeletons due to a variety of metabolic events including neurotransmitter synthesis. In brain, particularly pyruvate carboxylation has been shown to be active in the astrocytic compartment (Cerdan et al. 1990; Künnecke et al. 1993). Using glucose as substrate, PC in rats was shown to contribute more than 30% to the total glial TCA cycle flux (Künnecke et al 1993; Serres et al. 2008). On the other hand, acetate is not a substrate for anaplerosis and hence PC activity, though of glial location, has not been thoroughly assessed using acetate as probe. Notwithstanding, isotopomer analysis would even under these conditions potentially allow PC activity to be detected by the "absence" of specific labelling patterns. However, the mathematical models used for interpreting the isotopomer distributions turned out to be too instable for providing trustworthy values for PC activity with acetate being the sole labelled substrate (Künnecke et al. 1993). It may be argued that anaplerosis via PC is low under conditions of high acetate availability. Serres et al. (2008), however, demonstrated with labelled glucose high PC/PDH activity in rats co-administered non-tracer amounts of acetate.

26.4.5 *Substrate Trafficking and Pyruvate Cycling*

Pyruvate recycling in essence is the opposite of pyruvate carboxylation (see Fig. 26.2). Oxaloacetate or malate may be decarboxylated to form pyruvate through the combined action of phosphoenolpyruvate carboxykinase (PEPCK, EC 4.1.1.32) and pyruvate kinase (PK, EC 2.7.1.40) or through NADP-linked malic enzyme (ME, EC 1.1.1.40). Thus formed pyruvate may then re-enter the TCA cycle via PDH. Cerdan et al. (1990) were the first to describe the presence of this metabolic cycle in adult brain. They unexpectedly observed equimolar amounts of [5-^{13}C]- and [4-^{13}C]-glutamate in labelling studies with [1,2-$^{13}C_2$]-acetate as substrate. Under these conditions, decarboxylation in the pyruvate cycle leads to selective loss of one of the two adjacent labels introduced by [1,2-$^{13}C_2$]-acetate. This pyruvate can be used for lactate and alanine production, or for re-entry into the TCA via PDH, which will lead to mono-labelled acetyl-CoA. In terms of labelling patterns, this latter step ultimately leads to the formation of mono-labelled glutamate which shows up in the NMR spectrum as distinct singlet resonances among the typical doublet resonances introduced by the original double labelled substrate. The cellular location of this pyruvate recycling has been controversial. Cerdan et al. (1990) initially proposed pyruvate recycling to occur in the neuronal compartment. Their reasoning was that the mono-labelled species were predominantly observed in glutamate and not in glial glutamine. Mathematical modelling of labelling data obtained for acetate, 3-hydroxybutyrate and glucose as substrates supported this notion and also excluded the possibility that pyruvate recycling was a process occurring extra-cerebrally, e.g. in liver, which provided mono-labelled substrate to the brain (contributions of labelled glucose were indeed detected under conditions where labelled acetate or 3-hydroxybutyrate were provided as substrate, but were taken into account accordingly) (Serres et al. 2007; Künnecke et al. 1993). The enzymatic machinery for pyruvate recycling is potentially available in both the neuronal and glial compartments. Cruz et al. (1998) investigated the ontogeny of enzymatic activities and the extent of pyruvate recycling. The activities of ME and PEPCK were found to change their cellular localisation with postnatal development from mainly astrocytic to neuronal dominance in the adult brain. PEPCK activity was generally low after birth and further decreased toward adulthood, whereas ME activity increased to adult levels with the same time course as the activity of the pyruvate recycling did. Hence, their data further support the notion of a neuronal localisation of the pyruvate recycling. However, this localisation of pyruvate recycling has been challenged. Hassel et al. (1994) established that astrocytes in cultures release lactate formed via pyruvate from TCA cycle intermediates. While this apparent discrepancy might be explained by ontogenic differences between adult brain *in vivo* and primary cell cultures derived from developing brain, difference in the time course and extent of lactate labelling from [2-^{13}C]-acetate and [1-^{13}C]-glucose in *in vivo* studies were interpreted as being suggestive for a glial localisation of pyruvate recycling (Hassel and Sonnewald 1995). More convincing is the observation that [1,2-$^{13}C_2$]-acetate labelled lactate in control animals but not upon inhibition of glial acetate metabolism by fluoroacetate (Hassel et al. 1997). Melo et al. (2006a) reported more pyruvate recycling into glutamine than glutamate, yet a distinctly increased recycling was observed in glutamate in animals fed a ketotic diet. Sophisticated labelling experiments with [2-^{13}C,2-$^{3}H_3$]-acetate suggested that in addition to the neuronal pyruvate recycling, glial recycling may be active *in vivo* during acetate metabolism (Chapa et al. 2000). Similarly, Haberg et al. (1998a) came to a partially reconciling finding that, though pyruvate was assumed to be formed in the glial compartment, pyruvate/lactate produced in this manner were fuels for both the neuronal and glial TCA cycles. More recent investigations by Olstad et al (2007) on cultured neurons from mouse cerebellum demonstrated that apart from its previously implied glial location, neuronal pyruvate recycling is occurring extensively in neurons as originally postulated by Cerdan et al. (1990). It should be noted, however, that up to our current understanding pyruvate recycling and anaplerosis through PC are active simultaneously and thus the two pathways, if located in the same compartment, would potentially lead to extensive futile cycling. Proposal of further intra-cellular compartmentation of pyruvate recycling and PC may remedy this apparent contradiction. Data obtained from primary cultures of cortical astrocytes indeed showed differences in labelling ratios between glial glutamine C2/C4 and its precursor citrate which were taken as evidence for the existence of several TCA cycles within astrocytes (Sonnewald et al. 1994).

Irrespective of its location, speculations on the putative function of cerebral pyruvate recycling have been brought forward. Pyruvate recycling may provide essential substrate for neuronal TCA cycle metabolism under conditions of low glucose supply or impeded glucose oxidation (Pascual et al. 1998). Findings that pyruvate recycling was increased in animals fed a ketotic diet (very low in carbohydrates) are in support of this notion (Melo et al. 2006a). Moreover, quantitative analyses of pyruvate recycling in brains supplied with different substrates indicated that the activity of pyruvate recycling correlated with the preference of the substrates to enter the glial compartment (Künnecke et al. 1993). Alternatively it has been proposed that formation of pyruvate and lactate from glutamate or GABA released into the synaptic cleft may be used to deactivate these neurotransmitters. In turn, the production of lactate with its vasodilatory property may then provide a mechanism for mediating the increase in cerebral blood flow during neuronal activation (Laptook et al. 1988; Hassel and Sonnewald 1995).

26.4.6 N-Acetylaspartate and Lipids

N-Acetylaspartate (NAA) is an amino acid present in mammalian brain at high concentration and with almost exclusive localisation in the neuronal compartment (for a review see Baslow 2003). NAA had long been thought of being rather inert in the adult brain. Newer data has evidenced that NAA and in particular its acetyl moiety is in a dynamic turnover (Künnecke et al. 1993; Baslow 2003). Labelling experiments with acetate, however, strongly indicated that in contrast to the acetyl moiety in acetylcholine glial metabolism contributed only negligibly to NAA synthesis (Mukherji and Sloviter 1973; Künnecke et al. 1993). Notwithstanding, speculation about NAA's significance in cerebral tissue have been extensive and have included among others function as osmolyte, storage form of acetate or aspartate, intercellular signalling and participation in myelin synthesis in oligodendrocytes (Baslow 2003). The putative involvement of NAA in the synthesis of cerebral lipids is intriguing but also disputed. Radio-labelling studies demonstrated that in adult brain label from acetate is readily and rapidly incorporated into cerebral lipids but presumably without detour via NAA (Dhopeshwarkar et al. 1971). In the developing brain, however, a double labelling approach with [^{3}H]-acetate and [^{14}C]-acetate generated a ^{3}H/^{14}C ratio in fatty acids reminiscent of a distinct loss of hydrogen-3 in the pathway of acetate incorporation into lipids (D'Adamo and Yatsu 1966). A similar differential loss of label was also observed for the incorporation of acetate into NAA and thus the authors inferred that NAA was implicated in brain lipid synthesis during myelinogenesis.

26.5 Acetate's Role in Cerebral Pathologies

26.5.1 Brain Tumours

As outlined above, innumerous studies have demonstrated that acetate is readily taken up and metabolised in the mammalian brain. Brain tumours, in particular those of glial origin, retain these metabolic capabilities. Yet in such proliferating tissue, acetate is mainly destined for lipid- and protein biosynthesis that supports membrane expansion and tumour growth. Thus, in contrast to non-tumour brain tissue where acetate is chiefly metabolised in the tricarboxylic acid cycle(s) and rapidly egresses as CO_2, tumour tissue will incorporate the carbon skeleton of acetate and (radio-)labels introduced via this route in long-lived metabolites. Yoshimoto et al. (2004) successfully tested this concept of radio-labelled acetate being a proliferation marker in a variety of tumour cell lines including glioblastoma-derived cells. Dienel et al. (2001a) demonstrated with autoradiography that [2-^{14}C] acetate readily labels glial and also meningial brain tumours grown in rat brain *in vivo* and in human tumour explants. In their work, these authors noted that the rate of acetate metabolism in normal brain is lower but also more uniform than that of glucose, thus providing a consistent and low background for tumour detection. This paradigm has been taken up in several human PET studies in which [11-C] acetate was used to support detection of brain tumours (Liu et al. 2006; Tsuchida et al. 2008; Yamamoto et al. 2008). Despite the limited number of patients assessed to date, the collective of studies with acetate PET show promise in particular for the detection of gliomas. Tsuchida et al. (2008) and Yamamoto et al. (2008) reported that apart from mere detection, acetate PET enabled differentiation between low-grade and high grade glioma based on the extent of acetate retention.

26.5.2 Cerebral Ischemia

Increased interest has been focused on the role of neuronal-glial interaction and compartmentation during pathophysiological conditions of the brain. In an ischemic stroke, blood supply to part of the brain is decreased, leading to transient or permanent dysfunction of the brain tissue in that area. Numerous studies addressed the implications of cerebral ischemia on brain metabolism and it was concluded that astrocyte metabolism plays a pivotal role for neuronal survival. Using [1,2-$^{13}C_2$]-acetate as tracer, Pascual et al. (1998) demonstrated that focal cerebral ischemia in a time dependent manner decreased the oxidative metabolism of glucose in the affected area. Reduced glucose oxidation, in turn, was compensated for with an increased oxidation of glutamate and GABA in the astrocytes. Twenty-four hours after the insult oxidative metabolism of glutamine and the extent of pyruvate recycling were increased in the neuronal compartment. With the combined use of [1-^{13}C]-glucose and [1,2-$^{13}C_2$]-acetate, Haberg et al. (1998b, 2001, 2006) drew a differentiated picture of the metabolic ramifications in the core and penumbra of cerebral ischemia which essentially confirmed and extended findings reported previously. Penumbra was characterised by improved astrocytic metabolism as evidenced by increased acetate oxidation and glutamine formation. Conversely, neuronal mitochondrial activity was maintained despite a strongly reduced flux of glucose into the TCA cycle. During ischemia, glutamine was found to contribute increasingly to the synthesis of glutamate while neuronal TCA cycle activity gradually declined. This condition of gradually declining neuronal metabolic activity in connection with a lasting replenishment of glutamate was thought of rendering neurons susceptible to glutamate excitotoxicity (Haberg et al. 2001).

GABAergic neurons in the striatum were found to be especially sensitive to ischemia. This hypersensitivity was associated with a complete cessation of the use of astrocytic glutamine for the synthesis of GABA in these neurons (Haberg et al. 2001). In the hypoxic but non-ischaemic rat brain also a clear trend to reduced glutamine trafficking from the glial to neuronal compartment was observed (Chateil et al. 2001). Primary cultures of murine cortical astrocytes and cerebellar granule neurons subjected to prolonged hypoxia revealed that neuronal glycolysis and lactate production was increased whereas label incorporation from acetate into glial glutamine was greatly reduced (Sonnewald et al. 1994). In the light of major regional metabolic differences, conclusive interpretation of these data, however, is difficult since the cultured astrocytes and neurons were derived from very different brain regions and thus their individual metabolism may not reflect their natural interaction in intact brain. Yet, time dependent reduction of acetate uptake and metabolism was demonstrated in the ischemic rat brain (Hosoi et al 2007).

The effect of putative therapies on hypoxic and ischemic damage were assessed (Haberg et al. 2007; Sonnewald et al. 1998). U-101033E, a pyrrolopyrimidine lipid peroxidation inhibitor, primarily lead to a reduced mitochondrial lactate metabolism and substrate accumulation of glucose and acetate. The latter were interpreted as a drug-induced higher availability of those substrates (Haberg et al. 2007). Orotic acid, an intermediate in the pyrimidine biosynthesis with putative beneficial effects on purine uptake, supplied to hypoxic cultured astrocytes was capable of reverting impaired acetate utilization to normoxic levels (Sonnewald et al. 1998). On the other hand, Dienel et al. (2003) investigated the effect of behavioural training on the outcome of mild transient ischemia using [^{14}C]-acetate. They found that training indeed increased post-ischemic astrocytic metabolism in regions involved in motor activities but had no effect on functional neurological outcome.

26.5.3 Diabetes and Thyroid Hormone Deficits

Dysfunction of glucose homeostasis is one of the hallmarks of diabetes. Hypoglycaemia as well as severe hyperglycaemia lead to neurological malfunction including coma and even death. Mason et al. (2006) investigated the alterations of cerebral substrate transport and metabolism in patients with well-controlled diabetes type I. They found that during hypoglycaemia monocarboxylic acid transport in the brain was twofold increased in diabetic patients as compared to healthy subjects as was reflected by higher brain concentrations and rates of [2-^{13}C]-acetate oxidation. Since MCTs not only confer the transport of acetate but also that of lactate and ketone bodies, this finding is in line with the notion that during hypoglycaemia substrates other than glucose contribute significantly to cerebral metabolism in the adult brain (Cerdan et al. 1990; Künnecke et al. 1993; Mason et al. 2006). Upregulation of MCT thus is likely to contribute to the maintenance of cerebral energy supply during periods of hypoglycaemia. Garcia-Espinosa et al. (2003) assessed the effect of hyperglycemia in a rat model with streptozotocin-induced diabetes. Incorporation of acetate into cerebral metabolites was consistently higher in diabetic animals than in normoglycaemic controls. This metabolic alteration was ascribed to a stimulation of the glutamine synthesis in diabetic rats as a consequence of increased cortisol levels. Chapa et al. (1995) demonstrated that adult-onset hypothyroidism also has a marked influence on cerebral metabolism. Thyroid hormones act on the body among others to increase the basal metabolic rate and increase the body's sensitivity to catecholamines. Their absence after thyroidectomy was found to decrease the contribution of [1,2-$^{13}C_2$]-acetate to glial metabolism. Moreover, the glial-neuronal exchange was impaired as evidenced by the accumulation of glutamate and GABA.

Although indirect metabolic effects on cerebral metabolism could not entirely be excluded, these results were interpreted as glial and neuronal metabolism both being directly affected by adult-onset hypothyroidism.

26.5.4 Psychiatric and Neurological Disorders

Among the neurological disorders has epilepsy a particular connection with acetate metabolism. Back in the 1920s, the so-called ketogenic diet was developed to remedy difficult-to-treat seizures in children (see Freeman et al. 2007). The ketogenic diet is a high-fat/low-carbohydrate diet that leads to an increased supply of the brain with ketone bodies (including acetate) produced in the liver. Melo et al. (2006a) and Yudkoff et al. (2005) investigated the effect of such ketogenic diet on brain metabolism in rodents. Astrocytic metabolism was increased in response to ketosis, as was evidenced by an increased incorporation of label derived from acetate into glutamine, glutamate and GABA. Moreover, pyruvate carboxylation was increased in ketotic animals and paradoxically the contribution of pyruvate recycling was found to be higher in glutamine than in glutamate. Ross and co-workers (Ross et al. 2003; Shic and Ross 2003) found in human epilepsy patients increased glutamine synthesis and pyruvate carboxylase activity. Several different rodent models of epilepsy were used to further investigate the metabolic etiology of this disease. Genetic Absence Epilepsy Rats from Strasboug (GAERS) showed an increased glutamate-glutamine metabolism and PC/PDH activity in cortical areas together with decreased levels of GABA (Melo et al. 2006b, 2007). These findings are in line with the hypothesis of disturbances in the GABAergic and glutamatergic neurotransmission in the thalamocortical loop being intimately involved in absence seizures. Changes in cerebral metabolism observed during the development of young presymptomatic animals to adult GAERS involved an increase in astrocytic metabolism, thus suggesting that astrocytic alterations may play a pivotal role in triggering seizures (Melo et al. 2007). Results from labelling studies in lithium pilocarpine-induced and kainate-induced chronically epileptic rats are in some contradiction to these former data. Labelling of cerebral metabolites from [1,2-$^{13}C_2$]-acetate was similar in diseased and control animals but in the cortex of chronically epileptic animals the flow of neuronal glutamate to astrocytes was decreased (Melo et al. 2005; Müller et al. 2000). Yet another model of epilepsy investigated was based on pentylenetetrazole-induced seizures. Pentylenetetrazole is a GABA antagonist binding to the postsynaptic $GABA_A$ receptor and thus decreases the inhibitory tone and favours depolarisation of neurons (Bradford 1989). Eloqayli et al. (2001, 2004) found in the pentylenetetrazole model that chiefly changes in cerebellar metabolism were implicated in epileptic seizures. In cerebellum, the turnover of glutamate was decreased and from differential labelling in glutamate and glutamine the authors concluded that mainly the metabolic pool of glutamate was affected whereas the releasable neurotransmitter pool was unaffected (Eloqayli et al. 2001). Sodium valproate, in turn, is an anticonvulsant that has found clinical use in the treatment of epilepsy and bipolar disorder, as well as other psychiatric conditions requiring the administration of a mood stabilizer. Johannessen et al. (2002) investigated the acute effect of valproate on cerebral metabolism in mice. They concluded from their data that the TCA cycle activity was reduced and uptake of acetate and possibly other monocarboxylic acids was inhibited upon drug treatment. Blüml et al. (2002) had one human epilepsy patient on valproate treatment assessed by *in vivo* ^{13}C-MRS. In this single patient, carbon-13 accumulation was found to be increased in glutamine but markedly reduced in bicarbonate, and acetate usage was diminished as compared to healthy controls. These data were interpreted in view of valproate's putative mode of action of partially inhibiting hepatic urea cycle activity with resulting increased ammonia availability in the brain. In turn, this may have driven the conversion of cerebral glutamate to glutamine by means of glutamine synthase located in the glial compartment.

Schizophrenia is a mental disorders often described in terms of positive (e.g. delusions, hallucinations), negative (loss of normal traits and abilities, e.g. blunted affection and emotion, anhedonia) and cognitive symptoms. Numerous animal models have been developed and in particular pharmacologically induced models have found widespread application in schizophrenia research. Among these is MK801, a glutamate receptor antagonist acting on the N-methyl-D-aspartate (NMDA) subtype, which exhibits psychotogenic properties mimicking many of the traits found in schizophrenia. Sonnewald and co-workers (Brenner et al. 2005; Eyjolfsson et al. 2006; Kondziella et al. 2006) investigated the effect of MK801 administration on cerebral metabolism in rats. With their well proven approach of using [1-^{13}C]-glucose in combination with [1,2-$^{13}C_2$]-acetate to simultaneously probe neuronal and glial metabolism, respectively, they found that acute administration of MK801 did not affect metabolite labelling from acetate (Brenner et al. 2005). Conversely, subchronic treatment with MK801 (0.5 mg/kg/day i.p.) lead to significantly decreased contribution of glutamate and glutamine derived from [1,2-$^{13}C_2$]-acetate in the frontal lobes of the cortex (Kondziella et al. 2006). At lower doses (0.1 mg/kg/day i.p.), changes to glutamate and glutamine were restricted to the temporal lobe where their concentration as well as the labelling from glucose and acetate were increased as compared to controls (Eyjolfsson et al. 2006). Mice with a deletion of the STOP (stable tubule only peptide) protein show hyperdopaminergia with associated behavioural dysfunction and may serve as an alternative model of schizophrenia. In such knockout mice, Brenner et al. (2007) found an increased cerebral glutamine concentration but decreased glutamine synthesis from [1,2-$^{13}C_2$]-acetate. Together, these data suggest a major disturbance of the glutamate-glutamine cycle in these models of schizophrenia and are in line with the current "glutamate hypothesis" (as opposed to the dopamine hypothesis) which is thought to be pointing more at the root of the disease and is implying a hypofunction of glutamatergic signalling.

Several animal studies were carried out with acetate as tool for investigating effects on brain metabolism brought about by dysfunctional adenosinergic neuromodulation and dopaminergic neurotransmission. Haberg and co-workers (Hammer et al. 2001; Haberg et al. 2000) investigated the consequences of adenosine A_1 and

A_2 receptor agonists and antagonists using [1-^{13}C]-glucose or [1,2-$^{13}C_2$]-acetate as probes. Adenosine A_1 agonist and antagonist induced complementary changes. De novo synthesis of GABA was reduced and glutamate degradation, anaplerotic and glial TCA cycle activity were increased upon antagonist administration. Agonism of the A_1 receptor did not alter GABA synthesis and the contribution of the anaplerotic pathway to glutamate was decreased but no further effect on glial metabolism was detected (Haberg et al. 2000). A_2 receptor agonism and antagonism both had little effect on overall glial metabolism except for a shift to a reduced PC/PDH ratio in the cortex (Hammer et al. 2001). The metabolic interaction between the glutamatergic and dopaminergic system was investigated in transgenic mice lacking either the dopamine D_1 or D_2 receptor subtypes (Rodrigues et al. 2007). D_1 receptor-deficient mice showed a pronounced increase in label incorporation from [1,2-$^{13}C_2$]-acetate into glutamine. Conversely, no such effect was observed in the brains of D_2 receptor-deficient animals. The authors thus concluded that D_1 receptor-mediated dopamine signalling alters glial glutamine metabolism which is consistent with an increased activity of the glutamine-glutamate cycle and of glutamatergic neurotransmission. These findings were further substantiated by further pharmacologic modulation of the dopaminergic system. Wild-type mice treated with the dopamine depleting drug reserpine presented glutamine labelling similar to those found in D_1 knock-out mice. Subsequent administration of L-DOPA, a blood-brain barrier-penetrting precursor which in brain is rapidly converted to dopamine, reverted glutamine labelling in these animals to the original wild-type levels. Neither reserpine nor L-DOPA had an effect in D_1-receptor deficient mice, thus confirming the notion that metabolic interactions between the glutamatergic and dopaminergic system are intricately linked to D_1 receptor-mediated neurotransmission.

26.6 Conclusion and Outlook

This review has highlighted numerous studies and experimental approaches from cells to man which addressed a wide range of aspects of cerebral acetate metabolism in health and disease. Yet; this list is by far not exhaustive and many captivating studies have found no mention. These include among others work on enzyme deficiencies and inborn errors, intoxications and neurodegenerative diseases. All work described herein has in common that acetate was used as genuine substrate and especially as a tool for elucidating cerebral metabolism. Acetate's unique feature of being a cerebral fuel which almost exclusively enters the astroglial cells has made it possible to investigate cerebral metabolism from a distinct "glial" stance. The sheer number of studies is overwhelming, though only few were designed to provide truly quantitative data. The vast majority of the studies were semi-quantitative in that clear differences between different states of cerebral metabolism could be readily demonstrated, but often the precise nature of the underlying mechanisms has remained speculative. Investigations of cerebral metabolism using acetate have mainly focused on the metabolic interaction of glutamine, glutamate and GABA.

Since the early days after Lajhta et al. (1959) postulated the existence of two distinct glutamate compartments in mammalian brain, analysis of cerebral metabolism has experienced two major booms. With radiotracer methods the foundation of today's understanding of cerebral compartmentation were laid. With the advent of NMR applications in biochemistry, metabolic pathways and their complex interplay in the mammalian brain have been studied in more detail. The non-invasive nature of MRS has then enabled such studies to be carried out in situ in animals and ultimately in man. Judging from the number of publications, more recently the interest in the biochemical assessment of cerebral metabolism has somewhat cooled down. Thus, the question may be raised whether all aspects of cerebral metabolism are sufficiently clear? The answer is definitely no and more investigations have to be performed. Metabolic interactions in the living brain are currently best elucidated with MRS in combination with carbon-13 tracers but their assessment is limited by sensitivity issues especially for less- and lowly abundant neurochemicals. More recent developments and the introduction of hyperpolarised carbon-13 have been shown to be capable of increasing sensitivity orders of magnitude even under *in vivo* conditions. To date efficacious application of hyperpolarised carbon-13 is restricted to slowly relaxing carbon entities such as carboxylic groups. Acetate falls into this category and it remains to be seen whether cerebral biochemistry will soon experience another boost along these lines.

References

Badar-Goffer RS, Bachelard HS, Morris PG (1990) Cerebral metabolism of acetate and glucose studied by ^{13}C-n.m.r. spectroscopy: a technique for investigating metabolic compartmentation in the brain. Biochem J 266:133–139

Badar-Goffer RS, Ben-Yoseph O, Bachelard HS, Morris PG (1992) Neuronal-glial metabolism under depolarizing conditions: a ^{13}C-n.m.r study. Biochem J 282:225–230

Baslow MH (2003) N-Acetylaspartate in the vertebrate brain: metabolism and function. Neurochem Res 28:941–953

Berl S, Clarke DD (1983) The metabolic compartmentation concept. In: Hertz L, Kyamne E, McGeer G, Schousboe A (eds) Glutamine, glutamate and GABA in the central nervous system. Alan R. Liss Inc, New York, pp 205–217

Berl S, Nicklas WJ, Clarke DD (1968) Compartmentation of glutamic acid and metabilism in brain slices. J Neurochem 15:131–140

Blüml S, Moreno-Torres A, Shic F, Nguy C-H, Ross BD (2002) Tricarboxylic acid cycle of glia in the in vivo human brain. NMR Biomed 15:1–5

Bradford HF (1989) Glutamate, GABA and epilepsy. Epilepsia 30:17–25

Brand A, Richter-Landsberg C, Leibfritz D (1997) Metabolism of acetate in rat brain neurons, astrocytes and coculture: metabolic interactions between neurons and glia cells, monitored by NMR spectroscopy. Cell Mol Biol 43:645–657

Brenner E, Kondziella D, Haberg A, Sonnewald U (2005) Impaired glutamine metabolism in NMDA receptor hypofunction induced by MK801. J Neurochem 94:1594–1603

Brenner E, Sonnewald U, Schweitzer A, Andrieux A, Nehling A (2007) Hypoglutamatergic activity in the STOP knockout mouse: a potential model for chronic untreated schizophrenia. J Neurosci Res 85:3487–3493

Carroll PT (1997) Evidence to suggest that extracellular acetate is accumulated by rat hippocampal cholinergic nerve terminals for acetylcholine formation and release. Brain Res 753:47–55

Cerdan S, Künnecke B, Seelig J (1990) Cerebral metabolism of [1,2-$^{13}C_2$] acetate as determined by in vivo and in vitro ^{13}C NMR. J Biol Chem 265:12916–12926

Chapa F, Künnecke B, Calvo R, del Rey FE, de Escobar GM, Cerdán S (1995) Adult-onset hypothyroidism and the cerebral metabolism of (1,2-$^{13}C_2$) acetate as detected by ^{13}C nuclear magnetic resonance. Endocrinology 136:296–305

Chapa F, Cruz F, Garcia-Martin ML, Garcia-Espinosa MA, Cerdán S (2000) Metabolism of (1–13C) glucose and (2–13C, 2–2H3) acetate in the neuronal and glial compartments of the adult rat brain as detected by 13C, 2H NMR spectroscopy. Neurochem Int 37:217–228

Chateil J-F, Biran M, Thiaudière E, Canioni P, Merle M (2001) Metabolism of [1-^{13}C] glucose and [2-^{13}C] acetate in the hypoxic rat brain. Neurochem Int 38:399–407

Chowdhury GMI, Patel AB, Mason GF, Rothman DL, Behar KL (2007) Glutamatergic and GABAergic neurotransmitter cycling and energy metabolism in rat cerebral cortex during postnatal development. J Cereb Blood Flow Metab 27:1895–1907

Cremer JR, Braun LD, Oldendorf WH (1976) Changes duringdevelopment in transport processes of the blood-brain barrier. Biochim Biophys Acta 448:633–637

Cruz F, Scott SR, Brroso I, Santisteban P, Cerdán S (1998) Ontogeny and cellular location of the pyruvate recycling system in rat brain. J Neurochem 70:2613–2619

Cruz NF, Lasater A, Zielke HR, Dienel GA (2005) Activation of astrocytes in brain of conscious rats during acoustic stimulation: acetate utilization in working brain. J Neurochem 92:934–947

D'Adamo AF Jr, Yatsu FM (1966) Acetate metabolism in the nervous system. N-acetyl-L-aspartic acid and the biosynthesis of brain lipids. J Neurochem 13:961–965

de Graaf RA, Patel AB, Rothman DL, Behar KL (2006) Acute regulation of steady-state GABA levels following GABA-transaminase inhibition in rat cerebral cortex. Neurochem Int 48:508–514

De Koning-Verest IF (1980) Glutamate metabolism in ageing rat brain. Mech Ageing Develop 13:83–92

Dhopeshwarkar GA, Subramanian C, Mead JF (1971) Rapid uptake of [1-^{14}C] acetate by the adult brain 15 seconds after carotid injection. Biochim Biophys Acta 248:41–47

Dienel GA, Popp D, Drew PD, Ball K, Krisht A, Cruz NF (2001a) Preferential labelling of glial and meningial brain tumors with [2-^{14}C] acetate. J Nucl Med 42:1243–1250

Dienel GA, Liu K, Cruz NF (2001b) Local uptake of ^{14}C-labeled acetate and butyrate in rat brain in vivo during spreading cortical depression. J Neurosci Res 66:812–820

Dienel GA, Cruz NF, Ball K, Popp D, Gokden M, Baron S, Wright D, Wenger GR (2003) Behavioural training increases local astrocytic metabolic activity but does not alter outcome of mild transient ischemia. Brain Res 961:201–212

Dolezal V, Tucek S (1981) Utilization of citrate, acetylcarnitine, acetate, pyruvate and glucose for the synthesis of acetylcholine in rat brain slices. J Neurochem 36:1323–1330

Eloqayli H, Dahl CB, Götestam KG, Unsgard G, Hadidi H, Sonnewald U (2001) Pentylenetetrazole decreases metabolic glutamate turnover in rat brain. J Neurochem 85:1200–1207

Eloqayli H, Dahl CB, Götestam KG, Unsgard G, Sonnewald U (2004) Changes of glial-neuronal interaction and metabolism after a subconvolsive dose of pentylenetetrazole. Neurochem Int 45:739–745

Eyjolfsson EM, Brenner E, Kondziella D, Sonnewald U (2006) Repeated injection of MK801: an animal model of schizophrenia. Neurochem Int 48:540–546

Freeman JM, Kossoff EH, Hartman AL (2007) The ketogenic diet: one decade later. Pediatrics 119:535–543

Garcia-Espinosa MA, Garcia-Martin ML, Cerdán S (2003) Role of glial metabolism in diabetic encephalopathy as detected by high resolution ^{13}C NMR. NMR Biomed 16:440–449

Gonda O, Quastel JH (1966) Transport and metabolism of acetate in rat brain cortex in vitro. Biochem J 100:83–94

Haberg A, Qu H, Bakken IJ, Sande LM, White LR, Haraldseth O, Unsgard G, Aasly J, Sonnewald U (1998a) In vitro and ex vivo ^{13}C-NMR spectroscopy studies of pyruvate recycling in brain. Dev Neurosci 20:389–398

Haberg A, Qu H, Haraldseth O, Unsgard G, Sonnewald U (1998b) In vivo injection of [1-^{13}C] glucose and [1,2-^{13}C] acetate combined with ex vivo ^{13}C nuclear magnetic resonance spectroscopy: A novel approach to the study of middle cerebral artery occlusion in the rat. J Cereb Blood Flow Metab 18:1223–1232

Haberg A, Qu H, Haraldseth O, Unsgard G, Sonnewald U (2000) In vivo effects on adenosine A_1 receptor agonist and antagonist on neuronal and astrocytic intermediary metabolism studied with ex vivo ^{13}C NMR spectroscopy. J Neurochem 74:327–333

Haberg A, Qu H, Saether O, Unsgard G, Haraldseth O, Sonnewald U (2001) Differences in neurotransmitter synthesis and intermediary metabolism between glutamatergic and GABAergic neurons during 4 hours of middle cerebral artery occlusion in the rat: The role of astrocytes in neuronal survival. J Cereb Blood Flow Metab 21:1451–1463

Haberg A, Qu H, Sonnewald U (2006) Glutamate and GABA metabolism in transient and permanent middle cerebral artery occlusion in rat: importance of astrocytes for neuronal sutvival. Neurochem Int 48:531–540

Haberg A, Qu H, Hjelstuen MH, Sonnewald U (2007) Effect of the pyrrolopyrimidine lipid peroxidation inhibitor U-101033E on neuronal and astrocytic metabolism and infarct volume in rats with transient middle cerebral artery occlusion. Neurochem Int 50:932–940

Hammer J, Qu H, Haberg A, Sonnewald U (2001) In vivo effects of adenosine A2 receptor agonist and antagonist on neuronal and astrocytic intermediary metabolism studied with ex vivo ^{13}C MR spectroscopy. J Neurochem 79:885–892

Hassel B, Sonnewald U (1995) Glial formation of pyruvate and lactate from TCA cycle intermediates: implications for the inactivation of transmitter amino acids? J Neurochem 65:2227–2234

Hassel B, Sonnewald U, Unsgard G, Fonnum F (1994) NMR spectroscopy of cultured astrocytes: effects of glutamine and the gliotoxin fluorocitrate. J Neurochem 62:2187–2194

Hassel B, Bachelard H, Jones P, Fonnum F, Sonnewald U (1997) Trafficking of amino acids between neurons and glia in vivo. Effects of inhibition of glial metabolism by fluoroacetate. J Cereb Blood Flow Metab 17:1230–1238

Hirose S, Momosaki S, Hosoi R, Abe K, Gee A, Inoue O (2009) Role of NMDA receptor upon [^{14}C] acetate uptake into intact brain. Ann Nucl Med 23:143–147

Hosoi R, Okada M, Hatazawa J, Gee A, Inoue O (2004) Effect of astrocytic energy metabolism depressant on 14C-acetate uptake in intact rat brain. J Cereb Blood Flow Metab 24:188–190

Hosoi R, Kashiwagi Y, Tokumura M, Abe K, Hatazawa J, Inoue O (2007) Sensitive reduction in ^{14}C-acetate uptake in a short-term ischemic rat brain. J Stroke Cerebrovasc Dis 16:77–81

Hosoi R, Matsuyama Y, Hirose S-I, Koyama Y, Matsuda T, Gee A, Inoue O (2009) Characterization of ^{14}C-acetate uptake in cultured rat astrocytes. Brain Res 1253:69–79

Inoue O, Hosoi R, Momosaki S, Yamamoto K, Amitani M, Yamaguchi M, Gee A (2006) Evaluation of [^{14}C] phenylacetate as a prototype tracer for the measurement of glial metabolism in the rat brain. Nucl Med Biol 33:985–989

Johannessen CU, Petersen D, Fonnum F, Hassel B (2002) The acute effect of valproate on cerebral energy metabolism in mice. Epilepsy Res 47:147–256

Kondziella D, Qu H, Lüdemann W, Brinker T, Sletvold O, Sonnewald U (2003) Astrocyte metabolism is disturbed in the early development of experimental hydrocephalus. J Neurochem 85:274–281

Kondziella D, Brenner E, Eyjolfsson EM, Markinhuhta KR, Carlsson ML, Sonnewald U (2006) Glial-neuronal interactions are impaired in the schizophrenia model of repeated MK801 exposure. Neuropsychopharmacology 31:1880–1887

Kreis R, Hofmann L, Kuhlmann B, Boesch C, Bossi E, Hüppi PS (2002) Brain metabolite composition during early human brain development as measured by quantitative in vivo 1H magnetic resonance spectroscopy. Magn Reson Med 48:949–958

Künnecke B (1995) Application of ^{13}C NMR spectroscopy to metabolic studies on animals. In: Beckmann N (ed) Carbon-13 NMR spectroscopy of biological systems. Academic, San Diego, pp 159–267

Künnecke B, Cerdan S (1989) Multilabeled ^{13}C substrates as probes in in vivo ^{13}C and ^{1}H NMR spectroscopy. NMR Biomed 2:274–277

Künnecke B, Cerdan S, Seelig J (1993) Cerebral metabolism of [1,2-$^{13}C_2$] glucose and [U-$^{13}C_4$] 3-hydroxybutyrate in rat brain as detected by ^{13}C NMR spectroscopy. NMR Biomed 6:264–277

Lajhta A, Berl S, Waelsch H (1959) Amino acids and protein metabolism of the brain. IV. The metabolism of glutamic acid. J Neurochem 3:322–332

Laptook AR, Peterson J, Porter AM (1988) Effects of lactic acid infusions and pH on cerebral blood flow and metabolism. J Cereb Blood Flow Metab 8:193–200

Lear JL, Ackermann RF (1990) Evaluation of radiolabeled acetate and fluoroacetate as potential tracers of cerebral oxidative metabolism. Metab Brain Dis 5:45–56

Lebon V, Petersen KF, Cline GW, Shen J, Mason GF, Dufour S, Behar KL, Shulman GI, Rothman DL (2002) Astroglial contribution to brain energy metabolism in humans revealed by ^{13}C nuclear magnetic resonance spectroscopy: elucidation of the dominant pathway for neurotransmitter glutamate rpletion and measurement of astrocytic oxidative metabolism. J Neurosci 22:1523–1531

Li S, Chen Z, Zhang Y, Lizak M, Bacher J, Innis RB, Shen J (2005) In vivo single-shot, proton-localized ^{13}C MRS of rhesus monkey brain. NMR Biomed 18:560–569

Liu R-S, Chang C-P, Chu L-S, Chu Y-K, Hsieh H-J, Chang C-W, Yang B-H, Yen S-H, Huamg M-C, Liao S-Q, Yeh S-H (2006) PET imaging of brain astrocytoma with 1-11C-acetate. Eur J Nucl Med Mol Imag 33:420–424

Martinez-Hernandez A, Bell KP, Norenberg MD (1977) Glutamine synthase: glial localization in brain. Science 195:1356–1357

Mason GF, Petersen KF, Lebon V, Rothman DL, Shulman GI (2006) Increased brain monocarboxylic acid transport and utilization in type 1 diabetes. Diabetes 55:929–934

Melo TM, Nehling A, Sonnewald U (2005) Metabolism is normal in astrocytes in chronically epileptic rats: a ^{13}C NMR study of neuronal-glial interactions in a model of temporal lobe epilepsy. J Cereb Blood Flow Metab 25:1254–1264

Melo TM, Nehling A, Sonnewald U (2006a) Neuronal-glial interactions in rats fed a ketogenic diet. Neurochem Int 48:498–507

Melo TM, Sonnewald U, Touret M, Nehling A (2006b) Cortical glutamate metabolism is enhanced in a genetic model of absence epilepsy. J Cereb Blood Flow Metab 26:1496–1506

Melo TM, Sonnewald U, Bastholm IA, Nehling A (2007) Astrocytes may play a role in the etiology of absence epilepsy: a comparison between immature GAERS not yet expressing seizures and adults. Neurobiol Dis 28:227–235

Momosaki S, Hosoi R, Sanuki T, Todoroki K, Yamaguchi M, Gee A, Inoue O (2007) [^{14}C] Benzyl acetate is a potential radiotracer for the measurement of glial metabolism in the rat brain. Nucl Med Biol 34:939–944

Muir D, Berl S, Clarke DD (1986) Acetate and fluoroacetate as possible markers for glial metabolism in vivo. Brain Res 380:336–340

Mukherji B, Sloviter HA (1973) Metabolism of acetate and N-acetylaspartate in isolated perfused rat brain. J Neurochem 20:633–636

Müller B, Qu H, Garseth M, White LR, Aasly J, Sonnewald U (2000) Amino acid neurotransmitter metabolism in neurons and glia following kainate injection in rats. Neurosci Lett 279:169–172

Nakamura R, Cheng S-C, Naruse H (1970) A study on the precursors of the acetyl moiety of acetylcholine in brain slices. Biochem J 118:443–450

Nishima M, Suzuki M, Matsushita K (2004) Trichinella spiralis: activity of the cerebral pyruvate recycling pathway of the host (mouse) in hypoglycemia induced by the infection. Exp Parasitol 106:62–65

Norenberg MD, Martinez-Hernandez A (1979) Fine structural localization of glutamine synthetase in astrocytes in brain. Brain Res 161:303–310

O'Neal RM, Koeppe RE, Williams EI (1966) Utilization in vivo of glucose and volatile fatty acids by sheep brain for the synthesis of acidic amino acids. Biochem J 101:591–597

Olstad E, Olsen GM, Sonnewald U (2007) Pyruvate recycling in cultured neurons from cerebellum. J Neurosci Res 85:3318–3325

Pascual JM, Carceller F, Roda JM, Cerdán S (1998) Glutamate, glutamine, and GABA as substrates for the neuronal and glial compartments after focal cerebral ischemia in rats. Stroke 29: 1048–1057

Patel AB, de Graaf RA, Mason GF, Rothman DL, Shulman RG, Behar KL (2005) The contribution of GABA to the glutamate/glutamine cycle and energy metabolism in the rat cortex in vivo. Proc Natl Acad Sci USA 102:5588–5593

Pierre K, Pellerin L (2005) Monocarboxylate transporters in the central nervous system: distribution, regulation and function. J Neurochem 94:1–14

Pierre K, Pellerin L, Debernardi R, Riederer BM, Magistretti PJ (2000) Cell-specific lacalization of monocarboxylate transporters, MCT1 and MCT2, in the adult mouse brain revealed by double immunohistochemical labeling and confocal microscopy. Neuroscience 100:617–627

Rae C, Hare N, Bubb WA, McEwan SR, Bröer A, McQuillan JA, Balcar VJ, Conigrave AD, Böer S (2003) Inhibition of glutamine transport depletes glutamate and GABA neurotransmitter pools: further evidence for metabolic compartmentation. J Neurochem 85:503–514

Rodrigues TB, Granado N, Ortiz O, Cerdán S, Moratella R (2007) Metabolic interaction between glutamatergic and dopaminergic neurotransmitter systems are mediated through D_1 dopamine receptors. J Neurosci Res 85:3284–3293

Ross B, Lin A, Harris K, Bhattacharya P, Schweinsburg B (2003) Clinical experience with ^{13}C MRS in vivo. NMR Biomed 16:358–369

Schousboe A (2003) Role of astrocytes in the maintenance and modulation of glutamatergic and GABAergic neurotransmission. Neurochem Res 28:347–352

Serres S, Bezancon E, Franconi J-M, Merle M (2007) Brain pyruvate recycling and peripheral metabolism: an NMR analysis ex vivo of acetate and glucose metabolism in the rat. J Neurochem 101:1428–1440

Serres S, Raffard G, Franconi J-M, Merle M (2008) Close coupling between astrocytic and neuronal metabolism to fulfill anaplerotic and energy needs in the rat brain. J Cereb Blood Flow Metab 28:712–724

Shank RP, Bennett GS, Freytag SO, Campbell GL (1985) Pyruvate carboxylase: astrocyte-specific enzyme implicated in the replenishment of amino acid neurotransmitter pools. Brain Res 329:364–367

Shen J, Petersen KF, Behar KL, Brown P, Nixon TW, Mason GF, Petroff OA, Shulman GI, Shulman RG, Rothman DL (1999) Determination of the rate of the glutamate/glutamine cycle in the human brain by in vivo 13C NMR. Proc Natl Acad Sci USA 97:8235–8240

Shic F, Ross BD (2003) Automated data processing of 1H-decoupled 13C MR spectra acquired from human brain in vivo. J Magn Reson 162:259–268

Sollenberg J, Sörbo B (1970) On the origin of the acetyl moiety of acetylcholine in brain studied with differential labelling technique using ^{3}H-^{14}C-mixed labelled glucose and acetate. J Neurochem 17:201–207

Sonnewald U, Westergaard N, Schousboe A, Svendsen JS, Unsgard G, Petersen SB (1993) Direct demonstration by [^{13}C]NMR spectroscopy that glutamine from astrocytes is a precursor for GABA synthesis in neurons. Neurochem Int 22:19–29

Sonnewald U, Müller TB, Westergaard N, Unsgard G, Petersen SB, Schousboe A (1994) NMR spectroscopic study of cell cultures of astrocytes and neurons exposed to hypoxia: compartmentation of astrocyte metabolism. Neurochem Int 24:473–483

Sonnewald U, Akiho H, Koshiya K, Iwai A (1998) Effect of orotic acid on the metabolism of cortical astrocytes during hypoxia and reoxygenation: an NMR spectroscopy study. J Neurosci Res 51:103–108

Spence AM, Mankoff DA, Muzi M (2003) Positron emission tomography imaging of brain tumors. Neuroimag Clin N Am 13:717–739

Sze PY (1979) L-Glutamate decarboxylase. Adv Exp Med Biol 123:59–78

Tsuchida T, Takeuchi H, Okazawa H, Tsujikawa T, Fujibayashi Y (2008) Grading of brain glioma with 1-^{11}C-acetate PET: comparison with ^{18}F-FDG PET. Nucl Med Biol 35:171–176

Tucek S, Cheng S-C (1974) Provenance of the acetyl group of acetylcholine and compartmentation of acetyl-CoA and Krebs cycle intermediates in the brain in vivo. J Neurochem 22:893–914

Tyce GM, Ogg J, Owen CA Jr (1981) Metabolism of acetate to amino acids in brains of rats after complete hepatectomy. J Neurochem 36:640–650

Tyson RL, Gallagher C, Sutherland GR (2003) ^{13}C-labelled substrates and the cerebral metabolic compartmentalization of acetate and lactate. Brain Res 992:43–52

Van den Berg CJ (1970) Compartmentation of glutamate metabolism in the developing brain: experiments with labelled glucose, acetate, phenylalanine, tyrosine and proline. J Neurochem 17:973–983

Van den Berg CJ, Garfinkel D (1971) A simulation study of brain compartments. Biochem J 123:211–218

Van den Berg CJ, Ronda G (1972) The incorporation of double-labelled acetate into glutamate and related amino acids from adult mouse brain: compartmentation of amino acid metabolism in brain. J Neurochem 27:1443–1448

Van den Berg CJ, Mela P, Waelsch H (1966) On the contribution of the tricarboxylic acid cycle to the synthesis of glutamate, glutamine and aspartate in brain. Biochem Biophys Res Commun 23:479–484

Van den Berg CJ, Krzalic LJ, Mela P, Waelsch H (1969) Compartmentation of glutamate metabolism in brain: evidence for the existence of two different tricarboxylic acid cycles in brain. Biochem J 113:281–290

Waniewski RA, Martin DL (1998) Preferential utilisation of acetate by astrocytes is attributable to transprt. J Neurosci 18:5225–5233

Wyss MT, Weber B, Treyer V, Heer S, Pellerin L, Magistretti PJ, Buck A (2009) Stimulation-induced increase of astrocytic oxidative metabolism in rats and humans investigated with 1-^{11}C-acetate. J Cereb Blood Flow Metab 29:44–56

Yamamoto Y, Nishiyama Y, Kimura N, Kameyama R, Kawi N, Hatakeyama T, Kaji M, Ohkawa M (2008) ^{11}C-Acetate PET in the evaluation of brain glioma: comparison with ^{11}C-methionine and ^{18}F-FDG-PET. Mol Imag Biol 10:281–287

Yang J, Li SS, Bacher J, Shen J (2007) Quantification of cortical GABA-glutamine cycling rate using in vivo magnetic resonance signal of [2-^{13}C] GABA derived from glia-specific substrate [2-^{13}C] acetate. Neurochem Int 50:371–378

Yoshimoto M, Waki A, Obata A, Furukawa T, Yonekura Y, Fujibayashi Y (2004) Radiolabeled choline as a proliferation marker: comparison with radiolabeled acetate. Nucl Med Biol 31:859–865

Yu ACH, Drejer J, Hertz L, Schousboe A (1983) Pyruvate carboxylase activity in primary cultures of astrocytes and neurons. J Neurochem 41:1484–1487

Yudkoff M, Daikhin Y, Nissim I, Horyn O, Lazarow A, Luhovyy B, Wehrli S, Nissim I (2005) Response of brain amino acid metabolism to ketosis. Neurochem Int 47:119–128

Chapter 27
Brain Fatty Acid Uptake

James A. Hamilton, Kellen Brunaldi, Richard P. Bazinet, and Paul A. Watkins

Abstract The brain is a highly lipid-rich organ. Fatty acids are the primary constituents of brain lipids, and are provided either by exogenous uptake, or by de novo synthesis. Brain lipids are rich in polyunsaturated fatty acids containing double bonds at the ω3 or ω6 positions, which cannot be synthesized and therefore must be obtained from the circulation. The precise mechanism(s) by which the brain takes up these and other exogenous fatty acids have not been completely resolved. Interpretation of experiments designed to measure fatty acid transport can be problematical unless effects on subsequent metabolism of the fatty acid are also considered. Furthermore, brain fatty acid uptake is complicated by the presence of the blood-brain barrier. In this chapter, prevailing thought on how fatty acids enter the brain is presented. In the first section, brain uptake of fatty acids via unfacilitated diffusion is discussed. In the second section, we discuss the potential role(s) of candidate transport proteins and fatty acid binding proteins in the fatty acid uptake process. In the third section, current knowledge of *in vivo* brain fatty acid uptake is summarized.

Keywords Brain lipids • Fatty acid uptake – brain • Polyunsaturated fatty acids • Fatty acid transport proteins • Fatty acid diffusion

---

J.A. Hamilton • K. Brunaldi
Department of Physiology and Biophysics, Boston University School of Medicine, 700 Albany Street, Boston, MA 02118-2526, USA
e-mail: jhamilt@bu.edu; kbrunaldi@gmail.com

R.P. Bazinet, Ph.D.
Department of Nutritional Sciences, University of Toronto, FitzGerald Building 150 College Street, Room 306 Toronto, ON M5S 3E2, Canada
e-mail: richard.bazinet@utoronto.ca

P.A. Watkins, M.D., Ph.D. (✉)
Kennedy Krieger Institute and Department of Neurology, Johns Hopkins University School of Medicine, 707 N. Broadway, Baltimore, MD 21205, USA
e-mail: watkins@kennedykrieger.org

I.-Y. Choi and R. Gruetter (eds.), *Neural Metabolism In Vivo*, Advances in Neurobiology 4, DOI 10.1007/978-1-4614-1788-0_27,

Abbreviations Used

AA	arachidonic acid
AAT	aspartate aminotransferase
ACS	acyl-CoA synthetase
BBB	blood brain barrier
DHA	docosahexaenoic acid
FA	fatty acid
FABP	fatty acid binding protein
FABPpm	plasma membrane fatty acid binding protein
FAT	fatty acid translocase
FATP	fatty acid transport protein
HDL	high-density lipoprotein
LCFA	long-chain fatty acid
LDL	low-density lipoprotein
LDLr	low-density lipoprotein receptor
VLCFA	very long-chain fatty acid
VLDL	very low-density lipoprotein

27.1 Introduction

The brain is an extremely lipid-rich organ, second only to adipose tissue in its lipid content. Nearly 50% of brain dry weight is comprised of lipids (Morell and Toews 1996). Whereas adipose tissue stores triacylglycerol to meet tissue energy needs, brain lipids are primarily found as phospholipids, sphingolipids, and glycolipids. These lipids, which are rich in highly unsaturated fatty acids (FA), perform essential structural and signaling roles to support normal neural function. The FA constituents of brain lipids are derived from both uptake and *de novo* synthesis. The capacity of the brain for FA synthesis is limited. Polyunsaturated FA (PUFA) such as arachidonic (AA; C20:4ω6), docosahexaenoic (DHA; C22:6ω3), and eicosapentaenoic (C20:5ω3) acids and their precursors, the essential FA linoleic (C18:2ω6) and α-linolenic (C18:3ω3) acids, cannot be synthesized and must be taken up from the external milieu. Although the brain does not utilize FA for energy production, recent studies have demonstrated that FA β-oxidation for the generation of metabolic intermediates is robust in brain. Thus, there is a need for a constant supply of FA to the brain.

Transport of FA into the brain has been the subject of considerable controversy. A fundamental question is whether FA traverse the lipid bilayer of the plasma membrane via unfacilitated diffusion, by a protein-mediated process, or by a combination of both processes. The presence of the blood-brain barrier likely imposes additional constraints on FA transport. In the diffusion model, FA dissolve into the outer membrane leaflet, and the protonated FA can flip across to the inner leaflet to provide a net influx. FA can then be extracted into the intracellular environment. A simple equilibrium is formed across the membrane, and the direction of transport is influenced by concentration gradients and the physical properties of membranes. Although binding proteins and metabolic enzymes influence the concentration gradient, thereby affecting FA flux across the membrane, no membrane protein participates directly in the mass transport of FA across the membrane in this model.

The role for plasma membrane proteins in FA transport has been advocated by a number of investigators. Several candidate transporter proteins have been identified, including FA translocase (FAT) (Abumrad et al. 1993), plasma membrane FA binding protein (FABPpm) (Stremmel et al. 1985), and members of the FA transport protein (FATP) family (Schaffer and Lodish 1994; Gimeno 2007). An assumption of this model is that the plasma membrane presents a kinetic or thermodynamic permeability barrier to FA that impedes rapid uptake. Protein mediators would facilitate either bulk or selective transport of FA.

This chapter will be divided into three sections. In the first, the diffusion model of FA uptake will be reviewed. The role of membrane protein mediators, and of intracellular proteins, in FA transport into the brain will be discussed in the second section. Finally, the current state of knowledge regarding *in vivo* uptake of FA by the brain, independent of mechanism, will be presented.

27.2 Diffusion-Mediated Uptake of FA by the Brain

27.2.1 FA: Biological and Physical Properties

Unesterified FA are key nutrients in human physiology that enter cells mainly from the circulation (Neely and Morgan 1974). In heart muscle they serve as the primary cellular fuel, and in adipocytes they are stored as triglyceride for future release as supply fuel for other cell types. The exchange of FA between cells and plasma must occur rapidly to support life. FA in the circulation, which is bound mostly to albumin, has a turnover time of ~1 min, compared to about 1 month for the albumin molecule (Peter 1996). In a single pass through the brain, which lasts only seconds, 5% of albumin-bound FA is taken up into cells (Rapoport 2001), despite the fact that the FA must pass across several cellular membrane bilayers because of the tight junctions in the blood brain barrier (Drewes 2001).

The biological activities of FA are often dependent on the FA structure, which necessitates an understanding of both their chemical properties and their biophysical properties, such as interactions with binding proteins and membranes. PUFA are not only important structural components of membranes but can be oxidized to form eicosanoids such as prostaglandins, 13-HETE, and 15-HODE, which have very potent signaling and gene regulation activities (Nagy et al. 1998). Saturated very long chain (>18 carbons) FA (VLCFA), both unesterified and esterified, promote neurological dysfunction in human peroxisomal diseases by rigidifying membranes in a deleterious way (Moser et al. 2004, 2005). In contrast, the very long-chain ω3 PUFA are abundant constituents of cell membranes in the brain and neurological system, where they have beneficial membrane “disordering” effects. A variety of

FA, including saturated FA with 14–18 carbons and unsaturated FA, bind to all peroxisome proliferators-activated receptors (PPARs) *in vitro* (Xu et al. 1999), implying that may exert biological effects by this interaction.

FA properties are complex and very different from those of the other major nutrients, such as sugars and amino acids. The predominant FA in human diets, long chain FA (LCFA) with 14–18 carbons, have a low solubility (in the low μM range) in water or buffer at neutral pH. A prevalent misconception (Richieri et al. 1992) is that LCFA form micelles at physiological pH; instead they form acid-soaps with a bilayer structure similar to phospholipids (Cistola et al. 1988), aggregates that would be disadvantageous for FA transport and in fact rarely occur in human physiology. In the circulation and in cells, LCFA are present at concentrations of nM-μM as the unbound species, and up to mM bound to albumin, intracellular FABP (Richieri et al. 1994), and cell membranes (Hamilton and Kamp 1999). LCFA have very low solubility in water compared to glucose, but they are able to desorb rapidly from proteins and membranes to reach intracellular sites of their metabolism. Biophysical studies of the saturated VLCFA have illuminated their unique properties, such as their much slower rates of desorption compared to LCFA and their potential for rigidifying membranes (Zhang et al. 1996).

In the environment of cell membranes, FA can bind to the lipid bilayer, and possibly to proteins or other membrane components. FA can diffuse laterally through the lipid matrix of the membrane into regions of differing physical organization, such as rafts or caveolae. They may also enter these regions directly from the aqueous compartment. Numerous proteins have been proposed as catalysts for movement of the FA across the lipid bilayer in mammalian membranes, and the major postulated membrane-bound transporters for FA are discussed below. Some investigators support protein-mediated membrane transport on the basis of conventional methods with radio-labeled FA (Abumrad et al. 1998; Stump et al. 2001), but these methods generally do not distinguish between transport and metabolism and are not able to define a mechanism (Cooper et al. 1989; Hamilton 1998; Hamilton and Kamp 1999).

27.2.2 FA Transport in Membranes

If proteins are required for translocation of FA within membranes, the membrane must be impermeable to FA, or FA movement across the lipid bilayer would be too slow to support metabolism. To address the complicated issue of transport processes in a membrane, Hamilton and colleagues have used novel biophysical approaches centered on real-time fluorescence measurements of FA transport in protein-free phospholipid bilayers. The key steps of FA diffusion through the lipid bilayer (Fig. 27.1) begin with FA movement from the water phase into the lipid membrane (adsorption). The FA inserts with its tail aligned with the phospholipid fatty acyl chains and the head group at the aqueous interface, a positioning that results in equal populations of unionized and ionized FA at pH 7.4. This ionization property could allow FA as well as many other small amphipathic molecules to cross the

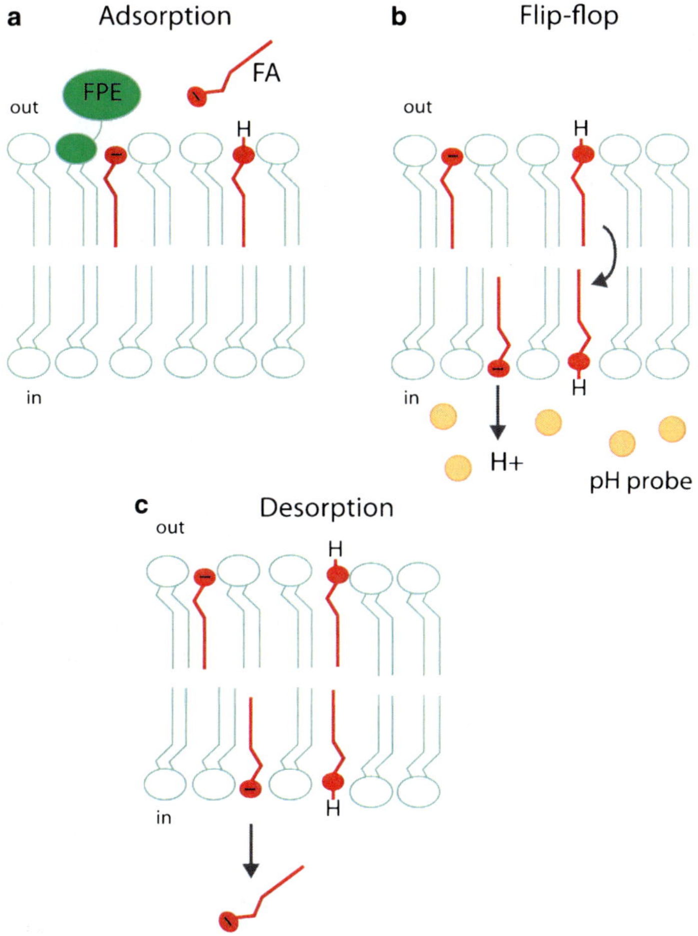

Fig. 27.1 Diffusion model of FA flip-flop through membranes. (**a**) Adsorption: FA molecules bind at the outer surface of the phospholipid bilayer membrane with the acyl chain embedded in the hydrocarbon core and the carboxyl group at the aqueous interface. The charge-sensitive probe directly measures adsorption (FPE – fluorescein phosphatidylethanolamine; (Brunaldi et al. 2007; Simard et al. 2008a)). (**b**) Flip-flop: The un-ionized FA flip to the inner leaflet, releasing H^+ to the intravesicular/intracellular compartment, which is measured as a drop in pH with a water-soluble pH-sensitive fluorescent probe (Kamp and Hamilton 1992). (**c**) Desorption: FA leave the lipid bilayer and partition into the intravesicular compartment at a slower rate than the combined steps of binding and flip-flop (Zhang et al. 1996; Simard et al. 2008b)

bilayer by diffusion (flip-flop) of the uncharged form with its relatively low-energy barrier for movement of the polar group through the hydrocarbon core.

Early FA diffusion experiments used a pH-sensitive fluorescent probe, pyranin, to detect intravesicular pH decreases after the inward flux of unionized FA (Kamp and Hamilton 1992). Similarly, the outward movement of FA was detected by a pH increase inside the vesicle. Hamilton and coworkers' research subsequently added other fluorescent probes, such as fluorescein phosphatidylethanolamine (FPE), that measures the adsorption of the FA to the bilayer (Brunaldi et al. 2007; Simard et al. 2008a), to distinguish this process from the transmembrane step of transport, as illustrated in Fig. 27.1. The very fast flip-flop of FA in these model membranes (half-time of ms) allowed them to conclude that there is no need for a protein to transport the ionized form of the FA to achieve net flux in either direction, or to accelerate an intrinsically slow rate of diffusion. After FA arrive at the inner leaflet, a very small fraction will desorb into the inner aqueous volume because of the well-recognized high partitioning in the lipid membrane. In cells, the amount of FA that desorbs will be enhanced by binding to intracellular membranes and FA binding proteins (FABP) and by utilization (metabolism).

There are now many studies demonstrating rapid diffusion of FA (half-time of sec or less) across the lipid membrane, both in model membranes and in biological membranes. Rapid binding of oleic acid (C18:1ω9), the most abundant FA in human plasma, and its diffusion across the plasma membrane have been documented in adipocytes (Kamp et al. 2003), HepG2 cells (Guo et al. 2006) and HEK cells (Meshulam et al. 2006; Simard et al. 2010). Although some FA are then activated on the cytosolic side of the plasma membrane by acyl CoA synthetases (e.g. FATPs), most exogenous FA entering cells are metabolized at intracellular sites after their desorption from the plasma membrane. Kinetic and thermodynamic data for a wide range of FA (14–26 carbons) (Zhang et al. 1996; Simard et al. 2008b) suggest that desorption is slower than flip-flop in model membranes, and this step could be the rate-limiting step of diffusion of FA into cells. The kinetics of desorption of typical dietary FA, including ω3 and ω6 PUFA, are fast (sec or less), whereas the desorption of VLCFA occurs in minutes to hours.

27.2.3 FA Transport Across the Blood Brain Barrier (BBB)

Nutrients supplied to all organs must pass through microvascular endothelium, but the brain has a special endothelial barrier, the blood brain barrier (BBB) with tight junctions between endothelial cells (Fig. 27.2) that prevent leakage of nutrients between these cells (Drewes 2001). The endothelial cell is thereby demarcated into luminal and transluminal surfaces, with membrane transporters at each membrane. As illustrated in Fig. 27.2, for several ions, these transporters might be asymmetrically distributed to fine tune control of ion transport in and out of the brain. Indeed, if the endothelial cell membranes were impermeable to FA, their transport could be quite complex. For example, if transporters on the luminal and transluminal sides were unidirectional for inward flux, additional transporters would be needed on each surface for the efflux of FA that are not utilized or that are generated in the adjacent brain cells.

Hamilton and colleagues have proposed a general mechanism for FA transport into brain cells which takes into account our current knowledge of the roles of the lipid bilayer and proteins (Hamilton and Brunaldi 2007) (Fig. 27.3). FA in its unesterifed

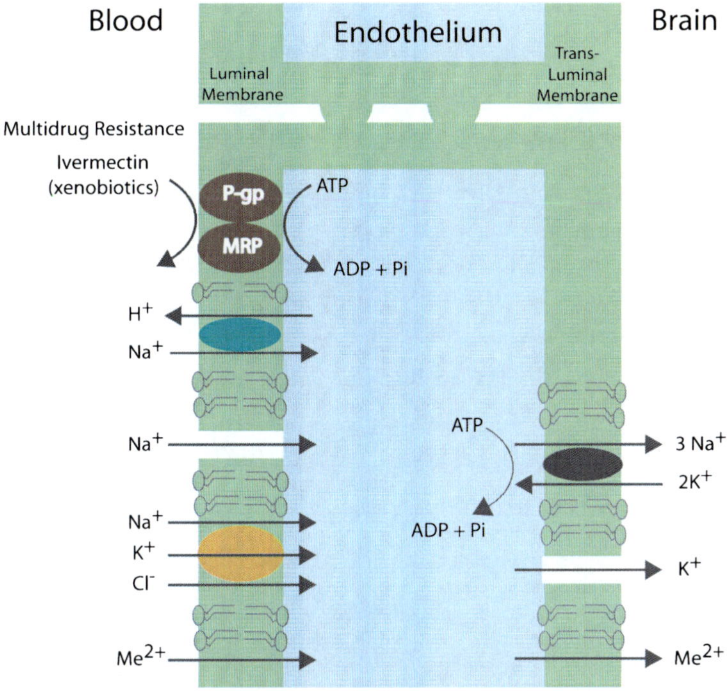

Fig. 27.2 Transport across the blood brain barrier (BBB). Tight junctions between each endothelial cell constitute the major obstacle to diffusion of ions, glucose, and amino acids. Membrane transporters asymmetrically distributed between the luminal and trans-luminal membranes provide a route for ions (Adapted from Drewes, (2001))

form equilibrates into the plasma from its binding sites on albumin. The FA binds to the luminal endothelial cell membrane and diffuses across the lipid bilayer in a non-energy dependent manner without a transporter. After desorption from the inner leaflet of the membrane into the cytosol, FA could diffuse through the cytosol with or without the aid of intracellular FABP. When the FA reaches the transluminal membrane, the same processes as in the luminal membrane are repeated. After FA diffuse across the membrane of the brain cell to reach the inner leaflet, some are thioesterified to coenzyme A by FATP (see below). The acyl CoA formed is highly charged, cannot freely diffuse across a lipid bilayer, and is trapped in the cell. Selectivity of FA begins at this step of uptake since all types of FA can diffuse across the lipid membrane. However, the different forms of FATP (see below) show preference for certain FA. Selectivity of FA then could continue according to the specificities of enzymes in the metabolic pathways.

The model proposed by Hamilton and colleagues is consistent with the very rapid appearance of FA in the brain. FA transport can occur in seconds or less in each step in the scheme, making the overall process very fast without postulating enhancement by membrane transporters. While not excluding any FA from entry

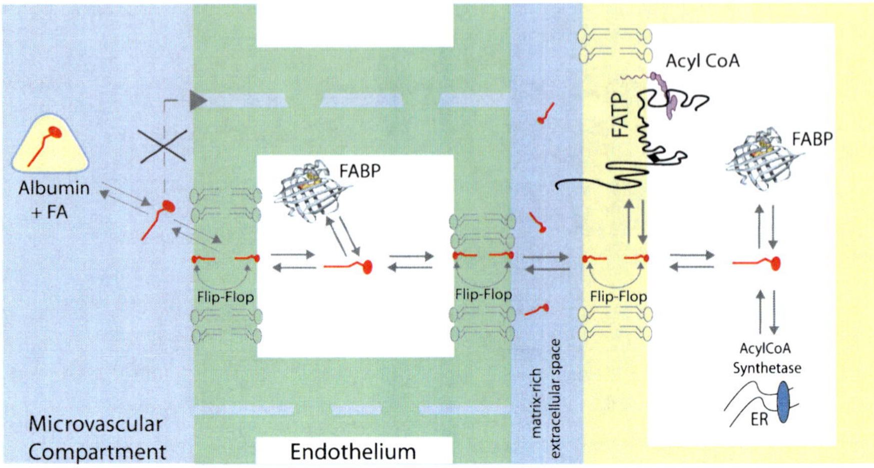

Fig. 27.3 Model for the transport of FA into brain cells. FA crossing the BBB is derived mainly from FA/albumin complexes. FA in the blood in equilibrium with the pool of FA bound to albumin first bind to the luminal surface of the endothelial cell. They diffuse across the luminal and trans-luminal leaflets of the endothelial cells and the plasma membrane of neural cells by reversible flip-flop. Within the cytoplasm of the endothelial cell and the neural cell, FA can diffuse through the aqueous medium unbound or bound to FABP. Acyl CoA synthetases trap FA by forming the acyl CoA, both at the inner leaflet with the membrane bound FATP and at intracellular sites in the endoplasmic reticulum. Unlike the unesterified FA, acyl-CoA cannot diffuse out of the cell. In this model, selection of FA is controlled largely by enzymes in the pathways of intracellular metabolism, beginning with ACSs, which are required for activation of FA for incorporation into lipids or oxidation (Adapted from Hamilton and Brunaldi (2007))

into the brain, the model explains the specificity of uptake. Small differences in the affinity of different long chain dietary FA and their rate of desorption for the membrane into the aqueous phase could affect the overall kinetics of uptake of FA into the brain. However, the model predicts that selectivity and retention of specific FA in the brain is determined largely by metabolism.

27.3 Proteins and the Uptake of Fatty Acids by the Brain

A role for membrane proteins in the uptake of FA by tissues, including the brain, has been the subject of investigation by many laboratories. It has been argued that the rate of diffusion across the plasma membrane is too slow to support the needs of tissues such as cardiac and skeletal muscle that rely on FA oxidation for ATP to support contraction. Several studies have indicated that FA transport is a saturable process (Bonen et al. 1998; Luiken et al. 1999a). These observations have led to the conclusion that FA transport must therefore involve protein carriers. However, dissecting pure transport from transport plus subsequent metabolism has often impeded the interpretation of experimental data that support, or oppose, a role for proteins in FA transport.

Proteins that have been investigated as FA transporters include FAT, FABPpm, and members of the FATP family. Other proteins that may also influence FA uptake include intracellular FABPs and α-synuclein. By binding and sequestering FA, the latter two proteins also affect the equilibrium distribution of FA across the plasma membrane.

It should be explicitly stated that very few studies of putative FA transport proteins have been conducted in brain, or brain-derived cell lines. Most studies described below have been done in skeletal and cardiac muscle, along with a few investigations of liver or adipose tissue. Nonetheless, several of these proteins are also expressed in brain and could be relevant for FA uptake in this organ.

27.3.1 Fatty Acid Translocase (FAT)

FAT was originally identified as CD36 (glycoprotein IV), a highly glycosylated integral membrane protein. CD36 is a class B scavenger receptor implicated in multiple biological processes including atherosclerosis, angiogenesis, and inflammation, as well as lipid metabolism (Febbraio et al. 2001). CD36 has two membrane-spanning domains that orient the protein with both the N- and C-termini facing the cytoplasm; a large, single extracellular loop constitutes the receptor domain (Gruarin et al. 2000). Numerous ligands bind to CD36, including thrombospondin-1, oxidized LDL, collagen, and free FA (Febbraio et al. 2001). This receptor is expressed in many tissues. In the brain, CD36 is expressed in microvascular endothelial cells, microglia, and perhaps neurons (Abumrad et al. 2005).

Abumrad and colleagues were the first to associate FAT/CD36 with LCFA transport (Abumrad et al. 1993; Harmon and Abumrad 1993). They showed that sulfo-succinimidyl FA inhibited the uptake of oleic acid by rat adipocytes in a dose-dependent manner by binding to an 88 kDa membrane protein (Harmon and Abumrad 1993). They cloned cDNA encoding this protein, which they named FAT, and found that its predicted amino acid sequence was 85% identical to that of human CD36 (Abumrad et al. 1993). Since that time, many studies have been published indicating an association between FAT/CD36 and LCFA uptake. However, the mechanism(s) by which FAT/CD36 promotes cellular FA uptake has not been established unequivocally.

Consistent with its receptor function, plasma membrane FAT/CD36 localizes to caveolae, membrane microdomains with well-characterized roles in lipid and lipoprotein metabolism (Lisanti et al. 1994). Significant amounts of this protein are also found in intracellular membrane compartments of heart (Luiken et al. 2002b) and skeletal muscle (Bonen et al. 2000). Intracellular FAT/CD36 is found associated with recycling endosomes (Luiken et al. 2004) and mitochondria (Campbell et al. 2004). In skeletal muscle, FAT/CD36 was observed to translocate to the plasma in response to insulin (Luiken et al. 2002a). Translocation was mediated by activation of phosphoinositide-3 kinase (Luiken et al. 2002b). In contrast, FAT/CD36 was found in an insulin-insensitive vesicle population in heart muscle (Muller et al. 2002). Muscular contraction also stimulated translocation of FAT/CD36 to the plasma membrane (Bonen et al. 2000) in a process mediated by AMP-activated protein kinase (Habets et al. 2007). Inhibition of mitochondrial FAT/CD36 with sulfo-*N*-succimidyloleate decreased mitochondrial β-oxidation of the LCFA, palmitate (C16:0), but not that of the medium-chain FA, octanoate (C8:0), suggesting a role for this protein in mitochondrial FA transport (Bezaire et al. 2006). Although the inhibitor did not affect carnitine-palmitoyltransferase-1 activity, oxidation of palmitoylcarnitine was significantly decreased. Thus, FAT/CD36 in mitochondria may function in concert with carnitine-palmitoyltransferase-1 to regulate entry of LCFA into the organelle, perhaps by facilitating the transfer of palmitoylcarnitine from carnitine-palmitoyltransferase-1 to carnitine:acylcarnitine translocase (Bezaire et al. 2006).

Some of the most relevant insights into the physiologic functions of FAT/CD36 have come from studies with genetic mouse models. Abumrad and coworkers produced transgenic mice with muscle-specific overexpression of FAT/CD36 (Ibrahimi et al. 1999). Increased FA oxidation was measured in soleus muscle during contraction stimulation. Very low-density lipoprotein (VLDL) triacylglycerol and blood free FA were lower in transgenic mice, and the FAT/CD36 overexpressing mice were more lean than wild-type mice. Cholesterol levels in blood were slightly lower, but blood glucose levels were significantly increased. Conversely, FAT/CD36-null mice had a significant increase in fasting levels of cholesterol, nonesterified free FA, and triacylglycerol, and lower fasting serum glucose levels than wild-type mice (Febbraio et al. 1999). Adipocytes from knockout mice took up less radiolabeled oleic acid than did control cells, and peritoneal macrophages showed significantly decreased binding and uptake of oxidized low density lipoprotein (LDL).

Drover and colleagues recently investigated intestinal absorption of FA in FAT/CD36-null mice (Drover et al. 2008). They found that absorption of LCFA was normal in knockout mice, but absorption of saturated VLCFA was severely impaired. Based on their results, a mechanism was proposed in which FAT/CD36 binds mixed micelles of bile salts and phospholipids. This complex spans across the unstirred water layer present at the luminal surface of the enterocyte plasma membrane. VLCFA, which cannot penetrate the unstirred water layer, adsorb to the hydrophobic mixed micelles, allowing them to migrate to the plasma membrane, which they enter via diffusion. This mechanism seems to be a plausible explanation for the general ability of FAT/CD36 to facilitate FA uptake across the plasma membrane.

27.3.2 Plasma Membrane Fatty Acid Binding Protein (FABPpm)

FABPpm is a ubiquitously expressed peripheral membrane protein; highest expression was found in heart, skeletal muscle, brain, liver, and kidney (Luiken et al. 1999b). Stremmel and coworkers initially isolated FABPpm from rat hepatocytes and characterized it as a 43 kDa protein with a high affinity for oleic acid (Stremmel et al. 1985). The protein is located in several subcellular compartments (Stump et al. 1993). In the plasma membrane, it is found in the outer leaflet. However, FABPpm is also found in the inner mitochondrial membrane, and is identical to the enzyme aspartate aminotransferase (AAT).

FAPBpm/AAT was perhaps the first protein proposed as a potential FA transporter, primarily because of its plasma membrane location and affinity for FA (Stremmel et al. 1985). Support for this notion came from studies showing that anti-FABPpm antibodies decreased FA uptake by liver and heart (Stremmel et al. 1986). However, although human $CaCo_2$ cells contained a protein immunoreactive with the anti-rat FABPpm antibody, LCFA uptake was not inhibited by pre-treatment of cells with this antibody (Trotter et al. 1996).

Because of the difficulties in separating pure transport of a FA from transport followed by subsequent metabolism, the potential role of FABPpm/AAT in FA transport was also investigated in giant sarcolemmal vesicles (Bonen et al. 1998). Bonen and coworkers have advocated this model system because the vesicles were devoid of FA metabolism (either oxidation or esterification). In addition, vesicles contained cytoplasmic FABP (FABPc), which can act as an intravesicular sink to bind transported FA. Using this system, these investigators found that FA uptake was decreased by both anti-FABPpm antibody and sulfosuccinimidyl oleate, the inhibitor of FAT/CD36. However, the decrease in FA uptake by both antibody and inhibitor were not additive, suggesting that FABPpm/AAT and FAT/CD36 may act in a cooperative fashion.

In addition to mitochondria, there is evidence that FABPpm/AAT localizes to intracellular membranes (Cechetto et al. 2002), and translocates to the plasma membrane in response to acute stimuli (Chabowski et al. 2005, 2007). Like FAT/CD36, FABPpm/AAT translocates in response to muscle contraction and AMP kinase activation; in contrast, insulin had no effect on the subcellular location of FABPpm.

Overexpression of FABPpm/AAT in skeletal muscle resulted in an increase in FA transport; however, the increase in uptake was modest in comparison to the large increase in quantity of FABPpm/AAT protein (Chabowski et al. 2007). Thus a direct role for FABPpm/AAT in FA uptake has been questioned. Neither mice transgenic for FABPpm expression, nor FABPpm knockout mice, have been reported.

27.3.3 The Fatty Acid Transport Protein (FATP) Family

The mammalian FATP family consists of six related proteins (Watkins et al. 2007; Watkins 2008). The original protein designated "FATP", now known as FATP1, was identified in a 1994 screen of a cDNA expression library for proteins that facilitated the accumulation of a fluorescent FA by 3T3-L1 adipocytes (Schaffer and Lodish 1994). In addition to FATP1, the other protein identified in this study as facilitating FA uptake was the long-chain acyl-CoA synthetase (ACS), ACSL1. Based on its amino acid sequence, FATP1 was found to be homologous to ACSL1, suggesting that it too was an ACS (Coe et al. 1999). However, it has been established that FATP1 belongs to a subfamily of ACSs that is distinct from the "long-chain" ACS subfamily. Amino acid sequence homology studies have subsequently identified five other members of the FATP family. FATP1-6 proteins are encoded by genes designated *SLC27A1-6*, respectively.

FATPs differ widely in their tissue expression patterns and subcellular locations, and their ability to facilitate cellular FA uptake. Only three members are found in brain. FATP1 is expressed in muscle, heart, brain, and adipose (Schaffer and Lodish 1994). FATP3 is found at low levels in brain, with higher expression in adrenal gland, testis, and ovary (Pei et al. 2004). FATP4 is perhaps the most abundant FATP present in brain, but it can also be found in adipose, skeletal muscle, heart, and small intestine (Stahl 2004; Gimeno 2007). FATP2, 5, and 6 have not been detected in brain. FATP2 is primarily present in liver and kidney; FATP 5 and 6 are found almost exclusively in liver and heart, respectively (Watkins et al. 2007; Watkins 2008).

Early reports suggested that murine FATP1-5 were all present in the plasma membrane (Schaffer and Lodish 1994; Stahl et al. 1999, 2001). Subsequent studies have clearly demonstrated that FATP2 and 3 are present in intracellular organelles, but not the plasma membrane, suggesting that they are incapable of functioning as direct mediators of FA uptake by cells (Steinberg et al. 1999a; Pei et al. 2004). FATP1 is also present in intracellular perinuclear vesicles, and translocation to the plasma membrane in response to insulin has been reported in skeletal muscle cells and adipocytes, but not in cardiomyocytes (Stahl et al. 2002; Wu et al. 2006b; Gimeno 2007). FATP4 has been found in several subcellular organelles, including mitochondria, peroxisomes, and endoplasmic reticulum (Jia et al. 2007), and some investigators have questioned whether FATP4 is actually present in the plasma membrane (Milger et al. 2006; Lobo et al. 2007). The liver-specific protein FATP5 was detected in the basal plasma membrane of hepatocytes, adjacent to sinusoids (Doege et al. 2006). The heart-specific protein, FATP6, was localized to the sarcolemma of cardiac myocytes in mice and rhesus monkeys (Gimeno et al. 2003b). When murine FATPs were expressed in yeast engineered to lack the ability to take up a fluorescent LCFA, uptake was restored by expression of FATP1, FATP2, and FATP4, but not by FATP3, FATP5, or FATP6 (DiRusso et al. 2005).

The amino acid sequences of FATP1-6 from mice and humans exhibit significant homology (Watkins et al. 2007). Independent of investigations of the properties of FATPs, an ACS that could activate the VLCFA, lignoceric acid (C24:0), to its CoA derivative was purified to homogeneity from rat liver peroxisomes (Uchida et al. 1996). cDNA encoding this enzyme was cloned, and the predicted amino acid sequence of the rat protein was nearly identical to that of murine FATP2, suggesting that they were the same protein (Uchiyama et al. 1996). Indeed, all six FATPs contain highly conserved, family-specific domains that are characteristic of all ACSs (Watkins et al. 2007). While several laboratories have investigated the FATP family for their role(s) in transport, others have studied their function(s) as ACSs. Although there was early disagreement as to whether FATPs were enzymes, there is now general consensus that all six proteins have ACS activity (Watkins et al. 2007; Watkins 2008). In general, all members of the FATP family are capable of activating long- to very long-chain FA (Uchida et al. 1996; Steinberg et al. 1999b; Herrmann et al. 2001; Hall et al. 2003, 2005; Pei et al. 2004; DiRusso et al. 2005). These enzymes are considered to be *very long-chain* ACSs because they are *capable* of activating

FA containing 24 or more carbons; however, activity with long-chain substrates is typically higher than with VLCFA. FATP5 also converts primary and secondary bile acids to their CoA thioesters (Steinberg et al. 2000; Mihalik et al. 2002); this activation step is necessary for the conjugation of bile acids to glycine or taurine.

The properties of individual FATPs have been investigated. Studies of cells that either overexpress FATP1, or are depleted of FATP1 by RNA interference, indicate that exogenous FA activated by this enzyme are channeled primarily into the triacylglycerol biosynthetic pathway (Hatch et al. 2002; Lobo et al. 2007). The relevance of this observation to brain lipid metabolism is not known. Mice transgenically overexpressing FATP1, as well as FATP1 knockout mice, have been produced, but no phenotypic abnormalities related to the brain were mentioned, and no studies addressing the role of this protein in brain were carried out. Heart-specific FATP1 overexpression in mice causes a lipotoxic cardiomyopathy (Chiu et al. 2005). FATP1-null mice fed a high-fat diet were protected from insulin resistance in skeletal muscle (Kim et al. 2004), but exhibited significant cold intolerance due to lack of this protein in brown adipose tissue (Wu et al. 2006a). Insulin did not induce LCFA uptake in adipocytes or skeletal muscle isolated from FATP1-null mice (Wu et al. 2006a). Knockout mice were resistant to diet-induced obesity, with decreased accumulation of triacylglycerol in muscle and adipose; in contrast, these mice accumulated lipids in the liver (Wu et al. 2006a).

FATP2 is identical to the very long-chain ACS described by Hashimoto and colleagues (Uchida et al. 1996; Uchiyama et al. 1996). It is expressed primarily in liver and kidney, but is not known to be present in brain. A knockout mouse was produced that was phenotypically normal (Heinzer et al. 2003). Knockout mouse liver and kidney peroxisomes had reduced capacity to activate VLCFA to their CoA derivatives and to degrade these FA via β-oxidation. Knockout did not lead to accumulation of VLCFA in liver, kidney, or brain, and rates of brain VLCFA β-oxidation were unaffected.

FATP3 is mainly found in adrenal gland, testis, and ovary (Pei et al. 2004). FATP3 mRNA expression is also high in embryonic mouse brain, but rapidly decreases to barely detectable levels by postnatal day 15. The role of this enzyme, which is found in subcellular vesicles, in FA metabolism is currently unknown. However, knockdown of endogenous FATP3 by RNA interference in mammalian cells did not decrease FA uptake (Pei et al. 2004). Neither transgenic nor knockout mice have been produced.

Mice with a spontaneous mutation in FATP4, as well as mice with targeted disruption of the FATP4 gene, have been reported (Gimeno et al. 2003a; Herrmann et al. 2003; Moulson et al. 2003). In two reports, FATP4-null mice were born with a rapidly fatal restrictive dermopathy, while the third group reported embryonic lethality. Activation of VLCFA, but not LCFA, to their CoA derivatives was severely impaired in brain, intestine, and skin of FATP4-null mice (Hall et al. 2005), and skin fibroblasts from knockout mice had reduced incorporation of VLCFA into both neutral lipids and phospholipids (Jia et al. 2007). Despite these biochemical deficiencies, brains of FATP4 knockout mice were examined and found to be histologically normal, and the lipid composition of knockout brain was indistinguishable from normal mouse brain (Herrmann et al. 2003). While the role of FATP4 in brain FA uptake has not been studied directly, uptake of a fluorescent LCFA by skin fibroblasts from FATP4-null mice was severely impaired (Jia et al. 2007).

FATP5 overexpression studies revealed that this enzyme functions primarily in the hepatic synthesis and recycling of bile acids (Mihalik et al. 2002). In addition, hepatocytes from FATP5 knockout mice had reduced uptake of LCFA (Doege et al. 2006). No effects of depletion of this liver-specific enzyme on brain FA metabolism have been reported. FATP6, like FATP2 and FATP5, is not expressed in brain. FATP6-overexpressing HEK293 cells has a higher rate of accumulation of fluorescent and radiolabeled LCFA (Gimeno et al. 2003b). Neither FATP6 transgenic nor FATP6 knockout mice have been reported.

27.3.4 Intracellular Fatty Acid Binding Proteins (FABP) and Synucleins

Several small, soluble, cytoplasmic proteins are known to bind FA, including FABPs and synucleins. By binding and sequestering FA that enter cells by any mechanism, these proteins thereby affect the equilibrium distribution of FA across cell membranes. The primary FABPs expressed in brain are FABP3 (heart FABP), FABP5 (keratinocyte or epidermal FABP), and FABP7 (brain FABP) (Storch and Corsico 2008). Structurally, FABPs have a β-barrel structure and typically bind a single FA with the carboxylic acid function directed inward. FABP5 binds primarily saturated FA, while FABP3 and FABP7 show preference for ω6 and ω3 FA, respectively (Murphy et al. 2005). Transfer of FA bound to FABP3 and FABP5 to intracellular membranes is thought to be via a collision-mediated process rather than by dissociation and diffusion; the mechanism for FABP7 is not known. Recycling of "unloaded" FABP thus promotes additional cellular FA uptake.

FABP3 is expressed in neurons. Mice with targeted disruption of FABP3, in which no compensatory upregulation of other FABP isoforms was observed, had reduced uptake of the ω6 FA, AA, but the uptake of the saturated 16-carbon FA, palmitate, was unaffected (Murphy et al. 2005). Brains from these mice had a reduced phospholipid mass, as well as reduced phospholipid AA content. These observations suggest that FABP3 may be specifically involved in ω6 FA transport and metabolism in the brain.

FABP5 is also expressed in neurons. It has been suggested that FABP5 participates in the transport of LCFA during neuronal differentiation (Storch and Corsico 2008). High levels of this protein are associated with neurogenesis in the prenatal and early postnatal periods, and cell culture studies show a role for this protein in neurite outgrowth.

FABP7 is present in both neurons and glia, with expression higher in fetal than adult brain; this protein is thought to be important for differentiation of both cell types (Storch and Corsico 2008). Although a role for FABP7 in establishing radial glial fibers along which neurons migrate, brains from FABP7-null mice had normal morphology and histology. Like FABP3 knockout mice, there was no compensatory upregulation of other FABPs in mice lacking FABP7. FABP7-null mice displayed behavioral changes such as increased memory of fear and increased levels of anxiety, and fetal mice had decreased brain content of the ω3 FA, DHA. Because of the normally high levels of DHA in brain and its importance in brain development, it has been speculated that FABP7 may play a specific role in transport of this FA into brain.

α-Synuclein (SNCA) is a small cytoplasmic protein that is highly abundant in neurons and astrocytes (Golovko et al. 2009). Although it bears some structural relationship to FABPs, SNCA's affinity for monomeric FA is significantly lower than for FABPs, casting doubt on its ability to function as a "classical" FABP. Nonetheless, brains of SNCA knockout mice have reduced uptake of palmitate and AA. Consistent with these observations, incorporation of radiolabeled palmitate and AA into brain phospholipids is reduced in knockout mice. Although lack of SNCA does not appear to affect brain uptake of DHA, incorporation of this FA into phospholipids is increased in null mice.

27.4 *In Vivo* Uptake of FA by the Brain

27.4.1 The Contribution of Plasma Fatty Acid Pools to Brain Uptake In Vivo

Whereas the brain can synthesize saturated and monounsaturated FA and cholesterol *de novo* (Marbois 1992; Quan 2003; Dietschy and Turley 2004), rates of brain synthesis of PUFA such as AA and DHA, from linoleic acid and α-linolenic acid, respectively, are <1% of their uptake rates from the unesterified plasma pool (Demar et al. 2005, 2006). During chronic ω3 PUFA deprivation, liver expression of desaturase/elongase enzymes and the synthesis rate of DHA increase (Igarashi et al. 2007b). However, expression of desaturase/elongase enzymes and the synthesis rate of DHA within the brain do not increase and remain low (Igarashi et al. 2007a). Therefore it is important that the brain maintains a constant supply of PUFA from the plasma.

Several plasma pools, including LDL (Edmond 2001), VLDL and high-density lipoprotein (Polozova et al. 2006), along with lysophosphatidlycholine (Thies et al. 1992) and unesterified FA (Robinson et al. 1992; Rapoport et al. 2001; Hamilton and Brunaldi 2007) have been proposed to contribute to brain FA uptake *in vivo*. Although it does appear that these pools can enter the brain, quantitative rates are needed for direct comparison. In a first approach to test if LDL is a major contributor to brain PUFA, Bazinet and colleagues measured PUFA concentrations within the brain of LDL receptor (LDLr) −/− mice and their wildtype controls (Chen et al. 2008a). Because the LDLr is selectively expressed in anterior part of the brain and not the cortex, these two regions were analyzed separately. Upon not detecting a

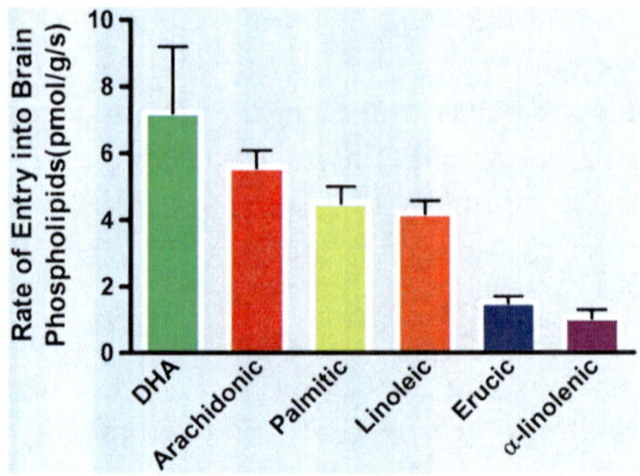

Fig. 27.4 *In vivo* incorporation rates of various FA, from the plasma unesterified FA pool, into rat brain phospholipids (From Chen et al. (2008b). With permission)

difference in any measured PUFA concentration, between LDLr−/− and wildtype mice, it was concluded that the LDLr is not necessary for maintaining brain PUFA levels and other mechanisms must exist that are sufficient to do so. The possibility that other lipoprotein receptors are necessary for maintaining brain PUFA concentrations is still under investigation. One limitation of this approach is that when a specific receptor pathway is knocked out, other pathways may compensate, thus maintaining brain PUFA concentrations.

In order to quantify the plasma uptake of unesterified FA, Rapoport and colleagues developed an intravenous infusion model in the rat. In this model, high-specific activity radiolabeled FA in a saline containing albumin solution are infused into the femoral vein at a variable rate of 0.223 $(1-e^{-19.2t})$ ml/min reaching steady state within 1 min and maintained for up to 5 min. In these studies, typically 170 μCi/kg of tracer is infused leading to a total infusion between ~200 pmol and 1 μmol of FA, depending on the specific activity of the FA, over the 5 min infusion. Steady state is confirmed by collection of arterial blood samples during the 5 min infusion. For studies in which individual lipid pools can be chemically separated, the experiment is terminated at 5 min either with high-energy head-focused fixation to rapidly denature brain proteins. For imaging studies, the radiolabeled FA is infused and the experiment is stopped 15 min after the end of the infusion, after plasma radioactivity levels return to baseline so as to minimize the noise of the tracer within brain capillaries. Importantly, the brain uptake from these infusion is more rapid than liver output of phospholipid or triglycerides with the radiolabeled FA; thus, it is possible to attribute FA esterification in brain phospholipids to the unesterified FA itself.

Using this methodology in rats and mice, the rates of plasma unesterified DHA, AA, palmitic acid, linoleic acid, erucic acid (C22:1ω9) and α-linolenic acid incorporation into brain phospholipids have been reported to range from ~2 to 10 pmol/g brain/s (Fig. 27.4) (Robinson et al. 1992; Chang et al. 1996, 1999; Contreras et al. 2000, 2001; Rapoport et al. 2001; Giovacchini et al. 2004; Bazinet et al. 2005, 2006a; Demar et al. 2005, 2006; Golovko and Murphy 2006; Igarashi et al. 2007a). These rates vary with experimental conditions, for instance the net rate of DHA entry into the brain is decreased 40-fold upon chronic ω3 PUFA deprivation in rats

(Contreras et al. 2000). Also, FA with lower rates of esterification into brain phospholipids (palmitic, linoleic, erucic and α-linolenic acids), upon entry into the brain are heavily β-oxidized (Robinson et al. 1992; DeMar et al. 2004, 2005; Golovko and Murphy 2006). Thus their net incorporation rate into the brain is more similar than their esterification rate into brain phospholipids suggests. Furthermore, adapting the Takasato method in the mouse (Takasato et al. 1984), Calon and colleagues have shown that the diffusion of DHA and EPA was not saturable up to 100 μM (Ouellet et al. 2009), approximately 10-fold above what is considered physiologically relevant (Bazinet et al. 2006a, b), consistent with a diffusion model. In this model, net uptake rate of eicosapentaenoic acid was similar to that of DHA; however, eicosapentaenoic acid was rapidly β-oxidized, possibly explaining its relatively low concentration within brain phospholipids (Chen et al. 2009; Ouellet et al. 2009).

In order to quantify the contribution of the unesterified FA pool to brain phospholipids, a half-life based on rate of incorporation from the plasma unesterified pool into brain phospholipids can be calculated according to equation (A) $T_{1/2}=0.693C_{PUFA}/$ rate of incorporation into brain phospholipids, where C_{PUFA} is the concentration of the PUFA in brain phospholipid (DeMar et al. 2004). The replacement of brain PUFA from the brain unesterified pool can then be compared to the direct half-life as measured upon i.c.v. injection of a radiotracer according to equation (B) $T_{1/2}=\log_{10}$ 2/(slope of the regression line) (DeMar et al. 2004). Using this approach, Rapoport and colleagues have shown that the half-life predicted by the rate of DHA entry from the plasma unesterified pool into the brain approximates the half-life measured directly with radiolabeled DHA injected intracerebrovascularly (Rapoport et al. 2007). Furthermore, based on the values provided in Contreras et al. the half-lives of unesterified AA in brain phospholipids as predicted by replacement from the plasma unesterified AA pool in rats consuming a diet deemed adequate in ω3 PUFA or deprived of ω3 PUFA are 42 and 48 days, respectively (Contreras et al. 2001). To test if the half-lives calculated from the unesterified AA pool are major contributors to the net half-life of AA in brain phospholipids, Bazinet and coworkers administered radiolabeled AA intracerebrovascularly to rats undergoing similar dietary manipulation. Similar to the half-lives predicted from uptake from the plasma unesterified pool, it was found that the half-lives in brain phospholipids were 44 ± 5 (ω3 PUFA adequate diet) and 46 ± 5 (ω3 PUFA deprived diet) days, suggesting that the plasma unesterified AA is the major pool contributing to brain phospholipids (Green et al. 2010).

27.4.2 Fatty Acid Uptake in Humans

Soon after methods for synthesis of ^{11}C-labelled PUFA were developed (Channing and Simpson 1993), Rapoport and Colleagues extended their FA infusion model in rodents to non-human primates (Arai et al. 1995; Chang et al. 1997) and eventually to humans with AA (Giovacchini et al. 2002, 2004; Esposito et al. 2007, 2008b) and

DHA (Umhau et al. 2009). Using this novel method, the net whole brain uptake of unesterified AA was estimated to be 17.8 mg/day whereas the uptake of unesterified DHA was 3.8 mg/day. If the assumption that plasma unesterified FA are major pools for brain FA uptake is made, then these rates likely approximate net brain uptake and consumption. However, if other pools contribute to brain PUFA uptake, these values would underestimate the net brain uptake and consumption of these two PUFA. Being that whole brain AA and DHA concentrations are approximately 3.8 and 5.1 g (Rapoport et al. 2007), respectively, the half-life of replacement can be calculated it can be calculated to be ($t_{1/2}=0.693\ c_{brain}$/net rate of entry into the brain) 148 and 930 days for AA and DHA, respectively. In regards to nutrition, this suggests that chronic low intakes may be necessary to deplete brain DHA concentrations, especially since DHA is conserved upon dietary deprivation in rats (DeMar et al. 2004). Conversely a slow rate of DHA uptake by brain could translate to equally long periods of time required to replenish brain DHA levels that might be decreased due to chronic low intakes or disease. A better understanding of the pools from which FA are taken up by the brain could improve strategies designed to manipulate brain PUFA levels and will aid in our understanding of positron emission tomography studies in which the rate of unesterified PUFA entry into the brain is measured (Esposito et al. 2007, 2008a).

27.5 Summary

The brain contains substantial quantities of FA, mainly found as constituents of membrane phospholipids, sphingolipids, and glycolipids. Although this organ does not metabolize bulk quantities of FA for energy production, there is significant daily turnover of brain FA. Thus, the brain is highly dependent upon uptake of FA, particularly ω3 and ω6 PUFA, from the circulation. Despite the importance of FA uptake by the brain, the precise mechanisms involved remain unresolved, and in some cases, the subject of much debate.

It is clear that *in vivo*, plasma unesterified FA are taken up by the brain. The extent to which other pools, such as triglycerides or phospholipids, contribute to *in vivo* uptake remains to be determined. Furthermore, the mechanism by which the brain maintains low levels of certain FA such as eicosapentaenoic acid requires further investigation. The proposed mechanism of FA transport across the BBB, passive diffusion, has sound biophysical foundations, and *in vitro* studies of several non-neural cell types demonstrating passive diffusion are consistent with *in vivo* uptake studies in the brain. However, mechanisms of FA uptake into the brain have not been studied in the same detail as in other cells, and thus the mechanism proposed in Fig. 27.3 should be considered a working hypothesis. Several proteins have been implicated in the cellular uptake of FA, but their importance in the brain has not been directly evaluated. Contributions of the FATP family to brain FA uptake can be hypothesized to reside in their ACS enzyme activity. Similarly, further downstream FA metabolism likely stimulates FA uptake. FABPs and synucleins participate in this process by sequestering FA and delivering them to intracellular sites of metabolism. In contrast, the contributions of FAT/CD36 to brain FA uptake, if any, will be difficult to assess because of the multiple ligands that FAT/CD36 binds and the multiple functions this membrane protein appears to have. Further studies of FABPpm/AAT will also be needed to determine whether this multifunctional protein contributes to brain FA uptake.

References

Abumrad N, Harmon C, Ibrahimi A (1998) Membrane transport of long-chain fatty acids: evidence for a facilitated process. J Lipid Res 39:2309–2318

Abumrad NA, Ajmal M, Pothakos K, Robinson JK (2005) CD36 expression and brain function: does CD36 deficiency impact learning ability? Prostaglandins Other Lipid Mediat 77:77–83

Abumrad NA, el-Maghrabi MR, Amri EZ, Lopez E, Grimaldi PA (1993) Cloning of a rat adipocyte membrane protein implicated in binding or transport of long-chain fatty acids that is induced during preadipocyte differentiation. Homology with human CD36. J Biol Chem 268: 17665–17668

Arai T, Wakabayashi S, Channing MA, Dunn BB, Der MG, Bell JM, Herscovitch P, Eckelman WC, Rapoport SI, Chang MC (1995) Incorporation of [1-carbon-11]palmitate in monkey brain using PET. J Nucl Med 36:2261–2267

Bazinet RP, Rao JS, Chang L, Rapoport SI, Lee HJ (2005) Chronic valproate does not alter the kinetics of docosahexaenoic acid within brain phospholipids of the unanesthetized rat. Psychopharmacology (Berl) 182:180–185

Bazinet RP, Rao JS, Chang L, Rapoport SI, Lee HJ (2006a) Chronic carbamazepine decreases the incorporation rate and turnover of arachidonic acid but not docosahexaenoic acid in brain phospholipids of the unanesthetized rat: relevance to bipolar disorder. Biol Psychiatry 59:401–407

Bazinet RP, Weis MT, Rapoport SI, Rosenberger TA (2006b) Valproic acid selectively inhibits conversion of arachidonic acid to arachidonoyl-CoA by brain microsomal long-chain fatty acyl-CoA synthetases: relevance to bipolar disorder. Psychopharmacology (Berl) 184:122–129

Bezaire V, Bruce CR, Heigenhauser GJ, Tandon NN, Glatz JF, Luiken JJ, Bonen A, Spriet LL (2006) Identification of fatty acid translocase on human skeletal muscle mitochondrial membranes: essential role in fatty acid oxidation. Am J Physiol Endocrinol Metab 290:E509–E515

Bonen A, Luiken JJ, Arumugam Y, Glatz JF, Tandon NN (2000) Acute regulation of fatty acid uptake involves the cellular redistribution of fatty acid translocase. J Biol Chem 275: 14501–14508

Bonen A, Luiken JJ, Liu S, Dyck DJ, Kiens B, Kristiansen S, Turcotte LP, Van Der Vusse GJ, Glatz JF (1998) Palmitate transport and fatty acid transporters in red and white muscles. Am J Physiol 275:E471–E478

Brunaldi K, Simard JR, Kamp F, Rewal C, Asawakarn T, O'Shea P, Hamilton JA (2007) Fluorescence assays for measuring fatty acid binding and transport through membranes. Methods Mol Biol 400:237–255

Campbell SE, Tandon NN, Woldegiorgis G, Luiken JJ, Glatz JF, Bonen A (2004) A novel function for fatty acid translocase (FAT)/CD36: involvement in long chain fatty acid transfer into the mitochondria. J Biol Chem 279:36235–36241

Cechetto JD, Sadacharan SK, Berk PD, Gupta RS (2002) Immunogold localization of mitochondrial aspartate aminotransferase in mitochondria and on the cell surface in normal rat tissues. Histol Histopathol 17:353–364

Chabowski A, Coort SL, Calles-Escandon J, Tandon NN, Glatz JF, Luiken JJ, Bonen A (2005) The subcellular compartmentation of fatty acid transporters is regulated differently by insulin and by AICAR. FEBS Lett 579:2428–2432

Chabowski A, Gorski J, Luiken JJ, Glatz JF, Bonen A (2007) Evidence for concerted action of FAT/CD36 and FABPpm to increase fatty acid transport across the plasma membrane. Prostaglandins Leukot Essent Fatty Acids 77:345–353

Chang MC, Arai T, Freed LM, Wakabayashi S, Channing MA, Dunn BB, Der MG, Bell JM, Sasaki T, Herscovitch P, Eckelman WC, Rapoport SI (1997) Brain incorporation of [1-11C] arachidonate in normocapnic and hypercapnic monkeys, measured with positron emission tomography. Brain Res 755:74–83

Chang MC, Bell JM, Purdon AD, Chikhale EG, Grange E (1999) Dynamics of docosahexaenoic acid metabolism in the central nervous system: lack of effect of chronic lithium treatment. Neurochem Res 24:399–406

Chang MC, Grange E, Rabin O, Bell JM, Allen DD, Rapoport SI (1996) Lithium decreases turnover of arachidonate in several brain phospholipids. Neurosci Lett 220:171–174

Channing MA, Simpson N (1993) Radiosynthesis of 1-[11C]polyhomoallylic fatty acids. J Label Compds Radiopharm 3:541–546

Chen CT, Liu Z, Ouellet M, Calon F, Bazinet RP (2009) Rapid beta-oxidation of eicosapentaenoic acid in mouse brain: an in situ study. Prostaglandins Leukot Essent Fatty Acids 80:157–163

Chen CT, Ma DW, Kim JH, Mount HT, Bazinet RP (2008a) The low density lipoprotein receptor is not necessary for maintaining mouse brain polyunsaturated fatty acid concentrations. J Lipid Res 49:147–152

Chen CT, Green JT, Orr SK, Bazinet RP (2008b) Regulation of brain polyunsaturated fatty acid uptake and turnover. Prostaglandins Leukot Essent Fatty Acids 79:85–91

Chiu HC, Kovacs A, Blanton RM, Han X, Courtois M, Weinheimer CJ, Yamada KA, Brunet S, Xu H, Nerbonne JM, Welch MJ, Fettig NM, Sharp TL, Sambandam N, Olson KM, Ory DS, Schaffer JE (2005) Transgenic expression of fatty acid transport protein 1 in the heart causes lipotoxic cardiomyopathy. Circ Res 96:225–233

Cistola DP, Hamilton JA, Jackson D, Small DM (1988) Ionization and phase behavior of fatty acids in water: application of the Gibbs phase rule. Biochemistry 27:1881–1888

Coe NR, Smith AJ, Frohnert BI, Watkins PA, Bernlohr DA (1999) The fatty acid transport protein (FATP1) is a very long chain acyl-CoA synthetase. J Biol Chem 274:36300–36304

Contreras MA, Chang MC, Rosenberger TA, Greiner RS, Myers CS, Salem N Jr, Rapoport SI (2001) Chronic nutritional deprivation of n-3 alpha-linolenic acid does not affect n-6 arachidonic acid recycling within brain phospholipids of awake rats. J Neurochem 79:1090–1099

Contreras MA, Greiner RS, Chang MC, Myers CS, Salem N Jr, Rapoport SI (2000) Nutritional deprivation of alpha-linolenic acid decreases but does not abolish turnover and availability of unacylated docosahexaenoic acid and docosahexaenoyl-CoA in rat brain. J Neurochem 75:2392–2400

Cooper RB, Noy N, Zakim D (1989) Mechanism for binding of fatty acids to hepatocyte plasma membranes. J Lipid Res 30:1719–1726

DeMar JC Jr, Lee HJ, Ma K, Chang L, Bell JM, Rapoport SI, Bazinet RP (2006) Brain elongation of linoleic acid is a negligible source of the arachidonate in brain phospholipids of adult rats. Biochim Biophys Acta 1761:1050–1059

DeMar JC Jr, Ma K, Bell JM, Rapoport SI (2004) Half-lives of docosahexaenoic acid in rat brain phospholipids are prolonged by 15 weeks of nutritional deprivation of n-3 polyunsaturated fatty acids. J Neurochem 91:1125–1137

Demar JC Jr, Ma K, Chang L, Bell JM, Rapoport SI (2005) alpha-Linolenic acid does not contribute appreciably to docosahexaenoic acid within brain phospholipids of adult rats fed a diet enriched in docosahexaenoic acid. J Neurochem 94:1063–1076

Dietschy JM, Turley SD (2004) Thematic review series: brain Lipids Cholesterol metabolism in the central nervous system during early development and in the mature animal. J Lipid Res 45:1375–1397

DiRusso CC, Li H, Darwis D, Watkins PA, Berger J, Black PN (2005) Comparative biochemical studies of the murine fatty acid transport proteins (FATP) expressed in yeast. J Biol Chem 280:16829–16837

Doege H, Baillie RA, Ortegon AM, Tsang B, Wu Q, Punreddy S, Hirsch D, Watson N, Gimeno RE, Stahl A (2006) Targeted deletion of FATP5 reveals multiple functions in liver metabolism: alterations in hepatic lipid homeostasis. Gastroenterology 130:1245–1258

Drewes LR (2001) Molecular architecture of the brain microvasculature: perspective on blood-brain transport. J Mol Neurosci 16:93–98, discussion 151–157

Drover VA, Nguyen DV, Bastie CC, Darlington YF, Abumrad NA, Pessin JE, London E, Sahoo D, Phillips MC (2008) CD36 mediates both cellular uptake of very long chain fatty acids and their intestinal absorption in mice. J Biol Chem 283:13108–13115

Edmond J (2001) Essential polyunsaturated fatty acids and the barrier to the brain: the components of a model for transport. J Mol Neurosci 16:181–193 discussion 215–121

Esposito G, Giovacchini G, Der M, Liow JS, Bhattacharjee AK, Ma K, Herscovitch P, Channing M, Eckelman WC, Hallett M, Carson RE, Rapoport SI (2007) Imaging signal transduction via arachidonic acid in the human brain during visual stimulation, by means of positron emission tomography. NeuroImage 34:1342–1351

Esposito G, Giovacchini G, Liow JS, Bhattacharjee AK, Greenstein D, Schapiro M, Hallett M, Herscovitch P, Eckelman WC, Carson RE, Rapoport SI (2008a) Imaging neuroinflammation in alzheimer's disease with radiolabeled arachidonic acid and PET. J Nucl Med 49:1414–1421

Esposito G, Giovacchini G, Liow JS, Bhattacharjee AK, Greenstein D, Schapiro M, Hallett M, Herscovitch P, Eckelman WC, Carson RE, Rapoport SI (2008b) Imaging neuroinflammation in Alzheimer's disease with radiolabeled arachidonic acid and PET. J Nucl Med 49:1414–1421

Febbraio M, Abumrad NA, Hajjar DP, Sharma K, Cheng W, Pearce SF, Silverstein RL (1999) A null mutation in murine CD36 reveals an important role in fatty acid and lipoprotein metabolism. J Biol Chem 274:19055–19062

Febbraio M, Hajjar DP, Silverstein RL (2001) CD36: a class B scavenger receptor involved in angiogenesis, atherosclerosis, inflammation, and lipid metabolism. J Clin Invest 108:785–791

Gimeno RE (2007) Fatty acid transport proteins. Curr Opin Lipidol 18:271–276

Gimeno RE, Hirsch DJ, Punreddy S, Sun Y, Ortegon AM, Wu H, Daniels T, Stricker-Krongrad A, Lodish HF, Stahl A (2003a) Targeted deletion of fatty acid transport protein-4 results in early embryonic lethality. J Biol Chem 278:49512–49516

Gimeno RE, Ortegon AM, Patel S, Punreddy S, Ge P, Sun Y, Lodish HF, Stahl A (2003b) Characterization of a heart-specific fatty acid transport protein. J Biol Chem 278:16039–16044

Giovacchini G, Chang MC, Channing MA, Toczek M, Mason A, Bokde AL, Connolly C, Vuong BK, Ma Y, Der MG, Doudet DJ, Herscovitch P, Eckelman WC, Rapoport SI, Carson RE (2002) Brain incorporation of [11C]arachidonic acid in young healthy humans measured with positron emission tomography. J Cereb Blood Flow Metab 22:1453–1462

Giovacchini G, Lerner A, Toczek MT, Fraser C, Ma K, DeMar JC, Herscovitch P, Eckelman WC, Rapoport SI, Carson RE (2004) Brain incorporation of 11C-arachidonic acid, blood volume, and blood flow in healthy aging: a study with partial-volume correction. J Nucl Med 45:1471–1479

Golovko MY, Barcelo-Coblijn G, Castagnet PI, Austin S, Combs CK, Murphy EJ (2009) The role of alpha-synuclein in brain lipid metabolism: a downstream impact on brain inflammatory response. Mol Cell Biochem 326:55–66

Golovko MY, Murphy EJ (2006) Uptake and metabolism of plasma-derived erucic acid by rat brain. J Lipid Res 47:1289–1297

Green JT, Liu Z, Bazinet RP (2010) Brain phospholipid arachidonic acid half lives are not altered following 15 weeks of n-3 polyunsaturated fatty acid adequate or deprived diet. J Lipid Res 51:535–543

Gruarin P, Thorne RF, Dorahy DJ, Burns GF, Sitia R, Alessio M (2000) CD36 is a ditopic glycoprotein with the N-terminal domain implicated in intracellular transport. Biochem Biophys Res Commun 275:446–454

Guo W, Huang N, Cai J, Xie W, Hamilton JA (2006) Fatty acid transport and metabolism in HepG2 cells. Am J Physiol Gastrointest Liver Physiol 290:G528–G534

Habets DD, Coumans WA, Voshol PJ, den Boer MA, Febbraio M, Bonen A, Glatz JF, Luiken JJ (2007) AMPK-mediated increase in myocardial long-chain fatty acid uptake critically depends on sarcolemmal CD36. Biochem Biophys Res Commun 355:204–210

Hall AM, Smith AJ, Bernlohr DA (2003) Characterization of the Acyl CoA synthetase activity of purified murine fatty acid transport protein 1. J Biol Chem 278:43008–43013

Hall AM, Wiczer BM, Herrmann T, Stremmel W, Bernlohr DA (2005) enzymatic properties of purified murine fatty acid transport protein 4 and analysis of acyl-CoA synthetase activities in tissues from FATP4 null mice. J Biol Chem 280:11948–11954

Hamilton JA (1998) Fatty acid transport: difficult or easy? J Lipid Res 39:467–481

Hamilton JA, Brunaldi K (2007) A model for fatty acid transport into the brain. J Mol Neurosci 33:146–150

Hamilton JA, Kamp F (1999) How are free fatty acids transported in membranes? Is it by proteins or by free diffusion through the lipids? Diabetes 48:2255–2269

Harmon CM, Abumrad NA (1993) Binding of sulfosuccinimidyl fatty acids to adipocyte membrane proteins: isolation and amino-terminal sequence of an 88-kD protein implicated in transport of long-chain fatty acids. J Membr Biol 133:43–49

Hatch GM, Smith AJ, Xu FY, Hall AM, Bernlohr DA (2002) FATP1 channels exogenous FA into 1,2,3-triacyl-sn-glycerol and down- regulates sphingomyelin and cholesterol metabolism in growing 293 cells. J Lipid Res 43:1380–1389

Heinzer AK, Watkins PA, Lu JF, Kemp S, Moser AB, Li YY, Mihalik S, Powers JM, Smith KD (2003) A very long-chain acyl-CoA synthetase-deficient mouse and its relevance to X-linked adrenoleukodystrophy. Hum Mol Genet 12:1145–1154

Herrmann T, Buchkremer F, Gosch I, Hall AM, Bernlohr DA, Stremmel W (2001) Mouse fatty acid transport protein 4 (FATP4): characterization of the gene and functional assessment as a very long chain acyl-CoA synthetase. Gene 270:31–40

Herrmann T, Van Der Hoeven F, Grone HJ, Stewart AF, Langbein L, Kaiser I, Liebisch G, Gosch I, Buchkremer F, Drobnik W, Schmitz G, Stremmel W (2003) Mice with targeted disruption of the fatty acid transport protein 4 (Fatp 4, Slc27a4) gene show features of lethal restrictive dermopathy. J Cell Biol 161:1105–1115

Ibrahimi A, Bonen A, Blinn WD, Hajri T, Li X, Zhong K, Cameron R, Abumrad NA (1999) Muscle-specific overexpression of FAT/CD36 enhances fatty acid oxidation by contracting muscle, reduces plasma triglycerides and fatty acids, and increases plasma glucose and insulin. J Biol Chem 274:26761–26766

Igarashi M, DeMar JC Jr, Ma K, Chang L, Bell JM, Rapoport SI (2007a) Docosahexaenoic acid synthesis from alpha-linolenic acid by rat brain is unaffected by dietary n-3 PUFA deprivation. J Lipid Res 48:1150–1158

Igarashi M, DeMar JC Jr, Ma K, Chang L, Bell JM, Rapoport SI (2007b) Upregulated liver conversion of alpha-linolenic acid to docosahexaenoic acid in rats on a 15 week n-3 PUFA-deficient diet. J Lipid Res 48:152–164

Jia Z, Moulson CL, Pei Z, Miner JH, Watkins PA (2007) Fatty acid transport protein 4 is the principal very long chain fatty acyl-CoA synthetase in skin fibroblasts. J Biol Chem 282:20573–20583

Kamp F, Guo W, Souto R, Pilch PF, Corkey BE, Hamilton JA (2003) Rapid flip-flop of oleic acid across the plasma membrane of adipocytes. J Biol Chem 278:7988–7995

Kamp F, Hamilton JA (1992) pH gradients across phospholipid membranes caused by fast flip-flop of un-ionized fatty acids. Proc Natl Acad Sci USA 89:11367–11370

Kim JK, Gimeno RE, Higashimori T, Kim HJ, Choi H, Punreddy S, Mozell RL, Tan G, Stricker-Krongrad A, Hirsch DJ, Fillmore JJ, Liu ZX, Dong J, Cline G, Stahl A, Lodish HF, Shulman GI (2004) Inactivation of fatty acid transport protein 1 prevents fat-induced insulin resistance in skeletal muscle. J Clin Invest 113:756–763

Lisanti MP, Scherer PE, Vidugiriene J, Tang Z, Hermanowski-Vosatka A, Tu YH, Cook RF, Sargiacomo M (1994) Characterization of caveolin-rich membrane domains isolated from an endothelial-rich source: implications for human disease. J Cell Biol 126:111–126

Lobo S, Wiczer BM, Smith AJ, Hall AM, Bernlohr DA (2007) Fatty acid metabolism in adipocytes: functional analysis of fatty acid transport proteins 1 and 4. J Lipid Res 48:609–620

Luiken JJ, Coort SL, Koonen DP, van der Horst DJ, Bonen A, Zorzano A, Glatz JF (2004) Regulation of cardiac long-chain fatty acid and glucose uptake by translocation of substrate transporters. Pflugers Arch 448:1–15

Luiken JJ, Dyck DJ, Han XX, Tandon NN, Arumugam Y, Glatz JF, Bonen A (2002a) Insulin induces the translocation of the fatty acid transporter FAT/CD36 to the plasma membrane. Am J Physiol Endocrinol Metab 282:E491–E495

Luiken JJ, Koonen DP, Willems J, Zorzano A, Becker C, Fischer Y, Tandon NN, Van Der Vusse GJ, Bonen A, Glatz JF (2002b) Insulin stimulates long-chain fatty acid utilization by rat cardiac myocytes through cellular redistribution of FAT/CD36. Diabetes 51:3113–3119

Luiken JJ, Schaap FG, van Nieuwenhoven FA, van der Vusse GJ, Bonen A, Glatz JF (1999a) Cellular fatty acid transport in heart and skeletal muscle as facilitated by proteins. Lipids 34:S169–S175

Luiken JJ, Turcotte LP, Bonen A (1999b) Protein-mediated palmitate uptake and expression of fatty acid transport proteins in heart giant vesicles. J Lipid Res 40:1007–1016

Marbois BN, Ajie HO, Korsak RA, Sensharma DK, Edmond J (1992) The origin of palmitic acid in brain of the developing rat. Lipids 27:587–592

Meshulam T, Simard JR, Wharton J, Hamilton JA, Pilch PF (2006) Role of caveolin-1 and cholesterol in transmembrane fatty acid movement. Biochemistry 45:2882–2893

Mihalik SJ, Steinberg SJ, Pei Z, Park J, Kim DG, Heinzer AK, Dacremont G, Wanders RJ, Cuebas DA, Smith KD, Watkins PA (2002) Participation of two members of the very long-chain acyl-CoA synthetase family in bile acid synthesis and recycling. J Biol Chem 277:24771–24779

Milger K, Herrmann T, Becker C, Gotthardt D, Zickwolf J, Ehehalt R, Watkins PA, Stremmel W, Fullekrug J (2006) Cellular uptake of fatty acids driven by the ER-localized acyl-CoA synthetase FATP4. J Cell Sci 119:4678–4688

Morell P, Toews AD (1996) Biochemistry of lipids. In: Moser HW (ed) Handbook of clinical neurology. Neurodystrophies and neurolipidoses, vol 22 (66). Elsevier, Amsterdam, pp 33–49

Moser H, Dubey P, Fatemi A (2004) Progress in X-linked adrenoleukodystrophy. Curr Opin Neurol 17:263–269

Moser HW, Raymond GV, Lu SE, Muenz LR, Moser AB, Xu J, Jones RO, Loes DJ, Melhem ER, Dubey P, Bezman L, Brereton NH, Odone A (2005) Follow-up of 89 asymptomatic patients with adrenoleukodystrophy treated with Lorenzo's oil. Arch Neurol 62:1073–1080

Moulson CL, Martin DR, Lugus JJ, Schaffer JE, Lind AC, Miner JH (2003) Cloning of wrinkle-free, a previously uncharacterized mouse mutation, reveals crucial roles for fatty acid transport protein 4 in skin and hair development. Proc Natl Acad Sci USA 100:5274–5279

Muller H, Deckers K, Eckel J (2002) The fatty acid translocase (FAT)/CD36 and the glucose transporter GLUT4 are localized in different cellular compartments in rat cardiac muscle. Biochem Biophys Res Commun 293:665–669

Murphy EJ, Owada Y, Kitanaka N, Kondo H, Glatz JF (2005) Brain arachidonic acid incorporation is decreased in heart fatty acid binding protein gene-ablated mice. Biochemistry 44:6350–6360

Nagy L, Tontonoz P, Alvarez JG, Chen H, Evans RM (1998) Oxidized LDL regulates macrophage gene expression through ligand activation of PPARgamma. Cell 93:229–240

Neely JR, Morgan HE (1974) Relationship between carbohydrate and lipid metabolism and the energy balance of heart muscle. Annu Rev Physiol 36:413–459

Ouellet M, Emond V, Chen CT, Julien C, Bourasset F, Oddo S, LaFerla F, Bazinet RP, Calon F (2009) Diffusion of docosahexaenoic and eicosapentaenoic acids through the blood-brain barrier: an in situ cerebral perfusion study. Neurochem Int 55:476–482

Pei Z, Fraisl P, Berger J, Jia Z, Forss-Petter S, Watkins PA (2004) Mouse very long-chain Acyl-CoA synthetase 3/fatty acid transport protein 3 catalyzes fatty acid activation but not fatty acid transport in MA-10 cells. J Biol Chem 279:54454–54462

Peter T Jr (1996) All about albumin. Academic, San Diego

Polozova A, Gionfriddo E, Salem N Jr (2006) Effect of docosahexaenoic acid on tissue targeting and metabolism of plasma lipoproteins. Prostaglandins Leukot Essent Fatty Acids 75:183–190

Quan G, Xie C, Dietschy JM, Turley SD (2003) Ontogenesis and regulation of cholesterol metabolism in the central nervous system of the mouse. Brain Res Dev Brain Res 146:87–98

Rapoport SI (2001) In vivo fatty acid incorporation into brain phosholipids in relation to plasma availability, signal transduction and membrane remodeling. J Mol Neurosci 16:243–261, discussion 279–284

Rapoport SI, Chang MC, Spector AA (2001) Delivery and turnover of plasma-derived essential PUFAs in mammalian brain. J Lipid Res 42:678–685

Rapoport SI, Rao JS, Igarashi M (2007) Brain metabolism of nutritionally essential polyunsaturated fatty acids depends on both the diet and the liver. Prostaglandins Leukot Essent Fatty Acids 77:251–261

Richieri GV, Ogata RT, Kleinfeld AM (1992) A fluorescently labeled intestinal fatty acid binding protein – interactions with fatty acids and its use in monitoring free fatty acids. J Biol Chem 267:23495–23501

Richieri GV, Ogata RT, Kleinfeld AM (1994) Equilibrium constants for the binding of fatty acids with fatty acid-binding proteins from adipocyte, intestine, heart, and liver measured with the fluorescent probe ADIFAB. J Biol Chem 269:23918–23930

Robinson PJ, Noronha J, DeGeorge JJ, Freed LM, Nariai T, Rapoport SI (1992) A quantitative method for measuring regional *in vivo* fatty-acid incorporation into and turnover within brain phospholipids: review and critical analysis. Brain Res Brain Res Rev 17:187–214

Schaffer JE, Lodish HF (1994) Expression cloning and characterization of a novel adipocyte long chain fatty acid transport protein. Cell 79:427–436

Simard JR, Kamp F, Hamilton JA (2008a) Measuring the adsorption of Fatty acids to phospholipid vesicles by multiple fluorescence probes. Biophys J 94:4493–4503

Simard JR, Meshulam T, Pillai BK, Kirber MT, Brunaldi K, Xu S, Pilch PF, Hamilton JA (2010) Caveolins sequester fatty acids on the cytoplasmic leaflet of the plasma membrane, augment triglyceride formation and protect cells from lipotoxicity. J Lipid Res 51:914–922

Simard JR, Pillai BK, Hamilton JA (2008b) Fatty acid flip-flop in a model membrane is faster than desorption into the aqueous phase. Biochemistry 47:9081–9089

Stahl A (2004) A current review of fatty acid transport proteins (SLC27). Pflugers Arch 447:722–727

Stahl A, Evans JG, Pattel S, Hirsch D, Lodish HF (2002) Insulin causes fatty acid transport protein translocation and enhanced fatty acid uptake in adipocytes. Dev Cell 2:477–488

Stahl A, Gimeno RE, Tartaglia LA, Lodish HF (2001) Fatty acid transport proteins: a current view of a growing family. Trends Endocrinol Metab 12:266–273

Stahl A, Hirsch DJ, Gimeno RE, Punreddy S, Ge P, Watson N, Patel S, Kotler M, Raimondi A, Tartaglia LA, Lodish HF (1999) Identification of the major intestinal fatty acid transport protein. Mol Cell 4:299–308

Steinberg SJ, Mihalik SJ, Kim DG, Cuebas DA, Watkins PA (2000) The human liver-specific homolog of very long-chain acyl-CoA synthetase Is cholate:CoA ligase. J Biol Chem 275: 15605–15608

Steinberg SJ, Wang SJ, Kim DG, Mihalik SJ, Watkins PA (1999a) Human very-long-chain acyl-CoA synthetase: cloning, topography, and relevance to branched-chain fatty acid metabolism [In Process Citation]. Biochem Biophys Res Commun 257:615–621

Steinberg SJ, Wang SJ, McGuinness MC, Watkins PA (1999b) Human liver-specific very-long-chain acyl-coenzyme A synthetase: cDNA cloning and characterization of a second enzymatically active protein. Mol Genet Metab 68:32–42

Storch J, Corsico B (2008) The emerging functions and mechanisms of mammalian fatty acid-binding proteins. Annu Rev Nutr 28:73–95

Stremmel W, Strohmeyer G, Berk PD (1986) Hepatocellular uptake of oleate is energy dependent, sodium linked, and inhibited by an antibody to a hepatocyte plasma membrane fatty acid binding protein. Proc Natl Acad Sci USA 83:3584–3588

Stremmel W, Strohmeyer G, Borchard F, Kochwa S, Berk PD (1985) Isolation and partial characterization of a fatty acid binding protein in rat liver plasma membranes. Proc Natl Acad Sci USA 82:4–8

Stump DD, Fan X, Berk PD (2001) Oleic acid uptake and binding by rat adipocytes define dual pathways for cellular fatty acid uptake. J Lipid Res 42:509–520

Stump DD, Zhou SL, Berk PD (1993) Comparison of plasma membrane FABP and mitochondrial isoform of aspartate aminotransferase from rat liver. Am J Physiol 265:G894–G902

Takasato Y, Rapoport SI, Smith QR (1984) An in situ brain perfusion technique to study cerebrovascular transport in the rat. Am J Physiol 247:H484–H493

Thies F, Delachambre MC, Bentejac M, Lagarde M, Lecerf J (1992) Unsaturated fatty acids esterified in 2-acyl-l-lysophosphatidylcholine bound to albumin are more efficiently taken up by the young rat brain than the unesterified form. J Neurochem 59:1110–1116

Trotter PJ, Ho SY, Storch J (1996) Fatty acid uptake by Caco-2 human intestinal cells. J Lipid Res 37:336–346

Uchida Y, Kondo N, Orii T, Hashimoto T (1996) Purification and properties of rat liver peroxisomal very-long-chain acyl-CoA synthetase. J Biochem (Tokyo) 119:565–571

Uchiyama A, Aoyama T, Kamijo K, Uchida Y, Kondo N, Orii T, Hashimoto T (1996) Molecular cloning of cDNA encoding rat very long-chain acyl-CoA synthetase. J Biol Chem 271: 30360–30365

Umhau JC, Zhou W, Carson RE, Rapoport SI, Polozova A, Demar J, Hussein N, Bhattacharjee AK, Ma K, Esposito G, Majchrzak S, Herscovitch P, Eckelman WC, Kurdziel KA, Salem N Jr (2009) Imaging incorporation of circulating docosahexaenoic acid into the human brain using positron emission tomography. J Lipid Res 50:1259–1268

Watkins PA (2008) Very-long-chain Acyl-CoA synthetases. J Biol Chem 283:1773–1777

Watkins PA, Maiguel D, Jia Z, Pevsner J (2007) Evidence for 26 distinct acyl-coenzyme A synthetase genes in the human genome. J Lipid Res 48:2736–2750

Wu Q, Kazantzis M, Doege H, Ortegon AM, Tsang B, Falcon A, Stahl A (2006a) Fatty acid transport protein 1 is required for nonshivering thermogenesis in brown adipose tissue. Diabetes 55:3229–3237

Wu Q, Ortegon AM, Tsang B, Doege H, Feingold KR, Stahl A (2006b) FATP1 is an insulin-sensitive fatty acid transporter involved in diet-induced obesity. Mol Cell Biol 26:3455–3467

Xu HE, Lambert MH, Montana VG, Parks DJ, Blanchard SG, Brown PJ, Sternbach DD, Lehmann JM, Wisely GB, Willson TM, Kliewer SA, Milburn MV (1999) Molecular recognition of fatty acids by peroxisome proliferator-activated receptors. Mol Cell 3:397–403

Zhang F, Kamp F, Hamilton JA (1996) Dissociation of long and very long chain fatty acids from phospholipid bilayers. Biochemistry 35:16055–16060

Chapter 28
The Ketogenic Diet and Brain Metabolism *

Marc Yudkoff, Yevgeny Daikhin, Oksana Horyn, Ilana Nissim, and Itzhak Nissim

Abstract Glucose ordinarily is the major brain fuel. However, the consumption of a diet high in fat evokes a brisk ketonemia (1–2 mmol/l) and provides brain with substrates (3-OH-butyrate and acetoacetate) that can furnish almost half of cerebral energy requirements. Such a diet also confers a potent anti-epileptic effect, even in patients whose epilepsy has proved refractory to anti-convulsant drugs. The precise basis of the therapeutic effect is not clear, but a ketogenic diet alters brain metabolism of many compounds, including the handling of neurotransmitter amino acids such as glutamate, aspartate, glutamine and GABA. This review summarizes some of these changes and considers how such adaptations might attenuate or even prevent a seizure diathesis.

Key words Anti-convulsant therapy • Brain amino acid metabolism • Epilepsy • Ketosis

* Supported by grants HD269711, RR00240, U54RR019453, U54RR023567, DK047870 from NIH

M. Yudkoff, M.D. (✉)
Children's Hospital of Philadelphia and Department of Pediatrics,
University of Pennsylvania School of Medicine, Philadelphia, PA, USA

Division of Child Development, Rehabilitation and Metabolic Disease,
Children's Hospital of Philadelphia, Philadelphia, PA, USA
e-mail: yudkoff@email.chop

Y. Daikhin, M.D., • O. Horyn • I. Nissim • I. Nissim
Children's Hospital of Philadelphia and Department of Pediatrics,
University School of Medicine, Philadelphia, PA, USA
e-mail: ssitz@mail.med.upenn.edu

I.-Y. Choi and R. Gruetter (eds.), *Neural Metabolism In Vivo*,
Advances in Neurobiology 4, DOI 10.1007/978-1-4614-1788-0_28,

28.1 Introduction: Overview of Brain Metabolism

A fundamental goal of all intermediary metabolism is to extract energy from the three primary nutrients of life – amino acids, fatty acids and carbohydrates. Cells achieve this goal by converting these relatively complex molecules into simpler species such as CO_2, H_2O and, in the case of amino acids, into NH_3. Much of the metabolic apparatus in any cell focuses on severing the molecular bonds of the major nutrients and capturing the energy formerly resident in them in the high energy phosphate bonds of compounds such as ATP and creatine phosphate.

The oxidation of these parent nutrients yields appreciable energy: ~4 kcal/g from either amino acids or carbohydrates and ~8 kcal/g from fatty acids. We might anticipate, given the ceaseless energy requirement of all cells, that individual organs would indiscriminantly oxidize any nutrient, the choice of fuel being limited only by the momentary accessibility of a particular substrate. However, the biochemical reality is much more complex, with many tissues exhibiting marked fastidiousness with regard to fuel selection.

Human brain metabolism highlights this phenomenon. Brain energy requirements are prodigious, with a mean value in excess of 300 kcal/kg/day, a parameter that exceeds by more than ten-fold the resting energy expenditure of a healthy adult (20–30 kcal/kg/day). The brain satisfies essentially all of this demand from the oxidation of glucose. A normal human brain comprises only 2% of body weight, but it extracts about 10% of the glucose in arterial blood (~310 μmol/kg/min) and almost 50% of arterial blood oxygen (1,560 μmol/kg/min) (Kety and Schmidt 1948). Whole body oxygen consumption is about 8,000 μmol/kg/min. Hence, brain oxygen utilization is almost 20% of the body total.

As the above data imply, the ratio of brain consumption of oxygen to that of glucose is 5:1 (1,560/310 μmol/kg/min). However, if glucose oxidation is complete this parameter ought to be 6:1, since the oxidation of 1 mole of glucose obliges consumption of 6 moles of oxygen: $C_6H_{12}O_6 + 6O_2 \rightarrow 6\ CO_2 + 6\ H_2O$. The discrepancy is attributable to the facts that: (a) some consumed glucose is "incompletely" combusted to lactate, not CO_2, and (b) a portion of overall oxygen consumption is applied not to glucose oxidation but to the synthesis of macromolecules (Clarke and Sokoloff 1999; Magistretti et al. 1999; Siesjo 1997).

The extraordinary intensity of cerebral metabolism reflects the prodigious work that brain cells are obliged to accomplish. Foremost among these functions is the maintenance of very high cross-membrane gradients of ions and neurotransmitters. Essential to all neurotransmission is the action potential, or the sudden dissipation of the potential difference that prevails between the interior of the neuron and the extracellular environment. Maintenance of the integrity of neurotransmission requires that this potential difference must be restored rapidly in order to prepare the neuron for yet another cycle of depolarization. Restoration of the membrane potential obliges the consumption of ATP by both astrocytes and neurons, which rapidly pump ions in order to re-establish the cross-membrane difference in electrical charge.

Propagation of the action potential along the axon typically culminates in the release of neurotransmitters into the synaptic space. Brain must expend considerable energy in swiftly clearing neurotransmitters from the synapse, often by uptake into astrocytes. Glial transport of a transmitter such as glutamate is a sodium-dependent process (see below), and astrocytes must consume ATP in order to restore the Na^+ gradient that enables such transport (Erecinska and Dagani 1990).

Finally, brain expends a great deal of energy to support its intense anabolic needs, including the synthesis of relatively large quantities of proteins and lipids. Brain also consumes substantial ATP in order to support the synthesis of glutamine via glutamine synthetase which, as we will see, is essential to cerebral amino acid metabolism.

As noted above, glucose is virtually the sole metabolic substrate to support these intense energy requirements. All brain cells probably can oxidize glucose to completion via the mitochondrial tricarboxylic acid cycle, but there are cels and cell compartments in which glycolysis alone is prominent. Selected astroglia filopodia may be too narrow to accommodate mitochondria. Glycolysis (and glycogenolysis) may predominate in such compartments in order to yield the energy needed to remove from the synapse the glutamate and K^+ released by pre-synaptic terminals during depolarization (Danbolt 2001; Hertz et al. 2007). It is unclear what becomes of the pyruvate that is generated when glycolysis is so active, but it seems likely that astrocytes convert at least a portion to lactate, which they then release to neurons for further oxidation. Neurons are able to respire on lactate alone (Schurr et al. 1999), but they probably must supplement this substrate with glucose in order to maintain large internal pools of glutamate and aspartate (Waagepetersen et al. 1998). A metabolic "dialogue" in which neurons oxidize glial-generated lactate would effectively couple overall brain metabolism to neuronal activity (Escartin et al. 2006; Magistretti et al. 1999).

Adaptation to an altered environment is the hallmark of any successful biologic system. Glucose is virtually the sole fuel of human brain in the basal state, but the human nervous system can consume other substrates when the supply of glucose is limiting and/or when an alternate fuel is abundantly present. The best known example of this phenomenon is cerebral consumption of ketone bodies in an individual who consumes a high-fat and low-carbohydrate diet. As we will see in the next section, the so-called ketogenic diet is of great therapeutic value in individuals with otherwise uncontrollable epilepsy.

28.2 Efficacy of the Ketogenic Diet in Controlling Epilepsy

Current interest in the metabolism of brain during ketosis has been sparked by the efficacy of a ketogenic diet as an adjunct therapy in controlling epilepsy. Recent decades have seen major advances in the pharmacology of the anti-epileptic medications, but such drugs either fail to completely control convulsions in many patients or they give rise to unacceptable side effects such as excessive lethargy, memory loss, changes in appetite or toxicity to major organ systems (Camfield and Camfield 1996).

A ketogenic diet has been found to be highly effective treatment, even in patients whose seizures are refractory to anti-epileptic medications (Freeman et al. 2006; Wilder 1921; Wilkins 1937; Yudkoff et al. 2001). About half of patients enjoy a 50% reduction in seizure frequency, with many manifesting a near-total remission (Freeman et al. 1998; Kossoff et al. 2002; Lefevre and Aronson 2000; Nordli et al. 2001; Thiele 2003; Vining 1999; Neal et al. 2008). The diet has been deployed primarily in children, but even adolescents and older patients may benefit (Kossoff et al. 2003; Kossoff et al. 2005; Mady et al. 2003). In most instances, the goal of diet therapy is to administer over 80% of calories as lipid, but some data suggests that even a relatively low-carbohydrate diet ("Atkins diet") could be therapeutically effective (Kossoff et al. 2002).

A remarkable aspect of the diet is that it controls seizures without inducing the undesirable side-effects that typify medications, including obtundation, memory loss and cognitive deficits. Growth and weight do not suffer profoundly (Vining et al. 2002). Hyperlipidemia can occur (Kwiterovich et al. 2003), but a recent review (Bravata et al. 2003) reports no major adverse effects on blood pressure, blood glucose, serum insulin or blood lipids.

The diet induces ketosis by providing 80–90% of calories in the form of long-chain triglycerides and the remaining 10–20% primarily as protein (Lefevre and Aronson 2000; Swink et al. 1997). The administration of medium-chain triglycerides also induces ketosis, but in some instances such dietary intervention has been associated with diarrhea and abdominal cramping. Clinicians traditionally initiate the diet with a 24–48 h period of fasting, but the gradual introduction of a high-fat diet also induces ketosis (>1.5 mmol/l) without apparent sacrifice of the therapeutic effect (Bergqvist et al. 2005). A diet that includes >60% of calories as polyunsaturated or monounsaturated fat induces an even more intense ketosis (Fuehrlein et al. 2004).

It remains uncertain whether ketone bodies directly confer the anti-epileptic effect or whether the therapeutic benefit results from restriction of dietary calories and/or carbohydrate. Both a high-fat diet and caloric restriction tend to diminish the blood glucose concentration. Caloric restriction alone affords seizure control in EL mice, a model of human multifactorial epilepsy (Greene et al. 2001; Mantis et al. 2004). Similarly, limiting calories lowers neuronal activity in the dentate gyrus (Bough et al. 2003) and attenuates the proconvulsant activity of pentylenetetrazole (Eagles et al. 2003). Restricting dietary calories by 50% does not significantly increase blood ketones but may be as effective as the ketogenic diet in diminishing seizure frequency (Bough et al. 2003; Eagles et al. 2003).

A series of scientific and ethical obstacles has made it quite difficult to study the efficacy of the ketogenic diet in a rigorous manner. However a recent (Neal et al. 2008) study randomized 145 children (2–16 years) with daily and drug-resistant seizures into either a control group or a group that received a ketogenic diet. After allowing for withdrawals and non-compliance, data were available for 103 children, 54 on the diet and 49 controls. The experimental group manifested a significantly greater decline in the frequency of seizures (62.0% vs 136.9% of the baseline value). More than one-third (38%) of the diet group enjoyed a greater than 50% reduction in seizure frequency compared with only 6% of the controls. Near complete (>90%)

reduction in seizure frequency was observed in 7% of the treated children and none of the controls. The type of seizure (generalized or focal) did not seem to affect results. Major side-effects in the treated group were constipation, vomiting, lack of energy and hunger.

28.3 Mechanisms of Action of the Anti-Epileptic Effect

Clinicians have known for centuries that starvation and its attendant ketosis can limit the severity and frequency of seizures. The precise mechanism of action remains elusive, although several theories have been propounded: (a) Ketosis entails the accumulation of several organic acids, most notably 3-OH-butyric acid and acetoacetic acid. This phenomenon might lead to acidification of the brain and consequent inhibition of neuronal sensitive H^+-sensitive ion channels. There is no good evidence for such acidification at the level of the whole brain (Al-Mudallal et al. 1996), although the possibility always exists that changes in H^+ concentration could occur in localized regions. (b) Ketone bodies or long-chain fatty acids may cause a hyperpolarization of ATP-sensitive K^+ channels (Vamecq et al. 2005), thereby making neuronal discharge less probable. (c) Fatty acids can inhibit neuronal function (Cunnane et al. 2002). Furthermore acetone, which forms from the decarboxylation of acetoacetate and can enter the central nervous system (Gerasimov et al. 2005), has been shown to raise the seizure threshold and attenuate the severity of induced seizures (Likhodii et al. 2003). (d) Glucose can increase neuronal excitability, even in physiologic concentrations (Burdakov et al. 2005). Similarly, frank hyperglycemia lowers the seizure threshold (Schwechter et al. 2003). It might be thought that the reduced blood glucose concentration that accompanies administration of a ketogenic diet would lower neuronal excitability and favor seizures, but this effect is neutralized by the increased brain glucose uptake that a ketogenic diet evokes (Cheng et al. 2003), in the process maintaining brain glucose levels even as the peripheral concentration declines. (e) Studies of mice that lack dopamine-β-hydroxylase indicate that the ketogenic diet recruits its therapeutic effect through the noradrenergic system (Szot et al. 2001), perhaps by favoring the synthesis and/or function of the biogenic amines. (f) Fatty acids at increased concentration evoke increased synthesis and activity of mitochondrial uncoupling protein (Mattson and Liu 2003; Sullivan et al. 2003; Sullivan et al. 2004). This process may provide an explanation for the beneficial effect of the ketogenic diet in other neurologic disorders, including glutamate toxicity (Noh et al. 2006), traumatic brain injury (Prins et al. 2004) and Parkinson's disease (Tieu et al. 2003). (g) There is abundant evidence that the ketogenic diet can change brain energy metabolism (Daikhin and Yudkoff 1998; DeVivo et al. 1978; Erecinska and Silver 1990; Lapidot and Haber 2002; Pan et al. 1999; Sato et al. 1995; Yudkoff et al. 1994, 2001, 2004, 2005' 2006). Indeed, it may be that 3-OH-butyrate is a more efficient metabolic substrate than glucose in terms of the energy formed per mole of oxygen consumed (Sato et al. 1995). Thus, studies in isolated heart show that infusion of 3-OH-butyrate increased as much as ten-fold the mitochondrial NADH/NAD ratio and lowered (2–4x) the coenzyme Q couple. This resetting of the redox potential increased energy obtained from mitochondrial oxidation of NADH from 53 kJ to 60 kJ/mole of oxygen consumed. Heightened efficiency of mitochondrial energy production might favor more rapid restoration of membrane potential and lessen free radical production following depolarization (Veech 2004; Veech et al. 2001).

28.4 Ketosis and Brain Metabolism of Neurotransmitter Amino Acids

Brain contains very large pools of metabolically active amino acids such as glutamate, aspartate, glutamine, alanine and GABA. The term "metabolically active" denotes the critical fact that, prior to final oxidation to CO_2, an appreciable fraction of glucose carbon traverses these amino acid pools. This intimate relationship between glucose and amino acid metabolism suggests the possibility that a ketogenic diet, which involves the partial substitution of glucose as a metabolic substrate in favor of the ketone bodies, might change the handling of neurotransmitter amino acids such as glutamate and GABA (Daikhin and Yudkoff 1998; Erecinska and Silver 1990; Melo et al. 2006; Yudkoff et al. 2005; 2001; 2004; 2006; 1994). It seemed plausible that one cogent factor to explain the antiepileptic effect might be an adaptation by brain in the metabolism of glutamate and GABA, respectively the major excitatory and inhibitory amino acids (Meldrum 2000; Petroff 2002).

In order to understand this hypothesis, it is necessary to explore in more detail brain handling of glutamate and GABA (Fig. 28.1). Brain maintains the intra-synaptic glutamate concentration at an extremely low level. Doing so is essential to neuronal function, since a low concentration of this neurotransmitter in the synapse both maximizes the signal-to-noise ratio upon release of additional glutamate from pre-synaptic terminals and also minimizes the risk of "excitotoxicity" that occurs when synaptic glutamate rises to an excessive concentration. Recent research clearly points to a pathogenic role of excessive neuronal excitation in many forms of brain injury, including hypoxia, traumatic insults and epilepsy (Olney 2003; Schousboe and Waagepetersen 2005; Waxman and Lynch 2005).

Brain astrocytes remove most intra-synaptic glutamate (Danbolt 2001; Gegelashvili and Schousboe 1998; Hertz et al. 2007; Takahashi et al. 1997). (In cerebellum neuronal uptake also may be important [Olstad et al. 2006]). Astrocytes are well-adapted for this role because of the large number of excitatory amino acid transporters they bear on their plasma membrane (Amara and Fontana 2002). In addition, they maintain a high membrane potential and very efficiently pump sodium, which facilitates Na^+-dependent transport of glutamate (Erecinska et al. 1996).

Once resident in astrocytes, glutamate must be restored to neurons in order to replace glutamate released from pre-synaptic terminals. Replenishment of the neuronal pool could not occur simply by transporting glutamate into neurons because this process would expose the system to the risk of unwanted depolarization as the

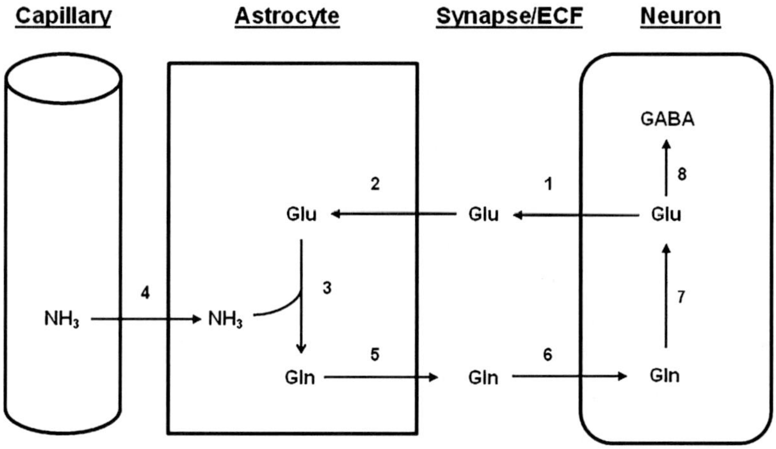

Fig. 28.1 The Glutamate-Glutamine Cycle. The cycle "begins" with the release of glutamate (1) into the synapse. Removal of synaptic glutamate occurs (2) when high-affinity, sodium-dependent transport systems on the surface membrane of astrocytes transport this neurotransmitter into the glia. The astrocytes convert glutamate to glutamine (3) via the glutamine synthetase pathway, which in brain is restricted to the glia. The glutamine synthetase pathway also utilizes ammonia that diffuses into brain from blood (4). The astrocytes release glutamine (5) via a specific carrier (5). Neurons take up glutamine (6), again via a specific carrier for neutral amino acids. The neurons hydrolyze glutamine to glutamate via mitochondrial phosphate-dependent glutaminase (7). In GABA-ergic neurons the glutamic acid is decarboxylated to GABA via the glutamate decarboxylase pathway (8)

glia release glutamate into the extracellular fluid. The brain solves the problem by first converting glutamate to glutamine via the mediation of glutamine synthetase (glutamate + NH_3 + ATP → glutamine + ADP + phosphate), an enzyme greatly enriched in astrocytes (Martinez-Hernandez et al. 1977; Norenberg and Martinez-Hernandez 1979). Ammonia necessary to the glutamine synthetase reaction derives in large measure from the plasma pool (Fig. 28.1) (Cooper et al. 1979), although oxidation of glutamate via glutamate dehydrogenase yields ammonia when neuronal depolarization is very active (Plaitakis and Zaganas 2001).

The plasma membrane of astrocytes also is rich in transport systems that mediate the release of glutamine to neurons (Chaudhry et al. 2002; Kanamori et al. 1998). In the latter cells the action of mitochondrial phosphate-dependent glutaminase (Sonnewald et al. 2004) hydrolyzes this amino acid to glutamate and ammonia.

It is evident that the interplay of glutamate and glutamine handling in neurons and astrocytes constitutes a cycle, the so-called "Glutamate Glutamine Cycle" (GGC) (Fig. 28.1), that remains the central conceptualization in all contemporary thinking about brain amino acid metabolism (Shank and Aprison 1977; Hertz et al. 1992; Waagepetersen et al. 2003). This model accommodates the two central requirements of brain glutamatergic neurotransmission: (a) Maintenance of very low intrasynaptic glutamate levels in order to maximize the signal-to-noise ratio with depolarization and to avoid excitotoxicity; (b) Restoration of glutamate carbon to neurons in a non-neuroactive form (glutamine) to avoid the risk of uncontrolled neuronal depolarization.

The model has been invaluable in accounting for experimental observations, but, like any heuristic model, it necessarily oversimplifies important aspects of brain glutamate handling. For example, it fails to specify the external sources of nitrogen that brain imports to replenish the inevitable losses to the circulation. Relatively little glutamate or glutamine is taken up from the peripheral blood (Grill et al. 1992; Smith et al. 1987), so an alternative reserve must be used.

Work in our laboratory and others suggests leucine as an excellent external source of amino nitrogen to replace that which is lost from oxidative metabolism. Neutral amino acid transporters at the brain:capillary interface very rapidly transport leucine into brain, which quickly transaminate this amino acid to yield the cognate ketoacid (2-oxo-caproic acid) and glutamate (Bixel et al. 1997; Hutson et al. 1992; Yudkoff et al. 1997). *In vivo* studies with magnetic resonance spectroscopy and [^{15}N]leucine indicate that the brain derives as much as one-third of all nitrogen from leucine (Kanamori and Ross 2004). If we add to this figure the contribution of valine and isoleucine, the other branched-chain amino acids, the total approximates 50%.

As emphasized above, the conceptualization implicit in the Glutamate Glutamine Cycle (Fig. 28.1) ignores the fact that glutamate, aspartate, and GABA are not only neurotransmitters, but metabolic intermediates that are in intimate articulation with the tricarboxylic acid cycle via very efficient transamination reactions such as aspartate aminotransferase (glutamate + oxaloacetate ↔ 2-oxo-glutarate + aspartate) or GABA transaminase (GABA + 2-oxo-glutarate ↔ succinic semialdehyde + glutamate). Hence, a metabolic substrate such as glucose (or 3-OH-butyrate) must, at least in part, pass through these pools prior to its final oxidation.

It is this factor that underlies the hypothesis that changes in neurotransmitter handling likely accompany fundamental changes in brain metabolism, including the shift from glucose as the sole cerebral fuel to the inclusion of the ketone bodies as important substrates.

28.5 Interplay of Brain Metabolism of Ketone Bodies and Glutamate

Evolution clearly has prepared humans to withstand the stress of caloric deprivation. The initial metabolic response to starvation is the breakdown of hepatic glycogen in order to maintain euglycemia. Once glycogen stores are nearly depleted, the catabolism of tissue protein, primarily from skeletal muscle and liver, provides key gluconeogenic substrate such as alanine (Felig et al. 1970). Ultimately, as starvation continues, the hydrolysis of triglycerides from adipose tissue will yield glycerol, a gluconeogenic substrate, and free fatty acids. The human liver converts the latter into ketone bodies, which liver does not consume but exports to peripheral tissues, particularly skeletal muscle and brain, which readily utilize these organic acids as metabolic substrates. Indeed, at a blood concentration of 2–4 mmol/l, a level that is typical of the ketotic state, the ketone bodies will furnish as much as 70% of the caloric requirements of the brain (Cahill 1998; Owen et al. 1967), in the process replacing glucose as the sole cerebral fuel.

In addition to starvation, a state of ketosis is inducible by a diet that provides 80–90% of calories as fatty acids. In this circumstance, the upregulation of moncarboxylate transporters at the level of the blood-brain barrier facilitate the entry into brain of 3-OH-butyrate and acetoacetate (Chiry et al. 2006; Pierre and Pellerin 2005). The rate of importation appears to be directly related to the ambient concentration of these compounds in blood, with both neurons and astrocytes demonstrating a capacity to switch from glucose to ketone bodies as a surrogate metabolic substrate. It is of interest that early in life, during the period of development of the human brain, the ketone bodies serve not only as a metabolic substrate, but as an important source of the acetyl-CoA that brain cells summon for the synthesis of myelin and other lipids (Gerhart et al. 1997; Koper et al. 1981; Lopes-Cardozo et al. 1984). In this regard, it is probably of teleologic significance that human milk is rich in fat, a circumstance that tends to favor ketosis (Dombrowski et al. 1989; Nehlig and Pereira de Vasconcelos 1993).

Oxidation of the ketone bodies involves the metabolism of 3-OH-butyrate to acetoacetate via 3-OH-butyrate dehydrogenase, a NAD-dependent enzyme. The succinyl-CoA transferase enzyme, which is abundant in brain, converts acetoacetate to acetoacetyl-CoA and succinate, the latter to be oxidized via succinate dehydrogenase in the tricarboxylic acid cycle (Fukao et al. 1997). A series of thiolases then hydrolyze acetoacetyl-CoA to acetyl-CoA, which enters the TCA cycle as citrate after condensation with oxaloacetate.

The carbon of either glucose or 3-OH-butyrate enters the tricarboxylic acid cycle as acetyl-CoA. Therefore, it might be thought that the choice of fuel should make no conspicuous difference to brain biochemistry and physiology. However, as shown in Fig. 28.2, whether brain respires on glucose or on 3-OH-butyrate might have a major influence on brain glutamate metabolism. Glucose handling initially must proceed through glycolysis, which yields NADH that the malate-aspartate shuttle transports to mitochondria. Glycolysis converts glucose carbon to pyruvate, which becomes decarboxylated to acetyl-CoA in the pyruvate dehydrogenase reaction. Glycolysis is restricted to the cytosol. In contrast, ketone body oxidation can occur *only* in mitochondria and *directly* to acetyl-CoA, the level of which is increased in brain of ketotic animals (Yudkoff et al. 2006). No intermediate comparable to pyruvate is interposed between ketone body oxidation and entry into the tricarboxylic acid cycle (Fig. 28.2). As noted above, when glucose is the sole metabolic substrate, astrocytes probably export to neurons a portion of glucose carbon as lactate (Magistretti et al. 1999). Hence, respiration on ketone bodies compared with respiration on glucose entails salient metabolic differences (Fig. 28.2): (a) More intense production of acetyl-CoA; (b) Greater flux through the citrate synthetase reaction (acetyl-CoA + oxaloacetate → citrate); and (c) Greater flux through the tricarboxylic acid cycle itself. A recent study (Melo et al. 2006) reflects this shift in metabolic organization. Rats received [1-^{13}C]glucose and [1,2-^{13}C]acetate in order to trace neuronal and glial metabolism, respectively. Comparisons were made between animals that consumed a control diet and those that received a ketogenic diet for a

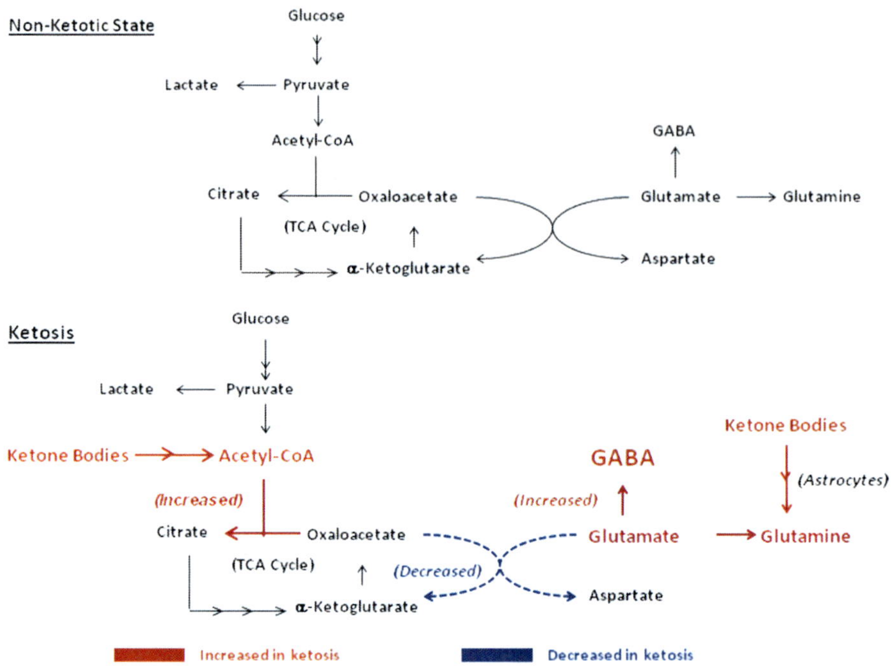

Fig. 28.2 This figure illustrates interactions of the metabolism of ketone bodies and that of amino acids and glucose. In the non-ketotic state (upper segment of the figure) the primary fuel of brain is glucose, which supplies almost all of cerebral energy requirements. The glycolytic pathway results in the production of pyruvate from glucose. Pyruvate dehydrogenase converts pyruvate to acetyl-CoA, which then enters the tricarboxylic acid cycle at the citrate synthetase reaction, involving condensation of acetyl-CoA with oxaloacetate. Glutamic acid forms from glucose carbon via the transamination α-ketoglutarate, an intermediate of the tricarboxylic acid cycle. With the consumption of a high–fat (ketogenic) diet (lower segment of Fig. 28.2) the levels of the ketone bodies increase both in blood and brain. Unlike glucose, the metabolism of which can proceed via glycolysis in order to derive energy, for these species there is no alternative to the tricarboxylic acid cycle. The ketone bodies *must* be converted to acetyl-CoA, which, as noted above, reacts with oxaloacetate to yield citrate via the citrate synthetase reaction (acetyl-CoA + oxaloacetate → citrate + CoA). In ketosis flow through the citrate synthetase reaction increases, thereby limiting the availability of oxaloacetate to the aspartate aminotransferase reaction (oxaloacetate + glutamate ↔ aspartate + α-ketoglutarate). Aspartate aminotransferase is a major pathway of glutamate metabolism. If less glutamate is so disposed because of reduced availability of oxaloacetate, more glutamate becomes accessible to glutamate decarboxylase for the production of GABA in GABA-ergic neurons. Furthermore, ketosis increases blood acetate and acetylcarnitine, which is metabolized in brain primarily in astrocytes. In glia the carbon of acetate becomes converted to glutamine, an excellent precursor to GABA in GABA-ergic neurons (see Fig. 28.1). GABA is the major inhibitory neurotransmitter. The red color denotes pathays that are intensified in ketotic brain. The blue color denotes pathways that are relatively less active

period of 21 days. The results indicated that the ketogenic diet was associated with diminished flux through glycolysis in neurons but increased consumption of acetate in the astrocytes.

Astrocytes have long been known to be a major site in brain for the oxidation of acetate as well as the synthesis of glutamine (Berl et al. 1962; Cerdan et al. 1990;

Martinez-Hernandez et al. 1977; Waniewski and Martin 1998). Both blood acetate and acetylcarnitine increase in the ketotic state (Scheppach et al. 1991). Hence, as indicated by studies with [1-^{13}C]glucose and [1,2-^{13}C]acetate, ketotic brain forms relatively more glutamine from acetate (Melo et al. 2006; Yudkoff et al. 2006).

In order to produce glutamine from acetate carbon, the latter must enter the tricarboxylic acid cycle via the citrate synthetase reaction, which is quite active in brain, the maximal velocity being more than tenfold greater than comparable flux through pyruvate dehydrogenase (Ratnakumari and Murthy 1989). Indeed, in synaptosomes the rate of flux through pyruvate dehydrogenase is barely equal to the metabolic rate (Erecinska and Dagani 1990). Hence, as the rate of flow of acetyl-CoA through citrate synthetase increases, there occurs a relative diminution in the level of oxaloacetate and a concomitant limitation of transamination of glutamate to aspartate, a key pathway of brain glutamate handling (Escartin et al. 2006; Mason et al. 1995; McKenna et al. 1993; Yudkoff et al. 1996, 2005) (Fig. 28.2). Aspartate aminotransferase is an equilibrium enzyme, the rate of flux through which must depend in part upon the concentration of the reactants. If oxaloacetate is lower in the brain during ketosis, then we anticipate a lowering of aspartate synthesis (DeVivo et al. 1978; Yudkoff et al. 2005).

Some experimental data support this hypothesis. We traced nitrogen metabolism in cultured astrocytes that we incubated with [^{15}N]glutamate. In the presence of acetoacetate (5 mmol/l) we observed diminished synthesis of [^{15}N]aspartate and no change of internal glutamate (Yudkoff et al. 2005). The latter observation indicated that a change of astrocytic glutamate transport was not likely. Injection of ketone bodies into suckling mice reduces brain aspartate levels (Thurston et al. 1986). Similarly, feeding a high-fat diet to rats lowers the concentration of this amino acid (DeVivo et al. 1978). Making animals ketotic for 21 days reduced production of [2-^{13}C]aspartate from [1-^{13}C]glucose, although there was no frank reduction in total brain aspartate (Melo et al. 2006).

It is possible that lessened transamination of glutamate to aspartate could exert an anti-epileptic effect. Depolarization of neurons favors glutamate transamination (McKenna et al. 1993; Peng and Hertz 1993; Waagepetersen et al. 2000a). Levels of aspartate also increase following a seizure (Carlson et al. 1992; Do et al. 1991). Aspartate itself is an excitatory neurotransmitter that may be of pathogenic significance in hippocampal epilepsy (Fleck et al. 1993; Martin et al. 1991; Meldrum 1994; Raiteri et al. 1990).

If less glutamate is transaminated to aspartate, it is possible that more glutamate could be decarboxylated to GABA (glutamate $\rightarrow$ GABA + CO_2). It is clear that GABA is the major inhibitory neurotransmitter and probably is important in the anti-epileptic effect of many anti-convulsants (DeDeyn et al. 1990; Lasley and Yan 1994; Gould et al. 1995; Hovanics et al. 1997; Lefevre and Aronson 2000; Loscher and Swark 1985; Meldrum 1994; Meldrum 2000; Olsen and Avoli 1997; Petroff 1996, 2002; White 1997). The glutamate decarboxylase reaction (K_m 0.1–1.2 mmol/l) may appear to be saturated at typical brain glutamate concentrations (8–10 mmol/l), but the level may be appreciably lower in GABA-ergic neurons (Storm-Mathisen et al. 1983; Storm-Mathisen and Ottersen 1986). Furthermore, less than 5% of total brain glutamate is a precursor to GABA (Patel et al. 1970; Patel et al. 1974) in GABA-ergic neurons (Storm-Mathisen et al. 1983; Storm-Mathisen and Ottersen 1986), which are less than one-fifth of all neurons. In addition, the flux through glutamate decarboxylase (~0.5 nmol/min.mg protein^{-1}) is tightly controlled and is far less than the Vmax for the enzyme (Battaglioli and Martin 1990). Hence, even small changes in the concentration of glutamate might affect GABA synthesis (Paulsen and Fonnum 1989). Finally, there are many factors that affect the binding of glutamate to the decarboxylase, including aspartate and Cl^- (Wu 1976; Porter and Martin 1988).

We noted a very rapid and significant increase of GABA in synaptosomes when this system was incubated with acetoacetate (5 mmol/l). Exposure of synaptosomes to either [^{15}N]glutamate or [2,3,3,4,4-d_5]glutamine (0.5 mmol/l) sharply increased the synthesis of either [^{15}N]GABA or [$^2H_4$]GABA (Erecinska et al. 1996). Intra-synaptosomal concentrations of aspartate were reduced by exposure to acetoacetate, again indicating a diversion of glutamate away from transamination to aspartate and toward decarboxylation to GABA (Erecinska et al. 1996). In these studies the actual source of GABA was glutamine, which GABA-ergic neurons efficiently transport and then hydrolyze to glutamate via phosphate-dependent glutaminase, a mitochondrial enzyme (Sonnewald et al. 2004). Indeed, it may be that glutamine is a more effective external precursor to GABA than is glutamate itself (Sonnewald et al. 1993). We have found (Yudkoff et al. 2006) that ketosis increases brain [^{13}C]glutamine synthesis from [1-^{13}C]acetate. Similarly, Melo et al. observed (2006) that administration of [1,2-^{13}C]acetate to ketotic epilepsy-prone rats (GAER rats) resulted in increased brain synthesis of [4,5-^{13}C]glutamate and [4,5-^{13}C]glutamine. As noted above, brain acetate metabolism is primarily of glial origin. The augmented ^{13}C-glutamine synthesis denotes in part more intense activity of pyruvate carboxylase, a glial enzyme (Shank et al. 1985) that restores to the tricarboxylic acid cycle a portion of the carbon that is consumed by dint of transamination of 2-oxo-glutarate to glutamate.

The picture that we can draw from these findings is that ketosis appears to accentuate overall astrocyte metabolism. The result is increased synthesis of glutamine, an excellent precursor to GABA. It is intuitively likely that the conversion of glutamine to GABA will be most active in the presence of increased neuronal activity, the most extreme example of which would be expected to occur consequent to a seizure. It may be that ketosis does not evoke increased steady-state levels of GABA but an increased capacity to synthesize GABA when the need arises, a demand that should become most acute during a seizure. A capacity for greater glutamine synthesis also would confer an anti-epileptic effect by favoring the conversion of excitatory glutamate into glutamine. Thus, ketosis theoretically could exert a dual action: (a) enhanced disposal of an excitatory factor (glutamate) and (b) enhanced formation of an inhibitory influence (GABA). This formulation is consistent with the finding (Melo et al. 2006) that ketosis evokes a modest (~ 13%) (although statistically insignificant) increase in the GABA concentration in the cortex. Ketosis did recruit a greater flow of glutamate carbon into GABA, as we noted previously in synaptosomes (Erecinska et al. 1996).

Investigations in humans also are consistent with the formulation that one component in the anti-epileptic effect of the ketogenic diet might be an influence on brain GABA handling. Thus, a recent investigation (Wang et al. 2003) with magnetic resonance spectroscopy in three patients revealed that, after a few months of therapy, the concentration of brain GABA increased by 52% and 34% in two patients. The third individual did not manifest such a change and even may have experienced a small (12%) decline in the concentration of this inhibitory transmitter.

Dahlin et al. (2005) studied the concentration of amino acids in the cerebrospinal fluid of children at 4 months following initiation of the diet. They noted a significant (~11%) increase in the GABA level and no significant change of either glutamate or aspartate. GABA levels in cerebrospinal fluid were highest in those patients who were categorized as "very good" responders to the ketogenic diet. For this group the pre- and post-treatment concentrations of CSF GABA were 4.47 and 4.8 μmol/l. In "good responders" the comparable values were 3.24 and 3.77 μmol/l. Interestingly, the concentration of alanine was diminished by nearly 25%. Patients who failed to respond to the ketogenic regimen had far lower CSF GABA levels both before and after treatment: 2.55 and 2.79 μmol/l, respectively.

28.6 Effects of Ketosis on Brain Amino Acid Transport

A complex skein of transport mechanisms subserves amino acid uptake into the brain (Broer and Brookes 2001; Hawkins et al. 2006). A key component of this transport system is the neutral amino acid transporter (L transporter), which mediates the uptake of most brain amino acids, including the branched-chain amino acids, via exchange with brain glutamine (Boado et al. 1999; Broer and Brookes 2001; Huang et al. 1996). A high brain glutamine level tends to favor such transport, although the affinity of the transporter for glutamine is not great (Yanagida et al. 2001). Ketosis tends to favor the entry of leucine into the brain (Yudkoff et al. 2006). A corollary of this phenomenon would be increased release of glutamine in the ketotic state, but this has not yet been directly tested. Obviously, the putative "loss" of nitrogen as glutamine would then be compensated by more intense accumulation of leucine (Hutson et al. 1992; Yudkoff et al. 1997). A heightened release of glutamine from brain would confer on the system another mechanism for the transient removal of glutamate if synaptic levels of this amino acid are elevated. It is tempting to speculate that this is the case, although there is no experimental data to support such a formulation.

Regrettably little is known of the effect(s) of ketosis on brain amino acid transport. However, it is clear that the ketotic state induces substantial alterations in the distribution ratio of individual amino acids between the brain and the blood. A notable example is alanine, for which fasting and ketosis causes a large decline in the blood concentration and little change in the level in brain, thereby resulting in a sharply increased brain:blood ratio (Yudkoff et al. 2006). Alanine is a key gluconeogenic precursor that is released by skeletal muscle and utilized by liver. Hence, when the organism fasts or consumes a low-carbohydrate diet, blood levels of this amino acid decline. Alanine also is a potentially important shuttle of $-NH_2$ groups in the brain (Waagepetersen et al. 2000b). It is tempting to speculate that the increased brain:blood ratio for alanine (and, conceivably, for other amino acids) in the ketotic state affords a gradient that favors the release of this amino acid to the blood, in the process effectively removing from brain nitrogen formerly resident in glutamate and favoring the oxidation of glutamate carbon:

1. Glutamate + Pyruvate → Alanine + 2-oxo-glutarate
2. Alanine (BRAIN) → Alanine (BLOOD)
3. 2-oxo-glutarate → → $CO_2 + H_2O$
4. Sum: Glutamate + Pyruvate → Alanine (BLOOD) + $CO_2 + H_2O$

It is thought that the direct export of glutamate from brain may occur (Hawkins et al. 2006; O'Kane et al. 1999, 2004). Whether a similar motif is applicable to other amino acids is uncertain, but if in ketosis these are present in high amount relative to the blood concentration, then the system might enjoy yet another mechanism for the rapid removal of excitatory glutamate from the synapse.

28.7 Summary

Even after decades of assiduous research, we still do not understand precisely how the ketogenic diet so remarkably attenuates a seizure diathesis in patients whose convulsions have proved otherwise refractory to aggressive intervention with anti-epileptic drugs. It is highly likely that there is no single mechanism to account for the therapeutic effect, but a substantial body of research points to adaptive changes in brain handling of neurotransmitter amino acids as operative factors. A shift in cerebral metabolism from glucose as the sole fuel to an energy economy based in part on ketone bodies (and, probably, acetate) appears to occasion important changes in handling of glutamate, glutamine and GABA. Many of these adaptations seem to occur in astrocytes, in which flux through glutamine synthetase becomes more intense, thereby providing more precursor for the synthesis of GABA in GABA-ergic neurons and facilitating the removal from the system of excitatory glutamate (Fig. 28.1). In addition, changes in the relative distribution of amino acids between brain and blood may enable the direct exportation of amino acids such as glutamine from the central nervous system.

In thinking about the biochemial mechanisms that underlie the efficacy of the ketogenic diet, it is important to bear in mind a crucial difference between this nutritional intervention and the anti-epileptic drugs. The latter often control seizures, but in most instances they also tend to "dampen" overall neurologic function, a phenomenon that becomes clinically evident as lethargy, changes in overall cognitive competence and alterations of memory and mood. It is not surprising that medications should induce such side-effects, since most of them either increase the tone of inhibitory receptors or inhibit the action of ion channels. In contrast, the ketogenic diet confers a therapeutic effect without causing such unwanted side-effects. A possible reason for this desirable outcome is that the diet does not alter the steady-state concentration of pertinent neurotransmitters such as glutamate and GABA. Hence, the diet does not change the basal inhibitory or excitatory "tone". Instead, the diet evokes the desired metabolic changes only when neuronal depolarization is most intense and the release of glutamate into the synaptic space is most active. In this circumstance the diet enables the overall competence of the system effectively to dispose of glutamic acid and augment availability of GABA. Additional research is needed in order to scrutinize this concept in finer detail.

References

Al-Mudallal AS, LaManna JC, Lust WD, Harik SI (1996) Diet-induced ketosis does not cause cerebral acidosis. Epilepsia 37:258–261

Amara SG, Fontana AC (2002) Excitatory amino acid transporters: keeping up with glutamate. Neurochem Int 41:313–318

Battaglioli G, Martin DL (1990) Stimulation of synaptosomal γ-aminobutyric acid synthesis by glutamate and glutamine. J Neurochem 54:1179–1187

Bergqvist AG, Schall JI, Gallagher PR, Cnaan A, Stallings VA (2005) Fasting versus gradual initiation of the ketogenic diet: a prospective, randomized clinical trial of efficacy. Epilepsia 46:1810–1819

Berl S, Takagaki G, Clarke DD, Waelsch H (1962) Metabolic compartments in vivo. Ammonia and glutamic acid metabolism in brain and liver. J Biol Chem 237:2562–2569

Bixel MG, Hutson SM, Hamprecht B (1997) Cellular distribution of branched-chain amino acid aminotransferase isoenzymes among rat brain glial cells in culture. J Histochem Cytochem 45:685–694

Boado RJ, Li JY, Nagaya M, Zhang C, Pardridge WM (1999) Selective expression of the large neutral amino acid transporter at the blood-brain barrier. Proc Natl Acad Sci 96:12079–12084

Bough KJ, Schwartzkroin PA, Rho JM (2003) Calorie restriction and ketogenic diet diminish neuronal excitability in rat dentate gyrus in vivo. Epilepsia 44:752–760

Bravata DM, Sanders L, Huang J, Krumholz HM, Olkin I, Gardner CD, Bravata DM (2003) Efficacy and safety of low-carbohydrate diets: a systematic review. JAMA 289:1837–1850

Broer S, Brookes N (2001) Transfer of glutamine between astrocytes and neurons. J Neurochem 77:705–719

Burdakov D, Gerasimenko O, Verkhratsky A (2005) Physiological changes in glucose differentially modulate the excitability of hypothalamic melanin-concentrating hormone and orexin neurons in situ. J Neurosci 25:2429–2433

Cahill GF Jr (1998) Survival in starvation. Am J Clin Nutr 68:1–2

Camfield CR, Camfield CS (1996) Antiepileptic drug therapy: when is epilepsy truly intractable? Epilepsia 37:S60–S65

Carlson H, Ronne-Engstrum E, Ungerstedt U, Hillered L (1992) Seizure-related elevations of extracellular amino acids in human focal epilepsy. Neurosci Lett 140:30–32

Cerdan S, Kunnecke B, Seelig J (1990) Cerebral metabolism of [1,2-^{13}C2]acetate as detected by in vivo and in vitro ^{13}C NMR. J Biol Chem 265:12916–12926

Chaudhry FA, Reimber RJ, Edward RH (2002) The glutamine commute: take the N line and transfer to the A. J Cell Biol 157:349–355

Cheng CM, Kelley B, Wang J, Strauss D, Eagles DA, Bondy CA (2003) A ketogenic diet increases brain insulin-like growth factor receptor and glucose transporter gene expression. Endocrinology 144:2676–2682

Chiry O, Pellerin L, Monnet-Tschudi F, Fishbein WN, Merezhinskaya N, Magistretti PJ, Clarke S (2006) Expression of the monocarboxylate transporter MCT1 in the adult human brain cortex. Brain Res 1070:65–70

Clarke DD, Sokoloff L (1999) Circulation and energy metabolism of the brain. In: Siegel GJ, Agranoff BW, Albers RW, Fisher SK, Uhler MD (eds) Basic Neurochemistry: Molecular, Cellular and Medical Aspects, 6th edn. Lippincott-Raven, Philadelphia, pp 637–669

Cooper AJL, McDonald JM, Gelbard AS, Gledhill RF, Duffy TE (1979) The metabolic fate of ^{13}N-labelled ammonia in rat brain. J Biol Chem 254:4982–4992

Cunnane SC, Musa K, Ryan MA, Whiting S, Fraser DD (2002) Potential role of polyunsaturates in seizure protection achieved with the ketogenic diet. Prostaglandins Leukot Essent Fatty Acids 67:131–135

Dahlin M, Elfving A, Ungerstedt U, Amark P (2005) The ketogenic diet influences the levels of excitatory and inhibitory amino acids in the CSF in children with refractory epilepsy. Epilepsy Res 64:115–125

Daikhin Y, Yudkoff M (1998) Ketone bodies and brain glutamate and GABA metabolism. Dev Neurosci 20:358–364

Danbolt NC (2001) Glutamate uptake. Prog Neurobiol 65:1–105

DeDeyn PP, Marescau B, MacDonald RL (1990) Epilepsy and the GABA-hypothesis: a brief review and some samples. Acta Neurol Belg 90:65–81

DeVivo DC, Leckie MP, Ferrendelli JS, McDougal DB Jr (1978) Chronic ketosis and cerebral metabolism. Ann Neurol 3:331–337

Do KQ, Klancnik J, Gahwiler BH et al (1991) Release of EAA: animal studies and epileptic foci studies in humans. In: Meldrum BS, Moroni F, Simon RP (eds) Excitatory Amino Acids. Raven, New York, pp 677–685

Dombrowski GJ Jr, Swiatek KR, Chao K-L (1989) Lactate, 3-hydroxybutyrate and glucose as substrates for early postnatal rat brain. Neurochem Res 14:667–675

Eagles DA, Boyd SJ, Kotak A, Allan F (2003) Calorie restriction of a high-carbohydrate diet elevates the threshold of PTZ-induced seizures to values equal to those seen with a ketogenic diet. Epilepsy Res 54:41–52

Erecinska M, Dagani F (1990) Relationships between the neuronal sodium/potassium pump and energy metabolism. Effects of K^+, Na^+ and adenosine triphosphate in isolated brain synaptosomes. J Gen Physiol 95:591–616

Erecinska M, Silver IA (1990) Metabolism and role of glutamate in mammalian brain. Progr Neurobiol 35:245–296

Erecinska M, Nelson D, Daikhin Y, Yudkoff M (1996) Regulation of GABA level in rat brain synaptosomes: fluxes through enzymes of the GABA shunt and effects of glutamate, calcium and ketone bodies. J Neurochem 67:2325–2334

Escartin C, Valette J, Lebon V, Bonvento G (2006) Neuron–astrocyte interactions in the regulation of brain energy metabolism: a focus on NMR Spectroscopy. J Neurochem 99:393–401

Felig P, Pozefsk T, Marlis E, Cahill GF (1970) Alanine: key role in gluconeogenesis. Science 167:1003–1004

Fleck MW, Henze DA, Barrionuevo G, Palmer AM (1993) Aspartate and glutamate mediate excitatory synaptic transmission in area CA1 of the hippocampus. J Neurosci 13:3944–3955

Freeman J, Veggiotti P, Lanzi G, Tagliabue A, Perucca E (2006) The ketogenic diet: From molecular mechanisms to clinical effects. Epilepsy Res 68:145–180

Freeman JM, Vining EP, Pillas DJ et al (1998) The efficacy of the ketogenic diet-1998: a prospective evaluation of intervention in 150 children. Pediatrics 102:1358–1363

Fuehrlein BS, Rutenberg MS, Silver JN, Warren MW, Theriaque DW, Duncan GE, Stacpoole PW, Brantly ML (2004) Differential metabolic effects of saturated versus polyunsaturated fats in ketogenic diets. J Clin Endocrinol Metab 89:1641–1645

Fukao T, Song XQ, Mitchell GA, Yamaguchi S, Sukegawa K, Orii T, Kondo N (1997) Enzymes of ketone body utilization in human tissues: protein and messenger RNA levels of succinyl-coenzyme A. CoA.:3-ketoacid CoA transferase and mitochondrial and cytosolic acetoacetyl-CoA thiolases. Pediatr Res 42:498–502

Gegelashvili G, Schousboe A (1998) Cellular distribution and kinetic properties of high-affinity glutamate transporters. Brain Res Bull 45:233–238

Gerasimov MR, Ferrieri RA, Pareto D, Logan J, Alexoff D, Ding YS (2005) Synthesis and evaluation of inhaled [^{11}C]butane and intravenously injected [^{11}C]acetone as potential radiotracers for studying inhalant abuse. Nucl Med Biol 32:201–208

Gerhart DZ, Enerson BE, Zhdankina OY, Leino RL, Drewes LR (1997) Expression of monocarboxylate transporter MCT1 by brain endothelium and glia in adult and suckling rats. Am J Physiol 273:E207–E213

Gould EM, Curto KA, Craig CR, Fleming WW, Taylor DA (1995) The role of GABA-A receptors in the subsensitivity of Purkinje neurons to GABA in genetic epilepsy prone rats. Brain Res 698:62–68

Greene AE, Todorova MT, McGowan R, Seyfried TN (2001) Caloric restriction inhibits seizure susceptibility in epileptic EL mice by reducing blood glucose. Epilepsia 42:1371–1378

Grill V, Björkhem M, Gutniak M, Lindqvist M (1992) Brain uptake and release of amino acids in nondiabetic and insulin-dependent diabetic subjects: important role of glutamine release for nitrogen balance. Metabolism 41:28–32

Hawkins RA, O'Kane RL, Simpson IA, Vina JR (2006) Structure of the blood-brain barrier and its role in the transport of amino acids. J Nutr 136:218S–226S

Hertz L, Peng L, Westergaard N, Yudkoff M, Schousboe A (1992) Neuronal-astrocytic interactions in metabolism of transmitter amino acids of the glutamate family. In: Schousboe A, Diemer NH, Kofod H (eds) Drug Research Related to Neuroactive Amino Acids, Alfred Benzon Symposium 32. Munksgaard, Copenhagen, pp 30–48

Hertz L, Peng L, Dienel GA 2007 Energy metabolism in astrocytes: high rate of oxidative metabolism and spatiotemporal dependence on glycolysis/glycogenolysis. J Cereb Blood Flow Metab [Epub ahead of print]

Hovanics GE, DeLorey TM, Firestone LL, Quinlan JJ, Handforth A, Harrison NL, Krasowski MD, Rick CE, Korpi ER, Makela R, Brilliant MH, Hagiwara N, Ferguson C, Snyder K, Olsen RW (1997) Mice devoid of gamma-aminobutyrate type A receptor beta3 subunit have epilepsy, cleft palate, and hypersensitive behavior. Proc Natl Acad Sci USA 94:4143–4148

Huang Y, Zielke HR, Tildon JT, Zielke CL, Baab PJ (1996) Elevation of amino acids in the interstitial space of the rat brain following infusion of large neutral amino and keto acids by microdialysis: leucine infusion. Dev Neurosci 18:415–419

Hutson SM, Wallin R, Hall TR (1992) Identification of mitochondrial branched chain aminotransferase and its isoforms in rat tissues. J Biol Chem 267:15681–15686

Kanamori K, Ross BD, Kondrat RW (1998) Rate of glutamate synthesis from leucine in rat brain measured in vivo by ^{15}N NMR. J Neurochem 70:1304–1315

Kanamori K, Ross BD (2004) Quantitative determination of extracellular glutamine concentration in rat brain, and its elevation in vivo by system A transport inhibitor, alpha-(methylamino) isobutyrate. J Neurochem 90:203–210

Kety SS, Schmidt CF (1948) The nitrous oxide method for the quantitative determination of cerebral blood flow in man: theory, procedure, and normal values. J Clin Invest 27:476–483

Koper JW, Lopes-Cardozo M, Van Golde LM (1981) Preferential utilization of ketone bodies for the synthesis of myelin cholesterol in vivo. Biochem Biophys Acta 666:411–417

Kossoff EH, Pyzik PL, McGrogan JR, Vining EP, Freeman JM (2002) Efficacy of the ketogenic diet for infantile spasms. Pediatrics 109:780–783

Kossoff EH, Krauss GL, McGrogan JR, Freeman JM (2003) Efficacy of the Atkins diet as therapy for intractable epilepsy. Neurology 61:1789–1791

Kossoff EH, Thiele EA, Pfeifer HH, McGrogan JR, Freeman JM (2005) Tuberous sclerosis complex and the ketogenic diet. Epilepsia 46:1684–1686

Kwiterovich PO Jr, Vining EP, Pyzik P, Skolasky R Jr, Freeman JM (2003) Effect of a high-fat ketogenic diet on plasma levels of lipids, lipoproteins, and apolipoproteins in children. JAMA 290:912–920

Lapidot A, Haber S (2002) Effect of endogenous β-hydroxybutyrate on brain glucose metabolism in fetuses of diabetic rabbits, studied by 13 C magnetic resonance spectroscopy. Devel Brain Res 135:87–99

Lasley SM, Yan QS (1994) Diminished potassium-stimulated GABA release in vivo in genetically epilepsy-prone rats. Neurosci Lett 175:145–148

Lefevre F, Aronson A (2000) Ketogenic diet for the treatment of refractory epilepsy in children: a systematic review of efficacy. Pediatrics 105:E46

Likhodii SS, Serbanescu I, Cortez MA, Murphy P, Snead OC 3rd, Burnham WM (2003) Anticonvulsant properties of acetone, a brain ketone elevated by the ketogenic diet. Ann Neurol 54:219–226

Lopes-Cardozo M, Koper JW, Klein W, Van Golde LM (1984) Acetoacetate is a cholesterogenic precursor for myelinating rat brain and spinal cord. Incorporation of label from [3-14 C]acetoacetate, [14 C]glucose and 3H2O. Biochem Biophys Acta 794:350–352

Loscher W, Swark WS (1985) Evidence for impaired GABAergic activity in the substantia nigra of amygdaloid kindled rats. Brain Res 339:146–150

Mady MA, Mc Kossoff EH, Gregor AL, Wheless JW, Pyzik PL, Freeman JM (2003) The ketogenic diet: adolescents can do it, too. Epilepsia 44:847–851

Magistretti PJ, Pellerin L, Rothman DL, Shulman RG (1999) Energy on demand. Science 283:496–497

Mantis JG, Centeno NA, Todorova MT, McGowan R, Seyfried TN (2004) Management of multifactorial idiopathic epilepsy in EL mice with caloric restriction and the ketogenic diet: role of glucose and ketone bodies. Nutr Metab 1:1–11

Martin D, Bustos GA, Bowe MA, Bray SD, Nadler JV (1991) Autoreceptor regulation of glutamate and aspartate release from slices of the hippocampal CA1 area. J Neurochem 56:1647–1655

Martinez-Hernandez A, Bell KP, Norenberg MD (1977) Glutamine synthetase: glial localization in brain. Science 195:1356–1358

Mason GF, Gruetter R, Rothman DL, Behar KL, Shulman RG, Novotny EJ (1995) Simultaneous determination of the rates of the TCA cycle, glucose utilization, alpha-ketoglutarate/glutamate exchange, and glutamine synthesis in human brain by NMR. J Cereb Blood Flow Metab 15:12–25

Mattson MP, Liu D (2003) Mitochondrial potassium channels and uncoupling proteins in synaptic plasticity and neuronal cell death. Biochem Biophys Res Comm 304:539–549

McKenna MC, Tildon JT, Stevenson JH, Boatright R, Huang S (1993) Regulation of energy metabolism in synaptic terminals and cultured rat brain astrocytes: Differences revealed using aminooxyacetate. Dev Neurosci 15:320–329

Meldrum BS (1994) The role of glutamate in epilepsy and other CNS disorders. Neurology 44:S14–S23

Meldrum BS (2000) Glutamate as a neurotransmitter in the brain: review of physiology and pathology. J Nutr 130:1007S–1015S

Melo TM, Nehlig A, Sonnewald U (2006) Neuronal-glial interactions in rats fed a ketogenic diet. Neurochem Int 48:498–507

Neal EG, Schwartz RH, Lawson MS, Edward N, Fitzsimmons G, Whitney A, Cross JH (2008) The ketogenic diet for the treatment of childhood epilepsy: a randomised controlled trial. Lancet Neurol 7:500–506

Nehlig A, Pereira de Vasconcelos A (1993) Glucose and ketone body utilization by the brain of neonatal rats. Progr Neurobiol 40:163–221

Noh HS, Kim DW, Cho GJ, Choi WS, Kang SS (2006) Increased nitric oxide caused by the ketogenic diet reduces the onset time of kainic acid-induced seizures in ICR mice. Brain Res 1075:193–200

Nordli DR Jr, Kuroda MM, Carroll J, Koenigsberger DY, Hirsch LJ, Bruner HJ, Seidel WT, De Vivo DC (2001) Experience with the ketogenic diet in infants. Pediatrics 108:129–133

Norenberg MD, Martinez-Hernandez A (1979) Fine structural localization of glutamine synthetase in astrocytes of rat brain. Brain Res 161:303–310

O'Kane RL, Martinez-Lopez I, DeJoseph MR, Vina JR, Hawkins RA (1999) Na(+)-dependent glutamate transporters (EAAT1, EAAT2, and EAAT3) of the blood-brain barrier. A mechanism for glutamate removal. J Biol Chem 274:31891–31895

O'Kane RL, Vina JR, Simpson I, Hawkins RA (2004) Na+–dependent neutral amino acid transporters A, ASC, and N of the blood-brain barrier: mechanisms for neutral amino acid removal. Am J Physiol Endocrinol Metab 287:E622–E629

Olney JW (2003) Excitotoxicity, apoptosis and neuropsychiatric disorders. Curr Opin Pharmacol 3:101–109

Olsen RW, Avoli M (1997) GABA and epileptogenesis. Epilepsia 38:399–407

Olstad E, Qu H, Sonnewald U 2006 Glutamate is preferred over glutamine for intermediary metabolism in cultured cerebellar neurons. J Cereb Blood Flow Metab [Epub ahead of print]

Owen OE, Morgan AP, Kemp HG, Sullivan JM, Herrera MG, Cahill GF (1967) Brain metabolism during fasting. J Clin Invest 46:1589–1595

Pan JW, Bebin EM, Chu WJ, Hetherington HP (1999) Ketosis and epilepsy: ^{31}P spectroscopic imaging at 4.1 T. Epilepsia 40:703–707

Patel AJ, Balazs R, Richter D (1970) Contribution of the GABA bypath to glucose oxidation, and the development of compartmentation in the brain. Nature 226:1160–1161

Patel AJ, Johnson AL, Balazs R (1974) Metabolic compartmentation of glutamate associated with the formation of γ-aminobutyrate. J Neurochem 23:1271–1279

Paulsen RE, Fonnum F (1989) Role of glial cells for the basal and Ca^{2+}-dependent K^{+}-evoked release of transmitter amino acids investigated by microdialysis. J Neurochem 52:1823–1829

Peng L, Hertz L (1993) Potassium-induced stimulation of oxidative metabolism of glucose in cultures of intact cerebellar granule cells but not in corresponding cells with dendritic degeneration. Brain Res 629:331–334

Petroff OA, Rothman D, Behar KL, Mattson RH (1996) Low brain GABA level is associated with poor seizure control. Ann Neurol 40:908–911

Petroff OA (2002) GABA and glutamate in the human brain. Neuroscientist 8:562–573

Pierre K, Pellerin L (2005) Monocarboxylate transporters in the central nervous system: distribution, regulation and function. J Neurochem 94:1–14

Plaitakis A, Zaganas I (2001) Regulation of human glutamate dehydrogenases: implications for glutamate, ammonia and energy metabolism in brain. J Neurosci Res 66:899–908

Porter TG, Martin DL (1988) Stability and activation of glutamate apodecarboxylase from pig brain. J Neurochem 51:1886–1891

Prins ML, Lee SM, Fujima LS, Hovda DA (2004) Increased cerebral uptake and oxidation of exogenous bHB improves ATP following traumatic brain injury in adult rats J. Neurochem 90:666–672

Raiteri M, Marchi M, Costi A, Volpe G (1990) Endogenous aspartate release in the rat hippocampus inhibited by M2 "cardiac" muscarinic receptors. Eur J Pharmacol 177:181–187

Ratnakumari L, Murthy CRK (1989) Activities of pyruvate dehydrogenase, enzymes of citric acid cycle, and aminotransferases in the subcellular fractions of cerebral cortex in normal and hyperammonemic rats. Neurochem Res 14:221–228

Sato K, Kashiwaya Y, Keon CA, Tsuchiya N, King MT, Radda GK, Chance B, Clarke K, Veech RL (1995) Insulin, ketone bodies, and mitochondrial energy transduction. FASEB J 9:651–658

Scheppach W, Pomare EW, Elia M, Cummings JH (1991) The contribution of the large intestine to blood acetate in man. Clin Sci 80:177–182

Schousboe A, Waagepetersen HS (2005) Role of astrocytes in glutamate homeostasis: implications for excitotoxicity. Neurotox Res 8:221–225

Schurr A, Miller JJ, Payne RS, Rigor BM (1999) An increase in lactate output by brain tissue serves to meet the energy needs of glutamate-activated neurons. J Neurosci 19:34–39

Schwechter EM, Veliskova J, Velisek L (2003) Correlation between extracellular glucose and seizure susceptibility in adult rats. Ann Neurol 53:91–101

Shank RP, Aprison MH (1977) Present status and significance of the glutamine cycle in neural tissues. Life Sci 28:837–842

Shank RP, Bennett GS, Freytag SO, Campbell GL (1985) Pyruvate carboxylase: an astrocyte-specific enzyme implicated in the replenishment of amino acid neurotransmitter pools. Brain Res 329:364–367

Siesjo BK (1997) Brain Energy Metabolism. Wiley, New York

Smith QR, Momma S, Aoyagi M, Rapoport SI (1987) Kinetics of neutral amino acid transport across the blood-brain barrier. J Neurochem 49:1651–1658

Sonnewald U, Westergaard N, Schousboe A, Svendsen JS, Unsgard G, Petersen SB (1993) Direct demonstration by [^{13}C]NMR spectroscopy that glutamine from astrocytes is a precursor for GABA synthesis in neurons. Neurochem Int 22:19–29

Sonnewald U, Schousboe A, Qu H, Waagepetersen HS (2004) Intracellular metabolic compartmentation assessed by 13 C magnetic resonance spectroscopy. Neurochem Int 45:305–310

Storm-Mathisen J, Leknes AK, Bore AT, Vaaland JL, Edminson P, Haug FM, Ottersen OP (1983) First visualization of glutamate and GABA in neurones by immunocytochemistry. Nature 301:517–520

Storm-Mathisen J, Ottersen OP (1986) Antibodies against amino acid neurotrnasmitters. In: Paunula P, Paivarinta H, Soinila S (eds) Neurohistochemistry: Modern Methods and Applications. New York, Alan R. Liss, New York, pp 107–136

Sullivan PG, Dube C, Dorenbos K, Steward O, Baram TZ (2003) Mitochondrial uncoupling protein-2 protects the immature brain from excitotoxic neuronal death. Ann Neurol 53:711–717

Sullivan PG, Rippy NA, Dorenbos K, Concepcion RC, Agarwal AK, Rho JM (2004) The ketogenic diet increases mitochondrial uncoupling protein levels and activity. Ann Neurol 55:576–580

Swink TD, Vining EP, Freeman JM, Freeman JM (1997) The ketogenic diet. Adv Pediatr 44:297–329

Szot P, Weinshenker D, Rho JM, Storey TW, Schwartzkroin PA (2001) Norepinephrine is required for the anticonvulsant effect of the ketogenic diet. Brain Res Dev Brain Res 129:211–214

Takahashi M, Billups B, Rossi D, Sarantis M, Hamann M, Attwell D (1997) The role of glutamate transporters in glutamate homeostasis in the brain. J Exper Biol 200:401–409

Thiele EA (2003) Assessing the efficacy of antiepileptic treatments: the ketogenic diet. Epilepsia 44:26–29

Thurston JH, Hauhart RE, Schiro JA (1986) Beta-hydroxybutyrate reverses insulin-induced hypoglycemic coma in suckling-weanling mice despite low blood and brain glucose levels. Metab Brain Dis 1:63–82

Tieu K, Perier C, Caspersen C, Teismann P, Wu DC, Yan SD, Naini A, Vila M, Jackson-Lewis V, Ramasamy R, Przedborski S (2003) D-beta-hydroxybutyrate rescues mitochondrial respiration and mitigates features of Parkinson disease. J Clin Invest 112:892–901

Vamecq J, Vallee L, Lesage F, Gressens P, Stables JP (2005) Antiepileptic popular ketogenic diet: emerging twists in an ancient story. Prog Neurobiol 75:1–28

Veech RL, Chance B, Kashiwaya Y, Lardy HA, Cahill GF Jr (2001) Ketone bodies, potential therapeutic uses. IUBMB Life 51:241–247

Veech RL (2004) The therapeutic implications of ketone bodies: the effects of ketone bodies in pathological conditions: ketosis, ketogenic diet, redox states, insulin resistance, and mitochondrial metabolism. Prostaglandins Leukot Essent Fatty Acids 70:309–319

Vining EP (1999) Clinical efficacy of the ketogenic diet. Epilepsy Res 37:181–190

Vining EPG, Pyzik P, McGrogan J, Hladky H, Anand A, Kriegler S, Freeman JM (2002) Growth of children on the ketogenic diet. Dev Med Child Neurol 44:796–802

Waagepetersen HS, Bakken IJ, Larsson OM, Sonnewald U, Schousboe A (1998) Comparison of lactate and glucose metabolism in cultured neocortical neurons and astrocytes using ^{13}C-NMR spectroscopy. Dev Neurosci 20:310–320

Waagepetersen HS, Sonnewald U, Larsson OM, Schousboe A (2000a) Compartmentation of TCA cycle metabolism in cultured neocortical neurons revealed by ^{13}C MR spectroscopy. Neurochem Int 36:349–358

Waagepetersen HS, Sonnewald U, Larsson OM, Schousboe A (2000b) A possible role of alanine for ammonia transfer between astrocytes and glutamatergic neurons. J Neurochem 75:471–479

Waagepetersen HS, Sonnewald U, Schousboe A (2003) Compartmentation of glutamine, glutamate, and GABA metabolism in neurons and astrocytes: functional implications. Neuroscientist 9:398–403

Wang ZJ, Bergqvist C, Hunter JV, Jin D, Wang DJ, Wehrli S, Zimmerman RA (2003) In vivo measurement of brain metabolites using two-dimensional double-quantum MR spectroscopy-exploration of GABA levels in a ketogenic diet. Magn Reson Med 49:615–619

Waniewski RA, Martin DL (1998) Preferential utilization of acetate by astrocytes is attributable to transport. J Neurosci 18:5225–5233

Waxman EA, Lynch DR (2005) N-methyl-D-aspartate receptor subtypes: multiple roles in excitotoxicity and neurological disease. Neuroscientist 11:37–49

White HS (1997) Clinical significance of animal seizure models and mechanism of action studies of potential antiepileptic drugs. Epilepsia 38(Suppl 1):S9–S17

Wilder RM (1921) Effects of ketonuria on the course of epilepsy. Mayo Clin Bull 2:307–310

Wilkins L (1937) Epilepsy in childhood. III. Results with the ketogenic diet. J Pediatr 10:341–357

Wu J-Y (1976) Purification, characterization and kinetic studies of GAD and GABA-T from mouse brain. In: Roberts E, Chase TN, Tower DB (eds) GABA in Nervous System Function. Raven, New York, pp 7–55

Yanagida O, Kanai Y, Chairoungdua A, Kim DK, Segawa Hi, Nii T, Cha SH, Matsuo H, Fukushima J, Fukasawa Y, Tani Y, Taketani Y, Uchino H, Kim JY, Inatomi J, Okayasu I, Miyamoto K, Takeda E, Goya T, Endou (2001) Human L-type amino acid transporter 1. LAT1.: characterization of function and expression in tumor cell lines. Biochem Biophys Acta 1514:291–302

Yudkoff M, Nelson D, Daikhin Y, Erecinska M (1994) Tricarboxylic acid cycle in rat brain synaptosomes. Fluxes and interactions with aspartate aminotransferase and malate/aspartate shuttle. J Biol Chem 269:27414–27420

Yudkoff M, Daikhin Y, Nelson D, Nissim I, Erecinska M (1996) Neuronal metabolism of branched-chain amino acids: Flux through the aminotransferase pathway in synaptosomes. J Neurochem 66:2136–2145

Yudkoff M, Daikhin Y, Nissim I, Grunstein R, Nissim I (1997) Effects of ketone bodies on astrocyte amino acid metabolism. J Neurochem 69:682–692

Yudkoff M, Daikhin Y, Nissim I, Lazarow A, Nissim I (2001) Ketogenic diet, amino acid metabolism and seizure control. J Neurosci Res 66:931–940

Yudkoff M, Daikhin Y, Nissim I, Lazarow A, Nissim I (2004) Ketogenic diet, brain glutamate metabolism and seizure control. Prostaglandins, Leukot. Essential Fatty Acids 70:277–285

Yudkoff M, Daikhin Y, Nissim I, Horyn O, Lazarow A, Luhovyy B, Wehrli S, Nissim I (2005) Response of brain amino acid metabolism to ketosis. Neurochem Int 47:119–128

Yudkoff M, Daikhin Y, Nissim I, Horyn O, Luhovyy B, Lazarow A, Nissim I (2006) Short-term fasting, seizure control and brain amino acid metabolism. Neurochem Int 48:650–656

Chapter 29
Redox Shuttles in the Brain

Beatriz Pardo and Laura Contreras

Abstract Herein, we summarize the most relevant systems operating in brain for the transfer of reducing equivalents from NAD^+/NADH or $NADP^+$/NADPH across the inner mitochondrial membrane. The kinetic properties, the transport characteristics and the calcium activation patterns of the carriers and enzymes involved in shuttling process are described. The involvement of redox shuttles in the physiology and the pathological states of the nervous system are also discussed.

The malate-aspartate shuttle (MAS) is considered to be the major redox shuttle system transferring reducing equivalents from cytosolic NAD^+/NADH to mitochondria that functions in brain, mainly important in neuronal cells. The function of MAS in brain is essential for maintaining a NAD^+/NADH ratio favourable for the oxidative metabolism of glucose. MAS deficiency in mice has been shown to produce severe neurological deficits and growth retardation. These mice exhibit pronounced motor coordination defects along with an impairment in myelination in the central nervous system. The existence of Aralar/AGC1 (aralar; aspartate-glutamate carrier)-knockout mice (lacking MAS activity in brain) has revealed new functions for MAS. Particularly, aralar was found to be essential in the supply of brain aspartate and N-acetylaspartate for myelin lipid synthesis and in the transmission of small Ca^{2+} signals to neuronal mitochondria. Interestingly, the first human patient with aralar deficiency associated to a severe hypomyelination has been recently reported.

B. Pardo, Ph.D. (✉)
Departamento de Biología Molecular, Centro de Biología Molecular "Severo Ochoa" UAM-CSIC, Universidad Autónoma, Centro de Investigación Biomédica en Red en Enfermedades Raras (CIBERER), ISCIII, Cantoblanco, Madrid, Spain
e-mail: bpardo@cbm.uam.es

L. Contreras
Departamento de Biología Molecular, Centro de Biología Molecular "Severo Ochoa" UAM-CSIC, Universidad Autónoma, Centro de Investigación Biomédica en Red en Enfermedades Raras (CIBERER), ISCIII, Cantoblanco, Madrid, Spain
e-mail: lcontreras@cbm.uam.es

I.-Y. Choi and R. Gruetter (eds.), *Neural Metabolism In Vivo*,
Advances in Neurobiology 4, DOI 10.1007/978-1-4614-1788-0_29,

The activity of the other major redox shuttle (glycerol-3-phosphate shuttle, GPS) in brain has been long questioned. Although indirect metabolic evidence might suggest that GPS is functional in the brain; for a long time the two enzymatic constituents of GPS (cytosolic and mitochondrial glycerol 3-phosphate dehydrogenases, cGPDH and mGPDH) were not found to colocalize in the same cell type, a fact that would be required for the GPS to be functional. Besides this, no neurological disturbances have been reported in GPS-deficient mice. Up to now, cGPDH had been exclusively found in oligodendrocytes, where it could provide glycerol phosphate for phospholipid synthesis and mGPDH had been found in neurons and astrocytes, where glycerol phosphate could be used as respiratory substrate. However, recent transcriptome studies in brain have revealed the coexistence of cGPDH and mGPDH in neurons and astrocytes, supporting the possible function of GPS in these cell types.

Other alternative NADH shuttles are described, particularly the citrate shuttles, malate-oxaloacetate and lactate shuttles. The citrate shuttles provide acetyl-CoA for lipid synthesis, and the citrate-pyruvate cycle, which is supposed to operate in glial cells provides also NADPH in the cytosol. These shuttles may be required when MAS is impaired; and/or in glial cells, where MAS has lower activity than in neurons. Finally, two redox systems thought to transfer reducing equivalents from mitochondria to cytosolic $NADP^+$/NADPH are described. These shuttles may be important for biosynthetic reactions in the cytosol and may play a critical role in the cellular defense against oxidative stress. However, the existence of these alternative shuttles has not been demonstrated by reconstitution assays, and because of this, should be regarded with caution.

Key words AGC1 • Aralar • Brain metabolism • Citrate shuttles • Mitochondria • Malate-aspartate shuttle • Glycerol 3-P shuttle • Lactate shuttles • Malate-oxaloacetate shuttle • NAD^+/NADH balance

Abbreviations

AAT	aspartate aminotransferase
Ac-CoA	acetyl-coenzyme A
ACS	acetyl CoA synthetase-1
ACL	ATP-Citrate lyase
AGC	aspartate-glutamate carrier
Asp	aspartate
AOAA	aminooxyacetic acid
ASPA	aspartoacylase
Asp-NAT	L-aspartate N-acetyltransferase
CaMCs	calcium binding mitochondrial carriers
CaU	calcium uniporter
CIC	tricarboxylate carrier
CS	citrate synthetase

DHAP	dihydroxyacetone phosphate
DIC	dicarboxylate carrier
GABA	α-aminobutyric acid
GAPDH	glyceraldehyde 3-phosphate dehydrogenase
Glt	glutamate
G3P	glycerol 3-phosphate
G6PDH	glucose 6-phosphate dehydrogenase
Gln	glutamine
GPx	glutathione peroxidase
GR	glutathione reductase
cGPDH	cytosolic glycerol 3-phosphate dehydrogenase
mGPDH	mitochondrial glycerol 3-phosphate dehydrogenase
GPS	glycerol 3-P shuttle
c/m ICDH	cytosolic/mitochondrial $NADP^+$-dependent isocitrate dehydrogenase
IDH	NAD^+-dependent isocitrate dehydrogenase
α-KG	α-ketoglutarate
LDH	lactate dehydrogenase
Mal	malate
MAS	malate-aspartate shuttle
MC	mitochondrial carrier
MCT	monocarboxylate carrier
MDH	malate dehydrogenase
c/m ME	cytosolic/mitochondrial malic enzyme
NAA	N-acetyl aspartate
3-NPA	3-nitropropionic acid
OAA	oxalacetate
OGC	α-ketoglutarate/malate carrier
Oxodic	oxodicarboxylate carrier
PC	pyruvate carboxylase
Pyr	pyruvate
ROS	radical oxygen species
SNP	single nucleotide polymorphism
TCA	tricarboxilic acid cycle
TIM	inner membrane translocase
TOM	outer membrane translocase
VOCC	voltage operated calcium channel.

29.1 Introduction

Most of our brain function is devoted to recognizing, understanding, and responding to the very complex external space and time challenges we continuously confront. These often-unpredictable challenges require rapid access to potentially useful information gained through personal experience or provided by others.

Efficient access to energy for these fast processing requirements constitutes a major necessity for the accurate functioning of brain.

Energy metabolism in cells is compartimentalized to either the cytoplasm or the mitochondria. Mitochondria performs a variety of energy-generating processes such as oxidative phosphorylation, the citric acid cycle and, in higher eucaryotes, oxidation of fatty acids. Parts of urea cycle, the biosynthesis of heme and certain amino acids are also carried out in mitochondria (Ernster and Gottfried 1981). All of these metabolic activities require the rapid and highly specific exchange of molecules between the cytosol and the mitochondrial matrix space. For energetic metabolism to continue, the NADH generated during glycolysis in the cytosol must be reoxidized to NAD^+. If this does not ocurr, the $NADH/NAD^+$ ratio increases and pyruvate is converted to lactate in order to reoxidize NADH to NAD^+. Such conversion of pyruvate to lactate might prevent the oxidative metabolism of pyruvate derived from glucose, via the tricarboxylic acid (TCA) cycle. But in brain, mitochondrial oxidative metabolism of pyruvate is essential for physiological function. Pyruvate is transported by specific carrier proteins, into the mitochondria where it is metabolized. Lactate, in contrast, is converted to pyruvate within the cytosol with the concomitant conversion of NAD^+ to NADH. The role of lactate in brain energy metabolism is controversial (for review, Pellerin 2003; but see Chih et al. 2001). Lactate has been considered for a long time a metabolic end-product in the CNS. However, a growing body of evidence supports that lactate may constitute an alternative cerebral oxidative energy substrate; even an obligatory substrate to recover after hypoxia-ischaemia (Maran et al. 1994; Schurr et al. 1997a, b; Cater et al. 2001, 2003). Certainly, an inability to maintain the balance of metabolic intermediates between the cytosol and mitochondria results in impairment of metabolism. In addition to many enzymes existing in either the cytoplasm or mitochondria, the pyridine nucleotides (NAD^+, NADH, $NADP^+$, and NADPH) are also compartmentalized between the cytoplasm and the mitochondria. Given that intact mitochondria are impermeable to NADH, an indirect pathway involving the movement of electrons from NADH, as opposed to the NADH itself, across the mitochondrial membranes must operate (Lehningher 1951; Purvis and Lowenstein 1961; Dawson 1979).

Redox shuttles provide a pathway for the transfer of reducing equivalents from cytosolic $NAD^+/NADH$ to mitochondria. Their significance has been extensively studied in brain tissue and they are crucial for the synthesis of neurotransmitters in the brain. Two redox shuttles have been described in brain: the malate-aspartate shuttle (MAS) and glycerol 3-phosphate shuttle (GPS). MAS is considered to be the most important shuttle in brain (Fitzpatrick et al. 1983; Cheeseman and Clark 1988) while the participation of glycerol 3-phosphate shuttle is controversial. However, other redox shuttles may participate in brain metabolism. Herein, we describe the shuttles that operate in brain, the kinetic properties, calcium activation and transport characteristics of the carriers and enzymes involved in shuttling process, as well as their roles in both the physiologic conditions of nervous system and in neuropathologic states.

29.2 The Malate-Aspartate NADH Shuttle (MAS)

29.2.1 General Description

The transfer of reducing equivalents from cytosolic NADH/NAD$^+$ to mitochondrial NADH/NAD$^+$ is energetically unfavorable, as the NADH/NAD$^+$ ratio is about two orders of magnitude larger in mitochondria than in the cytosol (Williamson et al. 1967; Gumaa et al. 1971). The malate aspartate shuttle, MAS (also known as the "Borst cycle") is an ubiquitous metabolic shuttle that operates by transfering reducing equivalents from cytosolic to mitochondrial NADH/NAD$^+$. To overcome the energetic barrier of this process, reducing equivalent transfer is coupled to H$^+$ entry in mitochondria, a process that uses of the proton electrochemical gradient. Six proteins, two mitochondrial carriers and four enzymes, participate in MAS. These are the aspartate-glutamate carrier (AGC, SLC25A12 and SLC25A13), the malate/α-ketoglutarate carrier (OGC, SLC25A11) and both the cytosolic and mitochondrial forms of malate dehydrogenase (cMDH, mMDH; EC 1.1.1.37) and aspartate aminotransferase (cAAT and mAAT; EC 2.6.1.1). MAS has been proved to be the most important shuttle in brain and is particularly important in neurons, especially during lactate utilization. The activity of the shuttle is extremely low in glial cells (Ramos et al. 2003; Xu et al. 2007; Berkich et al. 2007; Cahoy et al. 2008; but see Lovatt et al. 2007).

NADH (mainly generated during glycolysis by the enzyme glyceraldehyde 3-phosphate dehydrogenase (GAPDH, E.C.1.2.1.12)) is oxidized by cMDH while oxaloacetate is converted to malate (Fig. 29.1). Malate enters the mitochondria, in exchange for α-ketoglutarate via the OGC (SLC25A11). The mitochondrial

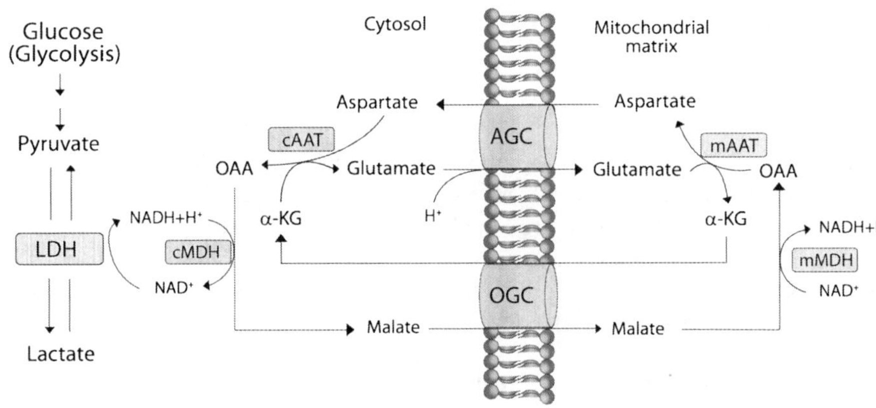

Fig. 29.1 The malate-aspartate shuttle (MAS) operates to transfer reducing equivalents from the cytosol to mitochondria. MAS is made up of two mitochondrial carriers, the aspartate-glutamate carrier (AGC) and the α-ketoglutarate-malate carrier (OGC) and four enzymes, mitochondrial and cytosolic aspartate aminotransferase (mAAT, cAAT) and mitochondrial and cytosolic malate dehydrogenase (mMDH, cMDH). LDH, lactate dehydrogenase; α-KG, α-ketoglutarate; OAA, oxalacetic acid

isoform mMDH, an enzyme of TCA cycle, converts malate to oxaloacetate forming NADH. Oxaloacetate is transaminated to aspartate by the mitochondrial aminotransferase (mAAT, EC 2.6.1.1) coupled to the conversion of glutamate to α-ketoglutarate. Aspartate leaves the mitochondria via the AGC in exchange for cytosolic glutamate, which enters the mitochondria together with a proton. To complete the shuttle, cAAT converts aspartate to oxaloacetate, simultaneously forming glutamate from α-ketoglutarate. By coupling aspartate-glutamate exchange to H^+ entry, the AGC becomes the only irreversible step of MAS under normal resting conditions (LaNoue and Tischler 1974, and references herein), and drives net redox transfer from the cytosol to mitochondria. The exchange of aspartate for glutamate via AGC appears to be the rate-limiting step of the shuttle (LaNoue et al. 2006).

29.2.2 *Isoforms of the Aspartate-Glutamate Carrier (AGC)*

The existence of glutamate transport in mitochondria was reported a long time ago by Azzi et al. (1967). But it was thanks to the pioneering research carried out by Kay LaNoue and colleagues that the activity of AGCs was characterized (LaNoue et al. 1974a, b; LaNoue and Tischler 1974; LaNoue and Schoolwerth 1979).

There are two AGCs in mammals, Aralar (also called aralar1 or AGC1; [del Arco and Satrústegui 1998]) and citrin/AGC2 (Kobayashi et al. 1999; del Arco et al. 2000) encoded by genes SLC25A12 and SLC25A13 present on chromosomes 2q31 (Sanz et al. 2000) and 7q21 (Sinasac et al. 1999), respectively. Del Arco and Satrústegui (1998) identified Aralar as the first known member of a subfamily of Ca^{2+} binding members of the mitochondrial carrier superfamily, the Calcium-binding mitochondrial carriers (CaMCs) with N-terminal extensions harboring the archetypical EF-hand calcium-binding motifs (Fig. 29.2a). Aralar and citrin are both constituents of MAS and AGC2/citrin, the liver isoform, is also a part of the urea cycle (LaNoue and Schoolwerth 1979; Palmieri et al. 2001). AGC2/citrin (del Arco et al. 2000) is the product of the gene mutated in type II-citrullinaemia (CTLN2; Kobayashi et al. 1999). Both aralar and citrin have a bipartite structure: a carboxyl-terminal half with the characteristic features of the mitochondrial carrier (MC) superfamily and an amino-terminal extension. Structurally, the MC family is characterized by the presence of three repeated regions, each about 100 amino acids long, composed of two hydrophobic transmembrane helices joined by an extensive hydrophilic region (Saraste and Walker 1982; Walker and Runswick 1993; Kunji, 2004; Fig. 29.2a). MCs may exist in two conformational states, the cytosol-state (c) that accepts substrate from cytosol and the matrix-state (m) that accepts substrates from mitochondrial matrix. Given the homology of MCs, the transition between states might constitute a general mechanism. The "single binding centre gated pore" model has been proposed to be operating for MCs. Substrate binding to a single site produces a conformational rearrangement with the movement of the transmembrane α–helices which permits translocation of the substrate (Kunji and Robinson 2006).

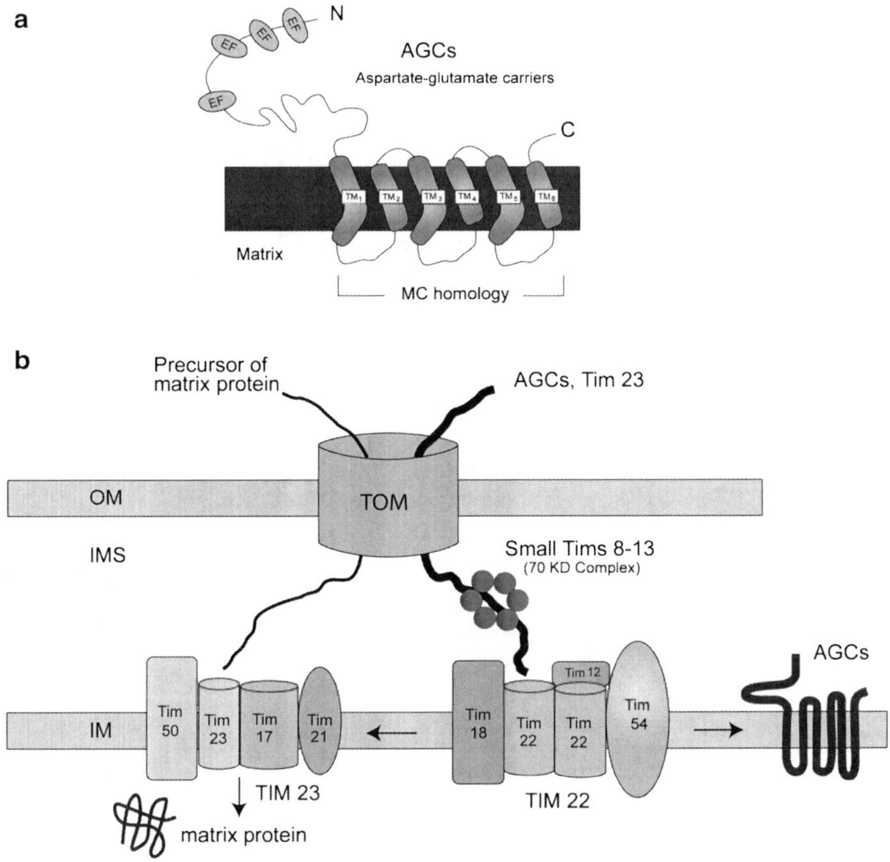

Fig. 29.2 Secondary structure of calcium-binding mitochondrial carriers (CaMCs) (A) and mitochondrial protein import machinery implicated in transport of CaMCs to mitochondrial inner membrane (IM; B). (a) Aralar1 has a typical structure of a mitochondrial carrier presenting six transmembrane α–helices and a large additional N-terminal domain facing the intermembrane space (IMS). **(b)** The outer membrane translocase (TOM) system transports mitochondrial proteins from the cytosol (where they are synthesized), to the IMS. The small Tim proteins form different 70-KDa complexes that escort inner membrane precursors across the IMS. The complex Tim8p-Tim13p in the IMS binds typically to hydrophilic N-extensions in aralar1. Once arrived to the IM, the TIM22 (inner membrane translocase) is the carrier translocase that mediates the import of these inner membrane proteins. Other pathway of import for mitochondrial matrix proteins, requiring TIM23, may be also implicated in Mohr-Tranebjaerg pathology. In this case, the transport of Tim23 (a constituent of TIM23) to the inner membrane is decreased due to defective Tim8p-Tim13p complex. However, Tim23 may be also scorted by the small Tims 9–10 (complexes Tim9p-Tim10p) in what constitutes an alternative route for its import

AGCs catalyze the exchange of matrix aspartate for external glutamate plus a proton. Transport activity of AGCs is driven by the proton electrochemical gradient, and constitutes the main point of regulation for MAS activity. Given that aspartate-glutamate exchange is strongly dependent on the mitochondrial inner membrane potential, in energized mitochondria this reaction is essentially unidirectional and favours the release of aspartate to the cytosol in exchange of glutamate entry into mitochondria. The effect of the membrane potential is a pronounced increase in transport activity (that depends on an electrogenic modulation of the transport rate constant of the negatively charged carrier-aspartate complex), and not to a change in the substrate affinity to the carrier (Palmieri et al. 2001). The latter observation was proved by comparing kinetic characteristics of AGCs in energized versus de-energized proteoliposomes; de-energization was induced by collapse of both electrical and chemical potential gradients ($\Delta\Psi$ and ΔpH) with ionophores such as valinomycin and nigericin (Palmieri et al. 2001). Under both conditions, transport affinities (Km values) of [1^4 C]aspartate and [1^4 C]glutamate for AGCs were the same, but the maximum transport rate (Vmax) of the [1^4 C]aspartate/aspartate and [1^4 C]glutamate/aspartate exchanges were different under energization and identical in the presence of ionophores (Palmieri et al. 2001). These properties of the AGC have been demonstrated in intact mitochondria (LaNoue et al. 1974a, b; LaNoue and Tischler 1974) and in proteoliposomes reconstituted with the native purified protein (Dierks and Krämer 1988; Dierks et al. 1988; Bisaccia et al. 1992) and in recombinant AGCs (Palmieri et al. 2001). The substrate affinities were clearly different for the two membrane sides both for aspartate (external 50 μM, internal 3 mM) and glutamate (external about 200 μM, internal 3 mM; Dierks et al. 1988).

29.2.3 Expression of AGC Isoforms

During embryogenesis, aralar and citrin have a widely overlapping expression pattern, and the full expression of aralar is only attained postnatally, a time at which citrin is only expressed at low levels in a few restricted neuronal clusters (Begum et al. 2002; Ramos et al. 2003; Contreras et al. 2010). Aralar is expressed in almost every tissue except the adult liver (del Arco and Satrústegui 1998; Begum et al. 2002; del Arco et al. 2002), and it is the major AGC isoform present in the adult CNS. Aralar is enriched in skeletal muscle and brain, whereas citrin (AGC2) is preferentially expressed in liver (del Arco el al. 2002). However, both isoforms are expressed in heart, kidney and other adult tissues (Begum et al. 2002; del Arco et al. 2002).

Aralar is the predominant isoform expressed in neurons and its levels undergo a marked increase during *in vitro* maturation at a time where there is also a prominent rise in MAS function (Ramos et al. 2003). In this work, aralar expression was evaluated at E18 (embryonic day 18), P0 (postnatal day 0), P7, P17 and P90 in mitochondria from rodent brain, where it was found to be slightly increased throught postnatal development up to P7 in rodents (Ramos et al. 2003). Also, aralar is one of the more enriched proteins during differentiation of P19 cells to a neuronal phenotype (Watkins et al. 2008). On the other hand, citrin was confined to neuronal cells within discrete clusters in the adult rodent brain. Particularly, it was detected in the deep cerebellar nuclei and a few midbrain nuclei, where it was present at very low levels in adult mouse brain (Contreras et al. 2010).

Aralar staining was found in all CNS neurons, with high levels associated to large cell body neurons scattered throughout the brain. Aralar mRNA show clear enrichment in sets of neuronal nuclei in the brainstem (Ramos et al. 2003; Berkich et al. 2007), in photoreceptor cells in the retina (Xu et al. 2007) and particularly in the ventral horn of the spinal cord (Ramos et al. 2003). The presence of aralar in a subset of cytochrome oxidase-rich neurons in the brain and spinal cord could thus reflect a higher energy demand of these neurons, which is met by the combination of high MAS and respiratory chain activities (Ramos et al. 2003).

In contrast, aralar has no detectable expression in white matter (Ramos et al. 2003; Xu et al. 2007; Berkich et al. 2007) and low levels in cultured glial cells (Ramos et al. 2003). These results have been challenged in a recent study of the transcriptome of acutely dissociated and purified astrocytes and neurons (Lovatt et al. 2007). However, in a second transcriptome study by Cahoy et al. (2008) this was not confirmed; on the contrary, aralar mRNA was found to be enriched in neurons with respect to astrocytes. This agrees with the preferential distribution of aralar in neurons obtained by in situ hybridization and immunocytochemistry studies. Remarkably, at present, in the vast majority of studies no detectable aralar expression in astrocytes has been reported (Ramos et al. 2003; Xu et al. 2007; Berkich et al. 2007; Cahoy et al. 2008) with the exception of Lovatt´s work (2007) that describes its expression in a subpopulation of protoplasmic astrocytes isolated from the adult murine cortex. Surprinsingly, oxidative metabolism seems to be a prominent part of energy metabolism in protoplasmic astrocytes (Lovatt et al. 2007). In glial cells, expression of the citrin isoform was also not found (Contreras et al. 2010). But more studies will be necessary to further clarify this issue.

29.2.4 Calcium Activation of Aralar-MAS in Brain

The most novel finding in the sequence of the AGCs was the presence of a number of EF-hand pairs in the amino terminal domain (del Arco and Satrustegui 1998; Contreras et al. 2007; Satrustegui et al. 2007a, b). EF-hands are Ca^{2+} binding domains consisting of a 12-residue loop flanked on both sides by α-helices (Nakayama et al. 1992; Zhou et al. 1997), and the two-EF-hand domain is considered to be the functional unit (Grabarek 2006; Kawasaki et al. 1998). Aralar has 4 EF-hand pairs, some of which are not canonical, and a single nonfunctional hypothetical EF9 in its N-terminus (Contreras et al. 2007). These Ca^{2+} binding domains face the intermembrane space (Palmieri et al. 2001). Since Aralar amino-terminal half was shown to bind calcium (del Arco and Satrústegui 1998), it may be concluded that some of these EF-hand pairs, satisfying six of the seven coordination bonds of a bona fide EF-hand, are operational calcium binding domains. In fact, recent studies by Marmol et al. (2009) have shown the requirement of EF-hands for aralar activation by calcium in insulinoma beta-cell lines.

Pardo et al. (2006) tested whether MAS activity in brain mitochondria, which contain aralar essentially as the only AGC isoform, was regulated by extramitochondrial Ca^{2+}. To this end, Ca^{2+} activation of MAS was studied in the presence of ruthenium red (RR) to block Ca^{2+} uptake in mitochondria. It was observed that MAS was indeed regulated by extramitochondrial Ca^{2+}, with an $S_{0.5}$ for Ca^{2+} stimulation of MAS of about 300 nM (Pardo et al. 2006); similar values were obtained for MAS activation in mitochondria from tissues expressing aralar as the only AGC isoform (Contreras et al. 2007). This value is lower than those at which the calcium uniporter is active (2–20 μM; Gunter et al. 2000; Moreau et al. 2006), suggesting that MAS activation could take place *in vivo* at Ca^{2+} concentrations that do not activate the Ca^{2+} uniporter. However, lower values are also cited for set points for mitochondrial Ca^{2+} uptake (0.3–1 μM, reviewed in Nicholls and Budd 2000).

To study the role of these small calcium signals in the activation of MAS in neuronal mitochondria, MAS activation was analyzed in cultured neurons by two-photon microscopic imaging of mitochondrial NADH. Neurons were incubated in Ca^{2+}-free media in the presence of 100 μM EGTA and exposed to 20 mM lactate, to boost MAS activity, together with 100 μM ATP or different Ca^{2+}-mobilizing agonists. ATP-induced cytosolic Ca^{2+} ($[Ca^{2+}]_i$) transients were small, with no detectable increase in mitochondrial Ca^{2+} concentration ($[Ca^{2+}]_{mit}$) in control or aralar-deficient neurons. The small Ca^{2+} signal induced by ATP resulted in a remarkable potentiation of lactate-dependent increase in mitochondrial NAD(P)H fluorescence in control neurons, which was notably absent in aralar-deficient neurons (Pardo et al. 2006); indicating that these small Ca^{2+} signals activate the aralar-MAS pathway.

Large calcium signals were also investigated by using K^+ stimuli to induce depolarization-dependent activation of voltage-operated Ca^{2+} channels (VOCC), and applied together with 20 mM lactate and millimolar external calcium concentrations. These signals clearly reach mitochondria through the entry of Ca^{2+} across the inner mitochondrial membrane Ca^{2+} uniporter (CaU), and subsequently the Ca^{2+} signal arising in mitochondria activates three Ca^{2+}-dependent dehydrogenases present in the mitochondrial matrix (pyruvate, α-ketoglutarate and isocitrate dehydrogenase), resulting in increased mitochondrial NADH. Indeed, a robust increase in cytosolic and mitochondrial Ca^{2+} was observed, with no difference between the increases in $[Ca^{2+}]_i$ and $[Ca^{2+}]_{mit}$ in control and aralar-deficient neurons (Pardo et al. 2006). Paradoxically, in this situation, the lactate-induced increase in the mitochondrial/cytosolic NAD(P)H fluorescence ratio potentiated by the large $[Ca^{2+}]_i$ signals was the same in control and aralar-deficient neurons (Pardo et al. 2006). As the increase in mitochondrial NAD(P)H obtained with high K^+ was expected to result from the additive effects of Ca^{2+} activation on MAS activity and Ca^{2+} activation of mitochondrial dehydrogenases, the lack of differences between control and aralar-deficient neurons indicated that MAS does not contribute to the increase in mitochondrial NAD(P)H under these conditions. It also suggests that MAS activity may be inhibited when mitochondrial Ca^{2+} uptake takes place. Indeed, Ca^{2+} stimulation of MAS activity in isolated brain mitochondria was not observed when RR was not present and Ca^{2+} uptake in mitochondria was allowed to take place (Pardo et al. 2006). The lack of activation of MAS in these conditions has been attributed to a competition between the carrier OGC and the enzyme α-KGDH for the shared substrate α-KG. In basal conditions, both have similar affinity for α-KG; calcium activation of

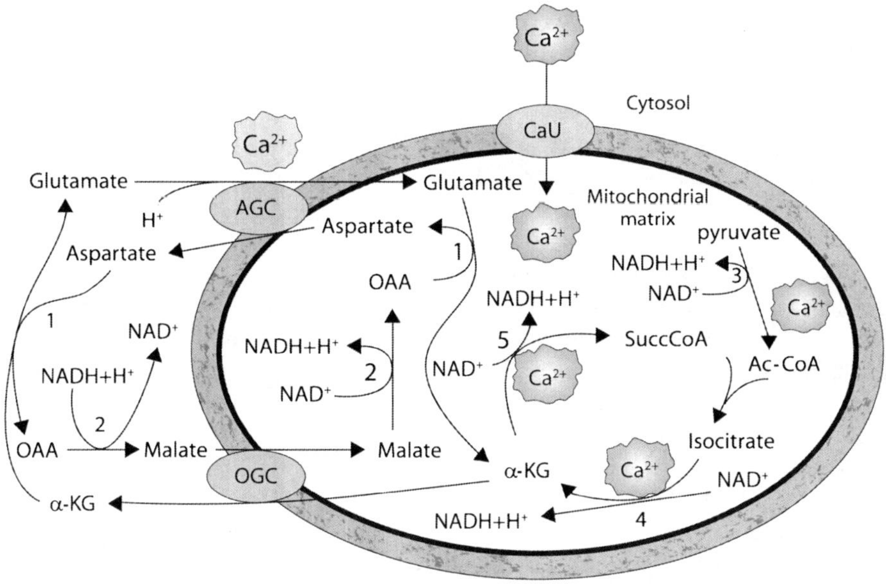

Fig. 29.3 Calcium activates mitochondrial metabolism via aralar-MAS or calcium uniporter. Extramitochondrial calcium can activate the aralar-MAS pathway, resulting in increased mitochondrial NADH. Mitochondrial matrix calcium is taken up through the calcium uniporter (CaU) resulting in the activation of 3 intramitochondrial dehydrogenases: pyruvate dehydrogenase (PDH), isocitrate dehydrogenase (IDH) and α-ketoglutarate dehydrogenase (α-KGDH) generating an increased supply of NADH to the electron transport chain. (1), cytosolic and mitochondrial aspartate aminotransferase (AAT); (2), cytosolic and mitochondrial malate dehydrogenase (MDH); (3), pyruvate dehydrogenase (PDH); (4), isocitrate dehydrogenase (IDH); (5), α-ketoglutarate dehydrogenase (α-KGDH). Ac-CoA, acetyl CoA; AGC, aspartate-glutamate carrier; α-KG, α-ketoglutarate; OAA, oxalacetic acid; OGC, α-ketoglutarate-malate carrier; SuccCoA, succinyl CoA

α-KGDH lowers its Km, and therefore, efflux of α-KG through the OGC becomes reduced with the consequent impairment of MAS activity (Pardo et al. 2006; Satrústegui et al. 2007b; Contreras and Satrústegui 2009).

In conclusion, large Ca^{2+} signals generate an increase in NAD(P)H arising mainly from the activation of mitochondrial dehydrogenases, whereas small Ca^{2+} signals effectively recruit the aralar-MAS pathway (for review, Satrústegui et al. 2007b; Fig. 29.3).

29.2.5 Role of MAS in the CNS

29.2.5.1 Physiological Implications

The importance of MAS for cerebral transfer of reducing equivalents from the cytosol to the mitochondrion and the maintenance of ATP levels has been inferred from experiments performed with inhibitors of MAS activity i.e. inhibitors of transamination, like aminooxyacetic acid (AOAA; Fitzpatrick et al. 1983; Cheeseman and Clark 1988; McKenna et al. 1993), or other drugs that impair MAS activity. Fitzpatrick et al. (1983) investigated the effects of several inhibitors of AAT, a key enzyme of MAS, on cerebral metabolism *in vitro*, showing that all of them reduced the rate of oxygen consumption in parallel to AAT activity. The content of ATP and phosphocreatine declined in slices incubated with glucose and AAT inhibitors. These data support that the operation of MAS provides a link between cerebral glycolysis (a continued need for NAD^+) and the TCA cycle that is vulnerable to several metabolic disturbances that impair brain function. Moreover, the role of MAS has been proved to be quite different in astrocytes and neurons or synaptic terminals. Studies carried out by McKenna et al. (1993) showed that AOAA leads to more than a 50% decrease in the rate of oxidation of lactate by synaptic terminals, in contrast to an increase of lactate oxidation in astrocytes. Cheeseman and Clark (1988) demonstrated that inhibition of MAS with another inhibitor, BMA (β-methyleneaspartate), led to decreased oxidation of glucose via the pyruvate dehydrogenase (PDH) reaction and the TCA cycle in synaptosomes. Other MAS inhibitors such as phenylsuccinate (blocks the OGC carrier) and 3-nitropropionic acid (3-NPA; impairs MAS by blocking the TCA cycle enzyme succinate dehydrogenase) have also been used to study the functional importance of the shuttle in the brain and brain cells (Palailogos et al. 1988; Gramsberger et al. 2000). Taken together, these data support the relevance of MAS mainly in neuronal cells and its crucial role for lactate utilization. Nonetheless, it should be noted that all of these pharmacological inhibitors have been demonstrated to be non-specific for MAS activity. AOAA not only inhibits aspartate aminotransferase but other aminotransferases are likely to be affected; BMA potently inhibits high affinity glutamate transport, and 3-nitropropionic acid also decreases mitochondrial respiration (for review see McKenna et al. 2006).

Because pharmacological inhibitors have off-target effects, genetic deficiencies of the components of MAS provide an additional and more specific means to study the role of MAS in brain metabolism. Disrupting the gene of aralar (slc25a12) in mice (by the method by Zambrowicz et al. 1998) has confirmed the role of MAS as the main redox shuttle in neurons and has enabled the discovery of new roles of aralar-MAS pathway (Jalil et al. 2005). On the one hand, it is required for synthesis of brain aspartate and N-acetylaspartate, a neuron-born metabolite that supplies acetate for myelin lipid synthesis; and on the other hand, it is essential for the transmission of "small Ca^{2+} signals" to mitochondria via an increase in mitochondrial NADH, as described above. Recently, AGC1 deficiency in humans has been described to be associated with severe psychomotor developmental arrest, hypotonia, seizures, and a global cerebral hypomyelination (Wibom et al. 2009). A phenotype quite similar to that described for Slc25a12-knockout mice (Jalil et al. 2005). Herein, we summarize some of the most important roles known at present for aralar in neurons i.e., lactate utilization, aspartate and N-acetylaspartate synthesis.

Role of Aralar-MAS in Lactate Utilization

Lactate appears to occupy a central position in intercellular metabolic trafficking within the brain. Lactate, produced in glial cells, could be supplied by astrocytes to neurons as a supplementary energy substrate, especially during periods of high activity (Tildon et al. 1993; Pellerin et al. 1998; Kasischke et al. 2004; Pellerin and Magistretti 2004). This process constitutes the "astrocyte-neuron lactate shuttle": astrocytes with endfeet bordering capillaries take up glucose from the circulation (or the extracellular space), converting it to lactate for export to neurons (for review, Pellerin 2003). This lactate shuttle is thought to be stimulated by increased concentrations of glutamate in the extracellular space and during excitotoxicity (Pellerin and Magistretti 1994; Schurr et al. 1999). Other studies (Caesar et al. 2008) provide evidence that lactate production in the cerebelar cortex is mediated by AMPA receptor activation present on both neuronal and glial cells; but do not answer whether lactate production has a neuronal or astrocytic origin. In astrocytes, the majority of the glucose taken up is utilized for their robust synthesis of lactate. However, although astrocytic cells can utilize lactate for energy, these cells synthesize and release considerably more lactate than they use for energy (Walz and Mukerji 1988). An important fact involved in lactate production by astrocytes is their limited redox shuttle activity as indicated above (see also LaNoue et al. 2006).

Commensurate with this model, specific monocarboxylate (including lactate) transporters have been found on neurons and astrocytes both *in vitro* and *in vivo*. Neurons primarily express the monocarboxylate transporter MCT2 (Bergensen et al. 2001; Pierre et al. 2002; Debernardi et al. 2003), the kinetics of which favour lactate uptake, whereas astrocytes express either MCT1 or MCT4 (Pierre et al. 2000; Bergensen et al. 2001; Debernardi et al. 2003), the kinetics of which favour lactate release, which is consistent with the lactate shuttle model. In support of this concept, Nissen and Schousboe (1979) and Tholey et al. (1981) have shown that the predominant isozyme of lactate dehydrogenase (LDH) in astrocytes is the anaerobic (M4) type involved in lactate production (LDH5). The LDH5 isoenzyme is composed of four identical A subunits, while at the other extreme LDH1 is formed by B subunits only. Specific kinetic properties are linked with the subunit composition of each isoenzyme. Isoenzymes enriched in A subunits (like LDH5) tend to be found predominantly in tissues producing lactate while tissues containing more B subunits (like LDH1) are more oxidative and consume lactate as energy substrate. In the rat brain, in situ hybridization of both LDH A and B subunit mRNAs revealed that while LDH A mRNA had a uniform distribution suggesting a glial localization, LDH B mRNA had a much more discrete expression reminiscent of a neuronal distribution (Laughton et al. 2000). Using immunohistochemistry, a selective expression of LDH5 in astrocytes and the prominent expression of LDH1 by neurons in both the hippocampus and occipital cortex of human brain was described (Bittar et al. 1996). These observations support the idea that astrocytes might act as lactate sources in the brain in response to synaptic activity, while neurons might represent lactate "sinks", consuming oxidatively this substrate. However, a huge debate is open in relation to the existence of the astrocyte-neuron lactate shuttle in which glial-produced lactate fuels neurons. Several studies rather support the conventional view that considers glucose as the main substrate for neuronal energy metabolism in the brain (Mangia et al. 2003, 2007; Chih et al. 2001).

Lactate constitutes an ideal substrate for synaptic terminals because it immediately yields NADH for energy, and synaptic terminals have an abundance of the isoform of LDH required for the aerobic utilization of lactate (Kull 1978). It is well established that lactate is an important substrate for developing brain (Medina 1985; Dombrowski et al. 1989; Vicario et al. 1991; McKenna et al. 1994) as well as for adult brain. Schurr et al. (1988, 1989) demonstrated that lactate can support the electrical activity in hippocampal slices from adult rat brain. Moreover, changes in lactate concentrations occur with no detectable change in oxygen content, suggesting that lactate generation is not merely secondary to anaerobic metabolism (Prichard et al. 1991; Hu and Wilson 1997).

The aralar-MAS pathway has been proved to be necessary for lactate utilization in synaptic terminals and neurons (McKenna et al. 1993; Pardo et al. 2006). As shown by two-photon microscopy imaging of mitochondrial NADH, addition of 20 mM lactate caused a substantial increase in mitochondrial NAD(P)H in control, but not in aralar-deficient neurons (Pardo et al. 2006), indicating that MAS is the main NADH shuttle present in neurons. The importance of aralar in glucose and lactate utilization was also confirmed by the marked increase of the lactate-to-pyruvate ratio observed in the culture medium of aralar-deficient neurons (Pardo et al. 2011). All these data together support the necessity of the aralar-MAS pathway for lactate utilization by these cells.

Neurons not only use lactate but recover more effectively from excitotoxic insults when using lactate rather than glucose alone (Schurr et al. 1997a, b; Cater et al. 2001, 2003; Ros et al. 2001). This is probably because fewer biochemical steps are required for oxidative phosphorylation from lactate than from glucose, and glucose, unlike lactate, requires an initial investiment of ATP to yield energy. More recently, Bliss et al. (2004) showed in cell cultures that overexpression of the glucose transporter (GLUT1) in glia together with an overexpression of the lactate transporter (MCT2) in neurons enhanced lactate uptake and neuronal resistance to excitotoxicity. Lactate prevented glutamate-induced neuronal cell death *in vitro*, even in the presence of glucose. Lactate protected against glutamate excitotoxicity in aralar-expressing but not aralar-deficient cultured neurons, showing that the aralar-MAS pathway was required (Llorente et al. 2007).

Role of Aralar-MAS in Brain Aspartate, N-Acetylaspartate (NAA) and Myelin Lipid Synthesis

Aralar deficiency resulted in a large drop in NAA and aspartate in the brain and in primary neuronal cultures derived from Aralar$^{-/-}$ mice. The fall in brain aspartate levels and the lack of aspartate production in the brain mitochondria of Aralar$^{-/-}$ mice (Jalil et al. 2005) suggests that a major route of aspartate production in the CNS is mitochondrial and that it depends on aralar for aspartate efflux to the cytosol.

The CNS pathology observed in aralar-null mice appears at about the time when the blood-brain barrier develops, preventing the entry of aspartate and other amino acids into most regions of the adult brain (Price et al. 1984). From this moment on, brain cells have to rely on their own endogenous production of aspartate, which takes place mostly in neurons (Urenjack et al. 1992, 1993). MAS is required in neuronal mitochondria for aspartate synthesis by providing the precursor malate, and for its efflux to the cytosol. The "partial TCA cycle" in neuronal mitochondria might be implicated in aspartate synthesis from glutamate, especially when glucose supply is low. In this cycle, the consumption of glutamate derived from glutamine (through the glutaminase reaction) is mediated in large part by a sequence of reactions involving aspartate aminotransferase (producing 2-oxoglutarate) and the segment of the TCA between 2-oxoglutarate and oxalacetate (Yudkoff et al. 1994). In this case, mitochondrial AAT and the AGC are also involved in the mitochondrial production and efflux of aspartate, respectively. Obviously, another possible pathway of aspartate production is by cytosolic AATs. However, when the cytosolic NADH/NAD$^+$ ratio is increased, as it is the case for aralar-deficient neurons, the cytosolic oxaloacetate level is expected to decrease, limiting aspartate production in the cytosol via AAT, which rapidly equilibrates aspartate and oxaloacetate (Shen 2005). The drastic fall in the concentration of aspartate in brain from Aralar$^{-/-}$ mice, representing about 11 and 6% of those of Aralar$^{+/+}$ mice at 10 and 20 days, respectively (Jalil et al. 2005) suggests that both mitochondrial and cytosolic aspartate formation are blocked by aralar deficiency.

In aralar-deficient mice, the fall in aspartate levels is associated with decreases in NAA levels (Jalil et al. 2005). NAA is derived from neuronal aspartate, and its synthesis is catalyzed by aspartate N-acetyl transferase (Asp-NAT) in brain mitochondria (Patel and Clark 1979; Madhavarao et al. 2003) and microsomes, which have four to five times higher activity than mitochondria (Lu et al. 2004) (Fig. 29.4). Aspartate and acetyl-CoA (Ac-CoA) are the substrates of Asp-NAT. The fall in NAA in aralar-KO mice is possibly caused by the lack of cytosolic aspartate for NAA synthesis by cytosolic Asp-NAT. However, if the mitochondrial Asp-NAT is the major source of NAA, as was initially reported (Knizley 1967; Goldstein 1969) then a possible explanation for the fall in NAA levels in aralar-null mice would be a drop of mitochondrial aspartate caused by a fall in mitochondrial glutamate, the amino donor for aspartate synthesis, as a result of the lack of AGC-driven glutamate uptake into mitochondria (Fig. 29.4).

NAA plays a role in myelin lipid formation through the supply of acetyl groups (D'Adamo and Yatsu 1966; Burri et al. 1991; Mehta and Namboodiri 1995; Chakraborty et al. 2001; Baslow 2003). In the adult brain, neurons are the main cell type able to produce NAA (Urenjak et al. 1992, 1993). The cytosolic pathway for NAA synthesis (Fig. 29.4) requires Ac-CoA which is formed from citrate in the ATP citrate lyase (ACL) reaction. Citrate is transported out of mitochondria through the citrate carrier (CIC) or the tricarboxylate carrier (TCC), which is highly expressed in neurons (Miyake et al. 2002a). The full operation of this pathway requires the citrate-malate shuttle (Fig. 29.4), to be described latter. NAA undergoes transaxonal transfer to oligodendrocytes (Chakraborty et al. 2001) where it supplies acetyl

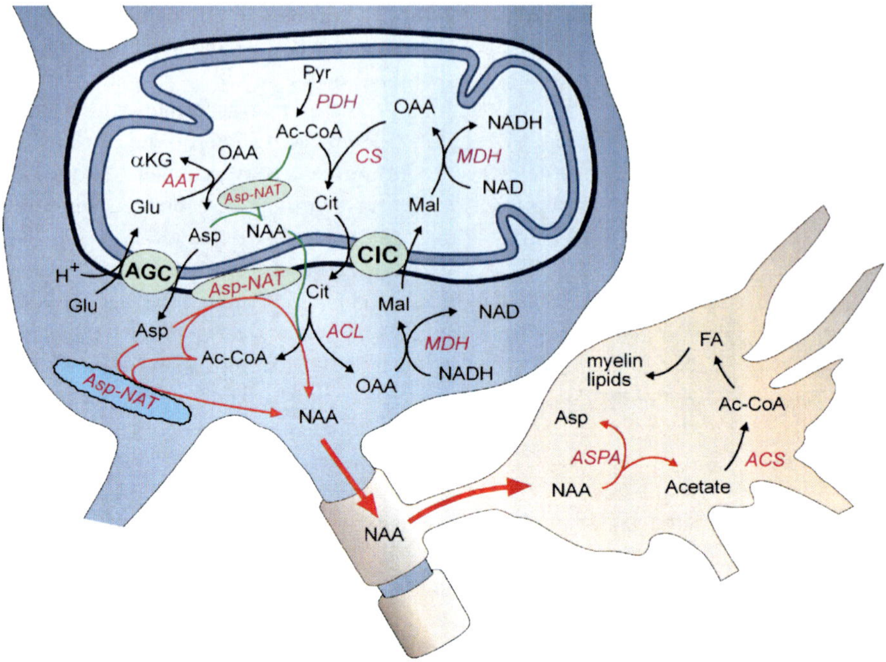

Fig. 29.4 Citrate/malate shuttle and aralar-MAS. Coordinated function for myelin synthesis. Acetyl CoA formed in the matrix effluxes as citrate through the citrate carrier, (SLC25A1, CIC) or the tricarboxylate carrier, TCC, with high expression in neurons (see text). Then it is cleaved by ATP citrate lyase (ACL) to Acetyl CoA plus OAA. The OAA returns to the mitochondria as malate, where it is oxidized again to regenerate OAA for condensation with newly produced Acetyl CoA. In the cytosol and/or mitochondria, Aspartate-N-acetyltransferase (Asp-NAT) utilizes acetyl CoA and aspartate to generate N-Acetyl aspartate (NAA), which would be transported transaxonally to the oligodendrocytes for myelin synthesis. If Asp-NAT is exclusively located in the cytosol or mitochondrial membrane, the generation of NAA by Asp-NAT would require the efflux of mitochondrial generated aspartate through the neuronal aspartate-glutamate carrier (AGC), aralar. However, if Aspartate-N-acetyltransferase (Asp-NAT) is present inside the neuronal mitochondria, NAA would be formed intramitochondrially. Ac-CoA, acetyl CoA; mAAT, mitochondrial aspartate aminotransferase; CS, citrate synthase; α-KG, α-ketoglutarate; cMDH, cytosolic malate dehydrogenase; mMDH, mitochondrial malate dehydrogenase; OAA, oxalacetic acid; PDH, pyruvate dehydrogenase (Modified from Satrústegui et al. 2007b)

groups for the synthesis of myelin lipids. In this study, radiolabeled NAA is injected into the posterior chamber of the eye showing that NAA is transported down axons and is passed from neurons to oligodendrocytes. The NAA cleavage enzyme, aspartoacylase (ASPA), is restricted primarily to oligodendrocytes (Bhakoo et al. 2001; Kirmani et al. 2003). A number of lipid components of myelin including ceramide and cerebrosides, but not cholesterol, are preferentially labeled by ^{14}C-NAA rather than from ^{14}C-acetate in the optic nerve, cerebrosides showing a particular high preference toward ^{14}C-NAA in the brain (Burri et al. 1991; Chakraborty et al. 2001). Mutations in ASPA cause Canavan´s disease (Matalon et al. 1988; Kaul et al. 1994,

1996; Zeng et al. 2002). This disease is characterized by a spongy degeneration of the CNS, increased levels of NAA in the brain and body fluids and an extensive loss of myelin. ASPA-deficient mice, generated as a model for Canavan´s disease (Matalon et al. 2000), show decreased myelin lipid synthesis and a very drastic reduction in brain acetate concentrations (Madhavarao et al. 2005), showing that the lack of NAA results in a lack of acetyl groups for myelin lipid synthesis in CNS. Interestingly, Aralar$^{-/-}$ mice show clear hypomyelination and a fall in myelin-specific lipids from brain, particularly galactocerebrosides, suggesting that the drop in NAA for myelin lipid synthesis causes this defect (Jalil et al. 2005; Satrústegui et al. 2007a, b). Besides, the first human patient presenting Aralar deficiency has been recently reported and a drastic lack of myelination in the cerebral hemispheres was found in association with a severe reduction in NAA (Wibom et al. 2009).

29.2.5.2 Pathological States Related to Aralar-MAS Defects

Aralar and Mohr-Tranebjaerg Syndrome

Mohr-Tranebjaerg syndrome (MTS/DFN-1) or deafness/dystonia syndrome is a recessive, X-linked neurodegenerative disorder characterized by progressive sensorineural deafness; resulting in cortical blindness, dystonia, dysphagia and paranoia (Tranebjaerg et al. 1995; Jin et al. 1996). Mohr-Tranebjaerg syndrome results from a mutation in deafness/dystonia protein 1/translocase of mitochondrial inner membrane 8a (DDP1/TIMM8a), which is implicated in deficient protein import into mitochondria.

As for most mitochondrial proteins, AGC precursor proteins are synthesized in the cytosol and then imported to mitochondrial inner membrane. Targeting information for localization in mitochondria in CaMCs is contained in their carboxy-terminal half, and all amino-truncated variants are targeted into mitochondrial membranes (del Arco et al. 2000; del Arco and Satrústegui 2004, 2005; del Arco 2005; Palmieri et al. 2001). However, the N-extensions appear to be important in selecting the mechanism for translocation across the mitochondrial intermembrane space of the CaMC proteins, and aralar. The biogenesis of the mitochondrial inner membrane is dependent on two distinct 70 KDa protein complexes. Tim8p partners with Tim13p to form a soluble 70 KDa complex in the intermembrane space (Koehler et al. 1999; Davis et al. 2000; Paschen et al. 2000); and Tim9p partners with Tim10p to form a similar 70 kDa complex in a 3:3 ratio (Curran et al. 2002a, b; Koehler et al. 1999). These small Tim proteins (Tim8p-Tim13p and Tim9p-Tim10p) escort inner membrane proteins across intermembrane space. Once into the inner membrane several translocases (TIMs) are operative; the TIM22 that mediates the import of inner membrane proteins, including Tim23p and AGCs; and TIM23, responsible for the import of mitochondrial matrix proteins (Fig. 29.2b).

These 70-KDa complexes have different substrate specificities, with the classical MC precursor proteins binding predominantly to the Tim9p-Tim10p complex (Paschen et al. 2000). In contrast, the Tim8p-Tim13p complex binds to the human and yeast AGC precursor proteins through their long hydrophilic amino-terminal domains (Curran et al. 2002b; Roesch et al. 2004). Pathway selection seems to be determined by the presence of hydrophilic amino-terminal extension of AGCs; thus the import of full-length aralar decreases 80% in mitochondria that lack the Tim8p-Tim13p complex, and shortening of the aralar amino-terminal domain results in preferential binding of aralar to the Tim9p-Tim10p complex (Roesch et al. 2004). In addition to citrin and aralar, the DDP1/TIMM8a-TIMM13 complex mediates also the import of Tim23p and the import of Tim23p was decreased by 75% in mitochondria lacking the Tim8p-Tim13p complex (Roesch et al. 2002).

Neuronal cell loss was shown in the CNS in two unrelated families affected by this disease. In lymphoblasts from MTS/DFN-1 patients, the mutations in Tim8p appear to result in impaired import of aralar. This impairment resulted in decreased mitochondrial NADH levels due to defects in MAS activity (Roesch et al. 2004). Specific defect in Mohr-Tranebjaerg might be localized to the large neurons in the cerebellum and basal ganglia, which are metabolically active and are enriched in expression of DDP1, TIMM13 and aralar (Ramos et al. 2003; Roesch et al. 2004). However, defects in mitochondrial import of other proteins that use the Tim8p-Tim13p, such as Tim23p or possibly the ATP-Mg/Pi carriers (small CaMCs) may also be involved.

Aralar and Autism

Autism is a neurodevelopmental disorder which manifests itself in early childhood. It is characterized by a deficit in verbal and non verbal communication, impairment in reciprocal social interactions, and patterns of repetitive or stereotyped behaviors and interests (Folstein and Rosen-Sheidley 2001). Cumulative evidence from family and twin studies suggests that genetic factors play an important role in the etiology of autism. Autism is thought to result from disruption of postnatal or experience-dependent synaptic plasticity (Zoghbi 2003). At birth, the brains of autistic children tend to be smaller than those of healthy infants, but between 6 and 14 months of age they undergo abnormally accelerated growth. Autistic brains are larger; the white matter is more prominent, although the cerebral cortex, hippocampus and amigdala are smaller than normal. Neuropathological studies show fewer Purkinje cells than normal, small neuronal size, and increased packing density in several nuclei of the amygdale; small neurons with decreased dendritic branches have ocasionally been observed in CA1 and CA4. It is suggested that autism could be a disorder of synaptic modulation or maintenance.

A strong linkage and association of the mitochondrial SLC25A12 gene, Aralar, with autism has been reported by several independent groups (Ramoz et al. 2004; Gallagher et al. 2003; Segurado et al. 2005; Turunen et al. 2007). Autism was associated with two single nucleotide polymorphisms (SNPs), rs2056202 and rs2292813, within the gene coding for aralar (SLC25A12) (Ramoz et al. 2004; Segurado et al. 2005); as well as a small delection of chromosome 2 in heterozygosis, in the region 2q32.1q32.2–2q32.2q32.3 which contains the aralar gene, was also observed in a patient with severe autism (Gallagher et al. 2003). Recently, Silverman et al. (2008)

reported that the rs2056202 polymorphism might be associated in patients with autism and related disorders; those with high levels of routines and rituals. These authors examined four repetitive behavior traits, "preoccupations/circumscribed interests" and "compulsive adherence to nonfunctional routine and rituals", in relation to the two SLC25A12 polymorphisms in a large series of multiply affected siblings with autism or related disorders. However, others found a lack of association between autism and SLC25A12 (Rabionet et al. 2006; Blasi et al. 2006).

Two polymorphisms located in flanking intronic sequences of exons 4 and 16 of SLC25A12 were associated with autism. The expression of a number of genes is regulated by cis-acting elements, and inherited variations in the SLC25A12 sequence could affect cellular functions when levels of the gene product are limiting or in excess. In fact, overexpression of this protein in beta cells increases MAS function (Rubi et al. 2004). However, the functional relevance of these SNPs or their impact on protein expression is unknown (Ramoz et al. 2004; Segurado et al. 2005). A recent report by Lepagnol-Bestel et al. (2008) showed an enhancement of aralar staining in prefrontal cortex, but not in other cerebral regions, of autistic children. Neuropathological findings and functional abnormalities in autism have been reported for Brodmann´s area (BA) 46 and the cerebellum. These authors found that SLC25A12 was expressed more strongly in the post-mortem brain tissues of autistic subjects than in those of controls, in the BA46 prefrontal cortex but not in cerebellar granule cells. SLC25A12 overexpression or silencing in mouse embryonic cortical neurons also modified dendrite length and the mobility of dendritic mitochondria (Lepagnol-Bestel et al. 2008). Expression of genes encoding for proteins involved in energy demand may be important for the development of cortical networks. The previous study carried out by Lepagnol-Bestel and colleagues shows the involvement of aralar in key mechanisms of normal brain development (sulcus formation in brain, dendritic outgrowth and synaptic plasticity) which might be particularly important in a neurodevelopmental disorder as autism.

Support for a role for mitochondrial dysfunction in autism comes from a study of Filipek et al. (2003), where a modest lactic acidosis is observed. Chugani et al. (1999) found that autistic children showed lower brain NAA levels and higher lactate levels both in plasma and brain. A decrease in NAA in different brain regions of autistic patients, similar to what was found in aralar-deficient mice, has been extensively reported (Friedman et al. 2003; Otsuka et al. 1999; Oliveira et al. 2005; Endo et al. 2007; Kleinhans et al. 2007).

29.3 The Glycerol 3-Phosphate Shuttle (GPS)

29.3.1 General Description

Cleavage of fructose 1,6 diphosphate by aldolase into two 3-carbon molecules, dihydroxyacetone phosphate (DHAP) and glyceraldehyde-3-phosphate, constitutes a critical point in the glycolytic pathway. The GPS interconverts L-glycerol 3-phosphate (G3P) and DHAP to transfer reducing equivalents to the mitochondrial

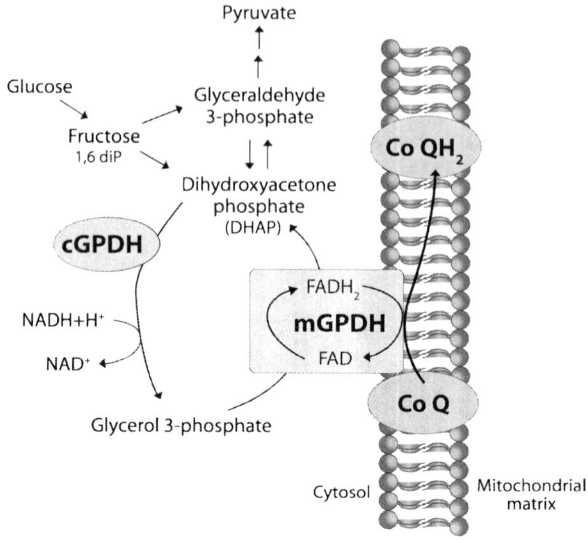

Fig. 29.5 Glycerol-3-phosphate shuttle (GPS). GPS is made up of two enzymes, cytosolic and mitochondrial glycerol 3-phosphate dehydrogenase (cGPDH and mGPDH, respectively). Glucose is converted into fructose 1,6 diphosphate, which is cleaved by aldolase into dihydroxyacetone phosphate and glyceraldehyde 3-phosphate; these trioses are interconvertible through the triosa phosphate isomerase activity. Electrons are transferred from NADH when dihydroxyacetone phosphate is reduced to glycerol 3-phosphate by cGPDH; and the former reoxidized to dihydroxyacetone phosphate by mGPDH. This enzyme is bound to an FAD prosthetic group on the outer side of the inner mitochondrial membrane. Finally, electrons are transferred to coenzyme Q (CoQ) and enter the electron transport chain at this point

electron transport chain in order to reoxidize cytosolic NADH generated from glycolysis (Fig. 29.5). In this shuttle, the NADH-dependent cytosolic glycerol-3-phosphate dehydrogenase (cGPDH, EC 1.1.1.8) reduces DHAP to G3P at the expense of NADH and H^+. This reaction is reversible, with a strong preference under physiologic conditions for the production of G3P. The FAD-dependent mitochondrial glycerol-3-phosphate dehydrogenase (FAD-mGDPH; EC 1.1.99.5) is encoded by a single gene, Gdm1, on chromosome 2 (Koza et al. 1996) and is located on the outer surface of the inner mitochondrial membrane. FAD-mGPDH catalyzes the irreversible conversion of G3P to DHAP, leading to intramitochondrial formation of $FADH_2$. Then, electrons are transferred from bound FAD through ubiquinone to complex III of the electron transport chain. In GPS, the electrons from cytosolic NAD^+/NADH are transferred to FAD, and skip mitochondrial complex I. This makes redox transfer by GPS energetically favorable. The two enzymes (cGPDH and mGPDH) that constitute the GPS are present in tissues capable of rapid oxidative metabolism of glucose, i.e. brain, skeletal muscle and pancreatic insulin β–cells (Brown et al. 1994).

Mice deficient in each of the isoforms (either cGPDH or mGPDH) have been generated. Mice lacking cGPDH are viable and phenotypically normal (although presenting high DHAP and reduced ATP levels in exercized muscle); however, mice lacking mGPDH have reduced body weight and viability. mGPDH$^{-/-}$ mice presented alterations in glucose, glycerol and triglycerides levels and a reduction in thermogenesis by GPS pathway. This deficit is compensated by a chronic stimulation of brown adipose tissue and increased expression of uncoupling protein-3, with a more pronounced phenotype in females than males (Brown et al. 2002a; DosSantos et al. 2003; Alfadda et al. 2004). Crosses were made between these two strains, and mice were obtained that lacked GPS (Brown et al. 2002b). They present hypoglycemic ketosis, glyceroluria, and very high G3P whereas reduced ATP, ADP and AMP levels. These mice, although active and nursing well for several days, failed to grow, and usually die within the first week.

29.3.2 Controversial Aspects

It has been assumed that continued oxidative metabolism of glucose in the presence of drugs that inhibit MAS provides evidence of GPS activity in brain cells (McKenna et al. 1993; Waagepetersen et al. 2001). However the existence of GPS in brain cells has not been proved since recently (Lovatt et al. 2007; Cahoy et al. 2008). Previous current research suggested that none of the major brain cell types coexpressed the two enzymatic isoforms (cGPDH and mGPDH), a fact that would be required for the GPS to be functional. Cytosolic GPDH was found to be highly enriched in cultured oligodendroglial cells and in white matter in brain (Leveille et al. 1980; Montz et al. 1985; Cheng et al. 1997), and the mitochondrial form was found in neurons and astrocytes (Nguyen et al. 2003). Accordingly, the latter authors suggested that the GPS would be of little importance in brain. Although the two isoforms could fulfill diferent roles in their host cells, like providing either glucose-derived glycerol phosphate for phospholipid synthesis (cGPDH) in oligodendrocytes, or lipid-derived glycerol-phosphate as a respiratory substrate (mGPDH) in neurons and astrocytes (Tretter et al. 2007). However, recent transcriptome studies demonstrated that cGPDH and mGPDH transcrits were coexpresed by both astrocytes and neurons, suggesting that GPS could be indeed active in these cell types (Lovatt et al. 2007; Cahoy et al. 2008), a matter that deserves further attention.

29.3.3 Calcium Activation of Glycerol 3P Dehydrogenase

The GPS is activated by calcium, which decreases the K_m of the enzyme mGPDH for the substrate G3P with no changes in the apparent V_{max} (Rasschaert and Malaisse 1991; Rutter et al. 1992). Ca^{2+} exerts its effects on the extramitochondrial side of the inner mitochondrial membrane. Activation by Ca^{2+} of FAD-GPDH is hardly modified by calmodulin inhibitors, suggesting that the enzyme reacts directly with Ca^{2+} (Wernette et al. 1981; MacDonald and Brown 1996). Half-maximal effects of Ca^{2+}

were observed in the range of 30–130 nM, depending on the concentration of glycerol phosphate (Rutter et al. 1992). In the protein sequence of mammalian FAD-GPDH, two EF-hand Ca^{2+}-binding motifs are found to be close to the carboxy terminus of the enzyme. The second of these domains appears to have features compatible with considerable affinity for Ca^{2+}, whereas the first does not (Brown et al. 1994). Overlay experiments with $^{45}Ca^{2+}$ suggest that EF-hand motifs present in the mammalian enzyme bind Ca^{2+} and are responsible for Ca^{2+} activation (MacDonald and Brown 1996). Interestingly, in humans, a mutation in the Ca^{2+}-binding domain associated with a reduced Ca^{2+} activation was found in a family of diabetic subjects (Novials et al. 1997).

29.4 Other Redox Shuttles in the Brain

A number of alternative shuttles for the transfer of redox equivalents across the inner mitochondrial membrane have been proposed to be functional in the brain, and may be of importance in case of MAS inhibition or impairment. Furthermore these alternative shuttles provide Acetyl-CoA and NADPH for important biosynthetic reactions in the cytosol and may play a role in the defense of the cell against oxidative stress.

Of them, just the malate-oxalacetate shuttle has been successfully reconstituted *in vitro* with brain mitochondria (Passarella et al. 1984), although based on the expression of the involved mitochondrial carriers and enzymes, the effective presence of these alternative shuttles must be taken into account in the general redox balance of the brain.

29.4.1 Alternative Mechanisms of NADH Shuttling to Mitochondria

29.4.1.1 Citrate Shuttles

Acetyl-CoA formed in mitochondria cannot cross freely the inner mitochondrial membrane, and it gains access to the cytosol in the form of citrate, which is effluxed in exchange for malate (see ahead). In the cytosol, cleavage of citrate renders back the OAA and the acetyl-CoA moiety, which will then be used for biosynthetic purposes such as acetylcholine, NAA or lipid synthesis. To close the cycle, OAA has to be transferred back to mitochondria, by either of two mechanisms: (1) the citrate-pyruvate shuttle, and, (2) the citrate-malate shuttle. In both pathways, there may be regeneration of glycolitic NAD^+, in the step catalyzed by cMDH and thus both may contribute to the overall redox state of the cell (Palmieri 2004).

Citrate-Pyruvate Shuttle

The citrate-pyruvate shuttle was originally proposed by Farfari et al. (2000) in the pancretic β-cell as an alternative to the pyruvate-malate shuttle (see Sect. 29.4.2.2). This cycle starts in the mitochondrial matrix with the carboxylation of pyruvate to OAA (pyruvate carboxylase, PC, see section "Pyruvate Carboxylase"), and the condensation with pyruvate-derived Acetyl-CoA to form citrate in the reaction catalyzed by citrate synthetase (CS, EC 2.3.3.1) (Watson and Lowenstein 1970). The citrate thus formed is transported to the cytoplasm (citrate carrier, CIC or tricarboxylate carrier, TCC, see section "Citrate-Malate Shuttle") in exchange for cytosolic malate and there undergoes cleavage by the citrate lyase (ACL, EC 2.3.3.8) to OAA and Acetyl-CoA. Then OAA is subsequently reduced to malate by cMDH (EC 1.1.1.37) and further transformed into pyruvate by cytosolic malic enzyme (cME, EC 1.1.1.40; see section "Malic Enzyme"). Transport of pyruvate back to the matrix closes the cycle (Fig. 29.6a, Farfari et al. 2000; Palmieri 2004). The overall outcome of the shuttle is one Acetyl-CoA transferred from the matrix to the cytosol and the conversion of one NADH generated in the glycolysis into cytosolic NADPH. Thus, there is no actual shuttling of redox equivalents between compartments, although NAD^+ is being regenerated in the cytosol. This cycle generates both the basic moiety and redox equivalents utilized for lipid synthesis. The shuttle is irreversible, because some of the enzymes (PC, CS, ACL) and the pyruvate carrier (pyruvate is transported together with a H^+; Halestrap 1975) mediate irreversible processes.

Because the expresion of PC and cME in the CNS is restricted to glial cells (that will be discussed as part of the pyruvate-malate shuttle as restricted to glial cells within the central nervous system, Sect. 29.4.2.2), the citrate-pyruvate shuttle will be accordingly restricted to glial cells within brain, although the other enzymes and carrier proteins involved have a more extended presence in the cells of the brain.

There are four transporters involved in the shuttle: the tricarboxylate carrier (see section "Citrate-Malate Shuttle"), the dicarboxylate carrier (DIC), the pyruvate carrier (PyC) and the phosphate carrier (PiC). The last one is encoded by only one gene in man (SLC25A3), with two isoforms arising from alternative splicing. Of them, PiC-A is present only in muscles, whereas PiC-B is ubiquitous (Fiermonte et al. 1998a; Palmieri 2004)

The transport activity of DIC (Slc25a10) has been demonstrated to be important for *de novo* fatty acid synthesis, as it is involved in supplying cytosolic malate for citrate efflux from mitochondria (Mizuarai et al. 2005). Thus, this member of the mitochondrial carrier superfamily is expressed mostly in lipogenic tissues like liver, although it has also been detected in heart and brain (Fiermonte et al. 1998b; Das et al. 1999). In the latter, the electroneutral exchange of malate and inorganic phosphate would play an important role as part of the citrate shuttles in the efflux of intramitochondrially generated acetyl-CoA necessary for myelin lipid synthesis (Fiermonte et al. 1998b; Palmieri 2004). On the other hand, in brain mitochondria, the activity of the pyruvate carrier is relatively low, compared with that of heart, kidney and liver (Hildyard et al. 2005).

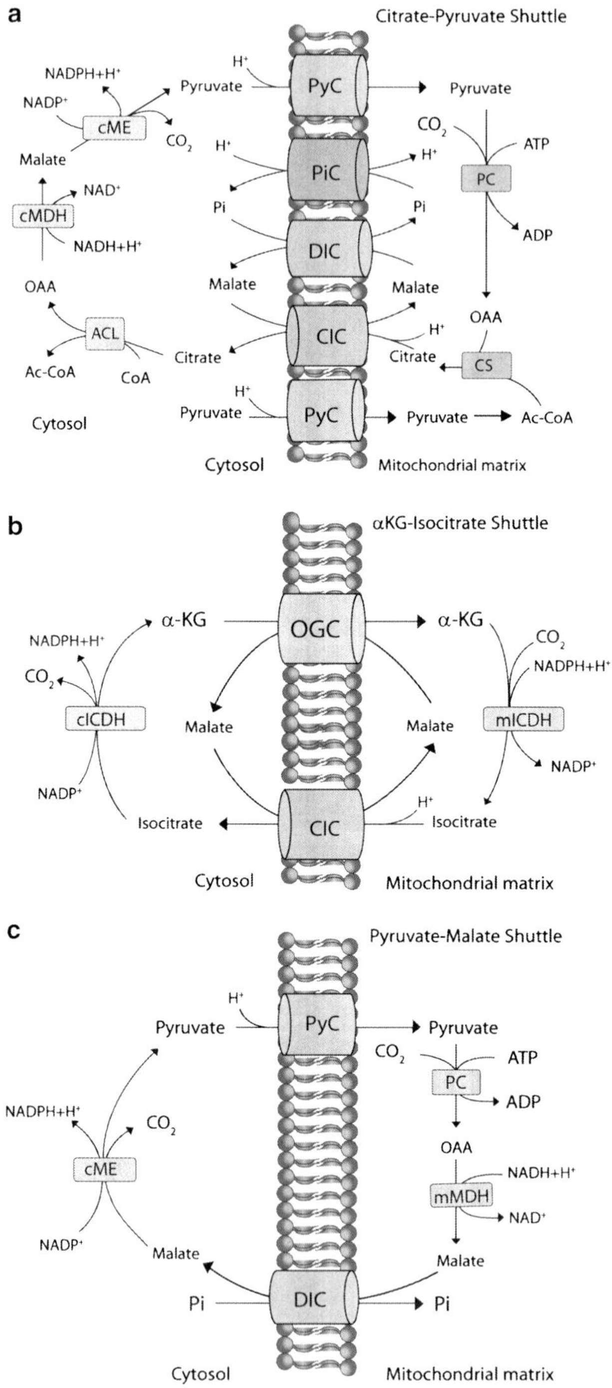
a
Citrate-Pyruvate Shuttle
NADPH+H⁺
NADP⁺
cME
Pyruvate
CO2
Malate
cMDH
NAD⁺
NADH+H⁺
OAA
ACL
Ac-CoA
CoA
Citrate
Cytosol
H⁺
PyC
PiC
Pi
DIC
Malate
CIC
Pyruvate
CO2
ATP
PC
ADP
OAA
H⁺
Pi
Malate
Citrate
CS
Pyruvate
Ac-CoA
Cytosol
Mitochondrial matrix
b
αKG-Isocitrate Shuttle
α-KG
OGC
NADPH+H⁺
CO2
cICDH
NADP⁺
Malate
Isocitrate
CIC
mICDH
NADP⁺
Isocitrate
Cytosol
Mitochondrial matrix
c
Pyruvate-Malate Shuttle
Pyruvate
PyC
NADPH+H⁺
CO2
cME
NADP⁺
Malate
Pi
DIC
PC
ATP
ADP
OAA
mMDH
NADH+H⁺
NAD⁺
Malate
Pi
Cytosol
Mitochondrial matrix

Citrate-Malate Shuttle

The citrate-pyruvate shuttle is limited to glial cells, because the enzymes pyruvate carboxylase and cME are not present in neurons. However, a variation of that shuttle may be present in neurons, and it could to be of particular importance in Acetyl-CoA efflux from mitochondria to generate NAA (see section "Role of Aralar-MAS in Brain Aspartate, N-Acetylaspartate (NAA) and Myelin Lipid Synthesis").

The citrate-malate shuttle, besides transferring the Acetyl-CoA moiety from the mitochondrial compartment to the cytosol, contributes also to the shuttling of glycolytic-born redox equivalents, as the conversion of OAA to malate in the cytosol regenerates NAD^+, and the opposite reaction in the matrix increases the mitochondrial NADH (Fig. 29.4; Patel and Clark 1979; Palmieri 2004).

The electroneutral exchange of mitochondrial citrate plus H^+ with cytosolic malate (Palmieri 2004) can be performed by two possible transporters of citrate. Initially, a transporter of molecular mass of 32.5 kDa was identified and purified from rat liver mitochondria (Bisaccia et al. 1989). Then the transporter was sequenced (Kaplan et al. 1993; Iacobazzi et al. 1997), mapped to chromosome 22q11 (Heisterkamp et al. 1995; Iacobazzi et al. 1997) and recognized as part of the mitochondrial carrier superfamily (Walker and Runswick 1993; Palmieri 2004). The presence of this transporter termed SLC25A1 or CIC was examined by mRNA expression pattern in human tissues, and found at high steady state levels in liver, kidney and pancreas, lower levels in heart, skeletal muscle and placenta, and no detectable mRNA in brain and lung (Huizing et al. 1998). However, citrate transport has been reported in the brain mitochondria (Rafalowska 1979), maybe because CIC is present in the brain albeit expressed at low levels not detected in the study by Huizing and coworkers. The existence of this carrier in the brain may be also backed up by the neurological symptoms of patients with velocardiofacial and DiGeorge syndromes, which are due to deletions on the 22q11 region coding, among others, the SLC25A1 citrate transporter (Heisterkamp et al. 1995; Goldmuntz et al. 1996).

Fig. 29.6 Alternative redox shuttles in the brain. (a). Citrate-pyruvate shuttle. The principal aim of the cycle is to export mitochondrial acetyl CoA to the cytosol to be used for lipid synthesis. Efflux of citrate through the citrate carrier (CIC), and requires previous exit of malate through the dicarboxylate carrier (DIC) in exchange of Pi. In the cytosol, the enzyme ATP-citrate lyase (ACL) cleaves it to render acetyl CoA and OAA. The latter is transformed to malate, regenerating NAD^+ for glycolysis, and the next reaction forms NADPH necessary for the synthesis pathways and pyruvate. The resulting pyruvate reenters the mitochondria through the pyruvate carrier (PyC) and is carboxylated to OAA. **(b). α – Ketoglurate-Isocitrate shuttle.** It is involved in the shuttling of NADPH between the mitochondrial and cytosolic compartments, by the interconversion of isocitrate to α-KG catalyzed by the mitochondrial and cytosolic isoforms of $NADP^+$-dependent isocitrate dehydrogenase (mICDH and cICDH). The flux across the inner mitochondrial membrane is accomplished by the oxoglutarate-malate (OGC) and the citrate (CIC) carriers. **(c). Pyruvate-malate shuttle.** This shuttle restricted to glia uses the mitochondrial carboxylation of pyruvate to generate OAA for reduction to malate, which will be decarboxylated in the cytosol to pyruvate by the malic enzyme thus generating NADPH for biosynthetic purposes. Both transporters (DIC, PyC) have been described to be present in the brain

However, a new family of tricarboxylate carriers with expression in the brain has been identified (Glerum et al. 1990). These transporters have higher molecular mass (37–38 kDa), and do not share the structural characteristics of the mitochondrial superfamily: they have only five putative transmembrane spans and no tripartite structure (Azzi et al. 1993). Two members of the family have been described: TCC is present in liver as well as in brain (especially in neurons), whereas BBC-TCC is expressed mainly in brain, particularly in Bergmann glial cells (Miyake et al. 2002a, b). Whether the members of this new family of mitochondrial carriers or the citrate transporter slc25a1 are involved in the actual citrate efflux from mitochondria in the brain remains to be determined.

In the citrate shuttles, the final product Acetyl-CoA is generated in the cytosol by the enzyme ATP-dependent citrate lyase (ACL, EC 2.3.3.8) from citrate with the hydrolysis of ATP to ADP. This enzyme is considered to be the primary enzyme responsible for acetyl-CoA synthesis, although another cytosolic enzyme, Acetyl-CoA synthetase 1 (ACS1, EC 6.2.1.1) may generate it from acetate and CoA, with hydrolysis of ATP to AMP (Beigneux et al. 2004). The tetrameric enzyme ACL is highly expressed in the developing brain (Beigneux et al. 2004) and has been reported to be associated with mitochondrial membranes, although it is generally regarded as cytosolic. Its activity in the brain has been related to the synthesis of acetyl-CoA for lipid synthesis, mainly in glial cells, as well as in the synthesis of the acetyl moiety of acetylcholine in cholinergic neurons and NAA for myelin synthesis (Janski and Cornell 1980; Szutowicz and Srere 1983; Elshourbagy et al. 1990).

Thus, impairments of the citrate-malate shuttle may result in a decrease of acetylcholine synthesis similar to that observed in the event of MAS activity impairment by Cheeseman and Clark, (1988), as aspartate efflux from matrix is necessary for this neurotransmitter synthesis (see section "Role of Aralar-MAS in Brain Aspartate, N-Acetylaspartate (NAA) and Myelin Lipid Synthesis").

29.4.1.2 Malate-Oxalacetate and Lactate Shuttles

As an alternative to MAS, Passarella et al. (1984) proposed a shuttle in which the redox pair is malate/OAA. The shuttle is composed of both cytosolic and mitochondrial MDH (EC 1.1.1.37, encoded by different genes and different regulatory properties; Malik et al. 1993), and the corresponding dicarboxylate carrier (DiC) and oxodicarboxylate carrier. The existence of this shuttle was based on the observed permeability of brain mitochondria to OAA (Passarella et al. 1985), and the incomplete block of glucose usage in the presence of MAS inhibitors like AOAA observed in brain slices (Fitzpatrick et al. 1983).

Thus, a malate-oxaloacetate shuttle was reconstituted *in vitro*, first with brain mitochondria (Passarella et al. 1984) and then with homogenates from cerebellar granule cells, where it was shown to be a target for glutamate excitotoxicity (Atlante et al. 1999); and also with heart mitochondria, where it is increased with hypertrophy and hypertension (Atlante et al. 2006). The limiting step would be the transport of OAA by the oxodicarboxylate carrier in exchange with cytosolic malate (SLC25A11) (Passarella et al. 1984; Fiermonte et al. 2001; Passarella et al. 2003), as the OAA transporter identified in yeast mitochondria (Palmieri et al. 1999), has no known homolog in mammals yet (del Arco and Satrustegui 2005). The mechanism of the shuttle is quite simple: reduction of OAA by cMDH (EC 1.1.1.37) regenerates NAD^+ in the cytosol, and malate is transported to the matrix in exchange with Pi through its carrier (DiC, Fiermonte et al. 1998b). In the matrix, oxidation of malate elevates NADH inside the mitochondria and produces again OAA, which returns to the cytosol through the oxodicarboxylate carrier in exchange with further malate (Passarella et al. 1984; Atlante et al. 1999).

In principle, all the processes involved are reversible, and on that account the shuttle could work in either direction. However, *in vivo* the relative ratios of $NADH/NAD^+$ in the matrix and cytosol do not favour the transfer of reducing equivalents from cytosol to mitochondria, if no other energy source is involved in the process (Williamson et al. 1967; Gumaa et al. 1971; Dawson 1979). Furthermore, the great difference of OAA concentrations between mitochondria and cytoplasm (0.0001 mM vs. 0.01 mM, respectively) implies a concentration gradient against non-energy driven OAA transport out of the mitochondria (Williamson et al. 1967). Thus, in an *in vivo* situation this mechanism would hardly function to transfer redox equivalents from cytosol to mitochondria, although in situations of elevated unbalanced redox state or MAS impairment redox transfer to mitochondria through this mechanism cannot be excluded.

Recently, a mitochondrial lactate shuttle has been proposed on the basis of the presence of monocarboxylate carriers (MCT) and lactate dehydrogenase (LDH) in mitochondria. This mechanism was proposed initially in muscle mitochondria (Brooks et al. 1999), but its participation in redox transfer to brain mitochondria has also been reported (Schurr 2006; Atlante et al. 2007; Schurr and Payne 2007). Basically, glycolysis would continue to complete oxidation of glucose to lactate, which would then be transported to the mitochondria by either MCT1 (Brooks et al. 1999; Schurr 2006; Schurr and Payne 2007) or in symport with a proton (Atlante et al. 2007), where it would be oxidized to pyruvate by mitochondrial LDH. Then, the pyruvate may either exit in exchange with further lactate (Atlante et al. 2007) or fuel the TCA (Brooks et al. 1999; Schurr 2006; Schurr and Payne 2007). In the latter case, as no cycling of a redox pair takes place, the process should not be considered a shuttle system. In any event, the localization of the mitochondrial LDH is not clear, as it has been reported both at the intermembrane/outer mitochondrial membrane (Brandt et al. 1987) and in the inner mitochondrial membrane/matrix fractions, where a complex with MCT1 and COX has been proposed for lactate oxidation (Brooks et al. 1999; Hashimoto et al. 2006). In any event, NADH produced in the intermembrane space would require a shuttle mechanism to transport the redox equivalents through the inner mitochondrial membrane. The fact that the lactate/pyruvate ratio is about 100-fold higher in the mitochondria than in the cytosol (Atlante et al. 2007), together with the more favourable kinetics of the MCT1 for pyruvate uptake (Yoshida et al. 2007), and the highly reduced state of the mitochondrial $NAD^+/NADH$ ratio (Williamson et al. 1967) argue against a prominent role for such a shuttle in the physiology of the cell. Consequently, a number of studies have been unable to observe lactate oxidation with mitochondria from muscle (Yoshida et al. 2007; and references therein) and thus the actual existence of this mechanism awaits further confirmation.

29.4.2 NADPH Shuttles Between Cytosol and Mitochondria in the Brain

NADPH is an important source of reducing equivalents for biosynthetic pathways, such as fatty acid, myelin and acetylcholine synthesis, hydroxylation reactions and in the regeneration of reduced glutathione (GSH), a critical player in cellular defense against oxidative damage (Lee et al. 2002; Minich et al. 2003; Dringen et al. 2005; Jezek and Hlavata 2005).

After oxidation, NADPH needs to be regenerated. This is accomplished mainly by cytosolic enzymes of the pentose phosphate pathway such as glucose-6-phosphate dehydrogenase (G6PDH, EC 1.1.1.49) and 6-phosphogluconate dehydrogenase (EC 1.1.1.44). However, other enzymes like the mitochondrial transhydrogenase (EC 1.6.1.1) take part also in the regeneration of NADPH, as well as the cytosolic and mitochondrial isoforms of the $NADP^+$-dependent isocitrate dehydrogenases and malic enzymes. These may regenerate NADPH independently or integrated in a mitochondrial shuttle system which would transfer redox equivalents between both compartments (Dringen et al. 2005; Ido 2007).

29.4.2.1 The α-Ketoglutarate-Isocitrate Shuttle

In this shuttle, the interconversion of isocitrate and α-KG by either the cytosolic or mitochondrial $NADP^+$-dependent isocitrate dehydrogenases (ICDH, EC 1.1.1.42) will allow the transfer of redox equivalents from the pair NADPH/$NADP^+$ between both cellular compartments.

Besides the enzymes, the α-KG-isocitrate shuttle is composed of two transporters, which allow the movement of the involved metabolites across the inner mitochondrial membrane: the tricarboxylate carrier (see section "Citrate-Malate Shuttle") and the oxoglutarate-malate antiporter (OGC) (Runswick et al. 1990; Palmieri 2004) (Fig. 29.6b).

The NADPH generating reaction of the cycle is the oxidative decarboxylation of isocitrate to $\alpha-$KG, which is catalyzed by the isocitrate dehydrogenases. Three different isoforms of the enzyme have been described. The first one plays a pivotal role in the tricarboxylic acid cycle, has NAD^+ as cofactor and is regulated by calcium (IDH, EC 1.1.1.41) (McCormack et al. 1988; McCormack et al. 1990). The other two are $NADP^+$-dependent enzymes and are involved in the αKG-isocitrate shuttle. Both mitochondrial (mICDH or ICDH1) and cytosolic (cICDH or ICDH2) isoforms have been described: mICDH is the predominant isoform in the heart, whereas cICDH is the main one in liver (Andrés et al. 1980; Plaut et al. 1983; Jennings et al. 1994; Yang et al. 1996; Minich et al. 2003).

In the brain, both mitochondrial and cytosolic isoforms have been reported (Loverde and Lehrer 1973; Bajo et al. 2002; Minich et al. 2003). The highest isocitrate dehydrogenase specific activity was found in astrocytic cultures (65 nmols min^{-1} mg^{-1} versus 30 nmols min^{-1} mg^{-1} in neuronal cultures), and was equally distributed in the mitochondrial and cytosolic fractions, in contrast with the elevated activity of the neuronal cytosolic fraction of about 75% of total activity (Loverde and Lehrer 1973; Juurlink 1993; Minich et al. 2003). However, it should be noted that *in vivo* kinetics may differ from that indicated, as a consequence of a distinct expression of enzymes in cultured cells, and particularly in astrocytes (Cahoy et al. 2008).

The main role of the α-KG-isocitrate shuttle would be NADPH production for GSH regeneration. The tripeptide GSH (γ-glutamyl-cysteinyl-glycine) is the electron donor to reduce peroxides to water or the corresponding alcohols in the reaction catalyzed by glutathione peroxidase (GPx, EC 1.11.19). The oxidation product is GSSG (glutathione disulfide) which can be reduced to GSH in a NADPH-dependent reaction catalyzed by the flavoenzyme glutathione reductase (GR, EC 1.8.1.7). Therefore, the ultimate antioxidant capacity of a tissue is determined by the supply of reducing potentials, and the detoxification system depends on the continuous supply of NADPH by the NADPH regenerating enzymes (Dringen et al. 2005). Accordingly, the α-KG-isocitrate shuttle plays a role in balancing the general redox state of both matrix and cytosol. This role has been backed up by reports showing protection by over expression of cICDH and conversely, a greater vulnerability to oxidative stress (elevated ratio GSSG/GSH) in cells with reduced activity of cICDH (Lee et al. 2002; Maeng et al. 2004). Also in conditions of G6PDH inactivity, cICDH could play an important role in addition to the pentose phosphate enzymes in cytosolic NADPH regeneration as demonstrated in the retina (Winkler et al. 1986). Similarly, in the regeneration of intramitochondrial GSH, a role for mICDH along with mitochondrial transhydrogenase has been proposed (Sazanov and Jackson 1994; Vogel et al. 1999; Jo et al. 2001).

It is usually considered that the α-KG-isocitrate shuttle renders net transfer of NADPH from the mitochondria (where $NADP^+$/NADPH ratio is maintained very low by the energy-linked transhydrogenase) to the cytosol in response to a stimulation of cytosolic NADPH utilizing enzymes (Williamson and Cooper 1980; Juurlink 1993; Yang et al. 1996). However, the processes involved are reversible, and therefore, shuttling of NADPH redox equivalents across the inner mitochondrial membrane can take place bidirectionally.

29.4.2.2 Pyruvate-Malate Shuttle

A pyruvate-malate shuttle has been demonstrated to operate across the mitochondrial membrane in pancreatic islets. This shuttle would provide cytosolic NADPH for lipid synthesis in the β-cell and is related to insulin secretion (MacDonald 1995). The cycle is initiated by the carboxylation of pyruvate to OAA by the enzyme pyruvate carboxylase (PC, EC 6.4.1.1). OAA is then converted to malate in the mitochondria by malate dehydrogenase (mMDH) and exported to the cytosol in exchange for Pi (dicarboxylate carrier, DIC), where it is decarboxylated by the cytosolic malic enzyme (cME, EC 1.1.1.40) to pyruvate that enters the mitochondria through its carrier (PyC) closing the cycle (Fig. 29.6c; Jitrapakdee and Wallace 1999). This cycle allows the shuttling of redox equivalents (NADH) generated in the mitochondria by the mMDH reaction to the cytosol in the form of malate, and there generates

NADPH in the cME reaction. The NADPH may then be used for lipid synthesis or glutathione regeneration (Ido 2007).

To our knowledge, the pyruvate-malate shuttle has not been reconstituted in brain mitochondria, although all the proteins involved have been described to be present in the central nervous system, at least in glial cells. As part of the Krebs cycle, mMDH is ubiquitously expressed. The two other enzymes involved in the shuttle have a more restricted expression pattern.

Pyruvate Carboxylase

PC is a biotin-containing enzyme found in astrocytes in the central nervous system (Cesar and Hamprecht 1995) and in dorsal root ganglion satellite cells and Schwann cells in the peripheral nervous system (Miller et al. 2002; Jitrapakdee et al. 2006). PC activity was found in cultured astrocytes (Shank et al. 1985; Shank et al. 1993) but could not be detected in neurons cultured from either cerebral cortex or cerebellum (Yu et al. 1983), where pyruvate is carboxylated mainly by ME (Hassel 2001). The enzyme has a leader sequence for its translocation to the mitochondrial matrix (Rohde et al. 1991; Jitrapakdee and Wallace 1999) where interactions with other mitochondrial proteins are likely to influence their characteristics and kinetic properties (Wallace et al. 1998; Jitrapakdee and Wallace 1999).

Malic Enzyme

The reversible oxidative carboxylation of pyruvate to malate takes place either in the cytosol or the mitochondrial matrix by the malic enzyme. In mammals, there are three different isoforms, encoded by different genes: one cytosolic isoform (ME1; EC 1.1.1.40) and two mitochondrial isoforms (ME2 (EC 1.1.1.38) and ME3 (EC 1.1.1.40)) (Loeber et al. 1994). ME1 and ME3 are specific for $NADP^+$ as a pyridine nucleotide cofactor, have a low *km* for malate and are present in human and rodent brain (Bukato et al. 1995a, b). ME2 uses either NAD^+ or $NADP^+$ as a cofactor, has a high Km for malate and is present in pancreatic islets of humans and rodents (MacDonald et al. 2009). Accordingly, cytosolic (cME) or mitochondrial (mME) malic enzyme activities in the brain exhibit different regulatory, structural and kinetic properties (McKenna et al. 1995). For example, the cytosolic enzyme exhibits a greater capability to catalyze the reductive carboxylation of pyruvate to malate than the mitochondrial enzyme (Bukato et al. 1995a, b). Thus the isozymes may play different metabolic roles in the brain. The mitochondrial isoform represents about 50–75% of total ME activity in the brain and it is expressed mostly in neurons, where it is highly enriched in the synaptosomal fracion; but also in astrocytes (Bukato et al. 1994; McKenna et al. 2000). mME would provide NADPH for reduction of intramitochondrial glutathione or play a role in the pyruvate recycling pathway in the brain (McKenna et al. 2000). On the other hand, the cytosolic isoform is primarily, if not exclusively a glial enzyme (25–50% of total ME in the brain, whereas up to 95% in cultured astrocytes) and hence may play a pivotal role in glial metabolism providing cytosolic NADPH for lipid synthesis or reduction of oxidized glutathione. It has also been suggested that cME is involved in the degradation of excessive Krebs cycle intermediates derived from glutamate or other amino acids to generate pyruvate as a fuel for the production of energy (Kurz et al. 1993; Bukato et al. 1994; McKenna et al. 1995; Vogel et al. 1998).

Therefore, although the other components of the shuttle are more broadly expressed, the biased presence of cME and PC restricts the possibility of a functional pyruvate-malate shuttle within the brain to the glial cells, and does not support a role for this shuttle in the redox balance and metabolism of neurons.

29.5 Concluding Remarks

Redox shuttles for redox transfer from NAD^+/NADH to mitochondria mainly participate in brain metabolism and neurotransmission. A robust body of evidence supports the critical role of MAS in brain neurons but its actual existence in glial cells is under investigation. Regarding GPS, controversial data about cellular localization of GPS constituents has cast doubt, until recently, on its relative importance in brain. In the past, the lack of pharmacological drugs to completely and specifically inhibit GPS and MAS activities did not facilitate the study of these redox shuttles. At present, important advances have been attained using genetic approaches and microscopic techniques with high spatio-temporal resolution in tissues. Some of the physiological roles of these redox shuttles have been unveiled thanks to the use of transgenic mouse models as is the case for the aralar-deficient mice, lacking MAS activity in brain. The actual existence of alternative NADH shuttles, particularly the citrate shuttles, has not been demonstrated by reconstitution assays, and should be thus regarded with caution. Interestingly, MAS and citrate-malate shuttles might function together in neuronal cells to generate NAA for myelin synthesis.

The NADPH shuttles are regarded as systems for redox transfer from mitochondrial NADP/NADPH to the cytosol, and are thought to be important in cells undergoing biosynthetic processes, mainly lipid synthesis, a process which is important in neurones, and especially glial cells. They include the α-ketoglutarate-isocitrate shuttle and the pyruvate-malate shuttle which is mainly represented in glial cells. Further work is required to clarify the actual role of these cycles *in vivo*.

Acknowledgments We are indebted to Prof. J. Satrústegui for critical reading of the manuscript, for helpful suggestions and discussions and for support. The authors would like to thank funding from the Ministerio de Educación y Ciencia (contract grant BFU2005-C02-01, GEN2003-20235-C05-03/NAC to JS), Instituto de Salud Carlos III del Ministerio de Sanidad (PI042457 to JS), European Union (contract grant number: LSHM-CT-2006-518153 to JS); Mutua Madrileña (to BP), and the institucional support of the Ramón Areces Foundation to the Centro de Biología Molecular Severo Ochoa. The CIBER de Enfermedades Raras is an initiative of the ISCIII.

References

Alfadda A, DosSantos RA, Stepanyan Z, Marrif H, Silva JE (2004) Mice with deletion of the mitochondrial glycerol-3-phosphate dehydrogenase gene exhibit a thrifty phenotype: effect of gender. Am J Physiol Regul Integr Comp Physiol 287(1):147–156

Andrés A, Satrústegui J, Machado A (1980) Development of NADPH-producing pathways in rat heart. Biochem J 186:799–803

Atlante A, Gagliardi S, Marra E, Calissano P, Passarella S (1999) Glutamate neurotoxicity in rat cerebellar granule cells involves cytochrome c release from mitochondria and mitochondrial shuttle impairment. J Neurochem 73:237–246

Atlante A, Seccia TM, De Bari L, Marra E, Passarella S (2006) Mitochondria from the left heart ventricles of both normotensive and spontaneously hypertensive rats oxidize externally added NADH mostly via a novel malate/oxaloacetate shuttle as reconstructed in vitro. Int J Mol Med 18:177–186

Atlante A, de Bari L, Bobba A, Marra E, Passarella S (2007) Transport and metabolism of l-lactate occur in mitochondria from cerebellar granule cells and are modified in cells undergoing low potassium dependent apoptosis. Biochim Biophys Acta 1767:1285–1299

Azzi A, Chappell JB, Robinson BH (1967) Penetration of the mitochondrial membrane by glutamate and aspartate. Biochem Biophys Res Commun 29:148–152

Azzi A, Glerum M, Koller R, Mertens W, Spycher S (1993) The mitochondrial tricarboxylate carrier. J Bioenerg Biomembr 25:515–524

Bajo M, Fruehauf J, Kim SH, Fountoulakis M, Lubec G (2002) Proteomic evaluation of intermediary metabolism enzyme proteins in fetal Down's syndrome cerebral cortex. Proteomics 2:1539–1546

Baslow MH (2003) N-acetylaspartate in the vertebrate brain: metabolism and function. Neurochem Res 28:941–953

Begum L, Jalil MA, Kobayashi K, Iijima M, Li MX, Yasuda T, Horiuchi M, del Arco A, Satrústegui J, Saheki T (2002) Expression of three mitochondrial solute carriers, citrin, aralar1 and ornithine transporter, in relation to urea cycle in mice. Biochim Biophys Acta 1574:283–292

Beigneux AP, Kosinski C, Gavino B, Horton JD, Skarnes WC, Young SG (2004) ATP-citrate lyase deficiency in the mouse. J Biol Chem 279:9557–9564

Bergersen L, Waerhaug O, Helm J, Thomas M, Laake P, Davies AJ, Wilson MC, Halestrap AP, Ottersen OP (2001) A novel postsynaptic density protein: the monocarboxylate transporter MCT2 is co-localized with delta-glutamate receptors in postsynaptic densities of parallel fiber-Purkinje cell synapses. Exp Brain Res 136:523–534

Berkich DA, Ola MS, Cole J, Sweatt AJ, Hutson SM, LaNoue KF (2007) Mitochondrial transport proteins of the brain. J Neurosci Res 85(15):3367–3377

Bhakoo KK, Craig TJ, Styles P (2001) Developmental and regional distribution of aspartoacylase in rat brain tissue. J Neurochem 79:211–220

Bisaccia F, De Palma A, Palmieri F (1989) Identification and purification of the tricarboxylate carrier from rat liver mitochondria. Biochim Biophys Acta 977:171–176

Bisaccia F, De Palma A, Palmieri F (1992) Identification and purification of the aspartate/glutamate carrier from bovine heart mitochondria. Biochim Biophys Acta 1106:291–296

Bittar PG, Charnay Y, Pellerin L, Bouras C, Magistretti PJ (1996) Selective distribution of lactate dehydrogenase isoenzymes in neurons and astrocytes of human brain. J Cereb Blood Flow Metab 16:1079–1089

Blasi F, Bacchelli E, Carone S, Toma C, Monaco AP, Bailey AJ, Maestrini E (2006) SLC25A12 and CMYA3 gene variants are not associated with autism in the IMGSAC multiplex family sample. Eur J Hum Genet 14:123–126

Bliss TM, Ip M, Cheng E, Minami M, Pellerin L, Magistretti P, Sapolsky RM (2004) Dual-gene, dual-cell type therapy against an excitotoxic insult by bolstering neuroenergetics. J Neurosci 24:6202–6208

Brandt RB, Laux JE, Spainhour SE, Kline ES (1987) Lactate dehydrogenase in rat mitochondria. Arch Biochem Biophys 259(2):412–422

Brooks GA, Dubouchaud H, Brown M, Sicurello JP, Butz CE (1999) Role of mitochondrial lactate dehydrogenase and lactate oxidation in the intracellular lactate shuttle. Proc Natl Acad Sci USA 96:1129–1134

Brown LJ, MacDonald MJ, Lehn DA, Moran SM (1994) Sequence of rat mitochondrial glycerol-3-phosphate dehydrogenase cDNA. Evidence for EF-hand calcium-binding domains. J Biol Chem 269:14363–14366

Brown LJ, Koza RA, Everett C, Reitman ML, Marshall L, Fahien LA, Kozak LP, MacDonald MJ (2002a) Normal thyroid thermogenesis but reduced viability and adiposity in mice lacking the mitochondrial glycerol phosphate dehydrogenase. J Biol Chem 277(36):32892–32898

Brown LJ, Koza RA, Marshall L, Kozak LP, MacDonald MJ (2002b) Lethal hypoglycemic ketosis and glyceroluria in mice lacking both the mitochondrial and the cytosolic glycerol phosphate dehydrogenases. J Biol Chem 277(36):32899–32904

Bukato G, Kochan Z, Swierczynski J (1994) Subregional and intracellular distribution of NADP-linked malic enzyme in human brain. Biochem Med Metab Biol 51:43–50

Bukato G, Kochan Z, Swierczynski J (1995a) Different regulatory properties of the cytosolic and mitochondrial forms of malic enzyme isolated from human brain. Int J Biochem Cell Biol 27:1003–1008

Bukato G, Kochan Z, Swierczynski J (1995b) Purification and properties of cytosolic and mitochondrial malic enzyme isolated from human brain. Int J Biochem Cell Biol 27:47–54

Burri R, Steffen C, Herschkowitz N (1991) N-acetyl-L-aspartate is a major source of acetyl groups for lipid synthesis during rat brain development. Dev Neurosci 13:403–411

Caesar K, Hashemi P, Douhou A, Bonvento G, Boutelle MG, Walls AB, Lauritzen M (2008) Glutamate receptor-dependent increments in lactate, glucose and oxygen metabolism evoked in rat cerebellum in vitro. J Physiol 586(5):1337–1349

Cahoy JD, Emery B, Kaushal A, Foo LC, Zamanian JL, Christopherson KS, Xing Y, Lubischer JL, Krieg PA, Krupenko SA, Thompson WJ, Barres BA (2008) A transcriptome database for astrocytes, neurons, and oligodendrocytes: a new resource for understanding brain development and function. J Neurosci 28(1):264–278

Cater HL, Benham CD, Sundstrom LE (2001) Neuroprotective role of monocarboxylate transport during glucose deprivation in slice cultures of rat hippocampus. J Physiol 531:459–466

Cater HL, Chandratheva A, Benham CD, Morrison B, Sundstrom LE (2003) Lactate and glucose as energy during, and after, oxygen deprivation in rat hippocampal and cultured slices. J Neurochem 87(6):1381–1390

Cesar M, Hamprecht B (1995) Immunocytochemical examination of neural rat and mouse primary cultures using monoclonal antibodies raised against pyruvate carboxylase. J Neurochem 64:2312–2318

Chakraborty G, Mekala P, Yahya D, Wu G, Ledeen RW (2001) Intraneuronal N-acetylaspartate supplies acetyl groups for myelin lipid synthesis: evidence for myelin-associated aspartoacylase. J Neurochem 78:736–745

Cheeseman AJ, Clark JB (1988) Influence of the malate-aspartate shuttle on oxidative metabolism in synaptosomes. J Neurochem 50:1559–1565

Cheng JD, Espinosa de los Monteros A, de Vellis J (1997) Glial- and fat-specific expression of the rat glycerol phosphate dehydrogenase-luciferase fusion gene in transgenic mice. J Neurosci Res 50:300–311

Chih ChP, Lipton P, Roberts EL (2001) Do active cerebral neurons really use lactate rather than glucose? Trends in Neuroscience 24(10):573–578

Chugani DC, Sundram BS, Behen M, Lee ML, Moore GJ (1999) Evidence of altered energy metabolism in autistic children. Prog Neuropsychopharmacol Biol Psychiatry 23:635–641

Contreras L, Gomez-Puertas P, Iijima M, Kobayashi K, Saheki T, Satrústegui J (2007) Ca^{2+} Activation kinetics of the two aspartate-glutamate mitochondrial carriers, aralar and citrin: role in the heart malate-aspartate NADH shuttle. J Biol Chem 282:7098–7106

Contreras L, Satrústegui J (2009) Calcium signaling in brain mitochondria: interplay of malate aspartate NADH shuttle and calcium uniporter/mitochondrial dehydrogenase pathways. J Biol Chem 284:7091–7099

Contreras L, Urbieta A, Kobayashi K, Saheki T, Satrústegui J (2010) Low levels of citrin (*SLC25A13*) expression in adult mouse brain restricted to neuronal clusters. J Neurosci Res 88 (5):1009–1016

Curran SP, Leuenberger D, Oppliger W, Koehler CM (2002a) The Tim9p-Tim10p complex binds to the transmembrane domains of the ADP/ATP carrier. EMBO J 21:942–953

Curran SP, Leuenberger D, Schmidt E, Koehler CM (2002b) The role of the Tim8p-Tim13p complex in a conserved import pathway for mitochondrial polytopic inner membrane proteins. J Cell Biol 158:1017–1027

D'Adamo AF Jr, Yatsu FM (1966) Acetate metabolism in the nervous system. N-acetyl-L-aspartic acid and the biosynthesis of brain lipids. J Neurochem 13:961–965

Das K, Lewis RY, Combatsiaris TP, Lin Y, Shapiro L, Charron MJ, Scherer PE (1999) Predominant expression of the mitochondrial dicarboxylate carrier in white adipose tissue. Biochem J 344(Pt 2): 313–320

Davis AJ, Sepuri NB, Holder J, Johnson AE, Jensen RE (2000) Two intermembrane space TIM complexes interact with different domains of Tim23p during its import into mitochondria. J Cell Biol 150:1271–1282

Dawson AG (1979) Oxidation of cytosolic NADH formed during aerobic metabolism in mammalian cells. Trends Biochem Sci 4:171–176

Debernardi R, Pierre K, Lengacher S, Magistretti PJ, Pellerin L (2003) Cell-specific expression pattern of monocarboxylate transporters in astrocytes and neurons observed in different mouse brain cortical cell cultures. J Neurosci Res 73:141–155

del Arco A, Satrústegu J (1998) Molecular cloning of Aralar, a new member of the mitochondrial carrier superfamily that binds calcium and is present in human muscle and brain. J Biol Chem 273:23327–23334

del Arco A, Agudo M, Satrústegui J (2000) Characterization of a second member of the subfamily of calcium-binding mitochondrial carriers expressed in human non-excitable tissues. Biochem J 345(Pt 3):725–732

del Arco A, Morcillo J, Martinez-Morales JR, Galian C, Martos V, Bovolenta P, Satrústegui J (2002) Expression of the aspartate/glutamate mitochondrial carriers aralar1 and citrin during development and in adult rat tissues. Eur J Biochem 269:3313–3320

del Arco A, Satrústegui J (2004) Identification of a novel human subfamily of mitochondrial carriers with calcium-binding domains. J Biol Chem 279:24701–24713

del Arco A (2005) Novel variants of human SCaMC-3, an isoform of the ATP-Mg/P(i) mitochondrial carrier, generated by alternative splicing from 3'-flanking transposable elements. Biochem J 389:647–655

del Arco A, Satrústegui J (2005) New mitochondrial carriers: an overview. Cell Mol Life Sci 62:2204–2227

Dierks T, Kramer R (1988) Asymmetric orientation of the reconstituted aspartate/glutamate carrier from mitochondria. Biochim Biophys Acta 937:112–126

Dierks T, Riemer E, Kramer R (1988) Reaction mechanism of the reconstituted aspartate/glutamate carrier from bovine heart mitochondria. Biochim Biophys Acta 943:231–244

Dombrowski GJ Jr, Swiatek KR, Chao KL (1989) Lactate, 3-hydroxybutyrate, and glucose as substrates for the early postnatal rat brain. Neurochem Res 14:667–675

DosSantos RA, Alfadda A, Eto K, Kadowaki T, Silva JE (2003) Evidence for a compensated thermogenic defect in transgenic mice lacking the mitochondrial glycerol-3-phosphate dehydrogenase gene. Endocrinology 144(12):5469–5479

Dringen R, Pawlowski PG, Hirrlinger J (2005) Peroxide detoxification by brain cells. J Neurosci Res 79:157–165

Elshourbagy NA, Near JC, Kmetz PJ, Sathe GM, Southan C, Strickler JE, Gross M, Young JF, Wells TN, Groot PH (1990) Rat ATP citrate-lyase. Molecular cloning and sequence analysis of a full-length cDNA and mRNA abundance as a function of diet, organ, and age. J Biol Chem 265:1430–1435

Endo T, Shioiri T, Kitamura H, Kimura T, Endo S, Masuzawa N, Someya T (2007) Altered chemical metabolites in the amygdale-hippocampus region contribute to autistic symptoms of autism spectrum disorders. Biol Psychiatry 62(9):1030–1037

Ernster L, Gottfried S (1981) Mitochondria: A historical review. J Cell Biol 91(3):227–255

Farfari S, Schulz V, Corkey B, Prentki M (2000) Glucose-regulated anaplerosis and cataplerosis in pancreatic beta-cells: possible implication of a pyruvate/citrate shuttle in insulin secretion. Diabetes 49:718–726

Fiermonte G, Dolce V, Palmieri F (1998a) Expression in Escherichia coli, functional characterization, and tissue distribution of isoforms A and B of the phosphate carrier from bovine mitochondria. J Biol Chem 273:22782–22787

Fiermonte G, Palmieri L, Dolce V, Lasorsa FM, Palmieri F, Runswick MJ, Walker JE (1998b) The sequence, bacterial expression, and functional reconstitution of the rat mitochondrial dicarboxylate transporter cloned via distant homologs in yeast and Caenorhabditis elegans. J Biol Chem 273:24754–24759

Fiermonte G, Dolce V, Palmieri L, Ventura M, Runswick MJ, Palmieri F, Walker JE (2001) Identification of the human mitochondrial oxodicarboxylate carrier. Bacterial expression, reconstitution, functional characterization, tissue distribution, and chromosomal location. J Biol Chem 276:8225–8230

Filipek PA, Juranek J, Smith M, Mays LZ, Ramos ER, Bocian M, Masser-Frye D, Laulhere TM, Modahl C, Spence MA, Gargus JJ (2003) Mitochondrial dysfunction in autistic patients with 15q inverted duplication. Ann Neurol 53:801–804

Fitzpatrick SM, Cooper AJ, Duffy TE (1983) Use of beta-methylene-D, L-aspartate to assess the role of aspartate aminotransferase in cerebral oxidative metabolism. J Neurochem 41:1370–1383

Folstein SE, Rosen-Sheidley B (2001) Genetics of autism: complex aetiology for a heterogeneous disorder. Nat Rev Genet 2:943–955

Friedman SD, Shaw DW, Artru AA, Richards TL, Gardner J, Dawson G, Posse S, Dager SR (2003) Regional brain chemical alterations in young children with autism spectrum disorder. Neurology 60:100–107

Gallagher L, Becker K, Kearney G, Dunlop A, Stallings R, Green A, Fitzgerald M, Gill M (2003) Brief report: A case of autism associated with del(2)(q32.1q32.2) or (q32.2q32.3). J Autism Dev Disord 33:105–108

Glerum DM, Claeys D, Mertens W, Azzi A (1990) The tricarboxylate carrier from rat liver mitochondria. Purification, reconstitution and kinetic characterization. Eur J Biochem 194:681–684

Goldmuntz E, Wang Z, Roe BA, Budarf ML (1996) Cloning, genomic organization, and chromosomal localization of human citrate transport protein to the DiGeorge/velocardiofacial syndrome minimal critical region. Genomics 33:271–276

Goldstein FB (1969) The enzymatic synthesis of N-acetyl-L-aspartic acid by subcellular preparations of rat brain. J Biol Chem 244:4257–4260

Grabarek Z (2006) Structural basis for diversity of the EF-hand calcium-binding proteins. J Mol Biol 359:509–525

Gramsbergen JB, Sandberg M, Kornblit B, Zimmer J (2000) Pyruvate protects against 3-nitropropionic acid neurotoxicity in corticostriatal slice cultures. Neuroreport 11:2743–2747

Gumaa KA, McLean P, Greenbaum AL (1971) Compartmentation in relation to metabolic control in liver. Essays Biochem 7:39–86

Gunter TE, Buntinas L, Sparagna G, Eliseev R, Gunter K (2000) Mitochondrial calcium transport: mechanisms and functions. Cell Calcium 28:285–296

Halestrap AP (1975) The mitochondrial pyruvate carrier. Biochem J 148:85–96

Hashimoto T, Hussien R, Brooks GA (2006) Colocalization of MCT1, CD147, and LDH in mitochondrial inner membrane of L6 muscle cells: evidence of a mitochondrial lactate oxidation complex. Am J Physiol Endocrinol Metab 290:E1237–E1244

Hassel B (2001) Pyruvate carboxylation in neurons. J Neurosci Res 66:755–762

Heisterkamp N, Mulder MP, Langeveld A, ten Hoeve J, Wang Z, Roe BA, Groffen J (1995) Localization of the human mitochondrial citrate transporter protein gene to chromosome 22Q11 in the DiGeorge syndrome critical region. Genomics 29:451–456

Hildyard JC, Ammala C, Dukes ID, Thomson SA, Halestrap AP (2005) Identification and characterisation of a new class of highly specific and potent inhibitors of the mitochondrial pyruvate carrier. Biochim Biophys Acta 1707:221–230

Hu Y, Wilson GS (1997) Rapid changes in local extracellular rat brain glucose observed with an in vivo glucose sensor. J Neurochem 68:1745–1752

Huizing M, Ruitenbeek W, van den Heuvel LP, Dolce V, Iacobazzi V, Smeitink JA, Palmieri F, Trijbels JM (1998) Human mitochondrial transmembrane metabolite carriers: tissue distribution and its implication for mitochondrial disorders. J Bioenerg Biomembr 30:277–284

Iacobazzi V, Lauria G, Palmieri F (1997) Organization and sequence of the human gene for the mitochondrial citrate transport protein. DNA Seq 7:127–139

Ido Y (2007) Pyridine nucleotide redox abnormalities in diabetes. Antioxid Redox Signal 9:931–942

Jalil MA, Begum L, Contreras L, Pardo B, Iijima M, Li MX, Ramos M, Marmol P, Horiuchi M, Shimotsu K, Nakagawa S, Okubo A, Sameshima M, Isashiki Y, Del Arco A, Kobayashi K, Satrústegui J, Saheki T (2005) Reduced N-Acetylaspartate Levels in Mice Lacking Aralar, a Brain- and Muscle-type Mitochondrial Aspartate-glutamate Carrier. J Biol Chem 280: 31333–31339

Janski AM, Cornell NW (1980) Association of ATP citrate lyase with mitochondria. Biochem Biophys Res Commun 92:305–312

Jennings GT, Sechi S, Stevenson PM, Tuckey RC, Parmelee D, McAlister-Henn L (1994) Cytosolic NADP(+)-dependent isocitrate dehydrogenase. Isolation of rat cDNA and study of tissue-specific and developmental expression of mRNA. J Biol Chem 269:23128–23134

Jezek P, Huavata L (2005) Mitochondria in homeostasis of reactive oxygen species in cell tisúes and organism. Int J Biochem Cell Biol 37(12):2478–2503

Jin H, May M, Tranebjaerg L, Kendall E, Fontan G, Jackson J, Subramony SH, Arena F, Lubs H, Smith S, Stevenson R, Schwartz C, Vetrie D (1996) A novel X-linked gene, DDP, shows mutations in families with deafness (DFN-1), dystonia, mental deficiency and blindness. Nat Genet 14:177–180

Jitrapakdee S, Wallace JC (1999) Structure, function and regulation of pyruvate carboxylase. Biochem J 340(Pt 1):1–16

Jitrapakdee S, Vidal-Puig A, Wallace JC (2006) Anaplerotic roles of pyruvate carboxylase in mammalian tissues. Cell Mol Life Sci 63:843–854

Jo SH, Son MK, Koh HJ, Lee SM, Song IH, Kim YO, Lee YS, Jeong KS, Kim WB, Park JW, Song BJ, Huh TL (2001) Control of mitochondrial redox balance and cellular defense against oxidative damage by mitochondrial NADP+–dependent isocitrate dehydrogenase. J Biol Chem 276: 16168–16176

Juurlink BH (1993) NADP-linked isozymes are the major forms of isocitrate dehydrogenase in mouse type-1-like astrocytes. Life Sci 52:1087–1090

Kaplan RS, Mayor JA, Wood DO (1993) The mitochondrial tricarboxylate transport protein. cDNA cloning, primary structure, and comparison with other mitochondrial transport proteins. J Biol Chem 268:13682–13690

Kasischke KA, Vishwasrao HD, Fisher PJ, Zipfel WR, Webb WW (2004) Neural activity triggers neuronal oxidative metabolism followed by astrocytic glycolysis. Science 305:99–103

Kaul R, Gao GP, Aloya M, Balamurugan K, Petrosky A, Michals K, Matalon R (1994) Canavan disease: mutations among Jewish and non-jewish patients. Am J Hum Genet 55:34–41

Kaul R, Gao GP, Matalon R, Aloya M, Su Q, Jin M, Jonson AB, Schutgens RB, Clarke JT (1996) Identification and expression of eight novel mutations among non-Jewish patients with Canavan disease. Am J Hum Genet 59:95–102

Kawasaki H, Nakayama S, Kretsinger RH (1998) Classification and evolution of EF-hand proteins. Biometals 11:277–295

Kirmani BF, Jacobowitz DM, Namboodiri MA (2003) Developmental increase of aspartoacylase in oligodendrocytes parallels CNS myelination. Brain Res Dev Brain Res 140:105–115

Kleinhans NM, Schweinsburg BC, Cohen DN, Müller R-A, Courchesne E (2007) N-acetyl aspartate in autism spectrum disorders: Regional effects and relationship to fMRI activation. Brain Res 1162:85–97

Knizley H Jr (1967) The enzymatic synthesis of N-acetyl-L-aspartic acid by a water-insoluble preparation of a cat brain acetone powder. J Biol Chem 242:4619–4622

Kobayashi K, Sinasac DS, Iijima M, Boright AP, Begum L, Lee JR, Yasuda T, Ikeda S, Hirano R, Terazono H, Crackower MA, Kondo I, Tsui LC, Scherer SW, Saheki T (1999) The gene mutated in adult-onset type II citrullinaemia encodes a putative mitochondrial carrier protein. Nat Genet 22:159–163

Koehler CM, Leuenberger D, Merchant S, Renold A, Junne T, Schatz G (1999) Human deafness dystonia syndrome is a mitochondrial disease. Proc Natl Acad Sci USA 96:2141–2146

Koza RA, Kozak UC, Brown LJ, Leiter EH, MacDonald MJ, Kozak LP (1996) Sequence and tissue-dependent RNA expression of Mouse FAD-linked glycerol-3-phosphate dehydrogenase. Arch Biochem Biophys 336(1):97–104

Kull HR (1978) Association of glycolytic enzymes with particulate fractions from nerve endings. Biochem Biophys Acta 522:1–19

Kunji ER (2004) The role and structure of mitochondrial carriers. FEBS Lett 564:239–244

Kunji ER, Robinson AJ (2006) The conserved substrate binding site of mitochondrial carriers. Biochim Biophys Acta 1757:1237–1248

Kurz GM, Wiesinger H, Hamprecht B (1993) Purification of cytosolic malic enzyme from bovine brain, generation of monoclonal antibodies, and immunocytochemical localization of the enzyme in glial cells of neural primary cultures. J Neurochem 60:1467–1474

LaNoue KF, Bryla J, Bassett DJ (1974a) Energy-driven aspartate efflux from heart and liver mitochondria. J Biol Chem 249:7514–7521

LaNoue KF, Meijer AJ, Brouwer A (1974b) Evidence for electrogenic aspartate transport in rat liver mitochondria. Arch Biochem Biophys 161:544–550

LaNoue KF, Tischler ME (1974) Electrogenic characteristics of the mitochondrial glutamate-aspartate antiporter. J Biol Chem 249:7522–7528

LaNoue KF, Schoolwerth AC (1979) Metabolite transport in mitochondria. Annu Rev Biochem 48:871–922

LaNoue KF, Carson V, Berkich DA, Hutson S (2006) Mitochondrial/cytosolic interactions via metabolite shuttles and transporters. Pages 589–617, in Gibson G and Dienel G, (eds.). Handbook of Neurochemistry and Molecular Neurobiology: Brain energetics. Integration of Molecular and cellular processes, 3rd edn. Springer, New York

Laughton JD, Charnay Y, Belloir B, Pellerin L, Magistretti PJ, Bouras C (2000) Differential messenger RNA distribution of lactate dehydrogenase LDH-1 and LDH-5 isoforms in the rat brain. Neuroscience 96:619–625

Lee SM, Koh HJ, Park DC, Song BJ, Huh TL, Park JW (2002) Cytosolic NADP(+)-dependent isocitrate dehydrogenase status modulates oxidative damage to cells. Free Radic Biol Med 32:1185–1196

Lehninger AL (1951) Phosphorylation coupled to oxidation of dihydrodiphosphopyridine nucleotide. J Biol Chem 190:345–359

Lepagnol-Bestel AM, Maussion G, Boda B, Cardona A, Iwayama Y, Delezoide AL, Moalic JM, Muller D, Dean B, Yoshikawa T, Gorwood P, Buxbaum JD, Ramoz N, Simonneau M (2008) SLC25A12 expression is associated with neurite outgrowth and is upregulated in the prefrontal cortex of autistic subjects. Mol Psychiatry 13:385–397

Leveille PJ, McGinnis JF, Maxwell DS, de Vellis J (1980) Immunocytochemical localization of glycerol-3-phosphate dehydrogenase in rat oligodendrocytes. Brain Res 196:287–305

Loeber G, Dworkin MB, Infante A, Ahorn H (1994) Characterization of cytosolic malic enzyme in human tumor cells. FEBS Lett 344:181–186

Lovatt D, Sonnewald U, Waagepetersen HS, Schousboe A, He W, Lin JH, Han X, Takano T, Wang S, Sim FJ, Goldman SA, Nedergaard M (2007) The transcriptome and metabolic gene signature of protoplasmic astrocytes in the adult murine cortex. J Neurosci 27(45):12255–12266

Loverde AW, Lehrer GM (1973) Subcellular distribution of isocitrate dehydrogenases in neonatal and adult mouse brain. J Neurochem 20:441–448

Lu ZH, Chakraborty G, Ledeen RW, Yahya D, Wu G (2004) N-Acetylaspartate synthase is bimodally expressed in microsomes and mitochondria of brain. Brain Res Mol Brain Res 122: 71–78

Llorente I, del Arco A, Satrústegui J, Pardo B (2007) Involvement of aralar and early ROS production in lactate-mediated neuroprotection against excitotoxicity 3rd European Society of Neuroscience Conference on Advances in molecular mechanisms of neurological disorders. Salamanca 19–22 May. J Neurochem 101(suppl):31

MacDonald MJ (1995) Feasibility of a mitochondrial pyruvate malate shuttle in pancreatic islets. Further implication of cytosolic NADPH in insulin secretion. J Biol Chem 270:20051–20058

MacDonald MJ, Brown LJ (1996) Calcium activation of mitochondrial glycerol phosphate dehydrogenase restudied. Arch Biochem Biophys 326(1):79–84

MacDonald MJ, Longacre MJ, Kendrick MA (2009) Mitochondrial malic enzyme (ME2) in pancreatic islets of the human, rat and mouse and clonal insulinoma cells. Simple enzyme assay for mitochondrial malic enzyme 2. Arch Biochem Biophys 488:100–104

Madhavarao CN, Chinopoulos C, Chandrasekaran K, Namboidiri MA (2003) Characterization of the N-acetylaspartate biosynthetic enzyme from rat brain. J Neurochem 86:824–835

Madhavarao CN, Arun P, Moffett JR, Szucs S, Surendran S, Matalon R, Garbern J, Hristova D, Johnson A, Jiang W, Namboodiri MA (2005) Defective N-acetylaspartate catabolism reduces brain acetate levels and myelin lipid synthesis in Canavan's disease. Proc Natl Acad Sci USA 102:5221–5226

Maeng O, Kim YC, Shin HJ, Lee JO, Huh TL, Kang KI, Kim YS, Paik SG, Lee H (2004) Cytosolic NADP(+)-dependent isocitrate dehydrogenase protects macrophages from LPS-induced nitric oxide and reactive oxygen species. Biochem Biophys Res Commun 317:558–564

Malik P, McKenna MC, Tildon JT (1993) Regulation of malate dehydrogenases from neonatal, adolescent, and mature rat brain. Neurochem Res 18:247–257

Mangia S, Garreffa G, Bianciardi M, Giove F, di Salle F, Maraviglia B (2003) The aerobic brain: lactate decrease at the onset of neural activity. Neuroscience 118:7–10

Mangia S, Tkác I, Gruetter R, Van del Moortele PF, Maraviglia B, Ugurbil K (2007) Sustained neuronal activation raises oxidative metabolism to a new steady-state level: evidence from 1 H NMR spectroscopy in the human visual cortex. J Cereb Blood Flow Metab 27:1055–1063

Maran A, Cranston I, Lomas J, Macdonald I, Amiel SA (1994) Protection by lactate of cerebral function during hypoglycaemia. Lancet 343:16–20

Mármol P, Pardo B, Wiederkehr A, del Arco A, Wollheim CB, Satrústegui J (2009) Requirement for aralar and its Ca^{2+}-binding sites in Ca^{2+} signal transduction in mitochondria from INS-1 clonal beta-cells. J Biol Chem 284:515–524

Matalon R, Michals K, Sebesta D, Deanching M, Gashkoff P, Casanova J (1988) Aspartoacylase deficiency and N-acetylaspartic aciduria in patients with Canavan disease. Am J Med Genet 29: 463–471

Matalon R, Rady PL, Platt KA, Skinner HB, Quast MJ, Campbell GA, Matalon K, Ceci JD, Tyring SK, Nehls M, Surendran S, Wei J, Ezell EL, Szucs S (2000) Knock-out mouse for Canavan disease: a model for gene transfer to the central nervous system. J Gene Med 2:165–175

McCormack JG, Bromidge ES, Dawes NJ (1988) Characterization of the effects of Ca2+ on the intramitochondrial Ca^{2+}-sensitive dehydrogenases within intact rat-kidney mitochondria. Biochim Biophys Acta 934:282–292

McCormack JG, Halestrap AP, Denton RM (1990) Role of calcium ions in regulation of mammalian intramitochondrial metabolism. Physiol Rev 70:391–425

McKenna MC, Tildon JT, Stevenson JH, Boatrigh TR, Huang S (1993) Regulation of energy metabolism in synaptic terminals and cultured rat brain astrocytes: differences revealed using aminooxyacetate. Dev Neurosci 15:320–329

McKenna MC, Tildon JT, Stevenson JH, Hopkins IB (1994) Energy metabolism in cortical synaptic terminals from weanling and mature rat brain: evidence for multiple compartments of tricarboxylic acid cycle activity. Dev Neurosci 16:291–300

McKenna MC, Tildon JT, Stevenson JH, Huang X, Kingwell KG (1995) Regulation of mitochondrial and cytosolic malic enzymes from cultured rat brain astrocytes. Neurochem Res 20:1491–1501

McKenna MC, Stevenson JH, Huang X, Tildon JT, Zielke CL, Hopkins IB (2000) Mitochondrial malic enzyme activity is much higher in mitochondria from cortical synaptic terminals compared with mitochondria from primary cultures of cortical neurons or cerebellar granule cells. Neurochem Int 36:451–459

McKenna MC, Waagepetersen HS, Schousboe A, Sonnewald U (2006) Neuronal and astrocytic shuttle mechanisms for cytosolic-mitochondrial transfer of reducing equivalents: current evidence and pharmacological tools. Biochem Pharmacol 71:399–407

Medina JM (1985) The role of lactate as an energy substrate for the brain during the early neonatal period. Biol Neonate 48:237–244

Mehta V, Namboodiri MA (1995) N-acetylaspartate as an acetyl source in the nervous system. Brain Res Mol Brain Res 31:151–157

Miller KE, Richards BA, Kriebel RM (2002) Glutamine-, glutamine synthetase-, glutamate dehydrogenase- and pyruvate carboxylase-immunoreactivities in the rat dorsal root ganglion and peripheral nerve. Brain Res 945:202–211

Minich T, Yokota S, Dringen R (2003) Cytosolic and mitochondrial isoforms of NADP+–dependent isocitrate dehydrogenases are expressed in cultured rat neurons, astrocytes, oligodendrocytes and microglial cells. J Neurochem 86:605–614

Miyake S, Yamashita T, Taniguchi M, Tamatani M, Sato K, Hawai Y, Senba E, Mitsuda N, Hori O, Yamaguchi A, Tohyama M (2002a) Expression of mitochondrial tricarboxylate carrier TCC mRNA and protein in the rat brain. Brain Res Mol Brain Res 100:67–73

Miyake S, Yamashita T, Taniguchi M, Tamatani M, Sato K, Tohyama M (2002b) Identification and characterization of a novel mitochondrial tricarboxylate carrier. Biochem Biophys Res Commun 295:463–468

Mizuarai S, Miki S, Araki H, Takahashi K, Kotani H (2005) Identification of dicarboxylate carrier Slc25a10 as malate transporter in de novo fatty acid synthesis. J Biol Chem 280:32434–32441

Montz HP, Althaus HH, Gebicke-Haerter PJ, Neuhoff V (1985) Glycerol phosphate dehydrogenase activity of oligodendrocytes isolated from adult pig brain: its inducibility by hydrocortisone. J Neurochem 45:1201–1204

Moreau B, Nelson C, Parekh AB (2006) Biphasic regulation of mitochondrial Ca^{2+} uptake by cytosolic Ca^{2+} concentration. Curr Biol 16(16):1672–1677

Nakayama S, Moncrief ND, Kretsinger RH (1992) Evolution of EF-hand calcium-modulated proteins. II. Domains of several subfamilies have diverse evolutionary histories. J Mol Evol 34:416–448

Nguyen NH, Brathe A, Hassel B (2003) Neuronal uptake and metabolism of glycerol and the neuronal expression of mitochondrial glycerol-3-phosphate dehydrogenase. J Neurochem 85:831–842

Nicholls DG, Budd SL (2000) Mitochondria and neuronal survival. Physiol Rev 80(1):315–360

Nissen C, Schousboe A (1979) Activity and isoenzyme pattern of lactate dehydrogenase in astroblasts cultured from brains of newborn mice. J Neurochem 32:1787–1792

Novials A, Vidal J, Franco C, Ribera F, Sener A, Malaisse WJ, Gomis R (1997) Mutation in the calcium-binding domain of the mitochondrial glycerophosphate dehydrogenase gene in a family of diabetic subjects. Biochem Biophys Res Commun 231:570–572

Oliveira G, Diogo L, Grazina M, Garcia P, Borges L, Vicente AM, Oliveira CR (2005) Mitochondrial dysfunction in autism spectrum disorders. Dev Med Child Neurol 47(3):185–189

Otsuka H, Harada M, Mori K, Hisaoka S, Nishitani H (1999) Brain metabolites in the hippocampus-amygdala region and cerebellum in autism: an 1 H-MR spectroscopy study. Neuroradiology 41:517–519

Palaiologos G, Hertz L, Schousboe A (1988) Evidence that aspartate aminotransferase activity and ketodicarboxylate carrier function are essential for biosynthesis of transmitter glutamate. J Neurochem 51:317–320

Palmieri L, Vozza A, Agrimi G, De Marco V, Runswick MJ, Palmieri F, Walker JE (1999) Identification of the yeast mitochondrial transporter for oxaloacetate and sulfate. J Biol Chem 274:22184–22190

Palmieri L, Pardo B, Lasorsa FM, del Arco A, Kobayashi K, Iijima M, Runswick MJ, Walker JE, Saheki T, Satrústegui J, Palmieri F (2001) Citrin and aralar1 are Ca^{2+}-stimulated aspartate/glutamate transporters in mitochondria. EMBO J 20:5060–5069

Palmieri F (2004) The mitochondrial transporter family (SLC25): physiological and pathological implications. Pflugers Arch 447:689–709

Pardo B, Contreras L, Serrano A, Ramos M, Kobayashi K, Iijima M, Saheki T, Satrústegui J (2006) Essential role of aralar in the transduction of small Ca^{2+} signals to neuronal mitochondria. J Biol Chem 281:1039–1047

Pardo B, Rodrigues TB, Contreras L, Garzón M, Llorente-Folch I, Kobayashi K, Saheki T, Cerdán S, Satrústegui J (2011) Brain glutamine synthesis requires neuronal-born aspartate as amino donor for glial glutamate formation. J Cereb Blood Flow Metab 31(1):90–101

Paschen SA, Rothbauer U, Kaldi K, Bauer MF, Neupert W, Brunner M (2000) The role of the TIM8-13 complex in the import of Tim23 into mitochondria. EMBO J 19:6392–6400

Passarella S, Barile M, Atlante A, Quagliariello E (1984) Oxaloacetate uptake into rat brain mitochondria and reconstruction of the malate/oxaloacetate shuttle. Biochem Biophys Res Commun 119:1039–1046

Passarella S, Atlante A, Quagliariello E (1985) Oxaloacetate permeation in rat kidney mitochondria: pyruvate/oxaloacetate and malate/oxaloacetate translocators. Biochem Biophys Res Commun 129:1–10

Passarella S, Atlante A, Valenti D, de Bari L (2003) The role of mitochondrial transport in energy metabolism. Mitochondrion 2:319–343

Patel TB, Clark JB (1979) Synthesis of N-acetyl-L-aspartate by rat brain mitochondria and its involvement in mitochondrial/cytosolic carbon transport. Biochem J 184:539–546

Pellerin L, Magistretti PJ (1994) Glutamate uptake into astrocytes stimulates aerobic glycolysis: a mechanism coupling neuronal activity to glucose utilization. Proc Natl Acad Sci USA 91:10625–10629

Pellerin L, Pellegri G, Bittar PG, Charnay Y, Bouras C, Martin JL, Stella N, Magistretti PJ (1998) Evidence supporting the existence of an activity-dependent astrocyte-neuron lactate shuttle. Dev Neurosci 20:291–299

Pellerin L (2003) Lactate as a pivotal element in neuron-glia metabolic cooperation. Neurochem Int 43:331–338

Pellerin L, Magistretti PJ (2004) Neuroscience. Let there be (NADH) light. Science 305:50–52

Pierre K, Pellerin L, Debernardi R, Riederer BM, Magistretti PJ (2000) Cell-specific localization of monocarboxylate transporters, MCT1 and MCT2, in the adult mouse brain revealed by double immunohistochemical labeling and confocal microscopy. Neuroscience 100: 617–627

Pierre K, Magistretti PJ, Pellerin L (2002) MCT2 is a major neuronal monocarboxylate transporter in the adult mouse brain. J Cereb Blood Flow Metab 22:586–595

Plaut GW, Cook M, Aogaichi T (1983) The subcellular location of isozymes of NADP-isocitrate dehydrogenase in tissues from pig, ox and rat. Biochim Biophys Acta 760:300–308

Price MT, Pusateri ME, Crow SE, Buchsbaum S, Olney JW, Lowry OH (1984) Uptake of exogenous aspartate into circumventricular organs but not other regions of adult mouse brain. J Neurochem 42:740–744

Prichard J, Rothman D, Novotny E, Petroff O, Kuwabara T, Avison M, Howseman A, Hanstock C, Shulman R (1991) Lactate rise detected by 1 H NMR in human visual cortex during physiologic stimulation. Proc Natl Acad Sci USA 88:5829–5831

Purvis JL, Lowenstein JM (1961) The relation between intra- and extramitochondrial pyridine nucleotides. J Biol Chem 236:2794–2803

Rabionet R, McCauley JL, Jaworski JM, Ashley-Koch AE, Martin ER, Sutcliffe JS, Haines JL, DeLong GR, Abramson RK, Wright HH, Cuccaro ML, Gilbert JR, Pericak-Vance MA (2006) Lack of association between autism and SLC25A12. Am J Psychiatry 163:929–931

Rafalowska U (1979) Transport of malate and citrate into rat brain mitochondria under hypoxia and anesthesia. Neurochem Res 4:355–364

Ramos M, del Arco A, Pardo B, Martinez-Serrano A, Martinez-Morales JR, Kobayashi K, Yasuda T, Bogonez E, Bovolenta P, Saheki T, Satrústegui J (2003) Developmental changes in the

Ca2+–regulated mitochondrial aspartate-glutamate carrier aralar1 in brain and prominent expression in the spinal cord. Brain Res Dev Brain Res 143:33–46
Ramoz N, Reichert JG, Smith CJ, Silverman JM, Bespalova IN, Davis KL, Buxbaum JD (2004) Linkage and association of the mitochondrial aspartate/glutamate carrier SLC25A12 gene with autism. Am J Psychiatry 161:662–669
Raschaert J, Malaisse WJ (1991) Hexose metabolism in pancreatic islets. Glucose-induced and Ca^{2+} -dependent activation of FAD-glycerophosphate dehydrogenase. Biochem J 278: 335–340
Roesch K, Curran SP, Tranebjaerg L, Koehler CM (2002) Human deafness dystonia syndrome is caused by a defect in assembly of the DDP1/TIMM8a-TIMM13 complex. Hum Mol Genet 11:477–486
Roesch K, Hynds PJ, Varga R, Tranebjaerg L, Koehler CM (2004) The calcium-binding aspartate/glutamate carriers, citrin and aralar1, are new substrates for the DDP1/TIMM8a-TIMM13 complex. Hum Mol Genet 13:2101–2111
Rohde M, Lim F, Wallace JC (1991) Electron microscopic localization of pyruvate carboxylase in rat liver and Saccharomyces cerevisiae by immunogold procedures. Arch Biochem Biophys 290:197–201
Ros J, Pecinska N, Alessandri B, Landolt H, Fillenz M (2001) Lactate reduces glutamate-induced neurotoxicity in rat cortex. J Neurosci Res 66:790–794
Rubi B, del Arco A, Bartley C, Satrústegui J, Maechler P (2004) The malate-aspartate NADH shuttle member Aralar1 determines glucose metabolic fate, mitochondrial activity, and insulin secretion in beta cells. J Biol Chem 279:55659–55666
Runswick M, Walker J, Bisaccia F, Lacobazzi V, Palmieri F (1990) Sequence of the bovine 2-oxo-glutarate/malate carrier protein: structural relationship to other mitochondrial transport proteins. Biochemistry 29:11033–11040
Rutter GA, Pralong WF, Wollheim CB (1992) Regulation of mitochondrial glycerol-phosphate dehydrogenase by Ca^{2+} within electropermeabilized insulin-secreting cells (INS-1). Biochim Biophys Acta 1175:107–113
Sanz R, del Arco A, Ayuso C, Ramos C, Satrústegui J (2000) Assignment of the calcium-binding mitochondrial carrier Aralar1 gene (SLC25A12) to human chromosome band 2q31 by in situ hybridization. Cytogenet Cell Genet 89:143–144
Saraste M, Walker JE (1982) Internal sequence repeats and the path of polypeptide in mitochondrial ADP/ATP translocase. FEBS Lett 144:250–254
Satrústegui J, Contreras L, Ramos M, Mármol P, Del Arco A, Saheki T, Pardo B (2007a) Role of aralar, the mitochondrial transporter of aspartate-glutamate, in brain N-acetylaspartate formation and Ca^{2+} signaling in neuronal mitochondria. J Neurosci Res 85:3359–3366
Satrústegui J, Pardo B, Del Arco A (2007b) Mitochondrial transporters as novel targets for intracellular calcium signaling. Physiol Rev 87:29–67
Sazanov LA, Jackson JB (1994) Proton-translocating transhydrogenase and NAD- and NADP-linked isocitrate dehydrogenases operate in a substrate cycle which contributes to fine regulation of the tricarboxylic acid cycle activity in mitochondria. FEBS Lett 344:109–116
Schurr A, West CA, Rigor BM (1988) Lactate-supported synaptic function in the rat hippocampal slice preparation. Science 240:1326–1328
Schurr A, West CA, Rigor BM (1989) Electrophysiology of energy metabolism and neuronal function in the hippocampal slice preparation. J Neurosci Methods 28:7–13
Schurr A, Payne RS, Miller JJ, Rigor BM (1997a) Brain lactate, not glucose, fuels the recovery of synaptic function from hypoxia upon reoxygenation: an in vitro study. Brain Res 744: 105–111
Schurr A, Payne RS, Miller JJ, Rigor BM (1997b) Glia are the main source of lactate utilized by neurons for recovery of function posthypoxia. Brain Res 774:221–224
Schurr A, Miller JJ, Payne RS, Rigor BM (1999) An increase in lactate output by brain tissue serves to meet the energy needs of glutamate-activated neurons. J Neurosci 19:34–39
Schurr A (2006) Lactate: the ultimate cerebral oxidative energy substrate? J Cereb Blood Flow Metab 26:142–152

Schurr A, Payne RS (2007) Lactate, not pyruvate, is neuronal aerobic glycolysis end product: an in vitro electrophysiological study. Neuroscience 147:613–619

Segurado R, Conroy J, Meally E, Fitzgerald M, Gill M, Gallagher L (2005) Confirmation of association between autism and the mitochondrial aspartate/glutamate carrier SLC25A12 gene on chromosome 2q31. Am J Psychiatry 162:2182–2184

Shank RP, Bennett GS, Freytag SO, Campbell GL (1985) Pyruvate carboxylase: an astrocyte-specific enzyme implicated in the replenishment of amino acid neurotransmitter pools. Brain Res. 329 (1–2): 364–367

Shank RP, Leo GC, Zielke HR (1993) Cerebral metabolic compartmentation as revealed by nuclear magnetic resonance analysis of D-[1-13 C]glucose metabolism. J Neurochem 61:315–323

Shen J (2005) In vivo carbon-13 magnetization transfer effect. Detection of aspartate aminotransferase reaction. Magn Reson Med 54:1321–1326

Silverman JM, Buxbaum JD, Ramoz N, Schmeidler J, Reichenberg A, Hollander E, Angelo G, Smith CJ, Kryzak LA (2008) Autism-related routines and rituals associated with a mitochondrial aspartate/glutamate carrier SLC25A12 polymorphism. Am J Med Genet B Neuropsychiatr Genet 147:408–410

Sinasac DS, Crackower MA, Lee JR, Kobayashi K, Saheki T, Scherer SW, Tsui LC (1999) Genomic structure of the adult-onset type II citrullinemia gene, SLC25A13, and cloning and expression of its mouse homologue. Genomics 62:289–292

Szutowicz A, Srere PA (1983) Purification and some properties of ATP-citrate lyase from rat brain. Arch Biochem Biophys 221(1):168–174

Tholey G, Roth-Schechter BF, Mandel P (1981) Activity and isoenzyme pattern of lactate dehydrogenase in neurons and astroblasts cultured from brains of chick embryos. J Neurochem 36:77–81

Tildon JT, McKenna MC, Stevenson J, Couto R (1993) Transport of L-lactate by cultured rat brain astrocytes. Neurochem Res 18(2):177–184

Tranebjaerg L, Schwartz C, Eriksen H, Andreasson S, Ponjavic V, Dahl A, Stevenson RE, May M, Arena F, Barker D et al (1995) A new X linked recessive syndrome with blindness, dystonia, fractures, and mental deficiency is linked to Xq22. J Med Genet 32:257–263

Tretter L, Takacs K, Hegedus V, Adam-Vizi V (2007) Characteristcs of α–glycerophosphate-evoked H_2O_2 generation in brain mitochondria. J Neurochem 100:650–663

Turunen J, Ylisaukko-oja T, Kilpinen H, Rehnstrom K, Kempas E, Vanhala R, Niemenin-von Wendt T, von Wendt L, Peltonen L (2007) Association analysis of SLC25A12 and EN2 in the finnish families with autism-spectrum disorders. Am J Med Genet B Neuropsychiatr Genet 141B(7):766, abstract

Urenjak J, Williams SR, Gadian DG, Noble M (1992) Specific expression of N-acetylaspartate in neurons, oligodendrocyte-type-2 astrocyte progenitors, and immature oligodendrocytes in vitro. J Neurochem 59:55–61

Urenjak J, Williams SR, Gadian DG, Noble M (1993) Proton nuclear magnetic resonance spectroscopy unambiguously identifies different neural cell types. J Neurosci 31(3):981–989

Vicario C, Arizmendi C, Malloch G, Clark JB, Medina JM (1991) Lactate utilization by isolated cells from early neonatal rat brain. J Neurochem 57:1700–1707

Vogel R, Hamprecht B, Wiesinger H (1998) Malic enzyme isoforms in astrocytes: comparative study on activities in rat brain tissue and astroglia-rich primary cultures. Neurosci Lett 247:123–126

Vogel R, Wiesinger H, Hamprecht B, Dringen R (1999) The regeneration of reduced glutathione in rat forebrain mitochondria identifies metabolic pathways providing the NADPH required. Neurosci Lett 275:97–100

Waagepetersen HS, Qu H, Schousboe A, Sonnewald U (2001) Elucidation of the quantitative significance of pyruvate carboxylation in cultured cerebellar neurons and astrocytes. J Neurosci Res 66:763–770

Walker JE, Runswick MJ (1993) The mitochondrial transport protein superfamily. J Bioenerg Biomembr 25:435–446

Wallace JC, Jitrapakdee S, Chapman-Smith A (1998) Pyruvate carboxylase. Int J Biochem Cell Bio l 30:1–5

Walz W, Mukerji S (1988) Lactate release from cultured astrocytes and neurons: a comparison. Glia 1:366–370

Watkins J, Basu S, Bogenhagen DF (2008) A quantitative proteomic analysis of mitochondrial participation in p19 cell neuronal differentiation. J Proteome Res 7(1):328–338

Watson JA, Lowenstein JM (1970) Citrate and the conversion of carbohydrate into fat. Fatty acid synthesis by a combination of cytoplasm and mitochondria. J Bio Chem 245:5993–6002

Wernette ME, Ochs RS, Lardy HA (1981) Ca^{2+} stimulation of rat liver mitochondrial glycerolphosphate dehydrogenase. J Biol Chem 256(24):12767–12771

Wibom R, Lasorsa FM, Töhönen V, Barbaro M, Sterky FH, Kucinski T, Naess K, Jonsonn M, Pierri CL, Palmieri F, Wedell A (2009) AGC1 deficiency associated with global cerebral hypomyelination. N Engl J Med 361:489–495

Williamson DH, Lund P, Krebs HA (1967) The redox state of free nicotinamide-adenine dinucleotide in the cytoplasm and mitochondria of rat liver. Biochem J 103:514–527

Williamson JR, Cooper RH (1980) Regulation of the citric acid cycle in mammalian systems. FEBS Lett 117(Suppl):K73–K85

Winkler BS, DeSantis N, Solomon F (1986) Multiple NADPH-producing pathways control glutathione (GSH) content in retina. Exp Eye Res 43:829–847

Xu Y, Ola MS, Berkich DA, Gardner TW, Barber AJ, Palmieri F, Hutson SM, LaNoue KF (2007) Energy sources for glutamate neurotransmission in the retina: absence of the aspartate/glutamate carrier produces reliance on glycolysis in glia. J Neurochem 101(1):120–131

Yang L, Luo H, Vinay P, Wu J (1996) Molecular cloning of the cDNA of mouse mitochondrial NADP-dependent isocitrate dehydrogenase and the expression of the gene during lymphocyte activation. J Cell Biochem 60:400–410

Yoshida Y, Holloway GP, Ljubicic V, Hatta H, Spriet LL, Hood DA, Bonen A (2007) Negligible direct lactate oxidation in subsarcolemmal and intermyofibrillar mitochondria obtained from red and white rat skeletal muscle. J Physiol 582:1317–1335

Yu AC, Drejer J, Hertz L, Schousboe A (1983) Pyruvate carboxylase activity in primary cultures of astrocytes and neurons. J Neurochem 41:1484–1487

Yudkoff M, Nelson D, Daikhin Y, Erecinska M (1994) Tricarboxylic acid cycle in rat brain synaptosomes. Fluxes and interactions with aspartate aminotransferase and malate/aspartate shuttle. J Biol Chem 269:27414–27420

Zambrowicz BP, Friedrich GA, Buxton EC, Lilleberg SL, Person Ch, Sands AT (1998) Disruption and sequence identification of 2,000 genes in mouse embryonic stem cells. Nature 392:608–611

Zhou J, Olcese R, Qin N, Noceti F, Birnbaumer L, Stefani E (1997) Feedback inhibition of Ca^{2+} channels by Ca^{2+} depends on a short sequence of the C terminus that does not include the Ca^{2+} -binding function of a motif with similarity to Ca^{2+} -binding domains. Proc Natl Acad Sci USA 94:2301–2305

Zeng BJ, Wang ZH, Ribeiro LA, Leone P, De Gasperi R, Kim SJ, Raghavan S, Ong E, Pastores GM, Kolodny EH (2002) Identification and characterization of novel mutations of the aspartoacylase gene in non-Jewish patients with Canavan disease. J Inherit Metab Dis 25:557–570

Zoghbi HY (2003) Postnatal neurodevelopmental disorders: meeting at the synapse? Science 302(5646):826–828

Chapter 30
Mitochondria in Neurodegeneration

Russell H. Swerdlow

Abstract In neurodegenerative diseases, particular parts of the brain, spinal cord, or peripheral nerves functionally fail and the neurons of the dysfunctional region die. Neuroanatomically localizable functional impairment and neurodegeneration associate with recognizable clinical syndromes. Mitochondrial dysfunction is found in numerous neurodegenerative diseases and justifies a "neurodegenerative mitochondriopathy" classification system. The role mitochondrial dysfunction plays in the etiology, pathophysiology, and histopathology of the neurodegenerative mitochondriopathies is not entirely clear at this time, but in different diseases mitochondria may represent a final common pathway, an intermediate step, or even primary driver of neurodegeneration. Mitochondria sit at the nexus of various hypotheses that attempt to explain particular neurodegenerative diseases, and are a therapeutic target.

Keywords Alzheimer's disease • Amyotrophic lateral sclerosis • Mitochondria • Mitochondrial DNA • Neurodegeneration • Neurodegenerative diseases • Parkinson's disease

30.1 Introduction

Many neurodegeneration researchers currently believe mitochondria are relevant to neurodegenerative disease development and progression. This acknowledgment has been long in coming. Investigators proposed impairment of oxidative metabolism may underlie dementia over 70 years ago without actually implicating mitochondria, the main source of cell oxidative metabolism (Quastel 1932). Approximately 40 years ago

R.H. Swerdlow, M.D. (✉)
Departments of Neurology, Molecular & Integrative Physiology, University of Kansas School of Medicine, 3901 Rainbow Boulevard, Kansas City, KS 66160, USA
e-mail: rswerdlow@kumc.edu

I.-Y. Choi and R. Gruetter (eds.), *Neural Metabolism In Vivo*,
Advances in Neurobiology 4, DOI 10.1007/978-1-4614-1788-0_30,

investigators observed oxidative metabolism was reduced in Alzheimer's disease (AD) brain and mitochondria looked abnormal in AD neurons (Friede 1965; Wisniewski et al. 1970). In the 1980s, data emerged directly implicating toxic inhibition of the mitochondrial enzyme NADH:oxidoreductase (complex I) in a clinical syndrome that resembled Parkinson's disease (PD) (Langston et al. 1983; Nicklas et al. 1985). The 1980s closed with the discovery that complex I activity is systemically reduced in PD patients (Bindoff et al. 1989; Parker et al. 1989; Schapira et al. 1989). This was followed in the 1990s by reports of mitochondrial dysfunction in several other neurodegenerative diseases. Reports focused both on abnormalities of specific mitochondrial enzymes, as well as more general abnormalities of mitochondria-associated phenomena such as oxidative stress. By the second half of the 1990s it was abundantly clear mitochondria play an important mechanistic role in cell death.

The significance of these findings has been hotly debated. The possibility an organelle so important to both the life and death of cells could fundamentally, indeed primarily, drive neurodegeneration seems obvious to some. Other neurodegenerative disease etiologic hypotheses, though, propose alternative causes and relegate mitochondria to a more secondary or even epiphenomenal position. While reviewing what is currently known about the mitochondria-neurodegeneration nexus, this chapter will attempt to capture the essence of this debate.

30.2 What Is a Neurodegenerative Disease, What Is a Mitochondrial Disease?

30.2.1 Neurodegenerative Diseases

Neurodegenerative diseases are characterized clinically by their insidious onset and chronic progression. Particular parts of the brain, spinal cord, or peripheral nerves functionally fail and the neurons of the dysfunctional region die. Neuroanatomically localizable functional impairment and neurodegeneration associate with recognizable syndromes that are ideally distinct, although in clinical and even neuropathologic practice substantial syndromic overlap exists. Neurodegenerative disease clinical syndromes are often categorized by whether they initially affect cognition, movement, strength, coordination, sensation, or autonomic control. Frequently, however, patients will present with symptoms and signs referable to more than one system. Either involvement of several systems can occur concomitantly, or else by the time the patient has functionally declined enough to seek medical attention multiple systems have become involved. Diagnosing neurodegenerative diseases can prove particularly intimidating to clinicians, because many times the diagnosis cannot be critically "confirmed" by a simple test, and because in most cases we are currently unable to effectively stop disease progression. There is often an air of finality when a neurodegenerative disease diagnosis is made, and many clinicians are understandably uncomfortable making such pronouncements based on limited interaction with a patient.

Neurodegenerative diseases are complicated on other levels. While the term "neurodegenerative" implies it is the loss of neurons that cause disease, it is possible neuronal demise is merely the final stage of a preceding period of neuron dysfunction. It is difficult to know whether clinical decline associates with actual neuron loss, or with a period of neuron dysfunction that precedes neuron loss. Also, particular neurodegenerative diseases are etiologically heterogeneous. The manifestations of human nervous system failure, not surprisingly, are limited. This brings up some very real questions of lumping versus splitting. Before the molecular revolution of the late twentieth century, lumping of clinical syndromes predominated. This began to change to some extent when it was learned mutations in different genes can cause essentially overlapping clinical syndromes. For example, in the 1980s a group of autosomal dominant cerebellar ataxias were syndromically lumped into four distinct categories (Harding 1983). As a consequence of advances in linkage analysis and positional cloning methodology, when diagnosing autosomal dominant ataxia syndromes clinicians can now often use genetic testing to provide a more specific diagnosis. This trend towards "splitting" should become more common as our etiologic understanding of the different neurodegenerative diseases increases.

In addition to syndromically defining neurodegenerative diseases by what neuroanatomical system is involved, these disorders are broken down along other clinical lines. Early (childhood, young adulthood, or middle aged adulthood) versus late (old age) onset is an important distinction. Some clinically similar neurodegenerative diseases are sub-categorized by their age of onset, despite the fact at the molecular level different forms of a particular disease may have very little in common. Sporadic onset versus Mendelian inheritance constitutes another important distinction, and most named neurodegenerative diseases have both sporadic (Mendelian inheritance is not recognizable) and Mendelian subtypes.

30.2.2 Mitochondrial Diseases

Compared to the mitochondrial disease field, the neurodegenerative diseases constitute a relatively mature construct. The first report of a "mitochondrial" disease did not occur until the 1960s (Luft et al. 1962). By the 1980s, a large number of rare metabolic disorders were recognized in which dysfunction of an enzyme or enzymes located within mitochondria occurred (Scholte 1988). During this period, advancing knowledge of mitochondrial genetics also allowed investigators to identify several rare disorders as likely arising from mutation of mitochondrial DNA (mtDNA). These disorders typically affected the central nervous system, causing encephalopathy, as well as muscle, causing weakness. In the late 1980s and in the 1990s, specific mutations of mtDNA were indeed shown to cause a number of these maternally inherited "encephalomyopathy" diseases (Holt et al. 1988; Lestienne and Ponsot 1988; Zeviani et al. 1988; Goto et al. 1990; Shoffner et al. 1990). In 1988, Leber's hereditary optic neuropathy (LHON), a maternally inherited neurodegenerative disease of the optic nerves, was shown to arise from mutation of mtDNA (Wallace et al. 1988).

Table 30.1 A brief list of well characterized neurodegenerative diseases, many of which qualify as neurodegenerative mitochondriopathies

I. Neurodegenerative disorders that can present as problems with:				
Cognition	*Movement*	*Strength*	*Coordination*	*PNS/ANS*
Alzheimer's disease	Parkinson's disease	Amyotrophic lateral sclerosis	Spinocerebellar atrophies	Amyloidoses
Frontotemporal dementia	Frontotemporal dementia	Frontotemporal dementia	Friedreich's ataxia	Friedreich's ataxia
Dementia with Lewy bodies	Dementia with Lewy bodies	Hereditary spastic paraparesis	Prion disorders	Multisystem atrophy
Corticobasal degeneration	Corticobasal degeneration		Progressive supranuclear palsy	Progressive supranuclear palsy
Prion disorders	Huntington's disease			Wilson's disease
II. Neurodegenerative disorders that currently qualify as neurodegenerative mitochondriopathies:				
ETC dysfunction observed				*Mutant protein locates to mitochondria*
Alzheimer's disease				Alzheimer's disease (autosomal dominant forms)
Parkinson's disease				Parkinson's disease (autosomal dominant and recessive forms)
Progressive supranuclear palsy				Huntington's disease
Huntington's disease				Amyotrophic lateral sclerosis (autosomal dominant form)
Amyotrophic lateral sclerosis				Friedreich's ataxia
Multisystem atrophy				Wilson's disease
				Hereditary spastic paraparesis (several forms)

This list is not inclusive, and diseases can present in ways in which they have not been listed. *ANS/PNS* autonomic nervous system/peripheral nervous system.

As is usually the case when a new disease construct is created, concepts used to define the mitochondrial disease field were soon challenged. In particular, maternal inheritance was initially thought to be a prerequisite for mtDNA disease. Genetic testing of individuals with optic nerve degeneration reminiscent of LHON but without a family history of that disorder soon showed the majority of those with a LHON mutation presented sporadically and not within the context of a recognizable matrilineal inheritance pattern (Johns et al. 1991; Newman 1993). Allelic and non-allelic heterogeneity were demonstrated to occur with suspected mitochondrial DNA diseases. Further, it became apparent neurodegenerative diseases with non-mitochondrial etiologies could also feature mitochondrial dysfunction. For example, in Huntington's disease (HD) mitochondrial dysfunction occurs, may be pathophysiologically relevant, but clearly is not the primary cause of the disorder (Parker et al. 1990a; Mann et al. 1990). For other neurodegenerative diseases mitochondrial dysfunction occurs, may be pathophysiologically relevant, and no clear upstream pathology has yet been elucidated. In light of the neurodegenerative disease-mitochondriopathy nexus, some have proposed creating a category of diseases called the "neurodegenerative mitochondriopathies" (Swerdlow 2009). In this construct, any neurodegenerative disease with demonstrable mitochondrial dysfunction is considered a neurodegenerative mitochondriopathy (Table 30.1).

30.3 Neurodegenerative Mitochondriopathies

30.3.1 Alzheimer's Disease

AD is the most common neurodegenerative disease of cognition, as well as the most common neurodegenerative disease overall. Advancing age is the greatest risk factor, and some studies suggest almost half of those over 85 meet diagnostic criteria for AD (Evans et al. 1989). The incredibly high prevalence in the elderly raises the question of whether AD in this group should be thought of more as a state of poor cognitive aging, as opposed to being considered a distinct disease entity (Swerdlow 2007b).

Clinically, AD is a heterogeneous disorder that may arise via several etiologic causes (Swerdlow 2007b). One of the more common AD schemas breaks it down into two basic forms, early and late onset. The border between the early and late onset forms is arbitrarily defined as occurring between 55 and 65 years of age. The early onset forms are further broken down into sporadic and autosomal dominant cases. Autosomal dominant AD typically presents before the age of 65, although very rare exceptions are reported in the literature. Mutations in genes encoding the amyloid precursor protein (APP), presenilin 1 (PS1), and presenilin 2 (PS2) have all been implicated in autosomal dominant AD kindreds. PS1 and PS2 play a role in APP processing, and the APP byproduct beta amyloid (Aβ) is the majority constituent of amyloid plaques found in the brains of elderly individuals with and without AD. Accordingly, an "amyloid cascade hypothesis" has proposed AD arises as a consequence of abnormal APP processing to particular Aβ variants, which then behave in a toxic fashion (Hardy and Allsop 1991; Hardy and Higgins 1992).

The major assumption of the amyloid cascade hypothesis is that Aβ causes neurodegeneration. However, it is vague as to how Aβ might actually induce neurodegeneration. One possibility is that Aβ interferes with mitochondrial function. This has indeed been shown under *in vitro* conditions. When maintained in the presence of Aβ, isolated mitochondria show diminished respiratory capacity in general, and specifically inhibition of cytochrome oxidase activity (Pereira et al. 1998; Canevari et al. 1999). In cell culture experiments, at adequate concentrations and in particular configurations Aβ is toxic. At concentrations insufficient to kill cells, Aβ appears to induce an increase in mtDNA levels and reduce the number of normal appearing mitochondria (Diana et al. 2008). Cells depleted of endogenous mtDNA (ρ0) cells, which lack functional electron transport chains (ETC), are impervious to Aβ (Cardoso et al. 2001). If data from *in vitro* experiments such as these is to be extrapolated to the *in vivo* condition, a case can be made that if Aβ is indeed driving neurodegeneration in AD, it is doing so through either direct or indirect effects on mitochondria.

Transgenic mice expressing mutant APP genes are commonly used in AD research. Mitochondrial function is altered in some of these mouse models (Reddy and Beal 2008; Eckert et al. 2010; Yao et al. 2009). Increased expression of genes that encode mitochondrial proteins and ETC subunits is an extremely early physiologic event in one AD transgenic mouse line (Reddy et al. 2004). As ETC enzyme activities are reduced in these transgenic mice, it is worth considering whether this represents a compensatory mitochondrial biogenesis that occurs in response to reduced mitochondrial function. Complementing this possibility are data that show brain mitochondrial content changes in aging and in AD. Data indicate mitochondrial number increases with advancing age (Barrientos et al. 1997; Hirai et al. 2001), which conceivably could represent an attempt to compensate for age-related mitochondrial impairment. Data regarding mitochondrial mass in AD brain are somewhat conflicting, since some studies show evidence of increased mitochondrial mass while others show decreased mitochondrial content (de la Monte et al. 2000; Hirai et al. 2001). Part of this discrepancy may be explained by the fact that all studies report a reduction in the number of morphologically normal mitochondria, and a concomitant increase in the number of morphologically abnormal mitochondria. There can be a marked increase of especially abnormal appearing mitochondria within autophagocytic lysosomes, indicating increased mitochondrial disposal rates. If mitochondrial disposal rates exceed an increase in mitochondrial biogenesis, and methods of mitochondrial ascertainment do not capture the mitochondrial discard pool, mitochondrial mass would appear reduced even though mitochondrial biogenesis is increased. This line of research could benefit from additional studies designed to firmly resolve this discrepancy.

Whether or not Aβ affects mitochondria through an indirect or direct effect in these AD models is unclear, but a number of investigators report transgenic mice show physical associations between mitochondria and Aβ as well as between mitochondria and APP. A physical association between mitochondria and Aβ also appears to occur in human AD (Anandatheerthavarada et al. 2003; Lustbader et al. 2004; Casperson et al. 2005; Crouch et al. 2005; ; Devi et al. 2006; Manczak et al. 2006). Physical interactions between presenilin 1 (PS1) and mitochondria are further reported (Hansson et al. 2004). In the case of Aβ, interference with normal mitochondrial physiology is suspected. The full physiologic spectrum of mitochondrial interactions with APP, Aβ, or PS1, though, has not been fully elucidated.

Mutations of APP, Aβ, or PS1 genes almost invariably result in AD (Swerdlow 2007b). Aside from these "deterministic" genes, the most studied AD genetic risk factor is the apolipoprotein E (APOE) gene on chromosome 19 (Corder et al. 1993). The mechanistic basis via which the APOE gene product influences AD risk is unknown, but mitochondrial interactions may account for the association. Apolipoprotein E degradation generates a fragment that contains a mitochondrial targeting amino acid sequence. This fragment has been localized to mitochondria and shown under experimental conditions to interfere with mitochondrial function (Chang et al. 2005).

Clear associations between mitochondrial function and AD are recognized. Vmax activities of several enzymes located within mitochondria are reduced (Swerdlow and Kish 2002). These include pyruvate dehydrogenase complex, alpha ketoglutarate dehydrogenase complex, and complex IV (cytochrome oxidase) of the ETC. These Vmax activity reductions are not limited to the brain, and at least for cytochrome oxidase the reduced Vmax appears systemic. Reduced cytochrome oxidase activity has been observed in brain, platelets, and fibroblasts. AD is therefore associated with systemic mitochondrial phenomena. It has been argued these systemic manifestations are unlikely to arise solely from interactions between mitochondria and Aβ, APP, or presenilins (Swerdlow and Khan 2009). Several lines of evidence further suggest reduced cytochrome oxidase activity in AD is at least partly mediated by mtDNA. As mentioned above, relative to persons without AD, platelet cytochrome oxidase activity is reduced in persons with AD (Parker et al. 1990b). When platelet mtDNA from AD subjects is expressed within neuronal cell lines grown in culture (cytoplasmic hybrid cell lines, or cybrids), the resulting cells continue to manifest reduced cytochrome oxidase activity over numerous passages (Swerdlow et al. 1997; Swerdlow 2007c) (Fig. 30.1). The exact mtDNA features that drive this are unknown. Polymorphic variations in protein coding genes, variation in the mtDNA control region (the "D loop"), low abundance heteroplasmic point mutations, deletions, and oxidative modifications of mtDNA nucleotides have all been proposed as potential culprits (Swerdlow and Khan 2009).

Whether or not a putatively causal mtDNA signature is inherited or acquired (somatic mutation) has been considered. Data from epidemiologic and neuropsychological studies are relevant to this debate, and suggest mtDNA inheritance may be more likely. It has been reported that for AD subjects who lack a demonstrable pattern of autosomal dominant inheritance, but who nevertheless have an AD-affected parent, the affected parent is more often the mother than the father (Edland et al. 1996). This relationship persists even after taking into consideration the fact women generally outlive men, placing them at an increased risk of developing AD. It makes sense to assume the converse relationship also holds, and that having a mother with

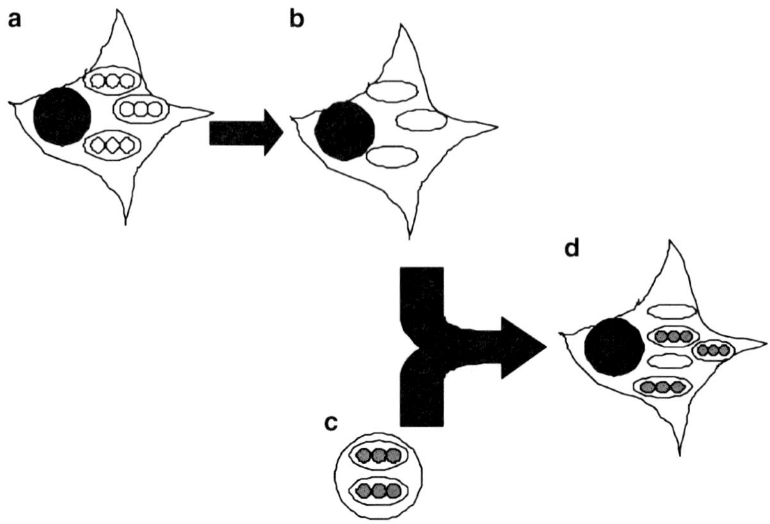

Fig. 30.1 The cybrid technique. (**a**) Shows a cell in which mtDNA is depicted as *empty circles* within oval mitochondria. It is possible to remove mtDNA from cells to create "ρ0" cells that lack mtDNA. This is depicted in (**b**) by the presence of empty mitochondria. Although ρ0 cells lack normal electron transport chain function, these cells will continue to grow and divide when provided adequate metabolic support. Exogenous mitochondria can be mixed with ρ0 cells; (**c**) shows a platelet, which lacks nuclei but contains mitochondria. The mtDNA in the platelet mitochondria is indicated by *shaded circles*. Mixing results in transfer of platelet mitochondria, and hence platelet mtDNA, to the ρ0 cell. The resulting cell is now a cytoplasmic hybrid, as it contains cytoplasmic components from the original cell line and from the platelet. This cytoplasmic hybrid, or cybrid, constitutes a unique cell line that has the nuclear DNA of the original cultured cell line and the mtDNA from the platelet donor. Introduction of exogenous mtDNA restores aerobic function to the previously ρ0 cell line

AD confers a slightly greater risk of AD to the offspring than having a father with AD. A large study of the offspring of individuals followed in the original Framingham study is consistent with this (Wolf et al. 2005). The investigators of the Framingham Offspring Study divided the middle aged individuals they were following into three groups. The first group was defined by having had both parents reach the age of 85 without dementing. The second group was defined by having had a father that developed AD, but a mother who reached the age of 85 without dementing. The third group was defined by having had a mother that developed AD, but a father who reached the age of 85 without dementing. The investigational endpoint was neuropsychological test performance. The middle aged individuals of the first group, as expected, performed well overall. The middle aged individuals of the second group (where the father but not the mother had AD) performed equivalently. The middle aged individuals of the third group (where the mother had AD), however, had statistically poorer performance on neuropsychological testing.

An obvious question arising from the Framingham Offspring Study cognitive analysis is whether the less robust cognitive testing results in the asymptomatic, middle aged children of AD mothers reflects an early manifestation of the AD process. Positron emission tomography (PET) studies are relevant in this regard. PET directly assesses glucose uptake by the brain across the blood brain barrier. This can further be used to calculate a "cerebral metabolic rate of glucose" (CMR_{glu}), which serves as a measure of brain glucose utilization. Persons with AD show reduced brain glucose uptake and CMR_{glu} in particular brain regions (Mosconi 2005). The temporal, parietal, and posterior cingulate regions are particularly affected. AD-like PET abnormalities can occur in cognitively asymptomatic persons at increased risk for AD as indicated by carriage of the APOE4 allele (Small et al. 1995; Reiman et al. 1996). Beyond this, one particularly relevant PET study divided cognitively intact middle aged individuals into three groups. The control group consisted of individuals for which neither parent developed AD prior to death. A second group consisted of individuals in which only the subject's father developed AD prior to death. A third group consisted of individuals in which only the subject's mother developed AD prior to death. The investigators found the control and paternal family history groups had normal PET studies, while the maternal family history group had PET scan changes typical of those seen in AD (Mosconi et al. 2007).

Taken together, epidemiologic (maternal inheritance bias), biochemical (systemic mitochondrial defects and perpetuation of mitochondrial defects in cybrid cell lines), neuropsychologic (differences in cognitive testing between children of AD-affected mothers and fathers), and PET (differences in CMR_{glu} between children of AD mothers and fathers) data suggest maternal inheritance-derived differences in mitochondrial function influence AD risk. As advancing age constitutes the single greatest risk factor for AD, it is interesting that longevity itself, while reflecting the longevity of both an individual's parents, is influenced more by maternal than by paternal longevity (Brand et al. 1992).

In summary, altered mitochondrial function in AD and rationales for such alterations in both early onset, autosomal dominant AD and sporadic AD are recognized. This recognition was previously used to justify a "mitochondrial cascade hypothesis" for AD (Swerdlow and Khan 2004) (Fig. 30.2). While some investigators feel the origin of and role this altered function plays in the development and progression of AD is upstream and critical, many investigators feel mitochondrial dysfunction in AD represents a downstream event and plays at most a minor role in the disease. At least one critical pro and con discussion of this controversial topic is available to those interested in this debate (Swerdlow and Kish 2002).

30.3.2 Parkinson's Disease

PD is the second most common neurodegenerative disease and the most common neurodegenerative movement disorder. The cardinal symptoms and signs of PD include resting tremor, bradykinesia, cogwheel rigidity, and postural instability. The clinical presentation is associated with the loss of substantia nigra dopaminergic neurons. Residual nigral neurons may contain intracytoplasmic, α synuclein-containing

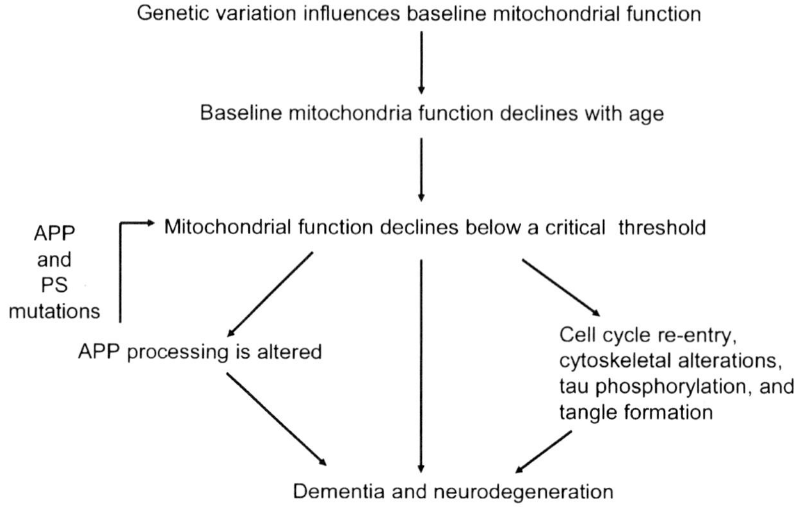

Fig. 30.2 The mitochondrial cascade hypothesis. In this hypothesis, mitochondrial failure activates a final common pathway of clinical symptoms, neurodegeneration, and histologic findings. In late-onset, sporadic AD mitochondrial failure is an aging-associated event that depends on an individual's inherent brain aging rate. In early-onset, autosomal dominant AD mitochondrial failure results from mutation-derived interference with mitochondrial function… *APP* amyloid precursor protein, *PS* presenilin

inclusions called Lewy bodies, and it is the presence of nigral Lewy bodies that establishes the histologic diagnosis. As is the case with AD, PD can be divided into early and late onset variants and Mendelian versus non-Mendelian forms exist. Although there is a greater age-of-onset overlap between the Mendelian and non-Mendelian forms than there is in AD, the percentage of cases caused by Mendelian gene mutations drops with advancing age, while the percentage of cases that do not show Mendelian inheritance rises.

Unlike AD, popular PD etiologic hypotheses have long entertained the possibility mitochondria may play an important role. Indeed, realization in the 1980s that mitochondrial toxins induce a form of subacute parkinsonism that resembles idiopathic PD catapulted mitochondria to the forefront of the PD research field (Langston et al. 1983; Nicklas et al. 1985). Early in that decade, recreational narcotic users in the San Francisco area developed parkinsonism after using meperidine contaminated with the chemical 1-methyl-4-phenyl-1,2,3,6-tetrahydropyridine (MPTP). By mid-decade it was established that MPTP is converted to a derivative, 1-methyl-4-phenylpyridinium (MPP+), which potently inhibits complex I of the mitochondrial ETC (Nicklas et al. 1985). Subsequently, investigators began using MPTP to generate experimental PD models. Since then, other complex I inhibitors such as rotenone have been used to produce popular cell culture and animal models of PD (Betarbet et al. 2000). Further, a number of epidemiologic studies suggest mitochondria-toxic pesticides may increase PD risk (Rajput et al. 1987; Hubble et al. 1993).

In 1989, persons with idiopathic PD were found to have reduced complex I activity (Bindoff et al. 1989; Parker et al. 1989; Schapira et al. 1989). This was initially shown using PD subject substantia nigra, platelets, and muscle as the mitochondrial source. It has since been shown that complex I activity is altered in fibroblasts and frontal cortex as well (Mytilineou et al. 1994; Parker et al. 2008). Persons with PD therefore appear to have a systemic reduction of complex I activity. Investigators have hypothesized the systemic complex I defect in PD arises as a consequence of exposure to exogenous inhibitors, systemic endogenous production of an inhibitory factor, or its mtDNA encoding (Swerdlow 2000).

Complex I is an extremely large multimeric enzyme that contains 46 known protein subunits. Seven of these proteins are encoded by genes on mtDNA. Since mtDNA makes such an important contribution to complex I structure and function and mtDNA abnormalities can produce sporadic disease, Parker hypothesized mtDNA was the most likely cause (Parker et al. 1989). Parker and colleagues decided to test this hypothesis using cybrids, and found repopulating mtDNA-depleted human neuroblastoma cells with mtDNA from PD subject mitochondria produced cell lines with reduced complex I activity (Swerdlow et al. 1996). This finding has since been replicated by other groups using human lung adenocarcinoma and teratocarcinoma ρ0 cell lines (Gu et al. 1998; Esteves et al. 2008). These cell lines also exhibit increased reactive oxygen species production, reduced mitochondrial calcium storage, and intracytoplasmic α synuclein accumulation. As is the case with AD cybrids, the actual nature of the mtDNA alteration that accounts for these findings remains unsettled. Several recent reports, though, claim mutations clustered within a particular stretch of the mtDNA ND5 gene are present in the brains of PD but not control subjects (Smigrodzki et al. 2004; Parker and Parks 2005). These mutations comprise an extremely small proportion of the total PD subject mtDNA pools, although the possibility mtDNA clones containing ND5 mutation expand in nigral neurons cannot be excluded. One of the more common PD ND5 mutations, when present in higher abundance, causes a devastating childhood neurodegeneration called Leigh's disease. In addition to ND5 mutations, expansion of deleted mtDNA molecules is seen in the brains of those afflicted with PD and in aging substantia nigra neurons (Kraytsberg et al. 2006; Bender et al. 2006).

Among those who believe mtDNA mutations influence PD risk, some would posit the mutations are acquired, while others would argue they are inherited. Several lines of investigation support the latter view. Epidemiologic studies suggest for non-Mendelian cases who nevertheless have a PD-affected parent, the affected parent is more likely to be the mother (Wooten et al. 1997; Swerdlow et al. 2001). Mitochondrial haplogroup and polymorphism association studies demonstrate mtDNA variations alter PD risk (Swerdlow 2000; van der Walt et al. 2003). The systemic nature of the PD complex I defect and PD cybrid studies also supports mtDNA inheritance over somatic mutation acquisition.

Enthusiasm for a mitochondrial hypothesis began to wane soon after α-synuclein gene mutation was shown to be a rare cause of autosomal dominant PD. Over the past decade a number of other genes have been implicated in both autosomal dominant and recessive PD. While this barrage of gene linkage and positional cloning studies initially moved interest in mitochondria further and further onto the etiologic backburner, Mendelian genetic research has paradoxically begun to rekindle interest in mitochondria. Under *in vitro* conditions and in animal models, complex I inhibition promotes α-synuclein aggregation (Lee et al. 2002; Sherer et al. 2002, 2003). Several of the genes implicated in Mendelian PD encode proteins (parkin, DJ1, PINK1, LRRK2) that are known or suspected to reside within or directly interact with mitochondria (Moore and Dawson 2008). Convergence of these mutant proteins on mitochondria suggests that in Mendelian PD mitochondria may give rise to a final common pathway of neurodegeneration.

30.3.3 Amyotrophic Lateral Sclerosis

Amyotrophic lateral sclerosis (ALS) is the most common neurodegenerative disease that primarily affects strength. Its course is characterized by progressive weakness arising from demise of both upper and lower motor neurons. As is the case with AD and PD, both sporadic and Mendelian forms exist, its incidence rises with increasing age, and the older the age of onset the less likely Mendelian inheritance is observed.

Abnormal mitochondrial morphology in ALS was observed prior to knowledge of specific functional alterations. These changes were observed in both the nervous system and in non-neuronal tissues (Swerdlow et al. 2000b). In the early and mid 1990s increased complex I activity was reported in an individual with familial ALS, and reduced cytochrome oxidase activity was shown in sporadic ALS patients (Bowling et al. 1993; Fujita et al. 1996). Soon after these reports, using Parker's hypothesis of sporadic disease risk transmission as justification ALS was studied using the cybrid technique. Cybrids produced on a neuroblastoma nuclear background and expressing mtDNA from sporadic ALS subject platelets showed a significant reduction in complex I activity and non-significant trends towards reduced complex III and IV activity (Swerdlow et al. 1998). The sporadic ALS reduction in complex I activity was corroborated by subsequent studies of ALS muscle. These studies also reported evidence of mtDNA perturbation (Wiedemann et al. 1998; Vielhaber et al. 2000).

The superoxide dismutase 1 (SOD1) gene was the first gene recognized to cause autosomal dominant ALS, and has been estimated to account for about 2% of ALS cases. SOD1 is classically thought of as an antioxidant enzyme localized to the cytoplasm. It was initially suspected SOD1 mutation would lead to oxidative stress and that oxidative stress would prove to be the cause of neurodegeneration in this disease. It was then shown that some causal SOD1 mutations do not reduce the ability of SOD1 to remove the superoxide radical. Combined with the fact that SOD1 mutation ALS is an autosomal dominant disorder, a rationale for some type of SOD1 gain of function mechanism was proposed. Mutant SOD1, and to a lesser extent wild type SOD1, has since been localized to the mitochondrial membrane (Hervias et al. 2006). Transgenic mice expressing human mutant SOD1 alleles certainly demonstrate substantially altered mitochondrial morphology and mitochondrial SOD1 accumulation (Dal Canto and Gurney 1994; Wong et al. 1995). Many investigators currently suspect the putative gain-of-function that occurs as a consequence of SOD1 mutation involves mitochondrial damage, which arises via a direct interaction between mitochondrial membranes and mutant SOD1 protein.

30.3.4 Strictly Mendelian Neurodegenerative Diseases

For some Mendelian neurodegenerative diseases mitochondrial dysfunction was recognized before any causal gene mutations were identified. For others, elucidating the responsible gene mutations implicated a potential role for mitochondrial dysfunction because the mutant protein is targeted to mitochondria.

HD is an example of the former case. HD is a degenerative movement disorder characterized by excessive characteristic movements called chorea. Neurodegeneration of the basal ganglia structures occurs. During the 1980s impaired ETC function was demonstrated in both cerebral and non-cerebral HD subject tissues and these findings were subsequently expanded upon (Swerdlow 2002). Toxic inhibition of the ETC enzyme complex II by either 3-nitropropionic acid or malonate was shown to induce neurodegeneration in mice, and this resulting pattern of neurodegeneration is anatomically reminiscent of that seen in human HD (Beal et al. 1993; Greene et al. 1993). Indeed, 3-nitropropionic acid and malonate administration were commonly used approaches for modeling HD, and are still used by some investigators. In 1993 HD was shown to associate with a polyglutamine repeat expansion in the first exon of the huntingtin gene on chromosome 4 (Huntington's Disease Collaborative Research Group 1993). This expansion causes a toxic huntingtin protein gain of function, but the exact nature of this gain of function is unknown. Published investigations claim mutant huntingtin affects mitochondrial function via both direct and indirect mechanisms (Browne and Beal 2004). Mutant huntingtin physically associates with mitochondrial membranes, where it appears to interfere with mitochondrial calcium handling (Panov et al. 2002). It can interact with other proteins that may mediate mitochondrial function. Finally, huntingtin may interfere with mitochondrial biogenesis by impeding transcriptional programs that mediate replication of the organelle. Currently, it is suspected this is accomplished through effects on peroxisome proliferator activated receptor γ coactivator 1 α (PGC-1a), a transcription co-activator that plays an important role in mitochondrial replication (Cui et al. 2006).

Friedreich's ataxia (FA) is an example of a Mendelian neurodegenerative disease in which a mitochondrial role was envisioned only after the responsible mutation was identified. FA is an autosomal recessive disorder, and it is one of the most common genetic neurodegenerative ataxias. FA is associated with homozygous mutation of the frataxin gene on chromosome 9, and mutations most commonly appear in the form of an intronic GAA repeat expansion. Frataxin is a mitochondrially targeted gene that plays a role in mitochondrial iron transport (Koutnikova et al. 1997). Other neurodegenerative diseases in which the mutant product is a mitochondrial protein include some forms of hereditary spastic paraplegia, Wilson's disease, and a deafness-dystonia syndrome (Swerdlow 2002).

30.4 Neurodegenerative Mitochondriopathy Controversies

So far this chapter has mostly considered data that support the view mitochondrial function is altered in several neurodegenerative diseases and likely plays an important role in the progression if not development of these diseases. It is only fair to point out that many feel mitochondria are not important in neurodegenerative disease progression and development, and that for many studies cited above that support a role for mitochondrial dysfunction, other studies report a lack of support. This is the case for epidemiologic studies of maternal inheritance bias in AD and PD, ETC dysfunction in AD and PD brains, ETC dysfunction in AD and PD non-cerebral tissues, studies of mtDNA mutation and polymorphism association in AD and PD, and cybrid studies of AD, PD, and ALS (Swerdlow 2000; Swerdlow and Kish 2002; Swerdlow et al. 2006; Swerdlow 2007c). It is also fair to point out, though, that negative studies have consistently failed to use methodologies equivalent to those used in the positive studies. Epidemiologic studies have varied in terms of the populations studied and methods of ascertainment. ETC enzyme analyses have varied in terms of mitochondrial sample preparation and enzyme rate determination. Methods for determining mtDNA sequence vary considerably, and many negative studies were not designed to detect heteroplasmic sequence deviations. The negative cybrid studies were done using ρ0 cells with non-neuronal nuclear backgrounds, different methods of mitochondrial acquisition, and functional assessments not comparable to those of the positive studies.

Published data that support a role for mitochondria in particular neurodegenerative diseases are sometimes non-committal in terms of where the observed mitochondria dysfunction falls in that disease's pathophysiologic hierarchy. For example, demonstrating a complex I defect in PD substantia nigra could potentially represent a downstream event or even a dying tissue epiphenomenon. It is harder to argue the complex I defect in PD platelets represents a dying tissue epiphenomenon. In general, the more upstream mitochondria are placed in a disease's pathophysiologic hierarchy, the more resistance there is. This tends to assure conservative views on the role of mitochondrial function in neurodegenerative diseases prevail. Also, the less mitochondria are accounted for within a prevailing disease paradigm, the more resistance there is to the idea that mitochondrial function plays an important role. For example, for many years the amyloid cascade hypothesis has managed to downplay any role for mitochondria in AD (Swerdlow 2007d).

30.5 Mitochondrial Tie-Ins to Other Neurodegeneration Hypotheses

Various mechanisms have been proposed to categorically explain neurodegeneration and neurodegenerative disease pathology. For example, oxidative stress is considered by many investigators to play an important role in neurodegeneration.

As mitochondria are both a source of free radical production and a target of free radical attack, oxidative stress hypotheses tend to conceptualize upstream, downstream, or dual mitochondrial involvement.

During the second half of the 1990s mitochondria were implicated in programmed cell death in general, and in apoptotic cell death specifically (Petit et al. 1997). Mitochondria are now known to sequester proteins that can activate various cell death pathways. In one popular model, induction of mitochondrial dysfunction initiates mitochondrial swelling and permeabilization of mitochondrial membranes through either rupture of the outer membrane or formation of a permeability pore. Some feel neuronal death in neurodegenerative diseases proceeds via such an "intrinsic" apoptotic pathway (Tsujimoto et al. 2006; Norenberg and Rao 2007). If so, this hypothesis could implicate a final common pathway role for mitochondria in multiple neurodegenerative diseases.

30.6 Treating Neurodegenerative Mitochondriopathies

Thus far this chapter has reviewed a spectrum of mitochondrial or mitochondria-related pathologies seen in neurodegenerative mitochondriopathies. Many of these specific pathologies have been therapeutically targeted. For example, antioxidants have been tried in most neurodegenerative diseases mentioned above (Swerdlow 2007a). The results of these trials have ranged from no observed benefit to at best slight and not easily replicated benefit. Antioxidant disappointments can be interpreted two ways. One interpretation is that the antioxidants evaluated were inappropriate, or were used at inadequate doses. Proponents of this view are readying a new generation of synthetic mitochondria-targeted antioxidants for clinical testing, or else exploring ways to activate endogenous cell antioxidant systems via upregulation of antioxidant enzyme expression. The other possible interpretation is oxidative stress does not drive neurodegeneration in the common neurodegenerative diseases. Some proponents of this view believe reactive oxygen and nitrogen species elevations are downstream of or peripheral to more critical pathologies. Other proponents suspect free radical elevations actually represent part of an intracellular signaling mechanism whose purpose is to activate compensatory adaptations (Finkel 2001; Schieke and Finkel 2006).

The idea of increasing cell energy supplies to benefit persons with a neurodegenerative disease has also been pursued (Swerdlow 2007a). Supplementing patients with natural or synthetic electron transfer-mediating molecules, such as coenzyme Q, have yielded results that range from no observed benefit to a slight benefit. If there is indeed a benefit, it is hard to know whether such agents are actually facilitating electron transfer. For some agents, the expected compartmentalization of the agent suggests it may simply be acting as an antioxidant. Dichloroacetate is an interesting drug that is used with the intent of enhancing oxidative phosphorylation (Bonnet et al. 2007). Dichloroacetate inhibits pyruvate dehydrogenase kinase, which reduces the inhibitory phosphorylation pyruvate dehydrogenase kinase places upon pyruvate

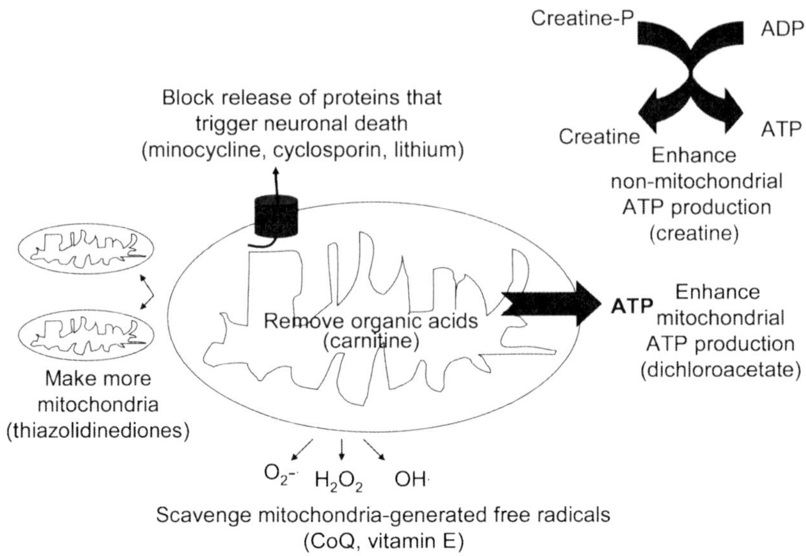

Fig. 30.3 Some attempts at "mitochondrial medicine"

dehydrogenase complex. This should theoretically direct more acetyl units into the Krebs cycle and reducing equivalents into the oxidative phosphorylation pathway. Unfortunately, a clinical trial of dichloroacetate in a rare mtDNA-derived mitochondriopathy, mitochondrial encephalopathy with lactic acidosis and strokelike episodes (MELAS), was disappointing from both efficacy and tolerability perspectives (Scaglia et al. 2006). Other agents, such as creatinine, are intended to help maintain cell energy supplies by enhancing cell energy storage (Tarnopolsky and Beal 2001).

Interfering with the mitochondrial permeability transition has been attempted with drugs such as cyclosporine and minocycline (Swerdlow 2007a). Results have been disappointing. Some pharmacologic interventions that probably affect mitochondria at other or at multiple levels are in the early stages of testing. An example is a recent small trial of lithium in ALS (Fornai et al. 2008). Lithium may reduce mitochondrial permeabilization, interfere with apoptotic cascades, alter energy homeostasis, and alter mitochondrial autophagy. Patients in this small trial were reported to do better than subjects randomized to placebo. More definitive trial results are awaited.

Pharmacologic induction of mitochondrial biogenesis for the treatment of neurodegenerative mitochondriopathies has been proposed (Swerdlow 2007a) (Fig. 30.3). Mitochondrial proliferation may represent a natural compensatory event in neurodegenerative mitochondriopathies. Attempting to facilitate this natural compensation makes sense, but is not without risk. Facilitating biogenesis of impaired mitochondria without fixing those mitochondria could prove detrimental. Preclinical testing of this strategy could help address such concerns or perhaps identify ways to maximize benefits while minimizing any adverse consequences.

The most ambitious strategies for treating mitochondrial dysfunction in human disease are those that focus on gene therapy. Techniques for replacing mutant mtDNA with non-mutant mtDNA in the intact, living organism are under development (Khan and Bennett 2004). Allotopic expression of mitochondrial genes has also been attempted *in vitro*, but with limited success (Khan et al. 2007).

30.7 Conclusions

Data accumulated over several decades indicate mitochondria function differently in persons with neurodegenerative diseases than they do in persons without such diseases. This is the case for very rare neurodegenerative disorders, and also for extremely common age-related ones such as PD and AD. The pervasive nature of altered mitochondrial function across the neurodegenerative disease spectrum justifies classifying them as "neurodegenerative mitochondriopathies." Given the likely role mitochondria play in both aging and neurodegeneration, a role for mitochondria in the pathophysiology and perhaps even etiologies of age-dependent neurodegenerative diseases should not be surprising.

While in some disorders mitochondria show generic patterns of altered function, in some cases very particular defects are demonstrable. Whether or not these defects represent upstream pathology is currently under debate. In the meantime, pharmacologically altering mitochondria or the consequences of mitochondrial dysfunction has been tried in an effort to ameliorate or reverse particular neurodegenerative disorders. Thus far no robust effects have been shown in clinical trials, but a greater understanding of the origin of and mechanisms through which mitochondria contribute to these diseases will hopefully facilitate development of more rational and efficacious therapies.

References

Anandatheerthavarada HK, Biswas G, Robin MA, Avadhani MG (2003) Mitochondrial targeting and a novel transmembrane arrest of Alzheimer's precursor protein impairs mitochondrial function in neuronal cells. J Cell Biol 16:41–54

Barrientos A, Casademont J, Cardellach F, Estivill X, Urbano-arquez A, Nunes V (1997) Reduced steady-state levels of mitochondrial RNA and increased mitochondrial DNA amount in human brain with aging. Brain Res Mol Brain Res 52:284–289

Beal MF, Brouillet E, Jenkins BG, Ferrante RJ, Kowall NW, Miller JM, Storey E, Srivastava R, Rosen BR, Hyman BT (1993) Neurochemical and histologic characterization of excitotoxic lesions produced by the mitochondrial toxin 3-nitropropionic acid. J Neurosci 13:4181–4192

Bender A, Krishnan KJ, Morris CM, Taylor GA, Reeve AK, Perry RH, Jaros E, Hersheson JT, Betts J, Klopstock T, Taylor RW, Turnbull DM (2006) High levels of mitochondrial DNA deletions in substantia nigra neurons in aging and Parkinson disease. Nat Genet 38:515–517

Betarbet R, Sherer TB, MacKenzie G, Garcia-Osuna M, Panov AV, Greenamyre JT (2000) Chronic systemic pesticide exposure reproduces features of Parkinson's disease. Nat Neurosci 3: 1301–1306

Bindoff LA, Birch-Machin M, Cartlidge NEF, Parker WD Jr, Turnbull DM (1989) Mitochondrial function in Parkinson's disease. Lancet 2:49

Bonnet S, Archer SL, Allalunis-Turner J, Haromy A, Beaulieu C, Thompson R, Lee CT, Lopaschuk GD, Puttagunta L, Bonnet S, Harry G, Hashimoto K, Porter CJ, Andrade MA, Thebaud B, Michelakis ED (2007) A mitochondria-K + channel axis is suppressed in cancer and its normalization promotes apoptosis and inhibits cancer growth. Cancer Cell 11:37–51

Bowling AC, Schulz JB, Brown RH Jr, Beal MF (1993) Superoxide dismutase activity, oxidative damage, and mitochondrial energy metabolism in familial and sporadic amyotrophic lateral sclerosis. J Neurochem 61:2322–2325

Brand FN, Kiely DK, Kannel WB, Myers RH (1992) Family patterns of coronary heart disease mortality: the Framingham longevity study. J Clin Epidemiol 45:169–174

Browne SE, Beal MF (2004) The energetics of Huntington's disease. Neurochem Res 29:531–546

Canevari L, Clark JB, Bates TE (1999) β-Amyloid fragment 25–35 selectively decreases complex IV activity in isolated mitochondria. FEBS Lett 457:131–134

Cardoso SM, Santos S, Swerdlow RH, Oliveira CR (2001) Functional mitochondria are required for amyloid beta-mediated neurotoxicity. FASEB J 15:1439–1441

Casperson C, Wang N, Yao J, Sosunov A, Chen X, Lustader JW, Xu WH, Stern D, McKhann G, Yan SD (2005) Mitochondrial Abeta: a potential focal point for neuronal metabolic dysfunction in Alzheimer's disease. FASEB J 19:2040–2041

Chang S, RanMa T, Miranda RD et al (2005) Lipid- and receptor-binding regions of apolipoprotein E4 fragments act in concert to cause mitochondrial dysfunction and neurotoxicity. Proc Natl Acad Sci USA 102:18694–18699

Corder EH, Saunders AM, Strittmatter WJ, Schmechel DE, Gaskell PC, Small GW et al (1993) Gene dose of apolipoprotein E type 4 allele and the risk of Alzheimer's disease in late onset families. Science 261:921–923

Crouch PJ, Blake R, Duce JA, Ciccotosto GD, Li QX, Barnham KJ, Curtain CC, Cherny RA, Cappai R, Dyrks T, Masters CL, Trounce IA (2005) Copper-dependent inhibition of human cytochrome c oxidase by a dimeric conformer of amyloid-beta1-42. J Neurosci 25:672–679

Cui L, Jeong H, Borovecki F, Parkhurst CA, Tanese N, Krainc D (2006) Transcriptional repression of PGC-1alpha by mutant huntingtin leads to mitochondrial dysfunction and neurodegeneration. Cell 127:59–69

Dal Canto MC, Gurney ME (1994) Development of central nervous system pathology in a murine transgenic model of human amyotrophic lateral sclerosis. Am J Pathol 145:1271–1280

de la Monte SM, Luong T, Neely TR, Robinson D, Wands JR (2000) Mitochondrial DNA damage as a mechanism of cell loss in Alzheimer's disease. Lab Invest 80:1323–1335

Devi L, Prabhy BM, Galati DF, Avadhani NG, Anandatheerthavarada HK (2006) Accumulation of amyloid precursor protein in the mitochondrial import channels of human Alzheimer's disease brain is associated with mitochondrial dysfunction. J Neurosci 26:9057–9068

Diana A, Simic G, Sinforiani E, Orru N, Pichiri G, Bono G (2008) Mitochondria morphology and DNA content upon sublethal exposure to beta-amyloid(1-42) peptide. Coll Antropol 32:51–58

Eckert A, Schulz KL, Rhein V, Götz J (2010) Convergence of amyloid-beta and tau pathologies on mitochondria in vivo. Mol Neurobiol 41:107–114, PMID: 20217279

Edland SD, Silverman JM, Peskind ER et al (1996) Increased risk of dementia in mothers of Alzheimer's disease cases: Evidence for maternal inheritance. Neurology 47:254–256

Esteves ARF, Domingues AF, Ferreira IL, Januário C, Swerdlow RH, Oliveira CR, Cardoso SM (2008) Mitochondrial dysfunction occurs in Parkinson's disease cybrids containing an NT2 neuron-like nuclear background. Mitochondrion 8:219–228

Evans DA, Funkenstein HH, Albert MS, Scherr PA, Cook NR, Chown MJ et al (1989) Prevalence of Alzheimer's disease in a community population of older persons. Higher than previously reported. JAMA 262:2551–2556

Finkel T (2001) Reactive oxygen species and signal transduction. IUBMB Life 52:3–6

Fornai F, Longone P, Cafaro L, Kastsiuchenka O, Ferrucci M, Manca ML, Lazzeri G, Spalloni A, Bellio N, Lenzi P, Modugno N, Siciliano G, Isidoro C, Murri L, Ruggieri S, Paparelli A (2008) Lithium delays progression of amyotrophic lateral sclerosis. Proc Natl Acad Sci USA 105: 2052–2057

Friede RI (1965) Enzyme histochemical studies of senile plaques. J Neuropathol Exp Neurol 24:477–491

Fujita K, Yamauchi M, Shibayama K et al (1996) Decreased cytochrome oxidase activity but unchanged superoxide dismutase and glutathione peroxidase activities in the spinal cords of patients with amyotrophic lateral sclerosis. J Neurosci Res 45:276–281

Goto Y, Nonaka I, Horai S (1990) A mutation in the tRNALeu(UUR) gene associated with MELAS subgroup of mitochondrial encephalomyopathies. Nature 348:651–653

Greene JG, Porter RH, Eller RV, Greenamyre JT (1993) Inhibition of succinate dehydrogenase by malonic acid produces an "excitotoxic" lesion in rat striatum. J Neurochem 61:1151–1154

Gu M, Cooper JM, Taanman JW, Schapira AHV (1998) Mitochondrial DNA transmission of the mitochondrial defect in Parkinson's disease. Ann Neurol 44:177–186

Hansson CA, Frykman S, Farmery MR et al (2004) Nicastrin, presenilin, APH-1, and PEN-2 form active gamma-secretase complexes in mitochondria. J Biol Chem 279:51654–51660

Harding AE (1983) Classification of the hereditary ataxias and paraplegias. Lancet 21: 1151–1155

Hardy J, Allsop D (1991) Amyloid deposition as the central event in the aetiology of Alzheimer's disease. Trends Pharmacol Sci 12:383–388

Hardy JA, Higgins GA (1992) Alzheimer's disease: the amyloid cascade hypothesis. Science 256: 184–185

Hervias I, Beal MF, Manfredi G (2006) Mitochondrial dysfunction and amyotrophic lateral sclerosis. Muscle Nerve 33:598–608

Hirai K, Aliev G, Nunomura A, Fujioka H, Russell RL, Atwood CS, Johnson AB, Kress Y, Vinters HV, Tabaton M, Shimohama S, Cash AD, Siedlak SL, Harris PL, Jones PK, Petersen RB, Perry G, Smith MA (2001) Mitochondrial abnormalities in Alzheimer's disease. J Neurosci 21: 3017–3023

Holt IJ, Harding AD, Morgan-Hughes JA (1988) Deletions of muscle mitochondrial DNA in patients with mitochondrial myopathies. Nature 331:717–719

Hubble JP, Cao T, Hassanein RE, Neuberger JS, Koller WC (1993) Risk factors for Parkinson's disease. Neurology 43:1693–1697

Huntington's Disease Collaborative Research Group (1993) A novel gene containing a trinucleotide repeat that is expanded and unstable on Huntington's disease chromosomes. Cell 72:971–983

Johns DR, Lessell S, Miller NR (1991) Molecularly confirmed Leber's hereditary optic neuropathy. Neurology 41(suppl 1):347

Khan SM, Bennett JP Jr (2004) Development of mitochondrial gene replacement therapy. J Bioenerg Biomembr 36:387–393

Khan SM, Smigrodzki RM, Swerdlow RH (2007) Cell and animal models of mtDNA biology: progress and prospects. Am J Physiol Cell Physiol 292:C658–C669

Koutnikova H, Campuzano V, Foury F, Dolle P, Cazzalini O, Koenig M (1997) Studies of human, mouse and yeast homologues indicate a mitochondrial function for frataxin. Nat Genet 16:345–351

Kraytsberg Y, Kudryavtseva E, McKee AC, Geula C, Kowall NW, Khrapko K (2006) Mitochondrial DNA deletions are abundant and cause functional impairment in aged human substantia nigra neurons. Nat Genet 38:518–520

Langston JW, Ballard PA, Tetrud JW, Irwin I (1983) Chronic parkinsonism in humans due to a product of meperidine-analog synthesis. Science 219:979–980

Lee HJ, Shin SY, Choi C, Lee YU, Lee SJ (2002) Formation and removal of alpha-synuclein aggregates in cells exposed to mitochondrial inhibitors. J Biol Chem 277(7):5411–5417

Lestienne P, Ponsot F (1988) Kearns-Sayre syndrome with muscle mitochondrial DNA deletion. Lancet 1:885

Luft R, Ikkos D, Palmieri G, Ernster L, Afzelius B (1962) A case of severe hypermetabolism of nonthyroid origin with a defect in the maintenance of mitochondrial respiratory control: a correlated clinical, biochemical, and morphological study. J Clin Invest 41:1776–1804

Lustbader JW, Cirilli M, Lin C, Xu HW, Takuma K, Wang N et al (2004) ABAD directly links Abeta to mitochondrial toxicity in Alzheimer's disease. Science 304:448–452

Manczak M, Anekonda TS, Henson E, Park BS, Quinn J, Reddy PH (2006) Mitochondria are a direct site of A beta accumulation in Alzheimer's disease neurons: implications for free radical generation and oxidative damage in disease progression. Hum Mol Genet 15:1437–1449

Mann VM, Cooper JM, Javoy-Agid F et al (1990) Mitochondrial function and parental sex effect in Huntington's disease. Lancet 336:749

Moore DJ, Dawson TM (2008) Value of genetic models in understanding the cause and mechanisms of Parkinson's disease. Curr Neurol Neurosci Rep 8:288–296

Mosconi L (2005) Brain glucose metabolism in the early and specific diagnosis of Alzheimer's disease. FDG-PET studies in MCI and AD. Eur J Nucl Med Mol Imaging 32:486–510

Mosconi L, Brys M, Switalski R, Mistur R, Glodzik L, Pirraglia E, Tsui W, De Santi S, de Leon MJ (2007) Maternal family history of Alzheimer's disease predisposes to reduced brain glucose metabolism. Proc Natl Acad Sci USA 104:19067–19072

Mytilineou C, Werner P, Molinari S et al (1994) Impaired oxidative decarboxylation of pyruvate in fibroblasts from patients with Parkinson's disease. J Neural Transm Park Dis Dement Sect 8:223–228

Newman NJ (1993) Leber's hereditary optic neuropathy. Arch Neurol 50:540–548

Nicklas WJ, Vyas I, Heikkila RE (1985) Inhibition of NADH-linked oxidation in brain mitochondria by 1-methyl-4-phenylpyridine, a metabolite of the neurotoxin, 1-methyl-4-phenyl-1,2,3, 6-tetrahydropyridine. Life Sci 36:2503–2508

Norenberg MD, Rao KV (2007) The mitochondrial permeability transition in neurologic disease. Neurochem Int 50:983–997

Panov AV, Gutekunst CA, Leavitt BR, Hayden MR, Burke JR, Strittmatter WJ, Greenamyre JT (2002) Early mitochondrial calcium defects in Huntington's disease are a direct effect of polyglutamines. Nat Neurosci 5:731–736

Parker WD Jr, Parks JK (2005) Mitochondrial ND5 mutations in idiopathic Parkinson's disease. Biochem Biophys Res Commun 326:667–669

Parker WD, Boyson SJ, Parks JK (1989) Electron transport chain abnormalities in idiopathic Parkinson's disease. Ann Neurol 26:719–723

Parker WD Jr, Boyson SJ, Luder AS, Parks JK (1990a) Evidence for a defect in NADH:ubiquinone oxidoreductase (complex I) in Huntington's disease. Neurology 40:231–1234

Parker WD, Filley CM, Parks JK (1990b) Cytochrome oxidase deficiency in Alzheimer's disease. Neurology 40:1302–1303

Parker WD, Parks JK, Swerdlow RH (2008) Complex I Deficiency in Parkinson's Disease Brain: Studies on Frontal Cortex. Brain Res 1189:215–218

Pereira C, Santos MS, Oliveira C (1998) Mitochondrial function impairment induced by amyloid beta-peptide on PC12 cells. NeuroReport 9:1749–1755

Petit PX, Zamzami N, Vayssiere JL et al (1997) Implication of mitochondria in apoptosis. Mol Cell Biochem 174:185–188

Quastel JH (1932) Biochemistry and mental disorders. Lancet 2:1417–1419

Rajput AH, Uitti RH, Stern W, Laverty W, O'Donnell K, O'Donnell D et al (1987) Geography, drinking water chemistry, pesticides and herbicides and the etiology of Parkinson's disease. Can J Neurol Sci 14:414–418

Reddy PH, Beal MF (2008) Amyloid beta, mitochondrial dysfunction and synaptic damage: implications for cognitive decline in aging and Alzheimer's disease. Trends Mol Med 14:45–53

Reddy PH, McWeeney S, Park BS, Manczak M, Gutala RV, Partovi D, Jung Y, Yau V, Searles R, Mori M, Quinn J (2004) Gene expression profiles of transcripts in amyloid precursor protein transgenic mice: up-regulation of mitochondrial metabolism and apoptotic genes is an early cellular change in Alzheimer's disease. Hum Mol Genet 13:1225–1240

Reiman EM, Caselli RJ, Yun LS et al (1996) Preclinical evidence of Alzheimer's disease in persons homozygous for the epsilon 4 allele for apolipoprotein E. NEJM 334:752–758

Scaglia F, Northrop JL, Kaufmann P, Englestad K, Wei Y, Jhung S, Sano MC, Shungu DC, Millar WS, Hong X, Gooch CL, Mao X, Pascual JM, Hirano M, Stacpoole PW, DiMauro S, De Vivo DC (2006) Dichloroacetate causes toxic neuropathy in MELAS: a randomized, controlled clinical trial. Neurology 66:324–330

Schapira AHV, Cooper JM, Dexter D et al (1989) Mitochondrial complex I deficiency in Parkinson's disease. Lancet 1:1289

Schieke SM, Finkel T (2006) Mitochondrial signaling, TOR, and life span. Biol Chem 387: 1357–1361

Scholte HR (1988) The biochemical basis of mitochondrial diseases. J Bioenerg Biomembr 20: 161–191

Sherer TB, Betarbet R, Stout AK, Lund S, Baptista M, Panov AV, Cookson MR, Greenamyre JT (2002) An in vitro model of Parkinson's disease: linking mitochondrial impairment to altered alpha-synuclein metabolism and oxidative damage. J Neurosci 22:7006–7015

Sherer TB, Kim JH, Betarbet R, Greenamyre JT (2003) Subcutaneous rotenone exposure causes highly selective dopaminergic degeneration and alpha-synuclein aggregation. Exp Neurol 179:9–16

Shoffner JM, Lott MT, Lezza AMS et al (1990) Myoclonic epilepsy and ragged red fibers disease (MERRF) is associated with a mitochondrial DNA tRNALys mutation. Cell 61:931–937

Small GW, Mazziotta JC, Collins MT et al (1995) Apolipoprotein E type 4 allele and cerebral glucose metabolism in relatives at risk for familial Alzheimer disease. JAMA 273:942–947

Smigrodzki R, Parks J, Parker WD (2004) High frequency of mitochondrial complex I mutations in Parkinson's disease and aging. Neurobiol Aging 25:1273–1281

Swerdlow RH (2000) Role of mitochondria in Parkinson's disease. In: Chesselet MF (ed) Molecular mechanisms of neurodegenerative diseases. Humana Press, New Jersey, pp 233–270

Swerdlow RH (2007a) Treating neurodegeneration by modifying mitochondria: potential solutions to a "complex" problem. Antioxid Redox Signal 9:1591–1603

Swerdlow RH (2007b) Is aging part of Alzheimer's disease, or is Alzheimer's disease part of aging? Neurobiol Aging 28:1465–1480

Swerdlow RH (2007c) Mitochondria in cybrids containing mtDNA from persons with mitochondriopathies. J Neurosci Res 85:3416–3428

Swerdlow RH (2007d) Pathogenesis of Alzheimer's disease. Clin Interv Aging 2:347–359

Swerdlow RH (2009) The neurodegenerative mitochondriopathies. JAD 17:737–751

Swerdlow RH, Khan S (2004) A "mitochondrial cascade hypothesis" for sporadic Alzheimer's disease. Med Hypoth 63:8–20

Swerdlow RH, Khan SM (2009) The Alzheimer's disease mitochondrial cascade hypothesis: an update. Exp Neurol 218:308–315

Swerdlow RH, Kish SJ (2002) Mitochondria in Alzheimer's disease. Int Rev Neurobiol 53:341–385

Swerdlow RH, Mitochondrial DNA (2002) and dysfunction in neurodegenerative diseases. Arch Path Lab Med 126:271–280

Swerdlow RH, Parks JK, Miller SW, Tuttle JB, Trimmer PA, Sheehan JP, Bennett JP, Davis RE, Parker WD (1996) Origin and functional consequences of the complex I defect in Parkinson's disease. Ann Neurol 40:663–670

Swerdlow RH, Parks JK, Cassarino DS et al (1997) Cybrids in Alzheimer's disease: a cellular model of the disease? Neurology 49:918–925

Swerdlow RH, Miller SW, Parks JK, Sheehan JP, Cassarino DS, Maguire DJ, Maguire RS, Bennett JP, Juel VC, Phillips LH, Trimmer PA, Pattee G, Tuttle JB, Davis RE, Parker WD (1998) Mitochondria in sporadic amyotrophic lateral sclerosis. Exp Neurol 153:135–142

Swerdlow RH, Parks JK, Pattee G, Parker WD Jr (2000) Role of mitochondria in amyotrophic lateral sclerosis. Amyotroph Lateral Scler Other Motor Neuron Disord 1:185–190

Swerdlow RH, Parker WD, Currie LJ, Bennett JP Jr, Harrison MB, Trugman JM, Wooten GF (2001) Gender ratio differences between Parkinson's disease patients and their affected parents. Parkinsonism Relat Disord 7:47–51

Swerdlow RH, Weaver B, Grawey A, Wenger C, Freed E, Worrall BB (2006) Complex I polymorphisms, bigenomic heterogeneity, and family history in Virginians with Parkinson's disease. J Neurol Sci 247:224–230

Tarnopolsky MA, Beal MF (2001) Potential for creatine and other therapies targeting cellular energy dysfunction in neurological disorders. Ann Neurol 49:561–574

Tsujimoto Y, Nakagwa T, Shimizu S (2006) Mitochondrial membrane permeability transition and cell death. Biochim Biophys Acta 1757:1297–1300

van der Walt JM, Nicodemus KK, Martin ER et al (2003) Mitochondrial polymorphisms significantly reduce the risk of Parkinson disease. Am J Hum Genet 72:804–811

Vielhaber S, Kunz D, Winkler K, Wiedemann FR, Kirches E, Feistner H, Heinze HJ, Elger CE, Schubert W, Kunz WS (2000) Mitochondrial DNA abnormalities in skeletal muscle of patients with sporadic amyotrophic lateral sclerosis. Brain 123:1339–1348

Wallace DC, Singh G, Lott MT et al (1988) Mitochondrial DNA mutation associated with Leber hereditary optic neuropathy. Science 242:1427–1430

Wiedemann FR, Winkler K, Kuznetsov AV et al (1998) Impairment of mitochondrial function in skeletal muscle of patients with amyotrophic lateral sclerosis. J Neurol Sci 156:65–72

Wisniewski H, Terry RD, Hirano A (1970) Neurofibrillary pathology. N Neuropathol Exp Neurol 29:163–176

Wolf PA, Beiser A, Au R, Auerbach S, DeCarli C (2005) Parental occurrence of dementia linked to lower cognitive function in the framingham offspring study. Neurology 64(Suppl 1): A267–A268

Wong PC, Pardo CA, Borchelt DR et al (1995) An adverse property of familial ALS-linked SOD1 mutation causes motor neuron disease characterized by vacuolar degeneration of mitochondria. Neuron 14:1105–1116

Wooten FG, Currie LJ, Bennett JP, Trugman JM, Harrison MB, Parker WD Jr (1997) Maternal inheritance in Parkinson's disease. Ann Neurol 41:265–268

Yao J, Irwin RW, Zhao L, Nilsen J, Hamilton RT, Brinton RD (2009) Mitochondrial bioenergetic deficit precedes Alzheimer's pathology in female mouse model of Alzheimer's disease. Proc Natl Acad Sci USA 106:14670–14675

Zeviani M, Moraes CT, DiMauro S et al (1988) Deletions of mitochondrial DNA in Kearns-Sayre syndrome. Neurology 38:1339–1346

Part IV
In Vivo Assessment of Metabolic Compartmentation Between Neurons and Astrocytes

Chapter 31
Studies of Brain Metabolism: A Historical Perspective

Arne Schousboe

Abstract Understanding of the complicated metabolic processes in the brain related to energy and amino acid neurotransmitter homeostasis is strictly dependent on the advancement of methodological technology both with regard to temporal and spatial resolution. The present review attempts to relate energy and amino acid metabolism at the cellular and inter-cellular level to use of radioactively labeled substrates and later ^{13}C-labeling combined with NMR and LC-MS technologies. The significance of cell culture techniques in this context is also discussed. The complexity of these metabolic processes is illustrated by a discussion of the compartmentation paradigm at the multicellular as well as the single cell level.

Keywords Amino acids • Astrocytes • Compartmentation • Glucose • Glycogen • Neurons • Oxygen

31.1 Introduction

It is well established that although the brain constitutes only 2% of the body weight in man, it accounts for 20% of the oxygen consumption of the body, a percentage that is even higher (50%) in young children (Clarke and Sokoloff 1994). The actual value for the oxygen consumption is about 160 μmol × min^{-1} per 100 g. As the respiratory quotient of the brain is close to unity this would translate into a glucose consumption of about 27 μmol × min^{-1} × 100 g. However, the actual value is higher than this leading to an oxygen/glucose ratio closer to 5 than 6, an observation leading to extensive discussions of the actual reason for this. One notion is that glycogen

A. Schousboe, D.Sc. (✉)
Department of Pharmacology and Pharmacotherapy, Faculty of Pharmaceutical Sciences, University of Copenhagen, Universitetsparken 2, DK-2100 Copenhagen, Denmark
e-mail: as@farma.ku.dk

I.-Y. Choi and R. Gruetter (eds.), *Neural Metabolism In Vivo*, Advances in Neurobiology 4, DOI 10.1007/978-1-4614-1788-0_31,

may play a role and moreover, a possible lactate production will influence this ratio (Clarke and Sokoloff 1994). The general concepts of cerebral metabolism as they appear today are based on classical studies on brain metabolism as summarized in the following.

31.2 Pioneering Studies of Brain Carbohydrate Metabolism

As mentioned above brain metabolism is under normal conditions almost exclusively based on oxidative metabolism of glucose. Therefore, in order to monitor this in the brain *in vivo* it was of great importance to develop reliable methods to determine rates of overall brain oxygen consumption ($CMRO_2$) and glucose utilization (CMR_{Gluc}).

$CMRO_2$ measurements are based on a determination of blood flow and a simultaneous determination of the arterious-venous difference in oxygen concentration in the blood. The measurement of blood flow was based on the use of a freely diffusible inert gas such as nitrous oxide or a radioactive inert gas as described by Kety and Schmidt (1948) and Freygang and Sokoloff (1958). This method is normally referred to as the Kety-Schmidt method and in the advent of modern PET technology it has been replaced by the use of $^{15}O_2$ (see Iversen et al. 2009).

Measurement of the cerebral rate of glucose consumption is based on the use of radioactively labeled (^{14}C) deoxyglucose, a method pioneered by Sokoloff and coworkers in the seventies (Sokoloff et al. 1977). Again, this method has been adapted to be used in PET studies by inserting ^{18}F into the deoxyglucose molecule allowing regional CMR_{Gluc} to be directly visualized (Reivich et al. 1979).

Understanding of regulatory principles in biochemical metabolic pathways does, however, require detailed knowledge about the individual, enzymatic reactions in such pathways. Our current knowledge in this regard is based upon a series of fundamental studies performed by pioneering neurochemists such as K.A.C. Elliott, J.H. Quastel, H.A. Krebs, O.H. Lowry, D.B. Tower, D. Richter, H.McIlwain and H.S. Bachelard. These basic studies of carbohydrate metabolism are comprehensively outlined in a highly useful chapter by Balazs (1970) in the first edition of Handbook of Neurochemistry edited by A. Lajtha.

In relation to glucose metabolism it appears pertinent to note that the glucose polymer, glycogen may be of interest. Compared to muscle and liver, the brain has only a limited glycogen store, which nevertheless could play an important role in brain function. Early studies by Lowry et al. (1964) have clearly shown that glycogen breakdown is a very rapid process. Moreover, its biosynthesis was also shown to be quite rapid (LeBaron 1955) suggesting a fast turnover. This notion may be in contrast to newer studies of glycogen turnover performed using ^{13}C-labeled glucose and NMR technology (Choi and Gruetter 2003). However, studies in isolated astrocytes which constitute the glycogen – containing cells in the brain (Cataldo and Broadwell 1986; Wender et al. 2000) have indicated a rapid turnover (Sickmann et al. 2005). The notion of glycogen as a dynamic molecule involved in the maintenance of neurotransmission processes is supported by studies in the optic nerve (Brown et al. 2005) as well as astrocytes (Walls et al. 2008, 2009) and the brain (Swanson 1992; Gibbs et al. 2006, 2008). The availability of pharmacological tools to specifically inhibit the glycogen phosphorylase, the enzyme responsible for mobilization of glucose-1-phosphate from glycogen has been of instrumental importance for a better understanding of a possible functional role of glycogen in relation to neurotransmission (Walls et al. 2008, 2009; Suh et al. 2007; Dienel et al. 2007; Sickmann et al. 2009). Based on these studies it has become evident that shunting of glucose through glycogen is crucial to maintain glutamatergic neurotransmission (Schousboe et al. 2010).

31.3 Pionering Studies of Amino Acid Metabolism

Glucose metabolism and that of neuroactive amino acids are closely associated *via* the tricarboxylic acid (TCA) cycle in combination with the high activity of aspartate aminotransferase catalyzing exchange between α-ketoglutarate/glutamate and oxaloacetate/aspartate, respectively (e.g. Waagepetersen et al. 2007) Thus, using radioactively labeled glucose it was very early on demonstrated that label could be rapidly found in neuronal glutamate (see Shank and Campbell 1983). This was originally taken as evidence that glucose serves as a precursor for glutamate. The subsequent finding that pyruvate carboxylase, the primary enzyme responsible for carboxylation and thus anaplerosis in the brain (Patel 1974) is only expressed in astrocytes (Yu et al. 1983; Shank et al. 1985) shows that these cells are obligatory for *de novo* synthesis of glutamate, a reaction requiring an anaplerotic process to replenish α-ketoglutarate after removal from the TCA cycle in the form of glutamate.

Pionering studies performed in the laboratories of H. Waelsch and J.H. Quastel led in the sixties to the notion that glutamate metabolism in the brain had to be extremely complex (e.g. Berl and Clarke 1969; Quastel 1975). Thus using radioactively labeled precursors such as glucose, aspartate, leucine, acetate and HCO_3^- it could be shown that while the specific radioactivity of glutamine did not exceed that of its precursor glutamate when glucose was the substrate, it always had a higher specific activity than glutamate in case of all the other substrates mentioned (for references see Berl and Clarke 1969; Quastel 1975). Interestingly, one study performed by Roberts et al. (1958) noted that when [^{14}C]GABA was given intracerebrally to mice, glutamine had a higher specific radioactivity than its precursor glutamate. However, these results were not associated with a compartmentalized glutamate metabolism. It must, however, be emphasized that subsequent studies by Roberts who originally discovered GABA in the brain (Roberts and Frankel 1950) and his co-workers have been instrumental for the current concept of GABA acting as the major inhibitory neurotransmitter in the brain (see Roberts 2007a). Altogether these fundamental studies led to the concept of metabolic compartmentation of glutamate metabolism in the brain with at least two compartments (Garfinkel 1966; Berl et al. 1968; van den Berg and Garfinkel 1971; Patel et al. 1974). Although these compartments were originally thought to represent nerve endings and neuronal cell bodies and possibly other cellular structures (Balazs et al. 1973) it was eventually realized that this reflected neurons and astrocytes, a notion compatible with the demonstration that anaplerosis by pyruvate carboxylase and glutamine synthesis by glutamine synthetase must be glial phenomena due to the exclusive localization of these enzymes in glia (Norenberg and Martinez-Hernandez 1979; Yu et al. 1983; Shank et al. 1985).

31.4 Metabolic Studies in Neurons and Astrocytes

Studies of metabolism related to oxygen consumption and energy turnover, glucose utilization and neuroactive amino acids in the brain *in vivo* or in brain tissue preparations with intact neuronal networks such as brain slices have provided extremely useful information about the relationship between brain functional activity (e.g. synaptic transmission) and energy requirements (see Attwell and Laughlin 2001). Such studies have, however, provided only indirect knowledge about the contribution of individual cell types such as neurons and astrocytes to these entities. Methods developed by S.P.R. Rose (Rose 1965, 1967, 1968) and A. Hamberger (Blomstrand and Hamberger 1969, 1970; Hamberger 1971) to isolate using differential centrifugation techniques viable neuronal perikarya and glial cells, facilitated greatly metabolic studies which showed characteristic properties of the individual cell types with regard to amino acid metabolism and transport. Thus, these preparations allowed the controversial conclusion to be made that glial cells exhibited high affinity transport of the neurotransmitter amino acids glutamate and GABA (Henn and Hamberger 1971). The differential centrifugation technique was also used by V. Whittaker (Gray and Whittaker 1960, 1962; Whittaker 1969) to prepare isolated nerve endings (synaptosomes) and synaptic vesicles. These preparations have been and are still extensively used to study processes such as transmitter biosynthesis, release and uptake.

While these preparations of cells and organelles have been instrumental for the understanding of basic metabolic functions at the cellular level it must be realized that the bulk-prepared neuronal and glial cells did not represent intact cells although many biochemical and physiological processes could be investigated (Hamberger and Sellström 1975). It was therefore of considerable interest to develop methods to maintain intact neurons and astrocytes in primary cell culture systems. Booher and Sensenbrenner (1972) succeeded in producing such cultures of astrocytes from neonatal brain cortical tissue and these cultures could subsequently be shown to exhibit a developmental pattern *in vitro* similar to that expected from astroglial cells *in vivo* (Schousboe et al. 1980). Moreover, these cultured astrocytes expressed high affinity transport systems for glutamate and GABA (Schousboe et al. 1977a, b; Hertz et al. 1978). During the next decades such cultures have proven extremely useful with regard to extensive studies of astrocytic metabolism (see below).

With regard to cell cultures of neurons, pioneering studies of Dichter (1978, 1980) and Messer (1977) led to establishment of primary cultures of cerebral cortical and cerebellar neurons, respectively. These culture systems subsequently have been utilized to generate a wealth of information concerning metabolic processes pertinent to GABAergic and glutamatergic neurotransmission (see below).

31.5 Metabolic Studies in Cultured Neurons and Astrocytes

Preparation of primary cultures of neurons and astrocytes as well as their basic properties have been summarized by Hertz et al. (1985). The progress in NMR technology with regard to better scanners with larger magnets provided the possibility of using this methodology in combination with ^{13}C-labeling to initiate studies of metabolism of glucose, lactate and neuroactive amino acids in cultured neurons and astrocytes. One of the pioneering studies led to the novel observation that astrocytes but not neurons could synthesize and release large amounts of citrate produced from glucose (Sonnewald et al. 1991). This indirectly confirmed that pyruvate carboxylation is an astrocyte specific reaction. It should, however, be mentioned that studies of Hassel and Brathe (2000a, b) in the brain as well as cultured cerebellar neurons using ^{14}C-labeled pyruvate and bicarbonate to study glutamate labeling were interpreted in favor of a neuronal pyruvate carboxylation likely occurring by the action of malic enzyme, the mitochondrial isoform of which is located in neurons (Vogel et al. 1998). That this issue is highly controversial is indicated by the fact that subsequent studies in cultured neurons and astrocytes utilizing 3-nitropropionic acid (3-NPA), a specific inhibitor of succinate dehydrogenase and hence TCA cycle activity (McKenna et al. 2006) could demonstrate by ^{13}C-labeling patterns of glutamate subsequent to incubation with [U-^{13}C]lactate that pyruvate carboxylation exclusively takes place in astrocytes and not in neurons (Waagepetersen et al. 2001b; Sonnewald et al. 2007).

Another controversial issue concerns possible differences between neurons and astrocytes with regard to oxidative metabolism. A comparison of TCA cycle activity estimated by modelling of experimental data from early studies of metabolic compartmentation (Garfinkel 1966; van den Berg and Garfinkel 1971) and recent studies of metabolism of ^{13}C-labeled glucose or acetate in rat or human brain (Preece and Cerdan 1996; Mason et al. 1992, 1995; Sibson et al. 1997; Shen et al. 1999; Lebon et al. 2002; Bluml et al. 2002) have resulted in the notion that neuronal TCA cycle flux and hence oxidative metabolism is much higher than that in astrocytes (Rodrigues and Cerdan 2007). Comparison of oxygen consumption in preparations of isolated neurons and astrocytes as summarized by Hertz and Schousboe (1975) shows, however, quite clearly that these two cell types have comparable oxidative metabolism, a notion strengthened in a recent review by Hertz et al. (2007).

The hypothesis put forward by Pellerin and Magistretti (1994) that glutamatergic activity is coupled to anaerobic glycolysis, i.e. conversion of glucose to lactate in astrocytes leading to transfer of lactate from these cells to neurons, a model named the astrocyte to neuron lactate shuttle hypothesis (ANLSH) further supported the view that neuronal oxidative metabolism is superior to that in astrocytes. Based on subsequent studies in different types of neurons and astrocytes in culture as well as modelling of *in vivo* data (see reviews by Hyder et al. 2006; Hertz and Zielke 2004; Hertz et al. 2007; Riera et al. 2008) it seems accepted that astrocytes do indeed have a pronounced oxidative metabolism. A study of the transcriptome in neurons and astrocytes separated from adult mouse brain using fluorescence or magnetic associated cell sorting technology (Lovatt et al. 2007) clearly has underlined that this is indeed the case. It should be noted in particular that even the fine processes of astrocytes contain mitochondria (Lovatt et al. 2007). A key element in the ANLSH is the ability of cells to oxidize lactate to pyruvate catalyzed by lactate dehydrogenase. It must be kept in mind that thermodynamically this reaction is highly favored in the direction towards lactate production (Williamson et al. 1967). It has been argued that since the LDH-1 isoenzyme is most prevalently expressed in neurons this would facilitate lactate utilization (Bittar et al. 1996). Even though LDH-1 has a lower K_m for its substrates compared to LDH-5 which has a higher expression in astrocytes (Bittar et al. 1996) it must be kept in mind that it is also more sensitive to inhibition by these substrates, i.e. in this case pyruvate inhibition (see Roberts 2007b). This means that when pyruvate increases from its average concentration close to the K_i value, conversion of lactate to pyruvate will be inhibited. Considering this together with the thermodynamics of the LDH reaction, there is no reason to believe that lactate should be a preferred substrate for neurons particularly not when glucose is available. Studies of the use of lactate and glucose as oxidative substrates to maintain activity in glutamatergic neurons in culture have supported this notion (Bak et al. 2006). Moreover, a recent study by Ghandi et al. (2009) has clearly demonstrated that lactate is transferred from neurons to astrocytes more vividly than in the opposite direction.

Another pathway which clearly is present in the brain (Cerdan et al. 1990) is pyruvate recycling, i.e. the series of reactions in which malate is converted to pyruvate which can enter the TCA cycle after oxidation to acetyl CoA. This requires malic enzyme to be expressed. Alternatively, oxaloacetate can be converted to phosphoenolpyruvate (PEP) by the action of PEP-carboxykinase (PEPCK) followed by conversion of PEP to pyruvate catalyzed by pyruvate kinase. This reaction was originally thought to take place in a compartment devoid of glutamine synthetase, i.e. being neuronal (Cerdan et al. 1990) since no ^{13}C-labeled glutamine could be detected after injection of [1,2-$^{13}C_2$]acetate. However, subsequent studies in cultured neurons and astrocytes have provided evidence that it occurs in astrocytes (Sonnewald et al. 1996; Waagepetersen et al. 2002). A recent study has shown that this process may also take place in neurons (Olstad et al. 2007).

The use of ^{13}C-labeling in combination with NMR and LC-MS technology in cultured neurons and astrocytes has been of instrumental importance in providing new insight into the metabolism and biosynthesis of neuroactive amino acids at the cellular level as well as the understanding of exchange of metabolites between neurons and astrocytes (for references see Hertz et al. 2000; Waagepetersen et al. 2007).

These studies have provided evidence that the glutamate-glutamine cycle is unlikely to operate in a stoichiometric manner since glutamate taken up into astrocytes is oxidatively metabolized (Westergaard et al. 1995; Sonnewald et al. 1997). Moreover, it has become clear that GABA biosynthesis requires participation of the TCA cycle, a notion not previously considered (Waagepetersen et al. 1999, 2001a). Additionally, such studies have led to an expansion of the concept of metabolic compartmentation emphasizing that this may even occur at the single cell level (Waagepetersen et al. 2003). This is at least partly due to the discovery of mitochondrial heterogeneity at the cellular level (Sonnewald et al. 1998; Waagepetersen et al. 2006).

Preparations of cultured neurons and astrocytes have also been of importance for the understanding of the homeostatic mechanisms regulating transfer of glutamate and GABA between neurons and astrocytes in relation to termination of the neurotransmitter action of these amino acids (see Schousboe et al. 2004). Hence, such studies provided evidence of a preferential astrocytic glutamate uptake decades before cloning of the pertinent transporters (Schousboe et al. 1977b; Hertz et al. 1978).

31.6 Concluding Remarks

This short historical review was intended to give some insight into the significance of development of methods to provide reliable brain tissue preparations to study events at the cellular level. This combined with advances in analytical technology both for brain imaging (PET) and the use of ^{13}C-labeling (NMR and LC-MS) has led us to realize that metabolic processes in the brain are highly complex and that communication between neurons and astrocytes with regard to exchange of metabolites is instrumental for brain function.

Acknowledgements The secretarial assistance of Ms Hanne Danø is cordially acknowledged.

References

Attwell D, Laughlin SB (2001) An energy budget for signalling in the grey matter of the brain. J Cereb Blood Flow Metab 21:1133–1145

Bak LK, Schousboe A, Sonnewald U, Waagepetersen HS (2006) Glucose is necessary to maintain neurotransmitter homeostasis during synaptic activity in cultured glutamatergic neurons. J Cereb Blood Flow Metab 26:1285–1297

Balazs R (1970) Carbohydrate metabolism. In: Lajtha A (ed) Handbook of neurochemistry. Plenum Press, New York, pp 1–36

Balazs R, Machiyama Y, Patel AJ (1973) Compartmentation and the metabolism of γ-aminobutyrate. In: Balazs R, Cremer JE (eds) Metabolic compartmentation in the brain. McMillan Press, Ltd, London, pp 57–70

Berl S, Clarke DD (1969) Compartmentation of amino acid metabolism. In: Lajtha A (ed) Handbook of neurochemistry. Plenum Press, New York, pp 447–472

Berl S, Nicklas WJ, Clarke DD (1968) Compartmentation of glutamic acid metabolism in brain slices. J Neurochem 15:131–140

Bittar PG, Charnay Y, Pellerin L, Bouras G, Magistretti PJ (1996) Selective distribution of lactate dehydrogenase isoenzymes in neurons and astrocytes of human brain. J Cereb Blood Flow Metab 16:1079–1089

Blomstrand C, Hamberger A (1969) Protein turnover in cell-enriched fractions from rabbit brain. J Neurochem 16:1401–1407

Blomstrand C, Hamberger A (1970) Amino acid incorporation *in vitro* in proteins of neuronal and glial cell enriched fractions. J Neurochem 17:1187–1195

Bluml S, Moreno-Torres A, Shic F, Nguy CH, Ross BD (2002) Tricarboxylic acid cycle of glia in the in vivo human brain. NMR Biomed 15:1–5

Booher J, Sensenbrenner M (1972) Growth and cultivation of dissociated neurons and glial cells from embryonic chick, rat and human brain in flask cultures. Neurobiology 2:97–105

Brown AM, Sickmann HM, Fosgerau K, Lund TM, Schousboe A, Waagepetersen HS, Ransom BR (2005) Astrocyte glycogen metabolism is required for neural activity during aglycaemia or intense stimulation in mouse white matter. J Neurosci Res 79:64–80

Cataldo AM, Broadwell RD (1986) Cytochemical identification of cerebral glycogen and glucose-6-phosphatase activity under normal and experimental conditions. II. Choroid plexus and ependymal epithelia, endothelia and pericytes. J Neurocytol 15:511–524

Cerdan S, Kunnecke B, Seelig J (1990) Cerebral metabolism of [1,2-$^{13}C_2$]acetate as detected by in vivo and in vitro ^{13}C NMR. J Biol Chem 265:12916–12926

Choi IY, Gruetter R (2003) In vivo ^{13}C NMR assessment of brain glycogen concentration and turnover in the awake rat. Neurochem Int 43:317–322

Clarke DD, Sokoloff L (1994) Circulation and energy metabolism of the brain. In: Siegel GJ, Agranoff BW, Albers RW, Molinoff PB (eds) Basic neurochemistry. Raven, New York, pp 645–680

Dichter MA (1978) Rat cortical-neurons in cell culture-culture methods, cell methods, cell morphology, electrophysiology and synapse formation. Brain Res 149:279–293

Dichter MA (1980) Physiological identification of GABA as the inhibitory transmitter for mammalian cortical neurons in cell culture. Brain Res 190:111–121

Dienel GA, Ball KK, Cruz NF (2007) A glycogen phosphorylase inhibitor selectively enhances local rates of glucose utilization in brain during sensory stimulation of conscious rats: implications for glycogen turnover. J Neurochem 102:466–478

Freygang WH, Sokoloff L (1958) Quantitative measurements of regional circulation in the central nervous system by the use of radioactive inert gas. Adv Biol Med Phys 6:263–279

Gandhi GK, Cruz NF, Ball KK, Dienel GA (2009) Astrocytes are poised for lactate trafficking and release from activated brain and for supply of glucose to neurons. J Neurochem 111: 522–536

Garfinkel D (1966) A simulation study of the metabolism and compartmentation in brain of glutamate, aspartate, Krebs cycle and related metabolites. J Biol Chem 241:3918–3929

Gibbs ME, O'Dowd BS, Hertz E, Hertz L (2006) Astrocyte energy metabolism consolidates memory in young chicks. Neuroscience 141:9–13

Gibbs ME, Hutchinson D, Hertz L (2008) Astrocytic involvement in learning and memory consolidation. Neurosci Biobehav Rev 32:927–944

Gray EG, Whittaker VP (1960) The isolation of synaptic vesicles from the central nervous system. J Physiol 153:35–37P

Gray EG, Whittaker VP (1962) The isolation of nerve endings from brain: an electron-microscopic study of cell fragments derived by homogenization and centrifugation. J Anat (Lond) 96:79–88

Hamberger A (1971) Amino acid uptake in neuronal and glial cell fractions from rabbit cortex. Brain Res 31:169–178

Hamberger A, Sellström A (1975) Techniques for separation of neurons and glia and their application to metabolic studies. In: Berl S, Clarke DD, Schneider D (eds) Metabolic compartmentation and neurotransmission. Plenum Publishing Corp, New York, pp 145–165

Hassel B, Brathe A (2000a) Neuronal pyruvate carboxylation supports formation of transmitter glutamate. J Neurosci 20:1342–1347

Hassel B, Brathe A (2000b) Cerebral metabolism of lactate in vivo: evidence for neuronal pyruvate and carboxylation. J Cereb Blood Flow Metab 20:327–336

Henn F, Hamberger A (1971) Glial cell function: uptake of transmitter substances. Proc Natl Acad Sci USA 68:2686–2690

Hertz L, Schousboe A (1975) Ion and energy metabolism of the brain at the cellular level. Int Rev Neurobiol 18:141–211

Hertz L, Zielke HR (2004) Astrocytic control of glutamatergic activity: astrocytes as stars of the show. Trends Neurosci 27:735–743

Hertz L, Schousboe A, Boechler N, Mukerji S, Fedoroff S (1978) Kinetic characteristics of the glutamate uptake into normal astrocytes in cultures. Neurochem Res 3:1–14

Hertz L, Juurlink BHJ, Szuchet S (1985) Cell cultures. In: Lajtha A (ed) Handbook of neurochemistry, 2nd edn. Plenum Press, New York, pp 603–661

Hertz L, Yu AC, Kala G, Schousboe A (2000) Neuronal-astrocytic and cytosolic-mitochondrial metabolite trafficking during brain activation, hyperammonemia and energy deprivation. Neurochem Int 37:83–102

Hertz L, Peng L, Dienel GA (2007) Energy metabolism in astrocytes: high rate of oxidative metabolism and spatiotemporal dependence on glycolysis/glycogenolysis. J Cereb Blood Flow Metab 27:129–149

Hyder F, Patel AB, Gjedde A, Rothman DL, Behar KB, Shulman RG (2006) Neuronal-glial glucose oxidation and glutamatergic-GABAergic function. J Cereb Blood Flow Metab 27:865–877

Iversen P, Sørensen M, Bak LK, Waagepetersen HS, Vafaee MS, Borghammer P, Mouridsen K, Jensen SB, Vilstrup H, Schousboe A, Ott P, Gjedde A, Keiding S (2009) Low cerebral oxygen consumption and blood flow in patients with cirrhosis and an acute episode of hepatic encephalopathy. Gastroenterology 136:863–871

Kety SS, Schmidt CF (1948) The nitrous oxide method for the quantitative determination of cerebral blood flow in man: theory procedure, and normal values. J Clin Invest 27:476–483

LeBaron FN (1955) The resynthesis of glycogen by guinea-pig cerebral cortex slices. Biochem J 61:80–85

Lebon V, Petersen KF, Cline GW, Shen J, Mason GF et al (2002) Astroglial contribution to brain energy metabolism in humans revealed by ^{13}C nuclear magnetic resonance spectroscopy: elucidation of the dominant pathway for neurotransmitter glutamate repletion and measurement of astrocytic oxidative metabolism. J Neurosci 22:1523–1531

Lovatt D, Sonnewald U, Waagepetersen HS, Schousboe A, He W, Lin JH, Han X, Takano T, Wang S, Sim FJ, Goldman SA, Nedergaard M (2007) The transcriptome and metabolic gene signature of protoplasmic astrocytes in the adult murine cortex. J Neurosci 27:12255–12266

Lowry OH, Passonneau JV, Hasselberger FX, Schulz DW (1964) Effect of ischaemia on known substrates and cofactors of the glycolytic pathway in brain. J Biol Chem 239:18–30

Mason GF, Rothman DL, Behar KL, Shulman RG (1992) NMR determination of the TCA cycle rate and alphaketoglutarate/glutamate exchange rate in rat brain. J Cereb Blood Flow Metab 12:434–447

Mason GF, Gruetter R, Rothman DL, Behar KL, Schulman RG, Novotny EJ (1995) Simultaneous determination of the rates of the TCA cycle, glucose utilization, alpha-ketoglutarate/glutamate exchange, and glutamine synthesis in human brain by NMR. J Cereb Blood Flow Metab 15:12–25

McKenna MC, Waagepetersen HS, Schousboe A, Sonnewald U (2006) Neuronal and astrocytic shuttle mechanisms for cytosolic-mitochondrial transfer of reducing equivalents: current evidence and pharmacological tools. Biochem Pharmacol 71:399–407

Messer A (1977) The maintenance and identification of mouse cerebellar granule cells in mono layer culture. Brain Res 130:1–12

Norenberg MD, Martinez-Hernandez A (1979) Fine structural localization of glutamine synthetase in astrocytes of rat brain. Brain Res 161:303–310

Olstad E, Olsen GM, Qu H, Sonnewald U (2007) Pyruvate recycling in cultured neurones from cerebellum. J Neurosci Res 85:3318–3325

Patel MS (1974) The effect of ketone bodies on pyruvate carboxylation by rat brain mitochondria. J Neurochem 23:865–867

Patel AJ, Johnson AL, Balazs R (1974) Metabolic compartmentation of glutamate associated with the formation of γ-aminobutyrate. J Neurochem 23:1271–1279

Pellerin L, Magistretti PJ (1994) Glutamate uptake into astrocytes stimulates aerobic glycolysis: a mechanism coupling neuronal activity to glucose utilization. Proc Natl Acad Sci USA 15:10625–10629

Preece NE, Cerdan S (1996) Metabolic precursors and compartmentation of cerebral GABA in vigabatrin-treated rats. J Neurochem 67:1718–1725

Quastel JH (1975) Metabolic compartmentation in the brain and effects of metabolic inhibitors. In: Berl S, Clarke DD, Schneider D (eds) Metabolic compartmentation and neurotransmission. Plenum Press, New York, pp 337–361

Reivich M, Kuhl D, Wolf A, Greenberg J, Phelps M, Ido T, Casella V, Fowler J, Hoffman E, Alavi A, Som P, Sokoloff L (1979) The [^{18}F]fluoro-deoxyglucose method for the measurement of local cerebral glucose utilization in man. Circ Res 44:127–137

Riera JJ, Schousboe A, Waagepetersen HS, Howarth C, Hyder F (2008) The micro-architecture of the cerebral cortex and its impact on functional neuroimaging. I-Metabolic demand. Neuroimage 40:1436–1459

Roberts E (2007a) Gamma-aminobutyric acid. Scholarpedia 2. doi:10.4249/scholarpedia.3356

Roberts EL Jr (2007b) The support of energy metabolism in the central nervous system with substrates other than glucose. In: Gibson GE, Dienel GA (eds) Handbook of neurochemistry and molecular neurobiology, 3rd edn. Springer Science, New York, pp 137–179

Roberts, E and Frankel S (1950) Gamma-aminobutyric acid in brain. Its formation from glutamic acid. J. Biol. Chem. 187:55–63

Roberts E, Rothstein M, Baxter CF (1958) Some metabolic studies of γ-aminobutyric acid. Proc Soc Exp Biol Med 97:796–802

Rodrigues TB, Cerdan S (2007) The cerebral tricarboxylic acid cycles. In: Gibson GE, Dienel GA (eds) Handbook of neurochemistry and molecular neurobiology. Springer, Berlin/Heidelberg, pp 63–91

Rose SPR (1965) Preparation of enriched fractions from cerebral cortex containing isolated metabolically active neuronal cells. Nature (Lond) 208:621–622

Rose SPR (1967) Preparation of enriched fractions from cerebral cortex containing isolated, metabolically active, neuronal and glial cells. Biochem J 102:33–43

Rose SPR (1968) Glucose and amino acid metabolism in isolated neuronal and glial cell fractions *in vitro*. J Neurochem 15:1415–1429

Schousboe A, Hertz L, Svenneby G (1977a) Uptake and metabolism of GABA in astrocytes cultured from dissociated mouse brain hemispheres. Neurochem Res 2:217–229

Schousboe A, Svenneby G, Hertz L (1977b) Uptake and metabolism of glutamate in astrocytes cultured from dissociated mouse brain hemispheres. J Neurochem 29:999–1005

Schousboe A, Nissen C, Bock E, Sapirstein VS, Juurlink BJH, Hertz L (1980) Biochemical development of rodent astrocytes in culture. In: Giacobini E, Vernadakis A, Shahar A (eds) Tissue culture in neurobiology. Raven, New York, pp 397–409

Schousboe A, Sarup A, Bak LK, Waagepetersen HS, Larsson OM (2004) Role of astrocytic transport processes in glutamatergic and GABAergic neurotransmission. Neurochem Int 45:512–527

Schousboe A, Sickmann HM, Walls AB, Bak LK, Waagepetersen HS (2010) Functional importance of the astrocytic glycogen-shunt and glycolysis for maintenance of an intact intra/extracellular glutamate gradient. Neurotox Res 18(1):94–99

Shank RP, Campbell GLeM (1983) Glutamate. In: Lajtha A (ed) Handbook of neurochemistry, 2nd edn. Plenum Press, New York, pp 381–404

Shank RP, Bennett GS, Freytag SO, Campbell GL (1985) Pyruvate carboxylase: astrocyte-specific enzyme implicated in the replenishment of amino acid neurotransmitter pools. Brain Res 329:364–367

Shen J, Petersen KF, Behar KL, Brown P, Nixon TW, Mason GF, Petroff OA, Shulman GI, Shulman RG, Rothman DL (1999) Determination of the rate of the glutamate/glutamine cycle in the human brain by in vivo ^{13}C NMR. Proc Natl Acad Sci USA 96:8235–8240

Sibson NR, Dhankhar A, Mason GF, Behar KL, Rothman DL, Shulman RG (1997) In vivo ^{13}C NMR measurements of cerebral glutamine synthesis as evidence for glutamate-glutamine cycling. Proc Natl Acad Sci USA 94:2699–2704

Sickmann HM, Schousboe A, Fosgerau K, Waagepetersen HS (2005) Compartmentation of lactate originating from glycogen and glucose in cultured astrocytes. Neurochem Res 30:1295–1304

Sickmann HM, Walls AB, Schousboe A, Bouman SD, Waagepetersen HS (2009) Functional significance of brain glycogen in sustaining glutamatergic neurotransmission. J Neurochem 109(suppl 1):80–86

Sokoloff L, Reivich M, Kennedy C, Des Rosiers MH, Patlak CS, Pettigrew KD, Sakurada O, Shinohara M (1977) The [^{14}C]deoxyglucose method for the measurement of local cerebral glucose utilization: theory, procedure, and normal values in the conscious and anesthetized albino rat. J Neurochem 28:897–916

Sonnewald U, Westergaard N, Krane J, Unsgård G, Petersen SB, Schousboe A (1991) First direct demonstration of preferential release of citrate from astrocytes using [^{13}C]NMR spectroscopy of cultured neurons and astrocytes. Neurosci Lett 128:235–239

Sonnewald U, Westergaard N, Jones P, Taylor A, Bachelard HS, Schousboe A (1996) Metabolism of [U-$^{13}C_5$]glutamine in cultured astrocytes studied by NMR spectroscopy: first evidence of astrocytic pyruvate recycling. J Neurochem 67:2566–2572

Sonnewald U, Westergaard N, Schousboe A (1997) Glutamate transport and metabolism in astrocytes. Glia 21:56–63

Sonnewald U, Hertz L, Schousboe A (1998) Mitochondrial heterogeneity in the brain at the cellular level. J Cereb Blood Flow Metab 18:231–237

Sonnewald U, Syversen T, Schousboe A, Waagepetersen HS, Aschner M (2007) Actions of toxins on cerebral metabolism at cellular level. In: Lajtha A, Gibson GE, Dienel GA (eds) Handbook of neurochemistry and molecular neurobiology: brain energetics. Integration of molecular and cellular processes, 3rd edn. Springer, Berlin/Heidelberg, pp 569–585

Suh SW, Bergher JP, Anderson CM, Treadway JL, Fosgerau K, Swanson RA (2007) Astrocyte glycogen sustains neuronal activity during hypoglycemia: studies with the glycogen phosphorylase inhibitor CP-316,819 ([R-R*, S*]-5-chloro-N-[2-hydroxy-3-(methoxymethylamino)-3-oxo-1-(phenylmethyl)propyl]-1H-indole-2-carboxamide). J Pharmacol Exp Ther 321:45–50

Swanson RA (1992) Physiologic coupling of glial glycogen metabolism to neuronal activity in brain. Can J Physiol Pharmacol 70(Suppl):S138–S144

Van den Berg CJ, Garfinkel D (1971) A stimulation study of brain compartments. Metabolism of glutamate and related substances in mouse brain. Biochem J 123:211–218

Vogel R, Jennemann G, Seitz J, Wiesinger H, Hamprecht B (1998) Mitochondrial malic enzyme: purification from bovine brain, generation of an antiserum, and immunocytochemical localization in neurons of rat brain. J Neurochem 71:844–852

Waagepetersen HS, Sonnewald U, Larsson OM, Schousboe A (1999) Synthesis of vesicular GABA from glutamine involves TCA cycle metabolism in neocortical neurons. J Neurosci Res 57:342–349

Waagepetersen HS, Sonnewald U, Gegelashvili G, Larsson OM, Schousboe A (2001a) Metabolic distinction between vesicular and cytosolic GABA in cultured GABAergic neurons using ^{13}C MRS. J Neurosci Res 63:347–355

Waagepetersen HS, Qu H, Schousboe A, Sonnewald U (2001b) Elucidation of the quantitative significance of pyruvate carboxylation in cultured cerebellar neurons and astrocytes. J Neurosci Res 66:763–770

Waagepetersen HS, Qu H, Hertz L, Sonnewald U, Schousboe A (2002) Demonstration of pyruvate recycling in primary cultures of neocortical astrocytes but not in neurons. Neurochem Res 27:1431–1437

Waagepetersen HS, Sonnewald U, Schousboe A (2003) Compartmentation of glutamine, glutamate and GABA metabolism in neurons and astrocytes: functional implications. Neuroscientist 9:398–403

Waagepetersen HS, Hansen GH, Fenger K, Lindsay JG, Gibson G, Schousboe A (2006) Cellular mitochondrial heterogeneity in cultured astrocytes as demonstrated by immunogold labeling of α-ketoglutarate dehydrogenase. Glia 53:225–231

Waagepetersen HS, Sonnewald U, Schousboe A (2007) Glutamine, glutamate and GABA: metabolic aspects. In: Lajtha A, Oja SS, Schousboe A, Saransaari P (eds) Handbook of neurochemistry and molecular neurobiology, 3rd edn, Amino acids and peptides in the nervous system. Springer, Berlin/Heidelberg, pp 1–21

Walls AB, Sickmann HM, Brown A, Bouman SD, Ransom B, Schousboe SA, Waagepetersen HS (2008) Characterization of 1,4-dideoxy-1,4-imino-D-arabinitol (DAB) as an inhibitor of brain glycogen shunt activity. J Neurochem 105:1462–1470

Walls AB, Heimbürger CM, Bouman SD, Schousboe A, Waagepetersen HS (2009) Robust glycogen shunt activity in astrocytes: effects of glutamatergic and adrenergic agents. Neuroscience 158:284–292

Wender R, Brown AM, Fern R, Swanson RA, Farrell K, Ransom BR (2000) Astrocytic glycogen influences axon function and survival during glucose deprivation in central white matter. J Neurosci 20:6804–6810

Westergaard N, Sonnewald U, Schousboe A (1995) Metabolic trafficking between neurons and astrocytes: the glutamate/glutamine cycle revisited. Dev Neurosci 17:203–211

Whittaker VP (1969) The synaptosome. In: Lajtha A (ed) Handbook of neurochemistry. Plenum Press, New York, pp 327–364

Williamson DHP, Lund P, Krebs HA (1967) The redox state of free nicotinamide-adenine dinucleotide in the cytoplasm and mitochondria of rat liver. Biochem J 103:514–527

Yu ACH, Drejer J, Hertz L, Schousboe A (1983) Pyruvate carboxylase activity in primary cultures of astrocytes and neurons. J Neurochem 41:1484–1487

Chapter 32
Glutamate-Glutamine Cycle and Anaplerosis

Gülin Öz, David A. Okar, and Pierre-Gilles Henry

Abstract Most excitatory neuronal activity in the brain is sustained by cycling the most prominent neurotransmitter, glutamate, and its metabolic product glutamine between neurons and astrocytes. Losses of these compounds during cycling are replenished by anaplerosis that primarily takes place in astrocytes. This chapter summarizes the methodological considerations in carbon tracer studies used to measure glutamate-glutamine cycling and anaplerosis, together with the estimated rates for these metabolic fluxes in the living brain. Questions regarding neuronal anaplerosis and the stoichiometry of the glutamate-glutamine cycle are examined. The possibility of a functional link between anaplerosis and the glutamate-glutamine cycle is discussed.

Keywords ^{13}C NMR spectroscopy • ^{14}C • Anaplerosis • Cataplerosis • Glucose • Glutamate-glutamine cycle • Neurotransmission • Synapse • TCA cycle

G. Öz, Ph.D. (✉) • P.-G. Henry, Ph.D.
Center for Magnetic Resonance Research, Department of Radiology, University of Minnesota Medical School, 2021 6th St. SE., Minneapolis, MN 55455, USA
e-mail: gulin@cmrr.umn.edu; henry@cmrr.umn.edu

D.A. Okar, Ph.D.
Department of Veterans Affairs Medical Center, Department of Biochemistry, Molecular Biology and Biophysics, University of Minnesota, 1 Veterans Drive, Minneapolis, MN 55417, USA
e-mail: David.Okar@va.gov

I.-Y. Choi and R. Gruetter (eds.), *Neural Metabolism In Vivo*, Advances in Neurobiology 4, DOI 10.1007/978-1-4614-1788-0_32,

32.1 Introduction

32.1.1 Glutamate and Metabolic Compartmentation

Despite its relatively small size (~2% of body weight), the human brain is the major consumer of oxygen (~20%) in the body (Rolfe and Brown 1997). The primary metabolic fuel for the brain is systemic glucose, whether derived from the diet, or synthesized in the liver and kidney. Not surprisingly, more than 50% of the energy produced in the brain is devoted to maintaining its defining function: Neurotransmission. This includes maintenance of resting membrane potentials, generation of action potentials, release of neurotransmitters, their postsynaptic action, reuptake, and refilling of the presynaptic vesicles, each of which requires energy (Attwell and Laughlin 2001; Lennie 2003). Neurotransmitters are a diverse family of compounds, not only on the basis of molecular size and chemical composition, but also in terms of their response to action potentials and frequency of release. The small molecule synapses have a relatively short recycling time, approximately 10 s (Danbolt 2001), and rapid firing frequency, 1–200 Hz (Korf and Gramsbergen 2007). Therefore, these synapses account for a large portion of the total energy consumption in the brain (Attwell and Gibb 2005). Among the small molecule neurotransmitters in the brain, glutamate is most prominent (Erecińska and Silver 1990). While it has been known for almost 50 years that glutamate stimulates neurons (Curtis et al. 1960), it was not fully accepted as a neurotransmitter for many years (Fonnum 1984). Although the neurotransmitter role of glutamate is essential for brain function, it is only one of the roles of this most intriguing amino acid. For example, astrocytic conversion of glutamate to glutamine not only comprises a central feature of the neurotransmitter cycle that is the topic of this chapter, but it also serves to clear neural ammonia. Thus, the brain is a net source of glutamine (Grill et al. 1992), which carries the nitrogen to the liver or kidney for elimination as uric acid or ammonia, respectively. As detailed in Fig. 32.1, glutamate sits at a critical metabolic juncture: It is in direct exchange with the TCA cycle intermediate α-ketoglutarate (αKG) and the precursor for γ-aminobutyric acid (GABA), the predominant inhibitory neurotransmitter. Glutamate is also metabolically related to aspartate, another small molecule excitatory neurotransmitter that is in direct exchange with the TCA cycle. The metabolism of glutamate and aspartate are also interrelated by the malate-aspartate shuttle (McKenna et al. 2006). Thereby, glutamate metabolism provides a functional link between the major consumers of chemical energy in the brain (small molecule neurotransmitters) and the major producers of that chemical energy (the TCA cycles in neurons and astrocytes).

Glutamate metabolism is compartmentalized in the central nervous system (CNS), especially with regard to glutamine synthetase and glutaminase, the activities of which are exclusively glial (Martinez-Hernandez et al. 1977) and primarily neuronal (Kaneko et al. 1987), respectively. Compartmentation supports the cycling of glutamate and glutamine between presynaptic neurons and astrocytes, which further fulfills the need to quickly remove glutamate from the synaptic cleft in order to

Fig. 32.1 Glutamate metabolism in the brain highlighting the close relationship of glutamate to other small molecule neurotransmitters, GABA and aspartate. Compartmentation is achieved by astrocytic localization of glutamine synthetase and neuronal localization of glutaminase

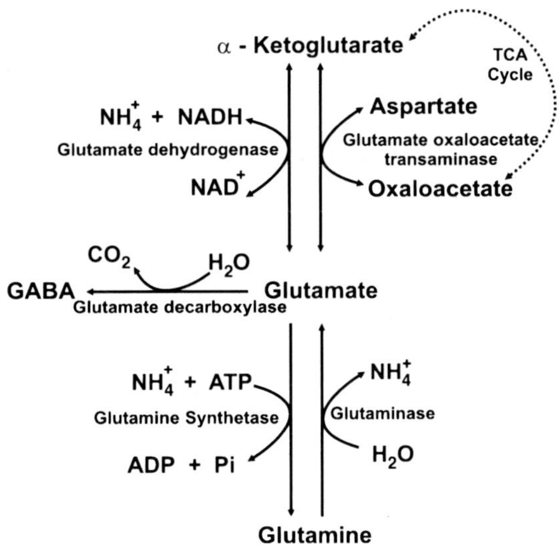

Fig. 32.2 The glutamate-glutamine cycle in the context of glutamatergic neurotransmission (NT) producing an action potential (AP) in the postsynaptic neuron. The trafficking of glutamate in the presynaptic neuron and the compartmentation of CO_2 fixation in the astrocyte provide functional context for the glutamate-glutamine cycle. *Glu* glutamate, *Gln* glutamine, *OAA* oxaloacetate, *αKG* α-ketoglutarate, *Cit* citrate

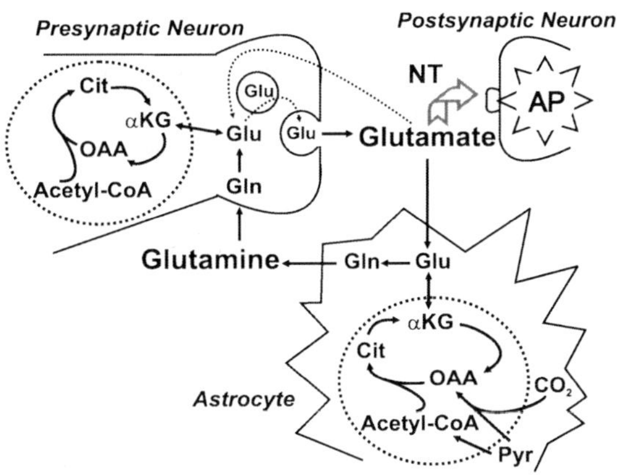

prevent excitotoxicity (Fig. 32.2). While the bulk of this chapter is devoted to measurements of the rates associated with this cycle in the living brain, here we consider it within the context of overall glutamate metabolism and brain structure. From this perspective, the glutamate-glutamine cycle is facilitated by the number, distribution and types of glutamate and glutamine transporters in the plasma membranes of astrocytes and neurons. Glutamate uptake into neurons and astrocytes is dominated by high affinity transporters, with similar K_m (3–80 μM) and V_{max} (1–50 nmol/mg protein/min) values in both cell types (Erecińska and Silver 1990; Schousboe et al. 2004). However, the expression of these transporters is very dense in astrocytes, but sparse in presynaptic neuronal membranes (Bergles et al. 1999; Danbolt 2001). Therefore, the glutamate uptake capacity of astrocytes exceeds the amount of the transmitter released by neurons by at least an order of magnitude, and reuptake of glutamate by neurons does not effectively compete with the astrocytic transporters (Danbolt 2001). On the other hand, glutamine uptake is mainly achieved by low affinity carriers (K_m ~ mM range) (Erecińska and Silver 1990; Broer and Brookes 2001). The glutamine transporters on astrocytes and neurons are functionally and structurally similar, except that the primary astrocytic carrier appears readily reversible and can mediate a net glutamine efflux, while the neuronal carriers favor glutamine uptake (Broer and Brookes 2001; Chaudhry et al. 2002). Clearly, the glutamine synthetase activity in astrocytes and glutaminase activity in neurons facilitate this astrocyte-to-neuron flux of glutamine by maintaining high and low intracellular glutamine concentrations in the respective cells (Erecińska and Silver 1990). Thereby, the distribution and activities of glutamate and glutamine transporters between neurons and astrocytes parallels the compartmentation of metabolic enzymes and both contribute to glutamate-glutamine cycling. Glutamate transport, metabolism, and cycling link the TCA cycles of the presynaptic neurons and astrocytes. Thus, the glutamate-glutamine cycle provides a functional link between glutamatergic neurotransmission, i.e. neuronal release of glutamate, and anaplerosis that is thought to be primarily associated with the astrocytic TCA cycle.

32.1.2 Glutamate-Glutamine Cycling and the Anaplerotic/Cataplerotic Balance

Nearly all excitatory neurons in the CNS are glutamatergic (Fonnum 1984). The glutamate-glutamine cycle (Fig. 32.2) and the conversion of glutamate to glutamine in astrocytes, account for the fact that glutamate is primarily localized in neurons (Chapa et al. 2000) despite its efficient uptake by astrocytes (Danbolt 2001). Upon excitation, presynaptic neurons release glutamate from vesicles in the nerve terminals into the synaptic cleft. After acting on its postsynaptic receptors, the majority of the glutamate is taken up by the neighboring astrocytes (Bergles et al. 1999) and converted to electrophysiologically inactive glutamine. Because glutamine synthesis is a glial process, most glutamine is found in astrocytes (Storm-Mathisen et al. 1992). Glutamine released from astrocytes is taken up by neurons and converted to glutamate, replenishing the presynaptic glutamate neurotransmitter pools and completing the cycle. The precise stoichiometry of this process is controversial, as will be discussed in detail below. Yet this does not change the fact that cycling of the glutamate/glutamine pool is not only central to glutamatergic neurotransmission, it also effectively links the TCA cycles of neurons and astrocytes. This is important because the TCA cycle is not an effective carbon sink. That is, the amount of TCA cycle intermediates is low relative to the flux of carbon through the cycle and, relative to the cytosolic metabolite pools, such as glutamate, glutamine and aspartate (Hertz and Dienel 2002). Thereby, any removal of TCA cycle intermediates for biosynthesis, e.g. net conversion of αKG to glutamate, must be closely matched by introduction of TCA cycle intermediates by some means. In other words, the "homeostatic" control of TCA cycle intermediates at concentrations far below the flux through the cycle implies a tight balance between anaplerosis and cataplerosis, the addition to and removal of carbon from the TCA cycle (Owen et al. 2002; Brunengraber and Roe 2006). Thus, because the glutamate-glutamine cycle links the TCA cycles between cells, it is possible for any cataplerotic loss of neuronal αKG to be compensated by astrocytic anaplerosis.

The concept of anaplerosis was developed by Kornberg during studies of microorganisms in which the TCA cycle is the source of biosynthetic carbon (Kornberg 1966). The focus of the early investigators was to understand how cells are able to acquire the carbon necessary for biosynthetic reactions. They coined the term "anaplerosis" to define reactions that "replenish" the TCA cycle intermediates removed for the purpose of biosynthesis. Given the relative lack of cell division and proliferation, the biosynthetic demands of the CNS are limited to maintaining the structural and functional integrity of the cells, e.g. synaptogenesis, and replenishing the neurotransmitter pools. The brain synthesizes few storage molecules such as triglycerides (Sastry 1985) or glycogen (Öz et al. 2007). Thus, neural biosynthesis is primarily devoted to the production of neurotransmitters and membrane lipids (Moore 2001; Pfrieger 2003; Satrustegui et al. 2007). Biosynthetic pathways account for a small proportion of the glucose taken up by the CNS because the oxygen-to-glucose ratio in the brain is approximately 5.5 in the resting state implying almost complete oxidation to CO_2 and H_2O (Complete oxidation of glucose would result in a ratio of 6 oxygen molecules per glucose molecule). Although other factors such as alternative fuel usage can also alter this ratio, anaplerotic entry of glucose carbons into the TCA cycle certainly contributes to lowering it. This does not imply that the anaplerotic activity in the brain, although relatively small, is in any way negligible, as de novo synthesis of lipid and small molecules is crucial to neurotransmission and brain function.

The adult, healthy brain does not oxidize significant amounts of lipid and ketone bodies, which would also necessitate anaplerotic/cataplerotic activity (Hakimi et al. 2005). Thus, cerebral anaplerotic/cataplerotic reactions primarily involve the exchange of aspartate and glutamate with oxaloacetate (OAA) and αKG, respectively, and CO_2 fixation. Traditionally, anaplerosis has been used interchangeably with CO_2 fixation in the brain because this has been the main anaplerotic flux that has been measured *in vivo*. Among the CO_2 fixing enzymes, pyruvate carboxylase (PC) is the most active one in the mammalian brain (Patel 1974). This enzyme combines one molecule of CO_2 with one molecule of pyruvate to make the TCA cycle intermediate OAA. In terms of metabolic compartmentation, PC is almost exclusively expressed in glial cells (Shank et al. 1985) and neuronal anaplerosis is not considered significant. Because continued glutamatergic neurotransmission requires the refilling of the presynaptic neuronal glutamate pool, any glutamate lost from the brain after synaptic release must be replaced by anaplerosis in the astrocyte, which returns it to the neuron as glutamine.

32.1.3 Questions and Controversies

Anaplerotic activity in the nervous system has been known since the 1960s, and identified in brain slices (Cheng et al. 1967), nerve preparations (Cheng and Waelsch 1962), and the living brain (Berl et al. 1962). However, the quantitative significance of anaplerotic activity relative to other cerebral metabolic fluxes, such as the neuronal and glial TCA cycles and the glutamate-glutamine cycle, remains a matter of debate. In addition to presenting a review of *in vivo* rate measurements of the glutamate-glutamine cycle and anaplerosis in the mammalian brain, this chapter will also highlight the following questions: Is the glutamate-glutamine cycle stoichiometric? Do neurons display significant anaplerotic activity? Does anaplerosis merely replenish TCA cycle intermediates or is it functionally linked to glutamate neurotransmission?

32.2 Methods to Measure Metabolic Fluxes *In Vivo*

32.2.1 Different Isotopic Labeling Methods: Continuous Infusion vs. Bolus Injections

There are multiple tracer methods to measure metabolic fluxes *in vivo*, such as radioactive tracer compounds in combination with positron emission tomography (PET) detection. This chapter will focus on the two main methods that can measure the anaplerotic and cataplerotic fluxes simultaneously with the glutamate-glutamine cycle activity in the living brain: Scintillation counting of ^{14}C-labeled metabolites in extracted brain tissue following injection of ^{14}C-labeled substrates, and NMR spectroscopy of ^{13}C-labeled metabolites either in extracted or intact brain tissue following or during injection of ^{13}C-labeled substrates. While this chapter focuses on carbon isotopes, ^{15}N-labeled substrates have also been utilized together with NMR to measure these fluxes (Kanamori and Ross 1993; Kanamori et al. 1995; Shen et al. 1998).

For these measurements one or more isotopically enriched precursor compounds, such as [^{13}C]glucose or [^{14}C]bicarbonate, are administered into the animal or human most commonly via the intravenous (IV) route, and less commonly via the subcutaneous, intracarotid, intraperitoneal or intracisternal route. Through the metabolism of the precursor compound the isotopic label flows into other metabolic pools. Labeled metabolites that are concentrated enough and are within the detection limit, such as glutamate, glutamine and aspartate, are quantified. If possible, the position of the label incorporated into the metabolites is determined, thus providing information about the fate of the precursor carbons, as well as the rates of the metabolic reactions that lie between the precursor compound and its metabolic product. The methods to quantify the labeled metabolites and to estimate metabolic rates using metabolic modeling are covered in detail in chapters 7 and 33 of this volume (Merle et al.; Henry et al.). Here we focus on the experiments (substrates, infusion protocols, outcomes) designed to specifically evaluate glutamate-glutamine cycling and anaplerosis.

The route of label administration makes a difference in the dynamics of the experiment, namely the rate at which the precursor compound reaches the brain. Clearly, the intracarotid, intracisternal and intravenous routes supply the labels quickly, while the intraperitoneal and subcutaneous routes are slower. Also, the potential contribution of peripheral metabolism (e.g. liver) to the measured brain fluxes depends on the route of administration, especially early in the experiment. For example, intracisternal injections are expected to result in minimal interference from peripheral metabolism, while label introduced via the tail vein will first traverse the liver. These differences need to be taken into account when analyzing the corresponding data. For example, when [1-^{13}C]glucose is perfused in the whole animal, gluconeogenesis in the liver scrambles the label, resulting in labeling that is different from the infusate, e.g. [1,2-^{13}C]glucose. This signal "pollution" can be estimated by measuring ^{13}C enrichments at all carbon positions in blood samples and can be overcome by co-infusion with two substrates (one labeled and one unlabeled) (Bouzier et al. 2000; Serres et al. 2008).

Even more important for the interpretation of the results however, is the time course of the label administration. When the labeled precursor is introduced as a bolus (single injection), the extent and time course of labeling of the products is very different from when the enriched precursor is supplied by continuous infusion or multiple injections. This was elegantly demonstrated in side-by-side experiments with the same tracer (^{14}C bicarbonate) and species (cat) where Otsuki et al. kept the specific activity (SA) of the blood $^{14}CO_2$ constant in one set of experiments (as in continuous infusions) while the SA of the blood $^{14}CO_2$ fell continuously in the other set (as in bolus infusions) (Otsuki et al. 1963). They obtained a constant ratio between the SA of the precursor $^{14}CO_2$ and its metabolic product aspartate with constant blood $^{14}CO_2$ SA, while this ratio varied when $^{14}CO_2$ SA was not constant. Furthermore, aspartate had the highest SA in the experiment with constant blood $^{14}CO_2$ SA, while the SA of glutamate was higher than that of aspartate in the second experiment, thereby suggesting different conclusions regarding the site and quantitative significance of CO_2 fixation. When conducted and interpreted with these caveats in mind, bolus and continuous infusion experiments each provide critical insight. The advantage of short time experiments with bolus injections is that they do not permit mixing of all the metabolic pools, thereby reflecting the initial fate of the administered compound (Lajtha et al. 1959; Berl et al. 1961). On the other hand, utilizing a continuous infusion protocol to achieve steady state isotopic enrichments throughout the experiment obviates many of the difficulties associated with data interpretation (Waelsch et al. 1964) and is advantageous for metabolic modeling of isotope enrichment time courses.

32.2.2 *^{14}C Labeling Methods and Substrates*

The main advantage of ^{14}C as a metabolic tracer is its high sensitivity such that small quantities can be administered in order to trace the metabolic pathways without

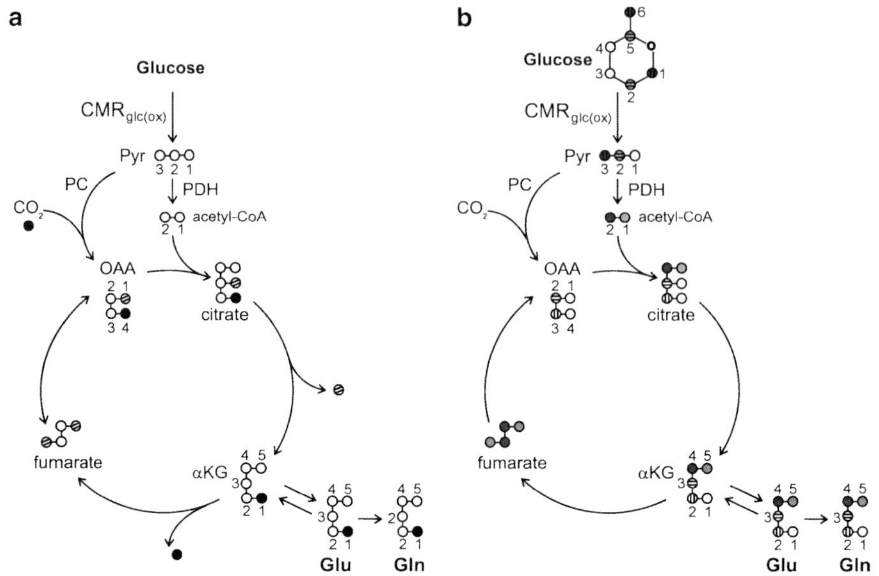

Fig. 32.3 (**a**) Label flow from bicarbonate, e.g. $^{14}CO_2$, to glutamate and glutamine through the TCA cycle. The numbering of the carbon positions is shown in pyruvate (*Pyr*), oxaloacetate (*OAA*), α-ketoglutarate (*αKG*), glutamate (*Glu*), and glutamine (*Gln*). The equilibration of the label in C-1 and C-4 positions of OAA via backflux to fumarate is displayed by *hatched bars*. *PC* pyruvate carboxylase, *PDH* pyruvate dehydrogenase. (**b**) Label flow from C-1 to C-4 positions of glucose to glutamate and glutamine. The label that enters the TCA cycle via PDH is shown by light and dark grey shades (acetyl-CoA C-1 and C-2), while label that enters the cycle via PC is shown by *hatched bars*

changing the metabolic status of the subject. However, since tissue extraction is necessary to follow incorporation of ^{14}C into individual metabolites, real time *in vivo* studies are not possible. The metabolites also need to be separated by chromatography to determine their specific activities. This may result in underestimation due to losses during sample handling. In addition, chemical derivation is necessary to measure labeling in specific carbon positions.

The most common substrates that have been utilized to follow cerebral metabolism are [^{14}C]bicarbonate to measure CO_2 fixation, [^{14}C]glucose to measure oxidative glucose utilization and label flow into amino acids, [^{14}C]glutamate, [^{14}C]glutamine or [^{14}C]aspartate to investigate precursor-product relationships and [^{14}C]acetate to monitor glial metabolism as acetate is preferentially taken up by astrocytes (Waniewski and Martin 1998). As an example, Fig. 32.3a displays the flow of label in the TCA cycle when $^{14}CO_2$ is used as the tracer. After entering the cycle at the OAA level via PC activity, it exits either via the αKG/glutamate exchange or in the following step in the TCA cycle catalyzed by α-ketoglutarate dehydrogenase. Therefore, each $^{14}CO_2$ molecule traverses the cycle only once making this a particularly desirable label for ease of data interpretation. Figure 32.3a also displays the backflux from OAA to fumarate and the resulting label scrambling between carbon positions 4 and 1 in

OAA. The label in C-1 of OAA is lost in the conversion of isocitrate to αKG and never reaches glutamate. Hence the extent of glutamate/glutamine labeling (from which the rate of CO_2 fixation can be extracted) depends on the extent of label equilibration via the fumarate backflux. Either complete or no scrambling of label via this flux has been assumed for *in vivo* studies of cerebral metabolism. Partial cycling (39%) between OAA and fumarate was shown in cerebellar astrocyte cultures (Merle et al. 1996). The astrocytic fumarate backflux was estimated in awake rats (Öz et al. 2004) and more studies are needed to measure this backflux under different physiological states as omission of this flux leads to underestimation of the PC flux.

32.2.3 ^{13}C Labeling Methods and Substrates

The advantage of ^{13}C as a metabolic tracer in combination with NMR is that it enables non-invasive, real time measurements of metabolism. Furthermore, the isotopic enrichment of metabolites at specific carbon positions can be measured without their chromatographic separation and chemical derivation. Recently, isotopomer analysis, which enables the quantitation of metabolites that are labeled at multiple carbon positions, has been demonstrated *in vivo* (Henry et al. 2003a). The main disadvantage of ^{13}C NMR is its low sensitivity, making high isotopic enrichments necessary. For example, [^{13}C]glucose studies are typically performed at hyperglycemia. This does not present a serious limitation for brain studies since CMR_{glc} (Cerebral Metabolic Rate of glucose utilization) is unchanged during acute hyperglycemia (Duckrow and Bryan 1987), indicating that ^{13}C NMR in conjunction with [^{13}C]glucose infusions reflects cerebral metabolic rates at normal physiology. Moreover, sensitivity in NMR experiments has been improving over the years. Using high-field magnets and new NMR pulse sequences and technologies, it is now possible to obtain ^{13}C-enrichments *in vivo* or ex vivo, even on small samples. This can be done for example by transferring magnetization from proton nuclei to ^{13}C nuclei (Pfeuffer et al. 1999) and high resolution magic angle spinning (HR-MAS) NMR spectroscopy (Risa et al. 2008).

A variety of substrates have been utilized to follow cerebral metabolism. For example, [1-^{13}C]glucose or [1,6-$^{13}C_2$]glucose can be used to measure neuronal and glial oxidative glucose utilization ($CMR_{glc,ox}$), TCA cycle, anaplerosis and glutamate-glutamine cycle rates. Use of [2-^{13}C]glucose allows specific investigation of CO_2 fixation by PC. Administering [3-^{13}C]lactate provides for studies of neuronal metabolism, while use of [^{13}C]acetate allows targeting of glial metabolism. Figure 32.3b outlines the flow of label from glucose to glutamate/glutamine. When carbon positions C-1 or C-6 of glucose are labeled, the ^{13}C label first reaches C-4 of glutamate if the ^{13}C-pyruvate enters the TCA cycle via pyruvate dehydrogenase (PDH), while C-2 of glutamate is labeled if the ^{13}C-pyruvate enters the cycle through PC. In the next turn of the TCA cycle the label from C-4 of αKG is distributed between C-2 and C-3. Therefore, equal C-2 and C-3 labeling indicates that there is no PC activity (as in neurons) and higher C-2 labeling than C-3 results from PC

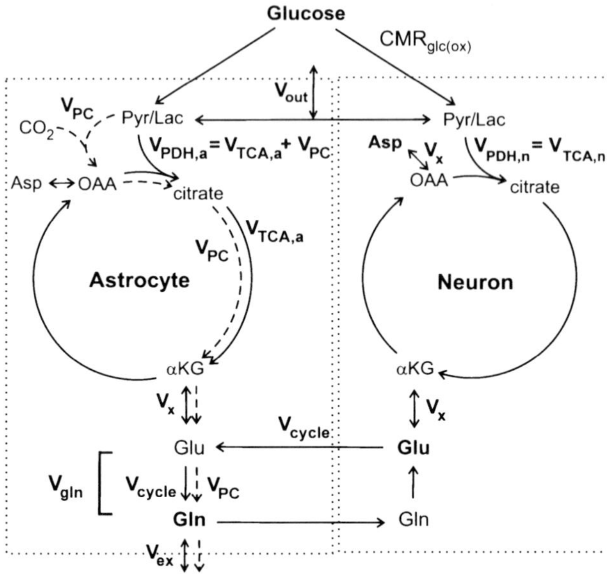

Fig. 32.4 Two-compartment model of astrocytic and neuronal oxidative glucose metabolism. *Glu* glutamate, *Gln* glutamine, *Asp* aspartate, *Pyr* pyruvate, *Lac* lactate, *OAA* oxaloacetate, *αKG* α-ketoglutarate; $CMR_{glc(ox)}$: cerebral metabolic rate of oxidative glucose utilization, $V_{PDH,n} = V_{TCA,n}$: neuronal pyruvate dehydrogenase rate; $V_{PDH,a}$: astrocytic pyruvate dehydrogenase rate; V_{PC}: pyruvate carboxylase rate; $V_{TCA,a}$: astrocytic TCA cycle rate leading to complete turning of the cycle; V_{cycle}: glutamate-glutamine cycle rate; V_{syn}: rate of glutamine synthetase; V_x: exchange rate between cytosolic amino acids and mitochondrial TCA cycle intermediates; V_{ex}: rate of label dilution of Gln; V_{out}: rate of label dilution of Lac

activity in astrocytes. Hence, the difference between glutamate/glutamine C-2 and C-3 labeling is used to deduce the PC rate. However, this difference is reduced by backflux from OAA to fumarate which complicates interpretation of the data. To overcome this, [2-^{13}C]glucose or [5-^{13}C]glucose can be used, which label glutamate C-5 upon entry into the TCA cycle via PDH and glutamate C-3 (and C-2 via fumarate backflux) upon entry via PC (Sibson et al. 2001). The label from C-5 of αKG is only scrambled to C-1 αKG on the next turn of the cycle, and therefore does not interfere with labeling in C-2 and C-3 positions, providing a less ambiguous estimate of the PC flux.

The two metabolic fluxes that are the focus of this chapter, those of PC (V_{PC}) and the glutamate-glutamine cycle (V_{cycle}) have typically been extracted from the rate of glutamine synthesis from glutamate (V_{gln}) in astrocytes in ^{13}C NMR studies combined with metabolic modeling. That is because carbon entering the TCA cycle via PC activity is assumed to leave the cycle at αKG and efflux in the form of glutamine (Fig. 32.4) because of the need to balance anaplerotic and cataplerotic fluxes. Hence in the metabolic models utilized thus far, V_{gln} has been assumed equal to the sum of V_{cycle} and V_{PC}. The assumption that anaplerotic carbon exits the system in the form of glutamine (Shen et al. 1999; Gruetter et al. 2001) is reasonable, considering that the brain is a net producer of glutamine (Grill et al. 1990). However, there is at least one other possible exit route for this carbon, namely glutamate oxidation in conjunction with pyruvate recycling which we will discuss below.

32.3 *In Vivo* Measurements of Glutamate-Glutamine Cycling

32.3.1 Discovery of the Glutamate-Glutamine Cycle

The earliest indications of compartmentation of glutamate and glutamine in the brain came from experiments where systemic administration of [^{14}C]glutamate to rodents led to higher plasma glutamine SA than that of its precursor tissue glutamate. These experiments indicated that the glutamine was not synthesized in the same compartment as the bulk of tissue glutamate (Lajtha et al. 1959). This "inverse product-precursor relationship" was also observed when [^{14}C]glutamate and [^{14}C]aspartate were administered intracisternally to minimize peripheral effects (Berl et al. 1961). Subsequent studies, comparing label incorporation from [^{14}C]glucose and [^{14}C]acetate into glutamate, aspartate and glutamine, demonstrated that the brain contained a large and a small glutamate pool (Van den Berg et al. 1969). Each pool was associated with a different TCA cycle, and the small glutamate pool was involved in glutamine synthesis. Furthermore, Patel et al. demonstrated that the development of compartmentation of glutamate metabolism coincides with the development and maturation of neuronal processes in the young rat brain (Patel and Balazs 1970). The subsequent discovery of the preferential localization of glutamine synthetase and glutaminase activities between astrocytes and neurons and the details of uptake and release of glutamate and glutamine to/from each cell type confirmed this compartmentation and lead to the current understanding of the glutamate-glutamine cycle.

32.3.2 Measurements of the Glutamate-Glutamine Cycle Rate

The first *in vivo* measurements of label transfer from [^{13}C]glucose to glutamate/glutamine in the mammalian brain by NMR were reported in the 1980s (Rothman et al. 1985; Behar et al. 1986) and were followed by the application of the same methodology in humans in 1990s (Beckmann et al. 1991; Gruetter et al. 1994). Despite the evidence of two glutamate pools, initially the cerebral TCA cycle rate (V_{TCA}) was estimated using a metabolic model with a single compartment (Fitzpatrick et al. 1990; Mason et al. 1992), which was accomplished by utilizing glutamate C-4 and C-3 labeling and multiple assumptions based on prior work. The single compartment model is generally recognized as adequate to fit glutamate labeling time courses since labeling of glutamate from [1-^{13}C]glucose reflects neuronal TCA cycle activity to a good approximation. Later studies took advantage of additional time courses measureable by *in vivo* NMR, starting with glutamine labeling. When infusing [^{13}C] glucose, astrocytic glutamine receives label primarily from neuronal glutamate, therefore glutamine labeling lags that of glutamate. This made it possible to deduce the glutamine synthesis rate (V_{gln}) from the difference in time courses of glutamate C-4 and glutamine C-4 labeling from [1-^{13}C]glucose in the human brain, although with a very large confidence interval (Mason et al. 1995). Sibson et al. were the first to report simultaneous measurements of the TCA cycle (without distinguishing between neurons and astrocytes) and the glutamate-glutamine cycle by simultaneously fitting glutamate and glutamine C-4 time courses (Sibson et al. 1997, 1998). These early measurements of the glutamate-glutamine cycle were based on a model that did not include a full astrocytic compartment, which led to various assumptions concerning the V_{PC} flux in particular. Initially, V_{PC} was assumed negligible. As a consequence, V_{gln} was assumed equal to the glutamate-glutamine cycle rate V_{cycle}, resulting in the claim that glutamatergic neurotransmission could be measured merely from glutamate C-4 and glutamine C-4 labeling (Sibson et al. 1997). This was later refined by measuring V_{gln} under pentobarbital anesthesia (isoelectricity), where $V_{cycle}=0$, and thus $V_{gln}=V_{PC}$ (Sibson et al. 1998). By assuming that the V_{PC} determined under isoelectricity did not change as electrical activity levels increased, V_{cycle} was then calculated from the difference between V_{gln} and V_{gln} at isoelectricity (Sibson et al. 1998).

A more accurate assessment of the cycling between the two cell types was possible with more sophisticated two-compartment models first introduced in the late 1990s (Gruetter et al. 1998, 2001; Shen et al. 1999). In contrast to previous models, these models included two distinct TCA cycles (one in neurons and one in astrocytes), as well as a pyruvate carboxylase flux specific to the astrocytic compartment. These models provided a more complete, although still simplified, representation of cerebral metabolic compartmentation between neurons and astrocytes and allowed the estimation of multiple metabolic fluxes in addition to V_{cycle}, including V_{PC} and the TCA cycle rates of neurons and astrocytes individually. In these studies, a ~5-fold higher neuronal V_{TCA} relative to that of astrocytes demonstrated that neurons are more oxidative than astrocytes (Shen et al. 1999; Gruetter et al. 2001; Lebon et al. 2002). A number of studies were subsequently published with similar two- and three-compartment models (Sibson et al. 2001; Lebon et al. 2002; Choi et al. 2002; de Graaf et al. 2003, 2004; Öz et al. 2004; Patel et al. 2004, 2005b; Mason et al. 2007; Boumezbeur et al. 2010; Patel et al. 2010; van Eijsden et al. 2010) although the details of the modeling and the assumptions used (e.g. on V_{PC}, astrocytic V_{TCA} or dilution fluxes) varied from study to study.

A crucial step in the direction of more accurate metabolic modeling was methodological developments that allowed the reliable measurement and inclusion of labeling in other carbon positions in glutamate and glutamine (C-2 and C-3), as well as aspartate labeling, in the modeling (Gruetter et al. 2001). The inclusion of C-3 time courses in modeling was particularly important for the simultaneous determination of V_{TCA} and the exchange rate between α-ketoglutarate and glutamate, a flux termed as V_x. With a one-compartment model, for example, the glutamate C-4 labeling curve was not sufficient to determine V_{TCA} and V_x reliably and both glutamate C-4 and C-3 curves were necessary for reliable V_{TCA} and V_x measurements (Henry et al. 2002). This is because the time course of glutamate C-3 labeling reflects the relative values of V_x and V_{TCA}. If V_x is fast compared to V_{TCA}, the glutamate pool is turned

Table 32.1 Glutamine synthesis, glutamate-glutamine cycle and pyruvate carboxylase rates (μmol/g/min) reported in the human brain

Label used	V_{gln}	V_{cycle}	V_{PC}	Reference
[1-^{13}C]glucose	0.47			Mason et al. (1995)
[1-^{13}C]glucose	0.40	0.32	0.08	Gruetter et al. (1998)
[1-^{13}C]glucose	0.36	0.32	0.04	Shen et al. (1999)
[1-^{13}C]glucose	0.26	0.17	0.09	Gruetter et al. (2001)
[2-^{13}C]acetate	0.36	(0.32[a])	(0.04[a])	Lebon et al. (2002)
[2-^{13}C]glucose and [1-^{13}C]glucose	0.37	0.34	0.02	Mason et al. (2007)
[1-^{13}C]glucose and [2-^{13}C]gcetate	(0.17, young subject) (0.14, elderly subjects)	0.16 (young subject) 0.13 (elderly subject)	(0.01[b], young subject) (0.01[b], elderly subject)	Boumezbeur et al. (2010)

Fluxes that were not fitted in the metabolic modeling but calculated from assumptions used in the paper are shown in parentheses

[a] V_{PC} assumed equal to 0.04 μmol/g/min based on Shen et al. (1999)

[b] V_{PC} assumed 6% of V_{cycle} based on Mason et al. (2007)

over before glutamate C-3 gets labeled, which results in an increased lag of the glutamate C-3 labeling curve relative to that of glutamate C-4. In contrast, if V_x is slow, the lag of the glutamate C-3 curve relative to C-4 is smaller. Similarly, with two-compartment models, simultaneous fitting of glutamate and glutamine C-4 and C-3 curves allows determination of V_x (Gruetter et al. 1998, 2001), whereas an assumption of V_x is necessary if only the C-4 curves of glutamate and glutamine are used (Shen et al. 1999).

Tables 32.1 and 32.2 list the values reported for V_{cycle} in the human and rat brain under a variety of physiological states. Although it is apparent from these data that, in general, V_{cycle} increases with increased electrical activity, there is considerable variation in the values reported even under the same physiological state by the same research group. Such variability may be partially explained by the difficulties associated with the reliable determination of V_{cycle} with the currently available *in vivo* data and metabolic models (Shestov et al. 2007). This underlines the need to further improve the reliability of metabolic modeling, e.g. by increasing the amount of experimental data used in the fitting procedure (use of labeling curves at multiple carbon positions, use of several different infused substrates simultaneously or in sequence and use of additional information available from isotopomers).

In any case, even a simple analysis of relative labeling patterns in glutamate, glutamine and aspartate demonstrates that the glutamate-glutamine cycle is a significant flux relative to other metabolic reactions, such as the TCA cycle in neurons and astrocytes. For example, the higher labeling in C-2 relative to C-3 of glutamine from [1-^{13}C] glucose (due to the PC flux in astrocytes) is transferred to glutamate in the awake (Aureli et al. 1997; Gruetter et al. 2001; Öz et al. 2004), but not anesthetized brain (Merle et al. 2002; Henry et al. 2003b). This demonstrates that the glutamate-glutamine cycle rate is fast enough to compete with the neuronal TCA cycle in labeling the large neuronal glutamate pool in the awake brain, but is slower under anesthesia when glutamate labeling is dominated by the contribution of the neuronal TCA cycle.

Table 32.2 Glutamine synthesis, glutamate-glutamine cycle and pyruvate carboxylase rates (μmol/g/min) reported in the rat brain under different anesthetics and physiological states

Label used	V_{gln}	V_{cycle}	V_{PC}	Reference
Under halothane with bicuculline induced seizures:				
[1,6-$^{13}C_2$]glucose	(0.625[a])	0.52	(0.105[a])	Patel et al. (2004)
Awake animals:				
$H^{14}CO_3^-$		0.46[b]		Lieth et al. (2001)
[1-^{13}C]glucose and $H^{14}CO_3^-$	0.62[c]	0.45	0.17	Öz et al. (2004)
Under morphine:				
[1-^{13}C]glucose or [1,6-$^{13}C_2$]glucose	0.44	(0.40[d])	(0.04[d])	Sibson et al. (1998)
Under halothane:				
[1,6-$^{13}C_2$]glucose	(0.346[e])	0.25	(0.096[e])	de Graaf et al. (2003)
[1,6-$^{13}C_2$]glucose	(0.406[e])	0.31[f]	(0.096[e])	de Graaf et al. (2004)
[1,6-$^{13}C_2$]glucose	(0.275[a])	0.22	(0.055[a])	Patel et al. (2004)
[1,6-$^{13}C_2$]glucose or [2-^{13}C]acetate	(0.725[a])	0.58	(0.145[a])	Patel et al. (2005b)
[2-^{13}C]acetate	(0.57)	(0.48)	(0.114[a])	Patel et al. (2010)
[U-^{13}C6]glucose or [2-^{13}C]acetate	(0.384[a])	(0.307)	(0.077[a])	van Eijsden et al. (2010)
Under a-chloralose:				
[1-^{13}C]glucose	0.21	(0.21[g])	(0[g])	Sibson et al. (1997)
[1-^{13}C]glucose or [1,6-$^{13}C_2$]glucose	0.17	(0.13[d])	(0.04[d])	Sibson et al. (1998)
Under a-chloralose and hyperammonemia:				
[^{15}N]ammonium acetate	0.20	0.135	0.065	Shen et al. (1998)
[1-^{13}C]glucose	0.297	0.20	0.096	Sibson et al. (2001)
Under pentobarbital:				
$^{15}NH_4Cl$	0.007 – 0.035			Kanamori et al. (1995)
[1-^{13}C]glucose or [1,6-$^{13}C_2$]glucose	0.04	(0[d])	(0.04[d])	Sibson et al. (1998)
[1-^{13}C]glucose	0.07	0.04	0.03	Choi et al. (2002)
[1,6-$^{13}C_2$]glucose, [2-^{13}C]acetate	(0.019[a])	0.015	(0.004[a])	Patel et al. (2005b)
Under pentobarbital and hyperammonemia:				
$^{15}NH_4Cl$	0.055 – 0.075			Kanamori et al. (1995)

Fluxes that were not fitted in the metabolic modeling but calculated from assumptions used in the paper are shown in parentheses

[a] Calculated based on the assumption that $V_{PC}/V_{gln}=0.2$ (Sibson et al. 2001)

[b] Calculated by assuming 0.075 mg protein/mg wet weight (Öz et al. 2004)

[c] Values determined from fitting ^{13}C and ^{14}C data simultaneously

[d] V_{PC} assumed equal to V_{gln} under pentobarbital (0.04 μmol/g/min), i.e. $V_{cycle,\,pento}=0$; V_{cycle} calculated from $V_{gln} - V_{PC,\,pento}$

[e] V_{PC} assumed equal to 0.096 μmol/g/min based on the value determined during hyperammonemia by Sibson et al. (2001); V_{gln} calculated from $V_{cycle}+V_{PC}$

[f] Value determined for pure gray matter

[g] V_{PC} assumed 0 at normoammonemia

Almost all metabolic models utilized thus far assumed a 1:1 stoichiometry for the glutamate-glutamine cycle; that is for each molecule of glutamate that is transferred from the presynaptic neuron to the astrocyte, a molecule of glutamine is returned to neurons. Other metabolites, such as lactate or the TCA cycle intermediates αKG and malate, have also been proposed to facilitate trafficking of glutamate carbons between astrocytes and neurons (Shank and Campbell 1984; Hassel and Sonnewald 1995;

Maciejewski and Rothman 2008), which would imply that the stoichiometry of the cycle is not strictly 1:1. However, *in vivo* evidence for a substantial contribution of such alternative cycles is currently lacking (Sibson et al. 2001; Lebon et al. 2002). Furthermore, comparable fluxes for transfer of neuronal glutamate to the glial compartment and glial glutamine to the neuronal compartment were estimated (Cruz and Cerdán 1999). Therefore further investigations are needed to substantiate the contribution of alternative cycles.

32.4 *In Vivo* Measurements of Cerebral Anaplerosis

32.4.1 Discovery of Anaplerotic Activity in the Brain

Carbon fixation in nervous tissue was first observed in the retina using $^{14}CO_2$ (Crane and Ball 1951), followed by the mammalian brain (Berl et al. 1962; Rossi et al. 1962; Siesjoe and Thompson 1965) and invertebrate (Cheng and Waelsch 1962) and amphibian nerves (Cote et al. 1966). Therefore, CO_2 fixation is a general property of the nervous tissue and not limited to mammalian CNS. In the mammalian brain studies, the higher SAs of glutamate, glutamine and aspartate isolated from the brain relative to those in blood indicated that $^{14}CO_2$ fixation must have occurred in the brain. In normoammonemic cats the highest SA was detected in aspartate, indicating that the CO_2 fixation occurred at the OAA level in the TCA cycle (Berl et al. 1962; Waelsch et al. 1964). Furthermore, the SA of glutamine was always higher than glutamate, implicating glia as the main site of anaplerotic activity in the brain (Berl et al. 1962; Rossi et al. 1962; Waelsch et al. 1964), although there has been some evidence that CO_2 fixation occurs in both metabolic compartments (Cheng 1971; Cheng and Nakamura 1972). Specifically, a lower glutamine-to-glutamate SA ratio obtained with $^{14}CO_2$ as the precursor than that obtained with glia-specific substrates, such as acetate, indicated that some CO_2 must also enter the neuronal compartment (Cheng and Nakamura 1972). On the other hand, the most active CO_2 fixing enzyme in the brain is PC (Patel 1974), a glial enzyme (Yu et al. 1983; Shank et al. 1985; Künnecke et al. 1993) whose activity increases with brain maturation (Yu et al. 1983) together with the development of compartmentation. This further supports the premise that glia are the primary site for CO_2 fixation.

Any CO_2 fixation in neurons must be facilitated by a different enzyme, such as malic enzyme. Intrastriatal injection of $NaH^{14}CO_3$ into rats lead to higher labeling of glutamate than glutamine (Hassel and Bråthe 2000). However, a bolus injection was utilized in this study in which case data interpretation is less straightforward because the labeling ratios in amino acids fluctuate over time as was discussed in more detail in section 2.1. Based on these data the authors claimed that neurons also fix CO_2 and compensate for any lost glutamate.

While there is evidence that the glutamate/glutamine pool and consequently TCA cycle intermediates are lost by glutamine efflux to the bloodstream, necessitating anaplerosis, another pathway in the brain may also contribute to the glutamate "loss". This pathway is glutamate oxidation, which starts with the entry of glutamate into the TCA cycle at αKG. It is thought that carbon entering the TCA cycle via this route exits as malate or OAA, which are converted to pyruvate via malic enzyme or phosphoenolpyruvate carboxykinase + pyruvate kinase, respectively. Pyruvate can then enter the TCA cycle via PDH and be fully oxidized, a pathway termed "pyruvate recycling". The conversion of malate/OAA to pyruvate within this pathway is a cataplerotic reaction and therefore needs to be counterbalanced by anaplerosis.

The compartmentation of pyruvate recycling and glutamate oxidation, as well as the quantitative significance of these pathways in the brain *in vivo*, is unclear. The pyruvate recycling pathway appears active in astrocyte cultures as these incorporate label from glutamate in the incubation medium into lactate (Sonnewald et al. 1993). However, the concentration of the extracellular glutamate determines the metabolic fate of the astrocytic glutamate, because higher levels in the medium upregulate glutamate oxidation in astrocytes (McKenna et al. 1996). Hence, it is unclear how these culture results translate to the *in vivo* situation, as extracellular glutamate is kept very low *in vivo* to prevent excitotoxicity. In fasted, awake mice C-2 of lactate was labeled from the glial specific substrate [2-^{13}C]acetate and not from [1-^{13}C] glucose, supporting the view that pyruvate recycling occurs in glia (Hassel and Sonnewald 1995). On the other hand, Cerdán and colleagues observed evidence for pyruvate recycling in glutamate but not glutamine labeling in the rat brain using [1,2-$^{13}C_2$]acetate and therefore concluded that the pathway was active in neurons (Cerdán et al. 1990; Cruz et al. 1998). This reasoning was later questioned, since glutamine was also not labeled in astrocyte cultures in the presence of glutamate oxidation and pyruvate recycling, indicating that the lack of glutamine labeling could not be taken as evidence for lack of pyruvate recycling/glutamate oxidation in astrocytes (Waagepetersen et al. 2002). Further experiments with deuterium (^{2}H) labeling in addition to ^{13}C lead to the conclusion that pyruvate recycling occurred in both cell types (Chapa et al. 2000). Interestingly, labeling patterns that implicate pyruvate recycling have been typically detected with [^{13}C]acetate, but not [^{13}C]glucose *in vivo* (Hassel and Sonnewald 1995; Öz et al. 2004). Even after 7 h of [1,6-^{13}C]glucose infusion in the anesthetized rat, isotopomers reflecting pyruvate recycling (labeling at the C-5 position) were only 3% of the total glutamate C-4 resonance (Henry et al. 2003a). Furthermore, Sibson et al. argued against significant glutamate oxidation and pyruvate recycling under hyperammonemia based on the close agreement between the anaplerotic flux they measured with [2-^{13}C]glucose and reported glutamine efflux rates under similar conditions (Sibson et al. 2001). They concluded that under these conditions anaplerosis had to be replenishing the glutamine lost to circulation, rather than the glutamate oxidized in the TCA cycle. Overall, acetate appeared to be a better label precursor for observing pyruvate recycling than glucose *in vivo* (Chapa et al. 2000). As discussed in the following paragraph, label scrambling via liver metabolism may reconcile these seemingly contradictory results.

Sibson et al. noted that the potential evidence for pyruvate recycling in their study, namely C-4 labeling in glutamate and glutamine from [2-^{13}C]glucose, could be due to hepatic label scrambling (Sibson et al. 2001). Indeed, comparison of metabolite labeling patterns in brain, blood and liver of rats demonstrated that the labeling patterns observed in brain metabolites indicative of pyruvate recycling mainly originated from metabolism of the ^{13}C-labeled precursor in the liver (Ebert et al. 2003; Serres et al. 2007). In these studies hepatic gluconeogenesis resulted in the redistribution of the ^{13}C label to blood glucose, making it available for oxidation in the brain. Hepatic label scrambling also explains why labeling patterns indicative of pyruvate recycling were typically observed with [^{13}C]acetate, but not [^{13}C]glucose: Hyperglycemia in studies with [^{13}C]glucose will reduce the rate of hepatic gluconeogenesis relative to normoglycemia. Under these conditions, the labeling pattern of the infused substrate, e.g. [1-^{13}C]glucose, will overwhelm the isotopomer distribution of blood glucose, making labeling from other glucose isotopomers, e.g. [1,2-^{13}C]glucose which results from pyruvate recycling, difficult to detect. In summary, unequivocal *in vivo* evidence for the existence of the pyruvate recycling and glutamate oxidation pathways in the brain is thus far lacking.

32.4.2 Measurements of Anaplerotic (Pyruvate Carboxylase) Activity in the Brain

The quantitative significance of CO_2 fixation can be assessed from the contribution of the PC and PDH pathways to the synthesis of amino acids that are in exchange with the TCA cycle intermediates. Early investigators concluded that pyruvate entry into the TCA cycle via CO_2 fixation vs. acetyl-CoA pathways occurred in a 1:10 ratio based on aspartate labeling from $^{14}CO_2$ in the cat brain (Otsuki et al. 1963). Aspartate enrichment reached ~ 10% of the SA of $^{14}CO_2$ suggesting that about 10% of OAA was formed via CO_2 fixation. However, this estimate does not take the compartmentation of aspartate into account. The $^{14}CO_2$ that enters the glial TCA cycle via PC does not readily reach the bulk of the aspartate pool that is localized in neurons. This is because the ^{14}C that is transferred to neuronal αKG via the glutamate-glutamine cycle is lost immediately after entering the neuronal TCA cycle through the conversion of αKG to succinate (Fig. 32.3).

Subsequently, a substantial PC flux relative to PDH was reported based on glutamine and glutamate labeling in awake and anesthetized mammals (Künnecke et al. 1993; Lapidot and Gopher 1994; Aureli et al. 1997; Lieth et al. 2001; Merle et al. 2002; Ebert et al. 2003; Serres et al. 2008). PC contributes more to the synthesis of glutamine than glutamate, which receives the label indirectly through the glutamate-glutamine cycle. Regarding the relative contributions of PC vs. PDH to glutamate and glutamine labeling, different studies can only be compared if values are reported at isotopic steady state since the % contributions change with the duration of label administration (Merle et al. 2002; Serres et al. 2008). This is because earlier during the infusion of the anaplerotic substrate the glutamine labeling mainly reflects the labeling pattern of the small astrocytic glutamate pool. The labeling pattern of the large neuronal glutamate pool is transferred to glutamine over time via the glutamate-glutamine cycle. At isotopic steady state the contribution of PC to glutamine labeling was 25-34% of that of PDH in both anesthetized (Merle et al. 2002; Serres et al. 2008) and awake animals (Lapidot and Gopher 1994; Lieth et al. 2001; Serres et al. 2008). Note that even these estimates of the relative contribution of PC vs. PDH to glutamine labeling at isotopic steady-state are only qualitative measures in the sense that they do not exactly report the ratio of the actual metabolic fluxes V_{PC} and V_{PDH}.

Absolute values for the PC flux estimated by metabolic modeling are listed in Tables 32.1 and 32.2. The PC flux constituted 20-30% of glutamine synthesis in awake rats (Öz et al. 2004) and rats under α-chloralose anesthesia (Sibson et al. 2001) during normoammonemia. This ratio increased to 43% under pentobarbital anesthesia where the glutamate-glutamine cycle is severely depressed (Choi et al. 2002). A larger range for the contribution of anaplerotic flux to glutamine synthesis was reported in humans (Table 32.1), ranging from 6% (Mason et al. 2007) to 35% (Gruetter et al. 2001).

Note that many of these studies, for example those that estimate the PC flux from the difference in ^{13}C labeling of C-2 and C-3 of glutamine from [1-^{13}C]glucose report a lower limit for this flux if the fumarate backflux is ignored (Öz et al. 2004). Furthermore, any anaplerotic activity in neurons (Hassel 2001) is also omitted. Label scrambling via fumarate backflux appears to be complete in neurons both in cultures (Merle et al. 1996) and *in vivo* based on equal C-2 and C3 labeling in aspartate in awake rats (Öz et al. 2004) and in glutamate in anesthetized rats (Merle et al. 2002; Henry et al. 2003b). Therefore any neuronal CO_2 fixation is not reflected in a difference in C-2 and C-3 labeling of aspartate and glutamate from [1-^{13}C]glucose.

In summary, there is solid evidence for the presence of substantial anaplerotic activity in the brain, primarily in astrocytes, compared to other metabolic fluxes, such as the glutamate-glutamine cycle and the TCA cycle.

32.4.3 Regulation of Anaplerosis in the Brain and Its Relationship to the Glutamate-Glutamine Cycle

The regulation of CO_2 fixation and its response to metabolic stress has been investigated since its discovery in the brain in the 1960s. Increased ammonia levels have been consistently shown to upregulate CO_2 fixation in the brain (Berl et al. 1962; Waelsch et al. 1964) leading to increased glutamine synthesis (Kanamori et al. 1995; Sibson et al. 1997, 2001; Shen et al. 1998). Hence there is consensus regarding a role for anaplerosis in cerebral ammonia detoxification.

A less conventional point of view concerns a link between CO_2 fixation and brain function, although the idea of glutamatergic activity as another regulatory factor of cerebral anaplerosis has also been around since the 1960s. For example, administration of ouabain, a drug that inactivates Na/K/ATPase and affects ionic movements and transmission of impulses, lead to reduced CO_2 fixation in the lobster nerve

(Cheng and Waelsch 1963). Similarly, a low sodium environment also decreased CO_2 fixation in the lobster nerve (Cheng and Mela 1966). Both the respiration rate and the flux of CO_2 into the TCA cycle increased upon stimulation of the frog nerve (Cote et al. 1966). Furthermore, conditions paralleling glutamate neurotransmission, such as increased extracellular potassium levels, were shown to increase CO_2 fixation (Kaufman and Driscoll 1992) and intracellular glutamine concentration (Brookes and Turner 1993; Brookes 2000) in astrocyte cultures. Consistently, sodium-dependent glutamate uptake was shown to activate astrocytic oxidative metabolism (Eriksson et al. 1995), which would be expected upon increased CO_2 fixation. *In vivo*, similar relative rates of PC to PDH were measured in anesthetized (Sibson et al. 2001; Merle et al. 2002) and awake animals (Lapidot and Gopher 1994; Lieth et al. 2001; Öz et al. 2004), indicating a corresponding increase of PC flux with brain activity. This was also reflected in studies where the PC flux was estimated assuming a fixed ratio to glutamine synthesis ($V_{PC}/V_{gln}=0.2$) under different levels of neuronal activity (Table 32.2) (Patel et al. 2004, 2005b). An increase in PC activity with increased neurotransmission was directly demonstrated recently by comparing V_{PC} and V_{cycle} in awake and deeply anesthetized rats using the same ^{13}C NMR methods under the two conditions. In the awake rat brain the PC flux was ~4-fold and the glutamate-glutamine cycle rate ~14-fold higher than under deep pentobarbital anesthesia (Choi et al. 2002; Öz et al. 2004). Similarly, a close coupling between PC flux and both astrocytic and neuronal TCA cycles at different brain activity levels was recently inferred from the PC-to-PDH ratios determined from glutamine labeling under anesthesia and in the awake state (Serres et al. 2008). In another study, the PC flux was unchanged during seizures, although the same study reported increased glutamine and TCA cycle intermediate levels implying increased anaplerosis with increased neuronal activity (Patel et al. 2005a).

Several potential mechanisms may increase anaplerosis with brain activity. Release of potassium from activated neurons increases the influx of bicarbonate into astrocytes (Brookes and Turner 1994), which in turn may increase anaplerosis (Gamberino et al. 1997). Also, depolarization in brain slices reduces glutamate levels (Badar-Goffer et al. 1992), which may trigger anaplerosis. It is likely that with increased cycling of glutamate and glutamine between neurons and astrocytes, the potential for loss of these amino acids by diffusion away from the active zone is increased as well, which in turn would need to be compensated by anaplerotic activity.

32.5 Concluding Remarks

Here we summarized the available data regarding *in vivo* measurements of glutamate-glutamine cycle and cerebral anaplerotic activity. We also discussed important considerations that must be taken into account when designing such experiments. The glutamate-glutamine cycle is a measure of excitatory neuronal activity in the brain and a major flux that facilitates carbon trafficking between neurons and astrocytes. Although a 1:1 stoichiometry between glutamate and glutamine has generally been assumed when studying this cycle, other metabolites involved in this trafficking have been proposed and need further investigation.

Anaplerotic activity, and specifically CO_2 fixation, is primarily localized to astrocytes. This activity compensates for any glutamate/glutamine losses, whether by efflux from the brain or by glutamate oxidation. The limited ability of neurons to adjust ana- and cataplerotic fluxes to maintain appropriate levels of cellular carbon in metabolic pools may also contribute to their observed fragility. Hence, although less active than in other tissues, astrocytic anaplerosis appears to play a critical role in neuroglial energetics. An increasing body of evidence supports a functional link between the anaplerotic activity in astrocytes and cycling of glutamate and glutamine between neurons and astrocytes, where increased cycling triggers increased CO_2 fixation. Further investigations are needed to elucidate the regulation of this process.

It is apparent from the reviewed data that further improvements are needed to increase the accuracy and precision of metabolic modeling. The glutamate-glutamine cycle flux appears to be particularly challenging to estimate (Shestov et al. 2007). Administering more than one labeled precursor simultaneously (Öz et al. 2004; Deelchand et al. 2009) or sequentially to constrain several fluxes in the final metabolic modeling (Lebon et al. 2002; Patel et al. 2005; Boumezbeur et al. 2010; van Eijsden et al. 2010) is expected to reduce the number of assumptions in these studies. However, it is important to remember that much information can be deduced in these tracer studies by comparing relative labeling levels of different carbons in metabolites (even though absolute metabolic rates cannot be measured this way) and that such information is invaluable for the understanding of the metabolic bases of brain function. Therefore both *in vivo* metabolic modeling and extract studies without metabolic modeling provide valuable insights. The progress in the field has been impressive and has increased our understanding of disease models and human brain diseases. Further development of both data acquisition and metabolic modeling methods is expected to continue to expand our understanding of brain function.

Acknowledgements This work was supported by the NIH grants R01 NS38672 (PGH), R21 NS060253 (GÖ) and VA Merit award (DAO). The Center for MR Research is supported by National Center for Research Resources (NCRR) biotechnology research resource grant P41RR08079 and Neuroscience Center Core Blueprint Award P30 NS057091. The Structural Biology NMR facility at the University of Minnesota is supported with funds from the NSF (BIR-961477), the University of Minnesota Medical School, and the Minnesota Medical Foundation.

References

Attwell D, Gibb A (2005) Neuroenergetics and the kinetic design of excitatory synapses. Nat Rev Neurosci 6:841–849

Attwell D, Laughlin SB (2001) An energy budget for signaling in the grey matter of the brain. J Cereb Blood Flow Metab 21:1133–1145

Aureli T, Di Cocco ME, Calvani M, Conti F (1997) The entry of [1-^{13}C]glucose into biochemical pathways reveals a complex compartmentation and metabolite trafficking between glia and neurons: a study by ^{13}C-NMR spectroscopy. Brain Res 765:218–227

Badar-Goffer RS, Ben-Yoseph O, Bachelard HS, Morris PG (1992) Neuronal-glial metabolism under depolarizing conditions. A ^{13}C-n.m.r. study. Biochem J 282(Pt 1):225–230

Beckmann N, Turkalj I, Seelig J, Keller U (1991) ^{13}C NMR for the assessment of human brain glucose metabolism in vivo. Biochemistry 30:6362–6366

Behar KL, Petroff OA, Prichard JW, Alger JR, Shulman RG (1986) Detection of metabolites in rabbit brain by ^{13}C NMR spectroscopy following administration of [1-^{13}C]glucose. Magn Reson Med 3:911–920

Bergles DE, Diamond JS, Jahr CE (1999) Clearance of glutamate inside the synapse and beyond. Curr Opin Neurobiol 9:293–298

Berl S, Lajtha A, Waelsch H (1961) Amino acid and protein metabolism – VI. Cerebral compartments of glutamic acid metabolism. J Neurochem 7:186–197

Berl S, Takagaki G, Clarke DD, Waelsch H (1962) Carbon dioxide fixation in the brain. J Biol Chem 237:2570–2573

Boumezbeur F, Mason GF, de Graaf RA, Behar KL, Cline GW, Shulman GI, Rothman DL, Petersen KF (2010) Altered brain mitochondrial metabolism in healthy aging as assessed by in vivo magnetic resonance spectroscopy. J Cereb Blood Flow Metab 30:211–221

Bouzier AK, Thiaudiere E, Biran M, Rouland R, Canioni P, Merle M (2000) The metabolism of [3-^{13}C]lactate in the rat brain is specific of a pyruvate carboxylase-deprived compartment. J Neurochem 75:480–486

Broer S, Brookes N (2001) Transfer of glutamine between astrocytes and neurons. J Neurochem 77:705–719

Brookes N (2000) Functional integration of the transport of ammonium, glutamate and glutamine in astrocytes. Neurochem Int 37:121–129

Brookes N, Turner RJ (1993) Extracellular potassium regulates the glutamine content of astrocytes: mediation by intracellular pH. Neurosci Lett 160:73–76

Brookes N, Turner RJ (1994) K^+-induced alkalinization in mouse cerebral astrocytes mediated by reversal of electrogenic Na^+-HCO_3^- cotransport. Am J Physiol 267:C1633–C1640

Brunengraber H, Roe CR (2006) Anaplerotic molecules: current and future. J Inherit Metab Dis 29:327–331

Cerdán S, Künnecke B, Seelig J (1990) Cerebral metabolism of [1,2-$^{13}C_2$]acetate as detected by in vivo and in vitro ^{13}C NMR. J Biol Chem 265:12916–12926

Chapa F, Cruz F, García-Martín ML, García-Espinosa MA, Cerdán S (2000) Metabolism of (1–^{13}C) glucose and (2–^{13}C, ^{2}–$^2H_3$) acetate in the neuronal and glial compartments of the adult rat brain as detected by ^{13}C, ^{2}H NMR spectroscopy. Neurochem Int 37:217–228

Chaudhry FA, Schmitz D, Reimer RJ, Larsson P, Gray AT, Nicoll R, Kavanaugh M, Edwards RH (2002) Glutamine uptake by neurons: interaction of protons with system a transporters. J Neurosci 22:62–72

Cheng SC (1971) CO_2 fixation in the nervous tissue. Int Rev Neurobiol 14:125–157

Cheng SC, Mela P (1966) CO_2 fixation in the nervous system – II. Environmental effects on CO_2 fixation in lobster nerve. J Neurochem 13:281–287

Cheng SC, Nakamura R (1972) Metabolism related to the tricarboxylic acid cycle in rat brain slices. Observations on CO_2 fixation and metabolic compartmentation. Brain Res 38:355–370

Cheng SC, Waelsch H (1962) Carbon dioxide fixation in lobster nerve. Science 136:782–783

Cheng SC, Waelsch H (1963) Some quantitative aspects of the fixation of carbon dioxide by the Lobster nerve. Biochem Z 338:643–653

Cheng SC, Nakamura R, Waelsch H (1967) Relative contribution of carbon dioxide fixation and acetyl-coA pathways in two nervous tissues. Nature 216:928–929

Choi IY, Lei H, Gruetter R (2002) Effect of deep pentobarbital anesthesia on neurotransmitter metabolism in vivo: on the correlation of total glucose consumption with glutamatergic action. J Cereb Blood Flow Metab 22:1343–1351

Cote LJ, Cheng SC, Waelsch H (1966) CO_2 fixation in the nervous system – I. CO_2 fixation in the sciatic nerve of the bullfrog. J Neurochem 13:271–279

Crane RK, Ball EG (1951) Relationship of $C^{14}O_2$ fixation to carbohydrate metabolism in retina. J Biol Chem 189:269–276

Cruz F, Cerdán S (1999) Quantitative ^{13}C NMR studies of metabolic compartmentation in the adult mammalian brain. NMR Biomed 12:451–462

Cruz F, Scott SR, Barroso I, Santisteban P, Cerdán S (1998) Ontogeny and cellular localization of the pyruvate recycling system in rat brain. J Neurochem 70:2613–2619

Curtis DR, Phillis JW, Watkins JC (1960) The chemical excitation of spinal neurones by certain acidic amino acids. J Physiol 150:656–682

Danbolt NC (2001) Glutamate uptake. Prog Neurobiol 65:1–105

Deelchand DK, Nelson C, Shestov AA, Uğurbil K, Henry PG (2009) Simultaneous measurement of neuronal and glial metabolism in rat brain in vivo using co-infusion of [1,6-$^{13}C_2$]glucose and [1,2-$^{13}C_2$]acetate. J Magn Reson 196:157–163

de Graaf RA, Brown PB, Mason GF, Rothman DL, Behar KL (2003) Detection of [1,6-$^{13}C_2$]-glucose metabolism in rat brain by in vivo ^{1}H-[^{13}C]-NMR spectroscopy. Magn Reson Med 49:37–46

de Graaf RA, Mason GF, Patel AB, Rothman DL, Behar KL (2004) Regional glucose metabolism and glutamatergic neurotransmission in rat brain in vivo. Proc Natl Acad Sci USA 101:12700–12705

Duckrow RB, Bryan RM Jr (1987) Regional cerebral glucose utilization during hyperglycemia. J Neurochem 48:989–993

Ebert D, Haller RG, Walton ME (2003) Energy contribution of octanoate to intact rat brain metabolism measured by ^{13}C nuclear magnetic resonance spectroscopy. J Neurosci 23:5928–5935

Erecińska M, Silver IA (1990) Metabolism and role of glutamate in mammalian brain. Prog Neurobiol 35:245–296

Eriksson G, Peterson A, Iverfeldt K, Walum E (1995) Sodium-dependent glutamate uptake as an activator of oxidative metabolism in primary astrocyte cultures from newborn rat. Glia 15:152–156

Fitzpatrick SM, Hetherington HP, Behar KL, Shulman RG (1990) The flux from glucose to glutamate in the rat brain in vivo as determined by ^{1}H-observed, ^{13}C-edited NMR spectroscopy. J Cereb Blood Flow Metab 10:170–179

Fonnum F (1984) Glutamate: a neurotransmitter in mammalian brain. J Neurochem 42:1–11

Gamberino WC, Berkich DA, Lynch CJ, Xu B, LaNoue KF (1997) Role of pyruvate carboxylase in facilitation of synthesis of glutamate and glutamine in cultured astrocytes. J Neurochem 69:2312–2325

Grill V, Gutniak M, Bjorkman O, Lindqvist M, Stone-Elander S, Seitz RJ, Blomqvist G, Reichard P, Widen L (1990) Cerebral blood flow and substrate utilization in insulin-treated diabetic subjects. Am J Physiol 258:E813–E820

Grill V, Bjorkman O, Gutniak M, Lindqvist M (1992) Brain uptake and release of amino acids in nondiabetic and insulin-dependent diabetic subjects: important role of glutamine release for nitrogen balance. Metabolism 41:28–32

Gruetter R, Novotny EJ, Boulware SD, Mason GF, Rothman DL, Shulman GI, Prichard JW, Shulman RG (1994) Localized ^{13}C NMR spectroscopy in the human brain of amino acid labeling from D-[1-^{13}C]glucose. J Neurochem 63:1377–1385

Gruetter R, Seaquist ER, Kim S, Ugurbil K (1998) Localized in vivo ^{13}C-NMR of glutamate metabolism in the human brain: initial results at 4 Tesla. Dev Neurosci 20:380–388

Gruetter R, Seaquist ER, Ugurbil K (2001) A mathematical model of compartmentalized neurotransmitter metabolism in the human brain. Am J Physiol Endocrinol Metab 281:E100–E112

Hakimi P, Johnson MT, Yang J, Lepage DF, Conlon RA, Kalhan SC, Reshef L, Tilghman SM, Hanson RW (2005) Phosphoenolpyruvate carboxykinase and the critical role of cataplerosis in the control of hepatic metabolism. Nutr Metab (Lond) 2:33

Hassel B (2001) Pyruvate carboxylation in neurons. J Neurosci Res 66:755–762

Hassel B, Bråthe A (2000) Neuronal pyruvate carboxylation supports formation of transmitter glutamate. J Neurosci 20:1342–1347

Hassel B, Sonnewald U (1995) Glial formation of pyruvate and lactate from TCA cycle intermediates: implications for the inactivation of transmitter amino acids? J Neurochem 65:2227–2234

Henry PG, Lebon V, Vaufrey F, Brouillet E, Hantraye P, Bloch G (2002) Decreased TCA cycle rate in the rat brain after acute 3-NP treatment measured by in vivo ^{1}H-{^{13}C} NMR spectroscopy. J Neurochem 82:857–866

Henry PG, Öz G, Provencher S, Gruetter R (2003a) Toward dynamic isotopomer analysis in the rat brain in vivo: automatic quantitation of ^{13}C NMR spectra using LCModel. NMR Biomed 16:400–412

Henry PG, Crawford S, Öz G, Ugurbil K, Gruetter R (2003b) Glucose and glial-neuronal metabolism in α-chloralose anesthetized rats measured by in vivo ^{13}C NMR spectroscopy. In: Proceedings of the 11th scientific meeting of the International Society for Magnetic Resonance in Medicine, Toronto, ON, p 1967

Hertz L, Dienel GA (2002) Energy metabolism in the brain. Int Rev Neurobiol 51:1–102

Kanamori K, Ross BD (1993) ^{15}N n.m.r. measurement of the in vivo rate of glutamine synthesis and utilization at steady state in the brain of the hyperammonaemic rat. Biochem J 293(Pt 2): 461–468

Kanamori K, Ross BD, Kuo EL (1995) Dependence of in vivo glutamine synthetase activity on ammonia concentration in rat brain studied by ^{1}H–^{15}N heteronuclear multiple-quantum coherence-transfer NMR. Biochem J 311(Pt 2):681–688

Kaneko T, Urade Y, Watanabe Y, Mizuno N (1987) Production, characterization, and immunohistochemical application of monoclonal antibodies to glutaminase purified from rat brain. J Neurosci 7:302–309

Kaufman EE, Driscoll BF (1992) Carbon dioxide fixation in neuronal and astroglial cells in culture. J Neurochem 58:258–262

Korf J, Gramsbergen JB (2007) Timing of potential and metabolic brain energy. J Neurochem 103:1697–1708

Kornberg HL (1966) The role and control of the glyoxylate cycle in Escherichia coli. Biochem J 99:1–11

Künnecke B, Cerdán S, Seelig J (1993) Cerebral metabolism of [1,2-$^{13}C_2$]glucose and [U-$^{13}C_4$]3-hydroxybutyrate in rat brain as detected by ^{13}C NMR spectroscopy. NMR Biomed 6:264–277

Lajtha A, Berl S, Waelsch H (1959) Amino acid and protein metabolism of the brain. IV. The metabolism of glutamic acid. J Neurochem 3:322–332

Lapidot A, Gopher A (1994) Cerebral metabolic compartmentation. Estimation of glucose flux via pyruvate carboxylase/pyruvate dehydrogenase by ^{13}C NMR isotopomer analysis of D-[U-^{13}C] glucose metabolites. J Biol Chem 269:27198–27208

Lebon V, Petersen KF, Cline GW, Shen J, Mason GF, Dufour S, Behar KL, Shulman GI, Rothman DL (2002) Astroglial contribution to brain energy metabolism in humans revealed by ^{13}C nuclear magnetic resonance spectroscopy: elucidation of the dominant pathway for neurotransmitter glutamate repletion and measurement of astrocytic oxidative metabolism. J Neurosci 22:1523–1531

Lennie P (2003) The cost of cortical computation. Curr Biol 13:493–497

Lieth E, LaNoue KF, Berkich DA, Xu B, Ratz M, Taylor C, Hutson SM (2001) Nitrogen shuttling between neurons and glial cells during glutamate synthesis. J Neurochem 76:1712–1723

Maciejewski PK, Rothman DL (2008) Proposed cycles for functional glutamate trafficking in synaptic neurotransmission. Neurochem Int 52:809–825

Martinez-Hernandez A, Bell KP, Norenberg MD (1977) Glutamine synthetase: glial localization in brain. Science 195:1356–1358

Mason GF, Rothman DL, Behar KL, Shulman RG (1992) NMR determination of the TCA cycle rate and alpha-ketoglutarate/glutamate exchange rate in rat brain. J Cereb Blood Flow Metab 12:434–447

Mason GF, Gruetter R, Rothman DL, Behar KL, Shulman RG, Novotny EJ (1995) Simultaneous determination of the rates of the TCA cycle, glucose utilization, alpha-ketoglutarate/glutamate exchange, and glutamine synthesis in human brain by NMR. J Cereb Blood Flow Metab 15:12–25

Mason GF, Petersen KF, de Graaf RA, Shulman GI, Rothman DL (2007) Measurements of the anaplerotic rate in the human cerebral cortex using ^{13}C magnetic resonance spectroscopy and [1-^{13}C] and [2-^{13}C] glucose. J Neurochem 100:73–86

McKenna MC, Sonnewald U, Huang X, Stevenson J, Zielke HR (1996) Exogenous glutamate concentration regulates the metabolic fate of glutamate in astrocytes. J Neurochem 66: 386–393

McKenna MC, Waagepetersen HS, Schousboe A, Sonnewald U (2006) Neuronal and astrocytic shuttle mechanisms for cytosolic-mitochondrial transfer of reducing equivalents: current evidence and pharmacological tools. Biochem Pharmacol 71:399–407

Merle M, Martin M, Villégier A, Canioni P (1996) Mathematical modelling of the citric acid cycle for the analysis of glutamine isotopomers from cerebellar astrocytes incubated with [1-^{13}C] glucose. Eur J Biochem 239:742–751

Merle M, Bouzier-Sore AK, Canioni P (2002) Time-dependence of the contribution of pyruvate carboxylase versus pyruvate dehydrogenase to rat brain glutamine labelling from [1-^{13}C]glucose metabolism. J Neurochem 82:47–57

Moore SA (2001) Polyunsaturated fatty acid synthesis and release by brain-derived cells in vitro. J Mol Neurosci 16:195–200, discussion 215-121

Otsuki S, Geiger A, Gombos G (1963) The metabolic pattern of the brain in brain perfusion experiments in vivo. I. The quantitative significance of CO_2 assimilation in the metabolism of the brain. J Neurochem 10:397–404

Owen OE, Kalhan SC, Hanson RW (2002) The key role of anaplerosis and cataplerosis for citric acid cycle function. J Biol Chem 277:30409–30412

Öz G, Berkich DA, Henry PG, Xu Y, LaNoue K, Hutson SM, Gruetter R (2004) Neuroglial metabolism in the awake rat brain: CO_2 fixation increases with brain activity. J Neurosci 24: 11273–11279

Öz G, Seaquist ER, Kumar A, Criego AB, Benedict LE, Rao JP, Henry PG, Van De Moortele PF, Gruetter R (2007) Human brain glycogen content and metabolism: implications on its role in brain energy metabolism. Am J Physiol Endocrinol Metab 292:E946–E951

Patel MS (1974) The relative significance of CO_2-fixing enzymes in the metabolism of rat brain. J Neurochem 22:717–724

Patel AJ, Balazs R (1970) Manifestation of metabolic compartmentation during the maturation of the rat brain. J Neurochem 17:955–971

Patel AB, de Graaf RA, Mason GF, Kanamatsu T, Rothman DL, Shulman RG, Behar KL (2004) Glutamatergic neurotransmission and neuronal glucose oxidation are coupled during intense neuronal activation. J Cereb Blood Flow Metab 24:972–985

Patel AB, Chowdhury GM, de Graaf RA, Rothman DL, Shulman RG, Behar KL (2005a) Cerebral pyruvate carboxylase flux is unaltered during bicuculline-seizures. J Neurosci Res 79:128–138

Patel AB, de Graaf RA, Mason GF, Rothman DL, Shulman RG, Behar KL (2005b) The contribution of GABA to glutamate/glutamine cycling and energy metabolism in the rat cortex in vivo. Proc Natl Acad Sci USA 102:5588–5593

Patel AB, de Graaf RA, Rothman DL, Behar KL, Mason GF (2010) Evaluation of cerebral acetate transport and metabolic rates in the rat brain in vivo using ^{1}H-[^{13}C]-NMR. J Cereb Blood Flow Metab 30:1200–1213

Pfeuffer J, Tkáč I, Choi I-Y, Merkle H, Ugurbil K, Garwood M, Gruetter R (1999) Localized in vivo ^{1}H NMR detection of neurotransmitter labeling in rat brain during infusion of [1-^{13}C] D-glucose. Magn Reson Med 41:1077–1083

Pfrieger FW (2003) Cholesterol homeostasis and function in neurons of the central nervous system. Cell Mol Life Sci 60:1158–1171

Risa O, Melo TM, Sonnewald U (2008) Quantification of amounts and ^{13}C content of metabolites in brain tissue using high- resolution magic angle spinning ^{13}C NMR spectroscopy. NMR Biomed 22(3):266–271

Rolfe DF, Brown GC (1997) Cellular energy utilization and molecular origin of standard metabolic rate in mammals. Physiol Rev 77:731–758

Rossi CA, Berl S, Clarke DD, Purpura DP, Waelsch H (1962) Rate of CO_2 fixation in brain and liver. Life Sci 1:533–539

Rothman DL, Behar KL, Hetherington HP, den Hollander JA, Bendall MR, Petroff OA, Shulman RG (1985) ^{1}H-Observe/^{13}C-decouple spectroscopic measurements of lactate and glutamate in the rat brain in vivo. Proc Natl Acad Sci USA 82:1633–1637

Sastry PS (1985) Lipids of nervous tissue: composition and metabolism. Prog Lipid Res 24:69–176

Satrustegui J, Contreras L, Ramos M, Marmol P, del Arco A, Saheki T, Pardo B (2007) Role of aralar, the mitochondrial transporter of aspartate-glutamate, in brain N-acetylaspartate formation and Ca^{2+} signaling in neuronal mitochondria. J Neurosci Res 85:3359–3366

Schousboe A, Sarup A, Bak LK, Waagepetersen HS, Larsson OM (2004) Role of astrocytic transport processes in glutamatergic and GABAergic neurotransmission. Neurochem Int 45:521–527

Serres S, Bezancon E, Franconi JM, Merle M (2007) Brain pyruvate recycling and peripheral metabolism: an NMR analysis ex vivo of acetate and glucose metabolism in the rat. J Neurochem 101:1428–1440

Serres S, Raffard G, Franconi JM, Merle M (2008) Close coupling between astrocytic and neuronal metabolisms to fulfill anaplerotic and energy needs in the rat brain. J Cereb Blood Flow Metab 28:712–724

Shank RP, Campbell GL (1984) Alpha-ketoglutarate and malate uptake and metabolism by synaptosomes: further evidence for an astrocyte-to-neuron metabolic shuttle. J Neurochem 42: 1153–1161

Shank RP, Bennett GS, Freytag SO, Campbell GL (1985) Pyruvate carboxylase: an astrocyte-specific enzyme implicated in the replenishment of amino acid neurotransmitter pools. Brain Res 329:364–367

Shen J, Sibson NR, Cline G, Behar KL, Rothman DL, Shulman RG (1998) ^{15}N-NMR spectroscopy studies of ammonia transport and glutamine synthesis in the hyperammonemic rat brain. Dev Neurosci 20:434–443

Shen J, Petersen KF, Behar KL, Brown P, Nixon TW, Mason GF, Petroff OA, Shulman GI, Shulman RG, Rothman DL (1999) Determination of the rate of the glutamate/glutamine cycle in the human brain by in vivo ^{13}C NMR. Proc Natl Acad Sci USA 96:8235–8240

Shestov AA, Valette J, Ugurbil K, Henry PG (2007) On the reliability of ^{13}C metabolic modeling with two-compartment neuronal-glial models. J Neurosci Res 85:3294–3303

Sibson NR, Dhankhar A, Mason GF, Behar KL, Rothman DL, Shulman RG (1997) In vivo ^{13}C NMR measurements of cerebral glutamine synthesis as evidence for glutamate-glutamine cycling. Proc Natl Acad Sci USA 94:2699–2704

Sibson NR, Dhankhar A, Mason GF, Rothman DL, Behar KL, Shulman RG (1998) Stoichiometric coupling of brain glucose metabolism and glutamatergic neuronal activity. Proc Natl Acad Sci USA 95:316–321

Sibson NR, Mason GF, Shen J, Cline GW, Herskovits AZ, Wall JE, Behar KL, Rothman DL, Shulman RG (2001) In vivo ^{13}C NMR measurement of neurotransmitter glutamate cycling, anaplerosis and TCA cycle flux in rat brain during [2-^{13}C]glucose infusion. J Neurochem 76:975–989

Siesjoe BK, Thompson WO (1965) The uptake of inspired $^{14}CO_2$ into the acid-labile, the acid-soluble, the lipid, the protein and the nucleic acid fractions of rat brain tissue. Acta Physiol Scand 64:182–192

Sonnewald U, Westergaard N, Petersen SB, Unsgard G, Schousboe A (1993) Metabolism of [U-^{13}C]glutamate in astrocytes studied by ^{13}C NMR spectroscopy: incorporation of more label into lactate than into glutamine demonstrates the importance of the tricarboxylic acid cycle. J Neurochem 61:1179–1182

Storm-Mathisen J, Danbolt NC, Rothe F, Torp R, Zhang N, Aas JE, Kanner BI, Langmoen I, Ottersen OP (1992) Ultrastructural immunocytochemical observations on the localization, metabolism and transport of glutamate in normal and ischemic brain tissue. Prog Brain Res 94:225–241

Van den Berg CJ, Krzalic L, Mela P, Waelsch H (1969) Compartmentation of glutamate metabolism in brain. Evidence for the existence of two different tricarboxylic acid cycles in brain. Biochem J 113:281–290

Van Eijsden P, Behar KL, Mason GF, Braun KP, de Graaf RA (2010) In vivo neurochemical profiling of rat brain by ^{1}H-[^{13}C] NMR spectroscopy: cerebral energetics and glutamatergic/GABAergic neurotransmission. J Neurochem 112:24–33

Waagepetersen HS, Qu H, Hertz L, Sonnewald U, Schousboe A (2002) Demonstration of pyruvate recycling in primary cultures of neocortical astrocytes but not in neurons. Neurochem Res 27:1431–1437

Waelsch H, Berl HW, Rossi CA, Clarke DD, Purpura DP (1964) Quantitative aspects of CO_2 fixation in mammalian brain in vivo. J Neurochem 11:717–728

Waniewski RA, Martin DL (1998) Preferential utilization of acetate by astrocytes is attributable to transport. J Neurosci 18:5225–5233

Yu AC, Drejer J, Hertz L, Schousboe A (1983) Pyruvate carboxylase activity in primary cultures of astrocytes and neurons. J Neurochem 41:1484–1487

Chapter 33
Brain Metabolic Compartmentalization, Metabolism Modeling, and Cerebral Activity-Metabolism Relationship

Michel Merle and Jean-Michel Franconi

Abstract Brain metabolism modeling is a powerful method for analyzing pre-steady-state and steady-state kinetics of ^{13}C incorporation in brain metabolites resulting from consumption of ^{13}C-labeled substrates. Studies on brain metabolism modeling were initially carried out by considering a simple one-compartment model to analyze glutamate C4 labeling from [1-^{13}C]glucose consumption. Thereafter, more sophisticated models including two or three compartments were used to analyze labeling of Glu, Gln, Asp and GABA at different carbon positions. In this chapter, after a brief recall of the cellular and molecular basis of brain metabolic compartmentalization, a survey of the metabolic pathways involved in the different models is presented. The discussion focuses on both the pathways generally considered in the modeling and those which are generally not considered, although evidence for their occurrence has been reported. Thereafter, the cerebral activity-metabolism relationship is analyzed through a review of the flux rates determined in rat brain under different cerebral activity status. More particularly, the relations between neuronal oxidative metabolism, glutamate-glutamine cycle, astrocytic oxidative metabolism and anaplerosis with cerebral activity are discussed. Some aspects concerning the reliability, the limitations and possible progress in brain metabolism modeling are finally discussed.

Keywords ^{13}C-NMR • Anaplerosis • Brain metabolism • Cerebral activity-metabolism relationship • Glutamate-glutamine cycle • Metabolic flux • Metabolism modeling • Pyruvate carboxylase

M. Merle, Ph.D. (✉) • J.-M. Franconi, Ph.D.
Unité de Résonance Magnétique des Systèmes Biologiques,
Université Victor Segalen, 146, rue Léo Saignat, 33076 Bordeaux cedex, France
e-mail: ftc.merle@gmail.com; franconi@rmsb.u-bordeaux2.fr

I.-Y. Choi and R. Gruetter (eds.), *Neural Metabolism In Vivo*,
Advances in Neurobiology 4, DOI 10.1007/978-1-4614-1788-0_33,

33.1 Introduction

Since the 1990s, the application of modern NMR in the neuroscience field has provided many data providing better understanding of brain functioning. In particular, detailed information about amino acid neurotransmitter metabolism and brain energetics have been obtained from NMR spectroscopy methods devoted to the detection of ^{13}C. The basic experimental approach was to analyze the incorporation of ^{13}C in metabolites after consumption of a ^{13}C-labeled substrate. Different experimental models were used: either isolated astrocytes and neurons in primary cultures, cocultures of these cells and brain slices incubated in a medium containing a ^{13}C-labeled substrate; or whole brain after administration of the labeled substrate to the animal or human subject. Studies on whole brain were initially carried out ex vivo from NMR analyses of small animal brain extracts. Thereafter, the experiments were conducted non-invasively *in vivo*, making it possible to extend the investigations to human brain.

Interpreting the data from experiments carried out on whole brain required the development of relevant analytical methods able to take into account the complexity of brain metabolism due to the occurrence of various interacting compartments. The fate of ^{13}C through the metabolic network was thus investigated using mathematical models devoted to the analysis of either isotopomers, specific enrichments at isotopic steady state or metabolite-labeling kinetics. As the amount and the quality of data from NMR spectroscopy increased with both the improvement of NMR technology and analytical methods and the availability of more substrates labeled at different carbon positions, data analysis required models of increasing complexity to obtain an integrated view of brain metabolism. The development of analytical tools was particularly critical for understanding the brain activity-metabolism relationship.

This chapter is focused on some aspects of the modeling developed for analyzing data from ^{13}C labeling experiments. In the first part, the emergence of the concept of brain compartmentalization as the fundamental brain metabolism characteristics is briefly recalled. In the second part, after presenting the basic principles of metabolism modeling and the current version of the two-compartment model related to glutamatergic neurotransmission, the occurrence or the absence of some pathways in the modeling are discussed in detail. In the third part, we discuss the cerebral activity-metabolism relationship, particularly the coupling between astrocytic and neuronal metabolism with cerebral activity. As a conclusion, the fourth part is devoted to concluding remarks about the limitations and the future of brain metabolism modeling.

33.2 Brain Metabolic Compartmentalization

The concept of brain metabolic compartmentalization between astrocytes and neurons emerged more than 40 years ago from the experimental demonstration with radioactive tracers of different glutamate pools in the brain. The basic result was that within a short period after intracisternal administration of [^{14}C]glutamate, glutamine in brain of rats or monkeys was more labeled than its precursor (Berl et al. 1961). Similarly, after intracarotidal infusion of [^{15}N]NH_3 in cats, labeling of the α-amino group was higher in brain glutamine than glutamate (Berl et al. 1962). These results demonstrated that there were two different glutamate pools in the brain and that newly formed glutamine derived from the metabolically very active and smallest pool, whose size was estimated to 2–20% of total brain glutamate. At the end of their pioneering work, the authors concluded: *"The results presented in this paper support the view that the concept of metabolic compartments or pools provides the operational terms by which structure and metabolism may be integrated and control of metabolism becomes in part a consequence of the structural heterogeneity of cells and tissue"* (Berl et al. 1961). Thereafter, comparing brain amino acid labeling from either labeled carbohydrates (particularly glucose) or ketogenic precursors (particularly acetate), O'Neal and Koeppe (1966) concluded in the occurrence of two different tricarboxylic acid (TCA) cycles in the brain. From their own results and those of other authors, they proposed that both TCA cycles were accessible to pyruvate, whereas only one was accessible to acetate and HCO_3^- and closely associated with glutamate dehydrogenase and glutamine synthetase activities. Their proposal was illustrated by a metabolic scheme which was the first rough model of brain metabolic compartmentalization (O'Neal and Koeppe 1966).

The nature of the compartmentalization was thereafter refined. Examining the time courses of amino acid labeling in mouse brain after intraperitoneal injection of [1-^{14}C]- and [2-^{14}C]glucose and [1-^{14}C]- and [2-^{14}C]acetate, Van den Berg et al. (1969) suggested that a large fraction of aspartate and glutamate pools were in direct connection with a fast turning TCA cycle. Their results revealed that glutamine was not formed from the large glutamate pool extensively labeled by glucose via the fast turning TCA cycle, but from the small pool in close connection with the less active TCA cycle. In their work, various important issues were evoked: (1) Are the two glutamate pools and related TCA cycles specific of the two different cell types, neurons and glial cells? (2) Is there rapid isotopic equilibrium between glutamate and α-ketoglutarate (and between aspartate and oxaloacetate)? At that time, no firm answer to these questions could be put forward owing to the lack of data and contradictory results. (3) Examining the labeling rate of aspartate and glutamate under different brain status, they made the link between metabolic and cerebral activities: *"The metabolic pathways leading to the production of energy are probably under direct control of "functional activity". It would therefore not be surprising if the metabolism of compounds not directly involved in energy production were separated from the reaction pathways involved in energy production. Metabolic pathways used for biosynthetic reactions or for the removal of waste material would then be separated from those pathways that should be critically and highly controlled by the need of energy for the functional activity of the brain."*. (4) Labeling of brain glucose from [^{14}C]acetate was observed, indicating the occurrence of gluconeogenesis and then the possibility of pyruvate recycling. On the other hand, from their studies on brain amino acid labeling (including that of γ-aminoacid butyrate, GABA) in brain slices incubated with [1-^{14}C]acetate, [^{14}C]glutamate, [^{14}C]

HCO_3^-, [1-^{14}C]GABA or [1-^{14}C]glutamate, and *in vivo* after administration of [1-^{14}C]- or [^{3}H]acetate, Berl et al. (1970) stated the fundamental elements of brain metabolic compartmentalization: (1) the compartment containing the small glutamate pool is associated with synthetic processes, (2) the other compartment containing the larger pool is associated with energy metabolism, (3) acetate obligatorily enters the TCA cycle devoted to synthesis, (4) CO_2 fixation for synthetic processes enters the TCA cycle at the oxaloacetate or malate stage and (5) GABA catabolism mainly occurs in the "synthetic cycle", whereas its synthesis appears closely associated with the "energy cycle". These points constituted (and largely remain) the foundation for building metabolic models.

The first operational model allowing quantitative simulations was described by Van den Berg and Garfinkel (1971). In the model, the two TCA cycles were linked via a metabolic pathway where the large glutamate pool in connection with the TCA cycle not consuming acetate was the precursor of GABA. GABA was catabolized through the TCA cycle consuming acetate from which the glutamate of the small pool was synthesized and further transformed to glutamine, the precursor of glutamate in the large pool. This model was designed as the simplest possible able to fit data from a unique source of experiments. From a formal point of view, it corresponded broadly to the GABA shunt. Simulations of the model under steady-state conditions gave interesting results in the light of our present knowledge about brain metabolism. Indeed, the ratios between the small and large glutamate pools, Glu_S/Glu_L ~0.15, and between the activity of the two TCA cycles, TCA_S/TCA_L~0.25, were in the range of recent estimations.

The relationship between compartments and cell types was rationalized owing to results of cellular biochemistry and enzymology works. The occurrence of high affinity transport systems specific for neurotransmitter amino acids (particularly glutamate and GABA) in glial cells, making possible their rapid uptake, raised the question of the role of glia in neurotransmission. Assigning the synthetic TCA cycle containing compartment to glia was ascertained by the discovery of the localization of the two key enzymes glutamine synthetase (Martinez-Hernandez et al. 1977) and pyruvate carboxylase (PC) (Shank et al. 1985; Yu et al. 1983) and by the specific metabolism of acetate (Waniewski and Martin 1998) in glia.

Recent studies on brain metabolism have all taken these pioneering findings as reference points for data analysis by mathematical modeling.

33.3 Metabolism Modeling

Metabolism modeling is primarily of great help in the comprehensive understanding of complex physiological processes involving numerous steps and components. As regards brain functioning, the aim of modeling is to give access to an integrated view on the cooperative behavior of astrocytes and neurons in both anaplerosis and energy metabolism, particularly in connection with cerebral activity. Also, this is a

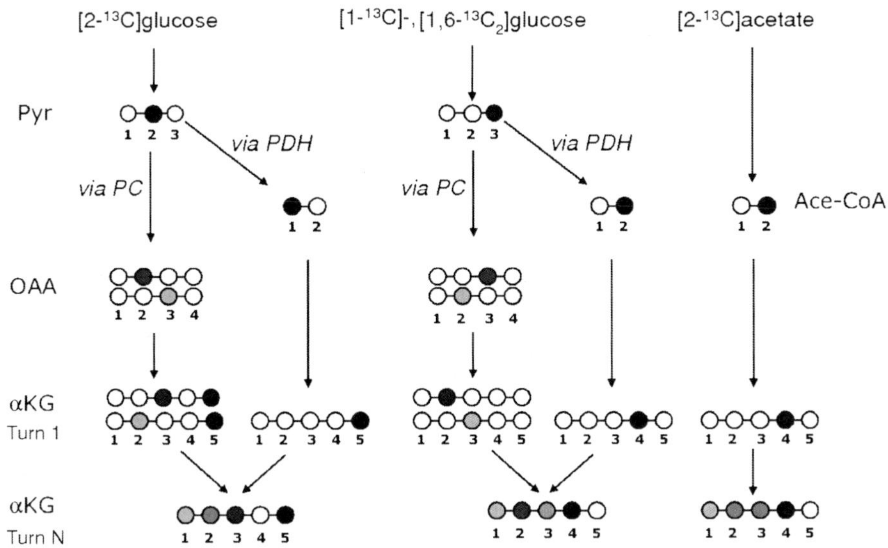

Fig. 33.1 *Fate of ^{13}C from different labeled substrates in metabolic pathways.* Labeling of α-ketoglutarate (αKG) from various substrates within the first TCA cycle turn (Turn 1) after the entry of the label into the TCA cycle, and under steady state (turn N), under the assumption of a glutamine efflux from the brain. From [2-^{13}C]glucose, the label entered via PDH from [2-^{13}C] pyruvate into the TCA cycle labels αKG at C5. Via PC, oxaloacetate (OAA) C2 is directly labeled. OAA C3 is thereafter labeled through partial back flow through malate and fumarase reactions, with C3 less enriched than C2. At steady state, αKG C1, C2, C3 and C5 are labeled with specific enrichments (ES): ES C5 > ES C3 > ES C2 > ES C1. From [1-^{13}C]- and [1,6-$^{13}C_2$]glucose, the label entered via PDH from [3-^{13}C]pyruvate labels αKG at C4. Via PC, OAA C3 is labeled, then OAA C2, with C2 less enriched than C3. At steady state, αKG C1, C2, C3 and C4 are labeled with ES C4 > ES C2 > ES C3 > ES C1. From [2-^{13}C]acetate, αKG C4 is labeled from [2-^{13}C]acetyl-CoA at the first turn and at C1, C2, C3 and C4 at subsequent turns with ES C4 > ES C2 = ES C3 > ES C1. Glutamate and aspartate are then labeled from their respective ketoacid. Symbols: •, •, • and ○, more or less enriched carbons

powerful tool for data analysis. It is particularly suited to analyze presteady- and/or steady-state data on compound labeling from metabolism of a ^{13}C-labeled substrate for determining metabolic fluxes. Nowadays, modeling is particularly useful for studying in situ brain metabolism by *in vivo* NMR spectroscopy.

The possibility of comprehensive modeling of brain metabolism was considerably improved owing to the availability of various ^{13}C-labeled substrates and developments of ^{1}H- and ^{13}C-NMR spectroscopy as analytical methods. The fate of ^{13}C in metabolic pathways is shown in Fig. 33.1, which illustrates the different labeling of α-ketoglutarate as a function of the labeled substrate used. Since the 1990s, many data have been collected from NMR studies thanks to the variety of experimental strategies offered: choice of a substrate specific for a particular cell type (e.g. Badar-Goffer et al. 1990; Hassel et al. 1995) or of the labeled carbon position (e.g. Sibson et al. 2001), use of ^{13}C- (e.g. Cerdan et al. 1990; Lapidot and Gopher 1994) or ^{13}C,

^{2}H-multi-labeled substrates (e.g. Chapa et al. 2000), competition between two substrates either labeled or not (e.g. Melø et al. 2005; Serres et al. 2005). In these studies, data analysis was based upon either direct comparison of metabolite enrichment levels, or on isotopomer analysis of labeled compounds and kinetic analysis of their labeling with sophisticated mathematical models.

33.3.1 Modeling of Isotopomer Data at Isotopic Steady-State

Isotopomer composition of labeled metabolites can be determined from the relative areas of ^{13}C multiplets in ^{13}C-NMR spectra arising from molecules labeled on adjacent carbons due to ^{13}C–^{13}C homonuclear J-couplings. The isotopomer composition of a given compound is potentially a very rich source of information about the metabolic pathways leading to its labeling. A model permitting the determination of substrate selection at the acetyl-CoA level and of relative fluxes of carbon on the oxidative (acetyl-CoA utilization) and nonoxidative (anaplerotic) pathways of the TCA cycle was developed by Malloy et al. (1988). This model was initially built to analyze heart metabolism where only one compartment was considered. The experimental data were the percent contributions of singlet, doublet(s) and triplet to C2, C3 and C4 glutamate carbon resonances under metabolic and isotopic steady state. Flux parameters were then evaluated by fitting the experimental data to the equation describing each of these contributions as a function of the parameters to be determined. The model was tentatively used to study the metabolism of cerebellar cells in primary cultures considered as homogeneous (one compartment) systems from the isotopomer composition of either glutamate from granule cells or glutamine from astrocytes (Martin et al. 1993). However, the original model of Malloy et al. (1988) assumed equal enrichments at C2 and C3 of the amino acid. It therefore appeared unsuitable for analyzing glutamine isotopomers from astrocytes incubated with [1-^{13}C]glucose because C2 was more enriched than C3, an evidence of pyruvate carboxylase activity (Martin et al. 1993). The model was therefore improved by considering the possibility of partial back flow in the TCA cycle from oxaloacetate to malate and fumarate (Fig. 33.2) through reversal of malate dehydrogenase and fumarase catalyzed reactions (Merle et al. 1996a). Analyzing the data with the improved model made it possible to determine an oxaloacetate-to-fumarate back flow of 39% in astrocytes, whereas 100% back flow was found in granule cells (Merle et al. 1996b).

Isotopomer analysis was applied to whole brain metabolism in rats infused with [2,4,6,8-$^{13}C_4$]octanoate (Ebert et al. 2003). As β-oxidation of fatty acids gives rise to acetyl-CoA without prior synthesis of pyruvate, brain metabolism of [2,4,6,8-$^{13}C_4$] octanoate could not give unequal labeling of glutamate/glutamine C2 and C3 via PC activity. Therefore, the authors used the method of Malloy et al. (1988) for data analysis. Although this method assumes only one metabolic compartment, discriminative

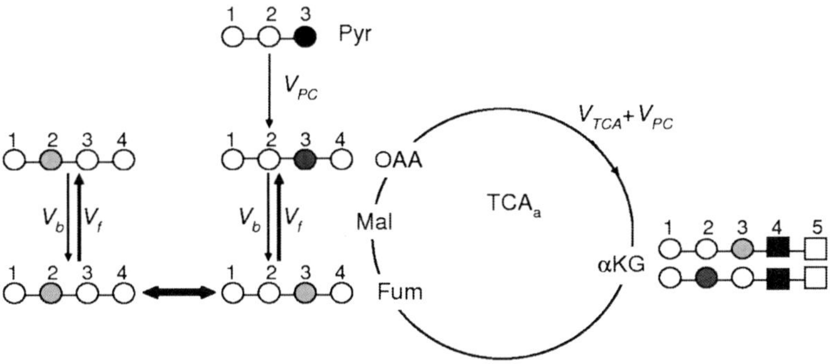

Fig. 33.2 *^{13}C-scrambling through back flow of oxaloacetate to malate and fumarate.* From [1-^{13}C]- or [1,6-$^{13}C_2$]glucose metabolism, carboxylation of [3-^{13}C]pyruvate via pyruvate carboxylase leads to C3- labeled oxaloacetate. Reversal of the malate dehydrogenase and fumarase catalyzed reactions leads to the formation of fumarate labeled either at C2 or C3, owing to the symmetry of the latter. Then, C2-labeled oxaloacetate is formed through the forward conversion of fumarate to oxaloacetate. Depending on the reverse yield (backward vs forward reaction rates, V_b and V_f, respectively), C2 and C3 labeled α-ketoglutarate isotopomers are formed in the next TCA cycle turn (with C2 more labeled than C3). In contrast, starting from [2-^{13}C]glucose leads to the initial formation of C2-labeled oxaloacetate and thus to α-ketoglutarate labeled at C2 and C3 (with C3 more labeled than C2). Symbols: •, •, • and ○ more or less enriched carbons, ■ and □ carbons from acetyl-CoA

information between the neuronal and glial compartments was obtained by analyzing glutamate and glutamine isotopomers, respectively.

Analytical solutions for isotopomer analysis of complex metabolic pathways including substrate oxidation, multiple pyruvate cycles and gluconeogenesis were recently described (Sherry et al. 2004). However, as this mathematic modeling assumes a mono-compartmental system, it is not suited for studying brain metabolism, which would require at least two compartments to be considered. Isotopomer analysis with a relevant model theoretically remains a potent analytical method to study brain metabolism (Mason and Rothman 2004). However the analysis requires data of very high quality, particularly from *in vivo* studies, making possible precise deconvolution of multiplets (as in Henry et al. 2003a).

33.3.2 Modeling of Dynamic Specific Enrichment Data

Basic principles of metabolic modeling applied to enrichment kinetics have already been described (e.g. Yu et al. 1997; Henry et al. 2006). From a defined metabolic network describing the flow of ^{13}C label, mathematic modeling relies on two types of equations describing mass and isotope balance for the metabolically connected compounds. For example, considering compound X synthesized from compounds

S_1, S_2 …and S_n at rates V_{in1}, V_{in2} and V_{inn} and precursor of products P_1, P_2 …and P_n at rates V_{out1}, V_{out2} and V_{outn}, as shown as follows:

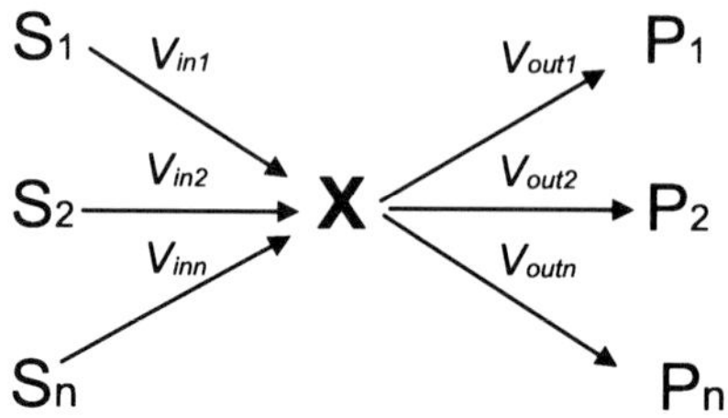

Mass balance equation is:

$$d[X]/dt = \sum V_{ini} - \sum V_{outi} \qquad (33.1)$$

and isotope balance equation is:

$$d[X^*]/dt = \sum V_{ini} \times \{S_i\} - \{X\} \times \sum V_{outi} \qquad (33.2)$$

where [X] is the concentration of compounds X, [X*] and {X} the concentration and specific enrichment of compound X labeled at a given carbon position, and $\{S_i\}$ the specific enrichment of compound S_i at the homologous carbon position, respectively.

At metabolic and isotopic steady state (d[X]/dt = d[X*]/dt = 0), the system of linear equations can be solved analytically, leading to the analytical expressions of parameters (rate constants) as a function of variables (enrichments). Under pre-steady-state conditions, the system is solved by numerical integration.

Current models for analyzing brain metabolism were derived from the two-compartment model initially proposed by Van den Berg and Garfinkel (1971) (Gruetter et al. 2001; Lebon et al. 2002; Öz et al. 2004; Patel et al. 2004; Shen et al. 1999; Sibson et al. 2001). The metabolic pathways for glutamatergic neurotransmission are shown in Fig. 33.3. From a formal point of view, there is a general consensus for the metabolic pathways represented in the figure by continuous arrows. In both astrocytic and neuronal compartments, these pathways include: (1) glycolysis as source of pyruvate to fuel the TCA cycle, (2) the efflux of lactate from the brain and the influx of blood lactate (V_{Lef} and V_{Ldil} in Fig. 33.3), and (3) the exchange between α−ketoglutarate and glutamate (V_x). Pathways specific of the glial compartment are (1) the formation of acetyl-CoA from acetate, (2) the formation of oxaloacetate from pyruvate via pyruvate carboxylase (V_{PC}), (3) the synthesis of glutamate from α-ketoglutarate through anaplerosis (V_{ana}), (4) the synthesis of glutamine from glutamate via glutamine synthetase and (5) the efflux of glutamine from the brain and the influx of glutamine from blood (V_{Gef} and V_{Gdil}). The conversion of glutamine to glutamate via glutaminase is specific to the neuronal compartment. The two compartments are

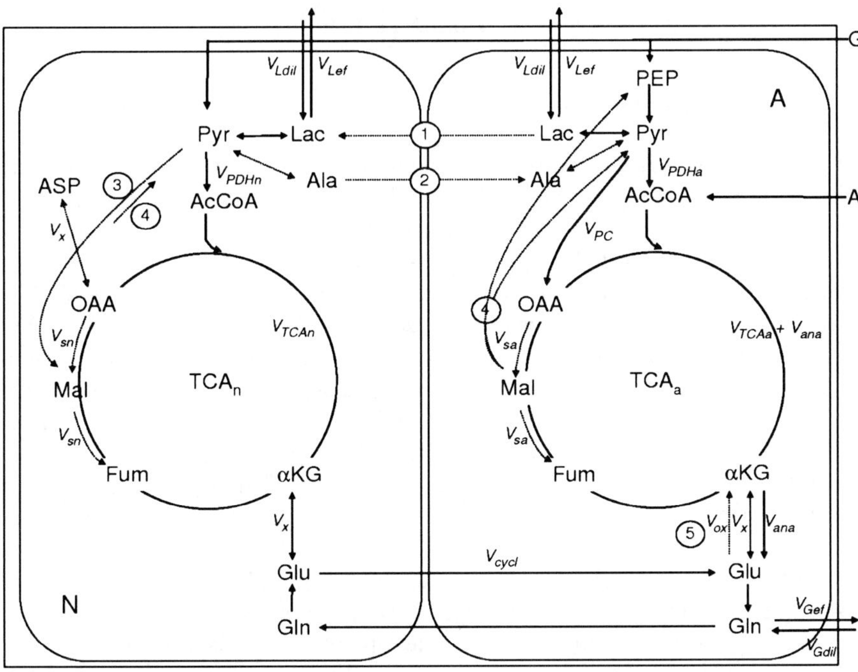

Fig. 33.3 *Astrocytic and neuronal metabolic pathways.* Metabolic pathways generally considered in astrocytes (*A*) and neurons (*N*) are defined by their rate constant: V_{Lef} and V_{Ldil}, efflux of lactate from the brain and influx of blood lactate; V_{PDHa} and V_{PDHn}, flux through astrocytic and neuronal pyruvate dehydrogenase; V_{PC}, flux through pyruvate carboxylase; V_{TCAa} and V_{TCAn}, flux through astrocytic and neuronal TCA cycle; V_{ana}, synthesis of glutamate from α-ketoglutarate through astrocytic anaplerosis; V_x, exchange between α-ketoglutarate and glutamate or between oxaloacetate and aspartate; V_{Gef} and V_{Gdil}, efflux of glutamine from the brain and influx of glutamine from blood; V_{cycl}, flux of neuronal glutamate to glia and reverse flux of glutamine to neurons; V_{sa} and V_{sn}, reverse flux from oxaloacetate to fumarate in astrocytes and neurons. Other pathways are: *1* transfer of lactate from astrocytes to neurons, *2* transfer of alanine from neurons to astrocytes, *3* neuronal pyruvate carboxylation, *4* pyruvate recycling and *5* glutamate oxidation

connected by the flux of neuronal glutamate to glia and the reverse flux of glutamine to neurons, constitutive of the glutamate-glutamine cycle (V_{cycl}).

Depending on the authors, back flow of oxaloacetate to malate and fumarate (V_{sa} and V_{sn}) and the exchange between oxaloacetate and aspartate (V_x) were not systematically considered. On the other hand, several metabolic pathways described in various works were generally not included in the modeling: (1) the transfer of astrocytic lactate to the neurons through the astrocyte-neuron lactate shuttle (ANLS) (Magistretti et al. 1999; Pellerin and Magistretti 1994), (2) the flux of alanine from neurons to astrocytes as component of a lactate-alanine cycle (Waagepetersen et al. 2000; Zwingmann et al. 2000), (3) neuronal pyruvate carboxylation via malic enzyme (Hassel and Brathe, 2000), (4) pyruvate recycling from malate via malic enzyme and/or oxaloacetate via phosphoenopyruvate carboxykinase and pyruvate kinase (Cerdan et al. 1990) and (5) glutamate oxidation in astrocytes (V_{ox}) (Qu et al. 2001).

33.3.3 Pathways Generally Considered in Modeling

33.3.3.1 Efflux of Lactate from the Brain and Influx of Blood Lactate

Using [1-^{13}C]glucose as labeled substrate, an important difference between half the specific enrichment of blood glucose C1 and the enrichment of brain glutamate C4 at steady state was always observed (e.g., Aurelli et al. 1997; Henry et al. 2002; Sibson et al. 1998; Tyson et al. 2003). This isotopic dilution may result from various processes: the hexose monophosphate shunt through which labeled glucose C1 is lost, dilution at the pyruvate/lactate node due to exchange with blood lactate and contribution of substrates other than glucose (endogenous ketone bodies, amino acids…) to brain metabolism, leading to a dilution at the acetyl-CoA node. Furthermore, entry of blood glutamine into brain tissue and its subsequent conversion into glutamate might also dilute glutamate labeling. Upstream the TCA cycle, dilution by blood lactate was generally included in modeling and considered as representing the various causes of isotopic dilution at the acetyl-CoA level (Gruetter et al. 2001; Mason et al. 2001; Öz et al. 2004; Patel et al. 2004, 2005a; Shen et al. 1999) (in the works by Patel et al. (2004, 2005a), [1,6-$^{13}C_2$]glucose was used instead of [1-^{13}C]glucose). In these works, the flux of lactate dilution (V_{Ldil} in Fig. 33.3) was determined by considering that blood lactate labeling corresponded to natural ^{13}C abundance (1.1%), i.e. the amount of lactate generated from glycolysis of the exogenous labeled glucose was negligible even after a long-lasting infusion (for example 3 h in Mason et al. 2001). In the study by Shen et al. (1999), the dilution flux was determined from the measured fractional enrichment difference between glucose C1 and glutamate C4 (taking into consideration the dilution by unlabeled pyruvate resulting from splitting of the 6 carbon molecule through glycolysis) at the end of the study, i.e. at isotopic steady state. In more recent studies, V_{Ldil} was determined by iteration. Values for V_{Ldil} are generally not given in result tables. However, when given, the values are in the same range as the glucose oxidation rate ($CMR_{Glc(ox)}$): $V_{Ldil} = 0.41 \pm 0.04$, $CMR_{Glc(ox)} = 0.41 \pm 0.04$ μmol g^{-1} min^{-1} in human (Gruetter et al. 2001) and $V_{Ldil} = 0.37 \pm 0.07$, $CMR_{Glc(ox)} = 0.34 \pm 0.03$ μmol g^{-1} min^{-1} in control rat (Patel et al. 2004), thus emphasizing the importance of the dilution. In the abovementioned studies, dilution at lactate level was considered as a true exchange between brain and blood lactate, i.e. influx = outflux ($V_{Lin} = V_{lout}$), except under intense neuronal activation where a net lactate efflux was considered (Patel et al. 2004).

In some studies, dilution through the hexose monophosphate pathway was considered independently of the dilution by blood lactate (Mason et al. 1992; Mason and Rothman 2004). In the modeling, removal of ^{13}C through the pentose phosphate pathway (PPP) was an input data based on the literature value ($V_{PPP} = 0.03 \times CMR_{Glc}$). Similarly, rates of brain-blood lactate exchange were calculated from literature data ($V_{Lin} = 0.05$ and $V_{Lout} = 0.085$ μmol g^{-1} min^{-1}, thus assuming net lactate efflux). These values are much lower than the values of V_{Ldil} determined by iteration.

In a recent study conducted with [1-^{13}C]glucose and [2-^{13}C]glucose, no dilution of the pyruvate/lactate level was considered and the fractional enrichment of pyruvate

C3 was assumed to be represented by that of glutamate C4 (Mason et al. 2007). The relevance of this assumption could be questioned given the importance of the isotopic dilution at the pyruvate/lactate/acetyl-CoA level (as noted above). However, the consequences of the assumption on the conclusions of the work are difficult to evaluate.

33.3.3.2 Efflux of Glutamine from the Brain and Influx of Blood Glutamine

The role of glutamine in cerebral nitrogen metabolism, and particularly in ammonia (and ammonium ion) detoxification, is widely acknowledged (Cooper 2001). This process relies on the high activity of the astrocytic glutamine synthetase which eliminates the excess of ammonia through glutamate amidation and thus makes brain detoxification possible via glutamine export through the blood-brain barrier (Grill et al. 1992). Net glutamine efflux was thus considered in most metabolism modeling studies.

Near isotopic steady-state (infusion time equal to or higher than 60 min in lightly anesthetized animals) labeling of glutamine C4 from [1-^{13}C]glucose metabolism was about 22% (mean value) less than that of glutamate C4 from ex vivo experiments (Aurelli et al. 1997; Bouzier et al. 1999; Chateil et al. 2001; Merle et al. 2002). Labeling of glutamine C4 was also 19% less than that of glutamate C4 from the ex vivo determination performed at the end of the infusion of awake rats with [1-^{13}C]glucose (Öz et al. 2004). A 26.5% drop in labeling in brain of rats infused with [1,6-$^{13}C_2$]glucose was also observed both in control animals and during intense neuronal activation (Patel et al. 2004). In their study, the authors showed the *in vivo* time courses of glutamate and glutamine C4 labeling determined on the basis of the measured end-point percent enrichments of the purified amino acids. In fact, there are generally no available data on glutamate and glutamine enrichments from purely *in vivo* studies. As a consequence, depending on the studies, isotopic exchange between brain and blood glutamine was more or less taken into consideration.

Neither glutamine efflux of brain glutamine nor dilution with blood glutamine was considered in the one-compartment model of Mason et al. (1992). This was also the case in their three-compartment model including GABAergic neurons, glutamatergic neurons and astrocytes (Mason et al. 2001).

Glutamine efflux was, however, considered in the recent update of their one-compartment model (Mason and Rothman 2004). In that model, it was assumed that, at isotopic steady state, glutamine and glutamate C4-specific enrichments resulting from [1-^{13}C]glucose metabolism were the same (the sole source of dilution between glucose and glutamate being upstream the TCA cycle) and that anaplerosis corresponded to glutamine efflux ($V_{ana} = V_{PC} = V_{Gef}$). Glutamine efflux was also considered in the study of [2-^{13}C]acetate metabolism in human brain by Lebon et al. (2002). The efflux rate was not determined in the study but its value from the literature was entered into the model as an input parameter to derive the astroglial TCA cycle flux (V_{TCAa}) and to assess its influence on the latter. The same assumption was made in the model of Gruetter et al. (2001) applied to human brain metabolism of [1-^{13}C]glucose.

Both net glutamine efflux from the brain and isotopic dilution by blood-borne glutamine were considered in some recent studies (Öz et al. 2004; Patel et al. 2004, 2005a; Sibson et al. 2001). In these studies, glutamine efflux was represented by V_{efflux}, but the meaning of this parameter was different. In the works by Sibson et al. (2001) and Patel et al. (2004; 2005a), V_{efflux} corresponded to the total glutamine efflux from the brain (V_{Gef} in Fig. 33.3), which is thus equal to the sum: dilution outflow + anaplerosis, or $V_{efflux} = V_{Gef} = V_{Gdil} + V_{ana}$. In the work by Öz et al. (2004), V_{efflux} corresponded to the net glutamine efflux equal to anaplerosis. Thus, under the assumption that both net brain glutamine efflux and exchange with blood glutamine occurred, $V_{efflux} = V_{ana} = V_{PC} = V_{Gef} - V_{gdil}$.

In the study by Sibson et al. (2001), rat brain metabolism was studied upon infusion of either [2-^{13}C]glucose or [1-^{13}C]glucose. V_{Gdil} was determined by fitting kinetic enrichments from [1-^{13}C]glucose and V_{ana} by fitting enrichments from [2-^{13}C] glucose. Öz et al. (2004) included a label dilution flux corresponding to an exchange with blood-borne glutamine in the glial compartment of the model previously used by Gruetter et al. (2001). This flux was considered to explain the lower glutamine than glutamate enrichment they observed by analyzing brain extracts 60 min after the start of glucose infusion (Öz et al. 2004). In the works by Patel et al. (2004, 2005a), the value for V_{Gdil} was iterated whereas that of anaplerosis was defined as a function of the TCA fluxes in glutamatergic and GABAergic neurons.

Label dilution flux and net glutamine efflux values were generally not given. The sole available values in the work of Öz et al. (2004) indicated similar rates for the two processes (~ 0.20 µmol gm^{-1}·min^{-1}) in the awake rat.

33.3.3.3 Exchange between α-Ketoglutarate and Glutamate

Labeling of brain glutamate from a labeled oxidative substrate requires transfer of the label from the mitochondrial TCA cycle intermediate α-ketoglutarate to cytosolic glutamate. This transfer proceeds mainly via α-ketoglutarate-glutamate and malate-aspartate shuttling through the mitochondrial membrane: (1) mitochondrial α-ketoglutarate is translocated into the cytosol via the malate/α-ketoglutarate exchanger and thereafter transaminated into glutamate by cytosolic aspartate aminotransferase and (2) cytosolic glutamate is translocated into mitochondria via the aspartate/glutamate exchanger and thereafter converted into α-ketoglutarate by the mitochondrial transaminase. This shuttling sustains the transfer of reducing equivalents from cytosol to mitochondria (Stryer 1995). The rate of label transfer from α-ketoglutarate to glutamate is an important issue for estimating TCA cycle flux from glutamate labeling. From a pioneering study with brain mitochondria, it was concluded that the exchange of label (^{14}C) between α-ketoglutarate and glutamate occurred at a more rapid rate than that of TCA cycle intermediate oxidation (Balazs and Haslam 1965). Rapid exchange (about tenfold the cycle flux) was also concluded from modeling of glutamate labeling by measuring the effects of different rates of conversion (Van den Berg and Garfinkel 1971).

The assumption of rapid exchange was retained by Fitzpatrick et al. (1990) who estimated the flux through TCA cycle from the *in vivo* kinetics of glutamate C4 labeling from [1-^{13}C]glucose. Regarding the very simple method they used to calculate the TCA cycle flux (a single exponential process), the assumption appeared validated by the delayed labeling of glutamate C3 (Fitzpatrick et al. 1990). Indeed, comparing the rates of glutamate C4 and glutamate C3 labeling from a labeled substrate leading to [2-^{13}C]acetyl-CoA is a-priori a good indicator of the exchange rate. As explained by Mason et al. (1992), if the exchange is rapid ($V_x >> V_{TCA}$), most of the ^{13}C entered into the TCA cycle leaves the cycle in the first turn to label the pool of glutamate (larger than the α-ketoglutarate pool). As a consequence, there is no more ^{13}C in the TCA cycle to label glutamate C3 at the subsequent turns. Glutamate C3 begins to be significantly labeled only after substantial C4 labeling, and therefore C3 labeling is delayed as compared to labeling at C4. On the other hand, if the exchange is not rapid ($V_x << V_{TCA}$), the label remains in the TCA cycle for several turns and thus labeling of glutamate C4 and C3 occurs at similar rates. Moreover, under this assumption, steady-state labeling of glutamate C4 and C3 are the same, independently of glutamate loss (for example, via glutamine efflux).

The data of Fitzpatrick et al. (1990) were analyzed by Mason et al. (1992) with a one-compartment model including both V_{TCA} and V_x as unknown parameters. The value for the V_x/V_{TCA} ratio was iterated to obtain the best fit of glutamate C3 labeling using as input data the labeling rate of glutamate C4. The results indicated that the exchange rate was around 100 times faster than the TCA cycle rate (Mason et al. 1992). As a consequence, they concluded that the TCA cycle rate could be simply determined from the kinetics of glutamate C4 labeling. However, in their analysis, they assumed no glutamate/glutamine loss (same glutamate C4 and C3 labeling at isotopic steady state), which could have biased their conclusion.

In many later studies with [1-^{13}C]glucose, fast exchange was assumed and the TCA cycle rate was estimated only from glutamate C4 labeling (Shen et al. 1999; Sibson et al. 1997, 1998; Mason et al. 2001). The same assumption was made in the study of [2-^{13}C]glucose metabolism by Sibson et al. (2001).

Fast exchange between the amino acids glutamate and aspartate and their homologous TCA cycle intermediates oxaloacetate and α-ketoglutarate was challenged by Gruetter et al. (2001) from the analysis of amino acid labeling from [1-^{13}C]glucose in human brain *in vivo*. They introduced the exchange rate V_x in their model, and considered the same rate for aspartate and glutamate. Simultaneous fitting of glutamate and glutamine C2, C3 and C4 and aspartate C2+C3 enrichment kinetics led to a value for V_x equivalent to the neuronal TCA cycle rate (0.57 μmol g^{-1} min^{-1}). In their study, they showed the profound effect of V_x on the estimation of V_{TCA} from glutamate C4 enrichment kinetics alone. Also, they showed the importance of the amino acid pool size on its enrichment kinetics (Gruetter et al. 2001). Their conclusion was strengthened by the finding that, under near physiological conditions, the rate of α-ketoglutarate synthesis in brain mitochondria was similar to that of α-ketoglutarate/glutamate efflux (Berkich et al. 2005). In this study, it was evaluated that, at each turn, 33% of mitochondrial α-ketoglutarate remained in the TCA cycle without equilibrating with cytosolic glutamate. As a consequence, the authors concluded that assuming fast equilibrium between α-ketoglutarate and glutamate could lead to underestimating the TCA cycle rate when the evaluation was based on the kinetics of glutamate C4 enrichment alone (Berkich et al. 2005).

The fact that V_x and V_{TCA} were in the same range suggested the coupling of the transfer of reducing equivalents from cytosol to mitochondria with oxidative metabolism through the energy demand linked to brain activation. This suggestion seemed strengthened by the studies by Choi et al. (2002) and Öz et al. (2004). Indeed, in deeply pentobarbital-anesthetized rats, the exchange rate was equal to the rate of neuronal pyruvate dehydrogenase flux, $V_x = V_{TCAn} = 0.35$ μmol g^{-1} min^{-1} (Choi et al. 2002), whereas higher values were found in awake rats, $V_x = 1.9$ and $V_{TCAn} = 1.13$ μmol g^{-1} min^{-1} (Öz et al. 2004). These results emphasized the need to consider parameter V_x in the model, as stated by Henry et al. (2002).

The issue of the exchange rate between α-ketoglutarate and glutamate was reappraised by Patel et al. (2004) from their study of [1,6-$^{13}C_2$]glucose metabolism in control and bicuculline-treated rat brain. As spectra acquired *in vivo* did not make possible the accurate determination of glutamate and glutamine C3 enrichments, they evaluated V_x from enrichments determined ex vivo at three different infusion times. In control rats, the value for V_x was 41 μmol g^{-1} min^{-1} (determined with a rather large error: 5–109 μmol g^{-1} min^{-1} 95% confidence interval) leading to a ratio $V_x/V_{TCAn} = 79$. Under seizures, V_x was 21 μmol g^{-1} min^{-1} (with a 4–51 μmol g^{-1} min^{-1} 95% confidence interval) leading to a ratio $V_x/V_{TCAn} = 37$. Thus, in both conditions, V_x appeared much higher than V_{TCAn}, in agreement with the assumption of fast exchange (Patel et al. 2004), and thus opposed to the conclusion of Öz et al. (2004). Moreover, there was no indication of the coupling of V_x with brain activity. In a more recent study, V_x was considered in the fitting of *in vivo* time courses of ^{13}C labeling of glutamate C3 and C4 and glutamine C4, and the enrichments of glutamate and glutamine C3 and C4 and GABA C2 and C3 determined ex vivo at three time points. However, the values found for V_x were not given (Patel et al. 2005a).

It seems unlikely that the abovementioned discrepancies about the exchange rate value can be explained by differences in modeling. The discrepancy originates rather from the fitting procedure and/or the difficulty of obtaining accurate data from *in vivo* experiments, particularly for glutamate C3 enrichment, which is crucial for V_x determination. At present, there is no consensus about the exchange rate between mitochondrial α-ketoacids and their homologous cytosolic amino acids.

33.3.3.4 Back Flow of Oxaloacetate to Malate and Fumarate

The issue of back flow of oxaloacetate to malate and fumarate is particularly important for evaluating astroglia anaplerosis. The pathway is shown in Fig. 33.2. Reversal of the malate dehydrogenase and fumarase reactions leads to the formation of the symmetrically labeled oxaloacetate isotopomer as regards the asymmetrical isotopomer initially formed (for example [2-^{13}C]oxaloacetate from [3-^{13}C]oxaloacetate). Back flow was evidenced in primary astrocytes by analyzing glutamine isotopomers obtained after cell incubation with [1-^{13}C]glucose. In that *in vitro* model and under the experimental conditions used, back flow amounted to 39% (Merle et al., 1996b).

Many *in vivo* studies were restricted to the analysis of glutamate/glutamine C4 labeling from a labeled substrate leading to [2-^{13}C]acetyl-CoA. Therefore, in these studies, modeling was not concerned by the issue of back flow. The situation is different regarding glutamate/glutamine C3 labeling from either [1-^{13}C]glucose, [2-^{13}C] glucose or [1,6-$^{13}C_2$]glucose. Concerning labeling at C2, the potential information is generally not accessible by *in vivo* spectrometry owing to signal overlapping in spectra (labeling at C2 was however determined in Gruetter et al. 2001). Therefore, depending on the authors, label scrambling through back flow of oxaloacetate to fumarate had a different status. In the recent update of their one-compartment metabolic model of [1-^{13}C] or [1,6-$^{13}C_2$]glucose metabolism, Mason and Rothman (2004) considered that complete equilibration (100% back flow) occurred.

In the two-compartment model used by Sibson et al. (2001) in their *in vivo* ^{13}C-NMR study of [2-^{13}C] or [5-^{13}C]glucose metabolism in the rat brain, back flow was considered as occurring in the astroglial TCA cycle. The rate of label scrambling between oxaloacetate C2 and C3 was introduced into the isotope balance equations. However, there was no detailed information about the fitting procedure used or the value found for this parameter (Sibson et al. 2001). The situation was slightly different in the *in vivo* study by Patel et al. (2005b) of [2-^{13}C]glucose metabolism by ^{1}H-[^{13}C]-NMR. Owing to overlapping, it was not possible to discriminate between glutamate and glutamine C2 or between glutamate and glutamine C3. However, potential back flow to fumarate of [2-^{13}C]oxaloacetate formed through pyruvate carboxylation was considered for determining the anaplerotic flux which was assessed from the sum of [2-^{13}C]- and [3-^{13}C]-glutamate and glutamine (Glx_{2+3}). However, no information about the yield of scrambling through back flow was available (Patel et al. 2005b).

Reverse flux from oxaloacetate to fumarate was neglected in the modeling by Gruetter et al. (2001) of [1-^{13}C]glucose metabolism in the human brain, although these authors were aware that estimation of pyruvate carboxylase activity could be underestimated when evaluated from the differential labeling of glutamate/glutamine C2 and C3. This was also the case in the *in vivo* ^{13}C-NMR study of [1-^{13}C] glucose metabolism in deeply pentobarbital-anesthetized rats (Choi et al. 2002).

In fact, the sole study where back flow was considered is that of Öz et al. (2004) on [1-^{13}C]glucose and $H^{14}CO_3^-$ metabolism in the awake rat. In this study, the model of Gruetter et al. (2001) was extended to allow for oxaloacetate-fumarate cycling. The rate of back flow (0.11 ± 0.12 μmol g^{-1} min^{-1}) was determined from the fit of ^{13}C and ^{14}C combined data. This back flow induced a 27% equilibration of the label entered into the TCA cycle via pyruvate carboxylase. Comparative analyses of the data when considering or not the occurrence of back flow revealed the slight underestimation of V_{PC} when back flow was neglected (Öz et al. 2004). This 27% equilibration was determined in awake rats, i.e. for a given metabolic status of the animals. One may suppose that the yield of back flow would depend on both TCA cycle and pyruvate carboxylase fluxes and also on concentrations of intermediates (three variables that are probably not independent). Therefore, it would be interesting to analyze this phenomenon as a function of brain activation.

33.3.4 Pathways Generally Not Considered in Modeling

33.3.4.1 Transfer of Astrocytic Lactate to Neurons

The astrocyte-neuron lactate shuttle (ANLS) has been proposed as the coupling mechanism between cerebral activity and energy metabolism (Pellerin and Magistretti 1994). The mechanism describes that after glutamatergic neurotransmission, (1) glutamate uptake (co-transported with 3 Na^+) into astrocytes induces Na^+-K^+ ATPase and glutamine synthetase activation, which results in ATP consumption, (2) this energy demand stimulates aerobic glycolysis, which leads to lactate synthesis and (3) the newly synthesized lactate is used as energy source in neuron oxidative metabolism (Pellerin 2003). According to this mechanism, two moles of ATP are required to recycle one mole of glutamate, which corresponds to the glycolysis of one glucose molecule per glutamate. Such a stoichiometry was reported to occur in brain (Sibson et al. 1998) but this does not constitute per se the ultimate evidence of the occurrence of the mechanism *in vivo*. Indeed, if the non-oxidative glucose consumption to support rapid neuronal firing was admitted by Shulman et al. (2001), the excess lactate production observed during intense neuronal activation was assumed to be totally effluxed (Patel et al. 2004). In fact, until recently, the relevance of the mechanism was hardly contested (Chih and Roberts 2003; Hertz et al. 2007). Although it has been demonstrated that exogenous lactate is more specifically consumed in neurons (Bouzier et al. 2000; Hassel and Brathe 2000a), a crucial point for the validation of the model was the demonstration of the transfer of glycolytic lactate from astrocytes to neurons and its subsequent oxidation in these cells. This demonstration required discrimination between the metabolism of glucose and that of glycolytic lactate when both compounds are processed via the same intermediate, pyruvate. As the metabolism of pyruvate is highly compartmentalized (Bak et al. 2007), this discrimination could not be achieved simply, for example, by using glucose as labeled substrate (as in most studies). However, the production of lactate from astrocytic glycolysis and its subsequent oxidation in neurons was demonstrated ex vivo using an indirect approach based on the analysis of the metabolic competition between endogenous and exogenous lactate (Serres et al. 2003, 2004, 2005). In addition, these studies showed unambiguously the relationship between brain activity and lactate metabolism. However, the conclusions of these works were essentially qualitative and the relative contributions of glucose and glycolytic lactate to neuronal metabolism under various cerebral activity status remains to be investigated (Pathway 1 in Fig. 33.3).

The absence of the pathway corresponding to a net transfer of lactate from astrocytes to neurons (pathway 1 in Fig. 33.3) in the published metabolic models can be explained by the fact that they were built according to experimental conditions not specifically chosen to explore the issue of the ANLS. Hence, they were unable to prove/disprove or to evaluate the flux on this pathway. From a theoretical point of view, considering this pathway in the model would imply adding new parameters describing different glycolytic fluxes and different lactate pool sizes in the two compartments, which per se does not constitute an insurmountable difficulty. However, the model would be operational only if adequate experimental data were available. Relevant data would be, for example, metabolite enrichment kinetics from parallel experiments carried out with either labeled glucose or labeled lactate (as in Serres et al. 2004) and particularly important data would be the specific enrichments of both blood and brain lactate. From a practical point of view, metabolic competition between exogenous and endogenous lactate would induce a significant isotopic dilution in metabolite labeling as compared to the enrichments obtained by using only one exogenous labeled substrate. However, it could be expected that this drawback would be balanced by the information gain resulting from the simultaneous analyses of data recorded with each labeled substrate.

Numerous arguments plead for the ANLS as a mechanism sustaining the critical role of astrocytes in brain metabolic regulations linked to cerebral activity (Pellerin et al. 2007). In particular, the conversion of glucose into lactate in these cells is strengthened by the fact that the Asp/Glu exchange carriers involved in the malate/aspartate shuttle for reoxidation of cytosolic NADH are predominantly expressed in neurons, not astrocytes (Ramos et al. 2003). As a consequence, although the glycerol 3-phosphate shuttle could be an alternative way to transfer electrons from cytosolic NADH into mitochondria (McKenna et al. 2006), regeneration of NAD^+ in the astrocyte cytosol mainly relies on reduction of pyruvate into lactate (Berkich et al. 2007).

In fact, the ANLS mechanism was recently explicitly acknowledged and considered in a revised model for coupling of brain glucose metabolism with neuronal activity (Hyder et al. 2006). However, considering that glucose (and other compounds from blood or released by astrocytes) imported from the extracellular medium remains available, it behaves as a competitor of glial lactate to fulfill the neuronal energy needs linked to cerebral activity. Actually, with cultured cerebellar neurons, it was suggested that synaptic activity does not induce an upregulation of lactate metabolism (Bak et al. 2006). Therefore, contrary to an extremely restrictive interpretation of the ANLS model, glial lactate might not be the unique substrate available for neuronal metabolism. The relative contributions of the two substrates to neuronal metabolism might be regulated by the redox switch/redox coupling mechanism proposed by Cerdan et al. (2006). As a consequence, in this framework, the ANLS model might not be the sole explanation for the reported stoichiometry of glycolysis of one glucose molecule per glutamate recycled (Sibson et al. 1998). The contribution of pyruvate/lactate produced from astrocytic glycolysis to neuronal oxidative metabolism was estimated to be around 70%, the complement coming from neuronal glucose oxidation (Hyder et al. 2006).

33.3.4.2 Lactate-Alanine Cycle

The occurrence of an active alanine-lactate cycle in brain was hypothesized from results of *in vitro* studies carried out with cultured astrocytes, neurons and cocultures (Waagepetersen et al. 2000; Zwingmann et al. 2000). As reviewed by Zwingmann and Leibfritz (2003), the basic findings were that (1) alanine synthesis from either

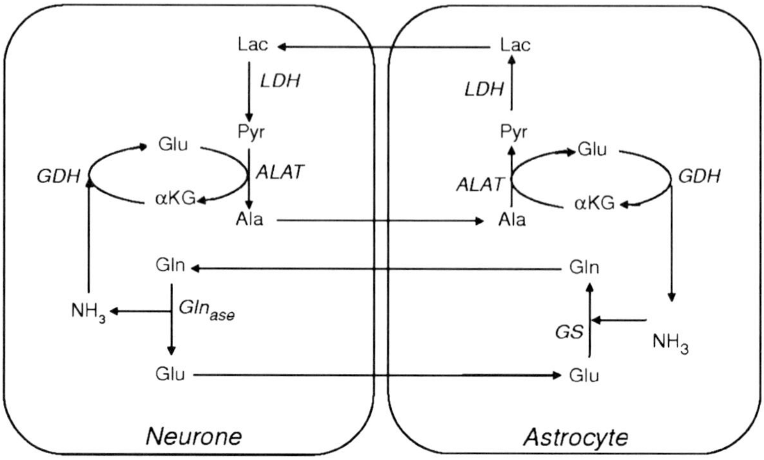

Fig. 33.4 *Alanine-lactate cycle between neurons and astrocytes.* The glutamate-glutamine cycle between astrocytes and neurons implies the release in neurons of one mole of ammonia/ammonium ion per mole of glutamate through glutaminase activity. Ammonia is recycled to astrocyte via (**a**) reductive amination of α-ketoglutarate to glutamate catalyzed by glutamate dehydrogenase (GDH), (**b**) synthesis of alanine catalyzed by alanine amino transferase (ALAT), (**c**) transfer of alanine to astrocyte and (**d**) release of ammonia in astrocytes via ALAT and GDH catalyzed reactions. The transfer of alanine from neurons to astrocytes is coupled to the transfer of lactate from astrocytes to neurons, thus defining the alanine-lactate cycle

glucose or lactate and its subsequent release in the culture medium were higher in the neuronal than the astrocytic cultures, and (2) when cells were incubated with alanine, the yield of conversion of the amino acid into lactate was much higher in the astrocytic than the neuronal cultures. These data suggested the possibility of a lactate-alanine cycle between astrocytes and neurons (more particularly glutamatergic neurons) *in vivo*, the primary function of which would be ammonia (and ammonium ion) detoxification of neurons. As regards ammonia metabolism, this mechanism could be the counterpart of the glutamate-glutamine cycle as illustrated in Fig. 33.4. Indeed, recycling of neurotransmitter glutamate in the astrocytic compartment via glutamine synthesis results in the mobilization of one mole of ammonia per mole of glutamate. This extra ammonia is released in the neuronal compartment through the glutaminase catalyzed conversion of glutamine to glutamate. In this compartment, the detoxification process could involve the reductive amination of α-ketoglutarate to glutamate catalyzed by glutamate dehydrogenase (GDH) and the subsequent transfer of the ammonia molecule from glutamate to alanine catalyzed by alanine amino transferase (ALAT). Then, alanine would be exported to the astrocytic compartment where recycling of ammonia to glutamine would proceed via the reverse reactions catalyzed by ALAT and GDH. The flux of carbon from neurons to astrocytes due to the export of alanine could be balanced by an import of lactate, thus completing the lactate-alanine cycle. From a formal point of view, this cycle is the perfect complement of the glutamate-glutamine cycle for the balance of carbon and ammonia. However, according to Fig. 33.4, the lactate-alanine cycle would specifically concern recycling of the nitrogen of the glutamine amido group. Indeed, the role of branched chain amino acids, and more particularly leucine, as nitrogen source of the α-amino group of glutamate and glutamine is well documented (Kanamori et al. 1998; Sakai et al. 2004). De novo synthesis of glutamate in the astroglial compartment for neurotransmitter glutamate pool replenishment involves both pyruvate carboxylase for the increase in α-ketoglutarate, and mitochondrial branched chain aminotransferase for glutamate synthesis by transamination with the branched chain amino acid (Hutson et al. 2005). Actually, a putative branched chain amino acid/branched chain α-keto acid (BCAA/BCKA) shuttle that carries nitrogen in the form of BCAA from the neurons to the astrocytes in connection with the transfer of nitrogen in the opposite direction through the glutamate/glutamine cycle has been proposed (Yudkoff et al. 1996; Hutson et al. 1998; Lieth et al. 2001). Moreover, there are alternative ways to recycle ammonia, in particular by simple NH_3 diffusion or active transport of NH_4^+ through ion channels (Marcaggi and Coles 2001). Therefore, the issue of the physiological significance and importance of the alanine flux from neurons to astrocytes *in vivo* remains to be clarified, all the more since no activity-dependent coupling of the glutamate-glutamine cycle with the lactate-alanine cycle was found *in vitro* (Bak et al. 2004) (Pathway 2 in Fig. 33.3).

Brain alanine metabolism was tentatively assessed upon infusion of [3-^{13}C]alanine (Serres and Merle, unpublished works). Although brain metabolite labeling was observed, no conclusion about the specific metabolism of alanine was possible owing to the degree of alanine peripheral metabolism (mainly in liver) which led to high enrichments of both blood glucose and lactate, even in the presence of an inhibitor of the gluconeogenesis pathway. Therefore, although the enzymatic equipment of both astrocytes and neurons supports the concept of the occurrence of a brain alanine-lactate cycle, there is no available data (especially NMR data) to test its relevance using metabolic modeling.

33.3.4.3 Neuronal Pyruvate Carboxylation

The concept of the occurrence in brain of two distinct compartments, one devoted to synthetic processes and the other to energy metabolism (Berl et al. 1970), and the astrocytic localization of pyruvate carboxylase (Shank et al. 1985; Yu et al. 1983) have led to the consensual idea that carbon flux through PC activity is the main (if not the sole) anaplerotic pathway via pyruvate carboxylation in the adult brain. In particular, this pathway is acknowledged as the main route for replenishment of the neurotransmitter glutamate and GABA pools. As a consequence, metabolic models do not allow for any alternative pathway. This idea was recently challenged by Hassel (2001a) who defended the thesis of the occurrence of neuronal pyruvate carboxylation in the adult brain. His argumentation was based on a previous finding that, after intrastriatal injection of [1-^{14}C]pyruvate or [^{14}C]HCO_3^- into mouse and rat brain, the specific enrichments of glutamate were higher than that of glutamine (Hassel and Brathe 2000a, 2000b). The activity of the pathway in neuronal cells was unambiguously evidenced by the release of labeled glutamate from cultured cerebellar granule neurons incubated with [1-^{14}C]pyruvate or [^{14}C]HCO_3^- (Hassel and Brathe 2000b). Moreover, comparison of aspartate enrichments obtained with [1-^{14}C]pyruvate or [2-^{14}C]pyruvate indicated that, under their experimental conditions, pyruvate carboxylation represented about one-third of pyruvate dehydrogenase (PDH) activity (Hassel and Brathe 2000b). The neuronal pyruvate carboxylation activity was attributed to malic enzyme (ME). In fact, mitochondrial malic enzyme activity was found to be much higher in cortical synaptosomes from adult rats than in primary cultures of cortical neurons or cerebellar granule cells (McKenna et al. 2000). In addition, immunocytochemical staining of mitochondrial ME in rat brain indicated a specific neuronal localization (Vogel et al. 1998b) (Pathway 3 in Fig. 33.3).

Neuronal pyruvate carboxylation was also deduced from NMR studies on primary cerebellar granule cells. With [2-^{13}C]pyruvate as labeled substrate, pyruvate carboxylation was evidenced by labeling of aspartate and glutamate C2 and C3 (Brand et al. 1992). Comparison of aspartate and glutamate labeling from either [2-^{13}C]- or [3-^{13}C]pyruvate led the authors to estimate that pyruvate carboxylation activity amounted to 28% of PDH activity (Brand et al. 1992). On the other hand, isotopomer analysis of glutamate from primary cerebellar granule cells incubated with [1-^{13}C]glucose revealed pyruvate carboxylation with a flux representing around 15% of the flux of citrate synthesis (Martin et al. 1993). However, in these two studies, pyruvate carboxylation was assumed to result from PC activity, but in both cases the results could also be explained by ME activity.

A major difference between neuronal and astrocytic pyruvate carboxylation (at least *in vitro*) is that equal enrichments of glutamate C2 and C3 were obtained in neurons, whereas glutamate/glutamine C2 was more enriched than C3 in astrocytes when primary cells were incubated with [1-^{13}C]glucose. This result was tentatively explained by the back flow of oxaloacetate to malate and fumarate in the TCA cycle, the consequence of which was that the prime asymmetrical labeling of oxaloacetate (at C3) resulting from carboxylation of [3-^{13}C]pyruvate via PC was lost at the level of the symmetrical intermediate fumarate (Merle et al. 1996a) (Fig. 33.2). Evaluating back flow yield led to values of 39% and 100% in cultured astrocytes and neurons, respectively (Merle et al. 1996b). This difference in back flow is in agreement with the proposal that neuronal pyruvate carboxylation involves ME (with a 100% back flow through reversal of only one step, the fumarase catalyzed reaction), whereas pyruvate carboxylation in astrocytes involves PC (with a 39% back flow through reversal of both malate dehydrogenase and fumarase catalyzed reactions) (Hassel 2001b).

With conventional labeled anaplerotic substrates (glucose, lactate, pyruvate) leading to either [2-^{13}C]- or [3-^{13}C]pyruvate, discriminating pyruvate carboxylation via either PC or ME in brain *in vivo* seems impossible. Indeed, even if it could be hypothesized that, in the astrocytic compartment, there is no (or at a given yield) back flow of oxaloacetate to fumarate - thus leading to no or partial equilibration of the oxaloacetate C2 (or C3) label entered via the PC pathway-, whereas there is total equilibration of malate labeling entered through ME activity in the neuronal compartment, data analysis using mathematical modeling would be an inextricable task owing to superposition of label scrambling through multiple TCA cycle turns and the interdependency of the neuronal and astrocytic TCA cycles through both the glutamate-glutamine cycle and the exchange between α-ketoglutarate and glutamate. Moreover, as the specific metabolite enrichments would be dependent on the relative activities on the pathways ME/PDH_n and PC/PDH_a, the fact that glutamine enrichment is higher than that of glutamate upon [2-^{13}C]glucose infusion (e.g., Sibson et al. 2001) does not constitute a proof that pyruvate carboxylation via ME does not occur or is negligible as compared to pyruvate carboxylation via PC. Indeed, with a flux ratio $PDH_n/PDH_a > 1$, the flux through ME could pass unnoticed, even if it is in the same range as the flux through PC. A final point is that reversibility of the malate reaction makes possible labeling of malate from pyruvate through a simple isotopic exchange, without any net anaplerotic flux (Kimmich et al. 2002).

The issue of neuronal pyruvate carboxylation therefore seems to remain an open question. The pathway is absent from current models of brain metabolism, although the occurrence of ME in adult neurons is well documented (Cruz et al. 1998; McKenna et al. 2000). The main reason is probably the experimental difficulty for evidencing the pathway, which cannot be distinguished from glial pyruvate carboxylation. As a consequence, only pyruvate carboxylation through PC is considered, which may constitute an oversimplification. Indeed, as noted by Öz et al. (2004), the possibility of significant replenishment of amino acid neurotransmitter pools by neuronal anaplerosis as a function of the demand would somewhat change the concept of the astrocyte-neuron metabolic relationship.

33.3.4.4 Pyruvate Recycling

The pathway known as pyruvate recycling corresponds to the net synthesis of pyruvate from malate/oxaloacetate via ME or via phosphoenolpyruvate carboxykinase (PEPCK) and pyruvate kinase, and the subsequent entry of the newly synthesized pyruvate into the TCA cycle via acetyl-CoA synthetase. The occurrence of pyruvate recycling in the brain was first reported by Cerdan et al. (1990) who found Glu isotopomers labeled at either C4 or C5 after metabolism of [1,2-$^{13}C_2$]acetate in the rat. As these isotopomers were not significantly transformed into the homologous Gln isotopomers, they concluded in the neuronal localization of the pathway. This was challenged by Hassel and Sonnewald (1995) who proposed rather a glial localization from their study in mice. The cellular localization and the ontogeny of the enzymes mediating pyruvate recycling were thus investigated by Cruz et al. (1998) who found that both ME and PEPCK were expressed in the rat brain, and more particularly in synaptosomes. They found that ME activity increased after birth to the adult level with the same time course as pyruvate recycling, whereas that of PEPCK remained low. From their results, they confirmed the neuronal localization of the pyruvate recycling pathway. In cultured cells, many studies substantiated the astrocytic localization of the recycling pathway (reviewed by Zwingmann and Leibfritz 2003). Its occurrence in glia was also reported from lactate labeling after incubation of freshly dissected rat retinas with $H^{14}CO_3^-$ (Lieth et al. 2001). However, the presence of the pathway was recently described in cultured cerebellar neurons (Olstad et al. 2007). Actually, the occurrence, the localization, the importance and the physiological significance of the pathway in the adult brain remain controversial (Pathways 4 in Fig. 33.3).

Brain pyruvate recycling in the adult rat brain was reported from NMR studies only when ^{13}C-labeled acetate was used. In most of these studies, involvement of peripheral metabolism was considered as very low and too small to question the conclusion of the cerebral origin of the detected pyruvate recycling. However, peripheral metabolism of acetate is known to interfere with its metabolism in brain. Van den berg et al. (1969) reported that ^{14}C-labeled glucose was present in brain of mice after intraperitoneal injection of ^{14}C-labeled acetate. They noticed that after 30 min, the specific radioactivity of glucose was in the same order of magnitude as the specific radioactivity in amino acids. More recently, it was found that a significant part of brain labeling from [2-^{14}C]acetate resulted from brain metabolism of blood [^{14}C]glucose (Cruz et al. 2005). These results are explained by the fact that although acetate (like fatty acids) is not a glucose precursor in animals, the transfer of the label to glucose proceeds via an isotopic exchange resulting in the labeling of TCA cycle intermediates that are gluconeogenesis precursors, particularly in liver (Kalderon et al. 1987).

The involvement of peripheral metabolism was not exhaustively investigated in NMR studies dealing with pyruvate recycling in the adult brain. In fed rats infused with [1,2-$^{13}C_2$]acetate, a significant contribution to brain metabolism of labeled glucose produced by gluconeogenesis was excluded because neither ^{13}C resonances from [1,2-$^{13}C_2$]- or [1-^{13}C]glucose in brain *in vivo* nor satellites for glucose resonance on ^{1}H-NMR spectra of plasma were detected (Cerdan et al. 1990). On the other hand, Hassel and Sonnewald (1995) concluded in the occurrence of pyruvate recycling in the brain of mice injected with [2-^{13}C]acetate from the presence of C2 or C3 labeled lactate in the brain. However, their results showed that blood lactate was more enriched than brain lactate, allowing the possibility that brain lactate (as well as blood lactate) could result from glycolysis of glucose labeled through liver metabolism. In fact, blood glucose enrichment from [1,2-$^{13}C_2$]acetate metabolism in rats was obvious in the work by Haberg et al. (1998) who found C1-glucose enrichment higher than that of GABA C2, glutamine C4 and glutamate C4 (after subtracting the labeling linked to the entry of [1,2-$^{13}C_2$]acetyl-CoA in the TCA cycle). Such a labeling indicated that the amino acid patterns could be fully explained by liver gluconeogenesis rather than by brain pyruvate recycling. With regard to the issue of pyruvate recycling, the authors noted the obvious influence of liver gluconeogenesis on cerebral metabolism. This was confirmed by Ebert et al. (2003) in a study with [2,4,6,8-^{13}C]octanoate as metabolic substrate, who suggested that the pyruvate recycling seen in the brain could be of hepatic origin. Recently, the issue of brain pyruvate recycling was specifically investigated by analyzing both the specific enrichments and NMR resonance patterns of brain and liver metabolites (glucose, lactate, alanine and glutamate) upon infusion of [2-^{13}C]acetate (Serres et al. 2007). The results demonstrated that the pyruvate recycling observed through brain metabolite labeling mainly originated from peripheral metabolism.

In most studies conducted with [1-^{13}C]- or [U-^{13}C]glucose, pyruvate recycling was never identified, likely because the enrichments resulting from metabolism of the endogenous glucose formed through neoglucogenesis was totally masked by the enrichments arising from metabolism of exogenous glucose. Moreover, as detection of the pathway relies on the occurrence and detection of specific lines in resonance signals (for example, the doublet due to glutamate C4-C5 coupling after [1-^{13}C] glucose metabolism), the demonstration of the pathways from *in vivo* studies remains practically impossible owing to the intrinsic limited resolution and sensitivity of *in vivo* NMR. In their *in vivo* study on [1,6-$^{13}C_2$]glucose metabolism in the rat brain, Henry et al. (2003b) found that the multiplets corresponding to glutamate C5 labeling possibly due to pyruvate recycling was around 3% of the glutamate C4 labeling after 7 h, thus suggesting minimal activity on the pathway. On the other hand, using [1-^{13}C]acetate or [2-^{13}C]glucose as metabolic substrates, the consequences of pyruvate recycling on metabolite labeling may become detectable. Significant enrichments of glucose carbons (in the range of 4% for C1 and C6 after 80 min) from metabolism of [2-^{13}C]acetate through hepatic gluconeogenesis were determined ex vivo in the rat (Serres et al. 2007). In the human brain, however, no significant contribution of pyruvate recycling to glutamate and glutamine C4 labeling was evidenced. An explanation for this was that the estimated contribution of the pathway to glutamate C4 labeling was in the range of the signal noise (Lebon et al. 2002). However, a difference between animal studies and human studies is that the rate of acetate infusion was usually higher in animals, thus leading to a more active pyruvate recycling (from either cerebral or hepatic metabolism). On the other hand, upon [2-^{13}C]glucose infusion in rats, significant labeling of glutamate and glutamine C4 were observed (Sibson et al. 2001). Assessing the possible source of this labeling (pentose phosphate pathway, metabolism of glucose labeled through gluconeogenesis and brain pyruvate recycling), the authors indicated that, in hyperammonemic rats, it could be fully explained by the hepatic metabolism and the phosphate shunt. In addition, they indicated that in normoammonemic rats, they were unable to distinguish glutamate/glutamine C4 labeling as a result of cerebral pyruvate recycling from labeling due to hepatic metabolism and phosphate shunt (Sibson et al. 2001). In a recent study, brain pyruvate recycling was concluded on the basis of the occurrence of labeled glutamate C4 (although the enrichment above natural abundance was very small, $0.16 \pm 0.22\%$) in the brain of rat infused with [2-^{13}C]glucose (Mason et al. 2007). However, in that study, the alternative pathways of glutamate C4 labeling were not considered and the labeling was explained solely by brain pyruvate recycling.

The occurrence and physiological significance of the pyruvate recycling pathway in the liver is well documented. In the brain, the enzymes defining the pathway are present, but it seems very difficult to obtain experimental data for the indisputable demonstration of its functionality. This is an important issue for understanding the cerebral activity linked neurone-glia relationship. Indeed, the pathway could be directly correlated with the oxidation of glutamate in the TCA cycle, as discussed below.

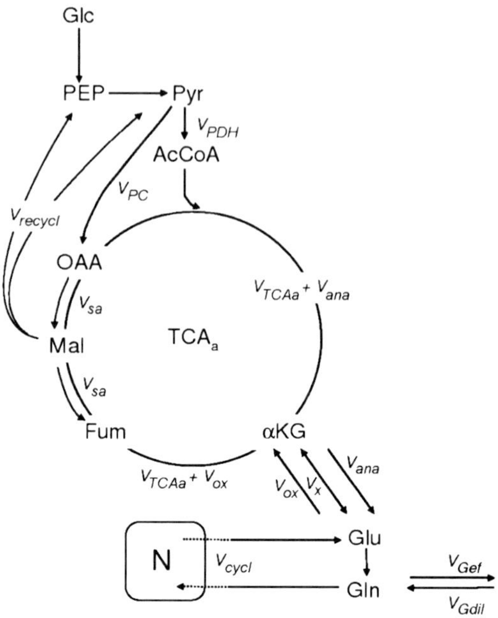

Fig. 33.5 *Astrocytic glutamate oxidation.* Neurotransmitter glutamate taken up by astrocytes can be either converted in glutamine by glutamine synthetase or oxidized via its entry in the TCA cycle at the rate V_{ox} after its oxidative deamination catalyzed by glutamate dehydrogenase. The four carbon intermediates formed from the glutamate carbon skeleton are then processed via two alternative pathways. (**a**) Oxaloacetate condenses with acetyl-CoA, thus leading to the formation of α-ketoglutarate from which new glutamate is formed by reductive amination. In this case, anaplerosis includes both glutamate oxidation and pyruvate carboxylation, $V_{ana} = V_{ox} + V_{PC}$. (**b**) The four carbon intermediates enter the pyruvate recycling pathways through malate enzyme and phosphoenol-pyruvate carboxykinase/pyruvate kinase activities at the rate $V_{recycl} = V_{ox}$. In this case, replenishment of the glutamate pool requires anaplerosis through pyruvate carboxylase activity and therefore, $V_{ana} = V_{PC}$. However, the two pathways are not exclusive and each of them may simultaneously be involved in glutamate oxidation according to complementary contributions

33.3.4.5 Glutamate Oxidation in Astrocytes

After release into the synaptic cleft, most of the neurotransmitter glutamate is taken up by astrocytes where the amino acid is further metabolized. Indeed, the jump in glutamate concentration due to the rapid influx of the amino acid activates the glutamate consuming pathways, thus leading to the buffering of the concentration increase. In astrocytes, there are two main pathways for glutamate metabolism: its amidation by glutamine synthetase and its oxidation in the TCA cycle. It is generally admitted that most of the glutamate is converted to glutamine and that the latter amino acid is thereafter exported to neurons, thus constituting the glutamate-glutamine cycle. Glutamate oxidation in the astrocytic TCA cycle implies the conversion of glutamate into α-ketoglutarate by oxidative deamination catalyzed by glutamate dehydrogenase and/or transamination. The 5-carbon atom skeleton of the amino acid is then oxidized to give 4-carbon atom TCA cycle intermediates. The glutamate oxidation pathway (V_{ox} in Figs. 33.3 and 33.5) must be distinguished from the exchange between α-ketoglutarate and glutamate (V_x). Indeed, the first two consequences of the entry of glutamate on the oxidation pathway are (1) the depletion of the neurotransmitter pool and (2) the increase in TCA cycle flux corresponding to the net entry of molecules. Therefore, to ensure correct brain functioning, the depletion in the glutamate pool must be compensated by a net glutamate synthesis (Pathway 5 in Fig. 33.3).

The simplest way for the recovery of the neurotransmitter glutamate pool is to build new molecules from the 4-carbon TCA cycle intermediates resulting from partial oxidation of the amino acid. As a consequence, to guarantee the continuity of the cycle, a stoichiometric increase in the entry of acetyl-CoA through citrate synthase is required. Thus, in the next TCA cycle turn, after condensation of oxaloacetate with acetyl-CoA, new α-ketoglutarate molecules are available to regenerate glutamate mainly by transamination. Thus, glutamate is its own anaplerotic substrate. With glucose as main brain energy substrate, this pathway would imply a link between astrocytic PDH activity and glutamatergic activation. From a formal point of view, this pathway is different from the exchange between α-ketoglutarate and glutamate, because the regeneration of one glutamate molecule requires a full TCA cycle turn. However, from an experimental point of view, the two pathways are indistinguishable when using labeled molecules because both occur via an isotopic exchange between α-ketoglutarate and glutamate (Fig. 33.5). In fact, as noted by Kimmich et al. (2002), the importance of isotopic exchange reactions would strongly moderate the conclusion of glutamate oxidation in studies about the fate of [U-$^{13}C_5$] glutamate or [U-$^{13}C_5$]glutamine in astrocytic cultures (Sonnewald et al. 1993; 1996; 1997). An important point is that through this pathway, glutamate regeneration does not require pyruvate carboxylase-mediated anaplerosis. Therefore, according to it, there would be no link between PC activity and glutamatergic activation.

The alternative way for net glutamate synthesis is anaplerosis through PC. In this case, PC activity would be coupled with glutamatergic activation. PC activity leads to the synthesis of oxaloacetate, and thus to an additional increase in the 4-carbon intermediate flux in the TCA cycle. As above, this pathway requires a stoichiometric increase in the entry of acetyl-CoA at the citrate level. In this second pathway and assuming that there is no loss of glutamate/glutamine through efflux to the blood, the mass balance for glutamate requires the flux through PC to match the flux of glutamate entry into the TCA cycle ($V_{pc} = V_{ox}$). However, this pathway requires the 4-carbon TCA cycle intermediates resulting from the entry of glutamate at the α-ketoglutarate level to leave the cycle before the next TCA cycle turn. The removal of intermediates occurs at the malate level. Malate is transported across the mitochondrial membrane to the cytosol, where it is converted to pyruvate by cytosolic ME, which is a specific glial enzyme (Vogel et al. 1998b). Then, pyruvate can be used to fuel the TCA cycle, allowing total oxidation of glutamate. In fact, this pathway corresponds to pyruvate recycling.

Under the assumption that a significant part of the neurotransmitter glutamate is oxidized in glia through the pathway involving pyruvate recycling, one would expect that brain pyruvate recycling would increase with cerebral activity. Although peripheral metabolism was demonstrated to be the main origin of the pyruvate recycling seen in brain, no difference in activity was found between pentobarbital-anesthetized and awake rats (Serres et al. 2007). This could appear surprising. In fact, a difference is evidenced only if the increase in activity is restricted to the pyruvate recycling pathway. Indeed, as pyruvate recycling is evidenced by specific resonance lines (for example, glutamate C4 singlet with [2-$^{13}C_2$]acetate (Cerdan et al. 1990) or glutamate C4–C5 doublet with [1-^{13}C]glucose (Serres et al. 2007)) resulting from pyruvate metabolism in the TCA cycle, the relative intensities of these lines remains the same if the activities on the different pathways leading to pyruvate synthesis are increased in the same way (thus leading to the same composition of pyruvate isotopomers). In other words, an increase in brain pyruvate recycling will not be detected (especially under steady state) if the synthesis of pyruvate via glycolysis is increased in the same way.

Modeling of glutamate pool replenishment would be simplified if glutamate oxidation in the astrocytic TCA cycle was the unique mode of glutamate/glutamine depletion. However, there is a net outflow of glutamine from the brain allowing detoxification of blood-borne ammonia (Cooper 2001). This outflow generates an additional depletion of the glutamate-glutamine pool, which must be filled up by PC mediated anaplerosis. Therefore, according to Fig. 33.5, the anaplerosis rate is $V_{ana} = V_{ox} + V_{Gef} - V_{gdil}$ and the pyruvate carboxylation rate through PC is $V_{PC} = V_{rec} + V_{Gef} - V_{gdil}$ (with V_{rec}, the rate of pyruvate recycling). In fact, determining which part of the PC activity is devoted to replenishment of oxidized glutamate and which part is devoted to replenishment of the lost glutamine is a daunting challenge.

At present, there is no satisfactory answer to the issue of glutamate oxidation. In fact, in most studies it is assumed that anaplerosis only involves PC activity and corresponds to the net efflux of glutamine into blood (Gruetter et al. 2001; Patel et al. 2005a; Sibson et al. 2001). According to this assumption, $V_{ana} = V_{PC} = V_{Gef} - V_{Gdil}$ (Fig. 33.5). Therefore, these authors implicitly assumed that there was no glutamate oxidation ($V_{ox} = V_{rec} = 0$). In a recent study, however, it was concluded that PC activity (which represented only about 6% of the rate of glutamine synthesis) was more related to glutamate oxidation in the glial TCA cycle than glutamine efflux, and therefore that the anaplerotic pathway was fully involved in neurotransmitter trafficking (Mason et al. 2007). This study was based on the analysis of glutamate and glutamine enrichments resulting from the metabolism of [1-^{13}C]- and [2-^{13}C]glucose in human cortex. The anaplerotic flux through PC was estimated by using a sophisticated mathematical model. However, its use required many assumptions and approximations which did not seem totally justified. For example, the rate of label scrambling between oxaloacetate C2 and C3 was chosen to be more than tenfold higher than the TCA cycle rate, thus allowing almost complete scrambling, whereas the data indicated that glutamate C3 was more enriched than glutamate C2. In the study, [2-^{13}C]glucose metabolism resulted in very low glutamate and glutamine enrichments in the range of natural abundance. Moreover, a considerable inter-subject variation induced very high standard errors for the enrichment, ranging from 49% for glutamine C3 to 218% for glutamine C4. Although the model used assumed that the anaplerosis rate was equal to the rate of glutamine efflux ($V_{ana} = V_{gef}$), the authors concluded that anaplerosis was mainly required for replacing oxidized glutamate by comparing the estimated value of the anaplerotic flux to that of glutamine efflux determined by the arterial-venous difference. Moreover, they took the occurrence of the small enrichment of glutamate C4 to be evidence of pyruvate recycling correlated with glutamate oxidation, without considering the possible interference of peripheral metabolism. This example illustrates how glutamate oxidation is difficult to prove and to evaluate, particularly from *in vivo* experiments. It also shows the limitations of mathematical modeling when adequate experimental data are not available.

33.4 Cerebral Activity-Metabolism Relationship

The most visible marker of oxidative brain metabolism detected by ^{13}C- or ^{1}H-NMR is glutamate. In human brain, the amino acid concentration is about 10 μmol g^{-1} (Gruetter et al., 1994; Mason et al. 2007). In the rat, glutamate content is also in the same range (Table 33.1). There are conflicting results about the changes in glutamate concentration as a function of brain electrical status. For example, either an increase (+ 38% in rats under morphine as compared to rats under pentobarbital (Sibson et al. 1998) or a decrease (–25% in rats under bicuculline as compared to controls (Patel et al. 2005b)) in glutamate concentration with cortical electrical activity have been reported. On the other hand, glutamate concentration under a given condition was not found always the same depending on the study (as reported in Table 33.1 for rats under halothane (Patel et al. 2004, 2005a, 2005b)). Actually, results from different studies show that the glutamate contents remained broadly very similar under the various conditions used (Table 33.1). Neuronal glutamate represents about 90% of the brain content (Cruz and Cerdan 1999; Öz et al. 2004), so it is generally considered a marker of the neuronal compartment. Furthermore, the distribution of other amino acids among different cell types is not definitely

Table 33.1 Glutamate concentration in rat brain under different status

Rat status	Glu concentration (μmol g^{-1})	References
Awake rat	10.5 ± 0.5	Öz et al. (2004)
Rat under bicuculline	11.5 ± 1.3	Patel et al. (2004)
	8.9 ± 1.3	Patel et al. (2005b)
Rat under morphine	11.2 ± 1.0	Sibson et al. (1998)
Rat under halothane	11.8 ± 1.8	Patel et al. (2004)
	14.4 ± 1.3	Patel et al. (2005a)
	8.9 ± 1.3	Patel et al. (2005b)
Rat under α-chloralose	9.7 ± 1.7	Sibson et al. 1997
	9.4 ± 1.7	Sibson et al. (1998)
	9.5 ± 0.7	Sibson et al. (2001)
Rat under pentobarbital	8.1 ± 0.9	Sibson et al. (1998)
	9.4 ± 0.9	Choi et al. (2002)
	12.1 ± 1.5	Patel et al. (2005a)

established: aspartate (~3 μmol g^{-1}) and GABA (~2 μmol g^{-1}) are also considered mainly neuronal whereas glutamine (~5 μmol g^{-1}) is considered mainly astrocytic (Öz et al. 2004; Patel et al. 2004, 2005a). The kinetics of glutamate C4 labeling from [1-^{13}C]glucose alone was used in the first evaluation of brain TCA cycle rate in an *in vivo* study using the simplest kinetic model for glutamate labeling (Fitzpatrick et al. 1990). In subsequent studies, discriminating the TCA cycle rate in neurons and astrocytes and evaluating other important fluxes by analyzing Glu, Gln, Asp and GABA labeling was possible by using various labeled substrates ([1-^{13}C], [1,6-$^{13}C_2$] or [2-^{13}C]glucose, [2-^{13}C]acetate…) and more sophisticated models. However, analyzing the relationship between cerebral activity and metabolism required working with different paradigms of brain electrical activity. In fact, there have been few *in vivo*/ex vivo studies where comparative metabolic analyzes were carried out under different cerebral activity status (Table 33.2).

In the work of Sibson et al. (1998), the TCA cycle rate and the rate of glutamine synthesis (V_{gln}) were evaluated in the brain of rats under either pentobarbital, α-chloralose or morphine by using the one-compartment model of Mason et al. (1995) to fit the kinetics of glutamate and glutamine C4 labeling from [1-^{13}C]glucose determined by ^{13}C-NMR at 7 T. In the fitting, there was considered to be a rapid exchange between mitochondrial α-ketoglutarate and cytosolic glutamate. The rate of neurotransmitter glutamate cycling (V_{cycl}) was then calculated from V_{gln} on the postulate that no glutamatergic neurotransmission occurred under pentobarbital.

In their work, Choi et al. (2002) and Öz et al. (2004) used the two-compartment model of Gruetter et al. (2001). Labeling kinetics from [1-^{13}C]glucose at glutamate and glutamine C3 and C4 and at aspartate C3 determined from *in vivo* ^{13}C-NMR at 9.4 T were used in Choi et al. (2002), whereas the enrichments at 5 time points of glutamate and glutamine C2, C3 and C4 and those at aspartate C2 and C3 from ex vivo analyses at 14.1 T were considered by Öz et al. (2004). Either pentobarbital-anesthetized or awake rats were used (Choi et al. 2002; Öz et al. 2004, respectively). The fitting parameters were neuronal TCA cycle rate ($V_{TCAn} = V_{PDHn}$), glial TCA cycle flux (V_{TCAa}), glial pyruvate carboxylation (V_{ana}), exchange rate between α-ketoglutarate and glutamate (V_x), label dilution at pyruvate/acetyl-CoA level (V_{Ldil}, V_{lef}, with $V_{Ldil} = V_{lef}$) and glutamate-glutamine cycling (V_{cycl}). The glial pyruvate dehydrogenase rate was thus calculated according to $V_{PDHa} = V_{TCAa} + V_{ana}$ (Choi et al. 2002). Additional parameters were considered by Öz et al. (2004), i.e. the influx of blood glutamine leading to isotopic dilution of the amino acid (V_{Gdil}) and oxaloacetate-fumarate back flow (V_{sa}) (however, the results shown in Table 33.2 were those obtained without considering oxaloacetate-fumarate back flow).

A two-compartment model similar to that of Gruetter et al. (2001) was used by Patel et al. (2004) to fit glutamate and glutamine C4 enrichment kinetics determined *in vivo* by spatially localized ^{1}H-(1-^{13}C)-NMR at 7 T, and the enrichments at 3 time points of glutamate and glutamine C3 and C4 and that of aspartate C3 determined ex vivo at 11.7 T. [1,6-$^{13}C_2$]glucose was used in the study. Control rats were under halothane anesthesia whereas intense neuronal activation was induced by bicuculline administration. Fitted parameters were neuronal TCA cycle ($V_{TCAn} = V_{PDHn}$) and

Table 33.2 Flux rates on pathways as defined in Fig. 33.3. Flux rate values are those found in the studies or calculated according to given information (as detailed in footnotes)

Flux rate (mmol min^{-1} g^{-1})	V_{PDHn} (= V_{TCAn})	V_{TCA}	V_{TCAa}	V_{PDHa}	V_{cycl}	V_{PC} (= V_{ana})	
Rat under pentobarbital		0.16 ± 0.08			0		Sibson et al. (1998)
Rat under α-chloralose		0.53 ± 0.08			0.13 ± 0.05		Sibson et al. (1998)
Rat under morphine		1.01 ± 0.23			0.40 ± 0.07		Sibson et al. (1998)
Rat under pentobarbital	0.35 ± 0.03		0.26 ± 0.06	0.29 ± 0.07[a]	0.04 ± 0.01	0.03 ± 0.01	Choi et al. (2002)
Awake rat	1.13 ± 0.14		0.40 ± 0.09	0.54 ± 0.12[a]	0.57 ± 0.21	0.14 ± 0.03	Öz et al. (2004)
Rat under halothane	0.52 ± 0.04[b]		0.09 ± 0.01[c]	0.15 ± 0.02[a]	0.22 ± 0.04	0.06 ± 0.01[d]	Patel et al. (2004)
Rat under halothane/bicuculline	1.14 ± 0.14[b]		0.20 ± 0.02[c]	0.33 ± 0.04[a]	0.52 ± 0.07	0.13 ± 0.02 [d]	Patel et al. (2004)
Rat under pentobarbital	0.35 ± 0.15[e]		0.06 ± 0.03[c]	0.07 ± 0.03[a]	0.02 ± 0.01	0.00 ± 0.00[d]	Patel et al. (2005a)
Rat under halothane	1.20 ± 0.05[e]		0.21 ± 0.01[c]	0.36 ± 0.02[a]	0.58 ± 0.03[f]	0.15 ± 0.01[d]	Patel et al. (2005a)
Rat under halothane						0.06 ± 0.01	Patel et al. (2005b)
Rat under halothane/bicuculline						0.07 ± 0.01	Patel et al. (2005b)

[a] Calculated from $V_{PDHa} = V_{TCAa} + V_{PC}$

[b] Calculated from $CMR_{glc(ox)N}$ with $V_{PDHn} = 2 \times CMR_{glc(ox)N}$

[c] Calculated with $V_{TCAa} = 0.15 \times V_{TCA(tot)}$ i.e. $V_{TCAa} = 0.176 \times V_{TCAn}$

[d] Calculated with $V_{PC} = V_{ana} = 0.2 \times V_{gln}$ with $V_{gln} = V_{cycl} + V_{ana}$ i.e. $V_{ana} = 0.25 \times V_{cycl}$

[e] Calculated with $V_{PDHn} = 2 \times CMR_{glc(tot)N} = 2 \times (CMR_{glc(ox)gaba} + CMR_{glc(ox)glu})$

[f] With $V_{cycl} = (V_{cyclNgaba} + V_{cyclNglu})$

glutamine synthesis rates ($V_{gln} = V_{cycl} + V_{ana.}$). Glial TCA cycle flux, glial PDH activity and anaplerosis (V_{TCAa}, V_{PDHa} and V_{ana}, respectively) were calculated assuming that $V_{ana} = 0.2 \times V_{gln}$ and $V_{TCAa} = 0.15 \times (V_{TCAa} + V_{TCAn})$. The exchange rate between α-ketoglutarate and glutamate (V_x) was also independently estimated from the glutamate C3 and C4 and aspartate C3 enrichment time courses determined ex vivo.

A three-compartment model including GABAergic neurons was considered by Patel et al. (2005a) in order to evaluate the contribution of GABA to glutamate/glutamine cycling. Metabolism in the brain of rats under pentobarbital or halothane was analyzed from the time courses of amino acid enrichments upon infusion of either [1,6-$^{13}C_2$]glucose + acetate or glucose + [2-^{13}C]acetate. Localized *in vivo* ^{13}C-NMR spectra were obtained at 7 T, whereas ex vivo analysis by ^{1}H-(1-^{13}C)-NMR was done at 11.7 T. Glutamate C3 and C4 and glutamine C4 labeling from [1,6-$^{13}C_2$] glucose was determined *in vivo*, whereas labeling of glutamate C3 and C4, GABA C2 and C3 and glutamine C3 and C4 from [1,6-$^{13}C_2$]glucose or [2-^{13}C]acetate was determined ex vivo at three different time points. In the model, the pools of GABA and glutamine were considered to be entirely neuronal and glial, respectively, whereas glutamate was assumed to be 10% in glia, 88% in glutamatergic neurons and 2% in GABAergic neurons. Seven parameters were iterated in the fitting procedure: flux through glutamatergic TCA cycle ($V_{TCA(Glu)}$) (as defined in Supporting Text of Patel et al. 2005a), net flux through GABAergic TCA cycle ($V_{TCA(Gaba)Net}$), flux through GABA shunt (V_{shunt}), diluting inflow flux of brain glutamine from blood glutamine (V_{dilGln}), diluting inflow of glial lactate/pyruvate from blood lactate ($V_{dil(a)}$), diluting inflow of GABAergic neuron lactate/pyruvate from blood lactate ($V_{dil(Gaba)}$) and diluting inflow of glutamatergic neuron lactate/pyruvate from blood lactate ($V_{dil(Glu)}$). Concerning the latter three parameters, there was no indication whether they were considered as independent parameters (no value given) and if the dilution at the acetyl-CoA level by metabolism of exogenous acetate was included in $V_{dil(a)}$. In the fitting, values of ratios ($V_{cyc(Glu/Gln)}$)/($V_{TCA(Glu)}$) and ($V_{cyc(Gaba/Gln)}$)/($V_{TCA(Gaba)}$) (with $V_{cyc(Glu/Gln)}$ and $V_{cyc(Gaba/Gln)}$, neurotransmitter glutamate and GABA cycle fluxes, respectively) determined from steady-state amino acid enrichments from [2-^{13}C] acetate in halothane-anesthetized rats were used as constraints. Pyruvate carboxylase flux was assumed to be 20% of the rate of total Gln synthesis flux and astroglial TCA cycle flux was assumed to be 15% of total TCA cycle flux.

The work by Patel et al. (2005b) concerned the estimation of pyruvate carboxylase flux in the brain of halothane-anesthetized control rats and under bicuculline seizures by analyzing [2-^{13}C]glucose metabolism. Amino acid enrichment kinetics were determined from localized *in vivo* ^{1}H-(1-^{13}C)-NMR spectra at 7 T using as reference the end point enrichment determined by ex vivo ^{1}H-and ^{13}C-NMR at 11.7 T. The flux through pyruvate carboxylase was estimated from the initial rate of glutamate and glutamine C2 and C3 labeling after correcting the enrichments for labeling due to the entry of [3-^{13}C]pyruvate into the TCA cycle (formed via pyruvate recycling in brain or peripheral metabolism and the pentose-phosphate pathway).

33.4.1 Neuronal Oxidative Metabolism

In the human brain, analyzing glutamate C4 labeling from [1-^{13}C]glucose during visual stimulation evidenced a large increase (up to 65%) in TCA cycle flux (Chen et al. 2001; Chhina et al. 2001). In the rat, the link between cerebral activity and rate of amino acid enrichment from [1-^{13}C]glucose was first noted by Shank et al. (1993) in their ex vivo experiments in awake and anesthetized animals. More recently, the neuronal TCA cycle rate (or V_{PDHn}) in rat brain was determined over a range of synaptic activity. As shown in Table 33.2, there were some discrepancies between values determined in the different studies, e.g. in rats under halothane, V_{PDHn} was more than twofold higher in Patel et al. (2005a) than in Patel et al. (2004) (also true for other rates). Therefore, the results in Table 33.2 are shown according to the different research groups for studies carried out with a given substrate. In this way, the differences in metabolic activity were self-consistent, i.e. independent of differences in experimental conditions or analytical procedures: NMR analyses, assumptions in modeling, constraints in fitting (however, data in Choi et al. (2001) were acquired *in vivo* whereas those in Öz et al. (2004) were from ex vivo analyses).

The results in Table 33.2 show that neuronal oxidative metabolism appears strongly correlated with cerebral activity. The same conclusion was drawn from studies carried out ex vivo with different combinations of labeled and unlabeled substrates (Serres et al. 2003, 2008). Actually, this is not surprising considering that brain activity induces an energy demand. This result is ascertained by the fact that, as a consequence of brain compartmentalization, neuronal oxidative metabolism is directly correlated with the kinetics of glutamate labeling from [1-^{13}C]- or [1,6-$^{13}C_2$] glucose. Moreover, as glutamate is the most concentrated brain amino acid, and because 90% of the neurotransmitter pool is neuronal (Cruz and Cerdan 1999), its enrichment at C4 from [3-^{13}C]pyruvate generates a strong easily detectable NMR signal both *in vivo* and *in vitro* with current NMR methods. A priori, a robust evaluation of the neuronal oxidation flux was therefore expected by analyzing glutamate C4 enrichment kinetics, even using a very simple model. However, a crucial parameter for evaluating the neuronal TCA cycle rate is the α-ketoglutarate/glutamate exchange rate. As there is no consensus about the value of V_x (as discussed above), the determination of the TCA cycle rate could involve some bias depending on the assumption of either a fast or less rapid exchange rate (as discussed by Henry et al. 2006). However, this would not invalidate the correlation between oxidative metabolism and cerebral activity.

In the work of Sibson et al. (1998), glial and neuronal TCA cycles were not distinguished. Therefore, the values in Table 33.2 were related to both TCA cycles. However, considering that glutamate is particularly a neuronal marker, the weight of the neuronal TCA cycle in the determination of V_{TCA} was much greater than that of the glial cycle. In the other studies, V_{PDHa} and V_{PDHn} were differentiated. In the works by Choi et al. (2002) and Öz et al. (2004), they were considered independent, whereas they were assumed dependent ($V_{TCAa} = 0.15 \times V_{TCAtotal}$, and thus $V_{TCAa} = 0.176 \times V_{TCAn}$) in the studies by Patel et al. (2004, 2005a). Depending on the relative weight of neurons and glia in brain metabolism, this constraint could have generated some bias in the determination of V_{PDHn}.

A linear relationship between cerebral oxidative glucose consumption ($CMR_{Glc(ox)}$) (the value of which was assumed to be half the TCA cycle rate) and the rate of glutamate-neurotransmitter cycling (V_{cycl}) was revealing in the work of Sibson et al. (1998). Its slope indicated a stoichiometry close to 1/1 between glucose oxidation and glutamatergic neuronal activity. Based on the coupling mechanism between glucose consumption and glutamatergic activity whereby glutamate uptake triggers aerobic glycolysis (Pellerin and Magistretti 1994), the authors proposed that the energy required for glutamate recycling was specifically provided by glycolysis (Sibson et al. 1998). Indeed, restoration of the ion balance after glutamate uptake (glutamate is co-transported with 3 Na^+) through Na^+/K^+-ATPase activation requires 1 ATP/glutamate mole, and the conversion of glutamate to glutamine by glutamine synthetase another ATP mole/glutamate. In fact, this need of 2 ATP/glutamate could be exactly covered by glycolysis, as 2 ATP/glucose are produced at the end of the glycolytic pathway. The observed 1/1 stoichiometry was thus interpreted as reflecting an activity-energy coupling between astrocytes and glutamatergic neurons where neurotransmitter recycling was energized by astrocytic glycolysis. The first consequence of this mechanism was that lactate, the end product of glycolysis, remained as the available substrate for neuronal oxidative metabolism. These results and the interpretation of Sibson et al. (1998)) were challenged by Gruetter et al. (2001) who used a more sophisticated metabolic model to analyze the kinetics of glutamate, glutamine and aspartate labeling at different carbon positions. The main critical point concerned the exchange rate between α-ketoglutarate and glutamate (V_x). They found that V_x was in the same range as V_{PDHn}. They considered therefore that assuming $V_x >> V_{PDHn}$ as in Sibson et al. (1998) led to an underestimation of the TCA cycle rate and thus, that the reported 1/1 stoichiometry between $CMR_{Glc(ox)}$ and V_{cycl} did not really hold.

In a recent review, the relationship between $CMR_{Glc(ox)}$ and V_{cycl} was reexamined taking into account all the published results (Hyder et al. 2006). The authors showed that all the data were in agreement with the linear relationship and the 1/1 stoichiometry of Sibson et al. (1998), although V_x was either similar (Choi et al. 2002; Öz et al. 2004) or faster (de Graaf et al. 2004; Patel et al. 2004) than V_{PDHn}. This could appear surprising given the incidence of V_x on the determination of V_{PDHn} (Henry et al. 2006). Therefore, the fact that the results remained compatible suggests the occurrence of some small but significant differences either in the processing of the NMR data or in the model. For example, although the two-compartment models used in the more recent studies were very similar, Choi et al. (2002) and Öz et al. (2004) considered V_{TCAa} and V_{TCAn} as two independent parameters and that no back flow of oxaloacetate to fumarate occurred, whereas dependent parameters and total back flow were considered by De Graaf et al. (2004) and Patel et al. (2004).

Neuronal oxidative metabolism is undoubtedly correlated with cerebral activity. The mechanism of coupling relies on astrocyte-neuron compartmentalization and on the occurrence of the glutamate-glutamine cycle. Much evidence pleads in favor of the ANLS mechanism initially proposed by Pellerin and Magistretti (1994), whereby astrocytic glycolytic lactate is used as oxidative fuel by neurons. However, a functioning according to a rigid interpretation of the mechanism in terms of stoichiometry seems unlikely. An amended version of the ANLS mechanism was recently proposed where glucose uptake and oxidation occur in both glia and neurons and where much but not all of glial lactate is transported to neurons (Hyder et al. 2006).

33.4.2 Glutamate-Glutamine Cycle

Recycling of glutamate neurotransmitter via the glutamate/glutamine cycle is the fundamental element in the cerebral activity-metabolism relationship. As this cycle involves glutamine synthesis, the kinetics of glutamine labeling from labeled glutamate was considered as the main marker of the activity on this pathway. However, as glutamine synthetase is specifically an astrocytic enzyme, there are two potential labeled glutamine precursors: the neurotransmitter glutamate labeled via the neuronal TCA cycle and taken up in astrocytes, and the astrocytic glutamate labeled via the astrocytic TCA cycle. As a consequence, from a theoretical point of view, determining the activity of the recycling pathway (V_{cycl}) requires a two-compartment model. Indeed, with the one-compartment model of Mason et al. (1992), only the glutamine synthesis rate was determined by analyzing the kinetics of glutamate and glutamine C4 labeling from [1-^{13}C]glucose (Mason et al. 1997). On the other hand, determining V_{cycl} independently of V_{TCAa} was possible with a two-compartment model (Gruetter et al. 1998; Shen et al. 1999).

In their study, Sibson et al. (1998) used the one-compartment model of Mason et al. (1995) to fit both glutamate and glutamine C4 enrichment kinetics. In this way, they determined the rate of glutamine synthesis (V_{gln} in their work). However, considering that in the group of rats under pentobarbital, there was no more synaptic activity, so no glutamate recycling, the value of V_{gln} in that condition was assumed to correspond to anaplerotic activity. Therefore, assuming that anaplerosis remained the same under the different brain activity status, this value was subtracted from other values determined for rats under α-chloralose or morphine to obtain V_{cycl}. (Table 33.2). From this study, the close activity/metabolism relationship was evidenced from the $CMR_{Glc(ox)}/V_{cycl}$ linear dependence.

Studies by Choi et al. (2002) and by Patel et al. (2005a) confirmed the very low synaptic activity (very small V_{cycl} value) under deep pentobarbital anesthesia (Table 33.2).

In the study of Patel et al. (2004), V_{TCAn} and V_{Gln} were determined by fitting with the two-compartment model. V_{cycl} was calculated from the equation $V_{Gln} = V_{cycl} + V_{ana}$ with $V_{ana} = 0.2 \times V_{gln}$. In this way, the parameters V_{cycl} and V_{ana} were not considered as independent. In the study by Patel et al. (2005a) where a three-compartment model was used, V_{cycl} was the sum of two terms $V_{cyc(GABA/Gln)}$ and $V_{cyc(Glu/Gln)}$ corresponding to cycling in GABAergic and glutamatergic neurons, respectively. The rates of GABA/Gln and Glu/Gln cycling were calculated from the iterated values of the TCA cycle rate in each type of neuron by using the relations $V_{cyc(GABA/Gln)}=0.63 \times V_{TCA(GABA)}$ and $V_{cyc(Glu/Gln)}=0.45 \times V_{TCA(Glu)}$ (determined from the specific amino acid labeling at steady state), considering that the contributions of $V_{TCA(GABA)}$ and $V_{TCA(Glu)}$ to the total oxidative metabolism in neurons were 18 and 82%, respectively.

For the different studies where a two- or three-compartment model was used, the ratio between cycling and neuronal TCA cycle rates appeared very similar : $V_{cycl}/V_{PDHn}=0.47\pm0.03$ (ratios under deep pentobarbital anesthesia were not included in the analysis owing to the very small neurotransmitter cycling in that condition). This could suggest that the determination of the V_{cycl}/V_{PDHn} ratio was more robust than that of each individual rate. This is for example particularly true for rats under halothane where V_{cycl} was either 0.22 or 0.58 mmol min^{-1} g^{-1} and V_{PDHn} was either 0.52 and 1.20 mmol min^{-1} g^{-1} in Patel et al. (2004) and (2005a), respectively. Taking into consideration the differences in the analytical procedures (modeling, assumptions, constraints between parameters), the reliability of the determination of V_{cycl} remains to be firmly established. However, all the data indicated a tight coupling between neurotransmitter recycling and neuronal oxidative metabolism.

33.4.3 Astrocytic Oxidative Metabolism

The strategic location of astrocytes between blood capillaries and neurons suggests an essential physiological role of these cells in brain functioning. At present, the involvement of astrocytes in various processes is well recognized, e.g. blood flow regulation, transport of nutrients, control of the extracellular environment, signaling (Nedergaard et al. 2003). In particular, in direct connection with cerebral activation via glutamatergic neurotransmission, the role of astrocytes in glutamate recycling and de novo synthesis is well documented. All these functions are energy-consuming but the role of glial oxidative metabolism in response to energy demand has not been clearly established, especially in NMR studies.

Indeed, as the glutamate content is about tenfold higher in neurons than in astrocytes (Cruz and Cerdan 1999), most NMR studies on metabolite labeling from [1-^{13}C]- or [1,6-$^{13}C_2$]glucose were focused on neuronal oxidative metabolism, especially since the early evaluation of the TCA cycle flux through the conversion of pyruvate in acetyl-CoA in the small and large glutamate compartments indicated a higher flux in this latter (0.30 and 1.25 μmol g^{-1} min^{-1}, respectively, in the rat) (Van den Berg and Garfinkel 1971). In human, glial TCA cycle rate (V_{TCAa}) determined upon [1-^{13}C]glucose infusion amounted to 0.06 ± 0.02 or 0.06 ± 0.04 μmol g^{-1} min^{-1} (Shen et al. 1999 or Gruetter et al. 2001, respectively). The lack of precision in these determinations illustrated the difficulty of evaluating this parameter. On the other hand, V_{TCAa} was determined upon [2-^{13}C]acetate infusion (Lebon et al. 2002) by taking advantage of the specific metabolism of acetate in astrocytes. The value found was 0.14 ± 0.04 μmol g^{-1} min^{-1}, i.e. about twice the values determined from [1-^{13}C] glucose, which illustrated the difficulty in determining this flux.

Concerning the relationship between astrocytic oxidative metabolism and brain activity, some values are given in Table 33.2. As already mentioned, V_{TCAa} and V_{TCAn} were not independent in the works by Patel et al. (2004, 2005a). This introduced a strong constraint in the fitting. The rationale of the constraint does not seem very clear. Indeed, Patel et al. (2004) assumed that V_{TCAa} amounted to 15% of the total TCA cycle flux ($V_{TCAa}=0.15 \times V_{TCAT}$) with reference to Sibson et al. (2001). In the work by de Graaf et al. (2004), V_{TCAa} amounted to 5% of the neuronal TCA cycle rate ($V_{TCAa}=0.05 \times V_{TCAn}$) using the same reference. However, only the assumption $V_{PDHa}=0.25 \times V_{PDHn}$ could be found in the work by Sibson et al. (2001). In the work on rats by Patel et al. (2005a), V_{TCAa} was assumed to represent 15% of V_{TCAT}, with reference to Lebon et al. (2002), whose work was carried out in human. In each of these studies, the constraint was applied in the same way independently of the animal status, i.e. when the rats were under pentobarbital, halothane or bicuculline. Obviously, owing to the constraint, oxidative metabolisms in astrocytes and neurons were considered as tightly coupled.

In the works by Choi et al. (2002) and Öz et al. (2004), V_{TCAa} and V_{TCAn} were considered as independent parameters in the fitting procedure. V_{TCAa} was found to be 1.5-fold higher in awake than pentobarbital-anesthetized animals, whereas V_{TCAn} increased threefold at the same time. This result thus indicated an increase in oxidative metabolism of astrocytes, although less than that determined for neurons. According to these studies, the TCA cycle rate in astrocytes was less than in neurons but remained in the same range (74 or 35%, under pentobarbital-anesthetized or awake rats, respectively).

A brain activity-dependent increase in glial oxidative metabolism was also evidenced by Serres et al. (2007) by analyzing the kinetics of glutamine labeling from [2-^{13}C]acetate metabolism in awake as compared to anesthetized rats. Moreover, the results indicated that the gain in oxidative metabolism resulted more from an increase in PDH than in acetyl-CoA synthetase activity. This conclusion was in agreement with the finding of a higher acoustic stimulus-induced increase in astrocytic oxidative metabolism than acetate utilization in the rat auditory pathway (Cruz et al. 2005; Hertz et al. 2007).

33.4.4 Anaplerosis

In all models, astrocytic anaplerosis via pyruvate carboxylase was considered as the sole pathway for replenishment of the glutamate/glutamine pool. The need for replenishment results from the loss of glutamine from the astrocytic compartment through its efflux to blood and/or the loss of glutamate through its oxidation (via the pathway including pyruvate recycling). As already mentioned, an important issue for evaluating V_{PC} (or V_{ana}) from [1-^{13}C]- or [1,6-$^{13}C_2$]glucose metabolism is the yield of oxaloacetate to fumarate back flow. To overcome the difficulty, [2-^{13}C]glucose can be used. Indeed, a priori, this glucose isotopomer seems particularly well suited for analyzing the anaplerotic pathway because, from [2-^{13}C]pyruvate, the label entered into the TCA cycle via PC leads to oxaloacetate and then α-ketoglutarate labeling at C2 and C3, whereas the label entered via PDH leads to α-ketoglutarate labeling at C5 (thereafter, the label is lost by decarboxylation in further TCA cycle turns). Therefore, the kinetics of amino acid labeling at C2 and C3 are mediated by V_{ana}, whatever the back flow yield.

This strategy has been used to specifically analyze PC activity in the brain of control rats and during bicuculline seizures (Patel et al. 2005b). However, using [2-^{13}C]glucose presents some drawbacks. (1) Labeling from [2-^{13}C]glucose through PC leads to much less carbon enrichment than labeling from [1-^{13}C]- or [1,6-$^{13}C_2$] glucose through PDH. For example, in control animals, glutamine C3 labeling from [2-^{13}C]glucose after 180 min (Patel et al. 2005b) was fourfold lower than after 85 min labeling from [1,6-$^{13}C_2$]glucose (Patel et al. 2004). As a consequence, the low amino acid-specific enrichments made their evaluation not easy considering the lack of resolution and the low signal/noise ratio of *in vivo* NMR. (2) An additional difficulty stems from the occurrence of [3-^{13}C]pyruvate. Indeed, this isotopomer arises from either glycolysis of glucose labeled at C1 and/or C6 through peripheral metabolism, pentose phosphate pathway or eventually brain pyruvate recycling. Its metabolism via PDH also leads to glutamate/glutamine C2 and C3 labeling from the second TCA cycle turn. Consequently, experimental labeling values determined under [2-^{13}C]glucose infusion must be corrected for both natural abundance and [3-^{13}C]pyruvate metabolism. These corrections were effectively done by Patel et al. (2005b). From their results, the authors concluded that the flow through pyruvate carboxylase was unaltered during seizures. This conclusion was drawn from the fact that the initial rates of glutamate + glutamine C2 + C3 labeling in control rats and under seizures were not different. The result was obtained by fitting the labeling at either three (rats under bicuculline, last point at ~70 min) or six time points (control rats, last point at ~ 160 min) by a simple exponential function. It was interpreted as an evidence that PC activity was the same under the two conditions. However, this interpretation may have been biased: owing to the very small number of points and their distribution on the time scale, the fitting procedure was unable to provide an accurate determination of the enrichment plateaus that were crucial for data interpretation. Indeed, the initial rate of an exponential process $y = A_\infty \times (1 - e^{-kt})$ depends on both the rate constant and the asymptotic value: initial rate = $k \times A_\infty$. Thus, the initial rates for two exponential processes $y_1 = A_1 1_\infty \times (1 - e^{-k1t})$ and $y_2 = A_{2\infty} \times (1 - e^{-k2t})$ are the same if $k_1/k_2 = A_{2\infty}/A_{1\infty}$. Thus, depending on the asymptotic values, the two processes may have different rate constants notwithstanding similar initial rates. More particularly, under seizures, an unchanged initial rate could be due to a higher PC activity coupled with a lower enrichment plateau. In fact, the authors noted the greater dilution of glutamate C3 labeling due to the increase in neuronal PDH. It therefore seems that the conclusion of unchanged cerebral pyruvate carboxylase flux with cerebral activation was not unambiguously demonstrated in the study of Patel et al. (2005b).

In studies carried out with [1-^{13}C]- or [1,6-$^{13}C_2$]glucose, no oxaloacetate to fumarate back flow was considered in the works of Choi et al. (2002) and Öz et al. (2004) (at least for data in Table 33.2), whereas total back flow was considered by Patel et al. (2004; 2005a) although asymmetrical labeling of glutamine C2 and C3 was observed (as discussed in Merle et al. 2002). Thus, in each case, a small bias was likely introduced into the fitting. More importantly, V_{ana} was considered to be an independent parameter by Choi et al. (2002) and Öz et al. (2004), whereas it was defined as dependent on V_{cycl} by Patel et al. (2004, 2005a) according to the relation $V_{ana}=0.2 \times V_{Gln}$ (Sibson et al. 2001). Implicitly, this meant that V_{ana} was coupled with V_{cycl} ($V_{ana}=0.25 \times V_{cycl}$), and thus coupled with cerebral activity. In fact, as shown in Table 33.2, V_{ana} was found to be much higher in awake (2004) than anesthetized rats (Choi et al. 2002). On the other hand, the same conclusion was drawn by Serres et al. (2007), who found that PC and astrocytic PDH activities increased simultaneously in awake as compared to anesthetized rats.

From these studies, it appears that PC activity is particularly difficult to assess from NMR studies owing to either the low contribution of PC relative to PDH in amino acid labeling and the oxaloacetate to fumarate back flow from [1-^{13}C]- or [1,6-$^{13}C_2$]glucose metabolism, or to the intrinsic low labeling from PC and the overlapping metabolism of [3-^{13}C]pyruvate when [2-^{13}C]glucose is used.

In modeling, anaplerosis was essentially related to the loss of glutamine through the efflux to blood: $V_{ana}=V_{efflux}$ (or $V_{ana}=V_{Gef} - V_{Gdil}$ when a diluting flux due to the entry of blood borne glutamine was considered). In fact, glutamine efflux is acknowledged as the main pathway for ammonium detoxification (Cooper 2001). On the other hand, brain activity has been shown to induce an increase in brain ammonium content (Tsukada 1966). As a consequence, an increase in glutamine efflux with cerebral activity would be expected. The loss in carbon and nitrogen would then be balanced by anaplerosis and the entry of blood ammonium (or ammonia) by diffusion and transport (Marcaggi and Coles 2001). Results indicating a cerebral activity-dependent increase in the flux through PC would be in agreement with this explanation, although the latter is probably not exclusive. Indeed, as above discussed, pyruvate recycling related to glutamate oxidation in glia also induces an increase in PC activity.

33.5 Conclusion

In metabolism modeling, the model is generally built on the basis of knowledge regarding metabolic characteristics previously demonstrated by using specific analytical methods. Thus, ideally, the model must contain this information. This is generally the case for fundamental characteristics whose occurrence is largely acknowledged. For example, concerning brain metabolism, modeling takes into account the occurrence of two glutamate pools and TCA cycles. Other information, however, is not (or not systematically) included in modeling: e.g., neuronal pyruvate carboxylation, brain pyruvate recycling, oxaloacetate-fumarate back flow, etc. This shortcoming may be for various reasons: either the information is considered as unreliable, contrary to other information taken as a dogma, or more likely and in most cases, the data are not sufficiently discriminative. Thus, estimating that a model is accurate (or as correct as possible) is not simple. For example, while there is much evidence that, from [3-^{13}C]pyruvate, the label entered into the TCA cycle via PC does not totally equilibrate at oxaloacetate C2 and C3 (Merle et al. 2002), this does not necessarily mean that models in which no or total equilibration was considered (Gruetter et al. 2001; Mason and Rothman 2004) are inadequate. Indeed, while from a theoretical and mathematical point of view, there is no fundamental reason for not considering a pathway (in the three-compartment model of Patel et al. (2005a), there are 30 isotope balance equations), the limit is dictated by their weight in the analysis when the metabolic network is compared with the experimental data. In fact, with complex networks, making the fitting procedure efficient requires reducing the number of degrees of freedom. This is obtained by making various assumptions, for example: amino acid pool sizes in the different compartments (e.g., the distribution of glutamate in glutamatergic neurons, GABAergic neurons and glia, Patel et al. (2005a)), interdependency between parameters or constraining parameter values. In that way, with a reduced number of parameters (seven parameters iterated in Patel et al. (2005a)), the fitting may be expected to give reliable parameter values.

The reliability of ^{13}C metabolic modeling was recently investigated using Monte-Carlo simulations (Shestov et al. 2007). Based on the two-compartment model of glutamatergic neurotransmission used in most studies (Gruetter et al. 2001; Patel et al. 2004…), the authors analyzed the influence of different parameters on results of the fitting procedure. They used as data the enrichment kinetics, from [3-^{13}C] pyruvate, of glutamate and glutamine C2, C3 and C4 and that of aspartate C2 and C3 calculated with the model, using pool sizes and flux rates corresponding to current published values as input parameters. Then, after adding a noise corresponding to a given standard deviation, they fitted the data with the same model in order to find the best value for V_{TCAn}, V_{TCAa}, V_{PC}, V_{cycl}, V_x and V_{Ldil} (with $V_{Ldil} = V_{lef}$) (as defined in Fig. 33.2). They then checked the influence of (1) the number of individual enrichment curves fitted (generally, with *in vivo* NMR, glutamate and glutamine C4 enrichments are accessible whereas those at C3 are more difficult to assess due to overlapping), (2) the number of points on individual enrichment kinetics, (3) the distribution of points throughout the kinetics and (4) the signal/noise ratio on the determination of parameters. Using all the data (seven enrichment curves), fitting provided a reliable determination for V_{TCAn}, V_{Ldil} (10-15% relative standard error) and V_{PC} (20-30% relative SD) and a large error for V_{TCAa} and V_x (50% and 100% relative SD, respectively), whereas V_{cycl} was poorly determined (358 and 670% relative SD with respect to the value found and the nominal value, respectively). Even when the fit was restricted to the estimation of V_{TCAn}, V_{Ldil} and V_{cycl} (i.e., using as input the nominal value for V_{TCAa}, V_{PC} and V_x), V_{cycl} was poorly determined (73% and 99% relative SD with respect to found and nominal values, respectively). On the other hand, reducing the data to three curves (glutamate C3 and C4 and glutamine C4, as in several studies) made V_{cycl} determination unreliable. Interestingly, when true experimental data were fitted, a rather good determination of V_{cycl} was obtained (29 ± 13% mean ± SD of the relative SD values from 12 different determinations) (Choi et al. 2002; Gruetter et al. 2001; Lebon et al. 2002; Öz et al. 2004; Patel et al. 2004, 2005a; Sibson et al. 1998, 2001), suggesting large differences in the methods used for error evaluation.

The study by Shestov et al. (2007) emphasized the need for the best possible fit between the experimental data and the metabolic network considered. Indeed, a sophisticated model involving numerous parameters is of limited use with data of insufficient quality. There are, however, several ways to improve the discriminative weight of the data. When possible, combining *in vivo* and ex vivo NMR data may be very useful, particularly for accurate determination of amino acid steady-state enrichments at the end of substrate infusion (Patel et al. 2004, 2005a). The constraint in the analysis could be increased by simultaneously fitting data from labeling experiments carried out with two different substrate isotopomers, e.g., [1-^{13}C]- or [2-^{13}C]glucose, taking advantage of the fact that labeling of α-ketoglutarate C2 and C3 from [2-^{13}C]glucose specifically results from ^{13}C entry into the TCA cycle through PC. Furthermore, two different substrates in combination could be used, e.g., [1-^{13}C]-/[1,6-$^{13}C_2$]glucose+acetate or glucose+[2-^{13}C]acetate (Patel et al. 2005a; Serres et al. 2008), taking advantage in this case that acetate is specifically consumed in astrocytes. On the other hand, a potential powerful information source specific to NMR spectroscopy that is not used in current modeling is the ^{13}C-^{13}C coupling figures on ^{13}C-NMR spectra, which are characteristic of metabolite labeling at adjacent carbon positions. Till now, isotopomer analysis was restricted to ex vivo experiments (e.g., Cerdan et al. 1990; Ebert et al. 2003; Kunnecke et al. 1993; Lapidot and Gopher 1994) owing to the lack of resolution of *in vivo* NMR. In the near future, the possibility of isotopomer analysis from data acquired *in vivo* may come about considering the improvements in both NMR spectroscopy (higher magnetic fields, better quality of detection probes, shimming, water elimination…) and methods for data analysis (automatic deconvolution, quantitation from a relevant reference) (Gruetter et al. 2003; Henry et al. 2003a, b). However, mathematic modeling previously developed for isotopomer analysis is not adapted for analyzing brain metabolism kinetics because it corresponds to a one-compartment system at isotopic steady state (Malloy et al. 1988; Merle et al. 1996a; Sherry et al. 2004). Therefore, there is a need for developing multi-compartment models dedicated to enrichment and isotopomer analysis at isotopic steady state and to the analysis of isotopomer enrichment dynamics.

Regarding the relevance of mathematical modeling, important issues are the temporal and spatial resolutions of *in vivo* NMR analyses. In human, at 2.1 T, ^{13}C-NMR spectra upon infusion of [1^{13}C]glucose were acquired every 5 min in a volume of 144 cm^3 (Mason et al. 2007; Shen et al. 1999), whereas spectra in a volume of 96 cm^3 were acquired every 15 min upon infusion of [2-^{13}C]acetate (which leads to a lower labeling owing to the astrocytic specificity of this substrate) (Lebon et al. 2002). At 4 T upon infusion of [1^{13}C]glucose, a 12-min acquisition time was used with a 45 cm^3 volume of interest (Gruetter et al. 2001). These acquisition conditions illustrate the low sensitivity of ^{13}C-NMR. In comparison, localized ^{1}H-NMR spectra of human brain were recorded every 2.7 min in a volume of 15 cm^3 at 7 T (Mangia, et al. 2006). In the rat, upon [1^{13}C]glucose infusion, localized ^{13}C-NMR spectra were acquired every 25 min at 9.4 T in a volume of 510 mm^3 (Choi et al. 2002). From [1,6-$^{13}C_2$]glucose metabolism, spectra in a volume of 360 mm^3 were acquired at 7 T every 4 min (Patel et al. 2005a). Using indirect localized ^{1}H-[^{13}C]-NMR, spectra in a ~350 mm^3 volume were acquired every 6 min at 3 T upon [1-^{13}C]glucose infusion, or at 7 T upon [1,6-$^{13}C_2$]glucose infusion (Henry et al. 2002; Patel et al. 2004). From the same volume, labeling from [2-^{13}C]glucose (via PC activity) was analyzed from spectra acquired every 30 min at 3 T (Patel et al. 2005b). On the other hand, spectra in 100 mm^3 were acquired at 7 T for 7 min (de Graaf et al. 2004). Although spectrum acquisition times in the same range as metabolite labeling half-times could be found in some of the above works (e.g., Gruetter et al. 2001; Patel et al. 2004), they were lower in most of them, suggesting rather good conditions for studying labeling dynamics. In contrast, with volumes ranging from either 50-150 cm^3 or 100-500 mm^3 in human or rat brain, respectively, spatial resolution was relatively low. In fact, the data collected corresponded to inhomogeneous brain regions. As significant regional differences in metabolism were demonstrated to occur, e.g. between gray matter, white matter, and subcortex (de Graaf et al. 2004; Mason et al. 1999), partial volume effects likely alter analyses of kinetic data acquired in a too large volume. Although discriminating the metabolism in gray and white matter was shown to be possible using spectroscopic imaging and tissue fractionation methods (Pan et al. 2000), the choice of the voxel remains crucial. Indeed, it depends at least on two contradictory requirements: the voxel must be small to minimize partial volume effects but large enough to provide a relevant NMR signal. The issue of partial volume effects is also particularly critical when ex vivo and *in vivo* NMR data analyses are coupled, owing to the difficulty in collecting the brain sample corresponding to the volume analyzed *in vivo*.

Owing to its low sensitivity, NMR spectroscopy requires working on a macroscopic scale (minimum voxel size in the 10 mm^3 range for ^{1}H-NMR). Notwithstanding the technical and methodological improvements that could be expected, the possibility of working at a microscopic range (at the cellular level) seems unlikely in the near future. Therefore, the current data represent the sum of the metabolism of millions of cells, whereas each individual or group of cells has its own metabolic characteristics depending on multiple factors: cell type, localization, activation status. As a consequence, the metabolic properties and parameter values assigned to a given group of cells from metabolism modeling are relative to their weight in the analysis. Therefore, they cannot be considered as representative of those of an individual cell.

Although it has intrinsic limits, *in vivo* NMR spectroscopy will likely remain the most powerful method to acquire data for the sophisticated analysis of brain metabolism using mathematical modeling.

References

Aurelli T, Di Cocco ME, Calvani M, Conti F (1997) The entry of [1–^{13}C]glucose into biochemical pathways reveals a complex compartmentation and metabolic trafficking between glia and neurons: a study by ^{13}C-NMR spectroscopy. Brain Res 765:218–227

Badar-Goffer RS, Bachelard HS, Morris PG (1990) Cerebral metabolism of acetate and glucose studied by ^{13}C-NMR spectroscopy. A technique for investigating metabolic compartmentation in the brain. Biochem J 266:133–139

Bak LK, Sickmann HM, Schousboe A, Waagepetersen HS (2004) Activity of the lactate-alanine shuttle is independent of glutamate-glutamine cycle activity in cerebellar neuronal-astrocytic cultures. J Neurosci 79:88–96

Bak LK, Schousboe A, Sonnewald U, Waagepetersen HS (2006) Glucose is necessary to maintain neurotransmitter homeostasis during synaptic activity in cultured glutamatergic neurons. J Cereb Blood Flow Metab 26:1285–1297

Bak LK, Waagepetersen HS, Melø TM, Schousboe A, Sonnewald U (2007) Complex glutamate labeling from [U-^{13}C]glucose or [U-^{13}C]lactate in co-cultures of cerebellar neurons and astrocytes. Neurochem Res 32:671–680

Balazs R, Haslam RJ (1965) Exchange transamination and the metabolism of glutamate in brain. Biochem J 94:131–141

Berkich DA, Xu Y, LaNoue KF, Gruetter R, Hutson SM (2005) Evaluation of brain mitochondrial glutamate and α-ketoglutarate transport under physiological conditions. J Neurosci Res 79:106–113

Berkich DA, Ola MS, Cole J, Sweatt AJ, Hutson SM, LaNoue KF (2007) Mitochondrial transport proteins of the brain. J Neurosci Res 85:3367–3377

Berl S, Lajtha A, Waelsch H (1961) Amino acid and protein metabolism-VI Cerebral compartments of glutamic acid metabolism. J Neurochem 7:186–197

Berl S, Takagaki G, Clarke D, Waelsch H (1962) Metabolic compartments in vivo: ammonia and glutamic acid metabolism in brain and liver. J Biol Chem 237:2562–2569

Berl S, Nicklas WJ, Clarke DD (1970) Compartmentation of citric acid cycle metabolism in the brain: labelling of glutamate, glutamine, aspartate and GABA by several radioactive tracer metabolites. J Neurochem 17:1009–1015

Bouzier A-K, Quesson B, Valeins H, Canioni P, Merle M (1999) [1-^{13}C]glucose metabolism in the tumoral and nontumoral cerebral tissue of a glioma-bearing rat. J Neurochem 72:2445–2455

Bouzier AK, Thiaudiere E, Biran M, Rouland R, Canioni P, Merle M (2000) The metabolism of [3-^{13}C]lactate in the rat brain is specific of a pyruvate carboxylase-deprived compartment. J Neurochem 75:480–486

Brand A, Engelmann J, Leibfritz D (1992) A ^{13}C NMR study on fluxes into the TCA cycle of neuronal and glial tumor cell lines and primary cells. Biochimie 74:941–948

Cerdan S, Kunnecke B, Seelig J (1990) Cerebral metabolism of [1,2-$^{13}C_2$]acetate as detected by in vivo and in vitro ^{13}C NMR. J Biol Chem 265:12916–12926

Cerdan S, Rodrigues TB, Sierra A, Benito M, Fonseca LL, Fonseca CP, Garcia-Martin ML (2006) The redox switch/redox coupling hypothesis. Neurochem Int 48:523–530

Chateil J-F, Biran M, Thiaudière E, Canioni P, Merle M (2001) Metabolism of [1-^{13}C] and [2-^{13}C] acetate in the hypoxic rat brain. Neurochem Int 38:399–407

Chapa F, Cruz F, Garcia-Martin ML, Garcia-Espinosa MA, Cerdan S (2000) Metabolism of [1-^{13}C] glucose and [2-^{13}C 2-^{2}CH$_3$]acetate in the neuronal and glial compartments of the adult rat brain as detected by [^{13}C, ^{2}H] NMR spectroscopy. Neurochem Int 37:217–228

Chen W, Zhu XH, Gruetter R, Seaquist ER, Adriany G, Ugurbil K (2001) Study of tricarboxylic acid cycle flux changes in human visual cortex during hemifield visual stimulation using (1) H-[(13)C] MRS and fMRI. Magn Reson Med 45:349–355

Chhina N, Kuestermann E, Halliday J, Simpson LJ, Macdonald IA, Bachelard HS, Morris P (2001) Measurement of human tricarboxylic acid cycle rates during visual activation by (13)C magnetic resonance spectroscopy. J Neurosci Res 66:737–746

Chih CP, Roberts EL Jr (2003) Energy substrates for neurons during neural activity: a critical review of the astrocyte-neuron lactate shuttle hypothesis. J Cereb Blood Flow Metab 23: 1263–1281

Choi IY, Lee SP, Kim SG, Gruetter R (2001) In vivo measurements of brain glucose transport using the reversible Michaelis-Menten model and simultaneous measurements of cerebral blood flow changes during hypoglycemia. J Cereb Blood Flow Metab 21:653–663

Choi IY, Lei H, Gruetter R (2002) Effect of deep pentobarbital anesthesia on neurotransmitter metabolism in vivo: on the correlation of total glucose consumption with glutamatergic action. J Cereb Blood Flow Metab 22:1343–1351

Cooper AJ (2001) Role of glutamine in cerebral nitrogen metabolism and ammonia neurotoxicity. Ment Retard Dev Disabil Res Rev 7:280–306
Cruz F, Scott SR, Barroso I, Santisteban P, Cerdan S (1998) Ontogeny and cellular localisation of the pyruvate recycling system in rat brain. J Neurochem 70:2613–2619
Cruz F, Cerdan S (1999) Quantitative ^{13}C NMR studies of metabolic compartmentation in the adult mammalian brain. NMR Biomed 12:454–462
Cruz NF, Lasater A, Zielke HR, Dienel A (2005) Activation of astrocytes in brain of conscious rats during acoustic stimulation: acetate utilization in working brain. J Neurochem 92:934–947
De Graaf RA, Mason GF, Patel AB, Rothman DL, Behar KL (2004) Regional glucose metabolism and glutamatergic neurotransmission in rat brain in vivo. Proc Natl Acad Sci USA 24:12700–12705
Ebert D, Haller RG, Walton ME (2003) Energy contribution of octanoate to intact rat brain metabolism measured by 13C nuclear magnetic resonance spectroscopy. J Neurosci 23:5928–5935
Fitzpatrick SM, Hetherington HP, Behar KL, Shulman RG (1990) The flux from glucose to glutamate in the rat brain in vivo as determined by ^{1}H-observed, ^{13}C-edited NMR spectroscopy. J Cereb Blood Flow Metab 10:170–179
Grill V, Bjorkman O, Gutniak M, Lindqvist M (1992) Brain uptake and release of amino acids in nondiabetic and insulin-dependent diabetic subjects: important role of glutamine release for nitrogen balance. Metabolism 41:28–32
Gruetter R, Novotny EJ, Boulware SD, Mason GF, Rothman DL, Shulman GI, Prichard JW (1994) Localized 13C NMR spectroscopy in the human brain of amino acid labeling from D-[1-13C] glucose. J Neurochem 63:1377–1385
Gruetter R, Seaquist ER, Kim S, Ugurbil K (1998) Localized in vivo ^{13}C-NMR of glutamate metabolism in the human brain: initial results at 4 tesla. Dev Neurosci 20:380–388
Gruetter R, Seaquist ER, Ugurbil K (2001) A mathematical model of compartmentalized neurotransmitter metabolism in the human brain. Am J Physiol Endocrinol Metab 281:E100–E112
Gruetter R, Adriany G, Choi IY, Henry P-G, Lei H, Oz G (2003) Localized in vivo ^{13}C NMR spectroscopy of the brain. NMR Biomed 16:313–338
Haberg A, Qu H, Bakken IJ, Sande LM, White LR, Haraldseth O, Unsgard G, Aasly J, Sonnewald U (1998) In vitro and ex vivo ^{13}C-NMR spectroscopy studies of pyruvate recycling in brain. Dev Neurosci 20:389–398
Hassel B (2001a) Pyruvate carboxylation in neurons. J Neurosci Res 66:755–762
Hassel B (2001b) Carboxylation and anaplerosis in neurons and glia. Mol Neurobiol 22:21–40
Hassel B, Sonnewald U, Fonnum F (1995) Glial-neuronal interactions as studied by cerebral metabolism of [2-^{13}C]acetate and [1-^{13}C]glucose. An ex vivo ^{13}C NMR study. J Neurochem 64:2773–2782
Hassel B, Sonnewald U (1995) Glial formation of pyruvate and lactate from TCA cycle intermediates: implications for the inactivation of transmitter amino acids. J Neurochem 65:2227–2234
Hassel B, Brathe A (2000a) Cerebral metabolism of lactate in vivo: evidence for neuronal pyruvate carboxylation. J Cereb Blood Flow Metab 20:327–336
Hassel B, Brathe A (2000b) Neuronal pyruvate carboxylation supports formation of transmitter glutamate. J Neurosci 20:1342–1347
Henry P-G, Lebon V, Vaufrey F, Brouillet E, Hantraye P, Bloch G (2002) Decreased TCA cycle rate in the rat brain after acute 3-NP treatment measured by in vivo ^{1}H-[^{13}C] NMR spectroscopy. J Neurochem 82:857–866
Henry P-G, Oz G, Provencher S, Gruetter R (2003a) Toward dynamic isotopomer analysis in the rat brain in vivo: automatic quantitation of ^{13}C NMR spectra using LCModel. NMR Biomed 16:400–412
Henry P-G, Tkac I, Gruetter R (2003b) ^{1}H-localized broadband ^{13}C NMR spectroscopy of the rat brain in vivo at 9.4 T. Magn Reson Med 50:684–692
Henry P-G, Adriany G, Deelchand D, Gruetter R, Marjanska M, Oz G, Seaquist ER, Hestov A, Ugurbil K (2006) In vivo ^{13}C NMR spectroscopy and metabolic modeling in the brain: a practical perspective. Mag Reson Imag 24:527–539

Hertz L, Peng L, Dienel GA (2007) Energy metabolism in astrocytes: high rate of oxidative metabolism and spatiotemporal dependence on glycolysis/glycogenolysis. J Cereb Blood Flow Metab 27:219–249

Hyder F, Patel AB, Gjedde A, Rothman DL, Behar KB, Shulman RG (2006) Neuronal-glial glucose oxidation and glutamatergic-GABAergic function. J Cereb Blood Flow Metab 26:865–877

Hutson SM, Berkich D, Drown P, Xu B, Aschner M, LaNoue KF (1998) Role of branched-chain aminotransferase isoenzymes and gabapentine in neurotransmitter metabolism. J Neurochem 71:863–874

Hutson SM, Sweatt AJ, LaNoue KF (2005) Branched-chain amino acid metabolism: implications for establishing safe intakes. J Nutr 135:1557S–1564S

Kalderon B, Gopher A, Lapidot A (1987) A quantitative analysis of the metabolic pathways of hepatic glucose synthesis in vivo with ^{13}C-labeled substrates. FEBS Lett 213:209–214

Kanamori K, Ross BD, Kondrat RW (1998) Rate of glutamate synthesis from leucine in rat brain measured in vivo by ^{15}N NMR. J Neurochem 70:1304–1315

Kimmich GE, Roussie JA, Randles J (2002) Aspartate amino transferase isotope exchange reactions: implications for glutamate/glutamine shuttle hypothesis. Am J Physiol Cell Physiol 282:C1404–C1413

Kunnecke B, Cerdan S, Seelig J (1993) Cerebral metabolism of [1,2-$^{13}C_2$]glucose and [U-$^{13}C_4$] β-hydroxybutyrate in rat brain as detected by ^{13}C NMR spectroscopy. NMR Biomed 6:264–277

Lapidot A, Gopher A (1994) Cerebral metabolic compartmentation. Estimation of glucose flux via pyruvate carboxylase/pyruvate dehydrogenase by ^{13}C NMR isotopomer analysis of D-[U-^{13}C] glucose metabolites. J Biol Chem 269:27198–27208

Lebon V, Petersen KF, Cline GW, Shen J, Mason GF, Dufour S, Behar KL, Shulman GI, Rothman DL (2002) Astroglial contribution to brain energy metabolism in humans revealed by ^{13}C nuclear magnetic resonance spectroscopy: elucidation of the dominant pathway for neurotransmitter glutamate repletion and measurement of astrocytic oxidative metabolism. J Neurosci 22:1523–1531

Lieth E, LaNoue KF, Berkich DA, Xu B, Ratz M, Taylor C, Hutson SM (2001) Nitrogen shuttling between neurons and glial cells during glutamate synthesis. J Neurochem 76:1712–1723

Magistretti PJ, Pellerin L, Rothman DL, Shulman RG (1999) Energy on demand. Science 283:496–497

Malloy CR, Sherry AD, Jeffrey MH (1988) Evaluation of carbon flux and substrate selection through alternate pathways involving the citric acid cycle of the heart by ^{13}C NMR spectroscopy. J Biol Chem 263:6964–6971

Mangia S, Tkac I, Gruetter R, Van de Moortele P-F, Maraviglia B, Ugurbil K (2006) Sustained neuronal activation raises oxidative metabolism to a new steady-state level: evidence from ^{1}H NMR spectroscopy in the human visual cortex. J Cereb Blood Flow Metab 27:1055–1063

Marcaggi P, Coles JA (2001) Ammonium in nervous tissue: transport across cell membranes, fluxes from neurons to glial cells, and role in signalling. Prog Neurobiol 64:157–183

Martin M, Portais J-C, Labouesse J, Canioni P, Merle M (1993) [1-^{13}C]glucose metabolism in rat cerebellar granule cells and astrocytes in primary culture. Evaluation of flux parameters by ^{13}C- and ^{1}H-NMR spectroscopy. Eur J Biochem 217:617–625

Martinez-Hernandez A, Bell KP, Norenberg MD (1977) Glutamine synthetase: glial localization in brain. Science 195:1356–1358

Mason GF, Rothman DL, Behar KL, Shulman RG (1992) NMR determination of the TCA cycle rate and alpha-ketoglutarate/glutamate exchange rate in rat brain. J Cereb Blood Flow Metab 12:434–447

Mason GF, Gruetter R, Rothman DL, Behar K, Shulman RG, Novotny FJ (1995) Simultaneous determination of the rates of the TCA cycle, glucose utilization, α-ketoglutarate/glutamate exchange, and glutamine synthesis in human brain by NMR. J Cereb Blood Flow Metab 15:12–25

Mason GF, Pan JW, Chu W-J, Newcomer BR, Zhang Y, Orr R, Hetherington HP (1999) Measurement of the tricarboxylic acid cycle rate in human grey and white matter in vivo by ^{1}H-[^{13}C] magnetic resonance spectroscopy at 4.1 T. J Cereb Blood Flow Metab 19:1179–1188

Mason GF, Martin DL, Martin SB, Manor D, Sibson NR, Patel A, Rothman DL, Behar K (2001) Decrease in GABA synthesis rate in rat cortex following GABA-transaminase inhibition correlates with the decrease in GAD_{67} protein. Brain Res 914:81–91

Mason GF, Rothman DL (2004) Basic principles of metabolic modeling of NMR ^{13}C isotopic turnover to determine rates of brain metabolism in vivo. Metab Engin 6:75–84

Mason FM, Petersen KF, de Graaf RA, Shulman GI, Rothman DL (2007) Measurements of the anaplerotic rate in the human cerebral cortex using ^{13}C magnetic resonance spectroscopy and [1-^{13}C] and [2-^{13}C]glucose. J Neurochem 100:73–86

McKenna MC, Stevenson JH, Huang X, Tildon JT, Zielke CL, Hopkins IB (2000) Mitochondrial malic enzyme activity is much higher in mitochodria from cortical synaptic terminals compared with mitochodria from prymary cultures of cortical neurons or cerebellar granule cells. Neurochem Int 36:451–459

McKenna MC, Waagepetersen HS, Shousboe A, Sonnewald U (2006) Neuronal and astrocytic shuttle mechanisms for cytosolic-mitochondrial transfer of reducing equivalents: current evidence and pharmacological tools. Biochem Pharmacol 71:399–407

Melø TM, Nehlig A, Sonnewald U (2005) Metabolism is normal in astrocytes in chronically epileptic rats: a ^{13}C NMR study of neuronal-glial interactions in a model of temporal lobe epilepsy. J Cereb Blood Flow Metab 25:1254–1264

Merle M, Martin M, Villégier A, Canioni P (1996a) Mathematical modelling of the citric acid cycle for the analysis of glutamine isotopomers from cerebellar astrocytes incubated with [1-^{13}C]glucose. Eur J Biochem 239:742–751

Merle M, Martin M, Villégier A, Canioni P (1996b) [1-^{13}C]glucose metabolism in brain cells: isotopomer analysis of glutamine from cerebellar astrocytes and glutamate from granule cells. Dev Neurosci 18:460–468

Merle M, Bouzier-Sore AK, Canioni P (2002) Time-dependence of the contribution of pyruvate carboxylase versus pyruvate dehydrogenase to rat brain glutamine labelling from [1-13C]glucose metabolism. J Neurochem 82:47–57

Nedergaard M, Ransom B, Goldman SA (2003) New roles for astrocytes: redefining the functional architecture of the brain. Trends Neurosci 26:523–530

Olstad E, Olsen GM, Qu H, Sonnewald U (2007) Pyruvate recycling in cultured neurones from cerebellum. J Neurosci Res 85:3318–3325

O'Neal RM, Koeppe RE (1966) Precursors in vivo of glutamate, aspartate and their derivatives of rat brain. J Neurochem 13:835–847

Öz G, Berkich DA, Henry PG, Xu Y, LaNoue K, Hutson SM, Gruetter R (2004) Neuroglial metabolism in the awake rat brain: CO_2 fixation increases with brain activity. J Neurosci 24:11273–11279

Pan JW, Stein DT, Telang F, Lee JH, Shen J, Brown P, Cline G, Mason GF, Shulman GI, Rothman DL, Hetherington HP (2000) Spectroscopic imaging of glutamate C4 turnover in human brain. Mag Reson Med 44:673–679

Patel AB, de Graaf RA, Mason GF, Kanamatsu T, Rothman DL, Shulman RG, Behar KL (2004) Glutamatergic neurotransmission and neuronal glucose oxidation are coupled during intense neuronal activation. J Cereb Blood Flow Metab 24:972–985

Patel AB, Chowdhury GMI, de Graaf RA, Rothman DL, Shulman RG, Behar KL (2005a) Cerebral pyruvate carboxylase flux is unaltered during bicuculline-seizures. J Neurosci Res 79: 128–138

Patel AB, de Graaf RA, Mason GF, Rothman DL, Shulman RG, Behar KL (2005b) The contribution of GABA to glutamate/glutamine cycling and energy metabolism in the rat cortex in vivo. Proc Natl Acad Sci USA 102:5588–5593

Pellerin L, Magistretti P (1994) Glutamate uptake into astrocytes stimulates aerobic glycolysis: a mechanism coupling neuronal activity to glucose utilization. Proc Natl Acad Sci USA 91:10625–10629

Pellerin L (2003) Lactate as a pivotal element in neuron-glia metabolic cooperation. Neurochem Int 43:331–338

Pellerin L, Bouzier-Sore A-K, Aubert A, Serres S, Merle M, Costalat R, Magistretti P (2007) Activity-dependent regulation of energy metabolism by astrocytes: an update. Glia 55:1251–1262

Qu H, Eloqayli H, Unsgard G, Sonnewald U (2001) Glutamate decreases pyruvate carboxylase activity and spares glucose as energy substrate in cultured cerebellar astrocytes. J Neurosci Res 66:1127–1132

Ramos M, del Arco A, Pardo B, Martinez-Serrano A, Martinez-Morales JR, Kobayashi K, Yasuda T, Bogonez E, Bovolenta P, Saheki R, Satrustegui J (2003) Developmental changes in the Ca^{2+}-regulated mitochondrial aspartate-glutamate carrier aralar1 in brain and prominent expression in the spinal cord. Dev Brain Res 143:33–46

Sakai R, Cohen DM, Henry JF, Burrin DG, Reeds PJ (2004) Leucine-nitrogen metabolism in the brain of conscious rats: its role as a nitrogen carrier in glutamate synthesis in glial and neuronal metabolic compartements. J Neurochem 88:612–622

Serres S, Bouyer J-J, Bezançon E, Canioni P, Merle M (2003) Involvement of brain lactate in brain metabolism. NMR Biomed 16:430–439

Serres S, Bezançon E, Franconi J-M, Merle M (2004) Ex vivo analysis of lactate and glucose metabolism in the rat brain under different states of depressed activity. J Biol Chem 279: 47881–47889

Serres S, Bezançon E, Franconi J-M, Merle M (2005) Ex vivo NMR study of lactate metabolism in rat brain under various depressed states. J Neurosci Res 79:19–25

Serres S, Bezançon E, Franconi J-M, Merle M (2007) Brain pyruvate recycling and peripheral metabolism: a NMR analysis ex vivo of acetate and glucose metabolism in the rat. J Neurochem 101:1428–1440

Serres S, Raffard G, Franconi J-M, Merle M (2008) Close coupling between astrocytic and neuronal metabolisms to fulfill anaplerotic and energy needs in the rat brain. J Cereb Blood Flow Metab (4):712–724

Shank RP, Bennett GS, Freytag SO, Campbell GL (1985) Pyruvate carboxylase: an astrocyte-specific enzyme implicated in the replenishment of amino acid neurotransmitter pools. Brain Res 329:364–367

Shank RP, Leo GC, Zielke HR (1993) Cerebral metabolic compartmentation as revealed by nuclear magnetic resonance analysis of D-[1-^{13}C]glucose metabolism. J Neurochem 61:315–323

Shen J, Petersen KF, Behar KL, Brown P, Nixon TW, Mason GF, Petroff OAC, Shulman GI, Shulman RG, Rothman DL (1999) Determination of the rate of the glutamate/glutamine cycle in the human brain by in vivo ^{13}C NMR. Proc Natl Acad Sci USA 96:8235–8240

Sherry AD, Jeffrey FMH, Malloy CR (2004) Analytical solutions for ^{13}C isotopomer analysis of complex metabolic conditions: substrate oxidation, multiple pyruvate cycles, and gluconeogenesis. Metab Engin 6:12–24

Shestov AA, Valette J, Ugurbil K, Henry P-G (2007) On the reliability of ^{13}C metabolic modeling with two-compartments neuronal-glial models. J Neurosci Res 85:3294–3303

Shulman RG, Hyder F, Rothman DL (2001) Cerebral energetics and the glycogen shunt: nurochemical basis of functional imaging. Proc Natl Acad Sci USA 98:6417–6422

Sibson NR, Dhankhar A, Mason GF, Behar KL, Rothman DL, Shulman RG (1997) In vivo ^{13}C NMR measurements of cerebral glutamine synthesis as evidence for glutamate-glutamine cycling. Proc Natl Acad Sci USA 94:2699–2704

Sibson NR, Dhankhar A, Mason GF, Rothman DL, Behar KL, Shulman RG (1998) Stoichiometric coupling of brain glucose metabolism and glutamatergic neuronal activity. Proc Natl Acad Sci USA 95:316–321

Sibson NR, Mason GF, Shen J, Cline GW, Herskovits AZ, Wall JE, Behar KL, Rothman DL, Shulman RG (2001) In vivo ^{13}C NMR measurement of neurotransmitter glutamate cycling, anaplerosis and TCA cycle flux in rat brain during [2-^{13}C]glucose infusion. Proc Natl Acad Sci USA 76:975–989

Sonnewald U, Westergaard N, Petersen SB, Unsgard G, Schousboe A (1993) Metabolism of [U-$^{13}C_5$]glutamate in astrocytes studied by 13C NMR spectroscopy: incorporation of more label into lactate than into glutamine demonstrates the importance of the tricarboxylic acid cycle. J Neurochem 61:1179–1182

Sonnewald U, Westergaard N, Jones P, Taylor A, Bachelard HS, Schousboe A (1996) Metabolism of [U-$^{13}C_5$]glutamine in cultured astrocytes studied by NMR spectroscopy: first evidence of astrocytic pyruvate recycling. J Neurochem 67:2566–2572

Sonnewald U, Westergaard N, Jones P, Taylor A, Bachelard HS, Schousboe A (1997) Glutamate transport and metabolism in astrocytes. Glia 21:56–63

Stryer L (1995) Biochemistry, 4th edn. Freeman, San Francisco

Tsukada Y (1966) Amino acid metabolism and its relation to brain functions. Prog Brain Res 21:268–291

Tyson RL, Gallagher C, Sutherland GR (2003) ^{13}C-labeled substrates and the cerebral metabolic compartmentalization of acetate and lactate. Brain Res 992:43–52

Van den Berg CJ, Krzalic LJ, Mela P, Waelsch H (1969) Compartmentation of glutamate metabolism in brain. Evidence for the existence of two different tricarboxylic acid cycles in brain. Biochem J 113:281–290

Van den Berg CJ, Garfinkel D (1971) A simulation study of brain compartments. Metabolism of glutamate and related substances in mouse brain. Biochem J 123:211–218

Vogel R, Hamprecht B, Wiesinger H (1998a) Malic enzyme isoforms in astrocytes: comparative study on activities in rat brain tissue and astrogia-rich primary cultures. Neurosci Lett 247: 123–126

Vogel R, Jennemann G, Seitz J, Wiesinger H, Hamprecht B (1998b) Mitochondrial malic enzyme: purification from bovine brain, generation of an antiserum, and immunocytochemical localization in neurons of rat brain. J Neurochem 71:844–852

Waniewski RA, Martin DL (1998) Preferential utilization of acetate by astrocytes is attributable to transport. J Neurosci 18:5225–5233

Yu ACH, Drejer J, Hertz L, Schousboe A (1983) Pyruvate carboxylase activity in primary cultures of astrocytes and neurons. J Neurochem 41:1484–1487

Yu X, Alpert NM, Lewandowski ED (1997) Modeling enrichment kinetics from dynamic ^{13}C-NMR spectra: theoretical analysis and practical considerations. Am J Physiol 272:C2037–C2048

Waagepetersen HS, Sonnewald U, Larsson OM, Shousboe A (2000) A possible role of alanine for ammonia transfer between astrocytes and glutamatergic neurons. J Neurochem 75:471–479

Yudkoff M, Daikhin Y, Nelson D, Nissim I, Erecińska M (1996) Neuronal metabolism of branched-chain amino acids: flux through the aminotransferase pathway in synaptosomes. J Neurochem 66:2136–2145

Zwingmann C, Richter-Landsberg C, Brand A, Leibfritz D (2000) NMR spectroscopy study on the metabolic fate of [3-^{13}C]alanine in astrocytes, neurons, and cocultures: implications for glial-neuron interactions in neurotransmitter metabolism. Glia 32:286–303

Zwingmann C, Leibfritz D (2003) Regulation of glial metabolism studied by ^{13}C-NMR. NMR Biomed 16:370–399

Chapter 34
Radiotracer Perspective on Cerebral Compartmentation

Bruno Weber, Renaud Jolivet, Matthias T. Wyss, and Alfred Buck

Abstract The most important compartments in neurobiology are the different types of cells present in cerebral tissue, as well as the interstitial space. In this chapter, the compartmentation of brain energy metabolism and its investigation with the use of compartmental models and radiotracer kinetics is introduced. Cerebral energy metabolism is a highly compartmentalized and complex system and the interaction between astrocytes and neurons plays a pivotal role. A current concept, known as the "astrocyte-neuron lactate shuttle", implies the production of lactate by astrocytes, which deliver it to neurons, to match their energy needs. The use of radiolabeled acetate serves as an example how to investigate brain energy compartmentation. The example further illustrates the importance to acquire dynamic data in order to appreciate meaningful readouts, such as compartment specific metabolic rates.

Keywords Brain energy metabolism • Neuron-astrocyte interactions • Lactate • Glucose • Positron emission tomography • Radiotracer

34.1 Introduction

In the field of pharmacology, a compartment is defined as a homogeneous volume of body fluids. The most often considered compartments are blood, interstitial fluid, fat and intracellular fluid. In the present chapter, a compartment is defined as a

B. Weber, Prof. Dr. • R. Jolivet, Ph.D.
Institute of Pharmacology and Toxicology, University of Zurich, Rämistrasse 100, 8091 Zurich, Switzerland
e-mail: bweber@pharma.unizh.ch; renaud.jolivet@a3.epfl.ch

M.T. Wyss, Dr. • A. Buck, M.D. (✉)
Department of Nuclear Medicine, University Hospital Zurich, Rämistrasse 100, 8091 Zurich, Switzerland
e-mail: matthias.wyss@usz.ch; fred.buck@usz.ch

I.-Y. Choi and R. Gruetter (eds.), *Neural Metabolism In Vivo*, Advances in Neurobiology 4, DOI 10.1007/978-1-4614-1788-0_34,

volume of the brain, in which all properties (concentrations, metabolism, transport characteristics) can be considered homogeneous. In most cases, semi-permeable membranes separate the compartments from each other. Depending on the used model, the definition of compartments can be as unspecific as "blood" and "cerebral tissue," but can also concern components with sub-cellular specificity, such as a specific type of receptor, or a cellular organelle.

In the field of neurophysiology, the most commonly used compartments are the different types of cells present in cerebral tissue, i.e. neurons and glial cells, as well as the interstitial space. Furthermore, when applying a compartmental model to investigate radiotracer kinetics, the delivery and drainage of the tracer occurs via the vasculature, so blood is yet another key compartment to be considered. Certain compounds are more or less freely diffusing between blood plasma and erythrocytes, so that blood can be considered as one single compartment. However, this is rather an exception so that blood plasma and erythrocytes need to be considered as two separate compartments.

For this chapter, we will focus on studies on brain metabolism. For this purpose, it seems mandatory to introduce the main concepts of cerebral metabolism. The most important blood-borne energy substrate is glucose that is transported from blood via glucose transporters located on the endothelial cells. Glucose is then either taken up by neurons (via glucose transporter GLUT 3) or by astrocytes (via GLUT 1) (Vannucci et al. 1997, 1998). Glucose taken up by neurons is normally fully metabolized to CO_2 and water via glycolysis and the tricarboxylic acid cycle (TCA cycle). The route via the astrocytes is the core of a more recent view on brain metabolism, called the astrocytes-neuron-lactate-shuttle (ANLS) hypothesis (Pellerin and Magistretti 1994; Magistretti et al. 1999). Astrocytes take up glucose, which is glycolytically metabolized to lactate that is then transferred to neurons via monocarboxylate transporter (astrocytic release of lactate via MCT 4 and MCT1 and neuronal uptake via MCT 2 (Pierre and Pellerin 2005)), where it is oxidatively metabolized for energy production. If we summarize these simplified pathways in terms of the involved compartments, we end up with the following ensemble: blood, astrocytes and neurons as well as the interstitial space. Moreover, within both the neurons and astrocytes a cytosolic and mitochondrial compartment can be distinguished for each cell type.

Every physiological data acquisition method has its inherent spatial resolution that – strictly speaking – determines the level with which compartments can be differentiated. As far as radiotracer methods are concerned, no available acquisition technology to date is able to reach cellular resolution. Nonetheless, radiotracer techniques allow the investigation of a specific cell type, if the appropriate tracers are used. It will be discussed later, how radiolabeled acetate allows to investigate astrocytic oxidative metabolism. In this chapter we will focus on autoradiography, positron emission tomography (PET) and beta probe studies.

One of the main points of the present chapter is to introduce the concept that the mere physical resolution of a method does not determine its physiological resolution. The investigator can apply several physiological tricks, to even reach subcellular resolution with imaging techniques, such as PET.

An example of investigating compartmentation with cellular – even subcellular – resolution is the recent work by Kasischke et al. (Kasischke et al. 2004). The authors

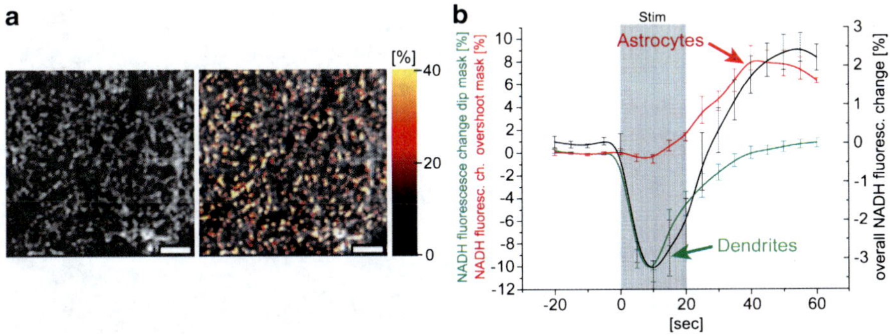

Fig. 34.1 Early oxidative metabolism in dendrites is eventually sustained by glycolysis in astrocytes. (**a**) Fluctuations of the NADH autofluorescence signal superimposed on the raw image (contrast enhanced) show that the NADH overshoot often colocalizes with astrocytes and their processes (scale bars 20 μm). (**b**) The biphasic NADH response is the sum of two spatially and temporally distinct monophasic metabolic responses. Early oxidative recovery metabolism in dendrites is followed by late activation of glycolysis in astrocytes ($n=8$, mean ± SEM) (Adapted with permission from Kasischke et al. (2004))

applied 2-photon microscopy (a fluorescence microscopy technique with subcellular resolution) on hippocampal slices of GFP-GFAP mutant mice and recorded NADH autofluorescence upon presynaptic stimulation of Schaffer collaterals. Slices were additionally stained with dyes specific for dendrites and mitochondria. By co-localizing the different channels, they could then allocate each pixel in the acquired images to either one of the four following compartments: the neuronal cytosol, neuronal mitochondria, the astrocytic cytosol and astrocytic mitochondria. NADH is particularly interesting in the context of brain energy metabolism because it is produced in the glycolytic pathway, is produced or consumed by the lactate dehydrogenase depending on the net direction of the reaction, is produced by the tricarboxylic acid cycle, consumed by the electron transport chain and cannot diffuse freely through the membrane of mitochondria. NADH autofluorescence can then be used to monitor the two major pathways involved in energy production in the brain, glycolysis and oxidative metabolism.

In their experiment, Kasischke et al. found a dip of the NADH signal located in mitochondria of dendrites at the onset of activation followed by a delayed overshoot of the NADH signal located in the cytosol of astrocytes (Fig. 34.1). However, they did not observe a coordinated increase of the oxidative metabolism in astrocytes.

These results are interpreted as a strong experimental validation of the astrocyte-neuron lactate shuttle hypothesis. Indeed, the initial dip signals the activation of oxidative metabolism in neurons while the delayed overshoot signals the activation of the glycolytic pathway in astrocytes with no coordinated increase of oxidative metabolism in these cells. It is then very likely that the glucose which enters the astrocytes is converted to lactate and released to the extracellular space from where it could be taken up by neurons as an energy substrate. This study is a perfect example in which

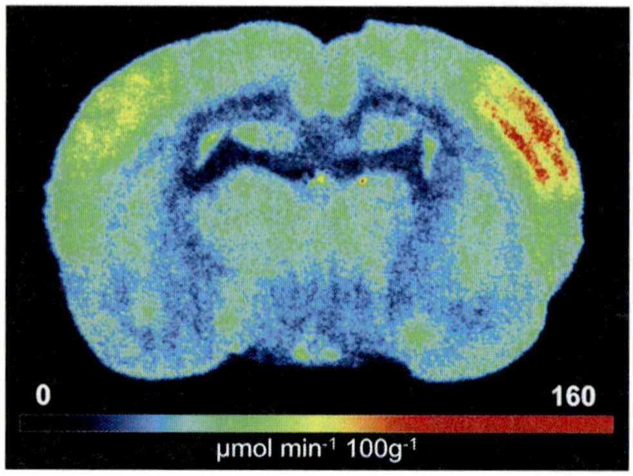

Fig. 34.2 Slice of an autoradiographic study using ^{18}F-fluorodeoxyglucose. A 2 Hz electrostimulation of the left infraorbital nerve was applied. Slices were put on a phosphor imaging plate for 4 h. Note the markedly increased glucose utilization in the stimulated somatosensory cortex (right hemisphere)

subcellular resolution of metabolic pathways could be achieved by combining a method with superb spatial resolution with various fluorescent dyes and the intrinsic properties of NADH, a key hub molecule in metabolic networks. The applicability of this approach in an *in vivo* setting is still unclear.

The aim of the present chapter is to introduce the most important radiotracer methods and to describe, how they can be used in compartment-specific investigations of cerebral metabolism. A study using 1-^{11}C-acetate is used as an example to explain some of the most relevant concepts.

34.2 Radiotracer Methods in Cerebral Metabolism Research

34.2.1 Autoradiography

The autoradiographical approach is the method of choice if a high spatial resolution is important. Briefly, after the tracer has circulated, the brain is removed and immediately frozen. Thereafter, sections have to be cut on a cryostat. The sections are then imaged using apposition exposure to a beta-sensitive film or a phosphor imaging system (Fig. 34.2). With ^{14}C or ^{3}H labeled compounds, a resolution below 100 μm can be achieved. The main disadvantages are its invasiveness and the fact that only one time point can be appreciated per animal, which obviously limits the applicability for investigations of tracer kinetics.

34.2.2 Positron Emission Tomography

Positron emission tomography (PET) is a three-dimensional imaging technique that detects gamma rays produced by positron emitting radioisotopes. With PET it is

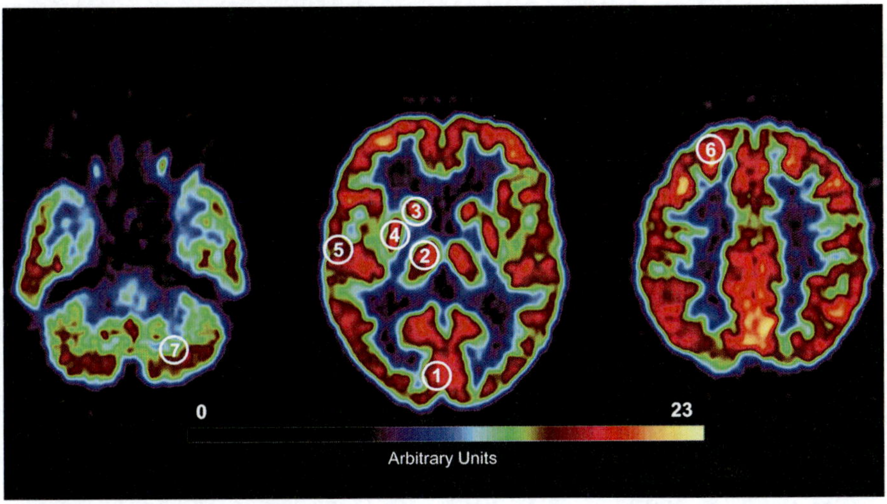

Fig. 34.3 ^{18}F-fluorodeoxyglucose PET of a normal human brain. The spatial resolution of the used scanner is 5 mm. This allows to differentiate various structures. (1) Visual cortex, (2) thalamus, (3) caudate nucleus, (4) putamen, (5) temporolateral cortex, (6) frontal cortex, and (7) cerebellum

possible to non-invasively obtain dynamic images of tracer concentration with excellent sensitivity, e.g. in the nanomolar range. The spatial resolution of human scanners is in the range of several millimeters (Fig. 34.3). Using state-of-the-art small animal PET scanners, the spatial resolution can be optimized to be even below 1 mm, although this comes at the cost of a reduced sensitivity.

34.2.3 Beta Probe

The big advantage of PET is that it allows to non-invasively acquire data of the whole brain. For certain applications, it is sufficient to determine the tracer kinetics at only a few points. An example is the study by Wyss et al., where a new tracer for the evaluation of glutamatergic receptors was validated using a beta probe system (Fig. 34.4), which allowed to measure the tracer kinetics at one point in the rat striatum (Wyss et al. 2007). Beta probes consist of a scintillating crystal, which is attached to a light-guiding fiber. Beta particles around the tip of the probe excite the crystal and the emerging photons are counted yielding a measure for the radioactivity concentration. The advantage of beta probes is the low price and the high temporal resolution. However, the use of the beta probe requires that the beta particles are emitted with a sufficiently high energy. This is the case for a lot of positron emitters such as ^{11}C, ^{18}F, ^{13}N and ^{15}O. For the commonly used isotopes ^{14}C and ^{3}H, the emission energy lies much below the detection level of beta probes.

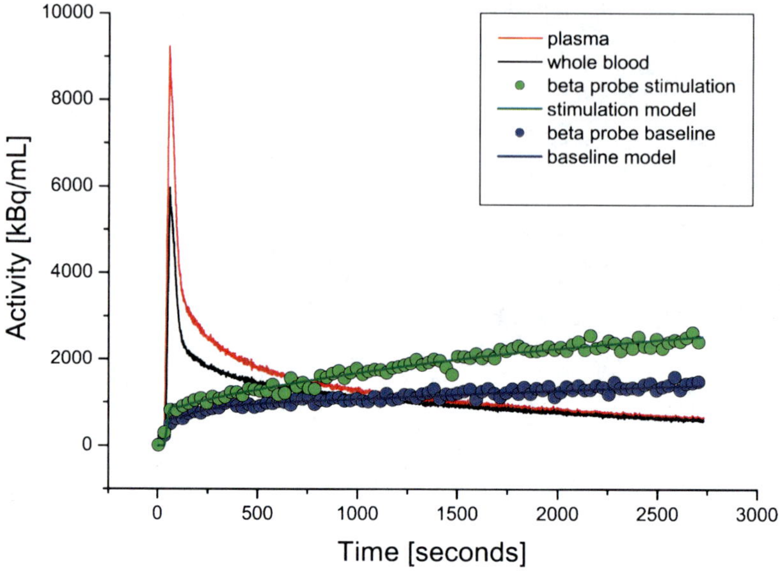

Fig. 34.4 Beta probe experiments in a rat. Shown are tissue time activity curves and blood curves of an ^{18}F-fluorodeoxyglucose study. On each hemisphere a beta probe was placed over the somatosensory cortex. The right infraorbital nerve was electrically stimulated with 2 Hz. This leads to a markedly faster increase of the FDG uptake

34.2.4 Data Analysis

The type of possible data analysis depends mainly on the method. The most complex data that can be derived from tracer experiments are fully quantitative values for physiological parameters such as cerebral blood flow, glucose metabolism or receptor densities. One of the most severe problems with all tracer methods is the dependence of the accumulated radioactivity on tracer delivery, e.g. cerebral blood flow. The elimination of flow effects requires in most cases some kind of kinetic modeling, which often mandates dynamic data acquisition.

34.2.5 Tracer Kinetic Modeling

At the base of tracer kinetic modeling is a compartmental model. An example of a typical one-tissue compartment model is demonstrated in Fig. 34.5. The meaning of the parameters depends on the application. In almost all models K_1 represents the delivery of tracer to tissue and is therefore closely related to cerebral blood flow (CBF). The exact relationship is $K_1 = CBF \times EF$, where EF is the first pass extraction fraction. Parameter k_2 describes back-transport of label from tissue to blood and its meaning completely depends on the tracer.

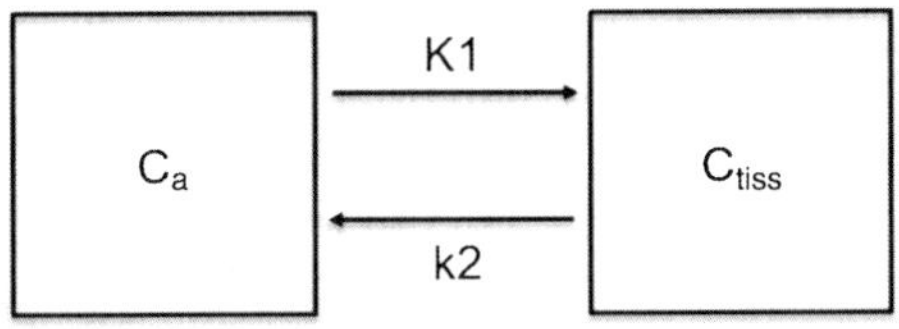

Fig. 34.5 One-tissue compartment model. K_1 describes the transfer of tracer from blood to tissue whereas k_2 corresponds to the flux of label from tissue into blood

34.2.5.1 Curve Fitting

In most cases the determination of the parameters of a model requires the measurement of the time-course of radioactivity in tissue (C_{tiss}) and in arterial blood (C_a). K_1 and k_2 are then fitted such that the calculated tissue time activity curve (TAC) fits the measured values as adequately as possible.

34.2.5.2 Autoradiographic Method

Autoradiography yields the tracer distribution at only one time-point. If the model has more than one free parameter, it is in most case not possible to calculate k values from autoradiographs without simplifying assumptions. For instance, the assumption that the ratio K_1/k_2 is the same in all gray matter is sometimes justified. The most famous example for such a case is the determination of the rate of glucose metabolism using labeled deoxyglucose or fluorodeoxyglucose. The autoradiographic method was developed and in detail described by Sokoloff et al. (Sokoloff et al. 1977).

34.2.5.3 Ratio Methods

In principle, the methods described above require the knowledge of thc time course of authentic tracer in arterial plasma, which is sometimes not so easy to get. One can then sometimes resort to simple ratios of tracer concentrations in different brain areas. For example, if neuroreceptors are of interest, and if there is a brain region devoid of receptors, the ratio of the tracer concentration in the region of interest to the region devoid of receptors at complete equilibrium is a measure of the receptor concentration.

34.3 Example

34.3.1 Investigation of Astrocytic Oxygen Consumption Using Radiolabeled Acetate

One open question in brain research concerns the relative contribution of astrocytes and neurons to total oxidative metabolism. This question is challenging to address *in vivo*, since it is extremely difficult to spatially resolve astroglia and neurons with

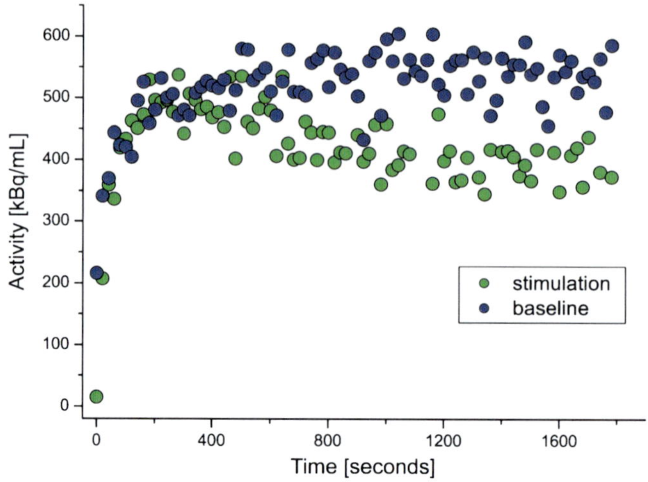

Fig. 34.6 Time activity curves from the rat somatosensory cortex of two subsequent 1-^{11}C-acetate injections. One injection was performed during baseline conditions the other during electrical infraorbital nerve stimulation leading to increased label washout rate. Note the faster washout of ^{11}C activity during stimulation

radiotracer methods (quite unlike the optical approach described in the introduction). However, radiolabeled acetate seems suitable to investigate astrocytic oxidative metabolism *in vivo*. This hypothesis is based on several observations. First, acetate is preferentially taken up by astrocytes. The reason is most probably due to a specific monocarboxylate transporter (MCT)-like acetate transporter which is absent in neurons (Waniewski and Martin 1998, 2004). Following intravenous injection, about 20% of the delivered substance crosses the blood brain barrier. In the astrocytes the labeled acetate enters the TCA cycle. A fraction will be oxidized to CO_2, another part leaves the TCA cycle at the stage of a-ketoglutarate and enters the glutamine/glutamate pool. From there it can re-enter the astrocytic TCA cycle and can get oxidized to CO_2. Another part of the labeled glutamine will leave the astrocytes and enter the neurons. However, in the first 20 min following injection this neuronal fraction is negligible (Lebon et al. 2002). We therefore hypothesized that in the first 20 min the label that leaves the brain tissue is labeled CO_2. If that is the case, the washout rate of the label from tissue might represent the production of CO_2 in the TCA cycle and therefore be a measure for total oxidative metabolism in the astrocyte.

The next question is what methodology would be suited to measure the washout rate of radioactive label. In animals one of the cost-effective methods is a beta probe, which is inserted into the brain structure of interest. In our experiments we used probes with 0.25 mm diameter (http://www.swisstrace.com) and performed experiments in rats using 1-^{11}C-acetate. The tissue kinetics of the label was measured in the somatosensory cortex at baseline and during electrical stimulation of the infraorbital branch of the trigeminal nerve. Tissue time-activity curves derived from the experiments are demonstrated in Fig. 34.6. The experiments demonstrated a significant 90% increase of the label washout rate from baseline to stimulation. From this result we concluded that the increase of oxidative metabolism in astrocytes during increased cerebral activity is substantial.

34.3.2 Autoradiography Versus Beta Probes

The usefulness of the autoradiographic method to derive quantitative values critically depends on the kinetic behavior of the tracer. If the tracer gets trapped in tissue and the trapping is not dependent on blood flow, one autoradiograph is sufficient to derive quantitative measures for physiological parameters. A typical example is the assessment of glucose metabolism with labeled deoxyglucose or fluorodeoxyglucose (^{14}C-deoxyglucose or ^{18}F-fluorodeoxyglucose) respectively. The methodology was developed and described in various publications of Louis Sokoloff (Sokoloff et al. 1977; Reivich et al. 1979). The top panel of Fig. 34.7 illustrates the change of the tissue radioactivity concentration of the tracer ^{18}F-fluorodeoxyglucose with a doubling of blood flow. At 45 min the doubling of blood flow leads only to a 10% increase in accumulated activity. For acetate the situation is different. The simulated time course of the ^{11}C activity following intravenous injection of 1-^{11}C-acetate for different flow and washout values is demonstrated in the bottom panel of Fig. 34.7. An increase in blood flow alone (K_1 change from 0.08 to 0.095 mL * min^{-1} * mL^{-1}) leads to a substantially different time course.

With a concomitant K_1 and k_2 increase as measured in our experiments, the activity is higher at the beginning but, due to the higher washout rate, lower after 18 min. What this means is that the autoradiographic method is not suited to evaluate acetate metabolism. Cruz et al. used 2-^{14}C-acetate to evaluate acetate metabolism during acoustic stimulation in rats (Cruz et al. 2005). The autoradiographs obtained 5 min following tracer injection demonstrate increased ^{14}C activity in the auditory structures. From this result the authors derived an increased acetate metabolism. However, as the simulation in Fig. 34.7 demonstrates, an increase in blood flow alone without associated increase in acetate metabolism (k_2) would already be sufficient to explain the result of increased uptake. In situations like these, the knowledge of the time course of the tracer distribution is a great advantage. In summary, autoradiographic studies for the quantification of physiological parameters are most useful, if tracer uptake is not or only minimally affected by blood flow. In situations where blood flow is important, the assessment of the full time-course of the tracer kinetics is advantageous.

34.4 Summary

The concept of compartmentation is central both in cerebral metabolism and in radiotracer methodology. In this chapter, we introduced the necessary background for understanding kinetic modeling of data obtained from radiotracer techniques,

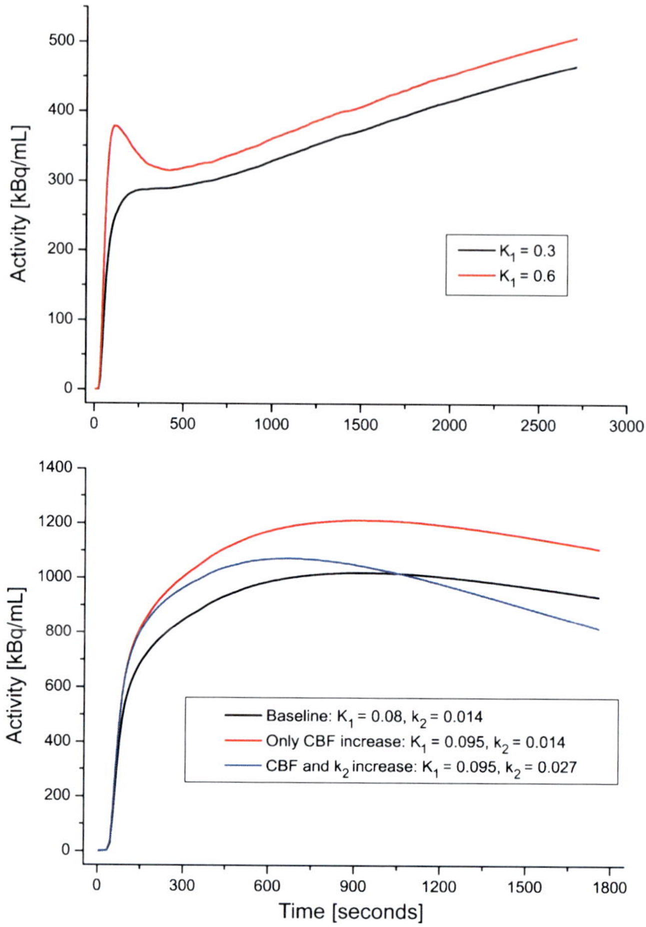

Fig. 34.7 Simulated tissue time-activity curves using ^{18}F-fluorodeoxyglucose (**a**) and 1-^{11}C-acetate (**b**). **a**: The *red curve* was simulated with double the blood flow of the *black curve*. The important point is the relatively small increase of tissue activity at 45 min, indicating that the flow dependence of the ^{18}F-fluorodeoxyglucose kinetics is small. **b**: Baseline tissue time-activity curves following injection of 1-^{11}C-acetate were generated with mean k values of our experiments (*black curve*). Note the markedly higher tissue activity when only blood flow is increased (*red curve*). The increase of blood flow and back-diffusion, as observed in our experiments, leads to higher tissue activity at the beginning but to a lower activity after 18 min (*blue curve*)

such as PET and beta-probes. The use of labeled acetate serves as an example to illustrate how important it is to acquire dynamic data in order to appreciate meaningful readouts, such as compartment specific metabolic rates.

References

Cruz NF, Lasater A, Zielke HR, Dienel GA (2005) Activation of astrocytes in brain of conscious rats during acoustic stimulation: acetate utilization in working brain. J Neurochem 92:934–947

Kasischke KA, Vishwasrao HD, Fisher PJ, Zipfel WR, Webb WW (2004) Neural activity triggers neuronal oxidative metabolism followed by astrocytic glycolysis. Science 305:99–103

Lebon V, Petersen KF, Cline GW, Shen J, Mason GF, Dufour S, Behar KL, Shulman GI, Rothman DL (2002) Astroglial contribution to brain energy metabolism in humans revealed by 13 C nuclear magnetic resonance spectroscopy: elucidation of the dominant pathway for neurotransmitter glutamate repletion and measurement of astrocytic oxidative metabolism. J Neurosci 22: 1523–1531

Magistretti PJ, Pellerin L, Rothman DL, Shulman RG (1999) Energy on demand. Science 283:496–497

Pellerin L, Magistretti PJ (1994) Glutamate uptake into astrocytes stimulates aerobic glycolysis: a mechanism coupling neuronal activity to glucose utilization. Proc Natl Acad Sci USA 91:10625–10629

Pierre K, Pellerin L (2005) Monocarboxylate transporters in the central nervous system: distribution, regulation and function. J Neurochem 94:1–14

Reivich M, Kuhl D, Wolf A, Greenberg J, Phelps M, Ido T, Casella V, Fowler J, Hoffman E, Alavi A, Som P, Sokoloff L (1979) The [18 F]fluorodeoxyglucose method for the measurement of local cerebral glucose utilization in man. Circ Res 44:127–137

Sokoloff L, Reivich M, Kennedy C, Des Rosiers MH, Patlak CS, Pettigrew KD, Sakurada O, Shinohara M (1977) The [14 C]deoxyglucose method for the measurement of local cerebral glucose utilization: theory, procedure, and normal values in the conscious and anesthetized albino rat. J Neurochem 28:897–916

Vannucci SJ, Maher F, Simpson IA (1997) Glucose transporter proteins in brain: delivery of glucose to neurons and glia. Glia 21:2–21

Vannucci SJ, Clark RR, Koehler-Stec E, Li K, Smith CB, Davies P, Maher F, Simpson IA (1998) Glucose transporter expression in brain: relationship to cerebral glucose utilization. Dev Neurosci 20:369–379

Waniewski RA, Martin DL (1998) Preferential utilization of acetate by astrocytes is attributable to transport. J Neurosci 18:5225–5233

Waniewski RA, Martin DL (2004) Astrocytes and synaptosomes transport and metabolize lactate and acetate differently. Neurochem Res 29:209–217

Wyss MT, Ametamey SM, Treyer V, Bettio A, Blagoev M, Kessler LJ, Burger C, Weber B, Schmidt M, Gasparini F, Buck A (2007) Quantitative evaluation of 11 C-ABP688 as PET ligand for the measurement of the metabotropic glutamate receptor subtype 5 using autoradiographic studies and a beta-scintillator. NeuroImage 35:1086–1092

Chapter 35
Nitrogen Metabolism *In Vivo*

Keiko Kanamori and Brian D. Ross

Abstract *In vivo* ^{15}N MRS has made a unique contribution to kinetic studies of the individual pathways that control the flux of glutamate in rat brain. Glutamate is the major excitatory neurotransmitter and is also a component of intermediary metabolism. Current evidence strongly suggests that the metabolic and neurotransmitter pools of glutamate are partly recycled through the glutamine/glutamate cycle, but other pathways such as transamination from leucine contribute to glutamate synthesis.

This chapter first describes the advantages and limitations of *in vivo* ^{15}N MRS and indirect detection of ^{15}N through coupled ^{1}H for kinetic study. We then describe major findings from our and other laboratories using these non-invasive methods in the following areas: (a) the uptake of neurotransmitter glutamate from the extracellular fluid into glia; (b) its metabolism to glutamine; (c) transport of glial glutamine to the extracellular fluid; (d) hydrolysis of neuronal glutamine to glutamate; and (e) contribution of transamination from leucine to replenishment of neuronal glutamate nitrogen. Comparison of the *in vivo* fluxes of these major pathways strongly suggests that approximately 75% of glutamate is recycled through the glutamine/glutamate cycle while leucine provides about 25% of neuronal glutamate nitrogen. *In vivo* glutamine synthetase activities measured at several levels of hyperammonemia showed that this enzyme becomes saturated when blood ammonia concentration exceeds 0.9 μmol/g, and causes elevation of brain ammonia. This *in vivo* ^{15}N MRS study, combined with measurement of brain ammonia and glutamine and observation of reversible neurobehavioral impairments in awake rats, strongly suggested that the elevation of brain ammonia, not the concomitant increase in brain glutamine, is the major factor in the pathogenesis of hyperammonemic encephalopathy.

K. Kanamori, Ph.D. (✉) • B.D. Ross, M.D., DPhil
Magnetic Resonance Spectroscopy Laboratory,
Huntington Medical Research Institutes, Pasadena, CA 91105, USA
e-mail: kkanamori@hmri.org; bdross.hmri@gmail.com

I.-Y. Choi and R. Gruetter (eds.), *Neural Metabolism In Vivo*,
Advances in Neurobiology 4, DOI 10.1007/978-1-4614-1788-0_35,

Keywords ^{1}H-^{15}N HMQC • Ammonia • Extracellular fluid • Glia • Glutamate • Glutamine • Glutamine synthetase • Hepatic encephalopathy • HSQC • *in vivo* • N-15 MRS • Neuron • Neurotransmitter • Rat brain • SAT • SN1 • Transport

35.1 Introduction

In vivo ^{15}N MRS is a useful technique that has made a unique contribution to our understanding of metabolic flux in the intact brain. Brain metabolites of neuropathological importance include glutamate (GLU), which mediates most of excitatory neurotransmission, and its precursor glutamine (GLN). As shown schematically in Fig. 35.1, neurotransmitter GLU is released from presynaptic vesicles in axon terminals to the extracellular fluid, binds to the receptor of postsynaptic neuron for neurotransmission. It is then taken up mainly into glia and metabolized to GLN by glia-specific glutamine synthetase (GS). The rapid clearance of GLU from the extracellular fluid prevents excessive stimulation of GLU receptor (GLU excitotoxicity) which is implicated in the initiation of epileptic seizures and neuronal death associated with a wide range of neurological disorders (Choi 1988; Danbolt 2001; Antonelli et al. 2007). GLN is then transported across the extracellular fluid mainly by transporters SN1 (Chaudhry et al. 1999; Bröer et al. 2002) and SAT (sodium-coupled amino acid transporter) (Varoqui et al. 2000) to the neuron and hydrolyzed, by phosphate-activated glutaminase (GLNase), to GLU. According to the concept of glutamine/glutamate cycle (Hertz 1979; Shank and Aprison 1981), this completes the recycling of GLU from GLN. An aminotransferase, viz. branched-chain aminotransferase (BCAT), also contributes to GLU synthesis (Kanamori et al. 1998; García-Espinosa et al. 2007). These major pathways that control the flux of the

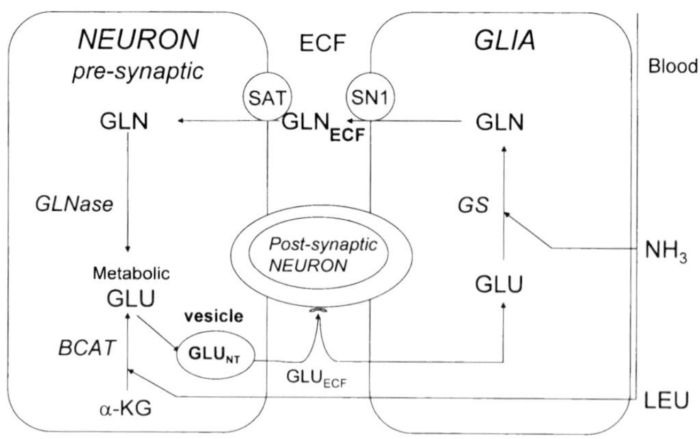

Fig. 35.1 The major pathways that contribute to glutamate flux in the brain. *α-KG* α-ketoglutarate, *BCAT* branched-chain aminotransferase, *ECF* extracellular fluid, *GLU* glutamate, GLU_{NT} neurotransmitter GLU, *GLN* glutamine, *GLNase* glutaminase, *GS* glutamine synthetase, *LEU* leucine

metabolic and neurotransmitter pools of GLU involve formation or transfer of amide and amine groups. Hence, non-invasive ^{15}N MRS makes a unique contribution to kinetic studies of these individual pathways *in vivo*. This chapter describes major findings from ^{15}N MRS studies of their fluxes in rat brain obtained in our and other laboratories, and discusses their neuropathological significance. First, we describe the characteristics of ^{15}N MRS that are important for *in vivo* studies.

35.2 Characteristics of *in Vivo* ^{15}N MRS

The most abundant isotope of nitrogen, ^{14}N (natural abundance 99.6%), has not been extensively used for *in vivo* MRS because it has nuclear spin quantum number I = 1, and hence a quadrupole moment which leads to rapid relaxation and broad signals, except in a few molecules with symmetrical electronic environments around ^{14}N such as choline (Witanowski et al. 1992). We will focus on ^{15}N which has nuclear spin I = ½ and provides a MRS signal with a narrow linewidth.

35.2.1 Direct ^{15}N MRS

^{15}N has a broad chemical shift range (500–900 ppm) and hence high spectral resolution (Levy and Lichter 1979). Its low natural-abundance (0.365%) is an advantage for *in vivo* MRS because only those metabolites that have been ^{15}N enriched by intravenous infusion of a labeled precursor are normally observed, without interference from naturally-occurring ^{15}N. This permits measurement of the fluxes of ^{15}N-metabolites upon a clean background. The sensitivity of ^{15}N is 1/1,000th of ^{1}H and 1/15th of ^{13}C for equal numbers of nuclei at constant field. However, proton-decoupling not only removes ^{15}N-^{1}H splitting, but can also lead to an enhancement of ^{15}N signal by nuclear Overhauser effect (NOE) for ^{15}N nuclei that relax via dipolar interaction with proton(s). Because ^{15}N has a negative gyromagnetic ratio, nuclear Overhauser enhancement factor (NOEF) is negative, with a theoretical maximum value of −4.93. This NOEF, when added to the original signal intensity (+1.0), results in an *inverted* and enhanced signal with a maximum NOE enhancement of −3.93 (Levy and Lichter 1979). Near-maximum NOE is attained in many neurochemicals of interest such as GLU and GLN. The spin-lattice relaxation times, T_1, of these nitrogens are reasonably short, < 5 s, permitting fairly rapid signal acquisition. Consequently, ^{15}N enriched brain metabolites in the 1–10 mM range can be observed with a time resolution of 5–30 min by proton-decoupled NOE-enhanced ^{15}N MRS *in vivo* at 20 MHz for ^{15}N. In Fig. 35.2, the top spectrum shows an *in vivo* proton-decoupled NOE-enhanced ^{15}N spectrum of [5-^{15}N]GLN and [2-^{15}N]GLU/GLN in rat brain acquired in 19 min during intravenous infusion of $^{15}NH_4Ac$ (Kanamori and Ross 1993, 1999). The time resolution permits measurement of reaction rates on the order of 0.1–10 μmol/g/h. The *in vivo* rates of many key reactions that control the flux of GLU are of this order of magnitude.

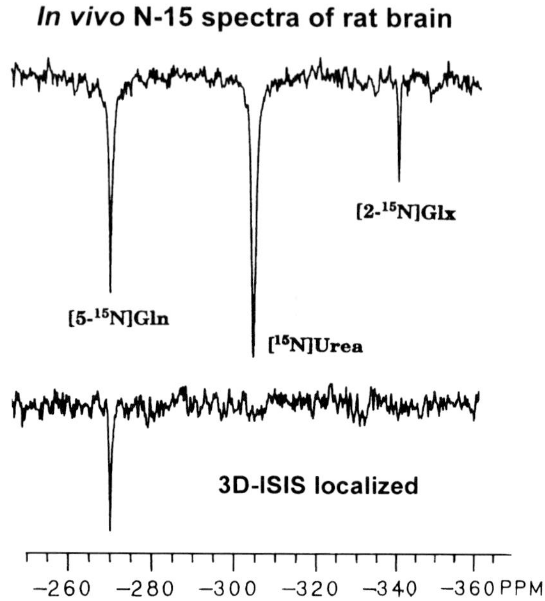

Fig. 35.2 *In vivo* ^{15}N spectra of rat brain, obtained with proton-decoupling and NOE-enhancement at 20 MHz for ^{15}N, without (*top*) and with (*bottom*) 3D ISIS localization. Glx: GLN/GLU

For localized observation of brain ^{15}N-metabolites, we have used image-selected *in vivo* spectroscopy (ISIS) developed by (Ordidge et al. 1986). In Fig. 35.2, the bottom spectrum shows an ^{15}N spectrum of [5-^{15}N]GLN in rat brain, acquired with three-dimensional ISIS in 28 min in a volume probe with a homogeneous B_1 field (Kanamori and Ross 1999). For localized observation of [2-^{15}N]amino acids, such as [2-^{15}N]GLU, which have short spin-spin relaxation time (T_2) due to exchange of the amine protons with water proton at physiological pH (Kanamori and Ross 1997b), ISIS has the advantage of entailing no inherent signal loss, whereas significant T_2-induced loss occurs in spin-echo localization methods such as PRESS (point-resolved spectroscopy) or STEAM (stimulated echo acquisition mode). The exchange causes change in ^{15}N-^{1}H coupling resulting in magnetic field fluctuation at ^{15}N which, in turn, causes shortening of T_2 due to scalar relaxation (Becker 1980). In the localized spectrum (Fig. 35.2 bottom), [2-^{15}N]GLX was not observed because the carrier frequency was set on [5-^{15}N]GLN to optimize observation of this peak. The resulting spatial displacement error for [2-^{15}N]GLX, $\Delta Z = \sigma B_0/G_z$ (with $\sigma = -71$ ppm, $B_0 = 4.7$ T and $G_z = 14$ mT/m), was 23 mm which is comparable to the dimension of rat brain. However, simultaneous observation of brain [5-^{15}N]GLN and [2-^{15}N]GLX in ISIS-localized spectrum with a spatial displacement error of <1.7 mm should be possible by setting the carrier frequency between the two peaks and using a gradient strength >95 mT/m. An alternative localization method is surface-coil localization, which has been used successfully in an ^{15}N NMR study of rat brain at 7 T (Shen et al. 1998).

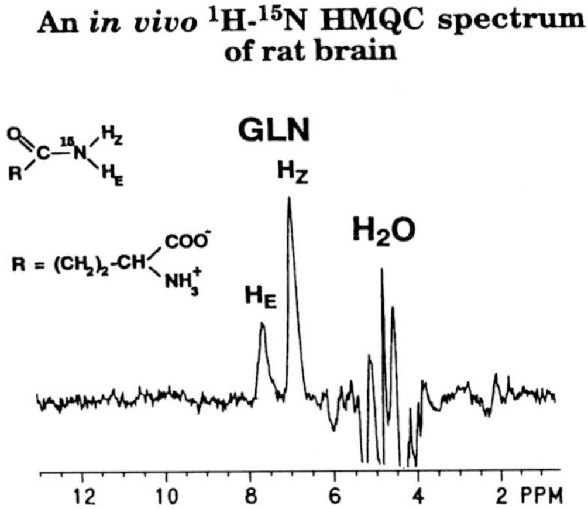

Fig. 35.3 An *in vivo* ^{1}H-^{15}N HMQC spectrum of rat brain acquired at 200 MHz for ^{1}H, with ^{15}N decoupling. The amide protons of [5-^{15}N]GLN are selectively observed. Protons not coupled to ^{15}N are suppressed by phase-cycling

35.2.2 *Indirect ^{15}N Detection through Coupled ^{1}H*

Sensitivity of ^{15}N detection is considerably enhanced by indirect detection through coupled ^{1}H by ^{1}H-^{15}N heteronuclear multiple-quantum coherence (HMQC) transfer MRS (Bax et al. 1983) and heteronuclear single-quantum correlation (HSQC) MRS (Kay et al. 1992). These methods permit selective detection of ^{1}H spin-coupled to ^{15}N provided that the proton is nonlabile at physiological pH and temperature. Among neurotransmitters and their precursors, [5-^{15}N]GLN and [$^{15}N_{indole}$] serotonin (Kanamori et al. 1991) are detectable *in vivo* and *in vitro* respectively by these methods. Figure 35.3 shows an *in vivo* ^{1}H spectrum obtained by this method from rat brain (Kanamori and Ross 1997a). Two amide protons of [5-^{15}N]glutamine, H_Z and H_E which are coupled to ^{15}N, are selectively observed, with ^{15}N decoupling. The signals from protons not coupled to ^{15}N that resonate upfield of water are suppressed by phase-cycling. Observation of line-broadening of H_Z, due to base-catalyzed exchange with water proton, has provided an unique assay of glial pH (~7.5) in hyperammonemia, as distinct from neuronal and whole-brain pH, as described in Kanamori and Ross (1997a). The natural linewidth of H_Z was measured in control non-hyperammonemic brain. A limitation of the indirect detection method is that the amine protons of GLU and GABA cannot be detected at physiological pH due to their rapid exchange with water protons Thus, while direct ^{15}N MRS is useful for *in vivo* study of a broad range of nitrogen-containing metabolites at 1–10 mM, the indirect ^{1}H-^{15}N method, which has intrinsically 50–100-fold higher sensitivity, is useful for study of low-concentration [^{15}N]-metabolites with non-labile protons, including [5-^{15}N]GLN in the extracellular fluid (the application section).

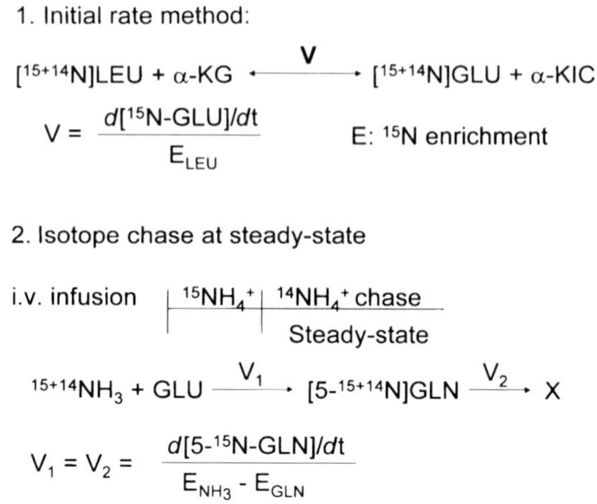

Fig. 35.4 Initial rate method and isotope chase method at steady-state. α-*KIC* α-ketoisocaproate. For other abbreviations, see Fig. 35.1 caption

35.2.3 ^{15}N-Enrichment and Kinetic Analysis

Brain metabolites are ^{15}N-enriched by intravenous infusion of ^{15}N-enriched precursors. In all *in vivo* ^{15}N MRS studies performed to date, the i.v. infused ^{15}N precursor is the actual substrate of the reaction under study. $^{15}NH_3$ is the substrate of glutamine synthetase to form [5-^{15}N]glutamine, and [^{15}N]leucine is the substrate of transamination to form [^{15}N]GLU. Hence, the rate of a single reaction is measured from the ^{15}N enrichment of the substrate and the time-course of ^{15}N-product formation. This simplifies kinetic analysis, and eliminates the need to make assumptions about the rates and metabolite pool sizes of intermediate pathways. We have used two kinetic analysis methods described below and shown in Fig. 35.4.

(a) *Initial rate method.* In this method, we measure the rate of increase of the ^{15}N-labeled product, e.g. [^{15}N]GLU *in vivo* (Kanamori et al. 1998). The rate of formation of total [^{15}N+^{14}N]GLU is then determined from the rate of initial linear increase divided by the steady-state ^{15}N enrichment of the substrate. This method permits accurate rate measurement when the efflux of the ^{15}N-product to other pathways is negligible.

(b) *Isotope chase method.* This permits accurate rate measurement when the efflux rate of the product is unknown. Infusion of the ^{15}N-labeled substrate is continued until total i.e. [^{14}N+^{15}N]concentrations of the substrate and the product reach steady-state. The infusate is then changed to ^{14}N and infusion continued to maintain steady-state while ^{15}N in the product is progressively replaced by ^{14}N. It has been shown that the rate of GLN synthesis is then given by the simple equation for V_1 from the rate of change of [5-^{15}N]GLN and E_{NH3} and E_{GLN} (the ^{15}N enrichments of the substrate and the product respectively) (Kanamori and Ross 1993).

35.2.4 Quantification

An ^{15}N metabolite is quantified by experimentally measuring the *in vivo* ^{15}N peak area per μmol of the metabolite. For brain [5-^{15}N]glutamine, for example, we measured its *in vivo* peak area per unit time as well as the actual quantity of [5-^{15}N]GLN found in the extract of the same brain in ten rats. The slope of the least-squares line through the plots gives the mean *in vivo* peak area per μmol of [5-^{15}N]GLN per g of brain. This quantity was used in subsequent time-course experiments to convert the observed *in vivo* peak area to [5-^{15}N]GLN concentration (Kanamori and Ross 1997b). The same method was used to quantify [5-^{15}N]GLN observed through coupled ^{1}H (Kanamori et al. 1995).

35.3 Applications

35.3.1 In Vivo Glutamine Synthetase Activity as a Function of Blood [NH_3]: Implication for Pathophysiology of Hyperammonemic Encephalopathy

Glutamine synthetase plays an important role in protecting the brain from a number of diseases. One is hepatic encephalopathy (HE) which is a reversible metabolic disorder of the brain caused by dysfunction of the liver, and by inborn errors of single genes which result in defects in the hepatic urea cycle (ornithine transcarbamylase deficiency). Mechanisms responsible for the pathogenesis of HE are unresolved, as most recently reviewed by (Zwingmann 2007). Decreased ammonia removal in the liver causes hyperammonemia and diffusion of excess ammonia into the brain. Ammonia is metabolized to GLN by glutamine synthetase in glia (Fig. 35.1), but ammonia can accumulate if the rate of its influx exceeds the rate of its removal by GS. Whether the neurotoxicity of ammonia itself causes encephalopathy or whether the concomitant accumulation of GLN, which is known to cause brain edema, contributes to pathogenesis (Albrecht and Dolinska 2001) has been a subject of considerable debate.

Most early investigations of possible correlation between severity of encephalopathy and brain ammonia and GLN concentrations in experimental animals used portacaval-shunted rats with hepatic artery ligation (Bosman et al. 1990; Swain et al. 1992; Mans et al. 1994). While they are good models of acute HE, complete loss of liver function results in coma and death in 6–10 h, and the effects of hyperammonemia are often difficult to distinguish from the other effects of irreversible liver failure. Hyperammonemia induced by administration of ammonia in rats with normal liver functions avoids this problem and has the additional advantage that reversal of encephalopathy can be studied. In our study, hyperammonemia was induced by intravenous infusion of NH_4Ac to attain steady-state blood [NH_3]. This permitted investigation of possible correlation between severity of encephalopathy,

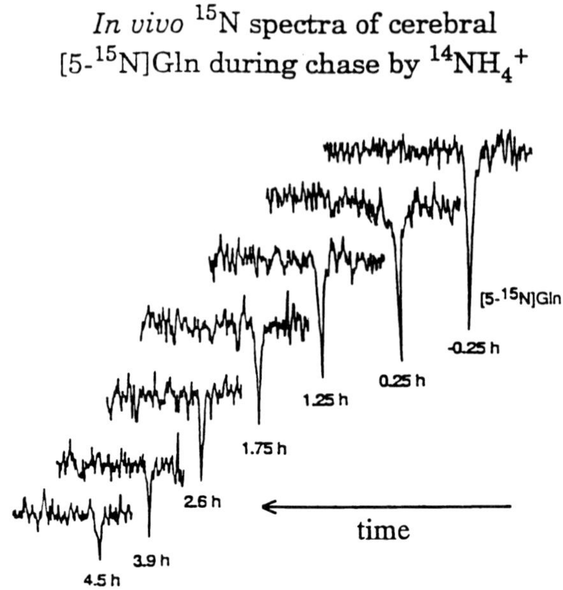

Fig. 35.5 *In vivo* ^{15}N spectra of brain [5-^{15}N]GLN before (t=–0.25 h) and during isotope chase by $^{14}NH_4^+$

as assessed by neurobehavioral impairments in awake rats, and brain ammonia and GLN concentrations. A unique feature of our study was the measurement of the rates of glutamine synthesis *in vivo* at several levels of hyperammonemia, to examine how the activity of this enzyme controls the concentrations of brain ammonia and GLN *in vivo* (Kanamori et al. 1996).

Figure 35.5 shows *in vivo* ^{15}N spectra of brain [5-^{15}N]GLN acquired during isotope chase by ^{14}N. Progressive decrease in the peak area is observed, as ^{15}N is replaced by ^{14}N while maintaining the total (i.e. ^{15}N + ^{14}N) concentration at steady-state. Figure 35.6 plots the change in brain [5-^{15}N]GLN concentration during the chase. The rate of glutamine synthesis was determined from the observed rate of decrease in [5-^{15}N]GLN and the ^{15}N enrichment of blood ammonia, 0.077 ± 0.017 and brain glutamine 0.24 ± 0.02 to be 4.8 ± 1.1 μmol/h/g (n = 5) at steady-state blood NH_3 concentration of 0.64 μmol/g (Kanamori and Ross 1993).

For the kinetic study at near-physiological blood ammonia concentration, we used the more sensitive indirect detection of ^{15}N through the coupled proton by heteronuclear multiple quantum coherence (HMQC) transfer MRS. By combining direct and indirect detection of [5-^{15}N]GLN, we measured *in vivo* GS activities at five different levels of hyperammonemia, ranging from mild, moderate to severe. As shown in Fig. 35.7, the *in vivo* GS activity increased with [NH_3] over 0.4–0.9 μmol/g and then leveled off. When *in vivo* GS activity (v) as a function of blood NH_3 concentration (s) was plotted as double-reciprocal Lineweaver-Burk plot, the linear increase of $1/v$ with $1/s$ permitted estimation of the *in vivo* glutamine synthetase activity at physiological blood ammonia concentration (0.17 ± 0.018 μmol/g) to be 1.3 ± 0.8 μmol/g/h (Kanamori et al. 1995). In hyperammonemia, the plateau reached

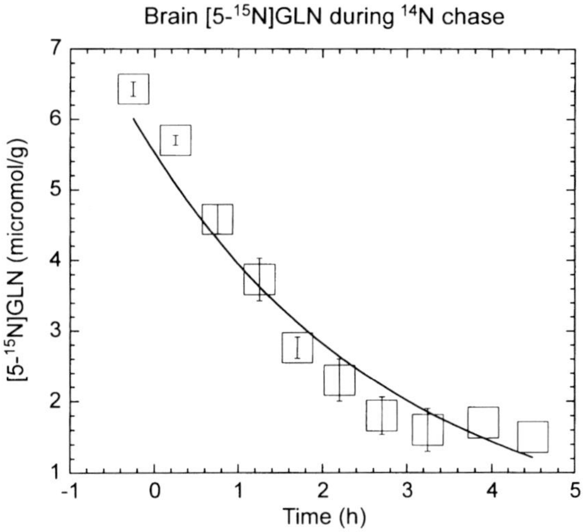

Fig. 35.6 Change in brain [5-^{15}N]GLN concentration during chase by ^{14}N (mean ± sem for $n = 5$)

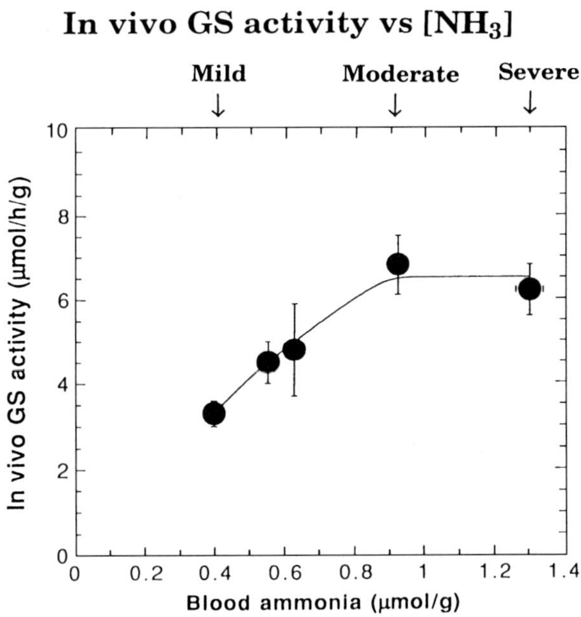

Fig. 35.7 *In vivo* glutamine synthetase activity (μmol/g/h) as function of the steady-state blood ammonia concentration. Hyperammonemia ranges from mild, moderate to severe

by *in vivo* GS activity (Fig. 35.7) strongly suggested that *in vivo* GS activity becomes saturated when [NH_3] exceeds 0.9 μmol/g. Figure 35.8 shows the effect of saturation of GS activity on brain [NH_3]. Also shown is the grade of encephalopathy observed in corresponding awake rats. Grades of encephalopathy, which parallel muscular incoordination progressing to stupor and coma in human HE patients (Zwingmann 2007), were assigned according to established criteria (Mans et al. 1994).

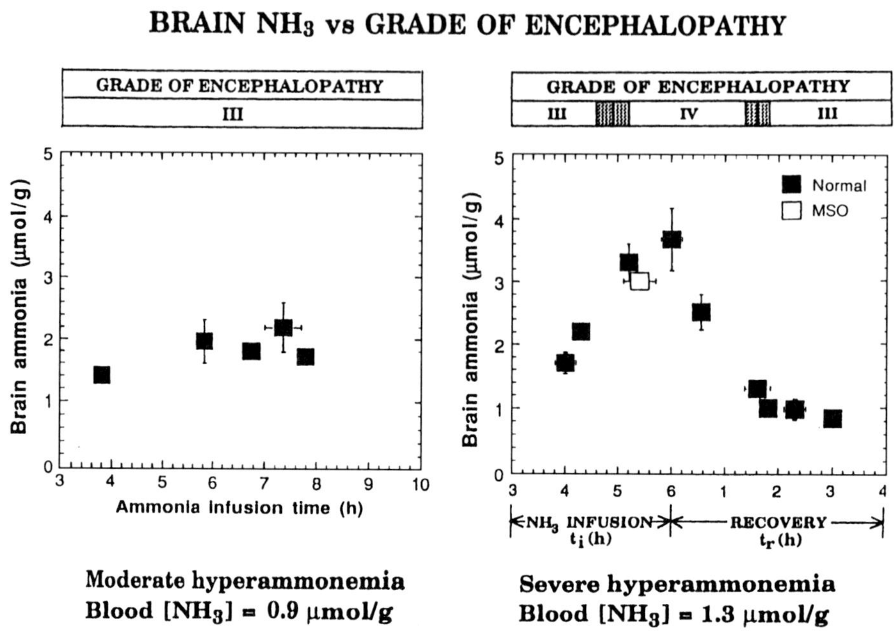

Fig. 35.8 Brain ammonia and grade of encephalopathy in moderate (*left*) and severe (*right*) hyperammonemia. Grade III is characterized by reluctance to move, but the righting reflex is present. Grade IV: motionless without righting reflex

In moderate hyperammonemia (left figure), ammonia diffusing into the brain was metabolized to GLN by GS and brain ammonia concentration remained around 2 μmol/g during 3–8 h of infusion. Corresponding awake rats showed only Grade III encephalopathy. By contrast, in severe hyperammonemia (right figure), brain ammonia was elevated to 3.5 μmol/g and the rats progressed from grade III to grade IV encephalopathy (For a brief description of Grades III and IV see Fig. 35.8 caption). When ammonia infusion was stopped at $t=6$ h (bottom x-axis), brain ammonia decreased and the rat recovered to grade III. The result strongly suggests that at this level of hyperammonemia, when *in vivo* GS activity became saturated (Fig. 35.7), the rate of ammonia influx into the brain exceeded the rate of metabolism to GLN and the resulting elevation of brain ammonia causes the onset of grade IV encephalopathy. Because GLN accumulation and the resulting brain edema have also been implicated as a possible cause of encephalopathy, we examined brain GLN concentration as well (Fig. 35.9). This was very similar in moderate and severe hyperammonemia and showed little correlation with grade of encephalopathy. Significantly, when glutamine synthesis was inhibited by preinjection of methionine sulfoximine (MSO), an irreversible inhibitor of GS, brain GLN was only 5.5 μmol/g (open circle in the right figure), but the rat showed grade IV encephalopathy when its brain $[NH_3]$ reached 3 μmol/g (open square in Fig. 35.8). This strongly suggested that elevation of brain $[NH_3]$, not accumulation of brain GLN, is the cause of encephalopathy in severe hyperammonemia.

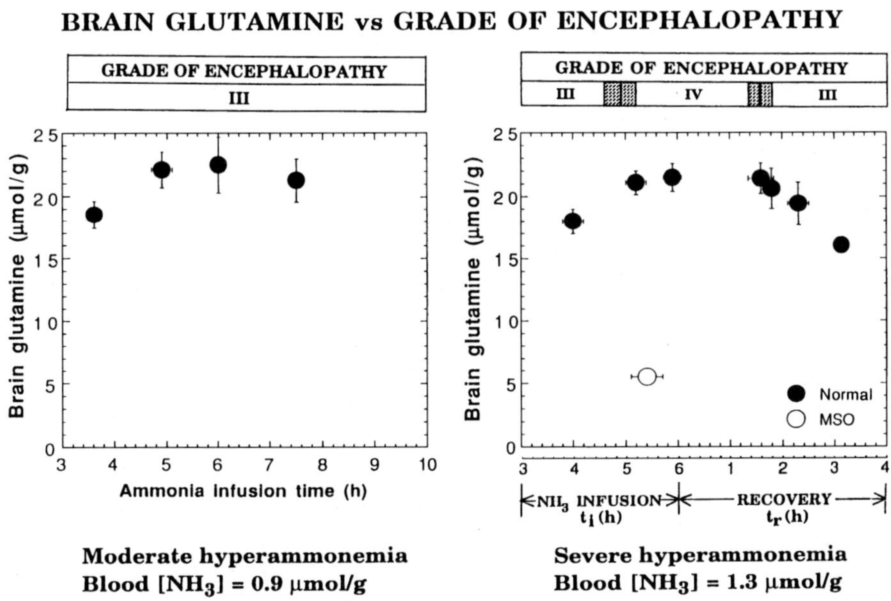

Fig. 35.9 Brain GLN and grade of encephalopathy in moderate (*left*) and severe (*right*) hyperammonemia. MSO: L-methionine-DL-sulfoximine, an irreversible inhibitor of glutamine synthetase

In conclusion, this study (Kanamori et al. 1996) resolved one of the controversial issues, viz. whether neurotoxicity of ammonia itself, or the concomitant accumulation of glutamine is the major factor in the pathogenesis of hyperammonemic encephalopathy. However, it should be noted that in hyperammonemia of human subjects both blood and brain ammonia levels and brain glutamine level can be much lower than those in the rat model. Then the effects of ammonia and glutamine become more difficult to resolve. This study also provided, for the first time, quantitative evidence for the level of hyperammonemia at which GS becomes saturated, resulting in accumulation of neurotoxic levels of brain ammonia.

35.3.2 Regulation of GLN Transport from Glia to the ECF in Vivo

In the glutamine/glutamate cycle, the pathway that had been the least well-understood was the mechanism of transport of GLN from the glia, through the extracellular fluid, to the neuron. Identification and characterization of two GLN transporters, SN1 (Chaudhry et al. 1999) (also termed SNAT3) and SAT1 (Varoqui et al. 2000) (also termed SNAT1) have significantly advanced our understanding of the process. The emerging picture of GLN transport in the adult brain is of GLN release from glia mainly mediated by the system N transporter, SN1, and of GLN uptake into neurons mainly mediated by system A transporter, SAT1 (reviewed by (Bröer and Brookes 2001)).

The glial transporter SN1 mediates the co-transport of GLN with Na^+ in exchange for H^+ (Chaudhry et al. 1999; Bröer et al. 2002). *In vitro*, SN1 can mediate the influx or efflux of GLN depending on the magnitude and direction of the driving forces (Chaudhry et al. 1999). A large transmembrane GLN gradient is expected to induce efflux despite the large inwardly directed driving force for Na^+ and the outwardly directed force for H^+ that would both tend to prevent efflux. These investigators proposed, on the basis of studies in PC12 cells expressing SN1 that extracellular GLN less than 0.4 mM induces GLN efflux.

Our recent study showed that GLN concentration in the ECF of normal rat brain, measured by the zero-flow method (Jacobson et al. 1985), is 0.385±0.016 mM (Kanamori and Ross 2004). Alexander et al. (1997) provided an estimate of 0.262±0.012 mM for GLN_{ECF} concentration in rat cortex, based on *in vivo* efficiency of their probe measured by the zero-flow method. Whole-brain [GLN], representing predominantly intracellular GLN, is ~8 mM (Erecińska and Silver 1990). At least 80% of intracellular GLN is estimated to be in glia on the basis of immunocytochemical studies (Ottersen et al. 1992; Nagelhus et al. 1996); also see (Kanamori and Ross 1997a). Hence, a reasonable estimate for GLN gradient across the glial plasma membrane is 17 (= 6.4/0.385 mM).

In view of the importance of glia-to-neuron GLN transport in the glutamine/glutamate cycle, we investigated the transport properties of SN1 *in vivo*, which had not been attempted previously. We studied the effect of transmembrane GLN gradient on the rate of glial GLN release through measurement of (a) GLN_{ECF} collected by microdialysis, (b) transmembrane GLN gradient, (c) the time-course of intracellular glial [5-^{15}N]GLN. This was measured by ^{15}N MRS and the time-course of extracellular [5-^{15}N]GLN by ^{1}H-^{15}N HSQC, during i.v. infusion of $^{14}NH_4Ac$ for 3 h to achieve steady-state glial [GLN], followed by $^{15}NH_4Ac$ infusion to ^{15}N-enrich brain GLN (Kanamori and Ross 2005).

As shown in Fig. 35.10, GLN_{ECF} was markedly elevated at t=4.5 h of NH_4Ac infusion. An intriguing possibility is that this elevation was caused by accumulation of GLN_{ECF} when its neuronal uptake mediated by SAT1 (with a K_m of 0.3–0.49 mM *in vitro* (Varoqui et al. 2000; Mackenzie et al. 2003) became saturated. The $[GLN]_i$/$[GLN]_{ECF}$ ratio was 9.6±0.9 compared with ~17 in normal brain. GLN_{ECF} then decreased substantially at t=4.9±0.1 h. To examine whether this decrease reflects reduction of glial GLN release due to decrease in transmembrane GLN gradient, we compared the time-courses of intracellular and extracellular [5-^{15}N]GLN during this period. As shown in Fig. 35.11a, intracellular, predominantly glial, [5-^{15}N]GLN increased progressively. By contrast, extracellular [5-^{15}N]GLN reached a plateau after T=1.9 h of $^{15}NH_4Ac$ infusion (Fig. 35.11b upper x-axis), corresponding to t=4.9 h of [^{14}N+^{15}N]ammonia infusion (lower x-axis). This strongly suggests that release of glial [5-^{15}N]GLN to ECF was reduced at this time. Significantly, this corresponds to total NH_4Ac infusion time of 4.9±0.1 h when GLN_{ECF} decreased substantially (Fig. 35.10). Taken together, the results strongly suggest that when the

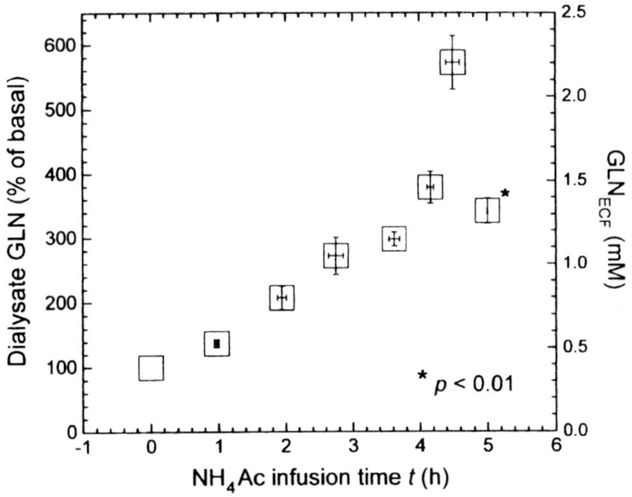

Fig. 35.10 Extracellular GLN concentration (mM) during NH_4Ac infusion (mean ± sem for $n = 5$). The mean value at $t = 5.0$ h is significantly different from that at $t = 4.5$ h ($p < 0.01$)

transmembrane GLN gradient decreased to 9.6 ± 0.9, the driving force for GLN release was reduced, resulting in suppression of glial GLN release to ECF *in vivo*. Other possible causes for the observed decrease in GLN_{ECF}, viz. osmotic disturbance and change in Na^+ gradient, have been examined and ruled out.

Partial suppression of glial GLN transport to ECF when GLN_{ECF} is elevated prevents excessive release, when its neuronal uptake by SAT has reached saturation and the rate of glia-to-neuron transport of GLN cannot be further increased This suggests that the bi-directional nature of the glial transporter SN1 permits fine control of the rate of glial GLN release to ECF.

In conclusion, this study showed for the first time that (a) combination of ^{15}N MRS to monitor glial intracellular [5-^{15}N]GLN and microdialysis/^{1}H-^{15}N HSQC to monitor extracellular [5-^{15}N]GLN permits study of how the GLN transporter SN1 operates *in vivo*; (b) the trans-membrane GLN gradient plays a key role in controlling glial GLN release to ECF at physiological Na^+ gradient.

35.3.3 In Vivo Glutaminase Activity

GLN released to ECF by transporter SN1, is taken up into neurons mainly by SAT1 and hydrolyzed by glutaminase (GLNase) to GLU. The *in vivo* rate of the GLNase pathway, measured by ^{15}N MRS at blood ammonia concentration of 0.4 μmol/g was 1.1 ± 0.2 μmol/g/h (Kanamori and Ross 1995).

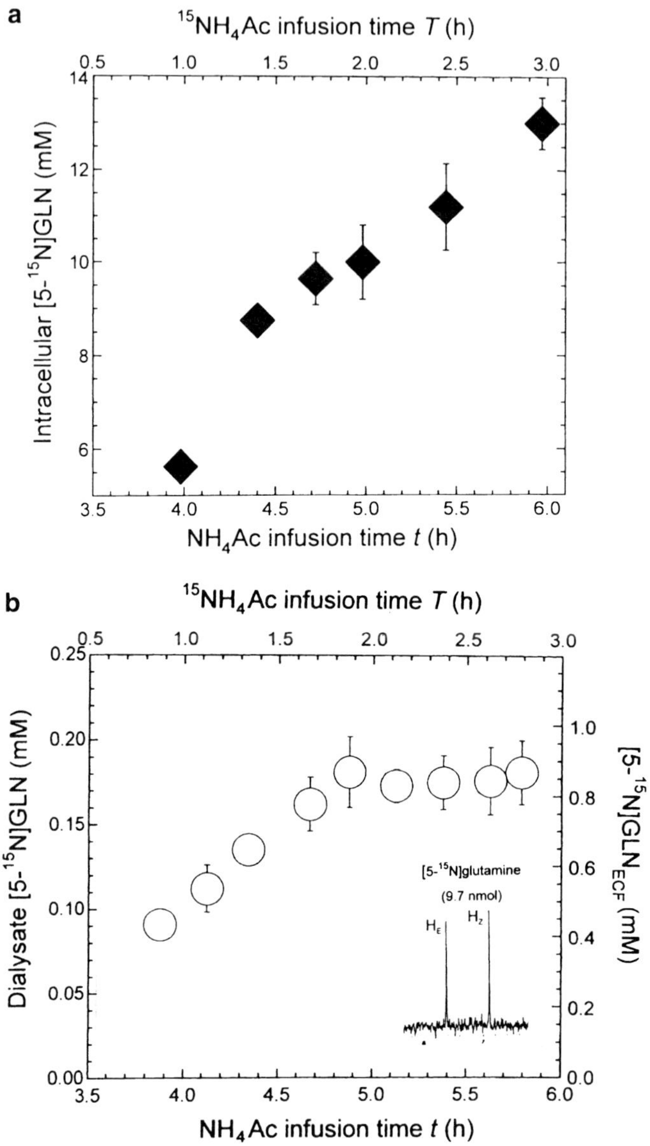

Fig. 35.11 (**a**) Time-course of intracellular [5-^{15}N]GLN (mean ± sem for $n=4$) measured by ^{15}N MRS. $^{15}NH_4Ac$ infusion was started after 3 h of $^{14}NH_4Ac$ infusion. Note that the total [^{14}N + ^{15}N] GLN was at steady-state during this period (Fig. 3b of Kanamori and Ross 2005). (**b**) The time course of extracellular [5-^{15}N]GLN (mean ± sem for $n=5$). The inset shows ^{1}H-^{15}N HSQC spectrum of extracellular [5-^{15}N]GLN (9.7 nmol) obtained at 600 MHz

35.3.4 The Rate of Glutamate Synthesis from Leucine

Another pathway that can contribute to GLU synthesis is transamination to α-ketoglutarate to form GLU. Transaminations from aspartate and alanine, although present at higher concentrations in the brain than leucine, are thought to operate in the direction of aspartate and alanine formation (Erecińska and Silver 1990) and do not contribute to de novo GLU synthesis because plasma aspartate and alanine have very low rates of uptake into the brain. Glutamate dehydrogenase has a very low affinity for NH_4^+ (K_m of 10–18 mM). Hence in normal brain with NH_4^+ concentration of 0.22 μmol/g, it is expected to make little contribution to GLU synthesis. The essential amino acid leucine (LEU) from dietary source, circulating in the blood, has a very high brain uptake index. In the brain, branched-chain aminotransferase (BCAT) catalyzes the transfer of the amine group from LEU to α-ketoglutarate to form GLU (Fig. 35.1). In cultured astrocytes, neurons and synaptosomes, LEU had been shown to be a good source of GLU nitrogen ((Yudkoff et al. 1996) and references cited therein), but its role in the intact brain was unknown. We measured the rate of GLU synthesis from LEU in rat brain *in vivo* by ^{15}N NMR (Kanamori et al. 1998), and evaluated the contribution of LEU to GLU synthesis, compared to that of GLN in glutamine/glutamate cycle.

Awake rats were fed exclusively by intravenous infusion of glucose, amino acids (including LEU) and minerals and vitamins for 2 days to attain metabolic steady-state. At the start of the third day, all plasma and brain amino acids were at steady-state and brain GLU concentration was normal. [^{14}N]LEU was then replaced by [^{15}N]LEU and the rate of GLU synthesis from LEU was observed *in vivo*. Figure 35.12 shows typical *in vivo* ^{15}N spectra of rat brain during [^{15}N]LEU infusion. Progressive increase in brain [^{15}N]GLU is observed. Figure 35.13 shows the increase in brain [^{15}N]GLU concentration, determined from the complete set of 16 *in vivo* time course spectra. Also shown are the concentrations determined at several time points in brain extracts from rats that were awake during [^{15}N]LEU infusion. These agreed well with the data determined from *in vivo* MRS in anesthetized rats. The rate of [^{15}N]GLU synthesis *in vivo* was determined, from the slope of the least-squares line during the initial linear increase to be 0.34 ± 0.01 μmol/g/h. From this rate and the observed ^{15}N enrichments of brain and blood LEU, the rate of GLU synthesis from LEU (designated as V_2) was determined to be 0.7–1.8 μmol/g/h at brain LEU concentration of 0.235 μmol/g (S_2). This concentration ensures that the observed rate of synthesis was not limited by the rate of LEU transport into the brain. At physiological brain LEU concentration (S_1) of 0.11 μmol/g, the *in vivo* rate (V_1) was estimated from the Michaelis-Menten equation, $V_2/V_1 = S_2[K_m + S_1]/S_1[K_m + S_2]$, the observed values of V_2 and S_2 and the known K_m (1.2 mM) of BCAT for LEU (Hall et al. 1993), to be 0.35–0.9 μmol/g/h, with LEU providing ~25% of glutamate nitrogen. A subsequent ^{13}C MRS study reports that LEU can provide as much as 50% of glutamate nitrogen (Sakai et al. 2004).

Fig. 35.12 Representative *in vivo* ^{15}N spectra of brain [^{15}N]GLU during [^{15}N]LEU infusion

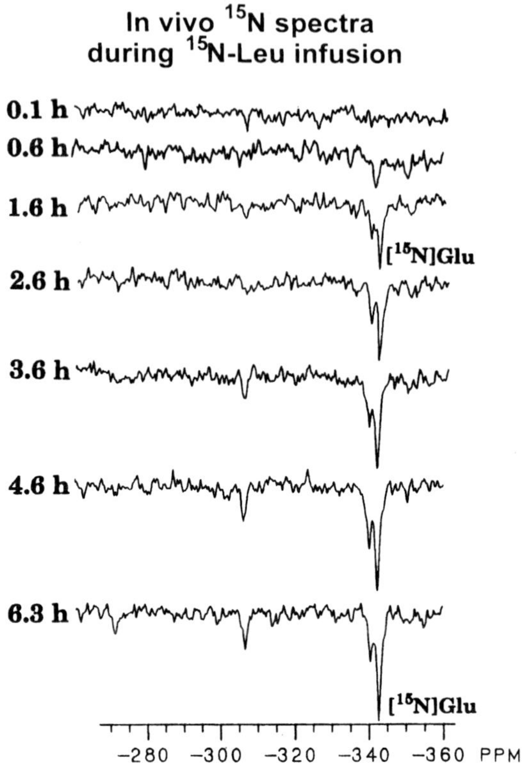

Fig. 35.13 Increase in brain [^{15}N]GLU during [^{15}N]LEU infusion. (mean ± sem for $n = 5$). *Filled circle*: anesthetized rats; *open triangle*: awake rats

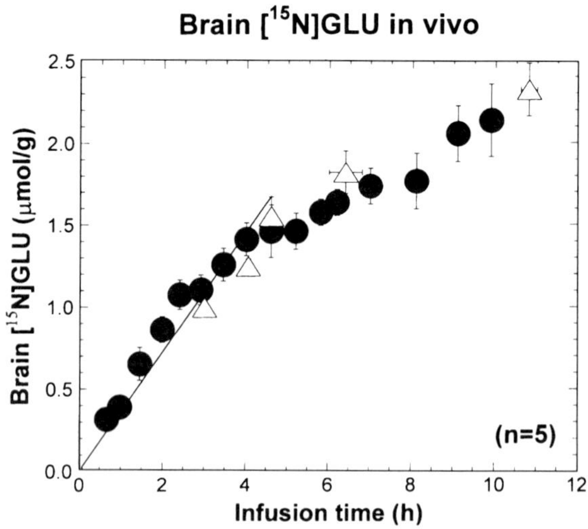

The branched-chain aminotransferase has two isoforms, the cytosolic $BCAT_c$ and the mitochondrial $BCAT_m$ (Hall et al. 1993). $BCAT_c$ represents 70% of total BCAT activity in the brain. At the time of the *in vivo* MRS studies, the distribution of BCAT expression in brain cell types (neuron vs glia) and subcellular compartments was unknown. A recent immunocytochemical study showed widespread expression of $BCAT_c$ in the cell bodies and axon terminals of glutamatergic, GABAergic and other types of neurons (García-Espinosa et al. 2007). Taken together, the results strongly suggest that $BCAT_c$-catalyzed transamination from LEU makes a significant contribution to replenishment of the metabolic pool of GLU in the neuronal compartment.

35.3.5 Glial Uptake of Neurotransmitter GLU

Metabolic GLU is taken up by a uni-directional transporter into presynaptic vesicles to form the neurotransmitter pool of GLU. It is released by exocytosis into synaptic fluid, binds to the receptor and is rapidly taken up mainly into glia by glutamate transporter GLT1 and is metabolized to GLN (Fig. 35.1). The uptake is facilitated by low glial GLU concentration (1–2 μmol/g) compared to whole brain (10 μmol/g). Our study of this process, using *in vivo* ^{13}C MRS and ^{13}C enrichment measurements of extracellular GLU collected by microdialysis and analyzed by gas-chromatography mass spectrometry (Kanamori et al. 2002), is beyond the scope of this chapter and will be described only briefly.

Brain GLU C5 was ^{13}C enriched by intravenous [2,5-^{13}C]glucose infusion, followed by [^{12}C]glucose infusion to chase ^{13}C from the small glial GLU pool. This leaves [5-^{13}C]GLU mainly in the large neuronal metabolic pool and the vesicular neurotransmitter pool. During the chase, the ^{13}C enrichment of whole-brain GLU C5 was significantly lower than that of extracellular GLU (GLU_{ECF}) derived from exocytosis of vesicular GLU. Glial uptake of neurotransmitter [5-^{13}C]GLU_{ECF} was monitored *in vivo* through the formation of [5-^{13}C,^{15}N]GLN during $^{15}NH_4Ac$ infusion. From the rate of [5-^{13}C,^{15}N]GLN synthesis (1.7 ± 0.03 μmol/g/h), the mean ^{13}C enrichment of extracellular GLU (0.304 ± 0.011) and the ^{15}N enrichment of precursor NH_3 (0.87 ± 0.014), the rate of synthesis of GLN (V'_{GLN}), derived from neurotransmitter GLU_{ECF}, was determined to be 6.4 ± 0.44 μmol/g/h. In an earlier ^{15}N NMR study, we determined the rate of synthesis of glutamine to be 6.8 ± 0.7 μmol/g/h under identical experimental conditions. This glutamine is synthesized from the metabolic glial GLU pool + neurotransmitter GLU taken up into glia. Comparison of the two rates strongly suggested that the neurotransmitter GLU provides 80–90% of the substrate GLU pool for GLN synthesis. Hence, under our experimental conditions, the rate of 6.4 ± 0.44 μmol/g/h also represents a reasonable estimate for the rate of glial uptake of GLU_{ECF}, a process that is crucial for protecting the brain from GLU excitotoxicity. This novel approach takes into account the vesicular neurotransmitter pool of GLU in addition to the metabolic glial and neuronal pools, requires ^{13}C enrichment analyses of extracellular GLU collected by microdialysis and is feasible in the animal brain. The significance of the kinetic studies of glutamine/glutamate cycle using strictly non-invasive ^{13}C NMR – the only feasible approach in normal human brain – is described in another chapter of this book.

35.4 Summary and Future Perspective

35.4.1 In Vivo Fluxes of the Major Pathways of GLU Synthesis

Figure 35.14 shows the *in vivo* rates of the major pathways that contribute to the synthesis and utilization of GLU, measured in spontaneously-breathing rats by non-invasive ^{15}N and/or ^{1}H-^{15}N HMQC MRS. Neurotransmitter GLU, taken up into glia, is the major source of substrate GLU for GLN synthesis. The *in vivo* rate of glutamine synthesis, catalyzed by glia-specific glutamine synthetase, is estimated to be 1.3 ± 0.8 μmol/g/h under physiological condition, while the rate of GLN hydrolysis to GLU, catalyzed by glutaminase which is present mainly in the neuron, is 1.1 ± 0.2 μmol/g/h. These fluxes through the glutamine/glutamate cycle are higher than the rate of GLU synthesis from LEU catalyzed by branched-chain aminotransferase (0.35–0.9 μmol/g/h). These results suggest that approximately 75% of brain GLU is recycled from GLN in the glutamine/glutamate cycle, while LEU provides about 25% of neuronal GLU nitrogen. Clearly, multiple pathways contribute to the flux of GLU in the intact brain. We hope that our studies, focusing on the *in vivo* rates of specific pathways, contribute, in combination with studies from other laboratories, to a clearer understanding of the complex mechanisms that regulate the neurotransmitter and metabolic pools of GLU *in vivo*. The significance of specific studies described in this chapter has been summarized at the end of the respective sections.

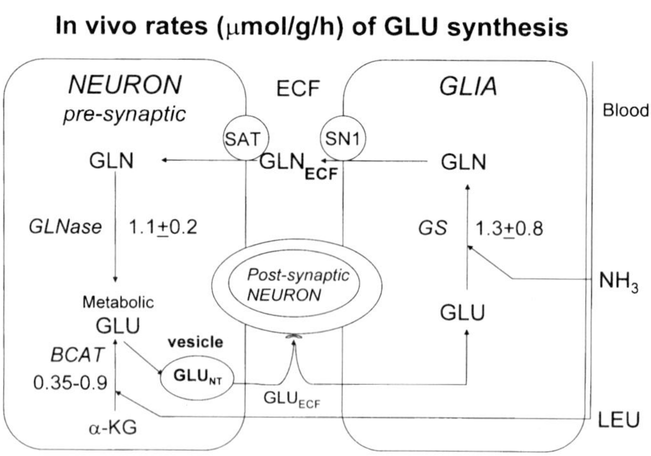

Fig. 35.14 *In vivo* rates (μmol/g/h) of major pathways that control the flux of glutamate in the brain

35.4.2 Future Perspective

35.4.2.1 Glutamine-Glutamate Cycle: A Missing Transporter

Glutamine transporters SAT1 and SAT2 are predominantly localized in glutamatergic and GABAergic neurons, and hence can play a major role in neuronal uptake of GLN_{ECF} (Varoqui et al. 2000; González-González et al. 2005). While this is well-established, recent detailed immunocytochemical studies in rat brain have shown that the neuronal SAT1 and SAT2 immunoreactivities are found mostly in somatodendritic compartments (perikarya and dendritic profiles) and are very scarce on axon terminals (Mackenzie and Erickson 2004; Melone et al. 2004, 2006; González-González et al. 2005). Hence, a transporter that takes up GLN_{ECF} into axon terminals to serve as precursor of neurotransmitter GLU remains to be identified.

35.4.2.2 Potential for Clinical Study

The sensitivity of detection of a brain ^{15}N-metabolite with a given concentration is expected to improve with increase in field strength from 4.7 T used in our study to higher fields now available for animal and clinical studies. The larger brain volume in human studies compared to that of rat brain should also significantly improve sensitivity of detection. The demonstrated clean background, due to lower abundance of nitrogen compared to carbon in biomolecules and the low-natural abundance of ^{15}N, is expected to facilitate localization to the brain, and can shorten exam time if there is no need to take pre-infusion natural-abundance spectra. Power deposition for proton-decoupling of ^{15}N is expected to be less than the level successfully used in clinical ^{13}C MRS because typical one-bond ^{1}H-^{15}N coupling (e.g. ~90 Hz in GLN amide) is smaller than the one-bond ^{1}H-^{13}C coupling (e.g. ~130 Hz in GLU C4). A major challenge in kinetic study is that accuracy of absolute flux measurement depends critically on the isotopic enrichment of the substrate in situ. In animal studies, we have measured the ^{15}N-enrichments experimentally in brain extracts (Sect. 3.4) or in extracellular fluid for study of GLN transport from the glia to the ECF (Sect. 3.2). In human studies, the isotopic enrichment of the substrate in the brain must either be quantifiable by MRS or a reasonable estimate made from its enrichment in plasma samples. We hope that the studies described in this chapter will help towards formulation of well-focused human studies with clinical significance.

Acknowledgments KK and BDR are visiting associates in the Division of Chemistry and Chemical Engineering at California Institute of Technology. The work described here was supported by research grants RO1-NS29048 and RO1-NS048589 from the National Institutes of Health, instrumentation grant from Rudi Schulte Research Institute, and institute funds from Huntington Medical Research Institutes.

We are grateful to all coauthors of the cited papers for their contribution.

References

Albrecht J, Dolinska M (2001) Glutamine as a pathogenic factor in hepatic encephalopathy. J Neurosci Res 65:1–5

Alexander GM, Grothusen JR, Gordon SW, Schwartzman RJ (1997) Intracerebral microdialysis study of glutamate reuptake in awake, behaving rats. Brain Res 766:1–10

Antonelli T, Fuxe K, Tomasini MC, Mazzoni E, Agnati LF, Tanganelli S, Ferraro L (2007) Neurotensin receptor mechanisms and its modulation of glutamate transmission in the brain; relevance for neurodegenerative diseases and their treatment. Prog Neurobiol 83:92–109

Bax A, Griffey RH, Hawkins BL (1983) Correlation of proton and nitrogen-15 chemical shifts by multiple quantum NMR. J Magn Reson 55:301–315

Becker ED (1980) High resolution NMR – theory and chemical applications, 2nd edn. Academic, New York

Bosman DK, Deutz NEP, de Graaf AA, van de Hulst RWN, van Eijk HMH, Bovée WMM, Maas MAW, Jorning GGA, Chamuleau RAFM (1990) Changes in brain metabolism during hyperammonemia and acute liver failure: results of a comparative ^{1}H NMR spectroscopy and biochemical investigation. Hepatology 12:281–290

Bröer S, Brookes N (2001) Transfer of glutamine between astrocytes and neurons. J Neurochem 77:705–719

Bröer A, Albers A, Setiawan I, Edwards RH, Chaudhry FA, Lang F, Wagner CA, Bröer S (2002) Regulation of the glutamine transporter SN1 by extracellular pH and intracellular sodium ions. J Physiol 539:3–14

Chaudhry FA, Reimer RJ, Krizaj D, Barber D, Storm-Mathisen J, Copenhagen DR, Edwards RH (1999) Molecular analysis of system N suggests novel physiological roles in nitrogen metabolism and synaptic transmission. Cell 99:769–780

Choi DW (1988) Glutamate neurotoxicity and diseases of the nervous system. Neuron 1:623–634

Danbolt NC (2001) Glutamate uptake. Prog Neurobiol 65:1–105

Erecińska M, Silver IA (1990) Metabolism and role of glutamate in mammalian brain. Prog Neurobiol 35:245–296

García-Espinosa MA, Wallin R, Hutson SM, Sweatt AJ (2007) Widespread neuronal expression of branched-chain aminotransferase in the CNS: implications for leucine/glutamate metabolism and for signaling by amino acids. J Neurochem 100:1458–1468

González-González IM, Cubelos B, Giménez C, Zafra F (2005) Immunohistochemical localization of the amino acid transporter SNAT2 in the rat brain. Neuroscience 130:61–73

Hall TR, Wallin R, Reinhart GD, Hutson SM (1993) Branched chain aminotransferase isoenzymes; purification and characterization of the rat brain isoenzyme. J Biol Chem 268: 3092–3098

Hertz L (1979) Functional interactions between neurons and astrocytes I. Turnover and metabolism of putative amino acid transmitters. Prog Neurobiol 13:277–323

Jacobson I, Sandberg M, Hamberger A (1985) Mass transfer in brain dialysis devices – a new method for the estimation of extracellular amino acids concentration. J Neurosci Methods 15:263–268

Kanamori K, Ross BD (1993) ^{15}N n.m.r. measurement of the in vivo rate of glutamine synthesis and utilization at steady state in the brain of the hyperammonaemic rat. Biochem J 293(Pt 2): 461–468

Kanamori K, Ross BD (1995) In vivo activity of glutaminase in the brain of hyperammonaemic rats measured by ^{15}N nuclear magnetic resonance. Biochem J 305(Pt 1):329–336

Kanamori K, Ross BD (1997a) Glial alkalinization detected in vivo by ^{1}H-^{15}N heteronuclear multiple-quantum coherence-transfer NMR in severely hyperammonemic rat. J Neurochem 68:1209–1220

Kanamori K, Ross BD (1997b) In vivo nitrogen MRS studies of rat brain metabolism. In: Bachelard H (ed) Advances in neurochemistry, vol 8. Plenum Publishing, New York, pp 65–90

Kanamori K, Ross BD (1999) Localized ^{15}N NMR spectroscopy of rat brain by ISIS. Magn Reson Med 41:456–463

Kanamori K, Ross BD (2004) Quantitative determination of extracellular glutamine concentration in rat brain, and its elevation in vivo by system A transport inhibitor, α-(methylamino)isobutyrate. J Neurochem 90:203–210

Kanamori K, Ross BD (2005) Suppression of glial glutamine release to the extracellular fluid studied in vivo by NMR and microdialysis in hyperammonemic rat brain. J Neurochem 94:74–85

Kanamori K, Ross BD, Parivar F (1991) Selective observation of biologically important ^{15}N-labeled metabolites in isolated rat brain and liver by ^{1}H-detected multiple-quantum coherence spectroscopy. J Magn Reson 93:319–328

Kanamori K, Ross BD, Kuo EL (1995) Dependence of in vivo glutamine synthetase activity on ammonia concentration in rat brain studied by ^{1}H-^{15}N heteronuclear multiple-quantum coherence-transfer NMR. Biochem J 311(Pt 2):681–688

Kanamori K, Ross BD, Chung JC, Kuo EL (1996) Severity of hyperammonemic encephalopathy correlates with brain ammonia level and saturation of glutamine synthetase in vivo. J Neurochem 67:1584–1594

Kanamori K, Ross BD, Kondrat RW (1998) Rate of glutamate synthesis from leucine in rat brain measured in vivo by ^{15}N NMR. J Neurochem 70:1304–1315

Kanamori K, Ross BD, Kondrat RW (2002) Glial uptake of neurotransmitter glutamate from the extracellular fluid studied in vivo by microdialysis and ^{13}C NMR. J Neurochem 83:682–695

Kay L, Keifer P, Saarinen T (1992) Pure absorption gradient enhanced heteronuclear single quantum correlation spectroscopy with improved sensitivity. J Am Chem Soc 114:10663–10665

Levy GC, Lichter RL (1979) Nitrogen-15 nuclear magnetic resonance spectroscopy. Wiley, New York

Mackenzie B, Erickson JD (2004) Sodium-coupled neutral amino acid (System N/A) transporters of the SLC38 gene family. Pflügers Arch Eur J Physiol 447:784–795

Mackenzie B, Schäfer MK, Erickson JD, Hediger MA, Weihe E, Varoqui H (2003) Functional properties and cellular distribution of the system A glutamine transporter SNAT1 support specialized roles in central neurons. J Biol Chem 278:23720–23730

Mans AM, DeJoseph MR, Hawkins RA (1994) Metabolic abnormalities and grade of encephalopathy in acute hepatic failure. J Neurochem 63:1829–1838

Melone M, Quagliano F, Barbaresi P, Varoqui H, Erickson JD, Conti F (2004) Localization of the glutamine transporter SNAT1 in rat cerebral cortex and neighboring structures, with a note on its localization in human cortex. Cereb Cortex 14:562–574

Melone M, Varoqui H, Erickson JD, Conti F (2006) Localization of the Na^+-coupled neutral amino acid transporter 2 in the cerebral cortex. Neuroscience 140:281–292

Nagelhus EA, Lehmann A, Ottersen OP (1996) Neuronal and glial handling of glutamate and glutamine during hypoosmotic stress: a biochemical and quantitative immunocytochemical analysis using the rat cerebellum as a model. Neuroscience 72:743–755

Ordidge RI, Connelly A, Lohman JAB (1986) Image-selected spectroscopy (ISIS). A new technique for spatially selective NMR spectroscopy. Magn Reson Med 66:283–294

Ottersen OP, Zhang N, Walberg F (1992) Metabolic compartmentation of glutamate and glutamine: morphological evidence obtained by quantitative immunocytochemistry in rat cerebellum. Neuroscience 46:519–534

Sakai R, Cohen DM, Henry JF, Burrin DG, Reeds PJ (2004) Leucine-nitrogen metabolism in the brain of conscious rats: its role as a nitrogen carrier in glutamate synthesis in glial and neuronal metabolic compartments. J Neurochem 88:612–622

Shank RP, Aprison MH (1981) Present status and significance of the glutamine cycle in neural tissues. Life Sci 28:837–842

Shen J, Sibson NR, Cline G, Behar KL, Rothman DL, Shulman RG (1998) ^{15}N NMR spectroscopy studies of ammonia transport and glutamine synthesis in the hyperammonemic rat brain. Dev Neurosci 20:434–443

Swain MS, Butterworth RF, Blei AT (1992) Ammonia and related amino acids in the pathogenesis of brain edema in acute ischemic liver failure in rats. Hepatology 15:449–453

Varoqui H, Zhu H, Yao D, Ming H, Erickson JD (2000) Cloning and functional identification of a neuronal glutamine transporter. J Biol Chem 275:4049–4054

Witanowski M, Stefaniak L, Webb A (1992) Nitrogen NMR spectroscopy. Academic, New York

Yudkoff M, Daikhin Y, Nelson D, Nissim I, Erecińska M (1996) Neuronal metabolism of branched-chain amino acids; flux through the aminotransferase pathway in synaptosomes. J Neurochem 66:2136–2145

Zwingmann C (2007) The anaplerotic flux and ammonia detoxification in hepatic encephalopathy. Metab Brain Dis 22:235–249

Part V
Cerebral Metabolism of Antioxidants, Osmolytes and Others *In Vivo*

Chapter 36
Glutathione (GSH) Synthesis and Metabolism

Maike M. Schmidt and Ralf Dringen

Abstract The tripeptide glutathione (GSH) is present in brain cells in millimolar concentrations. GSH has many important cellular and extracellular functions, of which the detoxification of reactive oxygen species and of xenobiotics are especially important. In the brain, astrocytes play a central role in the GSH metabolism. Due to their broad metabolic potential astrocytes can utilize various extracellular precursors to produce the amino acids glutamate, cysteine and glycine that are required as cellular substrates for GSH synthesis. In contrast, neurons rely for their GSH synthesis on the availability of extracellular cysteine that is provided by astrocytes. The mechanism involved in this metabolic supply of cysteine includes export of GSH from astrocytes, extracellular processing of the exported GSH by ectoenzymes, and uptake into neurons of the amino acid precursors for GSH synthesis. Disturbances of this metabolic interaction between astrocytes and neurons will affect GSH homeostasis in brain and could contribute to a compromised antioxidative defense in neurological diseases. In this article we will review the current knowledge on the metabolism of GSH in brain cells with a special focus on the mechanisms that are involved in the metabolic supply of cysteine by astrocytes for the synthesis of GSH in neurons.

M.M. Schmidt, Ph.D.
Center for Biomolecular Interactions, Center for Environmental Research and Sustainable Technologies, University of Bremen, PO. Box 330440, D-28334, Bremen, Germany
e-mail: maike.schmidt@uni-bremen.de

R. Dringen, Ph.D. (✉)
Center for Biomolecular Interactions Bremen, University of Bremen, PO. Box 330440, D-28334, Bremen, Germany

Center for Environmental Research and Sustainable Technologies, Leobener Strasse, Bremen, Germany

School of Psychology and Psychiatry, Monash University, Clayton, VIC, Australia
e-mail: ralf.dringen@uni-bremen.de

I.-Y. Choi and R. Gruetter (eds.), *Neural Metabolism In Vivo*,
Advances in Neurobiology 4, DOI 10.1007/978-1-4614-1788-0_36,

Keywords Astrocytes • Conjugation • Cysteine • Export • Glutamate • Glutamate cysteine ligase • γ-glutamyl transpeptidase • Glutathione • GSH • GSSG • Glycine • Mrp1 • Multidrug resistance proteins • Neurons • Neuron-glia interaction • PepT2 • Peptidases • Peptide transporter • Peroxidase • Oxidative stress • Metabolic cooperation • Transferase • Transport

36.1 Introduction

The tripeptide glutathione (GSH, γ-L-glutamyl-L-cysteinylglycine) is present in cells in millimolar concentrations (Cooper and Kristal 1997). GSH has essential cellular functions as important antioxidant. In addition, GSH serves as cofactor for some enzymes, is a storage and transport form of the amino acid cysteine and keeps the cellular thiol reduction potential in a strongly reduced form to protect essential thiol groups in enzymes against oxidation (Meister and Anderson 1983; Cooper and Kristal 1997; Janáky et al. 2007; Aoyama et al. 2008; Dringen 2009; Forman et al. 2009; Lu 2009). Recently, GSH has been demonstrated to be involved in cellular regulation, since the activity of many proteins can be modulated by glutathionylation (Dalle-Donne et al. 2007; Ghezzi and Di Simplicio 2007; Mieyal et al. 2008). Moreover, GSH is involved in processes such as cell proliferation (Pallardó et al. 2009), apoptosis (Circu and Aw 2008; Franco and Cidlowski 2009), autoimmunity (Perricone et al. 2009), and aging (Maher 2005; Rebrin and Sohal 2008). For all these cellular processes and functions, the availability of a sufficiently high concentration of GSH is essential. Thus, it is not surprising that a dysregulation of GSH metabolism has been connected with various human diseases (Franco et al. 2007; Ristoff and Larsson 2007; Ballatori et al. 2009b; Lu 2009).

The antioxidative function of GSH appears to be especially important for the brain, since the cells of this organ continuously generate reactive oxygen species (ROS) in substantial amounts as side product of an extensive oxidative metabolism (Clarke and Sokoloff 1999). This ROS generation has to be counter balanced by an appropriate antioxidative defense to prevent cell damage. Alterations in the homeostasis and the metabolism of the antioxidant GSH in brain have been connected with oxidative stress and oxidative damage. Evidence is growing that a compromised GSH metabolism contributes to the progression of neurological disorders such as Parkinson's disease (Chinta et al. 2007; Zeevalk et al. 2008; Cuadrado et al. 2009; Martin and Teismann 2009), Alzheimer's disease (Liu et al. 2004; Ballatori et al. 2009b; Fuller et al. 2010), as well as schizophrenia and bipolar disorders (Gysin et al. 2007; Matsuzawa et al. 2008; Andreazza et al. 2009; Behrens and Sejnowski 2009; Dean et al. 2009; Do et al. 2009; Wood et al. 2009).

Here we summarize the current knowledge on the cellular and extracellular metabolism of GSH in brain cells with the special focus on the mechanisms that are involved in the metabolic cooperation between astrocytes and neurons in the GSH metabolism of the brain.

36.2 GSH Content in Brain and Neural Cells

The specific GSH content of the brain depends on the species investigated and differs between brain regions. Forebrain and cortex seem to have the highest GSH content, followed by cerebellum, hippocampus, striatum, and substantia nigra (Kang et al. 1999; Calabrese et al. 2002), whereas brain stem, spinal cord, and the sciatic nerve have very low GSH contents (Cooper 1998; Janáky et al. 2007). Since the cellular GSH content is determined by the rates of GSH synthesis and GSH consumption, the observed differences in specific GSH contents of different brain regions may reflect regional variation in the availability of GSH precursors and/or in the demand of GSH for various cellular and extracellular functions. In addition, regional differences in the ratio of glial cells to neurons, are likely to contribute to the variation in GSH contents between different brain regions (Rice and Russo-Menna 1998; Srinivasan et al. 2010). For example, white matter contains significantly less GSH than gray matter (Srinivasan et al. 2010).

The most important functions of GSH in cells and in the extracellular space are listed in Table 36.1. It is important to consider that the concentrations of GSH in cells and in the extracellular space differ by three orders of magnitude. While GSH is present in the cytosol of cells in millimolar concentrations (Cooper and Kristal 1997; Dringen and Hamprecht 1998), the extracellular GSH in brain, as quantified by microdialysis, appears to reach just concentrations of up to 2 μM (Orwar et al. 1994; Yang et al. 1994; Lada and Kennedy 1997; Han et al. 1999). These differences in GSH concentrations have to be considered for the potential functions of GSH in brain cells and in the extracellular space (Table 36.1).

Brain cell cultures are frequently used to investigate the metabolism of GSH of the different types of brain cells. Substantial information is available on the GSH metabolism of cultured astrocytes and neurons, while less is known on the GSH metabolism of oligodendroglial and microglial cells (Dringen 2000, 2005, 2009; Dringen et al. 2005; Dringen and Hirrlinger 2010). Cultures that are enriched in astrocytes, neurons, oligodendroglial cells or microglial cells contain normally specific GSH contents between 20 and 40 nmol/mg protein (Dringen et al. 2005), although such values can differ strongly depending on the culturing conditions.

Table 36.1 Putative functions of GSH in brain

Intracellular functions (mM concentration range)	Extracellular functions (μM concentration range)
Radical scavenger	Substrate of enzymes (γGT)
Electron donor for non-enzymatic reduction	Radical scavenger
Substrate of enzymes (GPx, GST, isomerases)	Modulator of glutamate receptors
Leucotriene metabolism	Neurohormone
Detoxification of xenobiotics	Transport form of cysteine
Redox buffer	
Covalent modulator of protein functions via S-glutathionylation	
Storage form of cysteine	

Among the different types of cultured brain cells, microglial cells have the highest specific cellular GSH content, which is likely to contribute to the self-defense of microglial cells against the ROS that they produce and release (Dringen 2005; Dringen and Hirrlinger 2010).

36.3 Synthesis of GSH

GSH synthesis in brain cells follows the pathway that is well known for other tissues (Meister 1974). GSH is synthesized by the consecutive reactions of the two ATP-consuming enzymes glutamate cysteine ligase (GCL; γ-glutamylcysteine synthetase) and GSH synthetase. GCL generates the dipeptide γGluCys from glutamate and cysteine. GSH synthetase catalyzes the reaction of this dipeptide with glycine to GSH (Fig. 36.1). GCL is the rate limiting enzyme for GSH synthesis (Griffith 1999; Griffith and Mulcahy 1999) and is strongly regulated by GSH feedback inhibition (Richman and Meister 1975; Misra and Griffith 1998). Detailed information on structure, function and post-translational regulation of GCL have recently been summarized (Franklin et al. 2009). GCL is a heterodimeric protein that consists of a catalytic (GCLc) and a modulatory (GCLm) subunit. The catalytical subunit GCLc possesses γGluCys synthesis activity, while the modifier subunit GCLm modulates the activity of GCLc by decreasing the K_m value for the substrates ATP and glutamate, by increasing the K_{cat} value for γGluCys synthesis and by increasing the K_i value for the feedback inhibition of GCL by GSH (Chen et al. 2005). In the absence of the modifier subunit, the catalytic subunit is efficiently inhibited already by low concentrations of GSH, as indicated by the high sensitivity of the GCL for GSH inhibition in GCLm knockout mice (Yang et al. 2002). A competition by GSH with

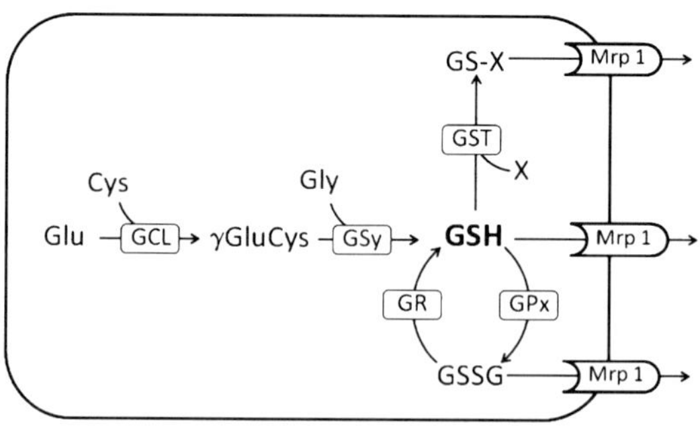

Fig. 36.1 Synthesis and metabolism of GSH. GSH is synthesized from the amino acid substrates glutamate, cysteine and glycine by the consecutive reactions of glutamate cysteine ligase (*GCL*) and GSH synthetase (*GSy*). GSH is oxidized during peroxide disposal by glutathione peroxidases (*GPx*) to GSSG and is regenerated by glutathione reductase (*GR*) from GSSG. GSH can also react with xenobiotics and endogenous compounds (X) in reactions that are catalyzed by glutathione-S-transferases (*GST*). GSH, GSSG and GSH conjugates can be exported from cells by multidrug resistance proteins such as Mrp1

glutamate on the substrate binding site and the reduction of an intersubunit disulfide bond are considered as molecular mechanisms of the GCL feedback inhibition by GSH (Huang et al. 1993; Fraser et al. 2002, 2003; Franklin et al. 2009). GCL activity is known to be modulated by a variety of chemical and physical agents (Soltaninassab et al. 2000). Recent data confirmed that GCL activity is increased by prostaglandins (Saito et al. 2007), thyroid hormones (Dasgupta et al. 2007), polyphenols, flavonoids and substances that activate the nuclear erythroid 2-related factor 2 (Nrf2) pathway (Lavoie et al. 2009).

The consequences of an impaired GSH synthesis become evident by the relative rare inborn errors of metabolism for the GSH synthesis enzymes GCL or GSH synthetase that can lead to dramatic phenotypes including neurological symptoms such as mental retardation and motor functional disturbances. GCL deficiency in humans has only been reported for nine patients that all suffered from rather mild hemolytic anemia. Only in four cases CNS dysfunction such as spinocerebellar degeneration and neuropathies belonged to the disease pattern (Ristoff and Larsson 2007). GSH synthetase deficiency leads to lower levels of cellular GSH and an accumulation of the dipeptide γGluCys in the cells. The patients suffer from hemolytic anemia, jaundice, metabolic acidosis and progressive CNS disorders (Dalton et al. 2004) with lesions in cortex and thalamus, cerebral necrosis and cerebral atrophy (Njalsson 2005).

Several mouse models have been developed that were used to study the consequences of alterations in the activity of the enzymes involved in GSH synthesis (Dalton et al. 2004; Botta et al. 2008; Ballatori et al. 2009b). The deletion of the catalytical subunit of GCL leads to an embryonic lethal phenotype, whereas heterozygous mice are viable and fertile but show a lowered tissue content of GSH (Dalton et al. 2000). In addition, the knockdown of GCLc using an RNAi technique resulted in mild oxidative stress in neurons (Diaz-Hernandez et al. 2007). GCLm knockout mice are viable and develop normally, but have tissue GSH levels that are below 20% of that of wild-type mice (Yang et al. 2002). Brain cells of GCLm deficient mice are more vulnerable to xenobiotics such as domoic acid (Giordano et al. 2006, 2007, 2008b, 2009) or environmental pollutants like flame retardants (Giordano et al. 2008a).

The transcription rate of the enzymes involved in GSH synthesis is controlled by the antioxidative response element (ARE) promoter, that is regulated by Nrf2. Both subunits of the GCL and the GSH synthetase are regulated by Nrf2 and are preferentially activated in astrocytes (Lee et al. 2003; Shih et al. 2003). Nrf2 activation or overexpression protects brain cells from oxidative or mitochondrial stress (Shih et al. 2003, 2005; Kraft et al. 2004) and leads to elevated levels of cellular GSH in cultured brain cells (Shih et al. 2003; Sun et al. 2005; Vargas et al. 2006, 2008). In contrast, Nrf2 deficient mice are more vulnerable to insults that are applied in animal models of neurodegeneration (Johnson et al. 2008; Nguyen et al. 2009; Vargas and Johnson 2009).

In addition to the enzymes responsible for GSH synthesis, the cellular GSH content depends strongly on the intracellular availability of the substrates glutamate, cysteine, and glycine (Fig. 36.1). Since glutamate and glycine are neurotransmitters and cysteine in high concentrations is toxic to neurons, the extracellular availability of these amino acids in the brain is limited (Dringen 2009). Thus, GSH synthesis in brain cells depends on their ability to take up suitable extracellular GSH precursors and/or to produce sufficient amounts of the substrates for GSH synthesis from such precursors. Cell culture experiments have demonstrated that uptake and metabolism of GSH precursors vary between different types of brain cells. GSH synthesis in neurons depends strongly on the availability of extracellular cysteine (Kranich et al. 1996; Dringen et al. 1999b), while astrocytes can use a wide range of amino acids and peptides that are taken up by the respective transporters and are subsequently converted to the substrate amino acids of GSH synthesis (Dringen and Hamprecht 1998). Cultured neurons do not use extracellular cystine as source for the GSH substrate cysteine and prefer glutamine as best extracellular precursor for the glutamate moiety of GSH (Kranich et al. 1996). In contrast, astrocytes prefer glutamate and cystine as extracellular GSH precursors (Kranich et al. 1998). Cultured oligodendrocytes rely on the presence of cysteine or cystine in the culture media. Absence of these amino acids leads to a decline in cellular GSH levels of cultured oligodendrocytes that is followed by cell death (Yonezawa et al. 1996; Back et al. 1998). These differential preferences for extracellular GSH precursors lower the competition between different types of brain cells for GSH precursors.

The availability of the sulfur-containing amino acids cystine and/or cysteine appears to be crucial for the GSH synthesis in brain cells. A variety of transporters have been described to mediate the cellular uptake of cysteine or cystine. Both astrocytes and neurons have been reported to take up cystine by the Xc^- transport system, an electro-neutral, sodium-independent cystine-glutamate exchanger (McBean 2002). In brain, the expression of this transporter varies between brain cell types and different brain regions (Shanker and Aschner 2001; Burdo et al. 2006). In addition, at least for astrocytes a contribution of the sodium-dependent X_{AG}- transport system (Bender et al. 2000) and of the γ-glutamyl transpeptidase (γGT) system (Shanker and Aschner 2001) have been described to contribute to cellular cystine uptake. Transport of cysteine into astrocytes and neurons involves sodium-independent as well as sodium-dependent transport systems (Shanker and Aschner 2001). For neurons, members of the sodium-dependent X_{AG}- transporter family represent the predominant mechanism for cysteine uptake (Shanker and Aschner 2001). Absence of the X_{AG}- transporter family member excitatory amino acid carrier 1 (EAAC1) causes neuronal GSH deficiency, oxidative stress and age-dependent neurodegeneration (Aoyama et al. 2006, 2008). Differences in the expression levels of cysteine and cystine transporters are likely to contribute to the observed preferences of brain cells for sulfur-containing amino acids as extracellular precursors for cellular GSH synthesis.

36.4 Redox Metabolism of GSH

GSH is an essential antioxidant for the cellular detoxification of ROS, since it directly reacts with radicals and is electron donor for the peroxide reduction by glutathione peroxidases (GPx). GSH is oxidized to GSSG during detoxification of peroxides (Fig. 36.1). For mammals eight GPx isoenzymes have been identified that differ in their tissue expression, substrate specificity and subcellular distribution (Ursini et al. 1995; Margis et al. 2008; Toppo et al. 2009). The cytosolic isoform GPx1 appears to be especially important for hydrogen peroxide reduction in brain cells (Dringen et al. 2005). The involvement of GPx1 in protecting brain cells against peroxide toxicity has been shown for GPx1-deficient neurons (de Haan et al. 1998; Taylor et al. 2005) and astrocytes (Liddell et al. 2006a, b). In addition, for several *in vivo* stress paradigms it was demonstrated that brains of GPx1-deficient mice are more seriously damaged than those of wild type mice (de Haan et al. 1998; Crack et al. 2001, 2006; Flentjar et al. 2002; Wong et al. 2008).

GSH is regenerated from GSSG (Fig. 36.1) by cellular glutathione reductase (GR) (Schirmer et al. 1989; Lopez-Barea et al. 1990). This enzyme uses NADPH as electron donor and has been shown to be present in all types of brain cells (Dringen et al. 2005). Reduction of GSSG by GR is highly efficient due to the low K_m-values of GR for GSSG and NADPH (Dringen and Gutterer 2002). Therefore, under unstressed conditions GSSG is hardly detectable in cells. However, during oxidative stress GR becomes rate limiting for GSH redox cycling as indicated by the cellular accumulation of GSSG (Dringen and Hamprecht 1997) and a corresponding shift in the thiol reduction potential of the stressed cells (Hirrlinger and Dringen 2010). For efficient GR-dependent reduction of GSSG to GSH, a suitable supply of reduction equivalents in the form of NADPH is required. Brain cells contain multiple pathways in cytosol and mitochondria that reduce $NADP^+$ to provide NADPH for the GR reaction (Dringen et al. 2005, 2007; Hirrlinger and Dringen 2010).

All types of brain cells contain GPx and GR (Dringen et al. 2005) and are thereby able to oxidize GSH to GSSG and to regenerate GSH. Among the different types of brain cells in culture, oligodendrocytes contain the highest specific activities of the enzymes GPx and GR and are highly efficient in peroxide detoxification (Hirrlinger et al. 2002a; Baud et al. 2004). Exposure of cultured brain cells to hydrogen peroxide caused transient accumulation of cellular GSSG (Dringen et al. 1999a; Hirrlinger et al. 2000, 2001, 2002a; Liddell et al. 2006a), demonstrating that the activity of GR is not sufficient to immediately reduce the GSSG that is formed by GPx. However, after removal of the peroxide all types of brain cells were able to rapidly restore their initial high GSH to GSSG ratio by GR (Dringen et al. 1998b, 1999a; Kussmaul et al. 1999; Hirrlinger et al. 2000; Liddell et al. 2006a, b). The GSH/GSSG redox cycling by GPx and GR does not deprive brain cells of their total glutathione (GSH plus GSSG). In contrast, GSH conjugation and export of GSH or GSSG lower the cellular GSH concentration and *de novo* synthesis of GSH is required to compensate for these types of GSH consumption.

36.5 Conjugation of GSH

Due to its reactive thiol group, GSH can rapidly react with various compounds (Fig. 36.1) in reactions that are either enzyme independent or are accelerated by glutathione-S-transferases (GSTs) (Commandeur et al. 1995; Salinas and Wong

1999; Hayes et al. 2005; Blair 2006). Four major groups of enzymes with GST activity have been identified, of which the large group of canonical soluble GSTs is best characterized. This group is subdivided in seven classes, of which the α, μ, and π classes are most prominent in mammals (Mannervik et al. 2005; Zimniak and Singh 2007). Although GSTs are generally considered as transferases that connect GSH to endogenous or xenobiotic acceptor substances, some of these enzymes act also as isomerases or show glutathione peroxidase activity (Zimniak 2007). GSTs are multifunctional enzymes with an impressive list of substrates (Sharma et al. 2007). GST isoenzymes are expressed in a tissue-dependent manner. In the brain, expression of GST isoenzymes of the α, μ, and π classes have been described (Dhanani and Awasthi 2007) and the expression of different GSTs depends strongly on the cell type as well as on the brain region (Johnson et al. 1993; Shang et al. 2008).

The physiological functions of individual GSTs in brain and in the different types of brain cells have not been elucidated in detail. Since compounds such as monochlorobimane, iodoacetate or iodoacetamide, and fumaric acid diesters rapidly lower the cellular GSH contents of cultured astrocytes or neurons (DeCory et al. 2001; Waak and Dringen 2006; Schmidt and Dringen 2009, 2010), a contribution of GSTs in the metabolism of such compounds has been postulated. In addition, GST-dependent detoxification in neurons has also been considered for various substances that cause selective necrosis to granule cells in the cerebellum (Fonnum and Lock 2004).

36.6 Export of GSH, GSSG, and Glutathione Conjugates

Export of GSH, GSSG, and GSH conjugates has been reported for various cell types and tissues including the brain (Borst and Elferink 2002; Dallas et al. 2006; Deeley and Cole 2006; Toyoda et al. 2008; Ballatori et al. 2009a). Such processes consume cellular GSH. Therefore, GSH synthesis is required to compensate for the cellular loss of GSH (Fig. 36.1).

Members of the family of multidrug resistance proteins (Mrps) are predominately responsible for the cellular export of GSH and its conjugates. Mrps are ATP-driven export pumps of organic anions and belong to the family of ATP-binding cassette transporters (Toyoda et al. 2008; Zhou et al. 2008; Ballatori et al. 2009a). Nine mammalian Mrps have been characterized that differ in their tissue distribution and in their substrate specificity. In addition to Mrps, also organic anion transporters (OATPs) or the cystic fibrosis transductance regulator (CFTR) have been reported to contribute to the transport of GSH (Ballatori et al. 2005). The current knowledge on the expression and functions of the different types of transporters that are involved in the export of GSH, GSSG and GSH conjugates in brain and neural cells has recently been summarized (Dallas et al. 2006; Ballatori et al. 2009a).

In addition to Mrps, the Ral-interacting protein RLIP76 has been described as novel GSH conjugate and multidrug transporter (Singhal et al. 2009). Although functional properties of RLIP76 have been predominantly obtained for cancer cells, at least the expression of RLIP76 in human brain tissue has been shown (Awasthi et al. 2007; Singhal et al. 2009). In epileptic brain, RLIP76 co-localizes exclusively with endothelial cells but not with glial cells or neurons (Awasthi et al. 2007). Whether this transporter is involved in the export of GSH conjugates from the brain remains to be elucidated.

Among cultured brain cells, only cultured astrocytes release substantial amounts of GSH (Hirrlinger et al. 2002b). GSH export from astrocytes depends strongly on the cellular content of GSH (Sagara et al. 1996) and is accelerated by glutamate (Frade et al. 2008). Under unstressed conditions these cells release around 10% of their cellular GSH per hour (Dringen et al. 1997), and even protect the GSH exported against oxidation (Stewart et al. 2002; Pope et al. 2008). Mrp1 has been shown to mediate about 60% of the GSH export from astrocytes (Hirrlinger et al. 2002b; Hirrlinger and Dringen 2005; Minich et al. 2006). The transporter(s) responsible for the residual 40% of GSH export from astrocytes has not been identified so far. However, pharmacological studies on cultured Mrp1-deficient astrocytes make a contribution of Mrps 4 and 5, OATPs 1 and 2 or CFTR in the GSH export from astrocytes unlikely (Minich et al. 2006). In addition to Mrp1, GAP junction hemichannels have been reported to mediate GSH release from astrocytes in culture (Rana and Dringen 2007) and brain slices (Stridh et al. 2008) under certain conditions.

Under unstressed conditions, GSSG is hardly detectable in brain cells (Hirrlinger and Dringen 2010). Thus, under such conditions astrocytes release exclusively GSH and only minute amounts of GSSG are detected in the media (Hirrlinger et al. 2002b; Minich et al. 2006). In contrast, during sustained peroxide stress, GSSG accumulates quickly in astrocytes and is efficiently exported (Hirrlinger et al. 2001; Hirrlinger et al. 2002c; Minich et al. 2006). The complete absence of any extracellular GSSG accumulation after exposure of Mrp1-deficient astrocytes to sustained peroxide stress (Minich et al. 2006) demonstrates that Mrp1 is exclusively responsible for GSSG export from astrocytes. Export of GSSG during oxidative stress has been discussed as protective mechanism that helps to at least partially prevent a shift in the thiol reduction potential of the cells (Akerboom and Sies 1990; Keppler 1999; Hirrlinger et al. 2001). However, it has to be considered that exported GSSG cannot serve anymore as substrate for cytosolic GSH regeneration by GR. Thus, ATP-dependent GSH *de novo* synthesis is required to compensate for the loss of cellular GSH due to GSSG export during oxidative stress.

36.7 Extracellular Metabolism of GSH

Extracellular GSH and GSH conjugates are substrates of the ectoenzyme γ-glutamyl transpeptidase (γGT). This enzyme catalyses the transfer of the γ-glutamyl moiety onto a suitable acceptor and the hydrolysis of the γ-glutamyl bond (Ikeda and Taniguchi 2005). In brain, γGT is strongly expressed in the endothelial cells of

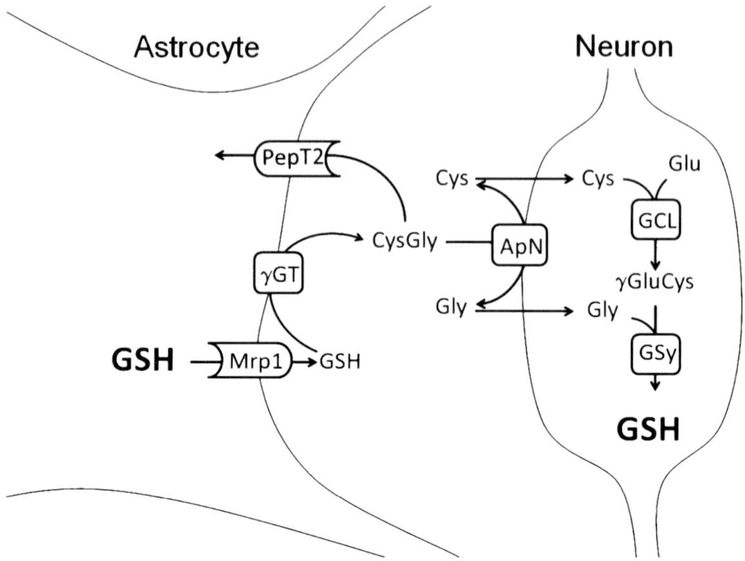

Fig. 36.2 Metabolic interaction between astrocytes and neurons in the GSH metabolism. The extracellular cysteine that is required for neuronal GSH synthesis is provided by astrocytes. This metabolic supply is initiated by GSH export from astrocytes via Mrp1 or other transporters. Extracellular GSH is used as substrate of the astrocytic ectoenzyme γ-glutamyl transpeptidase (γGT) that produces the dipeptide CysGly. This dipeptide can be hydrolyzed by neuronal aminopeptidase N (*ApN*) and the generated amino acids cysteine and glycine are taken up into neurons. Alternatively, CysGly can be taken up into astrocytes by the peptide transporter PepT2

capillaries and has been suggested to be involved in the ventilatory response to hypoxia (Lipton et al. 2001). However, γGT is also expressed in astrocytes. Presence of γGT activity (Dringen et al. 1997; Garcion et al. 1999; Gegg et al. 2003) and γGT immunoreactivity (Shine and Haber 1981; Philbert et al. 1995; Zhang et al. 1997) have been reported for astrocytes *in vivo* and in culture. γGT activity has also been reported for cultured oligodendrocytes, whereas hardly any activity of this enzyme was detectable for cultured microglial cells and neurons (Ruedig and Dringen 2004). The activity of γGT in cultured astrocytes is elevated after treatment of the cells with lipopolysaccharide (Garcion et al. 1999), NO donors (Gegg et al. 2003), tumor necrosis factor α (TNFα) (Ruedig and Dringen 2004), thyroid hormone (Dasgupta et al. 2005), 6-hydroxydopamine (Zhang et al. 2005) or ethanol (Rathinam et al. 2006). Such increases in γGT activity improve the ability of the cells to process the GSH that was released from astrocytes (Ruedig and Dringen 2004).

The product of the reaction of γGT with GSH is the dipeptide CysGly. This dipeptide is produced by γGT in equimolar concentrations to the GSH consumed. CysGly can be taken up into cells by peptide transporters or can be hydrolyzed by ectopeptidases (Fig. 36.2). Both pathways appear to be realized in brain. The high affinity proton-coupled oligopeptide transporter PepT2 is widely expressed in brain (Fujita et al. 2004; Groneberg et al. 2004; Shen et al. 2004). The highest level of expression was found in the choroid plexus (Shen et al. 2004). In addition to ependymal and choroid plexus cells, astrocytes and neurons have been reported to express PepT2 in brain (Berger and Hediger 1999; Shen et al. 2004). PepT2 expression has also been shown for cultured astrocytes (Dringen et al. 1998a; Xiang et al. 2006), whereas mRNA of PepT2 was not found in cultured neurons (Dieck et al. 1999). In cultured astrocytes, PeptT2 is responsible for the uptake of CysGly that is subsequently cleaved by a cellular peptidase to obtain cysteine and glycine for GSH synthesis (Dringen et al. 1998a). In contrast, cultured neurons appear not to take up intact CysGly. These cells express aminopeptidase N (ApN) (Dringen et al. 2001; Rathinam et al. 2006). Since the inactivation of this enzyme prevents the utilization of CysGly for GSH synthesis in neurons, CysGly is rather extracellularly hydrolyzed by neurons to deliver substrates for GSH synthesis than taken up as intact dipeptide (Dringen et al. 2001). ApN is widely expressed in various brain regions (Noble et al. 2001) and could therefore be responsible for the extracellular cleavage of the γGT-product CysGly in brain that is demonstrated by the observation that inhibition of γGT lowers the extracellular concentration of cysteine in brain (Han et al. 1999).

The uptake of cysteine and glycine into neurons is likely to be mediated by sodium dependent transporters. The sodium-dependent excitatory amino acid carrier 1 (EAAC1) is most likely responsible for the majority of cysteine uptake into neurons (Aoyama et al. 2008). Absence of EAAC1 causes neuronal GSH deficiency, oxidative stress and age-dependent neurodegeneration (Aoyama et al. 2006, 2008). The glutamate transporter-associated protein (GTRAP3-18) modulates the EAAC1-mediated cysteine uptake into neurons by interacting with EAAC1 in the plasma membrane. A decline of the GTRAP3-18 protein leads to an increase of the GSH content in neurons (Watabe et al. 2008).

The uptake of the amino acid glycine can be mediated by glycine transporters such as GlyT1 and GlyT2. These transporters differ in their expression patterns, stoichiometries, reversed uptake capacities and pre-steady-state kinetics (Roux and Supplisson 2000; Supplisson and Roux 2002; Eulenburg et al. 2005; Betz et al. 2006). GlyT1 appears to be expressed throughout the brain (Adams et al. 1995; Zafra et al. 1995; Jursky and Nelson 1996) and GlyT1 immunoreactivity has been shown for astrocytes and some subpopulations of neurons (Zafra et al. 1995; Cubelos et al. 2005), whereas GlyT2 immunoreactivity was found only for glycinergic neurons (Jursky and Nelson 1995). Studies on GlyT1 or GlyT2 knockout mice revealed that GlyT1 has an important function in the clearing of glycine from the synaptic cleft, whereas GlyT2 provides cytosolic glycine for vesicular release (Gomeza et al. 2003; Eulenburg et al. 2005; Betz et al. 2006). Whether the known glycine transporters or other transporters are involved in the uptake of the glycine that is required for GSH synthesis in neurons remains to be elucidated.

36.8 Metabolic Cooperation Between Astrocytes and Neurons in GSH Metabolism

In brain, especially neurons appear to be disadvantaged regarding GSH metabolism. Since neurons cannot use extracellular cystine as GSH precursor (Kranich et al. 1996), these cells rely on the supply of cysteine from neighboring cells (Dringen and Hirrlinger 2003; Dringen 2009). Indeed, presence of astrocytes maintains or even increases GSH levels in co-cultured neurons (Bolaños et al. 1996; Dringen et al. 1999b; Chen et al. 2001; Shih et al. 2003; de Bernardo et al. 2004; Gegg et al. 2005; Griffin et al. 2005; Watts et al. 2005; Rathinam et al. 2006; Vargas et al. 2006, 2008; Woehrling et al. 2007; Sandhu et al. 2009). The pathway involved in the supply of GSH precursors from astrocytes to neurons (Fig. 36.2) is initiated by the Mrp1-dependent export of GSH from astrocytes (Minich et al. 2006). Since neurons do not have the ability to take up intact GSH (Sagara et al. 1996), the extracellular breakdown of the GSH exported from astrocytes by the astroglial ectoenzyme γGT (Dringen et al. 1997, 1999b) and by a neuronal ectopeptidase (Dringen et al. 2001) is inevitable for the supply of the GSH precursor cysteine for neuronal GSH synthesis (Fig. 36.2). This scenario is strongly supported by the observation that inhibition of γGT in astrocyte-neuron co-cultures totally prevented the astrocyte induced increase of the neuronal GSH content (Dringen et al. 1999b). In addition, also the coordinated neuroprotective upregulation of GSH efflux from astrocytes, γGT activity in astrocytes, and ApN activity in neurons by ethanol (Rathinam et al. 2006) supports the view that the connection of these three metabolic steps are important for the cysteine supply from astrocytes for neuronal GSH synthesis. With the supply by astrocytes of glutamine and CysGly, precursors of all three amino acids that are needed for GSH synthesis are provided from astrocytes to neurons. Most evidences for the interaction of astrocytes and neurons in GSH metabolism of the brain have been deduced from data obtained on cultures and co-cultures of brain cells. However, γGT-dependent generation of cysteine from GSH by brain cells has been confirmed for brain slices (Li et al. 1996, 1999) and for the brain (Han et al. 1999), strongly suggesting that also *in vivo* astrocytes supply neurons with cysteine for GSH synthesis.

36.9 Conclusions

Metabolic cooperation between astrocytes and neurons is important for the homeostasis of GSH in brain. Neurons rely on the supply of GSH precursors, especially of cysteine, from neighboring astrocytes. The metabolic pathway that provides astrocyte-derived cysteine to neurons is initiated by the export of GSH from astrocytes. The exported GSH is subsequently processed by ectoenzymes to neuronal GSH precursors. Alterations of this metabolic cooperation between astrocytes and neurons will substantially disturb the availability of the GSH precursor cysteine for neurons. Such processes may contribute to a lowered antioxidative defense in neurons, to an increased susceptibility to oxidative stress and to neuronal death in neurodegenerative diseases. On the other hand, improvement of the ability of astrocytes to synthesize, export and/or process extracellular GSH is likely to enhance the supply of GSH precursors to neurons and could be beneficial to improve the antioxidative potential of neurons in brain.

References

Adams RH, Sato K, Shimada S, Tohyama M, Puschel AW, Betz H (1995) Gene structure and glial expression of the glycine transporter GlyT1 in embryonic and adult rodents. J Neurosci 15:2524–2532

Akerboom T, Sies H (1990) Glutathione transport and its significance in oxidative stress. In: Vina J (ed) Glutathione: metabolism and physiological functions. CRC, Boca Raton, pp 45–55

Andreazza AC, Kapczinski F, Kauer-Sant'Anna M, Walz JC, Bond DJ, Goncalves CA, Young LT, Yatham LN (2009) 3-Nitrotyrosine and glutathione antioxidant system in patients in the early and late stages of bipolar disorder. J Psychiatr Neurosci 34:263–271

Aoyama K, Suh SW, Hamby AM, Liu J, Chan WY, Chen Y, Swanson RA (2006) Neuronal glutathione deficiency and age-dependent neurodegeneration in the EAAC1 deficient mouse. Nat Neurosci 9:119–126

Aoyama K, Watabe M, Nakaki T (2008) Regulation of neuronal glutathione synthesis. J Pharmacol Sci 108:227–238

Awasthi YC, Sharma R, Yadav S, Dwivedi S, Sharma A, Awasthi S (2007) The non-ABC drug transporter RLIP76 (RALBP-1) plays a major role in the mechanisms of drug resistance. Curr Drug Metab 8:315–323

Back SA, Gan X, Li Y, Rosenberg PA, Volpe JJ (1998) Maturation-dependent vulnerability of oligodendrocytes to oxidative stress-induced death caused by glutathione depletion. J Neurosci 18:6241–6253

Ballatori N, Hammond CL, Cunningham JB, Krance SM, Marchan R (2005) Molecular mechanisms of reduced glutathione transport: role of the MRP/CFTR/ABCC and OATP/SLC21A families of membrane proteins. Toxicol Appl Pharmacol 204:238–255

Ballatori N, Krance SM, Marchan R, Hammond CL (2009a) Plasma membrane glutathione transporters and their roles in cell physiology and pathophysiology. Mol Aspects Med 30:13–28

Ballatori N, Krance SM, Notenboom S, Shi S, Tieu K, Hammond CL (2009b) Glutathione dysregulation and the etiology and progression of human diseases. Biol Chem 390:191–214

Baud O, Greene AE, Li J, Wang H, Volpe JJ, Rosenberg PA (2004) Glutathione peroxidase-catalase cooperativity is required for resistance to hydrogen peroxide by mature rat oligodendrocytes. J Neurosci 24:1531–1540

Behrens MM, Sejnowski TJ (2009) Does schizophrenia arise from oxidative dysregulation of parvalbumin-interneurons in the developing cortex? Neuropharmacology 57:193–200

Bender AS, Reichelt W, Norenberg MD (2000) Characterization of cystine uptake in cultured astrocytes. Neurochem Int 37:269–276

Berger UV, Hediger MA (1999) Distribution of peptide transporter PEPT2 mRNA in the rat nervous system. Anat Embryol (Berlin) 199:439–449

Betz H, Gomeza J, Armsen W, Scholze P, Eulenburg V (2006) Glycine transporters: essential regulators of synaptic transmission. Biochem Soc Trans 34:55–58

Blair IA (2006) Endogenous glutathione adducts. Curr Drug Metab 7:853–872

Bolaños JP, Heales SJR, Peuchen S, Barker JE, Land JM, Clark JB (1996) Nitric oxide-mediated mitochondrial damage: a potential neuroprotective role for glutathione. Free Radic Biol Med 21:995–1001

Borst P, Elferink RO (2002) Mammalian ABC transporters in health and disease. Annu Rev Biochem 71:537–592

Botta D, White CC, Vliet-Gregg P, Mohar I, Shi S, McGrath MB, McConnachie LA, Kavanagh TJ (2008) Modulating GSH synthesis using glutamate cysteine ligase transgenic and gene-targeted mice. Drug Metab Rev 40:465–477

Burdo J, Dargusch R, Schubert D (2006) Distribution of the cystine/glutamate antiporter system x_c^- in brain, kidney, and duodenum. J Histochem Cytochem 54:549–557

Calabrese V, Scapagnini G, Ravagna A, Fariello RG, Guiffrida Stella AM, Abraham NG (2002) Regional distribution of heme oxygenase, HSP70, and glutathione in brain: relevance for endogenous oxidant/antioxidant balance and stress tolerance. J Neurosci Res 68:65–75

Chen Y, Vartiainen NE, Ying W, Chan PH, Koistinaho J, Swanson RA (2001) Astrocytes protect neurons from nitric oxide toxicity by a glutathione-dependent mechanism. J Neurochem 77: 1601–1610

Chen Y, Shertzer HG, Schneider SN, Nebert DW, Dalton TP (2005) Glutamate cysteine ligase catalysis: dependence on ATP and modifier subunit for regulation of tissue glutathione levels. J Biol Chem 280:33766–33774

Chinta SJ, Kumar MJ, Hsu M, Rajagopalan S, Kaur D, Rane A, Nicholls DG, Choi J, Andersen JK (2007) Inducible alterations of glutathione levels in adult dopaminergic midbrain neurons result in nigrostriatal degeneration. J Neurosci 27:13997–14006

Circu ML, Aw TY (2008) Glutathione and apoptosis. Free Radic Res 42:689–706

Clarke DD, Sokoloff L (1999) Circulation and energy metabolism of the brain. In: Sigel GJ, Agranoff BW, Albers RW, Fisher SK, Uhler MD (eds) Basic neurochemistry: molecular, cellular and medical aspects. Lippincott-Raven, Philadelphia, pp 637–669

Commandeur JN, Stijntjes GJ, Vermeulen NP (1995) Enzymes and transport systems involved in the formation and disposition of glutathione S-conjugates. Role in bioactivation and detoxication mechanisms of xenobiotics. Pharmacol Rev 47:271–330

Cooper AJL (1998) Role of astrocytes in maintaining cerebral glutathione homeostasis and in protecting the brain against xenobiotics and oxidative stress. In: Shaw CA (ed) Glutathione in the nervous system. Taylor & Francis, Washington, DC, pp 91–115

Cooper AJ, Kristal BS (1997) Multiple roles of glutathione in the central nervous system. Biol Chem 378:793–802

Crack PJ, Taylor JM, Flentjar NJ, de Haan J, Hertzog P, Iannello RC, Kola I (2001) Increased infarct size and exacerbated apoptosis in the glutathione peroxidase-1 (Gpx-1) knockout mouse brain in response to ischemia/reperfusion injury. J Neurochem 78:1389–1399

Crack PJ, Taylor JM, Ali U, Mansell A, Hertzog PJ (2006) Potential contribution of NF-κB in neuronal cell death in the glutathione peroxidase-1 knockout mouse in response to ischemia-reperfusion injury. Stroke 37:1533–1538

Cuadrado A, Moreno-Murciano P, Pedraza-Chaverri J (2009) The transcription factor Nrf2 as a new therapeutic target in Parkinson's disease. Expert Opin Ther Targets 13:319–329

Cubelos B, Gimenez C, Zafra F (2005) Localization of the GLYT1 glycine transporter at glutamatergic synapses in the rat brain. Cereb Cortex 15:448–459

Dallas S, Miller DS, Bendayan R (2006) Multidrug resistance-associated proteins: expression and function in the central nervous system. Pharmacol Rev 58:140–161

Dalle-Donne I, Rossi R, Giustarini D, Colombo R, Milzani A (2007) S-glutathionylation in protein redox regulation. Free Radic Biol Med 43:883–898

Dalton TP, Dieter MZ, Yang Y, Shertzer HG, Nebert DW (2000) Knockout of the mouse glutamate cysteine ligase catalytic subunit (Gclc) gene: embryonic lethal when homozygous, and proposed model for moderate glutathione deficiency when heterozygous. Biochem Biophys Res Commun 279:324–329

Dalton TP, Chen Y, Schneider SN, Nebert DW, Shertzer HG (2004) Genetically altered mice to evaluate glutathione homeostasis in health and disease. Free Radic Biol Med 37:1511–1526

Dasgupta A, Das S, Sarkar PK (2005) Thyroid hormone stimulates γ-glutamyl transpeptidase in the developing rat cerebra and in astroglial cultures. J Neurosci Res 82:851–857

Dasgupta A, Das S, Sarkar PK (2007) Thyroid hormone promotes glutathione synthesis in astrocytes by up regulation of glutamate cysteine ligase through differential stimulation of its catalytic and modulator subunit mRNAs. Free Radic Biol Med 42:617–626 de Bernardo S, Canals S, Casarejos MJ, Solano RM, Menendez J, Mena MA (2004) Role of extracellular signal-regulated protein kinase in neuronal cell death induced by glutathione depletion in neuron/glia mesencephalic cultures. J Neurochem 91:667–682

de Haan JB, Bladier C, Griffiths P, Kelner M, O'Shea RD, Cheung NS, Bronson RT, Silvestro MJ, Wild S, Zheng SS, Beart PM, Hertzog PJ, Kola I (1998) Mice with a homozygous null mutation for the most abundant glutathione peroxidase, Gpx1, show increased susceptibility to the oxidative stress-inducing agents paraquat and hydrogen peroxide. J Biol Chem 273:22528–22536

Dean O, Bush AI, Berk M, Copolov DL, van den Buuse M (2009) Glutathione depletion in the brain disrupts short-term spatial memory in the Y-maze in rats and mice. Behav Brain Res 198:258–262

DeCory HH, Piech-Dumas KM, Sheu SS, Federoff HJ, Anders MW (2001) Efflux of glutathione conjugate of monochlorobimane from striatal and cortical neurons. Drug Metab Dispos 29:1256–1262

Deeley RG, Cole SP (2006) Substrate recognition and transport by multidrug resistance protein 1 (ABCC1). FEBS Lett 580:1103–1111

Dhanani S, Awasthi YC (2007) Glutathione S-transferase isoenzyme composition of human tissues. In: Awasthi YC (ed) Toxicology of glutathione transferases. Taylor & Francis, Boca Raton, pp 321–338

Diaz-Hernandez JI, Moncada S, Bolanos JP, Almeida A (2007) Poly(ADP-ribose) polymerase-1 protects neurons against apoptosis induced by oxidative stress. Cell Death Differ 14:1211–1221

Dieck ST, Heuer H, Ehrchen J, Otto C, Bauer K (1999) The peptide transporter PepT2 is expressed in rat brain and mediates the accumulation of the fluorescent dipeptide derivative beta-Ala-Lys-Nepsilon-AMCA in astrocytes. Glia 25:10–20

Do KQ, Cabungcal JH, Frank A, Steullet P, Cuenod M (2009) Redox dysregulation, neurodevelopment, and schizophrenia. Curr Opin Neurobiol 19:220–230

Dringen R (2000) Metabolism and functions of glutathione in brain. Prog Neurobiol 62:649–671

Dringen R (2005) Oxidative and antioxidative potential of brain microglial cells. Antiox Redox Signal 7:1223–1233

Dringen R (2009) Neuron-glia coupling in glutathione metabolism. In: Squire L, Albright T, Bloom F, Gage F, Spitzer N (eds) The new encyclopedia of neuroscience. Elsevier, Oxford, UK, pp 733–737

Dringen R, Gutterer JM (2002) Glutathione reductase from bovine brain. Methods Enzymol 348:281–288

Dringen R, Hamprecht B (1997) Involvement of glutathione peroxidase and catalase in the disposal of exogenous hydrogen peroxide by cultured astroglial cells. Brain Res 759:67–75

Dringen R, Hamprecht B (1998) Glutathione restoration as indicator for cellular metabolism of astroglial cells. Dev Neurosci 20:401–407

Dringen R, Hirrlinger J (2003) Glutathione pathways in the brain. Biol Chem 384:505–516

Dringen R, Hirrlinger J (2010) Anti-oxidative defense of brain microglial cells. In: Packer L, Sies H, Eggersdorf M, Cadenas E (eds) Micronutrients and brain health. Taylor & Francis, Boca Raton, pp 391–399

Dringen R, Kranich O, Hamprecht B (1997) The γ-glutamyl transpeptidase inhibitor acivicin preserves glutathione released by astroglial cells in culture. Neurochem Res 22:727–733

Dringen R, Hamprecht B, Broer S (1998a) The peptide transporter PepT2 mediates the uptake of the glutathione precursor CysGly in astroglia-rich primary cultures. J Neurochem 71:388–393

Dringen R, Kussmaul L, Hamprecht B (1998b) Rapid clearance of tertiary butyl hydroperoxide by cultured astroglial cells via oxidation of glutathione. Glia 23:139–145

Dringen R, Kussmaul L, Gutterer JM, Hirrlinger J, Hamprecht B (1999a) The glutathione system of peroxide detoxification is less efficient in neurons than in astroglial cells. J Neurochem 72:2523–2530

Dringen R, Pfeiffer B, Hamprecht B (1999b) Synthesis of the antioxidant glutathione in neurons: Supply by astrocytes of CysGly as precursor for neuronal glutathione. J Neurosci 19:562–569

Dringen R, Gutterer JM, Gros C, Hirrlinger J (2001) Aminopeptidase N mediates the utilization of the GSH precursor CysGly by cultured neurons. J Neurosci Res 66:1003–1008

Dringen R, Pawlowski PG, Hirrlinger J (2005) Peroxide detoxification by brain cells. J Neurosci Res 79:157–165

Dringen R, Hoepken HH, Minich T, Ruedig C (2007) Pentose phosphate pathway and NADPH metabolism. In: Dienel G, Gibson G (eds) Handbook of neurochemistry, 3rd edn, Neural energy utilisation. Springer Verlag, Heidelberg, pp 41–62

Eulenburg V, Armsen W, Betz H, Gomeza J (2005) Glycine transporters: essential regulators of neurotransmission. Trends Biochem Sci 30:325–333

Flentjar NJ, Crack PJ, Boyd R, Malin M, de Haan JB, Hertzog P, Kola I, Iannello R (2002) Mice lacking glutathione peroxidase-1 activity show increased tunel staining and an accelerated inflammatory response in brain following a cold-induced injury. Exp Neurol 177:9–20

Fonnum F, Lock EA (2004) The contributions of excitotoxicity, glutathione depletion and DNA repair in chemically induced injury to neurones: exemplified with toxic effects on cerebellar granule cells. J Neurochem 88:513–531

Forman HJ, Zhang H, Rinna A (2009) Glutathione: overview of its protective roles, measurement, and biosynthesis. Mol Aspects Med 30:1–12

Frade J, Pope S, Schmidt M, Dringen R, Barbosa R, Pocock J, Laranjinha J, Heales S (2008) Glutamate induces release of glutathione from cultured rat astrocytes - a possible neuroprotective mechanism? J Neurochem 105:1144–1152

Franco R, Cidlowski JA (2009) Apoptosis and glutathione: beyond an antioxidant. Cell Death Differ 16:1303–1314

Franco R, Schoneveld OJ, Pappa A, Panayiotidis MI (2007) The central role of glutathione in the pathophysiology of human diseases. Arch Physiol Biochem 113:234–258

Franklin CC, Backos DS, Mohar I, White CC, Forman HJ, Kavanagh TJ (2009) Structure, function, and post-translational regulation of the catalytic and modifier subunits of glutamate cysteine ligase. Mol Aspects Med 30:86–98

Fraser JA, Saunders RD, McLellan LI (2002) Drosophila melanogaster glutamate-cysteine ligase activity is regulated by a modifier subunit with a mechanism of action similar to that of the mammalian form. J Biol Chem 277:1158–1165

Fraser JA, Kansagra P, Kotecki C, Saunders RD, McLellan LI (2003) The modifier subunit of Drosophila glutamate-cysteine ligase regulates catalytic activity by covalent and noncovalent interactions and influences glutathione homeostasis in vivo. J Biol Chem 278: 46369–46377

Fujita T, Kishida T, Wada M, Okada N, Yamamoto A, Leibach FH, Ganapathy V (2004) Functional characterization of brain peptide transporter in rat cerebral cortex: identification of the high-affinity type H^+/peptide transporter PEPT2. Brain Res 997:52–61

Fuller S, Steele M, Imholz P, Münch G (2010) Activated astroglia during chronic inflammation in Alzheimer's disease - Do they neglect their neurosupportive roles? Mut Res Fundam Mol Mech Mutagen 690:40–49

Garcion E, Sindji L, Leblondel G, Brachet P, Darcy F (1999) 1,25-dihydroxyvitamin D3 regulates the synthesis of γ-glutamyl transpeptidase and glutathione levels in rat primary astrocytes. J Neurochem 73:859–866

Gegg ME, Beltran B, Salas-Pino S, Bolanos JP, Clark JB, Moncada S, Heales SJ (2003) Differential effect of nitric oxide on glutathione metabolism and mitochondrial function in astrocytes and neurones: implications for neuroprotection/neurodegeneration? J Neurochem 86:228–237

Gegg ME, Clark JB, Heales SJR (2005) Co-culture of neurones with glutathione deficient astrocytes leads to increased neuronal susceptibility to nitric oxide and increased glutamate-cysteine ligase activity. Brain Res 1036:1–6

Ghezzi P, Di Simplicio P (2007) Glutathionylation pathways in drug response. Curr Opin Pharmacol 7:398–403

Giordano G, White CC, McConnachie LA, Fernandez C, Kavanagh TJ, Costa LG (2006) Neurotoxicity of domoic acid in cerebellar granule neurons in a genetic model of glutathione deficiency. Mol Pharmacol 70:2116–2126

Giordano G, White CC, Mohar I, Kavanagh TJ, Costa LG (2007) Glutathione levels modulate domoic acid induced apoptosis in mouse cerebellar granule cells. Toxicol Sci 100:433–444

Giordano G, Kavanagh TJ, Costa LG (2008a) Neurotoxicity of a polybrominated diphenyl ether mixture (DE-71) in mouse neurons and astrocytes is modulated by intracellular glutathione levels. Toxicol Appl Pharmacol 232:161–168

Giordano G, Klintworth HM, Kavanagh TJ, Costa LG (2008b) Apoptosis induced by domoic acid in mouse cerebellar granule neurons involves activation of p38 and JNK MAP kinases. Neurochem Int 52:1100–1105

Giordano G, Li L, White CC, Farin FM, Wilkerson HW, Kavanagh TJ, Costa LG (2009) Muscarinic receptors prevent oxidative stress-mediated apoptosis induced by domoic acid in mouse cerebellar granule cells. J Neurochem 109:525–538

Gomeza J, Ohno K, Betz H (2003) Glycine transporter isoforms in the mammalian central nervous system: structures, functions and therapeutic promises. Curr Opin Drug Discov Dev 6:675–682

Griffin S, Clark JB, Canevari L (2005) Astrocyte-neurone communication following oxygen-glucose deprivation. J Neurochem 95:1015–1022

Griffith OW (1999) Biologic and pharmacologic regulation of mammalian glutathione synthesis. Free Radic Biol Med 27:922–935

Griffith OW, Mulcahy RT (1999) The enzymes of glutathione synthesis: γ-glutamylcysteine synthetase. Adv Enzymol Relat Areas Mol Biol 73:209–267

Groneberg DA, Rubio-Aliaga I, Nickolaus M, Doring F, Fischer A, Daniel H (2004) Direct visualization of peptide uptake activity in the central nervous system of the rat. Neurosci Lett 364:32–36

Gysin R, Kraftsik R, Sandell J, Bovet P, Chappuis C, Conus P, Deppen P, Preisig M, Ruiz V, Steullet P, Tosic M, Werge T, Cuenod M, Do KQ (2007) Impaired glutathione synthesis in schizophrenia: convergent genetic and functional evidence. Proc Natl Acad Sci USA 104:16621–16626

Han J, Cheng FC, Yang Z, Dryhurst G (1999) Inhibitors of mitochondrial respiration, iron (II), and hydroxyl radical evoke release and extracellular hydrolysis of glutathione in rat striatum and substantia nigra: potential implications to Parkinson's disease. J Neurochem 73: 1683–1695

Hayes JD, Flanagan JU, Jowsey IR (2005) Glutathione transferases. Annu Rev Pharmacol Toxicol 45:51–88

Hirrlinger J, Dringen R (2005) Multidrug resistance protein 1-mediated export of glutathione and glutathione disulfide from brain astrocytes. Methods Enzymol 400:395–400

Hirrlinger J, Dringen R (2010) The cytosolic redox state of astrocytes: maintenance, regulation and functional implications for metabolite trafficking. Brain Res Rev 63:177–188

Hirrlinger J, Gutterer JM, Kussmaul L, Hamprecht B, Dringen R (2000) Microglial cells in culture express a prominent glutathione system for the defense against reactive oxygen species. Dev Neurosci 22:384–392

Hirrlinger J, Konig J, Keppler D, Lindenau J, Schulz JB, Dringen R (2001) The multidrug resistance protein MRP1 mediates the release of glutathione disulfide from rat astrocytes during oxidative stress. J Neurochem 76:627–636

Hirrlinger J, Resch A, Gutterer JM, Dringen R (2002a) Oligodendroglial cells in culture effectively dispose of exogenous hydrogen peroxide: comparison with cultured neurones, astroglial and microglial cells. J Neurochem 82:635–644

Hirrlinger J, Schulz JB, Dringen R (2002b) Glutathione release from cultured brain cells: multidrug resistance protein 1 mediates the release of GSH from rat astroglial cells. J Neurosci Res 69:318–326

Hirrlinger J, Schulz JB, Dringen R (2002c) Effects of dopamine on the glutathione metabolism of cultured astroglial cells: implications for Parkinson's disease. J Neurochem 82:458–467

Huang CS, Chang LS, Anderson ME, Meister A (1993) Catalytic and regulatory properties of the heavy subunit of rat kidney γ-glutamylcysteine synthetase. J Biol Chem 268:19675–19680

Ikeda Y, Taniguchi N (2005) Gene expression of γ-glutamyltranspeptidase. Methods Enzymol 401:408–425

Janáky R, Cruz-Aguado R, Oja SS, Shaw CA (2007) Glutathione in the nervous system: roles in neural function and health and implications for neurological disease. In: Oja SS, Schousboe A,

Saransaari P (eds) Handbook of neurochemistry, 3rd edn, Amino acids and peptides in the nervous system. Springer Verlag, Heidelberg, pp 347–399

Johnson JA, el Barbary A, Kornguth SE, Brugge JF, Siegel FL (1993) Glutathione S-transferase isoenzymes in rat brain neurons and glia. J Neurosci 13:2013–2023

Johnson JA, Johnson DA, Kraft AD, Calkins MJ, Jakel RJ, Vargas MR, Chen PC (2008) The Nrf2-ARE pathway: an indicator and modulator of oxidative stress in neurodegeneration. Ann N Y Acad Sci 1147:61–69

Jursky F, Nelson N (1995) Localization of glycine neurotransmitter transporter (GLYT2) reveals correlation with the distribution of glycine receptor. J Neurochem 64:1026–1033

Jursky F, Nelson N (1996) Developmental expression of the glycine transporters GLYT1 and GLYT2 in mouse brain. J Neurochem 67:336–344

Kang Y, Viswanath V, Jha N, Qiao X, Mo JQ, Andersen J (1999) Brain γ-glutamyl cysteine synthetase (GCS) mRNA expression patterns correlate with regional-specific enzyme activities and glutathione levels. J Neurosci Res 58:436–441

Keppler D (1999) Export pumps for glutathione S-conjugates. Free Radic Biol Med 27:985–991

Kraft AD, Johnson DA, Johnson JA (2004) Nuclear factor E2-related factor 2-dependent antioxidant response element activation by tert-butylhydroquinone and sulforaphane occurring preferentially in astrocytes conditions neurons against oxidative insult. J Neurosci 24: 1101–1112

Kranich O, Hamprecht B, Dringen R (1996) Different preferences in the utilization of amino acids for glutathione synthesis in cultured neurons and astroglial cells derived from rat brain. Neurosci Lett 219:211–214

Kranich O, Dringen R, Sandberg M, Hamprecht B (1998) Utilization of cysteine and cysteine precursors for the synthesis of glutathione in astroglial cultures: preference for cystine. Glia 22:11–18

Kussmaul L, Hamprecht B, Dringen R (1999) The detoxification of cumene hydroperoxide by the glutathione system of cultured astroglial cells hinges on hexose availability for the regeneration of NADPH. J Neurochem 73:1246–1253

Lada MW, Kennedy RT (1997) In vivo monitoring of glutathione and cysteine in rat caudate nucleus using microdialysis on-line with capillary zone electrophoresis-laser induced fluorescence detection. J Neurosci Methods 72:153–159

Lavoie S, Chen Y, Dalton TP, Gysin R, Cuenod M, Steullet P, Do KQ (2009) Curcumin, quercetin, and tBHQ modulate glutathione levels in astrocytes and neurons: importance of the glutamate cysteine ligase modifier subunit. J Neurochem 108:1410–1422

Lee JM, Calkins MJ, Chan K, Kan YW, Johnson JA (2003) Identification of the NF-E2-related factor-2-dependent genes conferring protection against oxidative stress in primary cortical astrocytes using oligonucleotide microarray analysis. J Biol Chem 278:12029–12038

Li X, Orwar OWE, RevesjÖ C, Sandberg M (1996) γ-glutamyl peptides and related amino acds in rat hippocampus in vitro: effect of depolarisation and γ-glutamyl transpeptidase inhibition. Neurochem Int 29:121–128

Li X, Wallin C, Weber SG, Sandberg M (1999) Net efflux of cysteine, glutathione and related metabolites from rat hippocampal slices during oxygen/glucose deprivation: dependence on γ-glutamyl transpeptidase. Brain Res 815:81–88

Liddell JR, Dringen R, Crack PJ, Robinson SR (2006a) Glutathione peroxidase 1 and a high cellular glutathione concentration are essential for effective organic hydroperoxide detoxification in astrocytes. Glia 54:873–879

Liddell JR, Hoepken HH, Crack PJ, Robinson SR, Dringen R (2006b) Glutathione peroxidase 1 and glutathione are required to protect mouse astrocytes from iron-mediated hydrogen peroxide toxicity. J Neurosci Res 84:578–586

Lipton AJ, Johnson MA, Macdonald T, Lieberman MW, Gozal D, Gaston B (2001) S-nitrosothiols signal the ventilatory response to hypoxia. Nature 413:171–174

Liu H, Wang H, Shenvi S, Hagen TM, Liu RM (2004) Glutathione metabolism during aging and in Alzheimer disease. Ann N Y Acad Sci 1019:346–349

Lopez-Barea J, Barcena JA, Bocanegra JA, Martinez-Galisteo E, Peinado J (1990) Structure, mechanism, functions, and regulatory properties of glutathione reductase. In: Vina J (ed) Glutathione: metabolism and physiological functions. CRC, Boca Raton, pp 105–116

Lu SC (2009) Regulation of glutathione synthesis. Mol Aspects Med 30:42–59

Maher P (2005) The effects of stress and aging on glutathione metabolism. Ageing Res Rev 4:288–314

Mannervik B, Board PG, Hayes JD, Listowsky I, Pearson WR (2005) Nomenclature for mammalian soluble glutathione transferases. Methods Enzymol 401:1–8

Margis R, Dunand C, Teixeira FK, Margis-Pinheiro M (2008) Glutathione peroxidase family - an evolutionary overview. FEBS J 275:3959–3970

Martin HL, Teismann P (2009) Glutathione – a review on its role and significance in Parkinson's disease. FASEB J 23:3263–3272

Matsuzawa D, Obata T, Shirayama Y, Nonaka H, Kanazawa Y, Yoshitome E, Takanashi J, Matsuda T, Shimizu E, Ikehira H, Iyo M, Hashimoto K (2008) Negative correlation between brain glutathione level and negative symptoms in schizophrenia: a ^{3}T ^{1}H-MRS study. PLoS One 3:e1944

McBean G (2002) Cerebral cystine uptake: a tale of two transporters. Trends Pharmacol Sci 23:299–302

Meister A (1974) Glutathione, metabolism and function via the γ-glutamyl cycle. Life Sci 15:177–190

Meister A, Anderson ME (1983) Glutathione. Annu Rev Biochem 52:711–760

Mieyal JJ, Gallogly MM, Qanungo S, Sabens EA, Shelton MD (2008) Molecular mechanisms and clinical implications of reversible protein S-glutathionylation. Antiox Redox Signal 10:1941–1988

Minich T, Riemer J, Schulz JB, Wielinga P, Wijnholds J, Dringen R (2006) The multidrug resistance protein 1 (Mrp1), but not Mrp5, mediates export of glutathione and glutathione disulfide from brain astrocytes. J Neurochem 97:373–384

Misra I, Griffith OW (1998) Expression and purification of human γ-glutamylcysteine synthetase. Protein Expr Purif 13:268–276

Nguyen T, Nioi P, Pickett CB (2009) The Nrf2-antioxidant response element signaling pathway and its activation by oxidative stress. J Biol Chem 284:13291–13295

Njalsson R (2005) Glutathione synthetase deficiency. Cell Mol Life Sci 62:1938–1945

Noble F, Banisadr G, Jardinaud F, Popovici T, Lai-Kuen R, Chen H, Bischoff L, Parsadaniantz SM, Fournie-Zaluski MC, Roques BP (2001) First discrete autoradiographic distribution of aminopeptidase N in various structures of rat brain and spinal cord using the selective iodinated inhibitor [^{125}I]RB 129. Neuroscience 105:479–488

Orwar O, Li X, Andine P, Bergstrom CM, Hagberg H, Folestad S, Sandberg M (1994) Increased intra- and extracellular concentrations of γ-glutamylglutamate and related dipeptides in the ischemic rat striatum: involvement of glutamyl transpeptidase. J Neurochem 63:1371–1376

Pallardó FV, Markovic J, García JL, Viña J (2009) Role of nuclear glutathione as a key regulator of cell proliferation. Mol Aspects Med 30:77–85

Perricone C, De Carolis C, Perricone R (2009) Glutathione: a key player in autoimmunity. Autoimmun Rev 8:697–701

Philbert MA, Beiswanger CM, Manson MM, Green JA, Novak RF, Primiano T, Reuhl KR, Lowndes HE (1995) Glutathione S-transferases and γ-glutamyl transpeptidase in the rat nervous systems: a basis for differential susceptibility to neurotoxicants. Neurotoxicology 16:349–362

Pope SA, Milton R, Heales SJ (2008) Astrocytes protect against copper-catalysed loss of extracellular glutathione. Neurochem Res 33:1410–1418

Rana S, Dringen R (2007) Gap junction hemichannel-mediated release of glutathione from cultured rat astrocytes. Neurosci Lett 415:45–48

Rathinam ML, Watts LT, Stark AA, Mahimainathan L, Stewart J, Schenker S, Henderson GI (2006) Astrocyte control of fetal cortical neuron glutathione homeostasis: up-regulation by ethanol. J Neurochem 96:1289–1300

Rebrin I, Sohal RS (2008) Pro-oxidant shift in glutathione redox state during aging. Adv Drug Deliv Rev 60:1545–1552

Rice ME, Russo-Menna I (1998) Differential compartmentalization of brain ascorbate and glutathione between neurons and glia. Neuroscience 82:1213–1233

Richman PG, Meister A (1975) Regulation of γ-glutamyl-cysteine synthetase by nonallosteric feedback inhibition by glutathione. J Biol Chem 250:1422–1426

Ristoff E, Larsson A (2007) Inborn errors in the metabolism of glutathione. Orphanet J Rare Dis 2:16

Roux MJ, Supplisson S (2000) Neuronal and glial glycine transporters have different stoichiometries. Neuron 25:373–383

Ruedig C, Dringen R (2004) TNFα increases activity of γ-glutamyl transpeptidase in cultured rat astroglial cells. J Neurosci Res 75:536–543

Sagara J, Makino N, Bannai S (1996) Glutathione efflux from cultured astrocytes. J Neurochem 66:1876–1881

Saito Y, Nishio K, Numakawa Y, Ogawa Y, Yoshida Y, Noguchi N, Niki E (2007) Protective effects of 15-deoxy-Δ12,14-prostaglandin J2 against glutamate-induced cell death in primary cortical neuron cultures: induction of adaptive response and enhancement of cell tolerance primarily through up-regulation of cellular glutathione. J Neurochem 102:1625–1634

Salinas AE, Wong MG (1999) Glutathione S-transferases - a review. Curr Med Chem 6:279–309

Sandhu JK, Gardaneh M, Iwasiow R, Lanthier P, Gangaraju S, Ribecco-Lutkiewicz M, Tremblay R, Kiuchi K, Sikorska M (2009) Astrocyte-secreted GDNF and glutathione antioxidant system protect neurons against 6OHDA cytotoxicity. Neurobiol Dis 33:405–414

Schirmer RH, Krauth-Siegel RL, Schulz GE (1989) Glutathione reductase. In: Dolphin D, Poulson R, Avramovic O (eds) Glutathione: chemical, biochemical and medical aspects. Wiley, New York, pp 553–596

Schmidt MM, Dringen R (2009) Differential effects of iodoacetamide and iodoacetate on glycolysis and glutathione metabolism of cultured astrocytes. Front Neuroenergetics 1:1–10

Schmidt MM, Dringen R (2010) Fumaric acid diesters deprive cultured primary astrocytes rapidly of glutathione. Neurochem Int 57:460–467

Shang W, Liu WH, Zhao XH, Sun QJ, Bi JZ, Chi ZF (2008) Expressions of glutathione S-transferase α, μ, and π in brains of medically intractable epileptic patients. BMC Neurosci 9:67

Shanker G, Aschner M (2001) Identification and characterisation of uptake systems for cystine and cysteine in cultured astrocytes and neurons: evidence for methylmercury-targeted disruption of astrocyte transport. J Neurosci Res 66:998–1002

Sharma R, Ansari GAS, Awasthi YC (2007) Physiological substrates of glutathione S-transferases. In: Awasthi YC (ed) Toxicology of glutathione transferases. Taylor & Francis, Boca Raton, pp 179–203

Shen H, Smith DE, Keep RF, Brosius FC (2004) Immunolocalization of the proton-coupled oligopeptide transporter PEPT2 in developing rat brain. Mol Pharm 1:248–256

Shih AY, Johnson DA, Wong G, Kraft AD, Jiang L, Erb H, Johnson JA, Murphy TH (2003) Coordinate regulation of glutathione biosynthesis and release by Nrf2-expressing glia potently protects neurons from oxidative stress. J Neurosci 23:3394–3406

Shih AY, Imbeault S, Barakauskas V, Erb H, Jiang L, Li P, Murphy TH (2005) Induction of the Nrf2-driven antioxidant response confers neuroprotection during mitochondrial stress in vivo. J Biol Chem 280:22925–22936

Shine HD, Haber B (1981) Immunocytochemical localization of γ-glutamyl transpeptidase in the rat CNS. Brain Res 217:339–349

Singhal SS, Yadav S, Roth C, Singhal J (2009) RLIP76: A novel glutathione-conjugate and multidrug transporter. Biochem Pharmacol 77:761–769

Soltaninassab SR, Sekhar KR, Meredith MJ, Freeman ML (2000) Multi-faceted regulation of γ-glutamylcysteine synthetase. J Cell Physiol 182:163–170

Srinivasan R, Ratiney H, Hammond-Rosenbluth KE, Pelletier D, Nelson S (2010) MR spectroscopic imaging of glutathione in the white and gray matter at 7 T with an application to multiple sclerosis. Magn Reson Imaging 28:163–170

Stewart VC, Stone R, Gegg ME, Sharpe MA, Hurst RD, Clark JB, Heales SJ (2002) Preservation of extracellular glutathione by an astrocyte derived factor with properties comparable to extracellular superoxide dismutase. J Neurochem 83:984–991

Stridh MH, Tranberg M, Weber SG, Blomstrand F, Sandberg M (2008) Stimulated efflux of amino acids and glutathione from cultured hippocampal slices by omission of extracellular calcium: likely involvement of connexin hemichannels. J Biol Chem 283:10347–10356

Sun X, Erb H, Murphy TH (2005) Coordinate regulation of glutathione metabolism in astrocytes by Nrf2. Biochem Biophys Res Commun 326:371–377

Supplisson S, Roux MJ (2002) Why glycine transporters have different stoichiometries. FEBS Lett 529:93–101

Taylor JM, Ali U, Iannello RC, Hertzog P, Crack PJ (2005) Diminished Akt phosphorylation in neurons lacking glutathione peroxidase-1 (Gpx1) leads to increased susceptibility to oxidative stress-induced cell death. J Neurochem 92:283–293

Toppo S, Flohé L, Ursini F, Vanin S, Maiorino M (2009) Catalytic mechanisms and specificities of glutathione peroxidases: Variations of a basic scheme. Biochim Biophys Acta 1790:1486–1500

Toyoda Y, Hagiya Y, Adachi T, Hoshijima K, Kuo MT, Ishikawa T (2008) MRP class of human ATP binding cassette (ABC) transporters: historical background and new research directions. Xenobiotica 38:833–862

Ursini F, Maiorino M, Brigelius-Flohe R, Aumann KD, Roveri A, Schomburg D, Flohe L (1995) Diversity of glutathione peroxidases. Methods Enzymol 252:38–53

Vargas MR, Johnson JA (2009) The Nrf2-ARE cytoprotective pathway in astrocytes. Expert Rev Mol Med 11:e17

Vargas MR, Pehar M, Cassina P, Beckman JS, Barbeito L (2006) Increased glutathione biosynthesis by Nrf2 activation in astrocytes prevents $p75^{NTR}$-dependent motor neuron apoptosis. J Neurochem 97:687–696

Vargas MR, Johnson DA, Sirkis DW, Messing A, Johnson JA (2008) Nrf2 activation in astrocytes protects against neurodegeneration in mouse models of familial amyotrophic lateral sclerosis. J Neurosci 28:13574–13581

Waak J, Dringen R (2006) Formation and rapid export of the monochlorobimane–glutathione conjugate in cultured rat astrocytes. Neurochem Res 31:1409–1416

Watabe M, Aoyama K, Nakaki T (2008) A dominant role of GTRAP3-18 in neuronal glutathione synthesis. J Neurosci 28:9404–9413

Watts LT, Rathinam ML, Schenker S, Henderson GI (2005) Astrocytes protect neurons from ethanol-induced oxidative stress and apoptotic death. J Neurosci Res 80:655–666

Woehrling EK, Hill EJ, Coleman MD (2007) Development of a neurotoxicity test-system, using human post-mitotic, astrocytic and neuronal cell lines in co-culture. Toxicol in Vitro 21: 1241–1246

Wong CH, Bozinovski S, Hertzog PJ, Hickey MJ, Crack PJ (2008) Absence of glutathione peroxidase-1 exacerbates cerebral ischemia-reperfusion injury by reducing post-ischemic microvascular perfusion. J Neurochem 107:241–252

Wood SJ, Yucel M, Pantelis C, Berk M (2009) Neurobiology of schizophrenia spectrum disorders: the role of oxidative stress. Ann Acad Med Singapore 38:396–396

Xiang J, Chiang PP, Hu Y, Smith DE, Keep RF (2006) Role of PEPT2 in glycylsarcosine transport in astrocyte and glioma cultures. Neurosci Lett 396:225–229

Yang CS, Chou ST, Lin NN, Liu L, Tsai PJ, Kuo JS, Lai JS (1994) Determination of extracellular glutathione in rat brain by microdialysis and high-performance liquid chromatography with fluorescence detection. J Chromatogr B Biomed Appl 661:231–235

Yang Y, Dieter MZ, Chen Y, Shertzer HG, Nebert DW, Dalton TP (2002) Initial characterization of the glutamate-cysteine ligase modifier subunit Gclm(-/-) knockout mouse. Novel model system for a severely compromised oxidative stress response. J Biol Chem 277: 49446–49452

Yonezawa M, Back SA, Gan X, Rosenberg PA, Volpe JJ (1996) Cystine deprivation induces oligodendroglial death: rescue by free radical scavengers and by a diffusible glial factor. J Neurochem 67:566–573

Zafra F, Gomeza J, Olivares L, Aragon C, Gimenez C (1995) Regional distribution and developmental variation of the glycine transporters GLYT1 and GLYT2 in the rat CNS. Eur J Neurosci 7:1342–1352

Zeevalk GD, Razmpour R, Bernard LP (2008) Glutathione and Parkinson's disease: Is this the elephant in the room? Biomed Pharmacother 62:236–249

Zhang HF, Ong WY, Leong SK, Laperche Y (1997) Species differences in the localisation of γ-glutamyl transpeptidase immunopositive cells at the blood-brain interface. J Hirnforsch 38:323–330

Zhang J, Hu J, Ding J-H, Yao H-H, Hu G (2005) 6-Hydroxydopamine-induced glutathione alteration occurs via glutathione enzyme system in primary cultured astrocytes. Acta Pharmacol Sin 26:799–805

Zhou SF, Wang LL, Di YM, Xue CC, Duan W, Li CG, Li Y (2008) Substrates and inhibitors of human multidrug resistance associated proteins and the implications in drug development. Curr Med Chem 15:1981–2039

Zimniak P (2007) Substrates and reaction mechanisms of glutathione transferases. In: Awasthi YC (ed) Toxicology of glutathione transferases. Taylor & Francis, Boca Raton, pp 71–101

Zimniak P, Singh SP (2007) Families of glutathione transferases. In: Awasthi YC (ed) Toxicology of glutathione transferases. Taylor & Francis, Boca Raton, pp 11–26

Chapter 37
Brain Ascorbate: Protective, Yet Permissive for Redox Signaling

Margaret E. Rice

Abstract Ascorbic acid (vitamin C) occurs physiologically as the ascorbate anion. Ascorbate is a water-soluble antioxidant found throughout the body, but occurs in especially high concentrations (10 mM) in neurons. In the central nervous system (CNS), ascorbate is part of the antioxidant network, and as such is normally neuroprotective. Recent studies suggest that ascorbate may play a unique role in the CNS by permitting signaling by a novel diffusible messenger, hydrogen peroxide (H_2O_2), yet providing protection from damage from the reactive hydroxyl radical, which can be formed from H_2O_2 oxidation. In order to learn about the function of ascorbate in the CNS, it is necessary to understand how it is compartmentalized. Ascorbate enters the CNS primarily by active transport at the choroid plexus, but also by uptake of oxidized ascorbate, dehydroascorbate, via glucose transporters. After entry into the cerebrospinal fluid (CSF), it diffuses into brain extracellular fluid (ECF), where its concentration is regulated homeostatically, as well as dynamically. From ECF, ascorbate is taken up into brain cells, where it is further concentrated up to 20-fold. Thus, brain ascorbate is located predominantly in the intracellular compartment of brain tissue, with complementary distribution between neurons and glia that may contribute importantly to its role as a neuroprotective agent that also permits redox signaling.

Keywords Antioxidants • Ascorbic acid • Glia • Glutamate • Glutathione • Neurons • Neuroprotection • Oxidative stress • SVCT1 • SVCT2

M.E. Rice, Ph.D. (✉)
Department of Neurosurgery and Department of Physiology and Neuroscience,
New York University School of Medicine, 550 First Avenue, New York, NY 10016, USA
e-mail: margaret.rice@nyu.edu

I.-Y. Choi and R. Gruetter (eds.), *Neural Metabolism In Vivo*,
Advances in Neurobiology 4, DOI 10.1007/978-1-4614-1788-0_37,

37.1 Ascorbate Biosynthesis and Chemistry

Ascorbic acid is a water-soluble sugar acid, with a molecular weight of 176.13. It has two dissociable protons with pK_a values of 4.2 and 11.8 (Davies et al. 1991), so that it occurs as the monovalent anion, ascorbate, at physiological pH. The ene-diol structure of ascorbate enables it to be an electron donor: loss of two electrons leads to formation of dehydroascorbate (DHA), the final oxidation product. Most free radicals (molecules with one unpaired electron) generated by biological systems can cause a one-electron oxidation of ascorbate to form semi-dehydroascorbate, also known as the ascorbyl radical. Detection of ascorbyl radicals using electron-spin resonance has been used as an index of oxidative stress (Buettner and Jurkiewicz 1993). This radical intermediate is also formed in enzymatic reactions that involve ascorbate as an electron-donating co-factor (Diliberto et al. 1987).

The roles of ascorbate as an antioxidant and free radical scavenger are derived from its properties as an electron donor. Ascorbate has a relatively low redox potential and is therefore a broad-spectrum radical scavenger that is effective against peroxyl and hydroxyl radicals, superoxide, singlet oxygen, and peroxynitrite (Nishikimi 1975; Bodannes and Chan 1979; Machlin and Bendich 1987; Cohen 1994; Vatassery 1996). As part of the predominantly intracellular antioxidant network, ascorbate acts in concert with other low-molecular-weight substances, including glutathione (GSH) and vitamin E, as well as antioxidant enzymes, including superoxide dismutase and GSH peroxidase (Cohen 1994). Although redox reactions with ascorbate occur in the aqueous phase, this can prevent oxidation of lipid-soluble vitamin E (α-tocopherol), which can in turn stop peroxidation of cell membranes (Seregi et al. 1978; Niki 1991; Buettner 1993; Sato et al. 1993). Oxidized ascorbate, semi-dehydroascorbate and DHA, can be recycled by intracellular thiols, including GSH (Meister 1994; Winkler et al. 1994), and in some cells by a GSH-dependent DHA reductase (Rose 1993; Fornai et al. 1999). In the brain, immunocytochemical studies show regionally distinct patterns of localization for DHA reductase, with staining primarily in neurons (Fornai et al. 2001). Additionally, glutaredoxins, which are GSH-dependent reducing enzymes, also have DHA reductase activity (Washburn and Wells 1999; Ehrhart and Zeevalk 2003), as do thioredoxin reductase (May et al. 1997) and omega class GSH transferases (Board et al. 2000). Oxidized GSH, in turn, is recycled by GSH reductase (Meister 1994).

Another key function of ascorbate is to serve as an electron-donating enzyme co-factor. The most well-known ascorbate-dependent enzyme processes include collagen biosynthesis via hydroxylation reactions (Barnes 1975), noradrenaline/adrenaline synthesis by dopamine-β-hydroxylase (Diliberto et al. 1987), and myelin formation by enabling Schwann cells to lay down the basal lamina (Carey and Todd 1987; Eldridge et al. 1987). Promoting myelin formation may be the mechanism by which therapeutically administered ascorbate enhances myelination and reverses motor deficits in a mouse model of Charcot-Marie-Tooth disease, which is the most common hereditary peripheral neuropathy (Passage et al. 2004). More recently, it has been recognized that ascorbate is an essential factor for the prolyl 4-hydroxylases that control hypoxia inducible factor (HIF), a ubiquitous transcription factor that plays a key role in the cellular response to low oxygen conditions (Vissers et al. 2007; Pagé et al. 2008). Given the role of HIF in the transcriptional regulation of genes that can provide compensation for hypoxia, metabolic disruption, or oxidative stress (Siddiq et al. 2007), understanding the role of ascorbate, as well as ascorbate depletion, in HIF stabilization is an important emerging area in ascorbate research.

In most animals, ascorbic acid is synthesized from glucose in the following simplified pathway:

$$\text{D-glucose} \rightarrow \text{D-glucuronic acid} \rightarrow \text{L-gulonic acid} \rightarrow \text{L-gulono-}\gamma\text{-lactone} \rightarrow \text{L-ascorbic acid}$$

Humans, nonhuman primates, and guinea pigs cannot synthesize ascorbate because they carry a nonfunctional gene for the enzyme, L-gulono-γ-lactone oxidase, which catalyzes the last step of ascorbate biosynthesis (Nishikimi et al. 1994). Importantly, mutations in this gene were not fatal to the species that carry it for several reasons. First, ascorbate is available from dietary sources and is readily absorbed from the gut (Kallner et al. 1977). In addition, in animals that can synthesize ascorbate, synthesis occurs only in liver (mammals) or kidney (reptiles) (Chatterjee et al. 1975), with subsequent distribution to all other tissues via plasma. Consequently, all tissues have mechanisms for uptake and storage of ascorbate at cell-specific levels. In humans, the effects of vitamin C deficiency, which can lead to scurvy, were noticed first during long sea voyages, away from fresh fruits and vegetables, with primary symptoms caused by loss of collagen synthesis, although depression is also an early consequence of ascorbate deficiency (Carpenter 1986).

37.2 Ascorbate Transport and Transporters

For many years, understanding of ascorbate regulation in the CNS was impeded by the lack of molecular information about the transporters responsible for ascorbate uptake into the brain and into brain cells. This changed profoundly when Tsukaguchi et al. (1999) described the first members of a new family of Na^+-dependent vitamin C [ascorbate] transporters (SVCT1 and SVCT2), isolated from rat cDNA libraries. These isoforms share 65% amino acid homology and facilitate electrogenic, Na^+-dependent uptake of ascorbate with similar affinities (Tsukaguchi et al. 1999). The major difference between the transporter isoforms appears to be their complementary localization, with SVCT1 in kidney, liver and lung, for example, and SVCT2 in neural, neuroendocrine, exocrine and endothelial tissues. Only SVCT2 is found in the brain (Tsukaguchi et al. 1999). Shortly after this initial report, two human isoforms were cloned from human cDNA libraries (Rajan et al. 1999; Daruwala et al. 1999). Human SVCT2 shares a 95% sequence homology with rat SVCT2 and is also found in the brain (Rajan et al. 1999). SVCT1 and SVCT2 were subsequently classified as members of the SLC23 human gene family, which also contains two orphan members, SVCT3 and SVCT4 (for reviews, *see* Takanaga et al. 2004; Savini et al. 2008).

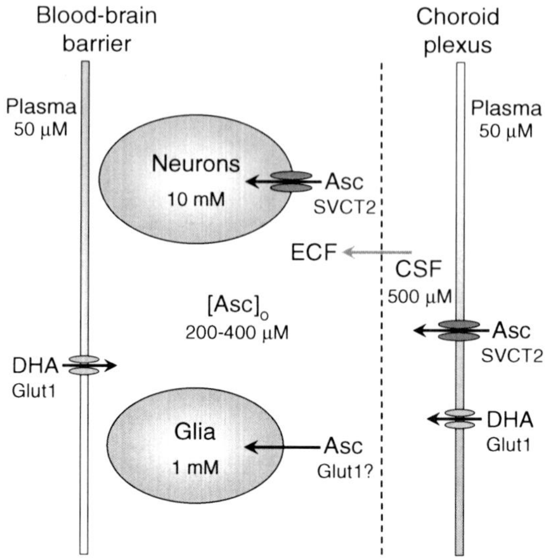

Fig. 37.1 Ascorbate transport, compartmentalization, and concentrations in the CNS. Whether ascorbate is synthesized in liver or kidney or acquired from the diet, it is distributed to all organs via plasma; plasma ascorbate concentration is typically 50 μM (see Sect. 3 for references). Active transport at the choroid plexus, via the SVCT2 isoform of the ascorbate transporter is the first step of brain entry of ascorbate. Transport via the choroid plexus, which is found in the cerebral ventricles, accumulates ascorbate in ventricular cerebrospinal fluid (*CSF*) to a concentration of 500 μM. This pool of ascorbate is contiguous with the extracellular fluid (*ECF*) compartment of the brain, so that ascorbate originating in the CSF is distributed throughout the CNS via ECF (*gray arrow*). Ascorbate can also enter brain ECF at the blood-brain barrier at the level of brain capillaries; however, this presumably occurs via GLUT1-facilitated transport of oxidized ascorbate, DHA, as brain capillaries do not express SVCT2. Existing evidence, discussed in Sect. 2, indicates that this is a minor route for ascorbate entry into the CNS. The extracellular concentration of ascorbate ($[Asc]_o$) is homeostatically regulated in a range of 200–400 μM, with dynamic modulation that is glutamate dependent, as discussed in Sect. 4. Ascorbate is taken up from ECF into neurons and glia. In neurons, ascorbate is transported via SVCT2; however, glia do not normally express SVCT2, leading to the suggestion that uptake of DHA via GLUT1 may be a primary route of entry for glial ascorbate

Consistent with the key role of the choroid plexus in transporting ascorbate into the CNS (Spector and Lorenzo 1973) (Fig. 37.1), *in situ* hybridization in rat brain indicates that this organ, which produces cerebrospinal fluid (CSF) from plasma, expresses high levels of SCVT2 (Tsukaguchi et al. 1999). Importantly, SVCT2 is also found in high levels in neurons, but not in glial cells (Tsukaguchi et al. 1999; Berger and Hediger 2000; Astuya et al. 2005), which is reflected in a ten-fold higher concentration of ascorbate in neurons than in glia, as discussed in Sect. 3. Transporter expression occurs in both excitatory and inhibitory neurons, including glutamatergic pyramidal cells of the hippocampus, glutamatergic granule cells of the cerebellum, and GABAergic cerebellar Purkinje cells (Tsukaguchi et al. 1999). Intriguingly, expression levels of SCVT2 are plastic, although the regulatory mechanisms are incompletely understood. For example, although apparently absent in glial cells *in situ*, SVCT2 can be expressed in astrocytes in culture (Berger and Hediger 2000); moreover, the transporters expressed are functional, as a number of studies have examined characteristics of ascorbate uptake in cultured astrocytes (e.g., Wilson 1989). This plasticity is physiologically relevant: SVCT2 is upregulated in both neurons and glia in the area surrounding a cerebral infarct during reperfusion in an experimental stroke model involving occlusion of the middle cerebral artery (Berger et al. 2003).

An additional mechanism for ascorbate entry across the blood-brain barrier (BBB) can occur through facilitated transport of oxidized ascorbate, DHA by a glucose transporter, GLUT1, with subsequent reduction of DHA back to ascorbate by thiols and DHA reductase activity (Agus et al. 1997) (Fig. 37.1). The fact that ascorbate is found almost exclusively in the reduced form in plasma (Dhariwal et al. 1991), however, suggests that transport of DHA into the CNS via GLUT1 is a minor route of entry for ascorbate (Tsukaguchi et al. 1999). The predominant role of SVCT2 in normal maintenance of brain ascorbate content was confirmed by studies in transgenic mice in which the mouse analogue of SVCT2 (Slc23a1) was knocked out (Sotiriou et al. 2002). Mice with this mutation died within a few minutes of birth from respiratory failure and widespread brain hemorrhage. Significantly, brain tissue from these animals was virtually devoid of ascorbate, demonstrating that active transport via SVCT2 is how ascorbate enters the CNS (Sotiriou et al. 2002).

37.3 Ascorbate Concentrations in Plasma, Cerebrospinal Fluid, Extracellular Fluid, and Brain Cells

As discussed in Sect. 2, the first step in providing ascorbate to the CNS is via stereospecific, Na^+-dependent transport at the choroid plexus via SVCT2 (Spector and Lorenzo 1973; Tsukaguchi et al. 1999). This uptake process transfers ascorbate from the blood into CSF (Fig. 37.1). In rat brain, this process accumulates ascorbate in CSF to a concentration of about 500 μM (Stamford et al. 1984), which is tenfold higher than the typically 50 μM concentration found in plasma in a variety of mammalian species (Hornig 1975; Chatterjee et al. 1975; Spector 1977; Drew et al. 1999) (Fig. 37.1). Although ascorbate can also enter the CNS by carrier-mediated uptake at the level of brain capillaries via GLUT1 (Agus et al. 1997) and by simple diffusion across brain capillaries (Lam and Daniel 1986) (Fig. 37.1), these capillaries do not express SVCT2 (Tsukaguchi et al. 1999; Qiao and May 2008), offering further evidence that the choroid plexus is the primary site of ascorbate entry.

Brain and spinal cord CSF is contiguous with the extracellular fluid (ECF) that surrounds brain cells, permitting equilibrium of CSF ascorbate concentration with that in ECF as seen with other extracellular molecules (e.g., Ichimura et al. 1991). Basal ascorbate concentration in ECF (or extracellular ascorbate concentration, $[Asc]_o$), is 200–400 μM (Fig. 37.1), with the lower limit determined in brain slices (McIlwain et al. 1956; Schenk et al. 1982) and the upper limit determined *in vivo* (Stamford et al. 1984; Miele and Fillenz 1996). From ECF, ascorbate is transported into brain cells (Spector and Lorenzo 1973; Kalir and Mytilenou 1992; Rice et al. 1994; Brahma et al. 2000; Rice and Russo-Menna 1998). The average concentration in neurons is estimated to be 10 mM, whereas the concentration in glia is only 1 mM (Rice and Russo-Menna 1998), consistent with expression of SVCT2 in neurons but not glia under normal conditions *in situ* (Tsukaguchi et al. 1999; Berger and Hediger 2000; Astuya et al. 2005) (Fig. 37.1). How do glia accumulate ascorbate? One hypothesis is that GLUT1 expressed by glia transport DHA, thereby providing an important mechanism for the recycling of ascorbate, should it become oxidized in the ECF through its role as an antioxidant (Astuya et al. 2005). Although ECF levels of DHA would be expected to be low, this process could contribute to maintenance of physiological levels of re-reduced ascorbate in glia. The mechanisms by which ascorbate concentrations are maintained at cell specific levels in either neurons or glia, however, are unknown.

Brain levels of ascorbate are region dependent, with higher tissue content in anterior structures like cortex and hippocampus, but progressively lower levels in more posterior regions like brainstem and spinal cord (Milby et al. 1982; Rice et al. 1995). This pattern largely reflects the increasing white matter content of posterior regions of the CNS, because the ascorbate content of glia-rich white matter is low compared to neuron-rich gray matter (Rice et al. 1995). Regional and sub-regional variation in neuron-to-glia ratio also contributes to tissue-specific levels. Difference in neuron density may also underlie gender differences in ascorbate content of several regions of rat brain, including hippocampus, with lower, estrogen-regulated levels in females than males (Kume-Kick and Rice 1998).

37.4 Homeostatic and Dynamic Regulation of Brain Extracellular Ascorbate

The brain, spinal cord, and adrenal glands have the highest ascorbate concentration in the body, as well as the greatest retention capacity (Hornig 1975). Under normal conditions, turnover of ascorbate in brain is about 2% per hour (Spector 1977). Under conditions of ascorbate deficiency, however, brain ascorbate content is retained tenaciously, with a loss of less than 2% per day in guinea pigs fed an ascorbate-free diet (Hughes et al. 1971). Thus, brain ascorbate levels are under strong homeostatic regulation (Spector 1977; Spector and Lorenzo 1973), which is largely due to the characteristics and location of SVCT2 in the CNS.

The Michaelis-Menten constant, K_m, for ascorbate uptake by SVCT2 expressed in most native cells or heterologous cell systems is generally 20–40 μM (Savini et al. 2008), which is similar to the K_m of 30–50 μM reported previously for ascorbate uptake by choroid plexus and brain slices (Spector and Lorenzo 1973; Spector 1977). However, cultured neurons have apparent K_m values outside this range, with both higher affinity uptake indicated by a K_m of ~10 μM (Castro et al. 2001) and lower affinity uptake indicated by a K_m of ~100 μM (Castro et al. 2001; Qiu et al. 2007). Assuming that typical K_m values of 20–40 μM reflect the state of SVCT2 *in vivo* rather than plastic changes from the *in vitro* environment required for analysis, these data indicate that the rate of SVCT2-mediated transport at the choroid plexus would be roughly half-maximal at normal plasma ascorbate concentrations (50 μM) and approach maximal rate at normal brain $[Asc]_o$ (200–400 μM). These conditions, coupled with the two-step uptake process involving transport from plasma to CSF then from CSF into neurons, contribute to the strong homeostasis of CNS ascorbate levels (Spector 1977).

Like overall brain ascorbate content, $[Asc]_o$ is also maintained homeostatically (Schenk et al. 1982). This was first shown in isolated mammalian brain slices, which rapidly lose ascorbate when incubated in ascorbate free-media. Up to 80% of tissue ascorbate content is lost from brain slices after only brief incubation (McIlwain et al. 1956; Schenk et al. 1982; Rice et al. 1994), possibly via reversal of the ascorbate transporter (Diliberto et al. 1987). In early studies of rat brain slices, McIlwain and colleagues found that the tissue lost ascorbate when incubated in media with ascorbate concentrations of less than 200 μM, but concentrated it in the presence of levels that were higher than this concentration (McIlwain et al. 1956). Subsequently, Schenk and colleagues used voltammetric microelectrodes to monitor ascorbate in the ECF of rat striatal slices during incubation in ascorbate-free medium and found that $[Asc]_o$ remained constant over an initial 30 min period while tissue content fell by 75% (Schenk et al. 1982). This suggested that $[Asc]_o$ was maintained homeostatically at the expense of intracellular stores. The mechanisms that determine the "set point" for $[Asc]_o$ are not yet understood.

Importantly, Meile and Fillenz discovered that $[Asc]_o$ is also regulated homeostatically *in vivo*, in awake, behaving animals (Miele and Fillenz 1996). In these studies, 100–1,000 μM ascorbate was perfused through a microdialysis probe, with voltammetric detection of $[Asc]_o$ in the tissue adjacent to the probe. Perfusion of ascorbate concentrations lower than 400 μM caused $[Asc]_o$ to fall, whereas higher concentrations caused $[Asc]_o$ to increase. After each perturbation, $[Asc]_o$ recovered within a few minutes, demonstrating efficient homeostasis. The concentration at which there was no net change in $[Asc]_o$ was 400 μM (Miele and Fillenz 1996), consistent with previous estimates for $[Asc]_o$ *in vivo* (Stamford et al. 1984). Homeostatic regulation of $[Asc]_o$ implies that the extracellular compartment of brain tissue is an important site of action for ascorbate.

Although brain $[Asc]_o$ is regulated homeostatically, dynamic, activity-dependent changes in $[Asc]_o$ also occur. Notably, $[Asc]_o$ in rat brain shows marked circadian variation (O'Neill et al. 1983); in these nocturnal animals, average $[Asc]_o$ is 20–60% higher throughout the brain during the dark cycle than in the light. Moreover, variation in $[Asc]_o$ is closely linked with motor behavior, with up to two-fold higher levels in motor striatum when an animal is active versus inactive (O'Neill et al. 1983). Consistent with these findings, the stimulant amphetamine causes an increase in striatal $[Asc]_o$, which is attenuated, albeit not abolished, in anesthetized compared to awake rats (Gonon et al. 1981). This amphetamine-induced increase in $[Asc]_o$ is not a generalized phenomenon, but is specific to motor striatum (for reviews, *see* Grünewald 1993; Rebec and Pierce 1994). The explanation is that activity-dependent increases in striatal $[Asc]_o$, including amphetamine-enhanced motor behavior, is a consequence of activation of glutamatergic input to the striatum, and involves glutamate release and uptake. The primary mechanism thought to underlie increases in $[Asc]_o$ is glutamate/ascorbate heteroexchange, in which ascorbate is released from cells as glutamate is taken up by a stereospecific transporter (O'Neill 1984; Grünewald and Fillenz 1984; Miele et al. 1994). Glutamate-dependent increases in $[Asc]_o$ *in vivo* can be prevented by glutamate uptake inhibitors, but not by glutamate-receptor blockers, further implicating glutamate transporters in the process (Cammack et al. 1991). Consistent with this mechanism, both circadian (O'Neill et al. 1983) and amphetamine-induced (Basse-Tomusk and Rebec 1990) increases in striatal $[Asc]_o$ are attenuated after cortical lesions that remove the cortical glutamatergic input to striatum. The amphetamine-induced response appears to require dopamine release, which then activates a loop that includes substantia nigra, thalamus, cortex and striatum (Rebec and Pierce 1994). Existing evidence implicates neurons (Grünewald and Fillenz 1984), rather than astrocytes (Wilson et al. 2000), as the cellular site of glutamate/ascorbate heteroexchange, but the specific glutamate transporters involved remain unresolved (for review, *see* Rice 2000).

37.5 Compartmentalization of Ascorbate Between Neurons and Glia

As discussed in Sect. 4, ascorbate is homeostatically regulated between the intra- and extracellular compartments of the brain, with additional compartmentalization between neurons and glia. How was the difference between neurons and glia determined? The most obvious approach would be to examine isolated populations of neurons and glia in culture. However, as already discussed, ascorbate is readily lost from mammalian cells *in vitro*. In fact, ascorbate loss from isolated cells is even greater than in brain slices, with ascorbate levels that are below detection limits in cells cultured in ascorbate-free media (Kalir and Mytilenou 1992; Makar et al. 1994; Siushansian and Wilson 1995).

Several lines of evidence from whole-tissue studies, however, pointed to preferential localization of ascorbate in neurons, rather than in glia. Histochemical studies identified ascorbate in the cytosol of neurons in the locus coeruleus, with lower levels in surrounding glia (Shimizu et al. 1960). This was of particular interest, because ascorbate is a co-factor for noradrenaline synthesis; however, later studies showed that there is little correlation between tissue levels of ascorbate and noradrenergic innervation (Milby et al. 1982). On the other hand, studies using HPLC to quantify tissue ascorbate content found that ascorbate levels in neuron-rich cerebral cortex from adult rats was roughly four-fold higher than in the essentially neuron-free optic nerve, consistent with predominant localization of ascorbate in neurons rather than glia (Rice et al. 1995).

Quantitative determination of intracellular ascorbate concentrations in neurons and glia was subsequently made through comparisons of the ascorbate contents of brain samples with known differences in neuron density or neuron-to-glial ratio. The first comparison examined the ascorbate content of samples of cerebral cortex from several mammalian species, based on the previous observation that neuron density varies across species with an inverse dependence on brain size (Tower and Elliott 1952). Consistent with primary neuronal localization, ascorbate content was found to increase linearly with increasing neuron density: human<rabbit<guinea pig<rat<mouse (Rice and Russo-Menna 1998). The y-axis intercept of the resulting plot of ascorbate content versus neuron density was used to estimate ascorbate levels in tissue with zero neuron density, i.e., a pure glial population. With appropriate values for the volume fractions of the intra- and extracellular compartments, an intracellular glial ascorbate concentration was calculated to be about 1 mM (Rice and Russo-Menna 1998). The second comparison examined ascorbate content in developing rat brain, in which regionally distinct patterns of changes in neuron-to-glia ratio occur over the first three postnatal weeks (Altman 1972; Parnavelas et al. 1983). In cerebral cortex, tissue ascorbate content is highest on postnatal day 3 (P3), when cortical tissue is nearly entirely composed of neurons with only few immature glial cells (Rice and Russo-Menna 1998). Over the next 3 weeks, ascorbate levels fall, consistent with the time course of cortical gliogenesis. In the cerebellum, by contrast, ascorbate levels are relatively constant until P9, when they begin to increase markedly, coincident with the onset of cerebellar granule-cell proliferation (Altman 1972); a maximum ascorbate content in developing cerebellum is reached by P15. The tissue ascorbate contents of rat cortex at P3 and cerebellum at P15 represent levels in relatively pure neuronal populations. Again using the appropriate intra- and extracellular volume fractions, intracellular ascorbate concentration in neurons was calculated to be 10 mM (Rice and Russo-Menna 1998). Because this concentration was obtained from two regionally and developmentally distinct neuron populations, high intracellular ascorbate levels appear to be a general characteristic of neurons.

37.6 Ascorbate and GSH

Ascorbate and GSH are the two most abundant low molecular weight antioxidants in the brain. Total tissue contents for each are 2–3 mM (Lyrer et al. 1991; Uemura et al. 1991; Rice et al. 1995), which are 10–20 times higher than that of cysteine and 500–1,000 times higher than that of uric acid (Lyrer et al. 1991; Uemura et al. 1991). Ascorbate and GSH have similarities and differences in their properties, which suggest that they have complementary, but distinct roles in the CNS (Orlowski and Karkowsky 1976; Makar et al. 1994; Rice 2000). GSH is an anionic tripeptide, γ-glutamyl-cysteinyl-glycine, synthesized in all cells in the CNS (Orlowski and Karkowsky 1976; Makar et al. 1994). In contrast to ascorbate, which is more highly concentrated in neurons than glia, as discussed in Sect. 5, GSH is more highly concentrated in glia, with average intracellular GSH levels that are 60% higher in glia (4 mM) than in neurons (2.5 mM) (Rice and Russo-Menna 1998). This distribution is consistent with higher levels of GSH synthesizing enzymes and GSH peroxidase in glia (Slivka et al. 1987; Raps et al. 1989; Makar et al. 1994; Desagher et al. 1996; Trépanier et al. 1996; Peuchen et al. 1997).

An apparently key function of both ascorbate and GSH is their ability to neutralize reactive hydroxyl radicals (•OH), because there are no enzymes analogous to superoxide dismutase or GSH peroxidase to scavenge •OH (Cohen 1994). In addition, ascorbate and GSH can interact as a redox couple (Meister 1994; Winkler et al. 1994); depletion of either ascorbate or GSH can be compensated for by the continued presence of the other. For example, ascorbate administration can prevent tissue damage after inhibition of GSH synthesis (Meister 1994; Avshalumov et al. 2004, 2007), as discussed further below, whereas administration of GSH ester can prevent the onset of scurvy in ascorbate-deficient guinea pigs (Meister 1994). On a cellular level, the distinct ascorbate:GSH ratio in neurons (4:1) and glia (1:4) (Rice and Russo-Menna 1998) might alter the dynamics of ascorbate/GSH interactions, with greater reliance on ascorbate in neurons and GSH in glia.

Brain slice studies again provided novel insight into dynamic consequences of an imbalance between oxidant production and antioxidant protection, i.e., oxidative stress, and how ascorbate and GSH prevent consequent oxidative damage. For example, *in vitro* studies using hippocampal slices from either rat or guinea pig showed a reversible inhibition of the population spike (PS) evoked by Schaffer collateral stimulation during brief exposure to exogenous H_2O_2 (1.0–3.0 mM) (Pellmar 1986, 1987, 1995; Avshalumov et al. 2000; Avshalumov and Rice 2002). In rat hippocampal slices, recovery of the primary PS following H_2O_2 washout is accompanied by mild epileptiform activity, indicated by an additional PS after the primary spike (Avshalumov and Rice 2002). When guinea pig hippocampal slices are exposed to H_2O_2 under identical conditions, the evoked PS is also reversibly depressed during H_2O_2 exposure. However, unlike rat slices, only a single PS is seen in guinea-pig hippocampal slices after H_2O_2 washout, indicating the absence of lasting pathophysiology (Avshalumov et al. 2004).

The higher tolerance of guinea-pig brain than rat brain tissue to H_2O_2 exposure, implies that the higher glia:neuron ratio in this species provides additional antioxidant protection, particularly from glia. The possible importance of GSH-based antioxidant protection in glia, was tested by examining the effect of GSH synthesis or GSH peroxidase inhibition on H_2O_2-induced pathophysiology in guinea-pig brain slices. Both GSH and ascorbate are readily lost from rat and guinea-pig brain slices *in vitro*, with slice contents that are at least 60% lower than those in intact tissue (McIlwain et al. 1956; Rice et al. 1994; Avshalumov et al. 2004). Nonetheless, in slices of guinea-pig hippocampus, an additional small, but significant, decrease in GSH (and ascorbate) content after inhibition of GSH synthesis by buthionine sulfoximine is accompanied by epileptiform activity in guinea-pig slices following H_2O_2 exposure (Avshalumov et al. 2004). The loss of H_2O_2 tolerance in guinea-pig hippocampal slices after GSH synthesis inhibition, despite a relatively small absolute decrease in GSH content, suggests that the protective effect of endogenous GSH is mediated through its role as a co-factor for GSH peroxidase, glutaredoxins, and GSH transferases, as well as by direct recycling of oxidized ascorbate (Meister 1994; Washburn and Wells 1999; Board et al. 2000; Rice 2000; Ehrhart and Zeevalk 2003; Avshalumov et al. 2004).

Consistent with this hypothesis, enhanced H_2O_2-induced pathophysiology in guinea-pig hippocampal slices also occurs when GSH peroxidase is inhibited by mercaptosuccinate (Avshalumov et al. 2004). Although GSH peroxidase inhibition alone has no effect on hippocampal PS amplitude, exposure to H_2O_2 during GSH peroxidase inhibition causes a larger suppression of the evoked PS than in ACSF alone. The recovery of PS amplitude during H_2O_2 washout under these conditions is not only accompanied by secondary H_2O_2-induced pathology, but is also delayed compared to that in ACSF alone. Importantly, inclusion of ascorbate at its normal extracellular concentration *in vivo* (400 μM) in the washout solution prevented these pathophysiological consequences of H_2O_2 exposure after GSH peroxidase inhibition.

Prolonged H_2O_2 exposure as a model of oxidative stress can also cause other pathophysiological effects in brain tissue, including edema (Brahma et al. 2000; Avshalumov et al. 2004). This consequence is also species dependent, however: significant water gain is seen in coronal slices of rat forebrain after 3 h exposure to H_2O_2 (Brahma et al. 2000), however no increase in edema is seen in guinea-pig slices incubated under identical conditions (Avshalumov et al. 2004). Once again, guinea-pig brain tissue behaves like rat tissue when GSH synthesis or GSH peroxidase is inhibited; importantly, H_2O_2-induced edema under these conditions was again prevented when ascorbate is present (Avshalumov et al. 2004). Studies in neuron-glia co-cultures have shown that the glial antioxidant network can protect neurons from H_2O_2 toxicity (Desagher et al. 1996; Drukarch et al. 1997, 1998; Tanaka et al. 1999; Dringen et al. 1999). The results from brain slices show that this is also true in tissue in which neuronal and glial integrity is maintained. Even more significantly, these data demonstrate that ascorbate can compensate for GSH depletion and GSH peroxidase inhibition to prevent pathophysiological consequences of peroxide-induced oxidative stress (Avshalumov et al. 2004).

37.7 Brain Ascorbate: Protective, Yet Permissive

37.7.1 Neuroprotection by Ascorbate

High levels of ascorbate and of SVCT2 in a variety of neuronal types imply a function for neuronal ascorbate beyond its actions as a cell-specific enzyme co-factor. Importantly, the tenfold difference in ascorbate levels between neurons and glia (Rice and Russo-Menna 1998) is consistent with the estimated tenfold higher rate of oxidative metabolism in neurons compared with glial cells (Siesjö 1980). Although ascorbate has been shown to be a pro-oxidant in some studies, albeit mostly *in vitro*, the occurrence of naturally high levels of ascorbate in neuronal cytosol together with other data discussed below argue strongly against normally pro-oxidant actions *in vivo* (Halliwell 1996; Rice 2000). For example, the importance of ascorbate as an intracellular antioxidant is supported by the finding that brain ascorbate (but not GSH) levels in pond turtles are two to three times higher than in mammals (Rice et al. 1995). These diving animals have a remarkable tolerance of hypoxia through a variety of mechanisms (Lutz 1992; Milton and Prentice 2007). Selective elevation of brain ascorbate levels, but not GSH or other antioxidant enzymes (Rice et al. 1995; Milton and Prentice 2007) could represent a specific adaptation to prevent oxidative damage during reoxygenation after a hypoxic dive (Rice et al. 1995; 2002). Interestingly, turtle brain tissue also maintains its ascorbate content *in vitro* (Rice and Cammack 1991), unlike isolated mammalian tissue, suggesting another neuroprotective adaptation. In hypoxia-tolerant turtles, whether normoxic or hypoxic, and in mammals under normal aerobic conditions, ascorbate is recycled as it is used (*see* Sect. 1), which provides on-going antioxidant protection *in vivo*.

When cerebral blood is compromised in mammals, for example, under conditions of ischemic or hemorrhagic stroke, energy failure and anoxic depolarization lead to efflux of ascorbate from cells to the ECF (Hillered et al. 1988). With continued ischemia, tissue levels of ascorbate and other low-molecular-weight antioxidants fall (Lyrer et al. 1991). Consequently, when aerobic metabolism resumes, intracellular stores of these agents are no longer adequate to quench reactive oxygen species (ROS), leaving cells vulnerable to oxidative damage. Increased detection of ·OH during post-ischemic reperfusion (Cao et al. 1988) is one consequence of antioxidant depletion. Can elevated ascorbate protect brain tissue from ischemic injury? The answer appears to be yes. Enhanced brain ascorbate content from dietary supplementation protects cortical mitochondria from *in vivo* ischemia/reperfusion injury in rats (Sciamanna and Lee 1993) and decreases focal ischemia-induced damage in primates (Ranjan et al. 1993; Henry and Chandry 1998). This may be due, at least in part to upregulation of SVCT2 in the region surrounding the infarct core, which should facilitate ascorbate uptake (Berger et al. 2003). Mechanisms of ascorbate protection suggested by *in vitro* studies include prevention of mitochondrial hyperoxidation and tissue edema formation during post-ischemic reoxygenation in brain slices (Pérez-Pinzón et al. 1997; MacGregor et al. 2003), as well as amelioration of excitoxicity from excessive glutamate receptors activation in cultured cells (for review, *see* Rice 2000). Notably, although transport of oxidized ascorbate, DHA, via GLUT1 is normally a minor route of ascorbate entry into the CNS, this pathway provides a much more efficient means of elevating brain ascorbate than transport via SVCT2 at the choroid plexus, which is already saturated at normal plasma ascorbate concentrations (Sect. 3). Indeed, administration of DHA provides potent neuroprotection *in vivo* in rodent stroke models, including decreased lipid peroxidation and a significant decrease in infarct volume (Huang et al. 2001; Mack et al. 2006).

Additional evidence for ascorbate neuroprotection via its properties as an antioxidant and ·OH scavenger comes from brain slice studies in which exogenous H_2O_2 is used to induce oxidative stress. Strikingly, H_2O_2 exposure, which normally leads to suppression of the evoked PS in hippocampal slices (Sect. 6) has no effect on the PS when applied in the presence of 400 μM ascorbate (Avshalumov et al. 2000). Similar prevention of PS suppression is provided by deferoxamine, a metal ion chelator that prevents iron or copper-catalyzed ·OH formation from H_2O_2, indicating that the effect of H_2O_2 on the PS is mediated by ·OH and that protection by ascorbate under these conditions is by ·OH scavenging (Avshalumov et al. 2000).

H_2O_2-dependent PS suppression is also prevented by isoascorbate (D-ascorbate) (Avshalumov et al. 2000), the non-biologically active stereoisomer of L-ascorbate that is not transported by stereoselective SVCT2 (Tsukaguchi et al. 1999), indicating an extracellular site of antioxidant action. As discussed in Sect. 6, ascorbate can also prevent long-lasting pathophysiological consequences of H_2O_2 exposure, indicated by the absence of epileptiform activity when ascorbate is present only during H_2O_2 washout in rat hippocampal slices (Avshalumov and Rice 2002) or in guinea-pig hippocampal slices in which the antioxidant network is compromised (Avshalumov et al. 2004). This secondary protection by ascorbate must occur intracellularly in neurons, however, because non-transported isoascorbate is not protective under these conditions (Avshalumov and Rice 2002). Taken together, these findings strongly support a neuroprotective role for both intra- and extracellular ascorbate.

A number of other actions of ascorbate, from alterations in neurotransmitter binding affinity to inhibition of Na^+/K^+-ATPase activity, have been reported in the literature. Because most of these studies were conducted *in vitro*, often in isolated cells or cell membranes that lack an intact antioxidant network, the potential for pro-oxidant effects of ascorbate is high, which casts doubt on the physiological relevance of some of the findings. This does not mean that ascorbate never acts as a pro-oxidant *in vivo*. Indeed, under pathological conditions such as ischemia, when ascorbate compartmentalization is disrupted, these actions could occur and contribute to CNS injury, as discussed above. Under normal conditions, however, ascorbate is neuroprotective.

37.7.2 Reactive Oxygen Species (ROS) as Signaling Molecules in the Brain

Demonstration of neuroprotection by ascorbate and other antioxidants is consistent with the traditional view that ROS are harmful agents and must be tightly controlled. This view is further supported by evidence of oxidative damage as a contributing factor to cell death following acute brain injury, including cerebral ischemia-reperfusion (Cao et al. 1988; Hyslop et al. 1995; Chan 2004; Starkov et al. 2004), as well as in slowly progressing neurodegenerative disorders like Parkinson disease (Cohen 1994; Ebadi et al. 1996; Olanow and Tatton 1999; Zhang et al. 2000; Greenamyre et al. 2001; Xu et al. 2002; Orth and Schapira 2002; Wood-Kaczmar et al. 2006; Fukae et al. 2007; Dodson and Guo 2007). This traditional perspective is evolving rapidly, however, in light of increasing evidence that ROS also act as cellular messengers that modulate processes from short-term ion-channel activation (Seutin et al. 1995; Krippeit-Drews et al. 1999; Avshalumov and Rice 2003; Tang et al. 2004; Avshalumov et al. 2005; Bao et al. 2005; Hidalgo et al. 2005) to gene transcription and cell proliferation (Suzuki et al. 1997; Sauer et al. 2001; Haddad 2002; Esposito et al. 2004; Rhee et al. 2005; Rhee 2006; Infanger et al. 2006; Bedard and Krause 2007).

Of the ROS generated by various cellular processes, H_2O_2 is a particularly strong candidate as a signaling agent because it is membrane permeable, possibly facilitated by aquaporins (Bienert et al. 2006, 2007). H_2O_2 is also relatively inert; in contrast to other ROS that are reactive free radicals, including superoxide and $^{\bullet}OH$, H_2O_2 is a mild oxidant (Cohen 1994; Rhee 2006). As discussed in Sects. 1 and 6, ascorbate works in concert with other components of the antioxidant network. The emerging role for H_2O_2 and other ROS as signaling agents, however, requires that oxidant regulation must be more subtle than previously thought.

The first evidence suggesting that H_2O_2 might regulate neurotransmission was the finding that exogenous H_2O_2 could suppress the evoked PS in hippocampal slices, possibly by inhibiting transmitter release (Pellmar 1986, 1987). The hypothesis that transmitter release can be inhibited by exogenous H_2O_2 was confirmed by the demonstration that evoked dopamine (DA) release in striatal slices is reversibly suppressed during brief H_2O_2 exposure (Chen et al. 2001). More significantly, subsequent studies showed that *endogenously generated* H_2O_2 inhibits axonal DA release in the striatum (Chen et al. 2001, 2002; Avshalumov et al. 2003; Avshalumov and Rice 2003) and suppresses the activity of dopaminergic neurons in the substantia nigra pars compacta (SNc) (Avshalumov et al. 2005). These studies were the first to show that endogenous H_2O_2 is an activity-dependent neuromodulator. Regulation of the nigrostriatal DA system is important because of the central role this pathway plays in the control of movement by the basal ganglia, with loss of dopaminergic neurons in the SNc leading to the motor deficits of Parkinson disease (Albin et al. 1989). In dopaminergic neurons, intracellular H_2O_2 modulates cell excitability via ATP-sensitive K^+ (K_{ATP}) channels (Avshalumov et al. 2005). By contrast, modulatory H_2O_2 in the striatum is generated downstream from the activation of glutamatergic AMPA receptors (Avshalumov et al. 2003), which are not found on dopaminergic axons (Chen et al. 1998; Bernard and Bolam 1998). This implies that modulatory H_2O_2 is not generated within dopaminergic axons, but rather must be produced postsynaptically in striatal neurons, then diffuse from these sites to modulate presynaptic DA release. Recent studies have confirmed this mode of intercellular communication by demonstrating AMPA-receptor dependent H_2O_2 generation in medium spiny neurons (MSNs) (Avshalumov et al. 2008), the principal output neurons of the striatum (Kemp and Powell 1971), but not dopaminergic axons.

Thus, H_2O_2 serves different functions in the striatum than in SNc. In striatum, diffusible H_2O_2 mediates regulation of axonal DA release by glutamate and GABA. When glutamatergic striatal afferents are activated, this has no direct effect on DA release, as there are no presynaptic glutamate receptors on dopaminergic axons, as already noted. However, this excitatory input leads to postsynaptic H_2O_2 generation in MSNs, which leads to inhibition of DA release. Thus, diffusible H_2O_2 mediates the effect of glutamate on DA release in the striatum. Local release of GABA in the striatum apparently opposes this process by suppressing the excitability of MSNs, resulting in *less* H_2O_2 generation and thus *enhancing* DA release. Overall, therefore, in striatum, H_2O_2-dependent inhibition via K_{ATP} channel activation on dopaminergic axons *reverses* the conventional effects of glutamate and GABA, with *inhibition* of

DA release following AMPA receptor activation and *enhancement* of DA release following $GABA_A$ receptor activation (Avshalumov et al. 2003). By contrast, spontaneous pacemaker activity in dopaminergic neurons of the SNc produces H_2O_2 that serves as an autoregulatory signal that leads to a decrease in cell firing rate, again via K_{ATP} channel activation (Avshalumov et al. 2005). Although the mechanism by which H_2O_2 activates K_{ATP} channels is not yet clear, this process circumvents the need for a catastrophic fall in cell ATP levels to activate K_{ATP} channels, thereby providing a subtle, but effective way to link metabolism to excitability.

Importantly, neuromodulation by H_2O_2 is not limited to the nigrostriatal system. Other studies have shown that H_2O_2 influences characteristics of LTP in the hippocampus (Auerbach and Segal 1997; Kamsler and Segal 2003, 2004), which has implications for memory formation. Diffusible H_2O_2 also plays a role in neuron-glia signaling in the hippocampus, in which neuronal activation leads to H_2O_2-dependent phosphorylation of myelin basic protein in oligodendrocytes (Atkins and Sweatt 1999). Overall, these findings indicate that H_2O_2 can act both as an intracellular signaling agent and as a diffusible messenger in the brain.

37.7.3 Ascorbate Is a Unique Neuronal Antioxidant

For H_2O_2 to act at both intracellular and potentially distant targets requires that the brain antioxidant network allows functional levels of H_2O_2 and other ROS, while still preventing oxidative stress, which adds a previously unrecognized dimension to oxidant management. Key features of such a permissive environment are the predominance of ascorbate in neurons and the predominance of cytosolic GSH and GSH peroxidase activity in glia (Cohen 1994; Rice and Russo-Menna 1998; Rice 2000; Avshalumov et al. 2004).

In striking contrast to the efficacy of ascorbate in preventing pathophysiological consequences of exogenous H_2O_2 exposure (Avshalumov et al. 2000, 2004; Avshalumov and Rice 2002), this antioxidant and ˙OH scavenger has *no effect* on dynamic modulation of axonal DA release by endogenous H_2O_2 generated downstream from AMPA receptor activation in the striatum (Avshalumov et al. 2003). This finding indicates that inhibition of DA release is a direct action of H_2O_2, rather than ˙OH. Equally important, however, it shows that ascorbate does not interfere with H_2O_2-dependent signaling in striatum. The lack of interference is consistent with *in vitro* studies showing little redox reactivity between ascorbate and H_2O_2: ascorbate, at 400 μM, is surprisingly stable in the presence of mM levels of H_2O_2, as is H_2O_2 in the presence of ascorbate, when precautions are taken to minimize the presence of trace metal ions (Avshalumov et al. 2000). By contrast, GSH is rapidly oxidized under similar conditions in the presence of H_2O_2 (Avshalumov et al. 2000). Thus, ascorbate is ideally suited as a key neuronal antioxidant because of its unique redox potential, which gives it the dual ability to permit H_2O_2 signaling, yet prevent pathological consequences that could occur from dysregulated H_2O_2 generation and ·OH production.

37.8 Conclusions and Future Directions

The high intracellular concentration of ascorbate in neurons suggests that it has a significant role in normal neuronal physiology. Given the well-established characteristics of ascorbate as an electron donor and free radical scavenger, it is likely that its primary role comes from neuroprotective actions as a component of the neuronal antioxidant network. In addition, ascorbate is one of the few antioxidants found in significant concentration in ECF. The fact that $[Asc]_o$ is regulated homeostatically, but modulated by glutamatergic activity suggests that the extracellular compartment is also an important site for ascorbate neuroprotection. Ongoing research on ascorbate and glutamate transporters should lay the groundwork for future understanding of ascorbate regulation in the brain. As new tools to manipulate ascorbate levels become available, new insights into its specific functions in the brain should also emerge.

It is also relevant to note that regulation of striatal DA release by endogenous H_2O_2 is similar in guinea-pig and rat, as well as in non-human primate (Rice et al. 2002). This not only demonstrates that H_2O_2 signaling is not impaired by the presence of the stronger antioxidant network afforded by the higher glia:neuron ratio of guinea-pig or primate versus rat brain, but also supports the hypothesis that precise neuron-glial antioxidant compartmentalization, regardless of species, is required for neuronal signaling, as well as neuronal protection. A key feature of this compartmentalization is the predominance of ascorbate in neurons, which is permissive for H_2O_2 signaling because it does not alter either intra- or extracellular concentrations of H_2O_2 or hinder the escape of diffusible H_2O_2 from the cells in which it is generated. On the other hand, homeostatically regulated levels of ascorbate in neurons, glia, and in ECF provides protection against $^{\bullet}OH$, as needed, in both the intra- and extracellular compartments of the CNS.

Acknowledgments The author gratefully acknowledges support from NIH/NINDS grant NS036362, the National Parkinson Foundation, and the NYU Ricciardi Research Fund, as well as constructive comments on this chapter from Dr. Jyoti C. Patel.

References

Agus DB, Gambhir SS, Pardridge WM, Spielholz C, Baselga J, Vera JC, Golde DW (1997) Vitamin C crosses the blood-brain barrier in the oxidized form through the glucose transporters. J Clin Invest 100:2842–2848

Albin RL, Young AB, Penney JB (1989) The functional anatomy of basal ganglia disorders. Trends Neurosci 12:366–375

Altman J (1972) Postnatal development of the cerebellar cortex of the rat: III. Maturation of the components of the granular layer. J Comp Neurol 145:465–514

Astuya A, Caprile T, Castro M, Salazar K, García Mde L, Reinicke K, Rodríguez F, Vera JC, Millán C, Ulloa V, Low M, Martínez F, Nualart F (2005) Vitamin C uptake and recycling among normal and tumor cells from the central nervous system. J Neurosci Res 79:146–156

Atkins CM, Sweatt JD (1999) Reactive oxygen species mediate activity-dependent neuron-glia signaling in output fibers of the hippocampus. J Neurosci 19:7241–7248

Auerbach JM, Segal M (1997) Peroxide modulation of slow onset potentiation in rat hippocampus. J Neurosci 17:8695–8701

Avshalumov MV, Rice ME (2002) NMDA-receptor activation mediates hydrogen peroxide-induced pathophysiology in rat hippocampal slices. J Neurophysiol 87:2896–2903

Avshalumov MV, Rice ME (2003) Activation of ATP-sensitive K^+ (K_{ATP}) channels by H_2O_2 underlies glutamate-dependent inhibition of striatal dopamine release. Proc Natl Acad Sci USA 100:11729–11734

Avshalumov MV, Chen BT, Rice ME (2000) Mechanisms underlying H_2O_2-mediated inhibition of synaptic transmission in rat hippocampal slices. Brain Res 882:86–94

Avshalumov MV, Chen BT, Marshall SP, Peña DM, Rice ME (2003) Glutamate-dependent inhibition of dopamine release in striatum is mediated by a new diffusible messenger, H_2O_2. J Neurosci 23:2744–2750

Avshalumov MV, MacGregor DG, Sehgal LM, Rice ME (2004) The glial antioxidant network and neuronal ascorbate: permissive yet protective for H_2O_2 signaling? Neuron Glia Biol 1:365–376

Avshalumov MV, Chen BT, Kóos T, Tepper JM, Rice ME (2005) Endogenous hydrogen peroxide regulates the excitability of midbrain dopamine neurons via ATP-sensitive potassium channels. J Neurosci 25:4222–4231

Avshalumov MV, Bao L, Patel JC, Rice ME (2007) H_2O_2 signaling in the nigrostriatal dopamine pathway via ATP-sensitive potassium channels: issues and answers. Antioxid Redox Signal 9:219–231

Avshalumov MV, Patel JC, Rice ME (2008) AMPA receptor-dependent H_2O_2 generation in striatal spiny neurons, but not dopamine axons: one source of a retrograde signal that can inhibit dopamine release. J Neurophysiol 100:1590–1601

Bao L, Avshalumov MV, Rice ME (2005) Mitochondrial inhibition causes functional dopamine denervation and striatal medium spiny neurons depolarization via increased H_2O_2, not decreased ATP. J Neurosci 25:10029–10040

Barnes MJ (1975) Function of ascorbic acid in collagen metabolism. Ann N Y Acad Sci 258:264–277

Basse-Tomusk A, Rebec GV (1990) Corticostriatal and thalamic regulation of amphetamine-induced ascorbate release in the neostriatum. Pharmacol Biochem Behav 35:55–60

Bedard K, Krause KH (2007) The NOX family of ROS-generating NADPH oxidases: physiology and pathophysiology. Physiol Rev 87:245–313

Berger UV, Hediger MA (2000) The vitamin C transporter SVCT2 is expressed by astrocytes in culture but not *in situ*. NeuroReport 11:1395–1399

Berger UV, Lu XC, Liu W, Tang Z, Slusher BS, Hediger MA (2003) Effect of middle cerebral artery occlusion on mRNA expression for the sodium-coupled vitamin C transporter SVCT2 in rat brain. J Neurochem 86:896–906

Bernard V, Bolam JP (1998) Subcellular and subsynaptic distribution of the NR1 subunit of the NMDA receptor in the neostriatum and globus pallidus of the rat: colocalization at synapses with the GluR2/3 subunit of the AMPA receptor. Eur J Neurosci 10:3721–3738

Bienert GP, Schjoerring JK, Jahn TP (2006) Membrane transport of hydrogen peroxide. Biochim Biophys Acta 1758:994–1003

Bienert GP, Møller AL, Kristiansen KA, Schulz A, Møller IM, Schjoerring JK, Jahn TP (2007) Specific aquaporins facilitate the diffusion of hydrogen peroxide across membranes. J Biol Chem 282:1183–1192

Board PG, Coggan M, Chelvanayagam G, Easteal S, Jermiin LS, Schulte GK, Danley DE, Hoth LR, Griffor MC, Kamath AV, Rosner MH, Chrunyk BA, Perregaux DE, Gabel CA, Geoghegan KF, Pandit J (2000) Identification, characterization, and crystal structure of the Omega class glutathione transferases. J Biol Chem 275:24798–24806

Bodannes RS, Chan PC (1979) Ascorbic acid as a scavenger of singlet oxygen. FEBS Lett 105:195–196

Brahma B, Forman RE, Stewart EE, Nicholson C, Rice ME (2000) Ascorbate inhibits edema in brain slices. J Neurochem 74:1263–1270

Buettner GR (1993) The pecking order of free radicals and antioxidants: lipid peroxidation, α-tocopherol, and ascorbate. Arch Biochem Biophys 300:535–543

Buettner GR, Jurkiewicz BA (1993) Ascorbate free radical as a marker of oxidative stress: an EPR study. Free Rad Biol Med 14:49–55

Cammack J, Ghasemzadeh B, Adams RN (1991) The pharmacological profile of glutamate-evoked ascorbic acid efflux measured by *in vivo* voltammetry. Brain Res 565:17–22

Cao W, Carney JM, Duchon A, Floyd RA, Chevion M (1988) Oxygen free radical involvement in ischemia and reperfusion injury to brain. Neurosci Lett 88:233–238

Carey DJ, Todd MS (1987) Schwann cell myelination in a chemically defined medium: demonstration of a requirement for additives that promote Schwann cell extracellular matrix formation. Brain Res 429:95–102

Carpenter KJ (1986) The history of scurvy and vitamin C. Cambridge University Press, New York

Castro M, Caprile T, Astuya A, Millán C, Reinicke K, Vera JC, Vásquez O, Aguayo LG, Nualart F (2001) High-affinity sodium-vitamin C co-transporters (SVCT) expression in embryonic mouse neurons. J Neurochem 78:815–823

Chan PH (2004) Mitochondria and neuronal death/survival signaling pathways in cerebral ischemia. Neurochem Res 29:1943–1949

Chatterjee IB, Majumbar AK, Nandi BK, Subramanian N (1975) Synthesis and some major functions of vitamin C in animals. Ann N Y Acad Sci 258:24–47

Chen Q, Veenman L, Knopp K, Yan Z, Medina L, Song WJ, Surmeier DJ, Reiner A (1998) Evidence for the preferential localization of glutamate receptor-1 subunits of AMPA receptors to the dendritic spines of medium spiny neurons in rat striatum. Neuroscience 83:749–761

Chen BT, Avshalumov MV, Rice ME (2001) H_2O_2 is a novel, endogenous modulator of synaptic dopamine release. J Neurophysiol 85:2468–2476

Chen BT, Avshalumov MV, Rice ME (2002) Modulation of somatodendritic dopamine release by endogenous H_2O_2: susceptibility in substantia nigra but resistance in the ventral tegmental area. J Neurophysiol 87:1155–1158

Cohen G (1994) Enzymatic/nonenzymatic sources of oxyradicals and regulation of antioxidant defenses. Ann N Y Acad Sci 738:8–14

Daruwala R, Song J, Koh WS, Rumsey SC, Levine M (1999) Cloning and functional characterization of the human sodium-dependent vitamin C transporters hSVCT1 and hSVCT2. FEBS Lett 460:480–484

Davies MB, Austin J, Partridge DA (1991) Vitamin C: its chemistry and biochemistry. Royal Society of Chemistry, Cambridge, UK

Desagher S, Glowinski J, Premont J (1996) Astrocytes protect neurons from hydrogen peroxide toxicity. J Neurosci 16:2553–2562

Dhariwal KR, Hartzell WO, Levine M (1991) Ascorbic acid and dehydroascorbic acid measurements in human plasma and serum. Am J Clin Nutr 54:712–716

Diliberto EJ Jr, Menniti FS, Knoth J, Daniels AJ, Kizer JS, Viveros OH (1987) Adrenomedullary chromaffin cells as a model to study the neurobiology of ascorbic acid: from monooxygenation to neuromodulation. Ann N Y Acad Sci 498:28–53

Dodson MW, Guo M (2007) Pink1, Parkin, DJ-1 and mitochondrial dysfunction in Parkinson's disease. Curr Opin Neurobiol 17:331–337

Drew KL, Osborne PG, Frerichs KU, Hu Y, Koren R, Hallenbeck JM, Rice ME (1999) Ascorbate and glutathione regulation in hibernating ground squirrels. Brain Res 851:1–8

Dringen R, Kussmaul L, Gutterer JM, Hirrlinger J, Hamprecht B (1999) The glutathione system of peroxide detoxification is less efficient in neurons than in astroglial cells. J Neurochem 72:2523–2530

Drukarch B, Schepens E, Jongenelen CA, Stoof JC, Langeveld CH (1997) Astrocyte-mediated enhancement of neuronal survival is abolished by glutathione deficiency. Brain Res 770:123–130

Drukarch B, Schepens E, Stoof JC, Langeveld CH, Van Muiswinkel FL (1998) Astrocyte-enhanced neuronal survival is mediated by scavenging of extracellular reactive oxygen species. Free Rad Biol Med 25:217–220

Ebadi M, Srinivasan SK, Baxi MD (1996) Oxidative stress and antioxidant therapy in Parkinson's disease. Prog Neurobiol 48:1–19

Ehrhart J, Zeevalk GD (2003) Cooperative interaction between ascorbate and glutathione during mitochondrial impairment in mesencephalic cultures. J Neurochem 86:1487–1497

Eldridge CF, Bunge MB, Bunge RP, Wood PM (1987) Differentiation of axon-related Schwann cells *in vitro*. I. Ascorbic acid regulates basal lamina assembly and myelin formation. J Cell Biol 105:1023–1034

Esposito F, Ammendola R, Faraonio R, Russo T, Cimino F (2004) Redox control of signal transduction, gene expression and cellular senescence. Neurochem Res 29:617–628

Fornai F, Saviozzi M, Piaggi S, Gesi M, Corsini GU, Malvaldi G, Casini AF (1999) Localization of a glutathione-dependent dehydroascorbate reductase within the central nervous system of the rat. Neuroscience 94:937–948

Fornai F, Piaggi S, Gesi M, Saviozzi M, Lenzi P, Paparelli A, Casini AF (2001) Subcellular localization of a glutathione-dependent dehydroascorbate reductase within specific rat brain regions. Neuroscience 104:15–31

Fukae J, Mizuno Y, Hattori N (2007) Mitochondrial dysfunction in Parkinson's disease. Mitochondrion 7:58–62

Gonon F, Buda M, Cespuglio R, Jouvet M, Pujol JF (1981) Voltammetry in the striatum of chronic freely moving rats: detection of catechols and ascorbic acid. Brain Res 223:69–80

Greenamyre JT, Sherer TB, Betarbet R, Panov AV (2001) Complex I and Parkinson's disease. IUBMB Life 52:135–141

Grünewald RA (1993) Ascorbic acid in the brain. Brain Res Rev 18:123–133

Grünewald RA, Fillenz M (1984) Release of ascorbate from a synaptosomal fraction of rat brain. Neurochem Int 6:491–500

Haddad JJ (2002) Antioxidant and prooxidant mechanisms in the regulation of redox(y)-sensitive transcription factors. Cellular Signal 14:879–897

Halliwell B (1996) Vitamin C: antioxidant or pro-oxidant *in vivo*? Free Rad Res 25:439–454

Henry PT, Chandry MJ (1998) Effect of ascorbic acid on infarct size in experimental focal cerebral ischemia and reperfusion in a primate model. Acta Neurochir (Wien) 140:977–980

Hidalgo C, Donoso P, Carrasco MA (2005) The ryanodine receptors Ca^{2+} release channels: cellular redox sensors? IUBMB Life 57:315–322

Hillered L, Persson L, Bolander HG, Hallstrom A, Ungerstedt U (1988) Increased extracellular levels of ascorbate in striatum after middle cerebral artery occlusion in the rat monitored by intracerebral microdialysis. Neurosci Lett 95:286–290

Hornig D (1975) Distribution of ascorbic acid, metabolites and analogues in man and animals. Ann N Y Acad Sci 258:103–117

Huang J, Agus DB, Winfree CJ, Kiss S, Mack WJ, McTaggart RA, Choudhri TF, Kim LJ, Mocco J, Pinsky DJ, Fox WD, Israel RJ, Boyd TA, Golde DW, Connolly ES Jr (2001) Dehydroascorbic acid, a blood-brain barrier transportable form of vitamin C, mediates potent cerebroprotection in experimental stroke. Proc Natl Acad Sci USA 98:11720–11724

Hughes RE, Hurley RJ, Jones PR (1971) The retention of ascorbic acid by guinea-pig tissues. Brit J Nutr 26:433–438

Hyslop PA, Zhang Z, Pearson DV, Phebus LA (1995) Measurement of striatal H_2O_2 by microdialysis following global forebrain ischemia and reperfusion in the rat: correlation with the cytotoxic potential of H_2O_2 *in vitro*. Brain Res 671:181–186

Ichimura T, Fraser PA, Cserr HF (1991) Distribution of extracellular tracers in perivascular spaces of the rat brain. Brain Res 545:103–113

Infanger DW, Sharma RV, Davisson RL (2006) NADPH oxidases of the brain: distribution, regulation, and function. Antioxid Redox Signal 8:1583–1596

Kalir HH, Mytilenou C (1992) Ascorbic acid in mesencephalic cultures: effects on dopaminergic neuron development. J Neurochem 57:458–464

Kallner A, Hartmann D, Hornig D (1977) On the absorption of ascorbic acid in man. Int J Vitam Nutr Res 47:383–388

Kamsler A, Segal M (2003) Hydrogen peroxide modulation of synaptic plasticity. J Neurosci 23:269–276

Kamsler A, Segal M (2004) Hydrogen peroxide as a diffusible signal molecule in synaptic plasticity. Mol Neurobiol 29:167–178

Kemp JM, Powell TP (1971) The structure of the caudate nucleus of the cat: light and electron microscopy. Philos Trans R Soc Lond B Biol Sci 262:383–401

Krippeit-Drews P, Kramer C, Welker S, Lang F, Ammon HP, Drews G (1999) Interference of H_2O_2 with stimulus-secretion coupling in mouse pancreatic beta-cells. J Physiol (Lond) 514:471–481

Kume-Kick J, Rice ME (1998) Estrogen-dependent modulation of rat brain ascorbate levels and ischemia-induced ascorbate loss. Brain Res 803:105–113

Lam DKC, Daniel PM (1986) The influx of ascorbic acid into the rat's brain. Quart J Exp Physiol 71:483–489

Lutz PL (1992) Mechanisms for anoxic survival in the vertebrate brain. Ann Rev Physiol 54: 601–618

Lyrer P, Landolt H, Kabiersch A, Langemann H, Kaeser H (1991) Levels of low molecular weight scavengers in the rat brain during focal ischemia. Brain Res 567:317–320

MacGregor DG, Avshalumov MV, Rice ME (2003) Brain edema induced by *in vitro* ischemia: causal factors and neuroprotection. J Neurochem 85:1402–1411

Machlin LJ, Bendich A (1987) Free radical tissue damage: protective role of antioxidant nutrients. FASEB J 1:441–445

Mack WJ, Mocco J, Ducruet AF, Laufer I, King RG, Zhang Y, Guo W, Pinsky DJ, Connolly ES Jr (2006) A cerebroprotective dose of intravenous citrate/sorbitol-stabilized dehydroascorbic acid is correlated with increased cerebral ascorbic acid and inhibited lipid peroxidation after murine reperfused stroke. Neurosurgery 59:383–388

Makar TK, Nedergaard M, Preuss A, Gebard AS, Perumal AS, Cooper AJL (1994) Vitamin E, ascorbate, glutathione, glutathione disulfide, and enzymes of oxidative metabolism in cultures of chick astrocytes and neurons: evidence that astrocytes play an important role in oxidative processes in the brain. J Neurochem 62:45–53

May JM, Mendiratta S, Hill KE, Burk RF (1997) Reduction of dehydroascorbate to ascorbate by the selenoenzyme thioredoxin reductase. J Biol Chem 272:22607–22610

McIlwain H, Thomas J, Bell JL (1956) The composition of isolated cerebral tissues: ascorbic acid and cozymase. Biochem J 64:332–335

Meister A (1994) Glutathione-ascorbic acid antioxidant system in animals. J Biol Chem 269: 9397–9400

Miele M, Fillenz M (1996) *In vivo* determination of extracellular brain ascorbate. J Neurosci Methods 70:15–19

Miele M, Boutelle MG, Fillenz M (1994) The physiologically induced release of ascorbate in rat brain is dependent on impulse traffic, calcium influx and glutamate uptake. Neuroscience 62:87–91

Milby K, Oke A, Adams RN (1982) Detailed mapping of ascorbate distribution in rat brain. Neurosci Lett 28:15–20

Milton SL, Prentice HM (2007) Beyond anoxia: the physiology of metabolic downregulation and recovery in the anoxia-tolerant turtle. Comp Biochem Physiol A Mol Integr Physiol 147:277–290

Niki E (1991) Action of ascorbic acid as a scavenger of active and stable oxygen radicals. Am J Clin Nutr 54:1119S–11124S

Nishikimi M (1975) Oxidation of ascorbic acid with superoxide anion generated by the xanthine-xanthine oxidase system. BBRC 63:463–468

Nishikimi M, Fukuyama R, Minoshima S, Shimizu N, Yagi K (1994) Cloning and chromosomal mapping of the human nonfunctional gene for L-gulono-γ-lactone oxidase, the enzyme for L-ascorbic acid biosynthesis missing in man. J Biol Chem 269:13685–13688

O'Neill RD (1984) Voltammetrically monitored brain ascorbate as an index of excitatory amino acid releases in the unrestrained rat. Neurosci Lett 52:227–233

O'Neill RD, Grunewald RA, Fillenz M, Albery WJ (1983) The effect of unilateral cortical lesions on the circadian changes in rat striatal ascorbate and homovanillic acid levels measured in vivo using voltammetry. Neurosci Lett 42:105–110

Olanow CW, Tatton WG (1999) Etiology and pathogenesis of Parkinson's disease. Annu Rev Neurosci 22:123–144

Orlowski M, Karkowsky A (1976) Glutathione metabolism and some possible functions of glutathione in the nervous system. Int Rev Neurobiol 19:75–121

Orth M, Schapira AH (2002) Mitochondrial involvement in Parkinson's disease. Neurochem Int 40:533–541

Pagé EL, Chan DA, Giaccia AJ, Levine M, Richard DE (2008) Hypoxia-inducible factor-1alpha stabilization in nonhypoxic conditions: role of oxidation and intracellular ascorbate depletion. Mol Biol Cell 19:86–94

Parnavelas JG, Luder R, Pollard SG, Sullivan K, Lieberman AR (1983) A quantitative and qualitative ultrastructural study of glial cells in the developing visual cortex of the rat. Phil Trans R Soc (Lond) 301:55–84

Passage E, Norreel JC, Noack-Fraissignes P, Sanguedolce V, Pizant J, Thirion X, Robaglia-Schlupp A, Pellissier JF, Fontés M (2004) Ascorbic acid treatment corrects the phenotype of a mouse model of Charcot-Marie-Tooth disease. Nat Med 10:396–401

Pellmar T (1986) Electrophysiological correlates of peroxide damage in guinea pig hippocampus *in vitro*. Brain Res 364:377–381

Pellmar TC (1987) Peroxide alters neuronal excitability in the CA1 region of guinea-pig hippocampus *in vitro*. Neuroscience 23:447–456

Pellmar TC (1995) Use of brain slices in the study of free-radical actions. J Neurosci Methods 59:93–98

Pérez-Pinzón MA, Mumford PL, Rosenthal M, Sick TJ (1997) Antioxidants, mitochondrial hyperoxidation and electrical recovery after anoxia in hippocampal slices. Brain Res 754: 163–170

Peuchen S, Bolanos JP, Heales SJ, Almeida A, Duchen MR, Clark JB (1997) Interrelationships between astrocyte function, oxidative stress and antioxidant status within the central nervous system. Prog Neurobiol 52:261–281

Qiao H, May JM (2008) Development of ascorbate transporters in brain cortical capillary endothelial cells in culture. Brain Res 1208:79–86

Qiu S, Li L, Weeber EJ, May JM (2007) Ascorbate transport by primary cultured neurons and its role in neuronal function and protection against excitotoxicity. J Neurosci Res 85: 1046–1056

Rajan DP, Huang W, Dutta B, Devoe LD, Leibach FH, Ganapathy V, Prasad PD (1999) Human placental sodium-dependent vitamin C transporter (SVCT2): molecular cloning and transport function. Biochem Biophys Res Comm 262:762–768

Ranjan A, Theodore D, Haran RP, Chandy MJ (1993) Ascorbic acid and focal ischaemia in a primate model. Acta Neurochir 123:87–91

Raps SP, Lai JCK, Hertz L, Cooper AJL (1989) Glutathione is present in high concentrations in cultured astrocytes but not in cultured neurons. Brain Res 493:398–401

Rebec GV, Pierce RC (1994) A vitamin as neuromodulator: Ascorbate release into the extracellular fluid of the brain regulates dopaminergic and glutamatergic transmission. Prog Neurobiol 43:537–565

Rhee SG (2006) H_2O_2, a necessary evil for cell signaling. Science 312:1882–1883

Rhee SG, Kang SW, Jeong W, Chang TS, Yang KS, Woo HA (2005) Intracellular messenger function of hydrogen peroxide and its regulation by peroxiredoxins. Curr Opin Cell Biol 17: 183–189

Rice ME (2000) Ascorbate regulation and its neuroprotective role in the brain. Trends Neurosci 23:209–216

Rice ME, Cammack J (1991) Anoxia-resistant turtle brain maintains ascorbic acid content *in vitro*. Neurosci Lett 132:141–145

Rice ME, Russo-Menna I (1998) Differential compartmentalization of brain ascorbate and glutathione between neurons and glia. Neuroscience 82:1213–1223

Rice ME, Pérez-Pinzón MA, Lee EJK (1994) Ascorbic acid, but not glutathione, is taken up by brain slices and preserves cell morphology. J Neurophysiol 71:1591–1596

Rice ME, Lee EJK, Choy Y (1995) High levels of ascorbic acid, not glutathione, in the CNS of anoxia-tolerant reptiles contrasted with levels in anoxia-intolerant species. J Neurochem 64:1790–1799

Rice ME, Forman RE, Chen BT, Avshalumov MV, Cragg SJ, Drew KL (2002) Brain antioxidant regulation in mammals and anoxia-tolerant reptiles: balanced for neuroprotection and neuromodulation. Comp Biochem Physiol (Part C) 133:515–525

Rose RC (1993) Cerebral metabolism of oxidized ascorbate. Brain Res 628:49–55

Sato K, Saito H, Katsuki H (1993) Synergism of tocopherol and ascorbate on the survival of cultured neurons. NeuroReport 4:1179–1182

Sauer H, Wartenberg M, Hescheler J (2001) Reactive oxygen species as intracellular messengers during cell growth and differentiation. Cell Physiol Biochem 11:173–186

Savini I, Rossi A, Pierro C, Avigliano L, Catani MV (2008) SVCT1 and SVCT2: key proteins for vitamin C uptake. Amino Acids 34:347–355

Schenk JO, Miller E, Gaddis R, Adams RN (1982) Homeostatic control of ascorbate concentration in CNS extracellular fluid. Brain Res 253:353–356

Sciamanna MA, Lee CP (1993) Ischemia/reperfusion-induced injury of forebrain mitochondria and protection by ascorbate. Arch Biochem Biophys 305:215–224

Seregi A, Schafer A, Komlos M (1978) Protective role of ascorbic acid content against lipid peroxidation. Experientia 34:1056–1057

Seutin V, Scuvee-Moreau J, Masotte L, Dresse A (1995) Hydrogen peroxide hyperpolarizes rat CA1 pyramidal neurons by inducing an increase in potassium conductance. Brain Res 683:275–278

Shimizu N, Matunami T, Onishi S (1960) Histochemical demonstration of ascorbic acid in the locus coeruleus of the mammalian brain. Nature 186:479–480

Siddiq A, Aminova LR, Ratan RR (2007) Hypoxia inducible factor prolyl 4-hydroxylase enzymes: center stage in the battle against hypoxia, metabolic compromise and oxidative stress. Neurochem Res 32:931–946

Siesjö BK (1980) Brain energy metabolism. Wiley, New York, pp 131–150

Siushansian R, Wilson JX (1995) Ascorbate transport and intracellular concentrations in cerebral astrocytes. J Neurochem 65:41–49

Slivka A, Mytilineou C, Cohen G (1987) Histochemical evaluation of glutathione in brain. Brain Res 409:275–284

Sotiriou S, Gispert S, Cheng J, Wang Y, Chen A, Hoogstraten-Miller S, Miller GF, Kwon O, Levine M, Guttentag SH, Nussbaum RL (2002) Ascorbic-acid transporter Slc23a1 is essential for vitamin C transport into the brain and for perinatal survival. Nat Med 8:514–517

Spector R (1977) Vitamin homeostasis in the central nervous system. N Engl J Med 296:1293–1398

Spector R, Lorenzo AV (1973) Ascorbic acid homeostasis in the central nervous system. Am J Physiol 225:757–763

Stamford JA, Kruk ZL, Millar J (1984) Regional differences in extracellular ascorbic acid levels in the rat brain determined by high speed cyclic voltammetry. Brain Res 299:289–295

Starkov AA, Chinopoulos C, Fiskum G (2004) Mitochondrial calcium and oxidative stress as mediators of ischemic brain injury. Cell Calcium 36:257–264

Suzuki YJ, Forman HJ, Sevanian A (1997) Oxidants as stimulators of signal transduction. Free Rad Biol Med 22:269–285

Takanaga H, Mackenzie B, Hediger MA (2004) Sodium-dependent ascorbic acid transporter family SLC23. Pflugers Arch 447:677–682

Tanaka J, Toku K, Zhang B, Ishihara K, Sakanaka M, Maeda N (1999) Astrocytes prevent neuronal death induced by reactive oxygen and nitrogen species. Glia 28:85–96

Tang XD, Santarelli LC, Heinemann SH, Hoshi T (2004) Metabolic regulation of potassium channels. Annu Rev Physiol 66:131–159

Tower DB, Elliott KAC (1952) Activity of the acetylcholine system in cerebral cortex of various unanesthetized animals. Am J Physiol 168:747–759

Trépanier G, Furling D, Puymirat J, Mirault ME (1996) Immunocytochemical localization of seleno-glutathione peroxidase in the adult mouse brain. Neuroscience 75:231–243

Tsukaguchi H, Tokui T, Mackenzie B, Berger UV, Chen X-Z, Wang Y, Brubaker RF, Hediger MA (1999) A family of mammalian Na^+-dependent L-ascorbic acid transporters. Nature 399:70–75

Uemura Y, Miller JM, Matson WR, Beal MF (1991) Neurochemical analysis of focal ischemia in rats. Stroke 22:1548–1553

Vatassery GT (1996) Oxidation of vitamin E, vitamin C, and thiols in rat brain synaptosomes by peroxynitrite. Biochem Pharmacol 52:579–586

Vissers MC, Gunningham SP, Morrison MJ, Dachs GU, Currie MJ (2007) Modulation of hypoxia-inducible factor-1 alpha in cultured primary cells by intracellular ascorbate. Free Radic Biol Med 42:765–772

Washburn MP, Wells WW (1999) The catalytic mechanism of the glutathione-dependent dehydroascorbate reductase activity of thioltransferase (glutaredoxin). Biochemistry 38:268–274

Wilson JX (1989) Ascorbic acid uptake by a high-affinity sodium-dependent mechanism in cultured rat astrocytes. J Neurochem 53:1064–1071

Wilson JX, Peters CE, Sitar SM, Daoust P, Gelb AW (2000) Glutamate stimulates ascorbate transport by astrocytes. Brain Res 858:61–66

Winkler BS, Orselli SM, Rex TS (1994) The redox couple between glutathione and ascorbic acid: a chemical and physiological perspective. Free Rad Biol Med 17:333–349

Wood-Kaczmar A, Gandhi S, Wood NW (2006) Understanding the molecular causes of Parkinson's disease. Trends Mol Med 12:521–528

Xu J, Kao SY, Lee FJ, Song W, Jin LW, Yankner BA (2002) Dopamine-dependent neurotoxicity of α-synuclein: A mechanism for selective neurodegeneration in Parkinson disease. Nat Med 8:600–606

Zhang Y, Dawson VL, Dawson TM (2000) Oxidative stress and genetics in the pathogenesis of Parkinson's disease. Neurobiol Dis 7:240–250

Chapter 38
N-Acetyl-Aspartate (NAA) Metabolism

Kishore K. Bhakoo

Abstract N-Acetyl-Aspartate (NAA) remains a unique molecule in medicine, where although its been associated with a number of brain pathologies, primarily due to its non-invasive visibility by magnetic resonance spectroscopy, its function/s remains an enigma. NAA is synthesized from aspartate and acetyl-coenzyme A in neurons and appears to be associated with early postnatal central nervous system (CNS) development. Perhaps the clearest indications as to its function comes from the observation that NAA is transported from neurons to the cytoplasm of oligodendrocytes, where the expression of lipogenic enzymes in oligodendrocytes, including the NAA-degrading enzyme aspartoacylase (ASPA) cleaves the acetate moiety for use in fatty acid and steroid synthesis, thus forming the building blocks for myelin lipid synthesis. Mutations in the gene for ASPA result in the fatal leukodystrophy Canavan disease, for which there is currently no effective treatment. In addition to postnatal myelination, in adult brain NAA also appears to adopt other roles, including a bioenergetic role in neuronal mitochondria and ion balance.

Keywords Brain • Glial • Ion balance • Myelination • NAAG • Neuronal

38.1 Introduction

N-Acetyl-Aspartate (NAA) is the second most abundant molecule (after glutamate) in the CNS (Tallan et al. 1956; Tallan 1957; Bluml 1999; Pan and Takahashi 2005). While, its presence is used increasingly in clinical and experimental magnetic resonance spectroscopy (MRS) studies as a putative marker of neuronal "well-being" (Fan

K.K. Bhakoo, Ph.D. (✉)
Translation Molecular Imaging, Singapore Bioimaging Consortium, Agency for Science, Technology and Research, #02-02 Helios, 11 Biopolis Way, 138667, Singapore
e-mail: kishore_bhakoo@sbic.a-star.edu.sg

I.-Y. Choi and R. Gruetter (eds.), *Neural Metabolism In Vivo*, Advances in Neurobiology 4, DOI 10.1007/978-1-4614-1788-0_38,

et al. 1986; Luyten and den Hollander 1986; Barany et al. 1987), this acceptance of neuronal distribution originates from a body of work suggesting that NAA is localized primarily within the nervous system (Tallan et al. 1956; Tallan 1957) and is synthesized by the synaptosomal mitochondria from aspartate and acetyl-coenzyme A. (Patel and Clark 1979; Bates et al. 1996). Although, the localization of NAA has been studied at length, its function(s) however, remain unclear and controversial. The proposed roles range from an involvement in carbon transfer as a donor of acetyl groups for lipogenesis (in myelination) in the developing brain, in the metabolism of specific brain fatty acids to ion balance and neuromodulation (Birken and Oldendorf 1989). This chapter will address the main aspects of NAA and its metabolism. For a more detailed analysis, the reader is also referred to the proceedings of the 1st International NAA Symposium (Moffett et al. 2006), with excellent contributions from the major researchers working in this field.

NAA has shown interesting correlations with several cerebral pathologies and a decrease in NAA is usually interpreted as an index of neuronal viability (Rigotti et al. 2007). Although, the decrease in NAA is associated with irreversible loss of neurons, there is growing evidence that these decreases in regional NAA levels can also represent reversible neuronal or mitochondrial dysfunction (De Stefano et al. 1995; Bates et al. 1996; Clark 1998; Gasparovic et al. 2001; Demougeot et al. 2003; Lee et al. 2010). Moreover, in neurodegenerative brain pathologies as diverse as multiple sclerosis, Alzheimer's disease, "mitochondrial encephalopathy with lactic acidosis and stroke-like episodes" (MELAS syndrome) (De Stefano et al. 1998) and hypothyroidism (Jagannathan et al. 1998), the loss or reduction of the NAA peak is related to neurological loss or dysfunction. Furthermore, decreases in NAA observed *in vivo* in several neurological conditions may be reversed – an observation that does not sit easily with the concept of NAA being a neuronal marker alone.

Although NAA is seen universally in all nervous tissues, there have a been a few reports where reduced myelination, psychomotor retardation, secondary microcephaly and seizures were associated with brain NAA levels being too low to be detected by MRS (Boltshauser et al. 2004; Burlina et al. 2006). This reduction is possibly due to genetic mutations in the gene for the NAA synthetic enzyme aspartate-N-acetyltransferase (Asp-NAT; EC 2.3.1.17).

In contrast, Canavan disease, which results in an inability to catabolize NAA, due to mutations in the gene for aspartoacylase (EC 3.5.1.15) (Kaul et al. 1993a), leads to high levels of NAA. The result is a progressive, fatal leukodystrophy in affected individuals (Canavan 1931; Hagenfeldt et al. 1987; Matalon et al. 1988; Bartalini et al. 1992), resulting in spongy degeneration of white matter tracts of the brain, and the characteristic NAA aciduria (Matalon and Michals-Matalon 1998), suggesting that excessive NAA levels may have detrimental effects in the CNS, particularly in white matter tracts.

Despite the presence of high levels of NAA in the CNS, and its correlation with a number of pathologies, there has been limited research into its functional role in the brain. Interestingly, the majority of research focus on NAA function in recent years has involved magnetic resonance spectroscopic based analyses of varying NAA levels in a number of neuropathological and psychopathological conditions (see Fig. 38.1).

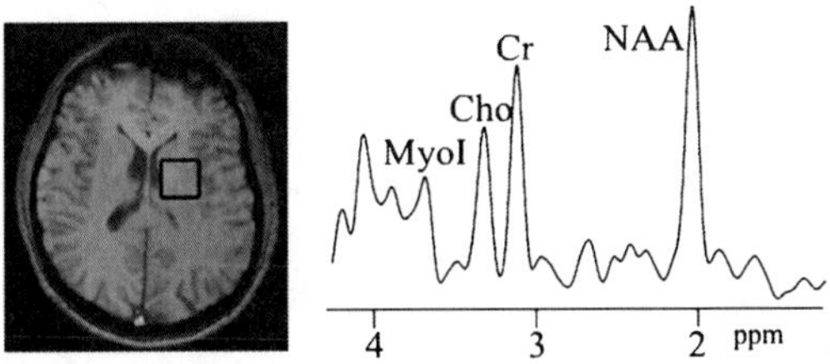

Fig. 38.1 Typical proton MR spectrum of normal human brain acquired using short-echo-stimulated echo mode (STEAM) spectroscopy. The spectrum is based on a voxel size of 3 × 3 × 2 cm. The major peaks of interest are NAA (2.02 ppm) creatine (Cr; 3.05 ppm), choline (Cho; 3.2 ppm) and myo-inositol (MyoI; 3.6 ppm)

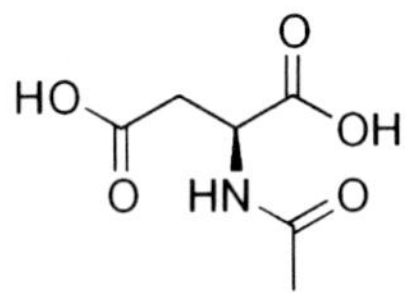

Fig. 38.2 N-acetylaspartate (NAA: (2 S)-2-Acetamidobutanedioic acid – mol mass 175)

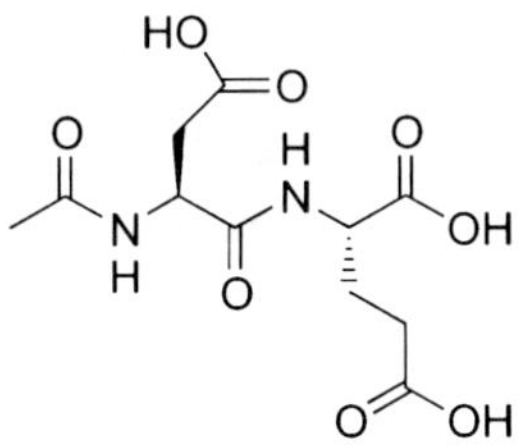

Fig. 38.3 *N*-acetylaspartylglutamate (NAAG: (2*S*)-2-[[(2*S*)-2-Acetamido-4-hydroxy-4-oxobutanoyl] amino]pentanedioic acid – mol mass 304)

The high concentration of NAA in the brain provides a dominant signal in water-suppressed proton MR spectra (Luyten and den Hollander 1986). The acetate moiety of NAA is coupled via the amine nitrogen of aspartate (see Fig. 38.2), resulting in a single prominent peak with a chemical shift of 2.02 ppm. Although the peak at 2.02 ppm is primarily due to NAA, this signal also includes contributions from other acetylated compounds, such as from the dipeptide, N-acetylaspartylglutamate (NAAG; see Fig. 38.3) (found predominately in the CNS) (Caramanos et al. 2005), N-acetylneuraminic acid (Varho et al. 1999), and underlying resonances of glutamate and glutamine.

38.2 NAA Synthesis

The earliest report of NAA synthesis demonstrated the acetylation of aspartate by a soluble enzyme, L-aspartate N-acetyltransferase (Asp-NAT) that could use acetyl CoA and aspartate to form NAA (Goldstein 1959), and was subsequently found to be specific to aspartate and detectable only in the neural tissues (Benuck and D'Adamo 1968). A subsequent report demonstrated that NAA is formed by acetylation of aspartate bound to tRNA rather than free aspartate (Clarke et al. 1975), and remains the only such report in the literature. Further studies confirmed the biosynthesis of NAA in the nervous system is via a membrane bound enzyme (Asp-NAT) that was only detectable in the neural tissues (Truckenmiller et al. 1985).

Using isolated rat brain mitochondria, Patel and Clark demonstrated that NAA is synthesized in the mitochondrion (Patel and Clark 1979), where a supply of both aspartate and acetyl CoA were required for the synthesis of NAA in these preparations. In a later study, it was proposed that NAA is transported out of mitochondria via a dicarboxylic acid transporter from the mitochondria to cytosol (Patel and Clark 1981).

Asp-NAT has been partially purified from rat brain, and that the active enzyme shown to be multimeric in structure (Madhavarao et al. 2003). Lu et al. have shown the enzyme to be distributed in two subcellular fractions, namely in the microsomal fraction and purified mitochondria (Lu et al. 2004).

The reports on mitochondrial synthesis of NAA provide tentative evidence for its role in neuronal energy metabolism (Patel and Clark 1979). This relationship was further confirmed when it was demonstrated that decrease in the activity of mitochondrial respiratory chain enzyme activities result in a decreased capacity of brain mitochondria to synthesize NAA. Hence, NAA production and mitochondrial metabolism are inextricably linked (Bates et al. 1996; Clark 1998).

[1-^{13}C]-glucose administration into the brain has provided a unique opportunity to study NAA biosynthesis using magnetic resonance spectroscopy (Tyson et al. 2003). These studies showed slow labeling of NAA from either labeled acetate (acetyl CoA) or aspartate.

Choi et al. examined the proportional incorporation of label from [1-^{13}C]-glucose in the rat brain, into both aspartyl and acetyl groups of NAA (Choi and Gruetter 2004). Label incorporation was detected in the acetyl group of NAA earlier than in the aspartyl group of NAA, suggesting that NAA is not likely to serve as a significant buffer of acetyl-CoA units, and due to its slow turnover, is unlikely to be involved as a major source of energy when the brain is in a resting state.

In contrast, using *in vivo* [1-^{13}C] glucose studies carried out in Canavan disease, and respective controls, have shown that NAA synthesis is decreased by about 60% (Moreno et al. 2001). NAA concentrations were increased by about 50% while aspartate levels were decreased to a similar extent in these patients. Therefore, decreased synthesis of NAA in the Canavan patients may have resulted from limited substrate availability of aspartate or from product inhibition by NAA, or both.

38.3 NAA Localization

The first report on the distribution of NAA in the brain demonstrated its presence in the brains of all vertebrates studied (Tallan 1957). Furthermore, it was found that the levels of NAA in cerebral gray matter, from bovine brain, were much lower then other regions, such as spinal roots, medulla, pons and cerebral white matter. However, the first non-neuronal localization was reported by Nadler and Cooper, where it was shown that NAA concentrations in glial tumors was much lower compared to normal brain tissue, and thus proposed that NAA was predominantly localized in neurons, as opposed to glia (Nadler and Cooper 1972).

Other evidence for the neuronal origin of NAA comes from a number of studies, where neuron-specific toxins resulted in depletion of NAA (Koller et al. 1984). Furthermore, immunohistochemical studies using antibodies to NAA display localization to a large number of neurons in the rat CNS, although many neurons remained unstained (Moffett et al. 1991; Simmons et al. 1991).

Furthermore, studies using purified cell populations derived from the CNS and high-resolution [^{1}H]-MRS confirmed earlier reports of the neuronal localization (cerebellar granule neurons) of NAA (Urenjak et al. 1992; Urenjak et al. 1993). Although this work also revealed its presence in the oligodendrocyte-type 2 astrocyte (O-2A) progenitor and immature oligodendrocyte, the metabolite was not detected in the mature oligodendrocyte. Similar *in vitro* findings were reported of O-2A cells and differentiated oligodendrocytes (Bhakoo and Pearce 2000), where highest levels of NAA were found in mature oligodendrocytes differentiated in the presence ciliary neurotrophic factor. Thus NAA, which represents a major chemical constituent of the CNS, has been interpreted as a marker of functional neurons in the adult brain.

Subsequent regional analysis of NAA in the developing rat brain demonstrated an increase in all areas of the brain (Florian et al. 1996), the most noticeable being after postnatal day 7. N-acetyl-aspartate levels were highest in gray matter (Florian et al. 1996).

It is important to note that there maybe subtle differences in NAA levels even within the same anatomical structures. Using three-dimensional multivoxel MRS of hippocampal tissue in healthy volunteers its been demonstrated that ratios of NAA:totalCr, NAA:Cho, and NAA:(Cho + totalCr) were significantly higher posteriorly than anteriorly (Ostojic et al. 2010). Thus changes in levels of NAA in pathology may relate to approximate voxel positioning within hippocampi.

38.4 NAA Catabolism

As discussed earlier the role of NAA hydrolysis in the adult brain is still uncertain, although there are several theories. One putative role of aspartoacylase may be to make the acetyl group available from NAA for lipid synthesis, i.e., N-acetyl-aspartate acts as an acetate shuttle from the mitochondrion to the cell membrane.

The enzyme that catalyzes the breakdown of NAA is known as aspartoacylase (EC 3.5.1.15). The enzyme is also known as acylase II, amidohydrolase II and N-acetyl-L-aspartate amidohydrolase. The presence of aspartoacylase was first reported in porcine kidney homogenates, with specific activity for NAA, with no activity for other acetylated amino acids (Birnbaum et al. 1952). Aspartoacylase has also been found in the mammalian brain, where it specifically hydrolyses NAA into aspartate and acetate (Birnbaum 1955). The first detailed studies of aspartoacylase involved the intracerebral injection of radiolabeled NAA, which was shown to be incorporated into brain lipid fractions with increased efficiency with age (D'Adamo and Yatsu 1966; Burri et al. 1991). The highest proportion of radiolabeled lipids was found in myelin and mitochondrial fractions. Aspartoacylase was purified from bovine brain and shown to be expressed in higher amounts in white matter than in gray matter (Kaul et al. 1991a), concomitant with the fact that immunostaining against aspartoacylase has shown that the enzyme is associated with myelin in the brain (Baslow and Resnik 1997).

Kaul and colleagues went on to clone the human aspartoacylase gene, with a predicted protein of 313 amino acids (Kaul et al. 1993b). Further investigations found a common missense mutation of the aspartoacylase gene in Canavan disease patients (Kaul et al. 1994b). Subsequently, a number of alternative mutations in the gene for aspartoacylase have been identified (Kaul et al. 1996; Zeng et al. 2002).

The tissue distribution of aspartoacylase activity has indicated that its expression profile in the CNS followed the time course of postnatal myelination (D'Adamo et al. 1973). Subsequent, *in vitro* studies using cell culture demonstrated that aspartoacylase enzymatic activity was present in oligodendrocytes, but not neurons or astrocytes (Baslow et al. 1999). Bhakoo and colleagues also found aspartoacylase activity in white matter, and the increase in activity during brain maturation correlated with the time course of myelination in the rat brain (Bhakoo et al. 2001). The authors also used several cell types, derived from the brain, to show lack of aspartoacylase activity in neurons, but with highest activity in mature oligodendrocytes and O-2A progenitors. Furthermore, Madhavarao and colleagues reported similar findings in oligodendrocytes but lower activity in O-2A progenitors (Madhavarao et al. 2002).

Aspartoacylase expression in oligodendrocytes and its relationship to the development of myelination was confirmed using *in situ* hybridization studies (Kirmani et al. 2002, 2003). The expression of aspartoacylase in oligodendrocytes was further confirmed with immunohistochemical studies (Klugmann et al. 2003).

38.5 NAAG

N-acetylaspartylglutamate (NAAG) is the most abundant and widely distributed peptide neurotransmitter/modulator in mammalian brain, where it is present in high micromolar to low millimolar concentrations (Williamson et al. 1991; Fuhrman et al. 1994; Coyle 1997; Neale et al. 2000). Its hypothesized that NAAG is synthesized by the enzymatic ligation of NAA and glutamate via NAAG synthetase (Cangro et al. 1987; Williamson and Neale 1988; Urazaev et al. 2001; Gehl et al. 2004; Arun et al. 2006). NAAG biosynthesis has been demonstrated in several tissues, including excised rat dorsal root ganglia (Cangro et al. 1987), excised frog retinas (Williamson and Neale 1988), crayfish nerve cord (Urazaev et al. 2001) and hemisected rat spinal cord (Gehl et al. 2004). In several recent reports, NAAG biosynthesis has been observed in cell culture, including primary rat astrocyte cell culture (Gehl et al. 2004), and in a continuous human neuroblastoma cell line (Arun et al. 2006).

The evidence to date indicates that NAAG is synthesized by peptide bond formation between NAA and glutamate, rather than by acetylation of aspartylglutamate. Tritiated NAA injected intracerebroventricularly has been reported to be incorporated at low levels into NAAG (Sinichkin et al. 1977), suggesting that NAA acts as a direct precursor for NAAG synthesis.

NAAG is released from synapses in a calcium dependent manner (Williamson et al. 1991), and acts through presynaptic metabotropic glutamate receptors to modulate the release of classical neurotransmitters (Zhao et al. 2001; Xi et al. 2002).

High levels of NAAG in the cerebrospinal fluid is also linked with leukodystrophy, characterized by severe hypomyelination, psychomotor impairment, and reduced cerebral myelination on MRI scans (Wolf et al. 2004).

38.6 NAA in Pathology

38.6.1 Canavan Disease

Canavan disease, also called Canavan-Van Bogaert-Bertrand disease, is the result of aspartoacylase deficiency or aminoacylase 2 deficiency, is an autosomal recessive (Namboodiri et al. 2006) degenerative disorder that causes progressive dystrophy of white matter tracts in the brain. Hence, this disease is one of a group of genetic disorders called leukodystrophies. The lack of aspartoacylase leads to the accumulation of NAA, and is highly specific, where elevated NAA level have been demonstrated by both biochemical as well as [^{1}H]-MR spectroscopic analysis (Matalon et al. 1989; Grodd et al. 1990; Kaul et al. 1991b).

This leukodystrophy, or spongy degenerative disease involving brain white matter was described in an infant by Canavan in 1931 (Canavan 1931). However, it took five decades until the link between elevated NAA levels and a progressive demyelinating cerebral atrophy was established in patients with progressive cerebral atrophy leading to NAA aciduria (Kvittingen et al. 1986). Matalon and colleagues were the first to connect N-acetylaspartic aciduria and aspartoacylase deficiency specifically to Canavan disease by showing high NAA levels in urine, and a lack of aspartoacylase activity in Canavan disease (Matalon and Michals-Matalon 1998). Canavan disease is a specific autosomal recessive genetic disorder where two mutations were prevalent among Ashkenazi Jewish patients (Kaul et al. 1994a). The incidences of mutations in the aspartoacylase gene are not only less common in non-Jewish patients, but the types of mutations are more divergent (Kaul et al. 1996; Sistermans et al. 2000).

Canavan disease highlights the neurotoxic effect of elevated NAA. Why should the failure to maintain normal levels of NAA in the brain produce the devastating effect on myelin, and why are they confined to white matter, with potential toxicity to neurons or oligodendrocytes in Canavan disease (Kitada et al. 2000; Leone et al. 2000)? In an interesting series of reports its been shown that administration of exogenous NAA can induce toxicity, where injection of NAA into the lateral cerebral induced profound seizures (Akimitsu et al. 2000).

38.6.2 Evidence for Function of NAA as an Osmolyte

Canavan disease results in a profound water imbalance in the brain. While there is no direct evidence to show that excessive NAA levels in the brain cause dysmyelination through osmotic damage, this hypothesis is often purported to underlie the pathogenesis of Canavan disease. This theory also suggests a reason why NAA, which appears on the whole to be metabolically inert, is present in such high concentration in the brain. This very metabolic inactivity makes it a perfect organic osmolyte. The hypothesis for the role of NAA as a molecular water pump in the brain has been reviewed extensively by Baslow (Baslow and Resnik 1997; Baslow 1998; Baslow 1999).

The function of NAA as a contributor of intracellular anion has been proposed earlier (McIlwain and Bachelard 1985; Cooper et al. 1991), and its role as an osmolyte both *in vitro* and *in vivo* has been demonstrated under pathological conditions. During cerebral ischemia, in which there is swelling of the brain cells, there is a fivefold increase in extracellular NAA levels (Sager et al. 1997). N-acetylaspartate release from neurons is also observed during hypo-osmotic swelling (Gotoh et al. 1997). Our recent work using cortical brain slices has shown that there is a 40% loss of NAA following a hypo-osmotic challenge (Bothwell et al. 2001).

An important question is whether the NAA itself is toxic to white matter, or whether it is the breakdown of a process requiring aspartoacylase that causes neurodegeneration. Affected brains show a profound water imbalance (Baslow 1999), but no defect in the initial myelination process. This supports the hypothesis that a NAA/aspartoacylase system is vital for water movement across cell membranes in the brain. Aspartoacylase deficiency stops the synthesis and degradation cycle of NAA, causing NAA and its hydration water shell to accumulate in the extracellular space. This increase in water pressure causes death of oligodendrocytes, and characteristic spongy degeneration of the white matter. It would also lead to the observed water imbalance in the brains of patients.

38.6.3 Multiple Sclerosis

A number of studies have indicated that NAA has a role in myelin maturation. The earliest indication of non-neuronal contributions to the changes seen in NAA levels came from the longitudinal studies on MS, which involves demyelination in white matter tracts of the brain. MS is principally a disease involving the destruction of oligodendrocytes, thus leading to demyelination. (McMorris and McKinnon 1996; Raine 1997; Lucchinetti et al. 1998; Zhou et al. 1998). Using *in vivo* [^{1}H]-MRS to follow the episodic changes in NAA during the relapse and remission phases of MS, Davie et al. demonstrated NAA, relative to creatine, was reduced in acute lesions, with a subsequent rise in the NAA/creatine ratio (Davie et al. 1994) This phenomenon had already been predicted earlier by Richards (Richards 1991), who hypothesized that serial *in vivo* [^{1}H]-MRS could be used to monitor progression and evaluate therapy using changes in the NAA/creatine ratio in MS. Reversible changes in brain NAA to creatine ratio have been shown to correlate strongly with clinical disability (De Stefano et al. 1995, 1997). Despite the overwhelming support for NAA as a neuronal marker, some workers have however ventured as far as stating that a decrease in NAA levels does not imply axonal loss alone (Davie et al. 1994; Narayana et al. 1998) Nevertheless the possibility of a reduction in NAA levels due to neuronal loss cannot be ruled out in MS lesions, especially in the chronic phase of the disease (Matthews et al. 1996; Fu et al. 1998). Hence, since oligodendrocytes have been shown to synthesize NAA in substantial quantities, it can therefore be surmised that a reduction in the NAA signal seen by [^{1}H]-MRS in MS lesions may be not solely due to neuronal pathology alone.

The relationship between NAA and oligodendrocytes has been further augmented by the studies of Jagannathan and colleagues (Jagannathan et al. 1998) on brains of hypothyroid patients by *in vivo* [^{1}H]-MRS, before and after thyroxine treatment. The authors reported lower levels of NAA in hypothyroid subjects compared to controls. Reversibility of NAA levels was seen even though thyroxine therapy was initiated at ages beyond which abnormalities in myelinogenesis are considered irreversible. Previously, proton NMR spectroscopic study on perchloric acid extract of neonatal hypothyroid rat brain also demonstrated a reduced level of NAA relative to control brain tissue (Sugie et al. 1990).

Cognitive dysfunction in multiple sclerosis (MS) is present in approximately 50% of the patients. More recently using MRSI (magnetic resonance spectroscopic imaging), to create "metabolic maps", a number of studies from Copenhagen have used whole-brain NAA "maps" in conjunction with cerebral glucose metabolism (^{18}FDG-PET) to improve the correlation between imaging techniques and cognitive dysfunction in MS (Sorensen et al. 2006; Mathiesen et al. 2006).

38.6.4 Epilepsy

[^{1}H]-MRS of the hippocampal regions of patients with epilepsy of different forms has revealed both focal and global reductions in the overall NAA and NAA/Cr ratio (Hugg et al. 1993; Briellmann et al. 2005; Aydin-Ozemir et al. 2010). Interestingly, while NAA/Cr is systematically reduced in the affected hemisphere, the contralateral hemisphere is also invariably involved (Vermathen et al. 2002), although these findings remain to be substantiated. Furthermore, following ablation of the lesion from the temporal lobe, NAA recovers both in the tissue around the resection, and in the contralateral hemisphere (Li et al. 2000; Serles et al. 2001).

More recently using MRSI it was demonstrated that, epileptic subjects with MRI-visible pathologies were associated with decreases in NAA/creatine, which extended beyond the volume indicated by the MRI (Riederer et al. 2007; Maudsley et al. 2010). Such NAA maps may provide additional landmarks where surgical intervention for the management of epileptic foci is indicated (Chernov et al. 2009).

38.6.5 Alzheimer's Disease

It is now widely accepted that a reduced regional brain NAA-creatine ratio is associated with dementia, with significant decreased ratio in the *cingulate gyrus* (Moats et al. 1994; Waldman et al. 2002). The finding that the posterior *cingulate gyrus* loses NAA early in the disease supports the hypothesis that this brain region is linked with attention and memory (Raichle et al. 2001). Moreover, similar reduction in the posterior *cingulate gyrus* may be the most reliable surrogate marker for the diagnosis for a number of dementias.

More specifically, in the familial form of Alzheimer disease, there is a marked reduction of NAA/Cr ratio in the posterior *cingulate gyrus* (Lin et al. 2003), where pre-symptomatic carriers have significantly reduced NAA/Cr and NAA-myo-inositol (NAA/MI) metabolite ratios (Godbolt et al. 2006).

38.6.6 Cerebral Oncology

[^{1}H]-MRS of cerebral tumors is consistently associated with reductions in NAA and creatine metabolites, and with increased signals from lactate, choline metabolites and lipids. However, in a number of reports, the MR spectra from brain tumors have shown a complete absence of NAA, whether benign or malignant (Bruhn et al. 1989b; Arnold et al. 1990; Fulham et al. 1992). This lack of NAA signal lies in the fact almost all brain tumors do not express the biosynthetic systems required for NAA production. However, in a rare instance of a neurocytoma, the presence of detectable levels of NAA have been reported in these cerebral masses, but with decreased NAA/Cr ratio (Jayasundar et al. 2003; Chuang et al. 2005). Subsequent studies by Amstutz and co-workers have shown that hypothalamic hamartomas, which can consist of either predominantly glial or neuronal populations, exhibit reduced NAA/Cr and increased mI/Cr ratios (Amstutz et al. 2006). Moreover, histological analyses showed that the NAA signal is greater in hamartomas with predominantly neuronal content, and lower in hamartomas with predominantly glial content.

38.6.7 Stroke

[^{1}H]-MRS studies have demonstrated large increases in lactate levels, with decreases in NAA levels in ischemic brain tissue several hours after onset (Bruhn et al. 1989a; Gideon et al. 1992; Wardlaw et al. 1998; Munoz Maniega et al. 2008). Localized decreases in NAA are seen within a few hours of ischemic onset in patients with clinical evidence of stroke, and low or absent in chronic infarcts (Gideon et al. 1994).

Furthermore, its been shown that decrease in the ratios of NAA to creatine (NAA/Cr) is of prognostic value in stroke patients (Parsons et al. 2000). There is now significant evidence, both from research and clinical studies that a reduction in NAA levels provide a useful marker of neuronal loss and viability, thus aiding in the long-term outcome following and ischemic insult. However, caution is required, as there is some evidence for the recovery of NAA in gray matter following the ischemic episode, suggesting some neuronal recovery (Walker et al. 2004). In contrast the recovery of NAA in white matter is well established, especially in the MELAS syndrome, which results in minor multiple stroke events characterized with a persistent expression of lactate and reversible loss of NAA (Li et al. 2000; Walker et al. 2004).

38.6.8 Psychiatric Disorders

Alterations in NAA concentrations, as measured by [^{1}H]-MRS, have been observed in the majority of neuropsychiatric disorders, especially in Schizophrenia. Most of the literature reports small, regional reductions in NAA or NAA/Cr (Bertolino et al. 1996; Callicott et al. 2000a; Deicken et al. 2001; Klar et al. 2010), although there is some variation between these studies (Deicken et al. 2000). In contrast, some studies have found no significant alterations in NAA or NAA/Cr in Schizophrenic patients.

Furthermore, correlations have been made in a number of studies between glutamatergic neurotransmission and NAA levels in Schizophrenia patients (Tsai et al. 1995, 1998; Moghaddam and Jackson 2003; Moghaddam 2003; Ohrmann et al. 2005). Moreover, NAA reductions in prefrontal cortex have been associated with dysregulation of dopamine release in the *striatum* of Schizophrenia patients (Bertolino et al. 2000). It is interesting to note that NAAG, which is synthesized from NAA, acts to regulate glutamate and dopamine release, most likely via activation of presynaptic metabotropic Glutamate R2/3 receptors.

In contrast a small number of studies have shown positive correlations, especially in the white matter tracts, between NAA levels and general measures of intellectual functioning (Jung et al. 2005; Yeo et al. 2006). These findings highlight the sensitivity and utility of MRS techniques for studying the role of NAA/NAAG in normal and pathological brain function, but also raise additional questions concerning the role of NAA in brain activity and cognition.

Although there is substantial evidence for the localization of NAA in neural tissues, and its status has been correlated with various types of lesions, there is limited evidence for its link with other metabolic markers. Some attempts have been made to correlate the levels of NAA with other metabolites using alternative imaging modalities such as MRI and PET (positron emitting tomography) imaging with a view to examining whether there is a relationship between cerebral metabolic rate of glucose (CMRglc) and NAA. The authors reported variation of CMRglc-to-NAA with cognitive status across subjects with cognitive dysfunctions (O'Neill et al. 2000). There was a correlation between CMRglc and [NAA] in cortical gray matter *in vivo*, accounting for the differing spatial resolutions of ^{18}FDG-PET and [^{1}H]-MRSI, where relative and/or absolute PET measures of gray matter CMRglc were found to increase linearly with increasing gray matter concentrations of NAA in 18 of 19 individual subjects. Moreover, the relationship of NAA to absolute CMRglc in gray matter was found to vary with cognitive status across subjects.

A number of studies have reported that NAA levels in the medial temporal lobe and the prefrontal cortex are reduced in schizophrenia (Abbott and Bustillo 2006; Szulc et al. 2007), and that these reductions mirror changes in cerebral blood flow measured with PET and functional MRI (Callicott et al. 2000b; Marenco et al. 2006).

Advances in MRS techniques have allowed more reliable relationships between anatomical changes in NAA levels with neuropsychological correlates of schizophrenia (Reid et al. 2010) and the effects of cognitive behavior therapy (Premkumar et al. 2010).

38.7 Summary

In summary, although NAA is one of the most prominent metabolites in the brain, and the measured levels (by MRS) are indicative of brain injury or disease, it nevertheless remains an ambiguous molecule. However, there has been renewed interest in elucidating its function/s. Some proposed functions for NAA, such as an osmoregulatory role, remains speculative, whereas there is substantive evidence for its function in synthesis of myelination. As more sensitive techniques for detecting *in vivo* levels of NAA become available, especially in the field of MRS, this will provide neuroscientists with new opportunities to elucidate possible roles of NAA in normal brain functions and pathologies.

References

Abbott C, Bustillo J (2006) What have we learned from proton magnetic resonance spectroscopy about schizophrenia? A critical update. Curr Opin Psychiatry 19:135–139

Akimitsu T, Kurisu K, Hanaya R et al (2000) Epileptic seizures induced by N-acetyl-L-aspartate in rats: *in vivo* and *in vitro* studies. Brain Res 861:143–150

Amstutz DR, Coons SW, Kerrigan JF, Rekate HL, Heiserman JE (2006) Hypothalamic hamartomas: Correlation of MR imaging and spectroscopic findings with tumor glial content. AJNR Am J Neuroradiol 27:794–798

Arnold DL, Shoubridge EA, Villemure JG, Feindel W (1990) Proton and phosphorus magnetic resonance spectroscopy of human astrocytomas *in vivo*. Preliminary observations on tumor grading. NMR Biomed 3:184–189

Arun P, Madhavarao CN, Moffett JR, Namboodiri MA (2006) Regulation of N-acetylaspartate and N-acetylaspartylglutamate biosynthesis by protein kinase activators. J Neurochem 98:2034–2042

Aydin-Ozemir Z, Terzibasioglu E, Altindag E, Sencer S, Baykan B (2010) Magnetic resonance spectroscopy findings in photosensitive idiopathic generalized epilepsy. Clin EEG Neurosci 41:42–49

Barany M, Spigos DG, Mok E, Venkatasubramanian PN, Wilbur AC, Langer BG (1987) High resolution proton magnetic resonance spectroscopy of human brain and liver. Magn Reson Imaging 5:393–398

Bartalini G, Margollicci M, Balestri P, Farnetani MA, Cioni M, Fois A (1992) Biochemical diagnosis of Canavan disease. Childs Nerv Syst 8:468–470

Baslow MH (1998) Function of the N-acetyl-L-histidine system in the vertebrate eye. Evidence in support of a role as a molecular water pump. J Mol Neurosci 10:193–208

Baslow MH (1999) Molecular water pumps and the aetiology of Canavan disease: a case of the sorcerer's apprentice. J Inherit Metab Dis 22:99–101

Baslow MH, Resnik TR (1997) Canavan disease. Analysis of the nature of the metabolic lesions responsible for development of the observed clinical symptoms. J Mol Neurosci 9:109–125

Baslow MH, Suckow RF, Sapirstein V, Hungund BL (1999) Expression of aspartoacylase activity in cultured rat macroglial cells is limited to oligodendrocytes. J Mol Neurosci 13:47–53

Bates TE, Strangward M, Keelan J, Davey GP, Munro PM, Clark JB (1996) Inhibition of N-acetylaspartate production: implications for ^{1}H-MRS studies *in vivo*. Neuroreport 7:1397–1400

Benuck M, D'Adamo AF J (1968) Acetyl transport mechanisms. Metabolism of N-acetyl-L-aspartic acid in the non-nervous tissues of the rat. Biochim Biophys Acta 152:611–618

Bertolino A, Nawroz S, Mattay VS, Barnett AS, Duyn JH, Moonen CT, Frank JA, Tedeschi G, Weinberger DR (1996) Regionally specific pattern of neurochemical pathology in schizophrenia as assessed by multislice proton magnetic resonance spectroscopic imaging. Am J Psychiatry 153:1554–1563

Bertolino A, Breier A, Callicott JH, Adler C, Mattay VS, Shapiro M, Frank JA, Pickar D, Weinberger DR (2000) The relationship between dorsolateral prefrontal neuronal N-acetylaspartate and evoked release of striatal dopamine in schizophrenia. Neuropsychopharmacology 22:125–132

Bhakoo KK, Pearce D (2000) *In vitro* expression of N-acetyl aspartate by oligodendrocytes: implications for proton magnetic resonance spectroscopy signal *in vivo*. J Neurochem 74:254–262

Bhakoo KK, Craig TJ, Styles P (2001) Developmental and regional distribution of aspartoacylase in rat brain tissue. J Neurochem 79:211–220

Birken DL, Oldendorf WH (1989) N-acetyl-L-aspartic acid: a literature review of a compound prominent in ^{1}H-NMR spectroscopic studies of brain. Neurosci Biobehav Rev 13:23–31

Birnbaum SM (1955) Amino acid acylases I and II from Hog Kidney. Methods Enzymol 2:115–119

Birnbaum SM, Levintow L, Kingsley RB, Greenstein JP (1952) Specificity of amino acid acylases. J Biol Chem 194:455–470

Bluml S (1999) *In vivo* quantitation of cerebral metabolite concentrations using natural abundance ^{13}C-MRS at 1.5 T. J Magn Reson 136:219–225

Boltshauser E, Schmitt B, Wevers RA, Engelke U, Burlina AB, Burlina AP (2004) Follow-up of a child with hypoacetylaspartia. Neuropediatrics 35:255–258

Bothwell JH, Rae C, Dixon RM, Styles P, Bhakoo KK (2001) Hypo-osmotic swelling-activated release of organic osmolytes in brain slices: implications for brain oedema *in vivo*. J Neurochem 77:1632–1640

Briellmann RS, Wellard RM, Jackson GD (2005) Seizure-associated abnormalities in epilepsy: evidence from MR imaging. Epilepsia 46:760–766

Bruhn H, Frahm J, Gyngell ML, Merboldt KD, Hanicke W, Sauter R (1989a) Cerebral metabolism in man after acute stroke: new observations using localized proton NMR spectroscopy. Magn Reson Med 9:126–131

Bruhn H, Frahm J, Gyngell ML, Merboldt KD, Hanicke W, Sauter R, Hamburger C (1989b) Noninvasive differentiation of tumors with use of localized [1]H-MR spectroscopy *in vivo*: initial experience in patients with cerebral tumors. Radiology 172:541–548

Burlina AP, Schmitt B, Engelke U, Wevers RA, Burlina AB, Boltshauser E (2006) Hypoacetylaspartia: clinical and biochemical follow-up of a patient. Adv Exp Med Biol 576:283–287, discussion 361–283

Burri R, Steffen C, Herschkowitz N (1991) N-acetyl-L-Aspartate is a major source of acetyl groups for lipid synthesis during rat brain development. Dev Neurosci 13:403–411

Callicott JH, Bertolino A, Egan MF, Mattay VS, Langheim FJ, Weinberger DR (2000a) Selective relationship between prefrontal N-acetylaspartate measures and negative symptoms in schizophrenia. Am J Psychiatry 157:1646–1651

Callicott JH, Bertolino A, Mattay VS, Langheim FJ, Duyn J, Coppola R, Goldberg TE, Weinberger DR (2000b) Physiological dysfunction of the dorsolateral prefrontal cortex in schizophrenia revisited. Cereb Cortex 10:1078–1092

Canavan MM (1931) Schildler's encephalitis periaxialsis diffusa. Arch Neurol Psychiatry 25:299–308

Cangro CB, Namboodiri MA, Sklar LA, Corigliano Murphy A, Neale JH (1987) Immunohistochemistry and biosynthesis of N-acetylaspartylglutamate in spinal sensory ganglia. J Neurochem 49:1579–1588

Caramanos Z, Narayanan S, Arnold DL (2005) ^{1}H-MRS quantification of tNA and tCr in patients with multiple sclerosis: a meta-analytic review. Brain 128:2483–2506

Chernov MF, Ochiai T, Ono Y et al (2009) Role of proton magnetic resonance spectroscopy in preoperative evaluation of patients with mesial temporal lobe epilepsy. J Neurol Sci 285:212–219

Choi IY, Gruetter R (2004) Dynamic or inert metabolism? Turnover of N-acetyl aspartate and glutathione from D-[1-^{13}C]glucose in the rat brain *in vivo*. J Neurochem 91:778–787

Chuang MT, Lin WC, Tsai HY, Liu GC, Hu SW, Chiang IC (2005) 3-T proton magnetic resonance spectroscopy of central neurocytoma: 3 case reports and review of the literature. J Comput Assist Tomogr 29:683–688

Clark JB (1998) N-acetyl aspartate: a marker for neuronal loss or mitochondrial dysfunction. Dev Neurosci 20:271–276

Clarke DD, Greenfield S, Dicker E, Tirri LJ, Ronan EJ (1975) A relationship of N-acetylaspartate biosynthesis to neuronal protein synthesis. J Neurochem 24:479–485

Cooper JR, Roth RH, Bloom FE (1991) The biochemical basis of neuropharmacology. Oxford University Press, New York; Oxford

Coyle JT (1997) The nagging question of the function of N-acetylaspartylglutamate. Neurobiol Dis 4:231–238

D'Adamo AFJ, Yatsu FM (1966) Acetate metabolism in the nervous system. N-acetyl-L-aspartic acid and the biosynthesis of brain lipids. J Neurochem 13:961–965

D'Adamo AFJ, Smith JC, Woiler C (1973) The occurrence of N-acetylaspartate amidohydrolase (aminoacylase II) in the developing rat. J Neurochem 20:1275–1278

Davie CA, Hawkins CP, Barker GJ, Brennan A, Tofts PS, Miller DH, McDonald WI (1994) Serial proton magnetic resonance spectroscopy in acute multiple sclerosis lesions. Brain 117:49–58

De Stefano N, Matthews PM, Arnold DL (1995) Reversible decreases in N-acetylaspartate after acute brain injury. Magn Reson Med 34:721–727

De Stefano N, Matthews PM, Narayanan S, Francis GS, Antel JP, Arnold DL (1997) Axonal dysfunction and disability in a relapse of multiple sclerosis: longitudinal study of a patient. Neurology 49:1138–1141

De Stefano N, Matthews PM, Fu L, Narayanan S, Stanley J, Francis GS, Antel JP, Arnold DL (1998) Axonal damage correlates with disability in patients with relapsing-remitting multiple sclerosis. Results of a longitudinal magnetic resonance spectroscopy study. Brain 121:1469–1477

Deicken RF, Johnson C, Pegues M (2000) Proton magnetic resonance spectroscopy of the human brain in schizophrenia. Rev Neurosci 11:147–158

Deicken RF, Eliaz Y, Feiwell R, Schuff N (2001) Increased thalamic N-acetylaspartate in male patients with familial bipolar I disorder. Psychiatry Res 106:35–45

Demougeot C, Bertrand N, Prigent-Tessier A, Garnier P, Mossiat C, Giroud M, Marie C, Beley A (2003) Reversible loss of N-acetyl-aspartate in rats subjected to long-term focal cerebral ischemia. J Cereb Blood Flow Metab 23:482–489

Fan TW, Higashi RM, Lane AN, Jardetzky O (1986) Combined use of ^{1}H-NMR and GC-MS for metabolite monitoring and *in vivo* ^{1}H-NMR assignments. Biochim Biophys Acta 882:154–167

Florian CL, Williams SR, Bhakoo KK, Noble MD (1996) Regional and developmental variations in metabolite concentration in the rat brain and eye: a study using ^{1}H-NMR spectroscopy and high performance liquid chromatography. Neurochem Res 21:1065–1074

Fu L, Matthews PM, De Stefano N, Worsley KJ, Narayanan S, Francis GS, Antel JP, Wolfson C, Arnold DL (1998) Imaging axonal damage of normal-appearing white matter in multiple sclerosis. Brain 121:103–113

Fuhrman S, Palkovits M, Cassidy M, Neale JH (1994) The regional distribution of N-acetylaspartylglutamate (NAAG) and peptidase activity against NAAG in the rat nervous system. J Neurochem 62:275–281

Fulham MJ, Bizzi A, Dietz MJ et al (1992) Mapping of brain-tumor metabolites with proton MR spectroscopic imaging - clinical relevance. Radiology 185:675–686

Gasparovic C, Arfai N, Smid N, Feeney DM (2001) Decrease and recovery of N-acetylaspartate/creatine in rat brain remote from focal injury. J Neurotrauma 18:241–246

Gehl LM, Saab OH, Bzdega T, Wroblewska B, Neale JH (2004) Biosynthesis of NAAG by an enzyme-mediated process in rat central nervous system neurons and glia. J Neurochem 90:989–997

Gideon P, Henriksen O, Sperling B, Christiansen P, Olsen TS, Jorgensen HS, Arliensoborg P (1992) Early time course of N-acetylaspartate, creatine and phosphocreatine, and compounds containing choline in the brain after acute stroke - a proton magnetic-resonance spectroscopy study. Stroke 23:1566–1572

Gideon P, Sperling B, Arlien-Soborg P, Olsen TS, Henriksen O (1994) Long-term follow-up of cerebral infarction patients with proton magnetic resonance spectroscopy. Stroke 25:967–973

Godbolt AK, Waldman AD, MacManus DG, Schott JM, Frost C, Cipolotti L, Fox NC, Rossor MN (2006) MRS shows abnormalities before symptoms in familial Alzheimer disease. Neurology 66:718–722

Goldstein FB (1959) Biosynthesis of N-acetyl-L-aspartic acid. Biochim Biophys Acta 33:583–584

Gotoh M, Davies SE, Obrenovitch TP (1997) Brain tissue acidosis: effects on the extracellular concentration of N-acetylaspartate. J Neurochem 69:655–661

Grodd W, Krageloh Mann I, Petersen D, Trefz FK, Harzer K (1990) *In vivo* assessment of N-acetylaspartate in brain in spongy degeneration (Canavan's disease) by proton spectroscopy [letter]. Lancet 336:437–438

Hagenfeldt L, Bollgren I, Venizelos N (1987) N-acetylaspartic aciduria due to aspartoacylase deficiency - a new aetiology of childhood leukodystrophy. J Inherit Metab Dis 10:135–141

Hugg JW, Laxer KD, Matson GB, Maudsley AA, Weiner MW (1993) Neuron loss localizes human temporal-lobe epilepsy by *in-vivo* proton magnetic-resonance spectroscopic imaging. Ann Neurol 34:788–794

Jagannathan NR, Tandon N, Raghunathan P, Kochupillai N (1998) Reversal of abnormalities of myelination by thyroxine therapy in congenital hypothyroidism: localized *in vivo* proton magnetic resonance spectroscopy (MRS) study. Brain Res Dev Brain Res 109:179–186

Jayasundar R, Shah T, Vaishya S, Singh VP, Sarkar C (2003) *In vivo* and *in vitro* MR spectroscopic profile of central neurocytomas. J Magn Reson Imaging 17:256–260

Jung RE, Haier RJ, Yeo RA, Rowland LM, Petropoulos H, Levine AS, Sibbitt WL, Brooks WM (2005) Sex differences in N-acetylaspartate correlates of general intelligence: an ^{1}H-MRS study of normal human brain. Neuroimage 26:965–972

Kaul R, Casanova J, Johnson AB, Tang P, Matalon R (1991a) Purification, characterization, and localization of aspartoacylase from bovine brain. J Neurochem 56:129–135

Kaul R, Michals K, Casanova J, Matalon R (1991b) The role of N-acetylaspartic acid in brain metabolism and the pathogenesis in Canavan disease. Int Pediatr 6:40–43

Kaul R, Gao GP, Aloya M, Michals K, Matalon R (1993a) Identification Of Mutations In Human Aspartoacylase (Hasp) Gene In Canavan Disease. Am J Human Genet 53:215–215

Kaul R, Gao GP, Balamurugan K, Matalon R (1993b) Cloning of the human aspartoacylase cDNA and a common missense mutation in Canavan disease. Nat Genet 5:118–123

Kaul R, Balamurugan K, Gao GP, Matalon R (1994a) Canavan disease: genomic organization and localization of human ASPA to 17p13-ter and conservation of the ASPA gene during evolution. Genomics 21:364–370

Kaul R, Gao GP, Aloya M, Balamurugan K, Petrosky A, Michals K, Matalon R (1994b) Canavan disease: mutations among Jewish and non-jewish patients. Am J Hum Genet 55:34–41

Kaul R, Gao GP, Matalon R, Aloya M, Su Q, Jin M, Johnson AB, Schutgens RB, Clarke JT (1996) Identification and expression of eight novel mutations among non-Jewish patients with Canavan disease. Am J Hum Genet 59:95–102

Kirmani BF, Jacobowitz DM, Kallarakal AT, Namboodiri MA (2002) Aspartoacylase is restricted primarily to myelin synthesizing cells in the CNS: therapeutic implications for Canavan disease. Brain Res Mol Brain Res 107:176–182

Kirmani BF, Jacobowitz DM, Namboodiri MA (2003) Developmental increase of aspartoacylase in oligodendrocytes parallels CNS myelination. Brain Res Dev Brain Res 140:105–115

Kitada K, Akimitsu T, Shigematsu Y, Kondo A, Maihara T, Yokoi N, Kuramoto T, Sasa M, Serikawa T (2000) Accumulation of N-acetyl-L-aspartate in the brain of the tremor rat, a mutant exhibiting absence-like seizure and spongiform degeneration in the central nervous system. J Neurochem 74:2512–2519

Klar AA, Ballmaier M, Leopold K, Hake I, Schaefer M, Bruhl R, Schubert F, Gallinat J (2010) Interaction of hippocampal volume and N-acetylaspartate concentration deficits in schizophrenia: A combined MRI and 1H-MRS study. Neuroimage 53:51–57

Klugmann M, Symes CW, Klaussner BK, Leichtlein CB, Serikawa T, Young D, During MJ (2003) Identification and distribution of aspartoacylase in the postnatal rat brain. Neuroreport 14:1837–1840

Koller KJ, Zaczek R, Coyle JT (1984) N-acetyl-aspartyl-glutamate: regional levels in rat brain and the effects of brain lesions as determined by a new HPLC method. J Neurochem 43:1136–1142

Kvittingen EA, Guldal G, Borsting S, Skalpe IO, Stokke O, Jellum E (1986) N-acetylaspartic aciduria in a child with a progressive cerebral atrophy. Clin Chim Acta 158:217–227

Lee SK, Kim J, Kim HD, Lee JS, Lee YM (2010) Initial experiences with proton MR spectroscopy in treatment monitoring of mitochondrial encephalopathy. Yonsei Med J 51:672–675

Leone P, Janson CG, Bilaniuk L et al (2000) Aspartoacylase gene transfer to the mammalian central nervous system with therapeutic implications for Canavan disease. Ann Neurol 48:27–38

Li LM, Cendes F, Antel SB, Andermann F, Serles W, Dubeau F, Olivier A, Arnold DL (2000) Prognostic value of proton magnetic resonance spectroscopic imaging for surgical outcome in patients with intractable temporal lobe epilepsy and bilateral hippocampal atrophy. Ann Neurol 47:195–200

Lin AP, Shic F, Enriquez C, Ross BD (2003) Reduced glutamate neurotransmission in patients with Alzheimer's disease - an *in vivo* ^{13}C-magnetic resonance spectroscopy study. Magma 16:29–42

Lu ZH, Chakraborty G, Ledeen RW, Yahya D, Wu G (2004) N-Acetylaspartate synthase is bimodally expressed in microsomes and mitochondria of brain. Brain Res Mol Brain Res 122:71–78

Lucchinetti CF, Brueck W, Rodriguez M, Lassmann H (1998) Multiple sclerosis: lessons from neuropathology. Semin Neurol 18:337–349

Luyten PR, den Hollander JA (1986) Observation of metabolites in the human brain by MR spectroscopy. Radiology 161:795–798

Madhavarao CN, Hammer JA, Quarles RH, Namboodiri MA (2002) A radiometric assay for aspartoacylase activity in cultured oligodendrocytes. Anal Biochem 308:314–319

Madhavarao CN, Chinopoulos C, Chandrasekaran K, Namboodiri MA (2003) Characterization of the N-acetylaspartate biosynthetic enzyme from rat brain. J Neurochem 86:824–835

Marenco S, Bertolino A, Weinberger DR (2006) *In vivo* NMR measures of NAA and the neurobiology of schizophrenia. Adv Exp Med Biol 576:227–240, discussion 361–223

Matalon R, Michals-Matalon K (1998) Molecular basis of Canavan disease. Eur J Paediatr Neurol 2:69–76

Matalon R, Michals K, Sebesta D, Deanching M, Gashkoff P, Casanova J (1988) Aspartoacylase deficiency and N-acetylaspartic aciduria in patients with Canavan disease. Am J Med Genet 29:463–471

Matalon R, Kaul R, Casanova J, Michals K, Johnson A, Rapin I, Gashkoff P, Deanching M (1989) Aspartoacylase deficiency: The enzyme defect in Canavan disease. J Inherit Metab Dis 12:329–331

Mathiesen HK, Jonsson A, Tscherning T, Hanson LG, Andresen J, Blinkenberg M, Paulson OB, Sorensen PS (2006) Correlation of global N-acetyl aspartate with cognitive impairment in multiple sclerosis. Arch Neurol 63:533–536

Matthews PM, Pioro E, Narayanan S, De Stefano N, Fu L, Francis G, Antel J, Wolfson C, Arnold DL (1996) Assessment of lesion pathology in multiple sclerosis using quantitative MRI morphometry and magnetic resonance spectroscopy. Brain 119:715–722

Maudsley AA, Domenig C, Ramsay RE, Bowen BC (2010) Application of volumetric MR spectroscopic imaging for localization of neocortical epilepsy. Epilepsy Res 88:127–138

McIlwain H, Bachelard HS (1985) Biochemistry and the central nervous system. Churchill Livingstone, Edinburgh

McMorris FA, McKinnon RD (1996) Regulation of oligodendrocyte development and CNS myelination by growth factors: prospects for therapy of demyelinating disease. Brain Pathol 6:313–329

Moats RA, Ernst T, Shonk TK, Ross BD (1994) Abnormal cerebral metabolite concentrations in patients with probable Alzheimer disease. Magn Reson Med 32:110–115

Moffett JR, Namboodiri MA, Cangro CB, Neale JH (1991) Immunohistochemical localization of N-acetylaspartate in rat brain. Neuroreport 2:131–134

Moffett JR, Tieman SB, Weinberger DR, Coyle JT, Namboodiri MA. 2006. N-Acetylaspartate-A unique neuronal molecule in the central nervous system. In: Advances in experimental medicine and biology, vol. 576. Springer, New York

Moghaddam B (2003) Bringing order to the glutamate chaos in schizophrenia. Neuron 40:881–884

Moghaddam B, Jackson ME (2003) Glutamatergic animal models of schizophrenia. Ann N Y Acad Sci 1003:131–137

Moreno A, Ross BD, Bluml S (2001) Direct determination of the N-acetyl-L-aspartate synthesis rate in the human brain by ^{13}C-MRS and [1-^{13}C]glucose infusion. J Neurochem 77:347–350

Munoz Maniega S, Cvoro V, Chappell FM, Armitage PA, Marshall I, Bastin ME, Wardlaw JM (2008) Changes in NAA and lactate following ischemic stroke: a serial MR spectroscopic imaging study. Neurology 71:1993–1999

Nadler JV, Cooper JR (1972) N-Acetyl-L-aspartic acid content of human neural tumours and bovine periperal nervous tissues. J Neurochem 19:313–319

Namboodiri AM, Peethambaran A, Mathew R, Sambhu PA, Hershfield J, Moffett JR, Madhavarao CN (2006) Canavan disease and the role of N-acetylaspartate in myelin synthesis. Mol Cell Endocrinol 252:216–223

Narayana PA, Doyle TJ, Lai D, Wolinsky JS (1998) Serial proton magnetic resonance spectroscopic imaging, contrast-enhanced magnetic resonance imaging, and quantitative lesion volumetry in multiple sclerosis. Ann Neurol 43:56–71

Neale JH, Bzdega T, Wroblewska B (2000) N-Acetylaspartylglutamate: the most abundant peptide neurotransmitter in the mammalian central nervous system. J Neurochem 75:443–452

Ohrmann P, Siegmund A, Suslow T, Spitzberg K, Kersting A, Arolt V, Heindel W, Pfleiderer B (2005) Evidence for glutamatergic neuronal dysfunction in the prefrontal cortex in chronic but not in first-episode patients with schizophrenia: a proton magnetic resonance spectroscopy study. Schizophr Res 73:153–157

O'Neill J, Eberling JL, Schuff N, Jagust W, Reed B, Soto G, Ezekiel F, Klein G, Weiner MW (2000) Method to correlate ^{1}H-MRSI and ^{18}FDG-PET. Magn Reson Med 43:244–250

Ostojic J, Kozic D, Konstantinovic J, Covickovic-Sternic N, Mijajlovic M, Koprivsek K, Semnic R (2010) Three-dimensional multivoxel spectroscopy of the healthy hippocampus-are the metabolic differences related to the location? Clin Radiol 65:302–307

Pan JW, Takahashi K (2005) Interdependence of N-acetyl aspartate and high-energy phosphates in healthy human brain. Ann Neurol 57:92–97

Parsons MW, Li T, Barber PA, Yang Q, Darby DG, Desmond PM, Gerraty RP, Tress BM, Davis SM (2000) Combined (1)H MR spectroscopy and diffusion-weighted MRI improves the prediction of stroke outcome. Neurology 55:498–505

Patel TB, Clark JB (1979) Synthesis of N-acetyl-L-aspartate by rat brain mitochondria and its involvement in mitochondrial/cytosolic carbon transport. Biochem J 184:539–546

Patel TB, Clark JB (1981) Mitochondrial/cytosolic carbon transfer in the developing rat brain. Biochim Biophys Acta 677:373–380

Premkumar P, Parbhakar VA, Fannon D, Lythgoe D, Williams SC, Kuipers E, Kumari V (2010) N-acetyl aspartate concentration in the anterior cingulate cortex in patients with schizophrenia: A study of clinical and neuropsychological correlates and preliminary exploration of cognitive behaviour therapy effects. Psychiatry Res Neuroimaging 182:251–260

Raichle ME, MacLeod AM, Snyder AZ, Powers WJ, Gusnard DA, Shulman GL (2001) A default mode of brain function. Proc Natl Acad Sci U S A 98:676–682

Raine CS (1997) The Norton Lecture: a review of the oligodendrocyte in the multiple sclerosis lesion. J Neuroimmunol 77:135–152

Reid MA, Stoeckel LE, White DM, Avsar KB, Bolding MS, Akella NS, Knowlton RC, Hollander JA, Lahti AC (2010) Assessments of Function and Biochemistry of the Anterior Cingulate Cortex in Schizophrenia. Biol Psychiatry 68:625–633

Richards TL (1991) Proton MR spectroscopy in multiple-sclerosis - value in establishing diagnosis, monitoring progression, and evaluating therapy. Am J Roentgenol 157:1073–1078

Riederer F, Bittsansky M, Lehner-Baumgartner E, Baumgartner C, Mlynarik V, Gruber S, Moser E, Kaya M, Serles W (2007) Decrease of NAA with aging outside the seizure focus in mesial temporal lobe epilepsy-a proton-MRS study at 3 Tesla. Brain Res 1179:131–139

Rigotti DJ, Inglese M, Gonen O (2007) Whole-brain N-acetylaspartate as a surrogate marker of neuronal damage in diffuse neurologic disorders. AJNR Am J Neuroradiol 28:1843–1849

Sager TN, Fink Jensen A, Hansen AJ (1997) Transient elevation of interstitial N-acetylaspartate in reversible global brain ischemia. J Neurochem 68:675–682

Serles W, Li LM, Antel SB, Cendes F, Gotman J, Olivier A, Andermann F, Dubeau F, Arnold DL (2001) Time course of postoperative recovery of N-acetyl-aspartate in temporal lobe epilepsy. Epilepsia 42:190–197

Simmons ML, Frondoza CG, Coyle JT (1991) Immunocytochemical localization of N-acetyl-aspartate with monoclonal antibodies. Neuroscience 45:37–45

Sinichkin A, Sterri S, Edminson PD, Reichelt KL, Kvamme E (1977) *In vivo* labelling of acetyl-aspartyl peptides in mouse brain from intracranially and intraperitoneally administered acetyl-L-[U-^{14}C]aspartate. J Neurochem 29:425–431

Sistermans EA, de Coo RF, van Beerendonk HM, Poll-The BT, Kleijer WJ, van Oost BA (2000) Mutation detection in the aspartoacylase gene in 17 patients with Canavan disease: four new mutations in the non-Jewish population. Eur J Hum Genet 8:557–560

Sorensen PS, Jonsson A, Mathiesen HK, Blinkenberg M, Andresen J, Hanson LG, Ravnborg M (2006) The relationship between MRI and PET changes and cognitive disturbances in MS. J Neurol Sci 245:99–102

Sugie H, Tsurui S, Ishikawa A, Matsuda F, Sugie Y, Igarashi Y, Fujise Y (1990) Effects of neonatal hypothyroidism on brain development: analysis of brain metabolites using high resolution phosphorus and proton magnetic resonance (NMR) spectroscopy. No To Hattatsu (Japanese) 22:166–172

Szulc A, Galinska B, Tarasow E et al (2007) N-acetylaspartate (NAA) levels in selected areas of the brain in patients with chronic schizophrenia treated with typical and atypical neuroleptics: a proton magnetic resonance spectroscopy (^{1}H-MRS) study. Med Sci Monit 13(Suppl 1):17–22

Tallan HH (1957) Studies on the distribution of *N*-acetyl-L-aspartic acid in brain. J Biol Chem 224:41–45

Tallan HH, Moore S, Stein WH (1956) N-acetyl-L-aspartic acid in brain. J Biol Chem 219: 257–264

Truckenmiller ME, Namboodiri MA, Brownstein MJ, Neale JH (1985) N-Acetylation of L-aspartate in the nervous system: differential distribution of a specific enzyme. J Neurochem 45:1658–1662

Tsai G, Passani LA, Slusher BS, Carter R, Baer L, Kleinman JE, Coyle JT (1995) Abnormal excitatory neurotransmitter metabolism in schizophrenic brains. Arch Gen Psychiatry 52:829–836

Tsai G, van Kammen DP, Chen S, Kelley ME, Grier A, Coyle JT (1998) Glutamatergic neurotransmission involves structural and clinical deficits of schizophrenia. Biol Psychiatry 44:667–674

Tyson RL, Gallagher C, Sutherland GR (2003) ^{13}C-Labeled substrates and the cerebral metabolic compartmentalization of acetate and lactate. Brain Res 992:43–52

Urazaev AK, Grossfeld RM, Fletcher PL, Speno H, Gafurov BS, Buttram JG, Lieberman EM (2001) Synthesis and release of N-acetylaspartylglutamate (NAAG) by crayfish nerve fibers: implications for axon-glia signaling. Neuroscience 106:237–247

Urenjak J, Williams SR, Gadian DG, Noble M (1992) Specific expression of N-acetylaspartate in neurons, oligodendrocyte-type-2 astrocyte progenitors, and immature oligodendrocytes *in vitro*. J Neurochem 59:55–61

Urenjak J, Williams SR, Gadian DG, Noble M (1993) Proton nuclear magnetic resonance spectroscopy unambiguously identifies different neural cell types. J Neurosci 13:981–989

Varho T, Komu M, Sonninen P, Holopainen I, Nyman S, Manner T, Sillanpaa M, Aula P, Lundbom N (1999) A new metabolite contributing to N-acetyl signal in ^{1}H-MRS of the brain in Salla disease. Neurology 52:1668–1672

Vermathen P, Ende G, Laxer KD, Walker JA, Knowlton RC, Barbaro NM, Matson GB, Weiner MW (2002) Temporal lobectomy for epilepsy: recovery of the contralateral hippocampus measured by ^{1}H-MRS. Neurology 59:633–636

Waldman AD, Rai GS, McConnell JR, Chaudry M, Grant D (2002) Clinical brain proton magnetic resonance spectroscopy for management of Alzheimer's and sub-cortical ischemic vascular dementia in older people. Arch Gerontol Geriatr 35:137–142

Walker PM, Ben Salem D, Lalande A, Giroud M, Brunotte F (2004) Time course of NAA T2 and ADC(w) in ischaemic stroke patients: ^{1}H-MRS imaging and diffusion-weighted MRI. J Neurol Sci 220:23–28

Wardlaw JM, Marshall I, Wild J, Dennis MS, Cannon J, Lewis SC (1998) Studies of acute ischemic stroke with proton magnetic resonance spectroscopy: relation between time from onset, neurological deficit, metabolite abnormalities in the infarct, blood flow, and clinical outcome. Stroke 29:1618–1624

Williamson LC, Neale JH (1988) Calcium-dependent release of N-acetylaspartylglutamate from retinal neurons upon depolarization. Brain Res 475:151–155

Williamson LC, Eagles DA, Brady MJ, Moffett JR, Namboodiri MA, Neale JH (1991) Localization and Synaptic Release of N-acetylaspartylglutamate in the Chick Retina and Optic Tectum. Eur J Neurosci 3:441–451

Wolf NI, Willemsen MA, Engelke UF et al (2004) Severe hypomyelination associated with increased levels of N-acetylaspartylglutamate in CSF. Neurology 62:1503–1508

Xi ZX, Baker DA, Shen H, Carson DS, Kalivas PW (2002) Group II metabotropic glutamate receptors modulate extracellular glutamate in the nucleus accumbens. J Pharmacol Exp Ther 300:162–171

Yeo RA, Brooks WM, Jung RE (2006) NAA and higher cognitive function in humans. Adv Exp Med Biol 576:215–226, discussion 361–213

Zeng BJ, Wang ZH, Ribeiro LA et al (2002) Identification and characterization of novel mutations of the aspartoacylase gene in non-Jewish patients with Canavan disease. J Inherit Metab Dis 25:557–570

Zhao J, Ramadan E, Cappiello M, Wroblewska B, Bzdega T, Neale JH (2001) NAAG inhibits KCl-induced [(3)H]-GABA release via mGluR3, cAMP, PKA and L-type calcium conductance. Eur J Neurosci 13:340–346

Zhou L, Trapp BD, Miller RH (1998) Demyelination in the central nervous system mediated by an anti-oligodendrocyte antibody. J Neurosci Res 54:158–168

Chapter 39
In Vivo GABA Metabolism

In-Young Choi and Jun Shen

Abstract γ-aminobutyric acid (GABA) is the major inhibitory neurotransmitter in the brain. In the central nervous system (CNS), significant concentration variations and gradients of GABA exist accompanied by its highly active metabolism and transport. To provide an understanding of GABA metabolism *in vivo*, this chapter covers the pathways of GABA metabolism, as well as methods for measuring GABA and GABAergic neurotransmission *in vivo*. GABA metabolism is compartmentalized and GABA turnover, its release from presynaptic terminals and reuptake into cellular cytoplasm, is linked to glucose and amino acid metabolism. GABAergic dysfunction involving GABA metabolism and/or neurotransmission has been implicated in several neurological and psychiatric disorders including epilepsy, depression and schizophrenia.

Keywords Degradation • Epilepsy • Metabolism • Neurotransmitter • NMR spectroscopy • Synthesis • Turnover • γ-aminobutyric acid (GABA)

39.1 Introduction

Since the identification of γ-aminobutyric acid (GABA) and its ability to decrease electrical activity in the brain in 1950s (Awapara et al. 1950; Roberts and Frankel 1950; Udenfriend 1950; Bazemore et al. 1956), GABA has been accepted as the

I.-Y. Choi, Ph.D. (✉)
Department of Neurology, Department of Molecular & Integrative Physiology, Hoglund Brain Imaging Center, University of Kansas Medical Center, 3901 Rainbow Blvd, Mail Stop 1052, Kansas City, KS 66160, USA
e-mail: ichoi@kumc.edu

J. Shen, Ph.D.
Section on Magnetic Resonance Spectroscopy, National Institute of Mental Health, National Institutes of Health, Molecular Imaging Branch, Mood & Anxiety Disorders Program, Bethesda, MD, USA
e-mail: shenj@intra.nimh.nih.gov

I.-Y. Choi and R. Gruetter (eds.), *Neural Metabolism In Vivo*,
Advances in Neurobiology 4, DOI 10.1007/978-1-4614-1788-0_39,

major inhibitory neurotransmitter in the mammalian central nervous system (Tapia et al. 1975). The important role of GABA has been suggested in connection with many neurological and psychiatric disorders including epilepsy, Huntington's disease, Parkinson's disease, alcoholism, schizophrenia and depression (Olsen and DeLorey 1999).

GABA is ubiquitously distributed in the brain and its inhibitory function is effective with nearly all neurons (Bloom and Iversen 1971). The concentration of GABA is reported to be about 1 μmol/g in the living human brain (Rothman et al. 1993a; Terpstra et al. 2002; Choi et al. 2004; Choi et al. 2005a; Jensen et al. 2005). However, GABA contents vary quite significantly based on brain regions (Banay-Schwartz et al. 1993) and may reach 10 μmol/g in substantia nigra (Okada et al. 1971) or even greater in other animals (Fonnum and Walberg 1973b). This significant regional variation is in accordance with the specific actions of GABAergic neurons in each region. However, the concentration of GABA in astroglia and interstitial fluid remains at a very low level due to highly active GABA metabolism via transaminase activity (Chan-Palay et al. 1979; Larsson and Schousboe 1990) and active transport, respectively. Therefore, a large concentration gradient of endogenous GABA exists between GABAergic neurons and the neighboring cells such as astroglia under normal physiological conditions, which is about 200 fold (Olsen and DeLorey 1999). Alteration of GABA concentration is often associated with an altered GABAergic function including the onset of convulsions as in seizure activity.

GABA is formed from the metabolism of glucose, the principal precursor for GABA formation *in vivo*. GABA metabolism consists of transamination and decarboxylation processes via GABA α-oxoglutarate transaminase (GABA-T, E.C. 2.6.1.19) and glutamate decarboxylase (GAD), respectively. GABA-T transfers the amino group of GABA to 2-oxoglutarate to yield glutamate and succinic semialdehyde while GAD decarboxylates glutamate to yield GABA. Like glutamate and aspartate, GABA differs from other transmitters in that its metabolism proceeds directly through substances that are important intermediates in the TCA cycle.

In this chapter, we focus on the metabolic pathways of GABA synthesis and catabolism, *in vivo* measurements of GABA metabolism and GABAergic neurotransmission. The compartmentation of various enzymes and metabolic pools participating in GABAergic neurotransmission is emphasized. GABA turnover, its release from presynaptic terminals, and their reuptake into cellular cytoplasm are discussed in relation to glucose and amino acid metabolism occurring within the various compartments involved. The actions of released transmitter GABA on the postsynaptic neuronal membrane and GABA receptors are also covered. The *in vivo* assessment of GABA metabolism can be provided by metabolic flux rate measurement using *in vivo* magnetic resonance spectroscopy (MRS).

39.2 GABAergic Pathways

39.2.1 GABA Synthesis

GABA is synthesized from the α-decarboxylation of glutamate via the action of glutamate decarboxylase (GAD), which is known to regulate the steady-state concentration of GABA (Martin and Rimvall 1993b). Both GABA and GAD are localized to GABAergic neurons although not exclusively. The synthesis of GABA in the central nervous system (CNS) takes place primarily in GABAergic neurons (Hertz 1979). In particular, axon terminals of GABAergic neurons possess a marked capacity for synthesizing GABA, which is then likely to be available for release into the synapse (Fonnum and Walberg 1973a; Martin and Rimvall 1993a).

GAD activity is the rate-limiting step in GABA synthesis and seems directly to participate in the synaptic GABA mechanisms. The regulation of GAD activity is linked to the inter-conversion between the cofactor-bound and free forms of GAD (Miller and Martin 1976). GAD exists in two different forms in the brain: GAD_{65} and GAD_{67} (Erlander and Tobin 1991; Martin and Rimvall 1993b). The two GAD isoforms show differing kinetic properties and cellular distributions suggesting different functional roles for cellular and vesicular GABA (Belhage et al. 1993; Soghomonian and Martin 1998). Within GABAergic neurons, GAD_{65} is highly concentrated in nerve terminals as demonstrated by immunocytochemical studies (McLaughlin et al. 1974; Pinal and Tobin 1998) and density gradient fractionation studies (van Kempen et al. 1965) whereas GAD_{67} is found throughout the neuron and exists mostly in a cofactor-bound active form (Kaufman et al. 1991b; Martin and Rimvall 1993a). GAD_{65} is associated with synaptic vesicles (Kaufman et al. 1991a; Martin and Rimvall 1993b), whereas GAD_{67} serves a nonsynaptic, intracellular GABA pool. In addition to its predominant localization in GABAergic neurons, GAD is also expressed in certain glutamatergic neurons (Sloviter et al. 1996). In comparison to neurons, an overwhelming body of evidence indicates that GAD is absent in astrocytes (Sellstrom et al. 1975; Schousboe et al. 1977; Houser et al. 1980; Hertz and Schouseboe 1986).

Alternate pathways of GABA synthesis from putrescine, which is formed via ornithine decarboxylation, and other polyamines have been suggested (Kremzner et al. 1975; De Mello et al. 1976; Sobue and Nakjima 1978; Murrin 1980; Seiler et al. 1980). However, GABA synthesis from polyamine pathways is quantitatively minor in the brain (Noto et al. 1986) even though polyamines appear to play a significant role in the developing brain (Zawia and Harry 1993) and retina (Yamasaki et al. 1999).

Although GABA is known to be produced exclusively in the neuronal compartment, the relative contributions of the two alternative substrates, neuronal glucose or glial glutamine has been controversial. Glutamine was proposed as an important precursor of the GABA pool (Bradford et al. 1978; Ward et al. 1983; Szerb and O'Regan 1984; Paulsen et al. 1988; Brainard et al. 1989; Fonnum and Paulsen 1990; Sonnewald et al. 1993). Early experiments compared radioactivity of GABA in the intact brain (Ward et al. 1983), brain slices (Szerb 1984), and synaptosomes (Bradford et al. 1978) after treatment by [^{3}H]glutamine and [^{14}C]glucose. However, the role of glucose as a direct precursor of GABA was difficult to assess in these early experiments because [^{14}C]glutamine produced during the glial metabolism of [^{14}C]glucose could also contribute to [^{14}C]GABA formation. Therefore, it was difficult to distinguish if [^{14}C]GABA was produced directly from the neuronal metabolism of glucose or after the formation of glial [^{14}C]glutamine. The conclusion of a more recent study claiming that GABA is formed from a glial pool of TCA cycle intermediates derived from the anaplerotic pathway (Brainard et al. 1989) was based on an erroneous assignment of the NMR signal of [2-^{13}C]GABA as [4-^{13}C]GABA. When [2-^{13}C]acetate was used, both [4-^{13}C]glutamine and [2-^{13}C]GABA were significantly labeled as assessed from ex vivo analysis of brain slices (Badar-Goffer et al. 1990) and co-cultures of cerebral astrocytes and neurons (Sonnewald et al. 1993), which has supported suggestions that glutamine is an important precursor for GABA (Yudkoff et al. 1989; Hassel et al. 1995; Waagepetersen et al. 1999). In contrast, recent *in vivo* MRS results have shown that the labeling of [2-^{13}C] GABA from [2-^{13}C]acetate is much less than observed ex vivo, indicating that the importance of glutamine as a precursor for GABA in the normal brain is less than previously believed. Many studies have also shown that neuronal glucose is quantitatively the major metabolic precursor of GABA in the intact brain (Cerdan et al. 1990; Bachelard and Badar-Goffer 1993; Chapa et al. 1995; Preece and Cerdan 1996). In particular, a recent *in vivo* isotope chase experiment showed that when [^{12}C]glutamine was a main precursor available to GABA, the turnover of GABA was extremely slow, further corroborating that glutamine is a minor metabolic precursor of GABA (Yang et al. 2009).

It has been suggested that GABA synthesis is regulated by product inhibition of GAD_{65} (Liden et al. 1987). In *in vitro* experiments, GABA was found to be a competitive inhibitor of GAD and to be able to convert holoGAD to apoGAD (Martin and Rimvall 1993a). Intracellular GABA levels, however, are not sufficiently high enough to regulate GAD (Neal and Shah 1990; Martin and Rimvall 1993a). There is no evidence of short-term regulation of GAD by second messengers. Most experimental evidence has shown that, unless the concentration of GABA is excessively high (Neal and Shah 1990; Yang and Shen 2005; de Graaf et al. 2006), GAD activity is not inhibited by GABA. In contrast to GAD_{65}, the expression of GAD_{67} is regulated by two mechanisms: (1) mRNA levels and (2) translation and protein stability. The latter mechanism is mediated by intracellular GABA levels (Rimvall and Martin 1994; Sheikh and Martin 1998). The down-regulation of GAD_{67} expression by elevated GABA concentration is also confirmed experimentally using chronic vigabatrin treatment, which causes reduced GAD_{67} protein content (Rimvall and Martin 1994; Sheikh and Martin 1998), development of tolerance to increased potassium-evoked GABA release from rat cerebral cortex (Neal and Shah 1990) as well as reduced GABA turnover rates (Manor et al. 1996; Mason et al. 2001; Patel et al. 2006).

39.2.2 GABA Degradation

GABA is catabolized in mitochondria to succinate in two-steps, which is mediated first by uptake into neurons and glia and second by transamination to succinic semialdehyde via GABA-T. The succinic semialdehyde is then oxidized to succinate via succinic semialdehyde dehydrogenase (SSADH). This metabolic pathway is known as the GABA shunt, which is the major pathway of GABA synthesis and catabolism (Balazs et al. 1970). When GABA-T is inhibited by its irreversible inhibitors such as vigabatrin (γ-vinyl GABA) or gabaculine, there is substantial accumulation of both cytosolic and vesicular GABA in the brain (Jung et al. 1977a; Mattson et al. 1994; Preece and Cerdan 1996). In cell culture studies, GABA-T inhibition has been shown to enhance depolarization-evoked release of endogenous GABA (Wood et al. 1988; Wu et al. 2001). Similarly, increased electrical stimulation-evoked GABA release has also been observed in brain slices treated with GABA-T inhibitors (Kihara et al. 1988). In *in vivo* studies, elevated GABA levels in the brain generated by administration of vigabatrin or gabaculine or by transplantation of genetically engineered GABA-producing cells have been found to correlate with protection against experimentally induced seizures in rodents (Gernert et al. 2002). Increased presynaptic availability and GABA release resulting from GABA-T inhibition are believed to cause the anticonvulsive effects of vigabatrin and gabaculine (Gale and Iadarola 1980). Recently, it has been reported that increasing GABA levels in the rostral agranular insular cortex using either vigabatrin or gene transfer mediated by a GAD_{67}-encoding viral vector leads to an elevated nociceptive threshold, producing long-term analgesia in rats (Jasmin et al. 2003).

39.3 *In Vivo* GABA Measurement and Metabolism

39.3.1 Magnetic Resonance Spectroscopy for Measuring GABA Metabolism

Rapid post-mortem accumulation of GABA in the brain is well known. Among many techniques that have been developed to study GABA metabolism, *in vivo* MRS has the unique ability to monitor GABA, its metabolic precursors and products *in vivo* in real time. Therefore, in principle, MRS is completely unaffected by post-mortem artifacts. For this reason, a brief description of MRS techniques for measuring GABA and its metabolism is provided below.

The concentration of GABA can be measured *in vivo* using proton (^{1}H) MRS techniques. Most MRS methods use spectral editing approaches to suppress the intense creatine methyl proton signal at 3.0 ppm and detect the gamma methylene protons of GABA at the same resonance frequency. Various techniques have been developed that include two-step J editing (Rothman et al. 1993b; Mescher et al. 1996; Mescher et al. 1998), double quantum filtering (Keltner et al. 1997; Shen et al. 2002; Choi et al. 2004), spectral editing by Hartmann-Hahn match (Choi et al. 2005b), and homonuclear polarization transfer (Shen et al. 2004) methods. $^{13}C\{^{1}H\}$ and $^{1}H\{^{13}C\}$ editing techniques, on the other hand, allow one to measure the kinetics of GABA labeling. Most efforts have focused on detecting the labeling (either directly using $^{13}C\{^{1}H\}$ MRS or indirectly using $^{1}H\{^{13}C\}$ MRS techniques) of GABA C2 (Pfeuffer et al. 1999; de Graaf et al. 2003; Yang et al. 2005), which is converted from glutamate C4 by GAD. The labeling of the carboxylic GABA C1 has also been used to study GABA metabolism *in vivo* by $^{13}C\{^{1}H\}$ MRS (Yang et al. 2009).

39.3.2 GABA Turnover

There has been a long history of controversies in the measurement of the GABA turnover rate since the initial isolation of GABA from the brain in 1950 (Awapara et al. 1950; Roberts and Frankel 1950). A large amount of data on the GABA turnover rate have been published using mostly *in vitro* techniques. These GABA turnover studies were based on the conceptual advances made by early studies of monoaminergic dynamics including development of kinetic models, mathematical equations, and procedural techniques (Cheney 1984). To date, the turnover rate of GABA in the brain has not been clearly documented in health and disease, in part due to rapid post-mortem accumulation of GABA associated with ex vivo studies. Due to the complexities of GABA metabolism, it is very challenging to design an optimal method for the measurement of the GABA turnover rate. Most GABA turnover rates refer to the GABA synthesis rate since the turnover rate equals the synthesis and degradation rates during steady state conditions.

39.3.2.1 GABA Turnover Rate by Labeled Precursors

When isotope-labeled glucose is administered to animals, a very quick labeling of GABA in the brain has been observed (Cremer et al. 1978), indicating a relatively high rate of GABA turnover, which has been well established. The rate of GABA turnover can be estimated using an infusion of labeled glucose to determine the rates of label incorporation into glutamate and subsequently GABA. The methodological differences (Neff et al. 1971) and the use of various mathematical models of neurotransmitter functions (Racagni et al. 1974) resulted in a range of GABA turnover rates reported. Early studies, e.g., (Cremer et al. 1978), assumed that GABA was formed from the total glutamate pool. Based on this erroneous assumption, the specific radioactivity of the total glutamate pool was used to determine the rate of GABA and these studies have yielded generally extremely high rates for GABA synthesis of approximately 0.8 μmol/g/min (Cremer et al. 1978) or even higher (Bertilsson et al. 1977). Later immunohistochemical studies have shown that glutamate concentration is quite low in GABAergic neurons (Storm-Mathisen et al.

1983; Ottersen et al. 1992). Since the precursor glutamate pool in GABAergic neurons is much smaller than the large glutamate pool in glutamatergic neurons, the small precursor glutamate pool turns over much faster than the large glutamate pool as reported by early radiolabeling studies (Patel et al. 1974). When the labeling of glucose or pyruvate is approximated to that of the GABA precursor pool, the calculated rate of GABA turnover is significantly reduced. Furthermore, another drawback of those early results is the requirement of an end product in the kinetic models, which are derived from the monoamine transmitter studies. In GABA catabolism, the products are assimilated into the TCA cycle and reused via the GABA shunt. To avoid errors in calculating the GABA turnover rate from changing glutamate specific activity, restricting the models to have constant glutamate labeling rates is required. The problem of the glutamate precursor can be ameliorated by collecting samples very early during the infusion of labeled glucose and by selecting brain regions where the pool of glutamate functions as the GABA precursor. To a certain extent, the normal product-precursor relationship between glutamate and GABA is fulfilled in the nucleus accumbens, nucleus caudatus, septum and tuberculum olfactorium. In those brain regions, the turnover rate of GABA could be reliably estimated using the labeled precursor pool methods (Cheney 1984). In contrast, in the cerebellar cortex, deep cerebellar nuclei, colliculus, hippocampus, or cortical areas, the turnover rate of GABA cannot be determined using similar methods (Revuelta et al. 1981).

In recent high-resolution ^{13}C NMR studies, the labeling of GABA C2 from infused [1-^{13}C]glucose was used to measure the GABA turnover rate (Manor et al. 1996; Hassel et al. 1998; Mason et al. 2001; Patel et al. 2005). In the study performed by Hassel and colleagues (Hassel et al. 1998), the ratio of percent ^{13}C enrichment of [2-^{13}C]GABA to that of [4-^{13}C]glutamate was used to estimate the GABA turnover rate to the overall TCA cycle rate (Hassel et al. 1998). The high GABA turnover rate of ~0.34 μmol/g/min or 17% of the overall TCA cycle flux, which is 2–3 times greater than those calculated on the basis of post-mortem formation of GABA in mice, was reported (Patel et al. 1970; Patel et al. 1974; Paulsen and Fonnum 1987). This high GABA turnover rate is probably due to the fact that the rate is considered to be proportional to ^{13}C enrichment, an assumption that is valid only at very early time points. Higher GABA synthesis rates were also reported based on ex vivo high-resolution MRS analysis of brain extracts (Manor et al. 1996; Mason et al. 2001; Patel et al. 2005). Using prolonged infusion of [1-^{13}C]glucose, ex vivo MRS analysis of brain extracts, and metabolic models of either two non-communicating compartments (glutamatergic and GABAergic neurons) or three communicating compartments (glutamatergic and GABAergic neurons, and astrocytes), the cerebral GABA turnover rate was estimated to be 0.10–0.15 μmol/g/min (20–30% of the total TCA cycle rate) in α-chloralose anesthetized rats (Manor et al. 1996; Mason et al. 2001). In contrast to the above studies which involves post-mortem processes, the GABA turnover rate was found to be 0.05–0.06 μmol/g/min (10–12% of the total TCA cycle rate) in anesthetized rats using *in vivo* ^{1}H[^{13}C] MRS to measure [2-^{13}C]GABA turnover from intravenously infused [1,6-^{13}C2]glucose (Yang and Shen 2005).

39.3.2.2 GABA Turnover Rate by GABA-T Inhibition

Since essentially all GABA is catabolized by GABA-T in the brain, inhibition of GABA-T has been used to measure the rate of GABA degradation and thus the rate of GABA turnover (Metcalf 1979; Fonnum 1985). Many inhibitors of GABA-T have been used for this purpose including aminooxyacetic acid (Wood 1975; Loscher and Frey 1978; Bernasconi et al. 1982; Loscher et al. 1989), cycloserine (Wood et al. 1978), gabaculine (Rando and Bangerter 1977; Bernasconi et al. 1982; Giorgi and Meek 1984; Behar et al. 1994; Yang et al. 2005; Patel et al. 2006), vigabatrin (Jung et al. 1977a; Casu and Gale 1981; Chapman and Evans 1983; Paulsen and Fonnum 1987; Paulsen and Fonnum 1988; Paulsen et al. 1988) and γ-acetylenic GABA (Jung et al. 1977b). Generally, the rate of GABA turnover measured using GABA-T inhibitors is all in the 0.03–0.07 μmol/g/min range in animals. Caution should be exercised when interpreting GABA degradation rates measured using inhibition of GABA-T because of (1) possible regionally dependent incomplete inhibition of GABA-T after systemic administration of GABA-T inhibitors; (2) potential limitation by the supply of precursors due to the relatively fast GABA turnover rate and (3) inhibition of GAD by certain GABA-T inhibitors such as aminooxyacetic acid and γ-acetylenic GABA (Bernasconi et al. 1988). All these conditions listed above could lead to underestimation of the rate of GABA turnover. However, the advantage of the GABA-T inhibition method is that the turnover rate of GABA can be measured independent of the type of GABA precursors, allowing measurement of turnover in brain regions where glutamate functions predominantly as an intermediate metabolite. Among the GABA-T inhibitors that have been investigated, gabaculine appears to be most suitable for measuring the GABA turnover rate because of its apparent ability to produce virtually complete inhibition of GABA-T within a short period of time (Rando and Bangerter 1977; Forchetti et al. 1982; Paulsen and Fonnum 1987). Local administration of vigabatrin by microinjection has also been a popular method for measuring GABA turnover in different brain regions (Paulsen and Fonnum 1987). Vigabatrin has a rapid onset of action and the effect is stable for an extended period of time (Jung et al. 1977a). The use of vigabatrin also avoids some systemic effects such as sedation (Casu and Gale 1981).

39.3.2.3 GABA Turnover Rate by Post Mortal Increase in GABA

The post-mortem GABA accumulation method (Patel et al. 1970, 1974; Paulsen and Fonnum 1987) is based on the fact that GABA catabolism is nearly completely inhibited because of anoxia after death, as a result of depletion of α-ketoglutarate and NAD. Post-mortem GABA accumulation was found to be nonlinear with the greatest rate of GABA accumulation occurring between 1 and 2 min post-mortem at 30 μg/g/min (Alderman and Shellenberger 1974). This nonlinearity is presumably due to compartmentalization of amino acids in the brain (Patel et al. 1974) and possibly rapid post mortal diffusion of GABA from regions with high GABA concentration to adjacent low concentration (Baxter et al. 1989). The usefulness of post-mortem methods is limited by two factors: (1) To simulate *in vivo* conditions, short incubation times are necessary. However, this short duration introduces experimental difficulties because a small increase in GABA levels will give a low degree of reproducibility. (2) GABA catabolism is not completely inhibited post-mortem. It was found that treatment with a GABA-T inhibitor before euthanasia increases the post-mortem accumulation of GABA by as much as 40–70% (Yamatsu et al. 1982). Using post-mortem methods, the GABA turnover rate was reported to be approximately 8% of the total cerebral TCA cycle activity (Paulsen and Fonnum 1987).

39.3.2.4 Factors Modulating GABA Turnover

When a high potassium ion concentration is present in the medium, GABA synthesis is significantly increased (Machiyama et al. 1970). Using isolated synaptosomes, electrical stimulation was shown to increase incorporation of glucose carbon into GABA (De Belleroche and Bradford 1972). In bicuculline-induced seizure models, the rate of accumulation of cortical GABA levels is approximately threefold greater than that in the control group. The results suggest that the GABA turnover rate increases in parallel with increases in the overall cortical metabolic rate observed during bicuculline seizure (Chapman and Evans 1983). GABA turnover is also known to be regulated by related receptor activity (Lindgren 1987). In general, increased activity of the neural pathway should be reflected in increased turnover of the released neurotransmitter including GABA. Electrical stimulation experiments have demonstrated that the turnover of GABA increases when the activity of the GABAergic pathway is enhanced and decreases when the activity is decreased (Mao et al. 1978). Stimulation of opiate receptors exerts an inhibitory influence on the turnover rate of GABA in nucleus caudatus (Moroni et al. 1978). Potentiation of postsynaptic GABAergic transmission by benzodiazepines and hypoglycemia also lead to downregulation of GABA turnover (Paulsen and Fonnum 1987).

39.3.3 GABAergic Neurotransmission

The functional activity in the brain is primarily composed of interplay between excitation and inhibition. In any brain region, the output is based on a complex processing of incoming signals that require both excitatory and inhibitory units. In each of these units or synapses, the neural activity depends on biosynthesis, release, interaction with receptors, and inactivation of the neurotransmitters. The GABAergic neurotransmission processes are initiated by the release of neurotransmitter GABA and its subsequent interactions with receptors on the postsynaptic neurons and in some cases on presynaptic neurons. GABA is cleared from the synaptic cleft by molecular diffusion and via the neurotrasmitter uptake, to a large extent by the presynaptic GABAergic neurons. The astrocytic processes surrounding the synapses also take up released neurotransmitter GABA. The components of the transmission of the primary inhibitory neurotransmitter GABA are reviewed below.

39.3.3.1 GABA Release and GABA Receptors

GABA is released exocytotically via a calcium-dependent vesicular mechanism (Sihra and Nicholls 1987) and via a calcium-independent nonvesicular mechanism by thermodynamic reversal of GABA transporters. $GABA_A$ receptors are ligand-gated ion channels (also known as ionotropic receptors) whereas $GABA_B$ receptor acts on membrane channels indirectly through a G-protein. $GABA_A$ receptors are the principal mediators of synaptic inhibition, and yet when intensely activated, dendritic $GABA_A$ receptors excite rather than inhibit neurons (Staley et al. 1995).

A potassium-induced release of GABA from primary cultures of cerebral cortical interneurons has been repeatedly demonstrated (Larsson et al. 1983). Astrocytes do not contribute to the potassium-induced release of GABA (Hertz and Schousboe 1980; Pearce et al. 1981; Hertz and Schousboe 1987). The synaptic action of GABA released from neurons by calcium-dependent vesicular release is terminated by uptake mainly into neurons (Isaacson et al. 1993). The uptake carrier derives its energy to accumulate GABA from the co-transport of two sodium ions and one chloride ion into the cell with each GABA molecule (Radian and Kanner 1983). A carrier with this stoichiometry transports net positive charge into the cell and must be able to run backward, releasing GABA into the extracellular space if intracellular [Na^+] rises high enough or if the membrane potential becomes sufficiently positive. In brain slice and synaptosomal preparations, GABA release can be evoked by depolarization with potassium or glutamate. This can occur even when vesicular release is inhibited by removal of extracellular calcium (Bernath 1992). Until recently, reversal of GABA transporters had been considered to be significant only during pathological conditions such as ischemia (Levi and Raiteri 1993). Recent studies have shown that GABA transporters reverse quite easily and are highly sensitive to changes in GABA concentration gradient and membrane potential (Attwell et al. 1993).

Carrier-mediated GABA release can occur in response to vigabatrin treatment (Wu et al. 2001). In an experiment using patch-clamp recording of rat hippocampal neurons, vigabatrin was shown to induce spontaneous GABA release via reversal of GABA transporters (Wu et al. 2001). Because this nonvesicular GABA release can inhibit surrounding neurons (Gaspary et al. 1998) and thus could contribute to the anticonvulsant properties of vigabatrin. GABA released via GABA transporter reversal may also make a large contribution to phasic GABAergic inhibition during high-frequency firing (Jackson et al. 2000; Wu et al. 2003). These results have established the functional importance of nonvesicular GABA release (During et al. 1995; Drew et al. 1997; Wu et al. 2001, 2003; Chen et al. 2005).

Many factors have been shown to regulate GABA release. However, our discussion is restricted to the effects of certain anticonvulsants in this chapter. Vigabatrin showed a preferential inhibition of neuronal GABA-T with IC50 values well below clinically relevant serum levels of vigabatrin (Larsson et al. 1986). Vigabatrin causes a pronounced increase in both cellular and releasable GABA pools. Acute administration of vigabatrin and other GABA-elevating drugs increases GABA efflux *in vivo*, spontaneous resting release of GABA and potassium-evoked GABA release from brain slices (Gram et al. 1988; Kihara et al. 1988; Neal and Shah 1990;

Jolkkonen et al. 1992; Qume et al. 1995; Smolders et al. 1997; Wu et al. 2001; Benturquia et al. 2004). Increased GABA release caused by acute inhibition of GABA-T has also been shown to quench the blood-oxgenation-level-dependent (BOLD) response to functional stimulation (Chen et al. 2005). In addition, anticonvulsant drugs which elevate brain GABA levels and which have little or no effect on GAD activity show a good correlation between the elevation in synatosomal GABA levels and an increase in whole brain GABA contents (Wood 1980).

39.3.3.2 GABA Uptake

The inactivation of GABA is primarily maintained by a highly efficient, high-affinity GABA transport system capable of maintaining a transmembrane concentration gradient of 10^4–10^5 fold. GABA transporters belong to a superfamily of Na^+/Cl^- dependent neurotransmitter transporters, with GAT1 and GAT3 being exclusively expressed in the brain. Among the GABA transporters, GAT1 is the most abundantly expressed and preferentially located on neurons (Gadea and Lopez-Colome 2001; Schousboe and Kanner 2002). GAT3 and GAT4 are primarily expressed in glia and other non-neuronal cells (De Biasi et al. 1998; Minelli et al. 2003). It is well established that GABAergic neurotransmission is based on recycling of neurotransmitter GABA (Hertz and Schousboe 1987), i.e., most released GABA is taken back into GABAergic nerve endings, thus enabling its incorporation into GABA vesicles and subsequent re-release (Schousboe 2003). Astroglial GABA transport has also been found to be of functional significance. A correlation between the anticonvulsant activity of GABA transporter inhibitors and their ability to selectively inhibit astroglial GABA uptake has been experimentally demonstrated. In comparison, no such correlation has been demonstrated for neuronal GABA uptake (White et al. 2002). Recent evidence has clearly shown that inhibitors of astroglial GABA transport can significantly increase the availability of GABA in the neurotransmitter pool (Sarup et al. 2003; Schousboe 2003).

The uptake of GABA is markedly Na^+-dependent. Experimental evidence of the energy requirement of GABA uptake or inhibition has been controversial. [^{14}C]2-deoxyglucose experiments have shown that inhibition is a metabolically active process (Ackermann et al. 1984; Nudo and Masterton 1986). However, fMRI of the inhibited motor cortex of healthy humans showed no measurable changes in the blood-oxygen-level-dependent signal in the motor cortex, indicating minimal metabolic cost associated with inhibition (Waldvogel et al. 2000). Glucose metabolism and GABA-glutamine cycling have been found to strongly correlate with anesthesia-induced changes in brain activity levels (Patel et al. 2005), suggesting significant contribution from GABAergic neurons and inhibition to cortical energy metabolism. In contrast, no metabolic cost associated with GABA uptake into astrocytes was measurable in cultured mouse cortical astrocytes (Chatton et al. 2003). Since the majority of released transmitter GABA is taken up back to GABAergic neurons, the energy cost associated with neuronal reuptake of GABA appears to be more relevant for assessment of the energy cost of inhibition.

39.3.3.3 GABA-Glutamine Cycling

Axon terminals of GABAergic neurons possess a high capacity for synthesizing GABA, which is then available for release (Fonnum and Walberg 1973a). Once released, GABA can be taken up into the GABAergic nerve endings, or into glial cells as described above (Sect. 39.3.3.2). Both neurons and astrocytes contain significant activity of GABA-T (Larsson et al. 1986). Astrocytic uptake of GABA, which is functionally significant, has to be compensated by a flow of precursors from astrocytes to neurons allowing replenishment of the neurotransmitter pool (Van den Berg and Garfinkel 1971; Tapia 1983; Cerdan et al. 1990; Lapidot and Gopher 1994; Preece and Cerdan 1996; Waagepetersen et al. 2003). This is due to the lack of a quantitatively important anaplerotic pathway in neurons (Cremer et al. 2010). The most prominent GABA precursor synthesized in astrocytes is glutamine (Patel et al. 2005; Yang et al. 2007). This GABA-glutamine cycle plays an important role in GABA neurotransmission. For example, the GABA-glutamine cycle has been found to regulate $GABA^{A}$ subunit composition (Cremer et al. 2010).

There are large disparities in the estimated GABA-glutamine cycling flux rate (Van den Berg and Garfinkel 1971; Yu et al. 1984a; Hassel et al. 1995; Patel et al. 2005; Yang et al. 2007). Based on computer modeling of data obtained using ^{14}C labeled glucose and acetate and assuming that the major metabolic flux connecting neurons and glia is the GABA-glutamine cycling flux, the GABA-glutamine cycling flux was estimated to be 0.14 μmol/g/min in awake mouse brain (9%) of total TCA cycle (Van den Berg and Garfinkel 1971), which is approximately twice of the value for the rat reported by Dzubow (Dzubow and Garfinkel 1970). Using intravenous infusion of [2-^{13}C]acetate into awake mice, the cerebral GABA-glutamine cycling rate was calculated to be 0.22 μmol/g/min (Hassel et al. 1995). An even higher value of 0.36 μmol/g/min was calculated by Cremer (Cremer et al. 1978; Cremer et al. 2010). Since GABA derived from astrocytic glutamine via the GABA-glutamine cycling flux has to be converted into GABA by GAD, which is absent in astrocytes, the rate of GABA-glutamine cycling flux cannot exceed that of GABA synthesis. Therefore, the high values of GABA-glutamine cycling rates should not be considered realistic. Since the activity in glutamatergic synapses expressed as units transmitter released and processed per time unit greatly exceeds that of GABAergic synapses (Hertz and Schousboe 1987; Schousboe et al. 2004), the GABA-glutamine cycling rate *in vivo* should be much smaller than the glutamate-glutamine cycling rate. The metabolic flux from glutamine toward GABA has also been measured in cerebral cortical neurons by incubating the cells with labeled glutamine and measuring incorporation of radioactivity into glutamate and GABA. Such experiments have shown a flux from glutamine via glutamate to GABA of about 0.03 μmol/g/min (Yu et al. 1984a). This flux is well below the measured GABA turnover rate and is comparable with the rate of GABA release (Yu et al. 1984b) and the GABA-glutamine cycling rate measured using *in vivo* MRS (Yang et al. 2007).

39.4 Disorders of GABA Metabolism

A variety of clinical disorders involve GABA metabolism and neurotransmission. It is well-established that GABAergic dysfunction plays a key role in several neurological disorders such as epilepsy (Pan et al. 2008). Clinical studies have demonstrated that vigabatrin, the highly specific GABA-T inhibitor, is an effective anticonvulsant drug in drug-resistant partial epilepsy with the elevated GABA concentration in the brain correlating with improved seizure control (Ben-Menachem et al. 1989; Riekkinen et al. 1989; Petroff et al. 1996). The inhibition of GABA-T by vigabatrin leads to augmented presynaptic availability and release of endogenous GABA and therefore reduced cortical excitability, which has been considered the mechanism by which vigabatrin exerts its anticonvulsant action in epileptic patients (Gram et al. 1989; Schechter and Sjoerdsma 1990). Vigabatrin has also been shown to depress the cerebral metabolic rate for glucose (CMR_{Glc}) and cerebral blood flow (CBF) in epileptic patients measured with ^{18}F-fluorodeoxyglucose and ^{15}O-water positron emission tomography (PET) with the decrease in CMR_{Glc} found to correlate with the increase in total cerebrospinal fluid (CSF) GABA concentration (Spanaki et al. 1999).

In contrast to vigabatrin which has a clear GABAergic mechanism of action, the possible GABAergic mechanism of action of valproate has been extensively debated ever since its introduction as an antileptic drug. Valporate influences a variety of other parameters which may be relevant to its mechanism of action as an anticonvulsant (Chapman et al. 1982). Studies investigating the influence of valporate on GABA synthesis have been conflicting, some demonstrated increased GAD levels (Loscher 1981) while others observing decreased GAD levels following long-term treatment with valporate (Emson 1976). Valporate may also inhibit GABA-T. However, the values of reported IC50 concentrations of valporate for the inhibition of GABA-T vary considerably with a general tendency towards concentrations being above clinically relevant levels (Larsson et al. 1986). A differential sensitivity of glial and neuronal GABA-T to valporate has also been demonstrated. Neuronal GABA-T was very sensitive to valporate with IC50 values approaching clinically relevant values (Larsson et al. 1986). Valporate has been found to be able to provoke a significant increase in evoked release of endogenous GABA (Gram et al. 1988).

Increasing evidence exists to suggest that major psychiatric disorders (e.g., schizophrenia and major depressive disorder) are associated with perturbations in the metabolism of GABA. Early studies reported that GABA level is abnormally low in the plasma of patients with unipolar depression. Post-mortem studies of cerebrospinal fluid GABA concentrations also identified abnormalities in GABAergic function in depression. Recently, *in vivo* MRS measurement of cerebral GABA concentrations in patients with major depressive disorders has consistently found reduced GABA levels in both occipital and prefrontal cortices (Sanacora et al. 1999; Hasler et al. 2007). This is consistent with the mood-stabilizing and antidepressant effects of some GABA-mimetic and anticonvulsant drugs.

39.5 Conclusions

The previous sections of this chapter have described major events and pathways involved in GABA metabolism. The role of GABAergic inhibition in the functions of the brain critically depends on its metabolic and transport mechanisms that are related to GABAergic neurotransmission. The knowledge of these biochemical processes of GABA is crucial to our understanding of various neuropsychiatric diseases and their pharmacological treatments.

References

Ackermann RF, Finch DM, Babb TL, Engel J Jr (1984) Increased glucose metabolism during long-duration recurrent inhibition of hippocampal pyramidal cells. J Neurosci 4:251–264

Alderman JL, Shellenberger MK (1974) Gamma-aminobutyric acid (GABA) in the rat brain: re-evaluation of sampling procedures and the post-mortem increase. J Neurochem 22:937–940

Attwell D, Barbour B, Szatkowski M (1993) Nonvesicular release of neurotransmitter. Neuron 11:401–407

Awapara J, Landua AJ, Fuerst R, Seale B (1950) Free gamma-aminobutyric acid in brain. J Biol Chem 187:35–39

Bachelard H, Badar-Goffer R (1993) NMR spectroscopy in neurochemistry. J Neurochem 61:412–429

Badar-Goffer RS, Bachelard HS, Morris PG (1990) Cerebral metabolism of acetate and glucose studied by 13C-n.m.r. spectroscopy. A technique for investigating metabolic compartmentation in the brain. Biochem J 266:133–139

Balazs R, Machiyama Y, Hammond BJ, Julian T, Richter D (1970) The operation of the gamma-aminobutyrate bypath of the tricarboxylic acid cycle in brain tissue in vitro. Biochem J 116:445–461

Banay-Schwartz M, Palkovits M, Lajtha A (1993) Heterogeneous distribution of functionally important amino acids in brain areas of adult and aging humans. Neurochem Res 18:417–423

Baxter CF, Parsons JE, Oh CC, Wasterlain CG, Baldwin RA (1989) Changes of amino acid gradients in brain tissues induced by microwave irradiation and other means. Neurochem Res 14: 909–913

Bazemore A, Elliott KA, Florey E (1956) Factor I and gamma-aminobutyric acid. Nature 178:1052–1053

Behar KL, Rothman DL, Spencer DD, Petroff OA (1994) Analysis of macromolecule resonances in 1H NMR spectra of human brain. Magn Reson Med 32:294–302

Belhage B, Hansen GH, Schousboe A (1993) Depolarization by K+and glutamate activates different neurotransmitter release mechanisms in GABAergic neurons: vesicular versus non-vesicular release of GABA. Neuroscience 54:1019–1034

Ben-Menachem E, Persson LI, Schechter PJ, Haegele KD, Huebert N, Hardenberg J, Dahlgren L, Mumford JP (1989) The effect of different vigabatrin treatment regimens on CSF biochemistry and seizure control in epileptic patients. Br J Clin Pharmacol 27(Suppl 1):79S–85S

Benturquia N, Parrot S, Sauvinet V, Renaud B, Denoroy L (2004) Simultaneous determination of vigabatrin and amino acid neurotransmitters in brain microdialysates by capillary electrophoresis with laser-induced fluorescence detection. J Chromatogr B Analyt Technol Biomed Life Sci 806:237–244

Bernasconi R, Klein M, Martin P, Christen P, Hafner T, Portet C, Schmutz M (1988) Gamma-vinyl GABA: comparison of neurochemical and anticonvulsant effects in mice. J Neural Transm 72:213–233

Bernasconi R, Maitre L, Martin P, Raschdorf F (1982) The use of inhibitors of GABA-transaminase for the determination of GABA turnover in mouse brain regions: an evaluation of aminooxyacetic acid and gabaculine. J Neurochem 38:57–66
Bernath S (1992) Calcium-independent release of amino acid neurotransmitters: fact or artifact? Prog Neurobiol 38:57–91
Bertilsson L, Mao CC, Costa E (1977) Application of principles of steady-state kinetics to the estimation of gamma-aminobutyric acid turnover rate in nuclei of rat brain. J Pharmacol Exp Ther 200:277–284
Bloom FE, Iversen LL (1971) Localizing 3H-GABA in nerve terminals of rat cerebral cortex by electron microscopic autoradiography. Nature 229:628–630
Bradford HF, Ward HK, Thomas AJ (1978) Glutamine–a major substrate for nerve endings. J Neurochem 30:1453–1459
Brainard JR, Kyner E, Rosenberg GA (1989) 13C nuclear magnetic resonance evidence for gamma-aminobutyric acid formation via pyruvate carboxylase in rat brain: a metabolic basis for compartmentation. J Neurochem 53:1285–1292
Casu M, Gale K (1981) Intracerebral injection of gamma vinyl GABA: method for measuring rates of GABA synthesis in specific brain regions in vivo. Life Sci 29:681–688
Cerdan S, Kunnecke B, Seelig J (1990) Cerebral metabolism of [1,2-13C2]acetate as detected by in vivo and in vitro 13C NMR. J Biol Chem 265:12916–12926
Chan-Palay V, Wu JY, Palay SL (1979) Immunocytochemical localization of gamma-aminobutyric acid transaminase at cellular and ultrastructural levels. Proc Natl Acad Sci USA 76: 2067–2071
Chapa F, Kunnecke B, Calvo R, Escobar del Rey F, Morreale de Escobar G, Cerdan S (1995) Adult-onset hypothyroidism and the cerebral metabolism of (1,2-13C2) acetate as detected by 13C nuclear magnetic resonance. Endocrinology 136:296–305
Chapman A, Keane PE, Meldrum BS, Simiand J, Vernieres JC (1982) Mechanism of anticonvulsant action of valproate. Prog Neurobiol 19:315–359
Chapman AG, Evans MC (1983) Cortical GABA turnover during bicuculline seizures in rats. J Neurochem 41:886–889
Chatton JY, Pellerin L, Magistretti PJ (2003) GABA uptake into astrocytes is not associated with significant metabolic cost: implications for brain imaging of inhibitory transmission. Proc Natl Acad Sci USA 100:12456–12461
Chen Z, Silva AC, Yang J, Shen J (2005) Elevated endogenous GABA level correlates with decreased fMRI signals in the rat brain during acute inhibition of GABA transaminase. J Neurosci Res 79:383–391
Cheney D (1984) Drug effects on transmitter dynamics: an overview. Dynamics of neurotransmitter function. Raven, New York, pp 63–80
Choi C, Coupland NJ, Hanstock CC, Ogilvie CJ, Higgins AC, Gheorghiu D, Allen PS (2005a) Brain gamma-aminobutyric acid measurement by proton double-quantum filtering with selective J rewinding. Magn Reson Med 54:272–279
Choi I-Y, Lee S-P, Merkle H, Shen J (2004) Single-shot two-echo technique for simultaneous measurement of GABA and creatine in the human brain in vivo. Magn Reson Med 51:1115–1121
Choi I-Y, Lee S-P, Shen J (2005b) Selective homonuclear Hartmann-Hahn transfer method for in vivo spectral editing in the human brain. Magn Reson Med 53:503–510
Cremer CM, Bidmon HJ, Gorg B, Palomero-Gallagher N, Escobar JL, Speckmann EJ, Zilles K (2010) Inhibition of glutamate/glutamine cycle in vivo results in decreased benzodiazepine binding and differentially regulated GABAergic subunit expression in the rat brain. Epilepsia 51:1446–1455
Cremer J, Sarna G, Teal H, Cunningham V (1978) Amino acid precursors: their transport into brain and initial metabolism, in Amino acids as chemical transmitters. In: Fonnum F (ed) Amino Acids as Chemical Transmitters. Plenum Press, New York, pp 669–689
De Belleroche JS, Bradford HF (1972) Metabolism of beds of mammalian cortical synaptosomes: response to depolarizing influences. J Neurochem 19:585–602

De Biasi S, Vitellaro-Zuccarello L, Brecha NC (1998) Immunoreactivity for the GABA transporter-1 and GABA transporter-3 is restricted to astrocytes in the rat thalamus. A light and electron-microscopic immunolocalization. Neuroscience 83:815–828

de Graaf RA, Mason GF, Patel AB, Behar KL, Rothman DL (2003) In vivo 1H-[13C]-NMR spectroscopy of cerebral metabolism. NMR Biomed 16:339–357

de Graaf RA, Patel AB, Rothman DL, Behar KL (2006) Acute regulation of steady-state GABA levels following GABA-transaminase inhibition in rat cerebral cortex. Neurochem Int 48:508–514

De Mello FG, Bachrach U, Nirenberg M (1976) Ornithine and glutamic acid decarboxylase activities in the developing chick retina. J Neurochem 27:847–851

Drew KL, Fitka T, Hu Y, Ungerstedt U (1997) Non-exocytotic GABA overflow in rat striatum inhibits gnawing. Life Sci 61:1593–1601

During MJ, Ryder KM, Spencer DD (1995) Hippocampal GABA transporter function in temporal-lobe epilepsy. Nature 376:174–177

Dzubow LM, Garfinkel D (1970) A simulation study of brain compartments. II. Atom-by-atom simulation of the metabolism of specifically labeled glucose and acetate. Brain Res 23:407–417

Emson PC (1976) Effects of chronic treatment with amino-oxyacetic acid or sodium n-dipropylacetate on brain GABA levels and the development and regression of cobalt epileptic foci in rats. J Neurochem 27:1489–1494

Erlander MG, Tobin AJ (1991) The structural and functional heterogeneity of glutamic acid decarboxylase: a review. Neurochem Res 16:215–226

Fonnum F (1985) Neuromethods. Humana Press, New Jersey, 3: Amino Acids:201

Fonnum F, Paulsen RE (1990) Comparison of transmitter amino acid levels in rat globus pallidus and neostriatum during hypoglycemia or after treatment with methionine sulfoximine or gamma-vinyl gamma-aminobutyric acid. J Neurochem 54:1253–1257

Fonnum F, Walberg F (1973a) The concentration of GABA within inhibitory nerve terminals. Brain Res 62:577–579

Fonnum F, Walberg F (1973b) An estimation of the concentration of g-aminobutyric acid and glutamate decarboxylase in the inhibitory Purkinje axon terminals in the cat. Brain Res 54:115–127

Forchetti CM, Marco EJ, Meek JL (1982) Serotonin and gamma-aminobutyric acid turnover after injection into the median raphe of substance P and D-ala-met-enkephalin amide. J Neurochem 38:1336–1341

Gadea A, Lopez-Colome AM (2001) Glial transporters for glutamate, glycine, and GABA: II. GABA transporters. J Neurosci Res 63:461–468

Gale K, Iadarola MJ (1980) Seizure protection and increased nerve-terminal GABA: delayed effects of GABA transaminase inhibition. Science 208:288–291

Gaspary HL, Wang W, Richerson GB (1998) Carrier-mediated GABA release activates GABA receptors on hippocampal neurons. J Neurophysiol 80:270–281

Gernert M, Thompson KW, Loscher W, Tobin AJ (2002) Genetically engineered GABA-producing cells demonstrate anticonvulsant effects and long-term transgene expression when transplanted into the central piriform cortex of rats. Exp Neurol 176:183–192

Giorgi O, Meek JL (1984) Gamma-aminobutyric acid turnover in rat striatum: effects of glutamate and kainic acid. J Neurochem 42:215–220

Gram L, Larsson OM, Johnsen A, Schousboe A (1989) Experimental studies of the influence of vigabatrin on the GABA system. Br J Clin Pharmacol 27(Suppl 1):13S–17S

Gram L, Larsson OM, Johnsen AH, Schousboe A (1988) Effects of valproate, vigabatrin and aminooxyacetic acid on release of endogenous and exogenous GABA from cultured neurons. Epilepsy Res 2:87–95

Hasler G, van der Veen JW, Tumonis T, Meyers N, Shen J, Drevets WC (2007) Reduced prefrontal glutamate/glutamine and gamma-aminobutyric acid levels in major depression determined using proton magnetic resonance spectroscopy. Arch Gen Psychiatry 64:193–200

Hassel B, Johannessen CU, Sonnewald U, Fonnum F (1998) Quantification of the GABA shunt and the importance of the GABA shunt versus the 2-oxoglutarate dehydrogenase pathway in GABAergic neurons. J Neurochem 71:1511–1518

Hassel B, Sonnewald U, Fonnum F (1995) Glial-neuronal interactions as studied by cerebral metabolism of [2-13C]acetate and [1-13C]glucose: an ex vivo 13C NMR spectroscopic study. J Neurochem 64:2773–2782

Hertz L (1979) Functional interactions between neurons and astrocytes I. Turnover and metabolism of putative amino acid transmitters. Prog Neurobiol 13:277–323

Hertz L, Schousboe A (1980) Interactions between neurons and astrocytes in the turnover of GABA and glutamate. Brain Res Bull 5:389–395

Hertz L, Schousboe A (1987) Primary cultures of GABAergic and glutamatergic neurons as model systems to study neurotransmitter functions. I. Differentiated cells. In: Vernadarkis A (ed) Model systems of development and aging of the nervous system. Martinus Nijhoff, Boston, pp 19–31

Hertz L, Schouseboe A (1986) Role of astrocytes in compartmentation of amino acid and energy metabolism. In: Fedoroff S and Vernadarkis A (eds) Astrocytes. Academic Press, New York

Houser CR, Vaughn JE, Barber RP, Roberts E (1980) GABA neurons are the major cell type of the nucleus reticularis thalami. Brain Res 200:341–354

Isaacson JS, Solis JM, Nicoll RA (1993) Local and diffuse synaptic actions of GABA in the hippocampus. Neuron 10:165–175

Jackson MF, Esplin B, Capek R (2000) Reversal of the activity-dependent suppression of GABA-mediated inhibition in hippocampal slices from gamma-vinyl GABA (vigabatrin)-pretreated rats. Neuropharmacology 39:65–74

Jasmin L, Rabkin SD, Granato A, Boudah A, Ohara PT (2003) Analgesia and hyperalgesia from GABA-mediated modulation of the cerebral cortex. Nature 424:316–320

Jensen JE, Frederick BD, Wang L, Brown J, Renshaw PF (2005) Two-dimensional, J-resolved spectroscopic imaging of GABA at 4 Tesla in the human brain. Magn Reson Med 54:783–788

Jolkkonen J, Mazurkiewicz M, Lahtinen H, Riekkinen P (1992) Acute effects of gamma-vinyl GABA on the GABAergic system in rats as studied by microdialysis. Eur J Pharmacol 229:269–272

Jung MJ, Lippert B, Metcalf BW, Bohlen P, Schechter PJ (1977a) gamma-Vinyl GABA (4-amino-hex-5-enoic acid), a new selective irreversible inhibitor of GABA-T: effects on brain GABA metabolism in mice. J Neurochem 29:797–802

Jung MJ, Lippert B, Metcalf BW, Schechter PJ, Bohlen P, Sjoerdsma A (1977b) The effect of 4-amino hex-5-ynoic acid (gamma-acetylenic GABA, gammma-ethynyl GABA) a catalytic inhibitor of GABA transaminase, on brain GABA metabolism in vivo. J Neurochem 28:717–723

Kaufman DL, Houser CR, Tobin AJ (1991a) Two forms of the gamma-aminobutyric acid synthetic enzyme glutamate decarboxylase have distinct intraneuronal distributions and cofactor interactions. J Neurochem 56:720–723

Kaufman DL, Houser CR, Tobin AJ (1991b) Two forms of the gamma-aminobutyric acid synthetic enzyme glutamate decarboxylase have distinct intraneuronal distributions and cofactor interactions. J Neurochem 56:720–723

Keltner JR, Wald LL, Frederick BD, Renshaw PF (1997) In vivo detection of GABA in human brain using a localized double-quantum filter technique. Magn Reson Med 37:366–371

Kihara M, Misu Y, Kubo T (1988) GABA transaminase inhibitors enhance the release of endogenous GABA but decrease the release of beta-alanine evoked by electrical stimulation of slices of the rat medulla oblongata. Life Sci 42:1817–1824

Kremzner LT, Hiller JM, Simon EJ (1975) Metabolism of polyamines in mouse neuroblastoma cells in culture: formation of GABA and putreanine. J Neurochem 25:889–894

Lapidot A, Gopher A (1994) Cerebral metabolic compartmentation. Estimation of glucose flux via pyruvate carboxylase/pyruvate dehydrogenase by 13C NMR isotopomer analysis of D-[U-13C] glucose metabolites. J Biol Chem 269:27198–27208

Larsson OM, Drejer J, Hertz L, Schousboe A (1983) Ion dependency of uptake and release of GABA and (RS)-nipecotic acid studied in cultured mouse brain cortex neurons. J Neurosci Res 9:291–302

Larsson OM, Gram L, Schousboe I, Schousboe A (1986) Differential effect of gamma-vinyl GABA and valproate on GABA-transaminase from cultured neurones and astrocytes. Neuropharmacology 25:617–625

Larsson OM, Schousboe A (1990) Kinetic characterization of GABA-transaminase from cultured neurons and astrocytes. Neurochem Res 15:1073–1077

Levi G, Raiteri M (1993) Carrier-mediated release of neurotransmitters. Trends Neurosci 16:415–419

Liden E, Karlsson L, Sellstrom A (1987) Is the concentration of gamma-aminobutyric acid in the nerve terminal regulated via product inhibition of glutamic acid decarboxylase? Neurochem Res 12:489–493

Lindgren S (1987) Effects of GABAergic drugs on the GABA turnover in the substantia nigra and the corpus striatum of the rat. J Neural Transm 69:33–46

Loscher W (1981) Valproate induced changes in GABA metabolism at the subcellular level. Biochem Pharmacol 30:1364–1366

Loscher W, Frey HH (1978) Aminooxyacetic acid: correlation between biochemical effects, anticonvulsant action and toxicity in mice. Biochem Pharmacol 27:103–108

Loscher W, Honack D, Gramer M (1989) Use of inhibitors of gamma-aminobutyric acid (GABA) transaminase for the estimation of GABA turnover in various brain regions of rats: a reevaluation of aminooxyacetic acid. J Neurochem 53:1737–1750

Machiyama Y, Balazs R, Hammond BJ, Julian T, Richter D (1970) The metabolism of gamma-aminobutyrate and glucose in potassium ion-stimulated brain tissue in vitro. Biochem J 116:469–481

Manor D, Rothman DL, Mason GF, Hyder F, Petroff OA, Behar KL (1996) The rate of turnover of cortical GABA from [1-13C]glucose is reduced in rats treated with the GABA-transaminase inhibitor vigabatrin (gamma- vinyl GABA). Neurochem Res 21:1031–1041

Mao CC, Peralta E, Morini F, Costa E (1978) The turnover rate of gamma-aminobutyric acid in the substantia nigra following electrical stimulation or lesioning of the strionigral pathways. Brain Res 155:147–152

Martin DL, Rimvall K (1993a) Regulation of gamma-aminobutyric acid synthesis in the brain. J Neurochem 60:395–407

Martin DL, Rimvall K (1993b) Regulation of gamma-aminobutyric acid synthesis in the brain. J Neurochem 60:395–407

Mason GF, Martin DL, Martin SB, Manor D, Sibson NR, Patel A, Rothman DL, Behar KL (2001) Decrease in GABA synthesis rate in rat cortex following GABA-transaminase inhibition correlates with the decrease in GAD(67) protein. Brain Res 914:81–91

Mattson RH, Petroff O, Rothman D, Behar K (1994) Vigabatrin: effects on human brain GABA levels by nuclear magnetic resonance spectroscopy. Epilepsia 35:S29–S32

McLaughlin PA, Nirmalan GP, Tam KH, Robinson GA (1974) Release of 131I from the ovary of the laying Japanese quail after injection of perchlorate, thiocyanate or iodide. J Endocrinol 63:337–342

Mescher M, Merkle H, Kirsch J, Garwood M, Gruetter R (1998) Simultaneous in vivo spectral editing and water suppression. NMR Biomed 11:266–272

Mescher M, Tannus A, Johnson MO, Garwood M (1996) Solvent suppression using selective echo dephasing. JMagn Reson Series A 123:226–229

Metcalf BW (1979) Inhibitors of GABA metabolism. Biochem Pharmacol 28:1705–1712

Miller LP, Martin DL (1976) A comparison of the kinetic properties of mouse brain glutamate decarboxylase activities in the mitochondrial and supernatant fractions. Life Sci 19:281–288

Minelli A, Barbaresi P, Conti F (2003) Postnatal development of high-affinity plasma membrane GABA transporters GAT-2 and GAT-3 in the rat cerebral cortex. Brain Res Dev Brain Res 142:7–18

Moroni F, Cheney DL, Peralta E, Costa E (1978) Opiate receptor agonists as modulators of gamma-aminobutyric acid turnover in the nucleus caudatus, globus pallidus and substantia nigra of the rat. J Pharmacol Exp Ther 207:870–877

Murrin LC (1980) Ornithine as a precursor for gamma-aminobutyric acid in mammalian brain. J Neurochem 34:1779–1781

Neal MJ, Shah MA (1990) Development of tolerance to the effects of vigabatrin (gamma-vinyl-GABA) on GABA release from rat cerebral cortex, spinal cord and retina. Br J Pharmacol 100:324–328

Neff NH, Spano PF, Groppetti A, Wang CT, Costa E (1971) A simple procedure for calculating the synthesis rate of norepinephrine, dopamine and serotonin in rat brain. J Pharmacol Exp Ther 176:701–710

Noto T, Hashimoto H, Nakao J, Kamimura H, Nakajima T (1986) Spontaneous release of gamma-aminobutyric acid formed from putrescine and its enhanced Ca2+–dependent release by high K+stimulation in the brains of freely moving rats. J Neurochem 46:1877–1880

Nudo RJ, Masterton RB (1986) Stimulation-induced [14C]2-deoxyglucose labeling of synaptic activity in the central auditory system. J Comp Neurol 245:553–565

Okada Y, Nitsch-Hassler C, Kim JS, Bak IJ, Hassler R (1971) Role of g-aminobutyric acid (GABA) in the extrapyramidal motor system. 1. Regional distribution of GABA in rabbit, rat, guinea pig and baboon CNS. Exp Brain Res 13:514–518

Olsen RW, DeLorey TM (1999) GABA and Glycine. In: Siegel GJ (ed) Basic neurochemistry: molecular, cellular and medical aspects. Lippincott-Raven Publishers, Philadelphia

Ottersen OP, Zhang N, Walberg F (1992) Metbaolic compartmentation of glutamate and glutamine: Morpolocial evidence obtained by quantitative immunocytochemistry in rat cerebellum. Neuroscience 46:519–534

Pan JW, Williamson A, Cavus I, Hetherington HP, Zaveri H, Petroff OA, Spencer DD (2008) Neurometabolism in human epilepsy. Epilepsia 49(Suppl 3):31–41

Patel AB, de Graaf RA, Martin DL, Battaglioli G, Behar KL (2006) Evidence that GAD65 mediates increased GABA synthesis during intense neuronal activity in vivo. J Neurochem 97:385–396

Patel AB, de Graaf RA, Mason GF, Rothman DL, Shulman RG, Behar KL (2005) The contribution of GABA to glutamate/glutamine cycling and energy metabolism in the rat cortex in vivo. Proc Natl Acad Sci USA 102:5588–5593

Patel AJ, Balazs R, Richter D (1970) Contribution of the GABA bypath to glucose oxidation, and the development of compartmentation in the brain. Nature 226:1160–1161

Patel AJ, Johnson AL, Balazs R (1974) Metabolic compartmentation of glutamate associated with the formation of gamma-aminobutyrate. J Neurochem 23:1271–1279

Paulsen R, Fonnum F (1987) Comparison of results obtained with different methods for estimating GABA turnover in rat neostriatum. Biochem Pharmacol 36:1539–1544

Paulsen RE, Fonnum F (1988) Regulation of transmitter gamma-aminobutyric acid (GABA) synthesis and metabolism illustrated by the effect of gamma-vinyl GABA and hypoglycemia. J Neurochem 50:1151–1157

Paulsen RE, Odden E, Fonnum F (1988) Importance of glutamine for gamma-aminobutyric acid synthesis in rat neostriatum in vivo. J Neurochem 51:1294–1299

Pearce BR, Currie DN, Beale R, Dutton GR (1981) Potassium-stimulated, calcium-dependent release of [3H]GABA from neuron- and glia-enriched cultures of cells dissociated from rat cerebellum. Brain Res 206:485–489

Petroff OA, Behar KL, Mattson RH, Rothman DL (1996) Human brain g-aminobutyric acid levels and seizure control following initiation of vigabatrin therapy. J Neurochem 67:2399–2404

Pfeuffer J, Tkac I, Choi IY, Merkle H, Ugurbil K, Garwood M, Gruetter R (1999) Localized in vivo 1H NMR detection of neurotransmitter labeling in rat brain during infusion of [1-13C] D-glucose. Magn Reson Med 41:1077–1083

Pinal CS, Tobin AJ (1998) Uniqueness and redundancy in GABA production. Perspect Dev Neurobiol 5:109–118

Preece NE, Cerdan S (1996) Metabolic precursors and compartmentation of cerebral GABA in vigabatrin-treated rats. J Neurochem 67:1718–1725

Qume M, Whitton PS, Fowler LJ (1995) The effect of chronic treatment with the GABA transaminase inhibitors gamma-vinyl-GABA and ethanolamine-O-sulphate on the in vivo release of GABA from rat hippocampus. J Neurochem 64:2256–2261

Racagni G, Cheney DL, Trabucchi M, Wang C, Costa E (1974) Measurement of acetylcholine turnover rate in discrete areas of rat brain. Life Sci 15:1961–1975

Radian R, Kanner BI (1983) Stoichiometry of sodium- and chloride-coupled gamma-aminobutyric acid transport by synaptic plasma membrane vesicles isolated from rat brain. Biochemistry 22:1236–1241

Rando RR, Bangerter FW (1977) The in vivo inhibition of GABA-transaminase by gabaculine. Biochem Biophys Res Commun 76:1276–1281

Revuelta AV, Cheney DL, Costa E (1981) Measurement of gamma-aminobutyric acid turnover rates in brain nuclei as an index of interactions between gamma-aminobutyric acid and other transmitters. Adv Biochem Psychopharmacol 27:169–181

Riekkinen PJ, Ylinen A, Halonen T, Sivenius J, Pitkanen A (1989) Cerebrospinal fluid GABA and seizure control with vigabatrin. Br J Clin Pharmacol 27(Suppl 1):87S–94S

Rimvall K, Martin DL (1994) The level of GAD67 protein is highly sensitive to small increases in intraneuronal gamma-aminobutyric acid levels. J Neurochem 62:1375–1381

Roberts E, Frankel S (1950) gamma-Aminobutyric acid in brain: its formation from glutamic acid. J Biol Chem 187:55–63

Rothman DL, Petroff OA, Behar KL, Mattson RH (1993a) Localized 1H NMR measurements of g-aminobutyric acid in human brain in vivo. Proc Natl Acad Sci USA 90:5662–5666

Rothman DL, Petroff OA, Behar KL, Mattson RH (1993b) Localized 1H NMR measurements of g-aminobutyric acid in human brain in vivo. Proc Natl Acad Sci USA 90:5662–5666

Sanacora G, Mason GF, Rothman DL, Behar KL, Hyder F, Petroff OA, Berman RM, Charney DS, Krystal JH (1999) Reduced cortical gamma-aminobutyric acid levels in depressed patients determined by proton magnetic resonance spectroscopy. Arch Gen Psychiatry 56:1043–1047

Sarup A, Larsson OM, Schousboe A (2003) GABA transporters and GABA-transaminase as drug targets. Curr Drug Targets CNS Neurol Disord 2:269–277

Schechter PJ, Sjoerdsma A (1990) Clinical relevance of measuring GABA concentrations in cerebrospinal fluid. Neurochem Res 15:419–423

Schousboe A (2003) Role of astrocytes in the maintenance and modulation of glutamatergic and GABAergic neurotransmission. Neurochem Res 28:347–352

Schousboe A, Kanner B (2002) GABA transporters: functional and pharmacological properties. Taylor & Francis Publisher, London

Schousboe A, Sarup A, Bak LK, Waagepetersen HS, Larsson OM (2004) Role of astrocytic transport processes in glutamatergic and GABAergic neurotransmission. Neurochem Int 45:521–527

Schousboe A, Svenneby G, Hertz L (1977) Uptake and metabolism of glutamate in astrocytes cultured from dissociated mouse brain hemispheres. J Neurochem 29:999–1005

Seiler N, Bink G, Grove J (1980) Relationships between GABA and polyamines in developing rat brain. Neuropharmacology 19:251–258

Sellstrom A, Sjoberg LB, Hamberger A (1975) Neuronal and glial systems for gamma-aminobutyric acid metabolism. J Neurochem 25:393–398

Sheikh SN, Martin DL (1998) Elevation of brain GABA levels with vigabatrin (gamma-vinyl-GABA) differentially affects GAD65 and GAD67 expression in various regions of rat brain. J Neurosci Res 52:736–741

Shen J, Rothman DL, Brown P (2002) In vivo GABA editing using a novel doubly selective multiple quantum filter. Magn Reson Med 47:447–454

Shen J, Yang J, Choi I-Y, Li SS, Chen Z (2004) A new strategy for in vivo spectral editing. Application to GABA editing using selective homonuclear polarization transfer spectroscopy. J Magn Reson 170:290–298

Sihra TS, Nicholls DG (1987) 4-Aminobutyrate can be released exocytotically from guinea-pig cerebral cortical synaptosomes. J Neurochem 49:261–267

Sloviter RS, Dichter MA, Rachinsky TL, Dean E, Goodman JH, Sollas AL, Martin DL (1996) Basal expression and induction of glutamate decarboxylase and GABA in excitatory granule cells of the rat and monkey hippocampal dentate gyrus. J Comp Neurol 373:593–618

Smolders I, Khan GM, Lindekens H, Prikken S, Marvin CA, Manil J, Ebinger G, Michotte Y (1997) Effectiveness of vigabatrin against focally evoked pilocarpine-induced seizures and concomitant changes in extracellular hippocampal and cerebellar glutamate, gamma-aminobutyric acid and dopamine levels, a microdialysis-electrocorticography study in freely moving rats. J Pharmacol Exp Ther 283:1239–1248

Sobue K, Nakjima T (1978) Changes in concentrations of polyamines and gamma-aminobutyric acid and their formation in chick embryo brain during development. J Neurochem 30:277–279

Soghomonian JJ, Martin DL (1998) Two isoforms of glutamate decarboxylase: why? Trends Pharmacol Sci 19:500–505

Sonnewald U, Westergaard N, Schousboe A, Svendsen JS, Unsgard G, Petersen SB (1993) Direct demonstration by [13C]NMR spectroscopy that glutamine from astrocytes is a precursor for GABA synthesis in neurons. Neurochem Int 22:19–29

Spanaki MV, Siegel H, Kopylev L, Fazilat S, Dean A, Liow K, Ben-Menachem E, Gaillard WD, Theodore WH (1999) The effect of vigabatrin (gamma-vinyl GABA) on cerebral blood flow and metabolism. Neurology 53:1518–1522

Staley KJ, Soldo BL, Proctor WR (1995) Ionic mechanisms of neuronal excitation by inhibitory GABAA receptors. Science 269:977–981

Storm-Mathisen J, Leknes AK, Bore AT, Vaaland JL, Edminson P, Haug FM, Ottersen OP (1983) First visualization of glutamate and GABA in neurones by immunocytochemistry. Nature 301:517–520

Szerb JC (1984) Storage and release of endogenous and labelled GABA formed from [3H]glutamine and [14C]glucose in hippocampal slices: effect of depolarization. Brain Res 293:293–303

Szerb JC, O'Regan PA (1984) Glutamine enhances glutamate release in preference to gamma-aminobutyrate release in hippocampal slices. Can J Physiol Pharmacol 62:919–923

Tapia R (1983) g-Aminobutyric acid: metabolism and biochemistry of synaptic transmission. In: Lajtha A (ed) Handbook of neurochemistry, vol 3. Plenum, New York, pp 423–466

Tapia R, Sandoval ME, Contreras P (1975) Evidence for a role of glutamate decarboxylase activity as a regulatory mechanism of cerebral excitability. J Neurochem 24:1283–1285

Terpstra M, Ugurbil K, Gruetter R (2002) Direct in vivo measurement of human cerebral GABA concentration using MEGA-editing at 7 Tesla. Magn Reson Med 47:1009–1012

Udenfriend S (1950) Identification of gamma-aminobutyric acid in brain by the isotope derivative method. J Biol Chem 187:65–69

Van den Berg CJ, Garfinkel D (1971) A simulation study of brain compartments. Metabolism of glutamate and related substances in mouse brain. Biochem J 123:211–218

van Kempen GM, van den Berg CJ, van der Helm HJ, Veldstra H (1965) Intracellular localization of glutamate decarboxylase, gamma-aminobutyrate transaminase and some other enzymes in brain tissue. J Neurochem 12:581–588

Waagepetersen HS, Sonnewald U, Schousboe A (1999) The GABA paradox: multiple roles as metabolite, neurotransmitter, and neurodifferentiative agent. J Neurochem 73:1335–1342

Waagepetersen HS, Sonnewald U, Schousboe A (2003) Compartmentation of glutamine, glutamate, and GABA metabolism in neurons and astrocytes: functional implications. Neuroscientist 9:398–403

Waldvogel D, van Gelderen P, Muellbacher W, Ziemann U, Immisch I, Hallett M (2000) The relative metabolic demand of inhibition and excitation. Nature 406:995–998

Ward HK, Thanki CM, Bradford HF (1983) Glutamine and glucose as precursors of transmitter amino acids: ex vivo studies. J Neurochem 40:855–860

White HS, Sarup A, Bolvig T, Kristensen AS, Petersen G, Nelson N, Pickering DS, Larsson OM, Frolund B, Krogsgaard-Larsen P, Schousboe A (2002) Correlation between anticonvulsant activity and inhibitory action on glial gamma-aminobutyric acid uptake of the highly selective mouse gamma-aminobutyric acid transporter 1 inhibitor 3-hydroxy-4-amino-4,5,6,7-tetrahydro-1,2-benzisoxazole and its N-alkylated analogs. J Pharmacol Exp Ther 302:636–644

Wood JD, Peesker SJ (1975) The anticonvulsant action of GABA-elevating agents: a re-evaluation. J Neurochem 25:277–282

Wood JD, Schousboe A, Krogsgaard-Larsen P (1980) In vivo changes in the GABA content of nerve endings (synaptosomes) induced by inhibitors of GABA uptake. Neuropharmacology 19:1145–1152

Wood JD, Kurylo E, Lane R (1988) gamma-Aminobutyric acid release from synaptosomes prepared from rats treated with isonicotinic acid hydrazide and gabaculine. J Neurochem 50:1839–1843

Wood JD, Peesker SJ, Gorecki DK, Tsui D (1978) Effect of L-cycloserine on brain GABA metabolism. Can J Physiol Pharmacol 56:62–68

Wu Y, Wang W, Richerson GB (2001) GABA transaminase inhibition induces spontaneous and enhances depolarization-evoked GABA efflux via reversal of the GABA transporter. J Neurosci 21:2630–2639

Wu Y, Wang W, Richerson GB (2003) Vigabatrin induces tonic inhibition via GABA transporter reversal without increasing vesicular GABA release. J Neurophysiol 89:2021–2034

Yamasaki EN, Barbosa VD, De Mello FG, Hokoc JN (1999) GABAergic system in the developing mammalian retina: dual sources of GABA at early stages of postnatal development. Int J Dev Neurosci 17:201–213

Yamatsu K, Yamanishi M, Ikeda M, Uzuo T, Okada Y (1982) Postmortem GABA increase in discrete regions of rat brain. In: Okada Y, Roberts E (eds) Problems in GABA research from brain to bacteria. Excerpta Medica, Amsterdam, pp 30–40

Yang J, Johnson C, Shen J (2009) Detection of reduced GABA synthesis following inhibition of GABA transaminase using in vivo magnetic resonance signal of [13C]GABA C1. J Neurosci Methods 182:236–243

Yang J, Li CQ, Shen J (2005) In vivo detection of cortical GABA turnover from intravenously infused [1-13C]D-glucose. Magn Reson Med 53:1258–1267

Yang J, Li SS, Bacher J, Shen J (2007) Quantification of cortical GABA-glutamine cycling rate using in vivo magnetic resonance signal of [2-13C]GABA derived from glia-specific substrate [2-13C]acetate. Neurochem Int 50:371–378

Yang J, Shen J (2005) In vivo evidence for reduced cortical glutamate-glutamine cycling in rats treated with the antidepressant/antipanic drug phenelzine. Neuroscience 135:927–937

Yu AC, Fisher TE, Hertz E, Tildon JT, Schousboe A, Hertz L (1984a) Metabolic fate of [14C]-glutamine in mouse cerebral neurons in primary cultures. J Neurosci Res 11:351–357

Yu AC, Hertz E, Hertz L (1984b) Alterations in uptake and release rates for GABA, glutamate, and glutamine during biochemical maturation of highly purified cultures of cerebral cortical neurons, a GABAergic preparation. J Neurochem 42:951–960

Yudkoff M, Zaleska MM, Nissim I, Nelson D, Erecinska M (1989) Neuronal glutamine utilization: pathways of nitrogen transfer studied with [15N]glutamine. J Neurochem 53:632–640

Zawia NH, Harry GJ (1993) Correlations between developmental ornithine decarboxylase gene expression and enzyme activity in the rat brain. Brain Res Dev Brain Res 71:53–57

Chapter 40
¹H-Magnetic Resonance Spectroscopy of Cerebral Phenylalanine Content and its Transport at the Blood-Brain Barrier

Roland Kreis

Abstract Proton magnetic resonance spectroscopy (¹H-MRS) has been used in multiple studies to quantify the cerebral phenylalanine (Phe) content in humans, mostly phenylketonuria (PKU) patients. PKU has also been extensively studied by magnetic resonance imaging (MRI). The major published findings about MRS and MRI in PKU are summarized. With different MRI techniques, the characteristic white matter lesions in PKU patients have been characterized, providing a picture that is in agreement with partly reversable dysmyelination, possibly in the form of spongiform alterations. The ¹H-MRS results from studies of PKU patients are summarized in a metaanalysis where major factors for differences between the published studies are identified and essential methodologic specifics needed for the determination of Phe concentrations are given. It was found that the steady state blood/brain Phe concentration ratio seems to be fairly stable for the large majority of PKU patients, leaving little room for individual dietary treatment, based on potential individual sensitivity to dietary Phe. Furthermore, ¹H-MRS has been used to investigate the dynamics of cerebral Phe uptake in humans with and without PKU. The theoretical background for the description of the blood-brain-barrier (BBB) dynamics is given and the main results from these studies described, among them the documentation of a BBB blockade for Phe inflow by oral supplementation with large neutral amino acids.

Keywords Blood brain barrier • Blood brain barrier transport • Brain • Magnetic resonance spectroscopy • Magnetic Resonance Imaging • Phenylalanine • Phenylalanine transport • Phenylketonuria • Quantitation • Transport • Transport phenylalanine • Treatment

R. Kreis, Ph.D. (✉)
Department of Clinical Research, University Bern, AMSM, DRNN/DKF, Pavilion 52, P.O. Box 35, Inselspital, CH-3010, Bern, Switzerland
e-mail: roland.kreis@insel.ch

I.-Y. Choi and R. Gruetter (eds.), *Neural Metabolism In Vivo*, Advances in Neurobiology 4, DOI 10.1007/978-1-4614-1788-0_40,

40.1 Introduction

Proton magnetic resonance spectroscopy (^{1}H-MRS) provides a means to determine local metabolite content in animal and human brain quantitatively and non-invasively (see detailed discussion in chapter 5 of this volume [de Graaf 2012]). Accuracy, precision, specificity, and limits of detection in terms of localization, size of investigated region, etc. strongly depend on experimental technique and instrumentation, as well as data evaluation and quantitation methods used. In standard clinical examinations, metabolites with tissue contents of 1–10 mmol/kg wet weight are normally targeted. With optimized data acquisition and evaluation schemes it is possible to aim at less abundant metabolites. In particular, the availability of very high magnetic fields for human investigations (Tkac and Gruetter 2005) and advances in receiver technology have led to a lowering of detection and quantitation thresholds. The quantitation of phenylalanine (Phe) in phenylketonuria (PKU), however, is already possible on standard clinical scanners with standard hardware, but definitely needs dedicated methodology (Bik-Multanowski and Pietrzyk 2007).

The ^{1}H-MR spectrum of Phe, depicted in Fig. 40.1a, contains contributions from the α and β protons of the alanine part of the molecule as well as the phenyl (φ) protons. The α and β protons resonate in the upfield part of the spectrum, overlapping with many resonance peaks of more abundant metabolites, such that selective detection of this part of the spectrum would be extremely difficult. However, the five aromatic protons resonate in the much less crowded downfield region of the spectrum. At very high fields they give rise to a coupled multiple line pattern around 7.37 ppm (chemical shifts between 7.32 and 7.43 ppm [Govindaraju et al. 2000]). Because of strong coupling, the resonances collapse into a single pseudo-singlet at the low clinical field strength of 1.5 T (Fig 40.1a). Because the peaks form a simple resonance pattern with the signal strength of five protons and because this peak appears in the downfield spectrum with no signal overlap with abundant metabolites, Phe can be detected and quantified fairly conveniently on clinical scanners at 1.5 T. Still, quantitation of steady state brain Phe content in healthy subjects, which is reported to be around 50 μM, is out of reach for clinical ^{1}H-MRS at 1.5 T. As illustration, a localized ^{1}H-MR spectrum of human brain is shown in Fig. 40.1c, demonstrating the relatively weak background signals in the downfield area (>5 ppm) compared to the upfield part, where only the largest signals have been labeled. With correct scaling to represent the Phe content in healthy brain, the Phe singlet at 7.37 ppm would be about the size of the peak-to-peak noise in the *in vivo* spectrum, i.e. ~ one fifth of the background peak at 7.3 ppm. Still, the main difficulty in determining Phe content in healthy controls is not noise, but the unknown nature of the underlying MR signals.

However, this has not been a major problem, since work on cerebral Phe content in humans has primarily focused on studies in phenylketonuria (PKU) patients, where plasma Phe content is usually 4–40 times higher than in healthy subjects and where there is no evidence to suggest that the underlying signals are altered with respect to healthy controls.

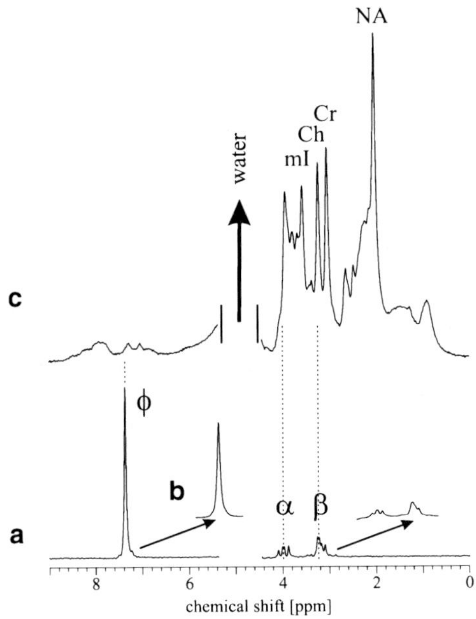

Fig. 40.1 ^{1}H-Magnetic resonance spectrum of an aqueous Phe solution in (**a**) and the same spectrum broadened to match the linewidth expected *in vivo* in (**b**). Part (**c**) shows an *in vivo* spectrum obtained from a 70 cm^3 supraventricular region of interest in a healthy subject (8.5 min acquisition time, clinical 1.5 T MR scanner, quadrature head coil). The resonances from α and β protons in the upfield part of the spectrum overlap with resonances from many abundant metabolites, while the phenyl protons (ϕ) lie in the less crowded downfield portion. However, because brain Phe concentration is only some 50 μM in healthy subjects and because the underlying resonances are of unknown size and shape, Phe cannot be measured in healthy subjects in steady state. Abbreviations: mI: myo inositol, Ch: cholines, Cr: creatines, NA: N-acetyl group containing metabolites, mainly N-acetylaspartate

PKU is an inherited disorder of amino acid metabolism, where blood levels of Phe are grossly elevated due to a deficiency of functional phenylalanine hydroxylase (PAH; Enzyme Commission 1.14.16.1) (Scriver and Kaufman 2001; National Institutes of Health Consensus Development Panel 2001). Untreated PKU leads to disturbed brain development with profound retardation, microcephaly, epilepsy and other neurologic symptoms, which can largely be prevented by institution of a Phe-restricted diet in the newborn. All babies are now screened at birth for excess Phe in urine and started on a low Phe diet if PKU is diagnosed. Nevertheless, there were two main motivations for the study of PKU patients with ^{1}H-MRS:

1. There is ample evidence that patients with apparently similar history, in particular a similar course of blood Phe levels, can have widely varying neurologic outcomes, suggesting that plasma Phe levels may be inadequate as prognostic and therapeutic parameter. It was hoped that brain Phe levels, as more direct measure of the toxic agent, can improve therapeutic guidance if the brain and blood Phe levels are not tightly linked (individual blood/brain Phe ratio).

2. Non-invasive access to brain Phe levels opens a direct avenue to study therapeutic measures dynamically in human patients. Blood brain barrier kinetics in response to oral Phe loads and dietary interventions was targeted. In particular, there was hope to observe potential blockade of Phe entry into the brain.

40.2 MRI in Early Treated PKU

The first MR investigations in PKU were done with MR imaging documenting hyperintensities in white matter on T_2-weighted images of early treated PKU patients. These alterations were primarily localized in posterior periventricular white matter, extending also anteriorly in more affected patients (Pearsen et al. 1990). These findings were corroborated and further elucidated by many authors (Bick et al. 1991; Bick et al. 1993; Thompson et al. 1993; Cleary et al. 1994; Pietz et al. 1996; Phillips et al. 2001; Dezortova et al. 2001; Kono et al. 2005; Vermathen et al. 2007; Sirrs et al. 2007; Leuzzi et al. 2007; Laule et al. 2007). Extent and severity of the lesions seem to depend on the recent history of dietary status, i.e. blood Phe levels (Thompson et al. 1993; Pietz et al. 1996). From relaxometry it became clear that an additional CSF-like water signal was causing the hyperintensities (Bick et al. 1993; Pietz et al. 1996; Vermathen et al. 2007; Sirrs et al. 2007; Laule et al. 2007) and that myelin composition is altered (Vermathen et al. 2007; Sirrs et al. 2007). Further characterization of these lesions was provided by diffusion weighted or diffusion tensor imaging (Phillips et al. 2001; Dezortova et al. 2001; Kono et al. 2005; Vermathen et al. 2007; Leuzzi et al. 2007) providing evidence for decreased apparent diffusion coefficients in the lesions, but also normal appearing white matter, consistent with bounded diffusion in the additional CSF-like spaces (Vermathen et al. 2007). These findings are consistent with dysmyelination possibly leading to the "spongiform alterations" in white matter as described in pathological post-mortem reports in un- or late-treated PKU (Malamud 1966). Whether or not and up to what stage these structural white matter changes are completely reversible by reinstitution of a strict diet (Thompson et al. 1993) and whether they are related to neuropsychological impairments (Anderson et al. 2007) is presently still unresolved.

40.3 Meta-Analysis for ^{1}H-MRS Results in PKU

Feasibility of Phe detection by *in vivo* ^{1}H-MRS in hyperphenylalaninemic animals and PKU patients was first demonstrated by the Yale group (Avison et al. 1990; Novotny et al. 1995). Other groups followed from the mid 1990s onward (Möller et al. 1995; Pietz et al. 1995; Kreis et al. 1995; Leuzzi et al. 2000; Koch et al. 2000). Quantitative results from studies performed by different research groups differed substantially (Moller et al. 2000; Kreis 2000), as also summarized recently in (Möller et al. 2006; Iles 2008). Since brain Phe levels depend on concurrent plasma Phe levels, the results obtained with different methods by different groups can be compared as concentration ratios found between plasma and brain Phe content. However, since for saturable blood-brain transport this ratio is also expected to depend on blood Phe levels, it is best to compare pairs of brain/plasma Phe ratios obtained with different methods. A direct comparison of results from the literature is provided in Fig. 40.2a. There is large scatter of brain Phe values for apparently similar plasma Phe levels within most studies. It is tempting to interpret these differences as individuality of the PKU patients. This would mean that there is an individual sensitivity to identical plasma Phe concentrations, which would have immediate consequences for individual treatment, in particular the strictness of the dietary regime needed. However, Fig. 40.2a also shows a very striking difference of brain Phe values between studies suggesting the presence of systematic methodological differences. Potential reasons for the large variance between and within these studies are listed below. If one accepts that there are probably technical reasons for the difference in results between studies and not systematic differences in the population of PKU patients studied, one can try to unify data from all studies into a single cohort. This was performed for Fig. 40.2b by appropriate scaling of the brain Phe values. The scaling factor for each study was found as follows. For all studies the brain to blood concentration ratio was calculated for patients with plasma values between 0.6 and 1.5 mM and the mean value of this ratio was calculated within studies, or rather uniform patient cohorts.[1] A calibration factor was then calculated for each study, such that the results for all reports were brought to the same ratio as determined as median study average [0.28 found for Ref. (Rupp et al. 2001)]. For clarity of presentation, data from Ref (Koch et al. 2000) was omitted to construct Fig. 40.2b since the variance within this particular study was more than two times higher than in all other reports. This unification scheme is supported by animal data showing a similar brain to blood ratio in hyperphenylalaninemia (Smith and Kang 2000). The unified data in Fig. 40.2b may give a more realistic representation of the distribution of brain Phe values in relation to plasma Phe levels. Interestingly almost all values below 1.5 mM plasma Phe now lie within a band of 0.1 mM around the linear regression line between brain and plasma values. 0.1 mM Phe can be estimated to correspond to more than two times the error estimate for individual data points – particularly in those studies that did not employ optimally specialized MRS methodology. Hence, this suggests that there is rather little room for individual sensitivity to plasma Phe among 'typical' PKU patients. Data from so-called atypical patients have been omitted from this meta-analysis. Atypical are those few reported PKU patients that have not been treated in early childhood, but appear to be free of substantial neurologic

[1] Data from the Münster Group (Möller et al. 1997) was first combined and then divided into two cohorts: data from the early papers (Möller et al. 1997; Moller et al. 1998) in one cohort, the newer data from a more recent publication (Weglage et al. 2001b) which seem to have been obtained with somewhat different methodology (Pietz et al. 2002) into another.

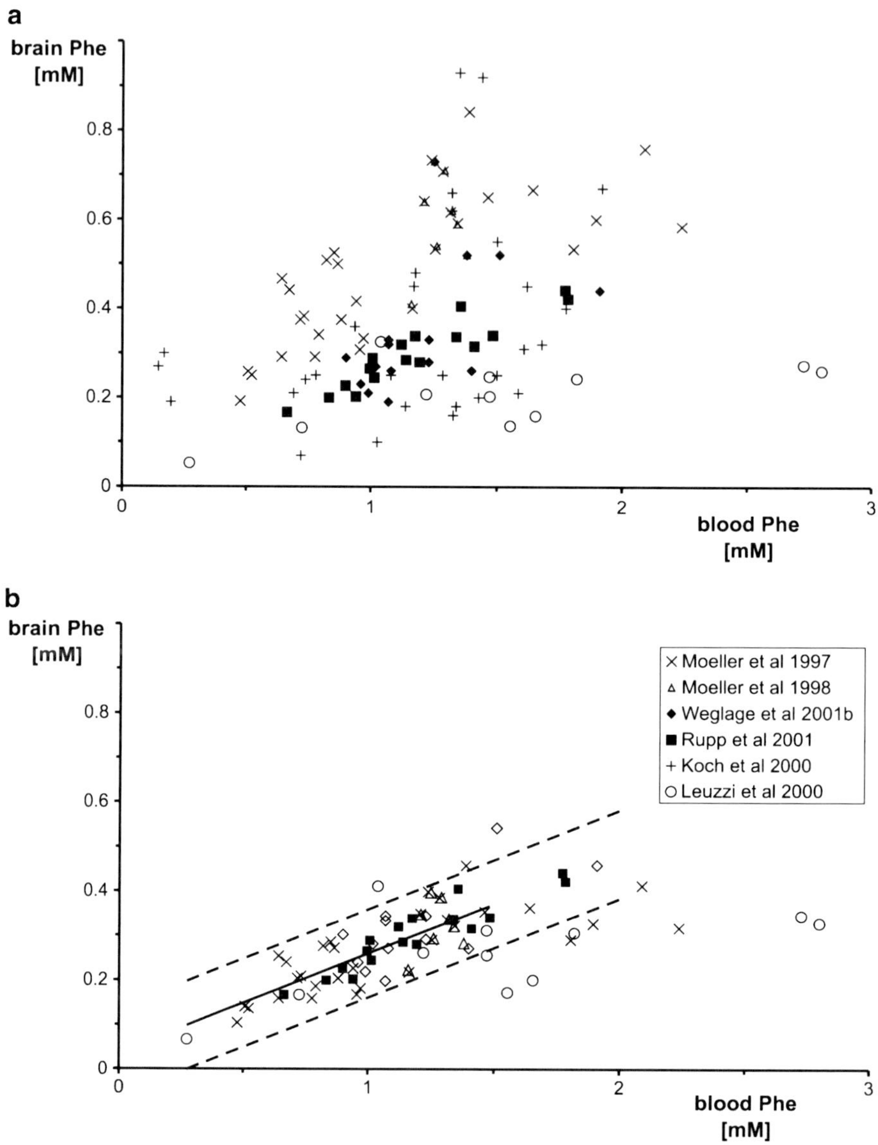

Fig. 40.2 Meta-analysis of cerebral Phe content as function of blood Phe level. Part (**a**) summarizes data from several research groups and several publications in their original form, but converted to identical units (mM). There is substantial scatter within and between studies. Part (**b**) shows data after scaling data to obtain identical mean brain to blood Phe ratio in all studies for PKU patients with blood Phe levels between 0.6 and 1.5 mM. With the between-study spread reduced, the data appears to confirm that brain Phe scales linearly with blood Phe up to 1.5 mM or even higher, that brain Phe content is considerably lower than blood Phe levels, and that there is fairly little interindividual spread for most "typical" PKU patients (Data from Ref. (Leuzzi et al. 2000) was converted from relative units to mM using creatine content and T_2 values from the literature and assuming equal sensitivity for Phe and creatine. Data from Ref. (Koch et al. 2000) have been left out of part (**b**) for clarity [see text])

sequelae expected for the plasma Phe values measured in these patients. Usually these patients would not be diagnosed, nor treated.

All these studies concerned adult or adolescent PKU patients, while age might of course be a determining factor. Recent data in newly diagnosed and still untreated neonates with PKU suggests that BBB protection against high levels of plasma Phe is not established to the same degree in developing brain, where substantially higher brain/blood Phe ratios have been measured than in adult patients (Kreis et al. 2007).

40.4 Methodology of ^{1}H-MRS for Quantitation of Brain Phe

In the first studies, detection and quantitation of Phe relied on direct difference spectroscopy, where an average spectrum from healthy controls was subtracted from a spectrum of a PKU patient in order to subtract the background signals and to visualize the signal from Phe. This is illustrated in Fig. 40.3. Elimination of the background signal is crucial, since there is a signal at ~7.3 ppm in healthy controls and PKU patients that is not related to Phe. It has recently been shown that this underlying peak is primarily caused by macromolecular compounds (Kreis et al. 2004) and may include contributions from phosphocreatine. The area under the curve of the Phe peak in the difference spectrum could be obtained by simple model fitting and the Phe tissue content, which is proportional to the area, could be calculated.

Optimization of the acquisition parameters at this stage included the evaluation of optimal voxel size, type of receive coil, quantitation method and echo time. Our conclusion at that time (Kreis et al. 1995) was to record signal from a large volume (~70 cm^3 above the ventricular system) using a homogeneous quadrature head coil,

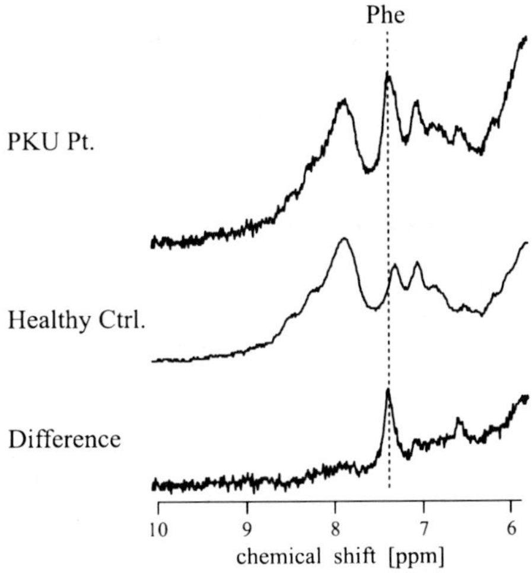

Fig. 40.3 Illustration of difference spectroscopy to visualize and quantitate the Phe resonance at 7.37 ppm, adapted from Ref. (Kreis 2000). The difference spectrum between spectra from a PKU patient and an averaged control spectrum shows a large singlet resonance at 7.37 ppm attributable to Phe. This does not coincide with the main background signal at 7.3 ppm in healthy control subjects

rather than smaller volumes (with better spectral resolution) using a more sensitive surface coil. This conclusion was mainly driven by the aim to obtain very reproducible repeat examinations for dynamic studies in the same subject. Today with the availability of multi-receiver array head coils this decision may be reconsidered. In particular, it could mean that Phe content might be evaluated in considerably smaller, cortical regions of interest, with a sensitivity similar to the one for the large ROI. A further specific issue that is not always taken into consideration is the potential for a large chemical shift offset in a large ROI. The chemical shift offset causes different resonance signals to originate from different (shifted) areas in the brain. This may be the cause for some of the variance found in the literature for the blood/brain ratio of Phe. In particular this is problematic, if the excitation signal is not applied on-resonance with the Phe signal at 7.37 ppm and if creatine is used as internal concentration reference. Recording of the unsuppressed water signal from the identical ROI appears to be a more precise and robust technique for Phe quantitation. Since the transverse relaxation time of the Phe signal is fairly short (Kreis et al. 1995) and the background signal does not decay quickly at 1.5 T (Vermathen et al. 2000), the use of a short echo time is recommended.

Further reasons for systematic differences between studies have been discussed in Ref. (Kreis 2000) and include possible corrections for Phe signal from CSF and blood and the concentration values assumed for the internal standard (water or creatine). The detectability of vascular Phe has been shown to depend on field strength and acquisition parameters (Kreis et al. 2005). The largest source for variation within a study is expected to arise from data analysis, especially if this is not performed automatically and may therefore be subject to operator bias. In particular, manual definition of the baseline is error prone. Modern data analysis methods in ^{1}H-MRS make use of as much prior knowledge as possible (Provencher 1993; Hofmann et al. 2002). The resonance patterns of Phe, relative frequency and phase, lineshape information, as well as a parameterized background signal can be invoked (Pietz et al. 1999). This is illustrated in Fig. 40.4. Further optimization of data processing and modelling yielded a very good reproducibility of cerebral Phe content. It was shown to vary by less than 10 μmol/kg upon immediate repetition of the MRS measurements in independent sessions (Kreis et al. 2009).

40.5 Dynamics at the Blood-Brain-Barrier

It has generally been accepted that Phe is primarily transported into brain through the blood-brain-barrier (BBB) by the L-system carrier ([Pardridge 1998a; Pardridge 1998b; Scriver and Kaufman 2001; Smith 2003], and refs therein), which seems to be encoded by LAT1 ([Moller et al. 2005] and refs therein). Transport via the L-carrier was shown to be bi-directional, sodium- and energy-independent, as well as saturable. The L-carrier system is shared for transport of all large neutral amino acids (LNAA's), in particular Phe, Trp, Leu, Met, Ile, Tyr, His, Val, and to some

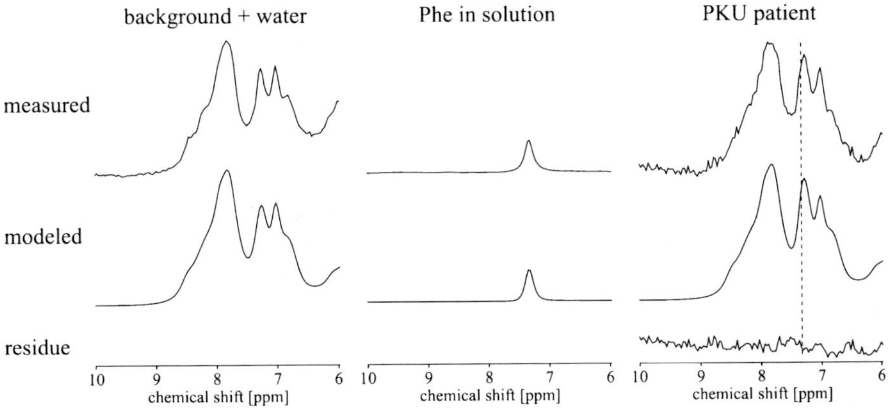

Fig. 40.4 Scheme for fully automated data fitting of the Phe MRS signal in PKU patients. The top row shows the experimental data used: an averaged spectrum from healthy controls, the ^{1}H-MRS spectrum of an aqueous solution of L-Phe, and the ^{1}H-MR spectrum of a PKU patient to be evaluated. Below are the modeled counterparts of these spectra, i.e. parameterized forms of the normal background (including variable signals from residual water). Further prior knowledge on the phase and lineshape are taken from reference scans of unsuppressed water. The lowest trace shows the residuals (measured - modeled) for this case (Adapted from Ref. [Pietz et al. 1999])

extent Thr and even Gln. The dynamics for Phe or any of the other LNAAs are usually modeled by Michaelis-Menten kinetics characterized by V_{max}, the maximal transport capacity, and K_m, the half-saturation concentration.

$$flux = \frac{V_{max}[Phe]}{K_m + [Phe]} \tag{40.1}$$

$1/K_m$ is a measure for the affinity of the carrier for the investigated LNAA. While K_m describes the transport of a single LNAA in absence of any other LNAA, competition for transport capacity between the LNAA's is handled by introducing an apparent kinetic constant K_m^{app}, that is influenced by concentrations and affinities of all other LNAAs.

$$K_m^{app} = K_m \left(1 + \sum_i \frac{[LNAA_i]}{K_{m,i}}\right) \tag{40.2}$$

On the plasma side of the BBB, the effective affinities are therefore much lower than the theoretical affinity for an isolated LNAA. K_m^{app} for Phe is 5–10 times higher than K_m, depending on the set of constants used. In diseases like PKU, where the content of one LNAA is drastically increased, the carrier can be swamped by this one LNAA, effectively shutting down transport for all other LNAAs. This could lead to a substantial decrease of the brain levels of the other LNAAs. This conclusion has led to the theory that imbalances in LNAA brain content are major factors in the etiology of PKU. They have also led to suggestions for therapy, where all or some of the LNAAs are given as oral supplements in addition or overproportionally to the standard AA replacement supplements (Ref. in [Scriver and Kaufman 2001]).

Recent *in vitro* data from membrane vesicle studies (O'Kane and Hawkins 2003) suggest the presence of a hitherto unknown sodium-dependent transport system for LNAA's at the abluminal membrane of cerebral endothelial cells, which would be of prime importance for the export of LNAA's from cerebral extracellular space into plasma. The authors of these studies (O'Kane and Hawkins 2003; O'Kane et al. 2004) propose that the four energy-dependent transport systems at the abluminal membrane are the cause for the large concentration gradient between plasma and extracellular fluid. These findings, if confirmed *in vivo*, will strongly influence the (re-)interpretation of the BBB transport studies by MRS and other methods described below.

Detailed analyses of the BBB dynamics of LNAA or Phe in particular, have mostly been carried out in model systems (e.g. isolated brain capillaries) or in invasive animal experiments (Pardridge 1998b). A few studies in humans have been performed using the invasive double indicator method (Knudsen et al. 1990; Knudsen et al. 1995; Knudsen and Paulson 1998) and most recently several studies were done with ^{1}H-MRS (Refs. (Pietz et al. 1995; Möller et al. 1997; Moller et al. 1998; Pietz et al. 1999; Rupp et al. 2001; Weglage et al. 2001a; Pietz et al. 2003a) for Phe; Ref. (Vermathen et al. 2000) for histidine). It should be noted that beside the non-invasiveness, there is another major difference between most invasive methods and ^{1}H-MRS. ^{1}H-MRS studies are performed on a much longer time scale than most of the other methods. Uptake studies of tracer compounds often reflect single transients of blood through the brain vasculature, whereas ^{1}H-MRS studies of Phe – for sensitivity reasons – can only measure changes over fractions of hours, or more typically even several hours. Steady state measurements are easiest to perform with MRS, they represent kinetics over many hours or days. This means that ^{1}H-MRS studies are more difficult to model, since mechanisms that may only have minor effects on the short time scale have time to become major determinants on the long run. Hence, features like brain efflux, equilibration among compartments, low affinity transport systems, passive diffusion, metabolic use, or even deviation from Michaelis Menten kinetics can be of major importance.

In analogy to earlier successful use in ^{1}H-MRS (Gruetter et al. 1996), Moeller et al (Möller et al. 1997; Moller et al. 1998) have suggested the use of a symmetric Michaelis Menten model for the description of Phe transport across the BBB. Both, steady state measurements (Möller et al. 1997) and kinetic data (Moller et al. 1998) have been neatly modeled in this framework. As a first approach, they used the simplest model possible with symmetric K_m^{app} and constant (maximal) metabolization rate V_{met}. It was found that the model parameters found were in keeping with animal and earlier human data from healthy subjects, but also that there was substantial variance between measurements that might point at large interindividual variation. It was correctly pointed out that (1) steady state measurements can only provide information on the ratio between V_{met} and $V_{\max}$, but not V_{met} or $V_{\max}$ alone, and that (2) determination of V_{met} and V_{max} through detection of deviations from steady state in kinetic experiments rely heavily on availability of brain Phe measurements early after Phe plasma increase, which is principally hard to get by with ^{1}H-MRS. In their most recent publication this group claimed that individual BBB Phe transport characteristics determine clinical outcome (Weglage et al. 2001b). This statement as applied to typical PKU patients was disputed by Pietz et al. (2002), leading to a consensus that more research into the BBB kinetics in PKU is warranted (Weglage et al. 2002).

The symmetric Michaelis Menten kinetics used for modeling in the work of Moeller et al was well justified as a first approach, but lacks the obvious asymmetry caused by differing competition for the carrier on the two sides of the BBB and of course does not take the recently proposed active export from extracellular space (O'Kane and Hawkins 2003) into account. Even leaving this new, but still unconfirmed transport pathway out of consideration for the moment, asymmetry has to be included in a refined model of BBB transport of Phe: K_m^{app} for facilitated transport by the L-system can be handled reasonably well on the plasma side, since plasma LNAA concentrations can be measured and K_m values from the published animal literature can be assumed. The brain side is more tricky for two reasons. (1) LNAA concentrations are not readily available in human subjects. (2) Assumptions on the kinetic model strongly influence the calculations. It is not clear a priori whether overall brain concentrations (tissue content from biopsies) or interstitial LNAA concentrations (extracellular concentrations from microdialysis studies) determine the calculation of K_m^{app} on the brain side. An extended model would include blood plasma, interstitial space, and cellular space as compartments, as well as the BBB at the capillary endothelium and brain cell membranes as barriers. If a single kinetic compartment with average brain concentrations is used on the brain side (fast equilibration across the brain cell membranes [Hommes and Lee 1990]), K_m^{app} differs by something like 30% only between the blood and brain side (plus asymmetry from possibly limited intracerebral distribution space). However, if interstitial concentrations, which are assumed to be considerably lower than cellular concentrations, are used to calculate K_m^{app} on the brain side of the BBB, considerable asymmetry results (Knudsen et al. 1990). K_m^{app} would be more than 3 times lower on the brain side invalidating the use of a two-compartment, symmetric Michaelis Menten model for evaluation of ^{1}H-MRS studies. Relevance of the interstitial space for LNAA dynamics in PKU is also suggested from a microdialysis study (Zielke et al. 2002). Deviations from symmetric Michaelis Menten kinetics have been discussed in (Möller et al. 2006) where the potential use of a reversible Michaelis Menten model has been suggested. Interestingly, the reversible Michaelis Menten model predicts a linear steady state relation between blood and brain Phe content, irrespective of blood Phe level, i.e. absence of carrier saturation. More experimental steady state data for blood Phe >2 mM would be needed to decide on the appropriateness of such a model.

A further complicating factor is the unknown rate of metabolization. Constant and maximum use of Phe for protein synthesis is the most basic assumption, but can be questioned in PKU because a constant net protein synthesis rate depends on availability of all AAs, which is not guaranteed at all in PKU since the protein synthesis rate was shown to decrease with a pathological increase in a single LNAA (Pardridge 1998a).

There would be less uncertainty if these ^{1}H-MRS studies could be done on control subjects, where BBB characteristics and metabolization rate are better characterized than in PKU patients. It has recently been shown that it is indeed possible to follow brain Phe changes as consequence of continued oral Phe supplementation and subsequent rise in blood Phe content over several hours (Pietz et al. 2003a). Comparison of BBB characteristics between PKU patients and healthy controls are currently under way.

A further MRS study on acute cerebral effects of a Phe load was conducted with ^{31}P-MRS, which is mostly used to study high energy metabolism. ^{31}P-MRS demonstrated subtle abnormalities of cerebral energy metabolism in PKU in steady-state conditions that were amplified by the Phe load (Pietz et al. 2003b). It may indicate that the challenge with Phe involves energy expenditure lowering the cerebral energy reserve. This would be in agreement with the proposition that the brain/plasma concentration gradient is maintained by active transport systems (O'Kane and Hawkins 2003; O'Kane et al. 2004). This may also hint at a link between Phe neurotoxicity and imbalances of cerebral energy metabolism.

40.6 Blockade of Phe Entry into Brain Through the BBB

Independent of quantitative details on kinetic constants, it has been shown by Pietz et al. (1999) that the competition among LNAAs at the BBB for transport capacity can be demonstrated by ^{1}H-MRS and possibly be put to good use in PKU as anticipated much earlier (Pratt 1982; Berry et al. 1990). The basic idea is that simultaneous increase of the plasma concentrations of all LNAAs other than Phe could block brain import of additional plasma Phe. This concept was proven in a crossover experiment where in one part of the study six PKU patients were given an oral load of 100 mg/kg$_{bw}$ L-Phe and subsequent increase in brain Phe levels was observed by ^{1}H-MRS (Fig. 40.5, solid lines). The same scheme was repeated in another measurement series, when additionally blood levels of seven LNAAs were simultaneously elevated by oral supplementation (150 mg/kg$_{bw}$ for each LNAA, distributed over five portions over 12 h). The net effect on BBB transport of a concentration variation of several LNAAs can be judged based on the simple construct of a K_m-normalized Phe to LNAA ratio (Pardridge 1988; Pietz et al. 1999)

$$\frac{[Phe]}{\sum_i \frac{K_{m,Phe}[LNAA_i]}{K_{m,i}}} \tag{40.3}$$

If this ratio increases in blood, it is expected that influx of Phe into brain is increased, while if it decreases, net flow of Phe is expected to go from brain to blood

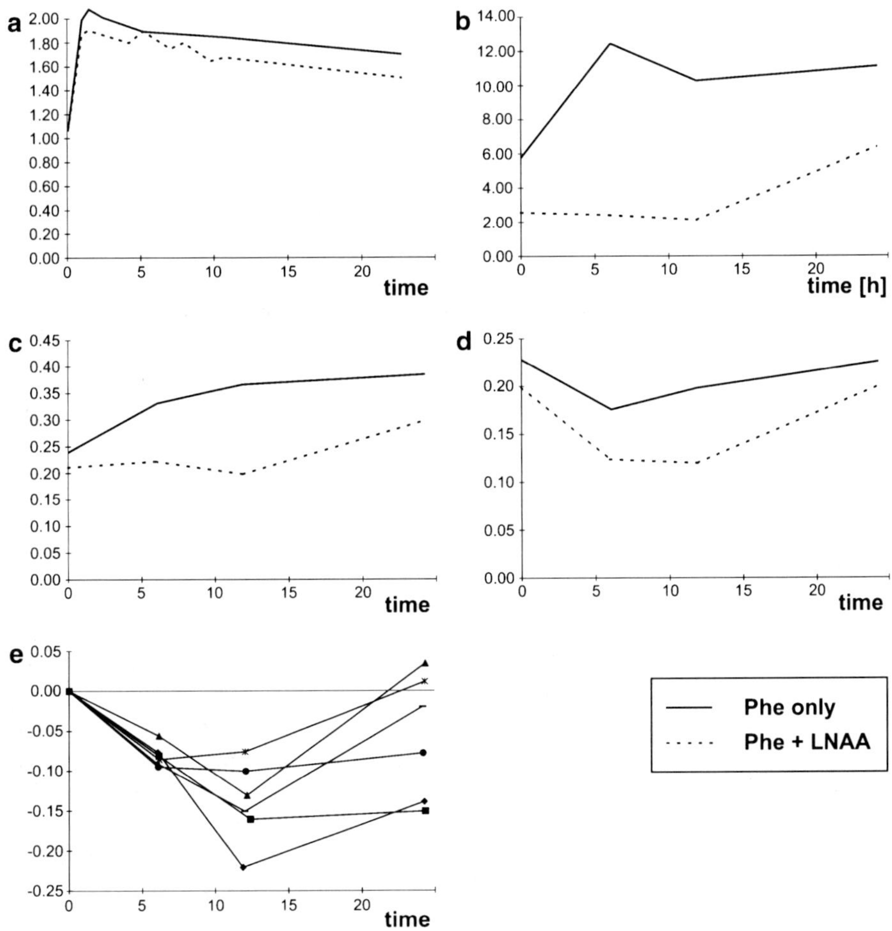

Fig. 40.5 Blocking the BBB for further entry of Phe in PKU subjects by supplementation with LNAAs. Time course data from Ref. (Pietz et al. 1999) demonstrating the average course of several parameters during both parts of the study, i.e. a single dose of oral Phe at time 0 (solid line.), vs. a single dose of Phe accompanied by multiple dose supplementation of all other LNAAs from −2 h to 10 h (dashed lines). Parts (**a**) to (**d**) contain the averaged values (six subjects) for plasma Phe content [mM] in (**a**); the K_m-normalized Phe/LNAA ratio [Eq. 40.3] in (**b**); the brain Phe content as determined by ¹H-MRS [mmol/kg$_{ww}$] in (**c**); and the resulting brain/plasma ratio for Phe in (**d**). Plot (**e**) illustrates the difference in how brain Phe content changes with time between the two series for each PKU subject [mmol/kg$_{ww}$]. The difference in brain Phe content at time zero was individually subtracted at all time points. Negative values indicate lower brain Phe content in the series with LNAA supplementation

even though the Phe content in blood may increase. The ^{1}H-MRS measurements proved the applicability of this simple picture, as indeed Phe content in brain did not increase over a time span of 12 h after an oral supplement with L-Phe, when LNAAs were given in addition. As depicted in Fig. 40.5a the plasma Phe content rose from about 1 to almost 2 mM with the specific dosage of Ref (Pietz et al. 1999), but the K_m-normalized Phe ratio decreased from 5.6 to 2.1 (Fig. 40.5b) and brain Phe levels did not increase until supplementation with LNAA was stopped (Fig. 40.5c). At the same time, EEG recordings showed that the normally occurring slowing of the EEG background activity under an acute Phe load did not occur with LNAA treatment.

A most recent workshop report suggests that this mechanism is of relevance even at the dosage that is clinically used for treatment of PKU and that traditional supplementation with a standard commercial amino acid powder mixture does indeed lead to lowering of the steady state brain Phe levels in PKU patients (Pietz and Kreis 2007).

40.7 Conclusions

^{1}H-MRS is a versatile tool to study Phe metabolism non-invasively in PKU patients and healthy controls. It allows dynamic determination of brain Phe content as response to challenges with oral Phe loads and hence elucidation of the Phe transport at the BBB. Systematic differences between results from different research teams hinder interpretation of published results. The existence of individual BBB kinetics that would lead to the need for individual therapy planning based on individually determined blood/brain ratio for Phe is not unequivocally documented. On the contrary, an attempt to unify the published data shows that most variation between typical patients may well be attributable to measurement uncertainty. Further research is needed to establish the proper model for interpretation of kinetic data and differences between healthy controls and PKU patients, as well as between individual PKU patients. Finally, blockade of Phe entry into brain has been shown to occur as consequence of carrier saturation by LNAAs.

Acknowledgement Many thanks to my collaborators, including Dr. Jean-Marc Nuoffer and in particular Dr. Joachim Pietz, without whose long-standing efforts and interest in improving our understanding of PKU our research would not have been possible. Support by the Swiss National Science Foundation is also gratefully acknowledged (3100A0-103938, 320000-120324).

References

Anderson PJ, Wood SJ, Francis DE, Coleman L, Anderson V, Boneh A (2007) Are neuropsychological impairments in children with early-treated phenylketonuria (PKU) related to white matter abnormalities or elevated phenylalanine levels? Dev Neuropsychol 32:645–668

Avison MJ, Herschkowitz N, Novotny EJ, Petroff OAC, Rothman DL, Colombo JP, Bachmann C, Shulman RG, Prichard JW (1990) Proton NMR observation of phenylalanine and an aromatic metabolite in the rabbit brain in vivo. Pediatr Res 27:566–570

Berry HK, Brunner RL, Hunt MM, White PP (1990) Valine, isoleucine, and leucine. A new treatment for phenylketonuria. Am J Dis Child 144:539–543

Bick U, Ullrich K, Stober U, Moller H, Schuierer G, Ludolph AC, Oberwittler C, Weglage J, Wendel U (1993) White matter abnormalities in patients with treated hyperphenylalaninaemia: magnetic resonance relaxometry and proton spectroscopy findings. Eur J Pediatr 152:1012–1020

Bick U, Ullrich K, Stöber U, Möller HE, Fahrendorf G, Ludolph AC, Wendel U (1991) MRI white matter abnormalities in patients with treated hyperphenylalaninemia: disturbed myelination or toxic edema? Neuropediatrics 22:174

Bik-Multanowski M, Pietrzyk JJ (2007) Brain phenylalanine measurement in patients with phenylketonuria: a serious diagnostic method or just reading tea leaves? Mol Genet Metab 91:297–298

Cleary MA, Walter JH, Wraith JE, Jenkins JPR, Alani SM, Tyler K, Whittle D (1994) Magnetic resonance imaging of the brain in phenylketonuria. Lancet 344:87–90

Dezortova M, Hajek M, Tintera J, Hejcmanova L, Sykova E (2001) MR in phenylketonuria-related brain lesions. Acta Radiol 42:459–466

de Graaf RA (2012) Principles of 1H NMR spectroscopy in vivo. In: Choi I-Y, Gruetter R (eds) Advances in Neurobiology, Vol 4, Chapter 5. Springer, New York, pp 133–148

Govindaraju V, Young K, Maudsley AA (2000) Proton NMR chemical shifts and coupling constants for brain metabolites. NMR Biomed 13:129–153

Gruetter R, Novotny EJ, Boulware SD, Rothman DL, Shulman RG (1996) 1H NMR studies of glucose transport in the human brain. J Cereb Blood Flow Metab 16:427–438

Hofmann L, Slotboom J, Jung B, Maloca P, Boesch C, Kreis R (2002) Quantitative 1H-magnetic resonance spectroscopy of human brain: Influence of composition and parameterization of the basis set in linear combination model fitting. Magn Reson Med 48:440–453

Hommes FA, Lee JS (1990) The control of 5-hydroxytryptamine and dopamine synthesis in the brain: a theoretical approach. J Inherit Metab Dis 13:37–57

Iles RA (2008) Nuclear magnetic resonance spectroscopy and genetic disorders. Curr Med Chem 15:15–36

Knudsen GM, Hasselbalch S, Toft PB, Christensen E, Paulson OB, Lou H (1995) Blood-brain barrier transport of amino acids in healthy controls and in patients with phenylketonuria. J Inherit Metab Dis 18:653–664

Knudsen GM, Paulson OB (1998) Measurement of blood-brain barrier in humans using indicator diffusion. In: Pardridge WM (ed) Introduction to the blood-brain barrier. Methodology, biology and pathology. University Press, Cambridge, pp 133–139

Knudsen GM, Pettigrew KD, Patlak CS, Hertz MM, Paulson OB (1990) Asymmetrical transport of amino acids across the blood-brain barrier in humans. J Cereb Blood Flow Metab 10:698–706

Koch R, Moats R, Guttler F, Guldberg P, Nelson M Jr (2000) Blood-brain phenylalanine relationships in persons with phenylketonuria. Pediatrics 106:1093–1096

Kono K, Okano Y, Nakayama K, Hase Y, Minamikawa S, Ozawa N, Yokote H, Inoue Y (2005) Diffusion-weighted MR imaging in patients with phenylketonuria: relationship between serum phenylalanine levels and ADC values in cerebral white matter. Radiology 236:630–636

Kreis R (2000) Comments on in vivo proton magnetic resonance spectroscopy in phenylketonuria. Eur J Pediatr 159(Suppl 2):S126–S128

Kreis R, Pietz J, Penzien J, Herschkowitz N, Boesch C (1995) Identification and quantitation of phenylalanine in the brain of patients with phenylketonuria by means of localized in vivo 1H magnetic resonance spectroscopy. J Magn Reson Series B 107:242–251

Kreis R, Salvisberg C, Lutz T, Boesch C, Pietz J (2005) Visibility of vascular phenylalanine in dynamic uptake studies in humans using magnetic resonance spectroscopy. Magn Reson Med 54:435–438

Kreis R, Vermathen P, Boesch C (2004) Metabolite and baseline characterization in the downfield region of the human cerebral 1H-MR spectrum in healthy subjects and patients with phenylketonuria. 12th Meeting of the International Society of Magnetic Resonance in Medicine. International Society of Magnetic Resonance in Medicine, Kyoto, p 303

Kreis R, Zwygart K, Boesch C, Nuoffer JM (2009) Reproducibility of cerebral phenylalanine levels in patients with phenylketonuria determined by 1H-MR spectroscopy. Magn Reson Med 62:11–16

Kreis R, Zwygart K, Boesch C, Pietz J, Nuoffer JM (2007) Probing the blood/brain barrier in neonates: 1H-MR spectroscopy shows low protection against high phenylalanine. 15th Meeting of the International Society of Magnetic Resonance in Medicine. International Society of Magnetic Resonance in Medicine, Berlin, p 2309

Laule C, Vavasour IM, Madler B, Kolind SH, Sirrs SM, Brief EE, Traboulsee AL, Moore GR, Li DK, MacKay AL (2007) MR evidence of long T2 water in pathological white matter. J Magn Reson Imaging 26:1117–1121

Leuzzi V, Bianchi MC, Tosetti M, Carducci CL, Carducci CA, Antonozzi I (2000) Clinical significance of brain phenylalanine concentration assessed by in vivo proton magnetic resonance spectroscopy in phenylketonuria. J Inherit Metab Dis 23:563–570

Leuzzi V, Tosetti M, Montanaro D, Carducci C, Artiola C, Carducci C, Antonozzi I, Burroni M, Carnevale F, Chiarotti F, Popolizio T, Giannatempo GM, D'Alesio V, Scarabino T (2007) The pathogenesis of the white matter abnormalities in phenylketonuria. A multimodal 3.0 tesla MRI and magnetic resonance spectroscopy (1H MRS) study. J Inherit Metab Dis 30: 209–216

Malamud N (1966) Neuropathology of phenylketonuria. J Neuropathol Exp Neurol 25:254–268

Möller HE, Feldmann R, Santer R, Ullrich K, Weglage J (2006) Phenylketonuria and blood-brain barrier competition investigated by magnetic resonance spectroscopy. In: Blau N. (Ed) Advances in Phenylketonuria and Tetrahydrobiopterin, SPS Verlagsgesellschaft. Heilbronn, pp 137–160

Moller HE, Ullrich K, Weglage J (2000) In vivo proton magnetic resonance spectroscopy in phenylketonuria. Eur J Pediatr 159(Suppl 2):S121–S125

Möller HE, Vermathen P, Ullrich K, Weglage J, Koch HG, Peters PE (1995) In vivo NMR spectroscopy in patients with phenylketonuria: changes of cerebral phenylalanine levels under dietary treatment. Neuropediatrics 26:199–202

Moller HE, Weglage J, Wiedermann D, Ullrich K (1998) Blood-brain barrier phenylalanine transport and individual vulnerability in phenylketonuria. J Cereb Blood Flow Metab 18:1184–1191

Möller HE, Weglage J, Wiedermann D, Vermathen P, Bick U, Ullrich K (1997) Kinetics of phenylalanine transport at the human blood-brain barrier investigated in vivo. Brain Res 778: 329–337

Moller LB, Paulsen M, Koch R, Moats R, Guldberg P, Güttler F (2005) Inter-individual variation in brain phenylalanine concentration in patients with PKU is not caused by genetic variation in the 4F2hc/LAT1 complex. Mol Genet Metab 86(Suppl 1):S119–S123

National Institutes of Health Consensus Development Panel (2001) National Institutes of Health Consensus Development Conference Statement: phenylketonuria: screening and management, October 16–18, 2000. Pediatrics 108:972–982

Novotny EJ Jr, Avison MJ, Herschkowitz N, Petroff OAC, Prichard JW, Seashore R, Rothman DL (1995) In vivo measurement of phenylalanine in human brain by proton nuclear magnetic resonance spectroscopy. Pediatr Res 37:244–249

O'Kane RL, Hawkins RA (2003) Na+-dependent transport of large neutral amino acids occurs at the abluminal membrane of the blood-brain barrier. Am J Physiol Endocrinol Metab 285: E1167–E1173

O'Kane RL, Vina JR, Simpson I, Hawkins RA (2004) Na+-dependent neutral amino acid transporters (A, ASC and N) of the blood-brain barrier: mechanisms for neutral amino acid removal. Am J Physiol Endocrinol Metab 287:E622–E629

Pardridge WM (1988) Phenylalanine transport at the human blood-brain barrier. In: Wurtman RJ, Ritter-Walker E (eds) Dietary phenylalanine and brain function. Birkhäuser, Boston, pp 56–62

Pardridge WM (1998a) Blood-brain barrier carrier-mediated transport and brain metabolism of amino acids. Neurochem Res 23:635–644

Pardridge WM (1998b) Introduction to the Blood-Brain Barrier (Methodology, biology and pathology). University Press, Cambridge

Pearsen KD, Gean-Marton AD, Levy HL, Davis KR (1990) Phenylketonuria: MR imaging of the brain with clinical correlation. Radiology 177:437–440

Phillips MD, McGraw P, Lowe MJ, Mathews VP, Hainline BE (2001) Diffusion-weighted imaging of white matter abnormalities in patients with phenylketonuria. AJNR Am J Neuroradiol 22:1583–1586

Pietz J, Kreis R (2007) MRS studies in PKU patients. International Symposium on "KU and other Hyperphenylalaninemias: where are we now?" Fulda, Germany

Pietz J, Kreis R, Boesch C, Penzien J, Rating D, Herschkowitz N (1995) The dynamics of brain concentrations of phenylalanine and its clinical significance in patients with phenylketonuria determined by in vivo 1H magnetic resonance spectroscopy. Pediatr Res 38:657–663

Pietz J, Kreis R, Rupp A, Mayatepek E, Rating D, Boesch C, Bremer HJ (1999) Large neutral amino acids block phenylalanine transport into brain tissue in patients with phenylketonuria. J Clin Invest 103:1169–1178

Pietz J, Kreis R, Schmidt H, Meyding-Lamade UK, Rupp A, Boesch C (1996) Phenylketonuria: findings at MR imaging and localized in vivo H-1 MR spectroscopy of the brain in patients with early treatment. Radiology 201:413–420

Pietz J, Lutz T, Zwygart K, Hoffmann GF, Ebinger F, Boesch C, Kreis R (2003a) Phenylalanine can be detected in brain tissue of healthy subjects by 1H magnetic resonance spectroscopy. J Inherit Metab Dis 26:683–691

Pietz J, Rupp A, Burgard P, Boesch C, Kreis R (2002) No evidence for individual blood-brain barrier phenylalanine transport to influence clinical outcome in typical phenylketonuria patients. Ann Neurol 52:382–383

Pietz J, Rupp A, Ebinger F, Rating D, Mayatepek E, Boesch C, Kreis R (2003b) Cerebral energy metabolism in phenylketonuria: findings by quantitative in vivo 31P MR spectroscopy. Pediatr Res 53:654–662

Pratt OE (1982) Transport inhibition in the pathology of phenylketonuria and other inherited metabolic diseases. J Inherit Metab Dis 5:75–81

Provencher SW (1993) Estimation of metabolite concentration from localized in vivo proton NMR spectra. Magn Reson Med 30:672–679

Rupp A, Kreis R, Zschocke J, Slotboom J, Boesch C, Rating D, Pietz J (2001) Variability of blood-brain ratios of phenylalanine in "ypical"patients with phenylketonuria. J Cereb Blood Flow Metab 21:276–284

Scriver CR, Kaufman S (2001) Hyperphenylalaninemia: phenylalanine hydroxylase deficiency. In: Scriver CR, Beaudet AL, Sly WS, Valle D (eds) The metabolic & molecular bases of inherited disease. Medical Publishing Division, McGraw-Hill, pp 1667–1724

Sirrs SM, Laule C, Madler B, Brief EE, Tahir SA, Bishop C, MacKay AL (2007) Normal-appearing white matter in patients with phenylketonuria: water content, myelin water fraction, and metabolite concentrations. Radiology 242:236–243

Smith CB, Kang J (2000) Cerebral protein synthesis in a genetic mouse model of phenylketonuria. Proc Natl Acad Sci USA 97:11014–11019

Smith QR (2003) A review of blood-brain barrier transport techniques. In: Nag S (ed) The Blood-Brain Barrier. Biology and Research Protocols. Humana Press, Totowa, pp 193–208

Thompson AJ, Tillotson S, Smith I, Kendall B, Moore SG, Brenton DP (1993) Brain MRI changes in phenylketonuria associations with dietary status. Brain 116:811–821

Tkac I, Gruetter R (2005) Methodology of H-1 NMR spectroscopy of the human brain at very high magnetic fields. Appl Magn Reson 29:139–157

Vermathen P, Capizzano AA, Maudsley AA (2000) Administration and (1)H MRS detection of histidine in human brain: application to in vivo pH measurement. Magn Reson Med 43: 665–675

Vermathen P, Robert-Tissot L, Pietz J, Lutz T, Boesch C, Kreis R (2007) Characterization of white matter alterations in phenylketonuria by magnetic resonance relaxometry and diffusion tensor imaging. Magn Reson Med 58:1145–1156

Weglage J, Pietsch M, Feldmann R, Koch HG, Zschocke J, Hoffmann G, Muntau-Heger A, Denecke J, Guldberg P, Guttler F, Moller H, Wendel U, Ullrich K, Harms E (2001a) Normal clinical outcome in untreated subjects with mild hyperphenylalaninemia. Pediatr Res 49: 532–536

Weglage J, Wiedermann D, Denecke J, Feldmann R, Koch HG, Ullrich K, Harms E, Moller HE (2001b) Individual blood-brain barrier phenylalanine transport determines clinical outcome in phenylketonuria. Ann Neurol 50:463–467

Weglage J, Wiedermann D, Feldmann R, Ullrich K, Moller HE (2002) Reply. Ann Neurol 52: 383–384

Zielke HR, Zielke CL, Baab PJ, Collins RM (2002) Large neutral amino acids auto exchange when infused by microdialysis into the rat brain: implication for maple syrup urine disease and phenylketonuria. Neurochem Int 40:347–354

Chapter 41
Measurement of Brain Creatine Metabolism *In Vivo*: Magnetic Resonance Spectroscopy Studies of Transgenic Animals

Arend Heerschap, Christine I.H.C. Nabuurs, Hermien E. Kan, Bé Wieringa, and Dirk Isbrandt

Abstract *In vivo* magnetic resonance spectroscopy (MRS) of transgenic mice with modified expression of enzymatically active proteins offers an attractive opportunity for non-invasive study of creatine metabolism in the brain as signals of substrates are particularly well-identifiable in MR spectra. This chapter reviews results of studies on mice with alterations in the expression of the enzymes creatine kinase and guanidinoacetate methyltransferase.

MRS studies on mice lacking guanidinoacetate methyltransferase have demonstrated metabolic changes comparable to those found in the deficiency of this enzyme in humans, which are (partly) reversible upon creatine feeding. Apart from being a model for creatine deficiency syndrome, these mice are of interest to study fundamental aspects of the biological role of creatine.

MRS studies on transgenic mice lacking creatine kinase isoenzymes have contributed to the view that these enzymes are involved in a large high-energy phosphate metabolic energy network, which is highly versatile and dynamically adapts to genotoxic or physiological challenges.

Keywords Creatine metabolism • MR spectroscopy • Transgenic mice

---

A. Heerschap, Ph.D. (✉) • C.I.H.C. Nabuurs, Ph.D. • H.E. Kan, Ph.D.
Department of Radiology, Radboud University Nijmegen Medical Centre, 6500 HB, Nijmegen, The Netherlands
e-mail: A.Heerschap@rad.umcn.nl; christine.nabuurs@maastrichtuniversity.nl; H.E.Kan@lumc.nl

B. Wieringa, Ph.D.
Department of Cell Biology, Radboud University Nijmegen Medical Centre, Geert Grooteplein 21, 6525 EZ, Nijmegen, The Netherlands
e-mail: b.Wieringa@ncmls.ru.nl

D. Isbrandt, Ph.D.
Zentrum für Molekulare Neurobiologie, University Medical Center, Hamburg-Eppendorf, ENP/ZMNH, Martinistrasse 52, 20246 Hamburg, Germany
e-mail: isbrandt@enp.org

I.-Y. Choi and R. Gruetter (eds.), *Neural Metabolism In Vivo*, Advances in Neurobiology 4, DOI 10.1007/978-1-4614-1788-0_41,

41.1 Introduction

Studies of genetically altered mice are important to understand the biological role of specific enzymes involved in cell signaling, metabolism, or structural organization of the brain. Non-invasive bio-imaging and mapping has been introduced for proper phenotyping of these animals (Budinger et al. 1999; Ntziachristos et al. 2003).

A particularly useful tool to investigate metabolism *in vivo* and *in vitro* is magnetic resonance spectroscopy (MRS). The MR sensitive nuclei mostly used in MRS of living beings are ^{1}H, ^{31}P and ^{13}C. The spectra of each of these nuclei offer unique information on distinct metabolic pathways, and also on tissue morphology and on some physiological processes. MRS of transgenic animals is most powerful when genetic modifications have an effect on the activity of enzymes that interact directly with MR-visible metabolites. This is the case for the enzymes described in this chapter, which are involved in creatine (Cr) synthesis and phosphorylation of creatine to phosphocreatine (PCr) and creatine-intermediates.

We first describe a few issues related to MRS of the mouse brain as this is more difficult than that of the rat brain because of its smaller size. For a detailed introduction of the basics of MRS and its localized application in living beings we refer to other textbooks, e.g. (de Graaf 2007). Subsequently, we give a brief description of the metabolic reactions and enzymes involved in creatine metabolism, before describing MRS findings.

41.2 *In Vivo* Magnetic Resonance Spectroscopy of the Mouse Brain

In general for MR spectroscopy, but in particular as applied to mice, two practical issues are of primary interest: sensitivity and field homogeneity. Therefore, much of the MR hardware components and software used in studies of mice have a dedicated design and these may differ for particular applications (Heerschap et al. 2004). Proportional spatial downscaling from rat to mice requires approximately a tenfold increase in sensitivity. To a large extent this can be achieved by the use of smaller radiofrequency coils for signal detection, although too small a coil may reduce coverage of the tissue of interest. Tissue coverage can be maintained by the application of coil arrays, which also has other advantages (Gareis et al. 2007). So-called cryoprobes can provide further increases in sensitivity for mice brain applications (Baltes et al. 2009). In all cases adequate probe building for mice requires skilled miniaturization. In principle, sensitivity can also be increased by going to higher magnetic fields. Common field strengths for mice MR studies range from 4.7 to 9.4 T, but higher fields may be desirable, also for a better separation of individual signals in spectra by increased chemical shift dispersion.

Proper imaging and localization requires a sufficiently strong magnetic field gradient set. For high spectral resolution, the optimization of the field homogeneity is

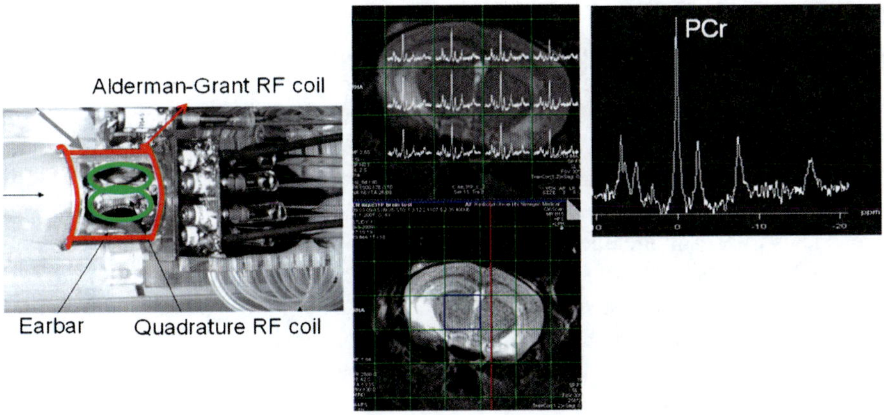

Fig. 41.1 ^{31}P MR spectroscopy of the mouse brain. At the left Radiofrequency probe for the mouse brain. The halve-volume Alderman-Grant coil, shown in red, is used for ^{1}H MR. A quadrature surface coil, shown in green, is used for ^{31}P MR spectroscopy and is positioned on top of the mouse head, which is fixed with earbars. At the middle MR images (T2 – weighted) with overlaid the grid with voxels from which the ^{31}P MR spectra (top) are obtained. 3D ^{31}P MR spectroscopic imaging was performed on a 7T Bruker Clinscan with a BIR-45° pulse for excitation, a repetition time of 1.5 s, a field of view of 24 mm^{3} and a matrix size of 8x8x8 and at the right ^{31}P MR spectrum of one of the voxels in the mouse brain showing a dominant peak for phosphocreatine (PCr) and to the right 3 peaks for ATP and to the left peaks for phosphomonoesters and diesters and inorganic phosphate

needed by shim coils of sufficient strength (Gruetter et al. 1998). Especially in the mouse, field in-homogeneities may be relatively severe since most tissue is in close vicinity of air due the small size of the animal. Therefore strong shim coils including second order terms are desirable (Tkac et al. 2004).

For reliable measurements of brain structures in the mouse, the animal has to be "immobilized" without suppressing respiratory motion or other basal physiological functions. Commonly, the animal is kept under anesthesia using inhalation or injection anesthetics. Possible ways of securing the mouse head, for instance, are the use of stereotactic earbars, teethbars and tight attachment of an RF coil on top of the head. As anesthetized mice are not able to maintain body temperature, a warming device, like a warm water bed, is needed. Furthermore, temperature has to be monitored continuously, just as breathing frequency, to adjust the depth of anesthesia, since this can influence the MR results.

In our MRS studies on the brain of mice with creatine or creatine kinase deficiency we have applied measurements with ^{1}H and ^{31}P nuclei. The RF probe used in most experiments is shown in Fig. 41.1. The ^{31}P MRS signal is acquired using a quadrature coil, while imaging, shimming and ^{1}H MRS were performed with an Alderman-Grant coil. More recently we also applied ^{13}C MRS to the mouse brain (Nabuurs et al. 2008) .

Of all nuclei applied in biomedical MRS, ^{1}H has the highest sensitivity and a natural abundance of nearly 100%, enabling MRS of relatively small volumes of interest or voxels (about a few μL) in mice or rats (Renema et al. 2003; in 't Zandt et al. 2004; Tkac et al. 2004). As the high signal for water (~45 M) may perturb the signal contributions from other MR visible molecules (~0.1–10 mM), it is commonly suppressed. In ^{1}H MR spectra of brain, signals of up to about 30 different molecules may be detected of which those of creatine, N-acetyl aspartate (NAA) and choline compounds (Cho) are usually dominating the spectrum.

The ^{31}P nucleus is applied extensively in MRS of bioenergetics, as important phosphates in energy metabolism are clearly visible in an *in vivo* ^{31}P MR spectrum (Fig. 41.1). ^{31}P has a natural abundance of 100%, but its sensitivity is only about ~7% of ^{1}H. Therefore, ^{31}P MR spectra are acquired from larger areas of the brain as compared to those for ^{1}H MRS. Apart from signals for inorganic phosphate (Pi), PCr and ATP, also the intracellular tissue pH can be calculated from the shift in resonance position between Pi and PCr (Brindle 1988; Moon and Richards 1973). In ^{31}P MR spectra of the brain also signals for phosphomonoesters (PME) and diesters (PDE) are present. And in addition often the level of "free" ADP is calculated from the creatine kinase equilibrium (ADP is not visible in the ^{31}P MR spectrum). By a method called magnetization transfer, flux rates in some biochemical pathways can be determined by ^{31}P MRS (Forsén and Hoffman 1963; Brindle 1988; Rudin and Sauter 1992). This technique has been applied to determine the flux rates in the creatine kinase (CK) reaction, involving the phosphorylation of Cr to PCr.

Additional assessments during an MR examination, for instance physiological measurements requiring the regular sampling of blood, are more difficult in mice than in rats.

41.3 Metabolic Reactions Catalyzed by Guanidinoacetate Methyltransferase and Creatine Kinase

Creatine is a small metabolite that fulfills a central role in brain energy metabolism. The two last steps in its biosynthesis are controlled by the enzymes arginine-glycine amidino transferase (AGAT) and guanidinoacetate methyltransferase (GAMT), of which the latter catalyzes the methylation of guanidinoacetate to form Cr (see Fig. 41.2). In mammals, this biosynthesis is assumed to take place mainly in the pancreas and liver. Subsequently, Cr is exported into the blood and taken up by tissues requiring Cr, such as muscle and brain (Wyss and Kaddurah-Daouk 2000). Intake of creatine-containing food also increases the blood level of creatine, and other tissues such as the brain may have the enzymatic machinery to synthesize creatine themselves (Braissant et al. 2001).

Tissues taking up creatine are generally characterized as having high and fluctuating energy demands. In these tissues, Cr is reversibly phosphorylated to PCr in the (near-equilibrium) reaction Cr + ATP ↔ PCr + ADP + H^+, which is catalyzed by the enzyme creatine kinase (CK). This reaction is thought to be important for the temporal buffering of the levels of so-called high-energy phosphates (~P) to balance

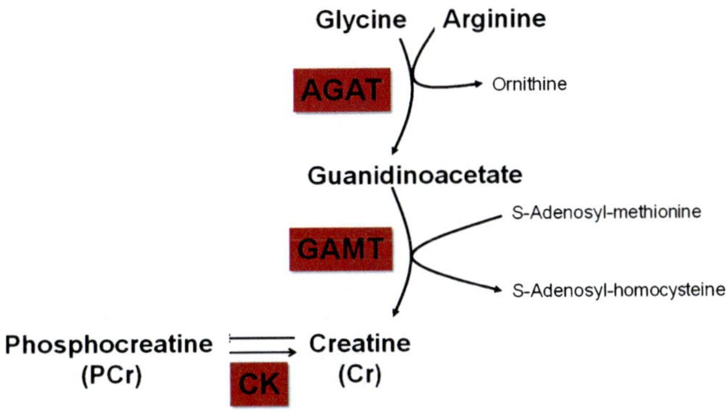

Fig. 41.2 Scheme of the synthesis of creatine starting from glycine and arginine. The enzymes involved are arginine-glycine amidino transferase (AGAT), guanidine methyltransferase (GAMT) and the enzyme creatine kinase (CK) for the conversion of creatine into phosphocreatine

(local) ATP/ADP ratios, and also for pH buffering (Wallimann et al. 1992; Wyss et al. 1992; Wallimann 1994; Wyss and Kaddurah-Daouk 2000). Over the years, discussion has arisen concerning the role of the Cr/CK system in the distribution and intracellular transport of high-energy phosphates, the importance of which is related to cellular diffusion distances. The presence and partitioning of different iso-enzymes of CK, apparently specifically associated with loci of energy production and consumption in the cell, has stimulated the idea of a structured uni-directional high-energy phosphate transport system, the so-called CK shuttle. This shuttle concept was originally proposed half a century ago (Bessman and Fonyo 1966; Bessman and Geiger 1981) and implicates cellular compartmentation of the reactants of the CK reaction. However, the spatial transport role of the Cr/CK system can also be understood in a simpler way, as a specific case of facilitated diffusion not necessitating the structural organization proposed for shuttle-type systems (Meyer et al. 1984).

In vertebrates, four CK genes have been identified, coding for altogether five different multimeric isoenzymes. Two genes for cytoplasmatic CK encode the muscle (M-CK) and brain (B-CK) isoforms, which assemble into active homo-dimeric (MM-CK, BB-CK) and hetero-dimeric (MB-CK) enzymes in the cytosol (Wallimann et al. 1992) BB-CK is expressed at high levels in brain and various other tissues in adult mice. Expression of the other class of CKs, ubiquitous (UbCKmit) and sarcomeric mitochondrial CK (ScCKmit), is also cell-type dependent. UbCKmit is mostly co-expressed with BB-CK. However, it appears that in the brain, BB-CK is preferentially expressed in astrocytes and inhibitory neurons, while mitochondrial CK seems to be restricted primarily to neurons (Tachikawa et al. 2004). UbCKmit occurs as dimeric or octameric enzyme and resides between the inner and outer membranes of mitochondria (Wallimann et al. 1992).

41.4 Transgenic Mice Lacking Guanidinoacetate Methyltransferase (GAMT)

41.4.1 General Phenotype of GAMT Deficient Mice

Guanidinoacetate N-methyltransferase (GAMT) knockout mice were generated by gene targeting in embryonic stem cells (Schmidt et al. 2004), as a model for the first-discovered Cr deficiency syndrome in humans (Stöckler et al. 1994). As these animals are not able to synthesize their own Cr, their tissues are completely deprived of Cr unless Cr is orally ingested. GAMT knockout mice have markedly increased guanidinoacetate (GAA) and reduced creatine and creatinine levels in brain, muscle, serum and urine, which are the typical abnormal metabolic findings also seen in GAMT-deficient patients. Biochemical alterations were highly reminiscent to those found in GAMT-deficient patients and can be attributed to the overall similar GAMT expression patterns in human and mouse tissues. GAMT deficiency is associated with increased neonatal mortality, muscular hypotonia, decreased male fertility and a non-leptin-mediated life-long reduction in body weight due to reduced body fat mass. Apart from lack of creatine, the genotoxic effects of this mutation may also be caused by accumulation of guanidinoacetate (Schmidt et al. 2004; Kan et al. 2005)

41.4.2 MRS Studies on GAMT-Deficient Mice

Localized ^{1}H MRS of mice lacking the GAMT enzyme (GAMT-/-) showed strongly reduced levels of Cr in the brain (Fig. 41.3), although residual Cr may still be present due to oral intake (Renema et al. 2003). GAA is phosphorylated in the body to PGAA, which was shown to be metabolically active in skeletal muscle (Renema et al. 2003). In ischemia-reperfusion experiments monitored by ^{31}P MRS, the PGAA signal decreased similarly to that of PCr. But recovery of PGAA was significantly delayed, also in the presence of Cr, indicating that CK has a lower affinity for GAA than for Cr (Kan et al. 2004). In agreement, saturation transfer experiments showed negligible transfer of the γ-phosphate of ATP to PGAA (Kan et al. 2004).

In confirmation with the ^{1}H MRS results localized ^{31}P MR spectra of the brain of GAMT-deficient mice showed very little PCr upon normal chow feeding (Fig. 41.4). In contrast to muscle this was not compensated for by the presence of high levels of PGAA for which only a small signal could be observed in brain spectra (Renema et al. 2003; Schmidt et al. 2004). Fluctuations of the PGAA and PCr signals, as seen in skeletal muscle, were not observed in the brain of different animals. Limited permeability of the blood-brain barrier for Cr (Wyss and Kaddurah-Daouk 2000) could explain these observations. This limited permeability was already suspected from the slow recovery of the Cr signal after long-term feeding of high amounts of Cr to GAMT-deficient patients (Wyss and Kaddurah-Daouk 2000; von Figura et al. 2001). Indeed, we observed a slower uptake of Cr in brain as compared to skeletal

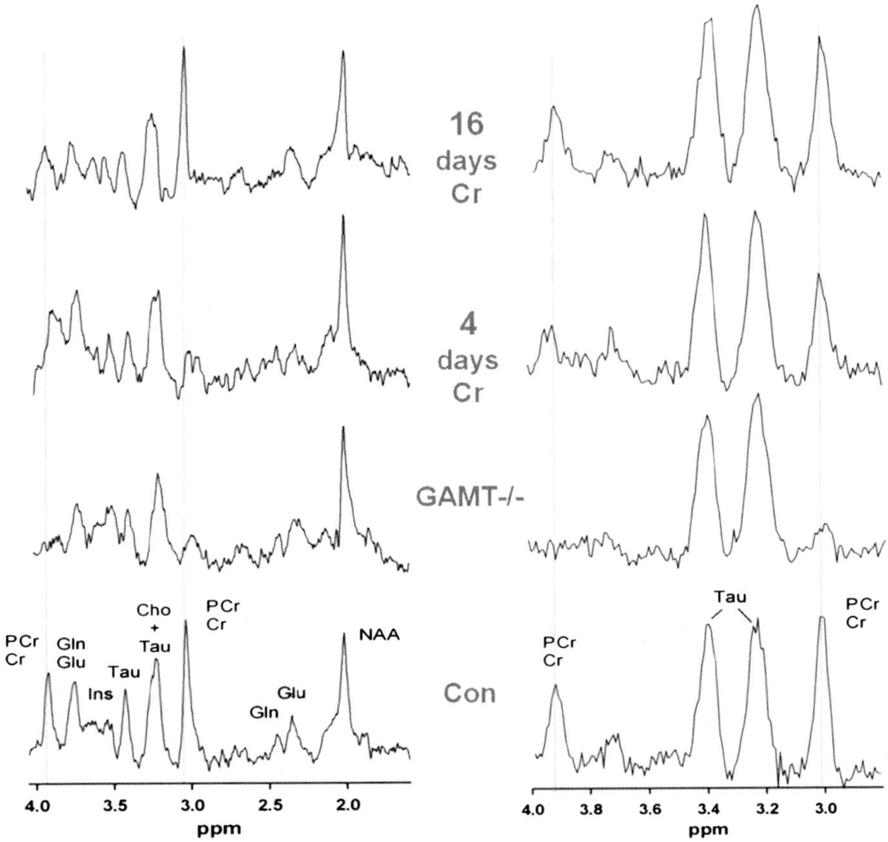

Fig. 41.3 Spectral changes observed in 1H MR spectra during the supplementation of creatine to GAMT deficient mice (GAMT-/-). At the left ^{1}H MR spectra of mouse brain and at the right for mouse leg skeletal muscle. The GAMT-/- mice clearly lacks peaks for the methyl and methylene protons of PCr and Cr in comparison to the control spectra (Con). During creatine supplementation peaks for creatine appear earlier in the spectra of muscle compared to the brain (e.g. compare the spectra at 4 days supplementation). NAA: N-acetylaspartate; Glu: glutamate; Gln: glutamine; Cho: choline, Tau: taurine; Ins:inositol

muscle of GAMT-deficient mice upon Cr feeding (Fig. 41.3). However the faster uptake in skeletal muscle only occurred during the first day of Cr supplementation, thereafter the uptake in brain and muscle was similar (Kan et al. 2007). This indicates the existence of a more subtle difference of Cr transport into these tissues of GAMT-deficient mice. NAA levels in the brain of GAMT-lacking mice were increased, but significantly decreased at higher Cr levels, possibly reflecting an osmolytic rearrangement. At the end of the supplementation period Cr levels were typically higher than in wild-type animals not supplemented with Cr.

It is of interest to compare findings in GAMT-deficient mice with those in β-guanidinopropionic acid (GPA) fed mice (Holtzman et al. 1989). As GPA can compete

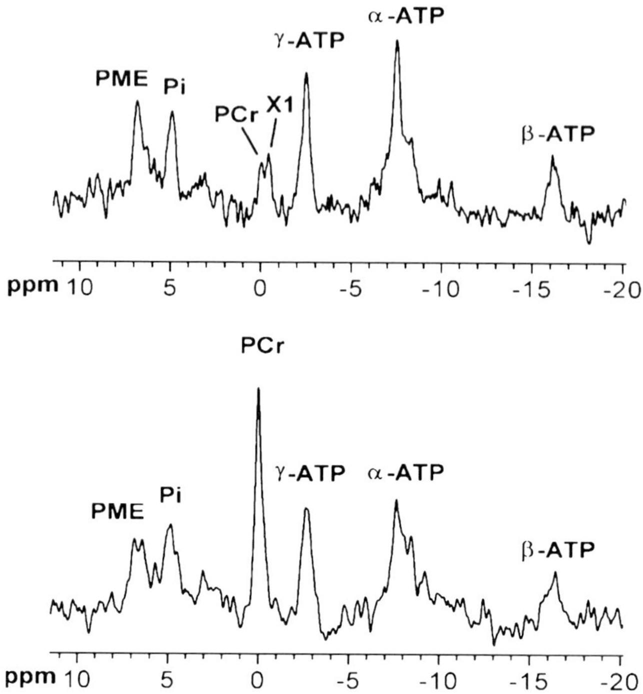

Fig. 41.4 ^{31}P MR spectra of the mouse brain. At the bottom a spectrum of a control mouse and at the top of a GAMT-/- mouse. The GAMT-/- mouse shows a strongly decreased PCr peak, while a new small peak appears (X1) at the position of phospho-guanidino-acetate. PME: phosphomonoesters; Pi: inorganic phosphate

with Cr uptake such feeding is aimed to deplete Cr. While Cr replacement appears to be possible to a large extent in murine skeletal muscle, in the mouse brain PCr is only partly replaced. From this result it was hypothesized that PCr is present in two different pools in the brain, one associated with neurons (the replaceable pool) and one with astrocytes (not replaceable by GPA). However, in the light of experiments with GAMT mice this partial replacement may also be caused by competition between GPA and Cr, still present in food, for transport over the blood brain barrier.

To study pure Cr depletion without any interference of (P) GAA a new mouse mutant with AGAT deficiency was recently generated and investigated by MRS (Nabuurs et al. 2009).

41.5 Brain Creatine Kinase Knock-Out Mice

In many MRS studies that have been performed in the past on ~ P transfer systems in rodent brain, main attention has been focused on the reversible CK-catalyzed reaction as levels of key substrates and products in this reaction (PCr, ATP, H^+) can be determined non-invasively from ^{31}P MR spectra of living tissue. In addition, the kinetics of the CK reaction and of ATPases can be determined from magnetization transfer experiments, although this has been little applied to the mouse brain (Holtzman et al. 1989; Jost et al. 2002).

Research on high-energy phosphoryl transfer in muscle and brain greatly benefited from the availability of mice with null mutations in CK genes (van Deursen et al. 1993; Steeghs et al. 1997a, b). Mice lacking either the cytosolic B-CK (B-CK-/- mice) (Jost et al. 2002), the ubiquitous mitochondrial CK isoenzyme (UbCKmit-/- mice) (Kekelidze et al. 2001) or both enzymes (B-CK/UbCKmit-/- mice) (in 't Zandt et al. 2004) have been generated to study the role of the CK reaction in the brain.

41.5.1 General Phenotype of Mice with Brain CK Deficiency

The double mutant B-CK/UbCKmit-/- mice, which were studied most intensively (Streijger et al. 2005), appeared viable, but their development was affected, resulting in an overall smaller body posture. Furthermore, the complete lack of CK in brain in these animals severely impaired spatial learning and was coupled to abnormal startle responses and hearing problems. In contrast, the visual and motor functions, exploration behavior, and anxiety-related responses were not changed, suggesting no global deficit in sensorimotor function or motivation. Various effects on behavior were consistent with underlying anatomical changes in the hippocampus. Others can best be explained by loss of functional integrity that may occur in development. Parallel to the anatomical and behavioral abnormalities, B-CK/UbCKmit-/- mice also had problems with thermoregulation, presumably due to an altered hypothalamic thermostat (Streijger et al. 2009). Furthermore, their response to chemical seizure induction appeared suppressed (Streijger et al. 2010).

41.5.2 MRS Studies of CK Isoenzyme Deficiency in the Mouse Brain

Localized ^{31}P MRS measurements of the brain of transgenic mice without cytosolic CK (B-CK-/- mice) showed normal PCr, Pi and ATP levels and pH values. The flux of ~P transfer through the CK reaction, measured by saturation transfer experiments, was dramatically lowered, although still measurable in these animals (Jost et al. 2002). Because the brain is an oxidative tissue, the mitochondrial content and fractional contribution of mitochondrial CK activity (up to 30% of total CK activity) is much higher than in muscle. Unlike in muscle, mitochondrial CK in brain may therefore be present at large enough quantities to support an MRS-visible exchange between the high-energy phosphates of γ-ATP and PCr.

It is not very likely that this exchange reflects an operational CK shuttle in the brain. As compared to muscle, the brain is composed of a much more heterogeneous mixture of cell types. Moreover, the cell-type distribution and co-expression of B-CK and UbCKmit isoforms in glial and neuronal cells appears to be rather variable and does not necessary seem to support the cellular co-existence of cytosolic and mitochondrial endpoints of the CK circuit in various brain regions (e.g. see Streijger et al. 2005 and references therein). Evidence from several studies showed close metabolic collaboration between glial and neuronal systems for brain energetics (Pellerin and Magistretti 2004), which makes it intrinsically difficult to understand how the PCr shuttle – if present – is integrated into the energy network of brain. In this respect it is of interest that Cr and the CK reaction have been proposed as players in a neuron-glial shuttle (Tachikawa et al. 2004) and that creatine has been suggested to act as a neuromodulator (Almeida et al. 2006).

Also mice lacking mitochondrial CK (UbCKmit-/-) appeared to have rather normal basal PCr, Pi, ATP and pH levels ('in 't Zandt et al. 2004), although a significantly lower PCr/ATP ratio in homozygous UbCKmit-/- mice was reported in one study (Kekelidze et al. 2001). During seizure induction by PTZ injection, UbCKmit-/- mice showed a more pronounced decrease in PCr and β-ATP signals while the PCr/β-ATP ratio remained lower than in wild-type mice (Kekelidze et al. 2001).

In the brains of double-knockout B-CK/UbCKmit-/- mice PCr appeared to be absent, while Pi and ATP levels and pH were virtually normal ('in 't Zandt et al. 2004). This picture is different from that in skeletal muscle, where PCr is still present in double-knockout animals despite the absence of cytosolic and mitochondrial muscle CK, presumably because of the presence of residual B-CK. In the brain of B-CK/UbCKmit double-knockout mutant, the total amount of creatine was decreased by 30% and levels of N-acetyl aspartate (NAA) increased, which could point to an osmolar redistribution or other adaptation mechanisms (in 't Zandt et al. 2004). Interestingly, a similar inverse relationship between Cr and NAA was observed during Cr feeding in GAMT deficient mice.

Saturation of the γATP peak in ^{31}P MR spectra of the mice brain results in a decrease of the PCr and Pi peak due to chemical exchange in CK and in ATP processing reactions respectively. In addition the α and βATP peaks are affected. Effects of similar size on α and βATP peaks seen in spectra of resting skeletal muscle of mice were assigned to a so-called NOE effect, which could only arise when a small fraction (<10%) of ATP is transiently bound (Nabuurs et al. 2010). As in spectra of mutant mice brain without any CK these effects still existed in the same magnitude, effects from exchange by the CK reaction can be excluded. Apparently, also in the brain a similar fraction of ATP is transiently immobilized. Moreover, as γATP and βADP co-resonate one would predict, upon their saturation, a substantial effect on the βATP resonance due to the CK reaction and no effect in the case of CK deletion. This is not observed and in analogy to our reasoning for skeletal muscle (Nabuurs et al. 2010) we conclude that ADP calculations from the CK equilibrium in the mouse brain are questionable.

The role of the PCr/CK circuit in the brain is far less well understood than in muscle. Apparently the circuit is not vital to basic brain physiology, but behavioral studies have demonstrated structural, sensory and cognitive deficiencies in brain CK mutants (Streijger et al. 2004, 2005). Testing brain function and metabolism under energy stress conditions in these mice is needed to uncover more specific functions of the circuit such as, for instance, in metabolic coupling between neuro-activation and hemodynamic response (Aubert et al. 2007).

41.6 Concluding Remarks

MRS provides unique contributions to the understanding of creatine-related bioenergetics and metabolism in the brain using transgenic animals. Typical properties of MRS, such as its non-destructive and non-invasive character, which enables dynamic and longitudinal studies, are major assets as compared to more invasive approaches in investigations of bioenergetic and metabolic changes in the brain. Because of its relatively low spatial resolution, the use of results from MRS examinations in the interpretation of processes at the (sub)cellular level may not be trivial and requires careful consideration of morphological, physiological, and biochemical knowledge of the brain. Cellular heterogeneity of the brain may be an issue when its composition changes due to the mutation introduced. Phenotypic consequences may also not be 100% penetrant, as was shown for ventricle enlargement by MR imaging of B-CK and other knockout mice (Auerbach et al. 2001; in 't Zandt et al. 2004). When only average values are evaluated, such changes may remain unnoticed.

Knockout mice often show a remarkable plasticity in coping with the effects of gene deficiency. Adaptations may occur with wider metabolic consequences than the direct effect of gene deletion. Hence, interpretation of a protein's function may not be straightforward (Bailey 1999), also due to partial redundancy or overlap with functions of other proteins. However, the adaptations that occur as a result of genetic manipulations are interesting in their own right and provide valuable information on the flexibility of the cellular network in which it is involved. For creatine and CK, which were long considered irreplaceable in the brain, a more subtle and dynamic role within such a network has emerged. Further work under different challenging conditions *in vivo* is needed to uncover more specific functions of ~ P-metabolizing iso-enzymes and the role(s) of creatine in the brain.

MRS of transgenic mice that represent models for disorders caused by defects in these enzymes, may have clinical relevance, as for example demonstrated in the studies of mice with GAMT and AGAT deficiency. These models provide unique possibilities to examine the potential reversibility of disease consequences, by following uptake of Cr in brain. Moreover, the effects of genetic lesions causing creatine deficiencies can be studied without developmental adaptations by suppressing these adaptations with Cr feeding until switching to a Cr free diet.

MRS of the brain of living mice has some technical challenges, but the studies described here and elsewhere demonstrate its feasibility. The development of new MR equipment, acquisition methods and practical approaches dedicated to mice are expected to uncover still more physiological details of energy transfer systems and creatine metabolism proper in the brain.

Acknowledgments This work was supported by NWO-ZONMW grants 901-01191, 902-37-088, and NWO-ALW grant 834-04-007,NWO-BIG (VISTA), PBF WAR08-15 and the DFG.

References

Almeida LS, Salomons GS, Hogenboom F, Jakobs C, Schoffelmeer AN (2006) Exocytotic release of creatine in rat brain. Synapse 60:5

Aubert A, Pellerin L, Magistretti PJ, Costalat R (2007) A coherent neurobiological framework for functional neuroimaging provided by a model integrating compartmentalized energy metabolism. Proc Natl Acad Sci U S A 104:4188–4193

Auerbach W, Hurlbert MS, Hilditch-Maguire P, Wadghiri YZ, Wheeler VC, Cohen SI, Joyner AL, MacDonald ME, Turnbull DH (2001) The HD mutation causes progressive lethal neurological disease in mice expressing reduced levels of huntingtin. Hum Mol Genet 10:2515–2523

Bailey JE (1999) Lessons from metabolic engineering for functional genomics and drug discovery. Nat Biotechnol 17:616–618

Baltes C, Radzwill N, Bosshard S, Marek D, Rudin M (2009) Micro MRI of the mouse brain using a novel 400 MHz cryogenic quadrature RF probe. NMR Biomed 22:8

Bessman SP, Fonyo A (1966) The possible role of the mitochondrial bound creatine kinase in regulation of mitochondrial respiration. Biochem Biophys Res Commun 22:597–602

Bessman SP, Geiger PJ (1981) Transport of energy in muscle: the phosphorylcreatine shuttle. Science 211:448–452

Braissant O, Henry H, Loup M, Eilers B, Bachmann C (2001) Endogenous synthesis and transport of creatine in the rat brain: an in situ hybridization study. Brain Res Mol Brain Res 86:193–201

Brindle KM (1988) NMR methods for measuring enzyme kinetics in vivo. Prog NMR Spectrosc 20:257–293

Budinger TF, Benaron DA, Koretsky AP (1999) Imaging transgenic animals. Annu Rev Biomed Eng 1:611–648

de Graaf RA (2007) *In vivo* NMR spectroscopy. Principles and techniques, 2nd edn. John Wiley, Chichester

Forsén S, Hoffman RA (1963) Study of moderately rapid chemical exchange reactions by means of nuclear magnetic double resonance. J Chem Phys 39:2892–2901

Gareis D, Wichmann T, Lanz T, Melkus G, Horn M, Jakob PM (2007) Mouse MRI using phased-array coils. NMR Biomed 20:8

Gruetter R, Weisdorf SA, Rajanayagan V, Terpstra M, Merkle H, Truwit CL, Garwood M, Nyberg SL, Ugurbil K (1998) Resolution improvements in in vivo 1H NMR spectra with increased magnetic field strength. J Magn Reson 135:260–264

Heerschap A, Sommers MG, in 't Zandt HJA, Renema WKJ, Veltien AA, Klomp DWJ (2004) In: conn PM (ed) NMR in laboratory animals Methods in Enzymology, vol 385, part A, Elzevier Inc., pp 41–63, ISBN: 0-12-182790-9

Holtzman D, McFarland E, Moerland T, Koutcher J, Kushmerick MJ, Neuringer LJ (1989) Brain creatine phosphate and creatine kinase in mice fed an analogue of creatine. Brain Res 483:9

In 't Zandt HJA, Renema WKJ, Streijger F, Jost C, Klomp DWJ, Oerlemans F, van der Zee CEEM, Wieringa B, Heerschap A (2004) Cerebral creatine kinase deficiency influences metabolite levels and morphology in the mouse brain: a quantitative in vivo 1H and 31P Magnetic Resonance study. J Neurochem 90:1321–1330

Jost CR, Van Der Zee CE, In 't Zandt HJ, Oerlemans F, Verheij M, Streijger F, Fransen J, Heerschap A, Cools AR, Wieringa B (2002) Creatine kinase B-driven energy transfer in the brain is important for habituation and spatial learning behaviour, mossy fibre field size and determination of seizure susceptibility. Eur J Neurosci 15:1692–1706

Kan HE, Renema WK, Isbrandt D, Heerschap A (2004) Phosphorylated guanidinoacetate partly compensates for the lack of phosphocreatine in skeletal muscle of mice lacking guanidinoacetate methyltransferase. J Physiol 560:219–229

Kan HE, Buse-Pot TE, Peco R, Isbrandt D, Heerschap A, de Haan A (2005) Lower force and impaired performance during high-intensity electrical stimulation in skeletal muscle of GAMT-deficient knockout mice. Am J Physiol Cell Physiol 289:C113–C119

Kan HE, Meeuwissen E, van Asten JJ, Veltien A, Isbrandt D, Heerschap A (2007) Creatine uptake in brain and skeletal muscle of mice lacking guanidinoacetate methyltransferase assessed by magnetic resonance spectroscopy. J Appl Physiol 102:2121–2127

Kekelidze T, Khait I, Togliatti A, Benzecry JM, Wieringa B, Holtzman D (2001) Altered brain phosphocreatine and ATP regulation when mitochondrial creatine kinase is absent. J Neurosci Res 66:866–872

Meyer RA, Sweeney HL, Kushmerick MJ (1984) A simple analysis of the "phosphocreatine shuttle". Am J Physiol 246:C365–C377

Moon RB, Richards JH (1973) Determination of intracellular pH by 31P magnetic resonance. J Biol Chem 248:7276–7278

Nabuurs CI, Klomp DW, Veltien A, Kan HE, Heerschap A (2008) Localized sensitivity enhanced in vivo 13C MRS to detect glucose metabolism in the mouse brain. Magn Reson Med 59:4

Nabuurs C, Romeijn M, Veltien A, Kan H, Isbrandt D, Heerschap A (2009) Creatine deficiency, uptake and breakdown studied in brain and muscle of Arginine:Glycine Amidinotransferase deficient mice. In: Proceedings of the International Society for Magnetic Resonance in Medicine 17, Honolulu

Nabuurs C, Huijbregts B, Wieringa B, Hilbers CW, Heerschap A (2010) 31P saturation transfer spectroscopy predicts differential intracellular macromolecular association of ATP and ADP in skeletal muscle. J Biol Chem 285:39588–39596

Ntziachristos V, Bremer C, Weissleder R (2003) Fluorescence imaging with near-infrared light: new technological advances that enable in vivo molecular imaging. Eur Radiol 13:195–208

Pellerin L, Magistretti PJ (2004) Neuroscience. Let there be (NADH) light. Science 305:50–52

Renema WK, Schmidt A, Van Asten JJ, Oerlemans F, Ullrich K, Wieringa B, Isbrandt D, Heerschap A (2003) MR spectroscopy of muscle and brain in guanidinoacetate methyltransferase (GAMT)-deficient mice: Validation of an animal model to study creatine deficiency. Magn Reson Med 50:936–943

Rudin M, Sauter A (1992) Measurement of reaction rates in vivo using magnetization transfer techniques. NMR Basic Pinciples Prog 27:257–293

Schmidt A, Marescau B, Boehm EA, Renema WK, Peco R, Das A, Steinfeld R, Chan S, Wallis J, Davidoff M, Ullrich K, Waldschutz R, Heerschap A, De Deyn PP, Neubauer S, Isbrandt D (2004) Severely altered guanidino compound levels, disturbed body weight homeostasis and impaired fertility in a mouse model of guanidinoacetate N-methyltransferase (GAMT) deficiency. Hum Mol Genet 13:905–921

Steeghs K, Benders A, Oerlemans F, de Haan A, Heerschap A, Ruitenbeek W, Jost C, van Deursen J, Perryman B, Pette D, Bruckwilder M, Koudijs J, Jap P, Veerkamp J, Wieringa B (1997a) Altered Ca2+ responses in muscles with combined mitochondrial and cytosolic creatine kinase deficiencies. Cell 89:93–103

Steeghs K, Heerschap A, de Haan A, Ruitenbeek W, Oerlemans F, van Deursen J, Perryman B, Pette D, Brueckwilder M, Koudijs J, Jap P, Wieringa B (1997b) Use of gene targeting for compromising energy homeostasis in neuro-muscular tissues: the role of sarcomeric mitochondrial creatine kinase. J Neurosci Methods 71:29–41

Stöckler S, Holzbach U, Hanefeld F, Marquardt I, Helms G, Requart M, Hanicke W, Frahm J (1994) Creatine deficiency in the brain: a new, treatable inborn error of metabolism. Pediatr Res 36:409–413

Streijger F, Jost CR, Oerlemans F, Ellenbroek BA, Cools AR, Wieringa B, Van der Zee CE (2004) Mice lacking the UbCKmit isoform of creatine kinase reveal slower spatial learning acquisition, diminished exploration and habituation, and reduced acoustic startle reflex responses. Mol Cell Biochem 256–257:305–318

Streijger F, Oerlemans F, Ellenbroek BA, Jost CR, Wieringa B, Van der Zee CE (2005) Structural and behavioural consequences of double deficiency for creatine kinases BCK and UbCKmit. Behav Brain Res 157:219–234

Streijger F, Pluk H, Oerlemans F, Beckers G, Bianco AC, Ribeiro MO, Wieringa B, Van der Zee CE (2009) Mice lacking brain-type creatine kinase activity show defective thermoregulation. Physiol Behav 97:10

Streijger F, Scheenen WJ, van Luijtelaar G, Oerlemans F, Wieringa B, Van der Zee CE (2010) Complete brain-type creatine kinase deficiency in mice blocks seizure activity and affects intracellular calcium kinetics. Epilepsia 51:9

Tachikawa M, Fukaya M, Terasaki T, Ohtsuki S, Watanabe M (2004) Distinct cellular expressions of creatine synthetic enzyme GAMT and creatine kinases uCK-Mi and CK-B suggest a novel neuron-glial relationship for brain energy homeostasis. Eur J Neurosci 20:144–160

Tkac I, Henry PG, Andersen P, Keene CD, Low WC, Gruetter R (2004) Highly resolved in vivo 1H NMR spectroscopy of the mouse brain at 9.4 T. Magn Reson Med 52:478–484

van Deursen J, Heerschap A, Oerlemans F, Ruitenbeek W, Jap P, ter Laak H, Wieringa B (1993) Skeletal muscles of mice deficient in muscle creatine kinase lack burst activity. Cell 74:621–631

von Figura K, Hanefeld F, Isbrandt D, Stöckler-Ipsiroglu S (2001) Guanidinoacetate methyltransferase deficiency. In: Scriver CR, Beaudet AL, Sly WS, Valle D, Childs V (eds) The metabolic and molecular bases of inherited disease, 8th edn. McGraw-Hill, New York, Chapter 84

Wallimann T (1994) Bioenergetics. Dissecting the role of creatine kinase. Curr Biol 4:42–46

Wallimann T, Wyss M, Brdiczka D, Nicolay K, Eppenberger HM (1992) Intracellular compartmentation, structure and function of creatine kinase isoenzymes in tissues with high and fluctuating.energy demands: the phosphocreatine circuit for cellular energy homeostatis. Biochem J 281:21–40

Wyss M, Kaddurah-Daouk R (2000) Creatine and creatinine metabolism. Physiol Rev 80: 1107–1213

Wyss M, Smeitink J, Wevers RA, Wallimann T (1992) Mitochondrial creatine kinase: a key enzyme of aerobic energy metabolism. Biochim Biophys Acta 1102:119–166

Index

A

AAT. *See* Aspartate aminotransferase
Absolute quantification, 198, 199, 201–204, 208
Ac-CoA. *See* Acetyl coenzyme A
Acetate, 96, 108, 182, 185–187, 203, 206, 211, 212, 224, 229, 231–237, 239–245, 399, 720, 735, 738, 740, 741, 763–786, 827–830, 832, 852, 856, 857, 866, 911, 913, 914, 928, 929, 931, 933–937, 949–951, 954, 957, 967–969, 971, 972, 974, 976, 980, 981, 985, 994, 995, 999–1002, 1077–1080, 1098, 1106, 1142
Acetoacetic acid, 823
Acetyl-CoA efflux from mitochondria, 292, 719, 732, 855, 862, 863, 865, 866, 959, 971, 974, 1078
Acetyl-CoA synthetase-1 (ACS), 803–805, 810, 866, 967, 981
Acetyl coenzyme A (Ac-CoA), 851, 855, 856, 1075, 1076
ACL. *See* ATP-citrate lyase
ACS. *See* Acetyl-CoA synthetase-1
Action potential, 398, 450, 452, 453, 456, 457, 686, 742, 821, 923
Activation by small cytosolic calcium signals, 850
Activation induced (AI) memri, 107
Activation of the brain, 418, 419, 422, 424, 427
Acyl-CoA synthetase fatty acid trapping, 803–805, 810
Adenosine diphosphate (ADP), 151, 170, 346, 348, 428, 623–626, 646, 725, 729, 732, 743, 859, 866, 1138, 1139
Adenosine triphosphate (ATP), 150–152, 154, 157, 160, 161, 164, 165, 168, 170, 171, 265, 287, 292, 335, 338, 339, 346, 350, 399, 400, 421–423, 426–428, 445, 624–627, 629, 630, 637–643, 645, 646, 678, 684, 685, 701, 703, 705, 719, 725, 729, 732, 737, 742, 743, 800, 820, 821, 825, 850–852, 854–856, 858, 861, 865, 866, 962, 978, 1032, 1036, 1037, 1064, 1065, 1137–1139, 1142–1144
ADP. *See* Adenosine diphosphate
Advanced method for accurate, robust, and efficient spectral fitting (AMARES), 160
Aerobic glycolysis, 282–285, 287, 292, 426, 719, 724, 962, 978
AGAT. *See* Arginine-glycine amidino transferase
AGC. *See* Aspartate-glutamate carrier
AGC isoform, 848–850
Aging, 21, 255, 285–286, 489, 570, 626, 889, 890, 894, 895, 899, 1030
Aglycemia, 395, 702
Alanine aminotransferase, 718, 719, 726
Alcoholism, 1096
Alpha ketoglutarate dehydrogenase, 891
Alpha-ketoglutarate-glutamate exchange, 718, 726, 729, 733, 735, 738, 928, 958, 977
Alpha-ketoglutarate-isocitrate shuttle, 731, 865, 868–869, 871
Alpha synuclein in Parkinson's disease, 893–896
Alzheimer's disease (AD), 169, 211, 223, 227, 245, 311, 607, 626, 627, 886, 888–893, 1030, 1076, 1084
AMARES. *See* Advanced method for accurate, robust, and efficient spectral fitting (AMARES)
Amino acid
  metabolism, 821, 824, 825, 911–912, 976, 1096, 1119
  transporters, 724, 824, 826

I.-Y. Choi and R. Gruetter (eds.), *Neural Metabolism In Vivo*, Advances in Neurobiology 4, DOI 10.1007/978-1-4614-1788-0,

Aminopeptidase N, 1038, 1039
Ammonia, 103, 726, 784, 825, 922, 938, 957, 964, 965, 972, 983, 1011–1017
AMPA receptor-dependent generation in striatum, 1064, 1065
Amphetamine, 258, 259, 387, 1057, 1058
Amyloid
  cascade hypothesis, 889, 890, 898
  imaging, 311
  β-peptide, 608, 610
Amyloid precursor protein (APP), 169, 608, 889, 894
Amyotrophic lateral sclerosis, 888, 896–897
Anaesthetics, 597, 666, 680
Anaplerosis, 719, 770, 777–779, 911, 912, 921–940, 950, 954, 955, 957, 958, 960, 967, 970–972, 976, 979, 981–983
Anaplerotic enzyme, 236
Angular momentum, 6, 66
Anisotropy
  cortical grey matter, 74
  white Matter, 72–74
ANLS. *See* Astrocyte-neurone lactate shuttle
Anticonvulsants, 829, 1104
Anti-epileptic effect, 822–824, 829–831
Antiepileptic therapy, 292
Antioxidants, 899, 1059, 1062, 1063, 1066
Antioxidative defense, 1030, 1040
Apolipoprotein E (APOE), 891
Apoptosis, 167, 1030
APP. *See* Amyloid precursor protein
Apparent diffusion coefficient, 28, 611
Apparent transverse relaxation time (T_2*), 138, 468
Aquaporins, 1064
Arachidonic acid, 428
Aralar, 841, 846, 848–859, 871
Arginine-glycine amidino transferase (AGAT), 1138, 1139, 1142, 1145
Arterial spin labeling (ASL), 29, 30, 419, 442, 469, 502, 516, 548–550, 567–570, 604
Arterial spin labeling RF coil, 569
Artifacts in autofluorescence, 342–345
Ascorbate
  brain slices, 1055–1058, 1060–1062
  concentration, 1054–1056, 1059
  enzyme co-factor, 1052, 1061
  homeostasis, 1057
  H_2O_2 signaling, 1060–1066
  molecular weight, 1052, 1059, 1062
  oxidation products, 1052
  p^Ka, 1052
  synthesis, 1052, 1053, 1058, 1060, 1061
  transport, 1053–1055
Ascorbate function
  antioxidant, 1052, 1056, 1060–1066
  edema prevention, 1061, 1062
  free radical scavenger, 1052, 1066
  HIF regulator, 1053
  hydroxyl radical scavenger, 1052, 1060
  neuroprotection in stroke, 1061–1063, 1066
Ascorbic acid. *See* Ascorbate
ASL. *See* Arterial spin labeling
Aspartate aminotransferase (AAT), 718, 733, 803, 826, 828, 829, 845, 851, 852, 855, 856, 911, 958
Aspartate-glutamate carrier (AGC), 845–848, 851, 856
  aralar and citrin, 730, 732, 846, 848
Aspartate, NAA and myelin synthesis, 780, 856, 866, 871
Aspartoacylase (ASPA), 856, 1076, 1079–1081
Asp-NAT. *See* L-Aspartate N-acetyltransferase
Astrocyte-neuronal lactate exchange, 420, 421
Astrocyte-neurone lactate shuttle (ANLS), 677–678, 684, 686, 955, 962, 963, 978, 979, 994
Astrocyte-neuron interaction, 1040
Astrocytes, 102, 182, 222, 281, 332, 386, 421, 435, 659, 678, 703, 708, 716, 756, 764, 807, 820, 842, 910, 922, 948, 994, 1019, 1030, 1055, 1080, 1097, 1139
Astrocytes metabolic cooperation with neurons, 182, 1030, 1040
Astrocyte-to-neurons lactate shuttle hypothesis (ANLSH), 237, 914
Astrocytic oxidative metabolism, 744, 775, 939, 980–981, 994, 999
Astrocytic tricarboxylic acid cycle, 210, 223, 235, 236, 238, 240, 743, 924, 930, 970, 972, 979, 999
Astrocytoma, 117, 172
Astroglia, 372, 373, 401, 732, 764, 765, 769, 771, 774–777, 821, 960, 999, 1096
ATP. *See* Adenosine triphosphate
α-ATP, 161, 638–642, 1143
ATP-citrate lyase (ACL), 855, 856, 863, 865, 866
Autism, 858–859
Autofluorescence, 333, 338–352, 356–361, 445, 994, 995
Autofluorescence signal, complications, 344, 360, 995
Autoradiography, 181, 256, 257, 282, 479, 480, 548, 576, 606, 680, 767–768, 775, 776, 780, 994–996, 998, 1000–1002
Axial diffusivity, 68, 73

B

Band pass filter, 345, 438, 447
Band pass RF filter, 190, 197
Bandwidth, 16, 59, 137, 143, 144, 197, 204
Bayesian framework, 449–450, 489
BBB. *See* Blood brain barrier
Beta-amyloid, 287
Beta probe system, 996
Binding potential, 267, 310–313, 315, 321, 580
Bioenergetics, 624–627, 630, 637, 642–645, 1138, 1145
Biosensors, 382–384, 386, 388, 391, 396, 401
Birth asphyxia,
Blood brain barrier (BBB)
  kinetics, 1120, 1127, 1130
  transport, 756, 1126–1128
Blood flow, 20, 29, 238, 241, 256, 333, 335, 336, 342, 352–354, 356, 358, 360, 372, 389, 394, 398, 400, 417–428, 434–440, 442, 444–447, 449, 459–462, 469–475, 478–480, 484, 488, 504, 505, 515–518, 526, 527, 595, 597, 757, 980, 998, 1000–1002, 1107
Blood oxygenation level dependent (BOLD), 23, 34, 433–489, 502–506, 510–516, 518–529, 534–535, 548–576, 578–585, 595, 597–603, 605–607, 612, 644, 645, 769, 1105
Blood oxygenation-level dependent contrast, 23, 504–506, 510, 514, 550, 554, 645
Blood pressure (arterial blood pressure), 478, 578, 579, 600, 601, 822
Blood steal, 471
Blood volume imaging, 29, 352–354, 358, 485, 504, 527, 548, 551, 552, 558, 567, 582, 595
BOLD. *See* Blood oxygenation level dependent
Bolus injections in isotopic labeling, 926–927
Brain activation, 23, 107, 113, 237, 285, 427, 436, 437, 460, 465, 550, 551, 571, 582, 602–604, 624, 644, 669, 682, 758, 960, 961
Brain content of phenylalanine, 1125
Brain development, 74, 163–172, 807, 859, 1119
Brain during starvation, 827–828
Brain uptake of fatty acids, 807, 810
Brain water, 160, 163, 165, 168, 506, 551, 552, 568, 660
B_1 shimming, 138, 191

C

^{11}C, 258, 260, 266, 267, 273, 287, 768, 781
^{13}C, 181, 198, 204, 766–769, 771–773, 784, 786
^{14}C, 207, 257, 272, 274, 374, 462, 502, 766, 768, 926–929, 931, 934, 937, 949, 950, 958, 961, 965, 966, 968, 1098, 1105, 1106
[1,2-$^{13}C_2$]acetate, 186, 203, 211, 936, 967, 968
[1-^{13}C]acetate, 186, 969
[2-^{13}C]acetate, 185, 186, 188, 224–225, 238, 933, 934, 936, 951, 957, 968, 969, 974, 976, 980, 981, 985, 1098, 1106
[^{14}C]acetate, 928, 931, 949
Ca^{2+}-dependent activation, 399, 850
Calcium activation of glycerol-3-phosphate shuttle, 861–862
Calcium activation of malate-aspartate shuttle, 849–851
Calcium binding mitochondrial carriers (CaMCs), 846, 847
Calcium waves in astrocytes, 488
Calibrated BOLD experiment, 466, 485
Caloric restriction, 822
Canavan disease, 1076, 1078, 1080–1082
Carbohydrates, 424, 779, 820, 949
Carbon-11. *See* ^{11}C
Carbon-13. *See* ^{13}C
Carbon-14. *See* ^{14}C
Carboxykinase, 719, 727, 778, 914, 936, 955, 967, 970
Carboxylase, 184–186, 206, 207, 210, 213, 222, 226, 234, 240, 718, 719, 737, 739, 758, 769, 770, 783, 830, 863, 865, 869, 870, 911, 912, 925, 928, 930, 932–934, 937–938, 950, 952–955, 961, 965, 970, 971, 976, 981, 982
[Ca^{2+}] reporters, 95–96
[^{14}C]aspartate, 928, 931
Catabolism, 716, 826, 950, 1079–1080, 1096, 1099, 1101–1103
Catalyzing enzyme, 266, 724, 735, 911
Cataplerosis, 925
Cataplerosis balance with anaplerosis, 925
CBF. *See* Cerebral blood flow
[^{14}C]bicarbonate, 926, 928
CBV. *See* Cerebral blood volume
^{13}C-^{13}C coupling, 184, 985
CD36, 801–803, 811
[^{11}C]DASB radioligand, 308, 310
[^{11}C]-deprenyl, 266, 267
Cell culture, 223, 708, 806, 890, 894, 912, 1034, 1080, 1081, 1099
Cellular distribution, 101, 104, 1035, 1097
Cellular mechanisms of swelling, 347
Central nervous system (CNS), 65–76, 80, 108, 117, 396, 400, 578, 595, 715–745, 823, 832, 863, 870, 887, 922, 1096, 1097

Cerebellar ataxias, 887
Cerebral activity-metabolism relationship, 947–986
Cerebral amyloid angiopathy (CAA), 608
Cerebral blood flow (CBF), 238, 256, 342, 353, 372, 389, 394, 398, 400, 417–428, 434, 440, 462, 502, 595, 624, 757, 998, 1107
Cerebral blood volume (CBV), 352, 353, 449, 460, 503, 547–585, 595, 658
Cerebral metabolic rate of ATP (CMR_{ATP}), 627, 642
Cerebral metabolic rate of glucose (CMR_{glc}), 266, 276, 277, 335, 417–428, 449, 462, 480, 662, 680, 893, 929, 1086, 1107
Cerebral metabolic rate of oxygen ($CMRO_2$), 283, 434, 549, 550, 760
Cerebral oncology, 1084
Cerebral pyruvate transport and metabolism, 745
[^{11}C]FLB radioligand, 308, 309
CFTR in glutathione transport, 1036, 1037
[2-^{13}C]GABA, 234, 239, 243, 1098, 1101
[1,6-$^{13}C_2$]glucose, 184, 186, 188, 192, 200–202, 206, 208, 209, 211, 929, 934, 951, 953, 956, 957, 960, 961, 969, 974, 976, 977, 980–983, 985, 986, 1101
[1-^{13}C]glucose, 182, 184, 186, 195, 202, 205, 206, 208, 209, 211, 222, 224–225, 233, 234, 237–240, 242–244, 927, 929, 931–934, 936–938, 952, 956–961, 966, 968, 969, 972, 974, 977, 979, 980, 985, 1101
[2-^{13}C]glucose, 184–186, 232, 929, 930, 933, 936, 951, 956, 958, 959, 961, 967, 969, 972, 974, 976, 981–983, 985, 986
[5-^{13}C]glucose, 930, 961
[^{13}C]Glucose, 926, 929, 931, 932, 936, 937
[^{14}C]glucose, 928, 931, 968, 1098
[^{14}C]glutamate, 928, 931, 949
[^{14}C]glutamine, 928, 1098
Chemical shift, 23, 24, 137, 139, 142, 150, 152, 159–161, 192, 195, 196, 629, 1007, 1077, 1124, 1136
Chemical shift imaging (CSI), 24, 154–157, 162, 165, 169, 172, 191, 196
Chemical shift reference, 152, 160
  of glycerophosphocholine, 160
Chemical shift selective (CHESS), 136, 139
Choline, 24, 134, 307, 1007, 1077, 1084, 1138
CIC. *See* Citrate carrier
Circular polarized B_1 field, 512
Circumscribing RF coil, 52–53
Citrate carrier (CIC), 732, 855, 856, 863, 865
Citrate-malate shuttle, 731, 855, 862, 863, 865–866, 868, 871
Citrate-pyruvate shuttle, 862–865
Citrate shuttles, 842, 862–866, 871
Citrate synthetase (CS), 827–829, 863
^{11}C-labeled fatty acid,
^{13}C-labeled substrates, 182, 183, 188, 198, 926, 951
^{14}C-labeled substrates, 926
^{13}C-labeling, 188, 189, 201, 209, 232, 738, 740, 741, 913–915, 929–931, 938, 948, 960
[3-^{13}C]lactate, 187, 188, 234, 929
Clinical indication
  angiography, 121
  cancer / metastases, 80–81, 121
  central nervous system (CNS), 117
  liver, 117–119
  lymphatic system, 116
Clinical microdialysis, 371–402
^{13}C NMR spectroscopy, 189–192, 221–246, 951
CNR. *See* Contrast-to-noise ratio
CNS. *See* Central nervous system
CO_2, 281, 389, 434, 462, 466, 468, 820, 824, 829, 832, 923, 925, 927–929, 935, 937–940, 950, 994, 999
  fixation, 923, 925, 927–929, 935, 937–940, 950
Co-cultures, 222–224, 227, 229, 703, 770, 776, 1040, 1061, 1098
Coenzyme Q (CoQ), 824, 860, 899
Comparison of MRI with PET and SPECT, 315
Compartmentalization, metabolic modeling, 947–986
Compartmental model, blood brain barrier, 264, 319
Compartmentation, 183, 206, 237, 717, 718, 726–728, 733–735, 740–745, 764, 765, 768–774, 779, 781, 786, 911, 913, 915, 922–925, 931, 932, 935–937, 993–1002, 1096, 1139
Compartment model, 182, 199, 205–210, 311, 313, 314, 320, 321, 461, 570, 775, 930–933, 948, 954, 957, 959, 961, 974, 976, 978–980, 984, 985, 998
Complex I defect, 895, 898
Compound action potential, 709
Computed tomography (CT), 258, 279–280, 548
Concentration reference, 134, 141, 161, 163, 1124
Confocal imaging, 444
Connectivity index (CnI), 111
Continuous infusion in isotopic labeling, 926–927
Contrast agent toxicity, 81, 82

Contrast-to-noise ratio (CNR), 522, 526–529, 531, 549–551, 557–560, 564, 567, 570, 579, 602, 604, 605, 607, 612
CoQ. *See* Coenzyme Q
Co-registration, 279–280, 419
Correction for reflectance changes, 344, 347
Cortex, 23, 98, 100, 103, 108, 111–113, 135, 159, 169, 170, 223, 238, 241–244, 279, 284–288, 290, 323, 333–335, 337, 341, 343–348, 353–357, 359–361, 372, 374–377, 386, 387, 398, 400, 436, 437, 440, 442, 443, 446, 447, 456–458, 471, 478, 481, 487, 502, 506, 512, 515, 518, 523, 525, 560–566, 582, 585, 605–607, 610, 612, 638, 644, 645, 667, 676, 678, 679, 683–685, 687, 708, 725, 736, 775–777, 783–785, 807, 849, 853, 858, 859, 870, 895, 972, 996, 997, 1000, 1016, 1031, 1033, 1056, 1058, 1059, 1085, 1086, 1098, 1099, 1101, 1105
Cortical column, 447, 448, 483
Coupling flow and metabolism
[^{11}C]PE2I radioligand, 308, 310
[^{11}C]Raclopride radioligand, 308
Creatine (Cr), 24, 134, 141–143, 170, 658, 820, 1077, 1083–1085, 1099, 1122, 1124, 1135–1145
Creatine kinase (CK)
  deficiency, 1143
  isoenzyme, 1143
  reaction, 158, 625, 637–641
  shuttle, 1139, 1144
Creatine phosphate, 820
Creatinine, 900, 1140
[^{11}C]SCH23390 radioligand, 308, 309
CSD. *See* Current source density
CSI. *See* Chemical shift imaging
Current source density (CSD), 337, 453–456
Current sources, 451
[^{11}C]WAY 100635 radioligand, 308, 310
Cybrids, 891, 895, 896
Cyclosporine, 900
Cytidine triphosphate, 150
Cytochrome oxidase defect in Alzheimer's disease, 626, 891
Cytosol, 240, 274, 280, 281, 339, 351, 444, 445, 462, 484, 624–626, 639, 719, 722, 723, 725, 727, 730–734, 740, 742–745, 799, 827, 842, 844–848, 851, 854–857, 862, 863, 865–871, 958, 960, 963, 971, 994, 1031, 1035, 1058, 1061, 1078
Cytosolic pyruvate
  compartmentation, 718
  metabolism, 718–719, 723–724

D
Data processing methods, 139, 272, 1124
DCE. *See* Dynamic contrast enhancement
Deafness-dystonia syndrome, 897
Decarboxylase, 222, 234–236, 239, 266, 718, 770, 825, 828, 830
Decarboxylation of glutamate, 738, 778, 823, 830, 1096, 1097
Dehydroascorbate (DHA), 794, 807–810, 1052, 1054–1056, 1062
Dementia, 262, 279, 287, 289, 322, 323, 885, 888, 1084
Dementia of Alzheimer's type (DAT), 286, 287, 289, 323
Dementia with Lewy bodies, 888
Deoxyglucose, 266, 272, 273, 462, 502, 577, 684, 700, 910, 999, 1000, 1105
2-Deoxyglucose (2DG), 104, 282, 479, 564, 576, 680, 717, 1105
2-Deoxyglucose (2DG) autoradiography, 479, 480
Deoxyhemoglobin, 23, 284, 352, 354, 434, 438, 449, 464–466, 468, 485, 501–535, 550, 557, 572, 574, 584, 602
Dephasing, 22, 119, 157, 195, 505, 513
Depolarizing agents, 386
Depression, 71, 105, 168–169, 290, 331–361, 445, 600, 605, 775, 1053, 1096, 1107
Detection of tissue at risk of ischemia, 359, 549
Detection sensitivity, 189, 437, 551, 629, 642, 645
Detuning, 42, 43, 48, 55, 57
Deuterium, 733, 734, 768, 771, 928. *See also* ^{2}H
DHA. *See* Dehydroascorbate; Docosahexaenoic acid
DHAP. *See* Dihydroxyacetone phosphate
DHA reductase, 1052, 1055
DHA transport via GLUT1, 1056
Diabetes, 280, 666, 668, 704, 774, 782–783
  type I, 171, 774, 782
DIC. *See* Dicarboxilate carrier
Dicarboxilate carrier (DIC), 729, 732, 863, 865–867, 869
Dichloroacetate, 899, 900
Differential subtype expression,
Diffusible messenger, 1065
Diffusion
  intercellular space, 372, 373, 381
  measurement, 70, 71, 75, 519
Diffusion coefficient, brain injury, 28, 70, 72, 87, 336–337, 355, 361, 377, 386, 389, 467, 611, 684, 824, 1063, 1086, 1120
Diffusion model, fatty acid transport, 795, 803–806, 818

Diffusion tensor imaging (DTI), 28, 29, 69, 72, 75, 611, 1120
Diffusion weighted imaging, 20, 28
Dihydroxyacetone phosphate (DHAP), 859–861
diphospho nicotinamide adenine dinucleotide (NAD), 629, 630
2,3-Diphosphoglycerate, 160
Direct and indirect sensitivity detection, 189
Distribution volume, 266, 316, 321, 659
Docosahexaenoic acid (DHA), 794, 807–810, 1052, 1054–1056, 1062
Dopamine
  receptor, 262, 309
  transporter, 308, 310
Dopamine-β-hydroxylase, 823, 1052
Dopamine-D_2-receptor, 612
Double-indicator method, 756
Double quantum filtering, 1099
Doublet, 151, 156, 202, 203, 237, 771, 778, 952, 969, 972
Double tuned RF coil, 59
Draining vein, 526, 585
DTI. *See* Diffusion tensor imaging
Dual tuned RF coil, 59
Dual-wavelength systems, 345
Dynamic (D)AIMEMRI, 107, 109
Dynamic contrast enhancement (DCE), 81
Dysmyelination, 1082, 1120

E

ECF. *See* Extracellular fluid
ECF glucose, 679, 681
Echo planar imaging (EPI), 507–510, 531, 533, 558
EEG. *See* Electro-encephalography
Effective transverse relaxation time $T_2^*(R_2^*)$, 527
Electrical stimulation, 334, 343–346, 350, 351, 386, 397, 399, 472, 606, 608, 678, 681, 683, 684, 687, 1000, 1103, 7091
Electrochemical gradients, 676
Electrode arrays, 447, 454
Electro-encephalography (EEG), 290, 291, 306, 346, 437, 487, 489, 645, 677, 1130
Encephalomyopathy diseases, 887
Endoplasmatic reticulum in neuron, 105
Endothelial cells in the blood brain barrier, 103, 659, 660, 756, 757, 798, 800, 1126
Energy consumption, 373, 374, 378, 394, 399, 480, 676, 677, 922
Energy metabolism, 24, 166, 167, 170, 171, 182, 183, 212, 256, 266, 373, 386–388, 390, 392, 398–400, 402, 434, 435, 462, 484, 488, 666, 676, 684, 685, 773, 774, 776, 823, 844, 849, 854, 950, 962, 965, 994, 1078, 1105, 1128, 1138
Energy requirement, 820, 1105
Enrichment dynamics, 985
Enzyme responsive contrast agents, 93–95
Enzymes, 115, 221, 229, 274, 383, 421, 629, 641, 642, 678, 716, 717, 724–727, 735–737, 768, 795, 799, 800, 804, 807, 844, 845, 860, 862, 863, 865, 868–870, 886, 887, 891, 912, 924, 925, 950, 967, 969, 1030–1033, 1035, 1036, 1052, 1059, 1060, 1062, 1096, 1136, 1138, 1139, 1145
Ependymoma, 172
EPI. *See* Echo planar imaging
Epilepsy, 109, 134, 168–169, 223, 227, 230, 231, 241–246, 280, 290–292, 377, 783, 784, 821–824, 829, 830, 1083–1084, 1096, 1107, 1119
Excitation, 8–16, 21, 24, 25, 27, 137–139, 143, 153, 307, 332, 340–342, 345, 346, 358, 360, 396, 444, 446, 453, 454, 458, 461, 471, 474, 476, 487, 505, 507, 509, 824, 924, 1103, 1124, 1137
Excitotoxicity, 115, 238–241, 688, 781, 824, 825, 853, 854, 866, 923, 936, 1006, 1021
Exercise, 170, 376, 386, 398, 402, 424, 425, 759
Experimental microdialysis, 372–374, 377, 378, 389, 395, 401
Extracellular brain compartment, 659, 758
Extracellular conductivity, 451, 454
Extracellular electrophysiological measurements, 401, 436, 447
Extracellular fluid (ECF), 378, 381, 825, 1006, 1009, 1015, 1023, 1054–1056, 1126
Extracellular inorganic phosphate, 151
Extraction of TACs, 264
Ex vivo nuclear magentic resonance, 985

F

^{18}F, 273
Facilitated diffusion, 264, 265, 660, 720, 755, 756, 1139
Fast axonal transport, 104
FAT. *See* Fatty acid translocase
FATP. *See* Fatty acid transport protein
Fatty acids
  adsorption, 796–798
  binding protein family, 795, 801, 803–806, 810
  desorption, 797–800

flip-flop, 797
fluorescent probes, 797–798
tracer in vivo, 808
uptake, 793–810
Fatty acid translocase (FAT), 801–802
Fatty acid transport protein (FATP), 795, 799–801, 803–806, 810
FDG. *See* Fluorodeoxyglucose
Feedback inhibition, 684, 1032, 1033
Feraheme contrast agent, 560, 579
[^{18}F]-fallypride, 261–264, 267
F_1F_0-ATP_{ase}, 624, 642
[^{18}F]-fluorodeoxyglucose, 256
[^{18}F]-fluorodopa, 259, 266, 267
Fick's principle, 418, 419, 676
Field dependence, 87, 88, 526–529
Filtered back projection, 263
Firing-rate models, 457, 458
Flavoprotein, 338, 344–345, 360
Flow-injection analysis, 389
Fluorimetry,
Fluorine-18. *See* ^{18}F
Fluoroacetate, 765, 776, 779
Fluorodeoxyglucose (FDG), 256, 271–292, 420, 996, 997, 999–1001, 1107
Flux, 5, 206, 207, 209, 210, 213, 225, 244, 264, 266, 338, 339, 379, 385, 393, 442, 467, 468, 637–643, 645, 741, 757–758, 827–830, 832, 924, 925, 929, 930, 932, 933, 936–940, 952, 955–962, 964–967, 971–977, 980–984, 998, 1006, 1007, 1022, 1023, 1096, 1101, 1106, 1125, 1138
fMRI. *See* functional magnetic resonance imaging
Forward flux, 639, 640
Forward modeling, 450–453, 455, 459
Forward rate constants, 637, 638, 640
Fourier Series Window CSI, 154
Fourier transformation, 20, 138, 140
Fractional anisotropy, 28, 29, 69
Fractional enrichment, 193, 775, 956
Frataxin in Friedreich's ataxia, 888, 897
Frequency domain, optical coherence tomography, 24, 139, 159, 437, 442–443
Frequency encoding, 15–19
Friedreich's ataxia, 888, 897
Fumarase, 718, 951–953, 960, 966
Fumarate, 738, 928–929, 938, 952, 953, 955, 960–961, 966, 974, 978, 981–983
Functional activation, 182, 196, 205, 422, 442, 443, 484, 549, 573, 576
Functional brain imaging, 306, 307, 551
functional magnetic resonance imaging (fMRI), 10, 11, 21–23, 26, 29, 245, 284, 306, 336, 433–489, 502–535, 547–585, 593–613, 644, 677, 679, 717, 1105

G

GAA. *See* Guanidinoacetate
GABA. *See* γ-aminobutyric acid (GABA)
$GABA_A$, 242, 399, 474, 598, 599, 604, 605, 608, 609, 613, 1065, 1104, 1106
$GABA_B$, 1104
Gabaculine, 1099, 1102
GABAergic function, 1096, 1107
GABAergic inhibition, 1104, 1108
GABAergic neurons, 222, 226, 228, 234–237, 239, 242–244, 726, 769, 775, 776, 781, 825, 828–830, 832, 957, 958, 976, 984, 1023, 1096, 1097, 1100, 1101, 1103, 1105, 1106
GABAergic neurotransmission, 1096, 1103–1106, 1108
GABA-glutamine cycling rate, 776, 1106
GABA transaminase (GABA-T), 239, 741, 776, 826, 1096, 1099, 1106, 1107
inhibitor, 1099, 1102–1105, 1107
GAD. *See* Glutamate decarboxylase
GAD_{65}, 1097, 1098
GAD_{67}, 1097–1099
γ-aminobutyric acid (GABA), 134, 604, 922, 1095
alternative pathways, 102, 969–970
concentration, 1096, 1098, 1102, 1104, 1107
degradation, 1099, 1102
irreversible inhibitors, 1099
metabolism, 1095–1108
modulation, 784, 785
precursor, 1101, 1106
receptors, 399, 477, 1096, 1104–1105
release, 1098, 1099, 1104–1106
synthesis, 221, 239, 242, 773, 776, 785, 830, 1096–1101, 1103, 1106, 1107
turnover rate, 1100–1103, 1106
uptake, 676, 1105
γ-amino butyric acid (GABA) type A receptor, 599
GAMT. *See* Guanidinoacetate N-methyltransferase
GAPDH. *See* Glyceraldehyde-3-phosphate dehydrogenase
Gap junction
in astrocytes, 478, 684, 687
hemichannels in glutathione transport, 1037

Gaussian distribution, 192, 200
Gd^{3+} chelates, 80, 82, 87, 91
Generalized Wiener estimate,
Genetic Absence Epilepsy Rats from Strasbourg (GAERS), 244, 783
Gene transfer, 1099
Germinoma, 172
g-factor, 509, 510
γ-Glutamyl transpeptidase, 1034, 1037
  in astrocytes, 1038
Glia, 76, 100, 103, 104, 169, 186, 208, 210, 211, 223, 246, 265, 286, 292, 372, 373, 395, 396, 398, 402, 703, 740, 756, 758, 769–771, 806, 825, 828, 854, 865, 912, 935, 936, 950, 955, 967, 969, 971, 976, 978, 979, 983, 984, 1006, 1011, 1015–1017, 1021–1023, 1054–1056, 1058–1061, 1065, 1066, 1079, 1099, 1105, 1106
Glial anaplerosis, 770, 777–779, 912, 921–940, 951, 954, 958, 967, 971, 972, 976, 981–983
Glial ascorbate concentration, 1059
Glioblastoma, 780
Glioma, 211, 283, 721, 781
Gliotransmitters, 477, 478
Glucose
  delivery, 676–677, 713
  metabolism, 170, 182, 232, 272, 274, 277, 278, 280, 282–292, 323, 337, 394, 399, 401, 418, 419, 421–427, 573, 575, 576, 606, 669, 675–689, 700, 704, 722, 740, 743, 827, 910, 911, 930, 953, 957, 959–961, 963, 969, 972, 976, 981, 983, 985, 998–1000, 1083, 1105
  oxidation, 282, 418, 421, 422, 425, 427, 676, 779, 781, 820, 956, 963, 978
  sensor, 374, 377, 679, 680, 682
  transport, 426, 427, 655–669, 701, 702
  uptake, 288, 291, 386, 400, 418, 421–424, 426, 427, 665, 668, 676, 688, 717, 742, 743, 758, 759, 823, 893, 979
  utilization, 239, 240, 283, 291, 292, 335, 394, 421, 426–428, 659, 893, 910, 912, 928–930, 996
Glucose comsumption (CMRGluc), 213, 288, 390, 399, 418, 422, 423, 426, 427, 470, 479, 660, 666, 668, 676, 678, 684, 701, 909, 910, 962, 978
Glucose levels in intercellular space, 386, 394, 401
Glucose-6-phosphate, 274, 282, 682, 868
Glucose transporter (GLUT1), 655, 659, 661, 663, 667, 668, 722, 744, 854, 1054–1056, 1062
  in ascorbate transport, 1054–1056
GLUT3, 659, 668, 721, 744
Glutamate
  dehydrogenase, 234–236, 718, 825, 949, 964, 970, 1019
  neurotransmission, 742, 926, 938
  neurotransmitter, 234–236, 238, 240, 241, 474, 676, 703, 741, 742, 776, 922, 924, 964, 965, 970, 971, 974, 976, 978, 979, 1006
  origin, 223, 239, 240, 683, 686, 777, 830, 853, 968, 971
  oxidation, 931, 935–937, 940, 955, 970–973, 983
  receptor agonists and antagonists, 241–242, 343, 386, 687, 784
  release, 398, 687, 688, 743, 825, 924, 1058
  spillover, 686
  uptake, 226, 386, 396, 399, 421, 426, 688, 703, 855, 915, 923, 939, 962, 978, 1058
Glutamate/ascorbate heteroexchange, 1058
Glutamate cysteine ligase (GCL), 1032, 1033
Glutamate decarboxylase (GAD), 222, 234–236, 239, 718, 770, 825, 828, 830, 1096–1100, 1102, 1105–1107
Glutamate-glutamine cycle, 186, 207, 209, 213, 244, 703, 784, 825, 915, 921–940, 955, 964, 965, 967, 978–980
Glutamate-glutamine cycling rate, 1106
Glutamatergic α-amino-3 hydroxy-5-methylisoxazole-4 propionic acid receptor (AMPA-R), 598, 599
Glutamatergic neurons, 225, 234, 236, 239, 243, 244, 726, 914, 957, 964, 976, 978, 979, 984, 1097, 1101
Glutamatergic N-methyl-D-aspartate receptor (NMDA-R), 598, 599
Glutamatergic synapses, 421, 709, 1106
Glutaminase, 236, 240, 703, 769, 770, 825, 830, 855, 922–924, 931, 955, 964, 1006, 1017–1018, 1022
Glutamine
  neurotransmission, 227, 737, 742, 783–785, 825, 922, 924, 932, 938, 939
  synthesis, 182, 206, 213, 223, 241, 686, 741, 770, 782–784, 830, 912, 924, 930–934, 938, 939, 964, 972, 974, 976, 979, 1012, 1014, 1022
  synthetase, 103, 115, 186, 207, 222, 223, 234–236, 244, 718, 737, 742, 776, 821, 825, 832, 912, 914, 922–924, 930, 931, 949, 950, 954, 957, 962, 970, 978, 979, 1006, 1010–1015, 1022
Glutathione (GSH)
  astrocytes, 1030, 1031, 1033–1041

concentrations, 1031, 1035
conjugation, 1035
export, 1037, 1038
extracellular processing, 1038, 1041
function, 1030, 1032, 1039
oxidation, 1030, 1037
peroxidase, 869, 1036
peroxidase inhibition, 1060, 1061
redox couple with ascorbate, 1060
reductase, 869, 1032, 1035
synthesis, 1029–1040, 1060, 1061
synthetase, 1032, 1033
transport, 1030, 1031, 1034, 1036
Glutathione-S-transferases, 1032, 1035
Glyceraldehyde-3-phosphate dehydrogenase (GAPDH), 642, 643, 727, 743, 845
Glycerol-3-phosphate dehydrogenase (GPDH), 861, 862
Glycerol-3-phosphate shuttle (GPS), 844, 859–862, 871
Glycerophosphocholine (GPC), 134, 151, 152, 156–158, 160, 163–165, 169, 172, 629, 630
Glycerophosphoethanolamine (GPE), 151, 152, 156–158, 163–165, 172, 629, 630
Glycogen
energy metabolism, 183, 386, 400, 684–686
metabolism, 629, 699–705, 708, 709, 712, 713
phosphorylase, 685, 702, 708, 911
turnover, 700, 704, 910
utilisation, 685
Glycogenolysis, 424, 681, 682, 684–686, 821
Glycogen-shunt activity, 685
Glycolysis, 171, 184, 234, 241, 274, 280–285, 287, 288, 291, 292, 350, 351, 418, 420–424, 426, 427, 444, 445, 484, 678, 682–685, 688, 701, 718, 719, 724, 725, 727, 738, 742–745, 769, 781, 821, 827, 828, 844, 845, 852, 860, 863, 865, 867, 913, 954, 956, 962, 963, 968, 972, 978, 982, 994, 995
Glycolytic enzymes, 421, 642, 724–726, 736, 737
GPC. *See* Glycerophosphocholine
GPE. *See* Glycerophosphoethanolamine
GPS. *See* Glycerol-3-phosphate shuttle
Gradient amplifiers, 36–39
Gradient coils, 14–15, 35, 36, 50, 68, 138
Gradient-echo (GE), 505, 507, 512, 523–526, 552–555, 559, 565, 567, 603
Gradient echo (GE) fMRI, 552–555, 559, 565, 567, 603
Gradient magnetic field, 14, 18, 28, 504, 512, 513, 1136
GSH synthesis, 1029–1041, 1060, 1061
inhibition, 1060
in neurons, 1034, 1039
GSSG, 869, 1032, 1034–1037
export, 1037
and NADPH, 869, 1035
Guanidinoacetate (GAA), 1138–1142
Guanidinoacetate N-methyltransferase (GAMT), 1138–1142, 1144, 1145
Guanosine triphosphate, 150, 151
Gulonolactone oxidase, 1053

H

^{2}H, 764, 766
^{3}H, 257, 764, 766, 780, 950
Half volume RF coil, 52
Hartmann-Hahn match, 1100
Hematocrit, 461, 469, 470, 483, 485, 573, 581, 582
Hemiparkinsonism-hemiatrophy syndrome, 289
Hemoglobin-oxyhemoglobin spectral shifts, 437–438
Hemoglobin saturation, 459, 467–469
Hepatic encephalopathy (HE), 211, 213, 245, 1011
Hereditary spastic paraplegia, 897
Heroin addiction, 578
Heteronuclear coupling, 156, 733
Hetero-nuclear RF coil, 56–60
Heteronuclear single-quantum correlation (HSQC), 1009, 1016–1018
Heteronuclear spectral editing,
Hexokinase, 266, 274, 277, 282, 283, 661, 667, 668, 680, 701, 704, 724
[^{3}H]Glutamine, 1098
High grade tumor, 288
High magnetic field, 189, 197, 198, 502, 505, 508, 517–519, 521, 534, 603, 607, 1118
High pass RF filter, 59, 263, 444
High reflection coefficient preamplifier, 44, 45
Hippocampal slice physiology, 488, 1060, 1061
Histocytoma, 172
^{1}H-^{15}N HMQC, 1009, 1022
Hollow fibers, 384, 385
Homonuclear coupling, 157
Homonuclear coupling A632 adenosine triphosphate, 157
Homonuclear polarization transfer, 1100
Homo-nuclear RF Coil, 59
Homonuclear spectral editing, 142–144
Homonuclear splitting pattern, 222

HSQC. *See* Heteronuclear single-quantum correlation
Huntingtin, 897
Huntington's disease, 170, 211, 626, 888, 889, 897, 1096
  skeletal muscle, 170
Hydrogen-2. *See* ^{2}H
Hydrogen-3. *See* ^{3}H
Hydrogen peroxide (H_2O_2), 383, 1035
Hydroxylase, 823, 1052, 1119
  basal lamina, 1052
  collagen biosynthesis, 1052
  myelin formation, 1052
5-Hydroxy-tryptamine (serotonergic) receptor, 310, 311, 387, 598, 605
Hypercapnia, 342, 466, 571, 574, 575, 578, 584, 585, 600, 608, 612, 757
Hyperdopaminergia, 784
Hypermetabolism, 288–290
Hyperphenylalaninemia, 1121
Hypoglycemia, 212, 504, 701–703, 705, 709, 711–712, 759, 774, 1103
Hypometabolism, 286, 287, 289–292
Hypothyroidism, 782, 783, 1076
Hypoxia, 70, 71, 151, 161, 166–168, 245, 278, 348, 356, 357, 371, 373, 374, 377, 445, 480, 488, 684, 704, 758, 781, 824, 844, 1038, 1053, 1061, 1062
Hypoxia inducible factor (HIF), 1053
Hypoxia-ischemia, 373
Hypoxia tolerant turtle brain, 1062

I

[^{123}I]-ADAM, 259
iCSD. *See* Inverse CSD (iCSD)
[^{123}I]-DATSCAN ([^{123}I]-FE-CIT), 258–260
[^{123}I]-IBZM, 258, 308, 322
Image alignment, 5, 272
Image reconstruction, 20, 256, 262–264
Image-selected in-vivo spectroscopy (ISIS), 137, 153, 154, 156, 162–164, 171, 195, 196, 1008
Imaging CMR_{ATP}, 646
  changes under brain stimulation, 107, 282, 284, 643
  changes under various brain states, 637
Imaging $CMRO_2$, 627, 629–637
  changes under brain stimulation, 643
  changes under various physiology conditions, 486, 489, 511, 642
Imaging progression of injury, 70, 71, 168, 337, 356
[^{123}I]-MIBG, 259
Inborn errors of metabolism, 1033
Incorporation fatty acids into phospholipids in vivo, 182, 183, 186, 187, 189, 196, 198, 200, 209, 232, 237, 238, 459, 700, 758, 767, 768, 772, 773, 776, 780, 786, 807, 808, 928, 931, 948, 1078, 1100, 1103, 1105, 1106
Increased relaxation for optimized neuroimaging (IRON), 549
Influence of ischemia on NADH signals, 238, 240, 349, 355, 375
Inhalation of $^{17}O_2$ gas, 632–635
Inheritance, 887, 889, 891, 893–896, 898
Inhibitory function, 1096
Inner-sphere contribution to relaxivity, 85–86
Inorganic phosphate (Pi), 151, 156, 164, 624, 630, 732, 733, 863, 1137, 1138, 1142
Intermediary metabolism, 134, 223, 820
Intracellular brain compartment, 659, 758
Intracellular electrophysiological measurements, 447
Intracellular Inorganic phosphate, 151
Intracellular [Mg] Adenosine triphosphate, 165
Intracellular pH, 150, 160–161, 720
Intrinsic optical signals, 332, 343, 352, 354, 355
Intrinsic optical signals in CCD-based imaging,
Inverse CSD (iCSD), 455
Ion balance, 978, 1076
Ion channels, 307, 334, 452, 688, 823, 832, 965, 1104
Ionic gradients, 334, 349, 372, 676
IRON fMRI, 549, 554–567, 571, 574, 575, 578–580, 585
Iron oxide, 79–122, 449, 549, 559, 560, 580, 602, 603
IRON (increased relaxation for optimized neuroimaging) signal, 549, 550, 555–557, 559–565, 567, 570, 573, 574, 576, 578, 581–584, 599
Irreversible trapping, 267, 1000
Ischemia, 72, 238–241, 245, 278, 336–338, 343, 349, 353, 355–359, 361, 373, 375, 377, 385, 397, 445, 549, 595, 596, 701, 703, 704, 758, 760, 763, 781–782, 1062, 1063, 1082, 1104, 1140
Ischemia-reperfusion, 1063, 1140
ISIS. *See* Image-selected in-vivo spectroscopy
Isoascorbate (d-ascorbate), 1063
Isocitrate dehydrogenases, 868
Isoform distribution, 725, 728, 732, 849, 853, 1035
Iso-orientation column, 517

Isotope, 66, 183, 259–261, 272, 273, 378, 568, 628, 631, 632, 700, 740, 766–768, 927, 953, 954, 961, 984, 1007, 1010, 1012, 1098, 1100
Isotope-isomer, 767
Isotopic enrichment, determination using isotopomers, 184, 186–189, 201–204, 206, 211, 230, 929, 1023
Isotopic labeling methods, 926–927
Isotopic steady-state, 187, 206, 209, 212, 938, 952–953, 957
Isotopomer, 200, 202, 203, 211, 212, 229, 738, 740, 741, 767, 771, 778, 929, 937, 952–953, 960, 966, 981, 982, 985

J

J-coupling, 160, 196
J-coupling constants, 143
J-difference editing, 143, 144

K

K_{ATP} channel activation, 1064, 1065
Ketogenic diet, 168, 169, 292, 783, 819–833
α-Ketoglutarate-malate carrier (OGC), 845, 851
Ketone bodies, 207, 292, 755, 758, 765, 782, 783, 821, 822, 824, 826–832, 925, 956
Ketosis, 783, 821–832, 861
Kety-Schmidt method, 418, 636, 910
Kinetics, 81, 103, 110, 213, 256, 259, 264–268, 309, 312, 313, 479, 488, 580, 627, 637, 656, 658, 660–666, 668, 701, 722, 732, 738, 741, 744, 775, 798, 800, 853, 867, 869, 948, 953, 959, 960, 963, 974, 976–979, 981, 982, 984, 985, 993, 996, 1000–1002, 1039, 1100, 1120, 1125–1127, 1130
  binding potential, 267, 310–313, 315, 321, 580
K_M in ascorbate transport, 1053–1055
Krebs cycle, 444, 870, 871, 900

L

Labelling pattern, 771
Lactate
  efflux in glycolysis, 759
  in glycolysis, 171, 282, 421, 423, 426, 444, 683, 995
  in head injury, 387, 760
  during hyperventilation, 760
  inflow of from blood, 568
  levels during hypercapnia, 466, 574
  levels in intercellular space, 372, 381
  in meningitis, 760
  origin, 134, 142, 224, 229, 239, 443, 683, 686, 777, 780, 853, 968, 1084
  outflow from brain, 391, 392, 825, 857, 968, 1084
  pool, 20, 208, 225, 226, 234, 236, 240, 390, 423, 426, 445, 682, 704, 727, 728, 743, 744, 764, 775, 783, 821, 826, 933, 962, 965, 967, 976, 977
  release, 391, 398, 426, 708, 779, 853
  shuttle, 396, 423, 677–678, 725, 758, 853, 866–867, 914, 955, 962, 994, 995
  shuttle in astrocyte-neuron interaction, 237, 396, 423, 677, 703, 758, 853, 955, 962, 967, 978, 995
  transport across blood brain barrier, 232, 239, 274, 291, 372, 394, 395, 397, 401, 424, 702, 718, 720, 756, 758, 827, 957
Lactate dehydrogenase (LDH), 167, 235, 339, 718, 719, 723–726, 728, 744, 845, 853, 867, 914, 994
Lactate utilisation
  in glycolysis, 677, 774
  by neurons, 372, 396, 401, 421, 423, 704, 743, 850, 853, 979, 994, 995
Laminar electrophysiological measurements, 446–448
Laminar optical tomography (LOT), 439, 482
Laminar population analysis (LPA), 456, 458, 487
Lamor frequency, 7–9, 12, 15, 40, 41, 46, 47, 55, 56, 86, 138, 511
Lamor precession, 40
Large neutral amino acids, 1124
LASER. *See* Localization by adiabatic spin-echo refocusing
Laser speckle contrast imaging, 436, 484
L-Aspartate N-acetyltransferase (Asp-NAT), 855, 856, 1076–1078
LCModel, 140, 141, 189, 199, 203, 658
LC-MS technology, 914
LDH. *See* Lactate dehydrogenase
L-DOPA, 310, 785
Learning & memory, 386
Leber's hereditary optic neuropathy, 887
Lewy bodies, 287, 888, 894
LFP. *See* Local field potential
Light scattering signals, 332, 343, 344, 352, 354, 439
Links
  to glutamate neurotransmission, 926
  metabolism and excitability, 34, 924

Lipid, 92, 135, 136, 142, 152, 186, 195, 240, 241, 257, 265, 281, 282, 292, 309, 627, 726, 780, 782, 794, 796–801, 805, 821, 822, 827, 852, 854–857, 861–863, 865, 866, 869–871, 925, 1052, 1062, 1079, 1080, 1084
 biosynthesis, 780, 782, 925, 1052, 1062, 1078, 1080
Liposomes, 91, 92
Lithium, 243, 783, 900
Lithium-pilocarpine, 243, 783
Local cerebral blood flow, 389, 394
Local field potential (LFP), 435, 447, 448, 450, 453–458, 470, 475–477
Localization
 with direct $_{13}$C detection, 165, 186
 with indirect ^{1}H[^{13}C] detection, 184, 190, 192, 194, 196, 199, 203, 210
Localization by adiabatic spin-echo refocusing (LASER), 136, 137, 196
Logan plot analysis, 321
Long-chain acyl-CoA synthetase (ACS), 800, 803–805, 810
Long-chain triglycerides, 822
Longitudinal recovery, 12, 13
Longitudinal relaxation time (T_1), 13, 14, 16, 20–22, 25, 26, 80, 82, 87, 98, 106, 107, 111, 116–120, 186, 196, 199, 595, 610, 628, 637, 640, 641, 1007
Low-grade tumor, 288
Low noise preamplifier, 45
Low pass RF filter, 59, 447, 458, 560
LPA. *See* Laminar population analysis
Lumped constant, 275, 277, 420

M

Macromolecules, 20, 25, 82, 140, 144, 382, 388, 505, 820
Magnetic field, 3, 35, 82, 188, 451, 501–535, 560, 602, 628, 985, 1008, 1118, 1136
Magnetic field inhomogeneity, 13, 22, 504, 512
Magnetic moment, 6, 84, 89, 524
Magnetic resonace spectroscopy (MRS)
Magnetic resonance (MR)
 diffusion, 68, 70, 72, 245
 relaxation, 66, 611
Magnetic resonance imaging (MRI), 3, 33, 65, 79, 138, 168, 246, 279–280, 306, 425, 434, 502, 548, 595, 628, 1081, 1120
Magnetic resonance spectroscopy (MRS), 24, 33, 149, 181, 377, 627, 676, 765, 1005, 1075, 1096, 1121, 1136
Magnetic susceptibility, 23, 24, 90, 150, 504, 506, 511, 558, 607
Magnetization, 3, 80, 136, 154, 195, 505, 549, 637, 929, 1138
Magnetization transfer (MT) method, 637, 639, 643, 1138
Magneto-encephalography (MEG), 306, 437, 450, 461, 469, 487, 489
Major depressive disorder, 1107
Malate-aspartate shuttle, 731, 732, 744, 827, 844, 845, 922
Malate dehydrogenase (MDH), 718, 845, 851, 856, 866, 869, 952, 953, 960, 966
Malate enzyme, 970
Malate-oxaloacetate shuttle, 866
Malic enzymes (ME), 719, 724, 726, 727, 732, 737, 740, 778, 868, 870, 871, 966, 967, 971
Malonate in Huntington's disease, 897
Mammalian brain regions, 142, 150, 152, 153, 157, 158, 163, 383, 595, 713, 741, 757, 764, 765, 769, 774, 777, 780, 786, 925, 926, 931, 935, 1057, 1080
Manganese
 focal brain injection, 101, 106
 intraperitoneal (i.p.) administration, 100
 intravenous (i.v.) administration, 100
 intra ventricular injection, 101
 transport, 101–104, 109
Manganese administration, 100–101, 103
 inhalation, 264
 sensory systems, 101
Manganese distribution, 101–105
 in glutamine synthetase, 103, 115
Manganese transfer index (MTI), 112, 113
Manic depressive illness, 169
MAP. *See* Maximum *a posteriori* probability (MAP)
Markov chain,
Martinotti cells, 474, 475, 487
Mass spectrometry (MS), 134, 157, 226, 768, 1021, 1083
Maximum *a posteriori* probability (MAP), 450
Maxwell's equations, 451
MCAO. *See* Middle cerebral artery occlusion
MCT. *See* Monocarboxilate carrier
MDH. *See* Malate dehydrogenase (MDH)
Measurement noise,
Medium-chain triglycerides, 822
MEG. *See* Magneto-encephalography
MELAS, 900, 1076, 1085
Membrane Mn-transport
 receptor, 102
 transferrin, 102, 103
 via Ca-Channel, 102
MEMRI, 79–122, 595

Mendelian inheritance, 887, 894, 896
Meningioma, 116–118, 171
M-EPI. *See* Multiplexed-EPI (M-EPI)
Mercaptosuccinate, 1061
Mesial temporal lobe epilepsy (MTLE), 223, 243–244
Metabolic compartmentation, 768–771, 911, 913, 915, 922–925, 932
  between neurons and glia, 932
Metabolic consequences of spreading depression, 331–361
Metabolic cooperation, 182, 1030, 1040
Metabolic energy, 280, 399, 400
Metabolic flux(es), 186, 199, 201, 210, 212, 230, 637, 639–643, 646, 926–930, 932, 938, 951, 1006, 1096, 1106
Metabolic interaction, 238, 717, 770, 785, 786, 1038
Metabolic modeling, 181–214, 926, 927, 930, 932–934, 938, 940, 953, 965, 984
  in disease, 211–212
Metabolic rate, 6, 182, 266, 275, 424, 434, 549, 627, 662, 680, 701, 710, 759, 782, 893, 926, 1002, 1086, 1103
Metabolic steady-state, 205, 212, 1019
Metabolic trafficking, 223, 393, 853
Metabolism, 24, 34, 101, 133, 163, 181, 222, 256, 272, 306, 332, 373, 418, 434, 504, 548, 597, 624, 656, 676, 700, 708, 715–745, 758, 764, 796, 820, 844, 885, 910, 922, 947–986, 993, 1014, 1029–1041, 1061, 1076, 1096
  of glutamate, 222, 765, 824, 911, 922
  during starvation, 826
Metabolite
  astrocytes, 1079
  correction, 268
  enrichment, 952, 963, 967
Metabolite concentrations
  brain tissue, 161
  extracellular space, 151
  plasma, 168
Metabotropic glutamate receptors, 687, 688, 1081
Methadone-maintenance, 171
Methionine, 223, 776, 1014
Methionine sulfoximine (MSO), 223, 224, 776, 1014, 1015
Michaelis Menten, 277, 656, 660–666, 668, 1019, 1056, 1125–1127
  asymmetric model, 663
  reversible model, 656, 661–666, 1127
  symmetric model, 1126, 1127
Microdialysis, 355, 371–402, 679, 680, 682, 756, 757, 760, 1016, 1017, 1021, 1031, 1057, 1127
Microfiltration, 371–402
  monitoring, 389–390
Microsensors, 383
MID. *See* Multi-infarct dementia (MID)
Middle cerebral artery occlusion (MCAO), 238–241, 246, 359, 596
Minocycline, 900
Mitochondria, 104, 151, 239, 281, 333, 375, 425, 444, 624, 683, 716, 781, 801, 821, 842, 885, 913, 930, 958, 994, 1062, 1076, 1099, 1139
  TCA cycle and electron transport, 339
Mitochondrial biogenesis, 890, 897, 900
Mitochondrial cascade hypothesis, 893, 894
Mitochondrial DNA, 627, 887, 889
Mitochondrial inorganic phosphate, 733
Mitochondrial pyruvate oxidation and anaplerosis, 719, 770, 777–779, 911, 912, 921, 940, 950, 954, 955, 957, 958, 960, 967, 970–973, 976, 979, 981–983
Mitochondrial pyruvate transport, 729–735
Mitochondrial redox potential, 333, 347–349
MK809, 375, 386, 784
Mn^{2+} chelates, 82, 97
Mn-distribution,
Mn-transport membrane, 102–104
Mohr-Tranejbaerg syndrome, 857–858
Monocarboxilate carrier (MCT), 711, 713, 720, 722, 724, 755, 756, 760, 765, 782, 867
Monocarboxylate, 426, 711, 718–723, 727, 728, 730, 743–745, 755–757, 760, 853, 867, 994
  Monocarboxylate transporter (MCT), 711, 718, 721, 722, 727, 730, 755–757, 760, 764, 765, 774, 782, 991. *See also* monocarboxylate carrierastrocyte-neuron interaction, 426, 711, 719, 722, 730, 742, 755
Monocarboxylic acid, 765, 773, 782, 784
Mono-crystalline iron oxide nanocompound (MION) contrast agent, 449, 450, 517, 560, 598, 599
Monte-Carlo simulations, 186, 204, 208–210, 984
Motion correction, 267–268
MPTP, 211, 894
MRI. *See* Magnetic resonance imaging
MRI/PET combination, 35
MRS. *See* Magnetic resonance spectroscopy (MRS)
MR spectra, modeling, 189
MR spectroscopic imaging (MRSI), 137, 138, 627, 633, 1083, 1084
MS. *See* Mass spectrometry (MS)

MSO. *See* Methionine sulfoximine (MSO)
MT. *See* Magnetization transfer (MT) method
[^{99m}Tc]-HMPAO, 264–266
[99mTc]-TRODAT-1, 260
MTI. *See* Manganese transfer index (MTI)
MTLE. *See* Mesial temporal lobe epilepsy (MTLE)
MUA. *See* Multiple unit activity
Multibanded pulses, 510
Multidrug resistance proteins (Mrps), 1036, 1037
Multidrug resistance proteins in glutathione transport, 1032, 1036
Multi-infarct dementia (MID), 287
Multilinear analysis, 319
Multiphoton imaging,
Multiple-labelled, 771
Multiple sclerosis, 26, 73, 81, 134, 1076, 1082–1083
Multiple single-site saturation (MSS) MT method, 638, 639, 641
Multiplet, 142, 143, 150, 151, 154, 156, 157, 160, 199, 211, 771, 772, 952, 953, 969
Multiple unit activity (MUA), 447, 448, 450, 456–458, 470, 472, 474–477, 479
Multiplexed-EPI (M-EPI), 510, 531
Multiplicities, 738
Multivariate machine learning algorithm, 534
Myelin, 25, 26, 73, 171, 780, 827, 841, 852, 854–857, 863, 865, 866, 868, 871, 1052, 1065, 1080, 1082, 1120
Myelination function, 1086

N

Na+, 91, 334, 336, 346, 349, 383, 400, 624, 676, 821, 824, 962, 978, 1016, 1017, 1053, 1055, 1063, 1104, 1105
NAAG. *See* N-acetylaspartylglutamat
N-acetyl aspartate (NAA)
  osmolyte in Canavan disease, 1082
  pathology, 773, 854, 1079, 1081, 1083
N-acetylaspartylglutamat (NAAG), 605–606, 1077, 1080–1081, 1085
NADH
  redox shuttles, 844, 862, 871
NADH/NAD ratio, 339, 340, 351, 445, 732, 824, 844, 845, 855, 867
NADPH
  redox shuttles, 844, 862
  regeneration, 869
  transfer, glutathione regeneration, 870
Na+ influx, 383, 1016, 1017
Na/K
  ATPase, 676, 678, 683, 688, 938
  pump, 676
N-[^{11}C]-methylspiperone, 266
Near infrared spectroscopy (NIRS), 310, 311, 353, 436–437, 442, 450
Negative BOLD, 470, 471, 475, 480, 550
Negative contrast agent, 80, 89, 550, 555, 564, 584, 612
Neural activity, 396, 397, 530, 613, 1103
Neural plasticity, 109, 121, 566
Neural tissue localization, 716, 1078, 1085
Neurodegeneration, 106, 115, 595, 605, 607–613, 885–901, 1033, 1034, 1039, 1082
Neurodegenerative diseases (disorders), 113, 361, 597, 607, 626, 637, 646, 785, 886–889, 897–899, 901, 1041
Neuroimaging, 279, 280, 292, 305–307, 324, 419, 420, 434, 485, 488, 548, 549, 608, 627, 629, 631, 635, 643, 644
Neuromodulator, nigrostriatal dopamine pathway, 1064, 1144
Neuron
  glia interaction in neurons, 1061, 1065, 1066
  mitochondrion, 851, 1078, 1079
Neuronal activation, 66, 338, 399–401, 463, 466, 517, 595, 609, 677–681, 683–686, 688, 689, 727, 758, 779, 956, 957, 962, 974, 1065
Neuronal anaplerosis, 925, 967
Neuronal ascorbate concentration, 1061
Neuronal excitability, 291, 292, 823
Neuronal oxidative metabolism, 914, 963, 977–980
Neuronal perikarya, 912
Neuronal tricarboxylic acid cycle, 234, 235, 281, 282, 288, 338, 624, 676, 678, 718, 735–737, 741, 764, 769–770, 774–776, 780, 821, 826–830, 844, 868, 911, 949, 994
Neuropeptide Y, 474, 480, 487
Neuroreceptors, 256, 268, 307, 999
Neurotransmission, 105, 211, 212, 227, 307, 309, 398–400, 548, 717, 719, 737, 740–743, 777, 783–785, 820, 825, 871, 910–911, 913, 922–926, 932, 938–939, 948, 950, 954, 962, 974, 980, 984, 1006, 1064, 1085, 1096, 1103–1108
Neurotransmitter, 134, 307, 420, 461, 600, 624, 676, 703, 709, 741, 765, 820, 844, 911, 922, 948, 1006, 1033, 1063, 1080, 1096

Neurovascular unit, 435
β-Nicotinamide adenine dinucleotide (NADH), 338, 425, 444, 624, 666–667, 683, 719, 824, 844, 963, 994
Nicotinergic acetyl-cholin receptor (nAChR), 599
NIRS. *See* Near infrared spectroscopy
Nitropropionic acid in mitochondrial biogenesis, 229, 852, 897, 913
NMDA. *See* N-methyl-D-aspartate
NMDA and AMPA receptors, 386
N-methyl-D-aspartate (NMDA), 105, 106, 282, 334, 374, 375, 386, 399, 401, 474, 487, 599, 687, 784
NMR. *See* Nuclear magnetic resonance
N-15 MRS, 1006–1011, 1016–1018
NMR technology, 910, 913, 948
Noninvasive imaging approach, 489, 548
Nuclear magentic resonance spectroscopy, 133–144, 189–192, 221–246, 374, 551, 569, 656, 658, 661–663, 667–668, 676, 685, 700, 705, 766, 767, 926, 929, 948, 951, 985, 986
Nuclear magnetic resonance (NMR), 65, 656, 676, 717, 766
Nucleotide triphosphate, 150, 151, 156, 159, 164
Nucleus in neuron, 1103

O

OATPs in glutathione transport, 1036, 1037
Obesity, 284, 805
Obsessive-compulsive disorder, 290
Occupancy, 258, 259, 323–324, 551, 570, 611, 767
OCT. *See* Optical coherence tomography
3-OH-butyrate, 823–824, 826, 827
3-OH-butyrate dehydrogenase, 827
3-OH-butyric acid, 823
50 ohm transmission, 51
OISI. *See* Optical Intrinsic Signal Imaging
^{17}O isotope, 631
Oligodendrocytes, 234–236, 780, 855–856, 861, 1034, 1035, 1038, 1065, 1079, 1080, 1082, 1083
Oligonucleotide reporters, 95
^{17}O MRS, 627–632, 635, 637, 642, 644
One compartment metabolic model, 961
One-compartment model, 199, 205–207, 209, 210, 932, 957, 959, 974, 979
On-line detection, 378, 382–383
Onset of convulsions, 1096
Ontogenesis, 765
Optical coherence tomography (OCT), 437, 439, 442–443, 483, 488
Optical imaging, 66, 81, 306, 324, 332, 333, 337, 352–356, 361, 436–437, 443–446, 461, 471, 479, 481, 483, 484, 502, 517, 582, 679
Optical intrinsic signal imaging (OISI), 436, 437, 463–466
Orotic acid, 782
Osmolyte, 780, 1082
Outer-sphere contribution to relaxivity, 86–87
[^{15}O]-water, 265
Oxaloacetate, 226–229, 234–237, 770, 778, 826–829, 845, 846, 855, 911, 914, 923, 925, 928, 930, 949–955, 959–961, 966, 967, 970–972, 974, 978, 981–984
Oxidative metabolism, 237, 239, 241, 244, 349, 350, 396, 397, 420, 422, 424, 624–627, 632, 635, 637, 642, 682, 683, 730, 735, 742, 743, 758, 775, 781, 826, 844, 849, 860, 861, 885, 886, 910, 913, 914, 939, 960, 962, 963, 977–981, 994, 995, 999, 1000, 1030, 1061
Oxidative phosphorylation, 281, 283, 292, 426, 484, 623–646, 683, 689, 733, 844, 854, 899, 900
  of ADP, 624
Oxidative stress, 115, 282, 862, 869, 886, 896, 898, 899, 1030, 1033–1035, 1037, 1039, 1041, 1052, 1053, 1060–1062, 1065
Oxodicarboxilate carrier (Oxodic), 866, 867
2-Oxoglutarate, 205–207, 210, 223, 228, 229, 826, 830, 832, 855, 1096
2-Oxoglutarate dehydrogenase, 223
Oxygen
  electrodes, 450, 483
  partial pressure, 91, 93, 334–335, 360, 443–444, 483
  utilization, 283, 549, 550, 563, 573, 583, 625, 627, 631, 632, 642, 646, 820
Oxygen consumption, 284, 399, 422, 424, 427, 445, 459–468, 470, 484, 505, 526, 550, 601, 626, 643, 676–679, 684, 686, 688, 775, 820, 852, 909, 910, 912, 913, 999–992 *see also* Cerebral metabolic rate of oxygen
Oxygen consumption rate, 505, 593 *see also* Cerebral metabolic rate of oxygen
Oxygen extraction fraction (OEF), 677
Oxygen/glucose ratio (O/G), 676, 677, 680, 685, 686
Oxygen/glucose turnover, 422, 685, 910
Oxyhemoglobin, 438, 461, 464

P

Parallel and antiparallel States, 8, 10
Parallel imaging, 507–510, 531, 559, 566, 585
Parkinson's disease, 169, 170, 211, 259, 286, 289, 322–323, 578, 580, 626, 823, 886, 888, 893–896, 1030, 1063, 1064, 1096
Parkinson's disease inheritance,
partial fourier sampling, 509
Partial volume RF coil, 52
Passive shimming,
Patlack graphical analysis, 276
PC. *See* Phosphocholine
PCA. *See* Principal component analysis
PCr↔ATP↔Pi exchange network, 637, 639, 642
Pentose phosphate pathway, 274, 422, 868, 956, 969, 976, 982
penumbra, 238, 240–241, 336, 337, 377, 388, 781
PepT2, 1038, 1039
Perfusion, 20, 48, 66, 116, 257, 290, 307, 336, 378, 419, 439, 516–518, 578, 602–603, 632, 682, 1057
Permeability-surface area, 756
Permeability transition, 900
Permissive for H_2O_2 signaling, 1066
Peroxide clearance in astrocytes, 1032–1035, 1037
PET/CT, 62, 293
PET/MRI, 60, 62, 63, 429
PGAA. *See* Phosphoguanidinoacetate (PGAA)
pH, 84, 86, 89–91, 121, 150, 152, 155, 159–161, 165, 168–172, 309, 335, 388, 600, 627, 720, 732, 757, 796, 797, 848, 1008, 1009, 1052, 1138, 1139, 1144
  homeostasis, 90
  responsive contrast agents, 91
Phased receive array, 585
Phase encoding, 15, 16, 18–20, 137, 154, 155, 509, 510, 526, 533
Phase modulation, 137, 154, 157
  in adenosine triphosphate, 137, 154, 157
Phenylalanine, 170, 1117–1130
  blood content, 1117–1130
Phenylketonuria, 170, 1118
  etiology, 1126
  treatment, 1121
Phosphatidylethanolamine (PtdE), 91, 164, 169
Phosphatidylinositol, 169
Phosphocholine (PC), 134, 151, 152, 156, 164, 165, 172, 629, 630
Phosphocreatine (PCr), 134, 143, 150, 151, 156, 159, 164, 335, 625, 630, 852, 1123, 1136, 1137
Phosphodiesters in membrane bilayer, 151
Phosphoenolpyruvate, 719, 725, 727, 732, 778, 914, 936, 967, 970
Phosphoethanolamine, 151, 629, 630
Phosphoethanolamine, intracellular pH, 150
Phosphoguanidinoacetate (PGAA), 1140
Phospholipids, 25, 101, 171, 794, 796, 802, 805, 807, 810
Phosphomonoesters (PME), 150, 156, 159, 164, 1137
Phosphorescence, 443–445, 483, 679
Phosphorescent lifetime, 444, 445, 484
Phosphorus, 47, 149–150
  intrinsic sensitivity, 149
  magnetic resonance properties, 149–172
2-Photon microscopy, 437, 439, 442, 445, 447, 471, 481, 485, 486, 488, 994
Photons, 8, 12, 256, 257, 260, 272, 340–342, 443, 996
Phototoxicity in autofluorescence, 340–341
Physiological stimulation, 283–285, 354, 604–606, 608, 612, 686
Pi. *See* Inorganic phosphate (Pi)
Plasmalogen, 169
plasma membrane fatty acid binding protein/ aspartate aminotransferase, 718, 733, 802–803, 826, 828, 829, 845, 851, 852, 855, 856, 911, 958
Plasma membrane monocarboxylate transport, 719
plasma tracers, 256, 266–268, 274–276, 311, 312, 314–317, 764, 765
PME. *See* Phosphomonoesters (PME)
^{31}P MRS, 627–643, 645, 646, 1128, 1137, 1138, 1140, 1143
PNET. *See* Primitive neuroectodermal (PNET)
Point resolved spectroscopy (PRESS), 24, 25, 33, 136, 137, 143, 153–154, 162, 163, 170, 195, 196, 1008
Point spread function (PSF), 523, 526, 564
Poisson's equation, 454
Polarization transfer, 157, 186, 189, 194–196, 199, 201, 1100
Polyamines, 1097
Polymerization, 94
Polyunsaturated fatty acid, 793
P:O ratio, 643
positional enrichment, 211
Positive contrast agent, 80
Positron emission tomography (PET), 34, 66, 170, 181, 255–268, 271–292, 305–307, 378, 418, 464, 504, 548, 610, 631, 677, 717, 765, 893, 910, 926, 994, 1083, 1107

Post-mortem accumulation, 1099, 1100, 1103
Potential energy, 10, 372, 400
Power, 4, 15, 189–191, 196–198, 279, 332, 338, 465, 530, 534, 549–551, 554, 556–561, 564, 569, 576, 579, 581, 768, 1023
Preamplifier, 40, 41, 43–46, 50, 54–57, 190
Precession, 6–9, 12–15, 38, 40, 52
Precursor, 167, 224, 266, 282, 608, 629, 682, 716, 765–766, 828, 855, 889, 911, 922, 949, 1006, 1034, 1081, 1096
Pre- post-synaptic neuronal signals, 351, 428
Presenilin, 889, 891, 894
PRESS. *See* Point resolved spectroscopy
Primitive neuroectodermal (PNET), 172
Principal component analysis (PCA), 456
Programmed cell death, 899
Prolyl 4-hydroxylases, 1052
Pro-oxidant effects of ascorbate, 1063
Protein-mediated fatty acid transport, 803–806
Proton data processing methods, 7, 8, 15
Proton decoupling, 1007, 1008, 1023
Proton density, 20–22
Proton (^{1}H) MRS techniques, 1099
pseudo-continuous arterial spin labeling,
PSF. *See* Point spread function (PSF)
Psychiatric disorders, 142, 289–290, 307, 607, 612, 1085–1086, 1096, 1107
PtdE. *See* Phosphatidylethanolamine (PtdE)
Putrescine, 610, 1097
Pyruvate
  carboxylation, 226, 227, 232, 777, 778, 783, 913, 955, 961, 965–967, 970, 972, 974, 983
  kinase, 719, 724, 725, 778, 936, 955, 967, 970, 1004
Pyruvate carboxylase (PC), 184–186, 206, 207, 210, 213, 234, 715, 718, 719, 737, 739, 758, 769, 770, 783, 830, 863, 865, 869, 870, 911, 912, 925, 928, 930, 932–934, 937–938, 950, 952–955, 961, 965, 970, 971, 976, 981, 982
Pyruvate carrier (PyC), 718, 729, 730, 863, 865, 869
Pyruvate dehydrogenase, 184–186, 281, 718, 735–737, 739, 827–829, 851, 852, 856, 891, 899, 928–930, 955, 960, 966, 974
  complex defect, 281, 718, 735, 891, 900
  and TCA cycle, 184, 281, 735–737, 852, 856, 928–930, 955, 960
pyruvate-malate shuttle, 731, 863, 865, 869–871

Q
Quadrature hybrid, 153, 168, 172, 190, 1123, 1137
Quantification, 183, 196–204, 208, 211, 373, 479, 607, 610, 613, 633, 639, 1002, 1011
  of direct ^{13}C spectra, 199–203
  of indirect 13C spectra, 203–204
  of isotopomers, 192, 237
  and modeling of CMR_{ATP}, 627, 642–646
  and modeling of $CMRO_2$, 198–204
Quantitative assessment, 277, 292, 602, 717
Quantitative (metaquant) microdialysis, 380
Quasistatic approximation, 451

R
Radial diffusivity, 68, 69, 73
Radiation, 26, 256–260, 272, 288, 289, 292, 419, 569, 766
Radiation necrosis, 288, 289
Radioactivity, 256, 258, 261, 262, 264, 267, 268, 276, 277, 313–318, 320, 321, 741, 766, 808, 911, 968, 996, 998, 1000, 1100, 1106
Radiofrequency (RF)
  chains, 46
  coils, 10–11, 50–51, 504, 509, 569, 1136
  decoupler amplifier, 190
  frequency synthesizer, 35
  power amplifier, 41, 136, 191, 196, 197
  receive amplifier, 36
  receive chain, 188, 190
  synthesizer, 36
  transmission, 34, 40–43, 45, 46, 51, 52, 55–57
  transmit chain, 36, 37
Radiofrequency magnetic field (B_1), 36, 40
Radioisotope modes of decay, 309, 662, 996
Radioisotope preparation, 309, 662, 996
Radio-label, 658, 764–766, 780, 796
Radiolabelled glycogen, 680
Radioligands, 307–315, 320, 322, 323
Radiopharmaceuticals, 256–258, 273, 306
Radio-tracer, 766
rat brain, 104, 133, 164, 182, 222, 378, 521, 552, 628, 657, 682, 704, 733, 758, 765, 808, 853, 931, 958, 1007, 1054, 1078, 1136
Rate of anaplerosis, 972
Rate of glutamate-glutamine cycle, 209, 930–935, 939
Reactive oxygen species (ROS), 1030, 1032, 1034, 1062–1065
  in Parkinson disease, 1063
  signaling agents, 1064

Receive only RF coil, 42
Receptor kinetic modeling, 998, 1002
Recording methods, 103, 256, 259, 261, 264, 266–268, 345–348, 355–358, 361, 393, 397, 446–448, 450, 455, 475, 502, 1130
Recurrent tumor, 288
Redox sensitive contrast agents, 344
Redox transitions in mitochondria, glycolysis, 867
Reduction of GSSG, 1035
Regeneration of GSH, 869, 1037
Regulation of CO2 fixation, 938
Regulation of dopamine release by H_2O_2 in striatum, 1085
Relative anisotropy, 69
relative cerebral blood volume (rCBV), 29
Relative quantification, 198–201
Relative specific activity, 764, 766
Relaxation, 10, 66, 80, 136, 154, 186, 257, 468, 505, 549, 595, 628, 700, 1007, 1124
Relaxation times, 66, 80, 158, 199, 468, 551, 554, 629
Relaxometric parameters, 93
Relevance of spreading depression for migraine, injury, 335–337, 343
Reliability of metabolic modeling, 207, 208, 933
Reproducibility and reliability, 182, 204, 207–210, 213, 613, 628, 635, 933, 980, 984, 1103, 1124
Resolution, 16, 75, 79, 133, 150, 188, 259, 280, 306, 332, 376, 419, 436, 502, 548, 602, 628, 680, 717, 871, 929, 969, 994, 1007, 1079, 1101, 1124, 1136
Response to therapy, 288
Resting state fMRI (r-fMRI), 526, 530–533
Retinotopy, 564
Reverse flux, 639, 641, 955, 961
Reverse rate constants, 637, 639, 642
Reversible two tissue compartment model, 979
r-fMRI. *See* resting state fMRI
RF transmit amplifier, 40–50. *See also* RF power amplifier
Ribose-5-phosphate, 152, 282
Rodent optic nerve, 709
Room temperature shim coils, 38
ROS. *See* Reactive oxygen species

S

SAR. *See* Specific absorption rate
SAT. *See* Sodium-coupled amino acid transporter
Saturation transfer, 157–159, 637, 1140
Scalar coupling, 137, 139, 142, 143, 631
Schizoaffective disorder, 169
Schizophrenia, 73, 168–169, 258, 259, 289, 323, 605, 611–612, 626, 784, 1030, 1085, 1086, 1096
Scintillation, 50, 306, 766, 768, 775, 926
Secondary (delayed) energy failure, 166, 167
Second-sphere contribution to relaxivity, 87–89
Seizure, 72, 109, 241–244, 289, 291, 292, 445, 704, 783, 822, 823, 832, 852, 934, 960, 976, 982, 1006, 1076, 1082, 1099
  activity, 290, 291, 728, 1096
  threshold, 336, 823
Semi-permeable membrane, 993
Semi-quantitative assessment, 278
Sensitivity, 11, 44, 71, 80, 133, 149, 182, 230, 257, 276, 309, 344, 382, 437, 507, 549, 602, 628, 657, 704, 728, 756, 768, 927, 969, 996, 1007, 1032, 1085, 1121, 1136
Sensitivity dependence upon T_1, T_2, T_2*, 13, 20, 90, 116, 117, 506, 509, 528, 628
Sensitivity gain at high-field, 628, 629
Sensitivity optimization for fMRI, 137, 1124, 1136
Serotonin receptor, 259, 310–311, 399
Serotonin transporter, 259, 308, 310–311
Shim amplifier, 40
Shim coils, room temperature, 38
Shimming
  automatic, 38
  B_1, 191
  passive, 612
Signaling, 282, 398, 434, 435, 474, 479–480, 488, 605, 624, 794, 795, 899, 980, 1051–1066, 1136
Signal-to-noise ratio (SNR), 45, 140, 150, 153–157, 161, 165, 168, 172, 446, 507, 509, 510, 514, 515, 526, 528, 531, 552, 557–560, 566, 569, 571, 585, 602, 628, 629
Simplified reference tissue model (SRTM), 317, 321
Single photon emission computed tomography (SPECT), 66, 255–268, 290, 305–309, 311, 312, 314, 322, 324, 419
Single-shot, inversion recovery (SIRENE), 154
Singlet, 134, 143, 152, 237, 723, 734, 767, 771, 773, 778, 952, 972, 1052, 1123
Singular value decomposition, 442
Sinogram, 261–263
Slice selection, 15–16, 136, 196, 199
Slow axonal transport, 104
Small animal positron emission tomography, 280, 996

Smooth muscle cell, 478, 479, 482, 609
SN1, 1006, 1015–1017
SNc degeneration in Parkinson disease, 1064
SNc regulation of DAergic neuron excitability by H_2O_2,
SNR. *See* Signal-to-noise ratio
Sodium-coupled amino acid transporter (SAT), 1006, 1017
Sodium waves in astrocytes, 688
Somatostatin, 474, 475, 477, 480, 487
Spatial localization, 135–139, 142, 186, 195, 444, 523, 561, 565
Spatial normalization, 261, 263
Spatial resolution, 16, 24, 75, 79, 168, 255, 259–262, 292, 332, 340, 341, 419, 436, 437, 439, 441, 488, 502, 509, 523, 525, 531, 533, 557, 564–566, 602, 609, 612, 680, 768, 985, 986, 994–997, 1086, 1145
  partial volume correction, 262
  relationship with detector properties, 260
Specific absorption rate (SAR), 197
Specificity, 114, 181, 246, 258, 259, 481–483, 514–527, 531, 534, 642
SPECT. *See* Single photon emission computed tomography
Spectral domain OCT,
Spectral imaging, 437–440, 471, 475, 483
Spectral quantification, 199–201, 203
Spectroscopy, 24, 76, 133, 149, 181, 221, 281, 374, 421, 437, 548, 627, 656, 676, 700, 756, 766, 826, 926, 948, 1008, 1075, 1096, 1117–1130
Spectrum analysis, 159–161
  prior knowledge, 159
Sphingomyelin, 169
Spin-echo (SE), 22, 137, 143, 153, 157, 196, 203, 505, 512, 513, 519–521, 553, 554, 565, 567, 572, 1008
Spin-echo (SE) fMRI, 505
Spin labeling, 30, 419, 548– 550, 567–570, 604
Spin-lattice (T_1) Relaxation times, 12, 505, 1007
Spin-spin (T_2), 12, 13, 154, 157, 1008
Spin-spin coupling, 225, 767
Splitting, 85, 184, 222, 767, 887, 956, 1007
Sporadic onset, 887
Spreading depression, 71, 331–361, 389, 445
SRTM. *See* Simplified reference tissue model (SRTM)
SSADH. *See* Succinic semialdehyde dehydrogenase (SSADH)
Standard uptake value (SUV), 278
Static magnetic field (B0), 4, 10, 14, 505, 512, 513, 522, 528
Steady-state, 187, 205, 206, 209, 212, 259, 391, 463, 464, 485, 486, 555, 577, 627, 635, 637, 640, 641, 656, 660–669, 775, 830, 833, 950, 951, 954, 959, 976, 985, 1010–1013, 1016, 1018, 1019, 1039, 1097, 1128
STEAM. *See* Stimulated echo acquisition mode (STEAM)
Stejskal-Tanner gradient pair, 519, 520
Stereoselectivity in ascorbate transport, 1061
Stimulated echo acquisition mode (STEAM), 24, 136, 137, 141, 154, 163, 193, 196, 1008, 1077
Stimuli and models, 121, 122, 333, 334, 345, 346, 376, 386, 566, 567, 575, 580
Stress, 66, 115, 241, 280, 282, 343, 375, 376, 386, 583, 600, 680, 682, 842, 862, 869, 886, 896, 898, 899, 938, 1030, 1033–1035, 1037, 1039, 1041, 1052, 1053, 1060–1062, 1065, 1145
Striatum, 100, 104, 108, 258, 259, 261, 309, 310, 323, 374–377, 381, 387, 390–396, 632, 679, 683, 781, 996, 1031, 1057, 1058, 1064, 1065, 1085
Stroke, 26, 70–72, 109, 134, 168, 227, 230, 231, 241, 245, 388, 389, 488, 566, 568, 596, 759, 781, 1055, 1062, 1076, 1085
Sub-cellular resolution, 994
Subcutaneous microfiltration, 100
Substantia nigra pars compacta (SNc), 1064, 1065
Substrate, 93, 134, 169, 183, 231, 273, 311, 350, 390, 418, 605, 663, 677, 703, 711, 722, 758, 820, 842, 911, 936, 948, 993, 1010, 1031, 1078
Substrate trafficking in astrocyte-neuron interaction, 778
Succinic semialdehyde dehydrogenase (SSADH), 1099
Superconducting magnet, 4
Superconducting shim coils, 37
Superoxide dismutase in amyotrophic lateral sclerosis, 896
Superoxide dismutase (MnSOD, SOD_2) in neuron mitochondria, 115
Surface coil, 10, 11, 52, 135, 136, 150–153, 158, 159, 163–167, 189, 190, 195, 198, 507, 601, 630, 1008, 1124, 1137
SUV. *See* Standard uptake value (SUV)
SVCT1, ascorbate transport, 1053
SVCT2, ascorbate transport, 1053
Swelling, 72, 167, 343, 344, 347, 359, 397, 443, 899, 1082

Synaptic activity, 332, 372, 426–428, 435, 446, 447, 450, 475, 487, 517, 676, 678, 853, 963, 977, 979
Synaptic excitation, 454
Synaptic inhibition, 473, 479, 1104
Synaptic potentials, 447, 676, 678, 683
Synaptic transmission, 105, 418, 420–422, 475, 566, 607, 689, 912
Synaptic vesicles, 105, 266, 372, 399, 912, 1097
Synaptosomes, 726, 829, 830, 852, 912, 966, 967, 1019, 1098, 1103
Synucleins, 806–810

T

TAC. *See* Time activity curve
Tail pinch, 375, 376, 386, 680–682
TCA cycle rate, 182, 186, 205–207, 209, 211, 773, 775, 776, 930, 931, 959, 960, 972, 974, 977, 978, 980, 981, 1101
Template-fitting technique, 457
Temporal lobe epilepsy, 109, 223, 243–244, 291
Temporal response, 559, 561–564, 580–583
Therapeutic intervention of birth asphyxia, 359, 611, 613
Thinned skull preparations, 436
Three compartment model, 314, 775, 932, 957, 976, 979, 980, 984
Three-spin chemical exchange model, 639
Thyroid, 258, 763, 782–783, 1033, 1038
Time activity curve (TAC), 264, 266, 267, 321, 998
Time-domain OCT, 442
Tomography, 66, 255–268, 271–292, 305, 378, 418, 442, 504, 548, 610, 631, 677, 717, 767–768, 810, 893, 926, 994, 1086
TORO. *See* Transmit Only, Receive Only
Tracers, 81, 207, 256, 259, 260, 264–268, 318, 378, 420, 548, 568, 656, 767, 768, 786, 948, 994
Tractography, 74–75
Tract tracing, 101, 106–113
Trafficking glutamate between astrocytes and neurons, 351
Transaminase, 229, 239, 741, 776, 826, 958, 1096
Transamination, 229, 234–236, 726, 770, 826, 828–830, 851, 965, 970, 971, 1005, 1010, 1019, 1021, 1096, 1099
Transcallosal, 475
Transcellular redox switch/Redox coupling hypothesis, 740–745
Transgenic animals, 1129–1145
Transgenic mice, 608, 609, 785, 802, 803, 805, 890, 896, 1140–1143, 1145
  with Alzheimer's disease, 608, 609, 785, 889, 890
Transmit array, 45
Transmit Only, Receive Only (TORO), 42–43, 45–51
Transmit only RF coil, 42
Transmit/receive mode, 42
Transmit/receive RF coil, 41
Transmit receive Switch (T/R-S), 41, 44
Transport, 101, 186, 233, 274, 309, 338, 373, 423, 445, 570, 624, 655–669, 686, 701, 711, 715–745, 755–760, 764, 794, 821, 844, 890, 912, 924, 950, 993, 1015, 1030, 1053, 1096, 1117–1130, 1139
Transporter(s), 103, 1037
  cysteine, 1033–1034, 1038–1040
  glycine, 1034, 1038–1040
  peptides, 1034, 1039, 1040
Transverse decay, 12, 21
Transverse relaxation time (T_2), 628, 1124
Trauma, 71, 373–375, 388, 389, 402
T_1 relaxation, 13, 16, 85, 98, 186, 199, 595
T_2 relaxation, 13, 16, 28, 468, 505
T_2 relaxation time, 468, 628, 1008
Tricarboxylic acid (TCA) cycle, 235, 281, 282, 338, 624, 718, 735–744, 769, 780, 821, 826–830, 844, 868, 911, 994
Trifluoroacetic acid, 223
Tritium, 257, 760. *See also* ^{3}H
TRUST, 469
Tumor, 74, 98, 121, 172, 278, 284, 288, 292, 580, 728, 1038
Turnover, 100, 134, 182, 229, 282, 373, 422, 681, 700, 727, 767, 910, 1056, 1096
Two-compartment model, 207, 570, 930, 948, 954, 961, 974, 979, 984
  metabolic modeling, 182, 206–210, 570, 930, 948, 954, 961, 974, 979, 984
Two-spin chemical exchange model, 640
Two-step J editing, 1099

U

[U-^{13}C]glucose, 184, 225–227, 233, 234, 237, 969
[U-$^{13}C_6$]glucose, 184, 186, 192, 200–202, 206, 208, 209, 211, 232, 274
[U-13C]lactate, 225–227, 232, 237–238
Ultrafiltration, 373, 379–381
Ultrahigh magnetic fields, 502, 534
Ultraslow microdialysis, 371, 393, 394

Ultra-small paramagnetic iron oxide (USPIO), 89, 116, 117, 579
Uncoupling, flow and metabolism, 265, 422, 423, 425, 427, 575, 682, 729, 823, 861
Unesterified fatty acid, 795, 800, 807–810
Uridine diphosphosugars (UDPS), 151, 152
Uridine triphosphate, 150
USPIO. *See* Ultra-small paramagnetic iron oxide

V
Validation, 213, 360, 389, 419, 461, 596, 635, 962, 995
Valproate, 292, 783, 784, 1107
VAN. *See* Vascular anatomical network
Vascular anatomical network (VAN), 459–461
Vascular coupling, 337
Vascular space occupancy (VASO), 551, 570–572, 584, 585
VASO. *See* Vascular space occupancy
Venous refocusing for volume estimation (VERVE), 551, 572–573, 582, 584, 585
VERVE. *See* Venous refocusing for volume estimation
Very long-chain acyl-CoA synthetase, 804–805
Very long-chain fatty acid, 795, 796, 798, 802, 804, 805
Vigabatrin, 741, 773, 776, 1098, 1099, 1102, 1104, 1107
Visual cortex, 23, 159, 170, 279, 284–286, 323, 335, 372, 400, 443, 478, 506, 512, 518, 525, 561–564, 585, 638, 644, 645, 678, 683, 684, 775, 997
Visual inspection, 278
Vitamin C, 1045. *See also* Ascorbate
Voltage sensitive dyes, 446, 479
Volume conductor theory, 451
Volume of distribution, 266, 313, 316, 318, 321, 659

W
Water suppression, 135, 136, 139, 204
White matte, cerebellum, 14, 21, 22, 27, 28, 72–76, 117, 165, 167–171, 196, 210, 262, 277, 288, 559, 658, 677, 709, 849, 858, 861, 986, 1031, 1056, 1076, 1079–1083, 1085, 1120
White matter lesions in magnetic resonance imaging, 117
Wilson's disease, 888, 897

X
[^{133}Xe], 257, 258, 264, 265
Xenon gas, 257, 504

Y
Yokoi-ito plot analysis, 318

Printed by Publishers' Graphics LLC USA
SO20120403-021
2012